REF
TS
156.8
.P764
1999

HWLCTC

Chicago Public Lit...

D1128727   d

BUSINESS/SCIENCE/TECHNOLOGY DIVISION
CHICAGO PUBLIC LIBRARY
400 SOUTH STATE STREET
CHICAGO, IL 60605

Chicago Public Library

REFERENCE

Form 178 rev. 1-94

PROCESS/ INDUSTRIAL INSTRUMENTS AND CONTROLS HANDBOOK

Chemical Engineering Books of Interest

Cascio • ISO 14000 GUIDE: THE NEW INTERNATIONAL ENVIRONMENTAL MANAGEMENT STANDARDS

Chopey • HANDBOOK OF CHEMICAL ENGINEERING CALCULATIONS, SECOND EDITION

Chopey • INSTRUMENTATION AND PROCESS CONTROL

Connell • PROCESS INSTRUMENTATION PROCESS MANUAL

Fitzgerld • CONTROL VALVES FOR THE CHEMICAL PROCESS INDUSTRIES

Fleming/Pillai • S88 IMPLEMENTATION GUIDE

Lieberman/Lieberman • A WORKING GUIDE TO PROCESS EQUIPMENT

Luyben/Tyreus/Luyben • PLANTWIDE PROCESS CONTROL

Meyers • PETROLEUM REFINING PROCESSES, SECOND EDITION

Miller • FLOW MEASUREMENT ENGINEERING HANDBOOK, THIRD EDITION

Perry/Green • PERRY'S CHEMICAL ENGINEERS' HANDBOOK, SEVENTH EDITION

Sawers/Eastman • PROCESS INDUSTRY PROCEDURES AND TRAINING MANUAL

Schweitzer • HANDBOOK OF SEPARATION TECHNIQUES FOR CHEMICAL ENGINEERS, THIRD EDITION

Shinskey • PROCESS CONTROLLERS FOR THE PROCESS INDUSTRIES

Shinskey • PROCESS CONTROL SYSTEMS, FOURTH EDITION

Woodside/Aurrichio/Yturri • ISO 14001 IMPLEMENTATION MANUAL

Yaws • CHEMICAL PROPERTIES HANDBOOK

PROCESS/ INDUSTRIAL INSTRUMENTS AND CONTROLS HANDBOOK

Gregory K. McMillan Editor-in-Chief

Douglas M. Considine Late Editor-in-Chief

Fifth Edition

McGRAW-HILL

New York   San Francisco   Washington, D.C.   Auckland   Bogotá
Caracas   Lisbon   London   Madrid   Mexico City   Milan
Montreal   New Delhi   San Juan   Singapore
Sydney   Tokyo   Toronto

Library of Congress Cataloging-in-Publication Data

Process/industrial instruments and controls handbook / Gregory K.
  McMillan. editor : Douglas M. Considine, late editor-in-chief. —
  5th ed.
      p.   cm.
  ISBN 0-07-012582-1
    1. Process control Handbooks, manuals, etc.   2. Automatic control
Handbooks, manuals, etc.   3. Engineering instruments Handbooks,
manuals, etc.   I. McMillan, Gregory K., 1946–   . II. Considine.
Douglas M.
TS156. 8. P764   1999
629.8—dc21   99-29591
                                                                CIP

McGraw-Hill

A Division of The McGraw·Hill Companies

Copyright © 1999, 1993, 1985, 1974, 1957 by The McGraw-Hill Companies, Inc.
All rights reserved. Printed in the United States of America. Except as permitted
under the United States Copyright Act of 1976, no part of this publication
may be reproduced or distributed in any form or by any means, or stored in
a data base or retrieval system, without the prior written permission of the
publisher.
    Figures 1 through 9 and portions of the text in "Batch Control: Applying the S.88.01
Standard" in Section 11 by Thomas G. Fisher have been reproduced courtesy of the ISA
(ANSI/ISA/S88.01/1995). (Copyright © ISA. Reprinted with permission. All rights reserved.)

    1 2 3 4 5 6 7 8 9 0   DOC/DOC   9 0 4 3 2 1 0 9

ISBN 0-07-012582-1

*The sponsoring editor for this book was Robert Esposito and the
production supervisor was Sherri Souffrance. This book was set
in Times Roman by TechBooks.*

Printed and bound by R. R. Donnelley & Sons.

McGraw-Hill books are available at special quantity discounts to use as premiums
and sales promotions, or for use in corporate training programs. For more information,
please write to the Director of Special Sales, McGraw-Hill, 11 West 19th Street,
New York, NY 10011. Or contact your local bookstore.

 This book is printed on recycled, acid-free paper containing a
minimum of 50% recycled, de-inked fiber.

Information contained in this work has been obtained by The McGraw-Hill
Companies. Inc. ("McGraw-Hill") from sources believed to be reliable.
However, neither McGraw-Hill nor its authors guarantee the accuracy or
completeness of any information published herein and neither McGraw-Hill
nor its authors shall be responsible for any errors, omissions, or damages
arising out of use of this information. This work is published with the understanding that McGraw-Hill and its authors are supplying information but are
not attempting to render engineering or other professional services. If such
services are required, the assistance of an appropriate professional should
be sought.

BUSINESS/SCIENCE/TECHNOLOGY DIVISION
CHICAGO PUBLIC LIBRARY
400 SOUTH STATE STREET
CHICAGO, IL 60605

CONTENTS

Section 7. CONTROL COMMUNICATIONS

Section 8. OPERATOR INTERFACE

Section 9. VALVES, SERVOS, MOTORS, AND ROBOTS

Section 10. PROCESS CONTROL IMPROVEMENT

Section 11. STANDARDS OVERVIEW

SECTION 1
INTRODUCTORY REVIEW

G. McMillan

From the mid-1970s to the mid-1990s, the hiring of engineers and technicians dwindled to new lows. Technical training was replaced with management training programs. Fast-track employees aspired to be managers rather than technical leaders. This, combined with the early retirement of most of the experienced engineers and technicians, led to a huge gap in age and experience in the area of process/industrial instrumentation and control. Companies have attempted to replenish their technical capability in the past few years, but the new hires are typically not given a mentor, access to a technical training program, and technical support. Nor are they given time to develop their skills on projects of incremental complexity. Often they are thrown immediately into some very difficult situations.

To help address this new need, the emphasis of the new material in the handbook has shifted from operating principles to application guidance. New features and process conditions that are important considerations for successful installations are discussed. Selection ratings, key points, and rules of thumb are offered. This update provides the reader with a perspective and appreciation for what is important for implementation from authors with decades of experience.

Plants have also suffered from neglect. In attempt to improve the return on equity, capital was not made available to replace old equipment. Meantime, the surge in the economy means plants are running at 200% or more of name-plate capacity. As a result, equipment is pushed beyond its normal operating region. This has increased the benefits from process control improvement to get the most out of a plant. Section 10 has been added to provide a comprehensive treatment of this important opportunity.

The biggest news, of course, is the move to smart instrumentation, the Windows NT platform, and Fieldbus. Distributed Control Systems and Field-Based Systems in Section 3, Knowledge-Based Operator Training in Section 8, Instrument Maintenance Cost Reduction in Section 10, and an Overview of the ISA/IEC Fieldbus Standard in Section 11 provide information essential to get the most out of these major shifts in technology.

Finally, standards have been recently developed to address safety, batch operation, and Fieldbus. Section 11 has been added to provide an overview of the important aspects of these new standards by authors who have played a key role in their development.

This handbook has been designed for the practitioner who needs to apply instrumentation and control systems in industry. The following is a walk-through of the technical articles.

SECTION 2: CONTROL SYSTEM FUNDAMENTALS

Control Principles

As was observed by readers of earlier editions, this has been one of the most widely used articles in this handbook. This article is intended not only for individual study, but also for use by groups of scholars in college, technical school, and in-plant training programs. The article commences with the nontheoretical analysis of a typical process or machine control system. Discussed are process reaction curves, transfer functions, control modes, and single-capacity and multicapacity processes—relating control characteristics with controller selection.

Techniques for Process Control

This article reviews from both the practical and the theoretical viewpoints the numerous advancements achieved in solving difficult control problems and in improving the performance of control systems where fractional gains in response and accuracy can be translated into major gains in yield and productivity. This article is the logical next step for the instrumentation and control engineer who understands the fundamentals of control, but who desires to approach this complex subject in a well-organized mathematical and theorectical manner. When astutely applied, this advanced knowledge translates into very practical solutions. The author proceeds in an orderly manner to describe state-space representation, transfer-operator representation, the mathematics of open-loop, feedback, feedforward, and multiple-loop control, followed by disturbance representation, modeling, the algebraics of PID (proportional-integral-derivative) design, adaptive control, pattern recognition, and expert systems. The techniques of least squares, batch parameters, the Kalman filter, recursive parameter identification, and projection also are described.

Basic Control Algorithms

Continuous process control and its counterpart in discrete-piece manufacturing control systems traditionally were developed on an analog base. Experience over the past few decades has shown that digital control provides many advantages over analog systems, including greater flexibility to create and change designs on line, a wider range of control functions, and newer functions, such as adaptation. But digital computation is not naturally continuous like the analog controller. The digital approach requires sophisticated support software.

This article addresses the basic issues of carrying out continuous control in the digital environment, emphasizing the characteristics that must be addressed in the design of operationally natural control algorithms. The author describes number systems and basic arithmetic approaches to algorithm design, including fixed-point and floating-point formats. Lag, lead/lag, and dead-time calculations required in the development of a basic control algorithm are presented. Also included are descriptions of quantization and saturation effects, the identification and matrix-oriented issues, and software and application issues. A closing appendix details the generalized floating-point normalization function.

Safety in Instrumentation and Control Systems

Never to be taken lightly are those features that must be engineered into control systems on behalf of protecting plant personnel and plant investment, and to meet legal and insurance standards. This is a major factor of concern to users and suppliers alike. Even with efforts made toward safety design perfection, accidents can happen.

The author of this article carefully defines the past actions and standards that have been set up by such organizations as the International Electrotechnical Commission (IEC). He gives descriptions of numerous techniques used to reduce explosion hazards, including design for intrinsic safety, the use of explosionproof housings, encapsulation, sealing, and pressurization systems. Obtaining certification approval by suppliers and users of intrinsically safe designs is discussed in some detail, along with factors pertaining to the installation of such equipment.

SECTION 3: CONTROLLERS

Distributed Control Systems

This article traces the evolution of the distributed control system (DCS). It provides an interesting perspective of how concerns and demands have been addressed. Of particular importance is the discussion of how the DCS is meeting the needs to be open and to take advantage of new market trends. The advantages of interfacing third-party software for advanced applications such as expert systems and production management control is highlighted. The effects of Fieldbus, Windows NT, and the Internet are analyzed. Finally, a comprehensive list of DCS selection criteria is offered to help the user make the most complex and far-reaching decisions in instrumentation and process control in that it sets the ease and degree of automation, maintenance, and operability of the plant.

Programmable Controllers

The author provides a perspective of the criteria for making the many choices of architecture, software, and hardware. The tables of the choices of networks and input/output (I/O) outline the essential issues. The discussion of controller size, modularity, and distribution addresses the key questions for any application. The article also provides a balanced view of alternatives such as PC-based soft control. Both the user and the supplier will benefit from this treatment of this important component of discrete manufacturing, safety interlock, and sequential control systems:

Stand-Alone Controllers

The continuing impressive role of these controllers, particularly in non-CIM environments, is emphasized. Descriptions include revamped and modernized versions of these decades-old workhorses. A potpourri of currently available stand-alone controllers is included, with emphasis on new features, such as self-tuning and diagnosis, in addition to design conformation with European DIN (Deutsche Industrie Norm) standards.

Hydraulic Controllers

The important niche for powerful hydraulic methods continues to exist in the industrial control field. The principles, which were established decades ago, are described, including jet pipe, flapper, spool, and two-stage valves. Contemporary servo valves are discussed. Hydraulic fluids, power considerations, and the selection criteria for servo or proportional valves are outlined. A tabular summary of the relative advantages and limitations of various hydraulic fluids, including the newer polyol esters, is included.

Batch Process Control

During the past few years much attention has been directed toward a better understanding of the dynamics of batch processes in an effort to achieve greater automation by applying advanced control knowledge gained from experience with continuous process controls and computers. This has proved to be more difficult and to require more time than had been anticipated. Standards organizations, such as the Instrument Society of America, continue to work up standards for a batch control model. In this article an attempt has been made to cut through some of the complexities and to concentrate on the basics rather than on the most complex model one can envision. Batching nomenclature is detailed, and definitions of the batch process are given in simplified, understandable terms. To distinguish among the many methods available for accomplishing batch control, a tabular summary of process types versus such factors as duration of process, size of lot or run, labor content, process efficiency, and the input/output system is given. Interfacing with distributed control system and overall networking are considered.

Automatic Blending Systems

Although the need to blend various ingredients in pots and crocks dates back to antiquity, contemporary blending systems are indeed quite sophisticated. The author contracts the control needs for batch versus continuous blending. A typical blend configuration is diagrammed in detail. Some of the detailed topical elements presented include liquid or powder blending, blending system sizing, blend controllers, stations, and master blend control systems. The application of automatic rate control, time control, and temperature compensation is delineated.

Distributed Numerical Control and Networking

An expert in the field redefines numerical control (NC) in the contemporary terms of distributed numerical control (DNC), tracing the developments that have occured since the days of paper-tape controlled machines. The elements of the basic DNC configuration are detailed in terms of application and functionality. Much stress is given to behind-the-tape readers (BTRs). The numerous additional features that have been brought to NC by sophisticated electronic and computer technology are described. The tactical advantages of the *new* NC are delineated. The manner in which numerical control can operate in a distributed personal computer (PC) network environment is outlined. UNIX-based networks, open architectures, and the Novell networks, for example, are described

Computers and Controls

This article, a compilation by several experts, commences by tracing the early developments of the main-frame computer, the 1960–1970 era of direct digital control (DDC), up to the contemporary period of personal computers (PCs) and distributed control system (DCs). Inasmuch as there is another article in this handbook on DCSs, primary attention in the article is on PCs. The basic PC is described in considerable detail, including its early acceptance, its major components (microprocessor, memory, power supply, keyboard, and I/O). The advantages and limitations of the PC's "connectability" in all directions, including networks, are discussed. Internal and external bus products are compared. PC software is discussed, with examples of specific languages and approaches. Software control techniques are presented in some detail. Progressive enhancement of the PC toward making it more applicable to process and factory floor needs is reviewed. In consideration of the fact that minicomputers and mainframe computers enter into some control situations, a few basic computer definitions are included in the form of an alphabetical glossary. This is not intended as a substitute for a basic text on computers, but is included as a convenient tutorial.

Manufacturing Message Specification

This article provides a detailed look into the structure and importance of an international standard for exchanging real-time data and supervisory control information among networked devices in a manner that is independent of the application function and the developer. The standard provides a rich set of services for peer-to-peer real-time communications over a network for many common control devices such as programmable logic controllers (PLCs), robots, remote terminal units (RTUs), energy management systems, intelligent electronic devices, and computers. The rigorous yet generic application services provide a level of interoperability, independence, and data access that minimizes the life-cycle cost (building, using, and maintaining) of automation systems.

Field-Based Systems

The concept and advantages of a field-based system are introduced. The importance of maximizing the utility of Fieldbus and the explosive trend of adding more and more intelligence in the field devices is emphasized by the citation of impressive benefits from the reduction in wiring, termination, calibration, configuration, commissioning, and maintenance costs. It is also apparent that since the field-based system uses the same graphical configuration and instruction set as foundation Fieldbus, the user can focus more on the application and make the location of functionality transparent. The embedding of more advanced functionality, such as self-tuning into the controller as a standard feature, promotes the integrity and use of these techniques. The process simulation links open up the possibility of knowledge-based training systems (see Section 8) and OPC connectivity enables value-added applications of third-party software.

SECTION 4: PROCESS VARIABLES—FIELD INSTRUMENTATION

Temperature Systems

Commencing with definitions of temperature and temperature scales and a very convenient chart of temperature equivalents, the article proceeds to review the important temperature measurement methodologies, such as thermocouples and resistance temperature detectors (RTDs), with a convenient tabular summary of each for selection purposes. Smart temperature transducers are illustrated and described. Other temperature measurement methods described include thermistors, solid-state temperature sensors, radiation thermometers, fiber-optic temperature sensors, acoustic pyrometers, and filled-system thermometers.

Pressure Systems

This article has been updated to reflect the use of ceramic differential-pressure transmitters and diaphragm seals. These are important topics since the proper application of these close-coupled ceramic d/ps, digital heads, or diaphragm seals can eliminate the installation of sensing lines, which are the source of most maintenance problems.

Flow Systems

The author provides an easy-to-read view of what is important to ensure the proper selection and installation of flow meters. The reader should appreciate the clear and concise comparison of the major types of in-line meters. The application matrix serves as a vital reference of performance parameters. From the discussion of how fluid conditions affect meters, the user realizes that the many supposed mass flow meters recently touted in the literature, such as temperature- and/or pressure-corrected pitot tubes, positive displacement (PD) pumps, vortex meters, magmeters, and thermal mass flow meters, are dependent on some stringent assumptions. These meters that compute mass flow from several measurements are based on a constant known composition, a user-defined equation between density, viscosity, and/or heat capacity and temperature and/or pressure, and a fixed velocity profile, except for the PD pump. Only the Coriolis mass flow meter is independent of the process fluid and velocity profile.

Level Systems

The author provides a good perspective of the effect of process conditions on the performance of level measurements. It becomes apparent that the only continuous level measurements essentially independent of the process fluid are radar measurements and floats since they detect the surface. Ultrasonic measurements also detect the surface but are affected by dust and the composition of the vapor. Hence a lot of discussion is devoted to the application and the installation of surface detection devices. Level measurements that use differential pressure or Nuclear devices are greatly affected by fluid density and hence on both fluid temperature and composition unless a second completely submersed measurement is used to compute density. Capacitance probes with coating rejection are affected by the dielectric constant unless a second completely submersed probe is used to measure the dielectric constant.

Industrial Weighing and Density Systems

Strain-gauge and pneumatic load cells for weighing various hopper and tank vessels as may be used in batching systems are described, as well as a microprocessor-based automatic drumfilling scale. Numerous fluid-density measuring systems are reviewed, including the photoelectric hydrometer and the inductance bridge hydrometer. Specific-gravity sensors described include the balanced-flow vessel, the displacement meter, and the chain-balanced float gauge. Several density and specific-gravity scales are defined.

Humidity and Moisture Systems

This is the most well-organized and comprehensive yet concise treatment of these measurements that can be found in any handbook or journal. This extensive discussion of features, advantages, and limitations of a wide variety of devices should eliminate much of the confusion about choices and help make these important measurements more commonly used. Diverse applications are summarized.

SECTION 5: GEOMETRIC AND MOTION SENSORS

Basic Metrology

Of fundamental interest to the discrete-piece manufacturing industries, this article includes the very basic instrumental tools used for the determination of dimension, such as the interferometer, optical gratings, clinometer, sine bar, optical comparator, and positioning tables.

Metrology, Position, Displacement, Thickness, and Surface Texture Measurement

Described are the fundamentals of metrology and rotary and linear motion and the instrumental means used to measure and control it, such as various kinds of encoders, resolvers, linear variable differential transformers, linear potentiometric, and the new magnetostrictive linear displacement transducers. Noncontacting thickness gauges, including the nuclear, x-ray, and ultrasonic types, are described. The importance and measurement of surface texture are described.

Quality Control and Production Gaging

The fundamentals of statistical quality control (SQC) are presented with definitions of common cause, control limits, histogram, kurtosis, median, normal distribution, paretochart, skewness, special cause, and standard deviation. The reader should see world class manufacturing in Section 10 to see how statistical indices are used for quantifying process control improvements. This article also illustrates typical installations of impedance-type dimension gauges and provides numerous examples of the applications.

Object Detectors and Machine Vision

This article starts with a description of the principles and features of inductive, capacitive, ultrasonic, and photoelectric proximity sensors. This is followed by an introduction to machine vision technology with an emphasis on data patterns and image processing. It concludes with a discussion of discrete-piece identification and bar coding.

Flat Web (Sheet) On-Line Measurement and Control

This article discusses important benefits and application considerations of on-line measurement and control of sheet thickness in both the cross direction (CD) and the machine direction (MD). The advantages of new modular, smarter, and more open Windows NT-based systems are discussed. Simple equations to predict the speed requirement and limits of CD and MD measurements are presented along with important application aspects of advanced profile control and constrained multivariable predictive control and real-time optimization of the sheet line.

Speed, Velocity, and Acceleration Instrumentation

Following definitions of terms, the many kinds of tachometers available are presented, including dc, ac, voltage-responsive, variable-reluctance, photoelectric, and eddy-current. The tachometerless regulation of servo speed is described as are governors. Air and gas velocity measurements, including air-speed indicators and anemometers, are delineated. Vibration measurement and numerous kinds of accelerometers, including piezoelectric, piezoresistive, and servoaccelerometers, are described. Velocity transducers for sensing relative motion are discussed.

Vibration Measurements

Vibration measurements and numerous kinds of accelerometers are described. The signal conditioning of piezoelectric and piezoresistive accelerometers are explored in greater detail. The effect of environmental conditions such as temperature, cable motion, mounting compliance, dynamic strain inputs, and electrostatic and electromagnetic fields are discussed along with the selection and the installation implications.

SECTION 6: REAL-TIME ANALYTICAL COMPOSITION MEASUREMENTS FOR INPUT TO PROCESS CONTROL

Introduction

The opening remarks to this section present a unique insightful viewpoint that can be gained only from decades of experience in designing and installing analyzers and sample systems. The list of common mistakes and then the steps that can lead to improved performance provide much-needed words of wisdom. This is followed by a discussion of practical considerations and trends.

Concentration Measurement Technology and Devices

This article starts with a description of the features of thermal conductivity and gas-density detectors. Next, the application options and considerations of conductivity analyzers are outlined. This is followed by an in-depth look at several different devices. A comprehensive look at pH measurement details the theory and reality, electrodes, problems and causes, and best practices for measurement, installation, and maintenance. An extensive list of key points summarizes the essential concepts and the rules of thumb summarize the important recommendations for pH measurement. The treatment of turbidity and refractive-index measurements is similarly complete in scope, addressing aspects of design, installation, calibration, problems and application data. Next, the features and capabilities of ultraviolet/visible absorption analysis and ionization concentration transducers are discussed. The article also provides a brief overview of a myriad of other techniques.

Sample Extraction, Conditioning, Preparation for On-Line Analysis

The success of a non-in-line analyzer depends on its sample system. The sample must present the right information in a form that maximizes analyzer reliability. This article provides a practical and extensive compilation of the principles of sample handling and transfer for continuous sampling and the advantages and sample preparation and multidimensional manipulation techniques for discrete sampling. It concludes with valve and device configurations and the benefits of trap and transfer techniques.

System Control and Managing Data

An analyzer system is often like a miniature chemical plant. This article addresses the many issues involving the control and programming of the system, digital signal processing, information display, storage, communication, and housing.

Calibration and Validation

This article discusses the aspects of calibration and validation necessary to ensure that the required performance is met and maintained. Details are provided on standards and methods and the decisions based on statistical process control (SPC) charts. Several examples are used to illustrate the use of SPC. Included are concept, maintenance cost evaluation, and performance monitoring.

Application Examples

Actual industry examples drive home the essential ideas and fill in the missing details needed for practical applications. This article lists informative successful analyzer applications. The system design is outlined and the results are plotted.

SECTION 7: CONTROL COMMUNICATIONS

Data Signal Handling in Computerized Systems

Networking, whether simple or complex, cannot succeed unless the raw data fed to the network are reliable, accurate, and free from competing signals. The author defines signal types, termination panels, field signals and transducers, sampled data systems, analog input systems, analog outputs, and digital inputs and outputs. Stressed are signal conditioning of common inputs, such as from the thermocouples, solid-state temperature sensors, and resistance temperature detectors (RTDs). Amplifiers, common-mode rejection, multiplexers, filtering, analog signal scaling, and analog-to-digital and digital-to-analog converters are among the numerous topics covered and profusely illustrated.

Noise and Wiring in Data Signal Handling

The basic problems that a control engineer must seek to correct or avoid in the first place, including grounding and shielding, are delineated. Troubleshooting for noise is highlighted. A tabular troubleshooting guide is included.

Industrial Control Networks

Early networking and data highway concepts are described as a basis for understanding the many more recent concepts. Network protocols, including CSMA/CD, token bus, and token ring, are defined. Communication models and layers are defined as well as open systems and Fieldbus. The important more recent roles of fiber-optic cables and networks are described, including the characteristics of optical fibers or cables and light sources and detectors. Note that this topic appears also in several other articles of the handbook.

SECTION 8: OPERATOR INTERFACE

Operator Interface—Design Rationale

The basics of good design are brought to the process and machine operator interface. There are discussions of the fundamental factors that determine good interface design, including human, environmental, and aesthetic considerations. Graphics used in panels are described as well as visual displays. The role of color is included. The article ends with a discussion of interface standards and regulations, maintainability, and miniaturization.

Cognitive Skills and Process Control

The author reports on special studies of the operator interface from an industrial engineering standpoint and explores in particular the cognitive skills required of an operator.

Distributed Display Architecture

This article essentially zeros in on the CRT and equivalent interfaces that do not enjoy the attributes of larger panels. Interactive graphics are described in some detail.

Operator Training

The need for these operator training systems has dramatically increased because of the decrease in resources, the push for more capacity from stressed equipment, and the advent of more complex automation strategies. This article describes the concept of a graphical Windows NT operator training system that uses a dynamic model and field-based system configuration as the knowledge bases for the plant and the control system, respectively. The incremental improvements and performance requirements are detailed.

Smart Alarms

The distributed control system (DCS) has increased the number of alarms by an order of magnitude. The operator becomes insensitive to frequent alarms and is subjected to a barrage of alarms during a crisis or a shutdown. This article describes how the alarm, instead of triggering alarms off of a high or a low measurement, should be built up to show the actual operating condition from information from diverse sources such as sensors, tasks, modes, outputs, and other alarms. When done properly, a single alarm is generated that identifies the root cause.

SECTION 9: VALVES, SERVOS, MOTORS, AND ROBOTS

Process Control Valves

This article describes not only both general and special types of control valves, actuators, and accessories in terms of features needed in a large variety of applications, it also describes the issues to be addressed for the best valve and material selection. Also offered are helpful hints on storage and protection and installation techniques. A new topic on control valve performance highlights the choices and the benefits associated with minimizing the dead band and the nonlinearity of the control valve characteristic. The new opportunity of using smart digital positioners to monitor and improve valve performance is outlined. Finally, an extensive troubleshooting chart lists the causes and the solutions for major problems and symptoms of erosion, leakage, and poor response.

Control Valve Cavitation

The author provides knowledge from years of study of the fundamentals of cavitation, emphasizing cavity behavior and its negative effects on valve and system performance. The importance of valve sizing and selection toward the avoidance of cavitation problems is stressed.

Control Valve Noise

This research specialist addresses the serious problem of valve noise. Noise terminology is defined. The kinds of noise encountered—mechanical, hydrodynamic, and aerodynamic—are delineated. Suggestions for reducing noise are given.

Servomotor Technology in Motion Control Systems

This rather exhaustive article, directed mainly to engineers in the discrete-piece manufacturing industries, also finds generous application in the process industries. It embraces factors in selecting a servomotor, describing the basic kinds of dc motors, hybrid servos, stepper motors, linear steppers, power transmission drives, stepper motor drives, emergency stop measures, machine motion control systems, and a potpourri of motion control systems.

Solid-State Variable-Speed Drives

There has been a profusion of solid-state variable-speed motor drives ranging from subfractional to multithousand horsepower rating. Semiconductor switching devices and their impact on the development of ac variable-frequency drives is described. There is an extensive review of the various types of medium-voltage variable-frequency drives such as the load commuted inverter, filter commuted thyristor inverter, current-fed GTO inverter, neutral-point-clamped inverter, multilevel series cell VFD, and the cycloconverter. The comparison table provides a useful aid for selecting the right drive.

Robots

The technology of robotics, after an amazing surge of activity, now has reached a reasonable stage of maturity and acceptance. In this article the basic format of the robot is described, that is, its characteristics, including axes of motion, degrees of freedom, load capacity, and power requirements, as well as its dynamic properties, including stability, resolution and repeatability, and compliance among other characteristics. End effectors or grippers are presented. Workplace configurations are analyzed. Robot programming and control are explained and numerous styles of robots are illustrated.

Current-to-Pressure Transducers for Control Valve Actuation

Diaphragm-motor valves (pneumatically operated) remain the principal choice as final controlling elements for fluid flow. Although the demand for pneumatic control generally has diminished over the past few decades, the process control valve is operated by pneumatic force. Thus modern electronic controllers with digital outputs must utilize some form of current-to-pressure (air) transducer at the valve site. Several forms are available, including the older flapper-nozzle principles. This article also describes the combination of the newer pressure sensors with electronic feedback control.

SECTION 10: PROCESS CONTROL IMPROVEMENT

World Class Manufacturing

This article documents the methodology for finding and quantifying the benefits of process control improvement that has proven successful in one of the largest chemical companies. The methodology is extremely powerful yet relatively simple. Indices are developed that quantify the performance of the process and the control system for key process variables. The difference between these two indices can be used to estimate the benefits from improved process control. These indices along with utilization numbers can also be put on line to monitor the health of new control systems implemented and document the benefits achieved, which is critical for operations motivation and justification for future improvements.

Plant Analysis, Design, and Tuning for Uniform Manufacturing

This article provides a technical overview of a comprehensive suite of concepts, tools, and techniques that have become the standard for process control improvement in the pulp and paper industry. These include plant analysis to measure process and product variability by use of time-series analysis techniques, plant auditing procedures designed to identify the causes of process variability, an interpretation of the results in both the time and the frequency domain, the use of spectral analysis for both diagnostics and design, the use of model-based controller tuning such as internal model control (IMC) concepts and lambda tuning for both plant design and controller tuning, the use of a tuning strategy to achieve coordinated dynamics of a process area by preselection of the closed-loop time constants for each control loop, and understanding the performance-robustness envelop of a control loop, the impact of actuator nonlinearities on control performance, and the variability propagation pathways through a complex process. These methods are applicable to all process industries, especially process control loops on liquid plug flow or unit operations involving gas and solid streams. For unit operations involving backmixed volumes, there is often a significant process time constant that attenuates the amplitude of the variability introduced by fast loops (pressure and flow) to the degree to which the effect on the uniformity of the final product is within the on-line or the lab measurement resolution and repeatability limits.

Control Valve Response

The author of this article is the technical leader in understanding how the shaft length and connections, packing, actuator construction, and positioner design affect the ability of the control valve to respond.

The article correctly focuses on the essential need of the control valve to move within a reasonably short time to a small change in controller output. The best way to determine this capability is to review the change in a low-noise highly sensitive flow measurement for a change in controller output of 0.5%. In many rotary control valves, the actuator shaft will respond, but because of loose connections and twisting of the shaft, the actual disk or ball will not move. Also, the common practice of making changes of 10% or more in controller output reveal few, if any, of the problems and make every control valve look alike.

Process Impact

This article provides definitive examples of how improvements in the regulatory control system can significantly reduce the variability in processes. These examples show how tuning and decoupling loops, adding feedforward, and reducing valve dead band can yield impressive results. A case is developed for dedicating the resources to analyze each loop methodically for process control improvement. Most truly successful constrained multivariable predictive control systems and real-time optimizations demand the improvement of the regulatory control system as a necessary prerequisite. The benefits from these critical basic improvements are a major part of the total benefits reported for the advanced control system.

Best Practices, Tools, and Techniques to Reduce the Maintenance Costs of Field Instrumentation

This article focuses on the source of maintenance problems and how to do predictive as opposed to preventative maintenance. Good engineering practices are itemized that can greatly reduce the magnitude of the problems to the point to which most problems originate not in the application but in the manufacturing of the instrument or in the wear-out phase of the instrument. The article discusses the practical value of specific types of instrument knowledge and process knowledge-based diagnostics. An important point is made that frequent calibration is counterproductive and that good practices and smart instrumentation combined with diagnostics provides the confidence needed to leave the instruments alone. Rules of thumb summarize the overall recommendations.

New Developments in Analytical Measurements

Advanced control systems need concentration measurements in order to truly optimize the process. This article is a compilation of new industrial methods, such as Near Infared, Fourier Transform Infrared, Mass Spectrometer, Raman Scattering, Nuclear Magnetic Resonance, X-Ray Fluorescence, Microwave, and Neutron Activation. Many offer the hope of less sample conditioning and recalibration and therefore less on-site support. They also open up the opportunity to measure new components reliably on-line. This article tempers the enthusiasm with a word of caution that newer analyzers use technologies that are sophisticated and require extensive engineering and setup. The discussion of application considerations reinforces a practical view of the opportunity.

The Improvement of Advanced Regulatory Control Systems

This article starts with an extensive summary of the problems and causes of poor loop performance. One of the key discoveries is that measurement and control valve resolution are the largest undocumented sources of loop dead time. The concepts presented help the user track down the cause of a control problem. It then summarizes good practices for implementing the solutions including instrumentation and control valve upgrades, a variety of strategies, such as feedforward, dead time compensation, and override control, and controller tuning techniques, such as the closed loop, open loop short cut, and simplified Lambda methods.

Multivariable Predictive Control and Real-time Optimization

The addition of constrained multivariable predictive control is becoming recognized as the largest proven general opportunity in process control. Unlike other advanced control systems, it has a recognized track record in increasing production capability on the average by 3%–5%. While improvements in yield have not been as consistently high because of a greater dependence on application specifics, they are still considerable. This article provides a concise yet thorough treatment of CMPC from the practitioner's viewpoint. It provides an understanding of the concepts and the implementation techniques. The rules of thumb and guidelines on corrections to the regulatory system, plant testing, controller construction and tuning, and the outlining of maintenance issues are the types of information that are greatly needed but not available in the open literature. The article concludes with a discussion of real-time optimization and future directions.

Neural Networks

The hype of being able to just dump and crank (mine) historical data has distracted users from the real benefits of neural networks. Neural networks can find nonlinear effects that cannot be computed based on first principles or even seen because of the number of variables and the complexity of the relationships. This article provides both a theoretical and a practical understanding of this potential. Applications as virtual analyzers and in supervisory control are discussed. Perhaps the most underrated opportunity is the one of knowledge discovery in which the neural network identifies previously unrecognized key relationships. The guidelines presented on building neural networks are based on years of industrial application experience.

SECTION 11: STANDARDS OVERVIEW

Safety-Instrumented (Interlock) Systems

This article provides an overview of ISA S84 and IEC 1508/1511 standards that define a detailed, systematic, methodical, well-documented design *process* for the design of safety-instrumented systems. It starts with a safety review of the process, implementation of other safety layers, systematic analysis, as well as detailed documentation and procedures. The steps are described as a safety-design life cycle. The intent is to leave a documented, auditable trail and make sure that nothing is neglected. While these procedures are time consuming, the improvements pay off in not only a safer but a better performing process (an in-depth study has shown that as safety increased, production also increased). The implications in terms of logic system, sensor, redundancy, final elements, and solenoid design and installation are discussed. Key points and rules of thumb summarize the recommendations.

An Overview of the ISA/IEC Fieldbus

Fieldbus promises to revolutionize instrument installations. This article documents the savings from reduced terminations, home run wiring, I/O cards, and control-room panel space and faster configuration, engineering, documentation, checkout, and commissioning. It also discusses the installation options, such as cable types, devices per spur, topology, and junction box details, and the execution and status of the basic and advanced function blocks. These function blocks are comprehensive enough to become the common control language that leads to standardization and a seamless transition between functionality in the control room and in the field. Rules of thumb for proper implementation are also offered.

Batch Control: Applying the S88.01 Standard

In this article, the S88.01 standard is discussed and a methodology is presented for applying the standard to the definition of control system requirements for batch processes. This methodology uses an object-oriented approach that fits well with batch control and the S88.01 standard, including the development of objects that can be reused from project to project. Significant savings from applying the S88.01 standard have been demonstrated in all phases of batch control projects. The separation of the recipe procedure from the equipment logic is emphasized. This separation is one of the major reasons that the S88.01 standard has been so successful. Recommendations for dealing with exception handling and detailed equipment logic, which are the major portions of a batch control project, is provided.

SECTION 2
CONTROL SYSTEM FUNDAMENTALS*

E. H. Bristol
Research Department, The Foxboro Company (a Siebe Company), Foxboro, Massachusetts. (Basic Control Algorithms)

G. A. Hall, Jr.
Westinghouse Electric Corporation, Pittsburgh, Pennsylvania. (Control Principles—prior edition)

Peter D. Hansen
Systems Development and Engineering, The Foxboro Company (a Siebe Company), Foxboro, Massachusetts. (Techniques for Process Control)

Stephen P. Higgins, Jr.
Retired Senior Principal Software Engineer, Honeywell Inc., Phoenix, Arizona. (Techniques for Process Control—prior edition)

Richard H. Kennedy
The Foxboro Company (a Siebe Company), Foxboro, Massachusetts. (Techniques for Process Control—prior edition)

E. C. Magison
Ambler, Pennsylvania. (Safety in Instrumentation and Control Systems)

Joe M. Nelson
Applications Systems, Honeywell Inc., Billerica, Massachusetts. (Techniques for Process Control—prior edition)

Robert L. Osborne
Manager, Diagnostics and Control, Westinghouse Electric Corporation, Orlando, Florida. (Techniques for Process Control—prior edition)

John Stevenson
West Instruments, East Greenwich, Rhode Island. (Control Principles)

---

* *Persons who authored complete articles or subsections of articles, or who otherwise cooperated in an outstanding manner in furnishing information and helpful counsel to the editorial staff.*

CONTROL PRINCIPLES

by John Stevenson*

In this article the commonly measured process variable temperature is used for the basis of discussion. The principles being discussed apply also to other process variables, although some may require additional sophisticated attention, as discussed in the subsequent article, which deals with techniques for process control.

In contrast with manual control, where an operator may periodically read the process temperature and adjust the heating or cooling input up or down in such a direction as to drive the temperature to its desired value, in automatic control, measurement and adjustment are made automatically on a continuous basis. Manual control may be used in noncritical applications, where major process upsets are unlikely to occur, where any process conditions occur slowly and in small increments, and where a minimum of operator attention is required. However, with the availability of reliable low-cost controllers, most users opt for the automatic mode. A manual control system is shown in Fig. 1.

In the more typical situation, changes may be too rapid for operator reaction, making automatic control mandatory (Fig. 2). The controlled variable (temperature) is measured by a suitable sensor, such as a thermocouple, a resistance temperature detector (RTD), a thermistor, or an infrared pyrometer. The measurement signal is converted to a signal that is compatible with the controller. The controller compares the temperature signal with the desired temperature (set point) and actuates the final control device. The latter alters the quantity of heat added to or removed from the process. Final control devices, or elements, may take the form of contactors, blowers, electric-motor or pneumatically operated valves, motor-operated variacs, time-proportioning or phase-fired silicon-controlled rectifiers (SCRs), or saturable core reactors. In the case of automatic temperature controllers, several types can be used for a given process. Achieving satisfactory temperature control, however, depends on (1) the process characteristics, (2) how much temperature variation from the set point is acceptable and under what conditions (such as start-up, running, idling), and (3) selecting the optimum controller type and tuning it properly.

PROCESS (LOAD) CHARACTERISTICS

In matching a controller with a process, the engineer will be concerned with process reaction curves and the process transfer function.

Process Reaction Curve

An indication of the ease with which a process may be controlled can be obtained by plotting the process reaction curve. This curve is constructed after having first stabilized the process temperature under manual control and then making a nominal change in heat input to the process, such as 10%. A temperature recorder then can be used to plot the temperature versus time curve of this change. A curve similar to one of those shown in Fig. 3 will result.

Two characteristics of these curves affect the process controllability, (a) the time interval before the temperature reaches the maximum rate of change, *A*, and (2) the slope of the maximum rate of change of the temperature after the change in heat input has occurred, *B*. The process controllability

* Engineering Manager, West Instruments, East Greenwich, Rhode Island.

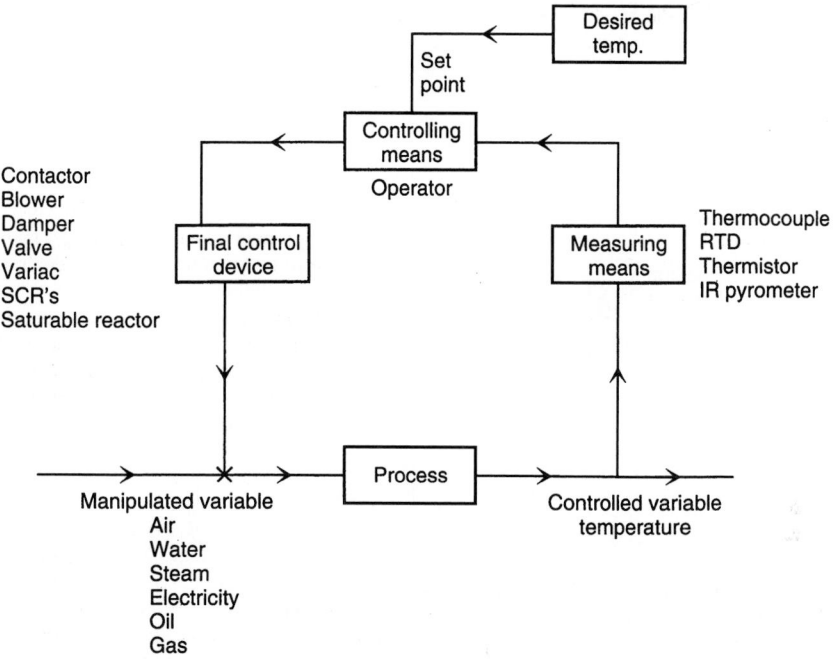

FIGURE 1   Manual temperature control of a process. (*West Instruments.*)

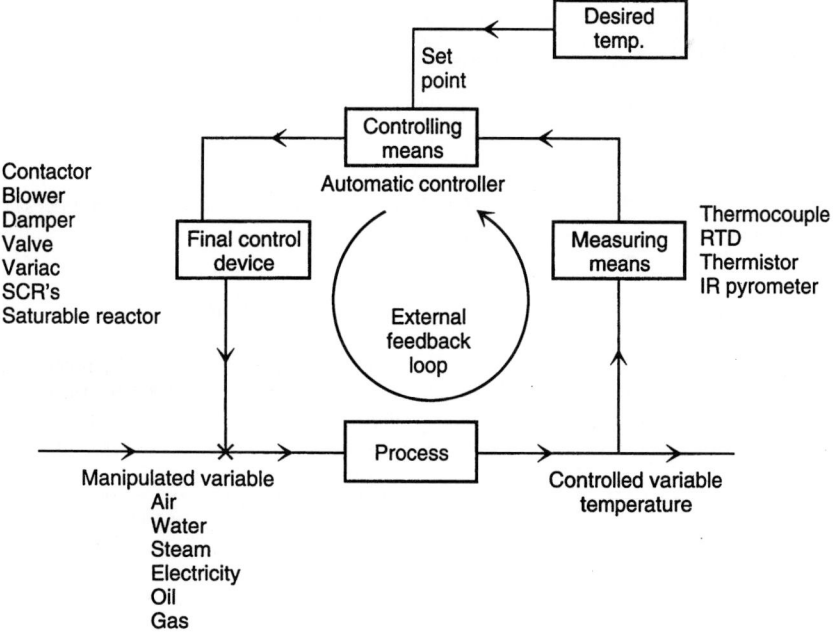

FIGURE 2   Automatic temperature control of a process. (*West Instruments.*)

decreases as the product of A and B increases. Such increases in the product AB appear as an increasingly pronounced S-shaped curve on the graph. Four representative curves are shown in Fig. 3.

The time interval A is caused by dead time, which is defined as the time between changes in heat input and the measurement of a perceptible temperature increase. The dead time includes two components, (1) propagation delay (material flow velocity delay) and (2) exponential lag (process thermal time constants). The curves of Fig. 3 can be related to various process time constants. A single time-constant process is referred to as a first-order lag condition, as illustrated in Fig. 4.

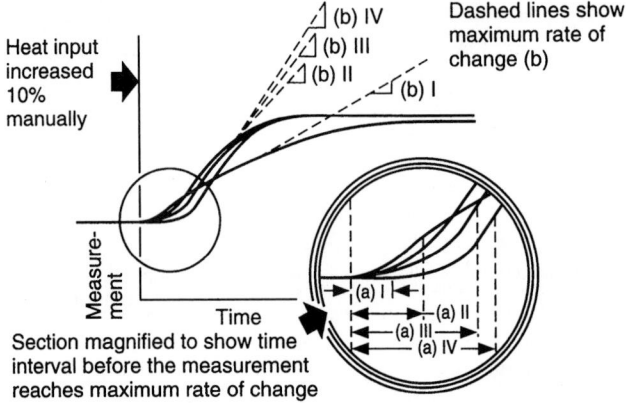

FIGURE 3   Process reaction curves. The maximum rate of temperature rise is shown by the dashed lines which are tangent to the curves. The tangents become progressively steeper from I to IV. The time interval before the temperature reaches the maximum rate of rise also becomes progressively greater from I to IV. As the S curve becomes steeper, the controllability of the process becomes increasingly more difficult. As the product of the two values of time interval A and maximum rate B increases, the process controllability goes from easy (I) to very difficult (IV). Response curve IV, the most difficult process to control, has the most pronounced S shape. Similar curves with decreasing temperature may be generated by decreasing the heat input by a nominal amount. This may result in different A and B values. (*West Instruments.*)

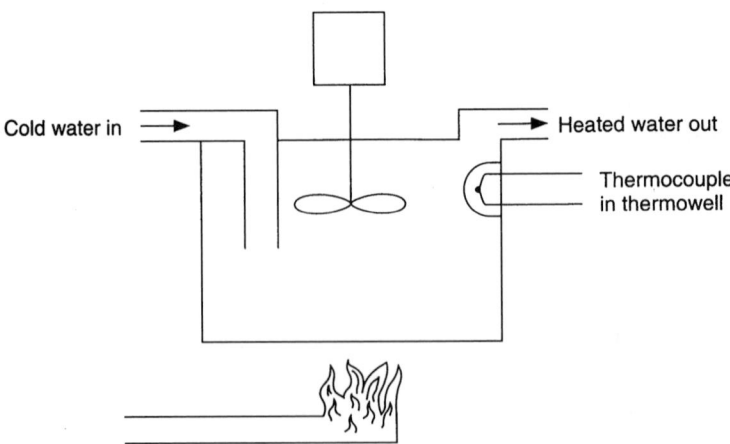

FIGURE 4   Single-capacity process. (*West Instruments.*)

This application depicts a water heater with constant flow, whereby the incoming water is at a constant temperature. A motor-driven stirrer circulates the water within the tank in order to maintain a uniform temperature throughout the tank. When the heat input is increased, the temperature within the entire tank starts to increase immediately. With this technique there is no perceptible dead time because the water is being well mixed. Ideally, the temperature should increase until the heat input just balances the heat taken out by the flowing water. The process reaction curve for this system is shown by Fig. 5.

The system is referred to as a single-capacity system. In effect, there is one quantity of thermal resistance R_1 from the heater to the water and one quantity of thermal capacity C_1, which is the quantity of water in the tank. This process can be represented by an electrical analog with two resistors and one capacitor, as shown in Fig. 6. R_{LOSS} represents the thermal loss by the flowing water plus other conduction, convection, and radiation losses.

It should be noted that since the dead time is zero, the product of dead time and maximum rate of rise is also zero, which indicates that the application would be an easy process to control. The same process would be somewhat more difficult to control if some dead time were introduced by placing the temperature sensor (thermocouple) some distance from the exit pipe, as illustrated in Fig. 7. This propagation time delay introduced into the system would be equal to the distance from the outlet of the tank to the thermocouple divided by the velocity of the exiting water. In this case the reaction curve would be as shown in Fig. 8. The product AB no longer is zero. Hence the process becomes increasingly more difficult to control since the thermocouple no longer is located in the tank.

A slightly different set of circumstances would exist if the water heater were modified by the addition of a large, thick metal plate or firebrick on the underside of the tank, between the heater and the tank bottom, but in contact with the bottom. This condition would introduce a second-order lag, which then represents a two-capacity system. The first time constant is generated by the thermal resistance from the heater to the plate and the plate heat capacity. The second time constant comes from the thermal resistance of the plate to the water and the heat capacity of the water. The system is shown in Fig. 9. The reaction curve for the system is given in Fig. 10. There is now a measurable

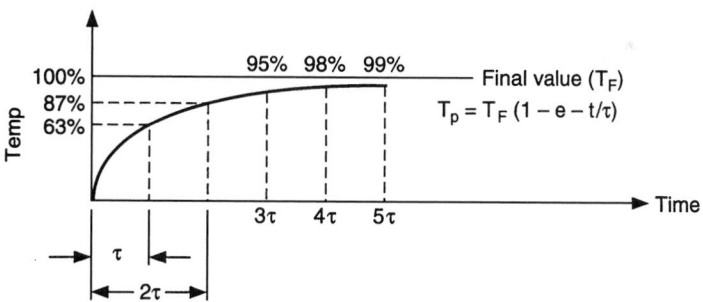

FIGURE 5   Reaction curve for single-capacity process. (*West Instruments.*)

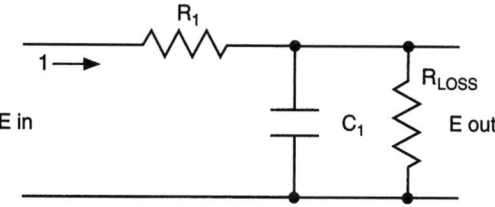

FIGURE 6   Electrical analog for single-capacity process. (*West Instruments.*)

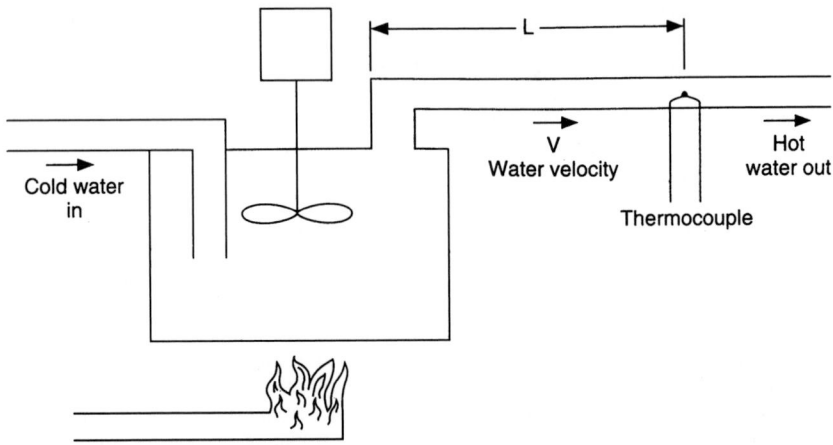

FIGURE 7   Single-capacity process with dead time. (*West Instruments.*)

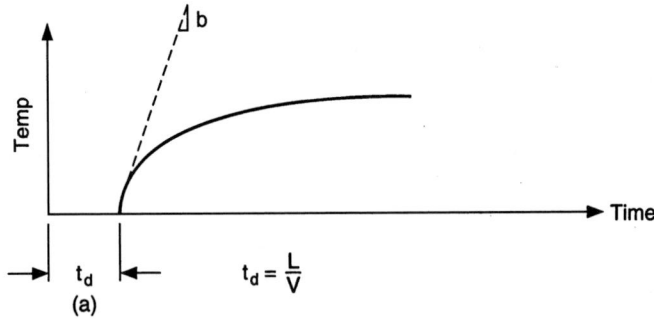

FIGURE 8   Reaction curve for single-capacity process with dead time. (*West Instruments.*)

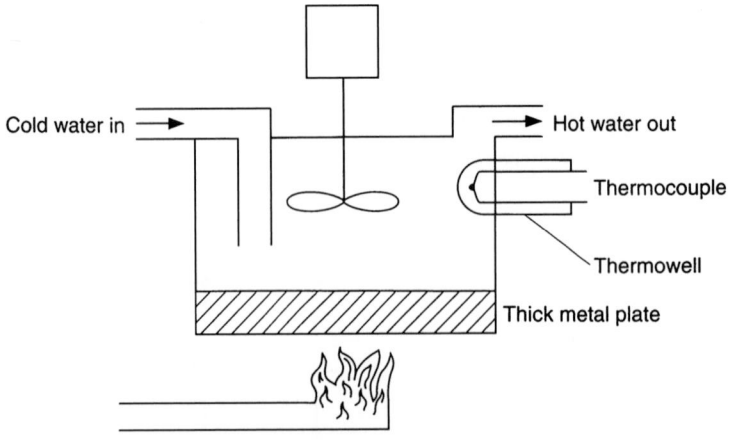

FIGURE 9   Two-capacity process. (*West Instruments.*)

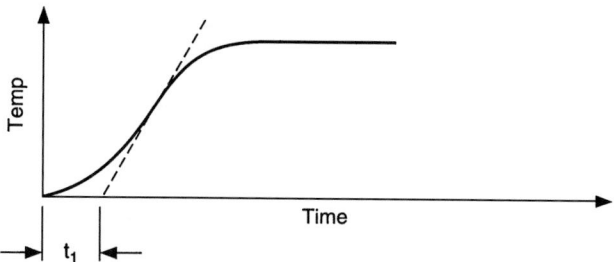

FIGURE 10   Reaction curve for two-capacity process. (*West Instruments.*)

time interval before the maximum rate of temperature rise, as shown in Fig. 10 by the intersection of the dashed vertical tangent line with the time axis. The electrical analog equivalent of this system is shown in Fig. 11. In the diagram the resistors and capacitors represent the appropriate thermal resistances and capacities of the two time constants. This system is more difficult to control than the single-capacity system since the product of time interval and maximum rate is greater.

The system shown in Fig. 9 could easily become a third-order lag or three-capacity system if there were an appreciable thermal resistance between the thermocouple and the thermowell. This could occur if the thermocouple were not properly seated against the inside tip of the well. Heat transfer from the thermowell to the thermocouple would, in this case, be through air, which is a relatively poor conductor. The temperature reaction curve for such a system is given in Fig. 12, and the electric analog for the system is shown in Fig. 13. This necessitates the addition of the R_3, C_3 time constant network.

Process Transfer Function

Another phenomenon associated with a process or system is identified as the steady-state transfer-function characteristic. Since many processes are nonlinear, equal increments of heat input do not

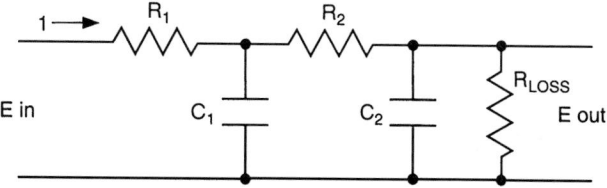

FIGURE 11   Electrical analog for two-capacity process. (*West Instruments.*)

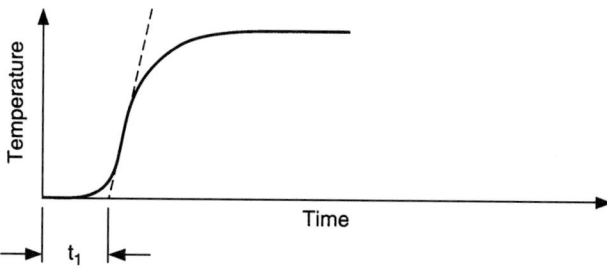

FIGURE 12   Reaction curve for three-capacity process. (*West Instruments.*)

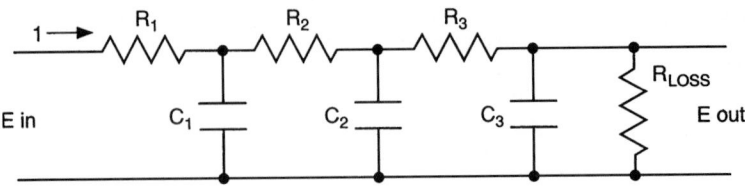

FIGURE 13   Electrical analog for three-capacity process. (*West Instruments.*)

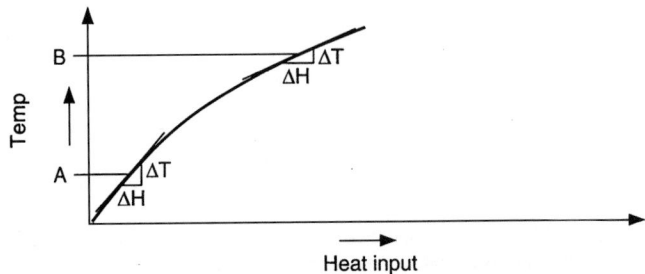

FIGURE 14   Transfer curve for endothermic process. As the temperature increases, the slope of the tangent line to the curve has a tendency to decrease. This usually occurs because of increased losses through convection and radiation as the temperature increases. This process *gain* at any temperature is the slope of the transfer function at that temperature. A steep slope (high $\Delta T / \Delta H$) is a high gain; a low slope (low $\Delta T / \Delta H$) is a low gain. (*West Instruments.*)

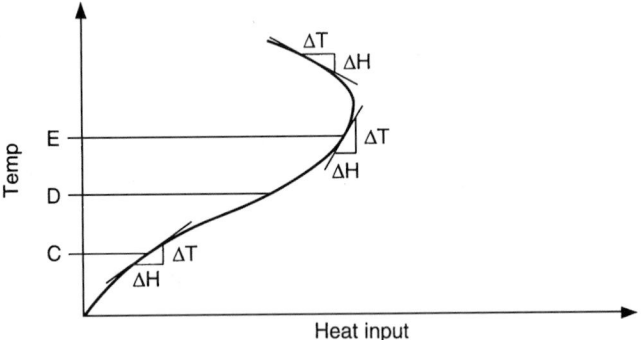

FIGURE 15   Transfer curve for exothermic process. This curve follows the endothermic curve up to the temperature level D. At this point the process has the ability to begin generating some heat of its own. The slope of the curve from this point on increases rapidly and may even reverse if the process has the ability to generate more heat than it loses. This is a negative gain since the slope $\Delta T / \Delta H$ is negative. This situation would actually require a negative heat input, or cooling action. This type of application is typical in a catalytic reaction process. If enough cooling is not supplied, the process could run away and result in an explosion. Production of plastics from the monomer is an example. Another application of this type is in plastics extrusion, where heat is required to melt the plastic material, after which the frictional forces of the screw action may provide more than enough process heat. Cooling is actually required to avoid overheating and destruction of the melt material. (*West Instruments.*)

necessarily produce equal increments in temperature rise. The characteristic transfer-function curve for a process is generated by plotting temperature against heat input under constant heat input conditions. Each point on the curve represents the temperature under stabilized conditions, as opposed to the reaction curve, which represents the temperature under dynamic conditions. For most processes this will not be a straight-line, or linear, function. The transfer-function curve for a typical endothermic process is shown in Fig. 14, that for an exothermic process in Fig. 15.

CONTROL MODES

Modern industrial controllers are usually made to produce one, or a combination of, control actions (modes of control). These include (1) on-off or two-position control, (2) proportional control, (3) proportional plus integral control, (4) proportional plus derivative (rate action) control, and (5) proportional plus integral plus derivative (PID) control.

On-Off Control Action

An on-off controller operates on the manipulated variable only when the temperature crosses the set point. The output has only two states, usually fully on and fully off. One state is used when the temperature is anywhere above the desired value (set point), and the other state is used when the temperature is anywhere below the set point.

Since the temperature must cross the set point to change the output state, the process temperature will be continually cycling. The peak-to-peak variation and the period of the cycling are mainly dependent on the process response and characteristics. The time-temperature response of an on-off controller in a heating application is shown in Fig. 16, the ideal transfer-function curve for an on-off controller in Fig. 17.

The ideal on-off controller is not practical because it is subject to process disturbances and electrical interference, which could cause the output to cycle rapidly as the temperature crosses the set point. This condition would be detrimental to most final control devices, such as contactors and valves. To

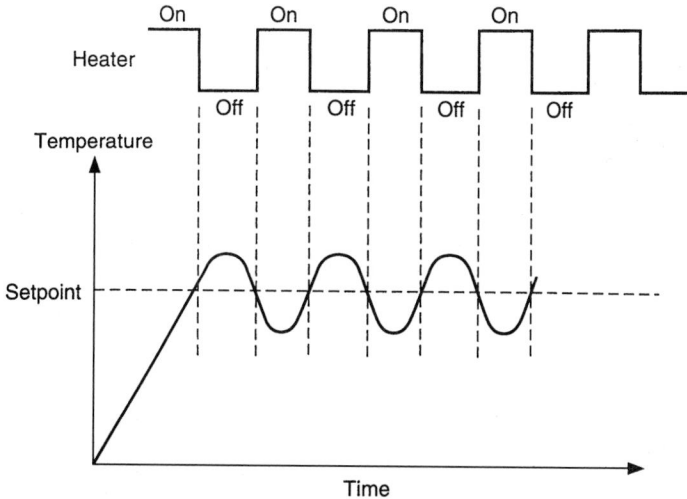

FIGURE 16   On-off temperature control action. (*West Instruments.*)

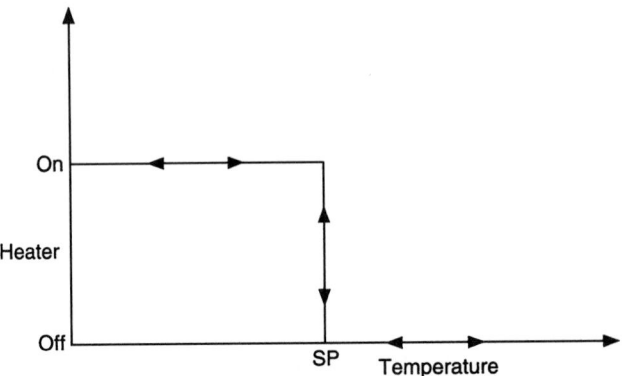

FIGURE 17 Ideal transfer curve for on-off control. (*West Instruments.*)

prevent this, an on-off differential or "hysteresis" is added to the controller function. This function requires that the temperature exceed the set point by a certain amount (half the differential) before the output will turn off again. Hysteresis will prevent the output from chattering if the peak-to-peak noise is less than the hysteresis. The amount of hysteresis determines the minimum temperature variation possible. However, process characteristics will usually add to the differential. The time-temperature diagram for an on-off controller with hysteresis is shown in Fig. 18. A different representation of the hysteresis curve is given in the transfer function of Fig. 19.

Proportional Control

A proportional controller continuously adjusts the manipulated variable so that the heat input to the process is approximately in balance with the process heat demand. In a process using electric heaters, the proportional controller adjusts the heater power to be approximately equal to the process

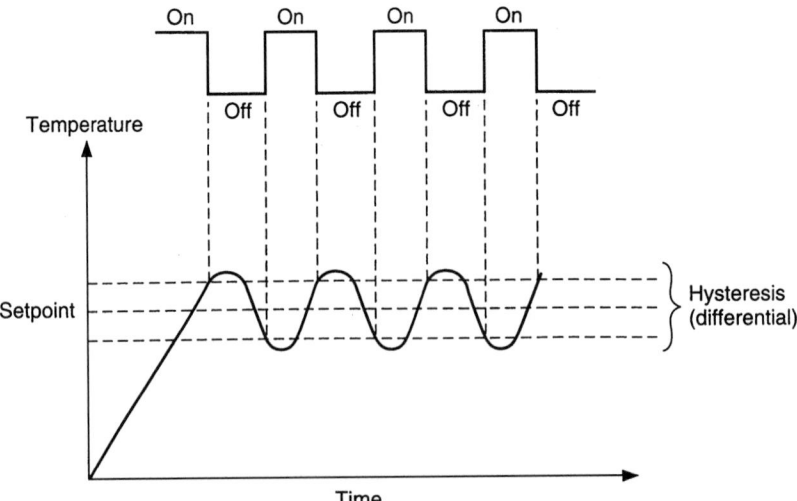

FIGURE 18 Time-temperature diagram for on-off controller with hysteresis. Note how the output changes state as the temperature crosses the hysteresis limits. The magnitude, period, and shape of the temperature curve are largely process-dependent. (*West Instruments.*)

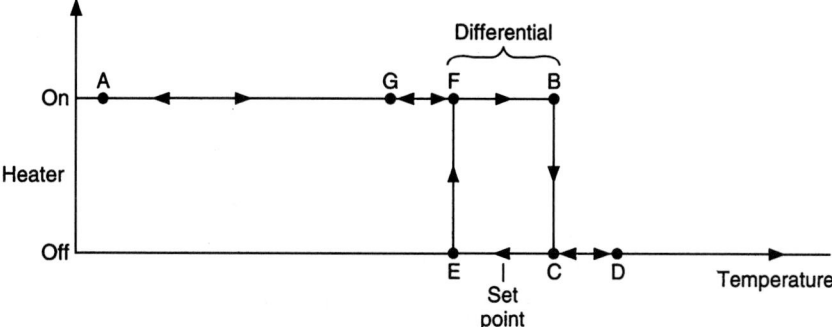

FIGURE 19 Another representation of the hysteresis curve—transfer function of on-off controller with hysteresis. Assuming that the process temperature is well below the set point at start-up, the system will be at *A*, the heat will be on. The heat will remain on as the temperature goes from *A* through *F* to *B*, the output turns off, dropping to point *C*. The temperature may continue to rise slightly to point *D* before decreasing to point *E*. At *E* the output once again turns on. The temperature may continue to drop slightly to point *G* before rising to *B* and repeating the cycle. (*West Instruments.*)

heat requirements to maintain a stable temperature. The range of temperature over which power is adjusted from 0 to 100% is called the proportional band. This band is usually expressed as a percentage of the instrument span and is centered about the set point. Thus in a controller with a 1000°C span, a 5% proportional band would be 50°C wide and extend 25°C below the set point to 25°C above the set point. A graphic illustration of the transfer function for a reverse-acting controller is given in Fig. 20.

The proportional band in general-purpose controllers is usually adjustable to obtain stable control under differing process conditions. The transfer curve of a wide-band proportional controller is shown in Fig. 21. Under these conditions a large change in temperature is required to produce a small change in output. The transfer curve of a narrow-band proportional controller is shown in Fig. 22. Here a

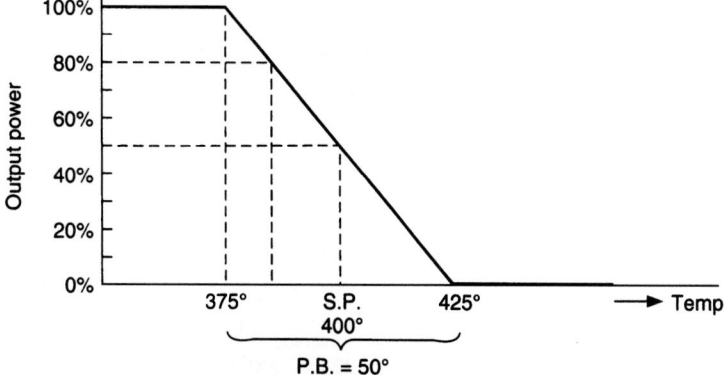

FIGURE 20 Transfer curve of reverse-acting controller. The unit is termed reverse-acting because the output decreases with increasing temperature. In this example, below 375°C, the lower edge of the proportional band, the output power is on 100%. Above 425°C the output power is off. Between these band edges the output power for any process temperature can be found by drawing a line vertically from the temperature axis until it intersects the transfer curve, then horizontally to the power axis. Note that 50% power occurs when the temperature is at the set point. The width of the proportional band changes the relationship between temperature deviation from set point and power output. (*West Instruments.*)

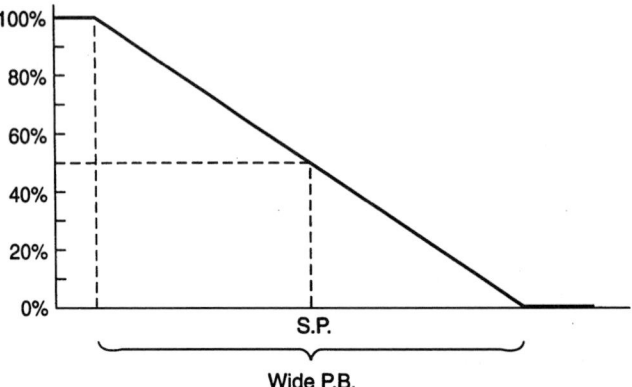

FIGURE 21   Transfer function for wide-band proportional controller. (*West Instruments.*)

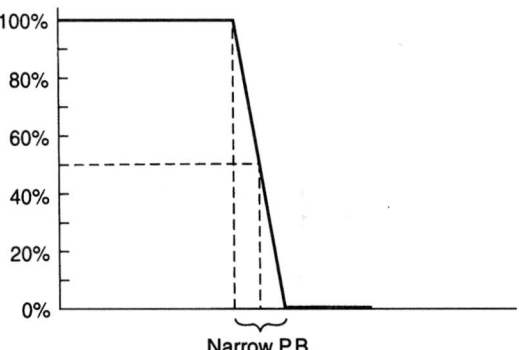

FIGURE 22   Transfer function for narrow-band proportional controller. (*West Instruments.*)

small change in temperature produces a large change in output. If the proportional band were reduced to zero, the result would be an on-off controller.

In industrial applications the proportional band is expressed as a percent of span, but it may also be expressed as controller gain in others. Proportional band and controller gain are related inversely by the equation

$$\text{Gain} = \frac{100\%}{\text{proportional band } (\%)}$$

Thus narrowing the proportional band increases the gain. For example, for a gain of 20 the proportional band is 5%. The block diagram of a proportional controller is given in Fig. 23. The temperature signal from the sensor is amplified and may be used to drive a full-scale indicator, either an analog meter or a digital display. If the sensor is a thermocouple, cold junction compensation circuitry is incorporated in the amplifier. The difference between the process measurement signal and the set point is taken in a summing circuit to produce the error or deviation signal. This signal is positive when the process is below the set point, zero when the process is at the set point, and negative when the process is above the set point. The error signal is applied to the proportioning circuit through a potentiometer gain control.

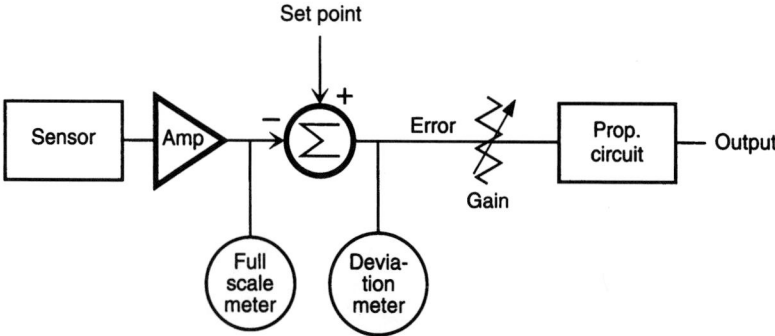

FIGURE 23   Block diagram of proportional controller. (*West Instruments.*)

The proportional output is 50% when the error signal is zero, that is, the process is at the set point.

Offset

It is rare in any process that the heat input to maintain the set-point temperature will be 50% of the maximum available. Therefore the temperature will increase or decrease from the set point, varying the output power until an equilibrium condition exists. The temperature difference between the stabilized temperature and the set point is called offset. Since the stabilized temperature must always be within the proportional band if the process is under control, the amount of offset can be reduced by narrowing the proportional band. However, the proportional band can be narrowed only so far before instability occurs. An illustration of a process coming up to temperature with an offset is shown in Fig. 24. The mechanism by which offset occurs with a proportional controller can be illustrated by superimposing the temperature controller transfer curve on the process transfer curve, as shown in Fig. 25.

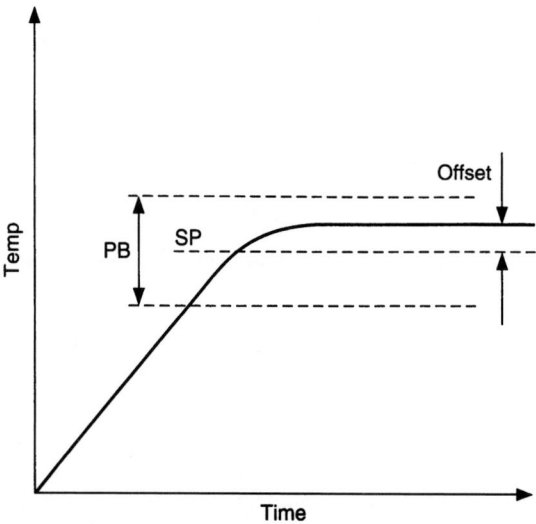

FIGURE 24   Process of coming up to temperature with an off-set (*West Instruments.*)

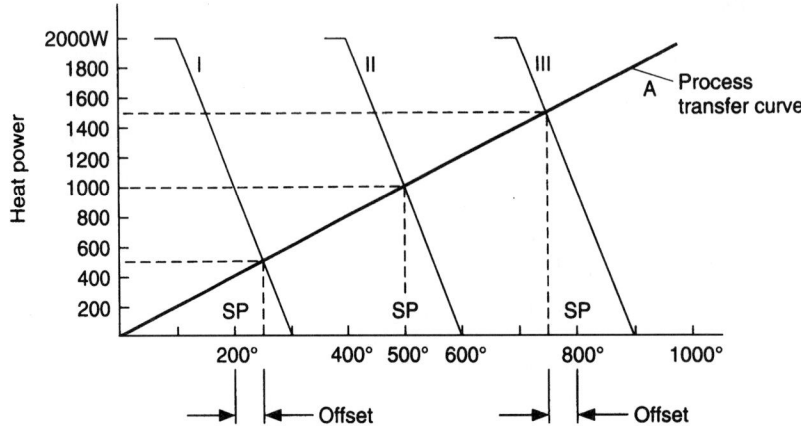

FIGURE 25 Mechanism by which offset occurs with a proportional controller. Assume that a process is heated with a 2000-watt heater. The relationship between heat input and process temperature, shown by curve A, is assumed to be linear for illustrative purposes. The transfer function for a controller with a 200°C proportional band is shown for three different set points in curves I, II, and III. Curve I with a set point of 200°C intersects the process curve at a power level of 500 watts, which corresponds to a process temperature of 250°C. The offset under these conditions is 250 to 200°C, or 50°C high. Curve II with a set point of 500°C intersects the process curve at 1000 watts, which corresponds to a process temperature of 500°C. There is no offset case since the temperature corresponds to the 50% power point. Curve III with a set point of 800°C intersects the process curve at 1500 watts, which corresponds to a temperature of 750°C. The off-set under these conditions is 750 to 800°C, or 50°C low. These examples show that the offset is dependent on the process transfer function, the proportional band (gain), and the set point. (*West Instruments.*)

Manual and Automatic Reset

Offset can be removed either manually or automatically. In analog instrumentation, manual reset uses a potentiometer to offset the proportional band electrically. The amount of proportional band shifting must be done by the operator in small increments over a period of time until the controller power output just matches the process heat demand at the set-point temperature (Fig. 26). A controller with manual reset is shown in the block diagram of Fig. 27.

Automatic reset uses an electronic integrator to perform the reset function. The deviation (error) signal is integrated with respect to time and the integral is summed with the deviation signal to move the proportional band. The output power is thus automatically increased or decreased to bring the process temperature back to the set point. The integrator keeps changing the output power, and thus the process temperature, until the deviation is zero. When the deviation is zero, the input to the integrator is zero and its output stops changing. The integrator has now stored the proper value of reset to hold the process at the set point. Once this condition is achieved, the correct amount of reset value is held by the integrator. Should process heat requirements change, there would once again be a deviation, which the integrator would integrate and apply corrective action to the output. The integral term of the controller acts continuously in an attempt to make the deviation zero. This corrective action has to be applied rather slowly, more slowly than the speed of response of the load. Otherwise oscillations will occur.

Automatic Reset—Proportional plus Integral Controllers

Automatic reset action is expressed as the integral time constant. Precisely defined, the reset time constant is the time interval in which the part of the output signal due to the integral action increases

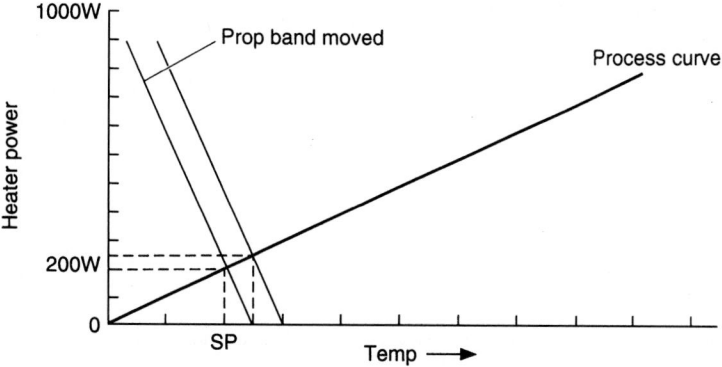

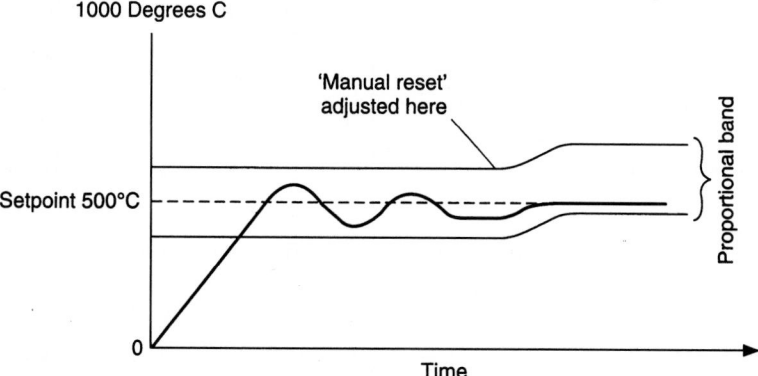

FIGURE 26 Manual reset of proportional controller. (*West Instruments.*)

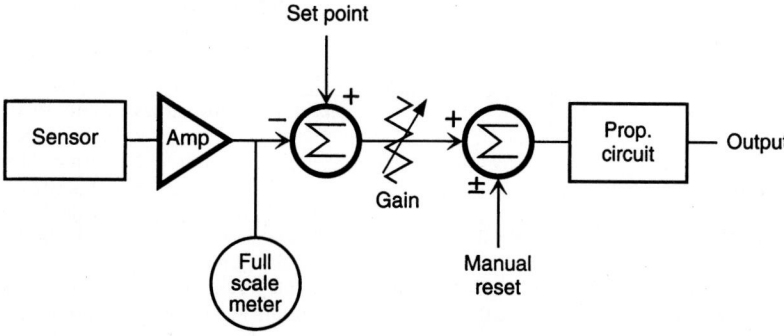

FIGURE 27 Block diagram of proportional controller with manual reset. (*West Instruments.*)

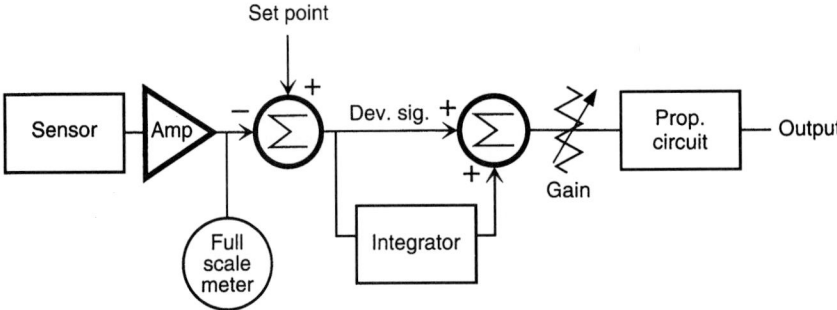

FIGURE 28 Block diagram of proportional plus integral controller. (*West Instruments.*)

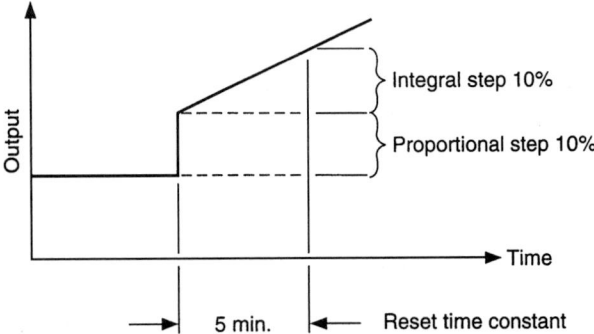

FIGURE 29 Reset time definition. (*West Instruments.*)

by an amount equal to the part of the output signal due to the proportional action, when the deviation is unchanging. A controller with automatic reset is shown in the block diagram of Fig. 28.

If a step change is made in the set point, the output will immediately increase, as shown in Fig. 29. This causes a deviation error, which is integrated and thus produces an increasing change in controller output. The time required for the output to increase by another 10% is the reset time—5 minutes in the example of Fig. 29.

Automatic reset action also may be expressed in repeats per minute and is related to the time constant by the inverse relationship

$$\text{Repeats per minute} = \frac{1}{\text{integral time constant (minutes)}}$$

Integral Saturation

A phenomenon called integral saturation is associated with automatic reset. Integral saturation refers to the case where the integrator has acted on the error signal when the temperature is outside the proportional band. The resulting large output of the integrator causes the proportional band to move so far that the set point is outside the band. The temperature must pass the set point before the controller output will change. As the temperature crosses the set point, the deviation signal polarity changes and the integrator output starts to decrease or desaturate. The result is a large temperature overshoot. This can be prevented by stopping the integrator from acting if the temperature is outside the proportional band. This function is called integral lockout or integral desaturation.

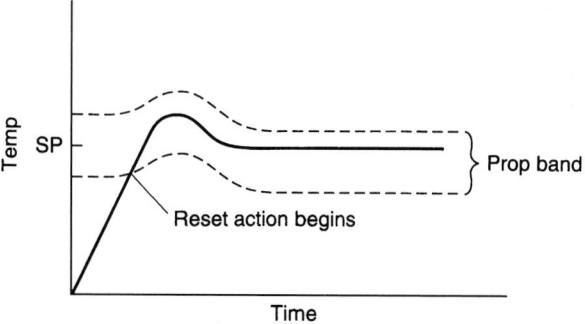

FIGURE 30   Proportional plus integral action. (*West Instruments.*)

One characteristic of all proportional plus integral controllers is that the temperature often over-shoots the set point on start-up. This occurs because the integrator begins acting when the temperature reaches the lower edge of the proportional band. As the temperature approaches the set point, the reset action already has moved the proportional band higher, causing excess heat output. As the temperature exceeds the set point, the sign of the deviation signal reverses and the integrator brings the proportional band back to the position required to eliminate the offset (Fig. 30).

Derivative Action (Rate Action)

The derivative function in a proportional plus derivative controller provides the controller with the ability to shift the proportional band either up or down to compensate for rapidly changing temperature. The amount of shift is proportional to the rate of temperature change. In modern instruments this is accomplished electronically by taking the derivative of the temperature signal and summing it with the deviation signal (Fig. 31(*a*)). [Some controllers take the derivative of the deviation signal, which has the side effect of producing upsets whenever the set point is changed (Fig. 31(*b*)).]

The amount of shift is also proportional to the derivative time constant. The derivative time constant may be defined as the time interval in which the part of the output signal due to proportional action increases by an amount equal to that part of the output signal due to derivative action when the deviation is changing at a constant rate (Fig. 32).

Derivative action functions to increase controller gain during temperature changes. This compensates for some of the lag in a process and allows the use of a narrower proportional band with its lesser offset. The derivative action can occur at any temperature, even outside the proportional band, and is not limited as is the integral action. Derivative action also can help to reduce overshoot on start-up.

Proportional plus Integral plus Derivative Controllers

A three-mode controller combines the proportional, integral, and derivative actions and is usually required to control difficult processes. The block diagram of a three-mode controller is given in Fig. 33. This system has a major advantage. In a properly tuned controller, the temperature will approach the set point smoothly without overshoot because the derivative plus deviation signal in the integrator input will be just sufficient for the integrator to store the required integral value by the time the temperature reaches the set point.

Time- and Current-Proportioning Controllers

In these controllers the controller proportional output may take one of several forms. The more common forms are time-proportioning and current-proportioning. In a time-proportioning output,

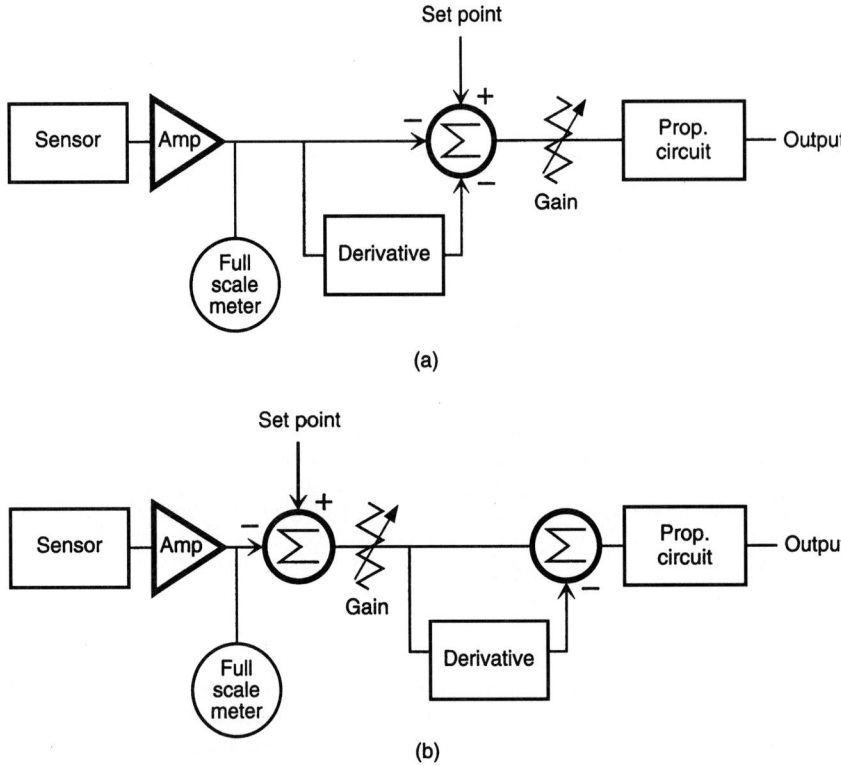

FIGURE 31   Block diagram or proportional plus rate controller. (*a*) The derivative of the sensor (temperature)
signal is taken and summed with the deviation signal. (*b*) The derivative of the deviation signal is taken. (*West Instruments.*)

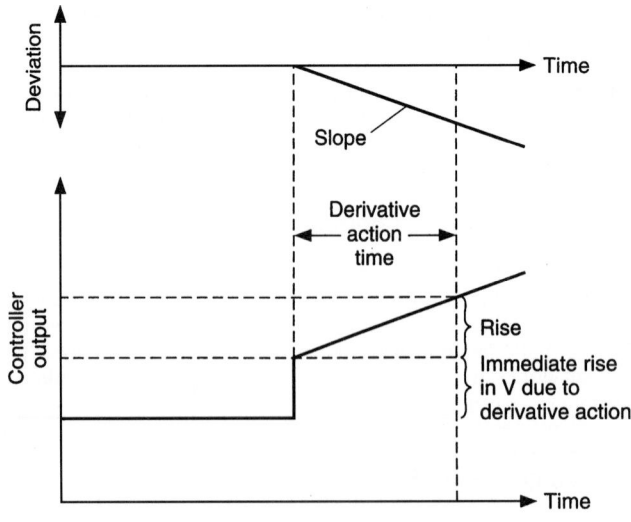

FIGURE 32   Derivative time definition. (*West Instruments.*)

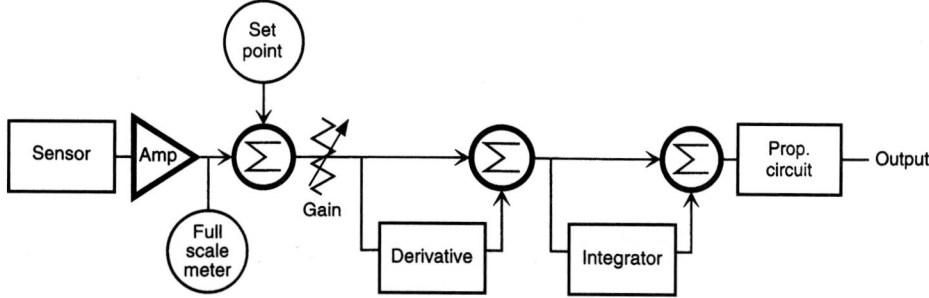

FIGURE 33 Proportional plus integral plus derivative controller. (*West Instruments.*)

power is applied to the load for a percentage of a fixed cycle time. Figure 34 shows the controller output at a 75% output level for a cycle time of 12 seconds.

This type of output is common with contractors and solid-state devices. An advantage of solid-state devices is that the cycle time may be reduced to 1 second or less. If the cycle time is reduced to one-half the line period (10 ms for 50 Hz), then the proportioning action is sometimes referred to as a stepless control, or phase-angle control. A phase-angle-fired output is shown in Fig. 35.

The current output, commonly 4 to 20 mA, is used to control a solid-state power device, a motor-operated valve positioner, a motor-operated damper, or a saturable core reactor. The relationship between controller current output and heat output is shown in Fig. 36.

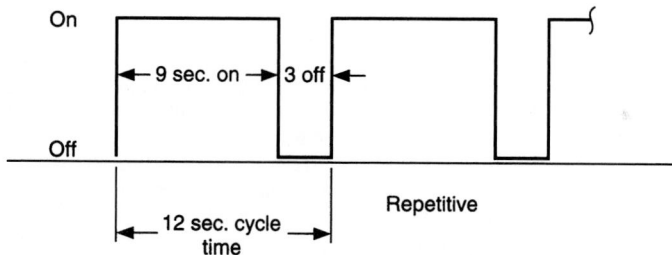

FIGURE 34 Time-proportioning controller at 75% level. (*West Instruments.*)

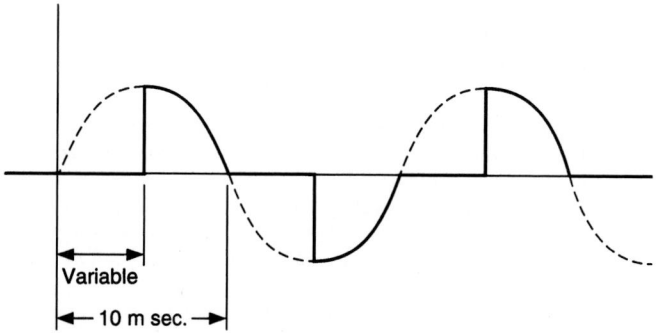

FIGURE 35 Phase-angle-fired stepless control ouput. (*West Instruments.*)

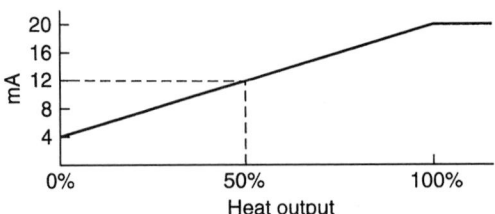

FIGURE 36 Current-proportioning controller. (*West Instruments.*)

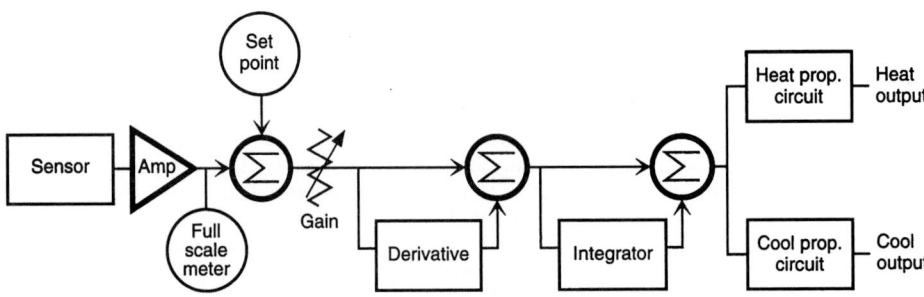

FIGURE 37 Heat-cool PID controller. (*West Instruments.*)

Heat-Cool PID Control

Certain applications that are partially exothermic demand the application of cooling as well as heating. To achieve this, the controller output is organized as shown in Fig. 37. The controller has two proportional outputs, one for heating and one for cooling.

The transfer function for this type of controller is shown in Fig. 38. Below the proportional band, full heating is applied; above the proportional band, full cooling is applied. Within the proportional band (X_{p1}) there is a linear reduction of heating to zero, followed by a linear increase in cooling with increasing temperature. Heating and cooling can be overlapped (X_{sh}) to ensure a smooth transition between heating and cooling. In addition, to optimize the gain between heating and cooling action, the cooling gain is made variable (X_{p2}).

PROCESS CONTROL CHARACTERISTICS AND CONTROLLER SELECTION

The selection of the most appropriate controller for a given application depends on several factors, as described in the introduction to this article. The process control characteristics are very important criteria and are given further attention here. Experience shows that for easier controller tuning and lowest initial cost, the simplest controller that will meet requirements is usually the best choice. In selecting a controller, the user should consider priorities. In some cases precise adherence to the control point is paramount. In other cases maintaining the temperature within a comparatively wide range is adequate.

In some difficult cases the required response cannot be obtained even with a sophisticated controller. This type of situation indicates that there is an inherent process thermal design problem. Thermal design should be analyzed and corrected before proceeding with controller selection. A good thermal

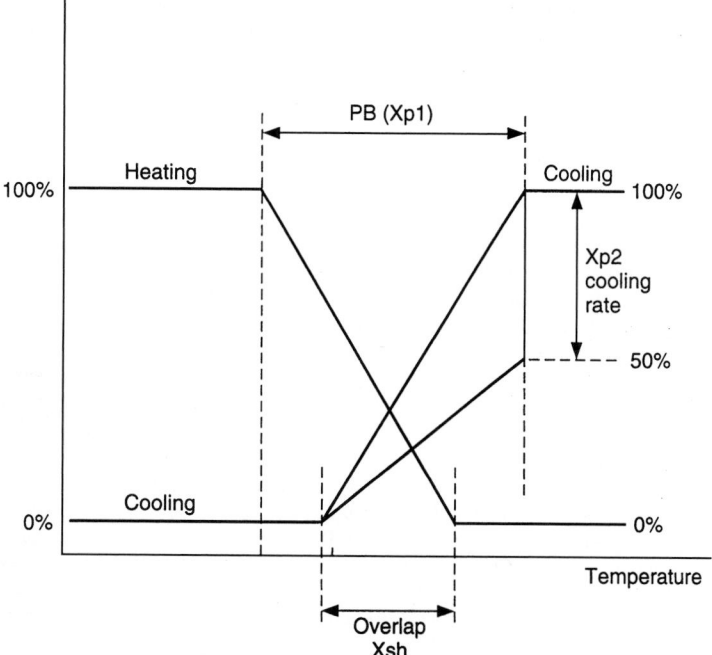

FIGURE 38 Transfer function for heat-cool PID controller. The controller has two proportional outputs, one for heating and one for cooling. (*West Instruments.*)

design will provide more stable control and allow the use of a less complicated and usually less expensive controller.

Controller Selection

Selection of the controller type may be approached from several directions:

1. Process reaction curve

2. Physical thermal system analysis

3. Previous experience

4. Experimental testing

The process reaction curve may be generated and observed to classify the process as easy or difficult to control, single capacity, or multicapacity. This knowledge should be compared with the process temperature stability requirements to indicate which type of controller to use.

The process controllability may be estimated by observing and analyzing the process thermal system. What is the relative heater power to load heat requirements? Are the heaters oversized or undersized? Oversized heaters lead to control stability problems. Undersized heaters produce slow response. Is the thermal mass large or small? What are the distance and the thermal resistance from the heaters to the sensor? Large distances and resistances cause lag and a less stable system. Comparing the controllability with the process temperature stability requirements will indicate which type of controller to use. This same system of analysis can be applied to process variables other than temperature.

Prior experience often is an important guideline, but because of process design changes, a new situation may require tighter or less stringent control. A method often used is to try a simple controller, such as proportional plus manual reset, and to note the results compared with the desired system response. This will suggest additional features or features that may be deleted.

Single-Capacity Processes

If the process reaction curve or system examination reveals that the process can be classified as single-capacity, it may be controlled by an on-off controller. However, two conditions must be met: (1) a cyclical peak-to-peak temperature variation equal to the controller hysteresis is acceptable, and (2) the process heating and cooling rates are long enough to prevent too rapid cycling of the final control devices. Controller hysteresis also has an effect on the period of temperature cycling. Wider hysteresis causes a longer period and greater temperature variation. A narrow hysteresis may be used with final control devices, such as solid-state relays, triacs, and SCRs, which can cycle rapidly without shortening their life. Typical system responses for oversized and undersized heater capacity are shown in Figs. 39 and 40, respectively.

If the previously mentioned two conditions are not acceptable, then the use of a proportional controller is indicated. A proportional controller would eliminate the temperature cycling. In a controller with adjustable proportional band, the band usually may be adjusted quite narrow and still maintain stability so that offset will not be a problem. If the controller has a fixed proportional band, at a value much larger than optimum, the resulting offset may be undesirable. Manual reset may be added to reduce the offset. A narrow proportional band will make the offset variations minimal with changes in process heat requirements so that automatic reset usually will not be required.

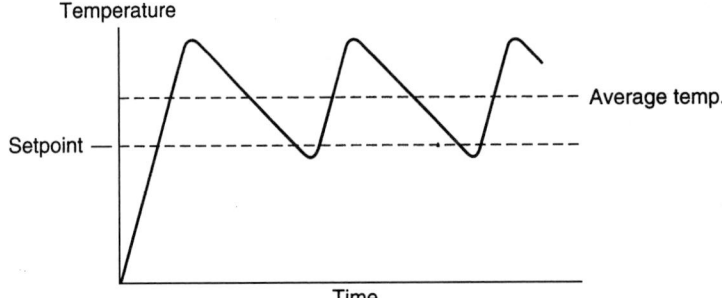

FIGURE 39   Typical system response for oversized heater condition. (*West Instruments.*)

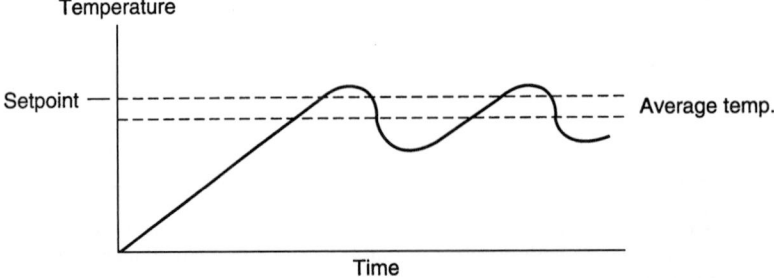

FIGURE 40   Typical system response for undersized heater condition. (*West Instruments.*)

A single-capacity process usually will not require derivative action. However, control action during process upsets may be improved by the addition of some derivative action. Adding too much derivative action (too long a derivative time constant) can cause instability with some controllers.

Multicapacity Processes

A multicapacity process or a single-capacity process with transport delay is generally not suited to on-off control because of the wide temperature cycling. These processes require proportional control. Depending on the process difficulty, as evidenced by the process reaction curve and the control precision requirements, a proportional controller or one with the addition of derivative and integral action will be required.

Proportional controllers must be "tuned" to the process for good temperature response. The question is—what is good response? Three possible temperature responses under cold start-up conditions are shown in Fig. 41.

Control systems usually are tuned under operating conditions rather than for start-up conditions. Tuning a controller requires first that the process temperature be stable near the operating point with the system in operation. Then a known process disturbance is caused and the resulting temperature response observed. The response is best observed on a recorder. The proper disturbance for tuning is one which is likely to occur during actual operation, such as product flow change or speed change. However, this may be impractical and thus a small set-point change usually is used as the disturbance. The optimum tuning for set-point changes may not produce optimum response for various process disturbances.

Process characteristics for a proportional-only controller are given in Fig. 42. The curves show the resulting temperature change after decreased process heat demand. Similar curves would result if the set point were decreased several degrees.

Processes with long time lags and large maximum rates of rise, such as a heat exchanger, require a wide proportional band to eliminate oscillation. A wide band means that large offsets can occur with changes in load. These offsets can be eliminated by the automatic reset function in a proportional plus integral controller. The system response curve will be similar to those shown in Fig. 43 for a decrease in heat demand.

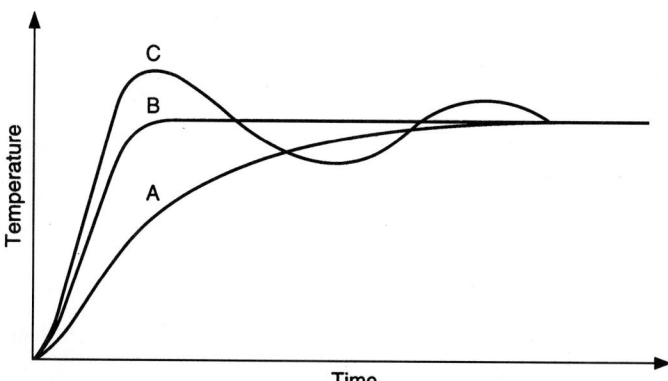

FIGURE 41   Three possible temperature responses of proportional controllers under cold start-up conditions. Curve *A* could be considered good response if a slow controlled heat-up is required. Curve *B* would be considered good response if the fastest heat-up without overshoot is required. Curve *C* could be good response if the fastest heat-up is required. The definition of "good response" varies with the process and operational requirements. (*West Instruments.*)

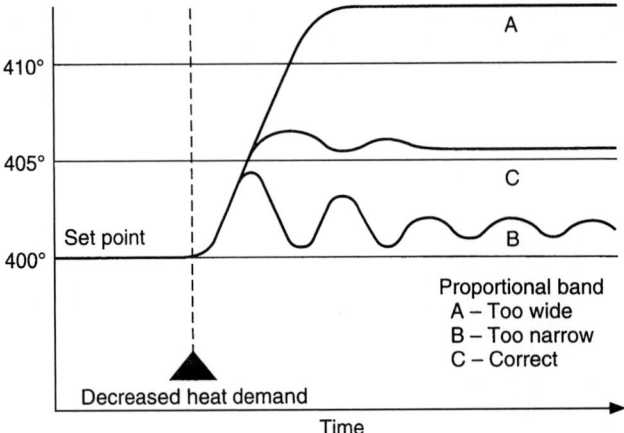

FIGURE 42 Process characteristics for a proportional-only controller. Curve *A* results when the proportional band is too wide. Note the large offset. The offset can be reduced by narrowing the proportional band. Instability results if the proportional band is too narrow, as shown by curve *B*. Optimum control, as shown by curve *C*, is achieved at a proportional band setting slightly wider than that which causes oscillation. If process parameters change with time or if operating conditions change, it will be necessary to retune the controller or avoid this by using a proportional band wider than optimum to prevent future instability. (*West Instruments.*)

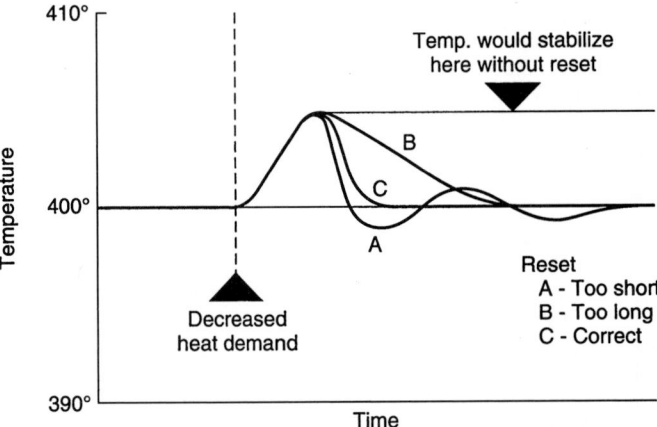

FIGURE 43 System response curves for proportional plus integral controller for an application such as a heat exchanger when there is a decrease in heat demand. An integral time constant which is too long for the process will take a long time to return the temperature to set point, as shown in curve *B*. An integral time constant that is too short will allow integral to outrun the process, causing the temperature to cross the set point with damped oscillation, as shown by curve *A*. If the integral time is much too long, continuous oscillation results. The integral time constant usually considered optimum is that which returns the temperature to set point as rapidly as possible without overshooting it, as shown by curve *C*. However, a damped oscillation (curve *A*) may be more desirable if the temperature must return to set point faster and some overshoot can be allowed. (*West Instruments.*)

Derivative (rate) action may be used to advantage on processes with long time delays, speeding recovery after a process disturbance. The derivative provides a phase lead function, which cancels some of the process lag and allows the use of a narrower proportional band without creating instability. A narrower proportional band results in less offset. The response of a proportional plus derivative controller in a system is dependent not only on the proportional band and the derivative time constant, but also on the method used to obtain the derivative signal. The curves of Fig. 44 show some typical results of the proportional plus derivative control algorithm as used in the West MC30 and previous controllers as well as most other brands of proportional plus derivative controllers. Figure 45 shows the response to a decrease in heat demand for the controllers described previously (Fig. 44).

As noted from the aforementioned illustrations, the problem of superimposed damping or continuous oscillation remains. This condition may be corrected by either decreasing the derivative time constant or widening the proportional band. The oscillations result from a loop gain that is too great at the frequency of oscillation. The total loop gain is the process gain times the proportional gain times the derivative gain. Decreasing any one of these gains will decrease the total loop gain and return stability.

The proportional plus derivative controller may be used to advantage on discontinuous processes such as batching operations involving periodic shutdown, emptying, and refiling. Here the proportional plus integral controller would not perform well because of the long time lags and intermittent operation. Derivative action also reduces the amount of overshoot on start-up of a batch operation.

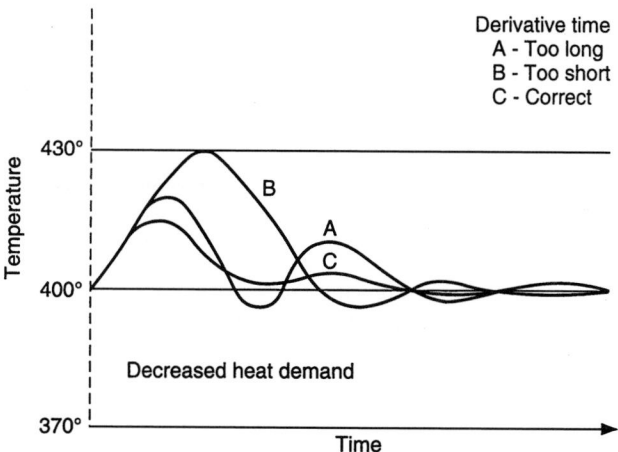

FIGURE 44 Response curves for controller using proportional plus derivative algorithm. A derivative time constant that is too long causes the temperature to change too rapidly and overshoot the set point with damped oscillation (curve *A*). A derivative time constant that is too short allows the temperature to remain away from the set point too long (curve *B*). The optimum derivative time returns the temperature to set point with a minimum of ringing (curve *C*). The damped oscillation about the final value in curve *A* can be due to excessive derivative gain at frequencies above the useful control range. Some controllers have an active derivative circuit which decreases the gain above the useful frequency range of the system and provides full phase lead in the useful range. The results of too short a derivative time constant remain the same as shown, not enough compensation for process lags. However, this method improves response at the optimun derivative time constant. It also produces two more possible responses if the time constant is longer then optimum. (*West Instruments.*)

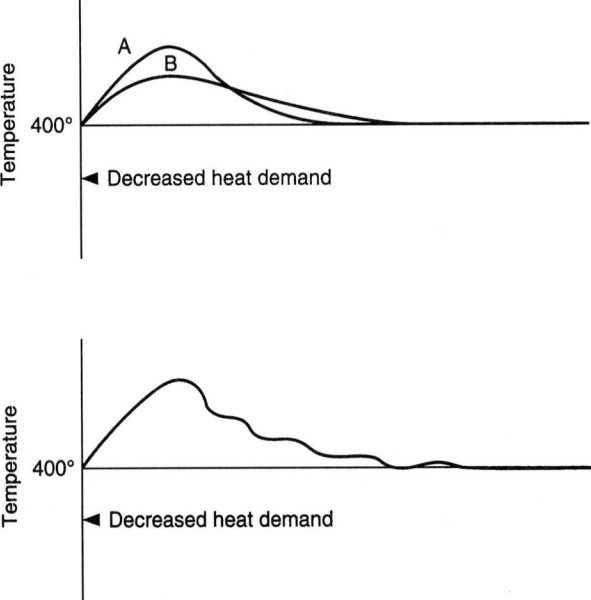

FIGURE 45 Response to a decrease in heat demand for the controller of
Fig. 44. Curve *A* (top diagram) is for the optimum derivative time constant.
The temperature returns smoothly to set point. Curve *B* shows one possibility
for a derivative time constant which is too long. The temperature deviation is
less, but the temperature returns to set point on the derivative time constant
curve. The curve in the bottom diagram shows another possibility for a derivative time constant which is too long. The temperature returns to set point on the
derivative time constant curve, but has either damped or continuous oscillation
superimposed. (*West Instruments.*)

   The most difficult processes to control, those with long time lags and large maximum rates of rise,
require three-mode or proportional plus integral plus derivative (PID) controllers. The fully adjustable
PID controller can be adjusted to produce a wide variety of system temperature responses from very
underdamped through critically damped to very overdamped.
   The tuning of a PID control system will depend on the response required and also on the process
disturbance to which it applies. Set-point changes will produce a different response from process
disturbances. The type of process disturbance will vary the type of response. For example, a product
flow rate change may produce an underdamped response while a change in power line voltage may
produce an overdamped response.
   Some of the response criteria include rise time, time to first peak, percent overshoot, settling time,
decay ratio, damping factor, integral of square error (ISE), integral of absolute error (IAE), and integral
of time and absolute error (ITAE). Figure 46 illustrates some of these criteria with the response to an
increase in set point.
   In many process applications a controller tuning that produces a decay ratio of ¼ is considered
good control. However, this tuning is not robust enough. Also the tuning parameters that produce
a decay ratio of ¼ are not unique and neither are the responses, as illustrated in Fig. 47. In some
applications the deviation from set point and the time away from set point are very important. This
leads to imposing one of the integral criteria, as shown in Fig. 48.

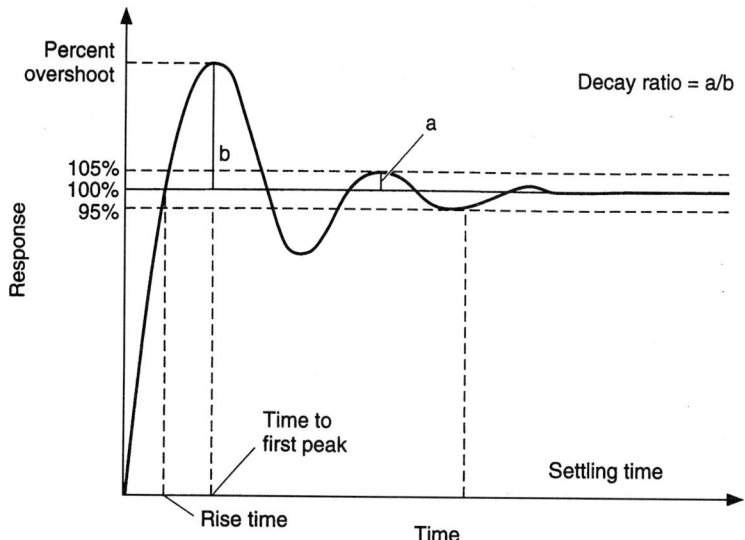

FIGURE 46 Tuning of PID control system depends on response required as well as on process disturbance to which it applies. (*West Instruments.*)

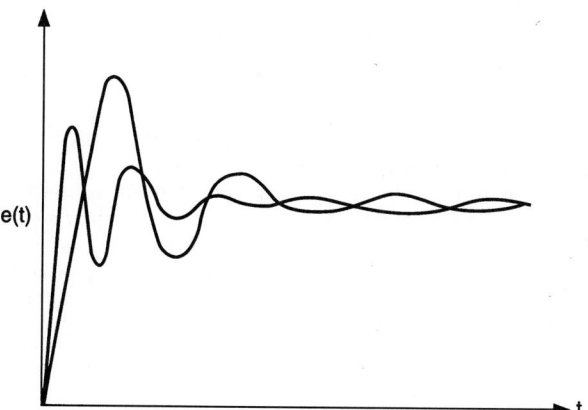

FIGURE 47 Nonunique nature of quarter decay ratio. (*West Instruments.*)

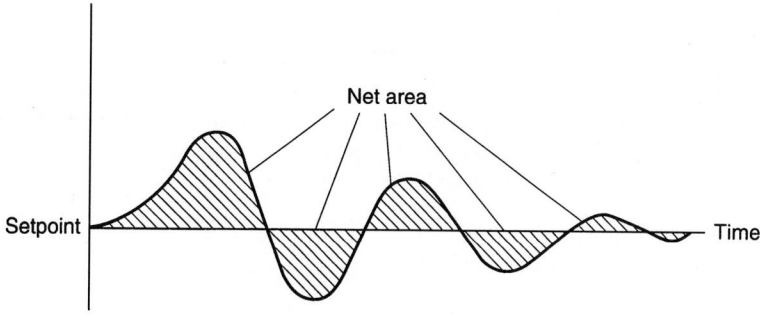

FIGURE 48 Applications of integral criteria. (*West Instruments.*)

REFERENCE

1. Astrom, K., and T. Hagglund, *PID Controllers: Theory, Design, and Tuning*, 2nd ed., Instrument Society of America, Research Triangle Park, North Carolina, 1995.

TECHNIQUES FOR PROCESS CONTROL

by Peter D. Hansen*

This article presents a number of techniques useful in the analysis and design of modern process control systems. Particular emphasis is given to transfer-function and adaptive methods, which lead to designs that cope with process delay (dead time), loop interaction, nonlinearity, and unmeasured disturbances.

The mathematical approaches described here can be used (1) by manufacturers of industrial controllers to achieve an improved and more versatile design, (2) by control engineers seeking a solution to difficult control applications, and (3) by students and researchers to achieve a more thorough understanding of control system dynamics.

An effort has been made to use consistent notation throughout this article. Uppercase letters represent transfer operators, functions of the differential operator s or the backward shift operator z^{-1}. Matrices are in boldface type. Scalar or vector variables and parameters are represented by lowercase letters. Key equations referenced throughout are the process equation [Eq. (5)], the target performance equation [Eq. (8)], the design equation [Eq. (10)], the open-loop controller equation [Eq. (11)], and the feedback controller equation [Eq. (13)].

DIGITAL CONTROL

A digital controller is generally considered to be superior to an analog controller. However, if it is used to emulate an analog controller, the digital device may be less effective because of phase (or delay) and resolution errors introduced by sampling and converting. The digital controller's advantage is its algorithmic flexibility and precision with respect to both calculations and logic, thereby facilitating on-line restructuring and parameter adaptation.

A digital control algorithm utilizes samples of its input signals which are discrete in both magnitude and time. Usually, continuous signals are sampled at a constant rate. Sampling the controlled variable introduces phase lag (effective delay) into the feedback loop because of

1. Low-pass filtering
2. Computation and transmission
3. Output holding between updates

---

* Bristol Fellow, Systems Development and Engineering, The Foxboro Company (a Siebe Company), Foxboro, Massachusetts.

Effective delay or parasitic lag, whether in the digital or the analog portion of a feedback loop, has an adverse effect on performance.

Loop delay, measurement noise, and output saturation determine the performance achievable with feedback control. Minimum integrated absolute error in response to an unmeasured load increases in proportion to the delay time for a dominant-delay process and in proportion to the square of delay for a dominant-lag process. Consequently the sampling-related delays should be made a small fraction of the total loop delay by using a small sampling interval.

State-Space Representation

Linear dynamic systems can be represented in terms of a state-variable vector x as a set of simultaneous first-order difference equations,

$$x\{t+h\} = \mathbf{M}x\{t\} + \mathbf{N}u\{t\}$$
$$y\{t\} = \mathbf{C}x\{t\} \tag{1}$$

where h is the computing interval, or as differential equations,

$$\frac{dx}{dt} = \mathbf{A}x + \mathbf{B}u$$
$$y = \mathbf{C}x \tag{2}$$

In these equations u is a vector of inputs, and y is a vector of measured variables. Matrices, but not vectors, are capitalized and in boldface type. $\mathbf{M}$ and $\mathbf{A}$ are square matrices and $\mathbf{N}$, $\mathbf{C}$, and $\mathbf{B}$ are rectangular (noninvertible). When a zero-order hold drives the continuous process, the representations are related at sampling instants by

$$\mathbf{M} = e^{\mathbf{A}h} = \sum_{n=0}^{\infty} \frac{(\mathbf{A}h)^n}{n!}$$

$$\mathbf{N} = (\mathbf{M} - \mathbf{I})\mathbf{A}^{-1}\mathbf{B} = \left[\sum_{n=0}^{\infty} \frac{(\mathbf{A}h)^n}{(n+1)!} \right] \mathbf{B}h \tag{3}$$

The inversion of $\mathbf{A}$ can be avoided by replacing $\mathbf{M}$ with its Taylor series, which converges (possibly slowly) for all $\mathbf{A}h$.

The state-space approach may be used to model multivariable systems whose characteristics are time-varying and whose controlled variables are not measured directly. However, the representation may be inefficient because the matrices are often sparse. The approach can be generalized to characterize nonlinear systems by considering the right-hand sides of Eqs. (1) or (2) to be vector functions of the state variables and inputs. However, a process with a time delay cannot be represented directly with a finite differential-equation form. The difference-equation form introduces an extra state variable for each time step of delay.

Methods of analyzing observability, controllability, and stability of state-space representations are discussed in many control texts [1]–[3], as are design methods for predictors and controllers. The state-space feedback-controller design procedures lead to inflexible global control structures, which are usually linear. All manipulated variables are used to control each controllable state variable, and all measured variables are used to calculate each observable state variable. Consequently an on-line redesign (adaptive) capability may be needed to retune for process nonlinearity and to restructure following either an override condition or a loss of a measurement or manipulator.

Transfer-Operator Representation

The state-space equations can be expressed in transfer-function form, using algebraic operators to represent forward shift z and differentiation s,

$$y = C(z\mathbf{I} - \mathbf{M})^{-1}\mathbf{N}u$$
$$y = C(s\mathbf{I} - \mathbf{A})^{-1}\mathbf{B}u \tag{4}$$

For a single-input, single-output time-invariant system, these equations can be expressed as

$$Ay = BDu + Ce \tag{5}$$

where y is the controlled or measured variable, u is the manipulated variable, and e is a load (or disturbance) variable. A, B, C, and D are polynomial functions of s or the backward shift $z^{-1} = e^{-hs}$. B contains stable zeros and is therefore cancelable. D may be noncancelable and has unity steady-state gain. A zero of a polynomial is a root, a value of its argument (s or z^{-1}) that causes the polynomial to be zero. Unstable (noncancelable, nonminimum phase) zeros are in the right half of the complex s plane or inside the unit circle in the complex z^{-1} plane. The delay operator, whose zeros are at the origin of the z^{-1} plane, is noncancelable and nonminimum phase. Its inverse, a time advance, is physically unrealizable.

For sinusoidal signals the differentiation operator becomes $s = j\omega$, and the backward shift becomes $z^{-1} = e^{-j\omega h}$. In steady state the radian frequency ω is zero, allowing s to be replaced with 0 and z^{-1} with 1 in the polynomial operators. The role of the C polynomial is played by an "observer" in state-space design. When e is dominated by measurement noise, e appears unfiltered at y; hence C is (almost) equal to A. When e is a load upset, e appears at y filtered by the process dynamics $1/A$; hence C is (nearly) 1. When e is considered an impulse, a process that would have a nonzero steady-state response to a steady-state e input has an additional zero at $s = 0$ or $z^{-1} = 1$ in its A and B polynomials.

An example of the conversion from the s domain to the z^{-1} domain is shown in [1]. A sampled lag $\{\tau_L\}$–delay $\{\tau_D\}$ process with gain $\{k\}$, whose input u is constant between sampling instants (because of a zero-order hold), is represented as

$$(1 - z^{-1}e^{-b})y = kz^{-n}[1 - e^{-a} + z^{-1}(e^{-a} - e^{-b})]u \tag{6}$$

where the delay is between n and $n + 1$ sampling intervals, $nh < \tau_D < (n + 1)h$, and

$$a = \frac{(n + 1)h - \tau_D}{\tau_L}$$

$$b = \frac{h}{\tau_L}$$

When $e^{-a} - e^{-b} < 1 - e^{-a}$, the first-order factor in parentheses on the right of Eq. (6) is cancelable and can be part of B. Otherwise it must be part of D.

When b is very small, because the sampling interval is very small, Eq. (6) becomes

$$[1 - z^{-1}(1 - b)]y = \left(\frac{k}{\tau_L}\right)z^{-n}[h(n + 1) - \tau_D + z^{-1}(\tau_D - nh)]u \tag{7}$$

Except for the b term on the left, this is indistinguishable from an integral $\{\tau_L/k\}$–delay$\{\tau_D\}$ process, signaling the likelihood of numerical difficulty in applications such as parameter (k and τ_L) identification.

Presuming that the desired behavior is a function of the measured variable y, the target closed-loop performance can be expressed as

$$Hy = Dr + Fe \tag{8}$$

where H is a (minimum-phase) polynomial with unity steady-state gain and stable zeros. H and F may be totally or partially specified polynomials. If e may have an arbitrary value in steady state, the steady-state value of F must be zero for y to converge to the set point (or reference input) r. Eliminating y from Eqs. (5) and (8) results in

$$ADr + AFe = HBDu + HCe \qquad (9)$$

Because D is not cancelable, this equation cannot be solved directly for u. However, the product HC may be separated into two parts, the term on the left-hand side AF and a remainder expressed as the product DG, so that D becomes a common factor of Eq. (9):

$$HC = AF + DG \qquad (10)$$

This equation is key to the controller design: selecting some and solving for other coefficients of H, F, and G when those of A, B, C, and D are known or estimated.

OPEN-LOOP CONTROL

The open-loop controller design results when Eq. (10) is used to eliminate AFe from Eq. (9), since D is not zero:

$$u = \frac{Ar - Ge}{BH} \qquad (11)$$

Of course, if e is unmeasured, open-loop control will not reduce e's effect on y. This causes F to be an infinite-degree polynomial HC/A and G to be zero. If e is measured, G is a feedforward operator. Substituting u from Eq. (11) back into the process equation, Eq. (5), and not canceling terms common to both numerator and denominator, results in the open-loop performance equation

$$y = \frac{BA(Dr + Fe)}{HBA} \qquad (12)$$

To avoid (imperfect) canceling of unstable roots, A as well as H and B must contain only stable zeros.

High-performance ($H \approx 1$) open-loop control applies the inverse of the process characteristic A/B to a set-point change. Because a dominant-lag process has low gain at high frequencies, its controller has high gain there. A rapid set-point change is likely to saturate the manipulated variable, but otherwise leaves its trajectory unchanged. The early return of this variable from its limit causes slower than optimal controlled variable response. This can be avoided by using nonlinear optimization (such as quadratic programming suggested in [4]) to compute the optimal controller-output trajectory, taking into account output limits, load level, and other process equality and inequality constraints.

The performance of an open-loop controller may be degraded by an unmeasured load or by mismatch between the process and the inverse controller at low frequencies. Mismatch at high frequency will not cause significant difficulty, however.

FEEDBACK CONTROL

Combining Eqs. (10) and (11) with the target equation, Eq. (8), to eliminate e results in the closed-loop control law

$$u = \frac{Cr - Gy}{BF} \qquad (13)$$

This equation also results when e is eliminated from the process and target equations, Eqs. (5) and (8), and D is made a common factor with the design equation, Eq. (10). From Eq. (13) it is clear that the disturbance polynomial C and its Eq. (10) decomposition terms F and G play a key role in the feedback-controller design. Various methods for determining these polynomials will be discussed. Except in the special case where $C = G$, the control output u does not depend exclusively on the control error $r - y$. However, the controller will provide integral action, eliminating steady-state error, if the steady-state values of G and C are equal and not zero, and either those of $e\{t\}$, A, and B are zero or that of F is zero.

Substituting u from Eq. (13) back into the process equation and not canceling terms common to both numerator and denominator results in the closed-loop performance equation

$$y = BC \frac{Dr + Fe}{HBC} \tag{14}$$

To avoid (imperfect) canceling of unstable roots, C as well as H and B must contain only stable zeros. However, it is not necessary that all of the zeros of A be stable when the control loop is closed. Zeros of C not common to A correspond to unobservable modes of the disturbance variable e. Zeros of B or D not common to A correspond to uncontrollable modes of y.

When the manipulated variable saturates, it is necessary to stop (or modify) the controller integral action to prevent "windup." If the integration were allowed to continue, the prelimited controller output would continue to rise (wind up) above the limit value, requiring a comparable period after the control error reverses sign before the manipulated variable could recover from saturation. This would cause a significant (avoidable) controlled-variable overshoot of the set point. A controller of a dominant-lag process, designed for good unmeasured-load rejection, employs significant proportional (and derivative) feedback G. When responding to a large set-point change, this feedback keeps the output saturated and the controlled variable rate limited longer than would a linear open-loop controller, resulting in a faster response [5]. Halting the integral action, while the manipulated variable remains limited, prevents appreciable overshoot.

The performance of the feedback loop is most sensitive to the process behavior in the frequency range where the absolute loop gain is near 1. Performance at significantly lower frequencies is often quite insensitive to the process characteristics, load, and controller tuning.

Robustness

The ability of a feedback loop to maintain stability, when the process parameters differ from their nominal values, is indicated with robustness measures. Denoting the locus of combined shifts of the process gain by the factor b and the process delay by the factor d, which make the loop marginally stable, is a useful indicator of robustness, providing, in a more physical form, the information contained in gain and phase margins. The use of the two parameters b and d is based on the idea that the process behavior, in the frequency range critical for stability, can be approximated with an n-integral-delay two-parameter model. The integrals, whose number n may range from 0 to 1 plus the number of measurement derivatives used in the controller, contribute to the gain shift b. The phase, in excess of the fixed contribution of the integrals, and the shift in phase $d\omega$ can be considered to be contributed by "effective" delay.

At marginal stability the return difference (1 plus the open-loop gain) is zero:

$$1 + \frac{G}{FB} \frac{bBD^d}{A} = 1 + \frac{bGD^d}{AF} = 0 \tag{15}$$

Here it is assumed that D is effective delay, which may include small lags not included in A and a $(1 - \tau s)/(1 + \tau s)$ factor for each nonminimum-phase zero.

Figure 1 is a plot of d versus b, using logarithmic scales, for PID control of pure-delay, integral-delay, and double-integral-delay processes. The proportional band (PB) and the integral time (IT) were determined for minimum overshoot using the algebraic PID design method described in a later section. In [6] a single number, characterizing robustness, is derived from a robustness plot. This robustness index is -1 plus the antilog of the length of the half-diagonal of the diamond-shaped box centered at the nominal design point ($d = b = 1$) that touches the d versus b curve at its closest point. A value of 1 indicates that the product or ratio of d and b can be as large as 2 or as small as 0.5 without instability.

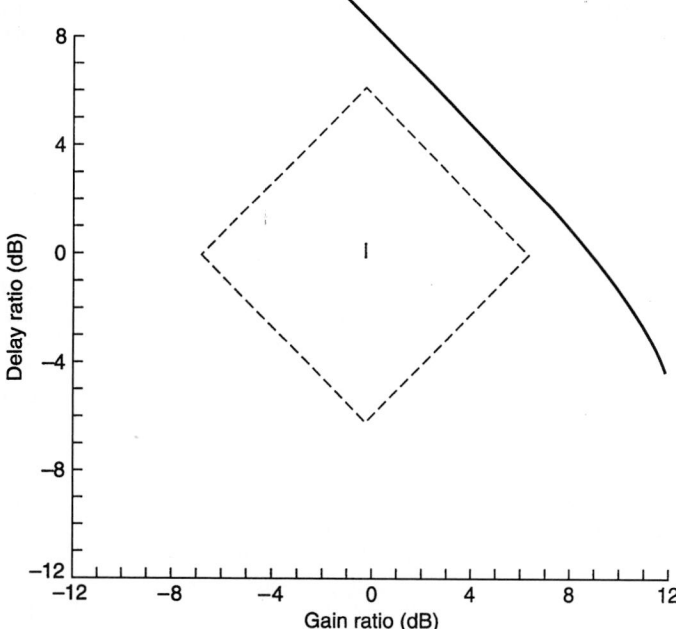

FIGURE 1(*a*)   Robustness plots. The diamond corresponds to a factor of 2 changes in the product or quotient of delay and gain from their nominal values. PI control of pure delay process. (*Foxboro.*)

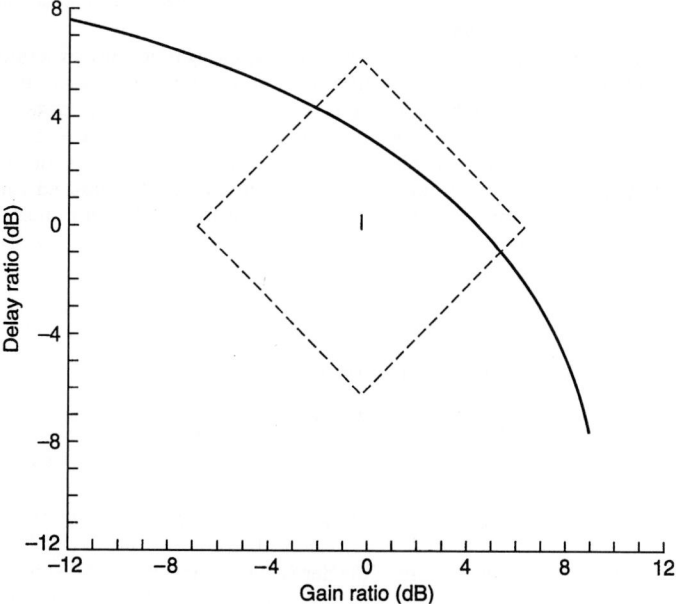

FIGURE 1(*b*)   PID control of integral-delay process. (*Foxboro.*)

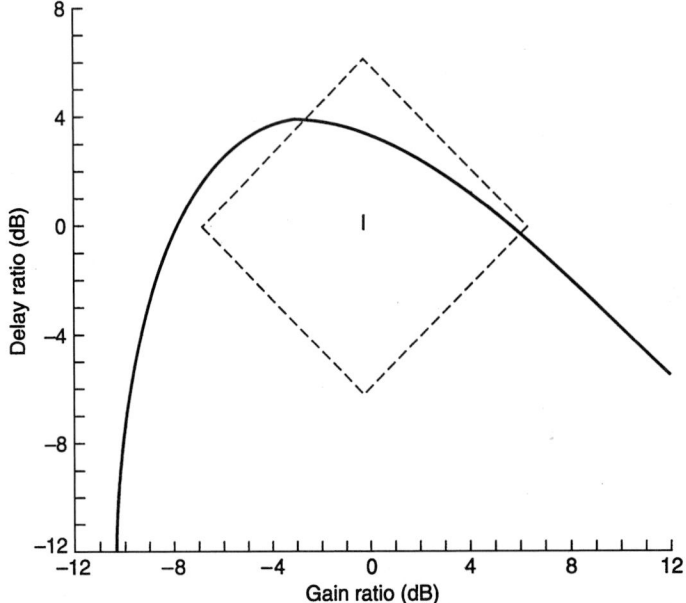

FIGURE 1(c)   PID control of double-integrator-delay process. (*Foxboro.*)

For the three cases of Fig. 1 the robustness index is 1.67, 0.47, and 0.30, each determined by sensitivity to delay shift. The diamond-shaped box in the figure would correspond to a robustness index of 1.

Most control schemes capable of providing high performance provide poor robustness (a robustness index near 0). Adaptive tuning may be required to keep a controller that is capable of high performance current with changing process conditions.

Digital simulation provides a useful means for exploring robustness experimentally. Nonlinearities can be included naturally in the time domain. It may not be necessary to use an exotic integration algorithm if the process can be modeled in real-factored form. The factored form can be much less sensitive to roundoff error than unfactored polynomial and state-space forms.

The simulation equations should be solved in causal sequence. Each equation's dependent variable should be updated based on the most current computed value of its independent variables (as is done in the Gauss-Seidel iterative solution of algebraic equations). A useful such model for a first-order factor, $y/x = 1/(1 + \tau s)$, is

$$y\{t\} = y\{t - h\} + \frac{h}{(\tau + h)}(x\{t\} - y\{t - h\}) \tag{16}$$

and for a damped second-order factor, $y/x = 1/(1 + Ts + T\tau s^2)$, is

$$v\{t\} = v\{t - h\} + \frac{h}{\tau + h}(x\{t\} - v\{t - h\} - y\{t - h\})$$

$$y\{t\} = y\{t - h\} + \frac{h}{T + h}v\{t\} \tag{17}$$

The internal variable v is a measure of the derivative of the output y,

$$\frac{dy}{dt} \approx \frac{v\{t\}}{T + h} \tag{18}$$

These models both give the correct result when $T = \tau = 0$: $y\{t\} = x\{t\}$. When the sampling interval h is very small compared with T and τ, it may be necessary to compute with double precision to avoid truncation, because the second term on the right of Eqs. (17) and (18) may become much smaller than the first before the true steady state is reached.

A fixed time delay may be modeled as an integer number of computing intervals, typically 20 to 40. At each time step an old data value is discarded and a new value added to a storage array. Incremented pointers can be used to keep track of the position of the delay input and output, as in a ring structure. This avoids shifting all the stored data each time step, as in a line structure.

FEEDFORWARD CONTROL

Feedforward control, to counteract the anticipated effect of a measured load e_M, combined with feedback control to mitigate the effect of an unmeasured load e_U, makes use of two design equations like Eq. (10), one for each load type,

$$
\begin{aligned}
HC_U &= AF_U + DG_U \\
HC_M &= AF_M + DG_M
\end{aligned}
\tag{19}
$$

C_M need not be cancelable and may include a delay factor. Combining the process equation, like Eq. (5), with the target equation, like Eq. (8), with the design equations (19), like Eq. (10), to eliminate e_U and D results in the combined feedback and feedforward control law, like Eqs. (11) and (13):

$$
u = \frac{C_U r - G_U y}{BF_U} - \left(G_M - \frac{F_M G_U}{F_U}\right)\frac{e_M}{BH}
\tag{20}
$$

The second (e_M) term is an additive feedforward correction. If $F_M G_U / F_U = G_M$, feedforward control is not capable of improving upon feedback performance. The $F_M G_U / F_U$ term represents the reduction, from the open-loop feedforward correction G_M, needed to prevent redundant (overcorrecting) contributions. F_M can be made (nearly) zero, at least at low frequencies, when there is no more effective delay in the manipulated-variable path D to the controlled variable y than in the measured disturbance path C_M. Then from Eqs. (19), $G_M = HC_M/D$ because D is a factor of C_M. The measured disturbance e_M is (almost) perfectly rejected with the feedforward correction $u_{FF} = -(C_M/BD)e_M$, provided the controller output does not limit.

Feedforward provides a means for this single-output transfer function approach to be applied to a process with interacting loops. Unlike the state-space approach, it is necessary to associate each controlled variable with a particular manipulated variable. Then the effect of other manipulated variables on that controlled variable can be removed or reduced with feedforward corrections. This approach has the advantage that the appropriate compensation can be applied to active loops even when other loops are saturated or under manual control.

Furthermore, feedforward compensation may be structured to multiply the feedback correction. Multiplicative compensation is particularly effective for a temperature or composition loop manipulated with a flow. This configuration is intended to make the process appear more linear, as seen from the feedback controller. Thus the feedforward can be considered to provide gain scheduling for the feedback controller. Alternatively, the feedback controller can be viewed as adaptively tuning the gain of the feedforward compensator.

Other nonlinearities, even though they may involve the feedback variable, may be removable with additive or multiplicative feedforwardlike corrections [7]. For example, consider a nonlinear dominantly second-order process, such as a robot arm with negligible actuator delay and linkage flexibility. The process

$$
g\{y\}\frac{d^2 y}{dt^2} + f\left\{y, \frac{dy}{dt}\right\} = u
\tag{21}
$$

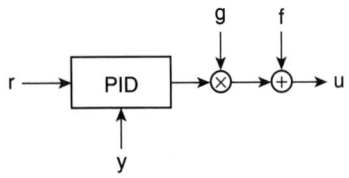

FIGURE 2 Feedback controller with multiplicative and additive compensation. (*Foxboro.*)

is controllable with u when the functions f and g are known,

$$u = f \left\{ y, \frac{dy}{dt} \right\} + g\{y\} \left[K(r - y) - D_M \frac{dy}{dt} \right] \quad (22)$$

This control, shown in Fig. 2, achieves linear closed-loop response to set point r,

$$y = \frac{Kr}{K + D_M s + s^2} \quad (23)$$

Here the proportional (K) and derivative (D_M) feedback terms should be chosen to achieve desired closed-loop performance, taking the neglected effective delay and high-frequency resonances into account, perhaps adaptively. If the K and D_M terms can be made large compared with f/g, the closed-loop performance may be quite insensitive to imperfect compensation for f. An integral term added to the proportional plus derivative controller would help to adapt the effective gain of the multiplicative g term. Dominantly first- or zero-order processes can be linearized similarly.

MULTIPLE-LOOP CONTROL

A cascade of control loops, where the output of a primary (outer-loop) controller is the set point of the secondary (inner-loop) controller, may improve performance of the outer loop, particularly when the primary measurement responds relatively slowly. Nonlinearity, such as results from a sticking valve, and disturbances within the fast inner loop can usually be made to have little effect on the slow outer loop. Limits on the primary output constrain the set point of the secondary loop. Typical secondary controlled variables are valve position and flow. Jacket temperature may be the secondary variable for a batch reactor. The design of the primary controller should provide means of preventing integrator windup when the secondary controller limits or is in manual [8].

Controllers also may be structured in parallel to provide a safety override of a normal control function. For example, the normal controlled variable may be a composition indicative of product quality. In an emergency, a pressure controller may take over its manipulated variable. This may be done by selecting the controller with the smaller (or larger) output or error to drive the manipulated variable. Means for preventing integrator windup of the unselected controller should be provided.

When there are multiple interacting controlled and manipulated variables, every controlled variable should be paired with a controller output in order to make each control loop as insensitive as possible to the status of the others. Bristol's relative gain array (RGA) [9] can help in the evaluation of potential pairs. A good pair choice may result by considering some controller outputs to be the sum (or ratio) of measurable variables. Control is then implemented with a cascade structure. One of the summed (or ratioed) variables, acting as a feedforward, subtracts from (or multiplies) the primary output to get the secondary set point for the other variable. Again, means to prevent integrator windup, when the secondary saturates, should be provided.

For example, in distillation column control (Fig. 3), the distillate (DF) and reflux (LF) flows may be manipulated to control the distillate (impurity) composition and condenser level. Normally LF is much larger than DF. If LF and DF were the controller outputs, an RGA would show that the larger flow LF could be paired with level and DF with composition [8]. However, if the composition controller adjusts [DF/(LF + DF)] and the level controller adjusts (LF + DF), an RGA would indicate minimal interaction since the ratio has no effect on level when the sum is constant. In this case the distillate set point is found by multiplying the composition controller output by the measured LF + DF, and the reflux set point is found by subtracting the measured DF from the level controller output. Dynamic compensation of the feedforward terms will not improve performance, since LF affects the composition with no more effective delay than DF.

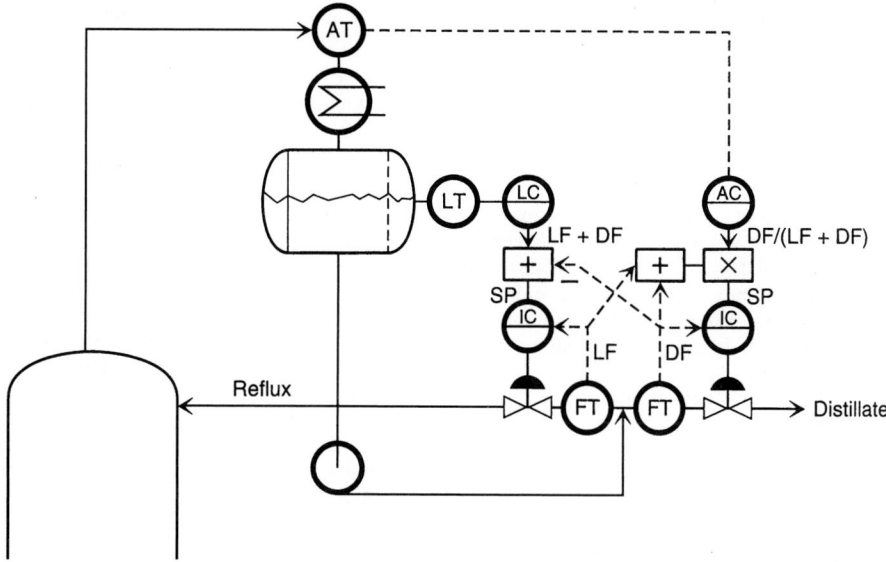

FIGURE 3   Level controller LC manipulates total flow and composition controller AC manipulates reflux ratio. (*Foxboro.*)

The RGA is an array where each element Γ_{ij} is a ratio of the sensitivities of a measurement y_i to an output u_j, the numerator having all other outputs fixed and the denominator having all other measurements fixed. For the process

$$
\begin{aligned}
A_1 y_1 &= b_{11} D_1 u_1 + b_{12} D_2 u_2 \\
A_2 y_2 &= b_{21} D_1 u_1 + b_{22} D_2 u_2
\end{aligned}
\tag{24}
$$

where A_i and D_j are dynamic operators and b_{ij} are constants, the RGA elements are also constants,

$$
\text{RGA} = \begin{bmatrix} \Gamma & 1-\Gamma \\ 1-\Gamma & \Gamma \end{bmatrix}
\tag{25}
$$

with only one interaction parameter Γ,

$$
\Gamma = \frac{b_{11} b_{22}}{b_{11} b_{22} - b_{12} b_{21}}
\tag{26}
$$

[The number of interaction parameters is $(n-1)^2$, where n is the number of interacting loops, because each RGA row and column sums to 1.]

When Γ is between 0.5 and 2, (u_1, y_1) and (u_2, y_2) could be pairs. When Γ is between -1 and 0.5, the opposite pairs could be used. Least interaction occurs when Γ is 1 or 0, which happens when one of the b_{ij} terms is zero. Values of Γ smaller than -1 or larger than 2 indicate that neither set of pairs should be used because the interaction is too severe. Saturating one of the loops, or placing it in manual, would change the gain in the other by more than a factor of 2.

When there is effective delay associated with each of the b_{ij} terms (here b_{ij} is not entirely cancelable), it is useful to compare the sum of the b_{11} and b_{22} delays with the sum of the b_{12} and b_{21} delays. If the interaction is significant (Γ not within 0.2 of either 0 or 1) and the combination with the smaller delay sum does not confirm the pairing based on Γ, a different choice of controller output variables (using a decoupling feedforward compensation) may be indicated.

DISTURBANCE REPRESENTATION

If the disturbance is a gaussian random variable, e is assumed to be a zero-mean white gaussian noise source. Colored noise is assumed to result from stably filtering the white noise e. The filter moving-average (numerator) characteristic is included in the C polynomial, and its autoregressive (denominator) characteristic is included in the A and B (or D) polynomials. The cross-correlation function of a linear filter's output with its input is equal to its impulse response convolved with the autocorrelation function of its input. When the input is white noise, its autocorrelation is an impulse function (the derivative of a step function). The cross correlation is then equal to the filter's impulse response [1]–[3].

Similarly, a deterministic disturbance may be considered to be the step or impulse response of a stable linear filter. A more complicated disturbance may be represented as the filter response to a sequence of steps (or integrated impulses). Any delay associated with an unmeasured disturbance is considered to determine the timing of the step in order that C_U be cancelable (stable zeros).

Pole-Placement Design

Pole-placement design [1] requires that A, B, C, and D be known and that a suitable H be selected. There may be difficulty in separating the BD product. As an expedient, B may be considered a constant and D allowed to include both stable and unstable zeros. If the degree of all of the polynomials is no greater than n, Eq. (10) provides $2n + 1$ equations, one for each power of s or z^{-1}, to solve for the coefficients of F and G. If the unmeasured disturbance e is considered an impulse, the degree of the G polynomial should be less than that of A, otherwise the degrees may be equal.

A design resulting in a B or F polynomial with a nearly unstable zero, other than one contributing integral action, should probably be rejected on the grounds that its robustness is likely to be poor. It may be necessary to augment H with additional stable factors, particularly if the degree of F or G is limited. The set point can be prefiltered to shift any undesired poles in the set-point function D/H to a higher frequency. However, the resulting uncompensatable unmeasured-load rejection function F/H may be far from optimal.

Linear-Quadratic Design

Linear-quadratic (LQ) design [1] provides a basis for calculating the H and F polynomials, but is otherwise like the pole-placement approach. In this case D contains both the stable and the unstable zeros and B is a constant. C/A is an impulse response function since e is specified to be an impulse. The H polynomial, which contains only stable zeros, is found from a spectral factorization of the steady-state Riccati equation,

$$\sigma H\{z\}H\{z^{-1}\} = \mu A\{z\}A\{z^{-1}\} + D\{z\}D\{z^{-1}\}$$

or

$$\sigma H\{s\}H\{-s\} = \mu A\{s\}A\{-s\} + D\{s\}D\{-s\} \tag{27}$$

The parameter σ is chosen to make the steady-state value of H unity and μ is an arbitrary parameter in the criterion function J,

$$J = E\{(r - y)^2 + \mu u^2\} \tag{28}$$

E is the expectation operator. The u term imposes a soft constraint on the manipulated variable with a penalty factor μ. A polynomial X, satisfying

$$X\{z\}A\{z^{-1}\} + \sigma H\{z\}G\{z^{-1}\} = D\{z\}C\{z^{-1}\}$$
$$X\{z\}H\{z^{-1}\} + \mu A\{z\}G\{z^{-1}\} = D\{z\}F\{z^{-1}\} \tag{29}$$

and Eq. (27) also satisfies Eq. (10) and minimizes Eq. (28) for an impulse disturbance e. The equations in s are similar. When the degree of A is n, the first equation of Eqs. (29) provides $2n$ equations, one for each power of z (or s). These can be solved for the $2n$ unknown coefficients of X and G, after H and σ are found from Eq. (27). G has no nth-degree coefficient and $X\{z\}$ and $D\{z\}$ have no zero-degree coefficient. [$X\{-s\}$ and $D\{-s\}$ have no nth-degree coefficient.] None of the polynomials is more than nth degree. F can then be found by polynomial division from Eq. (10) or the second of Eqs. (29).

For the optimization to be valid, the penalty factor μ must be large enough to prevent the manipulated variable u from exceeding its limits in responding to any input. However, if μ were chosen to be too large, the closed-loop response could be as sluggish as the (stabilized) open-loop response. The LQ problem may be solved leaving μ as a tuning parameter. Either experiments or simulations could be used to evaluate its effect on performance and robustness.

Despite the LQ controller being optimal with respect to J for a disturbance input, a switching (nonlinear) controller that takes into account the actual manipulated variable limits and the load can respond to a step change in set point r in less time and with less integrated absolute (or squared) error.

Minimum-Time Switching Control

The objective for the switching controller is to drive the controlled variable y of a dominant-lag process from an arbitrary initial value so that it settles at a distant target value in the shortest possible time. The optimal strategy is to maintain the manipulated variable u at its appropriate limit until y nears the target value r. If the process has a secondary lag, driving u to its opposite limit for a short time will optimally slow the approach of y to r, where it will settle after u is stepped to the intermediate value q needed to balance the load. Until the last output step, switching control is the same as "bang-bang" control. Determination of the output switching times is sufficient for open-loop control. The switching criteria must be related to y and its derivatives (or the state variables) in a feedback controller. Either requires solving a two-point boundary-value problem.

As an example, consider a linear integral $\{T\}$–lag$\{\tau_L\}$–delay$\{\tau_D\}$ process with constant manipulated variable (controller output) u. The general time-domain solution has the form

$$x\{t\} = at + b\exp\left\{-\frac{t}{\tau_L}\right\} + c$$

$$y\{t\} = x\{t - \tau_D\}$$

(30)

where x is an unmeasured internal variable and a, b, and c are constants that may have different values in each of the regimes. At time zero the controlled variable y is assumed to be approaching the target value r from below at maximum rate,

$$\frac{dy\{0\}}{dt} = \frac{dx\{0\}}{dt} = \frac{u_M - q}{T} = a - \frac{b}{\tau_L}$$

$$x\{0\} = b + c$$

$$y\{0\} = x\{0\} - \frac{(u_M - q)\tau_D}{T}$$

(31)

where u_M is the maximum output limit and q is the load. At that instant the output is switched to the minimum limit, assumed to be zero. If the next switching were suppressed, x and y would eventually achieve their negative rate limit,

$$\frac{dy\{\infty\}}{dt} = \frac{dx\{\infty\}}{dt} = -\frac{q}{T} = a$$

Combining with Eqs. (31) to eliminate a,

$$b = -\frac{u_M \tau_L}{T}$$

$$c = x\{0\} - b$$

(32)

At time t_1, x reaches the target value r with zero derivative,

$$\frac{dx\{t_1\}}{dt} = 0 = -\frac{q}{T} + \frac{u_M}{T} \exp\left\{-\frac{t_1}{\tau_L}\right\}$$

$$x\{t_1\} = r = x\{0\} - \frac{q(t_1 + \tau_L)}{T} + \frac{u_M \tau_L}{T}$$

(33)

Consequently,

$$t_1 = \tau_L \ln\left\{\frac{u_M}{q}\right\}$$

(34)

and switching from maximum to zero output occurred when

$$r - y\{0\} - \left(\tau_D + \tau_L - t_1 \frac{q}{u_M - q}\right)\frac{dy\{0\}}{dt} = 0$$

(35)

If q is zero, as it may be when charging a batch reactor, the qt_1 product is zero, even though t_1 is infinite.

Optimal switching, from zero output to that required to match the load q, occurs at time t_1. If t_1 is less than the delay time τ_D,

$$r - y\{t_1\} - \left(\tau_D + \tau_L - t_1 \frac{u_M}{u_M - q}\right)\frac{dy\{t_1\}}{dt} = 0$$

(36)

Otherwise

$$r - y\{t_1\} - \left(\tau_L - \frac{\tau_D}{\exp\left\{\dfrac{\tau_D}{\tau_L}\right\} - 1}\right)\frac{dy\{t_1\}}{dt} = 0$$

(37)

The controlled variable y will settle at the target value r at time $t_1 + \tau_D$. At this time the switching controller could be replaced by a linear feedback controller designed to reject unmeasured load disturbances and correct for modeling error. This combination of a switching controller with a linear controller, called dual mode, may be used to optimally start up and regulate a batch process [9].

Minimum-Variance Design

The minimum-variance design is a special case of the LQ approach where μ is zero. This may result in excessive controller output action with marginal improvement in performance and poor robustness, particularly when a small sampling interval h is used. Performance is assumed limited only by the nonminimum-phase (unstable) zeros and delay included in D. The H polynomial contains the stable zeros and the reflected (stable) versions of unstable zeros of D.

When D is a k time-step delay and e is considered an impulse, the minimum-variance solution for $F\{z^{-1}\}$ and $G\{z^{-1}\}$ can be found from Eq. (10) by polynomial division of HC by A. $F\{z^{-1}\}$ consists of the first $k - 1$ quotient terms. The remainder $(HC - AF)$ is $DG\{z^{-1}\}$. From Eq. (27) H is unity. However, if H is arbitrarily assigned, this design becomes pole placement.

As a minimum-variance example, consider D to be a delay. Again, H is unity. The disturbance e is assumed to be a step applied downstream of the delay. C is unity. B is a gain b. The A polynomial represents a kind of lag,

$$A = 1 - aD \tag{38}$$

In one delay time the controlled variable can be returned to the set point r, hence $F = 1 - D$. Equation (10) becomes

$$1 = (1 - aD)(1 - D) + DG \tag{39}$$

Solving for G,

$$G = 1 + a - aD \tag{40}$$

and the controller from Eq. (13) becomes

$$u = \frac{(r - y)/(1 - D) - ay}{b} \tag{41}$$

When the delay is one computing step h, the process can be considered a sampled first-order lag with time constant $h/\ln\{1/a\}$ and a zero-order hold. Equation (41) has the form of a digital proportional plus integral controller with proportional band PB $= b/a$ and integral time IT $= ah$. When $a = 1$, the process can be considered a sampled integral with time constant h/b and a zero-order hold. When a is zero, the process is a pure delay and the controller is "floating" (pure integral, PB IT $= bh$). The controlled variable y has dead-beat response in one time step for either a set point or a load step,

$$y = Dr + (1 - D)e \tag{42}$$

The response is also optimum with respect to the minimum largest absolute error and the minimum integrated absolute error (IAE) criteria.

When the delay has k time steps, the $F = 1 - D$ factor in the controller equation has k roots equally spaced on the unit circle. One provides infinite gain at zero frequency (integral action). The others make the controller gain infinite at frequencies that are integer multiples of $1/kh$, not exceeding the Nyquist frequency $1/2h$. These regions of high gain cause the loop stability to be very sensitive to mismatch between the actual process delay and kh used in the controller. As k approaches infinity, the robustness index approaches zero, indicating no tolerance of delay mismatch. A low-pass filter that improves the robustness by attenuating in these regions of high gain also degrades the nominal performance.

The control loop may be structured as two loops, the outer-loop integral controller providing the set point r_I to the inner-loop proportional controller,

$$r_I = \frac{r - y}{1 - D} = r - y + Dr_I \tag{43}$$

and

$$u = \frac{r_I - ay}{b} \tag{44}$$

The effect of closing only the inner loop is to create a delay process as seen by the outer-loop controller,

$$y = Dr_I + e \tag{45}$$

The outer-loop controller can be considered model feedback in relation to the closed inner loop, since the difference between the controlled variable's measured (y) and predicted (Dr_I) values is fed back as a correction to an open-loop (unity-gain) controller.

Model-Feedback Control

Model-feedback control, whose variations include, among others, Smith predictor, Dahlin, dynamic matrix, and model predictive (unless a special unmeasured disturbance model is used [10]), consists of an open-loop controller with a feedback correction equal to the model prediction error. If model-feedback control is applied without first closing a proportional inner loop, there results

$$u = \frac{A}{BH}\left[r - \left(y - \left(\frac{BD}{A}\right)u\right)\right]$$

$$y = \frac{D}{H}r + (1 - D)\left(\frac{C}{AH}\right)e$$

(46)

assuming no process-model mismatch and no unstable zeros of A. H is a filter chosen to improve robustness. For the above process, assuming H to be unity, the response to a step disturbance can be easily calculated by polynomial division,

$$F = (1 - D)\frac{C}{A} = \frac{1 - D}{1 - aD}$$

$$= 1 - (1 - a)D[1 + (aD) + (aD)^2 + (aD)^3 + \cdots]$$

(47)

The maximum error occurs during the first delay interval when $F\{0+\} = 1$. After n delay steps $F\{n+\}$ is reduced to a^n. Even though this result is optimum with respect to the minimum largest absolute error criterion, the recovery from a load upset can be very slow when a is close to 1 (or divergent, when a is greater than 1). The ratio of the IAE to the optimum is $1/(1-a)$. Consequently model-feedback control may not adequately reject an unmeasured load disturbance, when the process has a dominant lag (a near 1), unless well-chosen inner-loop feedback is applied or the model deliberately mismatches the process, as recommended in [6]. However, design of the inner-loop controller or the model mismatch, for near optimal load rejection, may require more detailed high-frequency knowledge (for example, a spectral factorization of process polynomials) than is necessary for selecting an output trajectory (open-loop controller) to achieve good set-point tracking.

The stability of a tightly tuned matched-model feedback loop is very sensitive to mismatch between the model and process delays. To achieve adequate robustness, it may be necessary to detune the controller with H, further sacrificing unmeasured-load rejection capability.

Without the inner loop or deliberate model mismatch, early return of the output from saturation may cause excessively slow controlled-variable response of a dominant-lag process to a large set-point step. As with open-loop control, an on-line nonlinear optimizer may be used to avoid this suboptimal behavior.

Algebraic Proportional plus Integral plus Derivative Design

In this section a two-phase method for applying Eq. (10) to the design of analog (or fast-sampling digital) proportional plus integral plus derivative (PID) controllers is described. Unlike Bode and root-locus design methods, this method allows all of the controller parameters as well as the closed-loop performance parameters (time scale and load sensitivity) to be found directly, without trial and error.

The process is represented with an A polynomial in s. B, D, and C are assumed 1. The inverse of a delay or small numerator zero factor (if representable as a convergent Taylor series up to the frequency range critical for stability) is included in the A polynomial,

$$(a_0 + a_1 s + a_2 s^2 + a_3 s^3 + \cdots)y = u + e$$

(48)

A large stable zero is unusual and requires special consideration. It should be approximately canceled by a controller or process pole. Such a pole is not available from a PID controller. However, an effective process cancellation, without a factorization, may result by disregarding the zero-order terms of both the process numerator and the denominator before determining A by polynomial division. When the process zero and pole are sufficiently dominant, mismatch in zero-order terms, which affects the very low-frequency behavior, will be corrected by high controller gain in that frequency range.

This two-phase design process implicitly imposes the performance limitation that would normally be imposed by including delay and nonminimum-phase zeros in D. The first design phase prevents inner-loop feedback when the open-loop process already approximates a pure delay. This is done by selecting the inner-loop gain and derivative terms,

$$G_I = K_M + D_M s \tag{49}$$

to make the closed inner loop H_I^{-1} approximate a delay at low and moderate frequencies. As many low-order terms of Eq. (10) are matched as are needed to determine the unknown controller and performance parameters.

The H_I polynomial is chosen to be the Taylor-series expansion of an inverse delay whose time τ_I and gain h_0 is to be determined,

$$H_I = h_0 \left[1 + \tau_I s + \frac{(\tau_I s)^2}{2} + \frac{(\tau_I s)^3}{6} + \cdots \right] \tag{50}$$

When F_I is chosen as 1 (instead of choosing h_0 as 1), Eq. (10) gives

$$H_I = A + G_I \tag{51}$$

The limited-complexity inner-loop proportional plus derivative controller can significantly influence only the low-order closed-loop terms and hence can shape only the low-frequency closed-loop behavior. Only the most dominant two poles (lowest in frequency) of the open-loop process may be unstable, since only they can be stabilized with proportional and derivative feedback. As a result the limiting closed inner-loop performance measures, the values of τ_I and h_0, are determined by a_2 and a_3, provided the latter have the same sign. Equating term by term and rearranging gives

$$\tau_I = \frac{3a_3}{a_2}$$

$$h_0 = \frac{2a_2}{\tau_I^2} \tag{52}$$

$$D_M = h_0 \tau_I - a_1$$

$$K_M = h_0 - a_0$$

When the sign of D_M is different from that of K_M, or derivative action is not desired, the parameters should be calculated with

$$\tau_I = \frac{2a_2}{a_1}$$

$$h_0 = \frac{a_1}{\tau_I} \tag{53}$$

$$D_M = 0$$

$$K_M = h_0 - a_0$$

For a pure delay process both K_M and D_M are zero. The closed inner loop becomes

$$y = \frac{r_I + e}{H_I} \tag{54}$$

The outer loop uses gain and integral terms applied to the error,

$$r_I = \left(\frac{1}{I_E s} + K_E\right)(r - y) \tag{55}$$

Using this equation to eliminate r_I from the previous gives

$$[1 + I_E s(K_E + H_I)]y = (1 + K_E I_E s)r + I_E se \tag{56}$$

The target closed-loop set-point behavior is chosen to approximate a nonovershooting delaylike model, n equal lags. The shape parameter n and the time constant τ_0 are to be determined, as are the controller parameters K_E and I_E,

$$\left[\left(1 + \frac{\tau_0 s}{n}\right)^n y\right] = r + \frac{I_E se}{1 + K_E I_E s} \tag{57}$$

Equating term by term and solving the four simultaneous equations gives

$$\begin{aligned}
n &= 10.4 \\
\tau_0 &= 1.54\tau_I \\
K_E &= 0.198h_0 \\
I_E &= \tau_0/h_0
\end{aligned} \tag{58}$$

A small value of I_E is desirable because its product with the output change is equal to the integrated error response to a load change. Since the controller is designed to achieve very small overshoot, the product is also nearly equal to the IAE for a step-load change. If the response shape parameter n were made infinite (corresponding to a pure delay target) instead of matching the fourth-degree terms, a faster but more oscillatory and less robust response would result (with near minimum IAE for a load step). Then the closed-loop time constant τ_0 would become $1.27\tau_I$ and K_E would equal $0.289h_0$. Equations (58) still would apply for I_E.

The resulting controller is a four-term noninteracting (sum of terms) type. The following equations may be used to convert these tuning values to those for a conventional three-term PID interacting (product of factors) type, adjusted for good load rejection,

$$K = K_E + K_M \tag{59}$$

If K^2 is greater than four times the ratio of D_M to I_E,

$$\frac{1}{PB} = 0.5\left[K + \left(K^2 - \frac{4D_M}{I_E}\right)^{0.5}\right]$$

$$IT = I_E/PB \tag{60}$$

$$DT = D_M PB$$

Otherwise,

$$PB = \frac{2}{K}$$

$$IT = DT = D_M\,PB \tag{61}$$

To achieve the designed set-point response also, the set-point input should be applied to the controller through a lead-lag filter. The lag time is matched to the integral time IT. The ratio of lead-to-lag time α is made equal to K_E PB. The resulting controller equation is

$$u = \frac{(1 + \alpha ITs)r - (1 + ITs)(1 + DTs)y}{PB\ ITs} \tag{62}$$

Derivative action is applied only to the controlled measurement, not to the set point. To prevent excessive valve activity at frequencies beyond the closed-loop bandwidth, it is customary to condition the controlled measurement y with a low-pass filter whose time constant is a small fraction (≈ 0.1) of DT. In order that the sampling process of a digital controller not further diminish the effectiveness of the derivative term, the sampling interval should be less than the effective filter time.

For example, consider a thermal process with a 300-second lag and a 10-second effective delay. The algebraic design gives an integral time IT $= 24$ seconds, derivative time DT $= 3.6$ seconds, and the closed-loop time constant $\tau_0 = 23.1$ seconds, all more sensitive to the delay than the lag. The sampling interval should not exceed 0.4 second in order not to compromise closed-loop performance. This is a surprisingly small interval, considering the relatively slow open-loop response.

Antialias Filtering

When an analog signal contains a component with frequency higher than the Nyquist frequency (half the sampling frequency f_S), the sampled signal component appears to have a frequency less than the Nyquist frequency, as shown in Fig. 4. If the analog signal component frequency f lies between odd integers $(2n - 1)$ and $(2n + 1)$ times the Nyquist frequency,

$$(n - 0.5)f_S \le f < (n + 0.5)f_S \tag{63}$$

the sampled signal component has the same amplitude but appears to be shifted to the frequency $|f - nf_S|$. This frequency shifting is called aliasing [1].

To achieve good performance with digital control, it is necessary to sample the controlled variable at a rate faster than twice the highest frequency significant for control and to attenuate, before sampling, components with higher than Nyquist frequency. This should be done with minimum attenuation or phase shifting of the lower-frequency components that are important for control.

It is particularly important to remove, before sampling, a signal component that is an integer multiple of the sampling frequency, because this component would shift to zero frequency, causing a steady-state offset error. An analog filter that averages between sampling instants removes such components completely. This filter can be realized with an analog-to-frequency converter and a

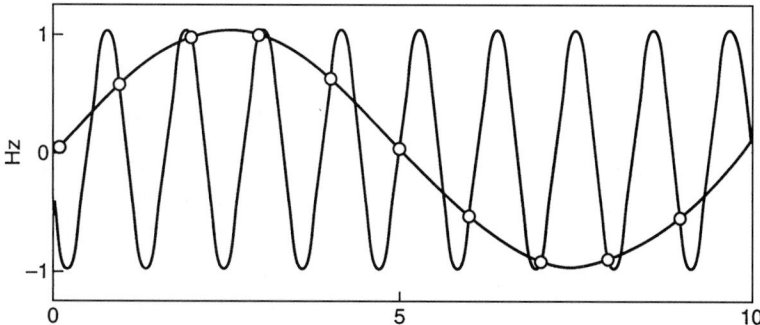

FIGURE 4 Two signals with different frequencies (0.1 and 0.9 Hz) have same values at 1-Hz sampling instants. The 0.9-Hz signal is aliased as 0.1 Hz after sampling. (*Foxboro.*)

sampled digital counter. The longer the sampling interval, the higher is the count and the greater is the digital signal resolution.

However, the analog averaging filter does not attenuate sufficiently near the Nyquist frequency. A stage of digital filtering, a two-sample average, completely removes a Nyquist-frequency component in the sampled signal. Its passband is flatter and its cutoff is sharper than those of a digital Butterworth (autoregressive) filter having the same low-frequency phase (effective delay).

This measurement antialias filter together with the manipulated variable zero-order hold adds an effective delay of $1.5h$ to the analog process. Calculation delay, which may be as large as the sampling interval h, is additional. Consequently the sampling interval should be small compared with the effective delay of the process, in order that feedback loop performance (unmeasured-load rejection) not be compromised. On the other hand, resolution is improved by averaging over a longer sampling interval.

An antialiasing filter removes or diminishes the effects of higher than Nyquist frequency process signals on the digital measurements. However, digital outputs apply a sequence of steps to the process that may excite a higher than Nyquist-frequency open-loop resonance (such as water hammer). This effect will be accentuated in high-performance loops with large output steps and in loops where the resonance is synchronized with a multiple (harmonic) of the output update frequency.

ADAPTIVE CONTROL

Time-varying or nonlinear process dynamics, variable operating conditions, slow response to upsets, dominant unmeasured disturbances, performance degradation resulting from deliberate upsets, and lack of tuning expertise are all reasons to consider adaptive (self-tuning) control. Economic incentives may result from improved control of product quality and yield, a higher production rate, less energy usage, improved plant safety, and less pollution.

Adaptive control schemes may be classified according to several design alternatives. Associated with each of these are concerns that should be addressed in order to assure robust adaptation.

First, the adaptor may be open- or closed-loop. An open-loop adaptor programs the controller tuning based on a model of the process. The model may be fixed and nonlinear (such as a neural net) or time-varying and linear, with parameters updated to reconcile measured process inputs and outputs. Mismatch between the model and the process may be caused by misidentification as a result of large nonstationary unmeasured disturbances, insufficiently rich inputs, or by model structural deficiencies. Mismatch may lead to inappropriate controller tuning that will not be corrected until (or unless) the model is revised. For example, even though the effective process delay limits how tightly the controller can be tuned, the model delay may be assigned arbitrarily (it is not easily identified) or the process delay masked by a larger sampling interval. On the other hand, a closed-loop adaptor monitors one or more performance measures for the closed control loop and adjusts controller parameters to drive these measures to target values. Desired performance is assured when the feedback adaptor converges. Its issues include the rate of convergence and the robustness of the adaptive loop.

Second, the adaptation may be based on observations of responses to deliberate or natural disturbances. For example, a performance-optimizing adaptor, using a hill-climbing strategy, requires the comparison of process performance measures for responses to identical, hence deliberate, disturbances. However, a deliberate disturbance degrades a well-tuned loop's performance. Therefore deliberate upsets should be applied infrequently and only when there is high likelihood that the controller is mistuned and is not likely to recover soon. A deliberate disturbance has the advantage that it is known and can be made large enough to dominate unmeasured disturbances and rich enough to excite important process modes. On the other hand, natural disturbances may be unmeasured and nonstationary, often consisting of significant isolated filtered steplike events and low-level noise. However, a response to a natural disturbance may contain incomplete information to observe all process modes and make an unambiguous adaptation.

Third, the target control-loop time scale may be fixed or optimal. One may choose to drive the closed-loop performance to that of a fixed model, as is done with pole-placement, fixed model-delay minimum-variance, or model-reference adaptors. A fixed time scale for all operating conditions must be user-selected to be as large as the largest minimum, and hence suboptimal for other operating conditions. Extensive knowledge of the process dynamics helps in choosing the time scale. Alternatively the control-loop time scale may be adapted in an open-loop minimum-variance adaptor by updating the effective delay time, or in a performance-feedback adaptor by choosing response shape parameters that are sensitive to the relative positions of the three (or more) most dominant closed-loop poles. On the other hand, a tightly tuned controller may not have an adequate stability margin, particularly if the (nonlinear) process may experience sudden large unmeasured load changes.

Fourth, the adaptations may be performed continuously, with updates each sampling interval, or aperiodically, following the response to each significant disturbance. When the controller tuning is updated each sampling instant, the update must recursively take into account a portion of past history. Over this time interval both the process parameters and the statistical measures of unmeasured disturbances are assumed to be stationary. However, real processes may be subjected to minimal load changes for extended intervals and to large unmeasured nonstationary load changes at other times. This makes choosing an adequate time interval difficult, particularly when effective adaptation requires that a time-varying linear model track a nonlinear process. On the other hand, an adaptive scheme designed to update controller parameters following significant isolated disturbance responses must also cope with cyclical and overlapping responses. It should distinguish among a loop instability, a stable limit cycle, and a cyclical disturbance.

Cycling can also be a problem for an adaptor based on model identification, because the signals may provide incomplete information for identifying more than two process parameters. An incorrect identification may lead to worse controller tuning. Furthermore, if the cycle amplitude is large, it may be impractical to make the identification unique by superimposing sufficiently significant deliberate disturbances.

An event-triggered adaptor provides an additional opportunity: the state and measured inputs, existing at the moment a new disturbance is sensed, can be used to select among several sets of stored tunings to determine the set to be used during, and updated following, the response interval. This capability (such as gain scheduling and multiplicative feedforward compensation) enables the controller to anticipate and compensate for the effect of process nonlinearity.

Three types of adaptive controllers will be discussed. First is the performance-feedback type, using expert system rules to cope with incomplete information, and using nonlinear dead-beat adaptation when information is complete. The second is an open-loop type that uses a recursive parameter identifier to update the parameters of a difference equation model. The controller is tuned as if the model were the process, thus invoking the "certainty equivalence" principle. Third is another open-loop type. This one identifies the low-order parameters of a differential equation model in order to update the coefficients in a feedforward compensator. It uses the moment-projection method on each isolated response.

Other types of adaptors, including the model reference type, may be more suitable for set-point tracking than for unmeasured load rejection. For example, the extended-horizon type uses a nonlinear optimizer to determine an open-loop controller's constrained output trajectory. The controlled-variable trajectory must be optimized over a prediction interval that exceeds that of the output by the process delay time. An extended-horizon optimizer can be used in conjunction with predictive-model feedback, useful (as explained earlier) for load rejection with a dominant-delay process and for model mismatch correction with set-point tracking. The nonlinear optimization calculations may require a large time step that compromises control loop performance with additional effective delay. Also, a linear moving-average process model may have so many degrees of freedom that on-line identification is impractical, both because the computational load is excessive and because a reasonable interval of natural-signal history does not persistently excite all of the model and process modes. A fixed nonlinear neural-net model may be more practical for a time-invariant process, even though its programming requires training with extensive signal records spanning the range of expected operating conditions.

PATTERN RECOGNITION AND EXPERT SYSTEMS, PERFORMANCE-FEEDBACK ADAPTOR

A performance-feedback adaptor monitors a single-loop variable control error [11]. Since the set point and process measurement are known, the controller output does not contain independent information, because it is determined by the controller equation. Pattern features are measured of the error response to a significant (typically unmeasured) disturbance. When enough of the response has been observed, the controller parameters are adjusted in order to make the feature values of the next response approach target values.

A significant error event is detected when the absolute control error exceeds a set threshold. The threshold value is chosen large enough so that an error event will not be triggered by low-level process or measurement noise. Since error peaks are the most prominent features of an oscillatory response, peak amplitudes and times are sought. Expert system (heuristic) rules may be used to distinguish response peaks from noise peaks.

Zero-to-peak $(-E_2/E_1)$ and peak-to-peak $[(E_3 - E_2)/(E_1 - E_2)]$ error amplitude ratios may be chosen as shape features. These are independent of the response amplitude and time scales and are called overshoot and decay, respectively. Target values for these ratios may be chosen to minimize a criterion function, such as minimum IAE, for a specific process and disturbance shape. The ratio of times between peaks is not recommended as a controlled-shape feature, because it is relatively sensitive to nonlinearity and noise and insensitive to the relative location of closed-loop poles.

The second and third peaks, E_2 and E_3, do not exist for an overdamped response. In this case a "knee" (or quasi-peak, defined by "expert" criteria) is sought to determine a response time scale used to terminate the peak search and to estimate the (negative) effective overshoot. Decay is zero in this case.

If the error response is a damped quadratic function, the first three error peaks E_i are related by

$$E_1 E_3 = E_2^2 \qquad \text{decay} = \text{overshoot} \tag{64}$$

This response, like the overdamped response, does not contain complete information for controller tuning [12]. A response containing both lag and underdamped quadratic terms provides complete information for tuning a PID controller. For the response

$$E\{t\} = \alpha e^{-at} + \beta e^{-bt} \cos\{\omega t\} \tag{65}$$

three shape ratios β/α, b/a, and b/ω provide sufficient information to update three controller parameters PB, IT, and DT. However, if either α or β were 0, values of a, or b and ω, would be unmeasurable and the information incomplete. It is desirable that the features used for adaptation reflect the relative pole positions indicated by the last two ratios, because the dominant error poles are usually closed-loop poles. Furthermore, the features should be insensitive to the first ratio, β/α, because this ratio is sensitive to the unmeasured disturbance shape and point of application, as indicated by the relative location of the error signal zeros.

When information is complete from the error response to a load step applied upstream of a dominant lag or delay,

$$E_1 E_3 > (E_2)^2 \qquad \text{decay} > \text{overshoot} \tag{66}$$

This corresponds to the lag (α) and quadratic (β) terms making contributions of the same sign to the first peak.

These inequalities can be reversed if the disturbance has a different shape or is applied at a different location. Reversal would result if the disturbance were a narrow pulse or if it were a step applied downstream of a dominant lag. Reversal also would result if the disturbance were statically compensated by a feedforward controller and could result if the disturbance were to affect interacting loops. The final part of this type of error response has a shape similar to the "usual" response so that, when the first peak, or peaks, is discarded and the remaining peaks are renumbered, decay becomes greater than overshoot.

Peak shifting desensitizes the pattern features to open-loop and disturbance-signal zeros, while maintaining sensitivity to the relative positions of the three most dominant error-signal poles. These poles usually are the closed-loop poles, but may include the poles of the disturbance signal if the disturbance is not applied suddenly.

The changes in the controller-parameter vector P are computed from the deviation of the measured-feature vector F from its target-value vector F_t, according to the adaptor's nonlinear gain function matrix $\mathbf{G}$,

$$P\{i+1\} = P\{i\} + \mathbf{G}(F_t - F\{i\}) \tag{67}$$

The response feature vector $F\{i\}$, measured after the ith response, is a nonlinear function of the controller parameter vector $P\{i\}$ (existing during the ith response), the process type, and the disturbance shape. For a given process and disturbance shape, simulation can be used to map feature deviations as a function of control parameter deviations $\delta\mathbf{F}/\delta\mathbf{P}$, allowing feature deviations to be predicted with

$$F\{i+1\} = F\{i\} + \frac{\delta\mathbf{F}}{\delta\mathbf{P}}(P\{i+1\} - P\{i\}) \tag{68}$$

If the unique inverse of the function matrix $\delta\mathbf{F}/\delta\mathbf{P}$ exists, the latest response contains complete information. Then a dead-beat adaptation ($F\{i+1\} = F_t$) is possible with

$$\mathbf{G} = \left(\frac{\delta\mathbf{F}}{\delta\mathbf{P}}\right)^{-1} \tag{69}$$

This multivariable adaptive loop is quite robust, being particularly tolerant of smaller than optimum $\mathbf{G}$. For example, if the optimum $\mathbf{G}$ is multiplied by a factor ranging between 0 and 2, the adaptive loop will remain stable. The eigenvalues of the $I - (\delta\mathbf{F}/\delta\mathbf{P})\mathbf{G}$ matrix, nominally located at the origin of the complex z plane, must stay within the unit circle.

If the process were a lag delay, a process-type variable, sensitive to the ratio of the lag-to-delay times, could be used to interpolate between the two extremes. The ratio of the controller IT to the response half-period is an indicator of the process type when decay and overshoot are fixed. This property may be used to identify the process type. The optimal IT-to-half-period ratio IT/T is smaller for the pure delay than for the integral delay.

When only two features, such as overshoot and decay, reliably characterize a response shape, only two controller parameters (PB and IT of a PID controller) are determined through performance feedback. However, because the optimal derivative-to-integral ratio is also a function of the process type, DT can be calculated after IT and the process type have been determined.

The process type, that is, the proportional-band ratio PB/PB$_t$ and the integral-time ratio IT/IT$_t$, is determined by interpolating stored data from performance maps for the process-type extremes, given overshoot, decay, IT/T, and DT/IT. The half-period T and the controller parameters PB, IT, and DT are values for the latest response. PB$_t$ and IT$_t$ are the newly interpolated values of the controller parameters predicted to produce the target features on the next response.

When the error-shape information is incomplete, as when the response is overdamped, quadratic, nonisolated, or nonlinear (because the measurement or controller output has exceeded its range), expert system rules are used to improve the controller tuning. These rules, of the if-then-else type, invoke a special strategy for each of these contingencies. Several retunings may be needed before a response shape contains sufficient information to achieve the desired performance on the next response. Even when the information is incomplete, robust tuning rules are possible, provided derivative action is not essential for stabilizing the control loop.

A nonisolated response is recognized if its start is detected while waiting for the last response to settle. A nonisolated response may be caused by the failure of the preceding response to damp quickly enough. If the decay of a nonisolated response is sufficiently small, even though it may be bigger than the target, the existing tuning is retained. Typically a continuing oscillation will be dominated by a quadratic factor, giving rise to incomplete information for retuning.

A nonisolated response may also be caused by a rapid sequence of load changes, the next occurring before the response to the last has settled. Peak shifting tends to desensitize the adaptor to a strange sequence of peaks, allowing detuning only when a conservative measure of decay is excessive.

A marginally stable loop is distinguished from a limit cycle or response to a cyclical load by observing the improvement in decay caused by an adaptive retuning. If retuning fails to reduce the decay measure, the last successful tunings may be restored and adaptation suspended until settling or an operator intervention occurs.

DISCRETE-MODEL IDENTIFICATION, OPEN-LOOP ADAPTATION

Adaptation of a feedback controller based on an identification of an input-output process model is most effective when the important process inputs are measured and (at least partially) uncorrelated with one another. An unmeasured disturbance is assumed to come from a stationary filtered white gaussian-noise source uncorrelated with the measured inputs. When a large unmeasured disturbance violates this assumption, the identified model may be a poor match for the process and poor controller tuning may result. Process-model mismatch may also result when the disturbance fails to independently excite process or model modes, a condition called nonpersistent excitation. A poor model structure, such as one having an incorrect unadapted delay or insufficient model degrees of freedom, may also cause mismatch that leads to poor control. Coefficients for a model having both linear and nonlinear terms may not be uniquely identifiable if the process input and output changes are small.

Two types of models may be identified, called explicit and implicit. An explicit model relates the process inputs and output with parameters natural to the process, such as Eq. (5). The explicit model is most useful for the design of an open-loop or model-feedback controller. A complicated design process involving Eq. (10) would be needed to compute the feedback controller parameters of Eq. (13). An implicit model combines the target equation, Eq. (8), and the feedback control equation, Eq. (13), so that the parameters needed for control are identified directly. In either case the identification model may be put in the prediction form

$$\Omega\{t + k\} = \Phi\{t\}^T \Theta + \varepsilon\{t + k\} \tag{70}$$

which predicts the value of Ω, k time steps ahead, given present and past values of the process inputs and outputs concatenated in the vector Φ. Θ is a corresponding vector of parameters determined by the identifier; ε is the identification error.

For the explicit model,

$$Ay = DBu + Ce + e_0 \tag{71}$$

Here e_0 is the steady-state offset and

$$\begin{aligned} A &= 1 + a_1 z^{-1} + \cdots \\ C &= 1 + c_1 z^{-1} + \cdots \\ B &= b_0 + b_1 z^{-1} + \cdots \end{aligned} \tag{72}$$

The time step h is assumed to be one time unit. The time step should be chosen small enough that the antialias filter, the digital computation, and the output hold do not dominate the effective delay, but large enough that roundoff or data storage do not cause difficulty. Here D is assumed to be a known and fixed k-time-step delay. The value of k must be large enough that B is stably cancelable. Of course, other choices for D are possible. The prediction model variables become

$$\begin{aligned} \Omega\{t\} &= y\{t\} \\ \Phi\{t - 1\}^T &= [u\{t - k\}, \ldots, -y\{t - 1\}, \ldots, \varepsilon\{t - 1\}, \ldots, 1] \\ \Theta^T &= [b_0, \ldots, a_1, \ldots, c_1, \ldots, e_0] \end{aligned} \tag{73}$$

If the model matched the process exactly, the prediction error $\varepsilon\{t\}$, which is uncorrelated with the variables in $\Phi\{t - 1\}$, would equal the white-noise disturbance $e\{t\}$. In order to identify C, past values of the prediction error are needed, but these cannot be found until Θ is identified. This difficulty can be overcome by solving for Θ recursively. When Θ is updated each time step, $\varepsilon\{t\}$ can be calculated using the most recent Θ. C must be constrained to have stable zeros. An algorithm that identifies C is said to be "extended." When Φ and Ω are prefiltered by C^{-1}, the algorithm is "maximum likelihood." If the identifier inputs are prefiltered by E^{-1}, $(E/C) - 0.5$ must be positive real in order to ensure that Θ can converge to its true value [13]. Convergence also requires no structural mismatch.

The model form for implicit identification uses the target equation, Eq. (8), to eliminate $r\{t - k\}$ from the controller equation, Eq. (13),

$$Hy\{t\} = BFu\{t - k\} + Gy\{t - k\} - C'r\{t - k - 1\} + e_0 + Fe\{t\} \tag{74}$$

where $C' = C - 1$ and D is a k-step delay $(h = 1)$. The prediction model variables become

$$\Omega\{t\} = Hy\{t\} = h_0 y\{t\} + h_1 y\{t - 1\} + \cdots \tag{75}$$

where H is specified.

$$
\begin{aligned}
BF &= \beta_0 + \beta_1 z^{-1} + \cdots \\
G &= \alpha_0 + \alpha_1 z^{-1} + \cdots \\
\Phi\{t - k\}^T &= [u\{t - k\}, \ldots, y\{t - k\}, \ldots, -r\{t - k - 1\}, \ldots, 1] \\
\Theta^T &= [\beta_0, \ldots, \alpha_0, \ldots, c_1, \ldots, e_0]
\end{aligned}
\tag{76}
$$

To identify the C polynomial, the set point r must be active, the updated control law implemented, and a recursive algorithm used. C must be constrained to have stable zeros. Also, k must be large enough that the zeros of BF are stable. If the model matched the controlled process exactly, the modeling error $\varepsilon\{t\}$, which is uncorrelated with any of the variables in $\Phi\{t - k\}$, would equal the closed-loop noise response $Fe\{t\}$. If $H = 1$, this implies a minimum-variance design, otherwise pole placement.

The same positive real and structural consistency requirements apply for convergence of an implicit parameter set as apply for the explicit set. The positive real condition ensures that the component of control error Fe, in phase with the model error ε, is positive and at least half as big.

Control based on either model will not have integral action when the identifier is turned off. Integral action depends on updating the offset e_0. The effective integral time constant depends on the quantity of past history taken into account in calculating Θ. Consequently it is likely to be significantly larger than optimal. On the other hand, integral action can be implemented explicitly in an outer-loop controller, such as

$$\delta r_I = \frac{r - y}{2k - 1} \tag{77}$$

without identifying e_0, if an incremental identification model is used to design the inner-loop controller,

$$\delta u = \frac{(C\delta r_I - G\delta y)}{BF} \tag{78}$$

For an incremental model, the values of variables in Ω and Φ are the changes δu and δy from one time step to the next. If C were 1 in such a model, the autocorrelation function of the unmeasured disturbance noise would be a step instead of an impulse. The inner-loop set point δr_I will be active, providing excitation to a mode (allowing identification of β_0 as well as α_0) that would not be excited in a single-loop structure when the set point r is fixed.

A restricted complexity model has fewer modes than the process. Consequently its modeling error will have components resulting from structural mismatch as well as unmeasured disturbances. It may have substantially fewer parameters than an "exact" model. For example, the C and β polynomials may be restricted to one term $(C = 1, BF = \beta_0)$ to be certain that they have no unstable zeros. Less past history is needed to reliably identify a small number of parameters since fewer equations are needed to solve for fewer unknowns. Therefore a restricted complexity model can be updated more quickly, allowing it to better track a changing or nonlinear process. The identifier inputs Φ and Ω should be filtered in order to make the process-model match best in the critical frequency range, where the open-loop absolute gain is near 1 for feedback control or near steady state for open-loop, or feedforward, control.

The implicit model form can be used to identify the delay time. The same Φ vector can be used for a number of predictor models, each predicting Ω a different number of time steps k into the future. If d is the largest possible value of k, the identifier equations can be time-shifted to yield a common Φ [14]. At time step t,

$$\Omega_k\{t - d + k\} = \Phi\{t - d\}^T \Theta_k \tag{79}$$

The prediction model with the largest β_0 coefficient can be chosen for the controller design, since this model indicates the greatest sensitivity of the controlled (predicted) variable to the present manipulated variable. Hence it will result in the smallest controller gain. Furthermore, if more than one β coefficient is identified, the model with the largest β_0 is most likely to have stable zeros. The model with the smallest prediction error is most likely the one with the smallest k because of the autoregressive α terms, but this model would not necessarily be best for control. The identification filter time constants can be made proportional to the identified k, since k determines the closed-loop performance and the critical frequency range.

CONTINUOUS-MODEL IDENTIFICATION, OPEN-LOOP ADAPTATION

A continuous (differential equation) model, in contrast to a difference equation model, is insensitive to the computing interval h, provided h is small. A restricted complexity identifier, for a process that includes delay, can be based on the method of moments [15].

The Laplace transform $X\{s\}$ of each signal's derivative $x\{t\}$ can be expanded into an infinite series of moments,

$$X\{s\} = \int_0^\infty e^{-st} x\{t\}\, dt = M_0\{x\} - s M_1\{x\} + \cdots \tag{80}$$

Signal derivatives are used so that the moment integrals, for an isolated response, converge to near final values in the finite time τ from the disturbance start,

$$M_n\{x\} \approx \int_0^\tau t^n x\{t\}\, dt \approx \sum_{k=1}^{\tau/h} (kh)^n \times \{k\}h \tag{81}$$

The signal transforms are related to the model polynomials on a term-by-term basis. Choosing the B and D polynomials to be 1 and

$$\begin{aligned} A\{s\} &= a_0 - s a_1 + \cdots \\ C\{s\} &= c_0 - s c_1 + \cdots \end{aligned} \tag{82}$$

in the process equation

$$u\{s\} = A\{s\} y\{s\} - C\{s\} e\{s\} \tag{83}$$

gives

$$\begin{aligned} M_0\{u\} &= a_0 M_0\{y\} - c_0 M_0\{e\} \\ M_1\{u\} &= a_1 M_0\{y\} + a_0 M_1\{y\} - c_1 M_0\{e\} - c_0 M_1\{e\} \\ &\ \ \vdots \end{aligned} \tag{84}$$

For each additional equation there are one plus the number of measured disturbances e of additional unknown parameters. The projection algorithm can be used to find the smallest sum of weighted squared parameter changes that will satisfy the equations. Using projection, only those parameters weighting signals that are significantly active are updated. Equations (84) expressed in vector and matrix form, for use in the projection algorithm, are

$$\Omega = \Phi^T \Theta$$

$$\Omega^T = [M_0\{u\}, M_1\{u\}, \ldots]$$

$$\Theta^T = [a_0, a_1, \ldots, c_0, c_1, \ldots] \tag{85}$$

$$\Phi^T = \begin{vmatrix} M_0\{y\}, & 0, \ldots, -m_0\{e\}, & 0, \ldots \\ M_1\{y\}, & M_0\{y\}, \ldots, -M_1\{e\}, & -M_0\{e\}, \ldots \end{vmatrix}$$

The moment-projection approach is particularly suited for adapting feedforward gain and delay compensators, because the inputs need not be persistently excited. Only two moments need be computed for each signal and two moment equations solved by projection. However, when signals are cycling or responses overlap, the moment integrals do not converge and the adaptation must be frozen. Since an adaptive feedback controller should be capable of stabilizing a stabilizable unstable loop, the moment-projection method is not suited for adaptation of a feedback controller.

LEAST-SQUARES METHOD, BATCH PARAMETER IDENTIFICATION

A batch identifier calculates the model parameters that best fit a block of measured data. The start may be triggered when a significant disturbance is sensed and the end may follow the settling of an isolated response or the detection of a preset number of peaks for a cycling response. A least-squares identifier finds the parameter vector Θ that minimizes the sum of squared prediction errors,

$$\varepsilon\{t\} = \Omega\{t\} - \Phi\{t - k\}^T \Theta \tag{86}$$

When the inverse of the matrix **P** exists,

$$\mathbf{P}^{-1} = \sum_i \Phi\{i - k\} \Phi\{i - k\}^T \tag{87}$$

the result is given by

$$\Theta = \mathbf{P} \sum_i \Phi\{i - k\} \Omega\{i\} \tag{88}$$

In order that $\mathbf{P}^{-1}$ not be dominated by steady-state components of Φ, it is customary to choose Φ and Ω to have nearly zero mean. If the means were completely removed, $\mathbf{P}^{-1}$ would be the covariance of Φ and **P** would be the covariance of Θ. **P** also appears in the recursive algorithm of the next section.

For Θ to be calculable, $\mathbf{P}^{-1}$ must not be singular. Nonsingularity is difficult to guarantee. If any of the process inputs were quiescent over the identification period, $\mathbf{P}^{-1}$ would be singular. This could happen if the controller output were limited or if the controller were in the manual mode. When **P** exists, the process is said to be persistently excited. It may be necessary to add otherwise undesirable probing signals to the normal process inputs to achieve persistent excitation.

KALMAN FILTER, RECURSIVE PARAMETER IDENTIFICATION

The Kalman filter [16] provides a one-step-ahead prediction of the parameter vector Θ (treated as a state variable) in a model equation,

$$\Theta\{t\} = \Theta\{t - h\} + v\{t\} \tag{89}$$

modified by observations of a related measured variable scalar (or vector) Ω,

$$\Omega\{t\} = \Phi\{t - k\}^T \Theta\{t\} + \varepsilon\{t\} \tag{90}$$

The model equation, Eq. (89), has a zero-mean white gaussian noise source vector v with covariance matrix $\mathbf{Q}$, which causes Θ to change randomly. The scalar (or vector) observation equation has a zero-mean white gaussian noise source scalar (or vector) ε, with covariance value (or matrix) R, in this case uncorrelated with v. Depending on which noise source dominates, the Kalman filter weights the other equation more heavily,

$$\Theta\{t\} = \Theta\{t - 1\} + K\{t\}(\Omega\{t\} - \Phi\{t - k\}^T \Theta\{t - 1\}) \tag{91}$$

Θ is the predicted parameter vector and K is the time-varying Kalman gain vector (or matrix) which can be precalculated using

$$K\{t\} = \mathbf{P}\{t - 1\}\Phi\{t - k\}(R + \Phi\{t - k\}^T \mathbf{P}\{t - 1\}\Phi\{t - k\})^{-1} \tag{92}$$

$$\mathbf{P}\{t\} = (\mathbf{I} - K\{t\}\Phi\{t - k\}^T)\mathbf{P}\{t - 1\} + \mathbf{Q} \tag{93}$$

When $\mathbf{Q}$ is a null matrix, this algorithm is the recursive least-squares algorithm, finding the parameter set Θ that minimizes the sum of squared model errors, equally weighted over all past observations. As time progresses the $\mathbf{P}$ matrix and gain K approach zero, so that eventually each new observation has almost no effect on the identified Θ. Therefore an unmodified recursive least-squares solver does not allow a model to adapt to a time-varying or nonlinear process.

Whereas R tends to reduce K and $\mathbf{P}$ by a factor each iteration, $\mathbf{Q}$ increases $\mathbf{P}$ by a fixed increment. Therefore $\mathbf{Q}$ has a relatively greater influence when $\mathbf{P}$ is small and vice versa when $\mathbf{P}$ is large. Thus when neither R nor $\mathbf{Q}$ is zero, $\mathbf{P}$ tends toward a midrange value. As a result, the Kalman gain K remains finite, so that the most recent observations affect the identified Θ, allowing the model to adapt to a time-varying or nonlinear process. In effect this method, in contrast to the variable forgetting factor approach, weights a different quantity of past history for each variable Φ_i, depending on its activity and the ratio of $\mathbf{Q}_{ii}$ to R.

When both R and $\mathbf{Q}$ are zero, the predictor becomes an orthogonal projection algorithm. Θ converges in a determinate number of iterations to a fixed vector, provided each of these observations contains some independent information. The number of iterations is the number of parameters in Θ divided by the number of components in Ω.

PROJECTION

If $\mathbf{P}$ is not updated and R is zero, the Kalman filter algorithm performs a projection each iteration, finding the smallest set of weighted squared parameter (state-variable) changes that satisfy the model equations exactly. The weighting matrix $\mathbf{P}$ is the a priori covariance of Θ.

REFERENCES

1. Astrom, K. J., and B. Wittenmark, *Computer Controlled Systems*, Prentice-Hall, Englewood Cliffs, New Jersey, 1984.

2. Takahshi, Y., M. J. Rabins, and D. M. Auslander, *Control Systems*, Addison-Wesley, Reading, Massachusetts, 1970.

3. Franklin, G. F., and J. D. Powell, *Digital Control*, Addison-Wesley, Reading, Massachusetts, 1980.

4. *Chemical Process Control Series*, American Institute of Chemical Engineers, New York, 1991.

5. Hansen, P. D., "Recovery from Controller Bounding," presented at the ASME Winter Annual Meeting, 1987.

6. Shinskey, F. G., "Putting Controllers to the Test," *Chemical Engineering*, pp. 96–106, Dec. 1990.

7. Slotine, J. J. E., and W. Li, *Applied Nonlinear Control*, Prentice-Hall, Englewood Cliffs, New Jersey, 1991.

8. Shinskey, F. G., *Process Control Systems*, 3rd ed., McGraw-Hill, New York, 1988.

9. Bristol, E. H., "On a New Measure of Interaction for Multivariable Process Control," *IEEE Trans. Automatic Control*, Jan. 1966.

10. Ricker, N. L., "Model Predictive Control: State of the Art," *Chemical Process Control Series*, American Institute of Chemical Engineers, New York, 1991.

11. Kraus, T. W., and T. J. Myron, "Self-Tuning PID Controller Uses Pattern Recognition Approach," *Control Engineering*, June 1984.

12. Hansen, P. D., "Recent Advances in Adaptive Control," presented at the 40th Chemical Engineering Conf., Halifax, Nova Scotia, Canada, July 1990.

13. Ljung, L., and T. Soderstrom, *Theory and Practice of Recursive Identification*, M.I.T. Press, Cambridge, Massachusetts, 1982.

14. Hansen, P. D., and T. W. Kraus, "Expert System and Model-Based Self-Tuning Controllers," in *Standard Handbook of Industrial Automation*, D. M. Considine, Ed., Chapman and Hall, New York, 1986, p. 216.

15. Bristol, E. H., and P. D. Hansen, "Moment Projection Feedforward Control Adaptation," in *Proc. 1987 American Control Conf.*, p. 1755.

16. Goodwin, G. C., and K. S. Sin, *Adaptive Filtering, Prediction and Control*, Prentice-Hall, Englewood Cliffs, New Jersey, 1983.

17. Ray, W. H., *Advanced Process Control*, New York, McGraw-Hill, 1981.

18. Hang, C. C. et al., *Adaptive Control*, Instrument Society of America, Research Triangle Park, North Carolina.

19. Wang, H. et al., *Advanced Adaptive Control*, Tarrytown, New York, Elsevier Science, 1995.

20. Luyben, W. L., and Luyben, M. L., *Essentials of Process Control*, McGraw-Hill, New York, 1997.

BASIC CONTROL ALGORITHMS

by E. H. Bristol*

INTRODUCTION

Continuous process control traditionally used analog control devices naturally related to the process. The modern era has largely replaced these with microprocessors and digital computations to reduce cost, but also for greater flexibility, supporting more diverse algorithmic forms. Initially digital control was implemented by the conversion of each analog control function into some (perhaps generalized) digital equivalent [1–4].

* The Foxboro Company, Dept. 0346, Bldg. C41–1H, Foxboro Massachusetts, 02035.

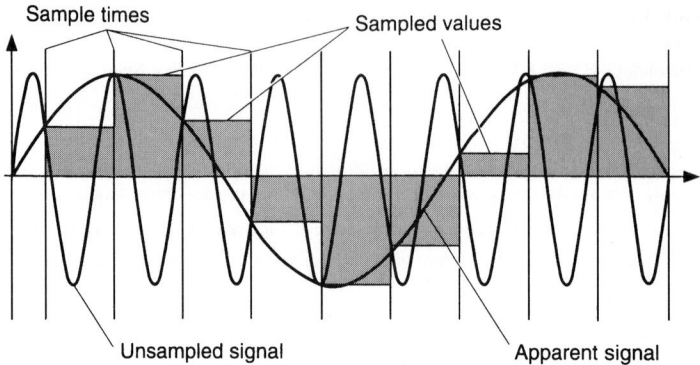

FIGURE 1 Aliasing of sampled to apparent signal.

This conversion requires an appropriate software architecture (more about that below) and the matching of the digital algorithms to analog behavior. Whereas the analog controller continuously sensed the process state and manipulated the actuators, the digital controller must repeatedly sample that state, convert it to a quantized number, use that number to compute control actions, and output those actions. Each of these steps involves its own problems and errors.

Standard digital control texts address the sampling problem thoroughly [5], but treat the broader control design in terms of neutral, computed parameters not clearly related to the process gains and time constants. Process control depends on standard control algorithms whose parameters may be set or tuned in terms of known process properties. The approach defined herein emphasizes control with traditional parameters, naturally related to the process and practice. This includes compensating sampled algorithms and their tuning properties for the sampling time, or even for irregular sampling.[1]

One aspect of sampling should be understood: aliasing. This is a bizarre effect in which sampled higher-frequency data appear to be at a much lower, more significant frequency (Fig. 1). The same effect limits the minimum apparent width in time of any disturbance pulse to one sampling period. A control loop attempting to control such a false disturbance causes a real one instead.

Aliasing is not normally a problem because most commercial controllers sample frequently and filter out the still higher frequencies. But users should be aware of aliasing if they construct home-brewed control algorithms of low sampling frequency.[2]

Except for aliasing, the choice of sampling time is not an issue today. Increasing the sample times faster than the dominant closed-loop time constants of the economically important process variables gives rapidly diminishing performance returns.[3]

For example, with end-product quality the sole criterion of a process that takes an hour to respond, even fast flow loops can be sampled once every 5 min. Of course, many "housekeeping functions," like flow control, may have constraining side effects whose violation would involve real costs if this logic were actually implemented. Sampling times faster than 1 s, well filtered, are rarely needed in continuous fluid process control, even when the local process dynamics are faster (as with flow loops).

The inexpensive microprocessor has caused a conservative design trade-off to favor faster sampling times, eliminating the issues for which the cost is so small. Nevertheless, a 10-to-1 reduction in

---

[1] The resulting parameter forms may look unnecessarily approximate. But recent standards efforts [6] have argued for even simpler control parameters. In either case it will always be possible to replace the proposed parameters by computations that support the more exact or simpler form. The chosen forms are based on the understanding that any use of digital control with analog process is inherently approximate and that intuitive tuning of parameters is the most important design consideration of these approximations. Apparently formal calculations of the parameters will always be misleading.

[2] Model-based control techniques, like internal model control and dynamic matrix control, are often constrained to operate at low sample times for reasons of modeling sensitivity.

[3] However, slow sampling frequency also causes tuning sensitivity, even for frequencies fast enough to give good control. Further doubling or tripling the frequency beyond this point should eliminate even this problem.

sampling time does correspond to a 10-to-1 reduction in computing resources if one has the tools and imagination to use the capability. Faster sampling times also require that internal dynamic calculations be carried out to a greater precision.

Process control algorithm design also differs from academic treatments in respect to accuracy considerations. Practice designs are for control and human operation, not for simulation or computation. When accuracy is important, a common thread through the discussion is the effect of differences among large numbers in exaggerating error and the role of multiplication in creating these large numbers.

Rarely is the parameter or performance precision (say, to better than 10%) important to control, even under adaptation. While there are sensitive high-performance, control situations, a 2:1 error in tuning is often unimportant. In contrast, there are situations within an algorithm in which a small error (say, a quantization error of 1 part in 100,000) will cause the control to fail. Computational perfection is unimportant; control efficacy is crucial.

A casual experimenter or designer of a one-use algorithm should be able to tailor a simple controller in FORTRAN floating point without any special considerations. The normal debugging and tuning should discover any serious deficiencies. A commercial design, applied in many applications, calls for a deeper understanding of algorithmic and fixed- versus floating-point issues and trade-offs. Fixed-point and machine language programming may become a lost art, even though their difficulties are overstated. The advantages, in speed and exactitude, are worth consideration, particularly on small control computing platforms, for which they may be essential. The problems are simply those of understanding and of effective specification and testing of the algorithm.

The discussion thus addresses the refined design of process-control-oriented continuous control algorithms and their software support. It emphases high-quality, linear, dynamic algorithms. This still includes representative examples of discrete computation's effects on modeled continuous control activities:

- imprecision, as it limits control performance,
- quantization, as it artificially disturbs the process,
- sampling, as it affects time continuous computations such as dead time.

In this environment, nonlinearity serves to compensate for a basically linear structure. One special concern, addressed below, is the backcalculation required for some aspects of these compensations.

The user interested in the more general computation of nonlinear functions should consult the general computing literature [7, 8]. The discussion briefly touches issues relevant to advanced forms like adaptive control, but their details are outside the intended scope.

STRUCTURE OF TRADITIONAL PROCESS CONTROL

The traditional process control structure is based on the combination of simple (PI/PID) controllers in simple single or cascaded loops, augmented with various nonlinear, feedforward, and dynamic compensators and constraint overrides.

Figure 2(a) illustrates the basic cascaded structure. In this structure, a primary controller, controlling a corresponding process variable, manipulates the set point to a secondary controller that acts to stabilize and control the corresponding secondary process variable.

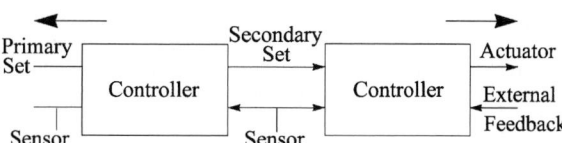

FIGURE 2(a)   Basic cascaded structure.

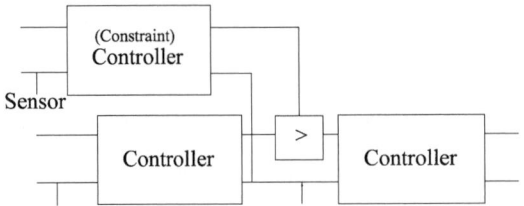

FIGURE 2(b) Constraint structure.

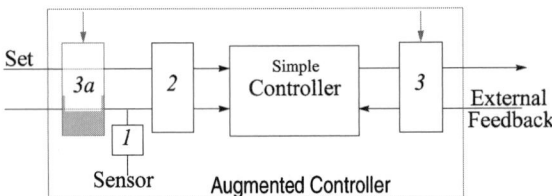

FIGURE 2(c) Augmented controller.

It will be noted that each controller has two connections on both the input side and the output side. On the input side these serve as the standard set point and measurement connection. The output side includes the normal output connection as well as the external and status feedback connections, described below, whose purpose is to adapt the control to loss of output control.

The structure can be continued with any number of controllers, defining a degree-of-freedom path of controlled variables. The control of the primary variable is then supported by successively more secondary variables until the ultimate control falls on the actuator. The single degree of freedom represented by that valve is then passed back along that path until it is the primary variable that is effectively free for manipulation.

The degree-of-freedom path can be overridden by any number of constraint controllers or limiters. Figure 2(b) illustrates the constraint structure. The selector symbol (>) represents a computation that outputs the greater of two inputs, thus imposing a low limit on the original degree-of-freedom path.

The constraint controller acts in feedback to set that limit value that will ensure that its measured value stays within the constraint limit defined by its set point. When the constraint controller takes over to enforce the constraint, it takes over the degree-of-freedom path, so that the path now runs from the constrained process variable through the secondary variable to the actuator. Any number of low- or high-limit or constraint override structures could similarly be included within the cascaded structure.

The traditional structure allows each controller in a cascade/constraint structure to be augmented in a variety of ways that enhance the control function without changing the underlying intent and structure. The figure shows the three distinct structural roles of this kind of compensation, with several variants:

1. Computation of the desired measurement form from the sensor signal.
2. Linearization compensation of the controller in terms of the controller input (applying the same computation to setpoint and measurement without changing their operationally displayed form).
3. Linearization compensation of the controller at its output, which requires the application of the compensation to the output, the inversion of that same compensation (as described below) into the external feedback, and the appropriate backpassage of any status information. A modified form of this structure is needed to support feedforward control, in which the disturbance signal (with any dynamic compensation) is brought into the compensating calculation as a kind of computing parameter.[4]

---

[4] A later distinction will be made about the bumpless application of tuning parameter changes. The feedforward signal will usually be intended to incorporate both any appropriate parameter change and the compensating disturbance bump!

3a. Partly for historical reasons, ratioing control is usually included as an operationally accessible set-point parameter multiplying a feedforward measurement to generate the true controller reference input.

A discussion below will show a branching out of the degree-of-freedom paths, matching the constraint branching together. The collective structures allow the implementation of all normal single-loop control. And each of the structures has multivariable counterparts.

NUMBER SYSTEMS AND BASIC ARITHMETIC

Operation on analog data involves not simply converting to decimal numbers, but also converting to the format that those numbers take in the digital computer. There are two such formats: fixed point for representing integers and floating point for representing real numbers. Even though analog data are usually conceived of in terms of real numbers, the most effective processing of continuous control data, particularly for small microprocessors, is in fixed-point arithmetic.

In this case, the data must be (painfully) scaled in the same way that analog control systems and simulations were scaled. Scaling and multiple-precision computations and conversions are the explicit responsibility of the control algorithm programmer. Apart from computational necessity, scaling is an art that is inherently significant to proper control design, since the process is itself fixed point. Meaningful control actions can be guaranteed only if the user is aware that the scale of the calculations is in fact appropriate to the process.

Floating-point data are automatically scaled, in that they have a fractional part f or mantissa, corresponding to the integer part of fixed-point data, and an exponent part e, which defines an automatically adjusted scaling factor equal to a base value b (usually 2) raised to that exponent. Although the fractional part is usually viewed as a fraction, the discussion will be less confusing if e is chosen so that f is an appropriate integer; any such value is then expressible entirely in terms of integers, as $f \times b^e$.

In the past, floating-point formats were not standardized [9]. Moreover, the floating-point format inherently involves arbitrary truncations and uncertain relationships between single- and double-precision computation. For this reason, fixed-point algorithms still permit the most precise designs.

Fixed-Point Format

A fixed-point format represents an integer as a binary number stored in a word containing a fixed number (n, below) of bits, usually in what is called two's complement format. This format represents positive numbers in a range of 0 to $2^n - 1$ and signed numbers in a range of -2^{n-1} to $2^{n-1} - 1$. In two's complement arithmetic, negative numbers are represented in binary notation as if they were larger positive integers. As a result, when the numbers are added, the natural additive truncation achieves the effect of signed addition.

Thus, in Fig. 3, (with $n = 3$) negative 2 is represented by a binary number corresponding to a positive 6. And 6 added to 2 becomes an 8 that, with truncation of the carry, corresponds to 0. Thus the 6 is a perfectly good negative 2.

Two's complement arithmetic is also related to modular, or around-the-clock, arithmetic. Because of this behavior, results too large to fit into a word must be taken care of in one of three ways:

- Data values can be distributed in a sufficiently large set of data words to represent them. The arithmetic hardware or software then utilizes carry bits to pass the data between words when a carry is required.
- Results too large to fit a word may be saturated.[5]

---

[5] That is, a value too large to fit the word is replaced with the largest number of the right sign that will fit.

Sign and Value	Two's Complement	
Decimal	Binary	Positive Decimal Equivalent
0	000	0
1	001	1
2	010	2
3	011	3
–4	100	4
–3	101	5
–2	110	6
–1	Carry bit 111	7
(2)+(–2)	1000	(3)+(5)=8

FIGURE 3   Two's complement arithmetic.

- One may ignore the problem in parts of the calculation for which it is clear that results will never overflow.[6]

The remaining discussion will be framed in decimal arithmetic, rather than in the unfamiliar binary arithmetic, for clarity's sake. Sufficient to say, actual designs are implemented by good fixed-point hardware including the basic binary operations and carry/overload bits to support effective single- and multiple-precision arithmetic. For this reason, multiplication and divisions of powers of 2 will be particularly efficient, to be preferred when there is a choice.

Fixed-Point Scaling

In the discussion of multiple precision and scaling, it is convenient to distinguish different tag ends to the variable name to represent different aspects of the variables under consideration. Thus if V is the name of a variable

- $V.S$ will become the scaled representation of the variable, taken as a whole.
- $V.0, V.1, V.2, \ldots, V.m$ (or $V.S.0, V.S.1, V.S.2, \ldots, V.S.m$) will be different storage words making up the multiple-precision representation of V (or $V.S$), with $V.0$ being the right-most word and $V.1$ being the next right-most word, etc., as shown in Fig. 4.
- When V is scaled by a fixed-point fraction, the numerator will be $V.N$ and the denominator will be $V.D$ [either of which may be itself in multiple-precision form (with $V.N.0, V.N.1, \ldots,$ or $V.D.0, V.D.1, \ldots$]:

$$V.S = \frac{V \times V.N}{V.D}$$

V	V.0	V.1	V.2
123456	12	34	56

FIGURE 4   Decimal multiple precision.

Normally the actual scaling computations will take place only when the data is processed for input/output (I/O) or display; control calculations will generally take place with respect to $V.S$. For this reason, when no confusion arises, the discussion will refer to V, substituting for $V.S$. Scaling conversions, like binary and decimal arithmetic and conversions, are straightforward though tedious; they need to be addressed further.

The special power of this notation is that it allows all scaling and multiple-precision computations to be expressed entirely in conventional algebra. In particular, for analytical purposes, a

---

[6] For which they will never become too large or too small to fit the word. This is a dangerous approach.

multiple-precision value can be expressed and operated on simply as a sum of the normal values $V.0, V.1, V.2, \ldots, V.m$, each with its own scaling factor:

$$V = V.0 \times B^m + V.1 \times B^{m-1} + V.2 \times B^{m-2} + \cdots + V.m$$

The value B is one greater than the largest positive number represented in a single-precision word ($B = 2^n$ for unsigned binary data; $B = 2^{n-1}$ for signed data).

Control algorithm parameter scaling is chosen according to need, computational convenience, and best use of the available data range, working with single-precision representation as much as possible. For a controller measurement or value on a 16-bit word machine, $1/2\%$ precision is normally minimally adequate. For positive data, a byte represents a data range of 0–255 ($2^8 - 1$), which is better than $1/2\%$ precision.

On the other hand, with a 16-bit word, there are 8 more bits in the word that could be used to give a smoother valve action and simplify calculations. When only positive values are worked with, 16 bits correspond to the normally unnecessary precision of 1 part in 65,535; with signed data, 1 part in 32,767; or, with a 1-bit safety margin, 1 part in 16,384. This scaling is well above the minimum process data precision, while still not forcing multiple-precision arithmetic in most cases. The corresponding 14-bit analog-to-digital (A/D) and digital-to-analog (D/A) converters for the process data are still reasonable.

Control parameters may have different natural data ranges or scalings. A controller gain might be scaled so that the minimum gain is $1/256$ and the maximum gain is 256. In this way the range of values equally spans high-gain and low-gain processes about a nominal unity gain. Time constants may require a certain resolution. When a minimum reset time of 1 s is adequate (1 h corresponds to 3600 s), 2^{16} corresponds to more than 18 h.

On the other hand, using the full range of data storage for the control parameters may require arithmetic routines that mix signed and unsigned arguments. At this level, there is a trade-off between the increased efficiency and the added complexity. Certainly it is more convenient to program uniformly in signed fixed-point (or even floating-point) arithmetic. But the costs of this convenience are also significant.

Range and Error in Fixed-Point Arithmetic

Good design for quantization and multiple-precision avoids poor control arising from inaccuracies not inherent in the real process. Normal control algorithms involve the standard combinations of additions, subtractions, multiplications, and divisions; rational functions of their data. Often the basic calculation allows many ways to order these calculations that are theoretically identical in their result as long as precision is indefinite.

Practical fixed-point programming requires a more careful understanding of the basic operations and the effect of ordering on the calculation. First, one should examine how each operation affects the worst-case range[7] and the error accumulation of the result. As will be seen, range effects are typically more important.

Error can be considered in terms of absolute error, i.e., the actual worst-case error, or relative error, i.e., the worst-case error as a percentage of the nominal value or scale. Usually the relative error is important in final results and products (or quotients), whereas the absolute error is important in determining how the error accumulates in a sum.

When two numbers are added or subtracted, their worst-case range doubles, either requiring a carry (into a multiple-precision result) or requiring a sign bit[8] [see Fig. 5(a)]. Additions and subtractions also cause the absolute error to increase. Addition of numbers of the same sign can never increase the relative error over the worst error of the added numbers. But their subtraction increases it, as does addition of numbers whose sign is uncertain. In Fig. 5(b),[9] two 10% relative error numbers, when

---

[7] The range of result for all possible combinations of the input data.

[8] When positive numbers are subtracted.

[9] In Fig. 5(b) the underlined numbers represent an error term being added or subtracted from the "ideal" value of a computational input or output.

$$(20 \pm 2) + (10 \pm 1) \;=\; (30 \pm 3)$$

$$99 + \;\;99 \;= 198$$
$$99 + (-99) = 0$$

$$\frac{(20 \pm 2) + (10 \pm 1)}{20 + 10} = \left(\frac{30}{30} \pm \frac{3}{30}\right)$$

$$99 - \;\;\;0 \;\;= 99$$
$$\;\;0 - \;\;99 \;= -99$$

$$\frac{(20 \pm 2) - (10 \pm 1)}{20 - 10} = \left(\frac{10}{10} \pm \frac{3}{10}\right)$$

(a) Range effects   (b) Error effects

FIGURE 5   Addition and subtraction.

$$\overline{99} \times \overline{99} = \overline{98}\;\overline{01}$$

$$\overline{98}\;\overline{01} \,/\, \overline{99} = \;\;\;\overline{99}$$

$$\overline{98}\;\overline{00} \,/\, \overline{99} = \;\;\;\overline{98}\;\text{Rem.: } \overline{98}$$

$$\text{but: } \overline{98}\;\overline{00} \,/\, \overline{1} = \overline{98}\;\overline{00}$$

$$(10 \pm 1) \times (10 \pm 1) = (100 \pm 21)$$

$$\frac{(100 \pm 8)}{(10 \pm 1)} \;=\; 10 \pm 2$$

(a) Range effects   (b) Error effects

FIGURE 6   Multiplication and division.

added, result in a 10% relative error. But when subtracted, the relative error can be arbitrarily large (30%) in this case. In this way, differences between large numbers explain most computing accuracy problems.

Fixed-point multiplication does more than just increase the range of the result; it doubles the required storage of the result [see Fig. 6(a)]. For this reason, most hardware implements single-precision multipliers to return a double-word result. By analogy, most hardware division divides a double-precision dividend by a single-precision divisor to obtain a single-precision quotient with a single-precision remainder. But, as Fig. 6(a) shows, such a division can still give rise to a double-precision quotient. The hardware expresses this as an overload (similar to division by zero).

Fixed-point hardware is designed to permit the effective programming of multiple-precision arithmetic. One of the strengths of fixed-point arithmetic, with its remainders, overflow and carry bits, and multiple-precision results is that no information is lost except by choice. The precision, at any point in the calculation, is entirely under the control of the programmer. This is not true of floating-point arithmetic. As shown in Fig. 6(b), multiplication and division always increase (e.g., double) relative error.

Fixed-Point Multiplication and Division

Multiplication's range explosion is its most problematic aspect. Among other consequences, it generates large numbers whose differences may cause large errors. It makes smaller numbers still smaller compared with errors caused by other large numbers. The important issue is the ratio of the large numbers (which make the errors) to the small numbers (which end up as the result). Multiplication's range expansion forces a choice: precision can be preserved or intermediate values can be rescaled and truncated back to their original data size.

Thus it is usually desirable to avoid repeated multiplications. Often multiplications and divisions occur together in a manner in which they can be alternated, the multiplication generating a double-precision value and the division returning that value to a single-precision quotient. The basic proportional controller calculation is a good example:

$$\text{Output} = \frac{100 \times \text{error}}{\text{Proportional Band}} + \text{bias}$$

Common combined operations such as this may usefully call for specially designed routines. In general, one should try to maintain a constant data range and size throughout the computation. Often a product is naturally intended to return a value in the same range as a process variable (output). There are two natural cases: either a value is multiplied by a gain or by a proper fraction (Fig. 7). In the controller calculation the 100/Proportional Band gain might be limited to the range of 1/256–256. If the gain is expressed as a single-precision integer (with a decimal point effectively in the middle of its digits) rather than as a fraction, one can still generate an appropriate single-precision result by taking the middle single-precision set of digits out of the double-precision scaled result [Fig. 7(a)], saturating and truncating the extraneous data.

(a): $\overbrace{\text{XX.XX}}^{\text{Gain}} \times \overbrace{\text{XXXX}}^{\text{Natural Range}} = \overbrace{\text{XXXXXX.XX}}^{\text{Natural Range}}$

(b): $\overbrace{\text{.XXXX}}^{\text{Fraction}} \times \overbrace{\text{XXXX}}^{\text{Natural Range}} = \overbrace{\text{XXXX.XXXX}}^{\text{Natural Range}}$

FIGURE 7 Scaled multiplication of (a) gain, (b) fraction.

As a common case of proper fractional multiplication, a lag calculation (time constant T) can calculate its output X as a weighted average of the past output Y' and the current input X:

$$Y = \frac{1}{T+1}X + \frac{T}{T+1}Y' = Y' + \frac{1}{T+1}(X - Y')$$

In this case, when the proper fractions are multiplied (scaled as integers), the proper result can nevertheless be returned as the left-most part of the double-precision result [Fig. 7(b)]. It is often appropriate to combine constants or parameters together into a common effective parameter that is the above kind of gain or proper fraction. This ensures that the linear operation with the process data is truncated by only the one final multiplication.

The combined parameters can be made to act consistently on the data even if in error as calculated; the errors can be reinterpreted as (presumably insignificant) errors in the original parameters. Note that the right-most expression within the above equation involves a feedback between Y and the difference between X and Y'. Often such a feedback can be included in the calculation to improve an otherwise error-prone algorithm, making it self-corrective, like any other feedback system.

There is a parallel between the above range discussion and traditional dimensional analysis [10]: sums must be of data elements with identical units and similar ranges. Multiplications change the units and also change the range. The ultimate purpose of any of our calculations is to convert data of limited range from the process to data with limited range to drive the process. Thus, whatever the internal gyrations, the process constrains the results to be reasonable. The challenge is to carry out the calculations so that the inherently limited practical range prevails throughout the calculation.

Digital Integration for Control

Figure 8 shows a different kind of division problem: process control integration. The object is to integrate the process error in a control (reset) calculation. This can be done by computation of a shared fractional multiplier $\Delta t/T$, which has been scaled to give a good range [which might also include the proportional band action PB, as in $100 \times \Delta t/(PB \times T)$]. This result would be multiplied by the error and added in double precision to the previous integrated (i.e., summed) value to get the current control value.

Double precision is essential here, since a large value of T corresponds to a small $\Delta t/T$ and a truncation loss of significant process errors. For example, suppose that the error and the sum are both scaled as signed single-precision integers, ranging from $-10,000$ to $+10,000$. With $\Delta t = 1$ and the minimum $T = 1$ (corresponding to 1 s), the corresponding maximum $\Delta t/T$, which equals 1, must be scaled to a value of 10,000. The natural scaling of the controller output can be achieved by dividing by 10,000 (see later scaling discussion below). That is, the scaled equations should be

$$\text{Sum}.S = \text{oldSum}.S + \left[\frac{(\Delta t/T).S \times \text{Error}.S}{10,000}\right]$$

$$\sum_{i=0}^{n} \frac{Error\,(i)\,\Delta t}{T} =$$

$$\frac{\sum_{i=0}^{n} Error\,(i)\,\Delta t}{T} = \frac{\sum_{i=0}^{n-1} Error\,(i)\,\Delta t}{T} + Error(n)\,\frac{\Delta t}{T}$$

$$= \sum_{i=0}^{n-1} \frac{Error\,(i)\,\Delta t}{T} + Quotient\left(\frac{Error(n)\,\Delta t\ +\ Remainder}{T}\right)$$

Note:

$$Error\,(n)\,\Delta t\ =\ T \cdot Quotient\left(\frac{Error(n)\,\Delta t + Remainder}{T}\right)$$

FIGURE 8   An integration trick.

In normal single-precision division, only the integer quotient is considered. Thus a product $[(\Delta t/T).S \times Error.S]$ less than 10,000 is truncated as zero: The error is ignored and the integration stops, causing a permanent offset. For T large, equal to 10,000 (which corresponds to $\sim$3 h) and $(\Delta t/T).S$ small (equal to 1.0 in this case), any error less than 100% (scaled to 10,000) will be lost. Under control the result is a 100% offset. Double precision alleviates the offset but still loses some information to truncation.[10]

The better method, shown in Fig. 8, preserves the division and achieves an exact result with the same storage as double precision. In this case, the product of the error and the sample time (the sample time may equal 1) is formed as a double-precision value, then added to the remainder from the previous sample calculation's division. This net value is divided by T to get back a single-precision quotient to be added to the old sum, and a remainder to be saved for the next sample time. Data results truncated from any quotient are never lost, but preserved and accumulated in the remainder, to show up in some later quotient.[11]

Floating-Point Format

Modern microprocessors are often supported with high-speed floating-point processors. Without these, floating-point calculations run an order of magnitude slower than fixed-point calculations. With the floating-point format, all remainders, carries, and multiple-precision results of single-precision computations disappear. Instead, the user chooses a fixed level of precision whose truncations show up in guard bits [9].

Floating-point errors introduced in small differences of large numbers become more severe, and at the same time less obvious because their processing is automated and invisible. For example, when a large number L (e.g., 10,000, stored to three-digit accuracy) is added to a small number S (e.g., 1), the smaller number totally disappears. The sum $L - L + S$ should equal S. If the calculation is carried out in the natural ordering, $(L - L) + S$, then S will result. But the nominally equivalent $(L + S) - L$ will generate zero.

---

[10] One alternative to multiple-precision integration uses random numbers (or dither, in mechanical engineering terms). The computation is carried out in normal double precision. However the higher-precision (lower-order) data word is ignored and not saved; it is replaced with a random number in the next sample time's calculation when it is needed again. This method works well practically and theoretically. A similar method was incorporated into the Digital Equipment Corp. PDP-1 floating-point package. But who would have the courage to use it on a real process?

[11] As pointed out in the next section, floating point lacks the support for such refined tricks. It is particularly subject to integration problems because a large sum may be big enough to prevent the addition of a small term, even though that term is rescaled to avoid its being truncated to zero.

There are a number of intricate ways of avoiding this problem:

- Convert all floating-point numbers to ratios of integers and operate in the fixed-point format. This is one way to study the properties of the algorithm.
- Use adequate precision. Practically this is often unpredictable, and theoretically it is impossible because repeating decimals require infinite data. This suggests also deferring division to the last operation.
- Reorder the calculation for the most favorable computation, either statically based on algorithm properties or by using an arithmetic package that continually reorders the operands. It is useful, in the following discussion, to consider every list of terms to be added (subtraction being taken as addition of a negated number) to be ordered by magnitude. The individual operations are carried out according to these rules.

Under addition or subtraction

- combine small numbers together before large numbers (to let them accumulate before being lost),
- take differences (subtractions or additions of numbers of opposite sign) before sums (to achieve all possible subtractive cancellation between larger numbers before these can lose the small numbers),
- from a list of numbers to be multiplied and divided, cancel or divide out most nearly equal numbers first (to minimize floating-point overload[12]),
- when all denominator terms are gone, multiply largest with smallest numbers (to minimize floating-point overflow).

One other advantage of following such a set of ordering rules is that it will give identical results to identical data even when they originally occurred in a different programmed order.

Generalized Multiple-Precision Floating-Point

Normally, the multiple-precision floating-point format is the same as the single precision with larger fraction and exponent data fields. The author has experimented with a more open-ended multiple-precision floating-point, illustrated in Fig. 9. The multiple-precision floating-point number is represented by a summed, ordered set of signed single-precision floating-point numbers, each mantissa having M digits.[13]

The numbers in the set are chosen so that the set can be truncated at any point to give the best possible truncated approximation. In this case, the value is considered to be made up of the truncated value and a signed ($\pm$) remainder that expresses the part cut off in truncation.[14] The generalized floating point would be supported by the following operations:

$$12340.36789 \Rightarrow 1234 \times 10^1 + 3679 \times 10^{-4} - 1000 \times 10^{-8}$$

Addition:
$$A = 3057 \times 10^6 \quad B = 4263 \times 10^4$$
$$A + B = 309963 \times 10^4 = 3100 \times 10^6 - 3700 \times 10^2$$

FIGURE 9   Generalized multiple-precision floating point ($M = 4$).

---

[12] The process of the floating-point exponent's getting too large for the provided data space.

[13] Analogous to multiple-precision fixed-point data represented as a summed scaled set of single-precision numbers. Each element is normalized (shifted), to use the full range of M digits. Recall that the general representation of a single-precision floating-point number is $f \times b^e$, where f, e, and b are integers. As above, the format is described in terms of a general b and illustrated with $b = 10$. In practice b will equal 2. Note that the different members in the summed set may have different signs! However, the sign of the total (set) value is still the sign of its initial and largest element.

[14] The remainder magnitude is always less than half the unit value of the next larger element in the set, since the best approximation requires that the remainder round to zero.

- Text conversion, converting to or from text general precision floating point to the internal format consisting of the summed set of single-precision values.
- (Set) Normalization, reprocessing the set of single-precision values so that they are ordered in magnitude, with largest first,[15] so that the magnitude of each mantissa is between b^{M-1} and b^M and so that consecutive values have exponents whose difference is greater than M (or, if the difference equals M,[16] then the magnitude of the second mantissa is less than or equal to $b^M/2$) [11].
- Addition and subtraction, merging entries into a final value set followed by a normalization of that set.
- Multiplication, multiplying of every pair of elements, one from each of the multiplicand and multiplier, to generate a double-precision result, followed by the merging and the normalization of the accumulated set of results.
- Division, dividing the largest elements in the dividend by the divisor, subtracting the divisor multiplied by the quotient from the dividend to obtain the remainder. This division is designed to select that quotient that returns the remainder with the smallest magnitude (of whatever sign). The remainder can be redivided by the divisor to compute any desired level of multiprecision quotients, with the final resulting quotient and remainder being normalized. The remainder so developed can be used in the earlier integration procedure.

Such a generalized floating point can be used to develop calculations whose precision expands indefinitely as needed. Such a system could give absolutely error-free results, without any special care.

SPECIFICATION OF FIXED-POINT ALGORITHMS

Clear fixed-point specification includes the proper statement of computation order and of scaling of intermediate and final results. This can be superposed on a conventional algebraic notation. For example, in the following control computation, the parentheses define any required ordering:

$$\text{Output}_5 = \left(\left(\frac{100 \times \text{Error}_1}{\text{Proportional Band}_2}\right)_4 + \text{Bias}_3\right)_5$$

Unparenthesized addition and subtraction are assumed to be from left to right and multiplication and division are assumed to alternate, as described above. Any constants that can be combined would certainly be combined in a working system. This should include any scaling constants. The subscripts refer to scaling specifications in the table of Fig. 10 below.

Such a table would completely specify all scalings, saturations, and conversions appropriate to the values within the calculation. Remembering that the internal scaling is reflected in the relation

$$V.S = \frac{V.N \times V}{V.D}$$

Subscript	I/O	Saturation Hi	Saturation Lo	V.N	V.D	Sign	Precision
1	IN	100	-100	16384	100	±	1
2	IN	16384	0	2	1	+	1
3	IN	100	-100	16384	100	±	1
4	–		YES	16384	200	±	1
5	OUT	100	0	16384	100	+	1

FIGURE 10   Table of variable scaling-related properties.

---

[15] In the form $f_o \times b^{e_o}$, $f_1 \times b^{e_1}$, $f_2 \times b^{e_2}$, ...
[16] Not possible after normalization if M equals 2.

we find that the conversion from functional algebraic equations to internal computational form is then carried out by the computation

$$V = \frac{V.D \times V.S}{V.N}$$

When all of these scalings are incorporated back into the original proportional controller calculations, its internal scaled form becomes

$$Output.S \times \frac{100}{16{,}384} = \frac{100 \times \dfrac{100}{16{,}384} Error.S}{\dfrac{1}{2} \times PropBand.S} + \frac{100}{16{,}384} \times Bias.S$$

or

$$Output.S = \left(\left(\frac{200 \times Error.S}{PropBand.S} \right) + Bias.S \right)$$

Such a simplification is to be expected in practical calculations as part of the combination of similar scaling terms and application constants. The final equation then becomes the programmed calculation, except that each operation would be saturated as specified. Saturation can, in fact, make the combination of terms invalid, but in this case, it may be worth considering whether or not the saturations might not better be left out, in the interests of a more perfect result.

Definition and implementation of an algorithm then have three parts:

- specification and programming of the necessary arithmetic and saturation routines,
- manual development of the scaled calculation, in terms of the routines,
- programming of the algorithm.

Ideally, an algorithm should be tried out first in a higher-level language (e.g., FORTRAN or C). Here it can first be expressed in floating point and then in scaled fixed point format. If it is later needed that the final form be in machine language, the three different forms can be run comparatively, greatly facilitating debugging.

OPERATIONAL ISSUES

The basic controls are normally expressed as linear algorithms, defined as if the process measurements and actuators were capable of perfect operation to whatever range might be needed. In fact, valves limit and sensors fail. The algorithms must be designed to accommodate temporary valve saturation or loss of sensor data. Moreover, they must be designed to allow the system to be restarted smoothly after longer-term failure or shutdowns.

The most common such problem is windup: the property of the controller that continues to integrate under error even after the actuator has limited, and is incapable of, further change. Windup requires recovery time even after the error reverses sign, blocking effective control, for that interval. The response to this problem requires that the algorithm be provided some indication of the limiting so that it can alter its behavior. The information can take the form of flags that inform the controller of the limiting actions being taken, propagated back to any affected controller. The flag then causes the integration to stop in some appropriate way.

A superior approach, called external feedback, senses the actual state of the manipulated variable and feeds it back, in comparison with its intended value. By working with the true state of the process, the algorithm can make much more refined accommodation strategies. Controller windup is not the only problem arising from actuator limiting; Blend pacing, split-range control, and multiple output control all relate to standard affects of valve limiting and its accommodation. The external-feedback strategy effectively unifies the handling of all of these issues.[17]

The problem is complicated because the effects of limiting must propagate back from the actual valve, through any control functions (e.g., cascaded controllers, ratio units), to the controller under consideration. Thus there is not only a need to notify controllers of valve limitings or sensor failures but to propagate this information to all affected control elements. Software must be designed to accomplish the propagation.

The external-feedback approach is particularly advantageous because it recognizes loss of control locally to the affected controller and responds only when the loss is material to it. Of course, the loss of control at any level may be a useful basis for alarming, independent of immediately recognized effects on control. The present discussion is largely limited to the algorithmic consequences of the problem, but the software consequences are just as important.

A common problem with ad hoc controller designs is that they bump the output whenever the parameters are changed for tuning. The PI controller computation below illustrates the problem:

$$O(t) = \left(\text{Error} + \frac{1}{\tau} \int_0^t \text{Error} \cdot dt \right) \frac{100}{PB}$$

$$= \frac{100}{PB} \text{Error} + \int_0^t \frac{100}{PB \cdot \tau} \text{Error} \cdot dt$$

The first of two nominally equivalent expressions computes the output with the tunings acting after the integration. The practical result is that any change in settings immediately bumps the output. In the second expression, the tunings act on the Error value, before integration. Any tuning changes affect only integrated errors occurring after the change.

Apart from output limiting and tuning bumps, there are a number of similar operational modalities that the controller should support, in relation to either manual or automatic operation. For example,

- Cold-start initialization: Sometimes it is desirable as part of the process start-up to start the controller in automatic mode, so that its output has no tendency to move, letting operators move the process on their own schedules. This is also the natural way to initialize secondary controllers when the primary controller goes to automatic mode. Cold start can be implemented if the controller set point is set to the present measurement value and the internal states of the controller are initialized so that the output matches the externally fedback or operator-set value of the output. Automatic control then continues naturally.

- Bumpless transfer: The purpose of this mode is to transfer control to automatic control in such a way that the process does not receive any immediate valve change, but moves from its prior position smoothly. In this case, the controller setpoint is left in place, but all internal states are initialized to be consistent with the current error and external feedback (and current controller output value).[18]

- Batch preload: In circumstances in which a known set-point change is to be applied (for example, in batch production), the controller may be set up to pick up on the set-point change with a preset initial internal integration value. The purpose is to give the process an initial kick, to get it to the set point in the fastest time. This strategy has a number of elaborations of varying sophistication.

- Ramped set-point change: The controller may be designed to limit the rate at which it responds to set-point changes to minimize the bumps to the process.

---

[17] A full accommodation of actuator or cascaded control loss would include both external feedback and the flags because some control functions do not lend themselves to the external-feedback solution. For example, some adaptive controllers depend on free response to their output to support meaningful adaptation.

[18] This mode can be applied to all controllers in a cascade, but because the secondary set points will then be matched by their primary controllers to their sensor, the result for them should be the same as if they had been initialized under a cold start. This is true only if the controller calculations are carried out in an appropriate order relative to each other. In lieu of this, it may be better to use the cold start on the secondaries anyway.

The variations on cold start and bumpless transfer depend on backcalculation: the recomputation of internal data[19] to be consistent with the unchanged output and external feedback. For example, a PID computation developed below computes its output O from the collective effect X of the proportional and the derivative effects and an internal bias B. The bias is computed in turn by application of a lag computation to the external feedback O_{FB}:

$$O(t) = X(t) + B(t)$$

$$B(t) = \frac{O_{FB}(t) \cdot \Delta t + B(t - \Delta t) \cdot \tau}{\tau + \Delta t}$$

We can accomplish the bumpless transfer by computing the new value of X and then rearranging the first equation above to backcalculate the bias from the new X and the old B:

$$B(t) = O(t) - X(t)$$

When the O is later calculated from this B, its value will remain initially at its old value, irrespective of manual changes in O, measurement, set point, or tuning changes reflected in X. The second equation might be used to backcalculate O_{FB}, but this value will be overridden by later computations anyway.

OUTPUT LIMITING: EXTERNAL FEEDBACK

As already indicated, external feedback represents a precise and smooth way of handling windup. However, implemented in a PID controller algorithm, it has the effect of including, in the integrated term, the difference between the controller output and the corresponding externally fedback measured state. The difference is added to oppose the normal integration. Thus, whenever the external feedback fails to follow the controller output, the difference builds up to stop the integration.

Feedback controllers are built about the processing of an error signal on their input. External feedback extends the principal to the output. It allows the control algorithm to be designed to alter its approach in the face of the output's failure to act. This same strategy can be generalized to apply to any control function:

- Blending: When several ingredient flows are ratioed to generate a blended product, two basic product properties are involved, the product quality and the product flow. Under the standard strategy of pacing, if one ingredient flow limits, the remaining flows are ratioed, not off their original target, but off the external feedback from the limited flow. Such a system extends the external-feedback concept to give up the control of product flow in favor of the more important product quality.

- Fuel/air ratio control: With certain liquid fuels, an excess accumulation of fuel in the burners constitutes a fire hazard. The fuel controller is designed to limit the fuel to be less than the combustible ratio to the measured air flow, at the same time limiting the air flow to be greater than the combustible ratio to the measured fuel flow. If either limits, the other is held to a safe value. Such a system extends the external-feedback concept to safety control.

- Multiple output control: In certain cases a manipulated resource may be duplicated so that several devices share a load. It is desirable that operating personnel be able to take one or more of such devices out of service in such a way that the others take up the load. In this case, a multiple output controller computes an output value such that when the value is ratioed as the set point to each device, the sum of the external feedbacks from all devices equals the net desired load. In this way, as one device is taken over in manual, the others will take up the slack.

---

[19] Particularly integration data. These guarantee that the integration will resume as if the controller had always been operating under the current error and output conditions.

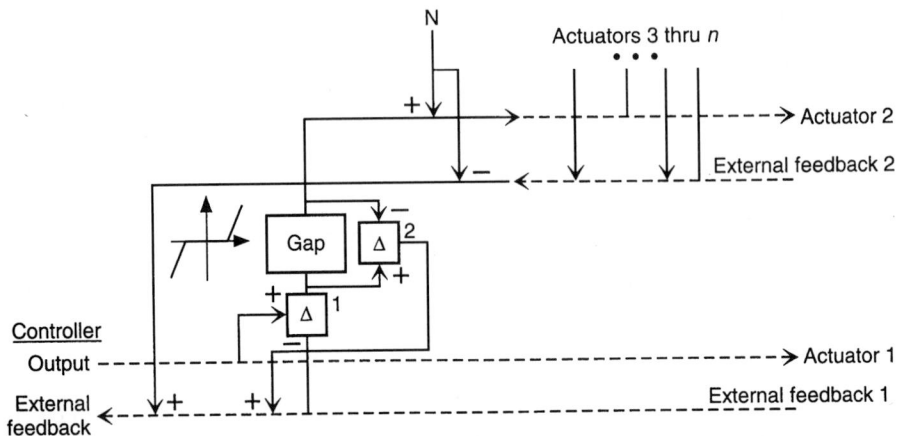

FIGURE 11 External-feedback-based backup.

- Backup (also split-range control): The normal external-feedback antiwindup action can be extended, by the recognized output/external-feedback error, to drive other actuators. Figure 11 illustrates a refined handling of this function.

  In this design, the controller output is normally connected to actuator 1. But if the calculated difference [the subtraction element (Δ) marked 1] between actuator 1 and its external feedback exceeds the gap parameter, the excess difference will serve to transfer the active control to actuator 2. In this way the controller can be backed up against any downstream failures or overrides. The nominal bias N reflects the preferred inactive value for actuator 2. (In more refined designs, N may be established dynamically by a separate controller.) Differences can further be passed on indefinitely to actuators 3 through n. Effectively this arrangement generalizes the behavior of split-range control, taking into account any downstream loss of control action.

  External feedback 2 is added into the controller external feedback to allow the controller to continue its full (integrating) control action, whatever actuators are in fact acting. The purpose of the gap is to ensure that actuator transfer does not give rise to chattering between actuators but acts only for significant loss of actuator control. However, to guarantee that the resulting temporary loss of control does not cause the controller to stop integrating, the actual amount of gapping action (the Δ marked 2) is added into the controller external feedback as well; the external feedback sees neither Gap nor actuator transfer.

  In actuality, the different actuators might call for different control dynamics and different compensation. This could be built into the control transferring paths. However, digital implementation allows the switching of controller tunings, as a function of the active actuator, to be carried out as part of the controller computation, a more natural arrangement. Digital implementation also allows the above structure to be black boxed flexibly, taking the confusing details out of the hands of the user.

- Linear programs and optimization: It has been argued that external feedback is incapable of dealing with connection to higher-level supervisory functions such as linear programs or optimizers. This position reflects higher-level functions not designed for operations, rather than any inherent problem with external feedback. The operationally correct optimizer will benefit from external-feedback data like any other control computation. In this case, each optimization target, with its external feedback, is associated with an implied output constraint. Whenever a difference develops between the two, the constraint limit and the violation become apparent.

  Thus the external-feedback value should be fed into the optimizer, parameterizing a corresponding optimization constraint. There are three special considerations:

- The control action actually implemented must push beyond the constraint so recognized. Otherwise, the constraint becomes a self-fulfilling prophecy that, once established, never gets retracted. Since the control action is presumed to be up against a real process constraint, it does not matter how much further into the constraint the target variable is set. However, it is probably better to exceed the constraint by some small number (e.g., 1%–5% of scale).
- The optimization computation is likely to be run infrequently compared with the normal regulatory dynamics. For this reason some lag filtering or averaging should be built into all external-feedback paths to minimize noise effects and increase their meaningful information content. The filter time constant should correspond to the optimization repetition interval.
- All of this assumes that the optimizer addresses the economic constraint dimensions only. Significant safety or quality constraint effects must always be separately addressed at the regulatory level.

With other control and operational nonlinearities, many issues come up, calling for many different kinds of thinking. Of course, these same differences must fit nonintrusively and naturally with the intentions and expectations of operating people. While operational users will normally not be aware of the technology behind these techniques, they will become intuitively aware of any inconsistencies between the handling of similar functions in different control elements. External feedback provides a powerful strategy for addressing many of these problems whose application uniformity the end user will appreciate.

EXTERNAL FEEDBACK IN NONLINEAR COMPENSATORS

In the introductory discussion of Fig. 2(c), the third form of compensation called for inversion of the associated nonlinear compensation. Traditionally this has been done with simple, analytically determined inverses. For example,

- A squaring of the direct compensation called for a square root to the external feedback; an exponential, a log.
- In feedforwards a subtraction of the feedforward signal inverted an addition; a division inverted a multiplication.

Historically, general inverting rules (like Newton's method) would not have been used for fear of failure to converge. Figure 12 shows a situation (the left graph) in which Newton's method, based on extrapolation of the derivative of the function, could successively overshoot the solution of the function (the intersection of the function graph and the shaded horizontal line) on each side. But if the interpolation is always based on prior guesses that bracket the solution (as on the right graph),

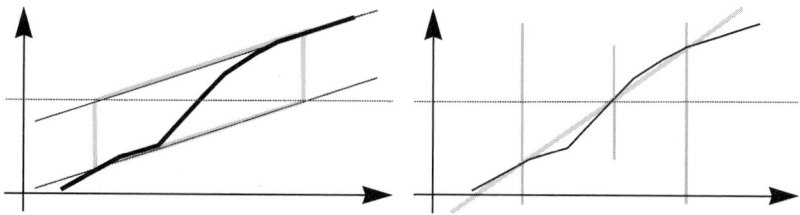

FIGURE 12   Newton's method vs. secant variant.

convergence is guaranteed. These kinds of methods can be generalized to invert multivariable functions, and find all solutions [12].

BASIC CONTROL ALGORITHMS

The Lag Calculation

The lag calculation corresponds to the following continuous transfer function:

$$\frac{L(s)}{I(s)} = \frac{1}{(\tau s + 1)}$$

where L is the output, I is the input, and τ is the lag time constant. The usual practice of going to the z transform for the corresponding sampled-data form should not be overemphasized. No exact approximation is possible. Instead, the algorithms are designed primarily to avoid operationally unnatural behavior. With this in mind, the best direct-sampled-data approximation of the above differential equation is

$$I(t) = \frac{\tau \Delta L}{\Delta t} + L(t) = \frac{\tau}{\Delta t}[L(t) - L(t - \Delta t)] + L(t)$$

or

$$L(t) = \frac{I(t) + \dfrac{\tau}{\Delta t}L(t - \Delta t)}{1 + \dfrac{\tau}{\Delta t}} = \frac{I(t)\Delta t + \tau L(t - \Delta t)}{\tau + \Delta t}$$

This approximation amounts to a weighted average of the new input and the old output. From the scaling point of view, each of the products has the same range as the sum; scaling is simple. The calculation is stable for all positive τ, accurate for large τ, and qualitatively natural as τ approaches 0, with $L(t)$ equaling $I(t)$ if $\tau = 0$, as intended.

The calculation is usable even in single precision if the term $I\Delta t$ is truncated up in magnitude (and τ is not too large). In this case, the product is never truncated to zero unless the product is truly zero. This guarantees that the output will always settle out at any steady-state input value. Normal truncation would leave the result below its theoretical steady-state value, a situation similar to the integral offset described above.

But a trick, similar to the one used with the integrators before, can be applied to calculating lags exactly:

$$L(t) = \text{quotient}\left[\frac{I(t)\Delta t + \tau L(t - \Delta t) + \text{remainder}}{\tau + \Delta t}\right]$$

with the remainder being saved for use in the next sampled calculation.

Lead/Lag Calculation

Filtered derivative and lead/lag calculations are most easily and reliably developed from the above lag calculation by analogy with transfer function calculations:

$$\frac{O(s)}{I(s)} = \frac{\tau \cdot s}{k \cdot \tau \cdot s + 1} = \frac{1}{k}\left[1 - \frac{1}{(k \cdot \tau)s + 1}\right]$$

Similarly,

$$\frac{O(s)}{I(s)} = \frac{\tau \cdot s + 1}{(k \cdot \tau)s + 1} = \frac{1}{k}\left[1 - \frac{1}{(k \cdot \tau)s + 1}\right] + \frac{1}{(k \cdot \tau)s + 1}$$

$$= \frac{1}{k} \times \left[1 + \frac{k - 1}{(k \cdot \tau)s + 1}\right]$$

or

$$\frac{O(s)}{I(s)} = \frac{\sigma \cdot s + 1}{\tau \cdot s + 1} = \frac{1}{\tau}\left(\sigma + \frac{\tau - \sigma}{\tau \cdot s + 1}\right)$$

The translation to digital form consists of carrying out all of the algebraic steps directly and replacing the lag transfer function with the digital lag algorithmic calculation described in the preceding section. Considering the last form and proceeding in reverse order, we would calculate the output of a lead/lag from the output of the lag calculation (with time constant τ):

$$O(t) = \frac{[\sigma I(t) + (\tau - \sigma)L(t)]}{\tau}$$

A basic filtered derivative can be calculated with a lag calculation, now indicated as L_D, and assuming, typically, $k = 0.1$ (In fixed point, k would be a power of 2:1/8 or 1/16.) and the lag time constant $0.1 \, \tau_D$:

$$D(t) = \frac{1}{k}[I(t) - L_D(t)] = 10[I(t) - L_D(t)]$$

PID Controller Calculation

PID controller designs are expressed in many forms:

$$O(s) = \left(\tau_D s + 1 + \frac{1}{\tau_I s}\right)\frac{100}{PB} \cdot \text{Error}$$

$$O(s) = (\tau_D s + 1)(\tau_{DI} s + 1)\frac{100}{PBs} \cdot \text{Error}$$

$$O(s) = (\tau_D s + 1)\left(1 + \frac{1}{\tau_I s}\right)\frac{100}{PB} \cdot \text{Error}$$

Each of these forms is capable of the same performance as the others, with one exception: the first form is capable of providing complex zeros. There is no generally argued requirement for complex zeros, but this is nonetheless a real distinction.

There is also some disagreement as to whether the $1/\tau$ terms should be replaced with gains (or whether the proportional band terms should be combined into independent proportional, integral, or derivative terms. The reason for giving all terms as gains is that this then places the most stable setting for all terms at zero (not entirely true of a derivative). This argument is pitched to operators. The reason for leaving the terms as above is that the time constants have process-related meaning for engineers who understand the control issues; the separate proportional band then becomes a single stabilizing setting for all terms.

Different implementations also apply different parts of the algorithm differently to the set-point and measurement terms within the error. This reflects that these terms have different effects within the process. Ideally one would provide separate tunings for load and set-point changes. A practical compromise is to apply all three actions to the measurement, but only the proportional and integrating action to the set point. The controller will then be tuned for load disturbances.

The above forms have a particular difficulty if the integrating calculation term is interpreted literally: The integrating term is most naturally translated digitally as

$$\sum_{i=0}^{[t/(\Delta t)]} \frac{\Delta t}{\tau} \cdot \text{Error}\,(i\,\Delta t)$$

However, the individual summed terms get unnaturally large when τ approaches zero. A practical way of bounding the value is to replace τ with $\tau + \Delta t$. When τ approaches zero, this still leaves an unnaturally large term for $i = t/\Delta t$, in competition with the proportional term. The solution is to replace $(t/\Delta t)$ with $(t/\Delta t) - 1$:

$$\sum_{i=0}^{[t/(\Delta t)]-1} \frac{\Delta t}{\tau + \Delta t} \cdot \text{Error}\,(i\,\Delta t)$$

[The summation (integration) can be carried out according to the above discussion.] The result can be justified in another way: Consider the last PID form introduced above:

$$O(s) = (\tau_D s + 1)\left(1 + \frac{1}{\tau_I s}\right)\frac{100}{PB}\cdot\text{Error}$$

The differentiation must be carried out lagged or filtered, as described above. If all but the integrating calculations are calculated as a combined result X (taking into account any separation of the treatment of set point and measurement), the result is:

$$O(s) = \left(1 + \frac{1}{\tau_I s}\right)X(s) = X(s) + \frac{X(s)}{\tau_I s}$$

This has an alternative formulation, which introduces a calculated bias B, particularly convenient for implementing the external feedback:

$$O(s) = X(s) + B(s)$$

$$B(s) = \frac{O_{FB}(s)}{\tau s + 1}$$

where O_{FB} is the external-feedback term, nominally equal to O. When this pair of transfer functions is translated to an algorithm, they become:

$$O(t) = X(t) + B(t)$$

$$B(t) = \text{quotient}\left[\frac{O_{FB}(t)\cdot\Delta t + B(t - \Delta t)\cdot\tau + \text{remainder}}{\tau + \Delta t}\right]$$

The collective effect of this calculation corresponds to the algorithmic expression of the more direct PID form with the modified integration proposed above:

$$O(t) = X(t) + \sum_{i=0}^{[t/(\Delta t)]-1} \frac{\Delta t}{\tau + \Delta t}X(i\,\Delta t)$$

As indicated above, external feedback must be added to this form by subtraction of the difference between output and external feedback from the $X(i\,\Delta t)$ term before integration.

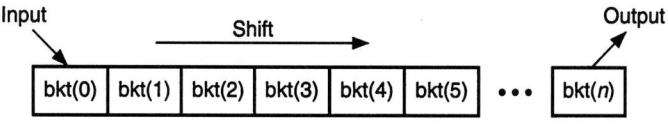

FIGURE 13   Bucket brigade.

Dead-Time Calculation

The dead-time calculation corresponds to the following continuous transfer function:

$$e^{-Ts}$$

Dead time represents a black box whose output exactly repeats the time form of its input, but is delayed by some amount in time. It represents the behavior of the state variables of product, input into a pipe or onto a conveyor belt, and output (delayed) at the other end. It is an essential modeling element for typical processes.

Dead time is conventionally approximated in two ways: through Padé (continued fraction) approximations of the above transfer function and through what is called a bucket brigade. The bucket brigade is implemented in an array of data cells, with the input entered at one end and then shifted down the array, one element per sample time until it comes out at the end, n sample times later, as in Fig. 13.

On the face of it, the Padé is capable of modeling a continuous range of dead times, whereas the bucket brigade is capable of representing only integral delays, for instance, by varying n. Either mechanism can represent a fixed dead time. A further problem arises if the goal is to represent changing dead times. In this case neither the Padé nor the bucket brigade with varying n really reflects the physical or the theoretical behavior of the process.

However, the bucket brigade more closely models the state behavior of product in a delay element; the internal bucket states do represent the internal propagation. The states in the Padé are unrelated to the internal product propagation. It turns out that the continuous Padé will model a changing dead time if the internal parameters are appropriately changed. However, mapping this into a discrete version, taking into account all of the earlier considerations, will be quite difficult.

Thus if the bucket brigade delay can be changed by speeding it up or slowing it down rather than by varying n (the output bucket), it can model changing delay times resulting from changes in flow rates. There are still design questions:

- How does one smoothly achieve dead times smaller than $n \Delta t$ (the number of buckets multiplied by the sample time)?

- How does one smooth the data between stored buckets [between shifts, as shown in Fig. 14(a)[20]]?

- How can one modify the discrete dead time (modeled by an integral number of sample time shifts) to represent a continuous range of (changing) dead times?

The first question is the easiest to answer: shift more than once at a time.[21] The second question is almost as easy to answer: On the input side, average the sampled inputs between shifts; on the output side interpolate between the last two buckets to smooth the effect of the shift [see Figs.14(a)[22] and 14(b)[23]]. The effect of all this is to create a dead time approximation whose effective dead time T corresponds to $(n + 0.5)$ shift times.[24]

---

[20] Figure 14(a) shows the process record (the solid curve), also as sampled (the thin, vertical dashed lines), as effectively sampled by the delay under infrequent shifts (the thicker vertical dashed lines), and as then sampled and held (the more deeply shaded histogram.)

[21] Also average the output values coming from such a multiple shift.

[22] The more lightly shaded histogram shows the effect of averaging the shifted values. The dashed record reconstruction shows the effect of the interpolation. Note the half shift-time delay due to the averaging and the interpolation.

[23] Figure 14(b) shows the modified bucket brigade with averaging at the input and the interpolation at the output.

[24] A more refined approximation would view it as a dead time of n shifts with a lag time of the shift time.

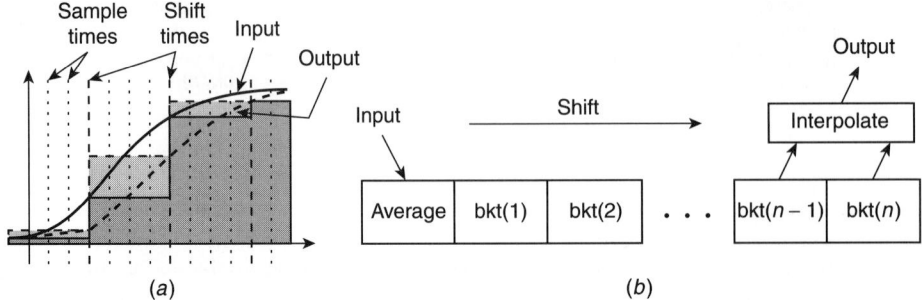

FIGURE 14 Bucket brigade for slow shifts.

The last question is the more subtle one and is answered by use of an irregular shifting frequency, whose immediate average value is equal to the current desired shifting frequency, corresponding to the desired (fractional) dead time. This solution is somewhat similar to the method used to achieve fine color variations on a CRT display that supports only a few basic colors: Mix a number of different pixels irregularly for the intended average affect.

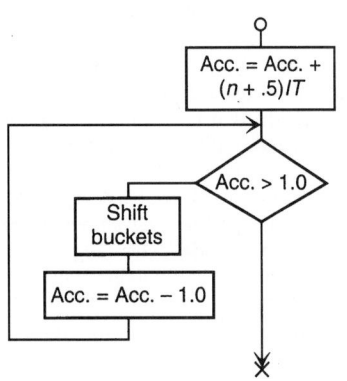

The flow chart (Fig. 15) shows the simplest way of achieving the desired irregular shift time. An accumulator variable (**Acc.,** initially zero) is incremented by $(n + 0.5)/T$ (corresponding to the desired fractional number of shifts per sample time[25]). If the accumulator variable has been incremented to 1 (indicating a net requirement for one full shift), a bucket brigade shift takes place and the accumulator is decremented by 1. Shifting and decrementing are repeated until the accumulator value drops back below 1.[26]

The result is a compensator algorithm capable of modeling dead time in fully time-varying situations.

FIGURE 15 Bucket brigade shift calculation flow chart.

Quantization and Saturation Effects

All of the calculations can now be carried out. The discussion has not addressed the effects of truncation on a derivative, but as developed here they are not more serious than for proportional control. In both cases truncation will cause a very small limit cycle, of the order of the minimum quantization of the D/A converter. However, if the derivative is not carefully filtered as part of its computation (as shown in integrated form above) severe problems arise. An approximate unfiltered derivative calculation has the form

$$D(t) = \tau_D \cdot \frac{\Delta\text{Measurement}}{\Delta t} = \frac{\tau_D}{\Delta t} \cdot \Delta\text{Measurement}$$

---

[25] If $(n + 0.5)\Delta t = T$, then $1/\Delta t = (n + 0.5)/T$.

[26] Actually the shift test value for the accumulator is irrelevant (and would be set to zero in practice) because the continual incrementing and decrementing balance it out to the same frequency, whatever the test point. A further simplification is to increment by $(n + 0.5)$ and decrement by T, further saving the need for floating point or the division.

Quantization has the awkward effect of forcing a minimum nonzero measurement change for any sampled calculation.[27] This value is multiplied by the gain $\tau_D/\Delta t$, which is usually very large. The result is very large pulses on the output of the controller. For ΔMeasurement quantized to 1 part in 1000 (0.1%), $\Delta t = 1$ s, and $\tau_D = 10$ min $= 600$ s, the minimum nonzero derivative is

$$D(t) = 600 \times 0.1 = 60\%$$

The use of filtering smears out the pulses and limits their height (The simple 0.1 time-constant-lagged derivative filter, developed above, limits the maximum pulse height to 10 times the quantization value. More refined algorithms can minimize the problem further by dynamically broadening the effective Δt in the calculation to get a better average derivative. The challenge is to get effective derivative action with a quantization that represents the realistic accuracy bounds of the data.

These derivative problems can be made far worse by internal saturations after integration in the PID algorithm, particularly in incremental algorithms (algorithms that output the desired change in valve position rather than the desired position). Such algorithms involve a second-derivative action. The problem arises because the saturation is likely to be unsymmetrical. When reintegrated later in the algorithm or system, the result is a significant offset. In the above case, the second differencing causes a doublet that extends 60% of scale in both directions. One-sided saturation, reintegrated, would create a 60% valve offset (bump).

IDENTIFICATION AND MATRIX-ORIENTED ISSUES

Theory-motivated control thinking emphasizes matrix-oriented formulations. These are becoming more common as engineers are trained in them. Properly understood, traditional methods are capable of equally good control, but there are aspects of normal control algorithm design for which these newer methods may be more appropriate. Adaptive control often suggests such methods. Space permits only an introduction to the problem but references [13–17] cover many important considerations.

A typical equation of this class defines least-squares data fitting of overdetermined parameters, as used in adaptation:

$$A^T A x = A^T y$$

In this equation, y is a data vector, x is a vector of parameters, and A is an $n \times m$ matrix $(n \geq m)$ of data vectors, occurring in a set of equations of the following form:

$$a_{i1}x_1 + a_{i2}x_2 + \cdots = y$$

The problem is to find the best fit for the parameters x, given the known A and y, solving the first equation:

$$x = (A^T A)^{-1}(A^T y)$$

Early direct and recursive solutions were unnecessarily sensitive to numerical rounding and truncation errors. The related eigenvalue problems were equally difficult until the problems were understood [13–16].

As above, the problem can be explained as an effect of differences of large numbers and the explosion of the digital data range under multiplication by numbers not close to 1. In matrix computations the concept of being close to 1 is formalized in several ways. Is $|A|$ close to 1? Because matrices

---

[27] Conventional resistor ladder A/Ds may be in substantial error in the calibration of their lowest-order bits to the extent that a constant slope measurement may appear to wander up and down as converted. This makes these quantization affects several times worse in practice.

involve several directions, the determinant is misleading. Another measure of a matrix is its norm:

$$\|A\| = \max_x \frac{\sqrt{x^T A^T A x}}{\sqrt{x^T x}}$$

Underlying the norm concept is the theory of orthogonal matrices and singular values [13]. An orthogonal matrix is one whose inverse equals its transpose: $Q^T Q = I$. (The letter Q will designate an orthogonal matrix.) In essence orthogonal matrices are equal to 1 in every possible way except for the identity:

4. Their determinant is equal to 1, but for sign.

5. They do not change the length of vectors that they multiply.

6. Therefore their norm equals 1.

7. Products of orthogonal matrices are orthogonal.

8. Every element of Q has magnitude ≤ 1.

9. The elements of the RGA [18] or interaction measure of an orthogonal matrix are all between 0 and 1. Thus easy computation corresponds to easy control.

The singular values σ_i of a matrix A are the square roots of the eigenvalues of $A^T A$. More interestingly, every matrix (square or not) obeys the singular-value decomposition theorem [13]. This theorem states that

$$A \equiv Q_1 \Sigma Q_2$$

for Σ, the diagonal matrix of the singular values σ_i, and some orthogonal Q_1, Q_2. Also,

10. The singular values of an orthogonal matrix are all equal to 1.

Under the singular-value decomposition, when A multiplies a vector, the immediate Q_1 or Q_2 twists the vector into an orientation in which each component is multiplied by one of the singular values. Thus the vector most amplified in length by A is the one oriented so that it is multiplied by the largest σ_i. For this reason the norm of A equals its largest σ_i. Also the vector most diminished in length by A is the one oriented so that it is multiplied by the smallest σ_i.

Generally a matrix is hard to compute with (is effectively much larger than 1) if there are significant off-diagonal terms and the ratio of largest σ_i to smallest σ_i is much larger than 1. (This also corresponds to large RGA elements.) Computing $A^T A$, as in the least-squares equation, squares this ratio, making computation that much worse. Effective algorithms minimize such operations.

As a simple example, consider the usual solution of the equation $Ax = b$, with A the matrix shown in Fig. 16. The conventional Gaussian solution[28] involves reducing the matrix to a triangular form

Gaussian (LU):

$$A = \begin{bmatrix} 0.1 & 1 \\ 1 & 1 \end{bmatrix} = LU = \begin{bmatrix} 1 & 0 \\ 10 & 1 \end{bmatrix} \times \begin{bmatrix} 0.1 & 1 \\ 0 & -9 \end{bmatrix}$$

Orthogonal (QR):

$$A = \begin{bmatrix} 0.1 & 1 \\ 1 & 1 \end{bmatrix} = QR$$

$$= \begin{bmatrix} -0.0995 & -0.995 \\ 0.995 & 0.0995 \end{bmatrix} \times \begin{bmatrix} -1.005 & -1.0945 \\ 0 & -0.8955 \end{bmatrix}$$

FIGURE 16   External-feedback-based backup.

---

[28] The example cheats a little, bypassing any pivot operation. These are imprecise compared with the orthogonal matrix methods.

that can then be backsolved. This is equivalent to factoring that same triangular form out, leaving a second triangular form (*L* and *U* in the figure). Note that the result involves large numbers (the 10 and −9).

   The same matrix can also be reduced to a triangular (backsolvable) form by multiplication by an orthogonal matrix. The corresponding factorization is called *QR* factorization, as shown. Note that all the calculations will now involve small numbers (0.995 and −1.005). The thrust of the newer matrix methods is to avoid matrix multiplication if possible (it expands the data range) and to try to restrict any multiplications to those involving orthogonal, diagonal, or triangular matrices.

SOFTWARE AND APPLICATION ISSUES

There was a brief reference above to software implications of control saturation. No modern discussion of control would be complete without observing the critical role of software [19–25]. Matrix-oriented control has been based on FORTRAN and purely mathematical approaches to control thinking. This has separated it from the major operational concerns of the industry. A control system should not only support the control computation, but the operational access to control data in a framework that is as easy to use as possible. It should include some model of sensible operator intervention.

   These considerations are accommodated naturally by the software for traditional control. Standard process regulatory control has been based on blocks, interpreted by the computer to carry out control.

   These blocks are blocks in two senses. They represent the digital equivalent of the old analog blocks in block diagrams. And they consist of data blocks whose data elements correspond to the signals and parameters of the analog block controller or are data pointers that make the connections between the blocks.

   But neither the existing systems nor the proposed standards offer attractive solutions, supporting the necessary flexibility and ease of use. The challenge is to provide flexible, future-oriented systems in which the goals and structure of the control system are transparent; new sensors, actuators, and algorithms are easily added; and sufficient standards included so control systems can include elements from many vendors [24, 25].

   And the block model is inefficient for the kinds of control that are now possible. For example, the iterated structure suggested in Fig.11 would be extremely inefficient if implemented in blocks, even as it would be quite simple programmed directly. And as normally interpreted, blocks have a predefined number of connections, limiting the use of structures with an indefinite number of inputs or outputs (e.g., feedforwards or overrides).

   Several references also show the limitations of the normal block diagram in representing the kinds of controls called for here [20–23, 26, 27]. Despite the universal belief in graphics, control diagrams are hard to understand. An improved notation and graphics separately list the loops making up the design, clearly distinguishing their controlled variables and purposes.

   For example, the Fig. 17 combination of two cascaded controllers with a pressure high constraint controller and a feedforward, according to the earlier structuring, would be expressed as

$$\text{T100}_{\substack{\text{REGULATE}\\\text{FEEDFWD F50}}} \quad \text{P100}_{\text{HICONSTRN}}\text{F100}_{\text{REGULATE}}\text{V100}$$

with the main degree-of-freedom path broken by the **P100** constraint expression.

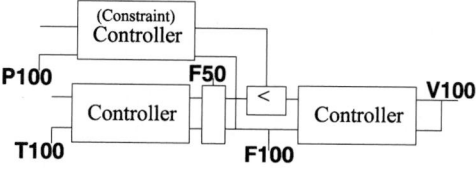

FIGURE 17   Constraint cascade.

Two operational issues are of special concern: smart field devices and interoperability. Modern sensors and actuators now include significant amounts of computational power themselves. The uses of this power will expand indefinitely. But both the field devices and the central controls must be developed in a standard model that permits control and its orderly operation to be supported without special or redundant programming of either.

There is a broader challenge: to solve the control software problem with solutions that capitalize on the computer as a truly intelligent control device, beyond the rigid scientific computations envisioned by the matrix control approaches or the programmable block diagram that mimics the dated analog controls.

SUMMARY

Digital control algorithms can be designed for experimental or single applications with the easiest tools available: FORTRAN, C, BASIC, and floating-point arithmetic. In this case, there is reasonable hope that normal commissioning debugging will weed out all the problems. But if a sense of workmanship prevails or if the algorithm is to be used in many applications, then attention to refinement and foolproofing are necessary:

1. Numerical effects of fixed- and floating-point arithmetic.
2. Documentation, control, and testing of detailed scaling, precision, and saturation within the algorithm.
3. Design for natural tuning and qualitative behavior, predictable from analog intuition.
4. Nasty quantization surprises.
5. Accommodations of windup and other control limiting effects.
6. Bumpless transfer and operational considerations.
7. Software, architectural, configuration considerations.

General-purpose digital programming languages and tools do not remotely address these issues.

REFERENCES

1. Bristol, E. H., "On the Design and Programming of Control Algorithms for DDC Systems," *Control Engineering*, Jan. 1977.
2. Clagget, E. H., "Keep a Notebook of Digital Control Algorithms," *Control Engineering*, pp. 81–84, Oct. 1980.
3. Fehervari, W., "Asymmetric Algorithm Tightens Compressor Surge Control," *Control Engineering*, pp. 63–66 Oct. 1977.
4. Tu, F. C. Y, and J. Y. H. Tsing, "Synthesizing a Digital Algorithm for Optimized Control," *Instr. Tech.*, pp. 52–56, May 1979.
5. Franklin, G. F., and J. D. Powell, *Digital Control of Dynamic Systems*, Addison-Wesley, Reading, Massachusetts, 1980.
6. ISA SP50 User Layer Technical Report.
7. Knuth, D. E., *The Art of Computer Programming*, vol. 2, Addison-Wesley, Reading, Massachusetts, 1981.
8. *Collected Algorithms from ACM*, 1960–1976.
9. ANSI/IEEE Standard 754–1985 for Binary Floating-Point Arithmetic.
10. *Chemical Engineers' Handbook*, 5th ed., McGraw-Hill, New York, pp. 2-81–2-85, 1973.
11. Considine, D. M., *Process/Industrial Instruments and Controls Handbook*, 4th ed., McGraw-Hill, New York, pp. 2.85–2.86, 1993.
12. Hansen, E., *Global Optimization Using Interval Analysis*, Marcel Dekker, New York, 1992.

13. Strang, G., *Linear Algebra and Its Applications*, Academic, New York, 1976.

14. Bierman, G. J., *Factorization Methods for Discrete Sequential Estimation*, Academic, New York, 1977.

15. Lawson, C. L., and R. L., Hansen, *Solving Least Squares Problems*, Prentice-Hall, Englewood Cliffs, New Jersey, 1974.

16. Laub, A. J., and V. C., Klema, "The Singular Value Decomposition: Its Computation and Some Applications," *IEEE Trans. Autom. Control*, **AC-25**, pp. 164–176, Apr. 1980.

17. MacFarlane, A. G. J., and Y. S. Hung, "A Quasi-Classical Approach to Multivariable Feedback Systems Design," *2nd IFAC Symposium, Computer Aided Design of Multivariable Technological Systems*, Purdue University, West Lafayette, Indiana, pp. 39–48, Sept. 1982.

18. Bristol, E. H., "On a New Measure of Interaction for Multivariable Control," *IEEE-PTGAC*, **AC-11**, pp. 133–134, Jan. 1966.

19. Shinskey, F. G., "An Expert System for the Design of Distillation Controls," presented at Chemical Process Control III, Asilomar, California, Jan. 12–17, 1986.

20. Bristol, E. H., "Strategic Design: A Practical Chapter in a Textbook on Control," presented at 1980 JACC, San Francisco, California, Aug. 1980.

21. Bristol, E. H., "After DDC–Idiomatic (Structured) Control," presented at 88th National Meeting of AIChE, Philadelphia, Pennsylvania, June 1980.

22. Prassinos, A., T. J. McAvoy, and E. H. Bristol, "A Method for the Analysis of Complex Control Schemes," presented at 1982 ACC, Arlington, Virginia, June 1982.

23. Bristol, E. H., "A Methodology and Algebra for Documenting Continuous Process Control Systems," presented at 1983 ACC, San Francisco, California, June 1983.

24. Bristol, E. H., "Super Variable Process Data Definition," ISA *SP50.4* Working Paper, Oct. 24, 1990 and updating.

25. Bristol, E. H., "An Interoperability Level Model: Super-Variable Categories," ISA *SP50.4* Continuing Working Paper.

26. Bristol, E. H. "A Language for Integrated Process Control Application," presented at the Retirement Symposium in Honor of Prof. Ted. Williams, Purdue University, West Lafayette, Indiana, Dec. 5–6, 1994.

27. Bristol, E. H. "Redesigned State Logic for an Easier Way to Use Control Language," World Batch Forum, Toronto, May 13–15, 1996; also *ISA Transactions*, **35**, pp. 245–257, 1996.

SAFETY IN INSTRUMENTATION AND CONTROL SYSTEMS

by E. C. Magison*

Extensive use of electrical and electronic control systems, computers, sensors, and analyzers in process control continues to focus attention on reducing the probability of fire or explosion due to electric instrument failure. At one time explosionproof housing was the common method of providing protection. Attention then turned to other means that provide the same or higher levels of safety, but with less weight and easier accessibility for maintenance and calibration and at equivalent or lower costs.

Because instrument manufacturers serve an international market, increased activity within national jurisdictions is being matched by recognition that standardization must be accomplished at the international level as well.

_____
* Consultant, Ambler, Pennsylvania.

AREA AND MATERIAL CLASSIFICATION

North America

In the United States, Articles 500 to 505 of the National Electrical Code (NEC) provide basic definitions of hazardous areas and the requirements for electrical installations. Articles 500 and 505 defines the classification of hazardous locations broadly in terms of kind and degree of hazard. The kind of hazard is specified by class and group. The degree of hazard is designated by division. Typical industrial locations, for example, may be classified as Class I, Group D, Division 1; or Class II, Group G, Division 2. The principal features of the NEC classification are summarized in Table 1. Many additional materials are listed in National Fire Protection Association publication NFPA 497 and NFPA 499. Similar definitions are given in the Canadian Electrical Code.

International Electrotechnical Commission

Most industrial nations have adopted or are adopting the area and material classification definitions of the International Electrotechnical Commission (IEC). Locations where a flammable concentration may be present are designated Zone 0, Zone 1, or Zone 2.

A Zone 0 location is a location where the atmosphere may be in the explosive range such a high percentage of the time (above 10%) that extraordinary measures must be taken to protect against ignition by electrical apparatus.

Zone 1 locations have a probability of hazard between Zone 2 and Zone 0. A Zone 2 location is similar to North American Division 2. Taken together, Zone 1 and Zone 0 equate North American Division 1.

In Zone 2, requirements analogous to those in North America are accepted in principle in many countries, but in practice Zone 1 types of protection are often used in Zone 2 because there is no accepted standard for Zone 2 apparatus. The advantage of distinguishing between the extraordinary hazards of Zone 0 and the lesser hazards of Zone 1 is that apparatus and installation requirements can be relaxed in Zone 1. For example, intrinsically safe systems in North America are judged on the basis of two faults because of the encompassing definition of Division 1. For use in Zone 1, consideration of only one fault is required, although two faults are assessed in Zone 0.

Material classification in most countries now uses IEC terminology. A Group I hazard is due to methane (firedamp) in the underground works of a mine. The presence of combustible dusts and other environmental aspects of mining works are assumed when preparing apparatus requirements for Group I.

Group II gases and vapors are flammable materials found in industrial aboveground premises. They are divided into Groups IIA, IIB, and IIC, which are similar although not identical to North American Groups D, C, and B, respectively.

Article 505, introduced in the 1996 NEC, defines Class I, Zone 0, Zone 1, and Zone 2 locations. It is the first step toward using the IEC method of area and material classification in the United States. This will allow eventual recognition of methods of protection such as increased safety and encapsulation which have been standardized for use in Zone 1 locations in Europe and in other industrial nations, but which have not been recognized for use in Division 1 in the United States. ISA, the International Society for Measurement and Control, is publishing a series of standards which mirror IEC requirements for types of protection useful in Zones 0, 1, and 2, modified to recognize North American standards and installation practices.

Presumably, when IEC agrees on definitions for zones in locations where combustible dust is the hazard, these will be proposed for addition to Article 505.

Classifying a Hazardous Location

The NEC definitions provide guidelines, but do not give a quantitative method for classifying a specific hazardous location. Factors to consider include the properties and quantity of hazardous material that may be released, the topography of the site, the construction of the plant or building, and

TABLE 1 National Electrical Code Area Classification System

Class I Gases and vapors	Class II Dusts	Class III Flyings
Group A—Acetylene Group B—Hydrogen or gases of similar hazardous nature, such as manufactured gas, butadiene, ethylene oxide, propylene oxide Group C—Ethyl ether, ethylene, cyclopropane, unsymmetrical dimethylhydrazine, acetaldehyde, isoprene Group D—Gasoline, hexane, naphtha, benzene, butane, propane, alcohol, acetone, benzol, lacquer solvent, natural gas, acrylonitrile, ethylene dichloride, propylene, styrene, vinyl acetate, vinyl chloride	Group E—Metal dusts Group F—Carbon black, coal, coke dusts Group G—Grain dust, flour, plastics, sugar	No group assigned. Typical materials are cotton, kapok, nylon, flax, wood chips—normally not in air suspension
Division 1*		
For heavier-than-air vapors, below-grade sumps, pits, etc., in Division 2 locations. Areas around packing glands; areas where flammable liquids are handled or transferred; areas adjacent to kettles, vats, mixers, etc. Where equipment failure releases gas or vapor and damages electrical equipment simultaneously.	Cloud of flammable concentration exists frequently, periodically, or intermittently—as near processing equipment. Any location where conducting dust may accumulate.	Areas where cotton, spanish moss, hemp, etc., are manufactured or processed.
Division 2*		
Areas adjacent to a Division 1 area. Pits, sumps containing piping, etc., in nonhazardous location. Areas where flammable liquids are stored or processed in completely closed piping or containers. Division 1 areas rendered nonhazardous by forced ventilation.	Failure of processing equipment may release cloud. Deposited dust layer on equipment, floor, or other horizontal surface	Areas where materials are stored or handled.

*In a Division 1 location there is a high probability that a flammable concentration of vapor, gas, or dust is present during normal plant operation, or because or frequent maintenance. In a Division 2 location there is only low probability that the atmosphere is hazardous—for example, because of equipment failure.

Until the 1971 revision, material classification in the NEC differed from the practice in almost all other countries except Canada. Material groupings were based on consideration of three parameters: autoignition temperature (AIT) (or, spontaneous ignition temperature SIT), maximum experimental safe gap (MESG), and the maximum pressure rise in an explosion test chamber. In Europe materials long have been grouped by AIT and separately by MESG. Pressure rise is not a material classification criterion. It is now recognized in the United States that there is no correlation between MESG and AIT. Hydrogen, for example, has a very samll MESG and a very high AIT. Many Group C and D materials have lower ignition temperatures but wider experimental safe gaps. Because the NEC classification was based on two uncorrelated parameters, United States experts could not use the results of experimental work on new material in other countries, or use other classification tables. The 1971 NEC revisions separate AIT from considerations of MESG. Explosionproof housings now can be designed for MESG typical of a group of materials. External surface temperatures shall not exceed the AIT of the hazardous gas or vapor of concern.

the past history of fire and explosion (of a particular location or plant as well as of an entire industry). Although authorities recognize the need, there are no concise rules for deciding whether a location is Division 1 or Division 2. The best guides to area classification known to the author are American Petroleum Institute publications API RP500A, B, and C, and API RP505 for petroleum installations, and NFPA 497 for installations in chemical plants. These documents are applicable to any industry. API RP505 specifies classification of Zones 0, 1, and 2.

Special Cases of Area Classification

It is common practice to pressurize instrument systems to reduce the area classification inside the enclosure. The inside of an instrument enclosure provided with a simple pressurization system, located in a Division 1 area, can be considered a Division 2 location because only by accidental failure of the pressurization system can the internal atmosphere become hazardous. If the pressurization system is designed to deenergize all equipment within the enclosure when pressurization fails, the interior can be considered a nonhazardous location. Two failures are required—(1) of the pressurization system and (2) of the interlock system—before an explosion can occur.

An important limitation of this philosophy is that if any single failure can make the enclosure hazardous, the interior of the enclosure must not be classified less hazardous than Division 2, regardless of the pressurization system design. Such is the case with bourdon-tube or diaphragm-actuated instruments where process fluid is separated from the instrument interior only by a single seal, namely, the bourdon or diaphragm. Unless the pressure is high enough or the enclosure air flow is great enough to prevent a combustible concentration inside the enclosure should the measuring element fail, the interior never should be classified less hazardous than Division 2. Such systems often are referred to as singly sealed systems.

In a doubly sealed system two seals are provided between the process fluid and the area being purged, and a vent to the atmosphere is provided between the seals. Failure of both seals is required to make the enclosure interior hazardous. Even so, pressurization can prevent the hazardous material from entering the compartment because the hazardous material is at atmospheric pressure. Article 501-5(f) of the NEC mandates a double seal wherever failure of a sealing element could force flammable material into the conduit system.

TECHNIQUES USED TO REDUCE EXPLOSION HAZARDS

The predisposing factors to fire or explosion are (1) the presence of a flammable liquid, vapor, gas, dust, or fiber in an ignitable concentration, (2) the presence of a source of ignition, and (3) contact of the source with ignitable material. The most obvious way to eliminate the possibility of ignition is to remove the source to a location where there is no combustible material. This is the first method recognized in the NEC, Article 500. Another method is to apply the principle of intrinsic safety. Equipment and wiring that are intrinsically safe are incapable, under normal or abnormal conditions, of igniting a specifically hazardous atmosphere mixture. For practical purposes there is no source of ignition.

Figure 1 summarizes the techniques used to reduce explosion hazards. Methods based on allowing ignition to occur force combustion under well-controlled conditions so that no significant damage results. A continuous source of ignition, such as the continuous pilot to localize combustion in gas appliances, is commonplace. Explosionproof enclosures contain an explosion so that it does not spread into the surrounding atmosphere. Historically this has been the most common technique. In Zone 2, enclosed break devices in which the enclosed volume is so small that a weak incipient explosion cannot escape the enclosure, are permitted in some countries.

There are several methods for reducing hazard by preventing the accumulation of combustible material in an explosive concentration or for isolating an ignition source from flammable material. Pressurization of instruments is common. Continuous dilution in which the interior of an enclosure is pressurized to exclude flammable material from entering and is also continuously purged to dilute any internal release of flammable material is applicable to analyzers and other devices in which flammable material may be released inside the enclosure. In hydrogen-annealing furnaces and hydrogen-cooled

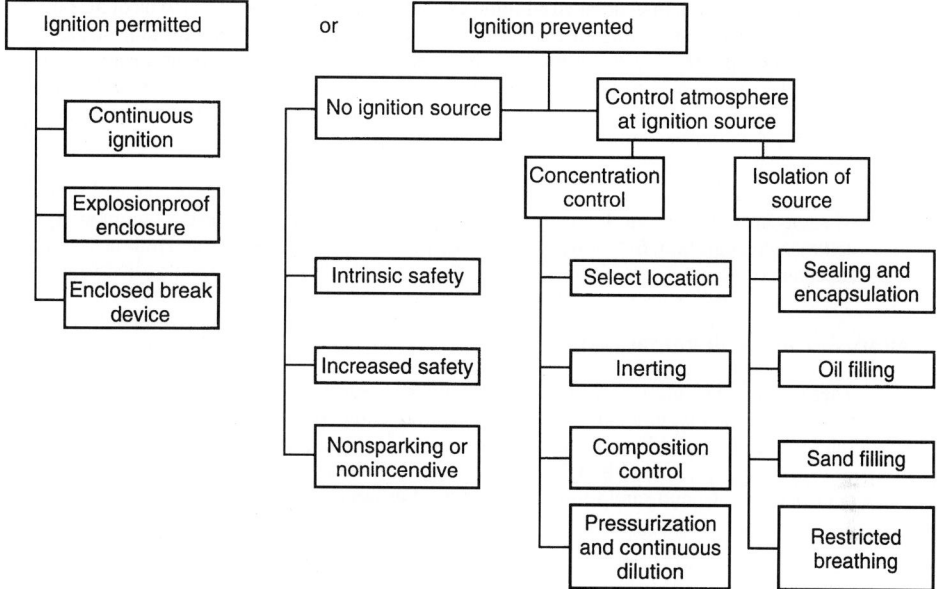

FIGURE 1   Techniques used to reduce explosion hazards.

electric generators the concentration is held above the upper explosive limit. Blanketing of tanks with nitrogen or carbon dioxide (CO_2) and rock dusting of coal mine galleries and shafts are examples of using inert materials to suppress a combustible mixture.

Several techniques are used to isolate the ignition source. Oil immersion prevents contact between the atmosphere and the ignition source. In Europe sand-filled equipment sometimes is used. Sealing and encapsulation both provide a barrier to impede contact.

Increased safety is a technique used for transformers, motors, cables, and so on, which are constructed with special attention to ruggedness, insulation, reliability, and protection against overtemperatures so that an ignition-capable failure is of very low probability. Increased safety, developed first in Germany and now accepted widely in Europe, may be used in Zone 1.

Nonsparking apparatus and nonincendive apparatus are suitable for use in Division 2 and Zone 2 because they have no normal source of ignition.

Restricted breathing is the technique of using a tight, but not sealed enclosure in Zone 2 to allow only slow access of flammable vapors and gases to the source of ignition. This technique, developed in Switzerland, is slowly achieving acceptance in Europe.

Explosionproof Housings

Termed flameproof enclosures in international English, explosionproof housings remain the most practical protection method for motor starters and other heavy equipment that produces sufficient energy in normal operation to ignite a flammable atmosphere. Explosionproof enclosures are not assumed to be vaportight; it is assumed that a flammable atmosphere will enter the enclosure.

A pressure rise of 100 to 150 lb/in² (690 to 1034 kPa) is typical for the mixture producing the highest explosion pressure. In small enclosures, loss of energy to the enclosure walls decreases the pressure rise. Because the enclosure must contain the explosion and also must cool escaping gases, cast or heavy metallic construction with wide, close-fitting flanges or threaded joints is typical. Nonmetallic construction is permitted.

For specific design criteria in the United States, reference should be made to the standards of the intended certifying agency. Although requirements of all agencies are similar, there are many differences in detail. In general, in addition to tests to ensure that an internal explosion is not transmitted to the outside, the enclosure must withstand a hydrostatic pressure of four times the maximum pressure observed during the explosion test and must not have an external case temperature high enough to ignite the surrounding atmosphere. In Canada the applicable standard is CSA C22.2, No. 30.

In Europe the applicable standards are CENELEC EN50014, "General Requirements," and CENELEC EN50018, "Flameproof Enclosure 'd'." These are available in English as British Standard BS 5501, Parts 1 and 5.

Requirements for flange gaps are less restrictive than North American standards. Routine testing of enclosures at lower pressures than those of the North American test is common. Type testing is achieving recognition. In North America, wider permissible flange gaps and routine testing are gaining acceptance.

Encapsulation, Sealing, and Immersion

These techniques seldom are applied to a complete instrument. They serve to reduce the hazard classification of the instrument by protecting sparking components or subassemblies. Oil immersion and sand filling are applied to power-handling apparatus, but neither technique has important applications in instrument systems, although oil immersion may be a convenient technique for some hydraulic control elements, and sand filling has been used in some portable devices.

Sealing

Article 501-3(b)(2) of the NEC states that general-purpose enclosures may be used in Division 2 locations if make-and-break contacts are sealed hermetically against the entrance of gases or vapors. The NEC provides no definition of a satisfactory hermetic seal, however. Seals obtained by fusion, welding, or soldering and, in some instances, plastic encapsulation are widely accepted. In reality, the leak rate of soldered or welded seals is lower than that required for protection against explosion in a Division 2 location.

The long-time average concentration inside a sealed enclosure approaches the average concentration outside. The function of a seal is to prevent transient excursions above the lower explosive limit (LEL) outside the device from raising the concentration inside the device to the LEL. Three mechanisms can force material through a seal, (1) changes in ambient temperature, (2) changes in barometric pressure—both effects tending to make the seal breathe, and (3) wind and strong air currents. The last named mechanism usually can be ignored because a sealed device must be installed in a general-purpose enclosure to protect it from such conditions.

Encapsulation involves enclosing a component or subassembly in a plastic material, a tar, or a grease, with or without the additional support of a can. If the encapsulated assembly is robust and has mechanical strength and chemical resistance adequate for the environment in which it is used, it can be considered the equivalent of a hermetic seal. An external hazardous atmosphere must diffuse through a long path between the encapsulating material and the device leads to reach the interior. Standards for sealed devices can be found in Instrument Society of America Standard ISA S12.12.

Pressurization Systems

Lowering the hazard classification of a location by providing positive-pressure ventilation from a source of clean air has long been recognized in the NEC, and instrument users have pressurized control room and instrument housings for many years. The first detailed standard for instrument purging (pressurizing) installations was ISA SP12.4 (now withdrawn). These requirements in essentially the same form make up the first section of NFPA 496, which also covers purging of large enclosures, ventilation of control rooms, Class II hazards, and continuous dilution. NFPA 496 defines three types of pressurized installation:

Type Z.   Pressurization to reduce the classification within an enclosure from Division 2 to nonhazardous

Type Y.   Pressurization to reduce the classification in an enclosure from Division 1 to Division 2
Type X.   Pressurization to reduce the classification within an enclosure from Division 1 to non-hazardous

Type Z Pressurization.   This permits the installation of ignition-capable equipment inside the enclosure. For an explosion to occur, the pressurized system must fail, and also, because the surrounding area is Division 2, there must be a process equipment failure which releases flammable material. Thus there must be two independent failures, and no additional safeguards in the pressurization system are necessary.

Figure 2 shows a typical installation for Type Z pressurization. Only a pressurization indicator is required. The probability that a process fault will make the location hazardous before any failure of

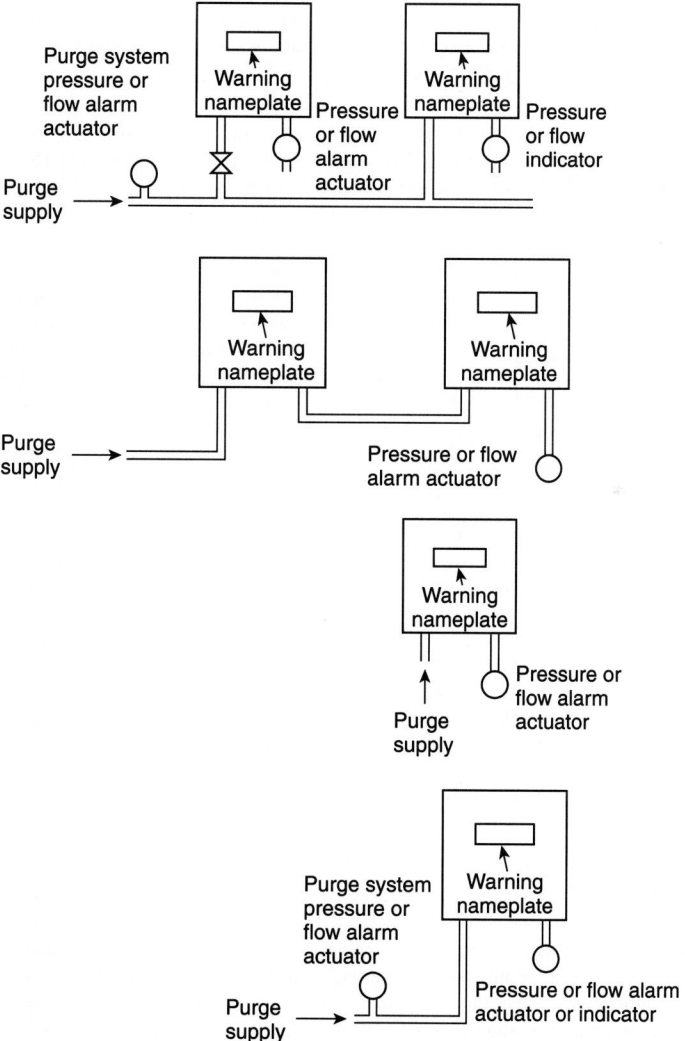

FIGURE 2   Typical installation for Type Y and Type Z pressurization.

the pressurization system is recognized and corrected is assumed to be extremely low. An electric indicator or alarm must meet the requirements of its location before pressurization. If a pressure indicator is used, no valve may be installed between the pressure indicator and the case. Any restriction between the case and the pressure device must be no smaller than the smallest restriction on the supply side of the pressure device. The case must be opened only if the area is known to be nonhazardous or if power has been removed. In normal operation the pressurization system must maintain a minimum pressure of 0.1 inch (2.5 mm) of water gage. The flow required to maintain this pressure is immaterial. The temperature of the pressurized enclosure must not exceed 80% of the ignition temperature of the gas or vapor involved when it is operated at 125% of rated voltage. A red warning nameplate must be placed on the instrument to be visible before the case is opened. Failure of the pressurization system must be alarmed.

Type Y Pressurization. In this case all requirements of Type Z pressurization must be met. Equipment inside the enclosure must be suitable for Division 2, that is, it is not an ignition source in normal operation. For an explosion to occur, the pressurization must fail, and the equipment inside must fail in a way to make it an ignition source.

Type X Pressurization. In this system the pressurization is the only safeguard. The atmosphere surrounding the enclosure is presumed to be frequently flammable. The equipment within the enclosure is ignition-capable. The pressurization system failure must automatically deenergize internal equipment.

All requirements for Type Z and Type Y pressurization must be met. The interlock switch may be pressure- or flow-actuated. The switch must be suitable for Division 1 locations, even if it is mounted within the instrument case, because it may be energized before purging has removed all flammable material. A door that can be opened with the use of a tool must be provided with an automatic disconnect switch suitable for Division 1. A timing device must prevent power from being applied until four enclosure volumes of purge gas can pass through the instrument case with an internal pressure of 0.1 inch (2.5 mm) of water gage. The timing device also must meet Division 1 requirements, even if inside the case (Fig. 3).

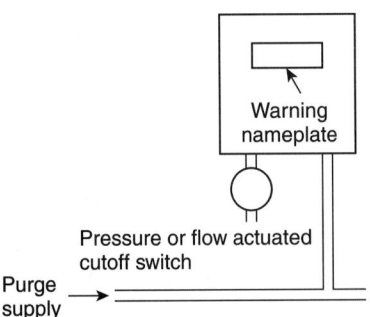

Purge supply

FIGURE 3 Typical installation for Type X pressurization.

IEC and CENELEC standards for pressurization systems are similar to those of NFPA, although the requirements are not phrased in terms of reduction in area classification. The minimum pressure is 0.2 inch (5.1 mm) of water gage.

The IEC and NFPA standards also cover continuous dilution. The hardware is similar to that required for pressurization, but the rationale for selecting the level of protection needed is based on the presence of a source of flammable material within the enclosure, as in an analyzer. The objectives are to prevent entry of an external flammable atmosphere (pressurization) and also to dilute any internal release to a low percentage of the lower flammable limit (continuous dilution).

INTRINSIC SAFETY

Experiment and theory show that a critical amount of energy must be injected into a combustible mixture to cause an explosion. If the energy provided is not greater than the critical ignition energy, some material will burn, but the flame will not propagate. An explosion occurs only when enough energy is injected into the mixture to ignite a critical minimum volume of material. The diameter of a sphere enclosing this minimum volume is called the quenching distance or quenching diameter. It is related to the maximum experimental safe gap (MESG), but is about twice as large. If the incipient flame sphere does not reach this diameter, it will not propagate.

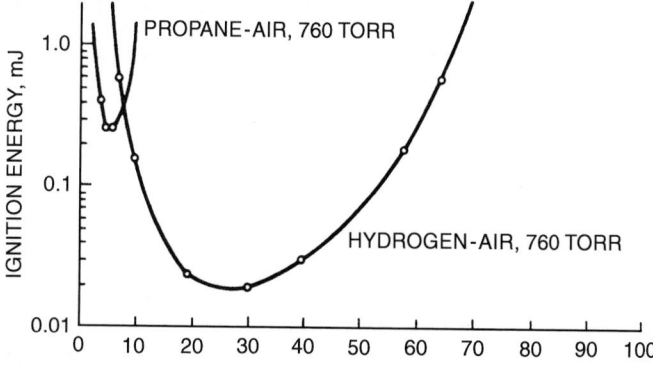

FIGURE 4   Effect of Concentration on ignition energy.

The energy required for ignition depends on the concentration of the combustible mixture. There is a concentration at which the ignition energy is minimum. The curve of ignition versus concentration is asymptotic to limits of concentration commonly called the lower explosive limit (LEL) and the upper explosive limit (UEL). Figure 4 illustrates the influence of concentration on the critical energy required to cause ignition. A hydrogen-air mixture, one of the most easily ignited atmospheric mixtures, supports combustion over a wide range of concentrations. A propane-air mixture, which is typical of many common hazardous materials, is flammable only over a narrow range of concentrations. The amount of energy required to ignite the most easily ignited concentration of a mixture under ideal conditions is the minimum ignition energy (MIE).

Definition

The NEC defines an intrinsically safe circuit as one in which any spark or thermal effect is incapable of causing ignition of a mixture of flammable or combustible material in air under prescribed test conditions, for example, those in ANS/UL913.

Early Developments

The British first applied intrinsic safety in direct-current signaling circuits, the first studies beginning about 1913. In 1936 the first certificate for intrinsically safe equipment for other than mining was issued. By the mid-1950s certification in Great Britain for industrial applications was common. At the U.S. Bureau of Mines work on intrinsically safe apparatus, although the term was not used, commenced about the same time as the British investigations. Rules for telephone and signaling devices were published in 1938.

During the 1950s increased use of electric equipment in hazardous locations stirred worldwide interest in intrinsic safety, and by the late 1960s almost every industrial country had either published a standard for intrinsically safe systems or drafted one. The major industrial countries also were active in the IEC Committee SC31G, which prepared an international standard for intrinsically safe systems.

The first standards for intrinsically safe equipment intended for use by the instrument industry were published as ISA RP12.2, issued in 1965. The NFPA used ISA RP12.2 as a basis for the 1967 edition of NFPA 493.

During the years following the publication of ISA RP12.2 and NFPA 493-1967 the certification of intrinsically safe systems by independent approval agencies, such as Factory Mutual and Underwriters Laboratories in the United States, CSA in Canada, BASEEFA in the United Kingdom, and PTB in Germany, became a legal or marketing necessity in most countries. All standards for intrinsic safety have therefore become much more detailed and definitive. Adherence to the standard is the objective, not a judgment of safety. The work of the IEC has served as the basis for later editions of NFPA 493 and for UL 913, which is now the U.S. standard for intrinsically safe equipment, as well as for Canadian Standard CSA C22.2-157 and CENELEC Standard EN50020. Any product marketed internationally must meet all these standards.

All the standards agree in principle, but differ in the details.

CSA and U.S. standards are based on safety after two faults, because in these cases Division 1 includes Zone 1 and Zone 0. European standards provide *ia* and *ib* levels of intrinsic safety for Zone 0 and Zone 1 application, based on consideration of two faults and one fault, respectively.

Standards for intrinsic safety can be less intimidating to the user if it is appreciated that most construction details are efforts to describe what can be considered a fault or what construction can be considered so reliable that the fault will never occur. When viewed in this light, creepage and clearance tables, transformer tests, and tests of protective components make much more sense. They are guidelines for making design decisions—not mandated values for design. They apply only if safety is affected.

Design of Intrinsically Safe Systems

The objective of any intrinsically safe design, whether produced by an equipment manufacturer or by a user attempting to assemble a safe system from commercially available devices, is the same—to ensure that the portion of system in the Division 1 location is unable to release sufficient energy to cause ignition, either by thermal or by electrical means, even after failures in system components. It is not necessary that the associated apparatus, that is, the apparatus located in Division 2 or a nonhazardous location connected to the intrinsically safe circuit, be itself intrinsically safe. It is only necessary that failures, in accordance with the accepted standard for intrinsic safety, do not raise the level of energy in the Division 1 location above the safe level.

BASIC TECHNIQUES USED BY MANUFACTURERS

Techniques used by manufacturers in the design of intrinsically safe apparatus are relatively few in number, and all manufacturers use the same fundamental techniques.

Mechanical and Electrical Isolation

The most important and most useful technique is mechanical isolation to prevent intrinsically safe circuits and nonintrinsically safe circuits from coming in contact. Often mechanical isolation is achieved solely by appropriate spacing between the intrinsically safe and nonintrinsically safe circuits. In other cases, especially at field connections or in marshaling panels, partitions or wireways ensure that the nonintrinsically safe wiring and intrinsically safe wiring are separate from one another. Encapsulation is sometimes used to prevent contact between the two types of circuits.

Related to mechanical isolation is what can be called electrical isolation. Except in battery-operated systems, intrinsically safe systems have some connection to the power line, usually through a power transformer. The designer must consider the possibility of a transformer fault that connects the line voltage primary winding to the low-voltage secondary winding. In many systems, if one must consider the presence of line voltage on secondary circuits, the value and power rating of limiting elements would make the design of an intrinsically safe system both functionally and economically impractical. Therefore in modern standards for intrinsically safe construction, several varieties of transformer construction are recognized to be so reliable that one can assume that a primary-to-secondary short circuit will never occur. In one such reliable construction, a grounded shield between primary and secondary ensures that any fault is from the primary to the grounded shield, so that the secondary winding potential is not raised to an unsafe voltage. In addition to special attention to transformer construction and testing, it is also necessary that the wiring layout prevent any accidental connection between wiring on the primary side of the transformer and wires connected to the transformer secondary.

Current and Voltage Limiting

Except in some portable apparatus, almost all intrinsically safe circuits require both current and voltage limiting to ensure that the amount of energy released under fault conditions does not exceed safe values. Voltage limiting is often achieved by use of redundant zener diodes to limit the voltage, but zener-triggered silicon-controlled rectifier (SCR) crowbar circuits are also. Redundancy is provided so that, in the case of the failure of a single diode or limiting circuit, the second device continues to provide voltage limiting. Current limiting in dc circuits and in most ac circuits is provided by film or wirewound resistors of high reliability (Fig. 5). Properly mounted resistors that meet the requirements for protective resistors in the applicable standard need not be redundant. They are of a level of quality that they will not fail in a way that allows current to increase to an unsafe level.

One common use of current and voltage limiting is in the zener diode barrier (Fig. 6). The unique feature of these barriers is the fuse in series with the zener diodes, so that when a fault causes current to flow through the zener diode, the fuse will open before the power in the zener diode reaches a level at which the diode may open. In the design shown in Fig. 6, the 20-ohm resistor does not perform a safety function. It allows testing of the barrier to determine that the diodes are still intact. The current limiting function is performed by the 300-ohm resistor.

Devices to be connected to terminals 3 and 4 must be approved as intrinsically safe, but any device incapable of applying voltage to terminal 1 higher than the barrier rating may be connected. If the barrier is designed to limit against full power line potential, then the equipment in the nonhazardous area may be selected, connected, and intermingled without regard to safety in the field circuits if no potential above power line voltage is present.

In use, terminals 2 and 4 are both connected to a busbar that is grounded through a very low (usually less than 1-ohm) ground resistance. The power supply also must be grounded. In operation, diodes D_1 and D_2 conduct only leakage current, which is small compared with the normal circuit flowing between terminals 1 and 3. When high voltage is applied to terminal 1, the diodes conduct and limit the voltage at terminal 3 to a safe value. R_3 limits the current into the hazardous area. Under fault conditions, the barrier looks like a low-voltage resistive source from the intrinsically safe side, terminals 3 and 4, and like a very low-impedance load at terminals 1 and 2.

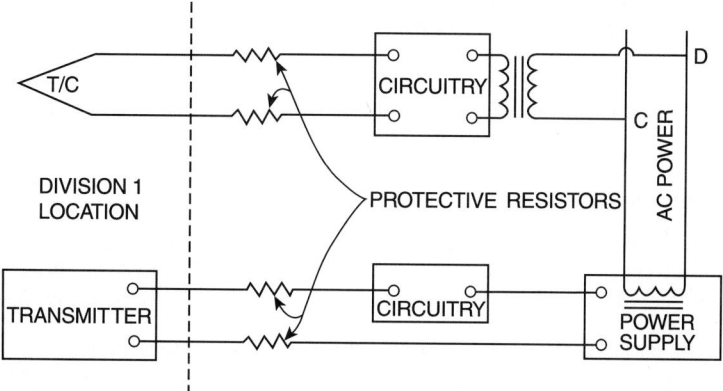

FIGURE 5 Use of resistors to limit current in hazardous location.

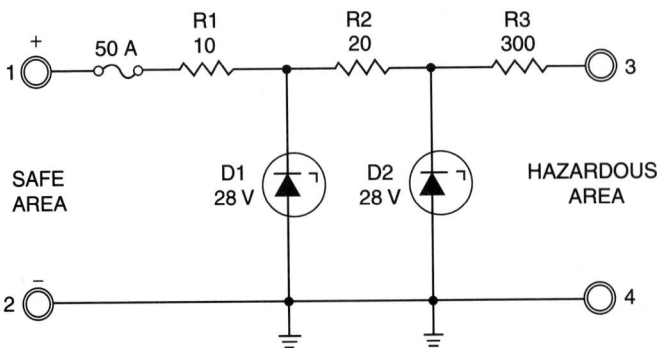

FIGURE 6   Schematic of zener diode barrier, positive type.

The values in Fig. 6 are for a 28-volt 93-mA barrier. Under fault conditions, the intrinsically safe circuit will appear to be driven from a nominal 28-volt source with a source resistance of 300 ohms. Safety is provided by the diodes and resistors. The resistors can be presumed not to fail. The diodes are redundant. Should one fail, limiting would still take place. The fuse serves no purpose regarding ignition and could be replaced by a resistor. Its function is to make the diode barrier economical. Should a fault occur, the zener diodes would connect heavily and, except for the fuse, would have to be impractically large and costly. The fuse is selected to blow at a current much lower than that which would damage the diode, permitting lower power, less costly diodes to be used.

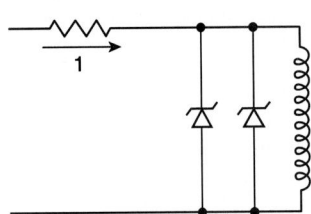

FIGURE 7   Shunt diodes reduce incendivity of inductor.

Shunt Elements

These devices are used to absorb the energy that would otherwise be released by an inductor to the arc. The function of a shunt diode is shown in Fig. 7. Although capacitors and resistors can be used to shunt inductors, in dc circuits diodes are placed so that in normal operation they are backbiased and draw no current. The shunt elements must be redundant so that if one fails, the other will continue to protect the inductor. Both must be connected close to the inductor being protected. Connection to the inductor must be especially reliable so that a fault between the inductor and the protective shunt diodes can be assumed not to occur. If such a fault occurs, ignition is possible because of the release of energy from the inductor. The purpose of the shunt diodes is to absorb energy stored in the inductor if the circuit external to the protected inductor opens.

Analytical Method for Circuit Design

The outline presented here can be considered both as a method of assessing a circuit and a mechanical design that already exists, or as a means for both analyzing the circuit and designing the layout. Only a slight difference in point of view is required. The steps in the analysis are essentially the same in both cases.

The first step is to identify the portion of the circuit that is to be intrinsically safe. Only the circuit in a Division 1 location need be intrinsically safe. A fault occurring in a nonhazardous or Division 2 location is of no consequence from the standpoint of the energy that may be released at that location. The fault is of concern if it affects the amount of energy that can be released in a Division 1 location.

Second, review the circuit or the hardware for isolating constructions that will allow one to assume that certain interconnections will not occur. If the hardware exists, review the mechanical layout to determine whether the use of adequate spacings or partitions allows one to avoid considering an interconnection between some nonintrinsically safe circuits and the intrinsically safe circuit. Usually the transformer must be one of the special protective construction. Otherwise it is unlikely that the circuits can be both functional and safe if one assumes a line voltage connection to the transformer secondary. If there is a transformer of special construction, then the wiring or the printed wiring board layout must ensure that the primary leads of the transformer are separated from the secondary leads by sufficient space (as defined in the relevant standard), or by a partition, so that connection between the two can be ignored. If the hardware has not been designed, one must determine where intrinsically safe portions of the circuit must be separated by appropriate spacing or partition from the higher-energy portion of the circuits. The specific nature of the spacings must be determined from further detailed analysis.

Having reviewed for isolating construction, one should assume normal operation of the circuit. Compute the current and voltage in the circuit and compare it with the reference curves to determine whether the appropriate test factor has been observed. If not, adjust the circuit constants until the requirements are met. In this step, and in subsequent steps, it is essential that an orderly approach to record keeping be adopted. Even a relatively simple circuit may require the consideration of many steps. It is essential to record each combination of open, short, or ground of field wiring (these are not counted as faults) and component failures, so that when the analysis is complete, one can verify that the worst-case situation has been analyzed. If one is submitting the circuit for later review by certifying agencies, the availability of a well-organized fault table will ease and expedite the assessment process.

Third, if the hardware already exists, one considers the layout and spacing to determine what circuits must be assumed to be shorted together, what connections can be considered to be a fault, and what connections can be considered never to occur. After identifying these, recompute the current and voltage under fault conditions. This must be done for a single fault in combination with opening, grounding, and shorting of the external wires—and also for two faults. One cannot assume that two-fault situations will be the most hazardous, because of the difference in test factor required when only a single fault is considered. Adjust the circuit constants until the voltages and currents are suitable.

After the analysis for arc ignition, consider whether current flowing in the circuit under fault conditions may produce a high surface temperature on resistors, transistors, and so on. The temperature rise of small components is typically 50 to 100°C/watt. If the hardware exists, of course, one can measure the temperature of components which are suspect under fault conditions. The limiting temperature to be considered must, of course, be determined from the standard being used as a criterion for design. Assessment is made based on the maximum voltage, current, and power ratings at the intrinsically safe input terminals.

Simplifying Assumptions

If one is analyzing a circuit, the validity of the analysis is only as good as the data with which the analysis can be compared. In general, the available data are limited to simple RL, RC, or resistive circuits. It may not be possible to analyze the effects of shunt elements included for some functional purpose. For example, an inductive force coil might be shunted with a variable resistor for a span adjustment. There is no reliable way to assess the additional safety provided by the shunt resistor. In general, one ignores such a parallel component in the analysis. Similarly, a capacitor might be wired in shunt with the force coil to provide damping of the electromechanical system. This, too, is ignored in the analysis, except with regard to the release of its own stored energy.

It is assumed by almost all experts that an iron-core inductor is less efficient in releasing stored energy than an air-core inductor, because some of the energy, rather than being released to the arc, is dissipated in eddy current and hysteresis losses. If one is analyzing a circuit in which the inductor has a ferromagnetic core and the circuit will be safe with an air-core inductor of the same value, then it will certainly be safe. The converse is not true. Inductance is a measure of the slope of the $B–H$ curve of the core material. Many small inductors, especially those with ferrite cores, have high inductance because the core material has a very high initial permeability. However, if the volume of material and the level at which it saturates are low, the amount of stored energy in the inductor may be considerably less than that calculated from a measured inductance value. Testing may verify safety.

Another simplifying assumption, which in many circuits reduces the amount of analysis considerably, is to determine the highest possible power supply voltage that may ever exist under fault conditions and with a high line voltage. This value is then used to determine the stored energy in all the capacitors or to determine the current and the resulting stored energy in all the inductors. If all the calculations are safe, one need not calculate the actual circuit currents and voltages.

Another simplification results from the need to prevent the discharge of a large capacitor in a hazardous location. Although curves giving the value of ignition voltage on a capacitor discharging through a resistor are available, one can assume that the capacitor is a battery charged to the fault voltage and select a series resistor based on the resistive circuit ignition characteristics. The resistor selected will be higher than that based on ignition tests of capacitors because the voltage on the capacitor decays when the current flows through the resistor. The connection between the resistor and the capacitor must be prevented from contacting any surrounding circuit. Therefore spacing, potting, or some other technique must be used to ensure that the capacitor cannot discharge except through the resistor.

Testing of Special Cases

It is not possible to determine the safety of all circuits by analysis alone. For example, some experts feel that one should verify the safety of diode-shunted inductors by conducting ignition tests. Inductors, especially small ones with ferromagnetic cores, may require testing to verify that they are safe despite high measured inductance.

Another common piece of apparatus that may have to be tested to determine safety is a regulated power supply. The reference curves of open-circuit voltage and short-circuit current for ignition in resistive circuits assume that the source impedance of the circuit, that is, the Thévenin equivalent impedance, is resistive. If the power supply is regulated, the voltage will remain essentially constant until a critical level of current is reached, beyond which the voltage will drop off with a further increase in current. The safety of maximum voltage and maximum current from such a supply cannot be determined from the reference curves for resistive circuits. In general, safety must be established by test, although some reference curves are available.

Transmission lines are another special case. The common method of assessing safety, that is, multiplying capacitance per foot (meter) or inductance per foot (meter) by the number of feet (meters) of cable and comparing these values with limit values for the voltage and current in the cable, yields conservative results. It is well known that because the resistance, capacitance, and inductance are distributed, the actual cable will be safer than this lumped constant analysis suggests. However, there are not sufficiently good reference data available to allow one to analyze on a more scientific basis. If answers from lumped constant approaches are not satisfactory, then the cable must be tested.

However, it is recognized that the L/R ratio of the cable, if it is sufficiently low, may be such that no hazard exists, even though the total inductance exceeds that which would be safe if the inductance were lumped. If one assumes a cable of resistance R and inductance L_x per foot (meter) operating from a circuit of maximum voltage V_{max} and source resistance R, then the maximum energy will be stored in the cable when the cable resistance is equal to the source impedance. The current will be $V_{max}/2R$. Therefore the maximum inductance permitted will be four times that permitted at the short-circuit current of the source. If the ratio L_x/R_x does not exceed $4L_{max}/R$, where L_{max} is the maximum permitted connected inductance, the cable will be safe regardless of length.

In summary, the design techniques used in all commercially available systems are similar. In this section we noted that the fundamental techniques are very few in number. Manufacturers may introduce variants of the basic techniques, some quite imaginative, but the fundamental design techniques are similar in all commercially available systems. Although some systems have used current limiting resistors and have put voltage limiting in the power supply, some have used active barrier isolators, and some have used zener diode barriers, there is no difference in safety among the various approaches. Any of the techniques properly applied will yield a safe system.

CERTIFICATION OF INTRINSICALLY SAFE APPARATUS

In the early years almost all intrinsically safe apparatus was certified as part of a complete loop, now called *loop certification*. The apparatus in the Division 2 or nonhazardous location, now called *associated apparatus*, was specified either by model number or, somewhat later, in the case of intrinsic safety barriers, by the electrical characteristics V_{max} and I_{max}. The intrinsically safe apparatus was also specified by model number.

In Germany a different scheme developed, now common worldwide. Associated apparatus is characterized by the maximum open-circuit voltage V_{oc}, the maximum short-circuit current I_{sc}, and the maximum permissible connected capacitance and inductance C_a and L_a.

Intrinsically safe apparatus is characterized by V_{max}, I_{max}, and P_{max}, the maximum voltage, current and power that can be safely applied at the terminals, and C_i and L_i, the effective capacitance and inductance seen at the terminals. (A large capacitor discharging into the external circuit through a current limiting resistor may appear at the terminals to be equivalent to a much smaller capacitor, and an inductor shunted by diodes may appear to be a small inductor.)

IEC and CENELEC standards are written around this kind of specification, known in North America as *entity approval*.

In current North American practice the manufacturer supplies a control drawing which details the interconnections of apparatus evaluated during the certification process. Apparatus may be defined broadly by specifying maximum parametric values, as in European practice, or very specifically by calling out model numbers. This drawing also specifies any special installation conditions, such as maximum load or power supply voltages, fusing, etc. for associated apparatus, or special grounding.

SYSTEM DESIGN USING COMMERCIALLY AVAILABLE INTRINSICALLY SAFE AND ASSOCIATED APPARATUS

General Principles

This section summarizes the principles underlying implementation of an intrinsically safe system using commercially available components. These components are the intrinsically safe apparatus in the Division 1/Zone 0/Zone 1 location and the associated apparatus in the Division 2/Zone 2 or nonhazardous location.

The controlling demand of all intrinsically safe system design is that every ungrounded conductor entering the Division 1 location (or Zone 1/Zone 0 location) must be protected against the inflow of nonintrinsically safe voltage, current, and power levels from the Zone 2/Division 2 or nonhazardous location.

If associated apparatus, such as a zener barrier, protects every ungrounded line so that application of line voltage to the associated apparatus does not cause nonintrinsically safe voltage and currents to flow in the protected lines then the apparatus on the nonhazardous side of the associated apparatus may be chosen with regard to function only. It need not be certified. The only restriction on this apparatus is that there be no voltage in it that exceeds the voltage rating of the associated apparatus. Most associated apparatus is rated for intrinsically safe outputs after application of line voltage, usually 250 V RMS.

The major obstacles to achieving an intrinsically safe system fall broadly into two categories.

1. The field-mounted equipment is not certified to be intrinsically safe.

2. The control house equipment is not certified as associated apparatus.

If the field-mounted apparatus and the control house interface apparatus have been certified intrinsically safe, separately or together, select equipment for the system which lies within the limits on their control drawings, choosing the alternatives that best fit the plant situation.

Field-mounted Apparatus not Certified.   If the field mounted device is simple apparatus, it need not be certified when connected to certified intrinsic safety barriers or other associated apparatus. Simple apparatus includes:

- passive devices such as switches, junction boxes, resistance temperature detectors, potentiometers
- simple semiconductor devices
- sources of stored energy, such as inductors or capacitors with well defined characteristics, which are to be taken into account when assessing intrinsic safety, or
- sources of generated energy such as solar cells and thermocouples. These sources shall not generate more than 1.5 V, 100 mA, or 25 mW.

In some standards the stored energy sources are not mentioned, and simple apparatus is defined as apparatus which neither generates nor stores more than 1.2 V, 100 mA, 25 mW, or 20 μJ. These standards tactly assume that inductive or capacitive storage elements will be taken into account as part of the L of C that may be connected to the associated apparatus.

Simple apparatus must be separated from nonintrinsically safe circuits and apparatus to ensure that nonintrinsically safe energy levels will not be injected into the intrinsically safe circuits. Most experts assume that 50-mm spacing or partitions provide sufficient protection against intermixing of the two circuits. Sometimes it has been stated that the simple apparatus shall not be in the same enclosure as nonintrinsically safe circuits or apparatus, but in the author's opinion, this is too restrictive a translation of reasonable intent into a rule. Always conform to limitations imposed by the control drawing for the associated apparatus.

Simple apparatus must be investigated for the applicable Temperature Code. If the barrier supplying it can deliver no more than 1.2 W maximum power then a T4 (135 C) rating is usually reasonable and defensible. If T5 (100 C) rating is desired it is necessary to investigate the temperature rise when exposed to the maximum power delivered from the barriers used in the system.

If the field-mounted apparatus is not certified, and does not conform to the limitations imposed on simple apparatus, the apparatus cannot easily be used in an intrinsically safe system. Few, if any, system designers can afford to make the investment required to self-certify a design, and because the user can't control the details of the design, the assessment applies only to the specific design of the piece of apparatus investigated. In the current regulatory environment few managers would risk using apparatus that has not been certified by a third party, so the need to consider self-certification is unlikely to arise. It may sometimes be necessary to permit temporary use of a product that has been submitted to a third party for certification. The supplier should provide a written assessment of the intrinsic safety of the device and a schedule for the availability of the certification. The device should not be installed unless the user has sufficient knowledge and resources to review and assess the vendor's self-certification document.

Control House Apparatus not Certified as Associated Apparatus.   For the same reasons that it is not practical to self-certify an item of field-mounted apparatus, it is not sensible to consider self-certifying or accepting a vendor's self-certification of control room apparatus. There is seldom any need to consider this option. In principle, one needs only to interpose between the control house apparatus and the field apparatus a certified associated apparatus, such as a barrier. The barrier must be compatible with the intrinsic safety parameters of the field device and must also allow the system to function as intended when it is connected between the field device and the control room device.

Uncertified Control Room Apparatus Connected to Uncertified Field Mounted Apparatus.   A system designer may occasionally encounter a special case of uncertified field-mounted apparatus connected to uncertified control room apparatus. One solution is to provide the field mounted device with another type of protection, usually an explosionproof enclosure or pressurization. Install barriers in its signal lines to the control room. These are usually mounted in the same enclosure as the apparatus generating the signal, but could be installed in a separate enclousre suitable for the location. Install barriers also at the control room apparatus. The two sets of barriers prevent ignition capable energy from entering the Division 1 location on the signal lines from either the field apparatus or from the control room apparatus. Select barriers so that the current drawn into a fault caused by shorting and grounding of the signal lines in the Division 1 location remains at a safe level, i.e. the sum of the short circuit currents of the barriers must be safe. This scheme is seldom applicable to a transmitter powered over the signal lines, as in two-wire ungrounded 4–20 mA transmitters. It is difficult to specify barriers that meet the functional requirements of the circuit, and also have safety descriptions with low enough values of current and voltage to permit summing four barrier currents into a ground fault in the line between the barriers. The technique is more frequently used with field mounted transmitters that have their own source of power, so that the signal from the transmitter to the control house is a voltage or a current of relatively low level, perhaps 1–5 V, 4–20 mA. Barriers with safety descriptions such as 9 V, 100 mA can be utilized in such signal lines to ensure both safety and function of the circuit.

Vendors of barriers make available a wealth of application information, including specific recommendations of barrier models for use in the application being discussed. The reader would be wise to use this free support.

INSTALLATION OF INTRINSICALLY SAFE SYSTEMS

Because an intrinsically safe system is incapable of igniting a flammable atmosphere even under fault conditions, cables of special construction or conduit are unnecessary. It is, however, necessary to install intrinsically safe systems so that ignition-capable energies will not intrude from another circuit.

An intrinsically safe system will be safe *after* installation if the installation

1. Conforms to the limiting parameters and installation requirements on which approval was based, as stated on the control drawing
2. Prevents intrusion of nonintrinsically safe energy on intrinsically safe circuits
3. Prevents power system faults or differences in ground potential from making the circuit ignition-capable

To prevent the intrusion of other circuits, intrinsically safe circuits must be run in separate cables, wireways, or conduits from other circuits. Terminals of intrinsically safe circuits must be separated from other circuits by spacing (50 mm) or partitions. The grounding of intrinsically safe circuits must be separate from the grounding of power systems, except at one point.

Nonincendive Equipment and Wiring

It is not necessary to provide intrinsically safe equipment in Division 2 locations. The equipment need only be nonincendive, that is, incapable in its normal operating conditions of releasing sufficient energy to ignite a specific hazardous atmospheric mixture. Such equipment has been recognized without specific definition in the NEC. In Division 2 locations, equipment without make-or-break or sliding contacts and without hot surfaces may be housed in general-purpose enclosures.

Requirements for apparatus suitable for use in Division 2 and Zone 2 locations have been published by ISA and IEC. These documents provide more detail than is found in the NEC, partially because of a trend toward certification of nonincendive circuits that may be normally sparking, but release insufficient energy in normal operation to cause ignition. These documents also define tests for sealed devices that are needed by industry. The documents are similar, except that the ISA standard does not cover restricted breathing and enclosed-break devices.

IGNITION BY OPTICAL SOURCES

Until the early 1980s it was assumed that transmitting instrumentation and control signals on optical fibers would avoid all the issues relative to ignition of flammable atmospheres by electrical signals.

Research showed that the power level required to ignite flammable vapors and gases directly is much higher than would be found in common measurement and control systems. However, if an optical beam irradiates a particle and raises its temperature sufficiently, ignition can occur. During 1990–1994 a collaborative program in the European Community undertook to bound the problem. The report of this work, EUR 16011 EN confirmed that ignition of a small particle occurs at low enough optical power levels to be of concern in measuring and control systems. Based on the data gathered during this investigation it was concluded that there is essentially no hazard of ignition by a continuous-wave optical beam in the visible or near infrared if the peak radiation flux is below 5 mW/mm^2 or the radiated power is less than 35 mW. This recommendation is about a factor of 2 lower than the lowest data points, which included materials with low AIT values, such as diethyl ether and carbon disulfide, and materials like hydrogen and carbon disulfide with low spark ignition energy.

Though this guidance is helpful because many uses of fiber optics operate below these limit values those using higher power levels in analytical instruments and those using laser power for other purposes are left without guidance.

Ignition by the heating of small particles is related to ignition by hot wires and components, and by AIT determinations in that it is a thermal process where the value of the critical power or power density measured depends on the amount of time lag before ignition occurs. However, the amount of power required does not correlate strongly with conventional combustible properties such as Minimum Ignition Energy, MESG, or AIT. Estimating the amount of power required for ignition of materials other than those tested. or for material for which there is sparse data is not easy.

A second series of investigations scheduled to be completed in 1999 is intended to provide more data. It is known that igniting power for many common industrial materials in Group D (Group IIA) is 200 mW or more depending on the size of the heated target.

Regulations and Standards

At the outset there was a tendency in some quarters, which persists still, to blindly rush to apply standards for mitigating hazards in electrical systems to this problem. These efforts ignored the fact that in many applications a fiber must break in order to illuminate a particle, and that many optical systems provide easy means of detecting this and other failures in time to take remedial action, such as shut-down. This author believes that the hazard of optical systems should be treated with the tools of risk assessment, fault-tree analyses, and similar tools used for assessing and mitigating other types of hazard in the process industries.

ACKNOWLEDGMENT

The cooperation of the Instrument Society of America in permitting extractions and adaptations of text and illustrations from *Electrical Instruments in Hazardous Locations* is greatly appreciated.

BIBLIOGRAPHY

ANSI/NFPA 70, "National Electrical Code (NEC)," National Fire Protection Association, Quincy, Massachusetts, Art. 500–505.

API RP 500A, "Classification of Areas for Electrical Installations in Petroleum Refineries," American Petroleum Institute, New York.

API RP 505, "Recommended Practice for Classification of Electrical Installations at Petroleum Facilities Classified as Class I, Zone 0, Zone 1, or Zone 2," American Petroleum Institute, New York.

BS 5501, "Electrical Apparatus for Potentially Explosive Atmospheres," British Standard Institution, London, 1978.

Part 1: "General Requirements" (EN 50014)
Part 2: "Oil Immersion 'o' " (EN 50015)
Part 3: "Pressurized Apparatus 'p' " (EN 50016)
Part 4: "Powder Filling 'q' " (EN 50011)
Part 5: "Flameproof Enclosures 'd' " (EN 50018)
Part 6: "Increased Safety 'e' " (EN 50019)
Part 7: "Intrinsic Safety 'i' " (EN 50020)

Dubaniewicz, T. H. Jr., K. L. Cashdollar, and G. M. Green, with appendix by R. F. Chalken, "Ignition of Methane-Air Mixtures by Laser Heated Small Particles," International Symposium on Hazards, Prevention, and Mitigation of Industrial Explosions, Schaumburg, Illinois, 1998.

ISA RP12.1, "Electrical Instruments in Hazardous Atmospheres," Instrument Society of America, Research Triangle Park, North Carolina.

ISA Monographs 110–113, "Electrical Safety Practices," Instrument Society of America, Research Triangle Park, North Carolina.

ISA RP12.6, "Installation of Intrinsically Safe Instrument Systems in Class 1 Hazardous Locations," Instrument Society of America, Research Triangle Park, North Carolina.

ISA S12.12, "Electrical Equipment for Use in Class 1, Division 2 Locations," Instrument Society of America, Research Triangle Park, North Carolina.

Magison, E. C., *Electrical Instruments in Hazardous Locations* 4th ed., Instrument Society of America, Research Triangle Park, North Carolina, 1998.

Magison, E. C., *Intrinsic Safety, an Independent Learning Module,* Instrument Society of America, Research Triangle Park, North Carolina, 1984.

Magison, E. C., "Are Optical Beams an Explosion Hazard?" *InTech.* pp. 61–66, Sept. 1997.

McGeehin, P., "Optical Techniques in Industrial Measurement: Safety in Hazardous Environments," Report EUR 16011 EN, Luxembourg, 1994.
    Annex I: Summary report, F. B. Carleton, and F. J. Weinberg, Imperial College
    Annex II: Final report, M. M. Welzel, H. Bothe, and H. K. Cammenga, PTB
    Annex III: Summary report, G. Tortoishell, SIRA Test and Certification Ltd.
    Annex IV: K. S. Scott, University of Leeds
    Annex V: Final report, C. Proust, INERIS

NFPA 496, "Purged and Pressurized Enclosures for Electrical Equipment," National Fire Protection Association, Quincy, Massachusetts.

NFPA 497, "Recommended Practice for Classification of Flammable Liquids, Gases or Vapors and of Hazardous (classified) Locations for Electrical Installations in Chemical Process Areas," National Fire Protection Association, Quincy, Massachusetts, 1997.

NFPA 499, "Recommended Practice for Classification of Combustible Dusts and of Class II Hazardous (Classified) Locations for Electrical Installations in Chemical Processing Plants," National Fire Protection Association, Quincy, Massachusetts, 1997.

IEC Pub. 79–15 "Electrical Apparatus with Type of Protection 'n,' " International Electrotechnical Commission, Geneva, Switzerland.

UL 913, "Intrinsically Safe Apparatus and Associated Apparatus for Use in Class I, II, and III, Division 1 Hazardous (Classified) Locations," Underwriters Laboratories, Northbook, Illinois.

Buschart, R. J., *Electrical and Instrumentation Safety for Chemical Processes*, Van Nostrand Reinhold, New York, 1991.

Guidelines for Hazard Evaluation Procedures, 2nd ed. Center for Chemical Process Safety, American Institute of Chemical Engineers, New York, 1992.

Welzel, M. M., S. Schenk, M. Hau, K. K. Commenga, and H. Bothe, "Einfluss des Brennstoff/Luft-Gemisches auf die minimale zuendfaehige Strahlungsleistung bei Strahlungsabsorption an einer Eisenoxideoberflaeche," PTB-Mittellungen No. 109, 2/99.

SECTION 3
CONTROLLERS*

L. Arnold
Johnson Yokogawa Corporation, Newnan, Georgia
(Stand-Alone Controllers)

Douglas L. Butler
Solutia Inc., St. Louis, Missouri (Computers and Controls,
Distributed Numerical Control and Networking)

Jim Cahill
Fisher-Rosemount Systems, Austin, Texas (Field-Based Systems)

Douglas M. Considine
Columbus, Georgia (Computers and Controls)

Terrence G. Cox
CAD/CAM Integration, Inc., Woburn, Massachusetts (Distributed
Numerical Control and Networking)

G. L. Dyke
Bailey Controls Company, Wickliffe, Ohio
(Stand-Alone Controllers)

Thomas J. Harrison
IBM Corporation, Boca Raton, Florida (Computers and Controls)

Tom Hudak
Rockwell Automation, Allen-Bradley, Mayfield Heights, Ohio
(Programmable Controllers)

John P. King
The Foxboro Company (a Siebe Company), Rahway, New Jersey
(Distributed Control Systems)

J. Kortwright
Leeds & Northrup (a Unit of General Signal), North Wales,
Pennsylvania (Stand-Alone Controllers)

Ralph Mackiewicz
SISCO Inc., Sterling Heights, Michigan (Programmable
Controllers, Manufacturing Message Specification)

C. L. Mamzic
Systems and Application Engineering, Moore Products Company,
Spring House, Pennsylvania (Stand-Alone Controllers)

---

* *Persons who authored complete articles or subsections of articles, or who otherwise cooperated in an outstanding manner in furnishing information and helpful counsel to the editorial staff.*

Donald McArthur
*Hughes Aircraft Company, Culver City, California
(Computers and Controls—prior edition)*

T. A. Morrow
*Omron Electronics, Inc., Schaumburg, Illinois (Timers and
Counters)*

Raymond G. Reip
*Sawyer, Michigan
(Hydraulic Controllers)*

B. R. Rusch
*Gould, Inc., Andover, Massachusetts (Programmable
Controllers—prior edition)*

H. L. Skolnik
*Intelligent Instrumentation (A Burr-Brown Company),
Tucson, Arizona (Computers and Controls)*

D. N. Snyder
*LFE Instruments, Clinton, Massachusetts
(Stand-Alone Controllers and Timers and Counters)*

J. Stevenson
*West Instruments, East Greenwich, Rhode Island
(Stand-Alone Controllers)*

Stan Stoddard
*Waugh Controls Corporation, Chatsworth, California
(Automatic Blending Systems)*

W. C. Thompson
*Fisher Controls International, Inc., Austin, Texas
(Batch Process Control)*

T. J. Williams
*Purdue University, West Lafayette, Indiana
(Computers and Controls—prior edition)*

Douglas Wilson
*Hughes Aircraft Company, Canoga Park, California
(Computers and Controls—prior edition)*

DISTRIBUTED CONTROL SYSTEMS

by John P. King*

INTRODUCTION

The definition of a distributed control system (DCS) has dramatically changed over its history. The DCS was originally conceived as a replacement for large panels, located in central control rooms and containing hundreds, sometimes thousands, of process instruments. The information-processing role of the DCS quickly expanded, adding advanced control such as model reference control and expert systems; information-analysis tools, such as statistical process control and intelligent alarming; decision support applications such as predictive maintenance and document management; and business system integration capabilities. Today the DCS is expected to be the real-time component of a manufacturing facility's enterprise management and control system, affecting virtually every aspect of the operation and profitability of both continuous and batch processes.

As its name implies, the DCS has three essential qualities. The first is to distribute its functions into relatively small sets of semiautonomous subsystems, which are interconnected via a high-speed communications network. These functions include data collection, process control, presentation of information, process analysis and supervision, storage and retrieval of archived information, and the presentation and reporting of information. The distribution of functions has several advantages over more classical process computer architectures, including:

- lower exposure to component or subsystem failure and better isolation to facilitate maintenance and upgrades
- better partitioning of application requirements for different process areas
- improved modularity for application development
- facilitation of geographical distribution, which reduces installation costs (reduced wiring) and provides more localized operational supervision while offering global access to information and control capabilities

The second is to automate the manufacturing process by integrating advanced regulatory control, logic and sequential control, and procedural languages into a cohesive application that is integrated with a variety of advanced application packages ranging from Batch to Expert Systems. The "control" aspect of the modern DCS has expanded to include information that is capable of supporting such manufacturing enterprise applications as

- activity-based cost accounting
- production scheduling and dispatching

---

* The Foxboro Company (a Siebe Company), Rahway, New Jersey.

- preventative or predictive maintenance scheduling
- validation of employee job certification and readiness
- information exchange with business, logistics, and transportation applications

The third characteristic is a system. Webster defines a system as "a set or arrangement of things so related or connected so as to form a unity or organic whole."[1] The systems aspect of a DCS organizes the command structure and information flow among its constituent parts so as to have it act as a single automation system unifying the various subsystems, including

- process signal input and conditioning
- process actuator signal output
- regulatory, combinatorial, and sequence logic and procedural and supervisory control
- human readable process displays of current values, alarms, trends, and calculations
- human actions including setpoint changes, manual overrides and alarm handling
- application subsystems such as process optimization and manufacturing support
- information-storage subsystems
- communications subsystems

EVOLUTION OF THE DISTRIBUTED CONTROL SYSTEM

The modern DCS has been constructed in layers, with each phase of its history contributing to a particular and essential quality. The following is a brief review of this history, including the contextual background of this evolution. The expectations and challenges associated with the various phases of evolution provide a perspective of the technology-driven market expectations and application challenges that had to be overcome for the DCS to displace then-current solutions.

In the 1940s ingenious mechanical and pneumatic devices controlled processes. These were located next to the process. The scope of control was typically limited to a single or at best a few process conditions, such as temperature, pressure, or flow. These devices made it too difficult to design and maintain coordinated process unit control systems. Processes were run conservatively to ensure stability rather than at peak economic performance.

In the late 1950s, electronic controllers were introduced. These devices could be wired together to create coordinated control strategies such as feedforward dynamic models. They could be centrally located in a control room by running wires to the measurements and actuators at the process. Operators could also be physically removed from potentially hazardous areas. Fewer human operators were required because a single operator could interact with multiple process units. Performance was improved because operators had access to more information. Engineering, constructing, and maintaining complex control systems continued to be cost prohibitive for most companies, which constrained their use to achieve optimal economic performance.

Process computers made their debut in the mid-1960s; however, their cost and the specialized skills required to effectively deploy them greatly limited their market acceptance. For those with the money, vision, and resources to deploy process computers, however, the journey to optimal economic performance had begun. Advanced control strategies no longer had to be simplified into models suitable for linear circuit technology. Advanced regulatory control, discrete logic, operational sequences, and procedural languages could now be combined into a cohesive strategy. The analysis and presentation of real-time process information could now be provided in a more intuitive and timely manner to improve the analysis skills of operators and supervisors.

By the early 1970s, analog systems architectures were introduced. Foxboro's Spec 200 is an example of this. Figure 1 illustrates the repackaging of individual analog controllers and auxiliary modules into a card, nest, and rack structure.

---

[1] *Webster's New World Dictionary of the American Language, Second College Edition*, The World Publishing Company, 1968.

FIGURE 1   Spec 200 Architecture (courtesy of The Foxboro company).

Spec 200 was designed by using linear circuit operational amplifier technology and was assembled into functioning control loops by selecting and wiring the appropriate input cards, dynamic characterizers, computing elements (summers, multipliers, etc.), controllers, displays, recorders, and output cards. This had the advantage of being able to select the components needed for the specific requirements of a process loop. The replacement of current transmission with voltage enabled components to be connected in parallel as well as in a series, thereby supporting much more complex control strategies. Complex control design and calibration maintenance were simplified. Specialized computer interface devices were introduced to facilitate communication with analog control loops, including the ability to read measurement, setpoint, output, auto/manual mode, remote/local mode, alarms, and controller status. In addition, these devices could write setpoint, output, computer/auto/manual output mode, and computer/remote/local setpoint mode.

This shift to a systems architecture from discrete instruments created a number of concerns in the control systems community, which had to be addressed to gain market acceptance. The most notable were

- a single component failure (e.g., power supply) would affect a large number of loops
- loops now comprised many components that had to be specified and wired
- the separation of control from the display made troubleshooting and maintenance more difficult
- most modules service multiple loops (e.g., multiple input card); functioning loops would be removed from operation during the replacement of a module with a failed component

Each of these concerns were addressed (e.g., parallel power supplies with auctioning circuitry), and the systems approach to instrumentation was born.

The synthesis of systematized instrumentation and process computers became the goal of distributed control systems in the 1970s.

By the mid-1970s, advances in microprocessor and software technology, coupled with the commercial availability of process control and operations software, made process computers affordable and useful tools for process control engineers.

Also by the mid-1970s, advances in microprocessor and software technology make it practical and affordable to replace operational amplifiers with digital circuitry. Honeywell's TDC 2000, Fig. 2, was an early example of this architecture. Digital circuitry had several advantages. First, the variety of component cards could now be dramatically reduced, as the function of the automatic control cards could reside in part as software programs. Several control calculations could be time multiplexed through a single central processing unit, thereby reducing the hardware required. Wiring could be replaced by "software referencing." Computations difficult to perform in electronic circuitry, such

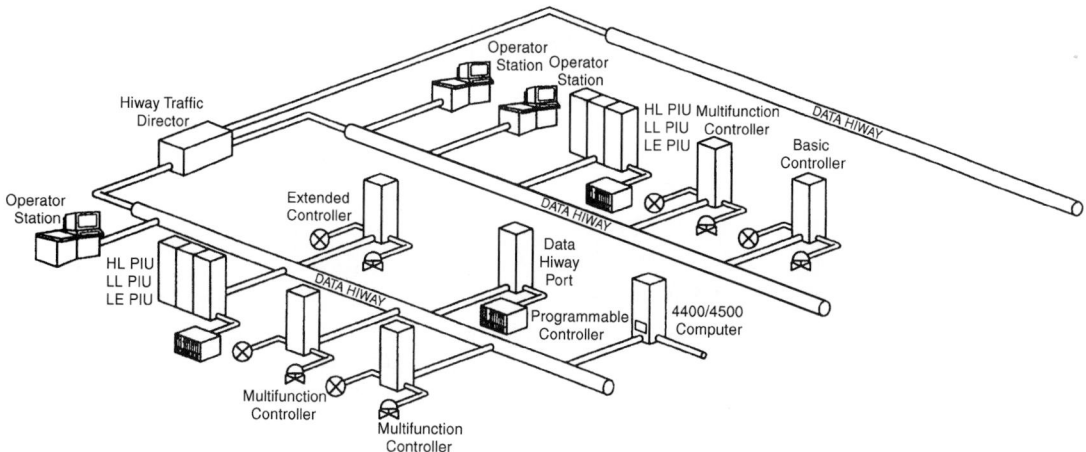

FIGURE 2   TDC 2000 architecture (courtesy of Honeywell).

as time delays (deadtime) or nonlinear relationships, could be easily accomplished with a digital computer. Automatic control systems were now digital.

Programmable logic controllers (PLC) also made their debut onto the market during this time frame. Their primary mission was to replace the large racks of relays found in both process and discrete manufacturing. Their acceptance was slow until a programming language was introduced that enabled electrical technicians to "program" these devices by using the same relay and coil designs they used to build relay racks.

The cathode ray tube (CRT)-based operator workstations made their debut in the mid-1970s as well. They replaced the display and operator entry components of panelboard instruments, including controller faceplates and auxiliary modules such as alarm annunciators, recorders, and discrete devices such as pushbuttons and switches. Hundreds of linear feet of panelboard containing instruments in the control rooms of large plants such as petroleum petrochemical plants could now be replaced by a few CRT terminals (Fig. 3). The economic case for these workstations was strong. Software applications for advanced navigation, alarming, and trending capabilities soon made these workstations the preferred operator interface even for smaller control rooms. Intelligent presentation and analysis of process information was the new tool for better economic performance.

Serious operational issues, however, had to be addressed to make these workstations acceptable. These included

- simultaneous access and contextual understanding of the information contained in hundreds of linear feet of panel when represented by a few consoles with their limited display surfaces
- ability to immediately identify, select, and interact with a single controller or group of controllers
- concurrent viewing of current values, trends, and alarms
- ability to intuitively navigate between summaries of a large number of loops and the details concerning a single loop or small group of loops.

Foxboro's Videospec, Figs. 4(*a*) and 4(*b*), introduced in 1976, represented an early solution to these concerns. It literally emulated the hundreds of feet of panel with different display views and offered a functional keyboard whose context changed with the screen view.

By the late 1970s to early 1980s, the second generation of distributed controllers greatly extended the use of software links between functional control elements. This enabled the control engineer to construct and modify precise dynamic models of processes. It also introduced new design considerations. These included the following.

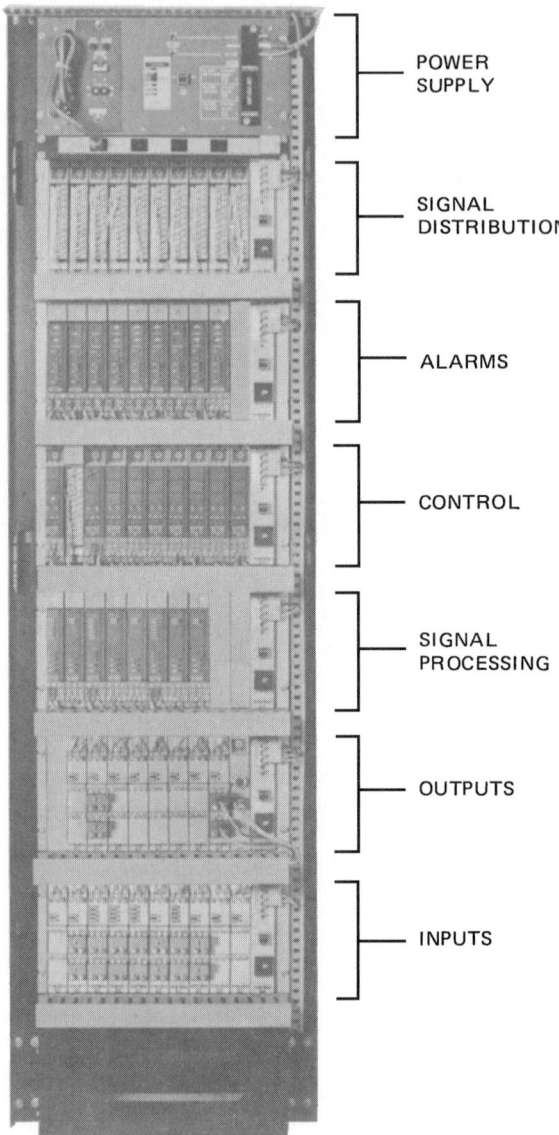

POWER
SUPPLY

SIGNAL
DISTRIBUTION

ALARMS

CONTROL

SIGNAL
PROCESSING

OUTPUTS

INPUTS

FIGURE 3(a)   Photo of a central control room, using panelboard instruments (courtesy of the Foxboro Company).

Sampled Data Interval Induced Deadtime

Software loops executed periodically instead of continuously like their linear circuit analogs. This equated to the introduction of deadtime in the control strategy. As long as the period of execution of the digital loop was significantly smaller than the primary time constant of the process, the impact was minimal to insignificant—unless the control strategy incorporated an inner loop that was isolated from the process.

FIGURE 3(*b*)   Photo of a CRT console as a panelboard replacement (courtesy of the Foxboro Company).

One example of this is the multiple-output control station, or MOCS, which enables multiple outputs to be manipulated to control a single measured variable. Examples of this include multiple pumps in parallel, which together comprise the energy needed to produce a single liquid flow rate, and multiple draft fans, which work together to produce a required air flow rate. The control problem is manifested by the effect of bringing one of the prime movers (pumps, fans) on or off line. In a simple control strategy the gain of the process would be radically altered by this event. The solution is to have the individual controller gains to each of the prime movers automatically compensate for the effect of one or more of them being brought on or off line.

This is accomplished in analog circuitry by creating an isolated inner loop, which automatically compensates for changes in dynamic response resulting from one or more of the prime movers being taken on or off line. This loop consisted of a high-gain proportional controller that received its setpoint from the output of the outer loop and its measurement from the weighted sum of the outputs to the prime movers (see Fig. 5(*a*)). Any discrepency between setpoint and measurement would instantaneously drive the weighted sum of the output to equal the demand setpoint. As there is no dynamic element in the inner loop, the control system is stable.

This would not be true with digital circuitry because the control algorithms are processed periodically. This time interval between computations introduces deadtime into the inner loop, making it unstable and driving it to oscillate wildly as shown in Fig. 5(*b*). Reducing the time period between executions would not dampen this effect; rather it would increase the frequency of oscillation.

The inherent oscillation is caused because of the time delay in the information being supplied to the measurement of the high-gain controller. It is always one sampled data interval older than the setpoint value, because the controller algorithm is calculated before the weighted average is computed and therefore the output of the summer is always the reading from the previous calculation.

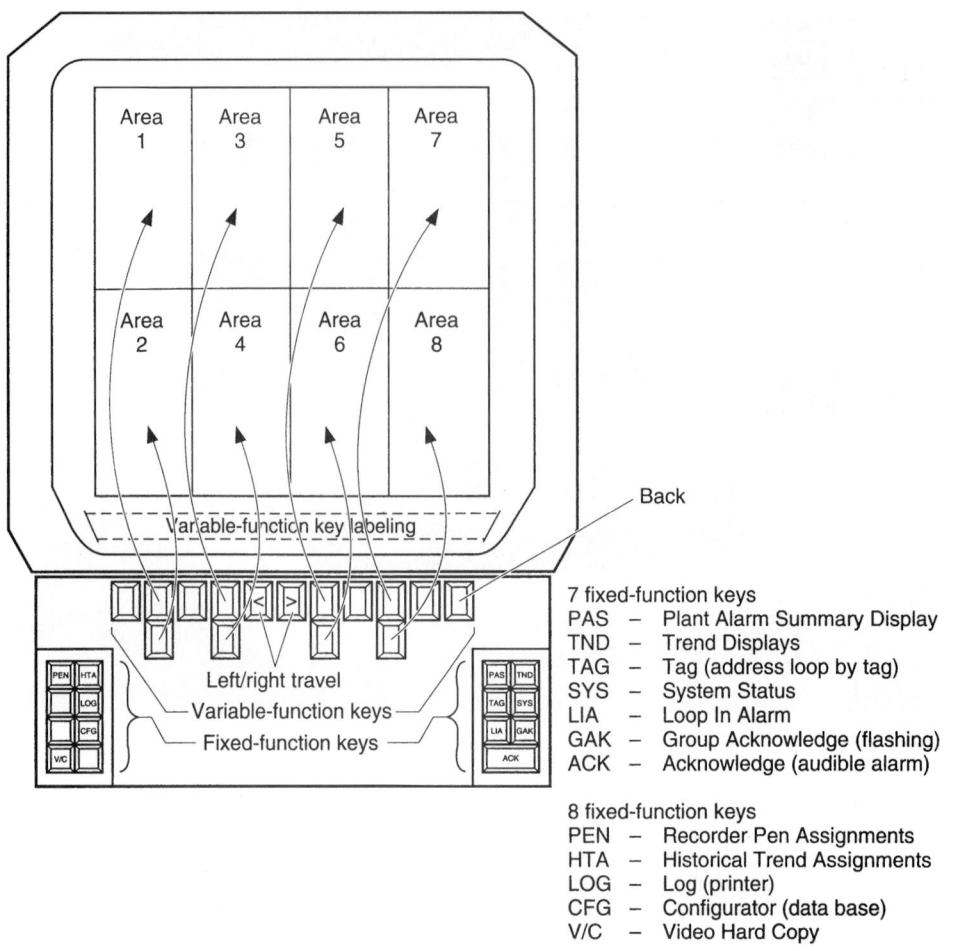

FIGURE 4(a)   Videospec workstation (courtesy of the Foxboro Company).

Figure 5(b) shows the effect of using a MOCS that uses high-gain proportional control on a digital (sampled data interval) controller. For purposes of this illustration it is assumed that the cycle time of the inner loop is sufficiently fast that any change to the output of the high-gain controller will have a negligible short-term (10 cycles) effect on the process and therefore the setpoint to the high-gain controller.

Several MOCS Controller designs are available for sampled data systems that provide the equivalent functionality and performance to analog circuitry. One solution is to replace the high-gain proportional controller with a high-gain (short integration time) integral-only controller, utilizing the formula

$$\text{Output} = \text{Previous Output} + (\text{Gain} \times \text{Error} + \text{Previous Error})/(\text{Gain} + 1),$$

where

$$\text{Error} = \text{Current Setpoint} - \text{Previous Output}$$

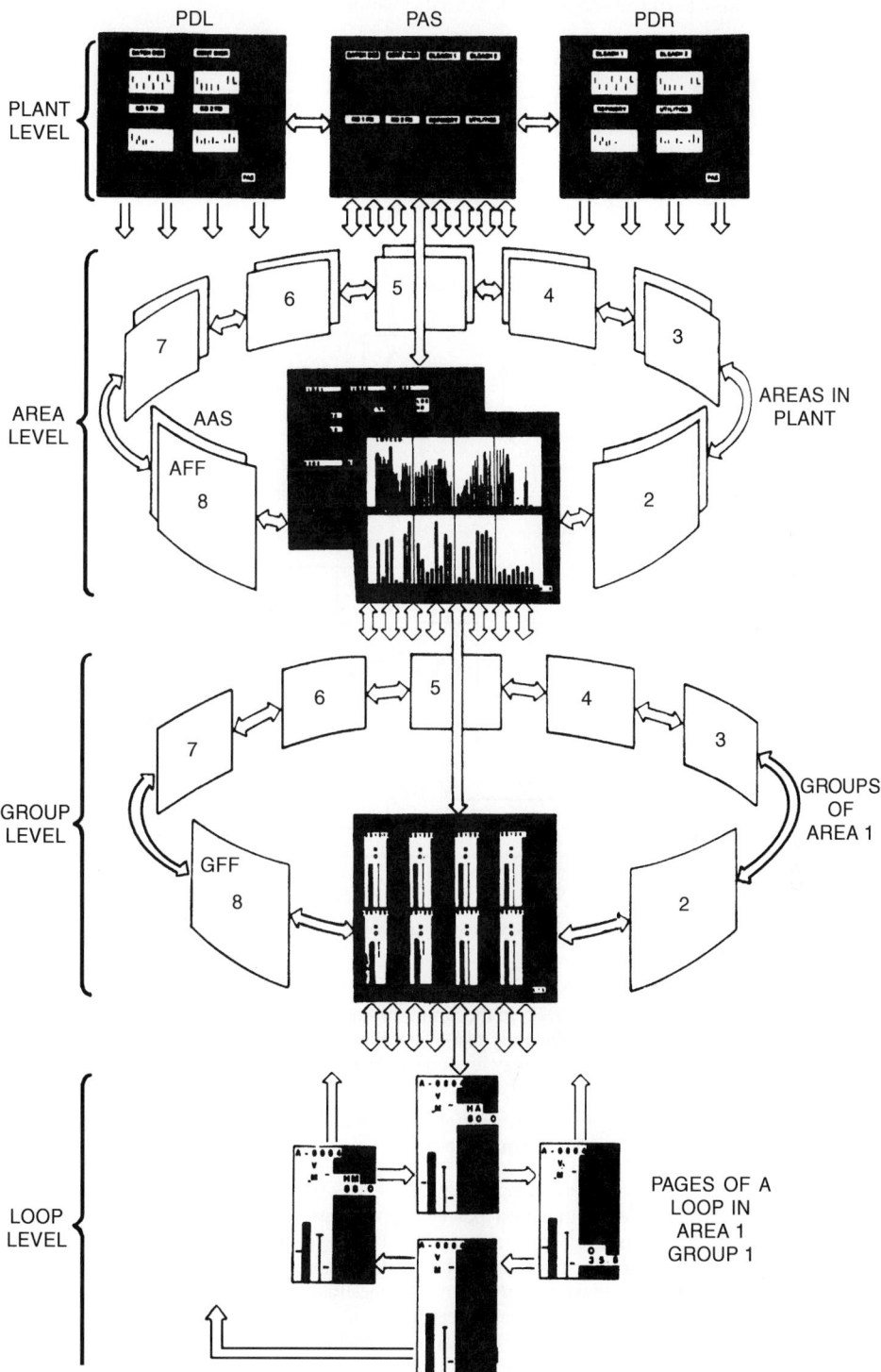

FIGURE 4(*b*)   Videospec screen navigation (courtesy of the Foxboro Company).

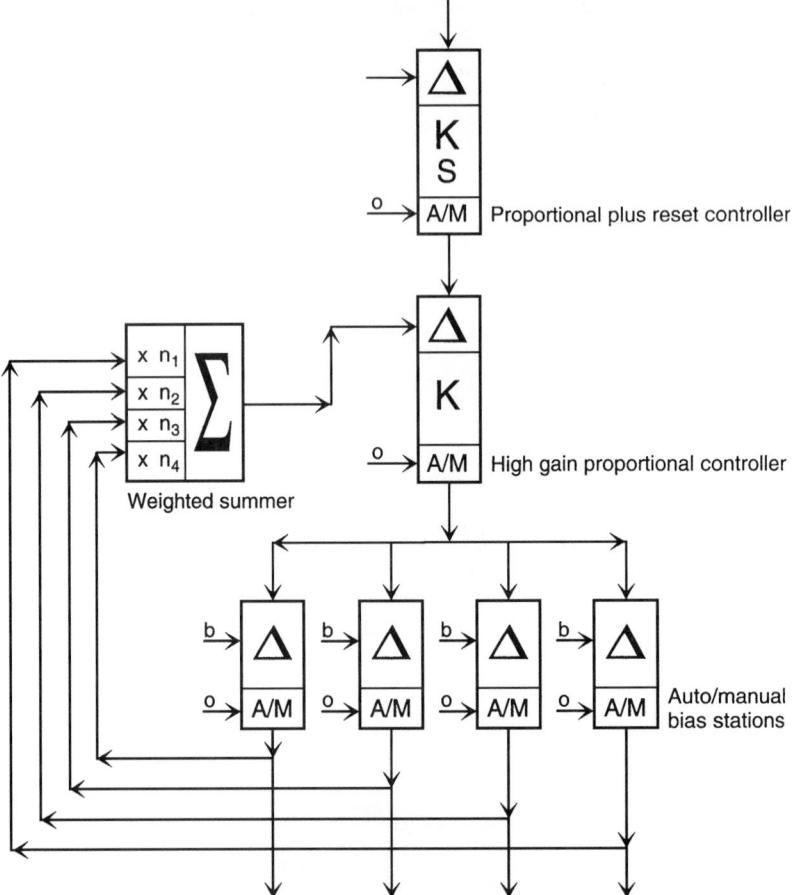

FIGURE 5(a)   SAMA drawing of MOCS, using a high-gain proportional analog controller.

The number of prime movers in the circuit has no effect on the convergence of the inner loop as long as the inner loop is sufficiently fast to stabilize before the outer loop is affected. The dynamic response of the new control strategy is shown in Fig. 6.

Loop Commissioning: Initialization Sequences and Bumpless–Balanceless Transfer

When a control loop is first turned on (powered up), it must establish its initial conditions by reading the measurement transmitter and assuming an initial value for the output. During initial commissioning this can be assumed (valve closed for instance). If the controller is commissioned in manual control, a careful transfer to automatic control can be orchestrated. If the control system is commissioned in automatic control, which may occur after a transient power outage of less than 1 s, the problem is more complex. The capacitance of the plant may well have enabled it to ride through this temporary loss of power, but the volatile memory of the digital controller may well have been lost. These include

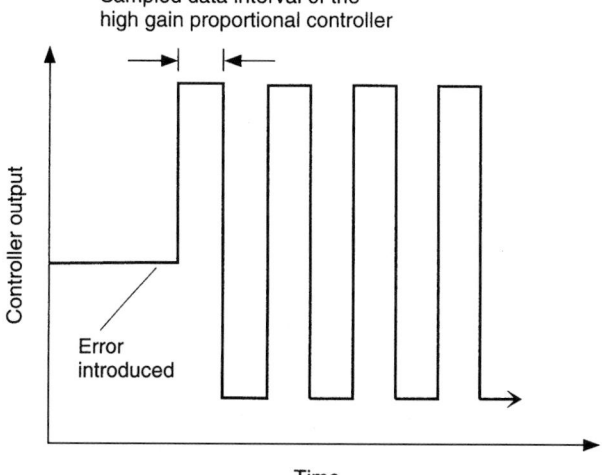

FIGURE 5(*b*) Dynamic response of MOCS, using a proportional-only sampled data controller.

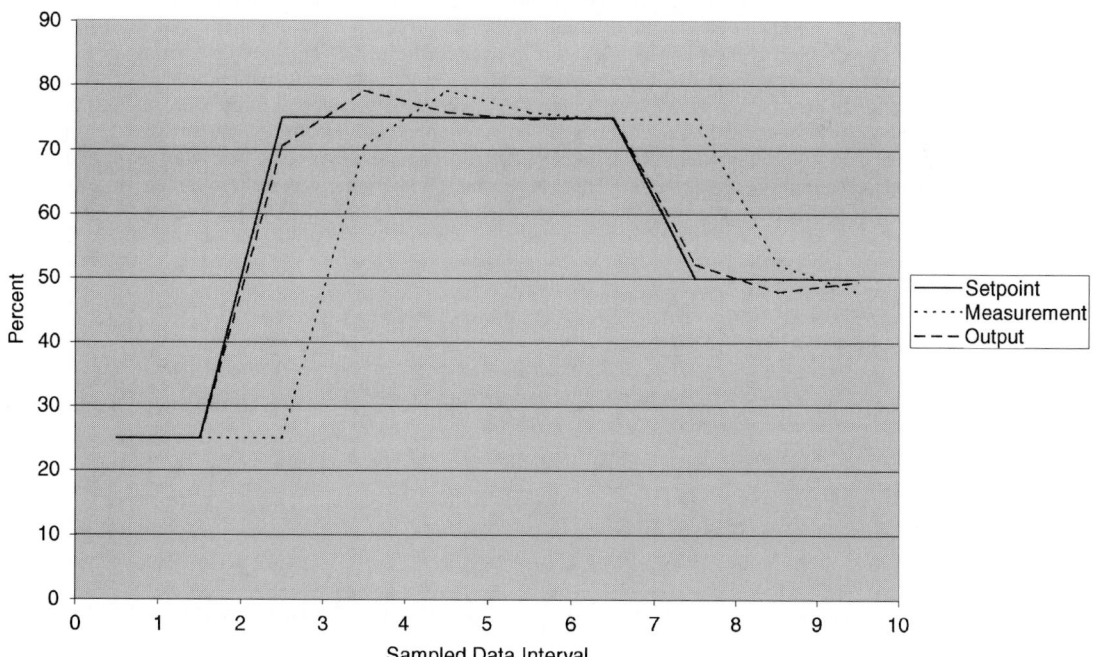

FIGURE 6 Dynamic response of MOCS, using an integral-only sampled data controller.

- providing some form of alternative power, e.g., battery backup
- frequently storing the values and parameters stored in volatile memory in nonvolatile storage

The issue then is how to set the initial conditions. The process measurement is not an issue—simply read the value from the transmitter. The current condition of the actuator is another issue. The value on the output register of the input/output (I/O) module is not necessarily representative of the actual output at initialization. There is no (unless provided as a separate input) feedback path from the actuator to the output module. Several alternatives exist to deal with this specific problem, such as forcing the output to a predetermined failsafe state, stored in nonvolatile memory on the I/O module, upon loss of control and assuming the actuator is in this position upon return. This would require the failsafe output value to coincide with the condition the actuator would assume if no power were provided to it.

This issue becomes more complex as more advanced control strategies and dynamic modeling are implemented. As was illustrated by the MOCS controller example, the serial nature of loop processing can create added complexities if information is required to flow from downstream blocks to those upstream. This is precisely what is required to propagate forward information about the state of final actuators that are at the end of a complex control block sequence.

Consider a temperature overflow cascade loop with a pressure override introduced via an autoselector, as shown in Fig. 7.

In this example the outputs of both the temperature and pressure controller connect to the input of the autoselector. The output of the autoselector connects to the setpoint of the flow controller. On return from a transient loss of power, the flow controller is able to determine both its measurement and its output based on the implementation of one of the schemes mentioned above. Its setpoint is another matter. The temperature and pressure controllers are able to determine their current measurement and read their last saved setpoint from persistent storage. Their output is another matter. Information regarding inputs or outputs associated with the autoselector are lost.

One solution to regaining automatic control without manual intervention is to institute a back calculation sequence. This method would utilize the primary controller's external reset feedback signal as a balancing mechanism. The strategy would assume that the control system was essentially in balance at the time of the memory loss. The first cycle of the calculation would set the output of both the pressure and temperature controllers as well as the setpoint of the flow controller equal to the measurement of the flow controller. The unfiltered reset feedback signal to the pressure and temperature controllers would be backcalculated to balance the proportional-gain amplified error against the initially set output value. This would enable both primary controllers to return to automatic control without "bumping" the flow controller setpoint during the next cycle.

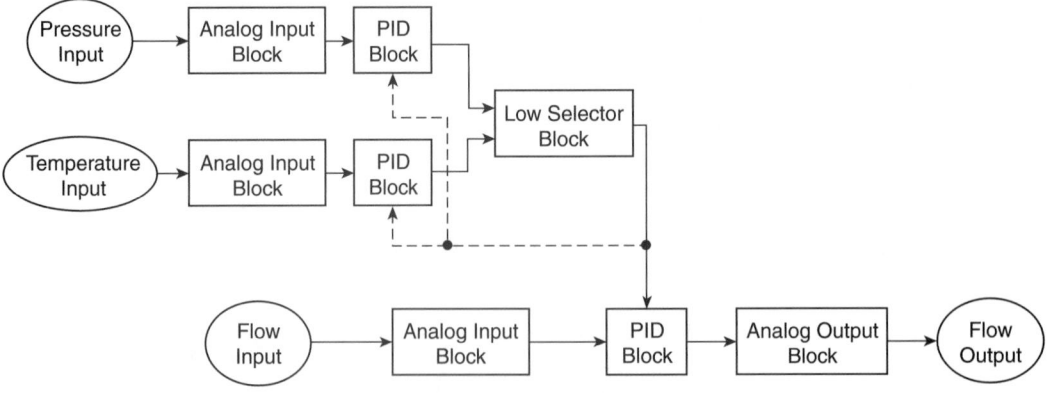

FIGURE 7 Temperature overflow cascade with pressure override, requiring backpropagation during an initialization sequence.

More complex strategies involving dynamic feedforward models with feedback trim would probably require several cycles to become rebalanced. If deadtime elements were included, portions of the control system might have to remain in a manual state until the control system regained the knowledge to perform correctly.

Hybrid Designs for Automatic or Manual Backup by Analog Techniques

There was also considerable distrust with the reliabiltiy and robustness of these new digital algorithms and controllers. This led to the creation of elaborate switching schemes between digital control strategies and simpler analog backup strategies, complicating the control engineer's responsibilities by forcing an analysis of the various permutations for commissioning hybrid strategies.

Transition from Automatic Control to Production Management

By the mid-1980s, computer hardware and software technology had reached the point where DCSs could address expanded issues concerning real-time information, and the emphasis shifted from automatic control to constructing tools to facilitate the production process:

* batch: product recipe management
* production scheduling and dispatching
* plantwide information collection, storage, retrieval, and reporting
* statistical process control
* manufacturing execution systems
* process accounting, including material and energy flow reconciliation
* process modeling and model reference control
* expert systems

To accommodate this new expectation, DCSs had to open their proprietary systems to accommodate third-party software from disciplines outside of the process control community by adopting standards-based designs. UNIX and the adaptation of communications standards such as ethernet (IEEE 802.3), token bus (IEEE 802.4), or token ring (IEEE 802.5) and variations of the seven-layer Open System Interconnect Model provided the communications framework in which great strides were made in transforming traditional DCSs into distributed computing environments in a relatively short time period.

One of the first systems to adopt this standards approach was Foxboro, with its Intelligent Automation Series introduced in 1987 (IA). It also hosted a myriad of third-party applications that were not directly associated with closed-loop automated control.

The Foxboro Intelligent Automation Series utilized the following:

* UNIX for its application-level operating system and a real-time operating system for mission-critical requirements, such as regulatory control
* ethernet and a proprietary communications messaging scheme called Object Management, used to symbolically reference information, remotely access file systems, and remotely execute programs and download binary software images to modules such as control stations and diskless workstations for its communications subsystem
* a commercially available relational database management system (RDBMS) for information storage and retrieval

In effect, Foxboro integrated a standards-based, open distributed computing environment with the DCS instead of offering the process computer as an appendage.

Once again this fundamental shift from conventional wisdom created a number of legitimate concerns in the industry. Among these were the following.

1. UNIX is essentially used for transactional processing and requires a high degree of systems administration, including being periodically taken off line for software maintenance. Can it really be used for mission critical control, which requires 7 days per week, 24 hours per day operation?

2. UNIX is a time-share system and not an interrupt driven, priority-based operating system. Can it adequately respond to the urgent real-time requirements of a DCS?

3. UNIX is very complicated and usually requires specialists not typically found in the process control community. Can these systems be adequately maintained?

4. RDBMSs are optimized to retrieve information efficiently. The demands of a real-time process information system are quite the opposite. They must correctly store hundreds to thousands of process values and events each second. RDBMS technology simply is not designed to address the real-time data-gathering requirements of DCSs.

5. Moving information technology applications onto a DCS unnecessarily complicates and therefore diminishes the reliability and performance of its primary goal—to control the process.

6. Issues such as quick and correct recovery from a system failure such as a loss of power are simply not designed into UNIX, increasing the overall system vulnerability and further compromising the primary mission–process control.

7. The complexity and time-share nature of UNIX makes it a poor candidate for redundant operations, and therefore highly available mission-critical operations cannot be achieved.

These issues were well understood by the suppliers of DCS systems and were for the most part successfully addressed in close cooperation with computer systems vendors such as SUN Microsystems. Issues such as system response time (workstation update time) and sluggish system "reboot" manifested themselves in the early versions of these hybrid computer–DCS offerings. However, since the performance and capacity of microprocessor technology essentially doubles for a given price point every 18 months, the conclusion was that these issues were transient. Considering the functionality and utility of the application-development environments available for these systems combined with the scope of the applications available for them, their choice was obvious. The complexity of these systems, however, has increased dramatically, making their overall operation increasingly vulnerable to "systems problems" beyond the capabilities of most users.

A key factor in the acceptance of general purpose operating systems such as UNIX was the exclusion of time-critical event processing such as process control and minimized dependency of safety related and mission-critical functions from their domain. The distributed nature of the DCS facilitated the utilization of the most appropriate technologies to achieve the primary goals of that subsystem. Therefore mission-critical components such as process control subsystems and I/O subsystems were designed with highly reliable, high-performance hardware and software while information storage, retrieval, presentation, and analysis and various forms of transactional processing including application-development environments were supported by information-technology products. Also, the distributed design enabled the use of highly reliable and optionally redundant subsystems in which appropriate and limited scope modules can survive the failure of a subsystem, albeit at a somewhat diminished capability. If an operator workstation were to fail, for instance, other workstations in the system could access the information of the failed subsystem. Fault tolerance was inherent in the distributed architecture.

By the early 1990s, most suppliers were repositioning their DCS offerings from panelboard replacement to being the real-time component of a manufacturing enterprise system utilizing distributed computing architectures. The issue was how to manage their early DCS customers' investments in what is euphemistically referred to in the industry as "legacy systems." That is, systems purchased from the mid-1970s until the early 1990s were basically proprietary designs focused on emulating the functionality of a panelboard, with most of the innovation focused on improved process operations and control features. Integrating with information technology (IT) systems was typically done via specialized gateways and custom programming. The introduction of high performance, high-capacity distributed computing environments enabled the use of a wide variety of third-party packages to be introduced as an integral part of the DCS. The emphasis in process control shifted from advanced regulatory control to model reference control and expert systems. Operator workstations began utilizing client/server technology by taking advantage of features such as X-Windows available in UNIX. Operator displays migrated from panelboard centric information to process performance information, taking advantage of information traditionally outside of the domain of process operators. More importantly, the market requirements for distributed control systems and the emphasis for innovation began shifting from control to information management. Achieving this new expectation required a massive increase in both the performance and capacity of the DCS and new paradigms in software technology. Any lingering doubt regarding the use of IT products in the DCS arena was now moot.

The performance, capacity, and software tools and utilities now availble within the DCS environment enabled much more sophisticated application software to be developed or adapted from other disciplines. As a result, a significant number of application products available in gateway systems were now mainstreamed; those available in proprietary DCSs were greatly enriched, and many available in other market segments but not utilized in process automation were now incorporated. Examples of these include

- product recipe management, dispatching and control
- statistical process control
- intelligent alarming
- model reference control
- expert systems
- production scheduling
- predictive maintenance
- dissemination and maintenance of operational and compliance documents
- data reconciliation: material and energy balances
- product conversion and yield accounting

Several corollary automation systems were experiencing a similar transformation because of the exponential performance improvements underway in computer and IT fields. These included:

- PLCs
- manufacturing execution systems (MESs)
- personal computer-based data-acquisition and control systems (SCADAs)
- plant information management systems (PIMSs)
- minicomputer-based information and control systems (DDCs)
- specialized control systems for various requirements, such as
  - safety override systems
  - batch weighing and blending systems
  - packaging line control
  - material transport systems
  - specialized equipment
- specialized information systems for various requirements, such as
  - plant maintenance
  - document management

- production scheduling
- inventory management
- cost accounting.

By the mid-1990s, it became apparent that Pandora's box had been opened. The market appetite for increasingly specialized application solutions for a widening diversity of market niches began growing exponentially. Considering the complexity and sophistication of addressing the requirements of each application solution and the reduced opportunity for amortization (volume sale) of specialized solutions coupled with the market expectation for "Microsoft pricing," the DCS vendor community would have to radically change their approach to product innovation and development. Constructing stand-alone solutions was no longer a profitable venture. Component based software architectures which served as the framework for a wide variety of niche applications would have to be constructed. This was both technically and culturally difficult for most of the DCS vendors, particularly the larger and more successful ones.

The primary technical issue centered around supporting the large installed base of customer systems in place and the commitment to maintain the viability of these investments for at least 10 years. This required the transition to component-based architectures to follow one of several paths. These included the following.

1. New component-based subsystems would be introduced and blended with the existing architecture, allowing for a gradual transition of existing systems.

2. A component-based DCS offering would be introduced in addition to the traditional offering, and funding for the traditional offering would be gradually diverted to the newer system based on market acceptance.

3. A hybrid offering would be created, supporting both traditional and component-based subsystems with investment continuing to support the traditional design in all aspects of the offering. This strategy was only available to those companies who were well on the way to converting their proprietary systems to a more standards-based approach in the late 1980s and early 1990s.

Culturally the issue centered around their large investment in proprietary or more recently standards-based designs. The large, successful DCS vendors invested considerable time, effort, and money to create and refine highly complex mission-critical systems and considerably more money promoting and gaining market preference. Their expectation was to recover these high initial costs over the lifetime of a product, which traditionally was 10 years of initial sale and an additional 10 years of existing sale expansions, modifications, and replacements. The high technology paradigm of the computer industry threatened to shorten this to 3 years. Also the component architecture threatened the sourcing of the addition (system enhancement) business and possibly the replacement and modification portions as well. Component-based architectures conforming strictly to industry standards also enabled users or third-party integrators to assemble complete systems from the foundational components.

Finally, although these large successful DCS corporations had a fundamental advantage of the requirements of DCSs, their expertise on the computer science of component technology would have to be acquired externally. Also many niche players with none of the DCS inertia described above were leveraging the extensive and highly integrated software architecture offered by Microsoft and PLC components to catapult themselves into the DCS market.

The choice of Microsoft as the software infrastructure provider is quite compelling. This company has a dominant (some say monopolistic) market position in the personal computer market, which has exceeded the power and capacity of most midrange computers of only a few years ago at a small fraction of the cost and is sold by the tens of millions each year. Second it has created a familiar desktop paradigm, Windows, for viewing and navigating through complex information, which is highly refined and accepted. Third and most importantly, it has created and integrated a complete operating, infrastructure support and application-development environment, including a large number of key applications, around a central framework, Microsoft's Distributed Component Object Model

(DCOM) software architecture. This provides a rich framework, which includes

- Windows NT multitasking, multithreaded operating systems, and infrastructure
- interprocess messaging and communications technologies
- software object libraries (ATL, MFC)
- application-development environments and languages (Visual Studio)
- database management systems (Access, FoxPro, MS SQL)
- a wide variety of applications
- a highly structured and supported application programming interface (ActiveX), which enables independent applications (and application developers) to communicate information and share application components

The combination of rapid application tools (RAD) such as Visual Basic coupled with ActiveX tools, such as the ability to "drag and drop" application objects previously developed or purchased from another party, makes the creation of complex and highly specialized applications for niche markets affordable. Independent application developers can create specialized applications or niche component products, using Microsoft's software infrastructure, at a small fraction of what it would otherwise cost.

Significant concerns persist, however, regarding the reliability, availability, and robustness of this approach, considering that Microsoft's products were not designed for the automation market and that manufacturing automation represents only a small fraction of the market being targeted by Microsoft. The size and complexity of these products coupled with the maturity of the product (as compared with, say, UNIX) and the rapid pace at which this product is evolving require careful scrutiny. The same challenge could, however, be made of any of the other basically commercial or IT products being used in this mission-critical arena, including UNIX.

Arguments have been made that the reliability, robustness, and high availability requirements for on-line transactional processing (OLTP), a major market for Microsoft and other IT manufacturers, have similar requirements to process automation and may be more demanding for some important criteria such as transactional response time.

Most DCSs at this point exclude from the IT domain those applications that directly affect equipment operation or could endanger the safety of process operators. Among these applications are process control or other critical event processing, such as safety shut-down systems and other mission-critical subsystems. These remain in the domain of more deterministic technologies.

MARKET TRENDS

The current initiatives in DCS development are focused around the construction of

- standards-based independent subsystems such as intelligent fieldbus devices
- incorporation of the Internet and Web technology into the DCS domain
- object-oriented software components and containers
- integrated process automation and manufacturing enterprise systems (see list above)

Fieldbus and Intelligent Devices

The move toward open standards of DCS subsystems is most obvious in the control and I/O arena, where intelligent devices and fieldbus technology are challenging multifunction controller subsystems and traditional I/O subsystems. This approach embeds the intelligence needed for device profiling and simple regulatory and (in some cases) discrete control into the primary sensing and actuating devices, thereby eliminating the need for the DCS to perform routine process and discrete control. The primary benefits for an open fieldbus standard for intelligent devices include the following.

- the sourcing of multiple measurements that are communicated in a time-multiplexed manner over a digital transmission line, thereby providing more information at a reduced cost
- the ability to source more information regarding the device at the source, thereby eliminating middleware that might pose ongoing maintenance problems
- the elimination of extraneous hardware such as I/O modules and controllers
- reduced wiring costs due to the multidrop nature of serial fieldbus and the multiplexed nature of digital transmission
- the ability to source intelligent devices from many vendors, thereby ensuring a wider selection of choice and competitive pricing

This has profound inplications for the DCS, as much of the control portion of its original mission, panelboard replacement, may no longer be required. Also, communication with these intelligent devices would be transactional rather than sampled data, thereby eliminating much of the distinction between DCSs and OLTPs. Quite a few open standards-based fieldbus technologies are available today or are in the final stages of development. Three stand out as having the strongest following as replacements for the control and I/O subsystems for DCSs. These are FOUNDATION Fieldbus, Profibus, and OPC. The following is a brief summary of these technologies.

The FOUNDATION Fieldbus organization was formed in 1994 to produce specifications for an all digital, serial two-way communications system that interconnects field-based intelligent devices such as sensors, actuators, and controllers to meet the ISA's SP50 requirements. It produced an early specification in 1995, and as of September of 1998 there were 12 devices that passed all required interoperabillity tests and 25 more in the queue for certification testing. FOUNDATION Fieldbus is designed specifically to meet the stringent, mission-critical demands for intrinsic safety and use in hazardous areas, volatile processes, and difficult regulatory environments. It is a detailed standard with strict compliance requirements to ensure interoperabillity among all devices and systems that conform. These includes the following.

1. Physical layer sends and receives messages to/from the communications stack and converts the messages to physical signals on the fieldbus transmission media.
2. Communications stack consists of three sublayers. The first is the data link layer, which controls the transmission of messages onto the fieldbus through a deterministic centralized bus scheduler called the link active scheduler. The fieldbus access sublayer manages the connections among devices in the various communications modes supported, including client/server, broadcast, and publish/subscribe. The fieldbus message specification supports the various types of messaging formats, which applications use to communicate across the fieldbus.
3. User application (blocks) represents the different types of functions supported within a given object. There are three types of blocks. These are resource blocks, which contain static information about the object (device), transducer blocks, which contain modifiable characteristics about the object, such as calibration data, and function blocks, which provide the functions that represent the object's behavior such as a PID function.
4. System management synchronizes the execution of function blocks and the communication of function-block parameters on the fieldbus.
5. Device descriptions contain extended descriptions of an object (device) to enable a manufacturer to extend the capability of a standard object by possibly adding additional features such as self-tuning or parameters such as special alarm or maintenance parameters.

Profibus, like FOUNDATION Fieldbus, is a vendor-independent, open fieldbus standard. Unlike FOUNDATION Fieldbus it is designed for a wide range of applications, including discrete manufacturing, process control, and building automation. Profibus has been available for quite a few years and is the fieldbus standard most popular in Europe. As of September of 1998 there were over two million devices installed that complied with the Profibus standard.

Profibus is a serial fieldbus system that supports the networking of intelligent field devices and supports both master and slave devices. Master devices determine the data communication on the bus. A master can send messages without an external request when it holds the bus access rights (the token). Slave devices do not have bus access rights and they can only acknowledge received messages or send messages to the master upon request. Typical slave devices include input/output devices, valves, drives, and measuring transmitters. There are actually three versions of Profibus.

1. Profibus-DP is optimized for high speed and inexpensive hookup. This Profibus version is designed especially for communication between automation control systems and distributed I/O at the device level. Profibus-DP can be used to replace parallel signal transmission with 24 V or 0–20 mA.

2. Profibus-PA is designed especially for process automation. It permits sensors and actuators to be connected on one common bus line even in intrinsically safe areas. Profibus-PA permits data communication and power over the bus by using two-wire technology according to the international standard IEC 1158-2.

3. Profibus-FMS is the general purpose solution for communication tasks at the cell level. Powerful FMS services open up a wide range of applications and provide great flexibility. Profibus-FMS can also be used for extensive and complex communication tasks.

OLE for Process Control is an open standard, supported by the OPC Foundation, a nonprofit organization, based upon the functional requirements of Microsoft's OLE/COM technology and intended to foster greater interoperabillity between automation/control applications, field systems/devices, and business/office applications in the process control industry. OPC defines standard objects, methods, and properties for servers of real-time information such as distributed process systems, programmable logic controllers, smart field devices, and analyzers in order to communicate the information that such servers contain to standard OLE/COM compliant technologies enabled devices. OPC's primary strength is that it is by design a plug and play software specification for Microsoft's DCOM.

Acceptance of control function enabled fieldbus technology into the process automation markets will be a very gradual process because of the inertia of the installed base of existing DCSs, transmitters, and actuators. The economics to retrofit fieldbus into an existing and viable facility is not compelling. Also, a limited form of intelligent field devices (excluding control) is already in the market.

Penetration of control function enabled fieldbus technology into the discrete markets will be much sooner because the retooling cycle in this sector is much quicker. This will probably initially take the form of a micro- or nano-PLC attached to a discrete device such as a motor starter, and it will eventually become an embedded controller associated with the device.

Internet Technology

The Internet (Intranet, Extranet) is another significant external technological influence affecting the direction of the automation market, including DCS. The Internet makes it practical to include distributed control systems as the real-time manufacturing component in a supply chain model of a corporate wide enterprise management program. In this role the DCS would furnish timely product availability and manufacturing capacity information to corporate-level planning and decision management systems so as to balance the flow of material and other resources throughout the corporation in the most efficient and profitable manner.

Internet technologies also make it economical to extend and further blend classical process automation functions with IT by adapting the internet appliance paradigm. Operator workstations, for instance, could adapt the Internet-portable appliance to create a personalized portable headset with the screen projected on a visor and wristband entry terminal. This device would be profiled to provide the information and services required by that specific operator or that class of operator job function. Information provided could be anything sourced by the Internet, including background music. This form of general purpose tool also be utilized by plant maintenance, engineering, and supervisory personnel and provide access to equipment drawings, material and safety data sheets, videoclips of setup procedures, video or voice communication, and so on. Communication among plant workers or in fact anyone logged on to the Internet is limited only by the permissions established in the profile.

We have previously discussed the use of Microsoft's DCOM software architecture as a means of constructing object-oriented software architectures. Another technology under serious investigation is JAVA, developed by SUN Microsystems. This technology is designed to be Internet centric and run as a virtual machine (VM). As such is can be utilized by any operating system supporting it. Application software suppliers can offer application components that run on a wide variety of operating systems in a highly integrated fashion. This may eventually offer significant competition to Microsoft's DCOM approach.

Object-Oriented Software Components and Containers

The confluence of market expectation, technological innovation, and the economies of open standards are driving the DCS toward adopting object-oriented software components and containers. Future process automation systems will increasingly adopt real-time software component architectures. Manufacturing enterprise solutions will merge with DCS functionality in a plug and play fashion, as will the application components of the PLC, safety shut-down system, manufacturing execution system, and PIMS, and well as other mission-critical real-time components.

Technology is now capable of absorbing the significant performance and capacity demands created by this heightened expectation at an affordable price. Object-oriented design frameworks provide the only software infrastructure available to achieve this expectation. Object-oriented design makes sophisticated application development an extensible process by building upon an extensive object class structure, thereby making the process affordable. The needed class libraries and extensive development tools to manage them are currently in place, as are a wide variety of components and applications that comply with these frameworks.

Presently there are three industrial strength frameworks with sufficient market strength and technical prowess to warrant consideration. These are Microsoft's Visual C++/COM/ATL/MFC, SUN Microsystem's JAVA/SWING, and the C++/STL ANSI Standard. Other object-oriented frameworks such as Small Talk are available and are quite robust but don't have the level of market acceptance or technological investment as those listed previously.

Of these, Microsoft's framework is the most mature and well developed, the ANSI Standard is the most open and available for any platform, and SUN's is the most compliant for network centric (the Internet) development, particularly Internet appliances.

JAVA is different from the other two in that it is not compiled in the native code of the hosting machine. Rather it executes in interpretive (bytecode) mode against a VM, which is rigorously defined and constructed within the operating system of the hosting machine. Therefore all services, including event handling, are coded identically, regardless of the differences in hosting machines. As long as the program is written in a form that does not depend upon any hosting system extensions, the identical program will execute correctly on any computing platform and (as an applet) in any internet browser that supports JAVA.

The issue is not whether to adopt an object-oriented software component and container framework—it is how to introduce it and in what time frame. This decision is especially crucial for process automation because object-oriented frameworks, particularly JAVA, are undergoing continuous and rapid evolution. This contrasts sharply with the capital investment philosophy for plant equipment, including process automation systems that typically are assumed to have an average depreciation term of 10 years. Rigorous change management formalisms will have to be introduced to maintain these installed systems to facilitate the frequent addition of hardware and software components. This continuously current maintenance philosophy will significantly extend the life of the DCS investment but will radically alter the manner in which these systems are supported and maintained. Validation methodologies will become essential to introduce these new components into mission-critical systems on the fly.

Expanded Architecture

Many of the traditional functions associated with the DCS such as I/O are becoming the purview of external systems, while many functions that were typically thought of as external to the traditional DCS, such as delivery of operator work instructions, are now being incorporated into the DCS. To accommodate this, the five subsystem (process I/O, control, workstation, applications, and communications) concept of the DCS must also now include the three software component categories for distributed computing environments. They are information, application, and presentation.

The purpose of the software component model is to create broad categories for the construction of component classes and containers. It is intended more for designers and developers than for users, but is important for users to understand. Information is concerned with the activities needed to capture, validate, and access all information required by either application or presentation activities regardless of its source, purpose, or user. The structure is complete and independent of the other two activities. Application is concerned with the activities that transform, translate, or otherwise represent the context of the information. Presentation is concerned with the activities needed to view information in its application context.

DCS SELECTION CRITERIA

The role of the DCS has expanded substantially from its original mission of panelboard replacement. The technological foundation has evolved even more dramatically, and the pace of change in both areas promises to quicken. Constructing a selection criteria for a DCS, however, relates more to its intended use than the scope of its capabilities. The requirements definition should consider all of the intended users and consumers of information produced by the system. The potential benefits for each department should be carefully explored in a series of meetings, surveys, and possible educational seminars. This is particularly important for potential information consumers.

Requirements Definition and Economic Benefits Analysis

There is an old saying, "If you don't know where you're going, any road will get you there." Thus the first step in any analysis is to completely define the economic incentives that justify the project. These should be quantified and described in as much detail as possible. Second is to identify any ancillary benefits expected, whether economic or otherwise. Finally, a survey should be taken of potential users of the system or consumers of the information produced by the system to determine their expectations and requirements. The total set of requirements should be carefully listed and described with an economic value placed on each item. Those requirements that don't necessarily lend themselves to an economic return, such as safety improvements, should be evaluated by generating the cost of the best alternative means to achieving the required result.

Complete and Detailed Functional Specification and System Requirements Definitions

There should be a clear association of a particular economic benefit to the set of functions required to achieve it. In most cases the relationship between economic benefits and functional requirements will be a many-to-many mapping. Therefore each functional capability should also identify the set of economic benefits that are dependent on it.

The functional specification and system requirements definition should be separate documents or, at least, separate, complete and independent sections of the same document. The following highly simplified example is intended to demonstrate the process of generating these documents.

1. **Economic Incentive and Defined Requirement:** generate an an annualized savings of $90,000 by preventing material losses of a particular ingredient by 10% caused by tank spillage.

2. **Functional Specification:** include high-level indicator on all material tanks with a high-level alarm set for 90% of capacity and automatic valve shut-off valves at 95% of capacity. Upon occurrence of the high-alarm condition, an audible alarm would sound and notification would appear on the screen currently being viewed by the process operator responsible for the tanks. Upon occurrence of the high-high alarm condition the inlet valve corresponding to the tank in alarm would automatically be driven to the off state by means of an override mechanism. The high-high alarm event would also be reported to the shift supervisor on his or her workstation and in the exceptional condition event log.

3. **System Requirement:** the system must be capable of generating the following alarm notifications on any analog signal connected to the system. The alarm condition may be used to notify process operators, supervisors, or both by posting a message at their workstations, triggering a control event, or generating an entry into an abnormal event log that includes the time, tag, alarm condition, and an attached message. The following alarm conditions are required: high absolute, high-high absolute, indication point is bad or taken out of service, and alarm point has been disabled (other conditions not required by this functional specification).

The systems requirement specification should also include the system configuration and equipment. Thus consideration must be given to the physical plant topology, process environment, and process equipment attached, including device specifications and I/O lists, operator workstation locations, and so on. The following general categories represent topics for consideration when the systems requirements specification is generated.

- Process I/O
  - signal accuracy, precision, validity, and sample interval
    - signal resolution in bits (10 bits = 0.1%)
    - sample conversion method (e.g., integrated voltage to frequency over time)
    - availability of status bits for measurement validation
    - signal conditioning when required (e.g., cold junction compensation for T/C's)
    - sampled data interval
  - signal termination and connection location
    - remote vs. local termination and remote vs. local I/O electronics
    - rack density and size
    - number of input and/or output points per card
    - ability to mix inputs and outputs of different types—granularity
    - output signal bypass capability to remove a card without disrupting the process
    - signal termination types
  - signal powering, grounding, electrical noise immunity, and isolation
    - location of signal ground and separation of signal from system ground
    - optional availability of field powering of transmitters
    - type of signal isolation: galvanic, optical, resistive
    - bundling of signal isolation: by point, by card, by group

- over voltage, over current, or reverse-polarity protection
- surge protection circuitry
- safety considerations for fire or explosion resulting from faulty operation
  - Compliance with applicable codes: FM, UL, CSA, PTB, BASEFA
  - intrinsic safety: energy levels too low to ignite a fire or explosion
  - explosion proof: enclosures that contain flame or heat to prevent ignition
- environmental considerations for ambient temperature, dust, water, chemicals, etc.
  - temperature ratings including thermal shock
  - moisture resistant enclosures: NEMA 12 (drip proof), NEMA 4 (water tight)
  - heat dissipation and heat removal considerations
  - corrosion and other chemical contaminant considerations
  - vibration or impulse shock
  - electrical: harmonic distortion (THD), electromagnetic (EMI), surges, or distortions
- control
  - control strategy construction schemes available, such as
    - preprogrammed function blocks
    - user-defined function blocks
    - sequential function charts
    - ladder logic
    - structured text
    - macro instruction lists
    - logic or timing diagrams
    - machine control languages
  - control algorithms available
    - device description: motor starter circuit, solenoid valve, variable speed drive, etc.
    - counters and totalizers
    - Regulation: PID, ratio, A/M station, self-tuning, lead/lag, deadtime, etc.
    - Logic: And, Or, Xor, Flip Flop, One Shot, Group Actions, etc.
    - Timer functions
    - calculation
    - sequencer
    - state controller
  - control modes
    - Auto-Manual-Initializing-Out of Service
    - Remote-Local-Tracking
  - execution paths, data flow, initialization schemes
  - alarm handling and exceptional condition processing
- presentation
  - navigation: area, tag, alarm or event
    - menu bar
    - tool bar
    - tree hierarchy

- graphical display pick point
- list box selection
- keyboard entry
- windows
  - multiple concurrent
  - drag and drop across applications
  - coordinated view sets, e.g., unit display, associated PV trends, engineering calcs
- displays
  - engineering detail of loop or device
  - graphical representation process unit or transfer line manifold
  - production chart
  - spreadsheets: engineering analysis
  - database: information extraction
  - email: supervisory or operational notifications
- trend
  - type: real time, historical, sequence of events
  - grouping: single variable or multivariable per trend
  - format: line, bar, scatter diagram
  - gridlines: variables vs. time, variables vs. an independent variable
- alarm notification
  - method: audible, annunciator, highlighted field on graphic, dialog box
  - type: absolute value, deviation from setpoint, rate of change, change of state, etc.
- interaction
  - method: fixed screen and field, action relative to pick, entry set (tag, param, value)
  - access protection: password, badge reader, etc.
  - validity checking
- reporting
  - printer speeds and capabilities
  - printer access from remote stations
  - printing formats
  - spooling and caching
- application
  - advanced and supervisory control
    - batch
    - statistical process control
    - model reference control
    - fuzzy logic
    - expert systems (forward and backward chaining)
    - neural networks
  - information management
    - continuous process historian
    - event historian

- tracking of product batch lot genealogy
- engineering configuration databases
- record keeping databases for maintenance, personnel, documents
- database access
  - access methods: SQL, ODBC, OLE DB
- database applications
  - process data and events
  - engineering configuration
  - product quality information
  - production management
  - business integration
- communications
  - request-reply, e.g., get/set/let
  - connected point, e.g., read a measurement every second
  - client-server and publish-subscribe, e.g., workstation display updates
  - file transfer, e.g., reload image
  - remote procedure call, e.g., execute a program using multiple processors
- systems issues
  - availability: reliability, redundancy, exposure to failure, maintainability
  - fault tolerance: fail-safe procedures, change control procedures, access procedures
  - performance, capacity, responsiveness
  - packaging issues: workstations, controllers, process I/O, application servers
  - power: regulation, distribution, protection
  - configuration limitations: workstations, controllers, other stations, independent nodes
  - documentation: installation, configuration, application, use, maintenance

System Evaluation: Budget, Schedule, and Risk Assessment

A formal procedure is usually required to budget, schedule, and evaluate a capital investment of this complexity. The first step in this process is to generate a predicted return on investment. This process can only be done correctly if both the economic return and the total cost are known with a high degree of certainty.

The potential economic return is quantified by assigning economic value to the requirements definition. The functional specification then identifies the set of functions needed for each of the requirements. These functions should be categorized in a format such as the following: (a) initial project justification—functions that must be provided to satisfy those requirements necessary for the project to go forward; (b) additional benefits—additional functions needed to achieve the additional set of quantifiable and desirable requirements; (c) other "users" of the system—additional capabilities needed to achieve unspecified requirements, which should be considered when evaluating the overall scope and capabilities of potential systems, especially if inclusion would possibly result in significant economic return; (d) other "consumers" of information—additional capabilities needed to achieve unspecified requirements, which should be considered when evaluating the scope of potential systems, especially if inclusion would possibly result in significant economic return; and (e) future needs assessment—functional and systems considerations for extensions of the system in size, functionality, or scope.

Systems should then be evaluated by assessing the confidence level the project team has to achieve each of the functions. If, after due diligence, it is not clear (or not known) that a particular system can accomplish a particular function, the system in question should be assigned a low confidence indicator for that function. Any system should be eliminated from further consideration if it does not indicate a high confidence for success for all functions in the initial justification category. The remaining categories should be used to reduce the selection process by using a weighting factor favoring those higher in the category list.

Cost is a major consideration in any purchase decision. However, cost should include the purchase, installation, project engineering, training, maintenance, operation, and administration for the time period considered for at least twice the return on investment period. After all, the objective is not simply to return your investment.

Technological innovation will accelerate the rate of depreciation of your investment in several ways. First, faster, cheaper, better (capacity, performance, functionality, ease of use) are almost guaranteed to appear on the market within 3 years of your purchase decision. This is true regardless of when you make the decision. This innovation has the potential to erode your ability to support, maintain, and enhance your system over time unless you plan for a continuous improvement investment program. This reality leads to several choices. These are as follows. First, plan your investment for a short payout with the strategy of replacing the system when it adversely affects the competitiveness of your business. Two considerations are required for this decision to be viable. First is the time delay associated with commissioning the system. Second is the total cost of replacement, including purchase, installation, project engineering, training, maintenance, operation, and administration. Next, plan your investment with the intention to upgrade to more technologically current versions when the economics justify this incremental investment. This strategy, of course, depends on your confidence in the commitment and ability of your suppliers to provide economically viable and compatible improvements that are competitive with future technologies. Finally, structure your primary requirements definition to be as insulated from technological innovation as possible by restricting the DCS requirements definition to process operations and control and support for standard application program interfaces. This, of course, doesn't solve the total problem—external systems are also subject to obsolescence—and probably makes the overall solution more expensive. It does, however, minimize the impact of technological innovation on the viability of the DCS.

Another major success factor is understanding that the resources available are more than adequate for both the initial project and the ongoing support of the system. A careful critical skills assessment should be conducted to ensure that the scope and complexity of the project does not exceed the skills of the committed resources. Available resources include those existing in house, planned as new hires, and those contracted or provided by the DCS supplier. Skill categories include engineering, operations, maintenance, ongoing training, administration and support. A common mistake made in many projects is to minimize the long-term continuity needed for ongoing engineering, training, and system administration (change control) by relying too heavily on contracted resources. Ongoing support in these vital disciplines is essential to the long-term viability of the system.

CONCLUSIONS

The demands for immediate, available, and reliable information in mainstream distributed (Internet) computing has created the opportunity for the distributed control systems to become the information provider to the manufacturing enterprise component of corporations. To accomplish this the DCS community has adopted mainstream distributed computing technology. Open computing standards and component-based architectures facilitate the seamless integration of applications and information among the various business, production management, operations, and process control systems. Distributed control systems, originally conceived as panelboard replacements, have dramatically improved their process information and management capabilities in recent years.

Essentially all DCS's on the market today offer a hybrid approach that uses open standards-based solutions such as UNIX or Microsoft Windows NT for supervisory and production management applications (information storage, retrieval and presentation, business application integration, and comprehensive development environments) while continuing to employ proprietary hardware and software for their embedded computing requirements such as process control.

New innovations for Internet appliances require a similar responsiveness, availability, and fault tolerance as distributed control systems. This technology promises to eventually displace the remaining proprietary solutions for applications such as process control. These, however, will have to address the mission-critical and safety concerns that differentiate real-time control from general purpose computing, particularly in those situations in which personal injury or property damage is at stake.

GLOSSARY

DCS	distributed control systems
PLC	programmable logic controller
CRT	cathode ray tube (references video-based operating view screens)
MOCS	multiple output control system having more than one actuator for a single controlled variable
IEEE 802.X	network communications protocols addressing the data link and network layer
UNIX	computer operating system usually associated with the Internet
Windows NT	new technology; Microsoft Windows most robust operating system
RDBMS	relational database management system for storage and retrieval of information
MES	manufacturing execution system for tracking and control of manual operations
PIMS	plant information management system for tracking and control of captured data
SCADA	scan, control, alarm, and data acquisition (usually a small PC-based system)
C++	object-oriented programming language
ATL	active template library; Microsoft's name for predefined C++ object classes
STL	standard template library; ANSI definition of predefined sofware object classes
MFC	Microsoft foundation class; C++ object classes for Windows visualization
JAVA	object-oriented programming (language designed for a VM)
Virtual machine	specification of generalized API for access to computer resources
API	application program interface to enable programs to access computer resources
SQL	structured query language for transacting with RDBMS's
RAD	rapid application development; software development tools and environments
OLTP	on-line tranactional processing; application execution concurrent with use
COM	Microsoft's component object model (foundation of their component architecture)
DCOM	Microsoft's distributed object model (network computing version of COM)

BIBLIOGRAPHY

Considine, D. M., *Process Industrial Instruments and Control Systems Handbook*, 4th ed., McGraw-Hill, New York, 1993.

Microsoft Home Page: http:/www.microsoft.com.

Lippman, and Lajoie, *C++ Primer*, 3rd ed., Addison-Wesley, Reading, Massachusetts, 1998.

Cappione, and Warlathz, *The JAVA Tutorial*, 2nd ed., Addison-Wesley, Reading, Massachusetts, 1998.

FOUNDATION Fieldbus home page: http://www.fieldbus.org/.

FOUNDATION Fieldbus Technical Overview, FD-043 Rev. 1.0, 1996.

Profibus Technical Overview: http://www.profibus.com/data/technic/index.html.

OPC Foundation Home Page: http://www.opcfoundation.org/.

JAVA Technology: http://java.sun.com.

PROGRAMMABLE CONTROLLERS

by Tom Hudak*

PROGRAMMABLE CONTROLLERS DEFINED

The first programmable controller, introduced in 1970, was developed in response to a demand from General Motors for a solid-state system that had the flexibility of a computer, yet could be programmed and maintained by plant engineers and technicians. These early programmable controllers took up less space than the relays, counters, timers, and other control components they replaced, and they offered much greater flexibility in terms of their re-programming capability (Fig. 1). The initial programming language, based on the ladder diagrams and electrical symbols commonly used by electricians, was key to industry acceptance of the programmable controller (Fig. 2).

Because programmable controllers can be programmed in relay ladder logic, it is relatively simple to convert electrical diagrams to the programmable controller program. This process involves defining the rules of operation for each control point, converting these rules to ladder logic, and identifying and labeling outputs (addressing). Today's work force has a mix of engineers—some of whom have been around for a while and are familiar with ladder logic as well as newer engineers more comfortable with computer-centric based programming and control. This has led to a mix of programming technologies that are applied based on user background and application need.

Programmable controllers can be place into two categories—fixed and modular. Fixed programmable controllers come as self-contained units with a processor, power supply, and a predetermined number of discrete and analog inputs and outputs. A fixed programmable controller may have separate, interconnectable components for expansion, and it is smaller, cheaper, and easier to install. However, modular controllers are more flexible, offering options for input/output (I/O) capacity, processor memory size, input voltage, and communication type and quantity. Originally, programmable controllers were used in control applications where I/O was primarily digital. They were ideal for applications that were more sequential and more discrete than continuous in nature. Over time, analog and process capabilities were added such that the programmable controller became a feasible solution for batch and process control applications. The evolution of programmable control has increased the options available for control systems that have traditionally relied on alternative technologies. Until the introduction of the micro-programmable logic controller (micro-PLC) in the mid-1980s, relays and single-board computers (SBCs) offered the most common means to increase automation on simple machines and less complex processes. Even though the functionality of the traditional programmable controller often benefited an application, the cost could not always be justified. If cost was not an issue, size often was. Sometimes even small programmable controllers were simply too large to fit in the space allocated for electrical controls.

It wasn't until micro-PLCs were introduced that programmable controllers could economically meet the demands of smaller machines in a more efficient manner than relays and SBCs. These fixed I/O controllers are generally designed to handle 10–32 I/Os in a package that costs less than $300 U.S., making them feasible replacements for even very small panels of relays. In addition, this low-cost controller option has opened the door for many small machine OEMs to apply automated control in places where it wasn't feasible in the past—for instance, one manufacturer uses a micro-PLC as the controller in lottery ticket counting machines. Modular programmable controllers are similar in function to fixed versions, but they physically separate the I/O from the controller. This allows the I/O racks to be distributed closer to the application, where they communicate back to the controller over an industrial network. Modular controllers allow a user to closely match the controller components to the specific application needs.

* Rockwell Automation, Allen-Bradley, Mayfield Heights, Ohio.

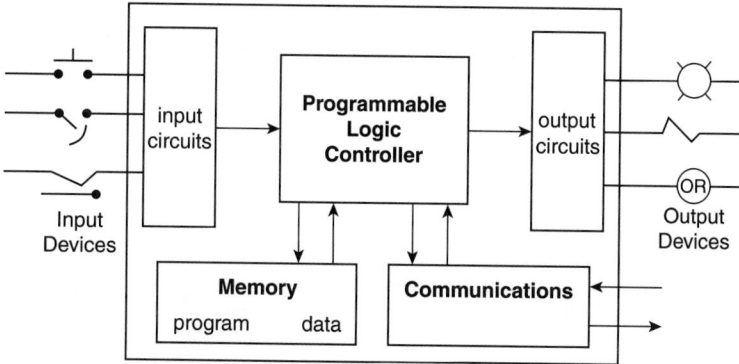

FIGURE 1 Basic PLC components.

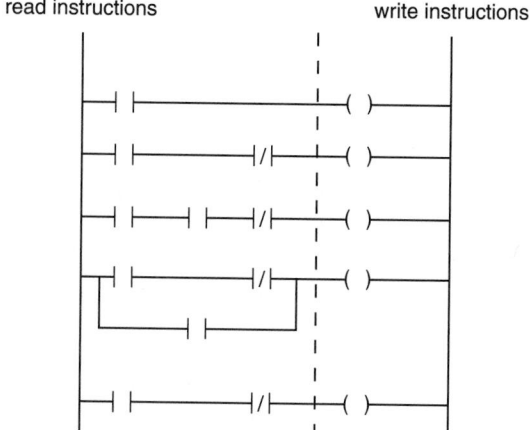

FIGURE 2 Relay ladder logic.

ALTERNATIVES TO PROGRAMMABLE CONTROLLERS

The success of the programmable controller has spurred innovation in a number of competing technologies. Below is a brief review of these technologies, which may in some cases be successfully applied as an alternative to programmable controllers.

PC-Based Control

Personal computer (PC)-based control is a broad term encompassing not only the controller but all aspects of the control system, including programming, operator interface, operating system, communication application programming interfaces, networking, and I/O. Soft control is the act of replacing traditional controllers with software that allows one to perform programmable controller functions on a personal computer. Soft control is an important development for individuals who have control applications with a high degree of information processing content. For hazardous processes, PC-based control is typically limited to supervisory control in which the PC sends set points for optimization. Regulatory control, sequencing, and interlocks are done in a PLC or distributed control system (DCS).

The operator determines whether the regulatory control system in the PLC or DCS accepts these set points. Typically, the supervisory set points are updated by pulses, so that a loss of the PC signals leaves the set points at last value.

Distributed Control System

A distributed control system is a technology that has evolved to meet the specific needs of process applications such as pulp and paper, utility, refining, and chemical processing. DCSs are generally used in applications in which the proportion of analog to digital is higher than a 60:40 ratio, and/or the control functions performed are more sophisticated. A DCS typically consists of unit controllers that can handle multiple loops, multiplexer units to handle a large amount of I/O, operator and engineering interface workstations, a historian, communication gateways, and an advanced control function is dedicated proprietary controllers. All these are fully integrated and usually connected by means of a communication network. A DCS typically takes a hierarchical approach to control, with the majority of the intelligence housed in microprocessor-based controllers that can each handle 10–1000 inputs and outputs. The advent of Fieldbus has resulted in the development of the field-based DCS (and PLC?) that facilitates both the movement of the control functions to the field and the use of additional measurements and diagnostics from smart instrumentation and control valves.

Relay-Based Control

Relays essentially serve as switching, timing, and multiplying mechanisms for input devices, such as push buttons, selector switches, and photoelectric sensors (Fig. 3). Relays are fairly intuitive; however, as mechanical devices, they do not offer the same programming and troubleshooting flexibility found in modern programmable control systems. Relays are also known for taking up considerable amounts of space, requiring extensive wiring, and needing regular maintenance.

Single-Board Controllers

The first electronic controls on circuit boards—or SBCs—appeared in the early 1960s. These "logic modules" were created by placing discrete components, such as transistors, capacitors, and resistors onto boards. These solid-state systems were inherently more reliable than relays because there were no

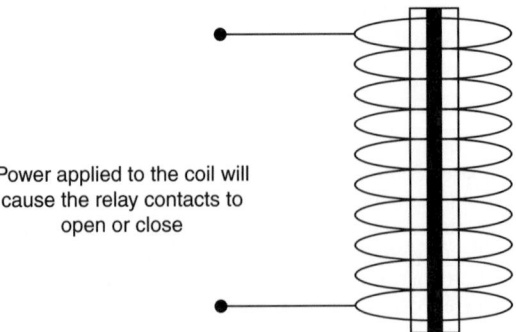

Electromagnet relays and a variety of special purpose components (timers, mechanical counters, cam switches, etc.) can be wired in unique configurations to achieve specific control actions.

Power applied to the coil will cause the relay contacts to open or close

FIGURE 3   Relay logic control.

Application Characteristic	Relay	Micro-PLC	SBC	Full-size Programmable Controllers
Inputs/outputs	1-12 max.	Up to 32	Varies	32-1,000s
Timers	Yes	Yes	Yes	Yes
Up/down counters	Yes	Yes	Yes	Yes
High-speed capabilities	No	Yes	Yes	Yes
Data calculations	No	Yes	Yes	Yes
Data acquisition	No	Yes	Yes	Yes
Communications	No	Limited	Yes	Yes
Operator interfaces	Primitive	Variable	Variable	Variable
Memory Size	N/A	1-10K	1-100K	up to 2 Meg

FIGURE 4   Evaluation of control methods for the application.

moving parts to wear out. In addition, there are numerous costs associated with installing, operating, and maintaining SBCs above and beyond the initial hardware cost. Because they are not typically available off the shelf, costs for SBCs involve securing the services of an electrical engineer to design the board and test for viability. Still, even today, many original equipment manufacturers choose to design and develop single-board controllers for their own unique machine applications. An SBC is usually very specific to a machine and can be cost effective initially. This is perfectly appropriate when the specific capabilities required by the machine cannot be achieved using standard, off-the-shelf control products.

Although there are many factors that must to be taken into consideration before determining whether a relay, SBC, micro-PLC, or full-size programmable controller is appropriate for the application, the following chart shows how each of the four meet some of the basic application requirements (Fig. 4).

PROGRAMMABLE CONTROLLER FUNCTIONS

Today's programmable controller is part of a system requiring the user to make choices concerning communication to other devices, I/O devices, and programming and other software tools. Understanding the choices available in each of these areas is essential to making a more informed decision about the proper control for the application.

NETWORKING CHOICES

Programmable controllers today are most often linked together in networks that together integrate plant operations. As a result, the choice of a programmable controller often depends upon where the programmable controller will fit within the plant's communication architecture. There are three main types of networks. An information network provides a link between the plant floor and manufacturing execution systems, connects to multiple vendors' host computers, has the capacity to transfer large data files, and supports standard network management and troubleshooting tools. A control network offers real-time performance, is deterministic and repeatable, supports peer-to-peer messaging, connects to programmable controllers, personal computers, man–machine interface devices, drives, and motion devices (and so on), and it supports programming and device configuration.

TABLE 1 Network Selection Criteria

If the Control Application Requires	Use This Network Type
High-speed data transfer between information systems and/or a large quantity of controllers	information network
Internet/Intranet connection	
Program maintenance	
Plantwide and cell-level data sharing with program maintenance	
High-speed transfer of time-critical data between controllers and I/O devices	control network
Deterministic and repeatable data delivery	
Program maintenance	
Media redundancy or intrinsic safety options	
Connections between controllers and I/O adapters	
Distributed controllers so that each controller has its own I/O and communicates with a supervisory controller	
Connections of low-level devices directly to plant floor controllers, without the need to interface them through I/O modules	device network
More diagnostics for improved data collection and fault detection	
Less wiring and reduced start-up time than a traditional, hard-wired system	
Modems	serial connection
Messages that send and receive ASCII characters to or from devices such as ASCII terminals, bar-code readers, message displays, weigh scales, or printers	
Supervisory control and data acquisition (SCADA)	

A device network reduces wiring costs (because devices do not need to be directly wired to a programmable controller), supports device-level diagnostics, and connects to multiple vendors' devices. In addition to, or in place of, these three main types of networks, many vendors offer a serial connection to a programmable controller.

Programmable controllers may be placed on any of these networks. Factors that determine the type of network they will be placed on include the type of information to send or receive, system performance, the distance or size of the application, available networks, and future expansion. Table 1 presents a guide to network type, based upon application requirements.

The following are examples of information, control, and device networks.

Information Network Example

The TCP/IP Ethernet network is a local area network designed for the high-speed exchange of information between computers and related devices (Fig. 5). With its high bandwidth (10–100 Mbps), an Ethernet network allows many computers, controllers, and other devices to communicate over vast distances. At the information layer, an Ethernet network provides enterprisewide systems access to plant-floor data. With an Ethernet network you have many possibilities because you can maximize communication between the great variety of equipment available from vendors. TCP/IP is the protocol used by the Internet.

Control Network Example

The ControlNet network is an open, high-speed deterministic network used for transmitting time-critical information (Fig. 6). It provides real-time control and messaging services for peer-to-peer communication. As a high-speed link between controllers and I/O devices, a ControlNet network combines the capabilities of existing Universal Remote I/O and DH+ networks. You can connect a variety of devices to a ControlNet network, including personal computers, controllers, operator interface devices, drives, I/O modules, and other devices with ControlNet connections. At the control

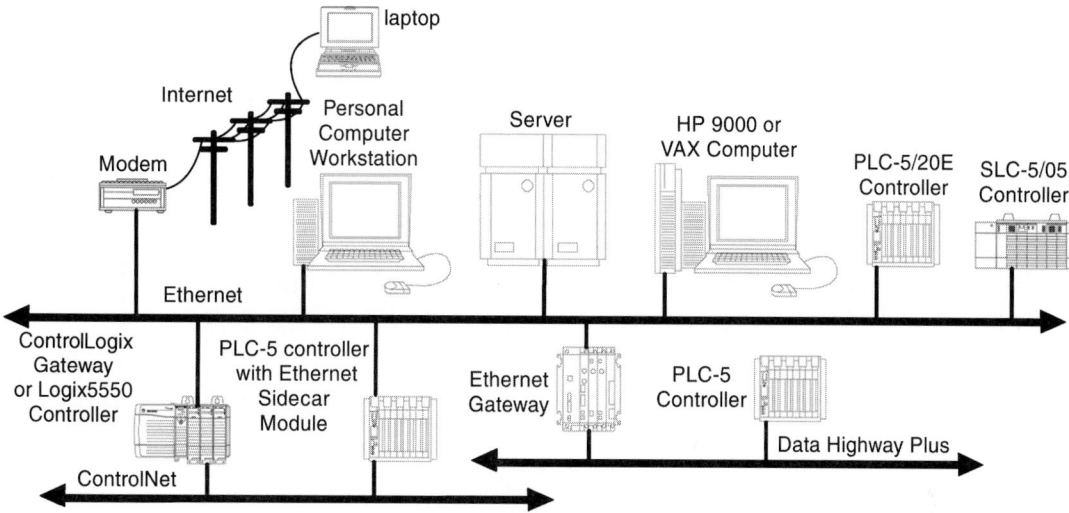

FIGURE 5   Information network example.

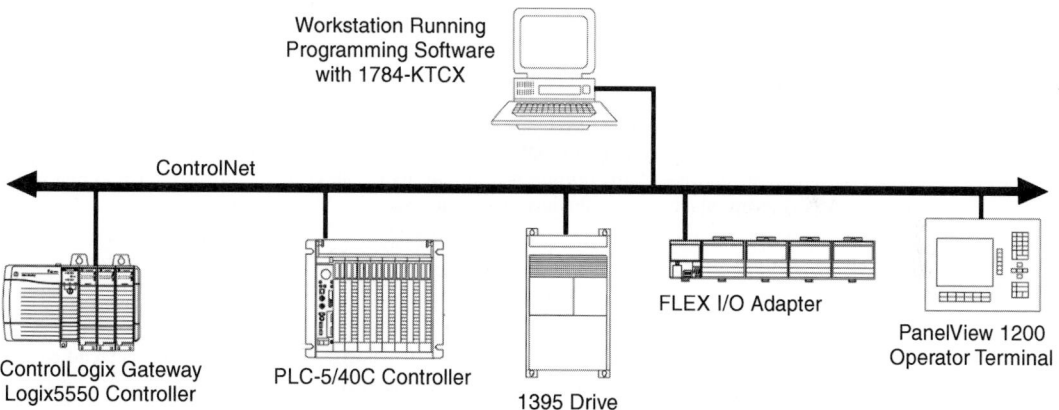

FIGURE 6   Control network example.

layer, a ControlNet network combines the functionality of an I/O network and a peer-to-peer messaging network. This open network provides the performance required for critical control data, such as I/O updates and controller-to-controller interlocking. ControlNet also supports transfers of noncritical data, such as program uploads, downloads, and messaging.

Device Network Example

A DeviceNet network (Fig. 7) is an open, low-level communication link that provides connections between simple, industrial devices (such as sensors and actuators) and high-level devices (such as controllers). Based on standard controller area network technology, this open network offers a level of interoperability between like devices from multiple vendors. A DeviceNet network reduces installation costs, start-up and commissioning time, and system and machine down time.

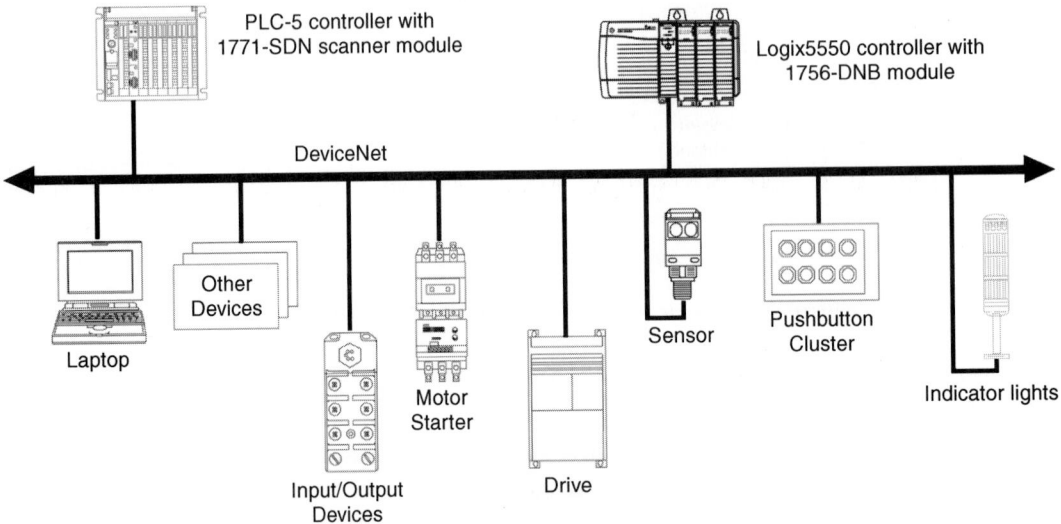

FIGURE 7 Device network example.

At the device layer, a DeviceNet network can connect low-level devices directly to plant-floor controllers without the need to interface them through I/O modules.

Serial Connection

The serial port found on many programmable controllers is used to connect devices (Fig. 8) that communicate by using vendor-specific communication protocols and send and receive ASCII characters, such as ASCII terminals, bar-code readers, and printers.

I/O CHOICES

The following questions attempt to provide some key decision parameters when one is faced with the wide array of choices available for I/O. Three key decisions are:

1. What physical form-factor will the I/O take, i.e., distributed or chassis based?

2. What types of devices will the I/O control?

3. What are the specific requirements of your application?

The solution, unfortunately, is seldom linear. Instead, a user typically must face trade-offs between the desired form-factor, the availability of an I/O interface to a particular device in that form-factor, and matching the specific application requirements with an existing I/O module. What follows is an overview of the choices the user will face.

Chassis-Based vs. Distributed I/O

The size of programmable controllers has been reduced over the past few years by several hundred percent. A key factor impacting the size of the controller has been the emergence of distributed I/O solutions. Previously, an I/O chassis was placed into an enclosure, wired to terminal strips, and then put on the machine before any interconnections were made. Now many users prefer a prepackaged

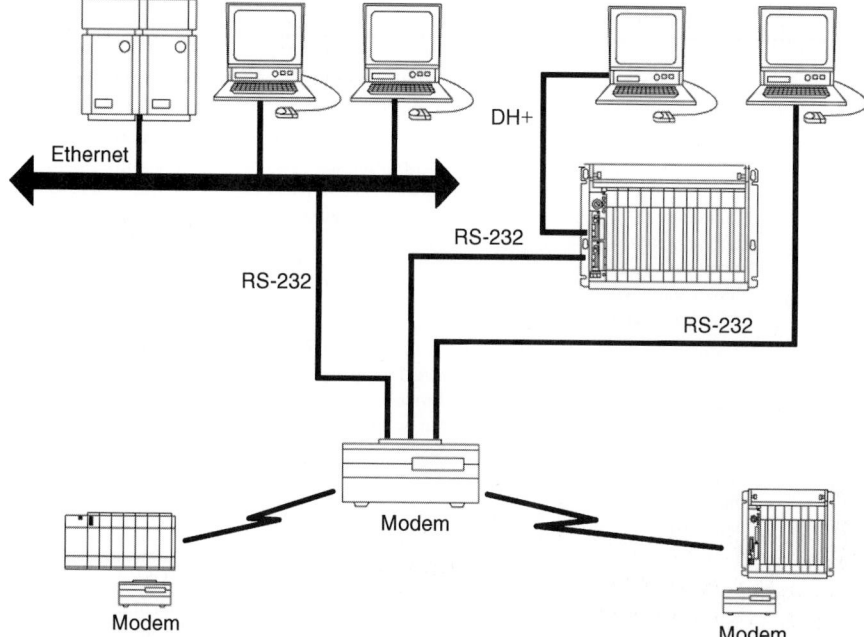

FIGURE 8   Serial connection example.

distributed I/O solution in which the physical packaging of the I/O is industrialized and now can be directly bolted to the machine. This can reduce wiring, provide the flexibility to continually add to the control system with maximum ease, and eliminate I/O from taking up panel space. On the other hand, chassis-based I/O makes it possible to place large clusters (100s to 1000s) of I/O either locally (together with the programmable controller) or remotely. In addition, chassis-based systems often allow the widest selection in terms of the types of devices that can be interfaced.

Types of I/O Devices

The types of devices that can be controlled by programmable controllers are digital or analog or require the use of so-called intelligent I/O modules (Table 2). The latter simply indicates the need to perform some processing on the I/O module independent of the programmable controller, and it varies from vendor to vendor.

Application-Specific Requirements

In addition to the types of devices being interfaced there are a number of I/O requirements based on the specific application. Some of these are listed below (Table 3).

SOFTWARE CHOICES

Today's programmable controller user has benefited from developments in the computer market. The software used to program today's programmable controllers typically runs on operating systems

TABLE 2   Types of I/O Modules and the Devices They Control

Choose This Type of I/O Module		For These Types of Field Devices or Processes (examples only)	With These Basic Functions	And with These Features for Special Requirements
Digital	input	selector, proximity, limit, or float switches pushbuttons thumbwheels circuit breakers photoelectric eyes motor starter contacts relay contacts	Sense input from on/off or opened/closed device signals. These signals can be either ac or dc.	isolation proximity switch source or sink (dc) fast response TTL
	output	alarms control relays fans lights horns valves motor starter solenoids	Provide output signals to on/off or opened/closed devices. These signals can be either ac or dc.	isolation protection (detection of failed triacs) TTL high current switching
Analog	input	temperature, pressure, load cell, humidity, and flow transducers potentiometers	Convert continuous analog signals into input values for the controller (A/D)	thermocouple RTD isolation accuracy resolution
	output	analog values actuators chart recorder electric motor drives analog meters	Interpret controller output to analog signals for the field device. This is generally done through a transducer (D/A).	PID isolation
Intelligent		encoders flow meters external device communication ASCII RF-type devices weight scales bar-code reader tag reader display devices	Used for specific applications such as position control, PID, and external device communication.	thermocouple RTD PID isolation

largely the same as those found in the home environment. The computer hardware also mimics that found in the home environment, though often it comes in an industrialized package to withstand the harsh environments where these programming devices reside.

As for programming software, the same trend found in the larger consumer software market is visible in the market for programmable controller software, with vendors offering greater consistency among programming packages at the same time that the number of features and user friendliness has increased.

The programming software interacts with the instructions embedded in the firmware of the programmable controller (Table 4). For this reason, most vendors offer both the hardware and the software package used for programming, though a number of software companies offer competitive programming software.

TABLE 3   Factors Affecting the Selection of I/O Modules

Selection Factors	Selection Criteria
Application-specific	voltage range and type
	level of diagnostic monitoring required
	speed required for passing I/O data between I/O circuit and controller
	backplane current
	output current
	number of I/O points
	isolation requirements
	sinking or sourcing inputs/outputs (dc)
	solid-state or relay outputs (dc)
Environmental	temperature
	space restrictions
	humidity
	proximity to wash-down area
	vibration
	shock restrictions
	noise restrictions
	distance restrictions
	atmospheric conditions (gases, corrosives)
Communication	networking requirements
	desired transmission communication rate
Termination	method of connection to sensors/actuators
	connection to (dc) sinking or sourcing sensor/actuator
Expansion	room for additional modules
	capacity of power supply to accommodate additional modules
	capacity of programmable controller memory and features to accommodate additional modules
Power	total current draw for all I/O modules in chassis
	total current draw for programmable controller in chassis
	total current draw for future I/O modules
	power supply internal or external to the chassis and input-line voltage

PROGRAMMING LANGUAGES

Five programming languages for writing the application program necessary to perform control are commonly found today. These are Ladder Diagram, Sequential Function Chart, Function Block, Structured Text, and Instruction List.

Ladder Diagram

For people who understand relay controls, ladder (Fig. 9) continues to be an advantage in terms of usability. Although it is possible to program all control logic in ladder, supplementing ladder with other languages allows users access to the language best suited for a particular control task.

Sequential Function Charts (SFCs)

SFC programming (Fig. 10) offers a graphical method of organizing the program. The three main components of an SFC are steps, actions, and transitions. Steps are merely chunks of logic, i.e., a unit of programming logic that accomplishes a particular control task. Actions are the individual aspects

TABLE 4 Common Instruction Types and the Functions They Execute

Instruction Family	Description and Common Symbolic Names
Relay-type	The bit (relay-type) instructions monitor and control the status of bits. XIC, XIO, OTE, OTL, OTU, IIN, IOT, IDI, IDO
Timer and counter	The timer and counter instructions control operations based on time or the number of events. TON, TOF, RTO, CTU, CTD, RES
Compare	The compare instructions compare values by using an expression or a specific compare instruction. CMP, EQU, GEQ, GRT, LEQ, LES, LIM, MEQ, NEQ
Compute	The compute/math instructions evaluate arithmetic operations using an expression or a specific arithmetic instruction. CPT, ACS, ADD, ASN, ATN, AVE, CLR, COS, DIV, LN, LOG, MUL, NEG, SIN, SQR, SRT, STD, SUB, TAN, XPY
Logical	The logical instructions perform logical operations on bits. AND, NOT, OR, XOR
Conversion	The conversion instructions convert integer and BCD values or convert radian and degree values. TOD, FRD, DEG, RAD
Bit modify move	The move instructions modify and move bits. BTD, MOV, MVM
File	The file instructions perform operations on file data and compare file data. FAL, FSC, COP, FLL
Diagnostic	The diagnostic instructions compare data to help you detect problems. FBC, DDT, DTR
Shift	Use the shift instructions to modify the location of data within files. BSL, BSR, FFL, FFU, LFL, LFU
Sequencer	Sequencer instructions monitor consistent and repeatable operations. SQO, SQI, SQL
Program control	Program flow instructions change the flow of ladder program execution. MCR, JMP, LBL, FOR, NXT, BRK, JSR, SBR, RET, TND, AFI, ONS, OSR, OSF, SFR, EOT, UIE, UID
Process control	The process control instruction provides closed-loop control. PID
Block-transfer	The block-transfer instructions transfer words to or from other devices. BTR, BTW, CIO
Message	The message instruction reads or writes a block of data to another station. MSG
ASCII	The ASCII instructions read, write, compare, and convert ASCII strings. ABL, ACB, ACI, ACN, AEX, AHL, AIC, ARD, ARL, ASC, ASR, AWA, AWT

of that task. Transitions are the mechanisms used to move from one task to another. Control logic for each step, action, and transition is programmed in one of the other languages such as Ladder Diagram or Structured Text.

As a graphical language, SFC programming offers you several choices for executing a program, each depicted in a visually distinct way. In a sequential configuration, the processor simply executes the actions in step 1 repeatedly, until the transition logic becomes true. The processor then proceeds to step 2. In a selection branch, only one branch is executed, depending on which transition is active. In a simultaneous branch, all branches are executed until the transition becomes active. In addition to various types of branches, the operation of individual actions within a step can be varied with the use of action qualifiers.

Action qualifiers determine how the action is scanned and allow actions to be controlled without additional logic. For example, one could use the L qualifier to limit the time that ingredient valve B is opened.

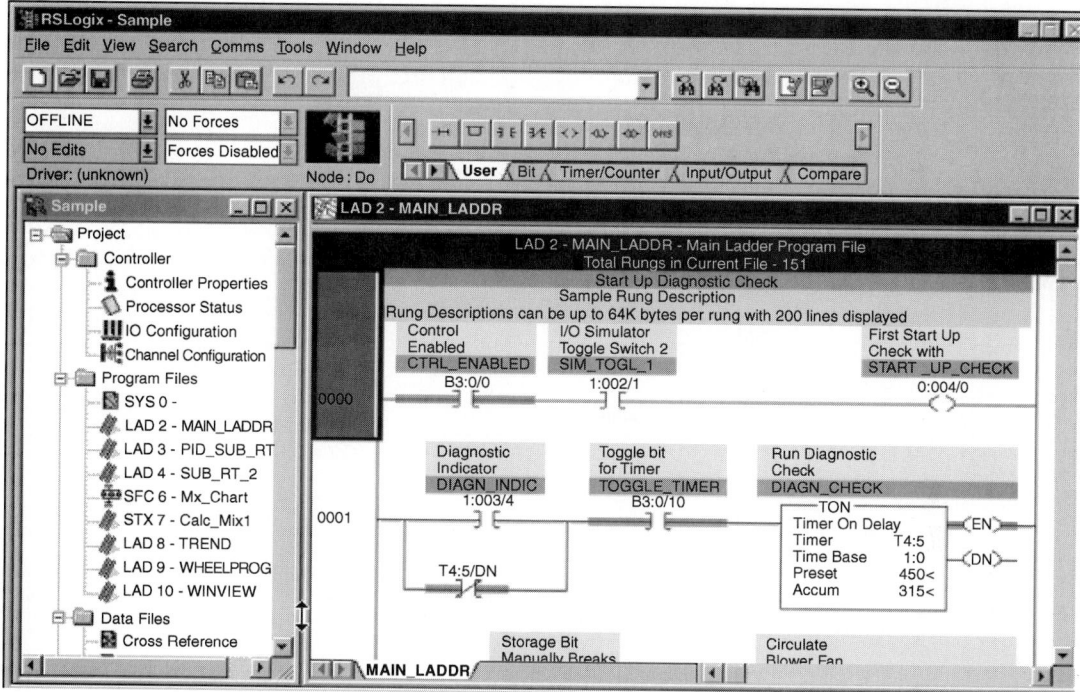

FIGURE 9   Ladder diagram example.

In practice, an active step is highlighted to signal to the programmer which part of the program is executing—a useful feature for troubleshooting. This highlighting is an example of the standard's extensibility—the ability of a vendor to add a feature not specified in the standard.

Note that the standard offers SFC programming as an organizing tool. The user chooses whether to use it or not, based on whether the process being controlled is sequential in nature. And even if SFC programming is used, the actions will be written in one of the four programming languages described below.

Function Block Diagram

Like SFC, Function Block (Fig. 11) is a graphical language that allows programming in other languages (Ladder, Instruction List, or Structured Text) to be nested within the function block. In Function Block, program elements appear as blocks that are "wired" together in a manner resembling a circuit diagram. Function Block is most useful in those applications involving a high degree of information or data flow between control components, such as process control.

Structured Text

This high-level language resembles Pascal or Basic, and, in fact, people trained in computer programming languages often find it the easiest language to use for programming control logic. When symbolic addressing is used, Structured Text programs (Fig. 12) resemble sentences, making it highly intelligible to the novice user as well. Structured text is ideal for tasks requiring complex math, algorithms, or decision making. Its concise format allows a large algorithm to be displayed on a single page (vs. multiple pages of Ladder logic).

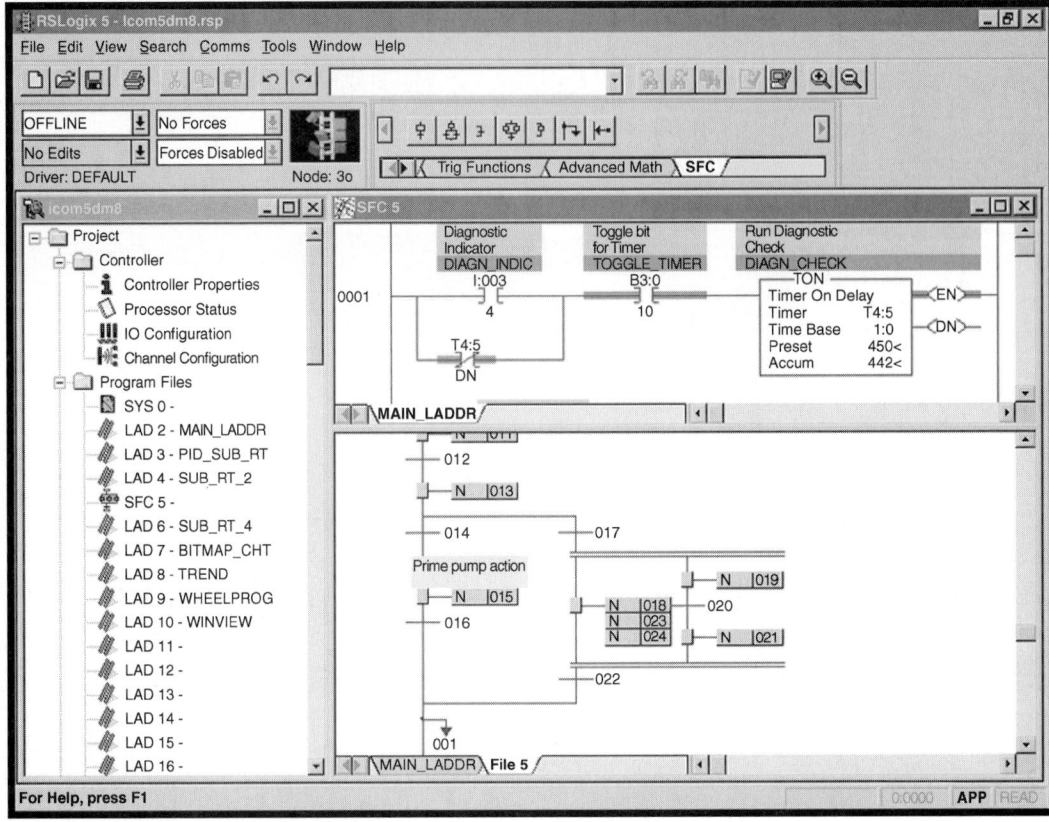

FIGURE 10 Sequential function chart example.

Benefits of Structured Text include the following: (a) people trained in computer languages can easily program control logic; (b) symbols make the programs easy to understand; (c) PowerText facilitates system debugging and application commissioning; (d) programs can be created in any text editor; and (e) it runs as fast as ladder logic.

Instruction List

This low-level language is similar to Assembly language and is useful in cases where small functions are repeated often. Although Instruction List (Fig. 13) is powerful, it is considered to be difficult to learn.

SOFTWARE STANDARDS

The popularity of the above as programming languages has led to the development of the IEC 1131-3 standard. IEC 1131-3 is an international standard for programmable controller programming languages. It specifies the syntax, semantics, and display for the previously mentioned suite of languages.

The availability of multiple languages for development of control programs allows developers to select the language best suited for a particular control problem. The IEC 1131-3 international standard makes it possible to use any one of the above languages with the confidence that it follows the same conventions all over the world, ensuring consistency and reducing error.

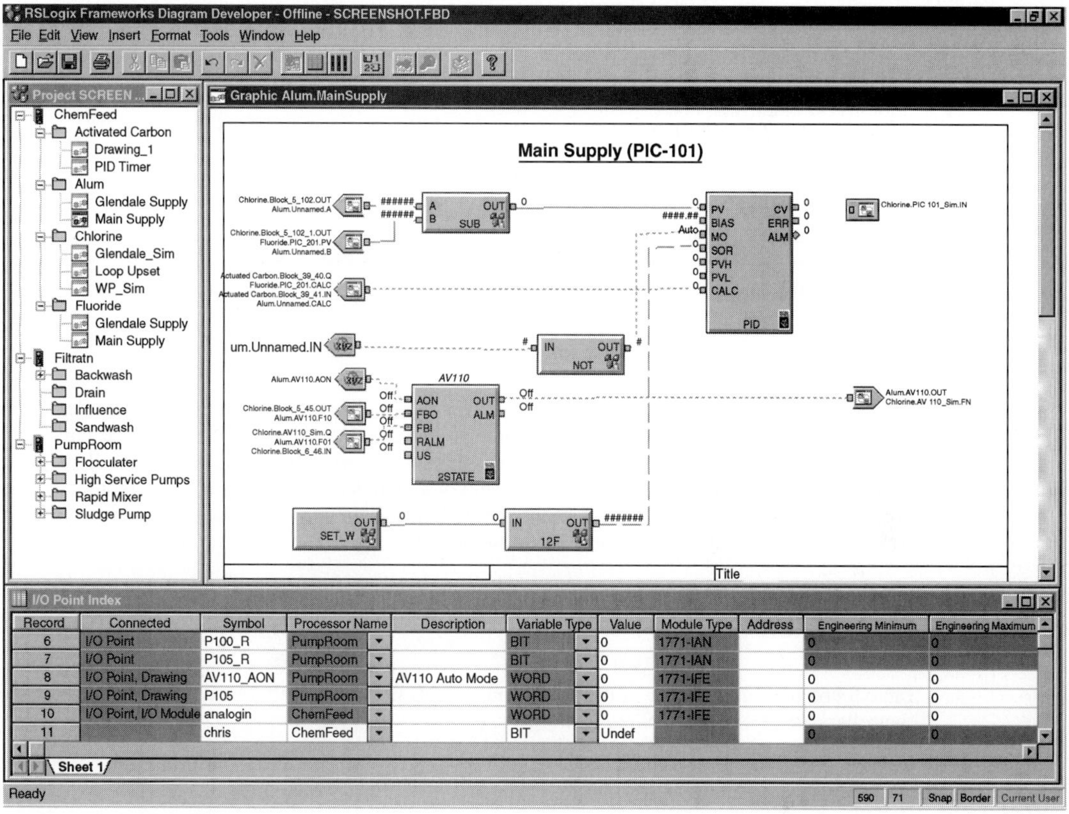

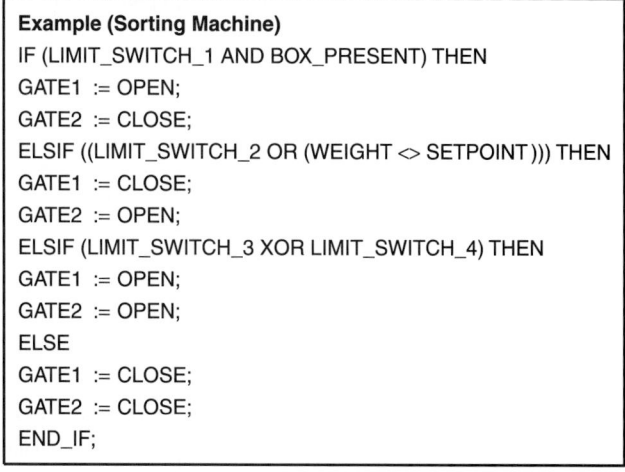

FIGURE 11   Function block example.

Example (Sorting Machine)

IF (LIMIT_SWITCH_1 AND BOX_PRESENT) THEN

GATE1 := OPEN;

GATE2 := CLOSE;

ELSIF ((LIMIT_SWITCH_2 OR (WEIGHT <> SETPOINT))) THEN

GATE1 := CLOSE;

GATE2 := OPEN;

ELSIF (LIMIT_SWITCH_3 XOR LIMIT_SWITCH_4) THEN

GATE1 := OPEN;

GATE2 := OPEN;

ELSE

GATE1 := CLOSE;

GATE2 := CLOSE;

END_IF;

FIGURE 12   Structured text example.

Instruction List Example (Calculate new weight by subtracting tare weight from net weight)

```
LD weigh_command
JMPC WEIGH_NOW
ST ENO
RET
WEIGH_NOW:  LD gross_weight
SUB tare-weight
```

FIGURE 13   Instruction list example.

SOFT CONTROL

As described in the introduction, soft control replaces traditional programmable controllers with software that allows you to perform programmable controller functions on a personal computer. While beyond the scope of our discussion here, soft control does have advantages, especially in those applications with large memory requirements and/or a high degree of information integration. In the case of the latter, soft control can provide the ability to combine man–machine interface (MMI), programming, and control on a single hardware platform.

OTHER SOFTWARE PACKAGES

Table 5 provides an overview of the other kinds of software commonly found where programmable controllers are also being applied.

CONTROLLER SELECTION

The selection of a programmable controller involves making the following selections.

1. Programmable controller.   Based on the number of I/O points required and the communication options, select the programmable controller. In a fixed controller, the I/O, chassis, and power supply are an integral part of the controller.

TABLE 5   Common Types of Industrial Software and the Functions They Perform Controller Selection

Software Type	Function Performed
MMI	monitor and control your production system
Data management	extend an MMI into data management for connectivity to industrial devices and data management
Trending	log and monitor programmable controller data in a wide variety of data-collection applications, including automatic data acquisition, data scaling, trending graphs, and log reports
Documentation	produce accurate, error-free schematics for simultaneous manufacturing and support documentation
Batch	create an object-oriented batch automation solution based on industry standards such as S88.01
Transaction processing	bridge the gap between a control system and the rest of the enterprise with a transaction-based system
Diagnostic	monitor timing relationships between I/O elements or bit addresses and diagnose real-time machine behavior that differs from accepted operation
Verification	automatically detect program and data changes and archive project files to a central location

2. Programming software.   Based on the controller, the software preferences you have, and the nature of your application, choose the programming software.

3. Communication modules.   Based on communication requirements (other that what is available on the controller), select the communication modules needed to connect to other networks.

4. I/O modules.   Select the I/O modules for the local chassis. This selection must be made for a modular controller.

5. Chassis.   Based on the total number of modules, select a chassis. This selection must be made for a modular controller.

6. Power supply.   Based on the total current draw of all modules in the chassis, select the power supply. This selection must be made for a modular controller.

7. Miscellaneous cables.   Based on the controller and the communication options, select the cables you need.

Today's typical PLC system involves a logic processor, programming software, power supply, and input/output cards. In a "fixed" PLC system these are integrated into a single housing, while in a "modular" PLC system these are independent units then housed in an industrial rack. Additionally, an operator interface, supervisory computer, specialty software, specialty modules, and additional I/O racks may be added. PLC systems are sized by the number of I/O they process, though other overlapping criteria play a factor, including memory, packaging, application, instruction set, networking, communication, and language capabilities. Micro (fewer than 32 I/O) PLC systems serve as a replacement for relays. Small (32–128 I/O) ones available in varying packaging and provide low-level communication. Medium (128–512 I/O) ones offer greater flexibility in type of I/O (analog and discrete), instruction set, and communications. Large (more than 512, and often more than 1000 I/O) ones are used in networked and distributed applications, as well as for stand-alone control with a variety of programming options.

Thus far, we've established that there are choices available for controller functions. There are also choices for controllers as well. As you determine which options are right for you, there are a number of factors to consider.

Openness

Choosing an "open" controller shouldn't by itself be the goal. A fundamental set of requirements—speed, scale, packaging, reliability, and peripherals—must be satisfied before any control system technology is successfully applied. Some of these requirements will be specific to industry segments (process, motion, or discrete), whereas others will be broadly applicable to all aspects of industrial control.

An operating system is at the heart of all open architecture solutions. The vast majority of the requirements for operator interface and program development have been satisfactorily met by the existing commercial operating systems products and have broad market acceptance. It is also a foregone conclusion that Windows-based solutions will be desired and accepted for these functions for the foreseeable future.

Speed

The term "real time" is used very freely in the control system market. Real time means that a transaction or event is processed as quickly and predictably as possible without it being saved to a file for batch processing at some later time. Thus, it is the application that determines how quickly real-time processing must be. Real time for an automated teller machine (ATM), for example, can be measured in seconds while users of motion control systems require a real-time response in fractions of milliseconds. This broad range of performance requirements is one of the first areas to examine when talking about programmable controllers within an operating environment. How fast must the controller be? How repeatable must it be? The key is to determine what speed means within the context of your own control application—and not to get involved in an exercise of evaluating specifications for specifications' sake.

Reliability/Environmental Operation

Programmable controllers are, in general, designed withstand the rigors of your manufacturing environment. The demands of a typical shop floor—airborne contaminants, vibration, shock, and extreme temperatures—require the programmable controller to exceed the operational requirements of a standard personal computer for reliability, flexibility, agility, ease of system upgrades, and rugged operation. If, however, your environment is clean, dust free, and low noise, is it worth it to purchase something that's really designed for rugged manufacturing? For this reason, many users have considered PC-based control.

Many vendors support the idea that an ordinary office PC can be used in a PC-based control system, leading end users to believe that all PCs provide the same "service." That statement isn't necessarily true; very few office PCs are robust enough to handle the industrial factory floor environment. The industrial computer is designed to withstand the assault of the shop floor. Many industrial PCs feature TFT flat panel displays, shock-mounted hard drives, solid-state flash hard drives with no moving parts, filtered ball-bearing cooling fans, internal temperature sensors, uninterruptible power supplies, redundant hard drives, and system control monitor cards. Additionally, many industrial PCs with integrated CRT (cathode ray tube) displays are designed with a high degree of EMI (electromagnetic interference) shielding to prevent electrical interference from causing display problems. These continuing improvements are designed to bring the robustness of dedicated PLC-based control systems to PC-based control systems.

But reliability and environmental operation are not hardware-only issues. From the standpoint of the operating system, it brings into question a number of practices that must be evaluated. It is a common practice for commercial operating systems to be upgraded every 18 months. It is also quite common to not be able to get a fix until the next release. Can the control solution live with this level of support? If so, you may want to use it. If not, then maybe a programmable controller is more appropriate.

The Application

Many vendors have come to the realization that process control isn't necessarily more difficult than discrete control—just different. Since the original programmable controllers were born in the discrete manufacturing industries (particularly automotive), it was only natural that the initial product developments catered to the needs of discrete manufacturers. However, the expanded capabilities of the controller and I/O modules—more memory, better math capabilities, and an expanded instruction set—have expanded the controller's usefulness well beyond discrete and into a wide array of process applications. In addition, the emergence of specialty I/O modules allows controllers to manage process functions such as temperature control, weigh scale load cells, plastic injection molding, and high-speed analog. The emergence of controller-compatible process function libraries, expanded OI capabilities (both hardware and software), and the debut of more capable networks also have increased the controller's prominence in process applications.

Programmable controllers are widely accepted in applications such as packaging and material handling, and their acceptance is growing in a number of others. Micro-PLCs, for instance, are expanding rapidly into the commercial markets and industrial applications because their compact size and low cost make them ideal for OEM and end-user applications. Programmable controllers, in general, are also being applied to a broader range of applications, especially process control and SCADA telemetry-type applications, such as food and chemical processing, water/wastewater, and oil and gas pipelines. This transition has been made possible because controllers now offer a complete process tool kit and the communication capabilities required by telemetry-type applications.

The evolution of the high-end and low-end controllers will be quite different. Many process manufacturers—especially in SCADA applications (water/wastewater, oil extraction, etc.) are looking to replace their dedicated remote terminal units (RTUs—small, relatively inexpensive SBCs) with small programmable controllers for cost, reliability and ease-of-use reasons. This has only been possible in recent years as networking capabilities of these systems improved.

On the high end, larger programmable controllers with more memory and I/O capabilities are becoming commonplace in small batch operations, places where the DCS was the system of choice in the past. DCS systems, in contrast, have evolved as well by improving scalability and lowering I/O costs. In addition, there are more opportunities to productively combine different functions in one control system. In effect, the lines have blurred between the DCS and the programmable controller.

Installation/Integration

The size of controllers has been reduced over the past few years by several hundred percent. Users have less panel space and, more importantly, are less inclined to pay for excess functionality. If they want the programmable controller to solve a given problem, they want the controller optimized to solve that problem, not problems primarily found in other applications. In other words, you don't want to buy a scientific calculator if you only want to do four-function math. Another key factor impacting the size of the controller has been the emergence of more distributed and "granular" I/O solutions.

Manufacturers are working closely with customers to understand the total requirement from a control solution perspective. In general, programmable control manufacturers do not supply all the products that comprise a total system. A system is more than just the control hardware, but also the integrating components. Robot controllers and weld control systems in automotive plants; weigh scales in process applications in food and drug industries; modems in telemetry applications; and software that is used in every application must all seamlessly integrate with existing plant floor controls and controllers. Often this is accomplished through a third party who develops network interfaces, allowing their products to communicate directly to the controller. Customers demand the same level of integration with their peripheral devices as they traditionally have had between the processor and the I/O, with the same degree of ease and security. Manufacturers are working closer with third-party companies to provide complementary products for controllers. By the development of a series of well-defined interfaces for the control architecture, the customers have access to a high degree of integration within their system.

Communications Performance

Control systems, and therefore programmable controllers, must provide the network functionality represented by the three layers that exist in most control systems: information, control, and device. Just as throughput in a control scheme should be measured from the input point back out to the output point, throughput from a system level should be thought of as time to get data out of a node on a device network up to a node on an information network.

End users are also asking manufacturers to provide an ideal way to combine control and information systems in a cost-efficient manner. Networking the systems that control inventory, maintenance, and other plant manufacturing operations is evolving as the demand for real-time information increases. Because of the distinct differences between how manufacturing and support tasks are performed and data are handled for various applications, more than one network will be needed in most facilities.

Safety/Data Integrity

Product and system-level diagnostics are becoming a big requirement so more information can be made available to the operator. Messages such as "limit switch one is broken" reduce the time needed to find malfunctions. Some of the diagnostic capability originates from peripheral products used with the programmable controller, such as sensors that are embedded with fuzzy logic. But some of the diagnostic capability has to inherently be shared among the I/O modules, the controller, and the software, to take advantage of the full benefits. Device networking within a plant allows for intelligence to be quickly shared between devices and the controller.

Engineers are using more efficient process design tools to improve development engineer productivity. Time-saving tools, such as Sequential Function Charts and Structured Text, are being used along with conventional relay Ladder logic design tools. Device level networks will be a significant trend in the market with direct benefits to the customer of improved diagnostics, life-cycle costs, and reduced wiring costs.

Cost

Controller manufacturers must meet or beat the price point that was set by the previous control solution. For example, most companies in SCADA applications used dedicated RTUs. In order to persuade companies to make the transition to programmable controllers, manufacturers must match the price point set by RTUs.

Customers must look beyond procurement price, however, and instead focus on the total life-cycle cost. Where automation investments were once viewed as one-time purchases, manufacturers are now examining the various stages of the automation investment life cycle and finding sophisticated ways of adding value to each stage. Manufacturers and customers must look beyond the initial purchase price and try to understand ways to reduce costs associated with justifying, purchasing, installing, operating, maintaining, and upgrading their automation systems. If the end user's process only involves collecting and managing non-time-critical information, PC-based control may be less expensive. However, if the process also involves time-critical information, plant-floor survival or other similar variables, PC-based control may be comparable in price to a traditional programmable controller. For instance, if the plant floor has radio frequency and electromagnetic interference, temperature swings and other environment factors, by the time the end user protects the PC with all of the component and system-level ruggedness that the process requires, the programmable controller may be less expensive.

Standard process runtime tools will result in lower training costs and more immediate operator acceptance of PC-based control. By the incorporation of autotuning into several of these packages, loop maintenance is simplified. And systems developed using standard tools such as Visual Basic, C++, ActiveX components, and others allow developers to take advantage of tools that are relatively inexpensive because of their use across an extremely cost-competitive PC marketplace.

With capabilities growing closer each day, cost is a major factor in deciding between the purchase of a DCS, personal computer, or programmable controller-based system. From the I/O to the user interface, the price differences can be staggering. While a typical DCS user interface (proprietary operating system) can cost $50,000, a typical industrial computer user interface running off-the-shelf OI software on an off-the-shelf Windows NT operating system can be purchased for less than $7,000. A micro-PLC, in contrast, can cost as little as $300.

With all the choices available for automation controllers today, making an informed decision requires looking beyond the box and into how the core functions solve your application. Look closely at your manufacturing process and the technologies driving today's innovations. While there is no single controller that is right for every application, taking a planned systematic approach to evaluating your control options will result in a system that adds value throughout the life cycle of your manufacturing system.

STAND-ALONE CONTROLLERS

A single-loop controller (SLC) may be defined as a controlling device that solves control algorithms to produce a single controlled output. SLCs often are microprocessor based and may be programmable or have fixed functionality.

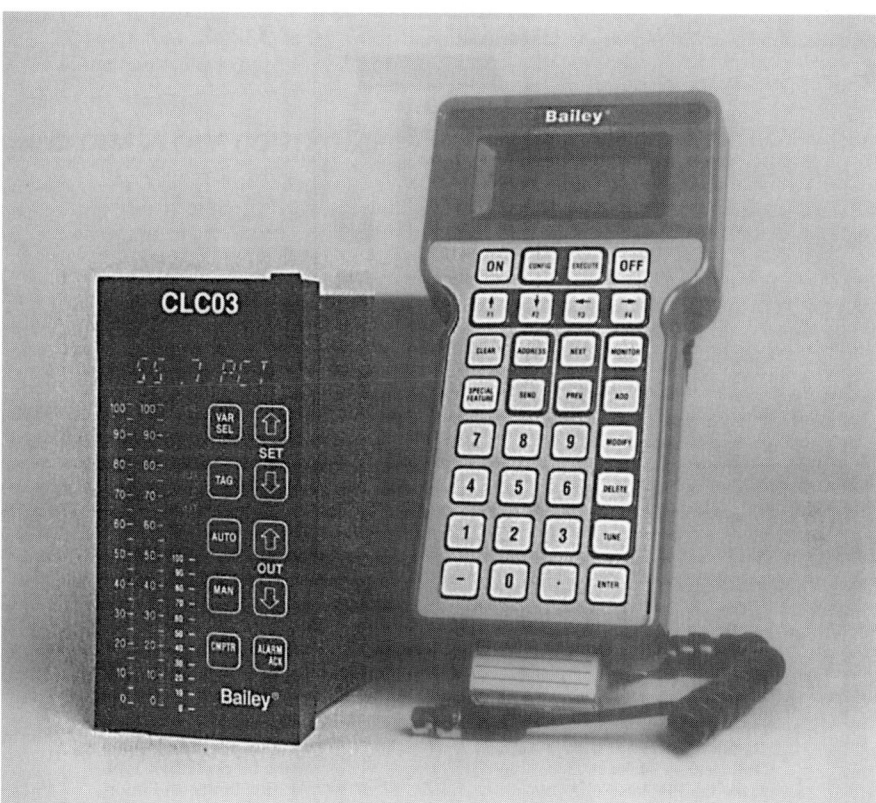

FIGURE 1 Stand-alone controller with dual-loop capability for controlling a variety of process variables. The instrument incorporates (*a*) gas plasma display for set point and control output, (*b*) on-board storage of over 75 proprietary function codes, (*c*) flexible input-output (4 analog, 3 digital inputs; 2 analog, 4 digital outputs), (*d*) optional loop bypass station permitting direct manual control of process outputs during maintenance periods, and (*e*) self-tunning. Configuration is obtained by way of a hand-held configuration and tuning terminal, which uses menu-driven prompts to "walk" the operator through "fill-in-the-blanks" configuration procedures. Full monitoring, control, and configuration capability for up to 1500 control points via a personal computer (PC) platform are available. Also, in the same product family there is a sequence controller that provides additional digital I/O for controlling up to three sequences, an RS-232C serial link for connection to external devices, such as printer or data logger. Common applications include flow, temperature, and pressure control of three-element boiler feedwater control and compressor surge control, as well as motor control, burner management, or other start-up/shut-down applications. (*Bailey Controls.*)

For many years, prior to the appearance of computer and advanced electronic technology, the SLC was the mainstay of industrial process control. It also found wide usage in the discrete-piece manufacturing industries, often where temperature control was of importance in connection with such operations as forging, heat treating, or drying ovens. By today's standards, huge centralized control panelboards, say, in a chemical plant or a petroleum refinery, would contain scores of analog SLCs for controlling temperature, pressure, flow, liquid level, and so on, in similar scores of places throughout the facility. The SLCs operated electric or pneumatic control elements such as valves, dampers, and conveyors, requiring large reels of wire or pneumatic tubing to connect them to the process. Frequently the SLCs of that day incorporated large indicating pointers, scales, and circular or strip-chart recorders.

Typically the contemporary SLC is digital. It is very much smaller (down to DIN sizes) and has electrical outputs (error signals) without recording or logging capability. It is built with standards in

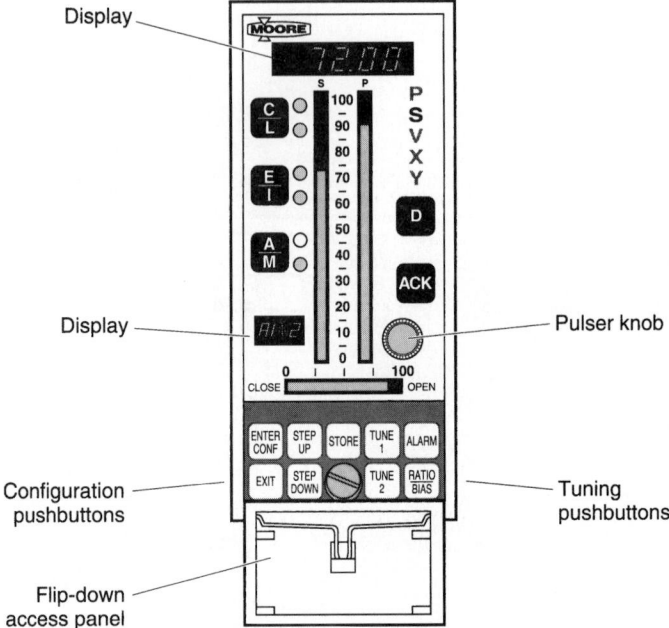

FIGURE 2 Microprocessor-based single-loop digital controller for general-purpose applications. Displays include analog liquid-crystal display (LCD) bar graphs for process, set point, and valve values; digital display for engineering unit values; alphanumeric display for status and alarm indication. Functions and operations, such as inputs, outputs, controls, and computations, are stored within the memory of the model as modular, easy to select function blocks (FBs). Typically the FBs have user-selected parameter values, calibration limits, and information specifying how a FB is linked to other FBs. The standard model includes FBs for single-loop, ratio-set, or external-set operation. An expanded version includes additional FBs for implementing advanced control strategies, namely, pressure-temperature compensation of a flow signal, dead-time compensation for transport lag, feedforward control, single-station cascade control, and override control. An additional third input option can be used to accommodate thermocouple, frequency, millivolt, RTD inputs, or computer pulse input. A serial data communications interface also is available for linking to communications networks (other controllers and computers). (*Moore Products Company.*)

mind for those situations in which a user may desire to make it part of a network, such as a distributed control system (DCS). SLCs continue to have advantages and to appeal to small- and medium-size plants that may not, within the foreseeable future, operate in a computer-integrated manufacturing (CIM) environment. There are tens of thousands of such situations. SLCs continue to enjoy a large share of the original equipment manufacturer (OEM) market for packaging with all types of equipment requiring some variable to be controlled (often temperature), which essentially is regarded by the user as isolated from other concerns with regard to mainline equipment. Even in some medium to large process and manufacturing situations, SLCs are considered as a control option, along with DCSs. Although CIM-type networking has numerous evident long-term cost-effective values (but which are sometimes difficult to justify), the cost of SLCs per loop of control needed is a handy criterion to use in making decisions on which way to go. With scores of SLC suppliers worldwide, it appears that the SLC is far from becoming obsolete. A potpourri of contemporary stand-alone controllers is presented in Figs. 1–6.

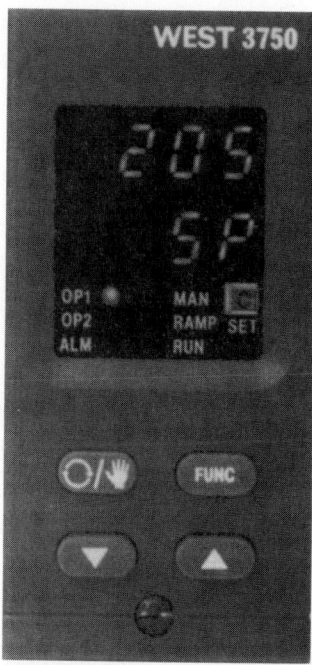

FIGURE 3   Stand-alone programmer controller in a ⅛
DIN case [96 by 48 mm (3.8 by 1.9 in.)] features a dual light-
emitting diode (LED) display, and seven dedicated LEDs
are used to show prompt legends during setup and the instrument status when a program is running. Full three-term
control output can be offered on output 1 (heat) and output
2 (cool) with the addition of one alarm output. RS 485 serial
communications option allows master-slave capabilities to
profile up to 32 other similar controllers. Other features
include soak hysteresis facility and dual-time-base capability to allow hour-minute or minute-second program rates.
Program parameters can be revised without interruption of
program. Auto-manual control allows the control of the process to be switched from automatic or closed-loop control to
manual or open-loop control. Pretune and self-tune may be
selected or deselected. Common applications include pot-
tery kilns, heattreating processes, food preparation, sterilization, and environmental chambers. (*West Instruments.*)

SINGLE- AND DUAL-LOOP CONTROLLERS

As is true of other instrumentation and control, stand-alone controllers have derived numerous technical advancements from digital electric circuitry, microprocessors, modern displays, and very creative
software. Contemporary stand-alone controllers differ markedly (perhaps even radically) from the
devices available as recently as a decade or so ago.

Size.   Not necessarily the most important, but one of the most noticeable features of the present
generation of SLCs is their smaller physical size. An impressive percentage of contemporary SLCs

FIGURE 4   Microprocessor-based ¼ DIN [96 by 96 mm (3.8 by 3.8 in.)] configurable controller. Instrument features blue vacuum fluorescent dot-matrix display (four lines of 10 characters per line), and self-tuning based on one-shot calculation of the optimum PID values based on a "cold" start. Calculated PID values are stored and can be displayed or changed on demand. Both direct- and reverse-acting outputs can be self-tuned. Four PID settings can be stored for each control output. PID set selection may be programmed as part of a profile, initiated by process variable value, internal trigger, keystroke, or logic input signal. Controller uses a dynamic memory rather than fixed allocation of memory. User can assign memory as needed during configuration for best memory utilization. Security by user ID number can be selected at any program level to prevent unauthorized access and program or data changes. Self-diagnostics are used on start-up and during normal operation by monitoring internal operations and power levels. Upon detection of an out-of-tolerance condition, controller shuts off output(s), activates alarm(s), and displays appropriate message. Controller can perform custom calculations using math programs. Device maintains a history file of event triggers as a diagnostic tool. Two process inputs and four outputs can be assigned to two separate control loops creating a dual-loop controller. Optional features include parallel printer output, logic I/O, digital communications, and PC interface. (*LFE Instruments.*)

follow the European DIN (*Deutsche Industrie Norm*) dimensions for panelboard openings. These are

⅛ DIN [96 by 96 mm (3.8 by 3.8 in.)]

⅛ DIN [96 by 48 mm (3.8 by 1.9 in.)]

¹⁄₁₆ DIN [48 by 48 mm (1.9 by 1.9 in.)]

By a substantial margin, the ¹⁄₁₆ DIN is the most popular size. The adoption of DIN sizes is not surprising since an estimated (1992) survey indicates that 30% of SLCs sold on the U.S. market are imported from Europe and Asia.

Control Functions.   It should be emphasized that numerous controllers are also capable of handling two control loops. By way of on-board microprocessors, most suppliers offer the fundamental on–off

FIGURE 5 Family of single-loop process controllers. These programmable controllers incorporate computational and control functions that can be combined in the same manner as programming a pocket calculator. The self-tuning function for optimizing control is particularly useful in multiproduct batch applications, where process characteristics can vary from product to product. It features a "one-shot" algorithm, triggered by set-point changes or on demand, and provides rapid response to process changes. Other features include feedforward control (with gain and bias computations), signal processing, analog inputs (4 points, 1–5 V dc), analog outputs (3 points, 1–5 V dc; 1 point, 4–20 mA dc). A companion programmable computing station provides data display, signal processing, and sequencing functions. The instrument features 10 status I/O points, each user-definable as input or output. Four programmable function keys on front of panel are used to start control sequences. Four associated lamps indicate sequence progress or serve as prompts. Control period is 0.1, 0.2, or 0.4 s; 500 program steps execute in 0.4 s. There are over 43 computational functions. A device of this kind can control several compressors, for example, as shown in Fig. 6. (*Johnson Yokogawa Corporation.*)

and PID (proportional–integral–derivative) formats. Some suppliers go much further in their design sophistication, offering many mathematical functions, either built into the SLC unit per se, or easily available in companion function modules. These functions, depending on brand and model, may include those listed in Table 1.

In addition to handling a variety of process variables, SLCs find application in the discrete-piece manufacturing industries.

Self-Tuning. This feature is available on nearly all the SLCs offered, except those in a low-cost category. The methodologies used in general are similar (Fig. 7).

Time Scheduling and Sequencing. Many available SLCs incorporate the ability to provide sequencing as well as time versus process variable scheduling. A number of processing operations, as found in the metallurgical industries in particular, require bringing a product up to temperature (as in annealing), holding the temperature for a given time interval, and then lowering it in several steps. This is described further in the following article in this handbook. Such schedules, of course, are not always limited to time and temperature, although this is the most common application.

Other Features. These include auto/manual, multi set point, self-diagnostics, and memory.

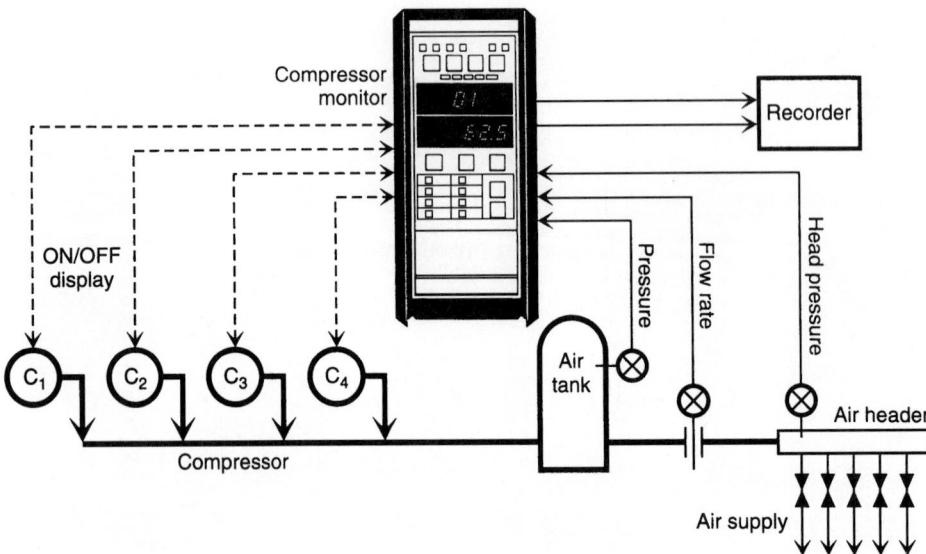

FIGURE 6   Control system for several compressors by using a stand-alone programmable computing station. (*Johnson Yokogawa Corporation.*)

TABLE 1   Mathematical Functions Offered in Stand-Alone Controllers (or Available in Computational Modules)*

Adder subtractor	$I_0 = \dfrac{G_0}{0.250}\left[G_A(V_A - b_A) \pm G_B(V_B - b_B) \pm G_C(V_C - b_C)\right] + 4\,\text{mA}$
Adjustable gain with bias	$I_0 = \dfrac{G_0}{0.250}\left[V_A - b_A\right] + K_0$
Deviation—adjustable gain with bias	$I_0 = \dfrac{1}{0.250}\left[G_0(V_A - V_B) + V_C\right]$

    Adjustable pots, 22 turn
    Gain and bias pots, 1 turn graduated dials

Multiplier/divider	$I_0 = \dfrac{G_0}{0.250}\left[\dfrac{(G_A V_A \pm b_A)(G_C V_C \pm b_C)}{(G_B V_B \pm b_B)}\right] + 4\,\text{mA}$

Other computation modules
    Lead/lag compensator
    Dual filter
    Ramp limiter
    Signal limiter
    Analog tracking
    Square-root extractor
    Signal selector (high/low)
    Signal selector (median)
    Peak picker
Signal converters
    Thermocouples: iron/constantan, Chromel/Alumel, copper/constantan, platinum/rhodium, Chromel/constantan
    Slidewire (nonisolated or isolated)
    Calculated relative humidity

* Examples are for inputs of 1.5 V dc and outputs of 4–20 mA.

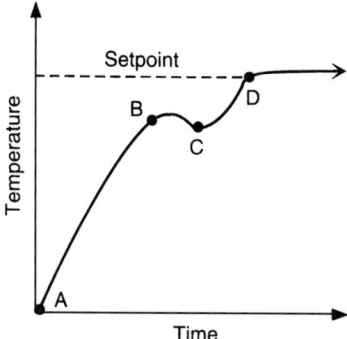

FIGURE 7   Self-tuning feature as available in one SLC. Starting from a cold start *A*, the controller provides 100% output power until (in this case) the temperature reaches *B*, where the output power is cut off. The controller then measures the system response *B* to *C* and calculates the appropriate PID constants for heat control. At point *C* the output is restored to bring the system to the desired set point *D* with minimum delay. Upon completion of the self-tuning routine, the controller automatically stores and displays the PID constants and starts using them in the control program. The operator can enter PID constants manually if it is decided not to utilize the self-tuning feature. (*LFE Instruments.*)

Networking.   Prior to the introduction of DCSs, single-loop controllers were commonly linked to a central control room. There may be occasional instances when a user may opt for networking the contemporary SLCs.

REFERENCE

1. Wade, H. L., *Regulatory and Advanced Regulatory Control: System Development*, Instrument Society of America, Research Triangle Park, North Carolina, 1994.

HYDRAULIC CONTROLLERS

by Raymond G. Reip*

A hydraulic controller is a device that uses a liquid control medium to provide an output signal that is a function of an input error signal. The error signal is the difference between a measured variable signal and a reference or set-point signal. Self-contained closed-loop hydraulic controllers continue to

* Shorewood Hills, Sawyer, Michigan 49125; formerly with L & J Engineering Inc., Crestwood, Illinois.

be used for certain types of process control problems, but as the use of the computer, with its electrical output, expands in process control applications, the electrohydraulic servo or proportional valve gains in usage. The combination adds the advantages of hydraulic control to the versatility of the computer. Also contributing to the expanding use of hydraulics is the steady improvement in fire-resistant fluids.

Since the servo or proportional valve does not accept low-level digital input directly, a digital-to-analog (D/A) converter and an amplifier are required. So where it can be used, a hydraulic controller that senses a controlled variable directly is preferred in the industrial environment for its easy maintainability.

ELEMENTS OF HYDRAULIC CONTROLLERS

The major elements include (1) an amplifier, (2) an error detector, and (3) a signal-sensing section. As in all control-loop applications, a measured variable must be sensed and compared with a set point, and the difference or error signal manipulated by the controller to cause some final control element to correct the process. Depending on the particular construction of the hydraulic controller, these functions often can be incorporated physically into the hydraulic controller per se.

The hydraulic relay is common to all hydraulic controllers. The relay allows a small mechanical displacement to control the output from a hydraulic power supply in a fashion that will drive a work load. The jet pipe and spool valve are examples of these relays (Fig. 1). If the process variable is being sensed directly, such as pressure by means of a diaphragm, the signal sensing and error detection can become an integral part of the hydraulic controller. As shown in Fig. 1c, a force is generated from the signal system and compared with a set-point force operated by the spring, and the resulting mechanical displacement (error signal) moves the hydraulic relay (in this case a jet pipe) to control the work cylinder.

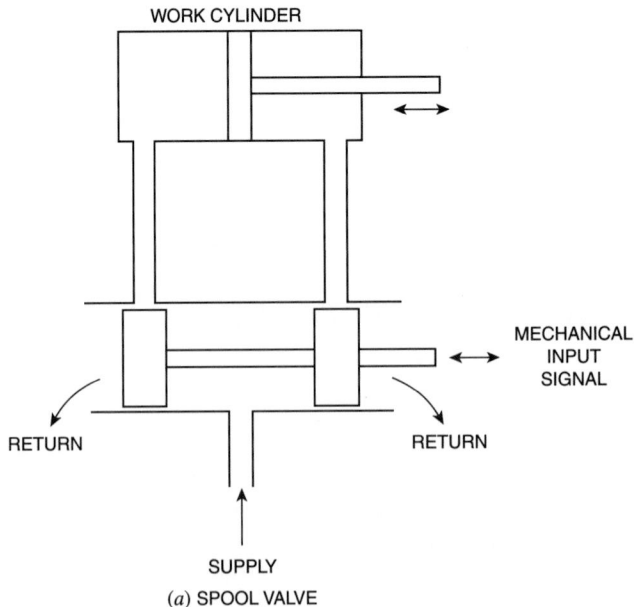

FIGURE 1(a)   Elements of hydraulic controllers.

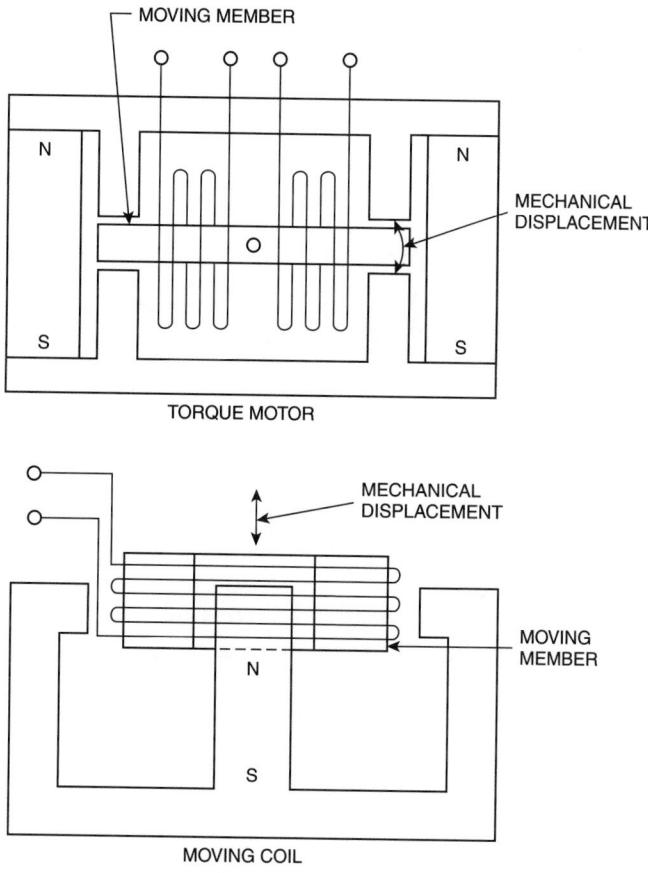

(b) ELECTROMECHANICAL TRANSDUCERS

FIGURE 1(b)   Elements of hydraulic controllers.

HYDRAULIC AMPLIFIERS

The three major traditional types of hydraulic amplifiers are (1) the jet pipe valve, (2) the flapper valve, and (3) the spool valve.

Jet Pipe Valve

By pivoting a jet pipe, as shown in Fig. 1(c), a fluid jet can be directed from one recovery port to another. The fluid energy is converted entirely to a velocity head as it leaves the jet pipe tip and then is reconverted to a pressure head as it is recovered by the recovery ports. The relationship between jet pipe motion and recovery pressure is approximately linear. The proportional operation of the jet pipe makes it very useful in proportional-speed floating systems (integral control), as shown in Fig. 2(a), or as the first stage of a servo or proportional valve.

Position feedback can be provided by rebalancing the jet pipe from the work cylinder, as shown in Fig. 2(b). A portional-plus-reset arrangement is shown in Fig. 2(c). In the latter case, the proportional

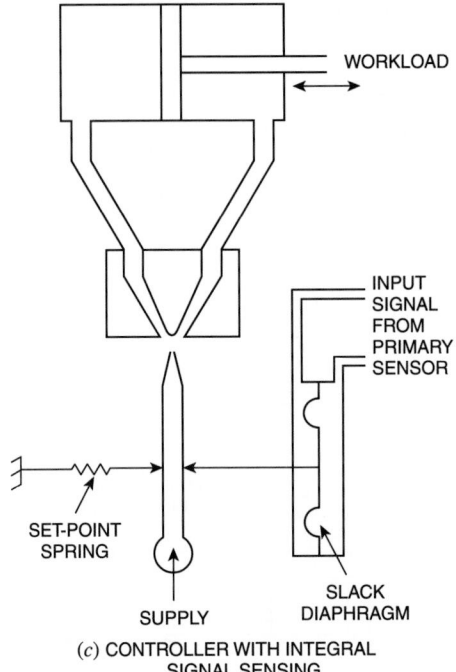

(c) CONTROLLER WITH INTEGRAL
SIGNAL SENSING

FIGURE 1(c)   Elements of hydraulic controllers.

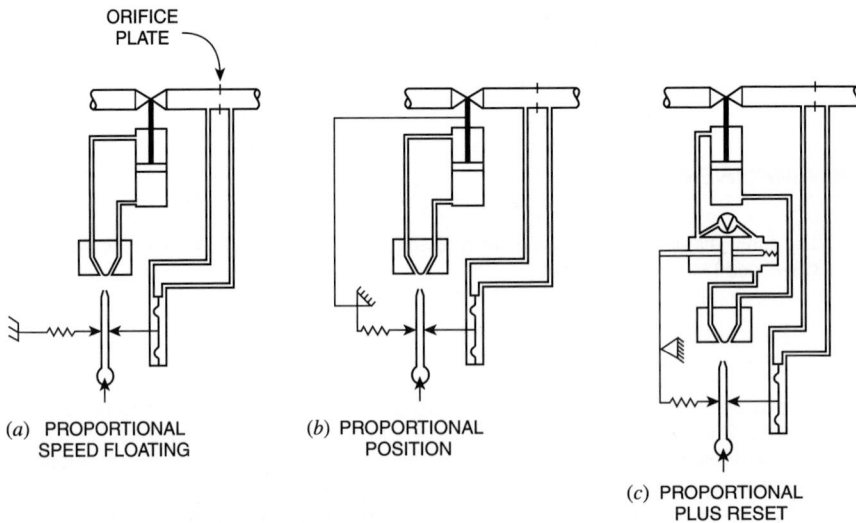

(a) PROPORTIONAL
SPEED FLOATING

(b) PROPORTIONAL
POSITION

(c) PROPORTIONAL
PLUS RESET

FIGURE 2   Hydraulic controllers arranged in different control modes.

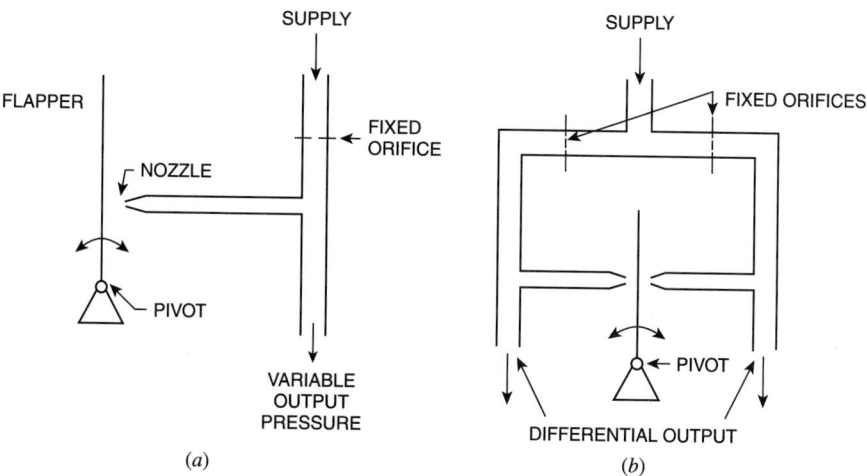

FIGURE 3 Flapper valves. (*a*) Single. (*b*) Double.

feedback is reduced to zero as the oil bleeds through the needle valve. The hydraulic flow obtainable from a jet pipe is a function of the pressure drop across the jet pipe.

Flapper Valve

This device makes use of two orifices in series, one of which is fixed and the other variable. The variable orifice consists of a flapper and a nozzle. The nozzle restriction is changed as the flapper is positioned closer to or farther from the nozzle. When the variable restriction is changed, the pressure drop across the fixed orifice changes, thus producing a variable output pressure with flapper position. Single- and double-flapper arrangements are available as diagramed in Fig. 3.

Spool Valve

This device, when used in automatic control or servo systems, usually is configured as a four-way valve, as previously shown in Fig. 1(*a*). A four-way valve can put full supply pressure on one side of the work cylinder and drain the other side, or vice versa. If the widths of the spool lands are greater than those of their respective ports, the valve is overlapped and is referred to as a closed-center valve. An underlapped valve is referred to as an open-center valve. A line-to-line valve is one in which the spool lands are the same width as the porting. Flow through the spool valve is proportional to the square root of the differential pressure across the valve, as is the case with a jet pipe.

Two-Stage Valves

When higher flows and pressures are required, a two-stage valve may be used. In these cases the second stage usually is a spool valve, while the first stage may be a jet pipe, a spool valve, or a flapper valve. Different combinations of two-stage valves are shown in Fig. 4. In two-stage valves the second

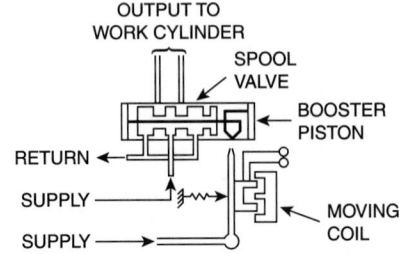

FIRST STAGE: JET PIPE
SECOND STAGE: SPOOL VALVE
FEEDBACK: HYDRAULIC

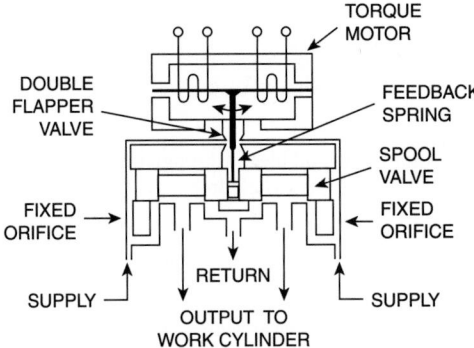

FIRST STAGE: DOUBLE FLAPPER
SECOND STAGE: SPOOL VALVE
FEEDBACK: MECHANICAL

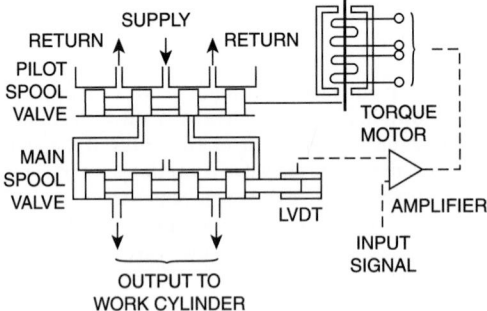

FIRST STAGE: SPOOL VALVE
SECOND STAGE: SPOOL VALVE
FEEDBACK: ELECTRIC

FIGURE 4   Various combinations of two-stage valves.

stage must be either positioned, such as with springs, or provided with a feedback to the first stage as required. Feedback may be mechanical, electrical, or hydraulic.

Electrohydraulic Amplifier

With reference to Fig. 4, frequently the process variable is converted to an electric signal by an appropriate transducer. In these cases generation of the set point and the error detection are performed electrically, and the error signal fed to the hydraulic amplifier is an electric signal. When this is done, an electromechanical transducer is required to generate the mechanical displacement. Two of the more popular schemes are shown in Fig. 4—the moving coil and the torque motor. The principal difference between these two transducers is in the direction of the mechanical displacement. The moving coil produces a linear mechanical displacement, while the torque motor output is rotational. Electromechanical transducers mechanically packaged together with a hydraulic amplifier commonly are referred to as servo or proportional valves.

Servo Valves

Figure 4 also illustrates one of the most widely used servo valve designs, a two-stage valve. One and three stages are also available. The single-stage design is used for the highest response rate; three-stage designs are used for the highest power output.

Typical dimensions and clearances for servo valves are given in Table 1. These figures apply to valves manufactured for aerospace or industrial use. The larger numbers shown apply to larger valves.

Most servo valve failures are due to sticking spools, usually directly traceable to contamination. Reliable servo valves depend on maintaining clean systems.

Proportional Valves

These valves grew out of a need for lower-cost servo valves capable of operating in a dirty environment. In many cases the basic designs are identical. However, there are some designs unique to proportional valves, such as proportional solenoids to drive the spool directly with a linear variable differential transformer (LVDT) feedback in a single stage. General frequency response of proportional valves is 2–10 Hz versus 10–300 Hz for servo valves.

The filtration requirement of servo valves is generally 3 μm, while 10 μm is specified for proportional valves. However, both figures are on the very clean side. A value of 40 μm generally is specified for industrial hydraulics.

TABLE 1   Typical Dimensions and Clearances for Servo Valves

Parameter	Dimension	
	Inches	Millimeters
Air-gap spacing (each gap)	0.010–0.020	0.254–0.508
Maximum armature motion in air gap	0.003–0.006	0.076–0.152
Inlet orifice diameter	0.005–0.015	0.127–0.381
Nozzle diameter	0.010–0.025	0.254–0.635
Nozzle-flapper maximum opening	0.002–0.006	0.051–0.152
Drain orifice diameter	0.010–0.022	0.254–0.559
Spool stroke	±0.010–±0.060	±0.254–±1.524
Spool or bushing radial clearance	0.0001–0.00015	0.0025–0.0038

In servo valves the spool and the sleeve or body are matched for very close fit and zero overlap. Spools and bodies are interchangeable in proportional valves, thus there is a significant cost differential. However, overlap results in dead zone, which can be a problem.

HYDRAULIC FLUIDS

Petroleum oil is still the optimum hydraulic fluid unless fire resistance is required. With additives, petroleum oil does not have major shortcomings of the magnitude of those found in all fire-resistant fluids. In designing a hydraulic control system, the choice of fluid must be made early because a number of features of the system must be tailored to the fluid used.

Although there are numerous petroleum-based hydraulic oils designed specifically for various types of systems, automatic transmission fluid has been found to be a good engineering compromise for many systems because (1) it is available worldwide to the same specifications; (2) it contains viscosity index improvers which reduce the effects of viscosity change with temperature (like 10W-30 motor oil); (3) like other quality hydraulic oils, it has rust, oxidation, and foam inhibitors; and (4) it contains an antiwear zinc compound additive, although not as much as is contained in extreme pressure (EP) hydraulic oils.

Synthetic hydrocarbon oils have all the advantages of natural oils, plus complete freedom from wax content. This permits synthetic oils to remain fluid and thus be pumpable down to −65°F (−54°C). Although the cost of synthetic oils is relatively high, their use makes it possible to install hydraulic controls in adverse outside environments.

With increasing environmental awareness, the new vegetable-oil-based hydraulic fluids must be given consideration for many applications. These oils are advertised as readily biodegradable and nontoxic. Tests indicate better lubricity (less pump wear) and a higher viscosity index (less viscosity change with temperature) as compared with good petroleum hydraulic fluids. Also, the vegetable oils have shown performance characteristics comparable to those of petroleum oil at normal system operating temperatures. The same seal materials and metals can be used for both types of oils. On the negative side, the newer vegetable-based oils cost roughly 2½ times more than the conventional petroleum oils and are sensitive to low temperature.

Fire-Resistant Fluids

The principal available fire-resistant fluids are described briefly in Table 2. Their characteristics are compared in Table 3 with those of petroleum oils.

TABLE 2 Representative Types of Fire-Resistant Hydraulic Fluids[a]

Letter Designation	Fluid Description
HF-A	High-water-content emulsions and solutions are composed of formulations containing high percentages or water, typically greater than 80%. They include oil-in-water emulsions and solutions which are blends of selected additives in water.
HF-B	Water-in-oil emulsion fluids consist of petroleum oil, water emulsifiers, and selected additives.
HF-C	Water-glycol fluids are solutions of water, glycols, thickeners, and additives to provide viscosity and other properties desirable for satisfactory use in hydraulic systems.
HF-D	Synthetic fluids are nonwater fluids, such as phosphate esters or polyol esters.

[a] Extracted from ANSI B93.5M.

TABLE 3 Comparison of Hydraulic Fluids

Characteristic	Petroleum Oil	HF-A, High Water Content, Low Viscosity	HF-B, Water Emulsion	HF-C, Water Glycol	HF-D Synthetic Phosphate Ester	Polyol Ester
Fire resistance	Poor	Excellent	Fair	Very good	Very good	Very good
Cost	1	0.1	1	2	5	5
Lubricity	Excellent	Fair	Good	Good	Excellent	Very good
High temperature	Excellent	Fair	Fair	Fair	Very good	Very good
Low temperature	Very good	Poor	Poor	Good	Fair	Good
Corrosion protection	Excellent	Good	Good	Good	Very good	Good
Standard hardware compatibility	Excellent	Fair	Fair	Fair	Good	Very good
Toxicity	Good	Very good	Good	Excellent	Fair	Very good

POWER CONSIDERATIONS

Hydraulic controllers are generally applied where either high-power or high-frequency response is required. The two requirements may or may not be needed for the same application. Frequency-response requirements are a function of the control loop dynamics and can be determined by appropriate automatic control theory techniques discussed elsewhere in this handbook. Power considerations will be discussed here.

Because hydraulic fluids can generally be considered incompressible, they can be handled under higher-pressure conditions than would be considered safe or practical with pneumatics. The amount of power delivered to the load is a function of the hydraulic power supply size, the pressure drop required by the hydraulic controller, and the pressure drop losses in the lines. With reference to Fig. 5 and neglecting any line losses, it can be said that the pressure drop available for the load is equal to the supply pressure less the hydraulic controller pressure drop. In equation form,

$$\Delta P_L = P_0 - \Delta P_c$$

If the oil flow under these conditions is Q, then

$$\text{Power available at load} = \Delta P_L Q$$

Q_{max} is determined by the hydraulic power supply capacity. As Q increases, the term $P_L Q$ tends to increase. However, increasing Q increases ΔP_c, thus reducing ΔP_L. It can be shown that for most hydraulic controllers the maximum $\Delta P_L Q$ occurs when $\Delta P_L = 2\Delta P_c$.

Figure 6 shows the force and power ranges normally encountered in hydraulics contrasted with the ranges commonly used with pneumatics. The load horsepower scales were calculated by selecting two representative load stroke lengths and three different full-stroke speeds, all corresponding to the force scale. The hydraulic power supply horsepower requirement is greater, however, since it must in addition supply the power for the hydraulic controller, line losses, and pump inefficiency. This can be

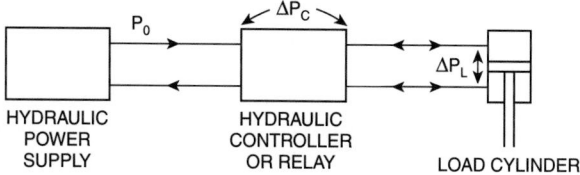

FIGURE 5 Hydraulic flow loop.

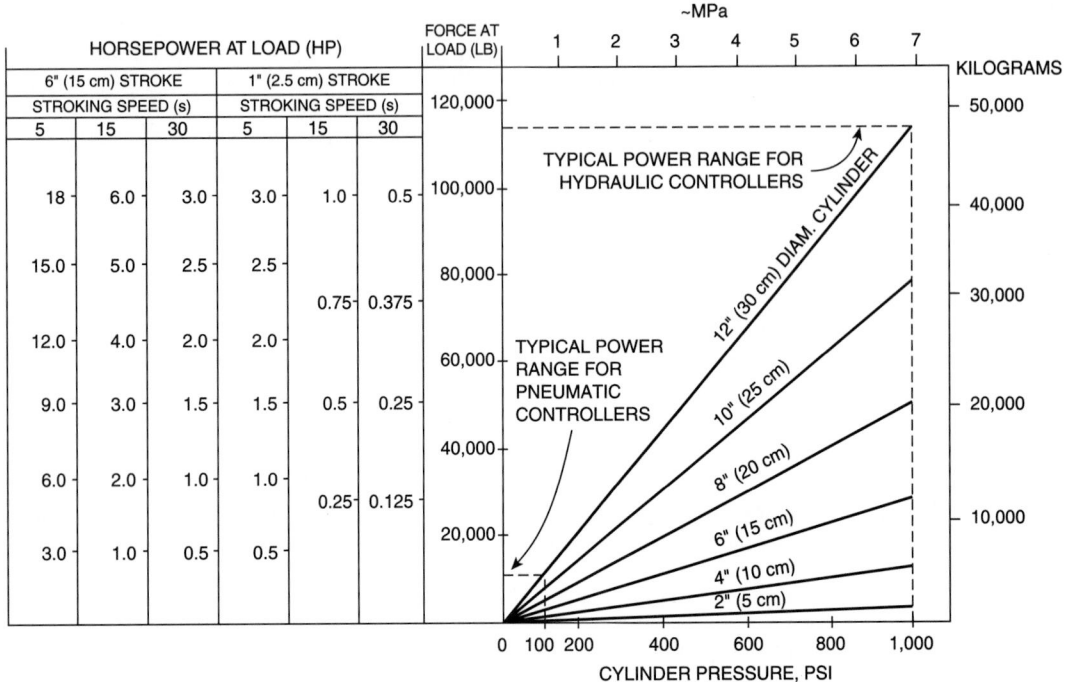

FIGURE 6   Force and power ranges normally encountered in hydraulic systems contrasted with ranges commonly used in pneumatic systems.

expressed as

$$HP_{ps} = \frac{HP_L + HP_c + HP_{LL}}{\text{pump efficiency}}$$

The power lost across the hydraulic controller is a function of the controller capacity. This is generally expressed in terms of a flow–pressure-drop curve.

Conservation

Increasing energy costs have resulted in greater use of systems that load the pump motor only when there is a demand for hydraulic control power. Methods include (1) open-center control valves that bypass flow at low pressure, and (2) pressure-compensated variable-volume pumps that pump only a volume sufficient to maintain pressure in the system.

APPLICATIONS

Generally hydraulic control systems are selected based on more demanding control requirements rather than on lower initial costs. However, as power requirements increase, even the initial cost may favor hydraulics. Three typical hydraulic controller applications are shown in Figs. 7 to 9. The approximate power and performance requirements versus applications of servo valve control systems are illustrated in Fig. 10.

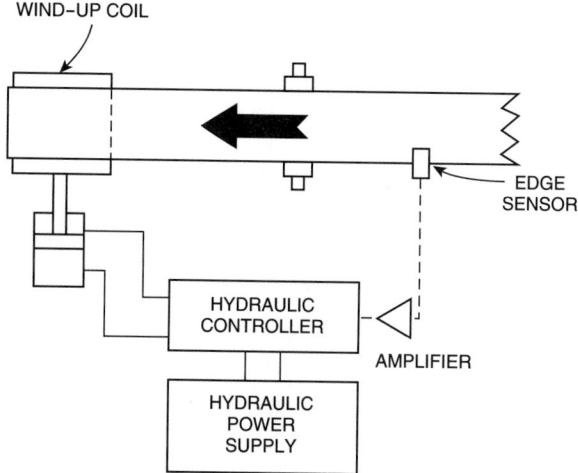

FIGURE 7 Hydraulic edge-guide control system. The windup coil is shifted in accordance with an edge sensor signal to provide an even coil.

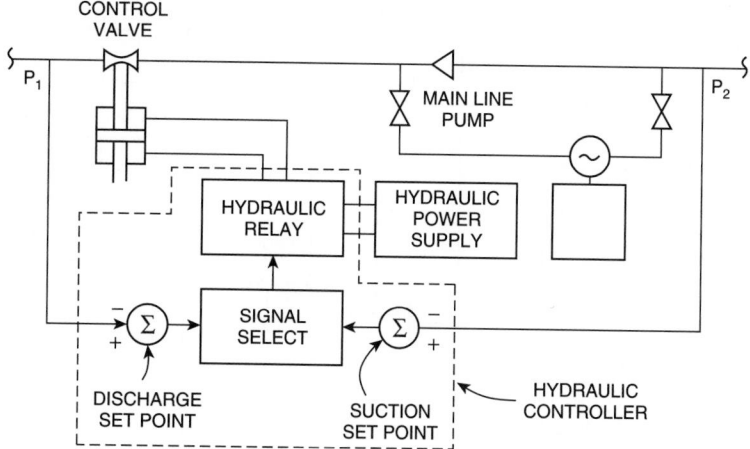

FIGURE 8 Hydraulic pipeline control system. If discharge pressure P_1 exceeds set point, or if suction pressure P_2 goes below set point, the control valve is throttled closed to correct the situation.

Size-Selection Procedures

Size selection for servo or proportional valves proceeds as follows:

1. Size the actuator area to produce a stall force 30% greater than the total force required at the supply pressure available:

$$A = \frac{1.3F_r}{P_s}$$

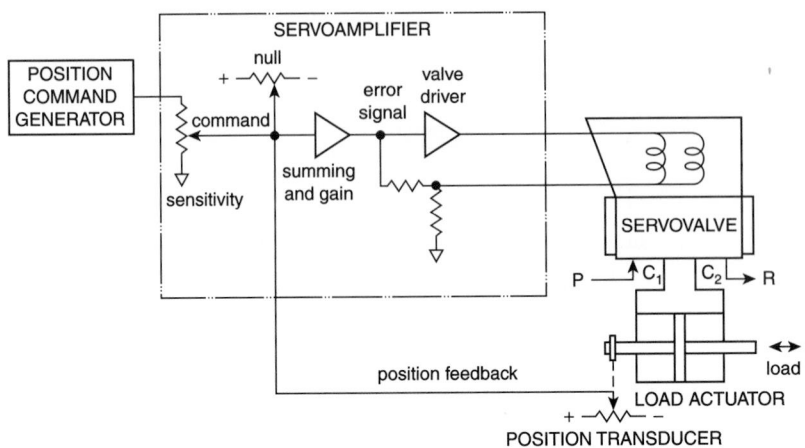

FIGURE 9 Hydraulic position control system, as used in machine tools.

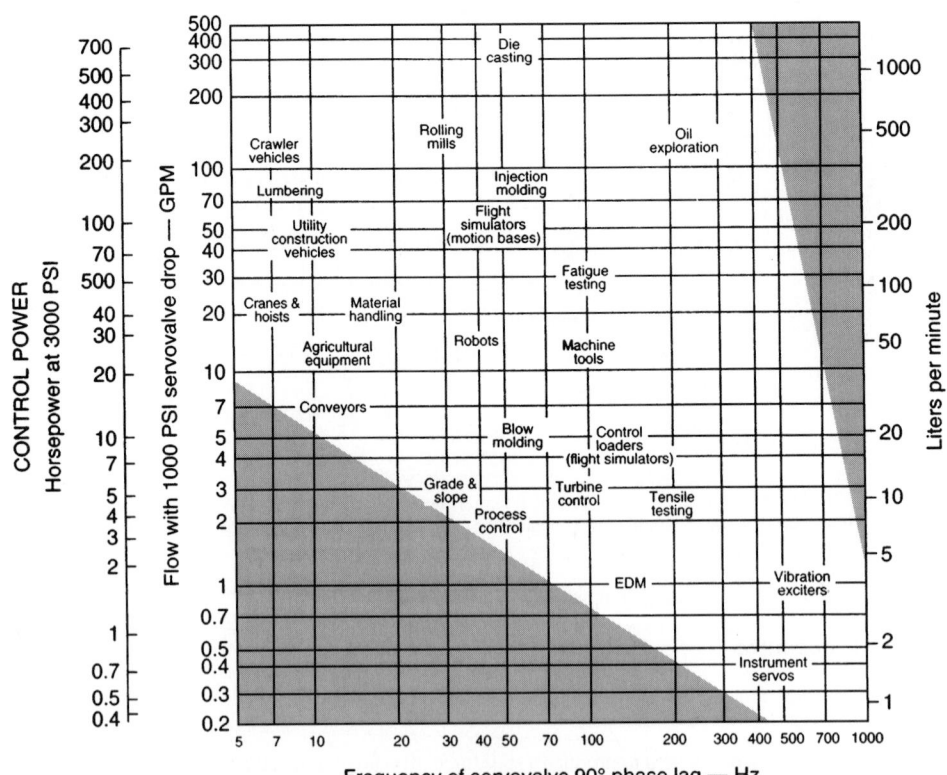

FIGURE 10 Spectrum of industrial applications for hydraulic servo valves. *Note:* 1000 psi = 6900 kPa.

where A = actuator area [in.^{2} (cm^2)]
F_r = total force required to move load [lb (kg)]
P_s = supply pressure [psi (kPa)]

2. From the maximum loaded velocity and the actuator force required at this velocity, determine the loaded flow:

$$Q_L = AV_L$$

where Q_L = loaded flow [in.^{2}/s (L/s)]
V_L = maximum loaded velocity [in./s (cm/s)]

3. Calculate no-load flow:

$$Q_{NL} = Q_L \sqrt{\frac{P_s}{P_s - P_L}}$$

where Q_{NL} = no-load flow [in.^{3}/s (L/s)]

4. Compute the rated flow required at the manufacturer's rated pressure drop. Increase 10% for margin.

$$Q_R = 1.1 \frac{Q_{NL}}{C} \sqrt{\frac{P_R}{P_s}}$$

where 1.1 = margin
Q_R = rated flow on manufacturer's data sheets
P_R = pressure drop at which manufacturer rates flow [generally 1000 psi (6900 kPa)
for servo valves; 150 psi (1000 kPa) for proportional valves]
C = conversion factor [in.^{3}/s (3.85 gal/min); L/s (60 L/min)]

Relative Advantages and Limitations

Advantages of hydraulic controllers include the following:

1. High speed of response. The liquid control medium, being effectively incompressible, makes it possible for a load, such as a work cylinder, to respond very quickly to output changes from the hydraulic controller.

2. High power gain. Since liquids can be readily converted to high pressures or flows through the use of various types of pumps, hydraulic controllers can be built to pilot this high-energy fluid.

3. Simplicity of the final actuator. Most hydraulic controller outputs are two hydraulic lines that can be connected directly to a straight-type cylinder to provide a linear mechanical output.

4. Long life. The self-lubricating properties of most hydraulic controls are conducive to a long, useful life.

5. Relatively easy maintainability.

Some limitations include the following:

1. Maintenance of the hydraulic fluids. Depending on the type of hydraulic hardware used, filtration is required to keep the fluid clean; the use of fire-resistant fluids, because of poorer lubrication and possible corrosive effects, requires more careful maintenance.

2. Leakage. Care must be taken with seals and connections to prevent leakage of the hydraulic fluid.

3. Power supply. New hydraulic power supplies usually must accompany new hydraulic controller installations. This is contrasted with pneumatic and electrical or electronic systems where such power normally is available.

REFERENCES

1. Henke, R., "Proportional Hydraulic Valves Offer Power, Flexibility," *Control Eng.*, vol. 28, no. 4, p. 68, 1981.

2. Maskrey, R. H., and W. J. Thayer, "A Brief History of Electrohydraulic Servomechanisms," *ASME J. Dyn. Syst. Meas. Control*, June 1978.

3. Niemas, F. J., Jr., "Understanding Servo Valves and Using Them Properly," *Hydraul. Pneum.*, vol. 31, no. 13, p. 152, October 1977.

4. Staff, "High Water Content Systems for Profit-Making Designs," *Hydraul. Pneum.*, vol. 35, no. 4, pp. HP1–HP18, April 1982.

5. Totten, G. E., "Novel Thickened Water-Glycol Hydraulic Fluids for Use at High Pressure," Paper No. 192-23.1, Proceedings of the International Fluid Power Applications Conference, March 1992.

6. Cheng, V. M., A. A. Wessol, and C. Wilks, "Environmentally Aware Hydraulic Oil," Paper No. 192-23.3, Proceedings of the International Fluid Power Applications Conference, March 1992.

7. McCloy, D., and H. R. Martin, *Control of Fluid Power, Analysis & Design,* Halsted Press, 1980.

8. Merritt, H. E., *Hydraulic Control Systems*, J. Wiley, 1967.

9. Holzbock, W. E., P. E, *Robotic Technology: Principles and Practice*, Van Nostrand Reinhold Co. Inc., 1986.

10. Parker Hannifin Corp., "Industrial Hydraulic Technology," Bulletin No. 0232-B1.

11. Moog Controls Inc., "Servovalve Selection Guide."

BATCH PROCESS CONTROL

During the last few years much attention has been directed toward a better understanding of the dynamics of batch processes in an effort to achieve greater automation by applying advanced control knowledge from experience with continuous process controls and computers. This has proved to be more difficult and to require more time than had been contemplated originally. This study is far from complete, as witnessed by the numerous papers and articles in contemporary literature. However, the Instrument Society of America (ISA) issued a batch control standard SP88 (see Section II for an overview).

BATCH MODE

Prior to the late 1930s practically all chemical and petroleum production processes essentially were of the batch mode. When crude oil, for example, was first cracked into lighter hydrocarbon fractions, thermal cracking was of the batch mode. Where inherently suited (chemically or physically) and where a large and continuous market demand for a single product existed, manufacturers found that continuous production warranted the costs of scaling up equipment and integrating the flow to and from various unit operations in a continuous production system. Because these prerequisites do not always exist, or the return on investment for going "continuous" may be insufficient, the batch mode continues to be quite extensive. In addition, the batch mode has certain innate values, including multiple use of equipment and greater safety with hazardous products because of small runs.

Because of tremendous variations in raw materials, final products, and operational equipment used, it is difficult to characterize a "typical" batch process. The opinions of a few experts are along these lines:

1. Every process is a batch process; the difference is how long the process runs between startup and shutdown. (While this may be true theoretically, it is indeed an oversimplification.)

2. A batch process is discontinuous, in which ingredients are sequentially prepared, mixed, reacted, cooked, heated, or cooled, finished, and packed—all in accordance with a time-sequenced schedule worked out by a process engineer.

3. A batch process seldom is purely batch; a continuous process seldom is purely continuous. Most batch processes require continuous control over one or more variables throughout the entire schedule, and thus the overall batch control system must be a form of hybrid. Many products made continuously will, at the end of the line, require batch packaging, inspection, handling, storing, and shipping.

Although precise definitions are difficult, some generalities may be stated. A batch process usually is completed within a matter of minutes to a few hours, as contrasted with a few seconds for what may be termed a discrete operation or weeks or months for a continuous process from start-up to shutdown. However, in the latter case, for any given molecule of substance entering the process, the interval of time within the process (residence time or throughput rate) before it exits as part of a finished product usually can be measured in seconds to very few minutes.

Simplistic Model.   For making soup prior to canning, the steps may be as follows:

1. Fill a kettle with a given volume of water.
2. Add ingredients A, B, and C.
3. Turn on the mixer and raise the kettle temperature to a near-boil and maintain for 1 hour.
4. Allow kettle and ingredients to cool (*a*) for a programmed length of time or (*b*) until a given temperature is reached—whichever criterion is shorter.
5. Add ingredient D.
6. Stir for 10 minutes.
7. Open dump valve to transfer contents.
8. Close dump valve (*a*) after programmed 3 minutes or (*b*) when kettle level indicates zero— whichever criterion is shorter.
9. Start rinse cycle, turning on water spray and actuating stirrer for 4 minutes. (A second rinse might be scheduled.)
10. Close dump valve.
11. Indicate "Ready for Next Batch" on display.

If self-diagnostics for the system were incorporated, all conditions would be checked for readiness for the next batch. In the foregoing, note the combined need for continuous and discrete control.

This description minimizes the number of actions that usually must be taken and the time-sequencing required for more complex processes, which may involve many more ingredients, intricate time relationships, and measurement of several additional variables that require continuous or discontinuous control, safety interlocks, and, in modern systems, the tabulation of historical and quality control parameters.

Packaged or customized batch control systems are available today that will handle up to hundreds of steps for a single recipe and memory for storing up to several thousand recipes that may use the same group of batching operations. On an average, however, a control system will need to accommodate only a few hundred recipes. Table 1 lists some characteristics of batch operations versus discrete and continuous operations.

BATCHING NOMENCLATURE

Recipe.   A list of ingredients, amounts, or proportions of each ingredient. Depending on the product, the recipe may be created by a chemist or other specialist as the result of formulating the desired product in a laboratory (possibly pilot plant) environment. The chemist specifies the types of equipment that will be used. Each type of equipment is referred to as a unit. A multipurpose, multiproduct batching system will have the capability of producing numerous products and using several pieces of equipment. Also, from historical records of equipment (units) available and of other recipes in storage, in modern procedures the chemist will configure the batch sequence interactively using a sequential function chart (SFC) editor to select the necessary unit sequence from among a group, while simultaneously entering parameters for that sequence in accordance with prompt messages. As pointed out by Adams [1], new recipes can be prepared easily by repeating this series of operations.

Grade.   A variation in a recipe or formula, usually achieved through the use of different product data parameters such as temperature, times, and amounts. This may constitute a new recipe.

Unit.   A vessel and its associated process equipment that acts independently of other process units.

Operation.   A time- and event-based sequence that defines the actions of a process unit in performing a process function. A typical operation may be the sequence to charge, heat, or dump a given unit.

Procedure.   A sequence of unit operations with specific product data that constitutes the batch cycle; also commonly referred to as a recipe or formula.

Discrete Control Device.   An on-off device, such as a solenoid valve, pump, or motor.

Regulatory Control.   A general term encompassing PID (proportional plus integral plus derivative) loops and continuous control functionality.

Sequenced Batch.   The most basic type of batching control. This can become quite complex because of the large number of operations in some cases. It may be considered as logic planning and incorporates all on–off functions and control of other discrete devices. In the pure form, sequence control involves no process feedback or regulatory control. It is quite amenable to ladder-logic-based programmable logic controllers (PLCs).

Programmed Batch.   Elements of sequenced batch plus the addition of some regulatory control. This system requires little operator intervention. Sophisticated regulatory controls (beyond PID) seldom are needed. The system is used frequently where a reactor or other batching equipment assembly is

TABLE 1 Process Types versus Process Characteristics*

Process type	Duration of process	Size of lot or run	Labor content	Process efficiency	Control type preference	Input-output content	
						Discrete	Analog
Discrete operations	Seconds	†	Medium to high	Low to medium	PLC	95%	5%
Batch operations	Minutes to hours	Small to medium	Medium	Medium	Various‡	60%	40%
Continuous operations	Weeks to months	Usually very large	Small	High	DCS	5%	95%

* Adapted from a chart originally presented by W. J. Loner, Allen-Bradley Co., Inc.

† Discrete operations to a large extent are encountered mainly in the discrete-piece manufacturing industries for the control of machines, conveyors, and so on. However, there are numerous discrete operations in process batching systems.

‡ In 1991 a survey indicated that programmable logic controllers (PLCs) were preferred by nearly 30% of users; approximately 45% preferred distributed control systems (DCSs); the remainder were various hybrid systems, some involving personal computers (PCs).

dedicated to a single product (or very few products) with limited process variations. The system also is amenable to PLCs in most cases. However, if there are several continuous control operations in proportion to the number of discrete operations, a hybrid, host, or distributed batch control system may be indicated.

High-Level Batch.   Elements of programmed batching with external and process feedback. Optimization routines may be used as, for example, in the production of polyvinylchloride (PVC) and other polymeric emulsions.

DEFINING AND OUTLINING THE BATCH PROCESS

Since the late 1980s considerable software has become available for assisting in creating batch control programs. ISA standard SP88 is geared toward a high degree of advanced control and will be described briefly later in this article. For those readers who may not be fully versed in these more advanced concepts, it may be in order to include the following "plain language" description.[1]

Implementing a batch control system can be very complex and difficult without the proper definition. It is very important to determine the operation of individual process units and the associated equipment that makes the unit work as an entity. Careful consideration must also be given to a unit's interaction with other units and all common resources that may be used concurrently by multiple units. Contention for common resources by multiple units, without the proper mechanism for ensuring their use, can lead to costly coordination problems when operating the process automatically from a procedure level.

The definition phase of a batch control process can be difficult. Ensuring the proper equipment states for any possible combination of process conditions generally requires multiple iterations until a final control strategy is reached. This usually continues well into the process start-up and must be updated periodically during plant operation, as process optimization takes place and the overall strategies change. Therefore a structured approach to defining and implementing batch control strategies is offered in the batch formalism (Fig. 1).

The batch formalism described here lays out a set of guidelines by which a process designer may follow in the route to the final batch control strategy. Each step of the formalism will build on the previous step, forcing the designer to complete each step before proceeding to the next.

Step 1.   The starting point is that of defining the control elements from the piping and instrumentation drawings (P&IDs). Using the P&ID, an equipment tag list can be constructed, including various regulatory and discrete devices.

Step 2.   Following the identification of the control elements, the individual strategies must be defined. Regulatory control strategies may be specified, using ISA symbology, or, for complex interactive loops, the Scientific Apparatus Makers Association (SAMA) standard RC22-11 may be useful. Due to the nature of batch control processes, attention should be given to loop mode changes, alarm status changes, and adaptive tuning requirements.

Traditionally the interlocking and sequencing of discrete control devices has been centered around ladder diagrams or in boolean logic form. These forms of diagraming, however accustomed designers are to them, are hard to follow and quite cumbersome when dealing with complex batch strategies. Time-sequence diagrams provide for a more organized time and event sequence and are considerably better for diagraming batch control processes.

While excellent for circuit wiring and fixed-sequence strategies, ladder logic does not indicate the time relationship required with most batch strategies. Time-sequence diagraming is essential in batch control due to the variable sequences, recipe changes, and common resource contentions, as well as the high degree of interaction between discrete and regulatory controls. This type of diagram is typified by Fig. 2. The symbology is shown later in Figs. 6 and 7.

---

[1] From information furnished by W. C. Thompson, Fisher Controls International, Inc., Austin, Texas.

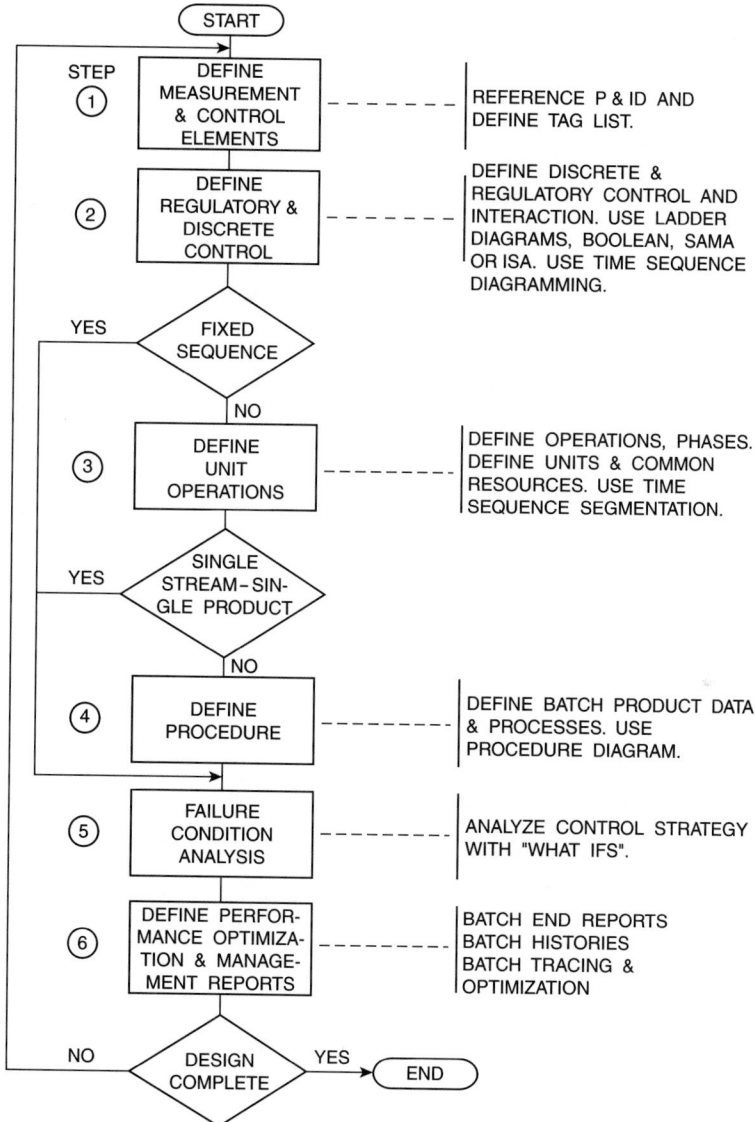

FIGURE 1 Batch formalism.

Additional considerations in Step 2 of the batch formalism are more complex strategies, such as split ranging, weighing and metering, and changing control strategies depending on the transient nature of the process.

If at this point in the procedure it is determined that the process sequence is fixed, the designer may proceed to Step 5, where the failure sequences based on analysis of the process conditions on control element failure may be defined. However, if there is more than one time sequence for a process unit and the order of those different time sequences may change from batch to batch, the designer should consider unit operations.

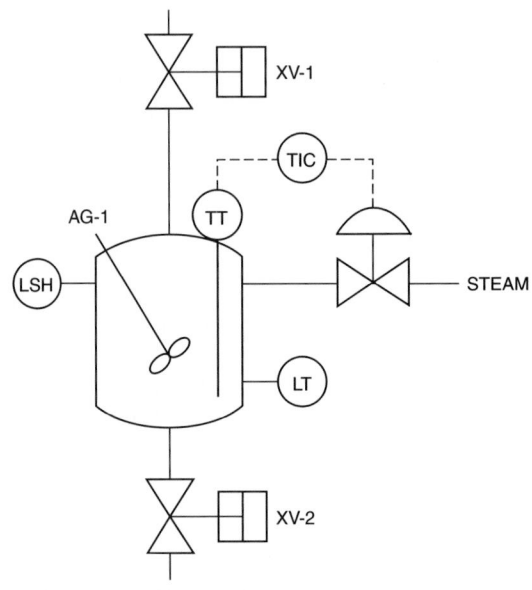

REACTOR AND ASSOCIATED INSTRUMENTATION

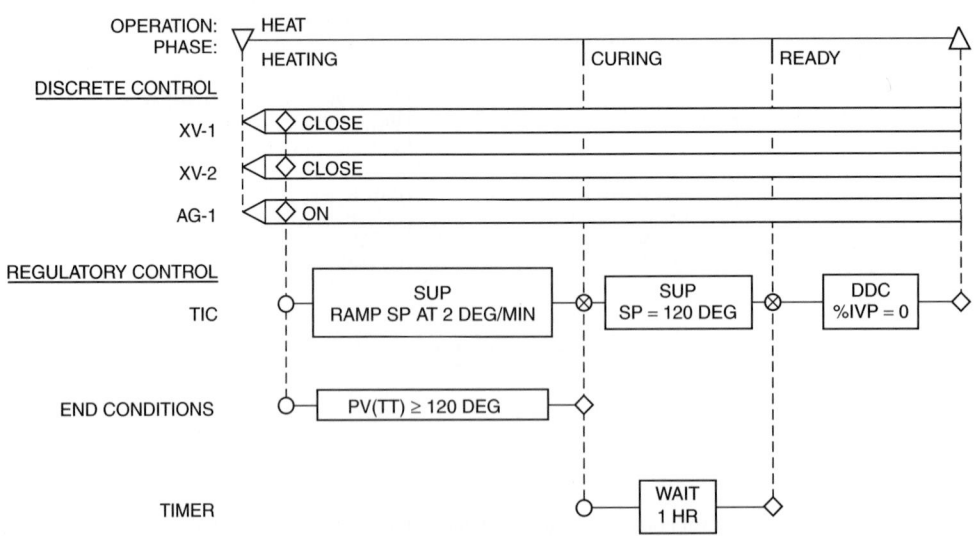

TIME SEQUENCE DIAGRAM

FIGURE 2   Representative reactor unit and associate time-sequence diagram.

Step 3.   Individual sets of time and event sequences are called operations. The actions within an operation include sequencing, interlocking, profiling, failure monitoring with emergency shutdown, calculations, integrators and timers, and parallel operations in addition to the discrete and regulatory functions. In addition the operation provides for manual operator entry and convenient entry and reentry points, all for coordinated control of the total unit.

In some operations, such as material transfers, unit equipment boundaries may overlap. In such cases, transfer headers may actually belong to different units at different times. This type of equipment is referred to as a common resource and must be coordinated between operations.

Figure 3 shows fill, heat, and dump operations for the unit shown in Fig. 2. It should be noted that the operations are subdivided into phases for satisfying operator interface and branching requirements.

Once the definition of the individual time sequences has been completed, the designer should define the operation or phase relationship and segment the process equipment by unit. Since this definition phase is highly application-dependent, defining the units and operation or phases is more easily done after the time-sequence diagrams are completed. Note that operations and phases are generally triggered at safe process conditions. When the designer is satisfied that the time sequences and units or operations are complete, the procedures should be defined.

Step 4: Procedure Definition.   As stated earlier, the procedure consists of a sequence of operations and sets of product parameters, because most batch processes are characterized by a series of small batches flowing sequentially through multiple units. Each batch follows some path, using some or all of the units or operations along that path. By treating each operation as a single control action in the procedure, a time-sequence-type diagram can be designed for the procedure (Fig. 4).

As with the operation or phase relationship, a procedure or process relationship is also established providing for safe hold points and better procedure definition for operator interface.

Just as common resources are contended for within operations, procedures may contend for the services of each unit in a multistream situation. The unit used in making the batch must be kept track of, so that cross contamination and improper mixes do not take place.

During procedure definition the designer should list all product parameters (values with engineering units), such as times, temperatures, ingredient amounts, and flow rates that are unique to the batch products and their various grades.

Step 5: Defining Failure Conditions.   After the first four steps of the formalism are complete, the designer should pass through the hierarchy once again to establish an analysis of the failure conditions of the process. Time sequences should then be defined for the application-dependent failure steps. These failure sequences should be incorporated by operation and should allow for reentry into the operation at a position that does not jeopardize personnel, equipment, or product.

Step 6: Optimization and Reports.   Finally, once all phases of the definition are complete, optimization of the performance of equipment and data measurement should be done and the determination of end report and batch history data completed. The format of batch end data should also be considered for printed reports.

The batch strategy is now complete and ready for implementation. Note that multiple iterations of this procedure may be necessary before implementation can be started.

SYSTEM SELECTION FACTORS

The actual implementation of the completed batch strategy is both application and control system dependent. The application and user preference usually dictate the architecture and selection of the control equipment. There are several different control system architectures available. One of the most popular utilizes distributed devices on local area networks, or highways. As stated previously, batch control packages are still available commercially. However, redundancy becomes an expensive issue compared with the selective redundancy of a distributed system. In addition, the total installed cost of centralized computers makes them less cost-effective due to wiring costs and the programming

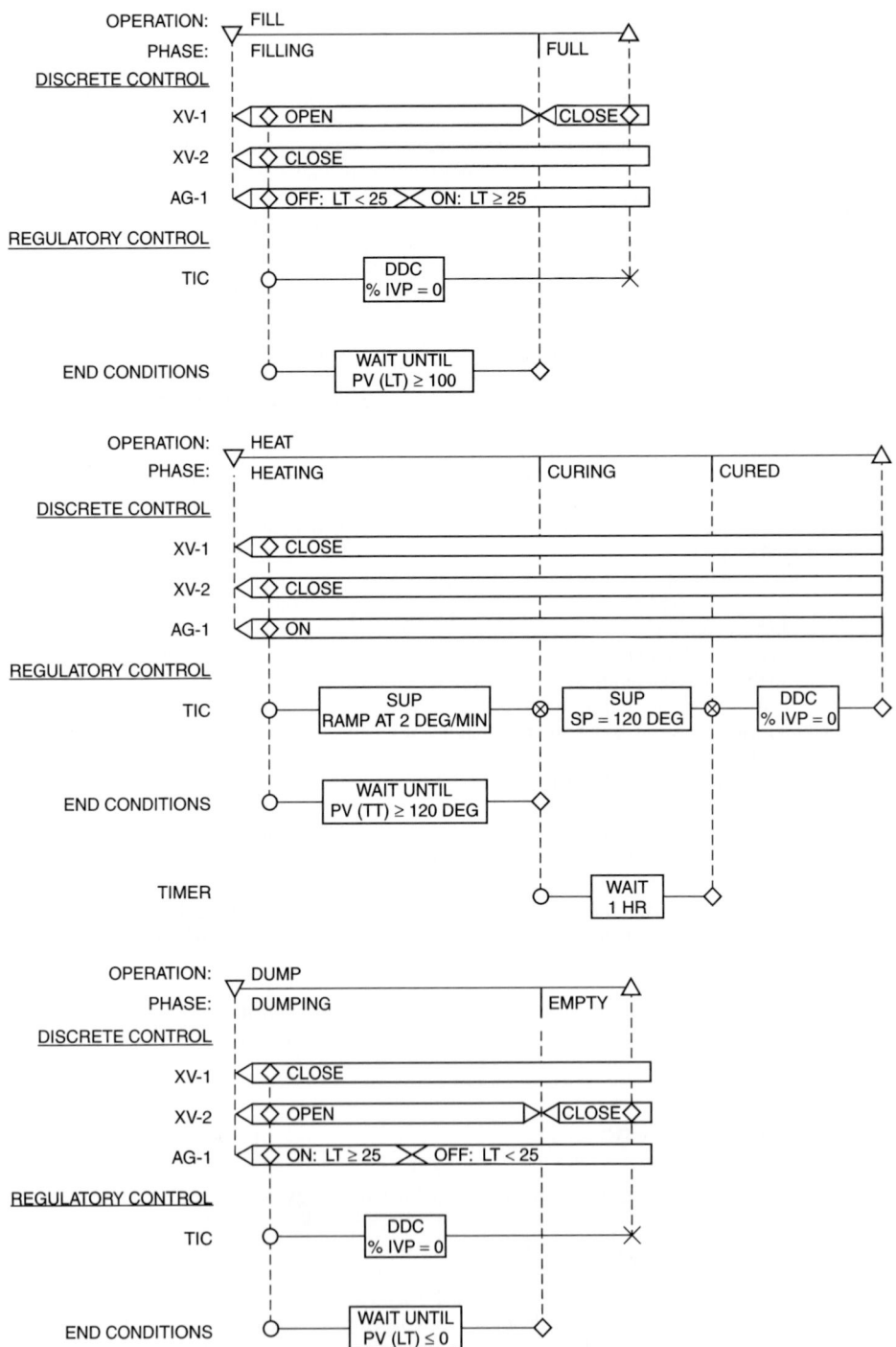

FIGURE 3   Time-sequence diagrams segmented into unit operations for unit of Fig. 2.

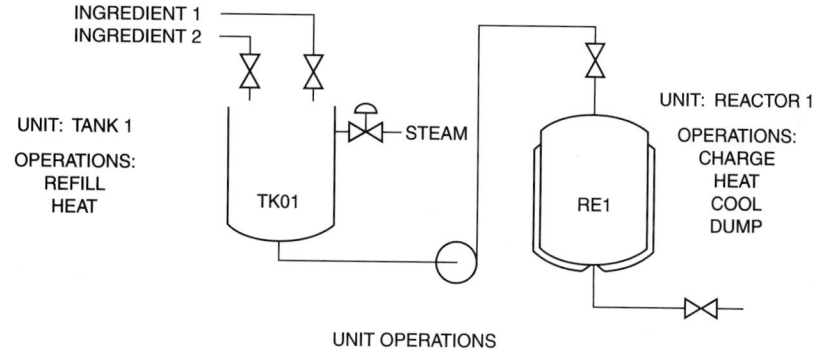

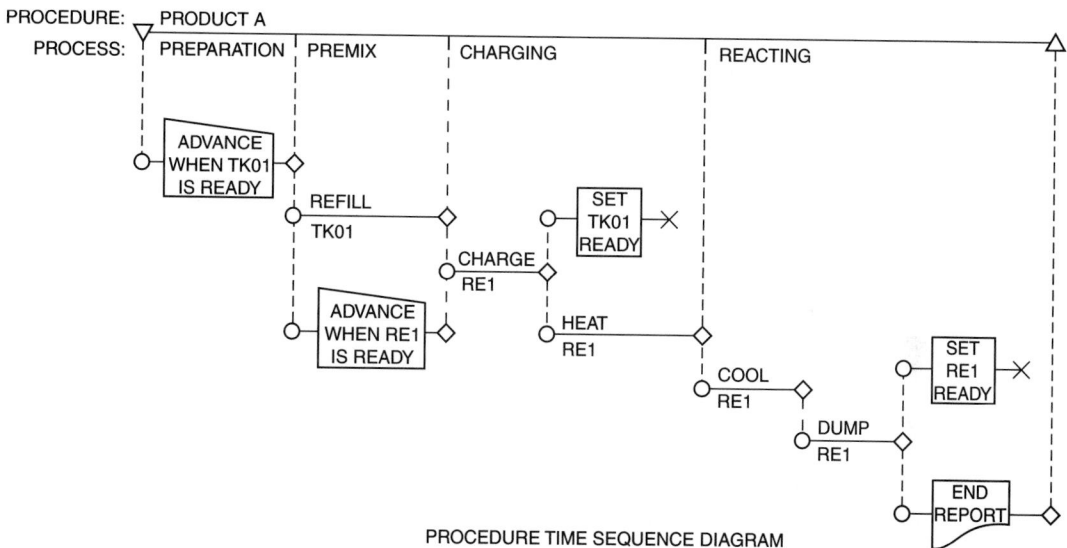

FIGURE 4 Time-sequence diagram for batch procedure.

efforts required with most centralized systems. Truly distributed control systems with advanced batch control software have been commercially available since the early 1980s. They incorporate sophisticated batch languages with hierarchical control that allow for easier implementation of unit operations, failure sequences, recipe level control, and effective operator interface. Depending on the process requirements, PCs, continuous controllers, and process management computers can easily be integrated with the batch (unit operations) controllers and consoles (for operator interface) to provide fully integrated plantwide control systems (Fig. 5). Subject to further confirmation, the symbology is shown in Figs. 6 and 7.

The system must meet certain minimum requirements.

1. The controllers must be able to handle the designer's control strategies. It is imperative in advanced batch control systems that the controller handle multiple unit operations for a number of units. Techniques such as tag aliasing (generating generic process tag names for use in unit operations) can greatly reduce configuration or programming time of the controllers.

Network traffic director
Continuous controllers
Unit operations controller
PLC interface
Unit operations controllers
Local traffic director 1
Operator consoles
Trend devices
Process computers
To other local highways and network devices

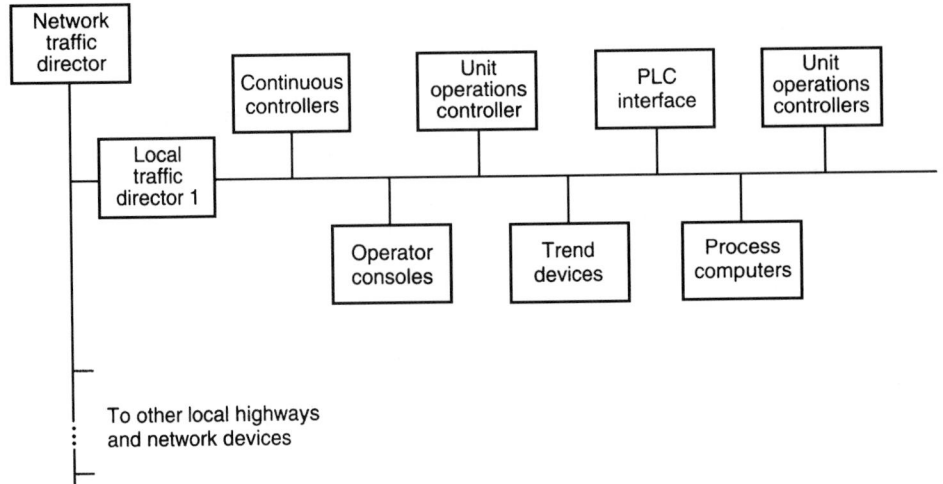

FIGURE 5 Generalized depiction of representative distributed control system (DCS) utilizing local area networks (LANs).

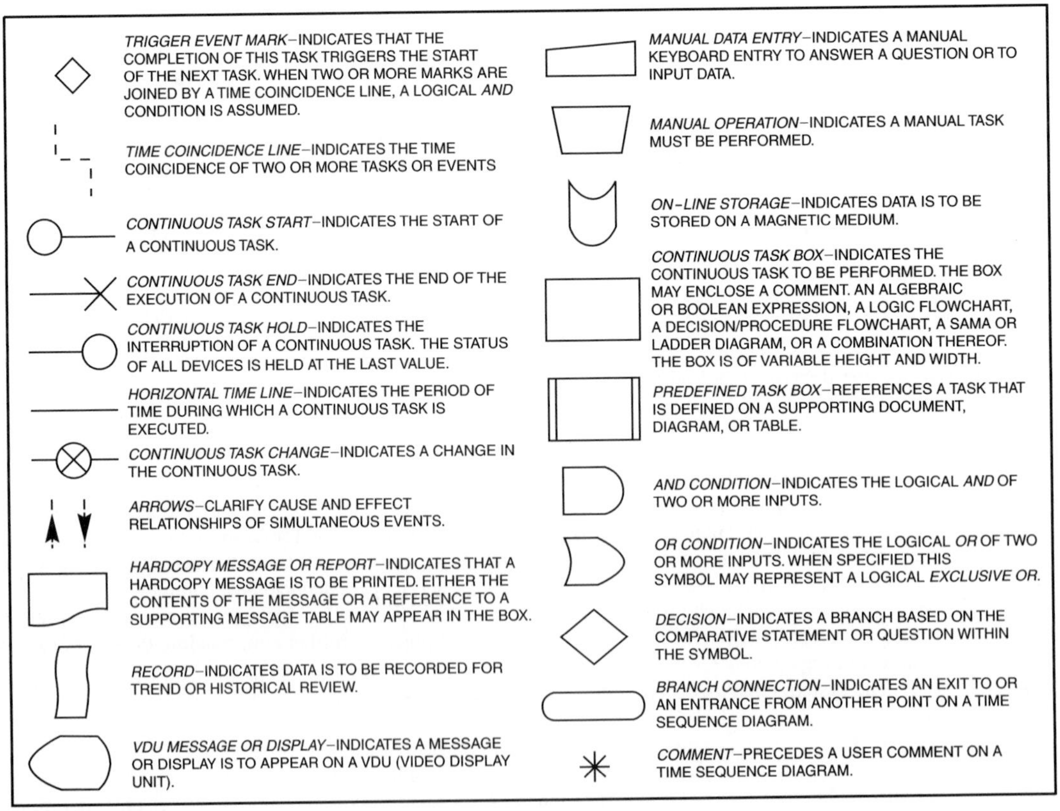

TRIGGER EVENT MARK–INDICATES THAT THE COMPLETION OF THIS TASK TRIGGERS THE START OF THE NEXT TASK. WHEN TWO OR MORE MARKS ARE JOINED BY A TIME COINCIDENCE LINE, A LOGICAL AND CONDITION IS ASSUMED.
TIME COINCIDENCE LINE–INDICATES THE TIME COINCIDENCE OF TWO OR MORE TASKS OR EVENTS
CONTINUOUS TASK START–INDICATES THE START OF A CONTINUOUS TASK.
CONTINUOUS TASK END–INDICATES THE END OF THE EXECUTION OF A CONTINUOUS TASK.
CONTINUOUS TASK HOLD–INDICATES THE INTERRUPTION OF A CONTINUOUS TASK. THE STATUS OF ALL DEVICES IS HELD AT THE LAST VALUE.
HORIZONTAL TIME LINE–INDICATES THE PERIOD OF TIME DURING WHICH A CONTINUOUS TASK IS EXECUTED.
CONTINUOUS TASK CHANGE–INDICATES A CHANGE IN THE CONTINUOUS TASK.
ARROWS–CLARIFY CAUSE AND EFFECT RELATIONSHIPS OF SIMULTANEOUS EVENTS.
HARDCOPY MESSAGE OR REPORT–INDICATES THAT A HARDCOPY MESSAGE IS TO BE PRINTED. EITHER THE CONTENTS OF THE MESSAGE OR A REFERENCE TO A SUPPORTING MESSAGE TABLE MAY APPEAR IN THE BOX.
RECORD–INDICATES DATA IS TO BE RECORDED FOR TREND OR HISTORICAL REVIEW.
VDU MESSAGE OR DISPLAY–INDICATES A MESSAGE OR DISPLAY IS TO APPEAR ON A VDU (VIDEO DISPLAY UNIT).
MANUAL DATA ENTRY–INDICATES A MANUAL KEYBOARD ENTRY TO ANSWER A QUESTION OR TO INPUT DATA.
MANUAL OPERATION–INDICATES A MANUAL TASK MUST BE PERFORMED.
ON-LINE STORAGE–INDICATES DATA IS TO BE STORED ON A MAGNETIC MEDIUM.
CONTINUOUS TASK BOX–INDICATES THE CONTINUOUS TASK TO BE PERFORMED. THE BOX MAY ENCLOSE A COMMENT. AN ALGEBRAIC OR BOOLEAN EXPRESSION, A LOGIC FLOWCHART, A DECISION/PROCEDURE FLOWCHART, A SAMA OR LADDER DIAGRAM, OR A COMBINATION THEREOF. THE BOX IS OF VARIABLE HEIGHT AND WIDTH.
PREDEFINED TASK BOX–REFERENCES A TASK THAT IS DEFINED ON A SUPPORTING DOCUMENT, DIAGRAM, OR TABLE.
AND CONDITION–INDICATES THE LOGICAL AND OF TWO OR MORE INPUTS.
OR CONDITION–INDICATES THE LOGICAL OR OF TWO OR MORE INPUTS. WHEN SPECIFIED THIS SYMBOL MAY REPRESENT A LOGICAL EXCLUSIVE OR.
DECISION–INDICATES A BRANCH BASED ON THE COMPARATIVE STATEMENT OR QUESTION WITHIN THE SYMBOL.
BRANCH CONNECTION–INDICATES AN EXIT TO OR AN ENTRANCE FROM ANOTHER POINT ON A TIME SEQUENCE DIAGRAM.
COMMENT–PRECEDES A USER COMMENT ON A TIME SEQUENCE DIAGRAM.

FIGURE 6 Subject to further confirmation, time-sequence diagram symbology developed during the 1980s.

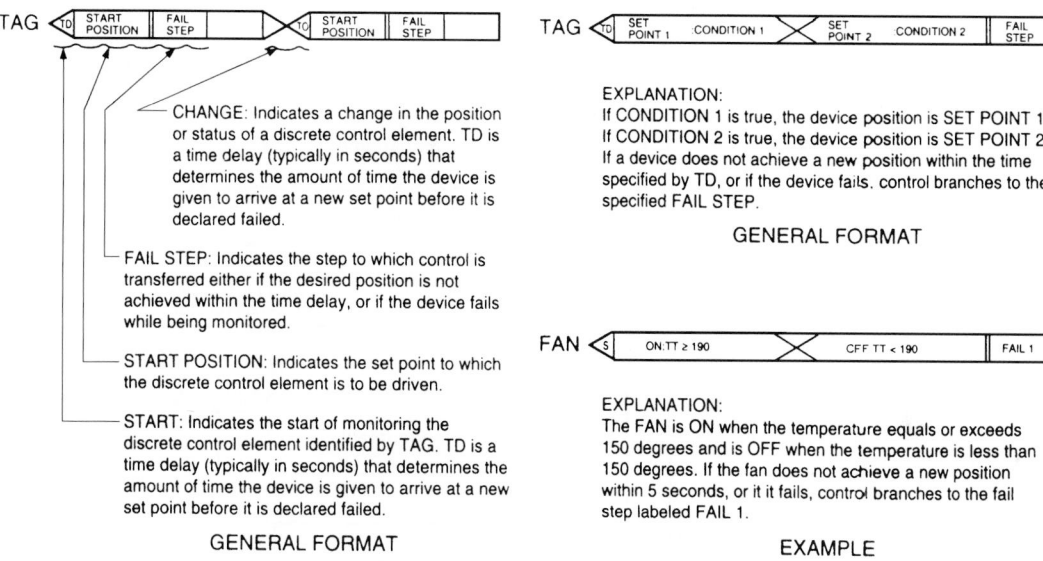

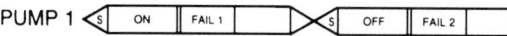

CHANGE: Indicates a change in the position or status of a discrete control element. TD is a time delay (typically in seconds) that determines the amount of time the device is given to arrive at a new set point before it is declared failed.

FAIL STEP: Indicates the step to which control is transfered either if the desired position is not achieved within the time delay, or if the device fails while being monitored.

START POSITION: Indicates the set point to which the discrete control element is to be driven.

START: Indicates the start of monitoring the discrete control element identified by TAG. TD is a time delay (typically in seconds) that determines the amount of time the device is given to arrive at a new set point before it is declared failed.

GENERAL FORMAT

EXPLANATION:
If CONDITION 1 is true, the device position is SET POINT 1.
If CONDITION 2 is true, the device position is SET POINT 2.
If a device does not achieve a new position within the time specified by TD, or if the device fails, control branches to the specified FAIL STEP.

GENERAL FORMAT

EXPLANATION:
The FAN is ON when the temperature equals or exceeds 150 degrees and is OFF when the temperature is less than 150 degrees. If the fan does not achieve a new position within 5 seconds, or it it fails, control branches to the fail step labeled FAIL 1.

EXAMPLE

EXPLANATION:
PUMP 1 is turned ON and is allowed 5 seconds to achieve this position. If the pump does not come on within this time, or if the pump later fails while being monitored, control branches to the fail step labeled FAIL 1. Later, PUMP 1 is turned OFF and is allowed 5 seconds to achieve this position. If the pump does not turn off within this time, or it the pump later fails while being monitored, control branches to the fail step labeled FAIL 2.

EXAMPLE

FIGURE 7 Time-sequence diagram symbology for discrete device control. Single condition shown at left, multiple conditions at right.

2. A clean operator interface with interactive color graphics is essential. Although continuous or steady-state control can be handled from panel-mounted operator stations or consoles with preformatted graphics, batch control generally requires custom graphic capability due to the complex interactions between units, loops, discrete devices, and the like.

3. Communications paths between controllers and consoles as well as peer-to-peer controller communication are usually essential. Because of the risk-distribution philosophy behind distributed control systems, most batch processes utilize more than one controller. It is generally required that most systems provide for standard interfaces to programmable controllers, continuous controllers, historical trend devices, and process or plant management computers, as well as providing a fully integrated package.

Control languages are available in many forms. Digital centralized computers generally require some knowledge of Fortran, Pascal, or other comparable languages by the designer. However, well-designed distributed control systems have taken the direction of configurability rather than programmability, allowing the designer of the process to perform the implementation. By using a high-level

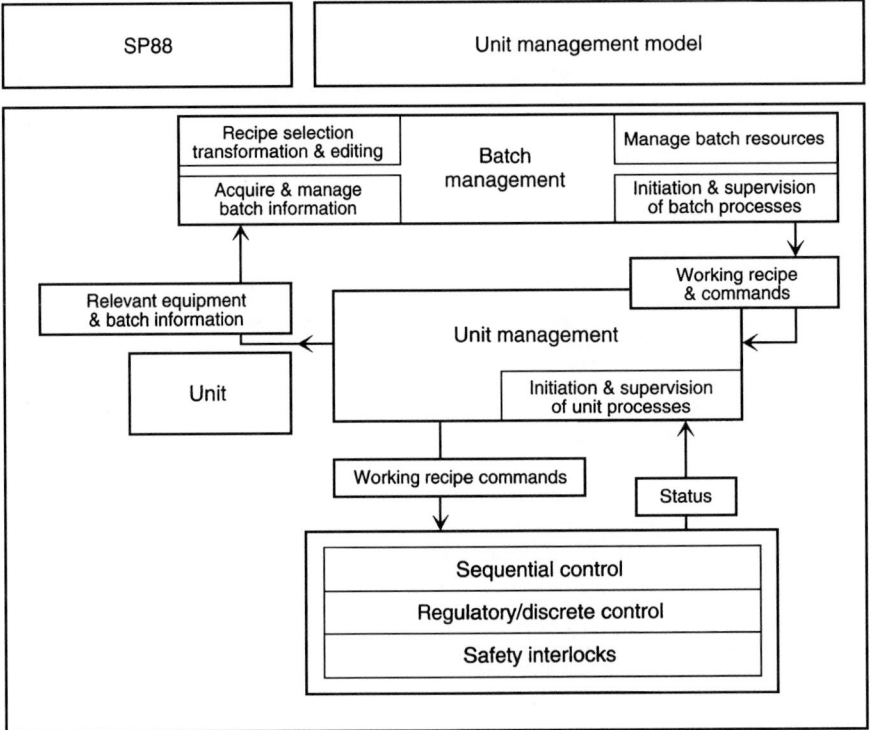

FIGURE 8   Batch management model as proposed in ISA SP88 draft.

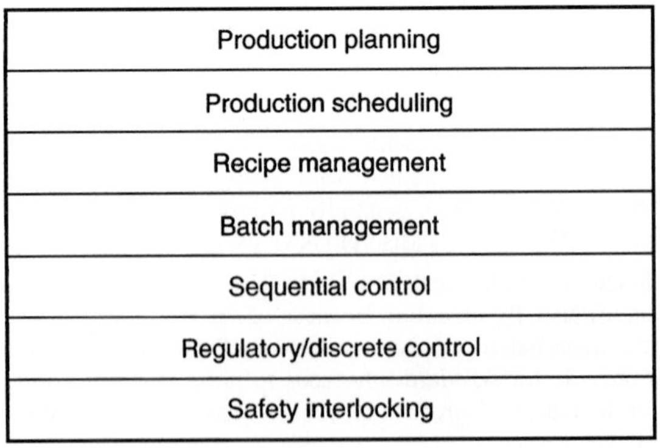

Production planning
Production scheduling
Recipe management
Batch management
Sequential control
Regulatory/discrete control
Safety interlocking

FIGURE 9   Control activity model.

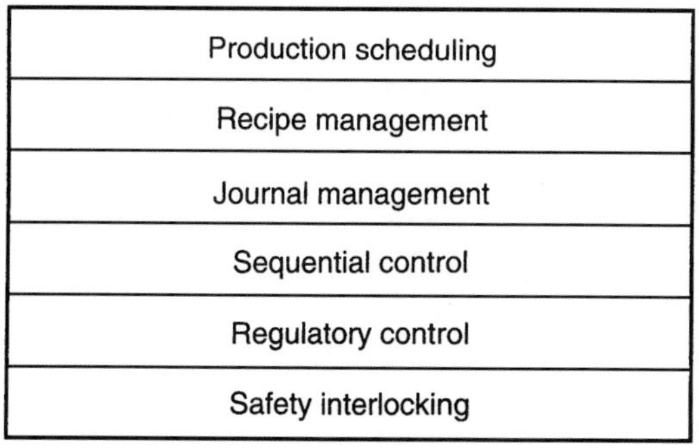

Production scheduling
Recipe management
Journal management
Sequential control
Regulatory control
Safety interlocking

FIGURE 10 Alternate control activity model.

Sequence controller model
Recipe selection transformation & editing
Acquire & manage batch information
Recipe management
Production scheduling
Initiation & supervision of batch processes
Relevant equipment & batch information
Working recipe & commands
Journal management
Sequential control
Working recipe commands
Status
Regulatory control
Safety interlocking

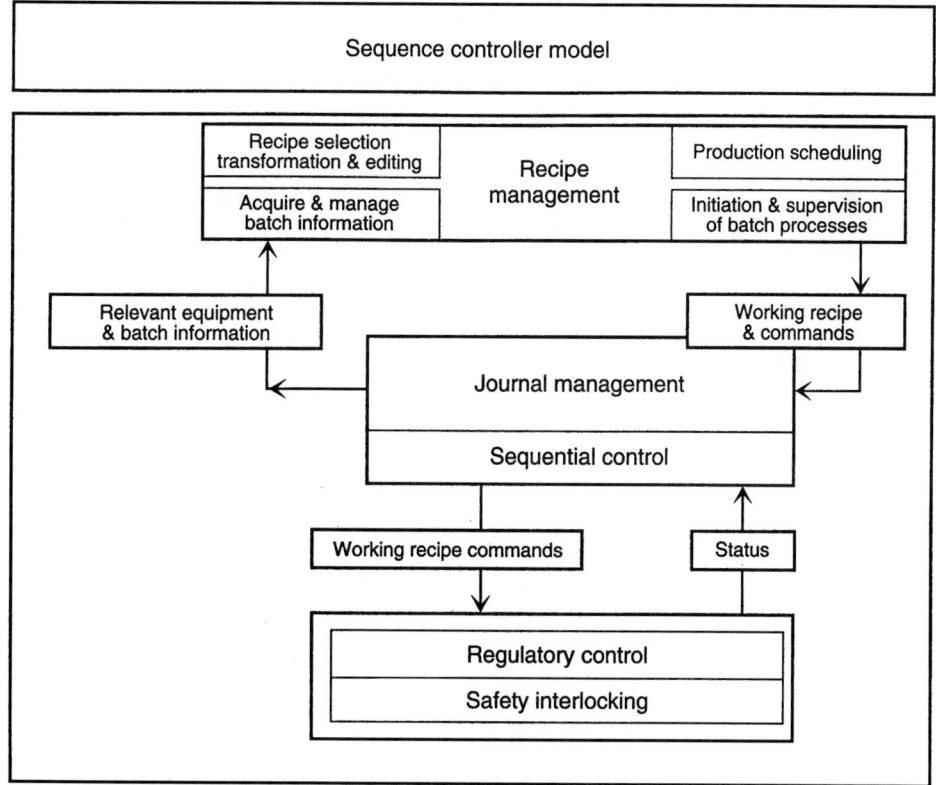

FIGURE 11 Controller characterization.

ACTIVATE	GO	RESTORE_RULES
BRANCH	HOLD	RESUME
CALL	IF	RULE
CAPTURE	INTERLOCK	SET_ABORT_KEY
CHAIN	LOG	SET_BIT
COMPARE	LOGIC	SET_HOLD_KEY
CONTINUE	MASK	STOP
CONTROL	MATH	SUSPEND_RULES
DEACTIVATE	MESSAGE	TREND
DELAY	ON_ABORT	UNLOCK
DEVICE	ON_ERROR	UNTIL
ELSE	OP_ENTRY	WAIT
END	PARALLEL	WHILE
END_WHILE	RAMP	COMMENT
FORMULA	REFER	DEFINE
FUNCTION	RELEASE	INCLUDE
GET_STATUS	REPEAT	NAME
	RESERVE	

FIGURE 12 Standardized names for sequence instruction steps.

engineering language with configurable expressions and mnemonics, the engineer can virtually configure the control system directly from the time-sequence diagrams. Consequently time–sequence diagramming, although it may increase the time for design phases, can substantially reduce the implementation phase of a project and reduce errors and redesign after start-up.

ISA SP88. This document [2] features a batch management model (Fig. 8), a control activity model (Figs. 9 and 10), and control characterization (Fig. 11). Tentative sequence instruction steps are tabulated in Fig. 12. The new standard is intended to provide flexibility for the user while including a comprehensive set of tools to accomplish a simple or a complex sequence control solution.

REFERENCES

1. Adams, A., "A Simplifying Batch Control and Recipe Management," *In-Tech*, January 1991, p. 41.
2. ISA SP88, "Batch Control Model," Instrument Society of America, final draft in process, 1993.
3. Wilkins, M. J., "Simplify Batch Automation Projects," *Chem. Eng. Progress*, 61, April 1992.
4. Doerr, W. W., and R. T. Hessian, Jr., "Control Toxic Emissions from Batch Operations," *Chem. Eng. Progress*, 57, September 1991.
5. Blankenstein, L. S., "Batch Tracking: A Key for Process Validation," *Chem. Eng. Progress*, 63, August 1991.
6. Concordia, J. J., "Batch Catalytic Gas/Liquid Reactions: Types and Performance Characteristics," *Chem. Eng. Progress*, 50, March 1990.
7. Musier, R. F. H., "Batch Process Management," *Chem. Eng. Progress*, 66, June 1990.
8. Fisher, T. G., "Batch Control Systems: Design, Application, and Instrumentation," Instrument Society of America, Research Triangle Park, North Carolina, 1990.
9. Staff: "Batch Control Service" (Videotapes), Instrument Society of America, Research Triangle Park, North Carolina, 1990.

AUTOMATIC BLENDING SYSTEMS

by Stan Stoddard*

The need to blend various ingredients in pots or vats dates back to antiquity. Over the years, various means of combining liquid or powder components in preprogrammed sequences and amounts were devised, including bucket trains, sprocket gears, water shells, and variable-speed gearing techniques. Later came pneumatic controllers, which could control the flow rate of a particular component. Proportioning techniques were developed that utilized mechanical and pneumatic devices. Electronics introduced analog amplifier comparators coupled with accurate metering devices and preset electromechanical counters for multiple-component flow control and metering. Further developments brought forth digital techniques with integrated circuits that performed precise measurement, comparisons, and control of multistream processes. Present microcomputer technologies and sophisticated programmable controllers find wide application in a variety of industries for bringing together multiple components in a precisely controlled manner. Applications are found in the petroleum, petrochemical, food and beverage, building materials, pharmaceutical, automotive, and chemical fields among others.

BATCH VERSUS CONTINUOUS BLENDING

In a batch-type process, a recipe is followed by adding specific amounts of ingredients in a predefined sequence with mixing, stirring, and brewing, or other processing times between the addition of each component. This practice, for example, was followed for many years in the manufacture of lubricating oils through programmed mixing, stirring, and heating of various hydrocarbon components and additives in large batch vats and allied equipment. This procedure has largely been replaced in many industries by in-line blending, which essentially refers to a process whereby component streams (liquids, gases, powders, or aggregates) are measured and controlled in a precise relationship or ratio to each other. All components flow together simultaneously to a central collection point, where they combine to form the finished product, such as a lubricating oil. The obvious advantage of in-line blending over the batch-type process is that large vessels for the mixing of components are eliminated. The blend heater, augmented by static or active in-line mixers, is all that is required to form the final product. The finished product can go directly to a canning line, to a finished product storage tank, or into a pipeline for distribution.

TYPICAL BLEND CONFIGURATION

A modern in-line blending scheme is shown in Fig. 1. The blend controller nominally utilizes microprocessor technology with a cathode-ray-tube (CRT) display. Each fluid component is pumped from a storage tank through a strainer and then through a flowmeter, with meters and valves selected for prevailing process conditions (viscosity, temperature, pressure, flow rates, and so on). The signal from the flowmeter is fed to the blend controller, which compares the actual flow rate to the desired flow rate. If the actual flow rate is incorrect, the blend controller will adjust the control valve via the 4- to 20-mA signal to the valve. In this way each component is controlled in a closed-loop fashion. Sometimes it is most practical to control the flow rate by means of proportioning pumps, which inject a precise amount of a specific fluid when a pulse signal from the blend controller is received. This

---

* Waugh Controls Corporation, Chatsworth, California.

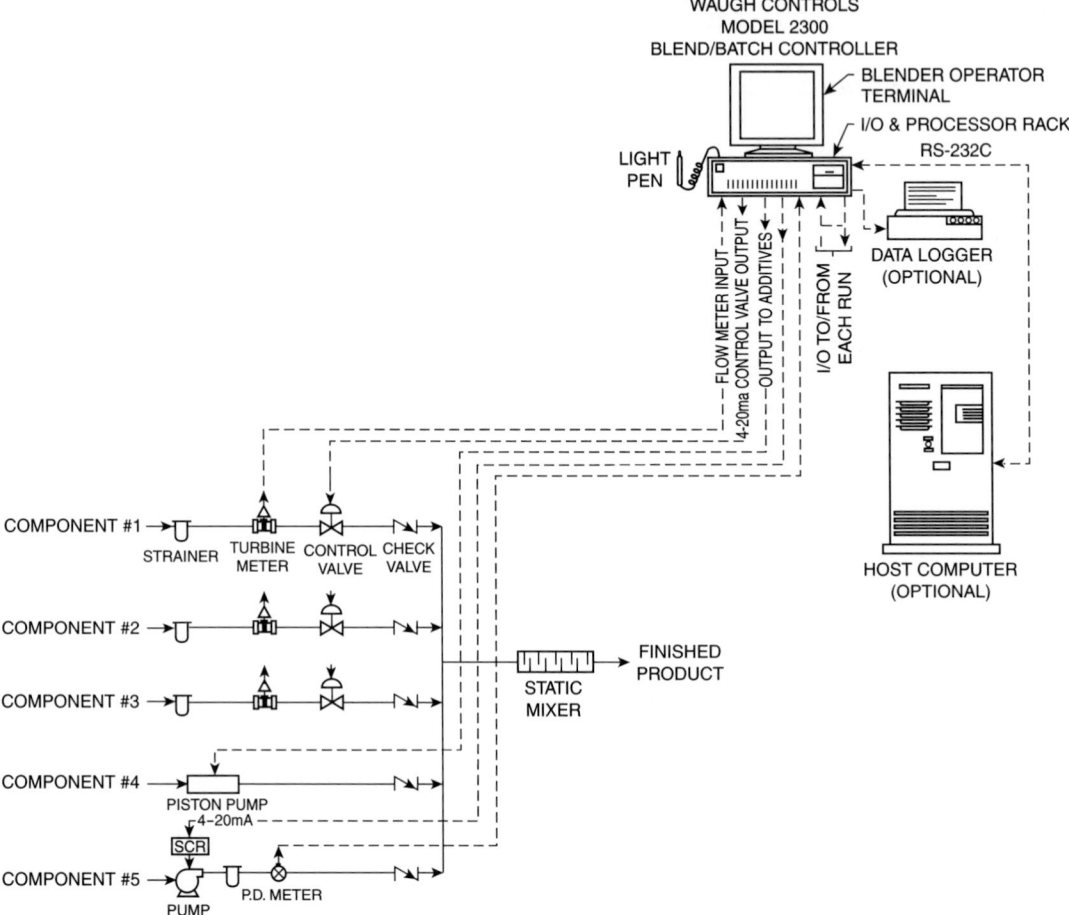

FIGURE 1   Typical blender configuration. (*Waugh Controls Corporation.*)

type of open-loop control is cost efficient, but some means for ensuring flow (not a dry line) should be considered inasmuch as no direct fluid measurement device is used.

Other variations of measurement and control involve the use of variable-speed pump motor controllers (silicon-controlled rectifiers) for flow control, adding a flowmeter in series with an injection pump, and the use of weigh belt feeders with variable feed-speed control and tachometer-load cell outputs (for powders and aggregates).

Liquid or Powder Blending

Solid materials or combinations of solid and fluid material, such as feeds to cement kilns, asphalt aggregates, and concrete aggregates, are readily controlled through the use of techniques similar to those used for fluids. Flow input, in the form of pulses representing the weight of the material, is obtained from weigh-belt feeders, while the 4- to 20-mA control output is used to operate a gate regulating the amount of material fed from a hopper onto the weigh belt. Many weigh-belt feeders require multiplication of belt speed by the mass of the belt in order to obtain the mass flow rate. This computation is performed by external computing modules in a rack. Many food and chemical products require a combination of both liquid and powder ingredients, blended together to form the finished product or an intermediate feed to additional processing. A combination-type blending system used to make bread or pastry dough is shown in Fig. 2.

Sizing a Blending System

The first step in designing a blending system is to construct a list of all the components required to form the various products. Next, after each component, list the ratio ranges for these fluids as they fulfill all daily production needs in an 8- to 9-hour shift, thus providing time for maintenance when required. Once the overall maximum delivery rate has been determined, each component stream can be sized to provide its percentage range of the total blend. The rangeability of the component streams then must be considered. This should not exceed 10:1 if turbine meters are used, or 20:1 in the case of positive-displacement (PD) meters. Possible future production rate increases should enter into sizing the component streams. However, caution should be exercised to avoid oversizing control valves excessively for the current flow rates.

BLEND CONTROLLER

Any number of blending systems of any type, or in any combination of types, are usually controlled by a microprocessor-based instrument system using one or more color monitors at the operator's control station. Such systems incorporate software to provide monitoring, control, data storage, and alarm functions, and input-output circuitry to accept signals from flowmeters and from weight and temperature transmitters, and to drive control devices, such as control valves, line-up valves, and pumps.

The system can use a distributed architecture, with monitoring and control instruments installed in the field near each blending unit and linked back to the central control system through a communications link, or a unit architecture, with all monitoring, control, and operator interfaces at the central control room. A typical modern blend control system is shown in Fig. 3.

Blend Stations

There will be a series of blend stations, usually skid mounted, each controlling the flow of a single component of the blend. Each skid-mounted station may include a pump, a pressure-regulating valve, a strainer, an air eliminator, a flowmeter, a control valve, and a check valve. A typical skid assembly is shown in Fig. 4.

Blend Setup

To set up a blend, the operator will enter into the blend controller the actual percentages to be controlled at each blend station, the rate at which the blend is to be run, and the quantity to be blended. If these formulations are repetitive, they can be stored in memory or on a hard disk as numbered recipes. The recipe storage feature provides quick blender setup by simple entry of the recipe number. A blend controller[1] can store hundreds of recipes. Alternatively, the blender parameters may be set from a supervisory computer system.

---

[1] For example, the Waugh 2300 blend controller.

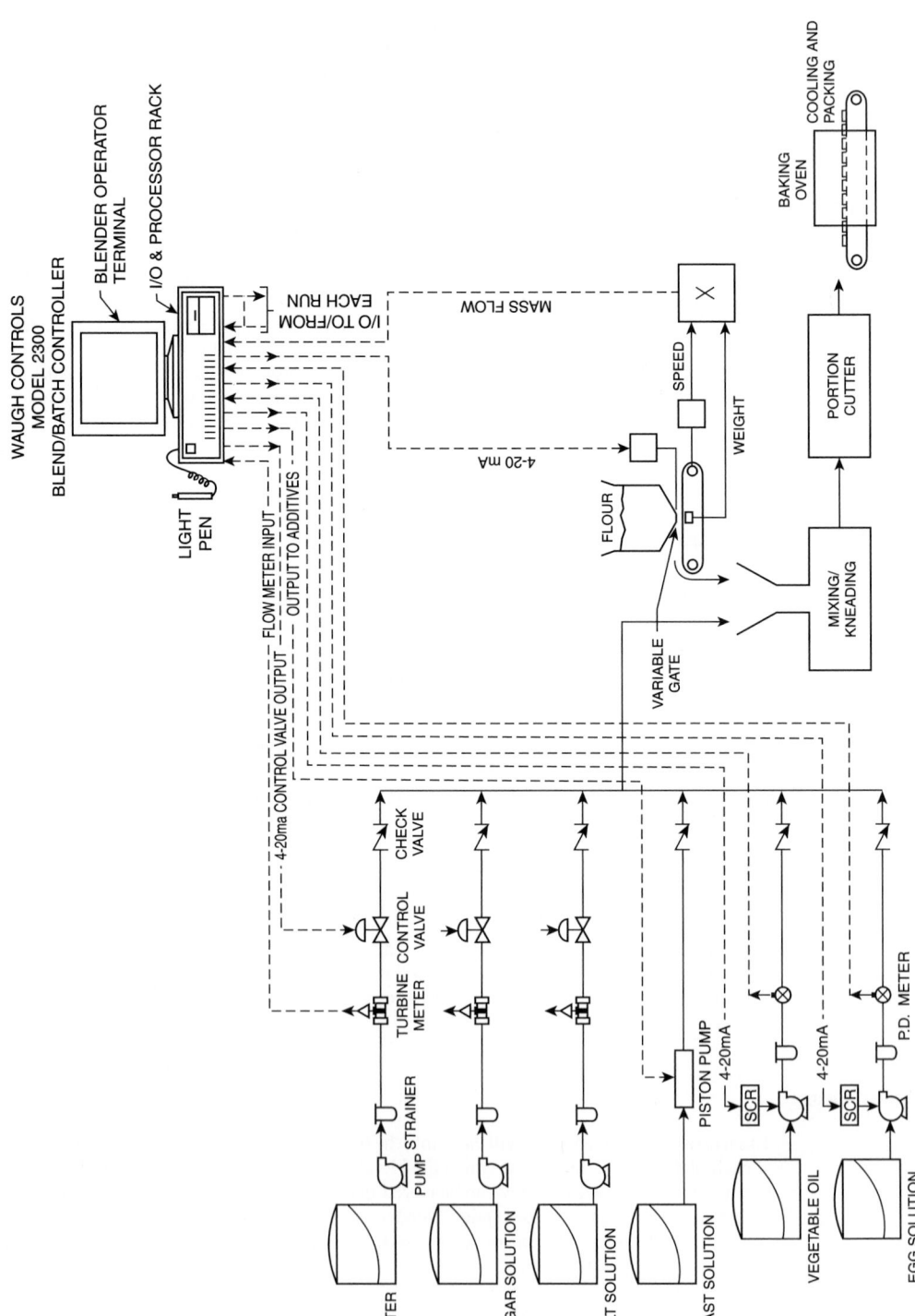

FIGURE 2 Blending system for preparing bread and pastry dough. (*Controls Corporation.*)

FIGURE 3 Blending process management system. (*Waugh Controls Corporation.*)

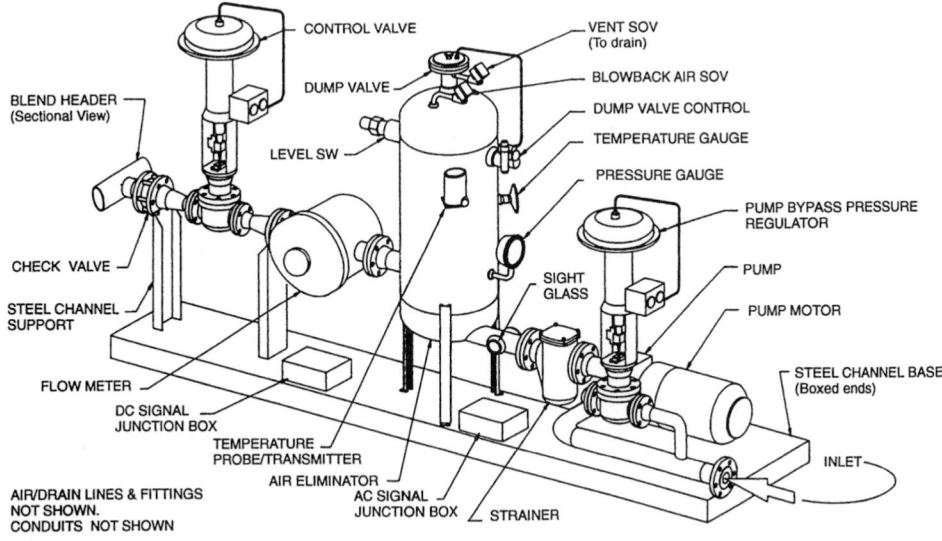

CONTROL VALVE

VENT SOV
(To drain)

DUMP VALVE

BLOWBACK AIR SOV

BLEND HEADER
(Sectional View)

DUMP VALVE CONTROL

LEVEL SW

TEMPERATURE GAUGE

PRESSURE GAUGE

PUMP BYPASS PRESSURE
REGULATOR

CHECK VALVE

SIGHT
GLASS

PUMP

STEEL CHANNEL
SUPPORT

PUMP MOTOR

STEEL CHANNEL BASE
(Boxed ends)

FLOW METER

DC SIGNAL
JUNCTION BOX

TEMPERATURE
PROBE/TRANSMITTER

INLET

AIR ELIMINATOR

AIR/DRAIN LINES & FITTINGS
NOT SHOWN.
CONDUITS NOT SHOWN

AC SIGNAL
JUNCTION BOX

STRAINER

FIGURE 4 Typical skid assembly.

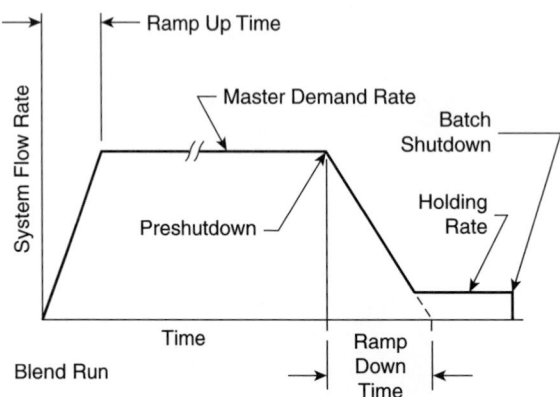

FIGURE 5   Master demand rate operation. (*Waugh Controls Corporation.*)

Master Blend Control

With all in readiness, the "blend start" button is pressed, from which point the operation is entirely automatic. Pumps are started, after which the control valves are opened and the blend flow rates increased gradually until the preselected flow rate is reached. The master demand rate (which represents the total system throughput) is maintained until the demand total or measured total reaches a preshutdown value. At that point the demand rate is ramped down to a new holding value until the total batch size is reached. When the batch size is attained, the master demand rate immediately goes to zero and all loop control outputs go to zero. A master demand rate operation is shown in Fig. 5.

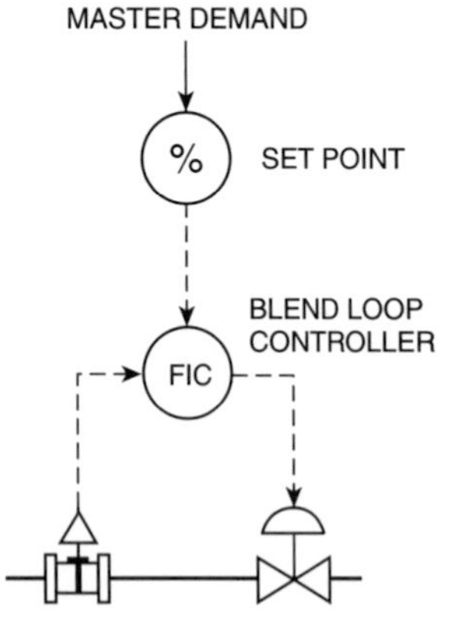

Blend Loop Control

For each component stream, closed-loop control is maintained by comparing the scaled pulse input from the flowmeter with the loop demand derived from the master demand total and the percent set point for that loop (Fig. 6). The difference between what is measured and what is demanded is the loop error or deviation error. The error from the present scan cycle is added to the previously accumulated errors. The blend loop controller then acts upon the integrated error and works to eliminate it. The output of the blend controller drives a final actuator to provide a fast response to error corrections. Blend loop control differs from conventional ratio control in that the accumulated loop error has a direct effect on the blend loop controller action. In contrast, ratio control uses only the error from the current cycle to bring the measurement back to the set point. It does not compensate for the error that accumulates while the controller is responding to the upset.

FIGURE 6   Blend loop controller. (*Waugh Controls Corporation.*)

Automatic Rate Control

If, during the course of a blending operation, one or more streams are unable to maintain a sufficient flow rate to achieve the desired blend ratios, the blender will automatically ramp down to a rate at which the selected ratios can be maintained. Upon restoration of flow in the lagging streams, the blender will automatically return to its originally set blend flow rate. The system determines that a loop is unable to maintain its required flow when its associated control valve is 95% open. Thus the blender will always operate either at its set master demand flow rate or at the maximum rate allowed by the lagging stream, but, in either case, the blend ratios will be maintained automatically.

Time Control

The quality of the blended product may be adjusted on line while the blend is in process via feedback from one or more analyzers that sample the blended stream. For example, in a lube oil blender an on-line viscometer may be used to control viscosity by adjusting the ratio of one of the lighter stocks. The analyzer signal acts as the process variable input to an internal PID controller function, whose set point is the desired product quality and whose output is then cascaded into the selected blend loop controller, serving to adjust the set point up or down. Changes in the flow of the selected stream affect the quality of the blend, thus closing the cascaded control loop.

Temperature Compensation

Each flow control loop possesses the capability to correct the volume as measured by the flowmeter to a standard volume at a selected reference temperature. Fluid temperature can be measured by a 100-ohm platinum probe with a 4- to 20-mA transmitter. Correction is made per American Petroleum Institute Standard 2540, Tables 6A, 6B, 6C, and 6D or a linear coefficient.

Programmable Logic Control

Blend controllers as described here contain integral PLC functions for interfacing with motor starters, solenoid valves, audible and visual alarms, valve limit switches, and other field equipment. For example, sequential starting of pumps to avoid sudden large increases in electrical load is readily accomplished by sequencing pump start relay closures at the beginning of each blend. A different time delay may be entered for each relay closure, thus permitting any desired time sequence of pump starting. Complex valve alignment procedures and interlocks can also be incorporated with these controllers.

DISTRIBUTED NUMERICAL CONTROL AND NETWORKING

by Terrence G. Cox*

Since the Industrial Revolution (around 1760) there has been a continuing search for more effective manufacturing methods. As part of this unrelenting process, numerically controlled (NC) machine tools were developed in the early 1960s and have been used widely through the present day.

* CAD/CAM Integration. Inc., Woburn, Massachusetts.

NCs have been used to cut down job lot sizes, reduce the setup time, and trim machining time and the skill level of the direct labor force. Punched paper or mylar tape was the early medium used to feed stored instructions to the controllers. Although a major technological development of its day, this methodology was found to be time consuming, error prone, and limiting because of the cumbersome long tapes. Tapes often were difficult to locate by an operator from an ever-expanding inventory, and it was hard to make certain that the latest revised tape was in hand. Loading and unloading the tape required care. Also, the operator had to document any changes (edits) made during the machining process and to make certain that these changes were received by engineering.

Paper tape proved to be a poor medium for the shop floor. Subject to exposure to oil and solvent spills, the tape information would sometimes be altered to the point where scrap parts would be produced. Mylar tapes alleviated some of the problems, but tape reels also became longer and unhandy as the part size and complexity increased. This required splitting a single long tape into several spools, a time-consuming operation. Special instruction sets were required at the start and end of each tape, not to mention the additional time required for loading and unloading each tape.

Direct numerical control (DNC) was later developed to overcome these problems. A further stage of development yielded what is now called distributed numerical control, also commonly abbreviated as DNC.

DIRECT NUMERICAL CONTROL

To overcome the limitations of the earlier, simple NC, several large firms undertook the design of DNC systems in order to establish an interface between a computer and a machine tool wherein a computer could feed instructions directly to a single machine tool and which, in turn, would also control the servos and cut the part. Unfortunately when DNC was developed, the only computers available were costly mainframes. A further drawback was that only *one* machine tool could be controlled at a time. Consequently the general result was that the direct numerical approach was not cost effective.

DISTRIBUTED NUMERICAL CONTROL

In DNC one computer feeds part programs to *multiple* machine tools. The advent of mini- and microcomputers, plus the ability to address multiple serial ports on a real-time basis, gained acceptance of this technology by manufacturers.

Basic DNC Configuration

In essence, a DNC system uploads and downloads part programs electronically to machine tools. In the initial use of this technology, the user does not need to have a CAD/CAM computer. Where a CAD/CAM system is not interfaced to the current DNC system, programs usually are entered through the machine tool controller in a manual data input (MDI) mode. Once in the controller, the program can be uploaded to the DNC system and, from that point, electronically downloaded to the required machine tool. The DNC system is used as (1) a file cabinet, (2) a traffic director, and (3) an interrupter.

File Cabinet.   It still was not unusual for a manufacturer to have thousands of part programs stored on punched tape. With a DNC system, these programs are conserved electronically and are placed in organized directories with well-documented revision levels.

Traffic Director.   Base systems usually have a number of machine tools associated with a single DNC management station. Distributing the needed programs to the appropriate machine tool at the required time in an organized method is the primary job of the system.

When a CAD/CAM computer is used, communications must be established and monitored between the host CAD/CAM and the DNC system regarding the system's ability to accept the transfer of jobs. The communications between the DNC system and each machine tool must be monitored in order for transmission to occur in a timely manner and to know when both systems are ready to accept information. The system also must take into account those instances in which transfer cannot occur because the machine tool's memory buffers are full. All of this requires management and coordination from the DNC management station.

Interrupter.   There is much concern among control engineers today over the importance of standards for control system networks and interconnectability. Over the years the machine tool control field has been a major offender in this regard. Several different protocols persist. Consequently the particular control for a given system frequently must be interrupted and matched by the DNC system so that part programs can be distributed. Many different protocols are spread over large numbers of machine tool builders, and, considering the introduction of new machine models periodically, the lack of standards has been a problem for several years.

In a modern DNC system the interrupter functionality must be of a user-friendly format so that end users can establish the needed communications as new machine tools are added.

DNC System Components

As shown in Fig. 1, a typical base DNC system will have four major facets: (1) a CAD/CAM system, (2) a DNC system, (3) the machine tools to be controlled, and (4) behind-the-tape readers (BTRs). The CAD/CAM system may be on a mainframe or on a personal computer (PC). The DNC system can run on any level of hardware, but usually will be PC based. There usually are many machine tools that are not uniformly based as to standards and distributed in many locations on the shop floor. The BTRs are used mainly for two reasons.

1. A machine tool does not have an active RS-232 input port. In this case there is only one way to load information, namely, through the tape reader. The BTR is a hardware and software combination device that is located between the tape reader and the controller and is interfaced to the DNC system. (As far as the controller is concerned, it is accepting information from the tape reader.)

2. A machine tool may or may not have an active RS-232 input port, but it does *not* have sufficient on-board memory to store the entire part program. In this case a BTR will interface in the same manner, but the revolving buffer within the BTR will compensate for the needed memory. In this case, as the machine tool uses information, the BTR is "spoon feeding" additional information to the DNC computer, which in turn is "spoon feeding" the BTR. Under these circumstances a part program is limited only by the capacity of the hard disk in the DNC computer.

Advantages of Basic DNC System

The benefits of a DNC system are direct and measurable and include the following.

Designer Efficiency.   Prior to direct numerical control it was the responsibility of the designer to punch out the tapes before they were run. If any edits were made after the tapes were punched, it was also the designer's responsibility to make the edits and to repunch the tapes. This process could interrupt the work of the designer by a factor ranging from 10 minutes to over an hour per day.

Reduced Machine Downtime.   On average the associated tool downtime for paper tapes exceeds 15 minutes per job run. This time is consumed by punching a tape at the CAD/CAM system, transporting

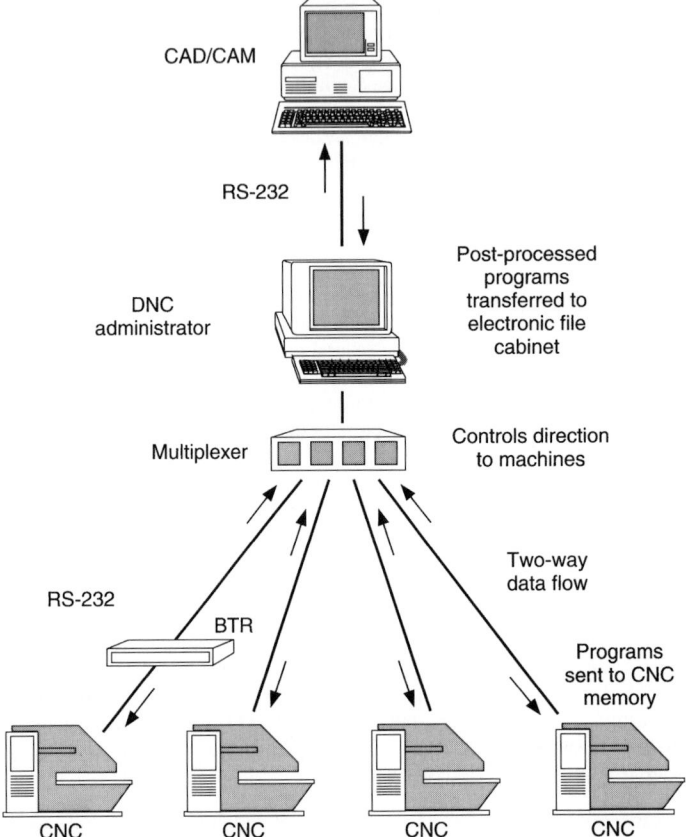

FIGURE 1   Direct numerical control (DNC) system that utilizes a standard disk operating system.

    Basic system: (1) Entry level, (2) configured in 4, 8, or 16 ports (up to 64 devices), (3) uses off-the-shelf 286 and 386 PCs.

    Features: (1) Centralized part programs, (2) eliminates paper tape, (3) full NC editor, (4) NC edit log, (5) electronic job transfer, (6) standard RS-232 or RS-422 communications.

    Advantages: (1) Reduced machine downtime—no loading or unloading of paper tapes, (2) programs transferred to and from machine tools in seconds, (3) automatic update of job transfer and activity log, (4) immediate printout or screen view of any job, (5) reduced network due to NC revision control.

    Hardware options: (1) Machine interface units (MIUs), (2) RS-422 communications, (3) behind-the-tape readers (BTRs), (4) communications box and queuing. (*CAD/CAM Integration, Inc.*)

the tape to a particular machine tool, loading the tape on the machine tool, unloading the tape at the end of the run, and punching a new tape at the machine tool if any revisions were made through the editor on the machine tool controller.

Easier Job Transfers.   Many DNC systems incorporate a job transfer and activity log. This log updates reports whenever a part program transfer is made to or from the shop floor. When used properly, this can be very effective for flagging management on production trends, thereby catching production problems before they are out of control. In turn, machine scheduling problems can be handled, thereby reducing machine downtime.

Viewing an Entire Job.   With DNC an entire program can be viewed on the screen should questions arise. Prior to DNC, only one line at a time could be viewed on the controller screen. Viewing the entire job allows the operator a far better frame of reference.

Reduction of Rework.   The organized archiving procedure for part programs and their revision levels, made possible by DNC, eliminates scrap due to incorrect tapes being loaded onto a machine tool.

Additional Executable Functions

Within the DNC management system, a number of executable functions are required.

NC Editor.   An editor, written specifically for NC code, should be available. It is used to make minor changes to the part program, such as feeds and speeds, or changing G codes from a different controller, all without going back to the CAD/CAM system. Common features include insert, overstrike, delete, resequencing, search and replace, cut and paste, and appending files. Advanced features found in some editors are mirroring, translate, and scaling.

Compare.   When a job is sent to a machine tool, the operator can use the controller's editor to alter the program. In order to manage these edits, the program designer must review the edits once uploaded from the machine tool and determine which should be kept, that is, either the original or the revised program. A DNC management station should have a compare feature to compare the two files efficiently and highlight the differences so that an intelligent decision can be made.

Activity Report.   This report should be available as a thumbnail sketch of shop floor activity. The report gives the manager the following information: what jobs have been downloaded or uploaded to which machine tools, at what time, and which jobs are waiting to be run at each machine tool.

DNC OPERATING SYSTEMS

There are several de facto standard operating systems available as of the early 1990s, including DOS (disk operating system), UNIX, VMS, OS/2, and various proprietary systems. In terms of trends. DOS, WINDOWS NT, and UNIX continue to have wide appeal. VMS appears to be giving way to ULTRAX (UNIX derivative). OS/2 is not receiving wide market acceptance. There is some consensus that proprietary systems will become virtually nonexistent by the early 2000s.

Disk Operating System.   A DOS environment provides a good platform. Because of the low cost of each license and minimum hardware requirements, a manufacturer can move into a tapeless environment for a modest cost. This is particularly important in connection with DNC because many thousands of small job shops constitute the major market. However, DOS is a single-tasking, single-user environment, and therefore only one task can be accomplished at a time. This is limiting because the computer will feed information to only one machine at a time, even if a number of them are multiplexed from one computer. As an example, if a program is attempting to transfer from the CAD/CAM system, it must wait until all other tasks of that computer are completed.

UNIX and WINDOWS NT Platforms.   UNIX/WINDOWS NT are much more powerful. They are true multitasking, environments. A part program can be transferred from the CAD/CAM system, transferred to all interfaced machine tools, and the operator can edit jobs on the DNC management station—all simultaneously.

In the larger manufacturing firms the foregoing become critical capabilities because more machine tools are interfaced to a system and, therefore, real-time resource sharing is necessary. Another niche where this becomes critical occurs when long part programs are generated and must be fed to memory-insufficient multiple machine tools. Multiplexed systems cannot handle this situation effectively, but the multitasking features of UNIX or WINDOWS NT can easily "drip feed" multiple machine tools.

Functionality Blur

As of the early 1990s a functionality "blur" or fuzziness is developing between the operating systems just described. Third-party developers are writing multitasking features for DOS. Examples include windows and time-slice drives for serial ports. Although these are not true multitasking capabilities, they are sufficiently good for handling the requirements of multitasking DNC for a large share of existing needs.

Networks are also blurring the multiuser issue (Fig. 2). Networks such as Novell (described later) allow a number of DOS nodes to access a file server simultaneously. This allows a DOS environment to become functionally multiuser at a very reasonable cost.

Because of the complexities of the computing environment, a manufacturer must look carefully before implementing a tapeless factory environment. To determine which operating system may be best suited to a given situation, manufacturing management must evaluate the firm's current computing environment, its networking strategic plan, and the required functionality of the proposed DNC system. PCs and cost-effective networks are major factors in contemporary automated manufacturing technology.

Personal Computer DNC Systems

There are key economic, tactical, and strategic reasons why PCs have grown in popularity not only for DNC, but for many other manufacturing systems.

Economic Advantages of PCs.   Unlike most other areas of a business, manufacturing requires a number of specialized software programs to create the optimal automation environment. In other departments a management resources planning (MRP) or CAD/CAM system can be procured that will handle the major share of the department's needs. Thus if these systems are on a mainframe and resources are being shared by a number of individuals, cost justification is easily defended. Furthermore, the software is written to function over a wide range of user applications, thereby reducing the need for customization.

Conversely, manufacturing applications tend to be more specialized, affecting a smaller realm of the corporate universe. Instead of having one MRP system to incorporate the needs of accounting planning, inventory control, traffic, and other departments, the manufacturing department requires many specialized programs, including DNC, statistical process control (SPC), scheduling, simulation, graphics and documentation, data collection, and cell control (Fig. 3). Each of these programs affects only a portion of manufacturing, each with its own specialized needs.

Because of this diversity of applications and specialization within each application, it is obviously not economically justifiable to procure a mainframe with, for example, an SPC application. The cost of a mainframe and the associated software would be too great in comparison to the payback.

A review of the market in the 1980s substantiates this point, as there was not much specialized manufacturing mainframe software available. What was available did not integrate easily. Melding of software applications could not justify the total purchase.

Tactical Advantages of PCs.   Of all the departments within a company that are dependent on computers, manufacturing would be the most seriously affected by computer downtime. It is evident that production lines would be shut down, inspection would be crippled, work in progress would not be trackable, and so on. In a worst-case scenario, if a mainframe controlled the manufacturing process and the mainframe malfunctioned, the entire manufacturing process would have to shut down. In

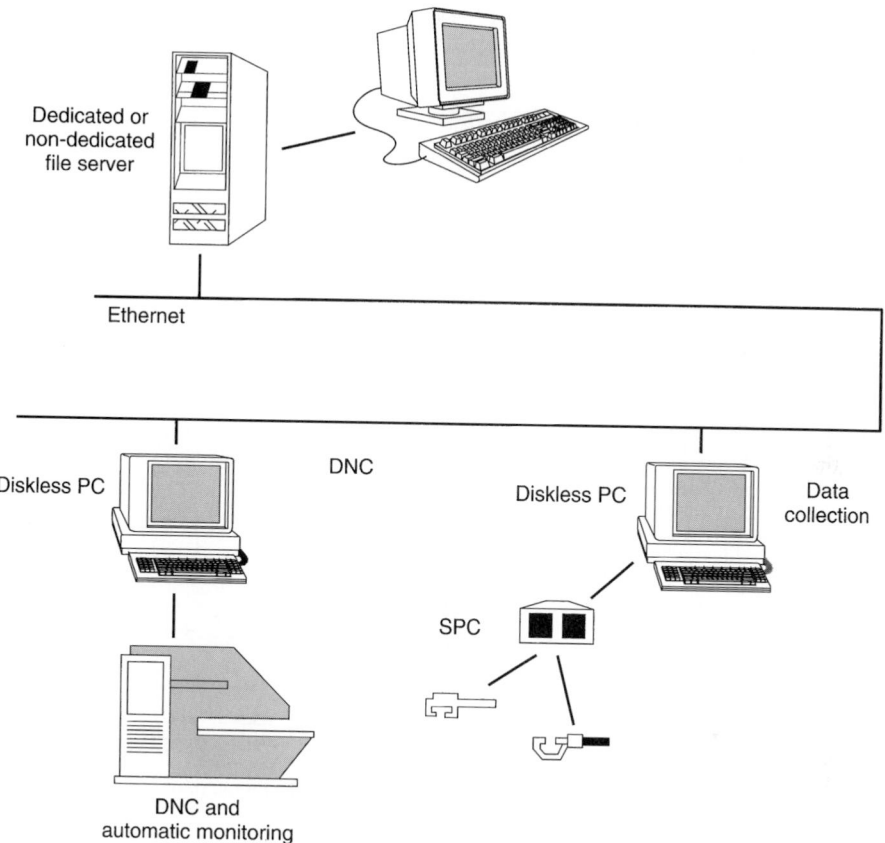

Dedicated or
non-dedicated
file server

Ethernet

Diskless PC

DNC

Diskless PC

Data
collection

SPC

DNC and
automatic monitoring

FIGURE 2   Entry-level network that is DOS based.

Basic system: (1) Entry-level standard network, (2) dedicated or nondedicated file server, (3) based on Intel 286 microprocessor, (4) supports up to five active users, (5) controls application information over network, (6) applications can be integrated in seamless environment through factory network control system.

Features: (1) High performance—files are transferred at 10 Mb/s, (2) easy-to-use menu-driven utilities put the network supervisor in control, (3) security—allows supervisor to restrict network access, (4) single-source database for complete control of all files, (5) cost-effective network uses diskless PCs, (6) virtually unlimited expansion capabilities.

Application: (1) Statistical process control (SPC), (2) direct numerical control (DNC), (3) data collection, (4) view graphics and documentation, (5) automatic monitoring. (*CAD/CAM Integration, Inc.*)

numerous instances the cost of such downtime is many thousands of dollars per minute. At best, repair time for the mainframe would be an hour or so, and at worst, a full day or more.

In a distributed PC network environment a hardware or software problem does not have the same rippling effect throughout manufacturing as would a mainframe. If a software application has a problem, a new node can be substituted by company personnel in a matter of minutes.

The majority of specialized manufacturing applications has been written for the PC-networked environment. Thus it is prudent for a manufacturer to establish a networked environment. PC interfaces tend to be more user friendly than the mainframe counterparts. Less training and better acceptance of PC systems by employees have been demonstrated.

Strategic Advantages of PCs.   After a long period of trial and error the corporate universe essentially has embraced the concept of distributed networking. Firms are now requiring that proprietary systems

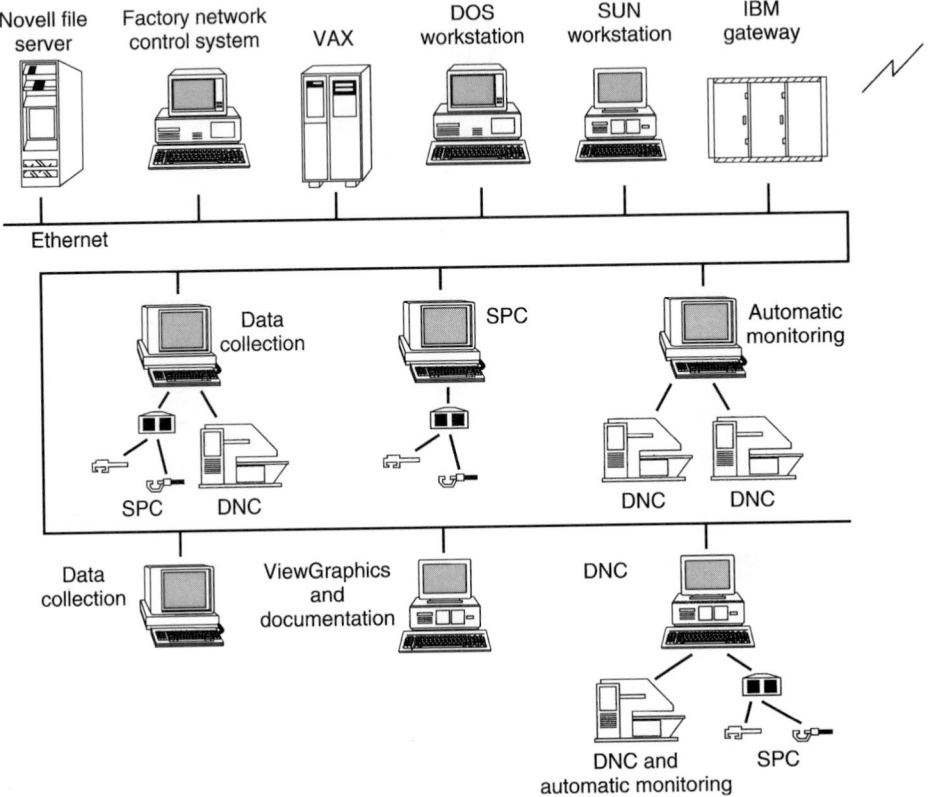

FIGURE 3 Advanced network that is DOS-based.

Basic system: (1) Controls application information of network, (2) applications can be integrated in seamless environment through factory network control system, (3) designed for medium to large business, (4) dedicated 386-based file server technology, (5) supports up to 100 active users on network, (6) high-performance network operating system.

Features: (1) Provides high functionality while maintaining high performance levels, (2) multiuser, multitasking architecture allows user to perform many operations simultaneously, (3) enhanced features for security, system reliability, and network management, (4) cell configurations decrease exposure to manufacturing downtime, (5) single-source database for complete control of files, (6) cost-effective network—uses diskless PCs, (7) simple-to-use menu-driven utilities make learning and adding applications easy.

Applications: (1) Statistical process control (SPC), (2) direct numerical control (DNC), (3) data collection, (4) view graphics and documentation, (5) automatic monitoring. (*CAD/CAM Integration, Inc.*)

must run over a single network and communicate with one another. The best hardware-software combinations are being purchased for specific applications, but with the proviso that the information generated must be capable of being transferred, accepted, and messaged by other applications of the network. (This strategy is better served through a distributed open architecture than a one-vendor mainframe mentality.)

UNIX-Based Network

The key feature of a UNIX-based network is that it allows the user to enter the realm of a true multitasking, multiuser environment. The power of UNIX is that it allows multiple users to perform multiple processes independently at each node, thereby reducing the network usage while simultaneously executing all needed tasks.

Because of the adherence to standards, a number of different UNIX machines can be interfaced to the same network, and any node can become a remote user of any other node. This facilitates the real-time exchange of information between departments that do not share the same application software. X-windows technology enhances this capability by standardizing the interface between all computers on the network.

Novell Network

The strong points of Novell are threefold: (1) a manufacturer can install a network at a more reasonable cost than a full UNIX system, (2) as a firm's networking strategy develops, a Novell network will make changes easier because of the wide variety of topologies that it supports, and (3) PC applications become available in a multiuser environment.

As of the early 1990s Novell networks can be procured in terms of a few hundred U.S. dollars. This permits users to justify the expense of installing a network and allows the network to expand as the firm becomes more comfortable with the networking concept. The justification argument, however, is not as great as in the one of Ethernet-based UNIX nodes.

When one couples lower cost with Novell's multiple topologies, there is a strong strategic case. As a firm develops the network strategy in both the manufacturing and the business areas, Novell can adapt and accommodate the different networks that emerge. Thus the investment is protected.

Considerable manufacturing software has been written for PC use. Novell allows single-user programs to share data and operate in a multiuser environment.

DNC System Trends

Standards-based PC and associated networks will become increasingly important. Communication standards, such as MMS and X-Windows, will play an integral part in the growing use of PCs and LANs in manufacturing. The degree of sophistication will continue to increase. Examples of contemporary DNC systems are shown in Figs. 4 and 5.

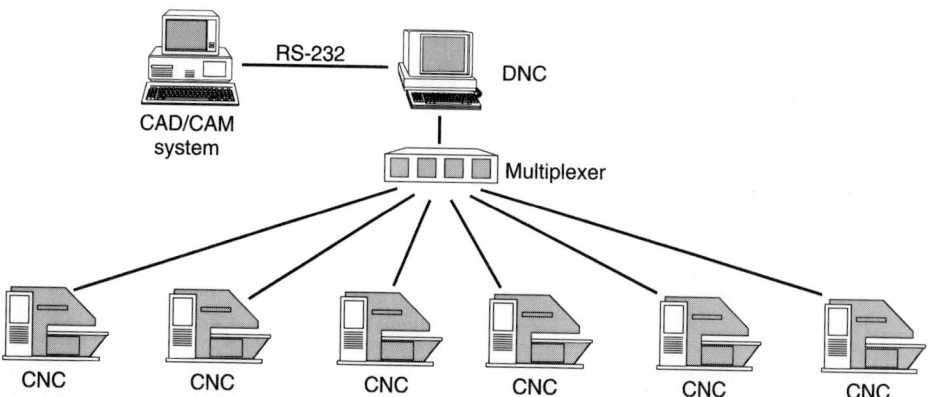

FIGURE 4   DNC system design for small workshop. Simple communications needed between CAM product and seven DNC machine tools with appropriate internal memory. There is a simple RS-232 interface between CAM product and DNC system. 8-port intelligent multiplexer interface to DNC system to communicate to seven machine tools via RS-232. This system allows job shop management to queue up jobs for each machine tool, compare original and revised part programs, and perform minor editing on part programs. System is fully upgradable to a LAN, and other third-party software packages can be added. (*CAD/CAM Integration, Inc.*)

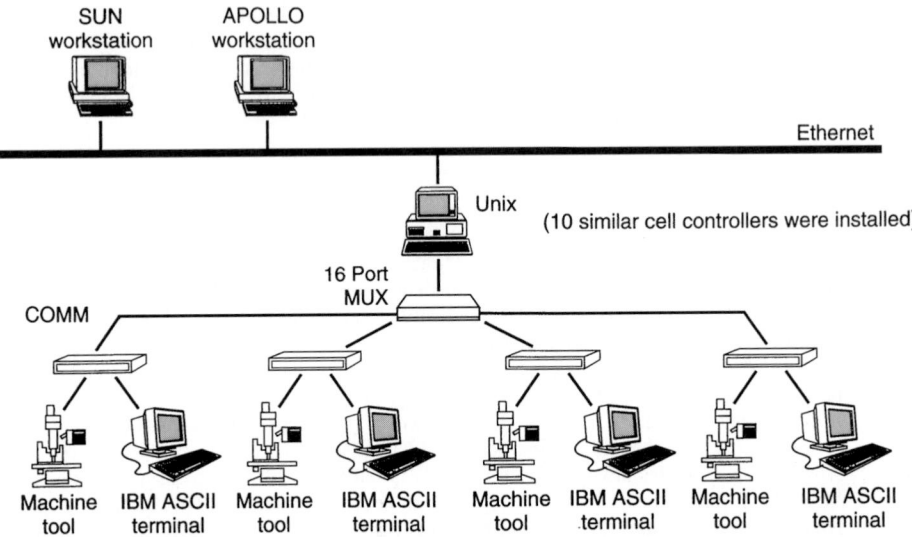

FIGURE 5 Large machine shop required communications to 10 machining cells with four machine tools in each cell. An existing Ethernet backbone was available. The management required that ASCII terminals be located at each machine tool, using a simple network that would minimize the number of wires running through the plant. It was desired to access the DNC system from the ASCII terminals to start job transfers to the machine tools. A UNIX node with star network configurations was installed at each machining cell. The star configuration provided machine tool interfacing to a central workstation hub, using a standard multiplexer. Each cell has additional multiplexer ports that could support additional machine tools if added at a later date. Each central hub attached to the Ethernet backbone provided closed -loop communications to existing workstations that already were making use of the network. A communications module was installed at each machine tool so that only one wire was drawn from the multiplexer to the machine tool. A communications module has two ports—one for interfacing to the machine tool controller and one or interfacing to the ASCII terminal. The installation is easily expandable both at the cell level and around the facility.

COMPUTERS AND CONTROLS

Ever since automatic controllers advanced from simple on-off devices, some form of computing has been required to implement algorithmic relationships in an effort to improve process and machine controllability. Over many decades, ingenious, essentially macroscale mechanical, pneumatic, hydraulic, and traditional electrical means were introduced and, considering their various vintage periods, operated quite satisfactorily. The computing capabilities required to achieve controllability that is considered acceptable in terms of the expectations of the early 1990s were enhanced by several factors, including the appearance of miniaturized solid-state electronic circuitry components with their high speed, their comparative simplicity, their progressively decreasing cost, and their inherent proclivity for handling instrumentation data in digital form, in contrast with the prior and essentially exclusive dependence on analog information.

The first electronic computer (although not solid-state) was developed by J. P. Eckert and J. W. Mauchly (Moore School of Electrical Engineering, University of Pennsylvania) with funds provided by the U.S. Army. The main purpose was that of calculating ballistic tables. Between 18,000 and

19,000 vacuum tubes replaced mechanical relays as switching elements. The machine weighed over 30 tons and occupied a very large room. Generally, the machine was not considered highly reliable. It existed until 1955 and logged some 80,000 hours of operation. The machine was known as the electronic numerical integrator and calculator (ENIAC).

Commencing essentially in the mid-1940s, digital computer development quickened in pace. IBM was active in the field and, in 1947, produced the selective sequence electronic calculator (SSEC). The Moore School developed a follow-up of the ENIAC, known as the electronic discrete variable computer (EDVAC), which was unveiled in 1952. Most historians now accept the fact that the EDVAC was the first stored-program computer. Unlike earlier machines, which were programmed at least partially by setting switches or using patch boards, the EDVAC and others to follow stored instructions and data in identical fashion. EDVAC used acoustical delay lines, which were simply columns of mercury through which data passed at the speed of sound, as the main memory. This type of storage soon gave way to magnetic-core memory, the antecedent of contemporary semiconductor memory.

In 1946 John Von Neumann, a mathematician at the Institute of Advanced Study in Princeton, New Jersey, prepared the paper, "Preliminary Discussion of the Logical Design of an Electronic Computing Instrument." The project, financed by the U.S. Army, suggested the principles of design for digital computers that were to be built over the next several decades. The principles outlined by Von Neumann included internal program storage, relocatable instructions, memory addressing, conditional transfer, parallel arithmetic, internal number-base conversion, synchronous internal timing, simultaneous computing while doing input-output, and magnetic tape for external storage. A computer built along these lines went into operation in the early 1950s. UNIVAC (universal automatic computer) was the first commercially available digital computer to use these principles. The UNIVAC was a descendent of ENIAC and EDVAC, having been built by Remington-Rand following the acquisition of the Eckert-Mauchly Computer Corporation. Eventually 48 UNIVACs were built, making Remington-Rand the number one computer manufacturer—until IBM commenced in earnest in 1954 with the introduction of the 700 line.

Early computer history would not be complete without mentioning the EDSAC (electronic data storage automatic calculator), developed by W. Renwick and M. V. Wilkes in 1949 at Cambridge, England; the Ferranti Mark I computer, developed at Manchester, United Kingdom, in 1950; the Pilot ACE (automatic computing engine), also developed at Manchester; the SEAC (standards eastern automatic computer), developed by the U.S. National Bureau of Standards (now NIST) in 1950; and WHIRLWIND I, developed by J. W. Forester at the Massachusetts Institute of Technology in 1950. It is interesting to note that WHIRLWIND I used an ultrasonic memory, later to be replaced by a Williams tube memory.

For nearly 3½ decades, commencing in the 1950s, computer architecture was largely based on the Von Neumann model. New concepts began to take hold in the 1980s, including parallelism, vector processing, multiprocessing, and memory access and interconnection schemes, such as the nearest-neighbor mesh, linear array, and hypercube schemes. Of course, a major event that has had a profound effect upon computers in instrumentation data acquisition and control systems was the introduction of the so-called personal computer (PC), described later in this article.

INITIAL CONSIDERATIONS OF COMPUTERS IN CONTROL SYSTEMS

A relatively few instrument and control engineers were early in their recognition of how digital computers could be used to advantage in control and data acquisition.

Direct Digital Control. In early systems a single mainframe computer accomplished data acquisition, control, reporting to the operator, and higher-level computation. In one of the first systems, the computer directly controlled the process, without intervening controllers. This architecture became known as direct digital control (DDC). Only one computer was used in the first system because of the

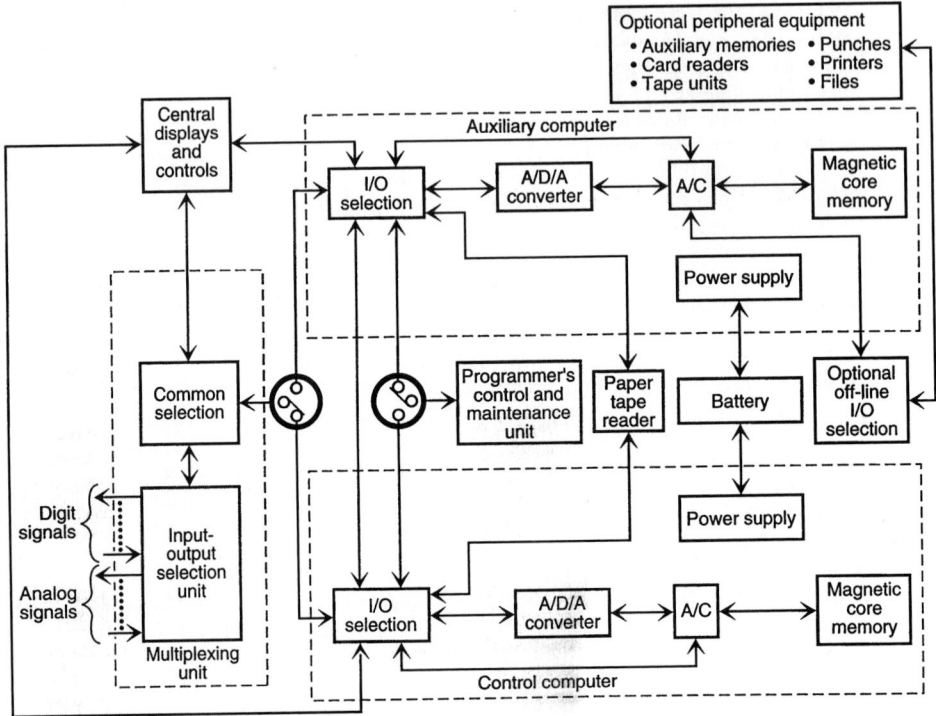

FIGURE 1   Direct digital control (DDC) system (with auxiliary computer) developed by D. Considine, D. McArthur, and D. Wilson. (*Hughes Aircraft Company.*)

high cost per processor and because of the general absence of computer-to-computer communications. Later systems used an auxiliary (redundant) computer. Early DDC system applications had a number of drawbacks. For example, if only one processor was used, a single failure could affect a large number of controlled variables and possibly disable an entire process. Although redundant processors were introduced to provide backup for hardware failure, additional processors for functional distribution did not appear until much later (Fig. 1).

Supervisory Control.   This concept, shown schematically in Fig. 2, developed over a period of years as an effort to use the many benefits available from the computer, but without the drawbacks of DDC. This concept had several advantages. In particular, it preserved the traditional panelboard control room while adding the capability of a digital computer and associated cathode-ray-tube (CRT) displays. This minimized the disruption of operations that could accompany a transition to full DDC. It also offered a buffer between the computer and the process, so that the operator had the option of selectively disabling the computer for troublesome loops. Although the impact of a computer failure was substantially reduced, it, too, used a few processors. Thus considerable functionality could still be disabled by a single failure.

Distributed Control Systems.   Introduced during the mid-1970s, distributed control systems (DCSs) combined three technologies, most of which either appeared for the first time or were much improved during the interim that dated back to DDC. These included (1) microprocessors, (2) data communications, and (3) CRT displays. Multiple microprocessors were used in some systems as dedicated control loop processors, CRT operating stations, and a variety of other control and communication functions.

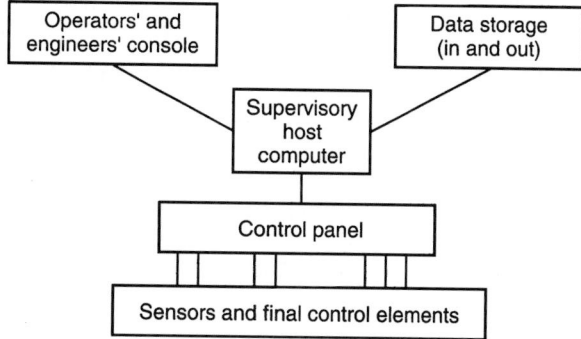

FIGURE 2   Schematic diagram of early supervisory control system.

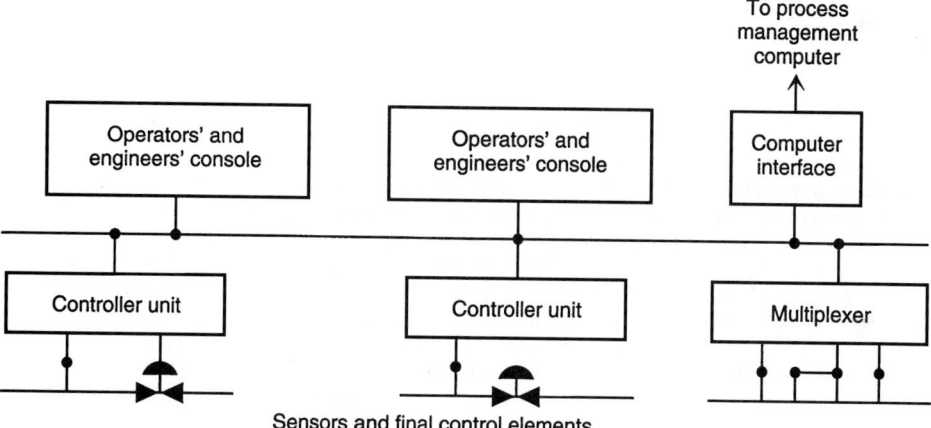

FIGURE 3   Schematic diagram of early distributed control system (DCS).

Although these systems could not accomplish all of the functions of the prior computer-based architectures in all instances, they did achieve considerable fault tolerance by minimizing the effect of a single failure. DCS is discussed in considerable detail in the first article of this handbook section (Fig. 3).

Hybrid Distributed Control Systems.   Because DCSs are limited in computer functionality and in logic or sequential control, they can be combined with supervisory host computers and with programmable controllers in hybrid architecture (Fig. 4). These systems use data communications channels between the host computer and both the programmable controllers and the DCS. Although these systems offer the strength of each subsystem, it is sometimes difficult to establish communication between the subsystems and confusing to provide software inasmuch as each subsystem may use different programming techniques and organizes data in separate databases.

In another version of the hybrid approach, the DCS is replaced by single-loop digital controllers. These offer some advantage, especially in some cases where a user may prefer a conventional panel as the primary operator interface rather than one or more CRTs.

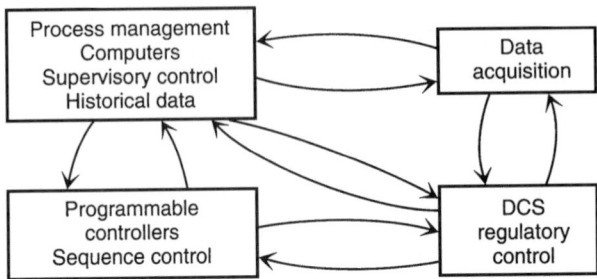

FIGURE 4   Schematic diagram of early hybrid distributed control system (DCS).

PERSONAL COMPUTERS IMPACT INDUSTRIAL INSTRUMENTATION[1]

Relatively early in the 1980s the PC enjoyed a multibillion dollar market and attracted the attention of pioneering professionals in the instrumentation and control field. It was recognized at an early date that one could develop a reasonably low-cost scan, control, alarm, and data acquisition (SCADA) platform and integrate it with general-purpose (shrink wrap[2]) software packages, such as spreadsheet and database management, to provide, in select cases, a cost-effective alternative to DCS. In a quantitative sense, prospective users of the PC tended to be cautious, and early applications occurred most often in cost-sensitive noncritical cases.

The initial acceptance of the PC was not universal, particularly with regard to using a PC as an industrial controller. Conservatives persisted in their opinions that PCs were limited to (1) off-line programming and downloading to real-time control computers, (2) for supervisor or operator advisory applications with artificial intelligence (AI) programs, (3) laboratory and small-scale pilot plant experiments, and (4) noncritical control applications, provided that programmable controller safety interlocks were also used. They were wary of the use of PCs in closed-loop situations but applauded their application for on-line engineering data acquisition and process performance analysis. Despite the conservative and limited acceptance by some, the PC has become very popular for a variety of control situations.

Basic PC

The advent of the modern PC makes it possible for virtually everyone to take advantage of the flexibility, power, and efficiency of computerized data acquisition and control. PCs offer high performance and low cost, coupled with an ease of use that is unprecedented. This contrasts with the implementation of data acquisition and control tasks that once required the power of a mainframe computer, the cost of which lead to timesharing among several users. The mainframe essentially was a centralized utility available to numerous parties. Consequently, small or remote jobs were often relegated to manual or, at best, simple electronic data-logging techniques. Thus smaller tasks could not benefit from the flexibility and power of a computerized solution.

Because of a significant (but not yet complete) degree of standardization among PC and data acquisition and control equipment manufacturers, a large selection of hardware and software tools and application packages has evolved. The end result is that an individual engineer or scientist can

---

[1] Personal computers are discussed in several other articles in this handbook. Check Index.

[2] Feature-laden high-quality inexpensive software spawned by enterprising software house and system integrators. This software was economically viable because the cost of the "intellectual property" could be spread over thousands of users.

now implement a custom data acquisition and control system within a fraction of the time and expense formerly required. It has become practical to tailor highly efficient solutions to unique applications. The nature of the PC encourages innovation, which in just a few years has revolutionized office automation and currently is producing similar results in manufacturing facilities, production lines, testing, laboratories, and pilot plants.

Major Components of the PC

A complete PC consists of a system unit containing (1) a microprocessor, (2) a memory, (3) a power supply, (4) a keyboard, and (5) one or more output devices, including monitors, printers, and plotters. The system unit usually is housed in an enclosure, separate from other major components such as the monitor.

Laboratory top and other miniature computers usually are *not* included in the true PC-compatible category because they lack expansion slots. However, growing numbers of small computers with expansion capabilities are appearing on the market.

An expansion slot is a physical and electrical space set aside for the connection of accessory hardware items to the PC. Electrical connection is made directly to the internal microcomputer bus. These accessory items usually take the form of a plug-in printed-circuit board (such as a graphics interface, a memory expansion module, or a data acquisition device).

Plug-in boards can be designed to be addressed by the microcomputer in two different ways, either as input–output (I/O) ports or as memory locations. Each method has its own advantages. However, memory addressing offers a higher level of performance that includes improved speed, extended address space, and the full use of the processor's instruction set.

Some of the computer I/O and memory addresses are reserved by the computer manufacturers for standard functions, such as graphics cards, RS-232 ports, memory, disk, and controllers.

PC Memory Types

Several distinct types of memory are used in PCs. These include RAM (random-access memory), ROM (read-only memory), floppy disk, and hard disk. Other memory technologies include magnetic tape and optical disk. RAM and ROM are semiconductor devices that offer a very high-speed operation. RAM has both read and write capabilities that are accessed by the microcomputer. ROM, in contrast, contains a fixed set of information that only can be read by the microcomputer. As with other computers, there is the central processing unit (CPU)—in this case, the microcomputer. ROM is preprogrammed at the factory and includes most of the basic CPU operating instructions, including the code that is required to start, or "boot," the computer. This is a special ROM or BIOS (basic I/O system). Other active program information is in RAM. The portion of RAM that can be used is a factor of the particular microprocessor chip used and, of course, the available software. The 20-bit address bus of the 8088 limits memory locations to about 1 Mbyte. The 80286 has a 24-bit bus which can address about 16 Mbytes. RAM is normally considered volatile, because in most systems its data will be destroyed with loss of power. Personal storage of data and programs usually is provided by the disk drives.

The complexities encountered with contemporary PCs usually discourage most users from attempting to "talk" directly with the CPU, BIOS, or disk drives. Interface software has been developed to bring the power of the PC within reach of nonspecialized users. Such software sometimes is referred to as the operating system.

PC Speed

The time required to run a program (that is, to process instructions) depends on several factors, including factors under the user's control, such as choice of language and the efficiency of the resulting code. Software efficiency partly refers to how many machine cycles are needed to execute the desired instructions. A so-called tight program will require the minimum number of machine cycles and thus, from a time standpoint, is most efficient. Other factors affecting speed include basic PC design factors, including the microprocessor chip selected, additional logic, circuit configuration, and clock frequency.

Data Conversion and Handling

Related topics, such as signal types, signal conditioning, sampled data systems, analog and digital I/O methodologies, and common-mode rejection are discussed in Section 7, Article 1 of this handbook.

Interfacing the PC

A PC may be connected to a data acquisition (and control) system in one of two basic ways: (1) by direct connection to the PC bus (internal bus products) or (2) by way of a standard communications channel (external bus products), such as RS-232, RS-422, or IEEE-488. There are advantages and limitations of each method.

Internal Bus Products.   Direct connection (Fig. 5) generally yields three advantages.

1. *Higher speed* results from avoiding the relatively slow, external communications-channel protocol. For example, the data acquisition rate using RS-232 communication at 9600 baud is limited to about 20 analog readings per second. By contrast, some direct PC bus products can receive data at a rate faster than one million samples per second.

2. *Lower cost* results because no separate enclosure or power supply is required. Power is obtained from the PC.

3. *Smaller size* results from the more efficient utilization of space.

Limitations of internal bus products include:

1. Lack of channel expansion capability.

2. Inability to add functions not originally procured.

Internal bus products are board-level systems that make direct connection to the computer expansion bus. These boards may have:

1. *Fixed arrangement* of analog and digital I/O, which is not conducive to changing for future needs, but which is usually available at a comparatively lower cost.

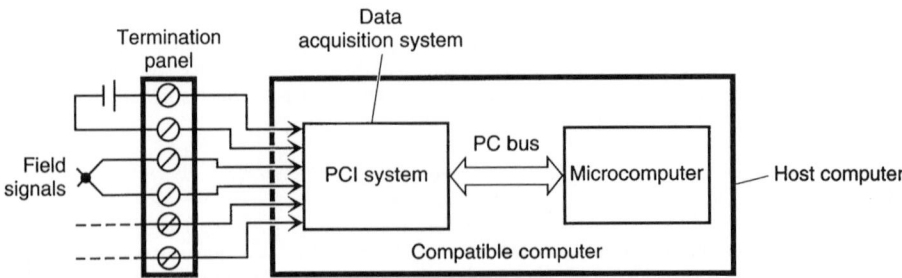

FIGURE 5   Block diagram of internal PC bus system. (*Intelligent Instrumentation.*)

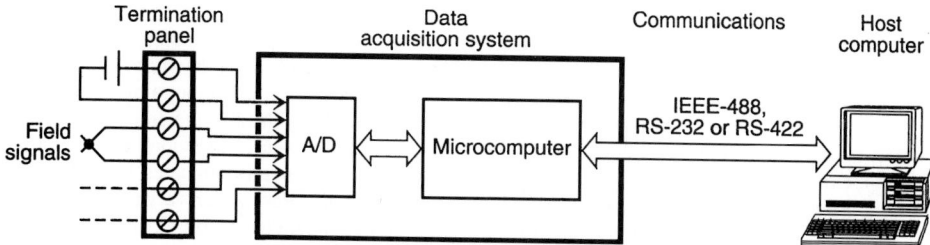

FIGURE 6 Block diagram of external bus data acquisition and control system. (*Intelligent Instrumentation.*)

2. *Modular design* that permits the user to select the number and configuration of the I/O functions needed. This is accomplished by using a grouping of function modules. (A modular board-level system offers some of the advantages of a box system.)

Fixed products include single-function and general-purpose configurations. The focused single-function boards often are effective in small applications or in well-defined instances where the board is embedded in a larger end product, as may be offered by an original equipment manufacturer (OEM).

Compromise is sometimes required in connection with general-purpose fixed I/O configurations. For example, the desired number of channels may not be obtainable (or the user may have to procure some unnecessary functions). Because of future developing needs, a mismatching may be inevitable. External add-on boards or boxes may be needed at a later date, frequently an undesirable situation because of space and cost considerations.

External Bus Products.   External PC connections (Fig. 6) frequently are advantageous for a number of reasons.

1. *Remote location* (from the host computer) permits the data acquisition portion of the system to be nearer to the sources of field signals and enables the construction of a distributed system. This allows many parameters to be monitored or controlled even if they are physically separated by an extensive distance from each other and from the host computer. For example, data originating from separate production lines in discrete-piece manufacturing or from separate units in a continuous chemical process—each with separate data acquisition and control systems—can be connected via RS-422. This permits monitoring by a single PC located several thousand feet (hundreds of meters) distant. For some systems this advantage can increase productivity and reduce overall costs.

2. *Offloading some data acquisition tasks* from the host computer through the use of data acquisition subsystems as described in 1.

3. *PC selection flexibility* because external bus products can be interfaced with virtually any kind of PC.

4. *System design flexibility* because the PC system need not be configured permanently from the start.

PC Software

In dealing with computers, including PCs, it is in order to frequently remind oneself that all communication with a computer must be in terms of digital 1s and 0s. At the outset of what has become computer science, only the machine language (a given voltage is present or is not present at an input) existed. Early computer programming was slow, error prone, difficult to maintain, and appreciated only by specialists.

Invention of the compiler was a great step forward in computer technology because a compiler understands (accepts) alphanumeric inputs and translates them into machine-readable code. Thus the compiler, which may be defined as an intermediate program, is a much better interface with human operators. Assembly language is the first step above machine code. An assembler is a low-level compiler that converts assembly language into machine code. The resulting code works with the computer's operating system to further simplify a specific programming task. Although more manageable than machine code, assembly language is not considered simple by the average computer user.

So-called high-level languages, such as BASIC, perform still more complex operations, while retaining a degree of recognizable user language, but still remain dominated by jargon and special syntax. C and Pascal are among several other high-level languages which vary in their ability in terms of execution speed and program maintainability. Compiled languages must be debugged as an entity. So-called interpreted languages are compiled incrementally, that is, line by line. This enables a single program line to be written and tested independently.

PC-based data acquisition systems have been designed so that users may write special programs for data acquisition, storage, display, logging, and control in high-level languages. It is the purpose of software to make these tasks as simple as possible. There are three classes of software available for PC-based data acquisition and control systems: (1) tutorial and program development tools, (2) function subroutine libraries, and (3) complete turnkey, menu-driven application packages.

In the case of program development tools, users may write their own unique application software, which usually includes "drivers" that provide the interface to the I/O hardware. Such packages ease the task of writing programs in high-level languages. This type of programming is considered quite flexible and generally useful.

Some application packages are designed for immediate system start-up, with little or no programming needed. These packages frequently are structured and less flexible than other types of software.

A large selection of third-party software is available, as prepared by independent software firms, or houses. Sometimes the term "genetic software" is used for such programs. LabVIEW[3] is an example of a third-party program. This is a SCADA (supervisory control and data acquisition) type of program for use in Windows NT. This software integrates process monitoring, data acquisition, output, test, measurement, analysis, and display. The program is compatible with most boards, carriers, modules, and IEEE-488 devices within the PC hardware system. The user interface is designed so that no programming is required. All features are selected with easy-to-use icons and menus. As a result, minimum computer skills are required to perform complex operations. The system includes fault-tolerant features, such as operating system error trapping and sensor voting. PID control and networking are optional capabilities. Custom color-graphic flow diagrams, including ISA symbols, provide display of real-time and trend information to the operator. Animation features can provide immediate visual feedback (such as the tank is filling or the door is open). Extensive alarm-processing capabilities are available. Alarms can be set to get the operator's attention, or they can automatically trigger a desired action. Data can be stored on floppy or hard disk, or written in a printer to provide historical records.

The LabVIEW[4] program is well suited for many real-world process control tasks in manufacturing and laboratory environments. It can automate experiments, control test sequences, perform calculations, display results in a graphic format, and generate reports. Applications include direct machine control, supervisory control, process monitoring, data logging, statistical process control (SPC), pilot plant production, and sequencing control. Specific areas of use include petrochemical and pharmaceutical processes, wave soldering, automotive machining, food processing, water treatment, and plastic extrusion. The program is menu-driven and icon-based. The conditions that define a current run are displayed on the screen and are readily modified. Conditions pertaining to a run can be stored and recalled as a group.

The program is flexible, permitting each channel to be set up with different characteristics. Sampling rates may vary from channel to channel and may also change at different times during a run.

---

[3] Name proprietary to National Instruments.

[4] If space permitted, numerous other PC software programs could be described. Programs like this typify the ingenious and rapid response of third-party software houses to the ever-increasing needs of the manufacturing and process industries.

Stored data can be played back as though they were being acquired in real time, thus permitting comparisons of current data with prior data.

With the program, channels can be used for purposes other than simple inputs or outputs. The program has the ability to derive channels from other channels as, for example, channels can "operate" on others by calculating averages, derivatives, integrals, and so on, in real-time. The group of mathematical, logical, statistical, and signal-processing functions also includes trigonometric functions, EXP, LN, LOG, OR, XOR, AND, filter, and FFT. These derived channels also can be used in determining triggers or as inputs to control loops.

Open- and closed-loop control algorithms may be implemented. In the open-loop mode, the user defines one period of any imaginable waveform, and the signal is then clocked out automatically during the run. For closed-loop control, both PID and on-off loops can be set up.

The program also includes a curve-fitting function. It uses an iterative routine to fit an arbitrarily complex model (up to 10 parameters) to the collected data. This and other routines, such as PID and thermocouple linearization, take advantage of the PC's optional coprocessor if available. This offers 80-bit real number processing, reduces round-off error, and permits faster computation.

Software Control Techniques

Many, if not most, data acquisition and control applications depend on the timely execution of read and write operations. When speed or timing is critical, three techniques for software control are available for consideration: (1) polling, (2) interrupts, and (3) direct memory access (DMA).

Polling. This is the simplest method for detecting a unique condition and then taking action. Polling involves a software loop that contains all of the required measurement, analysis, decision-making algorithms, and planned actions. The data acquisition program periodically tests the system's clock or external trigger input to sense a transition. Whenever a transition occurs, the program then samples each of the inputs and stores their values in a buffer. A buffer is simply a storage location that contains the values representing the specified inputs at a given time. The buffer can be stored in RAM, disk, or other types of memory. Each time the program senses a clock "tick," the inputs are scanned and converted, and a new value is added to the buffer.

Interrupts. These provide a means of tightly controlling the timing of events while allowing the processing of more than one task. Multitasking systems are also known as foreground/background systems. A method of putting data acquisition in the background is to relegate it to an interrupt routine. The clock or external timing signal, rather than being polled continuously, is used to generate an interrupt to the computer. Whenever the interrupt occurs, the computer suspends current activity and executes an interrupt service routine. The routine in this case might be a short program which acquires one frame of data and stores it in a buffer. The computer can perform other operations in the foreground while collecting data in the background. Whenever a clock tick or external interrupt occurs, the computer will automatically stop the foreground process, acquire the data, and then resume where it left off.

The reaction speed of the interrupt system is much slower than that of a well-prepared polling loop. This results because the interrupt mechanism in most computers involves a significant amount of software overhead. Also, the software complexity of interrupts can be significant. In most cases the programmer must be prepared to write assembly language code. In contrast, most polled systems can be written in a high-level language. Interrupts are useful in situations where the acquisition rate is slow, timing accuracy is not a priority, and background operation is important. When the time required to service an interrupt is small, compared to the rate at which the interrupts can occur, then this technique yields good results.

The foregoing factors illustrate that careful analysis is required before making a polling or interrupt decision.

Direct Memory Access.   This is a hardware technique that often allows the highest-speed transfer of data to or from RAM. Given the potentially more expensive hardware, DMA can provide the means to read or write data at precise times without significantly restricting the microprocessor's tasks. For example, one system,[5] under DMA control, can read or write any combination of analog, digital, or counter/timer data to or from RAM at a maximum rate of 300 kbyte/s. This is accomplished by taking minimal time from the other tasks of the microprocessor. The amount of time required to respond to a DMA request is very small compared with the time required to service an interrupt. This makes the goal of high-speed foreground/background operation possible.

Progressive Enhancement of the PC

In examining the progress that has been made in applying PCs to industrial data acquisition and control systems, one must recall that PC systems originally were designed for office automation, that is, for single-user, single-tasking, simple, comparatively low-cost machines. Thus PCs, as originally conceived, excel at word processing and spreadsheet applications. Often they are not capable of handling the real-time requirements of high-performance data acquisition and control. Thus it is not surprising that most of the technical advances found in the data acquisition and control boards available today are directed toward overcoming this initial limitation.

One excellent solution involves putting a high-performance processor right on the data acquisition board itself. This processor can assume control of the data acquisition functions, allowing the PC to act in a supervisory role. The PC can download programs to the data acquisition system and leave all the time-critical operations where they are most effectively handled. The architecture of one available system is exemplary of this approach inasmuch as the data acquisition functions already are on modules. This permits the limited available carrier board space to be dedicated to the on-board processor's functions.

Almost immediately after the IBM PC was introduced, data acquisition boards to plug into them became available. The early boards were basic, that is, they handled analog-to-digital (A/D) conversion, digital-to-analog (D/A) conversion, and digital I/O. This was followed by a period of increasing sophistication. Faster A/D converters with programmable gain instrumentation amplifier front ends were incorporated. Powerful features, such as on-board memory and exotic DMA techniques, appeared. The number of competitive sources increased; existing products often became obsolete within short periods as newer products with more powerful features became available. Thus a wide array of contemporary products has been developed, ranging from low-end digital I/O boards to array processing systems, the cost of which may be several times that of the PC itself.

Most manufacturers offer fixed-configuration multifunction data acquisition boards. A typical board will contain 16 single-ended or eight differential analog inputs, two analog outputs, 16 digital I/O bits, some sort of pacer clock or timer for generating A/D conversions, and often one or more event counters. The main differentiating factor among these boards generally is either the speed or the resolution of the A/D converter, or the mix of I/Os offered.

A problem with the fixed multifunction board approach is the difficulty of expanding or upgrading a system, as is frequently the case in research and development applications where a system may be used for numerous applications over a period of time. Even in fixed-configuration production systems, the number of channels provided by the board may be insufficient to handle a given problem. To expand the number of analog or digital I/O channels, the user is faced with purchasing a complete second board. This can be a significant penalty if all that is required is a few more digital I/O points. And to upgrade speed, a user must procure an entirely different board with a higher-performance A/D converter. This also may involve different software inasmuch as an earlier program written for a slower board may require rewriting to accommodate the different architecture of the faster board. For other applications there may be functions provided by the multifunction boards which are not needed.

---

[5] PCI system of Intelligent Instrumentation.

(Typically, only about 30% of data acquisition users require analog outputs. If they are provided, the user pays for them whether or not they are used.) Some manufacturers have provided solutions for most of the aforementioned problems.

TERMINOLOGY

During the last quarter century a massive transfer of computer and digital technology has infused into traditional instrumentation and control engineering. This was addressed in the mid-1980s by the editors of the third edition of this handbook by extensive discussions of computers and digital techniques. Because of the growing acceptance of the PC for data acquisition and control applications, and a deeper understanding gained by the profession during the last few years, there is less need for tutorial information. Thus the earlier glossary of terms has been reduced. However, some holders of the third edition may opt to retain their copy to augment this current, fourth edition.

Adder. A digital circuit which provides the sum of two or more input numbers as an output. A 1-bit binary adder is shown in Fig. 7. In this diagram, A and B are the input bits and C and $\bar{C}$ are the carry and no-carry bits from the previous position. There are both serial and parallel adders. In a *serial adder*, only one adder position is required and the bits to be added are sequentially gated to the input. The carry or no-carry from the prior position is remembered and provided as an input along with the bits from the next position. In a *parallel adder*, all the bits are added simultaneously, with the carry or no-carry from the lower-order position propagated to the higher position. In a parallel adder, there may be a delay due to the carry propagation time.

An adder may perform the subtraction as well as the addition of two numbers. Generally, this is effected by complementing one of the numbers and then adding the two factors. The following is an

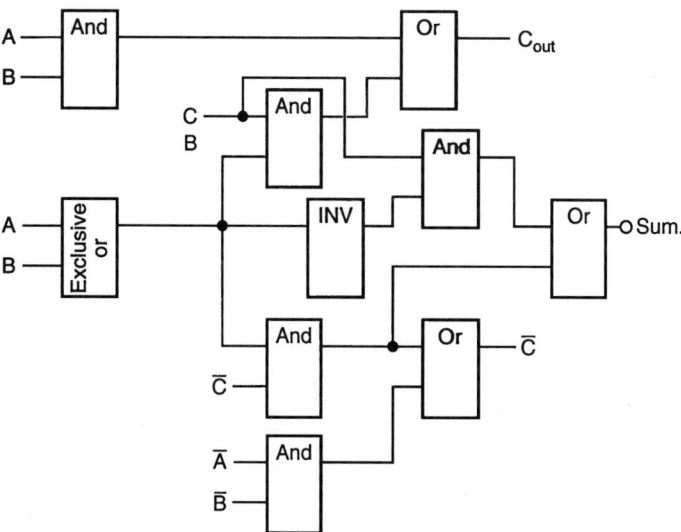

FIGURE 7   Binary adder.

example of a two's complement binary subtraction operation:

$$
\begin{array}{llll}
(a) & \quad 0110 & +6 & \text{(true)} \\
    & (+)\ 1010 & -6 & \text{(complement)} \\
\hline
    & \quad 10000 & \quad 0 & \text{(true)}
\end{array}
$$

$$
\begin{array}{llll}
(b) & \quad 0101 & +5 & \text{(true)} \\
    & (+)\ 1010 & -6 & \text{(complement)} \\
\hline
    & \quad 1111 & -1 & \text{(complement)}
\end{array}
$$

$$
(c) \qquad 1111\ \text{(complement)} = -0001\ \text{(true)}
$$

The two's complement of a binary number is obtained by replacing all 1s with 0s, replacing all 0s with 1s, and adding 1 to the units position. In (a), 6 is subtracted from 6, and the result is all 0s; the carry implies that the answer is in true form. In (b), 6 is subtracted from 5 and the result is all 1s with no carry. The no-carry indicates the result is in complement form and that the result must be recomplemented as shown in (c).

Address.   An identification, represented by a name, label, or number, for a digital computer register, device, or location in storage. Addresses are also parts of an instruction word along with commands, tags, and other symbols. The part of an instruction which specifies an operand for the instruction may be an address.

Absolute address or *specific address* indicates the exact physical storage location where the referenced operand is to be found or stored in the actual machine code address numbering system.

Direct address or *first-level address* indicates the location where the referenced operand is to be found or stored with no reference to an index register.

Indirect address or *second-level address* in a computer instruction indicates a location where the address of the referenced operand is to be found. In some computers, the machine address indicated can in itself be indirect. Such multiple levels of addressing are terminated either by prior control or by a termination symbol.

Machine address is an absolute, direct, unindexed address expressed as such or resulting after indexing and other processing has been completed.

Symbolic address is a label, alphabetic or alphanumeric, used to specify a storage location in the context of a particular program. Sometimes programs may be written using symbolic addresses in some convenient code, which then are translated into absolute addresses by an assembly program.

Base address permits derivation of an absolute address from a relative address.

Effective address is derived from applying specific indexing or indirect addressing rules to a specified address.

Four-plus-one address incorporates four operand addresses and a control address.

Immediate address incorporates the value of the operand in the address portion instead of the address of the operand.

N-level address is a multilevel address in which N levels of addressing are specified.

One-level address directly indicates the location of an instruction.

One-plus-one address contains two address portions. One address may indicate the operand required in the operation, and the other may indicate the following instruction to be executed.

Relative address is the numerical difference between a desired address and a known reference address.

Three-plus-one address incorporates an operation code, three operand address parts, and a control address.

Zero-level address permits immediate use of the operand.

AND *Circuit.*   A computer logical decision element which provides an output if and only if all the input functions are satisfied. A three-variable AND element is shown in Fig. 8. The function F is a binary 1 if and only if A and B and C are all 1s. When any of the input functions is 0, the output function is 0. This may be represented in boolean algebra by $F = A \cdot B \cdot C$ or $F = ABC$. Diode and transistor circuit schematics for two-variable AND functions also are shown in Fig. 8. In integrated

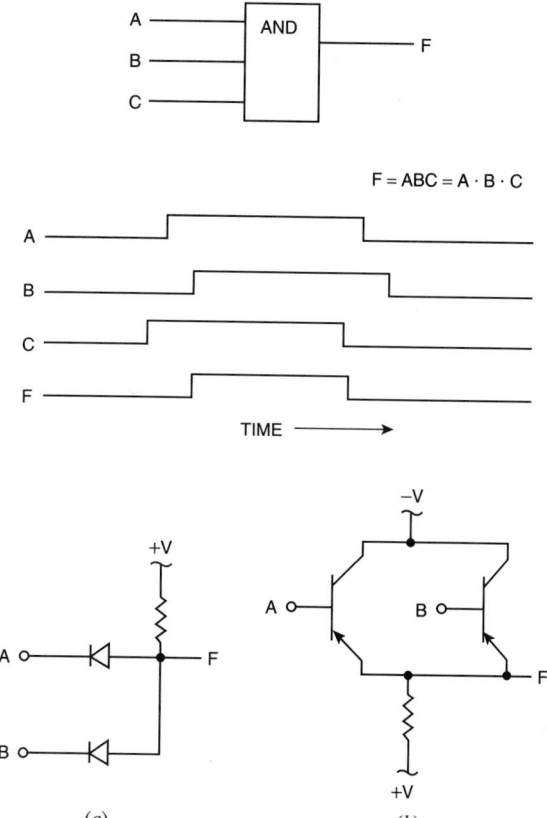

FIGURE 8   AND circuit: (*a*) diode-type circuit; (*b*) transistor-type circuit.

circuits, the function of the two transistors or diodes may be fabricated as a single active device. In the diode AND circuit, output F is positive only when both inputs A and B are positive. If one or both inputs are negative, one or both diodes will be forward-biased and the output will be negative. The transistor AND circuit operates in a similar manner, that is, if an input is negative, the associated transistor will be conducting and the output will be negative.

Generally referred to as "fan in," the maximum number of input functions of which a given circuit configuration is capable is determined by the leakage current of the active element. Termed "fan out," the number of circuits which can be driven by the output is a function of current that can be supplied by the AND circuit.

Assembler.   A computer program which operates on symbolic input data to produce machine instructions by carrying out such functions as (1) translating symbolic operation codes into computer instructions, (2) assigning locations in storage for successive instructions, and (3) assigning absolute addresses for symbolic addresses. An assembler generally translates input symbolic codes into machine instructions item for item and produces as output the same number of instructions or constants that were defined in the input symbolic codes.

Assembly language may be defined as computer language characterized by a one-to-one relationship between the statements written by the programmer and the actual machine instructions performed. The programmer thus has direct control over the efficiency and speed of the program. Usually the language allows the use of mnemonic names instead of numerical values for the operation codes of the instructions and similarly allows the user to assign symbolic names to the locations of the instructions and data. For the first feature, the assembler contains a table of the permissible mnemonic names and their numerical equivalents. For the second feature, the assembler builds such a table on a first pass through the program statements. Then the table is used to replace the symbolic names by their numerical values on a second pass through the program. Usually dummy operation codes (or pseudocodes) are needed by the assembler to pass control information to it. As an example, an origin statement is usually required as the first statement in the program. This gives the numerical value of the desired location of the first instruction or piece of data so that the assembler can, by counting the instructions and data, assign numerical values for their symbolic names.

The format of the program statements is usually rigidly specified, and only one statement per input record to the assembler is permitted. A representative statement is: symbolic name, operation code (or pseudocode), modifiers and/or register addresses, symbolic name of data. The mnemonic names used for the operation codes usually are defined uniquely for a particular computer type, with little standardization among computer manufacturers even for the most common operations. The programmer must learn a new language for each new machine.

Asynchronous.   A term used to designate the property of a device or action whose timing is not a direct function of the clock cycles in the system. In an asynchronous situation, the time of occurrence or duration of an event or operation is unpredictable because of factors such as variable signal propagation delay or a stimulus which is not under the control of the computer.

In terms of a computer channel, an asynchronous channel does not depend on computer clock pulses to control the transmission of information to and from the input or output device. Transmission of the information is under the control of interlocked control signals. Thus when a device has data to send to the channel, the device activates a service request signal. Responding to this signal, the channel activates a SERVICE OUT signal. The latter, in turn, activates a SECVICE IN signal in the device and also deactivates the request signal. Information is then transferred to the channel in coincidence with SERVICE IN, and the channel acknowledges receipt of the data by deactivating SERVICE OUT.

Asynchronous operation also occurs in the operation of analog-to-digital (A/D) subsystems. The systems may issue a command to the subsystem to read an analog point and then proceed to the next sequential operation. The analog subsystem carries out the A/D conversion. When the conversion is complete, the subsystem interrupts the system to signal the completion.

Asynchronous also has a broader meaning—specifically unexpected or unpredictable occurrences with respect to a program's instructions.

BASIC.   An acronym for *b*eginner's *a*ll-purpose *s*ymbolic *i*nstruction *c*ode. This language was developed at Dartmouth College by Kemeny and Kurz for timeshared use by students and other nonprofessional programmers. The language has been adopted by most computer firms and has been offered via timesharing services. BASIC is extensively used for small desktop computer systems and by computer hobbyists. The language is characterized by a very simple statement form. It normally requires that the first word in the statement be one of the small number of keywords and that a restricted naming convention for variables be used. An example of a program prepared in BASIC is

```
 05 LET E = -1
 10 READ A, B, C
 15 DATA 1, 2, 3
 20 LET D = A/B*C
 30 IF D = 4 GO TO 60
 40 IF D = 3/2 GO TO 80
 45 PRINT A, B, C, D, E
 50 STOP
 60 LET E = +1
 70 GO TO 45
 80 LET E = 0
 90 GO TO 45
100 END
```

The nucleus of the language is described by current ANSI specifications.

Baud.   A traditional unit of telegraph signaling speed derived from the duration of the shortest signaling pulse. A telegraphic speed of 1 baud is one pulse per second. The term *unit pulse* sometimes has the same meaning. A related term, *dot cycle*, refers to an on-off or mark-space cycle in which both mark and space intervals have the same length as the unit pulse.

Bit.   A contraction of *binary digit*. A single character in a binary numeral, that is, a 1 or 0. A single pulse in a group of pulses also may be referred to as a bit. The bit is a unit of information capacity of a storage device. The capacity in bits is the logarithm to the base 2 of the number of possible states of the device.

   Parity Bit.   A check bit that indicates whether the total number of binary 1 digits in a character or word (excluding the parity bit) is odd or even. If a 1 parity bit indicates an odd number of 1 digits, then a 0 bit indicates an even number of 1 digits. If the total number of 1 bits, including the parity bit, is always even, the system is called an even-parity system. In an odd-parity system, the total number of 1 bits, including the parity bit, is always odd.

   Zone Bit.   (1) One of the two leftmost bits in a system in which 6 bits are used for each character; related to overpunch. (2) Any bit in a group of bit positions used to indicate a specific class of items, such as numbers, letters, special signs, and commands.

Boolean Algebra.   Originated by George Boole (1815–1864), Boolean algebra is a mathematical method of manipulating logical relations in symbolic form. Boolean variables are restricted to two possible values or states. Possible pairs of values for the boolean algebra variable are YES and NO, ON and OFF, TRUE and FALSE, and so forth. It is common practice to use the symbols 1 and 0 as the Boolean variables. Since a digital computer typically uses signals having only two possible values or states, Boolean algebra enables the computer designer to combine mathematically these variables and manipulate them in order to obtain the minimum design which realizes a desired logical function. A table of definitions and symbols for some of the logical operations defined in Boolean algebra is shown in Fig. 9.

Branch.   A set of instructions that may be executed between a couple of successive decision instructions. Branching allows parts of a program to be worked on to the exclusion of other parts and provides a computer with considerable flexibility. The branch point is a junction in a computer routine where one or both of two choices are selected under control of the routine. Also refers to one instruction that controls branching.

Buffer.   An internal portion of a digital computer or data-processing system which serves as intermediary storage between two storage or data handling systems with different access times or formats. The word *buffer* is also used to describe a routine for compensating for a difference in data rate, or time of occurrence of events, when transferring data from one device or task to another. For example, a buffer usually is used to connect an input or output device with the main or internal high-speed storage. The term *buffer amplifier* applies to an amplifier which provides impedance transformation between two analog circuits or devices. A buffer amplifier may be used at the input of an analog-to-digital (A/D) converter to provide a low source impedance for the A/D converter and a high load impedance for the signal source.

Logical Operation	Symbol	Definition
AND	$\cdot$	$A \cdot A = A; A \cdot 0 = 0; A \cdot 1 = 1; A \cdot \bar{A} = 0$
OR	$+$	$A + A = A; A + 0 = A; A + 1 = 1; A + \bar{A} = 1$
NOT	$-$	$A\bar{A} = 0; A + \bar{A} = 1$
EXCLUSIVE OR	$\oplus$	$A \oplus B = \bar{A}B + A\bar{B} = \overline{A \odot B}$
COINCIDENCE	$\odot$	$A \odot B = A\bar{B} + AB = \overline{A \oplus B}$
NAND (or Sheffer stroke)	$/$	$A/B = \bar{A} + \bar{B} = \overline{AB}$
NOR (or Peirce)		$A\ B = \overline{A}\overline{B} = \overline{A + B}$

FIGURE 9   Boolean algebra symbols.

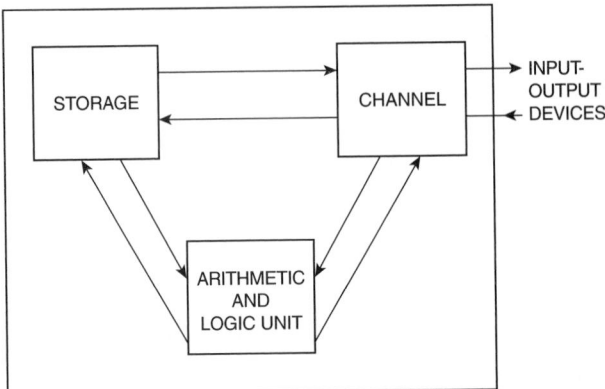

FIGURE 10   Computer central processing unit (CPU).

Byte.   A group of binary digits, usually shorter than a word and usually operated on as a unit. Through common usage, the word most often describes an 8-bit unit. This is a convenient information size to represent an alphanumeric character. Particularly in connection with communications, the trend has been toward the use of *octet* for an 8-bit byte, *sextet* for a 6-bit byte, and so on. Computer designs commonly provide for instructions that operate on bytes, as well as word-oriented instructions. The capacity of storage units is often specified in terms of bytes.

Central Processing Unit.   Also called the *mainframe*, the central processing unit (CPU) is the part of a computing system exclusive of input or output devices and, sometimes, main storage. As indicated in Fig. 10, the CPU includes the arithmetic and logic unit (ALU), channels, storage and associated registers, and controls. Information is transmitted to and from the input or output devices via the channel. Within the CPU, data are transmitted between storage and the channel and between each of these and the ALU. In some computing systems, if the channel and storage are separate assemblages of equipment, the CPU includes only the ALU and the instruction interpretation and execution controls.

A simplified example of CPU data flow was also given in Fig. 10. Generally, the operations performed in the CPU can be divided into (1) fetching instructions, (2) indexing (if appropriate to the instruction format), and (3) execution (Fig. 11). In the example shown, to initiate an operation, the contents of the instruction address register are transmitted to the storage address register (SAR), and the addressed instruction is fetched from storage and transferred to the storage data register. The operation code portion of the instruction is set into the operation code register (OP), and the data address part is set into the data address register. If indexing is defined in the instruction format, the indexing address is set into the index address register.

During the indexing phase of the operation, the index address register is transmitted to the SAR and the contents of the storage location are set into the storage data register. The contents of the storage data register and of the data address register are summed in the adder, and the result is placed in the data address register.

In connection with an arithmetic operation, the data address register is transmitted to the SAR, and the data in the referenced location are gated through the adder along with the contents of the data register. The result replaces the contents of the data register. If the arithmetic operation causes a carry-out of the high-order position of the word being operated on, the overflow indication is saved in a condition code trigger. This condition may be tested on a subsequent operation, such as a transfer on condition. On completion of the specified operation, the next instruction is fetched from storage and the previously described phases are repeated.

Channel.   The portion of the central processing unit (CPU) of a computer which connects input and output devices to the CPU. It may also execute instructions relating to the input or output devices. A channel also provides the interfaces and associated controls for the transfer of data between storage

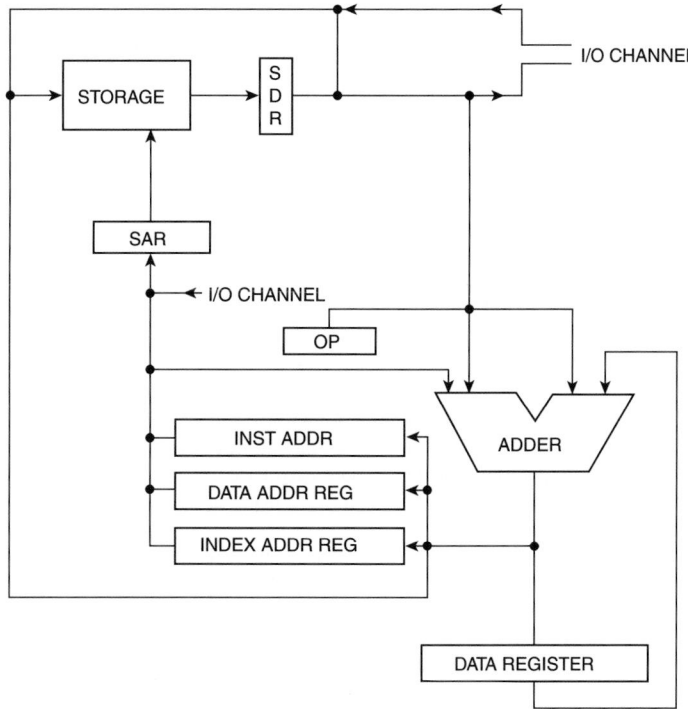

FIGURE 11   Block diagram of central processing unit of a computer.

and the input-output devices attached to the CPU. The channel generally maintains the storage address for the device in operation and includes the buffer registers needed for synchronizing with storage. A serial or selector channel may be used for slower devices.

A *multiplex channel* has the capability of concurrently servicing several devices. The storage address for each operating device is controlled by the channel and is maintained in main storage or in the logic of the channel. If the address is maintained in the logic, this increases the maximum data rate of the channel and reduces the number of storage references needed to service the device. The storage-access channel is an adaptation of the multiplex channel. With this type of channel, the device transmits both the data and the storage address to the central processor when it requires servicing.

Where a *series channel* is used, one or more devices may be physically attached to the channel interface, but only one of them is logically connected at any given time. Thus for the selected device, the full channel data rate capacity is available. For a given device data rate requirement, a series channel is less costly than either dedicated channels per device or a multiplex channel. In the latter instance, the saving is realized from the fact that the maximum required data rate is determined by the data rate of a single attached device. The maximum rate of a multiplex channel is the sum of the data rate requirements of several devices, plus the data rate required for device addressing.

Channel also may be used to describe other portions of some computers. For example, *analog input channel* refers to the path between the input terminals of an analog input subsystem and an analog-to-digital (A/D) converter. Similarly, it may describe the logical or physical path between the source and the destination of a message in a communications system. In some applications relating to data acquisition in physics experiments, it is synonymous with the quantization interval in an A/D converter. The tracks along the length of magnetic tape used for storing digital data are also referred to as channels.

Character.   One symbol of a set of elementary symbols, such as those corresponding to the keys of a typewriter, that is used for organization, representation, or control of data. Symbols may include the decimal digits 0 through 9, the letters A through Z, and any other symbol a computer can read, store, or write. Thus such symbols as @, #, $, and / are commonly used to expand character availability.

A *blank character* signifies an empty space on an output medium; or a lack of data on an input medium, such as an unpunched column on a punched card.

A *check character* signifies a checking operation. Such a character contains only the data needed to verify that a group of preceding characters is correct.

A *control character* controls an action rather than conveying information. A control character may initiate, modify, or stop a control operation. Actions may include the line spacing of a printer, the output hopper selection in a card punch, and so on.

An *escape character* indicates that the succeeding character is in a code that differs from the prior code in use.

A *special character* is not alphabetic, numeric, or blank. Thus @, #, . . . , are special characters.

Check.   A process of partial or complete testing of the correctness of computer or other data-processing machine operations. Checks also may be run to verify the existence of certain prescribed conditions within a computer or the correctness of the results produced by a program. A check of any of these conditions usually may be made automatically by the equipment; or checks may be programmed.

Automatic Check.   A procedure for detecting errors that is a built-in or integral part of the normal operation of a device. Until an error is detected, automatic checking normally does not require operating system or programmer attention. For example, if the product of a multiplication is too large for the space allocated, an error condition (overflow) will be signaled.

Built-in Check.   An error detecting mechanism that requires no program or operator attention until an error is detected. The mechanism is built into the computer hardware.

Checkpoint.   A point in time in a machine run at which processing is momentarily halted to perform a check or to make a magnetic tape or disk record (or equivalent) of the condition of all the variables of the machine run, such as the status of input and output devices and a copy of working storage. Checkpoints are used in conjunction with a restart routine to minimize reprocessing time occasioned by functional failures.

Duplication Check.   Two independent performances of the same task are completed and the results compared. This is illustrated by the following operations:

$$
\begin{array}{rrrr}
12 & 31 & 84 & 127 \\
9 & 14 & 43 & 66 \\
\hline
21 & 45 & 127 & 193 \\
\end{array}
$$

Parity Check.   A summation check in which the binary digits in a character or word are added, modulo 2, and the sum checked against a single, previously computed parity digit, that is, a check which tests whether the number of 1s in a word is odd or even.

Programmed Check.   A system of determining the correct program and machine functioning either by running a sample problem with similar programming and a known answer or by using mathematical or logic checks, such as comparing $A \times B$ with $B \times A$. Also, a check system built into the program for computers that do not have automatic checking.

Reasonableness Check.   Same as *validity check.* See below.

Residue Check.   An error detection system in which a number is divided by a quantity n and the remainder compared with the original computer remainder.

Sequence Check.   A data-processing operation designed to check the sequence of the items in a file assumed to be already in sequence.

Summation Check.   A check in which groups of digits are summed, usually without regard for overflow, and that sum is checked against a previously computed sum to verify that no digits have been changed since the last summation.

Validity Check. A check based on known limits or on given information or computer results; for example, a calendar month will not be numbered greater than 12, or a week does not have more than 168 hours.

Code. A system of symbols for representing data or instructions in a computer or data-processing machine. A machine language program sometimes is referred to as a *code*.

Alphanumeric Code. A set of symbols consisting of the alphabet characters A through Z and the digits 0 through 9. Sometimes the definition is extended to include special characters. A programming system commonly restricts user-defined symbols to only those using alphanumeric characters and for the system to take special action on the occurrence of a nonalphanumeric character, such as $, %, or &. For example, a job currently in progress may be stopped should a given special character be encountered.

Binary Code. (1) A coding system in which the encoding of any data is done through the use of bits, that is, 0 or 1. (2) A code for the 10 decimal digits 0, 1, . . . , 9 in which each is represented by its binary, radix 2, equivalent, that is, straight binary.

Biquinary Code. A two-part code in which each decimal digit is represented by the sum of the two parts, one of which has the value of decimal 0 or 5, and the other the values 0 through 4. The abacus and soroban both use biquinary codes. An example follows:

Decimal	Biquinary	Interpretation
0	0 000	0 + 0
1	0 001	0 + 1
2	0 010	0 + 2
3	0 011	0 + 3
4	0 100	0 + 4
5	1 000	5 + 0
6	1 001	5 + 1
7	1 010	5 + 2
8	1 011	5 + 3
9	1 100	5 + 4

Column-Binary Code. A code used with punch cards in which successive bits are represented by the presence or absence of punches in contiguous positions in successive columns as opposed to rows. Column binary code is widely used in connection with 36-bit-word computers, where each group of three columns is used to represent a single word.

Computer Code or Machine Language Code. A system of combinations of binary digits used by a given computer.

Excess-Three Code. A binary-coded decimal code in which each digit is represented by the binary equivalent of that number plus 3; for example:

Decimal Digit	Excess-3 Code	Binary Value
0	0011	3
1	0100	4
2	0101	5
3	0110	6
4	0111	7
5	1000	8
6	1001	9
7	1010	10
8	1011	11
9	1100	12

Gray Code.   A binary code in which sequential numbers are represented by expressions which are the same except in one place and in that place differ by one unit:

Decimal	Binary	Gray
0	000	000
1	001	001
2	010	011
3	011	010
4	100	110
5	101	111

Thus in going from one decimal digit to the next sequential digit, only one binary digit changes its value; synonymous with cyclic code.

Instruction Code.   The list of symbols, names, and definitions of the instructions that are intelligible to a given computer or computing system.

Mnemonic Operation Code.   An operation code in which the names of operations are abbreviated and expressed mnemonically to facilitate remembering the operations they represent. A mnemonic code normally needs to be converted to an actual operation code by an assembler before execution by the computer. Examples of mnemonic codes are ADD for addition, CLR for clear storage, and SQR for square root.

Numeric Code.   A system of numerical abbreviations used in the preparation of information for input into a machine; that is, all information is reduced to numerical quantities. Contrasted with *alphabetic code.*

Symbolic Code or Pseudocode.   A code that expresses programs in source language; that is, by referring to storage locations and machine operations by symbolic names and addresses which are independent of their hardware-determined names and addresses.

Two-out-of-Five Code.   A system of encoding the decimal digits $0, 1, \ldots, 9$ where each digit is represented by binary digits of which two are 0's and three are 1's, or vice versa.

Column.   A character or digit position in a positional information format, particularly one in which characters appear in rows and the rows are placed one above another; for example, the rightmost column in a five-decimal-place table or in a list of data.

Command.   The specification of an operation to be performed. In terms of a control signal, a command usually takes the form of YES (go ahead) or NO (do not proceed). A command should not be confused with an instruction. In most computers, an instruction is given to the central processing unit (CPU), as contrasted with a command, which is an instruction to be followed by a data channel. An input command, for example, may be READ and an output command, WRITE.

Compiler.   A program designed to translate a higher-level language into machine language. In addition to its translating function, which is similar to the process used in an assembler, a compiler program is able to replace certain items of input with a series of instructions, usually called subroutines. Thus where an assembler translates item for item and produces as output the same number of instructions or constants that were put into it, a compiler typically produces multiple output instructions for each input instruction or statement. The program which results from compiling is a translated and expanded version of the original.

Compiler language is characterized by a one-to-many relationship between the statements written by a programmer and the actual machine instructions executed. The programmer typically has little control over the number of machine instructions executed to perform a particular function and is dependent on the particular compiler implementation. Typically, the language is very nearly machine-independent and may be biased in its statements and features to a particular group of users with similar problems. Thus these languages sometimes are referred to as problem-oriented languages (POLs). Slightly different implementations of a given language are sometimes called *dialects.*

Computer Operating System.   Generally defined as a group of interrelated programs to be used on a computer system in order to increase the utility of the hardware and software. There is a wide range in the size, complexity, and application of operating systems. The need for operating systems arose from the desire to obtain the maximum amount of service from a computer. A first step was simple monitor systems providing a smooth job-to-job transition. Modern operating systems contain coordinate programs to control input-output scheduling, task scheduling, error detection and recovery, data management, debugging, multiprogramming, and on-line diagnostics.

Operating system programs fall into two main categories: (1) control programs and (2) processing programs:

1. Control programs

   a. Data management, including all input-output

   b. Job management

   c. Task management

2. Processing programs

   a. Language translators

   b. Service programs (linking programs, sort or merge, utilities)
      System library routines

   c. User-written programs

Control programs provide the structure and environment in which work may be accomplished more efficiently. They are the components of the supervisory portion of the system. Processing programs are programs that have some specific objective that is unrelated to controlling the system work. These programs use the services of supervisory programs rather than operating as part of them.

Data management is involved with the movement of data to and from all input-output devices and all storage facilities. This area of the system embraces the functions referred to as the input-output control system (IOCS), which frequently is segmented into two parts, the physical IOCS and the logical IOCS.

Physical IOCS is concerned with device and channel operations, error procedures, queue processing, and, generally, all operations concerned with transmitting physical data segments from storage to external devices. *Logical IOCS* is concerned with data organization, buffer handling, data referencing mechanisms, logical device reference, and device independence.

Job management involves the movement of control cards or commands through the system input device, their initial interpretation, and the scheduling of jobs so indicated. Other concerns are job queues, priority scheduling of jobs, and job accounting functions.

Task management is concerned with the order in which work is performed in the system. This includes management of the control facilities, namely, central processing unit, storage, input-output channels, and devices, in accordance with some task priority scheme.

Language translators allow users to concentrate on solving logical problems with a minimum of thought given to detailed hardware requirements. In this category are higher-level languages, report generators, and special translation programs.

Service programs are needed to facilitate system operation or to provide auxiliary functions. Sort programs, program linking functions, and file duplication, as well as system libraries are within this group.

User-written programs are programs prepared specifically to assist the user in the accomplishment of his or her objectives.

Concurrent Operation.   The performance of several actions during the same interval of time, although not necessarily simultaneously. Compare with *parallel operation*, described later in this section. Multiprogramming is a technique that provides concurrent execution of several tasks in a computer. See also *multiprogramming* in this section.

Conversion Time. The interval of time between initiation and completion of a single analog-to-digital (A/D) or digital-to-analog (D/A) conversion operation. Also, the reciprocal of the conversion rate. In practice, the term usually refers to A/D converters or digital voltmeters. The conversion time required by an A/D converter is comprised of (1) the time needed to reset and condition the logic, (2) a delay to allow for the settling time of the input buffer amplifier, (3) a polarity determination time, (4) the actual A/D conversion operation, and (5) any time required to transfer the digital result to an output register. Not all factors are always present. A unipolar A/D converter, for example, involves no polarity determination time.

Conversion time also is used in connection with data acquisition or analog input subsystems for process control computers. The more precise term in this case is *measurement time*. In this case the measurement rate may not be the reciprocal of the measurement time, inasmuch as some of the operations performed may overlap with other operations. With some types of A/D converters it is possible to select the next multiplexer point during the time the previous value is being converted from analog to digital form.

Conversion or measurement time also may include the time required for such operations as multiplexer address decoding, multiplexer switch selection, and settling time, and the time required to deactivate the multiplexer switches and permit the subsystem to return to an initial state.

Counter. A physical or logical device capable of maintaining numeric values which can be incremented or decremented by the value of another number. A counter, located in storage, may be incremented or decremented under control of the program. An example of such use is recording the number of times a program loop has been executed. The counter location is set to the value of the number of times the sequence of instructions is to be performed. On completion of the sequence, the counter is decremented by 1 and then tested for zero. If the answer is nonzero, the sequence of instructions may be repeated.

A counter also may be in the form of a circuit that records the number of times an event occurs. A counter which counts according to the binary number system is illustrated by the truth table shown in Fig. 12. Counters also may be used to accumulate the number of times an external event takes place. The counter may be a storage location or a counter circuit which is incremented as the result of an external stimulus. Specific counter definitions include the following.

A *binary counter* is (1) a counter which counts according to the binary number system, or (2) a counter capable of assuming one of two stable states.

A *control counter* records the storage location of the instruction word which is to be operated on following the instruction word in current use. The control counter may select storage locations in sequence, thus obtaining the next instruction word from the subsequent storage location, unless a transfer or special instruction is encountered.

A *location counter* or *instruction counter* is (1) the control section register which contains the address of the instruction currently being executed, or (2) a register in which the address of the current instruction is recorded. It is synonymous with *program address counter.*

	A_3	A_2	A_1
P_1	0	0	1
P_2	0	1	0
P_3	0	1	1
P_4	1	0	0
P_5	1	0	1
P_6	1	1	0
P_7	1	1	1
P_8	0	0	0

FIGURE 12 Truth table for binary counter. The situation illustrates a binary counter of three stages capable of counting up to eight pulses. Each trigger changes state when a pulse is gated to its input. In the instance of trigger A_2, it receives an input pulse only when trigger A_1 and the input are both 1. Subsequently, trigger A_3 changes its state only when both triggers A_1 and A_2 and the input pulse are all 1s. This table shows the values of each trigger after each input pulse. At the eighth pulse, the counter resets to zero.

Diagnostics. Programs provided for the maintenance engineer or operator to assist in discovering the source of a particular computer system malfunction. Diagnostics generally consist of programs which force extreme conditions (the worst patterns) on the suspected unit with the expectation of exaggerating the symptoms sufficiently for the engineer to readily discriminate among possible faults and to identify the particular fault. In addition, diagnostics may provide assistance in localizing the cause of a malfunction to a particular card or component in the system.

There are two basic types of diagnostic programs: (1) off-line and (2) on-line. Off-line diagnostic programs are those which require that there be no other program active in the computer system, sometimes requiring that there be no executive program in the system. Off-line diagnostics are typically used for central processing unit (CPU) malfunctions, very obscure and persistent peripheral device errors, or critically time-dependent testing. For example, it may be suspected or known that a harmonic frequency is contributing to the malfunction. Thus it may be desirable to drive the unit continuously at various precise frequencies close to the suspected frequency to confirm the diagnosis and then to confirm the cure. Interference from other activities may well make such a test meaningless. Thus all other activity on the system must cease.

On-line diagnostics are used mainly in a multiprogramming environment and are vital to the success of real-time systems. The basic concept is that of logically isolating the malfunctioning unit from all problem programs and allowing the diagnostic program to perform any and all functions on the unit. Many of the more common malfunctions can be isolated by such diagnostics, but there are limitations imposed by interference from other programs also using the CPU.

Digit.   A symbol used to convey a specific quantity either by itself or with other numbers of its set; for example, 2, 3, 4, and 5 are digits. The base or radix must be specified and each digit's value assigned.

Binary Digit.   A number on the binary scale of notation. This digit may be 0 or 1. It may be equivalent to an ON or an OFF condition or a YES or NO condition. Often abbreviated *bit*.

Check Digit.   One or more redundant digits carried along with a machine word and used in relation to the other digits in the word as a self-checking or error detecting code to detect malfunctions of equipment in data transfer operations.

Equivalent Binary Digits.   The number of binary positions needed to enumerate the elements of a specific set. In the case of a set with five elements, it will be found that three equivalent binary digits are needed to enumerate the five members of the set 1, 10, 11, 100, and 101. Where a word consists of three decimal digits and a plus or minus sign, 1999 different combinations are possible. This set would require 11 equivalent binary digits in order to enumerate all its elements.

Octal Digit.   The symbols 0, 1, 2, 3, 4, 5, 6, or 7 used as a digit in the system of notation which uses 8 as the base or radix.

Sign Digit.   A digit incorporating 1 to 4 binary bits, which is associated with a data item for the purpose of denoting an algebraic sign. In most binary, word-organized computers, a 1-bit sign is used: $0 = +$ (plus); and $1 = -$ (minus). Although not strictly a digit by the foregoing definition, it occupies the first digit position and it is common to consider it a digit.

EXCLUSIVE OR *Circuit.*   A logical element which has the properties that if either of the inputs is a binary 1, then the output is a binary 1. If both the inputs are a binary 1 or 0, the output is a binary 0. In terms of boolean algebra, this function is represented as $F = AB' + BA'$, where the prime denotes the NOT function. With reference to the transistor EXCLUSIVE OR circuit shown in Fig. 13, the output is positive when either transistor is in saturation. When input A is positive and B is negative, transistor T_2 is in saturation. When B is positive and A is negative, transistor T_1 is in saturation. When A and B are either both positive or both negative, then both transistors are cut off and the output F is negative. Although shown as discrete devices in the figure, fabrication using large-scale integrated circuit technology may utilize other circuit and design configurations.

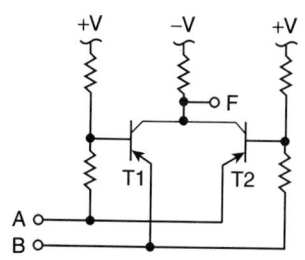

FIGURE 13   EXCLUSIVE OR circuit.

Fixed-Point Arithmetic.   A method of storing numeric data in a computer such that the data are all stored in integer form (or all in fractional form) and the user postulates a radix point between a certain pair of digits. Consider a computer whose basic arithmetic is in decimal and in which each computer word consists of seven decimal digits in integer form. If it is desired to add 2.796512 to 4.873214, the data are stored in the computer as 2796512 and 4873214, the sum of which is 7669726. It is recalled that a decimal point between digits 1 and 2 has been postulated. The result, therefore, represents 7.669726. Input and output conversion routines often are provided for convenience. These routines can add or delete the radix point in the external representation and align the data as required internally.

Fixed-point operations are fast and thus preferred over floating-point operations. It is important, of course, that the magnitude of the numbers be much better known than for floating-point numbers, since the absolute magnitude is limited by word size and the availability of double-length operations. For many applications, fixed-point calculations are practical and increase speed.

Fortran.   An acronym standing for *for*mula *trans*lation. It is a programming language designed for problems which can be expressed in algebraic notation, allowing for exponentiation, subscripting, and other mathematical functions.

Fortran was introduced by IBM in 1957 after development by a working group headed by John W. Backus. It was the first computer language to be used widely for solving numerical problems and was the first to become an American National Standard. Numerous enhancements have been added to the language. The current version is described in ANSI/ISA publications.

Index Register.   The contents of the index register of a computer are generally used to modify the data address of the instruction as the instruction is being read from storage. The modified address is called the *effective data address*. A particular index register is addressed by a specified field in the format of the instruction. The data address of the instructions would contain the address of the required data with reference to the start of the table. All instructions which reference table data are indexed by the specified index register which contains the address of the start of the table. Thus when the program is to perform these operations on another table of data, the value in the index register is changed to the start address of the new data table. This effectively modifies all the indexed instructions in the sequence.

Instruction.   (1) A set of characters which defines an operation together with one or more addresses, or no address, and which, as a unit, causes the computer to perform the operation on the indicated quantities. The term *instruction* is preferable to the terms *command* and *order*. *Command* is reserved for a specified portion of the instruction word, that is, the part which specifies the operation to be performed. *Order* is reserved for the ordering of the characters, implying sequence, or the order of the interpolation, or the order of the differential equation. (2) The operation or command to be executed by a computer, together with associated addresses, tags, and indices.

Interrupt.   A signal which causes the central processing unit (CPU) to change state as the result of a specified condition. An interrupt represents a temporary suspension of normal program execution and arises from an external condition, from an input or output device, or from the program currently being processed in the CPU. See earlier text in this article.

Language.   A communications means for transmitting information between human operators and computers. The human programmer describes how the problem is to be solved using the computer language. A computer language consists of a well-defined set of characters and words, coupled with a series of rules (*syntax*) for combining them into computer instructions or statements. There is a wide variety of computer languages, particularly in terms of flexibility and ease of use. There are three levels in the hierarchy of computer languages:

Machine Language.   (1) A language designed for interpretation and use by a machine without translation. (2) A system for expressing information which is intelligible to a specific machine, such as a computer or class of computers. Such a language may include instructions which define and direct machine operations, and information to be recorded by or acted on by these machine operations. (3) The set of instructions expressed in the number system basic to a computer, together with symbolic operation codes with absolute addresses, relative addresses, or symbolic addresses. In this case, it is known as an *assembly language*.

Problem-Oriented Language. A language designed for the convenience of program specification in a general problem area. The components of such a language may bear little resemblance to machine instructions and often incorporate terminology and functions unique to an application. This type of language is also known as an *applications language*.

Procedure-Oriented Language. A machine-independent language which describes how the process of solving the problem is to be carried out. For example, Fortran, Algol, PL/1, and Cobol.

Other computer languages include:

Algorithmic Language. An arithmetic language by which numerical procedures may be precisely presented to a computer in a standard form. The language is intended not only as a means of directly presenting any numerical procedure to any appropriate computer for which a compiler exists but also as a means of communicating numerical procedures among individuals.

Artificial Language. A language specifically designed for ease of communication in a particular area of endeavor, but one that is not yet "natural" to that area. This is contrasted with a natural language which has evolved through long usage.

Common Business-Oriented Language. A specific language by which business data-processing procedures may be precisely described in a standard form. The language is intended not only as a means for directly presenting a business program to any appropriate computer for which a compiler exists but also as a means of communicating such procedures among individuals.

Common Machine Language. A machine-sensible information representation that is common to a related group of data-processing machines.

Object Language. A language that is the output of an automatic coding routine. Usually object language and machine language are the same. However, a series of steps in an automatic coding system may involve the object language of one step serving as a source language for the next step, and so forth.

Logic. In hardware, a term referring to the circuits that perform the arithmetic and control operations in a computer. In designing digital computers, the principles of Boolean algebra are employed. The logical elements of AND, OR, INVERTER, EXCLUSIVE OR, NOR, NAND, NOT, and so forth, are combined to perform a specified function. Each of the logical elements is implemented as an electronic circuit which in turn is connected to other circuits to achieve the desired result. The word *logic* is also used in computer programming to refer to the procedure or algorithm necessary to achieve a result.

Logical Operation. (1) A logical or Boolean operation on N-state variables which yields a single N-state variable, such as a comparison of the three-state variables A and B, each represented by $-$, 0, or $+$, which yields $-$ when A is less than B, 0 when A equals B, and $+$ when A is greater than B. Specifically, operations such as AND, OR, and NOT on two-state variables which occur in the algebra of logic, that is, boolean algebra. (2) Logical shifting, masking, and other nonarithmetic operations of a computer.

Loop. A sequence of instructions that may be executed repeatedly while a certain condition prevails. The productive instructions in a loop generally manipulate the operands, while bookkeeping instructions may modify the productive instructions and keep count of the number of repetitions. A loop may contain any number of conditions for termination, such as the number of repetitions or the requirement that an operand be nonnegative. The equivalent of a loop can be achieved by the technique of straight-line coding, whereby the repetition of productive and bookkeeping operations is accomplished by explicitly writing the instructions for each repetition.

Macroassembler. An assembler that permits the user to define pseudocomputer instructions which may generate multiple computer instructions when assembled. Source statements which may generate multiple computer instructions are termed *macrostatements* or *macroinstructions*. With a process control digital computer, a macroassembler can be a most significant tool. By defining a set of macrostatements, for example, a process control engineer can define a process control programming language specifically oriented to the process.

Mask.   A pattern of digits used to control the retention or elimination of portions of another pattern of digits. Also, the use of such a pattern. For example, an 8-bit mask having a single i bit in the ith position, when added with another 8-bit pattern, can be used to determine whether the ith bit is a 1 or a 0; that is, the ith bit in the pattern will be retained and all other bits will be 0's.

As another example, there are situations where it is desirable to delay the recognition of a process interrupt by a digital computer. A mask instruction permits the recognition of specific interrupts to be inhibited until it is convenient to service them.

Memory.   Although the word *memory* is widely used, international standards list the word as a deprecated term and prefer the word *storage*. Memory or storage is a means of storing information. A complete memory system incorporates means of placing information into storage and of retrieving it. See earlier text in this article.

Microcomputer.   A computer that utilizes a microprocessor as its central processing unit (CPU). This CPU must perform two functions: (1) sequence through the instructions and (2) execute each instruction. A microcomputer requires two other fundamental elements: (1) memory for the program – a sequence of instructions to be performed, and (2) input–output circuits to tie it to external devices. A microcomputer is a digital logic device, that is, all input-output signals are at digital logic levels, either 0 or 5 V. A significant amount of interface circuitry is required between the microcomputer and external devices.

Microprocessor.   A program-controlled component, normally contained on a single crystal of silicon on which over one thousand transistors may be implemented. The microprocessor functions to control the sequence of instructions making up a program. It sequences through these instructions, examining each instruction in turn and executing it. This means that the microprocessor performs some manipulation of data. See text in earlier part of this article.

Microprogram.   Microprogramming is a technique of using a special set of instructions for an automatic computer. Elementary logical and control commands are used to simulate a higher-level instruction set for a digital computer. The basic machine-oriented instruction set in many computers is comprised of commands, such as ADD, DIVIDE, SUBTRACT, and MULTIPLY, which are executed directly by the hardware. The hardware actually implements each function as a combination of elementary logical functions, such as AND, OR, and EXCLUSIVE OR. The manner of exact implementation usually is not of concern to the programmer. Compared with the elementary logical functions, the ADD, SUBTRACT, MULTIPLY, and DIVIDE commands are a high-level language set in the same sense that a macrostatement is a higher-level instruction set when compared with the machine-oriented language instruction set.

In a microprogrammed computer, the executable (micro) instructions which may be used by the (micro) programmer are comprised of a logical function, such as AND and OR, and some elementary control functions, such as shift and branch. The (micro) programmer then defines microprograms which implement an instruction set analogous to the machine-oriented language instruction set in terms of these microinstructions. Using this derived instruction set, the systems or applications programmer writes programs for the solution of a problem. When the program is executed, each derived instruction is executed by transferring control to a microprogram. The microprogram is then executed in terms of the microinstructions of the computer. After execution of the microprogram, control is returned to the program written in the derived instruction set. The microprogram typically is stored in read-only storage (ROS) and thus is permanent and cannot be changed without physical replacement.

An advantage of microprogramming is increased flexibility. This can be realized by adapting the derived machine-oriented instruction set to a particular application. The technique enables the programmer to implement a function, such as a square root, directly without subroutines or macrostatements. Thus significant programming and execution efficiency is realized if the square root function is commonly required. Also, the instruction set of a character-oriented computer can be implemented on a word-oriented machine where adequate microprogramming is provided.

Minicomputer. There is no widely accepted definition of this term, although it is frequently used in the information processing industry, as are other computer classifications, such as microcomputer, supermini, supercomputer, and mainframe computer. Nevertheless, computers classified as minicomputers often have one or more of the following characteristics: (1) They are called minicomputers by their manufacturers; (2) they utilize a 16-bit instruction and data word, often divided into two separately addressable 8-bit bytes; the supermini utilizes a 32-bit word, usually consisting of four separately addressable bytes; (3) they are often packaged using a rack-and-panel construction, although some may consist of only a single printed circuit board; (4) they are highly modular, having a wide variety of optional features, peripheral equipment, and adapters for attaching specialized peripheral devices, such as data acquisition equipment; (5) they are often utilized by original equipment manufacturers (OEMs) to provide computing capability in specialized equipment, such as electronic assembly testers; and (6) they often provide as standard or optional features devices such as time-of-day clocks, interval timers, and hardware-implemented priority interrupts which facilitate their use in real-time applications. To a considerable extent, personal computers have markedly impacted on the use of the minicomputer. See text in earlier part of this article.

Module. In a computer system, particularly in connection with data acquisition and personal computers, a module or a series of modules may be used to expand or, in some cases, alter the ability and performance of a computer. See text in earlier portion of this article.

Multiprogramming. The essentials of a multiprogramming system in connection with digital computer operations are: (1) several programs are resident in main storage simultaneously; and (2) the central processing unit (CPU) is timeshared among the programs. This makes for better utilization of a computer system. Where only one program may reside in the main storage at any given time, inefficient use of the CPU time results when a program requests data from an input-output device. The operation is delayed until the requested information is received. In some applications, such delays can constitute a large portion of the program execution time. In the multiprogramming approach, other programs resident in storage may use the CPU while a preceding program is awaiting new information. Multiprogramming practically eliminates CPU lost time due to input-output delays. Multiprogramming is particularly useful in process control or interactive applications which involve large amounts of data input and output.

Input–output delay-time control is a basic method for controlling the interplay between multiple programs. Where multiprogramming is controlled in this manner, the various programs resident in storage are normally structured in a hierarchy. When a given program in the hierarchy initiates an input-output operation, the program is suspended until such time as the input-output operation is completed. A lower-priority program is permitted to execute during the delay time.

In a *time-slice* multiprogramming system, each program resident in storage is given a certain fixed interval of CPU time. Multiprogramming systems for applications where much more computation is done than input-output operation usually use the time-slice approach.

Multiprogramming systems allow multiple functions to be controlled simultaneously by a single process control digital computer. A multiprogramming system allows a portion of the storage to be dedicated to each type of function required in the control of the process and thus eliminates interference among the various types of functions. In addition, it provides the means whereby asynchronous external interrupts can be serviced effectively on a timely basis.

NAND Circuit. A computer logical decision element which has the characteristic that the output F is 0 if and only if all the inputs are 1's. Conversely, if any one of the input signals A or B or C or the three-input NAND element shown in Fig. 14 is not a 1, the output F is a binary 1. Although the NAND function can be achieved by inverting the output of an AND circuit, the specific NAND circuit requires fewer circuit elements. A two-input transistor NAND circuit is shown in Fig. 15. The output F is negative only when both transistors are cut off. This occurs when both inputs are positive. The number of inputs, or fan-in, is a function of the components and circuit design. NAND is a contraction of *not and.*

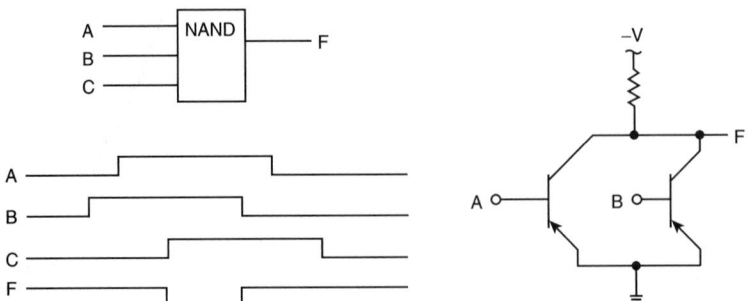

FIGURE 14   Schematic of NAND circuit.

FIGURE 15   Transistor-type NAND circuit.

NOR Circuit.   A computer logical decision element which provides a binary 1 output if all the input signals are a binary 0. This is the overall NOT of the logical OR operation. A Boolean algebra expression of the NOR circuit is $F = (AB)'$, where the prime denotes the NOT function. A two-input transistor NOR circuit is shown in Fig. 16. Output F is positive only when both transistors are cut off. This occurs when both inputs A and B are negative.

NOT Circuit.   Also known as an inverter circuit, this is a circuit which provides a logical NOT of the input signal. If the input signal is a binary 1, the output is a binary 0. If the input signal is in the 0 state, the output is in the 1 state. In reference to Fig. 17, if A is positive, the output F is at 0 V inasmuch as the transistor is biased into conduction. If A is at 0 V, the output is at $+V$ because the transistor is cut off. Expressed in boolean algebra, $F = A'$, where the prime denotes the NOT function.

OR Circuit.   A computer logical decision element which has the characteristic of providing a binary 1 output if any of the input signals are in a binary 1 state. This is expressed in terms of Boolean algebra by $F = A + B$. A diode and a transistor representation of this circuit is shown in Fig. 18. In a diode-type OR circuit, if either input signal A or B or both are positive, the respective diode is forward biased and the output F is positive. The number of allowable input signals to the diode OR gate is a function of the back resistance of the diodes. The input transistors of a transistor-type OR circuit are forced into higher conductivity when the respective input signal becomes positive. Thus the output signal becomes positive when either or both inputs are positive.

Parallel Operation.   The simultaneous performance of several actions, usually of a similar nature, through provision of individual similar or identical devices for each such action, particularly flow or processing of information. Parallel operation is performed to save time over serial operation. The decrease in the cost of multiple high-function integrated circuits has made parallel operation much more common in recent years. Because computer logic speeds already have approached a very high

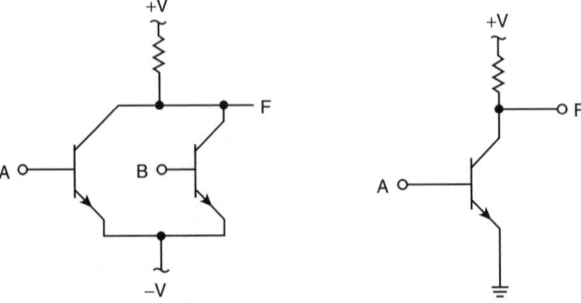

FIGURE 16   Transistor-type NOR circuit.

FIGURE 17   Inverter or NOT circuit.

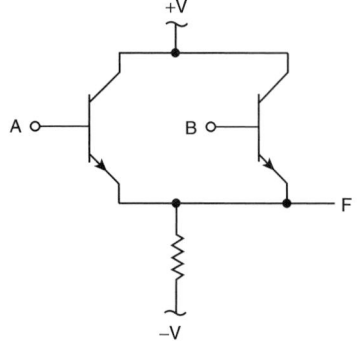

FIGURE 18 OR circuit.

limit, the use of parallel execution holds the best promise for increasing overall computing speeds. Multiprocessors represent a case where parallel operation is achieved by providing two or more complete processors capable of simultaneous operation. Compare with *concurrent operation.*

Pascal. Invented by N. Wirth in 1971, Pascal is a comparatively recent high-level programming language named in honor of Blaise Pascal, the seventeenth-century French philosopher who invented the first workable mechanical adding machine at age 19. The language was designed for the systematic teaching of programming as a discipline based on fundamental concepts of structure and integrity. Although taught widely, Pascal is increasingly being used outside the classroom in non-business-oriented applications. The language is characterized by strong typing and a syntax which encourages readable programs and good programming practices. It is a descendant of Algol, although not a strict superset of it. Pascal compilers are available on many computers, ranging from microcomputers to mainframes. Efforts to develop national and international standards for this language commenced in 1981.

Program. (1) The complete plan for the computer solution of a problem, more specifically the complete sequence of instructions and routines necessary to solve a problem. (2) To plan the procedures for solving a problem. This may involve, among other things, analysis of the problem, preparation of a flow diagram, preparing details, texting and developing subroutines, allocation of storage locations, specification of input and output formats, and the incorporation of a computer run into a complete data processing system.

Internally Stored Program. A sequence of instructions stored inside the computer in the same storage facilities as the computer data, as opposed to external storage on punched paper tape and pinboards.

Object Program. The program which is the output of an automatic coding system, such as an assembler or compiler. Often the object program is a machine-language program ready for execution, but it may well be in an intermediate language.

Source Program. A computer program written in a language designed for ease of expression of a class of problems or procedures by humans, such as symbolic or algebraic. A generator, assembler, translator, or compiler routine is used to perform the mechanics of translating the source program into an object program in machine language.

Program Generator. A program that permits a computer to write other programs automatically. Generators are of two types: (1) the *character-controlled generator*, which is like a compiler in that it takes entries from a library of functions but is unlike a simple compiler in that it examines control characters associated with each entry and alters instructions found in the library according to the directions contained in the control characteristics; (2) the *pure generator*, which is a program that writes another program. When associated with an assembler, a pure generator is usually a section of program which is called into storage by the assembler from a library and then writes one or more entries in another program. Most assemblers are also compilers and generators. In this case the entire system is usually referred to as an *assembly program.*

Programming Flowchart. A graphic representation of a program, in which symbols are used to represent data, flow, operations, equipment, and so forth. A digital computer program may be charted for two primary reasons: (1) ease of initial program design and (2) program documentation. By coding from a flowchart, instead of coding without any preliminary design, a programmer usually conserves time and effort in developing a program. In addition, a flowchart is an effective means of transmitting an understanding of the program to another person.

A programming flowchart is comprised of function blocks with connectors between the blocks. A specific function box may represent an input-output operation, a numerical computation, or a logic

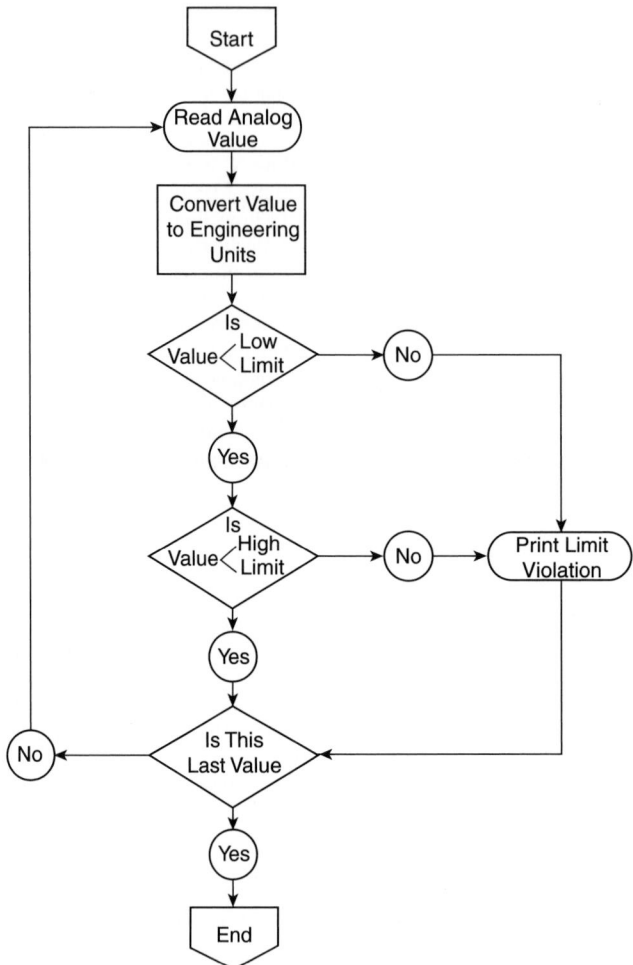

FIGURE 19   Representative computer programming flowchart.

decision. The program chart shown in Fig. 19 is of a program that reads values from the process, converts the values to engineering units, limit checks the converted values, and, if there is a violation, prints an alarm message on the process operator's display. A set of symbols used in flowcharting is described by the American National Standards Institute (ANSI).

Various levels of detail are presented in programming flowcharts. A functional block in a low-level flowchart may represent only a few computer instructions, whereas a functional block in a high-level flowchart may represent many computer instructions. A high-level flowchart is used mainly for initial program design and as a way of informing a nonprogrammer of what the program does. A low-level chart usually appears in the last stage of flowcharting before a program is actually coded. It is used for documentation that may be needed for later modifications and corrections.

Program Relocation.   A process of modifying the address in a unit of code, that is, a program subroutine, or procedure to operate from a different location in storage than that for which it was originally prepared. Program relocation is an integral part of nearly all programming systems. The technique allows a library of subroutines to be maintained in object form and made a part of a program by relocation and appropriate linking. When a program is prepared for execution by relocating the main routine and any included routines to occupy a certain part of the storage on a one-time basis, the process usually is termed *static relocation*. The resulting relocated program may be resident in a library and may be loaded into the same location in storage each time it is executed. Dynamic storage allocation schemes also may be set up so that each time a program is loaded, it is relocated to an available space in storage. This process is termed *dynamic relocation*. Timesharing systems may temporarily stop the execution of programs, store them in auxiliary storage, and later reload them into a different location for continued execution. This process is also termed dynamic relocation.

Software relocation commonly refers to a method whereby a program loader processes all the code of a program as it is loaded and modifies any required portions. Auxiliary information is carried in the code to indicate which parts must be altered. Inasmuch as all code must be examined, this method can be time-consuming.

In *hardware relocation* special machine components are used, as a base or relocation register, to alter addresses automatically at execution time in order to achieve the desired results. In dynamic relocation situations this is a fast method. Coding techniques also are used for relocation. The resulting code is caused to be self-relocating and executes in any storage location into which it is loaded. An index register may be used to make all references to storage. Double indexing is required to perform both the relocation and the normal indexing operations. Or the code may actually modify all addresses as it executes to provide the correct reference. This method is slow compared with other methods and requires additional storage.

Queue.　　When events occur at a faster rate than they can be handled in a computer system, a waiting line, or *queue*, must be formed. The elements of a queue typically are pointers (addresses) which refer to the items waiting for service. The items may be tasks to be executed or messages to be sent over communications facilities. The term is also used as a verb, meaning to place an item in a queue. Several methods of organizing queues are used. The *sequential queue* is common. As new elements arrive, they are placed at the end; as elements are processed, they are taken from the front. This is the first in–first out (FIFO) organization. In the *pushdown queue* the last one in is the first one out (LIFO). The *multipriority queue* processes from the front, but in terms of priority of the elements waiting. Essentially, this is a modified sequential queue containing subsequences and is sometimes referred to as priority in–first out (PIFO).

Resolution.　　In systems where either the input or the output of the subsystem is expressed in digital form, the resolution is determined by the number of digits used to express the numerical value. In a digital-to-analog converter, the output analog signal takes on a finite number of discrete values which correspond to the discrete numerical input. The output of an analog-to-digital converter is discrete, although the analog input signal is continuous.

In digital equipment, resolution is typically expressed in terms of the number of digits in the input or output digital representation. In the binary system, a typical specification is that "resolution is x bits." As an example, if V_{fs} is the full-scale input or output voltage range, this specification states that the resolution is $V_{fs}/2^x$. If $x = 10$ and $V_{fs} = 5$ V, the resolution is $5/2^{10}$, or 0.00488 V. It is also common to express resolution in terms of parts. A four-digit decimal converter may be said to have a resolution of 1 part in 10,000, and a 10-bit binary converter may be said to have a resolution of 1 part in 1024. The term *least-significant bit* (LSB) also is used. It may be stated, for example, that the binary resolution is $\pm\frac{1}{2}$ LSB. Also used is the term *least-significant digit* (LSD). This term is used in relation to decimal or other nonbinary digital equipment.

Routine.　　A set of coded instructions arranged in proper sequence to direct a computer to perform a desired operation or sequence of operations. A subdivision of a program consisting of two or more instructions that are functionally related, hence a program.

Diagnostic Routine.　　A routine used to locate a malfunction in a computer or to aid in locating mistakes in a computer program. Thus, in general, any routine specifically designed to aid in debugging or troubleshooting.

Executive Routine. A routine that controls loading and relocation of routines and in some cases makes use of instructions not available to the general programmer. Effectively, an executive routine is part of the machine itself. Synonymous with *monitor routine, supervisory routine*, and *supervisory program.*

Heuristic Routine. A routine by which the computer attacks a problem not by a direct algorithmic procedure but by a trial-and-error approach frequently associated with the act of learning.

Interpretive Routine. A routine that decodes and immediately executes instructions written as pseudocodes. This is contrasted with a compiler which decodes pseudocodes into a machine language routine to be executed at a later time. The essential characteristic of an interpretive routine is that a particular pseudocode operation must be decoded each time it is executed. Synonymous with interpretive code.

Service Routine. A broad class of routines provided at a particular installation for the purpose of assisting in the maintenance and operation of a computer as well as the preparation of programs as opposed to routines for the actual solution of production problems. This class includes monitoring or supervisory routines, assemblers, compilers, diagnostics for computer malfunctions, simulations of peripheral equipment, general diagnostics, and input data. The distinguishing quality of service routines is that they are generally tailored to meet the servicing needs at a particular installation, independent of any specific production-type routine requiring such services.

Tracing Routine. A diagnostic routine used to provide a time history of one or more machine registers and controls during execution of the object routine. A complete tracing routine reveals the status of all registers and locations affected by each instruction each time the instruction is executed. Since such a trace is prohibitive in machine time, traces which provide information only following the execution of certain types of instructions are more frequently used. Furthermore, a tracing routine may be under the control of the processor or may be called in by means of a trapping feature.

Scale Factor. In digital computing, an arbitrary factor which may be associated with numbers in a computer to adjust the position of the radix so that the significant digits occupy specified columns. In analog computing, a proportionality factor which relates the magnitude of a variable to its representation within a computer.

Serial Operation. The flow of information through a computer in time sequence using only one digit, word, line, or channel at a time. Serial addition in character-oriented computers permits the formation of sums with low-cost hardware. Addition occurs from right to left. Parallel addition is used in faster word- or byte-organized computers.

Magnetic disk and drum storage units may access and record data on a serial-by-bit basis. Conversion to (or from) the parallel form utilized in the central processing unit (CPU) is performed in the associated control unit. Except in the case of short distances, most communications between computers or between computers and many types of terminals take place on a serial-by-bit basis.

Software. The totality of programs, procedures, rules, and (possibly) documentation used in conjunction with computers, such as compilers, assemblers, narrators, routines, and subroutines. References are made to the software and hardware parts of a system, where the hardware comprises the physical (mechanical and electronic) components of the system. In some machines the instructions are microprogrammed in a special control storage section of the machine, using a more basic code actually wired into the machine. This is contrasted with the situation where the instructions are wired into the control unit. The microprogram technique permits the economic construction of various size machines which appear to have identical instruction sets. However, microprograms generally are not considered software and are sometimes called *firmware.* For additional information on software, see Index.

Storage. Any medium capable of storing information. As generally defined, however, a storage unit is a device on or in which data can be stored, read, and erased. The major classifications of storage devices associated with computer systems are (1) immediate access, (2) random access, and

(3) sequential access. As a general rule, the cost per bit of information is greater for immediate-access storage devices, but the access time is considerably faster than for the other two types.

Immediate-Access Storage. Devices in which information can be read in a microsecond or less. Usually an array of storage elements can be directly addressed, and thus all information in the array requires the same amount of time to be read.

Random-Access Storage. Devices in which the time required to obtain information is independent of the location of the information most recently obtained. This strict definition must be qualified by the observation that what is meant is *relatively* random. Thus magnetic drums are relatively nonrandom access when compared with monolithic storage, but are relatively random access when compared with magnetic tapes for file storage. Disk storage and drum storage units usually are referred to as random-access storage devices. The time required to read or write information on these units generally is in the 10- to 200-ms range, but is dependent on where the information is recorded with respect to the read-write head at the time the data are addressed.

Sequential-Access Storage. Devices in which the items of information stored become available only in a one-after-the-other sequence, whether or not all the information or only some of it is desired. Storage on magnetic tape is an example. Also see text in earlier portion of this article.

Storage Protect. Several methods are used to effect storage protection in digital computers. The objective is to protect certain areas of code from alteration by other areas of code. Storage may be protected by areas, each area being given a different program-settable code or key. A master key is used to permit the control program to refer to all areas. In another system, the storage is protected on an individual word basis, providing finer resolution but increasing the time required to protect a given area size. In both cases, instructions are provided to set the protect status by programming. Some systems provide for a disable of the feature from the console. This permits a preset protection pattern to be established, but with the program capability to disable selected protection status. Protection is required so that stored information will not be altered accidentally by store-type instructions. In single-word schemes, any storage modification is prevented. In the multikey area systems storage approach, modification is limited to operation in areas with the same key.

Synchronous. A synchronous operation takes place in a fixed time relation to another operation or event, such as a clock pulse. When a set of contacts is sampled at a fixed time interval, the operation is termed synchronous. This situation is to be contrasted with that where the contacts may be sampled randomly under the control of an external signal. Generally, the read operation of a main storage unit is synchronous. The turning on of the X and Y selection drivers and the sampling of the storage output on the sense line are controlled by a fixed frequency. Contrast with *asynchronous.*

Timesharing. The use of a device, particularly a computer or data-processing machine, for two or more purposes during the same overall time interval, accomplished by interspersing component and subsystem actions in time. In the case of a digital computer, timesharing generally connotes the process of using main storage for the concurrent execution of more than one job by temporarily storing all jobs in auxiliary storage except the one in control. This technique allows a computer to be used by several independent program users. The method most often is associated with a computer-controlled terminal system used in an interactive mode.

Time Slicing. A technique that allows several users to utilize a computer facility as though each had complete control of the machine. Several users can be serviced one at a time, unaware of each other, because of the relative speed between computer operation and human response. Essentially, time slicing is used by a software control system to control the allocation of facilities of a computer to tasks requesting service. The allocation basis is a fixed time interval. Each job is executed for the time period used in the time slice. The job is then temporarily stored on an auxiliary storage device (timesharing) or suspended (multiprogramming) while another job is being run. Each job, therefore, is run in an incremental fashion until complete.

Word. A character or bit string that traditionally has been an entity in computer technology. A word typically consists of one or more bytes. In small computers, there are usually 2 bytes per word. In large machines, there may be up to 8 bytes or more. Instructions are provided for manipulating words of data and, typically, most instructions occupy one word of storage. In addition, the internal data paths in a computer (parallel) are designed to transfer one word of data at a time.

MANUFACTURING MESSAGE SPECIFICATION

by Ralph Mackiewicz*

INTRODUCTION

What Is MMS?

MMS (Manufacturing Message Specification) is an internationally standardized application layer messaging system for exchanging real-time data and supervisory control information between networked devices and/or computer applications in a manner that is independent of the application function being performed or the developer of the devices or applications. MMS is an international standard (ISO 9506) that is developed and maintained by Technical Committee Number 184 (TC184), Industrial Automation, of the International Organization for Standardization (ISO).

The messaging services provided by MMS are generic enough to be appropriate for a wide variety of devices, applications, and industries. For instance, the MMS Read service allows an application or device to read a variable from another application or device. Whether the device is a Programmable Logic Controller (PLC) or a robot, the MMS services and messages are identical. Similarly, applications as diverse as material handling, fault annunciation, energy management, electrical power distribution control, inventory control, and deep-space antenna positioning in industries as varied as automotive, aerospace, petrochemical, electric/gas utility, office machinery, and space exploration have put MMS to useful work.

The History of MMS

In the early 1980s a group of numerical controller vendors, machine builders, and users, working under the auspices of committee IE31 of the Electronic Industries Association (EIA), developed draft standard proposal #1393A entitled "User Level Format and Protocol for Bi-directional Transfer of Digitally Encoded Information in a Manufacturing Environment." When the General Motors Corporation began its Manufacturing Automation Protocol (MAP) effort in 1980, they used the EIA-1393A draft standard proposal as the basis for a more generic messaging protocol that could be used for numerical controllers, PLCs, robots and other intelligent devices commonly used in a manufacturing environments. The result was the Manufacturing Message Format Standard (MMFS; sometimes pronounced as Memphis). MMFS was used in the MAP Version 2 specifications published in 1984.

During the initial usage of MMFS, it became apparent that a more rigorous messaging standard was needed. MMFS allowed too many choices for device and application developers. This resulted

---

* Vice President, SISCO, Sterling Heights, Michigan 48314.

in several mostly incompatible dialects of MMFS (one of which was called Knoxville, because it was close to Memphis). Furthermore, MMFS did not provide sufficient functionality to be useful for the process control systems found in continuous processing industries. With the objective of developing a generic and non-industry-specific messaging system for communications between intelligent manufacturing devices, the MMS effort was begun under the auspices of TC184, Industrial Automation, of ISO. The result was a standard based upon the Open Systems Interconnection (OSI) networking model, called the manufacturing message specification. In December 1988, the IS version of MMS (Version 1) was published as ISO 9506 parts 1 and 2.

Utility Communications Architecture. In 1990 the Electric Power Research Institute (EPRI) began an effort to develop a standard for electric utilities to use for building a modern real-time communications architecture. A key element of the utility communications architecture (UCA) is the use of MMS for real-time application-level communications. In addition to the use of MMS at the application level, UCA also defines a set of profiles for running MMS over both networks (TCP/IP and OSI) and serial-based systems for spread-spectrum or multiple-address system radios; a Common Application Service Model, which specifies how to use MMS to perform electric utility industry-specific functions such as select before operate and report by exception (called reporting); and device and object models for electric utility devices, called the Generic Object Models for Substation and Feeder Equipment (GOMSFE).

Further standards activity within IEC TC57 is being undertaken to develop a harmonized set of object models and communications profiles to develop a single international standard for substation communications. UCA and other selected networks such as Profibus are being used as the basis for much of this work. Although the work is incomplete at the time of this writing, international standard IEC 61850 should be the result of this work.

Previous IEC standardization efforts have also used MMS and UCA profiles as their basis. IEC 870-6 Telecontrol Application Service Element Number 2 (TASE.2) used EPRI's Intercontrol Center Communications Protocol (ICCP) as the basis for an international standard protocol for exchanging real-time data between utility control centers. ICCP was based upon MMS and UCA.

Also, the Gas Research Institute (GRI) under the auspices of its Remote Access Protocol Specification (RAPS) project are working on a gas industry version of UCA that builds upon the concepts of GOMSFE and IEC TC57 work to build gas industry-specific device and object models.

The MMS Standard

The MMS Standard (ISO 9506) itself is now jointly managed by TC184, Industrial Automation, of ISO and the International Electrotechnical Commission (IEC) and consists of two or more parts. Copies of the standard are available from ISO (www.iso.ch) or from your country's national standards organization. Parts 1 and 2 define what is referred to as the "core" of MMS. Part 1 is the service specification. The service specification contains definitions for a Virtual Manufacturing Device (VMD). The VMD represents devices on the network as an object (the VMD object) that has attributes (e.g., model number, etc.) and other subordinate objects (e.g., variables, etc.). The service specification also contains definitions for subordinate objects and their attributes that are contained with a VMD. Examples of these objects are variables, program invocations, events, and so on. It contains definitions for a set of communications *services* that client applications can use to manipulate and access the VMD and its objects. In general, MMS provides services that allow clients to perform the following functions over a network:

- determine and change the value or status of an object (e.g., Read, Write)
- determine and change the attributes of an object
- create and delete an object

Finally, the service specification gives a description of the behavior of the VMD to the service requests of client applications.

Part 2 is the protocol specification. The protocol specification defines the rules of communication, which include the sequencing of messages across the network and the format, or encoding, of the messages that are exchanged across the network. The protocol specification utilizes an ISO standard called the Abstract Syntax Notation Number One (ASN.1—ISO 8824 and ISO 8825) to specify the format of the MMS messages.

MMS provides a rich set of services for peer-to-peer real-time communications over a network. MMS has been used as a communication protocol for many common industrial control devices such as CNCs, PLCs, and robots. There are numerous MMS applications in the electrical/gas utility industry such as in remote terminal units (RTU), energy management systems (EMS), supervisory control and data acquisition (SCADA) systems, and other intelligent electronic devices (IEDs) such as reclosers, meters, and switches. Most popular computing platforms have MMS connectivity available either from the computer manufacturer or from a third party. MMS implementations support a variety of communications links, including Ethernet, RS-232C, OSI, and TCP/IP, and they can easily connect to the Internet by using commercially available bridges and routers. Gateways for translating between MMS and a host of other common device protocols are also available.

Benefits of MMS

MMS provides benefits by lowering the cost of building and using automated systems. In particular, MMS is appropriate for any application that requires a common communications mechanism for performing a diversity of communications functions related to real-time access and distribution of process data and supervisory control. When looking at how the use of a common communications service like MMS can benefit a particular system, it is important to evaluate the three major effects of using MMS that can contribute to cost savings: (1) interoperability, (2) independence, and (3) data access.

Interoperability is the ability of two or more networked applications to exchange useful supervisory control and process data information between them without the user of the applications having to create the communications environment. While many communication protocols can provide some level of interoperability, many of them are either too specific (to a brand/type of application or device, network connectivity, or function performed; see independence below) or not specific enough (provide too many choices for how a developer uses the network; see reference to MMFS above).

Independence allows interoperability to be achieved independent of the developer or supplier of the application, the network connectivity, the function performed, or data access.

1. **The Developer/Supplier of the Application.** Other communications schemes are usually specific to a particular brand (or even model in some cases) of application or device. MMS is defined by independent international standards bodies and is not specific to a particular vendor or type of equipment.

2. **Network Connectivity.** MMS becomes *the* interface to the network for applications, thereby isolating the application from most of the non-MMS aspects of the network and how the network transfers messages from one node to another. By use of MMS, the applications need not concern themselves with how messages are transferred across a network. MMS delivers the information in exactly the same form regardless of whether a serial link or a network is used.

3. **Function Performed.** MMS provides a common communications environment independent of the function performed. An inventory control application accesses production data contained in a control device in exactly the same manner as an energy management system would read energy consumption data from the same device.

Data Access is the ability of networked applications to obtain the information required by an application to provide a useful function. Although virtually any communications scheme can provide access to data at least in some minimal manner, they lack the other benefits of MMS, particularly independence (see above).

MMS is rigorous enough to minimize the differences between applications performing similar or complementary functions while still being generic enough for many different kinds of applications and devices. Communications schemes that are not specific enough can result in applications that all perform similar or complementary functions in different ways. The result is applications that cannot communicate with each other because the developers all made different choices when implementing.

While many communications schemes only provide a mechanism for transmitting a sequence of bytes (a message) across a network, MMS does much more. MMS also provides definition, structure, and meaning to the messages that significantly enhance the likelihood of two independently developed applications interoperating. MMS has a set of features that facilitate the real-time distribution of data and supervisory control functions across a network in a client/server environment that can be as simple or sophisticated as the application warrants.

Justifying MMS

The real challenge in trying to develop a business justification for MMS (or any network investment) is in assigning value to the benefits given a specific business goal. In order to do this properly, it is important to properly understand the relationship between the application functions, the connectivity functions, and the business functions of the network. In some cases, the benefit of the common communications infrastructure MMS provides is only realized as the system is used, maintained, modified, and expanded over time. Therefore, a justification for such a system must look at life-cycle costs versus just the purchase price (see Fig. 1).

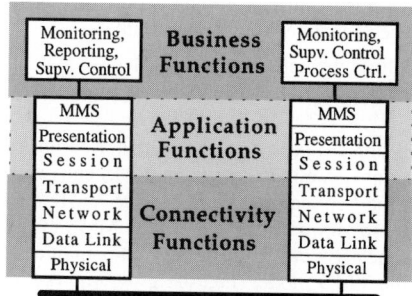

The MMS View of Network Applications

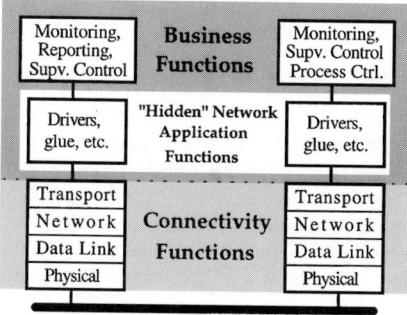

The Common View of Network Applications

FIGURE 1   MMS, as an application layer protocol, provides application services to the business functions, not connectivity services.

It is also important not to underestimate the cost associated with developing, maintaining, and expanding the application functions that have to be created if MMS is not used. A key element in assigning value is understanding that the business functions are what provide value to the enterprise. The cost of the custom network application functions directly reduces the amount of effort (i.e., cost) that can be placed on developing, maintaining, and expanding the business functions. With MMS, the communications infrastructure is built once and then reused by all the business functions. Justifying MMS requires that the user recognize the value provided by the network application functions in facilitating interoperability, independence, and access to data.

THE VMD MODEL

The primary goal of MMS was to specify a standard communications mechanism for devices and computer applications that would achieve a high level of interoperability. In order to achieve this goal, it would be necessary for MMS to define much more than just the format of the messages to be exchanged. A common message format, or protocol, is only one aspect of achieving interoperability. In addition to protocol, the MMS standard also provides definitions for:

- Objects. MMS defines a set of common objects (e.g., variables) and defines the network visible attributes of those objects (e.g., name, value, type).
- Services. MMS defines a set of communications *services* (e.g., read, write) for accessing and managing these objects in a networked environment.
- Behavior. MMS defines the network visible behavior that a device should exhibit when processing these services.

This definition of objects, services, and behavior comprises a comprehensive definition of how devices and applications communicate, which MMS calls the Virtual Manufacturing Device model. The VMD model is the key feature of MMS. The VMD model specifies how MMS devices, also called servers, behave as viewed from an external MMS client application point of view. The VMD model only specifies the network-visible aspects of communications. The internal detail of how a real device implements the VMD model (i.e., the programming language, operating system, CPU type, and input/output, or I/O, systems) are not specified by MMS. By focusing only on the network-visible aspects of a device, the VMD model is specific enough to provide a high level of interoperability. At the same time, the VMD model is still general enough to allow innovation in application/device implementation and making MMS suitable for applications across a range of industries and device types (see Fig. 2).

Client/Server Relationship

A key aspect of the VMD model is the client/server relationship between networked applications and/or devices. A server is a device or application that contains a VMD and its objects (e.g., variables). A client is a networked application (or device) that asks for data or an action from the server. In a very general sense, a client is a network entity that issues MMS service requests to a server. A server is a network entity that responds to the MMS requests of a client (see Fig. 2). While MMS defines the services for both clients and servers, the VMD model defines the network-visible behavior of servers only (see Fig. 3).

Many MMS applications and compatible devices provide both MMS client and server functions. The VMD model would only define the behavior of the server functions of those applications. Any MMS application or device that provides MMS server functions must follow the VMD model for all the network-visible aspects of the server application or device. MMS clients are only required to conform to rules governing message format or construction and sequencing of messages (the protocol).

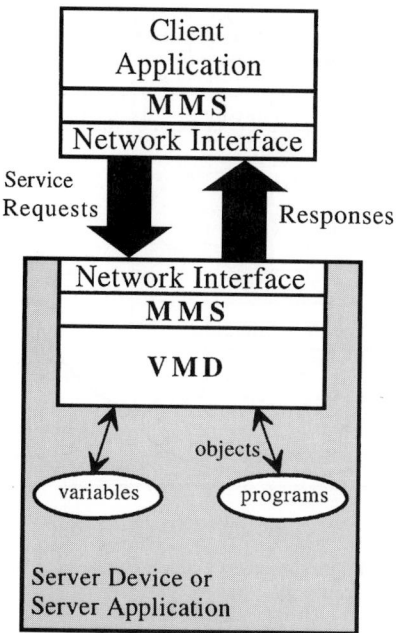

The Virtual Manufacturing Device (VMD) Model

FIGURE 2   The VMD model provides a consistent and well-defined view to client applications of the objects contained in the VMD.

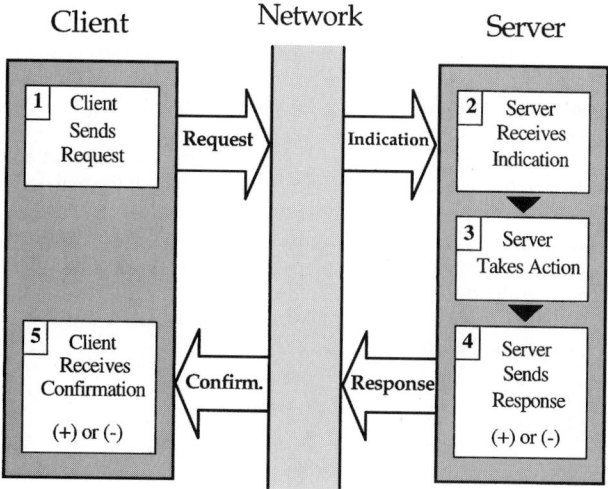

Client and Server Interactions

FIGURE 3   MMS clients and servers interact with each other by sending/receiving request, indication, response, and confirmation service primitives over the network.

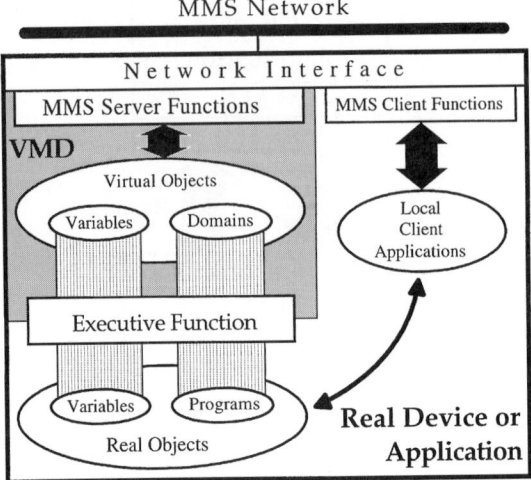

The MMS VMD Model

FIGURE 4   The executive function provides a translation, or "mapping" between the MMS defined virtual objects and the real-device objects used by the real device.

Real and Virtual Devices and Objects

There is a distinction between a real device and a real object (e.g., a PLC with a part counter) and the *virtual* device and objects (e.g., VMD, domain, variable, etc.) defined by the VMD model. Real devices and objects have peculiarities (product features) associated with them that are unique to each brand of device or application. Virtual devices and objects conform to the VMD model and are independent of brand, language, operating system, and so on. Each developer of a MMS server device or MMS server application is responsible for "hiding" the details of the real devices and objects, by providing an executive function. The executive function translates the real devices and objects into the virtual ones defined by the VMD model when communicating with MMS client applications and devices (see Fig. 4).

Because MMS clients always interact with the virtual device and objects defined by the VMD model, the client applications are isolated from the specifics of the real devices and objects. A properly designed MMS client application can communicate with many different brands and types of devices in the same manner. This is because the details of the real devices and objects are hidden from the MMS client by the executive function in each VMD. This virtual approach to describing server behavior does not constrain the development of innovative devices and product features and improvements. The MMS VMD model places constraints only on the network-visible aspects of the virtual devices and objects, not the real ones.

MMS Device and Object Modeling

The implementor of the executive function (the application or device developer) must decide how to "model" the real objects as virtual objects. The manner in which these objects are modeled is critical to achieving interoperability between clients and servers among many different developers. Inappropriate or incorrect modeling can lead to an implementation that is difficult to use or difficult to interoperate with. In some cases, it makes sense to represent the same real object with two different MMS objects. For instance, a large block of variables may also be modeled as a domain. This would provide the MMS client the choice of services to use when accessing the data. The *variable* services would give access to the individual elements in the block of data. The *domain* services would allow the entire block of data to be read (uploaded) or written (downloaded) as a single object.

MMS Objects

MMS defines a variety of objects that are found in many typical devices and applications requiring real-time communications. For each object there are corresponding MMS services that let client applications access and manipulate those objects. MMS defines the following objects:

- VMD: the device itself is an object
- domain: a resource (e.g., memory) represented by a block of untyped data
- program invocation: a runnable program consisting of one or more domains
- variable: a named or unnamed element of typed data
- type: a description of the format of a variable's data
- named variable list: a list of variables that is named as a list
- event: an object used to control the processing and notification of events
- semaphore: an object used to control access to a shared resource
- operator station: a display and keyboard for use by an operator
- file: data stored in files on a file server
- journal: an object used to keep a time-based record of variables and events

Object Attributes and Scope.   Associated with each object are a set of attributes that describe that object. MMS objects have a name attribute and other attributes that vary from object to object. Variables have attributes such as, name, value, and type. Other objects, program invocations for instance, have attributes such as name and current state.

Subordinate objects exist only within the scope of another object. For instance, all other objects are subordinate to, or contained within, the VMD itself. Some objects, such as the operator station object, may be subordinate only to the VMD. Some objects may be contained within other objects, such as variables contained within a *domain*. This attribute of an object is called its *scope*. The object's scope also reflects the lifetime of an object. An object's scope may be defined to be:

- VMD specific: the object has meaning and exists across the entire VMD (is subordinate to the VMD); the object exists as long as the VMD exists
- domain specific: the object is defined to be subordinate to a particular domain; the object will exist only as long as the domain exists
- application association specific: also referred to as AA specific; the object is defined by the client over a specific application association and can only be used by that specific client; the object exists as long as the association between the client and server exists on the network

The name of a MMS object must also reflect the scope of the object. For instance, the object name for a domain-specific variable must not only specify the name of the variable within that domain but also the name of the domain. Names of a given scope must be unique. For instance, the name of a variable specific to a given domain must be unique for all domain-specific variables in that domain. Some objects, such as variables, are capable of being defined with any of the scopes described on the preceding page. Others, like semaphores for example, cannot be defined to be AA specific. Still others, such as operator stations, are only defined as VMD specific. When an object like a domain is deleted, all the objects subordinate to that domain are also deleted.

VMD Object

The VMD itself can be viewed as an object to which all other MMS objects are subordinate (variables, domains, etc. are contained within the VMD). Because the VMD itself is also an object, it has attributes associated with it. Some of the network-visible attributes for a VMD are as follows.

1. Capabilities.   A capability of a VMD is a resource or capacity defined by the real device. There can be more than one capability to a VMD. The capabilities are represented by a sequence of character strings. The capabilities are defined by the implementor of the VMD and provide useful information about the real device or application. For instance, the capabilities of a PLC may contain information such as the amount of memory available.

2. LogicalStatus.   Logical status refers to the status of the MMS communication system for the VMD, which can be

   - state changes allowed
   - no state changes allowed
   - only support services allowed

3. PhysicalStatus.   Physical status refers to the status of all the capabilities taken as a whole, which can be equal to:

   - operational
   - partially operational
   - inoperable
   - needs commissioning.

4. LocalDetail.   This is a string of bits unique to that particular VMD that contains additional status information about the VMD. The meaning of these bits is determined by the manufacturer or developer of the MMS device or other MMS server application.

VMD Support Services

There are a number of communication services available to MMS clients to access the VMD object.

1. Identify.   This confirmed service allows the client to obtain information about the MMS implementation such as the vendor's name, model number, and revision level.

2. GetNameList.   This confirmed service allows a client to obtain a list of named objects that are defined within the VMD.

3. GetCapabilityList.   This confirmed service is used to obtain a list of the capabilities of a VMD.

4. Rename.   This confirmed service is used to rename a named MMS object at the server.

5. Status.   This confirmed service is used by a client to obtain the logical and physical status of the VMD. The client can optionally request that the VMD determine its status using an *extended derivation* (i.e., running a self-diagnostic routine).

6. UnsolicitedStatus.   This unconfirmed service is used by a server (VMD) to report its status to a client unsolicited by the client.

THE VMD EXECUTION MODEL

The VMD model has a flexible execution model that provides a definition of how the execution of programs by the MMS server can be controlled. Central to this execution model are the definitions of the domain and program invocation objects.

Domains

The MMS domain is a named MMS object that is a representation of some resource within the real device. This resource can be anything that is appropriately represented as a contiguous block of untyped data (referred to as load data). In many typical applications, domains are used to represent areas of memory in a device. For instance, a PLC's ladder program memory is typically represented as a domain. Some applications allow blocks of variable data to be represented as both domains and variables. MMS provides no definition for, and imposes no constraints on, the content of a domain. To do so would be equivalent to defining a "real" object (i.e., the ladder program). The content of the domain is left to the implementor of the VMD.

Domain Attributes.   In addition to the name, some of the attributes associated with MMS domains are as follows.

1. Capabilities.   Each domain can optionally have a list of capabilities associated with it that conveys information about memory allocation, I/O characteristics, and similar information about the real device. The capabilities of a domain are represented by a sequence of implementor-defined character strings.

2. State.   The state of a domain can be:

   - loading
   - complete
   - incomplete
   - ready
   - in use
   - several other intermediate states

3. Deletable.   This attribute indicates whether the domain is deletable by means of the DeleteDomain service. A domain that can be downloaded is always deletable. Nondeletable domains are pre-existing and predefined by the VMD and cannot be downloaded.

4. Sharable.   This attribute indicates if the domain can be shared by more than one program invocation (see the example of a batch controller).

Domain Services.   MMS provides a set of services that allow domains to be uploaded from the device or downloaded to the device. The MMS domain services do not provide for partial uploads or downloads (except as potential error conditions). Nor do they provide access to any subordinate objects within the domain. The set of services provided for domains is summarized below.

1. InitiateDownloadSequence, DownloadSegment, and TerminateDownloadSequence.   These services are used to download a domain. The InitiateDownloadSequence service commands the VMD to create a domain and prepare it to receive a download.

2. InitiateUploadSequence, UploadSegment, and TerminateUploadSequence.   These confirmed services are used to upload the contents of a domain to a MMS client.

3. DeleteDomain.   This confirmed service is used by a client to delete an existing domain, usually before initiating a download sequence.

4. GetDomainAttributes.   This confirmed service is used to obtain the attributes of a domain.

5. RequestDomainDownload and RequestDomainUpload.   These confirmed services are used by a VMD to request that a client perform an upload or download of a domain in the VMD.

6. LoadDomainContent and StoreDomainContent.   These confirmed services are used to tell a VMD to download (load) or upload (store) a domain from a file. The file may be local to the VMD or may be contained on an external file server.

UCA Usage of Domains.   Within UCA a domain is treated as a device. This allows a single VMD to represent multiple devices each containing their own *domain-specific* objects, particularly variable access objects. This is important for UCA because of the extremely large number of devices in a typical utility. For instance, a single substation could have many thousands of customers and therefore many thousands of meters. To require client applications to directly address each device using a unique network would require that the network infrastructure support many millions of addresses to accommodate all the potential devices. While this is possible to do using current networking technology, it would require higher levels of performance on the network, thereby increasing its cost. The utility industry has addressed this concern historically by using RTUs as data concentrators and gateways. Modeling data in devices as domain-specific objects allows UCA to easily support a data concentrator architecture while still conforming to the VMD model. By remaining within the bounds of the VMD model, generic non-UCA MMS clients are still able to communicate, to a large degree, with a UCA device without any prior knowledge of the UCA-specific object models (GOMSFE) and application service models (CASM).

Program Invocations

It is through the manipulation of program invocations that a MMS client controls the execution of programs in a VMD. A program invocation is an execution thread that consists of a collection of one or more domains. Simple devices with simple execution structures may only support a single program invocation containing only one domain. More sophisticated devices and applications may support multiple program invocations containing several domains each.

As an example, consider how the MMS execution model could be applied to a personal computer (PC). When the PC powers up, it downloads a domain called the operating system into memory. When you type the name of the program you want to run and hit the <return> key, the computer downloads another domain (the executable program) from a file and then creates and runs a program invocation consisting of the program and the operating system. The program by itself cannot be executed until it is bound to the operating system by the act of creating the program invocation.

Program Invocation Attributes.   In addition to the program invocation's name, the attributes of a program invocation are shown below.

1. State.   The state of a program invocation can be:
   - nonexistent
   - idle
   - running
   - stopped
   - unrunnable
   - other intermediate states

2. Deletable.   Indicates if the program invocation is deletable by means of the DeleteProgramInvocation service.

3. Reusable.   Reusable program invocations automatically re-enter the idle state when the program invocation arrives at the end of the program. Otherwise, a nonreusable program invocation in the running state must be stopped and then reset in order to bring it back to the idle state.

4. Monitored.   Monitored program invocations utilize the MMS event management model to inform the MMS client when the program invocation leaves the running state. Monitored program invocations have an event condition object defined with the same name as the program invocation.

5. List of domains.   The list of domains that comprise the program invocation.

6. Execution argument.   This is a character string passed to the program invocation using the Start or Resume service. The execution argument is used to pass data to the program invocation like parameters in a subroutine call.

Program Invocation Services.   The MMS services for program invocations allow clients to control the execution of VMD programs and to manage program invocations as follows.

1. CreateProgramInvocation and DeleteProgramInvocation.   These confirmed services are used to create and delete program invocations.

2. GetProgramInvocationAttributes.   This confirmed service returns the attributes of the program invocation to the requesting client.

3. Start, Stop, Reset, Resume and Kill.   These confirmed services are used by a client to cause the program invocation to change states (see Fig. 5).

Batch Controller Example

As an example of how the MMS execution model can be applied to a typical device, let us look at a VMD model for a simple batch controller (see Fig. 6). The figure depicts how the VMD model could be applied to define a set of objects (e.g., domains, program invocations, variables) appropriate for a

**Program Invocation State
Diagram**

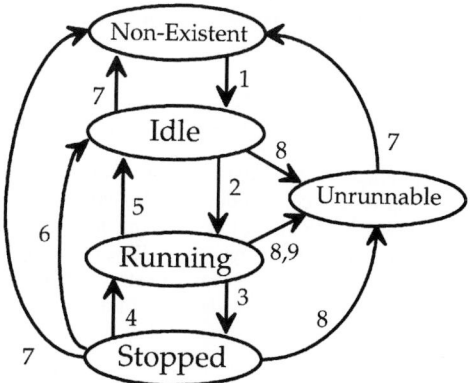

FIGURE 5   MMS Clients use MMS services to cause state transitions in the program invocation.

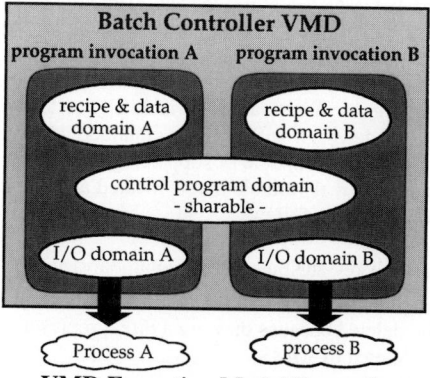

VMD Execution Model Example

FIGURE 6   The our example of a batch controller shows how to control two identical batch-oriented processes.

batch controller. This model will provide clients with an appropriate method of controlling the batch process using MMS services. In order to start up and control these two processes, a MMS client using this controller would perform the following actions.

1. Initiate and complete a domain download sequence for each domain: recipe and data domains A and B, I/O domains A and B, and the control program domain.
2. Create program invocation A consisting of I/O domain A, the control program domain, and recipe and data domain A.
3. Start program invocation A. Create program invocation B consisting of I/O domain B, the control program domain, and recipe and data domain B.
4. Start program invocation B.

The example above demonstrates the flexibility of the VMD execution model to accommodate a wide variety of real-world situations. Further examples might be a loop controller, in which each loop is represented by a single domain and in which the control loop algorithm (e.g., PID) is represented by a separate but common *sharable* domain. Process variables, setpoints, alarm thresholds, and so on could be represented by domain-specific (control-loop) variables. A program invocation would consist of the control-loop domains and their algorithm domains needed to control the process.

VARIABLE-ACCESS MODEL

MMS provides a comprehensive and flexible framework for exchanging variable information over a network. The MMS variable-access model includes capabilities for *named, unnamed* (addressed), and named lists of variables. MMS also allows the *type description* of the variables to be manipulated as a separate MMS object (named type object). MMS variables can be simple (e.g., integer, Boolean, floating point, string) or complex (e.g., arrays and structures). The services available to access and manage MMS variable objects support a wide range of data-access methods from simple to complex.

MMS Variables

A real variable is an element of typed data that is contained within a VMD. A MMS variable is a virtual object that represents a mechanism for MMS clients to access the real variable. The distinction between the real variable (which contains the value) and the virtual variable is that the virtual variable represents the *access path* to the variable, not the underlying real variable itself. MMS defines two types of virtual objects for variable access.

Unnamed Variable Object.   An unnamed variable object describes the access to the real variable by using a device-specific address. MMS includes unnamed variable objects primarily for compatibility with older devices that are not capable of supporting names. An unnamed variable object is a direct mapping to the underlying real variable that is located at the specified address. The attributes of the unnamed variable object are as follows.

1. Address.   This is the key attribute of the unnamed variable object.
2. MMS deletable.   This attribute is always *false* for an unnamed variable object. Unnamed variable objects cannot be deleted because they are a direct representation of the path to a "real" variable, which cannot be deleted by MMS.
3. Type description.   This attribute describes the type (format and range of values) of the variable.

Named Variable Object.   The named variable object describes the access to the real variable by using a MMS object name. MMS clients need only know the name of the object in order to access it.

Remember that the name of a MMS variable must also specify the *scope* of the variable (the scope can be VMD, domain—including the domain name, or association specific). In addition to the name, a named variable object has the following attributes.

1. MMS deletable. This attribute indicates if access to the variable can be deleted by means of the Delete-
VariableAccess service.

2. Type description. This attribute describes the type (format and range of values) of the variable.

3. Access method. If the access method is *public*, it means that the underlying address of the named variable object is visible to the MMS client. In this case, the same variable can be accessed as an unnamed variable object.

Addresses. A MMS variable address can take on one of several forms. Which specific form that is used by a specific VMD, and the specific conventions used by that address form, is determined by the implementor of the VMD based upon what is most appropriate for that device.

The possible forms for a MMS variable address are numeric, symbolic, and unconstrained. A numeric address is represented by an unsigned integer number (e.g., 103). A symbolic address is represented by a character string (e.g., "R001"). An unconstrained address is presented by a untyped string of bytes. In general, it is recommended that applications utilize named variable objects instead of the addresses of unnamed variable objects wherever feasible. Address formats vary widely from device to device. Furthermore, in some computing environments, the addresses of variables can change from one run time to the next. Names provide a more intuitive, descriptive, and device independent access method than addresses.

Named Variable Lists. MMS also defines a named variable list object that provides an access mechanism for grouping both named and unnamed variable objects into a single object for easier access. A named variable list is accessed by a MMS client by specifying the name (which also specifies its scope) of the named variable list. When the VMD receives a Read service request from a client, it reads all the individual objects in the list and returns their value within the individual elements of the named variable list.

Because the named variable list object contains independent subordinate objects, a positive confirmation to a Read request for a named variable list may indicate only partial success. The success/failure status of each individual element in the confirmation must be examined by the client to ensure that all of the underlying variable objects were accessed without error. In addition to its name and the list of underlying named and unnamed variable objects, named variable list objects also have a MMS deletable attribute that indicates whether or not the named variable list can be deleted by means of a DeleteNamedVariableList service request.

Named Type Object. The type of a variable indicates its format and the possible range of values that the variable can take. Examples of type descriptions include 16-bit signed integer, double-precision floating point, 16 character string, and so on. MMS allows the type of a variable to be either (1) *described* or (2) defined as a separate named object called a *named type*. A described type is not an object. It is a binary description of the type in a MMS service request (i.e., Read, Write, or InformationReport) that uses the *access by description* service option. The named type object allows the types of variables to be defined and managed separately. This can be particularly useful for systems that also use the DefineNamedVariable service to define names and types for unnamed variable objects. Other attributes of the named type object include the following.

1. Deletable. This parameter indicates if the named type can be deleted using a DeleteNamedType service request.

2. Type description. The MMS type description is very flexible and can describe virtually any data format in use today. The type description specifies the format and range of values that the variable can represent. MMS defines three basic formats for types: (1) simple, (2) array, and (3) structured. The nesting level is representative of the complexity of the variable's type (see Fig. 7).

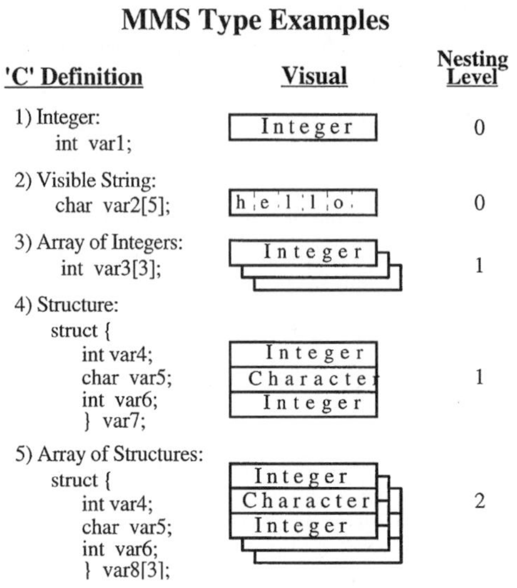

MMS Type Examples

'C' Definition	Visual	Nesting Level
1) Integer: int var1;	`Integer`	0
2) Visible String: char var2[5];	`h e l l o`	0
3) Array of Integers: int var3[3];	`Integer`	1

4) Structure:
    struct {
      int var4;
      char var5;
      int var6;
    } var7;
	`Integer`
	`Character`
	`Integer`
		1

5) Array of Structures:
    struct {
      int var4;
      char var5;
      int var6;
    } var8[3];
	`Integer`
	`Character`
	`Integer`
		2

FIGURE 7   The level of complexity that a VMD can support is defined by its nesting level.

Simple Types.   Simple types are the most basic types and cannot be broken down into a smaller unit by means of MMS. The other type forms (arrays and structures) are constructed types that can eventually be broken down into simple types. Simple type descriptions generally consist of the class and size of the type. The size parameter is usually defined in terms of the number of bits or bytes that a variable of that type would comprise in memory.

The various classes of simple types defined by MMS consist of the following.

1. Boolean.   Variables of this type can only have the values of true (value = non-zero) or false (value = zero). There is no size parameter for Boolean types and these are generally represented by a single byte or octet.

2. Bit string.   A Bit string is a sequence of bits. The size of a bit string indicates the number of bits in the bit string. The most significant bit of the most significant byte in the string is defined as Bit0 in MMS terminology.

3. Boolean array.   A Boolean array is also a sequence of bits in which each bit represents a true or false. It differs from an array of Booleans in that each element in a Boolean array is represented by a single bit, whereas each element in an array of Booleans is represented by a single byte. The size parameter specifies the number of Booleans (number of bits) in the Boolean array.

4. Integer.   MMS integer's are signed integers. The size parameter specifies the number of bits of the integer in 2's complement form.

5. Unsigned.   The unsigned type is identical to the integer type except that it is not allowed to take on a negative value. Because the most significant bit of an integer is essentially a sign bit, an unsigned with a size of 16 bits can only represent 15 bits of values or values of 0 through 32,767.

6. Floating point.   The MMS definition for floating point can accommodate any number of bits for the format and exponent width, including the IEEE 754 single- and double-precision floating point formats commonly in use today.

7. Octet string.   An octet string is a sequence of bytes (octet in ISO terminology) with no constraint on the value of the individual bytes. The size of an octet string is the number of bytes in the string.

8. Visible string.   The visible string type only allows each byte to contain a printable character. The character set is defined by ISO 10646 and is compatible with the common ASCII character set. The size of the visible string is the number of characters in the string.

9. Generalized time.   This is a representation of time specified by ISO 8824. It provides a millisecond resolution of date and time.

10. Binary time.   This is a time format that represents time as either (1) time of day by a value that is equal to the number of milliseconds from midnight or (2) date and time by a value that is equal to the time of day and the number of days since January 1, 1984.

11. BCD.   Binary coded decimal format is when four bits are used to hold a binary value of a single digit of zero to ten. The size parameter is the number of decimal digits that the BCD value can represent.

12. Object identifier.   This is a special class of object defined by ISO 8824 that is used to define ISO-registered network objects.

Array.   An array type defines a variable that consists of a sequence of multiple, identical (in format, not value) elements. Each element in an array can also be an array or even a structured or simple variable. MMS allows for arbitrarily complex nesting of arrays and structures.

Structures.   A structured type defines a variable that consists of a sequence of multiple, but not necessarily identical, elements. Each individual element in a structure can be of a simple type, an array, or another structure. MMS allows for arbitrarily complex nesting of structures and arrays. A structured variable consisting of individual simple elements requires a nesting level of 1. A structured variable consisting of one or more arrays of structures containing simple variables requires a nesting level of 3.

Variable-Access Services

These services are as follows.

1. Read.   Read is a confirmed service used by MMS clients to obtain the values of named, unnamed, or a named variable list objects.

2. Write.   Write is a confirmed service used by MMS clients to change the values of variable objects.

3. InformationReport.   This *unconfirmed* service is used by a VMD to report the values of variable objects to a MMS client in an unsolicited manner. It is roughly equivalent to sending a Read response to a client without the client having issued the Read request. The InformationReport service can be used to eliminate polling by client applications or as an alarm notification mechanism. The VMD could directly report changes in the value of variables, alarm conditions, or even changes in state of the VMD or program invocations to clients by using the InformationReport service.

4. GetVariableAccessAttributes.   This confirmed service is used by MMS clients to obtain the access attributes of a single named or unnamed variable object. The access attributes are the type (either a named type or type description), the MMS deletable attribute, and the address for unnamed variables. The address of named variables is optional and is only returned if the address is known and visible.

5. DefineNamedVariable.   This confirmed service allows MMS clients to create a named variable object by assigning a name to an unnamed variable object. Once defined, subsequent access to the unnamed variable object can be made by means of the named variable. This service also allows the MMS client to optionally specify a type definition for the named object that may be different from the type inherently defined for the unnamed variable object.

6. DeleteVariableAccess. This service is used to delete named variable objects when the MMS deletable attribute is *true*. The service provides options for deleting only specific named variable objects or all named variable objects of a specified scope.

7. DefineNamedVariableList, GetNamedVariableListAttributes, and DeleteNamedVariableList. These confirmed services are used to create, delete, and obtain the attributes (i.e., the list of underlying named and unnamed variable objects) of named variable list objects.

8. DefineNamedType. This confirmed service is used by a MMS client to create a new named type by specifying the type name (including scope), MMS deletable, and type description attributes.

9. DeleteNamedType. This confirmed service is used to delete an existing named type when the MMS deletable attribute is *true*. The service provides options for deleting only specific named type objects or all named type objects of a specified scope.

10. GetNamedTypeAttributes. This confirmed service is used by a MMS client to determine the MMS deletable and type description attributes of a named type object.

Variable-Access Features

The Read, Write, and InformationReport services provide several features for accessing Variables. The use of these service features, as described below, by MMS clients can provide enhanced performance and very flexible access to MMS variables.

Access by description is supported for unnamed variable objects only. It allows the MMS client to describe the variable by specifying both an address and a type description. These variables are called described variables. The described type may be different from the type inherently defined for the unnamed variable object. This can be useful for accessing data in devices in which the device's memory organization is simplistic. For example, many PLCs represent their data memory as a large block of 16-bit registers (essentially signed integers). Some applications may store ASCII string data in these registers. By using a described variable, a MMS client can have the data stored in these registers returned to it as a string instead of as a block of signed integers.

List of variables is a function that allows a list of named variable, unnamed variable, and named variable list objects to be accessed in a single MMS Read, Write, or InformationReport service. Care must be taken by the client to ensure that the resultant MMS service request message does not exceed the maximum message size (maximum segment size) supported by the VMD. This option also requires that a client examine the entire response for success/failure for each individual element in the list of variables.

Access specification in result is an option for the Read service that allows a MMS client to request that the variable's access specification be returned in the Read response. The access specification would consist of the same information that would be returned by a GetVariableAccessAttributes service request.

Alternate Access allows a MMS client to (1) partially access only specified elements contained in a larger arrayed and/or structured variable, and (2) rearrange the ordering of the elements contained in structured variables.

EVENT MANAGEMENT MODEL

In a real sense, an event or an alarm is easy to define. Most people have an intuitive feel for what can comprise an event within their own area of expertise. For instance, in a process control application, it is common for a control system to generate an alarm when the process variable (e.g., temperature) exceeds a certain preset limit called the high-alarm threshold. In a power distribution application, an alarm might be generated when the difference in the phase angle of the current and voltage waveforms of a power line exceeds a certain number of degrees.

The MMS event management model provides a framework for accessing and managing the network communication aspects of these kinds of events. This is accomplished by defining three named objects that represent (1) the state of an event (event condition), (2) who to notify about the occurrence of an event (event enrollment), and (3) the action that the VMD should take upon the occurrence of an event (event action).

For many applications, the communication of alarms can be implemented by using MMS services other than the event management services. For instance, a system can notify a MMS client about the fact that a process variable has exceeded some preset limit by sending the process variable's value to a MMS client using the InformationReport service. Other schemes using other MMS services are also possible. When the application is more complex and requires a more rigorous definition of the event environment in order to ensure interoperability, the MMS event management model can be used.

Event Condition Object

A MMS event condition object is a named object that represents the current state of some real condition within the VMD. It is important to note that MMS does not define the VMD action (or programming) that causes a change in state of the event condition. In the process control example given above, an event condition might reflect an *idle* state when the process variable was not exceeding the value of the high-alarm threshold and an *active* state when the process variable did exceed the limit. MMS does not explicitly define the mapping between the high-alarm limit and the state of the event condition. Even if the high-alarm limit is represented by a MMS variable, MMS does not define the necessary configuration or programming needed to create the mapping between the high-alarm limit and the state of the event condition. From the MMS point of view, the change in state of the event condition is caused by some autonomous action on the part of the VMD that is not defined by MMS.

The MMS event management model defines two classes of event conditions: network triggered and monitored. A network-triggered event condition is triggered when a MMS client specifically triggers it using the TriggerEvent service request. Network triggered events do not have a state (their state is always disabled). They are useful for allowing a MMS client to control the execution of event actions and the notifications of event enrollments. A monitored event condition has a state attribute that the VMD sets based upon some local autonomous action. Monitored event conditions can have a Boolean variable associated with them that is used by the VMD to evaluate the state. The VMD periodically evaluates this variable. If the variable is evaluated as *true*, the VMD sets the event condition state to *active*. When the Boolean variable is evaluated as *false*, the VMD sets the event condition state to *idle*. Event conditions that are created as a result of a CreateProgramInvocation request with the Monitored attribute *true* are monitored event conditions, but they do not have an associated Boolean variable.

In addition to the name of the event condition (an object name that also reflects its scope) and its class (network triggered or monitored), MMS defines the following attributes for both network-triggered and monitored event conditions.

1. **MMS deletable.**   This attribute indicates if the event condition can be deleted by using a Delete-EventCondition service request.

2. State. This attribute reflects the state of the event condition, which can be:
   - idle
   - active
   - disabled (network-triggered events are always disabled).

3. **Priority.**   This attribute reflects the relative priority of an event condition object with respect to all other defined event condition objects. Priority is a relative measure of the VMD's processing priority when it comes to evaluating the state of the event condition as well as the processing of event notification procedures that are invoked when the event condition changes state. Zero (0) is the highest priority, 127 is the lowest priority, and a value of 64 indicates "normal" priority.

4. Severity.   This attribute reflects the relative severity of an event condition object with respect to all other defined event condition objects. Severity is a relative measure of the effect that a change in state of the event condition can have on the VMD. Zero (0) is the highest severity, 127 is the lowest, and a value of 64 is for "normal" severity.

Additionally, MMS also defines the following attributes for monitored event conditions only.

1. Monitored variable.   This is a reference to the underlying Boolean variable whose value the VMD evaluates in determining the state of an event condition. It can be either a named or unnamed variable object. If it is a named object, it must be a variable with the same name (and scope) of the event condition. If the event condition object is locally defined or it was defined by means of a CreateProgramInvocation request with the monitored attribute *true*, then the value of the monitored variable reference would be equal to unspecified. If the monitored variable is deleted, then the value of this reference would be undefined and the VMD should disable its event notification procedures for this event condition.

2. Enabled.   This attribute reflects whether a change in the value of the monitored variable (or the state of the associated program invocation if applicable) should cause the VMD to process the event notification procedures for the event condition (*true*) or not (*false*). A client can disable an event condition by changing this attribute with an AlterEventConditionMonitoring service request.

3. Alarm summary reports.   This attribute indicates whether (*true*) or not (*false*) the event condition should be included in alarm summary reports in response to a GetAlarmSummaryReport service request.

4. Evaluation interval.   This attribute specifies the maximum amount of time, in milliseconds, between successive evaluations of the event condition by the VMD. The VMD may optionally allow clients to change the evaluation interval.

5. Time of last transition to active and time of last transition to idle.   These attributes contain either the time of day or a time sequence number of the last state transitions of the event condition. If the event condition has never been in the idle or active state, then the value of the corresponding attribute shall be undefined.

Event Condition Services.

1. DefineEventCondition, DeleteEventCondition, and GetEventConditionAttributes.   These confirmed services are used by MMS clients to create event condition objects, to delete event condition objects (if the MMS Deletable attribute is *true*), and to obtain the static attributes of an existing event condition object, respectively.

2. ReportEventConditionStatus.   This confirmed service allows a MMS client to obtain the dynamic status of the event condition object, including its state, the number of event enrollments enrolled in the event condition, whether it is enabled or disabled, and the time of the last transitions to the active and idle states.

3. AlterEventConditionMonitoring.   This confirmed service allows the MMS client to alter the priority, enable or disable the event condition, enable or disable alarm summary reports, and change the evaluation interval if the VMD allows the evaluation interval to be changed.

4. GetAlarmSummary.   This confirmed service allows a MMS client to obtain event condition status and attribute information about groups of event conditions. The client can specify several filters for determining which event conditions to include in an alarm summary.

Event Actions

An event action is a named MMS object that represents the action that the VMD will take when the state of an event condition changes. An event action is optional. When omitted, the VMD would execute its event notification procedures without processing an event action. An event action, when used, is always defined as a confirmed MMS service request. The event action is linked to an event condition when an event enrollment is defined. For example, an event action might be a MMS Read request. If this event action is attached to an event condition (by being referenced in an event enrollment), when the event condition changes state and is enabled, the VMD would execute this Read service request just as if it had been received from a client, except that the Read response (either positive or negative) is included in the EventNotification service request that is sent to the MMS client enrolled for that event. A confirmed service request must be used (i.e., Start, Stop, or Read). Unconfirmed services (e.g., InformationReport, UnsolicitedStatus, and EventNotification) and services that must be used in conjunction with other services (e.g., domain upload–download sequences) cannot be used as event actions. In addition to its name, an event action has the following attributes: MMS deletable—when *true*, it indicates that the event action can be deleted via a DeleteEventAction service request; Service request—this attribute is the MMS confirmed service request that the VMD will process when the event condition that the event action is linked with changes state.

Event Action Services.

1. DefineEventAction, DeleteEventAction, and GetEventActionAttributes.   These services are used by MMS clients to create, delete, and obtain the attributes of an event action object, respectively.

2. ReportEventActionStatus.   This service allows a MMS client to obtain a list of names of event enrollments that have referenced a given event action.

Event Enrollments

The event enrollment is a named MMS object that ties all the elements of the MMS event management model together. The event enrollment represents a request on the part of a MMS client to be notified about changes in the state of an event condition. When an event enrollment is defined, references are made to an event condition, an event action (optionally), and the MMS client to which Event Notification should be sent (see Fig. 8). In addition to its name, the attributes of an event enrollment

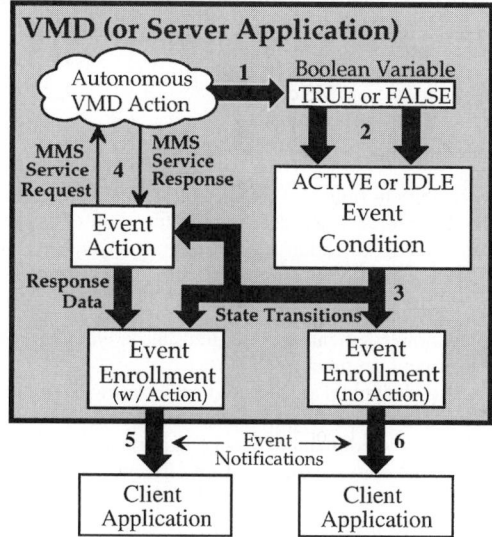

Monitored Event Conditions

FIGURE 8   A monitored event condition has a Boolean variable associated with it that the VMD sets.

are as follows.

1. MMS deletable. If *true*, this attribute indicates that the event enrollment can be deleted with a DeleteEventEnrollment service request.

2. Event condition. This attribute contains the name of the event condition about which the event enrollment will be notified of changes in state.

3. Transitions. This attribute indicates the state transitions of the event condition for which the VMD should execute its event notification procedures. The allowable state transitions are disabled to active, disabled to idle, idle to active, and idle to disabled.

4. Notification lost. If this attribute is *true*, it means that the VMD could not successfully complete its event notification procedures (1) because of some local problem or resource constraint or (2) because the VMD could not establish an association to the client specified in the event enrollment definition. Further transitions of the event condition will be ignored for this event enrollment as long as these problems persist.

5. Event action. This optional attribute is a reference to the event action that should be processed by the VMD for those state transitions of the event condition specified by the event enrollment's transitions attribute.

6. Client application. This attribute is a reference to the MMS client to which the EventNotification service requests should be sent for those transitions of the event condition specified by the event enrollment's transitions attribute. This attribute should only be defined if the VMD supports third-party services. This attribute is omitted if the duration of the event enrollment is current.

7. Duration. This attribute indicates the lifetime of the event enrollment. A duration of current means the event enrollment is only defined for the life of the association between the MMS client and the VMD (similar to an object with AA-specific scope). If the association between the VMD and the client is lost and there is no client application reference attribute for the event enrollment (duration = current), then the VMD will not reestablish the application association in order to send an EventNotification to the client in the event the association is lost. If the duration of the event enrollment is permanent, then the application association between the VMD and the client can be terminated without affecting the event enrollment. In this case, when a specified state transition occurs, the VMD will attempt to establish an application association with the specified MMS client.

8. State. The current state of the event notification procedure is as follows:

   • idle
   • active
   • disabled
   • a variety of other minor states

9. Alarm acknowledgment rules. This attribute specifies the rules of alarm acknowledgment that the VMD should enforce when determining the state of the event enrollment. If an acknowledgment to an EventNotification service request is required, the act of acknowledging or not acknowledging the Event Notification will affect the state of the event enrollment. The various alarm acknowledgment rules are summarized as follows:

   • none: no acknowledgments are required; these types of event enrollments are not included in alarm enrollment summaries
   • simple: acknowledgment are required, but acknowledgments of transitions to the active state will be reflected in the state of the event enrollment
   • ack-active: acknowledgment of event condition transitions to the active state will be required and will affect the state of the event enrollment (acknowledgments of other transitions are optional and will not affect the state of the event enrollment)
   • ack-all: acknowledgments are required for all transitions of the event condition to the active or idle state and will affect the state of the event enrollment

10. Time-active acked and time-idle acked. These attributes reflect the time of the last acknowledgment of the Event Notification for state transitions in the event condition to the active or idle state corresponding to the event enrollment.

Event Enrollment Services.

1. DefineEventEnrollment, DeleteEventEnrollment, and GetEventEnrollmentAttributes. These confirmed services are used by MMS clients to create, delete (if the MMS Deletable attribute is true), and to obtain the static attributes of an event enrollment object.

2. ReportEventEnrollmentStatus. This confirmed service allows the client to obtain the dynamic attributes of an event enrollment including the notification lost, duration, alarm acknowledgment rule, and state attributes.

3. AlterEventEnrollment. This service allows the client to alter the transitions and alarm acknowledgment rule attributes of an event enrollment.

4. GetAlarmEnrollmentSummary. This service allows a MMS client to obtain event enrollment and event condition information about groups of event enrollments. The client can specify several filters for determining which event enrollments to include in an alarm enrollment summary.

Event Notification Services. MMS provides several services for notifying clients of event condition transitions and acknowledging those event notifications as follows.

1. EventNotification. This is an unconfirmed service that is issued by the VMD to the MMS client to notify the client about event condition transitions that were specified in an event enrollment. There is no response from the client. The acknowledgment of the notification is handled separately by means of the AcknowledgeEventNotification service. The EventNotification service would include a MMS confirmed service response (positive or negative) if an event action was defined for the event enrollment.

2. AcknowledgeEventNotification. This confirmed service is used by a MMS client to acknowledge an EventNotification sent to it by the VMD. The client specifies the event enrollment name, the acknowledgment state, and the transition time parameters that were in the EventNotification request being acknowledged.

3. TriggerEvent. This service is used to trigger a network triggered event condition. It gives the client a mechanism by which it can invoke event action and event notification processing by the VMD. For instance, a client can define an event condition, event action, and event enrollments that refer to other MMS clients. When the defining client issues a TriggerEvent service request to the VMD, it will cause the VMD to execute a MMS service request (the event action) and send these results to other MMS clients via the EventNotification service.

SEMAPHORE MANAGEMENT MODEL

In many real-time systems there is a need for a mechanism by which an application can control access to a system resource. An example might be a workspace that is physically accessible to several robots. Some means to control which robot (or robots) can access the workspace is needed. MMS semaphores are named objects that can be used to control access to other resources and objects within the VMD. For instance, a VMD that controls access to a setpoint (a variable) for a control loop could use semaphores to only allow one client at a time to be able to change the setpoint (e.g., with the MMS Write service). The MMS semaphore model defines two kinds of semaphores. Token semaphores are used to represent a specific resource within the control of the VMD. Pool semaphores consist of one or more named tokens, each representing a set of similar but distinct resources under the control of the VMD.

Because semaphores are used solely for the purpose of coordinating activities between multiple MMS clients, the scope of a semaphore cannot be AA specific when the object exists only as an association between a single VMD and a single MMS client.

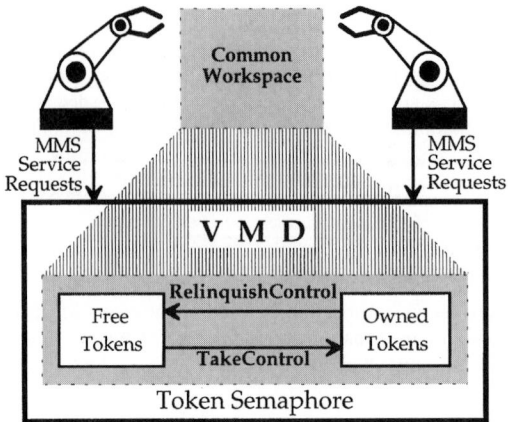

Token Semaphore Example

FIGURE 9 A token semaphore is modeled as a collection of free tokens and owned tokens.

Token Semaphores

A token semaphore is a named MMS object that can be a representation of some resource, within the control of the VMD, to which access must be controlled. A token semaphore is modeled as a collection of tokens that MMS clients take and relinquish control of using MMS services. When a client *owns* a token, the client may access the resource that the token represents. This allows both multiple or exclusive ownership of the semaphore. An example of a token semaphore might be when two users want to change a setpoint for the same control loop at the same time. In order for their access to the setpoint to be coordinated, a token semaphore, containing only one token, can be used to represent the control loop. When a user owns the token, he or she can change the setpoint. The other user would have to wait until ownership is relinquished (see Fig. 9).

A token semaphore can also be used for the sole purpose of coordinating the activities of two MMS clients without representing any real resource. This kind of "virtual" token semaphore looks and behaves the same except that it can be created and deleted by MMS clients using the DefineSemaphore service.

In addition to its name, the token semaphore has the following attributes.

1. Deletable.   If *true*, it means that the semaphore does not represent any real resource within the VMD. Therefore, it can be deleted by a MMS client by means of the DeleteSemaphore service.

2. Number of tokens.   This attribute indicates the total number of tokens contained in the token semaphore.

3. Owned tokens.   This attribute indicates the number of tokens whose associated semaphore entry state is owned.

4. Hung tokens.   This attribute indicates the number of tokens whose associated semaphore entry state is hung.

Pool Semaphores

A pool semaphore is similar to a token semaphore except that the individual tokens are identifiable and have a name associated with them. These named tokens can optionally be specified by the MMS client

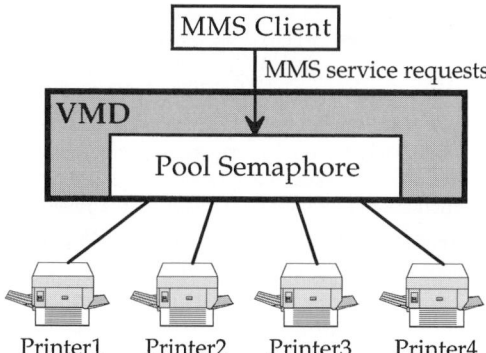

Pool Semaphore Example

FIGURE 10   A pool semaphore can be useful to control access to similar but distinguishable resources. In the example, each printer is represented as a separate named token.

when issuing TakeControl requests. The pool semaphore itself is a MMS object. The named tokens contained in the pool semaphore are not MMS objects. They are representations of a real resource in much the same way an unnamed variable object is. Pool semaphore objects are used when it is desired to represent a set of similar resources where clients that need control of such a resource may or may not care which specific resource they desire control over. For instance, the individual vehicles in an automated guided vehicle (AGV) system can be represented by a pool semaphore. MMS clients at individual work centers may desire to control an AGV to deliver new material but may not care which specific AGV is used. The AGV system VMD would decide which specific AGV, represented by a single named token, would be assigned to a given MMS client. The name of the pool semaphore is independent of the names of the named tokens (Fig. 10).

Pool semaphores can only be used to represent some real resource within the VMD. Therefore, pool semaphores cannot be created or deleted by using MMS service requests and cannot be AA specific in scope.

In addition to the name of a pool semaphore, the following attributes also are defined by MMS: (a) free named tokens—the list of named tokens that are not owned; (b) owned named tokens—the named tokens whose semaphore entries are owned; and (c) hung named tokens—the named tokens whose semaphore entries are hung.

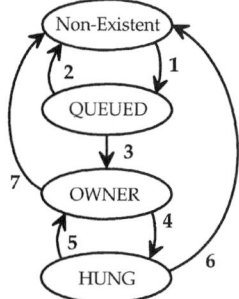

Semaphore Entry State Diagram

FIGURE 11   A semaphore entry is created each time a client attempts to take control of a semaphore. The semaphore entry reflects the state of the relationship between the client and the semaphore.

Semaphore Entry

When a MMS client issues a TakeControl request for a given semaphore, the VMD creates an entry in an internal queue that is maintained for each semaphore. Each entry in this queue is called a semaphore entry. The attributes of a semaphore entry are visible to MMS clients and provide information about the internal semaphore processing queue in the VMD. The semaphore entry is not a MMS object. It only exists from the receipt of the TakeControl indication by the VMD until the control of the semaphore is relinquished or if the VMD responds negatively to the TakeControl request (see Fig. 11).

The attributes of a semaphore entry are given below. Some of these attributes are specified by the client in the TakeControl request, whereas the others are set by the VMD.

1. Entry ID.   This is a number assigned by the VMD to distinguish one semaphore entry from another. The Entry ID is unique for a given semaphore.

2. Named token.   Valid only for pool semaphores. It contains the named token that was optionally requested by the client in a TakeControl request. If the semaphore entry is in the owned or hung state, it is the named token that the VMD assigned as a result of a TakeControl request.

3. Application reference.   This is a reference to the MMS client application that issued the Take-Control request that created the semaphore entry.

4. Priority.   This attribute indicates the priority of the semaphore entry with respect to other semaphore entries. Priority is used to decide which semaphore entry in the queued state will be granted a token (or named token) when multiple requests are outstanding. The value (0 = highest priority, 64 = normal priority, and 127 = lowest priority) is specified by the client in the TakeControl request.

5. Entry state.   The entry state attribute represents the relationship between the MMS client and the semaphore by one of the following values.

   • queued.   This means that a TakeControl request has been received but has not been responded to by the VMD. The client is waiting for control of the semaphore.
   • owned.   The VMD has responded positively to the TakeControl request and the client now owns the token (or named token).
   • hung.   This state means that the application association over which the MMS client issued the TakeControl request has been lost and the Relinquish if Connection Lost attribute is *false*. A MMS client can take control of a semaphore entry in the hung state by issuing a TakeControl request with the preempt option *true* and by specifying the MMS client to preempt (by the application reference attribute).

6. Relinquish if Connection Lost.   If this attribute is *true*, the VMD will relinquish control of the semaphore if the application association for the MMS client that owned the token is lost or aborted. If *false*, the semaphore entry will enter the hung state if the application association is lost or aborted.

7. Control timeout.   This attribute indicates how many milliseconds the client will be allowed to control the semaphore once control is granted. If the client has not relinquished control using a RelinquishControl request when the control timeout expires, the semaphore entry will be deleted and control of the semaphore will be relinquished. If the control timeout attribute is omitted in the TakeControl request, no control timeout will apply.

8. Abort on timeout.   This attribute indicates if the VMD should, in addition to relinquishing control of the semaphore, also abort the application association with the owner client upon the occurrence of a control timeout.

9. Acceptable delay.   This attribute indicates how many milliseconds the client is willing to wait for control of the semaphore. If control is not granted during this time, the VMD will respond negatively to the TakeControl request. If the acceptable delay attribute is omitted from the TakeControl request, it means that the client is willing to wait indefinitely.

Semaphore Services

1. TakeControl.   This confirmed service request is used by a MMS client to request control of a semaphore.

2. RelinquishControl.   This confirmed service request is used by a MMS client to release control over a semaphore that the client currently has control of.

3. DefineSemaphore and DeleteSemaphore. These confirmed services are used by clients to define and delete token semaphores that are used solely for coordinating the activities of two or more MMS clients.

4. ReportSemaphoreStatus and ReportPoolSemaphoreStatus. These confirmed services are used by MMS clients to obtain the status of semaphores.

5. ReportSemaphoreEntryStatus. This confirmed service is used by a MMS client to obtain the attributes of semaphore entries.

OTHER MMS OBJECTS

Operator Station Object

The operator station is an object that represents a means of communicating with the operator of the VMD by use of a keyboard and display. An operator station is modeled as character-based input and output devices that may be attached to the VMD for the purpose of communicating with an operator local to the VMD. MMS defines three types of operator stations: entry, display, and entry–display. An entry-only operator station consists of an alphanumeric input device such as a keyboard or perhaps a bar-code reader. A display-only operator station consists of an alphanumeric character-based output display (no graphics or control characters). The entry–display type of operator station consists of both an entry station and a display station.

Because the operator station is a representation of a physical feature of the VMD, it exists beyond the scope of any domain or application association. Therefore, MMS clients access the operator station by name without scope. There can be multiple operator stations for a given VMD. The services used by MMS clients to perform operator communications are input and output. MMS clients use the input service to obtain a single input string from an input device. The service has an option for displaying a sequence of prompts on the display if the operator station is an entry–display type. The output service is used to display a sequence of output strings on the display of the operator station.

Journal Objects

A MMS journal represents a log file that contains a collection of records (called a journal entry) that are organized by time stamps. Journals are used to store time-based records of tagged variable data, user-generated comments (called annotation), or a combination of events and tagged variable data. Journal entries contain a time stamp that indicates when the data in the entry were produced, not when the journal entry was made. This allows MMS journals to be used for applications in which a sample of a manufactured product is taken at one time, analyzed in a laboratory off line, and then placed into the journal at a later time. In this case, the journal entry time stamp would indicate when the sample was taken.

MMS clients read the journal entries by specifying the name of the journal (which can be VMD specific or AA specific only) and either (1) the date/time range of entries that the client wishes to read or (2) by referring to the entry ID of a particular entry. The entry ID is a unique binary identifier assigned by the VMD to the journal entry when it is placed into the journal. Each entry in a journal can be one of the following types.

1. Annotation. This type of entry contains a textual comment. This is typically used to enter a comment regarding some event or condition that had occurred in the system.

2. Data. This type of entry would contain a list of variable tags and the data associated with those tags at the time indicated by the time stamp. Each variable tag is a 32-character name that does not necessarily refer to a MMS variable, although it might.

3. Event data. This type of entry contains both variable tag data and event data. Each entry of this type would include the same list of variable tags and associated data as described above, along with a single event condition name and the state of that event condition at the time indicated by the time stamp.

The services available for MMS journals are as follows.

1. **ReadJournal.** This confirmed service is used by a client to read one or more entries from a journal.

2. **WriteJournal.** This confirmed service is used by a client to create new journal entries in a journal. A journal entry can also be created by local autonomous action by the VMD without a client using the WriteJournal service.

3. **CreateJournal and DeleteJournal.** These confirmed services are used by a client to create and delete (if the journal is *deletable*) journal objects. The CreateJournal service only creates the journal. It does not create any journal entries (see WriteJournal).

4. **InitializeJournal.** This confirmed service is used by a client to delete all or some of the journal entries that are in a journal.

Files

MMS also provides a set of simple file transfer services for devices that have a local file store but do not support a full set of file services by some other means. For instance, an electric meter may use the file services for transferring oscillography (waveform) files to an interested MMS client. The MMS file services support file transfer only, not file access. Although these file services are defined in an annex within the MMS standard, they are widely supported by most commercial MMS implementations. The services for files are described below.

1. **FileOpen.** This confirmed service is used by a MMS client to tell the VMD to open a file and prepare it for a transfer.

2. **FileRead.** This confirmed service is used to obtain a segment of the file's data from a VMD. The MMS client would continue to issue FileRead requests until the VMD indicates that all the data in the file have been returned. The number of bytes returned in each FileRead response is determined solely by the VMD and can vary from one FileRead response to the next up to the maximum negotiated at the time of establishing the application association.

3. **FileClose.** This confirmed service is used by a MMS client to close a previously opened file. It is used after all the data from the file have been read or can be used to discontinue a file transfer before it is completed.

4. **ObtainFile.** This confirmed service is used by a MMS client to tell the VMD to obtain a file. When a VMD receives an ObtainFile request it would then become an MMS client and issue FileOpen, FileRead(s), and FileClose service requests to the MMS application that issued the ObtainFile request. The original "client" would then become the *server* for the FileOpen, FileRead, and FileClose services. Once the transfer is complete, the VMD sends the ObtainFile response back to the original MMS client. A third-party option is available (if supported by the VMD) to tell the VMD to obtain the file from another node on the network, using some protocol (which may or may not be MMS).

5. **FileRename, FileDelete, and FileDirectory.** These confirmed services are used to rename, delete, and obtain a directory of files on the VMD, respectively.

CONTEXT MANAGEMENT

MMS provides services for managing the context of communications between two MMS nodes on a network. These services are used to establish and terminate application associations and for handling protocol errors between two MMS nodes. The terms *association* and *connection* are sometimes used interchangeably, although there is a distinction from a network technology point of view. The node that initiates the association with another node is referred to as the *calling* node. The responding node is referred to as the *called* node.

In a MMS environment, two MMS applications establish an application association between themselves by using the MMS Initiate service. This process of establishing an application association consists of an exchange of some parameters and a negotiation of other parameters. The exchanged parameters include information about restrictions that pertain to each node that are determined solely by that node (e.g., which MMS services are supported). The negotiated parameters are items for which the called node either accepts the parameter proposed by the calling node or adjusts it downward as it requires (e.g., the maximum message size).

The calling application issues an Initiate service request that contains information about the calling node's restrictions and a proposed set of the negotiated parameters. The called node examines the negotiated parameters and adjusts them as necessary to meet its requirements. It then returns the results of this negotiation and the information about its restrictions in the Initiate response. Once the calling node receives the Initiate confirmation, the application association is established and other MMS service requests can then be exchanged between the applications.

Once an application association is established, either node can assume the role of client or server, independent of which node was the calling or called node. For any given set of MMS services, one application assumes the client role while the other assumes the role of server or VMD. Whether or not a particular MMS application is a client, server (VMD), or both is determined solely by the developer of the application.

Associations vs. Connections

Although many people may refer to network connections and application associations interchangeably, there is a distinct difference. A connection is an attribute of the underlying network layers that represents a virtual circuit between two nodes. For instance, telephone networks require that two parties establish a connection between themselves (by dialing and answering) before they can communicate. An application association is an agreement between two networked applications governing their communications. It is analogous to the two telephone parties agreeing to use a particular language and to not speak about religion or politics over the telephone. Application associations exist independent of any underlying network connections (or lack thereof).

In a connection oriented environment (e.g., TCP/IP) the MMS Initiate service is used to signal to the lower layers that a connection must be established. The Initiate service request is carried by the network through the layers as each layer goes through its connection establishment procedure until the Initiate indication is received by the called node. The connection does not exist until after all the layers in both nodes have completed their connection establishment procedures and the calling node has received the Initiate confirmation. Because of this, the association and the connection are created concurrently in a connection-oriented environment.

In a connectionless environment (e.g., three layer), it is not strictly necessary to send the Initiate request before two nodes can actually communicate. In an environment where the Initiate service request is not used before other service requests are issued by a MMS client to a VMD, each application must have prior knowledge of the other application's exchanged and negotiated parameters by some local means (e.g., a configuration file). This foreknowledge of the other MMS application's restrictions is the application association from a MMS perspective. Whether an Initiate service request is used or not, application associations between two MMS applications must exist before communications can take place. In some connectionless environments such as the UCA three layer for serial link communications, MMS nodes still use the Initiate service to do the application association negotiation before communicating data or control information.

Context Management Services

These services include Initiate, Conclude, Abort, Cancel, and Reject.

1. Initiate.   This service is used to exchange and negotiate the parameters required for two MMS applications to have an application association.

2. Conclude.   This service is used by a client to request that a previously existing application association be terminated in a graceful manner. The conclude service allows the server to decline to terminate the association because of ongoing activities such as a download sequence or file transfer.

3. Abort.   This service is used to terminate an application association in an ungraceful way. The server does not have the opportunity to decline an abort. An abort may result in the loss of data.

4. Cancel.   This service is used by a client to cancel an outstanding MMS service request (e.g., TakeControl) that has not yet been responded to by the server.

5. Reject.   This service is used by either the client or server to notify the other MMS application that it had received an unsupported service request or a message that was not properly encoded.

SUMMARY

MMS provides a very flexible real-time messaging architecture for a wide range of applications. The concept of the VMD model of MMS is used in many lower level networking systems such as Foundation Fieldbus and Profibus, where a more limited subset of MMS is called the Fieldbus Messaging System. MMS has been used for years as a messaging system for Ethernet networks in automotive manufacturing, pulp-paper, aerospace, and other large and complex material handling systems. Recent technological innovations have allowed MMS to be applied into smaller and more resource-limited devices, such as small PLCs, RTUs, meters, breakers, and relays. With the backing of EPRI, GRI, and a large number of utility users, the use of MMS is gaining momentum in these industries as a considerable number of equipment suppliers now support MMS and UCA. MMS is an effective bridge between the plant floor and the management information systems as well as between the process control systems of the power plant and the distribution systems of the utility. With the additional refinements of GOMSFE and CASM, the level of interoperability between dissimilar equipment that is being achieved is welcome relief to an industry long plagued with numerous incompatible proprietary communication methods.

FIELD-BASED SYSTEMS

by Jim Cahill*

INTRODUCTION

Forty years ago, control relay systems were state of the art. Thirty years ago, analog controllers were the leading edge in control systems. Twenty years ago, microprocessors ushered in the distributed control system era. As microprocessors have become more powerful and able to go into rugged environments, a technology shift is again underway. Today's digital technology makes scalable, field-based systems a reality.

Welcome to the world of scalable, field-based systems. To fully appreciate the many benefits of a field-based system, we must understand what a field-based system is and what it can accomplish. A

* Fisher-Rosemount Systems, Inc., Austin, Texas 78759.

field-based system is a digital communications network that uses the power of microprocessor-based, field-device intelligence to improve plant performance. This system must be designed using open and interoperable standards to allow users to select the best solution for their application. Intelligent field devices share process variable information with a control system. In a field-based system, the current configuration, status, performance, and health of the device also are easily accessible from anywhere in the plant.

Intelligent field devices with HART and more recently FOUNDATION fieldbus technology are available, and many plants are currently using these devices. The ability to access the wealth of information contained within these devices is a recent technological advancement. Asset management software is software designed to extract the information contained in the intelligent field devices to allow predictive maintenance capabilities, as well as better process control capability. A field-based system integrates this information with process management to minimize downtime and maximize performance. This new found information allows users to make better decisions regarding the plant process (Fig. 1).

Currently, U.S. industries spend more than $200 billion each year on maintenance. Plant maintenance makes up 15–40% of the cost of goods produced in the West. In western markets, five times

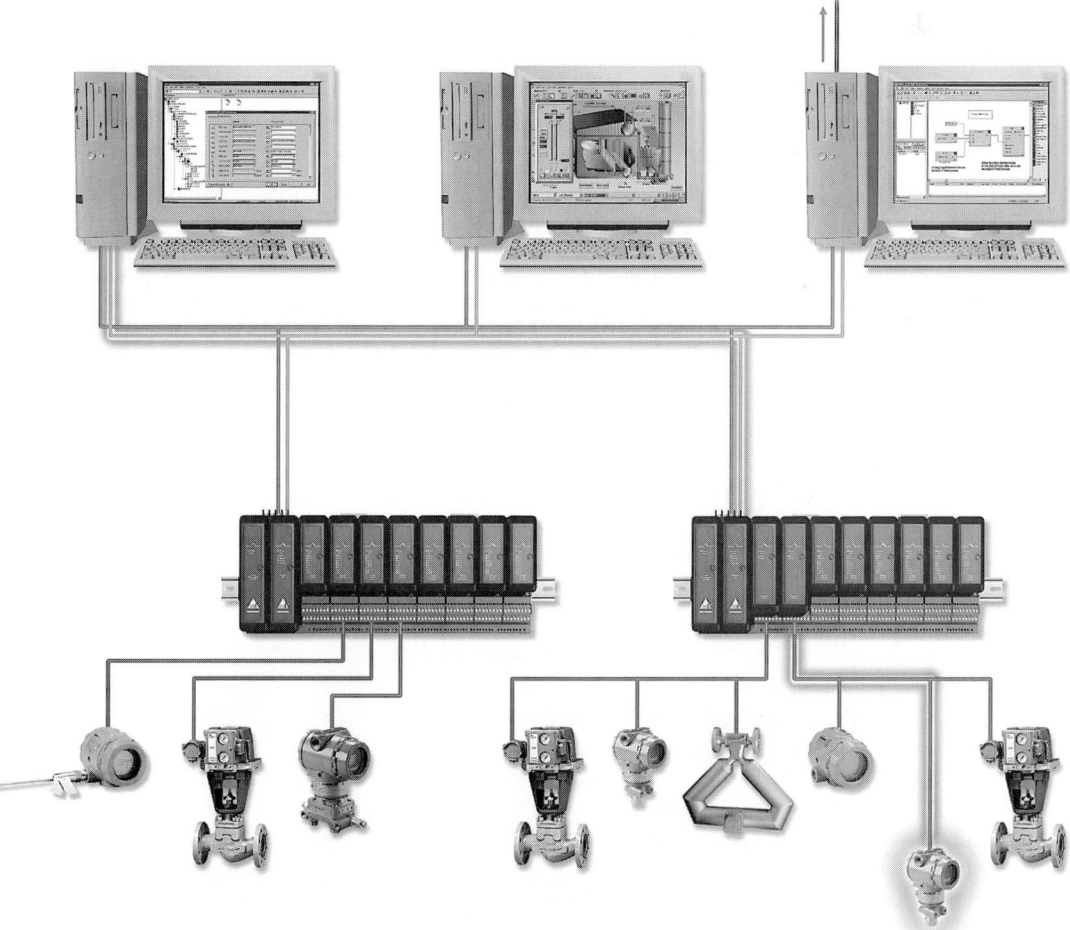

FIGURE 1   The power of intelligent field devices is the core of a field-based system.

more is spent on maintenance than on new plant construction. Also, more than $60 billion is lost on ineffective maintenance management each year [1]. Field-based systems with asset management software reduce this extensive use of time and money. Advanced diagnostics key in on devices that will soon need rework, before they affect product quality or shut down the process. This in turn allows scheduled maintenance to be done on time. This predictive method versus the old-fashioned reactive maintenance method is lowering plant operating and maintenance costs [2].

Time and cost savings are apparent in the areas of configuration and calibration using a field-based system with asset management software. Entire device configurations are downloaded using drag-and-drop techniques. Storing all calibration information in a database reduces calibration time. The user is prompted through the calibration process. The fact that all this can be done at a workstation adds convenience and value, especially if conditions such as weather, safety, and the location of field devices have inhibited configuration and calibration tasks in the past.

The field-based, built-for-bus architecture greatly reduces installation expenses (specifically in field wiring), the need for control room space, and field-device commissioning time. In April 1997, at the world's first FOUNDATION fieldbus installation in Fort Saskatchewan, Canada, Dow Chemical reported that there was a 74% reduction in field wiring, a 93% reduction in required control room space, and an 80% reduction in field-device commissioning time over previous installations done with conventional input/output (I/O) [3]. The economic advantages of a field-based system related to upfront engineering and installation are readily apparent here. Overall, Dow reported that by installing a field-based system they lowered their total project cost by 30% [4]. This is in comparison to a similar installation that used traditional devices. Ninety-six fieldbus devices were installed. Because of the success of this initial installation, 480 more fieldbus devices were installed 3 months later (Fig. 2).

Scalable in size and functionality, a field-based system has an intuitive look and feel. A scalable, field-based system can be applied from the smallest to largest process application because control capacity is added, intelligent device by intelligent device. In a field-based system, control can occur either in the intelligent field device or in the field-based system, wherever the application dictates. The layout of the process area relative to the control room might dictate where control should occur. For instance, a remote plant process with a single control loop is a logical spot to run control in the field device, either in the transmitter or valve positioner. Based on the user's process control requirements, it is cost effective to implement and grow control capability one loop at a time.

The current industry trend of producing and distributing *just in time* makes it imperative that the control process system seamlessly integrates with the enterprise with one common plug and play-based standard. Using OLE for process control (OPC), the field-based system does this. The field-based

FIGURE 2   Dow Chemical recorded significant installation savings.

system, OPC server delivers high-throughput information transfer across the plant network. Data rates of hundreds of thousands of data values per second throughput, per OPC server, is currently available. This standards-based information connection allows dependable and secure plug-and-play communications between many applications, including control, management execution, scheduling, and accounting systems.

HARDWARE

Field-based system hardware includes workstations, Ethernet control network, controller, I/O, and intelligent field devices. High-performance PCs running Windows NT bring the field-based system into clear view on the monitor. The Ethernet control network eliminates digital-to-analog and analog-to-digital signal conversion measurement inaccuracies. It facilitates multiple messages on the same wire, and because communications are bidirectional, peer-to-peer interactions between devices are easily accomplished. Since the hardware is built on a base of commercially available technologies, solutions can be rapidly designed and implemented based on the specifics of the application. For example, oil and gas production often have great distances between processing facilities. With the use of commercial Ethernet and satellite technology, solutions can be rapidly designed and implemented, whereas in the past, expensive, custom solutions were required.

The controller and I/O interfaces in a field-based system use FOUNDATION fieldbus, HART, serial, and conventional I/O standards for all monitoring and control. The application can dictate the I/O technology that makes the most sense. This I/O information is brought into a single engineering environment, also based on standards, including IEC 1131-3 and ISA S-88. It is not necessary for a user to have separate engineering environments for separate I/O technologies.

The FOUNDATION fieldbus and HART open-field communication standards use the power of field intelligence to reduce engineering, operating, and maintenance costs while improving process performance. As stated earlier, FOUNDATION fieldbus installations significantly decrease wiring, control room space, and commissioning time compared to a traditional installation. Table 1 summarizes the Canadian results [5].

These reductions are possible for the following reasons. Each fieldbus interface card supports two segments with a maximum of 16 fieldbus devices per segment. The fieldbus autosensing capability acknowledges the controller, I/O, devices, and tags, eliminating the need for dipswitch changing and address entering. Device calibration, configuration, and diagnostics are performed at the PC, which can significantly reduce trips to the field and other high-cost maintenance activities. These are all cost-saving benefits associated with a field-based system (Figs. 3 and 4).

HART communication protocols make it possible to communicate with conventional I/O without disturbing the 4/20 mA signal and still tap the wealth of information contained in the device. HART devices in a field-based system can present up to four multivariable signals from a single device and wire to the system. HART, similar to Fieldbus, reduces wiring, I/O count, and pipe intrusions.

If an intelligent HART device senses an internal malfunction, it sends a maintenance alert to the operator. Isolating equipment problems as they occur reduces process variability by enabling more effective control over the process. Process variability is reduced in a field-based system since it has access to this information, whereas a conventional system does not.

TABLE 1 Reductions Using FOUNDATION Fieldbus

Task	Percent Decrease
Field wiring	74
Control room space	93
Field-device commissioning time	80

FIGURE 3   Footprint is significantly reduced when FOUNDATION fieldbus is used. The full cabinet handles 320 inputs and outputs (I/O).

FIGURE 4   The nearly empty field-based system cabinet is shown with 384 I/O.

SOFTWARE

A field-based system requires software that is immediately familiar and intuitive for a first-time user. An object-oriented global database integrates operations, engineering, and multipurpose software applications in a field-based system. This object-based structure allows users to easily customize functions for their operations while still preserving the integrity of the original objects. Also, this object-based structure supports the ISA S-88 hierarchy and FOUNDATION fieldbus function block standard. All network communications between the software running in the PC and the controllers are automatic and transparent. A wide array of software applications provides full functionality to a field-based system (Fig. 5).

Operators have a window into their running process through the operator interface. The operator interface provides graphics, sophisticated alarm management and presentation, and real-time trending. Control information gathered from smart devices is used here to alert operators to a device malfunction. A combination of alarms and process conditions called smart alarms are built into the field-based

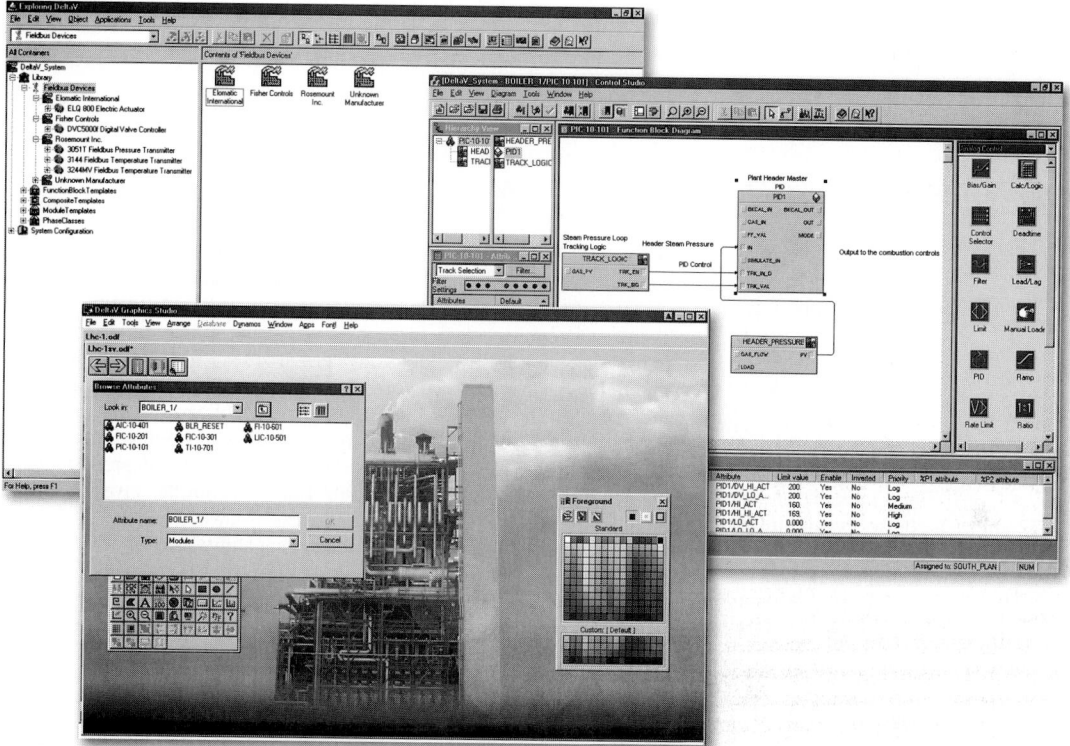

FIGURE 5 Field-based system software has the same look and feel as the Windows NT environment.

system. This application allows enabling or disabling alarms in real time, as the plant process requires both under user and control strategy control.

Operators and maintenance users also have access to top-to-bottom systemwide diagnostic views, including smart field devices. These views are accessible across the field-based control network. These data values are available for use in operator graphics, smart alarms, or OPC applications.

Field-based engineering software provides pre-engineered control strategies that drag and drop into configuration modules for FOUNDATION fieldbus, HART, and conventional devices. Other applications in the engineering environment provide:

- remote configuration, calibration, and diagnostics
- display and graphic symbol configuration
- configuration on-line viewing and debugging
- automatic process loop tuning
- batch recipe development and management
- process simulation

Additional software integrates the field-based system with the plant local area network for users across the plant and between plants. The following applications provide secure, reliable information to the right location in a field-based system.

- OPC server connectivity to client connectivity
- OPC server to server connectivity

- plantwide history integration
- optional field-based configuration, calibration, and diagnostic applications
- user-written and third-party applications

CHARACTERISTICS OF A FIELD-BASED SYSTEM

The power of a field-based system resides in its open communication standards. These standards allow HART, FOUNDATION fieldbus, and conventional devices to bidirectionally interact within the system. Fieldbus, HART, and classic I/O can be installed on the same controller. A 24-V bus power supply powers all devices. T junctions connect devices to the bus. Finally, this field-based system allows control to take place in the controller or the fieldbus device. Cascaded PID control using peer-to-peer communications on the digital bus or advanced control in the controller are both options in a field-based system. Deploying control in the field facilitates growing control capability systemwide in small manageable increments. It gives users the flexibility to deploy the control that best suits the application based upon physical location or process dynamics. This brings a higher degree of simple loop integrity and security to a control system. Deploying control in the controller gives advanced control capabilities not yet available from the devices. As microprocessor power increases both in the field-based system and in the intelligent field devices, more control capability is available at either location (Fig. 6).

Another technological advancement found in a field-based system is its capacity to integrate process control information with the rest of the enterprise. This is accomplished using the OPC

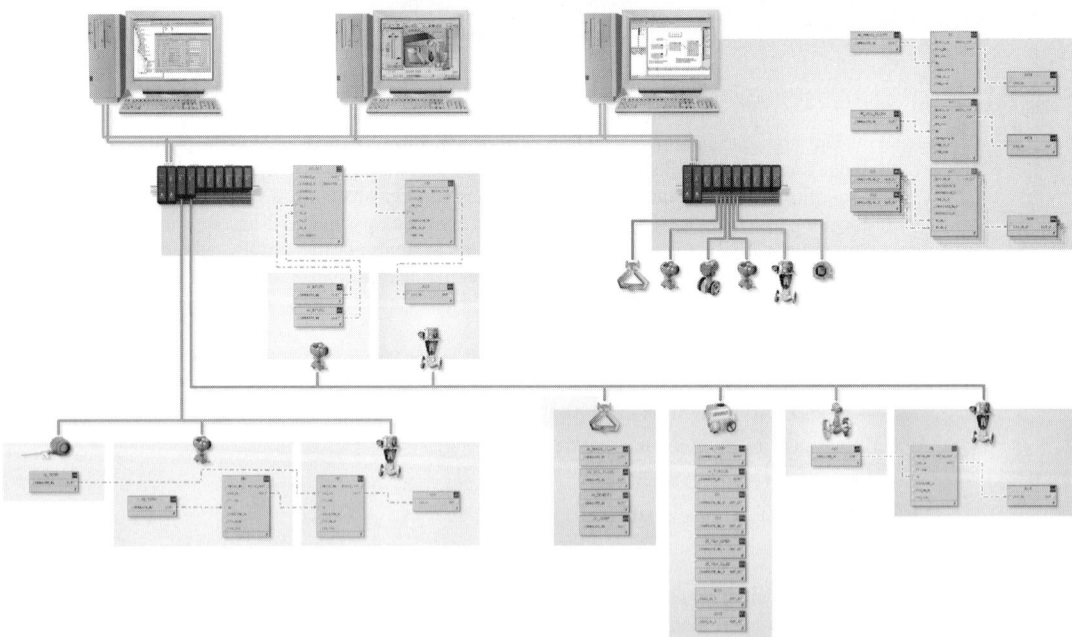

FIGURE 6   Peer-to-peer communication, as defined by the FOUNDATION fieldbus standard, facilitates the deployment of complex control strategies in intelligent field devices with no host interaction.

FIGURE 7 The OPC standard provides plug and play, high-performance connectivity.

standard. The OPC standard sustains transfer rates up to 60,000 values/s. OPC servers ensure that enterprisewide data are accurate and readily available. Applications such as Microsoft Excel receive up-to-the-second information from any other real-time data server using the OPC standard. The benefits of this technology are as follows (Fig. 7).

- OPC enterprisewide integration is simple to perform
- all applications resemble a Windows environment
- learning curve time is reduced
- exception reporting increases information throughput
- bidirectional communications increases information throughput
- no program interfaces are needed
- no custom interface maintenance is needed

Last, but far from least in importance, is the fact that a field-based system is easy to learn and use. HART and FOUNDATION fieldbus standards combined with a Windows-like environment greatly reduce a user's learning curve. Microsoft worldwide-accepted interface standards are used to produce an intuitive, user-friendly field-based system. Users are familiar with point-and-click and drag-and-drop techniques. Field-based systems use the latest technology to ensure that the power contained in the system is not lost to long learning curves and cryptic system design.

FIELD-BASED SYSTEM RESULTS

Increased functionality, performance, and interoperability leading to greater than ever plant efficiency and lower cost production continue to be the goals in designing process control systems. The open communications field-based standards architecture accomplishes this design goal. Tapping the wealth of intelligence within a field device and communicating these data with asset management software, enterprisewide, improves overall plant performance and significantly reduces associated costs.

Statistics from the first field-based installation at the Dow Chemical plant documented reductions in field wiring, control room space, and field-device commissioning time. Another field-based system installation at ARCO's West Sak field on Alaska's North Slope produced the following results (Fig. 8).

- 16% reduction in wellhead terminations
- 69% reduction in comparable wiring costs
- 98% reduction in home run wiring
- 83% reduction in instrument commissioning and checkout
- 90% reduced configuration time to add an expansion well
- 92% reduced engineering drawings to add or expand a well [6].

The option of control in the field or in the controller also adds to increased performance. Production costs decrease because device configuration, calibration, and diagnostics are performed at a PC instead of physically in the field.

The concept of seamless integration of the process with the rest of the enterprise is now a reality. Field-based systems using OPC make control information available plantwide. For complete enterprise integration, third-party applications communicate bidirectionally with the field-based system. In today's ultracompetitive environment, a field-based system is necessary to produce efficient and effective processes and products at the lowest possible production costs. The ability to disseminate process, plant, and asset information enterprisewide is another benefit of a field-based system. Additionally, the long-term diagnostic and maintenance benefits associated with a field-based system are currently being collected and documented.

FIGURE 8   Field-based system installation at ARCO's West Sak field on Alaska's North Slope.

CONCLUSIONS

The digital age is upon us. The process control industry can choose to embrace the new technological advances and receive the cost savings and improved performance of field-based systems or continue make due with mature technology. Although the technology that drives a field-based system appears to be a difficult change, upon careful consideration it becomes obvious that the short-term and long-term benefits of a scalable, field-based system are apparent.

Predictive maintenance capabilities that come from tapping the wealth of information hidden in intelligent field devices minimize downtime and maximize performance. This leads to an efficient, effective process control system. It makes the most sense to fix a device only when it needs maintenance. Neither reactive nor preventative maintenance provides this capability. Having to fix a broken device in the middle of running a process means reduced process availability and increased process variability, cost overruns, and reduced product quality. And furthermore, scheduled, preventative maintenance often causes unnecessary work and downtime.

Configuring and calibrating a field-based system at a PC by using asset management software provides convenience and value. Weather and location are no longer necessary considerations in a field-based system. The ability to download entire device configurations with drag-and-drop ease and to store all calibration data in a single database produce time and cost savings.

The vast improvement in the field-based built for bus architecture produces a major decrease in capital and installation expenses. Field wiring, control room space, and device commissioning time are all positively affected by this technology. The first FOUNDATION fieldbus installation at Dow Chemical produced a 30% decrease in total project cost compared to a traditional installation. It will be fascinating to see how much further total project costs decrease once the lessons learned from this first installation are incorporated into future installations.

The benefits of a field-based system far outweigh the comfort of the status quo. Increased plant capabilities, efficiency, and effectiveness combined with decreased installation and maintenance costs add up to an intelligent process control solution. Forty years ago, engineers dreamed of field-based systems. Today, they are a reality.

REFERENCES

1. Mobley, R., "An Introduction to Predictive Maintenance," in Lenz, G., *Nondestructive Evaluation and Diagnostics Needs for Industrial Impact*, hard copy version of keynote address at ISA 96.

2. Ibid.

3. Statistics compiled by Dow Chemical in Fort Saskatchewan, Alberta, Canada, 1997.

4. Ibid.

5. Ibid.

6. Figures compiled by Alaska Anvil Engineering for ARCO Alaska, 1997.

SECTION 4
PROCESS VARIABLES—FIELD INSTRUMENTATION*

L. Arnold
*Johnson Yokogawa Corporation, Newnan, Georgia
(Flow Systems)*

S. Barrows
*The Foxboro Company (a Siebe Company), Norcross, Georgia
(Temperature Systems)*

Gary Bugher
Endress & Hauser Inc., Greenwood, Indiana (Flow Systems)

W. H. Burtt
*The Foxboro Company (a Siebe Company), Foxboro,
Massachusetts (Resonant-Wire Pressure Transducers)*

W. A. Clayton
*Hy-Cal Engineering (a unit of General Signal), El Monte,
California (Temperature Systems—prior edition)*

R. Collier
*ABS-Kent-Taylor (ASEA Brown Boveri), Rochester, New York
(Temperature Systems—prior edition)*

Z. C. Dobrowolski
*(Deceased) Kinney Vacuum Company (a unit of General Signal),
Cannon, Massachusetts (High Vacuum Measurement—prior
edition)*

G. L. Dyke
*Bailey Controls Company, Wickliffe, Ohio (Pressure Transducers
and Transmitters)*

C. J. Easton
*Sensotec, Inc., Columbus, Ohio (Strain-Pressure
Transducers—prior edition)*

C. E. Fees
*Fischer & Porter Company, Warminster, Pennsylvania
(Rotameters—prior edition)*

H. Grekksa
*Infrared Measurement Division, Square D Company, Niles, Illinois
(Temperature Systems)*

---

* *Persons who authored complete articles or subsections of articles, or who otherwise cooperated in an outstanding manner in furnishing information and helpful counsel to the editorial staff.*

Jack Herring
Endress & Hauser Inc., Greenwood, Indiana (Humidity and Moisture Systems)

E. H. Higham
Foxboro Great Britain Limited, Redhill, Surrey, England (Magnetic Flowmeter—prior edition)

D. A. Jackson
Omega Engineering, Inc., Stamford, Connecticut (Temperature Systems—prior edition)

J. Kortright
Leeds & Northrup (a unit of General Signal), North Wales, Pennsylvania (Flow Systems; Temperature Systems)

A. J. Kurylchek
Wayne, New Jersey (Industrial Weighing and Density Systems)

R. W. Lally
Engineering Department, PCB Piezotronics, Inc., Depew, New York (Pressure Sensors)

G. Leavitt
ABB-Kent-Taylor (ASEA Brown Boveri), Rochester, New York (Temperature Systems—prior edition)

C. L. Mamzic
Systems and Application Engineering, Moore Products Company, Spring House, Pennsylvania (Flow Systems—prior edition)

G. R. McFarland
ABB-Kent-Taylor, Inc., Rochester, New York (Wedge-Type Flow Element)

Craig McIntyre
Endress & Hauser Inc., Greenwood, Indiana (Level Systems)

S. Milant
Ryan Instruments, Redmond, Washington (Thermistors)

R. W. Miller
Consultant, Foxboro, Massachusetts (Flow Differential Producers—prior edition)

A. E. Mushin
Omega Engineering, Inc., Stamford, Connecticut (Thermocouple Systems—prior edition)

R. Peacock
Land Instruments, Inc., Tullytown, Pennsylvania (Radiation Thermometers—prior edition)

B. Pelletier

Rosemount Inc., Measurement Division, Eden Prairie, Minnesota (Temperature Systems; Hydrostatic Level Gages)

G. Rebucci

Schenck Weighing Systems, Totowa, New Jersey (Weighing Systems—prior edition)

W. L. Ricketson

Toledo Scale Division, Reliance Electric Company, Atlanta, Georgia (Weighing Systems)

Bill Roeber

Great Lakes Instruments, Inc., Milwaukee, Wisconsin (Target Flowmeters)

Staff

Brooks Instrument Division, Emerson Electric Company, Hatfield, Pennsylvania (Positive-Displacement Meters)

Staff

Lucas Schaevitz, Pennsauken, New Jersey (Gage Pressure Transducers)

Staff

Raytek Incorporated, Santa Cruz, California (Temperature Systems)

Technical Staff

Moisture Systems Corporation, Hopkinstown, Massachusetts (IR Moisture Analyzer)

R. D. Thompson

ABB-Kent-Taylor (ASEA Brown Boveri), Rochester, New York (Temperature Systems—prior edition)

Ola Wesstrom

Endress & Hauser Inc., Greenwood, Indiana (Pressure Systems)

Peter E. Wiederhold

President, Wiederhold Associates, Boston, Massachusetts (Humidity—prior edition)

J. A. Wise

National Institute of Standards and Technology, Gaithersburg, Maryland (Temperature Systems—prior edition)

Gene Yazbak

MetriCor, Inc., Monument Beach, Massachusetts (Temperature Systems)

TEMPERATURE SYSTEMS

Over many decades the demand for temperature sensors and controllers has shown that temperature is the principal process variable of serious concern to the process industries, that is, those industries that handle and convert gases, liquids, and bulk solids into products and by-products. Chemical, petroleum, petrochemical, polymer, plastic, and large segments of metallurgical and food processors are examples. Temperature control is critical to such processes and operations as chemical reactions and in materials separations, such as distillation, drying, evaporation, absorbing, crystallizing, baking, and extruding. Temperature control also plays a critical role in the safe operation of such facilities.

Although critical temperature control applications occur less frequently in the discrete-piece manufacturing industries, there are numerous examples. The extensive demand for temperature controllers in the air-conditioning field is self-evident.

TEMPERATURE DEFINED

Although temperature fundamentally relates to the kinetic energy of the molecules of a substance (as reflected in the definition of the absolute, thermodynamic, or Kelvin temperature scale), temperature may be defined in a less academic fashion as "the condition of a body which determines the transfer of heat *to* or *from* other bodies," or even more practically, as "the degree of 'hotness' or 'coldness' as referenced to a specific scale of temperature measurement."

TEMPERATURE SCALES

Thermodynamic Kelvin Scale.   The currently accepted theoretical scale is named for Lord Kelvin, who first enunciated the principle on which it is based. Thermodynamic temperature is denoted by T and the unit is the kelvin (K) (no degree sign is used). The kelvin is the fraction $1/273.16$ of the thermodynamic temperature of the triple point of water. The triple point is realized when ice, water, and water vapor are in equilibrium. It is the sole defining fixed point of the thermodynamic Kelvin scale and has the assigned value of 273.16 K.

Celsius (Centigrade) Scale.   In 1742 Anders Celsius of Uppsala University, Sweden, reported on the use of thermometers in which the fundamental interval, ice point to steam point, was 100°. Celsius designated the ice point at 100° and the steam point at 0°. Subsequently Christin (1743) in Lyon, France, and Linnaeus (1745) at Uppsala independently interchanged the designations. For many years prior to 1948 it was known as the centigrade scale. In 1948, by international agreement, it was renamed in honor of its inventor. Used worldwide, temperatures are denoted as degrees Celsius (°C). By personal choice, the degree symbol is sometimes eliminated.

Fahrenheit Scale.   Daniel Gabriel Fahrenheit (1724) first defined the Fahrenheit scale, using the ice point (32°) and the human body temperature (96°) as the fixed points of the scale. The fundamental interval (ice point to steam point) turned out to be 180 degrees ($212 - 32 = 180$). Although very serious attempts have been and are being made to convert to the Celsius scale, the Fahrenheit scale remains popular in English-speaking countries. Scientific and engineering publications largely have converted to the Celsius scale, but because the conversion still is far from complete, many technical publications usually follow a value in °C by the equivalent value in °F. Again, by personal choice the degree symbol is sometimes eliminated, such as in 100 F.

Réaumur Scale.   Invented prior to 1730 by René-Antoine Ferchalt de Réaumur, the scale today is essentially limited to the brewing and liquor industries. The fundamental temperature interval is defined by the ice point (0°) and a steam-point designation of 80°. The symbol is °R.

Rankine Scale.   This scale is the equivalent of the thermodynamic Kelvin scale, but is expressed in terms of Fahrenheit degrees. Thus the temperature of the triple point of water on the Rankine scale, corresponding to 273.16 K, is very nearly 491.69° Rankine.

International Practical Temperature Scale.   For precision calibration needs, the concept of an international temperature scale with fixed reference points in addition to the ice point and the steam point was proposed as early as 1887. The last revisions to this scale of any note occurred with the publication of the fixed points for the International Practical Temperature Scale (IPTS) of 1968. In the usual applications of thermometry, this scale is not frequently used. Some of the intermediate reference points on the scale include the triple point of equilibrium of hydrogen, the boiling point of neon, the triple point of oxygen, and the freezing points of zinc, silver, and gold. The IPTS is reviewed periodically, as recently as 1990.

Temperature Scale Conversion

A convenient chart for converting degrees Celsius to degrees Fahrenheit and vice versa is given in Table 1.

TABLE 1 Temperature Conversion Table*,†

°C	°F	°F
-273.1	**-459.4**	
-268	**-450**	
-262	**-440**	
-257	**-430**	
-251	**-420**	
-246	**-410**	
-240	**-400**	
-234	**-390**	
-229	**-380**	
-223	**-370**	
-218	**-360**	
-212	**-350**	
-207	**-340**	
-201	**-330**	
-196	**-320**	
-190	**-310**	
-184	**-300**	
-179	**-290**	
-173	**-280**	
-169	**-273**	-459.4
-168	**-270**	-454
-162	**-260**	-436
-157	**-250**	-418
-151	**-240**	-400
-146	**-230**	-382
-140	**-220**	-364
-134	**-210**	-346
-129	**-200**	-328
-123	**-190**	-310
-118	**-180**	-292
-112	**-170**	-274
-107	**-160**	-256

°C	°F	°F
-17.8	**0**	32
-17.2	**1**	33.8
-16.7	**2**	35.6
-16.1	**3**	37.4
-15.6	**4**	39.2
-15.0	**5**	41.0
-14.4	**6**	42.8
-13.9	**7**	44.6
-13.3	**8**	46.4
-12.8	**9**	48.2
-12.2	**10**	50.0
-11.7	**11**	51.8
-11.1	**12**	53.6
-10.6	**13**	55.4
-10.0	**14**	57.2
-9.44	**15**	59.0
-8.89	**16**	60.8
-8.33	**17**	62.6
-7.78	**18**	64.4
-7.22	**19**	66.2
-6.67	**20**	68.0
-6.11	**21**	69.8
-5.56	**22**	71.6
-5.00	**23**	73.4
-4.44	**24**	75.2
-3.89	**25**	77.0
-3.33	**26**	78.8
-2.78	**27**	80.6
-2.22	**28**	82.4
-1.67	**29**	84.2
-1.11	**30**	86.0
-0.56	**31**	87.8

°C	°F	°F
10.0	**50**	122.0
10.6	**51**	123.8
11.1	**52**	125.6
11.7	**53**	127.4
12.2	**54**	129.2
12.8	**55**	131.0
13.3	**56**	132.8
13.9	**57**	134.6
14.4	**58**	136.4
15.0	**59**	138.2
15.6	**60**	140.0
16.1	**61**	141.8
16.7	**62**	143.6
17.2	**63**	145.4
17.8	**64**	147.2
18.3	**65**	149.0
18.9	**66**	150.8
19.4	**67**	152.6
20.0	**68**	154.4
20.6	**69**	156.2
21.1	**70**	158.0
21.7	**71**	159.8
22.2	**72**	161.6
22.8	**73**	163.4
23.3	**74**	165.2
23.9	**75**	167.0
24.4	**76**	168.8
25.0	**77**	170.6
25.6	**78**	172.4
26.1	**79**	174.2
26.7	**80**	176.0
27.2	**81**	177.8

°C	°F	°F
38	**100**	212
43	**110**	230
49	**120**	248
54	**130**	266
60	**140**	284
66	**150**	302
71	**160**	320
77	**170**	338
82	**180**	356
88	**190**	374
93	**200**	392
99	**210**	410
100	**212**	413
104	**220**	428
110	**230**	446
116	**240**	464
121	**250**	482
127	**260**	500
132	**270**	518
138	**280**	536
143	**290**	554
149	**300**	572
154	**310**	590
160	**320**	608
166	**330**	626
171	**340**	644
177	**350**	662
182	**360**	680
188	**370**	698
193	**380**	716
199	**390**	734
204	**400**	752

°C	°F	°F
260	**500**	932
266	**510**	950
271	**520**	968
277	**530**	986
282	**540**	1004
288	**550**	1022
293	**560**	1040
299	**570**	1058
304	**580**	1076
310	**590**	1094
316	**600**	1112
321	**610**	1130
327	**620**	1148
332	**630**	1166
338	**640**	1184
343	**650**	1202
349	**660**	1220
354	**670**	1238
360	**680**	1256
366	**690**	1274
371	**700**	1292
377	**710**	1310
382	**720**	1328
388	**730**	1346
393	**740**	1364
399	**750**	1382
404	**760**	1400
410	**770**	1418
416	**780**	1436
421	**790**	1454
427	**800**	1472
432	**810**	1490

°C	°F	°F
538	**1000**	1832
543	**1010**	1850
549	**1020**	1868
554	**1030**	1886
560	**1040**	1904
566	**1050**	1922
571	**1060**	1940
577	**1070**	1958
582	**1080**	1976
588	**1090**	1994
593	**1100**	2012
599	**1110**	2030
604	**1120**	2048
610	**1130**	2066
616	**1140**	2084
621	**1150**	2102
627	**1160**	2120
632	**1170**	2138
638	**1180**	2156
643	**1190**	2174
649	**1200**	2192
654	**1210**	2210
660	**1220**	2228
666	**1230**	2246
671	**1240**	2264
677	**1250**	2282
682	**1260**	2300
688	**1270**	2318
693	**1280**	2336
699	**1290**	2354
704	**1300**	2372
710	**1310**	2390

°C	°F	°F
816	**1500**	2732
821	**1510**	2750
827	**1520**	2768
832	**1530**	2786
838	**1540**	2804
843	**1550**	2822
849	**1560**	2840
854	**1570**	2858
860	**1580**	2876
866	**1590**	2894
871	**1600**	2912
877	**1610**	2930
882	**1620**	2948
888	**1630**	2966
893	**1640**	2984
899	**1650**	3002
904	**1660**	3020
910	**1670**	3038
916	**1680**	3056
921	**1690**	3074
927	**1700**	3092
932	**1710**	3110
938	**1720**	3128
943	**1730**	3146
949	**1740**	3164
954	**1750**	3182
960	**1760**	3200
966	**1770**	3218
971	**1780**	3236
977	**1790**	3254
982	**1800**	3272
988	**1810**	3290

°C	°F	°F
1093	**2000**	3632
1099	**2010**	3650
1104	**2020**	3668
1110	**2030**	3686
1116	**2040**	3704
1121	**2050**	3722
1127	**2060**	3740
1132	**2070**	3758
1138	**2080**	3776
1143	**2090**	3794
1149	**2100**	3812
1154	**2110**	3830
1160	**2120**	3848
1166	**2130**	3866
1171	**2140**	3884
1177	**2150**	3902
1182	**2160**	3920
1188	**2170**	3938
1193	**2180**	3956
1199	**2190**	3974
1204	**2200**	3992
1210	**2210**	4010
1216	**2220**	4028
1221	**2230**	4046
1227	**2240**	4064
1232	**2250**	4082
1238	**2260**	4100
1243	**2270**	4118
1249	**2280**	4136
1254	**2290**	4154
1260	**2300**	4172
1266	**2310**	4190

°C	°F	°F
1371	**2500**	4532
1377	**2510**	4550
1382	**2520**	4568
1388	**2530**	4586
1393	**2540**	4604
1399	**2550**	4622
1404	**2560**	4640
1410	**2570**	4658
1416	**2580**	4676
1421	**2590**	4694
1427	**2600**	4712
1432	**2610**	4730
1438	**2620**	4748
1443	**2630**	4766
1449	**2640**	4784
1454	**2650**	4802
1460	**2660**	4820
1466	**2670**	4838
1471	**2680**	4856
1477	**2690**	4874
1482	**2700**	4892
1488	**2710**	4910
1493	**2720**	4928
1499	**2730**	4946
1504	**2740**	4964
1510	**2750**	4982
1516	**2760**	5000
1521	**2770**	5018
1527	**2780**	5036
1532	**2790**	5054
1538	**2800**	5072
1543	**2810**	5090

°C	Temp	°F
-101	**-150**	-238
-95.6	**-140**	-220
-90.0	**-130**	-202
-84.4	**-120**	-184
-78.9	**-110**	-166
-73.3	**-100**	-148
-67.8	**-90**	-130
-62.2	**-80**	-112
-56.7	**-70**	-94
-51.1	**-60**	-76
-45.6	**-50**	-58
-40.0	**-40**	-40
-34.4	**-30**	-22
-28.9	**-20**	-4
-23.3	**-10**	14
-17.8	**0**	32

°C	Temp	°F
0	**32**	89.6
0.56	**33**	91.4
1.11	**34**	93.2
1.67	**35**	95.0
2.22	**36**	96.8
2.78	**37**	98.6
3.33	**38**	100.4
3.89	**39**	102.2
4.44	**40**	104.0
5.00	**41**	105.8
5.56	**42**	107.6
6.11	**43**	109.4
6.67	**44**	111.2
7.22	**45**	113.0
7.78	**46**	114.8
8.33	**47**	116.6
8.89	**48**	118.4
9.44	**49**	120.2

°C	Temp	°F
27.8	**82**	179.6
28.3	**83**	181.4
28.9	**84**	183.2
29.4	**85**	185.0
30.0	**86**	186.8
30.6	**87**	188.6
31.1	**88**	190.4
31.7	**89**	192.2
32.2	**90**	194.0
32.8	**91**	195.8
33.3	**92**	197.6
33.9	**93**	199.4
34.4	**94**	201.2
35.0	**95**	203.0
35.6	**96**	204.8
36.1	**97**	206.6
36.7	**98**	208.4
37.2	**99**	210.2

°C	Temp	°F
210	**410**	770
216	**420**	788
221	**430**	806
227	**440**	824
232	**450**	842
238	**460**	860
243	**470**	878
249	**480**	896
254	**490**	914

°C	Temp	°F
438	**820**	1508
443	**830**	1526
449	**840**	1544
454	**850**	1562
460	**860**	1580
466	**870**	1598
471	**880**	1616
477	**890**	1634
482	**900**	1652
488	**910**	1670
493	**920**	1688
499	**930**	1706
504	**940**	1724
510	**950**	1742
516	**960**	1760
521	**970**	1778
527	**980**	1796
532	**990**	1814

°C	Temp	°F
716	**1320**	2408
721	**1330**	2426
727	**1340**	2444
732	**1350**	2462
738	**1360**	2480
743	**1370**	2498
749	**1380**	2516
754	**1390**	2534
760	**1400**	2552
766	**1410**	2570
771	**1420**	2588
777	**1430**	2606
782	**1440**	2624
788	**1450**	2642
793	**1460**	2660
799	**1470**	2678
804	**1480**	2696
810	**1490**	2714

°C	Temp	°F
993	**1820**	3308
999	**1830**	3326
1004	**1840**	3344
1010	**1850**	3362
1016	**1860**	3380
1021	**1870**	3398
1027	**1880**	3416
1032	**1890**	3434
1038	**1900**	3452
1043	**1910**	3470
1049	**1920**	3488
1054	**1930**	3506
1060	**1940**	3524
1066	**1950**	3542
1071	**1960**	3560
1077	**1970**	3578
1082	**1980**	3596
1088	**1990**	3614

°C	Temp	°F
1271	**2320**	4208
1277	**2330**	4226
1282	**2340**	4244
1288	**2350**	4262
1293	**2360**	4280
1299	**2370**	4298
1304	**2380**	4316
1310	**2390**	4334
1316	**2400**	4352
1321	**2410**	4370
1327	**2420**	4388
1332	**2430**	4406
1338	**2440**	4424
1343	**2450**	4442
1349	**2460**	4460
1354	**2470**	4478
1360	**2480**	4496
1366	**2490**	4514

°C	Temp	°F
1549	**2820**	5108
1554	**2830**	5126
1560	**2840**	5144
1566	**2850**	5162
1571	**2860**	5180
1577	**2870**	5198
1582	**2880**	5216
1588	**2890**	5234
1593	**2900**	5252
1599	**2910**	5270
1604	**2920**	5288
1610	**2930**	5306
1616	**2940**	5324
1621	**2950**	5342
1627	**2960**	5360
1632	**2970**	5378
1638	**2980**	5396
1643	**2990**	5414

* General formula: $°F = (°C \times \frac{9}{5}) + 32$; $°C = (F - 32) \times \frac{5}{9}$

† The numbers in **boldface** type refer to the temperature (in either Celsius or Fahrenheit degrees) that it is desired to convert into the other scale.

If converting from degrees Farenheit to degrees Celsius, the equivalent temperature is in the left column, while if converting from degrees Celsius to degrees Fahrenheit, the equivalent temperature is in the column on the right.

Interpolation factors

°C		°F		°C		°F
0.56	**1**	1.8	**6**	3.33		10.8
1.11	**2**	3.6	**7**	3.89		12.6
1.67	**3**	5.4	**8**	4.44		14.4
2.22	**4**	7.2	**9**	5.00		16.2
2.78	**5**	9.0	**10**	5.56		18.0

TEMPERATURE SENSORS [1]

All materials are affected by temperature, and thus it is not surprising that there are so many means available for inferring temperature from some physical effect. Early thermometers depended on volumetric changes of gases and liquids with temperature change, and, of course, this principle still is exploited, as encountered in industrial gas- and liquid-filled thermal systems and in the familiar liquid-column fever thermometer. Although these instruments were accepted widely for many years, the filled-system thermometer has been significantly displaced by other simpler and more convenient approaches, including the thermocouple and the resistance temperature detector (RTD). The contraction and expansion of solids, notably metals, is a phenomenon that has been applied widely in thermometry as, for example, in bimetallic metallic temperature controllers commonly found in the air-conditioning field. Thermoelectric methods, such as the thermocouple, and thermoresistive effects, such as the change of electrical resistance with temperature change, as found in RTDs and thermistors, also have been known and applied for many decades. Thermal radiation of hot bodies has served as the basis for radiation thermometers [once commonly referred to as radiation pyrometers and now called infrared (IR) thermometers] and has also been known and practiced for many decades. Through technological advancements IR thermometers have grown in acceptance in recent years and displaced other measurement means for a number of temperature measurement applications.

Thus as is the case with other process variables, there is a wide selection of thermal sensors. An effort is made in this article to summarize the relative advantages and limitations of easily available temperature sensors, but for some uses, the justification of one method over another is sometimes difficult.

THERMOCOUPLES

For many years the thermocouple was the clear-cut choice of instrumentation and control engineers in the process industries, but in recent years the position of the thermocouple has been increasingly challenged by the RTD. Nevertheless, the thermocouple still is used widely.

Thermocouple Principles

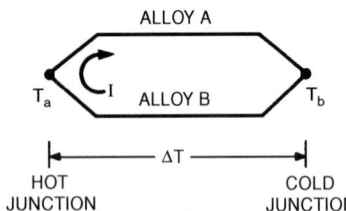

FIGURE 1 Seebeck's circuit. The direction and magnitude of the Seebeck voltage E_s depend on the temperature of the junctions and on the materials making up the thermocouple. For a particular combination of materials A and B over a small temperature difference, $dE_s = \alpha A, B \, dT$, where $\alpha A, B$ is a coefficient of proportionality called the Seebeck coefficient.

Seebeck Effect. As early as 1821, Seebeck observed the existence of thermoelectric circuits while studying the electromagnetic effects of metals. He found that bonding wires of two dissimilar metals together to form a closed circuit caused an electric current to flow in the circuit whenever a difference in temperature was imposed between the end junctions (Fig. 1).

Peltier Effect. Jean Peltier (1834) discovered that when an electric current flows across a junction of two dissimilar metals, heat is liberated or absorbed. When the electric current flows in the same direction as the Seebeck current, heat is absorbed at the hotter junction and liberated at the colder junction. The Peltier effect may be defined as the change in heat content when a quantity of charge (1 coulomb) crosses the junction (Fig. 2). The Peltier effect is the fundamental basis for thermoelectric cooling and heating.

---

[1] Temperature sensors are described in much more detail in *Industrial Sensors and Measurements Handbook*, D. M. Considine, Editor-in-Chief, McGraw-Hill, New York (1995).

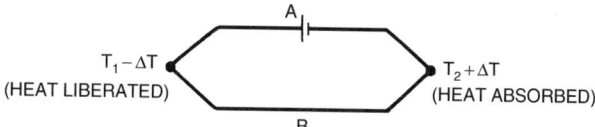

FIGURE 2   Peltier effect.

Thomson Effect.   Sir William Thomson (Lord Kelvin) discovered in 1851 that a temperature gradient in a metallic conductor is accompanied by a small voltage gradient whose magnitude and direction depend on the particular metal. When an electric current flows, there is an evolution or absorption of heat due to the presence of the thermoelectric gradient, with the net result that the heat evolved in an interval bounded by different temperatures is slightly greater or less than that accounted for by the resistance of the conductor. The Thomson effects are equal and opposite and thus cancel each other, thus allowing the use of extension wires with thermocouples because no electromotive force (EMF) is added to the circuit.

Practical Rules Applying to Thermocouples.   Based on decades of practical experience, the following rules apply:

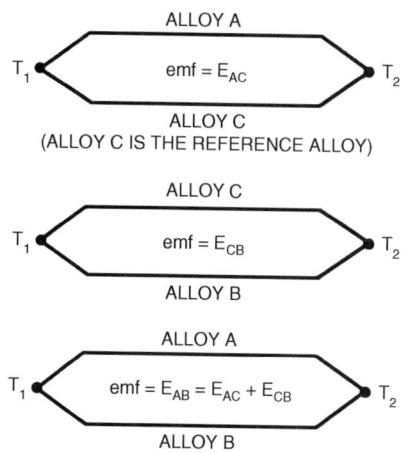

FIGURE 3   Thermocoule EMF algebraic sums.

- A thermocouple current cannot be sustained in a circuit of a single homogeneous material, however varying in cross section, by the application of heat alone.

- The algebraic sum of the thermoelectromotive forces in a circuit composed of any number of dissimilar materials is zero if all of the circuit is at a uniform temperature. This means that a third homogeneous material always can be added to a circuit with no effect on the net EMF of the circuit so long as its extremities are at the same temperature. Therefore a device for measuring the thermal EMF may be introduced into a circuit at any point without affecting the resultant EMF, provided all the junctions added to the circuit are at the same temperature. It also follows that any junction whose temperature is uniform and that makes a good electrical contact does not affect the EMF of the thermocouple circuit regardless of the method used in forming the junction (Fig. 3).

- If two dissimilar homogeneous metals produce a thermal EMF of E_1 when the junctions are at temperatures T_1 and T_2, and a thermal EMF of E_2 when the junctions are at T_2 and T_3, the EMF generated when the junctions are at T_1 and T_3 will be $E_1 + E_2$.

The application of this law permits a thermocouple calibrated for a given reference temperature to be used with any other reference temperature through the use of a suitable correction. Figure 4 shows a schematic example. Another example of this law is that extension wires having the same thermoelectric characteristics as those of the thermocouple wires can be introduced into the thermocouple circuit (from region T_2 to region T_3 in Fig. 4) without affecting the net EMF of the thermocouple.

Thermocouple Signal Conditioning

The thermocouple output voltage is quite nonlinear with respect to temperature. Further, changes in the reference junction temperature influence the output signal of thermocouples. Thermocouple signal conditioning is discussed in considerable detail in Section 7, Article 1, of this handbook.

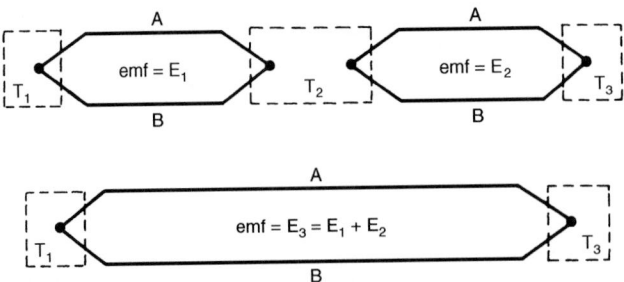

FIGURE 4   EMFs are additive for temperature intervals.

Types of Thermocouples

In present times thermocouple users have the advantage of applying a very mature technology. Over many decades of research into finding metal and alloy combinations that provide an ample millivoltage per degree of temperature and that resist corrosion in oxidizing and reducing atmospheres, a massive base of scientific and engineering information has developed. Much research also has gone into finding sheaths and protecting wells that will withstand both very high and very low temperatures. It should be pointed out, however, that the universal thermocouple has not been found. Thermocouple selection requires much experience to make the optimum choice for a given application.

This situation is evident from Table 2, which provides key specifying information on the five principally used base-metal thermocouples and the three precious- or noble-metal thermocouples that essentially are standard for high-temperature and high-accuracy needs of some applications, and illustrates the wide diversity of thermocouples easily obtainable from reliable manufacturers.

Thermocouple Fabrication and Protection

A very limited number of users prefer to construct their own thermocouple assemblies. Accepted methods for joining the wires to make a junction are shown in Fig. 5. The main components of a complete thermocouple assembly are presented in Fig. 6. Measurement junctions may be exposed, ungrounded, or grounded (Fig. 7).

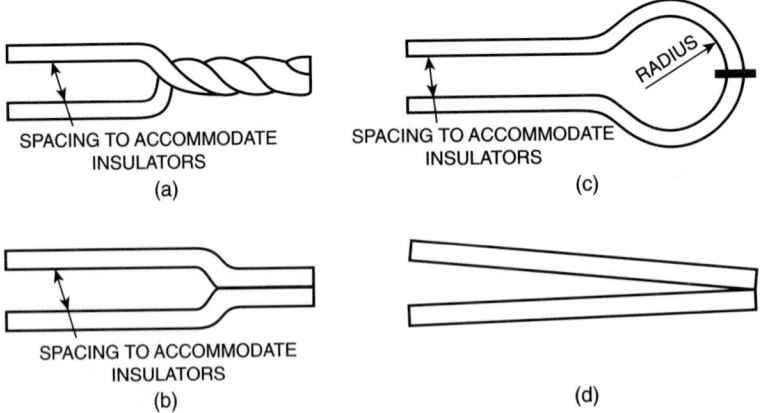

FIGURE 5   Methods for joining dissimilar wires in making thermocouples. (*a*) Twisting wires for gas and electric arc welding. (*b*) Forming wires for resistance welding. (*c*) Forming butt-welded thermocouple. (*d*) Forming wires for electric arc welding.

TABLE 2 Characteristics of Principal Industrial Thermocouples*

ANSI-ISA Types	Thermo-element Codes	Metal Combinations†	Chemical Composition	Common Temperature Ranges	Limits of Error — Standard Grade	Limits of Error — Premium Grade	Atmosphere Suitability
				Base-metal thermocouples			
J	JP	Iron/	Fe	−73°C to 427°C (−100°F to 800°F)	±2.2°C (±4°F)	±1/1°C (±2°F)	Oxidizing,
	JN	*constantan*	44Ni:55Cu	427°C to 760°C (800°F to 1400°F)	(±3/4%)	(±1/3%)	Reducing‡
K	KP	*Chromel*	90Ni:9Cr	0°C to 277°C (32°F to 530°F)	±2.2°C (±4°F)	±1.1°C (±2°F)	Oxidizing,
	KN	*Alumel*	94Ni:Al:Mn:Fe	277°C to 1149°C (530°F to 2100°F)	(±3/4%)	(±3/8%)	inert‖
T	TP	Copper/	Cu	−101°C to −60°C (−150°F to −75°F)	±1.7°C (±3°F)	(±1%)	Oxidizing,
	TN	*constantan*	44Ni:55Cu	−75°C to 93°C (−103°F to 200°F)	±0.8°C (±1.5°F)	(±3/4%)	reducing
E	EP	*Chromel*	90Ni:9Cr	99°C to 371°C (200°F to 700°F)	±1.7°C (±3°F)	±1.1°C (±2°F)	Oxidizing,
	EN	*constantan*	44Ni:55Cu	0°C to 316°C (32°F to 600°F)	(±1/2%)	(±3/8%)	inert
N	NP	*Nicrosil‖*	Ni:14.2Cr:1.4Si	316°C to 871°C (600°F to 1600°F)	±2.2°C (±4°F)	(±3/8%)	Oxidizing,
	NN	*Nisil*	Ni:4Si:0.15Mg	0°C to 277°C (32°F to 530°F)	(±3/4%)	—	inert
				277°C to 1149°C (530°F to 2100°F)			
				Precious-metal thermocouples			
R	RP	Platinum-rhodium/	87Pt:13Rh	Available up to 1480°C (2700°F), depending on sheath materials used	Check with supplier		Oxidizing,
	RN	*platinum*	Pt				inert
S	SP	Platinum-rhodium/	90Pt:10Rh	−18°C (0°F) to 538°C (1000°F) 538°C (1000°F) to 1149°C (2100°F)	±1.4°C (±2.5°F) (±1/4%)		Oxidizing, inert
	SN	*platinum*	Pt				
B	BP	Platinum-rhodium/	70Pt:30Rh	Available up to 1700°C (3100°F), depending on sheath materials used	Check with supplier		Oxidizing,
	BN	*platinum-rhodium*	94Pt:6Rh				inert, vacuum

* Specifications vary somewhat from one manufacturer to the next. Values in this table generally include temperature limitations of the type of sheath used.

† Terms in italics are registered trade names and proprietary compositions.

‡ Type J can be used in oxidizing and reducing atmospheres up to 760°C (1400°F), but above that point the iron oxidizes, causing accelerated deterioration of the accuracy of the sensor.

‖ Type K must not be used in reducing atmospheres, such as hydrogen, dissociated ammonia, carbon monoxide, and many reducing atmospheres as encountered in heat-treating applications. Under such conditions, the KP element, which contains chromium, forms green chromic oxide instead of a spinel-type nickel-chromium oxide. This condition sometimes is called "green rot."

General evaluation:

- Type J is the most widely used of all industrial thermocouples. It has high and fairly uniform sensitivity in millivolts per degree temperature change. Comparatively low-cost.
- Type K is a moderate-cost sensor, particularly for high-temperature measurements in oxidizing atmospheres.
- Type T is highly stable at subzero temperatures with a high conformity to published calibration data. These sensors are frequently the thermocouple of choice for cryogenic and ambient temperature conditions.
- Type E thermocouple provides the highest millivoltage per degree temperature change, providing maximum sensitivity, and is especially useful for short ranges or differential-temperature measurements.
- Type N provides superior thermal stability, longer life, and better accuracy for longer periods than type K. Depending upon manufacturer, this thermocouple can be somewhat limited to smaller outside diameters.
- Type S thermocouple is calibrated to IPTS 68 standard. Essentially, it is used for high-accuracy, high-temperature applications.

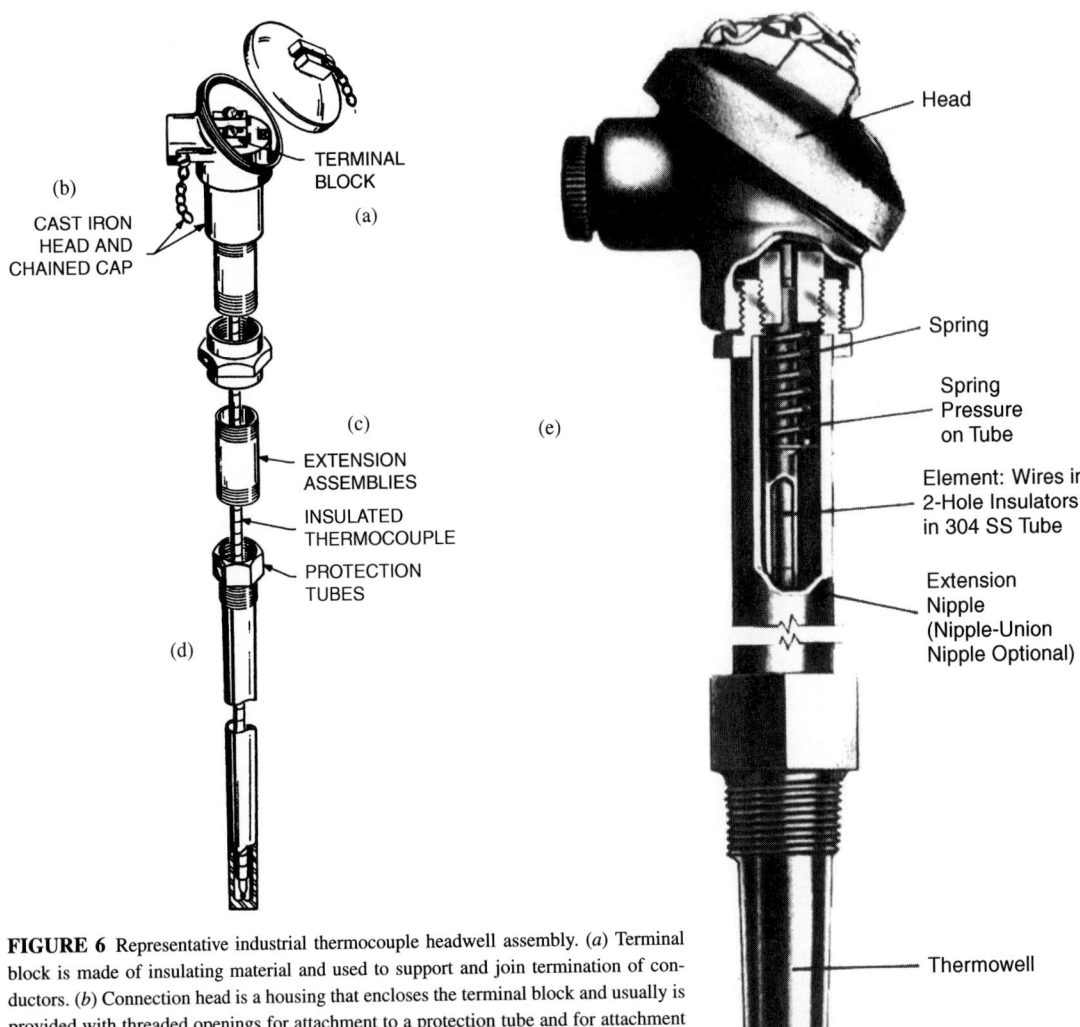

FIGURE 6 Representative industrial thermocouple headwell assembly. (*a*) Terminal block is made of insulating material and used to support and join termination of conductors. (*b*) Connection head is a housing that encloses the terminal block and usually is provided with threaded openings for attachment to a protection tube and for attachment of a conduit. (*c*) Connection head extension usually is a threaded fitting or an assembly of fittings extending between the thermowell or angle fitting and the connection head. Exact configuration depends on installation requirements. (*d*) Protection tube is used to protect sensor from damaging environmental effects. Ceramic materials, such as mullite, high-purity alumina, and some special ceramics, are used mainly in high-temperature applications. They also are used in lower-temperature applications for severe environmental protection. High-purity alumina tubes are required with platinum thermocouples above 1200°C (2200°F) because mullite contains impurities that can contaminate platinum above that temperature. (*e*) Spring-loaded thermocouple assemblies are particularly effective where a temperature measurement is made for control purposes. Spring loading not only improves response, but also protects the junction from the effects of severe vibration. In one design (*Leeds & Northrup*) a retaining ring is brazed to the tube or sheath close to the head end. A spring is compressed between the ring and a bushing assembly, forcing the thermocouple junction into contact with the tip of the well. A grounded junction is required. A silver plug contains the measuring junction. This results in superior response to a temperature change, with a time constant of about 12 seconds, including 3.5 seconds for initial response.

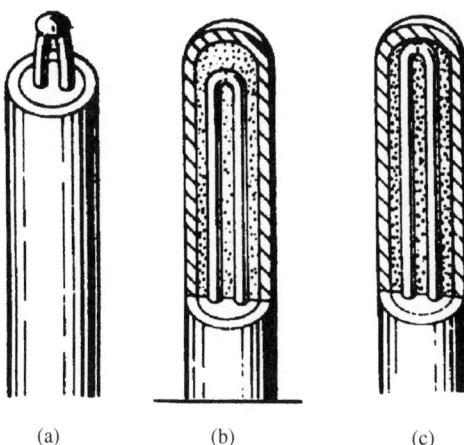

(a)   (b)   (c)

FIGURE 7 Thermocouple measuring junctions. (*a*) Exposed junction. (*b*) Ungrounded junction. (*c*) Grounded junction.

The exposed junction is often used for the measurement of static or flowing non-corrosive gas temperatures where the response time must be minimal. The junction extends beyond the protective metallic sheath to provide better resonse. The sheath insulation is sealed at the point of enry to prevent penetration of moisture or gas.

The ungrounded junction often is used for the measurement of static or flowing corrosive gas and liquid temperatures in critical electrical applications. The welded wire thermocouple is physically insulated from the thermocouple sheath by soft magnesium oxide (MgO) powder or equivalent.

The grounded junction often is used for the measurement of static or flowing corrosive gas and liquid temperatures and for high-pressure applications. The junction is welded to the protective sheath, providing faster response than an ungrounded junction.

Thermocouple Installation

Thermocouples with sheaths or placed in thermowells are subject to a temperature gradient along the length of the sheath. Such errors can be minimized by specifying long, small-diameter sensors, by using sheath materials with low thermal conductivity, and by providing high convective heat transfer coefficients between the fluid and the thermocouple. Mounting of the thermocouple also plays a large role in minimizing errors. Some users follow a long-regarded rule that the sensor immersion depth should be equal to 10 times the sheath diameter. But according to W. C. Behrmann of Exxon Research and Development Laboratories [1] (in an *InTech* article of August 1990, p. 36) the problem is more complex. In a mathematical study of thermocouple location geometry it has been found that 90° bend couples and curved couples are less prone to error than the common side-entering couples. Of significance, however, the installation difficulties are least with the side-entering geometry.

Thermocouple sheath materials fall into two major categories (Table 3):

1. Metals, such as Inconel 600 [maximum air temperature of 1150°C (2100°F)] and a number of stainless steels: 310SS [1150°C (2100°F)], 304SS, 316SS, and 347SS [900°C (1650°F)]

2. Ceramic materials, such as silicon carbide, Frystan, alumina, and porcelains of various types and brands

TABLE 3   High-Temperature Sheath Materials

Sheath Material	Maximum Operating Temperature	Workability	Working Environment	Approximate Melting Point	Remarks
Molybdenum*	2205°C (4000°F)	Brittle	Inert, vacuum, reducing	2610°C (4730°F)	Relatively good hot strength; sensitive to oxidation above 500°C (930°F); resists many liquid metals and most molten glasses
Tantalum†	2482°C (4500°F)	Malleable	Inert, vacuum	3000°C (5425°F)	Resists most acids and weak alkalies; very sensitive to oxidation above 300°C (570°F)
Platinum-rhodium alloy	1677°C (3050°F)	Malleable	Oxidizing, inert, vacuum	1875°C (3400°F)	No attack by SO_2 at 1093°C (2000°F); silica is detrimental; halogens attack at high temperatures
Inconel 600	1149°C (2100°F)	Malleable	Oxidizing, inert, vacuum	1410°C (2570°F)	Excellent resistance to oxidation at high temperature; do not use in presence of sulfur above 538°C (1000°F); hydrogen tends to embrittle

* Refractory metals are extremely sensitive to any trace of oxygen above approximately 260°C (500°F). They must be used in vacuum or in very pure inert gases such as helium and argon.
† Suitable for exposure to certain reducing atmospheres as well as inert gases and vacuum.

Thermocouple Wire Insulators

The dissimilar wires of a thermocouple must be insulated. Traditionally, ceramics have been used (Fig. 8).

Special Thermocouples

In addition to the traditional industrial thermocouples just described, there also are surface probes and cement-on styles. Thermocouples can be drawn in metal-sheathed form to as small as 0.25-mm (0.01-inch) OD. In wire form, 0.013-mm (0.0005-inch) thermocouples can be made.

Types K and E surface probes are commercially available. Type E usually is preferred because of its high accuracy in most low-temperature applications. Type K is used where high temperatures must be measured. Types J and T are not commonly used as probes.

Cement-on style thermocouples have grown in popularity in recent years. Special fast-responding techniques include thin-foil couples with plastic laminates for cementing directly on equipment. The full sensor is embedded between two thin glass-reinforced, high-temperature polymer laminates that both support and electrically insulate the foil section as well as provide a flat surface for cementing. The polymer-glass laminate, in general, determines the maximum temperature of the construction, which is 260°C (500°F) in continuous service (Fig. 9).

Thermocouple Circuit Flexibility

Normally one envisions the use of thermocouples one at a time for single temperature measurements. As shown in Fig. 10, thermocouples may be used in parallel, in series, and in switching and differential circuits.

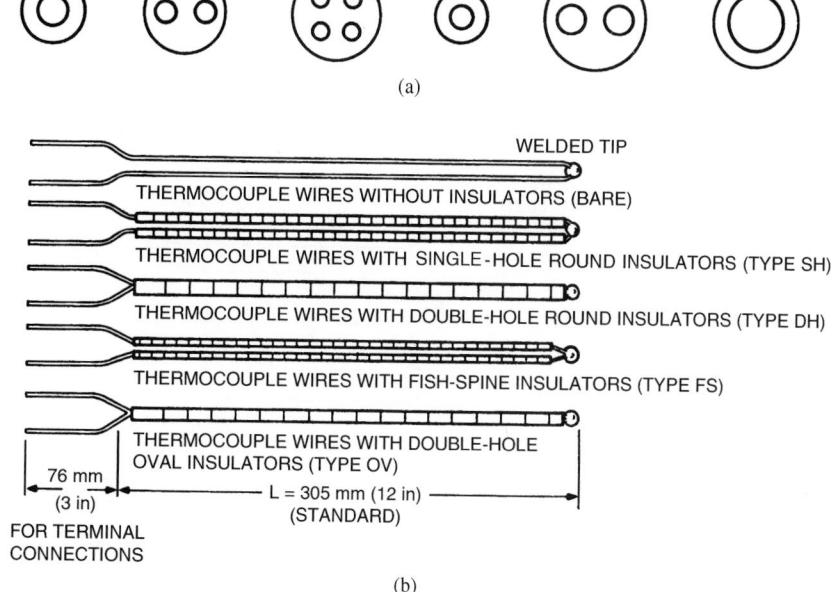

FIGURE 8 Thermocouple wire insulators. (*a*) Range of sizes, from left to right: 3.2, 2.4, 2.0, 1.6, 1.2, 0.8, and 0.4 mm ($\frac{1}{8}$, $\frac{3}{32}$, $\frac{5}{64}$, $\frac{1}{16}$, $\frac{3}{64}$, $\frac{1}{32}$, and $\frac{1}{64}$ inch) bore diameter. (*b*) Application of insulators to various styles of thermocouples.

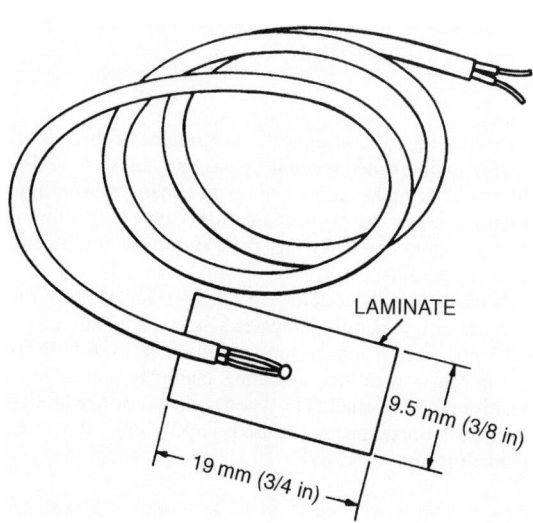

FIGURE 9   Cement-on-style thermocouples.

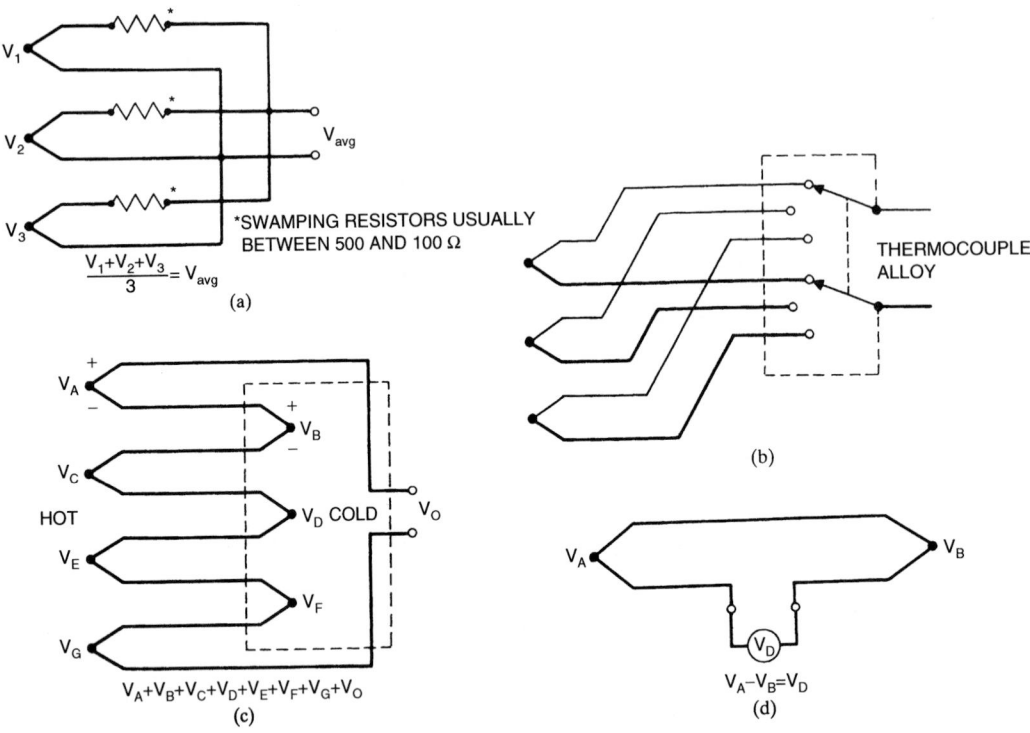

FIGURE 10   Use of thermocouples in multiples. (*a*) Thermocouples in parallel. (*b*) Thermocouples in switch circuit. Switch must be isothermal or made of the same thermocouple alloy material. (*c*) Thermocouples in series (thermopile). *Note:* V_B, V_D, and V_F are negative thermoelectric voltages compared with V_A, V_C, V_E, and V_G. However, the alloys are also reversed, thus creating a net voltage. (*d*) Differential circuit. *Note:* Output voltage cannot be accurately cold-junction compensated because of the nonlinearity of thermocouple EMF versus temperature. Approximations can be made if he absolute temperature is known.

Checking Thermocouples

Authorities suggest that the calibration of thermocouples be checked annually, or at least semiannually. In accordance with a maintenance schedule, past experience should be used for determining how often a thermocouple should be replaced prior to serious changes in calibration. This is particularly true of the base-metal sensors. Subsequent checking, that is, checking after initial installation, should be done in situ. Portable calibrators that can measure or simulate a thermocouple signal are readily available and simplify field calibration.

Resistance checks of thermocouples are effective for determining the condition of a sensor. For example, a low resistance generally indicates a satisfactory situation, whereas a high resistance may indicate that a thermocouple is approaching the end of its useful life. To determine correct installation, particularly in connection with thermocouples made "in house," a small magnet can be useful. For example, the positive thermoelement of a type J couple is iron, which is magnetic, whereas the negative element, constantan, is nonmagnetic. In the type K couple, the alumel element is magnetic and the other thermoelement is not.

RESISTANCE-TEMPERATURE DETECTORS

The science of measuring temperature by utilizing the characteristic relationship of electrical resistance to temperature has been advanced periodically since the early work of Faraday (circa 1835).

TABLE 4 Resistance versus Temperature for Various Metals

Metal	Resistivity, gΩ·cm	Relative Resistance R_t/R_0 at 0°C												
		−200	−100	0	100	200	300	400	500	600	700	800	900
Alumel*	28.1			1.000	1.239	1.428	1.537	1.637	1.726	1.814	1.899	1.982	2.066
Copper	1.56	0.117	0.557	1.000	1.431	0.862	2.299	2.747	3.210	3.695	4.208	4.752	5.334
Iron	8.57			1.000	1.650	2.464	3.485	4.716	6.162	7.839	9.790	12.009	12.790
Nickel	6.38			1.000	1.663	2.501	3.611	4.847	5.398	5.882	6.327	6.751	7.156
Platinum	9.83	0.177	0.599	1.000	1.392	1.773	2.142	2.499	3.178	3.178	3.500	3.810	4.109
Silver	1.50	0.176	0.596	1.000	1.408	1.827	2.256	2.698	3.616	3.616	4.094	5.586	5.091

* Registered trademark of Hoskins Manufacturing Co., Detroit, Michigan.

Certain suitably chosen and prepared materials that vary in resistance in a well-defined and calibrated manner with temperature became readily available around 1925, prompting the use of resistance thermometers as primary sensors for industrial applications where reproducibility and stability are of critical importance. Platinum resistance thermometers became the international standard for temperature measurements between the triple point of hydrogen at 13.81 K and the freezing point of antimony at 730.75°C. Since the 1970s RTDs have made very serious inroads on the thermocouple for very broad usage in industry—for practical industrial use, not just for applications requiring exceptional accuracy. The advantages and limitations of RTDs as compared with thermocouples in this present time span are presented later in this article.

Principles of Resistance Thermometry

For pure metals, the characteristic relationship that governs resistance thermometry is given by

$$R_t = R_0 (1 + at + bt^2 + ct^3 + \cdots)$$

where R_0 = resistance at reference temperature (usually at ice point, 0°C), Ω
R_t = resistance at temperature t, Ω
a = temperature coefficient of resistance, Ω/Ω (°C)
b, c = coefficients calculated on the basis of two or more known resistance-temperature (calibration) points

For alloys and semiconductors, the relationship follows a unique equation dependent on the specific material involved. Whereas most elements constructed from metal conductors generally display positive temperature coefficients, with an increase in temperature resulting in increased resistance, most semiconductors display a characteristic negative temperature coefficient of resistance.

Only a few pure metals have a characteristic relationship suitable for the fabrication of sensing elements used in resistance thermometers. The metal must have an extremely stable resistance-temperature relationship so that neither the absolute value of the resistance R_0 nor the coefficients a and b drift with repeated heating and cooling within the thermometer's specified temperature range of operation. The material's specific resistance in ohms per cubic centimeter must be within limits that will permit fabrication of practical-size elements. The material must exhibit relatively small resistance changes for nontemperature effects, such as strain and possible contamination which may not be totally eliminated from a controlled manufacturing environment. The material's change in resistance with temperature must be relatively large in order to produce a resultant thermometer with inherent sensitivity. The metal must not undergo any change of phase or state within a reasonable temperature range. Finally, the metal must be commercially available with essentially a consistent resistance-temperature relationship to provide reliable uniformity.

Industrial resistance thermometers, often referred to as RTDs, are commonly available with elements of platinum, nickel, 70% nickel–30% iron (Balco), or copper. The entire resistance thermometer is an assembly of parts, which include the sensing element, internal leadwires, internal supporting and insulating materials, and protection tube or case (Fig. 11 and Table 4).

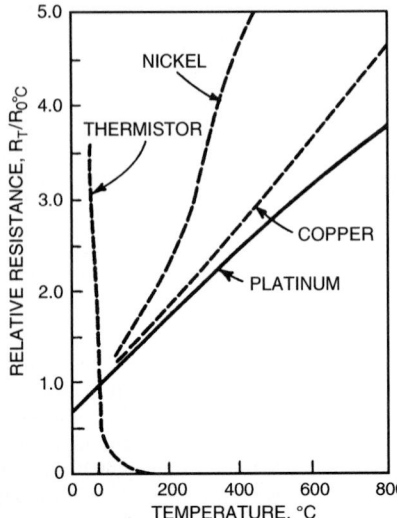

FIGURE 11   Resistance-temperature characteristics of thermoresistive materials at elevated temperatures. Platinum and nickel are the most commonly used metals for industrial applications.

Platinum RTDs

Of all materials currently utilized in the fabrication of thermoresistive elements, platinum has the optimum characteristics for service over a wide temperature range. Although platinum is a noble metal and does not oxidize, it is subject to contamination at elevated temperatures by some gases, such as carbon monoxide and other reducing atmospheres, and by metallic oxides.

The metal is available commercially in pure form, providing a reproducible resistance-temperature characteristic. Platinum with a temperature coefficient of resistance equal to 0.00385 Ω/Ω ($^\circ$C) (from 0 to 100°C) has been used as a standard for industrial thermometers throughout the United Kingdom and Western Europe since World War II and has gained prominence in recent years in the United States in the absence of a defined and commonly accepted standard coefficient. Platinum has a high melting point and does not volatilize appreciably at temperatures below 1200°C. It has a tensile strength of 18,000 psi (124 MPa) and a resistivity of 60.0 Ω/cmil · ft at 0°C (9.83 $\mu\Omega$ · cm).

Platinum is the material most generally used in the construction of precision laboratory standard thermometers for calibration work. In fact, the laboratory-grade platinum resistance thermometer (usually with a basic resistance equal to 25.5 Ω at 0°C) is the defining standard for the temperature range from the liquid oxygen point (-182.96°C) to the antimony point (630.74°C) as defined by the International Practical Temperature Scale.

The resistance-temperature relationship for platinum resistance elements is determined from the Callendar equation above 0°C,

$$t = \frac{100\,(R_t - R_0)}{R_{100} - R_0} + \delta \left(\frac{t}{100} - 1 \right) \frac{t}{100}$$

where
t = temperature, $^\circ$C
R_t = resistance at temperature t, Ω
R_0 = resistance at 0°C, Ω
R_{100} = resistance at 100°C, Ω
δ = Callendar constant (approximately 1.50)

The fundamental coefficient (temperature coefficient of resistance) α is defined over the fundamental interval of 0 to 100°C,

$$\alpha = \frac{R_{100} - R_0}{100 R_0}$$

Thin-Film Platinum RTDs.   Processing techniques developed during the past decade or two have provided the capability of producing thin-film platinum RTD elements that essentially are indistinguishable from wire-wound elements in reproduction and stability. An industrial RTD thermowell assembly is shown in Fig. 12. Thin-film platinum RTD elements are trimmed to the final desired resistance value, usually by automated equipment. These high-resistance elements have, in numerous instances, made obsolete the need to consider base-metal wire-wound sensors of nickel or nickel-iron. Thermistors, too, are sometimes replaced with a sensor of significantly greater stability and temperature range.

The use of sheaths and thermowells, as described previously for thermocouples, applies similarly to RTDs.

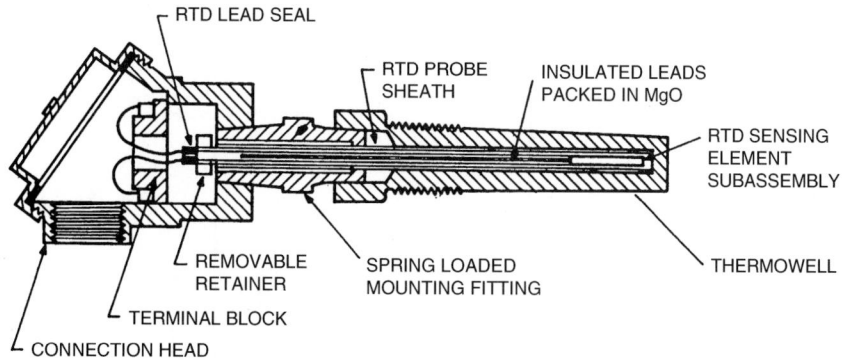

FIGURE 12   Representative industrial RTD thermowell assembly.

Wire-Wound Platinum RTDs.   Some users for certain applications prefer the wire-wound RTD. In the fully encapsulated RTD the platinum wire, usually 0.025-mm (0.001-inch) OD or less, is wound into a coil and inserted into a multibore high-purity ceramic tube, or may be wound directly on the outside of a ceramic tube. The most commonly used ceramic material is aluminum oxide (99.7% Al_2O_3). The winding is completely embedded and fused within or on the ceramic tube utilizing extremely fine granular powder. The resultant fully encapsulated element, with only two noble-metal leadwires exposed, provides maximum protection for the platinum resistance coil. Although special fusing techniques are used, such elements are not completely strain-free, but the effects of existing strains are fairly constant with resulting errors well within the permissible limits for industrial applications. The intimate contact between the platinum winding and the ceramic encapsulation permits a rapid speed of response with the thermal conductivity of ceramic adequate for heat transmission through the protecting layer. A platinum industrial RTD assembly with a maximum temperature range is shown in Fig. 13. For applications where high-temperature requirements are combined with high pressure, high flow, and high vibration, the assembly includes a thermowell protecting tube.

Nickel RTDs.   During recent years, widespread use of improved platinum RTDs with their superior performance characteristics, often at lower cost, has taken precedence over the use of nickel RTDs in a wide range of industrial applications. Current availability of nickel sensors has continued primarily as a component replacement for already existing industrial systems.

Copper RTDs.   The observations pertaining to nickel RTDs are also essentially applicable to copper RTDs. The straight-line characteristics of copper have in the past been useful in allowing two sensors to be applied directly for temperature-difference measurements.

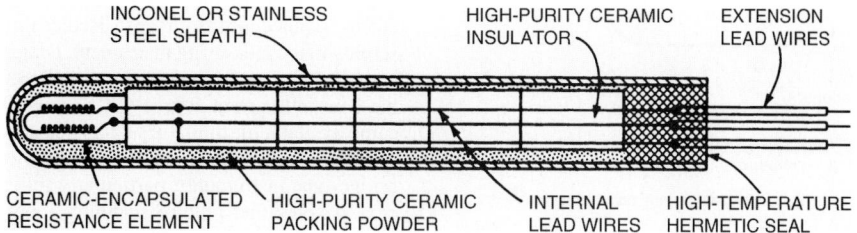

FIGURE 13   Representative platinum RTD assembly.

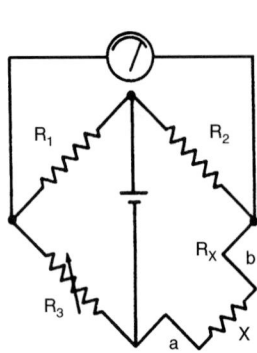

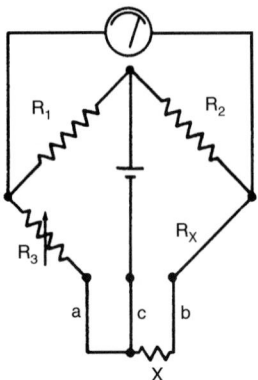

Circuit equations:

$$R_1 + R_3 = R_2 + a + b + X \quad (8)$$
$$R_1 = R_2$$
$$\therefore \quad R_3 = a + b + X \quad (9)$$

(a)

Circuit equations:

$$R_1 + R_3 + a + c = R_2 + b + X + c \quad (10)$$
$$R_1 = R_2$$

If a = b (lead resistance equal)

Then $R_3 = X$ \quad (11)

(b)

FIGURE 14 Traditional RTD circuits presented as backdrop to contemporary circuit designs. (*a*) Two-lead circuit permissible only when leadwire resistance can be kept to a minimum and only where a moderate degree of accuracy is required. (*b*) Three-lead circuit. Two leads are connected in close proximity to the resistance element at a common node. Third lead is connected to the opposite resistance leg of the element. Resistance of lead *a* is added to bridge arm R_3, while resistance of lead *b* remains on bridge arm R_X, thereby dividing the lead resistance and retaining a balance in the bridge circuit. Lead resistance *c* is common to both left and right loops of the bridge circuit. Although this method compensates for the effect of lead resistance, the ultimate accuracy of the circuit depends on leads *a* and *b* being of equal resistance. Special matching techniques must be used on leads *a* and *b*, particularly when distance between sensor and measuring equipment is realtively large. A four-lead circuit, not shown, is used only when the highest degree of accuracy, as in laboratory temperature standards, is required.

Ranges and Performance Characteristics of RTDs

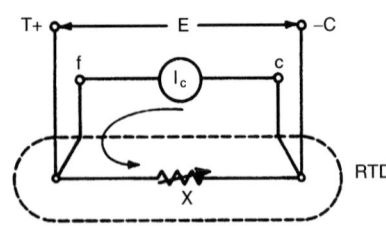

Circuit equations:

$$E = IX$$
$$I \cong constant$$
$$E = f(X) = f' \text{ (temperature)} \quad (18)$$

FIGURE 15 Four-lead constant-current measuring circuit.

The most common RTD temperature ranges commercially available and some additional performance characterisics are summarized in Table 5.

RTD Circuitry

Traditionally three methods have been used for making electric connections from the resistance thermometer assembly to the measuring instrument. Diagrams and their equations are presented in Fig. 14. When highly stable constant-current-source (CCS) power supplies became available in miniature packages at relatively low cost, these offered an effective alternative to null-balance bridge-type instruments, particularly for industrial process systems that require scanning (both manual and automatic) of up to 100 or more individual RTD points, often located at varying distances from a central point. The circuit shown in Fig. 15 is the basic

TABLE 5 Representative Standard RTDs Available Commercially

	Platinum	Platinum	Platinum	Platinum	Platinum	Nickel
Temperature	−200 to 200°C (−328 to 392°F)	−50 to 200°C (−58 to 392°F)	−100 to 260°C (−148 to 500°F)	−65 to 200°C (−85 to 392°F)	−200 to 650°C (−328 to 1200°F)	−130 to 315°C (−200 to 600°F)
Configuration	316 SS sheath 200°C (392°F) 17.2 MPa (2500 psig)	316 SS sheath 200°C (392°F) 15.2 MPa (2200 psig)	Surface mount	Surface mount	316 SS sheath 480°C (900°F); Inconel to 650°C (1200°F)	Stainless steel; other metals
Repeatability	±0.05% max ice-point resistance 0.13°C (0.23°F)	±0.025% max ice-point resistance	±0.04% max ice-point resistance 0.1°C (0.18°F)	±0.08% max ice-point resistance 0.2°C (0.36°F)	±0.26°C (0.47°F) up to 480°C (896°F); ±0.5% (reading) up to 650°C (1200°F)	Limit of error from ±0.3°C (±0.5°F) to ±1.7°C (3°F) at cryogenic temperatures
Stability	±0.08% max ice-point resistance	±0.035% max ice-point resistance	±0.05% max ice-point resistance	±0.15% max ice-point resistance	—	—
Time constant (to reach 63.2% of sensor response)	8 seconds	7 seconds	1.25 seconds	2.5 seconds	—	—
Immersion length	0.6 to 6 m (2 to 200 ft)	31, 61, 91, 122 cm (12, 24, 36, 48 in)	—	—	89 to 914 mm (3.5 to 36 in)	89 to 914 mm (3.5 to 36 in)
Leadwire	Teflon-insulated, nickel-coated 22-gauge standard copper wire	Teflon-insulated, nickel-coated 22-gauge standard copper wire	0.25-mm (0.01-in)-diameter platinum wire	Teflon-insulated, 24 AWG standard copper wire	—	—

four-lead circuit with two leads joined in close proximity to each side of the resistance element. The CCS can be connected across leads t and c supplying a constant current I_c across the resistance element X. The value for i_c may be kept to a minimum of 1 mA or less to avoid excessive self-heating errors. The voltage drop across the resistance element then is measured between T and C. The resultant voltage drops across the thermometer element, in the constant-current mode, then varies with resistance directly as a function of temperature. The advantage of CCS circuits becomes apparent to the user already versed in bridge measuring techniques. The CCS power supply continues to maintain the fixed constant current (within its compliance voltage limitations) across the thermometer element, thus making costly matching techniques associated with leadwires unnecessary. In addition, leadwire contact resistances associated with high-speed automatic switching are reduced to a minimum. An added feature of the CCS measuring circuit is its ability to interface directly with a wide variety of voltage measuring instruments. Digital linearizers can be applied to operate on the nonlinear platinum resistance thermometer function to display directly in engineering units with linearization conformities of a fraction of a degree (temperature). CCS measuring circuits provide added binary-coded decimal (BCD) output for interface with digital printers and process control computers.

COMPARISON OF THERMOCOUPLE AND RTD SELECTION

The choice of a thermocouple or a resistance temperature sensor for a given application is not always clear-cut. In making an optimum choice, the user must assign relative priorities to the most and the least important characteristics needed. As an example, for a specific application, vibration may be an outstanding problem and a thermocouple may be selected over an RTD because the thermocouple is less likely to be affected by vibration, even though most other criteria may favor the RTD. Or short-term reproducibility may be a key criterion that favors the RTD over the thermocouple.

Where special requirements are not a factor, then the principal criteria include accuracy, response time, size, original and installed cost (an important factor where many points of measurement are involved), sealing the detector from the process environment without significantly reducing response time, and the user's experience factor. When all other evaluating criteria are about equal, the user simply may prefer one type of detector over the other because in-house engineers, operators, and maintenance personnel have much more experience and know-how with a given type of temperature detector.

Numerous professionals have expressed their opinions. These have been distilled in Table 6. Thus a few inconsistencies may be expected.

THERMOCOUPLE AND RTD TRANSMITTERS

Data signal handling in computerized instrumentation systems is the topic of in Section 7, Article 1, of this handbook. Thermocouple, RTD, and solid-state temperature detectors are specifically mentioned in that article.

Although variously termed "smart" or "intelligent" temperature (and other process variables) transmitters have numerous electronically built-in advantages, the basic purpose of the transmitter must not be minimized: the ultimate and self-evident function of the transmitter is that of ensuring data integrity. This is particularly important in noisy electrical environments as encountered in numerous processing and manufacturing situations. Other transmitter capabilities, in essence, are extra dividends that can be derived conveniently and at reasonably low cost from associated electronics. These include, among other advantages, self-diagnostics, elimination of hardware changes or recalibration, and bidirectional communication with a distributed control system. Because so many industrial variable sensors, such as pressure, flow, and liquid level, also generate voltage and current signals, some available single units can be programmed electronically to replace any of the other units in a system, thus reducing replacement inventory costs. This is particularly true in the temperature field.

TABLE 6 Relative Advantages and Limitations of Thermocouples and Resistance Temperature Detectors

Advantages	Limitations
Thermocouples	
Cost: Although cost of RTDs is trending downward, thermocouples generally continue to be less expensive. Ruggedness: In terms of process environmental conditions, including high temperatures and vibration, thermocouples are regarded highly. Higher temperature range: Extends to about 1150°C (2100°F) or higher. However, the thermocouple's cryogeneic range is somewhat less than that of the RTD. Mounting cost: Generally considered to be somewhat lower than for RTDs.	Accuracy: Generally expected, after installation, ±4°C (±7.2°F), but there are exceptions. Thermocouples require a reference junction or special extension wire. The overall thermocouple system include the inaccuracies associated with two separate temperature measurements—the measuring junction and the cold or reference junction. Stability: Less than for the RTD. Estimated at 0.6°C (1°F) per year. Inherently small millivolt output: Can be affected by electrical noise. Calibration: Nonlinear over normal spans. Signal requires linearizing. Calibration can be changed by contamination.
Resistance Temperature Detectors	
Accuracy: Generally expected, after installation, ±0.5°C (±0.9°F). The platinum RTD, for example, is used to define the IPTS at the oxygen point (−182.97°C) and the antimony point (+630.74°C). No reference junction is required. Repeatability: Within a few hundredths of a degree; can often be achieved with platinum RTDs. Less than 0.1% drift in 5 years. Substantial output voltage (1 to 6 volts): This is an advantage in that the output can be controlled by adjusting the current and the bridge design in the RTD signal conversion module. Because a higher output voltage to the RTD usually can be achieved, the recording, monitoring, and controlling of temperature signals is simpler. This permits moe accurate measurements without requiring complex calculations for large spans. Short-term reproducibility: Superior to that of thermocouples. Reproducibility not affected by temperature changes. Relatively narrow spans: Some versions may have spans as narrow as 5.6°C (10°F). Compensation: Not required. Suppressed ranges: Available. Size: Generally smaller than thermocouples.	Cost: Generally higher than that of thermocouples. However, RTDs do not require compensation, special lead-wires, and special signal conditioning for long runs. With platinum-film technology, cost is trending downward. Less rugged: For adverse process environments, including high temperatures and vibration, RTDs are not regarded as highly as thermocouples. Lower temperature range: Limited to about 870°C (1600°F). Note, however, that RTDs extend to lower cryogenic ranges than do thermocouples. Self-heating errors: This may be a problem unless corrected in the transmitter's electronics.

Evolution of Temperature Transmitters

The traditional practice over many decades (still present in numerous systems) was that of using long wire runs to deliver low-level thermocouple and RTD signals to some operator-based central location. For reasons of cost and vulnerability to electrical noise, such systems were rather severely distance-limited, not to mention their high installation and maintenance costs.

Two-Wire Analog Temperature Transmitter.   This was widely accepted as a means for improving measurement stability and accuracy—accomplished by replacing low-level millivolts or ohms with high-level current signals. This enabled the replacement of high-cost shielded cables and conduits with twisted-pair copper wire in the case of RTDs, and of costly extension wires in the case of thermocouples.

TABLE 7   Comparison of Specifications of Main Types of Temperature Transmitters

Traditional Type	Smart or Intelligent Type	Throwaway Type (potted)
Analog	Digital	Analog
Thermocouple or RTD must be specified, changes require hardware reconfiguring	One model handles all thermocouple, RTD, mV, and ohm sensors	Limited to one input
Moderate cost	Premium cost, downward trend	Low cost
Variable span; moderate range	Variable span; wide range	Single span; limited range
Sometimes isolated	Isolated	Not isolated
Variety of applications	Almost every application	Only one application
Sometimes intrinsically safe	Nearly always intrinsically safe	Usually not intrinsically safe
Can be reconfigured on site or in shop	Can be reconfigured remotely	Very limited reconfiguration
	Remote diagnostics	
Moderate to good performance	Superior accuracy	Performance accuracy limited
RTD models easily linearized; thermocouples sometimes linearized	Linearization is selectable	Must linearize at receiver
Stability depends on manufacture and application	Quite stable with ambient temperature and time; infrequent needs to recalibrate	Stability depends on manufacture and application

Microprocessor-Based Transmitters

These smart devices were introduced in the late 1980s to take advantage of the large advances in digital electronics that had occurred just a few years previously. These transmitters have achieved wide acclaim because of their versatility. Generally the units are preprogrammed by the manufacturer with the information necessary to linearize a given sensor signal, including a given thermocouple, RTD, or other millivolt or ohm device. An important advantage of these transmitters is their potential for reconfiguration for accepting other sensor inputs without having to make hardware changes and recalibrate. Also, they offer a variety of self-diagnostics for power supply and loop problems (Fig. 16). The specifications of available transmitters are compared in Table 7.

THERMISTORS

The name thermistor is derived from thermally sensitive resistor, since the resistance of a thermistor varies as a function of temperature. Although the fundamental principles of the thermistor were established several decades ago, industrial and commercial utilization of thermistors in temperature measurement systems was very slow in developing. Since the early 1980s a proliferation of relatively low-cost, usually portable thermistor thermometers has become available. However, the thermistor to date has not been widely applied in the traditional industrial and process control temperature measurement field because of several inherent problems.

A thermistor is an electrical device made of a solid semiconductor with a high temperature coefficient of resistivity which would exhibit a linear voltage-current characteristic if its temperature were held constant. When a thermistor is used as a temperature sensing element, the relationship between resistance and temperature is of primary concern. The approximate relationship applying to most thermistors is

$$R_t = R_0 \exp B \left(\frac{1}{T} - \frac{1}{T_0} \right)$$

where R_0 = resistance value at reference temperature T_0 K, Ω
$\quad\quad R_T$ = resistance at temperature T K, Ω
$\quad\quad B$ = constant over temperature range, dependent on manufacturing process and construction characteristics (specified by supplier),

$$B \cong \frac{E}{K}$$

where E = electronvolt energy level
$\quad\quad K$ = Boltzmann's constant, = 8.625×10^{-5} eV/K

A second form of the approximate resistance-temperature relationship is written in the form

$$R_T = R_\infty e^{B/T}$$

where R_∞ is the thermistor resistance as the temperature approaches infinity, in ohms.

These equations are only best approximations and, therefore, are of limited use in making highly accurate temperature measurements. However, they do serve to compare thermistor characteristics and thermistor types.

The temperature coefficient usually is expressed as a percent change in resistance per degree of temperature change and is approximately related to B by

$$a = \frac{dR}{dT} \left(\frac{1}{R} \right) = \frac{-B}{T_0^2}$$

where T_0 is in kelvin. It should be noted that the resistance of the thermometer is solely a function of its absolute temperature. Furthermore, it is apparent that the thermistor's resistance-temperature function has a characteristic high negative coefficient as well as a high degree of nonlinearity. The value of the coefficient a for common commercial thermistors is of the order of 2 to 6 percent per kelvin at room temperature. This value is approximately 10 times that of metals used in the manufacture of resistance thermometers.

Resultant considerations due to the high coefficient characteristic of thermistors include inherent high sensitivity and high level of output, eliminating the need for extremely sensitive readout devices and leadwire matching techniques, respectively. However, limitations on interchangeability (particularly over wide temperature ranges), calibration, and stability—also inherent in thermistors—are quite restrictive. The high degree of nonlinearity in the resistance-temperature function usually limits the range of the readout instrumentation. In many applications, special prelinearization circuits must be used before interfacing with related system instrumentation. The negative temperature coefficient also may require an inversion (to positive form) when interfacing with some analog or digital instrumentation.

A number of metal oxides and their mixtures, including the oxides of cobalt, copper, iron, magnesium, manganese, nickel, tin, titanium, uranium, and zinc, are among the most common semiconducting materials used in the construction of thermistors. Usually compressed into the desired shape from the specially formulated powder, the oxides are then recrystallized by heat treatment, resulting in a dense ceramic body. The leadwires are then attached while electric contact is maintained, and the finished thermistor is then encapsulated. A thermistor can be made in numerous configurations, as illustrated in Fig. 17.

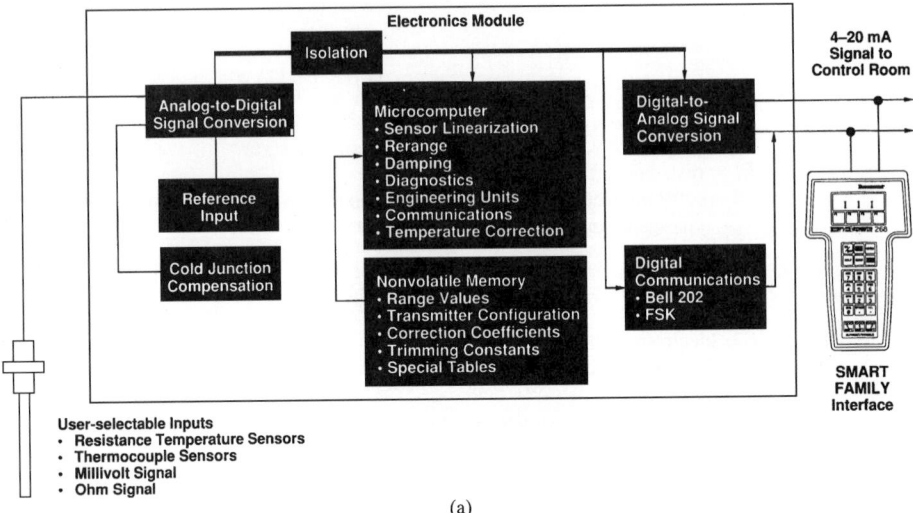

(a)

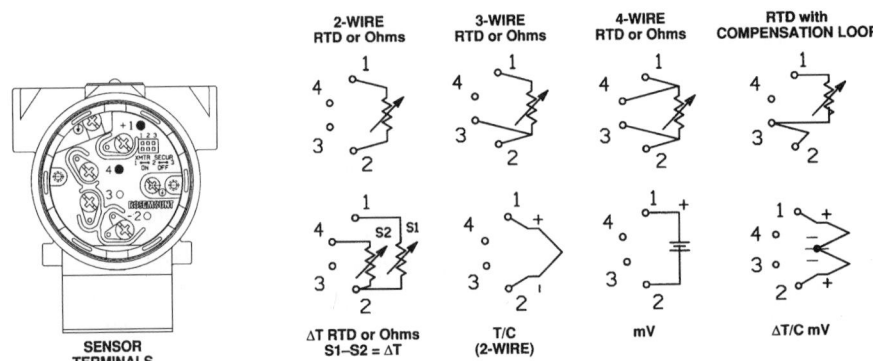

(b)

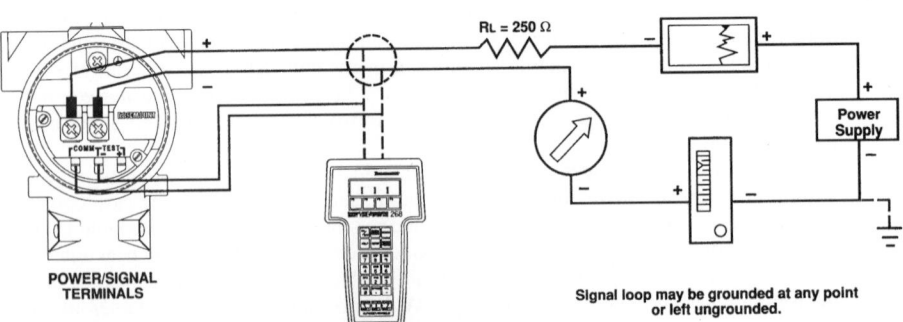

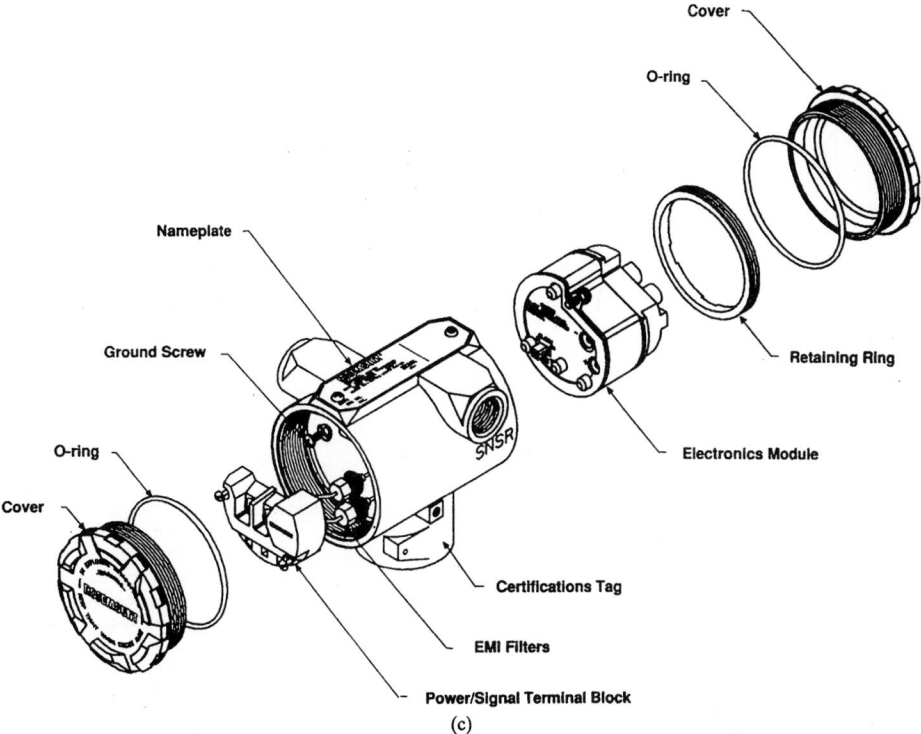

(c)

FIGURE 16 Example of contemporary smart temperature transmitter designed for use with a variety of sensors, such as RTDs, thermocouples, and other millivolt and ohm signals.

(*a*) Block diagram of electronics module. The interface can be connected at any termination point in the 4- to 20-mA signal loop. The electronics module consists of two circuit boards sealed in an enclosure with an integral sensor terminal block. The module utilizes digital ASIC (application-specific integrated circuit), microcomputer, and surface-mount technologies. The electronics digitize the input signal from the sensor and apply correction coefficients selected from nonvolatile memory. The output section of the electronics module converts the digital signal to a 4- to 20-mA output and handles the communication with other interfaces of the control system. An LCD (liquid-crystal display) meter may plug into the electronics module to display the digital output in user-configured units.

Configuration data are stored in nonvolatile EEPROM (electrically erasable programmable read-only memory) in the electronics module. These data are retained in the transmitter when power is interrupted—so the transmitter is functional immediately upon power-up.

The process variable (temperature) is stored as digital data and engineering unit conversion. The corrected data then are converted to a standard 4- to 20-mA current applied to the output loop. The control system can access the sensor reading directly as digital signal, thus bypassing the digital-to-analog conversion process for higher accuracy.

(*b*) Transmitter field wiring. The meter interface previously described can be connected at any termination point in the signal loop. In this system the signal loop must have 250-ohm minimum load for communications. In addition to the initial setting of the transmitter's operational parameters (sensor type, number of wires, 4- and 20-mA points, damping, and unit selection), informational data can be entered into the transmitter to allow identification and physical description of the transmitter. These data include Tag (8 alphanumeric characters), descriptor (16 alphanumeric characters), message (32 alphanumeric characters), date, and integral meter. In addition to the configurable parameters, the system's software contains several kinds of information that are *not* user-changeable, namely, transmitter type, sensor limits, and transmitter software revision levels.

The system performs continuous self-tests. In the event of a problem, the transmitter activates the user-selected analog output warning. Then the meter interface (or other designated control component) can interrogate the transmitter to determine the problem. The transmitter outputs specific information to the interface or control system component to identify the problem for fast and convenient corrective action. If an operator believes there may be a loop problem, the transmitter can be directed to provide specific outputs for loop testing.

During initial setup of the transmitter and for maintenance of the digital electronics, the format function is used. The top-level format menu offers two functions: characterize and digital trim. These functions allow the user to select the sensor type and adjust the transmitter's digital electronics to user plants standards.

(*c*) Exploded view of smart transmitter. (*Rosemount Inc., Measurement Division.*)

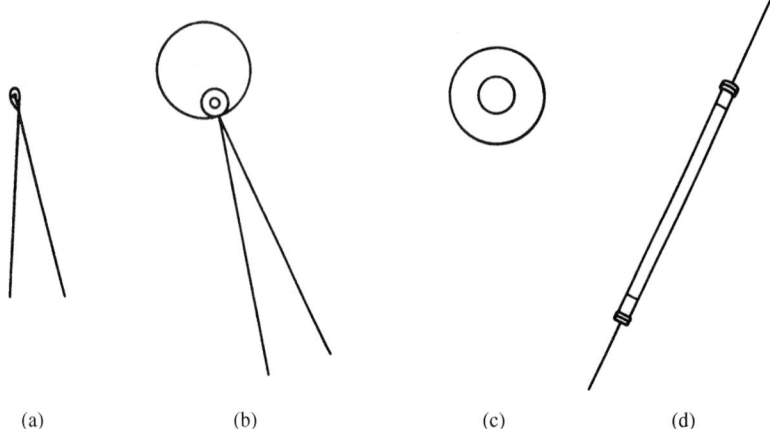

(a) (b) (c) (d)

FIGURE 17 Various configurations of thermistors. (*a*) Beads may be made by forming small ellipsoids of material suspended on two fine leadwires. The wires may be approximately 2.5 mm (0.1 inch) apart or even closer. If the material is sintered at an elevated temperature, the leadwires become tightly embedded within the bead, making electric contact with the thermistor material. For more rugged applications, the bead thermistor may be encapsulated or placed within a suitable metal sheath. (*b*) Disk configurations are manufactured by pressing the semiconductor material into a round die to produce a flat, circular probe. After sintering, the pieces may be silvered on the two flat surfaces. Thermistor disks may range from 2.5 to 25 mm (0.1 to 1 inch) or less in diameter, and from 0.5 to 13 mm (0.02 to 0.5 inch) or less in thickness. The disk configuration is well suited where a moderate degree of power dissipation is needed. (*c*) Washer-type thermistors are like the disk types, but provide a hole for bolt mounting. Often they are applied where high power dissipation is a primary requirement. (*d*) Rod-type thermistors are extruded through dies, resulting in long, cylindrical probes. Rod configurations generally are of high thermal resistance and are applied wherever power dissipation is not a major concern.

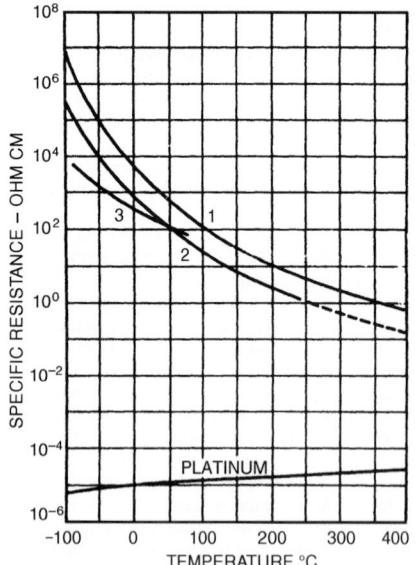

FIGURE 18 Resistance-temperature characteristics of three thermistors compared with platinum.

Thermistor Performance

The evaluation of thermistor performance characteristics is in many cases similar to that of resistancethermometers. Figure 18 shows the logarithm of the specific-resistance-versus-temperature relationship for three thermistor materials as compared with platinum metal. The specific resistance of the thermistor represented by curve 1 *decreases* by a factor of 50 as the temperature is increased from 0 to 100°C. Over the same temperature range, the resistivity of platinum will *increase* by a factor of approximately 1.39.

Thermistors range in terminal resistance at room temperature from about 1 ohm to the order of 10^8 ohms, depending on composition, shape, and size. Within a given type, they commonly vary 10 to 30 percent in resistance from the nominal at the reference temperature. Advanced manufacturing techniques and careful selection of suitable units can bring resistance values within closer limits.

Thermistor Applications

The application of thermistors as primary temperature elements follows the usual principle of

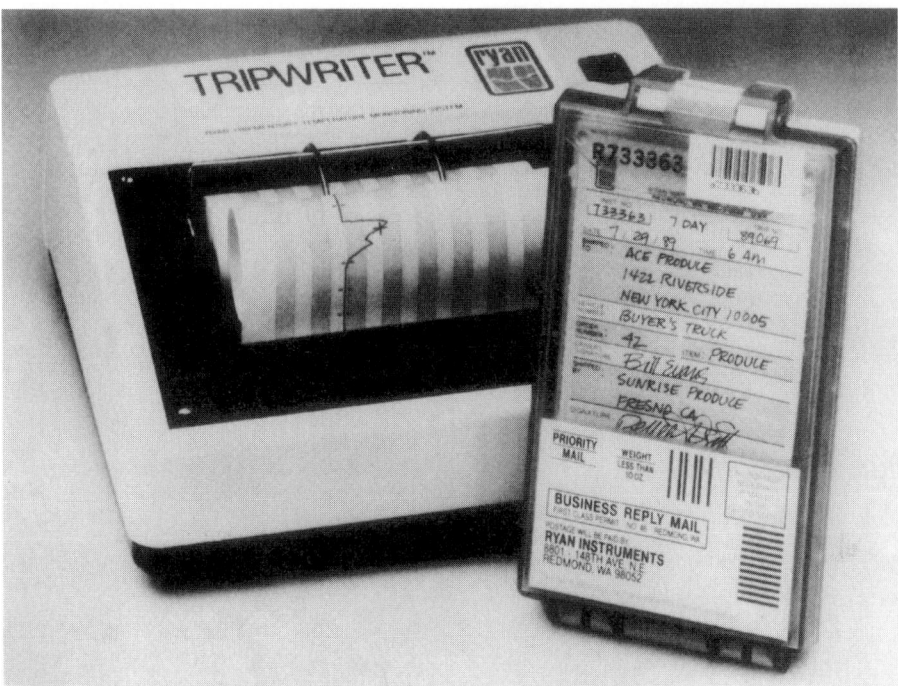

FIGURE 19   Thermistor-based temperature recorder (right) used for monitoring environmental conditions during the transport of various perishable foodstuffs, such as fruits and vegetables, for up to 16 days or more of travel time. These battery-operated units are packaged for mailing. Upon receipt of mailable units, such as *Tripmentor*™ and *Tripwriter*™, user may print out the information on the recorder (left). (*Ryan Instruments.*)

resistance thermometry. Conventional bridge or other resistance measuring circuits as well as constant-current circuits are used. Special application considerations must be given to the negative and highly nonlinear resistance-temperature relationship. Common to resistance thermometers, consideration must be given to keeping the measuring circuit small enough to avoid significant heating in order that the element resistance will be solely dependent on the measured medium.

A current very extensive use of thermistors is in trip thermometers, which have been well accepted in the food transportation industry. Small, portable recording-monitoring temperature instruments of the type shown in Fig. 19 initially were developed to provide a temperature history of foodstuffs (vegetables, fruits, seafoods) during transit from producer to final distributor by truck, rail, ship, and air. Such units not only provide information required by regulatory agencies, but also for the improvement of preparation and handling processes. A few examples of the more critical needs for such temperature information would include fresh lettuce, bananas, lobster, and pizza dough. Shipping time spans range from 4 to 16 shipping days. Temperature span covers -29 to $38°C$ (-20 to $100°F$), with a claimed accuracy of $±2$ percent (scale). Extended temperature ranges and longer shipment intervals (up to 75 days) are available. Both analog and digital versions are available. Some of the units also incorporate a humidity (capacitive) detector with scales ranging from 10 to 90 percent relative humidity.

Another rapidly developing market for thermistor thermometers where portability and convenience are important factors is found in laboratories, pilot plants, and for checking the temperature of flow and return pipes in heating and air-conditioning systems. One design makes a 3-second calibration from a special microprocessor each time it is turned on. A typical measuring range is from -51 to $150°C$ (-60 to $300°F$) with a resolution of $0.6°C$ ($1°F$) and an accuracy of $0.2°C$ ($0.4°F$). Power supply is a 9-volt battery. Depending on the model, the instrument weighs from 150 to 250 grams (5.3 to 7.8 oz).

SOLID-STATE TEMPERATURE SENSORS

Of limited, but expanding industrial use is the silicon or quantum detector. In one form, incident infrared radiation interacts with a bound electron within the semiconductor crystal lattice. The energy of a photon, if sufficiently large, is transferred to an electron to free it from its immobile state and permit it to move through the crystal. During the time it is free, the electron can produce a signal voltage in the detector. After a short interval, the electron will return to its bound state. This interval generally is far shorter than the thermal time constant of a thermal detector.

In essence, the quantum detector is a photon counter which is equally sensitive to all photons having the minimum energy to free a bound electron. A detector of this type will exhibit a fairly uniform response to all photons up to a particular wavelength. The practical advantage of these detectors lies in their ability to produce electrical signals that faithfully measure the incident photon flux. This permits a method of continuous temperature measurement without contact. Thus solid-state detectors are used in some infrared thermometer designs.

RADIATION THERMOMETERS

Radiation thermometry represents a practical application of the Planck law and Planck radiation formula (1900) and makes it possible to measure the temperature of an object without making physical contact with the object. Although these instruments are widely used because of the contactless feature, this is by no means the exclusive advantage of the method. Another important advantage is the wide useful temperature range—from subzero temperatures to extremely high, virtually unlimited values. Carefully constructed instruments have been used for many years to maintain laboratory primary and secondary standards above the gold point (1064.43°C) within an accuracy of ±0.01°C. Representative industrial designs generally have a precision of ±0.5 to ±1 percent, even under rather adverse conditions.

Radiation thermometers, although much improved during the past decade or two through the incorporation of technological advancements in electronic circuitry and optics, have been used for many decades. Originally called radiation pyrometers, then radiation thermometers, and, more recently, infrared (IR) thermometers, the initial applications of the method usually were in those processing and manufacturing applications where great amounts of heat were required, often in small spaces, thus creating very high temperatures. Often such materials were moving and could not be contacted by a temperature detector.

Thus early applications generally involved monitoring such operations as glass, metal, chemical, cement, lime, and refractory materials. During the past few decades, radiation thermometry has extended into lower temperature regions, including subzero measurements, as encountered in the foods, electronics, paper, pharmaceutical, plastics, rubber, and textile industries, among others. Portable IR thermometers also find wide application in industrial energy conservation strategies for checking large lines, vessels, steam traps, and so on, for faulty operation. Relatively recent increased usage of IR thermometers now gives it well over 10 percent of the industrial temperature measurement field.

IR Thermometry Principles

Planck's law predicts very accurately the radiant power emitted by a blackbody per unit area per unit wavelength, or complete radiation. It is written

$$M^b(\lambda, T) = \frac{C_1}{\lambda^5} \frac{1}{e^{C_2/\lambda T} - 1} \ \text{W} \cdot \text{m}^{-3}$$

where $C_1 = 2\pi hC^2 = 3.7415 \times 10^{16} \ \text{W} \cdot \text{m}^2$, called the first radiation constant
$C_2 = Ch/k = 1.43879 \times 10^{-2} \ \text{m} \cdot \text{k}$, called the second radiation constant

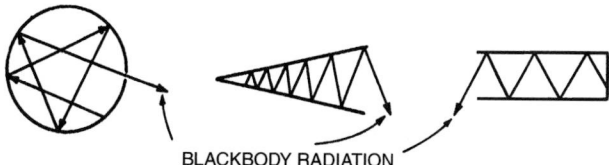

BLACKBODY RADIATION

FIGURE 20 Classical blackbody models. The radiation leaving a small hole in the sphere or the inside of the wedge, for example, will fulfill the definition of blackbody radiation, provided the walls of the cavity are opaque and the cavity is uniform in temperature. Because the geometry of these models promotes multiple internal reflection of the radiated energy, the wall material may be an imperfect radiator and thus a practical substance. Laboratory blackbody sources must be stable and uniform when used as reproducible sources in the calibration of secondary standards for radiation thermometers.

This radiation formula can be written in other forms, such as using wavenumbers instead of wavelengths. In this article, the above form expresses the radiant exitance[2] in terms of wavelength and absolute temperature. The units used are SI, and the nomenclature is that recommended by the Optical Society of America.

Blackbody Concept. This is central to radiation thermometer technology. The energy radiated by an object as a result of its temperature is quantitatively expressed in terms of a perfect radiating body, which is traditionally designated a blackbody. This concept has been described in several ways, such as a body that absorbs all the radiation it intercepts and a body that radiates more thermal energy for all wavelength intervals than any other body of the same area, at the same temperature. Physical realization of the blackbody, necessary for instrument calibration purposes, includes the spherical cavity and the wedge-shaped cavity as described by Mendenhall (Fig. 20). Numerous other blackbody source designs have been made and are available commercially (Fig. 21).

Emissivity. This is a measure of the ratio of thermal radiation emitted by a nonblackbody at the same temperature.

Total Emissivity. This is the ratio of the total amount of radiation emitted. By definition, a blackbody has maximum emissivity, the value of which is unity. Other bodies have an emissivity less than unity. The ratio is designated as total emissivity.

Gray Body. This refers to an object which has the same spectral emissivity at every wavelength, or one which has its spectral emissivity equal to its total emissivity.

Stefan-Boltzmann Law. As mentioned previously, Planck's radiation law predicts precise levels of radiation emitted per unit surface area of a blackbody at each wavelength. This may be written as

$$M^b(\lambda, T) = c_1 \lambda^{-5} \left(e^{c_2/\lambda T} - 1 \right)^{-1}$$

The total radiation emitted per unit surface area is the integral of this equation over all wavelengths,

$$M^b(T) = c_1 \int_0^{+\infty} \lambda^{-5} \left(e^{c_2/\lambda T} - 1 \right)^{-1} d\lambda$$

or, simply,

$$M^b(T) = \tau T^4$$

---

[2] A coined word not to be confused with excitance.

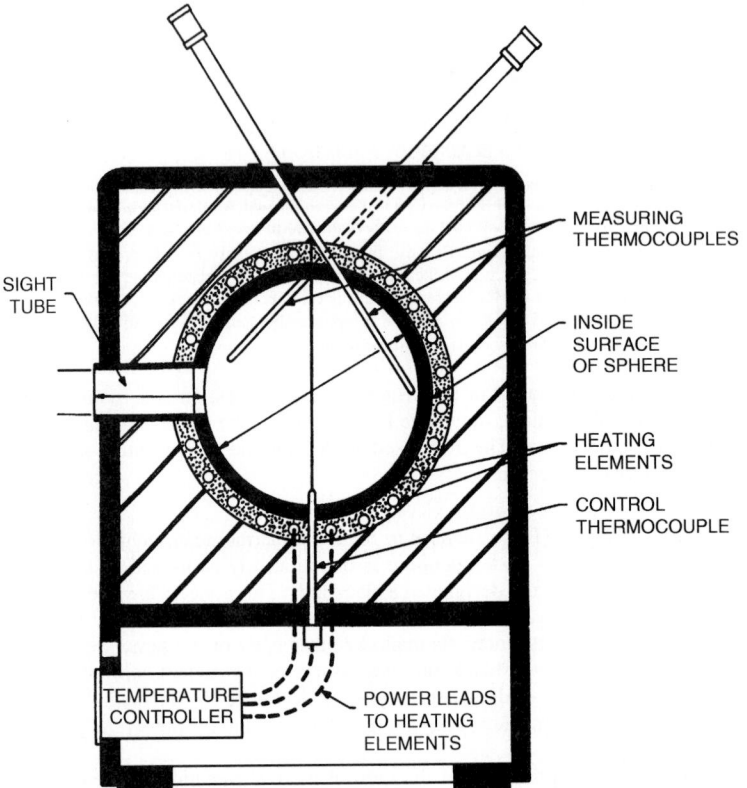

FIGURE 21 Schematic view of a commercial blackbody furnace. Entrance sight tube is shown to the left of the hollow sphere. One thermocouple is used to determine the sphere's temperature, another to determine temperature gradients. Blackbody source designs also may take the form of conical or cylindrical chambers where the length-to-diameter ratio is large. The emissivity of laboratory sources ranges rom 0.98 to 0.9998. In some sources a temperature uniformity of better than 1 K is obtainable. Thus, reproducible calibration conditions of ±1 percent are readily obtainable, and ±0.1 percent can be achieved for industrial instruments.

This is known as the Stefan-Boltzmann law. It is used extensively in the calculation of radiant heat transfer.

In situations where the exponential term in Planck's law is much greater than 1, that is, $e^{c_2/\lambda T} \gg 1$, Planck's equation can be approximated by

$$M^b(\lambda, T) = c_1 \lambda^{-5} e^{-c_2/\lambda T}$$

which is known as Wein's law.

A graphic representation of the radiant exitance as predicted by Planck's law, for several temperatures, is given in Fig. 22. Note that for each temperature there is a peak value of emission and that the peak value shifts to shorter wavelengths as the temperature is increased. This shift can be expressed as

$$\lambda_m T = b$$

which is known as Wein's displacement law.

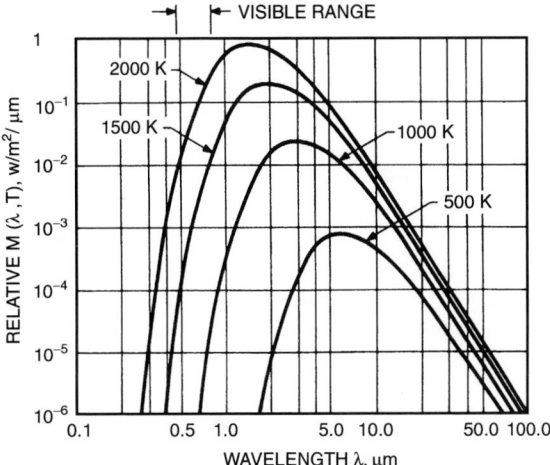

FIGURE 22 Radiant emission of thermal radiation from a blackbody at various temperatures (in kelvin).

Measuring Temperatures of Nonblackbodies

Temperature measurements of real nonblackbody objects can be complicated by three major factors:

1. Nonblackbodies emit less radiation than blackbodies, and often this difference is wavelength-dependent. Often, but not always, an emissivity correction must be made. Without the needed emissivity correction of the measured signal, the apparent temperature will be lower than the actual temperature.
2. Extra radiation from other radiant sources may be reflected from the object's surface, thus adding to the measured radiation and thereby increasing the apparent temperature. In certain cases the reflected amount may compensate for that needed to correct the object's emissivity.
3. The intensity of the emitted radiation may be modified in passing through media between the object and the instrument, thus resulting in a change in the apparent temperature. If radiation is lost, the apparent temperature will be low; if radiation is added, the temperature will read high.

In addition to these practical problems, there is often a need to measure an object that is transparent. Additional problems include added radiation produced by hot objects behind the object of measurement.

Effective Wavelength. The calibration function, or output, of a radiation thermometer is a nonlinear voltage or current. Mathematically it is an equation involving the spectral characteristics of the optical system and the detector response, integrated over all wavelengths. Once an instrument design is fixed, the relation between the thermometer output and a blackbody source temperature can be written,

$$V(T) = K \int_0^{+\infty} M^b(\lambda, T)S(\lambda)\, d\lambda$$

where $S(\lambda)$ = net thermometer wavelength sensitivity
         K = calibration constant

The calibration function is generated simply by aligning a unit with a blackbody source and measuring the output at different temperatures. Under conditions where Wein's law may be substituted for

Planck's law, this equation becomes

$$V(T) = Kc_1 \int_{\lambda_1}^{\lambda_2} \lambda^{-5} e^{-c_2/\lambda T} \, d\lambda$$

and a wavelength λ_e can be found such that

$$\lambda_e^{-5} e^{-c_2/\lambda_e T} = \int_{\lambda_1}^{\lambda_2} \lambda^{-5} e^{-c_2/\lambda T} \, d\lambda$$

The single wavelength λ_e, the effective wavelength, is representative of an instrument at temperature T. Thus the calibration function $V(T)$ can be written as

$$V(T) = K' e^{-c_2/\lambda_e T}$$

from which it can be shown that the effective wavelength can be expressed in terms of the rate of change of the thermometer calibration function at T as

$$\lambda_e = \frac{c_2}{T_2} \left[\frac{\Delta V(T)}{V(T) \Delta T} \right]^{-1}$$

N Values.   Over a range of temperatures, say from T_1 to T_2, the effective wavelength will change. Also, at a given temperature the calibration function can be approximated as a single-term power function of temperature:

$$V(T) = KT^N$$

Making use of the effective wavelength concept, it can also be shown that

$$N = \frac{c_2}{\lambda_e T}$$

where the power N is called the N value of the thermometer. A ratio thermometer can be described in terms of effective (or equivalent) wavelength, even though it has two distinct and possibly widely separated wavelength response bands.

At high temperatures, $c_2/\lambda T \gg 1$, the calibration function of a ratio thermometer is essentially the ratio of the two calibration functions of the individual "channels" $V_1(T)$ and $V_2(T)$,

$$V_R(T) = \frac{V_1(T)}{V_2(T)} = A \frac{e^{-c_2/\lambda_1 T}}{e^{-c_2/\lambda_2 T}}$$

where λ_1 and λ_2 are the effective wavelengths of the two channels. When a net effective or equivalent wavelength λ_E is defined as

$$\frac{1}{\lambda_E} = \frac{1}{\lambda_1} - \frac{1}{\lambda_2}$$

then the *form* of the calibration function $V_R(T)$ is nearly identical to that of a single-wavelength thermometer with an effective wavelength of λ_e. The nearer the two effective wavelengths of the ratio thermometer channels are, the longer the equivalent wavelength λ_E.

A mean equivalent wavelength and N value can also be used to characterize a given ratio thermometer. It should be noted that, in using the foregoing equations, the temperatures must be expressed as absolute temperatures.

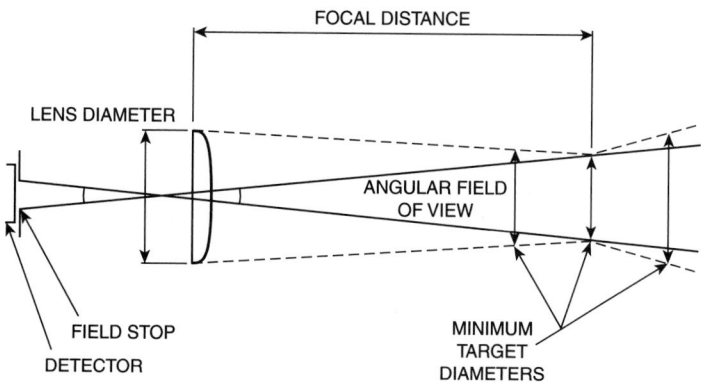

FIGURE 23 Schematic representation of the field of view of a radiation thermometer.

Optical Field of View. Inasmuch as Planck's law deals with the radiation emitted per unit surface area, radiation from a known area must be measured in order to establish a temperature measurement. This property, called the *field of view* of the instrument, is simply the target size that the thermometer "sees" at a given distance. Thus a target-size-versus-distance table, chart, or formula is essential for the correct use of a radiation thermometer. At a minimum, the object of measurement must fill the required target size at a given distance. The optical field of view of a radiation thermometer is shown in Fig. 23. Similarly, in calibrating a thermometer, the radiation source must fill the field of view in order to generate or check the calibration output. If the field of view is not filled, the thermometer will read low. If a thermometer does not have a well-defined field of view, the output of the instrument will increase if the object of measurement is larger than the minimum size.

Since a ratio thermometer essentially measures the ratio of two thermometer signals, anything that tends to reduce the actual target size will not upset the ratio. In principle, ratio thermometers should be immune to changes in target size or not seriously affected if the object does not fill the field of view. There are, however, two limitations to this: (1) the optical fields of view in the two wavebands must be the same, and (2) the net signals to be ratioed must be larger than any internal amplifier drifts, offsets, and noise.

Transparent Objects. The law of conservation of energy requires that, at every wavelength, the coefficients of transmission, reflection, and emission (absorption) of radiation add up to 1, as shown by

$$\varepsilon(\lambda) + r(\lambda) + t(\lambda) = 1$$

There are several ways to deal with transparent objects such as glass, plastic, semiconductor materials, and gases. The first and simplest is to select a waveband in which the object is opaque. For example, nearly all ordinary glasses are opaque at wavelengths longer than about 5.0 μm, provided the glass thickness is 3.0 mm or more. Similarly, most thin polyethylene plastics are opaque in a narrow wavelength band centered at 3.43 μm. As a simple guide in determining if an object is opaque, one must examine the spectral absorption coefficient that can be roughly deduced from spectral transmission curves using the relationship

$$\alpha = -\frac{1}{x} \ln \frac{T}{[1 - r(\lambda)]^2}$$

where $r(\lambda)$ = surface reflection coefficient that, at normal incidence, is obtained from the index of refraction n
         x = thickness
         T = transmission

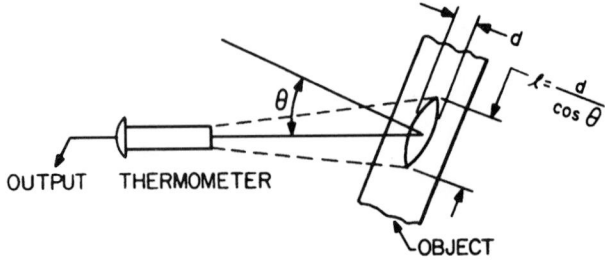

OUTPUT   THERMOMETER

OBJECT

FIGURE 24   Effect of angle of viewing on target size and shape.

$$r(\lambda) = \left(\frac{n-1}{n+1}\right)^2$$

Other Factors Affecting Performance.   These include factors in which (1) the object has unknown or varying emissivity and (2) attenuation or enhancement of radiation occurs in the instrument's sighting path. Still another factor is ambient temperature stability.

Emissivity data are obtainable from numerous handbooks on materials, but where these data cannot be located, as in the case of a brand new material for which data have not yet been published (or may be proprietary) or where the instrument supplier may not have any specific data, some laboratory work may be needed. In such cases the user either should measure the object's temperature, or at least that of a representative sample, by another accurate and reliable means and adjust the thermometer emissivity control to make the thermometer match; or should have such tests conducted by an approved outside laboratory.

It should be stressed that the angle of viewing the target affects the target size and shape, as shown in Fig. 24.

The user should determine which of the following situations may apply to a given application: (1) object and surroundings are at about the same temperature; (2) ambient conditions are cooler; or (3) ambient conditions are hotter. Measurements can be made that describe a given situation by using the following equation, which accounts for both the emissivity of the surface and the effect of thermal radiation from other sources reflected from the surface into the thermometer:

$$V(T_A) = \varepsilon(\lambda)V(T_O) + [1 - \varepsilon(\lambda)]V(T_B)$$

where $V(T_A)$ = indicated apparent blackbody temperature
$V(T_O)$ = blackbody temperature of object
$V(T_B)$ = blackbody temperature of net surroundings (assuming for simplicity that a reasonably uniform effective background exists)
$\varepsilon(\lambda)$ = spectral emissivity
$r(\lambda)$ = average reflectivity of object over same wavelength span, $= 1 - \varepsilon(\lambda)$

These problems are particularly important where the object is in hotter surroundings. What might be termed the specular solution to the problem can be used with many materials that have a relatively smooth surface, as, for example, flat glass in tempering furnaces. Rolled steel sheets for automotive and appliance uses are often processed in high-temperature ovens for annealing or cleaning. Such products often are highly specular, sometimes amounting to 90 percent or more, and can be measured successfully with very high N-value thermometers. Further, dual thermometers, as shown in Fig. 25, may be used.

In terms of attenuating the radiation received by the thermometer, water vapor is a major cause of error, particularly with wideband thermometers because of the overlapping wavelengths involved. Radiation also may be attenuated by dust, smoke, condensing steam, radiation-absorptive gases, and

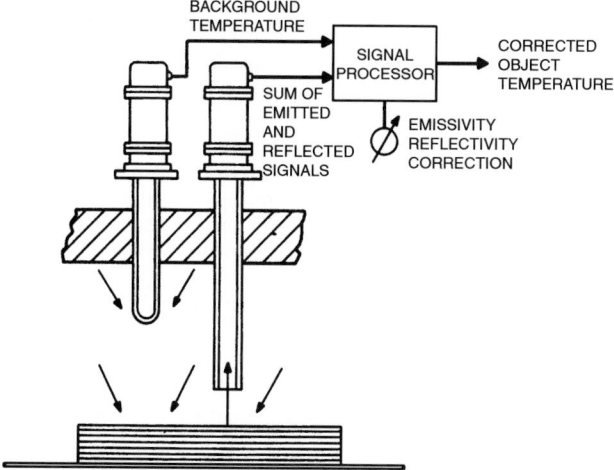

FIGURE 25 Dual-thermometer method for measuring the temperature of an object in surroundings that are hotter.

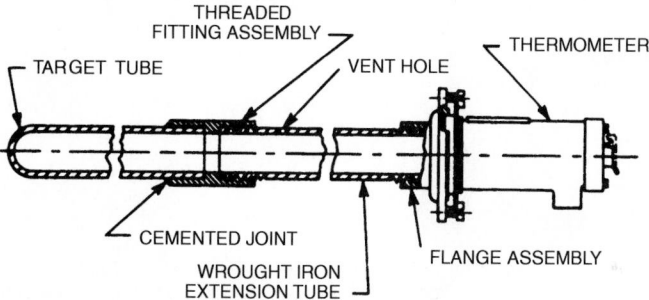

FIGURE 26 Radiation thermometer with closed-end target tube.

simply by any object that may block the field of view of the instrument. In difficult situations of this kind, a closed or open sighting tube may be used as part of the instrument (Fig. 26), or, for example, in the case of gas turbine measurements, a two-wavelength thermometer may be the solution.

IR Detectors

Thermal detectors are most frequently used in contemporary IR thermometers. These include thermocouples (types J, K, N, R, and S) and RTDs. Quantum detectors, which consist of a semiconductor crystal, are favored for some applications that require exceptional response speed. Both types of detectors have been described earlier in this article.

Classes of IR Thermometers

Generally IR thermometers are classified into five categories.

Wideband Instruments.   These devices are the simplest and of lowest cost. They respond to radiation with wavelengths from 0.3 μm to between 2.5 and 20 μm, depending on the lens or window material used. These instruments also have been called broadband or total radiation pyrometers because of their relatively wide wavelength response and the fact that they measure a significant fraction of the total radiation emitted by the object of measurement. Historically these devices were the earliest fixed or automatic units. Standard ranges include 0 to 1000°C (32 to 1832°F) and 600 to 1900°C (932 to 1652°F).

Narrowband Instruments.   These instruments usually have a carefully selected, relatively narrow wavelength response and are often selected to provide the particular wavelength response required by an application. Development of many different narrowband thermometers in recent years has been prompted by an increased need to match the IR thermometer with specific applications, as well as overcoming prior deficiencies of wideband instruments. Standard ranges are many, varying from one supplier to the next. Examples would include −50 to 600°C (−36 to 1112°F), 0 to 1000°C (32 to 1832°F), 600 to 3000°C (1112 to 5432°F), and 500 to 2000°C (932 to 3632°F).

Ratio (Two-Color) Instruments.   These instruments essentially consist of two radiation thermometers contained within a single housing. Some internal components, such as lens and detector, may be shared. The unique characteristic is that the output from the two thermometers, each having a

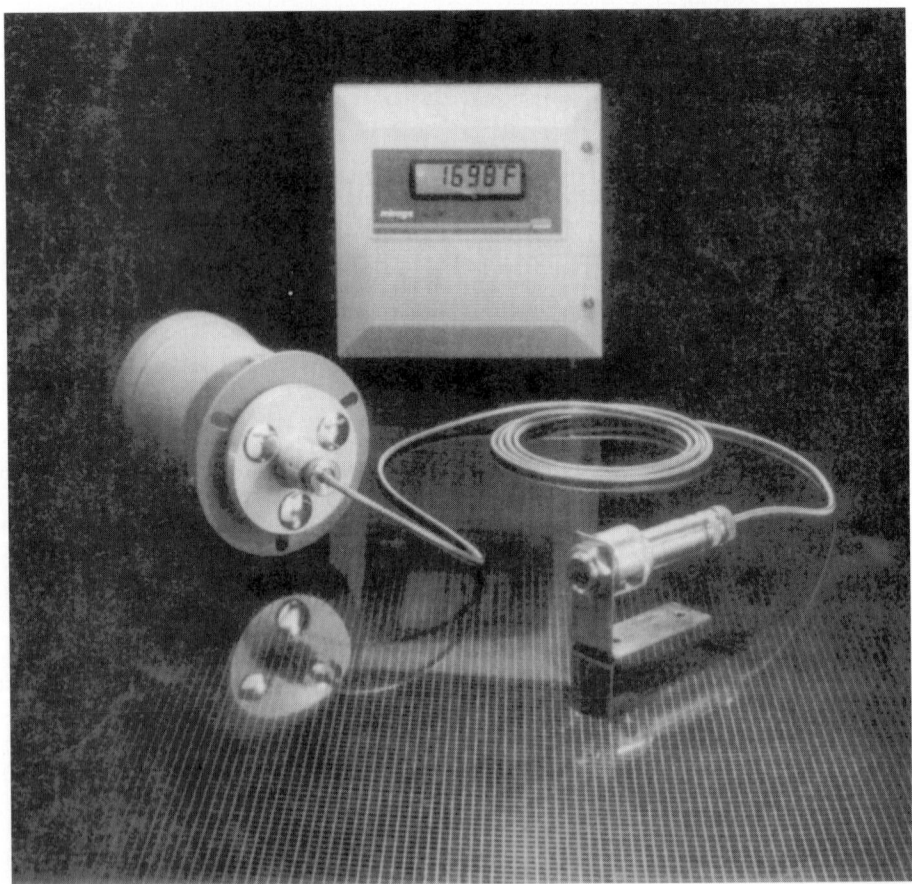

FIGURE 27   Fiber-optic IR thermometer. Instruments like this cover three spectral regions and can measure temperatures from 177 to 3500°C (350 to 6500°F). A two-color version is also available (*Fiber Optic Mirage*™). Claimed accuracy is within ±1 percent and repeatability, ±0.3 percent (full-scale). (*Square D. Company (Ircon), Infrared Measurement Division.*)

separate wavelength response, is ratioed. The concept behind the ratio thermometer is that the ratio signal is also a function of temperature, and so long as the ratio value is unchanged, the temperature measurement is accurate. Target, so long as it is sufficient, is not critical because the ratio of the signals from a small target is the same as that from a large target. These instruments cover wide temperature ranges. Examples would include 500 to 1800°C (932 to 3272°F), 0 to 1000°C (32 to 1832°F), and 825 to 1800°C (1441 to 3272°F).

Fiber-Optic Thermometers.   These instruments enable near-infrared and visible radiation to be transmitted around corners and away from hot, hazardous environments to locations more suitable for the electronics associated with contemporary radiation thermometers. Fiber optics also makes possible measurements in regions where access is restricted to larger instruments and where large electric or radio frequency fields would seriously affect an ordinary sensor. Conceptually, a fiber-optic system differs from an ordinary IR thermometer by the addition of a fiber-optic light guide, with or without a lens. The optics of the light guide defines the field of view of the instrument, while the optical transmission properties of the fiber-optic elements form an integral part of the thermometer spectral response (Fig. 27).

Disappearing-Filament IR Thermometer (Optical Pyrometer).   Also sometimes referred to as a brightness thermometer, this instrument makes a photometric match between the brightness of an object and an internal lamp. These pyrometers are sensitive only in a very narrow wavelength range. The instrument is particularly adapted to manual operation and portability. One of the earliest non-contact temperature-measuring devices, its reputation for high accuracy was established a number of decades ago. Optical pyrometers differ from other IR thermometers in both the type of reference source used and the method of achieving the brightness match between object and reference. Figure 28 is a schematic view of a typical visual instrument and the indicators for over, under, and matching conditions as seen by a manual operator. The temperature range is limited at the lower end by the

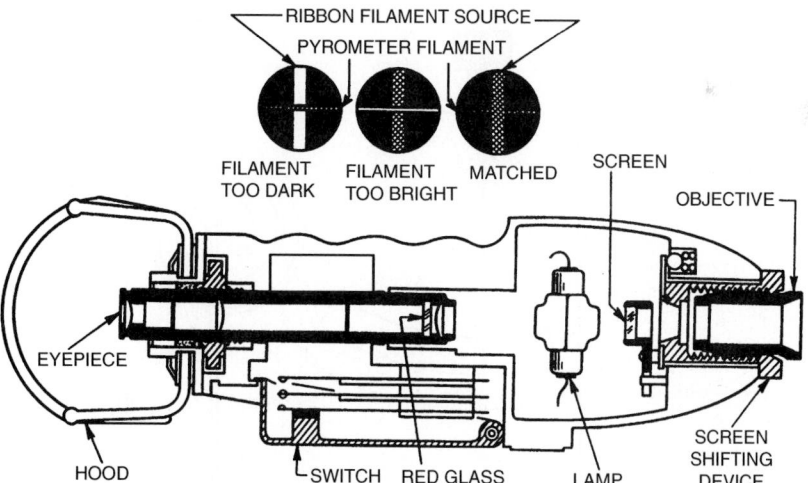

FIGURE 28   Optical pyrometer telescope and principle of operation. The reference source is an aged and calibrated tungsten strip lamp. In the manual version, the operator views the object to be measured and the lamp filament simultaneously through an optical filter. The combination of filter characteristics and the response of the average human eye produce a net instrument wave-length response band that is very narrow and centered near 0.65 μm. By adjusting the current flow through the lamp, or varying the intensity of the object radiation, the operator can produce a brightness match over at least a portion of the lamp filament, according to relative target size. Under matched conditions, the two "scenes" merge into one another, that is, the filament appears to vanish. Automatic optical pyrometers function similarly, in the same waveband, but utilize a photomultiplier tube or other detector element in place of the human eye. Advantage is taken of electronic feedback circuitry. (*Leeds & Northrup.*)

need for an incandescent image of the filament to about 800°C (1472°F). The upper temperatures are limited only by the needs of the application.

FILLED-SYSTEM THERMOMETERS

An industrial filled-system thermometer is shown schematically in Fig. 29. These instruments may be separated into two fundamental types. The measuring element (a bourdon, bellows, or helix) responds either to volume changes or to pressure changes with changes in temperature. The liquid expansivity with temperature is greater than that of the bulb metal, the net volume change being communicated to the bourdon. An internal-system pressure change is always associated with the bourdon volume change, but this effect is not of primary importance.

The following systems are generally available:

- *Class IA.*   The thermal system is completely filled with an incompressible liquid under pressure. The system is fully compensated for ambient temperature variations at the case and along the tubing (Fig. 30). Standard ranges, spans, and range limits are indicated in Table 8.
- *Class IIA.*   The thermal system is evacuated and partially filled with a volatile liquid, such as methyl chloride (CL), ether (ET), butane (BU), or toluene (T) (Table 8).
- *Class IIIB.*   The thermal system is filled under pressure with purified nitrogen (Table 8).

In all systems bulbs may be flange-mounted for direct insertion into the process, or contained in a thermowell. On special order, longer-capillary lengths, sanitary fittings, and other ranges and spans (within limitations) are commercially available.

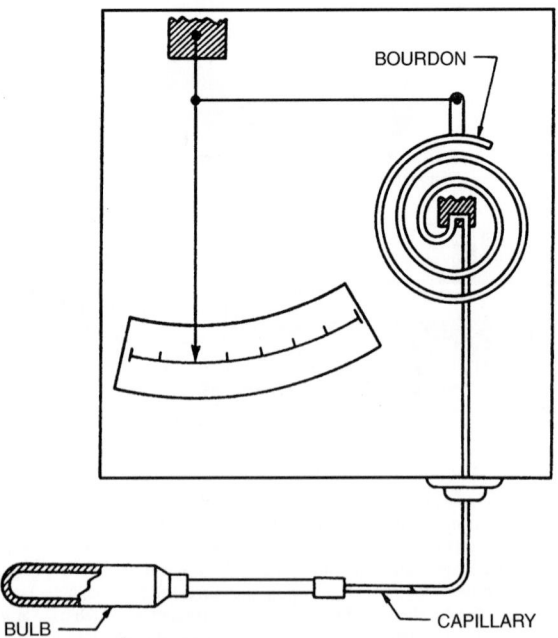

FIGURE 29   Operating principle of filled-system thermometer.

TABLE 8 General Specifications of Filled-System Thermometers

Class IA					Class IIA	Class IIIB		
	Bulb length		Range limits				Range limits	
Standard spans	mm	in	Minimum	Maximum	Standard ranges (fills)	Standard spans	Minimum	Maximum
°C			°C	°C	°C	°C	°C	°C
25	122	4.8	13	120	38 to 105 (BU)	100	−40	170
50	69	2.7	−15	200	105 to 150 (ET)	150	−15	400
75	53	2.1	−38	200	50 to 205 (T)	200	−15	500
100	43	1.7	−63	200	38 to 150 (ET)	250	−15	500
150	36	1.4	−73	250	50 to 250 (T)	300	−15	500
200	30	1.2	−73	260		400	−15	500
250	28	1.1	−73	260		500	−15	500
°F			°F	°F	°F	°F	°F	°F
50	112	4.4	50	250	100 to 180 (CL)	150	−50	240
100	69	2.7	0	400	100 to 220 (BU)	200	−50	425
150	51	2.0	−50	400	100 to 250 (BU)	300	0	825
200	43	1.7	−100	400	100 to 270 (BU)	400	0	1000
250	36	1.4	−100	400	100 to 300 (ET)	500	0	1000
300	33	1.3	−100	500	100 to 350 (ET)	600	0	1000
400	28	1.1	−100	500	100 to 400 (T)	700	0	1000
500	25	1.0	−100	500	200 to 500 (T)	800	0	1000
						1000	0	1000

Accuracy		
±0.5% of calibrated span for spans up to 215°C (400°F); ±0.75% of calibrated span for spans between 215 and 330°C (400 and 600°F).	±0.5% of calibrated span over upper two-thirds of scale.	±0.5% of calibrated span for spans up to 330°C (600°F); ±0.75% of upper range value for upper range value above 330°(600°F); ±0.75% of lower range value for lower range value below −45°C (−50°F).

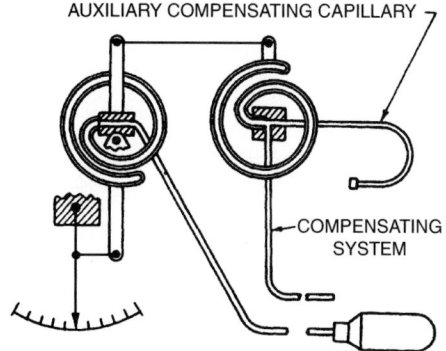

FIGURE 30 Fully compensated system as used in Class IA thermal systems.

AUXILIARY COMPENSATING CAPILLARY

COMPENSATING SYSTEM

Advantages and Limitations of Filled Systems. The fundamental simplicity of these systems favors lower cost, easy installation, and low maintenance. (However, in case of failure, the entire bulb and capillary assembly may have to be replaced.) The measuring system is self-contained, requiring no auxiliary power unless it is combined with an electronic or pneumatic transmission system. The capillary allows considerable separation between point of measurement and point of indication. [However, beyond about 120 meters (400 ft) it becomes more economic to use a transmission system.] In some cases the system can be designed to deliver significant power, if needed, to drive indicating or controlling mechanisms, including valves. Filled systems are limited to temperatures between −73 and 260°C (−100 and 500°F).

FIBER-OPTIC TEMPERATURE SENSOR
by Gene Yazbak[*]

A miniature fiber-optic sensor infers temperature from the refractive index (RI) of a thin layer of silicon that is "sandwiched" between two pieces of Pyrex glass. Rather than measuring the angle of refraction as is commonly done in many refractometers, an extrinsic Fabry-Perot interferometer spectrally modulates a light beam (wavefront) in proportion to temperature. Inasmuch as the units are based on spectral rather than amplitude modulation, they are not affected by such common system phenomena as fiber bending, connector losses, and source and detector aging (Fig. 31).

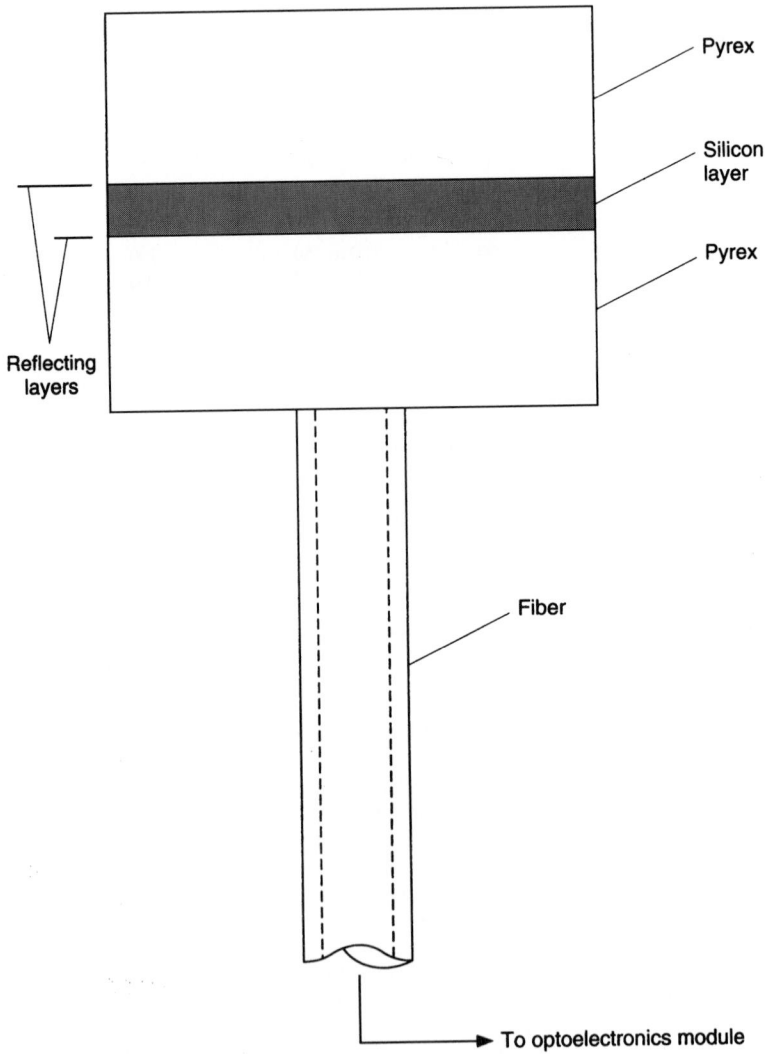

FIGURE 31   Fiber-optic temperature sensor. (*MetriCor, Inc.*)

[*] Senior Applications Engineer, MetriCor Inc., Monument Beach, Massachusetts.

Although the temperature probes can be made as small as 0.032 inch (0.8 mm), they usually are packaged as $1/8$-inch (3.2-mm) stainless-steel probe assemblies which are interfaced directly to process vessels or pipes, using off-the-shelf compression fittings. Claimed advantages include immunity to electromagnetic and radio frequency; installation and maintenance are simple.

The sensors have been found effective in chemical and food processing and notably in microwavable food packaging research for gathering data on bumping and splattering of foods during heating. Nonmetallic temperature probes available for such applications have been found to minimize error because of their low thermal conductivity and small size. Sensors can be located up to 914 meters (3000 ft) distant from the optoelectronics module.

Similarly conceived sensors are available for pressure measurement and direct refractive index measurement.

ACOUSTIC PYROMETER

The modulating effect of temperature on the speed of sound dates back to the early studies of Laplace over two centuries ago. Thus temperature can be measured inferentially by measuring the speed of an acoustic wave in a given environment. This principle has been applied to the measurement of furnace exit-gas temperatures. Traditionally such measurements are made by optical and radiation pyrometry or wall thermocouples, or estimated on the basis of heat balance calculations. Exit-gas temperatures are particularly important during boiler start-up so that superheater tubes will not be overheated prior to the establishment of steam flow.

If the speed of sound can be determined by measuring the flight time of an acoustic wave between two locations, then the temperature of the measured medium can be determined. Experiments with an acoustic pyrometer for furnace exit gas indicate that temperatures up to 1650°C (3000°F) may be encountered and that acoustic velocities may be in excess of 879 m/s (2880 ft/s) and wavelengths of the order of 1 meter (3.28 ft). As pointed out by some researchers, there still is much work to be done before a sound source is developed that will satisfy all of the conflicting requirements generated by the possible applications for acoustic pyrometers.

REFERENCE

1. McMillan, G. K. and Toarmina, C. M., *Advanced Temperature Measurement and Control,* Instrument Society of America, Research Triangle Park, North Carolina, 1995.

FLUID PRESSURE SYSTEMS

Pressure measurement not only is critical to the safe and optimum operation of such industrial processes as air and other gas compression; hydraulic equipment operation; separating operations, such as absorption, desorption, distillation, and filtration; steam generation; and vacuum processing—but other process variables, such as the contents level of tanks (hydrostatic pressure) and flow (differential pressure) can be inferred from pressure measurements. Pressure, of course, is the key to applying pneumatic transmitters and controllers. Further, pressure sometimes is selected over temperature as the variable to control. Air pressure also activates diaphragm motor valves.

Pressure instrumentation ranges widely from the comparative simplicity of low-cost bourdon- and bellows-actuated gages to some of the contemporary complex and sophisticated pressure sensors-transducers (transmitters) that have appeared in very recent years. Modern pressure transmitters differ from their historical counterparts mainly in two design respects:

1. Mechanical transducers utilizing links, levers, and pivots, having been replaced by electric and electro-optic transducers, which permits varying degrees of miniaturization of the force-receiving sensors

2. Introduction of what is commonly termed "smart" or "intelligent" electronics into transmitter design, notably the incorporation of a microprocessor along with other ingenious electronic circuitry

A 1991 survey of users indicates that about one-third of the modernized electronic pressure transmitters sold today are of the smart variety, and with an increasing rate of acceptance, even though costs run somewhat higher than for the simpler versions.

It is interesting to note that progress in the pressure transmitter field has stemmed mainly from discoveries and developments in the electronics and computer industries. Because of the dynamics of these industries, it would be unrealistic to assume that pressure transmitters or, in fact, any other areas of industrial instrumentation have reached a status that could be identified as mature. Thus the field will remain in a transitory phase for some time to come, which makes selection a continuing difficult task.

Because a substantial market remains for the historically less sophisticated pressure gages that incorporate simple mechanics and design, these devices are described in the early portions of this article.

MANOMETERS

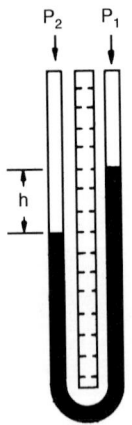

P_2 P_1

h

FIGURE 1 U-tube manometer.

Because of their inherent accuracy, manometers are used for the direct measurement of pressure and vacuum. Although some rugged designs can be used in the field and on-line, manometers largely serve as standards for calibrating other pressure-measuring instruments.

U-Tube Manometer. A glass U-tube is partially filled with liquid, and both ends are initially open to the atmosphere (Fig. 1).

When a gage pressure P_2 is to be measured, it is applied to the top of one of the columns and the top of the other column remains open. When the liquid in the tube is mercury, for example, the indicated pressure h is usually expressed in inches (or millimeters) of mercury. To convert to pounds per square inch (or kilograms per square centimeter),

$$P_2 = dh$$

where P_2 = pressure, psig (kg/cm^2)
d = density, lb/in^3 (kg/cm^3)
h = height, inches (cm)

For mercury, the density is 0.490 lb/in^3 at 60°F (15.6°C), and the conversion of inches of mercury to pounds per square inch becomes

$$P_2 = 0.490h$$

The density of water at 60°F (15.6°C) is 0.0361 lb/in³, and if water is used in a manometer, the conversion of inches of water to pounds per square inch becomes

$$P_2 = 0.0361h$$

The same principles apply when metric units are used. For example, the density of mercury at 15.6°C (60°F) may also be expressed as 0.0136 kg/cm³, and the conversion of centimeters of mercury to kilograms per square centimeters

$$P_2 = 0.0136h$$

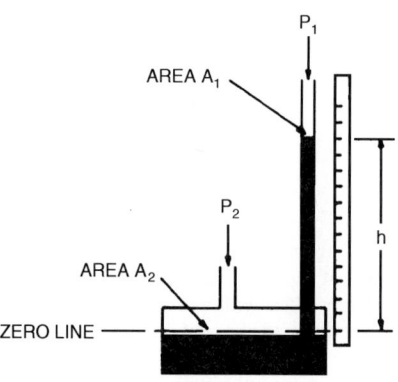

FIGURE 2   Well manometer.

For measuring differential pressure and for static balance,

$$P_2 - P_1 = dh$$

The U-tube manometer principle has also been utilized in industry in an instrument usually called a differential-pressure manometer. In this device the tubes are expanded into chambers and a float rides on top of the liquid in one of the chambers. The float positions an outside pointer through a pressuretight bearing or torque tube.

Well Manometer.   In this design one leg is replaced by a large-diameter well so that the pressure differential is indicated only by the height of the column in the single leg. The ratio of diameters is important and should be as great as possible to reduce the errors resulting from the change in level in the large-diameter well (Fig. 2).

The pressure difference can be read directly on a single scale. For static balance,

$$P_2 - P_1 = d \left(1 + \frac{A_1}{A_2} \right) h$$

where A_1 = area of smaller-diameter leg
          A_2 = area of well

If the ratio of A_1/A_2 is small compared with unity, then the error in neglecting this term becomes negligible, and the static balance relation becomes

$$P_2 - P_1 = dh$$

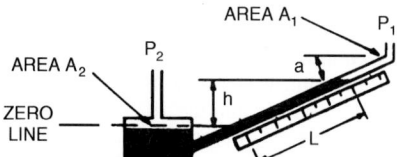

FIGURE 3   Inclined-tube manometer.

On some manometers this error is eliminated by reducing the spacing between scale graduations by the required amount.

Inclined-Tube Manometer.   In this device, so as to read small pressure differentials more accurately, the smaller-diameter leg is inclined (Fig. 3). This produces a longer scale so that

$$h = L \sin \alpha$$

Bell-Type Manometer.   This device utilizes a container immersed in a sealing liquid. The pressure to be measured is applied to the inside of the bell, the motion of which is opposed by a restricting spring (Fig. 4a). In the bell-type differential-pressure gage, pressures are applied to both the outside and the inside of the bell. Motion is restricted by an opposing spring (Fig. 4b).

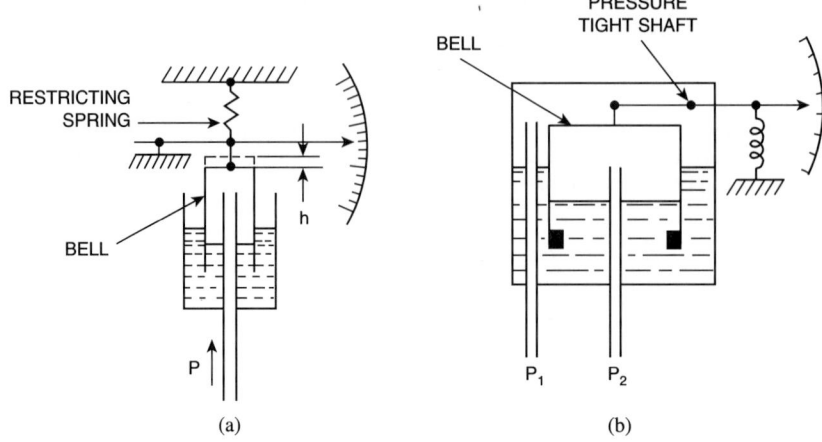

FIGURE 4   Bell-type manometers. (*a*) Liquid-sealed bell. (*b*) Differential-pressure gage.

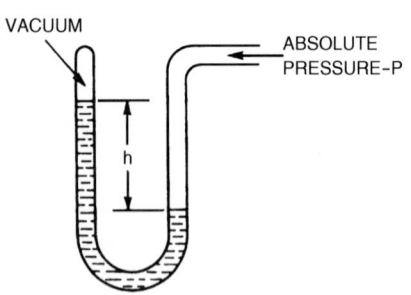

FIGURE 5   Liquid barometer.

Liquid Barometer.   A simple barometer may be constructed from a glass tube that is closed at one end and open at the other. The length of the tube must be greater than 30 inches (76.2 cm). The tube is first completely filled with mercury, the open end temporarily plugged, and then the plugged end placed in a container partially filled with mercury.

When the plug is removed, the mercury in the tube will drop by a certain amount, creating a vacuum at the top of the tube. The height of the column, as measured in Fig. 5 and expressed in inches or millimeters of mercury, will then be proportional to the atmospheric pressure.

Absolute-Pressure Manometer.   This type of gage comprises a glass U-tube partially filled with mercury, with the top of one leg evacuated and sealed (Fig. 6). The pressure to be measured is applied to the other leg, and *h* may be read in units of mercury absolute. To convert to pounds per square inch absolute (psia),

$$P = 0.490h$$

where *P* is absolute pressure in psia. If *h* is indicated in centimeters, this value may be converted to kilograms per square centimeter absolute by multiplying by 0.0136.

McLeod Gage (Liquid Manometer).   This device is designed for vacuum measurement. It functions essentially as a pressure amplifier. If, for example, 100 cm^3 (volume *V* in Fig. 7) of permanent gas is compressed into a section of capillary having a volume of 0.1 cm^3, the resulting pressure reading is amplified 1000 times. This principle allows pressure measurements into the 10^{-6}-Torr region, considerably below the 10^{-2}-Torr range of precision manometers.

If we assume that the volume *V* of gas trapped at the unknown pressure *p* (in centimeters of mercury for convenience) obeys Boyle's law, then $pV = (p+H)A$ where *A*, is the cross section of the closed capillary in square centimeters, and

$$P = \frac{AH^2}{V - HA}$$

FIGURE 6   Absolute-pressure gage (manometer).

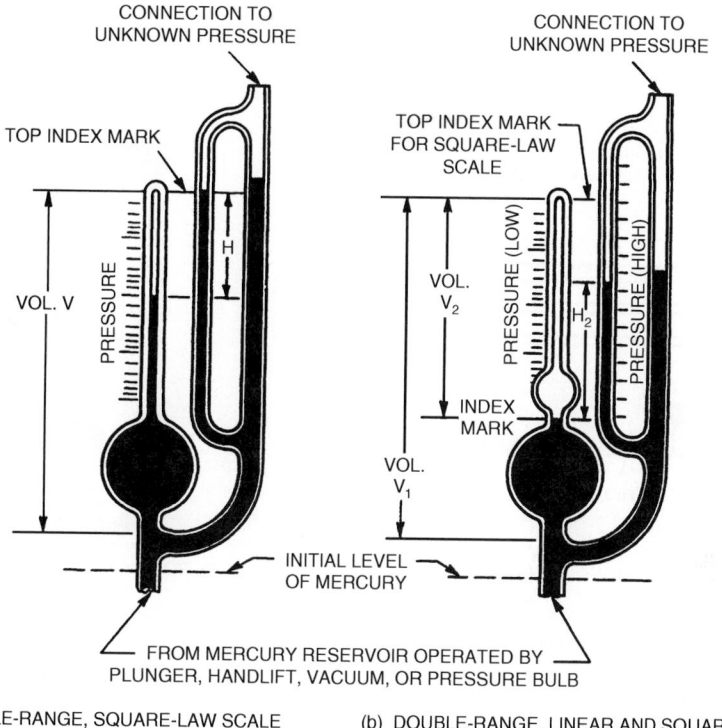

FIGURE 7 Two versions of McLeod gage.

In practice, HA is quite negligible when compared with volume V, with $p = 10AH^2/V$ Torr, and with other values expressed in centimeters.

A conveniently small McLeod gage may have a volume V of 200 cm³, with a capillary cross section A of 0.02 cm² and a length of 10 cm. Thus for $H = 0.1$ cm, $p = 5 \times 10^{-6}$ Torr, which would be the limit of unaided visual resolution and the reading could be wrong by 100 percent. At $H = 1$ cm, $p = 5 \times 10^{-4}$ Torr, the possible error becomes 10 percent. For various reasons, the only significant improvement in accuracy can be achieved by an increase in volume V.

A carefully constructed nonportable gage with a 1300-cm³ volume gives reproducible readings of ±0.5, ±0.6, ±2, and ±6 percent at pressures of 10^{-2}, 10^{-3}, 10^{-4}, and 10^{-5} Torr, respectively. The errors for other volumes can be estimated to be no lower than those based on volume proportionality. Thus in the previous example with $V = 200$ cm³ and $p = 10^{-4}$ Torr, percent error = $(1300/200) \times 2 = 13$ percent, which is in good agreement with the previous rough estimate of 10 percent.

Since the measured pressure in a McLeod gage is derived from basic (linear) dimensions, it is the industrial pressure standard with reference to which all other vacuum gages are calibrated. However, it should be emphasized that only the pressure of permanent gases is measured correctly. On account of the high compression ratio employed, vapor pressure of a substance of several tenths of a torr would not be detected with condensed liquid occupying negligible volume and not being visible to the eye. A highly portable version of the McLeod gage is shown in Fig. 8.

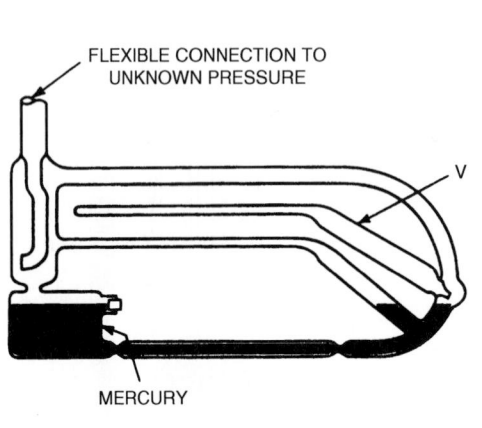

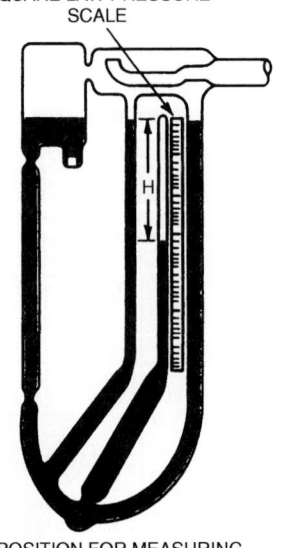

SQUARE LAW PRESSURE
SCALE

FLEXIBLE CONNECTION TO
UNKNOWN PRESSURE

V

MERCURY

(a) POSITION FOR REACHING PRESSURE
EQUILIBRIUM

(b) POSITION FOR MEASURING

FIGURE 8   Portable tilting-type McLeod gage.

ELASTIC-ELEMENT MECHANICAL PRESSURE GAGES

Dating back to the early years of steam power and compressed air and hydraulic technologies, this class of pressure sensors uses some form of elastic element whose geometry is altered by changes in pressure. These elements are of four principal types: bellows, bourdon tube, diaphragm, and capsule.

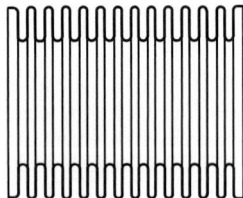

FIGURE 9   Common form of bellows used in pressure gage.

Bellows.   This is a thin-wall metal tube with deeply convoluted sidewalls that permit axial expansion and contraction (Fig. 9). Most bellows are made from seamless tubes, and the convolutions either are hydraulically formed or mechanically rolled. Materials used include brass, phosphor bronze, beryllium copper, Monel, stainless steel, and Inconel. Bellows elements are well adapted to use in applications that require long strokes and highly developed forces. They are well suited for input elements for large-case recorders and indicators and for feedback elements in pneumatic controllers.

Bourdon Tube.   In the 1852 patent its inventor E. Bourdon described the bourdon tube as a curved or twisted tube whose transfer section differs from a circular form. In principle, it is a tube closed at one end, with an internal cross section that is not a perfect circle and, if bent or distorted, has the property of changing its shape with internal pressure variations. An internal pressure increase causes the cross section to become more circular and the shape to straighten, resulting in motion of the closed end of the tube, a motion commonly called tip travel. Common forms of bourdon tubes are shown in Fig. 10.

A wide range of alloys can be used for making bourdon elements, including brass, phosphor bronze, beryllium copper, Monel, Ni-Span C, and various stainless-steel alloys.

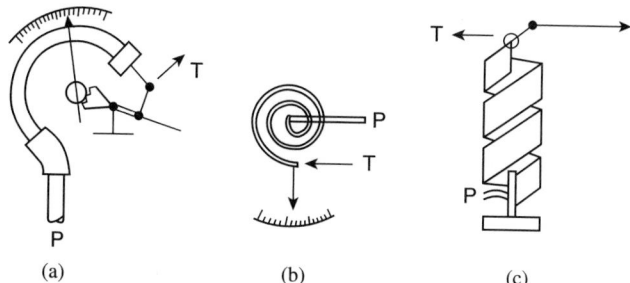

FIGURE 10 Types of bourdon springs. (*a*) C-type tube. (*b*) Spiral tube. (*c*) Helical tube.

Diaphragm. This is a flexible disk, usually with concentric corrugations, that is used to convert pressure to deflection. (In addition to use in pressure sensors, diaphragms can serve as fluid barriers in transmitters, as seal assemblies, and also as building blocks for capsules.) A diaphragm usually is designed so that the deflection-versus-pressure characteristics are as linear as possible over a specified pressure range, and with a minimum of hysteresis and minimum shift in the zero point. However, when required, as in the case of an altitude sensor, a diaphragm can be purposely designed to have a nonlinear characteristic.

Metals commonly used for making diaphragms are trumpet brass, phosphor bronze, beryllium copper, stainless steel, NiSpan C, Monel, Hastelloy, titanium, and tantalum. Both linearity and sensitivity are determined mainly by the depth and number of corrugations and by the angle of formation of the diaphragm face.

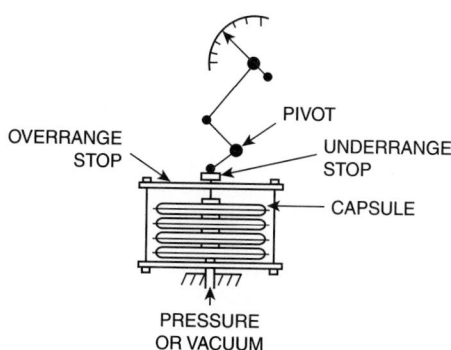

FIGURE 11 Use of capsule element in pressure gage.

In many pressure-measuring applications, the process fluid must not contact or seep into the pressure element in order to prevent errors due to effects of static head, to isolate the pressure element from corrosive and otherwise fouling fluids, to assist in cleaning (as in the food industry), and to prevent loss of costly or hazardous process fluids. Thus diaphragm seals are commonly used.

Capsule. A capsule is formed by joining the peripheries of two diaphragms through soldering or welding. Two or more capsules can be joined together (Fig. 11), and thus the total deflection of the assembly is equal to the sum of the deflections of the individual capsules. Such elements are used in some absolute pressure gages. These configurations also are used in aircraft applications.

Range of Measurement

The minimum and maximum pressure ranges of elastic-element mechanical pressure gages are given in Table 1.

FORCE OR PRESSURE SENSORS, TRANSDUCERS AND TRANSMITTERS

There is a marked distinction between a pressure sensor and a pressure transducer. The sensor provides the basis of measurement; the transducer converts energy from one form to another.

TABLE 1   Ranges of Elastic-Element Pressure Gages

Element	Application	Minimum Range	Maximum Range
Capsule	Pressure	0–0.2 in (0.5 cm) H_2O	0–1000 psig (70.3 kg/cm^2)
	Vacuum	0–0.2 in (0.5 cm) H_2O	0–30 in (76.2 cm) Hg vacuum
	Compound vacuum and pressure	Any span within pressure and vacuum ranges, with a total span of 0.2 in (0.5 cm) H_2O	—
Bellows	Pressure	0–5 in (12.7 cm) H_2O	0–2000 psig (141 kg/cm^2)
	Vacuum	0–5 in (12.7 cm) H_2O	0–30 in (76.2 cm) Hg vacuum
	Compound vacuum and pressure	Any span within pressure and vacuum ranges, with a total span of 5 in (12.7 cm) H_2O	—
Bourdon	Pressure	0–5 psig (0.35 kg/cm^2)	0–100,000 psig (7030 kg/cm^2)
	Vacuum	0–30 in (76.2 cm) Hg vacuum	—
	Compound vacuum and pressure	Any span within pressure and vacuum ranges, with a total span of 12 psi (0.84 kg/cm^2)	—

In the fully mechanical pressure instruments described previously, a spring may furnish the restoring force and, by means of links and lever, amplify and transmit the sensor value to a mechanically operated indicator, recorder, or controller.

In pneumatic pressure transducers, a counterpressure of air acts on the diaphragm, bellows, bourdon, or other elastic element to equalize the sensed (process) pressure. A force- or position-balance system may be used in pneumatic instruments, which are described in Section 3, Article 6, of this handbook. Current-to-pressure transducers used for the operation of pneumatic diaphragm control valves are described in Section 9 of this handbook.

In electronic or electro-optical transducers, sensor values are converted to electrical quantities—current, resistance, capacitance, reluctance, and alterations in piezoelectric and optical outputs.

Invention of the strain gage served as the initial impetus to use electrical transducers. There are numerous advantages for a large number of applications to be derived from some form of electronic transduction. Such units are quite small, they are easy to integrate into electrical networks, and numerous other electronic features can be added to transducers and transmitters, including built-in calibration checks, temperature compensation, self-diagnostics, signal conditioning, and other features, which may be derived from integrating a microprocessor in the sensor-transducer-transmitter unit.

Strain-Gage Transducers

These devices have been used extensively in pressure and weighing load cells for several years. Strain gages usually are mounted directly on the pressure sensor or force summing element. They may be supported directly by sensing diaphragms or bonded to cantilever springs, which act as a restoring force. The operating principle of a resistance-type strain gage is illustrated in Fig. 12; a contemporary unit is shown in Fig. 13. The performance and range characteristics of some commercially available units are summarized in Table 2.

The strain sensitivity is commonly called the gage factor when referring to a specific strain-gage material. Poisson's ratio for most wire is approximately 0.3. The strain sensitivity or gage factor is approximately 1.6 when considering only the dimensional change aspect. This means that a 0.1 percent increase in length within the elastic range should produce a resistance increase of 0.16 percent. When actual tests are performed, a metal or alloy exhibits different values of strain sensitivity for different temperatures

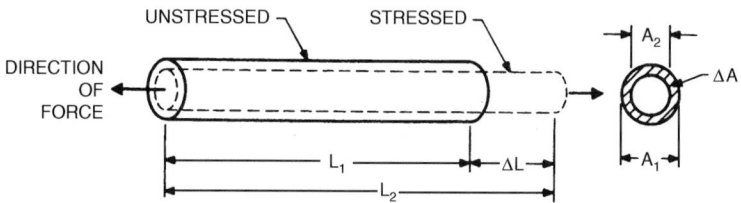

FIGURE 12 Basic relation between resistance change and strain in resistance-type strain gage. When under stress, the wire changes in length from L_1 to L_2 and in area from A_1 to A_2. The resistance is

$$R = p\frac{L}{A}$$

where L = conductor length
        A = cross section area
        p = resistivity constant

and the strain sensitivity (gage factor) is

$$S = \frac{\Delta R/R}{\Delta L/L}$$

where $\frac{\Delta R}{R}$ = resistance change
        $\frac{\Delta L}{L}$ = strain

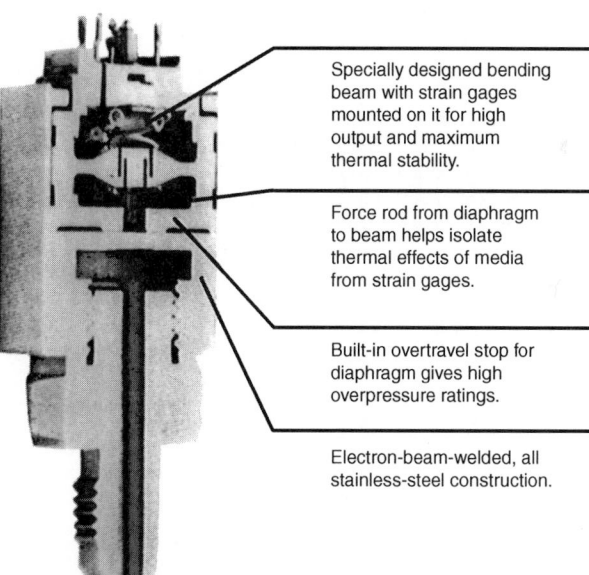

Specially designed bending beam with strain gages mounted on it for high output and maximum thermal stability.

Force rod from diaphragm to beam helps isolate thermal effects of media from strain gages.

Built-in overtravel stop for diaphragm gives high overpressure ratings.

Electron-beam-welded, all stainless-steel construction.

FIGURE 13 Strain-gage pressure transducer incorporating a force transfer rod between diaphragm and double cantilever beam with four foil strain gages on beam. Units are calibrated and temperature-compensated to ensure stability over specified ranges, as well as unit interchangeability. Units are available with required signal conditioning. (*Lucas Schaevitz.*)

TABLE 2 Representative Characteristics of Strain-Gage Pressure Transducers

Pressure Range	Pressure Reference	Performance, %FRO	Accuracy (Static Error Band)	Input	Output	Electrical Connection
Low pressure:						
0–2 in. H$_2$O to 0–100 psi (0–5 mbars to 0–7 bars)	Vented gage; absolute; differential; vacuum	Very low range; high performance	±0.50	10–32 V dc; ±15 V dc	5V; 4–20 mA	Cable
0.1.5 to 0–5 psi (0–100 to 0–350 mbars)	Vented gage; absolute	Low range; high performance	±0.35	10; 11–18; 18–32; 10–36 V dc; ±15 V dc	25 mV; 5 V; 2.5 V; 4–20 mA	Cable or connector.
0–1.5 to 0–5 psi (0–100 to 0–350 mbars)	Wet/dry unidirectional differential	Low range; high performance	±0.30	10; 11–18; 18–32; 10–36 V dc; ±15 V dc	25 mV; 5 V; 2.5 V; 4–20 mA	Cable or connector.
0–1.5 to 0–30 psi (0–100 mbars to 0–2bars)	Wet/wet unidirectional differential	Low range; differential measurement	±0.50	10; 11–18; 18–32; 9–36 V dc; ±15 V dc	25 mV; 5 V; 2.5 V; 4–20 mA	Cable or connector
Medium to high pressure:						
0–10 to 0–1000 psi (0–700 mbars, to 0–70 bars)	Unidirectional wet/dry differential	Medium range; high performance	±0.30	10; 11–18–32; 10–36 V dc; ± 15 V dc	25 mV; 5 V; 2.5 V; 4–20 mA	Cable or connector
0–10 to 0–3500 psi (0–700 mbars to 0–250 bars)	Wet/wet differential	High range; differential measurement	±0.35	10; 11–18; 18–32; 10–36 V dc; ±15 V dc	25 mV; 5 V; 2.5 V; 4–20 mA	Cable or connector
0–10 to 0–3500 psi (0–700 mbars to 0–250 bars)	Vented gage; sealed gage; absolute	Flush diaphragm	±0.30	10; 11–18;18–32; 10–36 V dc; ±15 V dc	25 mV; 5 V; 2.5 V; 4–20 mA	Cable or connector
0–10 to 0–10,000 psi (0–700 mbars to 0–700 bars)	Vented gage; sealed gage; absolute	Premium performance	±0.15	10; 11–18; 18–32; 10–36 V dc; ±15 V dc	25 mV; 5 V; 2.5 V; 4–20 mA	Cable or connector
0–75 to 0–10,000 psi (0–5 mbars to 0–700 bars)	Vented gage; sealed gage; absolute	High performance	±0.30	10; 10–36 V dc	24–32 mV; 5 V; 4–20 mA	Cable
0–15 to 0–5000 psi (0–1 bars to 0–350 bars)	Vented gage; sealed gage; absolute	Economical, general performance	±1.0	5; 9– 20;12–45 V dc	50 mV; 1–6 V; 4–20 mA	Cable or pins
Special configurations:						
13–25 to 13–80 psi (900–1700 to 900–5500 mbars)	Absolute	Configured for tank content	±0.20	10–36 V dc	4–20 mA	Connector
0–75 to 0–500 psi (0–5 to 0–35 bars)	Vented gage; sealed gage; absolute	Meets hygienic requirements for aseptic facilities	±0.50	10; 11–18; 18–32; 10–36 V dc; ±15 V dc	25 mV; 5 V; 4–20 mA	Cable or connector
0–150 to 0–6000 psi (0–10 to 0–400 bars)	Sealed gage; absolute	Meets long-term operations in subsea conditions	±0.30	10–36 V dc	4–20 mA	Cable or connector

Note: In addition to traditional process applications, strain-gage pressure transducers find wide application in other fields.

- *Aerospace:* Hydraulic pressure measurement on aircraft; in-flight pressure control; measurement of altitude (atmospheric pressure and air speed).
- *Energy applications:* Monitoring of nuclear reactor core pressure; measurement of water level behind hydroelectric dams; fossil-fuel power plants.
- *Marine applications:* Control of submersibles, such as measuring ballast pressure or each leg of an oil-rig platform as it is submersed to the ocean floor; content measurement on chemical tankers; monitoring hydraulic oil and fuel pressures of shipborne equipment.
- *Special process applications:* Measurement of wet/dry and wet/wet unidirectional differential pressures; tank contents pressure transmission.

Source: Lucas Schaevitz.

TABLE 3   Major Types of Resistance Strain Gages

Common Name	Basic Material of Strain Gage	Method of Attachment of Strain Gage to Surface	General Application
Unbonded	Wire	Connected at ends	Transducer
Bonded metallic	Wire or foil	Epoxy	Stress analysis and transducer
Flame spray	Wire	Spray-coated	Stress analysis
Welded	Foil	Spot-welded	Stress analysis
Bonded semiconductor	Silicon or germanium	Epoxy	Stress analysis and transducer
Diffused semiconductor	Silicon	Semiconductor diffusion	Transducer
Thin film	Metal alloy	Sputtering or deposition	Transducer

The ideal strain gage would change resistance in accordance with deformations of the surface to which it is bonded and for no other reason. However, gage resistance is affected by other factors, including temperature. Any resistive change in the gage not caused by strain is referred to as apparent strain. Apparent strain may be caused by a change in the gage factor due to temperature (thermal coefficient of the gage factor), by a change in resistance due to temperature (thermal coefficient of the resistance), by the stability of the metal, and even by the properties of the adhesive that bonds the gage to the surface being measured. Many improvements in strain-gage materials have been made in recent years, thus reducing the effects of apparent strain. Some common types of resistance strain gages and their characteristics are listed in Table 3.

Bonded Strain-Gage Systems.   Although increasing the gage factor makes the gage more sensitive to strain, this also increases the undesirable effects of temperature. Thus small size is preferred so that the gage can be placed close to the high-strain area. A high resistance permits larger input voltage excitation and thus a larger millivolt output with a lower power consumption.

Bonded foil strain gages are made using special metal alloy conductors with high resistivities, high gage factors, and low temperature coefficients. Wire strain gages are not used widely because in order to obtain 350 ohms by using no. 28 copper wire [0.000126 in^2 (0.08 mm^2) in cross section], 5360 ft (1633.7 meters) of wire would be needed [350 Ω/(65 Ω/1000 ft)]. The metal alloy of a bonded foil strain gage is formed into a back-and-forth grid to decrease the overall length of the strain gage system (Fig. 14). The length of the grid versus the width is designed to concentrate the strain-sensing grid over the high-strain area. Foil strain gages with gage resistance values of 120, 350, and 1000 ohms are common, with special gages for use with 4- to 20-mA electronic transmitters having resistances as high as 5000 ohms.

The sensing grid is tiny and fragile in comparison with the structure to which it is usually attached. Therefore pads for connecting leadwires must be manufactured as part of the strain gage. The strain gage is bonded to the specimen surface by a thin layer of epoxy adhesive (Fig. 15), and care must be

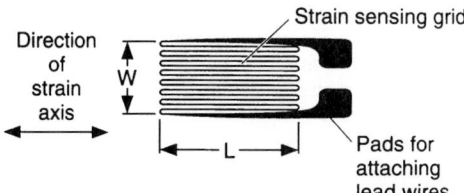

FIGURE 14   Representative single bonded-foil strain gage. Dimensions can be of the order of 0.021 × 0.062 in (0.79 × 1.57 mm). Thickness of a single gage may be the order of 0.0022 in (0.056 mm). With improved manufacturing techniques, the trend is toward smaller dimensions.

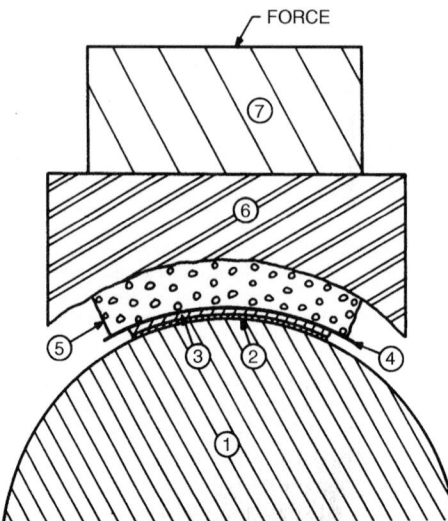

FIGURE 15   Installation of foil strain gage on non-planar surface. (1) Cylindrical specimen surface; (2) thin adhesive layer [0.001 inch (0.025 mm)]; (3) strain gage; (4) polymer sheet to prevent pressure pad from sticking; (5) pressure pad; (6) metal gaging block that conforms to specimen surface; (7) weight or clamp to apply pressure while adhesive is curing.

taken to ensure a thin, uniform, strong bond. A uniform bonding force applied by a contoured gaging block is used to exert a constant, even pressure against the strain gage. In summary, when installed and ready for use, the strain-gage system consists of the specimen surface, an effective bond between gage and specimen, the strain gage, appropriate leads and connectors, and, if needed, a protective waterproof coating.

Metallic Strain-Gage Materials.   All electrical conductors exhibit a strain-gage effect, but only a few meet the necessary requirements to be useful as strain gages. The major properties of concern are (1) gage factor, (2) resistance, (3) temperature coefficient of gage factor, (4) thermal coefficient of resistivity, and (5) stability. High-gage-factor materials tend to be more sensitive to temperature and less stable than the lower-gage-factor materials.

Strain-gage materials that have been commonly used in the past include the following:

Constantan.   Constantan or Advance (copper-nickel alloy) is primarily used for static strain measurement because of its low and controllable temperature coefficient. For static measurements, under ideal compensation conditions, or for dynamic measurements the alloy may be used from −100 to +460°F (−73.3 to +283°C). Conservative limits are 50 to 400°F (10 to 204°C).

Karma.   Karma (nickel-chrome alloy with precipitation-forming additives) provides a wider temperature compensation range than Constantan. Special treatment of this alloy gives minimum drift to 600°F (316°C) and excellent self-temperature compensation characteristics to ∼800°F (427°C).

Nichrome V.   Nichrome V (nickel-chrome alloy) is commonly used for high-temperature static and dynamic strain measurements. Under ideal conditions, this alloy may be used for static measurements to 1200°F (649°C) and for dynamic measurements to 1800°F (982°C).

Isoelastic.   Isoelastic (nickel-iron alloy plus other ingredients) is used for dynamic tests where its larger temperature coefficient is of no consequence. The higher gage factor is a distinct advantage where dynamic strains of small magnitude are measured.

479PT.   479PT (platinum-tungsten alloy) shows an unusually high stability at elevated temperatures. It also has a relatively high gage factor for an alloy. A gage of this material is recommended for dynamic tests to 1500°F (816°C) and static tests to 1200°F (649°C).

TABLE 4 Properties of Strain-Gage Materials

Material	Composition, %	Gage Factor	Thermal Coefficient of Resistivity, $°C^{-1} \times 10^{-5}$
Constantan (Advance)	Ni 45, Cu 55	2.1	±2
Isoelastic	Ni 36, Cr 8, Mn-Si-Mo 4, Fe 52	3.52 to 3.6	+17
Karma	Ni 74, Cr 20, Fe 3, Cu 3	2.1	+2
Manganin	Cu 84, Mn 12, Ni 4	0.3 to 0.47	±2
Alloy 479	Pt 92, W 8	3.6 to 4.4	+24
Nickel	Pure	−12 to −20	670
Nichrome V	Ni 80, Cr 20	2.1 to 2.63	10
Silicon	p-Type	100 to 170	70 to 700
Silicon	n-Type	−100 to −140	70 to 700

Semiconductor (Silicon) Strain-Gage Materials. Semiconductor material has an advantage over metals because its gage factor is approximately 50 to 70 times higher. However, the desirable increase in gage factor is partially offset by its greater thermal coefficient of resistivity (the common term is temperature effect). Comparatively recently, semiconductor strain gages are gaining in importance, particularly in the manufacture of miniature pressure and force transducers. Micromachined silicon assemblies also permit the integration of numerous other functions in a pressure transmitter. Representative properties of strain-gage materials, including semiconductors, are given in Table 4.

Strain-Gage Bonding Agents. The importance of the adhesive that bonds the strain gage to the metal structure under test or as part of a transducer cannot be overemphasized. An ideal adhesive should be suited to its intended environment, transmit all strain from the surface to the gage, have high mechanical strength, high electrical isolation, low thermal insulation, and be very thin. Also, it should not be affected by temperature changes. The adhesive must provide a strong bond while electrically isolating the gage from the surface to which it is attached. Electrical isolation is needed because most of the structures to which gages are bonded would electrically short out the elements if no electrical isolation existed. In a typical strain-gage installation, the electrical isolation between the gage and the specimen surface should be at least 1000 MΩ at room temperature and 50 volts dc. Electrical isolation (leakage) becomes a problem with bonding agents at high temperatures and in high-moisture environments. At high temperatures, even ceramic begin to exhibit a loss of electrical isolation. This is one of the most severe limitations on strain-gage performance at temperatures above 1200°F (649°C).

Because of the wide variation in properties obtainable with different resin and hardener combinations, epoxy resins are an important class of strain-gage adhesives. Alternate attachment methods, such as the flame spray technique, have been used.

Basic Strain-Gage Bridge Circuit. In order to make use of the basic operating principle of the bonded resistance strain gage (that is, change in resistance proportional to strain), the strain-gage input must be connected to an electric circuit capable of measuring small changes in resistance. Since the strain-induced resistance changes are small (typically 0.2 percent for full-scale output in one active gage), the gages are wired into a Wheatstone bridge. A Wheatstone bridge is a circuit designed to accurately measure small changes. It can be used to determine both dynamic and static strain-gage readings. The Wheatstone bridge also has certain compensation properties.

The Wheatstone bridge detects small changes in a variable by comparing its value with that of a similar variable and then measuring the difference in magnitude, instead of measuring the magnitude directly. For instance, if four equal-resistance gages are wired into the bridge (Fig. 16) and a voltage is applied

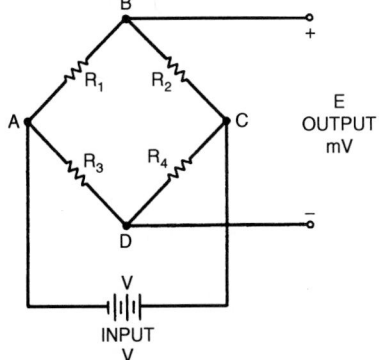

FIGURE 16 Four-arm Wheatstone bridge circuit. Strain gages are inserted at R_1, R_2, R_3, and R_4.

between points A and C (input), then there will be no potential difference between points B and D (output). However, any small change in any one of these resistances will cause the bridge to become unbalanced, and a voltage will exist at the output in proportion to the imbalance.

In the simplest Wheatstone bridge configuration, a strain-sensing grid is wired in as resistance R_1. For this circuit, the output voltage E_o can be derived easily. In reference to the circuit shown in Fig. 16, the voltage drop across R_1 is denoted by V_{ab} and given as

$$V_{ab} = \frac{R_1}{R_1 + R_2} \text{ volts} \tag{1}$$

Similarly, the voltage drop across R_4 is denoted by V_{ad} and given by

$$V_{ad} = \frac{R_4}{R_3 + R_4} \text{ volts} \tag{2}$$

The output voltage from the bridge E is equivalent to V_{bd}, which is given by

$$E = V_{bd} = V_{ab} - V_{ad} \tag{3}$$

Substituting Eqs. (1) and (2) into Eq. (3) and simplifying gives

$$\begin{aligned} E &= \frac{R_1}{R_1 + R_2} - \frac{R_4}{R_3 + R_4} \\ &= \frac{R_1 R_3 - R_2 R_4}{(R_1 + R_2)(R_3 + R_4)} \text{ volts} \end{aligned} \tag{4}$$

The voltage E will go to zero, and thus the bridge will be considered to be balanced when

$$R_1 R_3 - R_2 R_4 = 0$$

or

$$R_1 R_3 = R_2 R_4 \tag{5}$$

Therefore the general equation for bridge balance and zero potential difference between points B and D is

$$\frac{R_1}{R_4} = \frac{R_2}{R_3} \tag{6}$$

Any small change in the resistance of the sensing grid will throw the bridge out of balance and can be detected by a voltmeter.

When the bridge is set up so that the only source of imbalance is a resistance change in the gage (resulting from strain), then the output voltage becomes a measure of the strain. From Eq. (4), with a small change in R_1,

$$E = \frac{(R_1 + \Delta R_1)R_3 - R_2 R_4}{[(R_1 + \Delta R_1) + R_2](R_3 + R_4)} \text{ volts}$$

Most Wheatstone bridge circuits are produced with all four arms serving as active strain gages.

Types of Resistance Strain Gages. Many different types of resistance strain gages have been developed since the first bonded strain gage was introduced in 1936. Bonded gages have been used widely, but there have been numerous changes in technology, as described later, in the assembly of strain-gage sensors, notably as they are used in process-type pressure sensors and transducers. These alterations include bonded semiconductor strain gages and diffused semiconductor strain gages.

FIGURE 17 Strain-gage patterns that have been used successfully. (*a*) For measurement of strain in a diaphragm, elements 1 and 4 are subjected to compressive radial strains, while elements 2 and 4 are subjected to tensile tangential strains. (*b*) Rosette gage that measures srain in three directions simultaneously.

Bonded Foil Strain Gages.   Serious commercial attention to the strain gage commenced in the mid-1950s, at which time foil strain gages were produced by a printed-circuit process or by being stamped from selected alloys that had been rolled into a thin foil. Foil thicknesses ranged from 0.0001 to 0.002 inch (0.00254 to 0.00508 mm). The foil usually was heat-treated before use in order to optimize mechanical properties and the thermal coefficient of resistivity. For a given cross-sectional area, a foil conductor displays a large surface area. The large ratio of surface area to cross section provides superior mechanical stability under prolonged strain and high-temperature conditions. The large surface area also provides a good heat-transfer surface between grid and specimen and, therefore, high input-voltage levels are possible without developing severe temperature gradients across the insulating matrix.

Photoetching permits the manufacture of sensing grids in virtually any two-dimensional pattern. Good practice is to develop a geometric pattern that provides maximum electrical and mechanical efficiency from the sensing element. Some common configurations that have been used are shown in Fig. 17.

Bonded Semiconductor Strain Gages.   The principal difference between foil and semiconductor gages is the greater response of semiconductor gages to both strain and temperature. The large resistance-versus-strain characteristic of a properly oriented semiconductor crystal is due primarily to the piezoresistive effect. Gage behavior may be accurately described by

$$\frac{\Delta R}{R_0} = \left(\frac{T_0}{T}\right) E(GF) + \left(\frac{T_0}{T}\right)^2 E^2 C_2$$

where R_0 = unstressed gage resistance at T (changes as T changes)
ΔR = change in gage resistance from R_0
T = temperature, K
T_0 = 298 K (24.9°C)
E = strain
GF, C_2 = constants of particular gage in question

The resistance change due to strain is a parabola for high-resistivity *p*-type silicon. Pure material of this resistivity is not used to produce gages because of this severe nonlinearity. As can be seen in the equation, the linearity can be improved by reducing the nonlinearity constant C_2. Figure 18 shows the behavior of a typical *p*-type semiconductor strain gage for a material that has been doped

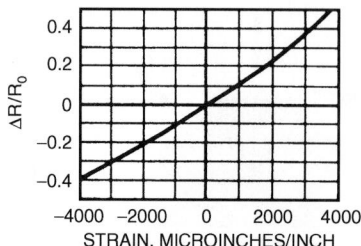

FIGURE 18 Gage sensitivity versus strain level for *p*-type semiconductor gage.

so that C_2 is low and the slope is more linear. The equation also shows that large tensile strains on the gage filament and higher temperatures increase gage linearity. As temperature T rises, the value of both terms on the right-hand side of the equation decrease, as does gage sensitivity. The nonlinearity coefficient, however, decreases faster than the gage factor coefficient, thus improving the linearity.

The *n*-type semiconductor strain gages are similar in behavior to *p*-type gages, except that the gage factor is negative.

The high output obtainable from semiconductor piezoresistive elements makes them particularly attractive for transducers that are $1/8$ inch (3.2 mm) and smaller. Miniature transducers are formed by attaching individual silicon strain gages to a force-collecting diaphragm or beam. When the diaphragm or beam deflects, the surface strains are sensed by the semiconducting elements. Output levels typically $\sim$100 mV full-scale are available from full bridge transducers with $\sim$10-volt input.

A typical semiconductor gage, unlike a bonded foil gage, is not provided with a backing or carrier. Therefore, bonding the gage to a surface requires extreme care in order to obtain a thin epoxy bond. The same epoxies used for foil gages are used for semiconductor gages.

Diffused Semiconductor Strain Gages.   A major advance in transducer technology was achieved with the introduction of diffused semiconductor strain gages. The gages are diffused directly into the surface of a diaphragm, utilizing photolithographic masking techniques and solid-state diffusion of an impurity element, such as boron. Since the bonding does not use an adhesive, no creep or hysteresis occurs.

The diffusion process does not lend itself to the production of individual strain gages and also requires that the strained member (diaphragm or beam) be made from silicon. Therefore, diffused semiconductors are used for manufacturing transducers (primarily pressure) instead of for stress analysis. Typically, a slice of silicon 2 to 3 inches (5 to 7.5 cm) in diameter is selected as the main substrate. From this substrate, hundreds of transducer diaphragms 0.1 to 0.5 inch (2.5 to 12.7 mm) in diameter with full four-arm Wheatstone bridges can be produced. A silicon pressure transducer diaphragm with a diffused semiconductor strain gage is shown in Fig. 19. The entire Wheatstone bridge circuitry is diffused into the diaphragm (strain gages and connection solder areas for leadwires).

A diffused silicon sensor is unsuitable for high-temperature measurements because the gage-to-gage resistance decreases sharply as a function of temperature. Two mechanisms combine to produce this effect. First, isolation from the substrate is accomplished by a *pn* junction, and its effectiveness is sensitive to increased heat. Second, the diaphragm (essentially an insulator) becomes increasingly conductive as the temperature is raised.

Refinements in the semiconductor diffusion process have allowed manufacturers to produce the entire strain-gage transducer diaphragm, strain gage, temperature compensation elements (that is, thermistors), and amplifier circuits with semiconductor technology. The introduction of very-high-volume, extremely low-cost transducers is now practical.

Thin-Film Strain Gages.   Another relatively recent development of strain-gage technology is the thin-film process. This technique controls the major strain-gage properties independently (that is, strain-gage and electrical isolation). It uses the inherent advantages of metal strain gages (low-temperature effects and high gage factors) and the process advantages available with the diffused semiconductor technique (no adhesive bonding). The thin-film strain gage is potentially capable of producing an ideal strain-gage system.

A thin-film strain gage is produced by depositing a thin layer of metal alloy on a metal specimen by means of vacuum deposition or sputtering. This technique produces a strain gage that is molecularly bonded to the specimen, so the disadvantages of the epoxy adhesive bond are eliminated. Like the diffused semiconductor process, the thin-film technique is used almost exclusively for transducer applications.

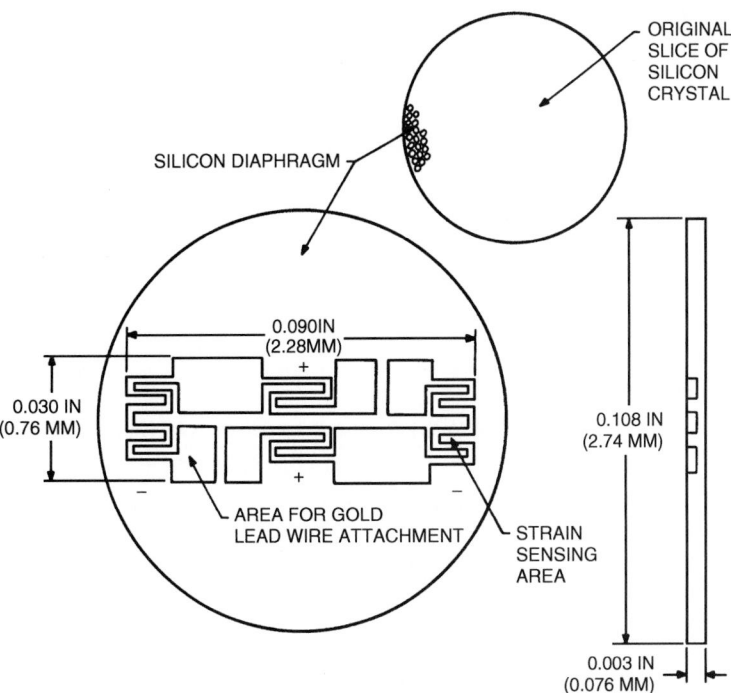

FIGURE 19   Representative silicon pressure transducer diaphragm with diffused Wheatstone bridge circuit. Many diaphragms are produced from a single slice of silicon crystal. The strain-gage elements are situated to measure compressive radial strains and tensile tangential strains.

To produce thin-film strain-gage transducers, first an electrical insulation (such as a ceramic) is deposited on the stressed metal member (diaphragm or beam). Next the strain-gage alloy is deposited on top of the isolation layer. Both layers may be deposited either by vacuum deposition or by sputtering.

In vacuum depositon, the material to be deposited is heated in a vacuum and vapor is emitted. The vapor deposits on the transducer diaphragm in a pattern determined by substrate masks.

The sputtering technique also employs a vacuum chamber. With this method, the gage or insulating material is held at a negative potential and the target (transducer diaphragm or beam) is held at a positive potential. Molecules of the gage or insulating material are ejected from the negative electrode by the impact of positive gas ions (argon) bombarding the surface. The ejected molecules are accelerated toward the transducer diaphragm or beam and strike the target area with kinetic energy several orders of magnitude greater than that possible with any other deposition method. This produces superior adherence to the specimen.

In order to obtain maximum bridge sensitivity (millivolt output) to minimize heating effects, and to obtain stability, the four strain gages, the wiring between the gages, and the balance and temperature compensation components are all integrally formed during the deposition process. This ensures the same composition and thickness throughout.

The thin-film strain-gage transducer has many advantages over other types of strain-gage transducers. The principal advantage is long-term stability. The thin-film strain-gage circuit is molecularly bonded to the specimen, and no organic adhesives are used which could cause drift with temperature or stress creep. The thin-film technique also allows control of the strain-gage resistance value. A resistance as high as 5000 ohms can be produced in order to allow increased input and output voltages with low power consumption.

Strain-Gage Bridge Correction Circuits.   When static strains or the static component of a varying strain are to be measured, the most convenient circuit is the Wheatstone bridge, previously shown in Fig. 16. The bridge is balanced ($E = 0$) when

$$\frac{R_1}{R_4} = \frac{R_2}{R_3}$$

Now consider a bridge in which all four arms are separate strain gages. Assume that the bridge is initially balanced, so that $R_1 R_3 = R_2 R_4$ and $E = 0$. A strain in the gages will cause a change in each value of resistance R_1, R_2, R_3, and R_4 by incremental amounts ΔR_1, ΔR_2, ΔR_3, and ΔR_4, respectively. The voltage output ΔE of the bridge can be obtained from

$$E = \frac{R_1 R_3 - R_2 R_4}{(R_1 + R_2)(R_3 + R_4)} \text{ volts}$$

which becomes

$$\Delta E = \frac{(R_1 + \Delta R_1)(R_3 + \Delta R_3) - (R_2 + \Delta R_2)(R_4 + \Delta R_4)}{[(R_1 + \Delta R_1) + (R_2 + \Delta R_2)][(R_3 + \Delta R_3) + (R_4 + \Delta R_4)]} \text{ volts}$$

After considerable simplification, this becomes

$$E = \frac{R_2/R_1}{(1 + R_2/R_1)^2}\left(\frac{\Delta R_1}{R_1} - \frac{\Delta R_2}{R_2} + \frac{\Delta R_3}{R_3} - \frac{\Delta R_4}{R_4}\right) \text{ volts}$$

This latter equation shows that if all four gages experience the same strain, the resistance changes will cancel out and the voltage change ΔE will equal zero. On the other hand, if gages R_1 and R_3 are in tension (ΔR_1 and ΔR_3 positive) and gages R_2 and R_4 are in compression (ΔR_2 and ΔR_4 negative), then the output will be proportional to the sum of all the strains measured separately. All four-arm Wheatstone bridge transducers are wired to give two gages in tension and two gages in compression. An example of a four-gage setup for the diaphragm of a pressure transducer is shown in Fig. 20. This design takes full advantage of the tensile tangential strains developed at the center of the diaphragm and the compressive radial strains present at the edge.

Another advantage of using a four-gage bridge, besides the increased output, is the effect on the temperature sensitivity. If the gages are located close together, as on a pressure transducer diaphragm, they will be subjected to the same temperature. Therefore the resistance change due to temperature will

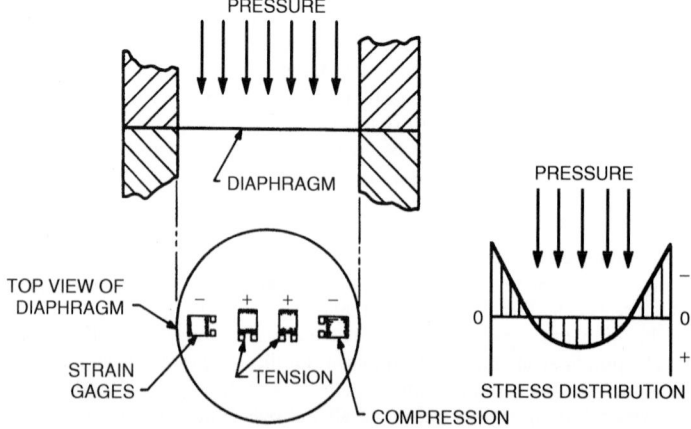

FIGURE 20   Representative strian-gage positions on a pressure diaphragm. Orientations take advantage of the stress distriubtion. The gages are wired into a Wheatstone bridge with two gages in tension and two in compression.

be the same for each arm of the Wheatstone bridge. If the gage resistance changes due to temperature are identical, the temperature effects will all cancel out and the output voltage of the circuit will not increase or decrease due to temperature.

The output voltage of the Wheatstone bridge is expressed in millivolts output per volt input. For example, a transducer rated at 3.0 mV/V at 500 psi (~73 kPa) will have an output signal of 30.00 mV for a 10-volt input at 500 psi (~73 kPa) or 36.00 mV for a 12-volt input. Any variation in the power supply will directly change the output of the bridge. Generally, power supply regulation should be 0.05 percent or better.

In production the four strain gages in a Wheatstone bridge never come out to be exactly equal for all conditions of strain and temperature (even in the diffused semiconductor process). Therefore various techniques have been developed to correct the differences in the individual strain gages and to make the strain-gage bridge easier to use with electronic instrumentation. Four main values normally need adjusting (Fig. 21): (1) electrical bridge imbalance, (2) balance shift with temperature, (3) span or sensitivity shift or bridge output with temperature, and (4) standardization of the bridge, output to a given millivolts-per-volt value. other transducer characteristics such as accuracy, linearity, hysteresis, acceleration effect, and drift are part of the transducer element design (beam or diaphragm) and cannot be corrected after the strain-gage bridge has been produced.

Figure 21 shows the circuit diagram of a Wheatstone bridge circuit with adjusting resistors. One corner of the bridge (points D and E) remains "open," so that the bridge can be adjusted electrically. This means that five leads come from the four gages. The zero-balance adjustment compensates for the electrical imbalance in the bridge caused by unequal resistances of the strain gages. Depending on which leg is unbalanced, R_{za} is placed between points E and F or between points D and F. The zero balance changes with temperature, and R_{zc} is inserted inside the bridge to correct for this change. A small piece of nickel wire is selected to provide a resistance change opposite the resistance change of

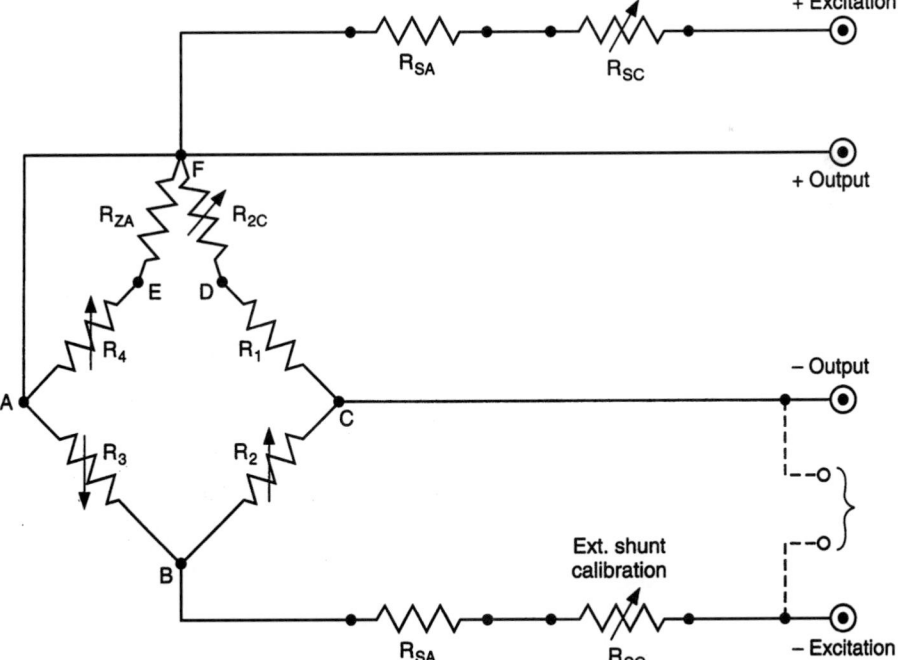

FIGURE 21 Strain-gage transducer circuit with four active strain-gage elements (R_1, R_2, R_3, and R_4). Balance, sensitivity, and thermal compensation resistors are also shown.

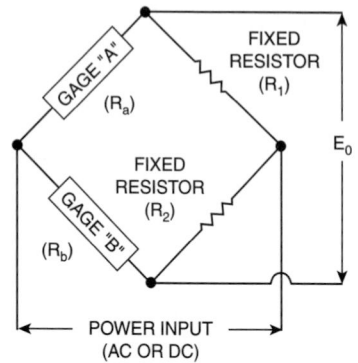

FIGURE 22   Wheatstone bridge circuit utilizing two strain gages.

the bridge. R_{sc} is also a temperature thermistor or sensor which changes resistance with temperature to adjust the excitation to the bridge. The values for R_{zc} and R_{sc} have to be selected by running each bridge over its desired temperature range [usually −65 to +300°F (−54 to +149°C)]. R_{sa} is a non-temperature-sensitive resistor, and it is used to adjust the output to a precise millivolts-per-volt value once all the balance and temperature-sensitive resistors have been inserted within the bridge. The user of transducers is not affected because all this circuitry is contained within the transducer and does not interfere with connections to amplifiers, power supplies, or computers.

A Wheatstone bridge can also be used in applications that require only one or two active strain gages. To compensate for temperature in two-gage applications, the gages must be located in adjacent arms of the bridge, as shown in Fig. 22. In placing gages, one must only recognize that the bridge is unbalanced in proportion to the difference in the strains of the gages located in adjacent arms and in proportion to the sum of the strains of gages located in opposite arms.

Electronics for Strain-Gage Transducers and Transmitters.   The full-scale output of a typical bonded-foil, four-active-element strain-gage bridge with all compensating and adjusting resistors connected is ~20 to 30 mV at 10-volt excitation. An amplifier must be used to obtain the 0- to 5-volt or 4- to 20-mA outputs used in control instrumentation. As a result of the advances in integrated circuitry, many transducers now have amplifiers that are internally installed within the transducer body (Fig. 23).

High-gain, low-noise, instrumentation-quality differential operational amplifiers such as the OP-07 make amplification of the strain-gage bridge output for standard 0- to 5-volt transducers and 4-

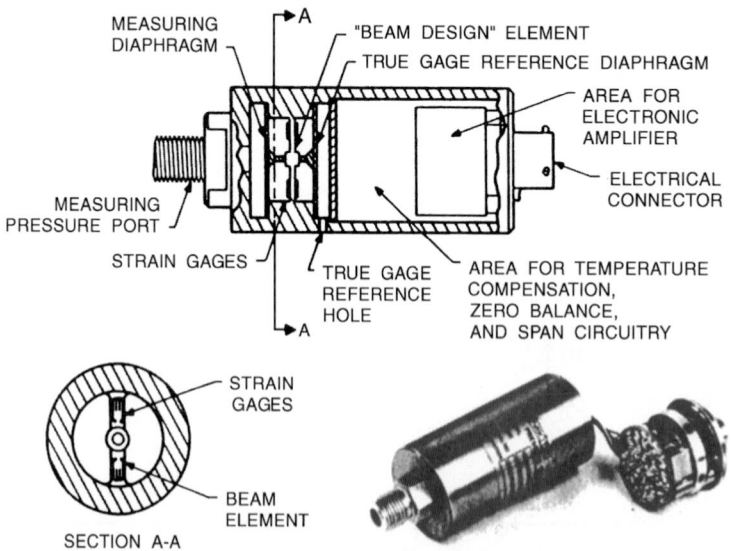

FIGURE 23   Pressure transducer (diaphragm-beam design) that compares measured pressure to atmospheric (reference) pressure. Welded stainless-steel diaphragms permit use in corrosive environments because strain gages are in environmentally protected chamber. (*Sensotec.*)

to 20-mA transmitters reliable and easy. These integrated-circuit amplifiers have a high common-mode rejection ratio and are thus well suited for use with Wheatstone bridge circuits. They are also inherently well compensated in order to deliver a constant output irrespective of temperature changes. The operational amplifiers used in instruments have controllable gains and zero-balance adjustments. Since the offset of the instrumentation channel's output is equal to the sum of the offsets in the bridge and in the amplifier, the combined offset can be adjusted at the amplifier so that the channel delivers 0 volts at zero stimulus (pressure, load, torque, and so on) for 0- to 5-volt output transducers or 4-mA at zero stimulus for 4- to 20-mA output transmitters.

The shear-web-element load cell is of a somewhat different configuration. This design is used in making high-capacity [50,000-lb (22,650-kg)] load cells for industrial weighing, crane scales, and so on. A shear web connects an outer, stationary hub to an inner, loaded hub. Strain gages detect the shear strain produced in the web. The large diameter of a very high-capacity shear element requires that the strain be measured at more than one location. Note the three webs in Fig. 24. This is common practice in all types of large transducer elements in order to obtain an average of the total strains on the element and to eliminate errors caused by minor off-center balancing. The strain gages are wired so that the user sees only one 350-ohm bridge.

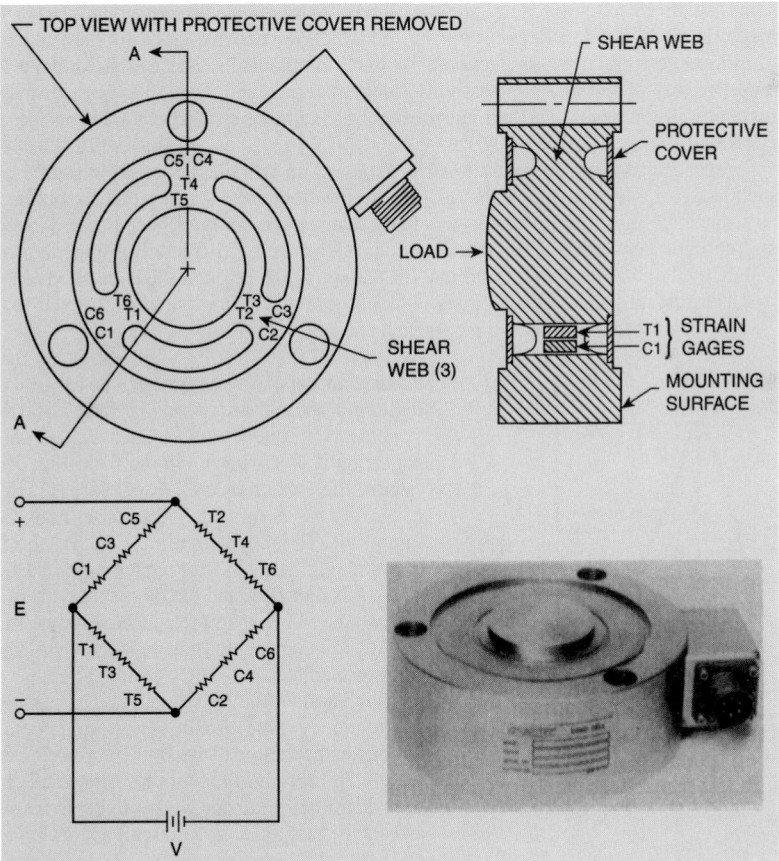

FIGURE 24   Pancake (shear-web) load cell. Each arm of the Wheatstone bridge circuit contains one strain gage from each of the three shear webs. The microstrains from the three webs are added together in one bridge circuit to determine the load. (*Sensotec.*)

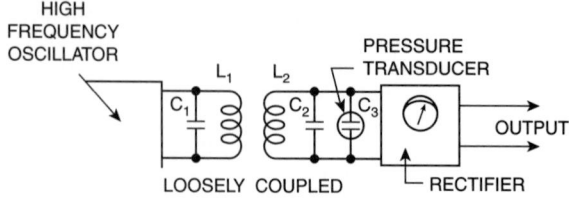

FIGURE 25 Schematic of tuned resonant circuit used in some capacitive pressure transducers.

Capacitive Pressure Transducers

In a traditional capacitance-type (capacitive) transducer, a measuring diaphragm (elastic element) moves relative to one or two fixed plates. Changes in capacitance are detected by an oscillator or bridge circuit. Generally, capacitive transducers are of low mass and high resolution, and they have good frequency response. Limitations have included a requirement for sophisticated signal conditioning, some sensitivity to temperature, and the effects of stray noise on sensor leads. As will be pointed out, much research and development during the past few years has gone forward to improve capacitive transducer performance—and generally with excellent results. These improvements largely have been made by way of testing and substituting new materials and through taking advantage of increasing ingenuity in the design and miniaturization of electronic circuitry, notably in the increasing use of microprocessors.

Transducer design concentration has centered on two classes of error sources. (1) Deficiencies such as in long-term stability, that is, essentially those error sources that cannot be corrected by the built-in electronics. Improved answers to these error sources have been derived essentially by testing and utilizing new materials as, for example, micromachined silicon, ceramics, quartz, and sapphire which, by their nature, exhibit minimal hysteresis. (2) Deficiencies that are amenable to improvement by electronic measures, including signal conditioning, calibration, and error self-diagnosis.

In a typical capacitive pressure transducer, as pressure is applied and changes, the distance between two parallel plates varies—hence altering the electric capacitance. This capacitive change can be amplified and used to operate into phase-, amplitude-, or frequency-modulated carrier systems. A frequency-modulated system using a tuned resonant circuit is shown in simple form Fig. 25. In this electric circuit, the capacitance C_3 is part of a tuned resonant circuit $L_2C_2C_3$. L_1C_1 forms part of a stable high-frequency oscillator circuit. The tuned circuit $L_2C_2C_3$ is loosely coupled to the circuit L_1C_1. The high-frequency potential induced in circuit $L_2C_2C_3$ is rectified, and the dc output current of the rectifier is indicated on a microammeter. The response of the tuned circuit $L_2C_2C_3$ to a constant frequency is shown in Fig. 26 as a function of the capacitance $C_2 + C_3$ of this circuit. Peak output occurs at point A when the circuit is tuned to resonate at the oscillator frequency. This circuit is tuned to its operating point B by increasing capacitor C_2 until the rectifier meter reads approximately 70 percent of maximum. Any small change in pressure transducer capacitance C_3, due to pressure on the diaphragm, affects the response of the circuit according to Fig. 26.

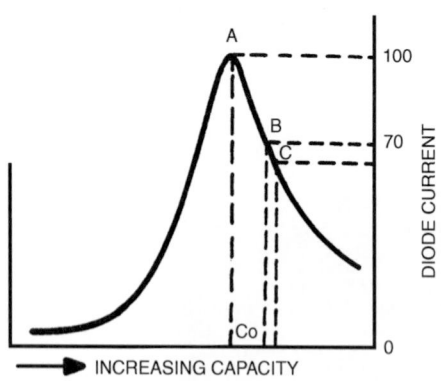

FIGURE 26 Response of resonant circuit to constant frequency shown as a function of circuit capacity.

In order to eliminate the effect of cable capacity between the transducer C_3 and the tuned circuit L_2C_2, a circuit of the type shown in Fig. 27 can be used. In this circuit, a coil L_3 is built as an integral part of the capacitive-transducer assembly. The coil L_3 is connected in parallel with the transducer capacity C_3 to form a tuned circuit with a resonant frequency (for example, 600 kHz). The tuned circuit L_3C_3

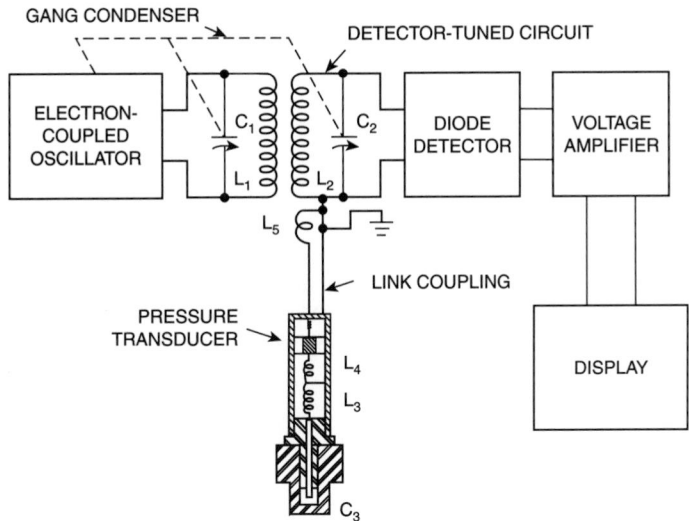

FIGURE 27 One type of circuit used to eliminate effect of cable capacity between transducer C_1 and tuned circuit L_2C_2.

is close-coupled to the tuned circuit L_2C_2 by means of the link coils L_4 and L_5, which are connected by means of a low-impedance (70-ohm) untuned cable. The change in cable capacity, such as that produced by vibration, is negligible when reflected into the high-impedance tuned circuit. In this way, long cables can be used between the transducer and the electronic unit. The tuning characteristics of a link-coupled circuit are shown in Fig. 28. The operating range is the linear section noted midway between the maximum and minimum readings obtained with changing capacity.

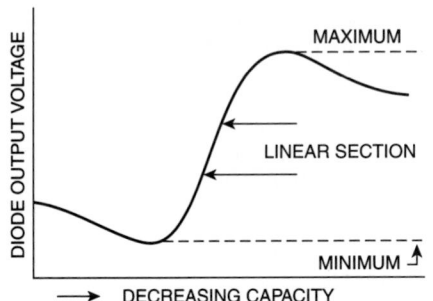

FIGURE 28 Tuning characteristic of a link-coupled circuit.

A phase-modulated carrier system can be used in combination with transducers that incorporate a radio-frequency matching transformer that is tuned with the fixed condenser plate and stray capacitances in the pickup to approximately the oscillator frequency and is properly matched to the transducer connecting coaxial cable. When pressure is applied to the diaphragm, the increase in capacity lowers the resonant frequency of the circuit. The resulting change in reactance is coupled back to the indicator by a suitable transmission line, producing a phase change in the discriminator. This, in turn, produces an output voltage that is a function of the pressure on the diaphragm. The voltage can be indicated or recorded by the usual methods.

The measuring circuit of a capacitive differential-pressure transducer transmitter is shown schematically in Fig. 29. In this system two process diaphragms (high side and low side) are mechanically attached to a connecting rod. In the middle of the connecting rod, the movable electrode is attached and held in position by a spring diaphragm. The differential pressure is balanced by the restoring force of the spring diaphragm. Hence the spring diaphragm represents the measuring element. When applying a differential pressure on the system, the movable electrode is shifted and the distances d_1 and d_2 to the fixed electrodes are changed simultaneously. As a result of the change in distance between the fixed and movable electrodes, the capacitances of the differential capacitor are also changed. This change is amplified electronically and transduced to a 4- to 20-mA dc output signal directly proportional to the differential pressure.

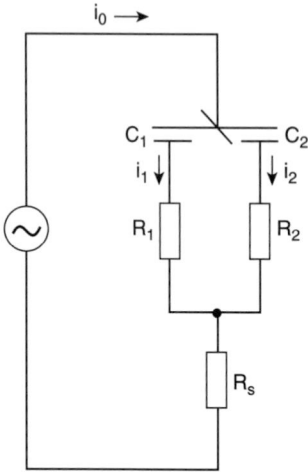

FIGURE 29   Schematic of a type of measuring circuit used in capacitive differential-pressure transducer transmitter.

Assume the gaps between the movable electrode and two fixed electrodes are both equal to d_0. When differential pressure $P_1 - P_2$ is applied, the connecting rod moves a distance of Δd. Then

$$d_1 = d_0 + \Delta d$$
$$d_2 = d_0 - \Delta d$$
$$\Delta d = K_1(P_1 - P_2)$$

where d_1 and d_2 represent the interelectrode gaps on the high and low sides, respectively; K_1 is a proportional constant. The capacitances C_1 and C_2 of the gaps are, respectively,

$$C_1 = \frac{K_2}{d_1} = \frac{K_2}{d_0 + \Delta d}$$
$$C_2 = \frac{K_2}{d_2} = \frac{K_2}{d_0 - \Delta d}$$

where K_2 is a proportional constant and depends on the electrode area and the dielectric constant of the material filling the gaps.

In the conventional capacitive pressure sensor, the compliance of the diaphragm is selected so that the device will produce about a 25 percent change in capacitance for a full-scale pressure change. This large change gives the device an advantage for the measurement of comparatively low pressures. The device also permits the designer to include backstops on either side of the diaphragm for overpressure protection. The relatively high sensitivity of the device also allows it to generate digital or analog outputs.

All-Silicon Capacitive Sensor.   In the recent past, performance of capacitive pressure sensors that utilize an all-silicon sandwich design provides better thermal stability because material mismatching (a major source of thermal effects) is eliminated. This provides an advantage over piezoresistive sensors because the latter normally use a sandwich constructed of different materials and hence different thermal coefficients of expansion.

As of early 1992 all-silicon sensors have been designed that displace only by 0.5 μm at a full-scale pressure of a 10-inch (25.4-cm) water column. The silicon diaphragm is sandwiched between two other pieces of silicon, the top piece serving as a mechanical overpressure stop, the bottom piece containing CMOS circuitry and the backplate of the capacitor, in addition to serving as an overpressure stop for pressure change in the opposite direction. The CMOS die contains both analog and digital circuitry, 13-bit digital output, an 8-bit parallel sensor interface, and an ASIC capability for local control or decision-point circuitry as well as for special communication protocols.

Piezoresistive Pressure Transducers

Stemming from research[1] in the 1950s on the piezoresistive properties of silicon-diffused layers and the development of a piezoresistive device for a solid-state accelerometer, the first piezoresistive pressure transducers were developed as pressure inputs for a commercial airliner in the 1960s. Although piezoresistive transducers have been available for other applications over an input pressure range of 1 to 680 atm, the principal application developed in the early 1970s was in the automotive field. Since

---

[1] Honeywell Inc.

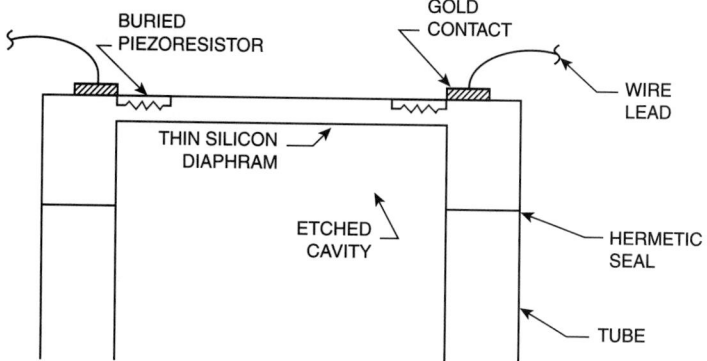

FIGURE 30   Cross section of piezoresistive sensing element with wire leads bonded to metal contacts.

that time, uses for piezoresistive pressure transducers in process control and industrial applications have increased.

The sensing element consists of four nearly identical piezoresistors buried in the surface of a thin, circular silicon diaphragm. Gold pads attached to the silicon diaphragm surface provide connections to the piezoresistors and serve as pads for probe-type resistance measurements or for bonding of wire leads. The thin diaphragm is formed by chemically etching a circular cavity into the surface opposite the piezoresistors. The unetched portion of the silicon slice provides a rigid boundary constraint for the diaphragm and a surface for mounting to some other member. A cross-sectional view of the sensing element with wire leads bonded to the metal contacts is shown in Fig. 30.

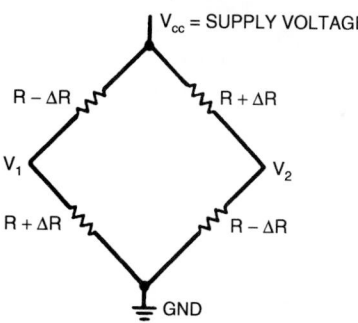

FIGURE 31   Full-bridge arrangement of piezoresistive transducer. $R + \Delta R$ and $R - \Delta R$ represent actual resistor values at applied pressure. R represents resistor value for undeflected diaphragm $(P = 0)$ where all four resistors are nearly equal in value. ΔR represents change in resistance due to applied pressure. All four resistors will change by approximately the same value. Note that two resistors increase and two decrease, depending on their orientation with respect to the crystalline direction of the silicon material.

Pressure causes the thin diaphragm to bend, inducing a stress or strain in the diaphragm and also in the buried resistor. The resistor values will change, depending on the amount of strain they undergo, which depends on the amount of pressure applied to the diaphragm. Hence a change in pressure (mechanical input) is converted to a change in resistance (electrical output). The sensing element converts energy from one form to another. The resistor can be connected to either a half-bridge or a full Wheatstone bridge arrangement. For pressure applied to the diaphragm using a full bridge, the resistors can theoretically be approximated as shown in Fig. 31 (nonamplified units). The signal voltage generated by the full-bridge arrangement is proportional to the amount of supply voltage V_{cc} and the amount of pressure applied, which generates the resistance change ΔR.

A half-bridge configuration used in a signal-conditioned version of the piezoresistive pressure transducer is shown in Fig. 32.

Among the pressure ranges of the transducers most frequently used are 0 to 1, 0 to 15, 3 to 15, 0 to 30, 0 to 100, and 0 to 250 psi (1 psi ≈ 6.9 kPa). Repeatability and hysteresis effects are typically less than 0.1 percent of full-scale, and combined linearity and hysteresis do not exceed ±1 percent of full-scale output. The operating temperature range for standard units is from −40 to 125°C (−40 to 252°F).

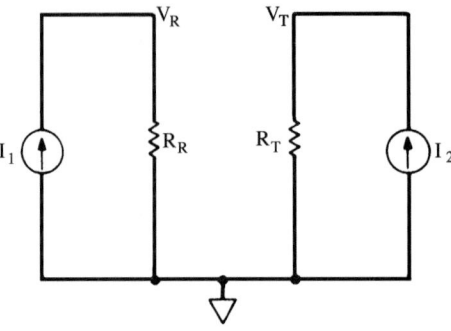

FIGURE 32 Half-bridge configuration used in signal-conditioned version of piezoresistive pressure transducer. Voltage across piezoresistors: $V_R = L_1 R_R = I_1(R_{RO} + kP)$ for radial resistor; and $V_T = I_2 R_T = I_2(R_{TO} = kP)$ across tangential resistor. I_1 and I_2 are adjusted at zero pressure to obtain $V_R = V_T$ or $I_1 R_{RO} = I_2 R_{TO}$. At other temperatures, when R_{RO} and R_{TO} vary, the equality will hold provided that the temperature coefficients of R_{RO} and R_{TO} are equal. I_1 and I_2 increase with temperature to compensate for the chip's negative temperature coefficient of span. The temperature coefficient of null, which may be of either polarity, is compensated for by summing a temperature-dependent voltage $V_N(T)$ with the piezoresistor voltage so that the output $V_O = V_R - V_T$ $\pm V_N(T)$, with the polarity of $V_N(T)$ selected to provide compensation.

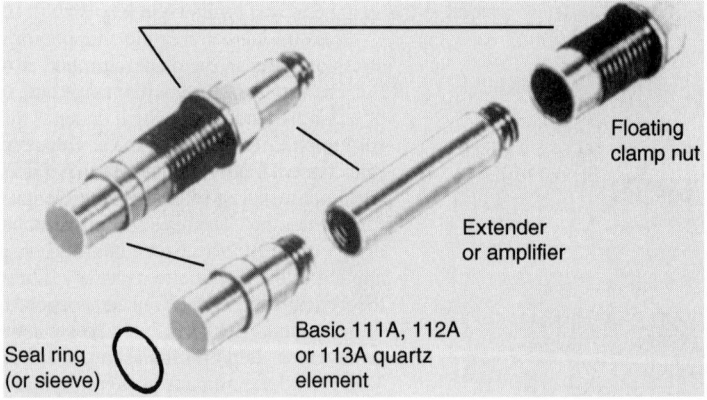

FIGURE 33 Modular assembly of piezoelectric dynamic pressure transducer. (*PCB Piezotronics, Inc.*)

Piezoelectric Pressure Transducers[2]

When certain asymmetrical crystals are elastically deformed along specific axes, an electric potential produced within the crystal causes a flow of electric charge in external circuits. Called the piezoelectric effect, this principle is widely used in transducers for measuring dynamic pressure, force, and shock or vibratory motion. In a piezoelectric pressure transducer, as shown in Fig. 33, the crystal elements

---

[2] Based on R. W. Lally, PCB Piezotronics, Inc., Depew, New York.

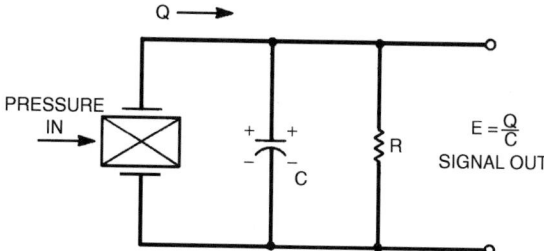

FIGURE 34   Piezoelectric crystal circuit. (*PCB Piezotronics, Inc.*)

form an elastic structure which functions to transfer displacement caused by force into an electric signal proportional to the pressure applied. Pressure acting on a flush diaphragm generates the force.

Piezoelectric pressure transducers historically have used two different types of crystals: (1) natural single crystals, such as quartz and tourmaline, and (2) synthetic polycrystalline ceramic materials, such as barium titanate and lead zirconate. With the relatively recent development of artificially cultured quartz crystals, the foregoing distinction is no longer clear-cut.

Cultured quartz has the advantage of being readily available and reasonably priced. Near-perfect elasticity and stability, combined with an insensitivity to temperature, make quartz an ideal transduction element. Ultrahigh insulation resistance and low leakage allow static calibration, accounting for the popularity of quartz in pressure transducers.

Natural tourmaline, because of its rigid, anisotropic nature, offers submicrosecond response in pressure-bar-type blast transducers. Artificial ceramic piezoelectric crystals and electret (permanently polarized dielectric material, the analog of a magnet) materials are readily formed into compliant transducer structures for generating and measuring sound pressure.

The charge signal from a piezoelectric pressure transducer is usually converted into a voltage-type signal by means of a capacitor, according to the law of electrostatics: $E = Q/C$, where E is the voltage signal, Q is the charge, and C is the capacitance. This circuit is shown in Fig. 34.

In response to a step-function input, the charge signal stored in the capacitor will exponentially leak off through the always finite insulation resistance of the circuit components, precluding static measurements. The initial leakage rate is set by the circuit discharge time constant $R \times C$, where R is the leakage resistance value, which can be as high as 10^8 MΩ in quartz crystals.

Because of the automatic rezeroing action of the discharge circuit, piezoelectric sensors measure relative pressure, sometimes denoted as psir. They measure pressure relative to the initial level for transient events and relative to the average level for repetitive phenomena. Sometimes the slow action of these circuits is mistaken for zero drift by impatient operators.

To prevent the charge signal from quickly leaking off through the recorder or oscilloscope input resistance, a special isolation amplifier is required between the crystal and the recorder. If the charge-converting capacitance is located at the input of this isolation amplifier, the amplifier is called a voltage amplifier. If the capacitor is in the feedback path, it is called a charge amplifier. Amplifiers are further classified as electrostatic (dc-coupled) or vibration (ac-coupled). The ac-coupling circuitry behaves similarly to the sensor discharge circuit.

The high-frequency response of piezoelectric sensor systems depends on the resonant behavior of the sensor's mechanical structure, or on electronic low-pass filters in the sensor, amplifier, or recorder.

The advent of microelectronics and charge-operated field-effect transistors (JFET and MOSFET) is continuing to change the design of piezoelectric sensors profoundly. The current practice is to package the isolation amplifier and signal-conditioning circuitry inside the sensor. These integrated circuit piezoelectric (ICP) sensors with built-in microelectronics, which operate over a simple two-wire cable, are called "smart" sensors.

To eliminate spurious signals caused by environmental effects, such as temperature and motion, the mechanical structures of some piezoelectric pressure sensors are quite sophisticated. A typical acceleration-compensated pressure sensor containing an integrated accelerometer to cancel out motion

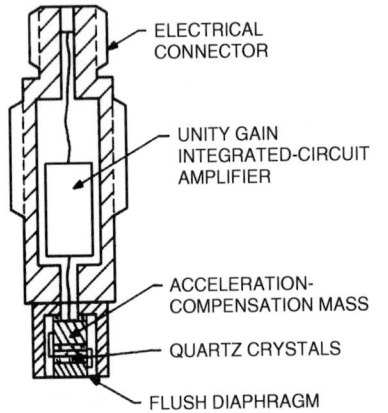

FIGURE 35 Acceleration-compensated quartz pressure sensor with built-in microelectronic unity-gain isolation amplifier. (*PCB Piezotronics, Inc.*)

signals is shown in Fig. 35. Durable conformal coatings of the sensor case and diaphragm provide electrical and thermal insulation. Hermetic seals are electron-beam welded.

Piezoelectric pressure sensors offer several advantages for measuring dynamic pressures. They are generally small in size, lightweight, and very rugged. One transducer may cover a measuring range of greater than 10,000:1 and a frequency range from less than 1 hertz to hundreds of kilohertz with little or no phase shift (time delay). As mentioned previously, piezoelectric sensors cannot measure static or absolute pressures for more than a few seconds, but this automatic elimination of static signal components allows unattended, drift-free operation.

Because of their unusual ruggedness, piezoelectric pressure sensors are widely used in difficult applications, such as ballistics, blasts, explosions, internal combustion, fuel injection, flow instabilities, high-intensity sound, and hydraulic or pneumatic pulsations—in connection with problems which may be encountered in connection with guns, shock tubes, closed bombs, rocket motors, internal-combustion engines, pumps, compressors, pipelines, mufflers, and oil exploration imploders.

Resonant-Wire Pressure Transducers[3]

A wire under tension is caused to oscillate at its resonant (or natural) frequency, and changes in pressure are converted to changes in this frequency. The application of an oscillating wire as a primary means of detecting force is based on fundamental principles initially outlined by Rayleigh's equations for a bar vibrating in a vacuum. Holst et al. (1979) modified Rayleigh's equations to fit an oscillating wire. Their approximation of the resonant frequency f_n of a wire in a vacuum is

$$f_n = \frac{1}{2l}\sqrt{\frac{T}{\rho A} + (12 + \pi^2)\frac{EK^2}{\rho l^2}} + \frac{1}{l}\sqrt{\frac{EK^2}{\rho}}$$

where $\rho =$ density of wire material
$A =$ cross-sectional area of wire
$T =$ tension in wire
$E =$ modulus of elasticity of wire material
$K =$ radius of gyration
$l =$ length of wire

Practical use of a resonant-wire pressure sensor requires that the wire be in a nonvacuum environment. Fairly rigorous modifications and refinements of the prior equation were made by Holst et al. (1979) to account for this. Simplifying and assuming that length, density, and area remain constant in the range of tension applied to the wire, the result of these modifications can be approximated as

$$f_n \propto T^2$$

A representative resonant-wire sensor for differential pressure or liquid-level measurement is shown schematically in Fig. 36; a resonant-wire sensor for measuring gage pressure is shown schematically in Fig. 37. This principle also is used in connection with liquid-level measurement. A block diagram of the electronic circuitry of a resonant-wire pressure transducer is presented in Fig. 38.

---

[3] Based on W. H. Burtt, The Foxboro Company (a Siebe Company), Foxboro, Massachusetts.

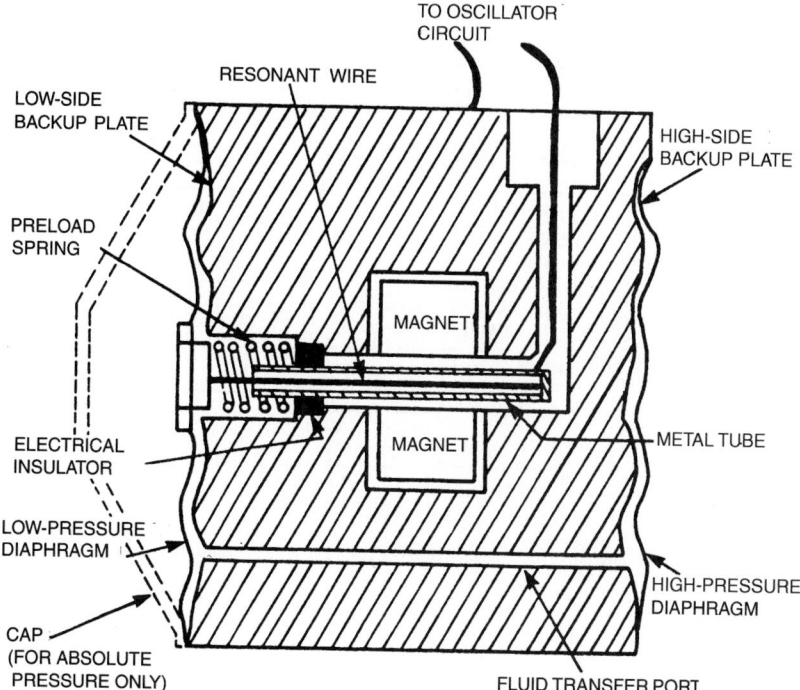

FIGURE 36 Schematic diagram of differential-pressure sensor that utilizes the resonant-wire principle. A wire under tension is located in the field of a permanent magnet. The wire is an integral part of an oscillator circuit that causes the wire to oscillate at its resonant frequency. One end of the wire is connected to the closed end of a metal tube, which is fixed to the sensor body by an electrical insulator. The other end of the wire is connected to the low-pressure diaphragm and loaded in tension by a preload spring. The spaces between the diaphragms and the backup plates, the fluid transfer port, and the metal tube are filled with fluid. An increasing pressure on the high-pressure diaphragm tends to move the diaphragm toward its backup plate. The fluid thus displaced moves through the fluid transfer port and tends to push the low-pressure diaphragm away from its backup plate. This increases the tension on the wire, raising its resonant frequency, and increasing the output signal of the transducer. Overrange protection for the wire is provided by an overrange spring (not shown) that is selected to limit the maximum tension on the wire to about two-thirds of its yield strength. The diaphragms are protected from overrange by the backup plates. (*Foxboro.*)

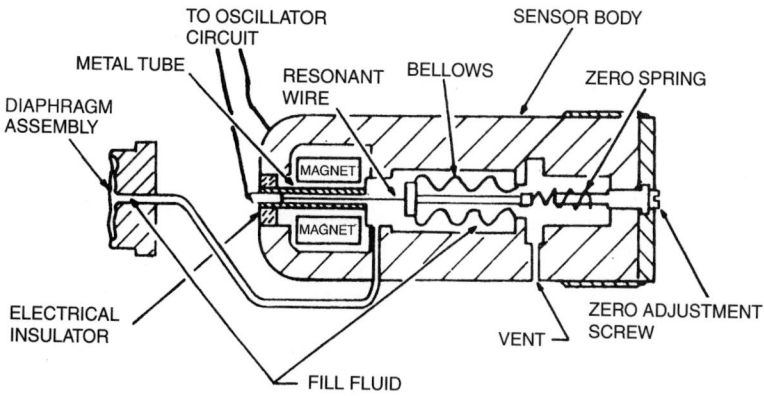

FIGURE 37 Schematic diagram of resonant-wire sensor for measuring gage pressure. (*Foxboro.*)

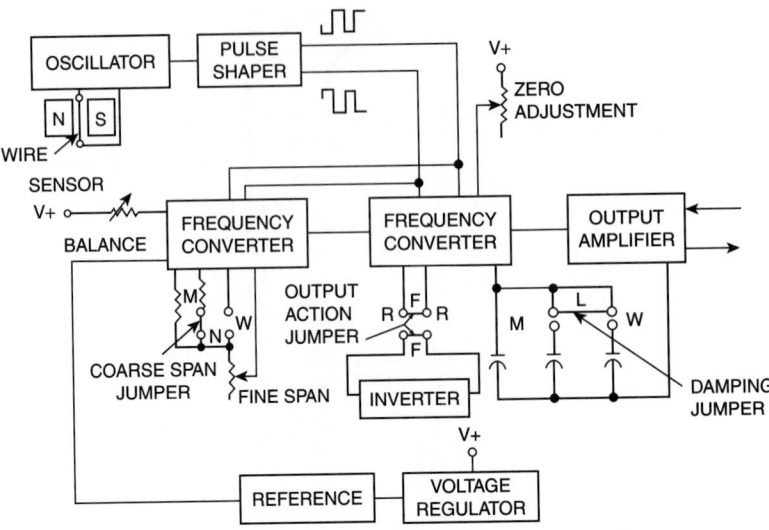

FIGURE 38   Block diagram of electronic circuitry of resonant-wire pressure transducer. (*Foxboro.*)

LVDT Pressure Transducers

Particularly for low-pressure measurement, some modern transducers utilize linear variable differential transformers (LVDTs) (Fig. 39). The LVDT measures the displacement of a pressure-carrying capsule. It is claimed that it is well suited for measuring low pressures accurately in absolute, vented gage, and differential-pressure applications. LVDTs are described in greater detail in Section 5, Article 1, of this handbook.

Carbon-Resistive Elements

These transducers no longer are popular, but should be mentioned to demonstrate the many transducer approaches used in pressure measurements. Some of these methods are described in Fig. 40.

Inductive Elements

These methods also have been reduced in terms of usage. They are described briefly in Fig. 41.

Reluctive Elements

Variable-reluctance pressure transducers are distinguished mainly by the manner in which the exciting electric energy enters the system. In these devices, energy is introduced as a magnetomotive force which may be derived from either a permanent magnet or an electromagnet assembly. Electromagnetic induction, roughly speaking, is the production of an electric current (or voltage) by the movement of a conductor through a magnetic field, or by changing the strength of the field while the conductor is within it. In either case it is the interaction of a magnetic field and an electric conductor in such a manner as to produce a current, a change in position, or a change in flux that is always involved (Fig. 42).

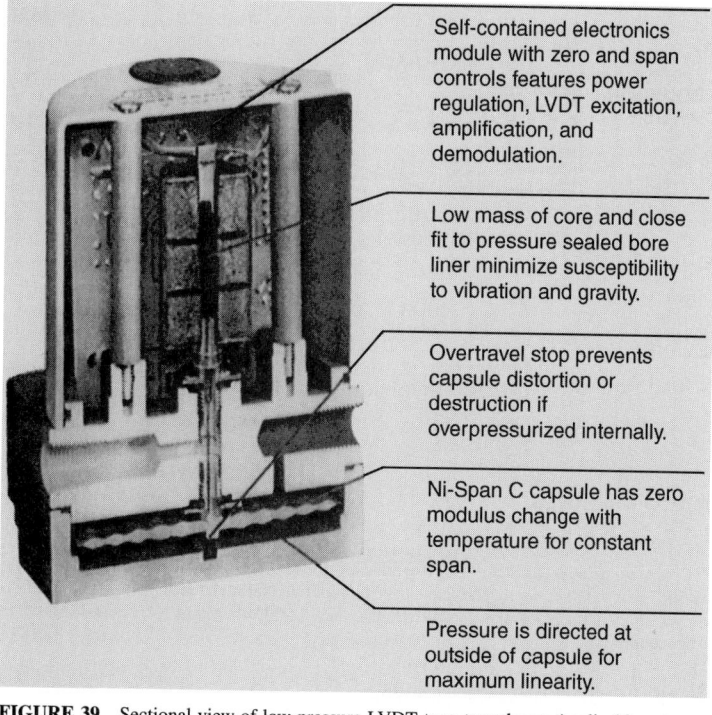

Self-contained electronics module with zero and span controls features power regulation, LVDT excitation, amplification, and demodulation.

Low mass of core and close fit to pressure sealed bore liner minimize susceptibility to vibration and gravity.

Overtravel stop prevents capsule distortion or destruction if overpressurized internally.

Ni-Span C capsule has zero modulus change with temperature for constant span.

Pressure is directed at outside of capsule for maximum linearity.

FIGURE 39   Sectional view of low-pressure LVDT-type transducer. Applicable pressure ranges are from as low as 2 inches (50 mm) of water to 100 psi (7 bars) full-scale. Typical applications include climate control and energy management systems, measurement of liquid levels in bulk storage tanks and closed pressure vessels. The unit incorporates a NiSpan C capsule, which offers low hysteresis and constant scale factor with temperature variation. Deflection of the capsule when pressurized is measured by an LVDT displacement sensor whose core is directly coupled to the capsule. The electrical output of the LVDT is directly proportional to core motion, which, in turn, is proportional to pressure applied to the capsule. (*Lucas Shaevitz.*)

Optical Pressure Transducers

Over many years, optical methods have been used to measure the movement of diaphragms, bellows, or other summing elements in pressure sensors (Figs. 43 and 44). A very recent (1992) fiber-optic pressure transducer is described in Section 7, Article 4, of this handbook.

VACUUM MEASUREMENT
by Z. C. Dobrowolski[*]

Subatmospheric pressure usually is expressed in reference to perfect vacuum or absolute zero pressure. Like absolute zero temperature (the concept is analogous), absolute zero pressure cannot be achieved, but it does provide a convenient reference datum. Standard atmospheric pressure is 14.695 psi absolute, 30 inches of mercury absolute, or 760 mmHg of density 13.595 g/cm^3 where acceleration due to gravity is $g = 980.665$ cm/s^2. 1 mmHg, which equals 1 Torr, is the most commonly used unit of absolute

[*] (Deceased). Formerly Chief Development Engineer, Kinney Vacuum Company (a Unit of General Signal), Cannon, Massachusetts.

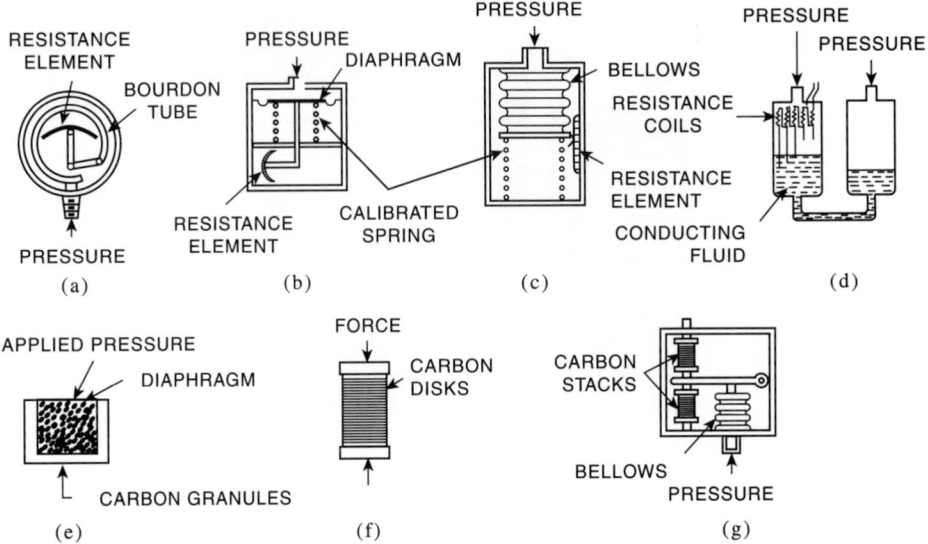

FIGURE 40 Group of resistive pressure transducers used for many years. Illustrations are highly schematic and essentially of historic interest. Other approaches have been miniaturized through the application of solid-state electronics. (*a*) Bourdon tube. (*b*) Diaphragm. (*c*) Bellows. (*d*) Differential coil. (*e*) Carbon pile. (*f*) Stacked carbon disk. (*g*) Carbon stacks with bellows coupling.

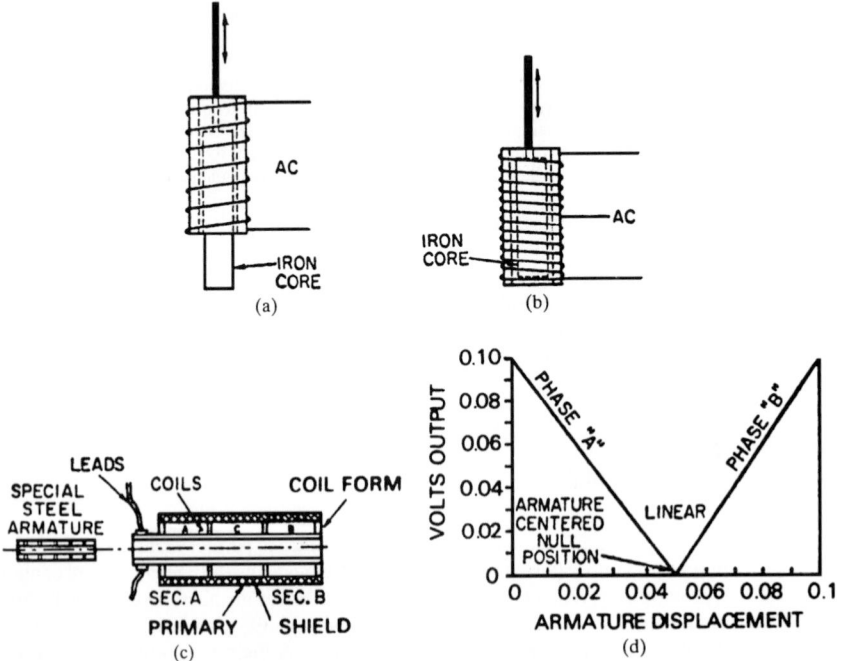

FIGURE 41 Various forms of inductive elements that are or have been used in pressure transducer designs. (*a*) Variable-inductance unit. (*b*) Inductance-ratio element. (*c*) Mutual-inductance element. (*d*) Phase relationship in mutual-inductance element.

TABLE 5   Range of Operation of Major Vacuum Gages

Principle	Gage Type		Range, Torr
Direct reading	Force measuring:		
	Bourdon, bellows, manometer (oil and mercury),		$760-10^{-6}$
	McLeod capacitance (diaphragm)		760×10^{-6}
Indirect reading	Thermal conductivity:		
	Thermocouple (thermopile)		$10-10^{-3}$
	Pirani (thermistor)		$10-10^{-4}$
	Molecular friction		$10^{-2}-10^{-7}$
	Ionization:		
	Hot filament		$10-10^{-10}$
	Cold cathode		$10^{-2}-10^{-15}$

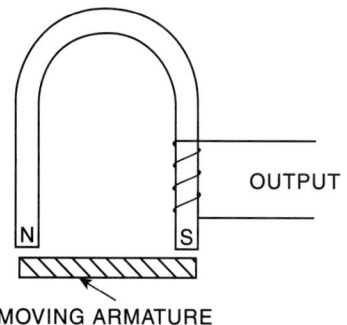

OUTPUT

N   S

MOVING ARMATURE

FIGURE 42   Schematic representation of variable-reluctance pressure transducer.

pressure. Derived units, the millitorr or micrometer, representing 1/1000 of 1 mmHg or 1 Torr, are also used for subtorr pressures.

In the MKS system of units, standard atmospheric pressure is 750 Torr and is expressed as 100,000 Pa (N/m^2) or 100 kPa. This means that 1 Pa is equivalent to 7.5 millitorr (1 Torr = 133.3 pascal). Vacuum, usually expressed in inches of mercury, is the depression of pressure below the atmospheric level, with absolute zero pressure corresponding to a vacuum of 30 inches of mercury.

When specifying and using vacuum gages, one must constantly keep in mind that atmospheric pressure is *not* constant and that it also varies with elevation above sea level.

Types of Vacuum Gages

Vacuum gages can be either direct or indirect reading. Those that measure pressure by calculating the force exerted by incident particles of gas are direct reading, while instruments that record pressure by measuring a gas property that changes in a predictable manner with gas density are indirect reading.

The range of operation for these two classes of vacuum instruments is given in Table 5. Since the pressure range of interest in present vacuum technology extends from 760 to 10^{-13} Torr (over 16 orders of magnitude), there is no single gage capable of covering such a wide range. The ranges of vacuum where specific types of gages are most applicable are shown in Fig. 45; pertinent characteristics of these gages are given in Fig. 46.

The operating principles of some vacuum gages, such as liquid manometers, bourdon, bellows, and diaphragm gages involving elastic members, were described earlier in this article. The remaining vacuum measurement devices include the thermal conductivity (or Pirani and thermocouple-type gages), the hot-filament ionization gage, the cold-cathode ionization gage (Philips), the spinning-rotor friction gage, and the partial-pressure analyzer.

Pirani or Thermocouple Vacuum Gage

Commercial thermal conductivity gages should not ordinarily be thought of as precision devices. Within their rather limited but industrially important pressure range they are outstandingly useful. The virtues of these gages include low cost, electrical indication readily adapted to remote readings, sturdiness, simplicity, and interchangeability of sensing elements. They are well adapted for uses where a single power supply and measuring circuit is used with several sensing elements located in different parts of the same vacuum system or in several different systems.

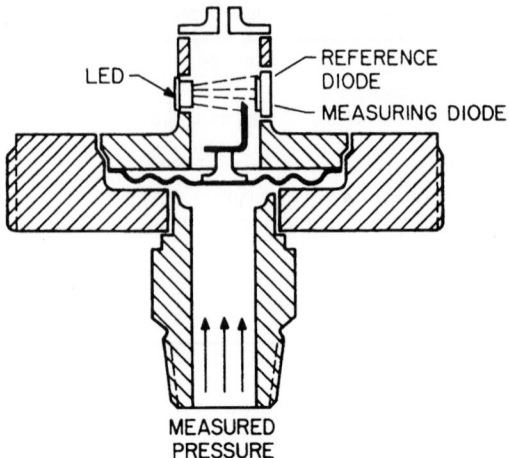

FIGURE 43 Cutaway view of Heise noncontacing optical sensor.

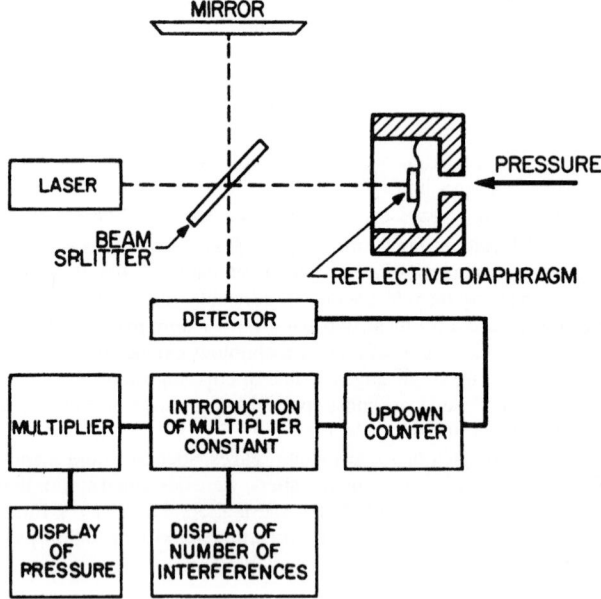

FIGURE 44 Supersensitive and accurate pressure sensor that utilizes optical interferometry.

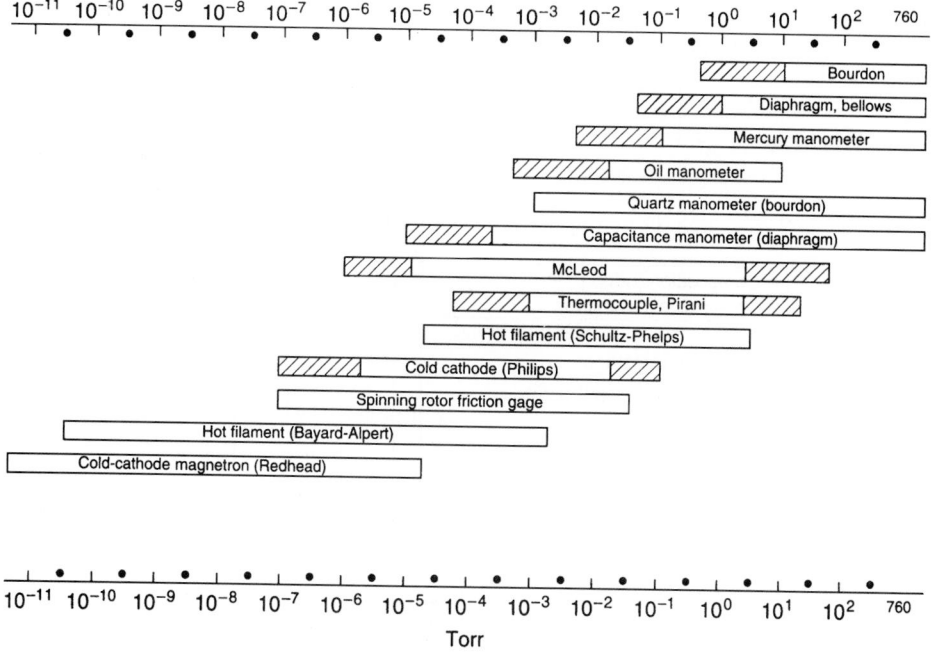

FIGURE 45 Ranges where certain vacuum gages are most suitable.

The working element of the gages consists of a metal wire or ribbon exposed to the unknown pressure and heated by an electric current (Fig. 47). The temperature attained by the heater is such that the total rate of heat loss by radiation, gas convection, gas thermal conduction, and thermal conduction through the supporting leads equals the electric power input to the element. Convection is unimportant and can be disregarded, but the heat loss by thermal conduction through the gas is a function of pressure. At pressures of approximately 10 Torr and higher, the thermal conductivity of a gas is high and roughly independent of further pressure increases. Below about 1 Torr, on the other hand, the thermal conductivity decreases with decreasing pressure, eventually in linear fashion, reaching zero at zero pressure. At pressures above a few torr, the cooling by thermal conduction limits the temperature attained by the heater to a relatively low value. As the pressure is reduced below a few hundred millitorr, the heater temperature rises, and at the lowest pressures, the heater temperature reaches an upper value established by heat radiation and by thermal conduction through the supporting leads.

Hot-Filament Ionization Vacuum Gage

The hot-filament ionization gage is the most widely used pressure-measuring device for the region from 10^{-2} to 10^{-11} Torr. The operating principle of this gage is illustrated in Fig. 48.

A regulated electron current (typically about 10 mA) is emitted from a heated filament. The electrons are attracted to the helical grid by a dc potential of about +150 volts. In their passage from filament to grid, the electrons collide with gas molecules in the gage envelope, causing a fraction of them to be ionized. The gas ions formed by electron collisions are attracted to the central ion collector wire by the negative voltage on the collector (typically −30 volts). Ion currents collected are on the order of 100 mA/Torr. This current is amplified and displayed using an electronic amplifier.

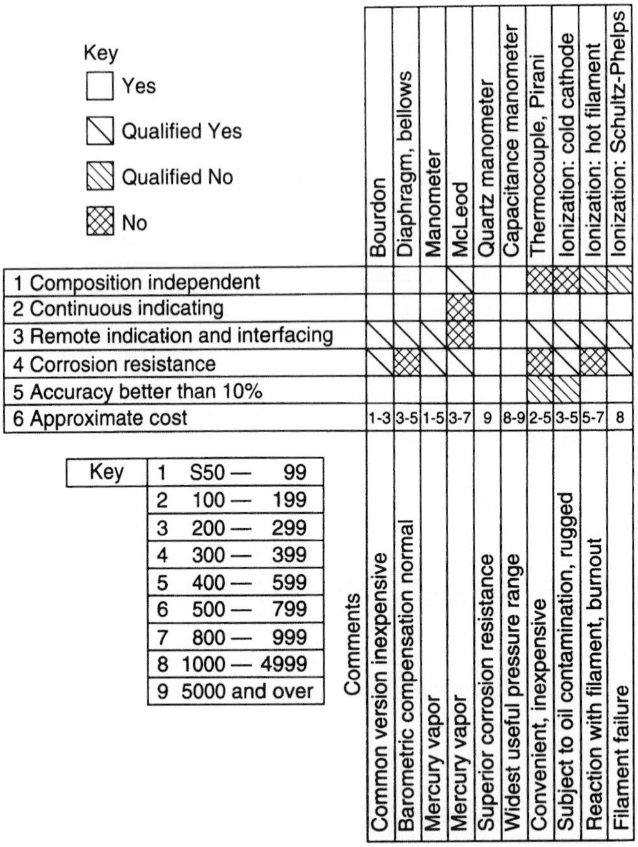
Key
Yes
Qualified Yes
Qualified No
No
Bourdon
Diaphragm, bellows
Manometer
McLeod
Quartz manometer
Capacitance manometer
Thermocouple, Pirani
Ionization: cold cathode
Ionization: hot filament
Ionization: Schultz-Phelps
1 Composition independent
2 Continuous indicating
3 Remote indication and interfacing
4 Corrosion resistance
5 Accuracy better than 10%
6 Approximate cost
1-3 3-5 1-5 3-7 9 8-9 2-5 3-5 5-7 8
Key
1 S50 — 99
2 100 — 199
3 200 — 299
4 300 — 399
5 400 — 599
6 500 — 799
7 800 — 999
8 1000 — 4999
9 5000 and over
Comments
Common version inexpensive
Barometric compensation normal
Mercury vapor
Mercury vapor
Superior corrosion resistance
Widest useful pressure range
Convenient, inexpensive
Subject to oil contamination, rugged
Reaction with filament, burnout
Filament failure

FIGURE 46 Vacuum-gage properties. Four different symbols are used in this chart, ranging from "yes" through "qualified yes" and "qualified no" to "no." For easy and uniform reference, these symbols are made to appear in shaded squares ranging from white (blank) to nearly black (heavy hash marks). This chart allows one to determine at a glance if the number of disadvantages or gage limitations is high or low. The assigned answers are unavoidably somewhat arbitrary. Reference to specific descriptions is suggested.

This ion current will differ for different gases at the same pressure, that is, a hot-filament ionization gage is composition-dependent. Over a wide range of molecular density, however, the ion current from a gas of constant composition will be directly proportional to the molecular density of the gas in the gage.

Cold-Cathode Ionization Vacuum Gage

This ingenious gage, invented by Penning, possesses many of the advantages of the hot-filament ionization gage without being susceptible to burnout. Ordinarily an electrical discharge between two electrodes in a gas cannot be sustained below a few millitorr pressure. To simplify a complicated set of relationships, this is because the "birthrate" of new electrons capable of sustaining ionization is smaller than the "death rate" of electrons and ions. In the Philips gage this difficulty is overcome by the use of a collimating magnetic field that forces the electrons to traverse a tremendously increased

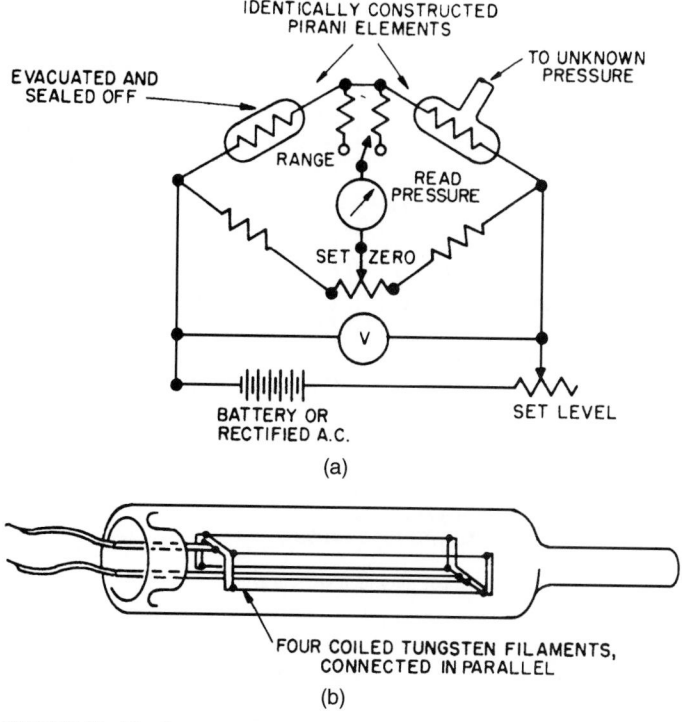

FIGURE 47   Pirani gage. (*a*) Gage in fixed-voltage Wheatstone bridge. (*b*) Sensing element.

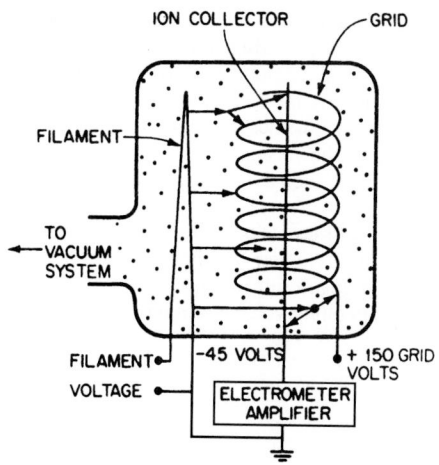

FIGURE 48   Hot-filament ionization gage (Bayard-Alpert type).

path length before they can reach the collecting electrode. In traversing this very long path, they have a correspondingly increased opportunity to encounter and ionize molecules of gas in the interelectrode region, even though this gas may be extremely rarefied. It has been found possible by this use of a magnetic field and appropriately designed electrodes, as indicated in Fig. 49, to maintain an electric discharge at pressures below 10^{-9} Torr.

Comparison with the hot-filament ionization gage reveals that, in the hot-filament gage, the source of the inherently linear relationship between gas pressure (more exactly molecular density) and gage reading is the fact that the ionizing current is established and regulated independently of the resulting ion current. In the Philips gage this situation does not hold. Maintenance of the gas discharge current involves a complicated set of interactions in which electrons, positive ions, and photoelectrically effective x-rays all play a significant part. It is thus not surprising that the output current of the Philips gage is not perfectly linear with respect to pressure. Slight discontinuities in the calibration are also sometimes found, since the magnetic fields customarily used are too low to stabilize the gas discharge completely. Despite these objections, a Philips gage is a highly useful device, particularly where accuracy better than 10 or 20 percent is not required.

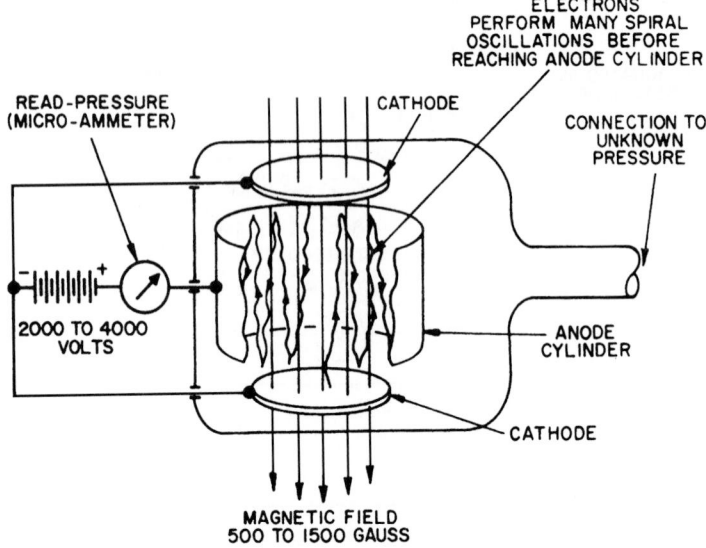

FIGURE 49   Philips cold-cathode ionization vacuum gage.

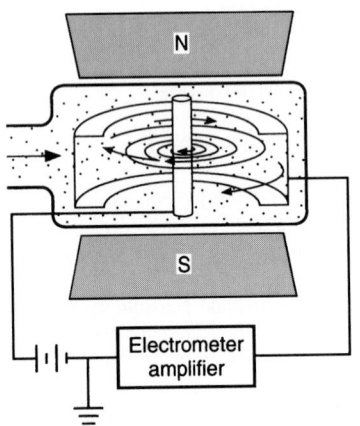

FIGURE 50   Inverted magnetron, a cold-cathode gage, produces electrons by applying a high voltage to unheated electrodes. Electrons spiraling in toward the central electrode ionize gas molecules, which are collected on the curved cathode.

The Philips gage is composition-sensitive but, unlike the situation with the hot-filament ionization gage, the sensitivity relative to some reference gas such as air or argon is not independent of pressure. Leak hunting with a Philips gage and a probe gas or liquid is a useful technique. Unlike the hot-filament ionization gage, the Philips gage does not involve the use of a high-temperature filament and consequently does not subject the gas to thermal stress. The voltages applied in the Philips gage are of the order of a few thousand volts, which is sufficient to cause some sputtering at the high-pressure end of the range, resulting in a certain amount of gettering or enforced take-up of the gas by the electrodes and other parts of the gage. Various design refinements have been used to facilitate periodic cleaning of the vacuum chamber and electrodes, since polymerized organic molecules are an ever-present contaminant.

The conventional cold-cathode (Philips) gage is used in the range from 10^{-2} to 10^{-7} Torr. Redhead has developed a modified cold-cathode gage useful in the 10^{-6} to 10^{-12} Torr range (Fig. 50). The operating voltage is about 5000 volts in a 1-kG magnetic field.

Spinning-Rotor Friction Vacuum Gage

While liquid manometers (U-tube, McLeod) serve as pressure standards for subatmospheric measurements (760 to 10^{-5} torr) and capacitance (also quartz) manometers duplicate this range as useful transfer standards, calibration at lower pressures depends on volume expansion techniques and presumed linearity of the measuring system. The friction gage allows extension of the calibration range

TABLE 6   Characteristics of Partial-Pressure Analyzers

Type	Minimum Partial Pressure, Torr	Resolution,* au	Magnetic Field
Magnetic sector	10^{-11}	20–150	Yes
Cycloidal	10^{-11}	100–150	Yes
Quadrupole	10^{-12}	100–300	No
Time of flight	10^{-12}	200	No

* Maximum mass number at which a mass number difference of 1 can be observed.

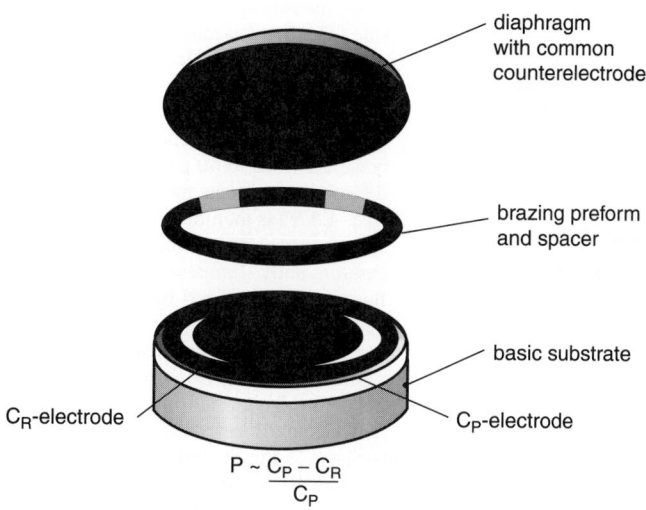

diaphragm with common counterelectrode

brazing preform and spacer

basic substrate

C_R-electrode

C_P-electrode

$$P \sim \frac{C_P - C_R}{C_P}$$

FIGURE 51   Construction of a ceramic sensor.

directly down to 10^{-7} Torr. It measures pressure in a vacuum system by sensing the deceleration of a rotating steel ball levitating in a magnetic field (Fig. 51).

Partial-Pressure Analyzers (Vacuum)[4]

Many applications of high-vacuum technology are more concerned with the partial pressure of particular gas species than with total pressure. Also, "total" pressure gages generally give accurate readings only in pure gases. For these reasons partial-pressure analyzers are finding increasing application. These are basically low-resolution, high-sensitivity mass spectrometers which ionize a gas sample in a manner similar to that of a hot-filament ionization gage. The resulting ions are then separated in an analyzer section, depending on the mass-to-charge ratio of each ion. The ion current corresponding to one ion type is then collected, amplified, and displayed. Partial-pressure gages are very valuable diagnostic tools in both research and production work. The major types are listed in Table 6.

[4] Also known as residual gas analyzers (RGAs).

CAPACITIVE CERAMIC (Al₂O₃) PRESSURE SENSORS

The capacitive sensor uses an AI_2O_3 ceramic with between 96 percent and 100 percent pureness. The 96-percent AI_2O_3 is a polycrystalline isotope material with high density and an average surface roughness of better than 0.4 μm (compared with metal membranes that are 0.6 μm). The remaining 4 percent consist of SiO_2 (silicon dioxide). The 100-percent ceramic is a pure single crystalline sapphire. It is absolutely resistant against chemicals, such as highly concentrated sulfuric acid, nitric acid, sodium hydroxide, or phosphoric acid.

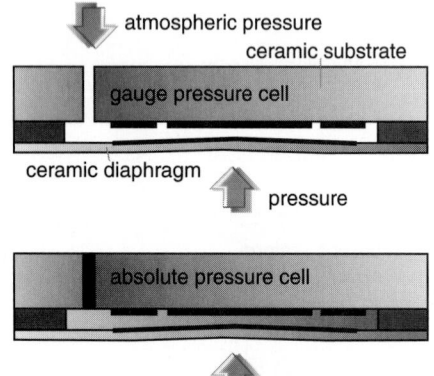

FIGURE 52   Function of the ceramic sensor.

The ceramic sensors are based on a capacitive principle and consist of a body, membrane, and electrodes. On the $\frac{1}{4}$-in-thick body, an inner and outer tantalum electrode is sputtered on. The other measuring electrode is sputtered on the membrane inner side. The inner electrode of the body is the sensitive measuring electrode; the outer electrode is the reference electrode. The reference electrode is very close to the joint part of the membrane and body and therefore has a constant capacitance in any case of varying pressure. The distance between membrane and body is well defined to 40 μm. This is accomplished by an active soldering that connects membrane and body. The raw sensor signal (0.5–4.5 V) is then measured and conditioned by an application-specific integrated circuit (ASIC) located on the back side of the ceramic body. Temperature compensation, calibration, and other functions are realized by a programmed E^2 ROM (see Figs. 51 and 52).

A ceramic sensor can withstand up to 800 times overpressure without suffering from zero drift or measurement errors. This behavior also results in excellent long-term stability, generally better than 0.1-percent/12-month stability.

In addition to universal corrosion resistance, this ceramic is absolutely resistant against temperature shocks, mechanical overloads, and vacuum. This is independent, whether 96-percent polycrystalline ceramic or 100-percent monocrystalline ceramic (sapphire) is used. The ceramic sensors are compatible with most known chemicals. For an extreme medium such as HF (hydroflouric acid), the 100-percent-pure ceramic is recommended.

The ceramic membrane ensures that no permeation of media takes place. AI_2O_3 has a 10^{-7} smaller diffusion coefficient than a metal diaphragm. The lack of any metal (nickel in particular) ensures that the ceramic sensors are also suitable for applications in which the risk of hydrogen diffusion may cause severe problems with ordinary metal diaphragms.

Self-monitoring of the diaphragm of a differential pressure cell ensures that an immediate alarm can be generated. A comparison of the total capacitance with an integrated temperature sensor allows for continuous self-monitoring of the sensor. For example, if the sensor is damaged and the fill fluid leaks, the total capacitance of the sensor, which is equal to a specific temperature because of the characteristics of the fill fluid, will no longer match with a preregistered value for temperature versus total capacitance (see Fig. 53).

Smart Pressure Transmitters

The description for a smart field device is generally any device that includes a microprocessor. Typically, this means that the device has the capability to provide better accuracy for parameters such as linearity correction and temperature compensation. The smart transmitter may also be able to work with more than one measured variable (for example, pressure, temperature).

Several other functions, such as self-diagnostics, semiautomatic calibration, and other beneficial functions, are often available. Smart pressure transmitters utilize a digital ASIC microcomputer. The

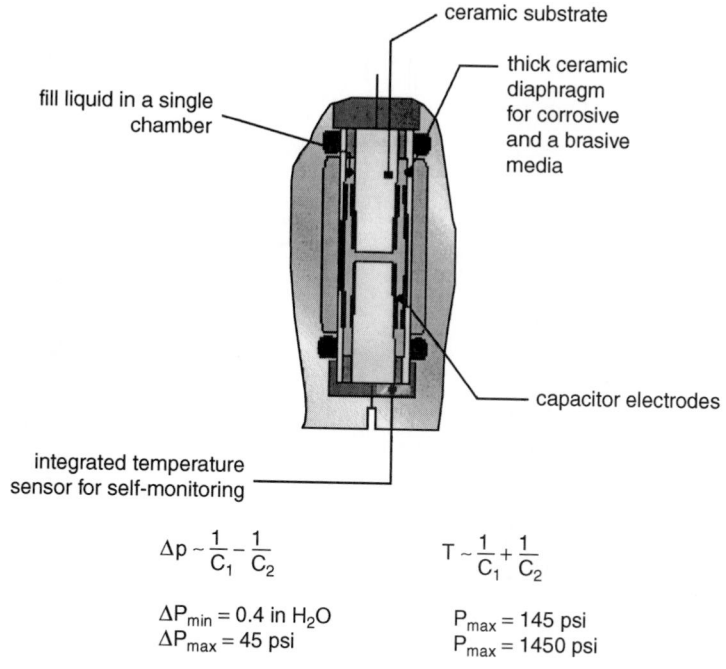

$$\Delta p \sim \frac{1}{C_1} - \frac{1}{C_2} \qquad\qquad T \sim \frac{1}{C_1} + \frac{1}{C_2}$$

$$\Delta P_{min} = 0.4 \text{ in } H_2O \qquad\qquad P_{max} = 145 \text{ psi}$$
$$\Delta P_{max} = 45 \text{ psi} \qquad\qquad P_{max} = 1450 \text{ psi}$$

FIGURE 53   Construction of ceramic differential pressure sensor.

electronics digitize the input signal from the sensor and apply correction coefficients selected from a nonvolatile memory. Configuration data are stored in a nonvolatile electrically erasable programmable read-only memory (EEPROM) in the electronics module. These data are retained in the transmitter when power is interrupted, so the transmitter is functional immediately on power-up. The output section of the electronics converts the data into a 4- to 20-mA signal with a superimposed digital protocol or into digital form for pure digital data transmission.

Remote configuration by means of a user interface such as a handheld terminal, personal computer is usually needed to configure a field-mounted smart device. Several communication protocols are in use; the most common protocol today is the HART® protocol (superimposed on the 4- to 20-mA signal), which offers an open solution for a number of manufacturers to use the same handheld terminal for their measuring devices. Another protocol that is gaining popularity is the pure digital protocol such as Foundation Fieldbus and PROFIBUS-PA. One of the advantages of fully digital signal transmission is that the digital/analog conversion is eliminated, resulting in improved performance. Another advantage is that more than one measured variable can be transmitted, together with diagnostic data at high speed (31.25 Kbytes/s) (see Figs. 54–56).

Application of Chemical Seals

Purpose and Theory.   A chemical seal is used as a barrier between the media being measured and the measuring device. The seal, by means of a fill fluid, will transmit the acting pressure to the sensing device, which in turn will give an output proportional to the acting pressure. A chemical seal system consists of one or two seals, a transmitter or sensing device, a connection between the transmitter and seal, and a fill fluid [see Figs. 57(*a*) and 57(*b*)].

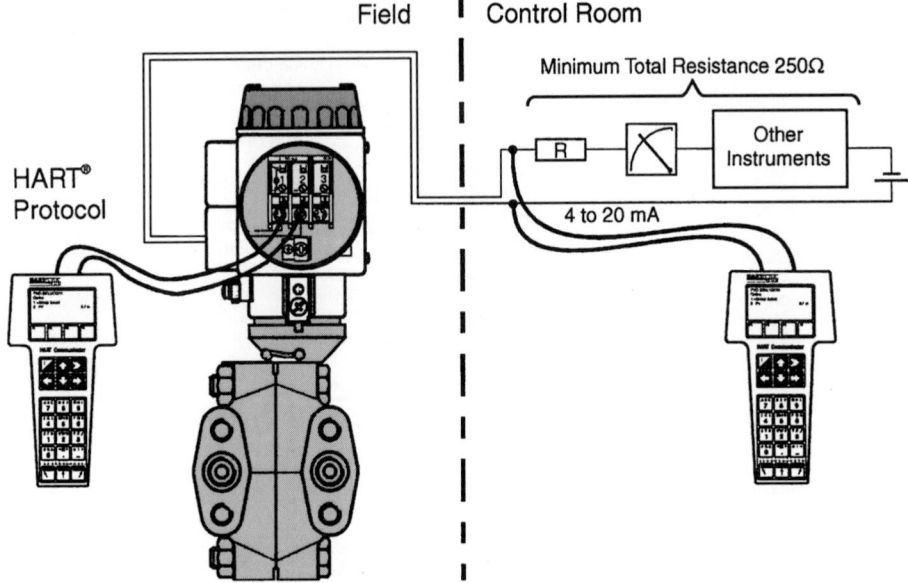

FIGURE 54   HART protocol (superimposed on the 4- to 20-mA signal).

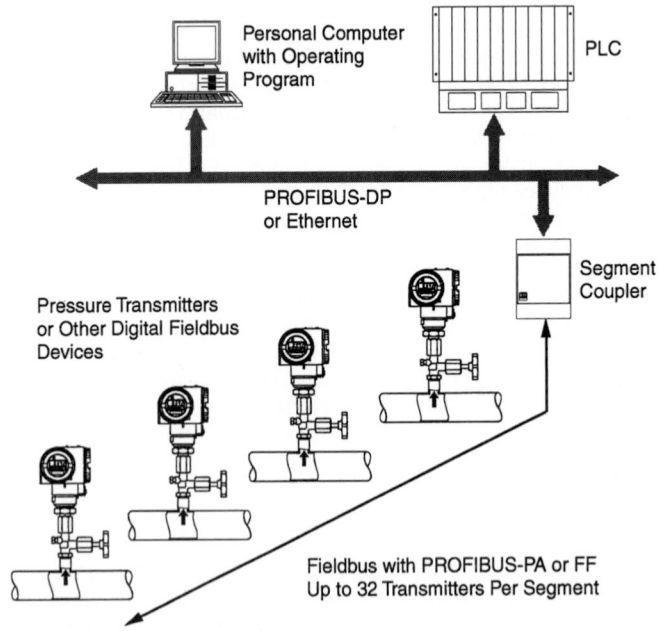

FIGURE 55   Fieldbus with PROFIBUS-PA or FF, up to 32 transmitters per segment.

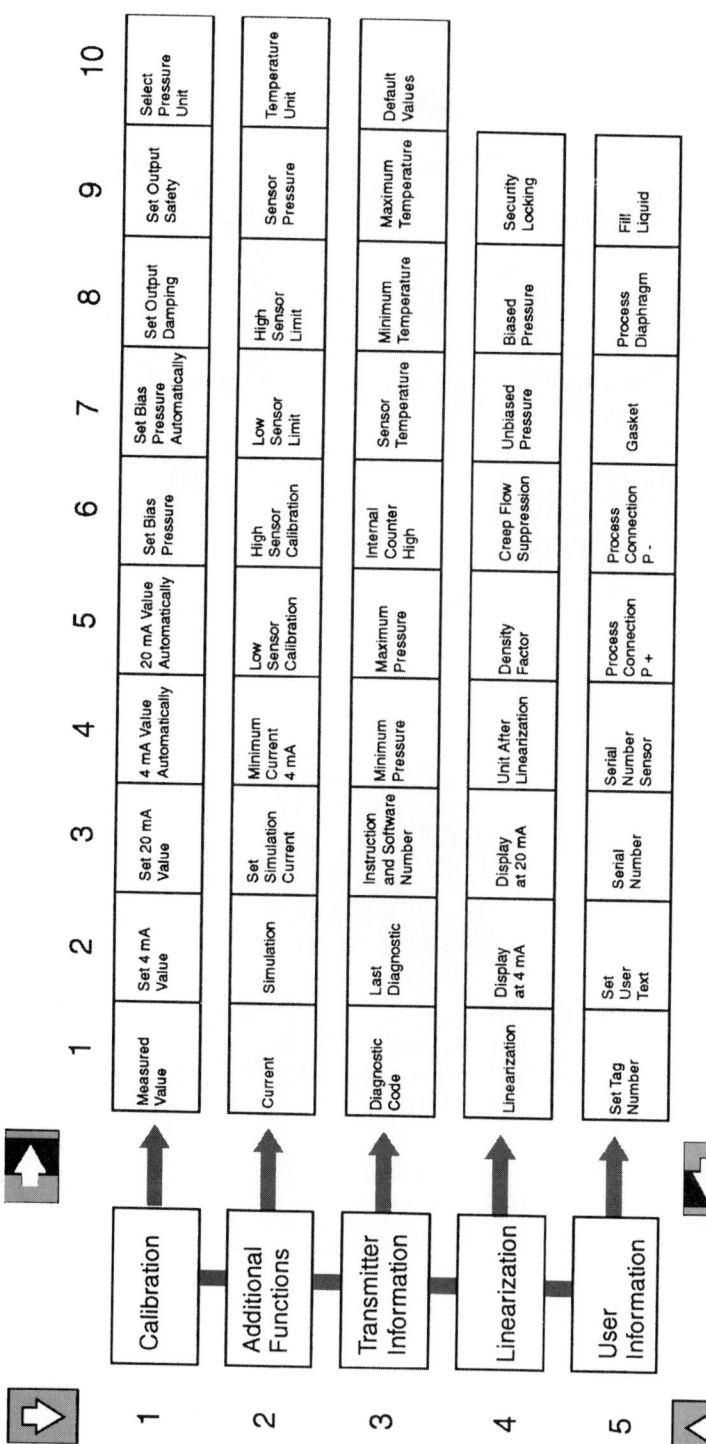

FIGURE 56 Smart transmitter programming data.

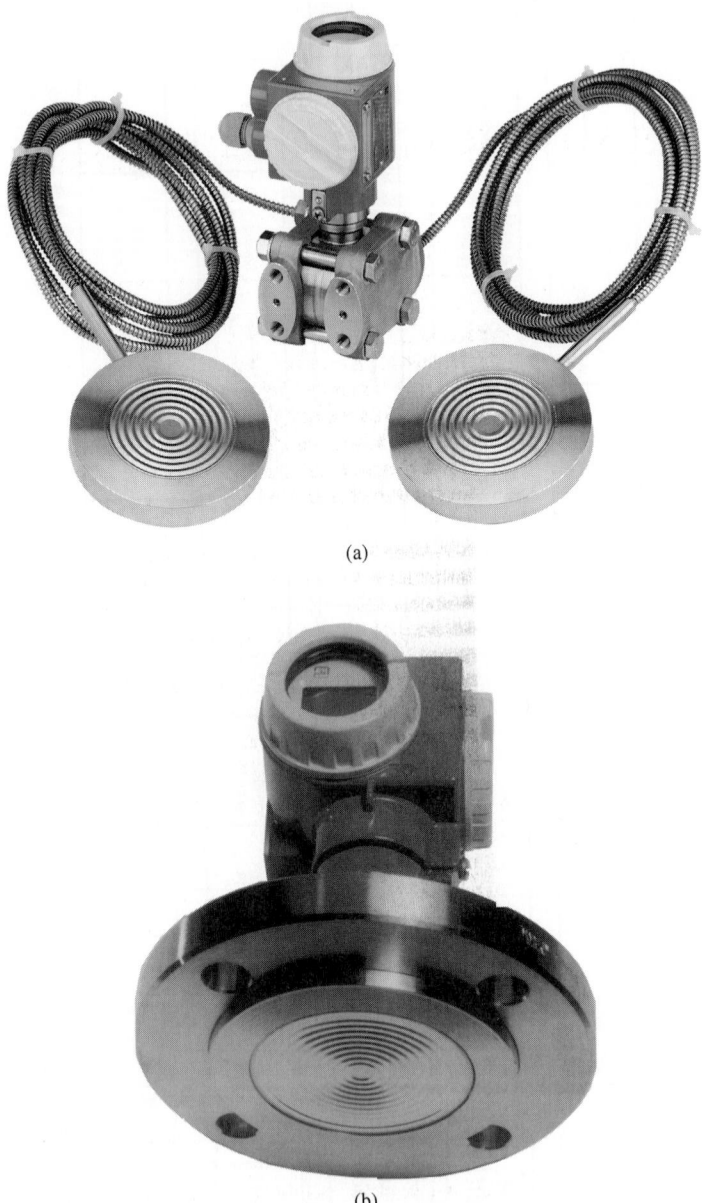

(a)

(b)

FIGURE 57   (*a*) Remote chemical seal system, (*b*) compact chemical seal systems.

There are many reasons why chemical seals have become so popular. When it is necessary to isolate the transmitter from the process chemical, seals can often solve the problem. The following conditions usually dictate the use of chemical seals:

- If the product is corrosive, it is usually necessary to use a chemical seal to act as a barrier between the corrosive media and the transmitter.
- When the temperature exceeds the transmitter's limitations, a seal can be the isolation device between the transmitter and the temperature.
- If the process contains particulates or is highly viscous, then a chemical seal may be necessary to prevent plugging.
- If wet or dry reference legs are not possible because of either condensation or evaporation, then it will be necessary to use a chemical seal to isolate the transmitter from the conditions within the vessel.
- If sanitary requirements exist, it is necessary to use chemical seals.
- If it is not possible to mount the transmitter integral to the measurement point, capillaries can be used to locate the transmitter at a remote location.

Chemical seals have been in service in industrial applications for years. It is a proven technology that fits a variety of applications in flow, level, and pressure.

Points to Consider when Choosing Chemical Seals.   There are multiple points that need to be addressed when choosing a chemical seal. These points, listed below, affect the performance, accuracy, and lifetime of the entire chemical seal system. Always keep in mind that the chemical seal system will only be as accurate as the chemical seals that are attached to the transmitter.

Materials of Compatibility

If the process is corrosive, the wetted material of the chemical must be taken into consideration. Seals have been made from practically every type of metal that can be machined and formed. Metals such as stainless steel and Hastelloy C, for aggressive media, tend to be most common. If an exotic metal is needed, it can usually be provided, but normally at a higher price. It should be understood that the diaphragm of the chemical seal (the measuring element) cannot withstand any corrosion because of its thickness. Therefore it is imperative that the diaphragm material can withstand the process.

Chemical Seal Hardware

There are normally either one or two basic pieces to a chemical seal. If it is a compact unit, then the only hardware is the chemical seal itself. If the unit needs to be remote for any number of reasons, then the seal will also have a capillary attached for the hydraulic transfer of the fill fluid.

The four types of basic chemical seals are flanged, extended, threaded, and pancake. There are also many types of special application seals such as in-line seals, saddle seals, and a full range of sanitary seals. When a remote chemical seal is used, capillaries transmit the pressure from the seal to the pressure sensor. Capillaries should always be kept at the shortest length possible to reduce temperature effects and response time.

When choosing a capillary, one needs to consider the inside diameter as it will have an effect on the measurement's response time and accuracy. The smaller the diameter, the less temperature effect you will see and the slower the response time. When choosing the correct capillary, decide whether reduced response time or reduced temperature effect is more important. If reduced temperature effect is more important, then a capillary with the shortest length and smaller diameter would be the best choice.

Fill Fluid

In any chemical seal system there is always a fill fluid. This fluid is perhaps one of the most important points to consider when assembling the system. The fill fluid will affect the safety of the process, accuracy and response time of the measurement, and quality of the product under fault conditions. The following four points need to be considered when choosing a fill fluid.

- *Temperature Limitations*: Every fill fluid has limitations as to the temperature under which it can operate. These temperature limitations also change under vacuum conditions. If a fill fluid is not operated under its limitations, then it will either solidify or vaporize and cause erroneous readings to occur. Always choose a fill fluid that will handle the most extreme conditions in the process.

- *Expansion Coefficient*: All fill fluids expand and contract with temperature fluctuations. But just as all materials have their own characteristics, all fill fluids have slightly different expansion coefficients. The smaller the expansion coefficient, the less effect you will see on the measurement because of temperature changes.

- *Process Compatibility*: Since a chemical seal has a very thin diaphragm in which to sense pressure changes, there is a great possibility that the seal could be damaged and fill fluid could leak into the process. Therefore it is important to make sure that the fill fluid being used is compatible with the process and will not affect the quality of it in any way.

- *Viscosity*: The viscosity of the fill fluid will have a large influence on the response time of the measurement. The less viscous the product, the better the response time. As the product becomes more viscous, the response time will also increase. It is also important to remember that as the temperature decreases, the viscosity will increase (see Fig. 58).

Temperature Effects

One of the most common problems in using a diaphragm seal is the inaccuracy that is caused by fluctuating temperature conditions. Every time the temperature changes, it will either expand or contract the fill fluid at a rate that is consistent with the fill-fluid thermal expansion coefficient. There are three variables that can be changed to reduce the amount of error caused by temperature effects:

- *Diameter of Capillary*: By reducing the diameter of the capillary, it is possible to decrease the volume of the fill fluid in the system, thereby reducing the temperature effect. When the diameter of the capillary is reduced it will increase the response time since it will take longer to move the same amount of volume through a smaller tube.

- *Choice of Fill Fluid*: If temperature effects will be a problem in the measurement, it is possible to choose a fill fluid that has a small thermal expansion coefficient. This will also help reduce temperature effects. When choosing a fill fluid, do not forget to look at the other key points such as temperature limitations and viscosity (see Fig. 59).

- *Diaphragm Size*: If a small-diameter diaphragm is used, a large temperature error could be seen because of the fill-fluid expansion or contraction. Smaller diaphragms have a much smaller displacement than larger diaphragms. The stiffness of the smaller diaphragms does not allow for temperature fluctuations without a significant error that is due to temperature changes. Larger diaphragms are not nearly as stiff and therefore are much more forgiving of expansion or contraction of the fill fluid.

Response Time

Response time can be a significant factor when chemical seal systems with capillaries are used. It takes a certain amount of time to transfer a volume of fill fluid from one end of a capillary to another.

Fill-Fluid Specifications

Fill Fluid	Suitable Temperature Range		Specific Gravity at Temperature		Viscosity at Temperature		Notes
	P_{abs} < 15 psi (°F)	$P_{abs} \geq$ 15 psi (°F)		(°F)	(cSt)	(°F)	
Silicone oil dc 200/50	N/A	-4 to +392	0.96	+77	50	+77	Standard
Silicone oil (4 cSt)	-130 to +176	-130 to +356	0.91	+68	4	+77	Low temperature
High-temperature oil	+14 to +212	-4 to +572	1.07	+68	39	+77	High temperature
High-temperature oil	+14 to +392	-4 to +750	1.07	+68	39	+77	High temperature and high vacuum
Halocarbon®	-40 to +176	-40 to +347	1.97	+68	14	+68	Oxygen and chlorine service
Flurolube®	N/A	-40 to +392	1.86	+77	5	+68	Oxygen and chlorine service
Glycerine	N/A	+60 to -462	1.26	+68	1110	+68	Food and beverage
Glycerine / water	N/A	+14 to +248	1.22	+68	88	+68	Food and beverage
Vegetable oil	+14 to +200	+14 to +400	0.94	+68	66	+68	Food and beverage
Food-grade silicone oil	N/A	0 to +572	0.97	+77	350	+77	Food and beverage

FIGURE 58   Fill-fluid specifications.

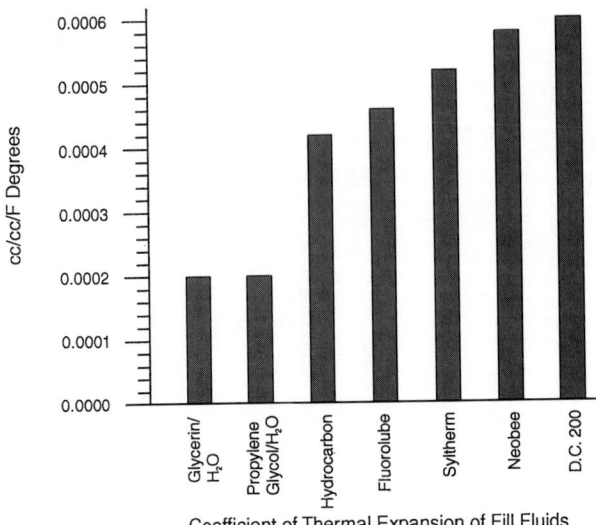

FIGURE 59 Thermal expansion coefficient of fill fluids.

The two major variables when response time on a chemical seal system is analyzed are capillary diameter and length and fill-fluid viscosity.

- *Capillary Diameter and Length*: The smaller the capillary, the longer it will take for the volume of fill fluid to be transferred through that capillary to the measuring device. When choosing a capillary, always be sure that the time response matches the demands of the process. Normal size capillaries are usually 2 mm. The length of the capillary should also be kept as short as possible to reduce the volume of fill fluid. If the fill-fluid volume is reduced, the response time will also be reduced.

- *Fill-Fluid Viscosity*: The higher the viscosity of the fill fluid, the longer the response time will be. If a fill fluid with a low viscosity is used, there will be a faster response time. But keep in mind that if a fill fluid with a low viscosity is used, it may or may not meet the requirements for thermal expansion and operating temperature range that the process demands.

Application Guidelines

Different applications require different installations. The most important point to remember is that the sensor range and zero point adjustment of a differential pressure (DP) transmitter allow for the offset of the negative leg of the chemical seal system. For this reason the transmitter may need to be installed at a different height than what is expected. The following diagrams should provide a guide to installing chemical seals in a variety of applications that could be seen in process (see Fig. 60).

Restriction Flow Measurement with Primary Devices and Differential Pressure Transmitters

The orifice plate is the oldest known method of metering flow. During the Roman Empire, water was brought to the homes of the city of Rome by viaducts. A tax was paid to the city and an orifice was installed in the line into the house. The size of the orifice in the line regulated the amount of water received. The tax paid was proportional to the size of the orifice. This was the earliest water meter. Today the orifice/differential pressure meter is still the most widely used flowmeter technology.

Level Measurement

Open Vessel

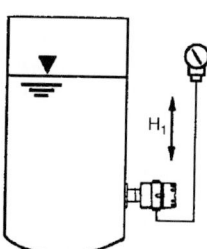

H_1 max. 22.75 ft.
with silicon oil fill.
H_1 max. 13 ft. with inert fill.
Instrument above point
or measurement.

Open Vessel

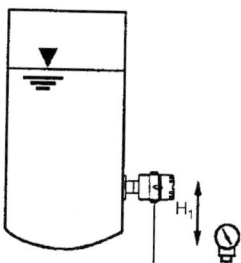

Instrument below point
of measurement.

Closed Vessel

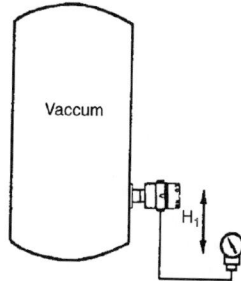

Vaccum

Absolute pressure instrument
always below point of measurement.

Closed Vessel

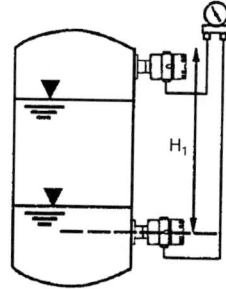

H_1 max. 22.75 ft.
with silicon fill.
H_1 max. 13 ft. with inert fill.

No vacuum.

Differential pressure instrument
for level measurament above
upper point of measurement

Closed Vessel

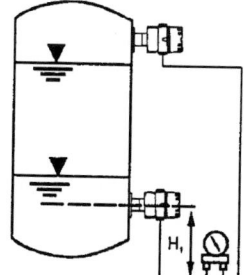

Must be used for vacuum.

Differential pressure instrument
for level measurement below
lower point of measurement.

Closed Vessel

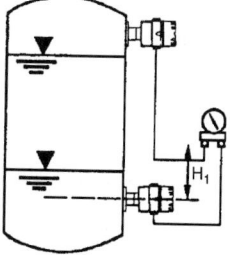

H_1 max. 22.75 ft.
with silicon oil fill.
H_1 max. 13 ft. with inert fill.

No vacuum

Differential pressure instrument
for level measurement between
points of measurement.

Flow Measurement

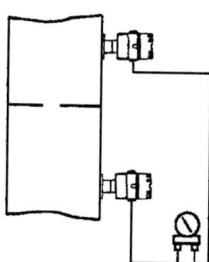

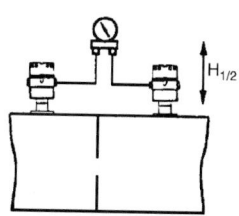

$H_{1/2}$ max. 22.75 ft.
with silicon oil fill.
$H_{1/2}$ max. 13 ft. with inert fill.

No Vaccum

FIGURE 60   Installation guidelines for chemical seals.

The pressure drop across an orifice, nozzle, venturi, or other device with a fixed passage diameter in a pipeline is a function of the flow rate. As the flow rate increases or decreases, the pressure drop varies exponentially with the change in flow rate. If the flow rate doubles, the pressure drop quadruples. Conversely, if the flow rate decreases by half, the pressure drop decreases to one fourth the drop. When the change in pressure drop is measured, the flow rate is implied. In the early days, an orifice plate was installed in the flow line and a simple U-tube manometer was installed to observe the difference in pressure across the orifice. The U-tube was typically filled with water. The difference in height of the water in one leg versus the other indicated the pressure differential. Most times it was measured in inches of water column (in. W.C.). The fill sometimes was mercury, before all the dangers of mercury were as well known as they are today. The U-tube naturally was very long and cumbersome. Differentials of 100 in are still common (see Fig. 61).

As instrumentation changed and improvements were made, a differential transmitter was substituted for the U-tube. Sensing lines conveyed the pressure to both sides of a diaphragm. The higher-pressure side, the upstream side, deflected the diaphragm toward the lower-pressure side. This caused the output of the transmitter to change its output, which remained proportional to the differential pressure. The output originally was pneumatic, 3–15 psi. This has now been almost entirely displaced by electronics by a 4–20-mA output. Since the relationship of pressure drop to flow is a geometric one, the pressure drop varies with the square of the flow. Either an internal or external square-root extractor is required.

Because this is a square relationship, the graduations on the resulting scale must also be spaced on a square relationship, which greatly limits the effective rangeability. This results in having the graduations at the low end of the scale too close together to be useful. For this reason, the rangeability of the differential pressure transmitter is normally considered to be 3:1 (33%–100%).

A simplified formula for determining pressure drop versus flow is

$$(Q_X/Q_K)^2 \times \Delta P_K = \Delta P_X$$

where Q_X is the unknown flow rate, Q_K is the known flow rate that generates a known pressure drop, ΔP_K is the known pressure drop produced at flow rate Q_K, and ΔP_X is the pressure drop that is produced at flow rate Q_X. This formula assumes that the operating conditions of temperature pressure, density, and viscosity are the same as the original conditions for which the orifice bore was selected. The following table shows the differential pressures for various flow rates and illustrates why the rangeability of the orifice plate meter is less than that of many other flowmeters.

Q (%)	100	90	80	70	60	50	40	30	20	10
ΔP (in)	100	81	64	49	36	25	16	9	4	1

DP transmitters are available in either pneumatic or electronic versions and are the most widely used communicating flowmeters today. The basic transmitter may be used to meter flow, level, or pressure variables, which makes it very versatile. The transmitter detects only the differential across the fixed restriction. It does not know what size the pipe is or what the fluid properties may be. This means that the DP may be mounted in any pipe size to meter any process fluid—liquid or gas—and not have to have any special features or special calibration. It has been calibrated to detect the expected pressure differential only. A pressure differential of 100 in W.C. is very common. If the transmitter is calibrated to have a 20-mA output at 100 in W. C., that is all that is necessary.

The key to making a DP transmitter work for a given application is the proper selection and installation of the differential producer—the orifice plate, most frequently. The calculation of the orifice bore is important. The wrong orifice bore cannot be expected to provide accurate flow measurement.

When the orifice bore is calculated, there are certain limits on how large the hole may be. You will hear the term beta or beta ratio used. This is only the ratio of the orifice bore (the hole) to the internal diameter of the pipe. Most users prefer to keep the beta to a maximum of 0.7, although some will go as high as 0.75. A beta of 0.75 means that if the pipe is a 2-in schedule 40 pipe, in which the pipe I.D. is 2.067 in, the hole will be 1.550 in. The beta may be much less than 0.7 and usually is for most installations. Pressure drop in the overall piping system is very often a large consideration in selecting the pipe size so users tend to keep flow velocity down whenever possible.

2-in schedule 40 pipe   *I.D. 2.067 in*

Beta ratio	0.2	0.3	0.4	0.5	0.6	0.7
Bore (in)	0.413	0.620	0.827	1.034	1.240	1.448

2-in schdule 40 pipe   *Beta ratio 0.7*

	50	100	150	200	400
Differential (in W.C.)	50	100	150	200	400
Flo wrate (gpm)	60	84	103	119	166
Perm press loss (in W.C.)	26	51	77	103	205

In addition to installing the orifice plate or other pressure differential producer and the transmitter itself, it is also necessary to install sensing lines to conduct the pressure to the transmitter. The sensing lines are frequently 1/4-in tubing, but may also be 1/2-in tubing or pipe. How long these sensing lines may be will vary with the application, but they should be kept to as short a length as reasonably possible. If the metering system is to installed outdoors, it may be necessary to protect the sensing lines from freezing in the wintertime. This can be done either by use of heat tape wrapped around the tubing or by heat tracing them with steam lines. For any application in which there may be solids or foreign matter present, provision should be made to permit the user to clean out the sensing lines as necessary. For those systems that are used to meter hot fluids such as steam, it may be necessary to keep the sensing lines filled with water or another liquid to isolate the high process temperature from the transmitter to protect the sensor and electronics.

Another element in the DP system is a three-valve manifold that is attached directly to the DP transmitter itself, providing a connection to the sensing lines. The purpose of the three-valve manifold is to permit the user to balance the pressure from one side of the orifice to the other for the purpose of zeroing the meter. Once the zeroing is done, the meter is ready for operation. Depending on how critical the application is, some users will recheck the meter zero monthly. Others will ignore it for long periods of time.

In some applications it may be desirable to shut down the transmitter and isolate it from the process without the need to shut down the process itself. In that instance, it is recommended to install two isolating valves, one on the high-pressure line and the other on the low-pressure line. If the transmitter must be isolated from the process, all that needs to be done is to close both valves. The process may continue uninterrupted.

The remaining large-cost item is the orifice flange. If the pipeline is small, most users will elect to use orifice flanges. These are typically special 300# flanges that have two holes bored through the side to permit the sensing lines to be attached through the side of the flanges and into the process. On very large pipe sizes, it is common to tap directly into the pipe wall itself. In that case the orifice plate is held in place by ordinary flanges. When orifice flanges are not used, the sensing lines are attached directly to the pipe wall. There are several types of taps used. The most common type is the radius tap or D, half D tap in which the upstream tap is one pipe diameter before the orifice plant and the downstream tap is 1/2 pipe diameter below the orifice plate. Others include the Vena Contracta and pipe taps. The latter seldom used because they are so far from the orifice plate and the permanent pressure loss is the highest. After the orifice plate there is some recovery of pressure as the flow slows and is converted back to pressure energy from velocity energy. See Fig. 62 for the locations of the various types of taps. Another type of tap that is seldom used in the U.S, but is sometimes used in Europe is the corner tap. This is a variation on the flange tap. It is cut into the flange so that the sensing is done right at the point where the orifice plate contacts the flange surface. It is done in such a manner that the pressure tap is from all sides of the pipe (the circumference). This type is, naturally, more expensive because of the extra machining required.

When selecting what type of flowmeter to use in any given application, it is important to look at the installed cost of the complete meter system as well as the cost of ownership. Many will overlook some of the secondary hardware necessary to the DP system. They look at the price of only the orifice plate and the DP transmitter itself. They forget to include the sensing lines, three-valve manifold, and the other large-cost item, the orifice flanges. The labor to install an orifice/DP system is higher because of the additional hardware required. It is also more expensive to maintain. The sensing lines have to be protected and kept free from foreign matter. If the installation is outside where

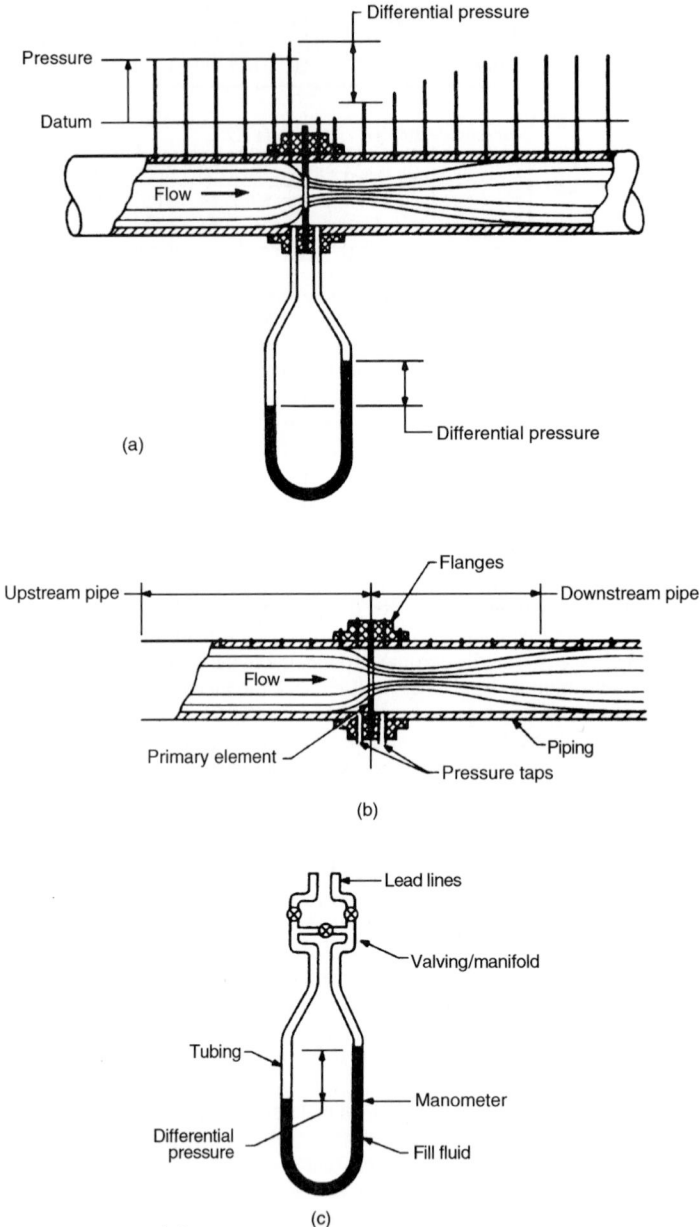

FIGURE 61   Differential pressure flowmeter. (*a*) Effect of orifice restriction in creating a pressure differential across plate and recovery of pressure loss downstream from plate is shown graphically in top view. A liquid manometer is used for illustrative purposes. Other DP transducers are used in practice. The pressure differential created depends on the type of restriction used. With an orifice plate, the flow contraction is abrupt; with a flow nozzle, it is more gradual. (*b*) Relationship between measured differential and flow rate is a function of tap locations, the particular type of restriction used, and the associated upstream and downstream piping. These factors are included in the discharge coefficient, which relates actual flow rate to theoretically calculated rate. (*c*) Commonly used piping arrangements used for installing a DP flowmeter.

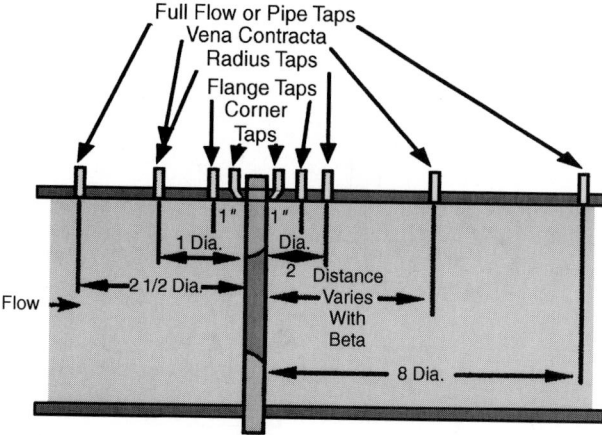

FIGURE 62 Various tap locations.

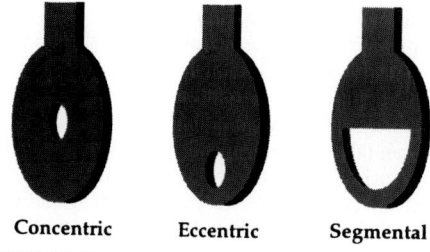

| Concentric | Eccentric | Segmental |

FIGURE 63 Various orifice configurations.

the ambient temperature may go below freezing in the winter, it may be necessary to protect the sensing lines from freezing. The orifice plate's sharp edge must be maintained. It some DP transmitters are overranged, it may become necessary to recalibrate it to be sure it is working properly.

When selecting the location on which to mount the orifice in the pipeline, it is very important to consider what is in the pipeline upstream of the meter. Any disturbance in the flow stream can have an adverse effect on the meter's performance. A single elbow upstream of the orifice may require 20 or more pipe diameters of straight pipe for allowing the flow time to settle down again, and establish a stable flow profile. If there are two or more elbows in multiple plains, it can take 100 or more diameters to regain some semblance of the proper flow profile. In some installations it may be necessary to install flow conditioning equipment after a pipe disturbance to reduce the straight run to something that will fit within the allowable space. There are tables and curves published by organizations such as ASME and AGA that will advise the user on just how much straight pipe is required for any given pipe configuration. These should be consulted before the orifice plate is installed to be sure the meter will perform as desired.

When there is a possibility that liquid will be present in a gas stream or gas may be present in a liquid stream, there is a possibility that there will be a buildup behind the orifice plate.

This is sometimes avoided by placing the orifice in a vertical run (see Fig. 63). If this is not feasible, special orifice plates are available that do not have the hole bored concentric with the pipe wall, but have an eccentric bore. An alternative is to bore a small weep hole in the plate so that it is adjacent to the pipe wall. This will allow the liquid or gas to pass on down the pipe without interrupting the flow measurement. There is a question, however, about how effective the weep hole may be.

Since the purpose of any flowmeter is to measure the flow accurately, it is necessary to maintain the meter properly to ensure its performance. The sharp edge of the orifice plate must be maintained. If the edge is permitted to become rounded, even slightly, it will cause the meter to become inaccurate. See below for the effects of degradation of the orifice plates and their subsequent loss of accuracy (see Fig. 64).

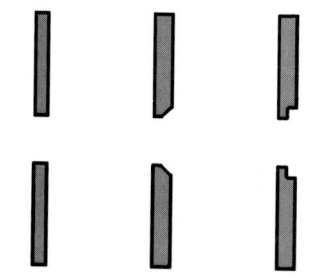

FIGURE 64 Typical orifice cross sections.

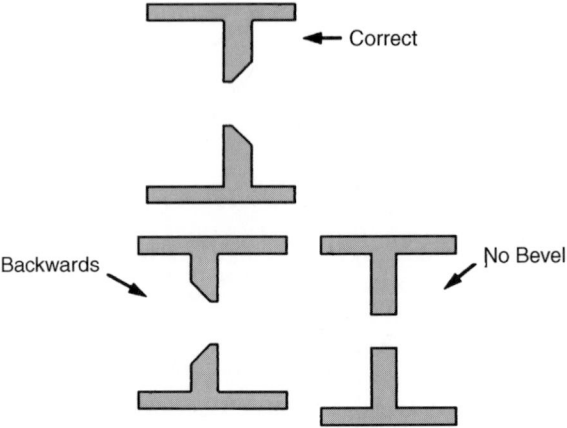

FIGURE 65   Orientation of orifice plates (flow from right to left).

The sharp edge of the orifice plate must butt against the flow, as indicated in Fig. 65. In small line sizes the thickness of the orifice plate, in relation to the pipe size, is too large and causes a shift in the orifice coefficient. To prevent this, either a chamfer or a step bore is included to reduce the land area of the bore itself. For large pipe sizes, over 3 in, there is no need for a chamfer on the sharp edge. If the orifice plate is installed in the wrong direction, the meter will not be accurate.

In some applications is may be desirable to use a differential producer other than an orifice plate. In large pipeline sizes, it is sometimes desirable to use a venturi to reduce the overall pressure loss. The costs associated with pumping large flows, particularly steam flows, can be very high. If, by installing a venturi instead of an orifice plate, the cost of pumping can be greatly reduced, the extra cost of the venturi may be well worth it. When a venturi tube is used, the high-pressure tap is installed just at the entrance of the venturi and the low-pressure tap is installed in the center of the throat of the venturi (the smallest passage in the venturi) (see Fig. 66).

Other types of differential producers are flow nozzles and flow tubes. These are sometimes fabricated on the job site, which is likely to require that they be calibrated on site. They are useful when there may be some solids present in the flow stream since they would allow the solids to pass on downstream and not clog up behind an orifice plate (see Fig. 67).

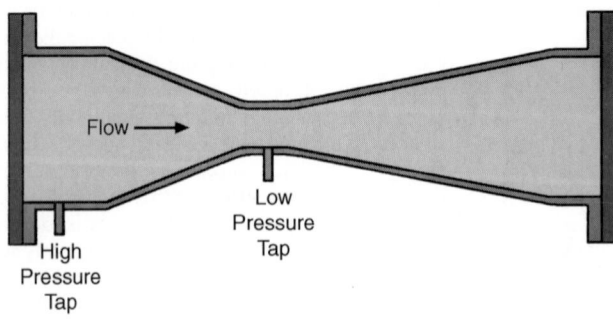

FIGURE 66   Venturi flowmeter.

Nozzles

• ISA 1932- Nozzle (ISO 5167)
- 2" to 20" (DN 50 to 500)
- Re ≥ 20,000
- Uncertainty around 1.1%
- Less dirt sensitive than orifices
- Smaller pressure loss
- Less tear at high velocities and temperatures
- Smaller influence of pipe roughness
- Smaller required straight pipes
- Welded nozzle is standard
- Higher price than orifices

• Quarter-circle Nozzle (VDI/VDE 2041)
- For very small Re (> 500)
- Uncertainty around 1 to 2%

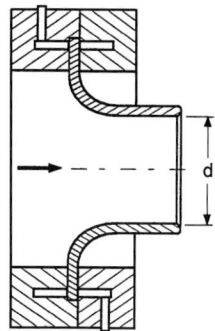

FIGURE 67   Example of nozzles.

Flow Tube

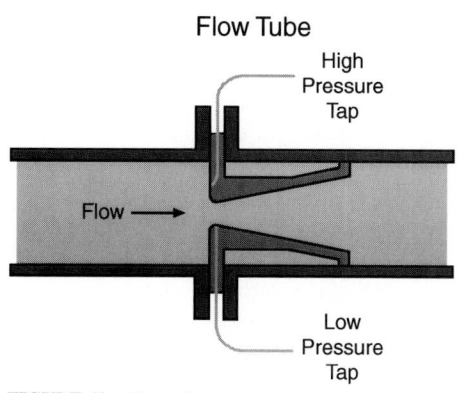

High
Pressure
Tap

Flow →

Low
Pressure
Tap

FIGURE 68   Flow tube.

The flow tube is very similar to the flow nozzle in that it permits foreign objects to pass on down the pipeline without clogging up behind the differential producer (see Fig. 68). Another differential producer is the pitot tube and the averaging pitot tube (see Fig. 69). These are inserted into the pipeline, often through a hot tap that permits the pitot tube to be inserted and removed without shutting down the process flow. They are often used in larger pipe sizes to reduce pressure loss and lower the cost of the installation. Note: the illustration of the averaging Pitot tube shows the high-pressure tap and the low-pressure tap separated for clarity. In actuality they are together in a single connnection through the pipe wall.

In some large line sizes, users will not install a special differentail producer, but instead will install taps on the inside and the outside radii of an existing pipe elbow. While this is not normally a high-accuracy installation, it does work adequately for noncritical measurements for which repeatability may be more important and the large size of the pipe makes the installation of an orifice plate very costly.

Errors in Orifice Plate System

For many years, the orifice plate/DP transmitter has been the method most often used to measure flow in pipelines. Even though there are now many new, and often better, flowmeters on the market, the old standby orifice plate is still most often used. Because the orifice plate has been widely used for so long, many flowmeter users believe that the orifice plate is always correct and accurate. Many orifice plates are installed and not checked again for years. Although many companies have a program to check the calibration of the DP transmitter regularly, they rarely check the orifice plate. On this basis, users are convinced they have a meter that is as accurate as the day it was installed. This is simply not true. Below is a list of some of the things that happen to orifice plates and the effects on the accuracy of the meter. This information was developed by testing orifice plate installations. The work was done by E. J. Burgin at the Florida Gas Transmission Company.

Condition	% Error
Orifice edge beveled 45° full circumference (machined):	
0.010 bevel width	−2.2
0.020 bevel width	−4.5
0.050 bevel width	−13.1
Turbulent gas stream:	
Upstream valve partially closed, straightening vanes in	−0.7
Upstream valve partially closed, straightening vanes out	−6.7
Liquid in meter tube 1 in deep in bottom of tube	−11.3
Grease and dirt deposits in meter tube	−11.1
Leaks around orifice plate:	
1. One clean cut through plate sealing unit	
a. cut on top side of plate	−3.3
b. cut next to tap holes	−6.1
2. Orifice plate carrier raised approximately 3/8 in	
from bottom (plate not centered)	−8.2
Valve lubricant on upstream side of plate:	
Bottom half of plate coated 1/16 in thick	−9.7
Three gob-type random deposits	0.0
Nine gob-type random deposits	−0.6
Orifice plate uniformly coated 1/16 in over full face	−15.8
Valve lubricant on both sides of plate:	
Plate coated 1/8 in both sides of full face	−17.9
Plate coated 1/4 in both sides full face	−24.4
Plate coated 1/8 in bottom half both sides	−10.1
Plate warp tests:	
Plate warped toward gas flow 1/8 in from flat	−2.8
Plate warped away from gas flow 1/8 in from flat	−0.6
Plate warped toward gas flow 1/4 in from flat	−9.1
Plate warped away from gas flow 1/4 in from flat	−6.1

As you can see, the errors introduces by orifice plate wear and tear are not trivial. Bear in mind that these potential occurrences are not just remote possibilities—they exist every day in thousands of orifice plated installations.

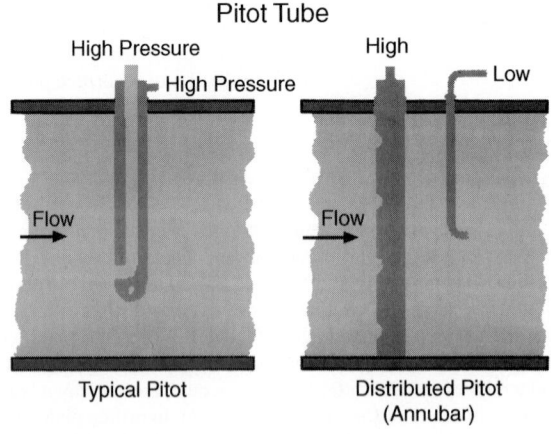

FIGURE 69   Pitot tube.

Advantages

1. The DP/orifice system has been around for many years; it is well understood by everyone from the process engineer to the instrument technician who must maintain it.

2. It is very versatile. You can change the orifice bore (within certain limits) to change the meter's flow range. The cost differential for one size pipe versus another is relatively small. A 100-in W. C. differential is the same for any pipe size.

3. The same transmitter can be used for measuring level and for measuring gage pressure and flow.

Disadvantages

1. The accuracy of the system is very dependent on how it is installed and maintained.

2. Long straight pipe runs are required, which limits where the meter may be located.

3. Solids in a flow stream may cause the meter to become clogged.

4. The sensing lines must be protected from freezing in winter if the meter is installed outside.

5. The accuracy is affected by wear and tear and damage to restriction.

Orifice Application, Device Selection

The main differences in the applications result from the different media states of aggregation, such as gas, liquid, or steam.

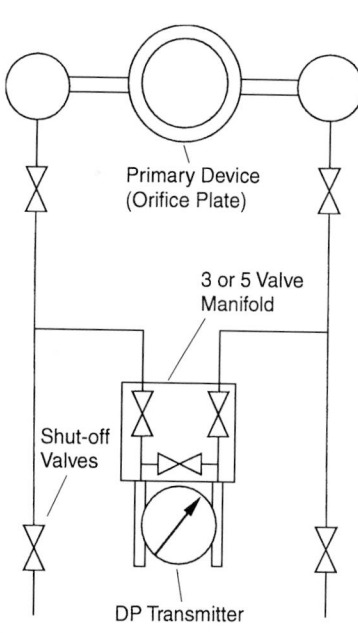

FIGURE 70   Steam installation.

Primary Device (Orifice Plate)

3 or 5 Valve Manifold

Shut-off Valves

DP Transmitter

The different states imply three different ways of installation. Depending on the purity of the fluid, variations are done accordingly. Generally, the impulse pipes should depart from the process pipe at the same height and should end at the transmitter at the same height. Therefore the fluid columns above the transmitter diaphragms have exactly the same weight and do not bias the DP measurement. Today's smart DP transmitters allow generally for a 100% bias adjustment to compensate for installation effects. To avoid permanent bubbles in liquid applications and liquid stoppers in gas applications, impulse pipes must not run horizontally—although a minimum of 4° angle is required, 10° is recommended.

Steam. With steam, the impulse pipe tappings should be at exactly the same heights. The transmitter should be installed below the process pipe. This also guarantees that sufficient steam can condense and that the impulse pipes and the transmitter are filled with only condensate (see Fig. 70).

Gas. With gases, the impulse pipes should depart to above the process pipe. The transmitter is installed above it. This prevents the impulse pipes and the transmitter from being filled with condensate, which would lead to a DP zero error.

For gas temperatures up to 175°F, the manifold block and transmitter can be mounted directly onto the orifice tappings. These compact versions lower the installation cost significantly (see Fig. 71).

Liquid. The impulse pipes must depart below the process pipe and orifice. This prevents gas bubbles from entering the impulse pipes and transmitter to avoid measurement errors (see Fig. 72).

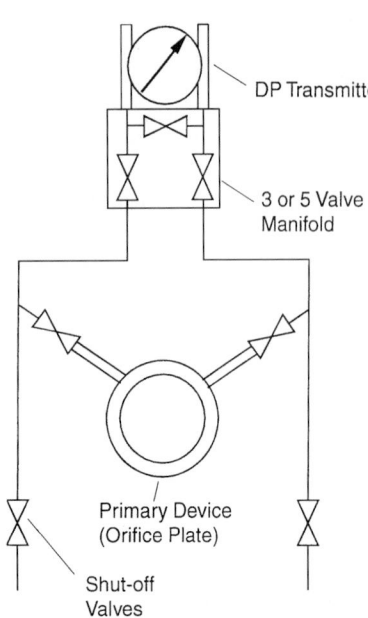

FIGURE 71   Gas installation.

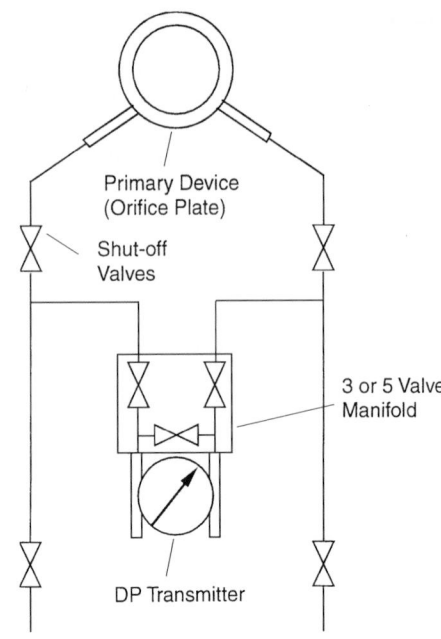

FIGURE 72   Liquid installation.

Wet Gas.   When a wet gas is measured, an installation according to Fig. 73 is recommended. Deposition chambers can be installed, which can be emptied regularly to prevent the impulse pipes from filling with condensate.

Liquids Carrying Gas.   Liquids carrying gas and cryogenic gases should have an installation as represented in Fig. 74. The impulse pipes are now filled with gas, and liquid cannot enter them. In cryogenic installations, natural or forced heating of the impulse lines must be guaranteed to avoid condensation (see Fig. 75).

Pitot Tube Application

The application of pitot tubes is based on the same principles as those of the orifices and they are installed in the same manner. Pitot tubes are typically used in gas and liquid flow rate measurements with larger pipe diameters and lower fluid pressures.

Gas.   The impulse lines should depart to above the process pipe and the pitot tube, and the transmitter should also be mounted above. In this way, the impulse pipes cannot fill with condensate and therefore cannot introduce an error in measurement.

For gas temperatures up to 175°F, the manifold block and transmitter can be mounted directly onto the flanged pitot tube tappings. These compact versions lower the installation cost significantly. However, the manifold cannot be dismounted for cleaning without interrupting the process (see Fig. 76).

Wet Gas.   If the flow rate of a wet gas is being measured, an installation like that of Fig. 77 is recommended. Deposition chambers can be installed that, when emptied regularly, prevent the impulse pipes from filling with condensate.

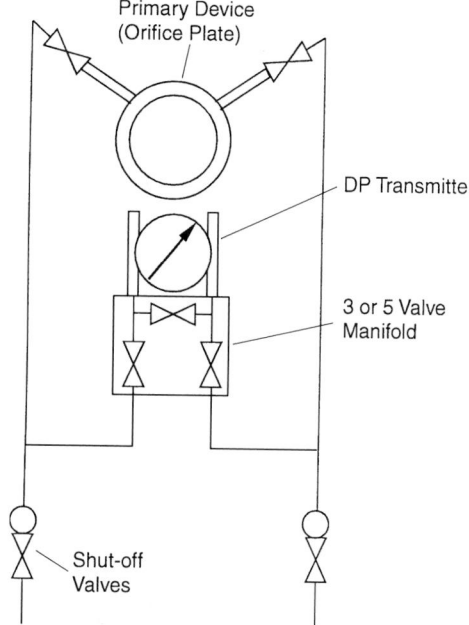

FIGURE 73   Wet gas installation.

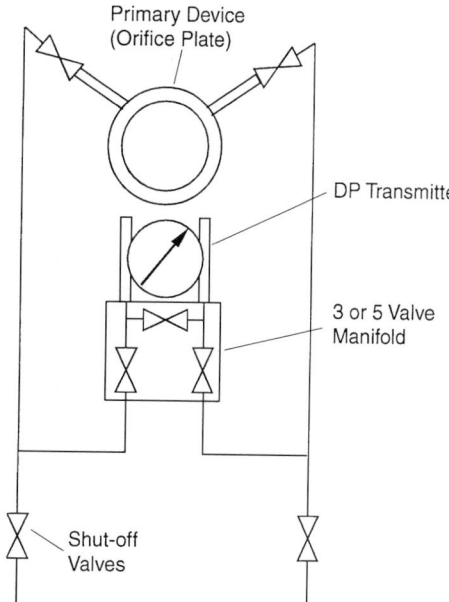

FIGURE 74   Installation of liquids carrying gas.

In addition, in installations above 25 in in inner pipe diameter, the pitot tube should have a counterbearing. With liquids or high velocities, such an installation is recommended from 20 in.

Pitot tubes should be operated at not more than 25 to 50 percent of their limit differential pressure to prevent them from vibrating under full load (see Fig. 78).

Steam and Gas—Flow Computer Applications

All steam, gas, and liquid applications are uncompensated volumetric and mass flow rate measurements for operating conditions. They are not compensated for temperature or pressure, which results in variations in density.

To compensate for these variations, a flow computer is necessary. To measure pressure and temperature the sensors must be selected according to the ranges and lengths of the application. A separate pressure and temperature sensor must be used.

Today, multivariable DP transmitters are also available. These transmitters have inputs for temperature and measure both pressure and DP. The flow computer may be integrated with these transmitters.

An installation for gas and saturated steam can include the parts specified in Fig. 79. For saturated steam, only the pressure compensation is required.

BASIC FLOW METERING CONSIDERATIONS

In selecting a flowmeter for any application, there are many different considerations to be made. One such consideration is the purpose the measurement is to serve. Is the measurement for accounting or custody transfer for which high accuracy and rangeability are very important? Or is the measurement for control of a process for which accuracy may not be as important as repeatability is the prime

Upstream/Downstream Straight Pipe Lengths
(ISO 5167)

| β | On upstream (inlet) side of primary device | | | | | | On down-stream (outlet) side |
	Single 90° bend or tee (flow from one branch only)	Two or more 90° bends in the same plane	Reducer (2 D to D over a length of 1.5 D to 3 D)	Expander (0.5 D to D over a length of 1 D to 2 D)	Globe valve fully open	Gate valve fully open	All fittings included in this table
< 0.20	10 (6)	14 (7)	34 (17)	5	18 (9)	12 (6)	4 (2)
0.25	10 (6)	14 (7)	34 (17)	5	18 (9)	12 (6)	4 (2)
0.30	10 (6)	16 (8)	34 (17)	5	18 (9)	12 (6)	5 (2.5)
0.35	12 (6)	16 (8)	36 (18)	5	18 (9)	12 (6)	5 (2.5)
0.40	14 (7)	18 (9)	– 36 (18)	5	20 (10)	12 (6)	6 (3)
0.45	14 (7)	18 (9)	38 (19)	5	20 (10)	12 (6)	6 (3)
0.50	14 (7)	20 (10)	40 (20)	6 (5)	22 (11)	12 (6)	6 (3)
0.55	16 (8)	22 (11)	44 (22)	8 (5)	24 (12)	14 (7)	6 (3)
0.66	18 (9)	26 (13)	48 (24)	9 (5)	26 (13)	14 (7)	7 (3.5)
0.65	22 (11)	32 (16)	54 (27)	11 (6)	28 (14)	16 (8)	7 (3.5)
0.70	28 (14)	36 (18)	62 (31)	14 (7)	32 (16)	20 (10)	7 (3.5)
0.75	36 (18)	42 (21)	70 (35)	22 (11)	36 (18)	24 (12)	8 (4)
0.80	46 (23)	50 (25)	80 (40)	30 (15)	44 (22)	30 (15)	8 (4)

	Fittings	Minimum upstream (inlet) straight length required
For all β values	Abrupt symmetrical reduction having a diameter ration > 0.5	30 (15)
	Thermometer pocket or well of diameter < 0.03 D	5 (3)
	Thermoter pocket or well of diameter between 0.03 D and 0.13 D	20 (10)

FIGURE 75   Upstream/downstream straight pipe lengths (ISO 5167) for orifice plate.

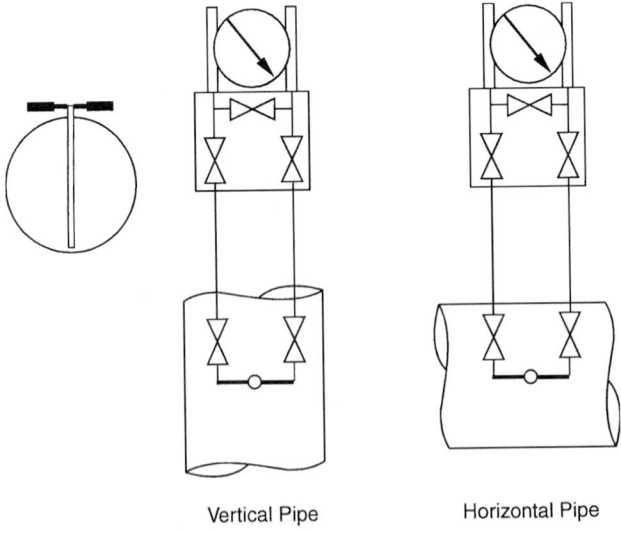

Vertical Pipe   Horizontal Pipe

FIGURE 76   Transmitter mounted above pitot tube.

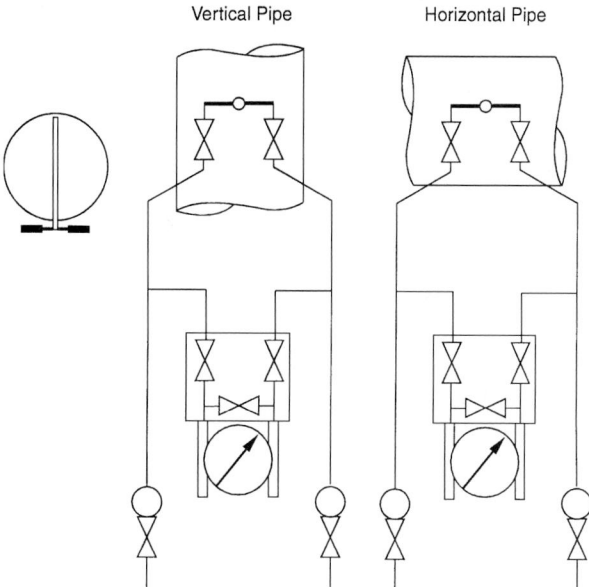

FIGURE 77   Installation for wet gas measurement.

concern? Do you require a meter for rate of flow or total flow? Is local indication or a remote signal required? If remote output is required, is it to be a proportional signal such as 4–20 mA or is it to be a contact closure to stop or start another device such as a pump or solenoid valve or even a contact to warn of a potential hazard?

Other parameters include the physical properties of the fluid to be metered. Is it a gas or a liquid at the point of measurement? Is it a clean liquid, or is it a slurry? Is it electrically conductive? What is the operating temperature and the operating pressure? What flow rate and range of flow rates are the meter to measure and expected to be able to withstand without failure?

All these parameters have a bearing on flowmeter selection. A great many types of flowmeters are available, but each type has some limitations. At this time, the perfect universal flowmeter has yet to be invented and probably never will be. The degree of satisfaction with the meter's performance is a function of how well the meter's capabilities and shortcomings are matched with the requirements and parameters of the application. Many meter technologies can be used or adapted to fit the purpose, but careful, informed meter selection pays off in performance.

DIFFERENTIAL-PRESSURE FLOWMETERS

Differential-pressure (DP) flowmeters have been used for many decades in numerous applications. In 1732 Henry Pitot invented the Pitot tube, and in 1791 Venturi did his basic work on the venturi tube. Venturi developed the mathematical basis and much of the theory now used in contemporary flowmeter computations. It was not until 1887 that Clemens Herschel developed a commercially satisfactory venturi tube.

Demands of the natural gas industry in 1903 led Thomas Weymouth to use a thin-plate, sharp-edged orifice to measure large volumes of gas. Throughout many years, DP flowmeters have enjoyed wide acceptance in the process and utility industries. Standards and vast amounts of engineering data have been developed by numerous technical organizations, including the American Gas Association,

90° elbow

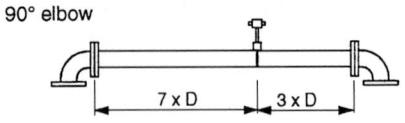

2 x 90° elbows

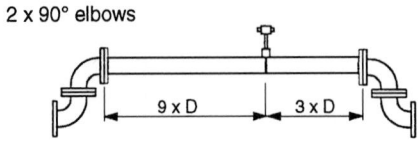

2 x 90° elbows
3-dimensional

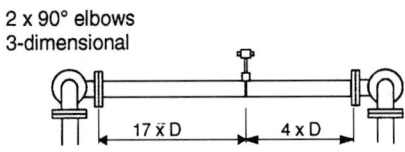

3 x 90° elbows
3-dimensional

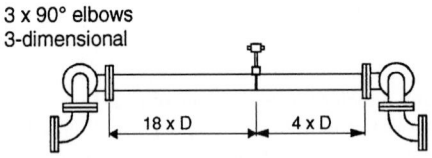

Reduction

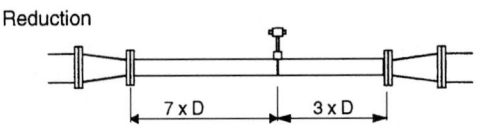

Expansion

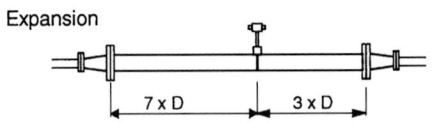

Control valve

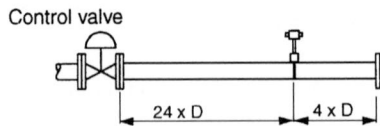

FIGURE 78   Upstream/downstream straight pipe lengths
(ISO 5167) for pitot tube.

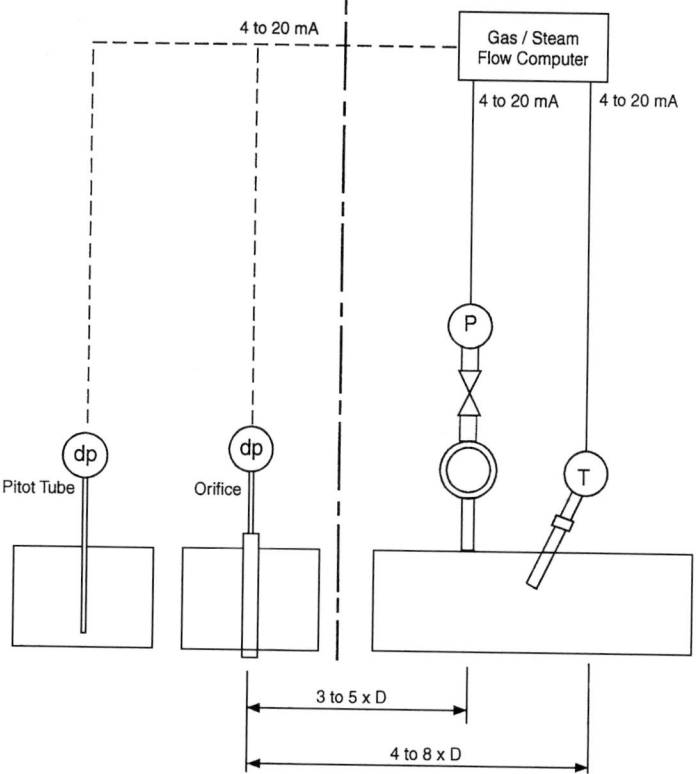

FIGURE 79   Installation for gas and saturated steam.

the American Society of Mechanical Engineers, the Instrument Society of America, the American National Standards Institute, and the National Engineering Laboratory (Germany). Some guidance for specifying DP flowmeters is given in Table 1, which lists the principal characteristics, including advantages and limitations.

Differential Producers

Flow is related to pressure by causing the flowing fluid to pass through some form of restriction in the transport pipe, which creates a momentary loss of pressure. As will be explained shortly, this pressure differential is related mathematically to the flow rate.

VARIABLE-AREA FLOWMETERS

The principle of using differential pressure as a measure of flow rate, as described in the case of the orifice, venturi, and other devices, is also applied in variable-area flowmeters, but with one important variation. In the orifice meter there is a fixed aperture (restriction) and the flow rate is indicated as a function of the differential pressure created. In the case of the variable-area meter there is a variable orifice and a relatively constant pressure drop. Thus the flow rate is indicated by the area of the annular

TABLE 1   Comparison of Differential Pressure and Target Flowmeters (*Continues*)

Applications:
    Liquid and gases:
        Orifice, venturi, flow nozzle, flow tube, elbow, pitot, target
    Steam:
        Orifice, venturi, flow nozzle, flow, tube, target

Flow range:
    Liquids (minimum):

Orifice	$0.1 \text{ cm}^3/\text{min}$
Venturi	$20 \text{ cm}^3/\text{min}$ (5 gal/min)
Flow nozzle	$20 \text{ cm}^3/\text{mi}$ (5 gal/min)
Flow tube	$20 \text{ cm}^2/\text{mi}$ (5 gal/min)
Elbow	Depends on pipe size
Pitot	Depends on pipe size
Target	0.25 L/min (0.07 gal/min)

    Liquids (maximum):

All types	Depends on pipe size

    Gases (minimum):

Orifice	Gas equivalent of liquids
Venturi	$35 \text{ m}^3/\text{h}$ (20 scfm)
Flow nozzle	$35 \text{ m}^3/\text{h}$ (20 scfm)
Flow tube	$35 \text{ m}^3/\text{h}$ (20 scfm)
Elbow	Depends on pipe size
Pitot	Depends on pipe size
Target	$0.5 \text{ m}^5/\text{h}$ (0.3 scfm)

    Gases (maximum):

All types	Depends on pipe size

Operating pressure (maximum):
    Generally depends on transmitter selected

Target meter	68 MPa (10,000 psig)

Operating temperature (maximum):
    Depends on construction material selected

Target meter	$398°C$ ($750°F$)

Scale:

All types	Square root

Nominal accuracy:

Orifice	$\pm0.6\%$ (maximum flow)
Venturi	$\pm1\%$ (maximum flow)
Flow nozzle	$\pm1\%$ (full-scale)
Flow tube	$\pm1\%$ (full-scale)
Elbow	$\pm\%5$ to $\pm10\%$ (full-scale)
Pitot	$\pm5\%$ or better (full-scale)
Target	±0.5 to $\pm5\%$ (full-scale)

Rangeability:

Target	3:1
Other types	4:1

Signal:

All types	Analog electric or pneumatic

Relative advantages and limitations:
    Comparatively high cost:
        Venturi

TABLE 1   (*Continued*)

Comparatively low cost:
   Orifice, flow nozzle, flow tube. Pitot, elbow, target
Comparatively easy to install:
   Orifice, elbow
Same transmitter can be used for variety of pipe sizes:
   Orifice, flow nozzle, flow tube. Pitot, elbow
Comparatively low pressure loss:
   Venturi, flow nozzle, flow tube
Good for dirty liquids and slurries:
   Orifice (if equipped with eccentric or segmental plates), venturi, target meter
   (especially for hot, tarry, and sediment-bearing materials)
Large and heavy:
   Venturi
Requires straight runs of piping (upstream, downstream):
   Orifice, target meter
Limited background engineering data:
   Flow nozzle, flow tube
Bidirectional flow measurement:
   Elbow
Minimum upstream straight piping:
   Elbow
Flow averaging:
   Pitot
Accuracy limited:
   Elbow

---

opening through which the flow must pass. This area is generally read out as the position of a float or obstruction in the orifice.

Frequently referred to as glass-tube flowmeters or rotameters, variable-area flowmeters are designed in three general configurations:

1. The *tapered-tube meter*, in which a float (usually of a density greater than the fluid being measured) is contained in an upright glass tube, which is tapered, that is, the bottom of the tube is of a somewhat smaller diameter than the top. Thus the area or concentric orifice through which the fluid flows is greater as the float rises in the tube. The term rotameter arose from early technology when the rotor was slotted and thus rotated by the flowing fluid. The float is lifted to a position of equilibrium between the downward (gravity) force and the upward force caused by the flowing medium. Later technology introduced rod- or rib-guided nonrotating floats.

2. The *orifice and tapered-plug meter*, which is equipped with a fixed orifice mounted in an upright chamber. The float has a tapered body with the small end at the bottom and is allowed to move vertically through the orifice. The fluid flow causes the float to seek an equilibrium position.

3. The *piston-type meter*, in which a piston is fitted accurately inside a sleeve and is uncovered to permit the passage of fluid. The flow rate is indicated by the position of the piston.

Schematic diagrams of the different types of variable-area flowmeters are shown in Fig. 1

Floats are made from numerous corrosion-resistant materials. For ordinary service, metering tubes are borosilicate glass. Metal tubes (partial or full) are used for high-pressure and hazardous applications, thus requiring a magnetic coupling to the float position to provide flow-rate indication. Float designs have varied over the years. Earlier meters were affected to some degree by viscosity changes

TABLE 2 Characteristics of Variable-Area Flowmeters

Service	Liquids, gases, steam
Flow rate:	
Liquids	0.2 to 23 gal/min (0.6 to 87 L/min)
Gases	0.9 to 106 scfm (26 to 2993 L/min)
Rangeability	10:1
Accuracy	Depends on model
	High accuracy designs:±1% of reading from 100 to 10% (full-scale)
	Purge meters: ±10% (full-scale)
Repeatability	Depends on model; average 0.5% (full-scale)
Configurations	Full view; armored for high pressure; glass and metal tubes (magnetically coupled indicator); alarms
Operating temperature (maximum)	Glass tube: 400°F (204°C)
	Metal tube: 1000°F (583°C)
Operating pressure (maximum)	Glass tube: 350 psig (2.4 MPa)
	Metal tube: 720 psig (4.9 MPa)
Floats	Rod-guided, rib guided, tube-guided glass or sapphire spheres
Construction materials	Metering tubes: borosilicate glass
	High pressure: stainless steel, Hastelloy, Mone
	Depending on service, floats are glass, sapphire, tantalum, Carboloy, stainless steel, aluminum, Hastelloy, CPVC, Teflon, PVC
Advantages	Relatively low cost; direct indicating; minimum piping required
Limitations	Must be vertically mounted; need minimum back pressure (gases); usually limited to direct viewing

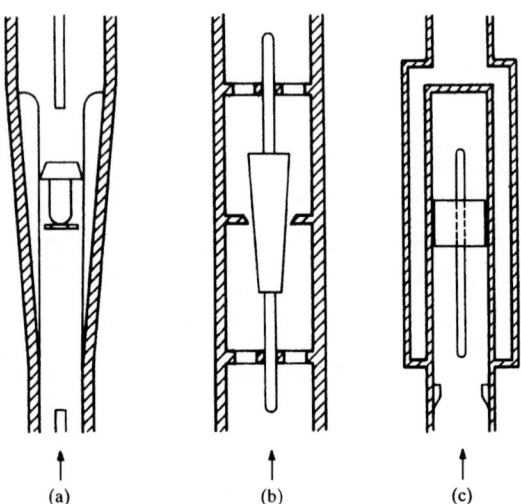

FIGURE 1 Schematic views of three basic types of variable-area flowmeters. (*a*) Tapered tube (rotameter). (*b*) Orifice and tapered plug. (*c*) Cylinder and piston.

of the metered fluid. Floats having sharp edges were found to be most immune to viscosity changes and are commonly used today for many applications. A majority of variable-area meters is used for direct reading, but they can be equipped with a variety of contemporary hardware for data transmission, alarming, and integration into sophisticated systems. The principal characteristics of variable-area flowmeters are summarized in Table 2.

MAGNETIC FLOW MEASUREMENT

General

The first magnetic flow-measuring system for closed pipes was developed in 1939 by a Swiss Father Bonaventura Thurlemann.

Magnetic flowmeters have been used for more than 50 years. Their construction and measuring features make them the ideal instruments for flow measurement of aqueous liquids.

The advantages of the instrument are now so well known that there is scarcely a branch of industry that does not utilize these advantages.

Because of continuing technical improvements in the electronics—advancing from valves through transistors to the microprocessor and to its construction (new electrodes and lining materials)—many versions are now available to the end user to fit specific requirements.

A conventional magnetic flowmeter comprises a sensor that receives a signal from the flowing liquid and a transmitter that converts this signal into a normalized current or pulse per unit of volume or time.

Magnetic flowmeter prices are comparable to that of orifice plates. Magnetic flowmeters are offered in both compact and conventional versions. In compact meters the sensor and transmitter form one mechanical unit; the measuring signal is converted into a current or pulse signal directly at the measuring point.

The latest generation of magnetic flowmeters has a microprocessor as a central control module. This microprocessor has taken over the functions performed in the past by discrete hardware. With the microprocessor changes in measuring ranges, time constants, and other parameters are easily programmable with specific limits.

For example, by push button it is possible to change a current output from 0–20 mA to 4–20 mA or pulse value from 1 to 20 gal per pulse. This technique is not only simple and reliable, but it also simplifies spare parts' stocking requirements, since the same electronics are used on all instruments.

Measuring Principle

Faraday's law of induction is the basis for magnetic flow measurement. When a conductor moves in a magnetic field, then a voltage is generated in this conductor (see Fig. 2);

$$U_e = B \cdot L \cdot v$$

where B = the strength of the magnetic field (induction)
L = the length of the conductor (distance of electrodes)
v = velocity of the conductor (average flow velocity)

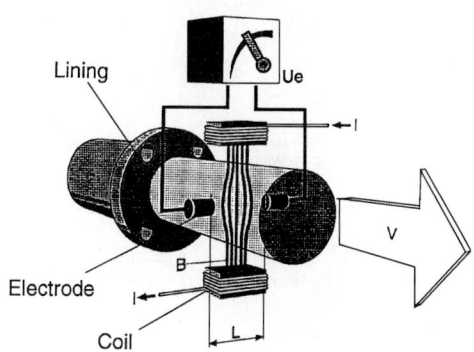

FIGURE 2   Magprin.

In magnetic flow measurement, the conductor is the flowing electrically conductive liquid. A cross section of liquid generally has no specific velocity. However, calculations show that the voltage generated between two opposite points on the inner pipe wall is largely independent of the velocity profile. Therefore the formula mentioned above also relates to liquids if the average flow velocity is substituted for v.

The magnetic field is generated by two coils that are energized by either an ac power line voltage or by pulsed dc voltage. The induced voltage U_e is received by two insulated electrodes. The insulation separation of the fluid being measured from the metal pipe is achieved by a lining material.

Electrode axis, magnetic field, and direction of flow are perpendicular to each other. The induced voltage at the electrodes (U) is proportional to the induction (B), to the flow velocity (v), and to the distance between the electrodes (L). Considering that the induction (B) and the distance between the electrodes (L) are constant values, the equation given in Fig. 3 (see also Fig. 4).

Since $\quad V = \dfrac{\text{Flow V}}{\text{Area A}}$

$$U_e = B \times \frac{V \times 4}{D^2 \times \pi}$$

Where D = Pipe internal diameter, therefore, U_e is proportional to V.

FIGURE 3 Formula.

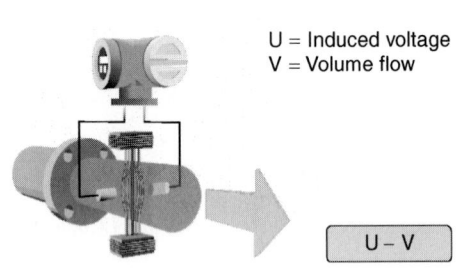

U = Induced voltage
V = Volume flow

U – V

FIGURE 4 Measuring system.

Sensor

A magnetic flowmeter always comprises a nonferromagnetic pipe (magnetically nonconductive) with a lining material, one or two coils with a magnetic core, and two measuring electrodes (see Fig. 5).

Construction

The measuring pipe is generally stainless steel (304 SS, 1.4301) with flanges to ANSI, DIN, SA, ASA, BS, or JIS standards. Plastic pipes can also be used.

The lining material separates the pipe from the liquid being measured. This is not required on plastic pipes. Standard lining materials, depending on application, temperature, and material being measured, are hard or soft rubber, polyurethane, Teflon, enamel, and ceramic (AI_2O_3) (see Table on next page).

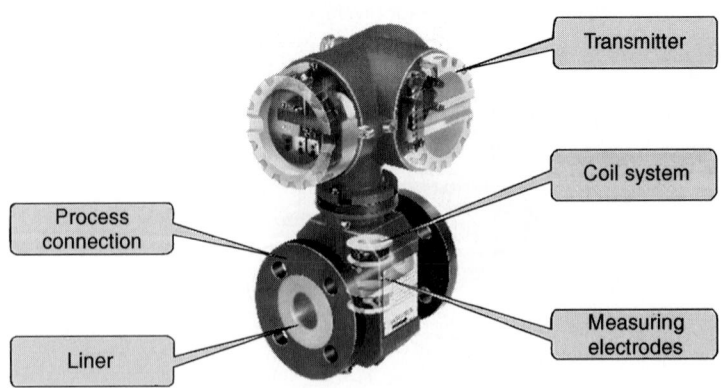

Transmitter

Coil system

Process connection

Measuring electrodes

Liner

FIGURE 5 Theory to practice.

Liner Materials for Magnetic Flowmeters—Common Types

Liner	Brand Name	Fluid Temp.	Industry
EPDM (soft rubber)	Keltan, Vistalon	$-58°$ to $+248°$F	Chemical, mining
Ebonite, SBR (hard rubber)		max. 176°F	Chemical, waste, paper, wastewater, Drinking water
PTFE	Teflon	$-40°$ to $+356°$F	Chemical, paper, food, heat measurement
PFA	Teflon	max. 392°F	Chemical, food, water, pharma, general purposes
FEP (bound)	Teflon	212°F	Chemical, food
Ceramic		$-40°$ to $+392°$F	Wastewater, biochemical, pharma
PU	Polyurethane	$-4°$ to $+176°$F	Chemical, mining, wastewater
CR	Neoprene, Bayprene	max. 176°F	Wastewater, chlorinated water
PP	Polypropylene	max. 140°F	Petrochemical, general purposes
NR	Natural rubber	176°F	Mining, aggregate, cement, concrete
PVDF		max. 392°F	Chemical, food, water, general purposes

The electrodes receive the signals and are therefore insulated from the pipe. The electrodes are in contact with the liquid being measured.

Consequently, corrosion-resistant electrode materials must be used such as stainless steel, hastelloy, tantalum, titanium, platinum/rhodium, and plastics or graphite. Electrode material selection can have a significant impact on the meter's price.

Application Areas

All aqueous solutions, such as water, effluent, clear sludge, pastes, juices, acids, and others can be measured with a magmeter. Select a version that is compatible with the minimum conductivity of the liquid being measured.

Petrochemicals, such as petroleum, benzene, and diesel fuel cannot be measured with the magnetic flow principle because of their low conductivity (for example, gasoline $= 0.00000001$ M^s/cm). For these substances, mechanical or other measuring techniques should be selected such as Coriolis or vortex flowmeters.

Accurate flow measurement is not affected by temperature, pressure, density, or viscosity of the liquid. The effect of the flow profile on the measurement is minimal, since the magnetic field is positioned in such a way that each point within the flow diameter in the area of the electrodes contributes to the measuring voltage (in millivolts). Measuring range full scale (100 percent) spans from 0.03 to 33 ft/s (1000:1).

The instrument can be installed at almost any position in the pipe, but vertical mounting is preferred. This prevents asymmetrical buildup on the electrodes and prevents air pockets from insulating the electrodes. The meter should be installed at a distance of at least 3 to 5 straight pipe diameters after equipment that generates turbulence (such as valves and restrictions).

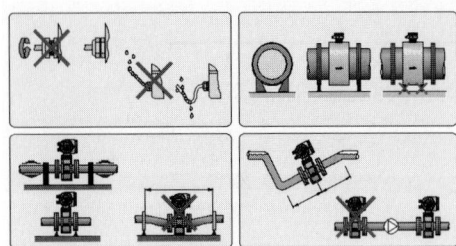

FIGURE 6   Installation hints.

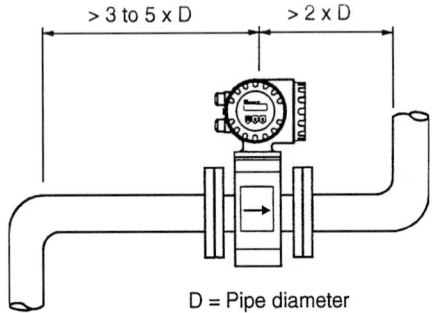

D = Pipe diameter

FIGURE 7   Inlet and outlet runs.

Magmeter Installation Tips and Suggestions

Inlet and outlet runs.   The sensor should be mounted upstream from fittings liable to generate turbulence; areas around valves, elbows, and T-junctions, for example (see Fig. 7).

Inlet run : $>3 \cdots 5 \times DN$
Outlet run : $>2 \times DN$

Mounting location.   Correct measurement is possible only when the pipe is full (see Fig. 8). The following locations should therefore be avoided:

- No installation at the highest point (air accumulation);
- No installation immediately before an open pipe outlet in a downward line. The alternative suggestion, however, permits such a location

Partly filled pipes.   For inclines, a mounting similar to a drain should be adopted. Do not mount the sensor at the lowest point (risk of solids collecting). Added security is offered by empty pipe detection (EPD). This option has an extra electrode in the measuring pipe. Note: Here too, the inlet and outlet lengths should be observed (see Fig. 9).

Download pipe.   With the installation suggested below, no partial vacuum is created with such downward pipe >5 m long (siphon, vent valve downstream of the sesor). (see Fig. 10).

Installation of pumps.   Do not mount the sensors on the suction side of pumps (see Fig. 11). There is a risk of vacuum!

Adapters.   The sensor can also be mounted in a pipe with a larger nominal diameter when suitable adapters (reducers and expanders) to DIN 28545 are fitted. The resultant increase in the rate of flow increases the accuracy of measurement with slowly moving fluids.

The nomogram given below can be used to determine the pressure loss caused.
Procedure:

1. Determine the ratio of the diameters d/D.

2. From the nomogram, read off the pressure loss at the flow velocity and d/D ratio.

Note: The nomogram applies to fluids with a viscosity similar to that of water (see Fig. 12).

What can be measured with a magnetic flow meter?

Chemicals	Acids, alkaline solutions, solvents, cooling liquids (glycol) additives
Food	Water, beer, wine, spirits, milk, yogurt, soft cheese, juices, molasses, sugar, and salt solutions, blood, sausage pulp
Metal industry	Pump control, cooling water, circulated water
Effluent	Effluent, raw sludge, purified sludge, neutralization chemicals, milk of lime, flocculants
Drinking water	Supply water, reservoir and pumping stations, consumption
Wood chip	Measurement and dosing of glue, pulp stock
Textiles	Water, chemicals, dyes, bleaching agents
Photography	Emulsions
Power station	Differential measurement of cooling circuits, heat quantity measurement
Animal feed	Water, molasses, liquid feed

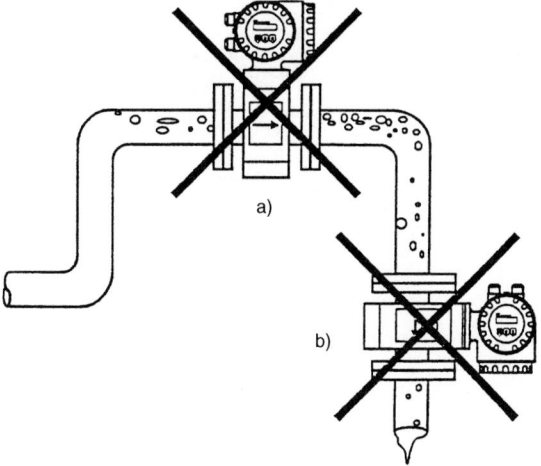

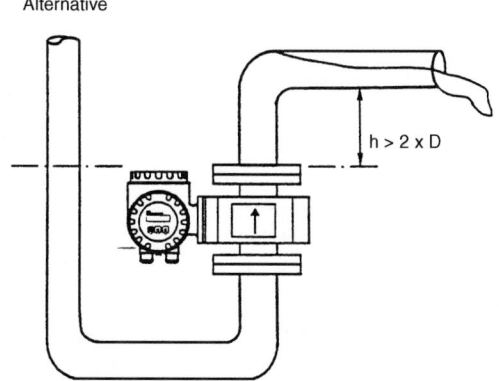

FIGURE 8   Mounting locations.

In conjunction with a heat calculator, the consumed or generated amount of thermal energy can be computed from the temperature difference between the forward and the reverse lines and the volumetric flow rate.

SELECTION CRITERIA: WHICH SYSTEM FOR WHICH APPLICATION?

Alternating (ac) Field System

Magmeters with ac magnetic fields are being replaced with instruments with pulsed direct fields (dc). Because of their simple electronics, AC systems had shown a small price advantage, but their versatility and capabilities are limited. These instruments operate at 50 or 60 Hz for magnetic-field generation and are large power consumers. These instruments are not completely maintenance free because of drifts of the sensor zero point that must be manually adjusted at no-flow conditions at frequent intervals.

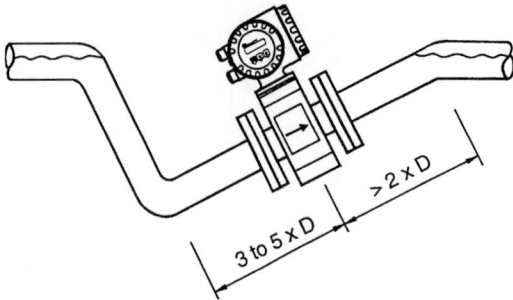

FIGURE 9   Partly filled pipes.

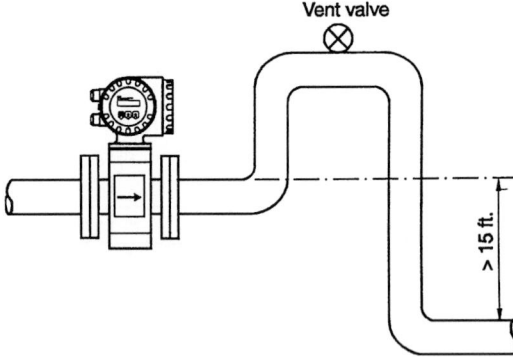

FIGURE 10   Downward pipe.

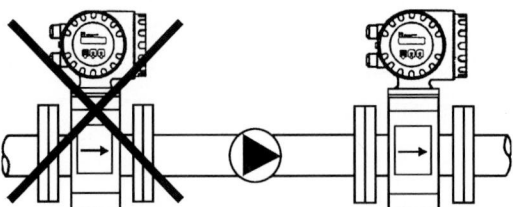

FIGURE 11   Installation of pumps.

Pulsed Direct (dc) Field Systems

Because of the automatic zero correction (AUTOZERO) feature, dc systems are particularly suitable for the following applications:

- liquids for materials with wide temperature and conductivity changes
- widely varying ambient temperatures
- liquids that tend to form a buildup inside the pipe

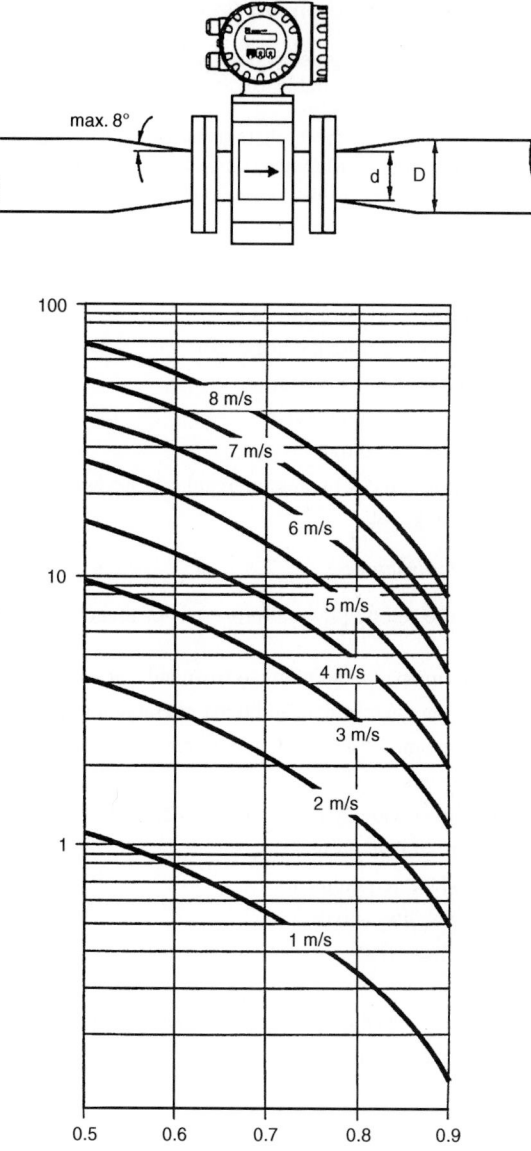

FIGURE 12   Nomogram.

Because of the high input impedance (10^{12} ohms) amplifier designs, even buildup of nonconductive material will have negligible effects on measuring performance. But too much buildup will cause the entire pipe's I.D. to be reduced, causing a permanent shift on the readings.

Power consumption is much lower. Therefore it is possible to drive these instruments for a low-voltage supply. Many of the microprocessor-controlled magmeters have a power consumption for all diameters of 15 VA.

The Inductive Flowmeter for Special Measurement Problems

Specific advantages of current magmeters have led to their increased use over traditional mechanical measuring devices. Designs are available to operate on very fast running events with short cycle times while achieving excellent repeatability.

The instrument is used in the food, pharmaceutical, and chemical industries in which a required volume fills a specific container. Magmeters are also used in applications in which the noise level is excessively high such as in slurry measurement or foods with solids, like yogurt with fruit, or soups containing vegetables.

Advantages:

- excellent repeatability: ±0.1 percent
- cleaning without removing from the line (CIP)
- no moving parts and therefore no wear
- ability to adjust for any desired volume
- internal preset counter (batching)
- measurement of tailoff, no costly pressure regulators necessary
- simple setup and installation

General magmeter advantages:

- Measurement not affected by physical properties of liquid being measured, such as temperature, pressure, viscosity
- No mechanical moving parts, therefore no wear and tear, maintenance free
- No diameter reduction, therefore no additional pressure loss
- Ideal for heavily contaminated liquids, sludge, entrained solids—but note velocity (see below)
- Wide range of diameters
- Excellent linearity over wide dynamic range and accuracies as high as ±0.2 percent
- Largely independent of flow profile
- Short straight in and out requirements
- High measuring reliability—no drift
- Can be cleaned in place (CIP)
- Bidirectional measurement
- Installed ground electrodes eliminate extraneous hardware like grounding rings
- Transmitter repair or replacement simple and convenient without shutting down the line, and requires no sensor recalibration

Limitations

- Only conductive liquids can be measured. Min. cond. 1 μS/cm (μmho), no gas or steam
- Buildup in pipes causes error because of reduction in pipe diameter
- Temperature and pressure limitations (technical specifications) of various liner materials

Recommended Flow Velocities	
Normal liquids	6 to 15 ft/s
Abrasives	<5 ft/s
Sludges	>5 ft/s
High percentage of suspended solids	>5 ft/s

TURBINE FLOWMETERS
by Bill Roeber[*]

Turbine flowmeters consist of a rotating device, called rotor, that is positioned in the fluid path of a known cross-sectional area, the body or pipe, in such a manner that the rotational velocity of the rotor is proportional to the fluid velocity. Since the cross-sectional area of the pipe is known, fluid velocity can be converted directly to volumetric flow rate by counting the number of turbine-wheel revolutions per unit of time. The following equation relates the conversion from fluid velocity (feet per second) to volumetric flow rate (gallons per minute):

$$Q = v \times A \times C$$

where Q = volumetric flow rate
   v = fluid velocity
   A = cross-sectional area
   C = constant

k Factor

A turbine flowmeter's K factor is determined by the manufacturer by displacing a known volume of fluid through the meter and summing the number of pulses generated by the meter. This summation of pulses, divided by the known volume of fluid that has passed through the turbine flowmeter, is the K factor. The turbine flowmeter's K factor is specified to be linear, within a tolerance, over a 10:1 range of flow rates. Extended flow ranges (greater than 10:1) normally restrict the user to specific fluid parameters. A representative graph of common turbine flowmeter characteristics is given in Fig. 13.

Reynolds Numbers

An important condition affecting the flow measurement is the Reynolds number of the fluid being measured. Reynolds numbers represent a unitless value that defines the ratio of a fluid's inertial forces to its drag forces. The Reynolds number equation relating these forces is

$$R = \frac{3160 \times Q \times G_t}{D \times \mu}$$

where R = Reynolds number
   Q = fluid flow rate, gal/min
   G_t = fluid specific gravity
   D = inside pipe diameter, inches
   μ = liquid viscosity, centipoise (cP)

The fluid flow rate and specific gravity represent inertial forces; the pipe diameter and viscosity are the drag forces. In a typical system the specific gravity and the pipe diameter are constants. Consequently low Reynolds numbers exist when the fluid flow rate is low or the fluid viscosity is high. The problem with measuring the flow of fluids with low Reynolds numbers using turbine flowmeters is that the fluid will have a higher velocity at the center of the pipe than along the pipe's inside diameter. This results in a K factor that is not constant over the flow rate range of the turbine flowmeter.

---

[*]Great Lakes Instruments, Inc., Milwaukee, Wisconsin.

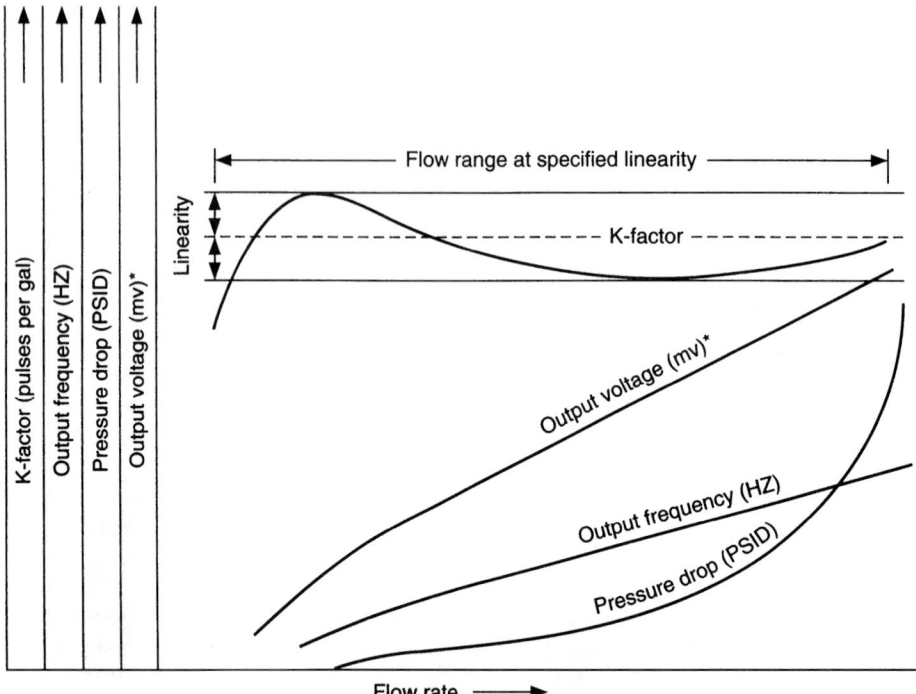

FIGURE 13 Turbine flowmeter performance characteristics. * Variable reluctance pickup only. (*Great Lakes Instruments, Inc.*)

Turbine Flowmeter Construction

Turbine flowmeters are constructed of a permeable metal rotor, housed in a nonmagnetic body, as depicted in Fig. 14. This allows proximity sensors to sense the rotational blades of the rotor directly through the body of the meter. Rotor designs vary from manufacturer to manufacturer, but in general, rotors with tangential axes are used for flow rates smaller than 5 gal/min (19 L /m) and axial rotors for higher flow rates. The rotor is designed to be of low mass with respect to the fluid momentum. This relationship shortens the response time of the turbine flowmeter by decreasing the rotational inertia required to accelerate the rotor. Turbine flowmeter bearings are of either the ball or the sleeve type. This provides long service life while minimizing frictional losses.

The mechanical losses in a turbine flowmeter can include bearing friction, viscosity shear effects, and magnetic or mechanical pickoff drag. These are illustrated in Fig. 15. These losses are most pronounced when turbine flowmeters are operated in the transitional or laminar flow region (at low fluid velocities or high fluid viscosities). Best results are obtained when turbines operate in media having Reynolds numbers in excess of 4000 or 5000. This varies with each manufacturer. Turbine flowmeter bodies have straightening vanes at the inlet and outlet to stabilize the fluid flow in the rotor area. A minimum of 10 pipe diameters of straight pipe upstream and 5 diameters downstream from the flowmeter installation is recommended for best measurement accuracy.

The rotor speed is sensed through the flowmeter body using several different methods. The most common for liquid service is the use of a variable-reluctance pickup. This device generates a magnetic field through a coil and the nonmagnetic flowmeter body into the rotor area. As each rotor blade enters the magnetic field, the discontinuity excites a voltage in the coil, producing an electrical sinusoidal wave current with frequency and voltage proportional to the rotor blade velocity.

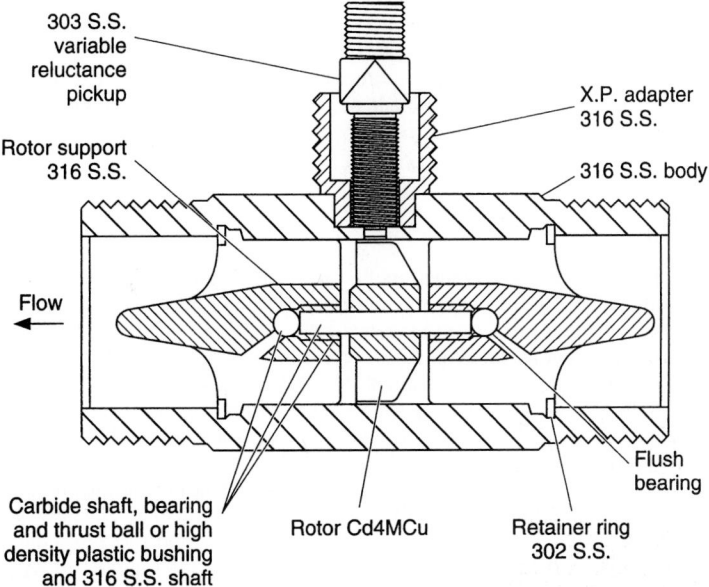

FIGURE 14 Typical axial-rotor [0.5- to 3-inch (12.7- to 76.2-mm)] turbine flowmeter construction. (*Great Lakes Instruments, Inc.*)

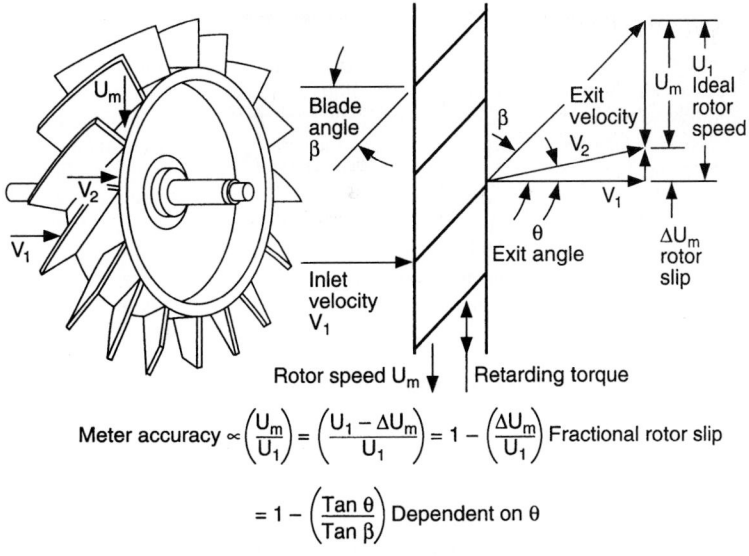

Meter accuracy $\propto \left(\dfrac{U_m}{U_1}\right) = \left(\dfrac{U_1 - \Delta U_m}{U_1}\right) = 1 - \left(\dfrac{\Delta U_m}{U_1}\right)$ Fractional rotor slip

$= 1 - \left(\dfrac{\text{Tan } \theta}{\text{Tan } \beta}\right)$ Dependent on θ

FIGURE 15 Standard turbine flowmeter rotor losses. (*Rockwell International.*)

Another popular method of sensing rotor speed is with a modulated-carrier electronic pickup. This type of pickup is designed to eliminate the physical loss caused by magnetic drag with the variable-reluctance pickup. This is especially important in low-velocity and gas flowmetering applications. The modulated-carrier pickup generates a "carrier" signal of a frequency much higher than that of the rotor blade. A coil, which is integral to the high-frequency oscillator, is positioned near the rotor. As each blade passes the coil, eddy-current losses in the coil increase, causing the amplitude of the carrier signal to decrease until the blade has passed. Filtering out the high-frequency component of the carrier retains the "modulated" signal, which is proportional to the fluid velocity.

Turbine Flowmeter Applications

Turbine flowmeters accurately measure flow over a large range of operating conditions. Primary uses include flow totalizing for inventory control and custody transfer, precision automatic batching for dispensing and batch mixing, and automatic flow control for lubrication and cooling applications. Typical measurement accuracy is ±0.5 percent of reading over a 10:1 flow range (turn-down ratio) and repeatability within ±0.05 percent of rate. A degradation in performance can be expected if there are variations in fluid viscosity, swirling of fluid within the pipes, or contamination which causes premature bearing wear. Turbine flowmeters are available for pipe sizes of 0.5 to 24 inches (12.7 to 610 mm) and flow ranges to 50,000 gal/min (11,358 m^3/h). Working pressures are generally determined by the inlet and outlet connections. Some designs are capable of working pressures in excess of 10,000 psig (68,950 kPa). Temperature limits are imposed by the pickup electronics. Most manufacturers specify 400°F (204°C) standard and extend to greater than 800°F (427°C) on special order. Popular end fitting connection styles include NPT, flared tube, flanged, sanitary, ACME, grooved, and wafer.

Paddle-Wheel Flowmeters

The paddle-wheel-type flowmeter, also referred to as an impeller flowmeter, is characterized in much the same way as the turbine flowmeter. The rotating portion of the flowmeter, called the rotor or impeller is positioned into the outside perimeter of the liquid path. As the liquid moves past the rotor, a moment is imposed on the blades of the rotor that are present in the stream. This moment causes the rotor to accelerate to a velocity equal to the liquid velocity. Since the cross-sectional area of the pipe is known, the volumetric flow rate can be easily obtained from the liquid velocity. The paddle-wheel flowmeter has inherently lower cost because the same sensing mechanism can be used with several pipe sizes and hardware configurations. Paddle-wheel flowmeters require a full pipe for operation and measure liquid velocities between 0 and 30 ft/s (9.1 m/s). This technology has typical linearity of +1 percent of full-scale and a repeatability of ±0.5 percent of full-scale. Operating conditions include pressures up to 2000 psig (139 kPa) and temperatures to 400°F (204°C).

Various types of mounting hardware adapt the paddle-wheel flowmeter to installation requirements. These include T-mounts, as shown in Fig. 16, insertion mounts for use with pipe saddles and weldments, spool pieces for mounting the sensor between pipe flanges, and hot tap hardware, which allows the insertion sensor to be serviced without depressurizing the system in which it is installed. The characteristics of turbine flowmeters are summarized in Table 3.

OSCILLATORY FLOWMETERS

As with the turbine flowmeter, the oscillatory meters sense a physical property of a moving fluid that is related to stream velocity. There are two principal meters in this classification, (1) vortex meters and (2) fluidic meters. Vortex meters are classified as vortex shedding or vortex precision types.

TABLE 3 Characterstics of Turbine Flowmeters*

Service	Liquids (clean), gases, and steam
Accuracy	$\pm$ 0.5% (reading) over 10:1 flow range; depends on specific designs; on average as quoted in literature, $\pm$ 0.25% (rate) for liquids, $\pm$ 1% (rate) for gases; factory calibration should include viscosity and lubricity factors for liquids
Repeatability	$\pm$ 0.05% (reading) over 10:1 flow range
Sizes	Generally 0.5 to 24 inches (12.7 to 610 mm), flanged or threaded connections
Rangeability	10:1 (generally); can be extended to 50:1, but with restricted flow parameters
Scale	Nominally linear, especially for Reynolds number higher than 10,000
Flow ranges	Extremely low flow rates claimed; depends on design; generally, for liquids, 5gal/min (19 L/min) to 50,000 gal/min (11.400 m^3/h); gases, 10 mil ft^3/h (285,000 m^3/h)
Pressure	Up to 3000 psig (20 MPa)
Temperature	Up to 400°F (204°C), but can be extended to 800°F (427°C) on special order; installations and maintenance relatively easy; can be used as bypass meter around pipeline orifice
Limitations	Problems may be encountered with two-phase flows; upstream straight piping required, including straighteners

* * Prepared by hand staff.*

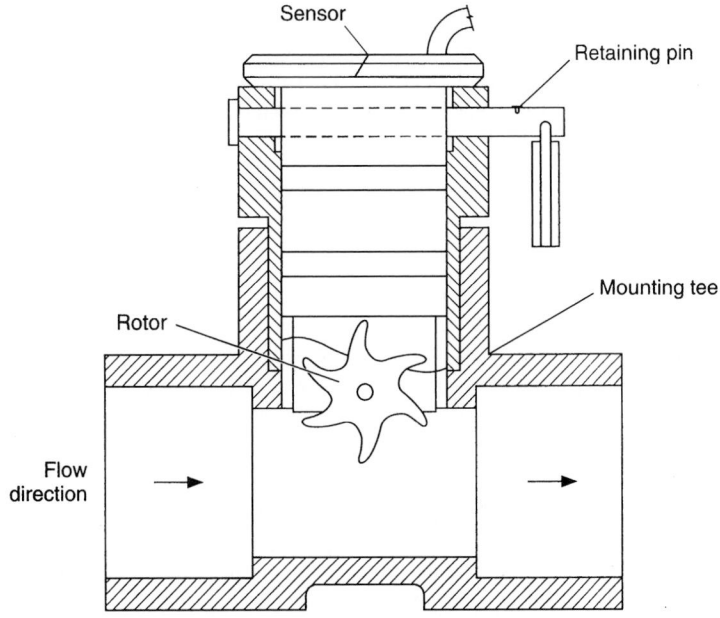

FIGURE 16 Typical paddle-wheel flowmeter construction. (*Great Lakes Instruments, Inc.*)

VORTEX SHEDDING FLOWMETERS

When a flowing medium strikes a nonstreamlined object or obstruction, it separates and moves around the object or bluff body before passing on downstream. At the point of the contact with the bluff body, vortex swirls or eddy currents separate from the bluff body on alternating sides.

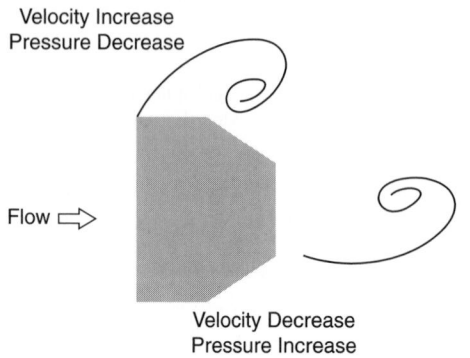

Velocity Increase
Pressure Decrease

Flow ⇨

Velocity Decrease
Pressure Increase

Net Pressure

FIGURE 17   Vortex diagram.

When this occurs, the separation or shedding causes an increase in pressure and a decrease in velocity on one side of the object. Simultaneously, a decrease in pressure with corresponding increase in velocity occurs on the other side of the object. After shedding from one side, the process is reversed, and a swirl or vortex is shed on the other side of the bluff body. The bluff body is sometimes referred to as a shedder bar (see Fig. 17).

In this way the vortex swirls are shed continuously, always 180 degrees out of phase with each other. The frequency of the shedding process is determined by the velocity of the material flowing past the bluff body or shedder bar. Each vortex swirl is of the same volume, without regard to the flowing medium.

The manufacturer performs a wet-flow calibration as part of the manufacturing process, during which the meter's K factor is determined. The K factor indicates how many pulses there are to a unit of volume, typically either gallon or liter. This K factor remains constant over the useful life of the meter and is determined by the meter body and the shedder bar dimensions. Unless corrosion or erosion drastically alters those dimensions, the meter will remain accurate. Therefore periodic recalibration is not required.

Typical vortex frequencies range from as low as one or two cycles per second to thousands of cycles per second, depending on flowing velocity, flowing medium, and meter size. Meters measuring gas normally have frequencies approximately 10 times the frequency found on meters in liquid applications. This is because gas-flowing velocities are normally much higher than liquid velocities in the same size pipeline. Smaller-sized meters normally have higher vortex shedding frequencies than large meters.

Vortex meter output is a linear function. The output signal may be either a 4–20-mA analog or a pulse/frequency as required. Accuracies of ±1 percent of rate or better are typical.

Application Guidelines

The vortex meter works well on relatively clean liquids, gas, and steam that do not contain a significant amount of solids. The liquid should also have a relatively low viscosity. The limiting factor for liquid is the Reynolds Number (Re). Typically the vortex meter is completely linear down to Re = 20,000. Below that Re the vortex shedding process becomes less regular. Vortex meters may be used below that down to as low as Re = 4000, at a slight loss in accuracy. Below Re = 4000 the vortex meter is not reliable.

In gas applications, the density of the gas is an important consideration. At a very low flowing velocity, a low-density gas has very little energy in the vortex swirl. At some point the energy present is too low for the sensor to detect the presence of flow and the meter goes to a zero flow condition.

For the most part, vortex meter application strengths fall within the same applications traditionally served by the orifice plate and differential pressure meter for many years—namely, clean liquids of low viscosity, gas, and steam.

Vortex Meter Advantages

Accuracy and rangeability are the main advantages of the vortex meter. The rangeability of the typical vortex meter is 20:1 and sometimes even higher. Ranges of 45:1 are possible, depending on the viscosity of the liquid or the density of the gas (see Figs. 18–20).

FIGURE 18   Prowirl 70 vortex meter.

FIGURE 19   Prowirl 77 vortex meter.

The proper installation of the vortex meter is important as it has an impact on the accuracy of the meter. Follow the manufacturer's instructions. Most of the time, when a vortex meter appears to be inaccurate, the cause is traceable to improper installation or the presence of a foreign object that causes unwanted turbulence in the flow stream.

Combating Potential Vortex Meter Problems

Vibration is one of the potential problems associated with vortex meters. Meters that use piezoelectric crystals for sensing the vortex shedding are subject to the stresses in the pipeline caused by vibration or pipeline noise—pulses in the flowing medium caused by valve cavitation or pump noise.

Some vendors attack the problem electronically by adding filters to the electronics. Another approach is to use mass-balanced differential switched capacitance (DSC) sensors that are immune to vibration and noise. When vibration is the cause of the output at a no-flow condition, the problem is sometimes solved by simple rotation of the meter in the pipeline, making the shedder bar parallel to the plane of the vibration.

Vortex meter electronics may be mounted directly on the meter body or mounted remotely where convenient. Contact the vendor for recommendations regarding the distance the electronics may be placed from the sensors.

FLUIDIC FLOWMETERS

Process Variables—Field Instrumentation

Fluidics is the technology of sensing, controlling, and information processing with devices that use a fluid medium. The operation of fluidic meters is based solely on the interaction between fluid streams. The particular function of each device, none of which has any moving parts, is dependent on its geometric shape.

Fluidics received considerable attention from the military in the 1960s as a control medium that could not be jammed, as is possible with unprotected electronic devices. The wall attachment fluid phenomenon (also called the Coanda effect) was first discovered in 1926 by Henri Coanda, a Rumanian engineer. The possible industrial application of fluidics was explored in the 1960s with the creation of amplifiers, valves, oscillators, and flowmeters, among other configurations.

Industrial fluidic flowmeters are currently available. These incorporate the principle of the fluidic oscillator (see Fig. 21).

In flowmeter applications, the fluidic oscillator has the advantages of

1. linear output with the frequency proportional to the flow rate,

2. rangeability up to 30:1,

3. being unaffected by shock, vibration, or field ambient-temperature changes,

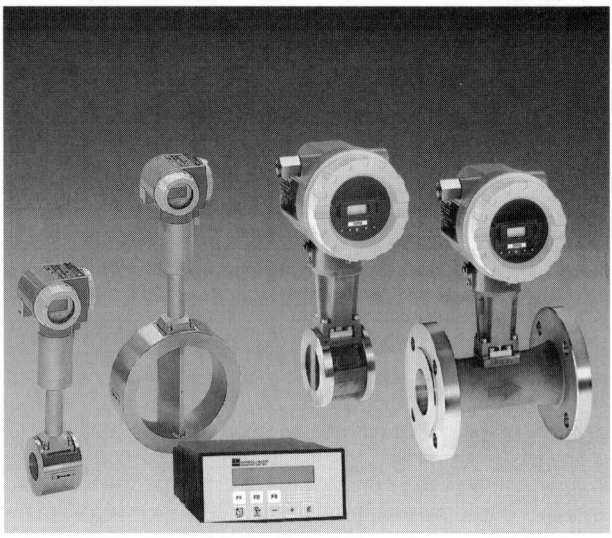

FIGURE 20   Prowirl family of vortex meters.

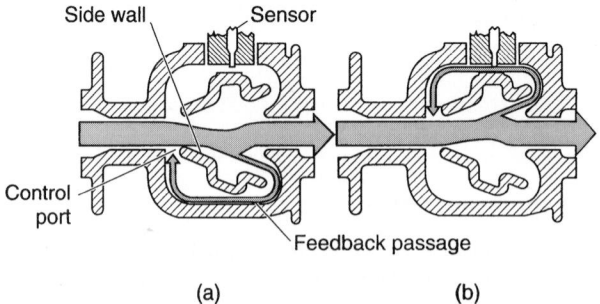

FIGURE 21   Operating principles of fluidic flowmeter.

4. calibration in terms of volume flow being unaffected by fluid density changes,

5. no moving points and no impulse lines.

A similar fluidic flowmeter, without diverging sidewalls, uses momentum exchange instead of the Coanda effect and has also proved effective in metering viscous fluids.

Fluidic Operating Principle

The geometric shape of the meter body is such that when flow is initiated, the flowing stream attaches to one of the sidewalls as a result of the Coanda effect. A portion of the main flow is diverted through a feedback passage to a control port.

The feedback flow increases the size of the separation bubble. This peels the main flow stream away from the wall, until it diverts and locks onto the opposite wall, where the feedback action is similar. The frequency of the self-induced oscillation is a function of the feedback flow rate. This rate is directly proportional to the flow rate of the mainstream (*Moore Products*).

Characteristics of Oscillatory Flow Meters

Flow range:	
Vortex shedding	3 to 5000 gal/min (11 to 19,000 L/min) for liquids
	10 mil scfh (283,200 m^3/h) for gases
Fluidic (Coanda)	1 to 1000 gal/min (4 to 4000 L/min) for liquids
Accuracy:	
Vortex shedding	±1% (rate) or better for liquids
	±2% (rate) for gases
Rangeability:	
Vortex shedding	20:1 or greater
Fluidic (Coanda)	up to 30:1
Scale:	
Vortex shedding	Linear (high Reynolds numbers)
Fluidic (Coanda)	Linear (high Reynolds numbers)
Operating pressure:	
Vortex shedding	Up to 3600 psig (25 MPa)
Fluidic (Coanda)	up to 600 psig (4 MPa)
Operating temperature:	
Vortex shedding	Up to 750°F (399°C)
Fluidic (Coanda)	0 to 250° (−18 to 121°C)
Advantages and limitations:	
Vortex shedding	Requires straight piping
	Sensitive to viscosity below minimum Reynolds number
	No moving parts
	Handles wide range of fluids, including steam
	Relatively good cost/performance ratio
Fluidic (Coanda)	Requires straight piping
	Sensitive to viscosity below minimum Reynolds number
	No moving parts
	Handles wide range of liquids
	Relatively good cost/performance ratio
	Bypass types available
Size ranges:	
Vortex shedding	0.5 to 12 in (15 to 300 mm)
Fluidic (Coanda)	1 to 4 in (25 to 100 mm)

MASS FLOW MEASUREMENT

Introduction

Traditionally flow measurement has been accomplished based on the volume of the flowing gas or fluid, even though the meter user may be more interested in the weight (mass) of the gas or liquid. Volumetric flowmeters are subject to ambient and process changes, such as density, that change with temperature and pressure. Viscosity changes may also affect certain types of volumetric flow sensors.

Mass flow measurement can be categorized as true mass flow measurement or inferential mass flow measurement. In the former, the mass flow is measured directly without regard to separately measured density or other physical properties. For inferential measurement, it is necessary to measure properties of the process such as temperature and pressure of a gas to infer the mass flow. For fluids with unknown properties, true mass flow measurement will be required.

Certain meters that use heat capacity (thermal mass flowmeters) are sometimes referred to as true mass flowmeters, since heat capacity is mass specific. But heat capacity is dependent on the fluid, and therefore the meter will read correctly if the fluid properties such as composition and heat transfer properties are known. To complicate the measurement, heat transfer is viscosity dependent. In contrast, a combination of a volumetric flowmeter and a density meter may be less dependent on the fluid properties.

For a number of years there has been a great interest in finding methods of measuring mass flow directly, rather than calculations alone to convert volume flow to mass flow. In today's industrial applications there are commonly three ways to determine mass flow.

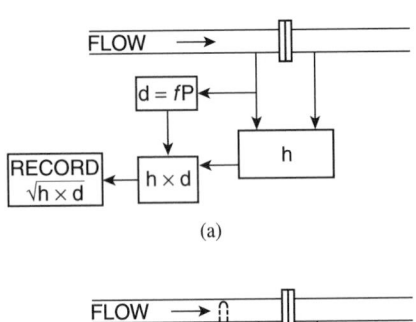

(a)

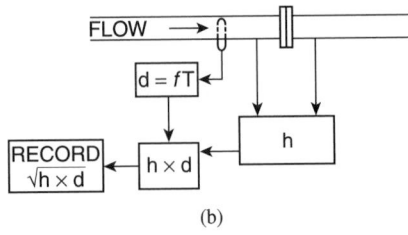

(b)

1. The inferential mass flow measurement—the application of microprocessor-based volumetric flowmeter technology to conventional volumetric meters. Separate sensors respond to velocity or momentum and pressure, temperature, etc.

2. The use of Coriolis flowmeters, which measure mass flow directly.

3. The use of thermal mass flowmeters, which determine mass flow by measuring heat dissipation between two points in the pipeline.

Inferential Mass Flow Measurement (Microprocessor-Based Volumetric Flowmeters)

As shown in Fig. 22 with a flow computer it is relatively simple to compensate for a volumetric flow meter with a density meter of the vibrating tube type. To compensate for temperature and pressure, it is helpful to know how well the gas follows the gas laws (how the gas expands or contracts with changes in temperature and pressure) for that particular gas.

The pressure- and the temperature-measuring points should be located close to the point of

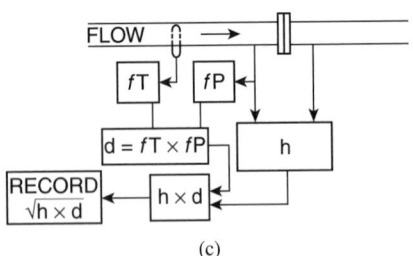

(c)

FIGURE 22   Method for compensating a volumetric flowmeter for mass flow.

volumetric flow measurements for the best accuracy. In some applications, this could be difficult, depending on the pipe configuration.

Examples of Mass Flow Measurement with Fluids of Known Properties

Flowmeters may have the ability to convert from volumetric flow to mass flow by obtaining additional information from additional measurements. For example, differential pressure devices require a value for actual density ρ for both volumetric and mass measurement. This may be obtained with a density cell (density measurement from a Coriolis mass flowmeter) of some sort or by measurement of the actual pressure p and actual temperature T for a gas and calculation of the actual density ρ when the gas law for that particular gas is known.

With the differential pressure Δp and additional sensor specific constants, this then provides a value for the mass flow rate Q_m. All values for the mass flow rate must be calculated in a flow computer or PLC that contains all these formulas. Today, with more powerful processors, more PLCs take over this function from flow computers.

Vortex Flowmeter Liquid Mass

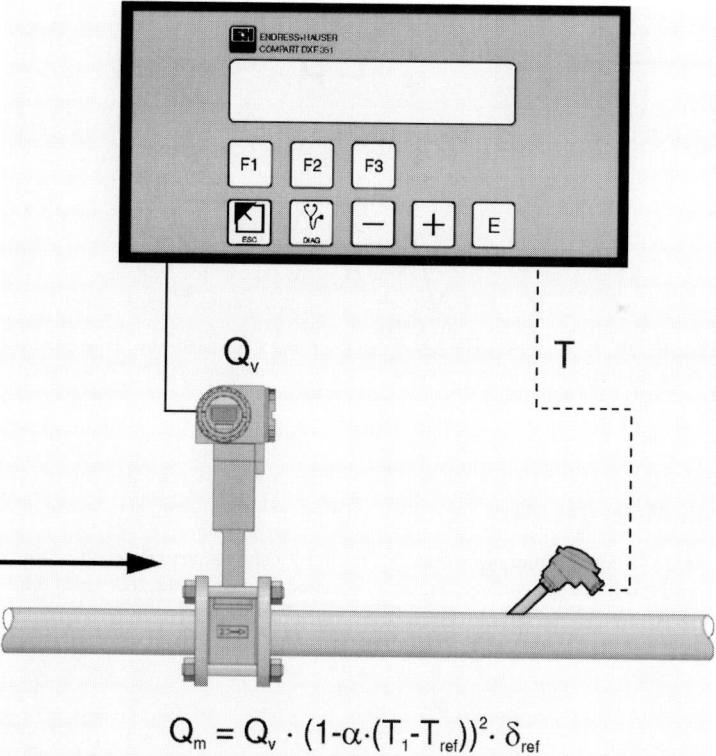

$$Q_m = Q_v \cdot (1 - \alpha \cdot (T_1 - T_{ref}))^2 \cdot \delta_{ref}$$

FIGURE 23 Required measurements Q_v and T; thermal expansion coefficient α, reference temperature T_{ref} and reference density δ_{ref} have to be programmed into the flow computer.

Vortex Flowmeter Gas Mass

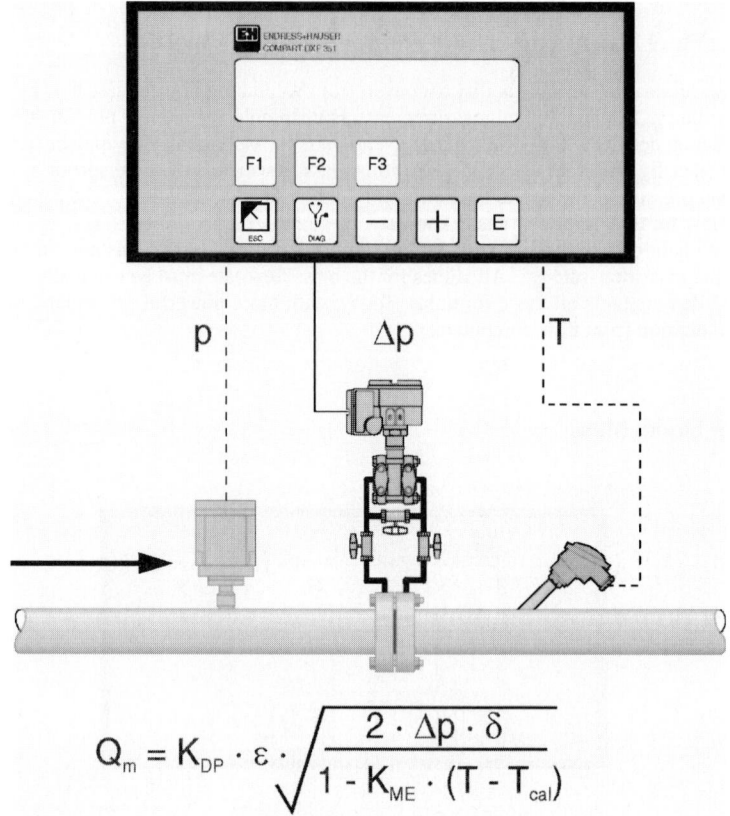

$$Q_m = K_{DP} \cdot \varepsilon \cdot \sqrt{\frac{2 \cdot \Delta p \cdot \delta}{1 - K_{ME} \cdot (T - T_{cal})}}$$

FIGURE 24   Required measurements Q_v and δ or P and T. Actual density can be calculated with the gas characteristics stored or programmed into the flow computer. The Z factor indicates how different a real gas behaves from an ideal gas under actual conditions that exactly obey the general gas law ($p \times V/T$ = constant; $2 = 1$).

Differential Pressure Meter Gas Mass

Figure 25 (Required measurements Q_v, P, and T) shows, actual density δ is calculated with steam tables stored in the flow computer, and the actual P and T saturated steam requires either a pressure P or temperature T measurement.

Vortex Flowmeter Steam Mass

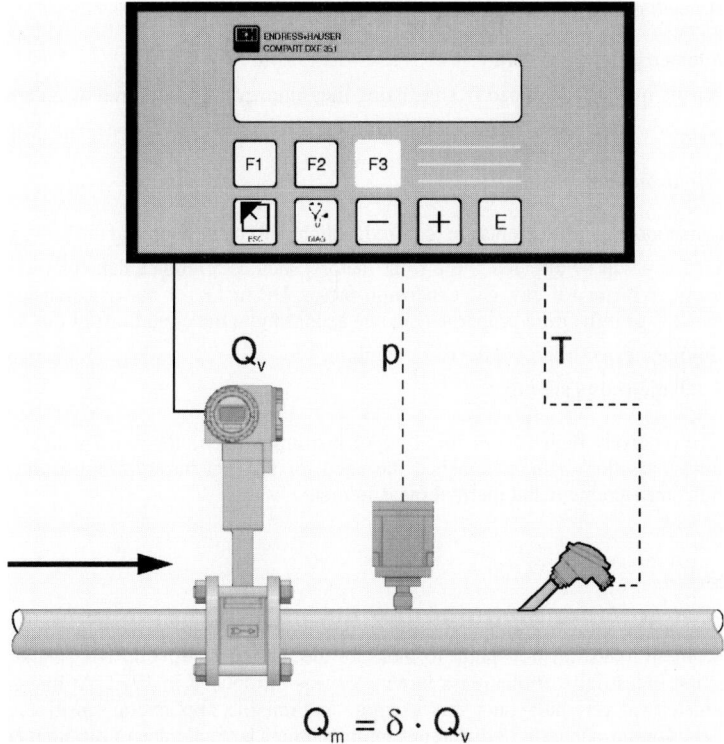

$$Q_m = \delta \cdot Q_v$$

FIGURE 25   Required measurements Q_v and δ and P and T. Actual density δ can be calculated with the gas characteristics stored or programmed into the flow computer (see also vortex meter). All other additional parameters describe the relationship between the flow rate and the measured differential pressure K_{DP}, E, K_{ME}, and T_{cal}.

Glossary

Q_m = Mass flow
Q_v = Uncorrected volume flow
α = Thermal expansion coefficient for fluid
δ = Actual density
δ_{ref} = Reference density
P = Actual pressure
P_{ref} = Reference pressure
T = Actual temperature
T_{ref} = Reference temperature
Z = Flowing Z factor
Z_{ref} = Reference Z factor
T_{cal} = Calibration temperature for DP cell, 294 K (20°F or 21°C)
E = Gas expansion factor
ΔP = Differential pressure
K_{DP} = DP factor
K_{ME} = Meter expansion coefficient

Many other types of volumetric or velocity meters may also be used in combination with density measurement.

Advantages:

- Volumetric flow is known.
- Density may be calculated from existing measurements or measured by a density meter.
- Mass flow may be programmed by powerful processors in PLCs or flow computers.

Disadvantages:

- Two or more measurements are required, which must be measured at the same location for accuracy.
- Accuracy will be affected if the right factors, such as reference density, thermal expansion coefficient, Z factor for gas, gas expansion factor, DP factor, or meter expansion coefficient are not known—or if the fluid properties change according to the conditions of the fluid.
- Multiple devices increase the risk of failure.
- Installation costs are higher.

The relatively high cost of this type of instrumentation, combined with an increasing need for reliable mass flow data, established the need for a direct mass flow instrument that incorporates Coriolis measurement and thermal measurement.

Coriolis Flowmeter

In comparison with differential pressure meters or positive displacement meters, the use of the Coriolis effect in an industrial flowmeter to measure the mass flow directly is a relatively new technology. The first industrial Coriolis mass flowmeter was introduced in 1972. At that time a Coriolis mass flowmeter was very bulky, not very accurate, not reliable, application sensitive, and very expensive.

Today, approximately 13 different suppliers offer a large number of different designs, from a single straight tube, to dual straight tubes, to all types of bent single tube or dual tubes. The latest generation of Coriolis mass flowmeters is accurate and reliable (typically 0.1% to 0.3%), and some of the designs are no longer bulky or application sensitive. Coriolis mass flowmeters have a very large installation base in liquid, slurry, and gas applications, and have been installed in some steam applications.

The success of the Coriolis mass flowmeter is based on the direct measurement of the mass flow and is unaffected by variations in fluid properties, such as viscosity, pressure, temperature, and density. In addition to the mass flow measurement, the Coriolis mass flowmeter is capable of obtaining measurements independent of mass flow, density, and temperature. The cost of a Coriolis mass flowmeter is very moderate when one considers the total cost of installation, maintenance, and actual output capabilities.

CORIOLIS FLOWMETER

Principle of Operation

The mass of a body on the Earth is usually determined by weight. Mass can also be determined by measurement of the acceleration, which is caused by an external force, as described in Newton's Second Law of Motion; force = mass × acceleration. The measuring principle is based on the controlled generation of Coriolis forces. These forces are always present when both translational (straight line) and rotational (revolving) movement occur simultaneously (see Fig. 26.)

$$\vec{F}_c = 2 \cdot \Delta m \, (\vec{\omega} \times \vec{v})$$

$\vec{F}_c$ = Coriolis force

Δm = Mass of moving body

$\vec{\omega}$ = Angular velocity

$\vec{v}$ = Radial velocity in a rotating oscillating system

The amplitude of the Coriolis force depends on the moving mass Δm, its velocity $\vec{v}$ in the system and therefore its mass flow.

FIGURE 26   Mass flow figure.

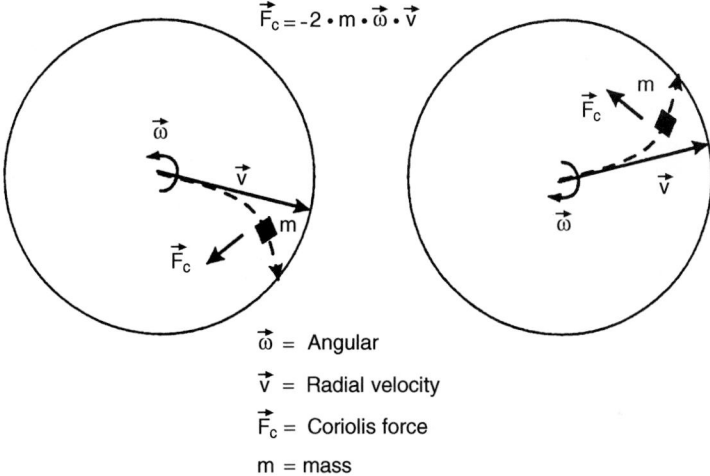

$$\vec{F}_c = -2 \cdot m \cdot \vec{\omega} \cdot \vec{v}$$

$\vec{\omega}$ = Angular

$\vec{v}$ = Radial velocity

$\vec{F}_c$ = Coriolis force

m = mass

FIGURE 27   Mass flow figure.

The amplitude of the Coriolis force depends on the moving mass Δm, its velocity v in the system, and therefore its mass flow.

The measuring principle is based on the controlled generation of Coriolis forces. The sensing meter contains a flow tube (or tubes) such that, in the absence of flow, the inlet and outlet sections vibrate in phase with each other. When fluid is flowing, inertial (Coriolis) forces cause a phase shift between inlet and outlet sections. Two sensors measure the phase difference, which is directly proportional to mass flow. The amplitude of the Coriolis force depends on the moving mass (process material), its velocity in the system, and therefore its mass flow (see Fig. 27).

The Promass (Coriolis mass flowmeter from Endress+Hauser) uses oscillation instead of constant angular velocity ω and one or two parallel measuring tubes. When process fluid is flowing through them, they are made to oscillate.

The Coriolis force produced at the measuring tube(s) causes a phase shift in the tube oscillation (see Fig. 28):

- When there is zero flow, i.e., with the fluid standing still, both phases are equal (1 in the figure), no phase difference.
- When there is flow (mass flow), the tube oscillation is decelerated at the inlet (2) and accelerated at the outlet (3).

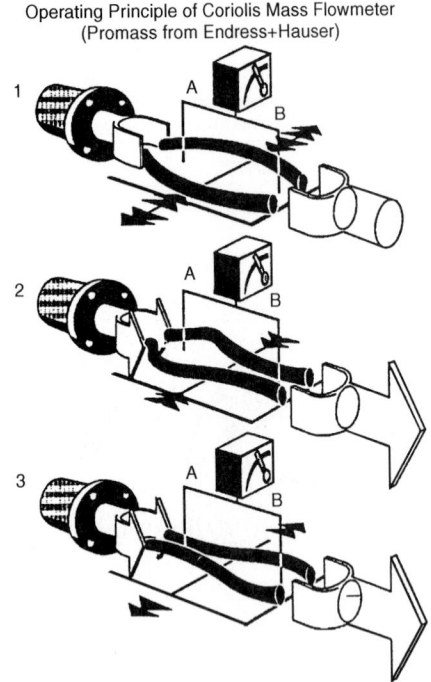

Operating Principle of Coriolis Mass Flowmeter
(Promass from Endress+Hauser)

FIGURE 28   Mass flow figure.

As the mass flow rate increases, the phase difference also increases ($A - B$). The oscillations of the measuring tube(s) are determined by electrodynamic sensors at the inlet and the outlet. The measuring principle operates independently of temperature, density, pressure, viscosity, conductivity, or flow profile.

Density Measurement

The measuring tubes are continuously excited at their resonant frequency. As the mass and therefore the density of the oscillating system change [measuring tube(s) and fluid], the vibrating frequency is readjusted. The resonant frequency is thus a function of the density of the fluid, enabling the processor to calculate density independently of the mass flow reading. This density information can be provided as an output signal.

Temperature Measurement

The temperature of the measuring tube(s) is determined to calculate the compensation factor that is due to temperature effects. This signal corresponds to the process temperature and is also an independently measured value as an available output.

Advantages

- Direct mass flow measurement unaffected by variations in fluid properties
- Multiple-variable measuring device (mass flow, density, and temperature). Some transmitters offer additional measuring variables based on calculation from the three independently measured values, such as percentage of mass flow, percentage volume flow, volume flow, standard volume flow, standard density, degrees Brix, degrees Baume.
- High accuracy (typical: $\pm 0.1\%$ to $\pm 0.3\%$)
- No moving parts; no rotating components
- Low maintenance
- Available in a range of corrosion-resistant measuring tube material

Disadvantages

- Initial cost for a Coriolis mass flowmeter ranges from moderate to high
- Limitation in size—6 in is the biggest size available

- Limitation in process temperature (typically 400°F or 200°C)
- High-pressure drop at full flow in some designs
- Some designs are very bulky
- Some designs are sensitive to vibration
- Some designs are installation sensitive

Application range

- Mass flow measurement for liquids, gases, and vapors are above the dew point (example superheated steam) as well as slurry or suspensions (mixture between liquid and solids as well as liquid and liquid).
- Do not use a Coriolis mass flowmeter to measure solid and gas, liquid and gas, or gas and liquid (saturated steam).

Installation hints

- Since vendors have their own sensor design, the performance of the Coriolis mass flowmeter can be affected by the selection of the installation location. Contact the vendor for specific installation requirements.
- Check corrosion compatibility between the fluid and the material used in the Coriolis mass flowmeter. Check what type of material of construction is used in the process.
- The measuring tubes of a Coriolis mass flowmeter have to be kept full; a partially filled pipe measurement is not possible.
- To ensure maximum performance of a Coriolis mass flowmeter, a zero point adjustment is recommended after installation and process startup. Contact vendor for specific instructions.
- Avoid entrained gas bubbles in a liquid because they affect the performance of the meter. If entrained gas is present, increase backpressure to dissolve the gas bubbles in the liquid. Some Coriolis mass flowmeters are capable of handling gas or air slugs by using smart process data filtering within the processor.
- Always install a Coriolis mass flowmeter on the high-pressure side of a pump and as close as possible to the pump.
- If possible, install the Coriolis mass flowmeter at the lowest point in the process. This guarantees the highest backpressure.
- If a Coriolis mass flowmeter is used to measure vapors or steam, ensure that the condensate will not collect in the measuring tubes.
- If a Coriolis mass flowmeter is used to measure slurry, be aware that some slurries are very abrasive. Since the measuring tubes (compared with the rest of the piping) are very thin, the life of the meter can be reduced. A single full-bore straight tube without a flow splitter will reduce abrasion and therefore the life of the meter. If the flow velocity is reduced by two, the abrasion factor will be reduced by four.
- If the ability to drain is important (contamination, safety, or to prevent buildup), straight tube meters have advantages over looped tube. Installing the sensor in a vertical pipe or in a horizontal pipe angled slightly upward will ensure that the sensor is completely drained.
- In general, elbows, valves, or pumps downstream of a Coriolis mass flowmeter do not affect the performance of the meter.
- If plugging is a potential problem, a straight tube sensor design is better because the sensor can be cleaned and inspected more easily.

BIBLIOGRAPHY

The book listed below was of particular value to the author of this section.

R. C. Baker, *An Introductory Guide to Flow Measurement* (Alden, Oxford, 1989).

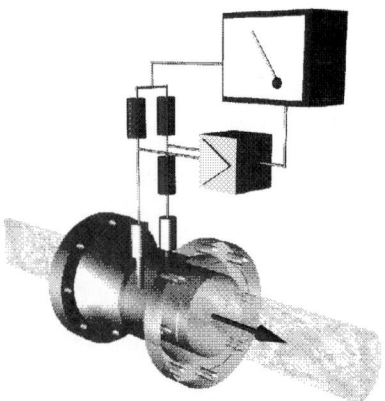

FIGURE 29   Thermal mass figure.

THERMAL MASS FLOWMETERS

There are two basic methods to sense the flow rate with a thermal mass flowmeter. In one type there are two heat sensors with a heat-producing element located between the two sensors. One sensor is upstream of the heating element, the other downstream. As the gas flows past the heating element, it increases very slightly in temperature (see Fig. 29).

The heat sensors detect the gas temperature before and after the gas passes the heating element. The heat acquired by the gas as it passes the heating element is a function of the mass flow of the gas. This type of sensing is often found in very small $1/8$- and $1/4$-in meters that are often used in gas chromatography.

In bigger meters ($1/2$ in and larger), it is more common to have just two RTDs located side by side. One of the RTDs detects the temperature of the gas. The electronics then increases the temperature of the other RTD so that there is a fixed temperature differential between the two—typically 30°F (see Fig. 30).

As the gas flows past the heated RTD, it is cooled slightly. The electronics then provides more power to the RTD to help maintain the temperature differential. The energy required for retaining the fixed temperature differential is a function of the mass flow of gas through the meter.

Thermal mass flowmeters are available in sizes from $1/8$ to 40 in. They are available as either insertion meters, in which the sensor is inserted into an existing pipe of duct, or as wafer or flanged meter bodies, depending on the pipe size (see Fig. 31).

While the meter is not significantly affected by changes in the gas pressure, it can be adversely affected by changes in gas temperature or gas composition. Since different gases have different thermal properties, the thermal mass flowmeter is often calibrated on the actual gas to be metered. If the flowing gas composition changes, these changing thermal properties will affect the rate at which the gas absorbs heat, rendering the meter inaccurate.

Another concern when thermal mass flowmeters are applied is the presence of moisture that may condense on the sensors. If condensate exists on one of the sensor RTDs, the affected sensor will not accurately measure either the temperature of the gas or the difference in the temperature of the gas from one RTD to the other.

There are minimum upstream straight pipe requirements that must be met to ensure that the flow profile is correct at the point of measurement. Consult the manufacturer's literature when selecting the location for the meter installation.

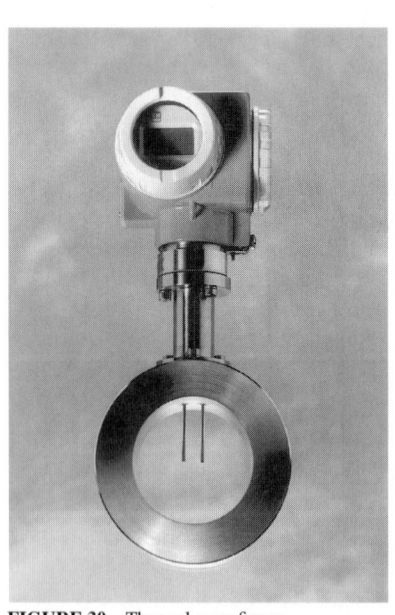

FIGURE 30   Thermal mass figure.

FLOW APPLICATION MATRIX

Element	Key Positive Points	Key Negative Points	Applications								Typical Features											
			Clean Liquid	Dirty Liquid	Viscous Liquid	Corrosive Liquid	Slurry	Clean Gas	Dirty Gas	Steam	Sizes available, in.	Accuracy, %	Flow Rangeability	Minimum Reynolds Number or Viscosity	Max. Pressure (psig)	Temperature Range (°F)	Straight piping required upstream	Pressure Loss	Purchase Cost	Installation Cost	Operational Cost	Maintenance Cost
Orifice Plate	Low cost, history of excellent service in many applications	High pressure loss, limited rangeability; accuracy affected by wear and fluid density	•	o		o		•		•	> 1	2-4 F	3:1	> 30,000	4000	≤ 1200	10-30	H	L	M-H	H	M-H
Flow Nozzle	Less expensive than venturi; more stable than orifice at higher temperatures and velocities	High pressure loss (similar to orifice); limited rangeability; accuracy affected by fluid density	•	o		o	o	•	o	o	> 2	1½ F	3:1	> 75,000	4000	≤ 1200	10-30	M	M	M	M	M
Venturi Tube	Low pressure loss; handles dirty fluids; short run of upstream piping	Limited rangeability; expensive; used at higher Reynolds numbers	•	o		o		•	o	o	> 2	1 F	3:1	> 100,000	4000	≤ 1200	5-10	L	H	M	L	M
Pitot Tube	Low cost, low pressure loss, can be used in irregular ducts; good for measuring velocity	Limited rangeability, low volumetric accuracy (3%); accuracy affected by fluid density; Low differential pressure, tends to clog	•			o		•		•	> 3	3 F	3:1	> 100,000	4000	≤ 1200	20-30	L	L	M	L	M
Wedge	Low pressure loss; handles dirty fluids	High cost; limited rangeability	•	•	•		o	•			1-4	½-2 F	4:1	> 500	4000	≤ 1200	10-30	M	H	H	M	H
Elbow	Low pressure loss; simple concept	Low accuracy; 2.5 ft/sec min. velocity	•	•	•	o	o	•			1-48	5-10 F	3:1	> 100,000	4000	≤ 1200	30	L	L	M	L	M
Variable Area	Low cost, direct reading, no power required; fluid can be clear or opaque; no upstream straight piping required	Expensive in larger sizes; accuracy affected by fluid properties	•	o	o	o		•			≤ 3	½-5 F	10:1	None	500	0-250	None	M	L	L	H	L
Target	Low cost; handles dirty fluids; suited for low Reynolds numbers; can be direct reading	Accuracy (½ – 5%) is affected by fluid properties	•	•	•	o		•	•	•	> 0.5	½-5 F	3:1	> 100	10,000	≤ 750	10-20	M	L	L	M	M

Table: Orientation / selection table for flowmeters (rotated landscape table).

Type	Characteristics	S1	S2	S3	S4	S5	S6	Size (in)	Inaccuracy	Rangeability	Viscosity (cSt)	Pressure (psig)	Temperature (°F)	Upstream diam.	R1	R2	R3	R4	R5
Turbine	Rangeability ≥ 10:1; measures low flow rates; easy to install	•			o		•	> 0.25	0.25 R	10-35:1	2-15 cSt	6000	-450-600	10-20	M	M	M	M	M-H
Positive-displacement	Most used for custody transfer; high accuracy; wide rangeability; good for viscous fluids; no upstream straight piping required	•	o	o		o	•	0.25-16	0.25 R	10:1	10 cSt	2000	-40-600	None	H	H	H	H	H
Vortex Shedding	Accuracy unaffected by fluid properties; rangeability 20-40:1; no moving parts; moderate pressure loss	•	o			o	•	> 0.5	1 R	20-40:1	> 4000	3000	-330-800	15-25	M	M	L-M	M	M
Fluidic	No moving parts; suitable for wide variety of liquids	•	•				•	1-4	1 F	30:1	> 3000	600	0-250	10-30	M	M	L	M	L
Magnetic	Can be used only with liquids having minimum threshold conductivity	•	•	•			•	> 0.1	½ R	30:1	None	5000	-40-350	5	N	M	L	L	L
Coriolis	Direct output in mass units without compensation; no obstructions to flow; high accuracy; wide rangeability; multi-variable; no upstream straight piping required	•	•	•		o	o	< 6	0.1 R	25:1	None	5700	-400-800	None	M	H	L	M	L
Thermal Mass	Direct output in mass units without compensation; negligible pressure loss; good for low velocity	•		o			•	> 1/8	1 F	50:1	None	4500	-40-150	10-30	L	H	L	L	L
Doppler	Nonintrusive; bidirectional; cost effective on large pipe sizes (clamp-on). Suspended particles needed by Doppler version; clean fluid needed by transit-time version, low accuracy	o	•	•		o		> 0.5	2-5 F	10:1	None	1000	-300-500	5-20	N	M	M	L	L
Transit Time	Nonintrusive; bidirectional; cost effective on large pipe sizes (clamp-on). Suspended particles needed by Doppler version; clean fluid needed by transit-time version, low accuracy	•	•	•		o		> 0.5	2 R	20:1	None	1000	-300-500	5-20	N	M	M	L	L

Legend:

• Recommended
o Limited application
☐ Not recommended

F Of Full scale
R of Rate

N None
L Low
M Medium
H High

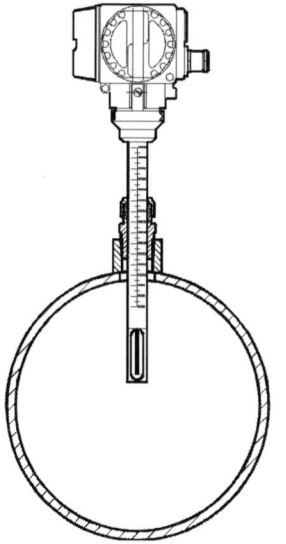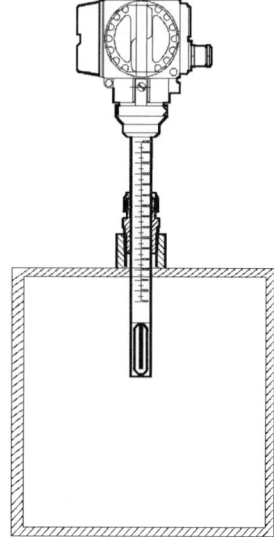

FIGURE 31   Thermal mass figure.

ULTRASONIC FLOWMETERS

Ultrasonic or acoustic flowmeters are of two principal types: (1) Doppler-effect meters and (2) transit-time meters. In both types the flow rate is deduced from the effect of the flowing process stream on sound waves introduced into the process stream. In clamp-on designs, these meters make it possible to measure the flow rate without intruding into the stream, and thus are classified as noninvasive. But even in configurations in which transducers are contained in shallow wells, the flowmeters are essentially nonintrusive.

The principles of ultrasonic flow measurement have been known for many years, but only within the past few decades have these meters made measurable penetration into the flowmeter field. This lag in acceptance has been variously explained, but the general consensus is that too many designs were introduced too soon—prematurely and without testing them against adverse plant environments—to the point that for several years ultrasonic flowmeters had somewhat of a tarnished image. Ultrasonic flowmeters have numerous innate advantages over most of the traditional metering methods: linearity, wide rangeability without an induced pressure drop or disturbance to the stream, achievable accuracy comparable to that of orifice or venturi meters, bidirectionality, ready attachment to the outside of existing pipes without shutdown, comparable if not overall lower costs, attractive features which can make a product prone to overselling. Within the last few years, ultrasonic flow measurement technology has found a firmer foundation, and most experts now agree that these meters will occupy a prominent place in the future of flowmetering.

Doppler-Effect (Frequency-Shift) Flowmeters

In 1842 Christian Doppler predicted that the frequencies of received waves were dependent on the motion of the source or observer relative to the propagating medium. His predictions were promptly checked for sound waves by placing the source or observer on one of the newly developed railroad trains. Over a century later, the concept was first considered for application in the measurement of flowing streams.

For the principle to work in a flowmeter, it is mandatory that the flowing stream contain sonically reflective materials, such as solid particles or entrained air bubbles. Without these reflectors, the

Doppler system will not operate. In contrast, the transit-time ultrasonic flowmeter does not depend on the presence of reflectors.

The basic equations of a Doppler flowmeter are

$$\Delta f = 2 f_T \sin \theta \frac{V_F}{V_S} \tag{1}$$

and, by Snell's law,

$$\frac{\sin \theta_T}{V_T} = \frac{\sin \theta}{V_S} \tag{2}$$

Simultaneously solving Eqs. (1) and (2) gives

$$V_F = \frac{\Delta f}{f_T} \frac{V_T}{\sin \theta_T} = k \Delta f \tag{3}$$

where
V_T = sonic velocity of transmitter material
θ_T = angle of transmitter sonic beam
K = calibration factor
V_F = flow velocity
ΔF = Doppler frequency change
V_S = sonic velocity of fluid
f_T = transmission frequency
θ = angle of f_T entry in liquid

Doppler-effect flowmeters use a transmitter that projects a continuous ultrasonic beam at about 0.5 MHz through the pipe wall into the flowing stream. Particles in the stream reflect the ultrasonic radiation that is detected by the receiver. The frequency of the radiation reaching the receiver is shifted in proportion to the stream velocity. The frequency difference is a measure of the flow rate. The configuration shown in Fig. 32 utilizes separated dual transducers mounted on opposite sides of the pipe. Other possible configurations are illustrated in Fig. 33. In essence the Doppler-effect meter measures the beat frequency of two signals. The beat frequency is the difference frequency obtained when two different frequencies (transmitted and reflected) are combined.

When the measured fluid contains a large concentration of particles or air bubbles, it is said to be sonically opaque. The more opaque the liquid, the greater the number of reflections that originate near the pipe wall, a situation exemplified by heavy slurries. It can be noted from the flow profile of Fig. 32 that these reflectors are in the low-flow-rate region. In contrast, the preponderance of particle reflectors will occur in the center of the pipe (where the flow rate is highest) when the fluid is less

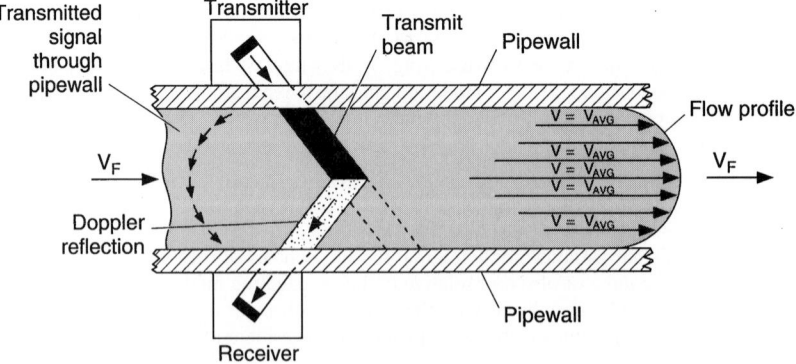

FIGURE 32   Principle of Doppler-effect ultrasonic flowmeter with separated opposite-side dual transducers.

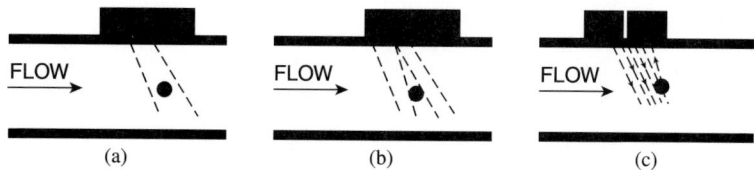

FIGURE 33   Configurations of Doppler-effect ultrasonic flowmeter. (*a*) Single transducer. (*b*) Tandem dual transducer. (*c*) Separate dual transducers installed on same side of pipe.

sonically opaque. Where there are relatively few reflective particles in a stream, there is a tendency for the ultrasonic beam to penetrate beyond the centerline of the pipe and to detect slow-moving particles on the opposite side of the pipe. Because the sonic opacity of the fluid may be difficult to predict in advance, factory calibration is difficult.

It will be noted from Fig. 32 that the fluid velocity is greatest near the center of the pipe and lowest near the pipe wall. An average velocity occurs somewhere between these two extremes. Thus there are numerous variables, characteristic of a given fluid and of a specific piping situation, that affect the interactions between the ultrasonic energy and the flowing stream. Should a measured fluid have a relatively consistent flow profile and include an ideal concentration and distribution of particles, these qualities add to the fundamental precision of measurement and thus simplify calibration. Various designs are used to minimize interaction inconsistencies. For example, separation of transmitters and receivers makes it possible to restrict the zone in the pipe at which the principal concentration of interaction occurs. This zone usually occurs in the central profile region of the pipe less affected by variations in sonic opacity than the region near the pipe wall.

Transit-Time Ultrasonic Flowmeters

With this type of meter, air bubbles and particles in the flowing stream are undesirable because their presence (as reflectors) interferes with the transmission and receipt of the ultrasonic radiation applied. However, the fluid must be a reasonable conductor of sonic energy (Fig. 34). At a given temperature and pressure, ultrasonic energy will travel at a specific velocity through a given liquid. Since the fluid is flowing at a certain velocity (to be measured), the sound will travel faster in the direction of flow and slower against the direction of flow. By measuring the differences in arrival time of pulses traveling in a downstream direction and pulses traveling in an upstream direction, this ΔT can serve as a measure of fluid velocity. Transit-time or ΔT flowmeters transmit alternately upstream and downstream and calculate this time difference. The operation is illustrated by the following equations:

$$V_F = \frac{(T_U - T_D)V_s}{\sin\theta}\frac{V_s\cos\theta}{d} = \frac{\Delta t\, V_s}{\sin\theta}\frac{1}{T_L} \tag{4}$$

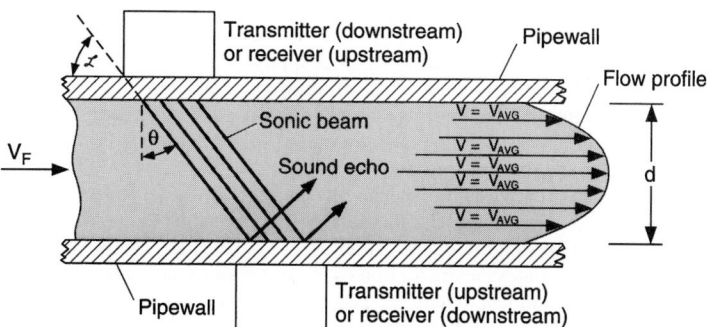

FIGURE 34   Principle of transit-time ultrasonic flowmeter, clamp-on type. Transducers alternately transmit and receive bursts of ultrasonic energy.

By Snell's law,

$$\frac{V_s}{\sin\theta} = \frac{V_C}{\sin\alpha} = K \tag{5}$$

$$V_F = \frac{K\,\Delta t}{T_L} \tag{6}$$

where T_U = upstream transit time
T_D = downstream transit time
T_L = zero-flow transit time
V_s = liquid sonic velocity
d = pipe inside diameter
V_F = liquid flow velocity
α = angle between transducer and pipe wall
V_C = transducer sonic velocity

In the clamp-on transit-time flowmeter, where the transducers are strapped to the outside of the pipe, the sonic echo is away from the receiver and thus the device can retransmit sooner and operate faster. Clamp-on meters are installed on a normal section of piping, which is particularly attractive for retrofit applications. Other designs make use of what are called wetted transducers, which are mounted within the pipe. In both designs, of course, the transducers are installed diagonally to the flow, that is, *not* directly across from each other. Wetted transducers are usually installed in a shallow well, but because there are no projections beyond the pipeline wall, they are still considered nonintrusive. There is no significant disturbance to the general flow profile. However, slight, localized eddy currents may form in the vicinity of the wells. To avoid eddy currents, at least one manufacturer puts the transducers within the wells, forming what is termed an epoxy window. This results in completely filling the stream side of the well, making it essentially flush with the inner pipe wall.

In another design (dual-path ultrasonic flowmeter), two pairs of transducers are installed in the piping. The upstream and downstream propagation times between each pair of transducers are integrated in a microprocessor-based electronics package to determine the flow rate.

POSITIVE-DISPLACEMENT FLOWMETERS

Positive-displacement (PD) flowmeters measure flow directly in quantity (volume) terms instead of indirectly or inferentially as rate meters do. In several rather ingenious ways, the PD meter separates the flow to be measured into different discrete portions or volumes (not weight units) and, in essence, counts these discrete volumes to arrive at a summation of total flow. Fundamentally, PD meters do not have a time reference. Their adaptation to rate indication requires the addition of a time base. Similarly, by making appropriate corrections for changes in density resulting from pressure, temperature, and composition changes, PD meters can serve to indicate flow in terms of mass or weight.

Although inferential rate-type flowmeters appear most frequently in process control loops, departments of facilities concerned with materials accounting (receiving, shipping, inter- and intraplant transfers, distribution, and marketing) largely depend on PD flowmeters. Also, in recent years there has been growing acceptance of PD flowmeters for use in automatic liquid batching and blending systems. Utilities and their consumers are among the largest users of PD flowmeters, with millions of units in use for distributing and dispensing water, gas, gasoline, and other commodities. Where PD meters are used to dispense these commodities, they are subject to periodic testing and inspection by various governmental weights and measures agencies. Accuracy requirements vary with type of meter and service. Usually water meters for domestic service and small industrial plants require an accuracy of only ± 2 percent and gas meters ± 1 percent, whereas meters associated with retail gasoline and diesel fluid pumping have a tolerance (when new or after repair) of only 3.5 cm^3 in 5 gal (19.9 L). Routine tests permit a tolerance of 7 cm^3 in some states.

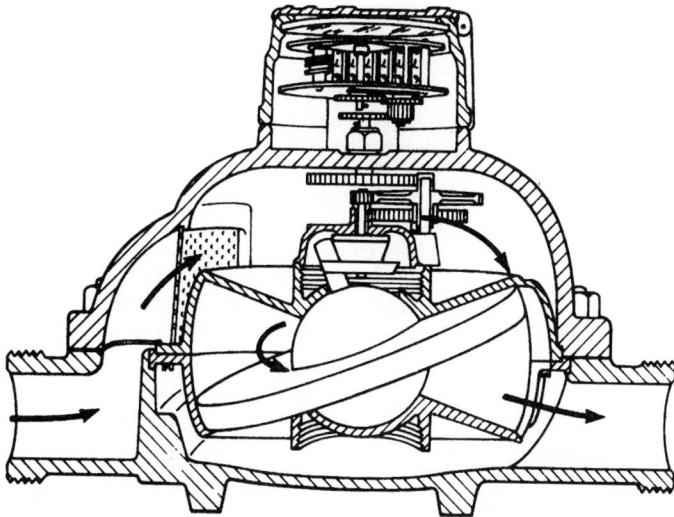

FIGURE 35 Sectional view of representative nutating-disk flowmeter. Each cycle (complete movement) of the measuring disk displaces a fixed volume of liquid. There is only one moving part in the measuring chamber, the disk. Liquid enters the inlet port and fills the space above and below the disk in a nutating motion until the liquid discharges from the outlet port. The motion of the disk is controlled by a cam which keeps the lower face in contact with the bottom of the measuring chamber on one side, while the upper face of the disk is in contact with the top of the chamber on the opposite side. Thus the measuring chamber is sealed off into separate compartments, which are successively filled and emptied, each compartment holding a definite volume. The motion is smooth and continuous with no pulsations. The liquid being measured form a seal between the disk and the chamber wall through capillary action, thus minimizing leakage or slip-page and providing accuracy even at low flow rates.

Nutating-Disk Meter

The nutating-disk meter is probably the most commonly encountered flowmeter found throughout the world for commercial, utility, and industrial applications. The meter is of particular importance in the measurement of commercial and domestic water. Although there are some proprietary design differences, the fundamental operation of the nutating-disk meter is shown in Fig. 35. In most water-metering situations the meter commonly incorporates a digital integrated readout of the "speedometer" style. Meters of this general type for industrial process use, such as in batching and blending, may incorporate a number of accessories. Among the common sizes are 0.5 inch (13 mm), which can deliver up to 20 gal/min (76 L/min), 0.75 inch (19 mm), 1 inch (25 mm), 1.5 inches (38 mm), and 2 inches (51 mm), which can deliver up to 160 gal/min (606 L/min).

Industrial nutating-disk meters provide accurate measurements for low-flow rates and are relatively easy to install and maintain. Depending on the construction materials used, the nutating-disk meter can handle a wide range of chemicals, including caustics. The useful temperature range is from -150 to $120°C$ (-238 to $248°F$).

Oscillating-Piston Meter

In principle, the oscillating-piston meter is similar to the nutating-disk meter, with the important difference that mechanical motion takes place in one plane only (no wobble). In one design* the

---
*Brooks Instrument Division, Emerson Electric Company, Hatfield, Pennsylvania.

measuring chamber consists of five basic parts: (1) top head, (2) bottom head, (3) cylinder, (4) division plate, and (5) piston. The only moving part in the measuring chamber is the piston, which oscillates smoothly in a circular motion between the two plane surfaces of the top and bottom heads. The division plate separates the inlet ports *A* and the outlet ports *B*. The piston is slotted to clear the division plate, which also guides the travel of the piston in its oscillating motion. A gear train transmits the piston motion to the register. The major components and operation of the meter are shown in Fig. 36.

The piston has a significantly smaller circumference than the chamber. This provides for maximum liquid area displacement for each oscillation. A small differential pressure across the meter produces motion of the piston within the measuring chamber. In order to obtain oscillating motion, two restrictions are placed on the movement of the piston. First, the piston is slotted vertically to match the size of a partition plate fixed to the chamber. This plate prevents the piston from spinning around its central axis and also acts as a seal between the inlet and outlet ports of the chamber. Second,

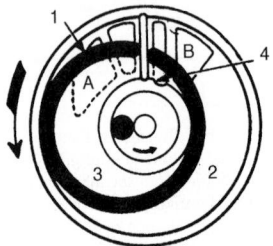

DIAGRAM 1

SPACES 1 AND 3 ARE RECEIVING LIQUID FROM THE INLET PORT, A, AND SPACES 2 AND 4 ARE DISCHARGING THROUGH THE OUTLET PORT B.

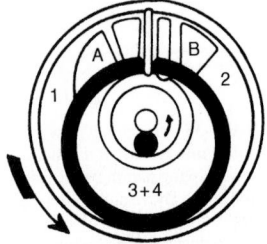

DIAGRAM 2

THE PISTON HAS ADVANCED AND SPACE 1, IN CONNECTION WITH THE INLET PORT, HAS ENLARGED, AND SPACE 2, IN CONNECTION WITH THE OUTLET PORT, HAS DECREASED, WHILE SPACES 3 AND 4, WHICH HAVE COMBINED, ARE ABOUT TO MOVE INTO POSITION TO DISCHARGE THROUGH THE OUTLET PORT.

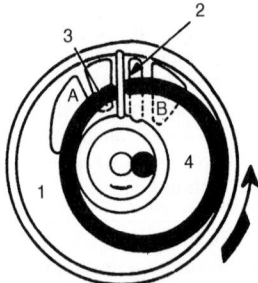

DIAGRAM 3

SPACE 1 IS STILL ADMITTING LIQUID FROM THE INLET PORT AND SPACE 3 IS JUST OPENING UP AGAIN TO THE INLET PORT, WHILE SPACES 2 AND 4 ARE DISCHARGING THROUGH THE OUTLET PORT.

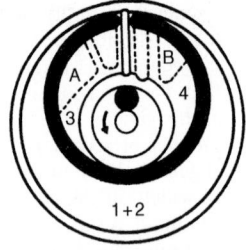

DIAGRAM 4

LIQUID IS BEING RECEIVED INTO SPACE 3 AND DISCHARGED FROM SPACE 4, WHILE SPACES I AND 2 HAVE COMBINED AND ARE ABOUT TO BEGIN DISCHARGING AS PISTON MOVES FORWARD AGAIN TO OCCUPY POSITION AS SHOWN IN DIAGRAM 1.

FIGURE 36 Principle of operation of oscillating-piston meter. (*Brooks Instrument Division, Emerson Electric Company.*)

the piston has a center vertical pin which is confined to a circular track that is part of the chamber. Differential pressure across the meter causes the piston to sweep the chamber wall in the direction of flow. This oscillating motion displaces liquid from the inlet to the outlet port in a continuous stream.

To further prevent unmeasured liquid from passing through the chamber, the piston has a horizontal partition or web. This web is perforated to promote balanced buoyancy of the piston within the chamber and a linear flow pattern of liquid through the meter. A drive bar and shaft are positioned through the top of the chamber so that, as the piston oscillates, the piston pin drives the bar and shaft in a circular or spinning motion. This rotating shaft is the driving link between the piston and the register or readout unit. Sizes range from 0.75 to 2 inches (19 to 51 mm), with capacity ranging from 5 to 150 gal/min (19 to 570 L/min).

Fluted Rotor Meter

Meters of this type are used in the flow measurement of crude and refined petroleum products and a variety of other commercial fluids. Frequently they are used on product loading racks and small pipelines. Usually the meters are equipped with direct readouts and often with ticket printers to provide authentically recorded documents for liquid-transfer transactions (Fig. 37).

Oval-Shaped Gear Flowmeters

In these meters, precision-matched oval-shaped gears are used as metering elements (Fig. 38). Meter sizes (connections) range from $\frac{1}{2}$ to 1, $1\frac{1}{2}$, 2, 3 and 4 inches ($\sim$1.3 to 2.5, 3.8, 5.1, 7.6, and 10.2 cm), with capacity ranges for different fluids as follows:

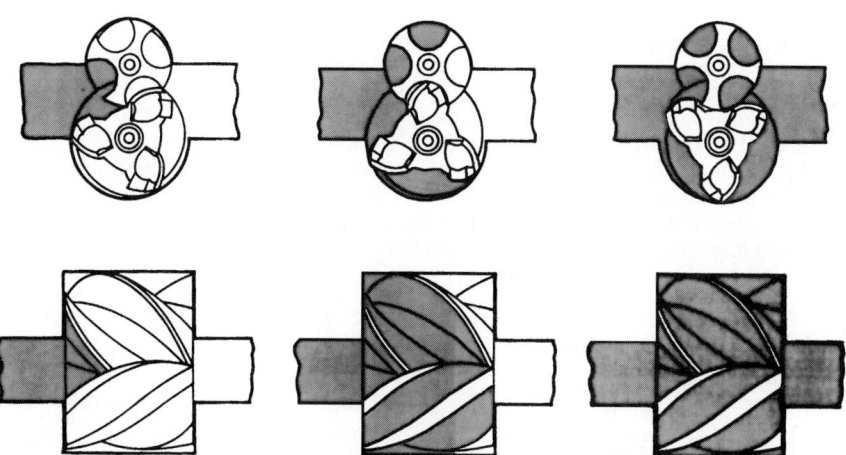

FIGURE 37 Operating principle of a Brooks *BiRotor* possitive-displacement flowmeter. As the product enters the intake of the measuring unit chamber, the two rotors divide the product into precise segments of volume momentarily and then return these segments to the outlet of the measuring unit chamber. During what may be referred to as "liquid transition," the rotation of the two rotors is directly proportional to the liquid throughput. A gear train, located outside the measuring unit chamber, conveys mechanical rotation of the rotors to a mechanical or electronic register for totalization of liquid throughput. (*Brooks Instrument Division, Emerson Electric Company.*)

Cold water	0.2—1.4 to 110—705 gal/min (U.S.) 0.8—5.3 to 416.4—2668.7 L/min
Hot water	0.3—1.0 to 132—484 gal/min 1.1—3.8 to 499.7—1832.1 L/min
Liquefied petroleum gas (LPG)	0.4—1.6 to 176—837 gal/min 1.5—6.1 to 666.2—3168.4 L/min
Gasoline (0.3—0.7 cP)	0.3—1.6 to 132—837 gal/min 1.1—6.1 to 499.7—3168.4 L/min
Kerosene (0.7—1.8 cP)	0.2—1.6 to 110—837 gal/min 0.8—6.1 to 416.4—3168.4 L/min
Light oil (2—4 cP)	0.1—1.9 to 705—1010 gal/min 0.4—7.2 to 2668.7—3826.3 L/min
Heavy oil (5—300 cP)	0.04—1.9 to 44—1010 gal/min 0.2—7.2 to 166.6—3826.3 L/min

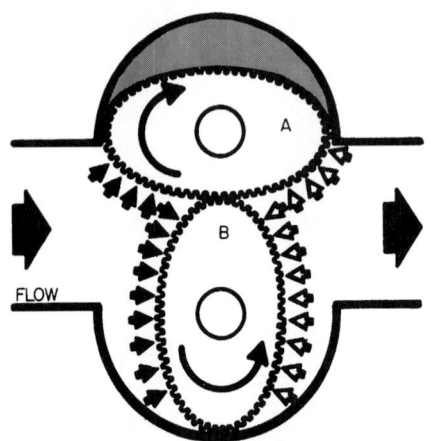

FIGURE 38 Sectional schematic of oval gear flowmeter, showing how a crescent-shaped gap captures the precise volume of liquid and carries it from inlet to outlet. (*Brooks Instrument Division, Emerson Electric Company.*)

In meter application engineering, three basic viscosity classifications are taken into consideration: (1) standard-viscosity class from 0.2 to 200 cP, (2) medium-viscosity class from 300 to 500 cP, and (3) high-viscosity class above 500 cP. Tests have shown that a meter calibrated on a 1-cP product and then applied to a 100-cP product does not shift more than 1.2 percent above the initial calibration. Normally, where the viscosity is 100 cP or greater, there is no significant shift in accuracy. Oval-gear PD meters are sized for maximum flow so that the pressure drop is less than 15 psi (103 kPa).

Normal operating pressure ranges from 255 to 710 psi (1758 to 4895 kPa), depending on whether steel or stainless-steel flanges are used. The maximum operating temperature is 230°F (110°C), but special meters are available for higher temperatures. The lower limit is 0°F (−18°C). Housings, rotors, and shafts are constructed of type 316 stainless steel or type Alloy 20 (CN-7M stainless steel). Bushings are of hard carbon.

Meters are available with numerous accessories, including an electric-impulse contactor, which is used to open and close an electric circuit at intervals proportional to the number of units measured.

Other Positive-Displacement Meters

In the lobed-impeller meter shown in Fig. 39 two rotors revolve with a fixed relative position inside a cylindrical housing. The measuring chamber is formed by the wall of the cylinder and the surface of one-half of one rotor. When the rotor is in vertical position, a definite volume of fluid is contained in the measuring compartment. As the impeller turns, owing to the slight pressure differential between inlet and outlet ports, the measured volume is discharged through the bottom of the meter. This action takes place four times for a complete revolution, the impeller rotating in opposite directions and at a speed proportional to the volume of fluid flow. Meters of this design can handle from 8 gal/min (30

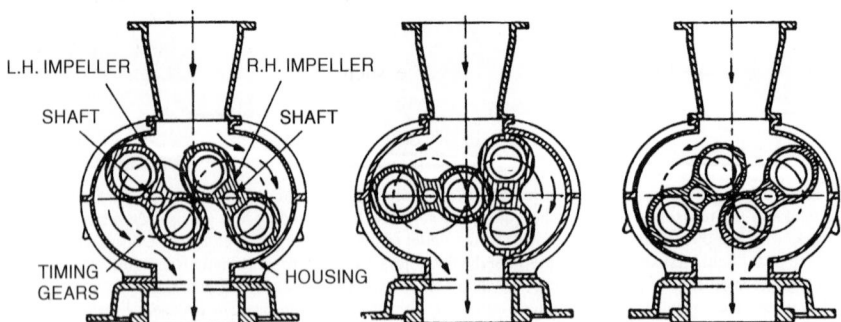

FIGURE 39 Lobed-impeller flowmeter.

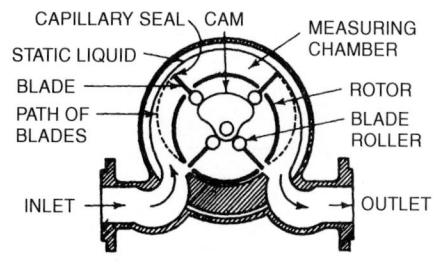

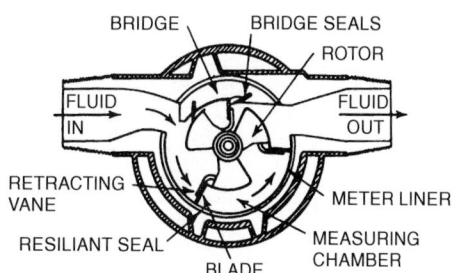

FIGURE 40   Sliding-vane rotary meter.   **FIGURE 41**   Retracting-vane rotary meter.

L/min) to 25,000 barrels/h. They perform well at temperatures up to 400°F (204°C) and pressures up to 1200 psi (8276 kPa). They are used for gases and a wide range of light to viscous fluids, including asphalts, and are best suited to high rates of flow.

In the sliding-vane rotary meter, vanes are moved radially as cam followers to form the measuring chamber (Fig. 40). Another version, the retracting-vane type, is shown in Fig. 41.

OPEN-CHANNEL FLOW MEASUREMENTS

An open channel is any conduit in which a liquid, such as water or wastes, flows with a free surface. Immediately evident examples are rivers, canals, flumes, and other uncovered conduits. Certain closed conduits, such as sewers and tunnels, when flowing partially full (not under pressure), also may be classified as open channels. Two measurement units are used in measuring open-channel flows: (1) units of discharge and (2) units of volume. The discharge (or rate of flow) is defined as the volume of liquid that passes a certain reference section in a unit of time. This may be expressed, for example, as cubic feet per second, gallons per minute, or millions of gallons per day. In irrigation the units used are the acre-foot [defined as the amount of water required to cover 1 acre to a depth of 1 foot (43,560 ft^3), or the hectare-meter (10,000 m^3)].

To determine open-channel flows, a calibrated device is inserted in the channel to relate the free-surface level of liquid directly to the discharge. These devices sometimes are called head-area meters and are used extensively for the allocation of irrigation water as well as in municipal sewage and industrial wastewater facilities.

Weirs

A weir is a barrier in an open channel over which liquid flows. The edge or surface over which the liquid flows is called the crest; the overflowing sheet of liquid is the nappe. The top of the weir has an opening, which may be of different dimensions and geometry (Fig. 42).

Of the various types of weirs, the three most commonly used are shown in Fig. 43. Weir calculations tend to be complex and reference to a source, such as *Stevens Hydrographic Data Book* (Leopolid & Stevens, Beaverton, Oregon—revised periodically), is suggested.

Parshall Flumes

These devices are available for measuring open-channel flows up to 1500 million gal/day (5.7 million m^3/day). The size of a flume is designated by the width W of the throat, which may range from 1 inch (2.5 cm) to 40 ft (12 meters) (Fig. 44). Generally economy of construction dictates that the smallest standard throat size be selected, provided it is consistent with the depth of the channel at maximum flow and permissible head loss. As a general rule, the throat size should be one-third to one-half the width of the channel. Prefabricated flumes often can be a cost advantage (Fig. 45).

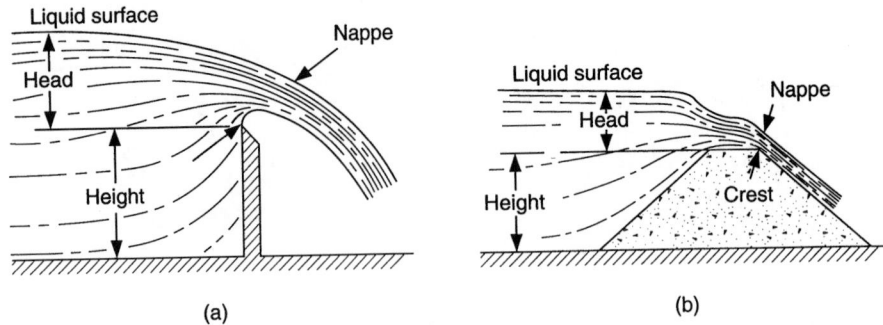

(a)   (b)

FIGURE 42   Basic forms of weirs. (*a*) Sharp-crested weirs, which are useful only as a means of measuring flowing water. (*b*) Non-sharp-crested weirs, which are incorporated into hydraulic structures as control or regulating devices, with measurement of flow as a secondary function. When the weir has a sharp upstream edge so that the nappe springs clear of the crest, it is called a sharp-crested weir. The nappe, immediately after leaving a sharp-crested weir, suffers a contraction along the horizontal crest, called crest contraction. If the sides of the notch have sharp upstream edges, the nappe also is contracted in width and the weir is said to have end contractions. With sufficient side and bottom clearance dimensions of the notch, the nappe undergoes maximum crest and end contractions, and the weir is said to be fully contracted.

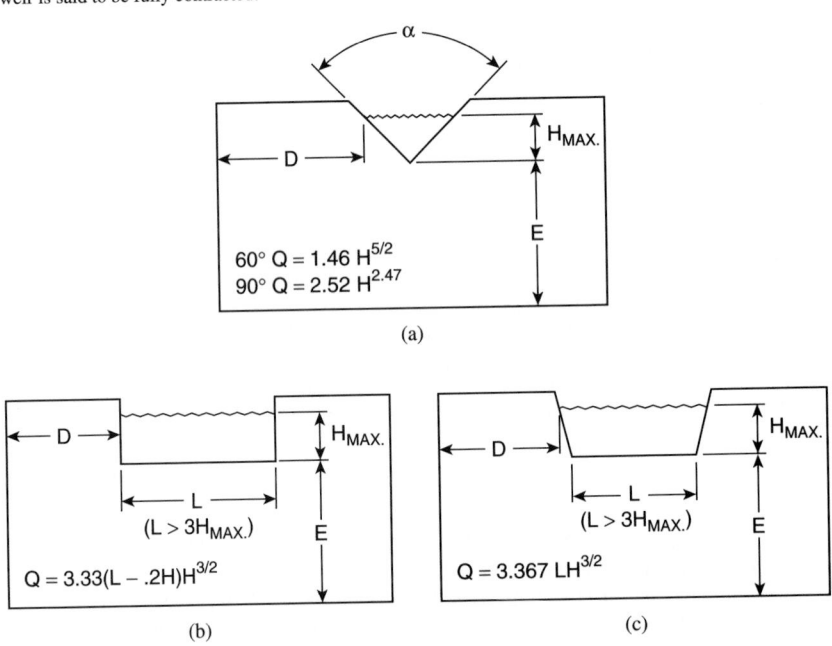

FIGURE 43   Weirs commonly used. (*a*) V-notch. (*b*) Rectangular. (*c*) Cipolleti. Occasionally, other weirs, such as hyperbolic (Sutro), broad-crested, and round-crested, are used.

Open-Flow Nozzle (Kennison)

The Kennison nozzle shown in Fig. 46 is an unusually simple device for measuring flows through partially filled pipes. The nozzle copes with low flows, wide flow ranges, and liquids containing suspended solids and debris. Because of its high-accuracy, nonclogging design and excellent head-versus-flow characteristics, the Kennison nozzle is well suited for the measurement of raw sewage,

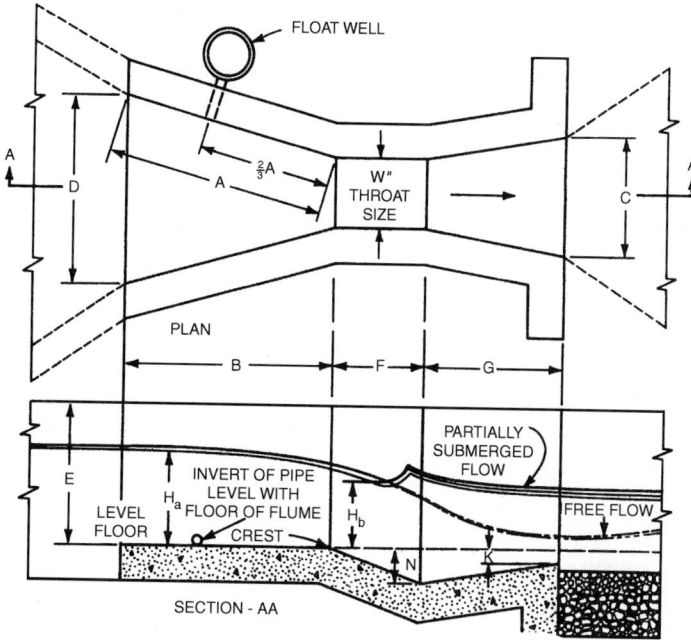

FIGURE 44 Parshall flume.

W	A	²/₃A	B	C	D	E	F	P	K	N	R	T
1″	1′2⁹/₃₂″	9¹⁷/₃₂″	1″2	3²¹/₃₂″	6¹⁹/₃₂″	1′8″	3″	2′1″	³/₄″	1¹/₈″	1³/₄″	³/₁₈″
2″	1′4⁵/₁₆″	10⁷/₈″	1′4″	5⁵/₁₆″	8¹³/₃₂″	9″	4¹/₂″	2′6¹/₂″	⁷/₈″	1¹¹/₁₆″	1³/₄″	³/₁₆″
3″	1′6³/₈″	1¹/₄″	1′6″	7″	10³/₁₆″	2′0″	6″	3′0″	1″	2¹/₄″	2¹/₂″	³/₁₆″
6″	2′7/₁₆″	1′4⁵/₁₆″	2′0″	1′3¹/₂″	1′3⁵/₈″	2′0″	1′0″	5′0″	3″	4¹/₂″	2¹/₂″	³/₁₆″
9″	2′10⁵/₈″	1′11¹/₈″	2′10″	1′3″	1′10⁵/₈″	2′6″	1′0″	5′4″	3″	4¹/₂″	2¹/₂″	³/₁₆″
1′0″	4′6″	3′0″	4′4⁷/₈″	2′0″	2′9¹/₄″	3′0″	2′0″	9′4⁷/₈″	3″	9″	2¹/₂″	¹/₄″
1′6″	4′9″	3′2″	4′7⁷/₈″	2′6″	3′4³/₈″	3′0″	2′0″	9′7⁷/₈″	3″	9″	2¹/₂″	¹/₄″
2′0″	5′0″	3′4″	4′10⁷/₈″	3′0″	3′11¹/₂″	3′0″	2′0″	9′10⁷/₈″	3″	9″	2¹/₂″	¹/₄″
3′0″	5′6″	3′8″	5′4³/₄″	4′0″	5′1⁷/₈″	3′0″	2′0″	10′4³/₄″	3″	9″	2¹/₂″	⁵/₁₆″
4′0″	6′0″	4′0″	5′10⁵/₈″	5′0″	6′4¹/₄″	3′0″	2′0″	10′10⁵/₈″	3″	9″	2¹/₂″	³/₈″

raw and digested sludge, final effluent, and trade wastes. Nozzle sizes range from 6 inches (15 cm) with a capacity of 90 gal/min (equivalent to 493 m³/day) up to 36 inches (91.4 cm) with a capacity of 14,000 gal/min (equivalent to 75,700 m³/day).

BULK-SOLIDS FLOW MEASUREMENT

In many industries, for the control of inventory, maintaining throughput rates, and materials balances, and in batching for recipe control, it is essential to measure the flow of bulk solids carefully. Although with some very uniform solids, flow may be measured in terms of volume, and volumetric feeders are available, but with some sacrifice of accuracy, in gravimetric devices, measurements are made on the basis of weight. Here flow measurement is expressed in weight (pounds or kilograms) per unit of time (minutes or hours). Thus, in essence, a belt conveyor scale is a mass flowmeter. For many years, prior to the introduction of strain-gage load cells and high-technology sensors, conveyor belts were coupled to mechanical weigh scales.

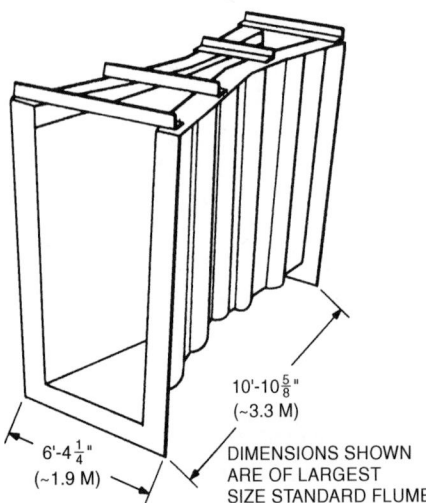

FIGURE 45 Rigid one-piece prefrabicated fiberglass Parshall flume. Advantages include smooth surface, dimensional accuracy and stability, with no external bracing needed, long life, and reduction of installation costs. Dimensions shown are for available large size.

FIGURE 46 Kennison nozzle. (*Leeds & Northrup, BIF Products.*)

Belt-Conveyor Weighers

Load-weighing devices can be installed on practically any size length and shape of conveyor—from 12-inch (30.5-cm)-wide belts, handling as little as 1 lb (0.45 kg)/min at a minimum belt speed of 1 ft (30.5 cm), to 120-inch (3-meter)-wide belts, handling as much as 20,000 tons (18,145 metric tons)/hour at speeds of up to 1000 ft/min (305 m/min). The basic design of a belt conveyor weigher is shown in Fig. 47.

Wherever possible, the scale should be installed where it will not be subjected to conveyor influences that cause a "lifting effect" of the belt off the weigh-platform section. Belt tensions and belt lift have more influence on accuracy than any other factor. It is important that the material (load) travel at the same velocity as the belt when crossing the weigh platform and not be in a turbulent state, as it is when leaving a feed hopper or chute. The best test for accuracy involves carefully weighing a batch of material and passing this exact weight several times over the belt scale. Another testing method is using a chain that is very uniform and matches closely the weight of the solid material normally weighed. Some manufacturers guarantee their product to weigh within ±0.5 percent of the totalized weight (when properly installed). The National Institute of Standards and Technology (NIST) has developed a standard certification code for conveyor scales.

In proportioning various solids and liquids, belt feeders and flow controllers, including controlled-volume pumps, can be integrated into a total digitized blending system. Microprocessor technology has been available for continuous weighing for several years. The microprocessor provides accuracy, automation, and flexibility features not feasible with previous analog and digital systems. The basic difference is that the microprocessor does things sequentially, using its memory and an instruction set (software), whereas analog and digital systems accomplish these tasks continuously by way of various electrical circuits (hardware).

In addition to belt-type weighers, the basic principle can be applied to screw feeders, sometimes used for very dry materials that tend to flush and flood. So-called solids flowmeters are also available. These are not true gravimetric devices, but rather, utilize impact or centrifugal force to generate a signal that is proportional to material flow. These devices are capable of providing long-term repeatability

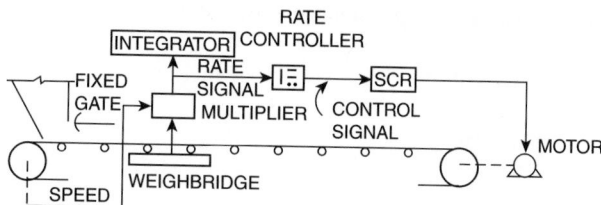

FIGURE 47   Gravimetric weigh-belt feeder with true rate control system where there are no inferred constants. Weight × speed = flow rate.

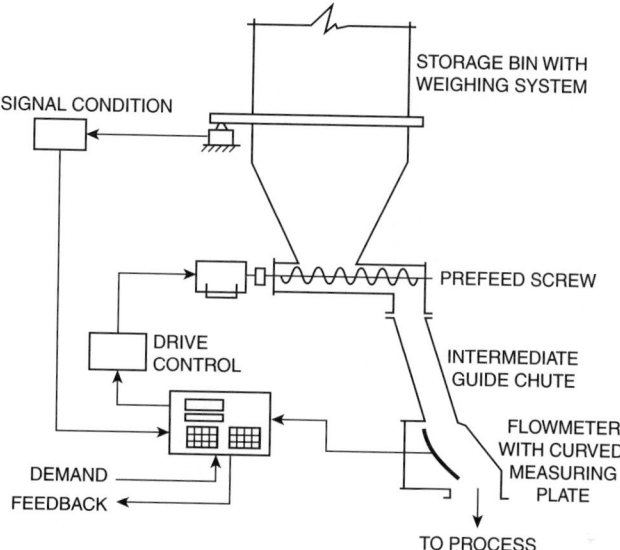

FIGURE 48   Flow-rate control system utilizing a solids flowmeter with an in-line calibration system. (*Schenck Weighing Systems.*)

in the 2 to 3 percent range (Fig. 48). In another gravimetric type (loss-in-weight feeder), a weigh hopper and metering screw are supported on load cells. As the metering screw delivers material from the system, the resultant load-cell signal decreases proportionately with time. The differential dG/dt produces the true feed rate F, as is shown graphically in Fig. 49.

Nuclear radiation methods also have been applied to weighing materials on a moving belt.

TYPES OF LEVEL MEASUREMENT SYSTEMS

A wide variety of level measurement systems are available to address the broad spectrum of applications, accuracy needs, installation requirements, and practices. Measurement technologies are made available in different versions to address a wide range of measurement needs or sometimes to address just one specific application. This subsection will attempt to define some of the general selection considerations of many available technologies, the general forms of these technologies, and some of their general advantages and disadvantages. As always, one must consult the specifications from the

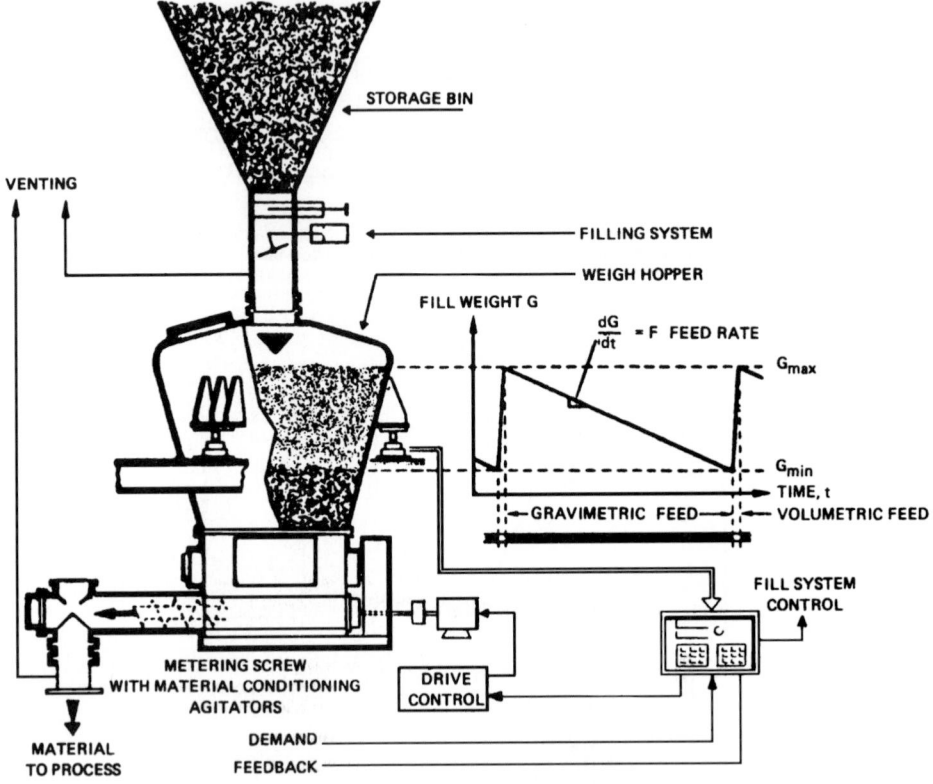

FIGURE 49   Gravimetric feeder using loss-in-weight principle. (*Schenck Weighing Systems.*)

various manufacturers for specific products and users' experiences in different installations to truly determine their applicability to measurement situations.

The family of level measurement systems can be divided into many categories: liquids or solids level measurement, point or continuous level measurement, electromechanical or electrical/electromagnetic level measurement, or contacting or noncontacting/nonintrusive level measurement.

For this subsection the focus is on the technologies of level measurement so the information will be presented in the following format:

- *General considerations in level measurement technology selection*

    Density and viscosity

    Chemical composition

    Ambient temperature

    Process temperature

    Process pressure

    Regulated environments

    Process agitation

    Vapor, mist, and dust

    Interfaces and gradients

Process conductivity and dielectric constants

Vibration

Material buildup or stickiness

Static charge

Humidity/moisture

Repeatability, stability and accuracy requirements

- *Electromechanical level measurement and detection systems*

  Floats for level detection and measurement of liquids

  Displacers for level detection and measurement of liquids

  Level detection of solids using rotating paddles

  Level measurement of liquids and solids using plumb bob

- *Electronic/electromagnetic energy level measurement and detection systems*

  Level detection of liquids by use of conductivity

  Level detection of liquids by use of vibrating forks resonance or rod attenuation

  Level detection of solids by use of vibrating fork or rod attenuation

  Level detection of liquids by use of ultrasonic gap

  Level detection of liquids by use of thermodispersion

  Level measurement of liquids by use of bubblers

  Level measurement of liquids by use of hydrostatic pressure

  Ultrasonic level detection and measurement of liquids and solids

  Capacitance level detection and measurement of liquids and solids

  Radar level detection and measurement of liquids and solids

  Level detection and measurement of liquids and solids by use of time-domain reflectometry

  Level measurement of liquids by use of magnetostrictive

  Level measurement of liquids by use of laser

  Level detection and measurement of liquids and solids by use of radiometric

  Level measurement of liquids and solids by use of weighing

  Level detection by use of optics

  Level detection in liquids by use of ultrasonic tank resonance

GENERAL CONSIDERATIONS IN LEVEL MEASUREMENT TECHNOLOGY SELECTION

The selection of the right level measurement solution is based on not only the choice of measurement technology but also the application and installation requirements. Unfortunately one universal technology that can address all level measurement requirements has yet to be discovered. So users and specifying engineers must address their level measurement needs from a broad array of technologies.

One has to review not only the capability of the measurement but also its survivability in a particular environment. Some measurement technologies must come into direct contact with the measured material.

Other technologies are noncontacting or even nonintrusive and do not touch the measured material. This can have direct bearing on how much maintenance may be required or the economic impact caused by failure of the measurement to provide correct information. Often the cost of a particular technology is much different when viewed from a total lifetime cost of ownership perspective. Purchase, installation, maintenance, and training costs all play a part in determining this.

The quality of level information required must also be reviewed. As in most measurement choices the quality required is determined by what is necessary to achieve targeted control capabilities or gain economically beneficial information. In other words the type of level measurement selected should produce the quality of level information required. The quality of level information is determined not only by the specifications of the measurement instrument but also by how the environment and the method of installation influence the measurement instrument.

Following are some general considerations one needs to take when selecting a level measurement device.

Density and Viscosity

Viscosity and density considerations affect the selection of many liquid level switches and continuous level transmitters. High-viscosity materials can cause mechanical technology switches to stick or hang up. The forces on probes and sensors from high-density materials can cause damage. An agitated liquid material may cause high lateral forces on a measurement probe when the viscosity or the density increases. The level information from some level measurement technologies is influenced by density changes. Bubblers, displacers, and hydrostatic instruments measure the hydrostatic head of a liquid. When the density of the measured liquid increases, given a fixed level, the indicated level will incorrectly show an increase.

The force from the shifting of solids may exceed the design of a measurement probe if the density increases from increased moisture or a change in composition. Noncontacting measurement devices generally are not affected by density and viscosity changes.

Chemical Composition

Measurement technologies must not only provide correct measurement of the particular material, given its chemical nature, but they need to be able to survive as instruments. The chemical characteristics of a material may change because of steps in the process. These characteristics, such as dielectric constant, conductivity, pH, etc., affect the different measurement technologies in different ways. Capacitance measurements are affected by changes of conductivity and/or dielectric constant, especially in the area between the conductive and the nonconductive states. Noncontacting ultrasonic measurement may not have to deal with direct contact with the measured material, but the transducer may have to contend with corrosive vapors or gases in the process vessel. A material compatibility table can be used to determine whether the measurement sensor materials of construction are suitable for the service. The effect of chemical composition on the measurement technology measurement capability may often be less with noncontacting technologies (i.e., radar, ultrasonic, gamma) but it should not be assumed that this is always the case.

Ambient Temperature

The temperature that the measurement instrument itself is exposed to must be reviewed. Cold or hot temperatures may damage, disable, or compromise the information from a measurement instrument. An optical level switch may be designed for proper operation up to 120°F ambient with a given limit of 150°F process temperature. If the level switch is mounted in an area where the outside temperature can exceed the 120°F ambient because of both radiated and transferred heat, the electronics in the switch may fail or report incorrect information. The quality of level information from a hydrostatic measurement may be compromised when remote seals are used to contact the process material. The expansion and the contraction of fill fluids within capillaries/impulse tubes connecting the process seal with the pressure sensor may cause level changes to be reported when ambient temperature changes take place.

Process Temperature

The process temperature that the measurement instrument is exposed to must also be reviewed. Naturally the measurement instrument materials of construction will dictate the maximum temperature allowed before damage occurs. The ambient temperatures may be fine for the instrument, but the process temperature may affect the level information. A hydrostatic sensor exposed to large change in process temperature may not be able to compensate immediately, thus giving a temporary offset to the level reported. The temperature of the fill fluid within the hydrostatic sensor can affect its density. If not compensated for, offsets in indicated pressure (level) can occur.

Process Pressure

The impact of static pressure and pressure changes (and vacuum) on the level information has to be taken into consideration. There are two aspects to be noted:

- The pressure rating of the instrument must at least match the process pressure in the tank. If a vacuum is present, the instrument needs a vacuum rating.
- The measuring principle must not be affected by the process pressure. Example: A hydrostatic measurement will not work properly in a tank with a head pressure. The obvious remedy, using differential pressure, will work only if the ratio between the hydrostatic pressure caused by the product and the head pressure does not exceed certain limits.

Regulated Environments

Measurements may often be required in environments where special regulations are in force. Installations where there is hazardous gas or dust often require a technology that is classified as meeting the requirements. National Electrical Code installation guidelines may have to be followed. Certificates from recognized agencies such as Factory Mutual (FM), Underwriters Laboratories (UL), and Canadian Standards Association (CSA) may be required. Some industries and even companies have specific requirements that the level technology used must fulfill. For example, the dairy industry requires measurement devices to meet 3A sanitary standards.

Process Agitation

Many process liquids must be agitated all or part of the time. Mixing blade movements may preclude using a contacting probe like capacitance. Some severe agitation may produce wavy or vortexlike surfaces. Noncontacting-type measurements like ultrasonic and radar may not get reliable signals back from these surfaces to provide reliable level information.

Vapor, Mist and Dust

The space above a liquid or solids material may contain some vapor, mist or dust. This becomes important when noncontacting measurements with laser, microwave, radar, and ultrasonic technologies are considered. The presence of enough dust and mist can attenuate the acoustic or electromagnetic signals to the point that measurement is unreliable. Ultrasonic measurements can be affected by the presence of vapor and/or changes in temperature of the vapor if the speed of sound is changed.

Interfaces or Gradients

Level measurements of interfaces between two dissimilar liquids, liquids and foam, and liquids and solids require careful analysis. Oil and water are a common example of a dissimilar liquids interface. Under some conditions, radar and time-domain reflectometry technologies can be used to detect the interface at the point of change from one low dielectric to another high dielectric liquid. Capacitance instruments can sometimes be used to measure one of the liquids and ignore the other if one is conductive/high dielectric and the other is nonconductive/low dielectric. Some types of ultrasonic technologies can detect the transition at which the speed of sound or reflection of acoustic energy is related to the interface location. Measurement technologies that are less affected by changes in physical properties (i.e., density only) like gamma and magnetostrictive sometimes can be used.

Often the transition between the two liquids is not well defined. There may be an emulsion layer or gradient that makes the definition of the interface more difficult to determine. The characteristics of the individual liquids may not be constant, making it difficult to use some level measurement technologies. Conductivity, dielectric constants, and acoustic, optical, and density variations may prove challenging to different measurement technologies. Liquid/foam interfaces can also be influenced by changes in these same parameters. Ultrasonic and radar measurements can be compromised by the presence of foam on a liquid if too little acoustic or microwave energy is reflected off the surface. Often level switch technologies are used that ignore the foam and react to the liquid. Vibration and ultrasonic gap switches can work here. Liquid/solid interfaces are also influenced by changes in the density of the solids.

Process Conductivity and Dielectric Constants

Conductive and capacitance measurements are dependent on predictable conductivity and dielectric constants of the measured materials. Selection of the proper probe is also affected. When conductivity and dielectric changes are unpredictable and cannot be compensated for, then another level technology should be used.

Vibration

Vibrations might be present for different reasons, from engine-borne vibrations caused by pumps or agitators to vibrating silo bottoms for easier material flow in solids. These vibrations usually do not cause major problems; nevertheless, a few considerations are to be made:

- Is there a chance for the vibration to interfere with the measurement itself? (Unlikely, since only few measuring principles depend on vibrations; only if the external vibrations occur in the same frequency and amplitude range as the device specific vibrations might an influence be noted.)
- Do the vibrations apply any unusual forces to the sensor element? If so, the instrument might be damaged over time.
- Are any strong vibrations applied to the sensor electronics? If so, a remote electronics might have to be considered. It has to be mounted at a point where no vibrations are present.

Material Buildup or Stickiness

Material buildup may make certain level systems fail or work less reliably. The question to be asked is always what kind of buildup can be expected in the process. Buildup must not always be a coating on the sensor (contacting principles); another form of appearance may be condensation (which can also affect noncontacting principles); material can also stick to the tank walls (causing wrong readings of level transmitters) or cause bridging between two elements of a sensor. Buildup can be addressed by different technologies in different ways (refer to the respective chapters).

Static Charge

Electrostatic charges/discharges are often generated during transport (pumping, pneumatic conveying, etc.) of nonconductive materials. These electrostatic charges can easily get a big as a couple of 10 kV. A sudden discharge through the sensor (for example, a capacitance probe) may apply an electrical energy to the sensor electronics high enough to destroy components, particularly semiconductors. One can avoid, or at least minimize, problems with electrostatic discharge (ESD) (1) by choosing a suitable measurement principle that is not subject to ESD or (2) by taking steps to reduce the electrostatic potential in the vessel and filling system (ground electrodes, etc.).

Humidity/Moisture

Some bulk solids can have variations in moisture content that are seasonal, composition related, or from different sources. Capacitance measurements, for example, can be affected as the change in moisture content also changes the dielectric constant of the solids. Depending on process temperatures, moisture may also cause condensation at the coldest point in the tank (typically the tank cover/ceiling). A measurement principle has to be selected that is not affected by condensation or changes in the material properties.

Repeatability, Stability, and Accuracy Requirements

In many cases, the repeatability of a measurement and the stability of the output signal is a bigger concern than the absolute accuracy. An accuracy of 0.2% under very specific laboratory conditions does not mean too much if the output drifts up and down in the real application. The main concern of an instrument manufacturer therefore is to design a product in a way that makes it extremely stable by itself and that makes it as insensitive as possible to external influences. Nevertheless, there is still the need for the user to select the right measuring principle, since a good instrument in a bad application will not perform as expected. Therefore any efforts put into the design of an instrument may be in vain if it ends up being used in a process that does not fit.

ELECTROMECHANICAL LEVEL MEASUREMENT AND DETECTION SYSTEMS

Although newer electronic level measurement technologies have displaced some of the electromechanical level measurement instruments, there are many applications for which they are best suited. Level measurement systems based on the conductivity or capacitance of the product or on its relative density that use displacer or pressure methods can be less reliable in bulk solids because of the effects of the variable conditions. These effects are as follows:

- Changes in the humidity content or composition of a product affect its electrical properties.
- Changes in relative density and bulk density that are due to the settling of the product affect the mechanical properties of the product.

The presence of foam, turbulence, air bubbles, and particles in suspension and high viscosity in liquid measurements should also be taken into consideration. Over the years, four type of mechanical measurement systems have been developed to cope with these disruptive factors. They are

- floats
- displacers
- rotating elements
- electromechanical plumb-bob systems

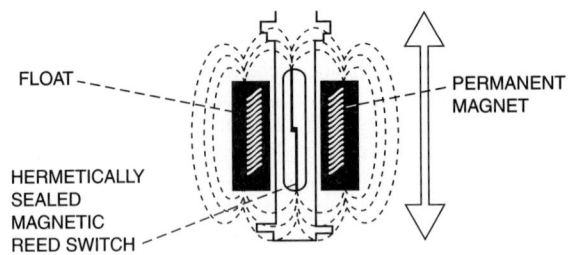

FIGURE 1 How a typical float switch works.

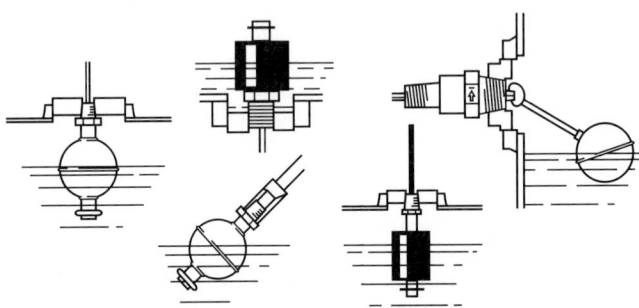

FIGURE 2 Typical float switches.

Level Detection and Measurement of Liquids by Use of Floats

Level detection of liquids is often done with a float-type level switch. The float moves on a mechanical arm or sliding pole and activates a switch when the level moves it upward. Sometimes the floats themselves contain a small magnet that changes the state of a reed switch when the liquid level moving up moves it into position (Figs. 1 and 2).

The simplest form of a float system for level measurement consists of a float, a small cable, two pulleys, and a weight that is suspended on the outside of the open tank. A scale is mounted on the outside of the tank, and the level of the tank contents is indicated by the position of the weight along the scale. In view of the requirements regarding mounting in sealed tanks, reading, and accuracy, the current industrial systems that use the float method are good examples of mechanical and electronic engineering, making them the most accurate level measuring systems for applications in very large storage tanks.

The advantages of this method are that it is relatively simple, suitable for various products, and very accurate. The disadvantages are that it requires a certain amount of mechanical equipment, especially in pressure vessels.

Level Detection and Measurement of Liquids by Use of Displacers

The displacement method is based on the difference between the weight of the displacement body and the upward force exerted by the medium on this body. The upward force depends on the volume of the displacement body, the relative density of the medium, and the level of the medium. For given volumes and relative densities, the upward force will depend on only the level of the medium. Obviously the displacement body must be heavier than the medium to be displaced.

The differential force is transmitted to a measuring transducer by means of a torque rod system in order to maintain a seal. The transmitter used with the displacement transducer or force transducer provides a 4–20-mA output. The displacement body is often mounted in a bypass pipe, which has the advantage of allowing any calibration and monitoring to be independent of the process involved. Also, the level of the product in the bypass pipe is somewhat more stable, resulting in a measurement that is not oversensitive. If the displacement body is mounted directly on the process tank, then it is advisable to mount it inside a guide tube or stilling well. The height of the separation layer between two products of different relative densities can also be measured with the displacement body method. The obvious advantage of this method is that the absolute level does not change.

The advantage of this method is accuracy. The disadvantages are that it is dependent on relative density and requires a significant amount of mechanical equipment.

Level Detection of Solids by Use of Rotating Paddles

Level is detected by the change in inertia of a rotating paddle depending on whether the paddle is in the air or in contact with a product.

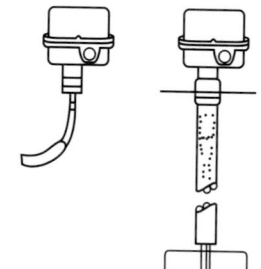

An example of a practical application is a paddle mounted on a shaft projecting into a silo, driven by a small synchronous motor through a reduction gear. When the paddle encounters resistance from the product, then the rotating drive system supported on bearings will move. Two microswitches, positioned to detect this movement, will change over. When the product no longer exerts a resistance against the paddle, the drive mechanism will return to its original position, and one contact will restart the motor, while the other will indicate the change in level. (Fig. 3).

The rotary level switch is used in simple processes and in cases in which product buildup is likely, since this type of switch is not generally sensitive to buildup. They are usually mounted horizontally in a silo at the desired switching point. The paddle is designed for mounting from the outside of the silo through a $1\frac{1}{4}$- or $1\frac{1}{2}$ in socket.

FIGURE 3   Paddle wheel switches.

The standard paddle may be too small for use with light products, which would not provide adequate resistance. In cases like this, a proportionally larger paddle should be used, but this would have to be mounted from inside the silo.

The product fill stream should not be directly on top of the shaft, and, when used as a minimum level limit switch, particular attention should be to the maximum load on the shaft. An appropriate solution to these problems may be to erect a protective screen above the paddle.

Level Measurement of Liquids and Solids by Use of a Plumb Bob

Level is measured when the surface of the product is plumbed from the top of the silo and, compared with the silo height, the difference is the product level. The plumb bob consists of a sensing weight, suspended inside the silo, on a tape or wire wound on a motor-driven reel. When the start is initiated manually or automatically by means of a time relay, the counter is automatically reset to the total silo height and the motor lowers the sensing weight into the silo. A measuring wheel, driven by the measuring tape, sends a pulse corresponding to the distance traveled. This is then subtracted from the reading on the counter. As the weight contacts the surface of the product, the tape will go slack. This is sensed, and the motor is reversed. The counter stops generating pulses and the measuring tape is wound back onto the reel. The weight then remains at the top of the silo, and the counter remains set to the last measured height until a signal is given for the measurement to be repeated.

This type of measurement system is suitable for level measurements of bulk materials and liquids in tall narrow silos and large storage tanks. The operating principle makes the plumb bob an efficient and inexpensive method for use with coarse-ground bulk materials, e.g., lime, stone, or coal. The accuracy is relatively high for tall silos and versions with accuracies of 1 mm are used in commodity storage tanks. Care must be taken to make sure the motor-driven wheel has the strength to pull up the plumb bob if it is buried or coated with product.

ELECTRONIC/ELECTROMAGNETIC LEVEL DETECTION AND MEASUREMENT SYSTEMS

Level measurement and detection systems for which electronic and electromagnetic principles are used have grown considerably. Advances in microprocessor/software, mechanical design, and measurement techniques have made it possible to develop measurement systems that can now do economically in a process plant what was possible only in the laboratory and at great expense.

Level Detection of Liquids by Use of Conductivity

This method is suitable only for level measurement in conductive liquids. The difference in conductivity of a partially insulated electrode is measured when the probe is covered and not covered with the conductive product. The metal wall of the tank can be used as the other electrode, or a second electrode can be inserted into the tank. Additional electrodes of different lengths can be added to detect other levels.

The advantages of this method are simple, inexpensive and suitable for dual or multiple point control. The disadvantages are probe cannot become contaminated with grease or other deposits and has limited suitability for products of varying conductivity (Fig. 4).

The principle for this form of level measurement is that the presence of a product will cause a change in the resistance between two conductors. A practical application of this principle is as follows. An indication of the level of electrically conductive products in a metal tank, or other container, can be obtained very simply by means of a probe insulated from the vessel and a conductivity amplifier. When the product is not in contact with the probe, the electrical resistance between the probe and the tank wall will be very high or even infinite. When the level of the product rises to complete the circuit with the probe and the tank wall, the resistance will be relatively low.

The calibration of conductivity switches is relatively simple. The production process should have been in operation, so there would be some natural contamination, or parallel resistance, caused by the buildup. The switch point is then chosen for a probe with an immersion of approximately 5 cm, then adjusted so the relay is energized. When the level subsequently drops and exposes the probe, the relay will deenergize.

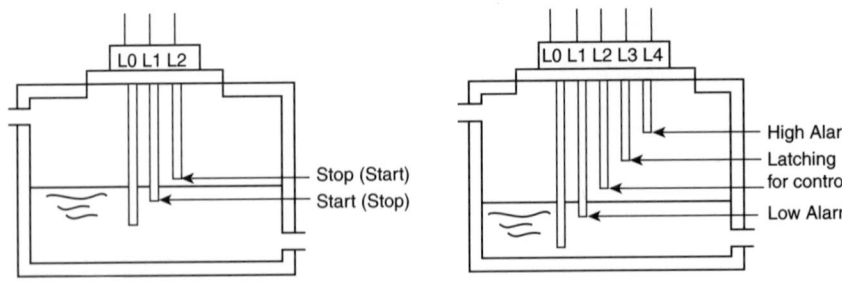

FIGURE 4   Conductivity switches.

Principle of operation

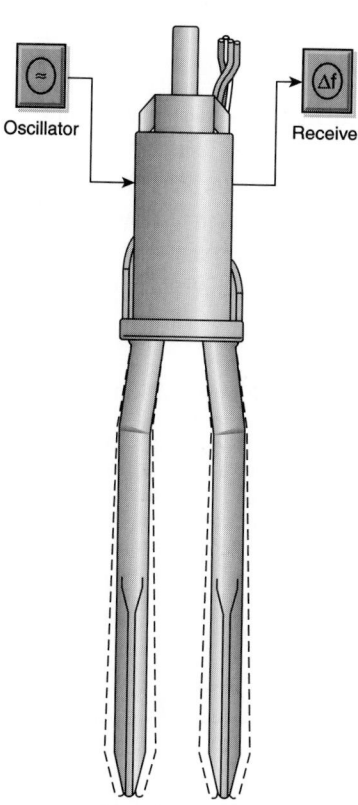

- Frequency shift tuning fork
- Piezo crystal oscillates forks
- Solution causes a reduction in resonant frequency
- Second crystal detects frequency change

FIGURE 5   Example of frequency shift tuning fork, used for measurement of liquids and slurries. (*Photo courtesy of Endress & Hauser.*)

Level Detection of Liquids by Use of Vibrating Forks Resonance or Rod Attenuation

The vibrating fork level limit switch consists of a symmetrical vibration fork with tines, which are wider at their ends. There is a thin membrane at the base of the vibrating fork which forms part of the process connection/mounting, so only a single stainless steel probe projects into the process. The vibrating fork is driven to its resonant frequency in air through the membrane by a piezoelectric crystal (Fig. 5).

The vibrating fork is designed so that when immersed in a liquid, there will be a shift in its resonant frequency of approximately 10% to 20%. The natural resonant frequency is picked up by a receiver crystal, and the shift in resonant frequency is detected by a reference circuit.

The vibrating tuning fork has a number of notable characteristics:

- no moving parts, thus no maintenance
- insensitive to appreciable buildup, strong flow, turbulence, air bubbles or particles in suspension
- can be mounted in any position
- small and can be mounted in a 1-in or smaller mounting boss

- no calibration required
- suitable for products with high viscosity

Some examples of applications are

- as dry-run safety devices in monopumps: fruit pulps, sauces, and syrups
- as minimum level monitoring system in processing plants
- as prelevel and high-level alarms in degassing tank for process water
- as minimum lubrication oil level safety devices in gear boxes and crank cases
- as condition indicators for the addition of subsequent components in mixing vessels
- as leak detectors in cellars, document stores, and computer rooms

Level Detection of Solids by Use of Vibrating Fork or Rod Attenuation

This type of system typically consists of two tines mounted on a membrane with a natural resonance of approximately 120 Hz (Fig. 6). Two piezoelectric crystals are mounted on the membrane. One of these crystals is driven by a 120-Hz oscillator, causing the system to resonate when the product does not cover the tines. When the product comes in contact with the tines or rod it causes an attenuation of the vibration that is detected by the second piezoelectric crystal (Fig. 7).

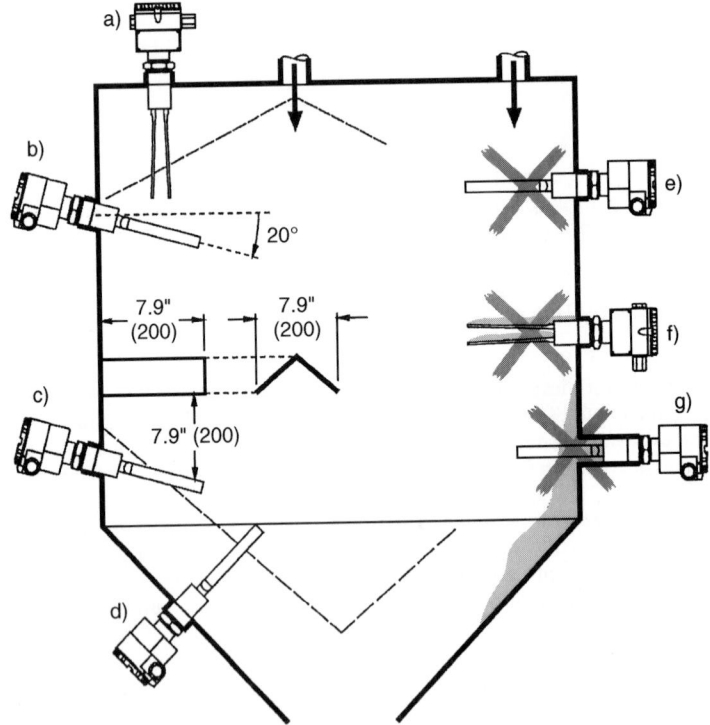

FIGURE 6   Correct installation of tuning fork.

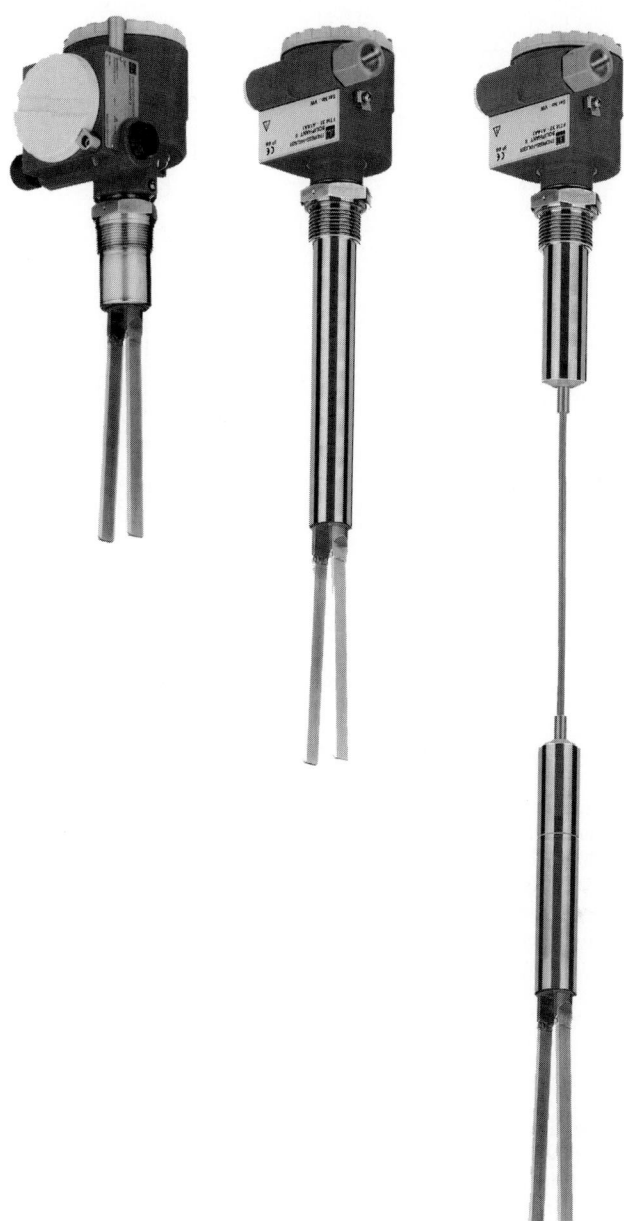

FIGURE 7   Example of amplitude shift tuning fork, used for measurement of solids. (*Photo courtesy of Endress & Hauser.*)

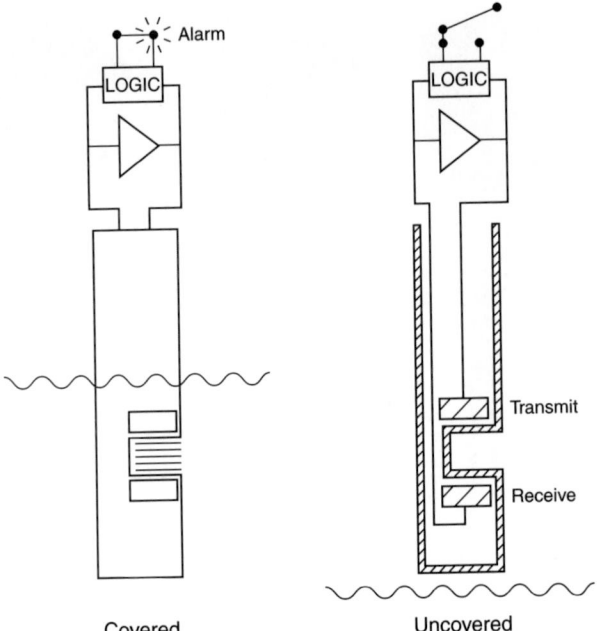

FIGURE 8   Covered and uncovered ultrasonic gap switches.

Level Detection of Liquids by Use of Ultrasonic Gap

An ultrasonic gap switch consists of transmit and receive piezoelectric crystals positioned in a probe where a liquid can move freely between them (Fig. 8). The transmit crystal emits acoustic pulses in the 500–1000-MHz range. When no liquid is in the gap this acoustic emission is not coupled to the receive crystal. As the liquid fills this gap the acoustic energy is conducted to the receive crystal and indicates level detection. Process temperatures exceeding 300°F and pressures over 2000 psi can be addressed by some versions. An advantage is that there are no moving parts to hang up or wear out and no routine maintenance requirements. A disadvantage is anything that can compromise the gap, such as plugging or crystalline buildup (Fig. 8).

Level Detection of Liquids by Use of Thermodispersion

This technique relies on the detection of the dispersion or removal of heat from a slightly heated element. A probe containing two temperature sensors and heating element is inserted in the vessel at the point level indication is desired (Fig. 9). A change in the rate of heat removal occurs when the liquid level makes contact with the liquid. The temperature sensors measure this change in heat transfer and provide an indication. An advantage of this approach is independence from material characteristics such as conductivity and dielectric constant. A disadvantage is susceptibility to coatings and buildup.

Level Measurement of Liquids by Use of Bubblers

In this method, the total hydrostatic pressure within a tank is measured by insertion of a narrow tube into the liquid and application of compressed air to the pipe so that the liquid column in the pipe

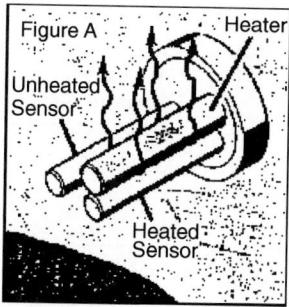

FIGURE 9 Thermal dispersion switch.

is pushed down and air bubbles are just formed in the liquid. These bubbles give the method its name. The pressure of the air in the pipe is now equal to the pressure of the liquid column and can be measured with a pressure transducer, which converts the pressure into an electrical signal. The compressed air can be supplied from the main air utility by means of a reducing valve or by using a miniature compressor.

The advantage of this method is simple assembly, suitable for corrosive substances. The disadvantages are that it requires air lines and air consumption, there is a danger of buildup of medium on the bubble pipe, and it is not suitable for use in pressurized vessels.

Level Measurement by Use of Pressure Transmitters

Principle. The principle for level measurement by use of pressure transmitters is based on the measurement of the hydrostatic pressure produced by a column of liquid at a given height. The pressure is calculated with the following formula:

$$P = h \cdot \rho \cdot g$$

where P = pressure
h = height of the liquid column
g = acceleration due to gravity
ρ = relative density

From this formula, it is evident that if the specific gravity of the medium does not vary, the only variable in the formula will be the height h. Therefore the pressure measured will be directly proportional to height h, the level of the liquid in the tank (see Fig. 10).

This diagram also illustrates that the pressure measured is equal to the pressure of the liquid column plus the surface pressure.

In tanks open to the atmosphere, the pressure on the surface of the liquids will be equal to the atmospheric pressure. Most pressure sensors compensate for atmospheric pressure. If the pressure on the surface of the liquid is greater or smaller than atmospheric pressure, then a differential pressure sensor can be used and, in this case, the pressure acting on the surface is subtracted from the total pressure, leaving only the pressure that is due to the column of liquid.

The SI unit of pressure is the Pascal (Pa), but pressure can also be expressed in psi or head of a liquid column.

Examples of pressure conversions:

1 Pascal $= 1$ N/m^2
14,504 psi $= 100$ kPA

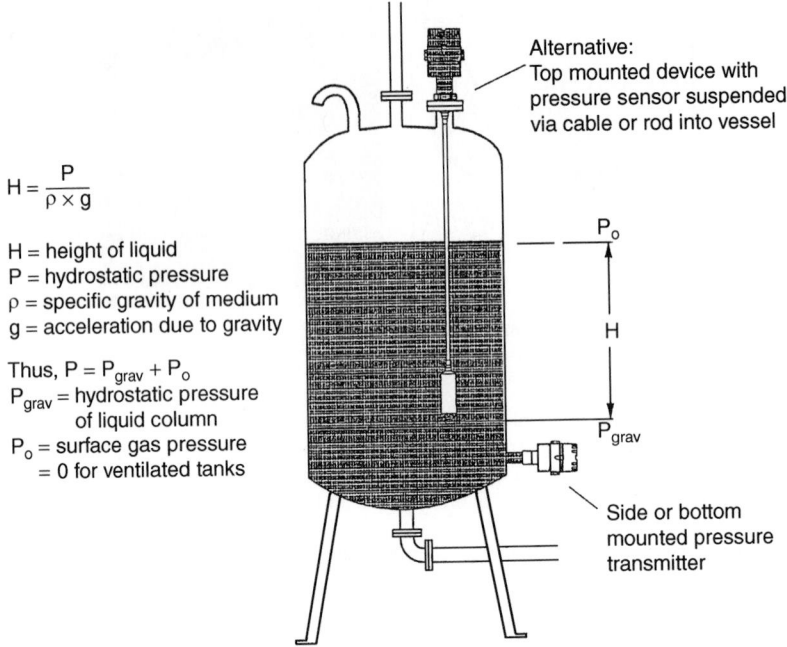

$$H = \frac{P}{\rho \times g}$$

H = height of liquid
P = hydrostatic pressure
ρ = specific gravity of medium
g = acceleration due to gravity

Thus, $P = P_{grav} + P_o$
P_{grav} = hydrostatic pressure
       of liquid column
P_o = surface gas pressure
    = 0 for ventilated tanks

Alternative:
Top mounted device with
pressure sensor suspended
via cable or rod into vessel

P_o

H

P_{grav}

Side or bottom
mounted pressure
transmitter

FIGURE 10 Pressure transmitters.

1 bar $= 10$ m head of water
14,504 psi $= 401.47$ in H_2O

Mounting. A pressure sensor is exposed to the pressure the system has to measure. Therefore it is usually mounted at or near the bottom of the tank. If the sensor cannot be mounted directly in the side of the tank at the appropriate depth, it can be mounted from the top of the tank and lowered to the appropriate depth on the end of a rod or cable (see Fig. 10). Cable-extended versions are also available for use in deep wells and reservoirs.

If the sensor is to be mounted on extension nozzles or long pipes, care must be taken to ensure that the medium will not crystallize or congeal in the pipe. If this is allowed to happen, the pressure will no longer be transmitted to the membrane and thus cannot be measured. To prevent this, either a different mounting system should be chosen or the nozzle or pipe could be heated.

Figures 11–17 show examples of recommended mounting of pressure and differential pressure (DP) transmitters for level measurement.

For applications in corrosive products in which a standard stainless-steel diaphragm is not sufficiently resistant, pressure transmitters are available with Monel, Hastelloy, tantalum, or ceramic membranes. The remaining part of the sensor body/flange can be made of virtually any corrosion-resistant material.

Should these special models still not meet corrosion resistance requirements, one solution would be to use the bubble tube principle (see Fig. 18). This involves connecting one end of a suitable plastic tube to a small compressor and reducing valve, while the other end is placed into the liquid by means of a standard pressure sensor. The compressor exerts sufficient enough pressure to exceed the pressure of the liquid column, causing bubbles to escape.

This creates a backpressure in the tube equal to the pressure exerted by the liquid column. The back pressure is measured by the pressure sensor. With this system the liquid level is measured without

- General recommendations for the routing of pressure piping can be taken from national or international standards, e.g. ISO 2186, ISO 5167, etc.
- Mount the transmitter directly on the tank.
- The negative side is open to atmospheric pressure.

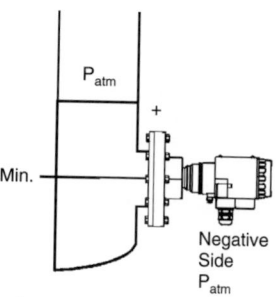

Flange Mounted Device
Open Tank

FIGURE 11   Flange-mounted device, open tank.

- Mount the transmitter below the lower tapping, so that the pressure piping is always filled with fluid.
- The negative side is open to atmospheric pressure.
- A trap prevents the build-up of dirt in the pressure piping.
- Install the pressure piping with a continuous fall of at least 10%

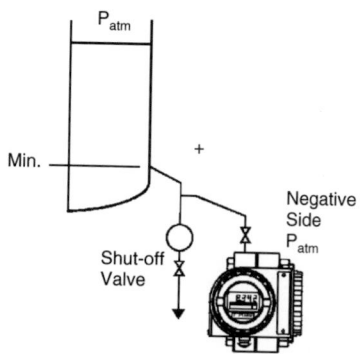

Impulse Pipe Connection
Open Tank

FIGURE 12   Impulse pipe connection, open tank.

- Mount the transmitter directly on the tank.
- The tapping for the negative side must be above the maximum level to be measured.
- A trap prevents the build-up of dirt in the pressure piping.
- Install the pressure piping with a continuous fall of at least 10%

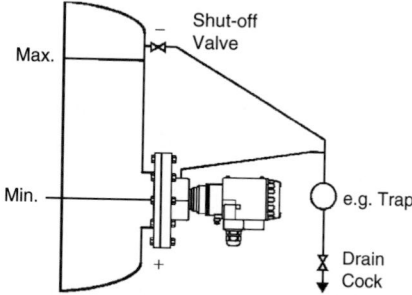

Flange Mounted Device
Closed Tank

FIGURE 13   Flange-mounted device, closed tank.

- Mount the transmitter below the lower tapping, so that the pressure piping is always filled with fluid.
- The tapping for the negative side must be above the maximum level to be measured.
- Traps prevent the build-up of dirt in the pressure piping.
- Install the pressure piping with a continuous fall of at least 10%.
- Use a three-way manifold for simple mounting without interruption of the process.

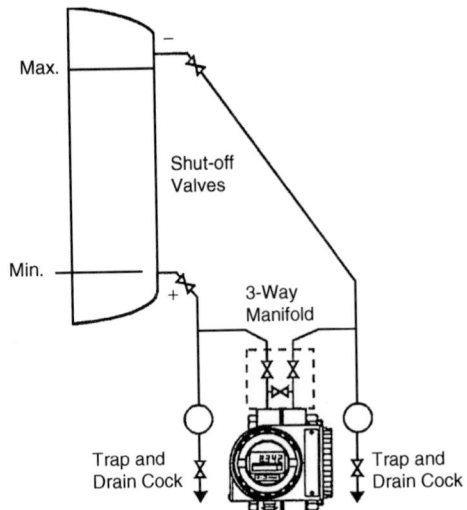

Impulse Pipe Connection
Closed Tank

FIGURE 14   Impulse pipe connection, closed tank.

- Mount the transmitter directly on the tank.
- The tapping for the negative side must be above the maximum level to be measured.
- Install the pressure piping with a continuous fall of at least 10%.
- A condensation trap ensures constant pressure on the negative side.

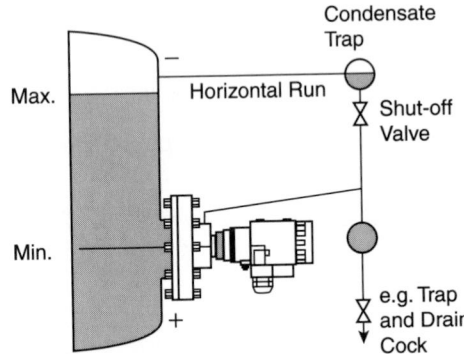

Flange Mounted Device
Closed Tank with Steaming Liquid

FIGURE 15   Flange-mounted device, closed tank with steaming liquid.

the equipment coming into contact with the liquid, eliminating product corrosiveness problems. Advantages of the bubble pipe method are its simple assembly and suitability for corrosive substances. Disadvantages include the requirement for air lines and air consumption and the danger of buildup of medium on the bubble pipe, which render the method as unsuitable for use in pressurized vessels.

Today, more and more bubbler systems are being replaced by top-mounted rod- or cable-suspended pressure sensors.

- Mount the transmitter below the lower tapping so that the pressure piping is always filled with fluid.
- The tapping for the negative side must be above the maximum level to be measured. The condensation trap ensures a constant pressure.
- Traps prevent the build-up of dirt in the pressure piping.
- Install the pressure piping with a continuous fall of at least 10%.
- Use a three-way manifold for simple mounting without interruption of the process.

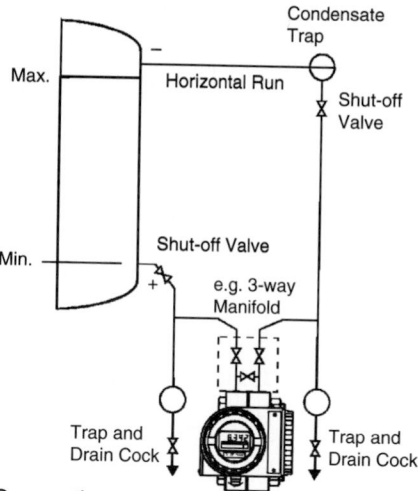

Impulse Pipe Connection
Closed Tank with Steaming Liquid

FIGURE 16   Impulse pipe connection, closed tank with steaming liquid.

- Mount the transmitter below the tapping point.
- Mount the remote seals on the tank.
- Install the pressure piping with a continuous fall of at least 10%.
- There should be no temperature difference between the capillaries.

(see also Diaphragm Seals section)

NOTE: If the material rises above the lower edge of the upper remote seal, no valid level measurement can be made.

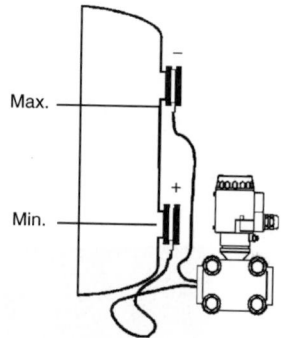

DP Transmitter with Remote
Diaphragm Seals
Closed Tank

FIGURE 17   DP transmitter with remote diaphragm seals, closed tank.

Electronic DP Measurement

To minimize potential measurement errors associated with the use of dry or wet impulse pipes or temperature-related errors on remote seals, in certain applications electronic DP measurement may be an alternative.

To obtain an electronic DP measurement, two gage pressure transmitters are used. One transmitter is placed at the bottom of the tank to measure the height of the liquid column plus pressure (see Fig. 19). The second transmitter is located at the top of the tank where it measures the head pressure

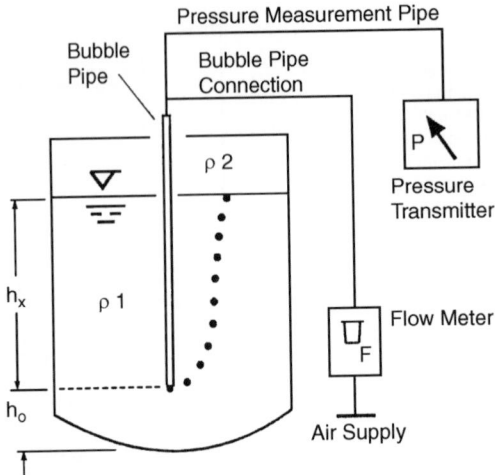

FIGURE 18 (Bubble pipe connection.)

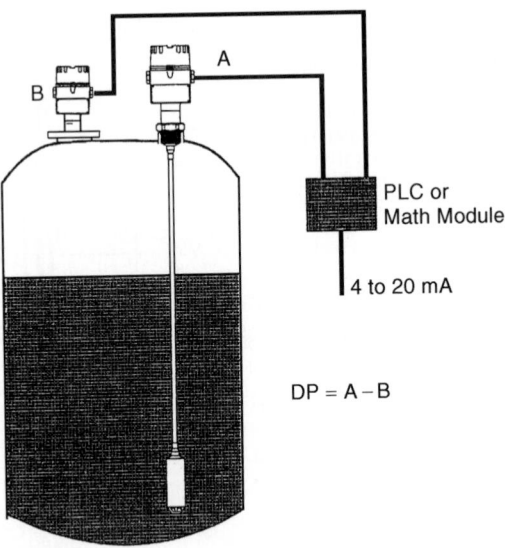

$$DP = A - B$$

FIGURE 19 Electronic DP measurement.

only (see Fig. 19). The two 4- to 20-mA signals can then simply be subtracted (level $= A - B$) to obtain the actual level measurement.

The formula is usually accomplished with a PLC or math module. To maintain accuracy and resolution of measurement, it is recommended only for applications for which there is a $1:6$ maximum ratio of hydrostatic pressure to head pressure.

Calibration

One of the practical advantages of pressure sensors is undoubtedly the fact that they are so easy to calibrate. The empty 0% reading is taken for an uncovered pressure sensor at atmospheric pressure and a certain level or x m head of water is chosen for the full 100% reading. This can be achieved with a pressure calibrator.

Level Detection and Measurement of Liquids and Solids by Use of Ultrasonic

Ultrasonic level measurement technology is based on the fact that sound travels through a medium with a known propagation speed, depending on the density and the temperature of that medium. The pulse time of flight (PTOF) method used in this kind of instrumentation utilizes a short ultrasonic pulse generated by a transducer. This pulse travels through the medium (typically air or a similar gas mix) and is reflected back to the transducer when it hits the surface of the material to be measured. The time it takes from emitting the signal to receiving the reflection is measured and then used (based on the known propagation speed of the sound wave) to calculate the distance to the level surface and the level height (based on the programmed calibration values of the system).

This noncontact form of measurement is ideal for many industrial applications for which corrosive conditions, changing product characteristics, and other factors make contacting level measurement devices less suitable. In water and wastewater applications, ultrasonic level measuring systems can be used to measure open channel flow, differential level across screens, and wet well level. Ultrasonic level systems are used in solids applications for accurate measurement of hopper level and bulk storage inventory.

The measurement system consists of a transducer or sensor, mounted on the tank or silo, and either a remote measurement transmitter or transmitter electronics built into the sensor (compact instruments) (Figs. 20 and 21). These compact systems offer many advantages, including simple installation, on-site calibration, and low prices. They can be very effective in applications like conveyor-belt monitoring, pump control, two-point control, and distance measurement (Fig. 22). However, the remainder of this discussion of ultrasonic level measurement will focus on traditional ultrasonic level measuring systems (sensor and remote transmitter). The basic considerations and physical facts apply to both types of instrumentation.

FIGURE 20   Ultrasonic sensors are available in a variety of sensors and constructed of various materials to suit the application.

FIGURE 21   Ultrasonic measuring systems are avaialble in two-wire versions for easy installation.

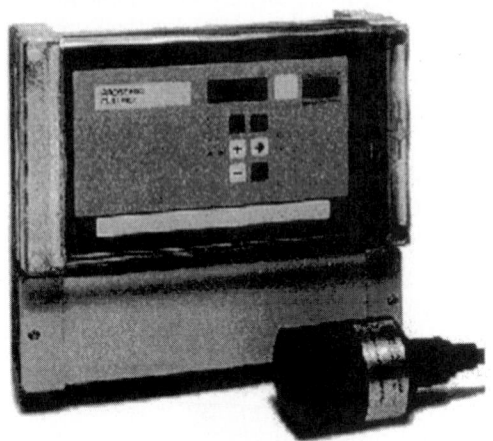

FIGURE 22   Photograph of an ultrasonic transmitter with one sensor.

Predicting all the conditions that affect an ultrasonic level measurement system can be a difficult task. Applications that appear to be simple may become very challenging because of filling noise, nuisance reflections, or variable material profiles. Even more challenging are applications for which these and other interfering conditions are changing constantly.

Echo Suppression or Tank Mapping.   When ultrasonic technology is used for level measurement, the signal can also be reflected by inlets, reinforcement beams, welding seams, etc. To prevent misinterpretations of these false echoes, the transmitter can be set up to ignore these nuisance reflections by automatically calculating a threshold curve, which overlays the actual reflected signal. This procedure has to be done while the tank is empty or at least partially empty down to the maximum desired mapping range. Every real level reflection will always add to the suppressed signal, thus generating a resulting signal above the threshold. All the transmitter has to do from there on is to ignore any signals below the threshold curve and evaluate only those signals that appear above it (Fig. 23).

Criteria for Selecting the Most Appropriate Ultrasonic System.   When selecting an ultrasonic level measuring system for a particular process, one must take several factors into consideration.

The first factor is the measuring distance. An ultrasonic system designed for very high accuracy and short distances may not be powerful enough for large distances.

The composition and properties of the product surface must also be considered. It is inherent in this type of system that at least a small part of the transmitted ultrasonic signal will be reflected from the product surface. In the case of a turbulent or very soft medium surface, a considerable part of the signal is either reflected away from the transducer or absorbed. Consequently, a higher signal transmission power might be required for a given distance compared with a smooth surface with a better reflectivity.

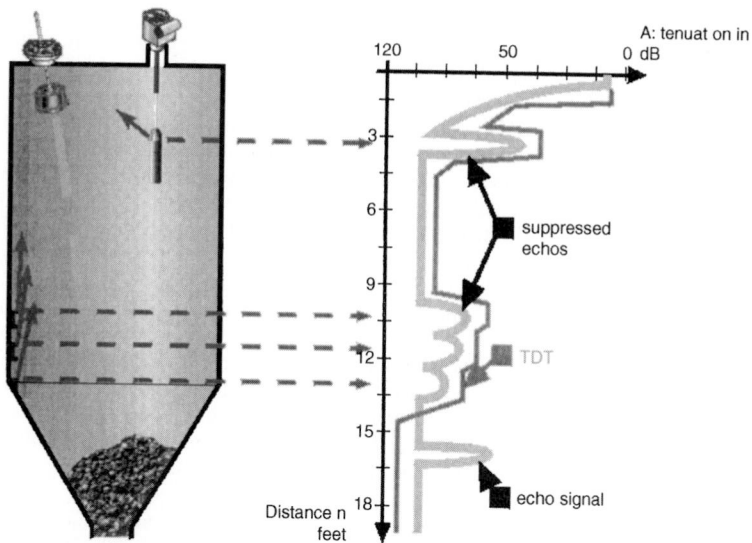

FIGURE 23   Tank mapping function illustrated. This powerful feature can solve many application dilemmas.

The presence or absence of foam has to be taken into account. A layer of foam on the liquid surface may (depending on its density) cause the signal to reflect off the foam surface (dense, thick foam), cause a complete signal absorption (soft, thick foam), or may not affect the signal in any significant way (light, thin foam).

In bulk solids, the grain size plays an important role. If the bulk material is composed of fine granules, then less energy will be reflected, so a greater transmission power is required for achieving the same distance measurement.

Both the signal transmitted from the sensor and the reflected signal must pass through the atmosphere in the tank or silo. All the factors, which are likely to weaken or interfere with the signal by absorption or reflection, must be carefully examined so the most appropriate system can be selected. These factors include the presence of dust, steam, pressure, temperature, and variations in the composition of gases.

Ultrasonic measurement systems are not sensitive to pressure variations. There will be no significant changes in the sound propagation speed. However, there are two limitations. First, the specified maximum pressure is purely a mechanical limitation. At very high pressures, the sensor may not be able to function because of the forces exerted on it. On the other hand, the system cannot operate at extremely low pressures because the air, or gas, will not be able to transmit the sound signal properly when a partial vacuum develops.

Variations in temperature will affect the sound velocity. Each transducer should be fitted with a temperature sensor that then electronically compensates for temperature variations of the air or gas. However, since temperature can be measured at only one point, certain variations may result because of temperature layers (different temperatures at the medium and the sensor).

The gas composition must also be considered. The propagation speed of sound waves will vary in air and in other gases or vapors. The transmitter will always process the signal on the basis of the speed of sound in air. Adjustments can be made when calibrating the system to compensate for different propagation speeds, but accurate readings can be achieved only based on the assumption that these gas properties are constant for the given application and do not change over the time.

Mounting.   For best performance and maximum range, the sensor must be mounted in the best possible position for the ultrasonic level measuring system to perform optimally. The sensor must be mounted in a position where there are no obstructions between the product surface and the sensor. Access ladders, immersion elements, stirrers, pump inlets, or a stream of falling product can all produce interference echoes. These can often be avoided by simply choosing a different position. The sensor should be in a location that maximizes its returned echo signal and minimizes vessel obstructions in its line of sight. Although an individual sensor has a constant, definable output beam, the more important consideration is the line-of-sight-reflected echo.

Minimizing vessel obstructions in the line of sight is accomplished by consideration of the geometry of both the vessel and the resulting reflected echoes. On vessels that have coned bottoms or dished (round) tops, positioning the sensor away from the vessel center is recommended to eliminate echo travel paths not related to the level.

Maximizing the returned level echo is generally accomplished by avoiding sensor mounts that either sight into the fill stream or position the sensor so that a large part of its output beam is lost into the vessel wall.

The position of the sensor must also comply with other requirements. The blocking distance (blind space, dead space, etc.) is the area directly below the sensor in which level cannot be sensed. Each time the sensor pulses, it must stop ringing before it can listen for a returned signal. This waiting period translates into a specific distance, the blocking distance, which is inherent to the sensor model specified. The product level must not be allowed within this distance.

Advantages/Disadvantages.   The main advantages of an ultrasonic system are that it has a noncontacting principle, it has no influences from changes of the material properties, its sensor is not subject to wear and tear, and it has a wide range of applications. The disadvantages are that it is sensitive to inconsistent gas concentrations in the free space between sensor and product, foam has to be taken into consideration, and it is not suitable for high temperatures and pressures.

Level Detection and Measurement of Liquids and Solids by Use of Capacitance

Capacitance level measurement systems measure the overall capacitance of a capacitor formed by (typically, but not necessarily) the tank wall and a probe (rod/cable) (Fig. 24). This capacitance changes depending on the degree of coverage of the probe with the material to be measured.

A capacitor is an electrical component capable of storing a certain electrical charge. It basically consists of two metal plates separated by an insulator known as a dielectric. The electrical size

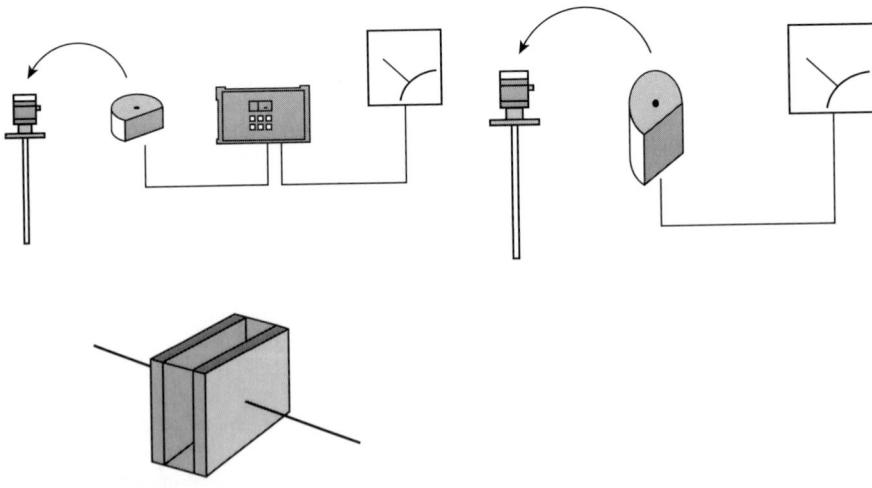

FIGURE 24   Capacitance level measuring principle.

(capacitance) of a capacitor depends on (1) the surface area of the plates, (2) the distance of the plates, and (3) the dielectric constant of the material between the plates. While the surface area and the distance of the plates will be individually different from measuring point to measuring point [depending on the desired indication range (probe length)], they will remain constant for a given application. The only variable left is the dielectric constant of the medium. When the process material replaces the empty space or air (dielectric constant 1) in the vessel, the capacitance of the capacitor increases (the higher the level, the bigger the capacitance). All liquids or solids have a dielectric constant higher than one (1). The transmitter electronics registers the change in capacitance. This change is converted to an electrical signal and used to provide an output for a switch point or a continuous level system.

To achieve a reliable measurement, the dielectric constant of the process material must be constant. If the material in a tank is changed against a material with a different dielectric constant, a recalibration is required for standard probes. More expensive self-compensating systems are available, but cannot always be applied.

The main applications for capacitance probes are level limit detection (point level) in liquids or solids and continuous measurement in liquids. Continuous measurement in solids is not recommended, since solids have the tendency to change their moisture contents, which affects the dielectric constant, which in turn leads to an unstable, drifting, nonreliable measurement.

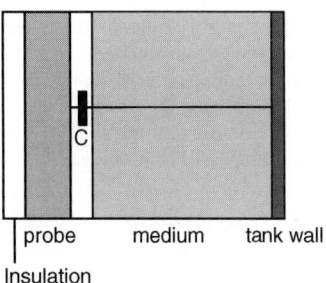

probe   medium   tank wall

Insulation

FIGURE 25   Example of function of a fully insulated probe.

Fully and Partially Insulated Probes

Two major versions of capacitance probes are available, regardless of the individual construction (rod, cable, or specially shaped sensor electrode). These two versions are (1) fully insulated probes, with the sensing element completely coated with an insulation material (Fig. 25), and (2) partially insulated probes, with the sensing element still isolated from ground (which is a requirement), but, besides that, is made of blank metal. The fully insulated probe provides a galvanic separation to the process material, while the partially insulated probe is in direct galvanic contact. The required probe type for a given application can be selected from Table 1.

TABLE 1   Required Probe Type for a Given Application

Application	Fully Insulated	Partially Insulated
Point level, tip sensitive, process material nonconductive	√	√
Point level, adjustable switch point, process material nonconductive	√	√
Point level, tip sensitive, process material conductive	√	√
Point level, adjustable switch point, process material conductive	√	Not suitable
Continuous level, process material nonconductive	√	√
Continuous level, process material conductive	√	Not suitable
Process material with extremely low dielectric constant	Limited suitability	√
Process material chemically aggressive	√	Only with special probe materials

Capacitance Probes and Ground References

A capacitance system always requires a reference potential (the second plate of the capacitor).

Depending on the probe design, this reference is already built into the probe, or a separate reference (typically the tank wall, but also an additionally mounted reference electrode) is used. Since a metal tank is usually connected to ground potential, this reference is often referred to as ground reference. Some manufacturers state that no ground reference is required, mainly for point level probes. These probes do use a ground reference as well, but it is already built into the probe; thus the statement should be "no additional ground reference required."

Because of the higher mechanical requirements for the manufacturing of a probe with a ground reference being built in or for making an additional ground reference electrode, these systems are more expensive than plain rod or cable versions. Hence the question arises of when the extra money should be spent for a probe with an integral or separate ground reference and when a standard probe can be used.

Here are some guidelines:

- In tanks made of nonconductive materials, a built-in or external ground reference is definitely required. Reason: Without a reference, the system will not respond properly and may give erratic readings.

- For a continuous measurement of nonconductive liquids in horizontal cylindrical tanks, tanks with a cone bottom or tanks with any irregular shape or with built-in items (like cooling or heating coils), a ground reference parallel to the probe is required, even if the tank is made from metal. Reason: The distance between the tank wall (or any internal installations) and the probe is not constant over the length of the probe, therefore the capacitance change per length unit (i.e., pF/in) varies depending on the level height. As a consequence, the measurement will be nonlinear.

- Level measurement (continuous) or detection (point level) in materials with very low dielectric constants can be considerably improved with built-in or external ground references mounted parallel to the probe. Reason: These references are much closer to the probe, and the capacitance change per length unit (i.e. pF/in) will be higher compared with that of a system without an additional reference. Consequently, the indication becomes more reliable; sometimes the presence or absence of a ground reference determines whether a system can be calibrated at all.

Capacitance Probes and Buildup

Buildup on a capacitance probe may be a serious source of trouble, depending on how thickly it coats the probe and how conductive it is. It can keep point level systems from properly switching and it might lead to a wrong level indication (offset) in continuous measurements. The influence of buildup on continuous systems can be minimized with two different approaches, which may be used in combination:

- Using a dead section at the upper couple of inches (between the sensing element and the process connection) prevents condensation, which typically occurs close to the tank ceiling or in nozzles, from influencing the measurement.

- If the sensing element of the probe is coated itself, particularly if the buildup is conductive, the selection of a higher measuring frequency considerably reduces the effect on the output signal (example: use a frequency of 1 MHz rather than 33 kHz). Systems with high frequencies are usually limited to shorter probes (rod probes rather than cable probes)

In applications that involve an extreme level of buildup, an alternative measurement technology should be considered.

For point level systems, buildup can again be handled in two different ways.

- Use a dead section (see above).

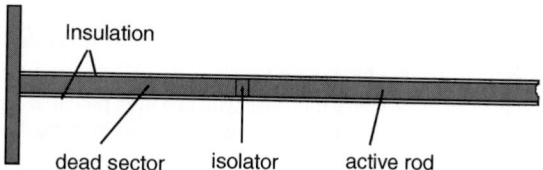

Insulation

dead sector   isolator   active rod

FIGURE 26   Example of capacitance probe with active guard.

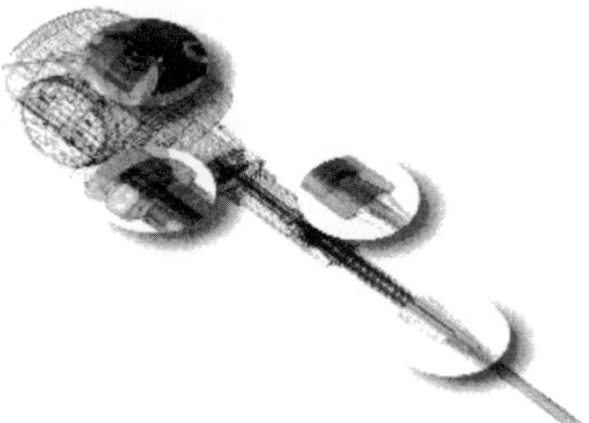

FIGURE 27   Illustration of a capacitance probe showing active guard.

- Use a compensation electrode that is an integral part of the probe (active guard, active compensation against buildup, driven shield). This electrode actively compensates for changing amounts of buildup, allowing the capacitance switch to work, even if the probe is completely coated (Fig. 26).

In a standard probe configuration (without active guard) the measuring loop consists of a high-frequency (HF) oscillator, probe, and impedance of the product. The switching of a level device is caused by the change of high impedance (no product) to low impedance (product at the probe). In case of heavy buildup that remains on the probe while the real level is gone, the level switch is short circuited and will still "see" the level.

A probe with an active guard system actually consists of two HF loops, one for the actual probe and one with a controlled signal for the guard (Fig. 27).

If any buildup is formed, the controller keeps the signal (amplitude and phase) of the guard on the same level as the signal at the probe. There is a current between the guard and ground, but no current between the probe and guard (since they are on equal potential); therefore there is no level indicated. If a real level is present, there is a current between the probe and ground, so the switch will be activated.

Installation and Application Considerations

Some important considerations have been mentioned above already (like dielectric constant and conductivity of the material to be measured, the tank material, buildup, etc.). Some other aspects should be taken into account as well:

- There must be a free, nonobstructed view from the mounting position of the probe down to the tank bottom in order to avoid the probe's touching (or getting too close to) either the tank wall or any internal installations (watch for agitators).

- When measuring materials with a high viscosity or a high potential for buildup in general, the probe must not be installed very close to the tank wall to avoid bridging between the probe and the tank, leading to a wrong indication.

- When installing the probe in a process tank with agitation, the lateral forces applied to the probe (depending on agitation speed, viscosity of the liquid, etc.) have to be considered. They might cause probe damage over time. Sometimes an improvement can be made by installation of the probe in a stilling well or use of a probe with a built-in ground tube.

- Special care should be taken when considering the installation of a capacitance probe into a plastic or glass tank, which cannot be used as a reference. These tanks require a probe with built-in or separate ground reference. Unfortunately, these kinds of tanks are typically used for chemically aggressive materials, like acids. Ground reference electrodes, on the other hand, should be in galvanic contact with the liquid to be measured—if even possible. This might pose a major problem with the chemical compatibility of the construction materials for the ground reference (expensive special materials, like tantalum, may need to be used). It is not recommended to use a ground reference with an insulation like the probe itself. Although a system like that would technically work, both electrodes (the probe and the reference) would form a large antenna, making the system sensitive to electromagnetic interference in the frequency range of the capacitance system. With no shield around the probe (grounded metal tank or grounded conductive liquid), the measurement is likely to become unreliable.

- For measurements in liquids with a high viscosity, particularly if these liquids are conductive, delays have to be taken into account for the level indication's being correct after draining a tank. It may take even hours before the material runs off the probe completely; as long as there is any buildup left, the level will be indicated with a certain offset (see above: "capacitance probes and buildup"). In applications like this, the suitability of a capacitance system should be questioned.

Advantages/Disadvantages

The main advantages of capacitance systems are easy installation, broad application range, many application-specific solutions available, good accuracy (in suitable applications), and well-known and proven technology.

The disadvantages are sensitivity to changes in the material properties (dielectric constant, conductivity), that buildup may be an issue, and it is an intrusive system.

Level Detection and Measurement of Liquids and Solids by Use of Radar

Radar is becoming a rapidly important method of measuring the level of liquids and in some cases solids. Once an expensive measurement solution, rapid improvements in technology have brought the cost into the range of other common level technologies. Even two-wire-powered versions are now available.

Point Level Measurement

The level in a tank can be indicated by microwave echo reflection for continuous, noncontact, level measurement of liquids, pastes, and slurries. For level detection, the microwave beam is projected horizontally from a transmitter on one side of a vessel through the air or gas to a receiver on the opposite side. The beam is attenuated by the dielectric constant of the measured product when it moves between the transmitter and receiver. Advantage is the measurement is able to detect very light-density solids material levels and is independent of vapor and temperature changes. The disadvantage

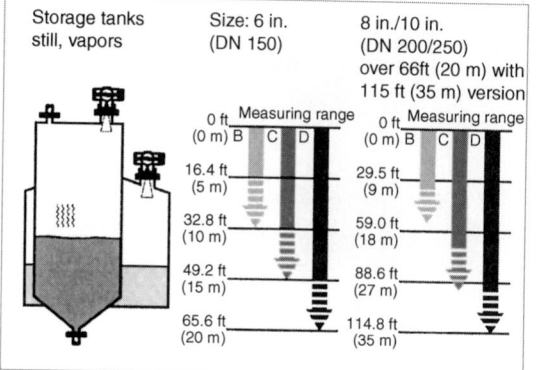

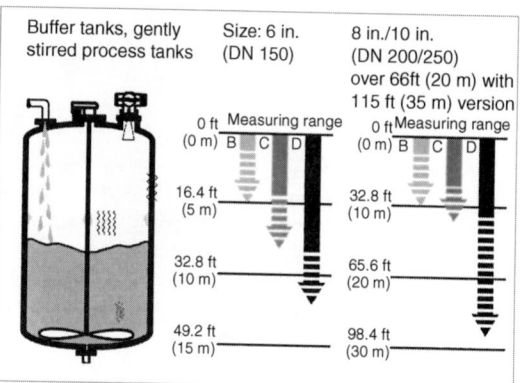

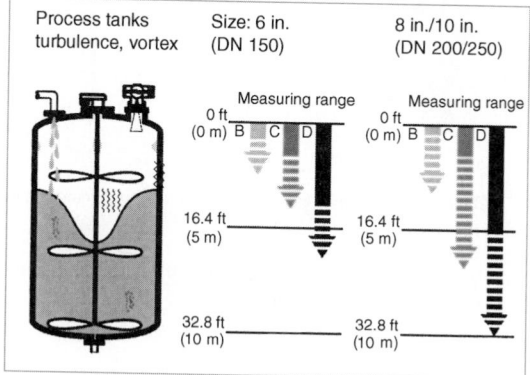

FIGURE 28   Radar device installed in a storage tank.

is the requirement for nonmetallic vessel walls or windows and vulnerability to buildup of material or moisture, which can cause false level indications.

Continuous Level Measurement

A microwave level transmitter, sometimes referred to as a radar level gage, uses a radar (PTOF) or frequency-modulated continuous-wave (FMCW) measurement methods and operates in a frequency band approved for industrial use. The low beam power used by the transmitter allows safe installation in metallic and nonmetallic vessels, with no risk to humans or the environment. The technical specifications of each radar level gage should be consulted carefully, as some units require a site license, from the FCC, before installation (Fig. 28).

Microwave or radar level measurement can be applied in storage, buffer, and process tanks in which temperature gradients, inert-gas blankets, and vapors are present. It can be used on bypass pipes and stilling wells as a substitute for mechanical systems. It provides reliable measurement under difficult conditions, such as when the dielectric constant of the medium is less than 1.9 or vortices are present.

Selection Criteria

Microwave measurements are virtually independent of temperature, pressure, and the presence of dust or vapor. Only the dielectric constant of the product must exhibit a minimum value, which is dependent on the measuring range. The selection of a suitable system is dependent on the application.

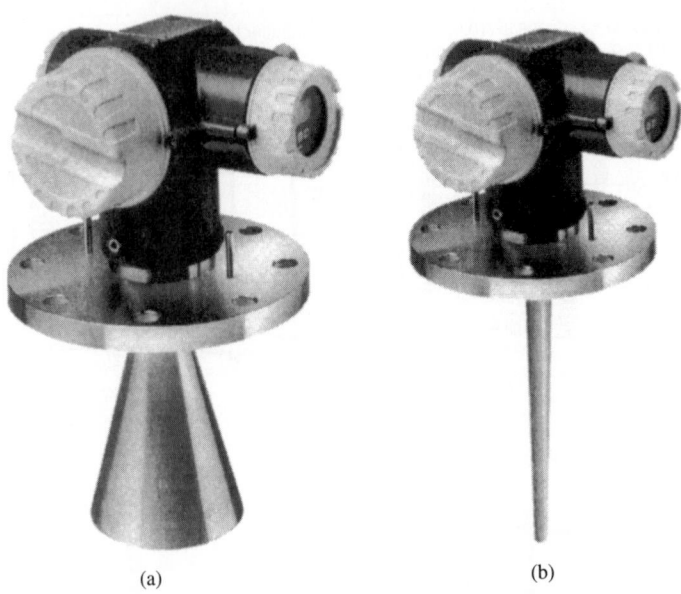

(a)   (b)

FIGURE 29   (a)Four-wire radar level gage with horn antenna, (b) four-wire radar level
gauge with Teflon rod antenna.

The antenna size and configuration are determined by the measuring range desired, process conditions,
dielectric constant, agitation, and foam (Fig. 29).

Installation Guidelines

The microwave pulse should, if possible, arrive unhindered at the product surface. The following
guidelines should be considered:

- The transmitter axis must lie perpendicular to the product surface.
- The front edge of the antenna horn must project into the tank.
- Avoid a central position on the tank, keeping away from fittings such as pipes when possible.
- Avoid measuring through the filling stream or into the vortex.
- For synthetic tanks like fiberglass, PVC, polypropylene, or unleaded glass, the transmitter can be
  mounted outside the tank, providing a totally nonintrusive measurement.

Where operating conditions require avoiding high temperatures or preventing buildup on the antenna,
it is recommended that the antenna be installed with a safety gap above the maximum product level.
Also, if the tank temperature is high, a high-temperature version with heat barrier can be used.

An advantage of radar is its broad applicability on most liquids and measurement independent of
pressure, temperature, vapor, and (to a degree) product dielectric constant. A disadvantage is that the
measurement may be lost because of heavy agitation of the liquid or the formation of foam.

Level Detection and Measurement of Liquids and Solids by Use
of Time-Domain Reflectometry

Microimpulse time-domain reflectometry, or MiTDR, is a technology only recently available for
industrial measurement and control. The principle of this technology is to launch an extremely short
microwave impulse on a waveguide, which can be a cable or rod. This waveguide, or probe, contains

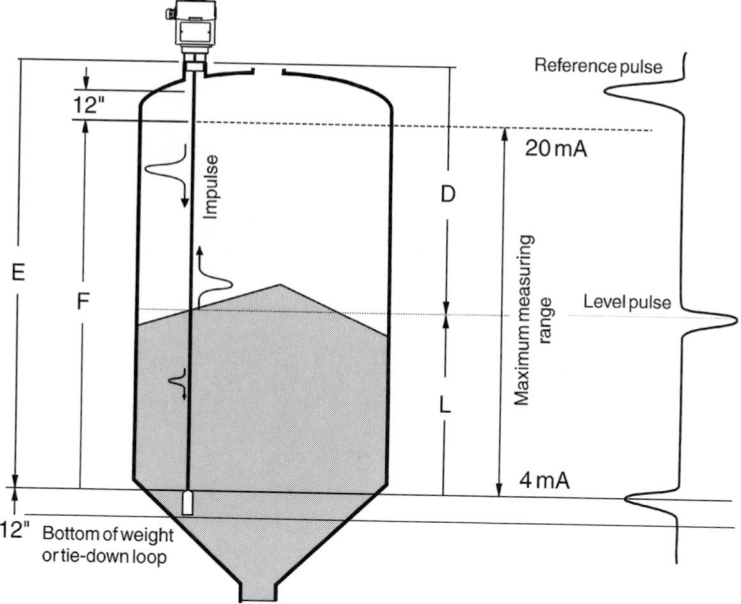

FIGURE 30 Time-domain reflectometry measuring principle.

the signal in a close area around the cable or rod. Since no real signal emission takes place, the signal is not affected by any adverse process conditions. The measurement effect is the time difference between launching the microimpulse and receiving a reflection from the level surface.

The above-mentioned reflection happens when the signal around the waveguide or probe hits an area with a dielectric constant different from the area where it came from (example: from air with a dielectric constant of 1 to the material to be measured with a dielectric constant of 1.8 or higher). The reflected pulse travels back up the probe to the pulse sampler where it is detected and timed. Any unreflected portion travels on to the end of the probe to provide an empty signal.

Each point along the probe is sampled for its pulse reflection behavior. The information accumulated over the sampling cycle is captured and passed onto signal processing. The signal produced by the change in dielectric constant at the air/product interface is identified (Fig. 30).

Typical applications of MiTDR level measurement in liquids are storage, buffer, or low-agitation process measurements. The system is not suitable for highly agitated liquids. Typical applications in solids are measurements in narrow or tall bins with very dusty and/or light materials. It can be used to measure a wide variety of bulk solids including minerals, plastics, agricultural products, foodstuffs, pharmaceuticals, and solid fuels.

The advantages of MiTDR level measurements are disregard of the presence of vapors, steam, dust, gas layers, buildup, temperature changes, pressure changes, acoustic noise, changing density of the material to be measured, changing dielectric constant of the material to be measured, and changing conductivity of the material to be measured.

Some disadvantages of this measuring system are that it does come in contact with the process and is an intrusive form of measurement.

Level Measurement of Liquids by Use of Magnetostrictive

A magnetostrictive system consists of a magnetostrictive wire in a probe and a donut-shaped float containing a permanent magnet. The float is the only moving part and travels up and down the probe. When a current pulse is induced from the end of the magnetostrictive wire, a donut-shaped magnetic field emanates down the wire. As this magnetic field interacts with the magnetic field of the permanent

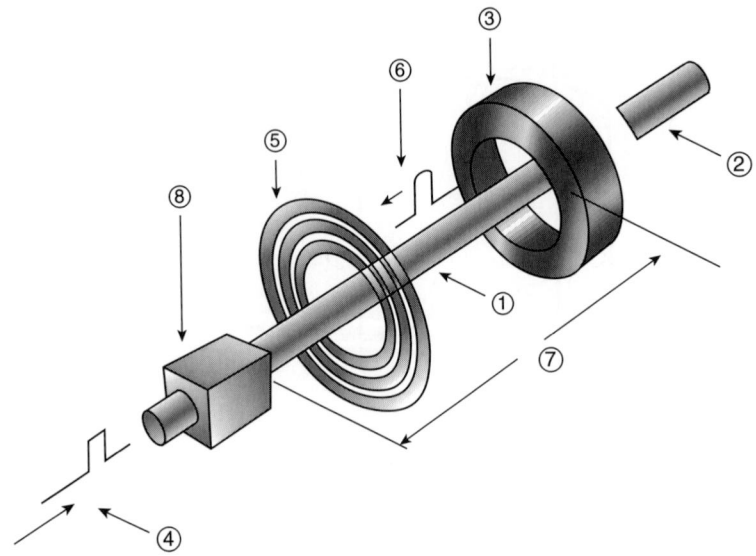

FIGURE 31 Magnetostrictive level measurement.

magnet in the float, a return pulse is sent down the wire. The electronics measures the time between the initial current pulse and the return pulse and calculates the location of the float and thus the level (Fig. 31).

Probes are typically available as rigid rods for tanks up to 25 ft and flexible cable for tanks up to 75 ft high. Magnetostrictive measurement systems claim accuracies of 0.02 in and repeatability of 0.0001. A variety of float materials and sizes are used for different applications. Advantages are high accuracy and independence from material characteristics such as conductivity and dielectric constants. Disadvantages are limitations to relatively clean liquids and anything that would cause the float to hang up on the probe.

Level Measurement of Liquids by Use of Laser

A laser can be directed at a shallow angle to the surface of a highly reflective liquid. A series of photocells are placed in the path of the reflected beam. As the level moves up and down, the reflected beam moves in step with it. A common application for laser level measurement system is on the forehearth of a molten glass furnace. The advantages are no contact with the process and very precise measurement of small level movements. A certain degree of dust and smoke can be tolerated. The disadvantages are limited applicability and sometimes a costly installation.

Level Detection and Measurement of Liquids and Solids by Use of Radiometric

Level measurement systems that use gamma or neutron radiation sources provide a means to obtain level information without any intrusion in the vessel. Gamma radiation is a form of electromagnetic radiation emitted from certain elements as a result of transitions in the nucleus of the atom.

Point Level Measurement. A beam of gamma rays is emitted at an exit angle of approximately 10°s from a source holder containing a radiation source. A gamma radiation detector is positioned on the vessel opposite the source with electrical connections to a level switch. The strength of the radiation source is calculated so that when the tank is empty, the detector transmits just enough pulses to the switching amplifier to energize the relay. As the level rises, the radiation will be attenuated, the detector will no longer transmit enough pulses, and the relay will deenergize. Thus the system provides level detection without coming into contact with the product and is external to the tank or reactor.

Gamma radiation level measurement systems are used in situations in which probes or other transducers cannot be placed inside a tank or reactor. Some examples are use with very corrosive or extremely adhesive products, in reactors or furnaces at very high pressures and/or temperatures, and with very coarse and abrasive bulk materials such as ores, limestone, coal, and basalt.

The four principal components of any level detection or measurement system that uses gamma radiation are:

1. radioactive source
2. source holder
3. detector
4. amplifier

The radioactive source, for industrial purposes, is usually cobalt 60 or cesium 137. These are artificial isotopes that emit pure gamma radiation, no alpha or beta particles or neutrons. The cobalt 60 source is used in the form of metal discs, while cesium 137 is in the form of small beads. In both cases, the radioactive material is fully enclosed in two sealed stainless-steel capsules to contain the radioactivity, preventing leakage in the event of mishandling the source. Neutron sources are used in very large or thick-walled vessels.

The source is placed in a source holder to ensure that the radiation is emitted only in the required direction. Like a light source, a radioactive source emits radiation in all directions, although for level detection only a fine beam of radiation is directed toward the detector. The holder consists of a hermetically welded steel spherical capsule, which is filled almost completely with lead. The high relative density of lead and the ease with which it can be processed make lead the best shielding material. The steel casing provides adequate protection not only against mechanical damage, but also in the event of fire. Lead fuses at a temperature as low as 325°C, but maintains its shielding properties, even when molten. The eccentric rotating mechanism enables the source to be turned on and off. This means that cleaning or maintenance operations can be carried out in the tank without the need to remove the radioactive source. The plug-in source housing can be locked in either the on or the off position. The complete locking mechanism is protected against dust and dirt ingress by cap and O-ring seal. As it is linked to the source carrier, the position of the cap indicates whether the source is on or off.

The detector used in gamma radiation point level measurement is usually a Geiger–Muller tube, more commonly called the Geiger counter. This consists of a glass tube with an anode and cathode. The tube is filled with an inert gas quenched with a small quantity of halogen and protected by a sealed metal casing. The counter is connected by means of a high-value resistor to a voltage of +400 V.

With no radioactivity present, the tube is a perfect insulator and no current flows in the circuit. When the tube is exposed to radioactive energy, the inert gas is ionized and, because of the high potential, there is a primary discharge of electrons. The speed of these primary electrons releases a secondary discharge known as an electron avalanche.

The current pulse is produced, and the number of pulses emitted is directly proportional to the strength of the radiation at the tube. As these low-current pulses cannot be transmitted over long distances, the detector casing also houses a preamplifier. The preamplifier circuit board also contains a high-voltage oscillator to supply the required anode voltage to the tube. The pulses produced are superimposed onto the supply current flowing in the two-core supply line to the amplifier. There are various models of detectors available, the differences being in the sensitivity and their use in flammable atmospheres.

Continuous Level Measurement.   Continuous level measurement that uses gamma radiation also uses the four main components mentioned in the preceding section. The main differences among these components will be noted here.

The standard source holder, as discussed above, is used, except that the exit angle used is wider, 20° or 40°. The radiation beam leaving the aperture is symmetrical, and thus the standard source holder must be mounted at an angle equivalent to half the exit angle. To prevent any premature attenuation, the source holder must be mounted directly opposite the highest measuring point. In the case of a tall tank with a small diameter, because of the limitation of the output angle, it may be necessary to use two identical sources and holders, the radiation from each covering a section of the detector.

The detector in analog measurement consists of a type of scintillation counter/photomultiplier or ion chamber, housed in a steel protective casing. The effective detection length obviously depends on the level range to be measured and the possible methods of installation. Detectors are available in lengths from 1 to 20 ft. If one detector is inadequate, then two detectors may be connected to one amplifier to extend the range.

The use of a rod-type scintillation counter has the advantage of allowing the use of a point source. It combines the advantage of the high sensitivity of scintillation crystals (compared with Geiger counters) with the safety and economy of the point source. It is, as the name implies, a rod of optically pure plastic within which scintillation crystals are uniformly distributed. In the presence of gamma radiation, the scintillation crystals emit flashes of light, which are then detected by a photomultiplier at the base of the rod and converted into electrical pulses.

The advantages of gamma are ease of installation outside of the vessel and independence from temperature, pressure, and material composition variations inside vessel. The disadvantages are high costs and the cost and liability of radioactive source disposal.

Level Measurement of Liquids and Solids by Use of Weighing

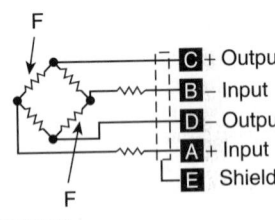

FIGURE 32

This indirect level measurement method is suitable for liquids and bulk solids. It consists of mounting the whole tank or silo on load cells or attaching strain transducers. The weight transducers are mostly based on the strain-gage principle (Fig. 32).

The advantages are very accurate level measurement for products with constant relative density and the fact that the system measures contents rather than level. The disadvantages are that installation cost is usually higher than other methods and the system requires large amounts of mechanical equipment to eliminate false weighing forces and can be very costly. If there is buildup on the vessel walls this can be falsely interpreted as an increased level.

Level Measurement of Liquids by Use of Optics

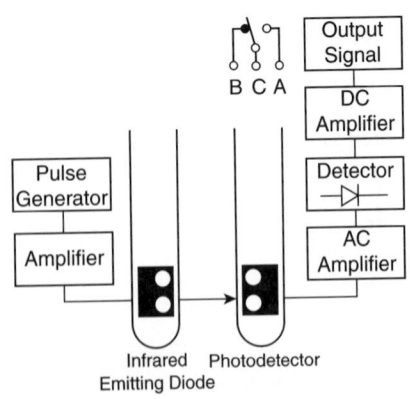

Infrared   Photodetector
Emitting Diode

FIGURE 33 Typical optical switch.

Single-probe optical technology can be applied to detect the level of clean liquids. A probe made of glass or clear plastic has a conical end that is positioned to make contact with the process liquid. The light source is directed from the other end of the probe. When liquid comes in contact with the probe, the index of reflection changes and the resultant change in light intensity is detected by a photocell. The advantages are low cost and low power operation. The disadvantage is that the system is limited to clean liquids that do not build up or coat the probe.

Another optical technology is the beam breaker. An infrared light source is positioned on one side of a vessel or container and a photocell on the other at the point detection of the level desired. When the material reaches this point, the light reaching the photocell is attenuated and an indication is given. Another approach is to mount the light source and detector next to each other in a probe and insert them into the vessel (Fig. 33).

The advantages are low cost and independence of conductivity and dielectric constant changes. The disadvantages are susceptibility to coatings and the requirement to put windows in the vessel wall in external installations (unless made of a transparent material).

Level Detection of Liquids by Use of Ultrasonic Tank Resonance

The tank resonance principle utilizes the fact that any object vibrating in its own resonance frequency will stop vibrating as soon as there are no more acceleration forces applied. This is not a sudden stop; it takes a certain time for the vibration to decay. This time again depends on the specific properties of the vibrating object and on the dampening force applied to it.

In a practical application, an ultrasonic transducer is attached to the wall of a tank at the position of the desired alarm point in a way that ensures a good acoustical coupling to that tank wall. A two-point calibration is performed then, with the tank being empty at the sensor position for the first calibration point, and the tank being filled above the sensor position for the second point. The transmitter electronics stores an acoustical profile of the tank wall (only looking at the area close to the sensor).

To generate this profile, the system first checks the resonance frequency of the tank wall, both for the free (empty) and the covered (full) situation. It then applies a short ultrasonic pulse with exactly this resonance frequency to the tank wall and measures the time it takes for the tank wall to stop vibrating after the sending of the pulse has been finished. This time will be longer for a free (not dampened) tank wall than for a covered (dampened) one. The time difference is the means for the system to determine the presence or absence of a liquid at the sensor position (Fig. 34).

Since the resonance frequency of a tank wall depends on the construction material as well as on the wall thickness, instruments for universal use are typically offered with different transducer types in order to accommodate all frequencies up into the megahertz range.

Installation of the transducers can be done by either bolting them to the tank (spot-welded bolts, rails), using tensioning bands around the tank, or using glue to hold the sensor in place. While the first option would be applicable to only metal tanks, the other two mounting methods can also be used on plastic and glass tanks (Fig. 35).

Certain application considerations are to be made:

- the process temperature (refer to instrument specifications)
- the viscosity of the liquid (if the viscosity is too high, the liquid will not properly run off the tank wall, thus leading to a false full indication)
- the overall buildup potential of the medium (crystalline buildup or any other hard residues are not allowed)

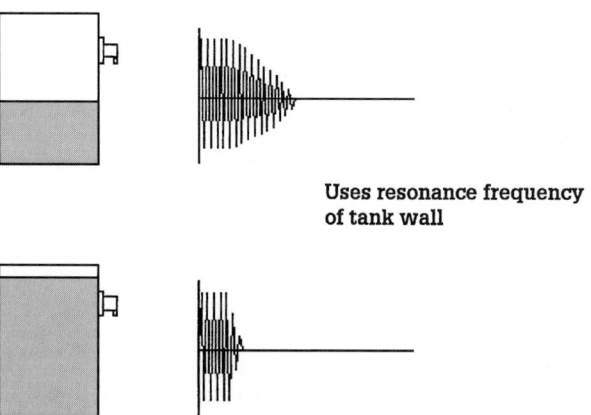

Uses resonance frequency of tank wall

FIGURE 34   Ultrasonic resonance principle.

FIGURE 35 Endress & Hauser Nivopuls can be strapped, bolted, or glued to the outside of the tank for nonintrusive level measurement.

- gas bubbles forming at the inside of the tank (decoupling the tank wall from the liquid)
- process pressure or changing product parameters have no influence on the measurement

The main advantages of the tank resonance principle are that it is nonintrusive (therefore, there are no issues with chemical compatibility), it has fast and easy installation and setup, and it has an excellent retrofit option for additional alarm points in existing applications for which no tank openings are available (certified pressurized tanks, glass-lined tanks, plastic or glass tanks, etc.). The disadvantages are that it has a relatively high instrument cost compared with that of a simple intrusive switch (but the cost savings for the installation have to be taken into account!), it is sensitive to buildup at the tank wall, and it may not work in highly aerated liquids (if gas bubbles collect at the tank wall).

INDUSTRIAL WEIGHING AND DENSITY SYSTEMS

INDUSTRIAL SCALES

For centuries the underlying principle of the steelyard, that is, determining weight through the use of levers and counterweights, persisted. Modern industrial and commercial scales as recently as the

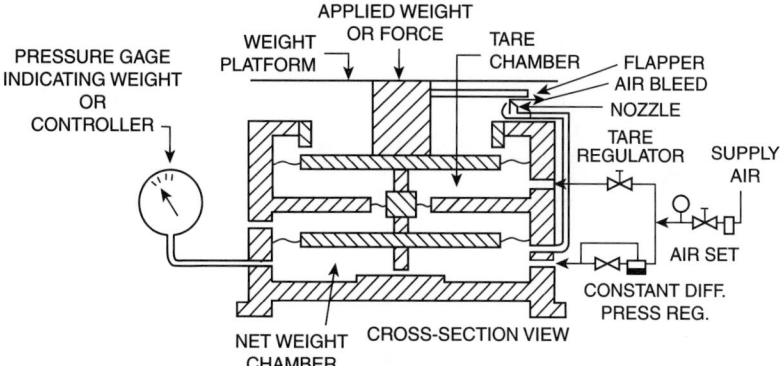

FIGURE 1 For some applications, a pneumatic load cell may be preferred over the widely used strain-gage cell.

midtwentieth century continued to rely on mechanical levers, pivots, bearings, and counterweights. Even without the benefits of advanced computing, electronics, and microprocessing techniques, a large truck or railroad platform scale could weigh 1 pound of butter or 1 part in 10,000. The principal design disadvantages of the former very rugged mechanical scales were their bulk, weight, and manual requirement to achieve balance. Self-balancing scales involving springs or counterbalancing pendulums essentially eliminated the need for manual operation. Invention of the strain gage and associated load cells revolutionized scale design. It should be pointed out that mechanical scales still are found in numerous isolated installations, although replacement parts are difficult to find.

Concurrent with the development of the strain gage, pneumatic and hydraulic load cells were developed and continue to be found in some installations (Fig. 1). However, the strain gage, by far, is the principal weight sensor used today. Refer to the article, "Fluid Pressure Systems" in this handbook section.

Industrial scales may be classified by several criteria.

1. Physical configurations, influenced by application needs, include (1) ordinary platform scales as found in warehousing, receiving, or shipping departments, (2) very large highway or railroad track scales, which today frequently are designed to weigh "on the move," (3) tank scales, in which entire tanks of liquids can be weighed periodically or continuously—for inventory control (actually used as flow controllers for some difficult-to-handle materials) and for feeding batch processes, (4) bulk-weighing or hopper scales, and (5) counting scales used in inventory and production control.

2. Scales also fall into a number of capacity categories, ranging from the most sensitive laboratory-type balances (not covered in this article) to platform scales from as low as 50 lb (23 kg) through 1000, 50,000, 200,000 up to 1 million lb (453,597 kg). The upper limit on platform scales usually is about 30,000 lb (13,607 kg). Scales also are available for weighing materials which travel along a monorail, such as in a meat-packing plant. Another specialized application for weight indication is that of a crane or derrick for lifting and lowering heavy loads.

A number of installations are shown in Figs. 2 through 5.

FLUID DENSITY

Density may be defined as the mass per unit volume and usually is expressed in units of grams per cubic centimeter, pounds per cubic foot, or pounds per gallon. Specific gravity is the ratio of the

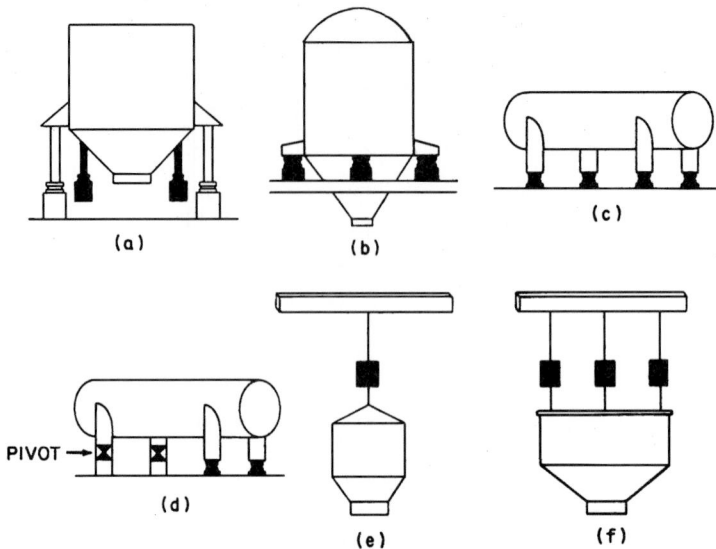

FIGURE 2 Arrangement of load cells and mounting assemblies (shown in black) for various hopper and tank configurations. (*a*) Four-load-cell compression arrangement (vertical tank built above floor) commonly used in batching and vehicle loading. (*b*) Three-cell compression arrangement (vertical tank built through floor or supported from beams) commonly used for storage of liquid or powders and for vehicle loading. (*c*) Four-cell compression arrangement (horizontal tank with pivot arrangement) for storage of liquids and powders. (*d*) Two-cell compression arrangement (horizontal tank with pivot arrangement) for storage of liquids. (*e*) Single-load-cell suspension arrangement (vertical hopper) for process weighing or batching of small loads. (*f*) Three-cell suspension arrangement (vertical hopper) for process weighing or batching of large loads.

FIGURE 3 Full-load-cell railway track scale. (*Reliance Electric.*)

FIGURE 4 Microprocessor-based automatic drum-filling scale. (*Reliance Electric.*)

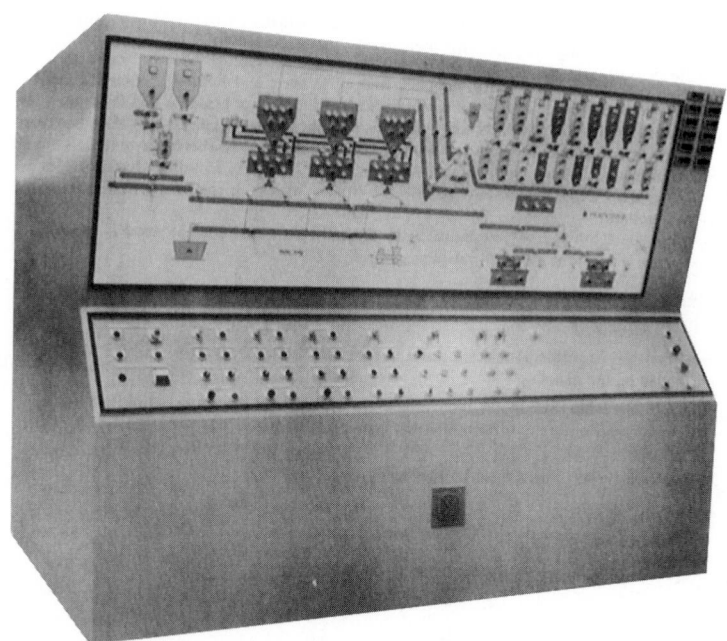

FIGURE 5 Some of the first industrial batching operations were centered about a sequence of weighing. Beyond the weight recipe handling, these batching systems need not be of the complexity and sophistication of some of the batching systems described in Section 3, Article 7, of this handbook. The batching center shown here involves several ingredient hoppers, collecting hoppers, and conveyors (*Reliance Electric.*)

density of the fluid to the density of water. For critical work, the reference is to double-distilled water at 4°C; in pratical applications, the reference commonly used is pure water at 60°F (15.6°C).

Several specific gravity scales are used in industry.

Balling.   Used in the brewing industry to estimate the percentage of wort, but also used to indicate percent (weight) of either dissolved solids or sugar liquors. Graduated in percent (weight) at 60°F (17.5°C).

Barkometer.   Used in tanning and the tanning-extract industry, where water equals zero and each scale degree equals a change of 0.001 in specific gravity, that is, specific gravity equals $1000 \pm 0.001n$, where n is in degrees Barkometer.

Brix.   Used almost exclusively in the sugar industry, the degrees represent percent sugar (pure sucrose) by weight in solution at 60°F (17.5°C).

Quevenne.   Used in milk testing, the scale represents a convenient abbreviation of specific gravity: 20° Quevenne means a specific gravity of 1.020. One lactometer unit is approximately equivalent to 0.29° Quevenne.

Richter, Sikes, and Tralles.   Three alcoholometer scales that read directly in percent ethyl alcohol by weight in water.

Twaddle .   This scale represents an attempt to simplify the measurement of industrial liquors heavier than water. The range of specific gravity from 1 to 2 is divided into 200 equal parts, so that 1 degree Twaddle equals 0.0005 specific gravity.

API.   Selected by the American Petroleum Institute and various standards institutions for petroleum products:

$$\text{Degrees (hydrometer scale at 60°F)} = \frac{141.5}{\text{specific gravity}} - 131.5$$

Baumé.   Widely used to measure acids and light and heavy liquids such as syrups, it was proposed by Antoine Baumé (1768), a French chemist. The scale is attractive because of the simplicity of the numbers representing liquid gravity. Two scales are used:

For light liquids $$°\text{Be} = \frac{140}{\text{specific gravity}} - 130$$

For heavy liquids $$°\text{Be} = 145 - \frac{145}{\text{specific gravity}}$$

The standard temperature for the formulas is 60°F (15.6°C).

A number of density and specific gravity measuring principles have developed over the years. Compared with the needs for sensing temperature, pressure, flow, and so on, the requirements for density instrumentation are relatively infrequent (Figs. 6 through 10).

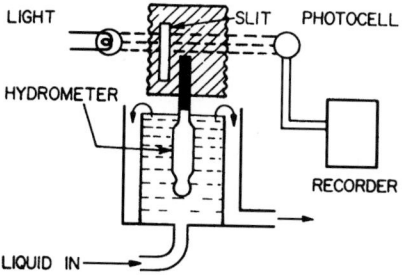

FIGURE 6   Photoelectric hydrometer. A glass hydyrometer, similar to the hand-held type, is placed in a continuous-flow vessel. The instrument stem is opaque, and as the stem rises and falls, so does the amount of light passing through a slit in the photocell. Thus photocell input is made proportional to specific gravity and can be recorded by a potentiometric instrument. The system is useful for most specific gravity recording applications not harmful to glass and is accurate to two or three decimal places.

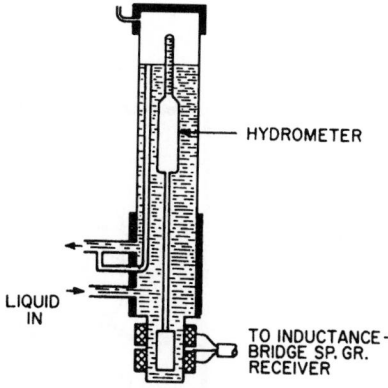

FIGURE 7   Inductance bridge hydrometer. The level of the measured liquid is held constant by an overflow tube. A glass hydrometer either rises or falls in the liquid as the specific gravity varies. The lower end of the hydrometer supports an armature in an inductance coil. Any movement of this armature is duplicated by a similar coil in a recording instrument. With this system, the temperature of the liquid usually is recorded along with the value of specific gravity, so that corrections can be made.

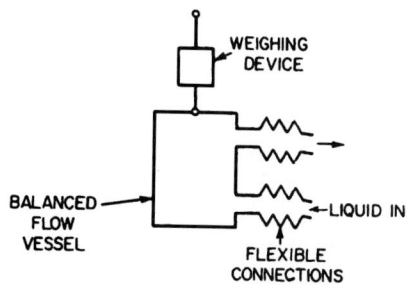

FIGURE 8   Balanced-flow vessel for measuring specific gravity and fluid density. A fixed-volume vessel is used, through which the measured liquid flows continuously. The vessel is weighed automatically by a scale, spring balance, or pneumatic force-balance transmitter. Since the weight of a definite volume of the liquid is known, the instrument can be calibrated to read directly in specific gravity or density units. Either open or closed vessels with flexible connections can be used. A high-accuracy measurement results, which is especially useful in automatic density control.

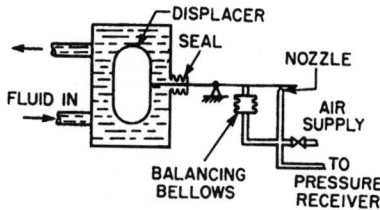

FIGURE 9   Displacement meter for measuring specific gravity and density. Liquid flows continuously through the displacer chamber. An upward force acts on the balance beam because of the volume of liquid displaced by the float. A pneumatic system, similar to the one shown, balances this upward force and transmits a signal proportional to the density of the liquid. Liquids with specific gravities 0.5 and higher can be measured with this equipment so long as suitable materials are used to prevent damage from corrosion. If the temperature of the flowing liquid changes, thermostatic heater may be used to maintain a constant temperature.

One of the most important developments in fluid density measurement over the last decade or so has been the Coriolis flowmeter, which measures flows in terms of their mass. See the article, "Flow Systems" in this handbook section.

Controlling the density of oil-well drilling mud is described by K. Zanker (*Sensors*, vol. 40, Oct. 1991).

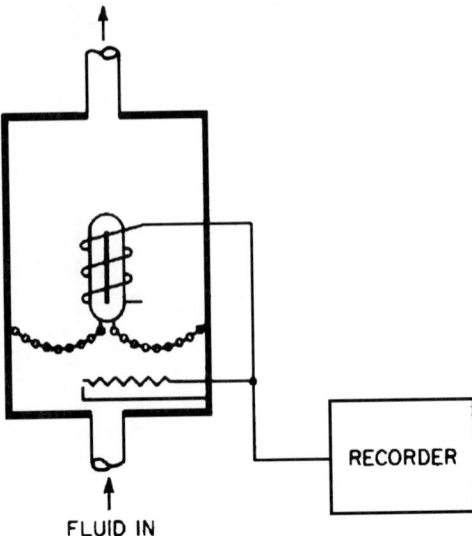

FLUID IN

FIGURE 10   Chain-balanced, float-type, density-sensitive element. This instrument uses a submerged plummet that is self-centering and operates essentially frictionless. It is not affected by minor surface contaminants. The volume of the body is fixed and remains entirely under the liquid surface. As the plummet moves up and down, the effective chain weight acting on it varies, and for each density within the range of the assembly, the plummet assumes a definite equilibrium point.

In order to transmit these changes in density, the plummet contains a metallic transmformer core whose position is measured by a pickup coil. This voltage differential, as a function of plummet displacement, is a measure of specific-gravity change. To compensate for a density change due to temperature, a resistance thermometer bridge notes the temperature change and impresses a voltage across the recorder, which is equal to and opposite the voltage transmitted to it by the pickup coil as a result of the temperature-induced density change.

HUMIDITY AND MOISTURE SYSTEMS

The presence of water affects all phases of life, and its control is vital to industrial, biological, and natural processes. Typically, water measurement is divided into two cases, humidity and moisture content. Humidity is the amount of water vapor in air or other gases. Moisture refers to the water content of a liquid or solid substance, such as paper, tobacco, textiles, films, candies, oils, solvents, etc. In general usage and in this discussion the two terms will be used interchangeably.

To be discussed are numerous options that are available to make these measurements, while attempting to point out the strengths and weaknesses of each technology. Firstly, the need to understand the fundamentals of the measurement and some common humidity parameters is addressed.

INTRODUCTION

The term humidity refers to water vapor, i.e., water in gaseous form. This is distinct from the term moisture, which relates to water in liquid form that may be present in solid materials or liquids. However, the term moisture is frequently used relating to measurements that are in fact water vapor measurements. For example, the term trace moisture is often used to express water vapor in the parts-per-million range, rather than trace humidity or trace water vapor.

Humidity is present everywhere in the Earth's atmosphere. Even in extremely dry areas and in boiloff from liquefied gases, there are traces of water vapor that in some applications could cause problems. Measurement of humidity is more difficult than measurement of most other properties such as flow, temperature, level, and pressure. One reason for this is the extremely broad dynamic range, which could start from 1 part per billion or less ($-112°C$ frost point), representing a partial vapor pressure of approximately 0.8×10^{-6} mm Hg, to steam at $100°C$ ($212°F$), representing a partial vapor pressure of 760 mm Hg. This amounts to a dynamic range of 10^9. Another reason is that measurements may have to be made in widely varying atmospheres, for example, from temperatures of $-80°C$ ($-112°F$) to $1000°C$ ($1832°F$), in the presence of a wide range of gases that could be corrosive or noncorrosive and in the presence of a variety of contaminants that are particulate or of a chemical nature.

Humidity measurements play an ever-increasing role in industrial, laboratory, and process control applications by allowing improvements in quality of product, reduction of cost, or increasing human comfort. In the tobacco industry, increasing proper humidity control greatly improves the quality of tobacco products. In warehouses, humidity control protects corrosive or humidity-sensitive materials, such as coils of steel, food, and dried milk. Cost-savings applications include industrial, textile, and paper dryers. If humidity in the dryer is monitored, the dryer can be turned off as soon as the humidity is below a specified level. This could save large amounts of money in energy costs compared with the traditional way of running the dryer for a sufficient length of time to ensure, with some safety margin, that the product is dry. Examples of human comfort and health are found in humidity-controlled hospital operating rooms, incubators, air conditioning, meteorological applications, auto emissions, air pollution, ozone depletion studies, and many other areas. These measurements are so important that the older and simpler humidity or dew point detectors such as dry-bulb and wet-bulb psychrometers, hair hygrometers, and dew cups are no longer considered sufficiently accurate or suitable for most industrial applications. This has led to the use of more sophisticated, often microprocessor-based, instrumentation to meet rigid requirements and government regulations of the Environmental Protection Agency (EPA), Food and Drug Administration (FDA), Federal Aviation Agency (FAA), and nuclear agencies.

HUMIDITY FUNDAMENTALS

One of the easiest ways to put humidity in its proper perspective is through Dalton's law of partial pressures. This law states that the total pressure P_T exerted by a mixture of gases or vapors is the sum of the pressure of each gas, if it were to occupy the same volume by itself. The pressure of each individual gas is called its partial pressure. For example, the total pressure of an air–water gas mixture, containing oxygen, nitrogen, and water is

$$P_T = P_{N_2} + P_{O_2} + P_{H_2O}$$

The measurement of humidity is directly related to the partial pressure of the water vapor in the gas being studied. This partial (vapor) pressure varies from less than 0.1 Pa (frost point $\approx -73°C$) to 1×10^5 Pa (dew point $\approx 100°C$). This covers a span of interest in industrial humidity measurement of more than one million to one.

Therefore the ideal humidity instrument would be a linear, extremely wide-range partial-pressure gage that is specific to water vapor and uses a fundamental (primary) measuring method. Since such an instrument, while possible, is inconveniently large and slow, nearly all humidity measurements are made by use of a secondary technique that responds to some humidity-related changes.

Common Humidity Parameters

As a basis for understanding humidity measurement, five terms need to be clearly defined. They are dry-bulb temperature, wet-bulb temperature, dew point temperature, relative humidity, and parts per million (ppm). These parameters are all interrelated and with only one humidity value, the dry-bulb temperature, and the total pressure, any other of these units of measure can be derived.

- Dry-bulb temperature: the ambient temperature, i.e., the temperature at the measurement point.
- Wet-bulb temperature: the temperature reached by a wetted thermometer as it is cooled by evaporation in an air stream. Typically an air stream flow rate of 4 to 10 m/s is required for proper cooling.
- Dew point temperature: the temperature when saturation occurs (100 percent relative humidity) or at which water begins to condense on a surface. When this temperature is below 0°C, the term frost point is proper.
- Relative humidity: the ratio of actual water vapor to the maximum water vapor possible in air at a given temperature, or the ratio of actual partial pressure of water vapor (partial pressure at dew point) to the saturation vapor pressure of water (partial pressure if the dew point were equal to the ambient temperature). For example, a dew point of 10°C at an ambient temperature of 25°C gives a relative humidity of 39.93 percent.

$$10°C = \text{vapor pressure of } 1227.76 \, \text{Pa}$$
$$25°C = \text{vapor pressure of } 3074.41 \, \text{Pa}$$
$$1227.76/3074.41 = .3993 = 39.93\% \, \text{RH}$$

Parts per million, ppm_v and ppm_w: the volume or mass ratio of water vapor to the total volume or mass of the carrier gas. The ppm by weight accounts for the ratio of the molecular weight of water to the background gas. Parts per billion (ppb) and parts per trillion (ppt) measurements are also ratio-based units that are of interest to the high-tech industry.

Table 1 shows these terms, their usual units of measurement, and some representative applications.

Other Important Terminology

Table 2 shows the units commonly used for humidity measurements in various industries to express the presence of water vapor. The unit of measure chosen depends on the use and type of industry involved and the unit that is the most meaningful for the application. Relative humidity (RH) is widely used in ambient environment monitoring since it represents the percentage of saturation of moisture in the gas at a given temperature. Dew point measurement is used as a direct indication of vapor pressure and absolute humidity. Conversion between relative humidity and dew point is possible if the ambient temperature (dry bulb) and total pressure are known. Measurement methods for dew point and relative humidity will be discussed in detail in the next section.

TABLE 1   Humidity Parameters and Representative Applications

Parameter	Description	Units	Typical Applications
Wet-bulb temperature	Minimum temperature reached by a wetted thermometer in an air stream	°F or °C	High-temperature dryers, air-conditioning, cooling tower control meteorological measurements and test chambers
Percent relative humidity	Ratio of the actual vapor pressure to saturation vapor pressure with respect to water at the prevailing dry-bulb temperature	0%–100%	Monitoring conditioning rooms, test chambers, pharmaceutical and food packaging
Dew/frost point temperature	Temperature to which the gas mixture must be cooled to achieve saturation; if the temperature is below 0°C, it is called the frost point	°F or °C	Heat treating, annealing atmospheres, compressed air dryers, meteorological and environmental measurements
Volume or mass ratio	Parts per million (ppm) by volume is the ratio of partial pressure of water vapor to the total pressure; ppm by weight is the same relationship as in ppm by volume multiplied by the ratio of the molecular weight of water to that of the carrier gas.	ppm_v, ppm_w	Monitoring of water as a contaminant primary in high-purity gases such as air, nitrogen, hydrogen, oxygen, methane, or argon

Psychrometric Charts

Psychrometric charts (see Fig. 1) provide a quick, though rather confusing, means for converting from one humidity unit to another. They are used to relate dew point, relative humidity, dry-bulb (ambient) temperature, and wet-bulb temperature to each other on one sheet of paper. Often enthalpy, entropy, and other units of measure are included in these charts. These charts must be used with some care since the pressure at the measurement may differ from that of the chart. This can cause significant errors if the pressure difference is large.

Computer Programs

There are now many computer programs that handle the unit conversion chore with ease. In addition, these programs allow the user to run other examples to predict moisture contents at elevated or reduced temperatures and pressures. These are most useful outside the normal range of psychrometric charts, the basic tool of air conditioning engineers.

Percent Relative Humidity

Percent relative humidity (%RH) is the best known and most widely used method for expressing the water vapor content of air. It is the ratio of the actual water vapor pressure to the water vapor pressure if the air were saturated. Percent relative humidity has derived from the invention of the hair hygrometer in the seventeenth century. The hair hygrometer operates on the principle that many organic filaments change length as a nearly linear function of the ratio of prevailing water vapor pressure to the saturation vapor pressure. A measurement of ambient (dry-bulb) temperature is also necessary to determine the water vapor content, e.g., 35% RH at 22°C.

TABLE 2   Terminology

Term	Description	Unit
For moisture (water in a solid or liquid)		
mass ratio, wet	$\dfrac{\text{mass of water}}{\text{total mass}}$	% or ppm_w
mass ratio, dry	$\dfrac{\text{mass of water}}{\text{mass of dry components}}$	% or ppm_w
water activity (a_w)	Relative vapor pressure of water in gas at equilibrium with bound water in solids or liquids, i.e., %RH over the sample under test	nondimensional ratio 0–1 or %
For humidity (water vapor in a gas)		
absolute humidity, volumetric (vapor concentration)	$\dfrac{\text{mass, vapor}}{\text{volume}}$	grams/m^3, grains/ft^3
absolute humidity, gravimetric (mixing ratio)	$\dfrac{\text{mass, actual vapor}}{\text{mass of dry gas}}$	% or ppm_w
relative humidity	$\dfrac{\text{partial vapor pressure}}{\text{saturation vapor pressure}}$	%
dew/frost point	Temperature at which saturation occurs	°F or °C
volume ratio	$\dfrac{\text{partial vapor pressure}}{\text{total pressure}}$	% by volume, or ppm_v
mass ratio	Same as mixing ratio	%, ppm by weight (ppm_w)
ppm by volume (ppm_v)	$\dfrac{\text{volume of water} \times 10^6}{\text{total volume}}$ or $\dfrac{\text{partial vapor pressure of water} \times 10^6}{\text{total pressure}}$	
ppm by weight (ppm_w)	$\dfrac{\text{mass of water} \times 10^6}{\text{total mass}} \times \dfrac{\text{molecular weight of water}}{\text{molecular weight of carrier gas}}$	

Over the years, other devices have evolved that permit a direct measurement of %RH. An example of these devices are electrochemical sensors that offers a degree of ruggedness, compactness, and remote readout capability. These electronic RH instruments essentially are derivatives of two basic designs, introduced in the 1940s, the Dunmore and the Pope cells. These and other technologies are described completely in the section on methods of measurement in gases. Please see Table 3 for typical technologies used to measure RH.

Dew Point Hygrometry

When more precise moisture measurement is needed, dew point measurements are used in many scientific and industrial applications. The dew point, the temperature at which water condensate begins to form on a surface, can be accurately measured from −5 to +100°C with a chilled-mirror hygrometer.

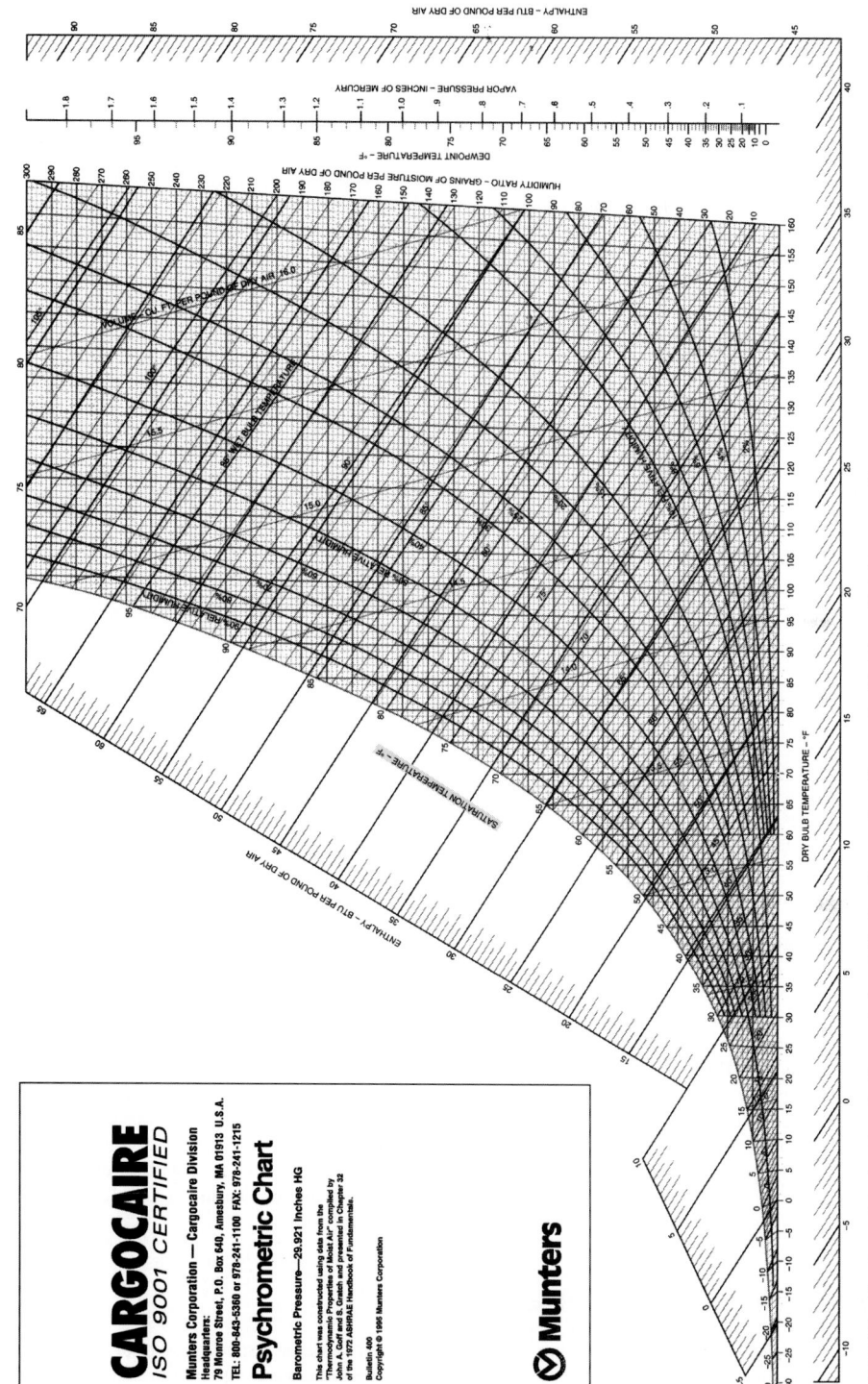

FIGURE 1 Psychrometric chart (*used with permission of Munters Corporation–Cargocaire Division; all rights reserved*).

TABLE 3   Instruments Giving Relative Humidity

Wet bulb/dry bulb	Evaporation cools a wetted thermometer to a minimum temperature and, by use of a psychrometric chart, relates the wet-bulb/dry-bulb temperature to %RH and dew point
Mechanical expansion	Expansion of element changes with %RH
Dunmore element	Bifilar wound wire on an insulating substrate coated with LiCl, a hygroscopic salt that takes up water, changing resistance of sensor (surface resistivity)
Pope cell	An ac resistance change of conducting polymer
Bulk effect	Electrodes in contact with a thin film of sulfonated polystyrene
Surface resistance	Resistance of many insulators depend on temperature and RH (adsorbed moisture)
Capacitance	Capacitance of polymer or metallic capacitors changes with RH
Strain gage	Resistance changes in a strain gage caused by mechanical stress in attached cellulose or polymer material relating to changes in RH

TABLE 4   Instruments Giving Dew or Frost Point

Chilled surface (manual or automatic)	Mirror or sensing surface cooled by chemical, electrical, or refrigerant until dew is detected
Adiabatic expansion	Gas under test is pressurized and released, causing a drop in temperature
Pneumatic bridge	Critical-flow nozzles from pneumatic bridge sensitive to mass flow with desiccant in one arm removing moisture from half the flow
Heated saturated salt	LiCl-coated bobbin bifilar wound with heater wires such that the current is determined by the surrounding vapor pressure (dew point)
Aluminum oxide	Capacitance of an aluminum oxide dielectric determined by the surrounding vapor pressure (dew/frost point)
Silicon oxide	Capacitance of an silicon oxide dielectric determined by the surrounding vapor pressure (dew/frost point)

Several types of instruments have received wide acceptance in dew point measurement: (1) condensation type hygrometer, (2) aluminum oxide sensor, (3) saturated salt sensor, and (4) adiabatic expansion hygrometer. Other instruments are used in specialized applications, including pressure ratio devices, dew cups, and silicon oxide sensors. Table 4 lists the technologies currently used to measure dew point directly. They are described in the methods of measurement in gases section.

Concentration Instruments

Knowing the ratio or concentration of moisture in a gas is especially useful when the water content is considered to be a contaminant. In high-moisture processes, this concentration percentage can give an indication of dryer efficiency. Direct concentration measurement is accomplished with the instruments listed in Table 5. They are also described in the methods of measurement in gases section.

Many manufacturers have developed circuitry and/or software to convert the primary unit of measure into many others to make the control interface as convenient as possible. However, there is obvious merit in selecting a technology that matches the unit of measure required. Conversion of units of measure nearly always means added errors that can be avoided with careful planning. Some technologies may not function under process conditions and may require a sample system. This could alter the selection of the technology used in any given process.

Direct Measurements Versus Sample Systems

It is always desirable to measure a process parameter *in situ* at a specific point of interest in a process. However, measuring moisture or humidity under process conditions is often impossible because they

TABLE 5   Instruments Giving Concentration (ppm$_v$ or %)

Electrolytic cell	The current according to Faraday's law, in bifilar windings controlled by electrolyzing water vapor attracted them by the phosphorous pentoxide coating
Quartz crystal	A shift in frequency of a hygroscopically coated crystal that is due to the mass change that is due to adsorbed water vapor
Sonic	Speed of sound is related to the percent of moisture in a high-humidity environment
Oxygen displacement	Measuring less than 21% oxygen content in air can indicate that this deficit was caused by the water vapor displacement
Infrared absorption	The decrease of only a specific wavelength of light caused by the concentration of water in its path
Lyman–alpha absorption	The decrease of only the Lyman–alpha wavelength of light caused by the concentration of water in its path
Titration	The quantity of reagent required for neutralizing a sample containing water

are too harsh for even the most robust of sensors. Adverse temperature, pressure, and contamination levels may not only interfere with the actual measurement, but in some cases are too severe for sensor survival. In these instances, installing a sample system close to the point of interest is the only alternative. The trade-off always includes added time lag in making the measurement and making decisions about controlling the process.

Sampling Systems.   When designing systems to make dew point measurements in a duct, room, process, or chamber, engineers often are not sure where to locate the sensor. The question is, should it be installed directly in the measurement area (*in situ*) or should a sampling system be used to carry a gas sample to a remote sensor? A sampling system is in many cases a must, for example, when the temperature of the gas being measured exceeds the maximum temperature rating of the sensor. Sampling is then needed to lower the gas temperature enough to prevent sensor damage. Because the dew point temperature is independent of the gas temperature (if not saturated), it can be measured at the other end of the sample line, after the gas has cooled. In many cases, either type of installation, if carefully designed, will provide reliable, continuous dew point measurements. However, there are several advantages to setting up a sampling system, even when an *in situ* measurement is possible. A sampling system often provides better, more reliable, long-term dew point measurements than other systems.

The advantages are that

- the incoming gas can be filtered and undesirable pollutants removed before contaminating the sensor

- the flow rate can be measured and controlled at the optimum rate for the sensor in use

- a remote external sensor located in a relatively constant ambient will provide more stable measurements when the dew point is measured in an environmental chamber that is programmed for wide and rapid temperature excursions

- a multiplexing sampling system can be made with manually operated valves or solenoid valves to select the points sequentially. Thus one hygrometer could take the place of several

- measurement errors will be minimal because all sampling points are measured with the same sensor

As in *in situ* measurements, there are generally two different purposes for the measurement of moisture/humidity—to verify that the process is operating as designed or actually to control the process to meet design parameters. The sampling techniques discussed here will be directed toward sample systems designed for continuous operation. However, they could also apply to systems primarily used for periodic or discontinuous monitoring or quality checks.

The purpose of a sample system is to modify *only* the process parameter(s) that are troublesome for the sensor, while retaining the sample integrity from the perspective of the water contained in it. This is often complicated by the behavior of the water molecule. A sample withdrawn from the process will immediately begin to be affected by the environment outside the process. Common sense and physical laws dictate that the measurement should be made at conditions as close to that of the process as possible, to retain the proper relationship of moisture to the background components. Sample system design must consider the temperature change caused by this transition as the sample as it is withdrawn. There will be dramatically different systems for gases, liquids, and solids, but all should change the parameters of the sample as little as possible.

All sample systems create a time lag between the extraction of the sample and the point when it is actually measured. This time is dependent solely on the flow characteristics of the sample system, and therefore the system should be designed to minimize this time delay.

There are several issues to consider in planning for a sample system installation. The sample system must not introduce errors in the measurement. It must not contain components that are incompatible with the process, and yet it must have a good speed of response to changes that occur in the process. The sample system has to operate in a cost-effective manner while being easy and safe to maintain.

The cost issues can be simplified when the bill of materials for the installation, along with the predicted annual maintenance, is compared with the projected value of the information. When the system is required for process control, this cost is usually capitalized as a part of the process and not calculated as an add-on. A good sample system is a balance of all of these issues considered together with the sole purpose of acquiring the data. Thus a sample system can be the great equalizer that allows options for the measurement that would not be available otherwise.

Sample Systems for Gases.   Gaseous sample system design must consider the ideal gas law effects, the physical behaviors of water, and the unique behavior of the sensor as the targeted process is modified. For example, if pressure is too great in the process for the sensor to operate properly, the pressure may be reduced with a pressure regulator to allow a measurement. This sounds simple, but there may be other effects that have an impact on the accuracy of the measurement.

Continuous Gas Sampling.   In continuous gaseous sampling, the following actions should be taken into consideration.

Pressure Issues

- When pressure is reduced by a regulator, consider the use of a fast-loop system to maximize response time (see Fig. 2). The pump shown here can also represent any pressure differential developed in the

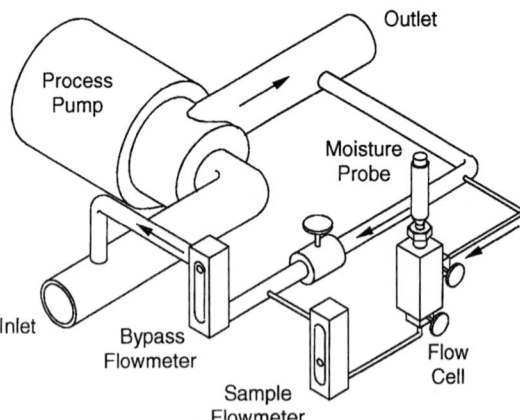

FIGURE 2   Fast-loop sample system.

process in which the sample circulation would not interfere with other measurements. Since sample systems require such a small fraction of the process for making the measurement, expanding this small quantity of gas through the regulator will slow the response of the system dramatically. The fast-loop allows the sample to be refreshed more often. This fast-loop system does not have to be complex and there are clever ways to create a fast loop by use of some sample system components, such as a coalescing filter.

- Care must be taken to reduce the pressure in a manner that does not cool the sample below the dew point of water or other condensable components, since this will change the integrity of the sample, compromising the measurement.

- When using a pump to extract the sample from the process, it is generally better to install the pump downstream of the sensor to reduce the wetted surface and volume between the sample extraction point and the sensor. This will increase the response speed of the system. Care should be taken in using a pump to simply move the sample, changing the pressure as little as possible from that of the process. Many dew point or specific humidity devices will be affected by any change in pressure.

Temperature Issues

- Temperature will always be different in the process than in the extracted sample. A variety of problems can result, such as creating condensation by extracting a sample from a process operating at high humidities and cooling it below the dew point of the gas. Water will condense at much higher temperatures than similar molecules. Very high relative humidity and/or condensation forming on the sensor can often damage sensors designed for measurement in the gas phase. This temperature change is not a concern when the humidity is very low.

- Heat tracing and temperature-controlled exchangers can be used to prevent condensation effects in all but the highest humidity processes. A sample taken at atmospheric pressure can support 100 percent water by volume (steam) at temperatures >100°C. If the sensing element can withstand this operating temperature, cooling a very hot sample with an unknown moisture content to +110°C, for example, will ensure the integrity of the sample from the perspective of the water component.

- By far the most common method of cooling a sample is through the heat exchange effect from the sample lines. In most applications, controlling the rate of flow through a given length of sample tube that is exposed to ambient conditions provides sufficient temperature stability. When cooling a sample, the overall flow rate of the sample will have an impact on the cooling rate through a given length of sample line. Keep in mind that the gas in the sample tubing will cool rapidly since its mass is low compared with the mass of the piping/tubing. The temperature of the gas is determined more by the piping temperature rather than the process temperature (see Fig. 3).

Other Issues

- *Contamination*: The sensor mechanism defines what constitutes a contaminant. In other words, contaminants are any components that interfere with the *measurement* of the moisture in the sample by use of a specific technique of measurement. These contaminants can be gases or aerosols or particulates. Their impact on the measurement must be taken into consideration. Aerosol or particulate contaminants can often be filtered out, but gases that interfere with a measurement technique will most often require selection of another measurement technique.

- *Flow Rate*: When optimizing flow rate for a specific sensor, consider the transit time required for refreshing the sample and if the sample is to be recaptured. When calculating transit time, include additional time for the sample to mix or dilute with the previous one during a change in the process moisture. One exchange of the sample over the sensor provides a dilution to half the new value. Thus a good rule of thumb is seven exchanges for 99.3% of the new value at the

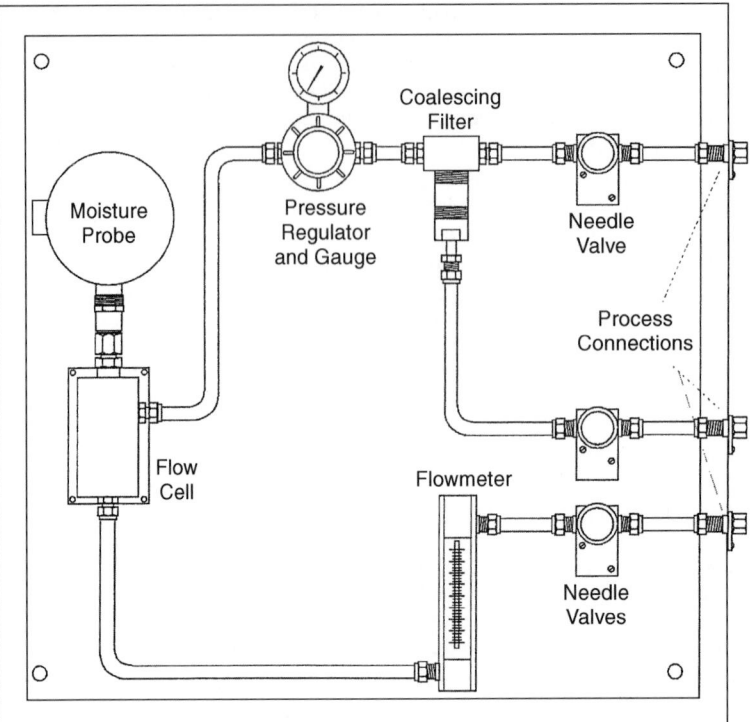

FIGURE 3   Sample system layout.

sensor. Sensor response time must be added to this to determine how the signal should be used for control.

- *Sample Site Selection*: Select the site for the sample extraction carefully, since process flow characteristics can cause moisture gradients across the piping. It is best to extract the sample from the middle third of the pipe in a straight run section of several feet.

- *System Site Selection*: Select the site for mounting the sample system panel so that the flow from the sample tap rises to intercept the panel. This will aid the flow of condensate back into the process if it should ever become entrained in the sample during an upset.

- *Leakage*: It is *very* important to have a tight system, since leaks will compromise the integrity of the sample. Test the plumbing/tubing without the sensor installed at a higher pressure than that at which the system will normally operate. For dew points $>-70°C$ this can be accomplished by application of a soapy solution over the external joints of the pressurized piping/tubing. For lower dew points, helium leak testing and high vacuum fittings are typically used.

- *Unit of Measure*: Sample system designers should keep in mind the principle of operation of a specific sensor. For example, a sample system should not be used for a %RH measurement, since the unit of measure is temperature dependent. However, a %RH sensor can be used with a sample system, if the measurement is converted from the moisture and temperature at the sensor into dew point or another nontemperature-dependent unit of measure.

- *Materials of Construction*: The water molecule is polar and will adsorb on internal component surfaces. The sample line length can be minimized by use of a small diameter tubing and reducing the number and size of components and filters ahead of the measurement point. This will enhance the performance of any sample system. Systems for measurements below −60°C should be constructed entirely of nonhygroscopic materials, such as electropolished stainless steel or specially coated electropolished stainless steel designed for fast dry down.

- *Sample Disposal*: Proper sample disposal can be very important. If a process leak would pollute the environment or be hazardous, then capturing the sample is essential. Putting the sample back into the process after modification for the convenience of the sensor can pose problems. This is especially true if the sensor cannot be operated at process pressure. If the process media are valuable, the waste of the small sample may become an argument for putting the sample back into the process.

Switched Gas Sampling.   If the sample is to be collected at intervals, special precautions must be taken, including the purging of the sample tap volume before data are taken. This volume can be significant if the sample lines are long and if the flow rate through them is low. An example of such a system might involve taking moisture measurements from a number of process points to a common sensor. Sample lines should be kept as short as possible, and the sensor should be located at the central hub of the points of extraction.

This spider arrangement can provide some economy when an expensive measurement technique is used. Good design can help minimize the effects of long sample transport time. If the sample incorporates a rotary selector valve for directing the different samples to the sensor, use a valve that allows all the samples to continue flowing to a common vent when not being selected for measurement. This will keep a fresh sample ready for measurement at the appropriate time. These valve systems often require an indexing system so that operators know which sample point is being measured. With less expensive measurement techniques this spider system might be more expensive than multiple instruments. Also consider that these measurements may gain in importance over time, and converting to a continuous measurement for a control loop would force individual systems to replace the spider extraction system plumbing.

Sample Systems for Liquids.   Liquid sample systems have many of the same problems as those used for gases. In addition, several issues should be considered when designing the system.

- Flow rates must be matched to the sensor and are generally lower because of dramatically higher density of liquid streams. This complicates response speed issues in control loops.

- Some liquid processes will change state if the temperature is changed. In these systems the sensor choices are very limited. The worst case would be designing and installing a system for a stream that polymerizes when the temperature drops to within the sensors' operating range.

- Because of their increased density, temperature drop in liquids is slower. If the sensor cannot withstand the process temperature, the cooling of the sample requires additional attention and is much more sensitive to the sample flow rate. The most common method of cooling is to use a length of tubing, usually coiled to save space, with ambient air convection currents providing the cooling. If additional cooling is required, tube and shell coolers connected to a small flow of municipal water will provide good results. These are inexpensive and easy to install.

- Sample disposal is usually more of an issue in liquid processes than in gases. When routing the sample lines, the sensor should continue to be as close to the sample point as possible. The sample return can be piped as far as necessary to get the pressure drop required for maintaining sufficient flow.

Sample Systems for Solids.   Solids moisture measurement is often accomplished on line for process control. However, if a sample must be collected from a solids process for process control or quality control documentation, their design must be approached systematically. Sample systems for solids are very specialized for the process they sample. These sampling operations are often manual, but whether manual or automated systems are being considered to monitor a solid's process, careful planning and design can pay big dividends.

If process control is the reason for the measurement, the overall process profile must be considered from beginning to end. Many of these systems are used in either intermediate or final drying operations. The process of drying a solid material is usually hot and can be relatively slow moving. In addition, the time delay for making changes in the process and seeing the results is much longer than in gaseous moisture applications.

Controlling these processes can often be aided by use of gaseous humidity sensors to track the areas of interest in and around the drying operation. Maintaining a consistent product recipe, along with the drying conditions of moisture and temperature, usually assists in sustaining uniform product performance. Care should be taken to avoid coating the sensor with condensable products driven off during the drying process.

There are many variables in the solids moisture problem relating to the sample, such as the location point, the quantity, the temperature, the method of containment, and the time needed for analysis. Whether a manual or automated system is in place, the following issues should be considered:

- The sample point is critical in solids. For example, in continuous drying or baking processes the product at the outer edges of the conveyor belt dries first, while the inner cross section is the last to dry. Since every product will release moisture at different rates, the location for obtaining the sample may need to be determined by trial and error.

- A sufficient sample must be removed to reduce the errors that can be introduced by use of smaller quantities. Small samples may test drier than the process because of the additional exposure to the ambient conditions during the time it takes to grab the sample, put it into a container, and seal it. If the sample temperature is very high with respect to the environment, this additional drying effect will likely be enhanced.

- Containers for transporting the sample to the lab should be clean and easy to seal quickly once the sample is obtained. They should be made of a material that does not change the sample's moisture content. Containers should also be easy to label, with the label containing all pertinent information needed by the lab for documentation.

- Samples should be transported to the lab in a timely manner and quickly analyzed to avoid errors caused by exposure to an environment other than that of the process. Another benefit of quick analysis is the ability to react to process changes in a timely manner.

Addressing these issues can easily translate into good process control and consistent quality.

METHODS OF MEASUREMENT IN GASES

There are several methods for measuring humidity in gaseous processes. Such a variety complicates the process of properly applying an instrument and expecting it to perform over the lifetime of the process equipment. Like almost every other parameter with many ways of measurement, most moisture sensor types were developed in response to a specific need. The sensing methods in the following section are divided by the primary sensing effect.

Thermometry Based

Wet-Bulb–Dry-Bulb Psychrometry.   Psychrometry has long been a popular method for monitoring humidity, primarily because of its simplicity and inherent low cost. A typical industrial psychrometer consists of a pair of matched thermometers, one of which is enclosed by a wick that is maintained in a wetted condition by distilled water. Air is passed over both thermometers, usually at a rate of approximately 215 m/min (700 ft/min) or more. The resultant evaporative cooling produces a wet-bulb temperature approximately equal to the thermodynamic wet-bulb temperature. The difference between the dry-bulb and the wet-bulb temperature is called the wet-bulb depression, which determines the humidity level. Many charts and curves converting wet bulb depression to relative humidity are shown in Ref. 1. Conversions can also easily be made with a computer disk like the one provided in Ref. 1. In a well-maintained psychrometer, with thermometer accuracies of $+/- 0.2°C$ ($+/- 0.38°F$) and within a temperature range of $+5$ to $+ 80°C$ (41 to 176°F), the relative humidity can be determined with an accuracy of approximately $+/- 3%RH$.

Accuracy of a psychrometer is strongly dependent on the accuracy of the thermometers used. For most precise measurements, platinum resistance thermometers are often chosen. Thermistors and thermocouples are also widely used. The absolute accuracy of the thermometers is of less importance than the ability to measure the wet-bulb depression accurately, i.e., the difference between dry- and wet-bulb measurements. This value is extremely small at very high humidities.

Generally speaking, the psychrometer is a fundamental measurement system and, if properly operated with accurately calibrated thermometers, a psychrometer such as the Assman laboratory type can provide very accurate, reliable, and repeatable measurements. For this reason the psychrometer was frequently used in the past as a fundamental calibration standard. However, most operators, especially in industry, do not take the time and necessary precautions and obtain far less accurate and less reliable results.

When the wet bulb reaches the maximum differential (depression), the relative humidity is determined on the psychrometric chart by following the wet-bulb temperature to the intersection with the dry-bulb temperature. For highest accuracy, both sensors are aspirated for proper cooling of the wet bulb and are thermally shielded from each other.

There are two types of psychrometers, a sling psychrometer (see Fig. 4) and an aspirated psychrometer. The sling psychrometer has the two thermometers mounted on a swivel, and they are whirled overhead manually. After the required number of revolutions, the wet-bulb temperature is read first and then the dry-bulb temperature. This information is used with the psychrometric chart to obtain the humidity value. The aspirated psychrometer uses a small battery-operated fan to draw air over the wet bulb. After the required time for the fan to circulate the air, the readings are taken as before.

A properly designed psychrometer has some inherent advantages:

- Its highest accuracy is near saturation (100% RH), making it superior to most other humidity sensors near saturation.

- If the wet bulb becomes contaminated or is improperly fitted, the simplicity makes it easily repaired at minimum cost.

- The psychrometer can be used at ambient temperatures above 100°C, and the wet bulb measurement is usable up to 100°C.

The limitations of this technique are as follows:

- When the RH drops below 20 percent, cooling the wet bulb to its full depression becomes difficult, resulting in reduced accuracy.

- Wet-bulb measurements below freezing are difficult to obtain with any degree of confidence.

- Since the wet bulb is a source of moisture, this technique can be used only when added water vapor is not a significant part of the total volume. Obviously, psychrometers cannot be used in small, closed volumes.

Chilled Surface Condensation Hygrometers.   The condensation-type hygrometer, often called a dew pointer, is one of the most accurate and reliable instruments, and it offers sensors that cover the widest range of all the technologies for humidity measurement. The trade-off for these benefits

FIGURE 4 Sling psychrometer (© *1994 Cole–Parmer Instrument Company; used with permission*).

is increased complexity and cost. The principle of operation is based on the behavior of water that allows for its condensation at normal ranges of pressure and temperature.

The most simplistic design for illustrating this measurement method is a shiny aluminum cup half full of cool water that can be stirred with a thermometer. As bits of crushed ice are added while stirring to cool the water further, dew formation will be eventually observed on the outside of the cup.

Reading the thermometer at the first sign of dew formation will produce surprisingly accurate atmospheric results with dew points above freezing. This crude method is obviously not for process measurements, but the principle of operation is adaptable to those conditions found in processes today.

One constant in all the condensation devices below is the change in the water molecule's state from a gas to a condensate as either dew or ice. In all of the optical detector types, the thickness of the condensate layer must be at least one-half the wavelength of light to be visible—many times greater than a monolayer of water molecules.

In very dry processes, it may take time to build enough condensate to become visible. For this reason, measuring extremely low dew points at the final temperature must be approached very slowly to avoid overshooting. A rapid approach could result in a reading well below the actual value.

Chilled Mirror Hygrometers.   A drawback of the chilled mirror hygrometer is that the instrument is more expensive than most other types, and requires considerable maintenance by skilled personnel. A new chilled mirror hygrometer, using a cycling mirror control technique, was developed during the 1980's. The method of mirror cycling reduces the sensor's sensitivity to contaminants, but at the expense of some of its fundamental measurement properties and accuracy.

The largest source of error in a condensation hygrometer is the difficulty of accurately measuring the condensation surface temperature. Special industrial and laboratory versions of the conventional instrument can be made with accuracies as high as $+/-0.1°C$ ($+/-0.18°F$) over a wide temperature span. Most instruments offer an accuracy of $+/-0.2°C$ ($+/-0.36°F$) which is better than what can be attained using other methods. Chilled mirror hygrometers can be made very compact by using solid state optics and thermoelectric cooling, and offer extensive capabilities through the use of microprocessors. There are three basic types of chilled mirror hygrometers:

Conventional Automatic Chilled-Mirror Hygrometer.   In its most fundamental form, dew point is detected by cooling a reflective condensation surface (mirror) until water begins to condense and by detecting condensed fine water droplets optically with an electro-optic detection system. The signal is fed into an electronic feedback control system to control the mirror temperature, which maintains a certain thickness of dew at all times.

Since the introduction of the chilled mirror hygrometer, this type of instrumentation has become widely accepted for precision laboratory calibrations and standards, despite its higher cost. Chilled-mirror hygrometers have also been used in many industrial applications in which high accuracy and traceability are required. This has resulted in many improvements, new developments, and several patents.

Operation of the basic optical dew point hygrometer is shown in Fig. 5. The surface temperature of a small gold- or rhodium-plated copper mirror is controlled by a thermoelectric cooler (heat pump). A high-intensity light-emitting diode (LED) illuminates the mirror. The quantity of reflected light from the mirror surface is detected by a phototransistor or optical detector. A separate LED and phototransistor pair are used to compensate for thermally caused errors in the optical components. The phototransistors are arranged in an electrical bridge circuit with adjustable balance that controls the current to the thermoelectric mirror cooler and therefore the mirror temperature.

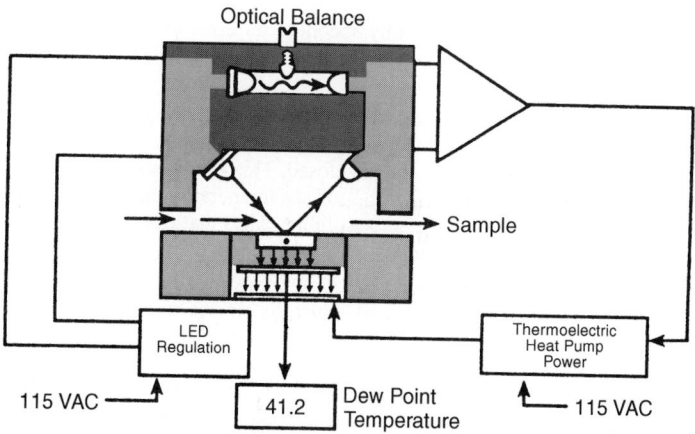

FIGURE 5   Schematic of conventional chilled-mirror sensor.

Reflectance is high when the mirror surface temperature is above the dew point (dry mirror) and maximum light is received by the optical detector. However, when the thermoelectric cooler reduces the mirror temperature below the dew or frost point (if below 0°C or 32°F), moisture condenses on the surface, causing light scattering, thereby reducing the amount of light received by the detector.

The system is designed so that the bridge is balanced and stable only when a predetermined layer of dew or frost is maintained on the mirror surface. Under these equilibrium conditions, the surface temperature is precisely at the dew point of the gas passing over the mirror. A precision NIST-traceable platinum or equivalent thermometer is embedded within the mirror surface to measure its surface temperature. The dew point temperature is displayed on the front panel of the instrument. When the mirror is clean, a perfect layer of condensation can be maintained, and high accuracy and repeatability results.

Some units use a second photodiode detector to measure the scattered light for better verification of the presence of dew or a second dry mirror to compensate for noncondensable contaminants in the sample. Temperature is continuously or cyclically controlled to achieve repeatable dew/frost layers during the temperature measurement of the mirror. This closed-loop control is inherently accurate since changes in the moisture content are detected and corrected for immediately. The mirror surface temperature measurement becomes the source of the signal used by the display and output circuitry.

Sensitivity to Contaminants.   Like any other humidity instrumentation, the main difficulty with chilled-mirror hygrometers is that the sensor is sensitive to contaminants. Even in clean applications, ultimately some contaminants will appear on the mirror surface, thereby influencing the optical detector and servobalancing functions of the instrument. A common practice to minimize this problem is periodically to open the servofeedback loop, causing the mirror surface to heat to a dry state and then readjusting the balance or reference of the optical circuit to compensate for the reduced reflectance of the mirror. In earlier models this procedure was performed manually. Subsequent improvements have been the so-called Automatic Balance and Continuous Balance by EG&G (now Edgetech), Automatic Balance by MBW and Michelle in Europe, and PACER by General Eastern. In all of these methods, the mirror is continuously operated at the dew point temperature and has at all times, except during the short balancing cycle, a dew or frost layer on it. For an extensive discussion of balancing methods, see Ref. 1.

Advantages of this technique include the following:

- It has a wide span; a properly designed condensation hygrometer can measure from dew points of 100°C down to frost points of −75°C.

- It has minimal errors because of its fundamental measuring technique, which essentially renders the instrument self-calibrating

- It has a good response time, usually specified in terms of its cooling or heating rate, typically 1.5°C/s, making it considerably faster than a saturated-salt dew point sensor and nearly as fast as most electrical %RH sensors.

- It uses simple verification that a dew/frost point has been detected by overriding the surface cooling control loop manually, causing the surface to cool further, and witness that the instrument returns to the same dew point when the loop is closed. This is a reasonable and valid check on the instrument's performance, assuming that the surface temperature measuring system is calibrated. If the instrument cannot cool further, the possibility exists that dew has not been detected and that the cooler has simply reached its limit.

- It is virtually indestructible because of its inert construction. Although the instrument can become contaminated, it is easy to wash and return to service without impairment of performance or calibration.

Actual calibration is similar to calibrating any electronic thermometer, i.e., by substituting a resistor and adjusting the readout electronics. Because of its fundamental nature and high accuracy and repeatability, this kind of instrument is used as a secondary standard for calibrating other lower-level humidity instruments.

The condensation (chilled-mirror) hygrometer measures the dew or frost point temperature. Unfortunately, many applications require measurement of %RH, water vapor in parts per million, or some other humidity parameter. In such cases the user must decide whether to use the fundamental high-accuracy condensation hygrometer and convert the dew or frost point measurement to the desired parameter, or to use lower-level instrumentation to measure the parameters directly. In recent years microprocessors have been developed that can be incorporated into the design of a condensation hygrometer, thus resulting in instrumentation that can offer accurate measurements of humidity in terms of almost any humidity parameter. An instrument of this type is shown in Fig. 8.

With the performance and range capabilities of this type of instrument, there are some limitations caused primarily by the optical detector system and the limitation of the cooling mechanism.

- Since the instrument is supposed to see the dew or frost on the mirror, the direct application of this instrument in industrial processes is limited. The light can be scattered by condensate on the mirror or by nonvolatile contamination.

- In many industrial applications it cannot be applied without sample systems. This additional expense, as well as the added cost to maintain and clean the mirror, should be considered. Cleaning the mirror may not be a time-consuming process in itself, but if the instrument is being used for controlling a process and the mirror becomes dirty, the error or runaway hunting of the unit can cause major process control problems. Thus these instruments have a relatively high installed cost in many processes.

- Another limitation is the cooling capacity of a specific instrument. The cooling function of a thermoelectric heat pump is specified in degrees of temperature depression, and it is limited. These coolers are often stacked to gain more depression capability, but eventually the lower temperature limit of the mirror is reached. If the sample is too dry, the optical detector never sees frost and never finds the actual frost point. Since the indication of frost point is a simple thermometer circuit, the instrument will indicate a false high reading in this case. Many automatic dew point devices offer a microscope option to allow the operator to verify the presence of condensate on the mirror.

A specialized dew point hygrometer for very low frost points operates with the same optical detection system as described above, but it uses a cryogenic source for cooling the mirror. One end of a thermally conductive rod is cooled in the cryogenic bath, and the other end is attached to the back of the mirror. The temperature control is made possible by a heater coil wrapped around the rod near the mirror. This method of cooling allows frost points to be measured down to $-110°C$. The control circuitry monitors the reflectance from the mirror and is designed to maintain a consistent layer of frost. Embedded in the mirror is an ultrastable precision thermistor that provides the frost point reading.

The cryogenic source can be either a cryogenic liquid, such as liquid nitrogen (LN_2), or a cryopump. These units offer the advantage of a faster response to very low frost points.

Cycling Chilled-Mirror Dew Point Hygrometer. A significant departure from the conventional chilled mirror system is the cycling chilled-mirror (CCM) control system, offering the user a sensor that requires considerably less mirror maintenance. The cycling chilled-mirror (CCM) instrument uses much of the same detection and chilling mechanisms in its operation. The primary difference is in the temperature control of the mirror (see Fig. 9). In addition, the new sensing technique makes it possible always to measure the dew point below 0°C (32°F), solving the well-known problem of dew versus frost point uncertainties that could cause interpretation errors of 1 to 3°C (1.8 to 5.4°F) in conventional hygrometers. The CCM hygrometer uses a cycling method. The mirror temperature is lowered at a precisely controlled rate until the formation of dew is detected. Before the dew sample is able to form a continuous layer on the mirror, the mirror is heated and the dew on the mirror surface is evaporated. Hence the mirror is almost always (95% of the time) in the dry state and contains a dew layer for only a short time (5% of the time) when a dew point measurement is made. Typically the measurement cycle is once every 20 s. Faster rates are possible. Another important benefit is that the dew layer is maintained for such a short time that it will never convert into frost, even when well below 0°C (32°F). This eliminates the problem with conventional optical hygrometes that one cannot

TABLE 6   Dew/Frost Point Conversion

Frost point (°C)	Dew Point (°C)	Deviation (°C)
0	0	0
−5	−5.6	0.6
−10	−11.6	1.2
−15	−16.7	1.7
−20	−22.2	2.2
−25	−27.7	2.7
−30	−33.1	3.1
−35	−38.4	3.4
−40	−43.7	3.7

be certain whether the instrument reads dew point or frost point when below 0°C (unless a mirror microscope is used). The reason is that, because of supercooling, a dew layer below 0°C (32°F) often remains for a long time, or even continuously, depending on flow rate, mirror condition, contaminants, and other factors.

Dew Point–Frost Point Conversion.   As pointed out above, misinterpretation of dew versus frost point could result in errors of up to 3.7°C (6.7°F) at −40°C (−40°F), as is shown in Table 6 above. In conventional continuous chilled-mirror hygrometers, when measuring below 0°C (32°F), initially a dew layer is established on the mirror and the instrument reads dew point. It is generally assumed that the dew layer converts to frost within a short time. However, this could in fact take several minutes, or even several hours, and is unpredictable. In some cases, the dew layer continues to exist indefinitely, even when the equilibrium temperature is well below 0°C (32°F). This is called supercooled dew, a condition that could exist because of certain types of mirror contaminants, a high flow rate, or a number of other reasons. In continuous chilled-mirror hygrometers, it is usually assumed that when the mirror temperature is below 0°C, the measurement represents frost point, but this is not at all certain. Even if this were the case, it could take a long time before the real frost point is measured. In the interim, the measurements could be in error without the user's realizing it. When such errors cannot be tolerated, it is customary to use a sensor microscope mounted on top of the sensor, making it possible to observe the mirror and determine whether dew or frost is measured. This is impractical and not an attractive industrial procedure. The CCM hygrometer therefore offers a significant advantage when measuring below 0°C because one can be sure that the measurement is dew point. If the user wants to read frost point, it is a simple matter to program the on-board microprocessor to display the frost point.

This system provides some distinct advantages over the continuous dew/frost layer control:

- The unit generally measures only dew point, even below 0°C, because of the short time the sensor is condensing moisture.
- This method has much reduced mirror contamination by allowing the mirror to remain dry for up to 95% of the measuring cycle and the addition of a bypass filter mechanism.

Although the maintenance of this device is relatively low, the accuracy is slightly less than that of continuous chilled-mirror units.

A specialized CCM hygrometer specifically designed for application in high temperature can measure dew points to 100°C. This device uses fiber-optic connections for the dew detector function. The upper temperature is limited by thermal limits of the thermoelectric heat pump or Peltier device. Applications are primarily limited to relatively clean processes since constructing and maintaining a heated sample system is costly.

Nonoptical.   There are four commercial methods of detecting the dew/frost layer that are nonoptical— capacitance, surface acoustic wave (SAW)/quartz crystal microbalance (QCM) devices, differential thermal, and differential capacitance detectors. Of these units, the capacitance and SAW devices are available and accepted.

Generally these sensors respond to dew/frost layers that are thinner than those that can be detected with optically based units. This does not change the theoretical accuracy of the reading, since if any detector were held even 0.01°C below the dew/frost point, the layer of condensate would continue to grow until the detector circuit recognized the thicker layer.

The capacitance sensor detects the formation of the condensate by measuring the change in surface capacitance of the chilled surface. The dielectric of water in the liquid phase is quite high and the detector works quite well for dew points above −15°C because of the thin dew layer required for detecting the onset of condensate formation. The sensor also works for frost points, but a thicker layer of ice is required for detection. This sensor is simple and fast since the mass of the detector is very low. Contamination does effect the measurement and the sensor can drift over time.

The SAW device is a mass detector that uses dual piezoelectric devices mounted on a common chip surface. This detector is cooled thermoelectrically. One of the piezoelectric crystals is excited, and the other is used as the receiver for the vibrations induced on the surface of the chip. As the chip reaches the dew/frost point, the mass of the condensate changes the frequency of the oscillations.

The slightest change of frequency is the detection threshold for the sensor. It responds quickly since the mass of the device is very low and it responds to very thin dew/frost layers. This detector requires high-frequency circuits that limit remote sensor locations, and it may suffer from adsorbed moisture in the chip components.

Since the mass change is the detection method, this sensor type is also subject to condensables and contamination other than water. The mass of these contaminants can cause errors and false dew point readings.

Manual Observation.   Manually operated dew point devices are used primarily for spot checking process gases. These units vary from the dew cup (see Fig. 6) to more elaborate devices with microscopes and electronic thermometry.

FIGURE 6   Manual dew cup. (*courtesy of Lectrodryer*).

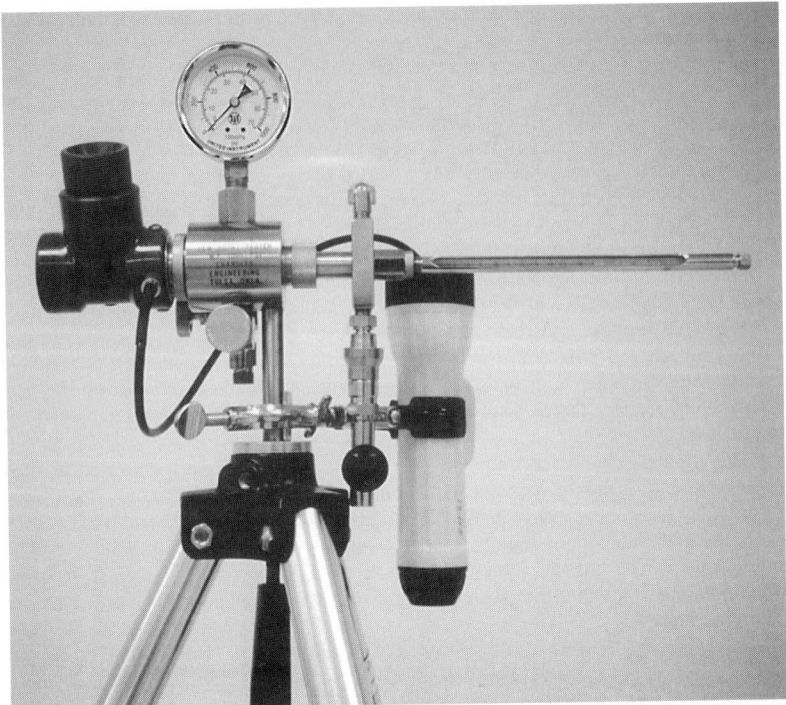

FIGURE 7 Manual optical dew point instrument (*courtesy of Chandler Engineering*).

The dew cup is a mirrored cup surrounded by an enclosure for conducting the sample across its outer surface. The enclosure has an observation window and is connected to the process by hoses. The dew point is taken in the same manner as above, by stirring a water and ice mixture, or more commonly by using acetone and dry ice until dew or frost is observed through the window. The thermometer reading is the dew point temperature.

Another manual device used primarily in natural gas and CO_2 measurements utilizes the expansion of a high-pressure gas to cool the backside of the mirror. Changing the flow rate of this gas controls the mirror temperature. The sample flow is directed to the observed mirror surface, and the condensate formation can be seen through the microscope. In newer units, internal optics allow the mirror temperature to be superimposed so it can be viewed in the same eyepiece. Operator training is essential in these applications, since hydrocarbon dew points can occur at the same range of temperatures and be confused with those of the water (Fig. 7).

Finally, when very low dew points are being measured, the cooling rate must be adjusted so the mirror temperature drops very slowly. This will prevent overshooting. Once the dew point is observed, the temperature should be cycled above and below that temperature to improve accuracy.

Adiabatic Expansion. The adiabatic expansion, or fog chamber, instrument is similar in one respect to the manual optical dew point instrument: An operator must see the moisture form (see Fig. 10). But in this case the visible moisture is in the form of a cloud or fog. The physical effect used here is the Joule–Thompson effect that describes the temperature change one observes while changing the pressure of a gas.

Rapidly changing the pressure of a gas easily demonstrates the basic operation of this instrument. If a gas mixture is pressurized, its temperature will rise. If this same gas mixture at the elevated pressure is allowed to cool to ambient and then released quickly, its temperature will drop rapidly.

FIGURE 8   Continuous temperature control optical dew point instrument.

TYPICAL CYCLING SEQUENCE OF THE CCM MIRROR

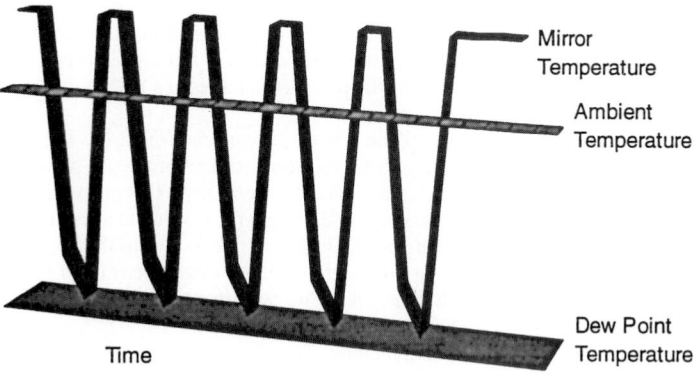

FIGURE 9   CCM temperature cycle.

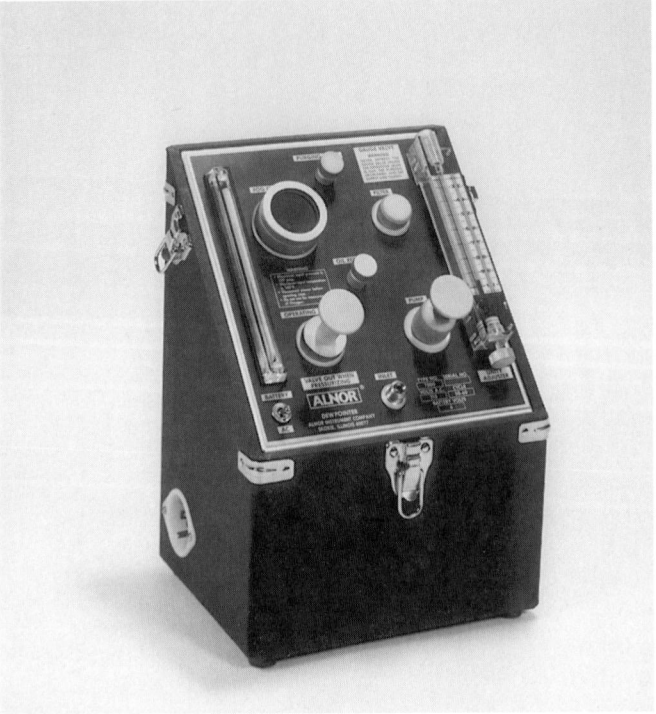

FIGURE 10   Adiabatic expansion or Fog chamber instrument (*courtesy of Alnov*).

If the gas mixture contains water vapor, it is possible to see a fog form momentarily. A relationship has been documented between the lowest pressure differential at which the fog is first observed, the temperature of the pressurized gas mixture, and the dew point of that gas mixture.

This type of instrument has been used primarily by the heat-treating industry because it is portable and simple to operate. The basic operation, after connecting to a sample tap, is to operate the manual plunger pump for several minutes to purge out the fog chamber, filling it with the sample gas. The fog chamber exhaust valve is shut and the chamber is pressurized with the pump.

The pressure is increased to an estimated value and read on the manometer on the front panel of the instrument. The temperature is observed and the sample is allowed to cool until it is stable again and is recorded. The operator observes the chamber through a port on the front of the panel and presses the pressure release valve. As the pressure is released a battery-powered light illuminates the chamber, and a fog may appear briefly.

The procedure is repeated until the lowest pressure at which the fog just appears is found. A circular slide rule is used to determine the moisture content of the gas mixture with the values of the pressure differential, the observed steady-state temperature at this pressure, and the process pressure. This trial and error method can be shortened if the typical value at a specific process sample point is known. Then a proper reading can be achieved in as few as 4–6 trials.

Even with its simple operating procedure, operators must be trained in observation skills. They must also be disciplined enough to repeat the measurement over a sufficient number of trials to get a good measurement.

Drawbacks to using this method include

- operator errors that can be reported as high as $\pm 5^{\circ}$C

- errors from using the unit with gases having Joule–Thompson coefficients that are very different from that of air. Such gases include hydrogen and helium with negative coefficients (hydrogen heats on expansion at normal temperatures and pressures) that produce false low readings or CO_2 with a very high cooling effect that produces false high readings.

- it is also a spot check method and not adaptable for continuous measurements.

Electrical Property Based

Capacitance

Aluminum Oxide Moisture Sensors.   This sensor is formed when a layer of porous, moisture sensitized aluminum oxide is deposited on a conductive substrate and the oxide is coated with gold.

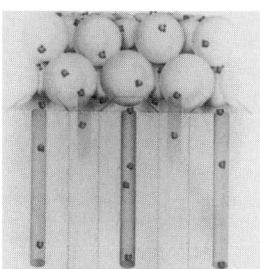

Recently, this technology has been refined to take advantage of semiconductor manufacturing techniques and higher purity substrates (such as ceramic) to produce sensors with consistent responses to moisture changes in the oxide (see Fig. 11).

The smaller size and surface area have led to the development of probe/transmitters as well as the classic probe and analyzer systems that have dominated this field for the past 30 years (see Fig. 12). The more predictable sensors allow simpler electronics and the possibility of on-line field calibrations for many common applications (see Fig. 13).

FIGURE 11   Aluminum oxide sensor construction.

Aluminum Oxide Hygrometer.   Aluminum oxide humidity analyzers are available in a variety of types. They range from low-cost, single-point systems, including portable battery-operated models, to multipoint, microprocessor-based systems with the capability to compute and display humidity information in different parameters, such as dew point, parts per million, and percent relative humidity. A typical aluminum oxide sensor is in essence a capacitor. The conductive base and the gold layer become the capacitor's electrodes, forming what is essentially an aluminum oxide capacitor. Water vapor penetrates the gold layer and is absorbed by the porous oxide layer. The number of water molecules absorbed determines the electrical impedance of the capacitor, which, in turn, is proportional to water vapor pressure. Construction of an aluminum oxide sensor, as shown in Figs. 11 and 14, is basically very simple. They do exhibit varying degrees of aging and hysteresis, and require periodic calibration. Over many years the companies that offer aluminum oxide sensors, such as Panametrics (USA), Endress & Hauser (Germany), and Shaw (England) have developed elaborate calibration systems and per-aging routines, resulting in sensors that maintain good accuracy for process applications over an extended period of time.

A typical aluminum oxide sensor is shown in Figs. 12 and 15.

Aluminum oxide hygrometers respond to the vapor pressure of water over a very wide range of vapor pressures. The impedance of the aluminum oxide layer between gold and aluminum electrodes is sensitive to water vapor from saturation vapor pressure down to below 10^{-5} mb (10^{-3} Pa), corresponding to $-110°C$ ($-166°F$) dew point. The strong affinity of water for this oxide, combined with the large dielectric constant of water, makes this device highly selective toward water. It does not respond to most other common gases or to numerous organic gases

FIGURE 12   Aluminum oxide probe-mounted sensor.

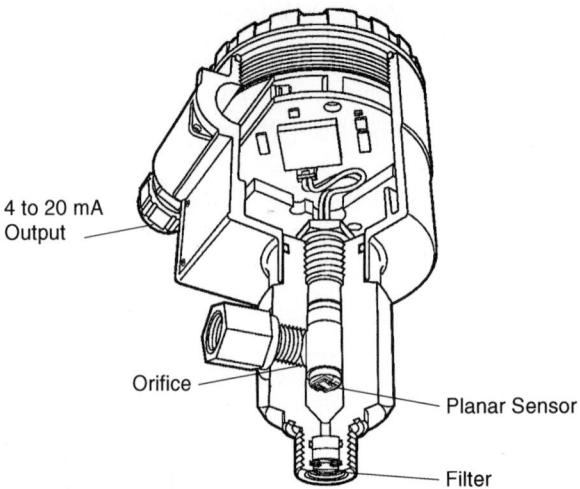

4 to 20 mA
Output

Orifice

Planar Sensor

Filter

FIGURE 13   Aluminum oxide dew point loop-powered transmitter.

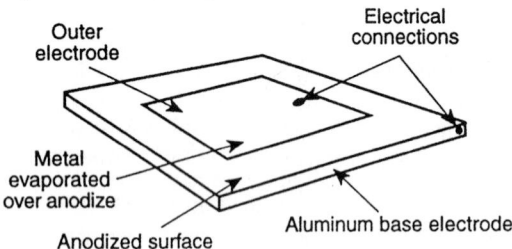

Outer
electrode

Electrical
connections

Metal
evaporated
over anodize

Aluminum base electrode

Anodized surface

FIGURE 14   Construction of aluminum oxide sensor. An aluminum oxide sensing element is a basically simple device consisting of an aluminum substrate, which is oxidized and on which a porous gold surface is deposited. Electrical connections are provided to the aluminum base and the gold surface and the device acts as a capacitor.

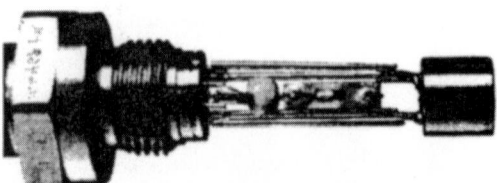

FIGURE 15   Aluminum oxide sensor (*courtesy of Panametrics*). The small aluminum oxide sensing element and often also a small temperature element are mounted in a rugged stainless-steel housing that can be threaded into an air duct or gas line.

and liquids. In situations in which aluminum can be chemically attacked, silicon sensors with a tantalum metal base could be used as an alternative. Such sensors are the most inert, although their sensitivity is somewhat less. The oxide layer is generally described in the form of a mass of tubular pores running up from the metal base to the exposed surface. Change in the size of these tubules with time is presumed to be the cause of the slow shifts in calibration often experienced with these sensors. Water is absorbed in these tubules in amounts directly related to the moisture content of the gas in contact with it. The amount of water is sensed electrically by measuring the change in capacitance or admittance produced by this water. Because of the radius of the pores in the aluminum oxide, the sensor is virtually specific for water molecules.

The dew point range that is customarily covered by standard sensors of this kind is between $-110°C$ ($-166°F$) and $+20°C$ ($68°F$), which corresponds to a range from approximately 1 ppb to 0.2 % by volume. Operation up to $100°C$ is theoretically possible but the shifts in calibration tend to be accentuated by higher temperature operation and very frequent calibration checks would be required. The sensor is therefore generally not recommended for operation above $70°C$ ($158°F$). Temperature coefficients are modest but need to be taken into account, especially in the higher gas temperature and dew point ranges. For accurate measurement in these ranges it is essential that probes are calibrated at their expected operating temperature. Response times for a 63% change are a few seconds at higher dew points, increasing as the dew point drops to a minute or more at $-50°C$ ($-58°F$) frost point.

Commercially available sensors can operate at pressures up to 34×10^6 Pa (4.9×10^3 psi) and at vacuum down to 0.7 Pa (1.015×10^{-4} psi). However, it is important to note that the reading obtained is the dew point of the gas at operating pressure. If line-pressure variations are common, it may be better to take a continuous sample of the gas and reduce its pressure to atmospheric or other controlled values. Accuracies are often specified in the range of $+/-1$ to $+/-2°C$ ($+/-1.8$ to $+/-3.6°F$) at higher dew points, increasing to $+/-2$ to $+/-3°C$ ($+/-3.6$ to $5.4°F$) at $-100°C$ ($-148°F$) frost point. However, these accuracies are obtained under laboratory and idealized conditions. In actual field use the accuracy will depend on the parameters of the application and the frequency of recalibration. These sensors can tolerate linear flow rates of up to 10 m (30 ft) per second at 10^5 Pa (14.4 psi). Calibrations do drift with time. The magnitude of these drifts for state-of-the-art sensors is modest. The stability varies from manufacturer to manufacturer.

Advantages

- Wide dynamic range from 1 ppm to 80% RH
- Probes can be installed remotely by 3000 ft (300 m) or more
- Probes can easily be used *in situ*
- Aluminum sensors are suitable for use in mutisensor configurations
- Relatively stable, with low hysteresis and temperature coefficients
- Independent of flow rate changes
- Suitable for intrinsically safe applications
- High selectivity for moisture
- Operation over a wide range of temperature and pressure
- Low maintenance
- Available in small sizes
- Very economical in multiple sensor arrangements
- Suitable for very low dew point levels without the need for cooling, with easy measurement of typical dew points down to below $-110°C$ ($-166°F$).

Limitations

- Slow drifts in calibration, which may accelerate at higher operating temperatures or in certain gases
- Probe affected by some corrosive gases. Analogous silicon or tantalum oxide sensors could be considered as alternatives.
- Temperature control often needed, especially at high temperature and dew points
- Offer only modest accuracy and repeatability
- Must be periodically calibrated to accommodate aging effects, hysteresis, and contamination (not a fundamental measurement)
- Nonlinear and nonuniform construction calls for separate calibration curves for each sensor.

Many improvements have been made over the past several years by various manufacturers of aluminum oxide sensors but it is important to remember that even though this sensor offers many advantages, it is a lower-accuracy device than any of the fundamental measurement types. It is a secondary measurement and therefore can provide reliable data only if kept in calibration and if incompatible contaminants are avoided.

As the sensor ages and fewer adsorption sites are available, it should be recalibrated to accommodate aging and contamination. Depending on the accuracy required by the specific application and the dew point and temperature extremes of the application, traceable calibrations may be suggested as often as every six months or up to several years.

In general, operation at low dew points, in clean gases, and at ambient temperature or below can reduce the need for recalibration. All aluminum oxide sensors require individual calibration curves. But computer technology allows the use of EPROM's, smart probes and digital data download for entry of calibration data. This removes the potential for error and reduces the inconvenience of recalibration.

Silicon Chip Sensors.   Hailed by the developers as the answer to the problems of long dry down, large temperature coefficients, and poor recovery times after saturation, the silicon oxide sensor offers the opportunity to be heated to assist in driving off excess adsorbed water from the sensor.

This sensor offers the following advantages:

- wide dynamic range
- independent of flow rate
- operational over a wide pressure range
- very fast response when used with the push-purge probe heater.

These instruments are similar in application to the aluminum oxide sensors. However, there are three major limitations to these types of sensors.

Silicon chip sensors are very temperature dependent and difficult to compensate for temperature variations. For this reason most commercial adaptations are kept at a controlled temperature, and accurate measurements are possible only if kept at this precise temperature.

These sensors require periodic calibration and are more expensive than aluminum oxide sensors. In general, silicon oxide sensors have not been able to displace aluminum oxide. However, they are designed for use in applications in which fast response is required, like spot checking with a portable instrument.

FIGURE 16   Polymer capacitance humidity sensor.

Polymer Sensors.   Polymer-based relative humidity sensors are widely used in applications ranging from environmental monitoring and control, food and materials storage, metrology, and a variety of processes for which humidity control is important. These sensors operate on a relative adsorption principle, producing a change in electrical capacitance that corresponds to a change in RH (see Fig. 16).

RH sensors measure water vapor, but only in relation to temperature. RH is the ratio of partial vapor pressure to the saturation pressure at the ambient temperature. Because the saturation pressure is dependent on only temperature, it is evident that moisture can be derived when both RH and temperature are known. Also, temperature compensation of the sensor's output signal is crucial to the accuracy of the final moisture value.

Polymer-based humidity sensors can work well in many gas applications in which frost points are higher than $-40°C$. An effective humidity sensor must be stable and have good resolution down to 1% RH. Factory calibration should be performed with a traceable reference such as a chilled-mirror instrument. The common advantages for this sensor are its compact size and often allowing *in situ* use and its low cost.

Hot and wet applications, such as industrial dryers, require a sensor that provides a water vapor measurement. This sensor can be invaluable for efficient process control. Surprisingly, measurement at high temperatures and high moisture levels requires a sensor that is stable and accurate at low RH ranges. For example, at a temperature of 150°C and 50°C dew point, the RH is only 2.6%.

With proper system design, polymer-based RH sensors offer an effective and reliable solution for many moisture applications. These are alternative solutions to those traditionally served by more expensive and complex moisture analyzers.

Resistance

Saturated-Salt/Heated Lithium Chloride.   The saturated-salt [lithium chloride (LiCl)] dew point sensor has been a widely used sensor because of its inherent simplicity and ruggedness (see Fig. 17).

The principle that the vapor pressure of water is reduced in the presence of a salt is used in the saturated-salt dew point sensor. When water vapor in the air condenses on a soluble salt, it forms a saturated layer on the surface of the salt. This saturated layer has a lower vapor pressure than water vapor in the surrounding air. If the salt is heated, its vapor pressure increases to a point at which it matches the water vapor pressure of the surrounding air and the evaporation–condensation process reaches equilibrium. The temperature at which equilibrium is reached is directly related to the dew point.

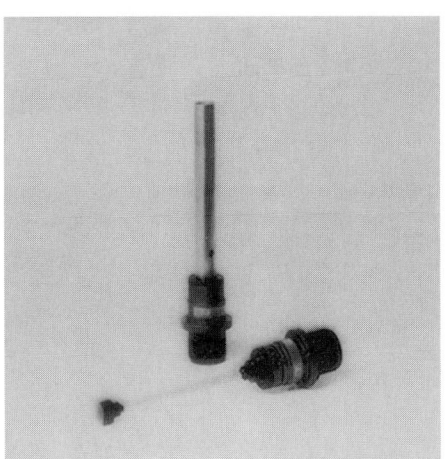

FIGURE 17   Heated Saturated-Salt (LiCl) sensor.

A saturated-salt sensor is constructed with an absorbent fabric bobbin covered with a bifilar winding of parallel fine-wire electrodes. This assembly is coated with a dilute solution of LiCl.

Lithium chloride is used as the saturating salt because of its hygroscopic nature, which permits application in relative humidities of between 11% and 100%.

An alternating current is passed through the winding and salt solution, causing resistive heating. As the bobbin heats up, water evaporates into the surrounding air from the diluted LiCl solution. The rate of evaporation is determined by the vapor pressure of water in the surrounding air. When the bobbin begins to dry out, as a result of the evaporation of water, the resistance of the salt solution increases. With less current passing through the winding, because of increased resistance, the bobbin cools and water begins to condense, forming a saturated solution on the bobbin surface. Eventually, equilibrium is reached and the bobbin neither takes on nor loses any water.

Properly used, a saturated-salt sensor is accurate to $\pm 1°C$ between dew point temperatures of $-12°C$ and $+38°C$.

Limitations of saturated-salt sensors include the following:

- a relatively show response time
- a lower limit to the measurement range imposed by the nature of LiCl.
- The active mechanism for this type of detector is still a salt. Therefore it can be washed away by liquids entrained in the sample or by losing power at high humidity. During such events, the sensor is rendered inactive and must be washed in ammonia and recharged with LiCl solution. The sensor can then be returned to service.

Dunmore Sensors.   The Dunmore cell [see Fig. 18(a)] also uses LiCl, but in an unheated resistive configuration around an insulator. This assembly is coated with a LiCl solution of a specific concentration. The resistance between the wires is a function of the %RH in the air surrounding the sensor.

The sensor is simple and, if applied properly, will give good repeatable service. The major drawback is the sensor's limited range. Specific sensors conditioned with different concentrations of LiCl solution can cover the entire range of RH, but a single sensor covers only approximately 20% of the 0%–100%RH scale. The other drawback is the same as for any of the salt-based sensors—they can be washed away by condensables, rendering the sensor inoperative.

Pope Cell—Sulfonated Polystyrene Sensor.   The Pope cell [see Fig. 18(b)] is an ion-exchange resistance device. The sensor is constructed by winding bifilar conductors over a polystyrene core.

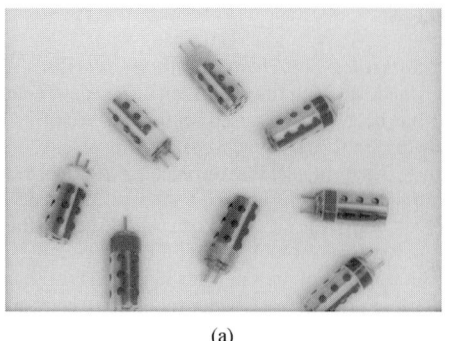

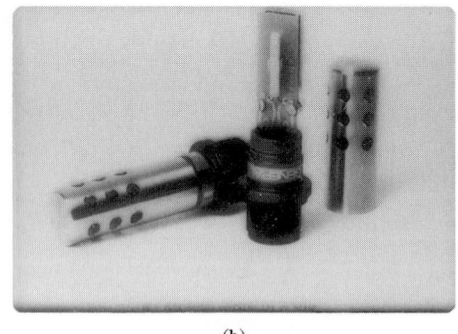

(a)   (b)

FIGURE 18   (a) Dunmore sensor, (b) Pope sensor.

This polystyrene core is treated with sulfuric acid. The H_2SO_4 in the matrix becomes very mobile and readily detaches from the polystyrene molecule to take on the H^+ ions from the water molecule. This alters the impedance of the cell dramatically and allows the RH to be measured.

This sensor responds rapidly and can be used over a wide range of relative humidity. It can be contaminated by salts and can lose calibration if wet. It also has a significant hysteresis.

Optical Property Based

Infrared Absorption.   Since the water molecule absorbs infrared radiation at well-defined wavelengths, this property can be used in both transmission and reflectance analyzers to measure moisture in gases, liquids, or solids. Typically, infrared radiation with alternating wavelengths of 1.94 μm (water absorption band) and 1.6 μm (no absorption) is passed through the gaseous or liquid sample by a windowed cell or reflected from the surface of a solid sample. The resulting signals are amplified, and the signal ratios are a function of moisture content. This system is immune to density changes, composition (because the wavelengths are specific for the moisture molecule), and other process variables. The useful range is from 0.1% to 95%. Since the materials of construction for this analyzer can handle corrosive gases and liquids, applications with the halogens and acids are possible with special flow cells. This sensor is very fast since the sample does not take long to equilibrate to a new value. The gas or liquid sample must be filtered so the windows remain as clean as possible. The temperature in the cell and the source for the light must be kept very constant. All this adds to the cost of the instrument. It is one of the most reliable methods to measure in corrosive gases. Although the instrument is normally off line with a sample drawn through the cell, process models have been used in which the window installations and path lengths are sufficient to measure on line.

Infrared Instruments.   Transmission of infrared and ultraviolet radiation through a gaseous medium does not occur uniformly at all wavelengths. Selective absorption of radiation occurs in absorption bands that are characteristic of the particular gas through which the radiation is passing. Operation of the infrared hygrometer is based on the absorption of infrared radiation at certain distinct frequency bands as a result of the presence of water vapor in the gas through which an infrared beam is passed. Water vapor strongly absorbs infrared radiation in bands centered at 2.7 and 6.3 μm. The feasibility of measuring water vapor by infrared absorption measurements was first demonstrated by Fowle in 1912, but did not result in practical instrumentation until the 1980s.

Infrared optical hygrometers are noncontact instruments. The sensing element is protected by a sapphire window and never contacts the air or gas that is being sampled. It is therefore a method suitable for monitoring moisture in highly contaminated fluid concentration. Operation of an infrared hygrometer is usually based on the dual-wavelength differential absorption technique. This requires identification of a primary wavelength within the most intense portion of the water vapor absorption band and a nearby reference wavelength at which water vapor absorption is negligible. The differential absorption technique involves computing the ratio of the transmission measured at the primary wavelength to that measured at the reference wavelength. This ratio can be considered a normalized transmission that is a direct measure of the water vapor absorption and is insensitive to drifts in electro-optical sensitivity, deposits on the windows, and haze or fog in the sample volume. It is based on the assumption that transmission losses that are due to gases other than water vapor are the same for both the primary and the reference wavelength bands. The two strongest infrared water vapor absorption bands are those centered at 6.3 μm. and 2.7 μm. The 2.7-μm band is used in most infrared hygrometers.

Advantages

- It is a noncontact instrument and is virtually unaffected by contaminants.
- The sensing element is protected by a sapphire window and never comes in contact with the fluid which is being sampled.
- Sensitivity can be increased if the path length is increased. Low-humidity level measurements require a long path length, sometimes more than 1 m. To prevent an odd size for the instrument, the folded-path version is customarily used. With this technique, the beam will be reflected and bounced back and forth one or more times by reflectance mirrors. Hence the beam will travel the same long distance, but the size of the instrument will be reduced.
- It offers a fast response at high and low absolute humidities.
- There is virtually no drift and no hysteresis.
- If properly calibrated, it can provide good long-term accuracy.
- Operation is available over a broad humidity range.

Limitations

- The infrared hygrometer is more expensive than the chilled-mirror hygrometer.
- As a secondary measurement device, it must be periodically calibrated.
- Optical windows can deteriorate over time or as a result of environmental conditions.
- The design is relatively complex.
- For low frost point measurements, the detector must be cryogenically cooled, adding further cost.

Lyman–Alpha Absorption.   The Lyman–α instrument method is very similar to the infrared technique, but uses a different wavelength of light in the vacuum ultraviolet region. It typically uses a hydrogen lamp to generate a monochromatic beam of light in the proper band to be sensitive to only water. This beam is passed through lithium fluoride windows to a nitric oxide ion chamber where the intensity of the beam is detected. The windows bracket a space of a couple of centimeters where the air can pass between them and the water content can be determined by the loss of light absorbed by it. This instrument is very fast and is most commonly used for high-altitude atmospheric research.

Applications of the infrared and the Lyman–alpha hygrometers have been limited to date, mainly because of their high cost and relative complexity.

Mechanical Effect Based

Vibrating Quartz Crystal (Piezoelectric).   Quartz-crystal moisture analyzers operate by monitoring the change in resonance frequency of a hygroscopic quartz crystal (Fig. 19). This frequency is directly related to the absorbed mass of moisture. This frequency can be calibrated in terms of moisture content.

FIGURE 19   Quartz-crystal oscillator instrument system (*courtesy of Ametek*).

A voltage potential is applied across the opposing faces of the disk, exciting a transverse oscillation within the crystal at a frequency in the 9-MHz range. Actually, the analysis procedure alternates the introduction of reference and sample gases into a moisture-sensitive sample chamber (cell). The sample cell consists of a small chamber through which the test sample and the reference sample gases alternately flow. Within the sampling chamber, a pair of electrodes support the thin quartz-crystal disk.

The sample gas is divided into two streams. One stream goes directly to the measuring crystal; the other passes through a dryer and then to the same measuring crystal, usually at 30-s intervals. Because the wet crystal is heavier than when it is dry, the vibration frequency decreases. Because the analyzers operate in this simple, physical way, taking on moisture molecules during the sample period and releasing them during the reference period, they are less subject to interference caused by the sample gas itself. The relative change in the sample cell behavior is registered, and absolute moisture content derived by real-time computation.

This technique is capable of measuring moisture levels in such gases as hydrogen, ethylene, refrigerants, and natural gas, with resolutions as low as 1ppb. Because this mode of operation always maximizes the partial pressure between wet and dry, it is possible to obtain readings quickly. For maximum accuracy, all sample handling must be done to high purity standards, with all leaks eliminated and sample systems should be designed for minimum volume.

Also, corrosive gases attack the quartz crystal, crystal coating, and/or sampling system. Ammonia, a molecule very similar to water, attaches to the hygroscopic coating and interferes with accurate moisture measurement.

The sensor head often also contains a calibrator that can be used to check the performance, although drift-free operation is usually possible for periods of a year or more. The instrument can cover a measurement range of approximately 20 ppb to 100,000 ppm by volume, although the sensor is normally used for measurements only up to 1000 ppm. Typical attainable accuracies are typically 5% of reading in most ranges and +/−10 ppb in the lowest range. Sensors are usually available in explosion-proof configurations and can be operated over a temperature range of −20°C to +55°C (−4°F to 131°F).

Advantages

- Wide operating range
- Low sensitivity to contamination
- Ability to measure very low frost points (down to ppb range)
- Autozeroing inherent in the design
- Built-in calibration checking
- Suitability for applications requiring explosion proofing
- Fast response time
- High accuracy at low frost point levels

Limitations

- More expensive than some other types
- Not suitable for *in situ* use
- Possible problems if the probe temperature is difficult to control

Applications

The quartz crystal sensor is especially useful in applications involving gases containing corrosive contaminants. Typical applications are

- Catalytic reforming
- Moisture in olefins
- Natural gas (cryogenic extraction, transmission stations, storage, distribution, LNG, production)
- Petroleum refining (reforming, alkylation, LPG, hydrocarbons)
- Petrochemicals (cracked gas, propylene, butadiene, ethylene)
- Chemicals (fluorocarbons gases, vinyl chloride, vinyl fluoride, refrigerants, reactor gas blankets, methyl chloride)
- Enhanced oil recovery (CO_2 pipelines)
- Electronics (doping gases, dielectric gases; soldering furnace atmospheres, blanketing gases)
- Metals (annealing furnace atmospheres)
- High-purity gas production
- Semiconductor manufacturing
- Cylinder gases

Pneumatic Bridge.   The pneumatic bridge device measures the mass difference between the sampled gas and a similar volume of the sample that has been dried. This difference is detected by a fluidic bridge that acts much like a Wheatstone bridge in electronics. The two streams are applied to opposing sides of the flow fixture. The added mass of the wet stream shifts the flow path inside to allow a differential to be determined.

When this flow rate is compared with the dry stream flow rate, a direct value of mixing ratio of water is provided. This unit of measure, along with a temperature measurement of the process, can be used to determine any other moisture value for the measurement.

The benefits of this device include its effective use at high temperatures and in pressurized streams. The drawbacks relate to the device's ability to balance the volumetric flow accurately and the maintenance of the desiccant canisters.

Bulk-Effect Sensors. Early polymer sensors operated on the bulk effect because of the property of many polymers to change geometry with RH. This effect translated into a change in capacitance or a change in resistance, depending on how the sensor was constructed.

The resistive bulk polymer sensor generally can be used over a temperature range $-10°C - +80°C$. Reasonable accuracy is achieved over the range 15%–99% RH.

These sensors are small, low cost, and capable of providing overall accuracies of $\pm 2\%$ RH and narrow range accuracies of $\pm 1\%$. Contamination directly on the sensor can offset the reading, but these sensors can be cleaned with few lasting effects. They must be periodically recalibrated and require temperature compensation. This type of sensor may not tolerate some contaminants. When in doubt, contact the vendor.

The capacitance bulk polymer sensor is slightly faster and often can be used to measure down to 2% RH. Use of the capacitance bulk polymer sensor to measure at RH levels greater than 95% is usually not recommended. Capacitance-based sensors must be temperature compensated and protected from conductive contaminants.

Strain Gage. It was observed that a certain variety of plants propagated by shooting their seeds during high-humidity conditions. This was caused as the seeds swelled in the humid conditions of their environment. Sections of these seeds impregnated with chemicals allowed this material to be used for sensors.

The treated material is bonded to a substrate, along with a strain-gage transducer. The sensors are calibrated to linearize their response to changes in RH. The sensors are inexpensive and small, but they can be contaminated and have a high-temperature coefficient that requires compensation.

Mechanical Expansion. This type of hygrometer is often called a hair hygrometer because of the material commonly used as the sensor. Hair length will increase slightly in the presence of high humidity, and it is this mechanical effect that is the basis for this technique. Human hair is still used, as is animal hair, but most instruments now use synthetic fibers. The hair or fiber is fastened to a fixed point on the chassis of the instrument, allowing the free end to be connected to and work against a spring mechanism or a weight. This has a geared mechanical advantage to drive a dial indicator. This is the typical RH meter in most walnut-framed combination thermometer–barometer–relative humidity instruments sold today. It is an inexpensive humidity device but not very adaptable to process conditions. The mechanical nature of this instrument requires a clean environment and that the instrument be protected from dust that could cause the mechanism to stick (Fig. 20).

Chemical Reaction Based

Electrolytic Sensor. A typical electrolytic hygrometer utilizes a cell coated with a thin film of phosphorus pentoxide (P_2O_5), which absorbs water from the sample gas (Fig. 21).

The cell has a bifilar winding of electrodes in a capillary tube. Direct current applied to the electrodes dissociates the water, which was absorbed by the P_2O_5 into hydrogen and oxygen. Two electrons are required for electrolyzing each water molecule, and thus the current in the cell represents the number of molecules dissociated. A further calculation based on flow rate, temperature, and current yields the parts-per-million concentration of water vapor.

The electrolytic hygrometer is desirable for dry gas measurements because it is one of the few methods that gives reliable performance for long periods of time in the low ppm and ppb regions. Electrolytic hygrometers require a relatively clean input gas, which usually necessitates the use of a suitable filter. The electrolytic cells should not be used with gases that react with P_2O_5, such as NH_3.

FIGURE 20 Mechanical expansion humidity/temperature indicator (*courtesy of Mitchell Instruments*).

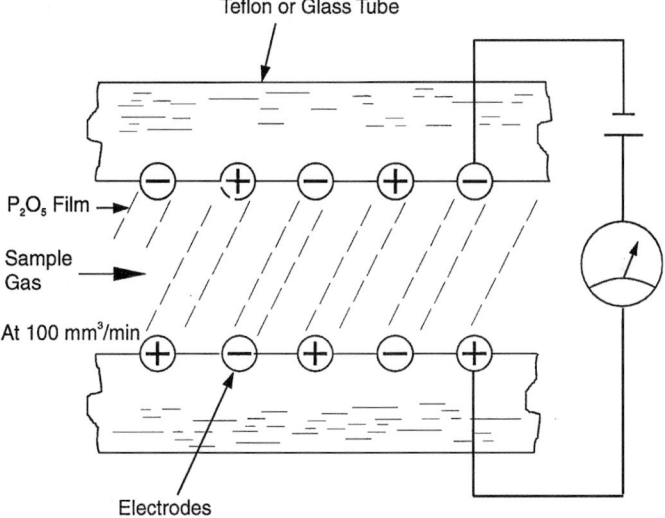

FIGURE 21 Electrolytic P_2O_5 sensor diagram.

   The ability of the electrolytic hygrometer to measure water vapor content of a gas or gaseous mixture is based on the fact that a water molecule can be electrolyzed into molecular hydrogen and oxygen by the application of a voltage greater than 2 V, the thermodynamic decomposition voltage of water. Since two electrons are required for the electrolysis of each water vapor molecule, the current in the electrolytic cell is proportional to the number of water molecules electrolyzed per second. A typical cell is shown in Fig. 22.

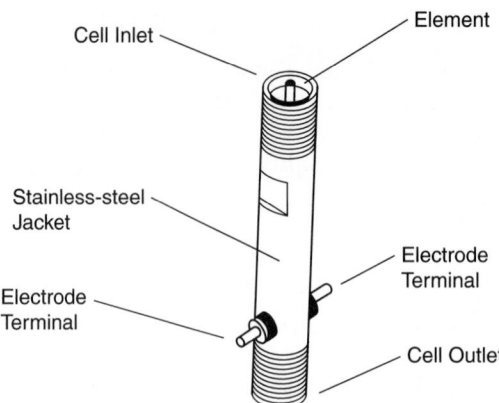

FIGURE 22   Electrolytic sensing cell.

It is a fundamental measurement technique, in that calibration of the electrolytic cell is not required. However, as with chilled-mirror hygrometers, annual recertification against a transfer standard is generally recommended.

- It is specific to water vapor.
- It provides good accuracy in the low ppm$_v$ and ppb$_v$ ranges at a reasonable cost.
- The instrument requires precise control of flow rate through the electrolytic cell. It is also highly sensitive to plugging and fouling from contaminants. The electrolytic cell sometimes requires factory rejuvenation after it is contaminated.
- Modern electrolytic cells are of epoxyless construction to minimize hygroscopic properties and improve response time. Their cost is not insignificant, but considerable progress and cost reductions have been accomplished in recent years resulting in many new applications.

To obtain accurate data, the flow rate of the sample gas through the cell must be known and properly controlled. The cell responds to the total amount of water vapor in the sample. It is most important that all components and sample line materials that come in contact with the sample gas before the sensor be made of an inert material, such as polished stainless steel. This is to minimize absorption/desorption of water. Most other materials are hygroscopic and act as a sponge at the low water vapor levels that are to be measured, causing outgassing, errors, and a slow response time.

Applications.   The typical minimum sample pressure that is required for this type of control is 68.9 to 689 kPa (10 to 100 psig). If the sample pressure is more than 689 kPa (100 psi), an external upstream regulator must be used. A filter is normally located upstream to prevent particulate contaminants from reaching the cell. A typical electrolytic hygrometer can cover a span from a few ppb to 2000 ppm$_v$, with an accuracy of $+/-5\%$ of reading or typically 1% of full scale. This is more than adequate for most industrial applications. The sensor is suitable for many applications involving clean, noncorrosive gases.

The electrolytic hygrometer can be used with most inert elemental gases and many organic and inorganic compounds that do not react with P_2O_5. Suitable gases include air, nitrogen, hydrogen, oxygen, argon, helium, neon, carbon monoxide, carbon dioxide, sulfur hexafluoride, methane, ethane, propane, butane, natural gas, and certain Freons. Unsuitable gases include corrosive gases and those that readily combine with P_2O_5 to form water (such as alcohols). Also to be avoided are certain acid gases, amines, and ammonia that can react with the P_2O_5 and certain unsaturated hydrocarbons (alkynes, alkadienes, and alkenes higher than propylene) that can polymerize and clog the electrolytic cell.

Some typical applications include

- Bottled gases, gas supplier industry—such as a QC instrument
- Plant, instrument, and process air
- Semiconductor industry—process gases
- Glove boxes for manufacture and encapsulation of electronic components
- Lightbulb manufacturing—inert gases
- Electric power industry—transformer gases (SF_6)
- Aerospace industry—argon and helium welding

When the electrolytic hygrometer is applied to clean gases in the low ppm_v region, the instrument offers long and reliable service. However, if misapplied, the electrolytic cell can be damaged. Although procedures are furnished by the manufacturers of these cells for unclogging the capillary and recoating with P_2O_5, this is not a trivial procedure. In some cases, a failed electrolytic cell must be replaced. On the other hand, since the electrolytic cell is essentially a primary device, no calibration is required when replacing the unit. Compared with the aluminum oxide sensor, the electrolytic cell, since it utilizes a primary method of measurement, has better long-term accuracy, exhibits no hysteresis, and is easier to keep in calibration. However, a sampling system is required that must be designed with special nonhygroscopic materials to prevent errors or a very slow response time. The aluminum oxide sensor is more convenient to use when in-line monitoring is desired or when a multiple sensor system is preferred.

A typical sampling system for ensuring constant flow is shown in Fig. 23. Constant pressure is maintained within the cell. The sample gas enters the inlet, passes through a stainless-steel filter, and

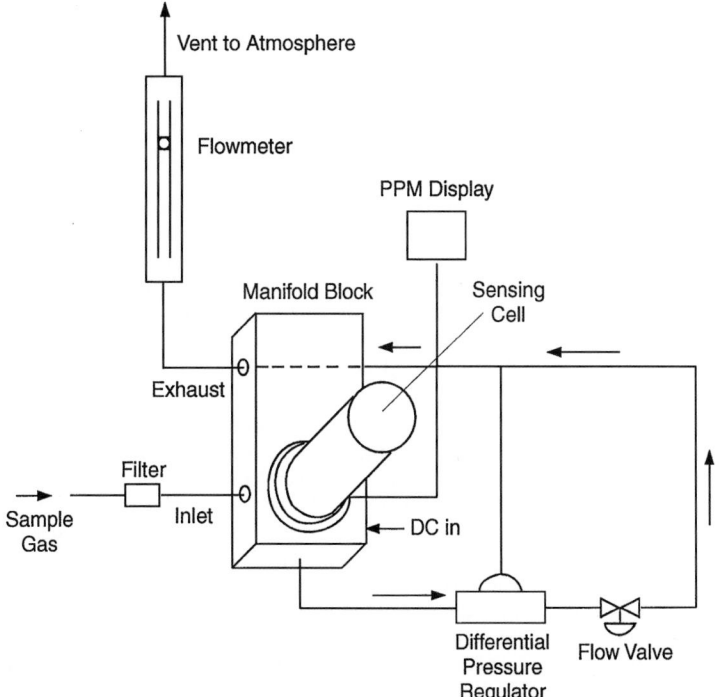

FIGURE 23   Electrolytic (P_2O_5) moisture analyzer flow schematic.

FIGURE 24   Electrolytic P_2O_5 moisture analyzer (*courtesy of Ametek*).

enters a stainless-steel manifold block. After the sample gas passes through the sensor, its pressure is controlled by a differential-pressure regulator that compares the pressure of the gas leaving the sensor with the pressure of the gas venting to the atmosphere through a preset valve and flowmeter. In this way, constant flow is maintained, even though there may be nominal pressure fluctuations at the inlet port. A typical instrument is shown in Fig. 24.

A typical electrolytic hygrometer can cover a span from 0 to 2000 ppm with an accuracy of ±5% of the reading, which is more than adequate for most industrial applications. The sensor is suitable for most inert elemental gases and organic and inorganic gas compounds that do not react with P_2O_5. This technique is not recommended for gases with high concentrations of hydrogen or oxygen because of the possibility of the recombination of the molecules to form additional water. Electrolytic hygrometers cannot be exposed to high water vapor levels for long periods of time because this may wash out the P_2O_5 salt and result in high cell currents.

Karl Fischer Titration.   Titration methods of moisture analysis have been used for many years in laboratories and are typically associated with liquid measurements. However, technicians found that whatever could be dissolved into the liquid, whether the sample started out as a solid or gas, could also be measured with the Karl Fischer titration method (Fig. 25). This method is a wet chemistry method that use the reaction of the water content in the sample with an iodine-based solution to determine its mass fraction. This reaction is based on the predictable nature of the chemistry that requires 10.71 C of electrical current for neutralizing every milligram of water in the solution. The method allows the determination of water in a wide concentration range from 1 ppm to 100%.

For measurement in gases, the mass flow of the gas sample is measured as it is bubbled for a time into the solution. The sensor current is proportional to the mass of the water contained in the sample. The method is repeatable and accurate with fresh solutions. The limitation is that there may not be sufficient time for the reaction to take place on all the water in the sample and it is an off-line device requiring a sample system.

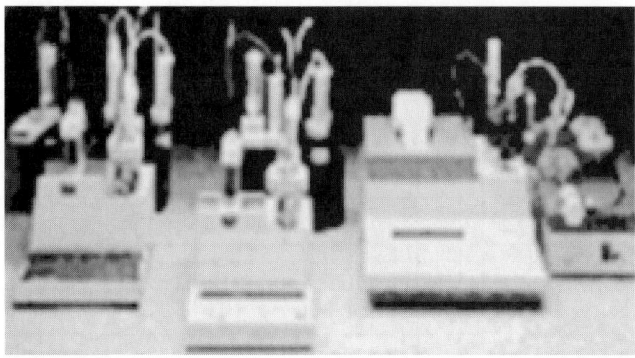

FIGURE 25   Karl Fischer titrator (*courtesy of Metler Toledo*).

Oxygen Displacement

Zirconium oxide sensors have long been known to provide accurate oxygen measurements from 1%–100% levels. They can also be applied to calculate the moisture content by measuring the oxygen content and assigning the depletion of the oxygen (from 20.9%) to increased in moisture.

Features of this technique include the following:

- no sampling required
- extremely fast response
- oxygen and moisture measurement with one sensor package.

Applications include flue gas or combustion-fired process gas because the sensor will operate in processes up to 1760°C. Since the oxygen accuracy is 0.1%, the moisture measurement accuracy is, at best, approximately 0.5% vol.

Sonic

High-temperature moisture measurements to 500°C can be made by comparing the sound velocity in air of a sample which contains a varying amount of steam. A fluidic oscillator, placed in an exhaust air stream from a paper machine hood or baking oven, for example, produces a tone. This frequency is measured very accurately by a sensor designed to handle the high temperatures that are encountered. Temperature compensation is handled by the electronics. This approach can eliminate the need for sampling, cooling, or other treatment of the gas for analysis. This is a relatively expensive approach, but is available as part of a complete control system.

APPLICATION GUIDE FOR GASES

Application of the proper instrument is a decision that requires that many issues be considered. Planning and preparation for the purchase and installation of a measurement system will save time and money. Below are suggested relationships among the types of instruments commercially available and the applications for which they can and cannot be used. Figure 26 is a basic guide for common process conditions for which moisture measurement is desired. Specific applications will have issues that cannot be addressed in this format. The instrument types have been assigned a letter in order to fit the chart to the page. The process condition illustrations are singular, and the instrument manufacturer

Moisture Selection for Gases

Y = Normally applicable
N = Not recommended
O = May be used, but more information or a sample system is normally used
NA = Not applicable

Process Condition	Wet Bulb/Dry Bulb	Optical Manual	CCM Optical (Optical Automatic)	Non-optical Dew Pointer	Fog Chamber	Aluminum Oxide	polymer Capacitance	Silicon Oxide	Heated Lithium Chloride	Dunmore	Pope	Infrared	Lyman-Alpha	Quartz Crystal	Bulk	Strain Gauge	Hair	Electrolytic	Titration	O₂ Displacement Sonic
Indoor Environmental Humidity	Y	N	Y	Y	Y	Y	Y	Y	Y	Y	Y	N	N	N	Y	Y	N	N	N	N
Outdoor Environmental Humidity	Y	Y	Y	Y	Y	Y	Y	N	N	N	N	N	N	Y	Y	Y	N	N	N	N
Vacuum from 0 - 1 Atm.	N	N	N	N	Y	Y	N	N	N	N	N	N	N	Y	Y	N	N	O	N	N
Pressurized from 1 - 7 Atm.	Y	Y	Y	Y	Y	Y	Y	O	O	O	O	Y	O	Y	Y	N	O	O	O	O
Pressurized from 7 - 340 Atm.	N	O	O	N	O	Y	O	N	N	N	N	Y	O	O	O	N	O	O	O	O
Temperature from - 40 to + 40°C	O	Y	Y	Y	Y	Y	Y	Y	Y	Y	Y	Y	Y	Y	Y	Y	Y	Y	Y	Y
Temperature from 40 to 100°C	Y	Y	Y	Y	Y	Y	Y	Y	Y	Y	Y	Y	O	Y	Y	N	O	N	Y	Y
Temperature from 100 to 200°C	O	N	N	O	O	Y	Y	O	O	O	O	Y	N	N	N	N	O	N	N	Y
Temperature from 200 to 350°C	N	N	N	N	N	Y	Y	N	N	N	N	Y	N	N	N	N	O	N	N	Y
Temperature above 350°C	N	N	N	N	N	N	N	N	N	N	N	Y	N	Y	N	N	N	N	N	Y
Corrosive gases	N	O	O	N	O	N	O	O	O	O	O	O	N	Y	Y	N	O	N	Y	Y
Dusty streams	N	N	N	O	O	Y	N	N	N	N	N	N	Y	N	Y	N	O	N	Y	N
Humidity <1 ppmv or dew points < -70°C non-condensing	N	Y	N	N	N	Y	N	N	N	N	N	Y	Y	Y	N	N	Y	Y	N	N
Humidity from 1 ppmv to 1000 ppmv or dew points < -20°C non-condensing	N	Y	Y	O	O	Y	Y	O	N	N	N	Y	Y	N	Y	N	Y	Y	N	N
Humidity from dew points of -20° to +50°C non-condensing	Y	Y	Y	O	O	Y	Y	Y	Y	O	O	Y	Y	N	Y	Y	O	Y	Y	Y
Humidity from dew points of 50° to 100°C non-condensing	Y	Y	Y	N	N	Y	Y	Y	N	N	O	Y	N	N	Y	N	N	N	Y	Y
Humidity to 100% RH condensing	O	Y	Y	N	N	Y	Y	N	N	N	N	N	N	N	N	Y	N	N	Y	O
No flow	N	Y	Y	Y	Y	Y	Y	Y	Y	O	O	Y	O	Y	Y	Y	O	N	Y	Y
Flow over 5 m/sec	Y	Y	Y	Y	Y	Y	Y	Y	O	N	O	Y	Y	Y	Y	N	Y	N	Y	Y
Measurement in hydrocarbons	N	Y	Y	Y	Y	Y	Y	Y	Y	Y	Y	Y	Y	Y	Y	Y	Y	Y	N	N
Measurement in H₂ or O₂ rich streams	O	Y	Y	N	Y	Y	Y	Y	Y	Y	Y	Y	Y	Y	Y	Y	N	Y	N	Y

FIGURE 26   Common process conditions for which moisture measurement is desired.

should be consulted before an instrument is installed that has a combination of these or other conditions. Within types of instruments there is often a wide variance in price and standard features from different manufacturers that may influence applicability of a specific instrument.

The legend letters for the instrument type are shown below the chart. The key is as follows: Y = normally applicable, N = not recommended for this application, NA = not applicable for the instrument type alone, O = may be used, but more information or a sample conditioning system is normally used. Relative cost is a factor when several techniques can work in the application and an instrument may work but be prohibitive in cost. This cost factor may cause an instrument type not to be recommended here.

METHODS OF MEASUREMENT IN LIQUIDS

Liquid applications use some of the same sensing types that are used for gases, with the exception of sensors that would be damaged or influenced by the liquid phase of the process. The techniques described below will only highlight the unique liquid issues if the technique has been described above.

Infrared Adsorption

This technique, as described above for gases, works with liquids as well—as long as the process connections and internal sample piping can maintain a seal. The liquid must also be transparent so the light beam is not obstructed. If the sample has a tint or color it is best if the color is constant since changes in color may interfere with the accuracy. This method is typically used off line and for applications for which the concentration of moisture is low to trace levels.

Capacitance

This type of capacitance sensor is of a rather large scale compared with that of the sensors discussed in the gases section. This device uses the fluid itself as the dielectric for determining the moisture content. Since the water in the fluid has a dielectric of ≈ 80, the carrier or process stream should have a very low and fairly stable dielectric of <2. If the carrier dielectric changes, a new calibration should be performed. These devices are often used in crude oil measurements to monitor flows to determine internal distribution for initial treatment—all the way to measuring for custody transfer. These devices have an insulated rod in a pipe carrying the sample. The capacitance is measured between these two conductors, and the dielectric is a function of the water content. Simple as this may seem, the thermal coefficient of this measurement, along with the salt and other contamination and buildup issues, can cause serious inaccuracies (see Fig. 27).

The common basic sand and water (BS&W) devices have been refined to reduce these effects. More sophisticated devices include dual sensors and mechanical and electronic refinements to achieve titration accuracies. These features do add cost to these systems, but when applied to processes that move thousands of barrels of oil per day this cost is amortized quickly. These units can measure from 0.05% to 30% water in oil and are made to handle the high pressures of the oil application.

Aluminum Oxide

Since this device uses the capacitance principle of operation, the dielectric of the water dissolved in the sample is the signature effect, as described in the gases section. The aluminum oxide sensors can be used in liquid hydrocarbons that have low dielectrics, are immiscible, and are noncorrosive with respect to gold or aluminum. The fluid must be immiscible in order to properly calibrate the sensor. The normal operating range for this type of application is from 0 to 5000 ppm_w, depending on the chemical

FIGURE 27   Precision water in oil analyzer and cell.

composition of the process fluid. Every hydrocarbon has a different affinity for water in solution, and this information must be obtained in order for the instrument to be configured to read out in ppm_w. This method allows readings on line as long as the flow rate is within manufacturer's specifications. This is advantageous for process and quality control. The major drawbacks for this sensor are the drift requiring recalibration every 6–12 months and the inability to handle strong corrosive contaminants.

Titration

As described in the section on gases, this technique is generally accepted as the definitive method for measuring moisture in liquids. It is not easily adapted to on-line measurements and thus has a limited role in process control. The sample is usually captured in the field or in the plant and taken to the lab for this analysis. This method does take time for the moisture dissolved in the sample to react with the reagent. Sample handling, as discussed above, is critical to getting accurate results. This technique has become more automated in the lab in recent years. However, trained technicians should operate this device to preserve sample integrity and maintain proper instrument practice for consistent results.

Conductivity

Conductivity-based instruments rely on the water content of the sample to change its ability to conduct electricity. Two conductors are insulated from each other and placed in contact with the sample stream. They are electrically excited, and the conductance is measured with the instrument electronics. These instruments must be used on streams where the baseline conductivity of the process is known and the only contaminant that might change that value is water. These devices need to be calibrated to the stream they are to measure in order to operate accurately. They require temperature compensation and are sensitive to buildup and contamination that can shift the baseline conductivity.

Centrifugal

When working with hydrocarbon liquids, one can easily see, for example, the difference in specific gravity between oil and water. This simple illustration can be expanded into the centrifugal method for water analysis by use of the artificial gravity created by spinning the collected sample. This method is used in many labs today to determine the water content of crude oil quickly. This technique does not lend itself to titration accuracies but can often determine if a process is repeatable or out

TABLE 7 Instrument Selection Guide

Moisture Selection for Liquids	Infrared Adsorption	Capacitance	Aluminum Oxide	Titration	Conductivity	Centrifugal	Distillation
Laboratory sample	Y	N	N	Y	N	Y	Y
Continuous measurement	Y	Y	Y	N	Y	N	N
0–5000 ppm$_w$ water	Y	N	Y	Y	N	N	Y
0%–30% water	Y	Y	N	Y	Y	Y	Y
Pressurized stream	Y	Y	Y	NA	Y	N	N
Corrosive contaminants	O	Y	N	O	Y	Y	Y

Y = normally applicable, N = not recommended for this application, NA = not applicable for the instrument type alone, O = may be used but more information or a sample conditioning system is normally used.

of specification. Small vials of the sample are measured out carefully, capped, and inserted into the holders along the circumference of the centrifuge. After the vials have spun the required time, the interface is plainly visible and compared with the graduations on the vial to obtain the concentration value. This technique cannot measure the dissolved moisture in the sample, and thus has a built-in offset of the saturation value for the hydrocarbon. Additives or contamination in the sample may affect the solubility and measurement of the mixture. This device is not adaptable to on-line measurements.

Distillation

This is another technique used for analyzing the water content from samples of liquid hydrocarbons. A sample is distilled, and the resulting product is separated in a miniature distillation column. Since the boiling points for hydrocarbons are dramatically different from those of water, the column is capable of providing a good separation. When a sufficient volume of distillate has been collected and measured, the ratios can be determined. Care must be taken in handling the sample throughout the analysis to ensure sample integrity in the beginning and prevent losses of gaseous products during the distillation.

APPLICATION GUIDE FOR LIQUIDS

Application of the proper instrument is a decision that requires consideration of many issues. Planning and preparation for the purchase and installation of a measurement system will save time and money. In Table 7 are suggested relationships among the types of instruments commercially available and the applications for which they can and cannot be used. This table is a basic guide for common process conditions for which moisture measurement is desired. Specific applications will have issues that cannot be addressed in this format. The process conditions' illustrations are singular and the instrument manufacturer should be consulted before an instrument that has a combination of conditions is installed. Within types of instruments there is often a wide variance in price and standard features from different manufacturers that may influence applicability of a specific instrument.

GENERAL OBSERVATIONS

Professionals stress that humidity remains one of the most difficult of the process variables to measure and that no universal methodology is on the horizon. Device selection is difficult. For each application there remain numerous trade-offs in selecting the optimum device from the standpoints of performance and cost, including maintenance. As the need for accuracy increases—as is encountered in various kinds of research involving humidity—the selection problem becomes more difficult. In some cases, user in-house engineering may be required in adapting commercially available sensors. Involving prospective vendors in the early planning phase of an installation is always cost effective. This discussion may uncover less expensive solutions or application-oriented risks not obvious in the engineering phase of the project.

Calibration

An essential part of humidity and moisture measurements is calibration against a standard. Without calibration, few instruments provide accurate or reliable information. The most fundamental standard that is used by national standards laboratories is the so-called gravimetric hygrometer. In this method, a certain amount of bone dry gas is weighed and compared with the weight of the test gas in exactly the same volume. From this, the amount of water is determined and vapor pressure calculated. The method is capable of providing the most accurate measurements possible, but such a system is very cumbersome, expensive and time consuming to use. Some national laboratories, such as NIST (National Institute for Standards Testing) in the United States, NPL (National Physics Laboratory) in the United Kingdom, and NRLM (National Research Laboratory of Metrology) in Japan, have the availability of a gravimetric hygrometer. However, these laboratories use the system only to calibrate other standards that are easier and faster to use for day-to-day calibrations, such as the two-pressure humidity generator, a precision chilled-mirror hygrometer, or a carefully designed psychrometer.

Traceability to National Standards.   Most commercial humidity measurement instruments are supplied with a calibration report showing the accuracy at the time of manufacture or shipment from the factory. In most cases, this does not truly reflect the way the instrument will perform in the field, nor does it need to in many of those cases. The user should know what to expect from the instrument in terms of performance in the field. Traceability means that the instrument has been calibrated against a primary or transfer standard.

Calibration Standards.   With regard to humidity measurements, it is commonly accepted that a standard is a system or device that can either produce a gas stream of known humidity by reference to fundamental base units, such as temperature, mass and pressure, or an instrument that can measure humidity in a gas in a fundamental way, by using similar base units. There are established standards for humidity in many countries, operating on various principles such as gravimetric systems, two-pressure generators, and other devices.

NIST (USA).   The primary responsibility of NIST (previously the National Bureau of Standards, or NBS) is to provide the central basis for the National Measurement System, to coordinate that system nationally and internationally, and to furnish essential services leading to accurate and uniform measurements throughout the USA. This section summarizes the activities of NIST and the humidity measurement systems used.

Standards used to calibrate humidity instruments fall into two classifications: primary standards and transfer standards. Below is a brief description of examples of the two classes of instruments.

Primary Standards.   These systems rely on fundamental principles and base units of measurement. A gravimetric hygrometer is such a device. It measures humidity in a gas stream by physically separating the moisture from the carrier gas and collecting it in a storage vessel. This is subsequently weighed to give the mixing ratio of the sample, and from this information other hygrometric parameters can be calculated. This method is extremely accurate but is cumbersome and time consuming to use. A gravimetric hygrometer is very expensive to build, and, at low humidity levels, can require many hours of operation to obtain a large enough sample. It is therefore not an attractive system for day-to-day use. At a lower level and at somewhat lower accuracies, the two-pressure and two-temperature generators and some other systems are customarily used as primary standards.

Transfer Standards. A number of instruments fall into this category. They operate on fundamental principles. These standards are capable of providing good, stable, and repeatable results but, if not properly used, can give incorrect results. Examples of commonly used instruments are the following:

1. Chilled-Mirror Hygrometer. This is probably the most widely used transfer standard. It has excellent accuracy and repeatability.

2. Electrolytic Hygrometer. This instrument operates on the principle of Faraday's law of electrolysis to determine the amount of moisture in a gas stream. Provided that the cell converts all the water in the gas stream into its component parts and that the sample flow is accurately monitored, the measurement of current represents an absolute measure of the moisture content.

3. Psychrometer. A fundamental instrument, but requiring a great deal of care in terms of operation and maintenance.

Applications

There are numerous requirements and applications for measuring humidity. Through the past 40 years the awareness and the need for these measurements have been rapidly increasing. When such requirements are to be made, the most important decision is to select the most suitable sensor and measurement method. Humidity can be measured in many ways, and many excellent humidity instruments can be found in the marketplace. However, no instrument is good for all applications and if an otherwise excellent instrument is used in the wrong application, unsatisfactory results will be obtained. It is highly advisable to carefully review each application and make a proper selection. It is impossible to review all conceivable applications; instead, a number of the more common applications will be discussed and the types of sensors typically used indicated.

Heat Treating. Atmosphere control in heat-treating furnaces is often accomplished by means of infrared analyzers capable of measuring and controlling CO, CO_2, and methane. Dew point instrumentation is used for control of water vapor, which can also provide an indirect measurement of other parameters.

The following humidity instruments are often used for heat treating.

1. Fog Chamber. The fog chamber is a manual and simple-to-operate device, but not very accurate. The fog chamber is popular and is widely used by heat treaters but is not suitable in applications for which continuous monitoring is needed. It usually does not meet requirements of regulatory agencies. The fog chamber measurements give one-time data only and are not suited for continuous monitoring or automatic process control functions.

2. Lithium Chloride Sensor. Although it is reliable, simple, and inexpensive, the lithium chloride sensor has a slow response time. In addition, although the sensor is rugged and able to operate in contaminated atmospheres, it is susceptible to contamination by ammonia, which makes it unsuitable for control of most carbon-nitriding atmospheres. Cleaning of the sensor, although not very difficult, is time consuming. This sensor is at present rarely used in heat-treating applications.

3. Chilled-Mirror Hygrometer. Although expensive and somewhat difficult to install, the chilled-mirror hygrometer is well suited for automatic, continuous dew point control of furnace atmospheres. It offers high accuracy, excellent repeatability, and is less expensive and easier to service than infrared analyzers. Compared with dew cups and fog chamber instruments, the condensation hygrometer provides much higher accuracy and repeatability and can easily be used in automatic control systems. The instrument can provide a continuous, 24-hour-a-day, record of the dew point in the furnace atmosphere, and therefore can allow the heat treater to do a much higher quality job. However, the cost of such an instrument is much higher and for measurements of very low dew points, sensor cooling is often required, which is an inconvenience and represents additional cost. Furthermore, the chilled-mirror sensor requires periodic cleaning and is difficult to maintain in heat-treating environments.

4. Aluminum Oxide Hygrometer. The aluminum oxide hygrometer is often used in heat-treating applications involving low frost points, i.e., below $-40°C$ $(-40°F)$, such as exothermic gas ovens. Although accuracies and repeatability for aluminum oxide hygrometers are considerably less than for chilled mirror systems, they are more convenient and less expensive to use at frost points below $-40°C$ $(-40°F)$. If the gas is relatively clean, acceptable results can be obtained using some of the more advanced aluminium oxide probes.

5. Infrared Analyzer. The infrared analyzer, often used in heat-treating applications, provides good accuracy and the capability of directly measuring carbon oxides and other gases. However, these instruments are more expensive than chilled-mirror hygrometers and require considerable maintenance and calibration. They also require maintenance and recalibration at modest to frequent intervals.

6. Electrolytic Hygrometer. Electrolytic cells have also found some applications in heat treating. The cells are useful only if the gas to be monitored is clean and filtered. These cells are primarily attractive when very low ppm levels are to be measured. Overall, heat-treating applications of electrolytic hygrometers have been limited.

Semiconductors. Sophisticated microcircuits (chips) involve extremely high packing densities of individual elements and are subject to damage due to contamination and moisture. The probability of failures resulting from a single defect increases as the distance between components on the chip decreases. Since large investments are required in capital equipment, processing times, and materials, the cost of failures has increased sharply. Hence it has become extremely important to control the atmospheres in which such sophisticated chips are manufactured. Among the contaminants that are most likely to cause semiconductor failures are metallic ions, other charged ions, oxygen, and water. Reliable humidity trace moisture instrumentation is often critically needed to ensure a high level of quality and reliability of the product and process. Levels of moisture that can cause damage are often very low (in the 1-ppm range and below). To measure and monitor such low levels of moisture requires highly specialized instrumentation. Instruments commonly used for semiconductor manufacturing are the electrolytic, aluminum oxide, and piezoelectric hygrometers.

Water Activity Measurements. Water activity (a_w), is defined as the free moisture available in a material as opposed to the chemically bound moisture. It is directly related to equilibrium relative humidity (%ERH). Quite simply, %ERH is expressed in terms of 0%–100%, and water activity in terms of 0–1. While water activity represents a very useful assessment of the free moisture of a material or substance for a wide variety of quality purposes, it does not necessarily reflect the total moisture content percentage, which is an entirely different measurement requiring the use of other principles. The total water content percentage equals the sum of bound water and free water. In simple terms, water activity is the equilibrium relative humidity created by a sample of material in a sealed air space, and expressed on a scale of 0–1 for 0%–100% ERH.

Measuring Water Activity. Since more and more quality restrictions are being placed on finished products, processing, and preservation of materials, applications for water activity measurement have increased rapidly and are becoming more diverse and more firmly established as part of documented quality procedures. Wherever there is a need to monitor or control free water in a material, an application to measure water activity exists. The free water is considered to be the water that is free to enter and leave the material by absorption/desorption. Generally, the main areas of need for water activity testing all have different reasons for making these measurements.

1. Food Processing

The physical–chemical binding of water, as measured by water activity (a_w) can be defined as

$$a_w = p/p_o = \%ERH/100$$

where $P =$ vapor pressure of water exerted by the food

           $p_o =$ vapor pressure of pure water

    $\%ERH =$ equilibrium relative humidity of the food

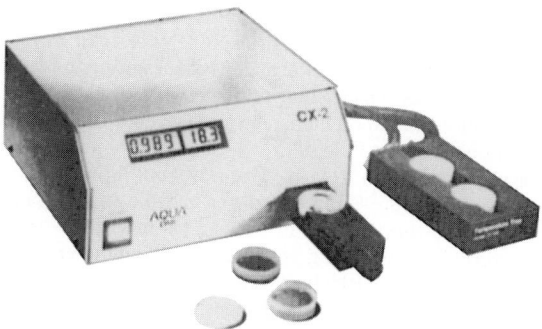

FIGURE 28   Chilled-mirror hygrometer for water activity measurements (*courtesy Decagon Devices, Inc.*). A chilled-mirror hygrometer
specifically designed for measurement of water activity. The instrument
includes special containers for the substances to be measured. Both dew
point and temperature of the vapor immediately above the substance
are measured and relative humidity (water activity) is calculated by a
microprocessor.

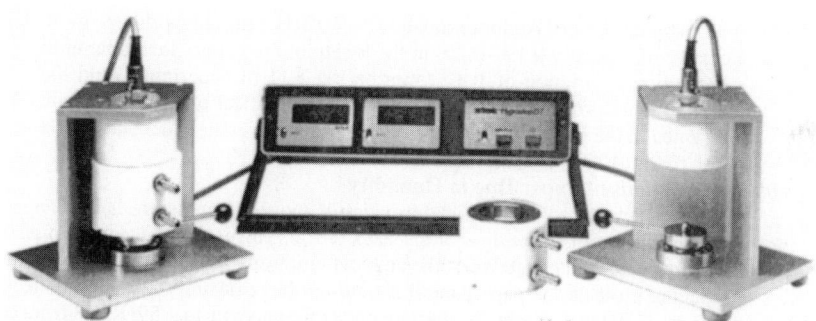

FIGURE 29   %RH instrument for measuring water activity (*courtesy Rotronics, Inc.*). A relative humidity
instrument to meausre %RH in the vapor immediately surrounding the specimen to be analyzed. Special
containers are included in the system.

In essence, the a_w of a food determines whether it will gain or lose moisture in a given environment.
For example, if a cracker and a slice of meat were independently transferred into an atmosphere
of 75% RH, we would find that over time the meat will lose water and the cracker will gain water.
They will come not to equal water content, but rather to equal a_w. The basis of protection by
packaging is to slow this process down as much as possible within economic as well as food safety
and quality limits. To do this, one must know both the sorption isotherms (moisture content versus
a_w) of the food product and the transport properties of the packaging film.

A typical chilled mirror hygrometer for water activity measurements is shown in Fig. 28 and a
%RH instrument in Fig. 29.

2. Paper Industry

Machinery for printing paper or cardboard, coating paper with aluminum, and other types of
unique applications are quite sensitive to the properties of the product to be transformed and to any
variation of some physical phenomena. One parameter that has long been recognized as influencing
the properties of paper and cardboard is moisture. Equilibrium relative humidity and the relative
humidity of the storage and work areas are responsible for changes occurring in the moisture
content of the product. Various studies have demonstrated the importance of %ERH control.

a. **Influence of Moisture Content on Paper**

Changes in moisture content result in swelling or shrinking of paper fibers. Humidity changes result in permanent dimensional changes because of the release of internal tensions in paper. Mechanical properties of paper and printing properties are also influenced by moisture. Moisture content is therefore an important parameter that should be monitored and controlled directly on the paper machines.

b. **Influence of Equilibrium Relative Humidity on Paper**

Differences between the %ERH of paper and the room relative humidity (%RH) result in changes in the moisture content of paper. This must be avoided to prevent problems during the paper conversion or printing process. It is difficult to eliminate any difference between the ambient %RH and the %ERH; however, controls or limits can be established. A %ERH of 50% is ideal for paper since any %RH changes, say in the range 40%–60% will have little effect on the moisture content. From time to time, printers experience difficulties that due to static electricity, such as paper sheets sticking together. This usually happens when the air and paper are too dry. When the paper and air are in the 40%–45% range this problem seldom occurs.

c. **Dimensional Changes**

Paper fibers absorb or desorb water, depending on the ambient relative humidity. This causes swelling or shrinking of the fibers, which affects the diameter of these fibers more than their length. In a sheet of paper most fibers run parallel to the running direction of the paper machine. Accordingly, dimensional changes that are the results of moisture variations are more important along the axis that is perpendicular to the running direction of the paper machine than along the axis parallel to it. At approximately 50% ERH a humidity change of 10% ERH results in a change of typically 0.1%–0.2% in the length of the paper. Such a humidity difference gives a dimensional variation of 1 to 2 mm on a 1 × 1 m paper and could therefore cause poor and inaccurate printing. Paper running through an offset press usually gains water since it is moistened in the process. The change in the moisture content depends not only on the % ERH of the paper (40%–60%), but also on the ambient %RH.

d. **Deformations of Paper Due to Humidity**

Paper in stacks or rolls shows deformation if too much moisture is exchanged with the surrounding air through the edges of the stack or roll. This is due to the uneven distribution of this moisture as it is exchanged with the ambient air during storage or transport. Water-vapor-tight packaging protects the paper, and it should not be removed without first checking %ERH of paper and %RH in the ambient environment. Differences up to +5% RH will not cause problems, while a differnce of 8%–10% RH could be critical.

e. **Temperature-Induced Deformation of Paper**

Temperature exerts a minimal influence on paper; however, any large temperature difference between the paper and the ambient air will have almost the same results as a humidity difference. This is due to the fact that %RH in the air layer in the immediate surroundings of the paper stack or roll is modified by the paper temperature. Assuming an ambient air of approximately 50% RH, a temperature difference of $+/-1°C$ ($+/-2°F$) will result in a humidity variation of $+3$% RH. Thus it is evident that when temperature differences approach 3–4°C, problems can occur.

f. **Curling of Paper Sheets**

Paper fibers do not all run exactly in the same direction across the thickness of a sheet of paper. Large moisture variations could result in unequal dimensional changes on both sides of the paper sheet, resulting in curling. This is a more prominent problem with coated stock, since both sides of the paper are not similar. When such paper is being used, humidity variations should be kept to less than 10% RH.

3. Pharmaceuticals

Some applications for a_w testing in pharmaceuticals relate to microbiological considerations, but many concern processing to achieve correct moisture levels that facilitate filling of capsules, tablet forming, and packaging. For example, gelatin capsules are often a_w maintained to allow them to be sufficiently supple for filling without cracking or breaking in the process. There are many effervescent powder manufacturers who control total moisture within specified limits by measuring a_w as a quality check because it is in some cases quicker and more practical.

4. Chemical Industry

The testing of a_w is a vital part of research and development with soap and hygiene products. Because of the different chemical components of these products, free water can migrate from within a product and component substances can be carried to the surface by using water as a vehicle. This process is known as efflovescence and can be a common problem with soap bars.

Natural Gas.   The presence of excessive amounts of water in natural gas presents many problems. For this reason, the maximum amount of water that the gas can contain is always specified and there is a great need for reliable instrumentation that can measure the trace water content in natural gas. Natural gas is almost always saturated with water when it leaves the ground so that slight changes in gas temperature or pressure can cause this water to condense or form hydrates. Liquid water can be very corrosive to steel gas pipelines, and hydrates, which resemble wet snow, must be avoided since they plug pipelines and valves, interrupting the steady flow of gas. Because of these problems, the allowable limit on water content is generally set at 80–96 gs/10^3 m^3 (5 to 6 pounds of water per million standard cubic feet).

Measurement Technology.   The measurement of water content in natural gas can be made on line or in the laboratory. Laboratory measurements offer the following advantages:

- Monitoring in one central test location
- Potential of using one instrument, such as a gas chromatograph, for several analysis

  The disadvantage is the following:

- Handling of the sample gas may affect the measurements and cause time delays because of the need to take samples to a laboratory at regular intervals.

  On-line measurements offer the advantages of

- real-time readings
- reduced sample-handling effects
- dedicated analysis
- possibility of process control

  The disadvantages of on-line measurements are

- such measurements are often required in remote locations
- requirement for well-designed sampling systems.

Contamination and the presence of hydrates in natural gas have created many problems with instrumentation used to measure water content or dew point in natural gas. As of this date, there are no instruments or sensors on the market that may be considered ideal for this application. Chilled-mirror hygrometers have been tried, but with little success because of contamination effects. They are therefore rarely used in natural gas applications. Aluminum oxide sensors have been used with some degree of success when used with a well-designed sampling system and if regular calibration and readjustment of sensors can be provided. The method is acceptable in some cases, but not ideal.

Electrolytic P$_2$O$_5$ hygrometers have been used with more success, although they require frequent maintenance and rejuvenation of the electrolytic cell with P$_2$O$_5$. A large number of P$_2$O$_5$ cells are in use for natural gas measurements, but they are not considered ideal for the application either. Piezoelectric quartz-crystal technology is well known and has been used for many years in communications and time-keeping equipment. This technology can be useful in natural gas applications, but is expensive. An ideal instrument for the natural gas market is not currently available.

ETO Sterilizers.   Many medical instruments, including one-time-use items for urological and blood work, are sterilized after packaging by exposure to ethylene oxide (ETO) gas. These medical devices are usually packaged in a sealed plastic or polymer envelope and boxed in cardboard cartons for shipment. Palletized cartons are then placed inside a humidity-controlled warehouse room. Humidity controlled in the range of 60–70% RH helps precondition the packaging and enhances penetration of the ETO gas.

Dryers.   A considerable amount of energy is expended in removing moisture from final and in-process products in a wide range of industries. On average, drying accounts for approximately 12% of total energy consumed in industrial areas such as food, agriculture, chemicals, textiles, paper, ceramics, building materials, and timber. In general, the operation of dryers, which come in a wide variety of shapes and sizes, is poorly understood. This frequently results in suboptimal performance. Humidity of the air within a convective dryer can have a significant effect on its operating performance and the characteristics of the dried product. In general, thermal efficiency of most industrial convective dryers is poor; a figure of 30% is not unusual. Most heat supplied to the dryer is lost in the exhaust gas. If the flow rate of the latter can be reduced and the humidity of the exhaust air correspondingly increased, there will be a reduction in the heat requirement of the dryer, even if the exhaust temperature has to be raised to avoid possible condensation in the gas cleaning equipment. In practice, there are practical constraints on the degree by which air flow can be reduced. For example, in fluidized-bed dryers, a sufficient air velocity is needed to maintain fluidization. Depending on the nature of the drying solids, the elevated humidity could also limit final moisture content of solids. The use of exhaust air recirculation, which also results in an elevated humidity within the dryer, suffers from similar limitations.

Hygrometers commonly used for drying applications include chilled-mirror instruments (good but expensive), and aluminum oxide hygrometers (easier to use but less accurate).

Gases.

1. Carbon Dioxide Gas. The presence of moisture in carbon dioxide gas can cause freeze ups in process lines at high pressures. Moisture sensors can be used at various locations in a carbon dioxide process line. One sensor may be placed in a bypass line that is vented to the atmosphere. Other sensors are usually operated directly in the gas line at pressures of approximately 2 MPa (300 psig). Results at both pressures are consistent and reliable. Nevertheless, operation at direct line pressure is generally more advantageous in terms of sensitivity and overall system response time. If an appropriate moisture-detecting system is used, maintenance costs and downtime can be virtually eliminated. In the case of high moisture levels, carbon dioxide gas can be recirculated through molecular sieve dryers before line freeze up occurs.

Typically water vapor detection is required in the 1–10-ppm$_v$ range, which corresponds to a dew point range of $-60°C$ to $-75°C$ ($-76°F$ to $-103°F$) at 1 atm. Operating pressures are in the range 0.1–2 MPa (15 to 300 psig). Aluminum oxide hygrometers are most commonly used for this application.

2. Gases Containing Hydrogen Sulfide. The presence of moisture in gas streams containing trace amounts of hydrogen sulfide produces corrosive byproducts. These byproducts can destroy expensive process pipes and fittings. In addition, high moisture content can render the process unusable or inefficient because of direct poisoning of the catalyst. Generally a number of sensors are used in various operating and regenerating process streams. The sensors are installed directly in the gas line. Since gas temperatures are in excess of $60°C$ ($14°F$), a cooler must be used in series with the sensor to decrease the sensor temperature to less than $60°C$ ($140°F$).

The aluminum oxide sensor appears to be ideally suited for this application. Phosphorous pentoxide hygrometers have also been used, but electrolytic cells are easily damaged, requiring frequent servicing and additional costs. When permanently installed, a suitable sensor, such as an aluminum oxide probe, has a wide dynamic range that allows for monitoring of both operation and regeneration processes. Multiple sensors with a single readout allow continuous monitoring of a number of process streams at modest cost. High maintenance costs are greatly reduced by effective moisture monitoring. The typical range to be monitored is 100 to 2000 ppm$_v$, corresponding to a dew point range of $0°C$ to $-45°C$ ($32°F$ to $-49°F$) at an operating temperature of $50°C$ ($122°F$) and pressure of 3.6×10^6 (500 psig). Because of its sensitivity to trace amounts of hydrogen sulfide, a chilled mirror hygrometer is not recommended for this application.

3. Steam Leaks in Process Gas Lines. Moisture measurements can be used to determine whether or not steam leaks occur within steam-jacketed air pipes. The presence of moisture indicates that leaks

in system gas lines that are heated by steam. The presence of steam in process lines causes corrosion of the metal and/or subsequent contamination of the product. A common procedure is to mount a moisture sensor in a sample cell. A sampling line, which includes a cooling coil, connects the sample cell inlet directly to the process stream. The cooling coil is necessary to reduce the temperature of the test gas to less than the maximum temperature the sensor can be operated at, i.e., $+90°C$ ($194°F$) for a chilled-mirror sensor, $+60°C$ ($140°F$) for an aluminum oxide probe. A steam leak can be immediately detected by an increase in the dew point, and an alarm system is used to provide an audible warning when a steam leak occurs.

Chilled-mirror and aluminum oxide hygrometers have been used for this application. This has prevented maintenance problems arising from corrosion of pipes by detecting a steam leak in the early stages of development. In this way, costly downtime and repairs can be kept to a minimum. The required water vapor range is approximately 1000 to 20,000 ppm_v, corresponding to a dew point range of $+20°C$ to $-20°C$ ($68°F$ to $-4°F$).

4. Cylinder Gases. Cylinder gases are generally supplied as a clean and uncontaminated product. The presence of water vapor is most undesirable and is a quality defect. Both the cylinder gas supplier and the user are interested in measuring the trace moisture content of the gas to determine its quality. Guarantees made and levels to be measured are normally in the low ppm and, in some cases, ppb range. A measurement of excess moisture is not necessarily caused by an inferior product; it can also be the result of problems with leaks or the gas-handling system. Particulate or chemical contamination is usually not a problem since such gases are generally very clean. Electrolytic and aluminum oxide hygrometers are frequently used for cylinder gas measurements. In some cases, the piezoelectric hygrometer may be used. If high accuracy and traceability are crucial, the chilled mirror hygrometer is selected, although the latter is expensive and difficult to use at the low ppm levels. Cylinder gas manufacturers obviously prefer fast-response instruments to cut test times and costs.

Industrial.

1. Instrument Air. Large amounts of instrument air are used in the power-generation industry to operate valves, dampers, and pneumatic instrumentation. Instrument air is also used in may other industries for a multitude of applications. Contaminants and moisture will cause wear on moving parts, leave deposits that can cause problems in operating various systems, deteriorate O-rings and seals, and cause freeze ups when exposed to wintertime temperatures. To eliminate excessive moisture in such instrument air lines, power plants often install refrigerative dryers. These use two towers (containers) filled with desiccant material that are alternated between being in operation or in refrigeration. They are regenerated by heating up the tower and purging the dry air. The regeneration cycle is timer controlled and not performance driven. When operating properly, the dew point of the output is approximately $-35°C$ ($-31°F$). The instrument is set to alarm and notifies the operators when the dew point approaches an unsafe condition. It is also customary to monitor changes in dew point to set the timer cycles on the dryer. When the dew point is too high after regeneration, the cycle needs to be lengthened or the desiccant must be replaced. Instrument air systems are also widely used in chemical plants, food and drug processing, pulp and paper mills, petrochemical processing, and many other industries. Frozen air lines can reduce production or shut down operations for many hours. The presence of moisture in the air used as a blanketing gas for sensitive instruments causes corrosion problems, rapid change of calibration, and loss of instrument sensitivity. Humidity sensors are usually installed at the source of the instrument air equipment. Alarm systems are always added to alert operators when the dew point deviates from a predetermined level. Close moisture control can greatly reduce the need for major repairs of delicate in-process instrumentation and is thus very cost effective.

Generally the water vapor pressure to be monitored for these applications is below 1000 ppm_v, corresponding to dew points below $-20°C$ ($-4°F$). Both chilled-mirror and aluminum oxide hygrometers are often used for instrument air applications.

2. Hydrogen-Cooled Electric Generators. Hydrogen is often used to cool large stationary generators because of its high heat capacity and low viscosity. It has a high heat capacity and therefore removes excess heat efficiently. Hydrogen also has a low viscosity, allowing higher-capacity operation of the generators while maintaining efficient cooling. The hydrogen must be kept dry to maintain both its heat capacity and viscosity. Ambient moisture is a contaminant that reduces the heat capacity and increases the viscosity of the cooling hydrogen. In addition, excess moisture increases the danger of arcing in the high-voltage high-current generators that could seriously damage the generators and cause ignition of the explosive hydrogen.

Cooling hydrogen is continuously circulated through the generator and through a molecular sieve dryer. To be certain that the hydrogen in the generator is dry, a probe or sample cell assembly is installed in the return line from the generator to the dryer. Detection of the moisture content above a predetermined acceptable level will trip the alarm relay and annunciator. Aluminum oxide systems with compatible moisture sensors are commonly used for this application. A special alarm relay, capable of carrying 125 VDC at 0.1 amp, is often available and may be required in some installations for interface with alarm annunciators. Lithium chloride hygrometers and in-line psychrometers have been used in the past. But they require a high degree of maintenance and have proven to be less reliable. Chilled-mirror hygrometers have also used in this application. The dew/frost point range to be covered is $+10°C$ to $-50°C$ ($50°F$ to $-58°F$) at operating temperatures of $30°C$ to $40°C$ ($86°F$ to $104°F$) and at operating pressures of 0.15 to 618 kPa (0.5 to 75 psi).

3. Steel Warehouses. After the manufacture of steel is completed, it is usually stored in a warehouse before being used in the next fabrication cycle. Many of these warehouses are expansive buildings with large doors for access by trucks, railroad trains, and vessels used to transport the steel. While in storage, the steel could be damaged by corrosion, which can easily happen if condensation occurs. Typically, the dew point of the outside air is lower than the surface temperature of the steel, so condensation will not occur. However, condensation happens when a rainstorm, the formation of fog, or an increase in outside air temperature, raises the dew point above the steel temperature. The steel should be protected against this condition, which can be accomplished by closing doors, turning on a dehumidifier, or heating the steel. To conserve energy, the warehouse would only be dehumidified or the steel be heated, if such weather conditions exist.

A chilled-mirror dew point hygrometer with a temperature sensor can be effectively used for this purpose. The temperature sensor must be installed in a small hole drilled into a large sample block of steel and is used to obtain the temperature of the steel in storage. The dew point of the surrounding air is then measured and compared with the steel temperature. Signal outputs of these temperatures are fed to a differential comparator that sounds an alarm whenever the dew point temperature approaches within a few degrees of the steel temperature. The dehumidifiers or heaters can then be turned on automatically to prevent damage and to operate in the most energy-efficient way.

4. Injection Molding. Small plastic parts and consumer products are often injection molded. The raw plastic is first heated and melted and is then forced under pressure into steel dies that shape the molten plastic into various commercial products. The raw plastic is purchased from the manufacturer in the form of small pellets. These are usually delivered in large containers, which are stored until needed. In some cases, the plastic pellets are too wet, either as received from the manufacturer or from condensation during storage. When this occurs, the plastic will not flow properly in the mold, which is a serious problem since many parts will then have to be rejected. Therefore the pellets are first directed into a hot air dryer just before molding, at a temperature of $80°C$ to $93°C$ ($175°$ to $200°F$). Although a large improvement in product uniformity can be obtained form drying the pellets, the operator has no way of knowing when they are dry enough to mold. Most dryers are automatically run for a fixed period of time. Some pellets that are very wet are not dried sufficiently, and these parts are rejected because of improper molding. Conversely, much money can be wasted by drying material that is already sufficiently dry. This problem can be solved by using a chilled-mirror or aluminum oxide hygrometer to measure the dew point at the dryer output duct. A high-temperature sampling system must also be used to reduce the air sample temperature to bring it into the measurement range of sensor. A sampling system allows the temperature of the air to be decreased by drawing the air through a certain length of tubing before measuring. Since the ambient temperature outside the tubing is much lower, the sample temperature equilibrates toward ambient. In effect, it is a heat exchanger. The dew point in the tubing remains constant and may be measured at the other end. Since the pellets are often dusty and dirty, a filter must be installed ahead of the sensor.

5. Battery Manufacturing. Many manufactured products are affected by ambient humidity. Some are processed in controlled chambers; others must be manufactured in humidity controlled rooms. In most cases, this can be handled by a properly designed air conditioning system, but in some cases, such as for the manufacture of lithium batteries, the entire manufacturing area must be controlled at a very low humidity level, sometimes well below $-20°C$ ($-4°F$). Substantial product losses could be suffered if this condition is not maintained. A complicating factor is that doors are occasionally opened and closed to allow personnel to enter or leave the room. The very dry conditions could also lead to discomfort on the part of the employees.

Dry rooms incorporating large air dryers are constructed to control humidity and temperature during the assembly process. Once the assembly process is completed and sealed, humidity control is no longer needed and the products may be removed safely from the controlled atmosphere.

In the past, dry room humidity measurements were often made with a dew cup. However, these are not very accurate, primarily because of operator inconsistencies and difficulties observing frost on the cooled surface. Chilled-mirror hygrometers, although much more expensive, have often been used in areas where humidity control is critical and product losses could be very substantial.

6. Gas Turbines. Gas turbines are usually connected to electrical generators and used in power plants and cogeneration installations. Most turbines are operated with natural gas. Normally steam or ammonia is injected into the intake air stream in order to control the flame temperature, which in turn determines the emission rate of oxides and nitrogen (NO_x). To properly set the correct rate of steam or ammonia injection and to prevent icing, which can cause catastrophic failures, the dew point of the upstream intake air must be closely monitored. EPA regulations limit the emissions of gas turbines, which are dependent on the relative humidity of the intake air. RH measurements can be used to make corrections in the emissions detected to establish compliance with EPA regulations. Such measurements must be accurate, traceable to humidity standards, and acceptable to the regulating agencies. Various types of dew point instruments have been used in the past, but with limited success, mostly because of the difficulty of frequent maintenance and/or need for calibration. At this time the cycling chilled-mirror (CCM) hygrometer is being successfully used in many installations. The use of certain polymer-type RH sensors have also been attempted.

7. Clean Rooms. Many products manufactured and processed in a clean room environment are moisture sensitive. For this reason, clean room specifications often include control levels of typically 35% relative humidity for year-round operation. The proper %RH level must generally be maintained within $+/-2$ %RH) at temperatures below $20°C$ ($68°F$). A relative humidity of 35% at $20°C$ corresponds to a dew point of about $4°C$ ($40°F$) dew point. The effects of higher humidity levels in close-tolerance environments can be detrimental to product quality and production schedules. In semiconductor manufacturing, when the humidity level fluctuates in a wafer fabrication area, many problems can occur. Bakeout times typically increase, and the entire process generally becomes harder to control. Humidity levels above 35%RH make the components vulnerable to corrosion. In pharmaceutical manufacturing facilities, high humidity causes fine powders to adsorb moisture, clogging the powder feed to the tablet press. Variations in humidity cause difficult adjustments in bed temperature and spraying rates, resulting in heat damage and moisture intrusion. Humidity in air ductwork creates moist areas for bacterial colonies to grow and cause process contamination. Two common approaches to humidity control are air conditioning and desiccants. Air conditioning lowers the temperature of a surface exposed to the clean room air stream below the dew point of that air stream. Excess water vapor condenses, and the resulting air is dehumidified. The air must then be reheated to the proper control temperature and routed to the clean room. Standard refrigeration equipment can produce dew points of $5°C$ ($+40°F$) on a reliable basis. In a desiccant system, the process air stream passes through a desiccant medium. The desiccant adsorbs moisture directly from the air stream, and the resulting dehumidified air is routed to the clean room. Measurement of %RH is not very difficult, and many approaches are possible. The most commonly used are the secondary RH elements like the ones used in air conditioning applications. In some cases in which accuracy and stability are of utmost importance, chilled-mirror hygrometers have been used to measure dew point and temperature and %RH is derived from a microprocessor calculation.

Automotive.

1. Automobile Exhaust Emissions. Nitric oxide emissions from automobile exhaust systems, known as NO_x are tightly regulated by the Environmental Protection Agency (EPA) in the USA and by similar institutions in several overseas countries. Aside from difficulties in measuring and controlling these emissions in trace amounts, they are also affected by, and dependent on, the relative humidity of the engine intake air. EPA specifications are based on a certain relative humidity, and corrections are allowed for measurements made at higher or lower relative humidity levels. To validate such corrections, accurate humidity measurements must be made and, more importantly, these measurements must be certified and traceable to national standards (such as NIST in the USA). For this reason, these measurements are customarily made by chilled-mirror hygrometers. The humidity range to be covered is usually within 20% to 80%RH, and measurements are not difficult to make. Most complexities are related to accuracy, repeatability, and traceability.

2. Engine Testing. Combustion engines require the proper mixture of ambient air and fuel. The efficiency and performance of such engines are affected by the moisture content of the air. Carburetor settings could be adjusted based on humidity measurements and thereby optimized for more fuel-efficient operation or better performance. Humidity measurements are therefore very important for engine testing and research. The most commonly used method for research in this field is the chilled-mirror hygrometer.

3. Paint Spray Booths. When automobile bodies are painted, conditions of humidity are critical. If water vapor levels in the air are too high, moisture can become trapped between the metal and the paint. When the painted shell is subsequently baked (or cured), the moisture migrates outward, and if the paint is still elastic, bubbles appear on the surface, which is most undesirable.

The solution to this problem is to maintain the ambient air at the appropriate humidity level before injecting it into the booths or to control the humidity in the paint area by using a humidity sensor situated in the return air duct. Typically, most manufacturers prefer to measure humidity in the return air ducts. In the return air, the sensor is not only subjected to the dust particles of the paints but also to the volatile vapors that are generated. For this reason, sensors must often meet intrinsic safety requirements.

Reliable humidity measurements in paint booths are very difficult because of the contaminants. Few RH sensors survive for long periods of time. Chilled-mirror sensors have also been tried, but with little success. The use of CCM-type sensors has the potential of providing an improvement, although such sensors are certainly not immune to the heavy contamination in spray booths.

Computers. Humidity measurements are important in computer rooms where large and expensive data storage systems are installed. It is a well-known fact that under extreme humidity and temperature conditions, magnetic devices used for data storage could be distorted and cause catastrophic failures to expensive equipment and the loss of important stored data. Certain types of computers are equipped with water-cooled storage systems, and in such cases humidity measurements are used to detect leaks by measuring humidity changes in the air surrounding the water-cooled devices. For these measurements, environmental conditions are usually right, i.e., air is clean and surrounding temperatures are close to ambient. Bulk polymer relative humidity sensors, capacitive or resistive, perform well under these conditions, and are often used. Such sensors should be periodically recalibrated, which is normally done on site. In cases in which humidity is extremely critical and in which very large losses could result from failure, the fundamental chilled-mirror hygrometer or transmitter types are used.

Data Communications through Telephone Cables. Modern telecommunications needs are placing a greater and greater burden on existing telephone cables. The use of these telephone cables for computer communications is often a problem because the integrity of data at high frequencies (high data rates) is critical. Older telephone cables may have paper or cotton wire insulation. This type of insulation can present large losses and become unreliable when moist. The amount of loss, and hence the quality of the line, depends on the amount of moisture in the cable. If the maximum moisture level can be accurately and reliably controlled, a higher data rate can be achieved. A solution to this problem is that when nitrogen or dry compressed air is used to dry cables that are covered in paper or cotton, the moisture content is monitored to maintain the required dryness at a level that is consistent with tolerable losses. Moisture levels are either periodically monitored or the standard hygrometer alarm wiring is used to activate the nitrogen or dry air supply if moisture exceeds a predetermined value.

Aluminum oxide sensors have often been used with compatible moisture sensors mounted inside the pressurized cable housing. Advantages are that if an existing cable can be made to handle more data, then the cost of installing additional cables can be eliminated. In cases in which it is impossible to install additional cables, the use of such a moisture system is the only way that existing cables can be made to handle the increased demands of modern telecommunications technology. The dew/frost point range to be covered is 0°C to −50°C (32°F to −58°F) at an operating temperature of 25°C to 30°C (77°F to 86°F), and pressure of 2.6 to 3 mPa (377 to 435 psia).

Buildings and Construction.   In the construction industry, the measurement of moisture in concrete is very important. For example, the presence of excess moisture prior to applying a water proof coating over the concrete, results in the following problems:

- The thermal effects of sealed-in moisture cause mechanical stresses that can result in fractures and subsequent loss of strength.

- Because of the various salts and other chemicals, moisture has a corrosive effect on reinforcing steel used in most concrete structures. This can again lead to a further loss of strength.

Moisture is often quantified as a percentage of water content by weight. This is usually not meaningful since the exact composition of the material is seldom known and a percentage moisture in one concrete mixture may differ significantly from another which has the same amount of water in it. Percent moisture is therefore not a reliable guide as to whether the structure is too wet or dry. For this reason, the technique of measuring relative humidity in a pocket of air that is in equilibrium with the material being tested is widely used. The measurement thus comes down to an equilibrium relative humidity measurement. The normal procedure is to drill a small hole in the concrete slab and place an RH probe inside. The measurement is made after sufficient time has lapsed so that the water vapor in the small hole is in equilibrium with the concrete. In many cases, typical polymer RH sensors have been used for this application. For precise measurements, such as for research purposes, a small chilled-mirror-type probe is also available, although at much higher cost.

Energy Management.   Several industrial and commercial processes require large amounts of energy. Energy requirements are in almost all cases very dependent on humidity and temperature control, and large energy savings can be obtained by optimizing such controls, which at today's state of technology can easily be done by programmed computers. Examples are large industrial dryers, air conditioning systems in large buildings, and climate control in museums and libraries. These applications normally require a large number of sensors mounted at various well-selected locations. These provide data to computers, which are then programmed to operate humidity and temperature controls in the building or dryer. There are many examples for which such energy management systems have yielded substantial energy savings, allowing such energy management systems, including humidity and temperature sensors, to be amortized within two years. Transmitters that use polymer % RH sensors are ideal for this application. Operational requirements are usually 20% to 70% RH at ambient temperatures of 0°C to 60°C (32°F to 140°F).

REFERENCE

1. P. R. Wiederhold, *Water Vapor Measurement, Methods and Instrumentation* (Marcel Dekker, New York, 1997).

SECTION 5
GEOMETRIC AND MOTION SENSORS*

J. N. Beach
*MICRO SWITCH, a Division of Honeywell Inc., Freeport,
Illinois (Electromechanical Switches), (Proximity Sensors)*

R. E. Gebelein
*Moore Products Company, Spring House, Pennsylvania
(Automatic Gaging—prior edition)*

S. Longren
Longren Parks, Chanhessen, Minnesota

Leslie Mantua
*MICRO SWITCH, a Division of Honeywell Inc., Freeport,
Illinois (Electromechanical Switches), (Proximity Sensors)*

Gregory K. McMillan
*Solutia Inc., Saint Louis. Missouri (Flat Web On-Line
Measurement and Control)*

T. A. Morrow
*Omron Electronics, Inc., Schaumberg, Illinois (Proximity
Sensors), (Geometric and Motion Sensors)*

Dave Nyce
*MTS Systems Corporation (Magnetorestrictive Linear
Position Sensors)*

Kathy Quillin
*MICRO SWITCH, a Division of Honeywell Inc., Freeport,
Illinois (Electromechanical Switches), (Proximity Sensors)*

Mark Rossi
*MTS Systems Corporation, Sensors Division,
Cary, North Carolina*

Robert D. Sill
*Endevco Corporation, San Juan Capristrano, California
(Vibration Measurement)*

Staff
*Compumotor Division, Parker Hannifin Corporation, Rohnert
Park, California (Encoders and Resolvers)*

---

* *Persons who authored complete articles or subsections of articles, or otherwise cooperated in an outstanding manner in furnishing information and helpful counsel to the editorial staff.*

Staff
Daedal Division, Parker Hannifin Corporation, Harrison City, Pennsylvania (Metrology)

Staff
Daytronic Corporation, Miamisburg, Ohio (Geometric Transducers, Accelerometers)

Staff
Lucas Ledex Inc., Vandalia, Ohio (Metrology)

Staff
Lucas Schaevitz, Pennsauken, New Jersey (Linear and Angular Displacement Transducers)

Staff
Lucas Sensing Systems Inc., Phoenix, Arizona

Staff
MTS Systems Corporation, Minneapolis, Minnesota

Staff
Parker Digiplan, Ltd., Poole Dorset, United Kingdom

Staff
TSI Incorporated, St. Paul, Minnesota (Constant-Temperature Thermal Anemometer)

L. Thompson
MTS Systems Corporation, Eden Prairie, Minnesota (Magnetostrictive Linear Displacement Transducers)

METROLOGY, POSITION, DISPLACEMENT, THICKNESS AND SURFACE TEXTURE MEASUREMENT[1]

Traditionally, the discrete-piece manufacturing and assembly industries, as represented by the appliance, land vehicle, aircraft, electronic components and instruments, business machine, fixture, jewelry, implement, and tool production, for example, have depended largely on the measurement and control of what may be termed the geometric variables for controlling product quality and production efficiency. Geometrically related variables of concern include such physical factors as dimension, straightness, shape, contour, displacement, position, and linear and rotary speed. Most of these variables, in essence, are related and can be derived from position determination.

When these variables are integrated into a total system, in contemporary parlance, they are commonly referred to as motion control systems.

BASIC METROLOGY

From a practical standpoint, industrial metrology falls into several main categories: (1) the basic instruments, tools, and standards normally found in the well-equipped metrology laboratory responsible for establishing and maintaining manufacturing quality standards; (2) instrumental methods for assisting inspectors in manually gaging parts, workpieces, jigs, fixtures, and final assemblies; (3) automated systems for gaging parts and pieces in high-volume production; (4) dimensionally oriented motion control systems for machine tools, assembly lines, welding, painting, fabricating, and testing, among scores of geometrically sensitive operations; and (5) thickness measurement and control of metal, plastic, wood, paper, glass, films, and plated and sprayed coatings. Closely allied position-sensitive operations include registration control, as found in printing and labeling, and object-detection systems, in which control actions are taken because an object occupies (even for an instant) a definite (or approximate) position (location).

DIMENSIONAL STANDARDS

Length is expressed in terms of the meter, abbreviated m. The meter is defined as 1,650,763.73 wavelengths in vacuum of the orange-red line of the spectrum of krypton 86. An interferometer is used to measure length by means of light waves.

A revised definition of the meter, based on very accurate measurement of the speed of light, was adopted by the General Conference on Weights and Measures in 1983. The meter has been defined as the length of path traveled by light in vacuum during a time interval of 1/299,792,458 of a second.

Practical realization of this definition is obtainable through time-of-flight measurements, by frequency comparisons with laser radiations of known wavelengths, or through interferometric comparisons with radiations of stated wavelengths, such as that from krypton 86.

Other SI length-related units include:

[1] The cooperation of the technical and engineering staffs of the following organizations in furnishing information for this article is gratefully acknowledged: *Daytronic Corporation,* Miamisburg, Ohio; *Lucas Ledex Inc.*, Vandalia, Ohio; *Lucas Sensing Systems Inc.,* Phoenix, Arizona; *Lucas Schaevitz,* Pennsauken, New Jersey; *MTS Systems Corporation,* Minneapolis, Minnesota; *Parker Digiplan, Ltd.*, Poole Dorset, BH17 7DX (U.K.); *Parker Hannifin, Compumotor Division,* Rohnert Park, California. Special appreciation is also extended to Steve Longren, Chanhassen, Minnesota.

TABLE 1 Common Equivalents and Conversions

Approximate common equivalents		Conversions accurate to parts per million	
1 inch	= 25 millimeters	inches × 25.4*	= millimeters
1 foot	= 0.3 meter	feet × 0.3048*	= meters
1 yard	= 0.9 meter	yards × 0.9144*	= meters
1 mile	= 1.6 kilometers	miles × 1.60934	= kilometers
1 square inch	= 6.5 square centimeters	square inches × 6.4516*	= square centimeters
1 square foot	= 0.09 square meter	square feet × 0.0929030	= square meters
1 square yard	= 0.8 square meter	square yards × 0.836127	= square meters
1 acre	= 0.4 hectare	acres × 0.404686	= hectares
1 cubic inch	= 16 cubic centimeters	cubic inches × 16.3871	= cubic centimeters
1 cubic foot	= 0.03 cubic meter	cubic feet × 0.0283168	= cubic meters
1 cubic yard	= 0.8 cubic meter	cubic yards × 0.764555	= cubic meters
1 quart (liq)	= 1 liter†	quarts (liq) × 0.946353	= liters
1 gallon	= 0.004 cubic meter	gallons × 0.00378541	= cubic meters
1 ounce (avdp)	= 28 grams	ounces (avdp) × 28.3495	= grams
1 pound (avdp)	= 0.45 kilogram	pounds (avdp) × 0.453592	= kilograms
1 horsepower	= 0.75 kilowatt	horsepower × 0.745700	= kilowatts
1 millimeter	= 0.04 inch	millimeters × 0.0393701	= inches
1 meter	= 3.3 feet	meters × 3.28084	= feet
1 meter	= 1.1 yards	meters × 1.09361	= yards
1 kilometer	= 0.6 mile	kilometers × 0.621371	= miles
1 square centimeter	= 0.16 square inch	square centimeters × 0.155000	= square inches
1 square meter	= 11 square feet	square meters × 10.7639	= square feet
1 square meter	= 1.2 square yards	square meters × 1.19599	= square yards
1 hectare†	= 2.5 acres	hectares × 2.47104†	= acres
1 cubic centimeter	= 0.06 cubic inch	cubic centimeters × 0.0610237	= cubic inches
1 cubic meter	= 35 cubic feet	cubic meters × 35.3147	= cubic feet
1 cubic meter	= 1.3 cubic yards	cubic meters × 1.30795	= cubic yards
1 liter	= 1.057 quarts (liq)	liters × 1.05669	= quarts (liq)
1 cubic meter	= 250 gallons	cubic meters × 264.172	= gallons
1 gram	= 0.035 ounces (avdp)	grams × 0.0352740	= ounces (avdp)
1 kilogram	= 2.2 pounds (avdp)	kilograms × 2.20462	= pounds (avdp)
1 kilowatt	= 1.3 horsepower	kilowatts × 1.34102	= horsepower

* Exact.
† Based on the U.S. survey foot (= 0.3048006 meter).

1. *Area*, expressed in terms of the square meter (m^2). Land is often measured by the hectare (10,000 m^2, or approximately 2.5 acres).

2. *Volume*, expressed in terms of the cubic meter (m^3). The liter is a special name for the cubic decimeter (0.001 m^3).

Common equivalents and conversions are given in Table 1. Prefixes that may be applied to all SI units are listed in Table 2.

Interferometer

An interferometer is a precision instrument that uses the interference of light waves as the basis of measurement. The optics of an interferometer are designed so that the variance of known wavelengths and path lengths within the instrument permits accurate measurement of distances. The French physicist Jacques Babinet suggested in 1827 the possibility of using the wavelength of light as a standard of length. However, it was not until 1960 that the meter was officially defined by the International

TABLE 2 Prefixes That May Be Applied to All SI Units

Symbol	Prefix	Pronunciation	Multiples and Submultiples
E	exa	ĕx'á	10^{18}
P	peta	pĕt'á	10^{15}
T	tera	tĕr'á	10^{12}
G	giga	ji'gá	10^{9}
M	mega	mĕg'á	10^{6}
k	kilo	kĭl'ŏ	10^{3}
h	hecto	hĕk'tŏ	10^{2}
da	deka	dĕk'á	10
d	deci	dĕs'ĭ	10^{-1}
c	centi	sĕn'tĭ	10^{-2}
m	milli	mĭl'ĭ	10^{-3}
μ	micro	mĭ'krŏ	10^{-6}
n	nano	năn'ŏ	10^{-9}
p	pico	pĕ'kŏ	10^{-12}
f	femto	fĕm'tŏ	10^{-15}
a	atto	ăt'tŏ	10^{-18}

Bureau of Weights and Measures in terms of interferometry, as described previously, thus replacing the former prime standard of length, namely, the length of an artifact, that is, a platinum-iridium bar, exact replicas of which were maintained at various international standards institutes. Now, with adequate skills, the standard of length can be duplicated anywhere in the world with equal precision.

The interferometer derives its name from the fact that it makes it possible to see light wave interference patterns as a series of bright and dark lines. If two light beams from a given source are directed optically over separate paths that may differ by as little as $^{1}/_{10}$ of a wavelength and then are recombined, detectable destructive interference will occur. Thus the interferometer may serve two roles: (1) to measure distances in terms of known wavelengths and (2) to make precise measurements of wavelengths. Currently laser interferometers are used industrially in the construction and assembly of precision machine tools where the laser can provide accuracies of better than 1.27×10^{-6} mm (20 millionths of an inch) in lengths up to 5080 mm (200 inches) (Fig. 1).

Optical Gratings

Diffraction gratings are used in optical instruments associated with precise dimensional measurements. Close, equidistant, and parallel grooves are ruled on a polished surface, commonly a glass base coated with aluminum. Gratings can be (1) reflection or (2) transmission types. The number of grooves range from several hundred to many thousand per inch (25 mm), depending on the dispersion required. The relative movement of two gratings produces optical patterns or fringes, sometimes referred to as moiré patterns (Fig. 2).

Gage Blocks

Usually furnished in sets of 81 blocks, ranging from 0.05 to 4.0 inches (1.27 to 101.6 mm), gage blocks serve as secondary standards traceable to the prime length standard. The blocks are rectangular pieces of hardened steel or carbide that are ground and polished flat with a square or oblong cross section. The length of a block is the perpendicular distance between the two opposite, polished faces. By carefully sliding the gaging surfaces of two blocks one over the other, they can be "wrung"

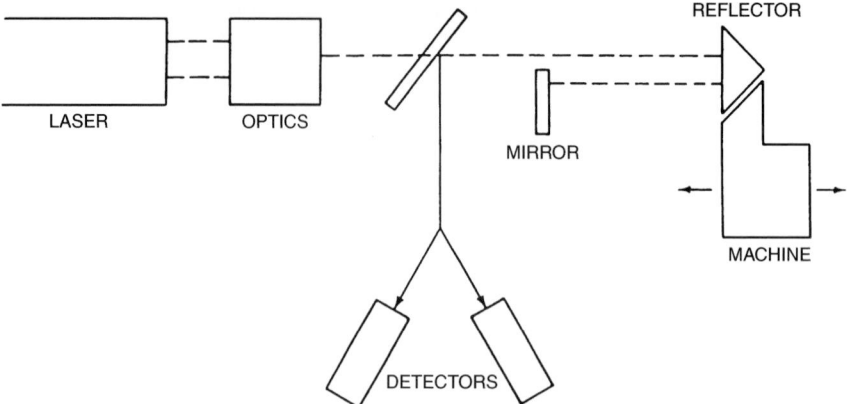

FIGURE 1   Laser interferometer capable of accuracies of better than 1.27×10^{-6} mm (20 millionths of an inch). Early designs required that the laser beam be reflected off a moving mirror that traveled along a track at least 1 meter long. This distance has been reduced to a few centimeters.

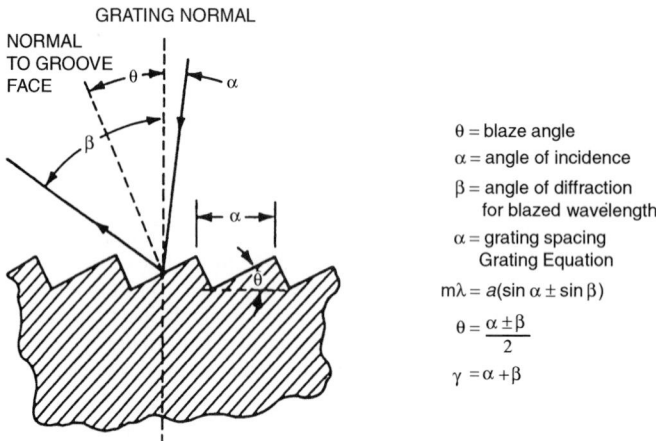

θ = blaze angle
α = angle of incidence
β = angle of diffraction
      for blazed wavelength
α = grating spacing
      Grating Equation
$$m\lambda = a(\sin \alpha \pm \sin \beta)$$
$$\theta = \frac{\alpha \pm \beta}{2}$$
$$\gamma = \alpha + \beta$$

FIGURE 2   Cross section of diffraction grating showing the "angles" of a single groove, which are microscopic in size on an actual grating.

together and thus built up to provide a useful range of standard lengths. The dimensions of individual blocks are established by interferometric methods with apparatus set up along the lines shown in Fig. 3. Class AA blocks have a tolerance of $\pm 2 \times 10^{-6}$ inch ($\pm 5.1 \times 10^{-5}$ mm) and are intended for reference purposes in temperature-controlled gage laboratories. These masters should be sent periodically to the National Institute of Standards and Technology (NIST), Gaithersburg, Maryland, for checking. Class A blocks are used in inspection departments and as masters. Blocks of class B and C quality are used throughout manufacturing for accurate measurements, tool setting, and instrument calibration.

Gage blocks are termed end standards. In addition there is need for line standards in the form of precision scales. Before the development of laser interferometric techniques, line standards were compared with master scales visually by use of a microscope. NIST now uses an automatic fringe-counting interferometer, which employs a laser light source to measure line standards directly in lengths up to 1 m with a precision of a few parts in 100 million.

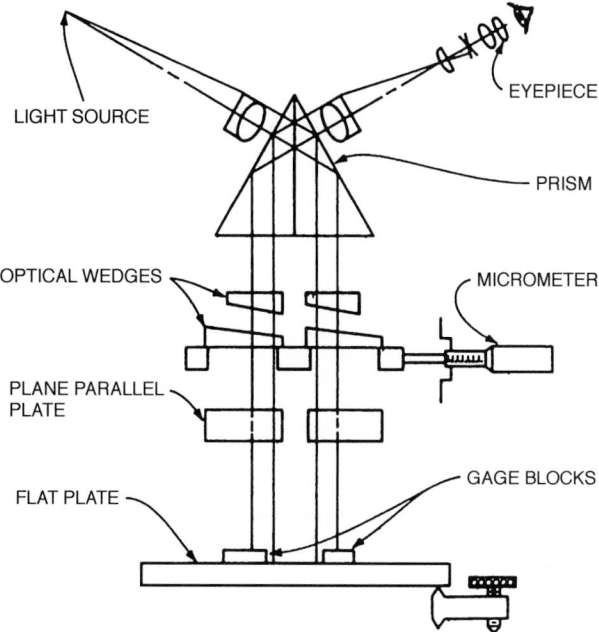

FIGURE 3   Interferometric comparator for comparing the lengths of two gage blocks. The difference in gage block lengths is measured by the horizontal distance through which optical wedges must be moved to bring the two interference patterns successively into coincidence with a cross hair in the eyepiece. Wedges and micrometer are rotated 90° for illustration.

Autocollimator

An autocollimator is an optical instrument for directly measuring small angles of tilt of a reflective surface. The instrument effectively combines the functions of a collimator and a viewing telescope into a single system.

Goniometer

A goniometer is an angular measuring device essentially similar to a circular table or a dividing head, except that autocollimating telescopes are used to establish the datum values for the readings taken from the divided circle, either directly or through a micrometer eyepiece. The instrument is used for determining circular divisions and utilizes the basic principle of the divided circle.

Clinometer

Essentially a divided-circle instrument, the clinometer simplifies the transfer of angles between planes. Bubbles are used to establish the null setting principle of a precision clinometer, while less precise instruments compare a measured angle to a datum surface (Fig. 4).

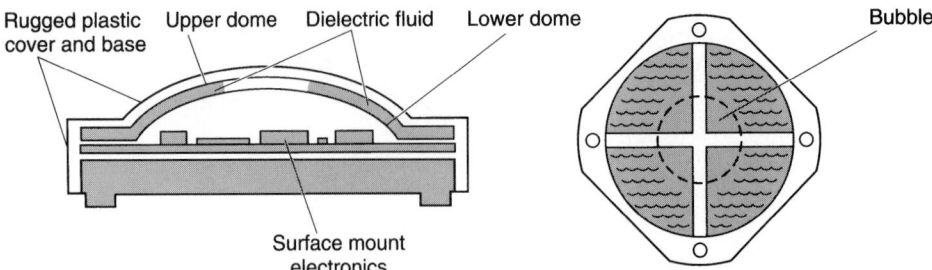

FIGURE 4   Dual-axis clinometer, which provides the function of two separate instruments. The sensor is composed of two hermetically sealed domes spaced about 3 mm ($\frac{1}{8}$ inch) apart. The lower, polyester plastic dome has four capacitance plates while an aluminum upper dome acts as a ground. A fluid with a high dielectric constant is sealed within the dome sandwich, leaving an air bubble about the size of a quarter. The bubble is centered at level position and will move from one side to the other as the device is tilted. When the sensor is tilted, the bubble (moving under the force of gravity) changes the capacitance. The resulting differential generates an output signal which reflects the relative tilt of the device in either the x or the y axis. The pin-selectable output signal provides either PWM or a dc analog voltage output. A fully electronic instrument is also available.

Clinometers are used in a number of industrial and commercial applications, including (1) level positioning of machine tools and other heavy machinery, manlifts, and cranes, and satellite and microwave antennas; (2) making camber, caster, and steering axis inclination adjustments in vehicle wheel alignment; (3) measuring pitch, roll, tilt, and angular position in construction equipment; (4) adjusting inclination of exercise equipment; (5) measuring heel angle or roll of sailboats and pitch and roll of undersea robot vehicles; and (6) leveling motor homes and RVs; automatic level control for private and commercial vehicles, among others. (*Lucas Sensing Systems Inc.*)

Theodolite

A theodolite is a surveying instrument that has been adopted for use in metrology. The instrument is comprised of a telescope in which rotation about the vertical axis is measurable. Theodolites are used for the alignment of large jigs and fixtures. Through triangulation, the instruments also can be used for length measurement.

Protractor

The familiar drafting protractor for measuring angles is now available in a digital electronic format (Fig. 5).

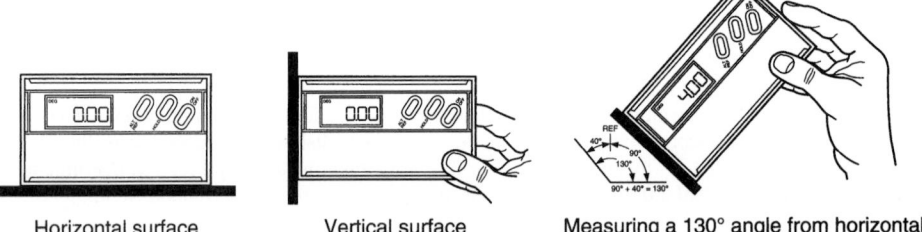

| Horizontal surface | Vertical surface | Measuring a 130° angle from horizontal |

FIGURE 5   Digital protractor (battery operated, portable) measures angles in degrees, mils, percent grade, millimeters per meter, and inches per foot. The range is ±60 percent with a resolution of ±0.01° (0 to 20°); ±0.1° (20 to 60°); or ±0.05° (3 min) (0 to 19.99°), and a repeatability of ±0.05°. The instrument incorporates digital statistical process control (SPC) output for data storage and processing. Applications include level positioning of machinery tables, positioning of aircraft runway lights, aircraft flight control surfaces and prop angle, automobile drive angle, and for the measurement of innumerable manufactured parts during various stages of production as part of SPC requirements. (*Lucas Sensing Systems Inc.*)

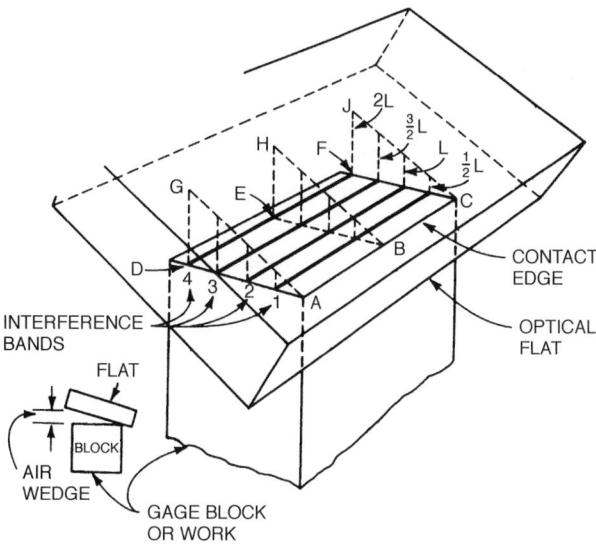

FIGURE 6   When an optical flat is manipulated so that a wedge of air is created, dark bands (interference bands) result at locations where the separation of flat and work equals a multiple of $L/2$, or one half-wavelength of the monochromatic light used.

Optical Flats

An optical flat is a transparent disk, usually of quartz, two sides of which are parallel. One side is polished for clear vision of the surface of the gage, part, or tool on which the flat is placed. The other side of the disk is ground optically flat. Under proper conditions, interference bands are created. Observation of these bands can be used for (1) determining the flatness of a surface, that is, to assess the amount of concavity or convexity, and (2) measuring the linear difference between a reference gage and an inspection gage, or between a gage block and a highly accurate part. Any linear difference can be detected between approximately 0.051 mm (0.002 inch) as a maximum and 2.5 to 1.3×10^{-5} mm (1 to 2 millionths of an inch) as a minimum (Fig. 6).

Spherometer

The spherometer is used for determining the radius of curvature of lenses and other spherical surfaces. A widely accepted design consists of a depth-measuring device, such as a micrometer screw, mounted at the center of a tripodlike support. The concave or convex surface to be measured is centered directly under the spherometer, and the micrometer screw is rotated until its tip just touches the spherical surface. The displacement of the tip of the micrometer screw above or below the plane of the three support points is a measure of the radius of curvature of the surface. A typical spherometer reads to 0.01 mm, although much greater precision can be obtained with highly refined instruments.

Sine Bar

A sine bar consists essentially of a bar serving as a straightedge and two cylindrical buttons, which may be on the side or on the undersurface. If the side-button type is used, one button rests on a gage block, and the thickness of that gage block is added to the height of the gage block stack used to set

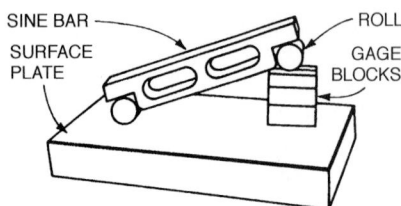

FIGURE 7 The sine bar allows work to be located in an accurate relationship to a plane or surface plate.

the second button. If a base-button sine bar is used, the first button rests directly on the surface plate or master flat. The second button rests on a stack of blocks equivalent to the sine of the wanted angle (Fig. 7).

Ellipsometer

The ellipsometer is used to determine the thickness of monomolecular dimensions. The instrument is basically a polarization interferometer utilizing a photometer as a readout device. A production-type ellipsometer can measure surface film thickness from a few angstroms to 1 μm and permits rapid determinations of thickness and refractive index. In one type, a laser source operates in a fully lighted room, providing a small intense spot that permits the study of very small samples. A solid-state detector is used. Automatic ellipsometers are available. In operation, a specimen is placed on the stage and the start button is depressed. A minicomputer displays and prints the film thickness in less than 4 s for most samples. High speed and simplicity make such an instrument ideal for monitoring rapidly changing films and for high-volume measurements in semiconductor manufacture.

Optical Comparator

Also known as a contour projector, the optical comparator is used for accurate visual inspection and measurement of a part in which a magnified image of the part or portion thereof is projected onto a viewing screen. Among their uses are the toolroom checking of cutter shapes and sizes and some delicate assembly operations. A contour projector can be conceived of as a variety of gages combined in a single unit.

Optical Bench

Although not compactly assembled in a black box, an optical bench (with accessories) may be considered a modular, highly flexible, disassembled instrument capable of convenient, customized assembly by a skilled user. An optical bench is a graduated support on which carriages for holding lenses, mirrors, and other components can be mounted and positioned for making precision optical measurements.

Measuring Machine

This is a highly instrumented device used to measure and record linear dimensions and xyz coordinates of holes and surfaces in parts and tools. Various styles operate under manual, motorized, or computer control. Direct-measuring machines with digital readout have been used widely. Some machines provide direct or absolute measurement of probe position or movement in three axes with a resolution ranging from 0.02540 to 0.00127 mm (0.001 to 0.00005 inch) (Fig. 8).

Miniature Positioning Tables (Stages)

Discrete-part manufacture and assembly is not confined to comparatively large items as encountered, for example, in automotive, aircraft, appliance, and machinery production, but the instrumentation technology also applies to very small parts and assemblies as may be encountered in the construction of instruments, controllers, electronic components, circuit systems, jewelry, and watchmaking, among

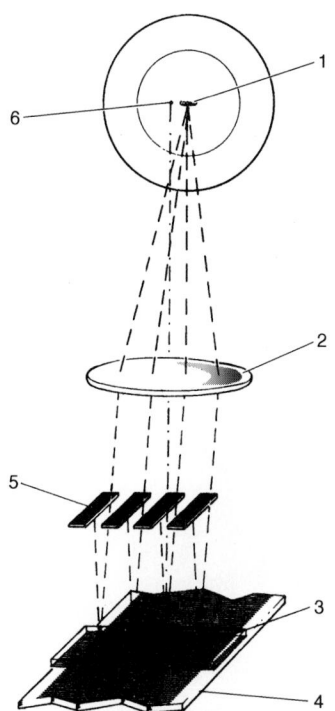

FIGURE 8 Optical system of one type of measuring machine. (1) Line filament; (2) collimating lens; (3) index grating; (4) scale grating; (5) photocell strips; (6) principal focus of lens.

numerous other products. Although, in some cases, these miniaturized production and testing setups are highly automated, in cases of low volume or in product development, miniature positioning tables (or stages) are used widely. As the size of the components and assemblies increases (in recent years by orders of magnitude), the need for precision and performance repeatability obviously increases.

Specific applications of miniature and small positioning tables include (1) frequent or one-time fine adjustments, (2) pinhole micrometer positioning (piggyback on a larger workstage), (3) fiber-optics research and development and alignment, (4) quality control testing and gaging, (5) laser scribing and cutting, (6) automated assembly and component insertion, (7) positioning probes and fine gas purges, (8) axial alignment of tubes and rods, and (9) individual positioning of elements of small gas lasers, among others.

Small and miniature positioning tables (stages) are available in a wide variety of configurations, including 1–2–3 (x–y–z) axes, manual and motorized drives, linear and rotary positioning, and ball and dovetail slides, and they can be equipped with a variety of instrumentation, including computer-controlled programmable systems (Figs. 9–11).

Positioning Table Geometry

Several key geometric factors that apply to large and small tables and stages affect their performance characteristics, such as the following.

Resolution.   Resolution is the smallest attainable increment of adjustment or positioning. With manually adjusted positioners, resolution is defined as the smallest movement achievable by controlled rotation of the adjustment screw or micrometer.

Repeatability.   This term defines how accurately a given position can be repeated or returned to an original location.

Positional Accuracy.   This is defined as the maximum achievable difference (error) between expected and actual travel distance. Pitch and yaw affect positional accuracy, as depicted in Fig. 12.

Straight-Line Accuracy.   Also called straightness and flatness of travel, it is, in theory, a linear slide or stage that moves along its axis of travel in a perfectly straight line. In reality the actual travel path deviates from the true straight line and flat line in both the horizontal and the vertical directions, respectively. Straight, flat line accuracy is defined as the maximum distance that the travel path deviates from the theoretical straight line in either plane, measured from the moving carriage surface center. Deviations result from the effects of yaw, pitch, and roll (Fig. 13).

Concentricity.   In theory, it requires that any point on the surface of a rotating table should travel along a path that forms a perfect circle. In reality, the actual path of travel will deviate from the perfect circle. Concentricity defines the maximum difference between a true circle and the actual circular path formed by the rotating point (Fig. 14).

Runout (Wobble).   In theory, it requires that any point on a rotary table should remain within a perfectly flat plane that is perpendicular to the axis of rotation. Runout describes the maximum distance that point will deviate from that plane (Fig. 15).

FIGURE 9(a) Linear stages for controlled, precise point-to-point positioning along a linear axis. Stages are comprised of (1) a precision linear ball slide which serves as a linear bearing and guide and (2) a drive mechanism which moves and positions the slide top along the linear axis. Three drive mechanisms are available, a fine screw, a micrometer, and a differential screw. The fine screw is used for fine-resolution positioning. The micrometer is used whenever a position readout is required. The differential screw is used for applications that require extremely fine-resolution positioning. Maximum mounting surface ranges from 32 by 32 to 152 by 152 mm; maximum load capacity is 3 to 45 kg (horizontal) and 0.5 to 27.5 kg (vertical); travel is 12.5 to 50 μm; straight-line accuracy is 2 μm per 25 mm of travel; and positional repeatability is 1.3 μm. (*Daedal Division, Parker Hannifin Corporation.*)

Other factors that must be considered when specifying a positioning table include (1) horizontal load capacity, the maximum load when the device is mounted or placed on a horizontal plane; (2) vertical load capacity, the maximum (pitch) load capacity (center of gravity, a specified design distance) of a positioning device when the device is mounted in the vertical plane (vertical load capacity is a measure of the load capacity effects of table bearing capacity and screw pitch); (3) overall load capacity, the maximum weight that a positioning device can support without causing excessive wear or damage to the device.

Machine Conditions and Requirements

The degree of positioning accuracy and repeatability required by various production machines differs widely: 0 ± 0.0025 mm (± 0.0001 inch) for boring machines, ± 0.025 mm (± 0.001 inch) for drilling and contour milling, ± 0.127 mm (± 0.005 inch) for armature insulator assembly, ± 0.254 mm (± 0.01 inch) for tube bending and frame welding, ± 0.381 (± 0.015 inch) for automobile seat cushion spring welding, and ± 0.1 percent of full-scale reading for asphalt batching machines. In the manufacture

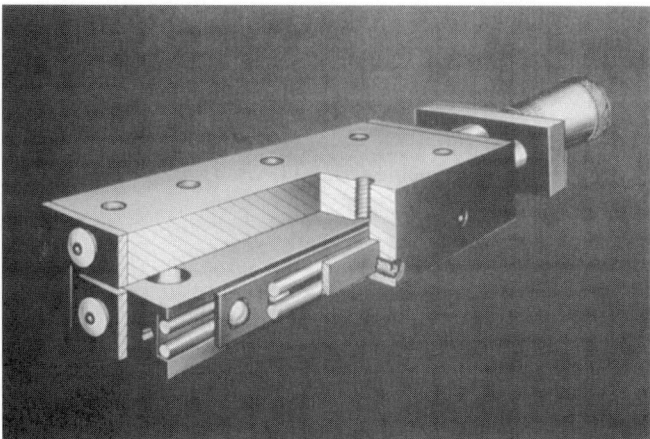

FIGURE 9b   Linear stages for controlled, precise point-to-point positioning along a linear axis. (*Daedal Division, Parker Hannifin Corporation.*)

FIGURE 10   Rotary stages and tables of varying sizes. They are used for controlled rotation and angular positioning. Each table comprises (1) a fixed housing (base), (2) a rotating member (shaft), (3) a bearing system, and (4) a control or drive mechanism. The bearing system rigidly supports the shaft and allows it to rotate freely within the housing. The drive mechanism controls shaft rotation, thereby converting the unit from a free rotating bearing assembly to a controllable rotary positioning device. In some cases the drive mechanism is a tangent arm drive. In other configurations a precision worm gear drive is used. This provides continuous angular positioning over a full 360° range. The stages range from 48 to 120 mm in diameter; load ranges from 4.5 to 22 kg (horizontal) and 1.5 to 9 kg (vertical); range is 360° continuous; vernier is 6 arc mintues. (*Daedal Division, Parker Hannifin Corporation.*)

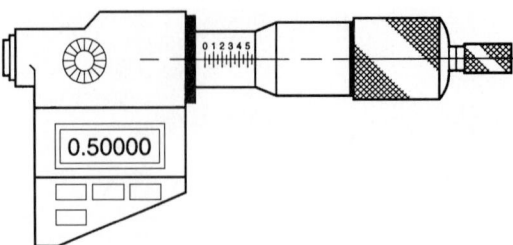

FIGURE 11   Precision electronic digital micrometer head provides a liquid crystal display (LCD) readout to 0.001-mm resolution. The micrometer features incremental or absolute positioning modes, zero set at any position, millimeter and inch readout, display hold, and automatic shutdown after 2 hours to conserve the integral battery. The micrometer travel is 25 mm. The battery will power the unit for 500 hours of use. (*Daedal Division, Parker Hannifin Corporation.*)

of solid-state circuits and components in the electronics industry, involving laser and electron beam manipulation, accuracy is in terms of a fraction of a micrometer. The complexity of the control system varies with the type of machine, particularly with the number of axes that must be controlled—ranging from two up to six or more axes. The speed of response needed is related to the total cycling time of the machine and ranges from seconds and fractions of seconds up to 2 minutes or greater.

Backlash.   To position a machine member with acceptable accuracy, it is necessary to establish the extent of the backlash or dead band region for the positioning mechanism used. The measuring transducer and its attendant dead band characteristics, when used with the machine member, determine the amount of dead band or backlash to be included in the total control system loop. For all types of repeat-back devices whose mechanical input is provided by rotating a shaft, an important consideration is the means of coupling the device to the positioned machine element. Frequently, a gear train is required to reduce member travel to one revolution or less of the transducer shaft. It is necessary to determine whether any backlash in this train is comparable, when expressed as an arc of the transducer shaft, with the positional accuracy requirement of the machine member. Satisfactory results on the basis of this comparison rest on the assumption that the electrical and mechanical error factors for the transducer are small ($^1/_{10}$ or less) compared with the machine member positioning tolerances.

Sometimes a transducer that does not require a mechanical input shaft should be considered. Where this is impractical, a separate rack and gear train, both exhibiting minimal backlash, may be used to position the mechanical input shaft of the associated transducer.

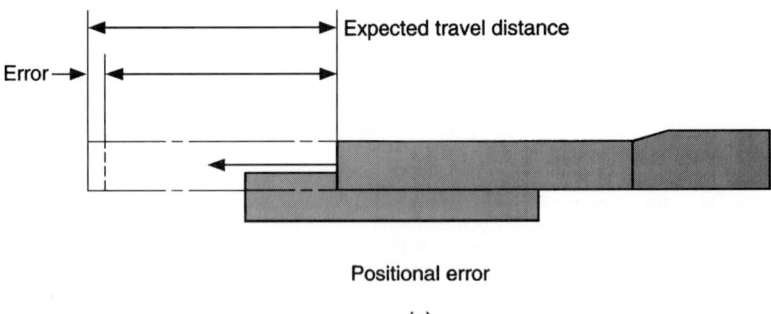

FIGURE 12   Positional error as affected by pitch error and yaw error. (*Daedal Division, Parker Hannifin Corporation.*) (*Continues*)

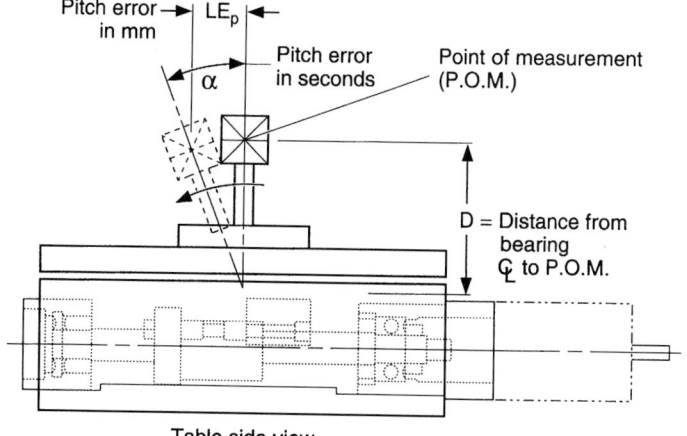

Table side view

Formula: $L_{EP} = [Tan (\alpha \div 3600)] \times D$
Example: $\alpha = 10$ sec; $D = 60$ mm
$L_{EP} = [Tan (10 \div 3600)] \times 60$ mm $= 2,9$ μm

Pitch error

(b)

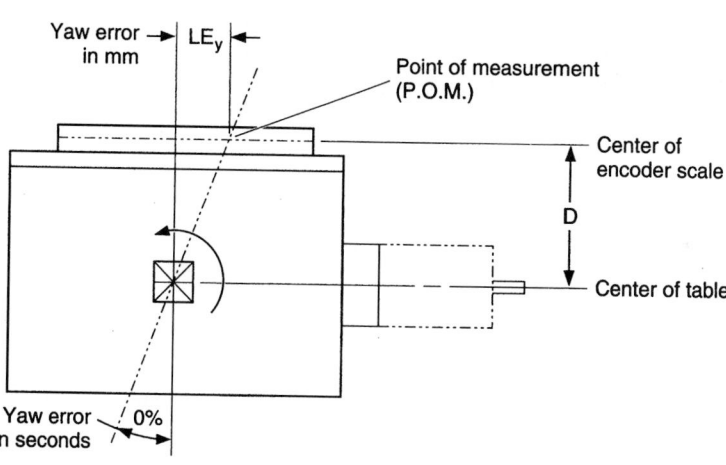

Table top view

Formula: $LE_Y = [Tan (\alpha \div 3600)] \times D$
Example: $\alpha = 10$ sec; $D = 85$ mm
$LE_Y = [Tan (10 \div 3600)] \times 85$ mm $= 4,1$ μm

Yaw error

(c)

FIGURE 12   (*Continued*).

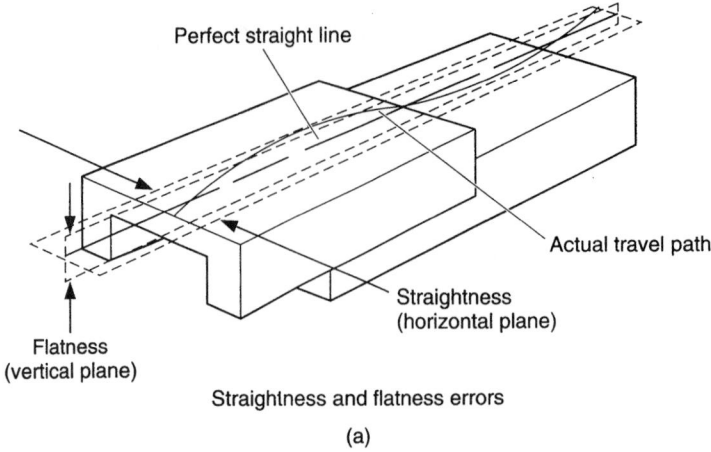

Straightness and flatness errors

(a)

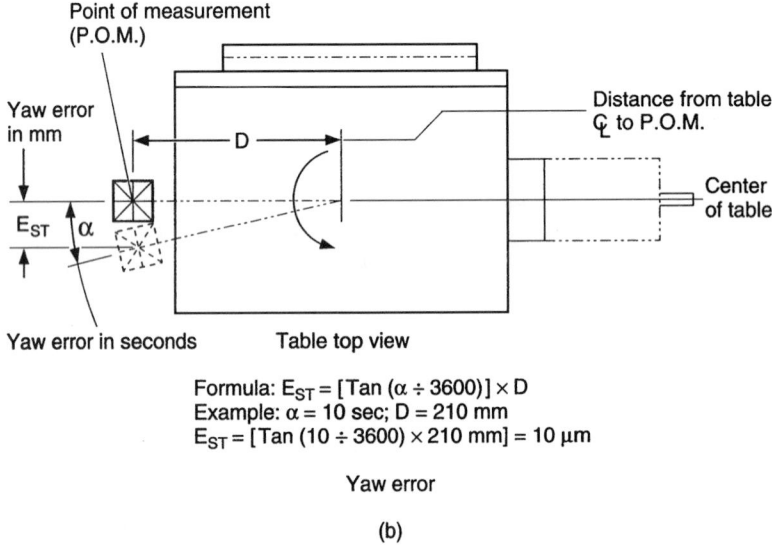

Formula: $E_{ST} = [\text{Tan}\,(\alpha \div 3600)] \times D$
Example: $\alpha = 10$ sec; $D = 210$ mm
$E_{ST} = [\text{Tan}\,(10 \div 3600) \times 210\ \text{mm}] = 10\ \mu\text{m}$

Yaw error

(b)

FIGURE 13   Straightness of travel error as affected by yaw error and roll error. (*Daedal Division, Parker Hannifin Corporation.*) (*Continues*)

Stable Machine Base.   The controlled member may have undesired movement with respect to the machine base in a direction transverse to the controlled axis of travel. If the slide and table ways wear nonuniformly, variation in the transverse position of a point on the table of a machine may cause a variation in the air gap of a magnetic slot transducer system, for example. The same problem will result in a misalignment of optical transducer systems if the table motion becomes crablike after wear of the slides has progressed.

Vibration.   Nondata components of both a cyclic and a random nature may be superimposed on the true data because of machine-induced vibration of the transducer. Thus every effort should be made

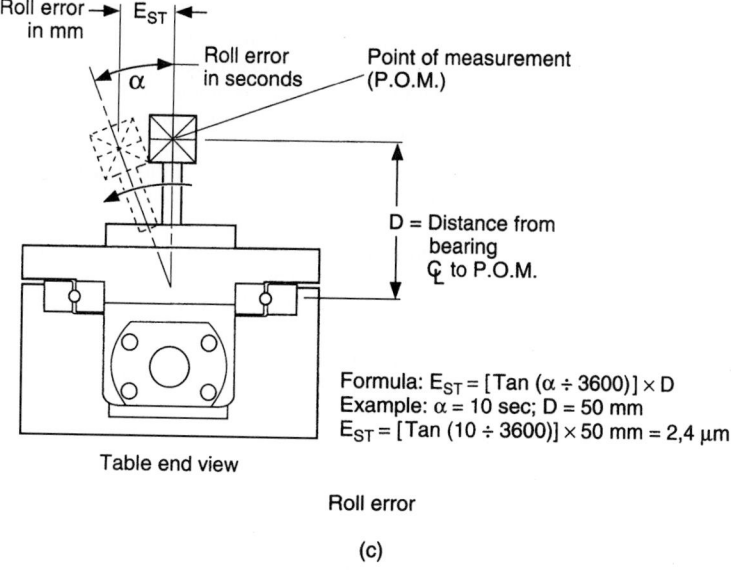

Formula: $E_{ST} = [\text{Tan} (\alpha \div 3600)] \times D$
Example: $\alpha = 10$ sec; $D = 50$ mm
$E_{ST} = [\text{Tan} (10 \div 3600)] \times 50$ mm $= 2,4$ μm

Table end view

Roll error

(c)

FIGURE 13   (*Continued*).

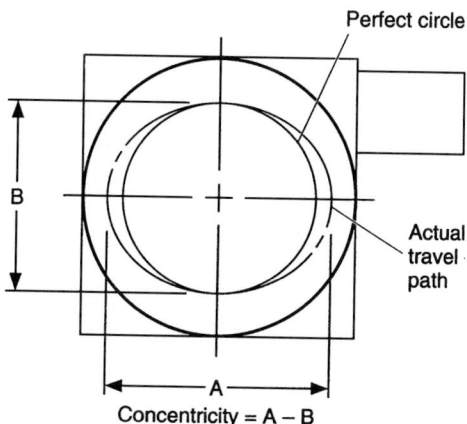

FIGURE 14   Demonstration of concentricity and deviation. (*Daedal Division, Parker Hannifin Corporation.*)

to reduce these effects and to take the residue effects fully into account when designing the total positioning system.

ROTARY MOTION

In general terms, an encoder may be defined as a device which translates mechanical motion into electronic signals used for monitoring position or velocity. Encoders are available for use in both rotary and linear motion systems.

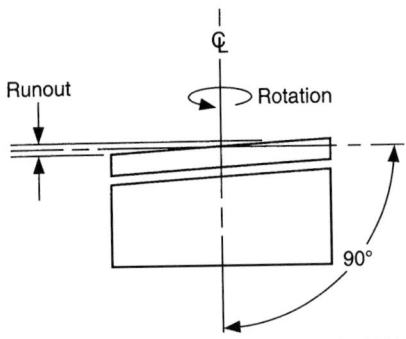

FIGURE 15 Demonstration of runout (wobble). (*Daedal Division, Parker Hannifin Corporation.*)

Encoders are of two basic types: (1) Absolute encoders provide a unique output signal for each single or multiple revolution of shaft gearing. An absolute encoder outputs a complete binary code (digital output) for each position. These devices are generally used in applications where position information rather than change in position is important. Absolute encoders have an individual digital address for each incremental move, and thus the position within a single revolution can be determined without a starting reference. By gearing two or more absolute encoders together, so that the second advances one increment for each complete revolution of the first (reminiscent of a mechanical counter), the range of absolute position can be extended. (2) Incremental encoders produce a symmetrical pulse for each incremental change in position. Pulses from the incremental encoder are counted for each incremental movement from a calibrated starting point in an up-down counter to track position.

Rotary Encoders

Over many years, a number of rotary encoders have been developed, including magnetic, contact, resistive, and optical devices. In recent years, the optical encoder has been preferred for many applications, including both absolute and incremental forms.

Rotary Incremental Optical Encoders

As shown in Fig. 16, optical encoders operate by means of a grating which moves between a light source and a detector. When light passes through the transparent areas of the grating, an output is seen from the detector. For increased resolution, the light source is collimated and a mask is placed between the grating and the detector. The grating and the mask produce a shuttering effect, so that only when their transparent sections are in alignment is light allowed to pass to the detector.

A rotary incremental optical encoder consists of five basic components: (1) a light source (LED or incandescent), (2) an encoder disk, (3) a grid assembly, (4) a photodetector, and (5) amplification electronics.

The disk shown in Fig. 17 is a key element of the encoder and is typically a glass material with imprinted marks of metal with slots precisely positioned. The number of marks or slots is equivalent to the number of pulses per turn. A typical resolution is 500 or 1000 pulses per revolution. (For example,

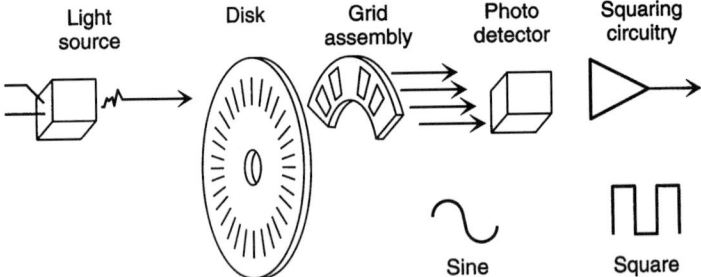

FIGURE 16 Rotary incremental optical encoder. Incremental encoders rely on a counter to determine position and a stable clock to determine velocity. These systems use a variety of techniques to separate "up" counts from "down" counts, and maximize the incoming pulse rate. Correct position information depends on accurate counting and the proper transmission of pulses. Use of a differential line drive can prevent false counting due to electrical noise developing stray electrical pulses. (*Parker Hannifin Corporation, Compumotor Division.*)

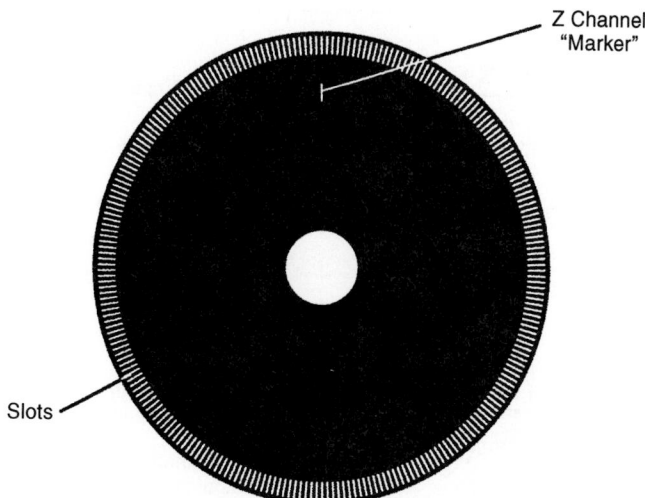

FIGURE 17   Disk used in rotary incremental optical encoder. (*Parker Hannifin Corporation, Compumotor Division.*)

a glass disk imprinted with 1000 marks would have moved 180° after 500 pulses. The maximum resolution of the encoder is limited by the number of marks or slots that can be physically located on the disk. Obviously, the larger the disk diameter, the greater the potential number of pulses per revolution. In general, commercially available rotary incremental encoders have an overall diameter of between 40 and 75 mm (1.5 and 3 inches).

Most rotary encoders also provide a single mark on the disk, called Z channel or marker. The pulse from this channel provides a reference once per revolution to detect error within a given revolution. The LED or other light source is constantly enabled. As the disk rotates, light reaches the photodetector at each slot or mark location. The detector is usually a phototransistor or, more commonly, a photovoltaic diode. This simple arrangement (Fig. 18), apart from its low output signal, has a dc offset which is temperature dependent, making the signal difficult to use.

Thus in practice, two photodiodes are used with two masks, arranged to produce signals with 180° phase difference for each channel, the two diode outputs being subtracted so as to cancel the dc offset (Fig. 19). This quasisinusoidal output may be used unprocessed, but more often it is either amplified or used to produce a square-wave output. Incremental rotary encoders thus may have sine-wave or square-wave outputs and usually have up to three output channels.

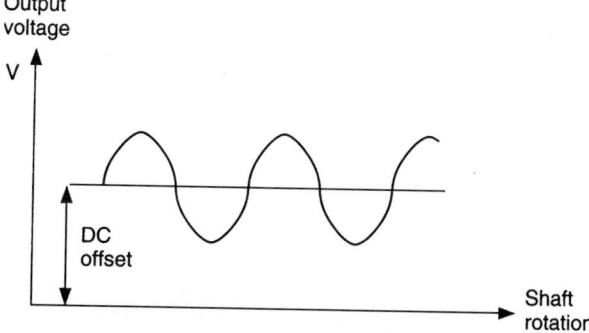

FIGURE 18   Output voltage (before conditioning) of rotary incremental optical encoder. (*Parker Hannifin Corporation, Compumotor Division.*)

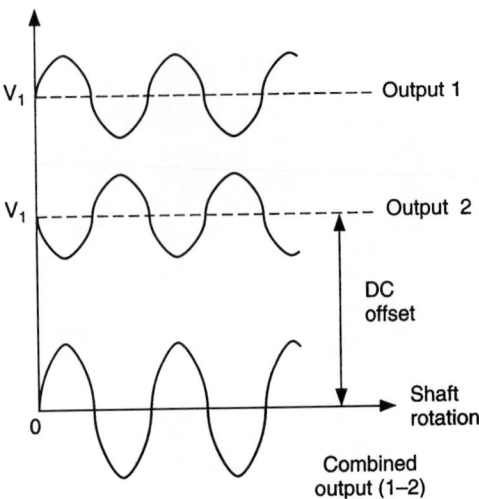

FIGURE 19   Output from dual-photodiode system used in rotary incremental optical encoder. (*Parker Hannifin Corporation, Compumotor Division.*)

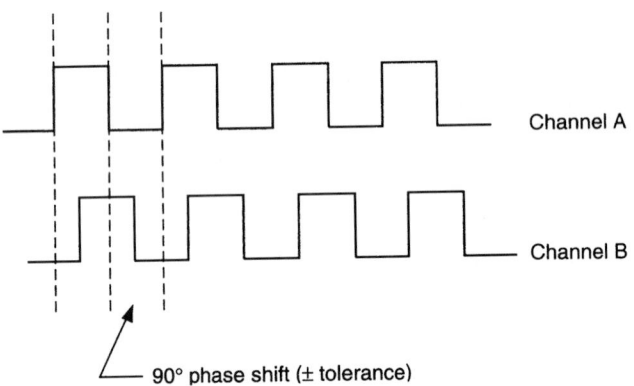

FIGURE 20   Quadrature output signals to rotary incremental optical encoder. (*Parker Hannifin Corporation, Compumotor Division.*)

A two-channel encoder, as well as giving the position of the encoder shaft, is also capable of providing information on the direction of rotation by examination of the signals to identify the leading channel. This is possible since the channels are normally arranged to be in quadrature, that is, 90° phase shifted (Fig. 20).

For most machine tool or positioning applications, a third channel, known as the index channel or Z channel, is also included. This gives a single output pulse per revolution and is used when establishing the zero position.

Figure 20 also shows that for each complete square wave from channel *A,* if channel *B* output is also considered during the same period, then four pulse edges may be seen to occur. This allows the resolution of the encoder to be quadrupled by processing the *A* and *B* outputs to produce a separate pulse for each square-wave edge. For this process to be effective, however, it is important that quadrature be maintained within the necessary tolerances so that the pulses do not run into one another.

Square-wave output encoders are generally available in a wide range of resolutions (up to about 5000 lines per revolution), and with a variety of different output configurations, such as the following.

Transistor-Transistor Logic. The encoder is commonly available for compatibility with transistor-transistor logic (TTL) levels and normally requires a 5-V supply. TTL outputs are also available in an open-collector configuration, allowing the system designer a choice of pull-up resistor value.

CMOS (Complementary Metal-Oxide Semiconductor). The encoder is available for compatibility with the higher logic levels normally used with CMOS devices.

Line Driver. A low-output impedance device, it is designed for driving signals over a long distance, and usually is used with a matched receiver.

Complementary Outputs. These outputs are derived from each channel, giving a pair of signals, 180° out of phase. These are useful where maximum immunity to interference is required.

Noise of Rotary Incremental Encoder

The control system for a machine is normally screened and protected within a metal cabinet, and an encoder may be housed similarly, but unless suitable precautions are taken, the cable connecting the two can be a source of trouble due to its picking up electrical noise. This noise may result in the loss or gain of signal counts, giving rise to incorrect data input and loss of position. Noise problems are briefly addressed in Figs. 21–23.

Accuracy of Rotary Incremental Encoder

In addition to noise problems, there may be machine-related sources of error, such as the following.

Slew Rate (Speed). An incremental rotary encoder will have a maximum frequency at which it will operate (typically 100 kHz), and the maximum rotational speed or slew rate will be determined by this frequency. Beyond this, the output will become unreliable and accuracy will be affected. Further, if an encoder is rotated at speeds higher than its design maximum, conditions may occur which will be detrimental to the mechanical components of the assembly.

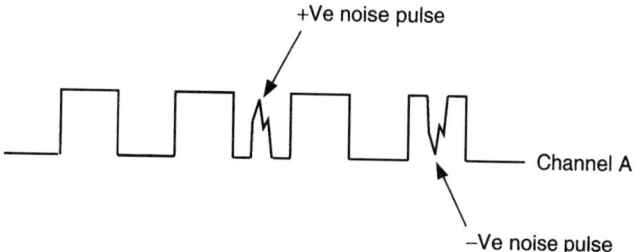

FIGURE 21 Example of effects of noise on encoder signal, showing how the introduction of two noise pulses has converted a four-pulse train into one of six pulses. Shielding of interconnecting cable may provide a solution, but inasmuch as the signals may be at low level (5 volts) and may be generated by a high-impedance source, further action may have to be taken. An effective way is to use an encoder with complementary outputs and connect this to the control system by means of shielded twisted-pair cable. (*Parker Hannifin Corporation, Compumotor Division.*)

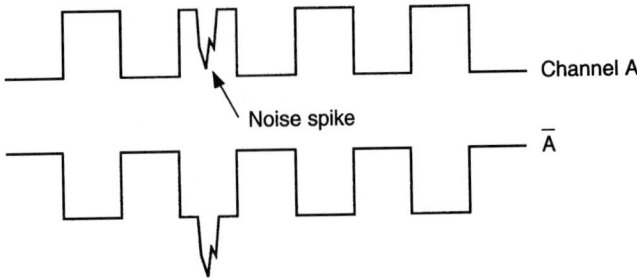

FIGURE 22   Encoder complementary output signals. The two outputs are processed by the control circuitry so that the required signal can be reconstituted without noise. (*Parker Hannifin Corporation, Compumotor Division.*)

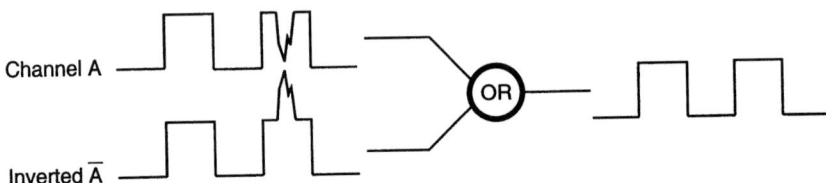

FIGURE 23   Complementary system. With reference to Fig. 22, if the *A* signal is inverted and is fed with the *A* signal into an OR gate (whose output depends on one signal or the other being present), the resultant output will be a square wave. (*Parker Hannifin Corporation, Compumotor Division.*)

Quantization Error.   In common with digital systems, it is not easy to interpolate between output pulses, so that knowledge of position is accurate only to the grating width (Fig. 24).

Eccentricity Error.   This error may be caused by bearing play, shaft run-out, incorrect assembly of the disk on its hub or of the hub on the shaft. Some of these error conditions include (1) amplitude modulation (Fig. 25(*a*)), (2) frequency modulation (as the encoder is rotated at constant speed, the frequency of the output will change at a regular rate, see Fig. 25(*b*)), and (3) interchannel jitter. The latter may occur if the optical detectors for the two encoder output channels are separated by an angular distance on the same radius. Then any "jitter" will appear at different times on the two channels, resulting in interchannel jitter.

Rotary Absolute Optical Encoders

An absolute encoder is a position verification device that provides unique position information for each shaft location. The location is independent of all other locations, whereas in the incremental encoder a count from a reference is required to determine position.

The disk used in the absolute optical encoder differs markedly from that used in the incremental encoder as previously shown in Fig. 17. In an absolute optical encoder there are several concentric tracks (unlike the incremental encoder with its single track). Each track has an independent light source (Fig. 26). As the light passes through a slot, a high state (true 1) is created. If light does not pass through the disk, a low state (false 0) is created. The position of the shaft can be identified through the pattern of 1s and 0s.

The tracks of an absolute encoder vary in slot size, moving from smaller at the outside edge to larger toward the center. The pattern of slots is also staggered with respect to preceding and succeeding tracks. The number of tracks determines the amount of position information that can be derived from

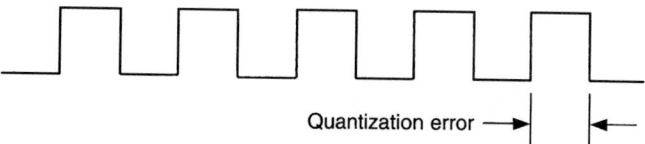

FIGURE 24   Demonstration of quantization error. (*Parker Hannifin Corporation, Compumotor Division.*)

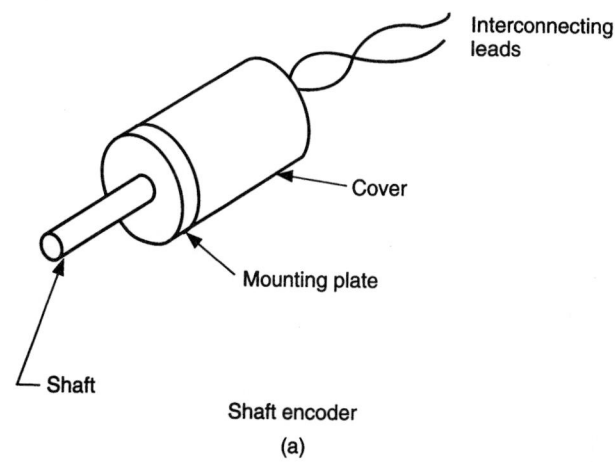

FIGURE 25   Errors caused by eccentricity problems. (*Parker Hannifin Corporation, Compumotor Division.*)

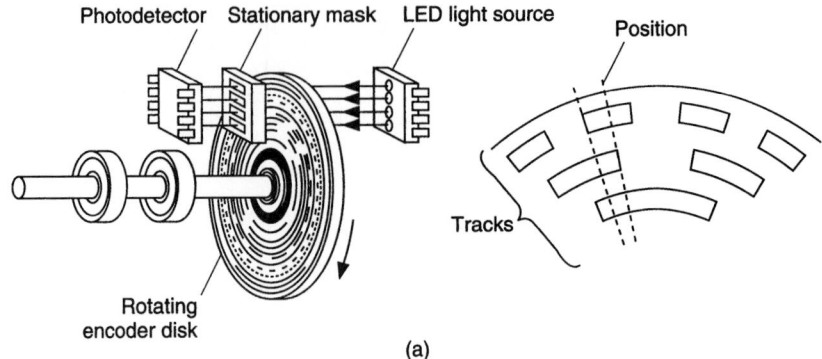

(a)

FIGURE 26(a) Representative patterning of disks in a rotary absolute optical encoder. *(a)* Disk and associated elements of system and schematic representation of tracks and position. *(Lucas Ledex Inc.)*

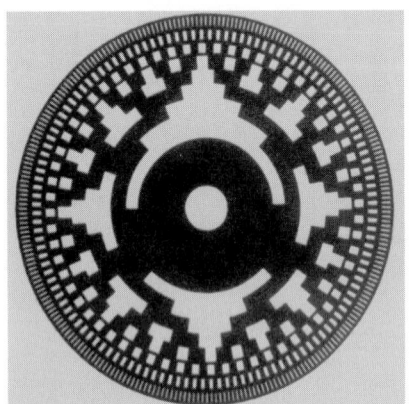

FIGURE 26(b) Another disk format. *(Parker Hannifin Corporation, Compumotor Division.)*

the encoder disk, that is, its resolution. For example, if the disk has 10 tracks, the resolution of the encoder usually would be 1024 positions per revolution, or 2^{10}.

For reliability it is desirable to have the disks constructed of metal rather than glass. A metal disk is not as fragile and has lower inertia.

The disk pattern of an absolute encoder is in machine-readable code, usually binary, gray code, or a variety of gray. Figure 27 represents a simple binary output with 4 bits of information.

Multiturn Absolute Rotary Encoders

Gearing an additional absolute disk to the primary high-resolution disk provides for turns counting, so that unique position information is available over multiple revolutions (Fig. 28).

Advantages of Absolute Encoders

Both rotary and linear absolute encoders offer several advantages in industrial motion control and process control applications. These include the following.

1. *No position loss on power down or loss of power.* An absolute encoder is not a counting device like an incremental encoder because an absolute system reads the actual shaft position. Lack of power does not cause the encoder to lose position information. Whenever power is supplied to an absolute system, it is capable of reading the current position immediately. In a facility where frequent power failures occur, an absolute encoder is necessary.

2. *Operates in electrically noisy environments.* Equipment, such as welders and motor starters, often generates electrical noise which can look like encoder pulses to an incremental encoder. Noise does not alter the discrete position that an absolute system reads.

3. *High-speed long-distance data transfer.* By using a serial interface, such as RS-422, absolute position data can be transmitted up to 1200 meters (4000 feet).

4. *Eliminates "go home" or referenced starting point.* An absolute system always "knows" its location. In numerous motion control applications it is difficult or impossible to find a "home" reference point. This situation may occur in multiaxis machines and on machines that cannot reverse

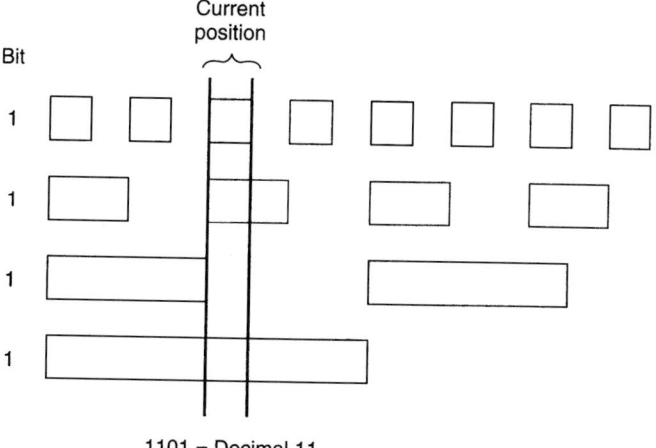

FIGURE 27 Disk pattern of absolute rotary encoder is in machine-readable code. Illustration represents a simple binary output with 4 bits of information. The current location is equivalent to the decimal number 11. Moving to the right from the current position, the next decimal number is 10 (0-1-0-1 binary). Moving to the left from the current position, the next position would be 12 (0-0-1-1). (*Parker Hannifin Corporation, Compumotor Division.*)

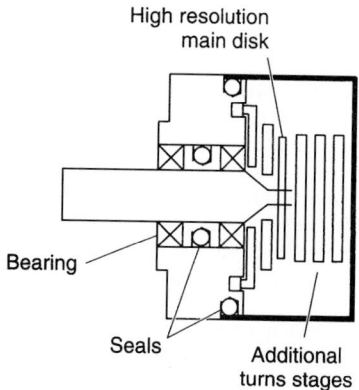

FIGURE 28 Multiturn absolute rotary optical encoder. The primary high-resolution disk has 1024 counts per revolution. A second disk with three tracks of information will be attached to the high-resolution disk geared 8:1. The absolute encoder now has eight complete turns of the shaft, or 8192 discrete positions. Adding a third disk geared 8:1 will provide for 64 turns of absolute positions. In theory, additional disks could continue to be incorporated but, in practice, do not exceed 512 turns. Encoders using this technique are called multiturn absolute encoders. This same technique can be incorporated in a rack and pinion style linear encoder, resulting in long length of discrete absolute locations. (*Parker Hannifin Corporation, Compumotor Division.*)

direction. This feature can be particularly important in a "lights-out" manufacturing facility. This reduces scrap and setup time resulting from power loss.

   5. *Reliable information in high-speed applications.*   The counting device often limits the use of incremental encoders because a counter is limited to a maximum pulse input of 100 kHz. An absolute encoder does not require a counting device. This is not a limitation of an absolute encoder.

Encoders for Stepping and dc Servomotors

Particularly designed for application on rotary shafts where standard encoders cannot be used, and intended primarily for direct shaft and face mounting on stepping and dc servomotors is the encoder shown in Fig. 29. The unit is designed for velocity and position sensing.

Resolvers

These devices have a long history of use in machine position control systems. Because of their fundamental analog output signal, conversion is required by a synchro-to-digital (S/D) converter.

   In principle, a resolver is a rotating transformer. The primary of the transformer is a winding on the shaft (rotor) much like a motor. The secondaries (stators) are wound in the case, again much like a motor. The difference in terminology of a resolver

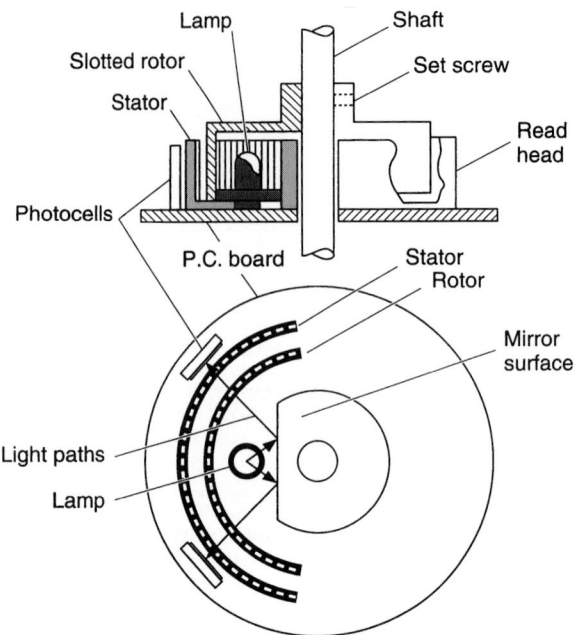

FIGURE 29   Position and velocity sensor intended primarily for direct shaft and face mounting on stepping and dc servomotors. Three types of disks are available—metal deposition on glass, photoemulsion on plastic, and etched metal. (*Lucas Ledex Inc.*)

and a synchro is the number of stator windings. A resolver has two stator windings 90° apart, whereas a synchro has three windings 120° apart. In both devices, as the shaft turns, the relative positions of the rotor and stator windings change, and the root-mean-square (rms) voltage output of the stator winding varies as the sine of the angle between them. Only the ratio of the outputs is used. It should be pointed out that although a synchro is a three-wire motorlike device, it is not a three-phase device. The ac outputs are either in phase or 180° out of phase. Phase shift does not change with angle except to reverse the phase at certain angles. The resolver or synchro is coupled to the shaft to be measured, such as a pinion gear, lead screw, or robot arm, and then wired directly to an S/D converter. The latter then outputs a digital word for further processing (Figs. 30–32).

The excitation voltage may be coupled to the rotating winding by slip rings and brushes, a disadvantage when used with a brushless motor. In such cases a brushless resolver may be used, as shown in Fig. 33.

Pancake Resolvers and Synchros

These transducers normally have large diameters and are of short length as compared with other resolvers. Diameters range from 50 to 90 mm (2 to 3.5 inches) (Fig. 34).

LINEAR MOTION

Linear Encoders

These sencoders, which may be absolute or incremental, are used to make direct measurements of linear movements. As shown in Fig. 35, many linear encoders utilize rack and pinion technology. A

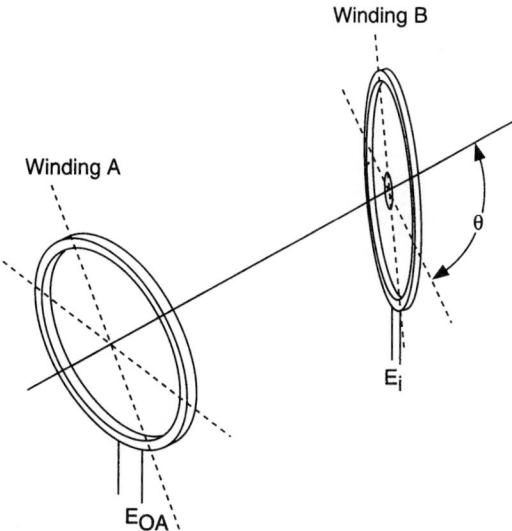

FIGURE 30   Resolver principle. Consider two windings A and B. If winding B is fed with a sinusoidal voltage, then a voltage will be induced into winding A. If winding B is rotated, the induced voltage will be a maximum when the planes of A and B are parallel and will be a minimum when the windings are at right angles. Also, the voltage induced into A will vary sinusoidally at the frequency of rotation of B so that $E_{OA} = E_i \sin \phi$. If a third winding C is positioned at right angles to winding A, then as B is rotated, a voltage will be induced into that winding, and that voltage will vary as the cosine of the angle ϕ, so that $E_{OC} = E_i \cos \phi$. (*Parker Hannifin Corporation, Compumotor Division.*)

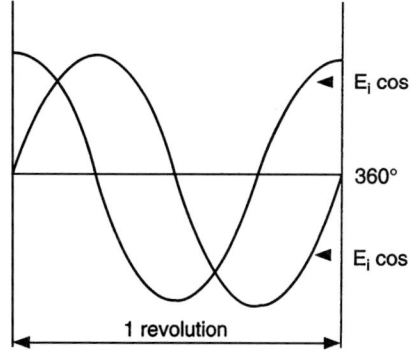

FIGURE 31   Resolver output. If the relative amplitudes of the two winding A and C can be measured at a particular point in the cycle, two outputs will be unique to that position. (*Parker Hannifin Corporation, Compumotor Division.*)

linear encoder comprises a linear scale, which in standard models ranges from about 200 mm to 6 m (8 inches to 20 feet), but considerably longer linear systems are constructed. Resolution is expressed in lines per unit length (normally, lines per centimeters or lines per inch). A linear absolute encoder is shown in Fig. 36. Typical mounting configurations are illustrated in Fig. 37.

Other Linear Position Transducers

Over the years, numerous methodologies have been developed to achieve linear positions and length measurements, some of which have been phased out while others are still in use. This limited description includes the following.

Inductive-Bridge Transducer.   This transducer type is used in production machines with restricted axis motion. Operation is based on the use of a fixed inductive member slightly longer than the axis to be measured (Fig. 38).

Inductive-Plate Transducer.   With this transducer type a slider moves across a scale, but with an air gap between the two. Since there is no physical contact, there is no apparent wear on the feedback device. The slider is attached to a movable push rod which can traverse up to 1270 cm

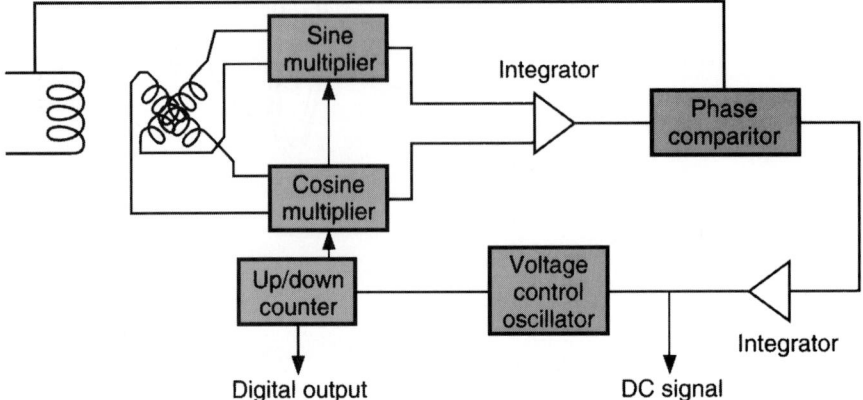

FIGURE 32 Resolver-to-digital converter. The information outputs (see Fig. 31) from the two phases are usually converted from analog to digital form for use in a digital positioning system. Resolutions up to 65,536 counts per revolution are typical. (*Parker Hannifin Corporation, Compumotor Division.*)

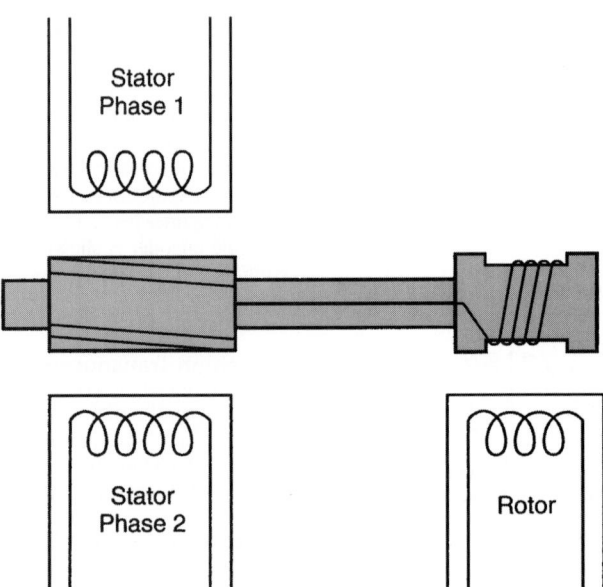

FIGURE 33 Brushless resolver used where it is desirable to avoid use of brushes and slip rings. (*Parker Hannifin Corporation, Compumotor Division.*)

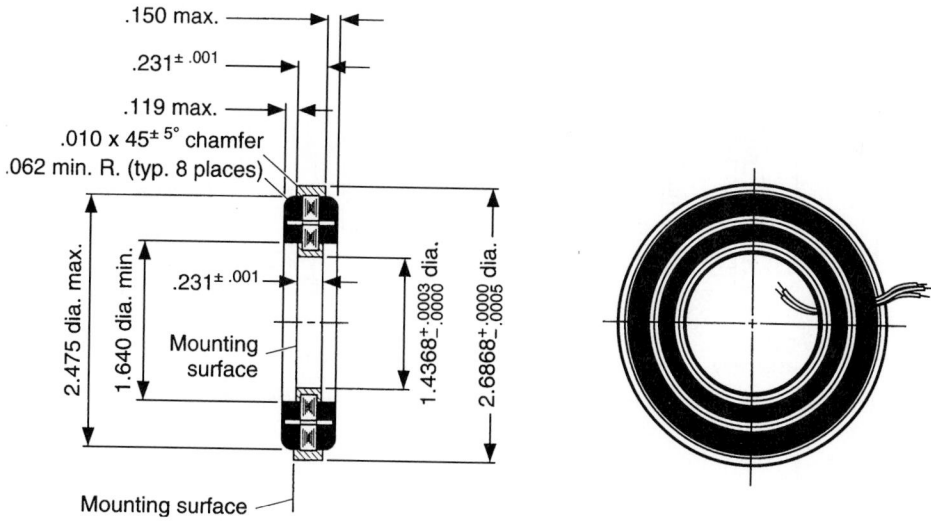

FIGURE 34   Representative pancake-type resolver, end and side views.

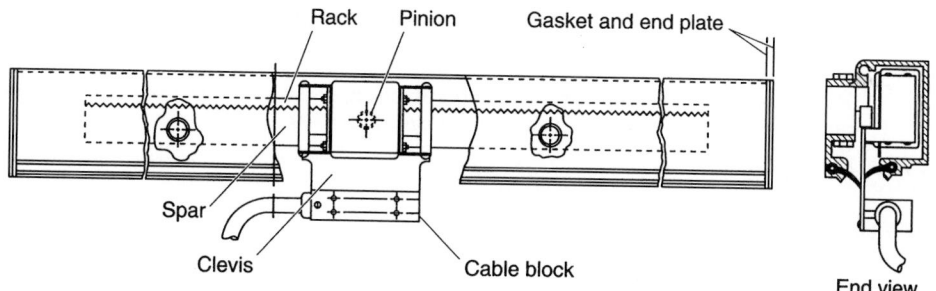

FIGURE 35   Linear encoder that utilizes rack and pinion technology. (*Parker Hannifin Corporation, Compumotor Division.*)

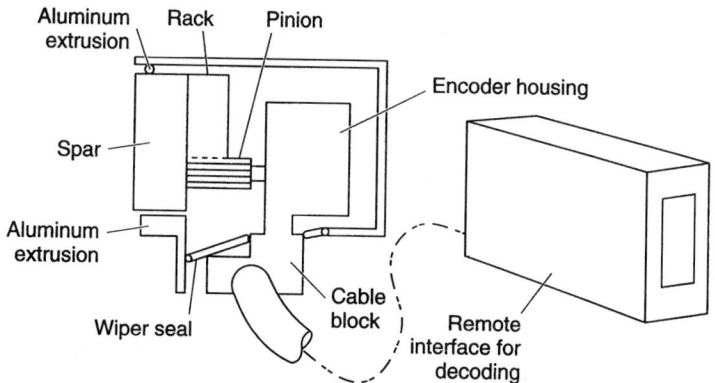

FIGURE 36   Linear absolute encoder. (*Parker Hannifin Corporation, Compumotor; Division.*)

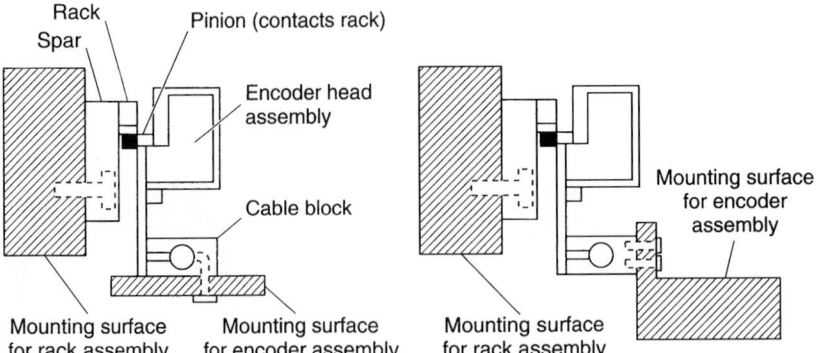

FIGURE 37   Typical mounting configurations of linear absolute encoder. (*Parker Hannifin Corporation, Compumotor Division.*)

(500 inches) per minute and attain a total of 2.5 mil travel cycles without replacing the seal. Scales are laser-checked to achieve an accuracy of ±0.0025 mm (±0.001 inch), are furnished in 60-meter (10-inch) lengths, and can be placed adjacent to one another for long travels of up to 60 meters (200 feet). Major applications have been on jig borers, horizontal boring machines, contouring machines, turning, milling, and drilling machines, as well as positioning tables, vertical turret lathes, and grinders (Fig. 39).

Magnetic Position Transducers.   A number of schemes have been developed that incorporate magnetic properties. They are shown diagrammatically in Fig. 40.

DISPLACEMENT TRANSDUCERS

Generally, displacement is thought of in terms of a motion of a few millimeters or less. Displacements as small as a wavelength of light can be measured. When displacements are quite large, the term distance may be preferred. Frequently a measurement of displacement is made to relate to some other measure, and hence displacement transducers are fundamental components of many instrumentation systems. A long established use is that of measuring the motion of the free end of a bourdon, the movement of a scale beam, or the deflection of an accelerometer. Displacement transducers are found in the measurement of thickness, among other variables.

Displacement implies motion from one point to another; and it also implies position, that is, a change from one position to the next. Displacement also implies the establishment of a new position as related to a stable, normal, or reference position.

Linear Variable Differential Transformers

The linear variable differential transformer (LVDT) is an electromechanical device that produces an electrical output proportional to the displacement of a separate nonmagnetic movable core. As shown in Fig. 41, one primary and two secondary coils are arranged symmetrically to form a hollow cylinder. A magnetic nickel-iron core, supported by a nonmagnetic push rod, moves axially within the cylinder in exact accordance with the mechanical displacement of the probe tip.

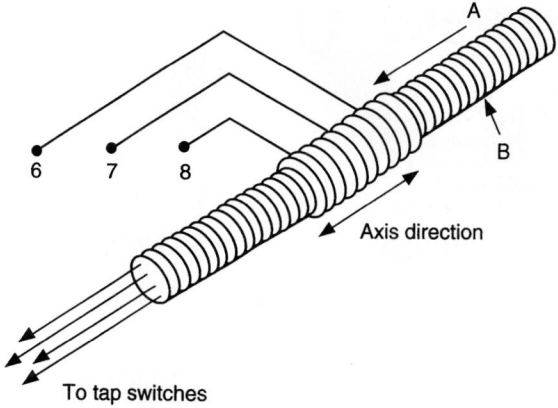

Position transducer

(a)

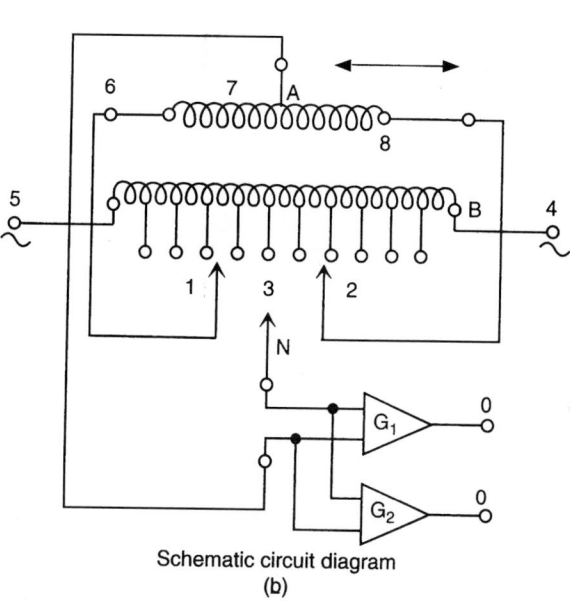

Schematic circuit diagram

(b)

FIGURE 38 Inductive-bridge transducer for position measurement. Operation is based on the use of a fixed inductive member B slightly longer than the axis to be measured, and a movable member A approximately half the length of B. Selectable taps are placed on B in a successive decade with externally located inductors to provide a bridge configuration that may be externally unbalanced by placing A (coil) and N (point) across a pair of tap points; then the coil is moved until equal voltage prevails between the two ends of the coil, as evidenced by the occurrence of a small voltage at O and O'. A disadvantage of the system is the relatively large number of wires that must be taken from the device through the machine to the control system. An advantage is the high output voltage per unit of displacement, 2 mV/0.01 mm (5 mV/0.001 inch). Supply frequency usually is between 400 and 1500 Hz.

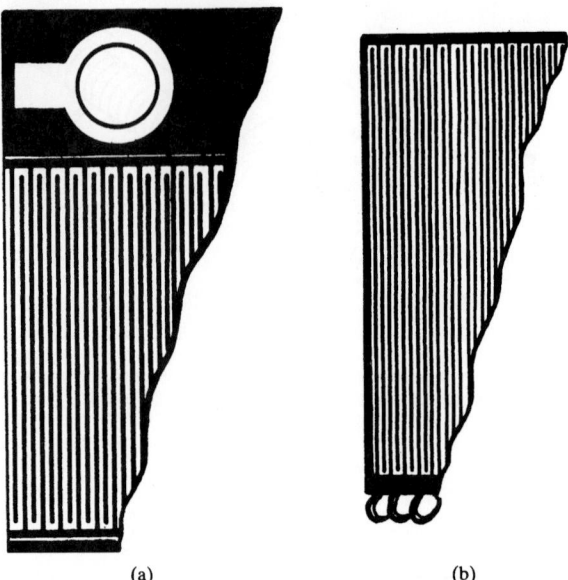

(a)   (b)

FIGURE 39   Inductive-plate position and motion transducer. (*a*) Section of scale made up of a copper pattern bonded to heat-treated steel. (*b*) Slider.

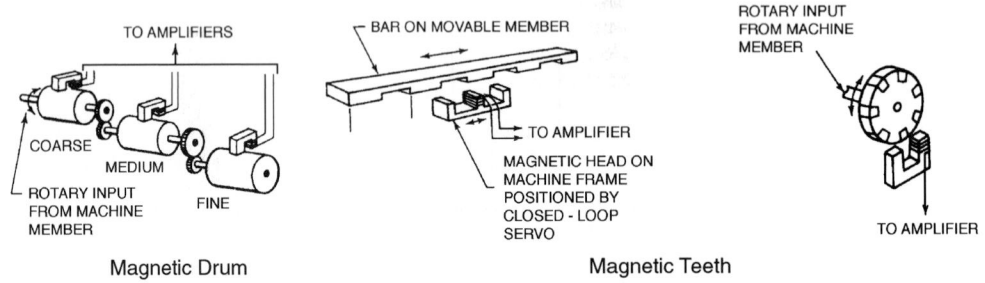

FIGURE 40   Types of magnetic configurations used in position transducers.

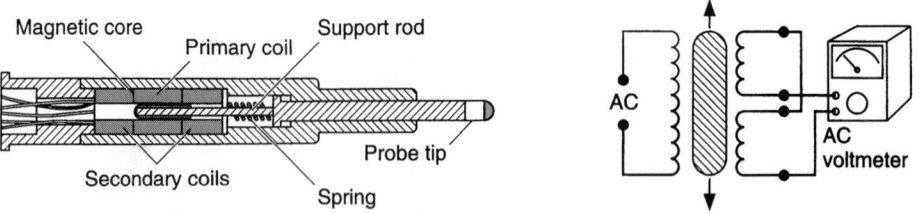

FIGURE 41   Typical linear variable differential transformer sensing probe and simplified circuit. (*Daytronic Corporation.*)

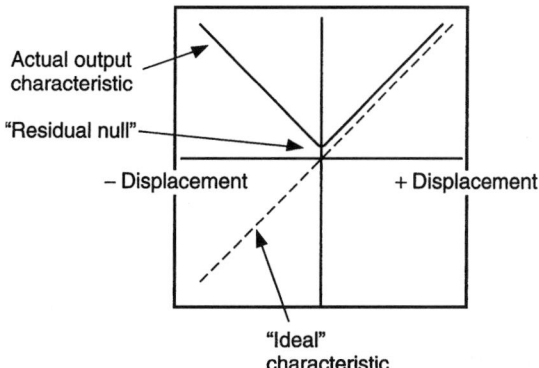

FIGURE 42   Output versus displacement in linear variable differential transformer. (*Daytronic Corporation.*)

With ac excitation of the primary coil, induced voltages will appear in the secondary coils. Because of the symmetry of magnetic coupling to the primary, these secondary induced voltages are equal when the core is in the center (null, or electric zero) position. When the secondary coils are connected in series opposition, as shown in Fig. 42, the secondary voltage will cancel and, ideally, there will be no net output voltage. If, however, the core is displaced from the null position, one secondary voltage will increase, while the other decreases. Since the two voltages no longer cancel, a net output voltage will now result. If the transducer has been properly designed, this output will be exactly proportional to the magnitude of the displacement, with a phase polarity corresponding to the direction of displacement. (See also Fig. 43.)

The LVDT enjoys wide diversity in its application as a displacement sensor, and, consequently, hundreds of configurations and sizes of the device are available from several manufacturers. The device, singly or in multiples, is used for sensing displacements from millimeters (microinches) to several meters (yards) and is useful in force, pressure, thickness, and other applications where a target variable can be converted to linear displacement. The following advantages account for the wide usage of the LVDT:

1. Essentially frictionless measurement because there is no physical contact between the movable core and the surrounding core structure. This permits its use in critical measurements that can tolerate the addition of the low-mass core, but cannot tolerate friction loading, as found in dynamic deflection or vibration tests of delicate materials and tensile or creep tests on fibers or other highly elastic materials.

2. Extremely long mechanical life as particularly required in high-reliability mechanisms and systems found in aircraft, missiles, space vehicles, as well as some industrial equipment.

3. Essentially infinite resolution, allowing the LVDT to respond to minute motion of the core.

4. Null repeatability, permitting the device to be used as a null-position indicator in high-gain closed-loop control systems.

5. Cross-axis rejection because the LVDT is sensitive to axial core motion, but relatively insensitive to radial core motion.

6. Environmental compatibility because, with the selection of proper construction materials, the LVDT can operate at cryogenic temperatures (immersed in liquid nitrogen or oxygen, for example, as well as in nuclear reactors with high radiation levels and in fluids at elevated temperatures and pressures).

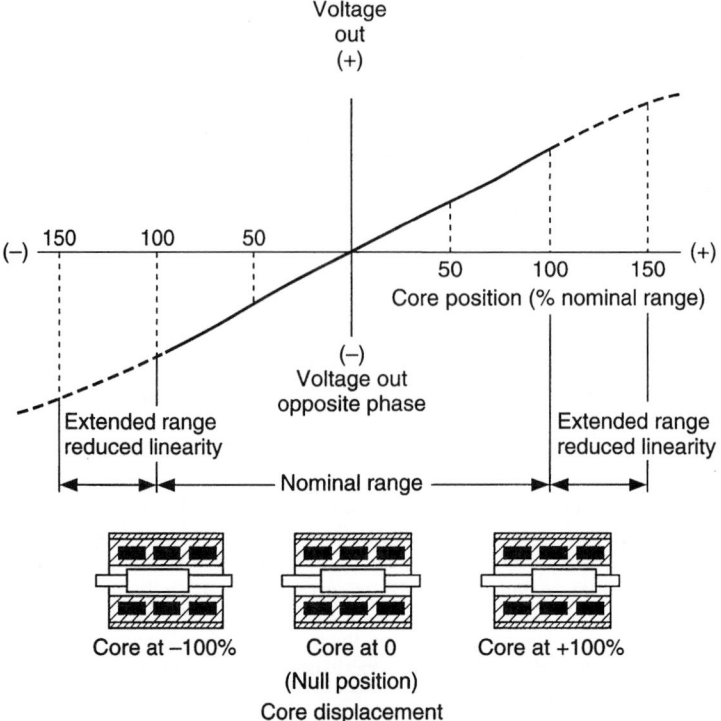

FIGURE 43   Linear variable differential transformer voltage and phase as a function of core position. (*Lucas Schaevitz*.)

The application versatility of the LVDT is illustrated in Fig. 44.
Other design variations of the LVDT include the following:

DC-LVDT.   It provides the characteristics of the AC-LVDT and the simplicity of dc operation. This design is based on the use of miniature high-performance solid-state components. Prior dc designs exhibited low sensitivity, poor stability, and output that varied with ambient temperature changes. The circuit of one contemporary dc unit is shown in Fig. 45.

Rotary Variable Differential Transformer.   The rotary variable differential transformer (RVDT) utilizes a specially shaped ferromagnetic rotor that simulates the linear displacement of the straight cylinder core of an LVDT. Although capable of continuous rotation, most RVDTs operate within a range of ±40°, with linearity better than ±0.5% of full-scale displacement. As with the LVDT, the RVDT's output voltage characteristically shifts 180° in phase around a null or zero shaft angle position.

Rotary Variable Inductance Transducer.   An inductive position sensor, the rotary variable inductance transducer (RVIT) combines a noncontacting variable inductance transducer with a proprietary (*Lucas Schaevitz*) digital Autoplex decoder. Operating on 5 V dc, the sensor can provide both digital and analog output signals.

Linear Potentiometers

These devices for the measurement of displacement and position take numerous forms. The simplest, least costly form is a single length of wire along which a slider or other form of moving device contacts

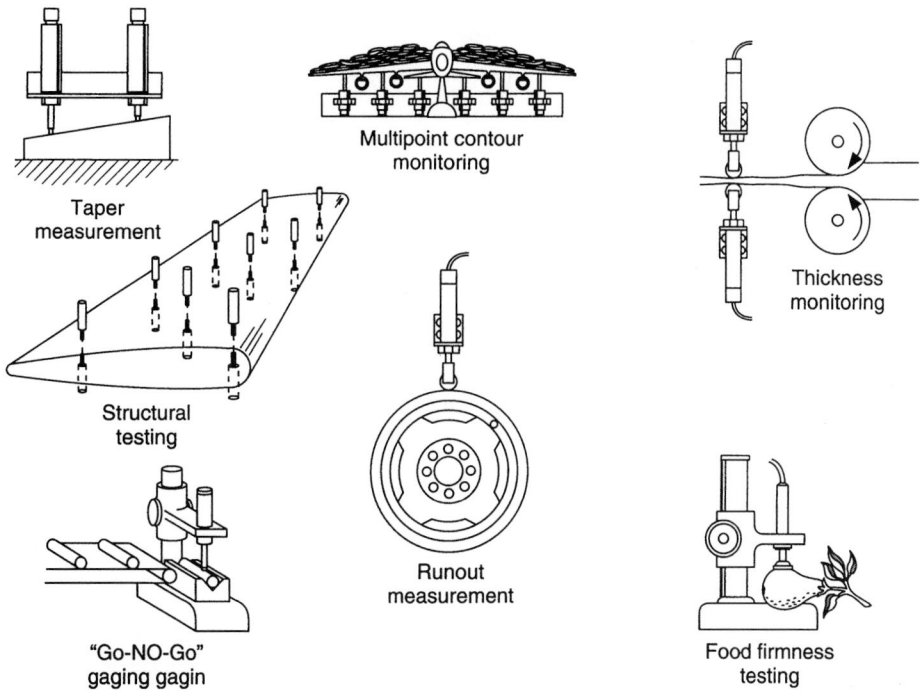

FIGURE 44   Examples of widely diverse applications of linear variable differential transformers. (*Daytronic Corporation.*)

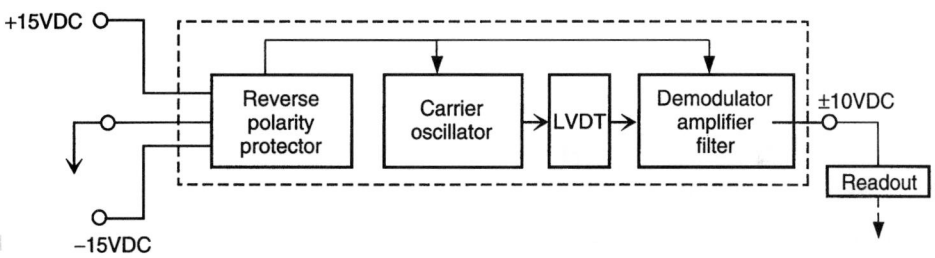

FIGURE 45   Block diagram of dc module (proprietary) for use with DC-LVDT. (*Lucas Schaevitz.*)

the wire. The position of the slider determines the effective length of the conductor. Hence a change in electrical resistance or a voltage drop is related to the position or displacement of the slider. This simple device is useful for laboratory demonstrations, but seldom is used industrially.

One example of a linear displacement transducer of the potentiometric type has been used in aircraft and missile production. As shown in Fig. 46, the device is comprised of two resistor elements (j), which are molded along with the slide bars (d) in the element block (s), which, in turn, is contained within an outer case (n). Wiper assemblies (h) are attached to wiper carrier (e) and are aligned to coincide with the resistance elements and contact bars immediately opposite each other. See sectional view. The wiper carrier is secured to the actuating shaft (a) in such a manner as to eliminate backlash and yet allow the shaft to rotate freely 360° when installed. O-ring sealings protect against sand, dust, and so on. Applications include aircraft control surface indication, landing gear retraction systems, missile stage separation, and various industrial uses.

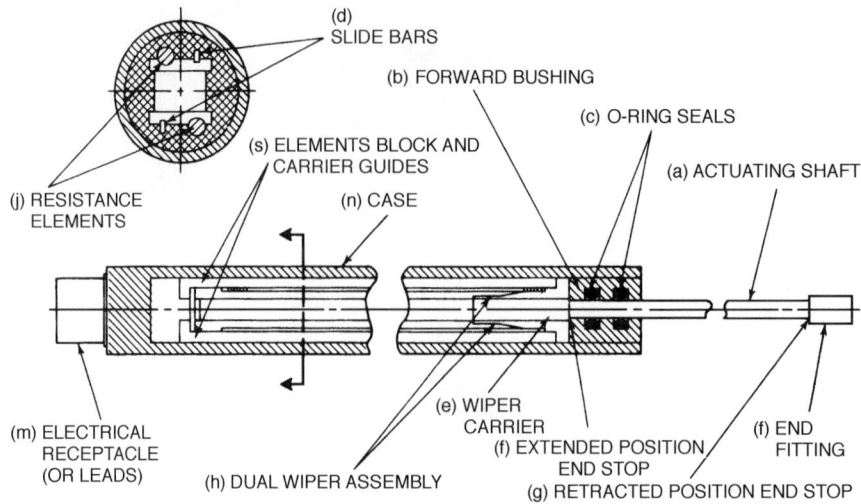

FIGURE 46   Cylindrical wire-wound potentiometric-type linear displacement transducer.

Linear Transformers

This device is a specialized synchro consisting of a salient-pole rotor and a single-phase stator, distributively wound. The winding on the stator is designed to produce an output voltage that varies linearly with the rotor position. This linear function is valid only within a restricted band about the zero position, generally $\pm 50°$ or $\pm 85°$, which is known as the excursion region. Past the excursion region, the output voltage bends to become sinusoidal.

MAGNETOSTRICTIVE LINEAR POSITION SENSORS

by Dave Nyce*

Linear position sensors are widely used as the feedback element for motion control in commercial and industrial products and systems. They can be incremental or absolute reading, contact or noncontact, and range through various levels of price and performance.

Magnetostrictive linear position sensors are noncontact and absolute reading. Noncontact means that the moving part of the sensor, attached to the member whose position is being measured, does not contact the stationary part of the sensor. The coupling between the moving and stationary sensor parts is achieved by means of a magnetic field; see Fig. 47. Therefore, any number of position changes can be made without causing wear of the sensor parts. This is in contrast to a contact sensor, like a

* MTS Systems Corporation.

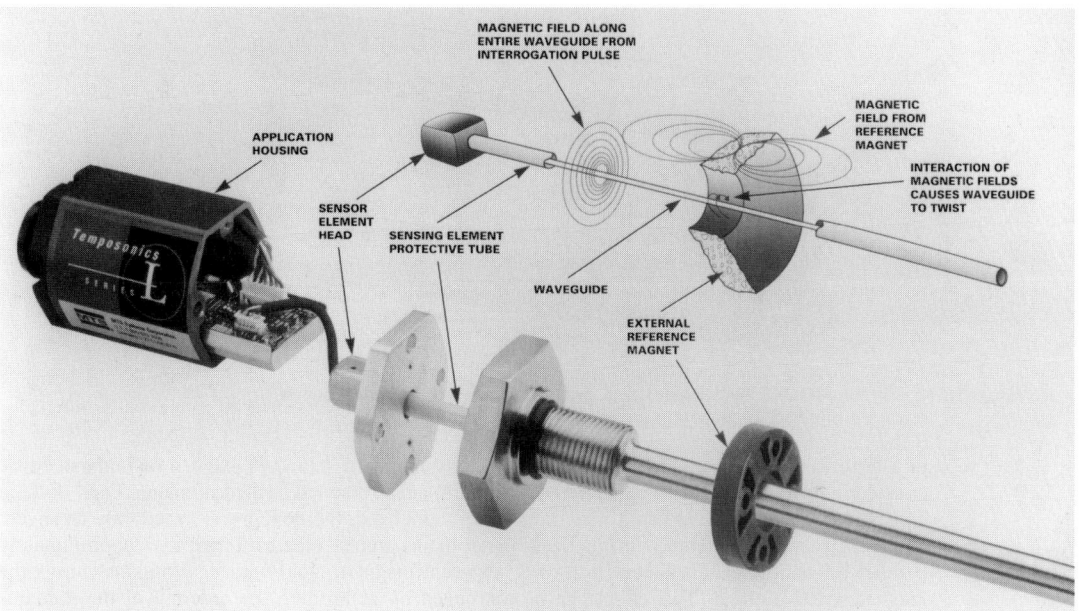

FIGURE 47   Coupling between moving and sensor parts by means of a magnetic field.

potentiometer, in which the wiper slides along the surface of a resistive element. This rubbing action is a source of noise, hysteresis (see Glossary), and limited lifetime. When wear significantly reduces the signal-to-noise ratio, or produces dead spots in the resistive element, replacement of the sensor is required. This can happen in a few months when the monitored parts have a constant dithering motion over the same area of the resistive element. For example, some high-quality potentiometers have a life rating of 100 million cycles. In industrial motion control applications, it is common to have a slight dithering of the measured part at 60 Hz. At this rate, 100 million cycles will be reached in 20 days, producing a dead spot if the mean position of the sensor has not changed. Normally, however, there would be several frequently used positions along the sensor measuring range, so the formation of dead spots would take a few months.

Since magnetostrictive sensors are absolute reading, the position is accurately known at power on, without the need for setting a zero position. Alternatively, an incremental sensor, like a linear optical encoder, indicates position changes from a set reference. On power up, or after a corruption of the count in memory by noise or other means, the system must drive to a reference location so that the sensor and the count can be rezeroed.

Magnetostrictive linear position sensors are manufactured in lengths from as short as 10 mm full stroke to more than 20 m long. Nonlinearity (without correction in software) is as low as 0.02% in sensors produced by MTS Systems Corp. Another popular absolute reading, noncontact, linear position sensor is the LVDT. This is an acronym for linear variable differential transformer. Standard LVDTs are manufactured with a full stroke measurement range as short as 1 mm and a nonlinearity of as low as 0.1%, but they have a nonlinearity of 0.2–1.0% at lengths over 25 mm and are difficult and expensive to produce with measurement ranges of over 100 mm.

Magnetostriction[1] is a property of ferromagnetic materials such as iron, nickel, and cobalt. When placed in a magnetic field, these materials change size and/or shape; see Fig. 48. The physical response

---

[1] Magnetostriction is a change in the size and/or shape of a ferromagnetic material caused by the application of a magnetic field.

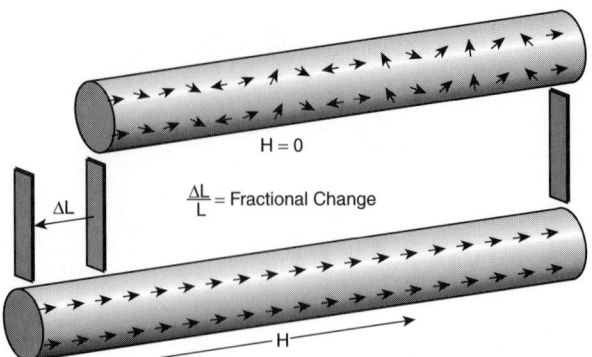

FIGURE 48   Ferromagnetic materials change size in a magnetic field.

of a ferromagnetic material is due to the presence of magnetic moments and can be understood by considering the material as a collection of tiny permanent magnets, called domains. Each domain consists of many atoms. When a material is not magnetized, the domains are randomly arranged. However, when the material is magnetized, the domains are oriented with their axes approximately parallel to each other. The interaction of an external magnetic field with the domains causes the magnetostrictive effect. This effect can be optimized by controlling the ordering of the domains through alloy selection, thermal annealing, cold working, and magnetic field strength [1].

The ferromagnetic materials used in magnetostrictive position sensors are transition metals, such as iron, nickel, and cobalt. In these metals, the 3d electron shell is not completely filled, which allows the formation of a *magnetic moment* (i.e., the shells closer to the nucleus than the 3d shell are complete, and they do not contribute to the magnetic moment). As electron spins are rotated by a magnetic field, coupling between the electron spin and electron orbit causes electron energies to change. The crystal then strains so that electrons at the surface can relax to states of lower energy [2].

When a material has positive magnetostriction, it enlarges when placed in a magnetic field; with negative magnetostriction, the material shrinks. The amount of magnetostriction in base elements and simple alloys is small, that is, of the order of 10^{-6} m/m. Since applying a magnetic field causes stress that changes the physical properties of a magnetostrictive material, it is interesting to note that the reverse is also true: applying stress to a magnetostrictive material changes its magnetic properties (e.g., magnetic permeability; see Glossary). This is called the Villari effect.[2] Normal magnetostriction and the Villari effect are both used in producing a magnetostrictive position sensor.

An important characteristic of a wire made of a magnetostrictive material is the Wiedemann effect[3]; see Fig. 49. When an axial magnetic field is applied to a magnetostrictive wire and a current is passed through the wire, a twisting occurs at the location of the axial magnetic field. The twisting is caused by an interaction of the axial magnetic field, usually from a permanent magnet, with the magnetic field along the magnetostrictive wire, which is present because of the current in the wire. The current is applied as a short-duration pulse, approximately 1 or 2 μm; the minimum current density is along the center of the wire with the maximum at the wire surface. This is caused by the skin effect.[4] The magnetic field intensity is also greatest at the wire surface. This aids in developing the waveguide twist. Since the current is applied as a pulse, the mechanical twisting travels in the wire as an ultrasonic wave. The magnetostrictive wire is therefore called the waveguide. The wave travels at the "speed of sound" in the waveguide material, approximately 3,000 m/s.

---

[2] The Villari effect is the change in magnetic properties of a ferromagnetic material in response to the presence of stress in the ferromagnetic material.

[3] The Wiedemann effect is the mechanical torsion that occurs when an electric current is passed along or through a long, thin ferromagnetic material while it is subjected to an axial magnetic .

[4] The Skin Effect is the non-uniform distribution of high-frequency current in a conductor, where current is concentrated in the outer layer rather than in the center of the conductor.

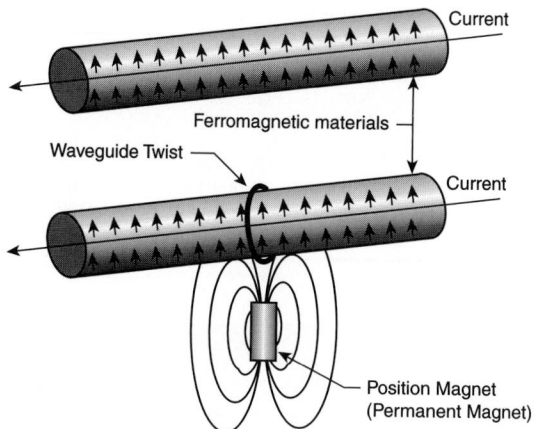

FIGURE 49   The Wiedemann effect.

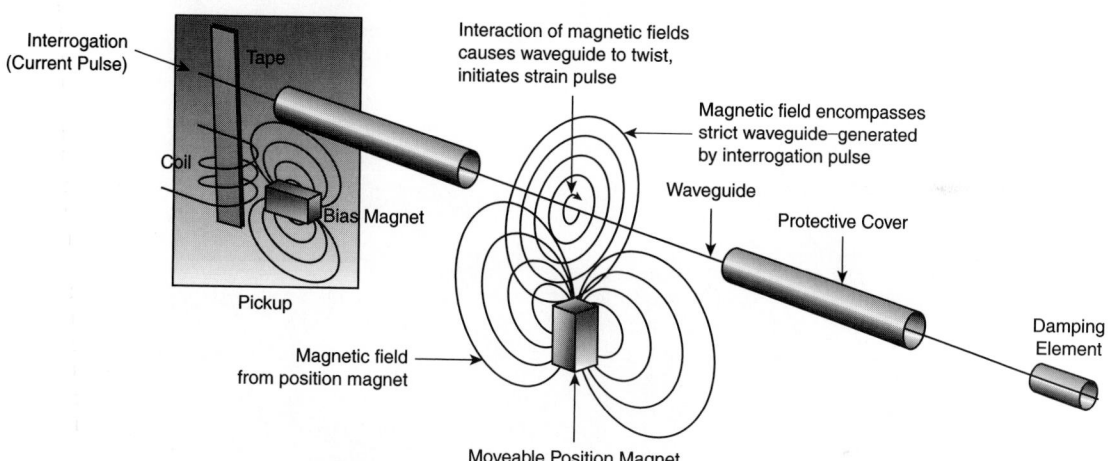

FIGURE 50   Operation of a magnetostrictive positive sensor.

The operation of a magnetostrictive position sensor is shown in Fig. 50. The axial magnetic field is provided by a position magnet. The position magnet is attached to the machine tool, hydraulic cylinder, or whatever is being measured. The waveguide wire is enclosed within a protective cover and is attached to the stationary part of the machine, hydraulic cylinder, and so on. The location of the position magnet is determined by first applying a current pulse to the waveguide. At the same time, a timer is started. The current pulse causes a sonic wave to be generated at the location of the position magnet (Wiedemann effect). The sonic wave travels along the waveguide until it is detected by the pickup. This stops the timer. The elapsed time indicated by the timer then represents the distance between the position magnet and the pickup.

The sonic wave also travels in the direction away from the pickup. In order to avoid an interfering signal from waves traveling in this direction, their energy is absorbed by a damping device (called the damp).

The pickup makes use of the Villari effect. A small piece of magnetostrictive material, called the tape, is welded to the waveguide near one end of the waveguide. This tape passes through a coil. The tape is magnetized by a small permanent magnet called the bias magnet. When a sonic wave

TABLE 3 Performance Feature Comparison

Parameter	LVDT	Magnetostrictive	Optical Encoder	Potentiometer	Auto-SE
Contact				×	
Noncontact	×	×	×		×
Absolute	×	×		×	×
Incremental			×		
Nonlinearity	0.1–1%	0.02%	0.01%	0.05%	0.1%
FS ranges	1–100 mm	10 mm – 20 m	50 mm–2 m	50 mm–1 m	80–250 mm
Cost	medium	medium	medium	low	low

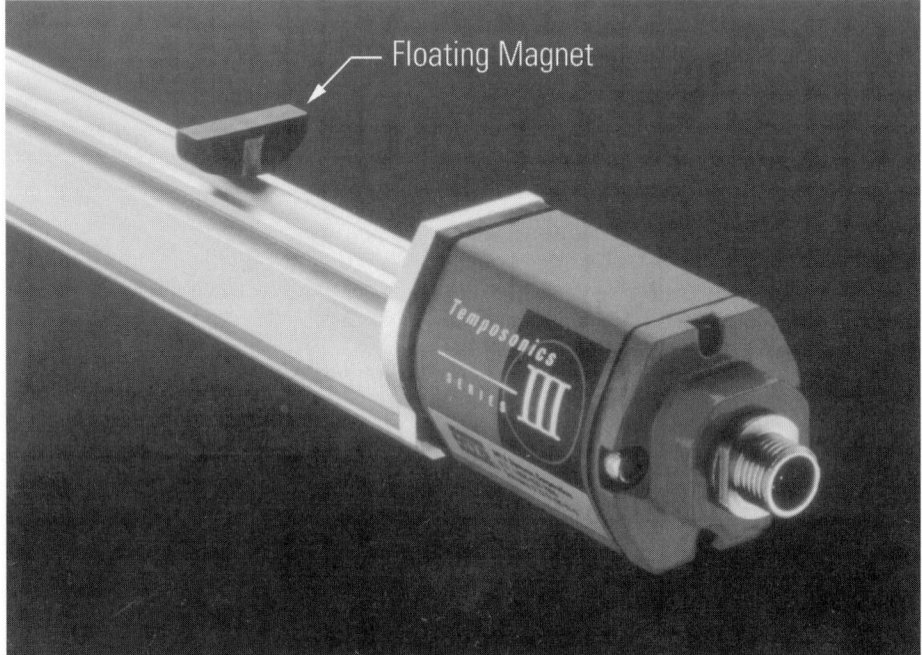

FIGURE 51   Floating magnet attached to suitable location on part to be measured.

propagates down the waveguide and then the tape, the stress induced by the wave causes a wave of changed permeability (Villari effect) in the tape. This causes a change in the tape magnetic flux density, and thus a voltage output pulse is produced from the coil (Faraday effect).[5] The voltage pulse is detected by the electronic circuitry and conditioned into the desired output. MTS magnetostrictive sensors are available with many outputs, including dc voltage, current, pulse width modulation, start–stop digital pulses, CANbus, Profibus, Serial Synchronous Interface, HART, and others.

Magnetostrictive position sensors have many different form factors. One important physical feature relates to the way in which the position magnet is applied. Figure 51 shows a floating magnet, which is attached by the user to a suitable location on the part to be measured. Figure 52 shows a captive magnet, which is enclosed in a sliding bearing so that it is more easily adapted to existing equipment.

Table 3 compares performance features of several types of linear position sensors, including standard Temposonics magnetostrictive sensors and the new Temposonics Auto-SE.

---

[5] The Faraday effect is the generation of a voltage by a coil of wire when the coil is subjected to a changing magnetic field.

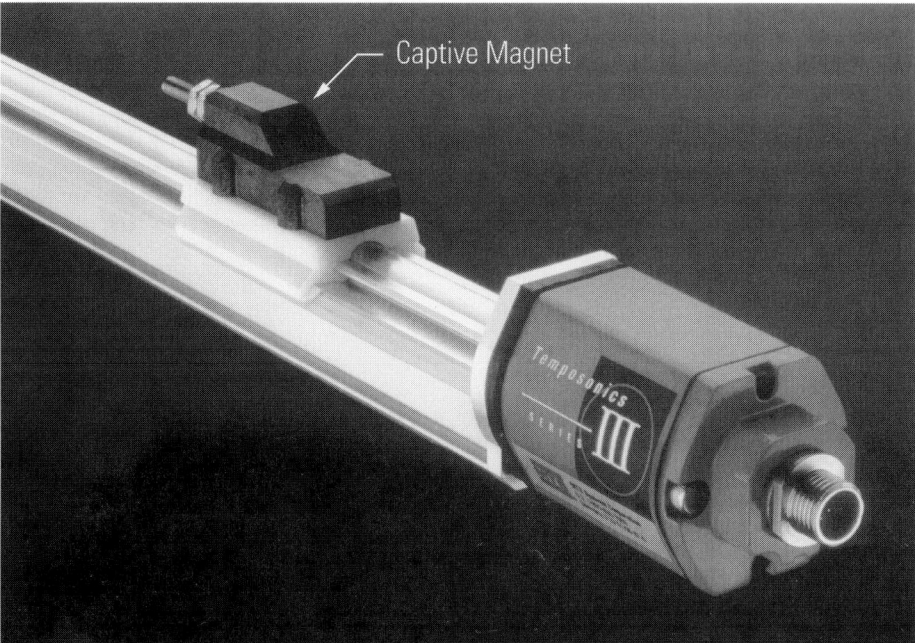

Captive Magnet

FIGURE 52   A captive magnet enclosed in sliding bearing.

Recently, there has been a major new development in magnetostrictive sensors: low-cost sensors for high-volume applications. Traditionally, magnetostrictive linear position sensors have required a substantial amount of manual assembly by skilled workers to produce the sensing element. Many jigs, fixtures, semiautomated tools and automated testers are used in manufacturing, but some operations still required manual assembly techniques.

Instigated by requests for this technology to be used in high-volume automotive applications, MTS has developed a fully automated manufacturing facility. An operator is responsible for making sure all parts feeders have a supply of parts. The fully automated production line assembles and tests all parts and final tests the completed sensors, and then packages them into the shipping pallets. Assembly techniques include many robotic operators, resistance welders, laser welding, vision systems, and computer analyses of assembly and performance data. The *auto*mated assembly *s*ensor *e*lement is called the *Temposonics Auto-SE.*

Sensors assembled on the MTS automated production line are presently not as accurate as those that are hand assembled for industrial use, but they are very low cost. Automotive and other high-volume applications use these sensors because of their unlimited lifetime, their stability over time and temperature, and especially because of their low cost. With automated assembly, a whole new field of application requirements can now be met with magnetostrictive position sensors!

SIDEBAR

The magnetostrictive linear position sensor was invented by Jacob (Jack) Tellerman in 1975. He was developing delay lines for use in computer memory devices when he had the idea to use similar technology to produce a position sensor. In the magnetostrictive memory device, digital ones and zeroes were represented by pulses of ultrasonic waves impressed onto one end of the waveguide (a magnetostrictive wire). Additional data could be pulsed into the waveguide until just before the first pulses would begin to arrive at the other end. Then that particular memory element was considered full. When the first pulses arrived at the other end, the data signals would be amplified and resent again as ultrasonic waves into the first end of the waveguide. This could continue indefinitely, keeping the data stored on the waveguide, until the data were needed again by the computer. Then, new data could be "written" to the waveguide.

Jack had the idea of generating the ultrasonic wave at locations along the waveguide by using a permanent magnet. Then the time taken until an ultrasonic pulse reached one end of the waveguide would indicate the position of the magnet. He co-founded a company called Temposonics, which was further developed after being acquired by MTS Systems Corp. in 1987.

MAGNETIC PROPERTIES GLOSSARY

1. Magnetic field intensity (H) is the force that drives the generation of magnetic flux in a material. It is also called magnetizing force and can be produced by the application of an electric current. H is measured in amperes/meter.

2. Magnetic flux density (B) is the amount of magnetic flux that results from the applied magnetizing force. B is measured in newtons/Ampere meter.

3. Magnetic permeability (μ) indicates the ability of a material to support magnetic lines of flux. The μ of a material is the product of the relative permeability of that material and the permeability of free space. The relative permeability of most nonferrous materials is near unity. In free space, magnetic flux density is related to magnetic field intensity by the formula

$$B = \mu_0 H,$$

where μ_0 is the permeability of free space, having the value $4\pi \times 10^{-7}$ henry/m. In other materials, the magnetic flux density at a point is related to the magnetic intensity at the same point by

$$B = \mu H,$$

where

$$\mu = \mu_0 \mu_r$$

and μ_r is the relative permeability [3].

4. Hysteresis is a phenomenon in which the state of a system does not reversibly follow changes in an external parameter [4]. In a linear position sensor, it is the difference in output readings obtained at a given point when approaching that point from upscale and downscale readings.

5. Magnetic hysteresis is represented in the hysteresis loop. When a ferromagnetic material is placed in an alternating magnetic field, the flux density (*B*) lags behind the magnetizing force (*H*) that causes it. The area under the hysteresis loop is the hysteresis loss per cycle, and this is high for permanent magnets and low for high permeability, low-loss magnetic materials [5].

6. Magnetic saturation is the upper limit of the ability of a ferromagnetic material to carry flux.

7. A magnetization curve shows the amount of magnetizing force required to saturate a ferromagnetic material. It is normally shown as a graph with *B* as the ordinate and *H* as the abscissa, and is known as the *B-H* curve. Figure 2 could indicate the magnetization curve for a specific material with the addition of calibration marks and curve adjustment to describe the characteristic of that material.

REFERENCES

1. Nyce, D. S., "Magnetostriction-Based Linear Position Sensors," *Sensors*, April 1994, vol. 11, no. 4, p. 22.

2. Philippe, R., *Electrical and Magnetic Properties of Materials*, Artech House, Norwood; Mass., 1988.

3. Esbach, O., *Handbook of Engineering Fundamentals*, Wiley, New York, 1975, p. 957.

4. Lerner, R., and G. Trigg, *Encyclopedia of Physics*, VCH Publishers, New York, 1990, p. 529.

5. Neelakanta, P., *Handbook of Electromagnetic Materials,* CRC Press, New York, 1995, p. 333.

BIBLIOGRAPHY

Burke, H., *Handbook of Magnetic Phenomena,* Van Nostrand Reinhold, New York, 1986.

Carstens, J. R., *Electrical Sensors and Transducers,* Regents/Prentice-Hall, Englewood Cliffs, N.J., 1992, p. 125.

Norton H., *Handbook of Transducers,* Prentice-Hall, Englewood Cliffs; N.J., 1989, pp. 106–112.

Cullity, B. D., *Introduction to Magnetic Materials,* Reading, Addison-Wesley, Mass., 1972.

Craik, D., *Magnetism Principles and Applications,* Wiley, New York, 1995.

Lorrain, P., and D. Corson, *Electromagnetic Fields and Waves,* W. H. Freeman, San Francisco; 1962.

Boll, R., *Soft Magnetic Materials,* Heyden and Son, London, 1977.

THICKNESS TRANSDUCERS

Thickness is the lesser of the three dimensions that define an object. All three dimensions, of course, are of major importance in the discrete-piece manufacturing industries. In terms of control engineering, thickness measurement and control are of particular importance to what may be called the flat-goods or continuous-length manufacturing industries that are involved in producing sheeted, webbed, and extruded end products, where thickness is of the utmost importance to the consumer. Examples include sheets of metal, plastic, paper, veneer, and plate glass. Thickness is also of paramount importance in the production of various films and coated or plated materials.

Most of the transducers for measuring position and displacement, as described earlier in this article, can measure thickness reliably for many applications. Figure 49 shown earlier, for example, illustrates the use of a pair of linear variable differential transformers (LVDTs) for thickness measurement. Further, some of the sensing principles of object detectors (described in the next article of this handbook section) can be adapted to thickness measurement. Covered in the following are systems that have been configured particularly for thickness measurement.

NONCONTACTING THICKNESS GAGES

Nuclear Radiation Thickness Gages

Both x-ray and nuclear-radiation principles have been applied to thickness measurement. The regulations imposed on instrumentation involving the use of ionizing radiation vary from country to country (Figs. 53 and 54).

In addition to beta radiation, nuclear fluorescence has been used in thickness gages for measuring the coatings on sheet steel or aluminum. Nuclear fluorescence is produced when gamma radiation

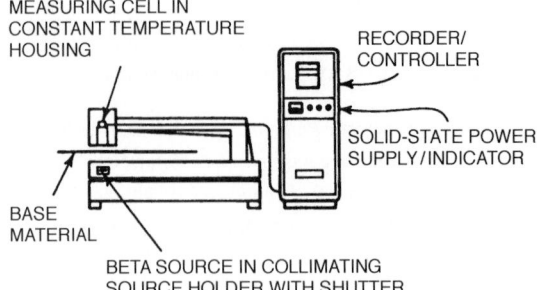

FIGURE 53 Operating principle of beta gaging system for measurement of continuous sheet materials. The gage can measure a single point or scan the sheet automatically. For paper, measurement is made in terms of basis weight (or other weight per unit area). The beta radiation source is located beneath the sheet. A detector cell is located above the sheet. Experienced precision is ±1 percent of range.

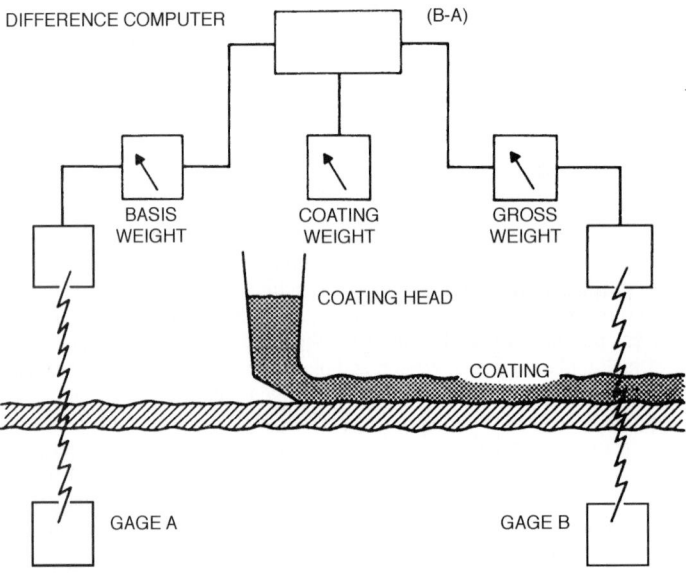

FIGURE 54 Double-gage beta-radiation system for controlling coating thickness on sheet materials.

excites electrons in the metal of the coated strip. Excitation is continuous, but each occurrence is only temporary. Electrons returning to the original unexcited state produce a characteristic low-energy radiation called fluorescence.

Continuous gaging can be an integral part of a computer-operated rolling mill. The gaging system may include an interface that receives commands for thickness settings and, in return, verify that such commands have been accomplished. A system also may include a display and record of the difference in thickness between the edge and the center, or "crown," of a metal sheet or bar, for example. A gaging system may be used to sense trends of change in thickness and may use these signals to actuate screw-down, speed, and tension controls to keep the material "on gage." Such control permits rolling

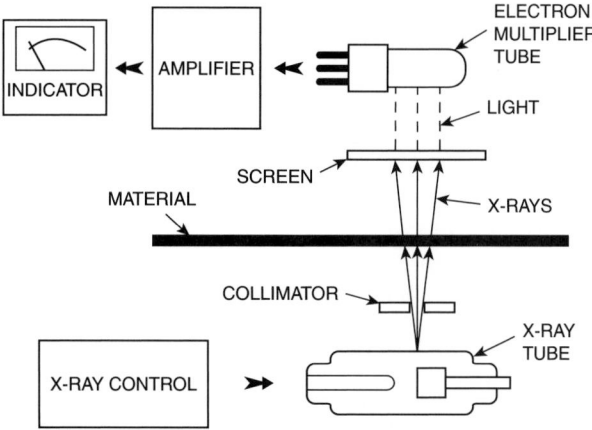

FIGURE 55 Principal elements of x-ray thickness gage.

to close tolerances, that is, maximum "on gage" lengths are being produced per ton of material, thus reducing scrap or "out-of-gage" materials at both ends of a coil or run.

X-Ray Thickness Gages

X-ray thickness gages can measure the thickness or density of hot or cold materials while the material is in motion. Steel, aluminum, brass, copper, glass, paper, rubber, plastic films, foils, and material coatings are amenable to such gaging. All materials absorb x rays to varying degrees, depending on thickness and density. Thickness can be determined by measuring the amount of x-ray energy absorbed by a material as it passes between an emitter and a receiver (Fig. 55). The gage is set to the desired thickness standard in mils or micrometers. A sample piece of material is used as a reference for calibration. An x-ray gaging system comprises three basic units: (1) a scanner that contains the x-ray generator and a detecting unit, (2) an operator control station, and (3) a power unit. The scanning unit generally is C- or O-frame mounted in a stationary position or on a traversing track, as on a rolling mill, process, or inspection line.

Ultrasonic Thickness Gages

The principle of the echo-type ultrasonic gage for presence/nonpresence object detectors is described in the following article.

In resonance-type ultrasonic thickness gages a frequency-modulated continuous-wave signal is produced. This provides a corresponding swept frequency of sound waves which are introduced into the part being measured. When the thickness of the part equals one half-wavelength, or multiples of half-wavelengths, standing-wave conditions or mechanical resonances occur. The frequency of the fundamental resonance, or the difference in frequency between two harmonic resonances, is determined by the instrument electronics. The thickness is calculated by the following formula:

$$\text{Th} = \frac{V}{2F}$$

where   Th = thickness of part
        V = speed of sound in material
        F = frequency, Hz

In the pulse-echo method, the following formula pertains to thickness measurement:

$$\text{Th} = \frac{VT}{2}$$

where   Th = thickness of part
V = speed of sound in material
T = transit time of sound pulse through one round trip in material

A common readout for ultrasonic gages is a cathode-ray tube (CRT). Frequency indications on the display are compared with an overlay scale calibrated in direct thickness readings. The readings are instantaneous, and thickness variations can be monitored as the transducer is scanned over the parts.

Direct-reading panel meters are available. For portability, small, battery-operated packages are available, with accuracies of between 0.5 and 1 percent of full scale. Both analog and digital versions are obtainable.

A limitation of ultrasonic gaging is the requirement for continuous coupling of the sound beam between the transducer and the part. Sound does not pass across an air-solid or air-liquid boundary. Liquid coupling, either a continuous thin film or some other type, is required. In some instances, complete immersion of the transducer and the material is feasible. In others it is possible to use a bubbler or partially contained water column to provide the continuous coupling path.

SURFACE TEXTURE MEASUREMENT

Continuous or periodic thickness measurement technology also can be applied in a customized manner for measuring the surface characteristics or texture of some materials. With increasing attention to quality control of in-process as well as finished products, this has become an important variable to measure and control. A version of thickness determination can be applied to some aspects of this problem.

For example, any surface produced by machining departs from the perfect form because of a variety of causes, such as inaccuracies in the machine tool, deformation of the work under the cutting force, and irregularities caused by vibration. Irregularities also may be caused by rupture of the material during separation of the chip. These factors, in turn, produce geometrical inaccuracies associated with errors of form, including surface texture—waviness and roughness. Roughness of surface is affected by both the size and the shape of the undulation. Wavelength spacing is just as important as height. What may seem to be a perfect surface can be changed to a very rough surface simply by changing the wavelength of the undulation. As the quantity of the undulation becomes smaller, the quality of the surface deteriorates. This deterioration becomes increasingly apparent to the eye as the wavelength becomes shorter, even though the height of the undulations remains the same.

The patterns formed on the surface by waviness often are varied. Some patterns, such as pronounced chatter marks and the coarse feed marks of a badly trued grinding wheel, can be identified at a glance. Others may require an instrument to reveal their presence. In the case of surfaces of revolution, the marks extending along the lay of the roughness often become the circumferential departures from roundness. Waviness, as seen in a profile graph, often can be appraised both as an undulation of the mean line and as an undulation of a line drawn through the more prominent crests. Thus the terms "crest line waviness" and "mean line waviness" (Fig. 56).

The most common specification used to control surface texture is average roughness expressed in micrometers or microinches. This value represents the actual vertical distance from a datum line of every point of the profile occurring in the length of the sampled surface. The position of the datum line or centerline is not a constant. The elevation varies with each specific length of profile being measured. By definition, the position of the centerline is such that the total areas of the profile lying above and below the line are equal. As shown in Fig. 57, surface texture includes roughness, waviness, lay, and flaws.

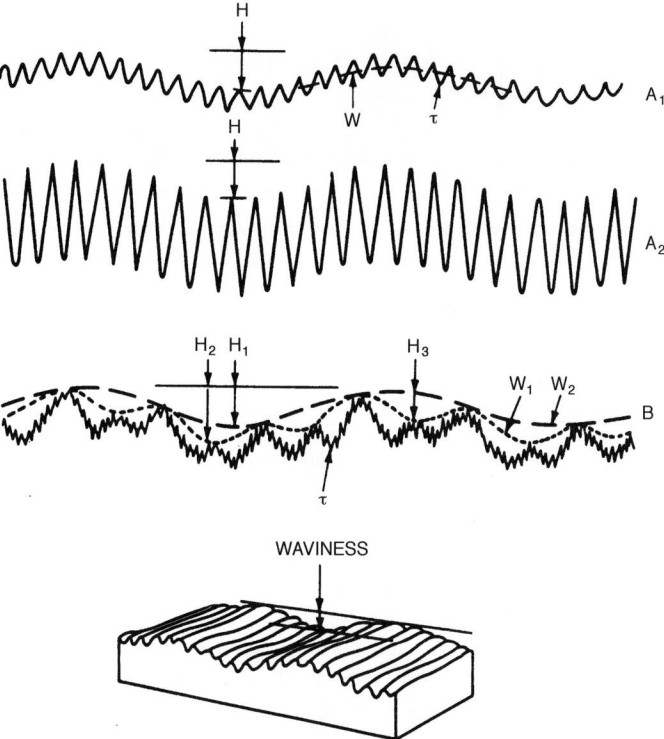

FIGURE 56 Examples of waviness. When measuring across the roughness lay, several types of waviness can be found. The profile A_1 has a wavy undulation W on which is superimposed roughness r of smaller amplitude. The profile A_2 is of the same kind, but the roughness is larger in amplitude than the waviness. In these cases, mean line and crest line waviness are substantially the same. The crest line is irregular and rarely identical with its mean line. On ground surfaces, the crest spacing of closely spaced waviness surfaces often is in the region from 0.02 to 0.1 inch (0.5 to 2.54 mm), and the height is generally less than half the overall height of the grinding texture. Surfaces also may be encountered where spacing and height are the same, yet the surfaces are quite different.

Since the measurement of surface roughness involves determination of the average linear deviation of the actual surface from the nominal surface, there is a direct relationship between the dimension tolerance for a part and the permissible surface roughness. It is evident that a requirement for the accurate measurement of a dimension is that the variations introduced by surface roughness not exceed the tolerance placed on a dimension. If this were not the case, measurement of the dimension would be subject to an uncertainty greater than the required tolerance.

Surface geometry has a fairly direct bearing on metrology and fits and limits. Measuring instruments ordinarily have anvils and gaging tips which, because of their size, make contact only with the highest points of surface irregularities. The intervening valleys, however, may have appreciable depth and, if this depth amounts to a large proportion of the tolerance, it may affect the size of the part. Subsequent removal of the high spots may differ from that indicated by the measuring instrument by an appreciable amount.

In order to accomplish measurements functionally and to an average numerical value, stylus-type instruments generally have been used. Noncontact methods, usually optical interference means, are effective for the interpretation of fringe values—hence interpretation of the surface. New methods employing scanning electron microscopy are now being used as a means of establishing a

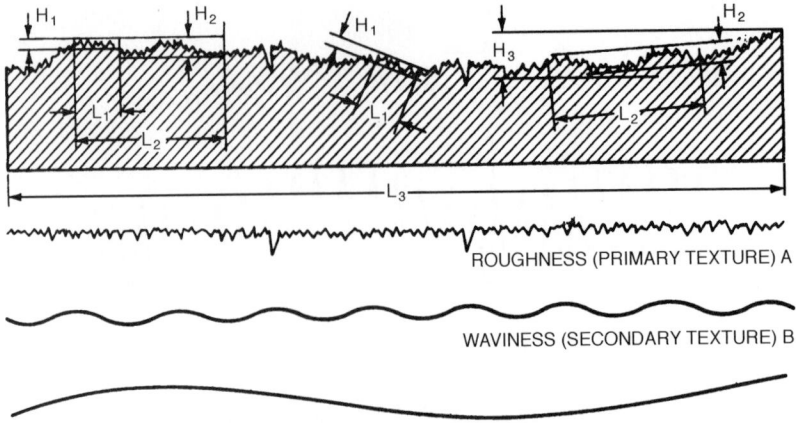

FIGURE 57   Surface texture representing the combined effects of several causes.

three-dimensional evaluation. The essential elements of a stylus-type instrument are (1) a sharply pointed stylus for tracing the profile at a cross section, (2) a means for generating a datum, and (3) a way of amplifying and indicating the stylus movement. Of the stylus-type instruments, two are in general use: (1) a carrier-modulated device in which the magnitude of a carrier current is controlled at every instant of time in accordance with the position of the stylus relative to the datum—regardless of how long the stylus remains in a given position, and (2) a device in which a current or potential is generated in accordance with the motion of the stylus as the stylus is displaced from one level to another.

Carrier-modulated instruments are useful for obtaining graphs because, in acting like simple levers, these devices faithfully reproduce every movement of the stylus relative to the datum, regardless of the spacing. These instruments are calibrated by using gage blocks or interferometrically. The instruments behave in a manner similar to a mechanical lever with magnification ratios of up to 1 million times. The generating instruments reproduce only if the stylus is rising and falling at a rate above the low-frequency limit. Therefore, widely spaced irregularities over which a stylus may be rising and falling only slowly will not be reproduced. This instrument is not desirable for a profile recording but is suitable for numerical evaluation. It is apparent that no matter how valuable and necessary the profile graph may be, some form of numerical assessment is required even if only for purposes of establishing a print value. However, a numerical value cannot be readily established until sufficient data concerning measurement of the component have been established.

Roughness width cutoffs utilized with average values are regarded as the greatest spacing of repetitive surface irregularities to be included in the measurement of average roughness height. A roughness width cutoff is rated in inches and must always be greater than the roughness width. When no value is specified, the value of 0.030 inch (0.8 mm) is assumed. The fundamental object of taking the length of the surface into consideration is based on the fact that different makes of instruments and different operators should obtain the same answer for any given surface. The profile graph on the other hand accurately defines and establishes all irregularities, giving values of height as well as width.

QUALITY CONTROL AND PRODUCTION GAGING

Statistical Quality Control

Maintaining product quality in accordance with acceptable standards has been a major role for industrial instrumentation since its inception decades ago. With the ever-growing interest in speeding up production, one becomes increasingly aware of the fact that rejects as well as acceptable products can be produced at very high rates. What constitutes product quality? Apparently, in the long run, it is that degree of excellence which the ultimate consumer demands for an affordable price. Obviously, manufacturers must compare cost versus quality acceptance in the marketplace. Based upon years of experience, the manufacturer learns what affordable quality really means in terms of market acceptance. Astute competitors also learn these basic facts.

With the foregoing knowledge, the manufacturer establishes quality standards, which then are reduced to engineering specifications for each product. But knowing in advance that there will be variations from exact specifications, acceptable variations must be established. These variations then are reduced to plus or minus tolerances.

The tolerable variations and the intolerable variations must be determined and, in an ideal situation, immediately fed back to production machines or processes. Adjustments made would again, ideally, affect every part or unit of substance being produced. But because the target of acceptance is bracketed ($\pm$), a strictly go/no-go type of sorting does not suffice. Thus at least as early as a century ago, the concept of statistical quality control (SQC) was introduced. The general intent of SQC is that of sampling units and parts being produced and essentially determining trends in deviation from production as continuously (affordable and achievable) as possible. Since the early part of this century, the literature on SQC theory has continued to grow, and because a full understanding of the concept involves rather intricate mathematics, most of the principles have been applied. But since the 1960s and continuing, the emphasis on SQC has shifted from essentially the exertion of manual controls and interpretation to the present semiautomation of data collection and interpretation. The mathematics have improved as the result of computer technology, including the development of algorithms and other shortcuts, and the interface between SQC and management has been streamlined by modern display technology, notably computer graphics, as well as by imbedded computer calculations.

Basic Assumptions of SQC

1. Variations are inherent in any process. Inherent variations are not the primary target of SQC. If there are *only* inherent variations, the process is said to be in statistical control. It is assumed that inherent variations affect all measurements and that, over a period of time, these variations will stabilize.

2. Other variations are designated as correctable, that is, they are in the realm of statistical control.

3. A key element in the SQC concept is the familiar normal distribution (bell curve), shown and explained later.

4. In some interpretations of SQC, a further objective is considered. The acceptable tolerance limits (+ and −) are not the target to seek, but those data within the + and − range should be further correlated so that the true "aim" of +0 is achieved as closely and as often as possible. For example, a manufacturer may specify that all products must fall within a given $\pm$ tolerance, but that the majority should be closer to the midrange of the tolerance, that is, as "near perfect" as possible. This differs from simply tightening the $\pm$ tolerance because it still allows less perfect units to escape full rejection.

SQC Glossary of Terms

An abridged glossary of terms used in SQC can be helpful. The following symbols are used in the next several paragraphs

where
N = number of data points
R = range
$\bar{R}$ = average range
s = standard deviation
x_i = value of specific data point
USL = upper specification limit
LSL = lower specification limit

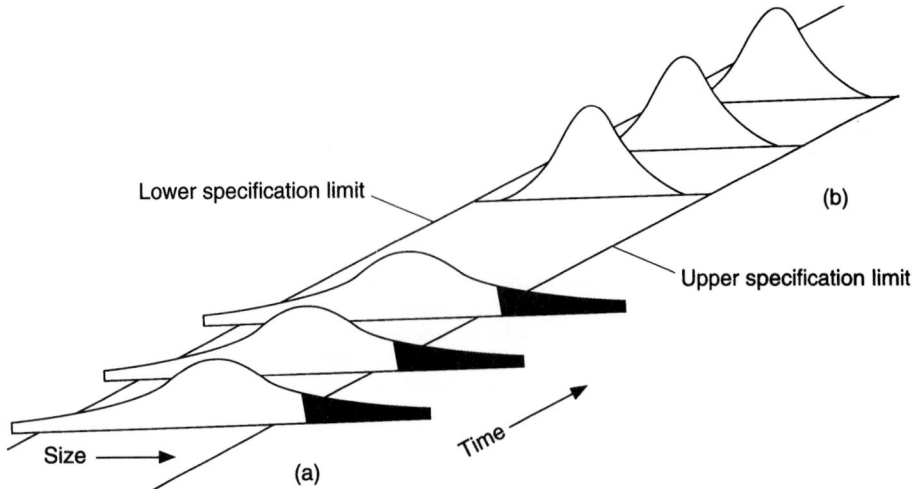

FIGURE 58   Demonstration of process variations. (*a*) In control, but not capable. (*b*) Variations from common causes are excessive. (*Moore Products.*)

Capability and Control (Concept of).   A process is in control if the only sources of variation are common causes. The mean spread of such a process will appear stable and predictable over time (Fig. 58).

The fact a process is in control does not imply that it will yield only good parts, that is, parts within specification limits. Control only denotes a stable process. The size variation due to the common causes may be so large that some parts are outside the specification limits. Under these conditions, the process is said to be *not* capable. Capability is the ability of the process to produce parts that conform with engineering specifications (Fig. 59).

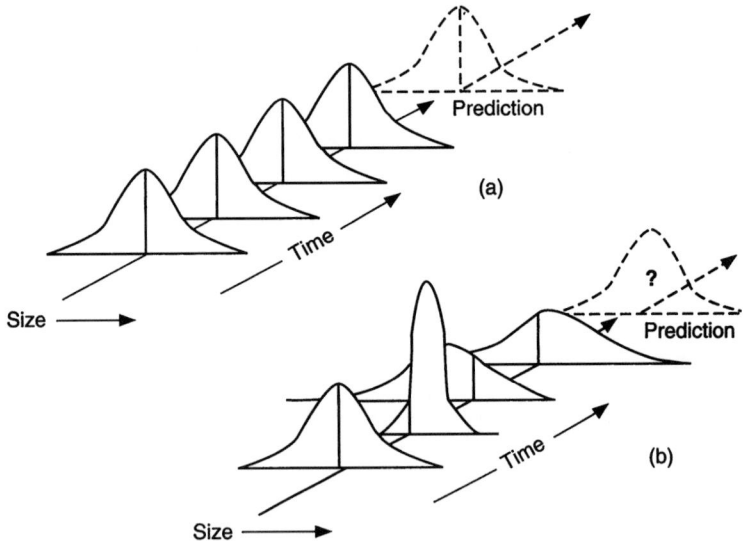

FIGURE 59   Concept of capability and control. (*a*) Process is in control. It is stable and predictable over time. (*b*) Process is neither stable nor predictable over time. (*Moore Products.*)

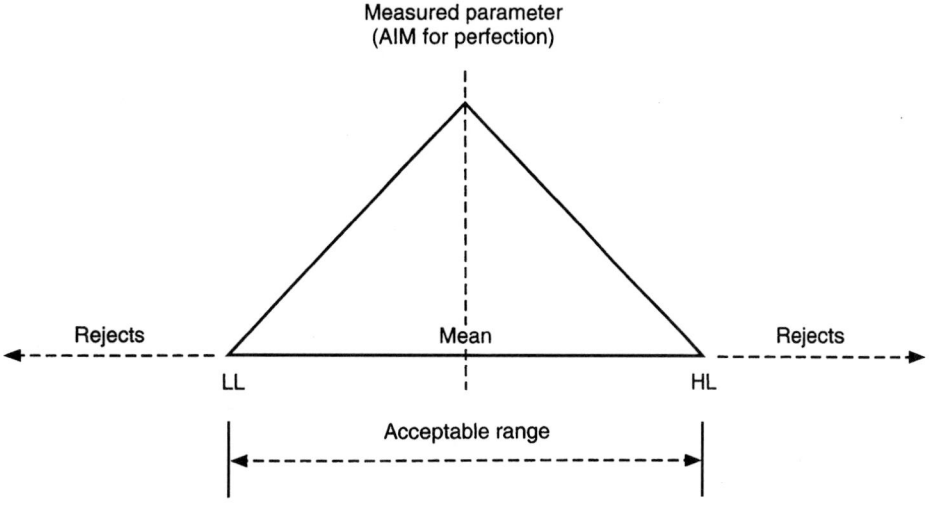

FIGURE 60 Parts or units are acceptable if they barely achieve either of the specified limits (Low, LL; High, HL). In the quality control program, the manufacturer can gear the operation such that a majority of units will lie in the middle of the acceptable range. This is the basis for the AIM mode of operation.

Cp (Inherent Capability of Process).   Cp is the ratio of the tolerance to 6 sigma. The formula is

$$Cp = \frac{USL - LSL}{6\sigma}$$

The Cp ratio is used to indicate whether a process is capable.

1.33 or greater   Process is capable.

1.0 to 1.3   Process is marginally capable, should be monitored.

1.0 or less   Process is *not* capable.

It should be noted that Cp does not relate the mean to the midpoint of the tolerances. If the mean is not at the midpoint, out-of-tolerance parts may still be probable, even if the process is capable (Fig. 60).

Cpk (Capability in Relation to Specification Limits).   Cpk relates the capability of a process to the specification limits, that is, Cpk equals the lesser of

$$\frac{USL - mean}{3\sigma} \quad \text{or} \quad \frac{mean - LSL}{3\sigma}$$

The Cpk value is useful in determining whether a process is capable and is producing good parts based on the specification limits. Values for the Cpk index have the following meaning:

Greater than 1.0   Both of the 6σ limits fall within the specification limits. The process is both capable and producing good parts (99.73 percent or greater).

1.0   At least one of the 6σ limits falls directly on the specification limits.

Less than 0.0   The mean is outside of the specification limits.

Capability Ratio. This ratio is the inverse of Cp, that is,

$$\frac{6\sigma}{\text{USL} - \text{LSL}}$$

The value of this ratio can be thought of as portion of the part tolerance consumed by 6σ. A common interpretation for the various values of the capability ratio is as follows:

50 or less	Desirable
51 to 70	Acceptable
71 to 90	Marginal
91 and greater	Unacceptable

Common Cause. A source of random variation that affects all of the individual measurements in a process. The distribution is stable and predictable. (See also *Special Cause* in this glossary.)

Control Limits (LCL and UCL). The upper and lower control limits are values used to determine whether or not a process is in statistical control. The values are inherent to the process and should not be confused with specification limits.

Histogram. A chart that plots individual values versus the frequency of occurrence and is used for statistical data analysis. Note that earlier, these diagrams were created manually. The diagrams now can be displayed on a CRT (computer graphics) and, of course, stored in memory (Fig. 61).

Individual. A single measurement of a particular process characteristic.

Kurtosis. An indication of whether the data in a histogram have a normal distribution. Specifically, it is a measure of the "flatness" or "peakness" of a curve. The formula for kurtosis is

$$\sum_{i=1}^{\eta} \frac{(X_i - \bar{X})^4}{4s^4}$$

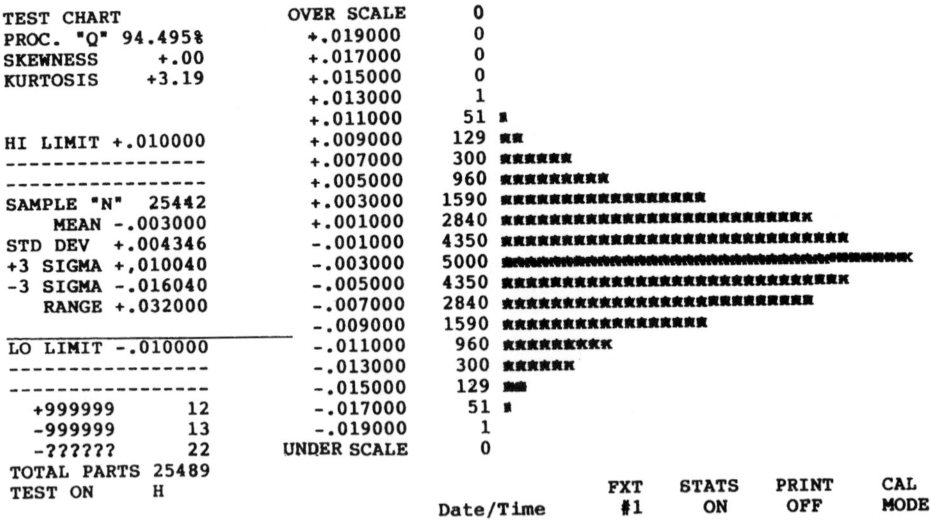

```
TEST CHART              OVER SCALE    0
PROC. "Q" 94.495%       +.019000      0
SKEWNESS      +.00      +.017000      0
KURTOSIS      +3.19     +.015000      0
                        +.013000      1
                        +.011000     51 ▮
HI LIMIT +.010000       +.009000    129 ▮▮
-----------------       +.007000    300 ▮▮▮▮▮
-----------------       +.005000    960 ▮▮▮▮▮▮▮▮▮
SAMPLE "N"   25442      +.003000   1590 ▮▮▮▮▮▮▮▮▮▮▮▮▮▮▮
     MEAN -.003000      +.001000   2840 ▮▮▮▮▮▮▮▮▮▮▮▮▮▮▮▮▮▮▮▮▮▮▮▮
STD DEV   +.004346      -.001000   4350 ▮▮▮▮▮▮▮▮▮▮▮▮▮▮▮▮▮▮▮▮▮▮▮▮▮▮▮▮▮
+3 SIGMA +.010040       -.003000   5000 ▮▮▮▮▮▮▮▮▮▮▮▮▮▮▮▮▮▮▮▮▮▮▮▮▮▮▮▮▮▮▮▮
-3 SIGMA -.016040       -.005000   4350 ▮▮▮▮▮▮▮▮▮▮▮▮▮▮▮▮▮▮▮▮▮▮▮▮▮▮▮▮▮
   RANGE +.032000       -.007000   2840 ▮▮▮▮▮▮▮▮▮▮▮▮▮▮▮▮▮▮▮▮▮▮▮▮
                        -.009000   1590 ▮▮▮▮▮▮▮▮▮▮▮▮▮▮▮
LO LIMIT -.010000       -.011000    960 ▮▮▮▮▮▮▮▮▮
-----------------       -.013000    300 ▮▮▮▮▮
-----------------       -.015000    129 ▮▮
  +999999     12        -.017000     51 ▮
  -999999     13        -.019000      1
  -??????     22      UNDER SCALE     0
TOTAL PARTS 25489
TEST ON      H                         FXT   STATS  PRINT  CAL
                        Date/Time      #1     ON     OFF   MODE
```

FIGURE 61 Facsimile of computer-generated histogram.

Values for kurtosis have the following meaning:

3	Normal distribution.
Less than 3	Leptokurtic curve, that is, the curve has high peak (data are concentrated close to the mean).
Greater than 3	Platykurtic curve, that is, the curve has low peak (data are disbursed from the mean).

Another version of the foregoing equation is

$$\frac{\eta^3 \sum_{i=1}^{\eta} X_i^4 - 4\eta^2 \sum_{i=1}^{\eta} X_i \sum_{i=1}^{\eta} X_i^3 + 6\eta \left(\sum_{i=1}^{\eta} X_i\right)^2 \sum_{i=1}^{\eta} X_i^2 + 3 \left(\sum_{i=1}^{\eta} X_i\right)^7}{\left[\eta \sum_{i=1}^{\eta} X_i^2 - \left(\sum_{i=1}^{\eta} X_i\right)^2\right]^2}$$

Mean X.   The value of the middle individual when the data are arranged in order from lowest to highest.

Mean skew 2$\bar{X}$.   The arithmetic average value of the data. The process mean is the average of all the process data. The subgroup mean averages just those values in the subgroup.

Median skew 2$\tilde{X}$.   The value of the middle individual when the data are arranged in order from lowest to highest. If the data have an even number of individuals, the median is the average of the two middle values.

Mode.   The most frequently occurring value, that is, the highest point on a histogram.

Normal Distribution.   Data often are summarized graphically as a means of better understanding and analyzing the variation. A plot of the frequency of occurrence of a particular variable is one of the common tools used for analysis. If a definite pattern emerges from the data, this plot is referred to as a distribution.

Many distributions have been identified and named as their pattern of variation is repeatable and certain mathematical characteristics can be defined for each distribution. One of the most commonly occurring patterns of distribution is the normal distribution. It describes many natural and other phenomena encountered in the field of statistics.

The mean and the standard deviation define a specific normal distribution. Knowing the values of the normal distribution, the total spread of expected outcomes can be predicted. It is important to note that 99.73 percent of the population lies between −3 standard deviations (often called −3σ) from the mean and +3σ standard deviations (+3σ) from the mean. This fact is the basis for many statistically calculated indications of the status of a process, including capability and control limits.

It is equally important to note that if the distribution of a process is not normal, a number other than 99.73 percent of produced parts or units will fall within 6 standard deviations. Because of this, many of the statistical control indicators calculated from a nonnormal distribution will not describe that distribution accurately. Often a distribution is sufficiently close to being normal, however, that is, the errors are insignificant (Fig. 62).

Pareto Chart.   This type of chart ranks events according to a specified parameter, such as rejected individuals by frequency of occurrence (Fig. 63).

Process Quality.   The process quality or process capability coefficient denotes the area under the normal curve that falls within the specification limits. It is expressed as a percentage. The calculation is a multistep procedure.

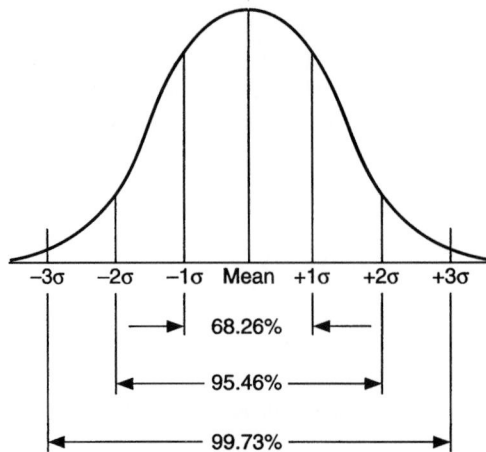

FIGURE 62 Normal distribution chart. This is basic to the formal statistical quality control (SQC) concept. The curve is useful for many other statistical purposes in other fields, such as variations in a population of people (clothing sizes, for example). The concept dates back to Laplace and Gauss and is sometimes referred to as Gaussian distribution.

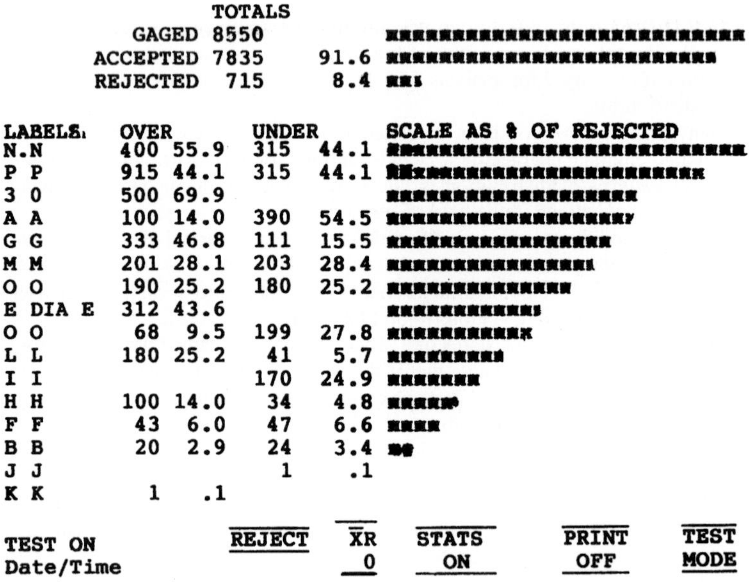

```
                      TOTALS
              GAGED   8550        ███████████████████████████████
           ACCEPTED   7835   91.6 █████████████████████████████
           REJECTED    715    8.4 ███

LABELS:   OVER          UNDER        SCALE AS % OF REJECTED
N.N       400  55.9     315   44.1   ██████████████████████████████████
P  P      915  44.1     315   44.1   █████████████████████████████
3  0      500  69.9                  ████████████████████████
A  A      100  14.0     390   54.5   ██████████████████████
G  G      333  46.8     111   15.5   █████████████████
M  M      201  28.1     203   28.4   ████████████████
O  O      190  25.2     180   25.2   ███████████████
E DIA E   312  43.6                  ████████████
O  O       68   9.5     199   27.8   ███████████
L  L      180  25.2      41    5.7   █████████
I  I                    170   24.9   ███████
H  H      100  14.0      34    4.8   ██████
F  F       43   6.0      47    6.6   ████
B  B       20   2.9      24    3.4   ██
J  J                      1    .1
K  K        1   .1
```

TEST ON	REJECT	X̄R	STATS	PRINT	TEST
Date/Time		0	ON	OFF	MODE

FIGURE 63 Facsimile of computer-generated pareto diagram. Listings include total parts gaged, parts accepted, and parts rejected, with the parameters for which parts are rejected in descendng order of frequency. This type of diagram stresses the relative importance of correction to be made.

1. Calculate the distance of the mean from the specification limits in terms of standard deviations:

$$L_U = \frac{USL - \bar{X}}{s} \qquad L_L = \frac{\bar{X} - LSL}{s}$$

(For this equation to be valid, the values for L_U and L_L must be greater than or equal to 0. If either is negative, the corresponding area is calculated using $A = 1 - A'$, where A' is obtained using $L' = -L$.)

2. Determine the area outside of the specification limits, expressed as a fraction of the normal curve:

$$A_{\text{upper}} = \frac{1}{\sqrt{2}} \cdot e^{-(L_U^2/2)} \cdot \left[B_1 Y_U + B_2 Y_U^2 + B_3 Y_U^3 + B_4 Y_U^4 + B_5 Y_U^5 \right]$$

$$A_{\text{lower}} = \frac{1}{\sqrt{2}} \cdot e^{-(L_L^2/2)} \cdot \left[B_1 Y_L + B_2 Y_L^2 + B_3 Y_L^3 + B_4 Y_L^4 + B_5 Y_L^5 \right]$$

where
$$Y_U = \frac{1}{1 + R \cdot L_U}$$
$$Y_L = \frac{1}{1 + R \cdot L_L}$$
$$R = 0.2316419$$
$$B_1 = 0.31938153$$
$$B_2 = -0.356563782$$
$$B_3 = 1.781477937$$
$$B_4 = -1.821255978$$
$$B_5 = 1.330274429$$

(When a tolerance limit is unused, that is, has a U prefix, the area under the normal curve that is outside that limit does not enter into the process quality calculation.)

3. Then the process quality is found from the equation:

$$\text{Process quality} = [1 - (A_{\text{upper}} + A_{\text{lower}})] \, 100\%$$

Range R.   The difference between the highest and lowest values in the group. The average of subgroup ranges is denoted by $\bar{R}$.

Run Chart (xR Chart).   A chart that displays the most recent individual measurement x and the absolute difference from the previous measure R on a consecutive basis.

Sample.   A synonym of subgroup in process control applications. However, sometimes "sample" is used to denote an individual reading within a subgroup. Because this can be confusing, "subgroup" is the preferred term.

3 Sigma. $\pm 3\sigma$ are the two specific points on a normal distribution centered about the mean. 99.73 percent of the population will fall between these values. Since this is essentially the entire population, $+3\sigma$ and -3σ represent the probable range of variation:

$$+3\sigma = X + 3s$$

$$-3\sigma = X - 3s$$

Skewness.   An indication of whether the data in a histogram have a normal distribution. Specifically, skewness is a measure of symmetry. Values for skewness have the following meaning:

0	Symmetrical distribution.
Greater than 1.0	Positive skewness, that is, the distribution has a "longer tail" to the positive side. The median is greater than the mode.
Less than 0	Negative skewness, that is, the distribution has a "longer tail" to the negative side. The median is less than the mode.

The formula for skewness is

$$\sum_{i=1}^{\eta} \frac{(X_i - \bar{X})^3}{3s^3}$$

Another version of this formula is

$$\frac{\eta^2 \sum_{i=1}^{\eta} X_i^3 - 3\eta \sum_{i=1}^{\eta} X_i \sum_{i=1}^{\eta} X_i^2 + 2\left(\sum_{i=1}^{\eta} X_i\right)^3}{\sqrt{\eta \sum_{i=1}^{\eta} X_i^2 - \left(\sum_{i=1}^{\eta} X_i\right)^2}}$$

Special Cause.   A source of nonrandom or intermittent variation in a process. The distribution is unstable and unpredictable. It is also referred to as an assignable cause. (See also *Common Cause* in this glossary.)

Specification Limits (LSL and USL).   The upper and lower specification limits (engineering-blueprint tolerances) are values that determine whether or not an individual measurement is acceptable. They are engineering tolerances established external from the process and should not be confused with control limits.

Standard Deviation.   A measure of the variation among the elements in a group. The formula for the standard deviation is

$$s = \sqrt{\sum_{i=1}^{\eta} \frac{(X_i - \bar{X})}{\eta - 1}}$$

Another version of this formula is

$$\sqrt{\frac{\sum_{i=1}^{\eta} X_i^2 - \left[\left(\sum_{i=1}^{\eta} X_i\right)^2 \Big/ \eta\right]}{\eta - 1}}$$

Subgroup.   A group of individual measurements, typically 2 to 10, used to analyze the performance of a process.

Variation Concept.   Variation occurs in manufacturing because the process conditions are never exactly identical for any two parts or units. If variation did not exist, quality would not be a problem.

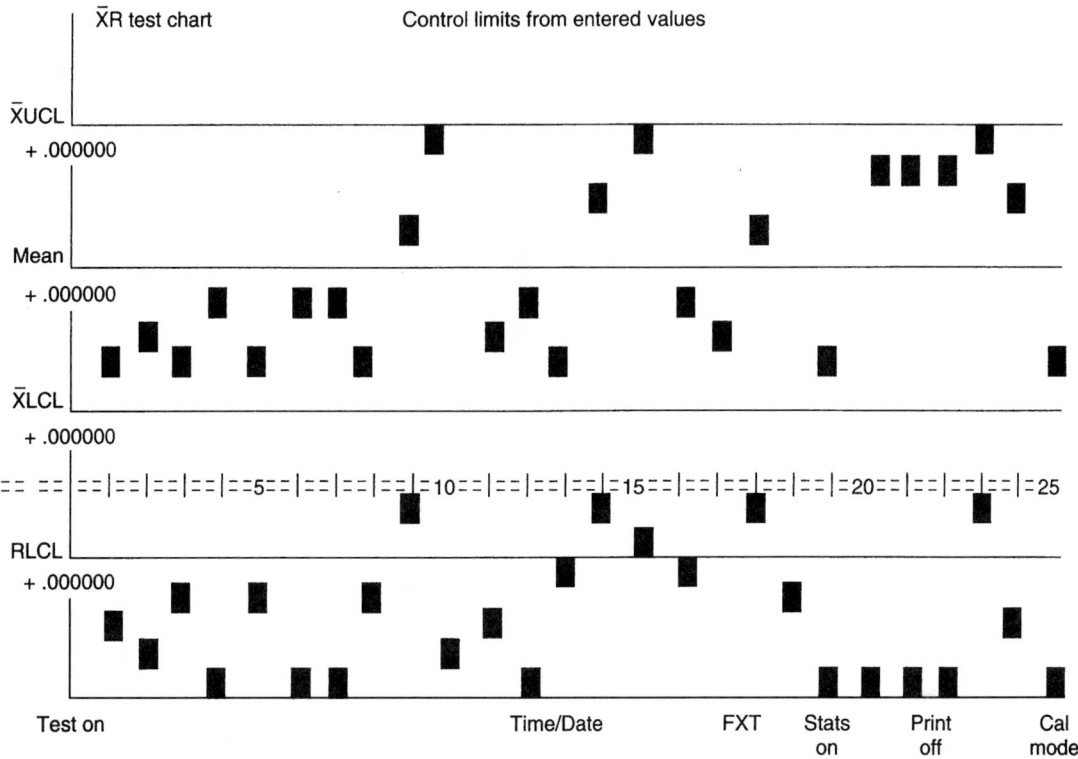

FIGURE 64   Facsimile of computer-generated $\bar{X}R$ chart.

Every piece or unit would be identical—so in the case of parts and pieces, all components would assemble and test correctly. Inasmuch as differences do exist, the goal of SQC is to reduce the variation as much as possible, thereby improving product consistency and quality.

skew2 $\bar{X}R$ Chart.   A chart that displays the subgroup mean and range on a consecutive basis. Upper and lower control limits are plotted to help analyze the process (Fig. 64).

Note that SQC data are plotted in various clumnar forms (Fig. 65).

System Approach to SQC

Modern instrumentation and computer technology have eliminated, or can eliminate, manually entered and manual charting by way of tremendously speeding up SQC data acquisition and the numerous computations required. SQC can be made more encompassing and penetrating. Histograms in electronic storage, and thus easily called up on graphic displays, essentially eliminate paperwork. The remaining areas, namely, the interpretation and corrective action aspects, can be further speeded up through the application of further automation technology. Much of the hardware is available to create SQC systems of varying extent and complexity. In applying advanced instrument and computer technology to SQC systems, there will be a tendency, as occurred in the past in attacking a new area for instrumentation, to overdo in some instances. The established guidelines still apply, that is, do *not* collect more data than are required. A flood of essentially useless information taxes the hardware, the software, and the personnel who are confronted with it.

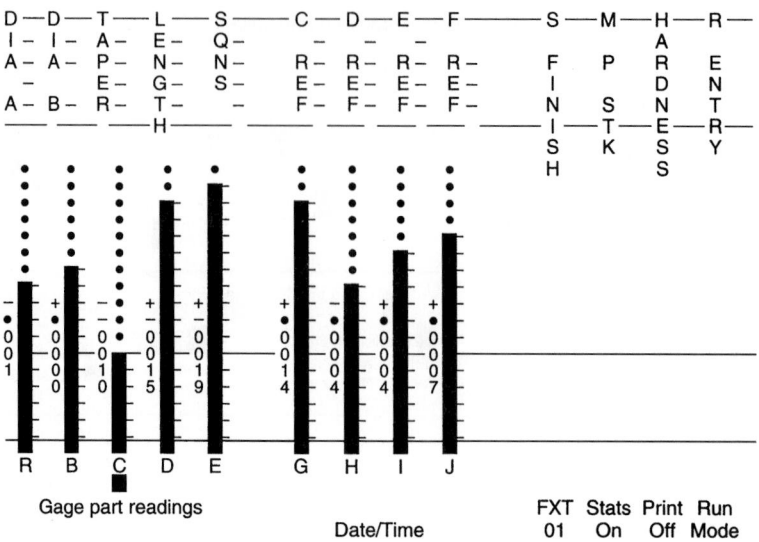

FIGURE 65   Facsimile of computer-generated SQC columnar data presentation.

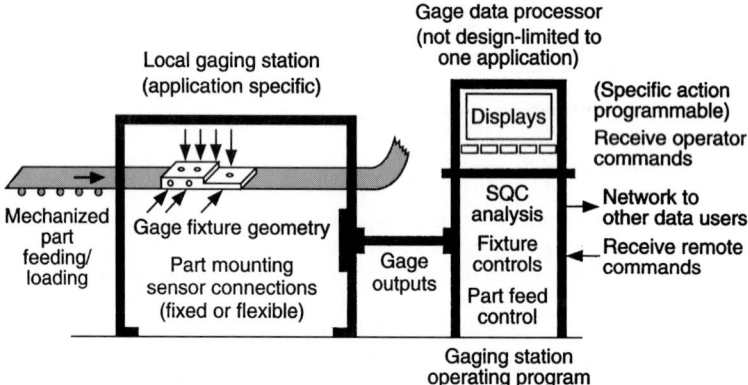

FIGURE 66   Concept of gage data processor that is programmable and need not be redesigned for each gaging station with which it will be used. Gage fixturing remains essentially very application specific, but gaging paperwork needs do not have to be incorporated through redesign of the data processor, but rather this can be done via reprogramming.

SQC is very easy to identify with controlling parts. It is somewhat more difficult to implement as the parts become subassemblies and the latter, in turn, become finished products ready for shipment. It is in the areas of subassembly and final assembly testing where much remains to be learned and achieved.

SQC in the Process Industries

The fundamentals of SQC also are applied in the continuous-process industries, but because of the marked differences in the characteristics of the end products and of the variables that are measured

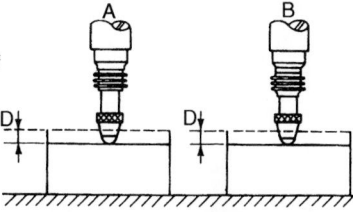

By placing two gage heads *A* and *B* parallel and on the same side of a workpiece and master with gage settings properly actuated, the indicator will show the difference between *A* and *B*. If both master and workpiece are affected by the same source of error *D* (such as temperature), the difference will remain the same. Such a setup can be used for measuring roundness, parallelism and flatness.

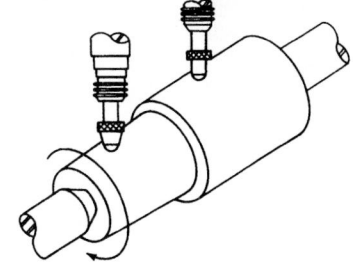

With this setup, only the difference in concentricity is shown on the indicator. If both parts are out of round in same amount, the reading is not affected.

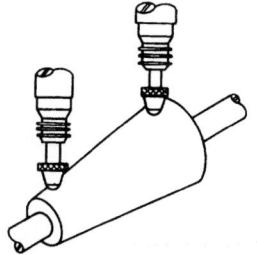

By placing two gage heads parallel on a tapered part, it is possible to chek the degree of taper as compared with a master part, without regard to its diameter.

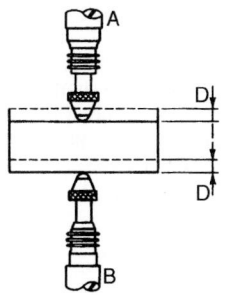

By placing two gage heads *A* and *B* perpendicular and opposite to the workpiece with gage settings properly actuated, the indicator will show the sum of *A* and *B*. If the workpiece is displaced either upward or downward, the indicator reading will not change. In this example, gage head *A* will move a distance +*D* while gage head *B* moves a distance –*D* so that the reading of the thickness of the part does not change because of such displacement of the part. Such a setup can be used to measure thickness or diameters without the need for precision fixtures.

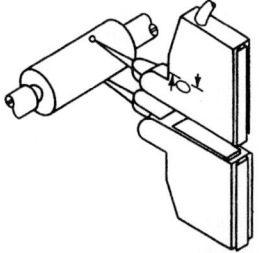

With this setup, the diameter of a workpiece is compared with that of a master without the necessity of placing the part in a precise fixture, since upward or downward displacement does not affect the accuracy of the reading.

FIGURE 67 Examples of the use of impedance-type dimension gages. The first three cases are concerned with "difference" measurements, whereas the last two cases deal with "sum" measurements. (*Brown & Sharpe.*)

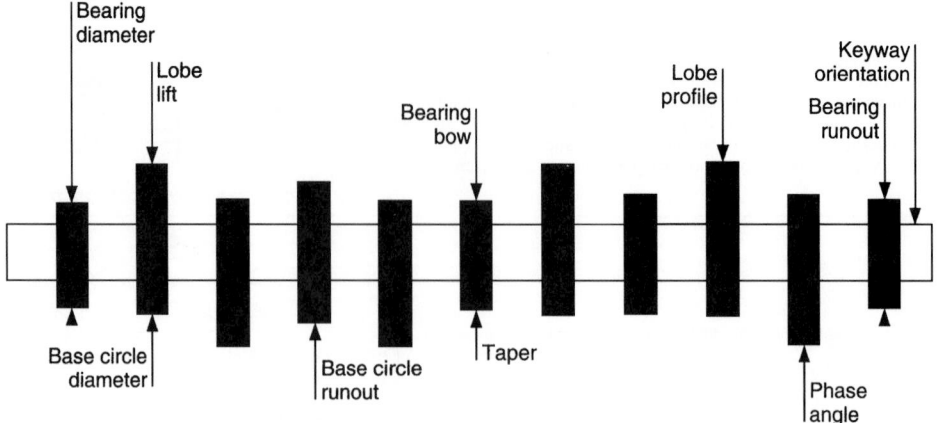

FIGURE 68 Multiple gaging of critical dimensions on the production floor. Difficult characteristics such as lobe profile no longer require sample checks in a laboratory environment. Tooling mounts the camshaft on centers. Gage tracking closely represents the actual usage in an engine.

and controlled, there are some differences in the application of SQC. This has led to the general acceptance of the term statistical process control (SPC) in these industries. The subject is addressed elsewhere in this handbook.

PRODUCTION GAGING SYSTEMS

The wide variety of dimension and position sensors previously described in this article comprises the basis for parts gaging systems, whether they are manual or semiautomated. The range of gaging needs is wide, involving different parts materials and sizes. Also, production runs, ranging from essentially continuous to short runs, add to the complexity and costs of automating gaging operations.

Prior to the availability of digital data processing, gaging operations depended on analog technology. The centralization of data gage information was not easy and SQC analyses were made manually. Depending on the size of the plant, gaging applications were approached on an application-by-application basis, with gage data display engineered along with the design of the fixtures for gaging.

Much progress has been made during the past decade toward streamlining and normalizing (not standardizing) the processing of gage data so that the display, analysis, and control aspects of localized gaging stations can be packaged in a manner such that a "gage data processor" can be used with most local gaging stations once the fixture is readied and the sensors are located and connected. Depending on a user's present and planned requirements, a data gage processor can be configured with a range of capacities and functions. Not having to reengineer the display, analysis, and control aspects of gaging each time a new gaging problem presents itself, is indeed a tremendous advantage. Gage data processors also can generate for and receive from computers and data processors at a higher level. Whereas the gage fixture is essentially inflexible, the gage data processor can have unlimited flexibility (Fig. 66).

Examples of the variety of properties that must be gaged in the discrete-piece manufacturing industries are given in Figs. 67 and 68. Typical applications where gage data processors have been used are listed in Table 4.

TABLE 4 Examples of Gage Data Processor Applications

Parameter	Valve Lifter Body
Measurements	Pin-hole location to open end
	Pin-hole location to seat
	Parallelism of flats to pin hole
	True position of flats to pin hole
	True position of flats to OD
	Pin-hole diameter size
	Pin-hole centrality to OD
	Pin-hole squareness to OD
	Slot width
	True position of slot to OD
Location	Plant floor; gage dedicated to a single production part
Probes	Back-pressure air circuitry transduced to electric LVDTs
	Leaf flexures
	D-plug locators
Gage data processor	Microprocessor based; connected to local area network

	Wheel analyzer
Measurements	Radial runout of each bead seat with respect to pilot, hole and bolt hole pattern
	Lateral runout of each bead seat with respect to pilot hole and bolt hole pattern
	Bolt hole pattern to pilot hole concentricity
Location	Plant floor; gage is adjustable with some change part tooling; designed to handle family of about 50 wheels per hour
Probes	Roller followers
	LVDTs
	Linear encoders
	System uses harmonic analyzer to determine runouts and concentricities
Gage data processor	Microprocessor based; connected to local area network

	Tapered roller bearing inner race
Measurements	Raceway diameter
	Roundness
	Width
	Undercut presence
Location	Plant floor; adjustability and change part tooling allow gage to handle wide range of "green" roller bearing inner races
Probes	LVDTs
	All measurements are dynamic as part is rotated during gaging cycle
Gaging rate	1800 parts per hour
Gage data processor	Microprocessor based; connected to local area network

	Tapered roller bearing outer race
Measurements	Upper, center, and lower IDs
	Taper, crown, and width
	Upper and lower ODs
	Taper OD
Location	Plant floor; dedicated change part tooling allows gage to handle wide range of bearing sizes while still maintaining required accuracy and repeatability
Probes	Tapered bore plugs with air circuits
	LVDTs
Gaging rate	1800 parts per hour
Gage data processor	Microprocessor based; connected to local area network

(Continues)

TABLE 4   (*Continued*)

	Piston
Measurements	Skirt diameter
	Ring groove diameters
	Ring groove width and location
	Wrist pin bore size, taper, and roundness
Location	Plant floor; gage is dedicated to single piston
Probes	LVDTs
	Bore plug with air circuits
Gage data processor	Microprocessor based; connected to local area network

	Single flank gear
Measurements	Relative angular velocities of two mating gears; frequency spectrum of driven gear is analyzed to locate source of transmission noise
Location	Laboratory
Probes	Rotary encoders
Gage data processor	Microprocessor based; connected to local area network

	Tight mesh gear
Measurements	Functional tooth thickness
	Pitch diameter runout
	Sectional runout
	Out-of-round
	Tooth-to-tooth
	Nicks
	Average lead
	Lead variation
	Average taper
Location	Plant floor; gage is dedicated to single part number
Probes	LVDTs
	Gear is run in tight mesh with two rolling master gears. The axis of the first master gear (called center-distance rolling master) is only allowed to translate. The second rolling master (lead and taper rolling master) is mounted on a two-axis gimbal. The positions of these rolling masters are monitored as they mesh with the part gear. These signals are analyzed to determine values for the various characteristics.
Gaging rate	About 250 per hour
Gage data processor	Microprocessor based; connected to local area network

	Transmission sprocket
Measurements	Thickness
	Tooth width
	Overall height
	Web thickness
	Sleeve diameter, size, straightness, and roundness
	Runout of spline to sleeve diameter
	Parallelism of hub to reference face
	Parallelism of thrust face to reference face
	Spline tooth—runout to sleeve diameter, pitch diameter, and nicks
Location	Plant floor; gage is dedicated to single transmission sprocket
Probes	Back-pressure air circuits transduced to electric LVDTs
Gaging rates	415 parts per hour
Gage data processor	Microprocessor based; connected to local area network

OBJECT DETECTORS AND MACHINE VISION

Control systems concerned with position measurement tend to fall into one of two main categories:

1. Systems as exemplified by various production machines, where achieving a position, measured in exact geometric coordinates, is the goal. These systems are described in Article 1 in this handbook section.
2. Systems where control action occurs because a given object occupies (even for an instant) a specific location, that is, a definite position within a manufacturing space. This article is concerned with that domain.

   Object detectors may be used to initiate motion, stop motion, and return motion, for example, in a pick-and-place robot or in the repetitive cycling of a machine. As the name implies, limit switches are used by the millions on machinery to limit a machine stroke (of vastly varying nature) in order to avoid injury to operators, machines, or materials in process. Object detectors (sometimes aided by machine vision) count and inspect products for filling level, label placement, and the absence (nonpresence) of too many or too few units in packaging, among many other applications.

   Object detectors have been so commonplace in the manufacturing scene for so many years that in a way they have "disappeared into the woodwork." Nevertheless, object detectors have been refined over generations of use and generally have kept pace with technology through the incorporation of scientific advancements, as found in Hall-effect and ultrasonic detectors. After many years of considering statistical quality control as an entity apart from instrumentation, the data inputs from object detectors now are incorporated into control system electronics for the purpose of making sometimes complex mathematical calculations and displaying the results (often right on the manufacturing floor) for rapid determinations of product quality (in-process and final) for both small and vast integrated manufacturing operations.

ELECTROMECHANICAL LIMIT SWITCHES

These switches are the "workhorses" of automated systems. Because of their extensive use, they usually are readily available out of stock. Depending on specific use conditions (temperature, hazardous atmosphere, light to heavy duty, and so on) most manufacturers offer a wide line of limit switches. In general, limit switches fall into three size (dimensional) categories. The switch box proper may range in height from 50 to 150 mm (2 to 6 inches), in width from 15 to 80 mm (0.6 to 3.2 inches), and in depth from 40 to 85 mm (1.6 to 3.4 inches). The upper operating temperature range, depending on the model selected, is 121°C (250°F); the lower temperature range is −32°C (−25°F). Housing material may be zinc, aluminum, or steel treated for corrosion and weather resistance. In some designs the inner portions of the switch circuitry are prewired and potted. Switches that meet European (DIN) specifications are also available.

Specifying Details.   The approximate specifying details, considering a range of models and brands, include the following.
   Control Actions.   Single-pole double-throw (SPDT) and double-pole double-throw (DPDT).
   Switching Capacity.   10 to 15 amperes 125 volts ac (inductive load); 5 amperes 125 volts ac (resistive load).

Standard. Overtravel 60° minimum, pretravel 15° maximum, differential travel 3° (single-pole) and 7° (double-pole) maximum.

Low-Differential-Travel Design. Overtravel 68° minimum, pretravel 7° maximum, differential travel 3° (single-pole) and 4°(double-pole) maximum.

Low-Operating-Torque Design. Overtravel 60° minimum, pretravel 15° maximum, operating torque 1.7 in · lb (0.19 N · m) maximum.

Low-Torque, Low-Differential-Travel Design. Overtravel 68° minimum, operating torque 1.7 in · lb (0.19 N · m) maximum, differential travel 3° (single-pole) and 4° (double-pole).

Sequence-Action Design. Delayed action between operation of two poles, in each direction; overtravel 48° minimum.

Center-Neutral Design. One set of contacts operates on clockwise rotation, and another set on counterclockwise rotation; overtravel 53° minimum.

Maintained-Contact Design. Operation maintained on counterclockwise rotation, reset on clockwise rotation, and vice versa; overtravel 20° minimum.

Actuators. Most problems of solving the geometry for matching the limit switch to the controlled machine have been solved with standard switch designs. These include (1) rotary operating heads, (2) plunger operating heads, and (3) wobble lever operating heads (Fig. 1). Gravity-return switches are also available. With these designs, the weight of the actuating lever must provide sufficient force to restore it to the free position. The very small 0.035-N · m (5-in · oz) operating torque is useful in some conveyor applications because it permits operation with small or lightweight objects.

Solid-State Limit Switches. Conventional limit switches incorporate Hall-effect principles and solid-state circuitry. The output is computer-compatible and requires no in-between electronics. The output interfaces directly with most electronic circuits, discrete transistor circuits, microprocessors, and integrated logic circuits. This switch cannot be used in areas where extremely high magnetic fields are present.

Fiber-Optic Limit Switches. A comparatively recent innovation in limit-switch design is the incorporation of fiber optics into the device's operation. Switching is accomplished with a cool light signal. The switch is not affected by electromagnetic radiation (EMI) or radio-frequency (RF) interference. The device is also intrinsically safe for use in hazardous locations. It is available for normally open or normally closed circuitry (in fiber-optic terminology). In the normally open mode, the light path is blocked by the shutter; in the normally closed mode, the light path is complete. Switch actuation blocks the light path.

PROXIMITY SENSORS

There are numerous instances in automated systems where an object (presence or nonpresence) must be detected, but not physically contacted (contactless). Several methods have been developed and refined over the years to achieve this task, including (1) electromagnetic inductive [radio-frequency (RF)] devices and other magnetic devices, such as variable-reluctance sensors, magnetically actuated dry-reed switches, Hall-effect switches, and Wiegand switches; (2) ultrasonic detectors; (3) photoelectric sensors; and (4) capacitive sensors. Machine vision also plays a role in some installations. Some devices are limited to ferrous materials, others, when properly calibrated, are applicable to other metals. Some proximity sensors are not affected by the target materials.

INDUCTIVE PROXIMITY SENSORS

The technology of inductive proximity sensors was established in the 1950s. Early sensors essentially were limited to the detection of iron objects, but as more was learned, their use expanded to the

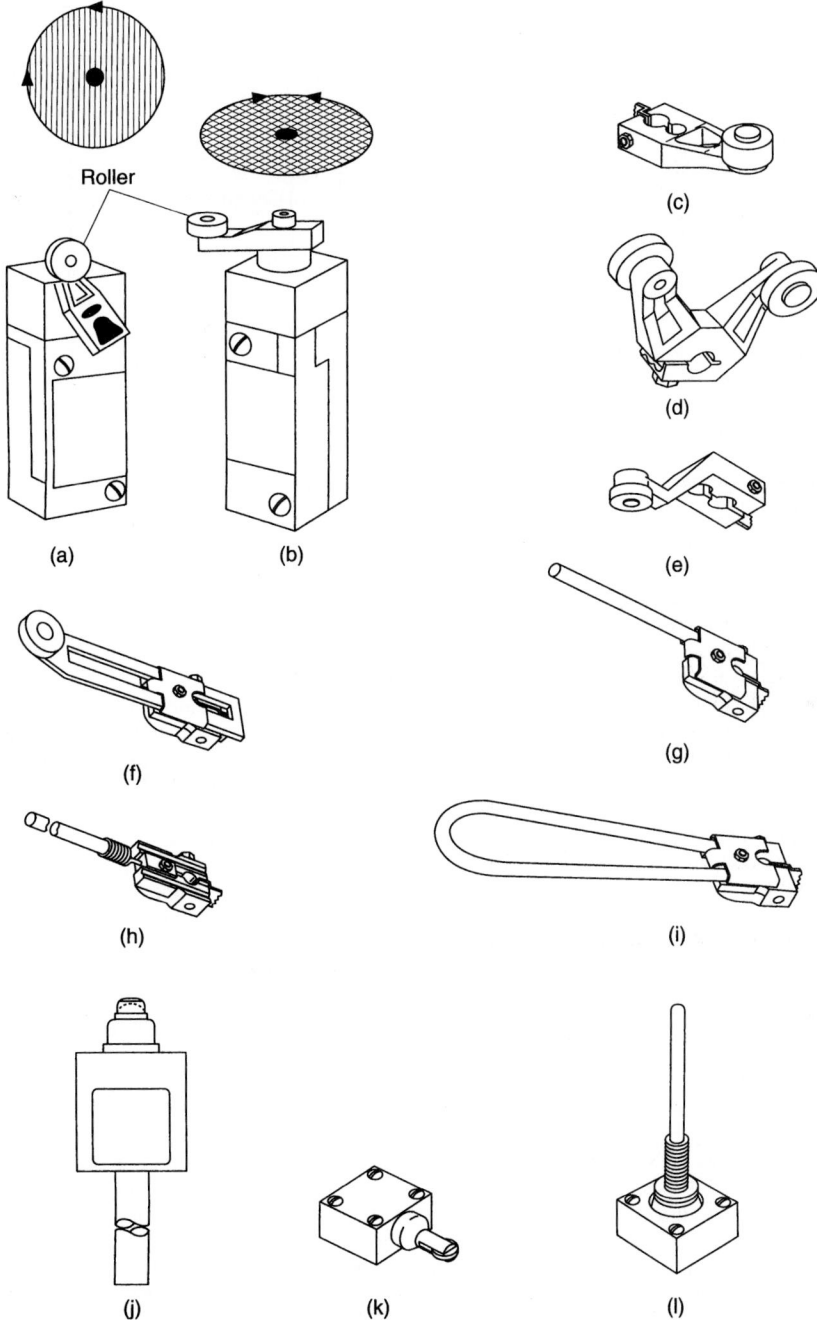

FIGURE 1 Various configurations of electromechanical limit switches. (*a*) Side-mounted roller. (*b*) Top-mounted roller. (*c*) Standard roller lever. (*d*) Yoke roller lever. (*e*) Offset roller lever. (*f*) Adjustable-radius roller lever. (*g*) Rod lever. (*h*) Spring-rod lever. (*i*) Flexible loop lever. (*j*) Top plunger. (*k*) Side roller plunger. (*l*) Wobbler switch. (*MICRO SWITCH.*)

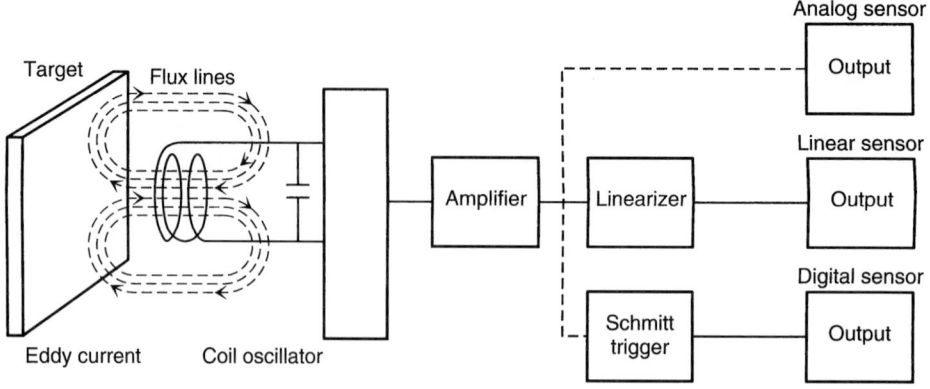

FIGURE 2   Inductive proximity sensor operating principle.

detection of other materials, including aluminum, brass, copper, heavily alloyed steels, and many others that make up industrial forms and parts production. It should be pointed out that sensors designed specifically for ferromagnetic materials are also available.

The key element of the detector is an oscillator which generates an RF field at the sensor face. The oscillator consists of an LC tank circuit (tuned circuit) and an amplifier circuit with positive feedback. The inductance and the capacitance of the LC network determines the oscillator frequency, which can range from 20 kHz to several megahertz (MHz).

The inductance portion L of the tuned circuit is formed by an air-wound or ferrite-core coil. The oscillator circuit has just sufficient positive feedback to sustain oscillation. This generates a sine wave that varies in amplitude, depending on the target's distance from the sensing face.

With a metal target present in front of the oscillator coil, the RF field generates eddy currents on the target's surface. This causes the amplitude to decrease or allows the oscillation to die off. This change in amplitude provides target presence or absence information that is sent either to the output (analog) sensor or through a level detector (Schmitt trigger) that produces a digital (square-wave) output. The digital output is either on or off. The analog version of the sensor is described later.

A schematic circuit of an inductive proximity sensor is shown in Fig. 2. A linearizer between the amplifier and its output provides a more accurate position of the target relative to the sensing face. The output signal has a direct linear relationship with the gap between the sensor and the target surface.

Certain variables must be considered when applying the sensor. These include (1) the nominal distance between sensor and target, (2) the target size, (3) the target material, (4) the target approach, (5) the switching frequency, and (6) the type of signal required, namely, digital, linear, or analog.

Target Distance and Size.   These are very closely related variables. Small targets decrease the nominal sensing distance. In contrast, however, beyond a point, targets larger than the standard do not increase the sensing distance. Suppliers, for each size and configuration of sensor, publish a value of "nominal sensing distance," which is the distance between target and sensor at which a given sensor will turn on. The actual sensing distance may be affected by a number of factors, including manufacturing and temperature tolerances. Variations of as much as ± 10 percent may be encountered, but they can be adjusted during calibration at the time of installation. Temperature drift tolerances also must be considered. Over a range of -25 to $85°C$ (-13 to $185°F$), the tolerance increases by ± 10 percent.

A shielded sensor will sense only the front of its face and ignores objects to the side. The presence of side materials, however, may cause a slight shift in operating characteristics.

A standard target is an object used for making comparative measurements of the operating distance. Usually a square of mild steel, 1 mm thick, is used. The length of the side of the square is equal to

TABLE 1 Correction Factors for Target Materials

Object material	Standard distance X
400 series stainless steel	1.15
Cast iron	1.10
Mild steel	1.00
Aluminum foil (0.05 mm)	0.9
300 series stainless steel	0.7
Brass MS63F38	0.4
Aluminum ALMG3F23	0.35
Copper CCUF30	0.3

either the diameter of the circle inscribed on the active surface of the sensitive face of the sensor, or three times the rated operating distance, whichever is greater.

The movement pattern of the target also affects the sensing distance capability and ultimately determines at which point the sensor will switch.

Target Material. When a target material differs from the standard, as described previously, a material correction must be made. The correction factors (multipliers of standard distance) listed in Table 1 can be used.

Switching Frequency. Defined as the actual number of targets to which the sensor can respond in a given time period, it is usually expressed in hertz (cycles per second).

General Sensor Characteristics. Cylindrically shaped sensors are used widely and have face diameters as small as 4 mm (0.16 inch) upward to 76 mm (3 inches). Standard sensing distances range from 0.8 to 32 mm (0.03 to 1.3 inches). Electrically there are several options, including 3-wire ac, 4-wire ac, 2-wire dc, 3- and 4-wire dc, and 2-wire ac/dc universal voltage sensors. External configurations also include limit-switch style and vane, rectangular, and ring self-contained noncylindrical sensors. Some designs are equipped with light-emitting diodes (LEDs) to indicate power on, short circuit, and so on (Fig. 3).

Analog Proximity Sensors. These devices provide a variable current output that is proportional to the distance between a metal target and the sensor face. The current also will vary if different metals or different numbers of small parts are placed in the sensing field. Consequently analog sensors are a convenient way of positioning, discriminating between assorted metals, and checking parts count. Historically the cost of the added electronics was a deterrent in the use of analog sensors. With the costs dropping, analog sensors are regaining acceptance (Fig. 4).

Hall-Effect Proximity Sensors

When a semiconductor, through which a current is flowing, is placed in a magnetic field, a difference in potential (voltage) is generated between the two opposed edges of the conductor in the direction mutually perpendicular to both the field and the conductor. This effect is utilized in some proximity sensors and has been incorporated into electromechanical limit switches, as shown in Fig. 5.

Wiegand-Effect Switches

A Wiegand wire is a small-diameter wire that has been selectively work-hardened so that the surface and the core of the wire differ in magnetic permeability. When subjected to a magnetic field, the wire

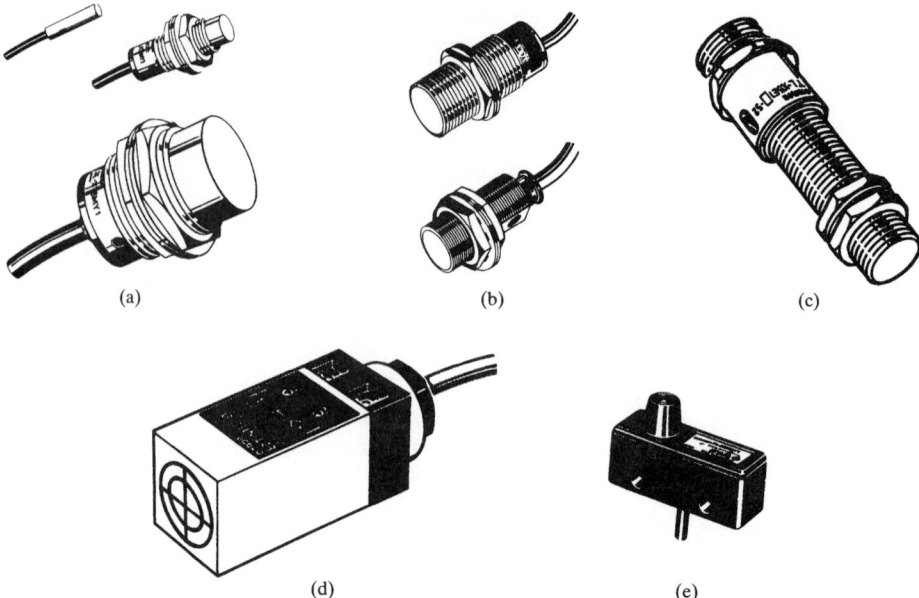

FIGURE 3 Representative inductive proximity sensors. (*a*) Short-length cylindrical sensor in metal housing, available in shielded or unshielded models. Ac and dc incorporate reverse polarity protection. Most feature short-circuit protection. Detecting distance 0.8 to 18 mm (0.03 to 0.71 inch); diameter 4 to 5.4 mm (0.15 to 0.21 inch); length 25 to 57 mm (0.98 to 2.24 inches). Incorporates operation indicators. Nickel-plated brass body. (*b*) Short-length cylindrical sensor. Wide operating voltage and extended service temperature ranges. Rugged construction. Shielded and unshielded models. Ac or dc incorporates reverse polarity protection. Detection distance 1 to 18 mm (0.04 to 0.71 inches); diameter 12 to 30 mm (0.47 to 1.2 inches); length 80 mm (3.15 inches). (*c*) Threaded cylindrical, shielded sensor. Accepts plug-in connection. Weld field immunity to 2k gauss. RFI immune. Short-circuit protection. (*d*) Short, limit-switch style sensor. Shielded for flush mounting in metal. Weld field immunity available. End- and side-sensing types. Detecting distance 12.5 mm (0.49 inches); height 77.7 m (3.06 inches); width 34.5 mm (1.36 inches); depth 34.5 mm (1.36 inches). (*e*) Basic switch size. Oil-tight epoxy housing. Operation indicator useful for retrofitting mechanical positioning switches. Detection distance 2 mm (0.08 inches); height 28.7 mm (1.13 inches); width 17.5 mm (0.69 inches); depth 49.2 mm (1.94 inches). (*Omron Electronics*, Inc.)

emits a well-defined pulse that requires little signal conditioning. This pulse induces a voltage in the surrounding sensing coil. The wire is insensitive to polarity and emits a pulse whether the magnetic field is flowing from north to south or vice versa. A Wiegand proximity sensor senses the presence or absence of ferromagnetic material.

Magnetically Actuated Dry-Reed Switches

Consisting of a thin reed (wire) contained in a hermetically sealed tubelike container, this type of switch is both inexpensive and rugged. Whenever an activating magnet approaches the critical range of the switch, a contact closure is made. Life expectancy generally is in excess of 20 million operations at contact ratings of about 15 VA. These switches generally can operate loads directly, thus avoiding the cost and complexity of comparable solid-state systems. Since the actuating magnet can be installe D counting, positioning, and synchronizing. Contact closure speeds can be up to 100 per second. Mercury switches with flexible electrodes that can be attracted by the proximity of a magnet also have been used.

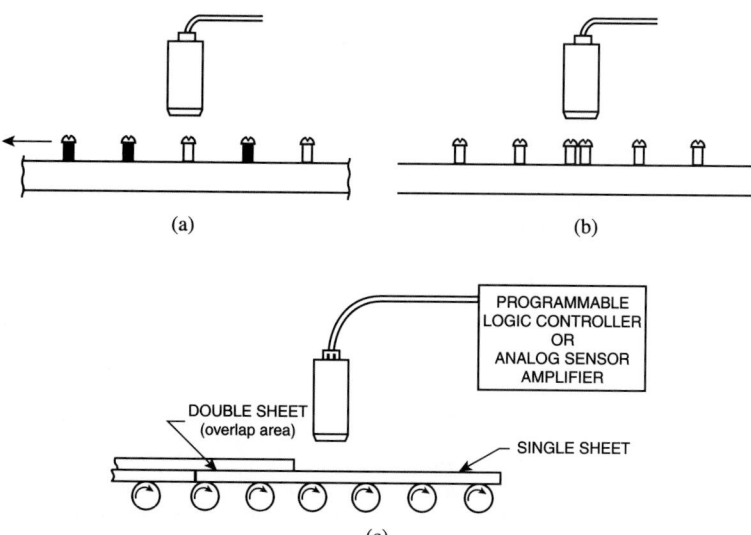

(a)   (b)

(c)

FIGURE 4   Some advantageous applications for analog inductive proximity sensor. (*a*) A stainless-steel screw mixed in with steel screws can easily be detected. The two metals yield two different current outputs, even without changing the sensing distance. (*b*) When an analog sensor detects two small parts instead of one, the change in the output is twice that of one part. The sensing distance remains constant. (*c*) Aluminum sheets slide down a conveyor, at a certain distance from the sensors, prior to a stamping operation. If two or more sheets are present, the surface passes closer to the sensor and causes the output curve to change. (*MICRO SWITCH.*)

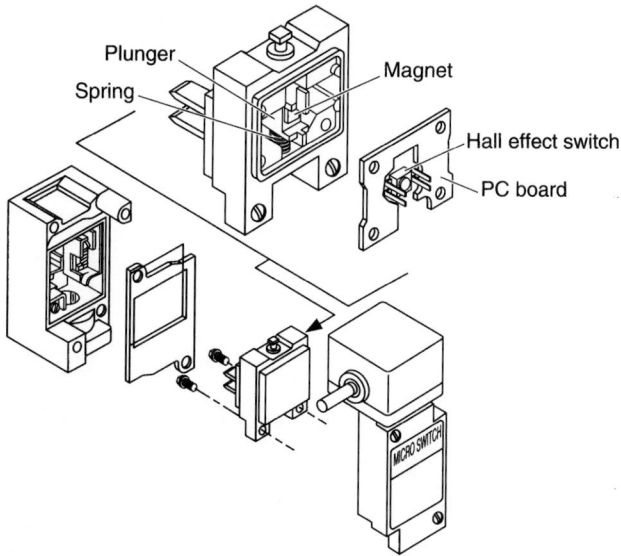

FIGURE 5   Hall effect is used in proximity detectors as well as in infrastructure of other devices, as shown here incorporated in a solid-state limit switch. (*MICRO SWITCH.*)

CAPACITIVE PROXIMITY SENSORS

Based on the fundamental phenomenon of electrical capacitance, these devices produce an oscillating electric field that is sensitive to (1) dielectric materials, such as glass, rubber, and oil, and (2) conductive materials, such as metals, salty fluids, and moist wood. The principle is shown schematically in Fig. 6.

Capacitance is a function of the size of the electrodes, the distance between the electrodes, and the dielectric constant D of the material between the electrodes. $D_{air} = 1$. The capacitance is given by

$$C = \frac{D \cdot A}{d}$$

where A = area
D = dielectric constant of material between electrodes
d = distance between electrodes

A simple capacitive sensor is shown in Fig. 7. The top electrode is the face of the sensor. A seal ring, the target, passes between it and the ground electrode (a metal conveyor belt). The sensor housing insulates the electrode from galvanic coupling to ground. The rubber seal ring has a dielectric constant D of 4.0. When it enters the electric field, the capacitance increases. The sensor detects the change in capacitance and provides an output signal.

Figure 8 illustrates a metal target, or some other conductive material, entering the electric field. The resulting increase in capacitance is detected and converted to an output signal. If the effective distance between electrodes is reduced, the result is an increase in capacitance.

Unshielded Capacitive Sensor.   This sensor is principally used to detect conductive materials at maximum distances. When detecting nonconductive materials, a path to ground is required. Unshielded sensors are designed to sense conductive materials through a nonconductive material, such as water in a glass or plastic container.

Conductive and nonconductive materials cause an increase in capacitance due to a dielectric change in comparison with air. The ground electrode, as shown previously in Figs. 6 through 8, is not required because a path to ground will serve equally well.

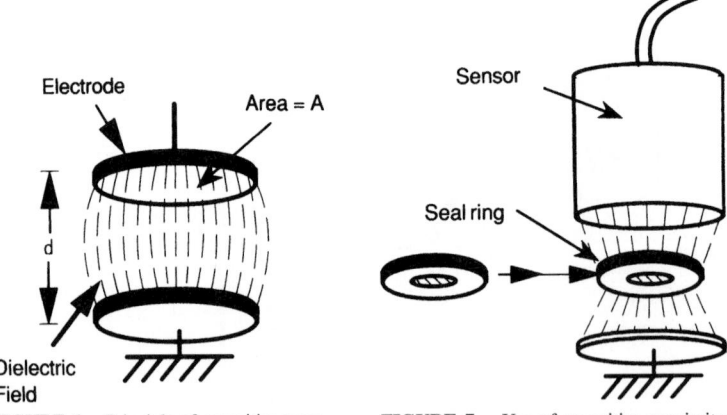

FIGURE 6   Principle of capacitive proximity sensor shown schematically. (*MICRO SWITCH.*)

FIGURE 7   Use of capacitive proximity sensor on metal belt conveying rubber seal rings. (*MICRO SWITCH.*)

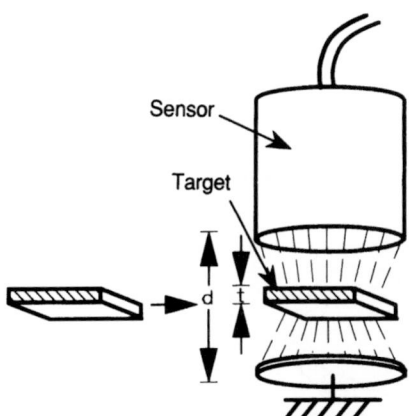

FIGURE 8 Capacitive proximity sensor detecting a metal target or some other conductive material entering the electrical field. Resulting increase in capacitance creates output signal. If the effective distance t is reduced, this will cause an increase in capacitance. (*MICRO SWITCH.*)

The sensor shown in Fig. 9 is generally referred to as unshielded. This type of device often is used for "looking through" a nonconductive material in order to control the level of the material. The device can also be used for controlling the level of a nonconductive material if a path to ground, such as a metal bin wall, is present.

Shielded (Metal-Body) Capacitive Sensor. A general-purpose device, it is used for sensing nonconductive materials, such as wood, plastic, cardboard, and glass. A path to ground does not always exist (Fig. 10).

Silo (Plastic-Body) Touching Sensor. This device is actuated by a material touching the sensor's detecting surface. A sensitivity adjustment is provided for tuning out certain materials as well as for setting detection for a specific mass of material.

The silo-type unit works on the same principle as the standard unshielded sensor. However, the silo is designed specifically for touching and detecting solid nonconductive materials, such as granular or powdered material in a bin or silo, and thus for determining the level of bulk materials.

Compensating for Contaminants. When using capacitive proximity sensors, any material entering the sensing field can cause an output signal. This includes water droplets, dirt, dust, and other contaminants on the sensor face. The use of a compensation electrode in the sensor can eliminate this problem (Fig. 11).

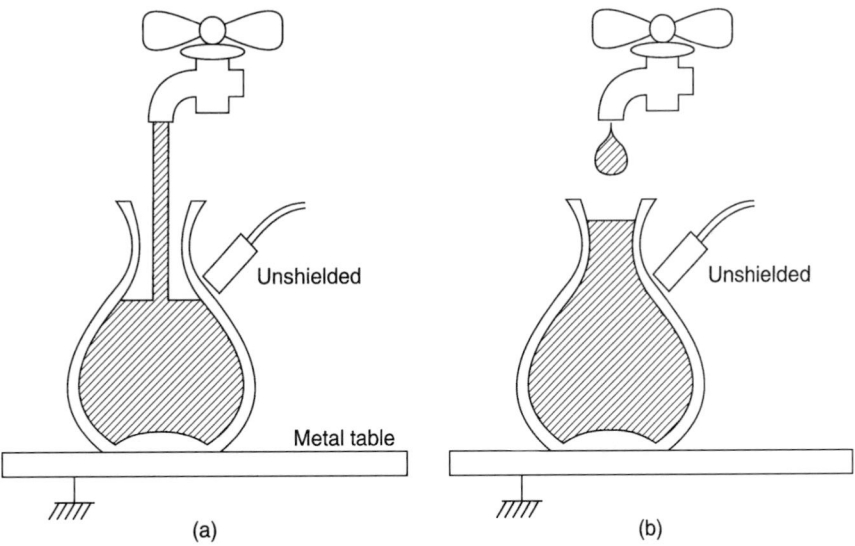

FIGURE 9 Use of unshielded capacitive sensor. (*a*) Level of conductive fluid pouring into glass bottle is below sensor. (*b*) Fluid has reached level of sensor, providing the ground electrode. This occurs even though fluid and metal table are separated by the glass of the bottle. The three materials form a capacitor. The alternating current provides a path to ground. With the ground electrode now in place, the circuit closes and a signal results. (*MICRO SWITCH.*)

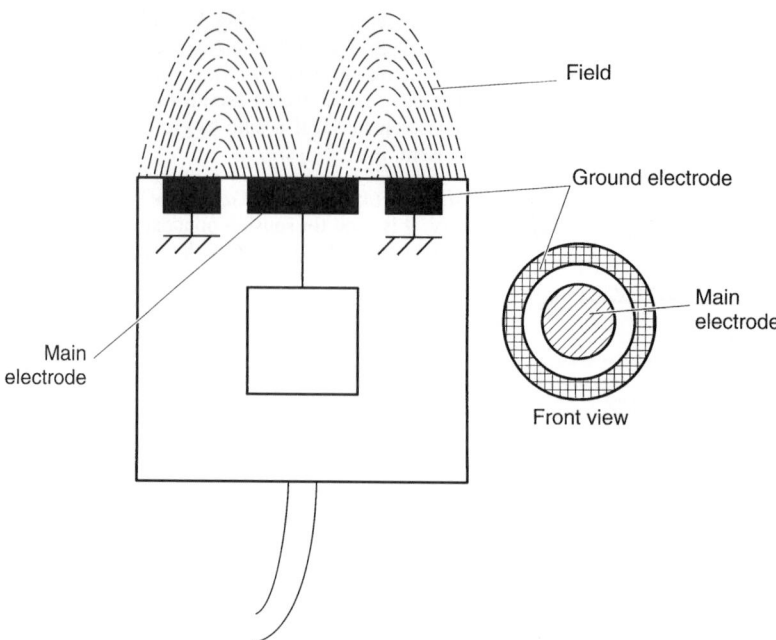

FIGURE 10   In a capacitive proximity sensor application, a path to ground does not always exist. By incorporating the ground electrode in the sensor, an electric field is created independently of any outside path to ground. The field functions the same as when the electrodes face each other. The shielded sensor shown can sense any material, grounded or ungrounded. (*MICRO SWITCH*.)

Combined Use of Shielded and Unshielded Sensors.   Some applications can take advantage of this combination, as illustrated in Fig. 12.

Physical Configurations of Capacitive Proximity Sensors.   These sensors commonly are of a cylindrical shape, as shown in Fig. 13*a*. Other formats are illustrated in Fig. 13*b* and *c*.

ULTRASONIC PROXIMITY SENSORS

Specific applications of ultrasonic measurement technology are associated with numerous processing and manufacturing variables and thus are mentioned in other sections of this handbook. The serious consideration of ultrasonic devices as proximity sensors is relatively recent as compared with their use in connection with other instrumental variables.

Proximity detection is one of the lesser demanding and sophisticated applications for ultrasonic technology because the application only requires the detection of presence or nonpresence, as contrasted, for example, accurately measuring the thickness of a part, the density of a fluid, or detecting exact locations of flaws in materials, among other demanding applications.

Ultrasonic gages generally fall into two categories: (1) resonance types, which produce a frequency-modulated continuous-wave signal, and (2) pulse-echo types, which operate somewhat like a sonar system. In the latter system, if a target is present, a small amount of the signal is reflected back to the transducer. Most ultrasonic sensors use a transducer which also serves as an emitter and receiver. Upon receipt of the echo, the amount of time required for the signal to travel between sensor and target is a measure of the target distance. Most sensors incorporate temperature compensation because of thermal influences on the speed of the ultrasonic wave. By definition, any sound exceeding

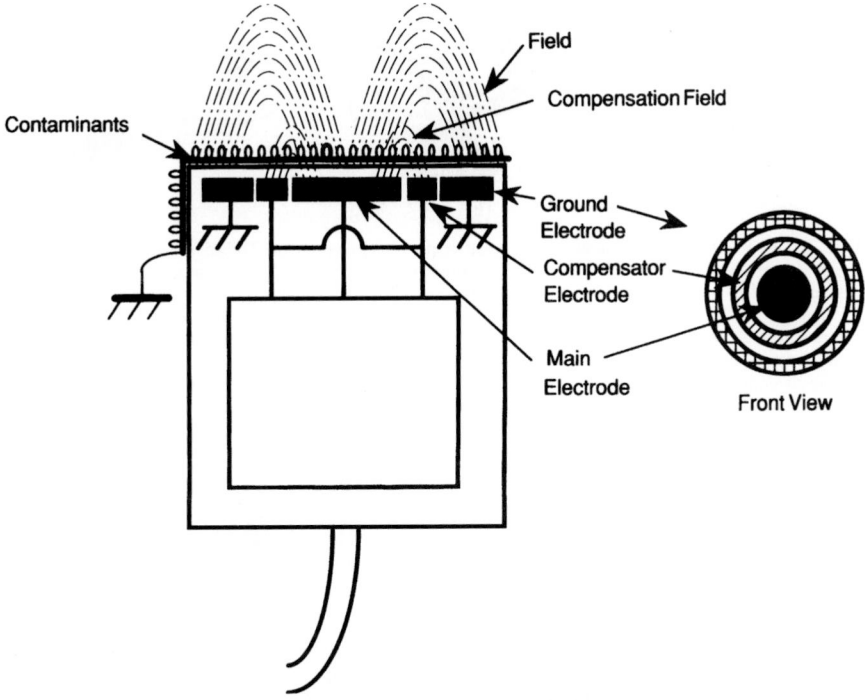

FIGURE 11   Shielded capacitive proximity sensor with two sensing fields, its own and a compensation field which the electrode creates. When contaminants lie directly on the sensor face, both fields are affected and the capacitance increases by the same ratio. The sensor does not "see" this as a change in capacitance, and an output is not produced. The compensation field is very small and does not extend very far from the sensor. When a target enters the sensing field, the compensation field is unchanged. The disproportionate change in the sensing field (with respect to the compensation field) is detected and converted to an output. (*MICRO SWITCH.*)

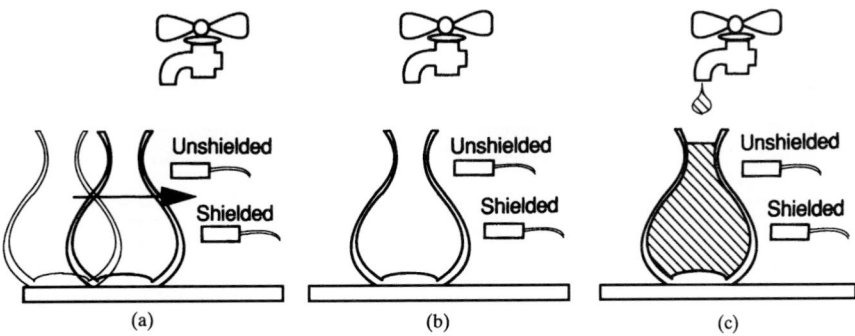

FIGURE 12   An unshielded sensor and a shielded sensor can work together. The shielded sensor locates the glass bottle so it can be filled with liquid. The unshielded sensor indicates that the fill level is reached and can be turned off. (*a*) Neither sensor switches as the bottle approaches. (*b*) The shielded sensor senses the entrance of the glass into its electric field, and it switches. (*c*) The fluid has reached the level of the unshielded sensor, and it switches. Shielded sensors may be flush-mounted in any solid material. (*MICRO SWITCH.*)

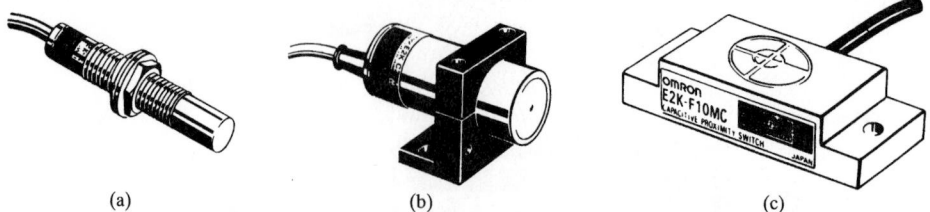

(a) (b) (c)

FIGURE 13 Representative configurations of capacitive proximity sensors. (*a*) Threaded cylindrical sensor. Detects glass, plastic, wood, liquids, and metallic objects. Detects materials inside nonmetallic containers. Fixed sensing distance (depending on model) 4 mm (0.16 inch), 8 mm (0.32 inch), 15 mm (0.59 inch); diameter (depending on model) 12 mm (0.47 inch), 18 mm (0.71 inch), 30 mm (1.2 inches); length 80 mm (3.15 inches). Plastic body. (*b*) Unshielded cylindrical sensor. Detects glass, plastic, wood, water, and metallic objects. Detects materials inside nonmetallic containers. Adjustable sensitivity for wide detection range. Furnished with mounting bracket. Detecting distance 3 to 25 mm (0.12 to 0.98 inch), adjustable; diameter 34 mm (1.34 inches); length 82 mm (3.23 inches). Plastic body. (*c*) Thin, flat-pack sensor. Detects metallic and nonmetallic objects. Compact size for mounting on conveyor walls or flush against metallic surfaces. Detecting distance 10 mm (0.39 inch); height 10 mm (0.39 inch); width 50 mm (1.97 inches); depth 20 mm (0.79 inch). Plastic body. (*Omron Electronics, Inc.*)

20,000 Hz is considered in the ultrasonic range. By using high frequencies, a device becomes less immune to environmental noise in its proximity. However, higher frequencies reduce the sensor's "seeing" distance, and thus the sensor designer must achieve a suitable compromise. Typically an industrial ultrasonic transducer will operate at 215,000 Hz.

From the standpoint of proximity sensing, ultrasonic devices have the advantage of working well in connection with nearly all materials when the devices are properly applied. Ultrasonic devices also have the advantage of detecting presence or nonpresence over longer distances than most other sensors, but some of the negative aspects include effects of temperature, surface finish (reflection), humidity, air turbulence, and inaccuracies that may arise from improper geometry between sensor and target.

The effect of distance between target and sensor is very significant. If a target inside the ultrasonic beam is positioned at 200 mm (7.9 inches) from the sensor, the received echo is 4 times stronger than if this distance were 400 mm (15.8 inches). Thus it is possible that a small object placed far from the sensor may not be detected.

Almost all materials and targets reflect sound and thus can be detected. Only sound-absorbing materials, such as textiles, weaken the echoes such that the maximum sensing distance must be greatly reduced. Thin-walled targets [below 0.005 mm (0.002 inch)] are also difficult to detect. The most effective sensor-target distance in such instances should be determined experimentally. Additional factors are illustrated and described in Figs. 14 through 17.

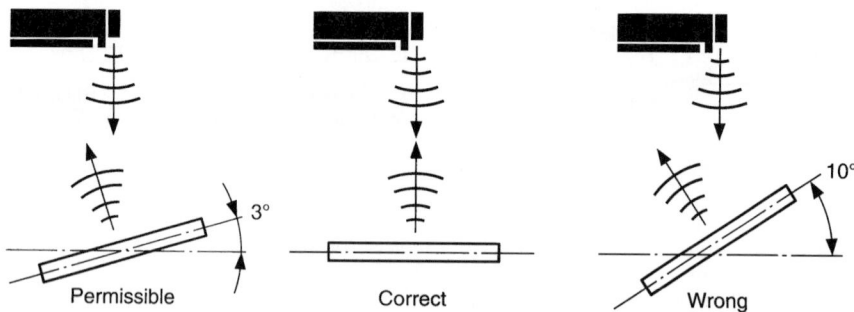

Permissible Correct Wrong

FIGURE 14 Effect of inclination on ultrasonic sensor beam. If a smooth, flat target is inclined more than $\pm 3°$ to the normal of the beam axis, part of the signal is deflected away from the sensor and the sensing distance is decreased. However, for small targets located close to the sensor, the deviation from normal may be increased to $+8°$. If the deflection angle exceeds $\pm 12°$, all of the signal is deflected and there is no signal response. (*MICRO SWITCH.*)

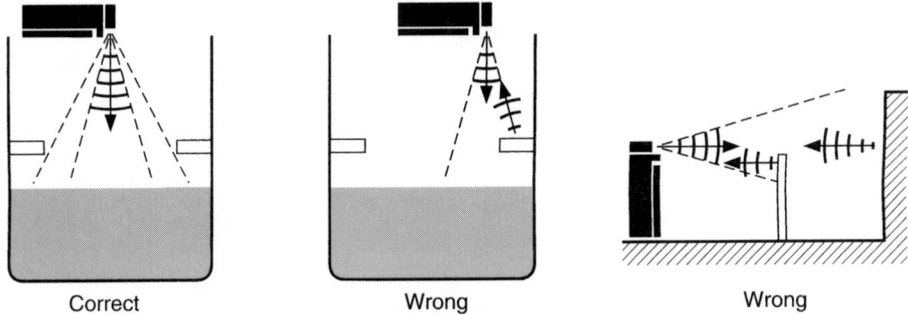

FIGURE 15   All sound-reflecting targets with sufficient surface situated inside the beam cone will be detected. To avoid false (parasitic) echoes, the interfering objects should be removed, or the ultrasonic sensor relocated. Similarly any sound-absorbing materials should not be in the beam path. (*MICRO SWITCH.*)

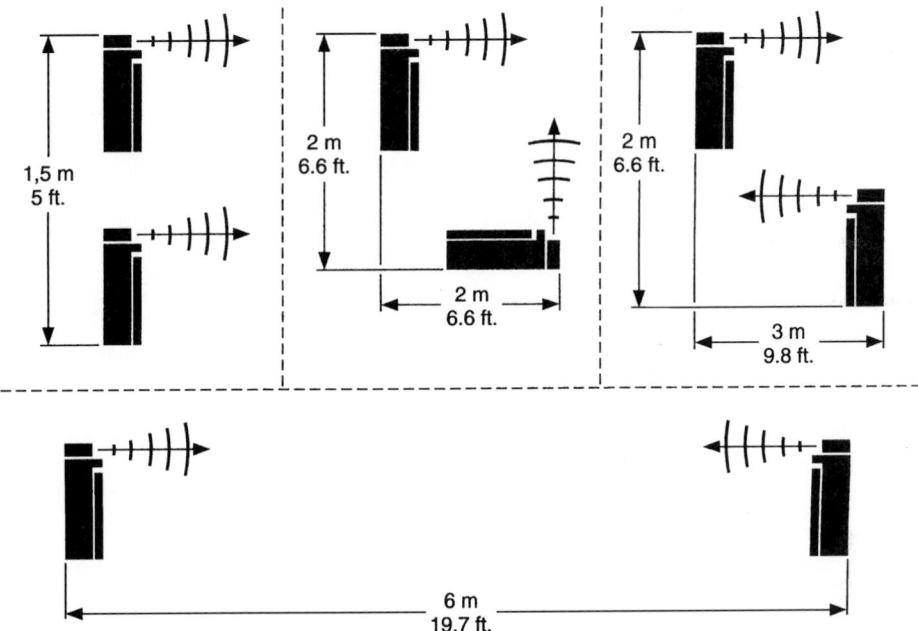

FIGURE 16   Where several ultrasonic detectors may be required within fairly close space confines, care must be taken to avoid acoustic interference between separate devices. The minimum practical sensor-to-sensor space is shown. (*MICRO SWITCH.*)

PHOTOELECTRIC PROXIMITY SENSORS

A cornerstone of early automated systems, practical applications of photoelectric devices date back several decades. They have profited over the years from advances in electronics, fiber optics, and practical application experience. Thus these devices continue to play very important roles in industrial control systems. The photoelectric effect, discovered by Hertz in 1887 and fully explained by Einstein in 1921, is the basis of numerous devices in a number of fields, including presence-nonpresence sensors.

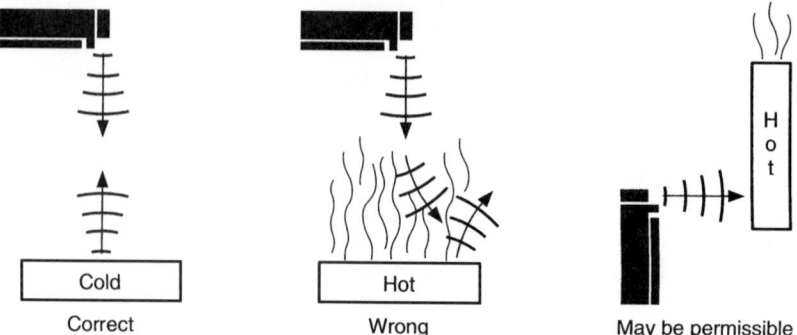

FIGURE 17   An internal temperature detector may be used within the ultrasonic device for adjusting the clock frequency of the elapsed time counter and the carrier frequency to compensate for air-temperature variations. However, large temperature fluctuations can cause dispersion and affect refraction of the ultrasonic signal adversely. If a hot object must be detected, the sensor should aim at the lower (cooler) portion of the target and thus avoid or minimize the effects of warm air currents. (*MICRO SWITCH.*)

Scope of Usage.   Photoelectric controls are found in many kinds of applications because they respond to the presence or absence of either opaque or translucent materials at distances from a fraction of an inch (a few millimeters) up to 100 or even 700 feet (30 to 210 meters). Photoelectric controls need no physical contact with the object to be triggered, which is very important in some cases, such as those involving delicate objects and freshly painted surfaces. Some of the more common applications include thread break detection, edge guidance, web break detection, registration control, parts-ejection monitoring, batch counting, sequential counting, security surveillance, elevator control, conveyor control, bin level control, feed or fill control, mail and package handling, and labeling. These applications are described later.

Photoelectric System Configurations.   A self-contained control includes a light source, a photoreceiver, and the control base function, which amplifies and imposes logic on the signal to transform it into a usable electrical output. A modular control uses a light source–photoreceiver combination separate from the control base. Self-contained retroreflective controls require less wiring and are less susceptible to alignment problems, while modular controls are more flexible in allowing remote positioning of the control base from the input components and hence are more easily customized.

Photoelectric controls are further classified as nonmodulated or modulated. Nonmodulated devices respond to the intensity of visible light. Thus, for reliability, such devices should not be used where the photosensor is subject to bright ambient light, such as sunlight. Modulating controls, employing LEDs, respond only to a narrow frequency band in the infrared. Consequently they do not recognize bright, visible ambient light.

Controls typically respond to a change in light intensity above or below a certain value of threshold response. However, certain plug-in amplifier-logic circuits cause controls to respond to the rate of light change (transition response) rather than to the intensity. Thus the control responds only if the change in intensity or brightness occurs very quickly, not gradually.

Operating Mode.   Both modulated and nonmodulated controls energize an output in response to either a light signal at the photosensor when the beam is not blocked (light-operated, LO), or a dark signal at the photosensor when the beam is blocked (dark-operated, DO). Although some controls have built-in circuitry that determines a fixed operating mode, most controls accept a plug-in logic card or module with a mode selector switch that permits either light or dark operation.

In addition to a light source, light sensor, amplifier (in the case of modulated LED devices), and power supply, a complete system includes an electrical output device (in direct interface with logic-level circuitry—the output transistor of a dc-powered modulated LED device or of an amplifier-logic card).

Scanning Techniques.   There are several ways to set up the light source and photoreceiver to detect objects. The best technique is the one that yields the highest signal ratio for the particular object to be detected, subject to scanning distance and mounting restrictions. Scanning techniques fall into two broad categories: (1) thru (through) scan and (2) reflective scan.

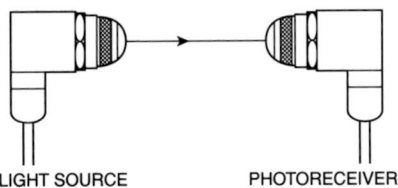

LIGHT SOURCE   PHOTORECEIVER

FIGURE 18   In direct, or thru, scan configuration, the light source is aimed directly at the photoreceiver. Sometimes the configuration is referred to as the transmitted-beam system. (*MICRO SWITCH.*)

Thru Scan.   In thru (direct) scanning the light source and photoreceiver are positioned opposite each other, so light from the source shines directly at the sensor. The object to be detected passes between the two. If the object is opaque, direct scanning will usually yield the highest signal ratio and should be the first choice (Fig. 18).

As long as an object blocks enough light as it interrupts the light beam, it may be skewed or tipped in any manner. As a rule, the object size should be at least 50 percent of the diameter of the receiver lens. To block enough light when detecting small objects, special converging lenses for the light source and photoreceiver can be used to focus the light at a small, bright spot (where the object should be made to pass), thereby eliminating the need for the object to be half the lens diameter. An alternative is to place an aperture over the photoreceiver lens in order to reduce its diameter. Detecting small objects typically requires direct scan.

Because direct scanning does not rely on the reflectiveness of the object to be detected (or a permanent reflector) for light to reach the photosensor, no light is lost at a reflecting surface. Therefore the direct scan technique permits scanning at farther distances than reflective scanning. Direct scanning, however, is not without limitations. Alignment is critical and difficult to maintain where vibration is a factor. Also, with a separate light source and photoreceiver, there is additional wiring, which may be inconvenient if the application is difficult to reach.

Reflective Scan.   In reflective scanning the light source and photoreceiver are placed on the same side of the object to be detected. Limited space or mounting restrictions may prevent aiming the light source directly at the photoreceiver, so the light beam is reflected either from a permanent reflective target or surface, or from the object to be detected, back to the photoreceiver. Polarized units require polarized reflectors and are used in applications where targets may be highly reflective or where sensing clear objects.

Retroreflective Scanning.   With retroreflective scanning the light source and photosensor occupy a common housing. The light beam is directed at a retroreflective target (acrylic disk, tape, or chalk), one that returns the light along the same path over which it was sent (Fig. 19). Perhaps the most commonly used retro target is the familiar bicycle-type reflector. A larger reflector returns more light to the photosensor and thus allows scanning at a further distance. With retro targets, alignment is not critical. The light source-photosensor can be as much as 15° to either side of the perpendicular to the target. Also, inasmuch as alignment need not be exact, retroreflective scanning is well suited to situations where vibration would otherwise be a problem.

Retroreflection from a stationary target normally provides a high signal ratio so long as the object passing between the scanner and the target is not highly reflective and passes very near the scanner. Retroreflective scanning is preferred for the detection of translucent objects and ensures a higher signal ratio than is obtainable with direct scanning. With direct scanning, the dark signal may not register very dark at the photosensor, because some light will pass through the object. With retroreflective scanning, however, any light that passes through the translucent object on the way to the reflector is diminished again as it returns from the reflector. The system is also useful where retroreflective tape or chalk coding can be placed on cartons for sorting. Retroreflective scanning normally can be used at distances up to 30 feet (9 meters) in clear air conditions. As the distance to the target increases, the retro target should be made larger so that it will intercept and return as much light as possible. Single-unit wiring and maintenance are secondary advantages of retroreflective scanning.

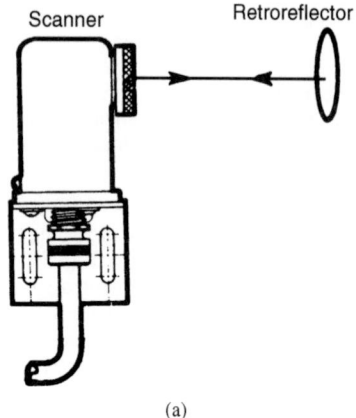

(a)

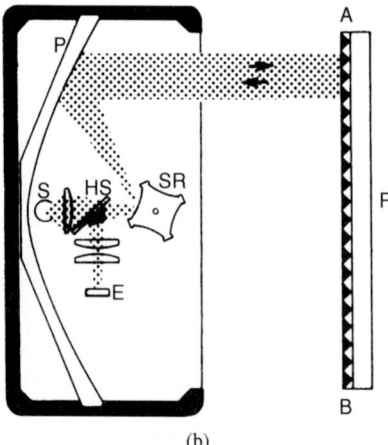

(b)

FIGURE 19 (a) Reflected beam (retroreflective scan) system in which light source and photoreceiver are contained in a single enclosure. This simplifies wiring and avoids critical alignment of source and sensor. (*MICRO SWITCH.*) (b) By adding a rotating-mirror wheel (SR), a parabolic reflector (P), and a semitransparent mirror (HS), a parallel-scanning beam can be obtained. This beam moves at high speed from A to B, thus forming a "light curtain," any interruption of which is detected and signaled by a relay. S—light source; E—photoreceiver. (*Sick Optik Elektronik.*)

Specular Scanning.   The specular scan technique uses a very shiny surface, such as rolled or polished metal, shiny plastic, or a mirror to reflect light to the photosensor (Fig. 20). With a shiny surface, the angle at which light strikes the reflecting surface equals the angle at which it is reflected from the surface. Positioning of the light source and photoreceiver must be precise (mounting brackets which fix the light source–photoreceiver relationship should be used), and the distance of the reflecting surface from the light source and photoreceiver must be consistently controlled. The size of the angle between the light source and photoreceiver determines the depth of the scanning field. With a narrower angle there is more depth of field; with a wider angle there is less depth of field. For a fill-level detection application, for example, this means that a wider angle between light source and photoreceiver allows detection of the fill level more precisely.

Specular scanning can provide a good signal ratio when required to distinguish between shiny and nonshiny (mat) surfaces, or when using depth of field to reflect selectively off shiny surfaces of a certain height. When monitoring a nonflat, shiny surface with high or low points that fall outside the depth of field, these points appear as dark signals to the photosensor.

Diffuse Scanning.   Nonshiny (mat) surfaces, such as kraft paper, rubber, and cork, absorb most incident light and reflect only a small amount. Light is reflected or scattered nearly equally in all directions. In diffuse scanning, the light source is positioned perpendicularly to a dull surface. Emitted light is reflected back from the target to operate the photoreceiver (Fig. 21). Because the light is scattered, only a small percentage returns. Therefore the scanning distance is limited (except with some high-intensity modulated LED controls), even with very bright light sources. It is often difficult to obtain a sufficient signal ratio with diffuse scanning when the surface to be detected is almost the same distance from the sensor as another surface (for instance, a nearly flat or low-profile cork liner moving along a conveyor belt). Contrasting colors can help in such situations.

Diffuse scanning is used in registration control and to detect material (corrugated metal, for example) with a slight vertical flutter—which may prevent a consistent signal with specular scanning. Alignment is not critical in picking up diffuse reflection.

Color Differentiation (Registration Control).   In distinguishing color, as in registration mark detection, contrast is the key. High contrast (dark color on light, or vice versa) provides the best signal ratio and control reliability. Therefore, if possible, bright, well-defined contrasting colors should be considered in the interest of the registration control system.

Diffuse scanning is normally used to detect color change. Table 2 gives some of the common color combinations that must be distinguished in registration control, plus the most suitable type of photosensor and scan technique.

When the background is clear (transparent), the best method is to detect any color mark with direct scanning. When the background is a second color, contrasts such as black against white usually ensure a sufficient signal ratio (difference between dark and light signals) to be handled routinely with diffuse scanning. Red, or a color that contains considerable red pigment (yellow, orange, brown), on a white or light background is a special case. In such instances, a photoreceiver with a

TABLE 2 Factors that Determine Selection of the Photosensor and Scan Technique in Registration Control Applications

Background	Mark	Photosensor	Scan technique
Clear film	Black, blue, red	Any	Thru scan
White (kraft paper, metal foil)	Black, blue	Phototransistor or CdSe photocell	Diffuse scan
	Red	CdS photocell with blue-green filter	Diffuse scan
Black, blue, or other dark colors	Red	Any	Diffuse scan
Red	Black, blue	Any	Diffuse scan

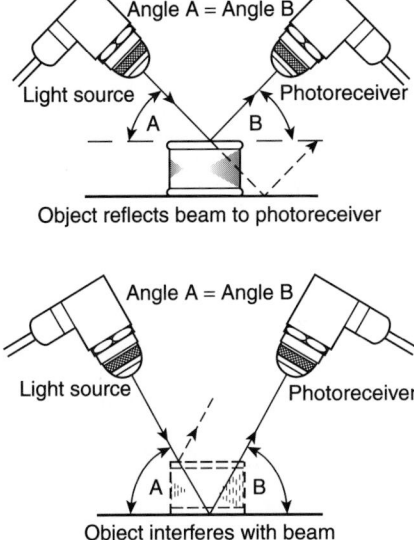

FIGURE 20 Specular scan technique uses a very shiny surface such as rolled or polished metal, shiny plastic, or a mirror to reflect light to the photosensor. (*MICRO SWITCH.*)

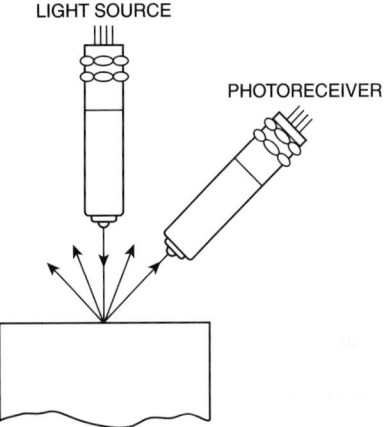

FIGURE 21 Diffuse scan is used in registration control and to detect material (corrugated metal, for example) with a slight vertical flutter—which might present a consistent signal with specular scan. Alignment is not critical in picking up diffuse reflection. (*MICRO SWITCH.*)

cadmium sulfide cell for detecting red marks is preferred because it makes red appear dark on a light background.

A retroreflective scanner with a short-focal-length lens (but without a retro target) can be used to detect registration marks. It is placed near the mark and is actually used in the diffuse scan technique. If a retroreflective scanner is employed to detect marks on a shiny surface, the scanner should be cocked somewhat off the perpendicular to make certain that only diffuse reflection will be picked up. Otherwise the shiny surface of the mark could mirror-reflect so brightly that it would overcome the dark signal a CdS cell normally receives from red. This would mean a light signal from both the background and the mark. In detecting colors, a rule of thumb is to use diffuse (weakened), rather than specular (mirror), reflection.

Sensitivity Adjustment. Most photoelectric controls have a sensitivity adjustment that determines the light level at which the control will respond. Conditions which may require an adjustment of sensitivity include (1) detecting translucent objects, (2) a high speed of response, (3) a high cyclic rate, (4) line voltage variation, and (5) a high electrical noise atmosphere.

Light Sources and Sensors. Early photoelectric control systems used incandescent light sources and traditional photocells—a combination still widely used. A photocell changes its electrical resistance with the amount of light that falls on it. A number of photocells have been used over the years for different applications (photoelectric controls, copying machines, television pickup tubes, and so on). Widely used for photoelectric controls are cadmium sulfide and cadmium selenide cells. During the relatively recent past, phototransistors and photodiodes have become available as sensors, and LEDs have been used as light (infrared) sources. The advantages of the newer components in certain types of applications are described shortly.

Photocells. The sensitivity of a photocell can be defined in two ways. Static sensitivity expresses the resistance of the cell at a given light intensity; the lower the resistance, the more sensitive the photocell. So long as the resistance falls within the range of the control unit with which the photocell is used, static sensitivity is usually not an important consideration. Dynamic sensitivity is an expression of the ratio of the photocell resistance at one light level to its resistance at a different level; the greater the ratio, the higher the sensitivity. This is a much more useful expression of photocell sensitivity.

The speed of response of a photocell is the time it requires to produce a change in resistance in response to a given change in light intensity. Although all photocells are fast, they require a finite amount of time to respond to changing light. And their speed of response depends on the amount of light falling on the cell; the greater the light intensity, the faster the response.

Light history effect refers to a characteristic of a photocell that has been kept dark or light for extended periods. Such a cell overresponds to a change in light before returning to its normal response (somewhat analogous to the response of the human eye to sudden changes in light). Although in some applications it is well to know about this effect, it normally is not a significant consideration in industrial uses.

The effect of temperature on a photocell is to increase its resistance and thus decrease its current with increases in temperature for a given level of illumination. The temperature effect is smaller when the level of illumination is high than when it is low.

Photocells also respond differently to different colors of light. Photocells generally used in industrial applications have a far greater response to colors in the red and infrared range than in the blue-violet range. Except in certain applications, as described previously for registration control, the color response is not a significant factor.

Phototransistors. A phototransistor produces a collector current that is a function of both base current and light. Since the base lead of a phototransistor is usually left unconnected, only variations in light intensity produce variations in current output. There are a number of differences between the phototransistor and the photocell. (1) The current output of a phototransistor is largely independent of the voltage across it, whereas that of a photocell is not. As a result, controls designed to work with photocells will not necessarily work well with phototransistors, and vice versa. (2) The response of phototransistors is affected by changes in temperature, but in a way opposite to that of photocells; the higher the temperature, the higher the current output. (3) Phototransistors have a polarity which must be observed, while photocells do not. (4) Phototransistors respond to light much faster than photocells, but typically have a lower sensitivity.

The photodiode response is narrower than that of the phototransistor, making the diode more effective in blocking stray light from incandescent, sun, or other light sources.

Light-Emitting Diodes. The introduction of LEDs as radiation sources brought a number of advantages for certain applications. The useful life of an LED is estimated at 100,000 hours, which is about 10 times that of an incandescent lamp. However, incandescent lamps are frequently used because they have a spectrum from the ultraviolet to the visible to the infrared, allowing a wide range of colored targets to be detected. LEDs have the advantage that they can be modulated directly, whereas incandescent lamps require a mechanical chopper. Silicon phototransistors and photodiodes are excellent matches for infrared LEDs because their greatest sensitivity peaks almost match precisely at the transmitter's (LED) wavelength. The role of LEDs in other instrumentation applications (displays, communications, and so on) is described in other parts of this handbook.

Photoelectric Sensor Configurations. Contemporary sensors tend to fall into three classifications: (1) general-purpose devices that handle a wide range of industrial applications, including packaging and materials handling; (2) special-purpose designs, including miniature heavy-duty models, some

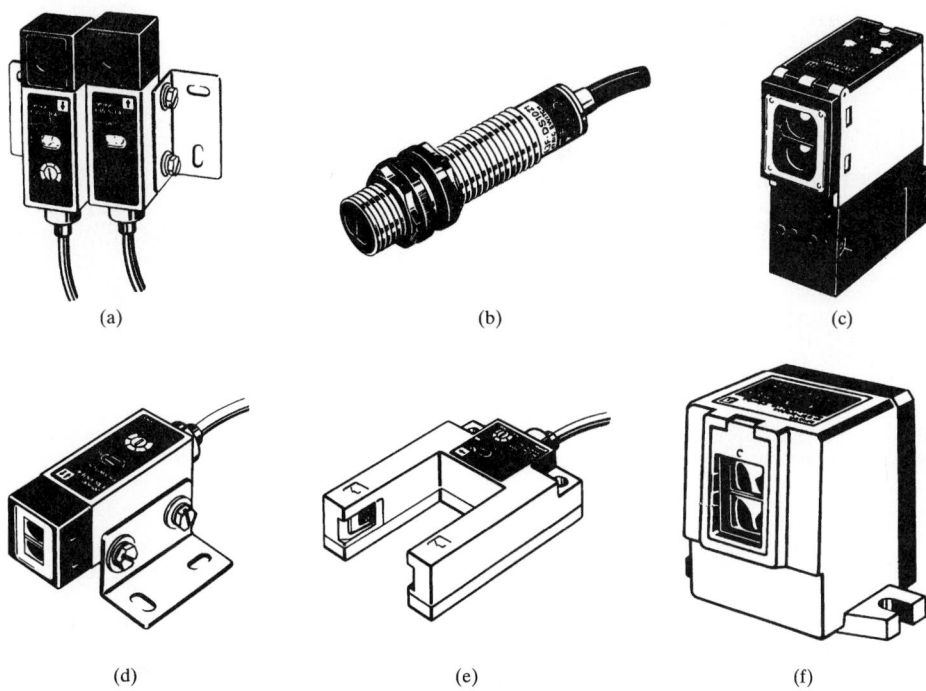

(a)   (b)   (c)

(d)   (e)   (f)

FIGURE 22   Representative configurations of photoelectric sensors. (*a*) Miniature sensor. Built-in dc amplifier offers fast response times. Horizontal or vertical mountings. Selectable light-on/dark-on operation. Adjustable sensitivity. Operation indicators. Sensing distance 2 meters (6.56 feet) or 5 meters (16.4 feet) separate type; 2 meters (6.56 feet) rectroreflective; 100 mm (3.94 inches) or 300 mm (11.8 inches) diffuse reflective. Fiber-optic versions available. (*b*) Threaded cylindrical. Built-in amplifier provides long sensing distance. Response time 30 ms (ac), 2.5 ms (dc). Polarized retroreflective model detects shiny objects. Operation indicators. Sensing distance 3 meters (9.84 feet) separate type; 0.1 to 2 meters (3.94 inches to 6.56 feet) retroreflective; 0.1 to 1.5 meters (3.94 inches to 4.92 feet) polarized retroreflective; 100 mm (3.94 inches) diffuse reflective. (*c*) Slim general-purpose. Universal power supply. Plug-in construction. Response time 30 ms (0.5 to 20 ms with built-in time delays). Operation indicators. Sensing distance 7 meters (22.97 feet) retroreflective; 5 meters (16.4 feet) polarized retroreflective; 2 meters (6.56 feet) diffuse reflective. (*d*) Color mark sensor. Self-contained dc amplifier for fast response and high-speed operations. Response time 1 ms. Horizontal or vertical mounting. Selective light-on/dark-on operation. Sensitivity adjustment. Sensing distance 12 mm (0.47 inch) or 50 mm (1.97 inches) diffuse reflectance. (*e*) Grooved head and slotted sensor for edge control, positioning, and mark detection. Response time 1 ms. Sensing distance 10 mm (0.39 inch) or 30 mm (1.18 inches). (*f*) Conventional conveyor sensor. Detects just the target, *not* the conveyor belt. Light source directed upward to avoid reflection from conveyor. Built-in time delays available. Sensing distance 10 m (32.8 feet) separate type; 0.3 to 3 meters (11.81 inches to 9.84 feet) retroreflective; 700 mm (27.56 inches) diffuse reflective. (*Omron Electronics, Inc.*)

of which include mark detection, transparent object detection, conveyor sensors that detect the object rather than the belt per se, belt grooved head and slotted sensors; and (3) designs that incorporate fiber optics, which are superior in long-distance applications, for detecting very small objects and shiny objects, and with an improved performance in hot environments (Fig. 22).

Fiber-Optic Sensors.   Separate-type fiber-optic cables detect opaque objects that break the beam for "thru scan." They require mounting space for separate emitter and receiver sensing heads. Diffuse reflective fiber-optic cables reflect the light off the object to be detected. Reflective sensors deliver and receive the light in a single sensing head. The detecting distance, however, is reduced.

Fiber-optic sensors in particular are preferred for hard-to-reach situations. Most plastic fiber-optic cables can be cut to custom lengths in the field from the original 2-meter (16.56-foot) length. When threaded heads are too large to reach the detection site, a cable with bendable steel tubing should be considered. The latter retains complex shapes and is well suited for multiple-sensor inspections

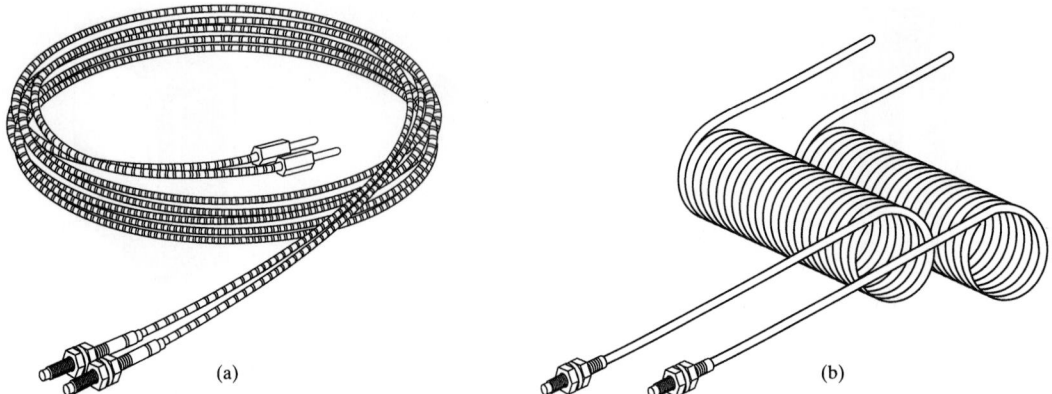

FIGURE 23   A wide selection of fiber-optic cables and accessories for photoelectric amplifiers makes it possible to solve difficult detection problems. (*a*) Armored through beam-sensing head. (*b*) Coiled cable for reciprocating/flexing equipment. (*Omron Electronics, Inc.*)

of minute assemblies and parts. Where the sensing distance of separate-type cables is required, an optional lens kit for increasing (by 7 times) the distance between emitter and receiver should be considered.

Needle probes may be selected to detect objects as small as 0.0006 inch (0.02 mm) passing flush by the fiber-optic cable lens. Side-view accessories and "periscope"-type needle probes provide space-efficient ways to achieve right-angle detection.

The use of convergent-beam sensing can be helpful when detecting highly polished reflective surfaces. The temperature-tolerant range for fiber-optic cables is −40 to 70°C (−40 to 158°F). For higher-temperature applications, plastic-sheathed or armored glass cables may be used; the latter are useful up to 400°C (750°F).

For equipments which involve flexure, such as a robot arm, fiber-optic cables with a retractable coiled section are usually well suited. Representative fiber-optic photoelectric sensors are shown in Fig. 23.

MACHINE VISION

Since its initial serious recognition in the late 1970s, machine vision (MV) has been variously defined:

> MV is the process of extracting information from *visual sensors* for the purpose of enabling machines to make intelligent decisions.

or

> MV is part of the larger technology of *artificial visual perception* that substitutes (partially or totally) the *human* visual capability by instruments which are backed up by electronic data processing (notably complex computations).

As will be noted from the foregoing, the real and practical objectives of MV tend to be nebulous and of far-reaching expectations.

MV became a "buzz word" of the 1980s—to the extent that MV, combined with robots, would bring about the most ambitious goals in terms of production automation. The fact is that as of the early 1990s, MV has lost much of its earlier luster. With relatively few exceptions, most contemporary MV systems are confined to sophisticated and demanding object-detection problems, with the majority of such detection problems currently being solved by the less exotic types of sensors previously described in this article. Because total MV systems generally still remain overly complex and costly, other nonvisual sensors have been greatly improved and applied more imaginatively to achieve many of the objectives that once appeared to lie within the province of MV. Cost, once again, has been the traditional motivating factor.

"Seeing" robots, once a major goal of MV, has proved disappointing from the standpoint of MV. To be true, in practice, robots, once programmed, do perform in many instances as though they could "see," but this has been brought about by other instrumental detectors (tactical, for example) than by literal "viewing" of parts, pieces, machines, and entire manufacturing scenes. For example, in the late 1980s a large automotive manufacturer originally invested rather heavily in MV systems in connection with robotic operations, but after a year or so of trials, abandoned them for possible consideration at some future date.

To be sure, MV developments will continue, and systems will be installed where other simpler, lower-cost approaches do not suffice. Coupled with trends toward reducing computation costs, MV may become more competitive.

Artificial Visual Perception

Industrial MV techniques initially stemmed from the interests of the military in a technology known as pattern recognition. This may be defined as seeing, analyzing, and interpreting patterns, as of scenery, juxtaposition, dimensional magnitudes, color, and other characteristics of the visual environment.

Human visual communication (information transfer) with the outside environment depends on the interactions (predominantly absorption, reflection, and refraction) of light (visual) radiation emanating from physical objects (in point, two, or three dimensions), thus enabling the human observer to cope with the surrounding environment in a safe and efficient manner. The vision activity (human or machine) must be complemented by some form of processor (human brain or electronic counterpart) to make a final identification of what has been seen. This process, defined by a few words, is pattern recognition.

The means that are applied to perform pattern recognition is known as the pattern processor. The human brain, in processing a pattern, does an amazing job of sorting out extraneous information in the input to quickly identify what really is present (the objects of interest that are in the pattern).

Interest in pattern recognition dates back some 70 years. The more the process is studied, the more one finds how fundamental the process really is to the human brain function. The amount of information (inputs) that the human system can observe and process is tremendous, but nevertheless remains poorly understood. This explains why MV, from a technological viewpoint, is closely associated with the study of artificial intelligence.

Elements of Pattern Recognition

Any pattern recognition system contains the same three basic elements: (1) sensing, (2) processing, and (3) implementing actions based upon input data.

The primary objective of pattern recognition is to classify a given unknown pattern as belonging to one of several classes of patterns. The applications of pattern recognition are many and varied. In the case of character recognition, for example, the patterns are easily generated and recognized by humans, and the basic goal has been to improve the human-machine communication. In other situations, the patterns are difficult for humans to recognize rapidly, as, for example, the very rapid interpretation of an electrocardiogram. The wide range of applications, as well as the relationship to diverse disciplines, including machine vision, communication and control, and the area of linguistics (word and sentence reading), have broadened the interest basis in pattern recognition.

Information Content of an Image

In terms of MV as used in industrial production situations, the information content of the image falls into three basic categories:

1. *Geometry*, which, in turn, portrays shape, position, dimension, and a number of other associated properties, including density and texture (which can be inferred from the known geometry in most cases).
2. *Color*, which is very helpful when present. There are, of course, color-blind persons and instruments that preclude its use.
3. *Movement*, which is present in two or more images of a dynamic process.

Extracting Information from Images

As with other forms of industrial instrumentation, the MV system senses (reads may be a better term) the image, but before the data obtained from the image can become meaningful, the sensed information must be compared with some form of standard, or prelearned, pattern. In more familiar terms, the combined actions of sensing and comparing constitute measurement.

As compared with the usual type of industrial sensor, such as a thermocouple, where measured data are conveniently available in the form of a ready-made electrical signal, in MV one deals not with just one or a very few points or locations of emanating data, but with thousands-plus bits of information—because what is being observed (measured) is comprised of a multitude of points (picture pixels[1]). In an artificial vision system, image data, as may be gathered by an electronic camera, must be compared electronically with information, that is, in some fashion with data that are stored in electronic memory.

Initially the application of a general-purpose computer to MV was the only choice available. The problem immediately encountered was the fact that the computer was designed to process computational data, *not* data patterns. The data from a video camera in an MV system are pattern data and, thus, are very different from computational data as exemplified by financial balance sheets or linear regression analysis. The situation is summarized as follows:

Image data quantity 484 × 320 pixel density

Hence, 154,880 pixels per frame

At 6 bits of gray-level data per pixel, 929,280 pixel values per frame

At a 30 times a second refresh rate, 27,878,400 bits per second

Since the early days of MV, a number of innovations have contributed to the simplification of this problem, including improvements in gray-scale systems, better algorithms, the pipelined image processing engine, the geometric arithmetic parallel processor, and associative pattern processing.

MV Sensors

Factors which usually must be considered are (1) optics and lighting, (2) field of view, (3) resolution, (4) signal-to-noise ratio, (5) time and temperature stability, and (6) cost. MV sensors which have been or are currently in use include the following.

---

[1] A pixel is a picture element, a small region of a scene within which variations of brightness are ignored.

Line Scanners. These include solid-state arrays, flying-spot scanners, and prism, mirror, or holographically deflected laser cameras. These scanners are fast high-resolution devices, which are relatively free of geometric distortion in one dimension. In order to capture a complete two-dimensional scene, the second dimension is obtained either by motion of the object past the scanner or by mirror or prism deflectors. Mechanical motion tends to slow down data acquisition and, in some cases, produces geometric distortion.

Area-Type Scanners. These scanners were used in early systems utilizing closed-circuit television cameras with vidicon image sensors. Advantages were comparatively low cost and relatively high resolution (300 to 500 television lines). They suffer geometric distortion, temperature instability, lag (requiring several television frames for full erasure), and sensitivity to nearby magnetic fields.

Solid-State Cameras. These cameras represent a major advance in image-sensing technology. Scenes can be digitized onto an array of photosensitive cells. Charge-coupled devices (CCDs) or charge injection devices (CIDs) have been used in these cameras. The arrays form a pixel grid containing the data currently appearing on camera. The solid-state camera is available in a variety of pixel densities, notably, 128×128, 256×256, 512×512, 1024×1024, and so on.

The limiting factor of cameras for imaging systems is not necessarily the resolution or speed. Camera technology essentially has kept up with the ability of computationally based processes to handle information, especially when multiple cameras are used. Traditionally, in MV systems sensors can detect information in real time a lot faster and in larger quantities than processors can handle.

MV Image Processing

From the beginning, a major task of MV engineering has been that of reducing the amount of information actually needed, and thus lessening and simplifying the processing task. Shortcuts are frequently taken to reduce data volume. Some processing systems that have been or are in current use include the following.

Binary MV System. The representation of all combinations possible in every pixel of every scene is unrealistic. Conversion of each pixel point into a binary value reduces the pixel data, but it also reduces the accuracy of the analysis. Binary systems evaluate each pixel as black or white. A threshold-adjusting capability can allow users to select what intensity of signal is to be the black-white border.

Gray-Level System. This system can interpret each pixel's value as a specific gray tone. These systems vary in precision, with each pixel being evaluated as 16, 64, or even 256 different values.

Windowing the Scene. This system can reduce the total data system load. Since analysis of an entire scene often is unnecessary for proper recognition, the use of a window can eliminate unneeded image data.

Segmentation. A common means of data reduction, this technique divides the image data into areas of interest and then interpolates surrounding pixels into those areas. There are several ways to segment an image, but all detract from accuracy. Four procedural options for segmentation may be considered:

1. *Algorithms.* In general, algorithms are a set of mathematical models that can be used to describe an image. A few years ago, the SRI algorithms were developed by the Stanford Research

Institute. These consisted of about 50 different features that are extracted from a binary image, such as the size of blank areas (holes) or objects (blobs) and their *centroids* or *perimeters*.

2. *Neighborhood Processing.*   This is an averaging method, achieved by treating each pixel value as if it were part of a group or neighborhood. This gray-level data-reduction technique can change an otherwise complicated image into areas with well-defined lines. Each pixel's point value is calculated by considering the value of a neighboring pixel. The process effectively averages the scene into regions or areas.

3. *Convolution.*   If an image is represented by the rate of light change per pixel instead of light density, the image will look like a line drawing because the greatest intensity of a pixel change is at the boundaries.

4. *String Encoding.*   In run-length, string, or connectivity encoding, the values of the first and last pixel positions of each scan line are compared to see if they are equal and, therefore, belong to the same region. The tables generated by this process also consider the vertical changes in pixel state and, as may be expected, are rather short because most simple binary images contain relatively few transition borders.

5. *Associative Pattern Processing.*   Rather than increasing the capability of processing large amounts of data in a shorter time span through the route of improving computer and data-handling techniques (hardware and software) as just described, a fundamentally different approach has three functional sections (as developed by APP (pattern processing technologies): (1) An image analyzer contains all the pixel data for the current image on the camera. (2) It converts the data into a unique statistical fingerprint, which is a unique set of response memory addresses. The response memory contains the statistical fingerprints for all trained images. (3) A matching histogram compares the fingerprints of the current image with those stored in the response memory. The theory of operation is based on a statistical phenomenon that reduces an image's data into a precise fingerprint, which is a set of values for an image that differs from other images and is a minute fraction of the total data of the image. Instead of one image being trained, all the images needed are trained and no programming is involved. One simply shows an image to the camera, allows the fingerprint memory locations to drive the response memory, and loads in a label for each image.

The Recognition Process

Once the user describes the image mathematically, the second half of the process step can occur. Data processing has been mentioned previously. The programming of a general-purpose computer requires programming expertise that is costly. Using this computational approach may entail performance levels that are too slow for real-time operation, unless the application only involves very basic procedures. Gains in processing power can be quickly taken up by application complexity with its increasing data demands.

MV Applications

Some form of actuation or control is the last step in implementing MV. Reasonably standardized methods of communication usually can be used with such common interfaces as RS-232 or IEEE-488.

MV suppliers in recent years have turned to specialized products oriented to specific industries. The advent of generic software has been possible because of numerous application similarities.

MV applications essentially fall into four categories: (1) inspection, (2) location, (3) measurement, and (4) recognition.

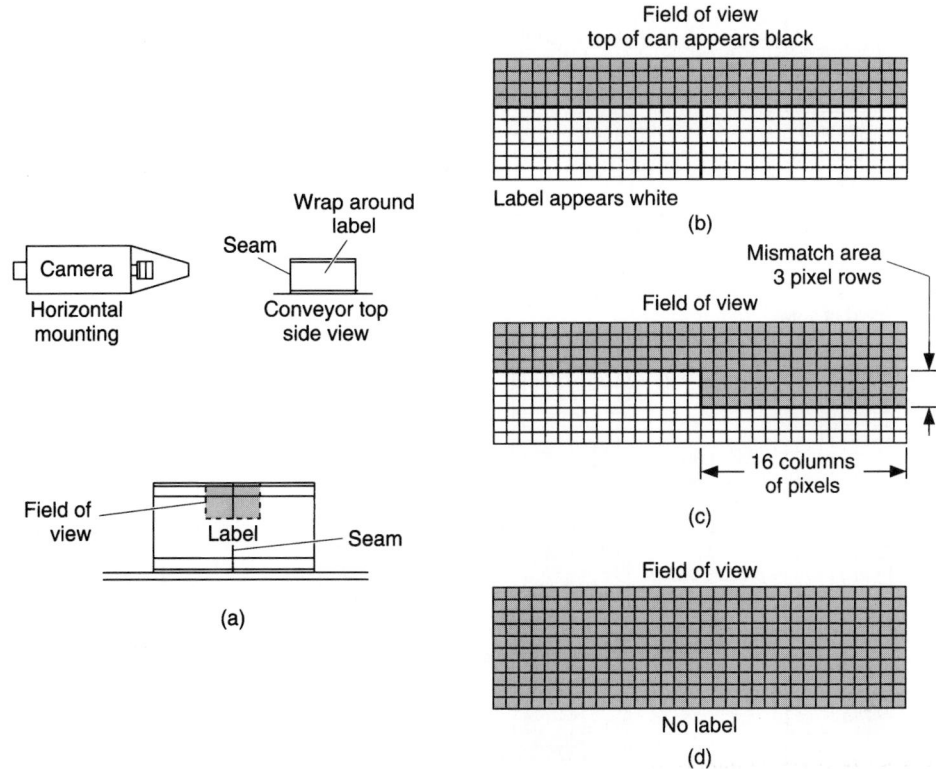

FIGURE 24 (*a*) Example of where a template or pattern match algorithm is used to check accuracy of a label placed on a can of food. The camera is located so that the field of view is horizontally across the area where the label ends touch, as shown in lower diagram. To set up the reference template, the operator adjusts the horizontal and vertical axes of the camera by viewing a display on the controller, during which time the operator also adjusts the mounted camera to obtain the kinds of views shown. (*b*) Label is properly placed. (*c*) Label is poorly placed, indicating a mismatch area of three pixels. (*d*) No label on can. (*Eaton.*)

MV can inspect to determine if parts are acceptable. Inspection tasks can be further categorized into one of four functions: (1) surface flaw detection, (2) presence or absence of certain features, (3) dimensional tolerance verification, and (4) shape verification. Three examples that use a gray scale are shown in Figs. 24 through 26.

DISCRETE-PIECE IDENTIFICATION—BAR CODING

The lack of uniformity of parts, pieces, subassemblies, and so on, that persists in the discrete-piece manufacturing industries imposes a myriad of complexities, as just alluded to in the preceding description of machine vision. Thus MV applications as generally defined must be highly customized to very specific applications. This imposes a crucial cost restraint.

If, however, pieces and parts could be "packaged" in a way to simplify their recognition and exact identification, most of the complexities of MV can be eliminated. That is, of course, exactly

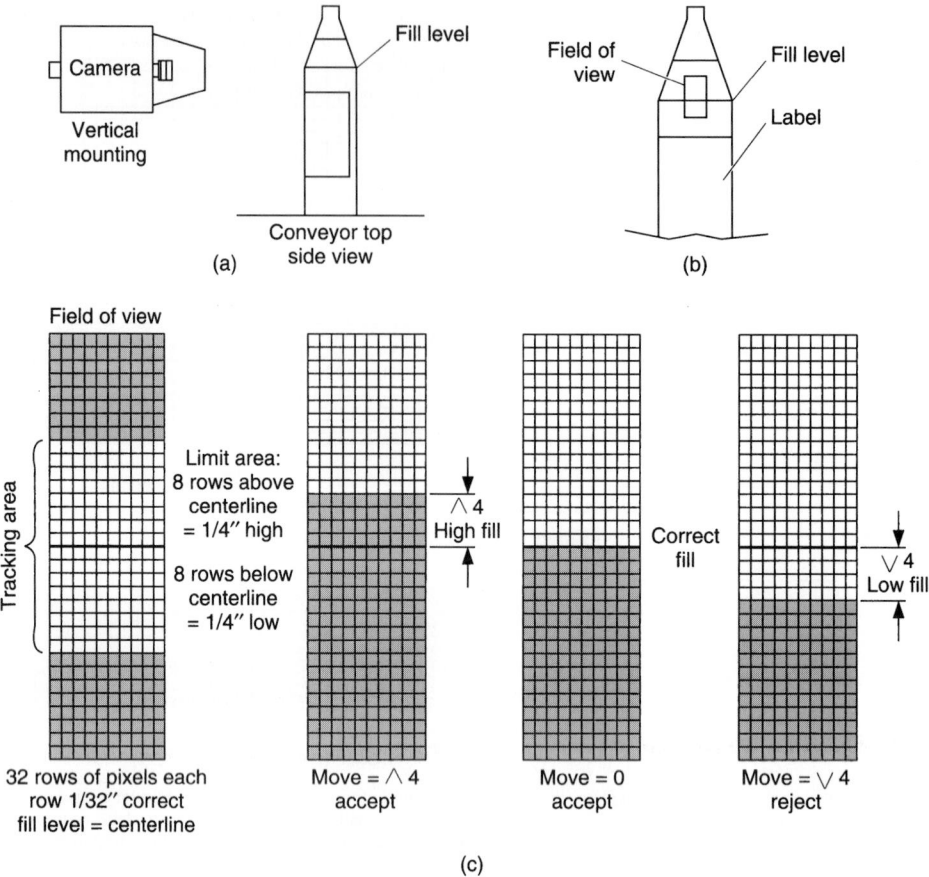

FIGURE 25   Example of use of *y*-axis algorithm to check bottle-filling operation. (*a*) Arrangement of camera and conveyor juxtapositions. (*b*) Location of field of view for given high and low levels. (*c*) Reference images showing high fill, correct fill, and low fill. (*Eaton.*)

what has been done for decades in terms of a large variety of manufactured end products, such as individual boxes of nuts and bolts, screws, cases of canned goods, boxed machine parts, containerized detergents, paints, hardware, plastic-encapsulated items, and even palletized groupings of identical merchandise.

The general concept of containerization, of course, originally was developed to simplify shipping and distribution to the end user. Later these packaging concepts became the basis for vastly improved inventory control and for automated warehousing by manufacturers, distributors, wholesalers, and, indeed, the final point of purchase.

With many millions of items produced, it became obvious that an extensive codification system was needed if piece identification were ever to be automated. Early trials with color coding proved that color-based systems were grossly inadequate. Entering from another direction (commercial and financial), the concepts of magnetic and optical coding were developed and rather quickly adopted by financial institutions. However, early magnetic and optical systems required a certain closeness of the object with the detecting means, as, for example, in identifying a check or reading the address on an envelope. Specially designed alphanumeric characters have served financial and commercial needs, but it is expected that bar coding ultimately may be used.

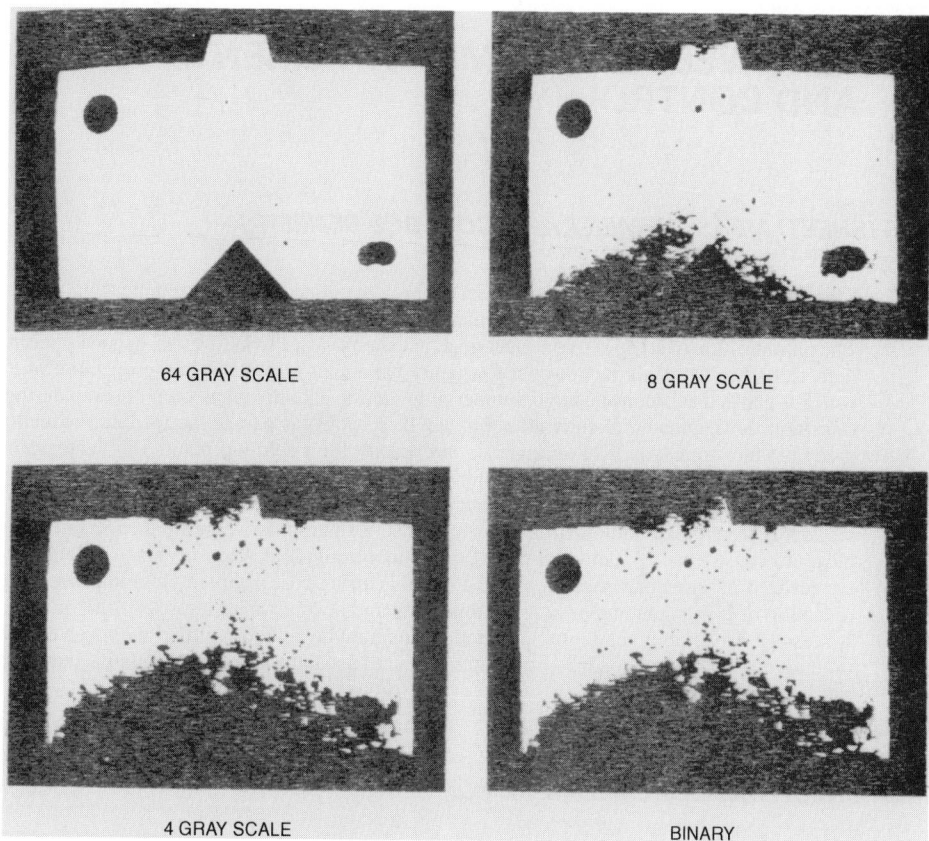

FIGURE 26   Illustration of advantages of high-intensity resolution. With 64 levels of gray, for example, part boundaries are distinguishable from background. Fine features, such as the shadow cast by the post in the lower right-hand corner of the part, are discernible. Using fewer levels of gray (8, 4, and 2, as shown) renders the scene more subject to the effects of shading. The slightly shaded portion of the part in the binary system has deteriorated, as have other subtle features. (*Analog Devices.*)

In the late 1960s the first supermarket checkout stand based on bar coding of merchandise was installed. The system did not get underway for several years, awaiting the willingness of manufacturers to include a bar code on each product. Ultimately the universal bar code, as shown in Fig. 27, was developed. During the late 1980s the bar code became a favorite of manufacturers in their finished goods operations, notably automated warehousing and inventory accounting. As the result, the bar code segment of the electronic industry has grown at an accelerated pace.

Invention of the laser scanner has made it possible to read bar codes on boxes in a warehouse as far as 10 feet (3 meters) away. Well-designed bar code readers are considered to be very close to error-free. Replacement of older style code readers with the hand-held wand has increased the acceptance and versatility of the system. Bar-coded tags for unpackaged materials, such as fabrics, have received wide acceptance.

FIGURE 27   Universal bar code.

FLAT WEB (SHEET) MEASUREMENT AND CONTROL

FLAT WEB (SHEET) MEASUREMENT AND CONTROL BENEFITS

The applications of on-line flat web (sheet) measurement and control are increasing in number and technology to improve sheet quality to ensure customer satisfaction and reduce the amount of scrap. The minimization of scrap is a continual goal, whether it is due to hard spots, gage variations, surface defects, or streaks. A single out-of-spec quality parameter may result in the rejection of the whole roll. For plants that produce a large number of grades, the additional amount of scrap during product transitions is significant. Even if the scrap can be recycled as part of the feed, the reduction in off-grade product corresponds to an increase in capacity for a sold-out plant. The decrease in recycle makes room for an increase in raw materials. The pounds of good-quality material produced per pound of raw materials (yield) and the pounds of raw material feed are both increased. A reduction in variability from greater automation also can translate to an increase in production rate by operating closer to constraints. If additional production is not needed, the feed is decreased and the benefits are realized in terms of a reduction in the manufacturing cost from the yield improvement and the reduction of lost orders and cost of handling of scrap and customer returns. If the product demand is seasonal or variable, the higher production rate may be able to build up the inventory enough to produce other products, conduct process tests, or shut down the line for preventative maintenance and process improvements.

FLAT WEB (SHEET) MEASUREMENTS

Besides the obvious noncontacting requirement, sheet measurements must have a fast response and both excellent short-term and long-term repeatability. Short-term repeatability is the variation (scatter) of successive measurements of the same true value. Long-term repeatability is the variation over a long period of time and is often described as drift. It can be corrected by either a zero adjustment to the measurement or a bias adjustment to the set point. This is not possible for short-term repeatability errors. The need for a rapid speed of response and short-term repeatability is extraordinary, especially at high line speeds and for sensors that transverse the sheet width. The response time (time to reach 98% of the final response) for a stationary sensor must be less than the width of the defect in the machine direction (MD) divided by the sheet speed as shown in Expression (1). For sensors that scan across the width of the sheet, the transverse time must satisfy this same relationship. The speed of the sensor is the width of the sheet divided by the transverse time. Since the width of the sheet is also divided up into data bins to detect defects in the cross direction (CD), the response time must be less than the transverse time divided by the number of bins as defined by Expression (2). For many transverse applications, the sensor must have a response time less than 25 mis to resolve CD variability. Additionally, the scan or update times of any digital devices, such as analog-to-digital converters and computers, must be faster than the these response times. This is outside the range of the typical distributed control system (DCS). Hence, special measurement subsystems and dedicated microprocessor-based controllers or personal computers are used for sheet measurement and control. It is important to realize that this speed requirement is for detection and not for control. These response times are negligible compared to the transport time (dead time) from the corrective mechanism, such as adjustable die bolts to the sensor. The detector response time contributes less than 0.1%, and a DCS scan time of 0.2 s contributes less than 1% to the total loop dead time. Thus, the control algorithms can be executed in the DCS. Also the DCS is fast enough for the detection of longer-term changes in sheet conditions such as temperature, moisture, composition, and MD gage and optimization. For CD control, the DCS to date is not used because of a lack of functional blocks for array handling and array math.

For the detection of MD defects,

$$T_{98} < W_d/S_s \tag{1}$$

For the detection of CD defects,

$$T_{98} < (W_d/S_s)/N_b \tag{2}$$

where N_b = number of CD data bins
 S_s = speed of the sheet (inches per second)
 T_{98} = maximum time to 98% response (seconds)
 W_d = minimum width of the MD defect (inches)

Some of the more common sensors are optical pyrometers for temperature, nuclear gages for basis weight, optical devices for percent transmission (clarity and color detection), infrared analyzers for moisture, and laser devices or CCD cameras for surface-defect detection. Nuclear gages that emit and measure beta particles have the response time and both the short- and long-term repeatability needed for CD and MD closed-loop control.

Since a nuclear gage measures basis weight from the amount of radiation attenuated by any mass between the source and the receiver, changes in the density of the sheet and air gap or dust particles can affect the reading. Thus, changes in air temperature and humidity and sheet temperature, moisture, and composition affect the accuracy of the inferred gage (thickness) of the sheet. Since the air gap is much larger than the sheet thickness, the attenuation of radiation by the air may be larger than that by the sheet. Changes in the air gap from changes in alignment of the source and receiver while scanning can be significant. Recent advances in sensor design have reduced the size of the air gap, helped maintain sensor alignment, and provided a curtain of constant-temperature clean air. The air curtain also helps regulate the position of the sheet relative to the sensor, and it reduces sheet flutter [1].

FLAT WEB (SHEET) CONTROL

As with all control systems, the performance of the control system is limited by the total dead time. For a proportional-plus-integral controller, the peak error occurs after 150% of the dead time [2]. After the defect passes by, the control system will continue its correction for at least another dead time. Small width defects such as streaks, bumps, and surface aberrations cannot be individually reduced by control action. Any attempted corrective action for an individual localized defect is liable to create a longer-term upset. Thus, the defect must have a width greater than 2.5 times the product of the overall dead time and line speed; otherwise the value of a control system reacting to the individual defect is debatable. Expression (3) defines this requirement in terms of the allowable total dead time in a dead-time-dominant loop in order to have some hope of an overall beneficial effect of CD or MD control action in reaction to a defect of width W_d in the machine direction.

For the control of defects,

$$TD_o < (W_d/S_s)/2.5 \tag{3}$$

where S_s = speed of the sheet (inches per second)
 TD_o = overall loop dead time (seconds)
 W_d = minimum width of the defect in the MD (inches)

Any variability associated with a sheet control loop basically is unattenuated and is seen in the final product since there is no backmixing from the extruder inlet to the finished roll. Oscillations

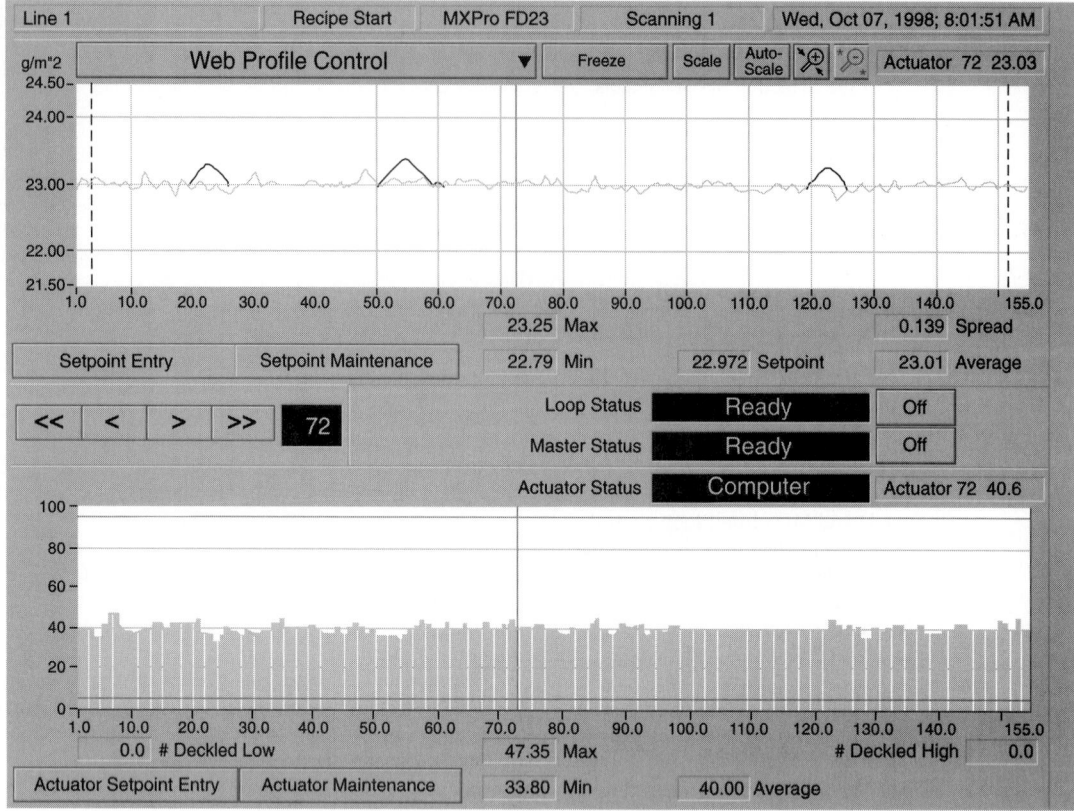

FIGURE 1   Sheet profile control screen on a Windows NT Platform (courtesy of Honeywell–Measurex).

in any loops, including those upstream of the sheet (such as feed concentration, extruder speed, and melt temperature and pressure), introduce variability into the final product. Also, the loops are typically dead-time dominant (overall loop dead time is larger than the largest time constant in the loop). It is essential to use lambda tuning to prevent oscillations and provide a smooth response with an adjustable closed-loop time constant. The use of Ziegler Nichols tuning and quarter amplitude response is a recipe for disaster in these applications. Power spectrum analysis and lambda tuning are essential tools for tracking down and eliminating variability in the sheet. The topics of plant analysis and advanced regulatory control in Section 10 on Process Control Improvement discuss these tools in more detail.

For profile control, there can be several hundred controlled variables (CD data bins) and manipulated variables (die bolts). Since the effect of a die-bolt adjustment affects more than one data bin, there is a mapping (association of data bins with die bolts) and interaction consideration. There is also considerable shrinkage of the sheet after emergence from the die. Several die bolts at each end of the sheet may be set to a fixed position because most of shrinkage occurs near the edges. Since line operating conditions such as melt pressure, sheet speed, moisture, and temperature affect shrinkage and mapping, there is need for inclusion of these as disturbance variables. While a constrained model predictive controller (CMPC) would normally be considered, the dimension of the matrix is larger and the execution speed of the algorithm is faster than what is used in a standard CMPC package such as Dynamic Matrix Control or Optimal Predictive Control. There is also some question whether the changes in mapping can be handled by the addition of disturbance variables. A proven successful approach is the use of individual lambda-tuned PID Smith predictors (dead-time compensators) for each manipulated variable (die bolt) with the tuning settings scheduled as a function of dead time (sheet speed), a user-specified process gain and time constant for the upward and downward deflection from set point, and finally, conventionally decoupled per a user-specified mapping [1]. Additionally, a beam-bending module is used to predict the lip in the sheet from die-bolt action, and a neural network can be employed to find the correct mapping [1]. The control system resides in a personal computer (PC) and uses a Windows NT platform to provide a friendly operator and engineering interface and an open database. Figure 1 shows the data bins (upper graph) and the die-bolt positions (lower graph) on a PC screen for CD profile control [1]. The architecture of a system for CD profile control including the scanner, measurement subsystem, interfaces, and local area networks is illustrated in Fig. 2.

The basis weight value of each data bin for the CD profile is the average of several fast samples. A high-frequency anti-aliasing filter is applied to the individual measurements. An exponential filter is then used between data bin values to remove the MD component from the CD profile.

If heaters are used instead of actuators for manipulation of the die bolts, a rather large nonlinear time constant is introduced because of the thermal response. Even though there is also additional dead time, the sheet CD response is no longer dead-time dominant and the user must be careful to enter the proper time constants for upward and downward deflections. Since the time constant is associated with

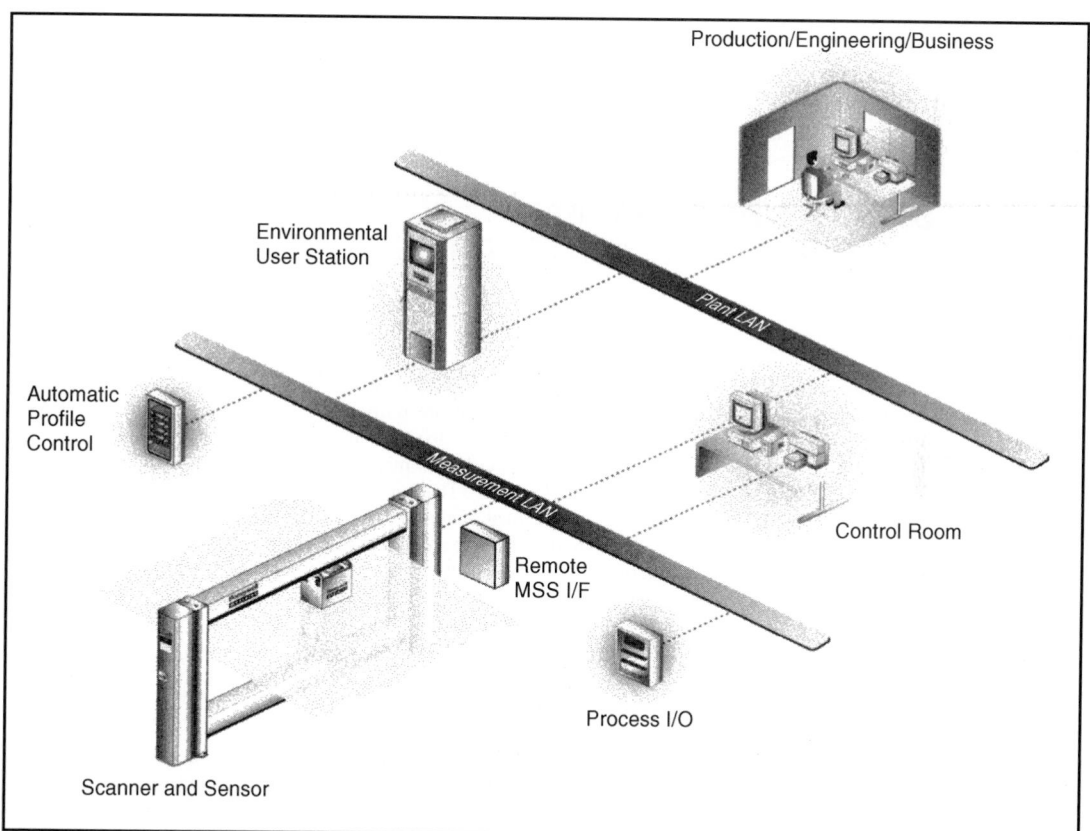

FIGURE 2   System architecture for sheet profile control (courtesy of Honeywell–Measurex).

the manipulated variable rather than in the process, the control system performance is significantly worse for die-bolt heaters because of the slowness of corrective action [2]. However, as a result of the large number of die bolts, the addition of actuators may be viewed as too expensive. An analysis of the cost of CD defects often will show that actuators will pay for themselves within 3 years.

There is an opportunity to use override control or a CMPC to optimize sheet speed to improve the production rate while honoring constraints such as sheet moisture, temperature, and average MD gage. These higher-level control systems also improve product quality by ensuring that constraints are not violated despite upsets in the extruder, melt, and sheet. Since most of the loops are dead-time dominant and interact, CMPC is a better choice than override control. Ultimately, there is the opportunity to use real-time optimization for data reconciliation, parameter identification, and optimization of operating conditions. A first principle model of the extruder, melt, and sheet processes is developed and used to iteratively search for the optimum. The topics of CMPC, real-time optimization, and advanced regulatory control in Section 10 on Process Control Improvement discuss these methods in more detail.

REFERENCES

1. Embleton, W., *MXProLine: On-line Measurement and Control, Engineered for Today's Producer*, October 22, Honeywell–Measurex White Paper, 1998.
2. McMillan, G. K., *Tuning and Control Loop Performance*, 3rd ed., Instrument Society of America Research Triangle Park, N.C., 1996.

SPEED, VELOCITY AND ACCELERATION INSTRUMENTATION

The more apparent applications for speed and velocity measurement and control relate to rotating electric motorized machines and equipment. There are numerous other applications, including the measurement of air (wind) and other effluent gas velocities, as may be needed, for example, in determining the pathways followed by air- or waterborne manufacturing pollutants for environmental detection, prevention, and enforcement purposes. The speed-related group of variables also is important in the research, design, and manufacture of a vast variety of aerodynamic and hydrodynamic machines, including vehicles. Central to the analysis and control of vibration, including its deleterious effects on equipment operation, useful life, and safety, is the measurement of acceleration. Numerous means have been developed over the years to detect the speed-related variables.

DEFINITION OF TERMS

Velocity.   Time rate of change of position. (Unless angular velocity is specified, this term generally is understood to refer to *linear* motion.) Strictly, the velocity of a moving point must specify both the speed and the direction of the motion and is, therefore, a *vector*, although the term sometimes is used more loosely as merely synonymous with speed. The velocity of a point is the time rate of the distance s from a fixed origin O, expressed as the vector derivative of s with respect to time, ds/dt

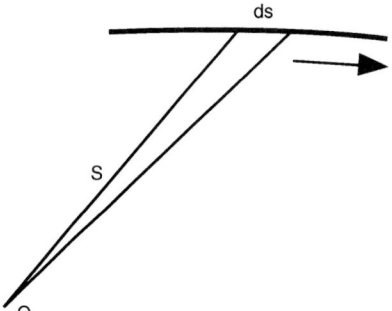

FIGURE 1 Velocity expressed as a vector derivative.

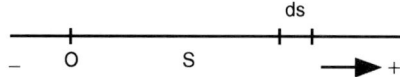

FIGURE 2 Treatment of distance and velocity as scalars.

(Fig. 1), while speed is the magnitude of the velocity and is not a vector. If the direction of motion is constant, so that the motion is in a straight line (but not necessarily with constant speed), and if the line of motion is clearly understood, it is convenient to treat the distance s and the velocity ds/dt as *scalars* with respect to some zero point on that line and with appropriate algebraic signs (Fig. 2). Otherwise they must be regarded as vectors. If the velocity is variable, account must be taken of the acceleration. Examples of both curved and rectilinear motion are approached from the mathematics of kinematics.

Angular Velocity. Quantity relating to rotational motion. While the term angular velocity may be extended to any motion of a point with respect to any axis, it is commonly applied to cases of rotation. Its instantaneous value is defined as the vector whose magnitude is the time rate of change of the angle Θ rotated through, as, for example, $d\Theta/dt$, and whose direction is arbitrarily defined as that direction of the rotation axis for which the rotation is clockwise. The usual symbol is ω or Ω.

The concept of angular velocity is most useful in the case of rigid body motion. If a rigid body rotates about a fixed axis and the position vector of any point P with respect to any point on the axis as origin is $\mathbf{r}$, the velocity $d\mathbf{r}/dt$ of P relative to this origin is $d\mathbf{r}/dt = \omega \times \mathbf{r}$, where ω is the instantaneous vector angular velocity. This indeed may serve as a definition of ω.

The average angular velocity may be defined as the ratio of the angular displacement divided by the time. In general, however, this is not a vector, since a finite angular displacement is not a vector. The instantaneous angular velocity is used more widely.

Angular velocities, like linear velocities, are vectorially added. For example, if a top is spinning about an axis which is simultaneously being tipped over toward the table, the resultant angular velocity is the vector sum of the angular velocities of spin and of tipping. This enters into the theory of precession. The derivatives of the eulerian angles are sometimes very useful in describing the angular motion of a rigid body which has components of angular velocity about all its principal axes.

Speed. Scalar quantity equal to the magnitude of velocity. Industrially, linear speeds are frequently inferred from rotational measurements simply because of the manner in which most machines are designed—with rotating shafts, wheels, and gears to which speed transducers can be conveniently attached. At one time, analog-type sensors of speed were used almost exclusively, and they still are in demand. Commencing on a small scale in the mid-1950s, digital speed sensors were developed; they are preferred for many applications because of the ease with which they can be integrated into otherwise digital systems.

Acceleration. Rate of change of velocity with respect to time. Acceleration is expressed mathematically by dv/dt, the vector derivative of velocity $\mathbf{v}$ with respect to time t. If the motion is in a straight line whose position is clearly understood, it is convenient to treat the velocity v and the acceleration dv/dt as scalars with appropriate algebraic signs; otherwise they must be treated by vector methods.

Acceleration may be rectilinear or curvilinear, depending on whether the path of motion is a straight line or a curved line. A body which moves along a curved path has acceleration components at every point. One component is in the direction of the tangent to the curve and is equal to the rate of change of the speed at the point. For uniform circular motion this component is zero. The second component is normal to the tangent and is equal to the square of the tangential speed divided by the radius of curvature at the point. This normal component, which is directed toward the center of curvature, also equals the square of the angular velocity multiplied by the radius of curvature. The acceleration due to gravity is equal to an increase in the velocity of about 32.2 feet (981.5 cm)/second/second at the earth's surface and is of prime importance since it is the ratio of the weight to the mass of a body.

TACHOMETERS

In some servo-controlled motion systems, feedback is required in terms of *position*, as furnished by encoders, resolvers, linear transformers, and so on, which are described in the first article of this handbook section. In other situations, feedback must be in terms of *velocity*.

In contemporary systems, a permanent-magnet dc motor may be used as a tachometer, because when driven mechanically, it generates an output voltage which is proportional to shaft speed (Fig. 3). A principal requirement of a tachometer for such control is that the output voltage be smooth over the operating range and that the output be stabilized against temperature variations.

Small permanent-magnet dc motors frequently are used as speed-sensing devices. These usually incorporate thermistor temperature compensation. They also use silver commutator and silver loaded brushes to improve commutator reliability at low speeds and at the low currents which are typical of the application. To combine high performance and low cost, dc servomotor designs often incorporate a tachometer mounted on the motor shaft and enclosed within the motor housing, as shown in Fig. 4.

There are two fundamental types of dc tachometers, (1) brush type and (2) brushless. Brush-type dc tachometers are of two constructions, (1) iron core and (2) moving coil. Fundamentally the ac tachometer is a three-phase electric generator with a three-phase rectifier on the output. Each type has relative advantages and limitations.

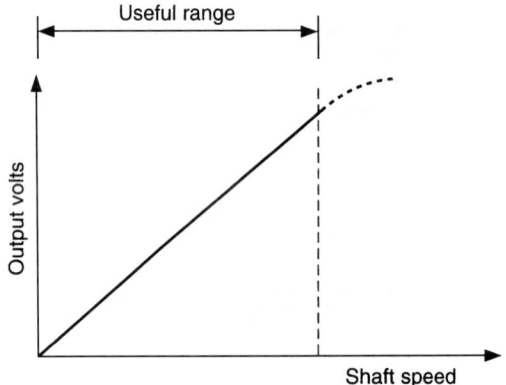

FIGURE 3   Desirable output characteristics of tachometer.

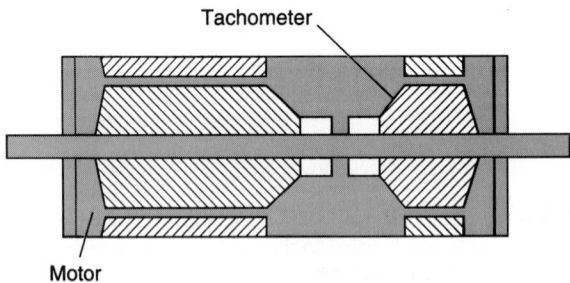

FIGURE 4   Motor with integral tachometer.

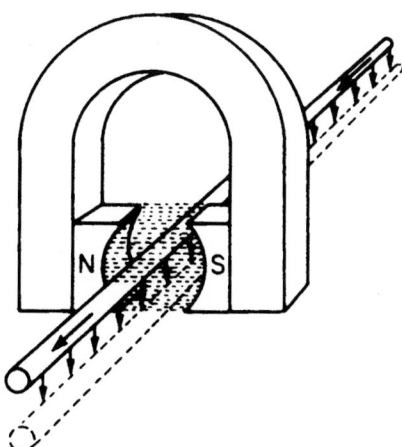

FIGURE 5 Operating principle of commutator-type dc tachometer generator. The magnitude of the voltage produced is a direct function of the strength of the magnetic field and the speed with which the conductor moves perpendicularly to it. Current will flow if the ends are connected to a load, such as an instrument. The polarity of the voltage and, therefore, the direction of current flow depend on the polarity of the field and the direction of conductor motion. This same effect can be obtained by rotating the magnet and holding the conductor still, that is, the principle of the ac or rotating-magnet tachometer.

DC Tachometers

As shown in Fig. 5, the dc tachometer depends on the relative perpendicular motion between a magnetic field and a conductor, which results in voltage generation in the conductor. A dc tachometer system consists of a dc generator and a dc indicator or recorder (Fig. 6). The composite characteristics of a representative dc tachometer generator are given in Table 1.

In the moving-coil brush-type dc tachometer, the winding is in the form of a shell or cup. In this construction there is a magnet on one side and an iron slug on the other. Thus the magnetic field passes through the cup-shaped winding. This overcomes much of the inertia because only the winding is rotating. The electrical inductance is also markedly reduced.

Brush-type dc tachometers are usually limited to relatively clean environments. Brush life is shortened in many cases because of particulate and erosive contaminants. Some airborne contaminants may also build up in the form of films on the commutator. Sealed enclosures can be used, but these create a thermal problem because of entrapped heat. This heat is not generated within the techometer per se because of the very low currents involved, but it can be conducted through the shaft. Magnets are sensitive to temperature (estimated to be 0.01 to 0.05 percent/°C) and therefore, if the stability of the output is critical, temperature compensation may be required.

Speed-Ratio Systems with dc Tachometers. By using two dc tachometer generators (Fig. 7) connected to a ratio meter mechanism, measurements that are depend on defferential processing speed, such as percent stretch and ratio of draw, can be taken and controlled through additional elements in the system. The system shown for a textile application has wide use in the paper and steel industries, as well as where it is important to know the ratio between two quantities expressible in terms of rotation. In these installations the minimum generator speed must be 400 r/min because of voltage requirements of the indicator. Full-scale range limits are from 10 to 100 percent shrink. Percent stretch = output – input × 100% input. If the input generator is 100 units/min and the output generator reads

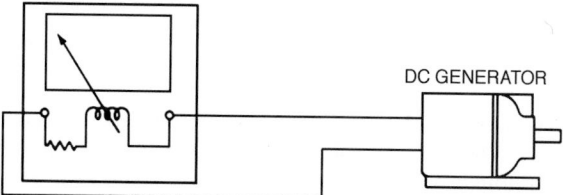

FIGURE 6 A dc tachometer system such as that shown may be used when the top operating generator speed is at least 100 and does not exceed 5000 r/min. Special indicators may be used for top operating generator speeds as low as 100 r/min, and special recorders are available for speeds as low as 10 r/min. Generally the characteristics of the tachometer can be matched to the equipment whose speed is being determined by using suitable gearing for effecting speed reduction or multiplication. Several types of dc tachometer generators are available.

TABLE 1   Composite Characteristics of Representative DC Tachometer Generator

Voltage output at 1000 r/min	6 V ±1%
Accuracy	±1%
EMF linearity	±0.15%
Permissible current drain	50 mA
Maximum rms value of ac ripple	2%
Allowable end play	0.005 in. (0.13 mm)
Maximum operating temperature	250°F (121°C)
Internal resistance at 77°F (25°C)	20 Ω ± 2%
Composition of brushes	Palladium-silver alloy
Armature	12 bars, 12 slots
High-potential test	500 V for 1 min
Bearings	Ball
Temperature compensation	$\frac{1}{10}$% per 10°C change
Normal continuous speed	2000 r/min
Minimum top speed	100 r/min
Maximum top speed	5000 r/min
Weight	~2 lb (0.9 kg) (dust-resistant models)
	~25 lb (11.3 kg) (weatherproof and explosion-proof models)
Shaft diameter	$\frac{3}{16}$ –$\frac{3}{4}$ in. (5–13 mm)
Adjustable magnetic shunt range	±4%
Length/width/height (approximate)	4½ –5 in./3 in./3 in. (114–127 mm/76 mm/76 mm) (dust-resistant models)
	12½ in./6 in./5 in. (318 mm/152 mm/127 mm) (weatherproof and explosion proof models)

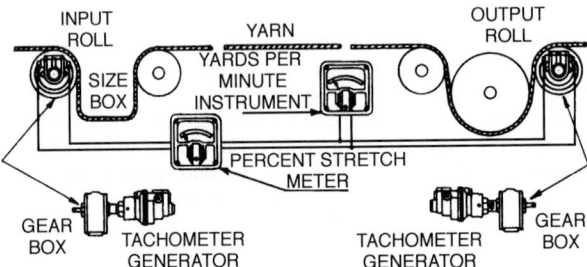

FIGURE 7   Speed-ratio System tachometer system used to control percent stretch in the textile industry or ratio of draw in other materials-processing industries.

125 units/min, the percent stretch is (125 − 100)/100 = 25 percent. Through suitable switching arrangements, the outputs of several pairs of generators may be selectively fed through the ratio and production rate instruments to provide readings from various sections of multistage machines.

AC Tachometers

There are (1) voltage-responsive tachometer systems and (2) frequency-responsive systems.

Voltage-Responsive Tachometer Systems.   Consisting of an ac generator and a rectifier-type indicator, as shown in Fig. 8, these systems may be used in any installation where the generator

TABLE 2   Composite Characteristics of Representative DC Tachometer Generator

Voltage output at 1000 r/min	10 V ±1% open circuit
Accuracy	±1%
Permissible current drain	150 mA
Frequency at 900 r/min	60-Hz sine wave
Allowable end play	0.005 in. (0.13 mm)
Maximum operating temperature	250°F (121°C)
Internal resistance at 77°F (25°C)	100 Ω ± 1% (voltage-responsive units)
	32 Ω ± 20% (frequency-responsive units)
	8 poles
Stator	
Rotor	Alnico V, 8 poles
Bearings	Ball
Temperature coefficient	$\frac{3}{10}$ % per 10°C change (voltage-responsive units)
	No temperature compensation (frequency-responsive units)
EMF linearity	Depends on load and speed
Minimum top speed	500 r/min
Maximum top speed	5000 r/min
Weight	About 3 lb (1.4 kg) (dust-resistant models)
	About 25 lb (11.3 kg) (spray-resistant and explosion-proof models)
Shaft diameter	$\frac{3}{16}$ –$\frac{3}{4}$ in. (5–13 mm)
Length/width/height (approximate)	4½ –5 in./3 in./3–3½ in. (114–127 mm/76 mm/76–89 mm) (dust-resistant models)
	12½ in./6 in./5 in. (318 mm/152 mm/127 mm) (spray-resistant and explosion-proof models)

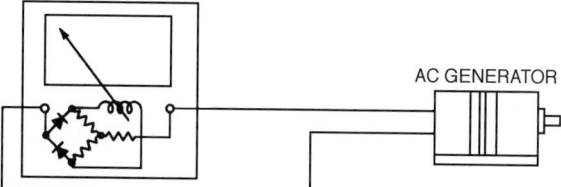

FIGURE 8   Highly schematic circuit of ac voltage-responsive tachometer system.

speed for full-scale is not less than 500 and not greater than 5000 r/min. With adequate attention given to bearings, conventional ac generators may be used at speeds up to 10,000 r/min. The ac tachometer generator embodies a stator surrounding a rotating permanent magnet. The output of the generator for voltage-response systems is temperature-compensated and is proportional to speed.

Frequency-Responsive Tachometer Systems. This type of system consists of a dc indicator or recorder, a frequency-responsive network which may be contained in the recorder or a separate transformer box, and an ac tachometer generator of either the conventional or the bearingless form. Several types of ac tachometer generators are available. The composite characteristics of a representative device are given in Table 2.

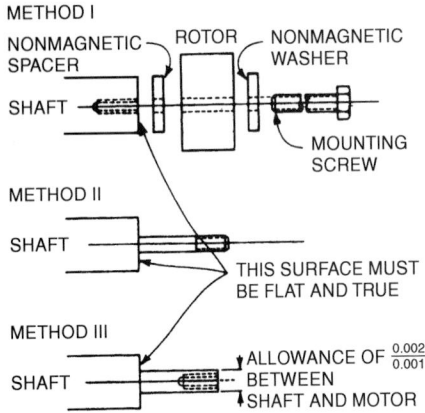

FIGURE 9   Examples of methods for mounting the rotor of a tachometer generator unit.

Bearingless Tachometer Generators

These devices are ac generators of the most basic form, consisting of only a permanent magnet rotor and a stator. The devices have no bearings or brushes. They are designed to be impervious to oil, grease, and relatively high temperatures and, consequently, may be installed in inaccessible areas, such as gearboxes, which permits saving of space. They have very low torque burdens of less than 1 oz·in and are capable of speeds up to 100,000 r/min.

In general, when a bearingless generator is used, the frequency-responsive approach is employed. Since the system is solely dependent on the frequency output of the generator, voltage variations caused by reductions in the magnetic strength of the rotor due to handling, poor alignment of stator and rotor, or axial travel of the rotor with respect to the stator will not affect the overall accuracy. The rotor of the generator unit should be mounted to the true center of the shaft with extreme care, particularly in high-speed installations. A few representative methods are illustrated in Fig. 9. The relatively low inertia of the rotor makes it possible to secure it to the shaft with a right-hand thread, regardless of the direction of rotation. It is recommended that a steel screw having an SAE thread be used to permit maximum tightening. The rotor should not be pressed onto the shaft because magnetic rotor material is brittle and may shatter.

Magnetic Speed Sensors

A magnetic pickup is essentially a coil wound around a permanently magnetized probe (Fig. 10). When discrete ferromagnetic objects, such as gear teeth, turbine rotor blades, slotted disks, or shafts with keyways, are passed through the probe's magnetic field, the flux density is modulated. This induces ac voltages in the coil. One complete cycle of voltage is generated for each object passed. If the objects are evenly spaced on a rotating shaft, the total number of cycles will be a measure of the total rotation, and the frequency of the ac voltage will be directly proportional to the rotational speed of the shaft. The magnetic pickup shown in Fig. 10 is used in conjunction with a 60-tooth gear to measure the revolutions per minute of a rotating shaft. Such a gear is often selected because the

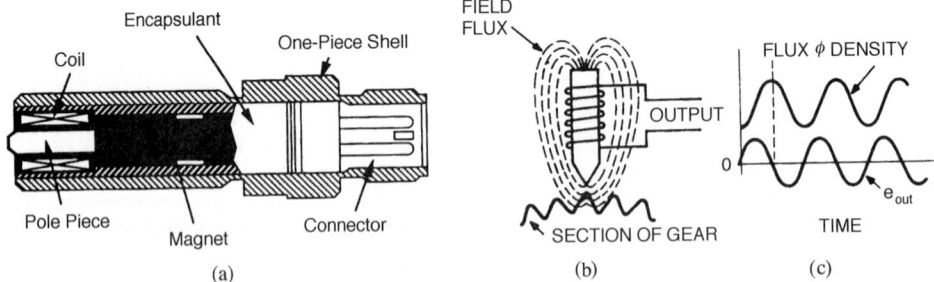

FIGURE 10   Magnetic speed sensor. (*a*) Sectional view. (*b*) Placement of probe, allowing a small air gap between pickup and gear teeth. (*c*) Output waveform, which is a function not only of rotational speed, but also of gear-tooth dimension and spacing, pole-piece diameter, and air gap. The pole-piece diameter should be less than or equal to both the gear width and the dimension of the tooth's top (flat) surface. The space between adjacent teeth should be approximately 3 times the diameter. Ideally, the air gap should be as small as possible, typically 0.13 mm (0.005 in.). A number of steel or cast-iron gears, precisely manufactured standards, are available. The standard solid gear comes with various dimensions and with 48, 60, 72, 96, and 120 teeth. (*Daytronic Corporation.*)

output frequency (in hertz) is numerically equal to revolutions per minute—a situation that allows frequency meters to be used without calibration. For very high rotational speeds, a smaller number of teeth may be better suited. This type of arrangement is used with some turbine-type flowmeters, as described in the article, "Flow Systems" in Section 4 of this handbook.

A magnetic pickup also can be used as a timing or synchronization device. Examples would include ignition timing of gasoline engines, angular positioning of rotating parts, or stroboscopic triggering of mechanical motion.

One commercially available unit is a passive or self-generating device, requiring no external excitation. When mounted in proximity to the teeth (or blades) of a conventional rotating gear (or turbine), it produces an approximately sinusoidal ac voltage signal output. The amplitude of the voltage also is generally proportional to the speed of rotation (Fig. 11).

The magnetic pickup circuit shown in Fig. 12 contains its own signal-conditioning circuitry for generating a clean square-wave output pulse for each ferrous discontinuity passing the head of the pickup. The output is either on or off, depending on the presence or absence of ferrous material.

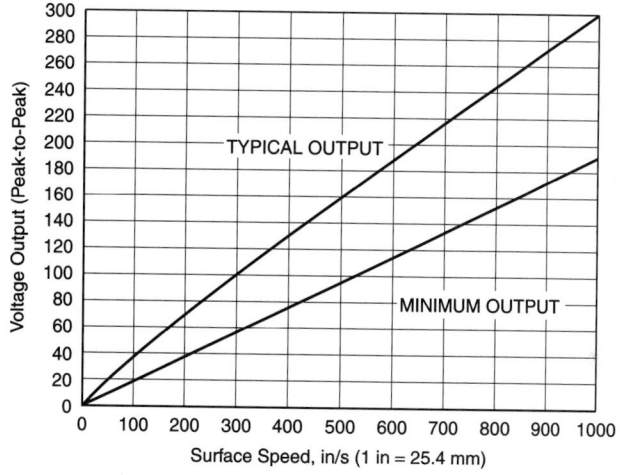

FIGURE 11 Output performance of magnetic pickup. (*Daytronic Corporation.*)

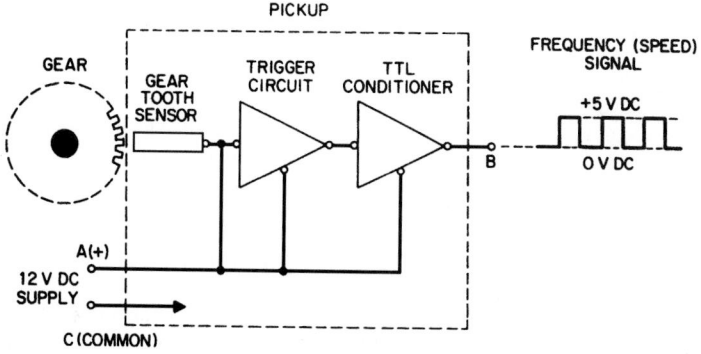

FIGURE 12 Circuit of zero-velocity magnetic pickup. (*Daytronic Corporation.*)

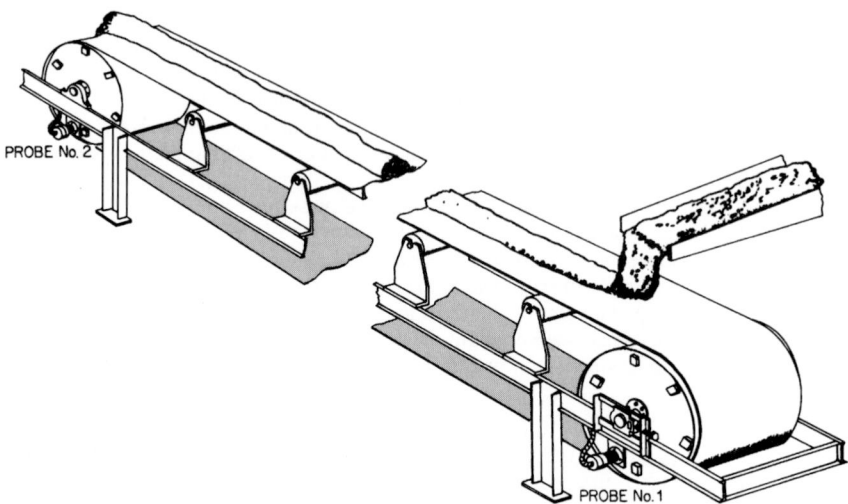

FIGURE 13 Conveyor loss-of-motion detection system that uses two probes, one on the head pulley and another on the tail pulley. By computing the plus or minus speed relation of the tail to the head pulley and comparing this value with percentage-slip set points, slip protection is ensured. A minimum percent feature of the system checks for mechanical failure between the motor and the tail pulley. Each of these features has its own time delay after start-up and one common delay to ignore nuisance alarms or shutdown. The four alarms are fed to a first-out annunciator and latch in the relay output circuit. Reset of the relay may be manual or automatic. (*Milltronics, Inc.*)

The unit senses motion down to "zero" velocity and continuously produces a pulse train of constant amplitude, irrespective of the rotational speed of the gear. Magnetoresistors are flux-responsive devices that detect magnetic field changes independent of the field's rate of change. Two magnetoresistors are used to cancel the effects of temperature drift and supply voltage variations. There are some restrictive mounting requirements.

Adaptations of proximity switches can be used for measuring speed and frequently are called speed switches. A description of these devices is given in a prior article in this handbook section. These units are popular for slowdown indication for conveyors and other process machinery (Fig. 13). On start-up, the incoming pulses from the probe are ignored for a brief, fixed time interval (such as 5 seconds) to permit equipment to accelerate to normal operating speed. This time delay is adjustable up to one full minute. The system is rugged and can be used for heavy-duty machinery in demanding atmospheres. The elements detected by the probe are ferrous pieces mounted directly on belt (or other) drive pulleys.

A capacitor-type or inductive proximity switch also can be used in speed measurement. A trigger cam is mounted on a rotating or reciprocating element of a machine so that it appears within the range of the proximity switch at every revolution or stroke. The distance between the cam and the switch must be no more than one-half the nominal detection range of the switch. The cam may be metallic or nonmetallic, depending on whether an inductive or a capacitive switch is used. The switch generates a pulse each time the cam appears within range. A controller measures the instantaneous rate between successive pulses and compares it with the set point. Through the use of appropriate electronics, the arrangement can be used for either overspeed or underspeed correction. Because these switches measure the time interval between two successive pulses, there is an inherent time lag in the response when the pulse rate decreases below the set point; and the size of the lag depends on the value of the set point. There is no time lag when the pulse rate increases above the set point.

These switches are described in more detail in the preceding article, dc, of this handbook.

Impulse Tachometers

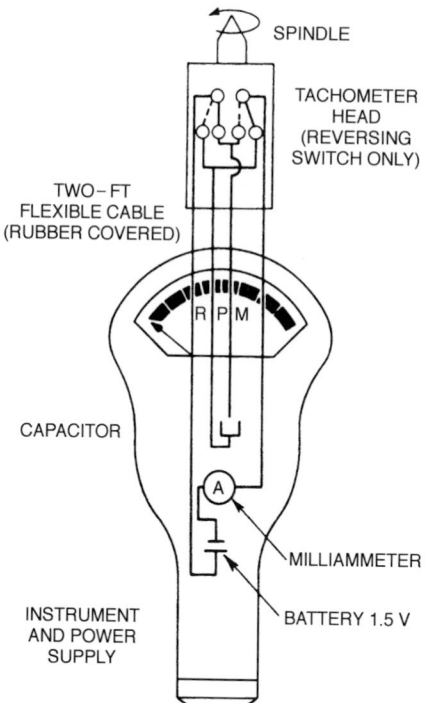

FIGURE 14   Capacitor-type impulse tachometer.

In the instrument shown in Fig. 14, the charging current of a capacitor is used. The pickup head usually contains a reversing switch, operated from a spindle, which reverses twice with each revolution. Thus battery potential is applied to the capacitor in each direction, and with each impulse a current is passed through the milliammeter. The indicator responds to the average value of these impulses. Therefore the indications are proportional to the rates of the pulses, which in turn are proportional to the rates of the spindle revolutions.

No current is drawn from the battery when the spindle is not revolving. The pulse current is approximately 1 mA. The spindle speed and battery voltage influence the indicator deflection. Thus it is important to check and correct the battery voltage at frequent intervals. This is accomplished by means of an adjustable resistor placed in the circuit.

The oscillating switch may be connected directly for speeds of 200 to 10,000 r/min and, with suitable gears, speeds below or above these values can be measured. The readings of a properly standardized instrument are not affected by temperature, humidity, vibration, or magnetic fields. The indicator and head may be separated up to a distance of 1000 feet (300 meters), and where suitably shielded connections are used, the distance may be increased. The indicator scale is uniform.

A high-accuracy instrument is also available wherein the capacitor and the reversing switch are connected to one leg of a bridge circuit. The pulses from the periodically charging capacitor upset the balance of the bridge and thus cause an indication on the milliammeter. Multiple ranges are obtained by using different capacitor values.

Optical Encoders

Many modern position control systems use an incremental optical encoder for position determination, as explained in the first article in this handbook section. By taking advantage of the calculating power of a microprocessor, the impulses from an optical encoder can be converted to a velocity measurement.

Optical encoders are available for handling very wide dynamic ranges, such as 10,000 to 20,000 to 1. Accuracies are claimed to be better than 0.01 percent per revolution.

Stroboscopic Tachometers

A stroboscope permits intermittent observation of a cyclically moving object in such a way as to produce an optical illusion of stopped or slowed motion. This phenomenon is readily apparent, for example, when rewinding a tape at many revolutions per minute when the tape deck is located under a 60-Hz incandescent lamp. Patterns on the reel tend to slow and then appear to stop before

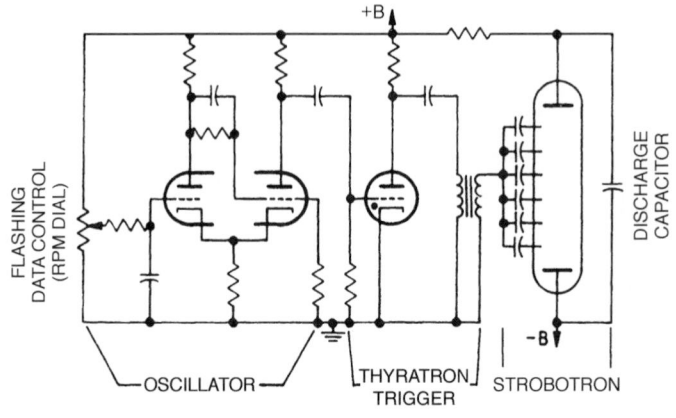

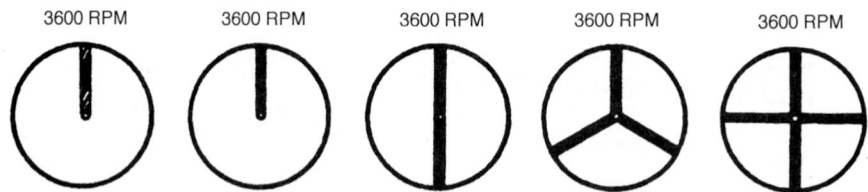

FIGURE 15 Electronic stroboscope. (*a*) Schematic representation of circuit. (*b*) Images obtained at harmonic and subharmonic flashing rates of a stroboscope. Even with an asymmetrical object, the correct fundamental image is repeated when the stroboscope is flashing at one-half, one-third, and so on, the speed of the object. The proper setting for a fundamental speed measurement is the highest setting at which a single stationary image can be achieved. This does not hold, however, if the fundamental is beyond the flashing rate of the stroboscope. There are several ways to distinguish fundamental from submultiple images. The flashing rate can be decreased until another single image appears. If this occurs at half the first reading, the first reading was the actual speed of the device. If it occurs at some other value, then the first reading was a submultiple. Or the user can double the flashing rate and check for a double image. Or the user can flip the range switch to the next higher range. Because of the 6:1 relationship between ranges, a 6:1 pattern should appear. The 6:1 relationship between ranges also makes it convenient to convert speed readings from revolutions per minute into cycles per second. One simply flips to the next lower range and divides the new reading by 10.

Because of their portability and easy setup, stroboscopes find a variety of applications, principally in machine and vehicle research, development, and testing. (*GenRad.*)

reversing their direction. Stroboscopic effects have been known for decades,[1] of course, one of the first scientific applications being found in very high-speed photography. Intermittency of observation can be provided by mechanical interruption of the line of sight (as with a motion picture camera) or by intermittent illumination of the object being viewed. The industrial stroboscope basically is a lamp plus the electronic circuits required to turn the lamp on and off very rapidly, at rates as high as 150,000 flashes per minute and higher.

The schematic diagram of an electronic stroboscope is shown in Fig. 15. The device includes a strobotron tube with its associated discharge capacitors, a triggering tube to fire the strobotron, an oscillator to determine the flashing rate, and a power supply. With the use of harmonic techniques, speeds up to 1 million r/min can be measured. Accuracy is nominally ±1 percent of the dial reading after calibration.

---

[1] Invented independently by Stampfer of Vienna and Plateau of Ghent in 1832. Stampfer chose the name "stroboscope," which is derived from the Greek words meaning "whirling watcher."

To serve as a tachometer, a stroboscope must have its own flashing-rate control circuits and calibrated dial. Stroboscope tachometer test disks are available. These disks can be cut out and mounted on light cardboard or metal. The center must be carefully located and fitted onto the drive shaft. Although nowadays more automatic means are available to measure belt slippage, as described previously, this was commonly accomplished by stroboscopes in earlier times.

Variable-Reluctance Tachometers

A pickup of this type produces pulses that are proportional to speed, are amplified and rectified, and control the direct current to a milliammeter. This type of instrument is rated at 10,000 to 50,000 r/min, with an accuracy of $\pm\frac{1}{2}$ percent of full-scale reading. The pickup is rated to withstand ambient temperatures from -60 to $500°F$ (-51 to $260°C$).

Photoelectric Tachometers

In one instrument of this type, designed to measure speeds up to 3 million r/min, the movable part subject to measurement is arranged to provide reflecting and absorbing areas. The interrupted reflected light produces, by means of a photocell, electric impulses which are applied to a frequency meter which generates a square wave from the pulse voltage and applies it to a discriminating circuit. A fixed current pulse at each half-cycle is produced. These pulses are rectified and applied to a dc milliammeter which indicates the average value. Thus the meter readings are proportional to the number of pulses per second, or the frequency.

Eddy-Current Tachometers

The eddy-current or drag-type tachometer has been widely used for certain types of speed measurements. A preponderance of these units has been employed in automobile speedometers, in which case a flexible shaft arrangement is used, but they also find industrial usage.

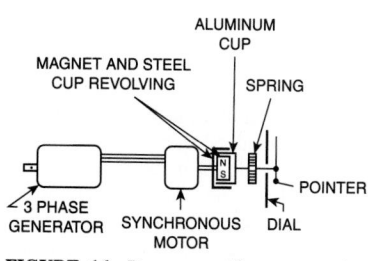

FIGURE 16 Drag-type eddy-current tachometer.

In its basic form, as shown in Fig. 16, the drag-type instrument uses a permanent magnet which is revolved by the source being measured. Close to the revolving magnet is an aluminum disk, pivoted so as to turn against a spring. A pointer attached to the pivoted disk is associated with a calibrated scale. As the permanent magnet is revolved, eddy currents are set up in the disk. The magnetic fields caused by these eddy currents produce a torque which acts in a direction to resist this action and turns the disk against the spring. The disk turns in the direction of the rotating magnetic field and turns (or is dragged) until the torque developed equals that of the spring. This torque is proportional to the speed of the rotating magnet. The instrument has a uniform scale.

The rotating field usually is produced by a permanent magnet but may be of any form which is steady. The disk may also take the form of a cup and may be of copper.

Remote indication is obtainable with one form of this tachometer. A three-phase generator is driven from the shaft whose speed is to be measured. The generator output is connected to a three-phase synchronous motor, attached to the indicator, which rotates the magnetic field. Several indicators, each with its own synchronous motor, may be connected to the three-phase generator and indicate in proportion to the speed of the generator. Since the synchronous motors keep in step with the generator frequency over a wide range, the indications are independent of the voltage developed by the three-phase generator.

Velocity Head or Hydraulic Tachometers

In devices of this type, advantage is taken of the fact that pumps or blowers produce a velocity which can be converted into a static pressure. The hydraulic tachometer incorporates a rotary pump as the transmitter and a piston as the receiver. The pump, usually contained in the indicator case, is driven by a flexible shaft and a gear train which automatically handles reversed speed, but the instrument normally is not equipped to show the direction of rotation. Pump displacement, which is positive and free of pulsations, raises or lowers a counterweight piston. The piston operates a pointer through a rack and gear. It also drives a tape-marking stylus for recording. The recorder tape is driven by the flexible shaft from the speed pickup.

The indicator may be read to 0.4 percent of full-scale value and is claimed to be accurate to within ±1 percent of full-scale reading. One application is in railroad locomotives, where the instrument accuracy is affected very little by ambient temperature changes.

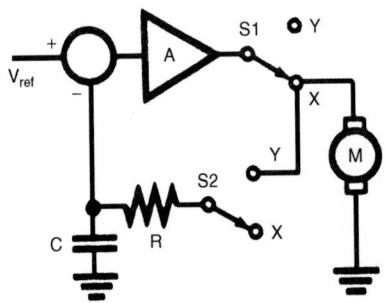

FIGURE 17 Motor-speed regulation without a tachometer. In the sample-and-hold diagram shown, the motor is switched from a free-running state, where it charges capacitor C, to a second state, where it is driven by an error voltage produced by the last stored sample. When $S1$ is in position Y, the motor is disconnected from the amplifier and is free running. At the same time, $S2$ connects the free-running motor to capacitor C.

TACHOMETERLESS REGULATION OF SERVO SPEED

Within the last few years, an interesting approach to regulating the speed of a motor without a tachometer has emerged. Basically, the arrangement consists of allowing a motor to coast for a very short interval, during which the back electromotive force (EMF) is measured. In one technique, sample and hold, the motor armature is time-shared. About 90 percent of the time it operates as a motor. During a 10 percent coasting period, the motor functions as a generator or dc tachometer. Thus it can provide an output voltage that is directly proportional to its speed. The applicable sample-and-hold block diagram is shown in Fig. 17. The motor inductance must be sufficiently small, as in the case of a printed-circuit motor, to qualify for this approach. Equations and more details are given by Geiger (1979) in the reference listed.

GOVERNORS

A governor is an automatic controller used to maintain the rotative speed of a machine at a desired value. It measures the speed, compares the measured value with the desired value, and acts to correct any error between the two values—usually by adjusting the flow of energy to the machine. Governors may be divided into two main types: (1) devices in which the speed-sensing element operates the energy metering device directly and (2) devices that use one or more stages of power amplification between the speed-sensing element and the energy control device. There is a natural distinction between these two types, arising from the fact that the first type usually gives stable control on an engine or other prime mover, whereas the second type requires the presence of some stabilizing factor to prevent continual oscillation of the speed (hunting).

AIR AND GAS VELOCITY MEASUREMENT

With accelerated interest in environmental measurements and control, there are numerous situations in which a determination of air, vapor, and effluent gas velocities must be made. Meteorological interest in air velocity has intensified, particularly in the operation of airport facilities for aircraft takeoffs and landings.

Pitot Tube Air-Speed Indicators

For many years, in testing the air speed of aircraft and in wind tunnel testing the Pitot tube has been used widely. The Pitot tube air-speed indicator consists of two elements: (1) a dynamic tube, which points upstream and determines the dynamic pressure, and (2) a static tube, which points normal to the air stream and determines the static pressure at the same point, as shown in Fig. 18. The tubes are connected to the two sides of a manometer or inclined gage so as to obtain a reading of velocity pressure, which is the algebraic difference between total pressure and static pressure.

The relationship between air velocity and velocity pressure is

$$V = \sqrt{2GH}$$

where V = velocity
G = acceleration due to gravity
H = velocity head or pressuret

The pressure differential created is quite small in relation to air velocity. At 110 ft/min (33.5 m/min) the velocity pressure is only 0.0625 in. (1.6 mm) water. Consequently the instrument is not generally used for measuring velocities less than 1000 ft/min (305 m/min). Turbulence in the air stream affects device accuracy, and the tubes are subject to clogging where dusty, unclean air is involved.

Venturi Air-Speed Indicators

Limitations of the Pitot tube led to the design of a venturi air-speed indicator in which a greater differential pressure is created. The device, shown in Fig. 19, requires individual calibration for best accuracy. Parts are not readily interchangeable.

Revolving-Vane Anemometers

This widely used device comprises a paddle wheel which is revolved by the moving air stream (Fig. 20). The wheel is attached to a counter, and by selection of the proper gear ratios and vane pitch the counter can be calibrated to read directly in feet of air. The air velocity can be determined by measuring the time interval. The device is supplied with a curve for correction of the nonlinear relation between air velocity and rotational speed of the vanes. Air density also should be considered when high accuracy is needed. The measurement tends toward the average air speed. The range of the device usually is 300 to 3000 ft/min (91.5 to 915 m/min).

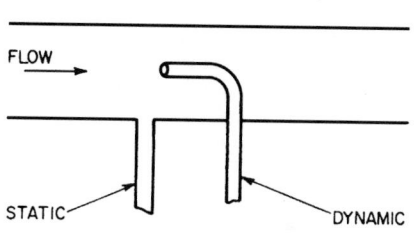

FIGURE 18   Pitot tube air-speed indicator.

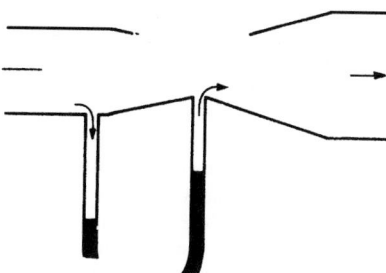

FIGURE 19   Venturi air-speed indicator.

FIGURE 20 Revolving-vane anemometer.

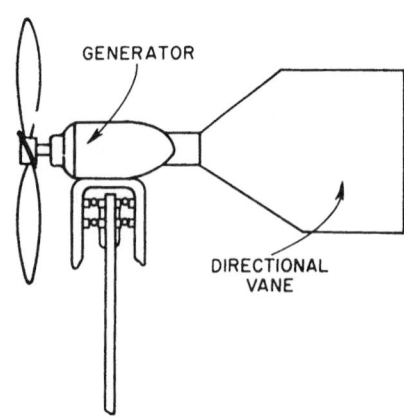

FIGURE 21 Propeller-type electric anemometer.

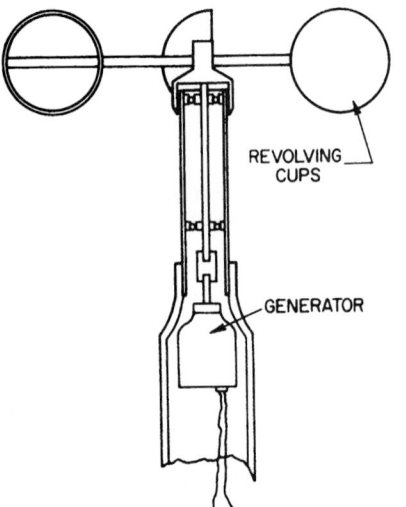

FIGURE 22 Revolving-cup electric anemometer.

Propeller-Type Electric Anemometers

This is a version of the basic rotating-vane device. Figure 21 shows the propeller type in which the blades are fastened to the shaft of an electric generator which develops an EMF or frequency proportional to speed. The EMF or frequency signal is fed to an indicator. The generator and propeller are pivoted so that the directional vane can keep the device headed directly into the direction of airflow. The device reads average air velocity. The direction of airflow can also be indicated mechanically or electrically.

Revolving-Cup Electric Anemometers

One type is shown in Fig. 22. The generator is mounted on a vertical axis, and, like the propeller type, its EMF or frequency output is proportional to the speed of the revolving cups. The speed readings are average, but the device is not directional.

Constant-Temperature Thermal Anemometers

In these devices, such as that shown in Fig. 23, the sensor element contains a heated wire. An electronic control circuit maintains the sensor element at a constant temperature regardless of the air velocity. The faster the air passes by the sensor, the more power is required to maintain the preset sensor temperature. The power dissipated by the sensor is directly related to air velocity.

An uncompensated constant-temperature hot-wire anemometer will measure accurately only if the temperature of the air flowing past the sensor remains constant. The sensor illustrated is compensated to allow accurate measurements over a wide range of air temperatures. Temperature compensation is achieved by incorporating an air temperature sensor into the control circuit. The sensors are calibrated to measure mass velocity at standard conditions [21°C (70°F) and 760 mmHg (14.7 psia)]. Built-in electronics also linearize the signal by using calibration data stored on a read-only memory (ROM). The sensors are available in three styles: (1) a general-purpose transducer, ruggedly constructed, with a protected probe tip; (2) a windowless transducer for less flow blockage, used for measurements in confined spaces, such as between circuit boards and ventilation slots; and (3) an omnidirectional transducer, also well suited to measurements in confined spaces.

Claimed accuracy for most sensors is ±1.5 percent of reading. Linear output is 0.5 V dc or 4 to 20 mA; minimum resolution is 0.1 percent full scale. Response time to flow is 0.2 second; response time to temperature ranges from 30 to 60 s. Temperature range is −45 to 93°C (−50 to 200°F); input power is 12 to 15 V dc, 250 mA maximum. There are five standard velocity ranges: 20 to 500, 20 to 1000, 20 to 2000, 50 to 5000, and 100 to 10,000 ft/min (0.1 to 2, 0.1 to 5, 0.1 to 10, 0.2 to 20, and 0.5 to 50 m/s).

VIBRATION MEASUREMENT

by Robert D. Sill[*]

VIBRATION MEASUREMENTS

Vibration measurements are used during all phases of development, construction, and operation of machines and systems. Vibration analysis can give insight to the design and integrity of a structure, and measurements during commissioning determine levels of vibration that could be destructive or cause excessive noise. Once a system is operational, monitoring vibration amplitude and trends are used as important indicators of mechanical health. Excessive or increasing vibration is often an early indicator of mechanical degradation.

Vibration is the variation of position over a period of time. It can be measured in many forms, such as displacement, velocity, or acceleration. For practical reasons described below, acceleration is the most common form measured.

Vibration sensing is required over a wide range of amplitudes and frequencies. For example, vibration from rotating machinery, such as pumps, motors, compressors, and turbines, occurs from ~1 Hz to over 20,000 Hz, the principal interest being from 10 to 2,000 Hz. Vibration amplitudes vary widely, depending on the equipment design. For example, a smooth-running motor may vibrate at 0.01g [1 g = 386.089 in./s^2 (980.665 cm/s^2)], but a high-speed gearbox can easily vibrate at more than 100 g at a frequency of more than 10,000 Hz.

Displacements associated with the accelerations will vary even more widely. A 1-g vibration at 10 Hz has an easily observed peak-to-peak displacement of 0.5 cm, whereas a displacement of 100 g at 10,000 Hz is a million times smaller, approximately the wavelength of visible light.

Ideally, a vibration sensor provides an output signal proportional to the vibrational input from the body to which it is attached. Sometimes it is not practical to attach a sensor directly to the surface of the moving body or to orient the sensor in the direction of interest (since most sensors are designed to respond to motion in only one direction, usually perpendicular to the mounting surface). In these cases, measurement is made by attaching the sensor to another body, such as a fixture, and making a measurement of the motion of that fixture. Clearly, the performance of the measurement would depend not only on the sensor but the structural characteristics of the fixture itself.

[*] Senior Project Engineer, Endevco Corp., 30700 Rancho Viejo Road, San Juan Capistrano, California 92675.

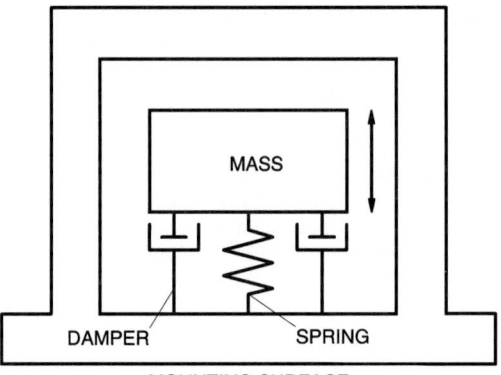

FIGURE 23 Mechanical schematic of inertial sensor. This is an idealized model that can closely match accelerometer performance. Most of the stiffness of the spring is provided by the sensor element, which deflects and supplies electrical output proportional to the inertial force ($F = \mathrm{ma}$) and therefore acceleration. Damping is associated with internal friction and interfaces between components.

One consequence of attaching a sensor (and fixture) to a structure is to change the vibration of the structure. For the effect to be small, the requirement is that the dynamic mass of the sensor be much less than the dynamic mass of the structure at the point of attachment.

Inertial Motion Sensing

Inertial motion sensors can be modeled and understood as a mass, a spring, a viscous damper, and a means of electrical detection. Figure 23 shows a simplified mechanical schematic of such a system. The response of this to vibration is well known. When the mass is small and the spring is stiff, the system can be used at frequencies below its resonant frequency, where its response is proportional to acceleration inputs. Thus it is an accelerometer. When the mass is large and the spring is flexible, the device can be used to sense relative displacement at a high frequency. Velocity transducers are used around or above the resonant frequency, although generally heavy damping is applied to completely suppress resonant amplification. However, damping does create phase shift. When damping is used for accelerometers or displacement transducers, it is desired to maintain its value at approximately seven-tenths of critical damping to achieve proportional or linear phase shift. When damping is not used in an accelerometer, the frequency range is limited to typically one-fifth of the resonant frequency.

Design Considerations

Accelerometer design is a matter of balancing tradeoffs between desired parameters, such as sensitivity and bandwidth, cost, transverse sensitivity, dynamic range, ruggedness, temperature limitations, and the cost of supporting cables and signal conditioning. Some tradeoffs are quantifiable; for instance, bandwidth is reduced by the square of the sensitivity. (That is, all other things equal, a transducer with four times the sensitivity will have a resonant frequency half as large, thus half the bandwidth. Full-scale range and ruggedness would be directly affected, decreased by a factor of 4.) Higher sensitivity can also come from gain from an internal electrical circuit, although this comes with the limitation due to the circuit, such as introduced noise or a decreased temperature range.

TABLE 3 Typical Accelerometer Performance Characteristics

Characteristics	Piezoelectric	PE with electronics	Piezoresistive	Variable Capacitance	Servo
Sensitivity	100 pC/g	100 mV/g	20 mV/g	200 mV/g	250 mV/g
Frequency range (Hz)	1–5,000	0–5,000	0–750	0–500	0–500
Resonance frequency (Hz)	20,000	25,000	2,500	1,000	1,000
Amplitude range (g)	1,000	80	25	10	15
Shock range (g)	10,000	5,000	2,000	5,000	250
Temperature range (°C)	−55 – +288	−55 – +120	0 – +95	−55 – +120	−40 – +85

ACCELEROMETERS

The most common type of vibration sensor is the accelerometer. It can be made small, lightweight, and rugged, although the size of the connectors in industrial applications increases mass significantly. Both self-generating accelerometers and those requiring electrical excitation are available. The most common is the piezoelectric (PE) device with electronics and industrial connectors. Typical performance characteristics for accelerometers are listed in Table 3.

Piezoelectric Accelerometers

A piezoelectric transducer is called self-generating because it converts a very small amount of mechanical energy into electrical charge (through transduction, hence the name transducer). These devices utilize a mass in direct contact with a PE element. When a varying motion is applied to the accelerometer, the sensor experiences a varying force excitation, causing a proportional electrical charge to be developed across it. Sensitivity is given in pC/g. The unit of charge, coulomb, is a very large value, so more practically the prefix pico (10^{-12}) is used. One pC is ~6 million electrons.

Two commonly used materials are lead zirconate titanate ceramic (PZT) and quartz. As self-generating materials, they both produce a large electric charge for their size, although the piezoelectric strain constant of PZT is ~150 times that of quartz. As a result, accelerometers using PZT are more sensitive or are much smaller. The mechanical spring constants for the piezoelectric components are high, and the inertial masses attached to them are small. Therefore, these accelerometers are useful to extremely high frequencies. Damping is rarely added to these devices, although electronic filtering is commonly added to associated electronics. Figure 24 shows a typical frequency response for such a device. PE accelerometers have a comparatively low mechanical impedance. Therefore, their effect on the motion of most structures is small. They are also rugged and have outputs that are stable with time and environment.

Two principal design configurations are used for PE accelerometers. One stresses the piezoelectric material in compression, whereas the other creates shear stresses. When the accelerometer is accelerated upward, the inertia of the mass causes it lag behind as the mounting base moves upward, squeezing the sensor between it and the base. Conversely, downward acceleration causes an upward relative motion. The resultant stress deforms the PE element. For a given geometry and force, more charge is induced in the compression mode rather than the shear. One disadvantage of the compression design is that its geometry allows other forces, such as strain at the mounting surface, to cause deformation in the element, resulting in output. If these strains are dynamic, the output that is due to base-strain sensitivity can be indistiguishable from the acceleration output, and so can be a significant source of error in accelerometers.

The most common design in industrial accelerometers is the annular shear, in which the PE element is a cylindrical ring mounted on a post that is perpendicular to the mounting surface, and the mass fits around the element. The components of an annular shear design are shown in Fig. 25. This arrangement is significantly less susceptible to errors caused by base strain compared to compression designs.

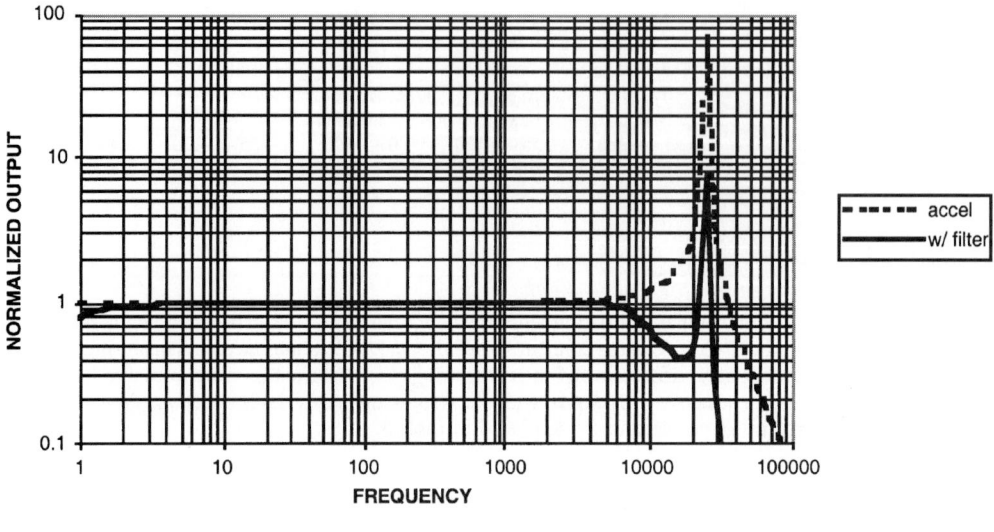

(a)

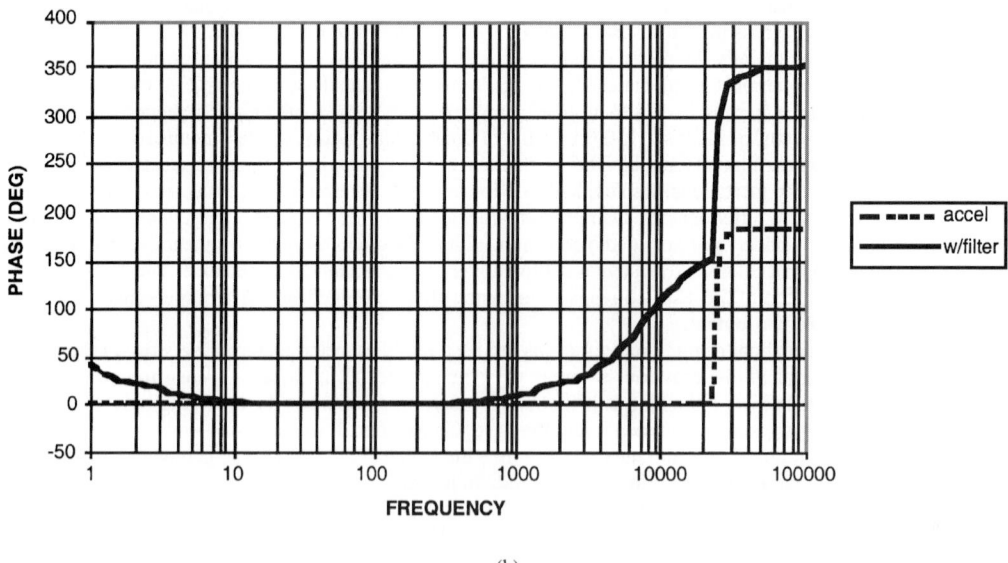

(b)

FIGURE 24   Typical responses of undamped and filtered accelerometers: (*a*) amplitude and (*b*) phase responses.

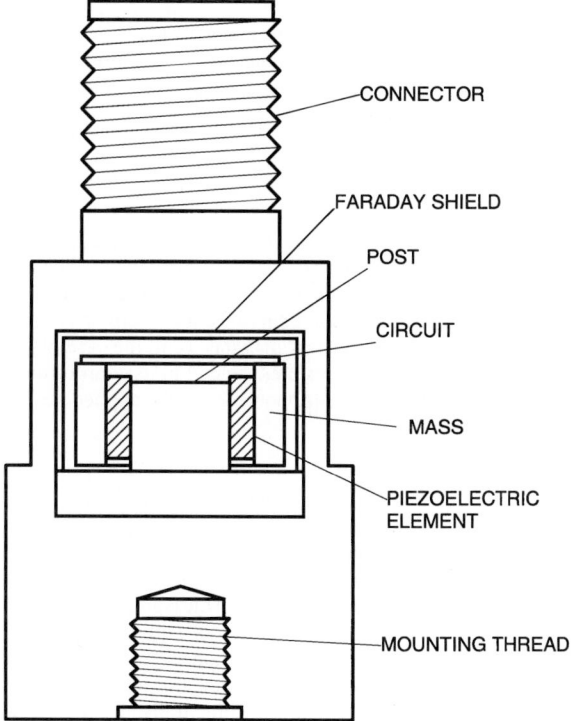

FIGURE 25   Components of an annular shear accelerometer.

Most industrial accelerometers also include integral electronics, providing low output impedance and in some cases filtering for the suppression of resonance. As described later in the section on Signal Conditioning, this simplifies signal conditioning and cabling requirements, although it limits the temperature range and reduces the flexibility allowed by an external signal conditioner in choosing the gain and filtering needed for a particular test condition.

Piezoresistive Accelerometers

Piezoresistive (PR) accelerometers are strain-gage sensors that use semiconductor strain gages in order to provide much greater gage factors than are possible with metallic gages. Higher gage factors are achieved because the material resistivity changes with stress, not just its dimensions. Significant miniaturization is possible by means of integrated circuit fabrication techniques used to manufacture the sensors, sometimes called microelectromechanical systems (MEMS).

A typical PR accelerometer uses either two or four active gages in a Wheatstone bridge. Designs can include overload stops to protect the gages from high-amplitude inputs, and oil to provide damping. Such an instrument is useful for acquiring vibration information at very low frequencies (for example, below 1 Hz), and the device can be used to sense static acceleration, unlike piezoelectric sensors.

Variable Capacitive

Another member of the MEMS family is the variable capacitive (VC) accelerometer. Also able to measure static acceleration, VC transducers generally have suppressed resonances for wider bandwidth by means of air damping, the viscosity of which changes much less with temperature than the oil sometimes used in PR accelerometers. This results in stable frequency response over temperature. With internal signal conditioning, VC accelerometers tend to have a higher output than PR devices, and usually they are more rugged, though in general neither PR nor VC accelerometers match the ruggedness of a PE transducer.

Servo Accelerometers

Servo accelerometers tend to be the most sensitive and least rugged, and they are also "dc coupled" so they can measure static acceleration or very low frequencies. Their design differs from previously described open-loop transducers in that the internal seismic mass does move freely relative to the base but is forced back into place by electronic servo feedback, either by current supplied to a coil attached to the mass in the presence of permanent magnets, or in the case of some MEMS implementations, by static charge. The resulting force restores the coil to its equilibrium position. The output signal is a measure of the forcing signal and is proportional to the applied acceleration.

SIGNAL CONDITIONING

Signal conditioners provide the interface between accelerometers and readout and processing instruments by (1) providing power to the accelerometer if it is not self-generating, (2) providing proper electrical load to the accelerometer, (3) amplifying the signal, and (4) providing an appropriate filtering and drive signal. The different classes of transducers require conditioners with distinct characteristics.

(Future trends are adding features to the conditioners, and changing their definition to "controllers." A new feature would be communication with "smart" transducers, which can transmit their identity and calibration characteristics to the conditioner. Some transducers contain analog-to-digital converters and communication circuitry to share a common bus with a controller, sending packets of digital data only as commanded. A bus could significantly reduce the length of cabling in an installation. Taking this trend to its extreme, elimination of wiring is the goal of "wireless" techniques, in which perhaps battery-powered sensors with radio receivers and transmitters could provide readings on demand to a centralized remote controller. Conventional conditioners, however, will still have a place, and these are described below.)

Conditioning Piezoelectric Accelerometers

The charge output of a piezoelectric accelerometer presents a very high source impedance to the conditioner. It may be regarded as a voltage source in series with a capacitance, or as a charge source in parallel with a capacitance. The signal conditioner determines how the transducer is treated in a given system. Both voltage and charge sensing can be used, although the charge converter is by far the most common approach. The charge converter is advantageous because the system gain and low-frequency response are well defined and are independent of the cable capacitance (determined by length) and accelerometer capacitance (which can change with temperature). The charge converter is essentially an operational amplifier with capacitive feedback.

The piezoelectric accelerometer and its cabling need an extremely high output impedance to prevent low-frequency noise. In addition, cables for PE transducers should be "noise treated" to minimize noise due to cable motion (see Environmental Effects, below).

Conditioning Piezoelectric Accelerometers with Internal Electronics

Piezoelectric accelerometers are available with simple electronic circuits internal to their cases to provide signal amplification, filtering, and low-impedance output. Most designs operate from fixed-dc-current supplies (typically 4 or 10 mA) and are designed to be intrinsically safe. The principal advantages of piezoelectric accelerometers with integral electronics are their relative immunity to cable-induced noise and spurious response, the ability to use lower-cost cable, and a lower signal conditioning cost.

These advantages do not come without compromise. Because the impedance matching circuitry is built into the transducer, gain cannot be adjusted to utilize the wide dynamic range of the basic transducer. Ambient temperature is limited to that which the circuit will withstand, and this is considerably lower than that of the piezoelectric sensor itself. In order to retain the advantages of small size, the integral electronics must be kept relatively simple. This precludes the choices possible in gain, filtering, and dynamic overload protection, and thus limits this application. But when conditions are relatively benign, these accelerometers can economically provide excellent noise immunity and signal fidelity.

Conditioning Piezoresistive Transducers

Piezoresistive transducers are relatively easy to condition. They generally have high-level output, low-output impedance, and very low intrinsic noise. These transducers require an external power supply. It must be a stable supply, since output is approximately proportional to the supply. This supply is usually dc voltage, but it may be alternating current, provided the carrier frequency is at least 5 to 10 times the maximum frequency of interest.

Most transducers are designed for constant voltage excitation and are used with relatively short cables. With long cables, wire resistance is not negligible. Moreover, resistance changes with temperature, and the voltage drop along the line varies as the transducer resistance of load changes. Any voltage drop will cause a direct reduction in transducer sensitivity. For these applications with long cables, transducers should be calibrated for constant current excitation so their output will be less dependent on external effects.

Many piezoresistive transducers are full-bridge devices. Some have four active arms to maximize sensitivity. Other have two active arms and two fixed precision resistor arms to permit shunt calibration by precision cablibration resistors in the signal conditioner.

Adjustment of the unbalanced output of an accelerometer can easily be performed in the signal conditioner. For full-bridge transducers, the balancing potentiometer is connected across the excitation terminals and a current-limiting resistor is connected between the wiper arm of the potentiometer and the bridge. For half-bridge transducers, a small balance potentiometer (typically 100 Ω) is connected between the bridge completion arms.

Environmental Effects

Temperature.   Accelerometers can be used over wide temperature ranges. Piezoelectric devices are available for use from cryogenic temperatures, $-270°C$ ($-454°F$), to over 650°C (1220°F). The sensitivity changes with ambient temperature, but the changes are systematic and can be calibrated. Some transducers include compensating capacitance elements that counteract the sensitivity deviations of the PE elements. If the ambient temperature changes suddenly so that strains develop within the accelerometer and within the time response of the measurement system, further errors can occur. These are evaluated by testing the response or accelerometers to step function changes in temperature per industry standard test procedures. Errors usually appear as a wandering signal or a low-frequency oscillation. These errors tend to be larger in designs using PE elements in compression than in those using shear.

Cable Motion.   Cabling from the accelerometer to the signal conditioner can generate spurious signals when it is subjected to movement or dynamic forces. This is usually significant only for systems using high-impedance piezoelectric accelerometers. The major noise-generating mechanism is triboelectric noise, which is caused by charge trapping due to relative motion, or localized separation between the cable dielectric and the outer shield around the dielectric. To reduce this effect, cabling is available that is "noise-treated." These cables have a conductive coating applied to the surface of the dielectric, which prevents charge trapping. Another way to eliminate this effect is to use a sensor that includes an electronic circuit for reducing the impedance.

Mounting Compliance.   Accelerometers are very stiff structures, generally with internal resonances of 10 kHz or higher. To correctly measure the motion of the surface to which they are attached, the stiffness of the attachment is ideally much greater than 10 kHz. The manufacturer's specified surface flatness, finish, and mounting torque are all required in attaining the necessary mounting stiffness. An overly flexible fixture or other compliant attachment introduces a lower resonance into the measurement system. Resonances by nature amplify the motion, so the reading would be inaccurate at those and nearby frequencies. Worse, rattling in a loose attachment would invalidate the measurement at all frequencies and possibly damage the transducer.

Dynamic Strain Inputs.   In vibration environments, some structures may dynamically flex, stretch, or bend at the mounting location of the accelerometer. Being in intimate contact with this strained area, the base of the accelerometer can also be strained. A portion of this base strain is transmitted to the sensing element. Any signals that result are an error source. In addition to strains in the structure, it is also possible to induce errors from forces or pressures onto the case or cable connector of the accelerometer. Outputs from these forces vary greatly, depending on the internal design of the accelerometer. The errors from these sources are usually checked against industry standard test procedures, and the results are included in specifications.

Electrostatic and Electromagnetic Fields.   Electrostatic noise can be generated by stray capacitance coupling into the measurement system. It is extremely important that the cabling between a high-impedance piezoelectric sensor and the signal conditioner be fully shielded. Using transducers with low-impedance output reduces this problem. Ground loops can be avoided by grounding the system at a single point, usually at the output of the signal conditioner. Most industrial transducers have cases that are isolated from both signal leads to prevent ground loops, and are protected from the effects of voltages on the mounting surfaces by an internal Faraday shield. Magnetically coupled noise can best be avoided by not placing signal cables in close proximity to power or high-current conductors and by avoiding electromagnetic sources when possible. Accelerometers should also be checked for their sensitivity to electromagnetic fields.

VELOCITY TRANSDUCERS

Electrodynamic velocity transducers are self-generating, creating voltage proportional to the velocity of the mounting surface, usually of sufficient amplitude that no voltage amplification is required. The disadvantages of velocity pickups are their large size and mass and their inability to be used for measurements at frequencies below ~10 Hz. Also, the output at high frequencies, above ~1,000 Hz, is quite small in most applications. Care must be taken in using these devices in strong magnetic field environments.

A typical velocity sensor consists of a damped magnetic core suspended in a housing rigidly attached to the vibrating surface. A coil of wire attached to the housing surrounds the core. Relative motion between the magnetic core and the housing causes magnetic lines of flux to cut the coil, inducing a voltage proportional to the velocity.

NONCONTACT (RELATIVE-MOTION) SENSORS

In some cases it is not practical to place a sensor in contact with the moving part. Relative-motion measurement approaches can be used, in which variations in the gap between the sensor and the moving surface are measured using one of several fields: magnetic (eddy current), electric (capacitance), or optic (reflectance or interferometery). Each technique can have specific requirements on the material or surface properties of the target, and a common problem is foreign objects getting into the gap and changing the properties of the field. An obvious requirement for any relative motion measurement is that the sensor must be mounted on a rigid structure that is not itself moving.

Eddy-Current Probe

An eddy-current displacement probe contains a small coil of fine wire at its tip that is excited by a remote RF oscillator to generate a magnetic field. As the tip of the probe is brought close to a conductive surface, such as a rotating shaft, eddy currents induced in the conductor by the probe's magnetic field oppose the field and reduce the amplitude of the carrier by a an amount proportional to the change in proximity. A demodulator, usually encapsulated in the same enclosure as the oscillator, converts the change in carrier amplitude to a low-impedance, calibrated voltage output.

An eddy-current displacement sensor and its companion oscillator-demodulator therefore constitute a gap-to-voltage measuring system. The average gap, or the distance between the probe tip and the conductive surface, is represented by a dc bias or offset on which is superimposed an ac analog of the surface's dynamic motion. A typical linear amplitude range is 1–2 mm (0.04–0.08 in) with a frequency-response capability from static to more than 2,000 Hz. The sensitivity changes for different target materials and with changes in cable length.

Optical Vibrometers

A large number of optical techniques exist for position or velocity measurement. Some are highly accurate and expensive, such as laser systems based on interferometery or Doppler effects. Others are much less expensive (based on variations in brightness of reflected light), very nonlinear, and subject to the optical properties of the surface. These generally are used for sensing presence rather than motion.

SECTION 6
REAL-TIME ANALYTICAL COMPOSITION MEASUREMENTS FOR INPUT TO PROCESS CONTROL*

C. H. Albright, Jr.
(Practical Considerations)

R. H. Cherry
Huntingdon Valley, Pennsylvania
(Thermal Conductivity Gas Analyzers)

Jimmy G. Converse
Sterling Chemicals, Inc., Texas City, Texas
(Sampling for On-Line Analyzers, Measurement Devices,
Calibration, Chapter Coordinator, Applications)

G. F. Erk
Philadelphia, Pennsylvania (Classification of Analysis
Instruments)

David M. Gray, Sr.
Leeds & Northrup (a Unit of General Signal),
North Wales, Pennsylvania (Thermal Conductivity
Gas Analyzers)

E. A. Houser
(Continuous Process Sampling)

David W. Howard
Brookfield Engineering Laboratories, Inc., Stoughton,
Massachusetts (Rheological Systems)

James A. Johnke
Source Technology Associates, Research Triangle Park
North Carolina (Continuous Emission Monitoring)

J. Kortright
Leeds & Northrup (a Unit of General Signal), North Wales,
Pennsylvania (Electrical Conductivity)

Gregory K. McMillan
Integrated Manufacturing Control Group, Solutia Inc.,
St. Louis, Missouri (pH Measurement)

Persons who authored complete articles or subsections of articles, or who otherwise cooperated in an outstanding manner in furnishing information and helpful counsel to the editorial staff.

Gregory Neeb

DeZurik (a Unit of General Signal), Sartell, Minnesota
(Rheological Systems)

Eugene Norman

Consultant, Green Bay, Wisconsin (Rheological Systems)

James Overall

Georgia-Pacific Corporation, Atlanta, Georgia
(Rheological Systems)

B. Pelletier

Rosemount Inc., Measurement Division,
Eden Prairie, Minnesota

J. G. Puls

DeZurik (a Unit of General Signal), Sartell, Minnesota
(Rheological Systems)

R. S. Saltzman

Saltzman & Associates, Wilmington, Delaware
(Ultraviolet Spectroscopy)

Don Soleta

Monsanto, Inc., St. Louis, Missouri
(Turbidity, Refractive Index)

E. Sperry

Beckman Industrial, Cedar Grove, New Jersey
(Electrolytic Conductivity—prior edition)

Geoff Wickens

Endress & Hauser, Inc., Greenwood, Indiana
(Electrical Conductivity)

INTRODUCTION

Analyzers have been identified by method of analysis, components measured, and specific application. There are numerous types of analyzers for each chemical component. It has become difficult to identify analyzer systems for lack of defined parts. We need to redefine things so that all analyzer systems can be conceived of as a group of the following modular functions: (1) sample extraction; (2) sample conditioning and clean up; (3) sample transfer and preparation; (4) component separation and isolation; (5) concentration/signal transducer; (6) system control and signal manipulation; and (7) information display and documentation. The configuration and number of modules may vary, but the functions of the system parts will be the same. The identification of all systems can be made by type of sampling, conditioning procedure, degree of component isolation, type and selectivity of the detector, and mode of control and signal handling.

CHANGING TIMES AND TECHNOLOGY

The Big Picture

We need to reach into the process at a point where a representative universe exists and extract the minimum quantity of material required to obtain the analytical concentration of the target analytes.

We need to transport this packet of sample to a safe and convenient location without adulteration or loss of components for chemical analysis.

We need to prepare this sample appropriately for the specified detection and measurement device in such a fashion that we can quantify the target analytes.

We need to isolate each target component so that other species present do not interfere with the measurement function.

If necessary, we should modify the chemistry of the target analyte to improve the stability or detection limit.

We need to make a ratio of the unknown quantity of analyte to a known standard to calibrate the system.

We need to provide an internal reference to monitor the performance of the analytical system in real time to validate our results.

We need to operate the system in a fashion that automatically keeps it clean and functional.

We need a simple, reliable control system to manage the system functions.

We need a simple, reliable communications scheme to report the results to the customer and to protect stored records for historical information.

SEPARATING SAMPLE PREPARATION FROM CONCENTRATION MEASUREMENT

There are many volumes of printed material describing sample extraction. The first rule is that the composition of the material extracted must be representative of the process-stream population. This does not mean that it has to be identical.

The main problem is that we have not addressed the "sample preparation" aspect of on-site automated chemical analysis adequately. Traditionally, we tapped the process stream and let it flow through the analyzer. As the applications progressed to dirtier and more complex materials, we added gadgets to the sample line to clean up the material. What evolved was miniature chemical processing units that became expensive to implement and difficult to maintain. They also became the main cause of analyzer system failure.

We developed better gadgets and automated system controls, but we did not improve reliability much. In 1978, remote discrete sampling was utilized to safely handle liquid hydrogen cyanide for gas chromatography (GC) analysis. It became obvious in 1983 that the GC column was technically part of the sample conditioning system. This realization immediately gave us new degrees of freedom in sample preparation. In 1985, discrete sampling was combined with automated cleanup of the sample conditioning system to reduce maintenance. We also added internal reference introduction to provide performance monitoring of the sample preparation system and the detector.

It has been standard practice to utilize valves for manipulation of the sample in process GC analyzers. Analytical chemists have coined the phrase "multidimensional gas chromatography" to describe such manipulations in laboratory instruments. This practice is multidimensional sample preparation and excludes the measurement part of chemical analysis. If we focus our attention on sample preparation and finally resolve this problem, the application of on-site automated chemical analyzers can take a giant step forward.

WHAT THE ANALYZER CUSTOMER SHOULD DEMAND FROM THE CONTRACTING ENGINEER

Avoid Common Mistakes

Successful sampling can best be achieved if one uses a modular approach to the task of sample preparation. First, distinguish between the roles of the measurement system and the sample preparation system. The prime objective of the analyzer is to determine the concentration of one or more species in a mixture within the required precision and accuracy.

The "primary task" in developing a sample preparation system is to maintain continuous unattended operation of the individual parts to provide a representative sample to the detector in a suitably conditioned form within the required precision. This can only be achieved if one focuses on the task of sample preparation by utilizing mechanical and electronic skills to apply the principles of physics, chemistry, and engineering.

In particular, beware if your proposed sample preparation system

1. is not designed for maintenance
2. contains complex miniature chemical processing units
3. generates large amounts of waste and disposal
4. is designed for ease of construction instead of use and maintenance
5. is based on cost and rushed to close out a project
6. requires excessive manual maintenance
7. has high emphasis on accessories (bells and whistles)
8. is overshadowed by measurement emphasis
9. lacks adherence to chemistry fundamentals.

Unfortunately, you more than likely will spot several of these problems. What can you do about them? First, you must understand the causes of the problems to ensure that you don't let them be designed into projects. Learn what is possible and then demand that project contractors address your concerns. Often, taking the following steps can lead to improved performance of the sample preparation systems.

1. Prior to designing the conditioning system, characterize the process material by extracting through filters and scrubbers and by analyzing the vapor, liquid, and solid phases.

2. Involve someone with an understanding of analytical chemistry and sample preparation fundamentals to weigh appropriate alternatives.

3. Minimize the quantity of material extracted from the process for conditioning by utilizing discrete sampling.

4. Demand as much automated system cleanup as is practical and prudent.

5. Keep the system simple and easy to maintain.

Nothing is Magic. There is a reason for everything. There are causes for every system failure.

PRACTICAL CONSIDERATIONS
by C. M. Albright, Jr.

Any practical appraisal of the merits of chemical-composition variables for process control purposes must recognize certain inherent physical limitations in their measurement. Generally speaking, these limitations are as follows.

1. Sample must be representative.   Although this requirement may appear obvious, it is a factor that is very frequently overlooked. In the first place, the sample must be gathered or drawn off in such a fashion that it will be of the same composition as the body of the processed material. Moreover, there must be assurance that any change in conditions, such as temperature or pressure, between the sampling and measuring points cannot influence the sample composition. In addition, in nearly all cases the probable compositions of the sample must be known ahead of time through some independent method before an analysis technique can be selected.

2. Physical state of sample.   The technique must provide for interaction between the applied energy and the entire sample, as well as for observation of the total result. This can seldom be accomplished. It is for this reason that a large majority of techniques are applicable to gases, where the molecules are widely spaced and free to react in a characteristic manner, and that fewer techniques are applicable to liquids and still fewer can be applied to solids.

3. Uniqueness of the specificity of the method.   The selection of the method must be tailored to the sample composition and to the information requirements.   Some methods or techniques involving atomic and molecular structure are rather universal in that they permit exact identification and measurement of every elemental or molecular constituent present in the sample.

These methods are usually the most complex and costly. They are sometimes considerably less sensitive than simple methods, whose only drawback is an inability to distinguish between related substances having similar gross interactions with energy, where the related substances are known not to be present in the sample. These simpler, less specific methods should always be considered.

TRENDS IN ANALYTICAL INSTRUMENTATION
by G. F. Erk

The trends in specific areas of analysis are brought out in the articles that follow in this handbook section. With the accelerated development of new chemical analytical techniques in the 1950s, a trend commenced that would move the formerly isolated chemical control laboratory on stream in terms of using continuous instrumental analyzers. A parallel trend occurred with the availability of improved nondestructive inspection techniques and semi-automated testing procedures in the discrete-piece manufacturing industries. These trends essentially started with the larger manufacturing firms and have been growing in acceptance and practicality with smaller manufacturers ever since. The driving forces for these actions have included the following.

1. Conservation of energy has increased efforts to use on-line analytical instruments. Fuel costs have risen over the past decade to a point where on-line instrumentation has become even more cost effective. One area immediately affected by this is the application of an oxygen and combustibles analyzer in determining optimum fuel-to-air ratios in the combustion process and the resulting heat generation. Likewise, the calorimetric analysis of fuel quality is becoming even more important, as are density and specific gravity analyses.

2. Requirements for monitoring pollution generate additional needs for high-quality analytical devices with the capability to provide good records of various pollutants and particulate emissions. Stack gas monitors are widely used for pollution control and monitoring and have design features permitting long-term unattended operations within stacks or through sampling systems.

3. There are pressing demands for more accurate, thorough, and rapid means of testing materials and product, from the receipt of raw materials and inspection throughout manufacturing to the completion of production, warehousing, and distribution.

Designers of contemporary analytical systems have taken advantage of all the technological amenities that have become available during the past decade or so, namely advanced communication and networking, improved data displays, microprocessors, and personal computers, self-diagnostics, and the miniaturization of sensors. Reliability and precision have progressed steadily. During the next decade, much more can be expected to come from the analytical instrument field. Scores of excellent analytical techniques still confined to the laboratory are waiting to be exploited by enterprising designers, who will convert them to configurations that will make them practical to use on the process and in factory floor environments.

OPPOSING APPROACHES TO A SIMILAR PROBLEM
by J. G. Converse*

There are two current schools of thought regarding how one designs analyzer systems. First, there are simple, reliable, single-component systems, which require low maintenance skills and low cost. Second, there are complex, reliable, multicomponent systems, which require high maintenance skills and high cost. Some expensive and complex analyzer systems are justified because they are the only way to achieve the needed results (i.e., NIR). Considering maintenance and total cost of ownership would dictate utilizing the simple approach wherever possible. One must utilize the simple approach if the facility does not have a competent and knowledgable analyzer technician, unless management is willing to pay the high price of contract maintenance. Justification for a complex and expensive analyzer system must be based on need and return on investment.

APPROACHES TO PROCURING ANALYTICAL SERVICES

There are different ways to acquire process analytical information:

- fully automated on-line analyzer systems
- automated at-line analyzers for operations to utilize
- central laboratory for low-frequency routine analyses
- outside laboratory to satisfy special needs when equipment and manpower are not practical.

The selection of one of these choices must be based on each sample and test required.

---

* Sterling Chemicals, Inc., Texas City, Texas.

CONCENTRATION MEASUREMENT TECHNOLOGY AND DEVICES

The following classification of physical and chemical properties of measurement devices may be useful.

MECHANICAL CONCENTRATION TRANSDUCERS

These can be classified by density (mass per unit volume); distillation (boiling point, distillation, gas chromatography, or GC, SimDis—measure boiling range distribution); vapor pressure (Reed System and Fowler/Trump method: log P vs. $1/T$); cloud point, pour point, or freezing point (temperature of phase change or flow); viscosity (rheological systems, consistency); and octane number (knock engines vs. NIR).

THERMAL CONCENTRATION TRANSDUCERS

Thermal Conductivity (and Gas Density)

This is for gas mixtures and column-effluent monitors. It is a nonspecific and nonabsolute method of gas-mixture analysis that depends on empirical calibration. It is generally used for binary mixtures and requires careful sample preparation. It is utilized as a detector for column chronography, in which components are isolated and many of the above problems are nonexistent.

Three specific application techniques are:

- diffusion (very low detector noise; mixing and exponential dilution; poor linearity)
- convection (radiative)
- flowthrough (retain peak symmetry; short time constant; high linearity; flow sensitive)

Electronic control of the detector sensors and bridge circuit can utilize three different modes of operation:

- constant voltage (simplest to operate)
- constant current (improved linearity)
- constant temperature (fastest return to baseline)

The assets are that it is a universal detector, which is simple to operate and maintain. The liability is an LDL of low parts per million.

The development of GC as a separation technique has given rise to a wide variety of general purpose and specialized detectors. The selection of the proper detector is dependent on the chromatographer's requirements for selectivity, sensitivity, response time, and type of sample. Secondary considerations such as destruction of the sample, corrosive reactions, gas phase, and so on can be important and must often be studied. The detector most widely used for universal analysis is the thermal conductivity detector (TCD). The gas density balance, which is a special case of the TCD, has been adopted for corrosive applications and molecular-weight determinations.

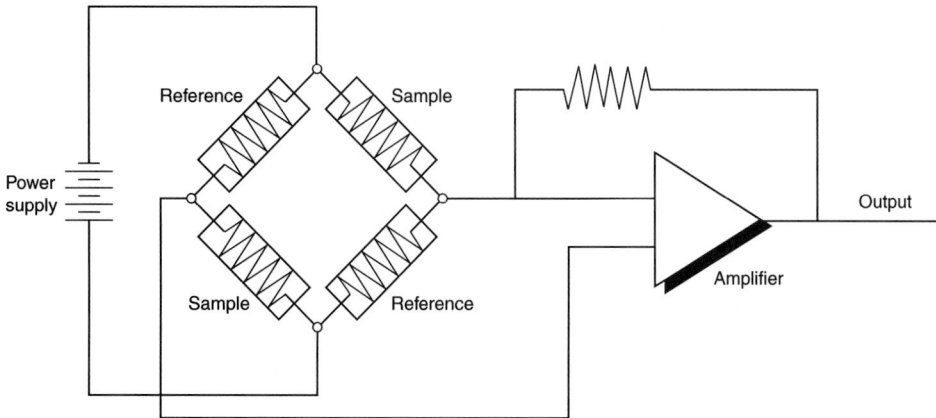

FIGURE 1   TCD Wheatstone bridge and readout connections.

Thermal Conductivity Detectors.   The TCD is one of the more widely used detectors for GC because it is simple in construction, rugged, versatile, sensitive, relatively linear over a wide range, nondestructive to the sample, and inexpensive. It is because of these qualities that it continues to be widely used, even though more sensitive and specialized detectors have been developed.

The TCD consists of a block (usually metallic) containing a cavity through which the gas flows. A heated element (resistance wire or thermistor) is positioned in the cavity and loses heat to the block, depending upon the thermal conductivity of the gas. For practical considerations, a differential method is usually used that requires two cavities and two heated elements. Only carrier gas is passed through one cavity, and the column effluent is passed through the other. For process analysis, a reference and sample gas are used (Fig. 1).

The two most commonly used detector transducers are metal resistance wires (filaments) and thermistors (beads of metallic oxide). Their operation is similar, except that filaments have a positive coefficient of resistance (resistance increases with temperature increase) and thermistors have a negative coefficient of resistance (resistance decreases as temperature increases). Also, the resistance change of filaments is smaller than that of thermistors, so filaments are sensitive over a wider temperature range.

The choice between thermistor or filament is usually based on temperature considerations: thermistors are used for ambient or subambient temperatures, and filaments are used for higher temperatures. Once the decision has been made to use filaments, the selection is based on the corrosiveness or oxidation characteristics of the materials to be analyzed.

Filaments are fabricated from a variety of metals; the most common of these is tungsten. Other materials used for filaments include nickel, platinum, rhenium tungsten, and tungsten with a plating of gold or rhodium. Normally the filament is made from wire that is ~0.001 in. (~0.002 cm) in diameter wound on a mandrel that is ~0.0012 in. in diameter.

Since the TCD is included in the class of detectors whose signal is dependent upon the concentration of the sample in the carrier gas, it is necessary to consider the internal volume of the cell (Fig. 2).

The size of tubing, its length, and the volume of the cavity, as well as any dead volume in the system, will affect the signal. The relative thermal conductivity of the sample constituents to that of helium (or another carrier gas such as hydrogen or argon) will also affect the signal strength. The heated elements are connected electrically into a Wheatstone bridge, as shown in Fig. 1. With the same gas passing through both cavities, the network is balanced by means of the balancing potentiometers so the electrical output is zero. The readout is usually a 1- or 10-mV potentiometric recorder or electronic integrator. When the thermal conductivity of the gas in one of the cavities changes as a result of the

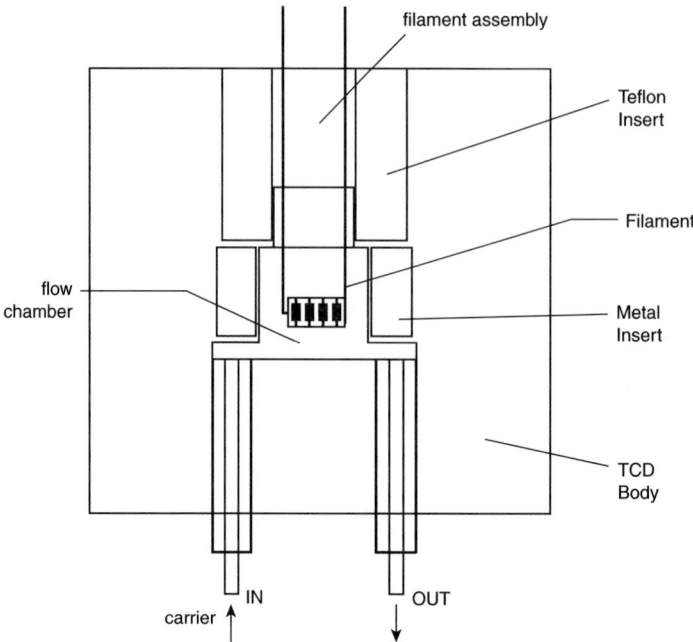

FIGURE 2   Small-volume TCD filament cell.

sample being *eluted* from the column, the temperature and resistance of the detector element in that cavity change and the imbalance of the bridge is recorded.

The ability of the TCD to follow peaks as they elute from the column is dependent on the speed of response, which in turn is dependent on the internal volume of the detector. The flow through the detector must remain constant for the entire analysis and is usually in the range of 5–100 ml/min.

Gas Density Detector.   The gas density detector operates on the principle that a sample fraction of molecular weight that is different from the reference/carrier gas will cause a flow imbalance between two symmetrical branched channels located in the same vertical plane (Fig. 3).

If the column effluent carries a sample component that is denser than the reference gas, the downward flow retards the flow in the lower branch and increases the flow in the upper branch. If the sample is less dense than the reference, the effect is reversed. The two measuring elements are thermal conductivity detectors connected as arms of a Wheatstone bridge. Here, however, *flow rate rather than molecular rate* is being sensed. Because the sensing elements are exposed only to the reference gas, the gas density detector is ideal for the detection of highly corrosive or reactive samples.

Detector Sensitivity.   The detector sensitivity depends on several factors. Most basic is the difference in thermal conductivity (or other characteristics) between the background (carrier, reference) gas and the sample.

Sensitivity of hot-wire TCDs depends mostly on the internal geometry, type of detector elements, and the temperature differential (delta T) between filaments and the cell block. Thermistor detectors are at their best in subambient and ambient operations. Experiments with filament and 8-K thermistor detectors with identical geometries demonstrate that a greater signal is obtained from the hot wires at cell temperatures of 75 °C and above.

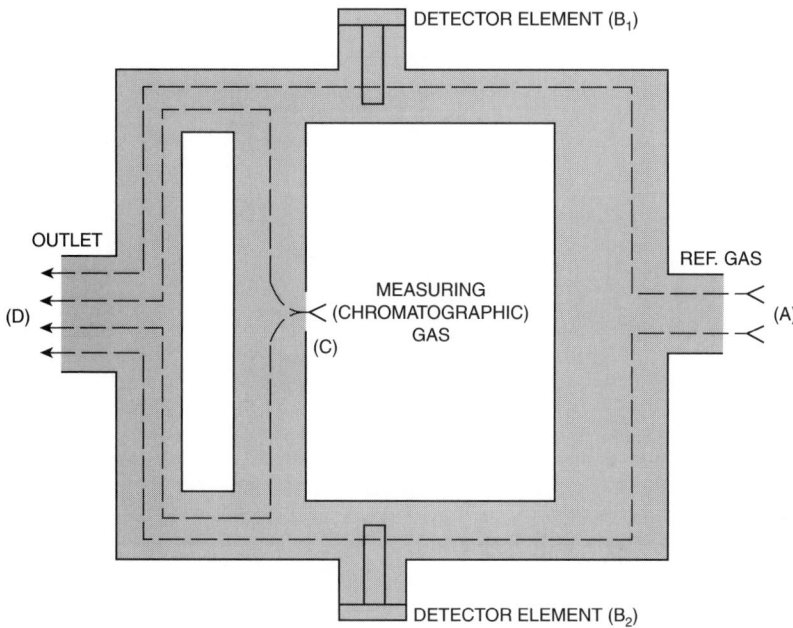

FIGURE 3   Gas density balance detector.

Gas density detectors vary in sensitivity directly with the difference in molecular weight, as well as with the number and type of elements.

Heat Flux (BTU)

This is for GC analysis and calculation from tables versus direct measurement. Fuel and air combusted in a chamber generate heat based on the fuel burned and flow rate. The heat flux sensor produces an electrical signal that is proportional to the quantity of energy passing through it. The burn chamber must be well insulated, and the fuel and air flow rates must be controlled. The chemical identity and percent of that species can be determined by its heat content compared to a table of reference materials. This is not a direct or highly accurate way to obtain the information desired. Alternatively, one can feed two or more reference materials to calibrate the heat-content analyzer and determine the heat content of an unknown material by comparison to the known materials. This provides a simple, accurate, and cost-effective heat-content analysis in real time (Fig. 4).

Catalytic Beads

These are combustibles and hydrocarbons in air. They are widely used in area safety monitors (organic vapor analyzers).

ELECTRICAL CONDUCTIVITY

This is in terms of ions in solution, mixtures, and column-effluent monitors.

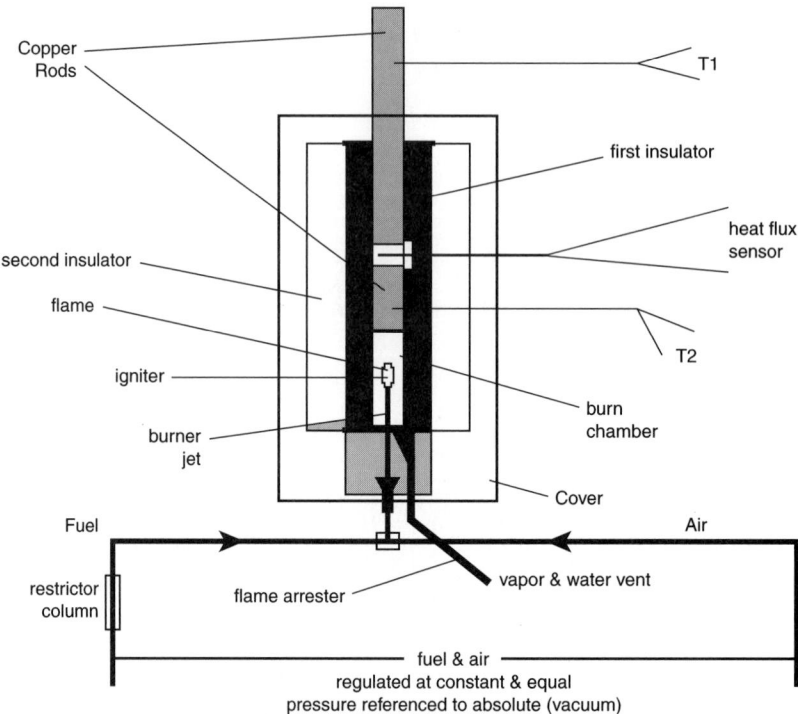

FIGURE 4   Direct heat-value measuring device, utilizing a heat flux sensor.

Electrical Conductivity Measurement

The conductivity measurement is used in many fields of process control and monitoring, since its applications and uses are many. Electrical conductivity is a measure of a liquid's ability to transport an electrical current—in other words, a measure of its resistance. Electrical currents are transported in liquids by positively and negatively charged particles (ions), typically resulting from salts, acids, and bases (see Fig. 5).

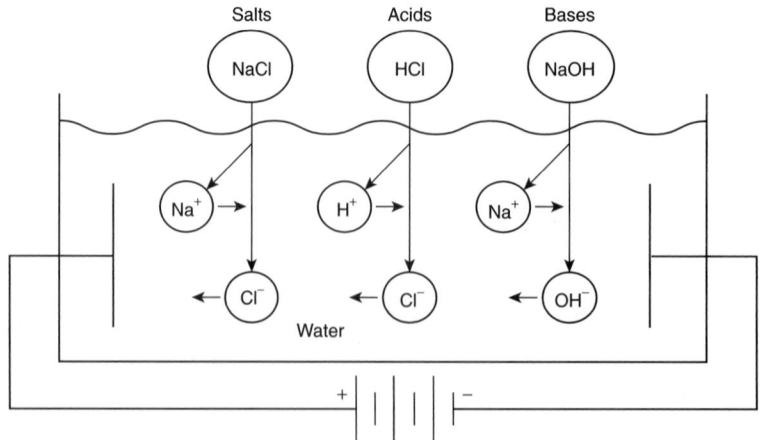

FIGURE 5   Dissociation.

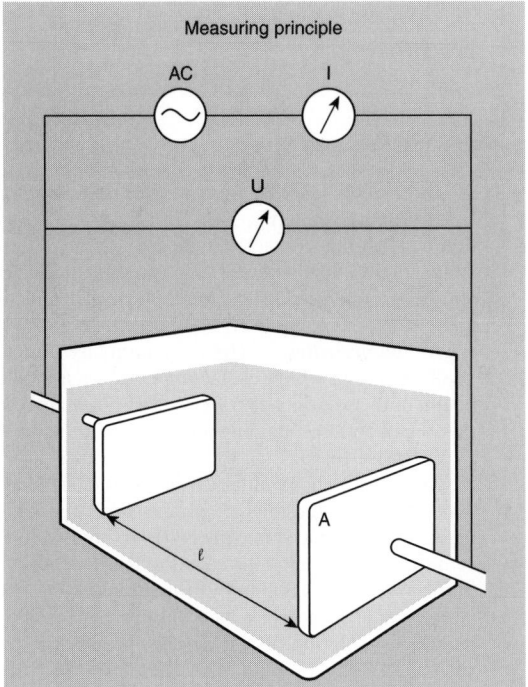

FIGURE 6   Two-electrode principle.

The unit of measure is siemens/cm = 1/resistance in ohms × cell constant.

For example, a solution has a resistance of 1.000.000 Ω, measured with a sensor with a cell constant of 0.1. The conductivity is 1/1.000.000 × 0.1 = 0.000001 S/cm. This could be better referenced as 1×10^{-6} or 1 μS/cm. Likewise, 2000 μS/cm = 2 mS/cm and 2000 mS/cm = 2 S/cm.

Cell Constant.   When two plates of fixed area and distance between them are placed in a solution, and potential is applied between them, a current will flow. Since the resistance is inversely proportional to the conductivity, with low conductivities the resistance is high. Therefore, in order to ensure a measurable current, the area and distance between the plates must be correctly matched to the desired measuring range. This is commonly referred to as the "cell constant." The cell constant is defined as c = distance between the plates/area of one of the plates (see Figs. 6 and 7).

The concept of using the Ohms law and dc circuit to explain the contacting conductivity measurement is a convenient one. In practice, though, measurements are made with an alternating current to ensure that cell polarization is greatly reduced.

Ions carry an electrical charge that is transferred to the surface of the cell. An ion cloud could form in front of the cell surface, preventing the flow of current. If direct current was to be used, the initial current would quickly go to zero.

Polarization also occurs when the frequency of oscillation is wrong or the selected cell constant is too low for the desired measuring range.

Contacting Conductivity Cells.   These cells are best used in low ranges, including typical water-based applications in which conductivities do no exceed 2000 μS/cm. However, with the correct

implementation, a contacting conductivity cell can be used for readings up to 200 mS/cm. Conductivity typically has a measuring scale from 0.05 μS/cm to 2000 mS/cm (Fig. 8).

Sensors, Cells and Probes

All three of these terms are sometimes used interchangeably to describe conductivity probes. To clarify, "cell" is used to describe contacting cells in which the liquid is in direct contact with the cell. "Sensor" is used when there is no electrical connection to the process stream, such as torodial conductivity sensors. The word "probe" can be used to describe both "cells" and "sensors."

High conductivities. These are normally found in concentrate, acid, and base applications, in which contacting cells are normally not practical because of the nature of the solutions to be measured. They are often chemically corrosive, and in the case of concentrates, higher densities and viscosities are commonplace (Fig. 9).

For these reasons the toroidal or electrodeless sensors have proven to be the most practical. The toroidal sensor is based on Faraday's law of magnetic induction. When two insulated coils are immersed into the solution to be measured, an oscillating voltage is induced into the first coil. This sets up a magnetic field, as the process solution flows through the center of the two coils. The amount of voltage induced into the second coil is directly proportional to the resistance of the solution (Fig. 10).

Toroidal sensors have an advantage because no metal comes into contact with the process solution. In addition, they can be produced from a variety of materials such as PEEK, TFE, and other chemically resistant plastics. Buildup on toroidal sensors does not cause problems and polarization is not an issue. Toroidal sensors can also be designed without cracks or crevices, making them suitable for food and pharmaceutical applications.

A disadvantage for toroidal sensors is their inability to reach much below 20 μS/cm without losing accuracy. In addition, most designs have temperature sensor elements that have a slow response time. Thus, where rapidly changing temperatures are occurring, external RTDs must be used (Fig. 11).

FIGURE 7 CLS15.

Temperature compensation

The effects of temperature on a solution are considerable. Unlike an electrical conductor, but similar to graphite liquids, solutions become more conductive as the temperature rises. Conductivity measurement, in most cases, is referenced back to 25 °C. For example, the conductivity of good drinking water is ~50 μS/cm at 25 °C. When the temperature rises to 50 °C, the conductivity will increase at a rate of 1.5%/K; thus (50–25 °C) × 1.5 = 37.5.

The conductivity without temperature compensation would be 87.5 μS/cm. Specific conductivity measurements are referenced back to 25 °C, so that this measurement can be used to detect real changes in ionic concentrations of process streams (Fig.12).

Recently the Water Quality Committee launched the new USP23 to replace the antiquated USP22 standards for purified water and water for injection that is used in pharmaceutical processes.

USP23 has the following benefits:

- permits on-line and off-line testing
- set the maximum allowable conductivity to detect the maximum allowable quantity of the ion that produces the lowest conductivity
- eliminates the handicap of using cold water

Areas of application

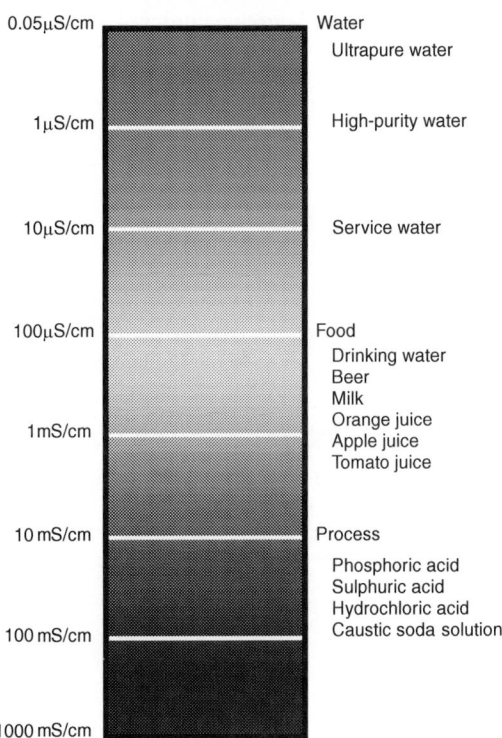

FIGURE 8 Application range.

- for off-line tests, allow for innocuous contamination that results from CO_2
- eliminates the variations of different temperature-compensation methods

USP23 Stage One refers to the on-line method that requires conductivity to be measured and displayed without temperature compensation. Instead, both variables are measured and the limit is set for $5\,°C$ steps. Some transmitters have this table implemented, and a dedicated relay contact will close when the limit is exceeded (also see Fig. 13).

Temperature ($°C$)	Conductivity ($\mu S/cm$)	Temperature ($°C$)	Conductivity ($\mu S/cm$)
0	0.6	55	2.1
5	0.8	60	2.2
10	0.9	65	2.4
15	1.0	70	2.5
20	1.1	75	2.7
25	1.3	80	2.7
30	1.4	85	2.7
35	1.5	90	2.7
40	1.7	95	2.9
45	1.8	100	3.1
50	1.9		

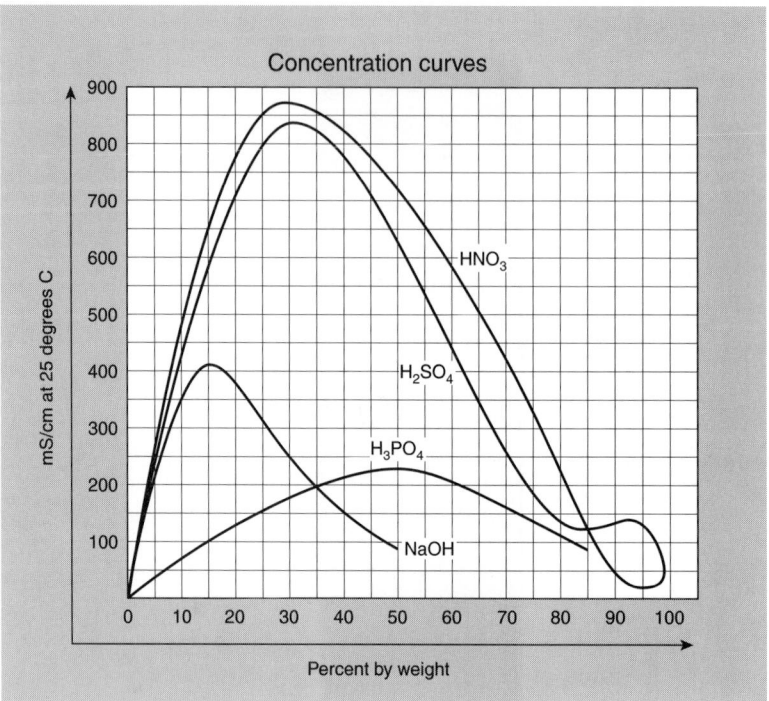

FIGURE 9   Concentration curves.

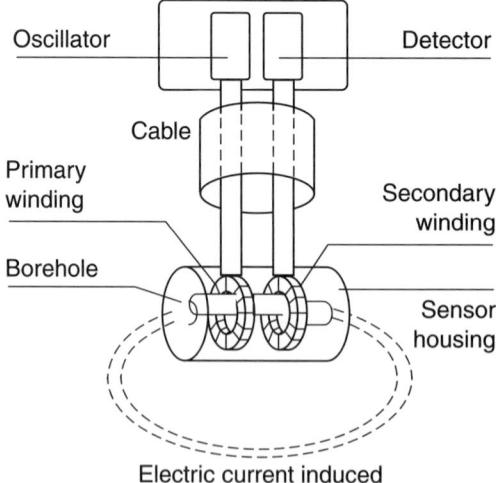

FIGURE 10   Toroidal measuring principle.

Acids and base concentrations are often determined by using conductivity. The amount of temperature compensation changes as the concentration changes. To make life easier for users, most modern transmitters have these temperature-compensation curves built into the software (Fig. 14).

Demineralized and ultrapure water also have nonlinear temperature-compensation characteristics and require the compensation to be done in the transmitter. The conductivities of these waters are often close to and below 1 μS/cm, to improve readability. Low conductivities are often displayed directly in ohms and are often referred to as resistivity.

Here are some typical values at 25 °C.

Pure water	0.048 μS/cm	18 MΩ
Power plant boiler feed water	0.05–1 μS/cm	18–1 MΩ
Deionized water	0.1–10 μs/cm	10–0.1 MΩ
Demineralized water	1–80 μS/cm	1–0.01 MΩ

Calibration Standards

The American Standard Test Method D1125–77 refers to conductivity standards made using ultrapure water and potassium chloride. All sensors should be referenced to this standard at one time or another (also see Fig. 15). Conductivity at 25 °C: 149.6 μS/cm, 1.406 mS/cm, 12.64 mS/cm, 107.00 mS/cm.

Generally systems do not need to be calibrated or checked on a regular basis, since they are solid-state sensors that are not subject to drift like electrochemical sensors. Cells can become coated with buildup, causing values to drift. Should buildup occur, remove the cell and clean with a suitable cleaning agent. Rinse well with a traceable standard, ensure temperature equilibrium, and calibrate in accordance with the manufacturer's recommendations. Transmitters can be easily checked by using fixed resistor values.

When calibration must be performed at values below 74 μS/cm, this must be done on line with a master/slave configuration.

The master cell measures the conductivity of the process stream and compares it to the reading of the installed cell (the slave). A closed environment must be maintained; otherwise, CO_2 from the atmosphere would change the liquid's conductivity (also see Fig. 16).

Sensor Selection and Installation

A system will consist of two basic components—sensor and transmitter. The sensor must always be considered first:

- cell constant for measuring range
- metallurgy of cell
- chemical resistance of plastics and seals
- pressure and temperature range
- process connection

FIGURE 11   Electrodeless sensor.

- surface finish
- speed of response of temperature sensor
- distance between sensor and transmitter

The transmitter or analyzer should have the following considered:

- remote or compact system
- two- or four-wire device

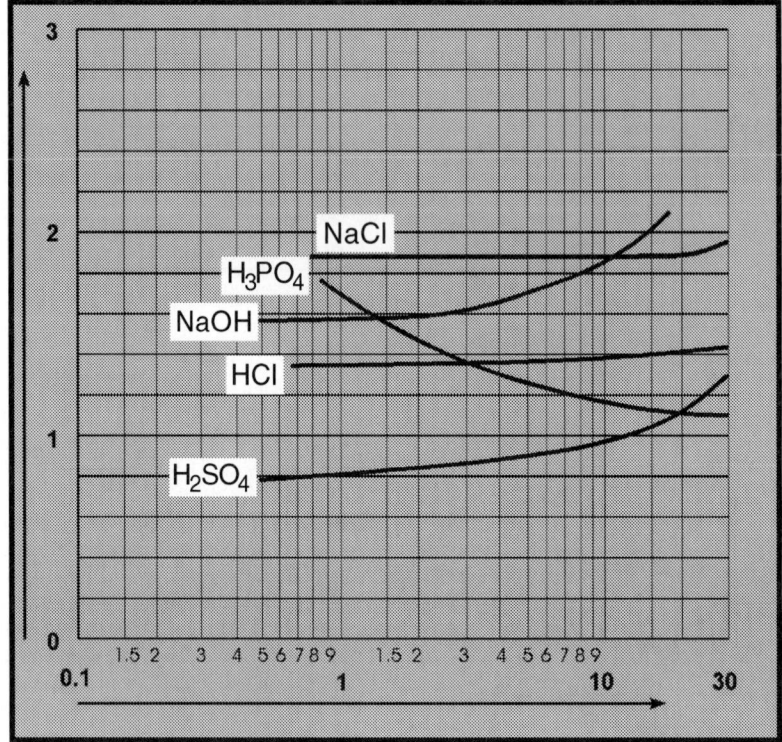

FIGURE 12   Temperature compensation.

- area classification
- types of outputs, i.e., dual current, relays, etc.
- analog or digital interface

In determining sensor installation, the user should consider the following:

1. It should be at a location where the sensor "sees" a representation of the solution to be measured.
2. It should be in a location where the sensor will always be covered—air gaps often collect at the top of horizontal pipe runs.
3. In ultrapure water streams, keep the sample line short and select material that will not absorb ions (avoid causing memory problems).
4. Conductivity cells used in ultrapure water applications must be mounted so that they are self-draining to facilitate cleanability.
5. Sensors should not be mounted where sediments could collect.
6. Toroidal sensors are sometimes bulky, so pressure drop and compatibility to the process stream should be considered.

ELECTRICAL RESISTANCE

This is in terms of purity and ion monitoring. In high-purity water, typically less than 1 μS/cm, the measurement is referred to as resistivity with units of megohm centimeters (MΩ cm). Pure water has a

FIGURE 13 Conductivity analyzer.

resistivity of $\sim$18.2 MΩ cm at 25 °C. One consideration that must be made when measuring solutions is the temperature coefficient of the conductivity of the water itself. To compensate accurately, a second temperature sensor and a compensation network are used. Specific sensors and analyzers are recommended for measurements in high-purity water.

Complex Impedance (H₂O)

The use of aluminum oxide sandwiched between two gold electrodes produces a sensor for water based on complex impedance (resistance, capacitance, impedance).

Piezoelectric Crystal (H₂O)

Placing a crystal in a tuned circuit will produce a resonant frequency that depends on mass. A hygroscopic coating (P_2O_5) on the crystal will cause it to absorb water from the surrounding environment, thus changing its mass and therefore the resonant frequency. The system can be calibrated to measure

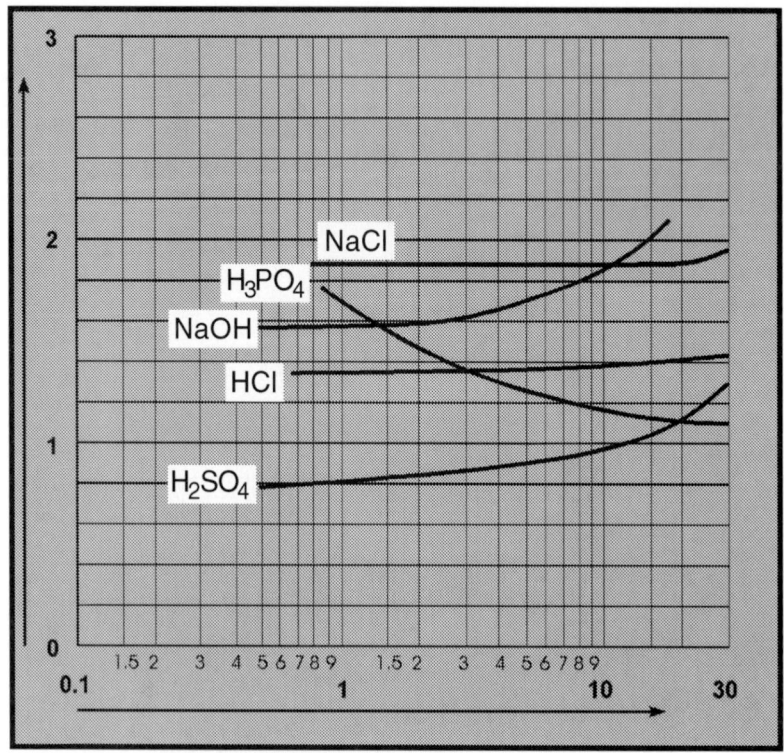

FIGURE 14 Temperature compensation.

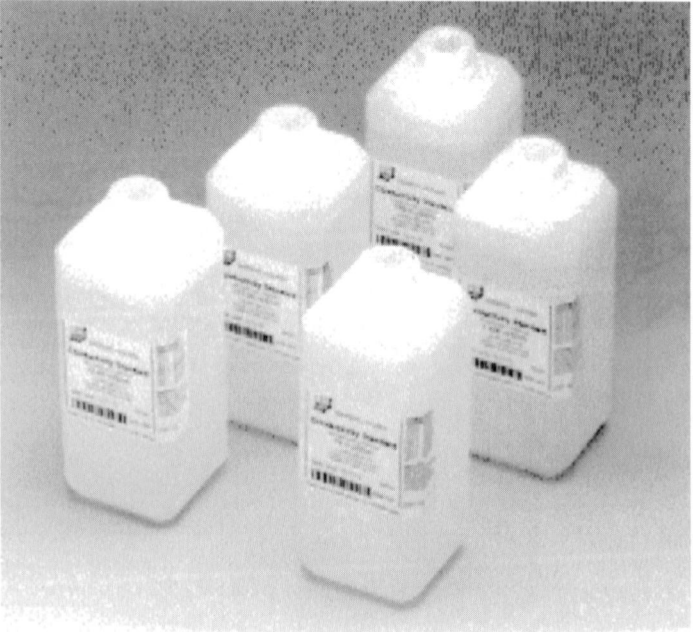

FIGURE 15 Calibration solutions.

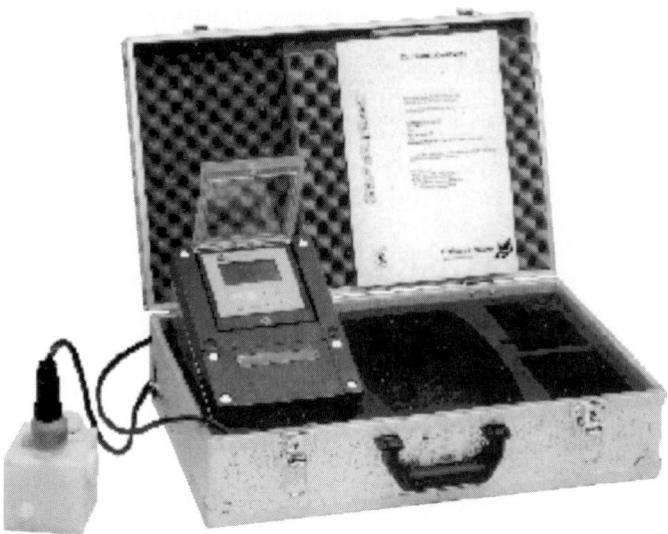

FIGURE 16   Con Cal.

ppb water in a gas mixture. Extreme care in operation and sample handling is required because of the abundance (1–3%) of water in our atmosphere.

pH MEASUREMENT: THEORY AND REALITY

by Gregory K. McMillan*

The pH measurement and reference electrodes each have an internal galvanic half-cell. The measurement and reference electrode half-cells typically both consist of a silver wire with a silver chloride coating at its end immersed in a solution with chloride ions. The reference electrode solution is usually potassium chloride, while the measurement electrode solution is usually a chloride buffer with a hydrogen activity approximately equivalent to 7 pH.

The actual sensing of pH is accomplished by having a pH-sensitive glass in contact with the internal fill, a 7-pH buffer, and the external sample or stream. The pH-sensitive glass develops potentials per Eq. (6.1a) and (6.2a), which are Nernst equations, by the hydrogen ion (proton) exchange between hydronium ions in the aqueous solutions and in the hydrated gel layer of the glass. The protons leave the hydronium ions in the aqueous solution, enter the sites vacated by positive alkali ions such as sodium and lithium, and recombine as hydronium ions in the hydrated gel layer. The potential developed is proportional to the difference in logarithms of the activity of hydronium ions in solution and in the gel layer on both sides of the glass membrane. If the gel layers have an equal number of sites for proton

* Integrated Manufacturing Control Group, Solutia Inc., St. Louis, Missouri.

exchange, the constants K_{g1} and K_{g2} will be equal. If all the original sodium ions at these sites in the gel are also replaced by protons, the activities a_{g1} and a_{g2} are equal. If these glass gel constants and activities are equal, Eqs. (6.1b) and (6.2b) can be combined to yield Eq. (6.3). By use of the definition of pH, the logarithms of hydrogen activity can be converted to pH, which yields Eq. (6.4), where the difference in potentials is proportional to the difference in pH. Also, since the internal fill has a hydrogen activity that corresponds to 7 pH, the equation for the potential difference is simplified to that shown in Eq. (6.4) [1]–[4]. An examination of this equation for the pH measurement electrode yields the following conclusions: (1) the millivolt output of the electrode decreases as the pH increases; (2) the millivolt output is zero at 7 pH; (3) the millivolt output is positive below 7 pH and negative above 7 pH; (4) the effect of temperature on the millivolt output approaches zero as the pH approaches 7; (5) at 25°C, the output changes by 59.2 mV per pH unit; (6) the activities of the gel layers on the outside and inside of the glass surface are assumed equal; and (7) the temperature at the *external* glass surface changes the Nernst potential (the sensor for automatic temperature compensation is buried inside the electrode).

$$E_1 = K_{g1} + 0.1984\,(T + 273.16)\log\frac{a_1}{a_{g1}} \tag{6.1a}$$

$$E_1 = K_{g1} + 0.1984\,(T + 273.16)\,[\log(a_1) - \log(a_{g1})] \tag{6.1b}$$

$$E_2 = K_{g2} + 0.1984\,(T + 273.16)\log\frac{a_2}{a_{g2}} \tag{6.2a}$$

$$E_2 = K_{g2} + 0.1984\,(T + 273.16)\,[\log(a_2) - \log(a_{g2})] \tag{6.2b}$$

If $K_{g1} = K_{g2}$ and $a_{g1} = a_{g2}$, then

$$E_1 - E_2 = 0.1984\,(T + 273.16)\,[\log(a_1) - \log(a_2)] \tag{6.3}$$

$$E_1 - E_2 = 0.1984\,(T + 273.16)\,(\mathrm{pH}_2 - \mathrm{pH}_1) \tag{6.4a}$$

$$E_1 - E_2 = 0.1984\,(T + 273.16)\,(7 - \mathrm{pH}_1) \tag{6.4b}$$

where $a_1 =$ activity of hydrogen ion in the external process fluid (normality)
$a_2 =$ activity of hydrogen ion in the internal fill fluid (normality)
$a_{g1} =$ activity of hydrogen ion in the outer gel surface layer (normality)
$a_{g2} =$ activity of hydrogen ion in the inner gel surface layer (normality)
$E_1 =$ potential developed at the external glass surface (millivolts)
$E_2 =$ potential developed at the internal glass surface (millivolts)
$K_{g1} =$ constant for potential for outer gel surface layer (millivolts)
$K_{g2} =$ constant for potential for inner gel surface layer (millivolts)
$\mathrm{pH}_1 =$ pH of external solution
$\mathrm{pH}_2 =$ pH of internal solution (typically 7 pH)
$T =$ solution temperature (degrees C)

The accuracy attainable with pH electrodes in a laboratory environment under ideal conditions is impressive. The short-term repeatability for a standard set of electrodes under ideal conditions is approximately ±0.01 mV. For a solution temperature of 25°C, Eq. (6.4) shows that the electrode potential changes by 59.2 mV/pH. Thus, a change of ± 0.01 mV corresponds to a change of ±0.0002 pH. The difference in internal and external potential at 7 pH, which is called the asymmetry potential, changes as the electrode membranes age. The drift in asymmetry potential for a standard set of electrodes sitting in a buffer solution is ~0.001 mV or 0.00002 pH units per day.

TABLE 1 Hydrogen and Hydroxyl Ion Concentrations for the pH Scale

ph	Example	Ion Concentration	
		Hydrogen	Hydroxyl
0	4% sulfuric	1.0	0.00000000000001
1		0.1	0.0000000000001
2		0.01	0.000000000001
3		0.001	0.00000000001
4		0.0001	0.0000000001
5		0.00001	0.000000001
6		0.000001	0.00000001
7		0.0000001	0.0000001
8		0.00000001	0.000001
9		0.000000001	0.00001
10		0.0000000001	0.0001
11		0.00000000001	0.001
12		0.000000000001	0.01
13		0.00000000000001	0.1
14	4% sodium hydroxide	0.00000000000001	1.0

The pH measurement error in process applications is an order of magnitude larger than the normally stated electrode error due to installation effects, such as varying glass surface conditions, dissociation constants, streaming potentials, concentration gradients, and diffusion or liquid junction potentials of actual installations [1]. Even under the best of industrial conditions (e.g., buffered systems between 4 and 10 pH at constant temperature), pH measurements are not more accurate than approximately ±0.02 pH. The titration curve slope is frequently flat at low pH and high pH so that even a relatively small pH measurement error for set points on the extremes of the curve may be larger than the pH error due to reagent delivery error from influent or reagent disturbances (the measurement error is greater than the control error). Table 1, which lists the hydrogen ion concentration for 0–14 pH, shows that a 0.02-pH error represents a relatively large hydrogen ion (acid concentration) and hydroxyl ion (base concentration) error at low pH and high pH, respectively.

Optimistic users and suppliers who provide one electrode per point in the process may believe they have achieved an accuracy of 0.02 pH or better. Results of installations with three electrodes per point in the process show that the error band over one a month at best is 0.05–0.1 pH for a flat titration curve and 0.25–0.5 for a steep titration curve. If all three electrodes agree continually within 0.01 pH for more than a few hours in an industrial application, it indicates that all of the electrodes are either coated, broken, or still have on protective caps [1].

The maintenance practice of removal and buffering of electrodes is costly and often counterproductive, because it reduces measurement accuracy as a result of damage to the fragile gel surface of the glass electrode, and the upset to the thermal and ionic equilibrium of the reference electrode. It is possible to improve the performance of pH electrodes and safety during pH maintenance and reduce the cost of pH maintenance by an order of magnitude through more realistic expectations, and the use of a better calibration policy.

Equation (6.4) shows that the potential difference will change at any pH other than 7 if the solution temperature changes. The magnitude of the pH error depends upon both the magnitude of the temperature error and the deviation of the pH reading from 7 pH. The size of the error is usually small because change in the process temperature is small compared to the 273.16 in Eq. (6.4) used to convert to degrees Kelvin and because most pH control system set points are near 7 pH.

The actual pH of the solution changes as a result of the change in dissociation constants of the ions with temperature. Until recently, compensation of only the change in millivolts generated per pH unit, or in other words, electrode temperature compensation, was offered. Microprocessor-based transmitters and receivers have enabled solution compensation. However, the relationship depends upon the composition and operating conditions of the process stream. For basic aqueous streams, the

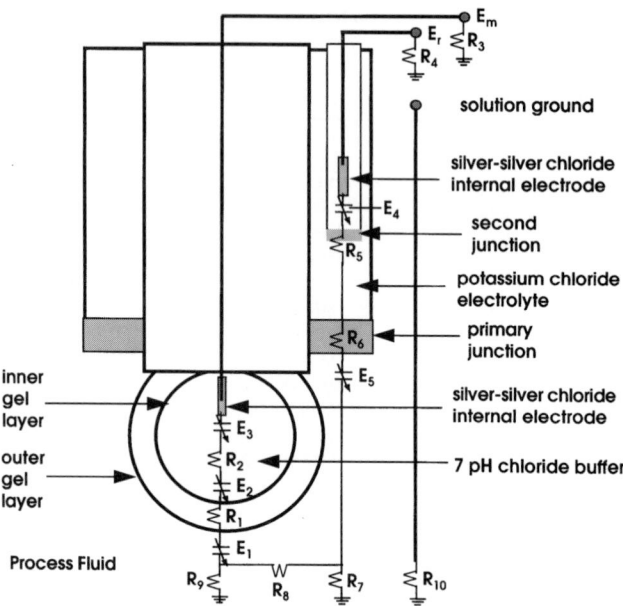

FIGURE 17   Schematic of a pH electrode.

change is often approximately −0.3 pH/10°C. Samples that cool down before measurement in the lab will indicate a higher pH than the electrodes installed in the process. The potential of interest in pH measurement is the difference between the potential developed at the outer and inner glass surfaces of the measurement electrode as defined by Eq. (6.4). Any other potential represents an error. Figure 17 shows the physical location of each potential and Eq. (6.5a) shows that the effect of these potentials is additive.

Whereas errors caused by changes in parameters in Eq. (6.1) and Eq. (6.2) result in either horizontal shifts of the isopotential point or changes in the slope of the line, all extraneous potentials in Eq. (6.5a) result in a vertical shift of the isopotential point:

$$E_i = E_1 - E_2 - E_3 + E_4 + E_5 - I_i \, (R_1 + R_2 + R_5 + R_6 + R_8) + E_8 \tag{6.5a}$$

$$E_o = M_t M_s E_i + M_z \tag{6.5b}$$

$$I_o = O_s E_o + O_z \tag{6.5c}$$

where E_1 = potential developed at external glass surface (millivolts)
E_2 = potential developed at internal glass surface (millivolts)
E_3 = half-cell potential of the measurement electrode (millivolts)
E_4 = half-cell potential of the reference electrode (millivolts)
E_5 = liquid junction potential of the reference electrode (millivolts)
E_i = transmitter input voltage (millivolts)
E_o = meter output voltage (millivolts)
E_s = electrode standardization potential (millivolts)
M_z = meter zero adjustment (bias)
M_s = meter span adjustment (gain)
M_t = meter temperature-compensation adjustment (gain)
O_z = transmitter output zero adjustment (bias)
O_s = transmitter output span adjustment (gain)
I_i = input leakage current of the meter amplifier (milliamps)
I_o = transmitter current output (milliamps)

R_i = measurement electrode glass resistance (ohms)
R_2 = measurement electrode internal fill resistance (ohms)
R_5 = reference electrode internal fill resistance (ohms)
R_6 = reference electrode liquid junction resistance (ohms)
R_8 = solution between measurement and reference electrode (ohms)

The measurement and reference electrode half-cell potentials in Eq. (6.5a), which are due to an electrochemical reaction between the internal electrode and fill, are of opposite sign and should ideally be equal so that their sum is zero. However, the half-cell potentials depend upon the internal electrode type (silver chloride or calomel), the internal fill concentration, and the electrode temperature. If the electrode type and fill are identical, then the change in half-cell potential with temperature will cancel out unless a temperature gradient exists between the reference and measurement electrode locations. The half-cell potential for a saturated calomel electrode ranges from 234 mV at 40° to 260 mV at 0°C, and for a saturated silver chloride electrode ranges from 193 mV at 40° to 237 mV at 0 °C.

The resistances in Eq. (6.4) are all relatively large. Fortunately, the input leakage current that flows through these resistances and the amplifier input is extremely small (~1 pA or one trillionth of a milliamp). The current flows from the positive measurement electrode terminal to the negative reference electrode terminal so that the sign of the potential drop is negative compared to the measurement electrode's outer potential and consequently causes the isopotential point to shift down and the pH measurement to go upscale.

The measurement electrode glass bulb resistance is normally the largest resistance. It ranges from about 50 MΩ to 500 MΩ at 25 °C depending upon the type of glass used. Electrodes whose gel layer has dehydrated will have an abnormally high resistance. The resistances of the internal electrode fills and external process fluid are normally only a few thousand ohms so that the potential drop due to leakage current flow is negligible. However, high-purity water (i.e., distilled water, deionized water, or steam condensate) and nonaqueous solutions have extremely high resistances.

Most glass measurement electrodes use bulbs. Various shroud designs are used and represent compromises for conflicting goals of bulb protection (from breakage and abrasion) and bulb exposure to flow (to ensure a representative reading and to prevent areas of stagnation and material buildup). The standard design has slots or holes to allow the process to flow around the bulb. Shrouds where the bulb is recessed without a slot or hole are used for pulp and very abrasive applications. The recessed area can become clogged or caked up, covering the bulb and reference junction. When material breaks loose, it can take the bulb with it. A bubble can also get trapped in the shroud. A tapered shroud is favored for coating applications, since it creates eddies to help keep the electrode surfaces clean [2], [3].

Most electrodes used today are a combination electrode, in which the measurement electrode is combined with the reference electrode and temperature compensator. The reference electrode forms a concentric ring around the measurement electrode (Fig. 18).

Flat-glass electrodes have been developed to minimize glass damage and maximize a sweeping action to prevent fouling. A small button, flat-glass electrode has a range of 0–10 pH, and large, flush, flat-glass electrodes have a stated range of 2–12 pH. High sodium ion concentrations and low hydrogen ion activity have a greater effect on flat glasses [2], [3]. Large flat-glass area electrodes have been reported to exhibit a significant undocumented and consequently uncompensated temperature effect (Fig. 19).

Some manufacturers offer several different glass formulations that are tailored to a process application. Higher glass impedance normally corresponds to a higher resistance to chemical attack from strong bases (sodium hydroxide) and strong acids (hydrochloric acid). Thick glass and flat glass are more abrasion resistant, but low ohmic glass is needed to reduce total impedance (reducing chemical resistance). In the late 1980s and early 1990s, progress was made is reducing the impedance of sodium ion and chemical resistant glass so that it could be used at temperatures as low as 5 °C [2], [3].

Pfaudler has patented a measurement electrode made of Glasteel, a unique material formed by the fusion of glass and steel. The probe has greater structural endurance, corrosion resistance, and a higher temperature rating of 140 °C below 5 pH. It is designed to withstand the rigors of heat, pressure, and dynamic agitation in reactors. It needs a modified or microprocessor-based transmitter, since its zero potential point lies between 1 and 3 pH. The cost of the Glasteel electrode is much higher than any other pH electrode [2], [3].

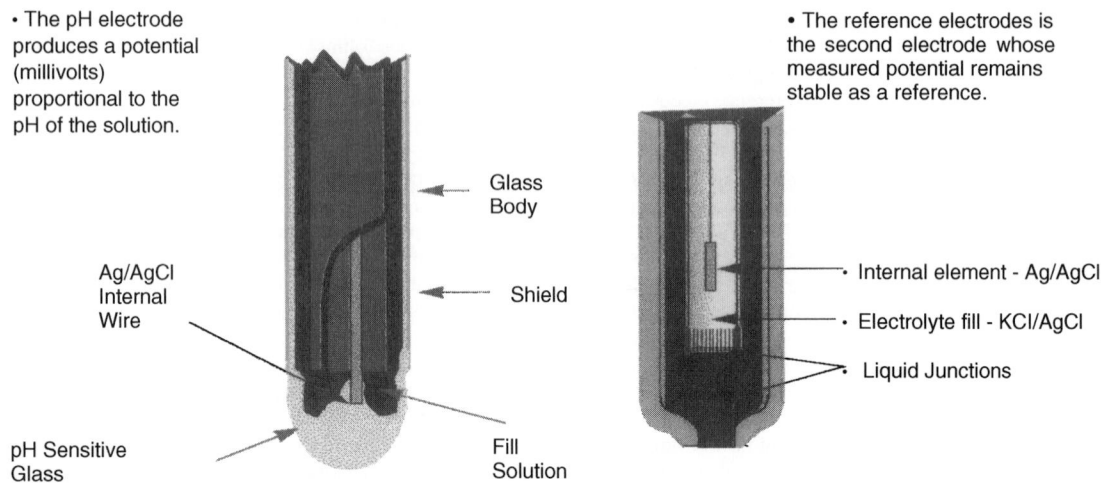

• The pH electrode produces a potential (millivolts) proportional to the pH of the solution.

Glass Body

Ag/AgCl Internal Wire

Shield

pH Sensitive Glass

Fill Solution

• The reference electrodes is the second electrode whose measured potential remains stable as a reference.

Internal element - Ag/AgCl

Electrolyte fill - KCl/AgCl

Liquid Junctions

FIGURE 18   Measurement and reference electrodes.

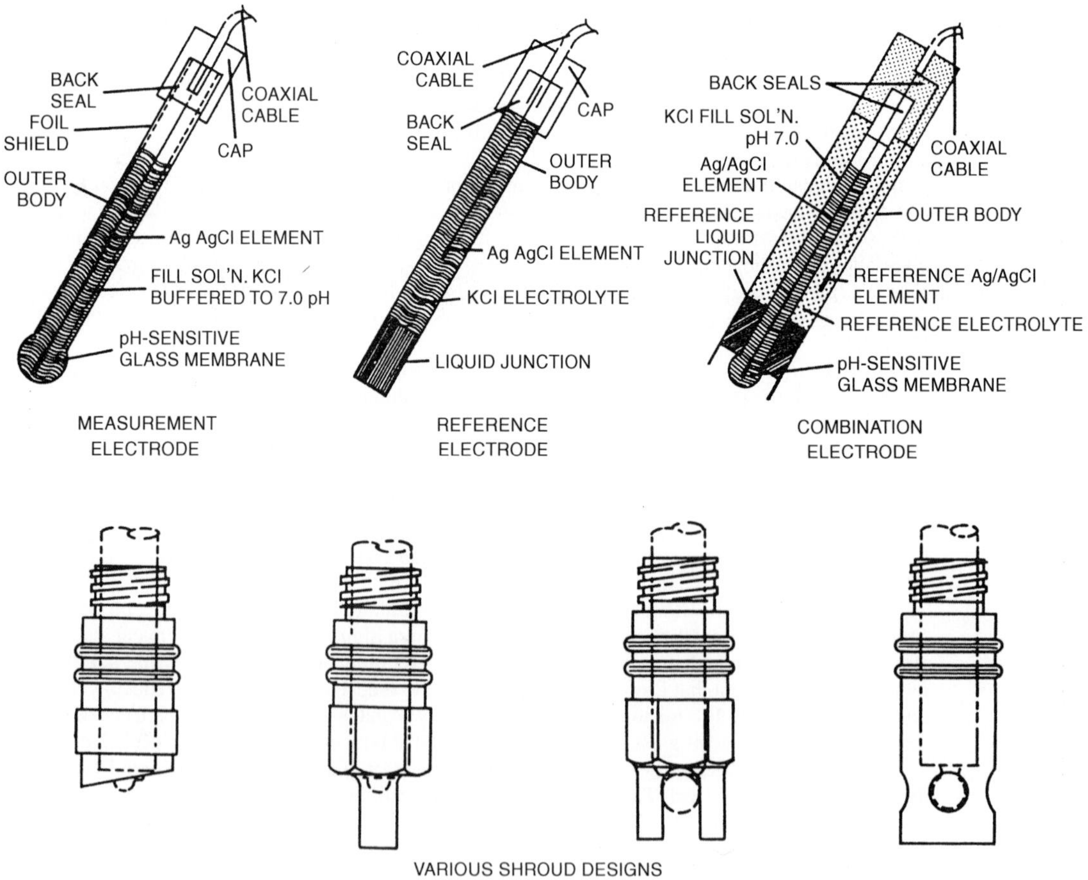

BACK SEAL
FOIL SHIELD
OUTER BODY
COAXIAL CABLE
CAP
Ag AgCl ELEMENT
FILL SOL'N. KCl BUFFERED TO 7.0 pH
pH-SENSITIVE GLASS MEMBRANE

MEASUREMENT ELECTRODE

COAXIAL CABLE
BACK SEAL
CAP
OUTER BODY
Ag AgCl ELEMENT
KCl ELECTROLYTE
LIQUID JUNCTION

REFERENCE ELECTRODE

BACK SEALS
KCl FILL SOL'N. pH 7.0
Ag/AgCl ELEMENT
REFERENCE LIQUID JUNCTION
COAXIAL CABLE
OUTER BODY
REFERENCE Ag/AgCl ELEMENT
REFERENCE ELECTROLYTE
pH-SENSITIVE GLASS MEMBRANE

COMBINATION ELECTRODE

VARIOUS SHROUD DESIGNS

FIGURE 19   Electrode shrouds.

Before the glass measurement electrode was developed for industrial measurement, antimony electrode was used. Antimony is a hard, brittle material. The electrode responds by oxidation of it surface. Since oxidant accumulation on its surface deteriorates the accuracy and repeatability of the measurement, periodic cleaning is necessary. The electrode's response is nonlinear and requires narrow calibration spans or special polynomials for adequate accuracy. The measurement range is reported to be 3–8 pH by some and 3–11 pH by others. It is very temperature sensitive. The approximate 50 mV output per pH unit increases from 1 to 3 mV/°C as the pH increases. The isopotential point is near 2 pH, so modified or microprocessor-based transmitters are required for temperature compensation. Finally, antimony is an oxidation reduction potential (ORP) electrode. It responds to oxidizing and reducing species, and its reading will be affected by a fraction of a ppm of chlorine. Antimony usually is used as a last resort—applications in which abrasion from slurries or etching from hydrofluoric acid causes glass measurement electrode failures. It cannot be used in food, beverage, or pharmaceutical applications because antimony is toxic [2], [3].

pH measurement electrodes have been developed to use iridium oxide. As with other metal oxides, this responds to oxidizing and reducing agents. The iridium oxide can be coated with a polymer to block negative ions and slow down positive ions, particularly those larger than the hydrogen ions. Without a sufficiently thick coating, reducing agents such as ascorbic acid decrease the millivolt potential developed and increase the pH reading. Oxidizing agents such as hydrochloride have the opposite effect. Increasing thickness decreases the speed of responses to pH, which already is slow in the neutral region. The electrode requires a modified or microprocessor-based transmitter, because the zero potential point is ~10 pH and the isopotential point is near 4 pH. It generates ~91 mV/pH unit at 25°C. Iridium oxide has better linearity and repeatability than antimony. It has a potential temperature rating of 200°C. It also eliminates problems of dehydration, etching sodium ions, and breakage associated with glass electrodes, and it can make repeatable measurements outside the 0–14 pH range [2], [3]. However, a high drift (e.g., 0.1 pH per day) makes frequent calibration an unfortunate necessity. Leeds & Northrup patented the DURAFET pH electrode shown in Fig. 20. It uses an

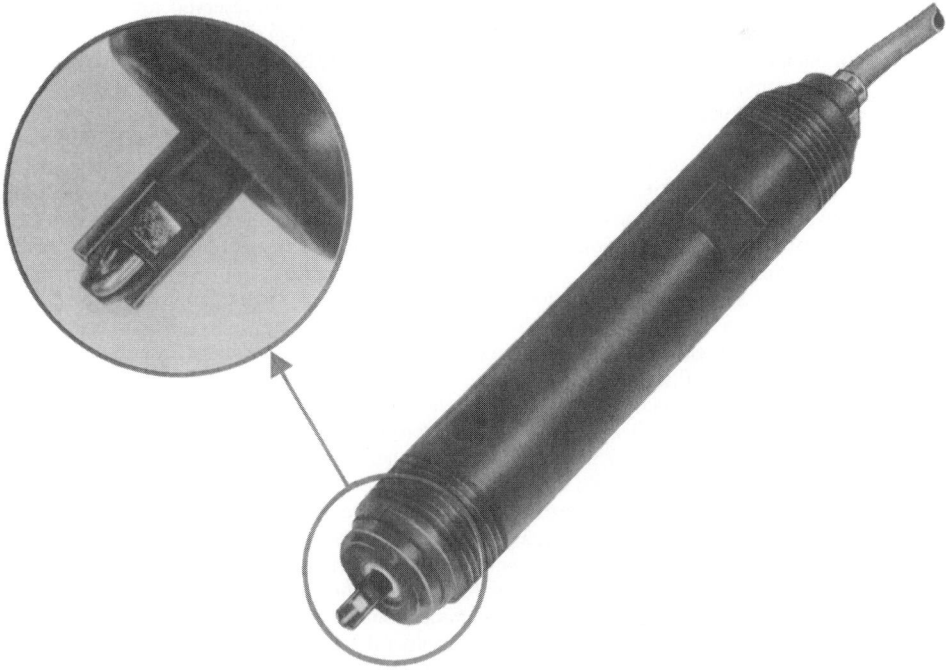

FIGURE 20   DURAFET field effect transistor electrode.

ion-sensitive field effect transistor (FET) instead of glass to provide a low impedance, breakage-resistant measurement electrode. The characteristic drift of laboratory versions of FET pH sensors has been reduced. Electrode life is expected to exceed that of glass. It is faster since the response is by field rather than by movement of hydrogen ions into a membrane. It has integral, automatic temperature compensation, and it has no sodium ion error or known ORP interference as long as the insulting silicon layer remains intact. It needs no hydration and uses a conventional reference electrode. The slope of ~59 mV/pH unit is close to that for glass, but the isopotential point shift necessitates the use of an adapter module to connect it to conventional pH transmitters [1]–[3]. The FET response has been reported by one user to be affected by ultraviolet light, and the thin silicon nitride crystal covering of the FET is susceptible to degradation if faced into the flow. Also, some units have suffered from manufacturing quality-control problems.

REFERENCE ELECTRODE

The reference electrode should provide a constant potential and electrical continuity between its internal electrode and the glass measurement electrode. The reference electrode has a junction where the internal electrolyte is in contact with the external process fluid. A liquid–liquid junction or diffusion potential develops as electrolyte ions and process ions migrate into the junction. When the charge accumulation becomes large enough to oppose further migration, the potential stops changing. This potential introduces an error, often overlooked in pH measurements. The junction potential magnitude and time to equilibrium generally increase for gel-filled or solid-state reference electrodes. The reference junction also is an entry path of process material into the internals of the reference electrode. Electrolyte contamination causes large shifts in the pH measurement. Flowing junctions minimize contamination and help establish a quicker, more constant reference junction but require electrolyte pressurization above the process pressure and periodic refilling of the electrolyte. Sometimes, the internal reference electrode is mounted in a pressurized external reservoir for added contamination protection and temperature stability. In the U.S., the combination electrode primarily is used to reduce the maintenance and replacement cost of electrodes. It has a sealed reference electrode that surrounds the measurement electrode and an annular reference junction [2].

The construction features of a particular reference design in a combination electrode present a tradeoff in the effort to quickly establish and minimize a constant junction potential, while resisting contamination and coating. The user must decide which is most important for a given application. A decrease in the area of the reference junction reduces contamination, but increases coating problems. A thickening or solidification of the internal fill, the addition of junctions, and a decrease in the junction porosity reduce contamination problems, but delay equilibrium and often increase the magnitude of the junction potential [1]–[3].

The potassium chloride solution is saturated and has a tendency to crystallize and reduce the diffusion rate and the associated potential at the junction. Also, silver from the silver–silver choloride internal element gets into the potassium chloride fill, reacts with sulfides, cyanides, and halides, and clogs the junction [2], [3].

In 1976, TBI patented a solid-state reference electrode that is nearly impervious to process contamination. The reference consists of four wooden dowels saturated with potassium chloride solution. Process material must work its way up into the innermost wooden dowel, where the internal silver–silver chloride element resides to create a drastic offset in the pH reading. Teflon is substituted for the dowel incontact with the process for applications requiring a sanitary surface or with chemicals that attack wood (hydrogen peroxide and hot, concentrated sodium hydroxide). Teflon is more difficult to keep saturated and slows down the time to equilibrium. This slowness is an advantage for continuous pH control, where it is desirable to ignore fluctuations in ionic strength from the steady state. However, it is difficult to get the electrode to match buffers within 1/10 of a pH for a two-point calibration. Since installation effects of more than 0.2 pH are normal, tight agreement with buffer solutions is not indicative of the electrode's installed performance. For high ionic strength process streams, high ionic buffers are recommended to precondition the electrode. A process calibration is essential for solidstate electrodes. For some batch applications, the reference does not have time to reach equilibrium, and pH errors of 0.5 pH and larger occur. In batch operations, it is important to use a wooden dowel for the reference junction [1]–[3].

Ingold had developed a Xerolyt stiff polymer reference electrode that contains potassium (not silver) chloride. It uses an aperture diaphragm to provide direct contact between process and reference electrodes. It has performance advantages in streams with high salt concentrations, emulsions, suspensions, proteins, sulfides, hot alkalides, and pressure fluctuations. However, it is not suitable for stream sterilization, large temperature changes, or acidic media (below 2pH) [2], [3].

The Rosemount R1 and the Amagruss refex are solid, nonporous references in which the whole lower outer sleeve of the reference is the active junction. The electrode is less susceptible to coatings and nearly impervious to contamination. Figures 21–24 show these various solid reference designs.

Great Lakes Instruments patented a differential pH electrode, using a second glass measurement electrode (behind a double junction) in a buffer solution for the reference electrode as shown in Fig. 25. The glass bulb isolates the silver–silver chloride element of the reference from sulfides, cyanides, and halides in the process. The buffer solution is chosen to be compatible with process fluids to ensure a constant buffer pH and constant reference potential. A ground rod eliminates errors and noise from ground potentials. The smaller positive ions in the buffer solution tend to move through the reference junction, out into organic and pure water streams. This creates an offset in the pH reading. The differential electrode has an excellent performance record in wastewater streams with inorganic acids and bases [2], [3].

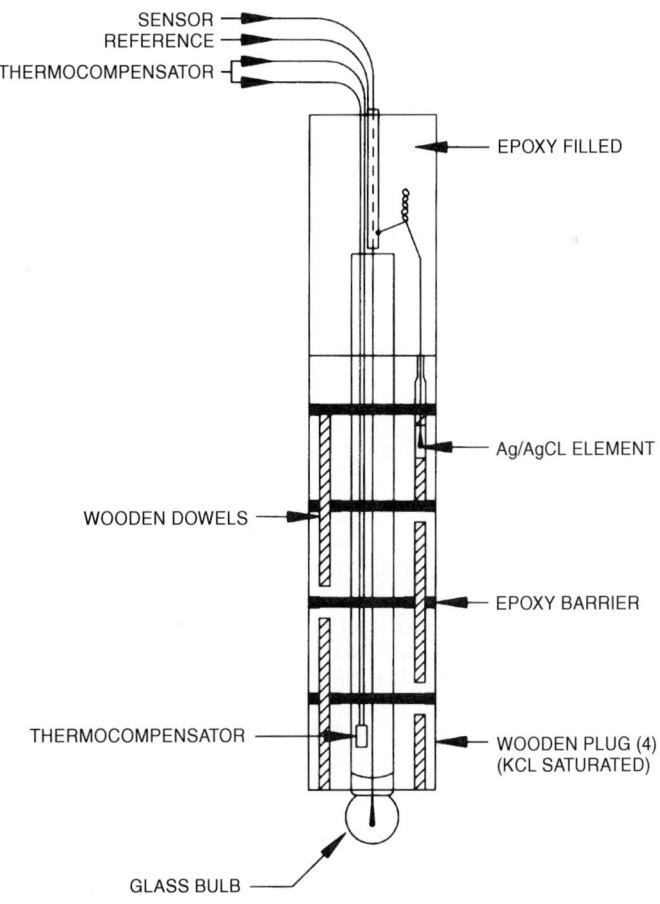

FIGURE 21   TBI solid-state reference electrode.

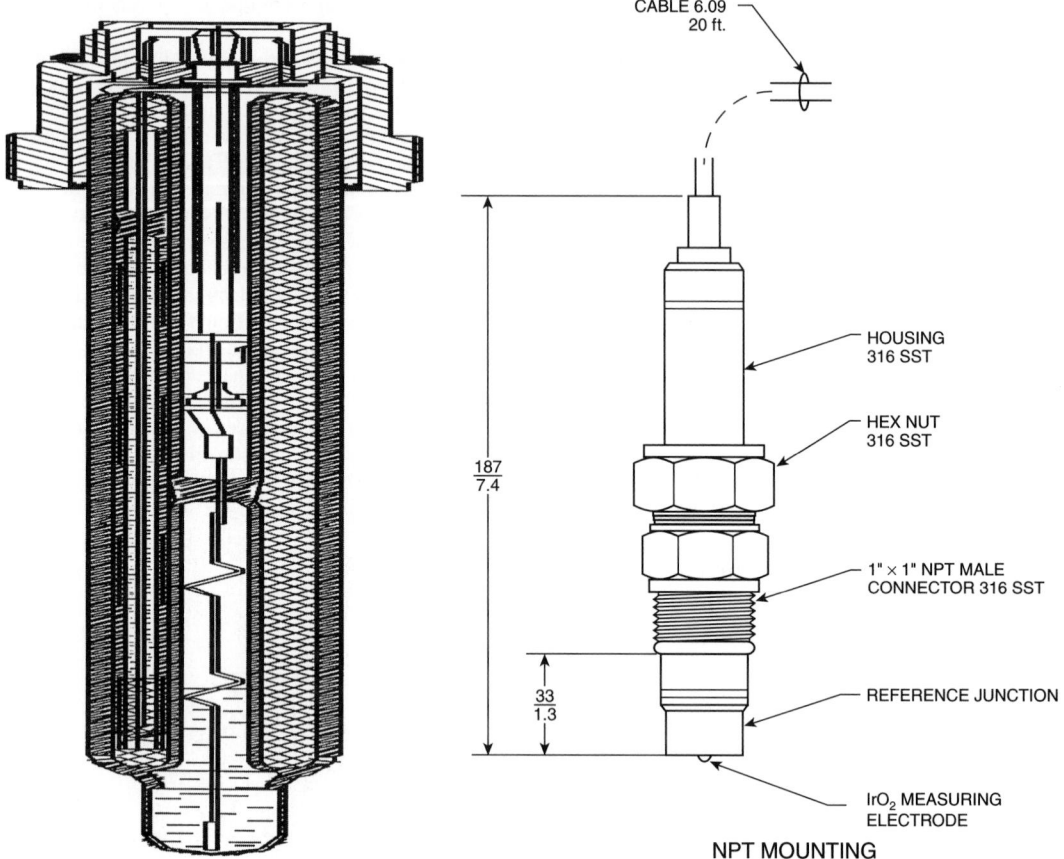

FIGURE 22 Mettler-Toleda (Ingold) Xerolyte solid-state reference electrode.

FIGURE 23 Rosemount R1 solid-state reference electrode.

CLEANING METHODS

Significant natural self-cleaning by turbulent eddies is achieved by a velocity of 5 or more ft/s past the electrode. A velocity of greater than 10 ft/s can cause excessive measurement noise and sensor wear. The obstructed area from the electrode must be subtracted from the total cross-sectional area when estimating the free area for flow around the electrode. The velocity is the volumetric flow in cubic feet per second divided by the free cross-sectional area in square feet. The pressure drop at the restricted cross section should be checked to ensure there is no cavitation [2], [3]. Some shroud designs, such as the tapered shroud, create more effective eddy action and could provide cleaning at a lower velocity. Flat surface electrodes get adequate cleaning at velocities of 1–2 ft/s. However, below velocities of 2–3 ft/s, the pH reading is more sensitive to velocity and can shift as much as 0.2 pH.

The addition of filters shifts maintenance from the electrode to the filter. The filter usually has to be changed more often that the electrode has to be cleaned. An extra filter is not recommended unless it can be back washed automatically. The filtrate master from TBI–Bailey (patent pending) is designed to provide a solids-free measurement for an assembly submerged in slurry. It reverses flow and pulses loose particles caught in the 10-μm metal filter [2], [3].

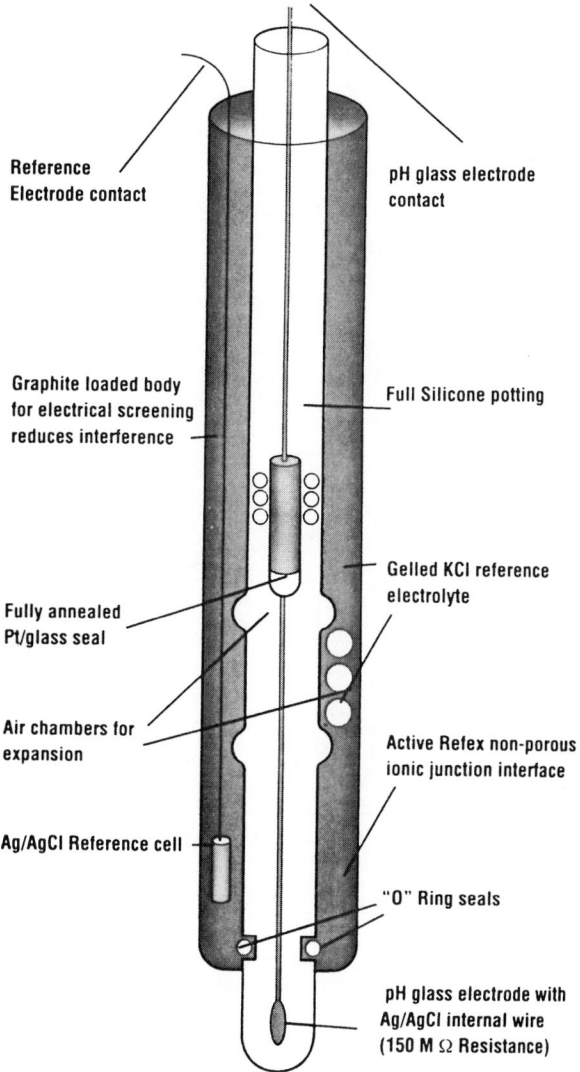

Reference Electrode contact

pH glass electrode contact

Graphite loaded body for electrical screening reduces interference

Full Silicone potting

Fully annealed Pt/glass seal

Gelled KCl reference electrolyte

Air chambers for expansion

Active Refex non-porous ionic junction interface

Ag/AgCl Reference cell

"O" Ring seals

pH glass electrode with Ag/AgCl internal wire (150 M Ω Resistance)

Cut-away illustration of the advanced Refex solid state pH/ORP combination electrode

FIGURE 24 Amagruss refex solid-state reference electrode.

The four major automatic types of cleaners are ultrasonic, brush, water, and chemical. The water cleaners have the same construction as the chemical cleaners. On-line electrode cleaners are shown in Fig. 26. These methods tend to concentrate on removing coatings from the measurement bulb. Particles and material clogged in the porous reference junction are more difficult to remove and can cause serious problems. The junction resistance can get so high that the pH reading goes offscale high [2], [3].

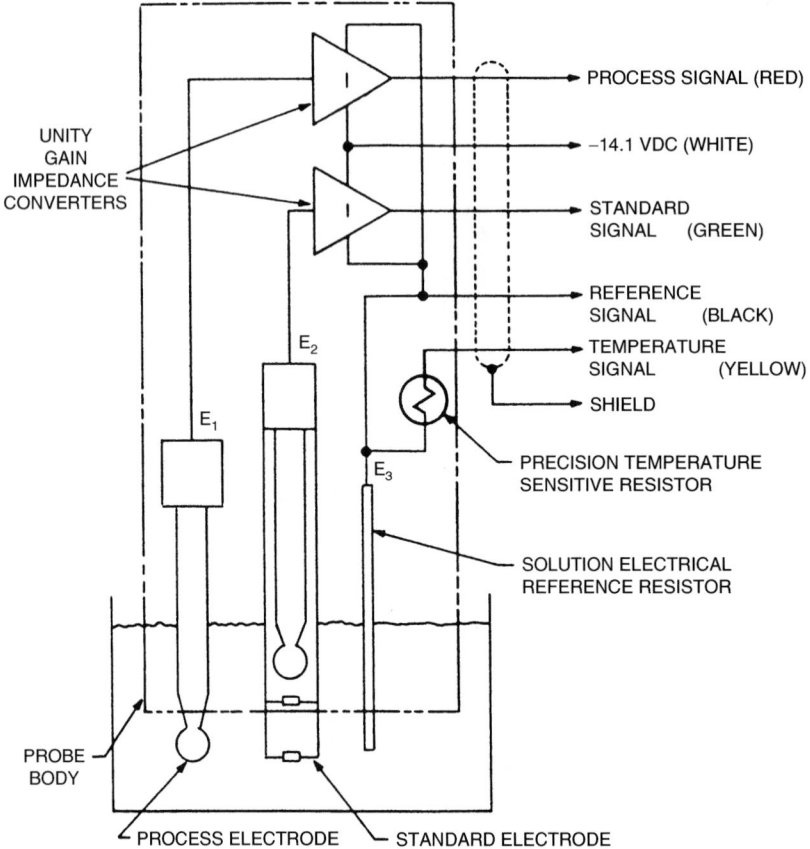

FIGURE 25   Differential pH electrode.

The ultrasonic cleaner uses ultrasonic waves to vibrate the liquid near the electrode surfaces. The effectiveness depends on the vibration energy and velocity past the electrodes. Heavy-duty electrodes are needed to withstand the ultrasonic energy. The ultrasonic cleaner works well in processes where fine particles and easily supersaturated sediments are formed or in suspension. It can remove loose and light particle and oil deposits. Most of the disappointment with ultrasonic cleaners comes from applications involving more difficult coatings [2], [3]. Because there are no brushes to replace or fluid to supply, users are lured into applying it where it is inappropriate. The brush cleaner tries to wipe off coatings by rotating a soft brush around the measurement bulb (but not reaching the reference junction). It has an adjustable height, a replaceable brush, and can be driven either electrically or pneumatically. Soft brushes or Teflon balls are used for glass electrodes; ceramic disks are used for antimony electrodes. Sticky material can gunk up the brush and get smeared on the bulb [2], [3].

The water jets direct a high-velocity spray to the measurement bulb. The reading will become erratic during the washing. A cycle timer that starts the jet also should freeze the pH reading and force the pH controller to manual during the wash, and for 2+ min after the wash for electrode recovery. The water jet works well for material easily dissolved in water [2], [3].

The chemical method uses a dilute acid or base, solvent, or bleach. It must be compatible with the process if it is used while the electrode is immersed. A base typically is used for resins and an acid is used for crystalline precipitation (carbonates) and amorphous precipitation (hydroxides). A dilute hydrochloric solution is used most often. Chemical cleaning tends to be most effective, but acid and base cleaners chemically attack the glass. Cleaning cycles that are too frequent or too long will cause

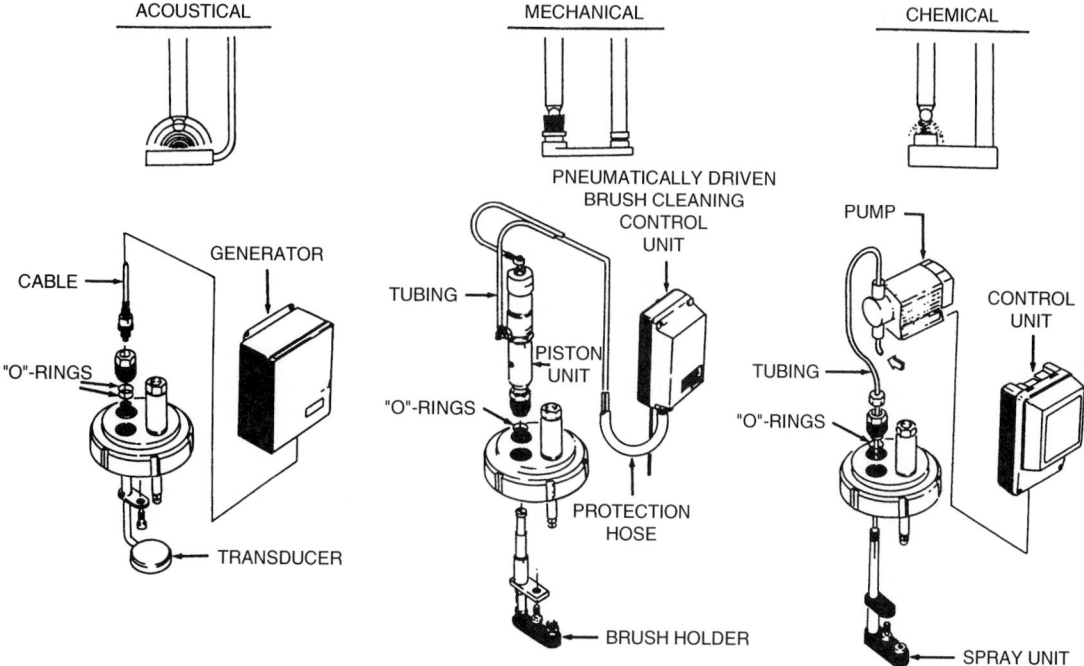

FIGURE 26   Major types of inline electrode cleaners.

premature failure of the glass electrode. As with the water jet, the cycle timer must provide a contact to suspend control action [2], [3].

PROBLEMS AND CAUSES

The pH electrode has the highest sensitivity to process conditions, and engineering and maintenance practices of the commonly used measurements. Since on-line pH measurements are important for improving yields, ensuring safety, and meeting environmental regulations, it is essential that the concepts and details of the application and maintenance issues be addressed. A discussion of the more frequently encountered problems and causes provides a background for understanding the recommendations.

Since the reference electrode electrolyte must be in contact with both the internal electrode (e.g., silver–silver chloride element) and the process as shown in Fig. 17, there is a pathway for process ions to move into the reference and electrolyte ions to move into the process. Ions migrate from areas of high to low concentration until a potential that is part of E5 in Fig. 17 builds up that opposes further movement [1]–[3]. Potassium and chloride ions have about the same mobility so they both move at about the same rate and the potential developed is rather small. The ions in other types of reference electrolytes and the process fluid tend to have a larger difference in ionic mobility and hence create a larger unbalance of charges in the junction. Thus, the potential in the junction E5 shifts until an ionic equilibrium is reached. Changes in process composition, temperature, and flow disrupt this ionic equilibrium. Plugging of the junction by the process reduces the electrolyte flow, increases the resistance of the junction R6, and can completely isolate the reference electrode from the process.

In order for a reference to have a nearly constant potential, the temperature and the flow would have to be constant, and reference would either have a clean consistent flowing or an immobilized electrolyte. If there is no positive flow of electrolyte out of the reference junction by design or by loss of pressurization, the process ions are free to move into a porous junction and from there into the reference electrode. The movement of process into the reference has a profound effect on first potential E5 and ultimately on E4 as process ions reaches the interface between the internal electrode and the electrolyte. Most severe shifts of pH measurements are due to contamination of the reference. The larger or more porous the junction and the more fluid the electrolyte, the faster the contamination. Also, since the electrolyte has silver ions from the internal electrode, sulfides, cyanides, bromides, and iodides will combine with them to form a black precipitate that plugs the junction and creates an offset in E5. Sulfides are common in biological systems.

For a flowing junction, important to periodically replenish the electrolyte so the fill doesn't drop to the internal element. For non flowing references with liquid or gel fills, fluctuations in pressure and temperature can cause a pumping action and depletion of the fill. If the liquid fill drops below the internal electrode or voids develop in a gel that spans the electrode, there is a loss in continuity and resistance R5 becomes infinite.

The glass electrode response is consistent only if the external gel layer thickness and structure is fixed and identical to the internal gel layer depicted in Fig. 17. The activity of the inside and outside gel layers must be identical for the Nernst equation that is the foundation of the pH measurement to be an accurate representation of the measurement [1][2][3][4]. This is not generally true, since the exterior surface is exposed to the harsh conditions of the process whereas the interior glass is in a protected environment of a buffer solution. Any difference between the layers shifts the isopotential point from the assumed 7.0 pH, which is the basis of the temperature compensation. This causes a small error in the temperature compensation in addition to the zero shift [5]. There is also a drift of about 0.0001 pH/day from the loss of active sites and the thickening of the gel layer associated with aging. Alternately drying and hydrating the glass and high temperatures can accelerate the aging by 50% and 1000%, respectively,. As compositions change and especially if the solution goes from being highly acidic to highly basic, there are considerable changes in the gel layer. There is a thinning and even a loss of the gel layer for exposure to high caustic or very high acid concentrations. It takes anywhere from a few minutes to a day to reestablish the gel layer depending upon the severity of the alteration. Until the gel layer reaches equilibrium where the outside loss to the process about equals the inside expansion into the glass, there is a shift in the pH measurement. High caustic concentrations also chemically attack the glass surface and cause a loss of active sites (silicon and alkaline oxides) from the glass. High temperatures greatly accelerate the chemical attack [6].

The gel layers shown in Fig. 17 are only 50–5000 Å thick and the glass bulb is typically less than 0.5 mm thick for a general purpose electrode to keep the measurement electrode resistance R1 less than 100 MΩ at 25°C [6]. Solids impinging upon the glass can easily cause a destruction of the gel layer and even the glass, depending upon the shape, hardness, and velocity of the particles. Streams below 3 pH containing hydrofluoric acid will etch the glass. Clumps of process material, and of course the usual assortment of nuts, bolts, and welding rods, can break the glass. The higher the resistance of the glass per unit thickness, the higher the chemical resistance. Rugged (thick) glass has been developed to reduce the chance of breakage. Vendors demonstrate this strength of these bulbs by pounding them on the table. However, in order to keep the total resistance from being more than three times as large as a general purpose electrode, a glass with a lower resistance per unit thickness, and hence a lower chemical resistance, is used. These rugged glasses are not suitable for use above 10 pH. Correspondingly, glasses designed to last longer in processes with high temperature, hydrofluoric acid, and sodium hydroxide have a higher chemical resistance per unit thickness. The total resistance R1 of these glass electrodes is dramatically higher than the general purpose electrode at lower temperatures, as shown in Fig. 27. The fill resistance R2 is negligible compared to the glass resistance R1 unless a large bubble is trapped inside the tip or around the internal electrode.

The resistance of the glass approximately doubles for every 10°C decrease in temperature. If an integral preamplifier is not used for process temperatures below 10°C for general purpose, 20°C for chemically resistant, and 40°C for rugged glass electrodes, the current signal becomes so low, especially near the zero potential point (7 pH), that the signal can become erratic. Also, the combination of a very high glass electrode resistance (e.g., 2000 MΩ) and the capacitance of a long run of coaxial cable (e.g., 100 ft) can cause a 98% response time for a change in potential as large as that for the glass electrode (e.g., 30 s) [7].

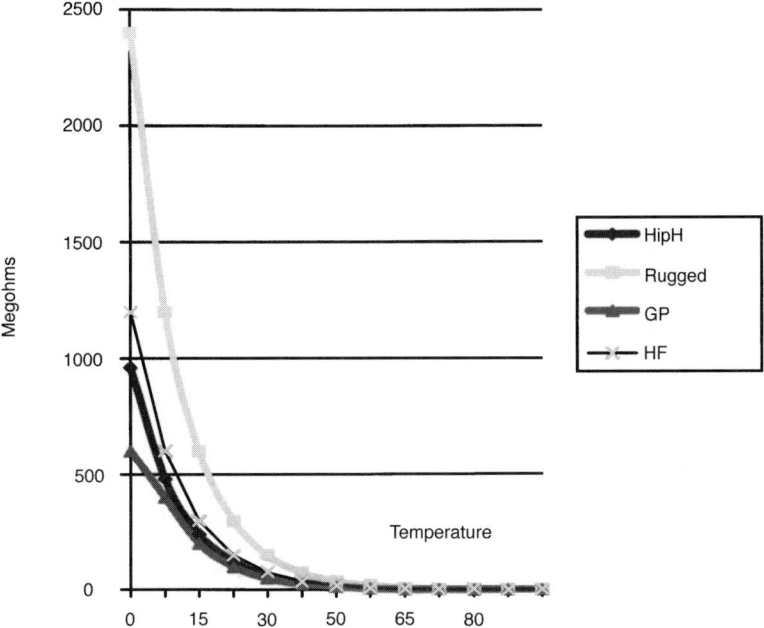

FIGURE 27   Exponential increase in resistance of various glasses with temperature.

The average spread in readings between adjacent electrodes averages is five times larger for measurements on a steep (e.g., 0.25 pH) versus a flat portion (e.g., 0.05 pH) of the titration curve [1]–[3], [8], [9]. This can be explained by small concentration gradients amplified by the incredible sensitivity of the pH measurement for a steep slope. These tend to be relatively fast errors (e.g., seconds). The attachment and release of gas bubbles at the electrode tip and spurious ground and streaming potentials can cause spikes in the measurement. Fluctuating ground potentials are often galvanic potentials that arise from the earthed metal parts (e.g., piping or vessel walls) in direct contact with a varying composition. The streaming potentials are the result of a moving conductive fluid's creating static charges on the glass bulb. There are slower errors (e.g., minutes or hours) due to upsets to the reference junction and gel layer equilibrium that manifest themselves after excursions to the extremes of the titration curve or after a reversal from acidic to basic conditions. These are all short-term errors in that a measurement that is high may be low just seconds, minutes, or hours later. Users with multiple electrodes at the same point in the process usually learn this, although some have driven themselves crazy by trying to make all of the measurements always agree. Users with just one measurement are baffled by the continual discrepancy between the lab and field.

There is a long-term error possible from a current flow between the solution ground and the instrument ground due to a difference in potentials, which can be as large as 1–2 V [10]. This continual unidirectional current flow can eventually cause a complete reduction of the reference electrode [4].

The biggest source of disagreement between the lab and field for processes at elevated temperatures is the temperature-induced changes in the process dissociation constants and in the measurement and reference electrode potentials not explained by the Nernst equation for the sample and on-line measurements. The nonideal changes in the electrode potentials, which can be as large as 0.5 pH, are not well documented and differ from one electrode design to another [11]. The use of different electrodes by the lab and field adds to the uncertainty. There are changes in actual pH due to changes in the dissociation constants. For basic solutions, these changes typically range between −0.015 and −0.035 pH per °C as shown in Fig. 28. For high ionic strength streams, the activity coefficient of the hydrogen ion, which affects the actual pH, and the activity coefficient of the process ions, which affects the junction potential, exhibit a significant change with temperature [5]. Often overlooked are the changes in the pH signal from changes in the electronic components of the field transmitter

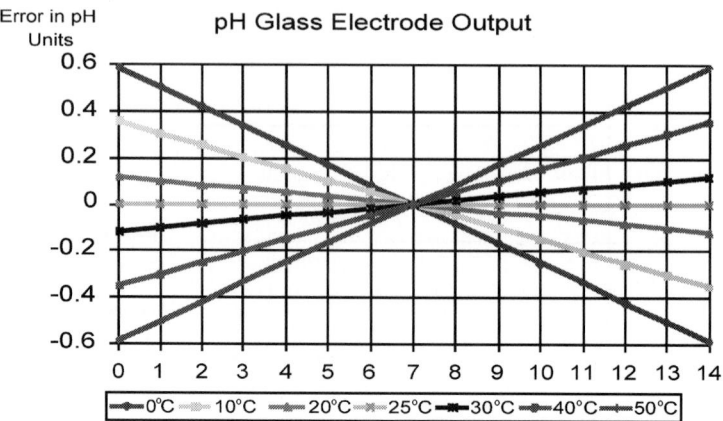

FIGURE 28   Effect of temperature on actual pH of basic solutions.

with ambient temperature for older nonsmart analog units. The standard method of temperature compensation does not deal with these issues. A pH measurement without a temperature measurement is meaningless [5]. For the effect that the compensator is designed to address, there is a dynamic error because the temperature sensor is inside the electrode and the Nernst potential change is at the exterior surface of the glass. The sensor temperature lags the process temperature by a minute or more since plastic and glass have a low thermal conductivity.

The largest most general source of discrepancy between the lab and the field is the difference in composition from the difference in location and time between the sample and measurement points. The variation in composition affects the hydrogen ion concentration, the ground potentials, the activity of the gel layer of the measurement electrode, and the junction potential of the reference electrode [4]. The discrepancy often results in a calibration request. If the electrodes are removed for inspection and buffer calibration, the thermal and ionic equilibrium is upset. If the electrodes are standardized against a sample, the correction for electrode A's being higher than B will make the error larger when electrode A is lower than B, as shown in Fig. 29. Either method of calibration introduces additional errors and confusion.

For process streams with water concentrations less than 60% or with high salt loading, changes in the concentration of the water for the former or salts for the latter case cause a change in the activity coefficient of the hydrogen ion, and consequently an unsuspected change in the pH reading [12]. High-purity water, on the other hand, causes an erratic signal due to the high resistance of the process fluid (R8 in Fig. 17) and due to the large changes caused by an inconsistent distribution of

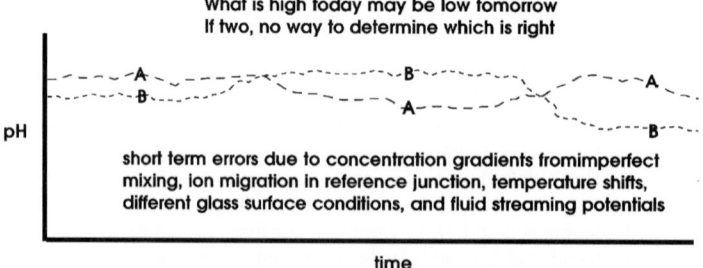

FIGURE 29   Short-term errors (what is high now may be low in a few minutes or hours).

reference electrode electrolyte in the circuit pathway through the process fluid from the reference to the measurement electrode. A low constant velocity, close proximity of the reference junction to the glass electrode, and either high flow or well-shielded nonporous references help sustain a more stable signal for high-purity water applications. The pH electrode assembly can be exposed to some high concentrations of acids and bases, particularly when there are excursions caused by control system failures or overshoot. O-ring failure is a common problem, but all of the wetted materials of construction must be able to handle the extremes of the pH range. High velocities and temperatures increase the corrosion rate. Control system upsets can deteriorate and even destroy electrodes. It is evident from the above list of errors and failures caused by upsets that the performances of the electrode and control system are interrelated. An erratic, slow, or inaccurate measurement can cause the control system to have a problem, and the consequential excursion in the pH can increase the problems with the electrode. The problems snowball. A pH control system can do quite well and then fall apart because of an escalating deterioration triggered by what may seem to be an insignificant change in a process condition, controller tuning, or maintenance [8].

Sources of pH Measurement Errors and Failures

The following are sources of pH measurement errors and failures.

- migration of process ions into the porous reference junction
- contamination of the reference electrode internals with process fluid
- silver precipitate from exposure to sulfides, cyanides, bromides, and iodides
- plugging of the porous reference junction
- upsets to the thermal equilibrium and ionic equilibrium of the reference electrode
- depletion of the reference electrolyte fill
- upsets to the gel layer equilibrium of the measurement electrode
- coating of the measurement electrode
- abrasion of the glass measurement electrode
- dehydration of the glass measurement electrode
- chemical attack of the glass measurement electrode
- aging of the glass measurement electrode
- breakage of the glass measurement electrode
- low process fluid temperature
- concentration gradients
- gas bubbles and spurious streaming and ground potentials
- process and ambient temperature changes
- lab samples at a different temperature and concentration than the on-line measurement
- lab electrode different than on-line electrode
- lab electrode not at thermal or ionic equilibrium
- frequent calibration adjustments
- changes in the water content for streams with water concentrations less than 60%
- changes in the salt concentration in high ionic strength streams
- low process fluid conductivity
- electrode holder and O-ring corrosion cracking and failure
- control system upsets

SELECTION

When the highest accuracy is needed, use pressurized flowing liquid reference electrodes to keep the junction potential E5 as small and as constant as possible, plus reduce the chance of contamination that drastically alters potential E4. The pressurization should be set to provide a positive flow of electrolyte into the process. When the electrode is immersed in a beaker for a sample or buffer measurement, the flow should be small enough so as not to alter the pH of the contents. The higher the pressure elevation above the process pressure, the larger the flow, and the smaller the potential E5 but the sooner the fill has to be replenished. Reference reservoirs are added to reduce the fill frequency. However this requires extra fittings and possibly freeze protection. If the internal electrode is in the reservoir, there is an offset in the measurement created by the glass and reference internal silver–silver chloride electrode elements' being at different temperatures. Unless the internal elements are identical and at the same temperature, there is a shift in the isopotential point.

For high ionic strength streams, use solid references or pressurized liquid reference electrodes to prevent contamination from the high ion concentrations. A flowing junction provides the best insurance against high and variable junction potentials, but the next best protection is offered by solid (e.g., solid state) references. The less porous the junction and the more torturous the internal electrolyte path, the slower the contamination as exhibited by reference R1 in Fig. 30. Solid nonporous junctions with immobilized electrolytes provide the greatest barrier to the ingress of process ions and reach ionic equilibrium the fastest of the solid-state electrodes.

For sulfides, cyanides, bromides, and iodides, isolate the process from the silver–silver chloride internal electrode to prevent the formation of silver precipitates that shift E5 and clog the junction. A second or even third internal reference junction with an intermediate electrolyte for the process junction that is free from silver ions and compatible with the process is used. A differential electrode in which the internal reference is a second glass electrode (patented by Great Lakes instruments) in a buffer solution compatible with the process provides the most isolation. However, the much faster migration of the hydrogen ions than the other ions in the buffer to the reference junction of the differential electrode creates a larger E5.

For batch operations, use reference junctions that are quick to reach ionic equilibrium; otherwise, the continual drift of the pH measurement will invalidate set points and end points. Flowing junctions

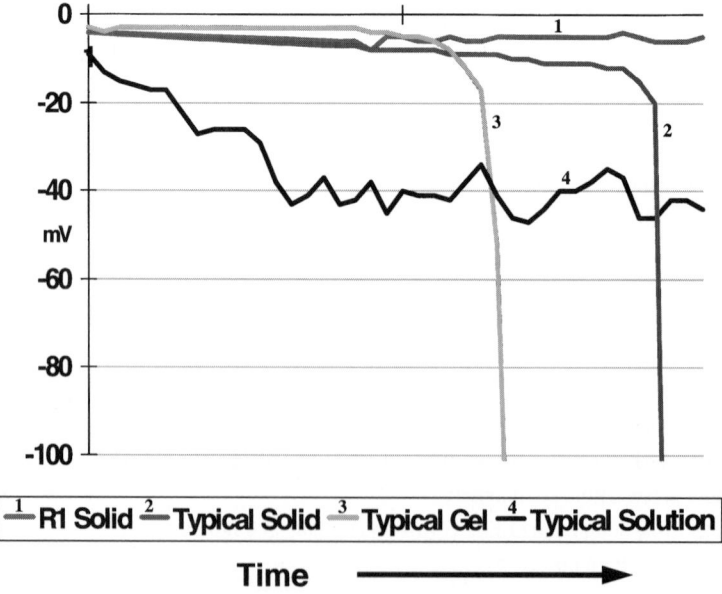

FIGURE 30   Time to contamination for various reference designs.

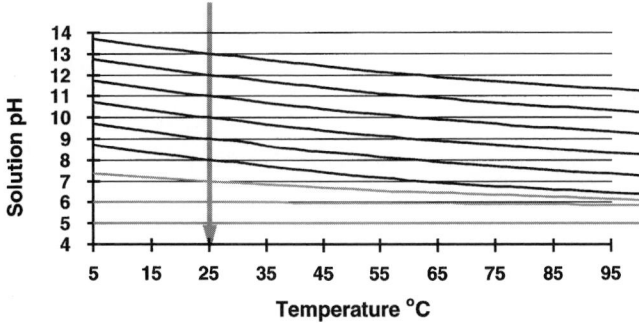

FIGURE 31 Error in Nernst potential corrected by standard temperature compensators.

are the quickest, and solid references with saturated (as opposed to immobilized) electrolyte are the slowest. For coating applications use flowing reference junctions or large nonporous reference junctions to keep the process from getting into the junction. It is extremely difficult to get the process out of the junction once it gets in. Mechanical and ultrasonic cleaning methods are ineffective. Chemical cleaners have a chance if the penetration is not too deep. Heating the reference electrode in a near-boiling 0.1-M KCL solution for 15 min, refilling the reference, air drying, and soaking may unclog the junction [13].

Use the electrode that minimizes maintenance. The previous selection practices are aimed at this result. If your application doesn't require exceptional accuracy and the process conditions do not have any of the difficulties just mentioned, it is best to use the electrode that instills the greatest confidence in operations, engineering, and maintenance so that the electrodes will be left alone. Many people develop a personal favorite, and the psychological advantage of using the favorite is important. Another consideration is the type of technical support and service available from the local supplier. Since pH applications can readily fail because of the wrong glass, materials of construction, and installation location and method, and because engineers and technicians today often don't have the time to develop special technical expertise, the quality and quickness of the on-site technical help you can get from the vendor can make or break the application. This includes how fast the vendor can deliver replacement electrodes and show up at your site to troubleshoot a rapidly deteriorating situation.

Compensate for changes in the Nernst potential with temperature. Automatic temperature compensation to correct the error shown in Fig. 31 is a standard feature that eliminates some of the error from changes in temperature. Even if the temperature is constant, the use of manual instead of automatic temperature compensation adds uncertainty because the settings are subject to either neglect or tweaking.

Compensate for changes in remaining electrode potentials and process pH with temperature. These often cause a larger measurement error than the Nernst potential effect addressed by standard automatic temperature compensators. Since these nonideal electrode and process effects are electrode and application specific, the only way to quantify them is to use the field electrode in a lab sample whose temperature is varied over the expected operating range. It is critical that the test only be done after the electrodes have reached thermal and ionic equilibrium. The change in pH reading with temperature is logged. The data are then fit with a either a straight line or a polynomial, depending upon the linearity and the options available in the transmitter or the distributed control system (DCS). Microprocessor-based pH transmitters offer a solution pH compensation for temperature, but the sensor buried in the electrode introduces a dynamic error from the thermal lag. A separate RTD installed close to the electrodes is preferable for the most accurate compensation, which means the calculation is done in the DCS.

For difficult, nasty service, use an automatically retractable electrode to limit exposure time. Assemblies with a pneumatic actuator have been developed, such as that shown in Fig. 32, which can be activated by a remote command [14]. If these are not permissible by the piping spec (e.g., high hazardous materials), a piping system with automated block valves can be used to block, drain,

ca. 1.4" (35mm)
ca. 18" (460mm)
INSERTED STATE
15°
2.5" (65mm)
RETRACTED STATE
CLEANING FLUID INLET
~23" (590mm)
CLEANING FLUID INLET

FIGURE 32 Automatically retractable pH assembly to limit exposure to harsh conditions.

flush, and soak the electrodes. However, the installed cost of the piping system is considerably more expensive than the automatically retractable assembly. The addition of a chemical cleaning step to the sequence adds considerable cost due to the piping and storage of a special cleaning fluid and should only be used if a high-velocity water flush is insufficient. Do not use distilled or deionized water or condensate for cleaning or soaking.

The temperature ratings of electrodes in the catalog are highly misleading. The life of an electrode, even though it is designed for higher temperatures, is exponentially reduced with temperature, as shown in Fig. 33. The effect of chemical attack and temperature is multiple rather than additive. Figure 34 provides a more realistic view of temperature ratings for glass electrodes. Any questionable or not

Months
12
10
8
6
4
2
0
12
6
3
1.5
25°C
50°C
75°C
100°C
Process Temperature

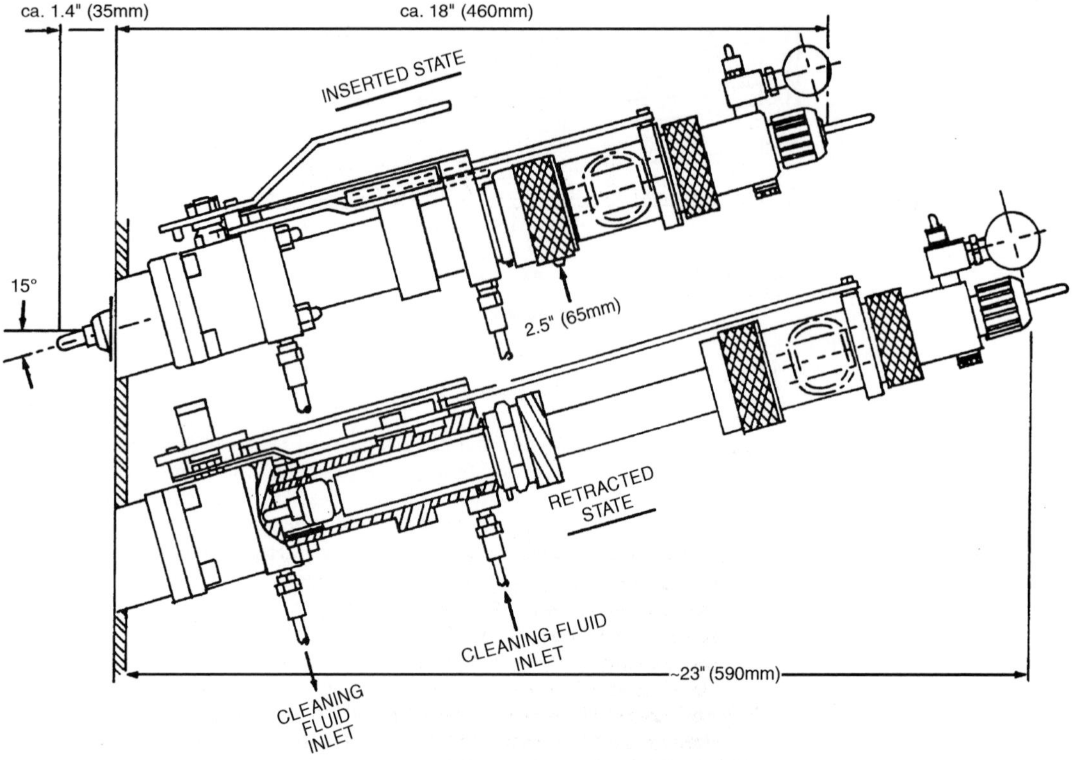

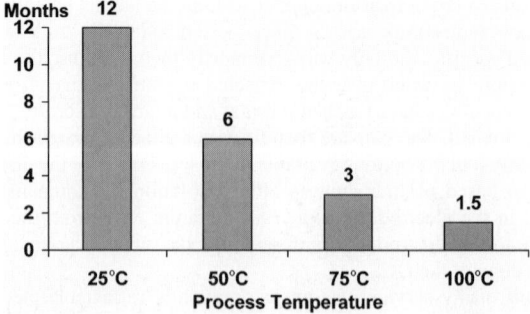

FIGURE 33 Life expectancy of glass electrodes vs. temperature for 2–10 pH.

good service	caution	not advisable

Chemical Attack Conditions	Temperature 5 < °C < 25	Temperature 25 < °C < 45	Temperature 45 < °C < 65	Temperature 65 < °C < 85	Temperature 85 < °C < 105
Hydrochloric Acid 0 < pH < 1		Limit Exposure			
Hydrochloric Acid 1 < pH < 2					
Hydrochloric Acid 2 < pH < 3					
Hydrochloric Acid 3 < pH < 7					
Sodium Hydroxide 7 < pH < 10					
Sodium Hydroxide 10 < pH < 11			Limit Exposure		
Sodium Hydroxide 11 < pH < 12		Limit Exposure			
Sodium Hydroxide 12 < pH < 13	Limit Exposure				
Sodium Hydroxide 13 < pH < 14					

FIGURE 34 Temperature ratings for glass electrodes to improve electrode life.

recommended rating for a glass electrode means the exposure to the high temperatures should be limited accordingly.

It is more appropriate for the DCS rather than the transmitter to schedule the insertion or commissioning and retraction or isolation of the electrodes so that it is coordinated with the mode of operation (e.g., position in a batch sequence or status of a continuous process equipment) and the process control strategy. Any controllers using the pH signal should be in output tracking or manual. Note that it is insufficient to freeze the pH signal because reset action in the controller will continue to ramp the loop output. The controller should not be released to use the signal until the reading has reached thermal equilibrium (e.g., 4 min after insertion). Ideally, it would be best to also wait for the signal to reach ionic equilibrium, but if the reference is slow, it is advantageous to get the reading before the migration of ions into the junction starts to shift the reading. Thus, for slow references there is a window of opportunity. The speed of recovery of the pH signal from the storage pH to the process pH provides a diagnostic of how clean or old the electrode is because a coated or thick gel layer will slow down the measurement electrode response. For easy to moderate service, use three electrodes and middle select for important measurements. This configuration provides the greatest reliability since it will inherently detect and ignore a single electrode failure of any type, even a failure to last value or at set point. Middle reading selection will also reduce the amplitude of both short-term and long-term errors without adding a measurement lag like a signal filter. It will also reduce maintenance if the user learns that short-term errors are normal and the electrodes should be left alone [1]–[3], [8], [9]. It facilitates a strategy to the keep the electrodes continually immersed to sustain their thermal and ionic equilibrium. Three electrodes allow the user to detect a fouled or noisy electrode by comparing the response to the other electrodes. With two electrodes,

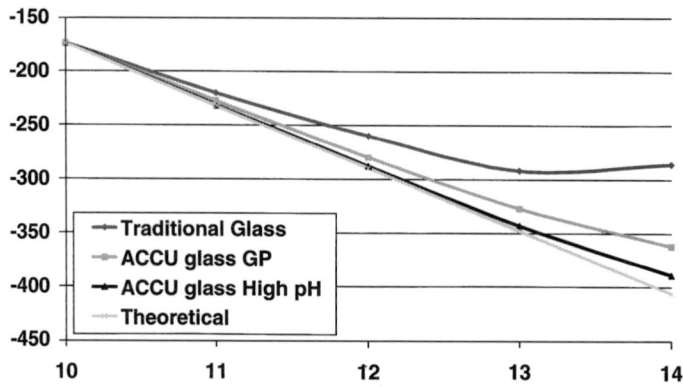

FIGURE 35   Sodium ion error for various types of glass electrodes.

you don't know which one is wrong. With three electrodes, the electrode furthest away from the middle reading is the suspect electrode. It is important to note that other selection logic schemes are not as effective as middle selection, which is a simple instruction often available in a DCS. However, for really nasty applications in which the electrode life expectancy is less than 3 months or in which on-line cleaning methods cannot keep the electrodes from getting coated in less than 2 weeks, the extra maintenance cost probably outweighs the process benefits from better accuracy and reliability of three electrodes and middle selection. For new difficult applications, it is better to start with one electrode to see if an automatically retractable assembly is a better choice from a total cost of ownership.

For pH >12 or high sodium ion concentrations, use high pH or wide range pH glass and keep the process temperature as close to 25 °C as possible to minimize sodium ion error and chemical attack of the glass electrode. The chemical attack and sodium ion error exponentially increase and the electrode resistance exponentially decreases with temperature. Thus, there is an optimum temperature around 25 °C. The high-pH glass uses the smaller lithium ion instead of the sodium ion for a tighter matrix that resists the penetration of the glass by sodium ions from the process. Sodium ions moving into the glass lowers the hydrogen activity of the gel layer compared to the process and lowers the pH reading [2], [3]. Figure 35 shows the errors caused by sodium ions at high pH for various types of glass.

For pH <2, use low pH or wide-range pH glass to minimize the acid error and disruption of the gel layer. The same glass used to minimize sodium ion error gel layer attack is effective in low-pH streams. However, some vendors offer a special glass designed for low pH service. In highly acidic streams (e.g., pH <0.5), protons moving into the glass increase the hydrogen activity of the gel layer compared to the process and cause the pH reading to be higher than actual [2], [3]. Some polymeric solid references are not designed for operation below a pH of 2.

The range of operating conditions for various types of electrodes is shown in Table 3. Note that while a high temperature and high pH is designated for a particular glass, it only means that the listed glass performs better than other types. It does not mean that the life expectancy will be acceptable.

For process temperatures <10 °C, avoid chemical resistant glass to prevent the electrical resistance from getting too large. However, if the life expectancy is unacceptable for general purpose glass, then the chemical resistant glass must be used and special measures taken, such as the use of an integral preamplifier and special electrode holder similar to what is used for high-purity water streams to prevent an erratic signal from electromagnetic interference (EMI) and spurious ground and streaming potentials.

For process temperatures <10 °C for general purpose, 20 °C for high pH, and 40 °C for rugged glass electrodes, use an integral preamplifier to immediately boost the signal so that it is less susceptible to EMI. Alternatively, the electrode cable run must be short and isolated from any possible EMI source.

For process temperatures >80 °C, limit exposure time and use a remote preamplifier to increase electrode life. As previously mentioned, the probe immersion can be scheduled by a DCS that sends

TABLE 2   Temperature Ratings for Glass Electrodes to Improve Electrode Life

Impedence @ 25°C	Operating Temperature Range	Continuous pH Operating Range	Application	Electrode Type
90 MΩ	32–212°F 0–100°C	0–10 pH	General water treatment	General purpose
250 MΩ	59–150°F 15–65°C	0–10 pH	Abrasive process	Ruggedized
130 MΩ	32–212°F 0–100°C	0–14 pH	pH >10 Na >1%	High pH
125 MΩ	32–212°F 0–100°C	0–10 pH	<100 ppm HF	HF resistant
140 MΩ	32–212°F 0–100°	0–10 pH	Steam sterilization to 130°C	High temperature

discrete outputs to an automatically retractable electrode assembly or to automated block valves for a piping system to block, flush, and drain the section of pipe for the electrode.

Use chemical and/or water jet washers to clean or hydrate probes on line. If chemical cleaning fluid cannot be put into the process, use an automatically retractable electrode. While other methods are effective for certain coatings, this is mostly from the viewpoint of keeping the measurement electrode clean. Chemical cleaning does a better job of cleaning the reference electrode, particularly if the junction is porous. The chemical cleaning method is more generally effective for even the measurement electrode [2], [3]. However, the installed cost of the chemical cleaning is higher because of the piping and storage needed for the special cleaning agent. A high-velocity water jet is preferable, if it can do the job by tapping into a process or utility water line. Do not use distilled or deionized water or condensate. Make sure the water pressure is always higher than the process pressure and be sure to use a check valve. Water has the additional advantage of hydrating electrodes, which is especially important for streams with very high acid or low water concentrations.

Use transmitters with the sensor diagnostic accurately corrected for glass impedance. The change in resistance of the glass electrode is so greatly dependent upon temperature and glass type that any diagnostic that does not take into account the glass type and temperature will either give false alarms or no alarms for measurement electrode coating diagnostics. A ground electrode is essential so that the much smaller resistance of the reference electrode can be deciphered from the glass electrode, as shown in Fig. 36. Properly done, resistance measurements can detect reference electrode plugging and contamination, nonimmersion, breakage of the glass, loss of continuity, and thick nonconductive coatings. The diagnostics should not be relied upon for alerting the user to coatings because a profound deterioration in the response starts for the very thin conductive coatings that commonly develop in industrial applications.

Best Practices for pH Measurement Selection

1. Where the highest accuracy is needed, use pressurized flowing liquid reference electrodes.
2. For high ionic strength streams, use solid references or pressurized flowing reference electrodes.
3. For sulfides, cyanides, bromides, and iodides, isolate from the silver–silver chloride internal electrode.
4. For batch operations, use reference junctions that are quick to reach ionic equilibrium.
5. For coating applications, use flowing reference junctions or large nonporous reference junction.
6. Use the electrode that minimizes maintenance.
7. Compensate for changes in the Nernst potential with temperature.
8. Compensate for changes in remaining electrode potentials with temperature.

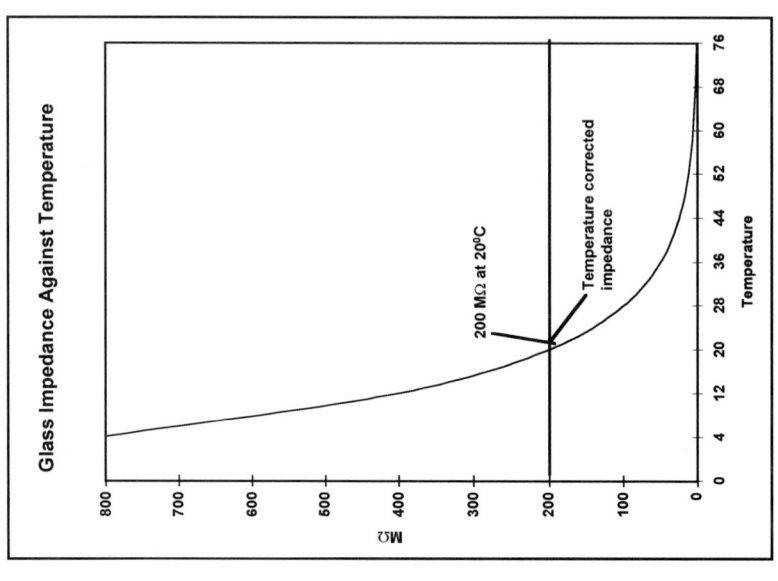

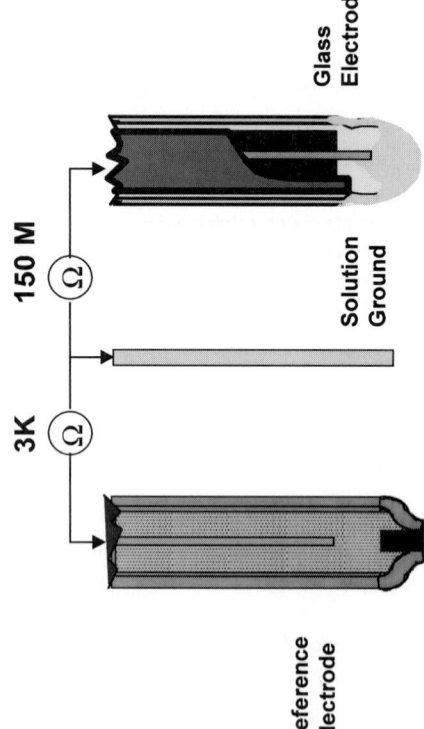

FIGURE 36 Sensor diagnostics with temperature-corrected glass impedence.

9. Compensate for changes in the process pH with temperature.

10. For difficult, nasty service, use an automatically retractable electrode to limit exposure time.

11. For easy to moderate service, use three electrodes and middle select for important measurements.

12. For pH >12 or high sodium ion concentrations, use high-pH or wide-range pH glass.

13. For pH >12, keep the temperature as close as possible to 25 °C.

14. For pH <2, use low-pH or wide-range pH glass.

15. For process temperatures <10 °C, avoid chemically resistant glass.

16. For process temperatures <10 °C for general purpose, 20 °C for high pH, and 40 °C for rugged glass electrodes, use an integral preamplifier.

17. For process temperatures <25 °C or $12 <$ pH <2, avoid thick (rugged) pH glass.

18. For process temperatures >80 °C, limit exposure time and use a remote preamplifier.

19. Use chemical and/or water jet washers to clean or hydrate probes on line.

20. If chemical cleaning fluid cannot be put into the process, use an automatically retractable electrode.

21. Use transmitters with sensor diagnostics accurately corrected for glass impedance.

INSTALLATION

For easy to moderate service, keep the electrode continually immersed in the process. The best results in terms of repeatability and reliability have been observed for electrodes kept in the process fluid so that they maintain an equilibrium with the process conditions [8]. If there are severe coating problems or chemical attack, it is better to limit their exposure and soak them in a solution as close to the same temperature and ionic makeup as the process as possible when not in use. The next best soaking solution is a 4-pH buffer.

Use a dedicated small-volume heat exchanger to reduce excessive temperatures. Since high temperature is the most frequent cause of premature electrode failure, the investment in a small exchanger for the flow past the electrode is worthwhile if a temperature reduction does not cause process problems or a coating. It is critical that the exchanger not add much volume and therefore transportation delay to the measurement. Also, the exchanger outlet temperature should not fluctuate more than a few degrees. For difficult nasty service, only insert the electrode deep enough to reach thermal equilibrium and get a full measurement electrode response (e.g., 4 min). Anything longer will lead to a shifting reading because of a drift from changes in the measurement glass gel layer and the reference junction potential. This is essentially what is done in the laboratory and explains why lab measurements seem to be more repeatable.

Where coating is a problem, keep the velocity between 5 and 9 f/s and have the fluid sweep past the electrode surfaces. This is the simplest and least expensive method of keeping electrodes clean. This mostly works in terms of inhibiting the start of a coating. If a sticky coating has already developed, the velocity won't remove it unless there is some solids, solvent, or cleaning agent in the fluid. Therefore, it is critical that the flow is never stopped and the electrode is never exposed to a stagnant process fluid. It also means that electrodes should be installed in pumped pipelines since the velocity can be achieved by changing the diameter of the pipe at the electrodes. In computing the velocity past the electrode, the occlusion by the electrode should be used in computing the net cross-sectional area. Velocities higher than 9 f/s can cause excessive wear and an erratic signal from abrasion and flashing. Ideally, the velocity should be kept as constant as possible to reduce the fluctuations in the reference potential. If the flow to the electrode stops, the line should be drained immediately to reduce the onset of a coating. The velocity in even highly agitated vessels is not much greater than 1 f/s unless you locate the electrodes perilously close to the agitator blades. The use of submersion assemblies in applications with potential coating problems is doubly troublesome since they are more likely to get coated and are more difficult to remove and service. Approximately 10–20 diameters downstream of a pump in a recirculation line sized for the right velocity is the optimum installation.

If the process fluid gets stickier as it cools, keep the electrodes hot. Hopefully this can be achieved by making sure the electrodes are soaking in a hot solution and the process fluid stays hot. In some cases, insulation and electrical heat tracing of the electrode assembly might be required. Since electrode life is severely reduced as a result of an elevated temperature, temperatures in excess of what is needed must be avoided.

Where abrasion or corrosion is a problem, keep the velocity below 1 ft/s and avoid direct impingement of the fluid on the glass surface. The line size is changed at the electrodes to ensure that the velocity is low enough. It is also critical that the solids can drain and there are no small cross-sectional areas (e.g., <1.5 in.) in the piping system.

For solids, vertical runs are favored with a the probe angle at a minimum (e.g., 20°) and at a maximum (e.g., 80°) to reduce coating and abrasion, respectively. Also, the slot or hole in the shroud should be faced away from the flow to protect the glass from direct impingement of abrasive solids and faced into the flow to provide a sweeping action across the glass for sticky solids. Figures 37(a) and (37b) show these probe angles and shroud orientations. The holes and slots in the shroud are designed to prevent a bubble from being trapped inside the shroud. A tapered shroud that is touted as creating vortices that help clean an electrode has no hole or slot, but the tail end is low enough to allow a bubble to escape.

Use a series installation of pH electrodes to help keep the concentration, velocity, and temperature the same for all of the electrodes and use a parallel installation to avoid breakage or blockage that simultaneously disables all electrodes. Flow and temperature can almost be made equal in the parallel paths if careful attention is paid to make the pressure drops and insulation identical for the branches. However, it is impossible to dictate an equal distribution of solids. This can then in turn affect pressure

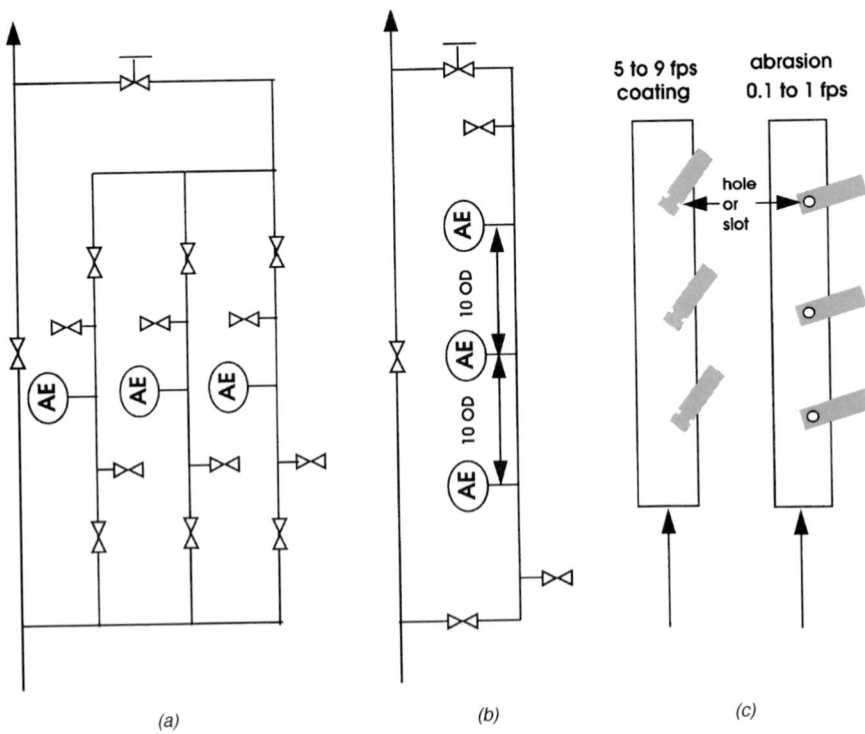

(a)   (b)   (c)

FIGURE 37   Vertical piping arrangements and probe orientations.

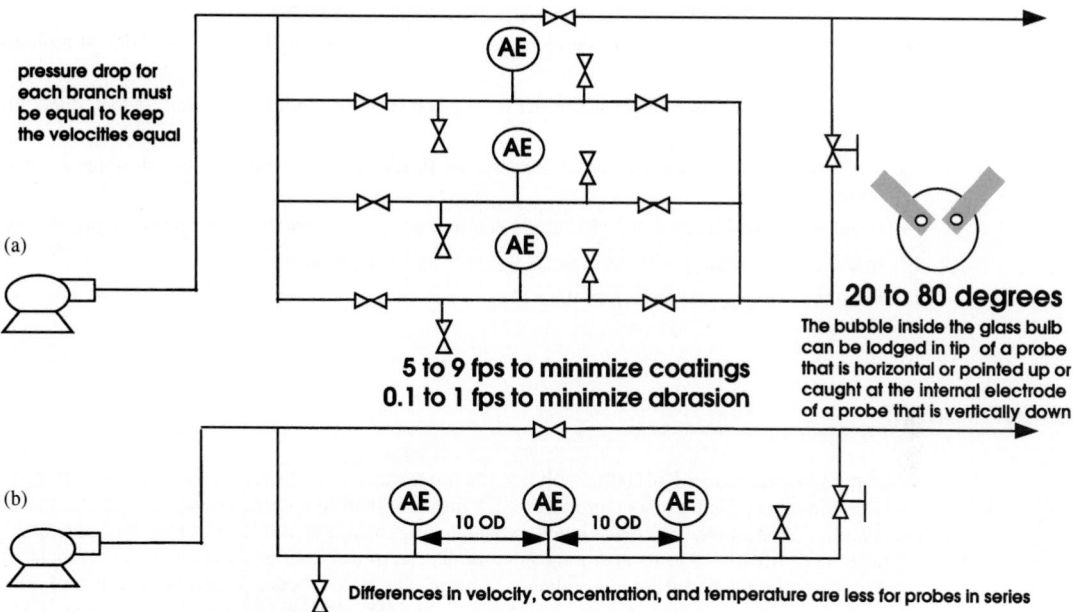

FIGURE 38   Horizontal piping arrangements.

drops through different fouling rates. A series installation is less expensive and keeps the velocities, temperatures, and solids the same for all of the electrodes if their insertion depths are identical. However, a blockage or large clumps, bolts, nuts, and welding rods could cause all three electrodes to fail. For a series installation, it is essential to have a pump strainer upstream and a line sized large enough that a plugged line is not possible. Figure 38 shows the alternate piping arrangements.

Install the electrodes at a 20°–80° angle from horizontal to keep the internal bubbles properly positioned. If the bubble resides in the electrode tip because of a horizontal installation or around the internal silver–silver chloride electrode because of a vertical installation, the signal can become erratic. The chance of a vertical installation causing a problem is not very probable and, in fact, many electrodes are installed at 90° with no reported problems.

Install a narrow range RTD as close as possible to the electrodes to get a fast accurate temperature indication. This should be used to compensate for the effect of temperature on the process (solution) pH and the electrode potentials not accounted for by the Nernst equation and the standard temperature compensator. For frequent, rapid temperature changes, the external RTD would be better than the internal temperature sensor for the Nernst compensation. Where possible, a bare RTD element should be used since a thermowell is the source of most of the dynamic and conduction error in the RTD measurement.

Keep the velocity variations less than 2 fps to minimize the changes in reference junction potential. This is essential for low conductivity (e.g. monomer and high purity water) streams.

Best Practices for pH Measurement Installation

1. For easy to moderate service, keep the electrode continually immersed in the process.

2. Use a dedicated small-volume heat exchanger to reduce excessive temperatures.

3. For difficult, nasty service, only insert the electrode deep enough to reach thermal equilibrium and get a full measurement electrode response (e.g., 4 min).

4. Where coating is a problem, keep the velocity between 5 and 9 f/s and have fluid sweep past the electrode surfaces.

5. If the process fluid gets stickier as it cools, keep the electrodes hot.

6. Where abrasion or corrosion is a problem, keep the velocity below 1 f/s and avoid direct impingement of the fluid on the glass surface.

7. Use a series installation of pH electrodes to help keep the concentration, velocity, and temperature the same for all of the electrodes.

8. Use a parallel installation to avoid breakage or blockage that simultaneously disables all electrodes.

9. Install the electrodes at a $20°$–$80°$ angle down to keep the internal bubbles properly positioned.

10. Install a narrow-range RTD as close as possible to the electrodes.

11. Keep the velocity variations less than 2 fps.

MAINTENANCE

Store spare electrodes in a 4-pH buffer solution (do not store or soak electrodes in distilled or deionized water or condensate) and do not store electrodes for more than 6 months. The 4-pH buffer keeps the gel layer hydrated and the reference junction from losing too much of its electrolyte. Since the inside gel layer has a life of ~3 years at storage conditions, electrodes lose an appreciable part of their life expectancy by being stored for more than 6 months. The best practice would be to use a supplier who delivers relatively new electrodes within a day. Middle selection of three electrodes should buy the user enough time to eliminate the need for on-site spares if the vendor is responsive. Electrodes should have expiration dates to protect the user.

Precondition electrodes for 2 days in a solution close in ionic composition and temperature to the process to reduce the amount of time it takes for electrode gel layer and reference junction potential to stabilize when the electrode is commissioned [4]. If middle selection is used and only one electrode is replaced at a time, this procedure is not as necessary.

Buffer calibrate the electrodes once just before installation using 7-pH buffer first to adjust the zero point. New electrodes need to have their span and efficiency (percent of slope) verified over at least a 3-pH range. The efficiency should be recorded for future reference to determine the degree of aging of the glass. The first buffer should be at 7 pH for glass electrodes since this is the zero potential and isopotential points. The next buffer is preferentially a 4-pH buffer. If the electrode is suspected to be nonlinear enough and the operating range in pH set points wide enough to adversely affect the degree of accuracy needed, then basic buffers may be needed. Make sure buffers above 7 pH have not drifted low as a result of carbon dioxide absorption. This can happen rapidly in open containers and slowly through closed plastic containers. Always check the expiration date and test a basic buffer before use.

Flush the buffer container and electrodes with tap water before filling the container with buffer and inserting the electrodes. Finally, ensure the buffer is stirred and the buffer value is corrected for temperature. Mixing helps speed up the response of the measurement electrode and helps disperse any foreign ions migrating out of the reference junction or coming off the electrode because of insufficient washing. If the buffer is used outside, there may be a significant shift due to outside temperature. For subzero temperatures, alcohols are added to the buffer for freeze protection. Special equations have been developed to predict the change in buffer pH with temperature and ethanol and glycol concentration [15]. If a buffer calibration is only done before the installation, the calibration can be done in the shop. Ideally, the buffer calibration should be at the same temperature as the process.

Don't install the electrodes until the system is filled with process fluid. gation of the electrodes during water batching is asking for problems in terms of broken and dirty electrodes. The readings during water batching are meaningless and often a source of concern because of rapid movement from lack of buffering. The periods in which the electrodes sit dry will cause dehydration [8]. Hydrate the glass measurement electrode 5 min, for each hour the electrode is left dry to reestablish the gel layer. The maximum amount of time needed to hydrate a dry electrode is ~1 day.

Don't remove the electrodes unless they are broken, worn out, or have a coating that cannot be cleaned on line. The removal of the electrodes upsets their equilibrium with the process and puts them at risk to being mishandled and overcalibrated. Buffer calibration should not be done to cover up the symptoms from a reference contamination, coating, or old gel layer. Since most systems have a single pH set point, a standardization near this set point is sufficient to maintain accuracy. In other words, changes in the span affect mostly the controller gain via changes in the measurement gain. These changes in measurement gain are insignificant compared to the changes in process gain and dynamics and control valve gain and dynamics. Frequent buffer calibrations recommended by most vendors and practiced by many users do more harm than good. The user ends up in a cycle of calibration adjustments, chasing the effects of previous calibration adjustments [8].

It is imperative to be able to detect coatings without removing the electrodes. A persistent coating can be diagnosed from the following symptoms:

- an electrode that is continually lagging other electrodes in a three-electrode installation
- a slow recovery during immersion in the process or 4-pH buffer after an on-line cleaning cycle
- a large increase in the control loop period

Just a 1-mm slime coating can increase the response time of the glass from 10 s to 7 min [16]. Therefore, the changes in the above symptoms are fairly dramatic. Only make a standardization adjustment after the reference has reached thermal and ionic equilibrium once after installation and every 6 months thereafter for a well-designed installation. Unless the electrode is preconditioned, it could take hours to a couple of days for some solid-state references to reach equilibrium [8]. Then a process sample should be taken when the process pH is near set point for making one standardization adjustment. For a well-designed installation, the electrode should not have to be standardized more frequently than every 6 months. A well-designed installation means the temperature and chemical attack is not extraordinary because of application conditions or exposure time.

Previously it was thought that frequent calibration adjustments (e.g., weekly) were needed to obtain an accuracy of 0.25 pH on a steep titration curves [1]–[3]. Now it is realized that the observed errors are short term and mostly due to concentration fluctuations, and the best policy is to leave the electrodes alone [8]. The standardization adjustment with the electrode in service is in conformance with the ASTME indirect method listed in Title 40 Code of the Federal Regulations Part 136 for pH and therefore should satisfy environmental requirements. Only make a standardization adjustment if three or more successive samples have the same error sign, and use the average rather than the last value unless there is a clear trend. When the sample deviation from the reading raises a red flag, at least two more samples should be taken over the next 2 days and a standardization made only if the deviation is consistent. The average of the deviations is most representative of the error unless there is a clear trend of an increasing deviation with time. An increasing deviation with time is a warning of a severe contamination problem, in which case the proper solution is removal and not recalibration. Patience is essential because of the predominance of short-term errors. Of course, a large deviation dictates immediate additional samples unless there is middle selection.

Use the same electrode for the lab samples that was used for the on-line measurement. Since there are unknown junction potentials and temperature effects that are peculiar to the type and options, electrodes used in the lab should be identical to those used in the process unless an accuracy of 0.5 pH is acceptable.

Make sure samples above 7 pH have not drifted because of carbon dioxide absorption, the samples with ammonia or hydrogen chloride have not drifted because of vaporization, sample containers will not contaminate the sample, and take the sample at the same point in the process and pipeline cross section as the on-line measurement. These practices are all aimed at making the sample composition as close as possible to the process fluid composition at the electrode when the reading was recorded [2], [3].

Change the temperature of sample, develop the correlation of pH reading to temperature, and compensate for changes in the pH reading of the sample and the process with temperature. This should be automated in both the lab and field for best repeatability. Lab meters to date only have the standard Nernst but not solution pH compensation, so the pH and temperature readings have to be brought into a personal computer and data-acquisition program for complete temperature compensation.

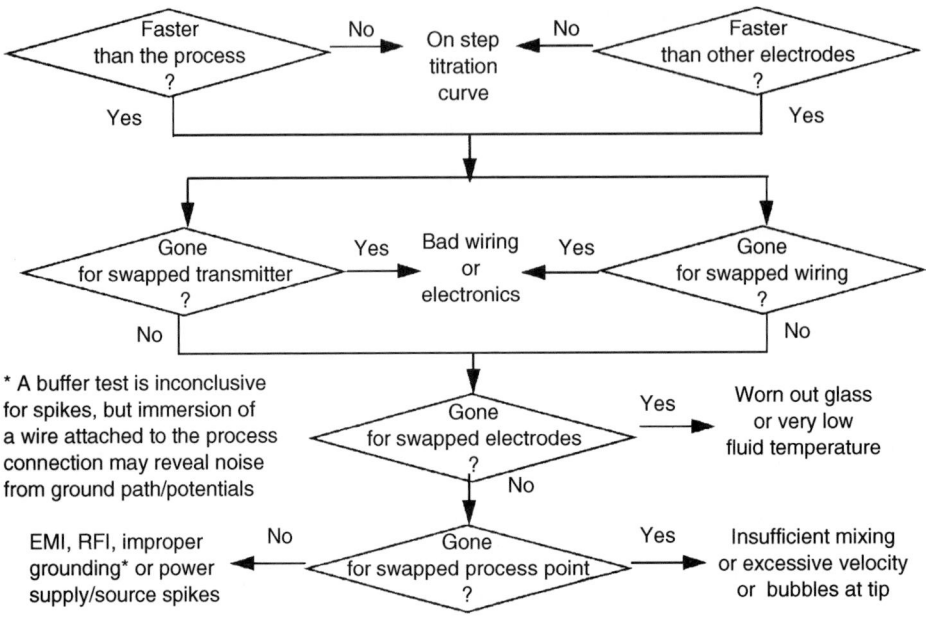

FIGURE 39 Troubleshooting logic for spikes in the pH reading.

To clean electrodes, soak the electrodes in stirred dilute HCL for 60 min; to rejuvenate electrodes, soak them in stirred very dilute HF for 1 min. Dilute 10% HCL is the most general purpose cleaner. For protein deposits, a combination 10% HCL and 1% pepsin solution is used. In some applications, polar solvents such as acetone or a bleach such as hypochlorite may be more effective [13]. The use of a very dilute HF solution will strip away the old gel layer. The electrode must then be soaked to create a new gel a layer. In theory, this could be done repeatedly to extend the life of the measurement electrode almost indefinitely if you could do it to the aging inside gel layer as well [7].

Figures 39 and 40 show the troubleshooting logic to find the cause of spikes in a pH reading or a sluggish response. Note that middle selection enables a simple immediate decision, the remaining logic requires much more analysis and is often inconclusive. The last step is removal or relocation of the electrode.

Best Practices for pH Maintenance

1. Store spare electrodes in a 4- or 7-pH buffer solution.
2. Do not store or soak electrodes in distilled or deionized water or condensate.
3. Do not store electrodes for more than 6 months.
4. Precondition electrodes for 2 days in a solution close in ionic composition and temperature to process.
5. Buffer calibrate the electrodes once just before installation, using 7-pH buffer first to adjust zero point.
6. Make sure buffers above 7 pH have not drifted low as a result of carbon dioxide absorption.
7. Flush the buffer container and electrodes with tap water.
8. Insure the buffer is mixed and the buffer value is corrected for temperature.
9. Don't install the electrodes until the system is filled with process fluid.
10. Hydrate the glass measurement electrode 5 min for each hour the electrode is dry.

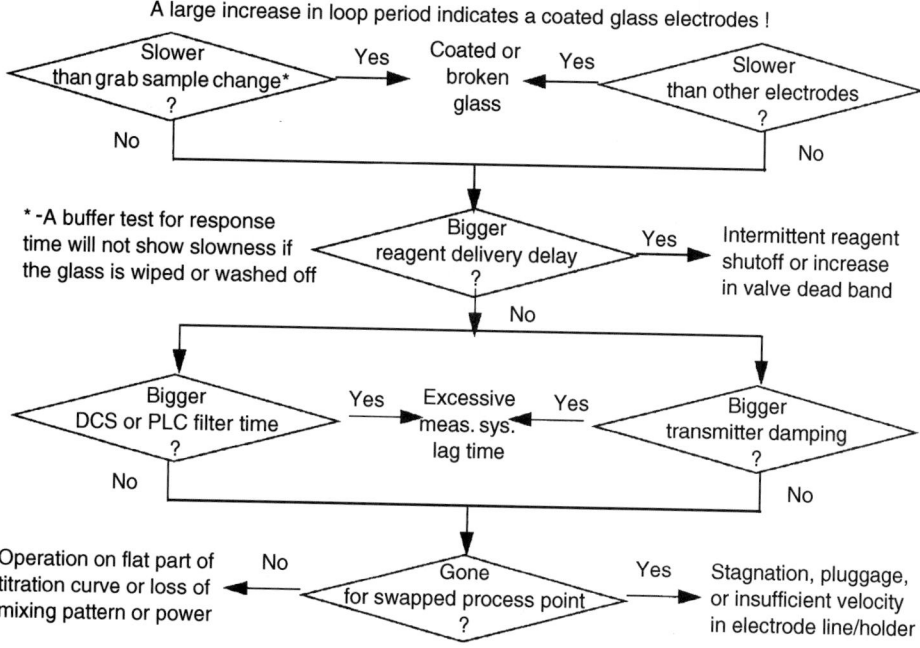

FIGURE 40   Troubleshooting logic for a sluggish pH reading.

11. Don't remove the electrodes unless they are broken, worn out, or have a coating that cannot be cleaned on line.

12. Only make a standardization adjustment after the reference has reached thermal and ionic equilibrium once after installation and every 6 months thereafter for a well-designed installation.

13. Only make a standardization adjustment if three or more successive samples have the same error sign and use the average rather than the last value unless there is a clear trend.

14. Use the same electrode for the lab samples that was used for the on-line measurement.

15. Make sure samples above 7 pH have not drifted as a result of carbon dioxide absorption.

16. Make sure samples with ammonia or hydrogen chloride have not drifted as a result of vaporization.

17. Make sure sample containers will not contaminate the sample.

18. Take the sample at the same point in the process and pipeline cross section as the on-line measurement.

19. Change the temperature of sample and develop the correlation of pH reading to temperature.

20. Compensate for changes in the pH reading of the sample and the process with temperature.

21. To clean electrodes, soak the electrodes in stirred dilute (e.g., 10%) HCL for 60 min.

22. To rejuvenate electrodes, soak them in stirred very dilute HF for 1 min.

Key Points

1. Selection, installation, and maintenance practices make or break the pH measurement.

2. The fragile gel layer of the glass electrode is easily disrupted by harsh or dry conditions.

3. The fragile gel layer has at best a 3-year life expectancy even when in storage.

4. Temperature and caustic have a multiple exponential effect on electrode life and accuracy.

5. The more significant effects of temperature on electrode accuracy have been ignored.

6. Unless the reference junction has either a high flow or an immobilized electrolyte, there is a significant shift in pH readings due to ion migration in and out of the reference junction.

7. Composition changes cause short-term errors by upsetting the gel layer and reference junction.

8. For steep titration curves, the short-term fluctuations are an order of magnitude larger.

9. Middle selection reduces errors, noise, and the need for preconditioning, calibration, and on-site spares for easy-to-moderate applications if the user leaves the electrodes alone.

10. There is an optimum velocity to minimize coatings and abrasion.

11. The electrode exposure in a harsh stream should be automatically limited.

12. Most long-term pH measurement errors are due to overcalibration and premature aging.

13. Most of the discrepancy between the field and lab is due to differences in process composition, electrode construction, and temperature.

14. Electrodes not prematurely aged by high temperature or caustic concentrations can be rejuvenated by a short exposure to a dilute HF solution to strip away the old gel layer.

Rules of Thumb

1. Unless the application has an extremely harsh operating condition or an exceptional accuracy requirement, the best electrode is the one favored by the maintenance technicians that has the most extensive and fastest technical assistance and service and spares.

2. Use special fills and internal barriers to prevent silver ions from combining with sulfides, cyanides, bromides, and iodides to form a black precipitate that plugs the junction and shifts the reading.

3. Keep the electrode continually exposed to a fluid with about the same temperature and ionic makeup.

4. For batch operation, use a reference that has the smallest and/or fastest junction potential.

5. For continuous operation, use a reference that has the smallest and/or most constant junction potential.

6. Use the right glass and wetted materials of construction for the most extreme operating conditions.

7. Document, correlate, and automatically correct the lab and field measurement for all (not just the Nernst) temperature effects.

8. Make sure lines for fluids with solids have no restrictions and will drain quickly and stay warm.

9. Keep the velocity at the electrode above 5 fps to prevent coatings and below 1 fps to prevent abrasion.

10. If velocity won't keep the electrodes clean, use a chemical or water jet or flush to clean the electrodes.

11. If electrode coating and aging rate is not excessive, use middle reading selection; otherwise, automatically isolate, flush, precondition (soak), and immerse the electrodes to limit the exposure.

12. For high temperatures, concentrations, and accuracy, use the same electrode in the lab as in the field.

13. Do not buffer calibrate electrodes after they have been installed, and only make a standardization adjustment if the lab and field show a consistent error near set point for three or more samples.

14. Make sure the equipment and control system design (e.g., excessive dead time) and controller tuning (e.g., excessive reset) do not cause excursions to harsh process operating conditions.

REFERENCES

1. McMillan, G. K., "Understand Some Basic Truths of pH Measurement," *Chemical Engineering Progress*, pp. 30–37, Oct. 1991.

2. McMillan, G. K., et. al., "pH Measurement," *Analytical Instrumentation,* Chilton, pp. 312–328, 1994.

3. McMillan, G. K., *pH Measurement and Control*, 2nd ed., Instrument Society of America, Research Triangle Park, N.C., 1994.

4. Ingold, W., "Principles and Problems of pH Measurement," Ingold Tech. Pub. E-TH 1-1-CH, pp. 1–23, 1980.

5. Ingold, W., "pH Measurement and Temperature Compensation," Ingold Tech. Pub. E-TH 8-2-CH, pp. 1–16, 1982.

6. Ingold, W., "pH Electrodes—Storage, Ageing, Testing, and Regeneration," Ingold Tech. Pub. E-TH 1-1-CH, pp. 1–13, 1982.

7. Ingold, W., "Calibration of pH Electrodes—Buffer Solutions, Electrolytes, Special Solutions and Accessories," Ingold Tech. Pub. E-TH 5- 1-CH, pp. 1–14, 1982.

8. McMillan, G. K., "Methods of Controlling pH," *Chemical Processing*, pp. 58–61, July 1997.

9. McMillan, G. K., "pH Measurement: The State of the Art," *InTech*, pp. 39–40, Feb. 1993.

10. Moore, R. L., *Environmental Protection by the Neutralization of Wastewater Using pH Control*, 2nd ed., Instrument Society of America, Research Triangle Park, N.C., 1995.

11. Hickey, D., conversations on test results (private communication).

12. "Measure pH in Mixed and Non Aqueous Solutions," *Today's Chemist At Work*, pp. 39–42, June 1993.

13. Boyle, P. F., "Some Cleaning Tips for Fouled Electrodes," *Control*, p. 108, Oct. 1993.

14. Loar, J., "Try Using Automatically Retractable Probes," *Control*, p. 106, Oct. 1992.

15. Moore, R. L., "Good pH Measurements in Bad Process Streams," *Instruments and Control Systems*, pp. 39–43, Dec. 1990.

16. Verhappen, I., "The Effects, of 'winterizing' pH Buffers," *ISA Transactions*, vol. 33, pp. 73–82, 1994.

OPTICAL CONCENTRATION TRANSDUCERS

by Don Soleta*

TURBIDITY

Turbidity instrumentation is used in many industries to measure the presence or concentration of particles[1] in liquid streams. Other terms used besides turbidity are clarity, haze, and transparency. They are all nontechnical appearance descriptors, and they describe the effect a sample, including suspended particles, has on light passing through it. Turbidity is most often used to describe the

---

* Analyzer Specialist, Monsanto, Inc., St. Louis, Missouri.
[1] Particles should be thought of as undissolved solids, liquids, or gases having a different refractive index than that of the carrier liquid.

measurement of low concentrations of particles ($<1\%$ v/v) that should not be in the process stream, like particles in drinking water or filtrate. Turbidimeters measure the scattering of light by particles at these concentrations and correlate it to particle concentration. A basic understanding of the principles of light scattering and the methods used to measure it will help in selecting and specifying the instrument needed for the required measurement.

LIGHT SCATTERING

Since turbidity is a visual phenomenon, it deals with the scattering of visible light, that is, electromagnetic radiation with wavelengths between 380 and 780 nms. When a focused beam of light is transmitted through a clear nonabsorbing liquid containing no particles, it passes through undisturbed in intensity and direction. However, these conditions rarely, if ever, exist. Even pure water will attenuate and scatter a beam of light, albeit at very low levels. If the beam of light encounters a particle, the intensity of the transmitted beam of light is decreased (see Fig. 41). Some of the light lost from the incident beam is absorbed, while the rest of the lost light is redirected or scattered by particles. Further discussion of scattered light theory can be found in the literature [1].

In a turbidity measurement, the intensity and direction of scattered light are determined by the properties of the particles suspended in the liquid and the properties of the liquid itself. Particle properties affecting scatter are the refractive index, size and shape, color, and concentration.

Refractive Index

The particle must have a different refractive index (RI) than that of the carrier liquid. This is usually not a problem with inorganic substances. The RI of organic substances can approach or equal that of the carrier liquid, making detection by light scattering difficult or impossible. The greater the difference in RI, the greater the intensity of the scattered light.

Size/Shape

Light scattered by a particle is not just in one direction as implied in Fig. 41. The direction of light scattered depends on a particle's size relative to the wavelength of the incident light and the particle's

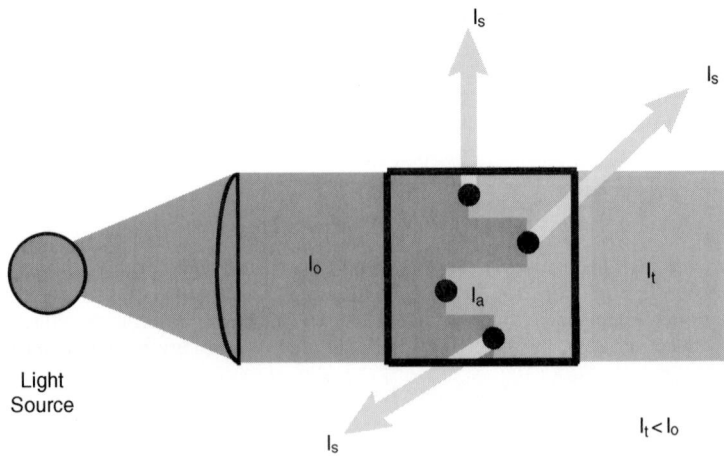

FIGURE 41 Basic Turbidimeter and Particle Influence on Light.

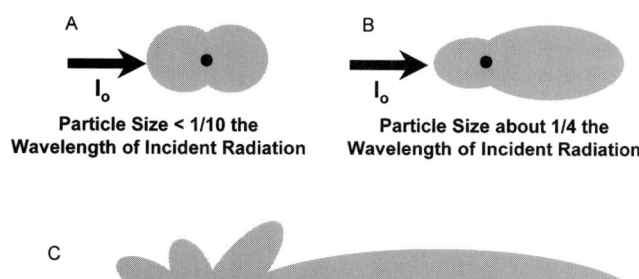

Particle Size Larger Than the Wavelength of Incident Radiation

FIGURE 42   Angular patterns of scattered intensity from particles of three sizes.

shape. When the size of a particle is less than one-tenth of the incident light wavelength, light is symmetrically scattered (Fig. 42A). More light is scattered in the forward direction as the size of the particle approaches the wavelength of the incident radiation (Fig. 42B). Two scattering characteristics are shown in Fig. 42C. When the size of the particle exceeds the wavelength of light, the forward-scattered light greatly intensifies. However, if the particle has an irregular shape, that is, it is not a sphere, light is again scattered in all directions, although not symmetrically as with small spherical particles.

Color

If a particle is colored, especially if it is black, part of the incident radiation interacting with the particle will be absorbed and not scattered. The absorbed light intensity, (I_a) in Fig. 41, decreases the intensity of light transmitted through the sample (I_t) and the intensity of scattered light (I_s). Ratioing the scattered light intensity to the transmitted light intensity (I_s/I_t) factors out the effect of absorbed light.

Concentration

As mentioned earlier, light scattering is used to measure low particle concentration levels. The upper limit of light scatter detection is reached when insufficient light is transmitted through the sample as a result of high levels of light absorption by the particles. At some concentration, dependent on particle color, the absorption of light exceeds the intensity of the scattered light (see Fig. 43). The measurement of particles by light scattering is then no longer possible. Analyzers based on light absorption are

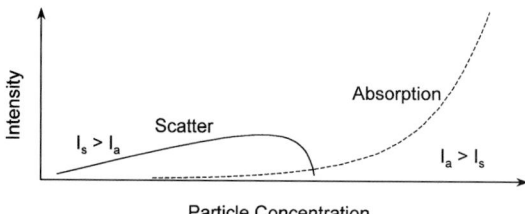

FIGURE 43   Relationship between scatter intensity and particle concentration.

used to measure these higher concentrations. However, the relationship between concentration and absorption (the loss of transmitted light) is nonlinear, and special compensation must be used to correlate particle concentration to absorbed light.

Properties of the liquid affecting light scattering are color, refractive index, and optical path length.

Color

A colored solution will also absorb part of the visible incident light, thus reducing the amount of light available for scattering. The intensity of the transmitted light will also be attenuated. Some of the scattered light will be absorbed by a colored solution and reduce the sensitivity of the method. Highly colored or dark solutions may reduce the amount of incident visible light to a level where turbidity measurement is not feasible.

Refractive Index

As mentioned before, the greater the difference between the RI of the particle and that of the carrier liquid, the greater the intensity of light scattering. If the carrier liquid is composed of two miscible liquids or if there is a dissolved compound in the carrier liquid, changes in the concentrations of any of these will change the RI of the carrier liquid. This will affect the angles at which light is scattered from the particles. Particle concentration would then appear to change under these special optical conditions, even though the actual concentration may be constant.

Optical Path Length

This parameter can be adjusted to compensate for highly colored solutions and/or high particle concentrations. Reducing the path length reduces the amount of sample between the light source and detector. This reduces the amount of absorbed incident light, allowing more light for scattering and transmission. However, very small path lengths may cause sample-plugging problems or not provide the required sensitivity. Long path lengths increase the amount of sample between the source and detector, increasing the intensity of scattered light. This aids in measuring very low particle concentrations.

METHODS OF MEASUREMENT

Turbidimeters consist basically of three components: a light source, one or more detectors, and a sample cell (see Fig. 44). Variations in the arrangement of these components account for the many types of turbidity instrumentation on the market. Forward scatter, 90° scatter, and backscatter are the main methods of scattered light measurement. Turbidity values between these methods should not be compared because there are no simple correlations relating the values between the methods. Figure 45 illustrates the difference between the methods.

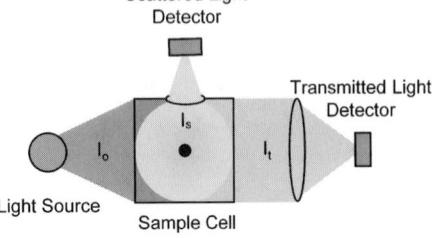

FIGURE 44   Basic turbidimeter.

90° Scatter

90° scatter or nephelometry is the method by which light scattered at right angles to the incident light beam is measured. Since smaller particles scatter more light at right angles to the incident beam, nephelometry is better suited to detect small particles

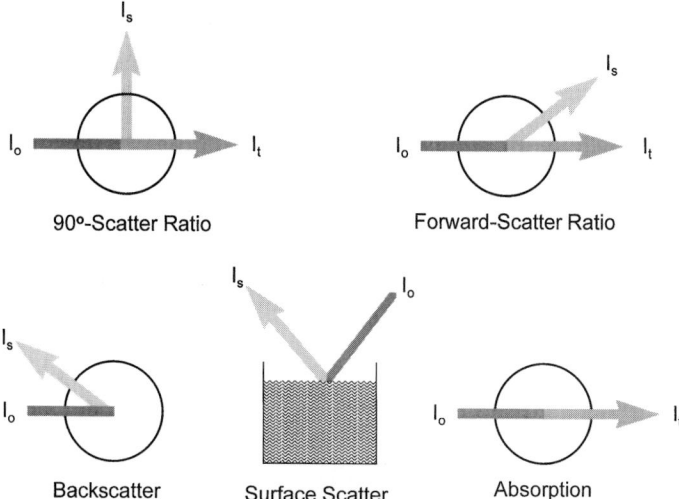

FIGURE 45   Methods of Light Scatter Measurement.

[2]. Anaspect of this method making it particularly suited for low concentrations is that the detector looks at a dark background, as opposed to the forward-scatter method, which has the detector looking at a background lighted from the incident beam. The advantage of this is analogous to the number of stars one can see at night in rural areas versus how many can be seen in a city. The light of the city makes it difficult to see the fainter stars in the sky, whereas they are visible in the absence of city lights. Sensitivity down to 0.001 nephelometric turbidity units (NTU) is achieved with this method.

Forward Scatter

As its name implies, this measures light scattered in the forward direction. Detectors are placed at angles, typically $10°-25°$, to the incident light beam. This method is better suited to the measurement of larger particles since they preferentially scatter light in the forward direction. This method may not be sensitive enough at very low particle concentrations because the small amount of forward-scattered light may be indistinguishable from the transmitted light. Sensitivity between 0.001 and 0.005 NTU is achieved with this method.

Backscatter

This places the light source and detector close to each other. Light scattered back near the source location is measured. This method is best suited for measuring high particle concentrations. The advantage of the method is that it can be performed with a single probe design. Most turbidity instruments require either all or part of the process stream to flow through the instrument. A probe can be inserted through a nozzle into a vessel or process line. The disadvantage to the probe is that there is no compensation for coatings on the probe lens or for changing color hue of the carrier liquid. Some manufacturers offer mechanical in-line cleaning apparatus to solve coating problems. This method has a sensitivity of 0.1 NTU.

Surface Scatter

A method used less frequently is surface scatter [3]. As shown in Fig. 45, a beam of light is focused on the surface of a sample, and light scattered above the sample is measured. The method requires the sample to be at atmospheric pressure.

The absorption method (Fig. 45), is not a true turbidimeter because transmitted light, not scattered light, is measured. Attenuation of the incident radiation may be caused by scatter, carrier liquid absorption, color, particle absorption, and fouled optics. Use of this method may be required for high particle concentrations in which light scatter methods are not suited. In some cases particle concentrations of 60% (v/v) have been be measured.

Other variations in the arrangement of light sources and detectors improve the accuracy and reproducibility of the turbidimeter. The most common variation is the addition of a second detector to measure the transmitted light in addition to the scattered light. The scattered light intensity is compared to the intensity of the transmitted light. This compensates for changes in the incident light intensity, carrier liquid color, and some fouling of the optics. Some manufacturers narrow the wavelength range of incident light to improve the reproducibility of the measurement due to the particle size–wavelength relationship. Descriptions of other light source–detector arrangements as well as additional methods of scattered light measurement can be found in the literature [4], [5].

UNITS OF MEASURE

There are many units of measure associated with the measurement of scattered light. Several of the units depend on the calibration standard used. Some are simply defined differently by various organizations. Table 4 shows one turbidity manufacturer's "rules of thumb" conversion factors between units [6]. Some conversion factors between units are different for various manufacturers. This exists because the correlation between different methods of scattered light measurement is usually not consistent.

Two of the most common standards used are diatomaceous (Fuller's) earth and formazin. The original turbidity unit of JTU, Jackson turbidity units, represents the turbidity caused by suspending specified diatomaceous earth in distilled water. Forward-scattering turbidimeters correlate well to this unit. The formazin standard is a polymeric suspension prepared by reacting hydrazine sulfate with examethylenetetramine [7]. Formazin standards are often used to calibrate instruments using the 90° scatter method.

APPLICATIONS

Turbidimeter applications number in the hundreds because the measurement is relatively simple and versatile. Turbidimeters are usually used to measure one of the following:

1. Particles that should not be in the process stream, such as water in a petroleum product or catalyst in filter effluent.

2. Concentration, of which examples are tomato juice concentration and the concentration of particles in a filter feed line.

3. The detection of changes, such as the onset of crystallization and immiscible interface zones.

4. Blending, such as adding the appropriate amount of filteraid to a filter feed stream.

Abridged lists of applications have been assembled in Table 5 [8].[2,3] These have been separated according to industry.

---

[2] HACH Company, Loveland, CO 80539.
[3] McNab, Inc., Mount Vernon, NY 10550.

TABLE 4 Conversion Table for Turbidity Units in Common Use[a]

Initial Unit	90° White Formazin EBC	90° White NTU	Absorption JTU	Absorption TE/F	90° White Formazin ASBC	Kieselguhr ppm SiO$_2$	APHA mod. ppm SiO$_2$	Mastix ppm
1 EBC corresponds to Formazin Turbidity Units acc. to European Brewery Convention	1	4	4	4	69	10	4	25
1NTU corresponds to Formazin Turbidity Units acc. to American regulations (EPH 90°)	0.25	1	1	1	17.25	2.5	1	6.3
1 JTU or JCU corresponds to Jackson Units acc. to Jackson candle turbidity measurement	0.25	1	1	1	17.25	2.5	1	6.3
1 TE/F or FE corresponds to Formazin Turbidity Units acc. to German standard[b]	0.25	1	1	1	17.25	2.5	1	6.3
1 ASBC corresponds to Formazin Turbidity Units–American Society of Brewing Chemists	0.014	0.058	0.058	0.058	1	0.14	0.058	0.36
1 Kieselguhr unit corresponds to 1 ppm SiO$_2$ acc. to German standard	0.1	0.4	0.4	0.4	6.9	1	0.4	2.5
1 APHA mod. corresponds to ppm SiO$_2$	0.25	1	1	1	17.25	2.5	1	6.3
1 mastix unit corresponds to ppm	0.04	0.16	0.16	0.16	2.75	0.4	0.16	1

[a] First look for initial turbidity unit in vertical column; then align it with the conversion turbidity unit in the horizontal row.
[b] Useful for concentrations below 40 NTU and calibrated with large particles (Formazin) without color effects considered.

TABLE 5 Turbidity Applications

Water and Waste Water Industry	Chemical and Petrochemical	Food and Beverage
Municipal water quality	sodium silicate after tank filter	concentration of tomato juice
Suspended solids in aeration basin	ppm filter aid in 95%	concentration of tea after centrifuge
Return activated sludge suspended solids	cyclohexane 5% polyethane solution	crude sugar syrup turbidity
Digested sludge suspended solids concentration	viscous polymer solutions	tomato pulp separation
	evaporator condensate line to check for carry over	clarity of instant tea
Final effluent clarity	brine clarity or filtration	turbidity of consommé
Filtered lake water clarity	water in liquid hydrocarbon stream	phosphorus content of corn and soybean oil
Monitor intake river water	turbidity of nylon water solution	fruit juice clarity for product uniformity
Drinking water turbidity	check for precipitate	fermentor cell count
Suspended solids in wastewater discharge	supended solids in stream from pressure leaf filter	product filter outlet
Sulfates in wastewater treatment	50% NaOH solution from filter	filter inlet to adjust addition of filter aid
Nonreactive silica in deionized water	hydrocarbon contamiantion in steam condensate	detection of bacterial growth
Influent and effluent streams in sedimentation, coagulation, flocculation processes	clarity of lube oils	
Dishwasher wastewater	monitor silver nitrate solution to control silver content in photographic solution	
	monitor the filtrate from the processing of clay slurries use to manufacture high voltage ceramic insulators	
	monitor paint resins to screen for contaminates	
	detection of iron contamination in dye baths	
	detection of undissolved additives in motor oil	
	iron in steam condensate	
	solids content of white liquor	
	pulp and paper stock consistency	
	lime slurry concentration	
	emulsified oil concentration in cooling lubricants	
	emulsion optical properties	
	catalyst concentration	
	iron content of post-scrubber steam condensate	
	carbon and sulfates in plating bath filtration systems	

SPECIFICATION

Turbidimeter selection is very application dependent, and not all models of turbidity instrumentation will work successfully on every application. Characteristics of the particles and carrier liquid influence which method to use, as was discussed. Most manufacturers require that an application data sheet be completed to help them specify the correct instrument. Table 6 lists application information needed to properly select the correct instrument.

TABLE 6 Turbidimeter Application Parameters

Parameter or Requirement	Requirement
Particle data	Mechanical
Solid, liquid, gas	wetted materials, i.e., metal, optics,
Size/shape	sealing materials compatible with
Refractive index	process
Color	sanitary clean in place (CIP)
Concentration range	required
	FDA approval required
Carrier liquid data	maximum & minimum pressure
Composition	and temperatures
Refractive index	sudden temperature changes such
Color	as thermal shcok aftr CIP and back
pH range	to process
	location of installation, i.e., vessel,
Electrical requirements	process line, sample line
Electrical classification of instrument	line size and pipe specification
location and electronics location	
Power source, i.e., 120 VAC,	Analytical
50/60 Hz, 230 VAC 50/60 Hz, 24 V dc	units of measure
Electronics enclosure requirements,	required reproducibility
i.e., NEMA 4X, air purge, matrial of	required accuracy
construction	satisfy EPA Method 180.1 or ISO
Distance between sensor and	7027
electronics	concentration alarm levels
Output signal, i.e., 4–20 mA	

DESIGN

A good installation design will also help ensure a successful application. The manufacturer's manual usually contains design and installation recommendations. General "rules of thumb" in turbidity instrument design are listed below and shown in Fig. 46.

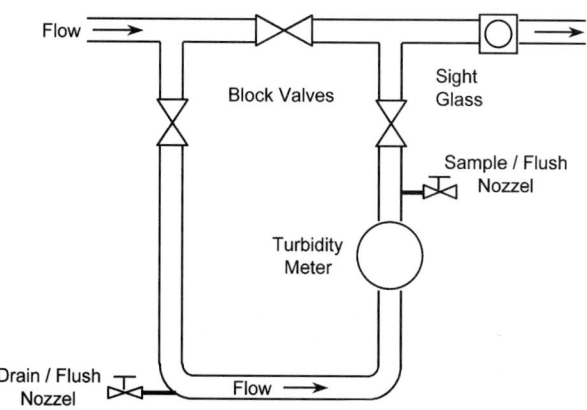

FIGURE 46 Turbidimeter installation.

1. Ensure the measuring sensor will stay liquid full at all times. Mount the sensor so the flow is in an upward vertical direction if possible.

2. Avoid the formation of bubbles in the sensor by keeping pressure on the sample. Locate the sensor before pressure letdown points such as a control valve to prevent degassing.

3. Avoid introducing stray light into the sensor. At low particle concentrations, very little stray light is needed to cause erroneous readings. Locate sight glasses away from the sensor by placing them several feet from the sensor or after pipe elbows.

4. Design for the ability to calibrate the instrument. A turbidimeter installed in a bypass stream with isolation valves allows for flushing the instrument with water for a close to zero calibration check. This also facilitates removal of the instrument for bench calibration or isolation of the instrument during maintenance without disrupting the process. Sample taps located downstream of the instrument provide a way to collect grab samples for lab analysis comparison.

CALIBRATION AND MAINTENANCE

Calibration

Manufacturers are very careful in making sure the instrument is calibrated before it leaves the factory. However, it is a good idea to verify calibration once the instrument has been installed. If a comparative method is used, make sure it uses the same method as that of the on-line meter. Measurements from forward-scatter instruments will not match nephelometric readings. Some of the ways to check calibration are as follows.

1. Compare the turbidity of a grab sample as measured with a lab meter to that of the on-line meter at the time of the sample. If this method is used it is also important that the integrity of the sample does not change. Changes in temperature and pressure may cause precipitation or formation of bubbles.

2. Flush the meter with water and compare the meter reading to that of the water as analyzed by a lab meter. Demineralized water will usually give low ppm readings.

3. Swap with a spare in shop. This method is more work and requires two on-line turbidimeters if continuous operation is required. However, calibration standards can be used with this method.

Calibration checks should be done as often as the user feels is required. Control charts go a long way in determining when the instrument is "out of control." Installation of a bypass loop around the meter will aid in performing the last two methods of calibration checks (Fig. 46). A manufacturer's representative should be consulted before adjusting the calibration.

Maintenance

Examine the output of the instrument daily to verify that the measured values make sense. The signal should be somewhat noisy and not "flat lined." The source light bulb will eventually burn out. Most instruments can sense when this happens, and it should be suspected if the reading suddenly drops to zero for no apparent process reason. Other common problems are listed below.

1. Bubbles are usually caused by the degassing of pressurized liquids. Most turbidimeter electronics can filter the noise caused by bubbles. If the signal noise from severe bubble formation cannot be filtered out, relocation of the instrument or the installation of a bubble trap may be required. The sample must be at atmospheric pressure to use a bubble trap.

2. Stray light is usually caused by a sight glass located too close to the instrument. This will show up as high readings. Covering up the sight glass may help, but relocation of the sight glass farther away from the instrument and preferably past a pipe elbow will solve the problem.

TABLE 7   Refractive Index–Medium Density Relationship

Medium	Density (g/cc)	RI
Air @ 20°C/1 atm	0.0012	1.0003
ethanol @ 20°C	0.7931	1.3290
Water @ 20°C	0.9982	1.3333
Acetic Acid @ 20°C	1.0496	1.3716
Glycerol @ 20°C	1.2633	1.4735

3. Window fouling by the process is compensated to some extent by turbidity instruments ratioing the scattered light to the transmitted light. However, accumulation of material on the process side of the optical window will attenuate the signal to the point where insufficient light is available to perform the measurement. This can be anticipated based on experience with the condition of the pipe wall for existing process lines. Ensuring a fast sample flow will usually prevent solids from building up on the optical windows. Aqueous samples containing high levels of oil and solids may cause problems that only mechanical cleaning can solve. The surface scatter method may solve this problem if the process stream can be open to the atmosphere. Abrasion and chemical attack will interfere with the measured signal by changing the light refraction and transmission characteristics of the optical windows. Properly specifying the window material to match the chemical properties of the process stream cannot be emphasized enough.

4. Moisture condensation on the nonprocess side of the optical windows in high-humidity environments and cold sample temperatures is the most common cause of significant measurement error. Even small amounts of moisture can cause an interference error of over 50% of the signal. Purging the optics with dry instrument air prevents condensation by removing the moisture from the optical components.

5. Low readings will occur if the particle concentration is too high for the method. As shown in Figure 3, absorption by too many particles causes the scattering of light to decrease even though the concentration of particles is increasing. Continuous operation in this range may require using the absorption method. Darkening of the liquid color can also affect the reading. Turbidimeters using the ratio measurement method compensate for this. However, if the liquid is too dark, there may be too little light available to reach the detectors.

6. Variations between instruments can be caused by comparing measurements from two different method turbidimeters. Also, using the incorrect conversion factors between units, that is, ppm, NTU, and JTU, can be a source of error. The manufacturers will know the correct conversion units and which methods can be compared to their instrument.

REFRACTIVE INDEX

The refractive index is defined as the ratio of the speed of light in a vacuum to the speed of light in a medium. Light travels slower in dense media, which increases the value of the RI. Table 7 shows this relationship for a few media.

When a light ray passes from one medium to another, not only will it change velocity based on the change in density, but its direction will change as well if the angle of incidence is other than 90°. Snell's law describes this change in direction at a single interface between two media with different refractive indices:

$$n_2/n_1 = \sin i / \sin r \qquad (6.6)$$

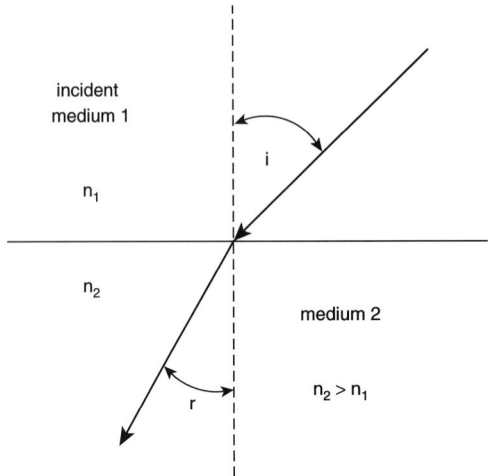

FIGURE 47   Refractive index.

where n_1 = refractive index of incident medium
n_2 = refractive index of second medium
i = angle of incidence
r = angle of refraction

The angle of incidence and refraction is measured from the normal to the interface. A ray passing from low to high RI media is bent toward the normal, as shown in Fig. 47. The ray is bent away from the normal when passing from high to low RI media. Since the incident medium is usually air, n_1 is very close to unity and the ratio of n_2/n_1 is simplified to n_2 or just n, the refractive index. The RI of a substance was one of the first physical methods used to identify a chemical substance and measure its concentration. Abbe and Pulfrich refractometers were among the first precision laboratory instruments utilizing the method. Most process refractometers measure the critical angle to determine the refractive index of a solution.

CRITICAL ANGLE MEASUREMENT

Figure 48 illustrates the principle of critical angle. When a light beam passes from a medium of higher RI to one of less RI at angles other than 90°, light will be either refracted or reflected at the medium

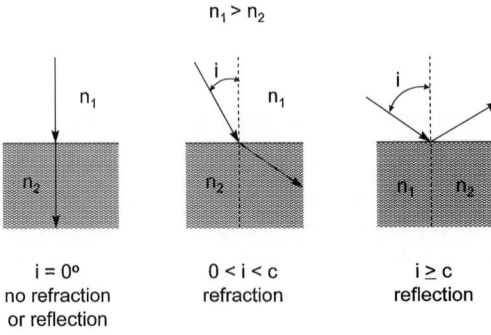

FIGURE 48   Critical angle.

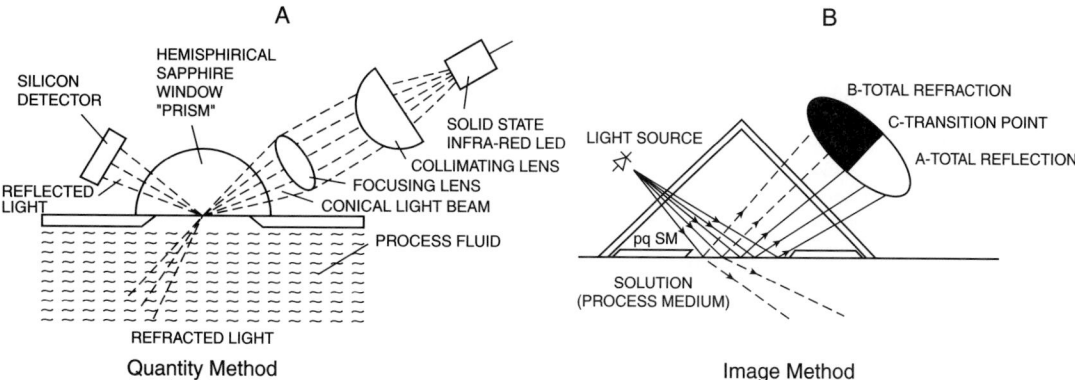

FIGURE 49   Methods of measuring critical angle.

interface. Light incident on the interface at angles smaller than the critical angle will pass through the interface and be refracted away from normal. Other light incident on the interface at angles equal to or greater than the critical angle will be reflected back into the incident medium.

Process refractometers measure the critical angle by one of two methods as shown in Fig. 49. The amount of reflected light is proportional to the RI. Measuring the quantity of the reflected light indirectly measures the RI (Fig. 49A) [9]. In Fig. 49B [10], the image of the refracted and reflected light is detected by a photosensitive diode array. The position of the transition point is a measure of the critical angle and thus the RI of the solution.

Another method of measuring RI uses a microchip interferometer permanently attached to the end of an optical fiber. The sample enters a cavity at the end of the optical fiber probe. As the sample's RI changes, so does the effective path length. The extent of spectral modulation is measured by the electronics and correlated to the refractive index of the sample.[4]

A method used more by laboratory refractometers is called the differential method. It is not an in line method and requires that a clean sample stream be presented to the analyzer. A light source is focused through the sample cell, which contains a reference material and the sample. The light beam passes through the cell twice as it is reflected back by a mirror. The change in the angle of the reflected light beam is a measure of the refractive index.

SPECIFICATION

The actual RI value is usually not of interest in an application, but the concentration of a component. Refractometers measure the RI and correlate it to a component's concentration. The RI of a solution is the sum of the RI of the individual components multiplied by their respective concentrations as expressed below.

$$RI_{solution} = RI_1 C_1 + RI_2 C_2 + RI_3 C_3 + \cdots$$

where $RI_1, RI_2, RI_3, \ldots$ = refractive indexes of components 1, 2, 3, $\ldots$
$\qquad$ $C_1, C_2, C_3, \ldots$ = molar concentrations of components 1, 2, 3, $\ldots$

As with any physical property measurement, RI only works with two-component mixtures. The presence of a third component will invalidate the correlation between RI and concentration of the component of interest. Exceptions are if the third component's concentration does not change within

---

[4] Photonetics, Peabody, Mass. 01960.

TABLE 8   Refractometer Specifications

Method	Max Press	Accuracy	Temperature Range
Critical angle (Quantity)	68 bars	0.5% of span	-25–$150\,°C$
Critical angle (Image)	25 bars	0.25% of span	0–$180\,°C^a$
Fiber optic	345 bars	0.1% of span	-270–$300^†\,°C$
Differential	8 bars	0.01% of span	30–$50\,°C$

[a] With cooling; w/o cooling max temp = 115 °C
[b] 6 to 8 pH continuous

TABLE 9   Refractometer Application Data

Analytical Data	Process Conditions	Electrical Requirements
Material to be measured	acceptable materials of construction	electrical classification of probe location and electronics location
Concentration range (span)	pipeline or vessel installation	
Units of measure	pipeline or vessel wall size	supply voltage and frequency
Required accuracy	sanitary clean in place required	electronics enclosure requirements, NEMA 4X, air purge, material of construction
RI data	FDA approval required	
pH range	max/min temperature	distance between probe and electronics
Concentrations of other stream components	max/min pressure	output signal, i.e., 4–20 mA
	flow rate past probe	

the required accuracy of the measured component or if the RI of the third component is the same as the other unmeasured component of the solution.

Bubbles and particulates can affect the RI measurement. If they adhere to the prism surface and cover it, the refractometer is isolated from the solution. Manufacturers offer washing options to clean the prism when these conditions exist. High concentrations of bubbles and particulates can cause refracted light to be scattered back into the prism. If the critical angle is measured by the quantity of reflected light (Fig. 49A), backscattered light may influence this method.

Solution density is a function of temperature, making RI temperature dependent. In general, the RI decreases 0.0002 units for every $1\,°C$ increase in temperature. Refractometers must be temperature compensated unless the solution temperature does not vary within very narrow limits. Most manufacturers will require a temperature range over which to compensate the measurement.

Process refractometers measure between 1.30 and 1.65 RI units. However, the RI range for an application will usually not be over the entire range. Process pressure, temperature, and composition all place restrictions on the application. Table 8 lists some limitations for the different types of refractometers. Of particular note is the need for cooling by using the critical angle image method when the process temperature exceeds $115\,°C$. Providing the manufacturer with accurate application data (Table 9) will help ensure a successful application.

APPLICATIONS

Use of the process refractometer spans all industries. Because of its rugged design and high degree of accuracy, applications range from the purity of water to the concentration of harsh chemicals. The ability to ignore undissolved solids and bubbles gives the refractometer an advantage over other optical methods. The probe design eliminates the need for a sample system and greatly reduces maintenance costs.

TABLE 10   Refractometer Applications

Food and Beverage	green syrup and molasses
juice	desugarization of molasses
juice concentrate	decolorization
alcohol distillation	separation
beer blending	vacuum pan
wort measurement	recovery pan
citric acid	Textile and Fibers
gelatine	textile sizing
monosodium glutamate	textile ultrafiltration
sucrose	dimethyl acetamide
fructose	polyamide
dextrose	Pulp and Paper
jams and jellies	washing control
soft drinks	evaporation
tomato paste	recovery boiler
salt concentration	green liquor
sugar concentration	paper sizing process
concentration of liquids	black liquor
purity of rinsing water in pipes	red liquor
Diary	white liquor
membrane filtration	starch
reverse osmosis	coatings
ultrafiltration	resin
ice cream	tall oil
Sugar Industry	Chemical
extraction	ammonia
evaporator control	formaldehyde
crystallization	phenol process
control	styrene in ethylbenzene

A common application of process refractometers in the Food and Beverage industry is the measurement of dissolved sugar concentration or % Brix. The Pulp and Paper industry uses refractometers to measure dissolved solids in many of its liquor streams. In the Chemical and Petroleum industries, refractometers are used to control product quality and measure the concentration of many binary solutions, especially from distillation columns. Table 10 lists more applications of process refractometers in the various industries.

INSTALLATION GUIDELINES

1. Mount the probe in a pipeline where the flow is highest and turbulent.
2. The fluid flow should be upward in a vertical pipeline installation.
3. The probe should be installed on the side of a pipe in horizontal pipeline applications.
4. Install the probe before a control valve or after a pump rather than after a control valve or before a pump to reduce sedimentation and air-trapping risks.
5. Installing the probe on a vessel recirculation line is preferred to mounting it in the vessel. If the probe must be mounted in a vessel, the location should be where maximum mixing is ensured.
6. Install a prism wash system if the possibility of a coating problem exists.
7. Mount the probe in a location of highest temperature to help keep the prism clean.
8. Mount the probe in an easily accessible location.
9. Install a sample point just downstream of the probe.

CALIBRATION AND MAINTENANCE

Process refractometers come precalibrated from the manufacturer. However, unless the application is well defined, field adjustments to the calibration will probably be necessary. Adjusting the zero of the refractometer to match lab measurements is almost always required. If the slope of the calibration curve is incorrect, recording the response of the refractometer, the process temperature, and the sample concentration at $\sim$7 concentration values is usually sufficient for the manufacturer to recalculate a new set of calibration coefficients.

Maintenance of process refractometers is usually minimal because of no moving parts. The instruction manual of the instrument describes procedures specific to a particular refractometer. Maintenance items common to probe-type critical angle refractometers are the coating of the prism, prism wash system failure, and light source failure.

REFERENCES

1. Clevett, K. J., *Process Analyzer Technology,* J. Wiley, New York, pp. 558–561, 1986.

2. Byrnes, J., and A. Valentine, *The Effect of Measurement Angle and Particle Size on Scattered Light in the Beer Line,* McNab Inc., Mount Vernon, New York 10550.

3. Sadar, M. J., *Turbidity Science Technical Information Series,* Booklet No. 11, Hach Company, Loveland, Col., p. 25, 1998.

4. Liptak, B. G., *Instrument Engineers' Handbook, Process Measurement and Analysis,* 3rd ed., Chilton, Radnor, Pa., pp. 1230–1235, 1995.

5. Lex, D., "Turbidimeter Technology Turns On the High Beam," *Intech,* pp. 36–38, June 1994.

6. HSB Instruction Manual, McNab, Incorporated, 20 North MacQuesten Parkway, Mount Vernon, New York 10550, pp. 2–4, copyright 1992.

7. ASTM Method D 1889–94, Section 10.

8. M. A. Betts, "Turbidimeters Solve Suspended Solids Monitoring Problems," *I&CS,* pp. 75–78, May 1998.

9. Model 725, *Process Refractometer Instruction Manual,* p. 2, Liquid Solids Control, Inc., P.O. Box 259, Upton, Mass. 01568, 1998.

10. *Instruction Manual for PR–01–S,* p. 5, K-Patents, Inc., 1804 Centre Point Circle, Suite 106, Naperville, Ill., 1996.

ULTRAVIOLET/VISIBLE ABSORPTION ANALYSIS

by R. S. Saltzman[*]

INTRODUCTION

Process stream analyzers based on the measurement of ultraviolet/visible (UV/Vis) radiation absorption are used throughout the process industries for quality control and for monitoring (and controlling)

[*] Saltzman & Associates, Wilmington, Delaware.

the concentrations of components in both gas and liquid streams. In contrast to a laboratory UV/Vis spectrophotometer, the plant analyzer is designed to be used in a plant environment, usually for the continuous or semicontinuous analysis of a specific component. The scanning spectrophotometer that reads absorbance (or transmittance) versus wavelength is a very useful laboratory tool. It can be utilized both to identify the components of a mixture and to determine the concentrations of some of the components, but it is not suited for reliable plant use (unless fiber optics are used and the spectrophotometer is located in a laboratory environment). However, process spectrophotometric analyzers based on photodiode array technology (described later) have been developed and are being successfully used for multicomponent analyses. The UV/Vis absorption patterns of compounds are generally not as distinctive as an infrared "fingerprint" absorption, and fewer compounds absorb in this region than in the infrared. However, several important classes of compounds absorb in the UV/Vis region, whereas water and the usual components of air do not absorb in this region. Hence, UV/Vis analyzers may be more selective and sensitive than infrared and other types of analyzers on many plant stream applications.

UV/VISIBLE ABSORPTION SPECTRUM

The region of the electromagnetic spectrum covered by UV/Vis absorption analyzers described here is from 190 to 690 nanometers (nm) and excludes the "vacuum" UV, below 190 nm. Absorption in this UV/Vis region, corresponding to photon energies between 6.44 and 1.77 eV, is due to the excitation of the more loosely bound valence electrons, including the "unsaturation" electrons of multiple bonds and the unpaired electrons of the free radicals.

The spectra of molecules generally are observed as broad bands when compared with the line spectra of atoms, because each electronic state has associated with it numerous vibrational and rotational energy levels. The vibrational-rotational fine structure of the bands is not resolved except for gas-phase spectra of light simple molecules at low total pressure. For example, ammonia and sulfur dioxide at low pressure show fine structure when studied at high resolution. Also, the absorption spectra of the simpler aromatics, such as benzene and cumene, show a distinct fine structure, even in the liquid phase. In general, however, pressure broadening and solvent interaction make the spectrum appear as the envelope of the otherwise expected fine-structure spectrum.

The source must provide the desired ultraviolet wavelength and may either be a line source, such as a mercury arc, or a continuous source, such as a hydrogen or deuterium arc. Tungsten and tungsten-iodine lamps also may be used at longer ultraviolet wavelengths. Optical filters or a spectral dispersing system are used for screening out unwanted radiation.

BEER'S LAW—CONCENTRATION VERSUS ABSORPTION

UV absorption data usually are tabulated and plotted in terms of absorptivity as a function of increasing wavelength (Fig. 50). The unit of wavelength in the UV/Vis region is the nanometer (nm) [1 nm $(10-9)$m $= 10$ Angstroms], which is the same as the millimicron $(m\mu)$ previously used extensively in analytical chemistry literature. The most common unit of absorptivity is liters per mole-centimeter. A material with absorptivity of unity at a specific wavelength has an absorbance of 1.0 in a path length of 1 cm when the sample concentration is 1.0 mole/liter.

The absorbance of a substance is directly proportional to the concentration of the material that causes the absorption in accordance with the Lambert–Beer law, more commonly referred to simply as Beer's law:

$$A = abc = \log I_o/I = \log 1/T \tag{6.7}$$

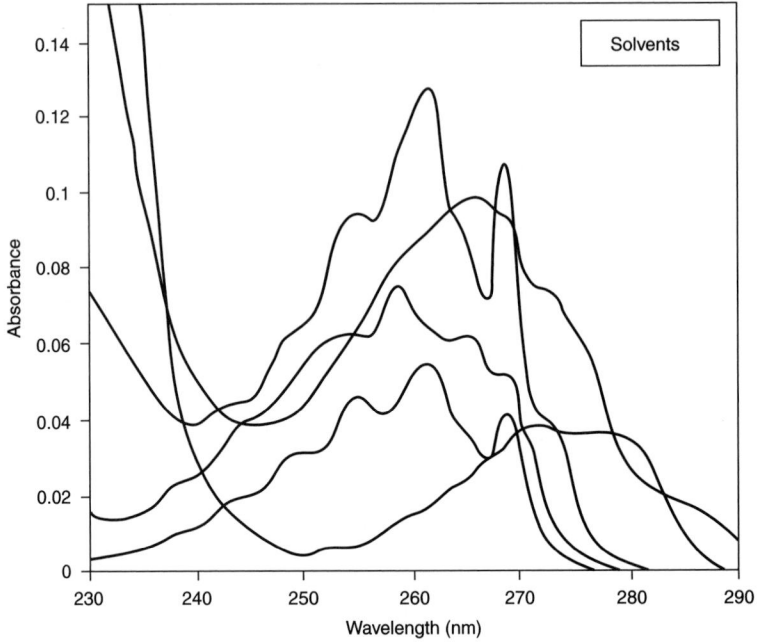

FIGURE 50   Absorbance vs. wavelength spectrum.

where A = absorbance
$\quad\quad a$ = molar absorptivity [liter/(mole) (cm)]
$\quad\quad b$ = path length [cm]
$\quad\quad c$ = concentration [mole/liter]
$\quad\quad I_o$ = intensity of radiation striking the detector with nonabsorbing sample in light path
$\quad\quad I$ = intensity of radiation striking the detector with concentration c of absorbing sample in light path b
$\quad\quad T$ = transmittance

For the vapor phase,

$$A = \frac{abc'}{2450} \quad \text{at 25°C and 760 Torr pressure} \tag{6.8}$$

where c' = volume percent or mole percent

$$A = \frac{abc'(P + 14.7)}{14.7} \times \frac{298}{(t + 273)} \tag{6.9}$$

at any temperature or pressure, where P is pressure in psig, and t is temperature in °C.
   For liquid phase,

$$c = \frac{(c''d) \times 10}{\text{M. W.}} = \text{mole/liter} \tag{6.10}$$

where c'' = weight percent in liquid
$\quad\quad d$ = density of liquid
$\quad$ M. W. = molecular weight of material to be measured

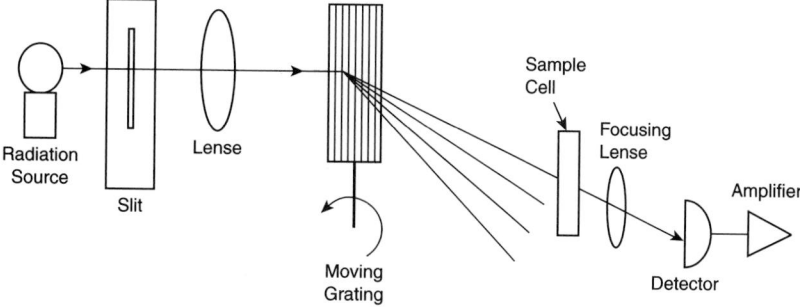

FIGURE 51   Absorption analyzer hardware configuration.

and hence

$$A = \frac{10 \times abc''d}{\text{M. W.}} \tag{6.11}$$

UV/VIS PROCESS ANALYZERS

The basic UV/Vis absorption analyzer consists of a radiation source, grating, sample cell, detector, and output meter (Fig. 51).

A transmittance measurement is made by calculating the ratio of the reading of the output with the sample in the cell to the reading with the cell empty (without UV/V is absorbing material). The concentration can be calculated from the known absorptivity of the substance by means of Beer's law [Eq. (6.7)] or may be obtained by comparison with known samples. Process diode array analyzers have computers with chemometric software programs and can be calibrated to provide concentration measurements of multiple components, even with overlapping absorbances. The source must provide the desired UV or visible wavelength and may be either line sources such as mercury discharge lamps, or continuous sources, such as deuterium or xenon discharge lamps. Tungsten and tungsten-iodine lamps also may be used at longer UV and visible wavelengths.

Optical filters or a spectral dispersing system are used for screening out wavelengths emitted by the source. The sample cell is equipped with windows that are transparent at the chosen wavelengths, and the path length between windows must be fixed. UV/Vis detectors may be photomultipliers, vacuum phototubes, photodiodes, or photodiode arrays. (In modern analyzers, microprocessors usually convert the detector outputs to digital signals and provide both digital and analog concentration outputs.)

Plant stream UV/Vis analyzers incorporate elaborate variations of these basic elements. Simple modifications of laboratory instruments are rarely adequate to meet the special requirements for continuous monitoring and control of process streams. A few basic design requirements must be met in nearly every case. For example, plant instruments must be rugged and serviceable. They usually cannot be, and are not, treated as carefully as laboratory types. Instruments in the field will be exposed to plant fumes and vibrations and must meet the electrical classification of the area. Maintenance periods, except for routine checks and adjustments, should be months apart, and the design should be as simple as possible with components readily accessible for maintenance. Concentration readings usually are accurate to ($\pm$)l% of the full-scale reading. However, for gas analyzers, without absolute pressure compensation, accuracy may be limited ($\pm$) 2%. If the analyzer is used in an automatic control loop, high reproducibility with a minimum of dead zone may be essential, and the instrument usually must rapidly respond to changes in concentration. UV/Vis analyzer response times are generally fast, so delays through long sampling lines must be avoided. The analyzer must be easy to calibrate. Calibration based on known values of absorptivity is easiest and accurate if the analytical radiation is monochromatic and unchanging in wavelength. Several types of process

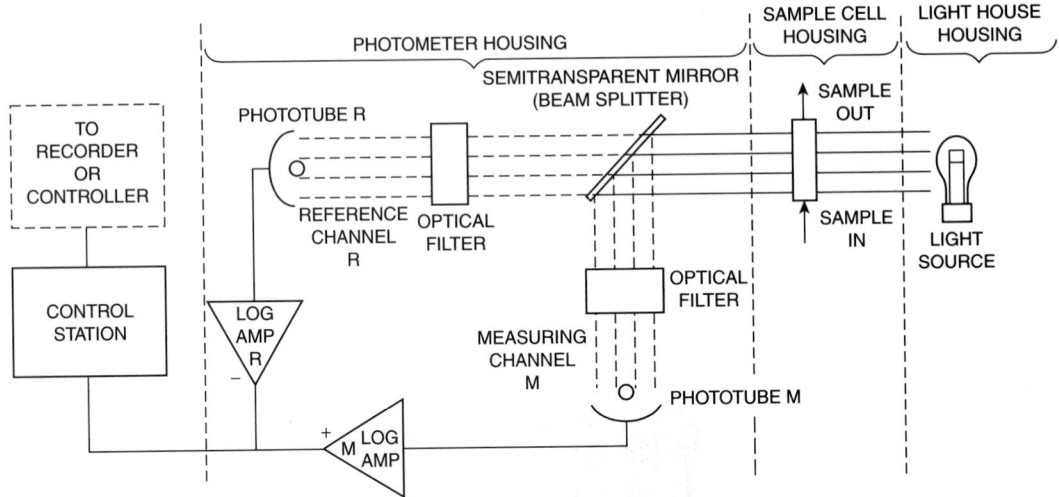

FIGURE 52 Functional diagram of split-beam UV/Vis analyzer.

analyzers based on UV/Vis absorption are commercially available. These instruments vary from very simple to highly sophisticated designs. The types of analyzers described in this section are not the only commercial UV absorption analyzers, but they are the most prevalent units used in industrial applications in the U.S.

Filter photometric analyzers are the most common of process UV/Vis analyzers, and the simplest is the single-beam type. The output of this type of instrument will be affected by fluctuations and drift of the light source, dirt or bubbles in the sample cell, and any drift in the detector or detector circuit. As a result, single-beam instruments must operate on relatively low sensitivity (high absorbance) levels to provide reasonably stable analyses. Stabilized light sources and improved detector circuits, in recent years, have improved stability to some extent and these simplified units have found extensive applications. A functional diagram of a typical split-beam analyzer is shown in Fig. 52.

This type of design is based on a simultaneous differential absorption measurement at two wavelengths. It has several distinct advantages, including compensation for dirt and bubbles in the sample cell. The most widely used split-beam analyzer is the AMETEK Model 4000 (previously the Du Pont 400) photometric analyzer. Radiation from the source, generally a gas discharge lamp, is partially absorbed in passing through the sample. The radiation leaving the sample is divided into two beams by a semitransparent mirror. Each beam passes through an optical filter to a photodiode. The filter removes radiation at all wavelengths except the one to be measured. Radiation at the analytical wavelength striking one photodiode is absorbed strongly by the component whose concentration is being measured. Radiation at the reference wavelength, directed to the second photodiode, is absorbed weakly or not at all by the component. Each photodiode develops a current directly proportional to the intensity of radiation striking the photodiode.

The photodiode current is converted to a dc voltage directly proportional to the negative logarithm of the photodiode current by a special logarithmic amplifier. The output voltages from the measuring and reference amplifiers are fed to a microprocessor, which provides digital and analog outputs calibrated in concentration units. This split-beam photometric analyzer has separate light source, sample, and photometer housings. The three housings of the analyzer are easily separated. Because of this, the analyzer is readily adapted for monitoring process gas and liquid streams in pipeline cells installed directly in the process line. The photometer unit may be mounted a considerable distance away from the light source. For example, a pipeline cell 1 m long has been used in some cases with this analyzer. This construction permits the use of special cells such as those used with high- or low-temperature gas and liquid samples. Figure 53 is a functional diagram of a filter-wheel chopper photometric analyzer.

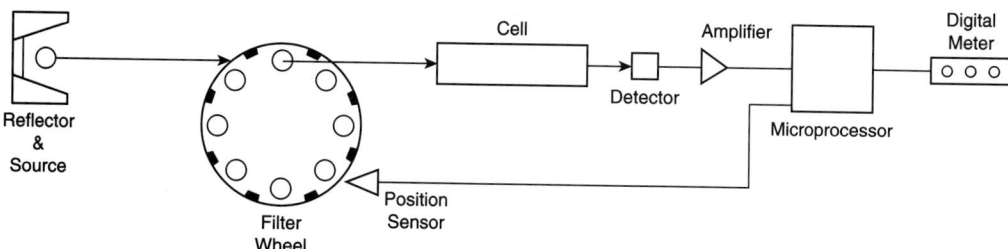

FIGURE 53 Functional diagram of a filter-wheel-chopper UV/Vis analyzer.

The filter wheel contains two or more interference filters (at different selected wavelengths) continuously rotated by the chopper motor sequentially into the optical path. The source light is collimated through the filters and the sample cell by a series of lenses, and a detector lens focuses the transmitted light on the detector. The detector provides electrical signals proportional to the intensities at each wavelength. As in the split-beam analyzer, two wavelengths are used for a single concentration measurement; however, if more filters at selected wavelengths are used, multiple concentrations may be measured. The analyzer microprocessor is programmed to compensate for overlapping absorbances. The functional diagram is typical of the ABB Process Analytics analyzer and several others.

The Western Research analyzer uses two separate hollow cathode sources with a system of mirrors and beam splitters to provide several discharge line wavelengths for multiple analyses (Fig. 54). Because of the low intensity emitted by these sources, a photomultiplier tube is used as the detector.

Process scanning UV/Vis spectrophotometers have not found appreciable applications in the process industries because of their relative complexity and sensitivity to vibrations and temperature variations. A few (such as the guided wave analyzer) have found use in the very near UV and visible regions, where fiber optics can be readily used, and the spectrophotometer is installed in a laboratory like environment.

Figure 55 is a functional diagram of a process diode array spectrophotometer. The light source, usually a deuterium or a pulsed xenon lamp, broadly emits UV and visible radiation. The light transmitted through the sample cell is collimated either directly with lenses and/or mirrors or through fiber optics. From the sample cell, the light is passed into the detector housing (directly or through fiber optics) and is focused onto the slit. From the slit, the light expands onto a holographic grating that disperses the light onto the linear photodiode array detector (LPDAD). The LPDAD is precisely positioned with respect to the grating to intercept a selected segment of the UV/Vis spectrum, resulting in each diode receiving light at a fixed number of, or fixed fraction of, nanometers away from adjacent diodes. The outputs of all of the diodes and thus all the wavelengths are read at essentially the same time. The photodiode array analyzer has no moving parts (except for a chopper when a continuous deuterium lamp is used), so it is well suited for process analyses. A computer is an integral part of the photodiode array analyzer and often can be calibrated, when programmed with chemometrics software, for accurate multiple analyses even with severely overlapping absorbances.

SAMPLING SYSTEM CONSIDERATIONS FOR UV/VIS ANALYZERS

Over 80% of the problems associated with analyzer installations are found in the sampling systems. With an analyzer designed to minimize sample handling problems, difficult analyzer applications can often be handled reliably with simplified sampling systems. Sample system considerations include the following.

1. Materials of Construction.   Components in sampling systems usually cannot be corroded to the same extent as is acceptable in the process. For long-term reliability, sample lines, critical valves,

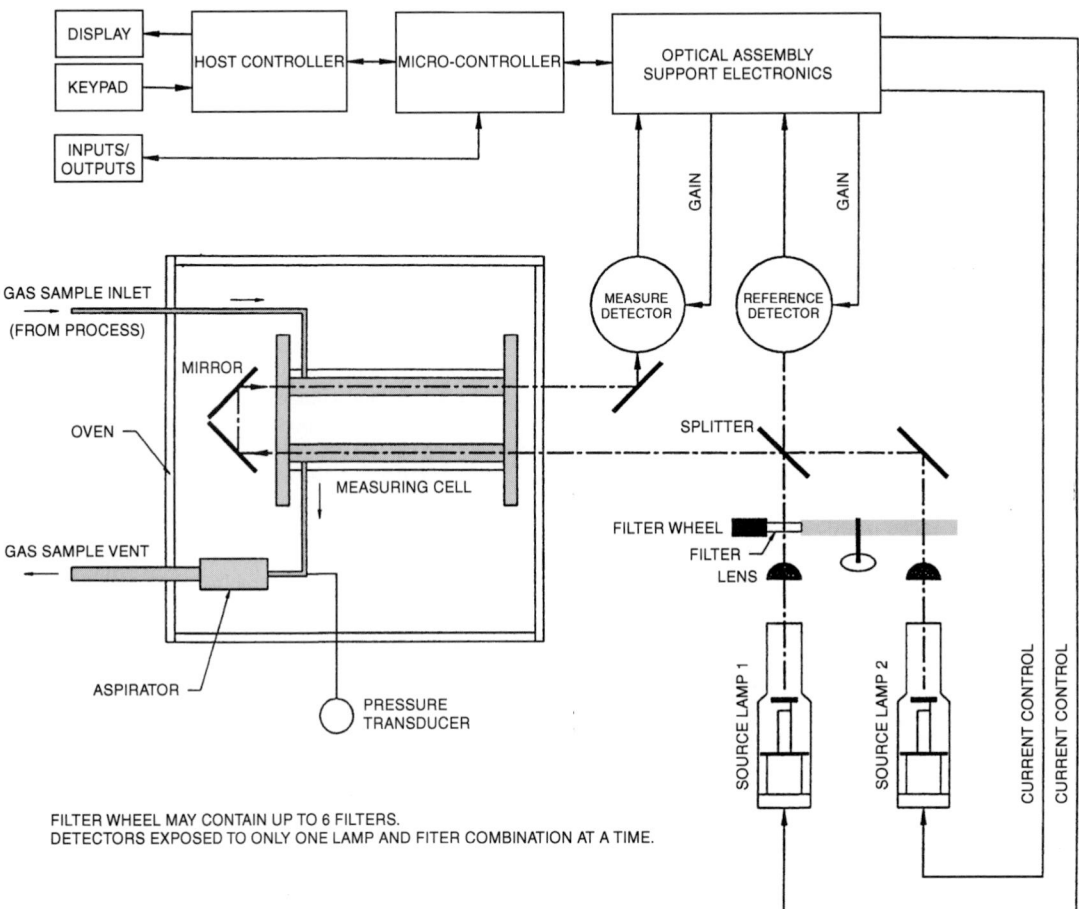

FIGURE 54   Hollow cathode source analyzer.

and other components of the sampling system should be made of highly corrosion-resistant material. Corrosion products can act similarly to a chromatographic column and adsorb the measured sample component and produce erroneous analyzer results.

2. Filtering.   Sample filtration must be adequate so that particulates in gas sample streams and suspended particles in liquid streams can cause no more than a 1% of full-scale deviation in the analyzer response. In analyzers using a split beam or two-wavelength measurement where compensation for particulates and haze is provided, filtering requirements are not nearly as stringent.

3. Traps.   Traps can be very useful in sampling systems to remove any condensate in gas samples and to remove gas bubbles in liquid samples. A trap in its simplest form consists of a "tee". In removing liquids and solids from gas streams, the sample enters the tee in a horizontal leg, and the sample to the analyzer flows up the vertical leg and the bypass stream down the vertical leg of the tee. In the removal of bubbles from liquid streams, the sample to the analyzer flows down from the vertical leg and the bypass stream up the vertical leg of the tee.

4. Sample Switching.   Automatic valves for sample switching are among the most common problems in sample system design. "Block-bleed-block" valving systems or the equivalent should be

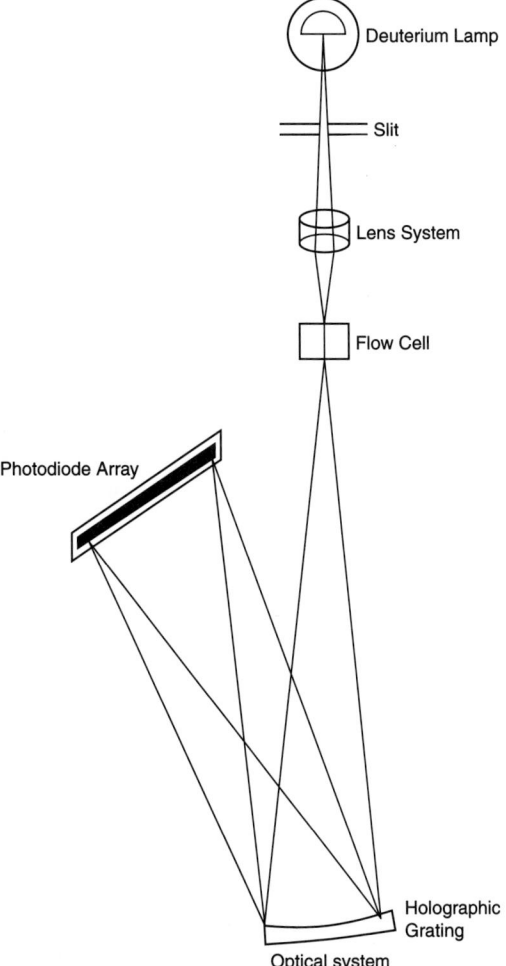

FIGURE 55 Process diode array analyzer.

used in any critical analyzer applications to ensure prevention of valve leakage and resultant erroneous analyses. One analyzer per stream is preferred for highest reliability.

   5. Sample Flow.   It is best to eliminate pumps for sample motive force. The most trouble-free method of moving the sample is to draw the sample from a point in the process at higher pressure than the discharge pressure of the analyzer. Generally, aspirators with vacuum break or pressure control systems have worked more reliably than pumps.

   6. Temperature.   When hot vapor samples are analyzed, the whole sample piping system must be kept at an appropriately elevated temperature to prevent condensation. The presence of condensate in the sample lines introduces response lags, which make the analyzer inaccurate or even useless. The condensate slowly dissolves and releases constituents of the sample to equilibrate with the instantaneous gas mixture. This condensate thus tends to level out rapid changes in concentration. When the sample is at high or low temperature, the cell temperature should be separately controlled to avoid temperature sensitivity dependence on flow rate.

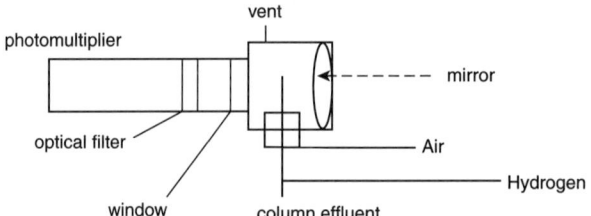

FIGURE 56   Flame photometric detector.

INFRARED ABSORPTION ANALYSIS

Infrared radiation stimulates molecular vibrations and produces a spectrum related to structure. [Scanning, selective detector (Luft type), filter photometer, Fourier transform, multiple detectors] The filter photometer has a much higher signal-to-noise ratio than FTIR or dispersive-type IR spectrophotometers. Proper design will produce a higher energy throughput and signal processing, which significantly reduces the noise (See Process Control Improvement Chapter, Section 10).

PULSED FLAME PHOTOMETRIC DETECTOR

The assets of this detector are high sulfur and phosphorus sensitivity, operational stability, and low maintenance. The liabilities are hydrocarbon quenching, hydrogen, and flame presence (Fig. 56).

Flame photometric detectors have been utilized primarily for sulfur measurement, but there is a tremendous potential for measurement of many atoms that can be excited by a hot flame. How hot must it be? Hydrogen/air, H_2/O_2, acetylene/air, and nitrous oxide/O_2 mixtures have increasingly higher temperatures (i.e., energy). The sample can be fed into the flame on a continuous basis, or it can be introduced in a discrete packet as is done for column separation and chronographic monitoring. One of the major reasons for expenditure of development funds was to provide a means for measuring the quantity of each different sulfur species in petroleum products. While the detector is selective for the target analyte, the large quantity of hydrocarbon caused quenching of the flame and degraded the operation of the detector. One approach to fixing this was to utilize a double flame. The first gets rid of the quenching and the second excites the sulfur atoms so they can be detected by a photomultiplier.

A very recent development in flame emission measurement is the pulsed flame photometric detector. It is based on a fuel concentration that cannot sustain continuous flame operation. The flame is reignited on a periodic basis, thus producing pulsed emissions. Since the combustion of the hydrocarbon molecule is very exothermic, fast, and irreversible, heteroatom species such as H_2S, HPO, HNO, and the like emit at the cooler yet reactive postpulsed flame conditions. Their flame emissions are electronically gated and separated in time from the hydrocarbon emission, thus providing elemental selectivity.

X-RAY FLUORESCENCE AND ABSORPTION

The assets are multiple element analysis, reconfigure application with software, 0.1–100%, it requires only electricity for utilities, and it has simple operation. The liabilities are that it is not so good for trace levels, not for light elements below sulfur (see Fig. 57).

X-Ray Fluorescence analyzers have been utilized to measure elemental concentration in mining ore slurries, but the analysis requires a computer to correct for complex matrix effects. This simple system works very well for monitoring one or two elements (catalysts) in water or an organic liquid, where the matrix effects are trivial. (See Process Control Improvement Chapter, Section 10, for details).

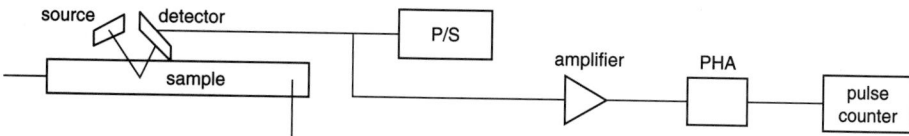

FIGURE 57   X-ray fluorescence and absorption.

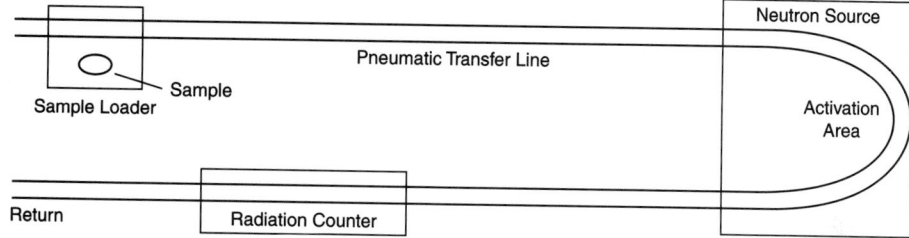

FIGURE 58   Neutron activation analyzer.

NEUTRON ACTIVATION EMISSION

The assets are low level detection limits, hard radiation free of interference, and a fast multielement. The liabilities are neutron source shielding, and the neutron source can be expensive (see Process Control Improvement Chapter, Section 10 for details). See Fig. 58.

NUCLEAR MAGNETIC RESONANCE ABSORPTION

The assets are that it is specific to proton [H+] molecular environment, it has a single response factor, and it is noncontact. The liabilities are the limited experience, and it is very expensive. (See Process Control Improvement Chapter, Section 10 for full details).

CHEMILLUMINESCENCE

This is light emitted from a chemical reaction. The assets are its very low detection limits and extreme specificity. The liabilities are that limited species apply, and there is complicated operation and maintenance. See Fig. 59.

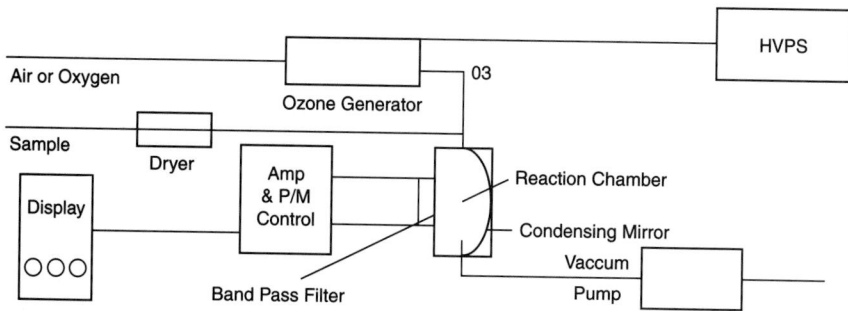

FIGURE 59   Chemilluminescence analyzer.

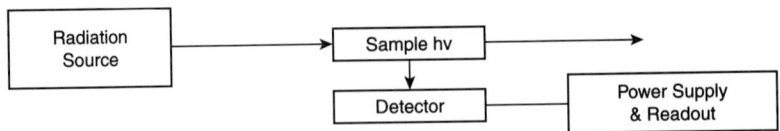

FIGURE 60   Fluorescence analyzer.

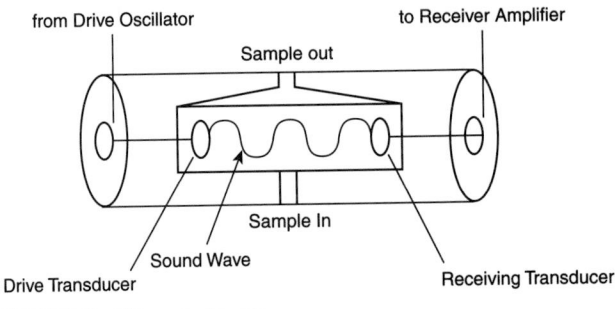

FIGURE 61   Ultrasonic detector.

RADIATION STIMULATED FLUORESCENCE AND PHOSPHORESCENCE

The assets are the very low detection levels; the liabilities are the limited number of molecules that respond without chemical reaction. See Fig. 60.

ULTRASONIC

The assets are the universal detector, 1 ppm, and linearity 10 (+6). The liabilities are that it requires a vapor sample, and it has a temperature limit of 200°C. See Fig. 61.

IONIZATION CONCENTRATION TRANSDUCERS

Flame

For these, hydrocarbons are burned in an air/hydrogen flame to form carbon ions; multiple H–C or C–C bonds are required to achieve high sensitivity. The assets are the primarily carbon counting selectivity, ppb sensitivity, and 10(+7) linearity. The liabilities are that it does not respond to most noncarbon compounds, and it requires clean air and hydrogen. See Fig. 62.

The operating mode is amperometric and the operating technique is direct current–high voltage. Hydrogen is mixed with the column effluent and directed through a narrow jet to the combustion chamber. Air flows up around the jet to support the burning of the hydrogen and any organic material eluting from the separation column. Carbon ions formed in the flame are attracted to the collector, causing a current to flow through an electrometer circuit. The electrometer is a very high impedance input amplifier, and the current is very small [10(−9) to 10(−12) amp].

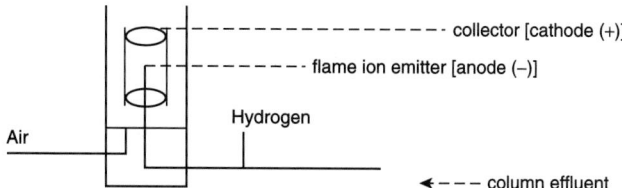

FIGURE 62   Flame ionization detector.

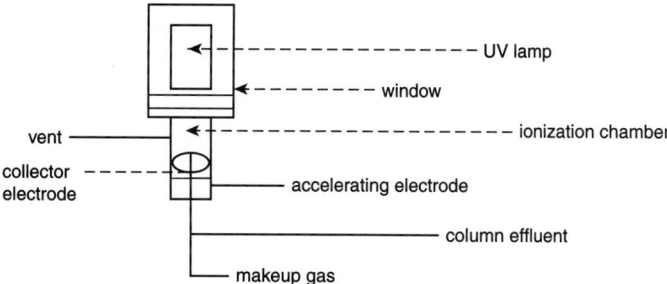

FIGURE 63   Photo ionization detector.

Photo

These are low ionization potential molecules (benzene, nitric oxide, H_2S, NH_3, etc.). The assets are no fire; no fuel and air required; it is more selective and sensitive than FID; it is simple, rugged, and safe; and it is low maintenance. The liabilities include drift due to source lam aging, lamp burn out, window fogging, and water reduces response. See Fig. 63.

The operating mode is light absorption and the operating technique is an ion-counting electrometer. A UV lamp (9.5, 10.2, or 11.8 eV) is utilized to ionize molecules in the separation column effluent as opposed to the flame method [R + hv → R(+) + e(−)]. Ions formed are collected by polarized electrodes. Only molecules having an ionization potential less than the energy of the photon will be ionized. Three different lamp energies are shown above, but only the 10.2-eV lamp has been stabilized well enough to produce a practical field application. Permanent gases, water, and C1-C4 alkanes have an ionization potential above 10.2 eV, so they do not respond to a PID. The detector response is related to the photoionization cross section by the equation $I = I(0)FN\,LC$, where $I(0)$ is the initial photon intensity, F is the Faraday, N is the Avogadro number, L is the light path, and C is the analyte concentration.

The PID is a concentration-sensitive detector, and flow rate is the only parameter that affects response. A small fraction of the analyte entering the radiation is destroyed by absorption, but the Beer–Lambert law is nearly followed. Typically the 10 to 50 times lower detection than that for FID is due to larger response factors and lower noise. (Good application: NH_3, PH_3, AsH_3, IH, I_2, SeH_2, and TeH_2 in the ppb range.)

Thermionic

This is for nitrogen and phosphorus containing organics. It is a mass-sensitive detector. The assets are that it is extremely selective and sensitive (ppb or ppt). The liabilities are that it requires hydrogen and flame, requires stable high-current supply, bead replacement, linearity ~10(4), and each compound must be calibrated frequently as response changes with time. See Fig. 64.

Sometimes called the "flameless alkali sensitized" detector, there is still no clear and undisputed explanation of how the thermionic detector works. Rubidium sulfate salt is placed on a ceramic heating element and suspended above the hydrogen/air mixture. The collector electrode placed above the metal

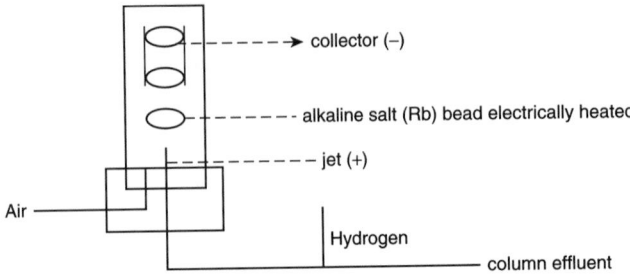

FIGURE 64   Thermionic detector.

atom (Rb) source is polarized negative (FID uses a positive collector). A low flow of hydrogen ($\sim$1-3 cc/min vs. $\sim$30 cc/min for FID) is mixed with the column effluent and directed through a jet (positive electrode) into the area containing metal atoms. By some gas-phase reaction, a very large current is generated across the detector electrodes, producing a strong signal from the electrometer amplifier. The low hydrogen-to-air ratio does not support normal combustion (thus the name *flameless*), but the hydrogen is necessary for the process to work. Currents of 2–3 A are required to heat the ceramic bead and generate the rubidium atom vapor. Sensitivity increases rapidly as the current passes a threshold value, and then a point is reached where the noise and instability of the signal dictate returning to a lower current value.

The hydrogen flow rate is very critical and must be controlled to a constant value. A 15-psig gas regulator connected to 10–15 ft of 0.005 in. ID capillary will allow stable control of a 1–3 cc/min H_2 flow. For a fixed-carrier gas flow through the column, H_2 flow is the critical parameter with a FID. The TID requires critical adjustment of the H_2 flow and the bead current, which makes it a little tricky to set up and optimize. Improvements in geometry of construction, fabrication of the rubidium bead, and current control have produced a field stable system that is reliable and maintainable (see HCN/AN area monitoring application).

Best response for phosphorus compounds is a He carrier; for the nitrogen compounds, it is a nitrogen carrier. Aging of the bead and contamination from the column or sample will change calibration.

Electron Capture

This is a nuclear electron source for multiple halogens, polynuclear aromatics, conjugated double bonds, and other molecules having a strong electron affinity. The assets are its extremely high sensitivity and selectivity. The liabilities are a dynamic linear range of only $\sim$10 (+4), tight control of operating parameters, and that extreme cleanliness is required. The carrier gas must be very dry and oxygen free, the concentration detector is very sensitive to the carrier gas flow rate. There is an extremely wide range of response factors for different analytes, and it is high maintenance. See Fig. 65.

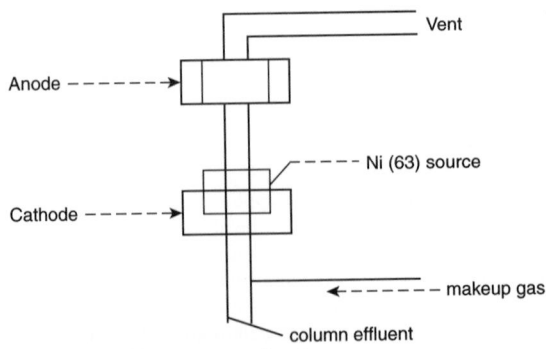

FIGURE 65   Electron capture detector.

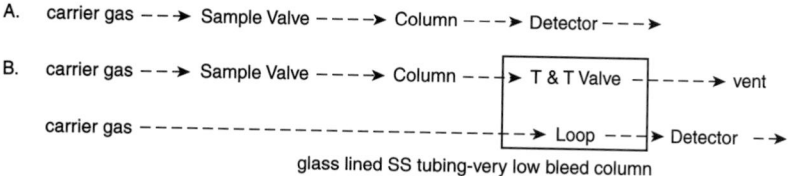

A.   carrier gas – – ➤ Sample Valve – – – ➤ Column – – – ➤ Detector – – – ➤

B.   carrier gas – – ➤ Sample Valve – – – ➤ Column – – ┤➤ T & T Valve –├ – – – ➤ vent

     carrier gas –├ ➤ Loop – –├➤ Detector – ➤

                         glass lined SS tubing-very low bleed column

FIGURE 66   System for protecting the electron capture detector.

The operating mode is amperometric and the operating technique is pulsed voltage constant current. The best applications are for polychlorinated pesticides. A radioactive beta source [63Ni] placed in the carrier gas emits electrons that collide with molecules of the carrier gas and ionize them. The charge carriers formed drift slowly in a weak electrical field, and after a number of collisions, the electrons lose most of their energy, until they retain only the thermal kinetic energy. Eventually, they are collected by the electrodes and a fixed steady background current is established. When a molecule with a significant electron affinity comes along, it captures an electron and forms a negative ion. Negative ions drift much more slowly in the electrical field and react much more rapidly with positive ions than electrons. The observed current decreases by the amount of electrons captured.

The detector principle is based on the huge difference between recombination rates of positive and negative ions on one hand and positive ions and electrons on the other. The detector current is related to the concentration of electron capturing analytes in a way similar to the Beer–Lambert absorption spectroscopy law. $I = I(0) \, e \, (-kCd)$, where I is the current observed when the analyte concentration is C. $I(0)$ is the current of the pure carrier gas with no analyte, d is the distance between the collecting electrodes, and k is the response factor that is a function of electrical field, temperature, carrier gas, and the analyte chemistry.

Contamination of the electron capture detector has probably been the major problem with its application. It is important in the laboratory, but is absolutely critical in an on-line analyzer. One way to keep column bleed and sample contaminants out of the detector internal region is to utilize the "trap and transfer" technique. Figure 66A shows the direct connection of the sample valve, column, and detector that allows unwanted materials to enter the detector. Figure 66B shows how trap and transfer vents them and sends the target analyte with minimal contaminants to the detector. Proper care in design and application can reduce maintenance and change failure to success.

SAMPLE EXTRACTION, CONDITIONING AND PREPARATION FOR ON-LINE ANALYSIS (CONTINUOUS SAMPLE FLOW VERSUS DISCRETE SAMPLING)

Sampling at the point of interest in a manufacturing process can be done in different ways. General practice has been to tap the process line and allow material under its own motive force (pressure) to flow to a point of least resistance (atmosphere) through a length of tubing. A measurement device is connected to the sample line such that material will flow past a sensor (concentration/voltage transducer) to provide a signal that can be set to represent a known amount of a component (analyte) in the mixture. Additional items were added to this configuration (sample conditioning system) to clean process material (filters), change phases (thermodynamic states), control physical properties (temperature, pressure, etc.), and isolate components. Probes were inserted into the process line to extract a more representative portion of the process material, and pumps or eductors were added to provide a motive force to move sampled material. Such systems can become very complex and difficult to maintain and can cause adulteration of a sample or loss of an analyte. No matter what operation is performed on a sample, it is essential that the concentration of an analyte at the sensor must be quantitatively related to its concentration in the process. Any changes must be constant and quantifiable.

Actual concentration of the analyte at the sensor is set by introducing a "standard" sample at the point of extraction of material from the process. This function is called "calibration of the analyzer system." Note that *system* is emphasized because we need a transfer function that relates the sensor concentration signal to the precise and accurate component population in the unadulterated process stream. This subject will be addressed in more detail under "Calibration and Validation."

CONTINUOUS PROCESS SAMPLING
by E. A. Houser

[The following excerpts are from Houser's book, "Principles of Sample Handling and Sampling Systems Design for Process Analyzers." A full copy of the book can be obtained from: ISA-Analysis Division; PO Box 12277; Research Triangle Park, NC, USA, 27709($35.00 US); (919)-549-8411]

Fundamental considerations for process sampling such as response time, sample tap location, materials of construction, etc. do not change for different methods of obtaining a sample.

Definition of a Process Analyzer Sampling System

A process analysis system includes equipment, the process, or required calibration sample, including necessary accessories to obtain the required operation exclusive of the analyzer itself. Note that this definition intentionally excludes any sample conditioning or sample processing that may be a part of the analyzer. This is consistent with the philosophy outlined previously. Materially, a sampling system consists of filters, pressure regulators, valves' pumps, flow-meter condensers, vaporizers, catch pots, pressure gages, and other pneumatic, hydraulic, and electric components.

A sampling system cannot be divorced from its plant environment, for example, packaging and installation. In this light, a sampling system also includes sample lines, steam tracing facilities, protective housings, utility supplies, and all the other physical equipment necessary to obtain proper sampling system operation. It is also important to recognize that a sampling system can be (and usually is) spread out over a wide area of the plant and is not limited to the neatly assembled cubicles and panels that can usually be obtained from the analyzer manufacturer.

Sampling System Functions

There are six overall functions of any sampling system.

1. To take a sample that is representative of the flowing process stream. The term *representative* is used in the most liberal sense, which is relative to the application needs of the particular system.

2. To transport the sample from the sample point to the analyzer.

3. To condition the sample so it is compatible with the analyzer and with application needs. Conditioning includes many operations; a few of these are cleaning, vaporizing, condensing, adjusting the pressure and temperature, and many others.

4. To switch from one stream to another where the analyzer is used on more than one sample stream. Additionally, to provide for proper introduction of calibration standards, where these are necessary. This function is vital to the correct operation of any sampling system because analysis accuracy is an important consideration.

5. To transport the sample from the analyzer to the desired point of rejection. Included here are venting arrangements, waste disposal systems, and methods for returning the sample to the process where necessary without adversely effecting analyzer operation.

6. To allow for the effects of corrosion and other reactions.

These functions should be accomplished with the timeliness, accuracy, and other needs of the analysis system. A well-designed sampling system optimizes the above performance factors with the cost of equipment, installation, operation, and maintenance that are a part of anything mechanical.

Importance of Good Sampling System Design

Most analyzers are designed to work with clean, dry, and noncorrosive, noninterfering samples at design conditions of temperature, pressure, and flow. It follows that use with samples in other conditions may adversely affect analyzer operation. Failure of the sampling system to supply a compatible sample may only affect analysis accuracy; or, it may prevent operation of the system beyond a few minutes time. A further danger exists in that the sampling system may introduce more problems of maintenance and/or accuracy than it solves. It is important, therefore, to consider all factors that influence the overall analysis system, as well as the performance required by the application.

This is not to say that such consideration will automatically result in a sampling-analysis system that runs by itself and in which there are no significant limitations on basic analyzer accuracy. Most sampling systems are compromises among cost, accuracy, maintenance requirements, and application performance. Therefore, do not be deluded into thinking that all of these areas can be obtained at maximum performance by the diligent application of sample handling theory and design. You will find that many problems require considerable field experimentation, revision, and cut and try. Even at the end of a lot of blood, sweat, and tears, the final system may still fall considerably short of what is actually desired. There is one best system for a particular analysis application, but a cooperative effort of several sources may be required to get to it.

A further observation is that the success of a sampling system is highly dependent upon what might seem to be insignificant or upon minor differences between one application and another.

Principles of Sample Handling

Extracting a Representative Sample.   We assume that those using this book already understand the fundamentals of physical chemistry, engineering stoichiometry, fluid flow, heat transfer, and so on, which are necessary to apply the techniques presented here. The principles of sample handling correspond approximately to the overall system functions:

- taking a representative sample
- transporting the sample
- conditioning the sample
- sample stream and calibration standards switching
- sample rejection or venting systems
- corrosion and other reactions

We will discuss some of the ways in which the principles learned are applied to the design of specific sampling systems. In some cases there is not a clear-cut separation between "principles" and "design." Although this is not surprising, it is hoped that the organization presented does not appear too arbitrary.

While it might seem logical that representative sampling should mean that the sample taken has exactly the same composition as the flowing process stream, this is a false assumption. For the purposes of sample handling, representativeness relates only to the measured variable and its significance. For example, it is frequently desirable to reject at the sampling point certain portions of the sample that may cause problems later on. A "true" sample may contain dirt, corrosive fractions, materials that may condense and foul the system, and so on, which are not wanted and which can be rejected to a large extent at the sample tap. Note also that representativeness can be obtained even though the actual concentration of the measured component changes, providing these changes are predictable or can be compensated. Furthermore, the degree of representativeness that is acceptable is a function of the overall purpose for making the analysis. For example, an application in which trends are much more important than absolute measurement can tolerate a rather large but relatively constant shift in actual sample concentration and still be representative in the sense of application suitability.

Representative sampling implies by its nature those activities that take place at the sample tap, usually by means of a sample probe or other device that affects the sample being extracted from the process stream.

Isokinetic and Nonisokinetic Sampling.   Isokinetic means "constant velocity." The basic principle is that the velocity (and direction) of the fluid flowing into the sampling opening is the same as the velocity (and direction) of the process stream passing around the opening. If the sample velocity is higher than the process stream, the heavier phase will be lower in the sample than it is in the process stream; if the sampling velocity is lower than the process stream, the heavier phase will be higher in the sample. This is because the heavier phase will not follow the flow lines as well as the lighter material.

The basic relationship between the sampling opening, process line size, and flow rate for isokinetic sampling is:

$$f = aF/A$$

where f = sample flow rate (consistent units)
F = process fluid flow rate
a = sampling port area
A = inside area of process line

Several methods have been worked out for controlling sample flow to maintain isokinetic conditions when a process flow varies. These methods are based on the use of a Pitot tube to measure process flow at the point of sampling. In general, isokinetic sampling is necessary only where two-phase samples are taken and where it is desirable to obtain a sample representative of the phase constituency of the process stream. The most common example is in sampling for dust or particulate matter. Nonisokinetic sampling in this case causes drastic analysis errors. With continuous process analyzers, the need for isokinetic sampling is uncommon. Usually, if the sample consists of more than one phase, such as traces of liquid droplets in a vapor, or solid particulates in a vapor or a liquid, it is desired to take a sample only of the main predominant phase and reject the other phases, which are usually regarded as contaminants. This can be a serious problem if the contaminants are a major percent of the process stream and vary in quantity. In these cases, it is desirable that the sampling be as nonisokinetic as possible so as to obtain the maximum rejection of the undesired phase.

There are some situations, however, in which isokinetic sampling is desirable, such as when there are traces of liquid droplets in a vapor and when the entire stream is to be either condensed or vaporized to obtain a single phase sample for the analyzer. An example of the former is the analysis of wet steam for conductivity or silica analysis, as in high-pressure boilers. A significant error would occur if the proportion of water droplets in the sample were either higher or lower than in the process stream. This is because the distribution of trace conductive components is considerably different in the two phases, and an overall analysis of the entire process stream is desired. Isokinetic sampling of wet steam is accomplished by properly designed sampling nozzles, which are installed at the sample taps at the time the plant is constructed.

As mentioned previously, extreme nonisokinetic sampling can be used to effectively reject contaminants at the sample tap. Sample probes and tap arrangements to accomplish this purpose are available.

Averaging of Multiple Samples.   When gases are sampled in large ducts, it frequently happens that there is a concentration gradient that exists across the duct diameter. These gradients can be caused by one or more of the following situations:

- poor mixing of two blended streams
- infiltration of diffusion of air (such as in stack gas sampling)
- stream-line flow distributions
- velocity gradients caused by abrupt changes in duct diameter

The rigorous procedure to determine the ideal location of sample probes is to make a velocity traverse across the duct, using a Pitot tube calibrated for the temperature and static pressure of the gases in the duct. A traverse can be plotted and will provide a velocity profile across the area of the duct. From this velocity profile, the required numbers of probes and positions can be determined for a given situation. A more simple approach is possible if uniform flow distribution across the duct can be assumed. For the velocity profile to be determined, it is advisable to make traverses in such a way that each position represents an equal area of circular sections of the duct (assuming a round duct). It is also desirable that a measurement be made for each area of 1 ft^2 or smaller. As a simplified method, a very approximate indication of the total flow can be made by measuring the flow at the center of the duct. It is usually ~1.1 times the average flow across the duct.

Examples of sample averaging include a separate probe and sample cleaning facility used for each sample to be averaged. This is necessary because individual flow meters and flow adjusting valves are used to set the flow rates. These three flows are then mixed and passed through the remainder of the system such as pump and analyzer. Samples must be cleaned before being metered for obvious reasons.

A method for averaging three samples from a river, ditch, or flume utilizes a peristaltic pump of the multiple-tube type, which simultaneously pumps three samples from the ditch. If equal sized tubes are used, the flow rates will be approximately equal. However, should it be desirable to have different flow rates because of a varying cross section of the ditch, then different sized tubes will provide different flow rates. After passing through the pump, a mixing chamber ensures that the three samples are thoroughly mixed.

Where the samples are relatively clean and would not clog orifices built into the probe, a single probe using several sample inlets can be used to obtain a cross section or averaged sample. The diameter of the sample inlets can be varied to take a flow rate that is proportional to the flow in that section of the duct. For gases, the flow rates can be made independent of pump inlet pressure by selecting a pump that pulls a vacuum of at least 0.5 atmosphere so that the orifices in the probe are operated under critical conditions.

Quenching Reactions.   When sampling a system in which the process gases are undergoing a continuous reaction, it is usually desirable to stop the reactions as soon as the sample enters the sampling probe. A probe specifically designed to accomplish this purpose is called a quench-type probe. There are three quenching methods, which may be used individually or in combination.

1. Heat exchange cooling.   A sample probe constructed with a water-cooling jacket will lower sample temperatures from several hundred degrees to cooling water temperature in the first few inches of the probe, depending upon the sample flow rate. Quenching is not instantaneous, but it is suitable for relatively slow reactions.

2. Expansion cooling.   If a restricting orifice or nozzle is placed at the inlet of the probe, some cooling will take place as a result of expanding the sample gas across the restriction. The amount of cooling is relatively limited. However, expansion cooling is usually used in conjunction with heat exchange cooling, that is, in a probe that is water jacketed, to obtain faster quenching than would be possible than either method alone.

3. Dilution with gas or liquid.   Process reactions can be quenched by mixing the sample with either a gas or a liquid at the inlet end of the probe. Of course, dilution with a gas requires that both the gas flow rate and sample flow rate be controlled or known accurately to prevent analysis errors. Quenching by liquid dilution means simply the mixing or spraying of water or other liquid into the sample at the inlet end of the probe. This arrangement is quite common with many probes designed for sampling oxygen flue gases. The primary reason for the spray is to clean the sample rather than quench reactions, although the latter is also accomplished.

Contaminant Removal.   The rejection or removal of unwanted contaminants from a sample at the sample probe is a good example of intentional sample modification while still retaining representative sampling from an analysis application standpoint. That is, we do not want to take a sample that is exactly the same in overall composition as the process fluid, but by proper design, we want to eliminate unwanted portions of the sample. Three types of probes that can be utilized for this purposes are:

- probes with filters
- water washing probes
- inertial separation probes

Techniques of Sample Transfer

Design of the sample transfer lines is just as important as all other aspects of the analyzer sample system. This may be because they are located in a vast area between the sample tap and the analyzer shelter. Pipefitters may be told to install lines between the two without specific instructions regarding the critical factors of sample transfer. Where adequate response time is an important requirement, it is essential that good design and installation practices be followed. The items that must be considered are as follows.

Sizing Liquid Sample Lines.

- supply pressure
- sample return
- actual distance to the analyzer and to the return point
- required response time
- fluid transport properties
- volume of and delay caused by components in the line (i.e., filters, traps, rotometers, etc.)

Sizing Gas and Vapor Lines.   An additional complicating factor is compressibility in addition to those for liquids. Tables of purge times for different flow rates in various size tubing and piping are available in the engineering literature. Spreadsheets are available for calculating these values and pressure-drop values as well as many other useful properties.

Phase Preservation.   It is important to keep a sample as all liquid or all vapor during transport, so heating or cooling may be required. This may be necessary because of some of the sample system components, such as rotometers, or it may be a result of maintaining true composition.

- dewpoint of vapors
- bubble point of liquids

Adsorption.   A frequent cause of analysis inaccuracy is the absorption sample components on tubing walls, gaskets, o rings, pipe dope, and so on. This is most prominent with trace level polar compounds. Smooth surfaces, washing, treating, and other techniques have produced lower losses. Utilization of stainless steel, Teflon, and especially the newer glass-lined steel tubing has shown much improvement. Water (H_2O) and hydrogen sulfide (H_2S) are very bad actors. Heating of the sample line and maintaining a high sample velocity are very important, also.

Diffusion. Use of plastics and rubber can contribute to loss of sample components through the walls of lines and vessels or intrusion of oxygen and water into a system. Permeability is different for different gases and materials, and it is dependent on pressure and temperature.

Basic Elements of Sample Conditioning

Sample conditioning in this work relates more to physical operations than to chemical processes, which we define as sample preparation. Items in this category consist of the following.

Cleaning (removing unwanted components).

- filtration
- centrifuging
- washing (must prohibit dilution and component losses)
- coalescing (small, finely divided liquid particles are caused to become larger)
- bubble removal

Condensing.

- air cooling
- water cooling (utilized to produce greater heat transfer)
- refrigeration cooling (subambient temperatures)

Vaporization

Pressure Reduction and Flow Control.

- orifice
- capillary
- regulator

Motive Force (don't use pumps unless absolutely necessary).

- mode of operation
- materials of construction
- packing and leakage
- pressure and temperature limits
- aspirators or eductors

Fundamentals of Sample Preparation

Component Isolation.

- column separation
- headspace
- membranes
- valves

Chemical Reaction.

- methanation (converting CO and CO_2 to CH_4 by hydrogen reduction)
- silanization (react silanes with reactive species to make stable volatile compounds)
- esterification (react acids with alcohols to make volatile esters)

DISCRETE SAMPLING
by J. G. Converse*

A "break from" conventional thinking is the concept of "extracting a minimum quantity of material from the process for analysis." A quote by Albert Einstein seems appropriate here: "The significant problems we face can not be solved at the same level of thinking we were at when we created them!" Instead of flowing a large quantity of material out of the process to be conditioned, we propose reaching into the process, extracting a small, discrete, representative sample, and utilizing it for analytical measurement. This provides the following list of benefits and savings, which are advantageous over the traditional continuous flowing out of gross amounts of unconditioned material.

Advantages of Discrete Sampling

1. Reduces sample waste and conditioning.
2. Reduces residues, emissions, and energy.
3. Reduces sample system duty cycle.
4. Allows automated system clean up.
5. Allows high velocity sample transport.
6. Allows autozero on carrier between injections.
7. Allows reference injection for diagnosis.
8. Provides real time validation.
9. Allows utilization of all preparation operators.
10. Sample size can be adjusted to fit concentration range.

Remote Discrete Sampling/Flow Injection/Multidimensional Sample Preparation

This includes column chromatography, as well as headspace, membrane separation, and valve switching techniques. Flow injection refers to the introduction of a discrete packet of material into a flowing carrier fluid. Different methods of injection will be presented and the nature of transport within the confined carrier will be discussed. While the carrier is generally inert or neutral to the injected sample, it can be or contain a reactive agent. The packet may be conditioned and manipulated to prepare it for chemical analysis, or it may transported in an unaltered state. Decoupling of the preparation and measurement systems can be evaluated for improved precision. Statistical quality control can be automatically internalized to verify performance.

The major difference between discrete and continuous sampling is the gross amount of material that must be conditioned. As an example, let us calculate some volumes.

$$100 \text{ cc/min} \times 60 \text{ min/hour} \times 24 \text{ hours/day} \times 365 \text{ days/year} = 52{,}568{,}000 \text{ cc/year}$$

$$0.001 \text{ cc/10 min} \times 60 \text{ min/hour} \times 24 \text{ hours/day} \times 365 \text{ days/year} = 52.6 \text{ cc/year}$$

---

* Sterling Chemicals, Inc., Texas City, Texas.

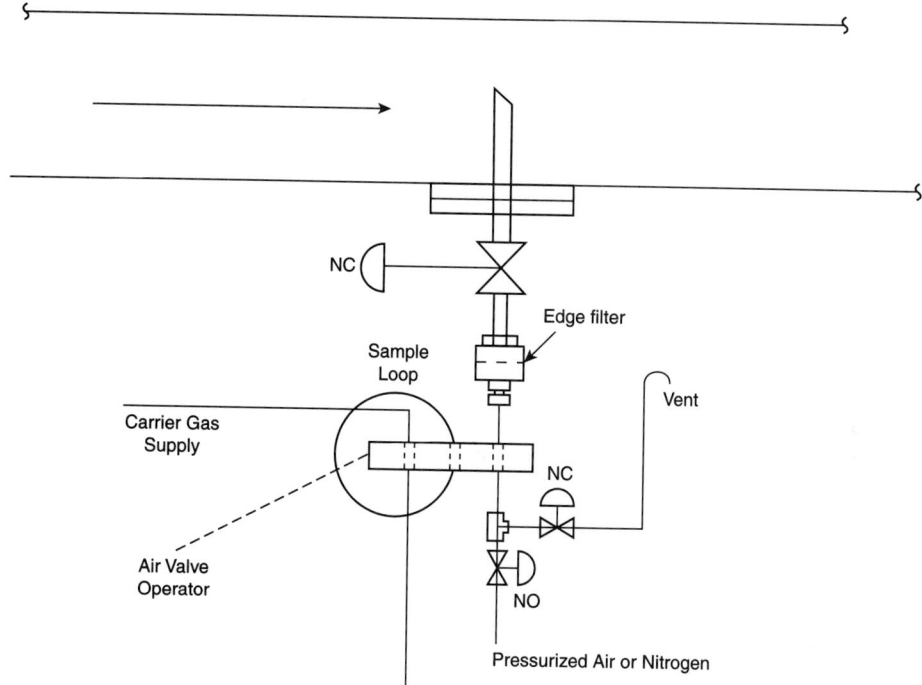

FIGURE 1 Extractive discrete sampling.

In the first instance, a conventional sample system would extract and condition 52M cc or 53K liters, or 13,742 gallons of material from the process each year. This material would then flow through a GC valve where 1 μl (0.001 cc) would be injected onto the column every 10-min. cycle. Depending on how dirty the process material was, a considerable amount of residue could be separated, which would then require disposal. At 10 ppm solids in the sample liquid, a filter removing it all would collect 524 g in one year (454 g = 1 lb)! This cleanup operation could require several pieces of hardware (filter, knockout pot, etc.) to condition the sample prior to injection. It can be seen from the above that either a very large filter or frequent cleaning of a small filter would be required to keep the flow path open. This is why continuous sampling practices are so maintenance intensive. *They are designed to fail.*

The second case defines the ideal case of discrete sampling. One μl is removed from the process every 10 min, rather than 1,000 ml. This may be cleaned with a filter, which can then be backpurged into the process during the 10-min cycle time before the next injection. Figures 1 and 2 show systems for extractive discrete and direct discrete sampling. The direct discrete system may not be achievable on a dirty stream, but the extractive configuration works very well. The extractive system can backpurge the valve and filter into the process or purge to sewer. In any case, much less material is extracted, much less residue is collected, and the system is cleaned each cycle. The actual duty cycle of the sample extraction line is only a small percentage of the total analysis time. The continuous sampling system only utilizes 52.6 cc of the 52,568,000 cc conditioned, which calculates to 0.0001%. *WHAT A WASTE!*

The number one advance in sample preparation has to be column capability and performance. Materials for column packing are available that achieve more selective separation and last longer in aggressive environments. Column life is critical to acceptable performance. Capillary columns were barely mentioned in the previous edition, but they now play a major role in process application. They produce significantly higher resolution of components and in many cases faster analysis cycles. Larger capacity capillaries have made them easier to apply in process instruments, and the outer

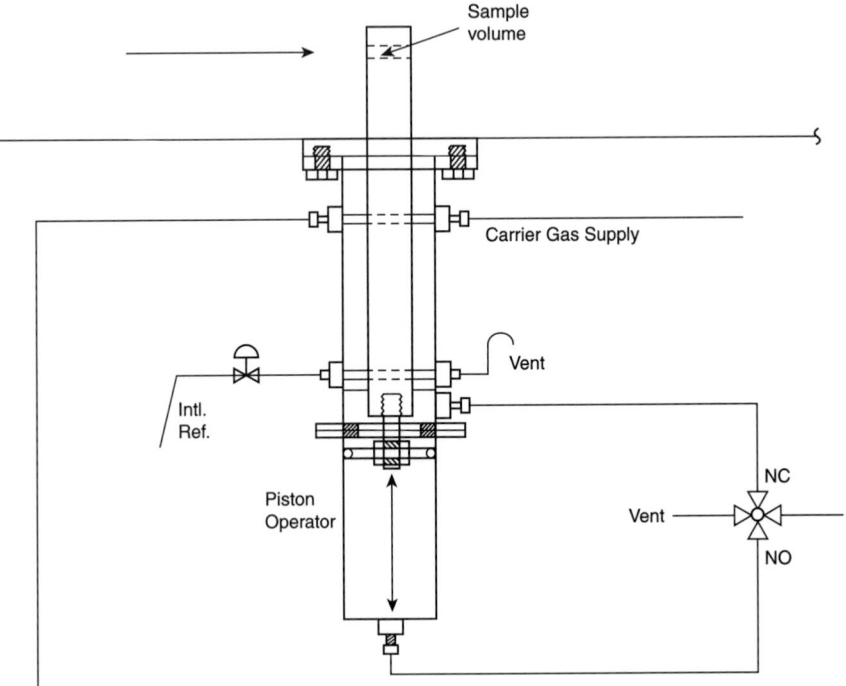

FIGURE 2 Direct discrete sampling.

surface clad-fused-silica columns are less fragile. Inert inner surface metal capillary columns with cross linked and chemically bonded stationary phases are now available for those who fear the "laboratory" glass columns. Micropacked capillary columns are once again finding many useful aplications. They provide a higher sample capacity than open capillaries with better resolution than larger packed columns. Literature and technical assistance are readily available to anyone wishing to apply this separation science technology. An analyzer practitioner doesn't have to develop his or her own columns now as the "ancients" did. Any buyers guide to analytical instruments will give a long list of column suppliers. These suppliers are also a very valuable source of books and training courses on applying chronography to chemical analysis for both laboratory and field systems.

Another major change is the technical information that is available to anyone wanting to enter the field. The free literature on column performance data, column selection procedures, sample introduction techniques, detector selection, performance, and application, and system operation and maintenance practices is unbelievable. You can become a near expert on the subject by some intensive reading. One still has to apply the knowledge and make a few mistakes to really become effective at utilizing this method of component separation.

In the beginning, the chromatograph was a single box with all the components packaged together. It was considered an analyzer unto itself and required a separate sample conditioning system. The process analyzer engineer began to develop techniques to modify the laboratory practice of using a single column. Multiple columns connected by switching valves were applied to backflush, foreflush, and cut components from the normal sequential elution scheme. Years later, the laboratory chemists started applying such techniques and coined the phrase "multi-dimensional gas chromatography." Our escape to new degrees of freedom came in 1983 when we realized that the column was part of the sample preparation system and not the measurement system. We developed "remote discrete sampling" in 1978, which allowed one to inject the small quantity of sample into a capillary hundreds of feet from the "analyzer" and transport it to the column as a packet. When we saw the transfer line heated jacket as a long oven, we were no longer constrained to a single box. The components of the original box could now be expanded into three-dimensional space by using several boxes with connecting capillaries.

Chronography (response versus time) is the correct title for what we practice and call "chromatography" (color versus time).

Fundamentals of Chronography

(See the Appendix in this section). Chronography is a physical/chemical method of separating the various components of a mixture into pure fractions or bands of each component. The carrier or moving phase may be a gas, liquid, or supercritical fluid. The separation is effected by distributing the mixture between the fixed or stationary phase in a column and the carrier. The stationary phase may be a solid or a liquid-coated solid packed into the column, or it may be attached to the walls of a capillary. Liquid samples that can be vaporized may be separated with a gas carrier. High-boiling and unstable compounds can be separated with a liquid carrier. Some materials behave better in a supercritical fluid or may be separated faster. This technology requires much more skill and is expensive; thus it has limited use at present.

Components of the sample are retained in the column for different lengths of time because of adsorption/desorption, solution/disolution, chemical affinity, size exclusion, and other mechanisms. Various components are continually washed from one part of the stationary phase and recaptured by another by the moving phase. Different components elute in groups from the column with respect to time from injection. Dispersion in the system causes the bands of components to emerge with a Gaussian distribution or a distorted peak shape. A simplified diagram of the process is shown in Fig. 3.

Only a superficial coverage of this vast separation technique can be presented in a work of this nature. Reference documents should be consulted and studied by anyone planning to apply this technology.

```
eluant                  peak      peak      peak      peak
position                #1        #2        #3        #4
  .     .     .     .    .         .         .         .
|     |     |     |    |aaaa|    |bbbb|    |cccc|    |dddd|
|     |     |     |aaaa|    |bbbb|    |cccc|    |dddd|     |
|     |     | aaa|    |bbbb|    |cccc|    |dddd|     |     |
|     |     | a  |bbbb|    |cccc|    |dddd|     |     |     |
|     |     | bbb|    |cccc|    |dddd|     |     |     |     |
|     |aabb|b    |cccc|    |dddd|     |     |     |     |     |
|     |aabb| ccc|    |dddd|     |     |     |     |     |     |
|abcd| ccd| c d| ddd|     |     |     |     |     |     |     |
|abcd| ccd|  dd|d    |     |     |     |     |     |     |     |
|abcd|  dd|d    |     |     |     |     |     |     |     |     |
|abcd|     |     |     |     |     |     |     |     |     |     |
+----+----+----+ ---- +----+----+----+----+----+----+----+
Inject                    ------> Time
```

FIGURE 3 Chronographic separation process; movement through column with time.

FAST GC

The major components of Fast gas chromatography (GC) are:

- very narrow, sharp-edged sample injection band
- small bore column (0.1–0.25 mm ID)
- optimization of column resolution

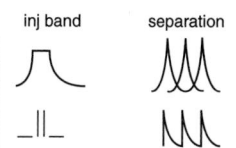

- carrier flow rate (faster than optimized flow)
- very low dead volume between the injector, column, and detector
- low dead-volume detector and/or carrier makeup

A very fast concentration change and narrow component peaks require an electronic signal sampling rate that is 10–100 times faster than conventional amplifiers.

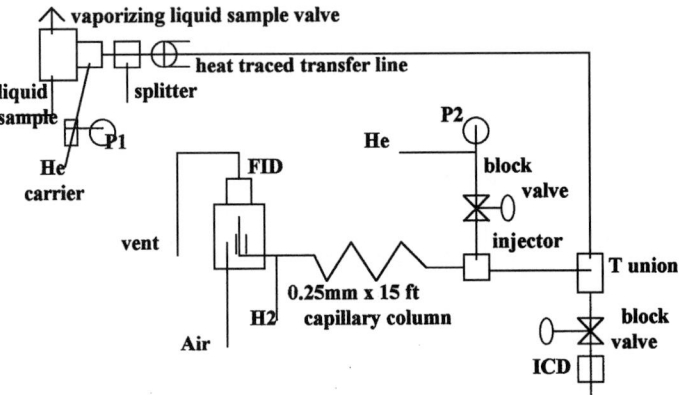

FIGURE 4   Remote discrete sampling with narrow band injection.

Sample Injection Techniques

Several techniques have been developed to obtain a very narrow injection band with sharply rising and falling edges. These range from stepping-motor driven valves to pneumatic capillary injection devices. Figure 4 shows a possible configuration for sampling and narrow band injection.

The dead volume of the flame ionization detector (FID) can be eliminated by passing the capillary column through the burner jet such that it extends ~1 mm past the end. A 5–10 μl liquid sample is injected into the helium carrier, utilizing a vaporizing liquid sample valve. A splitter can be used to reduce the vapor sample size and sweep the valve rapidly to maintain a tight sample packet in the heated transfer line. When the sample packet passes the injector, the block valves are actuated and sample is forced to flow toward the capillary column. As soon as the desired amount of sample passes the secondary carrier tee, the block valves are deactivated. Carrier at pressure P2 slices the sample forcing the front end onto the column and back flushing the other portion out to vent through the intercolumn detector. This produces a narrow, sharp-edged injection band based on the Besche injection process.

Examples of fast GC are as follows.

1. HCN and Acrylonitrile—area monitor—two components isolated in 30 s.

2. H_2O, methyl acetate, methyl iodide, acetic acid—reactor effluent—four components separated in 120 s.

Multidimensional Sample Manipulation Techniques

One-dimensional sample preparation uses one value of each variable. Multidimensional techniques use two or more values of any variable. Column switching is multidimensional because more than one separation device is used, as well as possibly different flow rates. Much effort has been made to develop very constant isothermal temperature zones in which to manipulate component isolation and measurement. There is no need for the temperature programming of sample preparation operations. Why should we force a critical parameter to be varied, try to control the rate of change tightly, and add extra hardware and software to heat and cool a system that is better off when held constant? Actually, temperature-programmed sample preparation is multidimensional, using many values of the temperature parameter while holding the other variables constant. Foreflush, backflush, and heart cutting are forms of multidimensional sample preparation. All of the techniques used by practitioners of process GC were developed before the laboratory folks coined the term "multidimensional." Switching valves are required to implement "traffic control" in order to effect changes of the critical variables. Constant temperature zones can be used instead of a continuously varyied temperature. The switching is done with mechanical valves directly or by the Deans pressure technique [1]. Components are manipulated by directional transfer, trap and store, or trap and transfer. This paper will discuss the utilization of all the sample preparation variables at different dimensional levels. Two or more values of a particular variable may be used, but none will be varied continuously.

Single Dimension.   Perhaps one should more accurately define a single-dimensional system before addressing multiple dimensions. Some systems using two sample injections for a single total analysis with two valves or two injections with one valve have been utilized. That surely is multidimensional analysis. It is essential that all the variables of a complete analysis be specified. If all of these have only one value, then the system is one dimensional. Development of a nonanalytical equation has been suggested as a way to itemize the variables of an analyzer system [2]. Such an equation for sample preparation by liquid/vapor equilibrium separation is shown in Fig. 5.

$$R = C_i \times f\,(f_g,\ f_L,\ T_L,\ L_L,\ P_v \ldots Z)$$

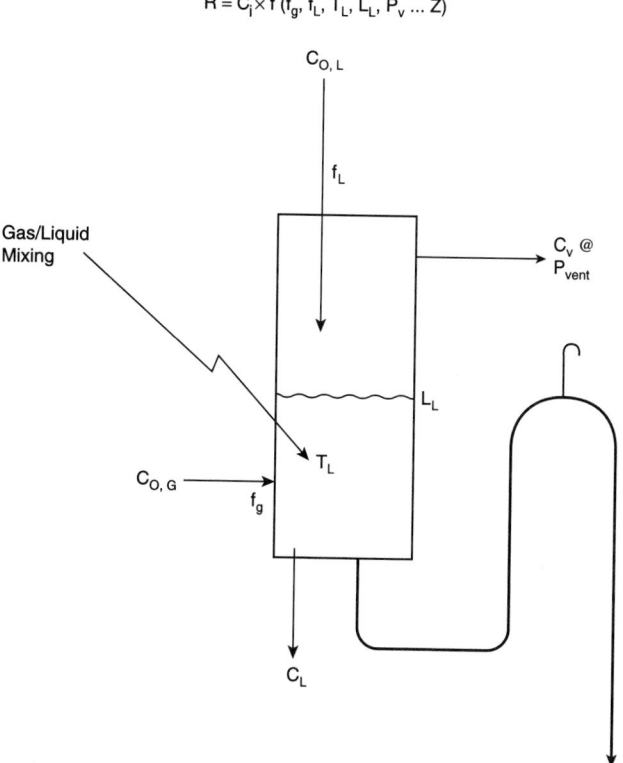

FIGURE 5   Variables for a dynamic gas/liquid sparger.

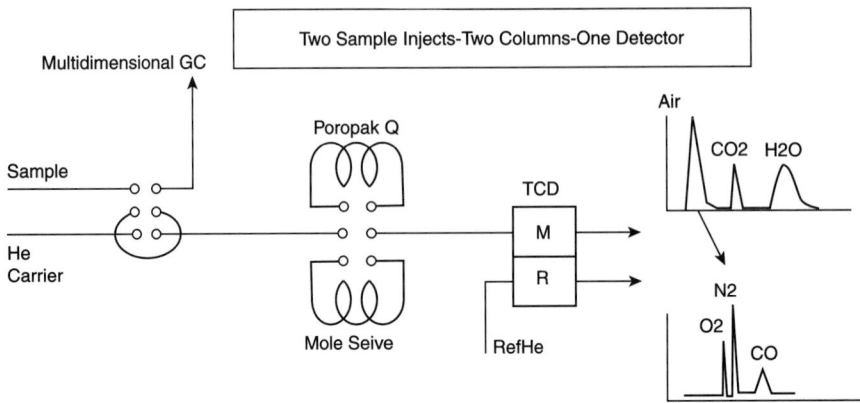

FIGURE 6 Multidimensional sample preparation.

It can be seen from this equation that even a simple one-dimensional system has many variables. Fortunately, we have learned how to hold most of these parameters constant, at least for the duration of a single analysis. It becomes immediately obvious that a multidimensional analysis requires the control of a large number of variables.

Multiple Dimensions. An example of two injections for one analysis is a mixture of O_2, N_2, CO, CO_2, and H_2O, directed first through a mole sieve column and then through a Poropak Q column into a thermal conductivity detector (TCD). We have utilized two samples, two columns, and one detector (Fig. 6).

Another case could use a small sample for major components and a large sample for trace components. The following parameters of possible analysis systems may have one or multiple values:

- number and size of sample injections
- number of temperature zones
- number and type of separation devices
- number of flow rates/pressures
- number and type of measurement devices
- number and type of carrier fluids
- number and type of valves and their configuration

Several examples will be presented to evaluate the merits of multidimensional sample preparation.

Temperature Dimension. When one is faced with separation of a wide boiling-point range sample, the obvious answer is temperature programming. There is, however, a better alternative. One can use multiple "isothermal temperature zones" and achieve better results because the temperature variable is held constant. The concept involves passing the lower boiling-point components through the highest temperature zone quickly to be separated at a lower appropriate temperature (Fig. 7).

A separation column of low resolution is used to split boiling range groups. A six-port valve is used to direct the groups to their appropriate destination. The high boilers are separated and measured in the highest temperature zone immediately. They are not held on the column until the temperature is programmed to a higher value. Total cycle time can be reduced significantly. The lower boiling components are sent to a lower temperature zone by directional switching or trap and transfer techniques. Additional separations and transfers can be used if required. The sample or portions thereof travel from a high to low temperature in all cases.

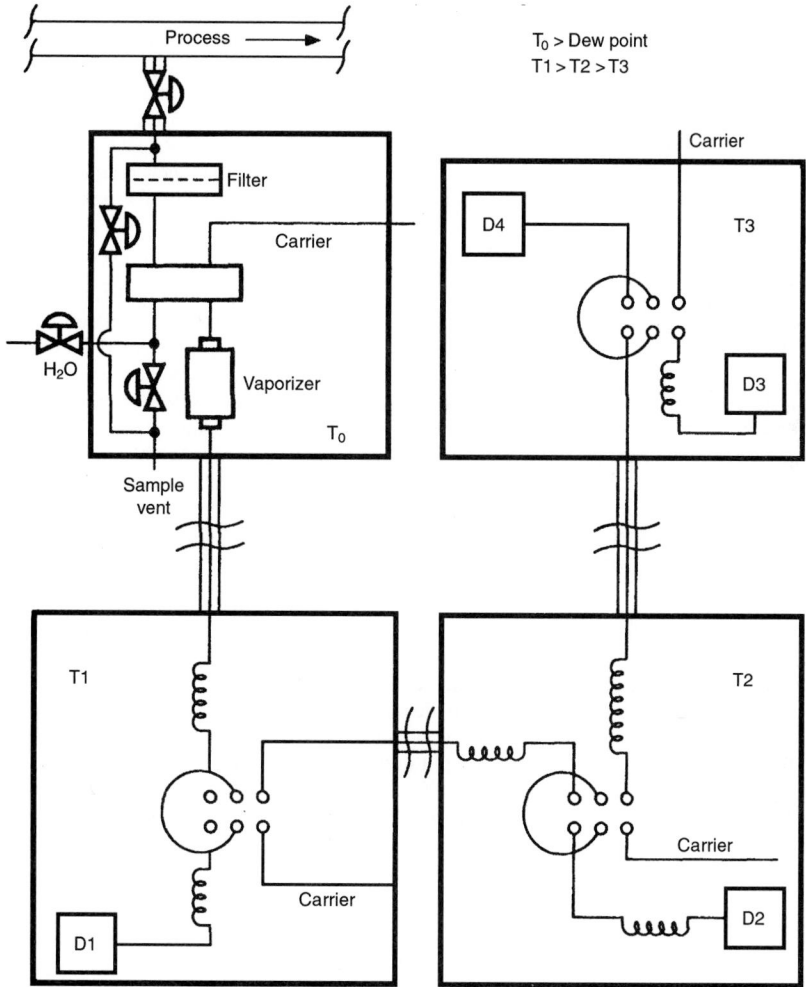

FIGURE 7   Temperature sequence separation.

Separation Device Dimension.   The magnificent discovery of column chromatography opened new vistas of component isolation made for chemical analysis. The first report of separating gas molecules on a column was made by Ramsey (1905), although the process was not named. Tsweet (1906) reported separating colored molecules in aqueous solution on a column of clay particles, which he called "chromatography." Martin and James developed practical applications of column separation techniques (1949–1952) but misnamed their process. Golay (1958) predicted column separation on theoretical principles and actually demonstrated the separation of gaseous molecules in an open metal capillary column. Tsweet really did demonstrate "chromatography." Golay also misnamed his process, which was actually "chronography," since elution of a component is marked with respect to time. Most of us practice gas chronography today.

We have struggled through many years of using packed columns for component separation developing a variety of selective surfaces to separate specific chemical species. Several diameters and lengths of column were used, and mesh size and liquid loading were varied. Capillary columns were then rediscovered and improved by coating the metal inner surface. Development of fused-silica capillary columns made further advances possible. The coating was crosslinked and bonded to the silica surface to produce high resolution, stable, and rugged separation devices. Highly selective separations were achieved by varying the two dimensions of coating and temperature. Other separation devices can also be utilized for sample preparation operations. Vapor/liquid equilibrium does not have to be performed on a column device. Special spargers, strippers, and separators have been used. Membranes are being developed to produce selective separations such as aqueous/organic and ion exchange. These can provide liquid/liquid extraction as well as gas/liquid separation sample preparation operations.

Flow Rate Dimension. Pressure/flow programming has been reported but not widely used. Better sensors and digital controllers may improve control, although flow has less effect on separation than temperature. Independent flow in different systems can be implemented by utilizing the trap and transfer technique. More use of pressure and flow rate parameters will occur as we learn to understand and control them better. Switching to a higher fixed pressure after the target peaks have eluted is useful in clearing the column of higher boiling components faster.

Significant new pressure programming equipment is now available that is primarily utilized to improve sample injection. These devices provide an opportunity to try new techniques. Supercritical fluid extraction and chronography are growing in application, which will enhance the pressure dimension performance. Pressure exerted on the sample and column not only changes the flow rate but may markedly change their behavior.

Measurement Device Dimension. Many detectors have been developed to measure chemical concentration of the isolated components. They vary in sensitivity and specificity. One application of two-dimensional detection is to place a nondestructive TCD in front of a higher sensitivity, more selective FID. The TCD measures all the components. The FID then provides adequate sensitivity for trace levels of hydrocarbon species. Detectors can be used in parallel operation, also. The fixed gases O_2, N_2, CO, CO_2, and H_2O can be split and transferred to a TCD while the organics are directed to a FID. Selective detectors such as ECD can be used without completely separating nonresponding species. *The choice of detector will markedly effect the sample preparation requirements.* Principles of operation and specific applications of the many chemical concentration/voltage transducers is beyond the scope of this paper. The reader is referred to [3]–[5].

Carrier Fluid Dimension. The carrier fluid can be varied to affect the behavior of the separation device or the detector. This has been most utilized in liquid chronography but can aid sample preparation in other ways. The use of an acid solution on one side and a base solution on the other side of an ion exchange membrane has been used to improve transfer of a selected species. Hydrophobic membranes can be utilized to reject water and concentrate the target species in an organic phase [6]. The exchange of nitrogen for helium in a column effluent is a means of preparing the sample for improved response in an electron capture detector. Our use of this parameter is only limited by our imagination.

Valves for Directing Path. The main types of valves currently used are spool, planar sliding, rotary sliding, diaphragm, and needle (see Fig. 8).

Primary connections are four, six, and eight ports. Valves with more ports can be considered combinations of these three. The valve is generally pneumatically operated, but electrical solenoids can be used at low pressure. They are placed in an isothermal temperature zone and operate by direct mechanical transfer. Another technique developed by Deans [1] uses mechanical valves outside the oven to produce pneumatic switching by changing the pressure differential across the separation columns. The common configurations for directing components of a sample to various locations utilize directional valves inside the oven. Pictures of the internal structure and principles of operation of these traffic directing devices are provided in [3].

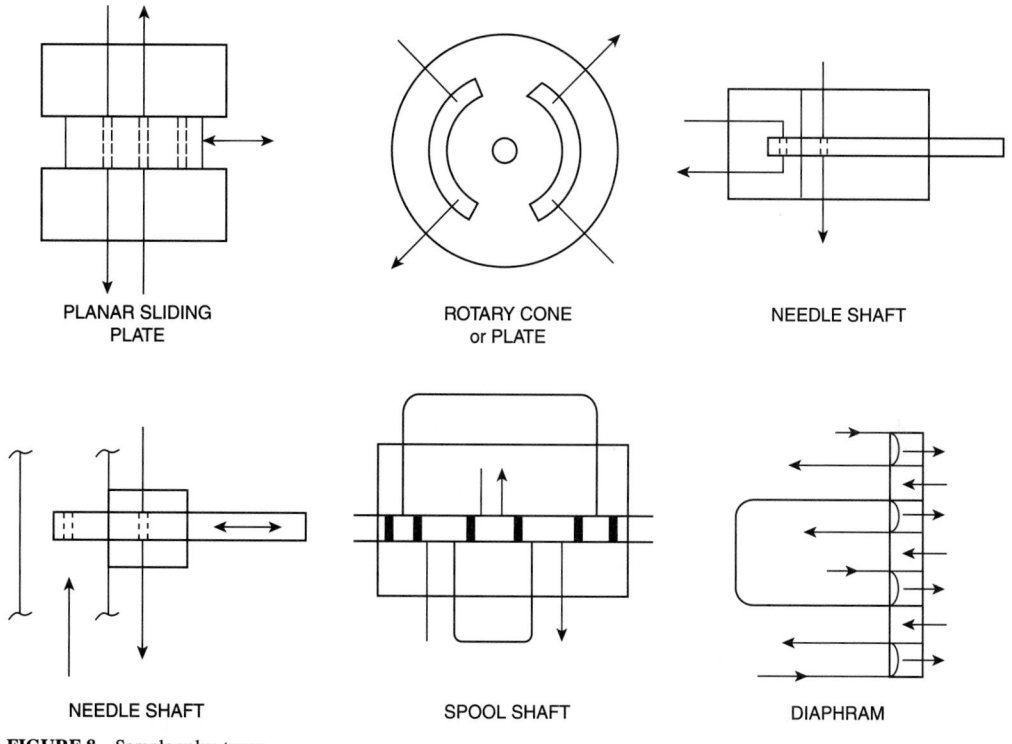

FIGURE 8 Sample valve types.

REFERENCES

1. Deans, D. R., and I. Scott, *Anal. Chem.,* vol. 45, p. 1137, 1973.

2. Converse, J. G., "Internal Reference Introduction for Performance Verification, Diagnostics, and Calibration Validation," *Proceedings of AID Symposium,* vol. 21, Boston, Mass., p. 185, 1985.

3. Clevett, K., *Process Analyzer Technology*, J. Wiley, New York, 1986.

4. McNair, H. M., and E. T. Bonelli, *Basic Chromatography,* Varian Ass., Consolidated Printers, Berkley, Calif., 1968.

5. Liptak, B. G., and K. Venczel, eds., *Instrument Engineers Handbook,* vol. 2, Process Measurement, Chilton Book, Radnor, Pa., 1982.

6. Lewis, J. J., and D. W. Yordy, "On-Line Multidimensional HPLC with Immiscible Mobile Phases," vol. 8, no. 1, pp. 52–56, 1989.

Valve/Device Configurations.   Techniques such as foreflush, backflush, and heart cut to the detector or vent have been developed to produce component isolation. Such manipulations are actually part of sample preparation prior to concentration measurement. A portion of the sample may be trapped and stored on a section of column for later separation and measurement. This author has developed a technique, called trap and transfer, utilizing a six-port external volume sampling vale (Fig. 9). Once a sample has been injected into the system, any portion can be quantitatively removed and directed to a specified destination by taking a subsample. The original sample maybe a vapor or a liquid that is subsequently vaporized. This technique can be utilized to capture a portion of a sample or to

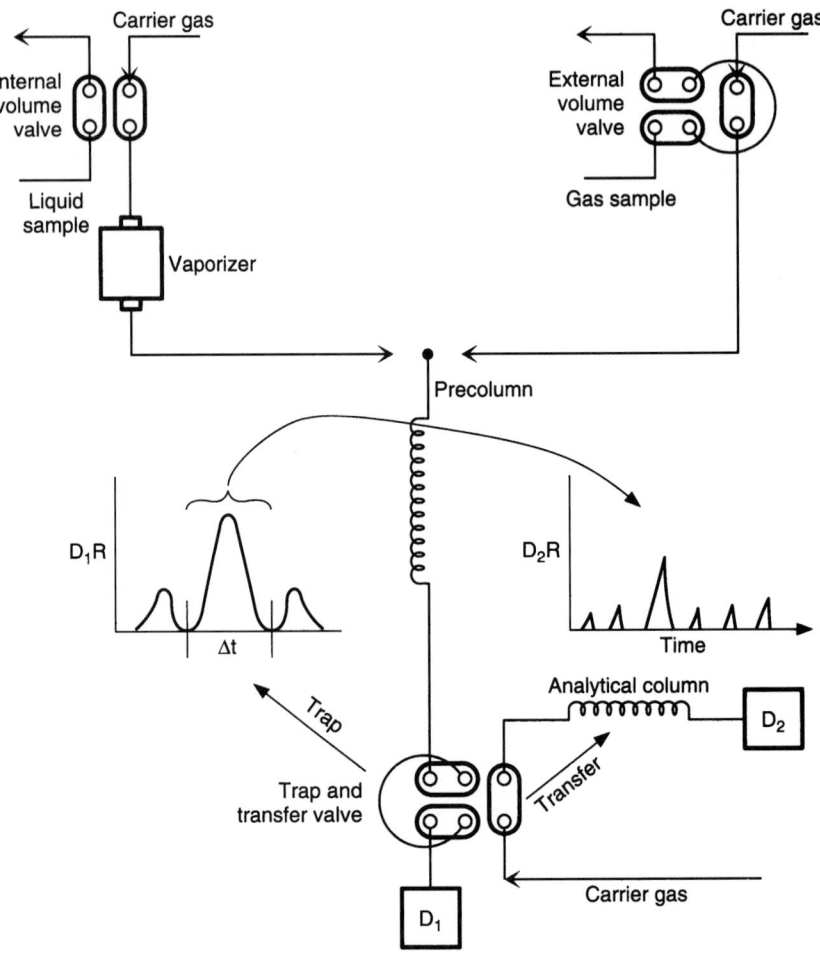

FIGURE 9 Trap and transfer valve operation.

dispose of it. The subsample can then be transported some distance from the tap without extensive heat tracing. Figure 10 shows a remote sampling arrangement that quantitatively disposes of the unwanted condensables, provides for automated cleanup of the sample conditioning devices, and ensures rapid transport of the remaining components to the measurement device. Other devices used for sample preparation (operators) can be configured in the same manner as columns.

Benefits of the Trap and Transfer Technique. The trap and transfer of portions of a sample has some major benefits:

1. Any peak broadening caused by diffusion in the precolumn is eliminated when the trapped portion is injected onto the second column.

2. The second column is an "independent analytical system" with temperature and flow rate isolated from the primary separation. There are no restriction valves to adjust and balance.

3. A large sample may be used on the precolumn and most of the material discarded. A larger quantity of the key component can be isolated without loading the analytical column and detector with unwanted material (concentration scheme).

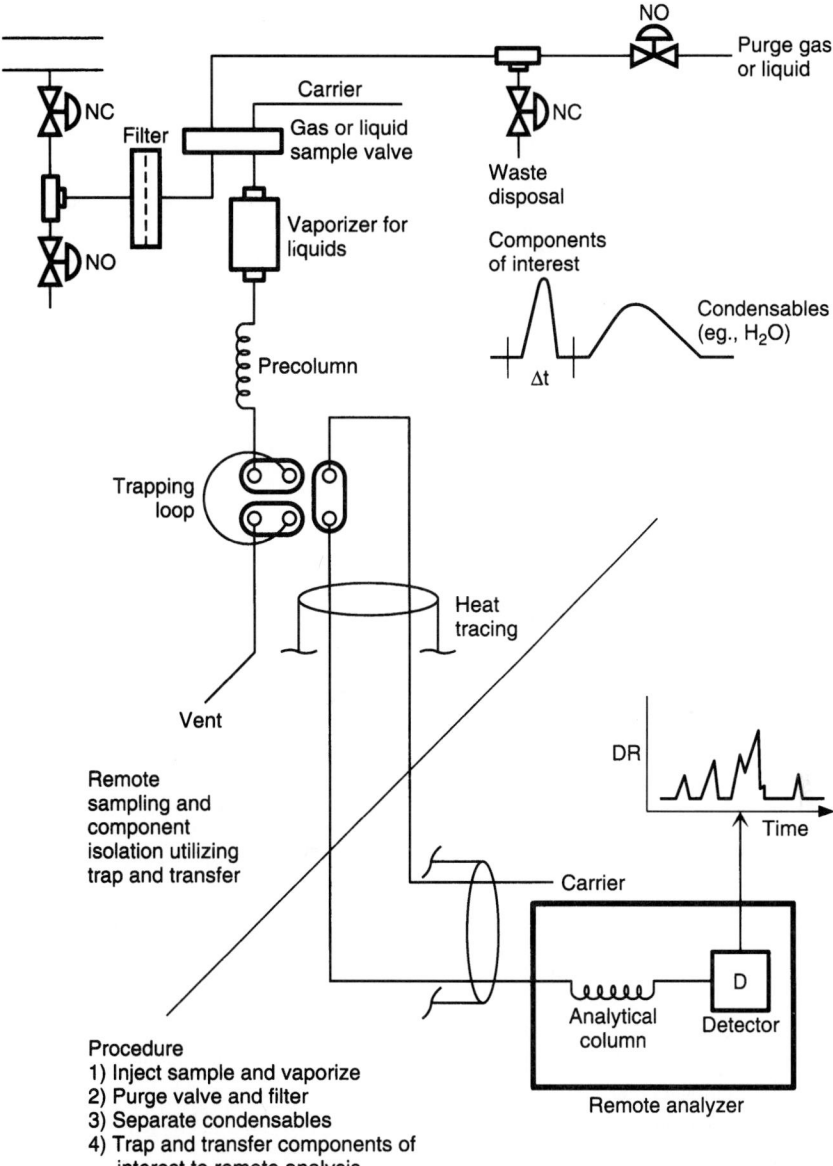

FIGURE 10   Condensables rejection.

4. A trace component on the tail of a major component can be trapped and transferred. The second independent system can better separate the two components because their concentration ratio has been reduced (improved resolution).

5. An uncoated column can be used to separate components of interest from high boiling or reactive residues, which cannot be backflushed from a reactive column material.

The trap and transfer technique promises to be an improved means of component isolation compared to the old directing traffic method.

CONCLUSIONS

Sample preparation may require many techniques to provide appropriate conditioning of a process stream. Reliable devices having controlled operation can be combined to produce a system that meets the overall needs of a specific application. *We do not need and cannot afford customization of each sample preparation system.*

APPENDIX A: LINEAR COLUMN COMPONENT SEPARATION PARAMETERS

   I. Introduction

   II. How the GC separation process works

     A. simple model—three steps

     B. stationary phase selectivity
       (1) dispersive interaction
       (2) dipole interaction
       (3) bacicity interaction
       (4) acidity interaction

   III. Type of column

     A. empty tube—An empty length of tubing will produce some component separation because of wall effects and analyte/carrier gas collisions.

     B. packed column—Packing the tube with a solid material will produce a greater surface area, which causes an increased interaction. The type of packing and the nature of the analytes passing over it will have various interactions. Coating the solid with a high boiling liquid will introduce additional types of interaction.

     C. capillary column—Coating the walls of small capillary tubing produces an increased resolution between analyte components, but this has several parameters which must be considered. Bonding of the coating to the tubing and crosslinking molecules of the coating material produces more stable and rugged columns.

     D. micropacked column—This is a tradeoff between larger packed columns and capillary columns. There are some good reasons to utilize these columns.

   IV. Stationary phase selectivity

     A. boiling point

     B. solubility

     C. polarity

     D. polarizability

     E. mixed phases

   V. Column diameter

     A. narrow bore (0.10, 0.20, 0.25, 0.32 mm)—lower capacity but higher resolution and shorter elution time.

     B. wide bore (0.53, 0.75 mm)—higher capacity, easier to work with and interface.

VI. Film thickness (0.1, 0.25, 0.5, 1.0, 2.0, 3.0, 5.0, 8.0, 12.0 μm)

    A. capacity (thicker film has a higher sample capacity but a higher bleed potential)

    B. resolution (thicker film has a higher resolution but a longer elution time)

VII. Column length

    A. resolution is improved by only the square root of length

    B. elution time is longer as column length is increased

    C. flow resistance increases (higher pressure is required to maintain the flow rate)

VIII. Temperature

    A. elution time decreases with higher temperature, but resolution is lost

    B. diffusion is greater at low temperature, causing band broadening

    C. higher temperature speeds up the separation process

IX. Carrier

    A. gas composition (H_2, He, N_2, Ar most common; some mixtures are utilized)

    B. linear velocity of carrier is a better measure than volume flow rate

    C. optimum velocity for best separation—tradeoff to reduce elution time

X. Internal pressure

    A. differential determines carrier velocity

    B. absolute can effect separation (supercritical fluid)

XI. Sample size (vapor and liquid)

    A. isolation

        (1) syringe

        (2) valve

        (3) pneumatic

    B. variability

    C. system loading

    D. detection limit

    E. special techniques

        (1) retention gap

        (2) partial venting

XII. Extracolumn effects on resolution

    A. sample injection

        (1) device

        (2) type

        (3) splitter

            a. repeatability

            b. discrimination

        (4) temperature

        (5) pressure

    B. tubing

        (1) size (volume)

        (2) material

    C. connections (dead volume and turbulence)

        (1) type of detector

        (2) makeup gas

(**3**) column termination
(**4**) detector restrictor connector
    a. flow stability
    b. performance

NOTE

It should be clearly understood that chromatography is solely a separation process. The chromatographic system accepts a mixture of substances and retains them to different extents so that, ideally, they are eluted as individual components from the system. However, if the eluent from the chromatographic system is monitored by an appropriate detection device that responds to solute mass or concentration, the quantity of each component present in the original mixture can be determined, thus producing an analysis. The determination, however, can *only* be obtained with the aid of the detector; the chromatographic system can *only achieve* separation.

SYSTEM CONTROL AND MANAGING DATA

by J. G. Converse*

SAMPLE SYSTEM AND ANALYZER CONTROL

The control of analyzer system operation progressed from electromechanical and electro-optical to fully digital electronic devices that also contained computational capability. We now utilize microprocessors, personal computers (PCs), minis, mainframes, and programmable logic controllers. One can still use cam and digital timers for simple operations. A big improvement in system control comes from digital electronics, which allows high-resolution timing (0.1 s or 0.01 s) and exact repeatability. Precise timing of sequential events is essential to multidimensional sample preparation that uses component isolation operators such as columns and switching valves.

Electronics for controlling instrument operating parameters such as temperature and pressure have contributed to more stable operation. This produces better repeatability of the system for multiple cycles, leading to a higher precision for component isolation. An advantage of improved measurement devices can be realized once the sample preparation function has been resolved. Computer hardware and software provide expanded capability for signal processing and data manipulation. Stable and precise sample preparation and measurement systems allow accurate analytical information, provided that adequate maintenance, calibration, and validation practices are applied.

The following Table 1 of timed events for injecting a discrete sample, performing conditioning operations, performing component isolation operations, directing analytes to different measurement

* Sterling Chemicals, Inc., Texas City, Texas.

TABLE 1 Example of Timing Table For System Control Device

Time (s)	Device	State (On/Off)	Function
000.0	timer	On	start analytical cycle
000.1	sample valves	On	open sample and vent valves to fill inject valve
005.0	inject valve	On	introduce sample
006.0	vent valve	Off	close the vent valve to direct purge air back through filter
010.0	inject valve	Off	return valve to purge and fill position
011.0	purge valve	On	clean inject valve and filter with purge air
080.0	T&T valve	On	trap and transfer selected analytes to analytical system
100.0	T&T valve	Off	trap and transfer return to normal position
220.0	gate 1	On	capture signal from component 1 and store
225.6	gate 1	Off	disconnect signal when component clears detector
237.2	gate 2	On	capture signal from component 2 and store
248.6	gate 2	Off	disconnect signal when component clears detector
271.0	gate 3	On	capture signal from component 3 and store
302.5	gate 3	On	disconnect signal when component clears detector
580.0	purge valve	Off	block purge air for next sample cycle
582.0	vent valve	Off	block vent line
585.0	sample valve	Off	block sample flow
600.0	timer	reset	start new cycle

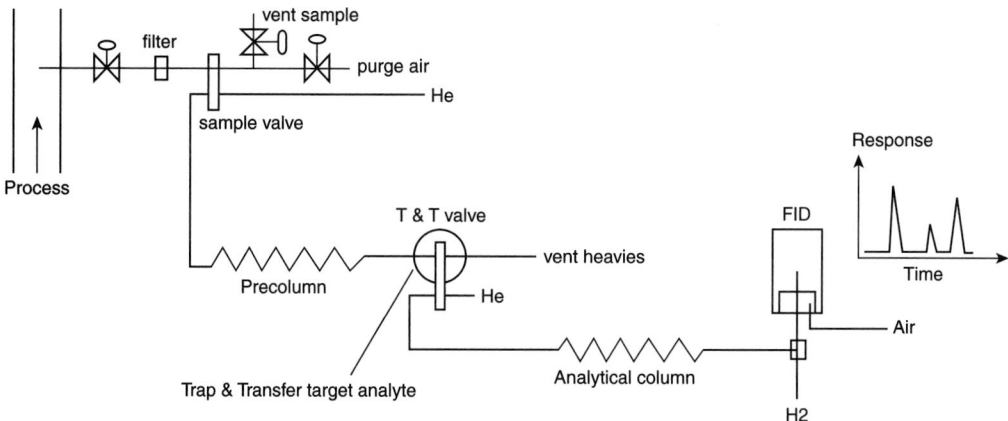

FIGURE 1 Diagram of valve system requiring control.

devices, and selecting the detector output when the analyte signal is ready for capture shows the general functions required. A diagram of the sample tap, sample valve, purge valves, precolumn, T and T valve, analytical column, and the detector with interconnecting transfer lines may be helpful in following the timed events in the table (Fig. 1). Several different kinds of devices have been utilized to control analyzer systems. The cam timer was used on early process chromatographs, and one optical programmer was developed by Mine Safety Appliances. The digital timer replaced the cam timer, which was then replaced by the digital microprocessor. This programming function is now being replaced by the PC.

The function of collecting the signal from the measurement device has also seen some evolution. The gated signal was originally fed to an analog strip chart recorder. The microprocessor digitized the peak area (or height if specified) and stored the information. It was sent to a printer along with other peaks and other information as a burst of digital information. Peaks moving out of a fixed time gate as a result of shifts in operational conditions caused the system to fail. Chromatographic software designed for PC operation in the laboratory is now being offered on process GC's. It can move with changes in a peaks elution time and do many more complex data-analysis functions. Digitized data are also sent directly to the process computer by means described in the section on communications. Two-way communication in the maintenance network allows the technician to evaluate the performance of the analyzer system and to make changes to the operating and data-processing parameters.

The programmer that has been developed so highly for process chronography can be utilized to control any discrete sample/flow injection system, including the sample preparation, component isolation, analyte measurement, signal manipulation, and information management.

Examples

1. Time of flight mass spectrometer—The very fast signal, "number of mass units with respect to time" can be digitized and stored in memory. Several samples can then be averaged to improve precision. The stored data can be fed to the programmer at a speed it can handle. Since the programmer is designed to accept "response vs. time" information, it can accept the data as though it were coming from a TCD, FID, or other detector measuring the effluent from a GC column. It has the ability to sort and store measurements, make calculations, and communicate with a process control computer.

2. Any scanning spectrophotometer—The absorbance with respect to wavelength or energy can be fed to the programmer at a rate that simulates the response of a signal coming from a GC detector monitoring a column effluent. Peak height corrected for tangent baseline would be proportional to concentration. Here again, the programmer can now perform all needed functions and communicate the results to the process computer.

3. Flow injection analyzers—The analyte flowing past a detector will generate a response vs. time signal. It will represent just one component, rather than several separated by column chronography, but the programmer can maintain a zero baseline by an autozero function between sample injections. It can integrate the area under the curve to measure all of the analyte present in the sample, thus producing a larger signal-to-noise ratio (see section on signal processing). Detectors utilized in flow injection analysis can encompass all of the measurement devices presented in Section 2.

The result of this approach to applying process analytical chemistry is that the number of different devices and the complexity of this technology could be reduced significantly.

SIGNAL MANIPULATION AND PROCESSING

There have been many significant changes in the hardware, system controls, and electronic data systems in recent years. It is hard to single out any one item that is the greatest, but microprocessors and PCs have certainly changed the way we do things. Smart sensors and analyzers are capable of bidirectional communication and can provide self-diagnostics. They can sense changes in operating conditions and correct for such changes.

A. Electronic devices for operating concentration/voltage transducers
  1. embedded diagnostic data acquisition sensors
  2. performance monitoring devices
B. area integrators vs. difference between voltages

Digital versus Analog Sensors

Future of Digital Signal Processing (DSP) Microprocessors in Detectors.   Generally, detectors have used analog signal processing techniques. For example, in most UV/VIS detectors, when the beams of ultraviolet light emerge from the beam splitter and pass through the two compartments of the flow cell (the reference cell and the active cell), a photodiode behind each cell monitors the energy level and emits a signal. An amplifier then takes this signal and sends a current that is directly proportional to the amount of light that passed through the cell. Detectors using analog signal processing have, for many years, proven to be effective, accurate, and dependable. However, at high sensitivity levels, problems caused by noise, interference, and electronic or ambient temperature fluctuations continue to be challenging, even in the best of systems.

When a relatively new technology—DSP—was discovered to possess extraordinary computational abilities, it seemed natural to explore its potential to overcome the challenges of high sensitivity detection. Consequently, instrument manufacturers are now exploring the use of digital filtering techniques in various types of detectors. Gilson, Inc. has integrated a DSP chip and custom software into its HPLC detectors.

High speed and sophisticated filtering capabilities are the two significant factors that make digital signal processing technology ideal for incorporation into detection instruments. Let's take a look at how a DSP-equipped detector works differently from a traditional detector with analog signal processing circuitry. Digital signal processing technology is application specific. Each DSP chip is individually programmed to perform a particular task, or series of tasks. As an example, a UV detector may incorporate a DSP microprocessor, programmed with custom-tailored, proprietary software designed specifically to provide a high level of digital filtering. Here's how it works: the detector takes ~50,000 "samples" or readings per second on two channels (active and reference). These "samples" are rapidly converted from analog to digital (A/D) format by means of high-precision, 16-bit resolution A/D chip. The DSP processes the 16-bit samples to approximately six decimal places. They are then logged, filtered, and transmitted to the host processor. The host then relays the samples (absorbance readings) by means of the serial input/output channel to the system controller.

DSP Provides Detectors with Greater Precision at High Sensitivity

Key to the success of digital signal processing is its ability to take vast quantities of data in an extremely short period of time, and modify, manipulate, and transmit it. But what does all of that mean in terms of real benefits to users of' detection instruments in laboratories? Here are some of the specific benefits of DSP technology over analog circuitry in detection devices.

1. DSP helps maintain the stability of the baseline at high sensitivity levels. In analog detectors, the circuitry itself contributes to baseline noise, with electronic noise and electronic temperature sensitivity, particularly at a high detection sensitivity. DSP allows the amount of analog circuitry to be minimized; thus the circuitry's noise level is held to an absolute minimum, significantly stabilizing the baseline.

2. DSP greatly improves detector performance at high sensitivity. Because the noise level is lower, detectors are capable of operation at higher sensitivities. For example, a sensitivity of 0.001 is usable because noise at that sensitivity is extremely low (less than 3%), and the baseline is virtually flat. With any detector, the noise determines the minimum detectable quantity.

3. DSP's temperature-resistant digital circuitry is more reliable and less vulnerable to outside stimuli. The electronics of DSP-equipped detectors are less sensitive to room temperature fluctuations, such as drafts, cold, or excessive humidity. Therefore, there is less effect on the signal from the electronics because they are inherently more stable. Also, the digital circuitry of DSP-equipped detectors is less vulnerable to external stimuli and less variable in structure than analog circuitry.

4. Superior readings from clearer, cleaner signals mean more accurate, reliable data. By taking in an extremely large number of signal samples in a short period of time and digitizing them, DSP ensures a higher degree of precision than analog technology can deliver.

5. Linear phase filtering eliminates high-frequency noise. With DSP, linear phase filtering techniques can be applied, which treat data equivalently. Linear phase filtering selectively and efficiently filters out high-frequency noise without introducing peak skewing. Nonlinear filtering, used in some detector designs, creates smooth peaks artificially by filtering out signals below a certain size, and this can result in small peaks being filtered out as noise. With linear phase filtering, the detector contribution to peak asymmetry is eliminated and small peaks are discriminated from noise.

From a manufacturing standpoint, the incorporation of DSP into detectors does not increase production costs. Although the DSP and A/D chips are rather expensive, higher material costs are offset by reductions in assembly and quality-control inspection times. The nature of the DSP chip is that it either works—or it doesn't. If it does, it is fully functional. If it doesn't, it is immediately replaced. There's never a need for "tweaking" or fine tuning the electronics. Given this parity in manufacturing costs between analog and digital circuitry, the dramatic and measurable improvements in detection sensitivity, accuracy, and performance of the newer technology ensure the growing significance of DSP detector technology.

DSP: A Refresher Course in the Technology

Ever since the powerful and rapidly developing computer technology called digital signal processing emerged a decade ago, it has brought about eye-opening changes into our homes, our automobiles, and our workplaces. The DSP chip has been hailed as the most exciting breakthrough in computer technology since the advent of the microprocessor. The potential inherent in DSP has only just begun to be tapped—and has given the world such time-saving, life-enhancing innovations as the compact disk player, high-speed modems and faxes, and voice mail. Soon to be perfected are high-definition TV, interactive TV, the videophone, and more.

DSP microprocessors are designed to perform very specific functions, and filtering is one of the functions they perform best. For example, DSP chips have been used to filter out engine and muffler noise in automobile operation. DSP-based shock absorbers allow cars to respond to bumps in the road with a counteraction that equalizes the ride. Compact disk players filter out electronic interference and ambient noise. And the soon-to-be marketed automobile "anti-collision" radar system is based on the DSP chip's sophisticated filtering capabilities.

Why is digital signal processing considered to be such a technological milestone? Let's take a look at how it works and what it can do.

Analog vs. Digital Signals

A signal is basically a continuous stream of information. Signal processing means acting on or modifying that information to make it understandable and usable. Analog information is represented by continuously variable values based on measurable quantities. Analog signals can be expressed in many ways—including music, images, voice, sounds, graphs, and light. Digital information, on the other hand, is expressed by discrete pulses that correspond to the patterns of binary digits—a sequence of zeroes and ones. Expressing information in this fashion places it in a logical domain, rendering it immune to environmental influences. How then can analog data be processed efficiently? Simply by digitizing it—that is, converting or translating it into digital code. An A/D converter chip is capable of receiving the signal and converting it into a sequence of numeric values that can then be processed by a digital computing device—the DSP chip.

The Many Advantages of DSP

Why is digital signal processing superior to analog? Here are some of the reasons.

1. It's more reliable. Digital signals are more reliable than analog since they work with logical values. They are not affected by temperature and are insensitive to changing environmental conditions. You can think of a DSP as a microcomputer with excellent mathematical computation skills. Just like your computer works day in and day out, so does a DSP.

2. It's flexible. This is the reason why DSP chips have been used to provide solutions to a wide range of mechanical, technical, and electronic problems. Each DSP chip can have its own set of signal processing commands built in, programming it to perform the function, or series of functions, for which it was designed. And if an improved method is developed, you simply need to replace the DSP chip; major hardware modifications are not needed for significant performance improvements.

In general, digital signal processors are dedicated instruments performing a specific function. The flexibility of a DSP allows it to be programmed for a range of functions—for example, sending a FAX, processing a document, printing out envelopes, playing music, and receiving voice mail on your personal computer.

3. It has greater dynamic range. There is a greater range of filters possible with DSP because it is a logical system that processes numbers mathematically. Filters that are virtually impossible in analog processing because of the need for high precision and low temperature sensitivity are simple to implement by means of DSP.

4. It's consistent. There are absolutely no variations from DSP to another. Whereas hardware components in analog circuits may have differences that result in varying performance, a DSP-based system always performs the same since it is working with logically identical mathematical values.

5. It's accurate. A properly programmed DSP will not degrade the signal of interest by adding noise or reducing its precision. In more concrete terms, the sound on the compact disk is more "real" to our ears than that on the plastic recording or magnetic audiotape because analog effects such as static or tape "hiss" are filtered out as the recording is made.

INFORMATION DISPLAY, STORAGE, AND COMMUNICATION

While it is a significant accomplishment to prepare the sample for analysis and measure the analyte concentration accurately, it is of little value if we cannot present this information to the customer in a clear, reliable, and useful way. Direct connection to the process computer for control purposes and information for maintenance technician are both important communication needs. That is what this section is all about.

1. Meters, charts, and control panels
2. Dataloggers, PC, DCS, and mainframe
3. Data highways (star, ring, bus)
4. Communication Protocols (superimpose digital on 4–20 mA analog signal, allow monitoring and downloading information, allow resetting and adjusting analyzer, use same menu driven system for reading and imputing information, remote diagnostics, maintenance, calibration, and record keeping)

- RS-485/RS-232
- Modbus
- Ethernet
- Token ring
- VistaNET(ABB Process Analytics)
- Fieldbus(ISA)
- HART (highway addressable remote transducer—Fisher-Rosemount)
- BRAIN (Johnson Yokagowa)
- Man–machine interface controller (Applied Automation, Inc.)
- (TCP/IP)
- Windows NT
- Winsock 1.1

One thing that standardization can provide is plug-and-play and true interconnectivity. Users want standardized communications protocols and data highways and plug-and-play connectivity and more bandwidth. They want yesterdays hardware and software to work with tomorrow's new technology.

They want tomorrow's instruments to work with today's controls systems and vice versa. Technology is always changing. A recent discussion on analyzer communication needs and interconnectivity revealed that protocol capability needs to move into the analyzer system. Ethernet, TCP/IP, Microsoft NT, and DCOM were mentioned as the vehicles of choice.

EXAMPLE OF CURRENT SYSTEM

Sterling Chemicals, Inc. has a network system that allows the engineer in his or her office to look at the data historian monitoring the analyzer performance.

The following diagram (Fig. 2) shows how anyone on the company computer network can monitor the performance of any analyzer in the system.

It also allows the onstream time to be measured accurately. Reliability and performance will no longer be a matter of conjecture or depend on who one talks to.

Figure 3 shows the output from a GC monitoring the quality of hydrogen cyanide. The system has an internal standard injected every cycle, so that it is obvious when changes in the analyte concentration are due to analyzer system failure. The target analytes in the process stream may vary with time, but the internal standard should always be constant. This graph shows the hydrogen cyanide analyzer concentration readout versus time for the internal standard and several critical components. The drop in the internal standard value occurred at ~6 a.m., but the analyzer was not checked until ~9 a.m. The column carrier flow had decreased, causing the peaks to move from their time gate. The appropriate maintenance was performed and the readout returned to correct values as demonstrated by the internal standard. The operators using the HCN feed material knew the analyzer was wrong and their feedstock was good. Operators focus on the internal standard so they know how to respond in real time and call analyzer maintenance instead of their raw-material supplier.

We have the hardware and the computer capability to analyze a process stream on line, introduce an internal standard, process the results, and display them to provide operators with the useful control tool of a real-time chemical concentration for critical components. Utilization of such information is

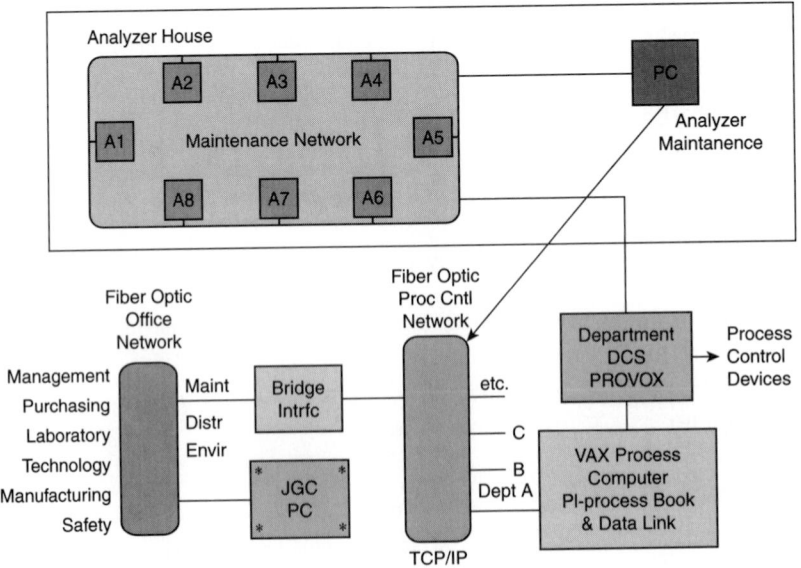

FIGURE 2   Communications network.

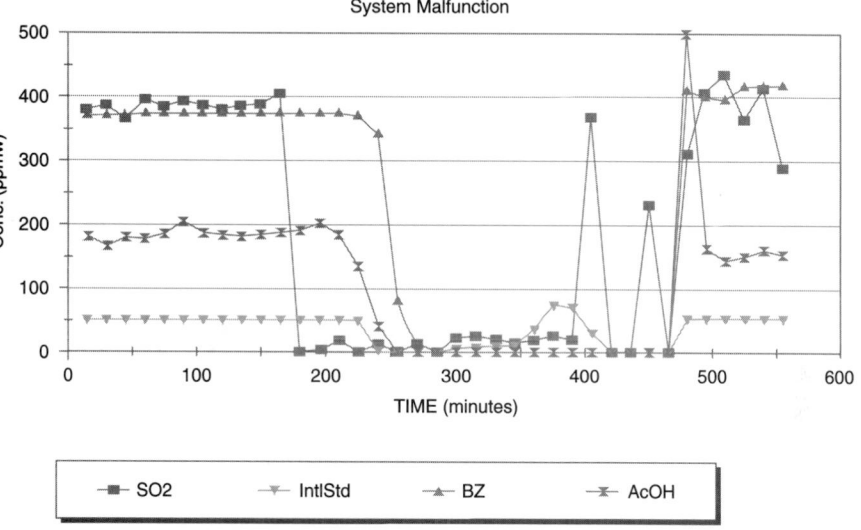

FIGURE 3   HCN analyzer output vs. time.

vital to the chemical process industries. The main problem with process analyzers is their reliability and credibility. Real-time validation can play a major role in overcoming this problem if it is properly utilized.

ENCLOSING THE HARDWARE

1. Power requirements
2. Utilities and sample lines
3. Signal transfer
4. Cabinets and houses

Housings and Ovens

Weather protection is necessary because of the electronic nature of most analyzers and because maintenance personnel have to work on the equipment. The degree of protection may range from a rain/sun protective roof to a fully environmentally conditioned building. Various enclosures may be utilized to isolate analyzer equipment for weather protection or safety in a hazardous location. Electrical codes usually decide the type of enclosure selected. IEC 1115 prescribes guidelines for Safety in Analyzer Houses, and wiring is regulated by IEC Code. Analyzer houses provide a good working environment for analyzers and technicians, but they can be a trap for hazardous vapors. Remote discrete sampling and multidimensional sample preparation allow small aggregates of equipment to be distributed in separate enclosures, thus avoiding the immense cost and potential hazards of houses.

Temperature is the most critical parameter in most sample preparation procedures and analytical measurements, so temperature control demands careful attention. Precision of temperature control will be dictated by the specifications of the measurement. Several methods of temperature control are available in different price ranges, but an excellent controller is of little use if wrongly applied. Temperature controllers are generally selected from specifications and then configured to meet the requirements of the current task. Modular units make design and fabrication of the total system much easier. Modules can be selected per need to avoid the expense of excessive features. Independent sample conditioning ovens are not as readily available as would be desired if the modular concept were adhered to. The oven must have adequate insulation for the internal temperature required and the external conditions. It must have the power for a range meeting the application requirements as well as control precision. A proper temperature sensor or sensors must be appropriately located to provide the appropriate feedback to the controller. As a minimum, 100 psig air pressure supply must be provided for valve actuation and a second regulator to provide heater air. Adequate space must be specified to contain one or more valves and separation columns as needed. The oven should be shallow to prevent hardware from being located in front of other components to simplify maintenance.

CALIBRATION AND VALIDATION

by J. G. Converse[*]

Appropriate sample system design will keep fresh sample flowing and present representative material to the measurement device. That is not sufficient to ensure reliable analysis unless maintenance and calibration are made an integral part of the total design. Regulatory agencies are dictating that appropriate calibration practices be followed for continuous emissions monitoring systems (CEMS). ISO-9000 standards will dictate further guidelines for the way we apply calibration and maintenance operations. Discrete sampling utilizing the flow injection analysis technique provides a sound technical approach to implementing automated maintenance, internal reference calibration, and real-time on-line statistical quality control. Examples will be presented to demonstrate applications of these functions.

Analyzer application professionals have implemented hardware and technology beyond the ability of many persons to maintain the systems we install. Management frequently does not fully understand maintenance requirements, and budget limitations prevent proper staffing of sufficiently qualified personnel. We have developed paradigms for process analytical chemistry that may be leading us into serious maintenance problems. We have combined sample preparation, measurement, and signal handling into a single, large, expensive package. Our customers do not understand why initial capital and continuing maintenance costs are so high for a simple chemical analysis. We need to examine our customers' needs and seek creative, cost-effective ways to serve those needs.

We need to change our paradigms, that is, our perceptions of how on-site automated chemical analysis should be performed. Possibly, customers would prefer a simple, reliable, and maintainable analyzer system that does not constitute a major capital project. Perhaps an analyzer system should concentrate on sample preparation and measurement utilizing modular components. Design should minimize sample usage, conditioning, response time, maintenance, and response variance. Only necessary functions and components should be included in the total package. A signal proportional to the target analyte concentration in the process should be the product of the analyzer system. It must represent a precise and accurate measure of the component population in the unadulterated process stream. Some people in the analysis instrumentation industry prefer a complex device that covers all needs for every situation. Others fail to realize that an excellent measurement on a bad sample is worse than no information at all. We need to direct our attention first to sample preparation, which presents a representative material to the measurement device. Such systems must also include consideration of maintenance and calibration needs. We need to stop talking about "keeping the system simple" and take action to honestly do it. To achieve such a goal, we have to be willing to accept change, listen to new ideas, and evaluate them carefully.

[*] Sterling Chemicals, Inc., Texas City, Texas.

STANDARDS PROCUREMENT

This is the procurement of "certified standards" to calibrate analyzers. The measurement is a comparison process in most cases. Absolute measurements are rare and difficult, so we generally utilize the practice of ratioing an unknown signal to that of a standard. In the case of chemical concentration measurements, the standard is prepared from first principles, such as weight, volume, pressure, and so on. Procedures must be developed for preparation of calibration samples, introduction of the material into the analyzer system, treatment of the measurement signal, and adjustment of the testing device. The purpose of these activities is to remove bias from the resulting measurement such that an accurate representation of true analyte concentration is produced.

Certified standards are required to demonstrate accuracy of our analyzer measurements. It is of no avail to calibrate an analyzer system that is "out of control" and "not capable." We must first demonstrate long-term stability, precision, and capability by means of real-time on-line control charts before calibration is meaningful.

Certified standards will be required for all product measurements related to customer quality specifications, both in the lab and for field measurements. Some will also be required to meet internal customer specifications. Procedures for good laboratory practices regarding the preparation of certified standards must be prepared and assembled into a calibration plan. A request form must be structured to provide adequate information to those preparing standards. An example is attached as a starting point. Equipment such as the balance used for weighing and all volumetric devices must be certified. Purity of solvents and analytes must be established and documented. Training on procedures will be required, along with records on training and actual standards preparation. Alternate analytical methods may sometimes be required for verification and capability for chemicals purification may be needed.

The program for certified standards preparation must be compared to cost and availability of purchased standards. This is a major effort and should not be undertaken without a thorough understanding of what it entails.

CERTIFIED STANDARD SAMPLE PREPARATION REQUEST FORM

Requestor:_____

Phone #_____ Instr. FI#_____ Date Required:_____

Phase: Liquid_____   Gas_____   Quantity:_____

Solvent:_____Purity:_____

Analytes:_____(Spec)_____(Spec)_____(Spec)
Calc. Conc._____(units)_____(units)_____(units)

Preparation Procedures:

Standard Prepared by:_____Date:___

Permeation Devices

Considerable development has produced some very reliable permeation devices for calibrating and validating on-line analyzers. A permeable tube (e.g., plastic) contained in a controlled environment is used to deliver a known quantity of analyte per unit time into a carrier gas to produce a known concentration. The permeation rate is determined by weighing the tube periodically. Temperature, pressure, and flow are critical variables that must be tightly controlled. The tube passing the carrier gas may, alternately, be immersed in the target analyte, but this provides only a reference unless the weight of the analyte can be determined. The standard material produced in this fashion may be exchanged for the unknown sample, or a discrete volume can be introduced into the system carrier gas. An example in which this is utilized is an area monitoring system. Twenty-nine sampling points are collected at a common manifold. Position number thirty is the permeation standard stream so that the analyzer is checked with an external standard once each full cycle.

METHODS FOR CALIBRATION

External Standards

One way of making comparison is to utilize the measurement of the process relative to an external known material. The external standard must be substituted for the process material within the analyzer. This places certain requirements on the system.

1. The system response must not change during the period of running the standard.
2. The sample size must remain the same for each test.
3. The measurement of the process must be interrupted during calibration, causing loss of information.
4. The cost of standard preparation and calibration adds to the system maintenance cost.
5. The system is perturbed by changing from unknown to standard and back to the unknown.
6. Several measurements of the standard must be made to establish stability and obtain an average value.

The quality of the results obtained using external standards is exceptionally good, considering all the possible sources of error. There are many arguments for and against using external standards. Quality and stability of standards will be discussed later.

Internal Standards

A comparison can be made by introducing a known amount of reference component into the unknown sample. The concentration of the target analyte is calculated from the ratio of the responses of the reference and target components. These response factors must first be determined from measurements on a sample of known analyte concentration. In the laboratory, the internal standard is added by weight or volume and mixed with the sample material. An automated on-line analyzer *does not* sense whether the materials are premixed. *It is only essential that the relative amounts are constant.* This allows the internal standard to be introduced at any appropriate time during an analysis. Several advantages are obvious.

1. System drift has less impact since the reference and target analyte are measured in close proximity.
2. The real-time measurement of the process is not interrupted.
3. Neat material can be used, thus avoiding preparation of standards, and calibration can be automated.
4. The system is not perturbed by switching samples.

The relative size of the process and reference samples is critical and must not change significantly. This technique should be considered, and ways to overcome objections should be sought.

Control Samples

This technique is similar to internal standard, except a known standard containing the target analyte is introduced intermittently between unknown samples. Ratioing the unknown to the known eliminates determining a response factor. It is generally utilized when the measurement system cannot be stabilized easily. In chronography, it has been termed "deferred standard." The injection of the standard is delayed such that none of the separated peaks interfere with the peaks of the unknown. Two valves can be used, but relative sample size is critical. An alternate flow of unknown and known through a single valve eliminates sample size consideration, but risks cross contamination. Preparation of standard is required but introduction is automated. Real-time measurement is not interrupted, but the time between unknown information update is lengthened. A single measurement on each of the materials alternately may degrade the precision of both. Care must be used when applying this technique.

Secondary Standard

Sometimes a stable and representative primary standard is difficult or impossible to obtain. Normal practice has been to draw a sample of the process stream and analyze it in the laboratory. The process analyzer is then adjusted to match the laboratory results. This technique has been highly criticized and should only be used when there is no other choice or the absolute value of the analyzer is not critical. The lab measurement is not infallible, nor is it certain that the primary standard used by the lab is correct. One should carefully study ASTM D 3764–92 on "Standard Practices for Validation of Process Stream Analyzers" to obtain guidance. Statistical practices must be utilized to obtain the best possible calibration accuracy, no matter what method is selected.

Auto Calibration

A calibration standard sample is prepared, placed in a vessel, connected to the analyzer system, and introduced periodically by automatic control. The process sample is bypassed during the period of calibration. The procedure is exactly the same as the manual calibration, except that a complex control scheme is required. Switching and identification add extra cost and complexity. Large amounts of standard must be prepared, and extreme care must be taken to prevent calibration when the standard runs out or changes. This means verification of the calibration measurement prior to changing analyzer settings.

DECISIONS TO CHANGE ANALYZER CALIBRATION BASED ON STATISTICAL QUALITY-CONTROL CHARTS

The periodic calibration of analyzers can lead to overadjustment, which introduces unnecessary error into the process measurement. Multiple analyses of reference samples are averaged and plotted on control charts to monitor analyzer performance. Upper and lower control limits are calculated for X and R. The analyzer calibration is not changed as long as all points fall within the control limits and pass additional tests. When a point falls outside the limits, a search for a probable cause is made and the problem corrected. Internal reference injection can be automated to provide data points for control charts. Analyzer capability can also be calculated. Once the analyzer system is operational, variables are evaluated and controlled to some acceptable degree. Calibration begins at some frequency and hopefully is required less often as the system is improved. No systematic procedure is followed to monitor "out-of-control" (OOC) results or determine the "measurement process" capability. Without control limits, the analyzer chugs along until the user complains about credibility. Without the use of statistical quality control (SQC), we have no valid measure of analyzer performance.

We hope to explain how SQC can be applied to process analyzer calibration and maintenance programs to improve the quality of the measurement process. A procedure will be developed to produce control charts, find probable cause for OOC results, and take action to bring the system back into control. A quantitative measure of control limits must be established to ensure that we don't overadjust the analyzer system. If the system is in control, don't fix it.

Statistical Quality Control

SQC is not just a fashionable subject or a short-term fad. It's the way we must demonstrate quality of our products if we are to remain in business. We must understand it, apply it, and really believe in it if we are to improve quality and productivity. SQC is based on the premise that the statistical treatment of data can help us separate systematic and random error. When that has been achieved, we can take action to remove the systematic error and improve our results. The subject of statistics and the details of implementing SQC are beyond the scope of this paper. We hope to present a cursory overview such that interested parties can pursue the practice further. Basically, averaging subsets of data points, quantifying control limits, and evaluating system capability are techniques to determine how good we are at doing whatever it is we do. First, one must make a large number of measurements and plot the results on a histogram to determine the distribution of our measurements. The size of the subset of data points for a control chart is established by that distribution. The averages of nonnormal distributions are essentially normally distributed if the subset size is large enough. A procedure is developed and communicated to those running the measurement process so that they can produce control charts and use them to control that process. Finding probable cause for OOC points and taking action to correct problems is not easy to implement. Collecting data and plotting charts are futile acts if nothing is done to improve the process. One just has a document showing how good or bad we are doing. There is no improvement without action.

Finding Probable Cause and Taking Action

Too often we think we see a change in calibration and adjust the analyzer response. Figure 1 demonstrates how this can cause overadjustment.

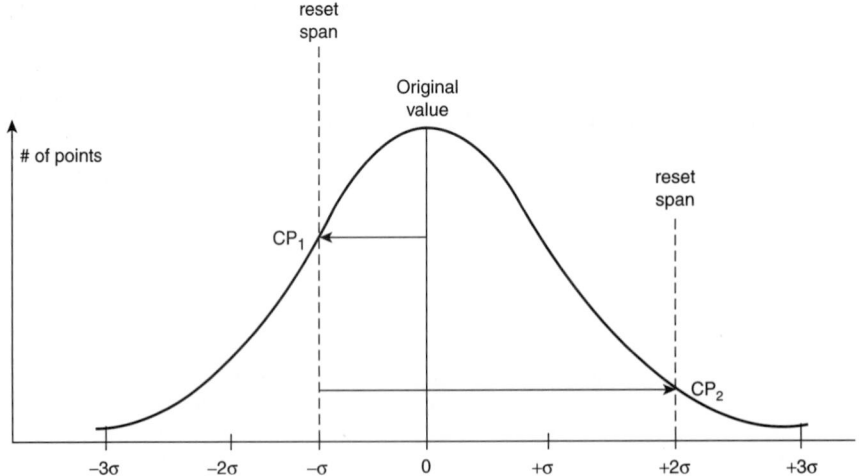

FIGURE 1   Overadjustment by resetting the analyzer span.

If the new calibration point, CP1, is within the control limits determined from the normal distribution, no change should be made. The next calibration point, CP2, may be on the other side of the distribution curve. Changing the span of the analyzer when the calibration points are within the established control limits is overadjustment. Leave the system alone when it is in statistical control.

When a calibration point is out of control, it does no good to recalibrate until the probable cause has been investigated. Once the problem is corrected through proper action, the system can be recalibrated. Many analyzer technicians spend a good portion of their careers recalibrating analyzers. If they find the cause and take proper action, they will ascend into the "Maytag" stage of maintenance grace. They will cease to calibrate when the system is in control. It is the responsibility of the analyzer technologist to teach them to perform their work in this way and to monitor maintenance practices.

The best approach to finding probable cause is to fully understand how an analyzer system works. All of the variables affecting system response must be defined and controlled [1]. Diagnostic procedures must be developed to identify which variables are out of control. Quoting Walter Trump, "Nothing is magic; there is a reason for everything." There is a cause when a system is out of control. When it is found, it can be fixed. It is generally true that most analyzer technicians can fix an instrument when they know what is broken. Their biggest problem is to diagnose what is causing the OOC results. Medical doctors have the same problem, but the systems they work on are much more complex. We can see from this why it is vitally important to keep analyzer systems simple. Reduce the number of variables and the cause of lost control can be more easily determined. Figure 2 shows the variables of a vapor/liquid sample conditioning system.

$$R = C_i \times f \left(f_g, f_L, T_L, L_L, P_v \dots Z \right)$$

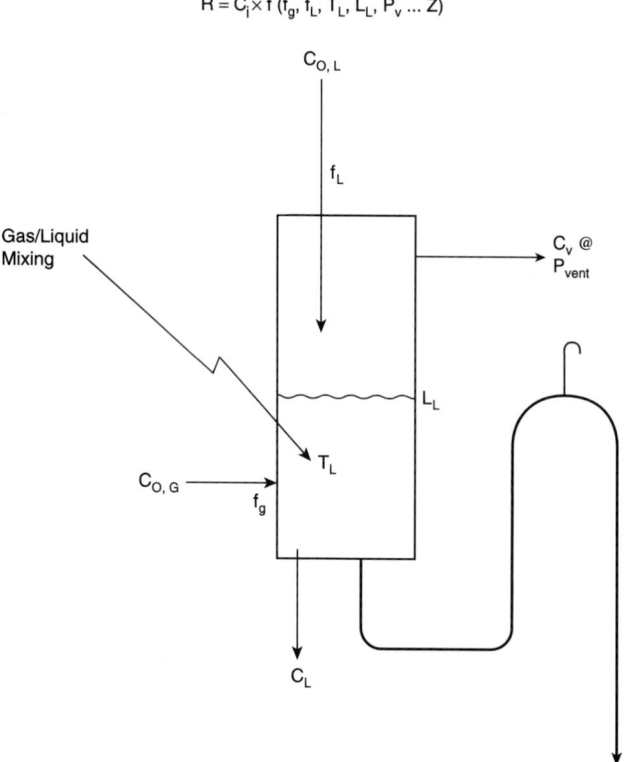

FIGURE 2   Vapor/liquid sample conditioning system.

All analyzer system variables including the measurement device and control and readout systems should be defined by a nonanalytical equation and evaluated. This provides a good starting point for problem diagnosis. Finding probable cause requires the action of breaking a complex system down into simpler subsystems. When we have a problem with a GC/computer system, we replace the computer with a chart recorder or a recording integrator. We then eliminate the sampling system and introduce a constant known reference sample. Continuing to reduce these subsystems into smaller cells, we ferret out the cause, changing one control device at a time. That is called troubleshooting. It is amazing that many people do not know about this procedure.

Statistical Process Control Utilizing Control Charts

A control chart can be constructed in several ways. We use run charts, X and R charts, and individual X and moving R charts, depending on the conditions of measurement [2]. A proper subset size is selected and data points are treated statistically by the appropriate procedure. The results are then plotted on graphs with respect to sequential time of data collection. After sufficient averages and ranges have been determined ($\sim$25), values of X, R, UCL, and LCL are calculated. Data points continue to be collected and added to the charts. When an average or range value exceeds the upper or lower control limit or fails one of several tests, we define the system as out of control. That is the time when we need to do something. Find the cause and fix it.

When we have the system in control, we can calculate capability and see if our measurements are acceptable within specification. The information from an analyzer that is in control and capable can then be used to apply statistical process control to manufacturing processes.

Two tests can be used to determine if the measurement process is adequate.

1. Adequate significant figures: $\dfrac{\text{MeasIncr}}{6(o\hat{\,})t} < 10\%$

2. Measurement adequate to detect real process changes: $\dfrac{\text{MeasIncr}}{6(o\hat{\,})t} < 20\%$

These indicators state that the measurement increment divided by six times the total standard deviation of the system must be less than or equal to 10%, and that the ratio of the measurement variance to the total variance must be less than or equal to 20%. Control charts on product quality can then be used to determine if the manufacturing process is in control and capable.

Automating Control Charting.   Control charting an analytical measurment manually is a lot of work and time consuming. A calibration sample must be analyzed *n* times (subset size). The values of the averages and ranges must be calculated and plotted. Values of X, R, UCL, and LCL must be determined periodically. A continuous analyzer must be taken out of service and alternately fed zero and span fluid. A conventional discrete sample analyzer (e.g., GC, LC, FIA) must also be disrupted from service to introduce a span fluid. The internal reference injection method [1] can provide control chart data on line without disrupting analyzer service. Computer treatment of the data can provide plotted charts of the calculated results. The proper selection of subset size and avoidance of auto correlation will produce a statistically sound evaluation of analyzer performance. The use of FIA for process monitoring has been demonstrated for years under the name of gas and liquid chromatography. Other detection methods employing flow injection analysis (acids and bases) [3] have been demonstrated. A general concept using FIA as a basis for automated on-site analytical chemical analyzers has been discussed [4]. The technology and the hardware are available. Only a little imagination and the will to do things differently are required to implement SQC for process analyzers. This must be done before we can apply SPC to manufacturing processes to improve quality and productivity.

We have generated a control chart of subset size one for our HCN analyzer to monitor performance continually. A second sample injection valve was installed ahead of the analytical column to introduce a discrete quantity of acrylonitrile (AN). The FIA analyzer separates and measures SO_2, AN, BZ, and ACOH in byproduct HCN. A second AN peak was generated by the discrete reference sample.

It should remain constant in area as long as the analyzer response does not change. The peak area is displayed on a trend recorder in the control room beside the four components being monitored. The operators use the reference AN trend to determine if the analyzer is functioning correctly. The measured components vary as the process changes. The reference trend provides the control chart to indicate analyzer problems immediately if they occur. A subset of n data points could be generated by programming the analyzer system to make the appropriate calculations and trend the averages or ranges. Calculating and marking the control limits would quantify performance.

These can be used to guide operations in making decisions about requesting analyzer calibration and maintenance.

Applications of SQC.   Diethylbenzene in ethylbenzene: Initial observations of our ethylbenzene analyzer indicated that the measurement process was OOC and not capable. The probable cause was determined to be sample flow rate sensitivity. When the sample flow rate was changed, the area of the component peaks changed. This was probably due to flashing of the sample in the injection valve.

Analyzer modifications: A new Amscor four-port liquid sample valve with a 1-μl slider volume was installed. The sample flows straight through this valve, and there is less chance of flashing. The electrometer electronics of the applied automation GC were modified so that the range could be programmed. This allowed the electrometer to measure a peak of 100% EB or a ppm peak of diethylbenzene (DIEB). A range control software program was installed to change range on command. The detector was tuned so that it could measure both the large EB peak and the small DIEB peak accurately (see Fig. 3).

Since the EB concentration does not change a significant amount during normal operations, it can be measured and used to correct for sample size variation. A printer interface card, a rotary switch, and software were added to allow printing of the analytical results. This allows the maintenance technician to monitor analyzer performance closely.

Initial testing: Initial testing was done to establish accuracy, precision and control limits. The tests showed a minimum detectable limit of 4 ppm for the DIEB measurement. This minimum is due to DIEB eluting on the tail of the large EB peak and therefore having a sloping baseline. Tests also show some offset of the benzene measurement. This is due to interference from some of the nonaromatic components. A column isolating the BZ could be selected if needed.

Calibration Procedure: A calibration procedure has been developed and documented for the maintenance personnel. Since absolutely pure EB was not available, correction was made for residual

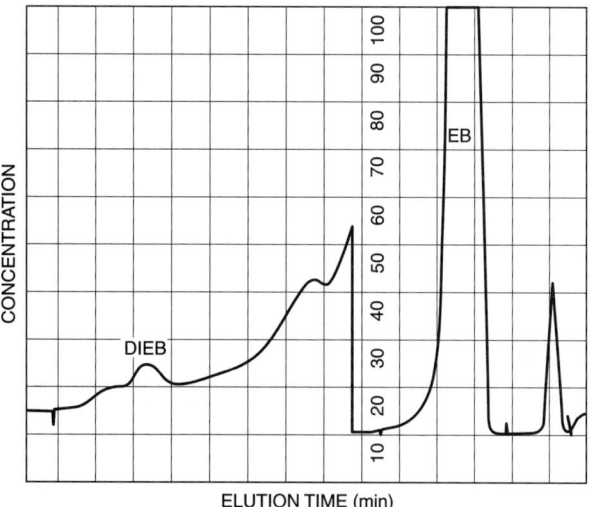

FIGURE 3   FID response to EB sample.

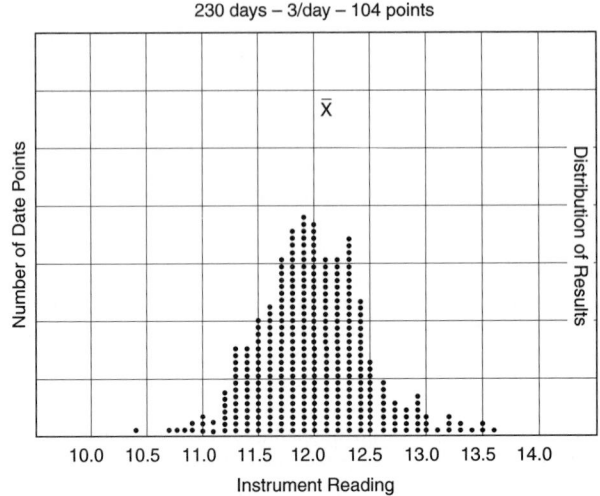

FIGURE 4 Distribution chart of measured results.

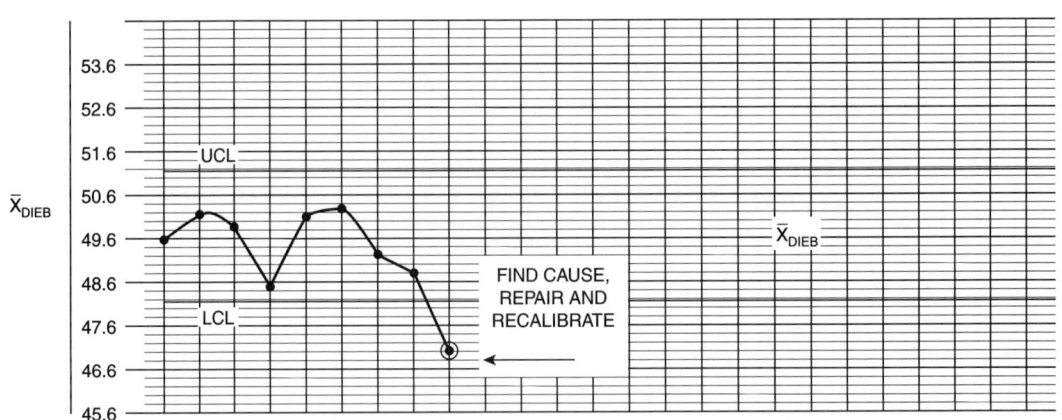

If $\bar{X}_{DIEB}$ falls outside the control range (48.1 to 51.1), find cause and correct it. Recalibrate the detector response factors as follows:

$$\frac{\text{Difference DIEB} * \text{Response Factor}}{\bar{X}_{DIEB}} = \text{New Response Factor}$$

FIGURE 5 Control chart for DIEB.

impurities. A sample of high-quality EB was analyzed and then spiked (standard addition). Concentrations of the key components were determined. A calibration sample was run for 24 h to generate a database. A distribution chart was prepared to determine the size of the subset (see Fig. 4).

The data indicated a normal distribution. A subset of five was selected for preparation of the average control chart (see Fig. 5).

Autocorrelation of the data caused the control limits to be too small. Three results averaged per point and two points taken >8 h apart averaged to form a subset worked better. Initially, the calibration

File:	44A102
Company:	PW Systems
Plant:	Syn–Gas
Department:	I/E
Machine:	44A102
Operation:	Oxygen
Characteristic:	span
Units:	percent

1	21.04	21.09	20.95	20.88	20.91	21.13	20.99	21.02	21.08	21.19	21.13	21.14	21.08	21.02	20.94	20.88	20.80	20.81	20.78	20.75
2	21.09	21.08	20.94	20.88	20.91	21.13	20.99	21.02	21.09	21.19	21.13	21.13	21.07	21.02	20.94	20.88	20.79	20.81	20.78	
3	21.09	21.06	20.93	20.88	20.93	21.13	20.98	20.99	21.10	21.17	21.12	21.13	21.06	21.02	20.93	20.85	20.77	20.79	20.77	
4	21.09	21.06	20.94	20.88	20.95	21.13	20.97	20.98	21.11	21.16	21.11	21.12	21.05	21.00	20.91	20.84	20.76	20.77	20.77	
5	21.09	21.06	20.93	20.88	20.98	21.15	20.98	20.97	21.11	21.14	21.11	21.10	21.04	20.97	20.91	20.81	20.77	20.76	20.76	
6	21.10	21.06	20.95	20.89	21.02	21.16	20.97	20.96	21.11	21.13	21.11	21.10	21.02	20.95	20.88	20.80	20.76	20.75	20.75	
7	21.11	21.04	20.93	20.91	21.05	20.95	20.97	20.97	21.13	21.13	21.09	21.07	21.02	20.95	20.87	20.78	20.74	20.75	20.75	
8	21.12	21.04	20.93	20.91	21.07	20.95	20.96	20.98	21.13	21.13	21.09	21.07	21.02	20.95	20.85	20.78	20.75	20.76	20.75	
9	21.11	21.03	20.91	20.93	21.09	20.96	20.97	21.00	21.14	21.13	21.09	21.07	21.02	20.95	20.86	20.79	20.75	20.76	20.74	
10	21.13	21.01	20.91	21.91	21.12	20.99	20.98	21.02	21.15	21.14	21.10	21.07	21.02	20.95	20.87	20.80	20.76	20.78	20.74	
11	21.12	20.99	20.90	20.91	21.11	20.99	21.00	21.04	21.16	21.13	21.12	21.08	21.02	20.95	20.87	20.80	20.78	20.78	20.74	
12	21.09	20.96	20.88	20.91	21.13	20.99	21.00	21.05	21.18	21.14	21.13	21.08	21.02	20.95	20.88	20.80	20.80	20.78	20.75	

Sample	1	2	3	4	5	6	7	8	9	10	11	12	13	14	15	16	17	18	19	20
X-bars:	21.100	21.040	20.923	20.892	21.022	21.054	20.980	21.000	21.124	21.148	21.111	21.098	21.037	20.973	20.892	20.817	20.769	20.775	20.757	
ranges:	0.098	0.130	0.070	0.050	0.220	0.210	0.040	0.090	0.100	0.060	0.040	0.080	0.060	0.070	0.090	0.100	0.060	0.060	0.040	

X-Bar Chart

LCL = 20.967 MEAN = 20.997 UCL = 21.026
using Tabular constants, based on samples 1.6

Range Chart

LCL = 0.032 MEAN = 0.112 UCL = 0.192

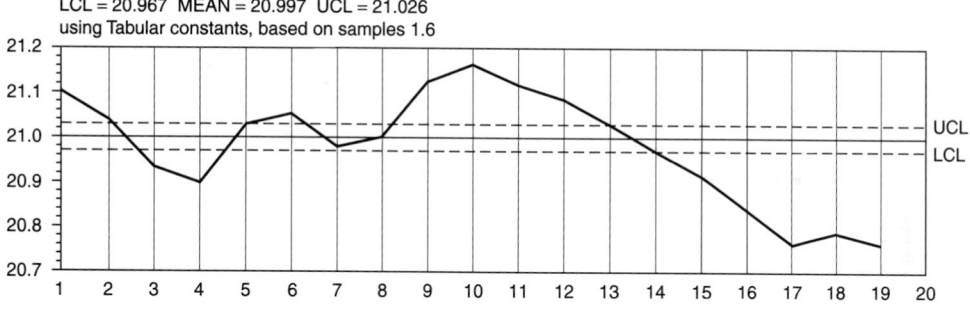

FIGURE 6   Control chart showing cross correlation for oxygen analyzer.

sample was run daily and the calculated values were plotted. UCL and LCL were set at 3σ from the distribution curve data. No change in calibration was made as long as the data points were in control. After average and range values were collected for 1 month, new values for the average of the averages and average of the ranges were calculated. Valid control limits were calculated from the average range and the standard deviation.

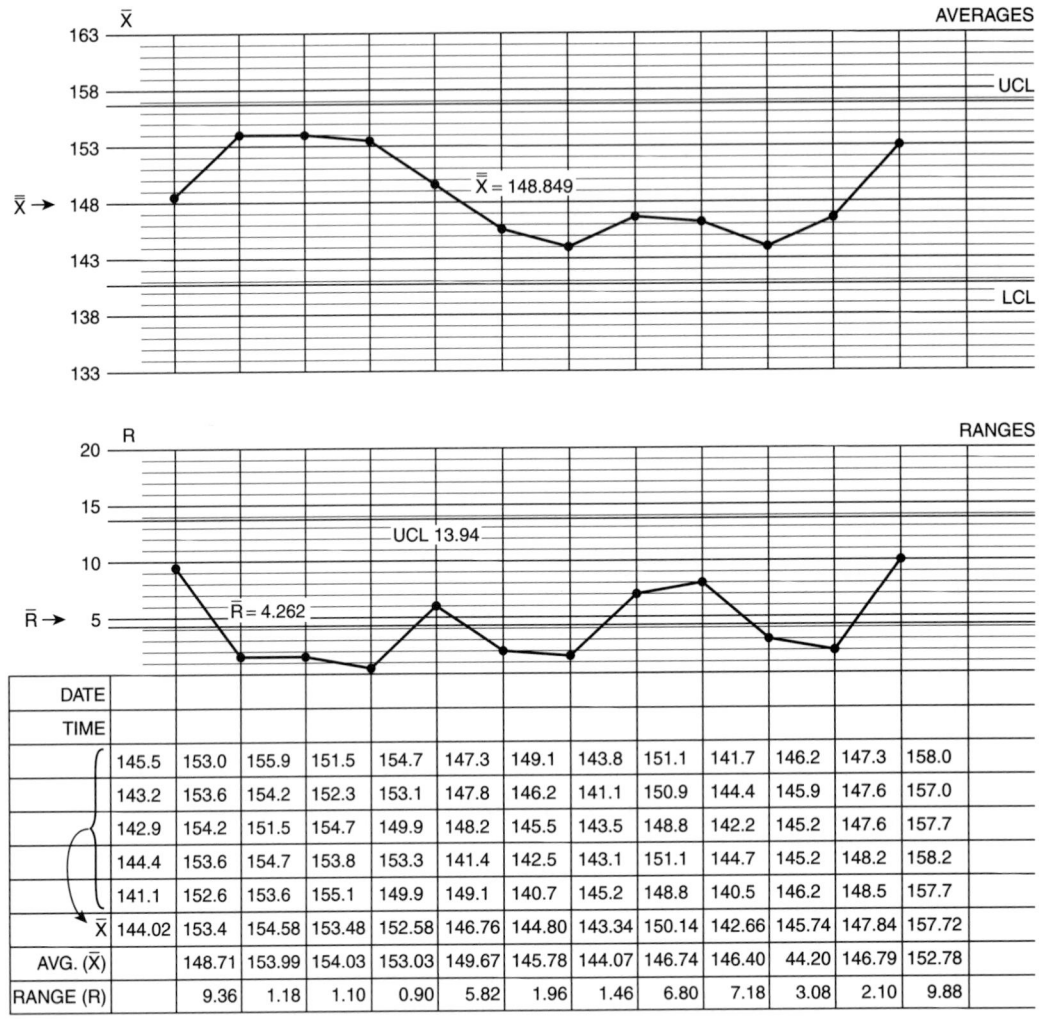

DATE													
TIME													
	145.5	153.0	155.9	151.5	154.7	147.3	149.1	143.8	151.1	141.7	146.2	147.3	158.0
	143.2	153.6	154.2	152.3	153.1	147.8	146.2	141.1	150.9	144.4	145.9	147.6	157.0
	142.9	154.2	151.5	154.7	149.9	148.2	145.5	143.5	148.8	142.2	145.2	147.6	157.7
	144.4	153.6	154.7	153.8	153.3	141.4	142.5	143.1	151.1	144.7	145.2	148.2	158.2
	141.1	152.6	153.6	155.1	149.9	149.1	140.7	145.2	148.8	140.5	146.2	148.5	157.7
$\bar{X}$	144.02	153.4	154.58	153.48	152.58	146.76	144.80	143.34	150.14	142.66	145.74	147.84	157.72
AVG. ($\bar{X}$)		148.71	153.99	154.03	153.03	149.67	145.78	144.07	146.74	146.40	44.20	146.79	152.78
RANGE (R)		9.36	1.18	1.10	0.90	5.82	1.96	1.46	6.80	7.18	3.08	2.10	9.88

FIGURE 7   Laboratory control chart for propionic acid.

These control limits are used to make decisions about analyzer calibration. If a data point falls outside the control limits, the cause must be determined and action taken to correct it. The analyzer can be recalibrated if it is deemed proper. A major change in the system requiring recalibration will require that new statistical limits be determined.

OXYGEN ANALYZER ON A FURNACE STACK GAS

Government regulations may require the documentation of analyzer performance, using control charts. Emission monitoring is a specific area. We have chosen oxygen measurement as an example of cross correlation that causes erroneous control limits. Figure 6 shows the control chart with narrow control limits.

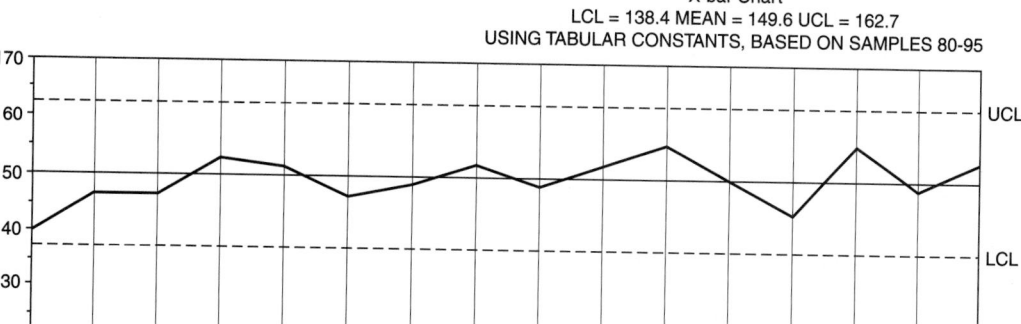

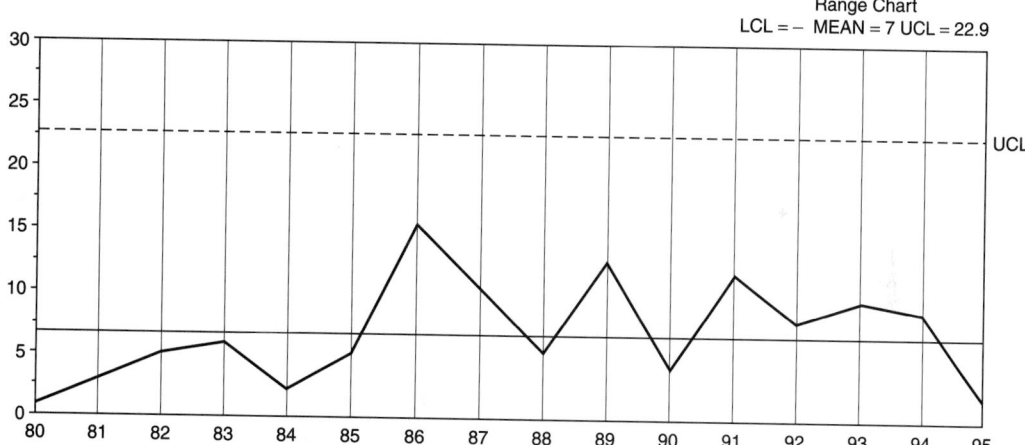

FIGURE 8   Field analyzer control chart with proper sampling rate.

The data obtained from periodic introduction of a standard sample are seen to be OOC most of the time. The frequency of testing was much higher than the long-term drift. Autocorrelation of the data points caused the control limits to be unrealistically small. When proper sampling frequency and subset size were selected, the results were inside reliable control limits.

PROPIONIC ACID IN ACETIC ACID

A furor erupted when on-line GC analyzer, QA Lab, and Technical Services Lab results continually failed to agree. One large standard sample was prepared and divided to calibrate the three different systems. Control charts were prepared for the field and QA Lab measurements. The average of subset averages was compared to determine bias between the two analyzer systems. The variance $(\sigma)2$ was compared to determine capability. It was found that the variance of the on-line field analyzer was 100 times lower than the QA Lab variance. The Lab found the causes and corrected them by using control charts (Fig. 7).

The two systems are now in much closer agreement. The TS Lab did not prepare run charts, so they could not document their performance capability. The furor has subsided and we live in closer harmony. Without control charts and solid numbers, the controversy could not have been resolved. Initially, the field analyzer control limits were too small. When we eliminated cross correlation, the measurements were shown to be in control (Fig. 8).

ACROLEIN IN ACRYLONITRILE

This analysis has become a very critical measurement in AN production. Specification levels for Acroleinc (ACR) have been lowered significantly, making the measurement requirements tougher. Both the on-line and lab measurements are made with gas chromatography. The Lab uses a single capillary column with a sample splitter and syringe injection. The field analyzer uses a packed column with a trap-and-transfer component isolation technique and automatic valve injection. Both use a flame ionization detector of equivalent sensitivity. The field analyzer delivers a much larger quantity of ACR to the detector per sample injection. The increment of measurement is 0.1 ppm for the field analyzer and 1 ppm for the lab. As we measure lower concentrations of ACR, the lab device does not provide enough significant figures. Rounding of the last digit prevents natural variation from being observed. Causes of variation due to syringe injection and low signal-to-noise ratio prevent adequate precision in the measurement. The lab device was found to be *not capable* at low ACR levels. Our customers insist on quality product, which we cannot demonstrate without a quality measurement process. Fortunately, the field analyzer has the required capability because of the measurement increment and the measurement variance (see Fig. 9).

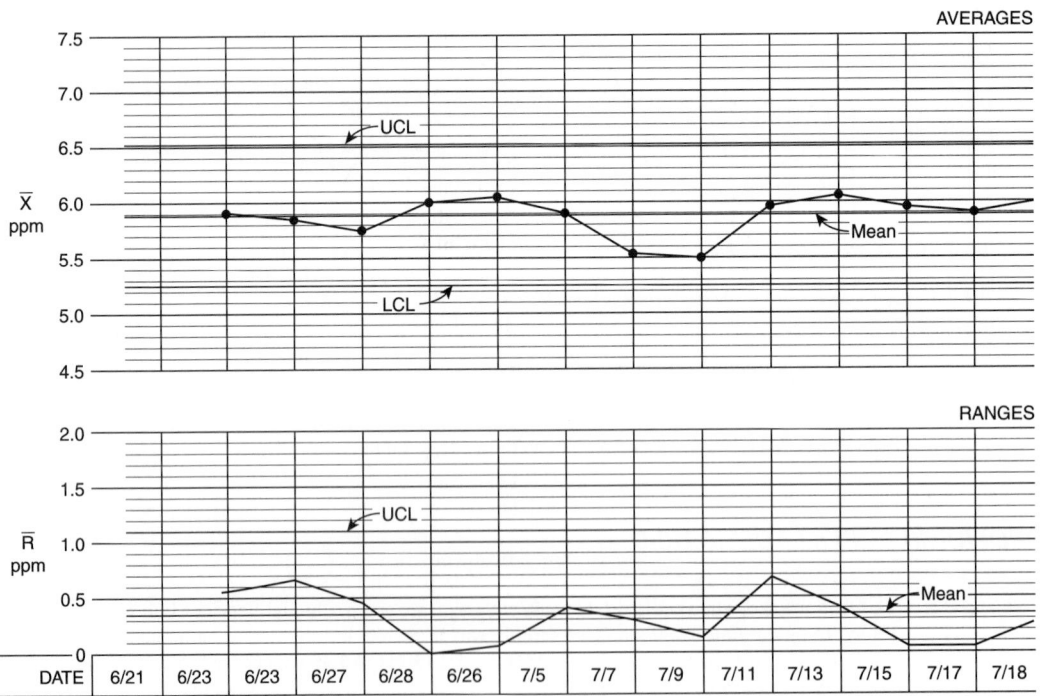

FIGURE 9   Control chart for ACR in AN.

REFERENCES

1. Converse, J. G., "Internal Reference Introduction for Performance Verification, Diagnostics, and Calibration Validation," *Proceedings of AID Symposium*, vol. 21, Boston, Mass., p. 185, 1985.

2. Taylor, J. K., *Quality Assurance of Chemical Measurements,* Lewis Publishers, Chelsea, Mich., 1987.

3. Schick, K. G., "High Speed On-Stream Acid-Base Titration Utilizing Flow Injection Analysis," presented at AID Symposium, Chicago, 1984.

4. Converse, J. G., "Flow Injection Analysis as a General Process Analyzer Concept," Presented at AID Symposium, Houston, 1987.

5. Mowery Jr., R. A., "The Potential of Flow Injection Analysis for On-Stream Process Applications," *Instrumentation & Control in the 80's,* p. 51, 1984.

Internal Reference Introduction for Performance Verification and Diagnostics

The continuous automated composition analyzer utilizing discrete sampling is ideal for internal standard calibration. The introduction of a pure reference material has several functions, such as indicating analyzer performance, providing malfunction diagnostics, and allowing relative response factor calibration. Monitoring the reference component provides a visible measure of confidence to both maintenance and operations personnel. The use of appropriate sample/reference valve volumes and a pure reference fluid eliminates the need for preparing standard samples repeatedly. Maintenance of composition analyzers has become very expensive. This cost can be reduced by innovative use of existing hardware and technology. The sample injection valve is a starting point.

My first introduction to a dual-valve sample injection system was through Jim Parks at Monsanto's Texas City plant. That concept for calibrating trace-level analyzers is explained in [1]. Briefly, a small volume valve is inserted into the loop of a larger external volume valve. An easily prepared standard sample with a higher than needed concentration (e.g., 1%) is passed through the small valve. The ratio of the two valve volumes is selected such that the detector will see an equivalent of 1/100th of the amount present if sampled by the large loop (i.e., 0.010%). This technique avoids trying to prepare a 100-ppm or lower concentration standard by the partial pressure technique.

Jim Franks [2] utilized this technique to calibrate analyzers that had components for which standard samples were difficult to prepare. All of us have had problems with erroneous standards, whether self-prepared or purchased. Proper use of the available technology should eliminate much of our frustration and wasted time caused by diagnosis and calibration. We have used a valve and column configuration to calibrate two detectors. The small valve is in series with the large valve. This would allow two internal volume valves to be used if necessary. A volume of the reference material can be sent to each detector by switching the detector select valve and double injecting the reference gas. Other workers [3, 4] have used nitrogen as an internal standard to correlate the results of several analyzers used to determine the composition of one sample. Peyron et al. [5] developed a "deferred internal standard" method in 1971 that is very similar to this work. Unfortunately, microprocessors were not readily available at that time, and their excellent publication apparently got lost in the literature. Crabtree and Blum [6] presented this technique in 1981, using a "key component" for microprocessor composition correction. J Osborne [7] describes the validation of analyzer data, using a computer prior to making process control changes based on analytical results. This concept may be simplified and used with internal reference samples to provide diagnosis as well as performance validation.

Maintenance Cost Evaluation. Some of the contributing factors to the high maintenance cost of composition analyzers are as follows.

1. Monitoring analyzer performance: Routine maintenance, consisting of checking temperatures, pressures, flows, voltages, currents, and so on, consumes considerable time. Data logging with electronic transducers can monitor these functions. Alarming by exception should be used, recording only points that are outside set limits. The cost of such equipment has to be weighed against man-hour charges.

2. Diagnosis of malfunction: Modern analyzers use "board level" changeout for electronic maintenance with diagnostics built into the microprocessor. The microprocessor can also be used to monitor the physical properties transducers previously mentioned. This device can do many other marvelous things if it receives information in the form of a proper electronic signal.

3. Verifying reliability: Most analyzer technicians can fix a malfunctioning analyzer once they diagnose the problem. Much time can be consumed in defining the problem, however. What is needed is a technique for monitoring the internal functions of the analytical system. Injection of a chemical internal reference material can produce a signal from the detector for performance verification, diagnostics, and calibration validation.

Performance Monitoring and Diagnostics.   Pure (100%) reference material should be used where possible to eliminate mixing of standards [5]. The reference chemical must be selected to match the detector response. Nitrogen can be used for thermal conductivity, but a hydrocarbon such as butane must be used for flame ionization. Constant mixtures such as air can be used for selective detectors such as oxygen. Initially, the purpose of the internal reference material was for calibration purposes. After installation and startup of a gas chromatographic analyzer system containing an internal standard valve, it became obvious that there were other benefits:

- monitoring column carrier flow and partial plugging
- monitoring losses in sample conditioning system
- monitoring detector response and system gain changes

The elution time of an inert material such as nitrogen reflects only the column transit time (i.e., carrier flow). Sharp drops in the reference peak area could indicate leaks. Slow changes may indicate detector aging or amplifier drift. The ratio of the original reference peak area to the current value produces a correction factor. This can be used to adjust the concentration values of the process sample each cycle without taking the analyzer out of service. Monitoring the correction factor provides an excellent indicator of analyzer changes or problems. Continuous analyzers that monitor zero-level impurities benefit greatly from an internal reference injection. Internal reference injection valves can be used to upgrade conventional continuous analyzers. An optional separation column can reduce interference, and a large sample loop can increase sensitivity.

Valve Volume Ratio.   The volume ratio of the two valves should be selected to match the concentration of the components in the sample. The standard 1 cc (1000 μl) sample loop of the external volume valve can be enlarged up to the point where resolution on the GC column is lost. (Note: 20 ft. of 1/4-in. tubing has a volume of $\sim$100 cc). The internal volume valve generally holds 1–10 μl. Ten μl of pure reference gas relative to a 1 cc sample produces a 1% equivalent reference peak. Two external or two internal volume valves can be used to acquire different volume ratios also. A sample volume of 100,000 μl and a reference volume of 1 μl produces a dynamic range of 10(−5) concentration units.

The exact volume ratio can be determined by sampling the same standard gas with both sampling valves. Reference [5] suggested filling each valve with a standard acid solution, flushing it into a flask with water, and titrating with a standard base solution to end point. This provides absolute volumes for the sample and reference valves. If the relative valve volumes are changed, the ratio must be determined in order to establish new relative response factors. If contaminants in the sample stream build up on the walls of the valve cavity, the sample volume and hence the volume ratio will change. This can also cause error when external standards are used and not checked periodically. The volume ratio should, therefore, be checked periodically on streams suspect of contamination.

Concept Evaluation.   In order to test the validity of this concept, a constant composition sample must be run. The parameters that affect the gain (span) of the analyzer system are then changed deliberately and systematically to see if the composition data remain unchanged. One should first write an equation containing all the known variables. A filament thermal conductivity detector response can be defined by the following nonanalytical equation:

$$\text{response} = \text{conc.} \times \text{function (fc, fr, Ib, Td, Gd, Tg, \%Tc, Ss, \ldots, Z)}$$

$$\text{Ss} = n = PVv/RT$$

where Ss = sample size, n = number of moles, P = pressure, Vv = valve volume, T = temperature (K); R = gas constant, fc = carrier flow rate, fr = ref. flow rate, Ib = bridge current, Td = detector filament temp., Gd = geometry of detector, Tg = temp. of gas in detector, %Tc = component loss on column, and Z = unknown variables yet to be identified. The easiest variables to change for checking correction are the bridge current (Ib) and the reference flow rate (fr).

A flame ionization detector response can be defined by the following nonanalytical equation:

$$\text{response} = \text{Conc.} \times \text{function (fc, fair, fh, Vb, Rf, Tg, Ss, \%Tc, \ldots, Z)}$$

where fair = air flow rate, fh = hydrogen flow rate, Vb = bias voltage, and Rf = feedback resistance for amplifier gain.

The easiest variables to change for checking correction are the H_2 flow rate (fh) and feedback resistance (Rf).

All detectors are dependent on sample size. Sample size is defined as the number of moles (n) of a component injected into the analyzer system. The sample volume is set by the valve. The amount of material in that valve volume is a function of temperature and pressure. It is approximated by the ideal gas law. Both valves are maintained at the same temperature and vent pressure [5]. A change in sample size caused by a buildup of material on the inside of the sample volume would cause the ratio to change and mislead the observer regarding the cause. It is also important to monitor the total area (sum) of all the peaks measured to ensure that a change in volume ratio is detected.

CONCLUSIONS

The introduction of an internal reference component has proven useful in monitoring analyzer performance. A trend readout of this component warns operations when calibration of the sample components is questionable. It is also useful in diagnosing analyzer malfunction. Special functions include verification that zero reading analyzers are alive, reduction in the number of blended standard samples, and calibration without removing the analyzer from service.

REFERENCES

1. Parks, J. C., and E. A. Hinkle, "A New Calibration Method for Process Chromatography," *Proceedings of AID Symposium*, vol. 8, 1962, p. 217.

2. Franks, J. C., "GC Calibration Method for High Temperature Vapor Components," *Proceedings of AID Symposium*, 1968.

3. Wainwright, M. S., and T.W. Hoffman, "GC Analysis of O-Xylene Oxidation Products," *J. Chrom. Sci.*, vol. 14, p. 159, March 1976.

4. VanCamp, C. E., et al; "On-Line Analysis of Thermal Cracker Effluent by GC," *J. Chrom. Sci.*, vol. 21, p. 259, June 1983.

5. Peyron, X., et al; "Specific Calibration Technique for Process Gas Chromatography," *J. Chrom. Sci.*, vol. 9, pp. 155–161, March 1971.

6. Crabtree, J. H., and D. Blum, "Cutting The Cost of Calibrating a Process Chromatograph," *Proceedings of AID Symposium*, vol. 19, 1981, pp. 13–22.

7. Osborne, J. E., "Computer-Aided Validation for Process GC and Continuous Analyzer Data," presented at AID Symposium, Houston, 1983.

TECHNIQUES FOR VALIDATION

The validation of analyzer performance in real time has become desirable and sometimes required. The use of a real-time internal standard for reference to a periodic calibration benchmark can produce valuable confidence for those depending on the analyzer to make important decisions. Just as the temperature, pressure, and flow monitoring devices provide an indication of external performance, the internal standard provides an indication of the system internal performance. It evaluates the sample preparation and measurement devices for stability and repeatability. Several methods of introducing internal standards to analyzer systems, ways to utilize the information gained, and results of several applications will be presented.

Internal Injection

Work on "internal reference injection" was presented in 1985 [1], and a thorough treatment of "deferred standards" was published by Guillemin [2] in 1988. Actual practices for the calibration of analyzers leaves something to be desired. United States Environmental Protection Agency protocol for Continuous Emissions Monitoring Systems provided the first clue that customers were not satisfied with our way of validating performance. Claims were, in fact, a bit better than reality. It has become necessary to prove performance in real time. We will review and evaluate some proposed ways of achieving real-time validation.

Having an analytical device on-line to monitor the concentration of critical components can be a valuable process control tool. Even batch processes utilizing discrete sampling and sample preparation techniques provide an updated analysis in 1–30 min. Many processes have long response times and do not require continuous measurement. No matter which technique, continuous or discrete sampling, the measurement must be valid if the resulting information is to be useful. Incorrect information is worse than no information. Not knowing, we proceed cautiously. Misinformed, we blindly charge into trouble.

Internal standards can be applied to *continuous* as well as *discrete* sampling systems. We calibrate analyzers with a known external standard to ensure that measurement of an unknown material produces accurate results. We need to verify the stability of the analytical system if we expect the calibration to be valid for the period following calibration. The instrument response cannot drift away from the calibrated value or the accuracy established at the time of calibration rapidly deteriorates as shown in Fig. 10. The standard is generally run at short intervals until stability is established. A periodic schedule for calibration is then set up to validate that the system is functioning within specifications. This schedule often degrades to calibration on demand when something appears to be wrong. Records of calibration and repair maintenance are often abandoned if they were kept in the beginning. If calibration is checked and

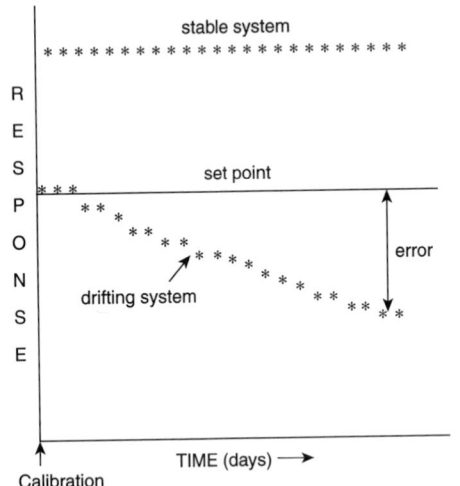

FIGURE 10   Analyzer stability.

an error is found as shown in Fig. 10, there is no way to know when the system drifted outside the specification limits. Regulatory agencies require frequent checks and good documentation for this very reason. Calibration is often put off or neglected because of some or all of the following reasons:

- a standard sample must be prepared
- the analyzer must be taken out of service
- technicians know there is nothing wrong with the analyzer
- technicians don't have time

- it takes a lot of effort to run a calibration
- the process results look good

Continuous sampling and measurement requires zero and span checks to determine that the system is in calibration. If the zero changes slowly, there is little indication that drift has occurred. The preparation of standard samples can be difficult and laborious, and purchased standards are expensive. They are occasionally inaccurate, so several different standards may be required to identify one that should be discarded.

High-quality calibration is not something that personnel want to perform frequently. Sometimes the customer does not want the measurement of a key parameter stopped at a time of process upset, even though the validity of the measurement may be in question. The argument that "correctness" is known without proof does not have any value. It must be demonstrated and proven to the customer. Is there time to produce inaccurate results? Should every effort be made to ensure that results are valid? Can looks be deceiving? Autocalibration that utilizes prepared external standards and an elaborate wiring/programming scheme is complex, expensive, and can put a control loop in a disastrous situation if something goes wrong.

The introduction of real-time analyzer system validation is not only desirable, it is becoming a necessity. This technology lends itself best to discrete sampling/flow injection techniques, but it can be applied to continuous sample flow, also. The *why* and *how to* details have been presented quite thoroughly by Converse [1] and Guillemin [2]. Actual applications that follow should demonstrate the value of "internal standard introduction." A classic example is the internal reference injected into a stack oxygen analyzer. (Figure 11). The control chart of the internal standard monitored the effect of pressure on the analyzer readout as the eye of hurricane Jerry passed Texas City, Texas on 15 October 1989 [3].

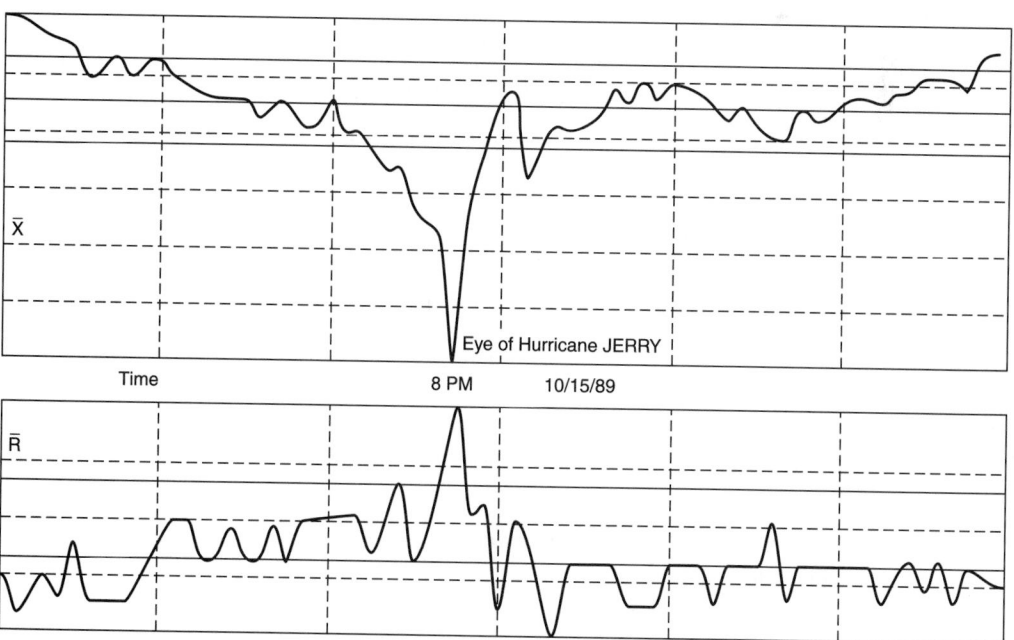

FIGURE 11   Internal reference in an oxygen analyzer.

CONCLUSIONS

Technology utilizing an internal standard introduction can be applied successfully to automated on-site analyzers, and indeed produces a means to monitor system performance. Requirement of proof that analytical data are correct means that real-time validation and control chart documentation become part of our job. Current computer capability will ease the burden of this massive task, but appropriate attention to application details is bound to remain with us a long time.

REFERENCES

1. Converse, J. G., "Internal Reference Introduction for Performance Verfication, Diagnostics, and Calibration Validation," ISA/*Proceedings of AID Symposium*, vol. 21, Boston, Mass., pp. 185–189, 1985.
2. Guillemin, C. L., and G. Guiochon, "Quantitative Gas Chromatography for Laboratory Analysis and On-Line Process Control," *J. of Chrom Library*, vol. 42, 1988.
3. Pevoto, L. F., and J. G. Converse, "Decisions to Change Analyzer Calibration Based on Statistical Quality Control Charts," *ISA Transactions*, vol. 30, No. 1, 1991.
4. Parks, J. C., and E. A. Hinkle, "A New Calibration Method for Process Chromatography," *Proceedings of AID Symposium*, vol. 8, p. 217, 1962.
5. Franks, J. C., "GC Calibration Method for High Temperature Vapor Components," *Proceedings of AID Symposium*, vol. 14, 1968.
6. Peyron, A. and C. L. Guillemin, et. al., "Specific Calibration Technique for Process Gas Chromatography," *J. Chrom. Sci.*, vol. 9, pp. 155–161, March 1971.

APPLICATION EXAMPLES

by J. G. Converse

WATER (H_2O)

What image is conjured up in your mind when someone asks you to specify a water analyzer? First, you probably scan some of the possible methods.

1. Electrochemical (hydrolysis)
2. Isolation/electrical (complex impedance)
3. Optical/thermal (dew point)
4. Spectroscopic (IR absorption)
5. Isolation/thermal conductivity (GC)
6. Isolation/refractive index (LC)

7. Chemical reagent (FIA)

8. Chemical reagent (titration)

9. Mass absorption/frequency shifts (piezoelectric)

10. Charge/mass separation (MS)

The point is that the title "water analyzer" tells one very little about the device. The specification of a method provides more information. However, the above case indicates ten different analyzers for measuring one component. We still haven't specified the sample phase (liquid or vapor), sampling technique (continuous or discrete), or conditioning procedures (phase separation, phase change, component isolation, etc.). The above list also confuses the issue by combining sample conditioning with detection method.

The detector definition should be restricted to the device that produces an electrical signal proportional to the chemical concentration of the specified component. It may utilize thermal, electrical, electrochemical, optical, spectrometric, or ionization properties. The selectivity of the detector must be defined since it specifies component isolation requirements. Each component should have an analyzer system specification even if the detector is multicomponent. The analyzer system should not be identified by the sample conditioning method. Examples can best be related to specific applications.

NITROGEN OXIDES

These are for UV, chemilluminescence, GC/PID, and PFPD.

ULTRAVIOLET ABSORPTION SPECTROSCOPIC METHOD FOR NITRIC OXIDE VAPOR IN INDUSTRIAL PROCESS VENT GASES

Applicability and Principle

Applicability: This method is applicable to measurement of nitric oxide (NO) vapor in process vent gases containing NO_2 and other species that do not absorb radiation in the region of 210–220 nm. A baseline correction is made for broad background absorption under the sharp peak at 214.08 nm. Ammonia would interfere with the measurement, but its presence in a quantity significant enough to interfere would be obvious. The method is calibrated against prepared standards of NO in nitrogen.

Principle: A grab sample is collected in a 10.0-cm path glass cell with quartz windows. It is immediately transported to the laboratory in a light-protected bag, and the peak and background absorbency at 214.08 nm are measured.

Apparatus

Sampling: The sample is collected from the process line into an 8-L steel cylinder. The cylinder is fitted with a valve at both ends in order to flush it with excess material for 3 min. The exit valve is closed first so the sample will be pressurized to the vent line pressure of 2–3 psig. The entrance valve is closed after ~1 min.

The sample cylinder is delivered to the laboratory within 15 min after collection, and spectroscopic measurements are performed within 30 min. Figure 1 shows measurement results of a NO concentration with respect to time delay of sample collection. It is apparent that long delays between sample

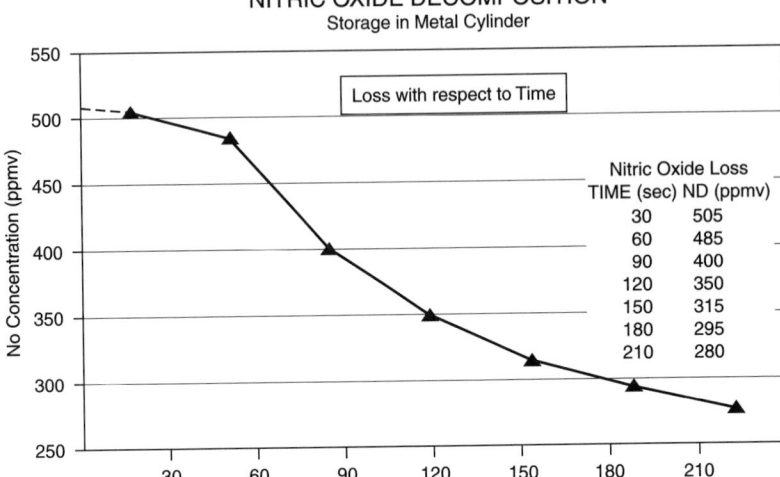

FIGURE 1 Disappearance of NO with respect to time.

collection and UV measurement should be avoided. The sample in the glass cell must be protected from exposure to light, which degrades NO.

An improvement of the sample collection was demonstrated by carrying the 10.0-cm quartz window–glass spectrophotometer cells to the field and filling them directly. The cells were not pressurized. Losses in the metal cylinder were avoided by eliminating it, as were secondary transfer losses. The glass cells were flushed thoroughly to be certain that sample replaced the nitrogen storage gas. The walls showed some fogging from moisture, but the NO spectrum was clear and without interference. An increase of $\sim$30 ppm NO was observed when the cells were directly filled, compared to filling them from a cylinder collected at the same time.

Analysis: The following equipment is needed for analysis: (a) two 10.0-cm glass absorption cells with quartz windows, and (b) a spectrophotometer with 0.01-nm resolution and PC hardware and software to process the absorption measurements and background correction.

Reagents

1. Methanol and water were used to wash glass cells.
2. Nitrogen was utilized to purge and dry glass cells.
3. Working standards were certified ppm NO in nitrogen purchased from Sigma Gas.
4. EPA protocol gas purchased to certify plant continuous emission monitoring systems (CEMS) systems was used to provide QA.

Procedures

The Sampling procedure is described in Section 2.3.

Measurement procedure: One 10.0-cm glass cell is filled with process gas, and the spectrum between 300.00 and 200.00 is collected. The absorbance at the peak maximum (214.08 nm) is measured by utilizing cross hairs on the PC monitor (Fig. 2). The baseline absorbance is determined by

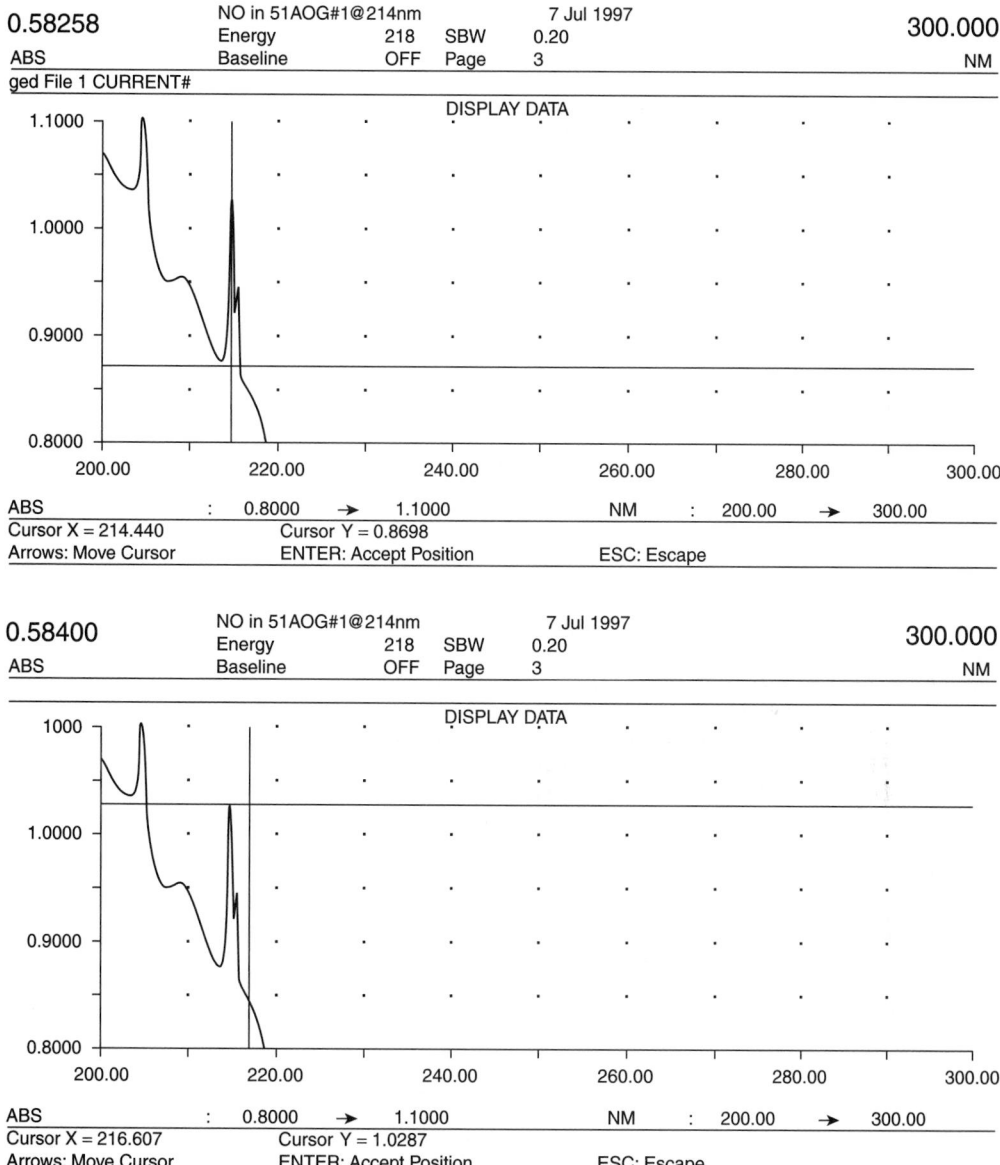

FIGURE 2 Spectrum of NO process sample and peak measurement.

positioning the cross hairs at 214.08 nm and adjusting the vertical scale to meet the visual baseline crossing at that wavelength (Fig. 2).

The baseline, peak maximum, and difference are recorded. The other cell filled with standard is measured the same way and the results are recorded (Fig. 3). The concentration of NO in the sample is calculated from the standard concentration and absorbance and the unknown absorbance.

Performance validation: The absorbance values of the standard gas and process sample are plotted daily on an X-bar chart to demonstrate instrument stability and measurement precision (Fig. 4).

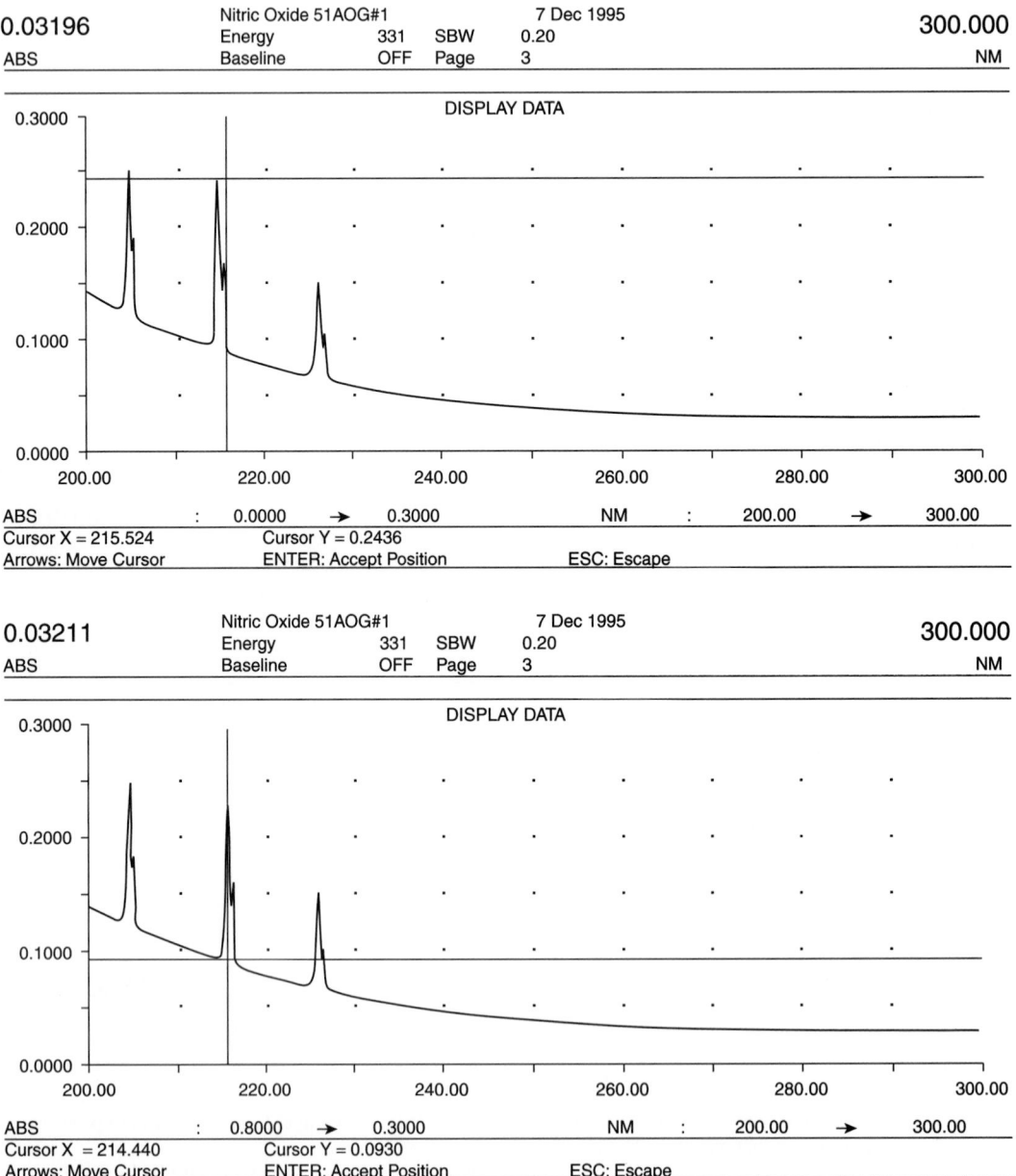

FIGURE 3 Spectrum of NO standard sample.

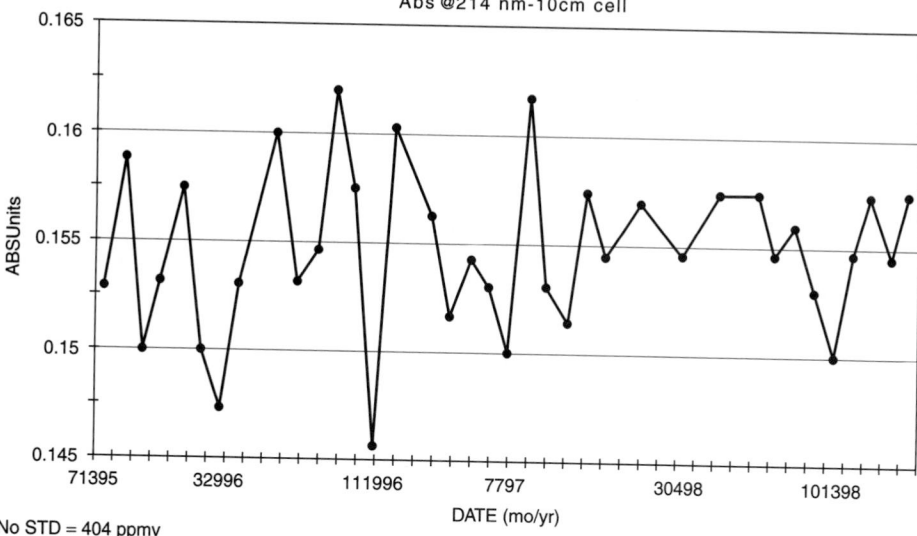

FIGURE 4 Control chart of NO standard.

Calibration

Calibration is performed for each unknown measurement by utilizing control samples.

Calculations

$$\text{process sample concentration} = \text{Std. Conc.} \times \frac{\text{Abs(sam)}}{\text{Abs(std)}} \tag{1}$$

$$\text{standard deviation } (\sigma) = \frac{(Xi - X)}{n - 1}, \qquad \text{RSD} = (+/- \sigma \times 100)/\text{mean} \tag{2}$$

REFERENCES

1. Saltzman, R. S., "Diode Array UV Analyzer for Source Monitoring," *ISA/AD Proceedings,* vol. 34, 1989.
2. Herman, B. E., and Klein, F., "Analysis of Combustion Stack Nitric Oxide Emissions with Non-Dispersive UV Detection," *ISA/AD Proceedings*, vol. 37, 1992.

This method can be applied on line directly by utilizing the Applied Analytics, Inc. diode array UV spectrometer (or equivalent) and a 100-cm cell.

STACK CONTINUOUS EMISSION MONITORING SYSTEMS (CEMS)

This is for conventional vs. RDS CEMS. At last count, there were more than a hundred companies listed as providing continuous emission monitoring systems for sale. There are thousands of governmental regulatory personnel, university researchers, and consultants sucking up the nation's wealth to guide and direct pollution monitoring efforts. All kinds of devices are being offered for sample extraction, conditioning, and component measurement, plus elaborate systems for data manipulation and storage. Conferences are held and papers are written, presented, and published. The cost of documenting the arguments on how we should deal with this problem is overwhelming. Cost to manufacturing companies required to install and maintain these systems threatens to put them out of business. The folks sucking up all that wealth don't seem to realize (or don't care) that when business shuts down, the source of the wealth dries up and we all wind up in the bread lines if there is any bread to hand out.

This example offers a lower cost higher-performance alternative to the many system now used.

What is the problem? There is a limit to how much pollution the Earth's atmosphere can tolerate. What is the solution? Pollution prevention! Change our processes to achieve prevention! Measuring the rate at which we are traveling to destruction doesn't offer much comfort. Devoting so much of our resources to CEMS is absurd. We must quantify emissions to determine the extent of prevention success. We should not build an industry for monitoring that destroys our industry of manufacturing. The desired information from a CEMS should be well defined and then obtained as simply and cost effectively as possible. We continue to design elaborate edifices to our brilliance rather than effective tools to achieve the desired results. Following is a system for sampling, conditioning, and measuring pollutants that will meet this need (see Fig. 5).

OPERATING PROCEDURE

1. Run ambient air.
2. Run stack sample.
3. Run standard gas sample.

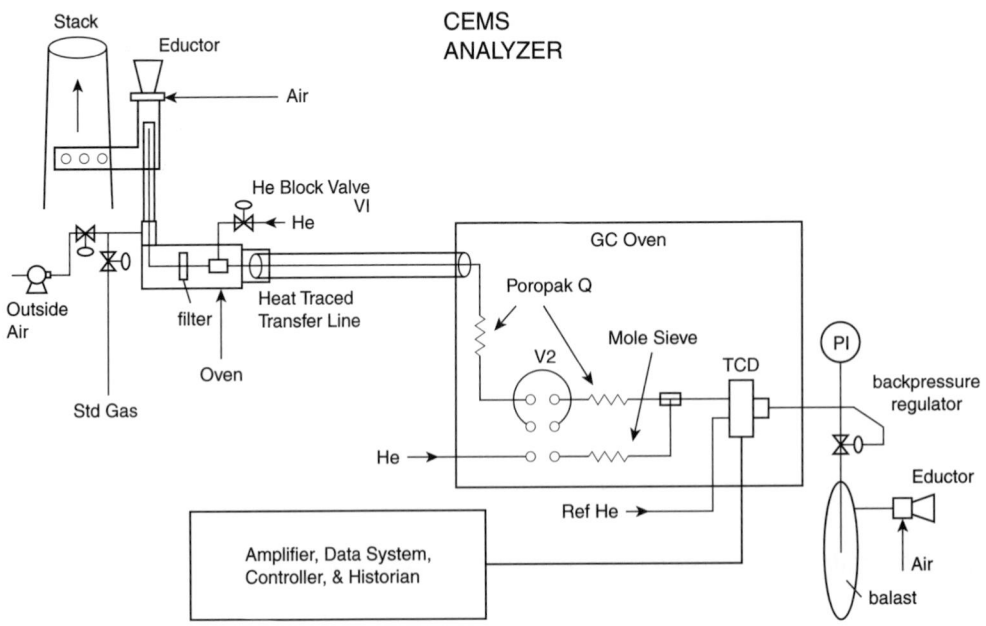

FIGURE 5   Simple discrete sampling CEMS.

(Step 2–Step 1) provides delta CO_2, delta H_2O, and delta O_2. Step 3 provides validation of analyzer system performance.

Continuous Emission Monitoring Systems required by the Clean Air Act have turned out to be very expensive projects. Members of the Analysis Division/ISA have requested proposals for lower-cost systems that can gain USEPA approval. A potentially less expensive approach has been suggested, but it will require demonstration of compliance. The system utilizes pneumatic remote discrete sampling, modular construction, component isolation, and simple thermal conductivity detection supported by PID and/or IRD if required. Cost savings will be significant. Some of this proposal has been demonstrated. These requirements can certainly be met.

The primary information being sought by a CEMS device is target analyte concentration and mass flow in the stack. The required value of lbs./h emission is on a dry basis. Reporting requirements are one analysis per 15 min for CEMS (BIF regulations require a 2 min response time). The required accuracy is $\pm 0.5\%$-oxygen and $\pm 20\%$ of span for CO and NO. An actual on-stream time of 95% is also required. The current practice is to use three different analyzers to measure O_2/magnetic, CO/IR, and NO/chemiluminescence. Sample extraction is continuous and water is removed from the sample under the assumption that none of the target analytes are lost in the process. Solids collected on the probe filter are blown back into the stack on a daily schedule.

Water washing or water knockout can cause loss of target analyte, as shown in the AAI Product Bulletin No. ASPB-93–3:

- CO_2, up to 45% removed
- CO, up to 5% removed
- O_2, up to 50% added
- SO_2, up to 99% removed
- N_2, up to 7% added   What about NO and NO_2?

Reference [1] discusses effects of moisture on measurement accuracy. Current systems use a continuous flowing sample. This adds cost, complexity, and maintenance effort to the CEMS. It also produces less accurate results because of component losses, correction calculations, and because total composition is not measured.

This proposal seeks to obtain quality concentration measurements on a discrete sample for all of the components in the stack, including water, utilizing 1–3 concentration/voltage transducers. Proposed gains are:

- more accurate measurement of all stack components
- more reliable sample acquisition and conditioning system
- reduced total cost of project

A major cost of current projects is an analyzer house. Modular construction of the system would allow placing components in less expensive enclosures. Modular discrete units could be replaced and the failed unit could be repaired at the instrument shop. We are now considering replacing the three analyzers for repair when they fail, rather than repair in place in order to meet on-stream time. It is also cheaper to purchase one analyzer versus three.

Figure 6 shows the component isolation demonstrated in our test facility and plant analyzer systems.

The NO peak location utilizing the thermal conductivity detector (TCD) is still unknown but has been isolated on mole sieve and MS-Q capillary columns with PID detection. Pneumatic sample extraction works as predicted, with an 8-s pulse producing an appropriate sample for the 1/8 in. separation column. The definition of additional operating parameters is continuing. The worst case may require an IRD for CO and a PID for NO, but they would be part of the GC system and function with the proposed sampling scheme.

Table 1 lists sampling technologies currently being evaluated and used. Several components are measured and their typical ranges are from ppm to percent levels. There are advantages and disadvantages to each measurement technique. Arguments for the various sampling and measurement

TABLE 1   CEMS Sampling Technologies

1. Cross stack (in situ)
2. Direct (wet with high-temperature heat trace)
3. Direct (dry with water removed)
4. Dilution (wet with lower-temperature heat trace)
5. Discrete (value vs. pneumatic)

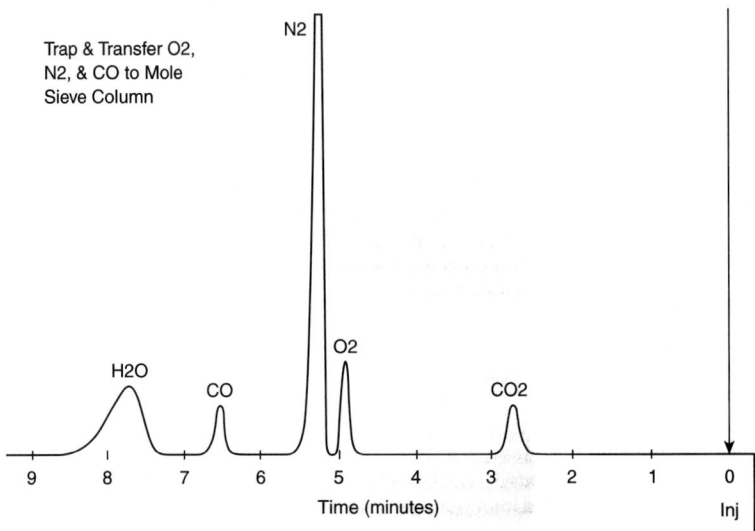

FIGURE 6   Response versus elution time for component separation.

techniques rage on, but personal preference and prejudice don't always lead to the best and most cost effective approach.

Reference [2] (attached as an Appendix) is an excerpt from a book "Continuous Emissions Monitoring" by Dr. James A. Jahnke, utilized in his A and WMA training course. Both the book and the course are recommended to those wanting details on the application of CEMS technology. Some real benefits can be gained by utilizing remote discrete sampling technology described below. The use of a TCD and a separation column for component isolation provide sufficient information to make accurate measurements.

Improved accuracy can be gained from normalization when all components are measured. The concentration of the individual components are determined by the following equation:

$$\text{normalized concentration area} = \frac{f(i) \times A(i)}{\sum_{I=1,\dots,n} [f(i) \times A(i)]} \times 100, \quad A(i) = \text{component } (i),$$

$$f(i) = \text{response factor}$$

Additional information can be gained from the proposed system if one alternately measures the ambient air sample and then the stack sample. The difference in CO_2 concentration (Fig. 5) allows one to determine the actual carbon burned, and H_2O concentration provides information on ambient moisture and total moisture for calculating components on a dry basis [2]. Use of the published F factors and Qh (heat input rate) allows one to calculate Qs, the volumetric flow rate.

Periodic measurement of the stack flow rate to check Qs calculation may eliminate the need for continuous in-line flow measurement. Reference [2] gives several equations for characterizing stack

TABLE 2 Comparison of Discrete vs. Continuous Sampling

RDS/Component Isolation/TCD	Continous Sampling
1. Measures N_2, CO_2, & H_2O as well as O_2, CO, & $(X)i$	1. Measures O_2, CO, & (X) i; lacks accurate total composition
2. Simple low-maintenance & low-cost sample system	2. Complex & high-cost, high-maintenance sample system
3. Component isolation removes interference	3. May have interfering components in measurement
4. No loss of components during sample preparation	4. May knock out components during water removal & conditioning
5. Simple, maintainable, & low-cost measurement device (TCD)	5. Complex & expensive multiple-measurement devices
6. Improved total accuracy from measuring all components & normalizing	6. Cannot normalize to reduce variations
7. Provides information to calculate other parameters	7. Limited information without adding more analyzers

emissions based on concentrations of O_2, CO, CO_2, and H_2O, measured by RDS CEMS. Table 2 lists my prejudices for using RDS versus continuous sampling.

CONCLUSION

Serious attention to developing this concept for emissions monitoring could lead to improved systems and reduced costs. The manufacturing community will have to push such an effort to make it go, because too much money is being spent by current practices. Unfortunately, many sample system components (gizmos and gadgets) suppliers will have to change their product lines.

APPENDIX I

1. Leung, D. Y. C., and F. S. K., Li, "Effect of Moisture on the Pollutant Concentration Measurement," *J. Air & Waste Management Association*, pp. 1224–1225, October 1994.

The effect of removing moisture from the sample gas on pollutant concentration measurements produces a high value whose percentage deviation of the measurement is a function of relative humidity and partial pressure of saturated water vapor only. The correction factor times ($\times$) the measured concentration gives the actual concentration.

$$\text{actual} = \text{measured} \times [1.0 - Pw]\text{, where } Pw \text{ is the partial pressure of water vapor}$$
$$\text{percent deviation from actual} = [100\ Pw\ /Ps]\exp(14.52 - 5358/T)$$
$$Ps = \text{saturated water vapor pressure@temperature } T$$

Under ambient conditions of 10–40°C and 40–100% RH, the percentage deviation varies 0.5–7.4%. This error is not related to analyte losses.

APPENDIX II

2. Jahnke, James A., "Continuous Emission Monitoring," Source Technology Associates, Research Triangle Park, NC.

F factor methods used for calculating flue-gas emissions simplify calculation by considering the known stoichiometry associated with combustion fuels. They are expressed in the general form of the following equation:

$$\text{emission rate } (E) = c(s) \times F \times D \,(\text{corr})$$

where $c(s)$ is the pollutant concentration and $D\,(\text{corr}) =$ a dilution correction factor.

$$F = \frac{\text{theoretical volume of gas generated by the complete combustion of a quantity of fuel}}{\text{amount of heat produced by the fuel upon combustion}}$$

Fd is for dry gases; Fw is for wet gases, and Fc is for volume of CO_2 from combustion.

Excess air is generally introduced to ensure complete combustion and water (H_2O) is present in the air and is generated during combustion. Carbon monoxide (CO) and Carbon dioxide (CO_2) are also produced.

$$V(t) = \text{volume of combustion products} = Q(s) \times D\,(\text{corr})$$

where $Q(s)$ is volumetric flow rate.

$$D(\text{corr}) \text{ for dilution air correction term} = \frac{20.9\% - \%O_2 d}{20.9\%} \text{ for dry air}$$

$$\frac{Q(s)}{Q(H)} = Fd\left(\frac{20.9\%}{20.9\% - \%O_2 d}\right)$$

where $Q(H) =$ combustion source heat input rate.

$$E = c(d) \times Fd \times D\,(\text{corr})$$

is used when pollutant and oxygen are measured on a dry basis. If the moisture content of the flue gas $B(ws)$ is measured, the emission rate is

$$E = c(w) \times Fd \times \left\{ \frac{20.9\%}{20.9\%[1 - B(ws)] - \%O_2 w} \right\}$$

where (w) denotes wet basis.

In situ and dilution CEMS systems measure flue gases on a wet basis. The Fd factor can be used if the changing moisture content is measured. If the wet factor Fw is used, the ambient moisture fraction $B(wa)$ must be included in the calculation:

$$E = c(w) \times Fw \times \left\{ \frac{20.9\%}{20.9\%[1 - B(wa)] - \%O_2 w} \right\}$$

This expression cannot be used if water is introduced or removed from the gas (e.g., scrubber, dryer, knockout). If carbon dioxide is measured Fc can used in the form

$$E = c(d) \times Fc \times \left(\frac{100}{\%CO_2 d}\right) \quad \text{or} \quad E = c(w) \times Fc \times \left\{\frac{100}{\%CO_2 w}\right\} \text{ also } \quad c(d) = c(w)/[1 - B(ws)]$$

If the pollutant is measured in ppm, it must be converted into units of (ng/scm) or (lb/scf). The F factors have been calculated and tabulated by the USEPA (see Tables in Ref. 2).

Since the factor $Fo = 20.9\% \, Fd/100 \, Fc = 20.9\% - \%O_2 d/\%CO_2 d$ is relatively constant, measured values of O_2 and CO_2 can be inserted into the equation to check consistency.

If the combustion source heat input rate $Q(H)$ is known, the flue-gas volumetric flow rate $Q(s)$ can be derived from the tabulated Fd and the measured $\%O_2$ by:

$$Q(s) = \frac{Fd \times Q(H) \times (20.9\%)}{(20.9\% - \%O_2 d)}$$

and checked against the flow meter. Since measuring the flow rate is a trouble and an expense, this might be a way to eliminate flow measurement, provided $Q(H)$ is available for a combustion system. Since we now measure $Q(s)$, we could build a database on $Q(H)$ for future applications. Many other equations can be derived from the basic wet and dry formulas, depending on what is known or measured and what is missing (see Ref. 2 for examples).

The adequacy of using this technique in any given application should certainly be cross checked by using a direct measurement for $B(ws)$ and $B(wa)$. Maximum deviations from the published F factors are estimated at $\sim 3\%$. Incomplete combustion can cause error, but this can be corrected if the CO concentration is determined.

AIR OXIDATION OF HYDROCARBONS

This is for a phthalic anhydride reactor effluent analyzer. A comparison of three laboratory and two on-line GC analyzers for phthalic anhydride reactor effluent samples demonstrated that no sets of relative response factors ever agreed. A single set of self-consistent factors was constant as long as no major changes were made to the hardware system. The calibration of an on-line analyzer for this application is very difficult and time consuming. Long-term stability of the sample preparation and measurement systems was also very difficult. In order to obtain reliable results for long periods of operation and without excessive tedious calibration, it became necessary to monitor changes in the system. The introduction of an internal standard into the flowing carrier gas of the process GC appeared to be the most practical solution to the problem. The readout of the analyzer could then be corrected relative to the value obtained for the standard. Bleeding a known quantity of standard into the reactor sample would be difficult because the flows would require very accurate control. The alternative would be to inject a fixed volume of internal standard and a fixed volume of reactor material into the carrier gas of the GC. Work by Parks [4], Franks [5], Peyron [6], and others indicated that the fixed volume method would produce the best results. These authors indicated that internal standard techniques would be impractical or difficult to apply to process analyzers. Executed properly, it turns out to be practical, simple, and very desirable to use an internal standard. The analyzer doesn't distinguish whether fixed quantities of material are mixed first and then introduced, or introduced unmixed into the flowing carrier. The essential item is that known relative quantities of each material are introduced and that their ratio remains constant. Butane was selected for this application according to the following criteria:

1. It is an easily commercially available source of pure material.

2. It is quickly eluted and easily separated from sample components.

3. It is responsive to flame ionization and TCD measurement devices.

The volume of the sample valve can be determined relative to the volume of the standard valve. The equivalent concentration of butane relative to the sample components and the relative response factors can then be determined. The relative volumes selected were $V(s) = 1000 \ \mu l$ and $V(r) = 10 \ \mu l$ to produce a volume ratio of $V(s)/V(r) = 100/1$. Peyron [6] suggested titrating the valve volumes filled with a known concentration of sodium hydroxide against standard hydrochloric acid to obtain absolute volumes. We used the 100% butane in both valves and reduced the FID sensitivity so that the detector was not saturated when measuring the 1-cc volume.

The following results were obtained: $V(s)$, 12,780,000 area counts, $\acute{o} = \pm 0.71\%$ relative for seven tests; $V(r)$, 118,600 area counts, $\acute{o} = \pm 0.96\%$ relative for seven tests; $V(s)/V(s) = 107.8/1$ $V(r) = 10.4 \ \mu l$ (volume specified from diameter and thickness), $V(s) = 1121 \ \mu l$ (1.121 ml).

A 10-μl volume of 100% butane produced an equivalent concentration of 0.928 vol.% in the 1.121-ml reactor sample volume. It is not necessary to do all this since several external standard samples must be run to calibrate the analyzer. Each time the system is calibrated, the value of the internal standard run each cycle is recorded along with the values of the individual components. An evaluation over several calibrations and extended periods of time will determine if the relative values

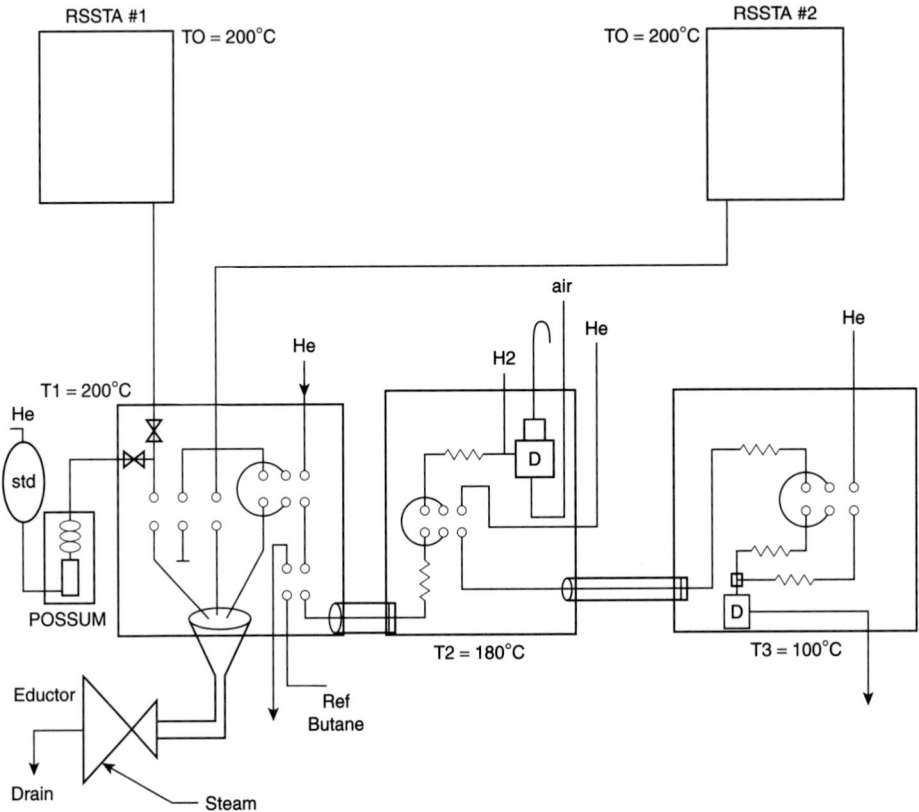

FIGURE 7 Phthalic anhydride GC analyzer.

are constant. If stability cannot be achieved, computer capability could be utilized to correct the analytical results each cycle, automatically. The primary purpose of this activity is to prove that the system is stable and that the analyzer is still accurate. If not, modification of the system to produce stable operation is recommended. The value of the internal standard should be displayed along with the values of the critical components. The operators will quickly learn when they cannot trust the reported results of the unknowns by watching the value of the known component. It is also advisable to set alarm limits on the internal standard to warn when the system is malfunctioning. Figure 7 shows the analyzer configuration.

In this case, it is obvious that drift was present and correction should be made to the results reported. A flat response will validate analyzer calibration. The response factors for the TCD portion of the system did not change with time, showing it was stable even though the FID was not. (Figures 8 & 9).

AIR/AMMOXIDATION OF HYDROCARBONS

Acrylonitrile (AN) has many impurities that have specification limits set by producers and customers through the American Society for the Testing of Materials, Committee E-13. Current practice utilizes a limited on-line capability combined with a slower, but more thorough, laboratory analysis. Better control could be achieved if all the critical components were measured on line in near real time. Since

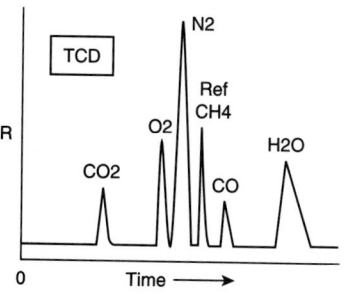
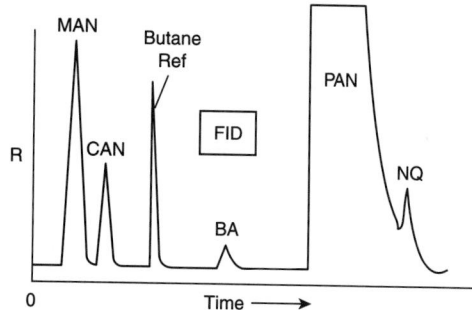

FIGURE 8   Response vs. elution time readout.

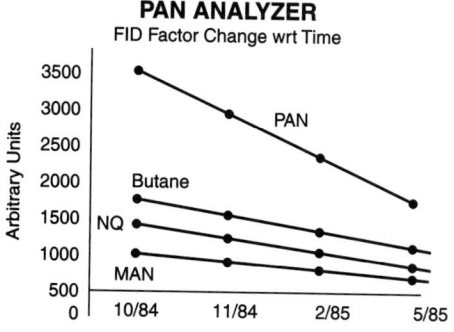
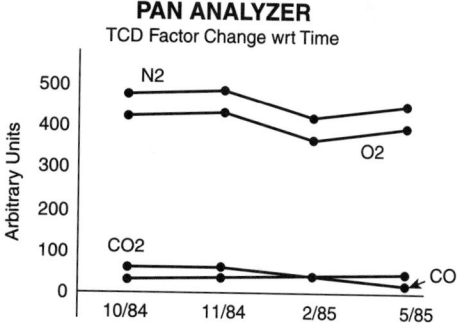

FIGURE 9   Change in FID component response factors with respect to time.

the analytical method is gas chromatography, a batch process update of quality occurs at the end of each analysis cycle. A cycle time of 10–20 min still produces a faster response to an off-specification product than laboratory testing. A choice must be made on how this enhanced capability will be achieved. Several simple analyzer systems may be utilized, or a single more complex system can be employed. Figure 10 shows a valve and column configuration for utilizing the Carbowax 20M column suggested by ASTM.

The first splitter allows a high carrier flow to sweep the liquid vaporizing sample valve and to introduce a discrete packet into the transfer line. A second splitter allows high flow through the transfer line and adjustment of loading on the capillary column. Two splitters in the system could lead to a variation in sample size introduced onto the column. In order to provide real-time monitoring of the system performance, an internal standard is introduced as a reference prior to the liquid sample introduction. Methane (100% CH_4) was selected as the internal standard, which was introduced by a 10-μl liquid slider valve. The elution time of methane is utilized to calculate the linear velocity of the helium carrier gas. This helps in setting up the column conditions. The area count of the CH_4 will indicate the variation in the split ratio of the two splitters. It should be recorded and displayed along with the analytes to be measured.

An equivalent concentration can be determined from the relative valve volumes if desired or an arbitrary response factor can be assigned. A change in methane concentration beyond set limits should alarm operations that the analyzer results are suspect. Figure 11 shows the response versus time chromatogram of the FID output. The sample and internal standard are introduced independently, so the methane peak can be placed anywhere on the chart.

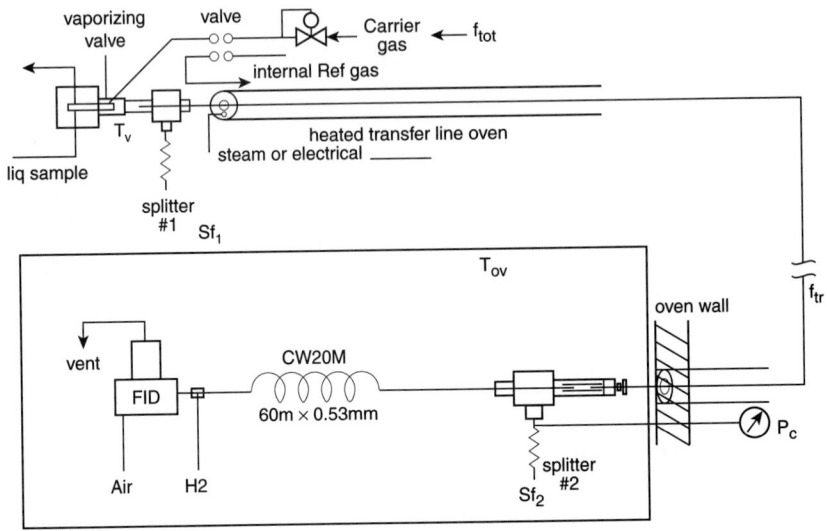

FIGURE 10   AN GC hardware connections layout.

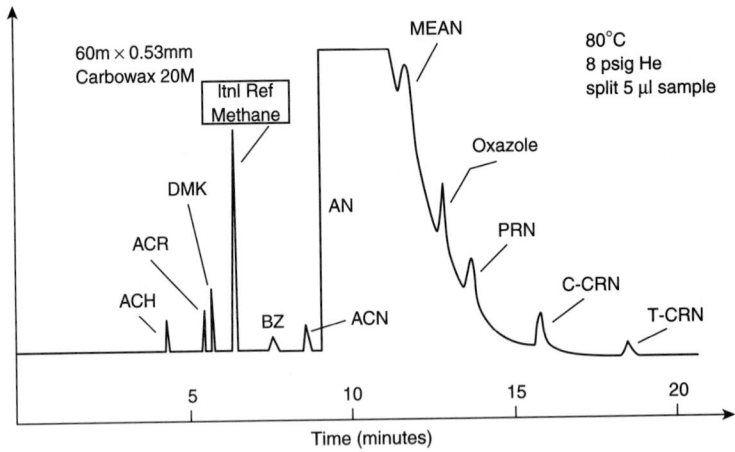

FIGURE 11   Chromatogram of AN.

WIDE BOILING RANGE LIQUID—OLEFINS AND ALCOHOLS

Sampling a liquid containing components having a wide boiling range presents a significant problem, with loss of light ends and residue. Continuous sample vaporizers tend to plug with the very high boiling residue and require large quantities of energy to generate excesses of vapor. Transporting liquid a long distance creates too much delay in sample transport, also.

The application of a remote, discrete, liquid sampling station provides a solution to many of the above problems, as well as others to be described in this project. We first started working on this

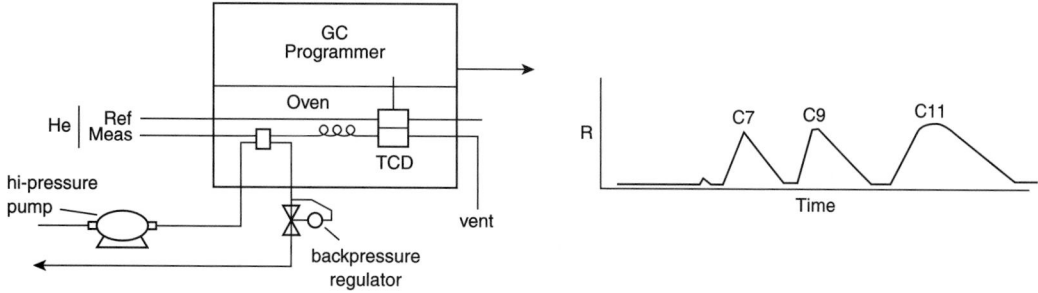

FIGURE 12   Wide boiling range analyzer.

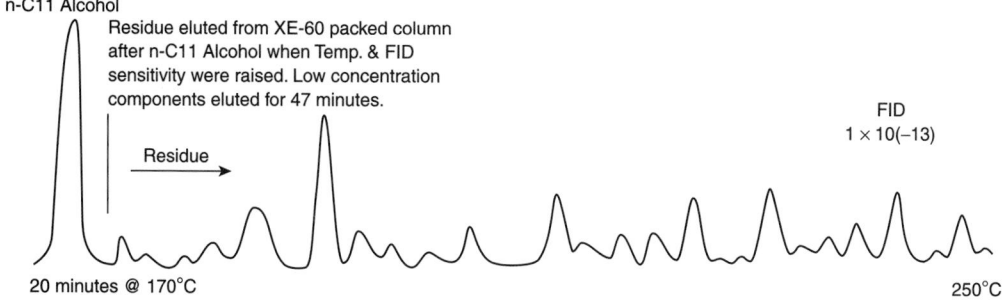

FIGURE 13   Residue from the process sample.

project in 1969, and much of the current technology, materials, and equipment were not available then. Our first approach to the sample injection was to utilize a high-pressure pump and a backpressure regulator. This prevented separation of the light ends (bubbles in the liquid) prior to liquid injection, but pressure was limited with the Beckman slider valve, which leaked above ~35 psig. The analyzer configuration was a gas chromatograph, shown in Fig. 12. This approach put most of the sample components onto the separation column, but the high boilers had a long elution time and the peak shapes were poor. The column life was short; we finally assigned the cause to very high boiling residue. This was identified by injecting a liquid sample into a temperature-programmed GC with a very sensitive FID (Fig. 13).

The resolution of the three major components was adequate, but capillary GC/MS indicated that ~210 components were present in this complex sample. The development of multidimensional sample preparation led the way to some significant improvements of the system. The design of a new analyzer system progressed over several years in very slow stages. The idea of partial separation followed by the trap and transfer of subgroups helped break the large problem down into smaller units. These could be operated on to produce better component isolation. Figure 14 shows how the light ends could be isolated and the very high boiling residue could be quantitatively excluded from the packed column.

The 1-μl liquid sample was flash vaporized in a heated block before it entered a bare, stainless-steel capillary (0.020 in. ID). The lower boiling components are trapped in the valve loop and transferred to the analytical packed column. The residue components that had previously stuck permanently on a reactive packed column were eluted from the more inert metal surface. The same principle can be utilized to isolate a midboiling group of components by using a short, less inert column packed with glass or silicon carbide particles (Fig. 15).

Applying this approach to the complex alcohol sample presented a surprising problem. The isolated group of components from the partial separation process are not homogeneous. Figure 16 shows that trapping a portion of the group can lead to gross error.

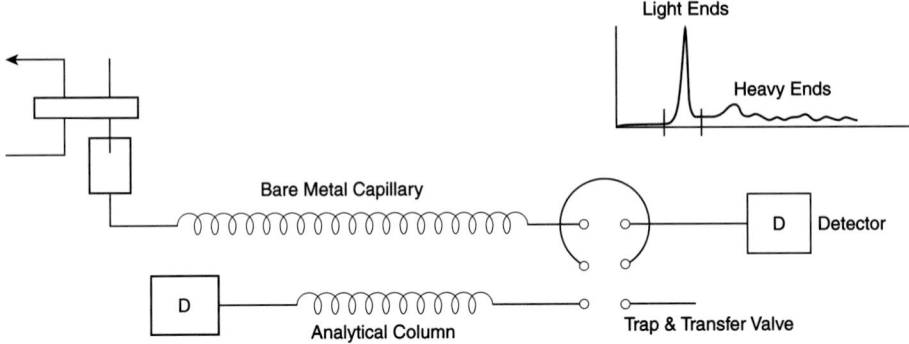

FIGURE 14   Isolating components of a mix with a bare metal column.

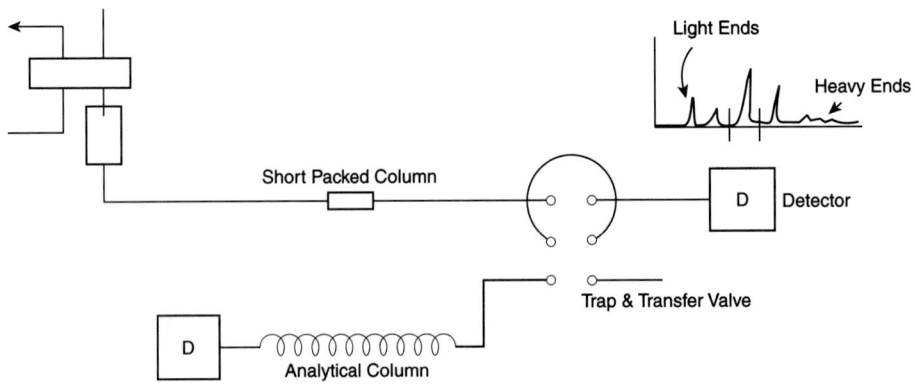

FIGURE 15   Isolating components of a mix with a short, inert, packed column.

TRAP AND TRANSFER CUTTING

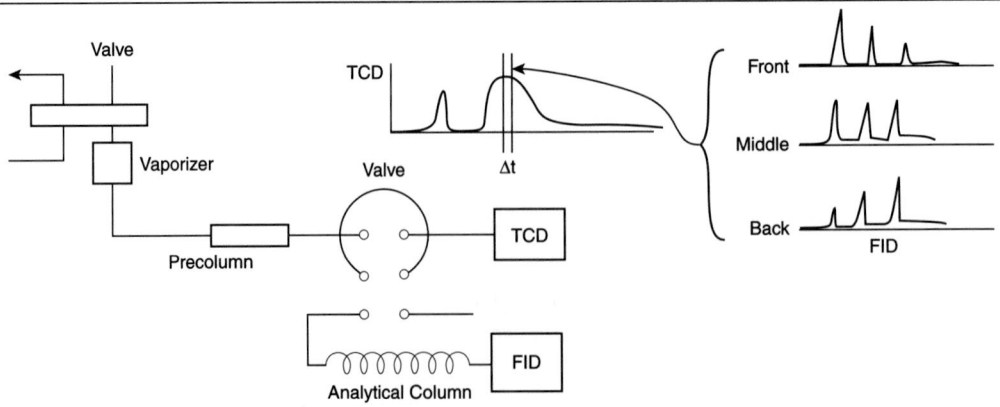

FIGURE 16   Possible discrimination with trap and transfer.

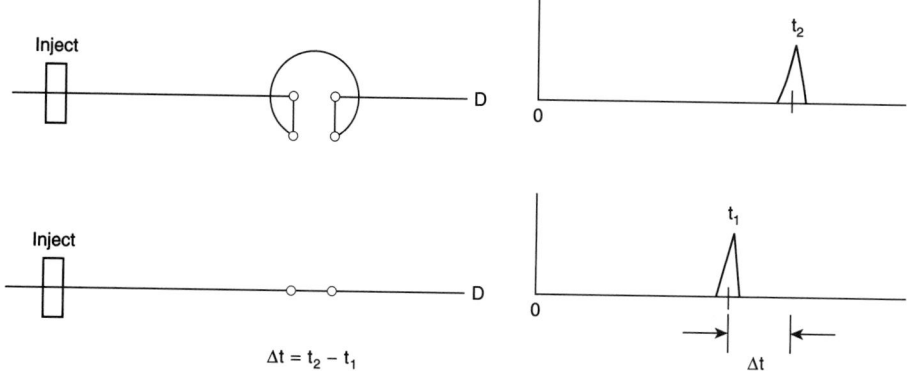

FIGURE 17 Determination of trapping time.

The size of the trapping loop will determine width of the cutting window delta t. This can be estimated from the cross-sectional area of the tubing times the loop length, divided by the carrier flow rate. The actual value can easily be measured by injecting a single component such as methane. The difference in elution time of the long path through the loop minus the short path with the loop out of the circuit is equal to delta t, as shown in Fig. 17.

One can use a very large loop in order to ensure that the entire group is transferred, or a splitter can be utilized to reduce the sample size. Another feature that is important to analyzing a wide boiling range sample is to have two or more temperature zones (ovens). We chose to purchase two process GC analyzers and connect them with a heated transfer tube. A simple way to do this is to insert a thick-wall aluminum tube into holes cut in the oven walls, and butt the ovens as close as possible. Insulation is wrapped around the tube, and the transfer channel is heated by the GC oven air. Using two analyzers gave extra valves and detectors at that time, but now one can purchase a second oven and control all its functions from the master analyzer. Application testing indicated that we needed to maintain the low temperature oven at 125°C and the high one at 170°C. Figure 18.

The vaporizing liquid valve allows the sample to be kept cool and maintained at sufficiently high pressure to prevent phase separation. A portion of the sample is removed with a splitter, and the remaining portion minus the residue is trapped and transferred to column 2 (C2). This separation device is selected to provide the desired resolution for the next subsampling process at V3. The low-boiling components are passed to the low-temperature oven, and the high-boiling components are trapped and transferred to C3. The analytical column C3 provides an adequate high resolution to allow interference free measurement with the individual components. Higher concentration components are monitored with the TCD, and the FID is utilized for trace components that pass through the nondestructive TCD. Selected components separated on C4 are transferred to C5 with V4. They separated on C5 and were measured with the FID (a TCD could be used if the concentrations were high enough or the components were not detected by the FID). A two-detector system could also be utilized as an oven, one if required it.

Figure 19 shows the complex mixture of the wide boiling range samples. Material from the first stage of reaction contains aldehydes, while a later stage of the process indicates that most of the aldehydes have been converted into alcohols.

The trap-and-transfer dual-temperature zone system developed for this application was able to effectively analyze material from any point in the process, from the olefin feed to the final alcohol product. A program was developed for each sample point and assigned a stream number. When the stream number was selected to match the material being sampled, appropriate operations of valves and component gates were carried out by the time-based program to produce a full analysis of the desired components. Capillary column methods have been developed to provide improved separation

Wide Boiling Range Dual Oven GC System

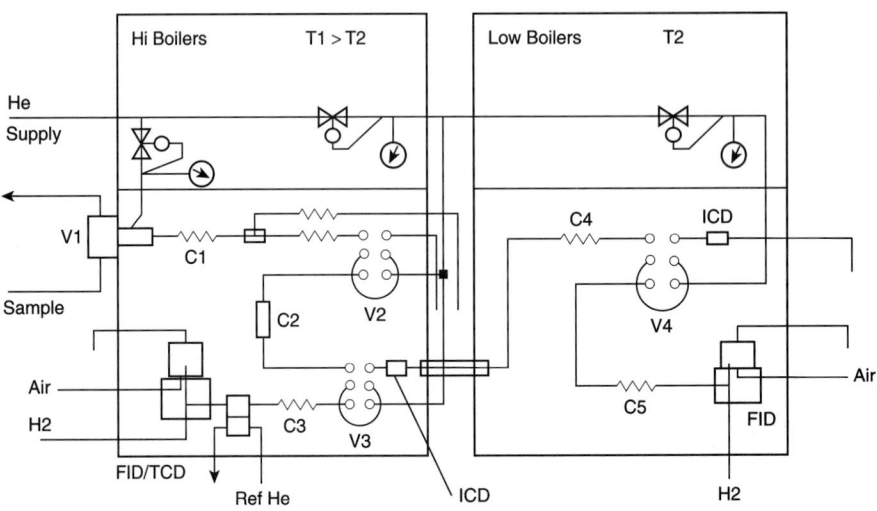

FIGURE 18   Configuration developed to achieve the required analysis.

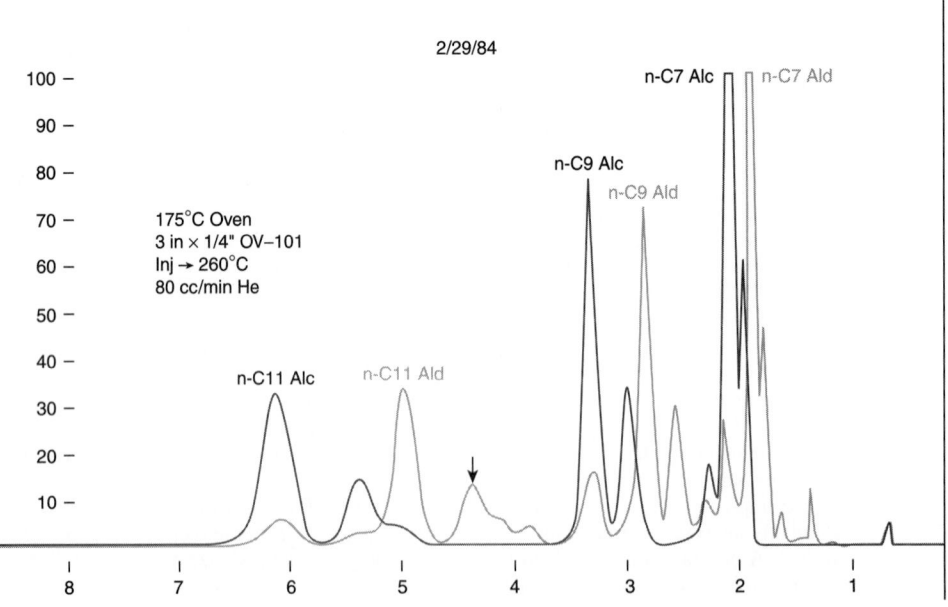

FIGURE 19   Separation of a complex mixture.

of the over 200 components identified in the process sample by GC/MS. Isolation of the required components has been achieved with our process analyzer system and provides adequate process control.

WASTE-WATER MONITORING

An at-line static headspace VOC analyzer developed to monitor Aromatics in a waste-water treatment facility utilized daily manual reference samples and control charts. Calibration samples were run at erratic and distant periods but showed the analyzer performance to be very stable ($\sigma = \pm 7$ ppbw/v over a 3-year period). Problems with preparation and handling of the reference samples caused their standard deviation to be $\sigma = \pm 20$ ppmw/v. This created problems with validation of performance for regulatory purposes. Figure 20 shows a proposed scheme to add an internal standard to a similar on-line system in order to avoid such problems with validation.

A quantity of water sample is drawn from the process stream by operating the composite sampling valve. A relative amount of internal standard is injected into the system by using a four-port liquid sample valve. Material from both valves is blown into the clean equilibrium chamber, and helium pressurizes the vessel to a set value. The chamber is isolated with an air-operated valve and is allowed to reach equilibrium. Pressurized vaporized content is then released to vent through a six-port vapor sample valve. A discrete sample is introduced into the carrier gas flowing to the separation column in the GC for analysis. The target analytes and the reference component are displayed and recorded. A control chart can be prepared from the reference values to demonstrate stability. Variations in the reference values outside a set limit can be utilized to alert operations of a potential problem.

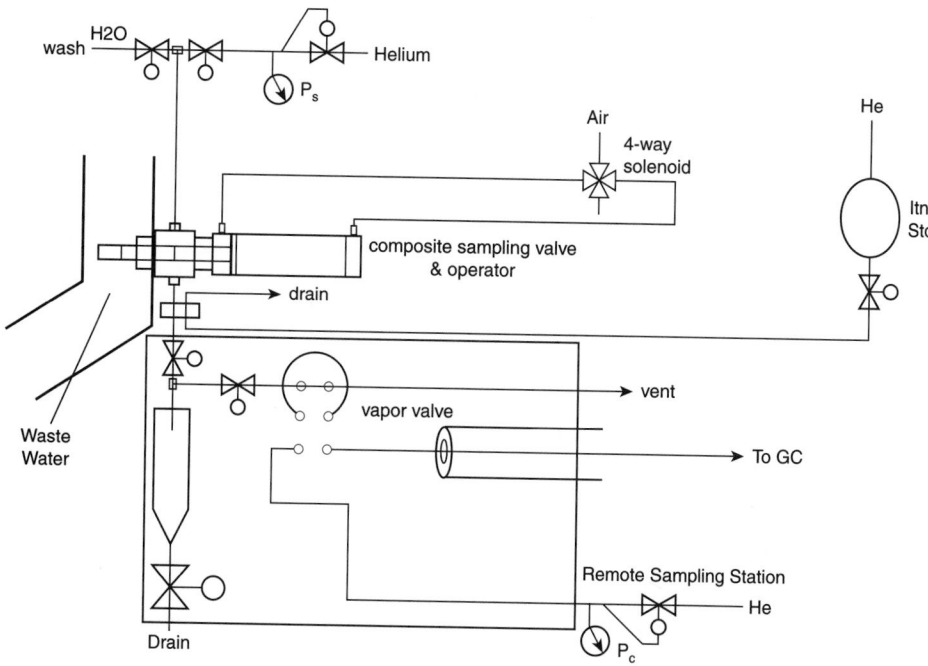

FIGURE 20   VOC in a water headspace analyzer.

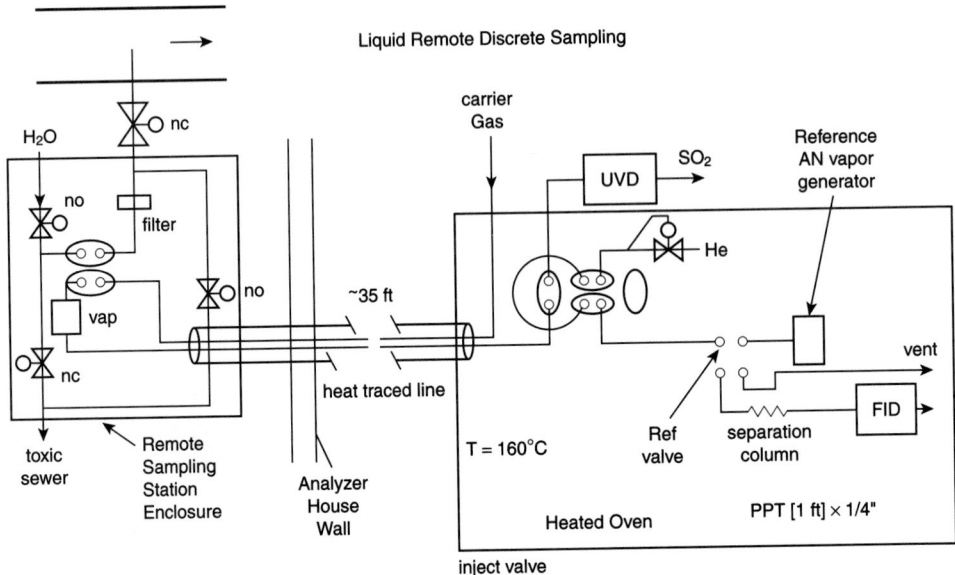

FIGURE 21 Hydrogen cyanide remote sampling station.

LETHAL SERVICE ON HYDROGEN CYANIDE

A remote discrete sampling system was developed for monitoring impurities in this highly toxic material, which had the sampling station located 250 ft from the analyzer. This minimized maintenance work in the very hazardous production area and kept the potential exposure in the analyzer house to a minimum. The impurities varied significantly with time, so an internal reference was introduced to notify the operations personnel of analyzer malfunction. Figure 21 shows the sampling system and detector indicating the location of the internal reference valve.

The internal reference selected was AN. The reference valve was timed to inject such that the second AN peak would be clear of other components. The operator would observe variability in the four analyte concentration trends, but the reference trend would be steady as long as the system was functioning properly. A change in reference trend indicated the analyzer was malfunctioning and that maintenance personnel should be contacted. This was a good start at real-time validation, but it lacked thoroughness. It is important that the internal reference be introduced such that it tests the entire system. We often introduce calibration sample into only part of the system or sometimes only to the detector. Taking our cue from the USEPA, we see that the standard must be introduced at the sample point so that we calibrate the "sample conditioning and preparation system" as well as the "measurement system." We should also note that certified standards are required. A lot of good analyzers have been calibrated with a lot of *bad* standards. One can better understand calibration of the entire system when vapor/liquid equilibrium techniques are utilized to separate volatile components from dirty water streams. A gas standard is used to measure the response of the detector, and a liquid standard is used to calibrate the efficiency of the vapor/liquid separator. It is apparent that in all cases we should introduce a standard that closely resembles the actual sample so that the sampling error as well as the measurement error can be determined.

One could actually introduce a small quantity of one of the target analytes into the sample on a periodic basis and test the analyzer system validation by treating the result as a standard addition.

The increase in concentration of the selected target analyte when a standard addition is introduced every few hours could be compared to the calculated increase. Failure to respond by specified value would warn of sample system or measurement device malfunction and send a warning to operations personnel.

SAFETY (OXYGEN MONITOR)

When hydrogen is compressed to several atmospheres, the presence of oxygen is not desirable. Oxygen analyzers on such an application will read zero, except for an alarm condition when oxygen is actually present. After weeks, months, and years of seeing the recorder pen hug the bottom of the chart, no one knows if the analyzer is working or not. A small, discrete air sample was injected into the hydrogen entering the analyzer to generate a response at 4-h intervals (Fig. 22).

The presence of the response gave confidence that the analyzer was functioning to detect oxygen. It turned out that the amplitude of the response indicated that the span of the analyzer was correct. One could inject a larger discrete sample of oxygen-free gas to indicate that the zero had not drifted. The valve, solenoids, and timer can be contained in a small enclosure. This same technique should be used for monitoring any zero or very low-level component to ensure that a dead system does not fool the people depending on a warning. Chemical species that poison the catalyst or are lethal to humans are examples of when real-time validation is important.

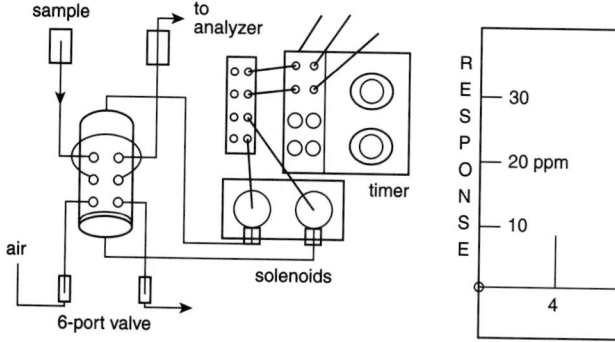 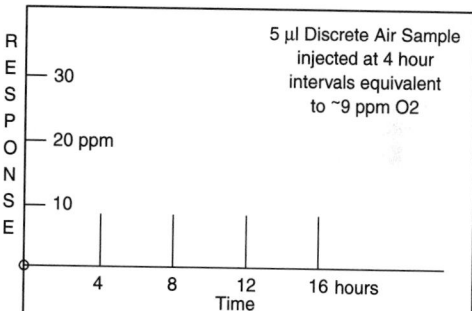

FIGURE 22 Oxygen analyzer.

REFINERY NUCLEAR MAGNETIC RESONANCE (NMR)

Sulfuric Acid Alkylation Process Control

1. Analyzer history: on line since April 1997; no NMR downtime.
2. Variables monitored include
   - acid strength
   - olefin-to-hydrocarbon ratio
   - hydrocarbon-to-acid ratio
   - acid soluble oil content.
3. Justification: 10% acid cost savings; unit optimization savings.

TABLE 3 Waste Gas Analysis

Parameter	Value		
Sample No.	a050198.19		
Date	05/01/98		
Time	06:43 PM		
Pressure (in psia)	51.6		
Temperature (in °C	23.2		
rms	460		
BTU/cu. ft	1398		
Specific gravity	0.814		
Vol.% methane	32.2	dch2	0.3117
Vol.% hydrogen	20.5	dch	0.2768
Vol.%C2– C5+	23.4	h2	0.6660
Vol.% olefin	3.0	ch4 peak	158
Vol.% nondetected	23.9	ch4	1.0447
ch	1.3715		
ch2	1.5312		
ch3	1.1417	sum noise	14.40

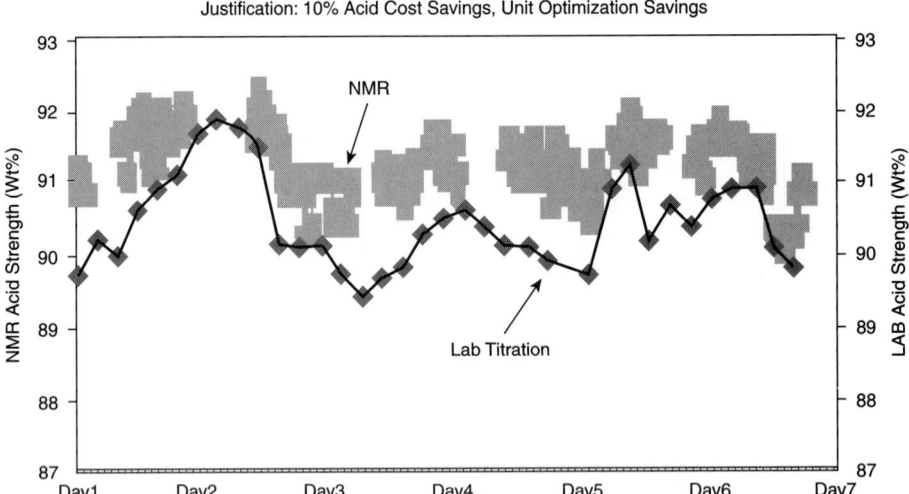

Justification: 10% Acid Cost Savings, Unit Optimization Savings

FIGURE 23 Comparison of 1 week of NMR and lab titration data.

Waste-Gas Analysis/Hydrgen Generation Control (Table 3.)

1. Analyzer History: on-line since October 1995; no NMR downtime.
2. Variables monitored: BTU content, CH bond type, olefin-type analysis, H_2 content, methane content.
3. Justification: rapid analysis; lower cost of ownership.

4. Present applications: environmental monitoring of refinery fuel gas.

5. Future applications: on-line control of fuel gas burners; feedforward control of hydrogen generation processes.

Figure 23 shows a comparison of Lab titration data with the NMR results. Figure 24 shows a typical NMR spectrum. Figure 25 shows the waste gas NMR spectrum obtained from the on-line analyzer.

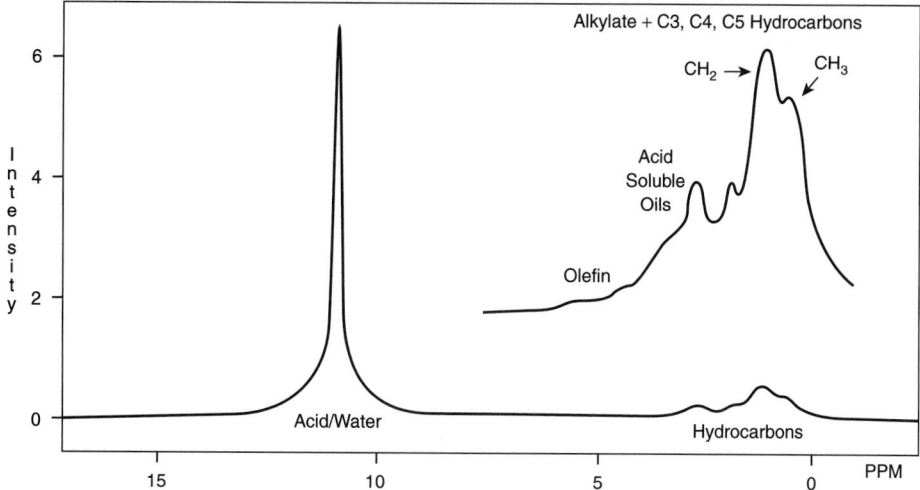

FIGURE 24   Typical[1] H NMR spectrum of intact emulsion process stream.

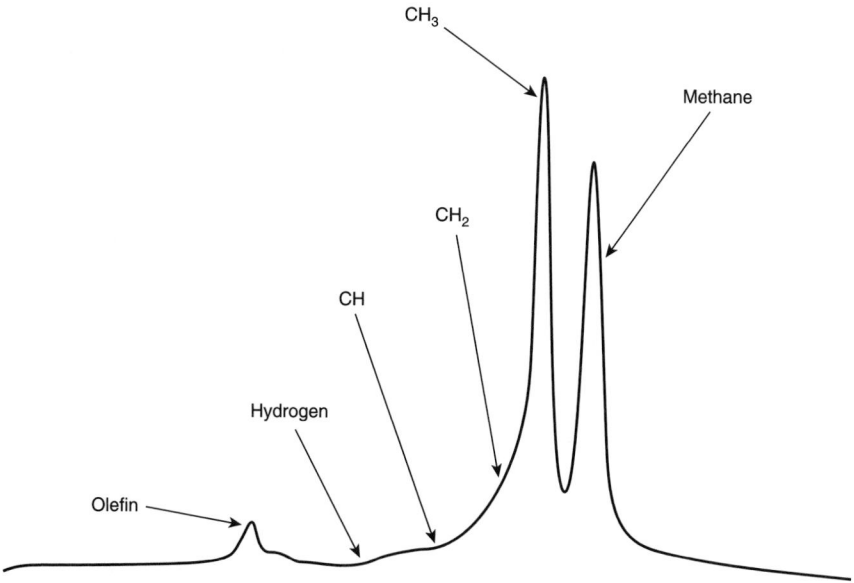

FIGURE 25   H NMR spectrum of waste-gas strem.

TABLE 4   Gasoline Blending Analysis

Parameter	R^2	RMSEP	Range
RON (D2699)	0.9805	0.35	87–100
MON (D2700)	0.9815	0.25	81–90
(R+M)/2	0.9854	0.26	84–94
FIA aromatics (D1319)	0.9638	1.80 vol.%	10–47 vol.%
FIA olefins (D1319)	0.9641	0.46 vol.%	1–10 vol.%
SFC aromatics	0.9813	0.51 vol.%	16–38 vol.%
SFC olefins	0.9804	0.33 vol.%	0.5–7.5 vol.%
Benzene (D3606)	0.9959	0.04 vol.%	0.4–2.5 vol.%
Oxygenates (D5599)	0.9969	0.29 vol.%	0.0–15.0 vol.%
Density (D1298)	0.9705	0.0015 g/ml	0.71–0.79 g/ml
Distillation (D86)			
IBP	0.8069	2.15° F	81–97° F
T10	0.8955	2.02° F	116–135° F
T50	0.9295	2.13° F	180–207° F
T90	0.9797	1.51° F	287–319° F
EP	0.9381	3.60° F	375–413° F
RVP (D5191)	0.987	0.21 psia	1.93–10.53 psia

Gasoline Blending (Table 4.)

1. Analyzer history: to be installed July/August 1998.

2. Variables monitored: RON, MON, aromatics, olefins, saturates, FIA, density, RVP, PNA content, benzene.

Figure 26 shows the comparison of oxygenated gasoline and regular gasoline as seen by the NMR analyzer.

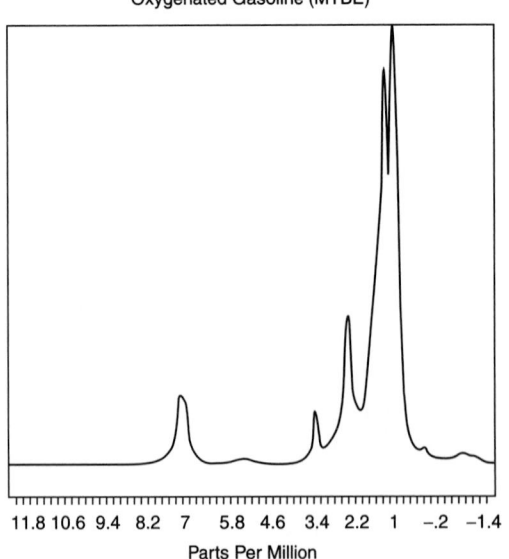

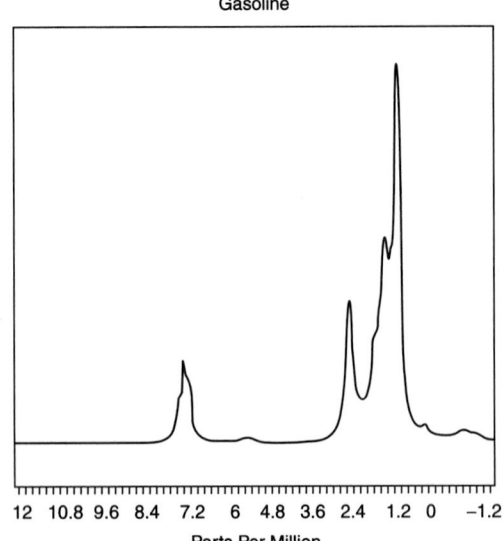

FIGURE 26   Gasoline NMR spectra.

SECTION 7
CONTROL COMMUNICATIONS*

B. A. Loyer
Systems Engineer, Motorola, Inc., Phoenix, Arizona (Local Area Networks—prior edition)

Howard L. Skolnik
Intelligent Instrumentation, Inc. (a Burr-Brown Company), Tucson, Arizona (Data Signal Handling in Computerized Systems; Noise and Wiring in Data Signal Handling)

---

* *Persons who authored complete articles or subsections of articles, or otherwise cooperated in an outstanding manner in furnishing information and helpful counsel to the editorial staff.*

DATA SIGNAL HANDLING IN COMPUTERIZED SYSTEMS

by Howard L. Skolnik*

Prior to the advent of the digital computer, industrial instrumentation and control systems, with comparatively few exceptions, involved analog, rather than digital, signals. This was true for both the outputs from sensors (input transducers—strain gages, thermocouples, and so on) and the inputs to controlling devices (output transducers—valves, motors, and so on). In modern systems many transducers are still inherently analog. This is important because computers can operate with only digital information. Therefore a majority of contemporary systems include analog-to-digital (A/D) and digital-to-analog (D/A) converters.

An important feature of data-acquisition products is how they bring together sophisticated functions in an integrated, easy-to-use system. Given the companion software that is available, the user can take advantage of the latest technology without being intimately familiar with the internal details of the hardware. When selecting a system, however, it is useful to have a basic understanding of data-acquisition principles. This article addresses how real-world signals are converted and otherwise conditioned so that they are compatible with modern digital computers, including personal computers

* Intelligent Instrumentation, Inc. (a Burr-Brown Company), Tucson, Arizona.

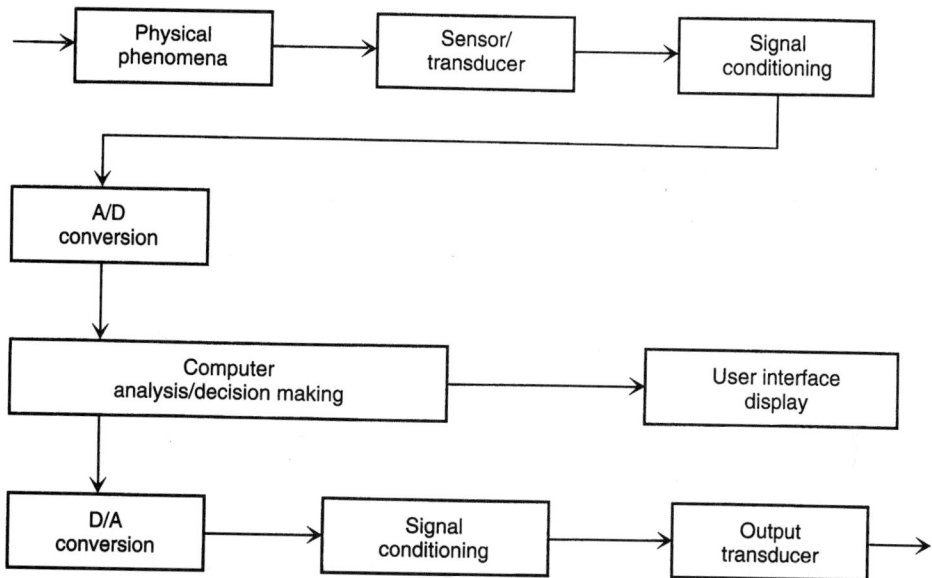

FIGURE 1   Flow diagram of modern computer-based data-acquisition and control system. The measured variable may be any physical or chemical parameter, such as temperature, pressure, flow, liquid level, chemical concentration, dimension, motion, or position. The sensor/transducer is the measuring device (primary element) that converts the measured variable into an electrical quantity. Common transducers include thermocouples, strain gages, resistance temperature devices, pH cells, and switches. The signal from a transducer can be in the form of a voltage, current, charge, resistance, and so on. Signal conditioning involves the manipulation of the raw transducer's output into a form suitable for accurate analog-to-digital (A/D) conversion. Signal conditioning can include filtering, amplification, linearization, and so on. Data conversion provides the translation between the real world (mostly analog) and the digital domain of the computer, where analysis, decision making, report generation, and user interface operations are easily accomplished. To produce analog output signals from the computer (for stimulus or control) digital-to-analog (D/A) conversion is used. Signal conditioning and output transducers provide an appropriate interface to the outside world via power amplifiers, valves, motors, and so on. (*Intelligent Instrumentation, Inc.*)

(PCs). The techniques suggested here are specifically aimed at PC-based measurement and control applications. These generally involve data-acquisition boards that plug directly into an expansion slot within a PC. References to specific capabilities and performance levels are intended to convey the current state of the art with respect to PC-based products (Fig. 1).

SIGNAL TYPES

Signals are often described as being either analog, digital, or pulse. They are defined by how they convey useful information (data). Attributes such as amplitude, state, frequency, pulse width, and phase can represent data. While all signals can be assumed to be changing with time, analog signals are the only ones to convey information within their incremental amplitude variations. In instrumentation and control applications most analog signals are in the range of -10 to $+10$ volts or 4 to 20 mA. Some of the differences between analog and digital signals are suggested in Fig. 2. Digital and pulse signals have binary amplitude values, that is, they are represented by only two possible states—low and high. While low and high states can be represented by any voltage level, transistor-transistor-logic (TTL) levels are most often used. TTL levels are approximately 0 and 5 volts. The actual allowable ranges

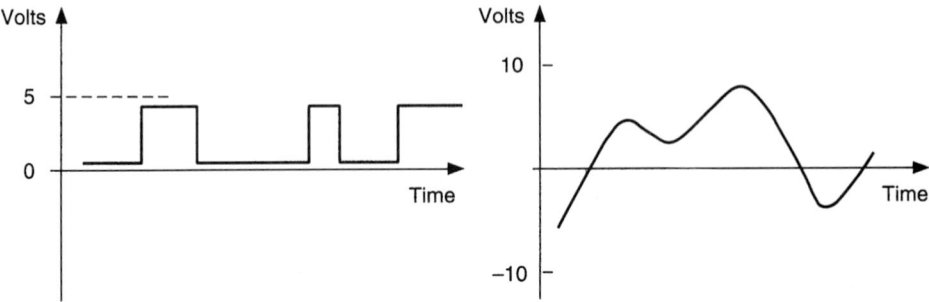

FIGURE 2 Comparison of digital signal (left) and analog signal (right). (*Intelligent Instrumentation, Inc.*)

for TTL signals are

Low level $= 0$ to 0.8 volt

High level $= 2.0$ to 5.0 volts

Thus with analog it is important how high the signal is, while with digital it matters only whether the signal is high or low (on or off, true or false). Digital signals are sometimes called discrete signals. While all digital signals have the potential of changing states at high speed, information is usually contained in their static state at a given point in time. Digital inputs can be used to indicate whether a door is open, a level is beyond a set point, or the power is on. Digital outputs can control (on or off) power to a motor, a lamp, an alarm, and the like. In contrast, analog inputs can indicate how high a level is or how fast a motor is turning. Analog outputs can incrementally adjust the position of a valve, the speed of a motor, the temperature of a chamber, and so on. Pulse signals are similar to digital signals in many respects. The distinction lies in their time-dependent characteristics. Information can be conveyed in the number of state transitions or in the rate at which transitions occur. Rate is referred to as frequency (pulses per second). Pulse signals can be used to measure or control speed, position, and so on.

TERMINATION PANELS

Termination panels are usually the gateway to a data-acquisition system. Screw terminals are provided to facilitate easy connection of the field wiring. Figure 3 suggests two of the many termination panel styles. Some models are intended for standard input or output functions, while others are designed to be tailored for unique customized applications. Mounting and interconnection provisions are provided for resistors, capacitors, inductors, diodes, transistors, integrated circuits, relays, isolators, filters, connectors, and the like. This supports a wide range of signal interface and conditioning capabilities. In most cases termination panels are located outside, but adjacent to, the data-acquisition system's host PC. Many mounting and enclosure options are available to suit different applications. Because the actual data-acquisition board is located inside the PC, short cables (normally shielded ribbon cables) are used to connect the termination panel's signals.

FIELD SIGNALS AND TRANSDUCERS

Whatever the phenomenon detected or the device controlled, transducers play a vital role in the data-acquisition system. It is the transducer that makes the transition between the physical and the electrical world. Data acquisition and control can involve both input and output signals. Input signals

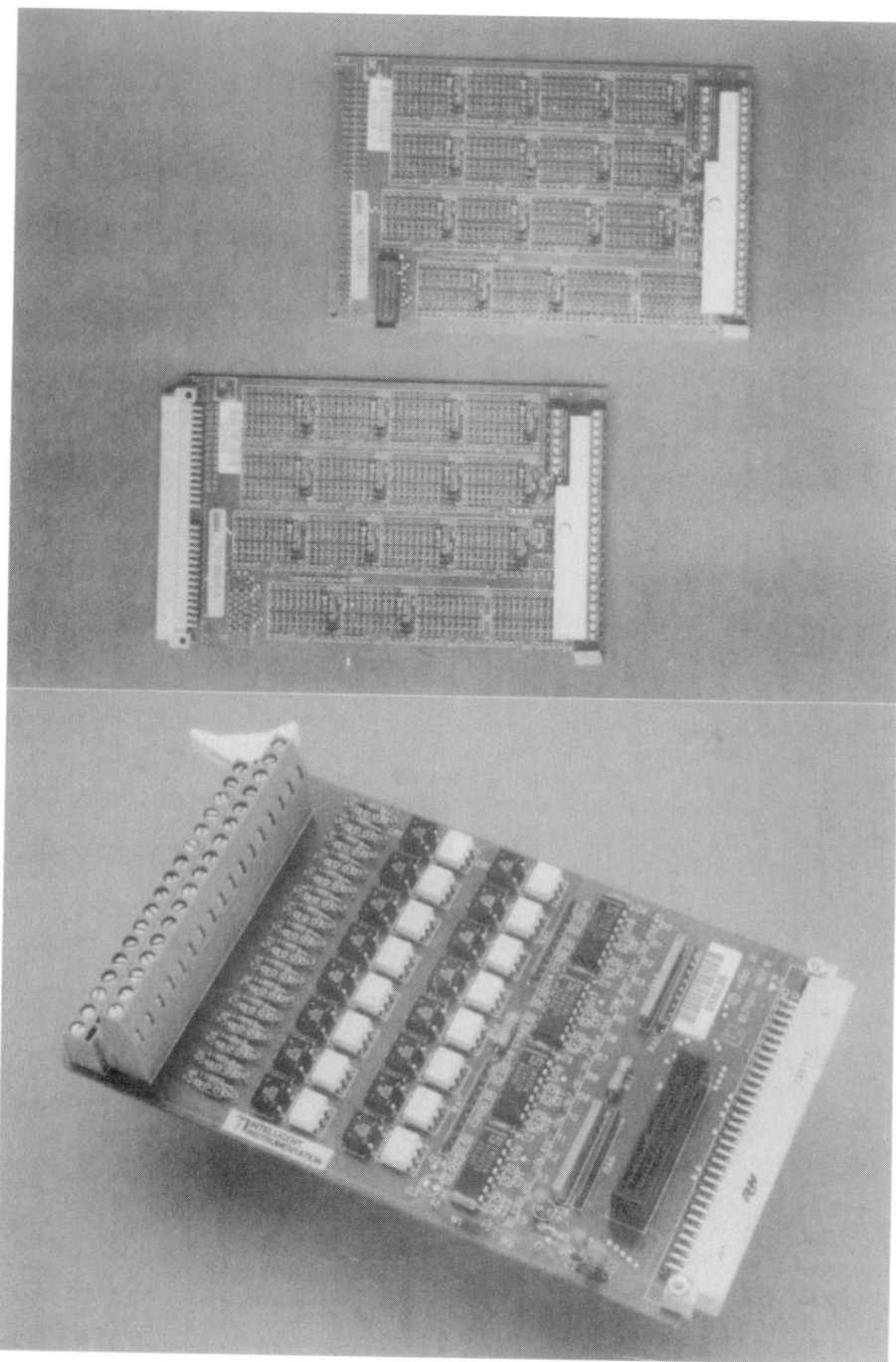

FIGURE 3   Representative termination panel styles. (*Intelligent Instrumentation, Inc.*)

can represent force, temperature, flow, displacement, count, speed, level, pH, light intensity, and so on. Output signals can control valves, relays, lamps, horns, and motors, to name a few. The electrical equivalents produced by input transducers are most commonly in the form of voltage, current, charge, resistance, or capacitance. As shown later, the process of signal conditioning will further convert these basic signals into voltage signals. This is important because the interior blocks of the data-acquisition system can only deal with voltage signals.

Thermocouples

Thermocouples are used widely to measure temperature in industry and science. Temperatures in the range of -200 to $+4000°C$ can be detected. Physically a thermocouple is formed by joining together wires made of two dissimilar metals. The resulting junction produces a voltage across the open ends of the wires that is proportional to temperature (the Seebeck effect). The output voltage is usually in the range of -10 to $+50$ mV and has an average sensitivity of 10 to 50 μV/°C, depending on the metals used. However, the output voltage is very nonlinear with respect to temperature. Many different thermocouple types are in wide use. For convenience, alphabetic letter designations have been given to the most common. These include the following:

J	Iron-constantan (Fe-C)
K	Chrome-Alumel (Ch-Al)
T	Copper-constantan (Cu-C)

Tungsten, rhodium, and platinum are also useful metals, particularly at very high temperatures.

Thermocouples are low in cost and very rugged. Still, they are not without their limitations and applications problems. In general, accuracy is limited to about 1 to 3 percent due to material and manufacturing variations. Response time is generally slow. While special thermocouples are available that can respond in 1 to 10 ms, most units require several seconds. In addition to the thermocouple's nonlinear output, compensation must also be made for the unavoidable extra junctions that are formed by the measuring circuit.

As mentioned previously, a single thermocouple junction generates a voltage proportional to temperature:

$$V = k(t) \tag{1}$$

where k is the Seebeck coefficient defining a particular metal-to-metal junction, and t is in degrees/kelvin.

Unfortunately the Seebeck voltage cannot be measured directly. When the thermocouple wires are connected to the terminals of a voltmeter or data-acquisition system, new thermoelectric junctions are created. For example, consider the copper-constantan (type T) thermocouple connected to a voltmeter shown in Fig. 4. It is desired that the voltmeter read only V_1 (of J_1), but the act of connecting the voltmeter creates two more metallic junctions, J_2 and J_3. Since J_3 is a copper-to-copper junction, it

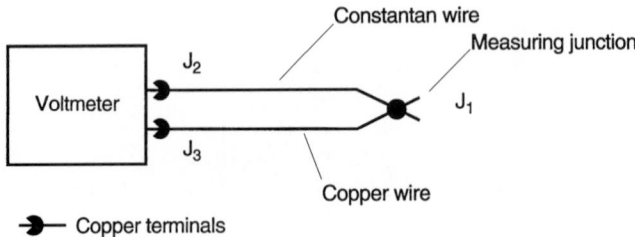

FIGURE 4   Essence of the thermocouple measurement problem. (*Intelligent Instrumentation, Inc.*)

creates no thermal voltage ($V_3 = 0$), but J_2 is a copper-to-constantan junction that will add a voltage V_2 in opposition to V_1. As a result, the voltmeter reading V_v will actually be proportional to the temperature difference between J_1 and J_2. This means that determining the temperature at J_1 requires a knowledge of the temperature at J_2. This junction is referred to as the *reference junction*, or *cold junction*. Its temperature is the reference temperature t_{ref}. Note that V_2 equals V_{ref}. Therefore it follows from Eq. (1),

$$V_v = V_1 - V_{ref} = k(t_1 - t_{ref}) \tag{2}$$

It is important to remember that k is highly nonlinear with respect to temperature. However, for measurement purposes it is not necessary to know the value of k. Tables have been compiled by the U.S. National Bureau of Standards (now the National Institute of Standards and Technology, NIST) that take variations in k into account and can provide t_1 directly in terms of V_v, assuming that t_{ref} is at 0°C. Separate tables were made for each thermocouple type. J_2 (and J_3) was physically placed in an ice bath, forcing its temperature to 0°C. Note that even under these conditions [see Eq. (1)], V_{ref} is not 0 volts. The Seebeck relationship is based on the kelvin (absolute zero) scale. In computer-based applications the thermocouple tables are transformed into polynomial equations for ease of use. Depending on the thermocouple type and the accuracy (compliance with the NIST tables) desired, between fifth- and ninth-order polynomials are used.

The copper-constantan thermocouple used in this example is a special case because the copper wire is the same metal as the voltmeter terminals. It is interesting to look at a more general example using iron-constantan (type J). The iron wire increases the number of dissimilar metal junctions in the circuit as J_3 becomes a Cu-Fe thermocouple junction. However, it can be shown that if the Cu-Fe and the Cu-C junctions (at the termination panel) are at the same temperature, the resulting voltage is equivalent to a single Fe-C junction. This allows the use of Eq. (2). Again, it is very important that both parasitic junctions be held at the same (reference) temperature. This can be aided by making all connections on an isothermal (same temperature) block.

Clearly, the requirement of an ice bath is undesirable for many practical reasons. Taking the analysis to the next logical step, Eq. (2) shows that t_{ref} need not be at any special temperature. It is only required that the reference temperature be accurately known. If the temperature of the isothermal block (the reference junction) can be measured independently, this information can be used to compute the unknown temperature t_1.

Devices such as thermistors, resistive temperature detectors, and semiconductor sensors can provide a means of independently measuring the reference junction. (Semiconductor sensors are the most popular for the reasons described hereafter.) A thermocouple temperature measurement, under computer control, could proceed as follows:

1. Measure t_{ref} and use the thermocouple polynomial to compute the equivalent thermocouple voltage V_{ref} for the parasitic junctions.
2. Measure V_v and *add* V_{ref} to find V_1.
3. Compute t_1 from V_1 using the thermocouple polynomial.

Solid-State Temperature Sensors

These devices are derived from modern silicon integrated-circuit technology, and are often referred to as Si sensors. They consist of electronic circuits that exploit the temperature characteristics of active semiconductor junctions. Versions are available with either current or voltage outputs. In both cases the outputs are directly proportional to temperature. Not only is the output linear, but it is of a relatively high level, making the signal interpretation very easy. The most common type generates 1 μA/K (298 μA at 25°C). This can be externally converted to a voltage by using a known resistor. The usable temperature range is −50 to 150°C. The stability and the accuracy of these devices are good enough to provide readings within ±0.5°C. It is easy to obtain 0.1°C resolution. Si sensors are ideal reference junction monitors for thermocouple measurements.

Resistance Temperature Detectors

Resistance temperature detectors (RTDs) exhibit a changing resistance with temperature. Additional detailed information on using RTDs can be found in the section on signal conditioning for resistive devices. Several different metals can be used to produce RTDs. Platinum is perhaps the most common for general applications. Yet at very high temperatures, tungsten is a good choice. Platinum RTDs have a positive temperature coefficient of about 0.004 Ω/°C. The relationship has a small nonlinearity that can be corrected with a third-order polynomial. Many data-acquisition systems include this capability. Platinum RTDs are usually built with 100-ohm elements. These units have sensitivities of about +0.4 Ω/°C. Their useful temperature range is about −200 to about +600°C.

Most RTDs are of either wire-wound or metal-film design. The film design offers faster response time, lower cost, and higher resistance values than the wire-wound type. The more massive wire-wound designs are more stable with time. High resistance is desirable because it tends to reduce lead-wire induced errors. To convert resistance into a voltage, an excitation current is required. Care must be taken to avoid current levels that will produce errors due to internal self-heating. An estimate of the temperature rise (in °C) can be found by dividing the internal power dissipation by 80 mW. This is a general rule that applies to small RTDs in a conductive fluid such as oil or water. In air the effects of self-heating can be 10 to 100 times higher.[1]

SAMPLED-DATA SYSTEMS

Modern data-acquisition systems use sampled-data techniques to convert between the analog and digital signal domains. This implies that while data may be recorded on a regular basis, they are not collected continuously, that is, there are gaps in time between successive data points. In general there is no knowledge of the missing information, and the amplitude of missing data points cannot be predicted. Yet under special circumstances it can be assumed that missing data fall on a straight line between known data points.

Fourier analysis reveals that signals, other than pure sine waves, consist of multiple frequencies. For example, a pulse waveform contains significant frequency components far beyond its fundamental or repetition rate. Frequencies extending to approximately $0.3/t_r$ are often important, where t_r is the pulse rise time. Step functions suggest that frequency components extend to infinity.

The Nyquist theorem defines the necessary relationship between the highest frequency contained in a waveform and the *minimum* required sampling speed. Nyquist states that the sample rate must be *greater* than two times the highest frequency component contained within the input signal. The danger of undersampling (sampling below the Nyquist rate) is erroneous results. It is not simply a matter of overlooking high-frequency information, but of reaching totally wrong conclusions about the basic makeup of the signal. See Fig. 5 for an example. Note that sampling a pure sine wave (containing only the fundamental frequency) at a rate in violation of the Nyquist criterion leads to meaningless results. This example suggests the presence of a totally nonexistent frequency. This phenomenon is known as aliasing.

If it were possible to sample at an infinite rate, aliasing would not be a problem. However, there are practical limits to the maximum sampling speed, as determined by the characteristics of the particular A/D converter used. Therefore action must be taken to ensure that the input signal does not contain frequency components that cause a violation of the Nyquist criterion. This involves the use of an input low-pass filter (antialiasing filter) *prior* to the A/D converter. Its purpose is to limit the measured waveform's frequency spectrum so that no *detectable* component equals or exceeds half of the sampling rate. Detectable levels are determined by the sensitivity of the A/D converter and the attenuation of the antialiasing filter (at a given frequency).

---

[1] Additional information on temperature sensors will be found in Section 3 of this handbook.

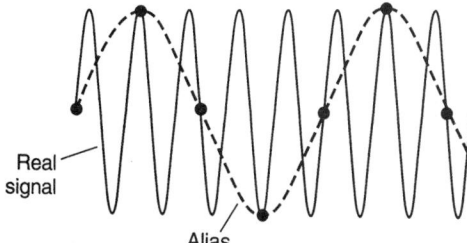

FIGURE 5   Aliasing because of insufficient sampling rate.
(*Intelligent Instrumentation, Inc.*)

Sequential Scanning

Systems are usually designed to collect data from more than one input channel. To reduce cost, most PC-based systems share significant components, including the A/D converter, with all of the channels. This is accomplished with a multiplexer (electronic scanning switch). However, when there is only one A/D converter, only one input channel can be acquired at a given point in time. Each channel is read sequentially, resulting in a time skew between readings. Techniques for minimizing time skew will be described.

ANALOG INPUT SYSTEMS

The fundamental function of an analog input system is to convert analog signals into a corresponding digital format. It is the A/D converter that transforms the original analog information into computer-readable data (a digital binary code). In addition to the A/D converter, several other components may be required to obtain optimum performance. These can include a sample/hold, an amplifier, a multiplexer, timing and synchronization circuits, and signal-conditioning elements.

A good starting point when choosing an A/D converter (or a system using an A/D converter) is to consider the characteristics of the input transducer. What is its *dynamic range*, maximum signal level, signal frequency content, source impedance, accuracy, and so on? A match in characteristics between the A/D converter and the transducer is usually desired. It is also important to consider possible sources of external interfering noise. This will have a bearing on the choice of A/D converter and the required signal conditioning. Some sensors have a very wide dynamic range. Dynamic range defines the span of input stimulus values that correspond to *detectable* output values. It is often expressed as the ratio of the maximum full-scale output signal to the lowest *meaningful* output signal. In this context, dynamic range and signal-to-noise ratio are the same. Caution! There is not necessarily a good correlation between sensor dynamic range and *accuracy*. Accuracy refers to how close the measured output corresponds to the *actual* input. For example, a transducer could have a dynamic range of 5000:1 and an accuracy of 1 percent. If the full-scale range is 100°C, a change as small as 0.02°C can be detected. Still, the actual temperature is only known, in absolute terms, to 1°C. 1°C is the accuracy; 0.02°C is the sensor's sensitivity. The proper choices of signal conditioning and A/D converter are essential to preserving the performance of the transducer. This example suggests the need for an A/D converter with more than 12 bits of resolution. Applications using other transducers may require 14-, 16-, or even 18-bit resolution.

Analog-to-Digital Converters

Many different types of A/D converters exist. Among these, a few stand out as the most widely used—successive-approximation, integrating, and flash (parallel) converters. Each converter has a set of unique characteristics that makes it better suited for a given application. These attributes include speed, resolution, accuracy, noise immunity, cost, and the like.

Industrial and laboratory data-acquisition tasks usually require a resolution of 12 to 16 bits. 12 bits is the most common. As a general rule, increasing resolution results either in increased cost or in reduced conversion speed. Therefore it makes sense to consider the application requirements carefully before making a resolution decision.

All A/D converters accomplish their task by partitioning the full analog input range into a fixed number of discrete digital steps. This is known as digitizing or quantizing. A different digital code corresponds to each of the assigned steps (analog values). Digital codes consist of N elements, or bits. Because each bit is binary, it can have one of two possible states. Thus the total number of possible steps is 2^N. N is often referred to as the converter's *resolution* (that is, an N-bit converter). Given the number of steps S, it follows that $N = \log S/\log 2$. Caution is required when using the term "resolution" because it can also be expressed in other ways. For example, a 12-bit system divides its input into 2^{12}, or 4096, steps. Thus if the A/D converter has a 10-volt range, it has a resolution of 2.4 mV (10 volts $\div$ 4096). This refers to the minimum detectable signal level. One part in 4096 can also be expressed as 0.024 percent of full-scale (FS). Thus the resolution is 0.024 percent FS. These definitions apply *only* to the ideal, internal capabilities of the A/D converter alone.

In contrast to this 12-bit example, 16 bits correspond to one part in 65,536 (2^{16}), or approximately 0.0015 percent FS. Therefore increasing the resolution has the *potential* to improve both dynamic range and overall accuracy. On the other hand, system performance (effective dynamic range and accuracy) can be limited by other factors, including noise and errors introduced by the amplifier, the sample/hold, and the A/D converter itself.

Flash-type A/D converters can offer very high-speed operation, extending to about 100 MHz. Conversion is accomplished by a string of comparators with appropriate reference voltages, operating in parallel. To define N quantizing steps requires $2^N - 1$ comparators (255 and 4095 for 8 and 12 bits, respectively). Construction is not practical beyond about 8 or 10 bits. In contrast, most data-acquisition and control applications require more than 10 bits.

12-bit devices that run in the 5- to 15-MHz range are available using a *subranging* or half-flash topology. This uses two 6-bit flash encoders, one for coarse and the other for detailed quantizing. They work in conjunction with differencing, amplifying, and digital logic circuits to achieve 12-bit resolution. It is essential that the input signal remain constant during the course of the conversion, or very significant errors can result. This requires the use of a sample/hold circuit, as described hereafter. There is also a speed trade-off compared with a full flash design, but the cost is much lower. Still, the size and power requirements of these flash converters usually prevent them from being used in PC-based internal plug-in products. These factors (cost, size, and power) generally limit flash converters to special applications.

Successive approximation converters are the most popular types for general applications. They are readily available in both 12- and 16-bit versions. Maximum sampling speeds in the range of 50 kHz to 1 MHz are attainable. These converters use a single comparator and are thus relatively low cost and simple to construct. An internal D/A converter (described below) systematically produces binary weighted guesses that are compared to the input signal until a "best" match is achieved. A sample/hold is required to maintain a constant input voltage during the course of the conversion.

Integrating converters can provide 12-, 16-, and even 18-bit resolution at low cost. Sampling speeds are typically in the range of 10 to 500 Hz. This converter integrates the unknown input voltage, V_x for a specific period of time T_1. The resulting output e_1 (from the integrator) is then integrated back down to zero by a known reference voltage V_{ref}. The time required T_2 is proportional to V_x:

$$V_x = V_{\text{ref}} \frac{T_2}{T_1}$$

Noise and other signal variations are effectively averaged during the integration process. This characteristic inherently smooths the input signal. The dominant noise source in many data-acquisition applications is the ac power line. By setting the T_1 integration period to a multiple of the ac line frequency, significant noise rejection is achieved. This is known as normal- or series-mode rejection. In addition, linearity and overall accuracy are generally better than with a successive approximation

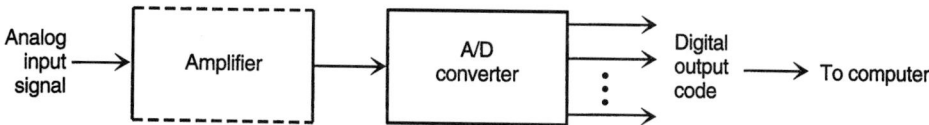

FIGURE 6 Simple analog input stage. *Note*: Amplifier may not be required in every application. (*Intelligent Instrumentation, Inc.*)

converter. These factors make the integrating converter an excellent choice for low-level signals such as thermocouples and strain gages.

Amplifiers

A simple analog input stage is shown in Fig. 6. This circuit can accommodate only one input channel. For multiple channels, several parallel stages can be used. However, the use of a multiplexer to share common resources can provide significant cost savings. This is suggested in Fig. 7. The amplitude of analog input signals can vary over a very wide range. Signals from common transducers are between 50 μV and 10 volts. Yet most A/D converters perform best when their inputs are in the range of 1 to 10 volts. Therefore many systems include an amplifier to boost possible low-level signals to the desired amplitude. Note that adding a fixed-gain amplifier increases sensitivity, but does not increase dynamic range. While it extends low-level sensitivity, it reduces the maximum allowable input level proportionally (the ratio remains constant). Ideally an input amplifier would have several choices of

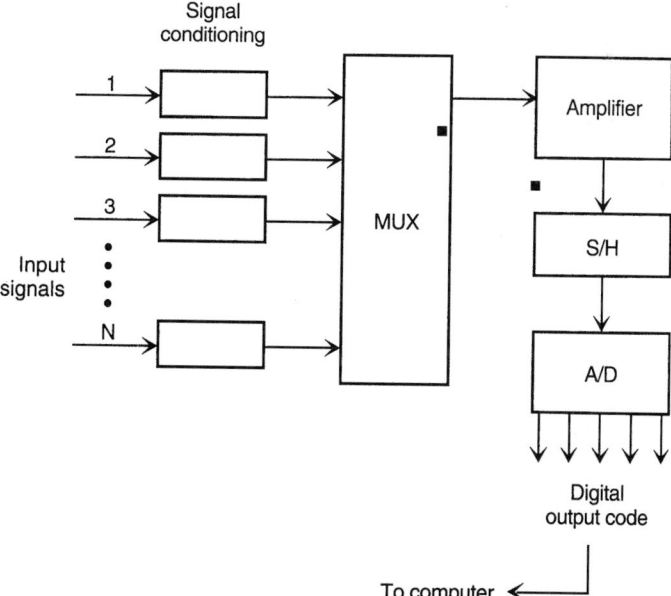

FIGURE 7 Complete analog input subsystem in which a multiplexer (MUX) is used to handle multiple signals. Software can control switches to select any one channel for processing at a given time. Since the amplifier and A/D converter are shared, the speed of acquisition is reduced. To a first approximation, the rated speed of the amplifier and A/D converter will be divided by the number of input channels serviced. Throughput rate is often defined as the per-channel speed multiplied by the number of channels. (*Intelligent Instrumentation, Inc.*)

gain, all under software control. Such a device is called a programmable gain amplifier (PGA). Gains of 1, 2, 4, and 8 or 1, 10, 100, and 200 are common. Unlike a fixed amplifier, a PGA can increase the system's effective dynamic range. For example, consider the following:

- A 12-bit converter (which can resolve one part in 4096) with a full-scale range of 10 volts
- A PGA with gains of 1, 10, and 100—initially set to 1
- An input signal that follows a ramp function from 10 volts down toward zero

Under these conditions, the minimum detectable signal will be 2.4 mV (10 volts ÷ 4096). However, software can be written to detect when the signal drops below 1 volt and to reprogram the PGA automatically for $G = 10$. This extends the minimum sensitivity to 240 μV while increasing a 0.999-volt signal to 9.99 volts. As the signal continues to drop (below 0.1 volt), the PGA can be set for $G = 100$, extending the sensitivity to 24 μV. The effect of making these corresponding changes in gain, to track signal variations, is to increase the *system's* dynamic range. Dynamically adding a gain of 100 is like adding close to 7 bits to the converter. Thus a 12-bit A/D converter can support a range of more than 18 bits. This technique is restricted to systems that have a PGA and to applications where the sample rate is slow enough to permit the time required for the autoranging function.

Most PGAs are differential input instrumentation amplifiers. They present a very high input impedance at both their + and − terminals. The common-mode rejection characteristic of this type of amplifier can attenuate the effects of ground loops, noise, and other error sources. Thus differential inputs are especially useful for measuring low-level signals. Most analog input systems have provisions for configuring the input multiplexer and amplifier for either single-ended or differential use.

Single-Ended versus Differential Signals

Single-ended inputs all share a common return or ground line. Only the high ends of the signals are connected through the multiplexer to the amplifier. The low ends of the signals return to the amplifier through the system ground connections, that is, both the signal sources and the input to the amplifier are referenced to ground. This arrangement works well for high-level signals when the difference in ground potential is relatively small. Problems arise when there is a large difference in ground potentials. This is usually caused by current flow (a ground loop) through the ground conductor. This is covered in further detail in the next article of this handbook section.

A differential arrangement allows both the noninverting (+) and the inverting (−) inputs of the amplifier to make direct connections to both ends of the actual signal source. In this way any ground-loop-induced voltage appears as a common-mode signal and is rejected by the differential properties of the amplifier. While differential connections can greatly reduce the effects of ground loops, they consume the equivalent of two single-ended inputs. Thus a 16-channel single-ended system can handle only eight differential inputs.

Ideally the input impedance, common-mode rejection, and bandwidth of the system's amplifiers would be infinite. In addition, the input currents and offset voltage would be zero. This would provide a measuring system that does not load or alter the signal sources. In contrast, real amplifiers are not perfect. Offset voltage V_{os} appears as an output voltage when the inputs are short-circuited (input voltage is zero). V_{os} can sometimes be compensated for in software. Input (bias) current can be more of a problem. This is the current that flows into (or out of) the amplifier's terminals. The current interacts with the signal source impedance to produce an additional V_{os} term that is not easy to correct. A resistive path must be provided to return this current to ground. It is necessary that the resistance of this path be small enough so that the resulting V_{os} (bias current × source resistance) does not degrade the system's performance significantly. In the extreme case where the inputs are left floating (no external return resistance), the amplifier is likely to reside in a nonlinear or otherwise unusable state. As a general rule, single-ended inputs do not require attention to the bias current return resistance. This is because the path is often provided by the signal source. In contrast, differential connections almost always require an external return resistor. Normally the system's termination panel will provide these resistors. Typically, values of 10 or 100 kΩ are used.

TABLE 1 Relationship between CMRR Expressed in dB and Common-Mode Error in LSB for Hypothetical Input System

Signal gain	Common-mode rejection ratio, dB	Error_{CM} absolute, LSB	Error_{CM}/V_{CM}, LSB/V
1	80	0.4	0.04
10	90	1.3	0.13
100	100	4.1	0.41
1000	110	13	1.3

Common-Mode Rejection

The ability of a differential-input amplifier to discriminate between a differential mode (desired input signal) and a common mode (undesired signal) is its common-mode rejection ratio (CMRR), expressed in decibels (dB). For a given amplifier the CMRR is determined by measuring the change in output that results from a change in common-mode input voltage. CMRR (dB) $= 20 \log(dV_{out}/dV_{CM})$. In a data-acquisition system the output signal from the input amplifier includes any error due to the finite CMRR. The A/D converter cannot discriminate between true and error portions of its input signal. Thus the relationship between the magnitude of the error and the sensitivity of the A/D converter is significant. This sensitivity is often referred to as the A/D converter's resolution, or the size of its least-significant bit (LSB). If the error exceeds 1 LSB, the A/D converter responds. As suggested, CMRR is used to measure the analog output error produced by a common-mode input. This makes sense because it is the ratio of two analog signals. However, a complete data-acquisition system (including an A/D converter) has a digital output. Therefore it is more meaningful to express the system's common-mode error in terms of LSBs. This is done by dividing the common-mode error voltage dV_{out} by the sensitivity of the A/D converter (1 LSB on the given range). Amplifier gain G_{diff} must be taken into account. Sensitivity is equal to the converter's full-scale range (FSR) divided by its resolution (number of steps):

$$\text{Error}_{CM}(\text{LSB}) = \frac{dV_{CM} \cdot G_{diff} \cdot 10^{-\text{CMRR}/20}}{\text{FSR/resolution}} \tag{3}$$

Table 1 shows the relationship between CMRR in decibels and common-mode error in least-significant bits for a hypothetical input system. A 12-bit A/D converter on a 10-volt range (0 to 10 volts or ±5 volts) is assumed in this comparison. A 10-volt common-mode signal is applied to the short-circuited (connected together) input terminals of the system. Dividing the common-mode error (in LSB) by the common-mode voltage yields a direct (useful) figure of merit for the complete data-acquisition system, not just the input amplifier.

Note that independently increasing (improving) CMRR at a *given* gain improves the system's performance. However, the increase in CMRR that accompanies an *increase* in gain actually results in a decrease of the system's overall accuracy. This is because the increase in CMRR (note the log relationship) has less effect than the direct increase in common-mode error ($dV_{CM} \cdot G_{diff}$).

Sample/Hold System

There is a distinct time interval required to complete a given A/D conversion. Only the integrating converter can tolerate input amplitude changes during this period. For the other converters a detectable change will result in significant errors. In general, an analog signal can have a continuously changing amplitude. Therefore, a sample/hold (S/H) is used as a means of "freezing" or holding the input constant during the conversion period. Fundamentally the S/H consists of a charge storage device (capacitor) with an input switch. When the switch is closed, the voltage on the capacitor tracks the input signal (sample mode). Before starting an A/D conversion, the switch is opened (hold mode), leaving the last instantaneous input value stored on the capacitor. This is maintained until the conversion is complete. In all practical applications, both successive approximation and subranging converters must use an S/H at their inputs. While a full flash A/D converter does not normally require an S/H, there are applications where it will improve its spurious-free dynamic range.

Multiplexers

The multiplexer (MUX) is simply a switch arrangement that allows many input channels to be serviced by one amplifier and A/D converter (Fig. 7). Software or auxiliary hardware can control this switch to select any one channel for processing at a given time. Because the amplifier and A/D converter are shared, the channels are read sequentially, causing the overall speed of the system to be reduced. To a first approximation, the rated speed of the amplifier and A/D converter will be divided by the number of input channels serviced. The throughput rate is defined as the sample rate (per-channel speed) multiplied by the total number of channels.

The user must be careful not to be misled by the speed specifications of the individual components in the system. Conversion time defines only the speed of a single A/D conversion. Software overhead, amplifier response time, and so on, can greatly reduce a system's performance when reading multiple channels.

In an ideal system all of the input channels would be read at the same instant in time. In contrast, multiplexing inherently generates a "skew," or time difference, between channels. In some cases the system may be fast enough to make it "appear" that the channels are being read at the same time. However, some applications are very sensitive to time skew. Given the fastest A/D converters available, there are still many applications that cannot tolerate the time difference between readings resulting from multiplexing. In critical applications the technique of *simultaneous* S/H can reduce time skew by a factor of 100 to 1000 (Fig. 8).

The simultaneous S/H architecture is ideal when the phase and time relationships of multiple input channels are critical to a given application. For example, assume the system in Fig. 7 is sequentially scanning four analog inputs at a throughput rate of 100 kHz. The elapsed time between conversions would be 10 μs. About 40 μs would be required to digitize all four channels. If the input signals are each 10 kHz sine waves, there will be an apparent phase shift of 144° between the first and fourth channels (40 μs/100 μs · 360°). In contrast, the simultaneous S/H system in Fig. 8 can capture all four channels within a few nanoseconds of each other. This represents a phase shift of about 0.01°.

This technique is particularly useful for preserving time and phase relationships in applications where cross-correlation functions must be calculated. Prime examples include speech research,

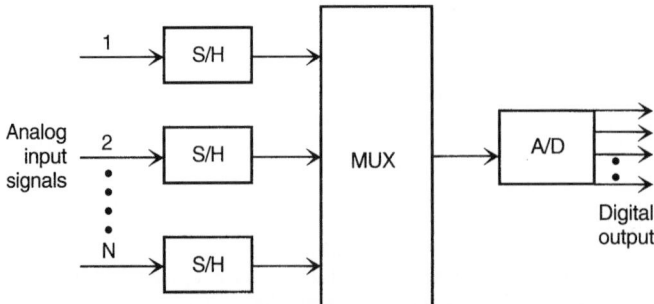

FIGURE 8   Simultaneous sample/hold system. The function of the S/H system is to grab the present value of the signal just before the beginning of an A/D conversion. This level is held constant, despite a changing input, until the A/D conversion is complete. This feature allows the accurate conversion of high-frequency signals. (*Intelligent Instrumentation, Inc.*)

materials and structural dynamics testing, electrical power measurements, geophysical signal analysis, and automatic test equipment (ATE) on production lines.

Analog Signal Conditioning

Analog input systems, based on the components described previously, are representative of most PC-based plug-in products. These boards are usually designed to accept voltage inputs (only) in the range of perhaps ± 1 mV to ± 10 V. Other signal ranges and signal types generally require preprocessing to make them compatible. This task is known as signal conditioning. The type of conditioning used can greatly affect the quality of the input signal and the ultimate performance of the system. Signal conditioning can include current-to-voltage conversion, surge protection, voltage division, bridge completion, excitation, filtering, isolation, and amplification. The required components can be physically located either remotely, at the signal source (the transducer), or locally, at the data-acquisition board (the host PC). Remote applications use *transmitters* that include the required components. They generally deliver high-level conditioned signals to the data-acquisition system by means of a twisted-pair cable. Local transducers usually connect directly to termination panels that include the required components.

Filtering

Of all the signal-conditioning categories, filtering is the most widely needed and most widely misunderstood. Simply stated, filtering is used to separate desired signals from undesired signals. Undesired signals can include ac line frequency pickup and radio or TV station interference. All such signals are referred to here as *noise*. Filtering can be performed, prior to the A/D conversion, using "physical" devices consisting of resistors, capacitors, inductors, and amplifiers. Filtering can also be accomplished, after conversion, using mathematical algorithms that operate on the digital data within the PC. This is known as digital signal processing (DSP).

Averaging is a simple example of DSP. It is a useful method for reducing unwanted data fluctuations. By averaging a series of incoming data points, the signal-to-noise ratio can be effectively increased. Averaging will be most effective in reducing the effects of random nonperiodic noise. It is less effective in dealing with 50- or 60-Hz or other periodic noise sources. When the desired signal has lower frequency components than the error sources, a low-pass filter can be used. This includes the case where the "real" input signal frequency components can equal, or exceed, half the sampling rate. Here the filter is used to prevent sampled-data aliasing. Aliasing results in the generation of spurious signals within the frequency range of interest that cannot be distinguished from real information. Hence serious errors in the interpretation of the data can occur. Noise-filtering techniques, whether implemented in hardware or software, are designed to filter specific types of noise. In addition to low-pass filters, high-pass and notch (band-reject) filters also can be used. For example, if the frequency band of interest includes the ac line frequency, a notch filter could be used to selectively remove this one component.

Signal termination panels are available that have provisions for the user to install a variety of filters. The most common types of filters are represented by the one- and two-pole *passive* filters shown in Fig. 9. The main difference between these passive filters and the *active* filters, mentioned hereafter, is the addition of amplifiers. Figure 9*b* is an example of an effective single-ended double-pole circuit to attenuate 50/60-Hz noise. The filter has a -6-dB cutoff at about 1 Hz while attenuating 60 Hz about 52 dB (380 times).

Figure 10 suggests a differential two-pole low-pass filter. In contrast to the circuits in Fig. 9, this can be used in balanced applications. Note that any mismatch of the attenuation in the top and bottom paths will result in the generation of a differential output signal that will degrade the system's common-mode rejection ratio. Therefore the resistors and capacitors should be matched carefully to each other. If it is given that all of the resistors and capacitors are of equal values, the pole position f_1

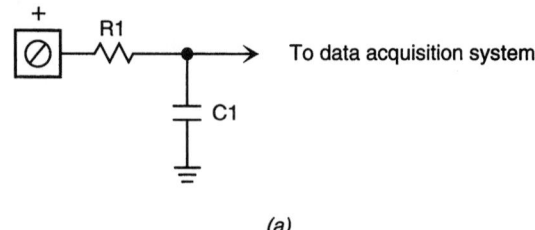

(a)

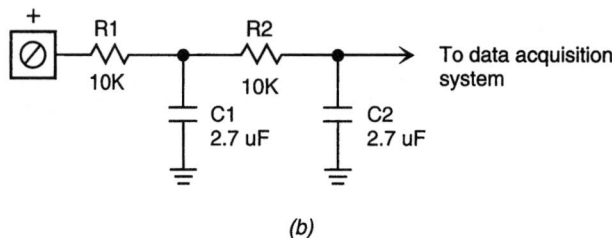

(b)

FIGURE 9   Low-pass filters. (*a*) One pole. (*b*) Two poles. (*Intelligent Instrumentation, Inc.*)

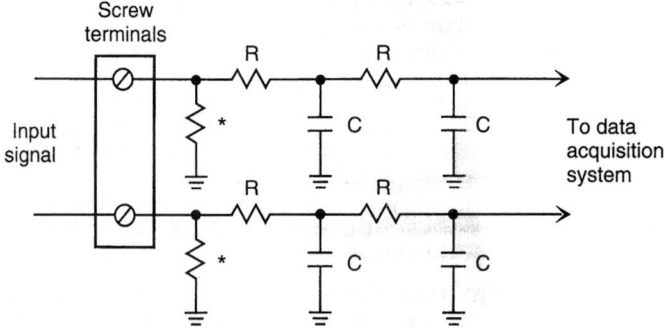

*Amplifier bias current return resistors

FIGURE 10   Two-pole differential low-pass filter. (*Intelligent Instrumentation, Inc.*)

for this *differential* two-section filter is

$$f_1 = \frac{0.03}{R \cdot C} \tag{4}$$

and the approximate attenuation ratio ($r = V_{in} / V_{out}$), at a given frequency f_x, is

$$r = \left(\frac{f_x \cdot R \cdot C}{0.088} + 1 \right)^2$$

$$= \left(\frac{0.3 f_x}{f_1} + 1 \right)^2 \tag{5}$$

$$\mathrm{dB} = 20 \log r$$

The equations for a single-ended single-pole filter are

$$f_1 = \frac{0.159}{R \cdot C}$$

$$r = \sqrt{\left(\frac{f_x \cdot R \cdot C}{0.159}\right)^2 + 1} \qquad (6)$$

and also
$$r = \sqrt{\left(\frac{f_x}{f_1}\right)^2 + 1}$$

The foregoing equations assume that the source impedance is much less than R and that the load impedance is much larger than R.

For filter applications, monolithic ceramic-type capacitors have been found to be very useful. They possess low leakage, have low series inductance, have very high density (small in size for a given capacitance), and are nonpolarized. Values up to 4.7 μF at 50 volts are commonly available.

In the ideal case a perfect low-pass filter could be built with infinite rejection beyond its cutoff frequency f_1. This would allow f_1 to be set just below one-half the sampling rate, providing maximum bandwidth without danger of aliasing. However, because perfect filters are not available, some unwanted frequencies will "leak" through the filter. Nyquist requires that the "2 times" rule be applied to the highest frequency that can be resolved by the A/D converter. This may not be the same as the highest frequency of interest f_1. Therefore the margin between the highest frequency of interest and the sampling rate must be adjusted. This could involve increasing the sampling rate or possibly forcing a reduction in signal bandwidth. In applications using simple passive filters, the attenuation rate might only be -20 to -40 dB per decade (one and two poles, respectively). This could require the sampling frequency to be 10 to 1000 times the filter corner frequency. The exact factor depends on the resolution of the A/D converter and the amplitude of the highest frequency component. Using high-order active filters (seven to nine poles) might require a factor of only 1.5 to 3 (relative to the original 2× Nyquist rule).

Complete ninth-order elliptic designs are available in a number of configurations. These filters have very steep rolloff (approximately -100 dB per octave), while maintaining nearly constant gain in the passband (±0.2 dB is common). In selecting elliptic filters care must be taken to choose a unit that has a stopband attenuation greater than the resolution of the system's A/D converter. For example, a 12-bit converter has a resolution of one part in 4096, which corresponds to 72 dB. The filter used should attenuate all undesired frequencies by more than 72 dB. Likewise, a 16-bit converter would require a 96-dB filter. Fixed-frequency filter modules, as well as switch- and software-programmable units, can be purchased from various manufacturers. Several of these modular filters can be installed directly on the system's input termination panels. Complete programmable filter subsystems are also available in the form of boards that plug directly into an expansion slot within the data-acquisition PC. The advantage of high-order active filters in antialiasing applications is now clear. Yet while offering excellent performance, they are physically large and expensive compared to simpler filters.

In summary, filtering is intended to attenuate *unavoidable* noise and to limit bandwidth to comply with the Nyquist sampling theorem which prevents aliasing. An antialiasing filter must be a physical analog filter. It cannot be a digital filter that operates on the data after A/D conversion. Noise suppression can often be accomplished, or assisted, with either an analog or a digital filter. Still, filters are not intended as substitutes for proper wiring and shielding techniques. Ground loops, along with capacitively or inductively coupled noise sources, require special attention.

Analog Signal Scaling

As indicated above, most A/D converters are designed to operate with high-level input signals. Common A/D conversion ranges include 0 to 10, ±5, and ±10 volts. When the maximum input signal

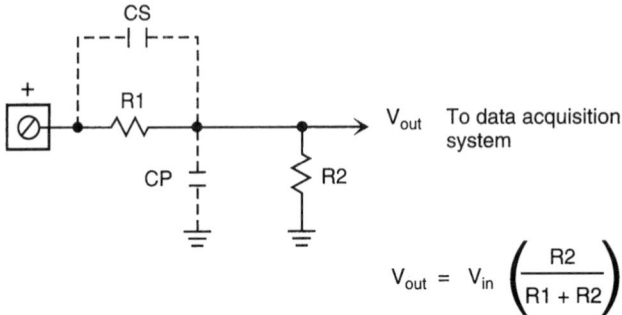

$$V_{out} = V_{in} \left(\frac{R2}{R1 + R2} \right)$$

FIGURE 11 Resistive voltage divider to reduce large analog input signals to below 10 volts. (*Intelligent Instrumentation, Inc.*)

is below 1 volt, accuracy is degraded. Under these circumstances it is often appropriate to amplify the signal before the A/D converter. Some A/D converter boards have amplifiers built in. If needed, external amplifiers can be added as part of the signal-conditioning circuitry on the termination panels. In addition to an input signal being too small, it is possible that it might be too large. Remember that most converters accept a maximum of 10 volts at their input. Signals could be 12, 48, or 100 volts (or more). Fortunately it is a simple matter to reduce excessive levels with a resistive voltage divider network. Figure 11 is appropriate for most analog signals. In selecting R_1 and R_2 there are practical factors to consider. Making R_1 large can introduce limitations on signal bandwidth, due to the low-pass filter produced by R_1 and the parasitic capacitance C_p in parallel with R_2. In some applications the network bandwidth can be extended by placing a capacitor C_s across R_1. The value should be selected to make the time constant $R_1 \cdot C_s$ equal to $R_2 \cdot C_p$. The equation assumes that the source (signal) impedance is very low compared with the series combination of R_1 and R_2, that is, $R_1 + R_2$. From this perspective, R_1 and R_2 should be as large as possible.

Input Buffering

The input characteristics of most data-acquisition boards are suitable for general applications. Yet in some cases the input resistance is too low or the bias current too high to allow accurate measurements. Input capacitance can also be an important factor. This is because some transducers, including piezo-electric and pH cells, exhibit a very high output impedance. Under these conditions, direct connection to the data-acquisition system can cause errors. These applications can be satisfied by adding a high input impedance buffer amplifier to the signal-conditioning circuitry. Figure 12 suggests one type of buffer circuit that can be used.

Resistance Signals

Resistance signals arrive at the data-acquisition system from primary sensors, such as strain gages and RTDs. Resistance is changed to a voltage by exciting it with a known current ($V_{out} = IR$). Figure 13 shows the simplest way to measure resistance. As suggested in Fig. 13a, the parasitic (unwanted) resistance of the two connecting lead wires can introduce significant errors. This is because the excitation current flows through the signal measurement wires. Figure 13b uses four connecting wires and a differential input connection to the data-acquisition system to minimize the lead-wire effects. This is known as a four-terminal (or kelvin) measurement. The extra wires allow the direct sensing of the unknown resistance. In both cases the resistance of the wires going to the data-acquisition system has little effect. This is because very little current flows in these leads. However, this technique is not well suited to RTD and strain-gage applications because of the very small change in measured voltage

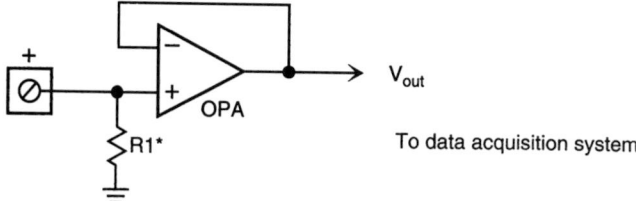

*R1 can be a very high impedance
resistor up to 10^9 ohms.

FIGURE 12   High-input impedance buffer circuit. (*Intelligent Instrumentation, Inc.*)

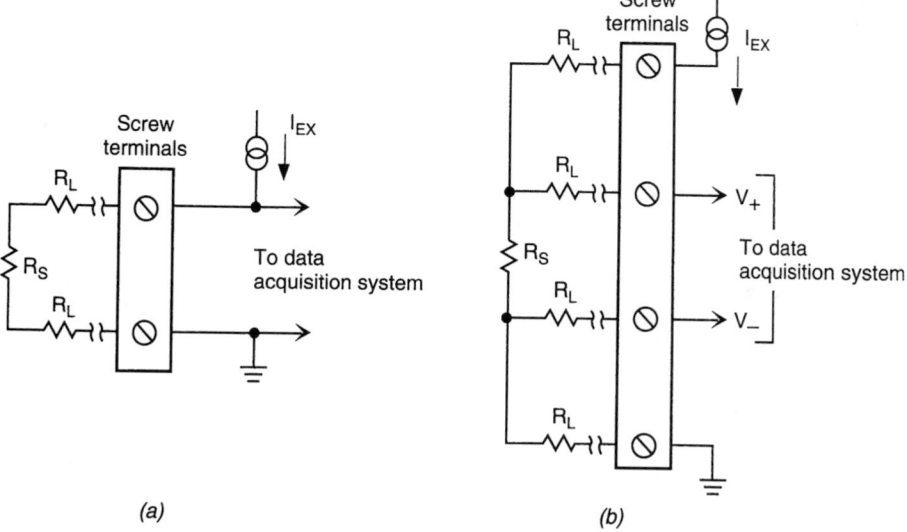

(a)   (b)

FIGURE 13   Measurement system for resistive device. R_L—lead-wire resistance. (*Intelligent Instrumentation, Inc.*)

compared to the steady-state (quiescent) voltage. The large quiescent voltage prevents the use of an amplifier to increase the measurement sensitivity.

It is usually better to measure resistive sensors as part of a Wheatstone bridge. A bridge is a symmetrical four-element circuit that enhances the system's ability to detect small changes in the sensor. In Fig. 14 the sensor occupies one arm of the bridge. The remaining arms are completed with fixed resistors equal to the nominal value of the sensor. In general, however, the sensor can occupy one, two, or four arms of the bridge, with any remaining arms being filled with fixed resistors. Note that the differential output from the bridge is zero when all of the resistors are equal. When the sensor changes, an easily amplified signal voltage is produced. Bridge-completion resistors should be of very high precision (typically 0.05 percent). However, stability is actually more important. Initial inaccuracies can be calibrated out, but instability always appears as an error.

A 100-ohm platinum RTD can be used to compare the merits of the approaches in Figs. 13 and 14. To control internal self-heating, the excitation level will be limited to 2 mA. Given that the sensitivity of this type of device is about +0.4 Ω/°C, the output will be about 0.8 mV/°C. This is indeed a small

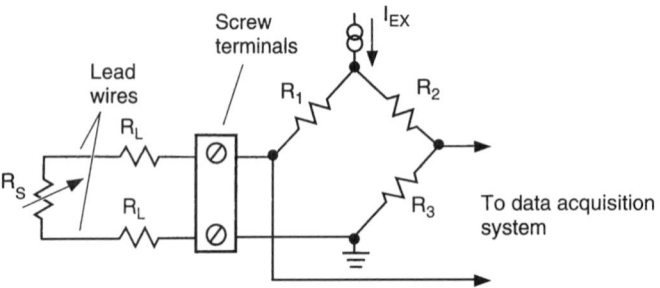

FIGURE 14   Two-wire bridge configuration. (*Intelligent Instrumentation, Inc.*)

signal that will require amplification. It would be useful to multiply the signal by 100 to 1000 to make best use of the A/D converter's full-scale range (typically 5 or 10 volts). However, the quiescent voltage across the RTD is 2 mA · 100 Ω = 0.2 volt. In Fig. 13 this limits the maximum gain to 10. Thus in a 12-bit system, the smallest detectable temperature change will be about 0.5°C. In contrast, the bridge circuit in Fig. 14 balances out the fixed or quiescent voltage drop, allowing greater magnification of the difference signal. This allows the detection of changes as small as 0.005°C.

Figure 14 has the same lead-wire resistance problem that the simple circuit had. The lead-wire resistances are indistinguishable from the transducer's resistance. So while this bridge circuit has high sensitivity, it is not suitable for precision applications.

Figure 15 shows a means of correcting for lead-wire effects. While the three-wire bridge requires an additional wire to be run to the sensor, several very important advantages are gained. If the (reasonable) assumption is made that the two wires bringing current to the sensor are of the same material and length, many of the potential error terms cancel. Comparing Figs. 14 and 15 shows that the additional signal wire has moved the measurement point directly to the top of the sensor. Again, the resistance of this wire is not important because current flow to the data-acquisition system is very low. Moving the measurement point has the effect of locating one current-carrying lead resistance in the top arm of the bridge while the other remains in the lower arm. The current in each is the same, so their voltage drops tend to cancel. While the compensation is not perfect, it does offer a significant performance improvement.

Transducer excitation and bridge-completion components are normally installed on the system's signal termination panels. While both voltage and current excitation can be used, current excitation is

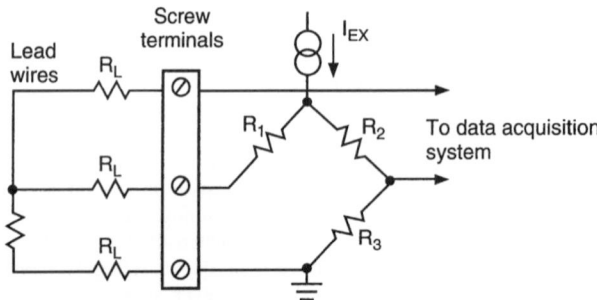

R_L is lead-wire resistance

FIGURE 15   Three-wire bridge configuration. (*Intelligent Instrumentation, Inc.*)

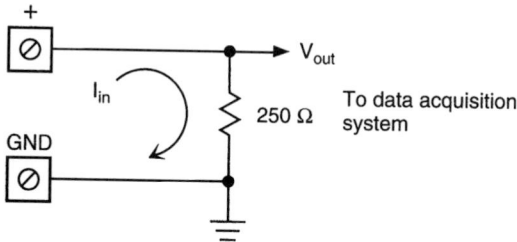

FIGURE 16   4- to 20-mA input conversion circuit (single-ended).
(*Intelligent Instrumentation, Inc.*)

generally more desirable. This is because current excitation provides a more linear output response, making the data interpretation easier.

Current Conversion

The need to measure signals in the form of currents is quite common in a data-acquisition system. The outputs from remote transducers are often converted to 4- to 20-mA signals by two-wire transmitters. At the data-acquisition system, current is easily converted back to a voltage with a simple resistor. Figure 16 shows how this is done. For a 4- to 20-mA signal a resistor value of 250 ohms can be used to provide a voltage output of 1 to 5 volts. As a general rule, the largest resistor that does not cause an overrange condition should be used. This ensures the maximum resolution. Stability of the resistor is essential, but the exact value is not important. Most systems have software provisions for calibrating the measurement sensitivity of each channel at the time of installation. Low-cost 0.1 percent metal-film resistors are usually adequate. For larger currents it is a simple matter to scale the resistor down to yield the desired full-scale voltage ($R = V/I$, where V is the full-scale voltage range and I is the maximum current to be read).

The technique of using only a resistor to convert from current to voltage does have limitations. If, for example, a 1-μA level is to be measured, a resistor of approximately 5 MΩ will be required. Unfortunately the use of high-value resistors leads to potentially large errors due to noise and measuring system loading. As suggested before, the data-acquisition system has a small but finite input current. This bias current (typically around 10 nA) will also flow through the conversion resistor and will be indistinguishable from the signal current. Therefore when very low currents must be measured, a different technique is used. Figure 17 suggests an active circuit that utilizes a precision field effect transistor (FET) amplifier to minimize the bias current problem. Both the simple resistor and the FET amplifier circuits require the same resistor value for a given current level. However, in the latter case the data-acquisition system's bias current is supplied by the amplifier and does not affect the measurement accuracy. A wide range of low-bias-current amplifiers are available for special applications. With the amplifier shown, currents as low as 10 pA can be read reliably.

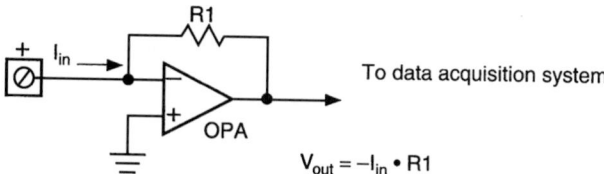

$$V_{out} = -I_{in} \cdot R1$$

FIGURE 17   Current-to-voltage converter circuit suitable for very low current levels. (*Intelligent Instrumentation, Inc.*)

Transmitters

When low-level signals (below 1 volt) are located remotely from the data-acquisition system, special precautions are suggested. Long wire runs with small signals usually result in poor performance. It is desirable to preamplify these signals first to preserve maximum signal-to-noise ratio. Two-wire transmitters provide an ideal way of packaging the desired signal-conditioning circuitry. In addition to signal amplification, transmitters can also provide filtering, isolation, linearization, cold-junction compensation, bridge completion, excitation, and conversion to a 4- to 20-mA current. Transmitters are ideal for thermocouples, RTDs, and strain gages. Current transmission allows signals to be sent up to several thousand feet (1500 meters) without significant loss of accuracy. While voltage signals are rapidly attenuated by the resistance of the connecting wires, current signals are not. In a current loop, the voltage drop due to wire resistance is compensated by the compliance of the current source, that is, the voltage across the current source automatically adjusts to maintain the desired current level. Note that power for the transmitter is conveyed from the data-acquisition system over the same two wires that are used for signal communications. No local power is required.

In addition to analog transmitters, there are also digital devices. These provide most of these features, except that the output signal is in a digital form instead of 4 to 20 mA. The output protocol is usually a serial data stream that is RS-232, RS-422, or RS-485 compatible. This is accomplished by including an A/D converter and a controller (computer) inside the transmitter. In many cases the output signal can be connected directly to a serial port on the PC without additional hardware. Two possible disadvantages of a digital transmitter are that it requires local power and, because of the added complexity, is generally more expensive.

Surge Protection

When a system can be subjected to unintentional high-voltage inputs, it is prudent to provide protection to avoid possible destruction of the equipment. High-voltage inputs can be induced from lightning, magnetic fields, static electricity, and accidental contact with power lines, among other causes.

Figure 18 suggests two different protection networks. Both circuits offer transient (short- duration) as well as steady-state protection. The circuit in Fig. 18a can tolerate continuous inputs of up to about 45 volts. When the overload disappears, the signal path automatically returns to normal. The circuit in Fig. 18b protects against continuous overloads of up to about 280 volts. In contrast, sustained

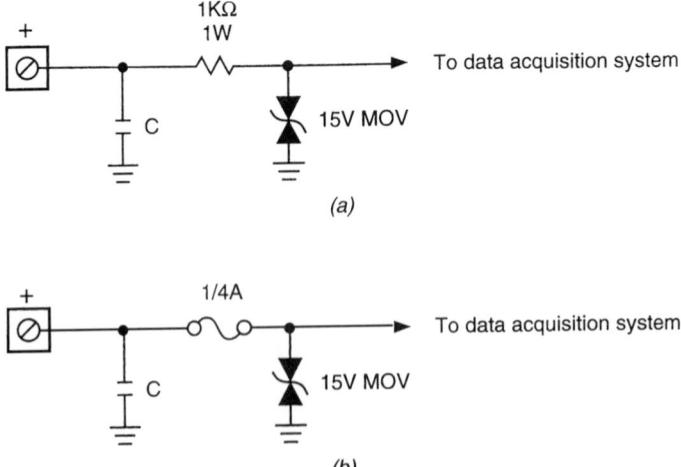

FIGURE 18   Representative input protection networks. (*Intelligent Instrumentation, Inc.*)

overloads to this circuit will cause the fuse to open (protecting the protection circuit). A disadvantage of this network is that the fuse must be replaced before the signal path is active again. In either case, signal flow is interrupted during the overload period. The resistor (or fuse) and the metal-oxide varistor (MOV) form a voltage clamp to ensure that transients will not get to the input of the data-acquisition system. MOVs are semiconductor devices that can react very quickly to absorb high-energy spikes. The 15-volt rating shown is high enough to pass all normal signals, but low enough to protect the data-acquisition system's input. Consideration should be given to the fact that even below the MOV's threshold voltage a small leakage current flows. If the series R is too large, the leakage could appear as a significant temperature-dependent error voltage (IR).

The optional capacitor can help suppress high-frequency transients. In some applications it must be rated for high voltage. For example, transients in power stations or other noisy environments can exceed 1000 volts. The capacitance value should be as large as physically possible, and the capacitor should be positioned as close as possible to the signal entry point of the system. Capacitors with low series impedance at high frequencies should be selected. This requirement eliminates electrolytic-type capacitors. If the input signal can change polarity, polarized capacitors must be avoided.

Analog Isolation

Isolators can be used to protect people and equipment from contact with high voltage. They usually provide the same protection as MOVs with the addition of one very important extra feature. Isolators can block overloads (protect) while simultaneously passing a desired signal. Applications include the breaking of ground loops, patient monitoring, and the removal of large common-mode signals. For example, if a thermocouple is connected to a motor winding, it could be in contact with 240 volts ac. Yet the thermocouple output voltage might be only 30 mV. The 30 mV (the actual signal) is seen as a differential signal while the 240 volts appears as a common-mode signal. The isolator operates in a way that is similar to a differential amplifier (described earlier). Its common-mode rejection capabilities block the effects of the unwanted portion of the signal. While standard differential amplifiers are generally limited to a ±10-volt common-mode signal, isolators are available with ratings beyond 5000 volts.

A family of industry-standard 5B signal-conditioning modules is available. These complete plug-in units are designed to provide a wide range of input and output capabilities. Each module supports a single channel, allowing the flexibility to mix the various types when configuring a system. Isolation, rated at 1500 volts, provides high-voltage separation between the signals and the data-acquisition system. Input modules are available for most voltage ranges, current ranges, thermocouples, RTDs, and strain gages. All of the required conditioning functions are included: protection, filtering, linearization, cold-junction compensation, bridge completion, and excitation. Output modules support 4- to 20-mA current loops. Standard termination panels accommodate up to 16 modules.

ANALOG OUTPUTS

Digital-to-Analog Converters

Analog outputs are required in many test and industrial automation applications. For example, they can be used to generate inputs (stimuli) to a device under test and to operate valves, motors, and heaters in closed-loop feedback control systems. A D/A converter is used to transform the binary instructions from the digital computer (PC) to a variable output level. Common analog output ranges include ±5, ±10, and 0 to 10 volts and 4 to 20 mA.

A popular type of D/A converter consists internally of N binary weighted current sources. The values (the levels can be scaled to suit speed and output requirements) correspond to 1, $1/2$, $1/4$, $1/8$, ..., $1/2^{N-1}$. N is also the number of digital input lines (bits). Each source can be turned on or off independently by the computer. By summing the outputs of the sources together, 2^N current combinations are produced. Thus a 12-bit converter can represent an analog output range with 4096 discrete steps. A current-to-voltage converter is included in voltage output models.

Faithful generation of a complex signal requires that the conversion rate (clock rate) of the D/A converter be very high compared with the repetition rate of the waveform. Ratios of 100 to 1000 points per cycle are common. This suggests that a "clean" 1-kHz output could require a 100-kHz to 1-MHz converter. This is pushing the current state of the art in PC-based data-acquisition products.

When operating in the voltage output mode, most D/A converters are limited to supplying around 5 mA of load current. This implies that most D/A converters will use some kind of signal conditioning when interfacing to real-world devices (transducers). When large loads such as positioners, valves, lamps, and motors are to be controlled, power amplifiers or current boosters need to be provided. Most data-acquisition systems do not include these high-power analog drivers internally.

Output Filtering

A D/A converter attempts to represent a continuous analog output with a series of small steps. The discontinuities inherent in a digitally produced waveform represent very high frequencies. This is seen as noise or distortion that can produce undesired effects. A low-pass filter is often used at the output of a D/A converter to attenuate high frequencies and, thus, "smooth" the steps.

DIGITAL INPUTS AND OUTPUTS

Most data-acquisition systems are able to accept and generate TTL level signals. These are binary signals that are either high or low (on or off). The low state is represented by a voltage near 0 volts (generally less than 0.8 volt), while a high state is indicated by a voltage near 5 volts (generally greater than 2 volts). Levels between 0.8 and 2 volts are not allowed. The output levels are intended to drive other "logic" circuits rather than industrial loads. As a result, drive capabilities are generally under 24 mA. Still, digital signals in many real-world applications are not TTL-compatible. It is common to encounter 24-volt, 48-volt, and 120/240-volt ac levels as digital input signals. High voltage and current outputs are often required to operate solenoids, contactors, motors, indicators, alarms, and relays.

Many types of digital signal termination panels are available to facilitate the connection of field wires to the data-acquisition system. In addition to screw terminals, the panels have provisions for signal conditioning, channel status indicators (such as light-emitting diodes), voltage dividers, and isolators. Thus the monitoring and the control of high dc levels, along with ac line voltage circuits, are readily accomplished.

Pulse and Frequency Inputs and Outputs

A variety of counting, timing, and frequency-measuring applications exists. Other applications require that devices be turned on and off for precise time periods. All of these functions can be provided by counter/timer circuits. The system's counter/timers are optimized for pulse applications, including frequency measurement and time-base generation. Counters are characterized by the number of input events that can be accumulated and by their maximum input frequency. Several independent counters are usually provided. They can be used to count events (accumulate), measure frequency, measure pulse width, or act as frequency dividers. Counting can be started from a defined initial value, and most counters can be configured to reset automatically to this value after it has been read. Software can easily interpret the counter's data as a sum or difference from an arbitrary starting point. Pulse generators (rate generators) are software programmable over a very wide range of frequencies and duty cycles. A rate generator is often used to provide the precise time base required for accurate data acquisition. Most systems use 16-bit counters that can accumulate pulses at frequencies up to 8 MHz. Up to 65,536 (2^{16}) events can be accumulated before the counter overflows. Two counters can generally be cascaded to provide 32-bit capability (more than 4 billion counts). Most counters accept only TTL level signals. Other levels require signal conditioning.

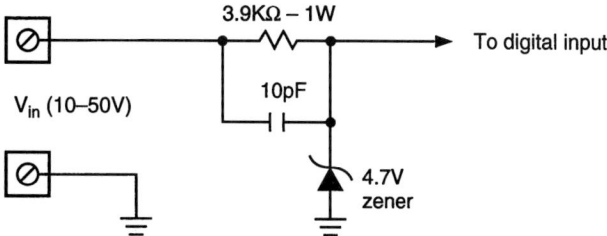

FIGURE 19 Circuit to convert large digital signals to TTL-compatible levels. (*Intelligent Instrumentation, Inc.*)

Frequency measurements using counters can be accomplished in different ways, depending on the application. When the unknown frequency is a TTL signal, it can be applied directly to the counter circuit. Analog signals can be converted to TTL levels with comparator circuits available from some manufacturers. Voltage dividers using resistors, zener diodes, or optoisolators can be used to scale down high-level signals. When using any kind of signal conditioner before a counter input, consideration should be given to possible speed limitations.

Two distinct options exist for measuring high or low frequencies. The first method counts a known clock generator for the period of the unknown input signal. This provides high resolution for low-frequency signals, while minimizing the time required for the measurement. Generally this is used for frequencies below 10 Hz. The second method counts cycles of the unknown input signal for a fixed time interval. The advantage of this technique is that it allows measurements up to the limit of the counter's speed (typically 8 MHz). It is easy to implement an auto-ranging software algorithm that optimizes resolution over a very wide frequency range.

Digital Signal Scaling

For large digital signals, the circuit in Fig. 19 can be used to produce TTL-compatible levels. Most digital circuits (digital input ports and counters) require fast input level transitions to ensure reliable operation. Steps faster than 10 μs are usually adequate. Parasitic capacitance at the input to the data-acquisition system can interact with the series resistor to degrade input steps. The 10-pF capacitor in Fig. 19 is included to help correct this problem. When the input is not fast enough, it can be made TTL-compatible with the Schmitt trigger circuit shown in Fig. 20.

Digital Isolation

A family of industry standard signal-conditioning modules is available. These complete plug-in units are designed to provide a wide range of input and output capabilities. Each module supports a single

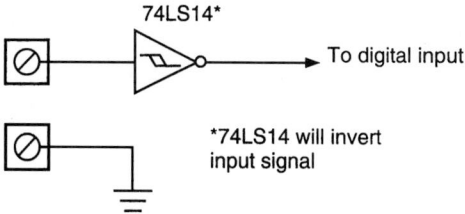

FIGURE 20 Schmitt trigger circuit to "speed up" slow input signals. Input levels must be TTL-compatible. (*Intelligent Instrumentation, Inc.*)

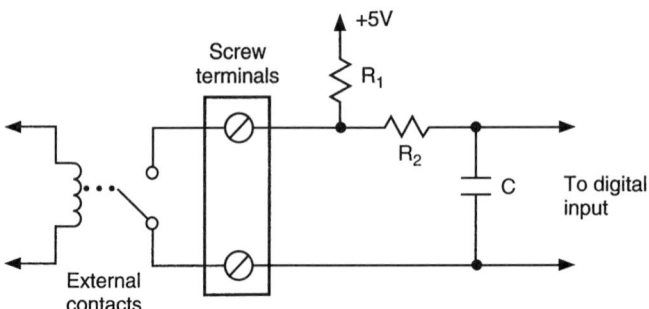

FIGURE 21   Contact sensing and wetting. (*Intelligent Instrumentation, Inc.*)

channel, allowing the flexibility to mix the various types when configuring a system. Optical isolation, rated at 4000 volts, provides high-voltage separation between the signals and the data-acquisition system. This is useful for safety, equipment protection, and ground-loop interruption. Output modules use power transistors or triacs to switch high-voltage high-current ac or dc loads. Loads up to 60 volts dc or 280 volts ac at 3 amperes can be accommodated. Input modules convert digital signals between 10 and 280 volts to TTL levels. Standard termination panels accommodate up to 16 modules.

Contact Sensing

As shown in Fig. 21, contact sensing can be implemented on a signal termination panel. When interfacing to relay or switch contacts, a pull-up current must be provided. The pull-up current converts the opening and closing of the contacts to TTL level voltages. Because all metal surfaces tend to oxidize with time, poor relay contacts can result. Both level generation and contact wetting can be accomplished by connecting a resistor between the input line and the +5-volt power supply. When the switch is open, the input system sees +5 volts. When the switch is closed, the input is 0 volts. This satisfies the TTL requirements of the data-acquisition system. A value of 250 ohms for R_1 will provide 20 mA of wetting current, which is usually enough to keep most contacts free of oxide buildup. R_2 and C_1 function as a very simple debounce filter to reduce erroneous inputs due to the mechanical bouncing of the contacts. Care must be taken to avoid slowing the signal transition so much that false triggering occurs. If needed, a Schmitt trigger can be added, as shown in Fig. 20. Digital filtering techniques can also be used to eliminate the effects of contact bounce.

Relay Driving

Figure 22 shows how a TTL output from a data-acquisition board can be connected to drive an external 5-volt relay coil. The digital output must be able to switch the coil current. The specifications of digital output ports vary considerably between models. However, most can support 16 to 24 mA. When large relays, contactors, solenoids, or motors are involved, an additional driver or intermediate switching network can be used. The diode D_1 protects the internal circuitry against the inductive kickback from the relay coil. Without the diode, the resulting high-voltage spikes will damage the digital port. Note that the direction (polarity) of the diode must be as shown in the diagram. Protection diodes must be able to respond very quickly and absorb the coil's energy safely. Most standard switching diodes fill these needs.

Motor Control

Many different types of motors are in common use today. When it comes to controlling these devices, specialized circuits are often required. Some applications, however, require only on-off operations.

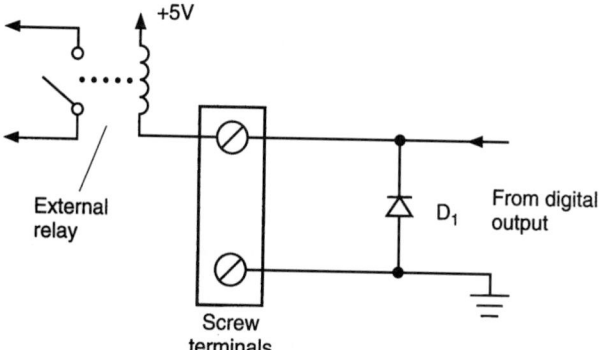

FIGURE 22   Relay driving circuit. (*Intelligent Instrumentation, Inc.*)

These can simply be driven by digital output ports, usually through optical isolators (loads of up to 3 amperes) or with various types of contactors (relays).

In general, when variable speed is desired, either analog or digital outputs from the data-acquisition system are used to manipulate the motor through an external controller. A wide range of both ac and dc controllers is available. Motor controls are discussed in more detail in Section 9 of this handbook.

Stepper-type motors are of particular interest in robotics, process control, instrumentation, and manufacturing. They allow precise control of rotation, angular position, speed, and direction. While several different types of stepping motors exist, the permanent-magnet design is perhaps the most common. The permanent magnets are attached to the rotor of the motor. Four separate windings are arranged around the stator. By pulsing direct current into the windings in a particular sequence, forces are generated to produce rotation. To continue rotation, current is switched to successive windings. When no coils are energized, the shaft is held in its last position by the magnets. In some applications these motors can be driven directly (via opto relays) by one of the data-acquisition system's digital output ports. The user provides the required software to produce the desired pulses in proper sequence. The software burden can be reduced by driving the motor with a specially designed interface device. These units accept a few digital input commands representing the desired speed, rotation, direction, and acceleration. A full range of motors and interfaces is available. Stepper motors are discussed in more detail in Section 9 of this handbook.

NOISE AND WIRING IN DATA SIGNAL HANDLING

by Howard L. Skolnik*

Signals entering a data-acquisition and control system include unwanted noise. Whether this noise is troublesome depends on the signal-to-noise ratio and the specific application. In general it is

* Intelligent Instrumentation, Inc. (a Burr-Brown Company), Tucson, Arizona.

desirable to minimize noise to achieve high accuracy. Digital signals are relatively immune to noise because of their discrete (and high-level) nature. In contrast, analog signals are directly influenced by relatively low-level disturbances. The major noise-transfer mechanisms include conductive, inductive (magnetic), and capacitive coupling. Examples include the following:

- Switching of high-current loads in nearby wiring can induce noise signals by magnetic coupling (transformer action).
- Signal wires running close to ac power cables can pick up 50- or 60-Hz noise by capacitive coupling.
- Allowing more than one power or signal return path can produce ground loops that inject errors by conduction.

Conductance involves current flowing through ohmic paths (direct contact), as opposed to inductance or capacitance.

Interference via capacitive or magnetic mechanisms usually requires that the disturbing source be close to the affected circuit. At high frequencies, however, radiated emissions (electromagnetic signals) can be propagated over long distances.

In all cases, the induced noise level will depend on several user-influenced factors:

- Signal source output impedance
- Signal source load impedance (input impedance to the data-acquisition system)
- Lead-wire length, shielding, and grounding
- Proximity to noise source or sources
- Signal and noise amplitude

Transducers that can be modeled by a current source are inherently less sensitive to magnetically induced noise pickup than are voltage-driven devices. An error voltage coupled magnetically into the connecting wires appears in series with the signal source. This has the effect of modulating the voltage across the transducer. However, if the transducer approaches ideal current-source characteristics, no significant change in the signal current will result. When the transducer appears as a voltage source (regardless of impedance), the magnetically induced errors add directly to the signal source without attenuation.

Errors also are caused by capacitive coupling in both current and voltage transducer circuits. With capacitive coupling, a voltage divider is formed by the coupling capacitor and the load impedance. The error signal induced is proportional to $2\pi f RC$, where R is the load resistor, C is the coupling capacitance, and f is the interfering frequency. Clearly, the smaller the capacitance (or frequency), the smaller is the induced error voltage. However, reducing the resistance only improves voltage-type transducer circuits.

Example. Assume that the interfering signal is a 110-volt ac 60-Hz power line, the equivalent coupling capacitance is 100 pF, and the terminating resistance is 250 ohms (typical for a 4- to 20-mA current loop). The resulting induced error voltage will be about 1 mV, which is less than 1 least-significant bit in a 12-bit 10-volt system.

If the load impedance were 100 kΩ, as it could be in a voltage input application, the induced error could be much larger. The equivalent R seen by the interfering source depends on not only the load impedance but also the source impedance and the distributed nature of the connecting wires. Under worst-case conditions, where the wire inductance separates the load and source impedances, the induced error could be as large as 0.4 volt. This represents about an 8-percent full-scale error.

Even though current-type signals are usually converted to a voltage at the input to the data-acquisition system, with a low-value resistor this does not improve noise performance. This is because both the noise and the transducer signals are proportional to the same load impedance.

It should be pointed out that this example does not take advantage of, or benefit from, shielding, grounding, and filtering techniques.

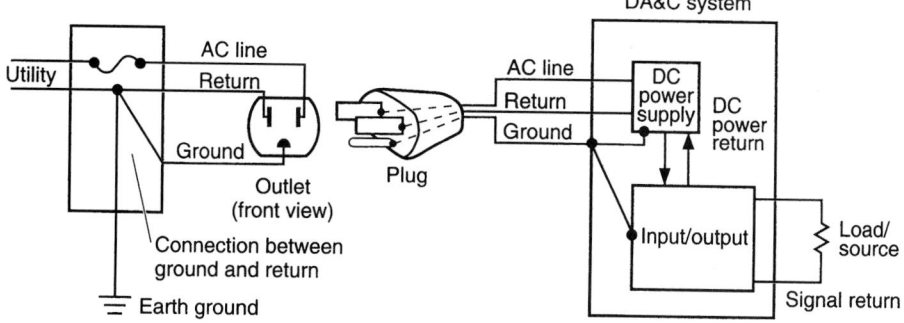

FIGURE 1   Differences between ground and return conductors. (*Intelligent Instrumentation, Inc.*)

GROUNDING AND SHIELDING PRINCIPLES

Most noise problems can be solved by giving close attention to a few grounding and shielding principles:

- Do not confuse the definitions of ground and return paths. Ground = safety; return = current-carrying.
- Minimize wiring inductance.
- Limit antennas.
- Maintain balanced networks wherever possible.

The foregoing directions appear simple, but what really is involved?

For a beginning, one should redefine some common terms. A ground is *not* a signal or power supply return path. A ground wire connects equipment to earth for safety reasons—to prevent accidental contact with dangerous voltages. Ground lines do not normally carry current. Return lines are an active part of a circuit—carrying power or signal currents (Fig. 1). Care should be taken to distinguish between grounds and returns and to avoid more than one connection between the two.

To be effective, return paths should have the lowest possible impedance. Someone once said that the shortest distance between two points is a straight line. But in geography this is not true, and it is not generally true in electronics either. Current does not take the shortest path; rather it takes the path of least resistance (really, of least impedance). Return impedance is usually dominated by the path inductance. Wiring inductance is proportional to the area inside the loop formed by the current-carrying path. Therefore impedance is minimized by providing a return path that matches or overlaps the forward signal path. Note that this may not be the shortest or most direct route. This concept is fundamental to ensuring proper system interconnections.

Three different grounding and connection techniques are suggested in Figs. 2, 3, and 4. The circuit in Fig. 2 allows the signal return line to be grounded at each chassis. This may look like a good idea from a safety standpoint. However, if a difference in potential exists between the two grounds, a ground current must flow. This current, multiplied by the wire impedance, results in an error voltage e_e. Thus the voltage applied to the amplifier is not V_1, but $V_1 + e_e$. This may be acceptable in those applications where the signal voltage is much greater than the difference in the ground potentials.

When the signal level is small and a significant difference in ground potentials exists, the connection in Fig. 3 is more desirable. Note that the return wire is not grounded at the amplifier and ground current cannot flow in the signal wires. Any difference in ground potential appears, to the amplifier, as a common-mode voltage. In most circuits the effects of common-mode voltage are very small, as long as the sum of signal voltage plus common-mode voltage is less than 10 volts. (Ten volts is the linear

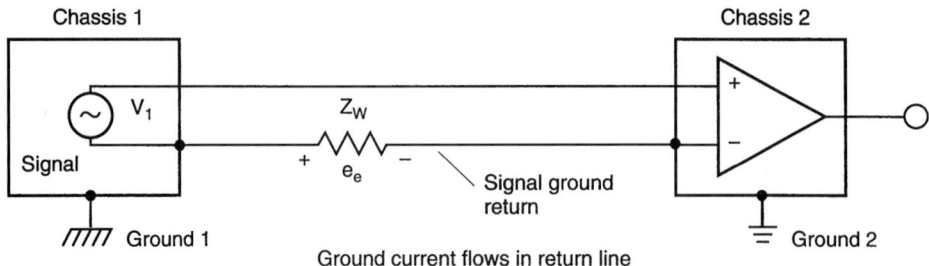

FIGURE 2   Single-ended connection. (*Intelligent Instrumentation, Inc.*)

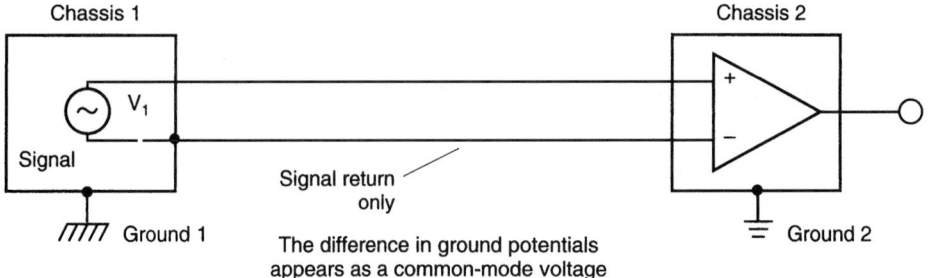

FIGURE 3   Differential connection. (*Intelligent Instrumentation, Inc.*)

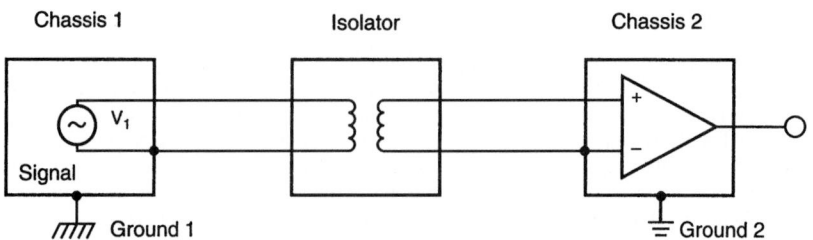

FIGURE 4   Isolated connection. (*Intelligent Instrumentation, Inc.*)

range for most amplifiers.) Additional information about common-mode rejection and single-ended versus differential amplifiers can be found in a prior article in this handbook.

If cost is not a limitation, Fig. 4 offers the highest performance under all conditions. Injecting an isolator into the signal path faithfully conveys V_1 to the amplifier while interrupting all direct paths. In this configuration multiple ground connections can be tolerated along with several hundred volts between input and output circuits. Additional information on both analog and digital isolators can be found in a prior article in this handbook.

Cable Types

What kind of wire should be used to interconnect a system? First, it must be emphasized that a single piece of wire is not generally useful. Circuits consist of complete paths, so pairs of wires are referred to in this discussion. Basically four kinds of wire are fundamental: (1) plain pair, (2) shielded pair,

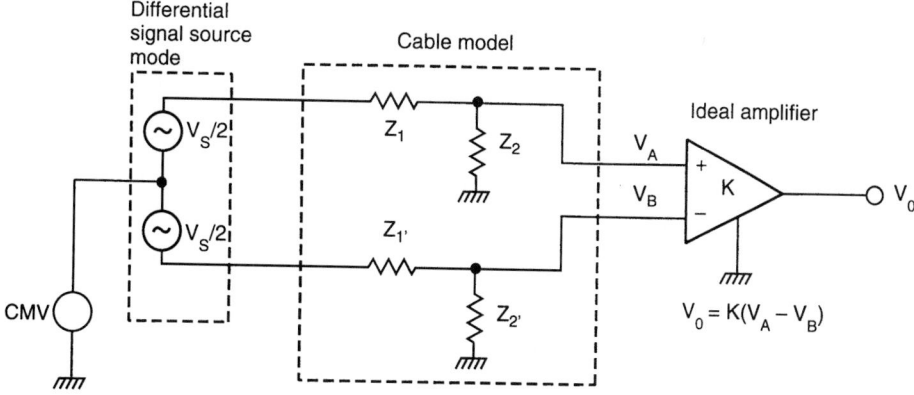

FIGURE 5   Influence of cable connectors on common-mode signal performance. (*Intelligent Instrumentation, Inc.*)

(3) twisted pair, and (4) coaxial cable. All but the coaxial wires are said to be balanced. Coaxial cable differs from the others in that the return line surrounds the central conductor.

Technically, the outer conductor should not be called a shield because it carries signal current. It is significant that the forward and return path conductors do not have exactly the same characteristics. In contrast, a shielded pair is surrounded by a separate conductor (properly called a shield) that does not carry signal current.

Figure 5 suggests a simple model for a differential signal connection. The attributes of the signal source have been split to model the influence of a common-mode voltage. Let us focus on the effect of forward and return path symmetry in the cable. Assuming that the amplifier is perfect, it will respond only to the difference between V_A and V_B. The technique of superposition allows us to analyze each half of the cable model separately and then to add the results. Z_1 is usually dominated by series inductance, while Z_2 is dominated by parallel capacitance. In any case, Z_1 and Z_2 form a voltage divider. If the dividers in both legs of the cable are identical, $V_A - V_B$ will not be influenced by common-mode voltage. If, however, the capacitance represented by Z_2 is different in the two paths, a differential voltage will result and the amplifier will be unable to distinguish the resulting common-mode error from a change in V_S.

Coaxial cable offers a very different capacitance between each of its conductors and ground. Not only does the outer conductor surround the inner, it is also connected to ground. Thus coaxial cable is intended for single-ended applications only. Note that even perfectly balanced cables can still attenuate differential signals.

Sometimes even a single-ended source is best measured with a differential amplifier. Refer again to Fig. 3. To maintain a high rejection of any ground difference potential, balanced cables are required.

TROUBLESHOOTING GUIDE FOR NOISE

One method of reducing errors due to capacitive coupling is to use a shield. Generally there is little that can be done to reduce the actual capacitance. (Wire length and physical location are factors, however.) Nevertheless, placing a conductive material (at ground potential) between the signal wires and the interference source is very useful. The shield blocks the interfering current and directs it to the ground. Depending on how complete the shield is, attenuations of more than 60 dB are attainable. When using shielded wire, it is very important to connect only one end of the shield to ground. The connection should be made at the data-acquisition system end of the cable (such as input amplifier). Connecting both ends of the shield can generate significant error by inducing ground-loop currents.

A shield can work in three different ways:

- Bypassing capacitively coupled electric fields
- Absorbing magnetic fields
- Reflecting radiated electromagnetic fields

Another approach is to use twisted pairs. Twisted-pair cables offer several advantages. Twisting of the wires ensures a homogeneous distribution of capacitances. Capacitances both to ground and to extraneous sources are balanced. This is effective in reducing capacitive coupling while maintaining high common-mode rejection. From the perspective of both capacitive and magnetic interference, errors are induced equally into both wires. The result is a significant error cancellation.

The use of shielded or twisted-pair wire is suggested whenever low-level signals are involved. With low-impedance sensors the largest gage-connecting wires that are practical should be used to reduce lead-wire resistance effects. On the other hand, large connecting wires that are physically near thermal sensing elements tend to carry heat away from the source, generating measurement errors. This is known as thermal shunting, and it can be very significant in some applications.

The previous discussion concentrated on cables making single interconnections. Multiconductor cables, for connecting several circuits, are available in similar forms (such as twisted pairs and shielded pairs). Both round and flat (ribbon) cables are used widely. Because of the close proximity of the different pairs in a multiconductor cable, they are more susceptible to crosstalk. Crosstalk is interference caused by the inadvertent coupling of internal signals via capacitive or inductive means.

Again, twisted pairs are very effective. Other methods include connecting alternate wires as return lines, running a ground plane under the conductors, or using a full shield around the cable.

Still another noise source, not yet mentioned, is that of triboelectric induction. This refers to the generation of noise voltage due to friction. All commonly used insulators can produce a static discharge when moved across a dissimilar material. Fortunately the effect is very slight in most cases. However, it should not be ignored as a possible source of noise when motion of the cables or vibration of the system is involved. Special low-noise cables are available that use graphite lubricants between the inner surfaces to reduce friction.

The key to designing low-noise circuits is recognizing potential interference sources and taking appropriate preventive measures. Table 1 can be useful when troubleshooting an existing system.

After proper wiring, shielding, and grounding techniques have been applied, input filtering can be used to further improve the signal-to-noise ratio. However, filtering should never be relied upon as a fix for improper wiring or installation.

Cable-Length Guidelines

What is the maximum allowable cable length? There is no direct answer to this question. The number of factors that relate to this subject is overwhelming. Signal source type, signal level, cable type, noise source types, noise intensity, distance between cable and noise source, noise frequency, signal frequency range, and required accuracy are just some of the variables to consider. However, experience can yield some "feel" for what often works, as per the following examples:

Analog Current Source Signals. Given 4- to 20-mA signal, shielded wire, bandwidth limited to 10 Hz, required accuracy 0.5 percent, and average industrial noise levels. Cable lengths of 1000 to 5000 feet (300 to 1500 meters) have been used successfully.

Analog Voltage Source Signals. Given ±1- to ±10-volt signal, shielded wire, bandwidth limited to 10 Hz, required accuracy 0.5 percent, and average industrial noise levels. Cable lengths of 50 to 300 feet (15 to 90 meters) have been used successfully.

Analog Voltage Source Signals. Given 10-mV to 1-volt signal, shielded wire, bandwidth limited to 10 Hz, required accuracy 0.5 percent, and average industrial noise levels. Cable lengths of 5 to 100 feet (1.5 to 30 meters) have been used successfully.

Digital TTL Signals. Given ground-plane-type cable and average industrial noise levels. Cable lengths of 10 to 100 feet (3 to 30 meters) have been used successfully.

TABLE 1 Troubleshooting Guide for Noise

Observation	Subject	Possible solution	Notes
Noise a function of cable location	Capacitive coupling	Use shielded or twisted pair.	a
	Inductive coupling	Reduce loop area; use twisted pair or metal shield.	b
Average value of noise:			
Not zero	Conductive paths or ground loops	Faulty cable or other leakage.	c
Zero	Capacitive coupling	Eliminate multiple ground connections. Use shielded or twisted pair.	
Shield inserted:			
Ground significant	Capacitive coupling	Use shielded or twisted pair.	a
Ground insignificant	Inductive coupling	Reduce loop area; use twisted pair or metal shield.	b
Increasing load:			
Reduces error	Capacitive coupling	Use shielded or twisted pair.	a
Increases error	Inductive coupling	Reduce loop area; use twisted pair or metal shield.	b
Dominant feature:			
Low frequency	60-Hz ac line, motor, etc.	1. Use shielded or twisted pair. 2. Reduce loop area; use twisted pair or metal shield. 3. Faulty cable or other leakage; eliminate multiple connections.	
High frequency		Complete shield.	d
Noise a function of cable movement	Triboelectric effect	Rigid or lubricated cable.	
Noise is white or $1/f$	Electronic amplifier, etc.	Not a cable problem.	

a. Complete shield to noise-return point and check for floating shields.

b. Nonferrous shields are good only at high frequencies. Use MuMetal shields at low frequencies.

c. Could be capacitive coupling with parasitic rectification, such as nonlinear effects.

d. Look for circuit element whose size is on the order of the noise wavelength (antennas). Openings or cracks in chassis or shields with a dimension bigger than the noise wavelength/20 should be eliminated.

Source: Intelligent Instrumentation, Inc.

Ground-plane cable reduces signal reflections, ringing, and RFI (radio frequency interference). Special termination networks may be required to maintain signal integrity and minimize RFI. If squaring circuits (e.g., Schmitt triggers) are used to restore the attenuated high-frequency signals, improved performance can be realized.

This information is given only as a typical example of what might be encountered. The actual length allowed in a particular application could be quite different.

The following relationships are offered as an aid to visualizing the influence of the most significant factors determining cable length. These relationships show how the various parameters affect cable length. These relationships are *not equations*, and will not allow the calculation of cable length.

For Current Source Signals:

Allowable length is proportional to

$$\frac{I_s D_n C_f}{f_n A N_i}$$

For Voltage Source Signals:

Allowable length is proportional to

$$\frac{V_s D_n C_f}{f_n A N_i R_L}$$

where I_s, V_s = signal level
$\qquad\quad C_f$ = coupling factor, which is inversely proportional to effectiveness of any shielding or twisting of wires
$\qquad\quad D_n$ = distance to noise source
$\qquad\quad f_n$ = noise frequency
$\qquad\quad A$ = required accuracy
$\qquad\quad N_i$ = noise source intensity
$\qquad\quad R_L$ = equivalent resistance to ground at signal input

INDUSTRIAL CONTROL NETWORKS

Commencing in the late 1960s and continuing through 1999 to the 2000s, industrial data and control networks will capture the ingenuity of control engineers, computer scientists, and, of course, communications specialists.[1] The well-justified thirst for information from top management down the industrial hierarchy will continue undiminished. To date, communication system designs have survived past difficult periods in an effort to find the best network systems for given applications, always with an eye toward finding the "universal" answer. Open systems hold some promise along these lines. Millions of hours of effort have gone into defining optimal protocols, improved communication media, practical and cost-effective bus configurations, and the reconfiguration of controls, such as programmable logic controllers (PLCs), distributed control systems, and personal computers (PCs), in an effort to make them increasingly "network friendly." There are other objectives and there have been tough decisions and roadblocks, but the leading technical societies have mustered strength through special committees in defining terms and establishing standards. This work will continue apace.

EARLY NETWORKING CONCEPTS

Early communication needs were served with point-to-point data links (Fig. 1). Very early standards, such as TTY (teletypewriter) current loops and RS-232, which allow different equipment to interface with one another, appeared and were accepted. From that, the star topology (Fig. 2) was developed so that multiple computers could communicate. The central, or master, node uses a communications port with multiple drops, as shown in Fig. 3. In this system the master is required to handle traffic from all the nodes attached, poll the other nodes for status, and, if necessary, accept data from one node to be routed to another. The heavy software burden on the master is also shared to a lesser degree among all the attached nodes. In addition, star topologies are inflexible as to the number of nodes that can

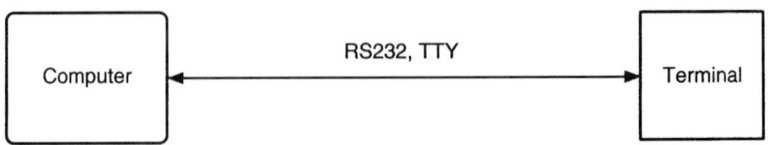

FIGURE 1   Point-to-point communication.

---

[1] Industrial control networks also are discussed in other portions of this handbook. See, in particular, the articles, "Distributed Control Systems," "Programmable Controllers," and "Distributed Numerical Control and Networking" in Section 3.

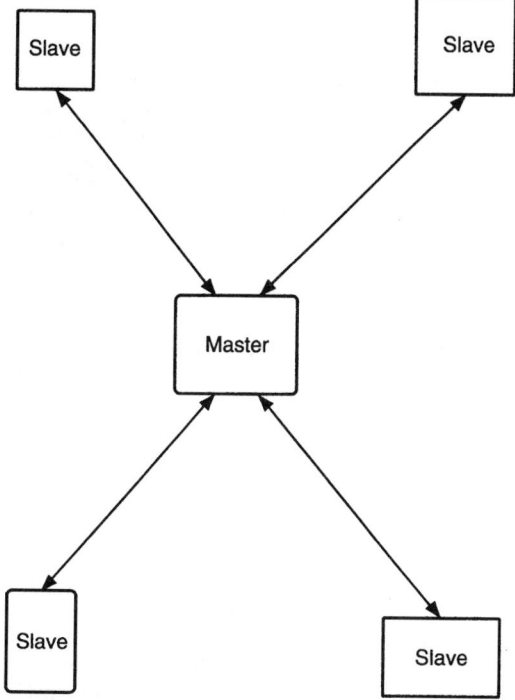

FIGURE 2   Star topology.

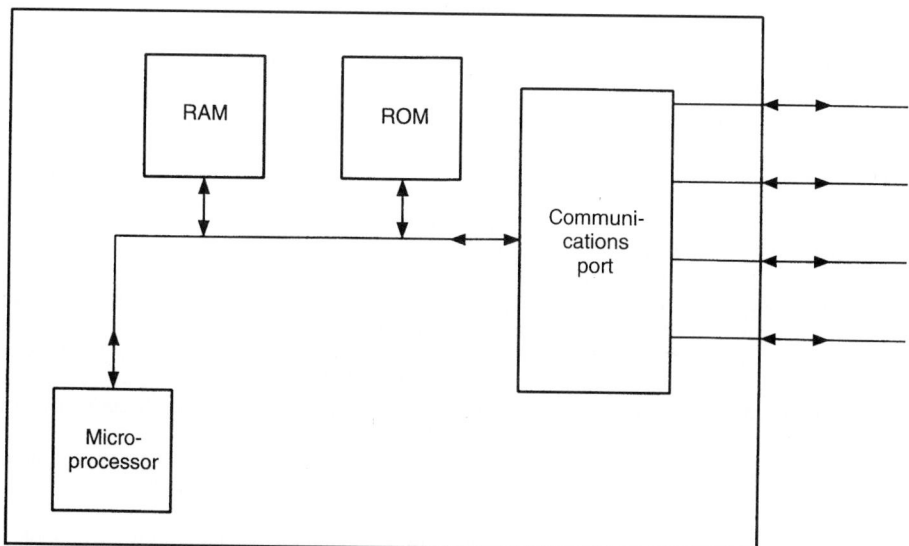

FIGURE 3   Master node for star topology.

be attached. Either one pays for unused connections (for future expansion), or a system results that cannot grow with demands.

To overcome some of these shortfalls, multidrop protocols were established and standardized. Data loop, such as SDIC (synchronous data link control), were developed as well as other topologies, including buses and rings (Fig. 4). The topology of these standards makes it easy to add (or subtract) nodes on the network. The wiring is also easier because a single wire is routed to all nodes. In the case of the ring and loop, the wire also is returned to the master. Inasmuch as wiring and maintenance are major costs of data communications, these topologies virtually replaced star networks. These systems, however, have a common weakness—one node is the master, with the task of determining which station may transmit at any given time. As the number of nodes increases, throughput becomes a problem because (1) a great deal of "overhead" activity may be required to determine which may transmit and (2) entire messages may have to be repeated because some protocols allow only master-slave communications, that is, a slave-to-slave message must be sent first to the master and then repeated by the master to the intended slave receiver. Reliability is another problem. If the master dies, communications come to a halt.

The need for multinode networks without these kinds of problems and restraints led to the development of the initial local area networks (LANs) using peer-to-peer communications. Here no one node is in charge; all nodes have an equal opportunity to transmit. An early LAN concept is shown schematically in Fig. 5.

In designing LAN architecture, due consideration had to be given to the harsh environment of some manufacturing and processing areas. Design objectives included the following.

Noise.   Inasmuch as a LAN will have long cables running throughout the manufacturing space, the amount of noise pickup can be large. Thus the LAN must be capable of working satisfactorily in an electrically noisy area. The physical interface must be defined to provide a significant degree of noise rejection, and the protocol must be robust to allow easy recovery from data errors. (See preceding article in this handbook section, "Noise and Wiring in Data Signal Handling.")

Response.   The LAN in an industrial situation should have an assured maximum response time, that is, the network must be able to transmit an urgent message within a specified time frame. The real-time aspect of industrial control demands this.

Priority Message.   On the factory floor, both control and status, when carried over the same network, should recognize the higher priority of the control message.

Early Data Highways

In 1972 the very first serial data communications highways were introduced. At that time the only purpose of the data highway was to allow host computers to adjust set points or, in some cases, perform direct digital control (DDC), while providing measurement data to the host computer. With such radically altered control concepts, in designing a data highway, great emphasis was placed on proposed data highways and their ability to operate at sufficiently high rates. There was concern that, during process upsets, many alarm conditions could suddenly change and these had to be reported to the entire system quickly so that remedial action could be taken. There also was major concern over start-up and shutdown procedures that cause heavy communication loads. Security also was a major concern.

In 1973 the distributed control system (DCS) appeared. It represented a major departure in control system architecture and impacted on the configuration of the data highway.

Ethernet.   The first LAN, developed by Xerox Corporation, has enjoyed years of application. The network uses CSMA/CD (carrier sense multiple access with collision) and is a baseband system with a bus architecture (Fig. 4b). Baseband is a term used to describe a system where the information being sent over the wire is not modulated.

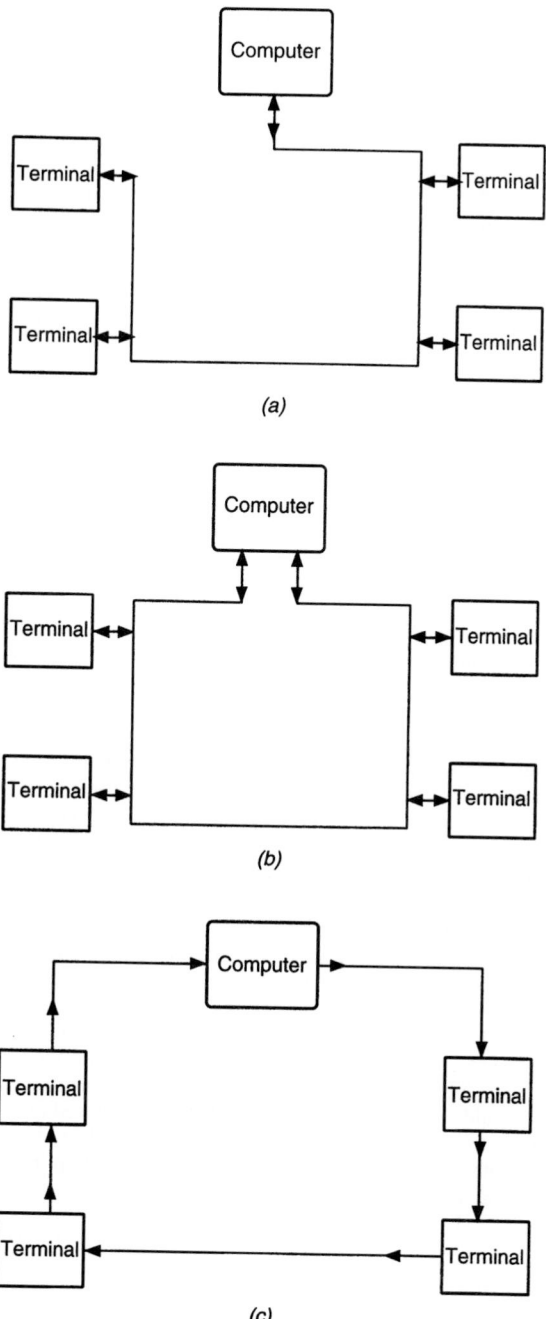

FIGURE 4   Basic communication standards. (*a*) Data-loop topology.
(*b*) Bus topology. (*c*) Ring topology.

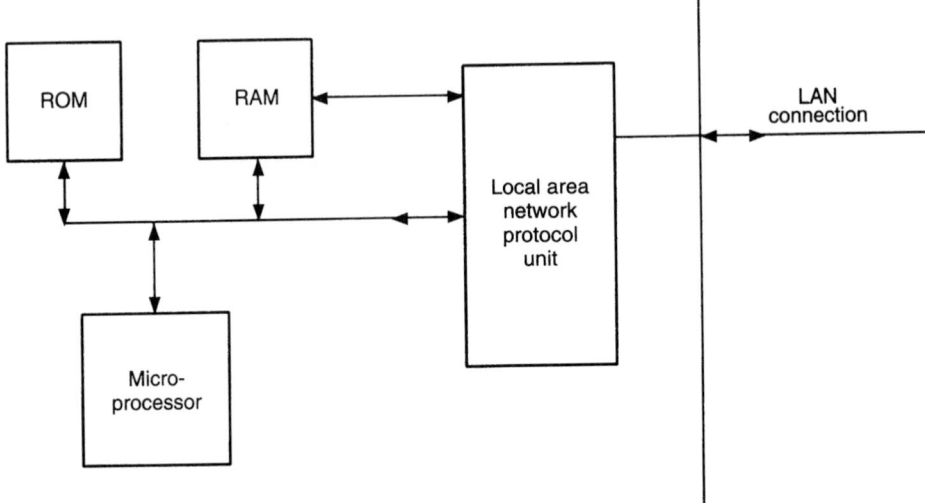

FIGURE 5   Early LAN concept shown schematically.

DECnet.   DEC (Digital Equipment Corporation) computers, operating on the factory floor in the early 1970s, were linked by DECnet. This was a token passing technique, described later under "Network Protocols."

DECnet/Ethernet.   In 1980, with an aim to support high-speed local LANs, DECnet and Ethernet were used together to form DECnet/Ethernet, a network that has been used widely over the years. One of the major advantages of combining the two concepts is that Ethernet's delay in one node's response to another's request is much shorter than that of a token passing protocol. Users generally found that these networks provide good real-time performance. Ethernet is inherently appropriate for transmitting short, frequent messages and it effectively handles the irregular data transfers typical of interactive terminal communications (Figs. 6 and 7).

CATV Cable.   In 1979 a CATV (community antenna television) cable system was announced. This system also had a central point of control and a multimaster protocol, but it used CATV cable at 1 Mbit/s. At these data rates, even in baseband, the transceiver design was based on radio-frequency (RF) technology. The topology of the network used a local star cluster with the CATV interconnecting all clusters. The data communication within the cluster was bit serial, byte parallel.

Later networks offered dual redundant mechanisms so that if one data highway failed, a second data highway would take over. The second highway was unused except for integrity diagnostics. To make certain that these data highways were in good order, elaborate mechanisms were implemented apart from the data communications to ensure cable and station integrity. CATV is mentioned later under MAP protocol.

NETWORK PROTOCOLS

A data communication protocol may be defined as "a set of conventions governing the format and relative timing of message exchange between two (or more) communications terminals," or, restated, "the means used to control the orderly communication of information between stations on a data link."

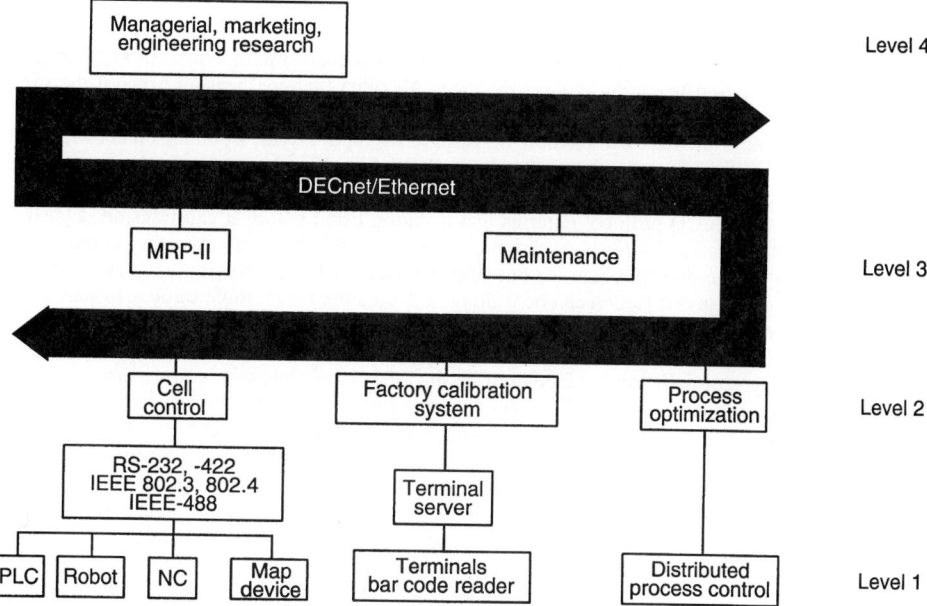

FIGURE 6 LAN for manufacturing applications combines features of DECnet and Ethernet. It may be defined as a multilevel functional model in which distributed computers can communicate over a DECnet/Ethernet backbone. Commencing in the 1980s, large numbers of these networks have been installed.

The communication protocol is vital to equipment design and must be defined accurately if all elements in a system are to work together in harmony in the absence of a lot of local "fixes." Thus much international effort has been made over several years by various technical society committees to accurately identify, define, and refine protocols. Because of the heavy investment involved, a protocol usually is debated for at least a few years before the various committees officially approve standards.

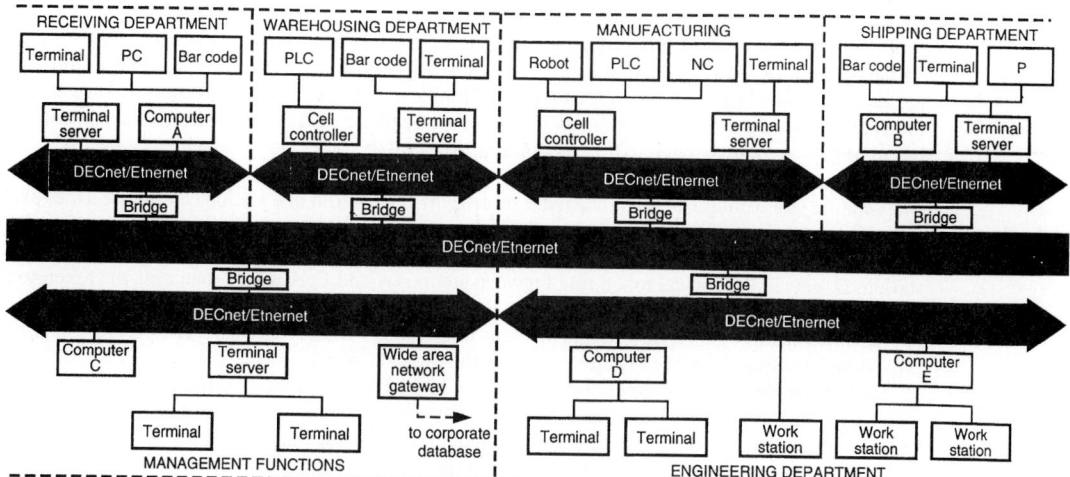

FIGURE 7 Schematic diagram shows how a DECnet/Ethernet baseband cable backbone can extend a manufacturing network to numerous plant areas, including a corporate database.

CSMA/CD Protocol (IEEE 802.3)

"Carrier sense multiple access with collision" is a baseband system with a bus architecture. Normally only one station transmits at any one time. All other stations hear and record the message. The receiving stations then compare the destination address of the message with their address. The one station with a "match" will pass the message to its upper layers, while the others will throw it away. Obviously, if the message is affected by noise (detected by the frame check sequence), all stations will throw the message away.

The CSMA/CD protocol requires that a station listen before it can transmit data. If the station hears another station already transmitting (carrier sense), the station wanting to transmit must wait. When a station does not hear anyone else transmitting (no carrier sense), it can start transmitting. Since more than one station can be waiting, it is possible for multiple stations to start transmitting at the same time. This causes the messages from both stations to become garbled (called a collision). A collision is not a freak accident, but a normal way of operation for networks using CSMA/CD. The chances of collision are increased by the fact that signals take a finite time to travel from one end of the cable to the other. If a station on one end of the cable starts transmitting, a station on the other end will "think" that no other station is transmitting during this travel time interval and that transmission can be resumed. After a station has started transmitting, it must detect when another station is also transmitting. If this happens (collision detection), the station must stop transmitting. Before quitting transmitting, however, the station must make sure that every other station is aware that the last frame is in error and must be ignored. To do this, the station sends out a "jam," which is simply an invalid signal. This jam guarantees that the other colliding station also detects the collision and quits transmitting. Each station that was transmitting must then wait before trying again. To make sure that the two (or more) stations that just collided do not collide again, each station picks a random time to wait. The first station to time out will look for silence on the cable and retransmit its message again.

Token Bus Protocol (IEEE 802.4)

This standard was developed with the joint cooperation of many firms on the IEEE 802 Committee. Since becoming a standard, it was selected by General Motors (GM) for use in its manufacturing automation protocol (MAP) as a local area network to interconnect GM factories.

This is also a bus topology, but differs in two major ways from the CSMA/CD protocol: (1) The right to talk is controlled by passing a "token," and (2) data on the bus are always carrier-modulated.

In the token bus system, one station is said to have an imaginary token. This station is the only one on the network that is allowed to transmit data. When this station has no more data to transmit (or it has held the token beyond the specific maximum limit), it "passes" the token to another station. This token pass is accomplished by sending a special message to the next station. After this second station has used the token, it passes it to the next station. After all the other stations have used the token, the original station receives the token again.

A station (for example, A) will normally receive the token from one station (B) and pass the token to the third station (C). The token ends up being passed around in a logical token ring (A to C to B to A to C to B . . .). The exception to this is when a station wakes up or dies. For example, if a fourth station, D, gets in the logical token ring between stations A and C, A would then pass the token to D so that the token would go A to D to C to B to A to D. . . . Only the station with the token can transmit, so that every station gets a turn to talk without interfering with anyone else. The protocol also has provisions that allow stations to enter and leave the logical token ring.

The second difference between token bus and CSMA/CD, previously mentioned, is that with the token bus, data are always modulated before being sent out. The data are not sent out as a level, but as a frequency. There are three different modulation schemes allowed. Two are single-channel and one is broadband. Single-channel modulation permits only the token bus data on the cable. The broadband method is similar to CATV and allows many different signals to exist on the same cable, including video and voice, in addition to the token bus data. The single-channel techniques are simpler, less costly, and easier to implement and install than broadband. Broadband is a higher-performance network, permitting much longer distances and, very important, satisfying the present and future communications needs by allowing as many channels as needed (within the bandwidth of the cable).

This protocol still is evolving. Considerations have been made to incorporate the standard of an earlier standards group, known as the PROWAY (process data highway), which was discussed in the mid-1970s by several groups worldwide, including the International Electrotechnical Committee (IEC), the International Purdue Workshop on Industrial Computer Systems (IPWICS), and the Instrument Society of America (ISA). It was also at about this time that the Institute of Electrical and Electronics Engineers (IEEE) became very active with its 802 Committee. PROWAY also is a token bus concept.

Technical committee standards efforts progress continuously, but often quite slowly, so that sometimes users with immediate requirements must proceed with approaches that have not been "officially" standardized. Association and society committees meet at regular intervals (monthly, quarterly, and annually). Thus the individual who desires to keep fully up to date must obtain reports and discussions directly from the organizations involved, or attend meetings and consult those periodicals which assiduously report on such matters.

Benefits to the token bus protocol gained from PROWAY are several.[2]

Token Ring Protocol (IEEE 802.5)

Originally token ring and token bus used the same protocol with different topologies. IBM suggested a different token ring protocol to IEEE. The proposal was accepted and became the basis for the token ring protocol, thereby forming two token protocols.

The topology is that of a ring (Fig. 4c). Any one node receives data only from the "upstream" node and sends data only to the "downstream" node. All communication is done on a baseband point-to-point basis. The "right to talk" for this network also is an imaginary token. Token ring has simplicity in that the station with the token simply sends it to the next downstream station. This station either uses the token or lets it go on to the next station.

A general summary of the three aforementioned protocols is given in Table 1.

COMMUNICATION MODELS AND LAYERS

Before describing more recent network protocols, such as MAP and MMS, it may be in order to mention tools and models that have been developed to assist in defining the various "layers" of data communications required from the factory floor or process to top-level corporate management. Computer-integrated manufacturing (CIM), for example, requires excellent communication at all levels before the promises of the concept can be achieved.

---

[2] PROWAY advantages include:

1. The immediate acknowledgment of a frame. If station A has the token and uses it to send a message to station B, station B can be requested to send an acknowledge message back to station A. Station B does not wait until it gets the token, but instead, station B "uses" station A's token for this one message. Station B, in essence, is telling station A that it received the message without any errors. This idea is allowed in 802.4.

2. Capability to initialize and control a station by sending a message over the network. (Every token bus protocol handler would have an input to control the rest of the node.)

3. Every station has some predefined data that are available upon demand. Any other station can request the data, and this station would send the information out immediately. (The station sending the response is using the other station's token.)

4. Access time of the network is deterministic and on upper bound must be predictable for any given system. This means that if a station on a network wants to send a message, one should be able to predict the maximum possible delay before it is sent out. (The original 802.4 specified a maximum access time per node, but no upper bound on the number of nodes.)

5. It provides a method of monitoring membership in the logical token ring. If a station died, every other station would be aware of it. Every station in the logical ring would have a list of all stations that are in the ring.

6. It provides for the accumulation of network performance statistics.

TABLE 1   General Characteristics of Basic Protocols

CSMA/CD protocol	Designed for a lot of short messages.
	Works well if there are not a lot of collisions.
	With heavy traffic there is a lot of overhead because of the increased number of collisions.
	Probabilistic in nature—every station has a finite chance of hitting some other station every time it tries to send. Thus there is no guarantee that a message may not be held up forever. This could be catastrophic.
	Baseband configuration means that digital data are represented as discrete levels on the cable, thus reducing noise effects and allowing longer cable lengths.
	Probably the most cost-effective for most applications.
Token bus protocol	Robust, able to recover from errors easily.
	Cable length is limited only by attenuation of signal in cable.
	Allows three different modulation schemes. Can carry multiple voice and video channels at the same time as data.
Token ring protocol	Deterministic access and priorities on messages.
	Can use fiber-optic cables. Maximum physical length of cable can be large.
	Requires more complex wiring than a bus because the last station must be connected to the first station to form a ring.
	Redundant cabling may be needed for fault tolerance because a single break or down station can stop all data transfer.

Experts have placed communication systems in levels ranging from three to seven in number. Examples of a three-level and a seven-level model are given.

Three-Level Concept.   This represented the state of the art prior to the development of the OSI reference model in the early 1970s.

Lowest Level.   Links groups of people and machines. The link involves the flow of information among different workstations at a department level. These local networks must have a fast payback because it is in this area where most manufacturing and processing alterations are made.

Middle Level.   Facilitywide networks that permit all departments within a facility to share data. Examples may include (1) obtaining employee vacation data at a moment's notice so that shift assignments can be made without delay, (2) tracing the history of a quality control problem immediately when a failure is noted, and (3) permitting a service supervisor to check the current readiness of production equipment.

Highest Level.   Corporatewide communications where all multifactories and departments can exchange information and report to a central information processing site. Manufacturing automation, such as CAD/CAM, materials management, and automated machine tools, can be linked to management information processing, including such functions as financial data, sales reporting, statistical quality control (SQC), statistical process control (SPC), and corporate planning, among many other functions. This arrangement enables managers to check the flow of production and product lead times as only one of many examples that could be given. This level of networking is more complex than the local area networks (LANs) or the facilitywide networks (WANs), in part because of the heavy information-exchange load.

OSI Reference Model.   The OSI (open system interconnections) model was developed in the 1970s as a joint effort of several groups, including the International Standards Organization (ISO), the American National Standards Institute (ANSI), the Computer and Business Equipment Manufacturers Association (CBEMA), and the National Institute of Standards and Technology (NIST), formerly the National Bureau of Standards (NBS). It has been estimated that this work represents tens of thousands of hours of effort by experts worldwide.

Initially the OSI reference model was oriented to telephony systems. Some of its achievements have included MAP (manufacturing automation protocol) and MMS (manufacturing message service). The seven distinct layers of OSI are shown in Table 2.

TABLE 2   Open-System Interconnections (OSI) Model

Layer	Name	Uses and applications
1	Physical	Electrical, mechanical, and packaging specifications of circuits. Functional control of data circuits.
2	Link	Transmission of data in local network-message framing, maintain and release data links, error and flow control.
3	Network	Routing, switching, sequencing, blocking, error recovery, flow control. System addressing and wide-area routing and relaying.
4	Transport	Transparent data transfer, end-to-end control, multiplexing, mapping. Provides functions for actual movement of data among network elements.
5	Session*	Communications and transaction management. Dialog coordination and synchronization. Administration and control of sessions between two entities.
6	Presentation†	Transformation of various types of information, such as file transfers; data interpretation, format, and code transformation.
7	Application‡	Common application service elements (CASE); manufacturing message service (MMS); file transfer and management (FTAM); network management; directory service.

*The session layer provides functions and services that may be used to establish and maintain connections among elements of the session, to maintain a dialog of requests and responses between the elements of a session, and to terminate the session.

†The presentation layer provides the functions, procedures, services, and protocol selected by the application layer. Functions may include data definition and control of data entry, data exchange, and data display. This layer comprises CASE (common application services), SAS (specific application services), and management protocols required to coordinate the management of OSI networks in conjunction with management capabilities that are embedded within each of the OSI layer protocols.

‡The application layer is directly accessible to, visible to, and usually explicitly defined by users. This layer provides all of the functions and services needed to execute user programs, processes, and data exchanges. For the most part, the user interacts with the application layer, which comprises the languages, tools (such as program development aids, file managers, and personal productivity tools), database management systems, and concurrent multiuser applications. These functions rely on the lower layers to perform the details of communications and network management. Traditionally, network vendors have provided a proprietary operating system for handling functions in the upper layers of the OSI model. These unique features have been the source of interconnection difficulties.

Manufacturing Automation Protocol

In the early 1970s the management of several industrial firms in the discrete-piece manufacturing industries realized that industrial computers and the networks that serve them were the key tools for achieving production automation on a grand scale as contrasted with the low-key efforts of the past, such as "isolated," or islands of, robotics and computerized numerical control of machines. Particularly in the United States the need to automate and pursue the ambitious goals of new production concepts, such as CIM, MRP I and II (materials requirement planning), FMSs (flexible manufacturing systems), and just in time (delivery), was precipitated by severely threatening competition from abroad. An excellent example of this was the decision by the management of GM to develop a program that would hasten the achievement of CIM, recognizing the pitfalls that existed then in attempting to utilize the control products of numerous manufacturers, that is, products which could not easily be connected and orchestrated in a practical operating network. Not necessarily the first, but certainly the most illuminated effort was the recommendation and demand of GM for simplification and implementation of control and communication components. Thus the MAP program was initiated. The U.S. government, also at about this time, created a special section of NBS (later NIST) to assist industry in the development of improved automation techniques, including networking. Again, the motive was a serious concern over the country's diminishing leadership in manufacturing. These actions placed great emphasis on LANs and the general concept of interconnectability.

GM, prior to the formation of the MAP plan, had for a number of years used a network of CATV cable (described previously) for closed-circuit television. To this had been added several channels of low-speed serial data communications, which involved RF modems to operate in the television broadband spectrum. To get a plantwide high-speed information network standard established for GM, the firm formed an internal committee called MAP. This well-publicized committee invited proposals from many vendors and circulated several papers on GM requirements.

The primary purpose of the plantwide ("backbone") network was not process control, but rather to allow the two-way flow of high-level production data. Otherwise the whole bandwidth of the backbone network would easily be consumed by local traffic. GM had defined a true hierarchical network environment and had clearly endorsed token bus and CATV, but not for process control. Also, during this period, a lot of interest was shown in TOP (technical office protocol).

The MAP concept developed at a good rate for several years, progressing to MAP 3.0. Scores of MAP networks have been installed in the United States and abroad, with the largest number located in large discrete-manufacturing firms, including, of course, GM, but also Ford Motor and Boeing. In June 1987 a 6-year freeze was imposed on the MAP 3.0 specification. This intended to allow manufacturers to build, and users to install, MAP networks without concern that the specifications continue to change over relatively short intervals of time. A leader in the MAP field indicated that the freeze would not affect the addition of functionality, compatible backward and forward, but that technology would not be added that would make obsolete what has been installed already.

Acceptance of MAP probably peaked in 1990. A large base, including Ethernet, DECnet, ARCNET, and others, remains in place. Thus, unfortunately, evaluations and guides to system selection are well beyond the province of a "permanent" handbook reference. However, a few general suggestions may be in order:

1. Are there severe noise conditions?

- YES—Seriously consider fiber-optic cables.
- NO—Hard-wired.

2. Are there time-critical throughputs—monitoring, data collection, control?

- YES—Provide equal access by all nodes.
- NO—If transmission distance is under 1 mile (1.9 km), consider twisted pair. If over this distance, consider telephone.

3. Is equal access by all nodes a requirement?

- YES—Use single high speed.
- NO—Consider combination of high-speed access and twisted pair.

4. Are there plans for future expansion?

- YES—Use standards-based network.
- NO—Use proprietary or standards-based network.

The general characteristics of PLC-based LANs are listed in Table 3 for various network types.

Open Systems

Although not necessarily fully accredited to the developers of the OSI model described previously, that group placed early emphasis on the concept of open-system architecture. The footnotes included with Table 2, which describes the OSI model, aptly define the presentation and application layers in open-system architecture. Also, the importance of MMS is listed under the applications layer of the OSI model. MMS is accepted internationally as a standard communications protocol for integrating mixtures of unlike devices that are used on the factory floor or by processing areas. The MMS reduces development costs for adding new devices to a network and diminishes the requirement for custom software and hardware for diverse device interfacing. Within the last few years it is estimated that some 40 major suppliers have recognized the MMS protocol.[3]

As of 1993 the topic of open-system architecture remains quite fluid. Some major firms are refining their most recent offerings of "open" networks that use the term as part of a proprietary trade name.

---

[3] It is interesting to note that 10 of these firms participated in an unusual demonstration at the ISA 1991 Exhibit in Anaheim, California, for the purpose of showing how MMS can handle data transfer between PLCs, PCs, NC, robots, and others on the plant floor.

TABLE 3 General Characteristics of PLC-Based LANs*

Very Good	Good	Fair	Poor
Noisy environments			
RHF FO	HC	TP	RTRL
Speed			
HC FO RHF		TP	RTRL
Throughput			
	HC FO RHF	TP	RTRL
Purchase price			
	TP RTRL	HC FO	RHF
Lifetime cost			
	RHF	FO	HC TP RTRL
Expandability			
HC FO RTRL RHF		TP	

*FO—fiber-optic; HC—hard-wired coaxial; RHF = redundant hard-wired/fiber; RTRL—remote telephone/radio link; TP = twisted pair.

Fieldbus

Since electronic measurement and control systems essentially replaced pneumatic systems several decades ago, industry has depended heavily on the 4- to 20-mA transmission standard, that is, until the *near* future! The ever-increasing use of microprocessor technology in sensors, transmitters, and control devices has created the need for a digital replacement of the 4- to 20-mA standard. Notably, this applies to "smart" transmitters.

A new fieldbus standard has been in preparation since 1985, sponsored essentially by the same society and institutional groups that have done admirable work in preparing other standards, network models, and so on. As of early 1992, completed portions of the new (SP 50) standard for both process control and factory automation include the physical layer and function block requirements. Field installations are under way in the United States and internationally. See Section 10 for an overview of the ISA SP50 fieldbus standard.

One portion of the standard (H1) specifies a low-speed powered link as a digital replacement for 4- to 20-mA transmission. When implemented, microprocessors embedded in smart transmitters will be able to communicate directly with digital control systems. Another portion (H2) specifies a high-speed unpowered link to operate at 1 Mbaud.

Other tests completed to date have included what have been described as the "worst of conditions," such as inserting a bad message, a missing terminator, and an open spur. Still other tests included adding and removing a device on line, adding crosstalk, adding a wide frequency range of white noise, and placing a walkie-talkie within 2 feet (30 cm) from the open cable for RF interference tests.

It can be safely forecast that the proposed field bus will be the subject of innumerable papers and discussions over the next few years.

FIBER-OPTIC CABLES AND NETWORKS

Fiber-optic technology has been used for well over a decade in telephony. The first large-scale demonstration was made by AT&T in 1980 in connection with the Olympic Games held at Lake Placid, New York. This test installation was only 4 km (2.5 miles) long, but tested out very successfully. The first actual commercial installations were made between Washington and New York and New York and Boston. As early as 1982 Leeds & Northrup offered an earlier version of the network, as shown in Fig. 8. This is a redundant electrical-optical highway which has been used in hundreds of installations worldwide.

Fiber-optics offers many advantages and relatively few limitations as a networking medium. Some of the advantages include the following:

- Not affected by electromagnetic radiation (RMI). For example, the cable can be installed in existing high-voltage wireways, near RF sources, and near large motors and generators. No shielding is required.

- With experience, fiber cable is easy to install and at less labor cost. Only a few special tools are needed.

- Not affected by lightning surges.

- Resists corrosion.

- Intrinsically safe.

- Compatibility makes fiber cables easy to integrate with existing platforms. Fiber cable is inherently suited to open systems.

- Increased security because of immunity to "bugging."

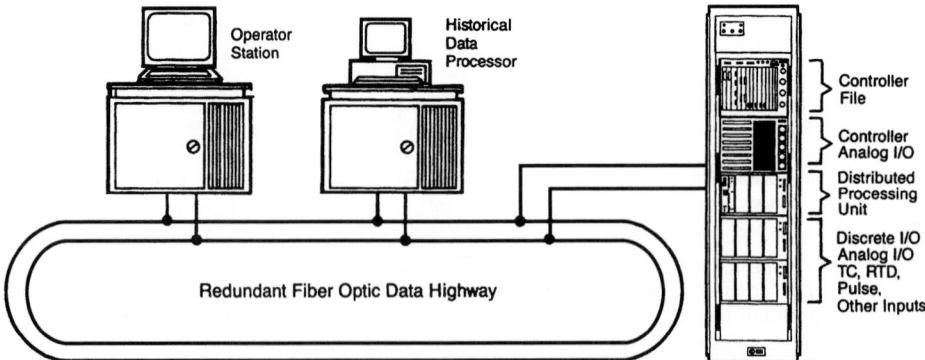

FIGURE 8   Redundant fiber-optic data highway for advanced data acquisition and plantwide control. This is an updated version of fiber-optic system first offered to industry in 1982, with ensuing hundreds of installations. Data highway consists of two redundant fiber-optic loops with repeaters to provide digital data communications among large numbers of multiloop controllers, operator stations, and computers. The optical data highway loops from one cluster to another, eventually returning to a control room with operator stations. One loop transmits digital data in a clockwise direction, while the other transmits counterclockwise. This ensures that communication between any two stations will be maintained no matter where a fault occurs. (*Leeds & Northrup.*)

- Higher data rates. The next generation of fiber-optic network protocol (FDDI) can transfer data at a 100-Mbit/s rate on the same cable fiber that now offers a 10-Mbit/s rate.

For reasons of caution and lack of better understanding of fiber technology, coupled with a continuing (but decreasing) cost differential with other media, optic cables still are in the lower portion of their growth curve. Predictions for increased use are very optimistic.

Characteristics of Optical Fibers and Cables

As in electrical transmission systems, the transmission sequence of a light-wave system begins with an electric signal. This signal is converted to a light signal by a light source, such as a light-emitting diode (LED) or a laser. The source couples the light into a glass fiber for transmission. Periodically, along the fiber, the light signal may be renewed or regenerated by a light-wave repeater unit. At its destination, the light is sensed by a special receiver and converted back to an electric signal. Then it is processed like any signal that has been transmitted in electrical form.

The system comprises transmitter circuitry that modulates or pulses in code the light from a light source. An optical fiber waveguide conducts the light signal over the prescribed distance, selected because of its particularly good transmission capability at the wavelength of the light source. The terminal end of the waveguide is attached to a detector, which may be a *pn* junction semiconductor diode or an avalanche photodiode, to accept the light and change the signal into an electromagnetic form for the receiver circuitry. The latter decodes the signal, making it available as useful electronic analog or digital output. When two-way communication is needed, the system is fully duplexed and two circuit links are needed.

Optical Fibers. Glasses of many compositions can be used for optical fibers, but for intermediate- and low-loss applications the options become increasingly limited. Multicomponent glasses containing a number of oxides are adequately suited for all but very low-loss fibers, which are usually made from pure fused silica doped with other minor constituents. Multicomponent glasses are prepared by fairly standard optical melting procedures, with special attention given to details for increasing transmission and controlling defects resulting from later fiber drawing steps. In contrast, doped fused silica glasses are produced by very special techniques that place them almost directly in a form from which fibers may be drawn.

Digital Light-Wave Systems. Much research has been directed toward light-wave systems that are digital. In a digital system, the light source emits pulses of light of equal intensity, rather than a continuous beam of varying intensity (analog approach). Each second is divided into millions of slices of time. The light source inserts 1 bit of information into each time slot, which flashes on briefly or remains off. The receiver looks for 1 bit in each slot. If the receiver senses a pulse, it registers a 1; if the absence of a pulse, a 0. Eight such bits of information make up a digital word. From a series of such words, other elements of the transmission system can reconstruct the original signal.

The capacity of a digital light-wave system is the maximum rate at which pulses can be sent and received. The maximum pulse rate is limited by how much the signal is distorted by dispersion as it travels along the fiber. *Dispersion* means that a pulse is spread out in time, so that some of the pulses arrive in the wrong time slot. If enough is lost from the proper slot, the receiver may not sense a pulse that was sent. If enough is received in an adjoining slot, the receiver may sense a pulse when none was sent. The greater the dispersion, the longer the time slots must be for the receiver to sense accurately.

Basic Fiber Types. Dispersion is of two kinds: (1) modal and (2) chromatic. Modal dispersion is the spreading of light as it traverses a length of fiber along different paths or modes (Fig. 9). Each path is a different length, and thus light takes a different time to travel through each. The highest-capacity fiber has only a single mode, so it has no modal dispersion. However, such fibers are much smaller, more difficult to couple light into, and harder to splice and connect with other types of fibers.

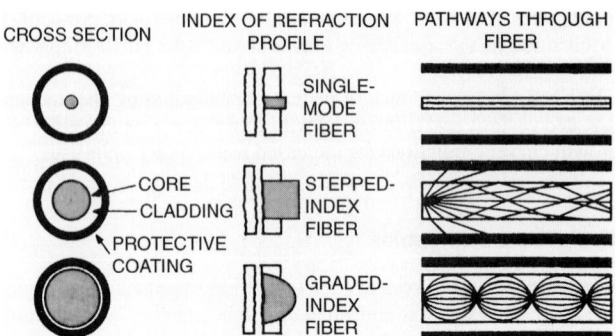

FIGURE 9   Schematic sectional views of fiber-optic cable. Not all layers are shown. Structure of the fiber determines whether and how the light signal is affected by modal dispersion. A single-mode fiber permits light to travel along only one path. Therefore there is no modal dispersion. In contrast, a step-index fiber provides a number of pathways of different lengths, but only one index of refraction boundary between layers, which bends the light back toward the center. Here modal dispersion is high. A graded-index fiber has many layers. The resulting series of graded boundaries bends the various possible light rays along paths of nominally equal delays, thus reducing modal dispersion.

The more common type of fiber is multimode, either step index or graded index. These fibers have wider-diameter cores than single-mode fibers and accept light at a variety of angles. As light enters at these different angles, it travels through the fiber along different paths. A light beam passing through a step-index fiber travels through its central glass core and in the process ricochets off the interface of the cladding adhering to and surrounding the core. The core-cladding interface acts as a cylindrical mirror that turns light back into the core by a process known as total internal reflection. To ensure that total internal reflection occurs, fibers are usually made from two glasses: core glass, which has a relatively higher refractive index, and clad glass, or possibly a plastic layer surrounding the core, which has a somewhat lower refractive index. When the seal interface between core glass and clad glass is essentially free of imperfections and the relative refractive indexes of the glasses used are correct, many millions of internal reflections are possible and light can travel through many kilometers of fiber and emerge from the far end with only a modest loss in brightness or intensity. A step-index fiber has just a single composition inside the cladding. Light must travel to this boundary before it is bent toward the center. The paths in this type of fiber disperse the pulse more than in a graded-index fiber.

In a graded-index fiber, light is guided through it by means of refraction or bending, which refocuses it about the center axis of the fiber core. Here each layer of glass from the center of the fiber to the outside has a *slightly* decreased refractive index compared to that of the layer preceding it. This type of fiber construction causes the light ray to move through it in the form of a sinusoidal curve rather than in the zigzag fashion of the step-index variety. With this type of fiber, when the physical design is correct and the glass flaws are limited, light can also be conducted over very long distances without severe loss because it is trapped inside and guided in an efficient manner. The fiber core is the portion of an optical fiber that conducts the light from one end of the fiber to the other. Fiber core diameters range from 6 to ~250 μm.

Fiber Cladding.   To help retain the light being conducted within the core, a layer surrounding the core of an optical fiber is required. Glass is the preferred material for the cladding, although plastic-clad silica fibers are common in less demanding applications. The cladding thickness may vary from 10 to ~150 μm, depending on the particular design.

Index of Refraction.   This is the ratio of the velocity of light passing through a transparent material to the velocity of light passing through a vacuum using light at the sodium D line as a reference.

The higher the refractive index of a material, the lower the velocity of the light passing through the material and the more the ray of light is bent on entering it from an air medium.

Numerical Aperture. For an optical fiber this is a measure of the light capture angle and describes the maximum core angle of light rays reflected down the fiber by total reflection. The formula from Snell's law governing the numerical aperture (NA) for a filter is

$$\text{NA} = \sin\theta = \sqrt{n_1^2 - n_2^2}$$

where n_1 is the refractive index of the core and n_2 the refractive index of the clad glass.

Most optical fibers have NAs between 0.15 and 0.4, and these correspond to light acceptance half-angles of about 8° and 23°. Typically, fibers having high NAs exhibit greater light losses and lower bandwidth capabilities.

Light Loss or Attenuation through a Fiber. This is expressed in decibels per kilometer. It is a relative power unit according to the formula

$$\text{dB} = 10\log\frac{I}{I_0}$$

where I/I_0 is the ratio of the light intensity at the source to that at the extremity of the fiber. A comparison of light transmission with light loss in decibels through 1 km of fiber is as follows:

80% transmission per kilometer $\simeq$ loss of $\sim$1 dB/km

10% transmission per kilometer $\simeq$ loss of $\sim$10 dB/km

1% transmission per kilometer $\simeq$ loss of $\sim$20 dB/km

Bandwidth. This is a rating of the information-carrying capacity of an optical fiber and is given either as pulse dispersion in nanoseconds per kilometer or as bandwidth length in megahertz-kilometers. Light pulses spread or broaden as they pass through a fiber, depending on the material used and its design. These factors limit the rate at which light carrier pulses can be transmitted and decoded without error at the terminal end of the optical fiber. In general, a large bandwidth and low losses favor optical fibers with a small core diameter and a low NA.

The longer the fiber, the more the dispersion. Thus modal dispersion limits the product of the pulse rate and distance. A step-index fiber can transmit a maximum of 20 Mbit of information per second for 1 km, and a graded-index fiber, more than 1000 Mbit. The process for making very low-loss fibers is essentially the same whether the fiber is step or graded index. Consequently nearly all multimode fibers presently used or contemplated for high-quality systems are of the higher-capacity graded-index type. Possibly for very high-capacity installations of the future, single-mode fibers may be attractive.

Cabling and Connections. Although optical fibers are very strong, having a tensile strength in excess of 500,000 psi (3450 MPa), a fiber with a diameter of 0.005 inch (0.1 mm), including the light-guide core and cladding, has a maximum tensile strength of only $\sim$10 psi (0.07 MPa). Unlike metallic conductors, which serve as their own strength members, fiber cables must contain added strength members to withstand the required forces. Also, pulling forces on unprotected fibers may increase their losses, as the result of bending or being under tension. Sometimes this is called microbending loss.

Imaging Requirements. In imaging, both ends of the group of fibers must maintain the exact same orientation one to another so that a coherent image is transmitted from the source to the receiver. Flexible coherent bundles of optical fibers having only their terminal ends secured in coherent arrays are used primarily in endoscopes to examine the inside of cavities with limited access, such as stomach, bowel, and urinary tract.

Rigid fiber bundles fused together tightly along their entire length can be made to form a solid glass block of parallel fibers. Slices from the block with polished surfaces are sometimes used as fiber-optic faceplates to transmit an image from inside a vacuum to the atmosphere. A typical application is the cathode-ray tube (CRT) used for photorecording. The requirement for this type of application is for both image coherence and vacuum integrity, so that when the fiber-optic array is sealed to the tube, the vacuum required for the tube's operation is maintained. However, any image formed electronically by phosphor films on the inside surface of the fiber-optic face is clearly transmitted to the outside surface of the tube's face. High-resolution CRT images can easily be captured on photographic film through fiber-optic faceplates.

Light Sources and Detectors

LEDs produce a relatively broad range of wavelengths, and in the 0.8-μm-wavelength range, this limits present systems to $\sim$140 Mbit/s for a 1-km path. Semiconductor lasers emit light with a much narrower range of wavelengths. Chromatic dispersion is comparatively low, so $\sim$2500 Mbit/s may be transmitted for a 1-km path.

Another factor affecting system capacity is the response time of the sources and detectors. In general it is possible to build sources and detectors with sufficiently short response times that the fiber, rather than the devices, becomes the capacity limiting factor. With single-mode fibers, lasers, and high-speed detectors, transmission rates of more than 10^9 bit/s have been achieved experimentally. This corresponds to more than 15,000 digital voice channels. Although it is interesting to learn how fast a rate can be achieved, in practice the system designer must balance other technical, operational, and economic constraints in deciding how much capacity to require of an individual fiber.

Present semiconductor lasers can couple $\sim$1 mW of optical power into a fiber. On the decibel scale, this is expressed as 0 dBm, meaning 0 dB above a reference power of 1 mW. Although some increase in power is possible, the small size and the temperature sensitivity of these lasers make them inherently low-power devices. LEDs can be made that emit as much power as lasers, but since they project light over a wide angle, much of it is lost just coupling it into the fiber. This loss is typically $\sim$10 to 20 dB. Lasers are more complex and require more control circuitry than LEDs, but they are the light source of choice when repeaters must be far apart and the desired capacity is high.

Light-wave receivers contain photodiodes which convert incoming light to an electric current. The receivers used in telecommunications system are avalanche photodiodes (APDs) made of silicon. They are called avalanche devices because the electric current is amplified inside the diode. This results in a more sensitive receiver than photodiodes without internal amplification. Again, this improved performance is achieved at the expense of added complexity. APDs require high-voltage power supplies, but they are the detectors of choice when high performance is desired.

Even with APDs, light-wave receivers are less sensitive than the best electrical ones; they require a larger minimum received power. This is a consequence of the random fluctuations in optical signal intensity known as shot noise. Light-wave systems can compensate for this. They can carry a much wider bandwidth than electrical systems, and bandwidth can be used to offset noise.

SECTION 8
OPERATOR INTERFACE*

J. N. Beach
*MICRO SWITCH Division, Honeywell Inc., Freeport, Illinois
(Operator Interface—Design Rationale)*

Kai-Chun Cheng
*School of Industrial Engineering, Purdue University, West
Lafayette, Indiana (Cognitive Skills and Process Control)*

Ray E. Eberts
*School of Industrial Engineering, Purdue University, West
Lafayette, Indiana (Adaptation of Manuscript on "Cognitive Skills
and Process Control)*

Gregory K. McMillan
*Solutia Inc., St. Louis, Missouri (Knowledge-Based Operator
Training, Intelligent Alarms)*

James A. Odom
*Corporate Industrial Design, Honeywell Inc., Minneapolis,
Minnesota (Operator Interface—Design Rationale)*

Staff
*Moore Products Company, Spring House, Pennsylvania
(CRT-Based Graphic Display Diagrams)*

*Persons who authored complete articles or subsections of articles, or otherwise cooperated in an outstanding manner in furnishing information and helpful counsel to the editorial staff.

OPERATOR INTERFACE—DESIGN RATIONALE*

The interface between a process or machine and the operator is the primary means for providing dialogue and for introducing human judgment into an otherwise automatic system. Although signals arrive at the interface and, once an operator judgment is made, leave the interface at electronic speeds, the operator responds at a much slower communication rate. Thus the operator, because of human limitations, is a major bottleneck in the overall system. The interface, whether it takes the form of

---

* Technical information furnished by *MICRO SWITCH*, a Division of Honeywell Inc., and James A. Odom, Corporate Industrial Design, Honeywell Inc., Minneapolis, Minnesota.

a console, a workstation, or other configuration, must be designed with the principal objective of shortening operator response time. The interface must be customized to the operator, and through a serious training program, the operator must become accustomized to the interface. The interface designer not only considers the hardware interface, but the software interface as well.

Inadequate interface design usually is a result of (1) considering the interface late in the overall system design process, (2) giving short shrift to human factors, and (3) procuring off-the-shelf interface configurations that have been compromised for generalized application rather than for specific needs. Cost cutting at the interface level of a system carries large risks of later problems and dissatisfaction.

HUMAN FACTORS

Interface design falls within the realm of human factors, which is sometimes referred to as human engineering or ergonomics—all of which pertain to the very specialized technology of designing products for efficient use by people. Human factors is concerned with *everything* from specific characteristics of interface components to the total working environment of the operator.

Importance of Initial Planning

An excellent starting point for the interface designer is that of providing a functional description of the interface and then a job description for the operator. These two descriptions should dovetail precisely. A pro forma questionnaire can be helpful (Fig. 1).

Operator Variables

The principal characteristics of the operator in designing an interface are (1) physical parameters, (2) experience, including trainability, and (3) long-established habit patterns. The physical aspects will be described shortly. In terms of experience, the amount of instruction required for efficiently using the interface as intended obviously depends on the complexity of the process or machine under control and of the interface per se. There are tremendous differences, ranging from the simplicity of operating an average copy machine to a complex machine tool or assembly line to a complex chemical process that incorporates many hundreds of control loops. When forecasting the amount of instruction that an operator will require for a given interface, the designer should establish the specific content of a training program as part of the overall interface design task.

Habit Patterns

People, as the result of past exposure, "expect" controls to move in certain ways. These expectations sometimes are called population stereotypes because they are so universally accepted. Where possible, component selection for an industrial control interface should be an extension of these stereotypes or habit patterns. For example, the wall-mounted toggle switch found in houses has established a habit pattern for turning on the lights. The upward flipping motion is associated with "on" and can be utilized with other instrumentation-type toggle-paddle switches for a natural transfer of a previously learned habit.

The clockwise motion of a rotary knob is frequently used to turn on a domestic appliance (television, range or oven, mixer). This same familiar action may be adapted to any control panel for an extension of a normal habit pattern. The scale of a slide switch or potentiometer should show an increase as the switch is moved upward or to the right. These control actions require the least amount of conscious effort to learn and are well established in our daily lives (Fig. 2).

Initial Planning Sheet—Interface Design

1. Functionality of interface:
   Variables to be indicated, recorded, adjusted, automatically alarmed...
   Provide instructions, ancillary and historical information—Printed or computer-stored................
   Extent of interactive graphics desired ....
   Special tasks of interface...
   Will interface satisfy only functional needs, or will it be used also as a "show piece" for VIP
   visitations? ....
2. What are cost constraints? space limitations? ....
3. Are there special environmental considerations?—Installed indoors (air-conditioned space?),
   outdoors, proximity to excessive noise, near moving machinery, vibration and shock, electrical
   interference, corrosive or explosive vapors ....
   Will design require guards, barriers or protective shields for components and user safety?...........
   What degree of "ruggedness" is required? ....
4. Is there a requirement for information confidentiality? ....
   Will interface be used constantly (three shifts) or be out of action parts of a 24-hour period or
   weekends?...
5. Will operator instructions be simple or complex? ....
   What will be the extent of operator training? ....
6. Who will install equipment and maintain it?—User technicians, supplier, outside service
   contractor?..
   How can installation and maintenance tasks be simplified?..
7. What are the choices in terms of overall physical configuration?—Console, workstation,
   pedestal, desk?..
8. What are hardware options?—Visual displays, audio alarms, voice reception and activation ....
   Operator's physical contact with interface, switches, touch screens, etc.?....................................
9. Are there any special procurement regulations, as for military installations? ....
10. Must the interface fit well into facility expansion plans? ....
11. What are the physical and mental profiles of your present very successful operators? ....
12. What have been common operator complaints in the past?..
13. Will complexity justify preparing a life-size mockup for achieving an interface that combines
    human factors with aesthetics?...
14. Functionality of operators—Complete job description, operator qualifications, physical size,
    agility, alertness, other factors (from past experience)..
    Is interface for a process or machine that could cause loss of life or equipment if permitted to go
    out of control? ....

FIGURE 1   Pro forma questionnaire for use by engineers and designers when initially considering new or revised operator interface.

When controls or control and display arrangements take advantage of these habit patterns, generally the following results can be expected:

1. Reaction time is reduced.

2. The first control movement by an operator is usually correct.

3. An operator can perform faster and can make adjustments with greater precision.

4. An operator can learn control procedures faster.

Operator-Interface Geometry

Extensive use of computer data terminals has generated a wide variety of interface architectures. However, there are certain common denominators. Typically, an operator will use a keyboard, a display unit, and a document station (a work holder for printed material). These elements require careful attention to their positioning with respect to the operator. A thorough ergonomic analysis

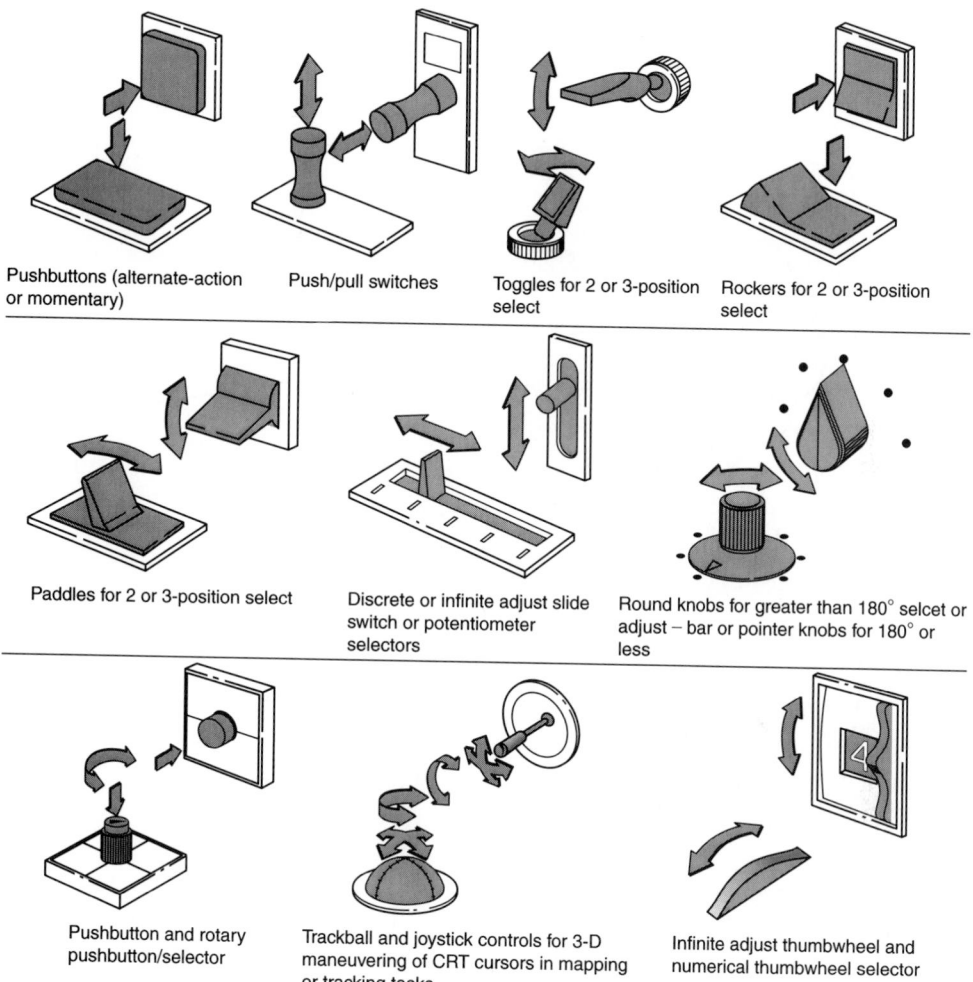

Pushbuttons (alternate-action or momentary)

Push/pull switches

Toggles for 2 or 3-position select

Rockers for 2 or 3-position select

Paddles for 2 or 3-position select

Discrete or infinite adjust slide switch or potentiometer selectors

Round knobs for greater than 180° selcet or adjust – bar or pointer knobs for 180° or less

Pushbutton and rotary pushbutton/selector

Trackball and joystick controls for 3-D maneuvering of CRT cursors in mapping or tracking tasks

Infinite adjust thumbwheel and numerical thumbwheel selector

FIGURE 2 Switch response control movements in accordance with habit patterns. (*MICRO SWITCH.*)

must be a part of interface design to ensure that lines of sight, reach, lighting, access, and angular relationships are addressed properly (Fig. 3).

The line of sight from the operator to the center of a display should be as near to perpendicular as possible. The top of the display should be at or below eye level.

Height, tilt, and swivel adjustment of seat, keyboard, display, supports, and footrests must be provided, since they are all parts of an interactive workplace system. Adjustments for operator posture should be easy to make and without requiring special tools, skills, or strength. If an operator cannot maintain a comfortable posture, there will be complaints of fatigue related to the eyes, neck, shoulders, arms, hands, trunk, legs, and feet.

Alphanumeric Displays

Data display terminals that use cathode-ray tubes (CRTs) are commonly referred to as video display terminals (VDTs) or video display units (VDUs). Data display systems also may use gas discharge,

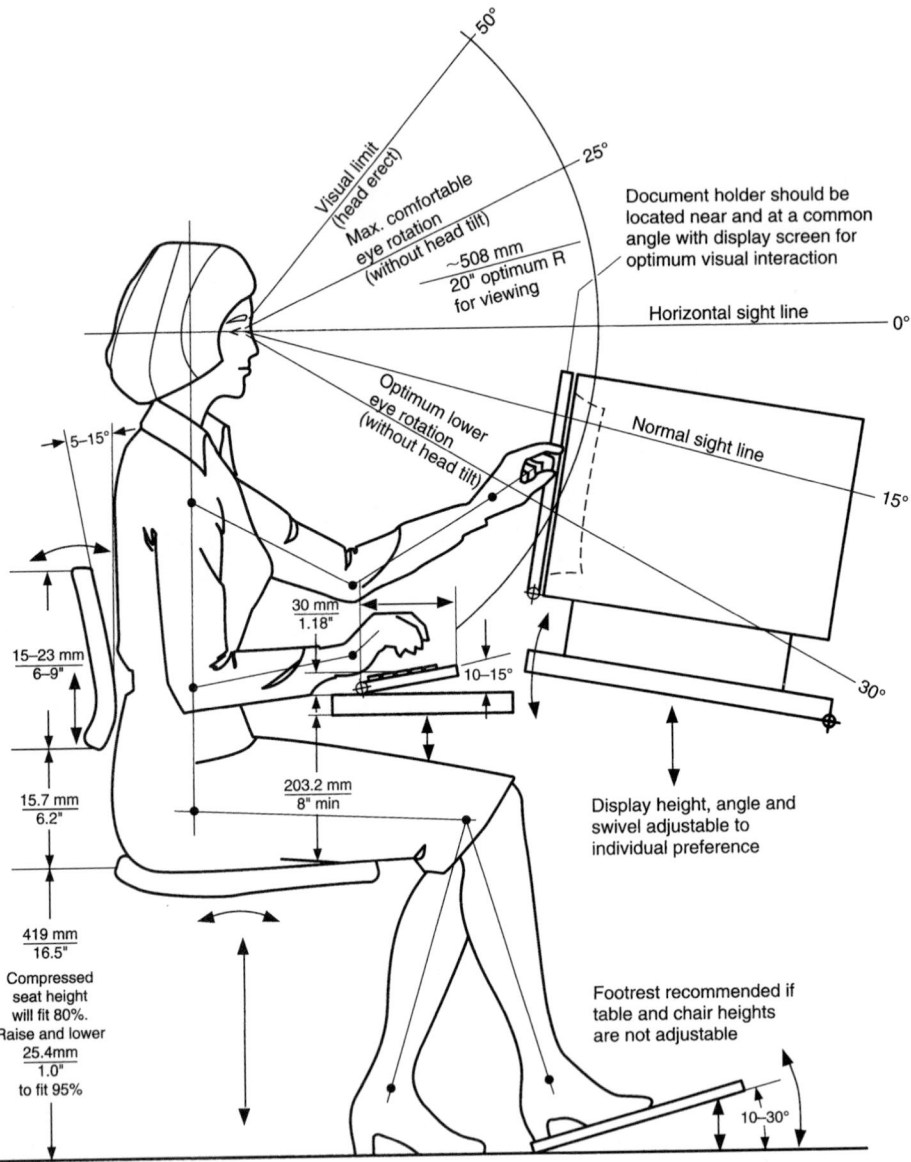

FIGURE 3 Geometry of operator-interface design. (*MICRO SWITCH.*)

light-emitting diodes (LEDs), and liquid-crystal displays (LCDs). Although non-CRT systems are often attractive because the interface screen can be relatively thin [2.5 to 5 cm (1 to 2 inches) front to back], the CRT enjoys wide usage because of its versatility in terms of graphics and use of color, and because of cost.

Although it is possible to display alphanumeric data on a single or abbreviated line display, as offered by some electronic typewriters, the full-page business-letter format usually is more desirable. If reference material is in the same format as the displayed information, the visual interaction between the two is compatible and operator perception problems are minimized. Many word-processing display

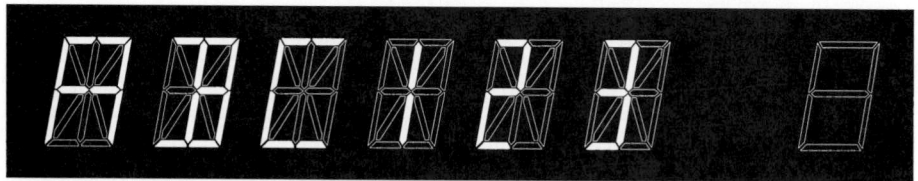

FIGURE 4 Types of alphanumeric matrix displays. (*Top*) 5 by 7 dot matrix. (*Middle*) 7 by 9 dot matrix. (*Bottom*) Bar matrix. (*MICRO SWITCH.*)

screens offered today do not have full vertical page capacity, but they move type upward (scrolling) when the lower line limit has been reached.

Alphanumeric displays on VDTs are typically dot-matrix construction because of the discrete addressing mode which they use. Both 5 by 7 and 7 by 9 rectangular dot groupings are used, with the larger sides vertical (Fig. 4). A practical working height for these characters is 2.5 mm (0.1 inch) minimum. With adequate spacing between characters and lines, a typical display screen [244 mm wide by 183 mm high (9.6 by 7.2 inches)] can reasonably accommodate 47 lines of 95 characters.

Although light characters on a dark background are most common in VDTs, testing has verified that there is improved legibility with dark characters on light backgrounds. By having the same contrast format for both reference document and display, there is considerably less eye strain as an operator shifts back and forth between two surfaces. It is good practice to have CRT displays include the capability of image reversal from positive to negative in full or selected areas. Dark characters on a CRT light background display (positive image) present less contrast than equivalent light characters on a dark background (negative display). In effect, backlighting of the positive image "washes" around the dark characters to make them appear narrower. To counteract this phenomenon, the stroke width of the positive image should be approximately 20 percent heavier than that of the equivalent negative image.

Glare on the face of CRTs is one of the most frequently cited problems associated with VDT operation (Fig. 5). Glare from uncontrolled ambient light reflects into the operator's eyes, making it difficult or impossible to distinguish images on the screen, as well as producing eye strain and fatigue. High ambient light also reduces the contrast between background and displayed characters, since it adds its energy to both. Antireflective coatings, tinted windows, louvered screens, and various other filter media can be used to minimize this problem. Some of these glare control solutions reduce ambient light bounce at the expense of contrast and image sharpness. The effectiveness of these

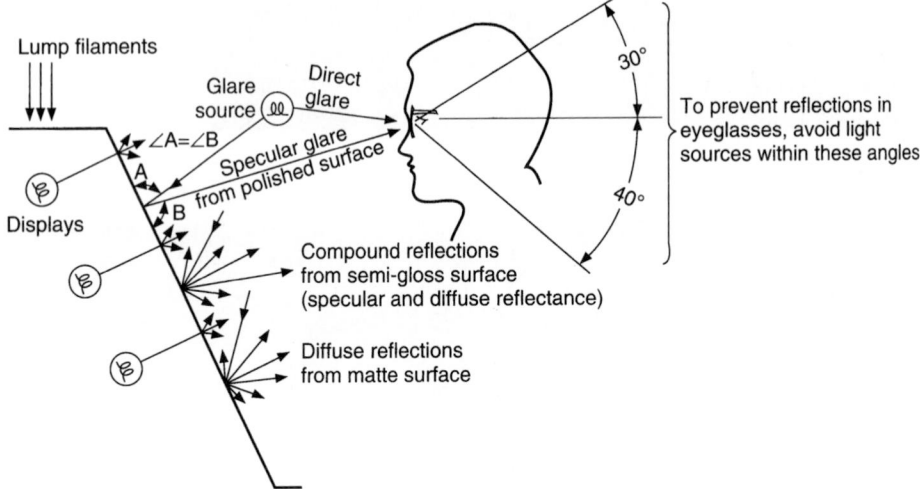

FIGURE 5 Glare from reflected ambient lighting can mask signals. (*MICRO SWITCH.*)

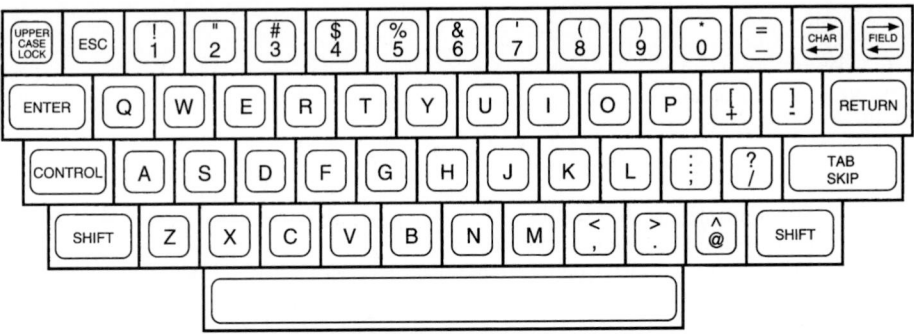

FIGURE 6 Keyboard with standard QWERTY layout.

various means should be evaluated carefully prior to making a commitment to use them. Further, the location and diffusion of ambient light sources should be investigated.

A wide range of image color is available on VDTs, but most use green, white, orange, or yellow contrasted with black. Any color image will satisfy the visual requirements of the display, so long as it does not fall at either end of the spectrum. Multicolor displays at these extreme wavelengths are also poorly perceived by the eye and should be avoided.

Keyboards

Current trends in programmable electronic data storage and retrieval systems have made the traditional typewriter keyboard all the more relevant as a critical interface between user and equipment (Fig. 6).

Key groupings for special functions and numeric entry have taken their place alongside the basic keyboard format. Full-function intelligent keyboards with integral microcomputer-based electronics may be encoded to meet user requirements.

FIGURE 7   Dvorak simplified keyboard input.

The physical effort required should be consistent with the speed and efficiency of the operator and the duration of the operation. Without attention to these needs, the operator's performance may suffer. Poorly conceived equipment standards have sometimes been established casually, or without meaningful studies of operator needs. With years of inertia, they become so ingrained in the user population that there is virtually no way of changing or even modifying their effect. The standard QWERTY keyboard (named by the arrangement of the first six keys on the upper row) is a prime example.

An alternative arrangement, the simplified keyboard (Fig. 7) was developed by Dvorak in 1936. It distributes the keys according to the comparative strength of the fingers and the frequency of letter occurrence in common English usage. This improved design has more than doubled typing speed and it is said to reduce the average typist's daily keyboard finger travel from 12 to 1 mile (19.3 to 1.5 km). Adopting of the system has been and most likely will continue to be difficult because of the number of QWERTY machines in existence and by the extensive retraining of operators that would be required.

Volumes have been written about the optimum keyboard angle for efficient use. There have been keyboard layouts designed to operate at every angle between horizontal and vertical. However, there is a general consensus that somewhere between 10 and 20° is a comfortable workplane. Most contemporary keyboards fall within this angular reference. Recent studies have shown that by providing a keyboard with an adjustable setting for use between 10° and 25°, the needs of individual operators can be better accommodated.

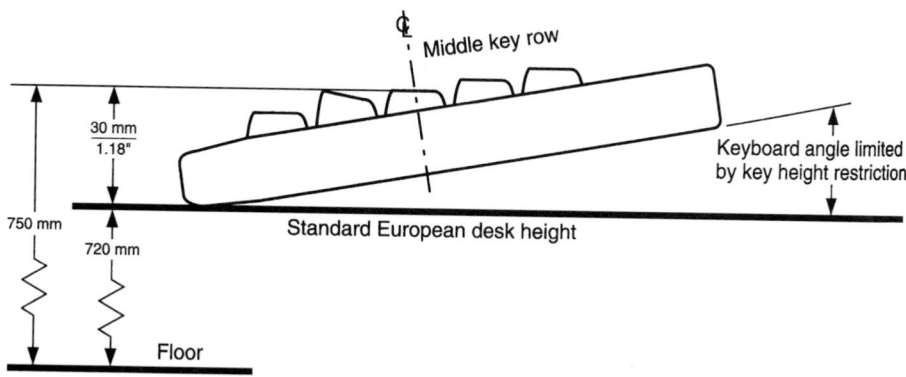

FIGURE 8   Basic DIN standard dimensions and angular restrictions. (*MICRO SWITCH.*)

The current European standard (Fig. 8) requires that the height of the keyboard measured at the middle key row should not exceed 30 mm (1.2 inches) above the counter top. Although there are a number of low-profile keyboards thin enough to allow meeting this requirement, the combination of the keyboard enclosure dimensions and the 30-mm DIN height standard may restrict the keyboard angle to 10° or less, and as stated previously, the preferred angle is between 10° and 20°.

DIN standards also recommend that if a keyboard is mounted at a height of more than 30 mm (1.2 inches), then a palm rest must be provided to reduce static muscle fatigue. This dictum, however, may be subject to question. A palm rest may be desirable for some low-speed data entry, but a high-speed typist does not rest palms on anything. The use of a palm rest could only promote inefficient and error-prone typing. In addition, a palm rest will require more desk space, which is usually at a premium.

For rapid keyboard entry, a 19.05-mm (0.75-inch) horizontal spacing between key centers is considered comfortable. For rapid entry it is desirable for traveling keys to have a displacement of between 1.3 and 6.4 mm (0.05 and 0.25 inch) along with an associated force of 25.5 and 150.3 grams (0.9 and 5.3 ounces). Traveling keys are preferred to low-travel touch panels, which are better suited for low-speed random entry situations. When used with a data display terminal, it helps to have the keyboard enclosure separable from the display, so the user can position it for maximum comfort.

Some of the more ambitious human factors efforts have created keyboards with a slight concavity, running from front to back on the surface of the combined key faces. This reflects the natural radii followed by the fingers moving up and down the key rows.

Other experimental keyboards have separated the left- and right-hand keys into two groups spaced slightly apart. The angle of each group parallels the natural angle of the hand and arm positioned in line and directly in front of a user. This minimizes the angular offset between hand and arm normally required for keyboard entry. These separated keyboards are not commercially available yet, but their invention indicates a continuing interest in improving the performance of keyboard operators.

Voice Recognition

Most voice recognition systems in current use are programmed to respond only to specific voices and accept a relatively limited command vocabulary. As speech processing technology continues to develop and as memory capabilities are expanded, these systems will be able to accept instructions from anyone. As the technology matures, voice recognition systems are expected to be able to recognize and identify both specific and nonspecific users. This process could eventually be used for authorized access to databases, and could supersede magnetic-card readers for clearance in security situations. More on voice recognition technology is given in a separate article in this handbook section.

ENVIRONMENTAL FACTORS

Numerous factors can affect the short- and long-term operating performance of the process or machine interface. Two of the most important of these are (1) the comparative gentleness or hostility of the local environment and (2) ambient lighting.

Local Environment

Industrial environments frequently pose a threat to the life and reliability of interface components. Often they are subjected to daily routines of abuse. Controls and displays may be splashed by water, oil, or solvents. They may be powdered with a layer of soot and dust, including metal particles, sticky vapors, and various granulated, gritty substances. However, even under these harsh circumstances, a resistant yet still attractive interface can be designed that will take advantage of oiltight manual controls, protective membrane touch panels, and ruggedized keyboards.

Ambient Light Conditions

External interface lighting almost always will either enhance or downgrade display visibility. As the ambient light level decreases, the visibility of self-illuminated displays will increase. For example, low-output lamps and projected color or "dead front" displays require low ambient light. Conversely, a brighter display is needed for recognition in high ambient light. By way of illustration, full indicator brilliance may be called for in the bubble of an aircraft cockpit exposed to direct sunlight, whereas a minimum glow from a display will be most appropriate at night. A situation like this will call for a brightness control for the indicator. Wherever possible, the designer should customize display brilliance to accommodate a wide range of brightness in any similar circumstances.

COGNITIVE SKILLS AND PROCESS CONTROL

Kai-Chun Cheng[*]

Ray E. Eberts[*]

INTRODUCTION

The process industry is a complex and dynamic system that involves the processing of multiple materials and energy to produce a new product. In addition, it often requires the integration of the efforts of many workers in the plant. Descriptions of conventional and modern process control system can be found in several references [1–4].

Because of the development of technology, automation has been widely incorporated into the design of process control systems. Process control is increasingly becoming a cognitive activity for the operator rather than a perceptual and control task. In a classification of the process operator's tasks by Lees [5], four of the five task categories are based on cognitive activities. Partly motivated by this, much research activity in human factors and cognitive psychology has been directed toward characterizing the cognitive skills developed by experts, how novices can be trained so that they develop those skills, and how equipment can be designed so that it fits in with the skills that operators have. An understanding of the cognitive skills possessed by operators can help in that design process. In addition, this understanding can also be used to train operators and to predict human performance in existing systems.

PROCESS CONTROL TASKS

The operator abilities needed for process control encompass the whole range of human abilities from perceptual motor tasks, such as tracking and manual control, to cognitive tasks, such as decision making and problem solving. Lees [5] provides an excellent review of the research on process control

[*] Purdue University, West Lafayette, IN 47907.

TABLE 8.1   Process Control Subtasks

Subtask	Description	References*
Monitor (vigilance, signal detection)	Monitor for seldom-occurring event; separate signal from noise	[7–9]
Control (tracking)	Keep system on optimal course	[10]
Interpret (categorization, quantization, estimation, filtering)	Separate random fluctuations of system from actual course; filter out noise	[11,12]
Plan (decisions, allocate resources, resource sharing)	Set goals and strategies; efficient use of resources; sequencing of tasks; heuristics; develop strategies	[13–15]
Diagnose	Identify the problem when a fault occurs	[16]

* Complete references can be found in Ref. 6.

operators. Several subtasks for process control in Table 8.1 have been extracted from the Lees article; this list of subtasks will be used throughout the article. The references for each of the subtasks describe characteristics of human performance and often suggest ways of improving performance. For example, the Peterson and Beach article [16] describes areas in which humans make poor decisions and therefore suggests ways to improve decision making by offering the right kind of information.

MODELS OF PROCESS CONTROL TASKS

The control technology had advanced from the conventional control techniques such as feedback and feedforward control to more advanced model-based predictive models and to recent developments such as adaptive control, artificial intelligence, and expert system [17, 18]. The goal was to develop models that can accurately capture the behavior of an actual system so that algorithms can be developed and used to design automatic process control systems.

Similarly, expert operators' behavior was studied and models were built to try to capture the characteristic of an expert. The goal was to learn how the experts could successfully control the processes and use this knowledge in the design of the control system. Thus novice operators can follow the expert operators' patterns and gain the expertise quickly. In addition, the experts' knowledge could help us design better training programs to help novice operators gain the desired expertise.

Quantitative Models

Model-based predictive controls are quantitative models that use dynamic mathematical formulas to describe the behavior of the system (references to several of the available models can be found in Ref. 6). In most cases, these models use physical system models and parameters to describe and understand human performance. Since many of the human tasks described in these models can now be performed automatically, the emphasis in modeling the operator has switched from these physical descriptions of performance to modeling the complex cognitive tasks required by current process control operators. These kinds of models are described in the following subsections.

GOMS and NGOMSL

GOMS (which stands for goals, operators, methods, and selection rules) and NGOMSL (natural GOMS language) models were formulated for human–computer interaction tasks. Since most of the complex cognitive tasks required by process control operators are performed through interaction with a computerized process control system, the human–computer interaction model is appropriate. For GOMS [19], human information processing is described by four parameters: cycle time, the representation code, storage capacity, and decay time. Each task is then analyzed to determine how the operator processes the task information. The cycle time parameter can be used to make time predictions for completion of the task. The other parameters (representation code, storage capacity, and decay time) indicate the limitations of the human operator. When these limitations are compared with the task demands, error predictions can be made.

The NGOMSL model, developed by Kieras [20], is an extension of the GOMS model that can be used to predict the usability of a product or system. An operator's interaction with a product or system is modeled by determination of the goals and the subgoals needed to accomplish tasks or subtasks, the methods required for accomplishing the goals, the steps (operators) to be executed in the methods, and the selection of rules when choosing between different methods. With the NGOMSL model, the steps that a person must go through to accomplish a task can be clearly listed, including both external and mental processes. Once the model is formulated, quantitative predictions can be made about the execution time, mental workload, learning time, and gains that are due to consistency. The operator–system interaction, when defined by the model, can be analyzed to find ways to improve performance of the operator, through system design changes, for the time needed to perform a task, the time needed to learn the system, the mental workload required for operating the system, and the consistency of procedures required of the operator.

GOMS and NGOMSL have been quite successful at accurately predicting human performance in computerized tasks. GOMS and NGOMSL can be applied to any process control tasks that require the operator to interact with a computer.

Expert Systems

Another approach to modeling the complex cognitive activity of process control operators is in terms of expert systems. At one level, expert systems can be differentiated from quantitative models in that the expert systems are heuristic and the quantitative models are algorithmic. An expert system is defined as an intelligent problem-solving computer program that uses knowledge and inference procedures to achieve a high level of performance in some specialized problem domain. This domain is considered to be difficult and requires specialized knowledge and skill. A common form of representing the knowledge in expert systems is to use production rules, which can be of the form

IF (premise) THEN (action)

The premise is a combination of predicates that, when evaluated by the program as true, lead to the specified action. The process control operator is modeled through the production rules used to control the process. KARL is an example [21] of an expert system developed for process control tasks.

Conclusions

The success of the applications of the above models has been mixed. The quantitative models are often based on models of physical systems and thus fail to capture abilities, such as flexible problem solving, that are uniquely human. These models are good at describing performance under ideal and optimal conditions (although not to the level found in most engineering models) and are also fairly good at performing control tasks; they fail if confronted with novel situations. The expert system models can deal with novel situations, although success of these models has been difficult to determine. GOMS and NGOMSL models have been shown to model human behavior accurately for those tasks in which the operator interacts with a computer.

EXTRACTING EXPERT KNOWLEDGE

Understanding expert process control operators is important in two ways. First, the purpose of a training program is to instruct a novice how to become more expert. Conveying how an expert performs tasks is important in training programs. Second, artificial intelligence techniques can be used, through computer programming, to make a process control system take on humanlike characteristics and intelligence. The knowledge or intelligence of experts must be determined in order to program such systems. In both cases, knowledge must be extracted from experts to use in training programs or artificial intelligence programs.

Unfortunately, extracting expert knowledge is often difficult. Experts generally store their knowledge and expertise implicitly in their mind rather than in a set of explicit rules to be followed. Thus expert operators generally have difficulty describing how they perform the control tasks. Sometimes, the expert's statement even contradicts his or her own behavior [22, 23]. The following techniques have been used for extracting expert knowledge.

Protocol Analysis

Following Newell and Simon's information processing theory [24], a technique for acquiring knowledge from experts was developed. This technique was referred to as the protocol analysis [25]. Protocol analysis requires an expert to solve a problem and at the same time verbalize the actions and/or thinking. The recorded protocols can then be analyzed to determine the thought process and information used by the expert. Another example of using protocol in extracting expert knowledge was an expertise transfer system [26] (ETS) developed based on Kelly's interview technique [27]. The ETS asks the expert series of "why" questions and uses the data to construct a knowledge base for the problem.

Scaling

Scaling techniques use statistical methods to calculate the spatial relationships between items of information and construct a hierarchical structure that represents an expert's knowledge base. Two methods are used often to scale the data, multidimensional scaling [28–30] (MDS) and cluster analysis [31, 32]. More detailed descriptions about these techniques can be found in the listed references.

Pattern Recognition

Expert knowledge can also be retrieved from the patterns in experts' behavior. These patterns could be how the experts interact with the system such as the when the display is checked, what type of information was checked, what action was taken, etc. When certain methods are used, the patterns of experts' behavior can be retrieved. Then, through the interpretation of these patterns, a model can be built that describes how experts control the system. Again, statistical tools could be used to retrieve the patterns in the recorded data. For example, Joyce and Gupta [33] and Gains, et al. [34] both had used the latencies between keystrokes to identify individual users. Another tool for identifying the patterns is by using neural networks. Past research in different areas has already shown that, when trained properly, neural networks have the ability to learn, memorize [35], and recognize patterns [36]. Kuespert and Mcavoy [22] had developed a framework for extracting expert knowledge through neural networks. A third tool for identifying the patterns in the data is to use filters. When a filter is used, patterns can be classified into different categories (e.g., Ref. 37).

EXPERTISE IN PROCESS CONTROL

Several characteristics can describe the expert process control operator. The first characteristic is that, after time, performing the task becomes automatic. Second, experts need accurate mental models of the system in order to control the system and make the right kinds of decisions. Third, operators must develop the ability to represent the system spatially.

Automatization

As a necessary, but not sufficient, condition, an expert can be characterized as someone who has been doing the task for a long time. Although it is difficult to determine when someone becomes an expert, one popular notion is that an expert has practiced the task for at least 5000 hours. The knowledge base accumulated by an expert is enormous. In a study of master chess players, Chase and Simon [38] estimated that these experts could recognize approximately 31,000 basic or primitive piece configurations. Brooks [39] has estimated that expert programmers have available to them between 10,000 and 100,000 rules that can be used to perform programming tasks. The expertise in a computerized expert system does not appear to be very expert in comparison. As a comparison, MYCIN, an expert medical diagnosis system and one of the most successful expert systems, contains 450 rules [40].

Researchers in expert systems claim that expert systems can, in essence, achieve the experience levels seen in true experts by acquiring expertise from several experts. The combined experience level would be greater than that of any one expert alone. Acquiring knowledge from an expert is difficult, as discussed above. Anecdotal evidence indicates that experts are often the least able to explain to novices their expertise. Lees [5] states that "it is well known, for example, that a pilot may be skilled in flying but may give a quite erroneous account of the control linkages in the aircraft." On a more empirical basis, Woodworth [41] cites evidence that conscious content of verbal reports disappeared with extended practice. Just at the time that, for example, a process control operator would be most useful to a knowledge engineer (one who gleans knowledge from an expert for an expert system), the expert operator may not be able to verbalize how to perform the task.

Accurate Mental Models

Experts are very good at knowing what to expect from the operation of the system. This ability is often referred to as the operator's having an accurate internal model of the system [42]. To have an accurate and effective internal model, the operator must be able to predict how the system will function. The events as they happen can then be compared with this prediction to see if anything unusual has occurred. Kragt and Landeweerd [43] characterized this ability as a "routine model" that the operators used when the process was performing in normal situations. Bainbridge [13] found that expert process control operators continually updated, in their heads, the current state of the system and where the system was moving. Without expectancies or without the internal model, the operator would have nothing on which to evaluate the current operation of the system.

Spatial Representations

Expert process control operators have available to them a spatial representation of the system. A plausible alternative to a spatial representation would be a propositionally based rule system similar to that used in expert systems. To differentiate between these two kinds of representations, consider a process in which the operator knows that the water level in a tank is dropping. If the operator stored a set of rules, he or she would mentally step through the rules to find one that fit; e.g., if the water level is dropping, an output valve is open. For a spatially based model, the operator would represent the task spatially as a physical system with locations and movements between the locations. To find the source of a problem, such as falling water level, the operator could mentally run a simulation of the process until a match is found. As an example, the operator could picture the water flowing out of the tank through an open valve. When expert steam plant operators were interviewed by Hollan et al. [44] the experts indicated that they did mentally run such simulations to solve system problems. Rasmussen [16] found a similar phenomenon in expert process control operators and characterized this kind of activity in fault diagnosis and troubleshooting as a topographical search. Spatial representations of process control tasks, especially fault diagnosis, were also found by Landweerd [45] and by Wickens and Weingartner [46].

A spatial representation of a task is an efficient way to store information. By spatially representing a process control task, the operator must represent physical locations, must know how the systems can interact with each other, and then well-learned and versatile reasoning and problem-solving strategies can be used to make inferences about the process. As a common example of how such a representation scheme could work, if someone inquires about the number of windows in your house, this number is usually not stored explicitly. Rather, the number can be arrived at by mentally picturing yourself moving through the rooms of the house, counting as you go. Propositionally based rules can then be generated from this spatial information as a secondary process. Storing and retrieving 100,000 rules from memory, as an example, would be difficult; a spatial representation of the information is more efficient. A computer model of this kind of expertise, however, is difficult to generate. To generate one, we would have to know how to incorporate the spatial information in computer code, how to model the reasoning that occurs on this spatial information, and how particular strategies (such as picturing yourself walking through rooms) are chosen from the vast amount of strategies, many fruitless, that must be available. Some research is being done in this area to design expert systems that capture this kind of expertise [47].

TRAINING COGNITIVE SKILLS

Novices can acquire expert behaviors in two ways: by training, thus receiving the experience necessary, or by design of the system that assists novices to exhibit behaviors that are more expertlike.

Training is costly but cost effective in the long run, as shown in the example of the Inland-Orange Mill, which adopted a well-planned training program [48]. An operator properly trained and skilled at the task can perform many functions that machines are not capable of yet, such as decision making and flexibility in problem-solving techniques. It reduces downtime for the machinery and costly errors. Also, as Younkin and Johnson [49] emphasized, training is the key to a successful automated computer control system.

Many available techniques, which were not available a few years ago, can bring down the cost of training. These techniques fall into two areas: advances in cognitive skill training and advances in computerized training methods.

Using a Predictor

Several techniques can be utilized in training novices to gain the desired cognitive skills that expert operators possess. Providing a predictor of the future state of the system is one way to increase cognitive abilities. Recall that an integral part of the operator's internal model was the ability to predict where the system should be so that this prediction could be compared with the current state of the system. Predictors can be used in either the training process or as a job performance aid. After training, predictors on an augmented display were shown to increase the cognitive skills of operators [50]. The operators had learned to internalize the predictor so that performance would be high even without the predictor on the screen; this is advantageous if a breakdown occurs in the computerized aid. If augmentation is used during training, some trials should be made without the predictor so that the operator does not form a dependency on this assistance.

Spatial Information

Another method to develop cognitive skills is to provide the operators with spatial information during training. Recall that the expert operators appeared to conceptually represent a process control task spatially, and thus enhancing this ability through training would be useful. Steamer is an example [51] of how spatial information can be used when training operators. Steamer is a computer-assisted instruction program developed for the U.S. Navy to train their steam plant operators. A steam plant is used to power large ships and it typically has approximately 1000 valves, 100 pumps, and various turbines, switches, gages, dams, and indicators. The operational procedures for these kinds of plants are contained in several volumes. In an analysis of experts performing the steam plant operations, Hollan et al. [44] concluded that the experts had a spatial representation of the plant that was used to solve problems. When problems occur, operators could run mental simulations by using this internal spatial representation of the plant.

Analogies and Metaphors

A third method to develop cognitive skills is to use analogies and metaphors to train novices about how a system works. With analogies and metaphors, novices learn how knowledge of a similar situation can be applied to the new situation [52]. As an example, one of the most used and successful metaphors has been the desktop metaphor. Our knowledge of how to manipulate objects on our own desk (such as pen, paper, files, etc.) was applied to the design of the computer desktop. Another example can be found in the article by Mayer and Bromage [53]. They proved that training subjects by using concrete models could help them perform better on tasks that require transfer of their knowledge to novel situations.

Active Learning

A fourth method for developing cognitive skills is to encourage active learning. Young [54] and Kessel and Wickens [55] showed that operators could detect the change in system control functions better if they had active control of the system instead of passively monitoring the system. Developments in learning theory also shows the importance of active learning. Constructivism [56] states that learners need to build their own knowledge instead of purely receiving it from the instructors. Building of the knowledge happens when learners actively interact with their environment and restructure their previous knowledge. Thus effective training must encourage learners to interact with the system actively.

Learning Styles

Finally, as indicated by Dixon [57], one of the important considerations when designing a training program is to determine the individual learning styles of the learners. Some people may have high spatial abilities and others may have high verbal abilities. In addition to these innate abilities, other factors that lead to different learning styles include personality, cultural background, and learning speed. An effective training program should be able to adapt to each individual's needs and provide appropriate help and directions.

IMPACT OF AUTOMATION

People make errors. Automation has been very successful at reducing these errors although it may just be relocating human error to another level: Errors can still occur in setup in manufacturing and maintenance and in programming. A person is still needed for performing cognitive-based tasks, as another system check, and to provide needed flexibility for unexpected events. A problem with automation is that it moves the human operator further away from the control loop. The issues associated with automating a process control task are considered further in the following subsections.

Vigilance Decrement

As more tasks become automated, the human operator is usually forced to monitor an automated system, and thus the monitoring load for the operator is increased. People are not very good monitors. This was first noticed when researchers studied radar operators during World War II [7] and found that the longer an operator continued at a task the more targets were missed; this became known as the vigilance decrement. With these long and monotonous tasks, the level of alertness decreases [58]. Solutions to this problem are to automate monitoring, provide special training, or provide effective feedback to the operators. Automating the monitoring is not a good solution because someone would then have to monitor the automatic monitoring. Special training is possible [59], and feedback, such as unexpected drills, should always be provided.

Out-of-Loop Familiarity

A second problem with automation in regard to cognitive skills is out-of-the-loop familiarity. As more inner-loop functions are automated, the operator is required to function more at the outer-loop stages. A problem occurs when an inner-loop function fails and the operator is required to jump in the loop and find the failure. Airlines have noticed that pilots have poor transfer from highly automated wide-body planes to smaller planes that require the inner-loop control [60]. Also, because much of the training of an operator is acquired on the job, automation can make learning more difficult. As mentioned above, Kessel and Wickens [55] and Young [54] found that operators who developed an internal model of the system from monitoring the system were not very good at detecting system faults. Operators who developed an internal model from controlling the system were much better at detecting system faults. This implies that skill maintenance of the operators may be required. Having the operators control the process in a simulator may be an important way to develop and maintain cognitive skills that may be at lost because of automation.

Overestimating the Intelligence of the Computers

A third problem with automation, especially with the incorporation of expert systems, is that the operator may think that the system is more intelligent than it actually is. Wickens [60] presents two examples of this occurring in aviation. In Detroit, Michigan, in 1975, a DC10 and an L-1011 were on a collision course; the air traffic controller saw this but did not act because he knew that if a collision became imminent the system would alert him. About this time, the air traffic controller took a break. The new controller saw the collision course and managed to contact the DC10 in time to just barely avert the crash. In another example, a crash of an L-1011 in 1972, the pilot put too much faith in the autopilot. When a fault occurred in the plane, the pilot put the plane on autopilot so he could diagnose the fault. He did not monitor the autopilot, the autopilot did not hold, and the plane crashed. Although many factors could contribute to the accidents, it appears that an important factor in both accidents was on overreliance on automation. The solution to these problems may not be less automation but could instead be to enable the operators to be aware of the limitations of the automation.

This problem also occurs in process control tasks. Jenkins [61] ran a simulation in which an operator along with an expert system controlled a nuclear power plant. He found that the operators assumed that the expert system knew more than it actually did; they thought that the system should have known when a failure in the cooling system occurred. In another computerized task, Rumelhart and Norman [62] found that novices often attribute humanlike characteristics to the computer and cannot understand when the computer makes nonhumanlike mistakes. If incorporating expert systems into process control, the human operator should be thoroughly trained on what the system does and does not know.

Information Overload

As systems become automated and processes are integrated, the information to be perceived and processed by each operator increases. One solution to this problem is a software program that would be capable of transforming the large amount of data into useful, manageable, and easily perceived and understandable formats. Graphical representation is a powerful tool for displaying large amounts of data in a meaningful way. It could enable operators to detect changes in the system status, interpret the recorded measurements, and achieve better retention [63–65].

In addition, a fault detection and analysis system could be used to detect process faults. A fault detection and analysis system can help operators diagnose the possible faults in the processes and thus reduces the operator's workload. Early detection of the process faults could help prevent accidents, improve safety, and increase efficiency and productivity. Shirley [66] and Sassen et al. [67] have provided examples of such fault detection systems.

CONCLUSIONS

Existing models of process control operators were investigated. Quantitative models were useful for modeling the physical responses of operators. GOMS and NGOMSL models, which were formulated for modeling human–computer interaction tasks, are useful for modeling the cognitive activity of operators required to interact with computers in process control systems. Expert systems can be used to capture some of the knowledge required of operators for performing complex and unanticipated tasks.

Methods to extract knowledge from experts were also summarized. This knowledge can be used for training novices to become more skilled or for making the process control system more intelligent by using human intelligence to program expert systems. One conclusion from this analysis is that a human expert operator has a huge database of knowledge that is used to operate a process control plant. Only a small subset of this knowledge can ever be extracted from the operator. Completely automating a process control system, in which the system performs the intelligent functions now performed by the operator, is not yet possible. An operator is still needed, although the role of the operator is changing.

Training of operators in process control is important. With a computer, we can copy an original disk to a new disk so that the new disk has the same information as that of the original disk. We cannot, however, merely copy the expert knowledge from an expert to a novice through knowledge extraction techniques and training programs. We can, on the other hand, study experts to find characteristics of how they store and process information about process control. This information for experts appears to be organized in a spatially based mental model that can be run, similar to running a simulation on a computer. This mental model is developed through active interaction with the system or a simulation of the system. Training programs should concentrate on methods that help develop these accurate mental models.

Finally, process control tasks are becoming more automated. Although automation can remove human error from some aspects of the system, automation provides other opportunities for human errors to occur. These potential problems with automating systems were outlined. Understanding these problems will help in avoiding them through proper training programs or further refinements in the design of process control systems.

REFERENCES

1. Edwards, E., and F. Lees, *The Human Operator in Process Control,* Taylor & Francis, London, 1974.

2. Patrick, D., and S. Fardo, *Industrial Process Control Systems,* Delmar, Albany, New York, 1997.

3. Murrill, P. W., *Fundamentals of Process Control Theory,* 2nd ed., Instrument Society of America, Research Triangle Park, North Carolina, 1991.

4. Polke, M., ed., *Process Control Engineering,* VCH, New York, 1994.

5. Lees, F. P., "Research on the process operator," in *The Human Operator in Process Control,* E. Edwards and F. P. Lees, eds., Halsted, New York, 1974, pp. 386–425.

6. Eberts, R. E., S. Y., Nof, B. Zimolong, and G. Salvendy, " Dynamic process control: cognitive requirements and expert systems," in *Human–Computer Interaction,* G. Salvendy, ed., Elsevier, Amsterdam, 1984, pp. 215–228.

7. Mackworth, N. H., *Q. J. Exp. Psychol.* **1**, 5–61, 1948.

8. Swets, J. A., Ed., *Signal Detection and Recognition by Human Observers: Comtemporary Readings,* Wiley, New York, 1964.

9. Mackie, R. R., *Vigilance: Relationships among Theories, Physiological Correlates, and Operational Performance,* Plenum, New York, 1977.

10. Poulton, E. C., *Tracking Skill and Manual Control,* Academic, New York, 1974.

11. Rouse, W. B., "A model of the human in a cognitive prediction task," *IEEE Trans. Syst. Man Cybern.* **SMC-3**, 473–477. 1973.

12. Rouse, W. B., "Human perception of the statistical properties of discrete time series: effects of interpolation methods," *IEEE Trans. Syst. Man, Cybern.* **SMC-6**, 466–472, 1976.

13. Bainbridge, L., "Analysis of verbal protocols from a process control task," in *The Human Operator in Process Control,* E. Edwards and F. P. Lees, eds., Halsted, New York, 1974, pp. 285–305.

14. Beishon, R. J., "An analysis and simulation of an operator's behavior in controlling continuous baking ovens," in *The Simulation of Human Behavior,* F. Bresson and M. deMontmollen, eds., Durod, Paris.

15. Peterson, C. R., and L. R. Beach, "Man as an intuitive statistician," *Psychol. Bull.* **68**, 29–46. 1967.

16. Rasmussen, J. "Models of mental strategies in process plant diagnosis," in *Human Detection and Diagnosis of System Failure,* J. Rasmussen and W. B. Rouse, eds., Plenum, New York, 1981, pp. 241–258.

17. Asbjonsen, O. A., "Control and operability of process plants," *Compu. Chem. Eng.* **13**, 351–364, 1984.

18. Fisher, D. G., "Process control: an overview and personal perspective," *Can. J. Chem. Eng.* **69**, 5–26. 1991.

19. Card, S. K., T. P. Moran, and A. L. Newell, *The Psychology of Human Computer Interaction,* Erlbaum, Hillsdale, New Jersey, 1983.

20. Kieras, D. E., "Towards a practical GOMS model methodology for user interface design," in *Handbook of Human–Computer Interaction,* M. Helander, ed., Elsevier, Amsterdam, 1988, pp. 67–85.

21. Knaeupper, A., and W. B. Rouse, "A model of human problem solving in dynamic environment," in *Proceedings of the 27th Annual Meeting of Human Factors Society,* Human Factors Society, Santa Monica, California, 1983, pp. 695–703.

22. Kuespert, D. R., and T. J. Mcavoy, "Knowledge extraction in chemical process control," *Chem. Eng. Commun.* **130**, 251–264, 1994.

23. Greenwell, M., *Knowledge Engineering for Expert System,* Ellis Horwood, Chichester, West Sussex, England.

24. Newell, A., and H. Simon, *Human Problem Solving,* Prentice-Hall, Englewood Cliffs, New Jersey, 1972.

25. Ericsson, K. A., and H. A. Simon, "Verbal reports as data," *Psycholog. Rev.* **87**, pp. 215–251, 1980.

26. Boose, J., "A framework for transferring human expertise," in *Human–Computer Interaction*, G. Salvendy, ed., Elsevier, Amsterdam, 1984, pp. 247–254.

27. Kelly, G. A., *The Psychology of Personal Constructs,* Norton, New York, 1955.

28. Kruskal, J. B., and M. Wish, "*Multidimensional Scaling,*" Sage, Beverly Hills, California, 1978.

29. Shepard, A., K. Romney, and S. B. Nerlove, eds., *Multidimensional Scaling: Theory and Applications in the Behavioral Sciences,* Seminar Press, New York, 1972.

30. Cox, T. F., and M. Cox, *Multidimensional Scaling,* Chapman & Hall, London, 1994.

31. Tryon, R. C., and D. E. Bailey, "*Cluster analysis,*" McGraw-Hill, New York, 1970.

32. Everitt, B., *Cluster Analysis,* Halsted, New York, 1993.

33. Joyce, R., G. Gupta, "Identify authentication based on keystroke latencies," *Commun. Assoc. Comput. Mach.* **33**, 68–176, 1990.

34. Gains, R., W. Lisowski, S. Press, and N. Shapiro, "Authentication by keystroke timing: some preliminary results," *Rand Rep. R2 56-NSF*, Rand Corporation, Santa Monica, California.

35. Chakravarthy, S. V., J. Ghosh, "Neural network-based associative memory for storing complex-valued patterns," *Proc. IEEE Int. Conf. Systems, Man Cybernetics* **3** 2213–2218 (1994).

36. Lo, S. B., J. S. Lin, M. T. Freedman, and S. K. Mun, "Application of artificial neural networks to medical image pattern recognition: detection of clustered microcalcifications on mammograms and lung cancer on chest radiographs," *J. VLSI Signal Process.* **18**, 263–274 (1998).

37. Siochi, A. C., and D. Hix, "A study of computer-supported user interface evaluation using maximal repeating pattern analysis," in *Human Factors in Computing Systems Conference Proceedings, CHI'91*, Association for Computing Machinery, New York, 1991, pp. 301–305.

38. Chase, W. G., and H. A. Simon, "Perception in chess," *Cognitive Psychol.* **4**, 55–81 (1973).

39. Brooks, R. E., "Towards a theory of the cognitive processes in computer programming," *Int. J. Man-Machine Stud.* **9**, 737–751, 1977.

40. Cendrowska, J., and M. A. Bramer, "A rational reconstruction of the MYCIN consultation system," *Int. J. Man-Machine Stud.* **20**, 229–317, 1984.

41. Woodworth, R. S., *Experimental Psychology,* Holt, New York, 1938.

42. Veldhuyzen, W., and H. G. Stassen, "The internal model—what does it mean in human control?" in *Monitoring Behavior and Supervisory Control*, T. B. Sheridan and G. Johannsen, eds., Plenum, New York, 1976, pp. 157–171.

43. Kragt, H., and J. A. Landeweerd, "Mental skills in process control," in *The Human Operator in Process Control*, E. Edwards and F. P. Lees, eds., Halsted, New York, 1974, pp. 135–145.

44. Hollan, J. D., A. Stevens, and N. Williams, "STEAMER: an advanced computer assisted instruction system for propulsion engineering," paper presented at the Summer Simulation Conference, Seattle, Washington, 1980.

45. Landeweerd, J. A., "Internal representation of a process fault diagnosis and fault correction," *J. Appl. Ergonom.* **22**, 1343–1351, 1979.

46. Wickens, C. D., and A. Weingartner, "Process control monitoring: the effects of spatial and verbal ability and current task demands," in *Trends in Ergonomics/Human Factors II*, R. E. Eberts and C. G. Eberts, eds., North-Holland, Amsterdam, 1985, pp. 25–32.

47. De Kleer, J., and J. S. Brown, "Mental models of physical mechanisms and their acquisition," in *Cognitive Skills and Their Acquisition,* J. R. Anderson ed., Erlbaum, Hillsdale, New Jersey, 1981, pp. 285–305.

48. Meadows, K., "The making of a training program," *Intech* **37**, 172–173, 1990.

49. Younkin, C. S., and F. B. Johnson, "Operator training: the key to successful computer control," *J. Water Pollut. Control Federation* **58**, 944–948, 1986.

50. Eberts, R. E., and W. Schneider, "Internalizing the system dynamics for a second order system," *Hum. Factors* **27**, 371–393, 1985.

51. Stevens, A., B. Roberts, L. Stead, "The use of sophisticated graphics interface in computer-assisted instruction," *IEEE Comput. Graphics Appl.* **3**(2), 25–31, 1983.

52. Gentner, D., "The mechanism of analogical learning," in *Similarity and Analogical Reasoning*, S. Vosniadou and A. Ortony, eds., Cambridge U. Press, New York.

53. Mayer, R. E., and B. K. Bromage, "Different recall protocols for technical texts due to advance organizers," *J. Education. Psychol.* **72**, 209–225, 1980.

54. Young, L. R., "On adaptive manual control," *IEEE Trans. Man-Mach. Syst.* **MMS-10**, 229–331, 1969.

55. Kessel, C. J., and C. D. Wickens, "The transfer of failure-detection skills between monitoring and controlling dynamic systems," *Hum. Factors* **24**, 49–60, 1982.

56. Blais, D. M., "Constructivism: a theoretical revolution in teaching," *J. Developmental Education* **11**, 2–7, 1988.

57. Dixon, N., "Incorporating learning style into training," *Training Development J.* **36**, 62–64, (1982).

58. Moore-Ede, M., S. Campbell, and T. Baker, "Effect of reduced operator alertness on the night shift on process control operator performance," *Adv. Instrumen. Proc.* **44**, 967–970, 1989.

59. Fisk, A. D., and W. Schneider, "Control and automatic processing during tasks requiring sustained attention: a new approach to vigilance," *Hum. Factors* **23**, 737–758, 1981.

60. Wickens, C. D., *Engineering Psychology and Human Performance,* Merrill, Columbus, Ohio, 1984.

61. Jenkins, J. P., "An application of an expert system to problem solving in process control displays," in *Human–Computer Interaction*, G. Salvendy, ed., Elsevier, Amsterdam, 1984, pp. 255–260.

62. Rumelhart, D. E., and D. A. Norman, "Analogical processes in learning," in *Cognitive Skills and Their Acquisition*, J. R. Anderson, ed., Erlbaum, Hillsdale, New Jersey, 1981.

63. Robertson, P. K., "A methodology for scientific data visualization: choosing representations based on a natural scene paradigm," in *Proceedings of Visualization '90*, IEEE Computer Society, Los Alamitos, California, 1990, pp. 114–123.

64. Schutz, H. G., "An evaluation of formats for graphic trend display—experiment II," *Hum. Factors* **3**, 99–107, 1961.

65. Standing, L., J. Conezio, and R. N. Harber, "Perception and memory for pictures: single trial learning of 2560 visual stimuli," *Psychonom. Sci.* **19**, 73–74, 1970.

66. Shirley, B. S., "An expert system for process control," *Tappi J.* **70**, 53–56. 1972.

67. Sassen, J. M., E. Buiel, and J. H. Hoegee, "A laboratory evaluation of a human operator support system," *Int. J. Hum. Comput. Stud.* **40**, 895–931. 1994.

DISTRIBUTED DISPLAY ARCHITECTURE

Since the early 1970s instrument displays have been undergoing numerous changes that reflect and incorporate the marked progress that has occurred throughout instrumentation and control technology. These would include development of the early minicomputers and the later introduction of the personal computer (PC); the advent of the microprocessor; advances in cathode-ray tube (CRT) technology and other display methodologies, such as liquid-crystal displays (LCDs), vacuum fluorescent displays (VFDs), and plasma (gas discharge); the decentralization of control centers by way of distributed control; the development of advanced networking concepts; a better understanding of human factors as they relate to the operation of machines and processes; and the engineering of interactive graphics among numerous other advancements that have seriously affected display practices, the general result of which is a severely altered interface. The contrast is dramatically apparent by comparing Figs. 1 and 2.

Contemporary systems are quite software-intensive. Installation and maintenance require different skills than in the case of prior-generation systems. Fewer electricians and hardware technicians are needed, whereas more expertise is required in the system engineering and software areas.

THE GRAPHICS INTERFACE

"Pictures" help immeasurably to improve an operator's comprehension of what is going on in a controlled process or machine. This, of course, is wisdom known since antiquity. It was first applied in industrial instrumentation in the 1940s, when elaborate process or manufacturing diagrams were put on permanent panels, with illuminated indicators placed directly on the diagrams, and where attempts were made to locate single-loop controllers to reflect the relative geometry of where the measurement sensors were located in the process. Such panels became plant "showpieces" and often

FIGURE 1   Long panelboard containing indicators and controllers used in large processing plant. Generally this type of panel was used in the early 1970s. The very beginnings of the cathode-ray tube (CRT) displays for annunciators are indicated by the CRTs in the upper right-hand portion of the view.

were constructed of costly enameled steel. Unfortunately changes were inevitably made in the process or instrumentation, and the beautiful initial panel soon was marked over with tape. A cluttered appearance progressively emerged, which hardly could be considered an inspiration to the operator's confidence in the system. Use of a CRT as the panel face essentially solved this problem of panel obsolescence. Later embellished through the use of color, it offered immense flexibility, not only at one location, but at several locations along a network. Currently these properties are being exploited further through the use of the X-window and X-network concepts. But the CRT has had some limitations too, notably the comparatively small space in which an entire control panel must be projected. Even though through menu techniques a process may be shown in detailed scenes, some loss of comprehension occurs, brought about by crowding and by the complexities introduced when expanding a scene into subscenes ad infinitum. Larger basic presentation techniques (such as flat screens) are constantly being refined.

Graphic displays are formed by placing a variety of symbolic forms, originally created by an artist and later reproduced electronically, on the display surface where they can be viewed by the user. These may enter the system by way of certain display standard protocols, such as GKS/PHIGS (graphical kernel system/programmer's hierarchical interactive graphics system), or otherwise software created. In some systems the objects are formed by using the basic output primitives, such as text, lines, markers, filled areas, and cell arrays.

The display software must provide additional higher-level graphic objects formed with the basic primitives. For graphic displays on process systems, the fundamental types of objects available should

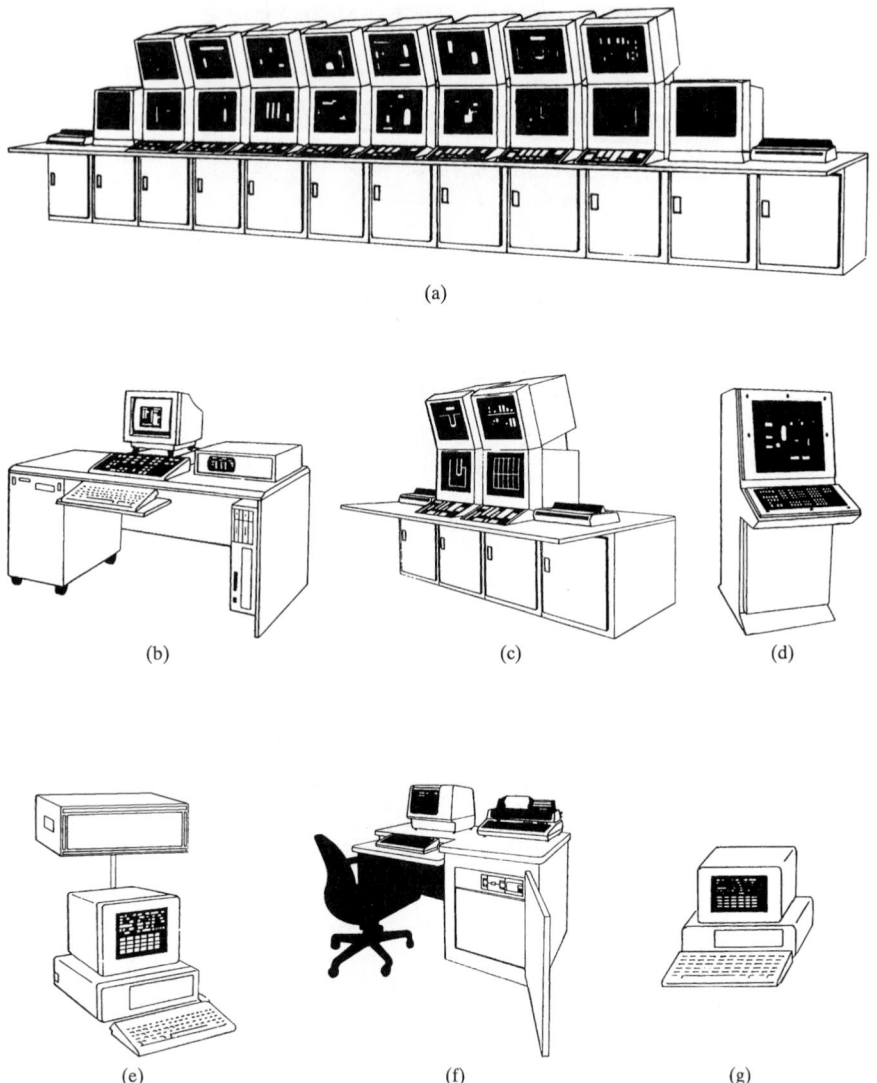

FIGURE 2 Examples of how the flexibility of CRT-based panels contributes to various interface configurations and distributed display architecture. (*a*) Central configurable CRT station with distributed historian. (*b*) Engineer's workstation. (*c*) Area CRT station. (*d*) Field-hardened console. (*e*) Independent computer interface (serial-interface–personal-computer). (*f*) Batch management station. (*g*) General-purpose computer.

include background text, bar graphs, trends, pipes, and symbols. It is desirable to allow each of the fundamental types to be freely scaled and positioned. It is also desirable for the trends to be displayable as either point plots or histograms, and for both text and trend fields to be scrollable together if there are more data available than can be reasonably displayed in the space allocated for the object. The system should provide libraries of symbols and facility for the user to edit.

Well-designed graphics displays can show large amounts of dynamic information against a static background without overwhelming the user. The objects used to form the static portion of the display may be as simple as textual headings for a table of dynamic textual data, or a graphic representation of a process flow diagram for a section of a process, with dynamic information positioned corresponding to the location of field sensors. In a GKS system the static portion of a display usually is stored in a device-independent form in a metafile; in PHIGS, an archive file may be used. From there it can be readily displayed on any graphics device in the system, including various terminals and printers.

Display software also can provide the ability to define "pickable" objects and a specific action to take if an object is picked. Picking an object may call up a different display, show some additional information on the same display, or initiate an interaction sequence with the user.

The graphic display software should provide the ability to update dynamic data objects at specifiable time intervals, upon significant changes, or on demand. The update specification should be on a per object rather than a per display basis. The updating of each object should involve the execution of arithmetic expressions with functions available to access the database of the process information system. Besides changing the values of the data to be represented, the updating could result in changes to the attributes of the primitives used to form the objects.

For some users, especially process operators, color can add an important dimension to graphic displays, expanding the amount of information communicated. Color graphics are used widely today. (One must make certain that there are no color-blind operators.) Making hard copies of color displays has improved quite a bit during recent years.

USER INTERACTIONS

The most important aspect of any interface is the means by which the operator can interact with the system. This is the mechanism that the user may employ when accessing, manipulating, and entering information in the system. It is the key point in any interface design and in specifying requirements prior to procurement.

It has been established that best acceptance occurs when an operator becomes confident that there is no harm in using an interactive graphics system. To achieve this, software should be provided and used that makes certain that all interactions are syntactically consistent and that all data entered will be checked for validity. The system should reject invalid inputs and provide a message that the user can readily understand without external reference material.

The user interface must be responsive. The goal should be that all system responses to user actions occur faster than it is physically possible for the operator to perform a subsequent action. Some situations require more responsive interfaces than others. In general, the number of interactions and the complexity of each interaction should be minimized, even though this may require additional sophisticated hardware. Most interactions with a process computer system can be performed in an easier fashion with an interactive graphics system than with a textual dialogue system. However, even with interactive graphics, some textual interaction sometimes may be necessary.

"CONVERSATIONAL" INTERACTIONS

Historically systems have interacted by having the user carry out a textual dialogue in a dedicated portion of the display screen, sometimes called a dialogue area. The flow of the dialogue is similar to a vocal conversation—with the system asking for information with a prompt that the user reads in the dialogue area, and the user providing answers by entering alphanumeric text strings or pressing dedicated function keys on a keyboard. The system provides feedback to each user keystroke by echoing the user's input to an appropriate position in the dialogue area, importantly of a different color than the system's prompt. This is a natural form of interaction for the operator. The tools used to carry out such interactions are easy to construct.

The person who has programmed the system and the user should fully understand the dialogue language that is being exchanged. This can be troublesome because natural language communication can be ambiguous, considering the multiple meanings of some words that arise from the level of education and sometimes the geographic background of the user. Programmers sometimes have difficulty in creating meaningful prompts that are short and easily accommodated in the dialogue area. Live voice systems also can be effective, but have not been accepted on a wide scale.

In process control systems the dialogue area frequently appears at the bottom of the display surface, allowing the user to view the displays while the interaction is progressing. In this way the user will not, in error, attempt to use old data left on the screen during an extended or interrupted interaction. If display updating is halted during interactions, the system should either obscure the data by performing the interaction in a dialogue box in the center of the screen, or in some way make it obvious that the displayed data are not reliable.

Dialogues should be designed to minimize the number of questions required and the length of user inputs needed. Prompts should be simple and easy to understand. They should list menus of possible alternatives when appropriate. Whenever possible, default values should be provided to facilitate the rapid execution of functions.

DATA INPUTS

An interactive graphic user's interface may consist of a bit-map graphic display device (usually with a dedicated processor), a keyboard, and one or more graphics input devices. Hence the electronic interface is essentially the same as the traditional interface, except that the video display unit has some additional capabilities and a graphics input device has been added. Normally, interactive graphics capabilities will be in addition to those of the conventional panel.

Graphics input devices are a combination of hardware and software that allows the positioning of a graphic pointer or cursor to a desired location on the screen and "triggering" an event to perform some action based on the selected pointer position. The event normally is triggered by pressing or releasing a button. Examples of physical input devices that have been used include the light pen, joystick, track ball, mouse, graphics tablet, touchpad, touchscreen, and keyboards. Once a device is selected, one should concentrate on how it can be used effectively. Among the most commonly used are the keyboard, a mouse, or a touchscreen.

Keyboard. A keyboard should be available with an alphanumeric section for entering text and an array of function buttons for certain specialized functions. In contrast to nongraphic interactions, the alphanumeric section will be used infrequently. On process systems, the function button section of the keyboard is important even in a graphic environment, because buttons can provide immediate generation of commands or random access to important information. Sometimes there is a need for keyboards with in excess of 100 function buttons on a single workstation. Also, there may be a need for lamps associated with keys that can be used to prompt the user. Keys should provide some kind of feedback when actuated. Tactile feedback is desirable, but not necessary. Many operators are prone to favor tactile feedback. Visual feedback from every keystroke is essential.

Touchscreen. This is probably the simplest of the graphic input devices to use because it requires no associated button for triggering events. This can be done by sensing a change in the presence of a finger or some other object. A touchscreen occupies no additional desk or console space and appears to the operator as an integral part of the screen. The touchscreen is useful to naive operators for selecting one of several widely spaced items that are displayed on the screen. However, because of calibration problems, the curvature of the CRT screen, and its coarse resolution (that is, between screen and human finger), implementations using touchscreens must be considered carefully in advance—probably involving prior experience to get the "feel" of the system. From a human factors standpoint, it is important that the system be designed to trigger the event upon the user's removal of a finger from the screen rather than upon the initial contact of the finger. This allows the user to correct the initial (frequently inaccurate) position of the cursor by moving the finger until the graphic pointer is properly positioned over the designated object on the screen, at which point the finger can be removed, thus triggering the selection of the object.

Touchscreens operate on several principles, but most require an overlay on the monitor. Touchscreen technologies include the following.

Resistive.   Two conductive sheets separated by an array of clear, tiny elastic dots. The first substrate layer may be glass or plastic. This has a conductive coating (such as indium or tin oxide) that possesses uniform and long-term stability. The second layer (cover sheet) may be of polycarbonate or oriented polyester (Mylar) plastic, which has a conductive inner coating on the surface facing the substrate conductive coating. A voltage measurement indicates where the circuit is completed. Resistive screens indicate only a single point to the computer at the center of the area of contact. In discrete resistive touchscreens, one of the layers includes equally spaced horizontal rows, which are etched into the surface; the others are in the form of vertical lines (columns). Thus the two sheets form a grid. Discrete screens usually are application-specific. Analog versions allow total software control for the touch zones. Characteristics of resistive touchscreens are 4000 × 4000 resolution, low cost, very rapid response, and no limitation on stylus used.

Capacitive.   These operate on the principle that high-frequency alternating current couples better between conductors than direct current. The screen system is controlled by a radio-frequency source (approximately 10 MHz), the outputs of which go to each corner of the screen. When the screen is touched, current passes from the screen through the operator's finger. The amount of current emanating from each corner of the screen relates to the location of the touch. Characteristics of capacitive touchscreens are good resolution (1000 × 1000 points per screen side), fairly high cost, very rapid response, and some limitations on stylus used. The screen must be shielded from internal electronics noise and external electromagnetic interference. Very good optical properties.

Surface Acoustic Wave.   This is a relatively new technology. When surface acoustic waves (SAWs) propagate across a rigid material, such as glass, they are efficient conductors of acoustic (sound) energy—at precise speeds and in straight lines. Special transducer-generated SAW signals travel along the outside edges of the screen, where they encounter reflective arrays on the screen. Each array reflects a small portion of the wave. Through complex logic circuitry, the SAWs determine the coordinates of any pliable device (including a finger) that may press the screen surface to form a transient detent. Some drawbacks solved by this technology include a combination of transparency, ruggedness, and ease of installation. Characteristics of SAW screens are a modest resolution (15–100 points per inch), high cost, and very rapid response.

Piezoelectric.   Pressure-sensitive electronics detect and determine touch location. A touch on the screen causes the screen glass to flex slightly, applying pressure to each of the stress transducers. This pressure is inversely proportional to the distance from the touch point. Each transducer generates a signal proportional to the amount of pressure applied. Signals are processed into X and Y coordinates for transmission. Sometimes vibrations in the environment may adversely trigger one or more transducers. Resolution is limited (80 to 90 touch points); response is very rapid.

Infrared.   The principle used is interruption of a beam of infrared (IR) light. The system consists of a series of IR light-emitting diodes (LEDs) and phototransistors arranged in an optomatrix frame. The frame is attached to the front of the display. A special bezel is used to block out ambient light. An IR controller scans the matrix, turning on LEDs and the corresponding phototransistors. This is done about 25 times per second. Characteristics of IR systems include reasonable cost, any type of stylus may be used, and response is very rapid.

Mouse and Track-Ball Input Devices.   These devices are used to drive a cursor on the graphics screen. To move a cursor, the mouse is moved across a hard surface, with its direction and speed of movement proportionally related to the cursor's movement on the display screen. A track ball operates on the same principle, but the movement of a ball mounted on bearings within the track ball positions the cursor. It can be thought of as an upside-down mouse. Both the mouse and the track ball have one or more buttons that are used for control. One button is used to indicate when the cursor has reached the proper location. One button is used to activate or deactivate the device. The screen cursor can be used to make menu selections from the display or to place graphic elements.

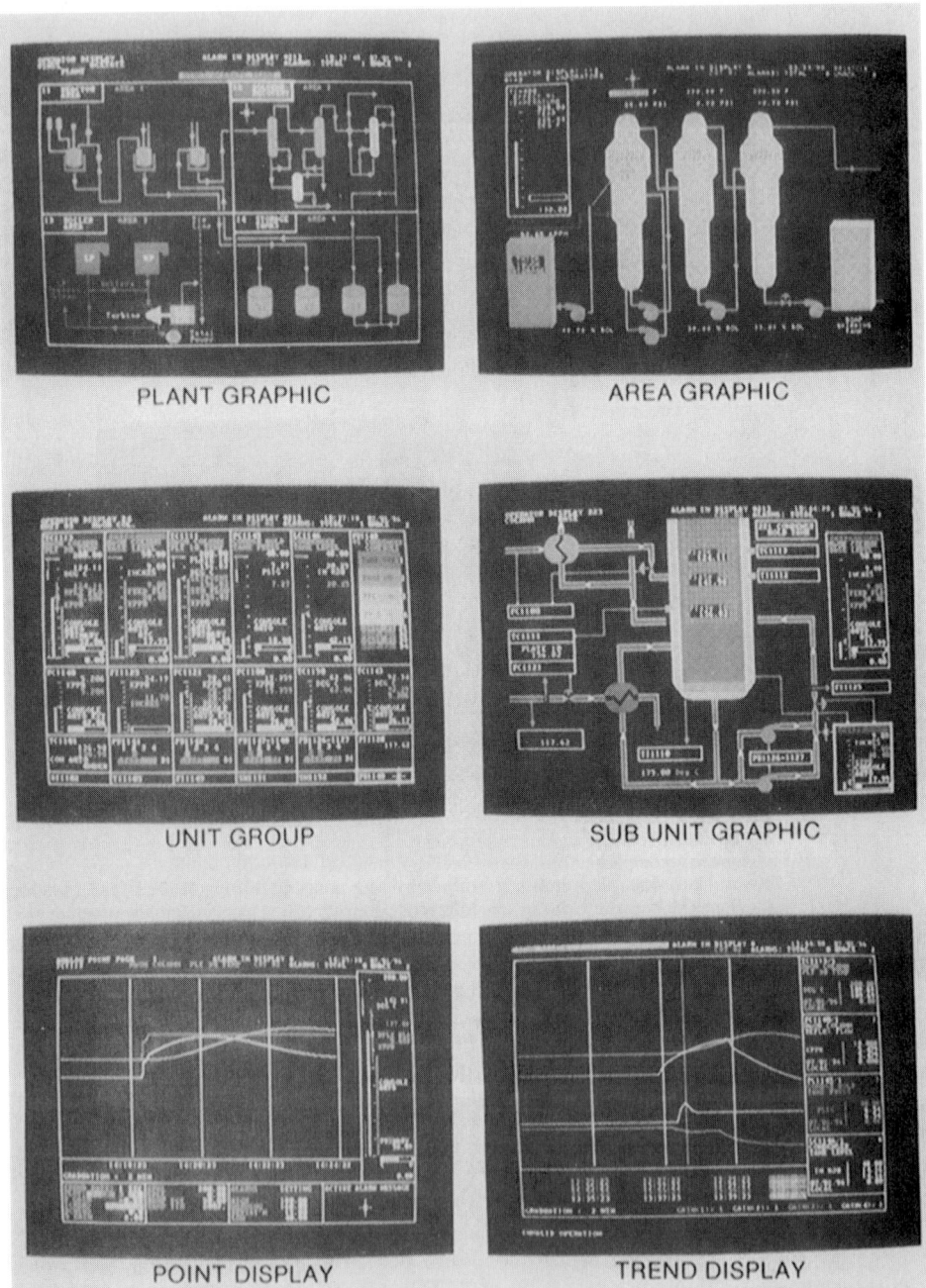

FIGURE 3   Depending on the operator's needs and preferences, the vailable area on a CRT screen can be used in total for a given "scene," or, more commonly, the area will be divided into windows. These arrangements are accomplished through the software program. Via menu-driven techniques the entire operation can be brought into view (plant graphic), followed by close-ups of areas, units, and subunits. Further, point and trend displays may be brought to the screen. Interactive graphics can be planned into the program wherever desired.

UNIT GRAPHIC

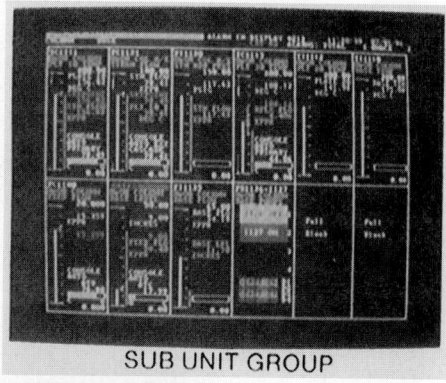

SUB UNIT GROUP

FREE FORM GROUP/GRAPHIC

FIGURE 3   (*continued*) Depending on the operator's needs and preferences, the available area on a CRT screen can be used in total for a given "scene," or, more commonly, the area will be divided into windows. These arrangements are accomplished through the software program. Via menu-driven techniques the entire operation can be brought into view (plant graphic), followed by close-ups of areas, units, and subunits. Further, point and trend displays may be brought to the screen. Interactive graphics can be planned into the program wherever desired.

Digitizer.   This has been a common input device, especially for computer-aided design (CAD) systems. Three main components of a digitizing tablet are pad, tablet, or surface; the positioning device that moves over the surface; and electronic circuitry for converting points into x–y coordinates and transmitting them to the computer. The device can be used to place a graphic element on the screen, make a menu selection from the screen, or make a menu or symbol selection from the digitizing tablet itself.

Joystick.   Identical in function to the joystick supplied with home computer games, these devices are sometimes preferred when working with three-dimensional drawings because of their free range of motion. The joystick works on the same general principles as the mouse and track ball.

Light Pen.   One of the very early means for interacting a narrow beam of light with photoreceptors behind the screen, it is similar in principle to some of the other systems described.

VISUAL DISPLAY DEVICES

Cathode-Ray Tube.   The operating principles of this ubiquitous display device were well developed and the device was thoroughly seasoned by the time of its first use in industrial instrumentation and

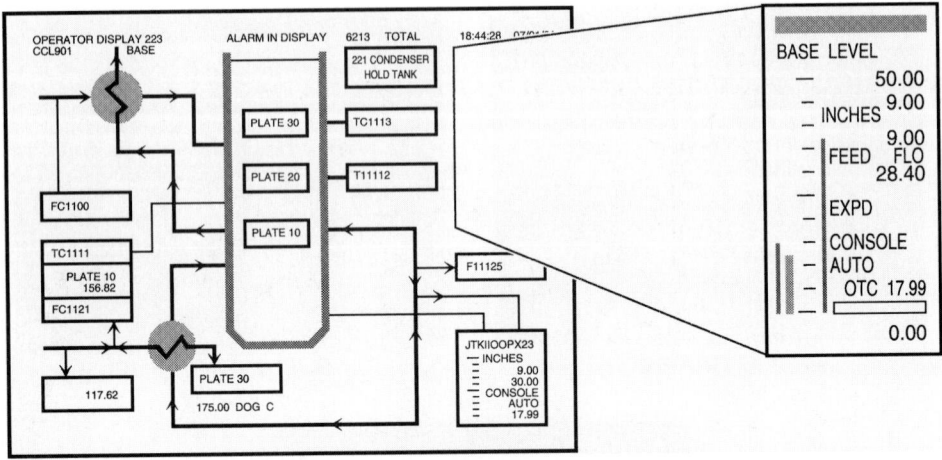

FIGURE 4   By way of software, a portion of a CRT display, namely, a window, can be singled out for detailed viewing. In the X-Window scheme, one or more windows could be designated as "dedicated" for use on an X-Network, making it possible for several stations along the network to obtain instant information that may be of interest to several operators in a manufacturing or processing network.

control systems. Of course, for many years it has been used in the form of the oscilloscope, where it enjoys a leading role in test instrumentation. In terms of process control and factory applications, it made its first appearance in CATV (closed-circuit television) applications, where it was used as a "seeing eye" for monitoring remote operations. Then it was later used in some of the earliest attempts to achieve several basic advantages of computer-integrated manufacturing.

The immense flexibility of the CRT is demonstrated by Figs. 3 and 4. Also as evident from other fields of CRT usage, such as animation and CAD applications, the CRT seems almost unlimited in its potential when in the hands of ingenious programmers and software engineers. See the articles, "Distributed Control Systems" in Section 3 and "Industrial Control Networks" in Section 7 of this handbook.

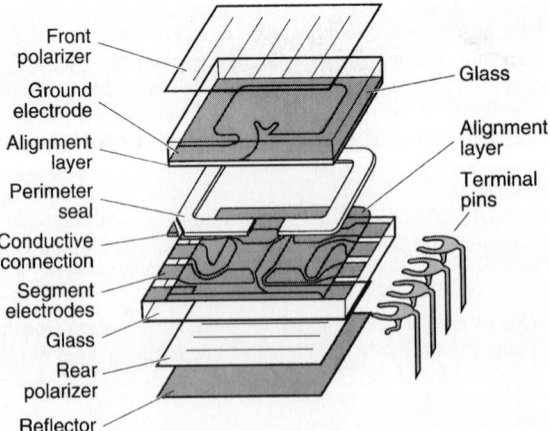

FIGURE 5   Cross-sectional view of typical TN (twisted nematic) liquid-crystal display with transparent indium—tin oxide electrodes patterned on glass substrates using photolithography techniques. (*Standish Industries, Inc.*)

Liquid-Crystal Displays.   Introduced several years ago, liquid-crystal displays (LCDs) are achieving a measurable degree of design maturity, particularly after the addition of numerous colors, as indicated by the many thousands of units that are installed and ordered each year. LCDs are nonemissive and therefore require good ambient light at the proper angle for good viewing. The physical chemistry of LCDs is a bit too complex for inclusion here. However, a cross-sectional view of a typical LCD unit is shown in Fig. 5.

Electroluminescent Panels.   These displays use a thin film of solids that emit light with an applied electric field. Because they are monochromatic, small, and expensive, they usually are applied when applications require a highly readable, lightweight, low-power-consuming visual display unit with very rapid update times. Design research in this field is continuing.

Vacuum Fluorescent Panels.   These display units create thin images by accelerating electrons in a glass envelope to strike a phosphor-coated surface. They are bright, available in multiple colors, and have fast update times. Usually they are small in size and, therefore, find applications that require low-information displays, such as reprogrammable control panels or video-control stations, where graphics are simple and space is a premium consideration.

KNOWLEDGE-BASED OPERATOR TRAINING

G. K. McMillan*

Operator training systems are the key to successful implementation of new systems and the safe and efficient operation of existing control systems. Often the training system also serves the essential purpose of testing the control system design and configuration. The need for these operator training systems has dramatically increased because of the following:

1. Early retirement of experienced operators
2. Reduction in the number of operators
3. Ever-changing proprietary computer interfaces
4. Increased information displayed
5. Barrages of spurious alarms
6. Increased economic and environmental pressures
7. Aged and stressed process equipment
8. The need to push processes to their constraints
9. Increased product grades and decreased inventory
10. Increased sophistication and complexity of control systems
11. Loss of operator involvement from advanced automation

* Senior Fellow, Solutia Inc., St. Louis, Missouri.

For new plants, there is obviously a need to develop operator expertise in the process. What is less apparent is the great need to do the same in existing plants because of the loss of the experience base from years of corporate downsizing. This has affected both the operators and the engineers who provide technical and production support. Also, many projects to replace panel board instruments with distributed control systems (DCSs) or programmable logic controllers (PLCs) were justified based on head-count reduction. In the meantime, there was a constant evolution of operator interfaces for the DCS and the PLC. The differences between manufactures and even within a given manufacturer from generation to generation are significant. Finally, there is a new movement toward a Windows environment. While the Instrument Society of America standard for Fieldbus promises to create a common control language, there is no similar effort to establish a common interface.

The DCS has also increased the number of alarms by an order of magnitude over panel board instruments since a DCS typically comes with three free alarms whereas a panel board alarm required an annunciator window to be engineered, purchased, and installed. Unfortunately, most of these new DCS alarms are spurious. They pose a serious distraction and make the operator insensitive to real alarms. This, coupled with the dramatic increase in information from the additional process variables and diagnostics from smart transmitters, can cause information overload.

At the same time there is a tremendous increase in economic pressures from worldwide competition and the need to constantly increase profits and the return on equity to meet stockholder expectations and an increase in environmental regulations and voluntary initiatives to reduce emissions. Spills or inefficient operation that results in a loss in yield and increase in waste treatment are intolerable and make the difference between a plant's running or being sold or leveled. The need for restraint in capital spending means that many plants cannot replace existing equipment even though production targets have increased. Plants are typically running from 20% to 200% more than name-plate capacities with equipment that is 20 or more years old. Operating conditions are pushed closer to constraints (equipment or process limits on operating temperatures, pressures, levels, and flows) so that production rate can be increased without a loss in yield, safety, or on-stream time. The need to increase the product spectrum and coincidentally reduce product inventory leads to an increase in product changeovers that places additional stress on equipment and operations to maintain quality, yield, and capacity.

This combination of fewer and less-experienced engineers and operators with older equipment and higher production requirements is a recipe for a performance that does not meet expectations, which shows up in a loss of on-stream time, production rate, and yield. It is particularly noticeable in the time it takes to resolve a customer complaint, solve a production problem, or restart after a shutdown.

Advanced control systems that ride but do not violate constraints and provide automatic startups, changeover, and shutdowns can force a plant to meet production targets. However, this is a double-edged sword because the operator becomes too dependent on the control system. The more successful the implementation, the fewer the chances for the operator to develop and exercise his or her skills. The result is operator disengagement and eventually an inability for the operator to take over and run the process when the control system can no longer do its job because of changes, deterioration, or failures in the process, equipment, control valves, measurements, or controllers [1].

The following incremental improvements to operator training systems to help meet this increasing need are listed in order of increasing complexity.

1. Tiebacks
2. Ramps, filters, gains, dead times, and dead bands.
3. Liquid-material balances
4. Energy balances
5. Phase equilibrium
6. Vapor-material balances
7. Component balances
8. Charge balances
9. Reaction kinetics
10. Mass transfer

11. Polymerization kinetics
12. Electrolyte effects

The operator training system should have the following features. It should

1. Use the actual operator interface
2. Require no modification of the control system configuration
3. Test the complete plant
4. Test the complete control system
5. Ensure enough fidelity to show interactions and actual operating conditions
6. Ensure enough fidelity to provide process and control system knowledge
7. Show true process conditions for startups, changeovers, failures, and shutdowns
8. Automatically be updated for changes in process and equipment and the control systems
9. Provide faster than real time test runs of both the plant and the control system
10. Offer a freeze and bumpless restart of the plant model
11. Be capable of being quickly developed, tested, and demonstrated on a laptop by most process and process control engineers as opposed to modeling specialists
12. Become the source and depository of knowledge on the process and the control system

The simplest form of model for training is the tieback, in which the DCS or the PLC inputs are tied to the outputs by a simple relationship. For discrete control devices such as pumps, agitators, fans, and on–off valves, this tieback provides the motor run contacts and valve limit switches feedback inputs in the expected pattern based on the discrete outputs after a specified transition time. For control loops, the feedback measurement is the loop output multiplied by a gain. In some models, filters, dead times, and dead bands are added to provide more realistic dynamics. These parameters can be adjusted so that the loop responds in a fashion that is reasonable to the operator and can use loop tuning settings that are in the ball park. However, dynamics is not typically the forte of operators. While you might be able to satisfy the operators by an iterative correction of the response, there is very little assurance that the resulting dynamics is realistic. Tieback models reenforce these preset notions on the process and control system behavior. The main purpose of these models is to familiarize the operator with the DCS or the PLC interface. They also are simple enough that the whole plant can be simulated, and if the connections are between the model and the controller inputs and outputs, the whole configuration can be tested. While ideally the logic for the tieback model should be done independently of that of the configuration, the time savings in automatically generating these tiebacks from the configuration generally outweigh the loss of separation of the test and control system software. For critical loops and safety interlocks, the test software should be constructed independently by a different person from the one doing the configuration of the control or safety interlock system. Thus automatically generated tiebacks should be relegated to discrete device control and simple loops for which there are no safety, performance, or operational issues.

The use of ramp rates tuned to simulate levels is absurd in that they require more effort than a simple liquid-material balance and rarely provide a good representation of actual level response. Thus liquid and solid levels should be based on inventory or accumulation, which is the integral of the difference in flow into and out of a volume (mass balance). Gas pressure can be computed by the ideal gas law from the density of the vapor by extension of the mass balance to the vapor space. It may involve vaporization and condensation rates based on heat transfer and boiling points based on compositions. This necessitates the addition of energy and component balances. More of an issue than the additional complexity of the process model is the fast dynamics that makes it necessary to use special stiff integration methods, small step sizes, and/or larger than actual volumes [2].

The simulation of pH can be done from a charge balance based on dissociation constants and concentrations of acids and bases. For solutions that are not dilute, the changes in activity coefficients based on electrolyte effects should be included. Interval halving is used to find the pH where the summation of positive and negative charges is nearly zero. This search routine converges rapidly. Alternatively, a polynomial fit to a titration curve can approximate pH from the ratio of reagent added to the vessel or stream. However, this does not take into account the shift of the titration curve from the change of the dissociation constants with temperature or the change in composition of the influent.

Reactors require kinetic expressions to determine the rate of component consumption and formation. Often an activation energy and preexponential factor are sufficient to show the dependence of the reaction rate on concentration and temperature. For vapor-phase reactions, pressure is also important. For polymerization, viscosity effects and chain initiation, building, and termination must be included. For copolymers and triploymers, cross linking requires complexity beyond the capability of most models.

The additional fidelity from the inclusion of energy and component balances, phase equilibrium, and kinetics is necessary to provide the proper process gains, dynamics, and interactions necessary to convey process knowledge and test the control system performance. It extends the utility of the training system from just training operators to training engineers and technicians for operations, maintenance, and projects. It also opens the door to using the model as an experimental tool for troubleshooting, debottlenecking, and new product and process research and development.

The model for training operators and engineers and testing the control system must not be just a process model. It needs to be a plant model. It must include models of instrumentation and control valves (everything outside of the control system). Lags and delays should be added as needed to simulate the dynamics of sensors and actuators. It is important to simulate the lags of thermowells and electrodes and the stroking time of control or on–off valves greater than 4 inches since these are relatively slow. However, most of the other lags associated with sensors and actuators are negligible. More often, the dead time from the sensitivity and/or resolution limits of sensors, digital components, mechanical components, and control valves, the dead time from the cycle time of analyzers, and the dead time from transportation delays of temperature and composition changes in pipelines are the limiting factors to control system and thus plant performance. The resolution limit of control valves and mechanical components can be simply simulated by a dead band (backlash) and the resolution limit of sensors and digital components by a quantizer. The dead time from cycle times can be simulated by a various types of sample and holds. The input (read) and the output (write) are done at the beginning of the cycle for digital devices, whereas the input is done at the beginning and the output is done at the end of the cycle for analyzers. The dead time of transportation delays for pipelines and ducts is the volume divided by the flow rate. The dead time of transportation delays for spin, conveyor, and sheet lines is the length divided by the speed.

The effect of noise is also significant and can be more of a problem than dead time. All measurements have noise (even temperature sensors); it is just a matter of how much and how slow. The control system can distinguish a load upset only if it is larger than the noise amplitude. If the noise frequency approaches the control system natural frequency, the noise is amplified. Also, if the control valve moves as the result of noise, the control system inflicts a disturbance on itself.

If the plant model has a friendly graphical user interface (GUI) and high-level functions and offers presentations suitable for design and documentation, as shown in Fig. 1, it can become the source and the depository of plant knowledge that is readily maintained and widely used. The same principles are true for the control system. The common control language established by the Fieldbus standard (see section 11) and used in field-based systems (see section 3) offers this capability. If these plant models and configurations can be interfaced and combined in a personal computer, the plant and control system knowledge is centered in the training system and the paradigm for operator training is shattered.

The major architectures used in computer-based operator training systems involve major distinctions as to modification, location, and interconnection of the major components (plant model, control system, trainer interface, and operator interface). Often these reside in different personal computers to allow the trainer, model support person, and trainees to have simultaneous and separate access to the training system, as shown in Fig. 2. When the operator interface uses special computers and screens and the control system can reside in only a proprietary microprocessor-based controller, the connection between the controller inputs and outputs and the model is done by special communication packages and serial interface cards. When the control system and/or interface can be emulated in

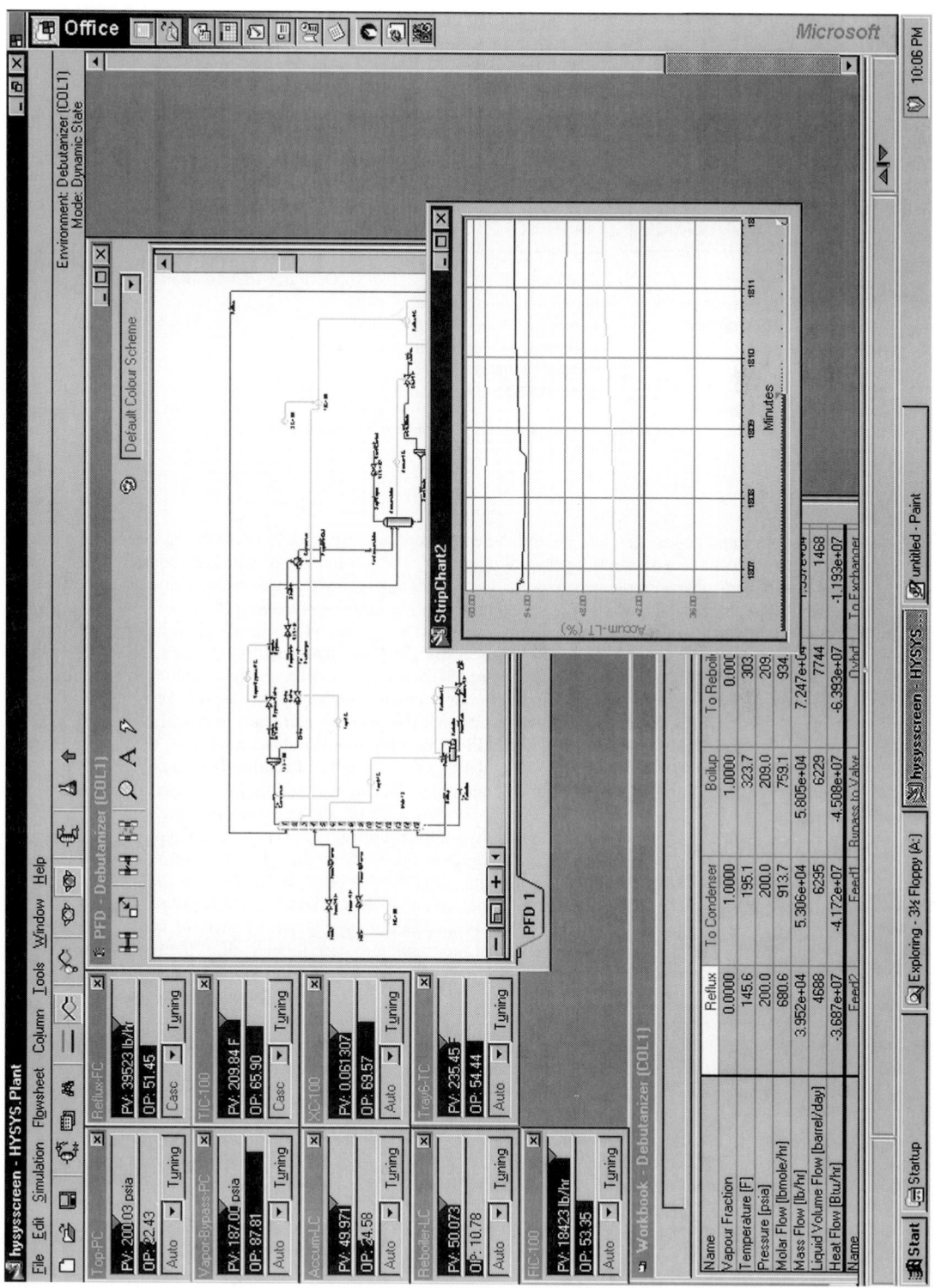

FIGURE 1 An operator training model that serves as the source and depository of plant knowledge. (© Hyprotech Ltd. 1998.)

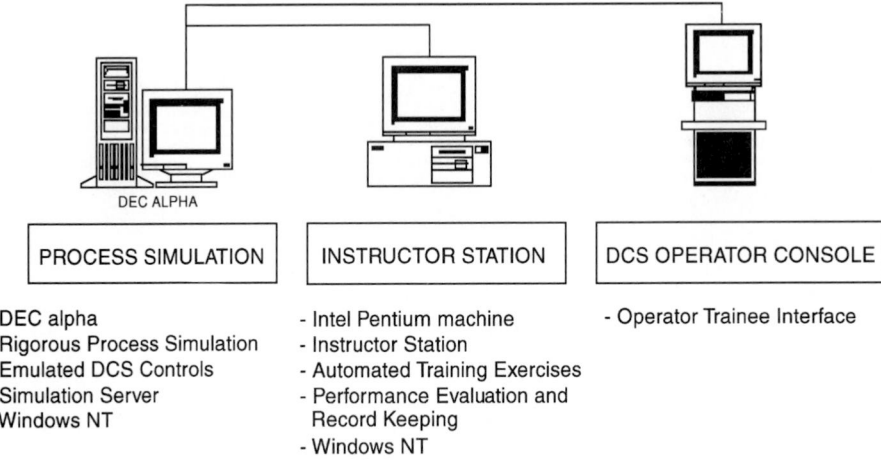

PROCESS SIMULATION	INSTRUCTOR STATION	DCS OPERATOR CONSOLE

- DEC alpha
- Rigorous Process Simulation
- Emulated DCS Controls
- Simulation Server
- Windows NT

- Intel Pentium machine
- Instructor Station
- Automated Training Exercises
- Performance Evaluation and
  Record Keeping
- Windows NT

- Operator Trainee Interface

FIGURE 2   An operator training system network with separate stations. (© Hyprotech Inc., 1998.)

the same computer as the plant model, there is an opportunity to develop, test, and demonstrate the model and control system on a personal computer. The advantages of emulated control systems in terms of being able to implement and use the trainer anywhere and share results with everyone cannot be overemphasized. The control system emulation has additional value in terms of opening up the possibility of running the control system faster than real time and inherently providing initialization of setpoints and outputs. If the control system for the faster than real time studies is distinct from the DCS or the PLC control system, there are extensive duplication and initialization requirements. However, the emulation of proprietary operator screens and keyboards as soft keys or new keys on a PC is undesirable since it fails to ensure that the operator can learn to quickly make changes during real operating crises. As DCSs and PLCs develop true Windows interfaces, the changes can be transparent to the operator and the emulation of the interface becomes both feasible and highly desirable. The Windows NT environment and object linking embedding for process control enables the communication between the simulation, control system, and operator interface without special software or hardware and enables the system to run completely on a personal computer.

The connection between the plant model and the DCS or the PLC must be at the input and the output blocks of the configuration. If the connection is done to the loop via the data highway, signal calculations (split ranging and compensation), characterization, and selection are bypassed. The configuration and the performance of these functions are generally more of an issue than the tuning settings of the loop. It is critical that the training system include all of the input and output signal processing in the configuration.

REFERENCES

1. Lauritsen, M., "Dynamic Simulation Applications and Benefits to Engineering and Operations," presented at the American Institute of Chemical Engineers CAST meeting, September, 1998.

2. McMillan, G., *Modeling and simulation of processes, in Instrument Engineer's Handbook*, 3rd ed., Chilton, Location, 1994, pp. 70–76.

3. HYSYS™ is a trademark of Hyprotech Ltd. (information and figures are reproduced by permission of Hyprotech Ltd.).

INTELLIGENT ALARMS

G. K. McMillan*

Smart alarms or the equivalent alarm filtering can eliminate 90% or more of the alarm presented to the operator [1]. The goal should be zero alarms for normal operation and a single alarm for an abnormal condition that pinpoints the root cause. This is especially important during upset conditions because the response of the operator rapidly deteriorates as the number of alarms increases. Figure 1 hows that the operator needs at least 1 min after one alarm, 10 min after two alarm, and 20 min after three alarms to have any chance at all of a proper diagnosis as to the cause of the alarm so that the probability of failure to diagnose the alarm is less than one. Figure 1 also shows that the operator needs 10 or more minutes to ensure that the probability of failure to diagnose a single alarm is low (<0.1) [2].

Instead of triggering alarms off of a high or low measurement, the alarm should be built up to show the actual operating condition from information from diverse sources such as sensors, tasks, modes, outputs, and other alarms. If done properly, a single alarm is generated when the root cause is identified. Alarm filtering can be made to accomplish this same goal after the fact by use of a similar logic to present to the operator only one alarm related to the root cause. However, the number of alarms generated is very large and the highway traffic is very high during critical periods. Thus it is better to build in the proper logic in the beginning to reduce the number of alarms generated rather than filters to screen out the extraneous alarms. Alarm analyzers that tell the operator which

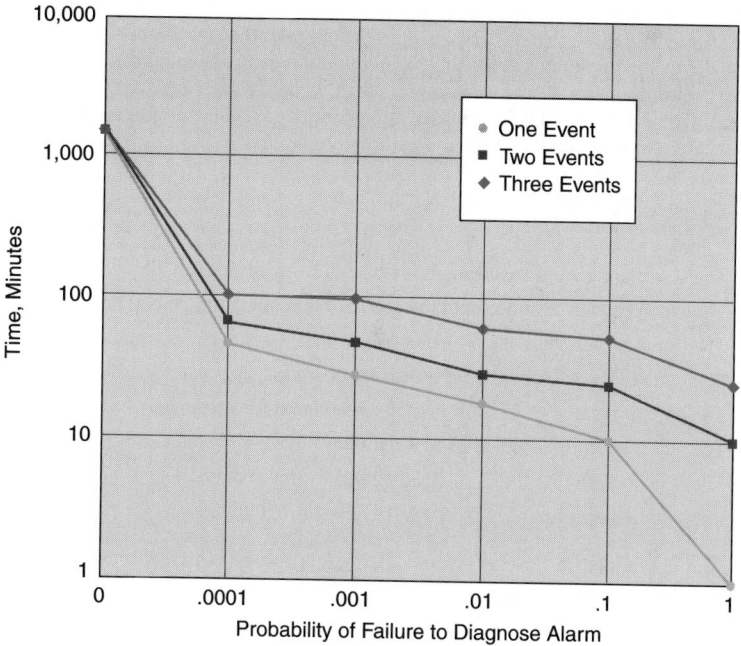

FIGURE 1

* Senior Fellow, Solutia Inc., St. Louis, Missouri.

one of the alarms was really important are too late because the damage has already been done by the barrage of alarms. Even if the extraneous alarms do not activate a horn or flash, they still pose a serious undesirable and unnecessary distraction to the operator. Thus, to minimize highway loading and operator distraction from extraneous information, the control systems should reduce the number of alarms generated rather than screen or classify alarms.

Each of the conditions that set a smart alarm needs to have the option of a time delay to ignore spurious signals and allow time for a valves, pumps, process, and instruments to respond. Also, the conditions to set an alarm should be based on analog and discrete measurements (sensors) so that the failure of motors or valves does not cause a false alarm. On the other hand, conditions to clear the alarm should be generally built up from action requests (set points for unit points, positioners, and discrete control devices) rather than feedback measurements or statuses to ensure conditions to deactivate the alarm are recognized early enough. Also, each analog value used to set and clear an alarm must have a dead band to eliminate alarm chatter from noise.

The basic features for build a simple smart alarm are illustrated by the following logic used to activate a smart alarm to alert an operator that a batch has gotten too thick (e.g., excessive polymerization). In this example, there is an agitator and a thickening task in the batch sequence.

The conditions to set a smart alarm to alert the operator that the batch is too thick are as follows:

1. The batch sequence unit point is in thickening task for more than 5 s.

2. The agitator amps are too high for more than 5 s.

The conditions to clear a smart alarm that indicates that the batch is to thick are as follows:

1. The batch sequence unit point requested task is anything but thickening.

2. The agitator amps have dropped below the alarm point minus the dead band.

The basic features required for building a more complex smart alarm are illustrated by the following logic used to activate a smart alarm to alert the operator that a feed nozzle or control valve is plugged to a fed batch reactor. In this example, consider that there is a feed tank, pump, solids filter, flow control loop, automated isolation valve, smart alarm to notify the operator that the feed filter is plugged, and a batch sequence. The smart alarm for the plugged filter might use pressure drop and would be considered to be a precursor to the low flow condition for the smart alarm for the plugged nozzle or control valve.

The conditions to set a smart alarm to alert the operator that a feed nozzle or control valve is plugged are as follows:

1. The feed tank level measurement is above 4% for more than 14 s.

2. The batch sequence unit point is in the feed task for more than 10 s.

3. The pump motor run contact is closed for more than 8 s.

4. The actual valve position from the smart positioner is greater than 40% for more than 6 s.

5. The isolation valve open limit switch is activated for more than 4 s.

6. The smart alarm that indicates that the feed filter is plugged is not activated for more than 2 s.

7. The flow measurement field status is good for more than 10 s.

8. The flow measurement is less than 10% for more than 4 s.

The conditions to clear the smart alarm that indicates that the feed nozzle or control valve is plugged are as follows:

1. The feed tank level measurement is below 2%.

2. The batch sequence unit point requested task is anything but the feed task.

3. The pump motor set point is stopped.

4. The flow controller output (smart positioner set point) is less than 20%.

5. The isolation valve set point is closed.

6. The smart alarm that the feed filter is plugged is activated.

7. The flow measurement field status is bad.

8. The flow measurement is above 20%.

In the above examples and for most applications, one can derive the conditions to clear the smart alarm from the conditions to set the alarm by substituting set points for measurements, by using dead bands for analog values, and by omitting time delays. However, for the more general case, it may be advantageous to offer the user the option of deriving the clear condition from the set condition and adding distinct or separate clear conditions. In other words, for each set condition, the user would fill in the tag of the loop, indicator, discrete control device, unit point, or precursor alarm followed by the associated trigger point, time delay, and dead band, and finally check a box as to whether to automatically derive the clear condition. Finally, an optional entry of specifically different clear conditions would be offered.

The alarm is activated (st) or deactivated (cleared) or bypassed by the operator by means of a discrete status rather than by changing trip points between designated and unreachable values. Newer DCS systems support this feature, which eliminates the need for housekeeping of valid and invalid analog alarm settings.

The above method of constructing smart alarms leads the user into more carefully thinking of the conditions that uniquely identify the abnormal operating condition. In contrast, most DCSs ask the user to fill in the high, high–high, low, and/or low–low settings for each analog point and to choose a deviation alarm for all points. In early DCS designs, a setting was mandatory and the only way to deactivate an alarm was to choose a setting that could never be reached. As a result, the DCS created an order of magnitude increase in alarms compared with panel board analog controllers. The DCS now has the opportunity to generate more meaningful and fewer alarms than those of the analog days by giving the user the ability to go exclusively to smart alarms. Additionally, the advent of the field-based DCS opens the ability to use diagnostics from smart instrumentation to help make alarms smarter.

REFERENCES

1. Bray, M., and D. Corsberg, "Practical alarm filtering," *InTech*, February, 34–36, 1994.

2. Heavner, L., L. Schulltz, and G. McMillan, "Dynamic alarming for DCS," *InTech*, November, 22–25, 1993.

SECTION 9
VALVES, SERVOS, MOTORS, AND ROBOTS*

Mark Adams
Fisher Controls International, Inc., Marshalltown, Iowa
(Process Control Valves)

Len Auer
Rosemount Inc., Eden Prairie, Minnesota (Current-to-Pressure
Transducers for Control-Valve Actuation)

Allen C. Fagerlund
R. A. Engel Technical Center, Fisher Controls International, Inc.,
Marshalltown, Iowa (Control Valve Noise)

Bill Fitzgerald
Fisher Controls Company International, McKinney, Texas
(Control Valve Troubleshooting)

David A. Kaiser
Compumotor Division, Parker Hannifin Corporation, Rohnert
Park, California (Servomotor
Technology in Motion Control Systems)

J. J. Kester
Bodine Electric Company, Chicago, Illinois (Stepper and
Other Servomotors)

S. Longren
Longren Parks, Chanhassan, Minnesota (Servomotors)

Richard H. Osman
AC Drives, Robicon Corporation, Pittsburgh, Pennsylvania
(Solid-State Variable-Speed Drives)

Bert J. Peterson
Fisher Controls International, Inc., Marshalltown, Iowa
(Process Control Valves)

C. Powell
GMFanuc Robotics Corporation, Auburn Hills, Michigan
(Robots)

---

* *Persons who authored complete articles or subsections of articles, or otherwise cooperated in an outstanding manner in furnishing information and helpful counsel to the editorial staff.*

Marc L. Riveland
Applied Research, Fisher Controls International, Inc.
Marshalltown, Iowa (Control-Valve Cavitation)

PROCESS CONTROL VALVES

by Bert J. Peterson*

INTRODUCTION

The control valve, or final control element, is the last device in the control loop. It takes a signal from the process instruments and acts directly to control the process fluid. Control valves maintain process variables such as pressure, flow, temperature, or level at their desired value, despite changes in process dynamics and load. Control valves must be designed to accommodate the needs and characteristics of the process fluid they control. Likewise, the control valve must react to the protocol and needs of the controlling devices in the process control system. The evolution of control valves is in response to the combined forces of the processes they handle and the systems that control them. Evidence of these factors exists in the design of valve bodies, actuators, valve controllers, and interface accessories.

This article covers control valves in their major subsegments. Following a brief history of control valves, a summary of valve bodies and of the criteria for valve applications, sizing, and selection is presented. These factors are strongly influenced by the needs and specifications of the process. The next section discusses the increasingly important consideration of control valve performance. A review of actuator styles, distinctions, and selection follows the performance section. This review centers on the needs of the control valve body and the needs of the process and power source. Finally, the article covers digital valve controllers, positioners, and other valve and actuator accessories.

CHRONOLOGY

The modern history of control valves is rooted in the industrial revolution and began with the invention of the industrial steam engine. Steam to power these engines built up pressures, which had to be contained and regulated. Instantly, valves (which had existed since the middle ages) took on new significance. The process control systems at that time were simple: human beings. Things changed

---

* Senior Publications Specialist, Fisher Controls International Inc., Marshalltown, Iowa.

TABLE 1 Principal Selection Criteria and Availability of Generic Valve Styles

Valve Style	Main Characteristics	Available Size Range		Typical Std. body Materials	Typical Std. End Connection	Typical Max. Pressure		Relative Flow Capacity
Regular sliding stem	Heavy duty, versatility	NPS 1/2–20	DN 15–500	Cast iron; carbon, alloy, or stainless steel	Flanged, welded screwed	Class 2500	PN 420	Moderate
Bar stock	Compact	NPS 1/2–3	DN 15–80	Stainless steel; nickel alloys	Flangeless, screwed	Class 600	PN 100	Low
Economy sliding stem	Light duty, inexpensive	NPS 1/2–4	DN 15–100	Bronze, cast iron, carbon steel	Screwed, flanged	Class 300	PN 50	Low
Through-bore ball	On–off service	NPS 1–24	DN 25–600	Carbon steel or stainless steel	Flangeless	Class 900	PN 150	High
Partial ball	Characterized for throttling	NPS 1–24	DN 25–600	Carbon steel or stainless steel	Flangeless, flanged	Class 600	PN 100	High
Eccentric plug	Erosion resistance, versatility	NPS 1–12	DN 25–300	Carbon steel or stainless steel	Flangeless, flanged	Class 600	PN 100	Moderate
Swing-through butterfly	No seal	NPS 2–36	DN 50–900	Cast iron; carbon or stainless steel	Flangeless, lugged, welded	Class 2500	PN 420	Moderate
Lined butterfly	Elastomer or PTFE lined	NPS 2–24	DN 50–600	Cast iron; carbon or stainless steel	Flangeless, lugged	Class 150	PN 20	High
High-performance butterfly	Offset disk, flexible seals	NPS 2–72	DN 50–1800	Carbon steel or stainless steel	Flangeless, lugged	Class 600	PN 100	High
Special	Custom to application	NPS 2–24	DN 50–600	Carbon, alloy, or stainless steel	Flanged, welded	Class 4500	PN 760	High

(Continues)

quickly, however. The invention of the flyball governor heralded a new era of "feedback control" and permanent linkage of process to control valves.

The next leap of technology took place in 1875, when William Fisher invented the self-contained automatic pump governor. This device used pump output pressure to control valves that throttle steam flow to the pump engines. It was the first process control device to achieve a set point by offsetting process pressure acting on a piston with the force of a mechanical spring. This combination of forces, linked to a valve body, was the basis on which control valve actuators and control valves later evolved.

For the next 50 years, control valves consisted of a variety of self-contained governors (now called regulators), float valves, and lever valves. The most common method of valve actuator was a spring-opposed piston or diaphragm motor (an actuator) that operated directly from the process fluid.

In the mid-1930s, pneumatic pressure control instruments began to emerge. Instrument companies coaxed the valve companies to make valve actuators that reacted to standardized pneumatic signals rather than process pressure. The new control instrumentation improved the fidelity of process control dramatically and required an upgrade of control valve components. Characterized valve plugs were developed, and the valve positioner made its debut.

As process pressure increased, high-pressure valves were developed. As flow rates increased, larger-capacity valves were developed. Valves changed to cope with process changes. The late 1970s witnessed a wholesale move to centralized electronic control, and control valves were modified to accept analog electronic signals. Today, control valves exist in an environment of distributed digital control, integral digital control, integral accessories, modularity, and two-way communication. The evolution continues. Control valves are adapting to change in the processes they control and to the instruments that control them.

TABLE 1 (*Continued*)

Rel. Shutoff Capability	Noise or Cavtitation Trim Option	Available Control Characteristic	Flow Rangeability	Application Temp. Range	Pressure Drop Capability	Best Economic Size Range	
Excellent	Yes	Equal percentage, linear, quick opening, special	Moderate	Quite low to very high	High	NPS 1–4	DN 25–100
Excellent	No	Equal percentage, linear	Moderate	Moderate	Moderate	NPS 1/2–1	DN 15–25
Good	Yes	Equal percentage, linear	Moderate	Moderate	Moderate	NPS 1–2	DN 25–50
Excellent	Yes	Equal percentage	Low	Moderate	Moderate	NPS 4–8	DN 100–200
Excellent	Yes	Equal percentage	High	Quite low to quite high	Moderate	NPS 4–8	DN 100–200
Excellent	No	Linear	Moderate	Quite low to quite high	High	NPS 4–8	DN 100–200
Poor	No	Equal percentage	Moderate	Very low to quite high	Moderate	NPS 6–36	DN 150–900
Good	No	Equal percentage	Low	Moderate	Low	NPS 6–24	DN 150–600
Excellent	No	Linear	Low	Very low to quite high	Moderate	NPS 6–72	DN 150–1800
Excellent	Yes	Custom	Moderate to high	Very low to quite high	High to very high	–	–

CONTROL VALVE BODIES

General Categories of Control Valves

Control valve here means any power-operated valve, whether used for throttling or on–off control. Varieties from which to select, as listed in Table 1, include sliding stem valves and rotary valves. Typical sliding-stem valves are straight-pattern valves (sometimes called globe valves) and angle-pattern valves. Rotary valves include ball and butterfly valves. Other varieties such as motorized gate valves, louvers, pinch valves, plug valves, and self-operated regulators are not considered here. These major types, sliding-stem and rotary, are further divided into ten subcategories according to relative performance and cost. Despite variations found within each category—such as cage guiding and stem guiding—all valves within a given subcategory can be considered very much alike in the early stages of the valve selection process. Selecting a valve involves narrowing your selection to one of these subcategories and then comparing specific valves in that group (Table 2).

Designations NPS and DN are used in Table 1 and throughout this section. NPS is a designation for nominal pipe size. It comprises the letters NPS followed by a dimensionless number, which is indirectly related to the physical size, in inches, of the end connections. DN is an international designation for nominal diameter. It comprises the letters DN followed by a dimensionless whole number, which is indirectly related to the physical size, in millimeters, of the end connections.

Sliding-Stem Valves

The most versatile of the control valves are the sliding-stem valves. Straight-pattern, angle-pattern, and three-way valves can be purchased in sizes ranging from NPS 1/2 to NPS 20 or from DN 15 to

TABLE 2 Control-Valve-Characteristic Recommendations for Liquid-Level, Pressure, and Flow Control*

Liquid-Level Systems

Control Valve Pressure Drop	Best Inherent Characteristic
Constant ΔP	Linear
Decreasing ΔP with increasing load, ΔP at maximum load >20% of minimum-load ΔP	Linear
Decreasing ΔP with increasing load, ΔP at maximum load <20% of minimum-load ΔP	Equal percentage
Increasing ΔP with increasing load, ΔP at maximum load <200% of minimum-load ΔP	Linear
Increasing ΔP with increasing load, ΔP at maximum load >200% of minimum-load ΔP	Quick opening

Pressure Control Systems

Application	Best Inherent Characteristic
Liquid process	Equal percentage
Gas process, small volume, less than 10 ft of pipe between control valve and load valve	Equal percentage
Gas process, large volume (process has receiver, distribution system, or transmission line exceeding 100 ft of nominal pipe volume), decreasing ΔP with increasing load, ΔP at maximum load >20% of minimum-load ΔP	Linear
Gas process, large volume, decreasing ΔP with increasing load, ΔP at maximum load <20% of minimum load ΔP	Equal percentage

Flow Control Processes

Flow measurement signal to controller	Location of control valve in relation to measuring element	Best Inherent Characteristic	
		Wide range of flow set point	Small range of flow but large ΔP change at valve with increasing load
Proportional to flow	In series	Linear	Equal percentage
	In bypass[†]	Linear	Equal percentage
Proportional to flow squared	In series	Linear	Equal percentage
	In bypass	Equal percentage	Equal percentage

* Based on a combination of applied control theory and actual experience. (*Fisher Controls International, Inc.*)
[†] When control valve closes, flow rate increases in measuring element.

DN 500. Examples of sliding-stem valves are shown in Figs. 1–5. More choices of materials, end connections, and control characteristics are available for sliding-stem valves than for any other product family. Sliding-stem valves are available in cage-guided, post-guided, and stem-guided designs with flanged, screwed, or welding ends. Economical cast iron as well as carbon steel, stainless steel, and other high-performance body materials are available. Pressure ratings up to and above Class 2500 or PN 420 are available. Their precise throttling capabilities, overall performance, and general sturdiness

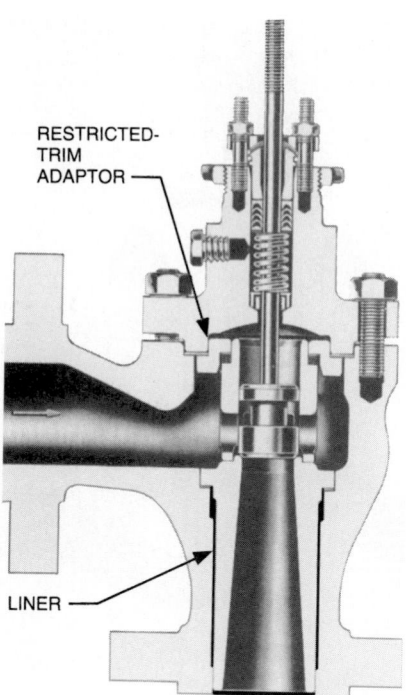

FIGURE 1   Reduced trim, angle-pattern sliding-stem valve shows the capability for trim reduction in a sliding-stem valve. The valve also features an outlet liner for resistance to erosion. The unbalanced plug provides tight shutoff but requires a larger actuator than balanced designs. (*Fisher Controls International, Inc.*)

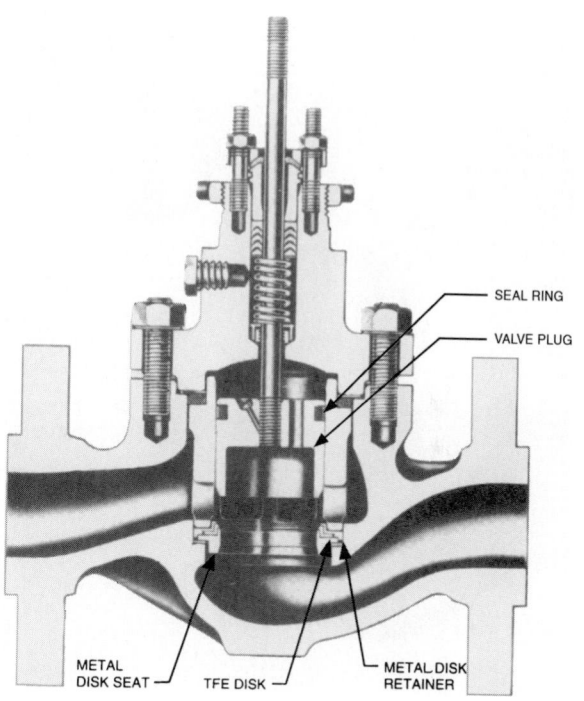

FIGURE 2   Standard straight-pattern sliding-stem valves are available in a broad range of sizes, materials, and end connections. The balanced plug shown reduces unbalance force and allows the use of smaller actuators. A soft seat provides tight shutoff. Valves such as this are the first choise for applications smaller than NPS 3 or DN 80. (*Fisher Controls International, Inc.*)

make sliding-stem valves a bargain, despite their slight cost premium. The buyer gets a rugged, dependable valve intended for long, trouble-free service. Sliding-stem valves are built ruggedly to handle conditions such as piping stress, vibration, and temperature changes. In sizes NPS 3 or DN 80, incremental costs over rotary valves are low in comparison to the increments in benefits received.

For many extreme applications, sliding stem valves are the only suitable choice. This includes valves for high pressure and temperature, antinoise valves, and anticavitation valves. Because of process demands, these products require the rugged construction design of sliding-stem products.

Bar stock valves are small, economical sliding-stem valves whose bodies are machined from bar stock (Fig. 6). Body sizes range from fractions of an inch up to NPS 3 or DN 80; flow capacities generally are lower than those of general-purpose valves. End connections usually are flangeless (for clamping between piping flanges) or screwed. The main advantage of this type of valve is that far more materials are readily available in bar stock form than in cast form. Consequently these valves are often used where there are special corrosion considerations. However, their compactness and general high-quality construction make them attractive for flow rates below the range of the regular sliding-stem subcategory, even when corrosion is not a consideration. Overall, they are an economical choice when they can be used.

The lowest-cost products among sliding-stem valves are called general-purpose or economy bodies. These valves are used for low-pressure steam, air, and water applications that are not demanding (Figs. 7 and 8). Available sizes range from NPS 1/2–4 or DN 15–DN 100. Body materials include bronze, cast iron, steel, and stainless steel (SST). Pressure classes generally stop at Class 300 or PN 50.

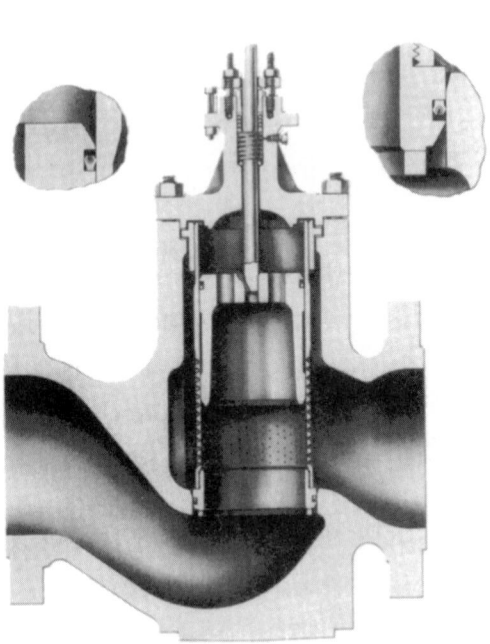

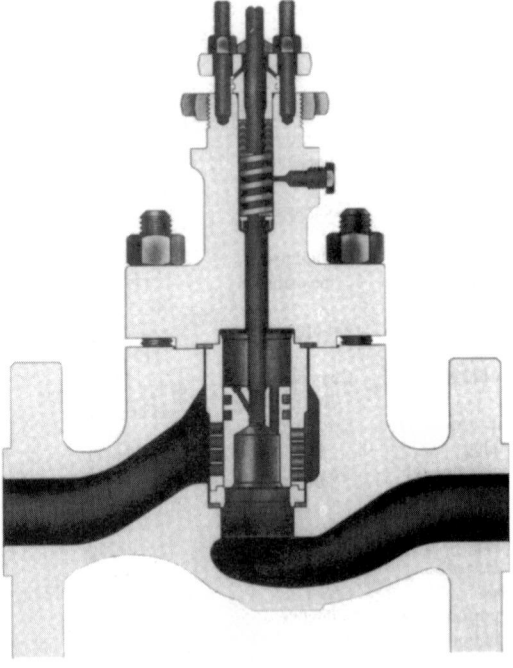

FIGURE 3   Severe-service capability in a large, straight-pattern valve. It features a drilled-hole cage for attenuation of flow noise by splitting the flow into multiple passages. Hole spacing is controlled carefully to eliminate jet interaction and high resultant noise levels. (*Fisher Controls International, Inc.*)

FIGURE 4   High-pressure globe valves are typically available in sizes NPS 1–20 (or DN 25–500) and Classes 900, 1500, and 2500 (or PN 150, PN 260, and PN 420). These valves provide throttling control of high-pressure steam and other fluids. Antinoise or anticavitation trim is often used to handle problems caused by high-pressure drops. (*Fisher Controls International, Inc.*)

Compared to regular sliding-stem valves, these units are very simple, their actuators are smaller, and their cost is normally three-quarters to one-half as much. Severe service trims for noise and cavitation service are not usually available in these products.

Ball Valves

There are two subcategories of ball valves. The through-bore or full-ball type shown in Fig. 9 is often used for high-pressure drop throttling and on–off applications in sizes to NPS 24 or DN 600. Full-port designs exhibit high flow capacity and low susceptibility to wear by erosive streams. However, sluggish flow throttling response in the first 20% of ball travel makes full-bore ball valves unsuitable for throttling applications. Newer designs in full-ball, reduced-bore valves provide better response. Pressure ratings up to Class 900 or PN 150 are available, as are a variety of end connections and body materials. Another popular kind of ball valve is the partial-ball style (Fig. 10). This subcategory is very much like the reduced-bore group, except that the edge of the ball segment has a contoured notch shape for better throttling control and higher rangeability. Intended primarily for modulating service, not merely for on–off control, partial-ball valves are generally higher in overall control performance than full-ball products. They are engineered to eliminate lost motion, which is detrimental to performance. The use of flexible or movable metallic and fluoroplastic sealing elements allows tight shutoff and wide temperature and fluid applicability. Their straight-through flow design achieves high capacity. Sizes range through NPS 24 or DN 600. Pressure ratings go to Class 600 or PN 100. Price is normally lower than that of globe valves.

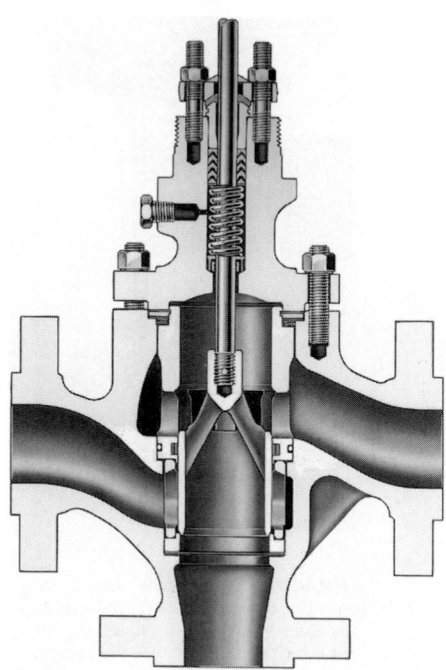

FIGURE 5   Three-way valves have three end connections to allow for converging (flow mixing) or diverging (flow splitting) operation. (*Fisher Controls International, Inc.*)

FIGURE 6   Bar-stock valves provide economical solutions to small flow requirements. Pressures to 1500 psig (104 bars) and temperatures to 450°F (232°C) can be handled. Compact spring-and-diaphragm actuators complement these small valve bodies. (*H. D. Baumann Inc.*)

Eccentric-Plug Valves

Eccentric-plug valves combine many features of sliding-stem and rotary products and use rotary actuators. These valves are available for different types of service. The valve in Fig. 11 is used for a variety of fluids in both industrial process and utility applications. The valve in Fig. 12 features oversized shafts and rigid seat design for severe service and erosion resistance. Both designs have excellent throttling capability and combine many of the good aspects of rotary and sliding-stem valves. Sizes are available through NPS 8 or DN 200 in ratings to Class 600 or PN 100. Flanged and flangeless constructions are usually available.

Butterfly Valves

Butterfly valves are divided into three subcategories: swing through, lined, and high performance. The most rudimentary is the swing-through design (Fig. 13). Rather like a stovepipe damper, but considerably more sophisticated, this kind of valve has no seals—the disk swings close to, but clear of, the body's inner wall. Such a valve is used for throttling applications that do not require shutoff tighter than ~1% of full flow. Sizes range from NPS 2 to NPS 96 or DN 50 to DN 400. Body materials

FIGURE 7   Screwed-end bronze valve body capable of handling many utility applications. It is complemented by a wide variety of seat ring sizes and control characteristics. (*H.D. Baumann Inc.*)

FIGURE 8 Valve style typical of general-purpose valves. Availability usually extends to size NPS 4 or DN 100 and to Class 300 or PN 50 ratings. These valves feature compact, reversible diaphragm actuators and inexpensive positioners. (*Fisher Controls International, Inc.*)

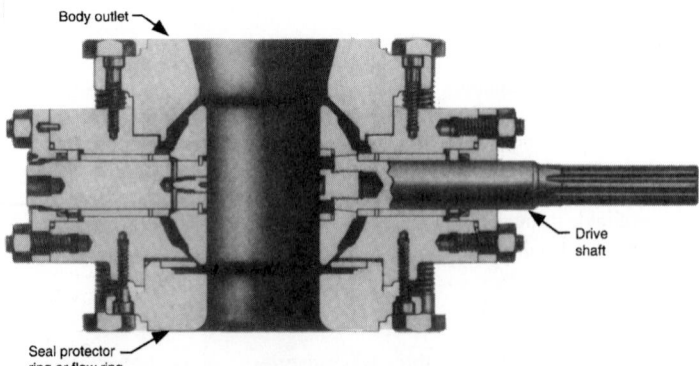

FIGURE 9   High-pressure ball valve, featuring heavy shafts and full-ball design. The valve shown can be used for pressure drops to 2220 psi (152 bars). Class 600 and 900 or PN 100 and 150 bodies are available; sizes range to NPS 24 or DN 600. (*Fisher Controls International, Inc.*)

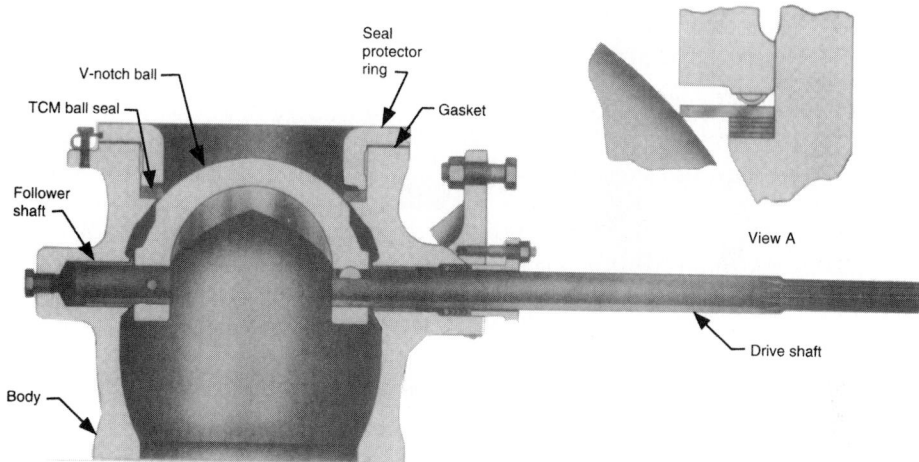

View A

FIGURE 10   Applications to Class 600 or PN 100 can be handled by this segmented or partial ball valve. The flangeless body incorporates many features to improve throttling performance and rangeability. Tight shutoff is achieved by either metal or composition seals. (*Fisher Controls International, Inc.*)

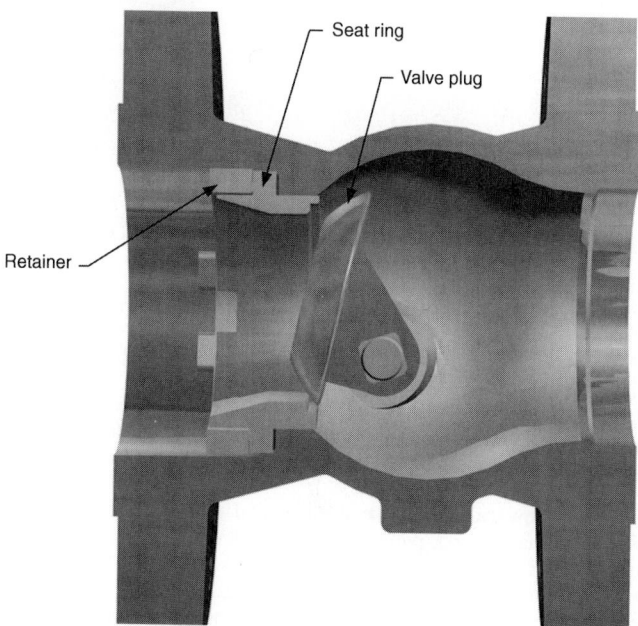

FIGURE 11   Rotary eccentric-plug valve for industrial process and utility applications. (*H. D. Baumann Inc.*)

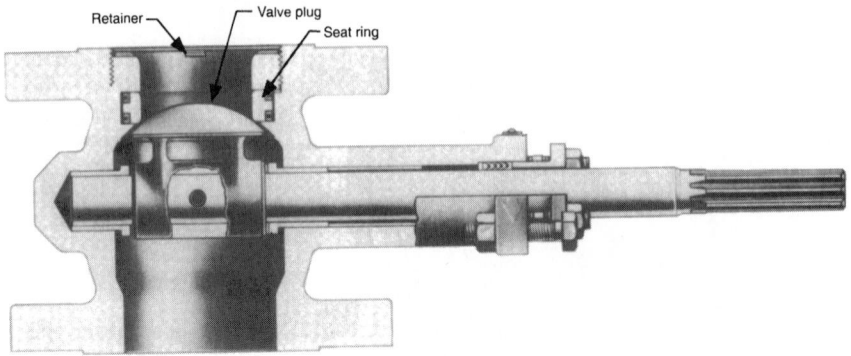

FIGURE 12   Rotary eccentric-plug valve for severe applications and highly erosive process fluids. (*Fisher Controls International, Inc.*)

FIGURE 13   Swing-through butterfly valve provides an economical solution to high-flow throttling applications. Leakage is higher than for other designs because no sealing mechanism is used. (*Fisher Controls International, Inc.*)

are cast iron, carbon steel, or stainless steel. Mounting is flangeless, lugged, or welded. Body pressure ratings up to Class 2500 or PN 420 are common, and wide temperature ranges are also available. While a very broad range of designs is available in these products, they are limited by lack of tight shutoff.

Need for no or low leakage requires the lined and high-performance butterfly valves. Lined butterfly valves feature an elastomer or polytetrafluoroethylene (PTFE) lining that contacts the disk to provide tight shutoff (Fig. 14). Because this seal depends on interference between disk and liner, these designs are more limited in pressure drop. Temperature ranges are also restricted considerably because of the use of elastomeric materials. A benefit, however, is that because of the liner, the process fluid never touches the metallic body. Thus these products can be used in many corrosive situations. Elastomer-lined butterfly valves are generally the lowest-priced products available as control valves in medium to large sizes.

High-performance butterfly valves such as the one shown in Fig. 15 are characterized by heavy shafts and disks, full pressure rated bodies, and sophisticated seals that provide tight shutoff. These

FIGURE 14   Lined butterfly valves offer tight shutoff but are limited to low temperatures. Liner material keeps process away from the metallic body, eliminating many corrosion problems. (*Fisher Controls International, Inc.*)

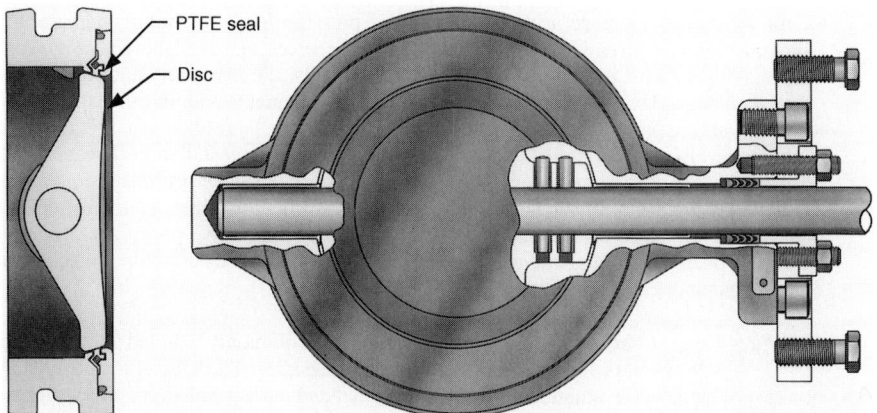

FIGURE 15   High-performance butterfly valves provide excellent performance and value. High-pressure capability, tight shutoff, and excellent control are standard. Designs are available in Classes 150, 300, or 600 (PN 20, PN 50, or PN 100) and sizes to NPS 72 or DN 1800. (*Fisher Controls International, Inc.*)

valves provide an excellent combination of performance features, light weight, and very reasonable pricing. Eccentric shaft mounting allows the disk to swing clear of the seal to minimize wear and torque. The offset disks used allow uninterrupted sealing and a seal ring that can be replaced without removing the disk. High-performance butterfly valves come in sizes from NPS 2 to NPS 72 or DN 50 to DN 1800 with flangeless or lugged connections. Bodies are carbon-steel or stainless steel, and pressure ratings are up to Class 600 or PN 100. With their very tight shutoff and heavy-duty construction, these valves are suitable for many process applications. Advanced metal-to-metal seals provide tight shutoff in applications that are too hot for elastomer-lined valves to handle.

Special Control Valves

Standard control valves can handle a wide range of control applications, which can be defined as:

- pressure ratings to Class 2500 or DN 420
- −150 to 450°F (−101 to 232°C)
- flow coefficient values of 1.0–25,000 C_v or 22,000 K_v
- within the limits imposed by common industrial standards

Perhaps the need for careful consideration of valve selection and the need for special valves become more critical for applications outside the standard limits mentioned above. However, corrosiveness and viscosity of the fluid, leakage rates, and many other factors demand consideration, even for standard applications.

Special valves can include body liners and seals to contain corrosive or toxic materials (Fig. 16) or valves with special trims for special purposes (Fig. 17). Some categories of special valves are discussed in the following sections.

High-Capacity Control Valves.   The following are often considered to be special valves:

- globe-style valves larger than size NPS 12 or DN 300
- ball valves larger than size NPS 24 or DN 600
- high-performance butterfly valves larger than size NPS 48 or DN 1200

As valve sizes increase arithmetically, static pressure loads at shutoff increase geometrically. Consequently, shaft strength, bearing loads, unbalance forces, and available actuator thrust all become more significant with an increasing valve size. Normally, the maximum allowable pressure drop is reduced on large valves to keep design and actuator requirements within reasonable limits. Even with lowered working pressure ratings, the flow capacity of some large-flow valves is tremendous.

Noise levels must be carefully considered in all large-flow installations because sound pressure levels increase in direct proportion to flow magnitude. To keep valve-originated noise within tolerable limits, large cast (Fig. 18) or fabricated (Fig. 19) valve body designs have been developed. These bodies, normally cage-style construction, use unusually long valve plug travel, a great number of small flow openings through the wall of the cage, and an expanded outlet line connection to minimize noise output and reduce fluid velocity.

Low-Flow Control Valves.   Low-flow applications are commonly handled in one of two ways. One method is with special trims in standard control valve bodies. The special trim is typically made up of a seat ring and valve plug that have been designed and machined to very close tolerances to allow accurate control of very small flows. These types of constructions can often handle C_V or K_V values as low as 0.03 or 0.025. Using these special trims in standard control valves provides economy by reducing the need for spare-parts inventory for special valves and actuators. Using this approach also makes future flow expansions easy by simply replacing the trim components in the standard control valve body.

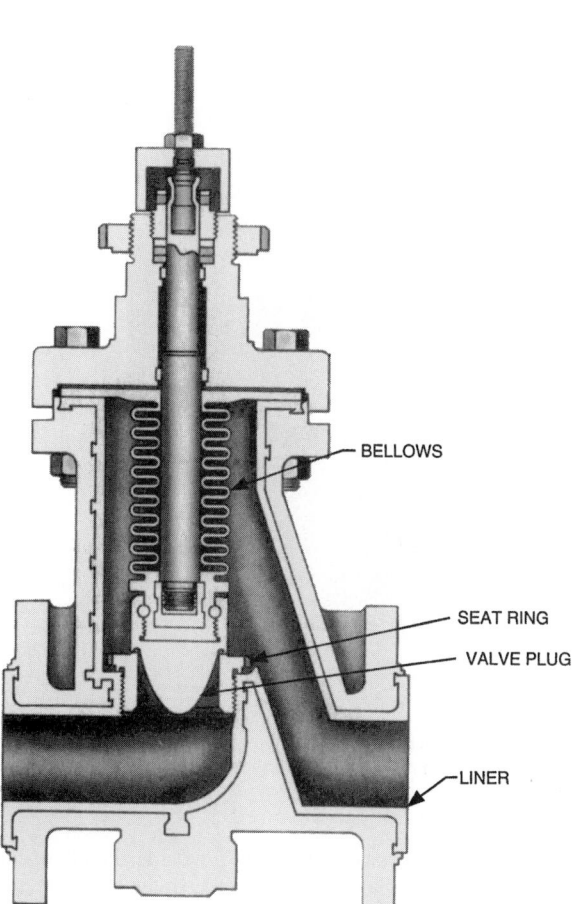

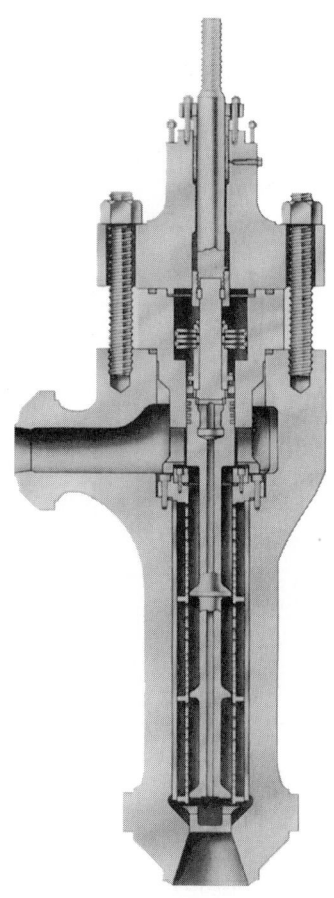

FIGURE 16   Special sliding-stem valve used for severely corrosive or toxic fluids. It has a full PTFE liner and PTFE trim. It also has a bellows seal to eliminate leakage. (*Fisher Controls International, Inc.*)

FIGURE 17   Special sliding-stem valve for super-heater bypass service in power plants. The application requires tight shutoff and flows that range from cold cavitating water to flashing water to superheated steam. (*Fisher Controls International, Inc.*)

Control valves specifically designed for very low-flow rates (Fig. 20) also handle these applications. These valves often handle C_V or K_V values as low as 0.000001. In addition to the very low flows, these control valves are compact and light weight.

Valves for Sanitary Service.   Valves for sanitary service (Fig. 21) are used in the food and beverage, pharmaceutical, biotechnical, and semiconductor industries. Sanitary valves have several design features specifically for the intended service: highly polished surfaces, self-draining features, and minimum areas that can hold process fluid that is not moving in the process stream. Diaphragm-type valves are often used for this service because this design minimizes or eliminates valve trim guiding surfaces that are in contact with process fluid. (Instead of a valve plug, a diaphragm seats against a seat ring to control the flow of process fluid.) Sanitary valves are often designed for CIP (clean in place) and SIP (sanitize in place) procedures. End connections are often weld ends or clamped connections.

High-Temperature Control Valves.   Control valves for service at temperatures above 450°F (232°C) must be designed and specified with the temperature conditions in mind. At elevated temperatures,

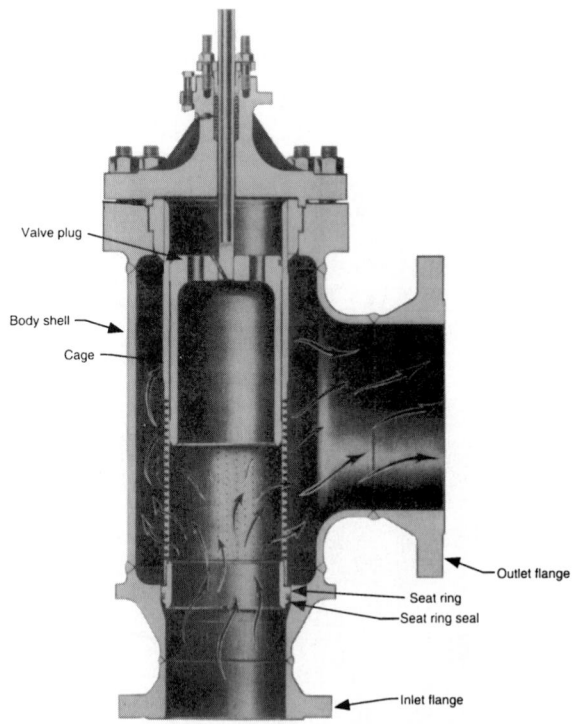

FIGURE 18   High-capacity cast globe valve for noise attenuation service. (*Fisher Controls International, Inc.*)

FIGURE 19   Extremely high-flow fabricated valve. The valve can be custom made to match the required flow and piping configuration, and it features a drilled-hole noise-reduction trim. (*Fisher Controls International, Inc.*)

such as with boiler feedwater systems and superheat bypass systems, the standard materials of control valve construction might be inadequate. For instance, plastics, elastomers, and standard gaskets often prove unsuitable and must be replaced by more durable materials. Metal-to-metal seating materials are always used. Semimetallic or laminated graphite packing materials are commonly used, and spiral-wound stainless steel and flexible graphite gaskets are necessary.

Chromium-molybdenum (Cr-Mo) steels are often used for valve body castings for temperatures above 1000°F (538°C). Chromium-molybdenum steel (such as WC9) is used up to 1100°F (593°C). For temperatures up to 1500°F (816°C), 316 stainless steel (such as CF8M) is often selected. For temperatures between 1000 and 1500°F (538 and 816°C), the carbon content must be controlled to the upper end of the range, that is, 0.04–0.08%.

On high-temperature service, extension bonnets help protect packing box parts from extremely high temperatures.

Cryogenic Service Valves.   Cryogenic service normally involves temperatures below −150°F (−101°C). Plastic and elastomeric components often cease to function appropriately at temperatures below 0°F (−18°C). At these temperatures, components such as packing and plug seals require special consideration. For plug seals, a standard soft seal will become very hard and less pliable, thus not providing the shutoff required from a soft seat. Special elastomers have been applied in these temperatures but require special loading to achieve a tight seal.

Packing is a concern in cryogenic applications because of the frost that can form on the valves. Moisture from the atmosphere condenses on colder surfaces and can freeze into a layer of frost. As the stem is stroked by the actuator, the layer of frost on the stem is drawn through the packing, causing

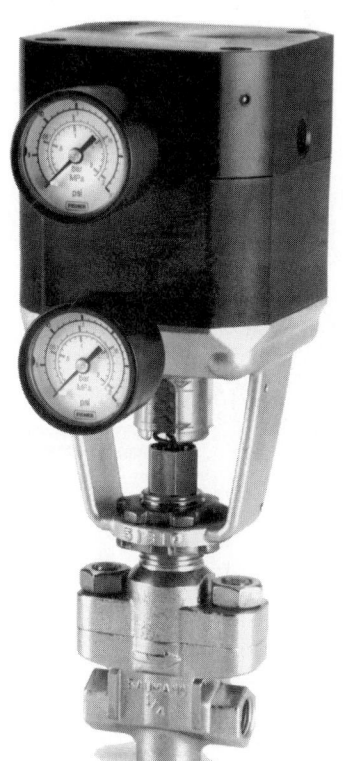

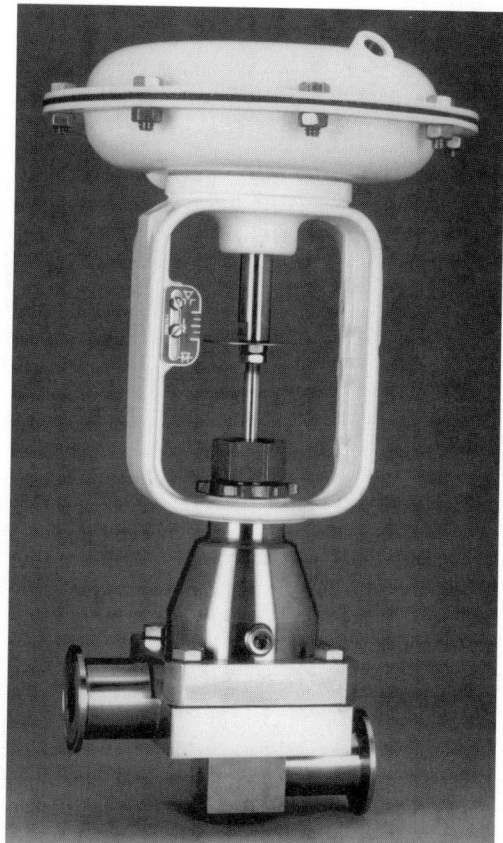

FIGURE 20 Control valve designed for very low flow rates. (*H. D. Baumann Inc.*)

FIGURE 21 Control valve for sanitary service; it has highly polished surfaces and clamp-type end connections. (*H. D. Baumann Inc.*)

tears and loss of seal. The solution is to use extension bonnets (Fig. 22). Extension bonnets allow the packing box area of the control valve to be warmed by ambient temperatures, thus preventing frost from forming on the stem and packing box areas. The length of the extension bonnet depends on the application temperature and insulation requirements. The colder the application, the longer the extension bonnet required.

Valves for Nuclear Service. Nuclear-service valves must meet many special requirements. Strict compliance with government regulations is required. Manufacturers in many areas must provide documented evidence that components were manufactured, inspected, and tested by proven techniques performed by qualified personnel according to documented procedures.

Valves Subject to Sulfide Stress Cracking. NACE International (National Association of Corrosion Engineers) is a technical society concerned with corrosion. NACE has a standard that provides guidelines for the selection of materials that are resistant to sulfide stress cracking. Sulfide stress cracking is a concern with oil and gas that contains hydrogen sulfide (commonly called sour gas service).

The NACE standard lists the types of materials and their heat-treating conditions that are most resistant to sulfide stress cracking. In some areas, conformance of portions of the standard is required by law.

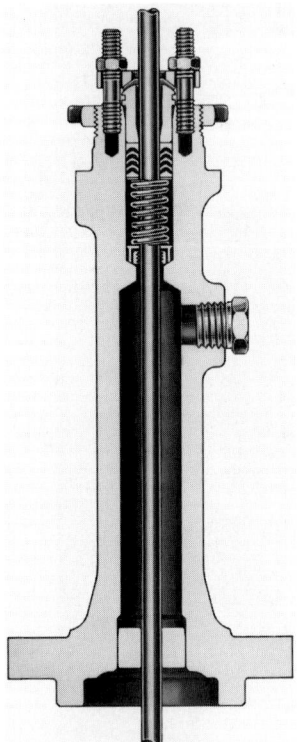

FIGURE 22 Typical extension bonnet to help isolate packing from process temperatures. (*Fisher Controls International, Inc.*)

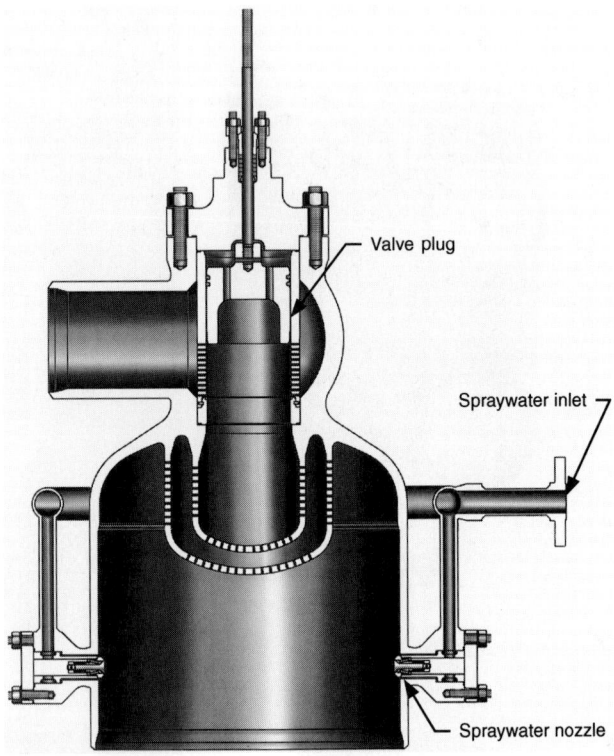

FIGURE 23 Steam conditioning valve that combines steam pressure reduction and desuperheating. (*Fisher Controls International, Inc.*)

Steam Conditioning Valves. A steam conditioning valve is used for the simultaneous reduction of steam pressure and temperature to the level required for a given application (Fig. 23). Frequently, these applications have high inlet pressures and temperatures and require significant reductions of both. Steam conditioning valves are best as forged and fabricated bodies that can better withstand steam loads at elevated pressures and temperatures. Forged materials permit higher design stresses, improved grain structure, and an inherent material integrity over cast valve bodies. The forged construction also allows pressure ratings to Class 4500 or PN 760.

Spraywater is provided to the valve to cool the steam. The spraywater nozzles are near the flow vena contracta below the main valve seat. The water is injected at a point of high velocity and turbulence, where it is distributed quickly and evenly throughout the flow stream.

The forged and fabricated design allows the manufacturer to provide different sizes and pressure class ratings for the inlet and outlet connections to more closely match the adjacent piping.

Other advantages of combining the pressure reduction and desuperheater function include:

- improved spraywater mixing because of the optimum utilization of the turbulent expansion zone downstream of the pressure reduction elements
- improved rangeability
- increased noise abatement; there is additional attenuation of noise as a result of the spraywater injection

- in some designs, improved response time because of an integrated feedforward capability
- ease of installing and servicing only one device

CONTROL VALVE PERFORMANCE [1]

Global competition in the process industries is putting increasing pressure on companies to provide the highest quality products and the maximum plant throughputs with fewer resources. While meeting these demands, companies also must meet ever-changing customer needs.

A company makes a profit through the production of a quality product that conforms to a set of specifications. Any deviation from the established specification means lost profit because of excessive material use, reprocessing costs, or wasted product. Reducing process variability through better process control allows optimization of the process and the production of products right the first time.

The control valve is often overlooked when process variability is reduced because its impact on dynamic performance is not realized. Extensive studies of control loops indicate that as many as 80% of the loops do not do an adequate job of reducing process variability. Furthermore, the control valve was found to be a major contributor to this problem for a variety of reasons.

To verify performance, manufacturers must test their products under dynamic process conditions. Evaluating control valve assemblies under closed-loop conditions provides the only true measure of variability performance (Fig. 24). Closed-loop performance data prove significant reductions in process variability can be achieved by choosing the right control valve for the application.

For best performance, valves must be optimized or developed as a unit. Some of the most important design considerations include:

- dead band
- actuator–positioner design
- valve response time
- valve type and sizing

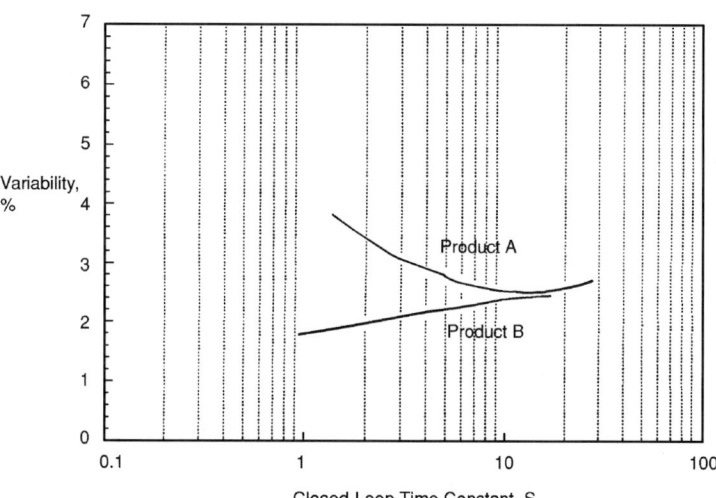

FIGURE 24   Testing under dynamic process conditions can demonstrate control valve performance. (*Fisher Controls International, Inc.*)

Dead Band

Dead band is a major contributor to excess process variability, and control valve assemblies can be a primary source of dead band in an instrumentation loop.

Dead band has many causes, but friction and backlash in the control valve, along with shaft wind-up in rotary valves, and relay dead zone are some of the more common forms. Because most control actions for regulatory control consist of small changes (1% or less), a control valve with excessive dead band might not even respond to many of these small changes. A well-engineered valve should respond to signals of 1% or smaller to provide an effective reduction in process variability.

Actuator–Positioner Design

Actuator and positioner design must be considered together. The combination of these two pieces of equipment greatly affects the static performance (dead band), as well as the dynamic response of the control valve assembly and the overall air consumption of the valve instrumentation.

The most important characteristic of a good positioner for process variability reduction is that it be a high-gain device.

Typical two-stage positioners use pneumatic relays at the power amplifier stage. Relays are preferred over spool valves for this purpose because relays can provide a high-power gain that gives excellent dynamic performance with minimal steady-state air consumption.

Positioner designs are changing dramatically, with microprocessor devices becoming increasingly popular. These microprocessor-based positioners provide dynamic performance equal to the best conventional two-stage pneumatic positioners. They also provide valve monitoring and diagnostic capabilities to help ensure that initial good performance does not degrade with use.

High-performance positioners with both high static and dynamic gain provide the best overall process variability performance for any given valve assembly.

Valve Response Time

For optimum control of many processes, it is important that the valve reach a specific position quickly. A quick response to small signal changes (1% or less) is one of the most important factors in providing optimum process control. In automatic, regulatory control, the bulk of the signal changes received from the controller is for small changes in position. If a control valve assembly can quickly respond to these small changes, process variability is improved.

Valve Type and Characterization

The style of valve used and the sizing of the valve can have a great impact on the performance of the control valve assembly in the system. Although a valve must be of sufficient size to pass the required flow under all possible contingencies, a valve that is too large for the application is a detriment to process optimization.

Flow capacity of the valve is also related to the style of valve through the inherent characteristic of the valve. The inherent characteristic is the relationship between the valve flow capacity and the valve travel when the differential pressure drop across the valve is held constant. Also see the discussion in the Valve Capabilities and Capacities section.

For process optimization, the installed flow characteristic of the entire process and the gain are more important.

The best process performance occurs when the required flow characteristic is obtained through changes in the valve trim rather than through the use of positioner cams or other methods. Proper selection of a control valve designed to produce a reasonably linear installed flow characteristic over the operating range of the system is a critical step in ensuring optimum process performance.

Valve Sizing

Process variability reduction efforts are limited by valves being oversized. The oversizing results from using line-size valves, especially with high-capacity rotary valves, as well as the conservative addition of multiple safety factors at different stages in the process design.

The second way oversized valves hurt process variability is that an oversized valve is likely to operate more frequently at lower valve openings, where seal friction can be greater, particularly in rotary valves. Because an oversized valve produces a disproportionately large flow change for a given increment of valve travel, this phenomenon can greatly exaggerate the process variability associated with dead band that is due to friction.

When selecting a valve, consider the valve style, inherent characteristic, and valve size that will provide the broadest possible control range for the application.

Economic Results

Consideration of the performance factors discussed can have a dramatic impact on the economic results of an operating plant. More and more control valve users focus on dynamic performance parameters such as dead band, response times, and installed gain (under actual process load conditions) as a means to improve process-loop performance. Although it is possible to measure many of these dynamic performance parameters in an open-loop situation, the impact these parameters have becomes clear when closed-loop performance is measured (Fig. 24).

Process industries have become increasingly aware that control valve assemblies play an important role in loop/unit/plant performance. They also have realized that traditional methods of specifying a valve assembly are no longer adequate to ensure the benefits of process optimization. Although important, such static performance indicators as flow capacity, leakage, materials compatibility, and bench performance data are not sufficiently adequate to deal with the dynamic characteristics of process control loops.

Parts of the loop cannot be treated individually to achieve coordinated loop performance. Likewise, performance of any part of the loop cannot be evaluated in isolation. Isolated tests under nonloaded, bench-type conditions will not provide performance information that is obtained from testing the hardware under actual process conditions.

VALVE SELECTION

Picking a control valve for a particular application once was simple. Usually only one general type of valve was considered—the sliding-stem valve. Each manufacturer offered a product suitable for the job, and the choice depended on obvious considerations such as cost, delivery, vendor relationships, and user preference. Selection is now considerably more complicated—especially for engineers with limited experience or for those who have not kept up with changes in the control valve industry. For many applications, an assortment of sliding-stem, ball, and butterfly valves is available. Some are touted as universal valves for almost any size and service, and others are claimed to be optimum solutions for narrowly defined needs. Like most decisions, the selection of a control valve involves a great number of variables. Presented here is an overview of the selection process.

General Selection Criteria

Most of the considerations that guide the selection of the valve type are rather basic. However, there are some matters that might be overlooked by users whose familiarity is limited. A checklist includes the following:

1. Body pressure rating and limits
2. Size and flow capacity
3. Flow characteristics and rangeability
4. Temperature limits
5. Shutoff leakage
6. Pressure drop (shutoff and flowing)
7. End connection requirements
8. Material compatibility and durability
9. Lifecycle cost

Pressure Ratings

The most common pressure ratings for steel and stainless-steel valves are Classes 150, 300, and 600 [2], [3] or PN 20, PN 50, or PN 100. For a given body material, each rating prescribes a profile of maximum pressure that decreases with temperature according to the strength of the material. Each material also has minimum and maximum service temperatures based on loss of ductility or loss of strength. For most applications, the required pressure rating is dictated by the application. However, because not all products are available for all ratings, it is an important consideration for selection.

Operating Temperature

Required temperature capabilities are usually also a foregone conclusion, but one that is likely to further narrow the range of selections. Considerations here include the strength or ductility of the body material as well as the relative thermal expansion of the valve internal parts. Temperature limits also might be imposed as a result of disintegration of soft parts at high temperatures or loss of resiliency at low temperatures. The soft materials under consideration include various elastomers, plastics, and PTFE. They might be found in parts such as seat rings, seal or piston rings, packing, rotary shaft bearings, and butterfly valve liners. Typical upper temperature limits for elastomers are in the 200–350°F (93–177°C) range, and the general limit for PTFE is 450°F (232°C).

Temperature affects valve selection by excluding certain valves that do not have high- or low-temperature options, such as lined butterfly valves. It also might have some effect on valve performance. For instance, going from PTFE to metal seals for high temperatures generally increases the shutoff leakage flow. Similarly, high-temperature metal bearing sleeves in rotary valves impose more friction load on the shaft than PTFE bearings do, so that the shaft cannot withstand as high a pressure-drop load at shutoff.

Selection of valve packing is very often based on the service temperature. Two packing types, PTFE V rings and graphite, meet most packing requirements. These materials have proven reliable, inexpensive, and effective.

PTFE V-ring packing is composed of solid rings of molded PTFE. Generally, in a given packing set, there are two or more packing rings with a V cross section, a male adaptor, and a female adaptor. The packing can be used over a temperature range of −40 to 450°F (−40 to 232°C) and for nearly all chemicals. PTFE packing can be used with a spring or as jam-type packing. Stem friction is low. This packing is the preferred packing material for most applications. Packing is discussed in greater detail in the Material Selection section.

Graphite packing systems are used mainly for temperatures above 450°F (232°C). They are composed of graphite ribbon rings, graphite filament rings, and sacrificial zinc washers. The graphite rings perform the sealing function and the zinc washers protect the valve stem from galvanic corrosion.

MATERIAL SELECTION

Material compatibility and durability are complex considerations. The consideration might be corrosion by the process fluid, erosion by abrasive material, flashing, cavitation, or simply one of process pressure and temperature. The material used for piping is a good predictor of control-valve body material. However, because the velocity is higher in valves, other factors must also be considered. When these items are included, often valve and piping materials will differ. Trim materials are usually a function of the body material, temperature range, and qualities of the fluid. When a body material other than carbon, alloy, or stainless steel is required, the use of alternate valve types, such as lined or bar stock, should be considered.

Control valves are required to function with precision in some very extreme environments. A number of factors must be considered to ensure that a material will perform properly in service. These factors fall primarily into two categories: suitability to function mechanically and compatibility with the process environment. These constraints conflict in many instances, making it difficult to satisfy all considerations with a single material. In these cases, the best compromise must be identified.

Carbon-Steel Bodies and Bonnets

The most standard material for control-valve bodies is carbon steel such as WCB or WCC steel. Carbon steel is easily cast, welded, and machined. It is used for a large majority of process applications because of its low cost and reliable performance. Its use is strongly recommended over any other material if possible because of its broad availability and low cost.

Alloy-Steel Bodies and Bonnets

When higher temperatures or pressures are involved, alloy steels are often specified. Most are steels with chromium or molybdenum added to enhance their resistance to tempering and graphitization at elevated temperatures. The chromium and molybdenum additions also increase their resistance to erosion in flashing applications. Among the more popular materials are WC9 and WC6 steel.

Stainless-Steel Bodies and Bonnets

The most common stainless steel used for bodies and bonnets is CF8M, which is the cast version of S31600. With its nominal 19½% Cr, 10½% Ni, 21½% Mo composition, CF8M is a relatively low-cost material with good high-temperature properties and excellent resistance to corrosion.

Selection of Materials

Comparing pressure-temperature ratings in ASME B16.34 is much simpler when the ratings are presented in graphic form. The first discovery that is made is that the Class 150 ratings for WCB, WCC, WC9, and C5 are identical over their common temperature ranges, and that CF8M is only rated slightly lower at temperatures below 550°F (288°C). The second discovery is that the rating plots for these materials in all other classes have the same shapes, and all that changes is the y-axis scale for the allowable pressures. Figure 25 is a plot of the pressure-temperature ratings, where the allowable pressure has been normalized to 100 percent. This curve is representative of the relative pressure-temperature ratings of the materials for ANSI Classes 300 through 4500. If the material providing the maximum allowable pressure at any temperature is determined from the plot, three material regimes can be established. From ambient temperature to 750°F (399°C), WCC has the highest pressure-temperature ratings of this group of materials. From 750°F to 950°F (399–510°C), WC9 has the highest ratings, and from 950°F (510°C) to higher temperatures, CF8M has the highest ratings.

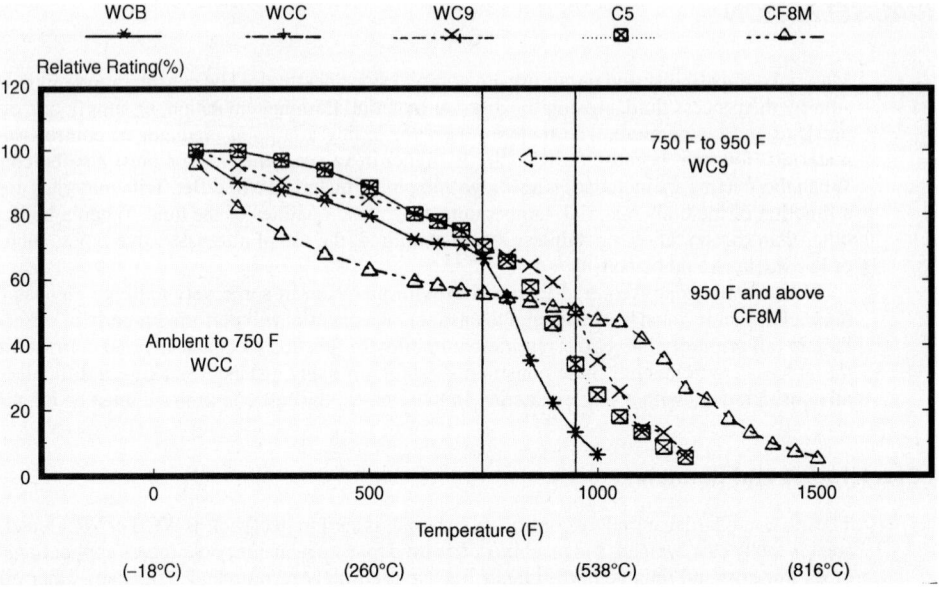

FIGURE 25   Control valve pressure ratings depend on temperature and material. This chart compares normalized ratings as a function of these factors. These are relative ASME B16.34 pressure-temperature ratings (ratings vs. WC9 steel at room temperature). (*Fisher Controls International, Inc.*)

Trim Parts

Valve trim components have much different material requirements than valve bodies and bonnets. Trim parts are not pressure retaining, so they are not directly safety related. However, because the trim components provide the flow control, they are very important to valve performance. Trim materials must usually have excellent resistance to corrosion by the process fluid. If not, adequate flow control and mechanical stability will not be maintained. Each component must have other characteristics, depending on the valve design, process fluid, and application.

Valve Packing [1]

Most control valves use packing boxes with the packing retained and adjusted by a flange and packing nuts (Fig. 26). PTFE and graphite materials are often used as packing rings.

PTFE V-Ring Packing.

- plastic material with the ability to minimize friction
- molded in V-shaped rings that are spring loaded and self-adjusting in the packing box; lubrication is not required
- resistant to most known chemicals, except molten alkali metals
- requires extremely smooth stem finish to seal properly; packing will leak if the stem or packing surface is damaged
- normally used from −40°F to +450°F (−40 to +232°C)
- not suitable for nuclear service because PTFE is easily destroyed by radiation.

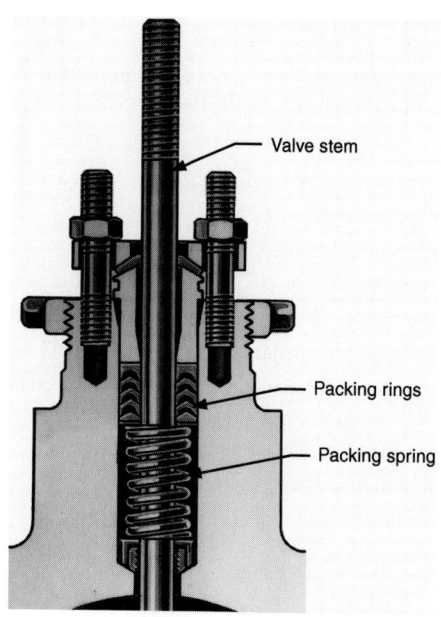

FIGURE 26 Typical spring-loaded PTFE V-ring packing. (*Fisher Controls International, Inc.*)

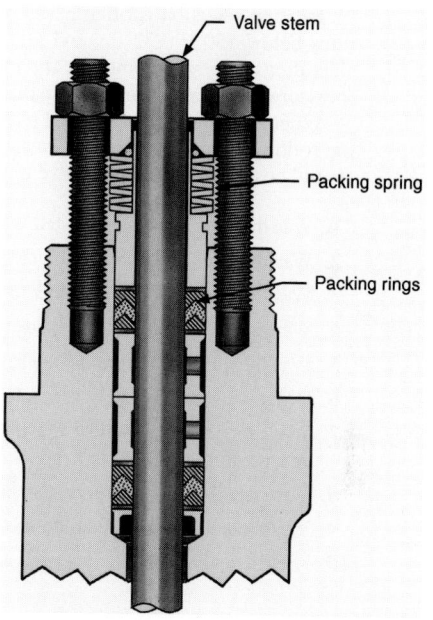

FIGURE 27 One example of the new packing technologies is the ENVIRO-SEAL packing system. (*ENVIRO-SEAL is a mark owned by Fisher Controls International, Inc.*)

Laminated and Filament Graphite Packing.

• suitable for high-temperature nuclear service or where low chloride content is desirable (Grade GTN)

• provides excellent sealing, high thermal conductivity, and long service life, but produces high stem friction and resultant hysteresis

• impervious to most hard-to-handle fluids and high radiation

• can be used for temperatures from cryogenic to 1200°F (649°C)

• lubrication is not required, but an extension bonnet or steel actuator yoke should be used when packing box temperature exceeds 800°F (427°C)

Improved Packing Technology.   Environmental concern over fugitive emissions has resulted in governmental regulations that restrict the amount of emissions of various fluids that can be permitted. This concern, along with the economic concern over the loss of valuable process fluids, made it necessary to improve the sealing of valve stems and shafts. New valve packing technologies meet these challenges. Also, the new technologies extend packing seal life and performance.

One example of the new packing technologies is the ENVIRO-SEAL packing system (Fig. 27). Improved sealing is made possible by:

• anti-extrusion rings that contain the pliable seal material

• proper alignment of the valve stem or shaft within the bonnet bore

• managed packing force provided through springs

• minimizing the number of seal rings to reduce consolidation, friction, and thermal expansion

Packing selection is often based on process temperature; that is, PTFE is selected for temperatures below 450°F (232°C), and graphite is selected for temperatures above 450°F (232°C). Considerations also should include the effect of packing friction on process control, seal performance (pressure/temperature/emission sealing capabilities), and service life.

When selecting a packing seal technology for fugitive emission service, it is important to answer the following questions to help ensure long-term performance. Detailed answers based on test data should be available from the valve manufacturer.

1. Was the packing system tested within the valve style to be used?
2. Was the packing system subjected to multiple operating cycles?
3. Was the packing system subjected to multiple thermal cycles?
4. Were packing adjustments made during the performance test?
5. Was the packing system tested at or above the service conditions of the planned application?
6. Did testing of packing systems for rotary valves include deflection of the valve shaft?
7. Was stem leakage monitored using a procedure covered by government regulations or an industry-accepted practice?
8. Were the packing components examined for wear after the completion of each test?
9. Was the compression load on the packing measured as the test progressed?
10. Are the test results documented and available for review?

VALVE CAPABILITIES AND CAPACITIES

Flow characteristic, rangeability, pressure drop capabilities, end connection style, shutoff, and capacity are very important to consider when you select a valve. Valve manufacturers publish these characteristics as specifications in sales literature or data sheets.

Flow Characteristic

The next selection criterion—inherent flow characteristic—refers to the pattern in which the flow at constant pressure drop changes according to valve position. Typical characteristics are quick opening, linear, and equal percentage. The choice of characteristic has a strong influence on the stability or controllability of the process, because it represents the change of valve gain relative to travel. Most control valves are carefully "characterized" to exhibit a certain flow characteristic by means of contours on a plug, cage, or ball element. Some valves are available in a variety of characteristics to suit the application, while others offer little or no choice.

To determine the best flow characteristic for a given application quantitatively, a dynamic analysis of the control loop can be performed. In most cases, however, this is unnecessary; reference to established rules will suffice. Figure 28 illustrates typical flow characteristic curves. The quick-opening flow characteristic provides for maximum change in flow rate at low valve travels with a fairly linear relationship. Additional increases in valve travel give sharply reduced changes in flow rate, and when the valve plug nears the wide open position, the change in flow rate approaches zero. In a control valve, the quick-opening valve plug is used primarily for on–off service, but it is also suitable for many applications where a linear valve plug would normally be specified.

The linear flow characteristic curve shows that the flow rate is directly proportional to the valve travel. This proportional relationship produces a characteristic with a constant slope so that with constant pressure drop, the valve gain will be the same at all flows. The linear valve plug is commonly specified for liquid-level control and for certain flow control applications requiring constant gain.

In the equal-percentage flow characteristic, equal increments of valve travel produce equal percentage changes in the existing flow. The change in flow rate is always proportional to the flow rate

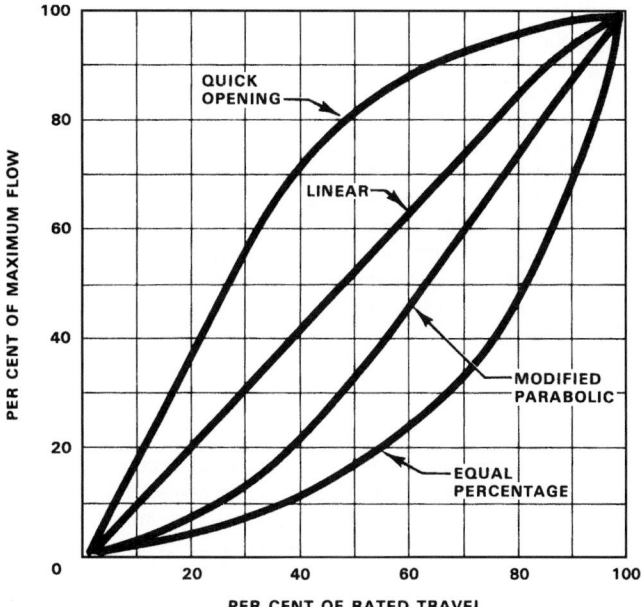

FIGURE 28 Many control valves offer a choice of control characteristics. The selection to match process requirements is guided by simple rules. Adherence to the guidelines will help ensure a stable process operation. (*Fisher Controls International, Inc.*)

just before the change in valve plug, disk, or ball position is made. When the valve plug, disk, or ball is near its seat and the flow is small, the change in flow rate will be small. With a large flow, the change in flow rate will be large. Valves with an equal-percentage flow characteristic are generally used on pressure control applications, and on other applications where a large percentage of the pressure drop is normally absorbed by the system itself, with only a relatively small percentage available at the control valve. Valves with an equal-percentage characteristic should also be considered where highly varying pressure drop conditions can be expected. Table 2 lists characteristic recommendations by process type.

Rangeability

One aspect of flow characteristic is its rangeability, which is the ratio of maximum and minimum controllable flow rates. Exceptionally wide rangeability might be required for certain applications to handle wide load swings or a combination of start-up, normal, and maximum working conditions. Rotary valves, especially partial ball valves, normally have greater rangeability than sliding-stem varieties.

Pressure Drop

The maximum pressure drop the valve can tolerate at shutoff and when partly or fully open is an important selection criterion. Sliding-stem valves are generally superior in both regards because of the rugged, well-supported design of their moving parts. Unlike most sliding-stem valves, many rotary valves are limited to pressure drops well below the body pressure rating, especially under flowing conditions, because of dynamic stresses imposed on the disk or ball segment by high-velocity flow.

Noise and cavitation are two considerations that, although unrelated, are often grouped together because they both usually accompany high pressure drops and flow rates. They are handled by special modifications of more or less standard valves. Cavitation is the noisy and potentially damaging implosion of bubbles formed when the pressure of a liquid momentarily dips below its vapor pressure through a constriction at high velocity. In controlling gases and vapors, noise results from the turbulence associated with high-velocity streams. When cavitation or noise is judged likely to be a problem, its severity must be predicted from the valve's specifications according to well-known techniques, and valves with better specifications must be sought if necessary. Cavitation-control and noise-control trims for various degrees of severity are widely available in regular sliding-stem valves—at a progressive penalty in terms of cost and flow capacity. Rotary valves have more limited noise- and cavitation-control options and are also much more susceptible to cavitation and noise at a given pressure drop. Please refer to subsequent articles in this handbook section concerning control-valve noise and cavitation.

End Connections

At some point in the selection process, the valve end connections must be considered. The question to be answered is simply whether the desired connection style is available in the valve style being considered. In some situations, end connections can quickly limit the selection or dramatically affect the price. For instance, if a piping specification calls for welded connections only, the choice might be limited to sliding-stem valves. The few weld-end butterfly and ball valves that are available are rather expensive.

Shutoff Capability

Some consideration usually must be given to a valve's shutoff capability, which ordinarily is rated in terms of classes specified in ANSI/FCI 70-2 [4] or IEC 534-4. In actual service, shutoff leakage depends on many factors, including pressure drop, temperature, the condition of the sealing surfaces, and—very importantly for sliding- stem valves—the force load on the seat. Because shutoff ratings are based on standard test conditions (Table 3), which might be very different from service conditions, service leakage cannot be predicted very well. However, the shutoff classes provide a good basis for comparisons among valves of similar configuration.

Tight shutoff is particularly important in high-pressure valves because leakage can cause seat damage, leading to ultimate destruction of the trim. Special precautions in seat materials, seat preparation, and seat load are necessary to ensure success. Valve users tend to overspecify shutoff requirements, incurring unnecessary cost. Actually, very few throttling valves really need to perform double duty as tight block valves. Since tight shutoff valves generally cost more initially and to maintain, serious consideration is warranted.

Flow Capacity

The criterion of capacity or size can be an overriding constraint on selection. For very large lines, sliding-stem valves are much more expensive than rotary types. On the other hand, for very small flows, a suitable rotary valve might not be available. If the same valve is desired to handle a significantly larger flow at a future time, a sliding-stem valve with replaceable, restricted trim might be indicated. Rotaries generally have much higher maximum capacity than sliding-stem valves for a given body size. This fact makes rotaries attractive in applications where the pressure drop available is rather small. But it is of little or no advantage in high pressure drop applications such as pressure regulation or letdown.

At the risk of overgeneralizing, you can simplify the process of selection roughly as follows. For most general applications it makes sense, both economically and technically, to use sliding-stem

TABLE 3 Maximum Leakage and Test Conditions for Control-Valve Leakage Classes

ANSI B16.104-1976*	Maximum Leakage*	Test Medium	Pressure and Temperature
Class II	0.5% valve capacity at full travel	Air	Service ΔP or 50 psid (3.4-bar differential), whichever is lower, at 50–125°F (10–52°C)
Class III	0.1% valve capacity at full travel	Air	Service ΔP or 50 psid (3.4-bar differential), whichever is lower, at 50–125°F (10–52°C)
Class IV	0.01% valve capacity at full travel	Air	Service ΔP or 50 psid (3.4-bar differential), whichever is lower, at 50–125°F (10–52°C)
Class V	5×10^4 mL/min/psid/in. port dia (5×10^{12} m^3/s/bar differential/mm port dia)	Water	Service ΔP at 50 to 125° F (10–52°C)

Class VI	Nominal port diameter		Bubbles/min	mL/min	Air	Service ΔP or 50 psid (3.4-bar differential), whichever is lower, at 50–125°F (10–52°C)
	in.	mm				
	1	25	1	0.15		
	1½	38	2	0.30		
	2	51	3	0.45		
	2½	64	4	0.60		
	3	76	6	0.90		
	4	102	11	1.70		
	6	152	27	4.00		
	8	203	45	6.75		

*Copyright 1976 Fluid Controls Institute, Inc. Reprinted with permission.

valves for the lower ranges, ball valves for intermediate capacities, and high-performance butterfly valves for the very largest sizes. For the very least demanding services, in which price is the dominant consideration, one might consider economy sliding-stem valves for the small-size applications and butterfly valves for the largest.

For sizes of NPS 1/2 to NPS 3 or DN 15 to DN 80, general-purpose sliding-stem valves provide an exceptional value. For a minimal price premium over rotary products, they offer unparalleled performance, flexibility, and service life. The premium for these devices over rotary products is warranted. For severe service applications, the most frequently used, and often the only available product is the sliding-stem valve.

Applications ranging from NPS 4 to NPS 6 or DN 100 to DN 150 are best served by such transitional valve styles as the eccentric plug valve or the ball valve. These products have excellent performance and lower cost. They also offer higher capacity levels than globe designs.

In sizes NPS 8 or DN 200 and larger, pressures and pressure drops are much lower. This gives rise to the possibility of using high-performance butterfly valves for most situations. These valves are economical, offer tight shutoff, and provide good control capability. They provide cost and capacity benefits well beyond those of globe and ball valves.

Special considerations require special valve solutions. There are valve designs and special trims available to handle high-noise applications, cavitation, high pressure, high temperature, and combinations of these conditions.

The obvious point here is different types of valves are appropriate for use in different size ranges, because they provide the most cost-effective solution in each given instance. If you stick with the same type of valve over a wide size range, you sacrifice either performance at the low end or economy at the high end, or both.

After going through all the other criteria for a given application, people who specify valves often find that they can use several types of valves. From there on, selection is a matter of price versus capability as discussed here—coupled with the inevitable personal and institutional preferences. Because no single control-valve package is cost effective over the full range of applications that are normally encountered, it is important to keep an open mind for alternative choices.

VALVE SIZING

It used to be common practice in the industry to select valve size strictly as a function of pipe size. Soon it became apparent that this practice contributed to very poor control and resulting process problems. The wide range of flow, pressure, and fluid conditions required a more in-depth selection methodology. With time, methods were developed and the days of selecting a valve based on pipe size are gone forever.

Selecting the correct valve size for a given application requires a knowledge of the flow and process conditions the valve will actually see in service as well as information on valve function and style. Sizing valves is based on a combination of theory and empirical data. The results are predictable, accurate, and consistent.

Early efforts in the development of valve sizing centered around liquid flow. Daniel Bernoulli was one of the early experimenters who applied theory to liquid flow. Subsequent experimental modifications to this theory produced a useful liquid-flow equation.

$$Q = C_v \sqrt{\frac{P_1 - P_2}{G}}$$

where Q = flow rate
   C_v = valve sizing coefficient, determined by testing
   P_1 = upstream pressure
   P_2 = downstream pressure
   G = liquid specific gravity

This equation rapidly became widely accepted for sizing valves on liquid service, and manufacturers of valves began testing and publishing C_v data in their catalogs.

It was inevitable that the good results obtained from the C_v equation would strongly tempt its use to predict the flow of gas. The results, however, were inaccurate. Modifications were made to the equations over time, with consequent improvement of results. Various companies used techniques they developed, but there was no common formulation until the Instrument Society of America (ISA) put forth its standardized guidelines.

In order to assure uniformity and accuracy, the procedures for measuring flow parameters and for valve sizing are addressed by ISA standards. Measurement of C_v and related flow parameters is covered extensively in ANSI/ISA S75.02, 1981 [5]. The basic test system and hardware installation are outlined so that coefficients can be tested to an accuracy of ± 5 percent. Water is circulated through the test valve at specified pressure differentials and inlet pressures. Flow rate, fluid temperature, inlet and differential pressure, valve travel, and barometric pressure are all measured and recorded. This yields sufficient information to calculate necessary sizing parameters. Numerous tests must be performed to arrive at the values published by the valve manufacturer for use in sizing. It is important, also, that these factors be based on tests, not estimates, because the results are not always predictable.

Basic Sizing Procedure

The procedure by which valves are sized for liquid flow is straightforward. Again, to ensure uniformity and consistency, a standard exists that delineates the equations and correction factors to be used for a given application (ANSI/ISA S75.01-1985 [6]).

The simplest case of liquid-flow application involves the basic equation developed earlier. Rearranging the equation so that all of the fluid and process-related variables are on the right-hand side, the expression for the valve C_v required for the particular application is:

$$C_v = \frac{Q}{\sqrt{(P_1 - P_2)/G}}$$

Based on a given flow rate and pressure drop, a required C_v value can be calculated. This C_v can then be compared to C_v values for a particular valve size and valve design. Generally, the required C_v should fall in a range of between 70% and 90% of the selected valve's C_v capability. Allowance for minimum and maximum flow pressure conditions should also be considered.

Once a valve has been selected and C_v is known, the flow rate for a given pressure drop, or the pressure drop for a given flow rate, can be predicted by substituting and solving for the appropriate quantities in the equation.

This basic liquid equation covers conditions governed by the test assumptions. Unfortunately many applications fall outside the bounds of these standards and therefore outside of the basic liquid-flow equation. Rather than develop special flow equations for all of the possible deviations, it is possible to account for different behavior with the use of simple correction factors. These factors, when incorporated, change the form of the equation to the following:

$$C_v = \frac{Q}{NF_p F_R \sqrt{P_1 - P_2/G}}$$

where $N =$ numerical coefficient for unit conversion
   $F_p, F_R =$ correction factors

Choked Flow

A plot of the basic equation (Fig. 29) implies that flow can be increased continually by simply increasing the pressure differential across the valve. In reality the relationship given by this equation

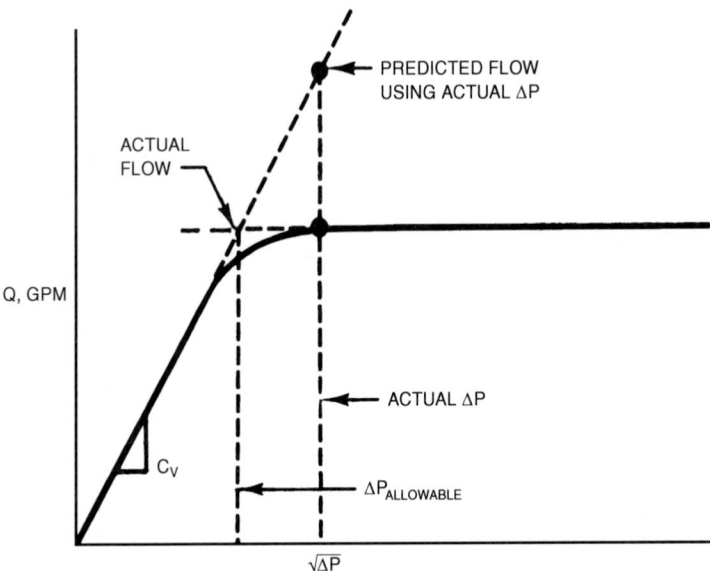

FIGURE 29 Sizing equations suggest that as the pressure drop is increased, flow will increase proportionally—forever. In reality, this relationship holds only for certain conditions. As the pressure drop is increased, choked flow caused by the formation of vapor bubbles in the flow stream imposes a limit on liquid flow. A similar limitation on flow of gases is realized when velocity at the valve vena contracta reaches sonic velocity. These choked-flow conditions must be considered in valve sizing. (*Fisher Controls International, Inc.*)

holds for only a limited range. As the pressure differential is increased, a point is reached where the realized flow increase is less than expected. This phenomenon continues until no additional flow increase occurs in spite of increasing the pressure differential. This condition of limited maximum flow is known as choked flow. This phenomenon occurs on both liquids and gases. It is necessary to account for the occurrence of choked flow during the sizing process to ensure against undersizing a valve. Predictions must be made using a valve recovery coefficient F_L for liquids and X_T for gases.

Viscous Flow

One of the assumptions implicit in the sizing procedures presented up to this point is that of fully developed, turbulent flow. In laminar flow, all fluid particles move parallel with one another in an orderly fashion with no mixing of the fluid. Conversely, turbulent flow is highly random in local velocity direction and magnitude. While there is certainly net flow in a particular direction, instantaneous velocity components in all directions exist within this net flow. Significant fluid mixing occurs in turbulent flow. The factor F_R is a function of the Reynolds number and describes the degree of turbulent flow. It can be determined by a simple nomograph procedure.

Piping Considerations

When a valve is installed in a field piping configuration which is different than the standard test section, it is necessary to account for the effect of the altered piping on flow through the valve. Recall that the standard test section consists of a prescribed length of straight pipe up- and downstream of the valve. Field installation might require elbows, reducers, and tees, which will induce additional pressure losses adjacent to the valve. To correct for this, the factor F_p is introduced.

Gas and Steam Sizing

Although most comments so far pertain to liquid sizing, they closely parallel the procedures used for air, gas, and steam valve sizing. The only additional steps involve correction for the physical properties of the particular gas and pressure ratio factors which determine the degree of compression and predict choked flow. The general form of the sizing equation for compressible fluids is:

$$C_v = \frac{Q}{N F_p P_1 Y \sqrt{X/GT_1 Z}}$$

where
$Y =$ expansion factor
$X = \Delta P/P_1$
$T_1 =$ temperature
$Z =$ compressibility factor

For additional information on valve sizing, consult the referenced ISA publications or the manufacturer's literature. Computer sizing programs are available, which alleviate the need to solve complex equations manually and which provide exceptional accuracy.

ACTUATORS

Actuators are the distinguishing elements between valves and control valves. The actuator industry has evolved to answer a wide variety of process needs and user desires. Actuators are available with many designs, power sources, and capabilities. Proper selection involves process knowledge, valve knowledge, and actuator knowledge. A control valve can perform its function only as well as the actuator can handle the static and dynamic loads placed on it by the valve. Proper selection and sizing are, therefore, very important. The actuator represents a significant portion of the total control-valve package price, and careful selection can minimize costs.

The range of actuator types and sizes on the market is so great that it seems the selection process might be highly complex. It is not. With a few rules in mind and knowledge of your fundamental needs, the selection process can be very simple.

The following parameters must be known at the beginning of the selection process. They are important because they quickly narrow the selection process.

1. Power source availability
2. Failure-mode requirements
3. Torque or thrust requirements (actuator capability)
4. Control functions
5. Economics
6. Size, modular construction, easy maintenance

Power Source

The power source available at the location of a valve can often determine what type of actuator to choose. Typically, valve actuators are powered either by compressed air or by electricity. However, in some cases water pressure, hydraulic fluid, or even pipeline pressure can be used. The majority of actuators sold today use compressed air for operation. They operate at supply pressures from as low as 15 psig (1.0 bar) to a maximum of about 150 psig (10.4 bars).

Since most plants have ready availability of both electricity and compressed air, the selection depends on the ease and cost of furnishing either power source to the actuator location. One must also consider reliability and maintenance requirements of the power system and their effect on subsequent valve operation. Consideration should be given to providing backup operating power to critical plant loops.

Failure Mode

The overall reliability of power sources is quite high. However, many loops demand specific valve action should the power source ever fail. Desired action on signal failure might be required for safety reasons or for the protection of equipment. Fail systems store energy, either mechanically in springs or pneumatically in volume tanks or hydraulic accumulators. When power fails, the fail systems are triggered to drive the valves to the required position and then maintain this position until resumption of normal operation. In many cases the process pressure is used to ensure or enhance this action.

Actuator designs are available that allow a choice of failure mode between failing open, failing closed, or holding in the last position. Many actuator systems incorporate failure modes at no extra cost. Spring and diaphragm types are inherently fail open or closed. Electric actuators nearly always hold in their last position.

Actuator Capability

An actuator must have sufficient thrust or torque for the application. In some cases this requirement can dictate actuator type as well as power-supply requirements. For instance, large valves requiring a high thrust might be limited to only electric or electrohydraulic actuators because of a lack of pneumatic actuators with sufficient torque capability. Conversely, electrohydraulic actuators would be a poor choice for valves with very low thrust requirements. The matching of actuator capability with valve-body requirement is best left to the control valve manufacturer, as there is considerable variation in frictional and fluid forces from valve to valve.

Control Functions

Knowledge of the required actuator functions will most clearly define the options available for selection. These functions include the actuator signal (such as pneumatic, electric, analog, frequency), signal range, ambient temperatures, vibration levels, operating speed, cycle frequency, and quality of control required.

Generally, signal types are grouped as being either two position (on–off) or analog (throttling). On–off actuators are controlled by two-position electric, electropneumatic, or pneumatic switches. This is the simplest type of automatic control and the least restrictive in terms of selection.

Throttling actuators have considerably higher demands put on them for both compatibility and performance. A throttling actuator receives its input from an electronic or pneumatic instrument that measures the controlled process variable. The actuator must then move the final control element (valve) in response to the instrument signal in an accurate and timely fashion to ensure effective control. The two primary additional requirements for throttling actuators are compatibility with instrument signal and better static and dynamic performance to ensure loop stability.

Compatibility with instrument signals is inherent in many actuator types, or it can be obtained with add-on equipment. But the high-performance characteristics required of a good throttling actuator cannot be bolted on. Low hysteresis and minimal dead band must be designed into actuators.

Stroking speed, vibration, and temperature resistance must also be considered if critical to the application. Stroking speed is generally not critical; however, flexibility to adjust it is desirable. With liquid service, fast stroking speeds can be detrimental because of the possibility of water hammer.

Vibration or mounting position can cause problems as the actuator weight, combined with the weight of the valve, might require bracing. If extremes of temperature or humidity are to be experienced by the control valve, this information is essential to the selection process. Many actuators contain either elastomeric or electronic components, which might be subject to degradation from high humidity or temperature.

Economics

An evaluation of the economics in actuator selection requires combining initial cost, installation, maintenance, and reliability factors. A simple actuator, such as a spring-and-diaphragm actuator, has few moving parts, is easy to service, and will normally cause fewer problems. Initial cost is low. Maintenance people understand and are comfortable working with them. An actuator made specifically for a control valve eliminates the chance for a costly performance mismatch. An actuator manufactured by the valve manufacturer and shipped with the valves eliminates separate mounting expenses and ensures easier coordination of spare parts purchases. Interchangeable parts also are important to minimize spare parts inventory.

Savings of installation and maintenance costs are available from packages that combine the valve, actuator, and accessories in a modular unit. The components are designed to work together, external piping is reduced, and complicated exposed linkages are eliminated.

Actuator Designs

There are many types of actuators for rotary and sliding-stem valves. There are five major categories:

1. Spring-and-diaphragm actuators (Figs. 30 and 31).
2. High-pressure spring-and-diaphragm actuators (Fig. 32).
3. Pneumatic piston actuators (Figs. 33–36).
4. Electric motor actuators (Fig. 37).
5. Electrohydraulic actuators (Figs. 38 and 39).

Each actuator has weaknesses, strong points, and optimum uses. Most actuator designs are available for either sliding stem or rotary valve bodies. They differ only by linkage or motion translators. The basic power sources are identical (Table 4).

Most rotary actuators (Figs. 31, 35, and 36) are similar to sliding-stem actuators. Rotary actuators use linkages, gears, or crank arms to convert direct linear motion of a diaphragm or piston into the 90° output rotation required by rotary valves. The most important consideration for control valve actuators is the requirement for a design that limits the amount of lost motion in the internal linkage and valve coupling. Rotary actuators are available that use tilting pistons or diaphragms. These designs eliminate most linkage points (and the resultant lost motion) and provide a safe, accurate, and enclosed package.

When considering an actuator design, consider the method by which it is coupled to the drive shaft of the control valve. On rotary valves, slotted connections mated to milled shaft flats generally are not satisfactory if throttling is required. Pinned connections, if constructed solidly, are suitable for nominal torque applications. The best connectors are clamped, splined shapes. This type of connection eliminates all lost motion, is easy to disassemble, and is capable of high torques.

Sliding-stem actuators are rigidly fixed to valve stems by threaded-and-clamped connections. Sliding stem actuators are very simple in design. Because they do not have any linkage points and their connections are rigid, they exhibit no lost motion and excellent inherent control characteristics.

Because rotary and sliding-stem actuators are so similar in concept and characteristics, they will not be further differentiated in this section unless necessary.

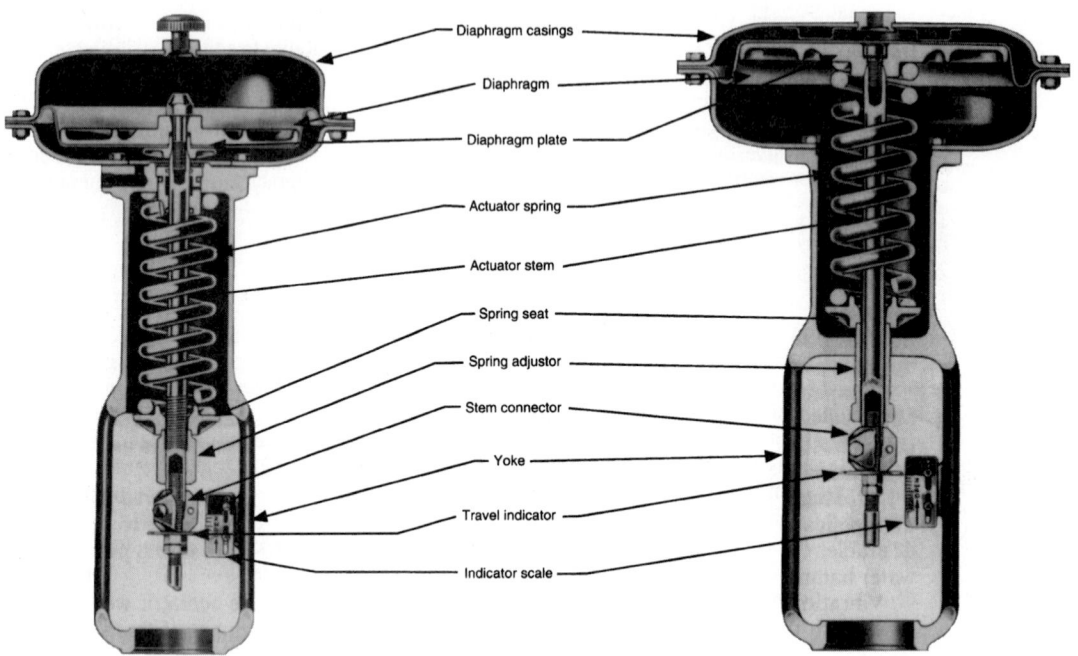

Diaphragm casings

Diaphragm

Diaphragm plate

Actuator spring

Actuator stem

Spring seat

Spring adjustor

Stem connector

Yoke

Travel indicator

Indicator scale

FIGURE 30 Spring-and-diaphragm actuators offer an excellent choice for most control valves. They are inexpensive and simple, and they have an ever-present, reliable spring fail action. Shown are two styles. On the left, air (operating pressure). opens the valve and the spring closes it (air to open; spring closes). On the right, air to close, spring opens. (*Fisher Controls International, Inc.*)

Diaphragm Actuators. The most popular and widely used control-valve actuator is the pneumatic spring-and-diaphragm style (Figs. 30 and 31). Diaphragm actuators are extremely simple and offer low cost and high reliability. Diaphragm actuators normally operate over the standard signal ranges of 3 to 15 psig (0.2–1 bar) or 6 to 30 psig (0.4–2 bars). Therefore they are often suitable for throttling service using instrument signals directly. Many designs offer either adjustable springs or wide spring selections to allow the actuator to be tailored to the particular application. Because diaphragm actuators have few moving parts that might contribute to failure, they are extremely reliable. Should they ever fail, maintenance is extremely simple. Improved designs include mechanisms to control the release of spring compression, reducing the possibility of injury to personnel during actuator disassembly.

The overwhelming advantage of the spring and diaphragm actuator is the ever-present provision for fail action. As pressure is loaded on the actuator casing, the diaphragm moves the valve and compresses the spring. The stored energy in the spring acts to move the valve back to its original position as pressure is released from the casing. Should there be a loss of signal pressure to the instrument or the actuator, the spring can move the valve to its initial (fail) position. Actuators are available for either fail-open or fail-closed action.

The only real drawback to the spring-and-diaphragm actuator is a relatively limited capability. Much of the thrust created by the diaphragm is taken up by the spring and thus does not result in output to the valve. Therefore the spring-and-diaphragm actuator is seldom used for high force requirements. It is not economical to build and use very large diaphragm actuators because the size, weight, and cost grow out of proportion to capability. This limitation is mitigated, however, by the fact that most valves are small and have low force requirements.

High-Pressure Spring-and-Diaphragm Actuators. High-pressure spring-and-diaphragm actuators (Fig. 32) share many of the advantages of standard spring-and-diaphragm actuators and offer additional advantages. The use of higher supply pressure allows the actuator to be smaller and lighter than typical

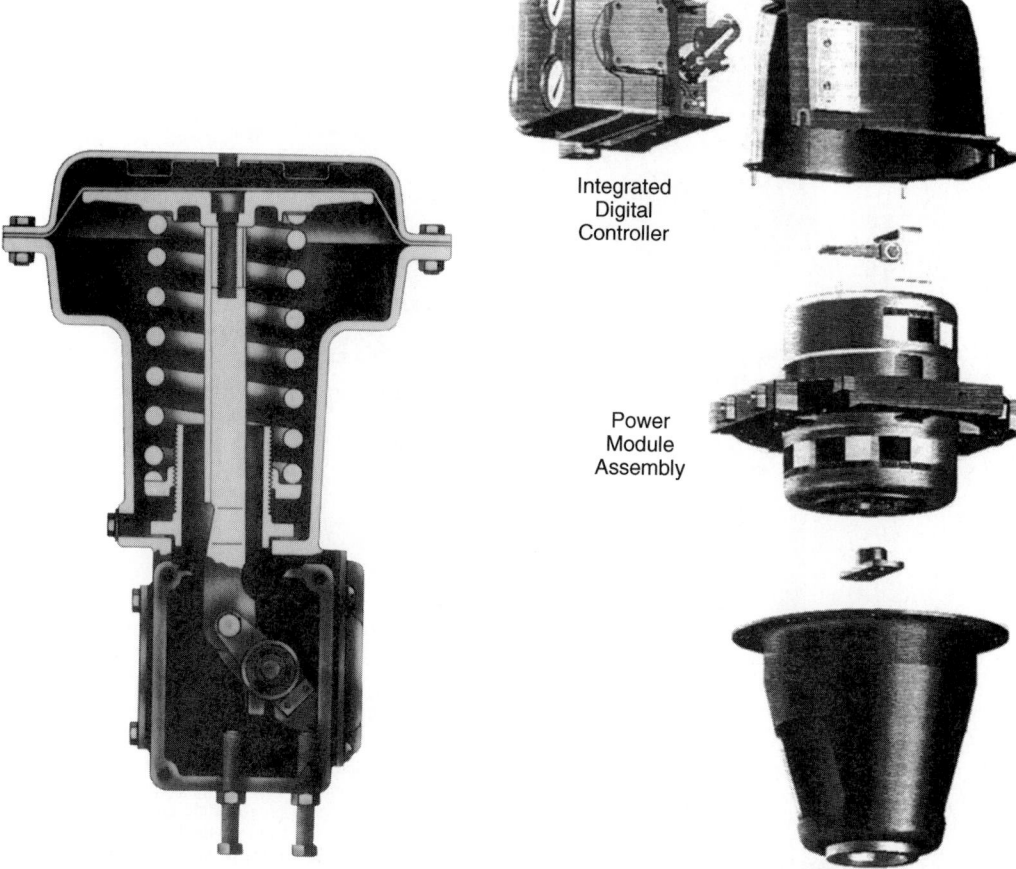

Integrated
Digital
Controller

Power
Module
Assembly

FIGURE 31 Spring-and-diaphragm actuators offer many features that provide precise control. The splined actuator connection, clamped lever, and single-joint linkage all contribute to low lost motion. (*Fisher Controls International, Inc.*)

FIGURE 32 High-pressure spring-and-diaphragm actuator, featuring integral control and accessories and modular construction. The spring and diaphragm are contained in the power module assembly. Tubing, linkage, and mounting brackets are either eliminated or enclosed. (*Fisher Controls International, Inc.*)

diaphragm actuators. The smaller size makes modular construction easier to provide. Modularity makes maintenance easier and allows complete integration of instruments and accessories.

Piston Actuators.   Piston actuators, such as those shown in Figs. 33–36, are the second most popular control-valve actuator style. They are generally more compact and provide higher torque or force outputs than spring-and-diaphragm actuators. Piston styles normally work with supply pressures of between 50 and 150 psig (3.5 and 10.4 bars). Although piston actuators can be equipped with spring returns, this construction has limits similar to those of the spring and diaphragm style.

Piston actuators used for throttling service must be furnished with double-acting positioners, which simultaneously load and unload opposite sides of the piston. The pressure differential created across the piston causes travel toward the lower pressure side. The positioner senses the motion of the output, and when the required position is reached, the positioner equalizes the pressure on both sides of the piston.

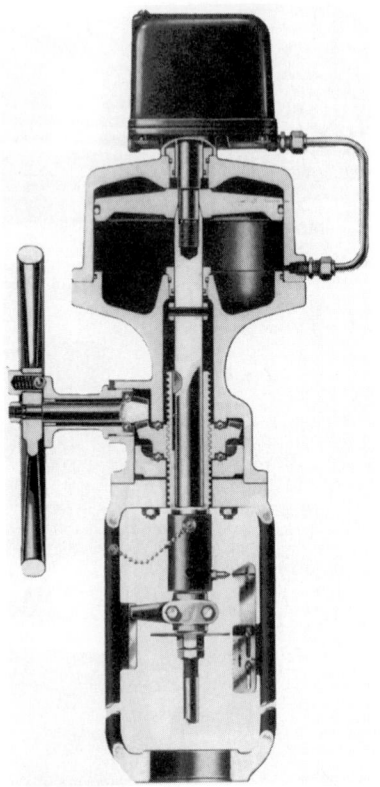

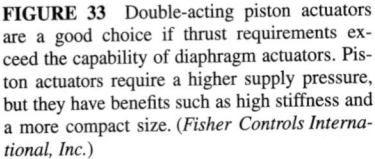

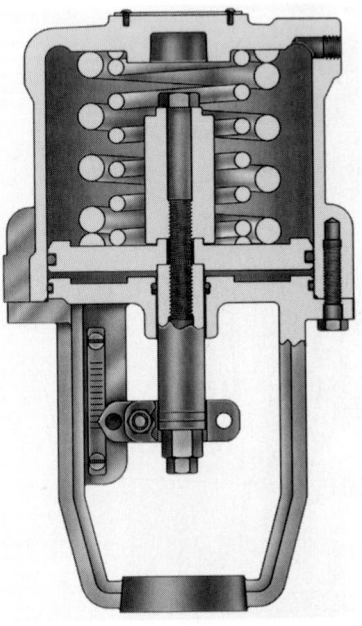

FIGURE 33 Double-acting piston actuators are a good choice if thrust requirements exceed the capability of diaphragm actuators. Piston actuators require a higher supply pressure, but they have benefits such as high stiffness and a more compact size. (*Fisher Controls International, Inc.*)

FIGURE 34 Spring fail action is available in this spring-biased piston actuator. Process pressure acting on the valve plug can aid fail action, or the actuator can be configured so that the spring alone closes or opens the valve on failure of operating pressure. (*Fisher Controls International, Inc.*)

The pneumatic piston actuator is an excellent choice when a compact high-power unit is required. It is also easily adapted to services where high ambient temperatures are involved.

The main disadvantages of piston actuators are the high supply pressures required, the requirement for positioners when used for throttling service, and the lack of inherent failure-mode systems. Two types of spring-return piston actuators are available. The variations are subtle, but significant. It is possible to add a spring to a piston actuator and operate it much like a spring and diaphragm. These designs use a single-acting positioner, which loads the piston chamber to move the actuator and compress the spring. As pressure is unloaded, the spring moves the piston back. These designs use large high-output springs, which are capable of overcoming the fluid forces in the valve.

The alternative design uses a much smaller spring and relies on valve fluid forces to help provide the fail action. In normal operation they act like a double-acting piston. In a fail situation the spring initiates movement and is helped by unbalance forces on the plug.

The only failure-mode alternative to springs are pressurized air volume tank pneumatic trip systems to move the piston actuator to its fail position. Although these systems are quite reliable, they add to overall system complexity, maintenance difficulty, and cost. Therefore for any failure-mode requirement prime consideration should be given to spring-return actuators if they are feasible.

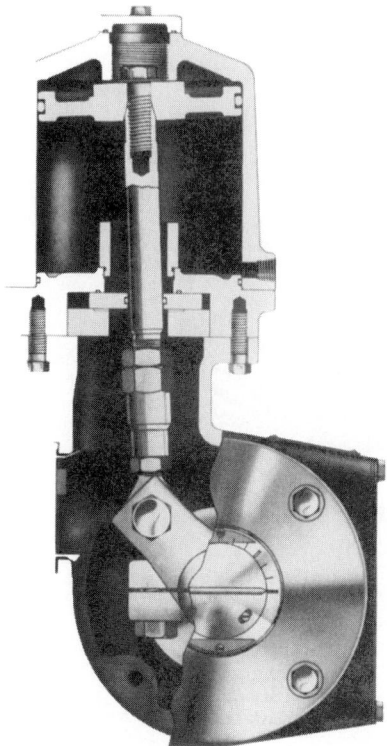

FIGURE 35 This piston actuator controls a rotary valve. The valve linkage and clamped connector eliminate lost motion and provide throttling accuracy. (*Fisher Controls International, Inc.*)

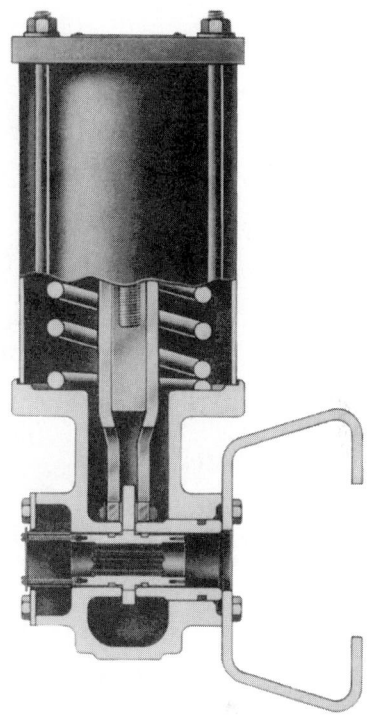

FIGURE 36 For on–off service and some throttling applications, requirements for accuracy and minimum lost motion are not necessary, and a simple design such as this can save money. The actuator shown features spring-return action. (*Fisher Controls International, Inc.*)

Use special care during the selection of throttling piston actuators to get one that has minimal hysteresis and dead band. As the number of linkage points in the actuator increases, so does the dead band. As the number of sliding parts increases, so does the hysteresis. An actuator with high hysteresis and dead band can be quite suitable for on-off service. However, caution is necessary when attempting to adapt this actuator to throttling service by simply bolting on a positioner.

The cost of a diaphragm actuator is generally less than that of a comparable-quality piston actuator. Part of this cost savings is in the ability to use instrument output air directly, thereby eliminating the need for a positioner. The inherent provision for fail action in the diaphragm actuator is also a consideration.

Electric Actuators. Electric actuators can be successfully applied in many situations. Most electric operators consist of motors and gear trains (Fig. 37). They are available in a wide range of torque outputs, travels, and capabilities. They are suited for remote mounting where no other power source is available or for use where there are specialized thrust or stiffness requirements. Electric actuators are economical, compared with pneumatic ones, for applications in small size ranges only. Larger units operate slowly, weigh considerably more than pneumatic equivalents, and are more costly. Precision throttling versions of electric motor actuators are quite limited in availability. One very important consideration in choosing an electric actuator is its capability for continuous closed-loop control. In applications where frequent changes are made in control valve position, the electric actuator must have a suitable duty cycle.

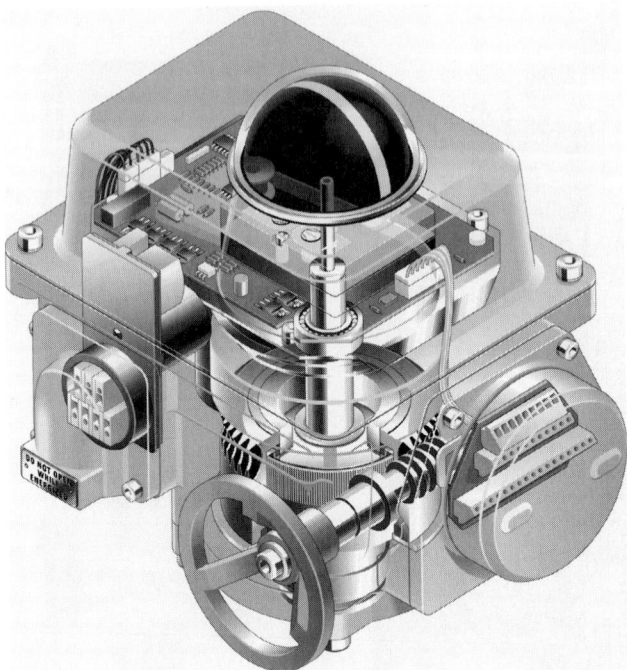

FIGURE 37   Technical improvements have made electric actuators practicable for control purposes. They offer high thrust or torque and high stiffness. (*El-O-Matic International.*)

While having many disadvantages, the electric actuator will generally provide the highest output available within a given package size. In addition electric actuators are very stiff, that is, resistant to valve forces. This makes them an excellent choice for good throttling control of large high-pressure valves.

Electrohydraulic Actuators.   Electrohydraulic actuators, like those in Figs. 38 and 39, are electric actuators in which motors pump oil at high pressure to a piston, which in turn creates the output force. The electrohydraulic actuator is an excellent choice for throttling because of its high stiffness, compatibility with analog signals, excellent frequency response, and positioning accuracy. Most electrohydraulic actuators are capable of very high outputs, but they are limited by high initial cost, complexity, and difficult maintenance. Failure-mode action on electrohydraulic actuators can be accomplished by the use of springs or hydraulic accumulators and shutdown systems.

Actuator Sizing

The last step in the selection process is the specification of the actuator size. The process of sizing is to match the actuator capabilities as closely as possible to the valve requirements. In practice, the mating of actuator and valve requires the consideration of many factors. Valve forces must be evaluated at the critical positions of valve travel (usually open and closed) and compared to actuator output. Valve force calculation varies considerably between valve styles and manufacturers. In most cases it is necessary to consider a complex summation of forces, including the following:

- Static fluid forces
- Dynamic fluid forces and force gradients

FIGURE 38   Self-contained electrohydraulic actuator. This single unit contains the electric motor, hydraulic pump, reservoir, hydraulic positioner, and actuator cylinder. (*Fisher Controls International, Inc.*)

FIGURE 39   Electrohydraulic actuators provide the ultimate in thrust, speed, frequency of response, and stiffness. The type shown is operated by an external hydraulic power supply. (*Fisher Controls International, Inc.*)

- Friction of seals, bearings, and packing
- Seat loading

Although actuator sizing is not difficult, the great variety of designs on the market and the ready availability of vendor expertise (normally at no cost) make detailed knowledge of the procedures unnecessary.

Summary of Actuator Selection Factors

In choosing an actuator type, the fundamental requirement is to know your application. Control signal, operating mode, power source available, torque required, and fail position can make many decisions

TABLE 4   Comparison of Valve Actuator Features

Advantages	Disadvantages
Spring and diaphragm	
Lowest cost	Limited output capability
Ability to throttle without positioner	Large size and weight
Simplicity	
Inherent failure-mode action	
Low supply-pressure requirement	
Adjustability to varying conditions	
Ease of maintenance	
High-pressure spring and diaphragm	
Compact, light weight	Requires high supply pressure—40
No spring adjustment needed	psig (2.8 bars) or higher
Costly cast components not needed	Positioner required for throtting
Inherent fall-safe action	
No dynamic stem seals or traditional stem connector block needed	
Design can include integral accessories	
Pneumatic piston	
High force or torque capability	Fall-safe requires accessories or
Compact, light weight	addition of a spring
Adaptable to high ambient temperatures	Positioner required for throttling
Fast stroking speed	Higher cost
Relatively high actuator stiffness	High supply-pressure requirement
Electric motor	
Compact	High cost
Very high stiffness	Lack of fail-safe action
High output capability	Limited duty cycle
Supply pressure piping not required	Slow stroking speed
Electrohydraulic	
High output capability	High cost
High actuator stiffness	Complexity and maintenance
Excellent throtting ability	difficulty
Fast stroking speed	Fail-safe action only with accessories

for you. Keep in mind simplicity, maintainability, and lifetime costs. Safety is another consideration that must never be overlooked. Enclosed linkages and controlled compression springs available in some designs are very important for safety reasons. The pros and cons of the various actuator styles are listed under "Summary Checklist."

The spring-and-diaphragm actuator is the most popular, versatile, and economical type. Try it first. If the limitations of available diaphragm actuators eliminate them, consider pistons or electric actuators, bearing in mind the capabilities and limitations of each.

VALVE CONTROLLERS AND ACCESSORIES

No study of control valves would be complete without a look at devices that augment the valve function and interface it to control systems. Included in this category are devices such as digital valve controllers as well as traditional valve positioners, electropneumatic transducers, limit switches, and manual actuator overrides. These devices assure controllability, provide information about valve operation, and also allow for operation or shutdown in emergency situations.

Valve Positioners and Controllers

Positioners are instruments that help improve control by accurately positioning a control valve actuator in response to a control signal. Positioners receive an input signal either pneumatically or electronically and provide output power, generally pneumatically, to an actuator to assure valve positioning. A feedback linkage between valve stem and positioner is established so that the stem position can be noted by the instrument and compared with the position dictated by the controller signal (Fig. 40).

Use of positioners is generally desirable to linearize the control valve plug position with a control signal. Positioners will often improve the performance of control valve systems. There are situations, however, where process dynamics eliminate the use of positioners. On very fast loops it has been found that the use of positioners will degrade performance because the response of the positioner might not be able to keep up with the system in which it is installed.

Positioners operate with a pneumatic input and output signal or with an electronic input signal and pneumatic output. Some of the electronic versions accept an analog input signal, and others accept a digital input signal. Digital positioners and digital valve controllers are discussed in the next section.

FIGURE 40 Electropneumatic positioners combine the function of a current-to-pressure transducer with those of a positioner. It receives an electronic input signal and ensures valve position by adjusting output pressure. (*Fisher Controls International, Inc.*)

Digital Positioners [1]

Digital valve positioners are of three types:

1. Digital Noncommunicating—A current signal (4–20 mA) is supplied to the positioner, which both powers the electronics and controls the output.

2. HART (highway addressable remote transducer) communications—This is the same as the digital noncommunicating but is also capable of two-way digital communication over the same wires used for the analog signal. (HART is a mark owned by HART Communications Foundation, Inc.)

3. Fieldbus—This type receives digital signals and positions the valve by using digital electronic circuitry coupled to mechanical components. An all-digital control signal replaces the analog control signal. Additionally, two-way digital communication is possible over the same wires. The shift in field communications technology toward a fieldbus technology benefits the user by enabling improved control architecture, product capability, and reduced wiring.

There is a general trend toward greater use of digital valve controllers on control valves (Fig. 41). There are several reasons for this trend:

1. There is a reduced cost of loop commissioning, including installation and calibration.

2. Diagnostics are used to maintain loop performance levels.

3. There is an improved process control through reduced process variability.

4. Offset the decreasing mechanical skill base of instrument technicians.

Two aspects of digital valve controllers make them particularly attractive:

1. Automatic calibration and configuration: Considerable time savings are realized over traditional zero and spanning.

2. Valve diagnostics: Through the DCS (distributed control system), PC software tools, or handheld communicators, users can diagnose the health of the valve while it is in the line (Fig. 42).

Electropneumatic Transducers

Electropneumatic transducers (Fig. 43) are devices that convert an electronic input into a pneumatic output signal that is proportional to the input signal. Electropneumatic transducers are used in

FIGURE 41   Rotary control valve with digital valve controller (*H. D. Baumann Inc.*).

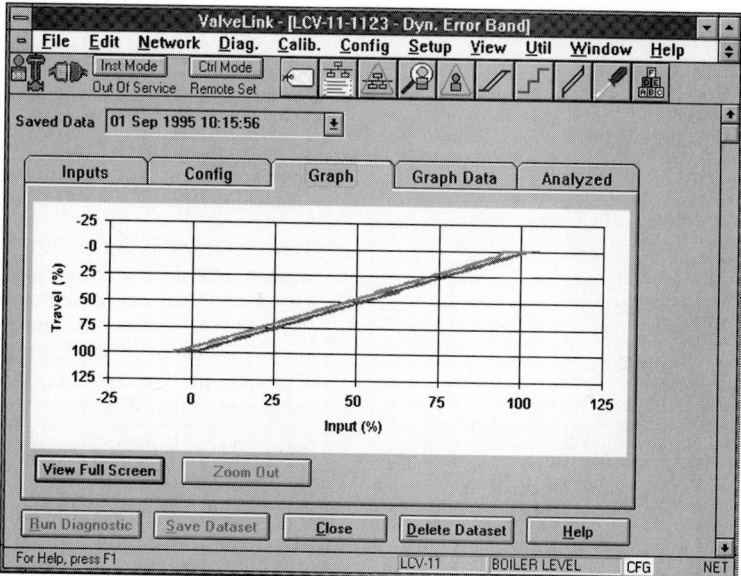

FIGURE 42   Diagnostics programs can provide information to help plan predictive maintenance. (*Fisher Controls International, Inc.*)

electronic control loops to help operate pneumatic control valves. Most transducers convert a standard
4- to 20-mA (milliampere) signal to a 3- to 15-psig (0.2–1.0 bar) pneumatic output. Devices also are
available that can respond to digital signals and nonstandard analog inputs. The transducer function
is sometimes included with the valve positioner. If the transducer is included, the device is known as
an electropneumatic positioner: the input is an electronic signal and the output is position.

Volume Booster.   The volume booster is normally used in control-valve actuators to increase the
stroking speed. These pneumatic devices have a separate supply pressure and deliver a higher-volume
output signal to move actuators rapidly to their desired positions. Special booster designs (Fig. 44) are
also available for use with positioners. These devices incorporate a dead-band feature to adjust their
response and eliminate instabilities. This booster, therefore, permits high actuator stroking speeds
without degrading the steady-state accuracy provided by positioners in the loop.

Trip Valves.   Pressure-sensing trip valves are available for control applications where a specific
actuator action is required when supply pressure fails or falls below a specific point. When supply
pressure falls below the preadjusted trip point, the trip valve causes the actuator to fail up, lock in
last position, or fail down. When supply pressure rises above the trip point, the valve automatically
resets, allowing the system to return to normal operation. Auxiliary power to provide for actuator
action in case of trip is provided by pneumatic volume tanks. Fig. 45 shows a system installed on a
valve actuator.

Limit Switches.   Electrical position switches are often incorporated on control valves to provide the
operation of alarms, signal lights, relays, or solenoid valves when the control-valve position reaches
a predetermined point. These switches can be either integrated, fully adjustable units with multiple
switches or stand-alone switches and trip equipment. Use special care in the selection of limit switches
for harsh environments to assure functionality over time (Figs. 46 and 47).

Solenoid Valves.   Small, solenoid-operated electric valves are often used in a variety of on-off or
switching applications with control valves. They provide equipment override, failure-mode interlock

FIGURE 43 Electropneumatic transducers are common accessories. They convert an analog electronic signal to a pneumatic output signal. The best transducers are compact and accurate and consume little supply air. (*Fisher Controls International, Inc.*)

of two valves, or switching from one instrument line to another. A typical application involves a normally open solenoid valve, which allows positioner output to pass directly to the actuator. On loss of electric power, the solenoid valve will close the port to the valve positioner and bleed pressure from the diaphragm case to the control valve, allowing it to achieve its fail position.

Position Transmitters. Electronic position transmitters are available that send either analog or digital electronic output signals to control-room devices. The instrument senses the position of the valve and provides a discrete or proportional output signal. Electrical position switches are often included in these transmitters (Fig. 48).

Manual Handwheels. A variety of actuator accessories are available which allow for manual override in the event of signal failure or lack of signal previous to start-up. Nearly all actuator styles have available either gear-style or screw-style manual override wheels. In many cases, in addition to providing override capability, these handwheels can be used as adjustable position or travel stops. Figure 49 shows the installation of a manual handwheel on a spring and diaphragm actuator.

CONTROL VALVE INSTALLATION [1]

Never install a valve where service conditions could exceed those for which the valve was intended. Contact the manufacturer if you have questions concerning applicable service conditions.

Storage and Protection

Consider storage and protection early in the selection process, before the valve is shipped. Typically, manufacturers have packaging standards that are dependent upon the destination and intended length of storage before installation. Because most valves arrive on site some time before installation, many problems can be averted by making sure the details of the installation schedule are known and discussed with the manufacturer at the time of valve selection. In addition, take special precautions when you receive the valve at the final destination. For example, store the valve in a clean, dry place away from any traffic or other activity that could damage the valve.

Installation Techniques

Always follow the control valve manufacturer's installation instructions and cautions. Typical instructions are summarized here.

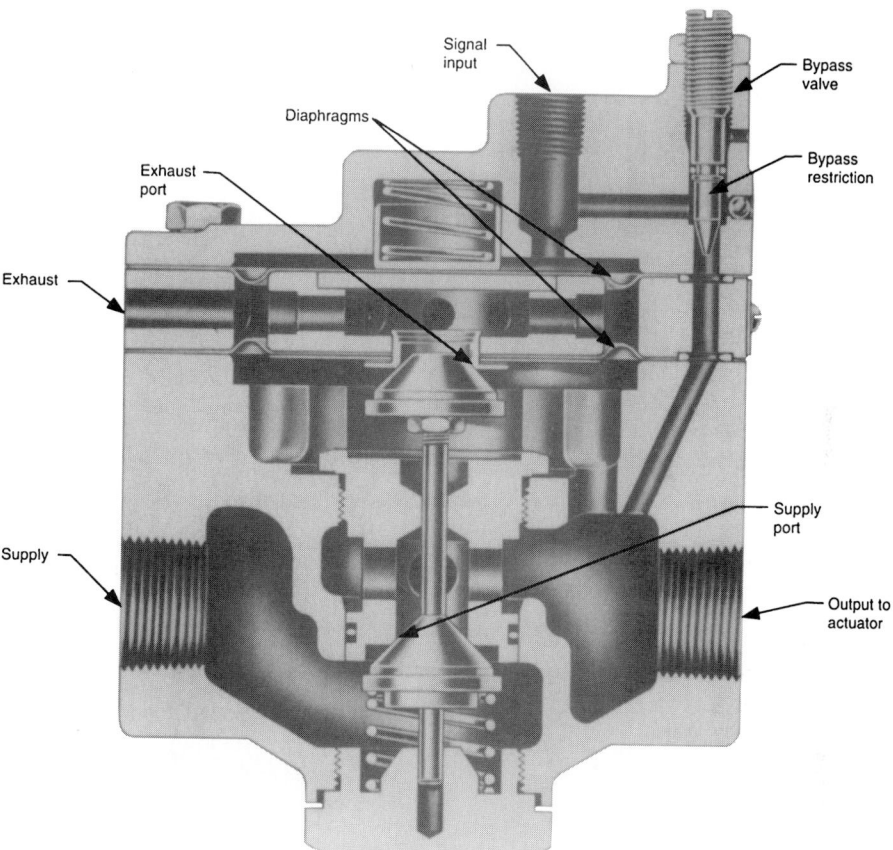

Signal
input

Bypass
valve

Diaphragms

Bypass
restriction

Exhaust
port

Exhaust

Supply
port

Supply

Output to
actuator

FIGURE 44   Volume booster that delivers added air volume for rapid actuator stroking. The booster shown is specifically made for valve positioners. It uses a bypass to allow small output changes to pass through, but, when the signal changes exceed preset dead-band limits, it delivers high-volume output pressure for rapid stoking. (*Fisher Controls International, Inc.*)

Read the Instruction Manual.   Before installing the valve, read the instruction manual. Instruction manuals describe the product and review safety precautions to take before and during installation. Following the manual helps ensure an easy and successful installation.

Be Sure the Pipeline Is Clean.   Foreign material in the pipeline could damage the seating surface of the valve or even obstruct the movement of the valve plug, ball, or disk so that the valve does not shut off properly. To help reduce the possibility of a dangerous situation from occurring, clean all pipelines before installing. Make sure pipe scale, metal chips, welding slag, and other foreign materials are removed. In addition, inspect pipe flanges to ensure a smooth gasket surface. If the valve has screwed end connections, apply a good grade of pipe sealant compound to the male pipeline threads. Do not use sealant on the female threads because excess compound on the female threads could be forced into the valve body. Excess compound could cause sticking in the valve plug or accumulation of dirt, which could prevent good valve shutoff.

Although valve manufacturers take steps to prevent shipment damage, such damage is possible and should be discovered and reported before the valve is installed.

FIGURE 45   Fail action for piston actuators can be accomplished by using pneumatic trip systems. A switching valve transfers stored pressure from volume tanks to the piston to stroke the valve and maintain the predetermined failure position. (*Fisher Controls International, Inc.*)

Do not install a control valve that has been damaged in any way.

Before installing, check for and remove all shipping stops and protective plugs or gasket surface covers. Check inside the valve body to make sure no foreign objects are present.

Use Good Piping Practices.   Most control valves can be installed in any position. However, the most common method is with the actuator vertical and above the valve body. If horizontal actuator mounting is necessary, consider additional vertical support for the actuator. Be sure the body is installed so that fluid flow will be in the direction indicated by the flow arrow (Fig. 50) or instruction manual.

Be sure to allow ample space above and below the valve to permit easy removal of the actuator or valve plug for inspection and maintenance. Clearance distances are normally available from the valve manufacturer as certified dimension drawings. For flanged valve bodies, be sure the flanges are properly aligned to provide uniform contact of the gasket surfaces. Gently tighten bolts after establishing proper flange alignment. Finish tightening them in a criss-cross pattern (Fig. 51). Proper tightening will avoid uneven gasket loading and will help prevent leaks. It also will avoid the possibility

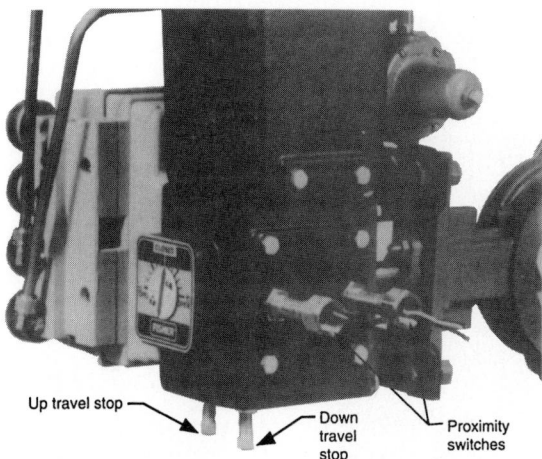

FIGURE 46 Actuator, featuring externally adjustable travel stops and integrally mounted cam-operated proximity limit switches. All linkages for switches and travel stops are fully enclosed. (*Fisher Controls International, Inc.*)

FIGURE 47 Limit switches are common actuator accessories. The unit shown has two limit switches, but similar designs can hold up to six switches and trip points can be adjusted to any point of travel. (*Fisher Controls International, Inc.*)

of damaging, or even breaking, the flange. This precaution is particularly important when connecting to flanges that are not the same material as the valve flanges.

Pressure taps installed upstream and downstream of the control valve are useful for checking flow capacity or pressure drop. Locate such taps in straight runs of pipe away from elbows, reducers, or expanders. This location minimizes inaccuracies resulting from fluid turbulence.

Use 1/4- or 3/8-in. (6–10 mm) tubing or pipe from the pressure connection on the actuator to the controller. Keep this distance relatively short and minimize the number of fittings and elbows to reduce system time lag. If the distance must be long, use a valve positioner or a booster with the control valve.

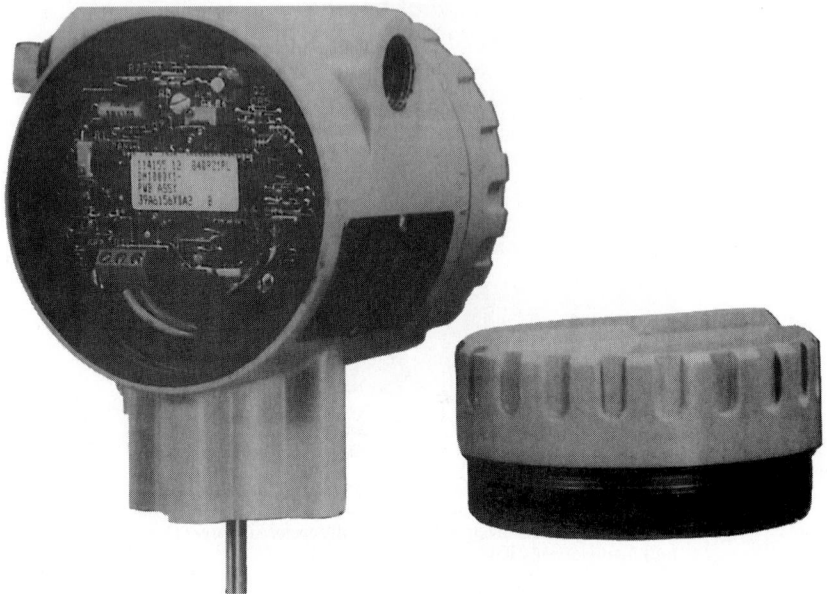

FIGURE 48 Stem position transmitters provide discrete or analog output of valve position for use by control-room instrumentation. (*Fisher Controls International, Inc.*)

SUMMARY CHECKLIST

The subject of control valves is complex and ever evolving. Valve styles are changing to meet changing process conditions and accessories and instrumentation continue to evolve to meet the requirements of the control systems. The key to the selection process is to understand both the needs of the process and the needs of the controlling instrumentation. Tips for valve selection, sizing, and actuator selection follow.

1. Valve Body Selection
   a. Sliding-stem valves provide the widest variety and best capability in the industry. Their performance and versatility make them very popular. In large sizes they may be expensive, but for sizes NPS 3 or DN 80 and less they are a first choice.
   b. Rotary-ball and eccentric-plug valves provide excellent control and are especially good values in sizes NPS 4 to 6 or DN 100 to DN 150. Erosion-resistant designs and trims are available to extend their life in many difficult applications.
   c. Butterfly and high-performance butterfly valves are most popular and economical in sizes above NPS 6 or DN 150. In many large-size cases they are the only available choice.
   d. Special requirements necessitate special valve solutions. Valve designs and special trims are available to handle high noise, cavitation, high pressure, high temperature, and combinations of these.
2. Sizing of Valve
   a. The liquid sizing equation is simple to use and based on empirically determined sizing coefficients.
   b. A valve size should be selected which gives the required application C_v at 70–90% of travel.
   c. Sizing and trim selection are influenced by choked flow and the presence of cavitation. These phenomena limit flow and may cause significant damage.

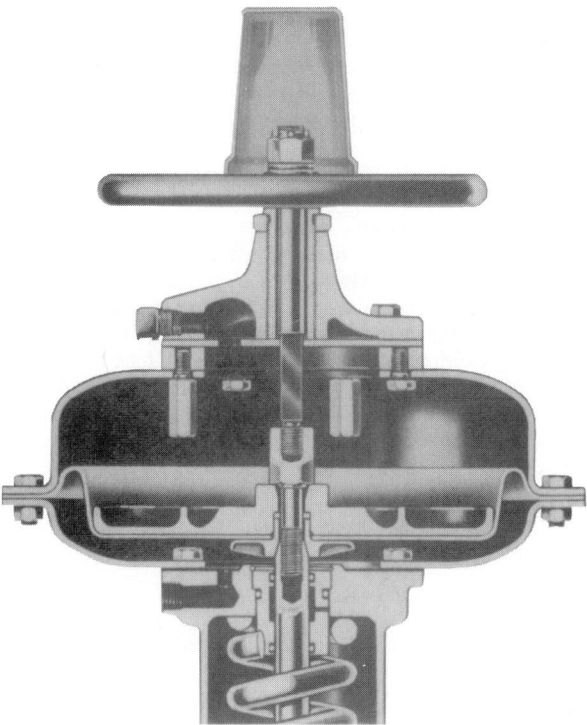

FIGURE 49   The handwheel on this actuator can act as a travel stop or for emergency operation of the valve. The actuator is an air-to-open, spring-close actuator. (*Fisher Controls International, Inc.*)

FIGURE 50   Be sure flow is in the same direction as the flow arrow on the valve. (*Fisher Controls International, Inc.*)

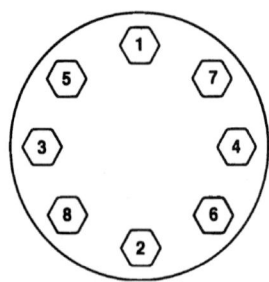

FIGURE 51 Tighten flange bolts or studs and nuts in a criss-cross pattern (*Fisher Controls International, Inc.*).

d. Viscosity and piping corrections must be made in many sizing situations. Piping considerations are especially important when high-recovery valves are specified.

e. Sizing valves for gas flow involves physical principles similar to liquid flow. However, effects of compressibility and critical flow factors must be considered.

3. Actuator selection

a. Actuator selection must be based on a balance of process requirements, valve requirements, and cost.

b. Spring and diaphragm actuators are simpler, less expensive, and easier to maintain. Consider them first in most situations.

c. Piston actuators offer many of the advantages of pneumatic actuators, with higher thrust capability than diaphragm styles. They are especially useful where compactness is desired or long travel is required.

d. Electric and electrohydraulic actuators provide excellent performance. They are, however, much more complex and difficult to maintain.

e. Actuator sizing is not difficult, but the wide variety of actuators and valves makes this difficult to master. Vendor expertise is widely available.

REFERENCES

1. *Control Valve Handbook*, 3rd ed., Fisher Controls International, Inc., Marshalltown, Iowa, 1998.

2. ANSI B16.34-19, "Steel Valves," American National Standards Institute, New York, 1988.

3. ANSI B16.1-19, "Cast Iron Pipe Flanges and Flanged Fittings," American National Standards Institute, New York, 1989.

4. ANSI/FCI 70-2-1976 (R1982), "Quality Control Standard for Control Valve Seat Leakage," Fluid Controls Institute, 1982.

5. ANSI/ISA S75.02-1988, "Control Valve Capacity Test Procedure," Instrument Society of America, Research Triangle Park, North Carolina, 1988.

6. ANSI/ISA S75.01-1985, "Flow Equations for Sizing Control Valves," Instrument Society of America, Research Triangle Park, North Carolina, 1985.

7. ANSI/ISA S75.05-1983, "Control Valve Terminology," Instrument Society of America, Research Triangle Park, North Carolina, 1983.

8. ANSI/ISA S75.11-1985, "Inherent Flow Characteristic and Rangeability of Control Valves,"Instrument Society of America, Research Triangle Park, North Carolina, 1985.

9. *Control Valve Sourcebook—Power and Severe Service,* Fisher Controls International, Inc., Marshalltown, Iowa, 1990.

10. Fitzgerald, W., *Control Valves for the Chemical Process Industries,* McGraw-Hill, New York, 1995.

11. Baumann, H. D., *Control Valve Primer—A Users Guide*, 3rd ed., Instrument Society of America, Research Triangle Park, North Carolina, 1998.

CONTROL VALVE TROUBLESHOOTING

Table 5 is a listing of a number of different valve problems or symptoms, with potential root causes, and recommended corrective action. It is presented in the form of a troubleshooting diagram that references certain common procedures used in valve maintenance. These procedures are explained in greater detail in the next section.

TABLE 5   Troubleshooting Diagram

Problems and Symptoms	Causes	Solutions
1. Body erosion.	1*a*. Velocity. 1*b*. Particulates in flowstream. 1*c*. Cavitation and flashing.	1*a*. Increase valve trim size to slow fluid. 1*b*. Switch to streamlined design to reduce fluid impingement. 1*c*. Switch to C5 body material. 1*d*. Switch to torturous-path trim to slow fluid. 1*e*. Switch to low-recovery valve and trim to conrol cavitation. 1*f*. Repair by welding up with stainless material.
2. Trim erosion.	2*a*. Velocity. 2*b*. Particulates in flowstream. 2*c*. Cavitation and flashing.	2*a*. Increase valve and trim size to slow fluid. 2*b*. Switch to hardened trim. 2*c*. Switch to torturous-path trim to slow fluid. 2*d*. Switch to low-recovery valve and trim to control cavitation. 2*e*. Switch to streamlined design to reduce impingement.
3. Seat ring-to-plug leakage.	3*a*. Low load (benchset, calibration, friction, etc.). 3*b*. Poor surface condition (lapping, materials).	3*a*. Use proper surface preparation (lapping). 3*b*. Correct actuator and valve setup (benchset, calibration, friction, etc.).
4. Seat ring-to-body leakage.	4*a*. Low load (inadequate torque, parts stack-up, improper gasketing). 4*b*. Surface condition (cleanlinness, finish). 4*c*. Porosity in body.	4*a*. Correct bolt load, parts stack-up, gasketing. 4*b*. Recut, clean up gasket face. 4*c*. Porosity in casting can sometimes result in leakage around gaskets. Check for porosity. Grind out and weld up.
5. Packing leakage.	5*a*. Stem finish/cleanliness. 5*b*. Bent stem. 5*c*. Low packing load. 5*d*. Wrong packing type or configuration. 5*e*. Excessive packing stack height (graphite.) 5*f*. Corrosion and pitting (graphite). 5*g*. Seized or cocked packing follower.	5*a*. Clean up and polish stem to 4 rms finish. 5*b*. Straighten stem to within 0.002 in. over stroking length. 5*c*. Retorque bolting or use live-loading. 5*d*. Check packing type and configuration against application. Repack as necessary. 5*e*. Install spacers to minimize packing height. Repack valve. 5*f*. Use sacrificial washers. Remove graphite packing if valve is to be inactive for more than 2 to 3 weeks. 5*g*. Inspect and replace any damaged parts such as flanges, nuts, and followers. 5*h*. Switch to high-performance packing system.
6. Sliding wear.	6*a*. High cycling (unstable loop?). 6*b*. Excessive contact stress. 6*c*. Misalignment. 6*d*. Surface finish not to specification. 6*e*. Incorrect materials choice	6*a*. Tune loop; reduce friction to reduce instability. 6*b*. Increase bearing size. 6*c*. Remachine parts to correct alignment. 6*d*. Polish surfaces. 6*e*. Review materials choice in light of application. 6*f*. Switch to sliding-stem globe-style valve because of better guiding.

(Continues)

TABLE 5   Troubleshooting Diagram

Problems and Symptoms	Causes	Solutions
7. Bonnet-to-body leakage	7a. Low load from bonnet bolting (torque, internal parts, stack-up, spring rate in gasket). 7b. Surface finish. 7c. Stud leaks.	7a. Retorque bolting. Check parts stack-up against drawings. 7b. Retouch, clean up gastket faces. 7c Sometimes casting porosity can let process fluid seep into bottom of stud holes. Leakage around studs looks like leakage past bonnet gasket (see Fig. 52). Grind out and weld up porosity.
8. Loose stem connection. or broken stem.	8a. Improper torque or pinning. 8b. Vibration or instability.	8a. Purchase the stem and plug as an assembly. 8b. Review trim-style application. 8c. Reduce clearances between cage and plug. 8d. Switch to a welded plug or stem connection.
9. Excessive leakage past piston seal.	9a. Cage finish too rough. Cage I.D. too large. 9b. Improper installation: graphite rings, omniseal. 9d. Is leakage normal for the type of seal? 9d. Exceeding temperature limitations for seal. 9e. Seal simply worn out due to cycling.	9a. Polish cage bore, check I.D. against drawings. 9b. Replace seal, follow installation instructions. 9c. For some seals such as graphite piston rings, high leakage is normal. 9d. Change to high temperature design. 9e. Replace seal. Address loop stability if cycling is caused by this.

(Continues)

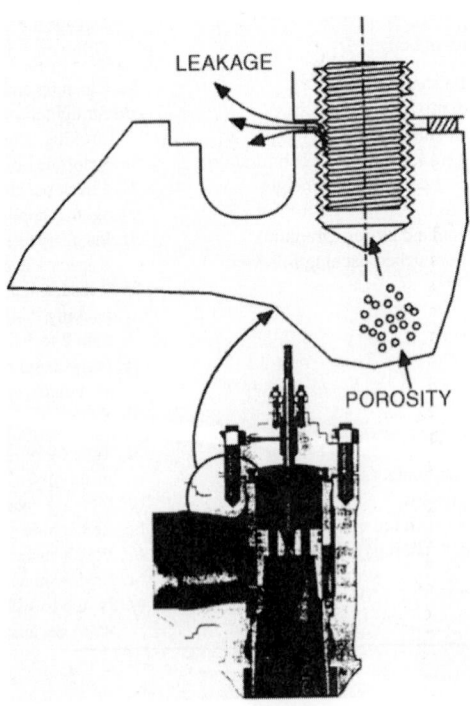

FIGURE 52   Porosity in body masquerading as bonnet joint leakage.

TABLE 5   Troubleshooting Diagram

Problems and Symptoms	Causes	Solutions
10. Valve will not respond to signal.	10a. No air supply or low air supply.	10a. Check the system in accordance with the P&IDs. Verify that all air supply valves are open.
	10b. Leaks in actuator.	10b. Measure and verify sufficient air supply pressure.
	10c. Solenoid closed on inlet lines.	10c. Listen for blowby at the seals or diaphragm. Repair or replace defective parts.
	10d. No controller input signal.	10d. Actuate solenoid valve. Replace if defective.
	10e. Crimped, broken air lines.	10e. No controller input may indicate a fuse has blown. Replace.
	10f. Leaking air fitting.	10f. Check all air lines to see they are not crimped or broken. Repair or replace.
	10g. Incorrect flow direction causing excessive loads on plug.	10g. Check fittings for leaks. Tighten or repalce
	10h. Incorrect air line connections.	10h. If the valve was just installed, check the flow arrow to ensure the process is flowing in the proper direction. Flow above the seat can add pressures the actuator may not be able to overcome. Reverse flowing direction, if appropriate.
	10i. Packing parts binding on stem or shaft.	10i. Check the air to and from a piston actuator to ensure the supply is not connected to the exhaust and vice versa. Check all connections.
	10j. Defective positioner or I/P.	10j. Check the packing gland. Improper gland configuration is a primary cause of rod binding. Replace parts and polish trim.
	10k. Packing overtightened.	10k. Check the positioner and/or the I/P to see if the output can be changed manually. If not, it is defective. Repair or replace.
	10l. Trim is seized.	10l. Overtightened packing or binding in guides can cause excessive friction that blocks valve. Loosen, lubricate, cycle, and retorque.
	10m. Plug stuck in seat.	10m. Replace or repair seized trim. Damage may be polished out.
		10n. Pull or machine plug out of seat. Repair or replace affected parts.
11. Valve will not open to rated travel.	11a. Insufficient supply pressure.	11a. Verify adequate supply pressure.
	11b. Leaks in the actuator or accessories.	11b. Stop all leaks in actuator, air lines fittings, and accessories.
	11c. Incorrect positioner or I/P calibration.	11c. Correct positioner and/or I/P calibration.
	11d. Incorrect travel adjustment.	11d. Readjust valve travel.
	11e. Incorrect actuator spring rate.	11e. Change actuator spring.
	11f. Incorrect benchset.	11f. Adjust benchset.
	11g. Bent stem or shaft.	11g. Replace bent stem or shaft.
	11h. Damaged valve trim.	11h. Replace damage trim.
	11i. Debris in trim.	11i. Clean out valve trim.
	11j. Incorrect flow direction.	11j. Reverse flowing direction.
	11k. Actuator is too small.	11k. Replace actuator.
	11l. Excessive packing friction.	11l. Loosen packing, cycle, lubricate and retorque.
	11m. Incorrect position of manual operator on travel stop.	11m. Readjust manual operator or travel stop.

(Continues)

TABLE 5 Troubleshooting Diagram (*Continued*)

Problems and Symptoms	Causes	Solutions
12. Valve travel sluggish or slow.	12a. Excesive packing friction. 12b. Stem or shaft bent. 12c. Inadequate supply pressure. 12d. Inadequate supply volume. 12e. Undersized accessories. 12f. Excessive friction in piston-type actuator. 12g. Bearing friction. 12h. Poor positioner response.	12a. Readjust or replace packing. 12b. Replace bent shaft or stem. 12c. Increase supply pressure. 12d. Go to bigger supply line or add capacity at valve. 12e. Increase flow capacity of accessories. 12f. Clean out, polish cylinder I.D., remove excess lubricant. 12g. Repair or replace defective bearings. 12h. Repair or replace positioner.
13. Valve travel is jumpy.	13a. Stick-slip action in packing seals or bearings. 13b. Volume booster bypass may need to be adjusted. 13c. Positioner may be defective. 13d. Positioner gain may be too high.	13a. Loosen, lubricate packing. Replace or repair seals and bearings. 13b. Adjust booster bypass. 13c. Repair or repalce positioners. 13d. Adjust positioner gain. Replace with lower gain model.
14. Rotary valve will not rotate	Rotary valves have some unique problems. In addition to those items already covered in items 10 and 11: 14a. Actuator stops set wrong, stopping the valve mechanically before it fully rotates. 14b. Broken shaft. 14c. Overtravel can cause severe damage to eccentric valves; valves can jam. 14d. Dirt or corroded valve seats can cause broken stems or valve can jam. 14e. Changing service conditions, higher pressures and greater pressure drops may stop the valve from rotating due to insufficient torque, high bearing loads. 14f. Overtightened line bolting can increase friction between the ball and seal.	14a. Readjust actuator stops. 14b. Replace shaft. 14c. Replace damaged parts, readjust travel. 14d. Replace or clean parts. 14e. Recheck actuator sizing and valve service limits. Change valve and/or actuator, as appropriate. 14f. Loosen line bolting.
15. Poor flow control and (rotary and sliding stem).	See items 12 and 13 relating to sluggish response "jumpy" travel. Other causes include: 15a. Deformed cage. 15b. Damaged piston rings. 15c. Erosion, corrosion, and cavitation can alter trim profile. 15d. A twisted shaft will indicate a position that is untrue in regard to disk and seat. The valve may indicate full open or full closed and may really be mid-range. 15e. Valve may be installed backward. 15f. Incorrect selection of flow characteristic. 15g. Low performance valve package.	15a. Replace cage. 15b. Replace piston rings. 15c. Resolve sources of damage. Replace parts. 15d. Replace Shaft. 15e. Reverse valve in line. 15f. Correct flow characteristic. 15g. Select valve assembly with conrol requirements taken into account.

Common Valve Maintenance Procedures

Packing Maintenance.   Valve packing is one of the more troublesome elements of control valve operation. As a result, the end user is often faced with the prospect of pulling it out and installing a new set. The best way to do this is to take the bonnet off of the valve and then push the old packing out from the bottom, using the following procedure. Note that this procedure covers a sliding-stem globe-style valve, and, as such, it can be done in line. For rotary valves, the procedure differs in that there is no bonnet, so the valve has to be taken from the line to extract the packing as indicated below:

1. Apply enough air pressure to the actuator to put the valve in an intermediate position so that there is no residual stem load. Disconnect the actuator and valve stems. Relieve the air pressure, and disconenct the actuator supply and any leakoff piping.

2. Remove the yoke coupling, yoke locknut, or the yoke bolting, and remove the actuator from the bonnet.

3. Loosen the packing flange nuts so that the packing is not tight on the valve plug stem. Remove any travel indicator disk and stem locknuts from the valve plug stem threads. *Safety note:* When lifting the bonnet, be sure that the valve plug and stem assembly remains on the seat ring. This avoids damage to the seating surfaces as a result of the assembly dropping from the bonnet after being lifted part way out. The parts are also easier to handle separately. Use care to avoid damaging gasket sealing surfaces. If the cage cannot be held in the body due to gasket adhesion, control it so that it will not cause equipment damage or personal injury should it fall unexpectedly.

4. Unscrew the bonnet bolting and carefully lift the bonnet off the valve stem. If the valve plug and stem assembly start to lift with the bonnet, use a brass or lead hammer on the end of the stem and tap them back down. Set the bonnet on a cardboard or wooden surface to prevent damage to the bonnet gasket surface.

5. Remove the valve plug, the seat ring, and the cage. *Note:* All residual gasket material must be removed from the cage gasket surfaces. If the gasket surfaces are scored or damaged during this process, smooth and polish them by hand, sanding with 360-grit paper and using long, sweeping strokes. Failure to remove all residual gasket material and/or burrs from the gasket surfaces will result in leakage.

6. Clean all gasket surfaces with a good-quality degreaser. Remove and residual tin or silver from all gasket surfaces.

7. Cover the opening in the valve body to protect the gasket surface and to prevent foreign material from getting into the body cavity.

8. Remove the packing flange nuts, packing flange, upper wiper, and packing follower. Carefully push out all the remaining packing parts from the body side of the bonnet using a rounded rod or other tool that will not scratch the packing box wall.

9. Clean the packing box and the related metal packing parts: packing follower, packing box ring, spring or lantern ring, special washers, etc.

10. Inspect the valve-stem threads for any sharp edges that might cut the packing. A whetstone or emery cloth may be used to smooth the threads if necessary. They can also be chased with a die.

11. Remove the protective covering from the body cavity, and install the cage using new top gaskets. Install the plug and then slide the bonnet over the stem and onto the studs. Lubricate the stud threads and the faces of the hex nuts. Replace hex nuts and torque the nuts in a crisscross pattern to no more than one-quarter of the nominal torque value specified. When all the nuts are tightened to that torque value, increase the torque by one-quarter of the specified nominal torque and repeat the crisscross pattern. Repeat this procedure until all the nuts are tightened to the specified nominal value. Apply the final torque value again and, if any nut still turns, tighten every nut again.

12. Install new packing and the metal packing box parts according to the appropriate arrangement in the instruction manual. If desired, packing parts may be prelubricated for easier installation.

Slip a smooth-edged pipe over the valve stem, and gently tamp each soft packing part into the packing box.

13. Slide the packing follower, wiper, and packing flange into position. Lubricate the packing flange studs and other related parts and the faces of the packing flange nuts. Replace the packing flange nuts. *For spring-loaded TFE V-ring packing,* tighten the packing flange nuts until the shoulder on the packing follower contacts the bonnet. *For other standard packing types,* tighten the packing flange nuts to the recommended torque. For high-performance packing sets, adjust the live-loading springs as indicated in the instruction manual.

14. Mount the actuator on the valve body assembly, and reconnect the actuator and valve stems accoding to the procedures in the appropriate instruction manual.

15. Cycle the valve 20 to 30 times and recheck packing load.

Packing can be replaced with the valve in the line, but it is not recommended due to the increased risk of stem or packing box damage. If it must be attempted, follow the above procedure with the changes noted below:

1. Remove the packing loading parts so that the top of the packing rings can be seen.

2. Very carefully insert a corkscrew packing extraction tool into the packing box and twist it into the top of the packing until it can be used to pull the top packing ring out.

3. Repeat this procedure until all the upper packing has been removed. If there is a spacer or bushing below the packing or between the upper and lower packing sets on a double arrangement, it usually has some type of slot or extraction hole. If it does not, it will have to be left in place. Assuming that it can be extracted, pull it out, and continue the above process with any packing left below the spacer.

4. Once all the packing and internal parts have been removed, do your best to clean the box out and inspect for any signs of damage. This cleaning and inspection will be very difficult to accomplish with the bonnet in place.

5. Normally you should remove the stem connector and the actuator so the rings can be slid down over the stem. If this is not possible, split rings can be used, and they can be forced onto the stem by twisting them until the opening is large enough to slide over the stem. Split rings are not recommended due to their propensity to leak. If they are used, make sure to stagger the splits to reduce the potential for leakage.

6. If any damage is found, the valve should be disassembled and the situation corrected at the first opportunity. Effective corrective action cannot be taken with the bonnet on the valve, and repacking with the bonnet on the valve will improve packing performance for a limited time, at best, if the stem or box is damaged in any way.

7. Repack and reassemble as noted above, using split rings if the stem connector was not removed.

Lapping the Seats

Lapping is procedure used to provide a better fit and surface finish between the valve plug and the mating seat. It applies only to metal-to-metal seating and is normally used for Class IV or V shutoff on control valves. Classes I, II, and III don't require it and Class VI nearly always requires soft seats. The plug and seat in their as-machined state do not always fit together perfectly around their circumference. Imperfections in fit result in excess leakage, so lapping is required to eliminate these imperfections and to make sure that the two parts fit together as closely as possible. Lapping should be carried out as follows:

1. Lapping should be done with the standard guiding in place to make sure that the parts are lapped in the positions that they will be in once the valve is fully assembled. For this reason, it is normally done with the bonnet in place.

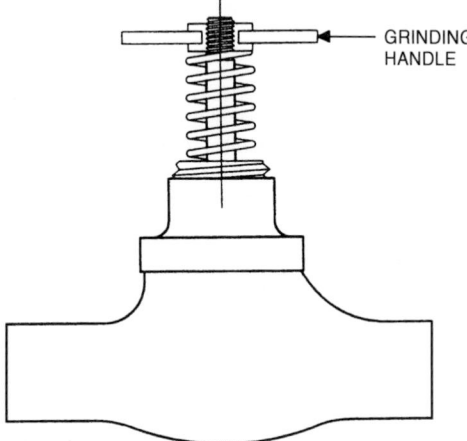

GRINDING
HANDLE

FIGURE 53 Lapping tool used with spring.

2. With the seat ring in place in the body, apply a light coating of corase grinding compound (600 grit) to both the seat ring and the plug. If the seating surfaces are made of stainless steel, use some white lead in the grinding compound to keep it from tearing or galling. Insert the plug and stem into the body and assemble the bonnet onto the body opening. The bonnet does not have to the bolted into place as long as the guiding simulates actual service.

3. Lapping requires that a very light load be applied to keep from tearing the metal, so if the plug is heavy, a spring should be used to support some of the load. The spring can be inserted over the stem and then a piece of strap iron can be locked into place on the stem and used as a grinding handle (Fig. 53).

4. Gently rotate the plug and stem four or five times, over about a 45° arc. Pick it up and move to a new position and repeat. Continue this procedure, lapping over the entire circumference at least once. Pull the assembly apart, clean he surfaces, and look for a fine continuous lap line on both the plug and the seat. Using a mirror will make the line easier to see on the seat ring inside the valve body.

5. If the lap lines look good, reassemble and repeat the procedure with a fine grit compound. If the lap lines are not continuous, repeat with the course compound. If they are still not continuous, repeat with the course compound. If they are still not continuous, try coining the surfaces by hitting the top of the stem two or three times with a heavy, but soft hammer, and lap again. If this still doesn't provide the desired results, the plug and seat should be remachined to provide a better initial fit and the process restarted.

6. When the fine grinding is done, thoroughly clean the surfaces and reassemble, torquing the bonnet in place. If possible, a seat-leak test should then be carried out to ensure tight shutoff.

7. High-temperature valves should be heated, if possible, before beginning this process to better duplicate actual guiding and fit in service.

8. Double-ported bodies will never seal as well as a single-port design, but they can still be lapped to improve shutoff. Special considerations for these valves include:

   The top seat grinds faster than the bottom. Use a coarser grit on the bottom ring to help correct for this.

   Never leave one seat dry while grinding the other one. This will tear the metal and hurt the shutoff.

   Heavy grinding on one seat may be required to get the two seats to contact at the same time.

9. Note that despite claims to the contrary, blue-lining to check for seat contact will not provide the same tight shutoff seen with lapping. Tests have shown that there can still be relatively

large imperfections present even though the blue-line shows continuous contact between the two surfaces.

Replacing the Actuator Diaphragm. After isolating the valve assembly from all pneumatic and/or fluid pressures, relieve *spring compression* in the spring, if possible. (On some spring and diaphragm actuators for use on rotary-shaft valve bodies, spring compression is not externally adjustable. Initial spring compression is set at the factory and does not need to be relieved in order to change the diaphragm.) Remove the upper diaphragm case. On direct-acting actuators, the diaphragm can be lifted out and replaced with a new one. On reverse-acting actuators, the diaphragm head assembly must be dismantled to change the diaphragm.

Most pneumatic spring and diaphragm actuators utilize a *molded diaphragm* for control valve service. The molded diaphragm facilitates installation, provides a relatively uniform effective area throughout the valve's range, and permits greater travel than could be possible if a *flat-sheet diaphragm* were used. If a flat-sheet diaphragm is used in an emergency repair situation, it should be replaced with a molded diaphragm as soon as possible.

When reassembling the diaphragm case, tighten the cap screws around the perimeter of the case firmly and evenly to prevent leakage. Be careful not to tear the diaphragm in the area of the bolt holes during reassembly. Avoid reusing a diaphragm since they are prone to leak if reused.

Replacing threaded-in seat rings. Threaded-in seat rings are no longer the preferred design for control valves in the chemical process industry. Nevertheless, this design is encountered fairly often due to its popularity in the past. The main reason this design has fallen from favor is that the seat rings can be very diffcult to get out. Adhering to the following recommended practice should help extract the seat ring with a minimum of effort and risk of personnel:

1. Before trying to remove the seat ring(s), check to see if it has been tack-welded into the body. If it has, grind out the weld.

2. To make disassembly easier, soak the ring and threads with penetrating oil and allow them to sit for some time so that the oil can do its job in loosening up the threads.

3. Insert a seat ring puller like that shown in Fig. 54 against the lugs or in the slots of the ring. Be careful to hold the puller down against the ring while applying torque, and any rounded edges on the lugs or slots should be corrected to keep the puller from slipping past the lugs or slots.

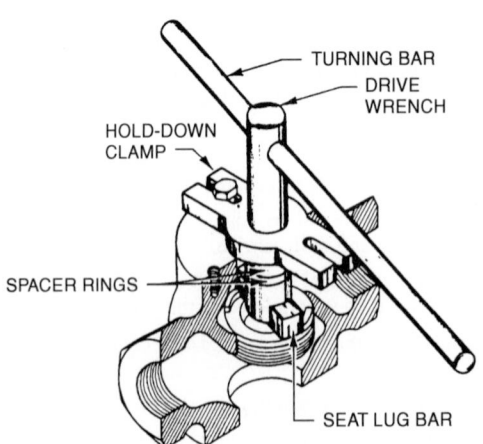

FIGURE 54 Seat ring puller. (*Fisher Controls International, Inc.*)

4. The torque can be applied manually or with the aid of a hydraulic torque wrench. If the power wrench is used, be extra careful to avoid slippage due to the high torques and the safety risk to personnel if something slips or breaks. If the valve has been pulled from the line, a lathe or boring mill may be the easiest way to apply the torque to back the ring out.

5. The bonnet bolting can be used as a reaction point for the torque and to hold the puller down into the body.

6. On particularly stubborn rings, using an impact wrench can help to break them loose.

7. As the ring starts to come out, the bolts holding the puller in the body must also be loosened to permit the ring to move up.

8. Once the ring is out, thoroughly clean and chase all threads.

9. Apply a heavy coat of lubricant or pipe compound to all threads and reinstall and torque to specified levels. The ring may be tackwelded in place, as necessary.

10. On double-ported valves, the port the farthest distance from the actuator is the smallest and needs to be installed first.

REFERENCES

1. Preckwinkle, S. E., *Maintenance Guide for Air Operated Valves, Pneumatic Actuators & Accessories*, Electric Power Research Institute, Palo Alto, Calif., 1991.

2. Ozol, J., "Experiences with Control Valve Cavitation Problems and Their Solutions," *Proceedings of EPRI Power Plant Valves Symposium EPRI*, Palo Alto, Calif., 1987.

3. McElroy, J. W., *Light Water Reactor Valve Performance Surveys Utilizing Acoustic Techniques,* Philadelphia Electric Co., Philadelphia, Pa. 1987.

4. Fitzgerald, W. V., "Automated Control Valve Troubleshooting: The Key to Optimum Valve Performance," *ISA, Proceedings*, ISA, Research Triangle Park, N.C., 1991.

5. Ferguson, Brian, "Air-Operated Valve—Preventive Maintenance Program," *Proceedings of the 2d NRC/ASME Symposium on Pump & Valve Testing*, Washington, D.C., 1992.

6. Hutchison, J. W., *ISA Handbook of Control Valves, ISA*, Research Triangle Park, N.C., 1971.

7. *Control Valve Handbook*, 1st ed., Fisher Controls, Marshalltown, Iowa, 1977.

8. *Instruction, Manual, EHD, EHS, & EHT, Form 5163*, Fisher Controls, Marshalltown, Iowa, 1985.

CONTROL VALVE CAVITATION: AN OVERVIEW

by Marc L. Riveland*

INTRODUCTION

Cavitation is of significant concern to the process control industry. It can be the source of unacceptable noise, vibration, material damage, and a decrease in the efficiency of hydraulic devices. Left

* Sr. Engineering Specialist, Applied Research, Fisher Controls International, Inc., Marshalltown, Iowa 50158.

unaddressed, cavitation can shorten the operating life of critical and expensive hardware, upset process control, and create hazardous or unsuitable work environments of plant personnel.

Theoretically, cavitation can occur in any process element or fitting that induces pressure changes. However, control valves, by virtue of their throttling function, are inherently problematic in this regard. Understanding the basic nature of cavitation and becoming familiar with available cavitation control products and techniques are the most effective means of avoiding these negative consequences in practice.

CAVITATION FUNDAMENTALS

Valve Hydrodynamics

Cavitation is a specific fluid behavior whereby pressure dynamics induce explosive growth and collapse of cavities within a liquid. Weaknesses in the fluid continuum allow rapid growth and vaporization of the liquid when the local fluid pressure decreases to near the liquid vapor pressure. Conversely, if the local pressure subsequently rises to a value above the vapor pressure, the reverse process occurs and the cavity collapses. Localized pressures and velocities associated with the collapse phase are the source of most of the aforementioned problems. This behavior can occur in cases in which the liquid is static (as in a propeller spinning in a large body of water) or in motion (as in the case of liquid flow through pipes and fittings).

Pressure dynamics conducive to cavitation within a control valve are related to the flow of the liquid through the restriction. The *mean* pressure profile associated with flow through a simple restriction, such as a control valve, is shown in Fig. 1.

The shape of this general curve is a consequence of fluid continuity and conservation of energy. The mean kinetic energy increases as the fluid accelerates through the restricted flow area of the valve throat. Correspondingly, the mean fluid pressure decreases to maintain the fluid energy balance. The fluid decelerates as it moves into the increased flow area of the valve outlet and downstream pipe, the mean kinetic energy decreases, and the pressure again increases. However, a small amount of fluid energy is irreversibly dissipated in this recovery process so that the mean fluid pressure is not completely restored to its original value. This results in a pressure differential across the device, often referred to as the *observed* pressure differential.

The point of minimum flow area is known as the *vena contracta* and, for all practical intents and purposes, corresponds to the minimum pressure condition. The relationship between these two key pressure differentials is embodied in the liquid pressure recovery factor, F_L, defined by the following equation:

$$F_L^2 = \frac{\Delta P_{\text{obs}}}{P_1 - P_{\text{vc}}} \tag{1}$$

Proper use of this factor is discussed later in the section on valve selection.

Incipient cavitation is the level of cavitation associated with discernible onset. This would *theoretically* occur when the vena contracta pressure is equal to the vapor pressure of the fluid and the corresponding outlet pressure is above the vapor pressure. According to the simple model just described, it would initially appear reasonable to utilize Eq. (1) to predict this condition. While this simple model is suitable for conceptualizing the cavitation process, and it does provide a first-order understanding of the forces that give rise to cavitation, it is not sufficiently complete enough to explain all observed behaviors. In fact, it is a particularly poor model for predicting the incipient cavitation condition.

Recall that this model depicts *mean* fluid pressure. Flow through the complex geometry of control valves gives rise to significant deviations from this mean pressure. These include instantaneous pressure fluctuations associated with fluid turbulence and low pressures in the cores of vortices and eddies associated with boundary-layer separation, free shear zones, stagnation regions, and re-entrant zones. These phenomena can produce local pressures significantly higher or lower than the mean pressure, sufficient to initiate cavitation in very localized regions. Typically, cavitation begins well before the

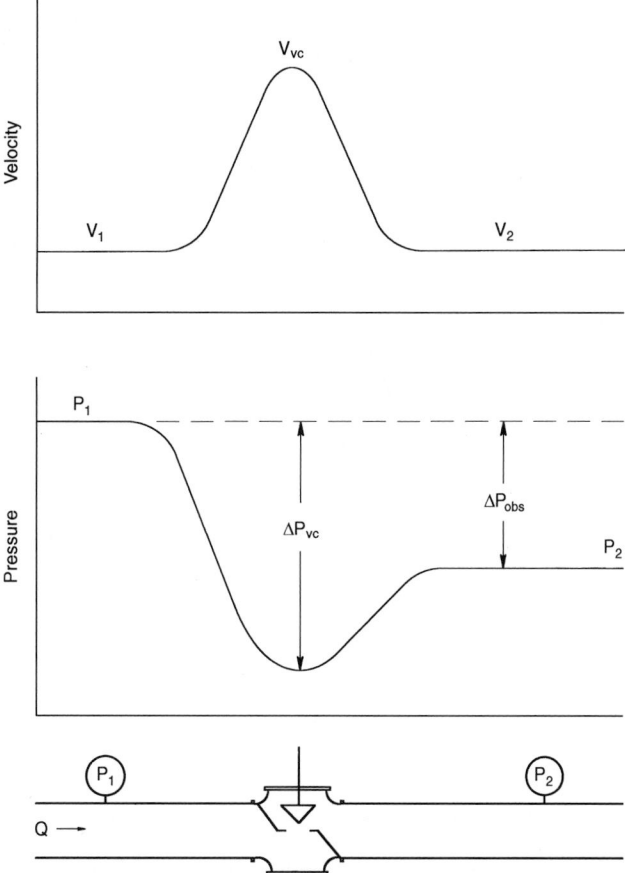

FIGURE 1 Idealized mean fluid pressure profile for a control valve. This model represents mean pressure changes associated with mean velocity changes. It does not account for local pressure fluctuations.

minimum mean pressure is reduced to the vapor pressure. Additional empirically based parameters are required to predict different levels of cavitation. These are discussed later.

Cavity Mechanics

Understanding cavity behavior is the key to developing both systemic and valve-based solutions to cavitation–related problems. Cavitation has been the focus of both academic and industrial research for decades, and much has been learned about individual and collective cavity behavior. A comprehensive review of this subject is outside the scope of the immediate article and will be discussed only in the context of cavitation control. By necessity, many of the details are omitted. The interested reader is referred to a very thorough coverage of the subject by Hammitt [1].

Cavitation is generally recognized to consist of four distinct events, that is, nucleation, growth, collapse, and rebound. All four events contribute to the overall extent of cavitation-related problems. However, the latter two events are the primary source of noise, vibration, and material damage.

When pressure conditions are right, cavities begin to grow from weak spots in the fluid continuum known as *nuclei*, undergoing an initial period of stable growth. Upon attaining a critical diameter, growth proceeds explosively with substantial vaporization of the liquid. Cavity growth ceases and the collapse process begins when the fluid pressure increases. The ultimate degree and extent of cavity growth will be determined by the number and size of entrained nuclei, the properties of the liquid, and the extent of the low-pressure region.

Unlike the comparatively symmetrical growth of the cavity, the initial collapse is very rapid and highly asymmetrical. This results from the inertial forces acting on the cavity as it moves into a lower-velocity, higher-pressure region. Experimental studies [2] have revealed the presence of a very small, high-velocity jet (referred to as a *microjet*) formed during such asymmetrical collapse.

Several additional growth–collapse cycles may follow the initial cycle in a phenomenon known as rebound. This occurs when the rate of cavity collapse exceeds the condensation rate, or if the cavity contains substantial amounts of a noncondensable gas. The cavity contents are compressed by the liquid rather than condensed to the liquid state. Mechanical energy is stored in the compressed gases and can be released to initiate another cycle. The total cavity volume decreases on each successive cycle until the process ceases. The collapse of a rebound cavity is generally more symmetrical than the initial collapse and is marked by the absence of the high-velocity microjet. However, the rapid movement of the liquid surrounding the cavity induces a shock wave, which propagates away from the cavity.

The collapsing cavity serves as the primary source of hydrodynamic noise and vibration. In fact, hydrodynamic noise is usually attributed entirely to cavitation; noise levels at subcavitating flow are not typically troublesome. The general subject of hydrodynamic noise and related prediction methods is very involved and is discussed elsewhere in this handbook. In practice, treatment of cavitation (the source) brings resultant hydrodynamic noise levels to within an acceptable level. A more detailed discussion is outside the immediate scope of this article.

Damage Mechanisms

Physical damage to the valve is probably the most frequent concern because of the associated cost, inconvenience, and unpredictable nature. Damage, as used in the context of this article, refers to any permanent deformation or loss of material. The collapsing cavity initiates an attack on adjacent material surfaces. This attack on the material surface comprises a mechanical component and, more selectively, a chemical component. The response or reaction of the material to the attack and the total time of exposure determine the extent of total damage to the material.

Mechanical attack may consist of high velocity, microjet impingement, shock-wave impingement, or most likely a combination of the two, on the material surface. Mechanical attack must always originate from a cavity collapsing *near the material surface* in order to impart damage to that surface. If the microjet (established during the asymmetrical cavity collapse) is close to the surface and impinges directly on the surface, a damaging attack will occur; otherwise no adverse material effect results. Interestingly, these microjets exhibit a preferred orientation toward rigid surfaces. Presumably the fluid resistance near the wall reduces the "supply" of fluid to that side of the cavity. Flow to the outboard side of the cavity is relatively unimpeded, so that the jet orientation is in the direction of the wall. The coupling dynamics of a highly compliant surface (such as an elastomeric material) effect the opposite behavior; that is, the orientation is *away* from the surface. Similarly, the intensity of the shock waves dissipates rapidly with the propagation distance so that shock waves originating from cavities far removed from the surface (that is, more than one bubble diameter or so) have insufficient strength to impart significant damage.

Corrosion can play a significant role in the damage process in that it interacts with the mechanical forms of attack in a synergistic manner. Protective coatings can be removed by the mechanical attack, allowing chemical attack to occur. The chemical attack in turn deteriorates the material, making it more susceptible to mechanical attack. The process continues, resulting in a more aggressive damage process than that associated with either of the constituent forms individually.

Repeated attack by millions of microjets and shock waves results in the characteristic appearance of cavitation—a very rough pitted surface.

Scale Effects

A number of factors can potentially affect the intensity of cavitation or, more importantly, the level of the associated negative effects of cavitation. Sometimes called *scale effects*, these factors can either intensify or diminish cavitation-related problems relative to equivalent installations under hydrodynamically similar operating conditions. The scale effects of most concern to control-valve applications are pressure- and velocity-related effects, size effects, and air-content effects. Other effects such as viscosity, surface tension, and various thermal property effects have also been investigated. However, in most industrial applications these are either of little significance or not sufficiently quantified to be able to adequately account for them.

In general, as the pressure, velocity, or valve size increase, the associated cavitation problems get worse. Numerous investigations (as documented by Hammitt) have borne out the fact that the degree of damage resulting from cavitation is very sensitive to the fluid velocity; that, in fact, the total damage imparted is an exponential function of the fluid velocity. The range of values reported for the exponent is very broad, usually between three and ten. However, there is some agreement that six is a representative number. Tullis [3] and Mousson [4] both provide data showing an increase in the damage rate as the upstream pressure increases. Investigations by Tullis [3] report an increase in the severity of the negative side effects of cavitation associated with an increase in the nominal size of the device.

Likewise, there are effects associated with the change in backpressure applied to a valve. For a monotonically decreasing outlet pressure at constant inlet conditions, two opposite trends can be rationalized: an intensifying effect (due to increasing vapor volume) and a diminishing effect (due to decreasing collapse intensity). The issue becomes a matter of determining which effect dominates at any given outlet pressure. Mousson's data [4] supply a partial answer to this by showing a maximum damage level existing roughly midway between the two extremes. Field experience with control valves is consistent with this. In some cases unacceptable levels of noise and vibration existing at flow conditions well below choked flow have been observed to diminish to satisfactory levels at choked flow conditions.

The presence of dissolved or entrained air (or any other noncondensable gas) has multiple effects on the cavitation process and associated problems. Increasing the amount of such gases has the effect of providing additional nucleation sites in the fluid. This contributes to an increase in the overall amount of cavitation and consequently the level of problems associated with cavitation. However, continued increase in the amount of air reduces the collapse velocities and disrupts the microjet and shock-wave attack mechanisms. This results in an overall attenuation of the negative effects of cavitation, even though cavitation is still occurring. Mousson [4] reports a significant reduction in the damage levels with only a few percent of air (by volume) entrained in water under otherwise constant conditions.

CAVITATION ABATEMENT STRATEGIES

There is no single best method for controlling the problems created by cavitation. A number of techniques are available, each with inherent advantages and shortcomings. Familiarity with these practices provides greater opportunity to implement the most technically satisfactory and cost-effective solution for a given application.

All cavitation abatement techniques are based on effecting control over one or more of the basic elements of cavitation problems, that is, cavitation intensity and attack, material of construction (response to attack), or duty cycle (time of exposure). Cavitation control should be considered on two levels—the system level and the control valve level. Whenever possible, it is desirable to contend with cavitation at the system level.

SYSTEM-LEVEL CAVITATION CONTROL

A primary and preferred strategy is to consider potentially problematic conditions at the time a process system is designed. Awareness and avoidance of conditions conducive to control-valve cavitation are a highly effective means of reducing the risk of cavitation-related problems.

Valve Placement

Cavitation-related problems can be averted by simply locating the valve in a region of high overall pressure. The flow rate and pressure differential across the valve remain the same regardless of location, but overall fluid pressure within the valve body is increased proportionately, as depicted in Fig. 2. This results in a greater margin between the minimum pressures throughout the valve and the vapor pressure, thus decreasing the likelihood that the fluid pressure will fall below the vapor pressure of the liquid.

This technique can be effected in practice by locating the valve as far upstream in the system as possible. When the piping losses are minimized upstream and shifted downstream of the valve, backpressure to the valve is increased. This, as noted above, increases the overall pressures internally in the valve.

Backpressure Devices

When placement of the valve within the system is not flexible, fluid pressures within the control valve may be increased by introducing additional resistance to flow downstream of the valve. When a restriction such as an orifice plate or a second valve is placed downstream of the valve, the backpressure is increased by the amount of the pressure differential across that restricting device.

Usually the pressure drop across the control valve is decreased by this amount, that is, the inlet pressure remains constant but the backpressure increases. Consequently, the valve will realize less pressure drop for the same flow rate and the required valve coefficient must increase accordingly. In addition to increasing the fluid pressure within the valve, the fluid velocities will generally be reduced since the valve will operate at a larger opening. The combined effects of increased fluid pressure and reduced velocity can be very effective in controlling cavitation. A word of caution is needed, however.

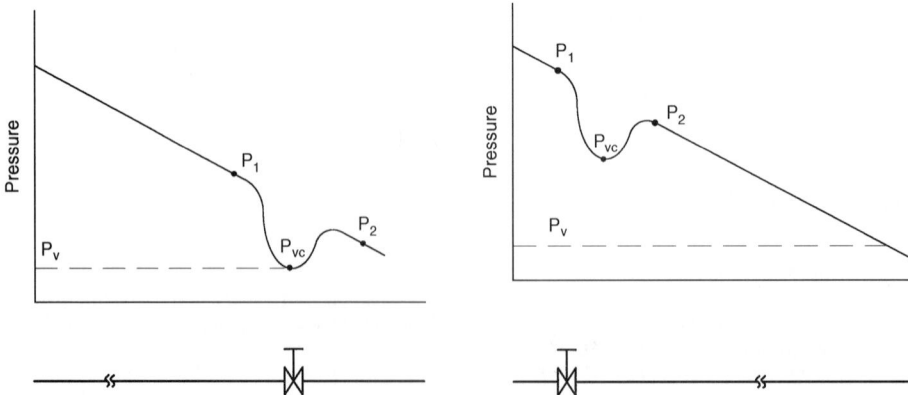

FIGURE 2   The effect of locating a valve in a region of higher system pressure is to increase the overall mean pressure within the valve.

Cavitation within the overall system may not always be controlled by this method, but rather merely displaced from the valve to another location within the system. It is important to account for the possibility and consequence of any resulting cavitation at the downstream restriction.

The fixed-restriction alternative is best suited to on–off service since the device can only be optimized for a single flow rate. If the restriction is sized for high load and the system is operating at lower load, the pressure drop across the restriction will be very low as a result of a decreased fluid velocity through the restriction. Consequently, the valve will realize an increased pressure drop and again be at risk of cavitation-related problems. At higher flow rates than designed for, the orifice may well become the primary restriction and in turn limit or choke the flow at a lower flow rate than desired. The use of a second control valve in series affords a greater effective range, but this is usually a more expensive solution and requires a more sophisticated control scheme for optimum performance.

Gas Injection

Another method of controlling the damage, noise, and vibration resulting from cavitation (but not totally eliminating cavitation) is through noncondensable gas injection. This method, while effective, is very selective since not all processes will tolerate the introduction of gases.

A gas that will not condense under the prevailing downstream conditions is injected (or aspirated) into the flowing fluid near the vena contracta. The continued presence of the gas phase during pressure recovery disrupts the cavity collapse process and limits the negative effects associated with cavity collapse. Caution must be exercised to introduce the gas at or downstream of the throat of the valve in order to avoid a reduction in flow from a two-phase mixture at the vena contracta. Individual valve suppliers should be consulted as to the practicability of this method for the valve style being considered, as well as to the minimum amount of gas required and the exact location of introduction to the flow stream.

CONTROL-VALVE SOLUTIONS

If it is not possible to avoid cavitating conditions in the process system, it is necessary to contend with it at the control-valve level. Control-valve solutions are predicated on the concepts of cavitation reduction, resistance to cavitation attack, and control of the collapse region of cavities.

Material Selection

First consideration should be given to material selection. If physical damage is of primary concern, it is sometimes possible to create a cavitation-tolerant environment by selecting materials more resistant to cavitation attack. Standard trims constructed of materials suited to the process fluid, and service conditions often provide a cost-effective solution.

Proper material selection is not a black-and-white issue, nor is there one "best" material. The characterization of a material's resistance to cavitation lacks rigorous quantification. Currently, qualitative force ranking of material resistance to cavitation attack in combination with empirical "rules of thumb" governs selection. As a general rule, a material's hardness and resistance to corrosion are the foremost properties considered. Other properties that have shown a correlation to cavitation damage resistance to varying degrees include the ultimate resilience and strain energy to failure. However, no single property offers a consistent numerical correlation.

It is important to base material selection on the total attack, that is, considering both the mechanical and the chemical components. A notable exception to the "harder is better" rule is the widespread use of cobalt alloy 6 in cavitating service. Its combined hardness and corrosion resistance makes it a preferred choice to harder materials currently available. However, even this material is not universally superior. It provides very poor protection in applications of boiler feedwater treated with hydrazine.

Even though the material is extremely hard, the amines attack the material and render it structurally inferior. Other materials that are chemically more resistant to the amines, such as S44004 (440C), are a preferred choice. Other popular materials frequently used in cavitating liquid service include other alloy steels, tool steels, certain stainless steels (such as the 300 series), and precipitation-hardened materials.

Ceramics is an emerging material category that shows promise of good cavitation damage resistance. Ceramics of practical interest to the control-valve industry consist of metals or metalloids combined with oxygen, carbon, nitrogen, or boron. Examples include aluminum oxide, zirconium dioxide, silicon carbide, and silicon nitride.

Other nonmetallic materials, such as elastomers and compliant materials in general, exhibit an ability to withstand levels of cavitation attack greater than standard structural indicators would suggest. This paradox apparently results from a dynamic interaction between the surface and the cavity, which orients the microjet away from the surface, thus eliminating the mechanical attack. While such behavior is appealing from a damage-control standpoint, Sanderson [5] points out that bonding difficulties, as well as potential pressure and temperature limitations, have curbed the widespread use of such materials in the industry.

It should be emphasized that all materials are vulnerable to cavitation attack. The rate of damage is a complicated function of the intensity of the cavitation attack, the total time of exposure, and the material characteristics. Material selection by itself can prolong the life of a component, but it will not completely eliminate the possibility of damage and therefore is best utilized in conjunction with other abatement strategies.

Special Trim Designs

If the protection offered by material selection is deemed inadequate by itself, or if noise and vibration are also of concern, it may be necessary to use special trim designs. A number of proprietary products are available from different valve manufacturers. These products and trims come in a wide variety of configurations, but they are all based on one or more fundamental operational strategies with different tradeoffs between cost and performance.

The most common design concepts embraced by different valve manufacturers parallel many of the techniques used on a larger scale in the context of system strategies. The foremost objective of good cavitation control product design is to control energy conversions within the valve. Pressure-recovery characteristics and trim velocities are favorably affected by strategically introducing resistance into the flow path. In general, overall fluid pressure recovery is reduced in such trims. Reduced-pressure recovery effectively reduces the tendency of the valve to cavitate by raising overall pressure in the valve compared to those in a high-recovery valve under the same conditions, as depicted in Fig. 3. Further benefit is realized in that, if cavitation does occur, the pressure differential driving cavity collapse $(P_2 - P_v)$ is reduced, which in turn tends to reduce the negative effects of cavitation.

This objective is commonly achieved in practice by forcing fluid flow through successive stages or tortuous paths. When the pressure drop across a valve is staged, a portion of the total pressure drop is taken across each of a series of restrictions, or stages. This creates a much less efficient hydrodynamic path than an equivalent single restriction and results in a lower pressure recovery. Furthermore, the decreased flow efficiency requires comparatively larger flow passages, hence lower velocities, under similar flow conditions. Tortuous-path treatment, in contrast, utilizes a labyrinthal flow path to induce irreversible energy conversions, which in turn have the same impact on pressure recovery and velocity that staging does.

Another fundamental design strategy consists of dividing the flow stream into multiple parallel flow paths. Whereas many of the cavitation-related problems tend to scale with the physical size of the flow stream, the reduced size of individual flow paths helps to reduce overall cavitation-related problems, particularly noise and vibration. To avoid potential plugging problems associated with restrictive flow passages, a compromise between degree of cavitation control and passage clearance must be reached.

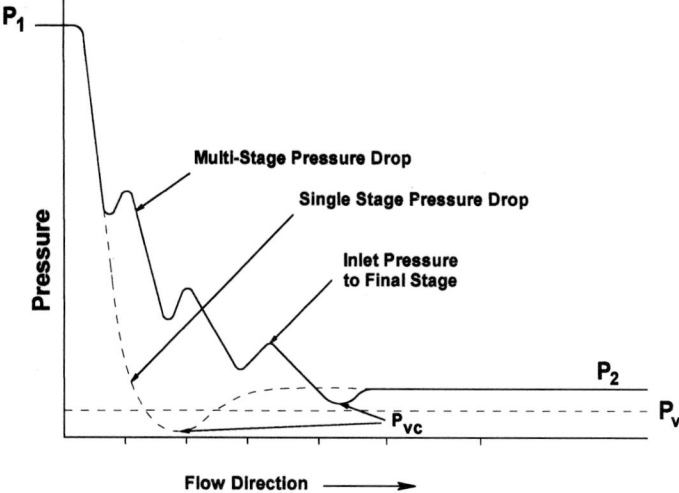

FIGURE 3   Successive contractions and expansions, and tortuous flow paths reduce caviation problems. Pressure is dissipated while maintaining higher overall fluid pressures.

Finally, it is possible to redirect the flow stream to avoid direct impingement of cavitating flows on critical control surfaces. As pointed out earlier, a cavity must collapse in close proximity to a surface in order to damage that surface. By separating collapsing cavities from the surface the threat of damage is minimized, even though the cavitation has not necessarily been eliminated or reduced.

The design of any control valve to minimize the effects of cavitation involves a tradeoff between higher comparative costs, lower relative capacity, the degree of protection required by a particular application, and the ability of the valve to tolerate dirty fluids. Highly optimized designs actually prevent the formation of any significant cavitation, whereas standard trim would cavitate heavily. This degree of protection is not warranted by all process control applications. Therefore a variety of trims generally designed to a reasonably specific set of conditions are available to meet the variety of process needs.

CONTROL-VALVE SELECTION

Correct process operation and maximum valve life are predicated on proper valve selection for specific application conditions. This is true not only for special cavitation control trims, but for so-called standard valves as well. The following discussion provides an overview of the generally accepted industry methods available at the time of this writing. This subject remains an active area of development within the industry, and changes are likely forthcoming.

Background

Recognizing that cavitation is a complex phenomenon, it comes as no surprise that a single, simple method to properly select a valve for cavitation remains elusive. Numerous "manufacturer specific" methods have evolved over the years, with varying degrees of success. A fairly recent initiative by the Instrument Society of America (ISA) has resulted in the first meaningful step toward bringing these diverse methods into a single, cohesive, and universally applicable framework. While the technology is still evolving and many questions remain, this effort goes a long ways toward providing a common vernacular and collection of parameters by which performance can potentially be quantified. This effort culminated with the release of a recommended practice [6] that provides background information on the cavitation phenomenon, nominal testing and evaluation methods, and a basic application framework. The discussion that follows is based on this information. The interested reader is referred to this document for more information. (Note: At this juncture, the scope of discussion is narrowed to exclude noise prediction technology. This is an equally involved topic and is covered elsewhere in this handbook.)

Cavitation Parameters and Coefficients

From the discussion presented earlier it is evident that pressure characteristics are relevant to cavitation behavior. It comes as no surprise, then, that control-valve service and performance are generally characterized by variant forms of the pressure coefficient or pressure ratios. Several different forms have evolved over time, the exact choice of which varied between valve manufacturers. The ISA method [6] has adopted the following basic form:

$$\sigma = \frac{P_1 - P_v}{P_1 - P_2} \tag{2}$$

where P_1 = absolute pressure upstream of control valve,
P_2 = absolute pressure downstream of control valve,
P_v = absolute thermodynamic vapor pressure.

The application method is built on the notion of comparing an operating condition expressed in terms of Eq. (2) to a meaningful limit expressed in the form of a valve cavitation coefficient, σ_v. If σ is greater than σ_v, then the valve satisfies the criterion for acceptable operation implied by the coefficient.

Clearly, the critical step of this method is the establishment of a meaningful and appropriate valve cavitation coefficient. Numerous benchmarks are defined, some of which are readily measurable and others that are desired but are more difficult to objectively quantify. Several principal values utilized in the recommended practice [6] include the following:

- incipient cavitation coefficient, σ_i: the value of σ at the onset of cavitation
- constant cavitation coefficient, σ_c: the value of σ at conditions of mild but steady cavitation
- incipient damage coefficient, σ_{id}: the value of σ at the onset of damage by cavitation
- manufacturer's recommend coefficient, σ_{mr}: the minimum value of σ recommended by the valve manufacturer for satisfactory operation of the control valve.

It is very difficult to observe and measure cavitation directly. No "scientifically pure," or completely objective, laboratory method exists for evaluating the cavitation that occurs in a control valve or all coefficients of interest. The usual approach is to monitor the effect of cavitation on characteristics such as noise levels, vibration levels, damage rate, or flow efficiency, and infer information about the behavior of other cavitation effects under field conditions.

Means of evaluating σ_i and σ_c are available [6]–[8]. In essence, pipe wall vibration (or in some cases control-valve noise) are measured in a prescribed laboratory installation. The measured quantity (i.e., vibration or noise) is plotted over a range of σ. Figure 4 depicts a typical plot of this information. The incipient level of cavitation and the constant level of cavitation are associated with the inflection points indicated on the plot.

The incipient level of cavitation is generally too conservative for most applications. Most control valves, especially those designed for cavitating service, can tolerate some degree of cavitation. This level of cavitation is an appropriate limit only for those applications in which no cavitation whatsoever can be tolerated by the application.

The constant level of cavitation is a limit defined primarily by the test. It is certainly a discernible level of cavitation; however, it is difficult to say whether it is a tolerable level of cavitation or not with

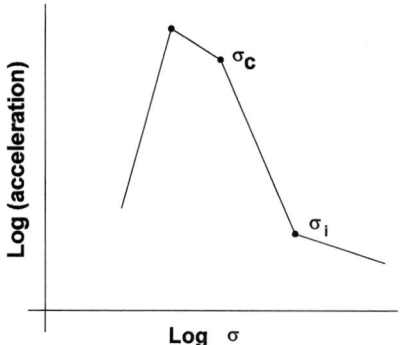

FIGURE 4 Typical pipe wall vibration curve, showing defined cavitation levels.

respect to any given application. In other words, constant cavitation is not necessarily an appropriate universal limit or threshold. Again, in many applications, it is likely to be too conservative, resulting in unnecessarily expensive solutions.

The incipient damage level of cavitation is a threshold of significant practical interest. If the onset of damage could be predicted, and damage subsequently avoided, many cavitation applications would be solved. However, the problem again lies in the objective evaluation of this condition. From previous discussion it is evident that the onset and extent of material damage is a function not only of the extent and intensity of cavitation, but also of the material of construction and the total duration of attack (duty cycle). No test method currently exists that consistently quantifies this process and accounts for all of these factors.

For the foreseeable future, the threshold described by σ_{mr} probably provides the greatest utility to the end user. The manufacturer's knowledge of valve design parameters, tested performance, and accumulated experience provide a strong basis for establishing the value of meaningful operational limits.

It should be noted that none of these parameters by themselves are complete similarity parameters. They do not, without modification, account for the numerous scale effects previously identified (such as pressure or size). Some scaling relationships have been established for certain parameters and are offered in the recommended practice [6]. Others, such as σ_{mr}, may or may not already account for one or more of the scale effects. They should not be subject to universal scaling laws without specific direction to do so by the manufacturer.

Whereas the pressure-recovery characteristics of a control valve are quantified in the F_L parameter, the question naturally arises as to the appropriate use of this parameter in the context of selecting a valve for cavitating service.

Universally, F_L has only one *quantitative* use, and that is to determine the choked flow rate through a specific valve under a given set of conditions [9]. However, because it is a pressure recovery term, it is *qualitatively* related to a valve's tendency to cavitate. Reflecting on the previous discussions regarding pressure recovery, it is apparent that if the same pressure differential is applied to both a high-recovery device and a low-recovery device, the high-recovery device will have the lower vena contracta pressure (see Fig. 5).

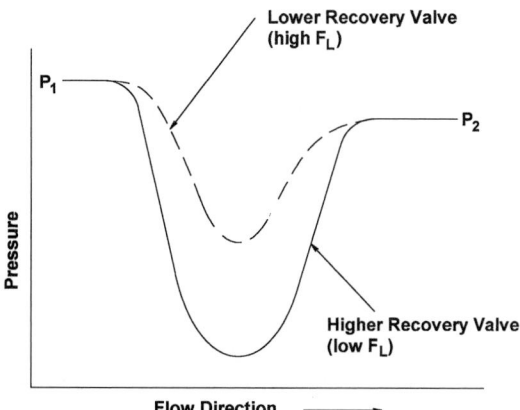

FIGURE 5 Low-recovery valves tend to have higher overall mean fluid pressure than high-recovery valves under equivalent flow conditions. Local pressure fluctuations may offset this positive effect.

Low-pressure recovery devices are characterized by large values of the pressure-recovery coefficient. Therefore, a valve with a high value of F_L is generally, but not universally, a better candidate for cavitating service than one with a low value insofar as overall fluid pressures within the valve are greater. Used in conjunction with other parameters and additional information, F_L can play a role in sizing valves for cavitation service. However, it should not be construed as *the* singular cavitation index. Many cavitation control hardware features are not reflected in the value of the pressure-recovery coefficient.

Example

Consider a cage-guided globe valve in the following service.
Water

$$T_1 = 26.7\,°C,$$
$$P_1 = 1034\ kPa,$$
$$P_2 = 470\ kPa.$$

The associated valve coefficients are:

$$F_L = 0.82,$$
$$\sigma_i = 2.89,$$
$$\sigma_{mr} = 1.57.$$

Water at the given temperature has a vapor pressure of 3.5 kPa; therefore by Eq. (2),

$$\sigma = \frac{1034 - 3.5}{10.34 - 470}$$
$$= 1.83$$

Since $\sigma < \sigma_i$, the valve will be cavitating. However, since $\sigma > \sigma_{mr}$, the valve will operate satisfactorily according to the manufacturer's specification.

CLOSURE

Cavitation is a complex phenomenon that can have an adverse impact on process performance, equipment service life, and operational behavior. Useful application technology is emerging but is not fully mature and is still evolving. This technology is most effective when used in combination with engineering judgment that is based on an understanding of the basic behaviors and experience.

REFERENCES

1. Hammitt, F. G., *Cavitation and Multiphase Flow Phenomena*, McGraw-Hill, New York, 1980.
2. Knapp, R. T., and A. Hollander, "Laboratory Investigations of the Mechanism of Cavitation," *Trans. ASME*, vol. 70, 1948.
3. Tullis, J. P., "Cavitation Scale Effects for Valves," *J. Hydraulics Div., ASCE*, vol. 99, p. 1109, 1973.
4. Mousson, J. M., "Pitting Resistance of Metals under Cavitation Conditions," *Trans. ASME*, vol. 59, pp. 399–408. 1937.

5. Sanderson, R. L., "Elastomers for Cavitation Damage Resistance," paper C. I. 82–908, presented at the ISA International Conference And Exhibit, Philadelphia, Pennsylvania, Oct. 1982.

6. ISA RP75.23–1995, "Considerations for Evaluating Control Valve Cavitation," Instrument Society of America, Research Triangle Park, North Carolina, 1985.

7. Riveland, M. L., "The Industrial Detection and Evaluation of Control Valve Cavitation," *ISA Trans.*, vol. 22, no. 3, 1983.

8. Ball, J. W., and J. P. Tullis, "Cavitation in Butterfly Valves," *J. Hydraulics Div., ASCE*, vol. 99, p. 1303, 1973.

9. ISA S75.01–1985 (R 1995),"Flow Equations for Sizing Control Valves," Instrument Society of America, Research Triangle Park, North Carolina, 1995.

CONTROL VALVE NOISE

by Allen C. Fagerlund*

INTRODUCTION

Fluid transmission systems are major sources of industrial noise. Elements within the systems that contribute to the noise are control valves, abrupt expansions of high-velocity flow streams, compressors, and pumps. Control-valve noise is a result of the turbulence introduced into the flow stream in producing the permanent head loss required to fulfill the basic function of the valve.

NOISE TERMINOLOGY

Noise is described or specified by the physical characteristics of sound. The definitive properties of sound are the magnitude of sound pressure and the frequency of pressure fluctuation, as illustrated in Fig. 1.

Sound pressure, P_S, measurements are normally root-mean-square (rms) values of sound pressure expressed in microbars. Because the range of the sound pressure of interest in noise measurements is ~10^8 to 1, it is customary to deal with the sound pressure level (L_P) instead of sound pressure. L_P is a logarithmic function of the amplitude of sound pressure and is expressed mathematically as

$$L_P = 20 \log_{10} \frac{P_s}{0.0002\,\mu\text{bar}} dB$$

The selected reference sound pressure of 0.0002 μbar is approximately the sound pressure required at 1,000 Hz to produce the faintest sound that the average young person with normal hearing can detect.

* Senior Research Specialist, R. A. Engel Technical Center, Fisher Controls International, Inc., Marshalltown, Iowa.

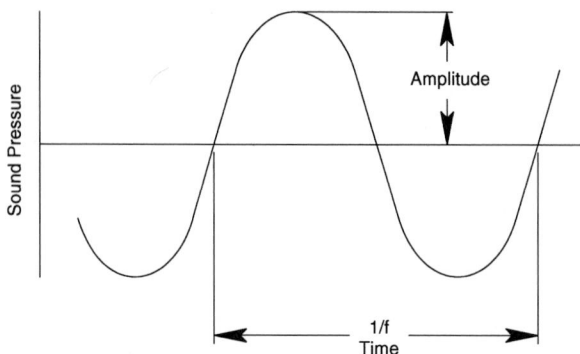

FIGURE 1 Properties of sound.

The characteristic of the L_P scale is such that each change of 6 dB in level represents a change in the amplitude of sound pressure by a factor of 2.

The apparent loudness of a sound varies not only with the amplitude of sound pressure but also as a function of frequency. The human ear responds to sounds in the frequency range between 20 and 18,000 Hz. The normal ear is most sensitive to pressure fluctuations in the neighborhood of 3,000 to 4,000 Hz. Therefore the degree of annoyance created by a specific sound is a function of both sound pressure and frequency.

L_P measurements are often weighted to adjust the frequency response. Weighting that attenuates various frequencies to approximate the response of the human ear is called A weighting. Figure 2 shows L_P correction as a function of frequency for A-weighted octave-band analysis.

Approximate overall sound levels of some familiar sound environments are shown in Table 1.

Any study of valve noise will evaluate the following basic phenomena. Acoustic power, generated by fluid flow through a control valve, propagates through the piping and creates a fluctuating pressure field, which forces the pipe wall to vibrate. These vibrations in turn cause pressure disturbances outside the pipe that radiate as sound. The difference in sound pressure levels from inside to outside the pipe is called the transmission loss.

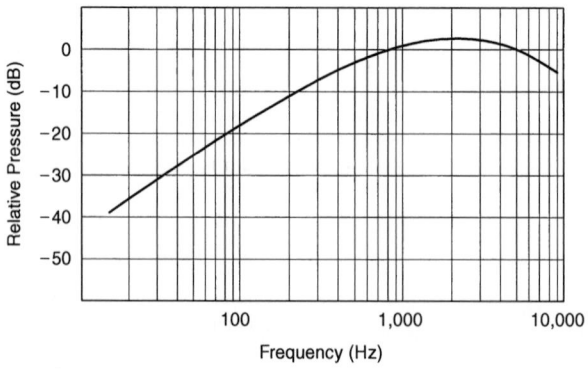

FIGURE 2 A-weighting curve.

TABLE 1   Approximate Sound Levels of Familiar Sounds

Sound	Sound Level (dB)
Pneumatic rock drill	130
Jet takeoff [at 200 ft (61 m)]	120
Boiler factory	110
Electric furnace (area)	100
Heavy street traffic	90
Tabulation machine room	80
Vacuum cleaner [at 10 ft (3 m)]	70
Conversation	60
Quiet residence	50
Electric clock	20

SOURCES OF VALVE NOISE

The major sources of control valve noise are (1) mechanical vibration of valve components and (2) fluid-generated noise, namely hydrodynamic noise and aerodynamic noise.

Mechanical Noise

The vibration of valve components is a result of random pressure fluctuations within the valve body or fluid impingement on movable or flexible parts. The most prevalent source of noise resulting from mechanical vibration is the lateral movement of the valve plug relative to the guide surfaces. Sound produced by this type of vibration normally has a frequency of less than 1,500 Hz and is often described as a metallic rattling. The physical damage incurred by the valve plug or associated surfaces is generally of more concern than the noise emitted.

A second source of mechanical noise is a valve component resonating at its natural frequency. Resonant vibration of valve components produces a single pitched tone, normally having a frequency between 3,000 and 7,000 Hz. This type of vibration produces high levels of stress that may ultimately produce fatigue failure of the vibrating part.

Noise resulting from mechanical vibration has, for the most part, been eliminated by improved valve design and is generally considered a structural problem rather than a noise problem.

Hydrodynamic Noise

Control valves handling liquid flow streams can be a substantial source of noise. The flow noise produced is referred to as hydrodynamic noise and may be categorized with respect to the specific flow classification or characteristic from which it is generated. Liquid flow can be divided into three general classifications: (1) noncavitating, (2) cavitating, and (3) flashing.

Noncavitating liquid flow generally results in very low ambient noise levels. It is generally accepted that the mechanism by which the noise is generated is a function of the turbulent velocity fluctuations of the fluid stream, which occur as a result of rapid deceleration of the fluid downstream of the vena contracta as the result of an abrupt area change.

The major source of hydrodynamic noise is cavitation. This noise is caused by the implosion of vapor bubbles formed in the cavitation process. Cavitation occurs in valves controlling liquids when the following two service conditions are met: (1) the static pressure downstream of the valve is greater than the vapor pressure and (2) at some point within the valve the local static pressure is less than or equal to the liquid vapor pressure because of either high velocity or intense turbulence.

The cavitation phenomena is discussed in detail in another section of this handbook. Vapor bubbles are formed in the region of minimum static pressure and subsequently are collapsed or imploded as they pass downstream into an area of higher static pressure. Noise produced by cavitation has a broad frequency range and is frequently described as a rattling sound similar to that which would be anticipated if gravel were in the fluid stream. Since cavitation may produce severe damage to the solid boundary surfaces that confine the cavitating fluid, noise produced by cavitation is generally of secondary concern.

Flashing is a phenomenon that occurs in liquid flow when the differential pressure across a restriction is greater than the differential between the absolute static and vapor pressures at the inlet to the restriction, that is, $\Delta P > P_1 - P_v$. The resulting flow stream is a mixture of the liquid and gas phases of the fluid, which causes large variations in fluid density and acoustic wave speed. Noise resulting from a valve's handling a flashing fluid is a result of the deceleration and expansion of the two-phase flow stream.

Test results supported by field experience indicate that noise levels in noncavitating and flashing liquid applications are quite low and generally are not considered a noise problem.

Aerodynamic Noise

Aerodynamic noise is created by turbulence in a flow stream as a result of deceleration or impingement. The principal area of noise generation in a control valve is the recovery region immediately downstream of the vena contracta, where the flow field is characterized by intense turbulence and mixing. Subcritical flow is similar to noncavitating liquid flow in that turbulence caused by shear is the primary source of noise. Once the compressible flow is critical with shock waves present, the character of the noise generation changes. The interaction of the turbulence with the shock waves becomes the predominant noise source.

NOISE PREDICTION

International standards now exist that provide methods for predicting valve noise under a wide variety of conditions. A reader wishing detailed methods of calculation should refer to the most current revision of the appropriate standards mentioned below.

Hydrodynamic Noise: IEC 534-8-4 [1]

Methods in the standard are for both the noncavitating and cavitating conditions of liquid flow. They are based on the premise that the acoustic power generated in side the pipe is proportional to the available stream power. The proportionality factor or efficiency is developed empirically and differs greatly between the two regimes. Transmission loss through the pipe wall is then calculated to yield an estimate of the external sound pressure level at 1 m from the outside surface of the pipe. Flashing conditions are not covered in the standard. The large variation in the fluid density and the acoustic wave speed must be accounted for in attempts to predict this noise, and to date a reliable method has not been developed.

Aerodynamic Noise: IEC 534-8-3 [2]

Methods for predicting aerodynamic noise are built on an understanding of the structure of open compressible jets over the full pressure ratio range. The acoustic power inside the pipe is defined by the stream power at the vena contracta multiplied by an efficiency term, which is dependent on valve geometry and the pressure ratio. Pipe-wall transmission loss terms are well defined to yield a prediction of external noise levels.

NOISE CONTROL

Either one or both of the following basic approaches can be applied for noise control:

1. Source treatment: prevention or attenuation of the acoustic power at the source (quiet valves and accessories).

2. Path treatment: reduction of noise transmitted from a source to a receiver.

Quiet Valves

Based on the preceding discussion, the parameters that determine the level of noise generated by compressible flow through a control valve are the geometry of the restrictions exposed to the flow stream, the total valve flow coefficient, the differential pressure across the valve, and the ratio of the differential pressure to the absolute inlet pressure.

It is conceivable that a valve could be designed that utilizes viscous losses to produce the permanent head loss required. Such an approach would require valve trim with a very high equivalent length, which becomes impractical from the standpoint of both economics and physical size.

The noise characteristic or noise potential of a regulator increases as a function of the differential pressure ΔP and the ratio of the differential pressure to the absolute static pressure at the inlet $\Delta P/P_1$. Thus for high-pressure-ratio applications ($\Delta P/P_1 > 0.7$), an appreciable reduction in noise can be effected by staging the pressure loss through a series of restrictions to produce the total pressure head loss required.

Generally in control valves, noise generation is reduced by dividing the flow area into a multiplicity of smaller restrictions. This is readily accomplished with a cage-style trim, as shown in Fig. 3: (a) slotted multipath, (b) drilled-hole multipath, (c) drilled-hole multipath, multistage, and (d) plate-style multipath, multistage. Both acoustic and manufacturing technology have improved to where individual stages in a trim can be custom designed for the conditions they see, as well as take advantage of three dimensional flow passages, as shown in Fig. 3(e).

Proper hole size and spacing is critical to the total noise reduction that can be derived from the utilization of many small restrictions versus a single or a few large restrictions. It has been found that optimum size and spacing are very sensitive to the pressure ratio $\Delta P/P_1$.

For control-valve applications operating at high-pressure ratios ($\Delta P/P_1 \geq 0.7$) the series restriction approach, splitting the total pressure drop between the control valve and a fixed restriction (diffuser) downstream of the valve, can also be very effective in minimizing the noise. In order to optimize the effectiveness of a diffuser, it must be designed (special shape and sizing) for each given installation so that the noise levels generated by both the valve and diffuser are minimized. Figure 4 depicts a typical valve-plus-diffuser installation.

Pertaining to the design of quiet valves for liquid applications, the problem resolves itself into one of designing to reduce cavitation. Service conditions that will produce cavitation can readily be calculated. The use of staged or series reductions provides a practicable solution to cavitation and hence hydrodynamic noise.

Path Treatment

A second approach to noise control is path treatment. Sound is transmitted through the medium that separates the source from the receiver. The speed and efficiency of sound transmission is dependent on the properties of the medium through which it is propagated. Path treatment consists of regulating the impedance of the transmission path to reduce the acoustic energy communicated to the receiver.

In any path-treatment approach to control valve noise abatement, consideration must be given to the amplitude of noise radiated by both the upstream and the downstream piping. Since, when all else is equal, an increase in static pressure reduces the noise transmitted through a pipe, the upstream

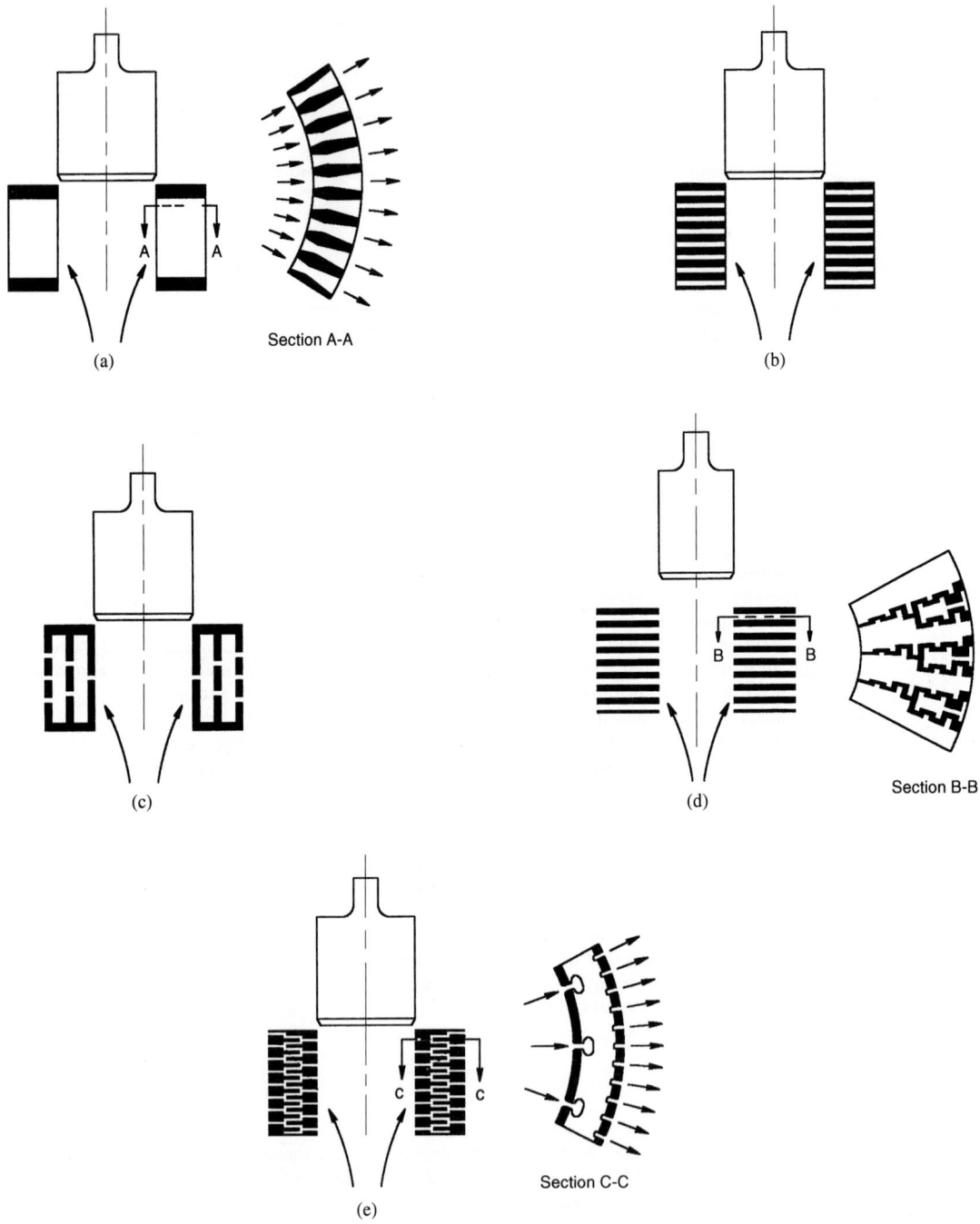

FIGURE 3   Valve cage designs for noise attenuation.

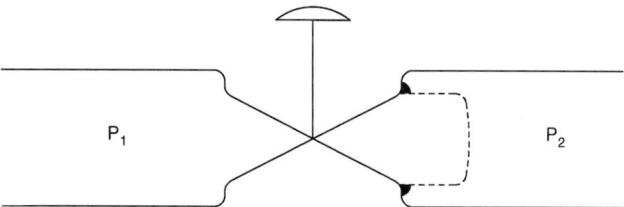

FIGURE 4 Two-stage pressure reduction with diffuser.

noise levels are always less than those downstream. Also, the fluid propagation path is less efficient moving back through the valve.

Dissipation of acoustic energy by the use of acoustical absorbent materials is one of the most effective methods of path treatment. Whenever possible, the acoustical material should be located in the flow stream either at or immediately downstream of the noise source. This approach to abatement of aerodynamic noise is accommodated by in-line silencers. In-line silencers effectively dissipate the noise within the fluid stream and attenuate the noise level transmitted to the solid boundaries. Where high mass-flow rates or high-pressure ratios across the valve exist, in-line silencers are often the most realistic and economical approach to noise control. The use of absorption-type in-line silencers can provide high levels of attenuation; however, economic considerations generally limit the insertion loss to ~25 dB.

Noise that cannot be eliminated within the boundaries of the flow stream must be eliminated by external treatment or isolation. This approach to the abatement of control-valve noise includes the use of heavy-walled piping, acoustical insulation of the exposed solid boundaries of the fluid stream, and the use of insulated boxes, rooms, and buildings to isolate the noise source.

In closed systems (not vented to the atmosphere) any noise produced in the process becomes airborne only by transmission through the solid boundaries that contain the flow stream. The sound field in the contained flow stream forces the solid boundaries to vibrate, which in turn causes pressure disturbances in the ambient atmosphere that are propagated as sound to the receiver. Because of the relative mass of most valve bodies, the primary surface of noise radiation to the atmosphere is the piping adjacent to the valve. An understanding of the relative noise transmission loss as a function of the physical properties of the solid boundaries of the flow stream is necessary in noise control for fluid transmission systems.

A detailed analysis of noise transmission loss is beyond the scope of this article. However, it should be recognized that the spectrum of the noise radiated by the pipe has been shaped by the transmission loss characteristic of the pipe and is not that of the noise field within the confined flow stream. For a comprehensive analysis of pipe transmission loss, see [1].

Acoustic insulation of the exposed solid boundaries of a fluid stream is an effective means of noise abatement for localized areas, if installed correctly. Specific applications should be discussed with insulation suppliers, since results can vary widely.

Path treatment such as the use of heavy-walled pipe or external acoustical insulation can be a very economical and effective technique for localized noise abatement. However, it should be pointed out that noise is propagated for long distances by means of fluid stream and that the effectiveness of heavy-walled pipe or external insulation terminates where the treatment is terminated. Path treatment effects are summarized in Fig. 5.

A simple sound survey of a given area will establish compliance or noncompliance with the governing noise criterion, but not necessarily either identify the primary source of noise or quantify the contribution of individual sources. Frequently piping systems are installed in environments where the background noise due to highly reflective surfaces and other sources of noise in the area make it impossible to use a sound survey to measure the contribution a single source makes to the overall ambient noise level.

A study of sound transmission loss through the walls of commercial piping indicated the feasibility of converting pipe-wall vibrations to sound levels. Further study resulted in a valid conversion technique as developed in [2].

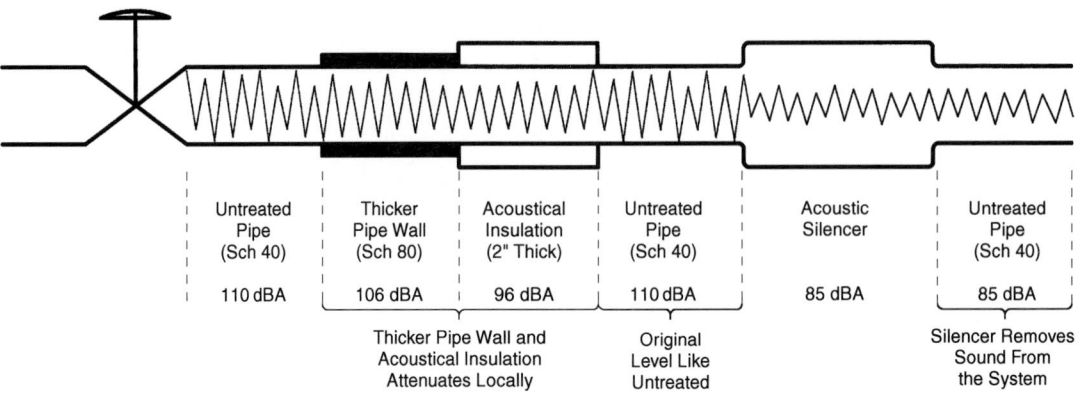

Untreated Pipe (Sch 40)	Thicker Pipe Wall (Sch 80)	Acoustical Insulation (2" Thick)	Untreated Pipe (Sch 40)	Acoustic Silencer	Untreated Pipe (Sch 40)
110 dBA	106 dBA	96 dBA	110 dBA	85 dBA	85 dBA
	Thicker Pipe Wall and Acoustical Insulation Attenuates Locally		Original Level Like Untreated		Silencer Removes Sound From the System

FIGURE 5   Comparison of path-treatment methods.

The vibration levels may be measured on the piping downstream of a control valve or other potential noise source. Sound pressure levels expected are then calculated based on the characteristics of the piping. Judgment can then be made as to the relative contribution of each source to the total sound field as measured with a microphone. The use of vibration measurements effectively isolates a source from its environment.

REFERENCES

1. "Prediction of Noise Generated by Hydrodynamic Flow," IEC-534-8-4-1994, 1994.
2. "Control Valve Aerodynamic Noise Prediction Method," IEC-534-8-3-1995, 1995.
3. Fagerlund, A. C., "Sound Transmission through a Cylindrical Pipe Wall," *J. Eng. Ind. Trans. ASME,* vol. 103, pp. 355–360, 1981.
4. Fagerlund, A. C., "Conversion of Vibration Measurements to Sound Pressure Levels," Publ. TM-33, Fisher Controls International, Inc. Marshalltown, Iowa.

SERVOMOTOR TECHNOLOGY IN MOTION CONTROL SYSTEMS

by David A. Kaiser*

INTRODUCTION

Motion control systems that employ servo motors generally have the topology shown in Fig. 1. The innermost servo loop is usually a torque or force loop consisting of a servo motor and a servo drive.

* Staff Engineer, Compumotor Division, Parker Hannifin Corporation, Rohnert Park, California.

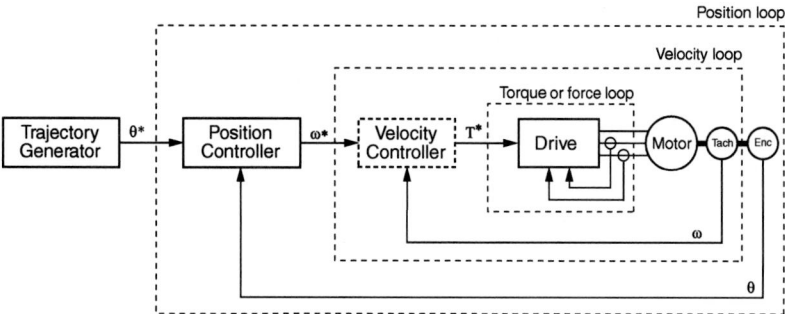

FIGURE 1   Basic servo loop topology.

The input command to this loop is either a torque or force signal and is designated by T* in Fig. 1. Bandwidths of these loops are of the order of 500 Hz and greater. Servo drives apply a switched voltage waveform to the servo motor's winding and sense the resultant current. Generally, controlling a servo motor's current directly controls the servo motor's shaft torque, or in the case of linear motors, carriage force.

The servo motor may have a tachometer connected to its shaft if a velocity loop is closed. The input command to this loop is a desired velocity and is designated in Fig. 1 by ω^*. Bandwidths of these loops are of the order of 50 Hz and higher; however, they must not be higher than the torque loop or instability is almost always ensured. In general, motion systems use a position device and velocity information is derived. Typical positioning devices are encoders and resolvers.

Servo systems might also employ a position loop controller and some type of motion trajectory mechanism commanding a desired position, θ^*. Bandwidths of these loops are of the order of 5 Hz and greater. Again, position loop bandwidths in this topology must be lower than the velocity loop for stable and predictable control.

TYPES OF SERVO MOTORS

Electric motors have been used in servo systems for over a century. The improvement in power electronic devices in the past 20 years has greatly increased the popularity of servo motors. In general, the type of servo motor used is highly dependent on the application.

Before selecting a servo motor technology, the user must consider the following:

- the environment in which the motor be placed (i.e., temperature, vibration, fluids, air borne particles),
- the required velocity, torque, and power profiles including the duty cycles
- the acceptable torque and/or velocity ripple
- the available operating voltages
- the need to comply with local or worldwide regulatory standards (i.e., CE, UL, etc.)
- the acceptable audible noise

The most popular servo motor technologies are summarized in Table 1, along with their respective power levels and typical advantages and disadvantages. This table is by no means a comprehensive list. Table 2 [1] defines the different types of servo applications along with specific examples. It is very typical that more than one type of servo motor will be adequate for a given application. Factors such as cost, availability, and user's experience will ultimately play the central role in selecting a motor technology.

TABLE 1   General Characterstics of Motors Used in Servo Applications

Type	Major Advantage	Major Disadvantage	Typical Power Range (hp)
DC brushed	low initial cost	maintenance	<20
DC brushless	low cost	torque ripple	<20
AC synchronous	high performance	high initial cost	<30
Induction	low cost, simple construction	possible problems with full torque at zero speed	>5
Reluctance	low cost	torque ripple, audible noise	<20

TABLE 2   Generic Motion Applications Using Servo Motors

Application	Definition	Examples
Feed to length	Applications in which a continuous web, strip, or strand of material is being indexed to length, most often with pinch rolls or some sort of gripping arrangement. The index stops and some process occurs (cutting, stamping, punching, labeling, etc.).	BBQ grill-making machine film advance on-the-fly welder
X/Y Point to point	Applications that deal with parts handling mechanisms that sort, route, or divert the flow of parts.	optical scanner circuit board scanning
Metering/dispensing	Applications in which controlling displacement and/or velocity are required to meter or dispense a precise amount of material.	telescope drive engine test stand capsule filling machine
Indexing/conveyor	Applications in which a conveyor is being driven in a repetitive fashion to index parts into or out of an auxiliary process.	indexing table rotary indexer conveyor
Contouring	Applications in which multiple axes of motion are used to create a controlled path, (e.g., linear or circular interpolation).	engraving machine fluted-bit cutting machine
Tool feed	Applications in which motion control is used to feed a cutting or grinding tool to the proper depth.	surface grinding machine transfer machine flute grinder disk burnisher
Winding	Controlling the process of winding material around a spindle or some other object.	monofilament winder capacitor winder
Following	Applications that require the coordination of motion to be in conjunction with an external speed or position sensor.	labeling machine window blind gluing moving positioning systems
Injection molding	Applications in which raw material is fed by gravity from a hopper into a pressure chamber (die or mold). The mold is filled rapidly and considerable pressure is applied to produce a molded product.	plastic injection molding
Flying cutoff	Applications in which a web of material is cut while the material is moving. Typically, the cutting device travels at an angle to the web and with a speed proportional to the web.	rotating tube cutting

General Characteristics and Comparison of Servo Motors

Regardless of the motor technology, all servo motors consist of a motor housing or stack, some type of position or velocity feedback device, and numerous options including connectorization, gearheads, brakes, flanges and shafts. Figure 2 shows the anatomy of a rotary servo motor with a typical list of the many options available.

In addition to mechanical options, servo motor manufacturers will often offer an assortment of different motor windings for a given frame size. This allows additional flexibility especially in matching a particular servo drive with the motor.

Since the mid-1990s, linear servo motors have increased in popularity in applications requiring direct linear motion. All of the subsequent motor technologies discussed here apply to linear as well as rotary motors.

dc Brushed Motor.   Arguably the most common servo motor is the dc brushed motor. The basic structure is shown in Fig. 3. The motor is made up of three main parts: the stator, containing the permanent magnets; the rotor, made up of coils of wire wound in slots in an iron core; and the commutator, which consists of a brush assembly that maintains the proper orientation of magnetic fields to produce maximum shaft torque for a given motor current.

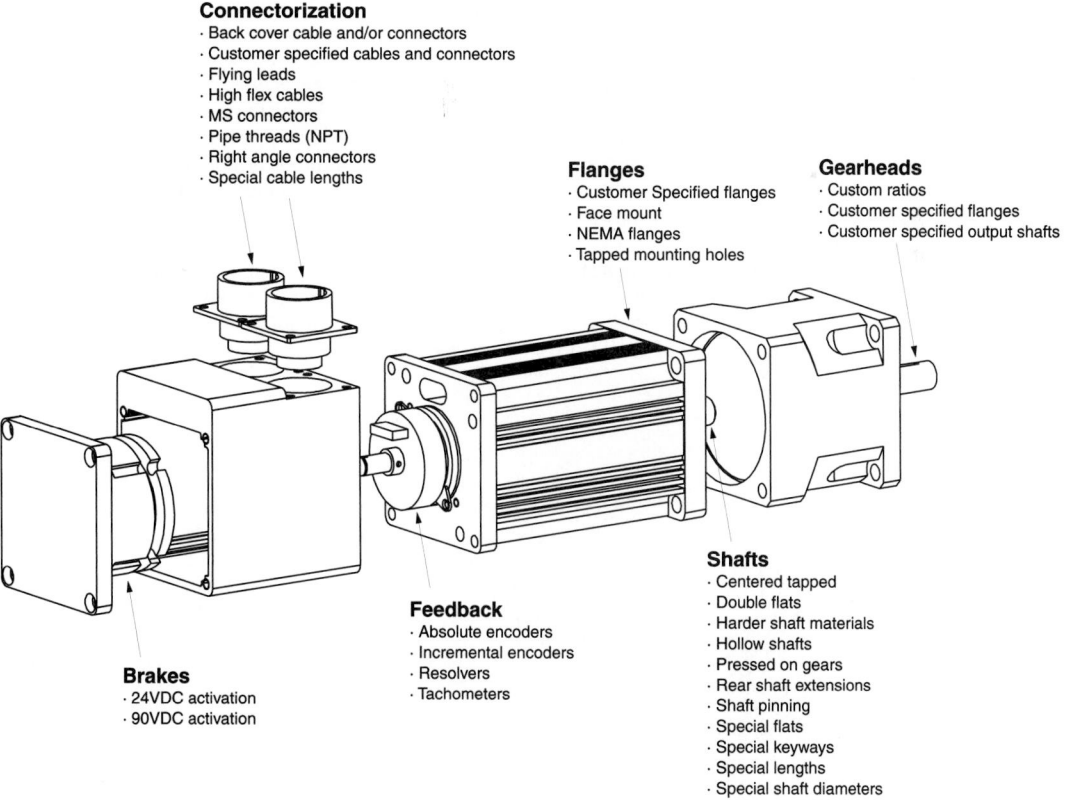

Connectorization
· Back cover cable and/or connectors
· Customer specified cables and connectors
· Flying leads
· High flex cables
· MS connectors
· Pipe threads (NPT)
· Right angle connectors
· Special cable lengths

Flanges
· Customer Specified flanges
· Face mount
· NEMA flanges
· Tapped mounting holes

Gearheads
· Custom ratios
· Customer specified flanges
· Customer specified output shafts

Brakes
· 24VDC activation
· 90VDC activation

Feedback
· Absolute encoders
· Incremental encoders
· Resolvers
· Tachometers

Shafts
· Centered tapped
· Double flats
· Harder shaft materials
· Hollow shafts
· Pressed on gears
· Rear shaft extensions
· Shaft pinning
· Special flats
· Special keyways
· Special lengths
· Special shaft diameters

FIGURE 2   Anatomy of a rotary servo motor.

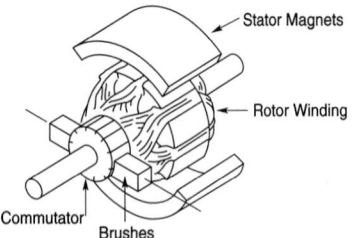

FIGURE 3 dc motor (courtesy of Robbins and Myers/Electrocraft).

These motors are generally the lowest in initial cost for a given power rating. Their long-term cost, however, might be higher because they require routine brush maintenance. If an application requires the motor to turn in both directions, a see-sawing action occurs on the brushes, and the wear-out time is even shorter. It therefore has the potential to be the minimal cost solution in cases in which the required motion is unidirectional and/or limited in duty.

In terms of power density, the dc brush motor suffers from three fundamental design constraints limiting its overall usage. First, because the current-carrying conductors are on the rotor, it is difficult to keep the winding temperatures cool because of the high thermal impedance between the rotor and the motor case. Second, the commutator assembly can take up significant space in the motor housing. Third, because mechanical contact is made by means of the brushes, there is a maximum speed limitation that is lower than typical brushless motors.

A key advantage for the this type of servo motor is the ease of driving and controlling it. The simplest of power amplifiers can be used keeping the overall system cost low.

dc brushed motors have enjoyed the advances made in permanent magnet materials, especially neodymium-iron-boron (NdFeB). Although these magnets are more expensive than the historical Alnico (*Al*uminum, *Ni*ckel and *Co*balt) types, the improvement in power density is substantial.

Brushless dc Motor. In the brushless dc motor, permanent magnets are mounted on the rotor and the wound iron core field is contained in the stator. This is depicted in the cross-sectional view and the equivalent three-phase winding model shown in Fig. 4. Since the current-carrying coils are now located in the housing, there is a short, efficient thermal path from the windings to the outside air. Cooling can further be improved by finning the outer casing as this is typicaly done.

The term "brushless dc" can mean one of two types of motors. Using the term "dc" in its description relates to the nature of the currents going to and from the motor. The first type of brushless dc motor has only two wires and the motor current is "dc." This type of motor is generally not used in servo applications because of the power requirements needed for the commutation mechanism mounted inside. This type of motor is shown in Fig. 5 and is commonly used in small fans to cool electronic enclosures.

The far more common type of dc brushless motor is a three-phase design with concentric windings. This type of motor operates with constant levels of dc current; however, the commutation of the currents is done externally in the servo drive, with Hall effect devices mounted inside the motor providing shaft-position information. This topology is shown in Fig. 6.

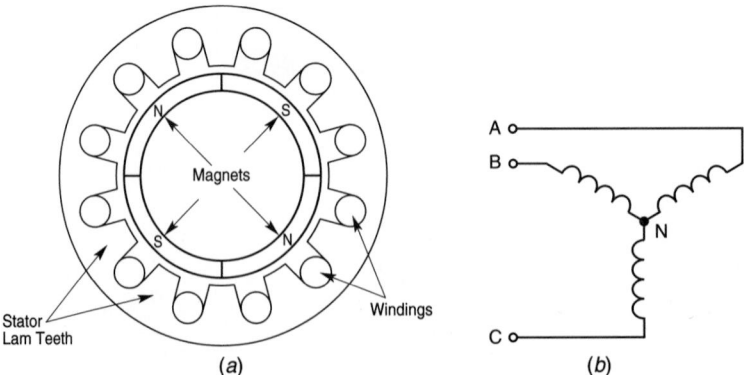

FIGURE 4 (a) Typical cutaway view of brushless dc motor; (b) three-phase winding representation.

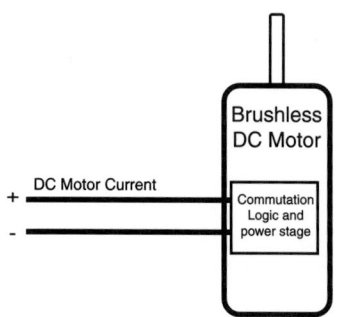

FIGURE 5   One type of brushless dc motor.

The typical motor voltage and current waveforms for this type of brushless motor are depicted in Fig. 7. The dotted lines represent a Hall state. These states are used to commutate the motor. These motors are also referred to as "trapezoidal back EMF motors," "trap" or "six-state" motors because of the nature of their phase voltages and currents.

The chief disadvantage of this type of motor is that it torque pulsations can be sensed as the Hall position boundaries are crossed. This will affect the overall velocity smoothness of the servo, especially in low-inertia or high-performance applications.

ac Synchronous Motor. Arguably the most popular motor used in general purpose motion control today is the ac synchronous motor. It is also commonly referred to as a permanent magnet ac (PMAC) motor. The cutaway view of this motor is similar to the brushless dc motor in Fig. 4 with one major difference: the stator winding pattern and the rotor magnet placement are done in such a way as to minimize the torque ripple. Mechanically, this is done by spatially distributing the stator windings in conjunction with either skewing the rotor magnets and/or the stator slots. Consequently, the servo drive must also be designed such that it can produce smooth sinusoidal ac voltages and currents to fully exploit this motor. This is shown in Fig. 8. Theoretically, the smoother the resultant flux wave, the lower the shaft torque ripple.

Some manufacturers in an effort to further reduce the torque ripple, resort to a "slotless" stator design in an attempt to reduce the electrical "cogging" torque to zero.

Depending on the mechanical constraints, these motors are designed in both a flat pancake and a long rectangular shape. In the <10 hp range, these motors by and large dominate the market.

As with the brushed dc motors, the ac synchronous motors today also predominately use high-energy magnets composed of NdFeB.

Induction Motor.   The induction motor has long been the work horse motor in the world. Figure 9 shows the common elements that make up the induction motor. This motor is typically used in three primary ways: line-start applications, constant velocity or spindle applications, and finally, servo applications.

The line start applications include general purpose machinery, pumps, fans, compresors, and conveyors. Line start refers to the method of control. Since induction motors tend to draw large currents when energized directly off line, line starters are typically circuit breakers with a high current inrush capability. These motors operate at fixed rotational frequencies slightly less than the applied electrical frequency/pole pair combination. This limits their use.

Spindle applications, in contrast, employ some type of motor inverter. In this case, the input line frequency no longer directly affects the motor speed. Spindle applications include more advance fan control, pumping, and general purpose turning machines. A classic example of this mode of control is a machining station that has one large spindle drive controlling the main turning axis or mandrel.

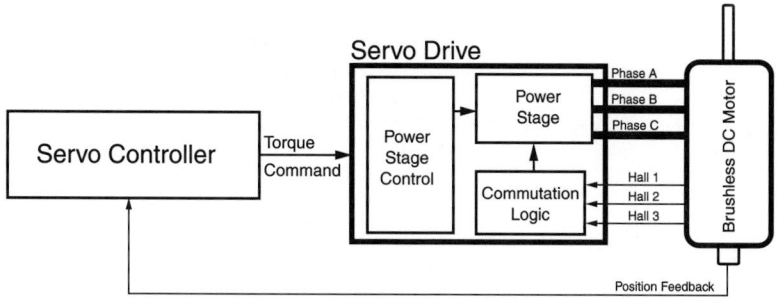

FIGURE 6   Typical topology of a brushless dc motor used in servo applications.

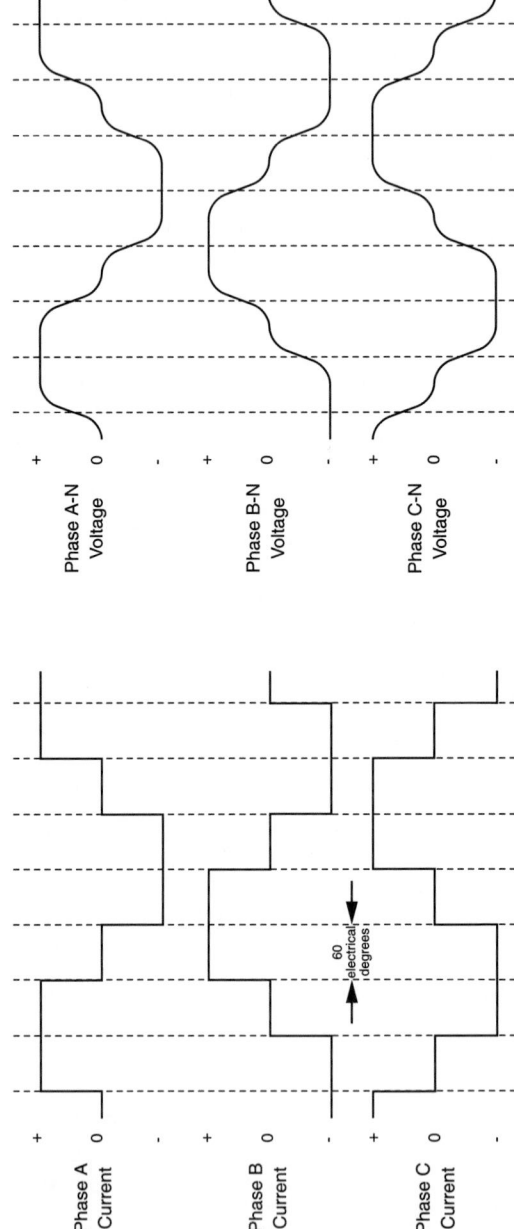

FIGURE 7 Typical phase voltages and currents of a brushless dc motor used in servo applications.

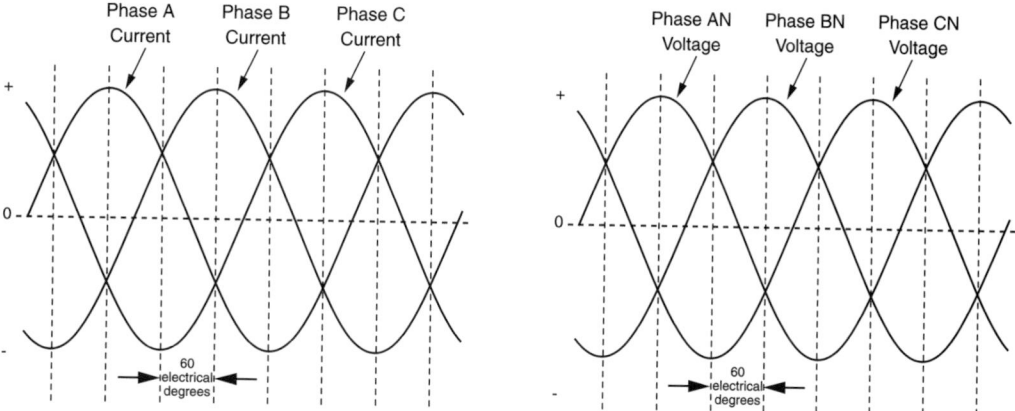

FIGURE 8   Typical phase voltages and currents of a PMAC synchronous motor.

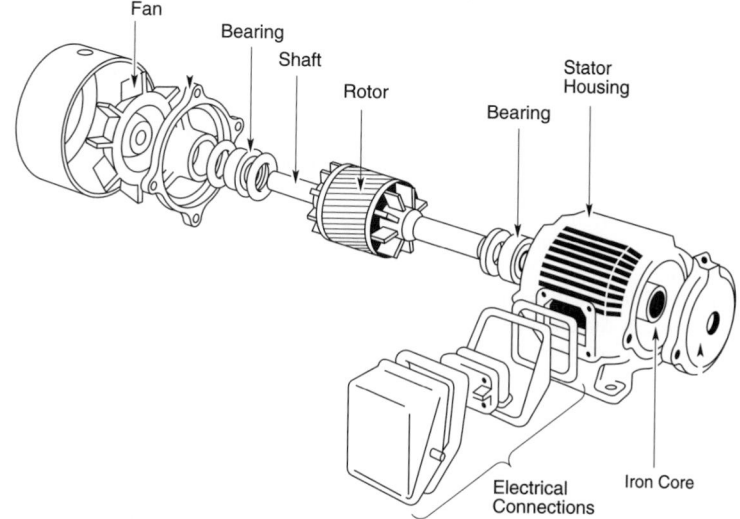

FIGURE 9   Induction motor (courtesy of Danfoss Drives) [2].

All cutting operations on the work piece are then controlled by smaller, higher-performance positioning servos that are synchronized with the main spindle axis position.

Finally, the third mode of induction motor control is servo position control. It is also sometimes called "vector" or "field orientation" control. These drives either require position sensing or some type of internal complex motor model to achieve field orientation. Torque production in induction motors requires a "slip" between the rotating stator field and the rotor. This results in rotor losses that are unavoidable and limit the ability to produce full rated torque at zero speed. In the fractional and integral horsepower range, these motors usually lose out to the ac synchronous type. In the high-power range, the cost of the induction motor is significantly less than the permanent magnet ac motor, and with the performance demands not as high, these motors are the preferred choice.

Variable Reluctance.   The final motor considered for use in servo applications is the variable reluctance motor. It is also commonly called either a "switched reluctance," "brushless reluctance,"

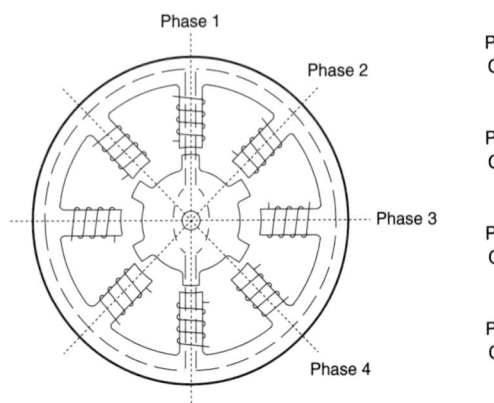

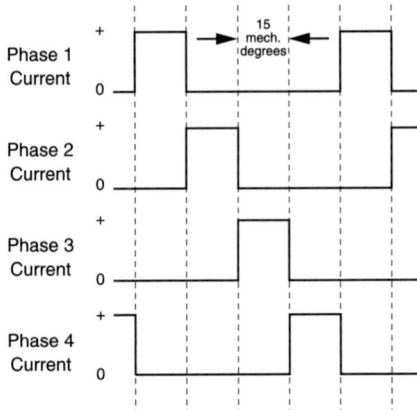

FIGURE 10   Cutway view of a four-phase reluctance motor.

FIGURE 11   Current waveforms of a reluctance motor.

or "commutated reluctance" motor. It origins can be traced to the early 1800s [3]. The fact that no permanent magnets are used not only keeps the manufacturing cost low, but also allows for high-temperature operation. Figure 10 shows a cut away view of a four-phase design. The stator teeth and rotor geometry are similar to permanent magnet stepper motors, but with the absence of magnets, the rotational losses are much smaller. One drawback of this motor is the relatively high audible noise it makes under normal operation.

The motor currents for a variable reluctance servo motor are generally switched on and off in a simple pattern, making the control relatively simple. Figure 11 shows what the ideal current waveforms look like.

The chief drawback in this servo motor design is the large torque ripple associated with the current pulses. This, in conjunction with the audible noise, ultimately limits the variable reluctance motor's use in high-performance servo applications.

GENERAL CONSIDERATIONS

There are numerous considerations to take into account when selecting a servo motor. Before a motor technology is chosen, the desired speed, torque, and duty cycle have to be approximated, the available mounting space estimated, and finally a target cost set. Refer to Table 1 in the previous section for a list of the common motors used for a given power range. After the motor technology is selected, there is a plethora of both electrical and mechanical details to consider. Fortunately, there are several regulatory agencies servo motor manufacturers conform to that ease in the selection of the proper motor configuration.

Motor Parameters, Definitions, and Terminology

Depending on the motor technology, motor parameter definitions can vary. Careful attention must be paid to the implied assumptions and conditions for each parameter, as well as its tolerance, units, and method of measurement. One area that is consistently overlooked is a servo motor's thermal rating. Motor parameters are typically given at ambient temperatures of 25°C or 40°C, with the assumption that the motor is mounted to a specified heatsink.

Historically, dc motors were the first widely accepted, general purpose servo motor, resulting in a strong tendency to put as many ac servo motor parameters as possible in their equivalent dc motor terms. This is a source of great confusion. For example, a servo motor's stall torque or its torque rating at zero speed is given by Eq. (1) and is straightforward for a dc motor. The current is dc and the units for the torque constant, K_T, are Newton meters/Amp (or Nm/A).

$$T_{\text{stall}} = I_{\text{stall}} K_T \qquad (1)$$

To calculate this for an ac motor, the units of K_T might be as follows: Nm/A, Nm/A rms, Nm/A peak. In making this calculation, we now must understand what kind of current is assumed for the ac servo motor.

Servo motors are compared to one another by either their current, voltage, power, and/or their speed ratings. For a given motor technology and frame size, there is a fixed amount of torque that can be produced before either magnetic saturation sets in or the motor losses are so high that the motor basically overheats. Equation (2) relates a servo motor's power rating to its torque and speed rating.

$$P_{\text{rated}} = T_{\text{rated}} \, n_{\text{rated}} \qquad (2)$$

A typical motor data sheet will include this information along with a host of other details. Table 3 is an example of the kind of information manufacturers will provide.

In situations in which the servo motor has an integral gearbox or brake, the manufacturer will provide additional derating information if necessary. Care must be taken to fully understand the thermal interface between the motor and its mounting to correctly apply any derating information provided.

Name-Plate Ratings

Name-plate information varies from manufacturer to manufacturer, although the basic information of make and model number is always provided. It is also typical that the dc winding resistance is given so that a field technician can quickly determine if the motor has a shorted phase winding. Figure 12 shows what a motor name plate might look like.

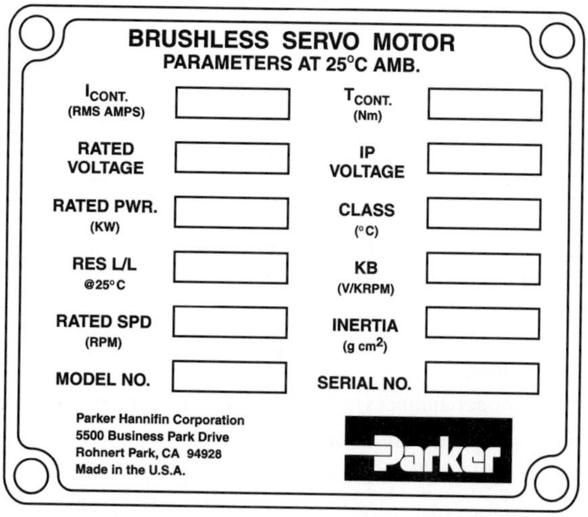

FIGURE 12 Typical servo motor name plate.

TABLE 3 Typical Servo Motor Parameters Provided by the Manufacturer

Typical Motor Data Sheet Parameters
1. Acceleration at rated torque
2. Bearing class, internal/external
3. Bearing grease
4. Constant(s): torque, voltage, motor, electrical time, mechanical time, thermal
5. Current: rated, peak
6. dc resistance terminal (line-line)
7. Damping
8. Dielectric strength (winding to frame)
9. Inductance: terminal (line-line)
10. Insulation class
11. IP classification
12. Maximum winding temperature
13. Mechanical dimensions with tolerances
14. Number of motor poles
15. Output power: rated
16. Rotor inertia
17. Shaft seal pressure
18. Shaft: radial play (front to back), radial loading, material, magnet type
19. Speed: rated, maximum
20. Speed vs. torque curves
21. Stator phase sequence
22. Thermal impedance
23. Thermostat reset temperature
24. Thermostat trip temperature
25. Torque(s): continuous stall at XX °C ambient, peak, static friction, % ripple, derating curves
26. Vendor/supplier
27. Voltage: rated, max
28. Weight
29. Winding capacitance to frame
30. Wiring diagrams of the motor and any of its optional feedback and brake devices

Options
    Brakes: release time, holding torque, operating voltage
    Gearbox: ratios, deratings
    Hall devices
    Incremental encoder: manufacturer, supply voltage, resolution, accuracy
    Keyway
    Resolver: manufacturer, electrical specs, accuracy, model no.
    Tachometer: manufacturer, electrical specs, model no.

NEMA (National Electrical Manufacturers Association) specifies (standard MG 7 [4]) that the minimum standard name-plate information for servo motors 3 in. in diameter or greater contain at least the following information:

1. Manufacturer's name

2. Manufacturer's model number (includes motor type, i.e., ac, dc, and so forth)

3. Manufacturer's serial number or date code

4. Maximum continuous stall torque at either 25°C or 40°C ambient

5. Maximum continuous rms current at either 25°C or 40°C ambient

6. Maximum continuous output power at either 25°C or 40°C ambient

7. Maximum allowable intermittent voltage (brush motors only)

8. Maximum allowable speed

9. Phase-phase resistance at 25°C

Note that the specifications are given at a certain operating temperature. This is important, as even the motor's resistance can increase significantly at elevated operating temperatures.

ELECTRICAL CONSIDERATIONS

There are numerous electrical considerations to take into account when selecting a servo motor. The available supply voltage is usually considered first. Invariably the servo motor will be connected to a servo drive, and therefore the supply voltage becomes more of a drive requirement than a motor. In order to minimize errors in the drive motor system interface, drive manufacturers either manufacture servo motors or provide guidelines on how to connect servo motors to them. This becomes extremely important in the cases in which regulatory acceptance of the entire system is required by the customer.

Another area of equal importance is understanding the acceptable power loss in a servo motor. Electrical parameters such as motor inductance and resistance play an important role in determining these losses, as well as the servo drive's current control methodology.

Regulatory Considerations

It is always important to determine where the servo motor will ultimately be used. This will affect the need for regulatory approval. The common regulatory agencies that affect servo motors are UL and CE.

UL.   The initials UL refer to Underwriters Laboratories, Inc., a nonprofit agency that has set forth a series of safety standards covering a wide range of products from consumer goods to industrial equipment. The primary goal of UL testing is to certify that a product will not be a source of a sustaining fire under any condition of use or misuse.

The term *UL Approved,* however is a misnomer. The most common designation, *UL Listed,* can be found on virtually any consumer good. This designation means that a stand-alone product complies with a particular UL safety standard appropriate for that product's intended use. A second designation is *UL Recognized Component.* This designation covers components intended to be used as part of an end product or system. Since servo motors are used as part of an end product or system, *UL Recognized Component* is the appropriate designation. Depending on the motor and the application, the following list of UL standards might apply: UL519, UL547, UL674, and UL1004 (see [5]).

In addition, products that are certified by UL to meet the Canadian National Standards and Codes display the UL Mark for Canada, known as *cUR.*

CE.   Since the 1990s, there has been a strong push to buy and sell freely in the common European markets. Although historically, the U.S. suppliers have met the regulatory requirements for a particular country they were doing business in, there is now a strong push to consolidate all the European countries into one set of "directives" mandated by the European Union laws. What makes this difficult is that servo motors are always connected to something. The whole system consisting of the motor and drive now must meet the regulatory requirements. This generally requires special attention paid to all wire connections and particularly shielding.

There are two general directives spelled out in the CE mark. The first directive is commonly referred to as the low voltage directive (LVD). It refers to the product's ability to withstand high-voltage surges without degrading. Another part of the low-voltage directive (EC Directive 73/23/EEC) requires a "technical file" be kept on servo motors produced by a given manufacturer. This file contains a wealth of information about the safety testing of the motor as well as information about manufacturing quality control. The second directive is referred to as the electromagnetic compatibility, or EMC, directive.

TABLE 4 Overall Required Test to claim CE Compliance

EC standard	Description
EN 55022 (1987)	radiated and conducted electromagnetic emissions
EN 61000-4-2	electrostatic discharge immunity
EN 50140 (1994)	radiated electromagnetic field immunity
EN 61000-4-4 (1995)	electrical fast transient burst immunity
EN 61000-4-8 (1993-06)	power frequency magnetic field immunity
ENV 50141 (1993-08)	radio frequency common mode immunity
TBD (Mains harmonic content)	Power line disturbance

This directive is broken down into two additional classes: emissions and susceptibility. These classes refer to the product's ability to limit the amount of electrical noise it emits and to reject external noise from affecting its operation. Table 4 summarizes the required tests to become CE compliant.

Speed Versus Torque Curves

Servo motors operate with full torque at zero speed. This torque is sometimes referred to as stall or continuous torque ($T_{continuous}$). Associated with this value is the peak stall torque ($T_{maximum}$). Depending on the type of motor and the expected life, this value can be 2–5 times the continuous value. Other useful values from the speed versus torque curve include rated and no-load speeds and rated torque. These points are all illustrated in a typical speed versus torque curve shown in Fig. 13.

Depending on the servo motor technology, feedback limitations, and the assumed servo drive connected to the motor, the actual speed torque curves from a servo motor manufacturer might look very different. It is therefore critical that both the drive and motor be considered together when the speed versus torque curve is analyzed.

To determine if the amount of time in the intermittent region is acceptable, the rms, or root-mean-square, torque must be calculated for a motion profile and be within the continuous torque region at the relevant speeds. To do that, first the required motion profile is used to determine the torque profile, and then that profile is used to determine the rms torque. Finally, a safety factor is included to provide

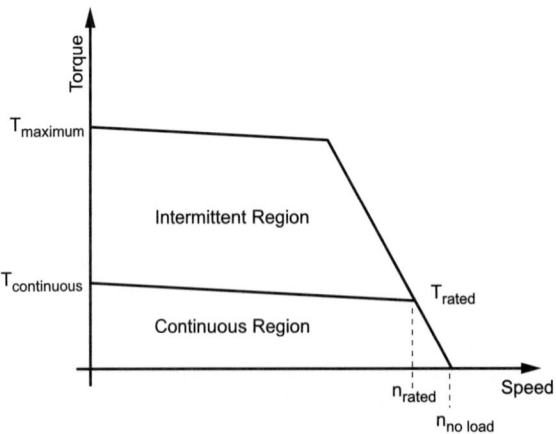

FIGURE 13 Typical speed vs. torque curve for a servo motor.

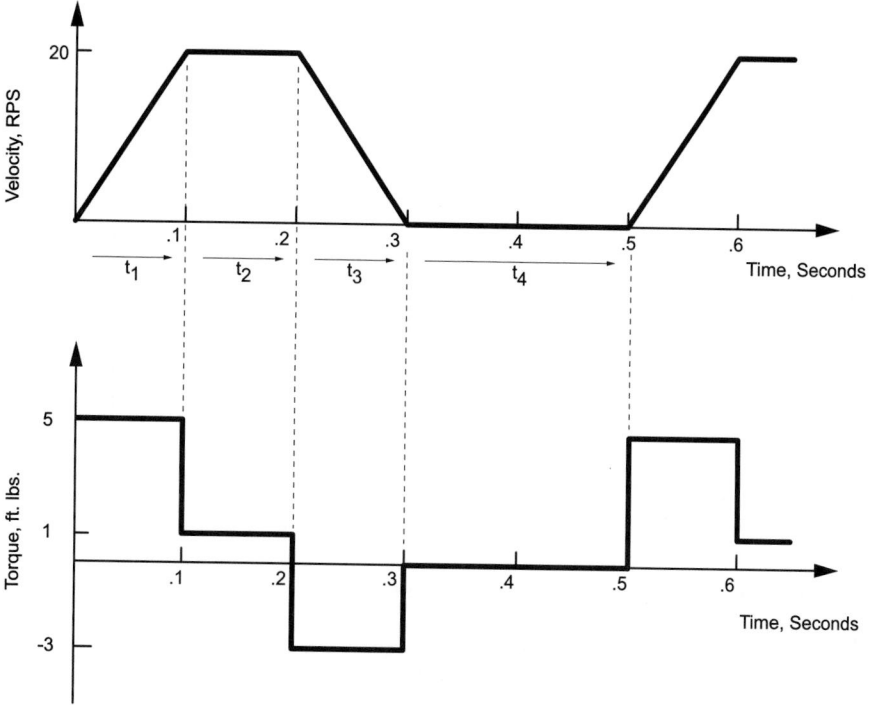

FIGURE 14 Typical servo motion profile.

additional headroom in sizing. The equation for rms torque is given in Eq. (3).

$$
\text{torque}_{\text{rms}} = \sqrt{\dfrac{\sum\limits_{i=1}^{n} \text{torque}_i^2 t_i}{\sum\limits_{i=1}^{n} t_i}}
\tag{3}
$$

An example of what a typical profile might look like is illustrated in Fig. 14.

Example. A required motion profile with its corresponding torque profile is given in Fig. 14. Calculate the rms torque.

The rms torque is calculated as follows:

$$
\text{torque}_{\text{rms}} = \sqrt{\dfrac{\text{torque}_1^2\, t_1 + \text{torque}_2^2\, t_2 + \text{torque}_3^2\, t_3 + \text{torque}_4^2\, t_4}{t_1 + t_2 + t_3 + t_4}}
$$

or

$$
\text{torque}_{\text{rms}} = \sqrt{\dfrac{5^2 \times 0.1 + 1^2 \times 0.1 + (-3)^2 \times 0.1 + 0^2 \times 2}{0.1 + 0.1 + 0.1 + 0.2}} = 2.6\,\text{ft. lbs.}
$$

It is also typical for servo motor vendors to provide a peak torque or current derating curve. An example of one is shown in Fig. 15.

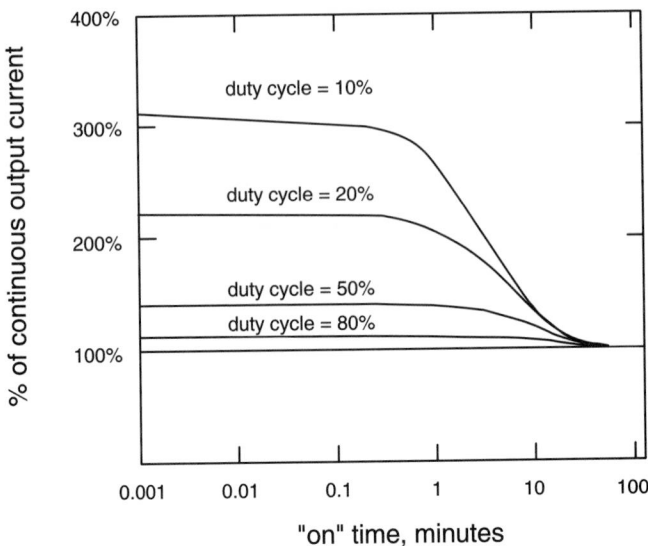

FIGURE 15   Typical servo motor derating curve.

In this case, the percent over continuous torque is determined and the amount of time that is needed is used to find the allowable amount of "on" time. As the time gets longer, the amount of intermittent torque approached the continuous value, as it should.

Torque ripple is another figure of merit used in comparing one servo type with another. Ideally, the exact shaft torque should not be affected by the shaft location. Motor characteristics, feedback resolution, and servo drive current regulation all affect the motor's torque ripple. In generally, Eq. (4) is used in comparing motors. Typical values are under 5%.

$$\text{torque ripple} = \frac{\text{torque (peak} - \text{peak)}}{\text{torque(continuous)}} \tag{4}$$

Finally, it is not uncommon that a servo motor's no-load speed is limited mechanically. Rotational speeds above 10,000 rpm invariably require special balancing, manufacturing, and, in some cases, special position feedback devices.

Thermal Ratings—Insulation Class

In general, the insulation class of the windings will dictate the maximum allowable winding temperature. In the case of permanent magnet machines, the magnets may limit the maximum operating temperature before the winding insulation does. Typical winding classes are shown in Fig. 16. A useful rule of thumb is that motor insulation life is reduced by approximately a factor of 2 for every 10° rise over the rated temperature. Altitude also plays a role in determining these numbers, as further deratings are needed for altitudes above 3300 ft (1000 m).

Servo motor manufacturers will generally put thermal cut-out switches inside the motor windings to protect the motor from permanent thermal damage. These switches are sensed by the servo drive connected to it and rely upon the drive to cease operation of the motor. The cut out or maximum allowable motor temperatures are summarized in Table 5. These maximum temperatures correspond to approximately 30,000 hs of operation. For the most part, servo motor windings are generally class H.

TABLE 5 Insulation Class Rating

Class	Max. Allow. Winding Temp. (°C)
A	105
B	130
F	155
H	180

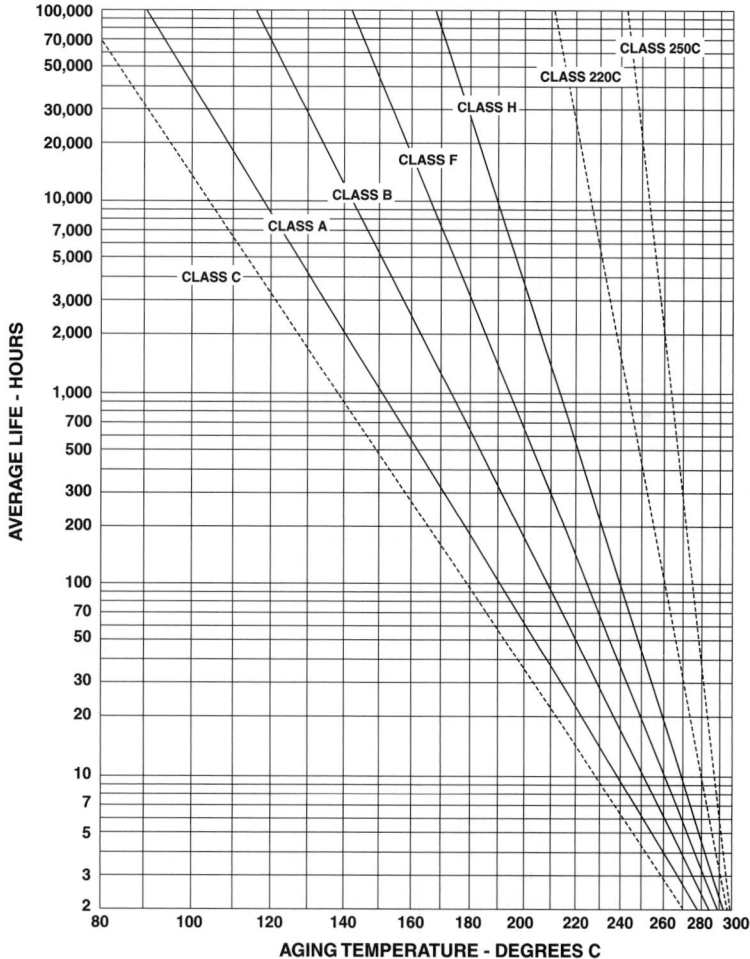

FIGURE 16 Life temperature lines of varnished twist-test specimens as determined by AIEE 57 test procedure. (Anaconda Wire and Cable Company.)

MECHANICAL CONSIDERATIONS

Mounting

Unfortunately, there is no real dimensional consistency from servo motor manufacturers; however, some vendors do follow the NEMA dimensional specification (MG7-1993 [4]) for the larger frame size motors. As of 1998, all dimensions are metric, with the English dimensions under consideration. Figure 17 and Table 6 show NEMA's servo motor dimensional specifications. Note that there are two standard NEMA mountings: the type "C" face mount with tapped mounting holes, and the type "D" flange mount with free mounting holes. It is not uncommon to specify only the mounting face. This gives the servo motor manufacturer the ability to adjust the other motor dimensions while maintaining the mounting. The chief advantage of following the NEMA guidelines is the compatibility of gearboxes and other transmission devices.

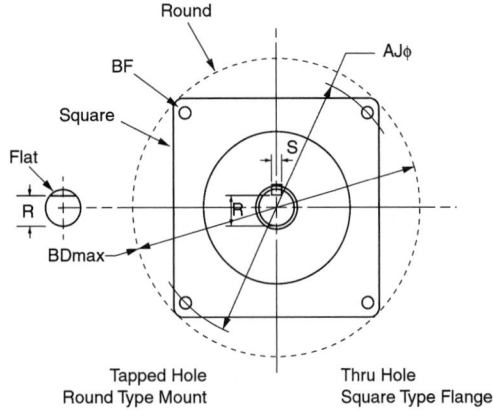

Tapped Hole
Round Type Mount

Thru Hole
Square Type Flange

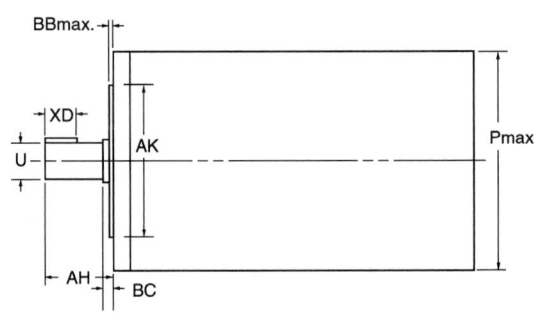

Legend:

AJ	Diameter of mounting bolt circle in face, flange or base of machine.	**BB$_{max}$**	Maximum depth of male or female pilot of mounting face, flange or base machine.
AK	Diameter of male or female pilot on face, flange or base of machine.	**BF**	Threaded or clearance hole in mounting face, flange or base of machine.
AH	Mounting surface or face, flange of base of machine to end of shaft.	**XD**	Usable length of keyseat.
U	Diameter of shaft extension.	**S**	Width of keyseat.
P$_{max}$	Maximum width of machine (end view) including pole bolts, fins, and such, but excluding terminal housing, lifting devices, feet and outside diameter of face or flange.	**R**	Bottom of keyseat or flat to opposite side of shaft or bore.
BD$_{max}$	Maximum outside diameter of mounting face, flange or base of machine.	**BC**	Distance between mounting surface of face, flange, or base of machine to shoulder on shaft.

FIGURE 17   Dimensions for servo motors (NEMA MG7-1993 2.4).

TABLE 6 Metric Mounting Dimensions for Servo Motors (NEMA MG7-1993 Section 2.4.1.3)

Flange[a]	Number[b]	AJ (mm)	AK (mm)	AK Tol[c] (mm)	AM (mm) Pri	AM (mm) Sec	U (mm) Pri	U (mm) Sec	U Tol[c] (mm) Pri	U Tol[c] (mm) Sec	BC (mm)	P_{max}/BD_{max} (mm)	BB_{max} (mm)	BF (mm)	BF Tol[c] (mm)	Thread	XD (mm) Pri	XD (mm) Sec	S (mm) Pri	S (mm) Sec
55CM	55DM	55	40	+0.011 / −0.005	20	30	9	14	+0.007 / −0.002	+0.008 / −0.003	0	70	2.5	5.8	+0.300 / −0	M5 × 0.8	15	20	3 × 3	5 × 5
65CM	65DM	65	50	+0.011 / −0.005	20	30	9	14	+0.007 / −0.002	+0.008 / −0.003	0	80	2.5	5.8	+0.300 / −0	M5 × 0.8	15	20	3 × 3	5 × 5
75CM	75DM	75	60	+0.012 / −0.007	23	40	11	16	+0.007 / −0.002	+0.008 / −0.003	0	91	2.5	5.8	+0.300 / −0	M5 × 0.8	18	30	4 × 4	5 × 5
85CM	85DM	85	70	+0.012 / −0.007	30	40	14	19	+0.008 / −0.003	+0.009 / −0.004	0	105	2.5	7	+0.360 / −0	M6 × 1.0	20	30	5 × 5	6 × 6
100CM	100DM	100	80	+0.012 / −0.007	30	40	14	19	+0.008 / −0.003	+0.009 / −0.004	0	120	3	7	+0.360 / −0	M6 × 1.0	20	30	5 × 5	6 × 6
115CM	115DM	115	95	+0.013 / −0.009	40	50	19	24	+0.009 / −0.004	+0.009 / −0.004	0	140	3	10	+0.360 / −0	M8 × 1.25	30	40	6 × 6	8 × 7
130CM	130DM	130	110	+0.013 / −0.009	40	50	19	24	+0.009 / −0.004	+0.009 / −0.004	0	160	3.5	10	+0.3.60 / −0	M8 × 1.2	30	40	6 × 6	8 × 7
145CM	145DM	145	110	+0.013 / −0.009	40	50	19	28	+0.009 / −0.004	+0.009 / −0.004	0	165	3.5	10	+0.360 / −0	M8 × 1.2	30	40	6 × 6	8 × 7
165CM	165DM	165	130	+0.014 / −0.011	50	50	24	28	+0.009 / −0.004	+0.009 / −0.004	0	200	3.5	12	+0.430 / −0	M10 × 1.5	40	40	8 × 7	8 × 7
200CM	200DM	200	114.3	+0.0 / −0.25	79	—	35	—	+0.01 / −0.00	—	0	235	4	13.5	+0.430 / −0	M12 × 1.75	70	—	10 × 8	—
215CM	215DM	215	180	+0.014 / −0.011	60	—	32	—	+0.018 / +0.002	—	0	250	4	15	+0.430 / −0	M12 × 1.75	50	—	10 × 8	—
265CM	265DM	265	230	+0.016 / −0.013	85	—	48	—	+0.018 / +0.002	—	0	300	4	15	+0.430 / −0	M12 × 1.75	60	—	14 × 9	—
300CM	300DM	300	250	+0.016 / −0.013	85	—	48	—	+0.018 / +0.002	—	0	350	5	19	+0.520 / −0	M16 × 2.0	60	—	14 × 9	—
350CM	350DM	350	300	+0.016 / −0.016	110	—	55	—	+0.030 / +0.011	—	0	400	5	19	+0.520 / −0	M16 × 2.0	90	—	16 × 10	—

a The suffix CM denotes C face-mounting motors with tapped mounting holes.
b The suffix DM denotes D flange-mounting motors with free mounting holes.
c The reference for the tolerances is ISO Recommendation 286, S.I. Metric System of Limits and Fits. If a stepper motor is designed to be a direct substitute for servo motors in servo motor applications, the stepper motor dimensions should comply with the dimensions specified for servo motors.

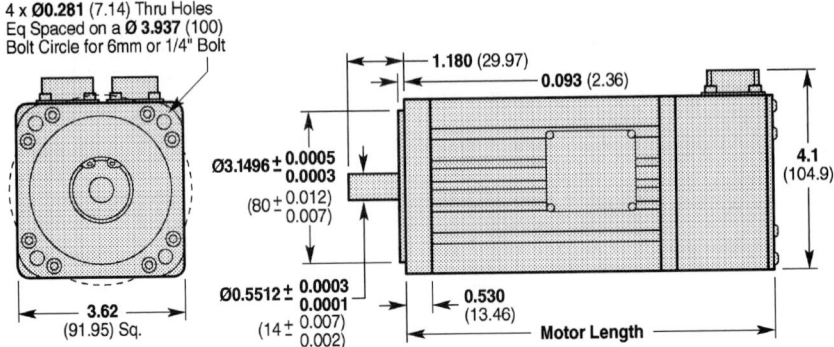

FIGURE 18   Example of NEMA 100DM.

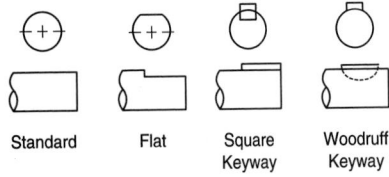

Standard   Flat   Square   Woodruff
                    Keyway   Keyway

FIGURE 19   Examples of different shaft options.

An example of an ac synchronous motor that conforms to the NEMA mounting standards is shown in Fig. 18. The length of the motor is generally not important provided that the user has space for it.

Modifications to motor shafts are another mechanical consideration to consider. The default standard is not to make any shaft modifications. At one time keyways were used, but with small, high-performance servo motors, they have two drawbacks. First, any sharp cuts made into a motor's shaft will be a point of structural weakness; second, since keyways are generally made from a softer material, this potentially causes a rotor imbalance. Figure 19 illustrates the common shafts options.

Motor connections can vary from flying leads or pig tails to full military-style waterproof connectors. Figure 20 illustrates some of the more typical arrangements.

One of the more common connections made to servo motors is a gearbox. Planetary styles are generally preferred because of both high speed constraints and balancing considerations. Some servo motor manufacturers will offer the gearbox as an integral part of the servo motor. This is shown in Fig. 21.

After all the options are selected for the servo motor, the general classification of the motor must be specified.

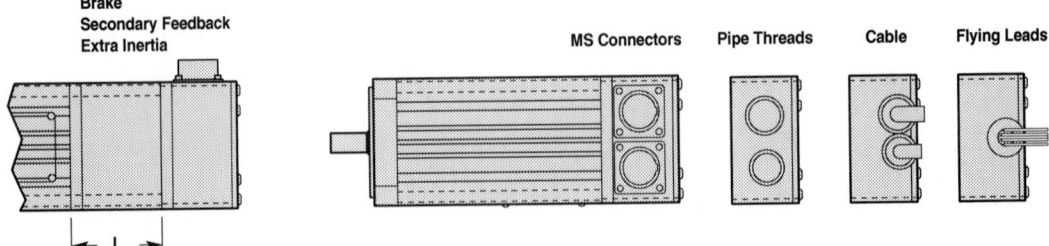

FIGURE 20   Examples of different connector options.

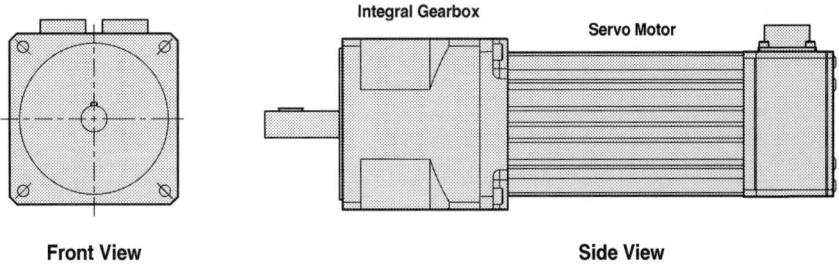

FIGURE 21   Example of an in-line gearbox attached to a servo motor.

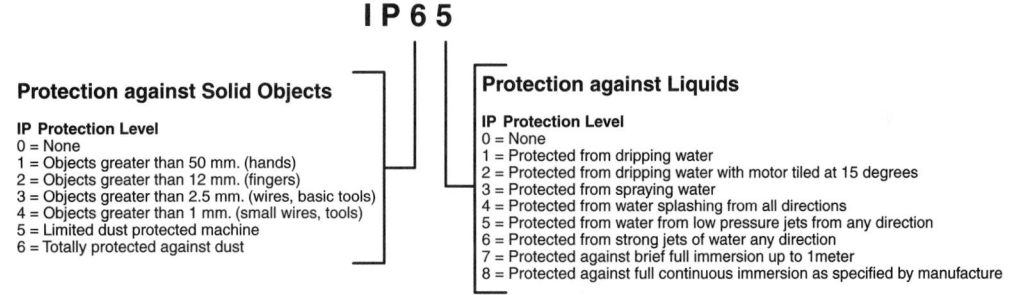

FIGURE 22   IP classification.

IP Classification

Servo motors use a standard classification system to distinguish different working environments. The IP classification comes from the IEC (International Electrotechnical Commission) standard (34-5) [14]. This rating system uses two numbers after the letters IP that relate to the degree of protection against solids and liquids. Figure 22 illustrates the different classifications.

In general, servo motors used for industrial applications are rated at IP 65. For metals-cutting and other harsher environments, IP67 is typical.

Couplers

The ideal coupler joining a servo motor and load would be completely rigid, allowing the torque produced in the servo motor to be transmitted directly to the load. Rigid couplers require perfect alignment, however, which can be difficult or even impossible to achieve. In real systems, some misalignment is inevitable, partly because of tolerance buildups of components.

The three misalignment conditions are:

- end float: a change in the relative distance between the ends of two shafts
- angular misalignment: the center lines of two shafts intersect at an angle other than 0°
- parallel misalignment: the offset of two mating shaft center lines, although the center lines remain parallel to each other

These conditions can exist in any combination. They are illustrated Fig. 23.

TABLE 7   Coupler Comparison

Type	Rigidity	Rel. Misalign. Allowance	Rel. Cost	Rel. Inertia
Beam	low	high	low	high
Bellows	medium	medium	medium	low
Metal Disk	high	low	high	medium

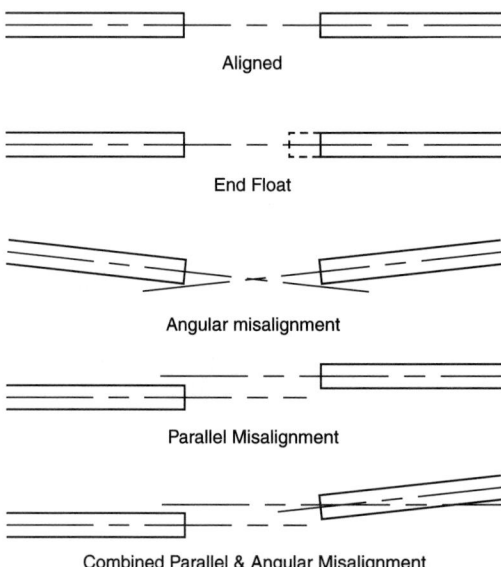

FIGURE 23   Coupler misalignments.

There are several types of couplers to consider. In addition to coupler dynamics, the coupler's inertia may also have to be considered. Table 7 summaries the characteristics of the three most common types [8].

Beam or helical style couplers are low cost and they handle large misalignments, but they can be extremely hard to tune for high-performance servo systems. Metal disk couplers offer the best torsional rigidity, but they allow very little angular alignment. Bellows couplers offer very good torsional rigidity, relatively low inertia, and some tolerance to misalignment.

Bearings

The most common type of bearing used in servo motors is the radial ball bearing. One useful classification for bearing tolerances is stated by the Annular Bearing Engineering Committee (ABEC) or Anti-Friction Bearing Manufacturers Association (AFBMA). Different classes or grades numbering odd from one to nine indicate increasing levels of tightness or tolerance. Grades 1 and 3 are the most common for general purpose servo motors.

Servo motor manufacturers make an assumption on the life expectancy of the bearings used in the motors they produce. Factors such as temperature, bearing ball speed, and axial and radial loading play an important part in estimating the useful life of a bearing.

TABLE 8 Relationship between Reliability (r) and a_1

Reliability (r) (%)	L_n	Life Adjustment Factor for Reliability (a_1)
90	L_{10}(rated life)	1
95	L_5	0.62
96	L_4	0.53
97	L_3	0.44
98	L_2	0.33
99	L_1	0.21

The most common method to determine the bearings reliability is to use a fatigue life equation (see [9]). An example of this is given in Eq. (5).

$$L_n = \frac{16667 a_1 a_2 a_{\text{safety}}}{N} \left(\frac{C_B}{P} \right)^3 \tag{5}$$

with the following definitions:

$L_n =$ number of hours for a given reliability rate "r" (see Table 8)
$a_1 =$ life adjustment factor based on reliability
$a_2 =$ life adjustment factor for material ($\sim$1–3, depending on the steel)
$a_{\text{safety}} =$ optional servo motor manufacturer's safety factor ($\sim$0.6–0.7)
$N =$ rotor speed (rpm)
$C_B =$ basic dynamic load rating from manufacturer (lb)
$P =$ equivalent radial load on bearings (lb)

To decide on the value for a_1, a reliability rate must be determined. The most common one for servo motors is the L_{10} ($n = 10$) rating. This means that 90% of the bearings will meet the number of hours calculated in Eq. (5). Table 8 illustrates the other values of a_1 for their respective reliability rates.

Since there are generally two bearings in a servo motor, the equivalent radial load P must be calculated. P is a function of the actual radial load, F_R, the distance the load is applied, x, and other geometric dimensions of the motor. Figure 24 shows a cutaway view of a typical bearing arrangement in a servo motor.

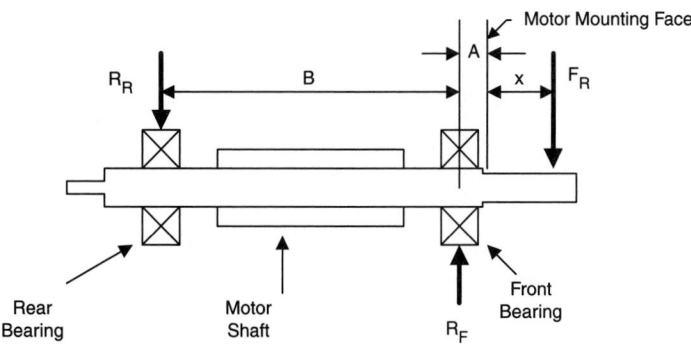

FIGURE 24 Bearing placement in servo motors.

The reaction forces are found by summing the moments around each bearing. For the front bearing:

$$R_F = F_R \left(\frac{A + x + B}{B} \right) \tag{6}$$

And for the rear bearing we have

$$R_R = F_R \left(\frac{x + A}{B} \right) \tag{7}$$

The front bearings generally have significantly more loading than the rear ones and therefore are used to determine the overall life of the motor.

$$P \approx R_F \tag{8}$$

If axial loads are applied, they are compensated for by modifying the effective radial load. This effective radial load is generally calculated by summing the real radial load with two times the axial load.

Equation (5) is usually shown graphically for a given reliability rate n. For a given reliability rate, the acceptable load as a function of rotor speed is given for radial load at a distance x. These curves are generated for different values of distance. Figure 25 is an example of the nature of the curve.

For a fixed radial loading distance x, as the loading is increased, the corresponding speed must be lowered to maintain the same number of reliable hours.

Lubrication

There are many different types of lubrication for servo motors. In selecting the lubrication, it is extremely important to understand the type of environment in which the servo motor will be placed.

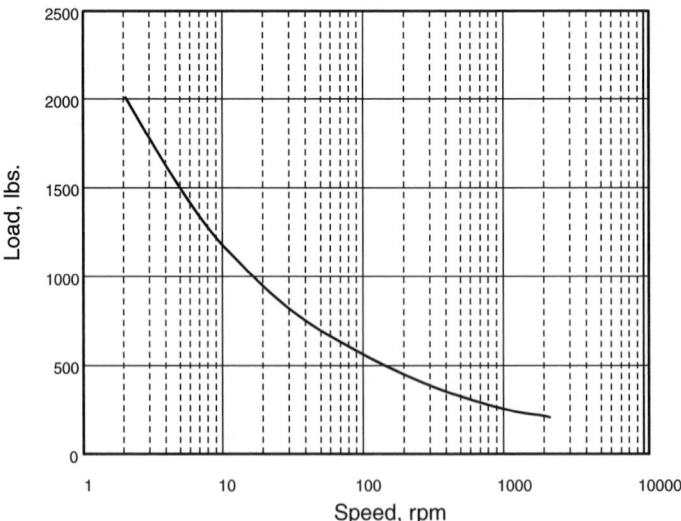

FIGURE 25   Typical radial loading curve for a fixed radial load of x.

TABLE 9 Lubricating Grease Components

Lubricating Grease =	Fluid+	Thickening Agents	+Special Ingredients
	mineral oils	Soaps	oxidation inhibitors
	esters	lithium	rust inhibitors
	organic esters	sodium	VI improver
	glycols	barium	tackiness
	silicones	calcium	perfumes
		strontium	dyes
			metal deactivator
		Nonsoaps (inorganic)	
		microgel (clay)	
		carbon black	
		silica gel	
		Nonsoaps (organic)	
		urea compounds	
		terepthlamate	
		organic dyes	

In general, lubrication grease can be described by the three components that make it up: fluid, thickener, and special ingredients. Table 9 illustrates the common ingredients for these three components.

Seals

Servo motors cover such a wide range of speeds and operating environments that no one seal design is preferable in all cases. The two basic types of dynamic seals are the face seal and the lip seal.

Face seals or axial mechanical seals are mounted between the rotor and stator housing. A spring mechanism provides the required axial pressure on the seal. These types of seals can be very lossy, and they are therefore limited in use.

Lip seals create a radial barrier between the rotating shaft and the motor housing. This type of seal looks like a U-shape O ring with a garter spring encompassing the entire shaft. Lip seals are self-adjusting but will wear out prematurely if operating at high speeds. Depending on the environment, many different types of materials are used. The most popular general purpose seals for servo motors are made with either Teflon or Viton [11]. Lip seals are by far the most common for the general purpose servo motor market.

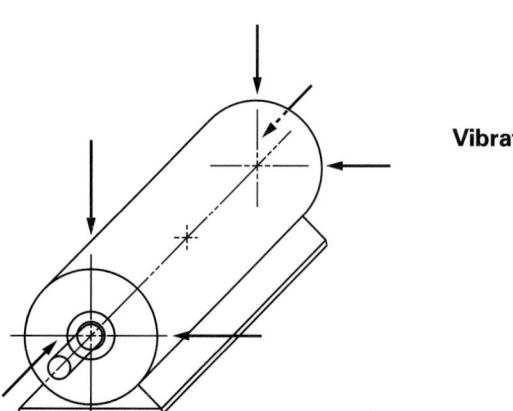

Vibration

Vibrations in servo motors are caused by either an unbalance in the shaft, misalign bearings, or slight deformations in the housing of the motor caused by an interaction between the rotor's magnetic field and the stator's. Despite the fact that the rotors of servo motors can be mechanically balanced, it is possible for other secondary effects to dominate.

One standard that servo motor manufacturers reference is ISO 2373 [10]. This standard provides a set of guidelines for maximum allowable vibration velocities as a function of shaft height. Figure 26 illustrates the points of vibration measurements, and Table 10 summarizes the specifications for the different quality grades.

FIGURE 26 Measuring points for a vibration test.

TABLE 10   Recommended Limits of Vibration Severity

Quality Grade	Speed *n* (rpm)	Maximum rms Values[a] of the Vibration Velocity for the Shaft Height, *H* (in mm)					
		80 < H < 132		132 < H < 225		225 < H < 400	
		mm/s	in/s	mm/s	in/s	mm/s	in/s
N	600 < *n* < 3600	1.8	0.071	2.8	0.110	4.5	0.177
(Normal)							
R	600 < *n* < 1800	0.71	0.028	0.12	0.044	1.8	0.071
(Reduce)	1800 < *n* < 3600	1.12	0.044	1.8	0.071	2.8	0.110
S	600 < *n* < 1800	0.45	0.018	0.71	0.028	1.12	0.044
(Special)	1800 < *n* < 3600	0.71	0.028	1.12	0.044	1.8	0.071

[a] A single set of values, e.g., those applicable to the 132- to 225-mm shaft height, may be used if shown by experience to be required.

REFERENCES

1. Parker Hannifin Engineering Technical Reference, Parker Hannifin Corporation, Compumotor Division, 1997.

2. AC Technical Reference, Danfoss Drives.

3. Miller, T. J. E., *Brushless Permanent-Magnet and Reluctance Motor Drives*, Oxford University Press, New York, 1989.

4. NEMA Standards Publication MG 7-1993 Motion/Position Control Motors and Controls: Covers all rotational and linear electric servo and stepper motors and their power requirements, feedback devices, and controls intended for use in a motion/position control system that provides precise positioning, speed control, torque control, or in any combination. Adopted by the U.S. Department of Defense.

5. Underwriters Laboratories, Inc.: UL 508C Power Conversion Equipment, UL519 Impedance Protected Motors, UL547 Thermal Protectors for Motors, UL674 Electric Motors and Generator sfor Use in Division 1 Hazardous Locations; UL1004 Electric Motors.

6. IEC 60034-1 (1996), Rotating electrical machines, Parts 1–12, 1996.

7. IEC Standard Publication 34-5, 2nd ed., Bureau Central de la Commission Electrotechnique Interntionale, Geneva, Switzerland, 1981.

8. Kaiser, D. A., J. Morris, and C. Durkin, "Dynamic Modeling of Couplers for Drivescrew Applications," *PCIM*, June 1997.

9. Torrington/FAFNIR/Kilian Engineering Reference Catalog 100-295-75M.

10. Teflon is a registered trademark of DuPont; Viton is a registered trademark of DuPont Dow Elastomers.

11. ISO 2373, 2nd ed., 1987-06-01 Mechanical Vibration of Certain Rotating Electrical Machinery with Shaft Heights between 80 and 400 mm. Measurement and Evaluation of the Vibration Severity, Switzerland, 1987.

SOLID-STATE VARIABLE SPEED DRIVES

by Richard H. Osman[1]

INTRODUCTION

The global power electronics industry continues the rapid pace of solid-state drive development. Over the years, many drive circuits have become virtually obsolete, and new ones have been introduced. The user is still confronted with a wide variety of drive types that are suitable for virtually every kind of electrical machine, from the subfractional to the multimegawatt rating. The integral horsepower standard polyphase ac motor and dc motor are major consumers of electric power in industrial applications, and they represent the opportunity for substantial improvement in the user's process. Both new installations and the retrofit of existing machines are possible.

Despite the diversity of power circuits, there are two common properties of these drives:

1. All of them accept commonly available ac input power of fixed voltage and frequency and, through switching power conversion, create an output of suitable characteristics to operate a particular type of electric machine; that is, they are machine specific.

2. All of them are based on solid-state switching devices. The development of new devices is the most important driver of the technology.

Figure 1 shows the basic structure of most common ac drives. There is an input conversion circuit, which converts the utility power into dc, and then an output inversion stage, which changes the dc back into variable ac. This type of drive is commonly called a "variable frequency drive," or VFD. (For dc drives, the motor is in the dc link, or a chopper is used in place of the inverter.)

REASONS FOR USING A VARIABLE SPEED DRIVE

There are a number of reasons to use a variable speed drive:

1. Energy savings where variable flow control is required. In any situation in which flow is controlled by a throttling device (valve or damper), there is the potential for energy savings by removing the throttle and slowing the fan or pump to regulate flow.

2. Optimizing the performance of rotating equipment; e.g., SAG mills, compressors, conveyors, pumps, and fans.

3. Elimination of belts and gears or other power transmission devices by matching the base speed of the motor to the driven load.

4. Automation of process control by using the VFD as the final control element, leading to more efficient part-load operation.

5. Reduction of the rating and cost of the electrical distribution system by eliminating motor starting inrush.

6. Extending the life of motors, bearings, seals, liners, and belts.

---

[1] Vice President of Technology, Robicon, 100 Sagamore Hill Road, Pittsburgh, Pennsylvania 15239.

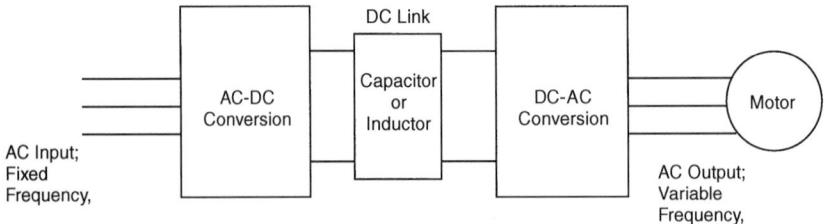

FIGURE 1   Basic circuit arrangement of modern commercial ac drives.

7. Reducing noise and environmental impact—electric drives are clean, nonpolluting, quiet, efficient, and easy to repair.

SEMICONDUCTOR SWITCHING DEVICES

Even though many of the basic power conversion principles were developed in the thirties, when the circuits were constructed with mercury arc rectifiers, it was not until the invention of the thyristor in 1957 that variable speed drives became truly practical. Figure 2 shows a comparison of the properties of devices commonly in use today.

The thyristor (SCR) is a four-layer semiconductor device that has many of the properties of an ideal switch. It has low leakage current in the off state, a small voltage drop in the on state, and takes only a small signal to initiate conduction (power gains of over 10^6 are common). When applied properly, the thyristor will last indefinitely. After its introduction, the current and voltage ratings increased rapidly. Today it has substantially higher power capability than any other solid-state device but no longer dominates power conversion in the medium- and higher-power ranges. The major drawback of the thyristor is that it cannot be turned off by a gate signal, but the anode current must be interrupted in order for it to regain the blocking state. The inconvenience of having to commutate the thyristor in its anode circuit at a very high energy level has encouraged the development of other related devices as power switches.

Transistors predate thyristors, but their use as high-power switches was relatively restricted (compared with thyristors) until the ratings reached 50 A and 1,000 V in the same device, during the early 1980s. These devices are three-layer semiconductors that exhibit linear behavior but are used only in saturation or cutoff. In order to reduce the base drive requirements, most transistors used in

Device Type	Maximum Voltage	Maximum Current	u sec Off Time	Peak Gate Power	"ON" Voltage	Cost
Diode	7000	10000	50	N/A	1.0	Low
Thyristor	7000	10000	10-300	2 W	1.25	Low
GTO	6000	4000	10-50	12kW	3.5	High
Transistor	1400	1800	3-5	20 W	2.5	Med.
IGBT	6000	2400	0.5	3 W	3.0	Med.
IGCT	6000	4000	2-3	45kW	2.0	Med.

FIGURE 2   Comparison table of modern semiconductor switching devices.

variable speed drives are Darlington types. Even so, they have higher conduction losses and greater drive power requirements than thyristors. Nevertheless, because they can be turned on or off quickly via base signals, transistors quickly displaced thyristors in lower drive ratings, and they were once widely used in pulse-width-modulated inverters. They in turn were displaced by insulated gate bipolar transistors (IGBTs) in the late 1980s. The IGBT is a combination of a power bipolar transistor and a MOSFET that combines the best properties of both devices. A most attractive feature is the very high input impedance that permits them to be driven directly from lower-power logic sources. Their power handling capability has increased dramatically, and they are now practical alternatives to thyristors, GTOs, and IGCTs in the largest drive ratings.

It has long been possible to modify thyristors to permit them to be turned off by a negative gate signal. These devices are four-layer types and are called gate-turn-off thyristors, or simply GTOs. These devices have been around since at least 1965, but only in the mid-1980s did their ratings increase to high power levels. Present-day GTOs have about the same forward drop as a Darlington transistor (twice that of a conventional thyristor). GTOs require a much more powerful gate drive, particularly for turn off, but the lack of external commutation circuit requirements gives them an advantage over thyristors. GTOs are available at much higher voltage and current ratings than power transistors. Unlike transistors, once a GTO has been turned on or off with a gate pulse, it is not necessary to continue the gate signal because of the internal positive feedback mechanism inherent in four-layer devices. Unfortunately, high cost and very large switching losses restricted the use of GTOs to only those applications in which space and weight were at a premium. In 1997 the IGCT (integrated gate controlled thyristor) was introduced. This is identical in construction to the GTO, but a new method of turn off and special metallurgy has resulted in a device considerably better than the GTO in forward drop and switching losses.

Today the thyristor, IGBT, GTO, and IGCT form the technological base on which the solid-state variable speed drive industry rests today. There are other device technologies and enhancements in various stages of development that may or may not become significant, depending on their cost and availability in large current (>50 A) and high voltage (1,000 V) ratings. These include: (1) trench gate construction for IGBTs, (2) silicon carbide semiconductors, and (3) variants of the four-layer switch, such as the MTO (MOS turn-off thyristor) and MCT (MOS controlled thyristor). We should expect new switches to come along and significantly improve on the devices currently in use. While the type of semiconductor device is not necessarily the most important issue to a user, in general the newer devices provide a better drive performance.

DRIVE CONTROL TECHNOLOGY

Parallel to the development of power switching devices, there have been very significant advances in hardware and software for controlling variable speed drives. These controls are a mixture of analog and digital signal processing. The advent of integrated circuit operational amplifiers and integrated circuit logic families made possible dramatic reductions in the size and cost of the drive control, while permitting more sophisticated and complex control algorithms without a reliability penalty. These developments occurred during the 1965–1975 period. Further consolidation of the control circuits occurred after that as large-scale integrated circuits (LSI) became available. In fact, the pulse-width modulation control technique was not practical until the appearance of LSI circuits because of the immense amount of combinatorial logic required.

Clearly, the most significant advance in drive control has been the introduction of microprocessors into drive control circuits. The introduction of cheap and powerful microprocessors continues to expand the capability of drive controls. A modern drive should have most of these features. The performance enhancements include:

1. More elaborate and detailed diagnostics, resulting from the ability to store data relating to drive internal variables, such as current, speed, firing angle, and so on; the ability to signal to the user if a component has failed.

2. The ability to communicate both ways over industry standard protocols with the user's central computers about drive status.

3. The ability to make drive tuning adjustments by means of keypads, with parameters such as loop gains, ramp rates, and current limits stored in memory rather than potentiometer settings.

4. Self-tuning and self-commissioning drive controls.

5. More adept techniques to overcome power circuit nonlinearities.

SOLID-STATE dc DRIVES

The introduction of the thyristor had the most immediate impact in the dc drive area. Ward Leonard (motor-generator) variable speed drives were quickly supplanted by thyristor dc drives of the type shown in Fig. 3 (six-thyristor full converter), for reasons of lower cost, higher efficiency, and lower maintenance cost. This circuit arrangement (the Graetz circuit) has become the workhorse of the electrical variable speed drive industry, as will be seen by the number of other drives that use it. By utilizing phase control of the thyristors, the converter behaves as a programmable voltage source. Therefore, speed variation is obtained by adjustment of the armature voltage of the dc machine. Because the phase control is fast and precise, critical features such as current limit are easily obtained. In fact, almost all thyristor dc drives today are configured as current regulators with a speed or voltage outer loop. The line-side characteristics are only fair in that input power factor is proportional to speed, and the input current has 31% harmonic content. The full converter offers low output ripple and the ability to regenerate or return energy to the ac line. The system can be made into a four-quadrant drive by the addition of a bidirectional field controller. Torque direction is determined by field current direction. Because of the large field inductance, torque reversals are fairly slow (100–500 ms) but adequate for many applications.

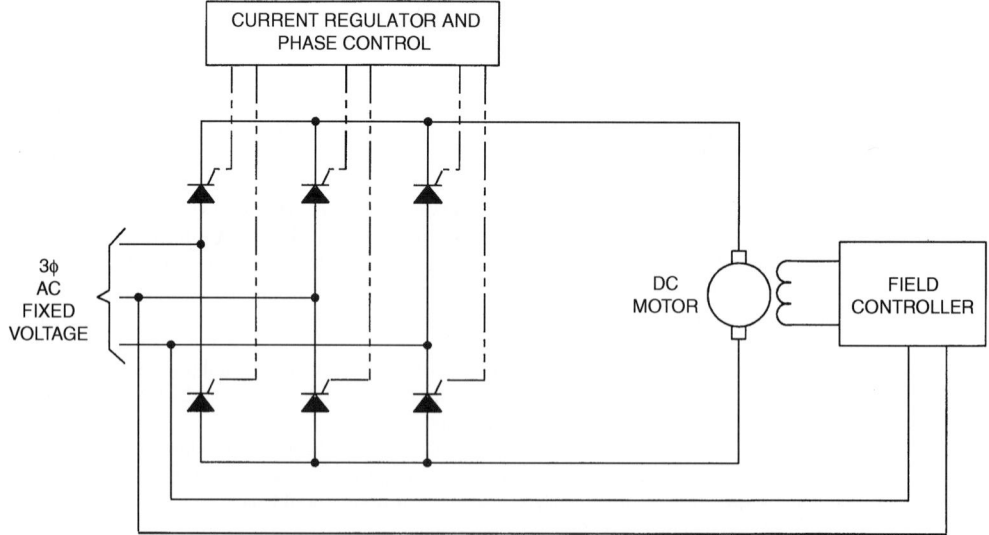

THYRISTOR DC DRIVE–3 PHASE FULL CONVERTER

FIGURE 3   Thyristor converter dc drive.

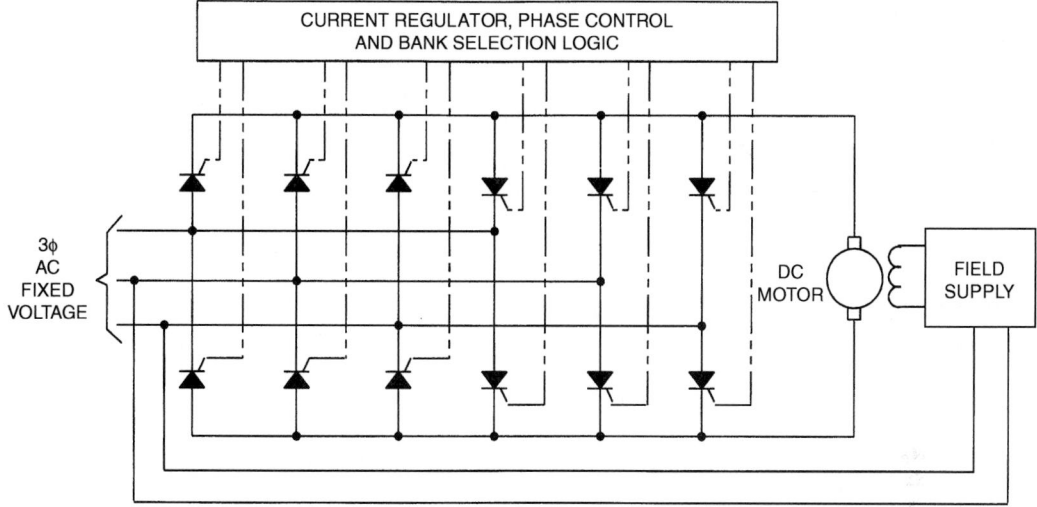

THYRISTOR DC DRIVE–ARMATURE DUAL CONVERTER

FIGURE 4   Dual thyristor converter dc drive.

For the best speed of response in four-quadrant thyristor dc drives, the dual armature converter of Fig. 4 is preferred. This is simply two converters (as shown in Fig. 3) connected back to back. Torque direction is determined by the direction of armature current, and since this is a low inductance circuit, reversal can be accomplished in 10 ms (typically). Obviously, only one converter is conducting at one time with the other group of six thyristors not being gated. This is called "bank selection."

SUMMARY OF THYRISTOR dc DRIVES

The two types of thyristor dc drives just described all share a common property in that the thyristors are turned off by the natural polarity reversal of the input line. This is called natural or line commutation. Thus, the inability to turn off a thyristor from the gate is no practical drawback in these circuits. Consequently, they are simple and very efficient (typically 98.5%) because the device forward drop is very small compared with the operating voltage. These drives can be manufactured to match a dc machine of any voltage (commonly 500 V) or horsepower (1/2 to 2500 HP, typically). The main drawback of dc drives is the machine, not the power electronics. The dc machine, although easy to control, is larger, heavier, less robust, incompatible with corrosive or hazardous environments, generally not available above 750 V, and much more expensive than its ac counterpart of the same rating. Today, the only remaining reason for selecting a dc drive is if an inexpensive dc motor is available, or a retrofit situation exists. ac variable frequency drives now offer better response and generally better overall performance.

ac VARIABLE FREQUENCY DRIVES

The impact of new solid-state switching devices was even more significant on ac variable speed drives, but it occurred somewhat later in time as compared with dc drives, and it shows no sign of stopping. ac drives are machine specific and more complex than dc drives, mostly because of the simplicity of the ac machine. Solid-state variable speed drives have been developed and marketed for wound-rotor induction motors (WRIMs), cage-type induction motors, and synchronous motors.

Historically, WRIM-based variable speed drives were in common use long before solid-state electronics. These drives operate on the principle of deliberately creating high-slip conditions in the machine and then disposing of the large rotor power that results. This is done by varying the effective resistance seen by the rotor windings. The WRIM is the most expensive ac machine. This has made WRIM-based variable speed drives noncompetitive as compared with cage induction motor (IM) drives or load commutated inverters using synchronous machines. The WRIM has become a casualty of the tremendous progress in ac variable speed drives as applied to cage induction motors.

INDUCTION MOTOR VARIABLE SPEED DRIVES

Because the squirrel cage induction motor is the least expensive, least complex, and most rugged electric machine, great effort has gone into drive development to exploit the machine's superior qualities. Because of its extreme simplicity, it is the least amenable to variable speed operation. Since it has only one electrical input port, the drive must control flux and torque simultaneously through this single input. As there is no access to the rotor, the power dissipation there raises its temperature; so very low-slip operation is essential. Induction motor variable speed drives in the past have had the greatest diversity of power circuits. Today, for drives rated 600 VAC or below (LV), there are essentially only two choices: (1) The IGBT pulse-width-modulated drive and (2) the autosequentially commutated current-fed inverter (ASCI; see Fig. 6).

1. The IGBT pulse-width modulated drive: In this type of voltage source inverter, both the frequency and amplitude are controlled by the output switches alone. A representative circuit based on IGBT's is shown in Fig. 5. The input converter is a diode bridge so that the dc link operates

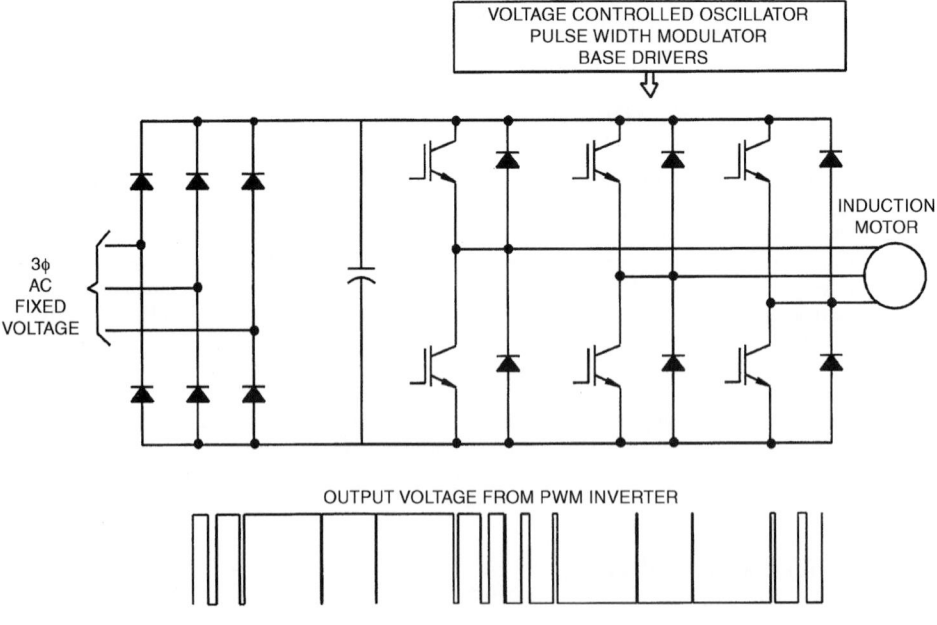

FIGURE 5   IGBT PWM variable frequency drive.

at a fixed unregulated voltage. The diode front end gives virtually unity power factor, independent of load and speed. This type of drive is called pulse-width modulated (PWM), because the output voltage waveform is synthesized from constant amplitude, variable-width pulses at a high (1–3 kHz) frequency so that a sinusoidal output is simulated; the lower harmonics (5, 7, 11, 13, 17, 19, . . .) are not present in modern PWM drives. One advantage is smooth torque, low harmonic currents, and no cogging.

Although this approach eliminates the phase control requirements of the thyristor converter, it requires fast output switches. Since every switching causes an energy loss in the output devices, fast switches are needed in order to support a high switching frequency without excessive losses in the output switches. Occasionally, high-frequency switching may cause objectionable acoustic noise in the motor, but that has been overcome with special modulation techniques and higher switching frequencies (3 kHz and up).

There are IGBT PWM VFD's on the market today in the range of 1 kW to 1 MW at 460, 600, and 690 VAC. As with all voltage source inverters fed by diode rectifiers, regeneration of power to the line is not possible.

2. The autosequentially commutated current fed inverter (ASCI; Fig. 6): An entirely different approach to an IM drive is to generate a smooth dc current and feed that into different pair of windings of the machine so as to create a discretely rotating magnetomotive force (MMF). This type of inverter is called the autosequentially commutated current-fed inverter. This circuit was invented later than voltage-fed inverters and is much more popular in Europe and Japan than in the United States. The input stage is a three-phase thyristor bridge, which is current regulated. A dc link choke smoothes the current going to the output stage. There a thyristor bridge distributes the current into the motor windings with the same switching function as the input bridge, except at variable frequency. (Notice

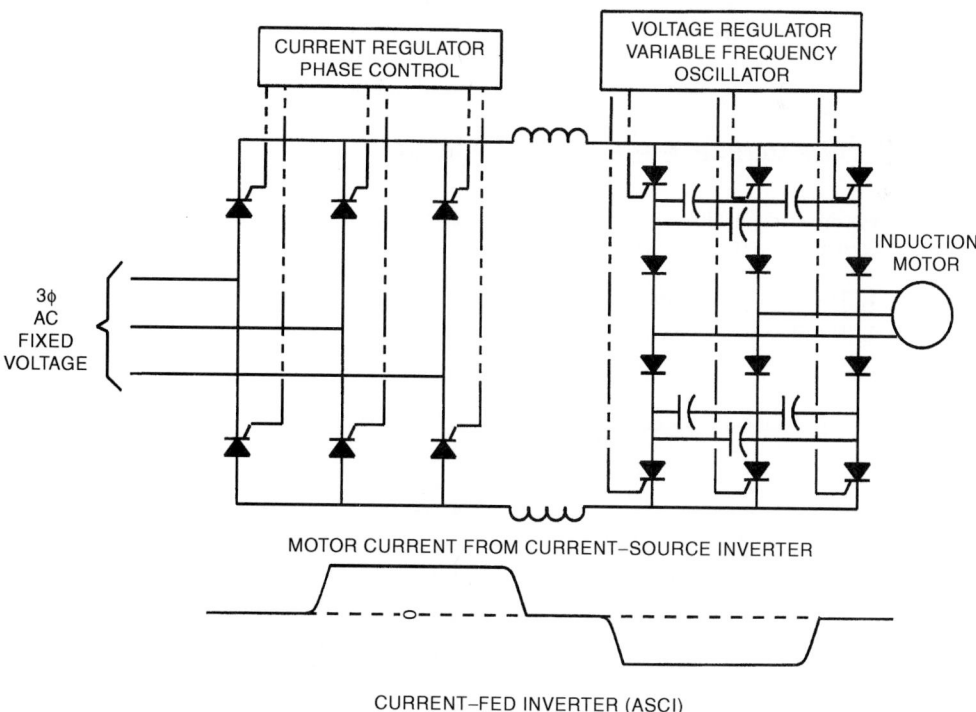

FIGURE 6   Thyristor Current-fed drive.

the similarity to the LCI). The current waveform is a quasi-square wave whose frequency is set by the output switching rate and whose amplitude is controlled by the current regulator. The capacitors and rectifiers are used to store energy to commutate the thyristors, since the induction motor cannot provide this energy and remain magnetized, in contrast to the synchronous motor. This type of drive has simplicity, good efficiency (95%), excellent reliability, and four-quadrant operation up to ~120 Hz. Harmonics in the output current are greater than in the PWM drive, but reasonably low, giving a form factor of 1.05 (same as the LCI). The input power factor is equal to the product of the load power factor times the PU speed. Above 500 kW, the ASCIs are very cost effective. Because of the controlled current properties, this drive is virtually immune to damage from ground faults, load shorts, and commutation failures. Since MMF (current) is directly controlled and the drive is regenerative, ASCIs can readily be equipped with field oriented controls for the most demanding four-quadrant operation.

CURRENT-FED VS. VOLTAGE-FED CIRCUITS: THE TWO BASIC TOPOLOGIES

Voltage-fed and current-fed refer to the two basic VFD strategies of applying power to the motor. In Europe, these are called voltage-impressed and current-impressed, which is a clearer description. In voltage-fed circuits, the output of the inverter is a voltage, usually the dc link voltage. The motor and its load determine the current that flows. The inverter doesn't care what the current is (within limits). Usually, these drives have diode rectifiers on the input. The main dc link filter is a capacitor. In current-fed circuits, the output of the inverter is a current, usually the dc link current. The motor and its load determine the voltage. The inverter doesn't care what the voltage is. Usually these VFD's have a thyristor converter input stage, and the dc link element is an inductor. See Fig. 7 for a summary comparison of the properties of the two types of VFDs in a six-pulse 600 VAC or smaller (LV) configuration.

MEDIUM-VOLTAGE VARIABLE FREQUENCY DRIVES

For drives rated 2300 VAC and above on the output, there are a number of choices of design of both current and voltage fed types.

1. The load commutated inverter (LCI; Fig. 8).
2. The filter commutated thyristor inverter (Fig. 9).

LV Current-fed Type VFD	LV Voltage-fed Type VFD
● Lower Cost at High HP, LV	● Lower Cost at Low HP, LV
● Four- Quadrant	● Two-Quadrant
● P.F. = P.U. Speed*Load P.F.	● 95% displacement P.F.
● 96.5% Efficiency	● 96-97.5% Efficiency
● Immune to short circuits	● Requires protection for S.C.
● More low-cost Components	● Few Higher-cost Components
● Large Magnetics	● Small or no Magnetics
● Lower motor noise	● Low to Medium Motor Noise
● Non-Critical layout	● Critical construction layout
● 30% Harmonic Current (6-P)	● 40% Harmonic current (6-P)
● Low dV/dt at output	● High dV/dt at output
● High Common-mode Voltage	● Low Common-mode Voltage
● Output filter required	● Output filter not needed

FIGURE 7   Comparison of low-voltage current-fed and voltage-fed drives.

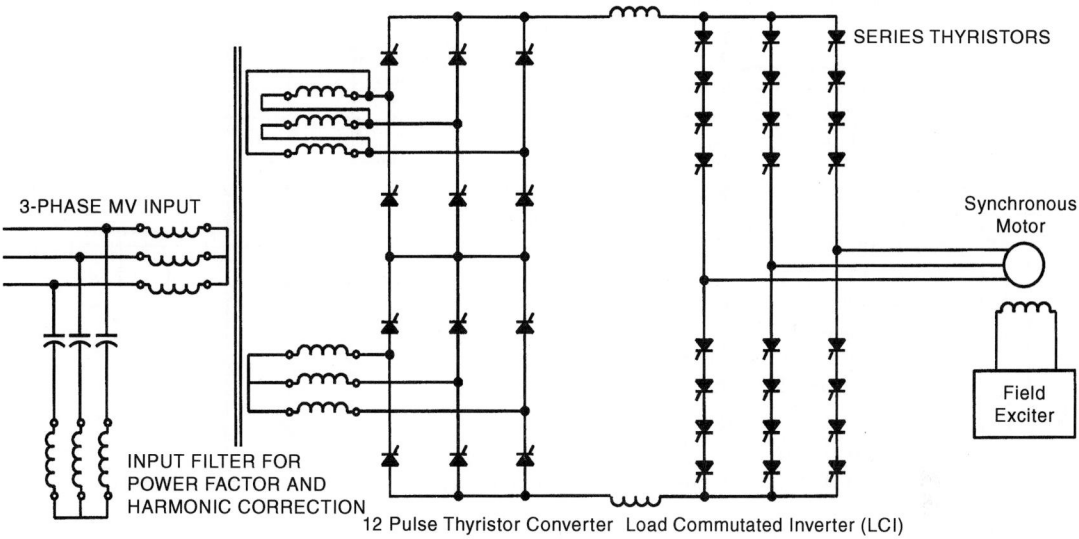

FIGURE 8 Load commutated inverter.

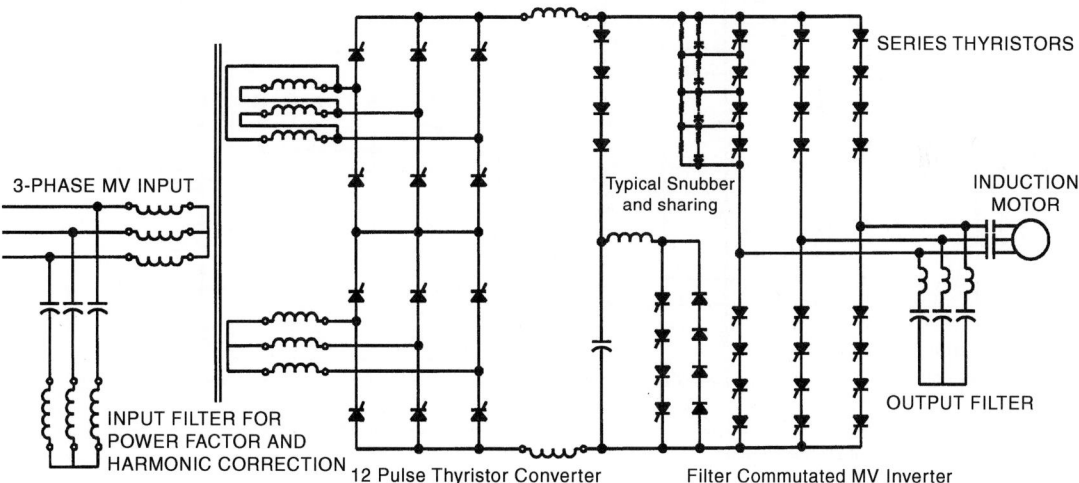

FIGURE 9 Filter-commutated thyristor drive.

3. The current-fed GTO inverter (Fig. 10.).

4. The neutral-point-clamped inverter (Fig. 11).

5. The multilevel series cell VFD (Figs. 12 and 13).

6. The cycloconverter (Fig. 14).

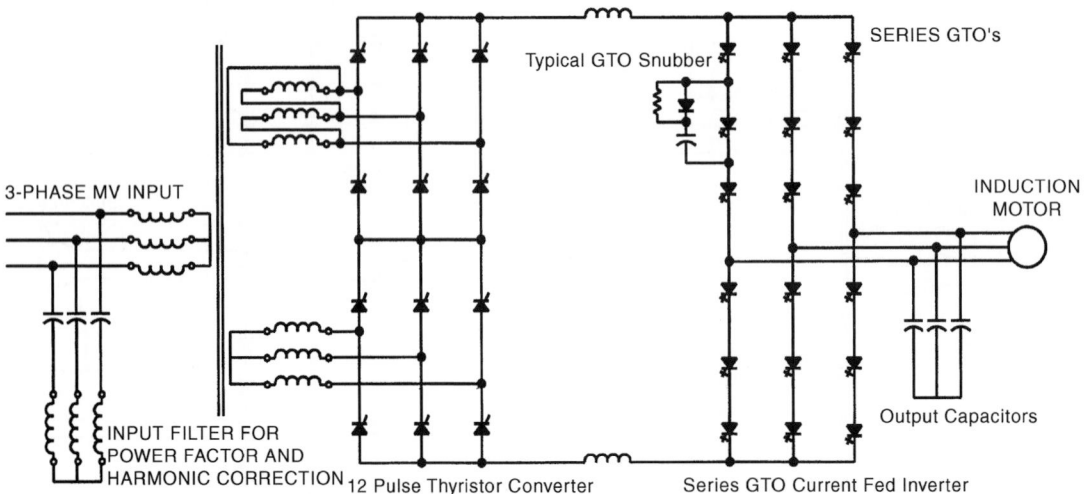

FIGURE 10   Current-fed GTO inverter.

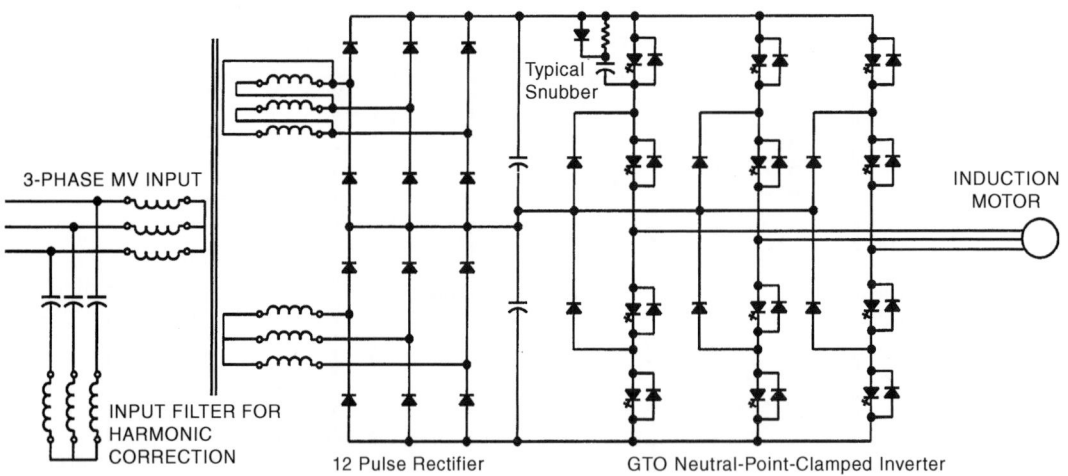

FIGURE 11   Neutral-point-clamped inverter.

THE LOAD COMMUTATED INVERTER

As shown in Fig. 8, the load-commutated inverter is based on a synchronous machine. All the thyristors are naturally commutated, because the back EMF of the machine commutates the load side converter. The machine side converter operates exactly like the line side converter, except the phase back angle is ~150°. The machine naturally applies reverse voltage to an off-going device before the next thyristor is gated. This imposes some special design criteria on the synchronous motor. It has to be able to operate at a substantial leading power factor over the speed range, it must have enough leakage inductance to limit the thyristor di/dt, and it has to be able to withstand harmonic currents in the

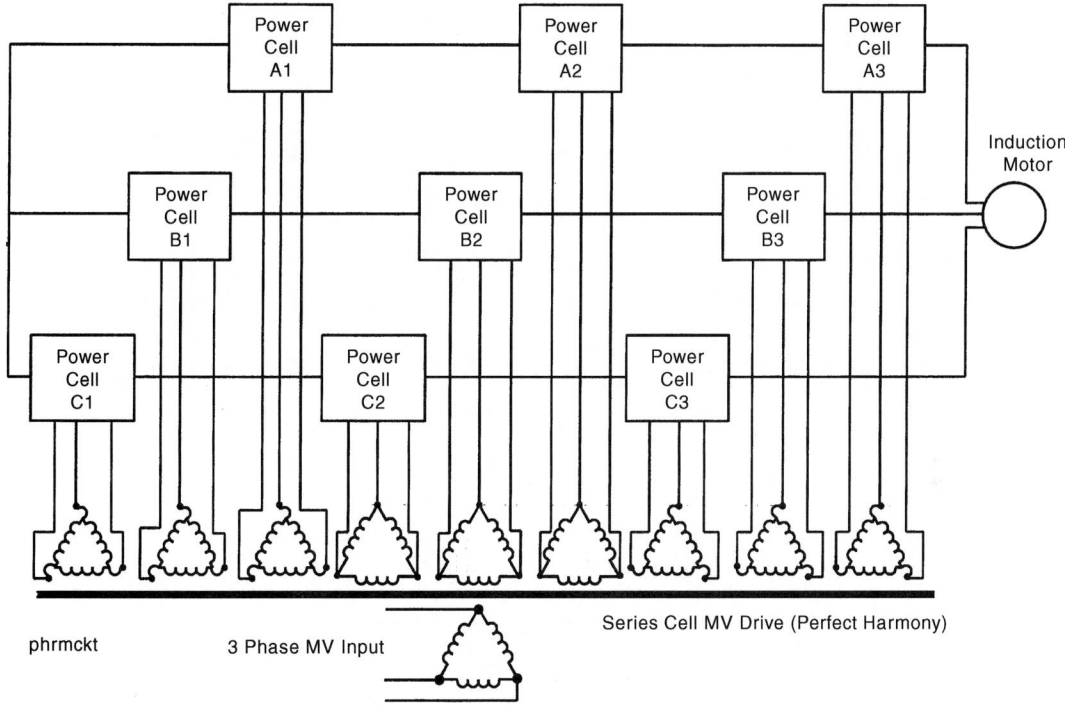

FIGURE 12   Multilevel series-cell drive.

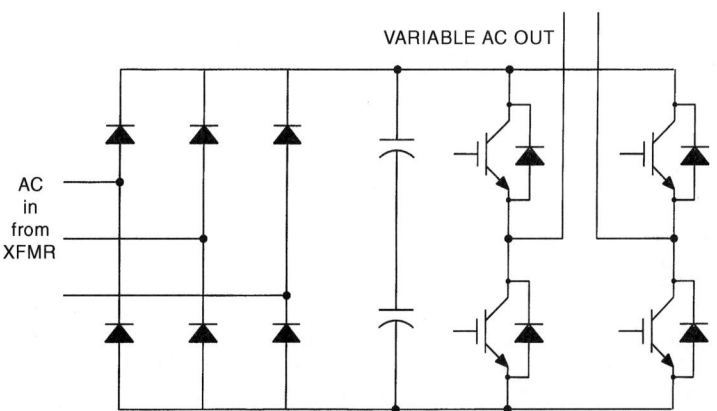

FIGURE 13   Conversion cell of Multilevel VFD.

damper windings. The LCI uses two thyristor bridges—one on the line side and one on the machine side. The requirement for the machine to operate with a leading power factor requires substantially more field excitation and a special exciter compared with that normally applied to synchronous motor. This also results in a reduction in the torque for a given current. The machine side devices are fired in exact synchronism with the rotation of the machine, so as to maintain constant torque angle and constant commutation margin. This is done either by rotor position feedback or by phase control

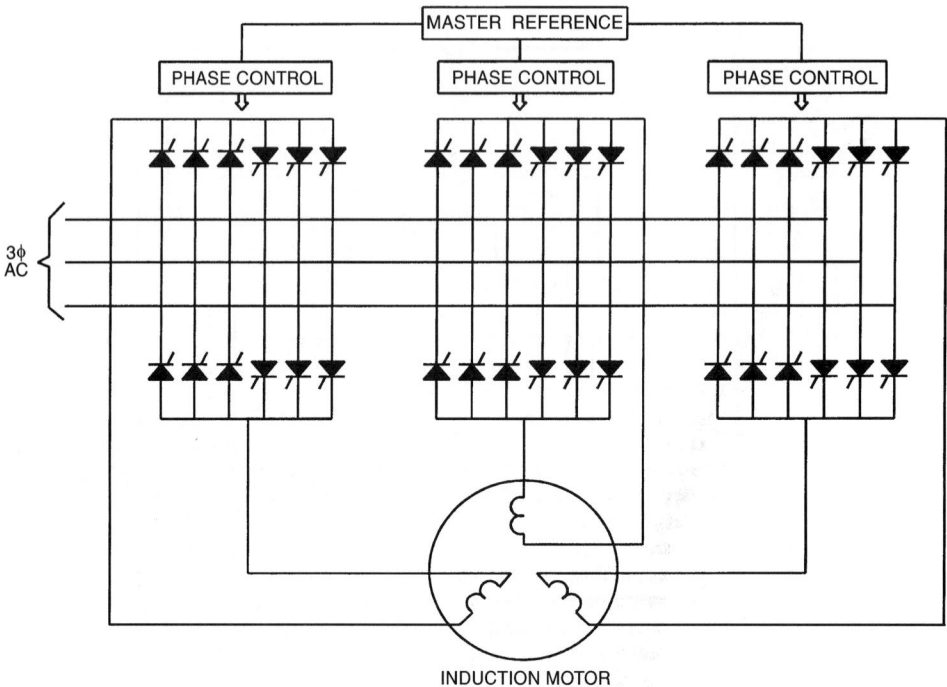

FIGURE 14   Cycloconverter.

circuits driven by the machine terminal voltage. Only RC networks for voltage sharing are necessary. The output current is very similar in shape to the input current, which means a substantial harmonic component. The harmonic currents cause extra losses in the damper bars, and they give rise to very significant torque pulsations. The drive is not self-starting because of the low machine voltage at low speeds.

Therefore, the drive is started by interrupting the dc link current with the line-side converter in order to commutate the inverter thyristors. The line-side converter is regulated to control torque. A choke is used between converters to smooth the link current. LCIs came into commercial use about 1980 and are used mainly on very large, medium-voltage drives (1–100 MW). At these power levels, multiple series devices are employed (typically four at 4 kV input), and conversion takes place directly at 2.4 or 4 kV or higher. The efficiency is excellent, and reliability has been very good. Although they are capable of regeneration, LCIs are rarely used in four-quadrant applications because of the difficulty in commutating at very low speeds where the machine voltage is negligible. Operation above line frequency is straightforward. Despite the need for special synchronous motors, the LCI drive has been very successful, particularly in very large sizes, where only thyristors can provide the current and voltage ratings necessary. Also, high-speed LCIs have been built. Now that self-commutated VFDs are available, the LCI is becoming less popular.

FILTER COMMUTATED THYRISTOR DRIVE

In the circuit shown in Fig. 9, the output capacitor filter is chosen to supply all the magnetizing current of the motor at ~50% speed. Above that point the load (machine and filter combined) power factor remains leading and the inverter thyristors are naturally commutated, that is, the voltage across the device is naturally reversed before the reapplication of forward voltage. In this mode of operation, the thyristor waveforms are similar to those in an LCI. The filter must supply (at a minimum) all of the reactive current requirements of the motor at full load, and is typically 1 PU of the drive kVA rating. In addition to the large ac capacitors, the filter has to have some series inductive reactance to limit the di/dt applied to the inverter thyristors. Since the filter is capable of self-exciting the motor, a contactor is required to isolate the filter from the motor when the drive is off line. The large filter has the added advantage of providing a path for the harmonic currents in the inverter output (which is a six-step current), so that the motor current waveform is good near rated frequency. As the output frequency decreases, the filter becomes less effective and the motor current waveform deteriorates. The fundamental current into the filter increases with the square of the frequency up to rated voltage, since the voltage is also increasing with the frequency. Maintaining control of the voltage requires that the inverter "drain off" more of the filter reactive current as the frequency increases, therefore making it difficult to achieve more than ~1.1 PU of base frequency.

Since the filter cannot provide commutation down to zero frequency, it is necessary to provide an auxiliary commutation means to get the drive started. This circuit acts on the dc link current and is commonly called the diverter. When it is time to switch inverter thyristors, the dc link current is temporarily interrupted (diverted), allowing the devices to recover. Then the next thyristor pair is gated, and dc link current is restored. The auxiliary circuit has to be able to withstand full link voltage, and interrupt the rated dc link current for several hundred microseconds to permit the inverter thyristors to recover. (High-voltage thyristors require long turn-off times as a consequence of design compromises in achieving a high blocking voltage). Thus, the auxiliary commutation circuit is quite significant in rating. It is not usually intended for continuous operation, but only to get the speed up to the point where the filter commutation commences. The drive controller has to be able to manage two modes of operation.

This circuit has been implemented with four 3-kV thyristors in series per leg of the output bridge. (It is possible to add additional thyristors for redundancy.) Each leg of the bridge experiences the peak motor line–line voltage of ~6000 V in both polarities, so the devices must have symmetrical blocking voltage. As in the input converter, the issue of voltage sharing during steady-state and switching arises. Combinations of device matching and/or RC snubbers are needed. Gate circuits for thyristors are simple and typically deliver 3–5 W of power, although they are designed for somewhat more. This approach has been most successful in those applications in which the drive operates more or less continuously and in the range of 60–100% of rated speed.

CURRENT-FED GTO INVERTER

Another medium-voltage bridge inverter circuit is shown in Fig. 10. Here the output devices are GTOs (three 4-kV units per leg will be required) that can be turned off with the gate. This reduces the size of the filter as compared to the filter-commutated inverter to perhaps 0.8 PU, but it does not eliminate it. Since the motor appears to be a voltage source behind the leakage reactance, it is not possible to commutate the current between motor phases without a voltage to change the current in the leakage inductance. When a GTO turns off, there must still be a path for the current trapped in the motor leakage inductance, which is provided by a capacitor bank. The capacitors resonate with the motor leakage inductance during the transfer of current. The choice of capacitor is determined by the permissible ring-up voltage during commutation. All current-fed VFDs have the need of a "buffering" capacitor between the impressed current of the inverter and the inductance of the motor. Furthermore, if the capacitor bank exceeds 0.2 PU, the possibility of self-excitation of the motor exists, necessitating a contactor between machine and drive. Voltage-fed circuits do not require these elements, because the voltage can arbitrarily be changed across the leakage inductance.

Since the capacitor bank is smaller than in the filter-commutated VFD, it does not provide as much filtering of the output current. Motor current improvements are made by harmonic elimination switching patterns for the GTOs. At low frequencies, many pulses per cycle are possible and harmonic elimination is quite effective, but the GTO frequency limit of a few hundred hertz restricts harmonic elimination at rated frequency to the fifth and maybe the seventh. This frequency limit comes about because of the nature of the GTO turn-off (and to a lesser extent, turn-on) mechanism. The device is turned off by extracting charge from the gate over a period of a few tens of microseconds and interrupting the regenerative turn-on mechanism. Near the end of the charge extraction period, the voltage across the GTO rises and the current begins to fall. During this time the device experiences extremely high internal power dissipation, which must be mitigated by the use of a large (1–5 μF compared to 0.1 μF for thyristors) polarized snubber located very close to the GTO. In that snubber, the capacitor is connected through a diode (the diode requires the same voltage rating as the GTO) to the GTO, so turn-off current can divert into the snubber, but the capacitor cannot discharge into the snubber at turn on. The energy transferred to the snubber capacitor must be disposed of in some way so that the capacitor is discharged before the next turn off. Thus GTOs typically have a minimum "on" time (say 10 μS) and a minimum "off" time (say 100 μS) to permit the internal switching heat to flow away from the junction and for the snubber to recover.

Violation of the minimum time limits, or an unsuccessful turn-off attempt can result in destruction of the GTO. This limits the the maximum switching rate with tolerable losses to a few hundred hertz. The GTO gate driver, in addition to a providing a turn-on pulse comparable to the thyristor driver, has to be able to deliver a peak negative current of 1/5 to 1/3 the anode current in order to turn-off the device. Thus, the GTO driver has a peak VA rating of 2–3 orders of magnitude higher than that for a thyristor, and perhaps ten times the average power requirement. This is an important factor in that the all the gate power must be delivered to a circuit floating at medium-voltage potential.

The snubber losses can have a noticeable effect on part-load efficiency for a GTO drive. Some circuit implementations use patented energy recovery techniques to avoid efficiency deterioration, but these add serious complexity. The snubber loss is proportional to the frequency and to the snubber capacitance, but to the square of the voltage. Those circuits have to use devices with a comparable voltage rating to the GTO.

The design compromises in the metallurgy of the GTO result in a noticeably higher forward drop (2.5–4 V) than that of the conventional thyristor. The device design is further complicated by the requirement for symmetrical voltage blocking in the current-fed topology.

NEUTRAL-POINT-CLAMPED INVERTER

Despite the obvious complications of series GTO designs, they have also been used successfully in voltage fed drives. Figure 10 illustrates such a circuit, the neutral-point-clamped inverter. There have been many of this type applied at a 3,300-V output with 4.5-kV GTOs, but the circuit has only recently been extended to 4-kV ac, probably because of the improved properties of the IGCT. In the new erversions of this drive, the GTOs are replaced with IGCTs. These devices are similar in construction to a GTO, but they are turned off quickly (1 μs) by drawing all the anode current out through the gate, so that the turn-off gain is unity. This requires a higher current gate driver, but lower average power requirements since the turn-off time is so short. The main claim of improvement is that the IGCT supposedly can operate with a very small or no snubber. In this 4- kV ac output design, the total dc link voltage is 6 kV, with a midpoint established at the center of the capacitor filter. Each leg of the bridge consists of two 6-kV IGCTs in series. There are diodes in reverse across each GTO to permit motor current to flow back to the link, and still more diodes (same voltage rating as the GTOs/IGCTs) connecting the midpoints of the inverter legs back to the midpoint of the dc link. The total device count is 12 GTOs and 18 diodes (plus 12 more diodes in the GTO snubbers, if GTOs are used). The neutral-point-clamped inverter offers several advantages in those cases in which series devices would be necessary anyway. First, the clamping diodes permit another voltage level, the dc link midpoint, at the output. This cuts the voltage step seen by the motor in half, and more importantly, creates another degree of freedom in eliminating output harmonics. Also, the clamping diode positively limits the voltage across any one device to half the link voltage, enforcing voltage sharing without additional RC networks.

It is still necessary to equip each GTO with a close-coupled snubber and manage the snubber losses. Since the switching devices in this circuit are never subjected to reverse voltage, it is preferable to use asymmetrical devices in which the absence of reverse blocking is traded off for lower conduction and switching losses.

Device protection during a short circuit is a problem, as the GTO/IGCT can carry almost unlimited fault current like a thyristor. Unlike the current-fed circuits in which fault current is limited, in the voltage fed circuit, the dc link capacitor can source very large fault currents in the event of a short or a commutation failure.

Protection schemes generally attempt to detect the onset of fault current and turn off the devices before it grows beyond the safe turn-off level. Another approach of turning on all the GTOs to distribute the fault current may protect the devices, but it applies a bolted fault to the motor, resulting in extremely large torques. It is possible to use the NPC topology with IGBTs as the switching devices. As IGBTs are currently limited to 3,300 V, the IGBT NPC cannot yet reach a 4-kV ac output, but a European manufacturer has announced a 6-kV IGBT to be available in the first quarter of 1999. The concept of NPC can be extended to M-level inverters, although the number of diodes grows rapidly. Since each device is topologically unique, adding redundant devices would require twice as many, instead of just one more.

MULTILEVEL SERIES-CELL INVERTER

The patented series cell arrangement of Figs. 11 and 12, also known as the Perfect Harmony drive, addresses the previously mentioned design issues in a unique way. Since there are no devices in series, only series cells, the problem of voltage sharing does not exist. The rectifier diodes and the IGBTs are both closely coupled to the dc link capacitor in the cell and thus cannot be exposed to more than the bus voltage, regardless of the load behavior. Since there is no dc link choke, a voltage transient on the ac mains is converted into a current pulse by the relatively high leakage reactance of the transformer secondary, and does not add to the voltage seen by the diodes.

Each cell generates the same ac output. The fundamentals are equal in magnitude and in phase, but the carrier frequency is staggered among the cells in a particular phase. Although an individual cell operates at 600 Hz, the effective switching frequency is 3 kHz, so the lowest harmonic is theoretically the 100th. This low switching frequency and the excellent high-frequency characteristics of the IGBT has the advantage that the IGBT switching losses are totally negligible. The devices can switch well above rated current without the need for snubbers which also helps in maintaining excellent efficiency. Waveform quality is unaffected by speed or load. For the 5 cell/phase VFD, there are ten 620-V steps between the negative and positive peaks. With this technique, the concern for high dv/dt on the motor windings is avoided entirely.

A major advantage of the IGBT over all other power switches is the extremely low gate power required. The peak power is about ~5 W, with an average of much less than 1 W. This dramatically simplifies the delivery of gate power compared to the GTO/IGCT. Although there are more active devices in the Perfect Harmony (60 IGBTs and 60 diodes in the inverter sections) than in the other circuits, the elimination of snubbers, voltage sharing networks, and high-power gate drivers compensates for the additional switching devices. The type of IGBTs employed are third and fourth-generation isolated base modules, generally the same mature product as those found in 460 VAC and 690 VAC PWM drives, and that are also used in traction applications. The IGBTs are protected by an out-of-saturation detector circuit, which augments the built-in current limiting behavior. Since the cells are assembled into a nonconducting framework and are electrically floating, the mounting and cooling of the IGBTs is no more complex than in a low-voltage PWM drive. It is possible to put redundant cells in the string and also to operate at reduced output with one cell inoperative.

CYCLOCONVERTER

Still another approach in an IM drive is to "synthesize" an ac voltage waveform from sections of the input voltage. This can be done with three dual converters, and the circuit is called a cycloconverter

	Load Commutated Inverter	Neutral Point Clamped Inverter	Filter Commutated Inverter	GTO Inverter	Multilevel Series-Cell Inverter
Input Harmonics	Fair (12-pulse) Poor (6-pulse)	Good (12-pulse) V Good (18pls)	Fair (12-pulse) Poor (6-pulse)	Fair (12-pulse) Poor (6-pulse)	Excellent
Input Power Factor (uncorrected)	Poor	Very Good	Fair	Fair	Very Good
Output Harmonics	Poor	Good	Good (at full speed)	Fair	Excellent
Output Common-mode Voltage	High (fair)	None (excellent)	High (poor)	High (poor)	None (excellent)
Output dV/dt (unfiltered)	High (poor)	High (poor)	Low (good)	Low (good)	Low (good)
Regeneration Capability	Yes	No	Yes	Yes	No
Torque Pulsations	High (poor)	Low (very good)	Low (good)	Low (fair)	None (excellent)
Special Motor Required?	Yes	No	Yes (for common-mode voltage)	Yes (for common-mode voltage)	No
Speed Range (PU)	.15 -- 2.0	0 – 2.0	0.5 – 1.1	0 – 1.1	0 – 2.0
Special Starting Mode?	Yes	No	Yes	No	No

FIGURE 15   Comparison of medium-voltage motor drives.

(see Fig. 13). The output voltage is rich in harmonics but of sufficient quality for IM drives as long as the output frequency does not exceed 1/3 of the input frequency. The thyristors are line commutated, but there are 36 of them. The cycloconverter is capable of very heavy overloads and four-quadrant operation, but it has a limited output frequency and poor input power factor. For special low-speed high-power (>10 MW) applications, such as cement-kiln drives, the cycloconverter has been used successfully.

COMPARISON OF MEDIUM-VOLTAGE MOTOR DRIVES

All the types of drives mentioned above are capable of providing highly reliable operation at a justifiable cost, and they have been proven in service. They all have efficiencies above 95%. The most significant differences among them have to do with power quality, that is, how close to a sinewave is the input current, and how well does the output resemble the sinusoidal utility voltage. Figure 15 compares them on a number of different factors. Note that the voltage-fed drives have an advantage in input harmonics and power factor, and the drives, which do not use thyristors, have a wider speed range.

BIBLIOGRAPHY

Bedford, B. D., and R. G. Hoft, *Principles of Inverter Circuits*, Wiley, New York, 1964.

Bose, B. K., *Adjustable Speed ac Drive Systems*, Wiley, New York, 1981.

Brichant, F., *Force-Commutated Inverters*, Macmillan, New York, 1984.

Ghandi, S. K., *Semiconductor Power Devices*, Wiley, New York, 1977.

Kosow, 1. L., *Control of Electric Machines*, Prentice-Hall, Englewood Cliffs, New Jersey, 1973.

Pelly, B. R., *Thyristor Phase-Controlled Converters and Cycloconverters,* Wiley, New York, 1971.

Schaefer, J., *Rectifier Circuits: Theory and Design*, Wiley, New York, 1965.

Scoles, G. J., *Handbook of Rectifier Circuits*, Wiley, New York, 1980.

Sen, P. C., *Thyristor dc Drives*, Wiley, New York, 1981.

ROBOTS

Early industrial robots date back several decades. Initially they were used mainly to assist or take over dangerous or difficult handling operations, essentially to protect manual laborers from undue exposure to harmful substances, temperature, radiation, and so on. Research and development expense in such situations was relatively easy to justify, but all along the ultimate objective was to design robots that could perform manual tasks better, more cheaply, and more quickly than people. As with numerous other specialized equipment technologies, the robot, in concept, was far ahead of the components needed to enhance its performance, as have later emanated from advancements in solid-state electronics, computer controls, and communications. Very large strides in robot development have been made since the mid-1970s, particularly as the result of some piece- and parts-handling industries (automotive being a major example) to improve their competitive position in terms of increased productivity and product quality.

BASIC FORMAT OF ROBOT

It is not always easy to distinguish a mechanized handling machine from what is generally considered to be a robot. For example, a modern, complex conveyor system would meet some of the general descriptive criteria of a robot, but in professional parlance, a conveyor by itself would seldom be considered a robot. However, in terms of total robotic technology, one or many conveyors could be involved. A definition, coined several years ago by the Robot Institute of America, still provides a good definitive foundation, even though some of the words used are rather general and perhaps superconclusive. The definition is:

> A robot is a *reprogrammable, multifunctional* manipulator designed to *move* materials, parts, tools, or specialized devices through variable programmed *motions* for the performance of a variety of tasks.

In terms of their classification, robots may be considered from a number of viewpoints:

1. Axes of motion, including type of motion, number of axes, and the parameters of axis travel
2. Load capacity and power required, namely, weight of load, and electrically, pneumatically, or hydraulically operated
3. Dynamic properties
4. End-effectors or grippers used
5. Programming and control system
6. General-purpose or special task

AXES OF MOTION

A robot may be movable from one factory location to another, as may be required by factory layout changes or by major alterations in job assignment. However, for any given task that will be repeated over and over for long periods, a robot will be firmly fastened to the operating floor (sometimes the ceiling). The firm location establishes a fixed geometrical location of reference, an unchangeable position that will geometrically relate precisely with an associated machine, or in the case of a work cell, involving several other machines and often other robots.

For relatively moderate changes in the robot's working envelope, the average "stock" robot will incorporate considerable flexibility within its design so that changes can be made without altering the location of reference. Sometimes, in the case of a "smart" robot, final very small changes in the positioning of an arm can be made by outputs from a machine vision or tactile system.

Less frequently, a robot will be intentionally designed for movability so that it can be transferred to the worksite, rather than grouping one or more robots about specific locations, as will be mentioned later.

Degrees of Freedom

Designed or built-in axes of motion essentially define the robot's ability to move parts and materials, sometimes referred to as degrees of freedom. The axis of motion refers to the separate motion a robot has in its manipulator, wrist, and base. The designer usually will select from one of four different geometric coordinates for any given robot.

Revolute (Jointed-Arm) Coordinates.   In this system the robot arm is constructed of several rigid members, which are connected by rotary joints. Three independent motions are permitted (Fig. 1). These members are analogous to the human upper arm, forearm, and hand, while the joints are equivalent to the human shoulder, elbow, and wrist, respectively. The arm incorporates a wrist assembly for orienting the end-effector, in accordance with the demands of the workpiece (Fig. 2). These three articulations are pitch (bend), yaw (swing), and roll (swivel). In some applications, fewer than six articulations may suffice, depending on the geometry of the workpiece and the machine which the robot is serving.

Cartesian Coordinates.   In this system all robot motions travel in right-angle lines to each other. There are no radial motions. Consequently the profile of a Cartesian-based robot will have a rectangularly shaped work envelope (Fig. 3).

Some systems utilize rotary actuators to control end-effector orientation. Robots of this type generally are limited to special applications. A robot may incorporate rectilinear cartesian coordinates as, for example, a continuous-path extended-reach robot gains much versatility through a bridge and trolley construction, which enables the robot to have a relatively larger rectangular work envelope. When ceiling mounted, this system may service many stations with several functions, thus leaving

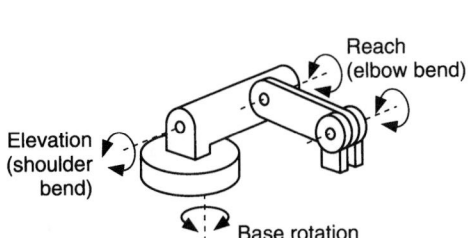

FIGURE 1 Jointed-arm manipulator, incorporating revolute coordinates.

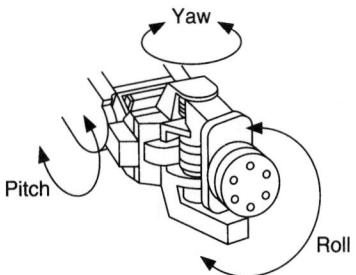

FIGURE 2 Wrist assembly on robot arm for orienting end-effector in accordance with requirements of workpiece.

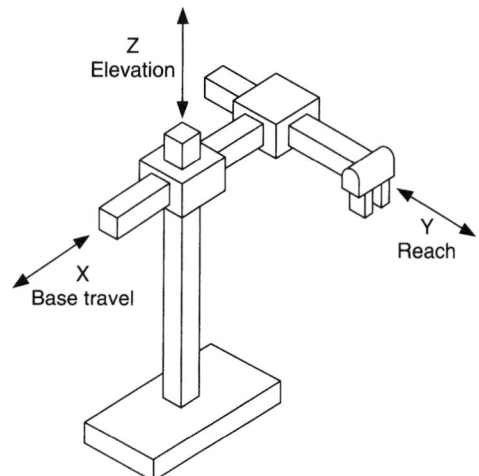

FIGURE 3 Manipulator incorporating Cartesian coordinates.

the floor clear. X and Y motions are performed by the bridge and trolley; the vertical motions are accomplished by using telescoping tubes.

In a Cartesian coordinate system the location of the center of the coordinate system is the center of the junction of the first two joints. Except for literally moving the robot to another factory location, the center does not move. In effect, it is tied to the "world" as if anchored in concrete. If the X measurement line points toward a column in the work area where the robot is placed, the X line will always point toward that same column, no matter what way the robot turns while performing its programs. These are known as the world coordinates for a given robot installation (Fig. 4).

In the operation of a robot, having an origin for a measurement reference is not sufficient. One also needs to know the point to which measurements will be made. This measurement is made from the origin of the coordinate system to a point that is exactly in the center of the circle, on which the tool (end-effector) is to be mounted. This system moves with the tool and is aptly called the tool coordinate system. In the tool coordinate system the X and Y lines lie at right angles flat on the tool-mounting surface. The Z line is the same as the axis of rotation for the point, that is, it points directly through the tool in one direction and through the wrist in the other direction. The system is *not* tied to the world. Instead it stays in position on the tool-mounting surface and moves wherever the tool moves.

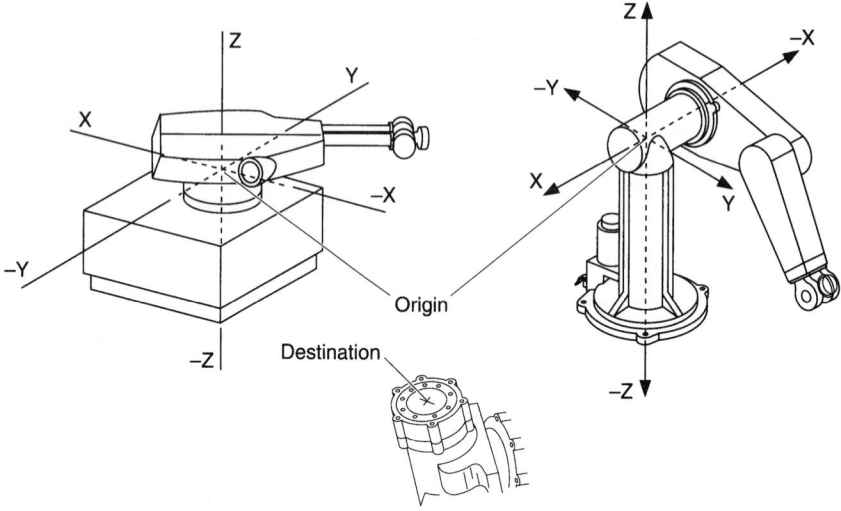

FIGURE 4   World coordinate system of robot using Cartesian coordinates.

While the origin of the system is thus allowed to move around, the destination (where it measures to) is left to the discretion of the user. Sometimes the tool coordinate system is actually used to measure where the tip of the tool lies relative to where it is mounted; sometimes it is used to measure where one position in space lies relative to some other point in space (Fig. 5).

Cylindrical Coordinates.   Robots designed with this system have a horizontal shaft that goes in and out and rides up and down on a vertical shaft. The latter rotates about the base (Fig. 6). Additional rotary axes are sometimes used to allow for end-effector orientation. Cylindrical-coordinate robots are often well suited for tasks to be performed on machines to be serviced that are located radially from

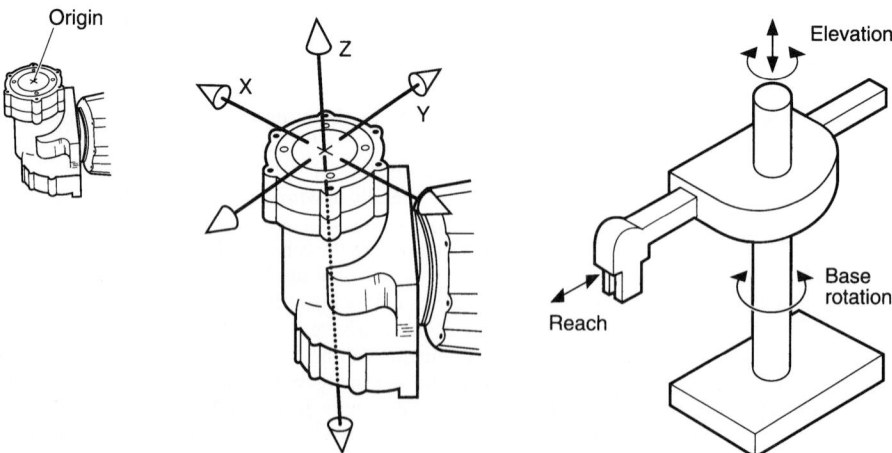

FIGURE 5   Tool coordinate system of robot using Cartesian coordinates.

FIGURE 6   Manipulator incorporating cylindrical coordinates.

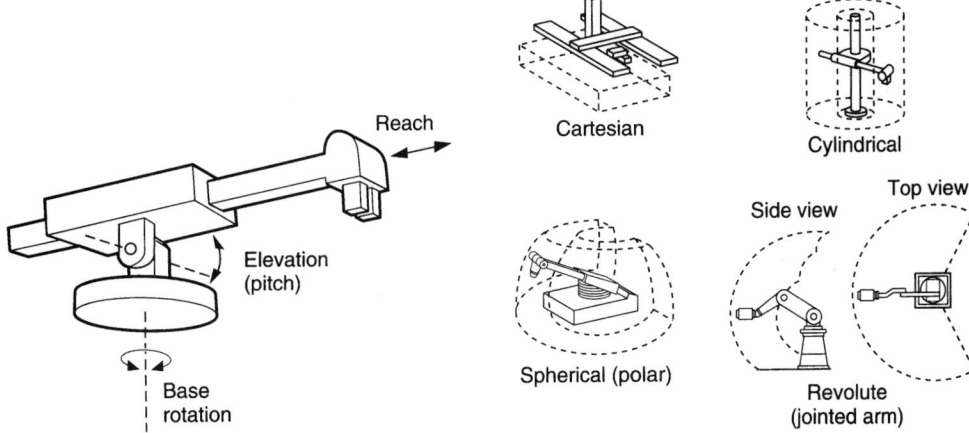

FIGURE 7   Spherical-coordinate manipulator, the operation of which is comparable to a tank turret.

FIGURE 8   Work envelope of a robot is that area in space which the robot can touch with the mounting plate on the end of its arm.

the robot and where no obstructions are present. A robot that incorporates cylindrical coordinates has a working area or envelope that is a portion of a cylinder.

Spherical (Polar) Coordinates.   Robots using this system may be likened to a tank turret, that is, they comprise a rotary base, an elevation, and a telescoping extend-and-reach boom axis. Up to three rotary wrist axes (pitch, yaw, and roll) may be used to control the orientation of the end effector (Fig. 7).

Work Envelope

The area in space that a robot can touch with the mounting plate on the end of its arm is known as its work envelope (Fig. 8).

LOAD CAPACITY AND POWER REQUIREMENTS

With need and proper design, robots can be designed to handle miniature (tiny) pieces that weigh a few ounces or grams, as found, for example, in electronics manufacture, up to heavy industrial loads ranging from 135 to 1045 kg (300 to 2300 lb) and even much greater loads where robotic equipment is used, for example, in earth-moving situations, as may be found in earthquake debris removal. A recent survey of user demand shows that robots lie within the range of 9 kg (20 lb) on the low side to 136 kg (300 lb) on the high side for the majority of applications. A majority of robots are electrically actuated by servomotors, particularly stepping and permanent-magnet dc motors, as described previously in detail in this handbook section. Less frequently used are pneumatic and hydraulic actuators.

DYNAMIC PROPERTIES OF ROBOTS

Important dynamic properties of robots include (1) stability, (2) resolution, (3) repeatability, and (4) compliance. Considering these factors, the design of a robot is innately complex because of the manner

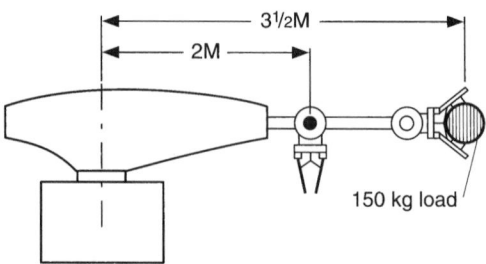

FIGURE 9 Consider a robot arm that has a retracted hand position of 2 m and an extended hand position of 3.5 m. Consider also that this arm might carry a load of 150 kg, and that the arm should go from position to position, with or without load, at any extension and without overshoot. For the configuration shown, the variation in moment of inertia is from 70 kg Msec^2 when tucked in and unloaded to 230 kg Msec^2 when fully extended and loaded. To achieve a critically damped servo with position repeatability of 0.5 mm under all operating conditions is difficult. Note that 0.5-mm resolution for an arm with 300° of rotation requires position encoding to an accuracy of 1 part in 33,000, or 2^{15}. The foregoing deals only with a major robot arm articulation. In a full arm the interactions among the various articulations complicate both dynamic performance and accuracy. For example, a robot arm designed to achieve an individual-articulation natural frequency of 50 Hz degenerates to an overall 17 Hz in a six-articulation arm. (*Westinghouse.*)

in which these properties interrelate. This also contributes to the difficulties of optimizing a design. Figures 9 through 12 illustrate specific examples of dynamic problems.

Stability

This characteristic is associated with oscillation in the motion of the tool. The fewer the oscillations that are present, obviously the more stable is the operation of the robot. Negative aspects of oscillations include the following. (1) Extra wear is imposed on the mechanical, hydraulic, and other parts of the robot arm. (2) The tool will follow different paths in space during successive repetitions of the same movement, thus requiring more distance between the intended trajectory and the surrounding parts. (3) The time required for the tool to stop at a precision position will be increased. (4) The tool may overshoot the intended stopping position, possibly causing a collision with some object in the system.

Oscillations may be damped or undamped. Damped (transient) oscillations will degrade and cease with time; undamped oscillations may persist or may grow in magnitude (runaway oscillation) and are the most serious because of the potential damage they may cause to the surroundings.

Variations of internal and gravitational loads on the individual joint servos (as the arm's posture changes) make the operation of the robot prone to oscillation.

In one approach to solving oscillation problems, the joint servos operate continuously. Some sophisticated servo designs (the result of experience from numerical control of machine tools) prevent oscillation from starting, regardless of the load carried. In another approach the robot controller locks each joint independently the first time it reaches its set point. Special circuitry also decelerates the joint after it comes within a prescribed distance of that position. When the joints are all locked (total coincidence), the arm is stationary, and it can then begin to move to the next position. If the position is held for more than a few seconds, the tool slowly creeps away from its programmed position.

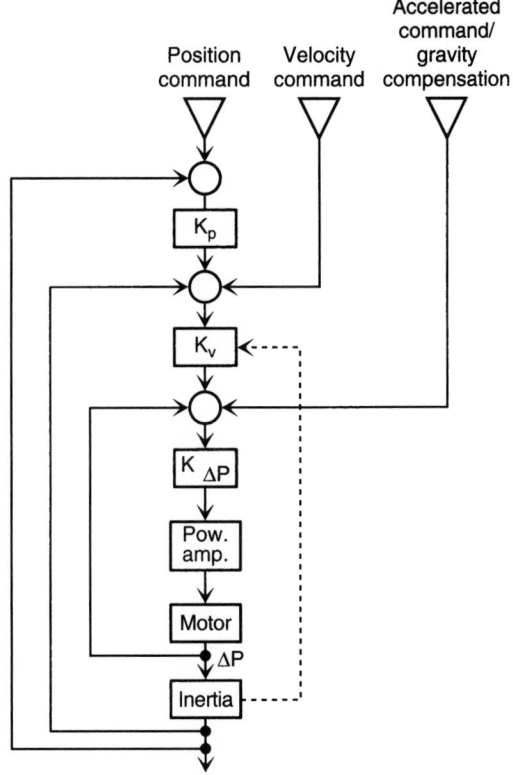

FIGURE 10   Block diagram for functionally describing key elements of a single-articulation servo system, including velocity and acceleration feedback and interarticulation bias signals. In estimating the time to complete a task (without actually simulating the entire process), the interface with the workplace complicates the process. Paths to avoid obstacles add program steps. Some steps must be very precise, calling for closing out to zero error before the program advances. Other steps may be the corners in a motion path, which can be passed through "on the fly" so to speak. The use of interlock switches may introduce transport lags. Simple programs often permit using a rule of thumb. For example, if one allows 0.8 second for each motion taught, short steps as well as long, a time for program completion can be estimated quite closely. However, if a program is complex, as in spot welding a car body, there are too many variables to permit the use of such methods. Other factors that must be included are weld-gun inertia, weld-gun operating time, metal thickness, proximity of spots to one another, among other critical variables.

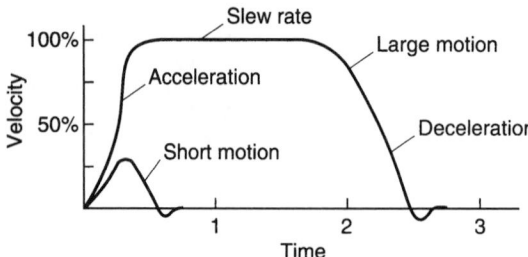

FIGURE 11   It is common for robots to be offered with abbreviated specifications that list the slew rates and the repeatability of each articulation. What is really needed is the total amount of time required to go from position to position and net accuracy of all articulations in consort. Shown here are two typical velocity traces for a short-arm motion and a large-arm motion. It is evident that the slew rate is no measure of elapsed time in making a motion, particularly a short motion in which the slew rate may not be attained at all.

Resolution and Repeatability

Repeatability is affected by resolution and component inaccuracy. Both short- and long-term repeatability exist in a robotic system. Long-term repeatability is of concern in robot systems that must perform tasks over a several-month period. During long-term repetitive use, components wear and age to the extent that repeatability must be checked periodically. Short-term repeatability is influenced most by temperature changes within the control and the environment as well as by transient conditions between shutdown and start-up of the system. These factors frequently are grouped under the umbrella term, drift. The accuracy of a robotic system can range from several hundredths of an inch for a simple robot to several thousandths of an inch for a robot doing precision assembly or handling small parts. In the case of a robot used in testing printed-circuit boards and other electronic manufacturing operations, the need for precision is paramount. Repeatability claims for standard or stock robots are listed in Table 1.

TABLE 1   Manufacturers' Claimed Repeatability of Randomly Selected Contemporary Robots

Load Capacity		Claimed Repeatability (±)	
lb	kg	inches	mm
5	2.2	0.004	0.1
14	6	0.004	0.1
22	10	0.008	0.2
35	16	0.001	0.03
66	30	0.002	0.05
110	50	0.020	0.5
132	60	0.020	0.5
150	68	0.020	0.5
176	80	0.020	0.5
200	90	0.010	0.25
264	120	0.04	1.0

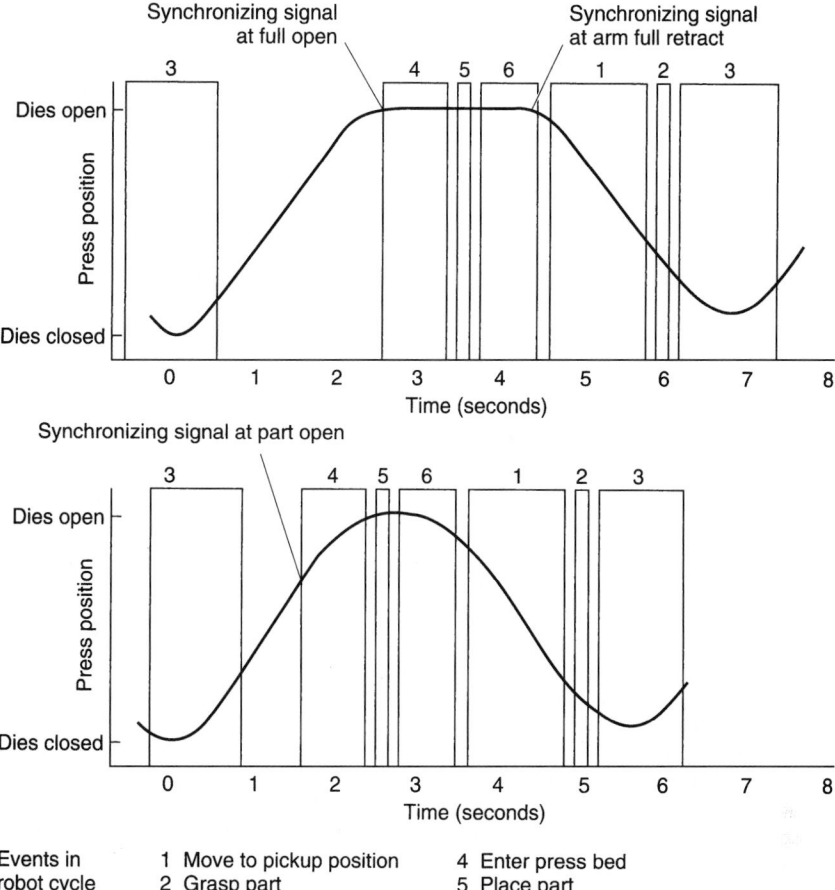

Events in   1 Move to pickup position   4 Enter press bed
robot cycle   2 Grasp part   5 Place part
                3 Transport part to press   6 Retract arm

FIGURE 12   For some operations, program time is critical, such as when a robot is serving heavy, expensive capital equipment. If the production rate is paced by the robot rather than the equipment, the project would not seem viable because of loss of throughput. Optimizing such a program may involve a range of techniques. A typical application might be press-to-press transfer of sheet-metal parts. A line of presses runs at a gross production rate of up to 700 parts/h. At this rate a robot must make a complete transfer and return for the next pickup in 5.16 s. With presses on center-to-center distances of 6 m, this is a demanding transfer speed. To meet this rate, a robot was modified by increasing the capacity of both hydraulic supply and servo valves. Acceleration and deceleration times were reduced at some sacrifice in damping and accuracy. This was compensated by providing the die nests with leads or strike bars. Finally, interlocks were refined so that the robot could make approaches and departures during the rise and fall of the moving platens of the press. The curves given here show how the time can be shortened by tight interlocks that do not wait for press-cycle operation. For safety, this approach cannot be used with human operators.

Compliance

The compliance of a manipulator is indicated by its displacement relative to a fixed frame in response to a force (torque) exerted on it. The force may be a reaction force (torque) that arises when the manipulator pushes (twists) the tool against an object, or it may be the result of the object pushing (twisting) the tool. High compliance means that the tool moves a lot in response to a small force, and the manipulator is then said to be spongy or springy. If it moves very little, the compliance is low and the manipulator is said to be stiff.

Compliance is a complex quantity to measure. In practice, a manipulator may be defined as a nonlinear, anisostropic tensor quantity that varies with time and with the manipulator's posture and motion. It is a tensor because a force in one direction can result in displacements in other directions and even rotation. A torque can result in rotation about any axis and displacement in any direction. A 6×6 matrix is a convenient representation for a compliance tensor. Time can affect compliance through changes in temperature and hence the viscosity of hydraulic fluid. Compliance will often be found to be a function of the frequency of the applied force or torque. A manipulator, for example, may be very compliant at frequencies around 2 Hz, but very stiff in response to slower disturbances. Compliance may exhibit hysteresis. For example, the servos in one design of hydraulic manipulator turn off when the arm stops moving. In this condition all the servo valves are closed, and the compliance has a value that is determined by the volume of incompressible hydraulic fluid trapped in the hydraulic hoses and the elasticity of the hoses. However, if an outside force on the tool should move any of the joints more than a given distance from the position at which they are supposed to remain, then the servos on all joints will turn on again. The compliance then changes to a completely different value (presumably stiffer in some sense).

Both electric and hydraulic manipulators have complex compliance properties. In an electric manipulator the motors generally connect to the joints through a mechanical coupling. The sticking and sliding friction in such a coupling and in the motor itself can cause strange effects on the compliance measured at the tool tip. In particular, some of these couplings are not very back-drivable. For example, if one pushes on the nut of a lead screw (backdrive), the lead screw will not turn unless the screw's pitch is very coarse and ball bearings are used between the threads to reduce friction. But one can turn the screw easily, and the nut will move.

For manipulators that (1) operate open-loop in the sense that (2) they go blindly to a given point in space, (3) without any regard to the actual position in the environment, or (4) without regard to any reaction forces (feedback) that those objects exert on the arm (or tool)—then less compliance than that of the surrounding objects would be an advantage. High-frequency oscillations can be filtered out without degrading the overall response. Such filtering actually requires no special effort inasmuch as the combinations of servo valves and actuators commonly used have relatively low bandwidths.

Tactile sensors, which measure forces and moments exerted on the tool, can allow the manipulator to track or locate objects. Even in such cases, however, oscillations may arise in the force-feedback control loop if the compliance at the point of sensing is too low (stiff). Examination of a particular servo design is needed to predict reliably whether it will provide the kind of compliance needed for a specific task. There is no substitute for an actual test with the real tool on the manipulator.

END-EFFECTORS (GRIPPERS)

The device that is fastened to the free end of a manipulator is known as an end-effector or gripper. The usual function of the device is to grasp an object or a tool and then hold it while the manipulator moves, thereby moving the object, and finally releasing the object. Many end-effectors are widely used and available from stock. Custom-designed end-effectors, however, are not uncommon where standard designs do not satisfy specific application needs. Normally the end-effector is not included in the price of a robot per se. Hence their cost often can create substantial additional expense. Mechanical clamps (grippers) are commonly used. Other forms include vacuum-operated holders and both permanent and electromagnetic holders (Fig. 13).

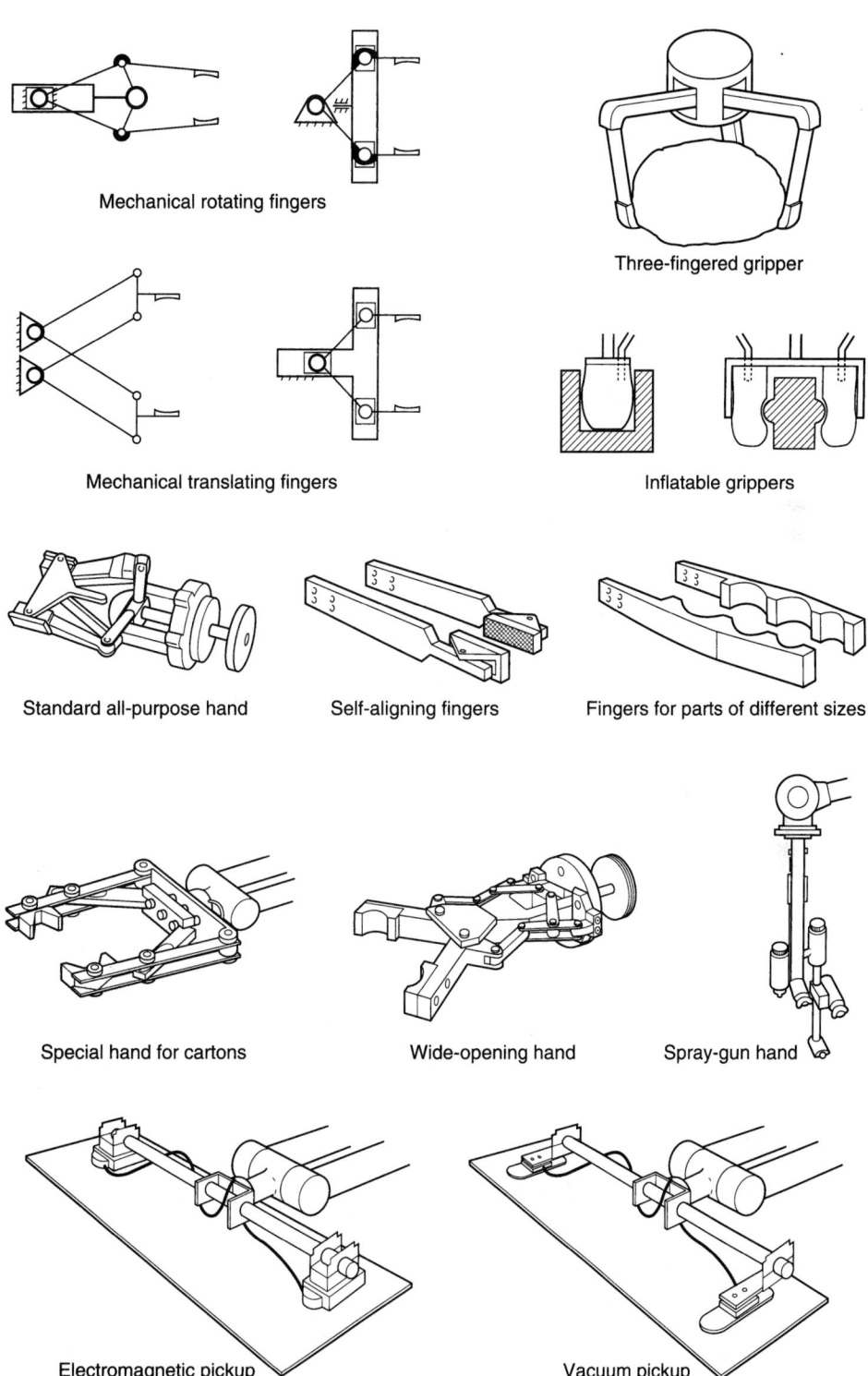

Mechanical rotating fingers

Three-fingered gripper

Mechanical translating fingers

Inflatable grippers

Standard all-purpose hand

Self-aligning fingers

Fingers for parts of different sizes

Special hand for cartons

Wide-opening hand

Spray-gun hand

Electromagnetic pickup

Vacuum pickup

FIGURE 13 Representative robot end-effectors.

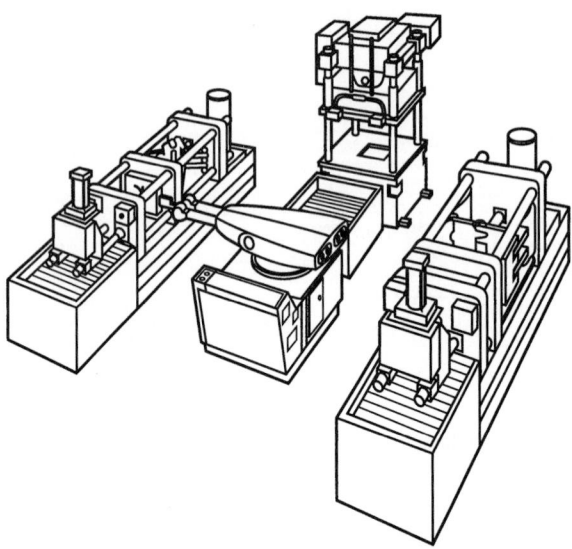

FIGURE 14 Die-casting installations to unload, quench, and dispose of part. In this installation, quite exemplary of earlier robot installations, the work is arranged around the robot.

WORKPLACE CONFIGURATIONS

There are four basic situations pertaining to the flow of work and the location of the robot:

1. Work may be arranged around the robot (Fig. 14).
2. Work is brought to the robot (Fig. 15).
3. Work may travel past the robot.
4. The robot travels to the work.

Work Cells

A robotic work cell may be defined as a cluster of two or more robots and several machine tools or transfer lines that are interconnected in such a way that they work in unison. All of the necessary accessory equipment is embraced within the work cell and, together, establishes a particular work environment. The cell level is one step higher than the station level in the hierarchy of control and command. Keeping very close supervision over statistical quality control has become a paramount consideration in instrumenting the cell-level concept (Fig. 16).

ROBOT PROGRAMMING AND CONTROL

The two general classifications of robots from a control standpoint are (1) nonservo robots and (2) servo-controlled robots. For noncritical, simple applications, nonservo robots may suffice, particularly where low cost is a major consideration. Most of the early designs were in this category. However, the wide acceptance of the servo-controlled robot, as technological advancements made them possible, are evident from the inspection of Table 2.

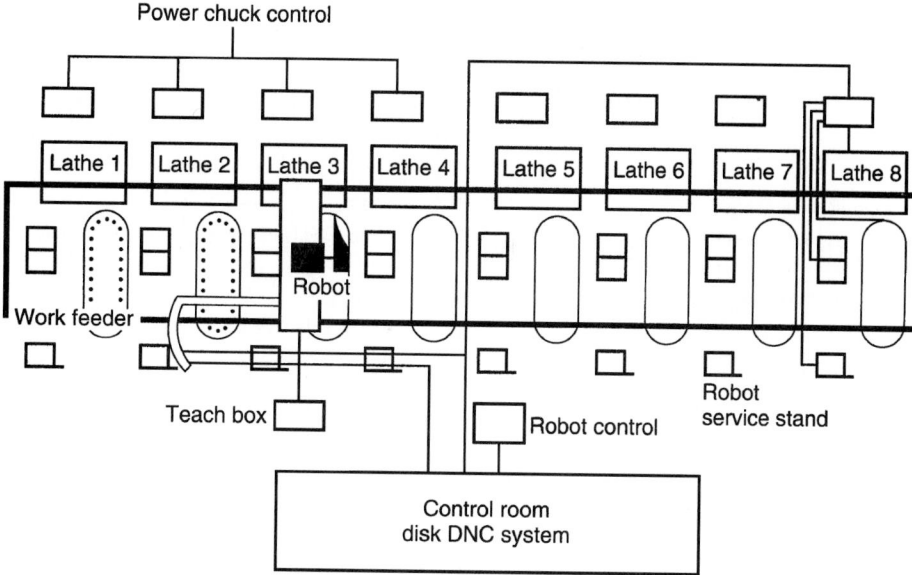

FIGURE 15   Overhead robot system, where the robot travels to the work. In this system an overhead robot system allows one robot to serve eight numerically controlled lathes.

Of historical interest, when robots and other automation techniques were largely associated with replicating the skills of human operators, the detailed steps and operations of the human operator were carefully studied and recorded. Initially this was the main source of robot programming information. Out of these studies the early "playback" concept was developed. In this method, the robot was "taught" by manually recording all of the movements that robot had to take to accomplish a given task. Obviously, at that time this method represented a "shortcut" because the path of the robot did not have to be measured or described in complex mathematical terms. Since those earlier years, of course, the techniques of mathematical simulation and appropriate motion algorithms have been developed. Much research along these lines has been conducted over the past decade or so by a combination effort made by robot designers and manufacturers, by large robot-using firms, and by academic institutions. For example, a pioneer in the field has been the Robotics Institute, Carnegie Mellon University, Pittsburgh, Pennsylvania, which initially developed a program known as VAST (versatile robot-arm dynamic simulation tool). As has become an accepted practice pertaining to instrumentation and control in the process industries, many robots have developed a high dependence on manufacturers and consultants for robot programming and software systems. In cases where applications from one user to the next may differ only in minor detail, packaged computer controls and software are now available and may be used with few, if any, alterations.

For example, large numbers of robot designs have been refined over several generations. Thus special controllers, visual operating displays, and software programs are available from robot suppliers. Standard applications that fit these criteria include those used in general materials handling, palletizing, arc welding, and, more recently, laser cutting, welding, etching, and surface-hardening applications.

Customized software for palletizing, for example, provides quick setup, easy modification of existing applications, and automatic calculation of all robot paths, eliminating the process of position teaching. In connection with robot welding applications, touch-sensing systems adaptively locate weld joints, and a through-arc seam-tracking system offers further enhancement by allowing the robot to compensate for weld-joint deviations and to correct the robot's path in real time.

TABLE 2 Nonservo versus Servo-Controlled Robots

Characteristic	Nonservo Robot	Servo-Controlled Robot
Flexibility	Limited in terms of program capacity and positioning capability. Arms can travel at only one speed and can stop only at end points of their axes.	Maximum flexibility provided by ability to program axes of manipulator to any position within limits of travel. Can vary speed at any point within envelope. Ability to move heavy loads in a controlled fashion.
Speed	Relatively high.	Relatively slow.
Repeatability	Approximately ±0.5 mm.	±0.1 to 0.5 mm and better, depending on design and application.
Cost	Comparatively low.	Comparatively high.
Complexity	Simple operation, programming, and maintenance.	Permits storage and execution of more than one program, with random selection of programs from memory via externally generated signals. Subrouting and branching capabilities may be available, permitting robot to take alternative actions within a program when commanded.

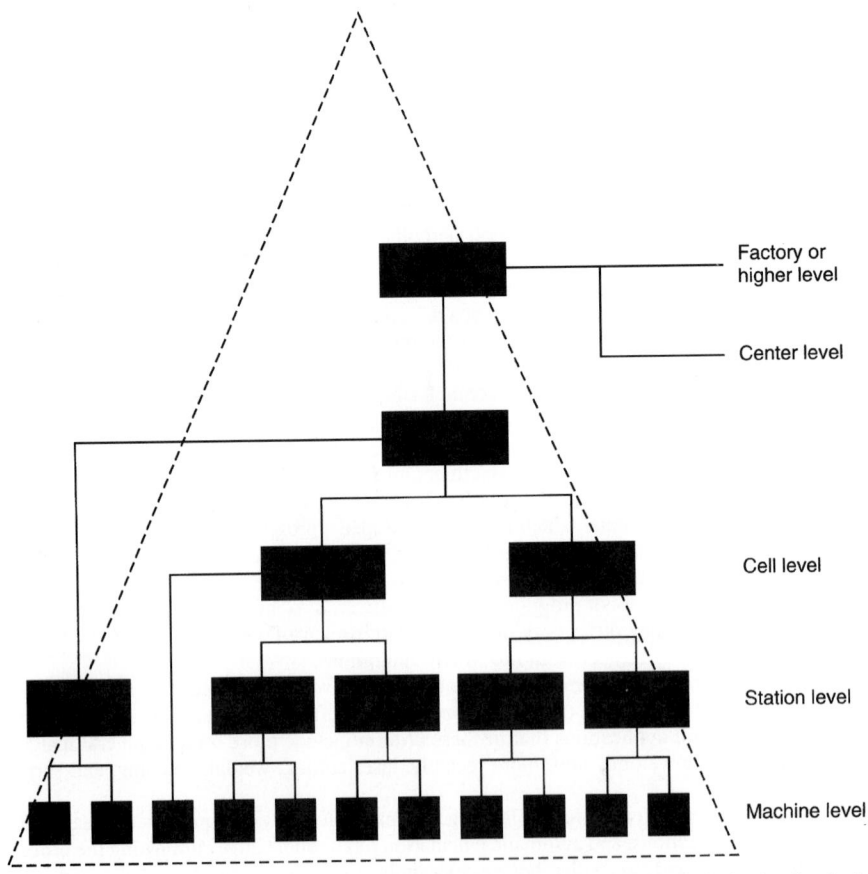

FIGURE 16 Pseudopyramidal hierarchy where communications are predominantly vertical rather than horizontal.

For parts and piece-dispensing operations, a software package minimizes the amount of code that is needed. For more complex operations, vision systems equip robots with advanced gray-scale vision capabilities. This is covered in more detail elsewhere in this handbook. Software also has been developed for painting applications and adds many teaching, editing, and programming capabilities. In these cases the robot can be controlled by a specially developed electronics package, or by using a teach pendant. A representative grouping of contemporary industrial robots is given in Fig. 17.

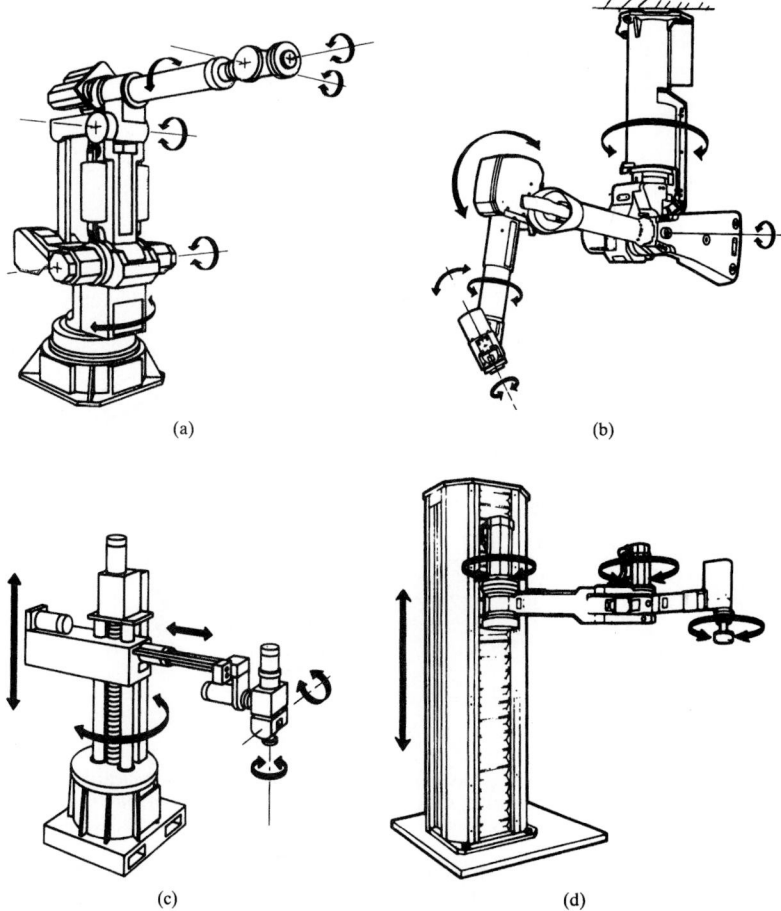

(a) (b)

(c) (d)

FIGURE 17 Representative industrial robots. (*Courtesy of GMFanuc Robotics Corporation.*)

(*a*) Spot welding, heavy part or tool handling, parts transfer, palletizing, material removal. Payload 120 kg (264 lb). Six axes of motion; floor- or wall-mounted; repeatability ±0.5 mm (0.02 in.); base rotation 300°; vertical travel 2731 mm (107.5 in.); reach 2413 mm (95 in.). (*S-420*)

(*b*) Arc welding of large parts on conveyors and fixtures. Payload 5 kg (11 lb). Six axes of motion; overhead-mounted; repeatability ±0.1 mm (0.004 in.); base rotation 300; reach 1309 mm (51.5 in.). (*ArcMate OH*)

(*c*) Material handling, machine loading, palletizing, mechanical assembly in severe environments. Payload 50 kg (110 lb). Three to five axes of motion; floor-mounted; repeatability 0.5 mm (0.02 in.); base rotation 300°; vertical travel 550 or 1300 mm (21.6 or 51.2 in.); horizontal travel 500 to 1100 mm (19.7 to 43.3 in.). (*M-100*)

(*d*) Palletizing and machine loading. Payload 50 kg (110 lb). Four to five axes of motion; floor-mounted; repeatability ±0.5 mm (0.02 in.); vertical travel 1850 mm (72.8 in.); radius reach 1930 mm (76 in.); access to two or more conveyors. (*M-400*) (*Continues*)

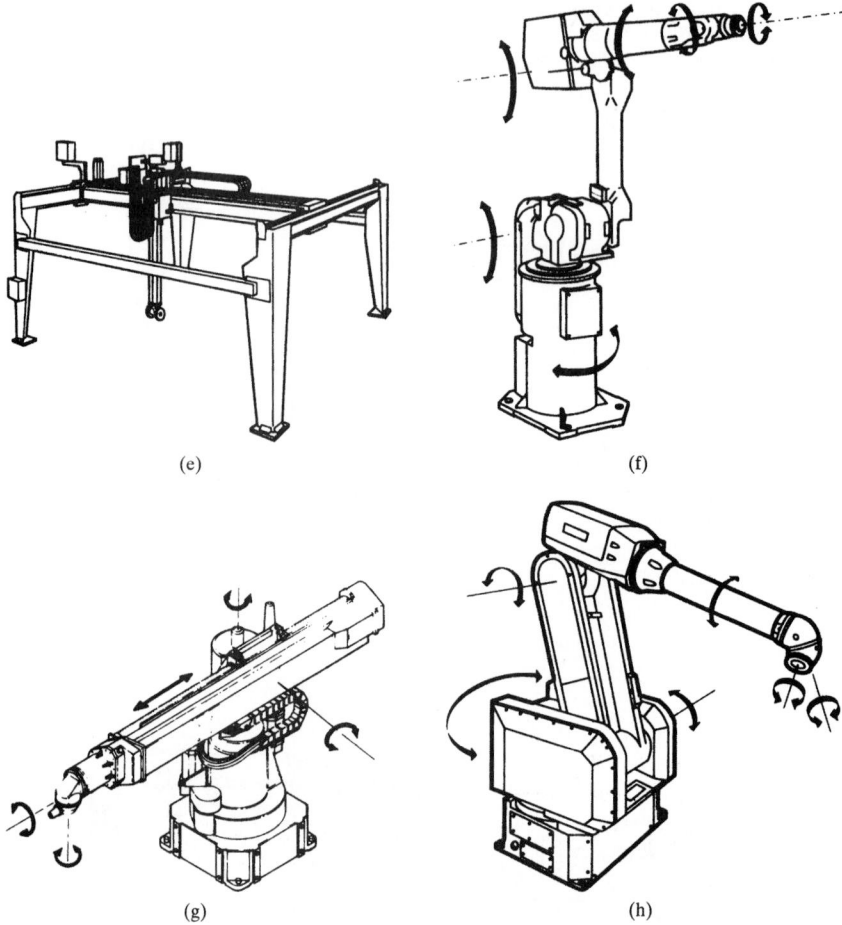

(e)   (f)

(g)   (h)

FIGURE 17   (*Continued*)

(*e*) Gantry robot for medium-to heavy-payload machine load and unload uses. Also palletizing, mechanical assembly, parts transfer. Cartesian coordinates. Area (shown) or linear configurations. Payload 50 kg (110 lb). Repeatability ±0.5 mm (0.02 in.); very large work envelope; two to four axes of motion for linear design; three to five for area design. (G-500)

(*f*) Multipurpose material handling; light-payload applications. Payload 10 kg (22 lb). Six axes of motion; floor-, ceiling-, or wall-mounted; repeatability 0.2 mm (0.008 in.); base rotation 300°; front reach 1529 mm (60.2 in.). (*S-10*)

(*g*) Laser robot for integration with a laser generator. For precision-path laser processing-welding, cutting, heat treating, and cladding. Payload 5 kg (11 lb). Five axes of motion; floor-mounted; antibacklash drive; repeatability ±0.05 mm (0.002 in.); base rotation 200; vertical travel 1968.5 mm (77.4 in.); horizontal travel 3964 mm (156 in.). Complete robotic laser cells available. (*L-100*)

(*h*) Industrial and automotive paint finishing of stationary or moving parts. Payload 7 kg (15.5 lb). Six or seven axes of motion; floor-or rail-mounted; repeatability 0.5 mm (0.02 in.); maximum reach 2613 mm (103 in.), large work envelope. Robot also used for dispensing and applying antichip sealers and underbody deadeners. (P-155)

CURRENT-TO-PRESSURE TRANSDUCERS FOR CONTROL-VALVE ACTUATION

by Len Auer[1]

Current-to-pressure transducers (I/Ps) are used primarily in process control to change a 4-to 20-mA electronic signal from a computer controller into a 3-to 15-psi (21-103 kPa) pneumatic signal. The output signal from the I/P is then used to fill a diaphragm or piston actuator, which, in turn, modulates a valve. An effective I/P must provide air to the receiver quickly, accurately, and in sufficient quantity. The I/P device also must be able to exhaust air quickly when the signal decreases, consume a minimum amount of supply air for operation, and be easy to repair. In most industrial applications the I/P also must be sufficiently rugged to withstand difficult environmental conditions, including vibration, dirty supply air, temperature extremes, and corrosive conditions.

Since the mid-1960s I/Ps have used the traditional flapper-nozzle design concept with relatively few alterations. However, in the late 1980s several I/P manufacturers introduced new technologies in an attempt to overcome some of the difficulties encounted with the traditional design. These technologies vary in approach, and application success often depends on the integration of several design features within the same I/P.

TRADITIONAL FLAPPER-NOZZLE DESIGN

The traditional flapper-nozzle design is shown in Fig. 1. The input current (4 to 20 mA) is applied to a coil-armature arrangement that acts on a beam. The beam ("flapper") positions itself against a nozzle that has air flowing through it. The gap between flapper and nozzle determines the backpressure, also called pilot pressure, that builds up in the nozzle. Other variations of the flapper-nozzle concept are shown in Fig. 2. A bellows sometimes is connected to the nozzle area to balance the forces on the

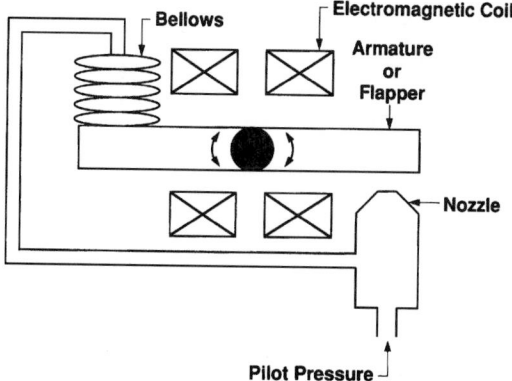

FIGURE 1   Traditional flapper-nozzle design. (*Rosemount Inc.*)

---

[1] Product Marketing Manager, Rosemount Inc., Eden Prairie, Minnesota.

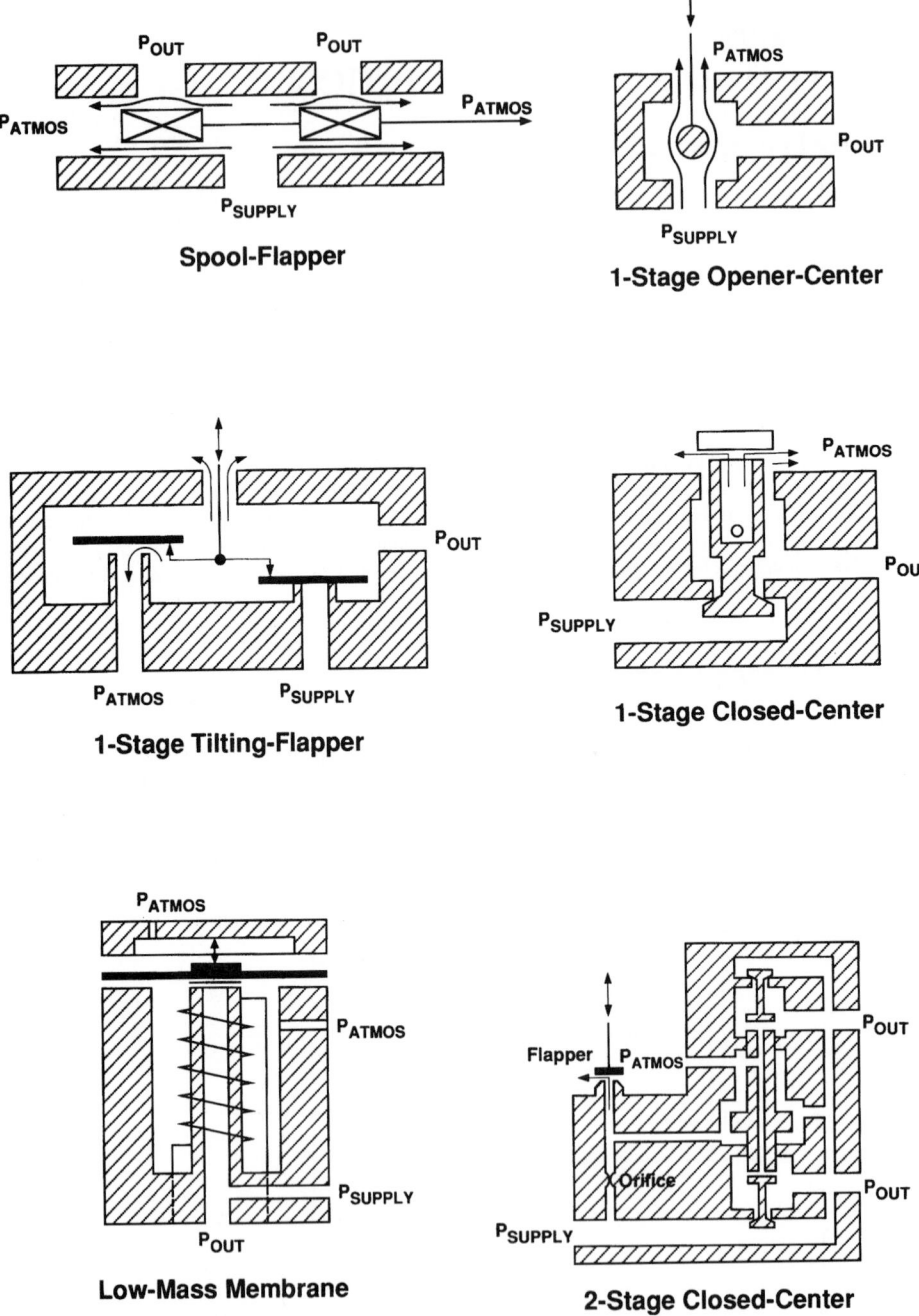

FIGURE 2 Variations of flapper-nozzle concept. (*Rosemount Inc.*)

armature-flapper. The pilot pressure usually is channeled to a pneumatic booster or relay. This booster translates the low pilot pressure into a higher output pressure and capacity, which typically is 3 to 15 psi (20 to 100 kPa) and 4 ft³/min (0.11 m³/min), respectively.

I/Ps that use the flapper-nozzle principle alone may have difficulty with some of the environmental factors faced in an application. Such I/Ps essentially are mechanical in nature and do not use electronic feedback sensors. The flapper is susceptible to vibration and traditionally has forced users to mount the I/P separately on a pipe or rack. This requires additional tubing to carry the I/P output signal to the valve. Output tubing installation costs offset any benefits from mounting the I/Ps together in a common location. The dead time and lag time introduced into the loop by longer output signal tubing can have a significant impact on loop performance. In addition to vibration, traditional I/Ps also can be adversely affected by fluctuations in air supply, downstream tubing leaks, temperature changes, and aging of the magnetic coil within the I/P. Periodic calibration checks are required in order to maintain the output of the I/P within the desired range.

Dirty supply air can be a major cause of I/P downtime. While mechanical I/Ps do not have electronic feedback to compensate for partial plugging, the nozzle opening traditionally has been designed to be at least 0.015 in. (0.4 mm) in diameter. This has reduced the likelihood of the nozzles plugging and is a strong point for those I/Ps that have maintained the larger nozzle diameters.

INTRODUCTION OF NEW I/P CONCEPTS

Since the late 1980s several new technologies have been introduced within the I/P. These new concepts have changed the nature of the pilot stage and incorporated sensor-based electronics. There are some inherent tradeoffs when combining different versions of these new technologies, and they have met with mixed results in the field. Two concepts are described here.

Piezoceramic Bender-Nozzle.   This device, another variation of the flapper concept, is shown in Fig. 3. The unit does not use the coil to move the flapper, but instead, the flapper itself is made of layers of different materials which are laminated together. These different materials flex or bend when a voltage is applied across them. The 4- to 20-mA input signal to the I/P must be converted to a voltage in the range of 20 to 30 V dc.

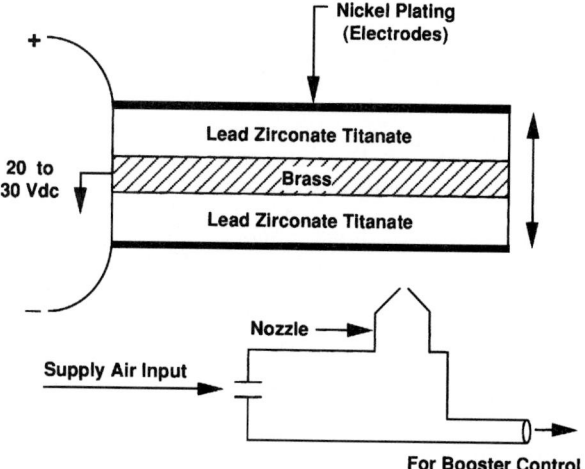

FIGURE 3   Piezoceramic bender-nozzle. (*Rosemount Inc.*)

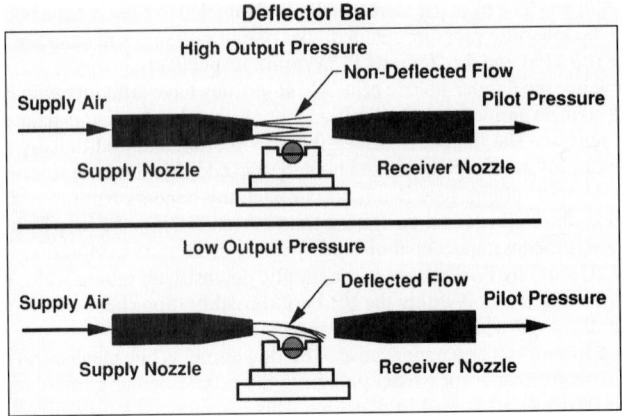

FIGURE 4 Deflector bar design. (*Rosemount Inc.*)

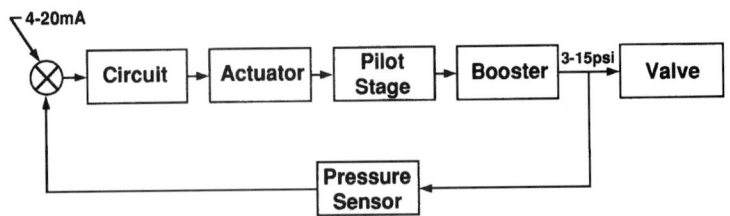

FIGURE 5 Electronic feedback control as applied to I/P device. (*Rosemount Inc.*)

This design tends to be more stable in vibration than the typical flapper armature, particularly when combined with an electronic feedback loop. Several drawbacks also have become evident from field applications. The bender does not have a very good "memory" and will tend to locate in a different position for the same input signal. This creep can be cumulative and eventually will exceed the adjustment range of the calibration mechanism. An electronic feedback sensor can be combined with the piezoceramic bender to compensate for the creep temporarily, but the feedback circuit typically uses much of the power available from the input signal. This leaves little power to energize the bender. The bender cannot balance against the force of the nozzle air, unless the nozzle is kept relatively small. Thus larger nozzles must be traded for improved bender control. Plugging of small nozzles or orifices typically is the leading cause of I/P field failure.

Deflector Bar Design.   Shown in Fig. 4, this pilot stage concept also was introduced in the late 1980s. This design uses an electromagnetic coil similar to the traditional flappers. The deflector bar, however, replaces the flapper as the main moving part in the assembly. The flapper no longer is used to block the airflow coming out of the nozzle. The deflector bar design is based on the Coanda effect, which may be defined as the tendency of an airstream to attach itself to a surface with which it makes oblique contact.

The actual hardware consists of two opposed 0.015-in. (0.4-mm)-diameter nozzles, fixed on the same centerline, spaced about 0.15 in. (0.4 mm) apart. One nozzle provides a high-velocity airstream from the air supply, the second nozzle recaptures the airstream and converts its kinetic energy to a pressure (potential energy).

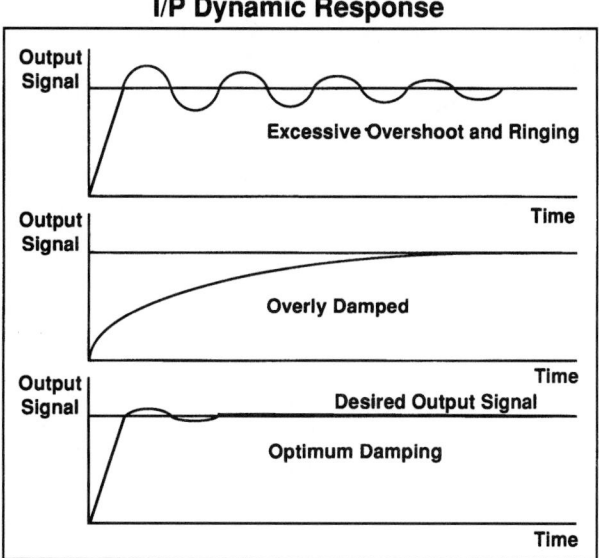

FIGURE 6   Possible variations in dynamic response of I/P devices. (*Rose-mount Inc.*)

To vary the pilot stage output pressure, a 0.019-in. (0.5-mm)-diameter solid deflector bar, which is positioned crosswise and midway between the nozzles, is caused to move into the airstream. The stream attaches to the surface of the bar and follows its curvature for some distance before separating. This deflection of the airstream away from the receiver nozzle results in a decrease in the pilot output pressure. To increase the output pressure again, the bar is simply pulled out of the airstream.

The deflector bar is low mass, which adds vibration stability, and only travels about 0.003 in. (0.08 mm) to produce a full-scale output change. This small travel requirement, coupled with the fact that the bar movement does not directly oppose the airflow, allows for a low-power magnetic actuator coil. The nozzle diameter is not limited by force-balance versus power tradeoffs, as is the case with flapper designs. The nozzle diameter is relatively large at 0.016 in. (0.15 mm), and is only limited by the desirable range for air consumption.

When combined with an electronic feedback sensor, this type of pilot stage is virtually unaffected by vibration and provides quick response to input changes. Lag time, the rate of change to reach a new output pressure, actually can be reduced by the quick response, as compared with other pilot stage designs.

ELECTRONIC FEEDBACK

Several I/Ps introduced since the late 1980s have incorporated pressure sensors and electronic feedback control. This feedback, shown in Fig. 5, detects the actual output pressure and is completely internal to the I/P. There are several types of sensors used, the most common of which is a solid-state silicon strain-gage type. The electronics in the I/P contain an error-correction circuit that continuously compares the output sensor reading with the input signal. The electronics then adjust the current to the pilot stage to make any needed corrections to the output of the I/P.

Electronic pressure sensor feedback is the foundation on which the installed advantages of recent I/Ps are based. These advantages include vibration immunity, calibration stability, repeatability, quick dynamic response, and reduced downtime. Performance of the electronic feedback circuits is relatively consistent across the I/Ps available as of the early 1990s. However, many of the advantages just given are contingent on integration of the electronic feedback with the optimum pilot stage design. If tradeoffs are made in order to incorporate electronics into the I/P, reliability can be adversely affected. In addition, dynamic response may vary considerably, as shown in Fig. 6. In particular this can be evident when the same I/P design is used to fill a wide range of output end volumes. Proper balance of damping and responsiveness is critical to the operation of the I/P into the full range of typical output volumes. The importance of the role played by the I/P in the performance of a loop cannot be overemphasized.

SECTION 10
PROCESS CONTROL IMPROVEMENT*

James F. Beall IV
Eastman Chemical Company, Longview, Texas

William L. Bialkowski
EnTech Control Engineering Inc., Toronto, Ontario, Canada

Jimmy G. Converse
Sterling Chemicals, Inc., Texas City, Texas

Mark T. Coughran
Fisher Controls International, Inc., Marshalltown, Iowa

John Edwards
Process NMR Associates, LLC (Foxboro), Danbury, CT 06810

Gregory Gervasio
Process Analytical Technology, Solutia Inc.,
St. Louis, Missouri

Tony Harding
Spectrace (Division of Thermo Instrument Systems),
Fort Collins, Colorado

J. B. Klahn
Applied Automation, Elsag-Bailey, Bartlesville, Oklahoma

K. K. Konrad
INTEK Corporation, Houston, Texas

Paul Luebbers
Solutia Inc., Cantonment, Florida

Gregory K. McMillan
Solutia Inc., St. Louis, Missouri

Michael J. Pelletier
Spectroscopy Products Group, Kaiser Optical Systems, Inc.,
Ann Arbor, Michigan

R. J. Proctor
GAMMA-METRICS, San Diego, California

Joseph P. Shunta, P. E.
E. I. du Port de Nemours, Wilmington, Delaware

Persons who authored complete articles or subsections of articles, or who otherwise cooperated in an outstanding manner in furnishing information and helpful counsel to the editorial staff.

Joseph Zente

Epsilon Industries, Austin, Texas

WORLD-CLASS MANUFACTURING

by Joseph P. Shunta, P. E.*

INTRODUCTION

Companies competing in a global market are constantly under pressure to reduce costs and improve quality. One of the ways to gain competitively is by increasing the productivity of manufacturing operations. This is an area to which process control can bring substantial benefits by improving quality of products, increasing yields, production rates, and uptime, and decreasing cycle time.

However, this does not happen automatically by simply installing the most modern control equipment as we experienced when distributed control systems became available in the 1980s. Many companies installed distributed control systems, expecting improved performance, only to find later that it was not significantly better than before. One of the reasons was that the control strategies had not been improved; they just duplicated the old analog systems. What is needed is to take advantage of the power of the new digital systems and upgrade the control strategies to gain concrete business benefits, not just to ensure stable operation of equipment. Companies are now going back and reevaluating how process control is applied and looking for ways to increase productivity through improved control (Fig. 1). This section presents a methodology for analyzing a process and identifying where process control can be improved in order to achieve world-class performance in manufacturing.

MODEL FOR IMPROVING PROCESS CONTROL TO ACHIEVE BUSINESS BENEFITS

Figure 2 is a model for how process control should be applied or improved to gain business benefits. Starting with a manufacturing process, product is made and shipped to a customer. The customer can be either external or internal. The customer uses the product and feeds back how well the product performed. The feedback may be in terms of the quality, price, or availability of the product.

Customer feedback is then transformed into internal business metrics that will be used to monitor and drive improvements in process performance. Examples of these metrics are process capability and process performance indices (C_p and P_p, respectively) for quality, first-pass, first-quality yield, throughput, percentage of uptime, and cycle time; these are defined in Figure 3. These metrics must be achieved in order to meet the customer and business requirements. Technical programs should be aimed at achieving these goals including process control improvement programs.

However, these metrics are at too high a level to apply directly to process control so another transformation must take place. This transformation identifies the key product properties and process variables along with their allowable ranges of variability. The idea is that, by controlling at the required targets and within specified ranges of variability, the business metrics are achieved and ultimately the customer needs are met as well. The measurements and control strategies are assessed to determine how well the controls achieve these operating goals, and improvements are identified when the controls fall short.

---

* Principal Consultant, E.I. DuPont de Nemours & Co., Wilmington, Delaware 19898.

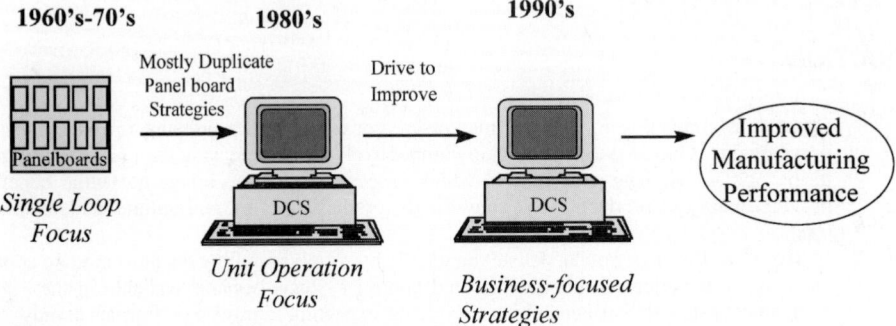

FIGURE 1   Evolution of control capability.

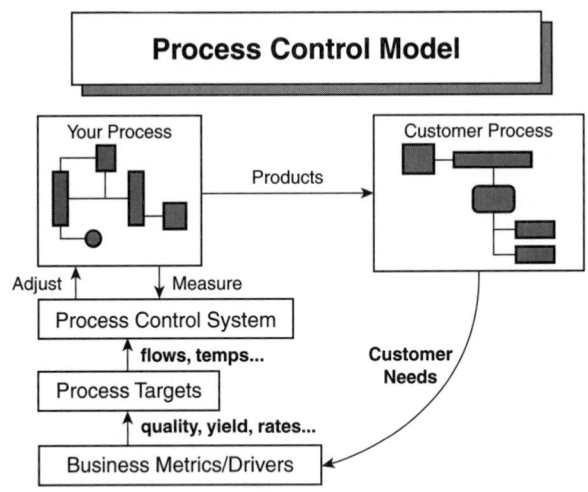

FIGURE 2   Process control model.

ANALYSIS TO IDENTIFY CONTROL IMPROVEMENTS

The first step in Fig. 2 is to receive feedback from the customer and transform it into specific business metrics or goals in terms of product quality, yield, throughput, uptime, and so forth. This task requires a close working relationship with the customer and falls largely on operations management but could include technical specialists as well. The subsequent tasks are, however, clearly within the realm of the control engineer. A methodology is shown in Fig. 4 that identifies what control improvements are necessary to achieve the business goals. The benefits of following a disciplined approach are shown in Fig. 5.

Key Business Drivers

- Use as metrics to assess performance
- Quality = product properties conform to customer wants, measured by C_p and P_p indices
- % First pass, first quality yield = make product right the first time
- Throughput = rate at which product is made
- Cycle time = elapsed time from receiving raw materials to shipping finished product
- % Uptime = time process is available to run at full rates

FIGURE 3   Key business drivers.

Process Control Analysis

- **Identify Business Drivers/Goals**
- **Identify Key Product Properties and Process Variables that Impact Business Drivers**
- **Identify Key Variables that have High Variability**
- **Assess Process Measurements and Control Strategies to Identify Improvement Opportunities**
- **Estimate Stakes and Feasibility and Prioritize**

FIGURE 4   Process control analysis.

Benefits Of Process Analysis

Identifies Business-focused Improvements
....ensuring results will impact business drivers

Provides Direction for Implementation
...through control strategy designs

Provides Basis for Economic Expectations
...so results can be clearly quantified

Provides Basis for Continuous Improvement
...to hold the gains

FIGURE 5   Benefits of process analysis.

Identify Key Product Properties and Process Variables

In the first step of the analysis, the business metrics are transformed into key product properties and process variables (flows, pressures, temperatures, etc.) that must be measured and controlled to ensure that

- product quality meets the customer specifications
- production rates, uptime, and cycle time are achieved so that product orders are filled on time
- first-pass, first-quality yields are achieved to minimize rework and costs

In other words, the key product properties and process variables are the ones that have the greatest impact on meeting the business goals. Prime examples of process variables are reactor temperature, reactor feed ratios, and distillation column temperatures. Product properties are things like denier and viscosity and may be measured and controlled on line or may be measured in the laboratory and not controlled directly. In the latter case, we have to identify the process variables that have the greatest impact on the product property. In addition to identifying the key variables, the associated steady-state targets, or set points, and the allowable ranges of variability must be determined. It is then the job of the controls to maintain the key variables at their respective targets within the prescribed ranges of variability.

Often the key variables are fairly obvious, but there are some tools that can be applied when the key variables are not so obvious. Two useful tools are modeling and design of experiments. Various types of steady-state modeling can be used:

- first-principle models
- empirical models such as statistical (regression) models or neural networks.

First-principle models describe mathematically the chemistry, physics, material, and energy balances of the process. Variables can be selectively varied to see the effect on the product, yield, etc. Empirical models are based on operating data and relate how changes in the variables affect the process outputs. Design of experiments is a procedure for testing the operating plant to see the actual effects of changing variables. This process is often complex and time consuming and may result in off-quality product being generated during the testing. However, the payback gained in increased process understanding and improved control is often well worth the effort.

Identify the Need for Improved Measurement and Control

The second step in the analysis identifies which of the key variables are in need of improved control. This is done by assessing variability since reducing variability is the basis on which we will achieve improved performance, as shown in Fig. 6. There are basically three ways we can reduce variability, as shown in Fig. 7; these are explained below.

First, an understanding of the types of variability and their sources is necessary to determine how to reduce it. Special causes of variability are gross upsets in the process. Examples are large changes in raw-material composition, operating blunders, and equipment malfunctions. It is the job of statistical process control (SPC) to identify when a special cause has occurred so it can be identified and removed. Automatic process control (APC), on the other hand, compensates for these upsets by adjusting a process variable to counteract the effect. What actually occurs is the controls transfer

---

Process Variability Impacts Business Goals

- Better consistency in product properties
  > product quality

- Tighter conformance to desired setpoints
  > Yield

- Run closer to constraints > throughput

- Faster transitions from one grade to
  another > cycle time

- Avoid process conditions which lead to
  fouling, etc. > uptime

FIGURE 6   Impact of process variability on business goal.

Ways to Reduce Variability

- Make frequent adjustments to process by
  automatic controls > APC

- Monitor key variables to signal special
  cause, identify cause and eliminate it > SPC

- Change the process

FIGURE 7   Ways to reduce variability.

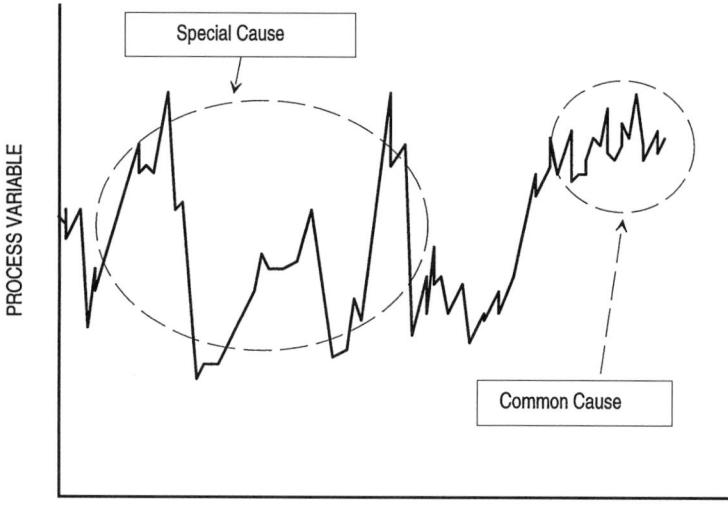

FIGURE 8   Special-cause and common-cause variability.

the variability from where it matters to where it does not matter. For example, to maintain a reactor temperature, the controls will transfer the variability to reactor coolant flow and temperature. If feasible, chronic special causes should be removed rather than rely on control to compensate for them. However, the analysis described here identifies the improvements in process control that are needed to compensate for the special causes.

The other type of variability is called common-cause variability and is inherent or built in to the process. This kind of variability is frequent, random, and generally small, so it is sometimes referred to as process noise. Some common causes are turbulence, machine vibration, steam supply pressure variations, and small random variations in raw-material composition. Process control is not recommended to compensate for common-cause variability because it may result in over-control and create instability. A better approach is to identify the cause if it cannot be ignored and make a process change to correct for it. For example, changing the process design, replacing equipment, and adding damping or filtering to the measurement to remove the variability are often remedies. Figure 8 graphically illustrates the difference between special-cause and common-cause variability.

Statistical Metrics

Identifying where process control can be improved is a matter of assessing the magnitude of the special causes of variability. The common statistical metric for variability is the standard deviation. One form of the standard deviation measures variability from all sources, special and common. This is called the total standard deviation, S_{tot} (Fig. 9). Another form measures only common-cause variability and is called the capability standard deviation, S_{cap} (Fig. 10). S_{cap} gets its name because the variability remaining after all the special-cause variability has been removed is the minimum variability the process can achieve, that is, the capability of the process. Note that S_{tot} takes differences between the mean and each data point whereas S_{cap} takes differences between adjacent data points. This form of S_{cap} is the mean-squared successive difference (MSSD) formula. If the process had no special causes occurring, S_{tot} would be equivalent to S_{cap}. Statisticians refer to a process in which only common-cause variability is present as being in the state of SPC or being stable. Note that the meaning of stability to

Standard Deviation and Mean

- Standard Deviation

$$S_{tot} = \sqrt{\frac{(X_1-\overline{X})^2 + (X_2-\overline{X})^2 + ... + (X_n-\overline{X})^2}{n-1}}$$

- Mean

$$\overline{X} = (X_1 + X_2 + ... + X_n)/n$$

FIGURE 9 Standard deviation and mean.

Mean Square Successive Difference MSSD

$$S_{cap} = \sqrt{\frac{(X_2-X_1)^2 + (X_3-X_2)^2 + ... + (X_n-X_{n-1})^2}{2n-1}}$$

FIGURE 10 Mean-square successive difference.

Process Capability Index

$$C_p = \frac{\text{Upper Specification - Lower Specification}}{6\,(S_{cap})}$$

For one-sided specification or the average is not midway between the specification range:

$$C_{pk} = \frac{|\text{average - nearer specification}|}{3\,(S_{cap})}$$

FIGURE 11 Process capability index.

Process Performance Index

$$P_p = \frac{\text{Upper Specification - Lower Specification}}{6\,(S_{tot})}$$

For one-sided specification or the average is not midway between the specification range:

$$P_{pk} = \frac{|\text{average - nearer specification}|}{3\,(S_{tot})}$$

FIGURE 12 Process performance index.

a control engineer is different: Stability means that variations in a process are not increasing without bounds.

Standard deviation has units of degrees, pounds per square inch, pounds per hour, etc., depending on the process variable being measured. To assess variability, it is more convenient to normalize the variability to remove engineering units. This allows us to compare variability of temperature, pressure, etc., on the same basis. The statistical metrics for normalizing variability are the process capability indices C_p and C_{pk} (Fig. 11) and the process performance indices P_p and P_{pk} (Fig. 12). The standard deviations are related to the allowable variability ranges (specifications) that are in the numerator of the index. The range may be a product specification range for a quality variable or the allowable variability range for a process variable. Again, the ranges are chosen to meet the business goals. C_p and P_p assume that the mean of the variable (also called average) is at the midpoint of the range. C_{pk} and P_{pk} are used when the variable is not midrange, and it is important to reflect that, or if the specification is one sided, that is, there is either a high specification or a low specification.

The desired value for the indices depends on the needs of the customer but as a rule, a value of 2.0 is considered world class. A value above 1.5 is also considered very good. Values of approximately 1.0 are just adequate and below 1.0 are poor. Figure 13 illustrates graphically the effect of the numerical

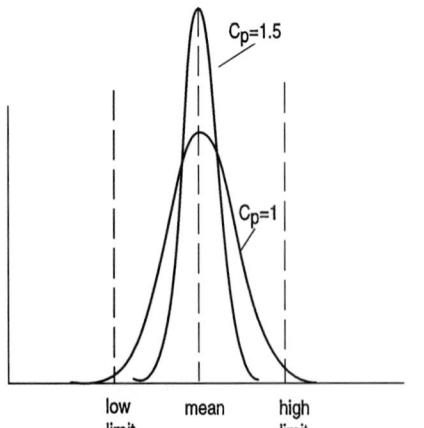

FIGURE 13 Histogram comparing C_p's.

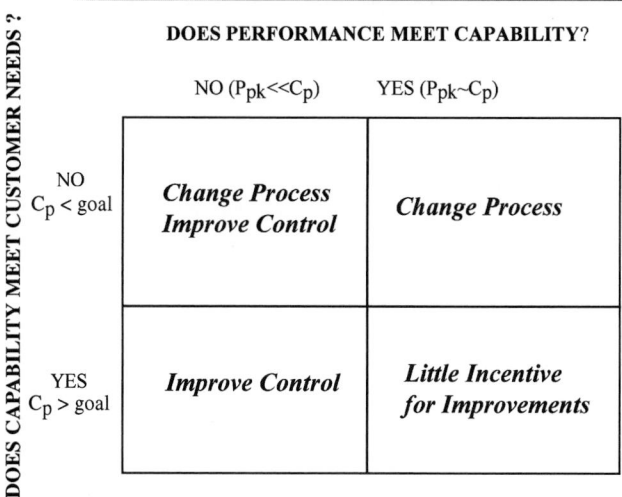

FIGURE 14   The capability and performance matrix.

value of the index. A value of 1.0 means that 99.73% of the data are within the specification if the mean is a midrange but there is no room for error in the process. A higher value allows some leeway in the position of the controller set point and allows some variation in the process without violating specifications.

The capability index is compared with the performance index to identify where control can be improved. The indices are calculated for each key product property and process variable and then compared, as per Fig. 14:

- If the capability index is below the goal, a process change is needed.
- If the performance index is less than capability, a process control improvement is needed.

Assessment of Measurement and Controls

The next step in the analysis is to assess the measurements and controls for the product properties and process variables to identify specific improvements or fixes. The kinds of questions to ask in making these assessments are shown in Figs. 15 and 16. This calls for a team effort, a team that includes process engineers, operators, instrument technicians, quality-control specialists, and process control engineers. By using process diagrams that show the main control loops, the team discusses measurement and controls for each of the key variables needing improved control and arrives at a conceptual design for the improvement. Detailed designs are worked out later.

Estimate Benefits for the Improvements

The final list of improvement opportunities may be large enough to require a prioritization to make sure funds and resources are used effectively. One method of prioritization is to rank candidate

Assessing Measurements

- accuracy, precision, sensitivity
- repeatability
- reliability
- frequency
- appropriate operating range
- suitable technique
- sampling procedures
- lab practices
- lack of measurement

FIGURE 15 Assessing measurements.

Assessing Control Strategies

- operates in intended mode
- handles upsets well
- tuning
- interactions with other loops
- turndown capability
- stable over whole range
- appropriate for control task
- complexity

FIGURE 16 Assessing control strategies.

Prioritizing Opportunities

- Identify process control opportunities
- Estimate % reduction in variability
- Estimate impact on business drivers and estimate savings
- Assess feasibility
- Prioritize

FIGURE 17 Prioritizing opportunities.

improvements according to stake and feasibility (Fig. 17). The control assessment just completed may suffice to assess cost feasibility and technical feasibility. What may be needed in addition is to estimate the economic benefits to the business for each improvement.

Quality Stake

Since quality is such a fundamental product parameter, one would expect that estimating the stake for improved quality in terms of increased sales would be easy to obtain. However, that is usually not the case. What is easier to estimate is the effect of quality on rework, yield, or throughput. The following is a generic stake calculation for improved quality in terms of increased earnings that can be applied in several possible cases:

$$Q = \text{(annual production rate in units/yr)} \times \text{(% increase in earnings/100%)}$$
$$\times \text{(earnings in \$/unit of product)}$$

Yield Stake

Yield is related to the amount of raw materials that get lost either through the formation of second-grade or waste byproducts in chemical reactions or lost physically in vents, cleanouts, waste streams, etc. First-pass, first-quality yield describes the amount of raw materials turned into first-grade products without blending or rework. Rework introduces yield losses and, in a sold-out market, has the additional penalty of tying up equipment that could be making new product instead of reworking old product.

The following two equations are used to estimate the stakes for reducing physical and chemical losses by improved control.

$$Y1 = \text{(flow rate of lost stream in units/h)} \times \text{(% valuable components/100%)}$$
$$\times \text{(% reduction in loss/100%)} \times \text{(\$/unit)} \times \text{(h/yr)}$$

$$Y2 = \text{(raw material flow rate in units/h)} \times \text{(% recoverable yield loss/100%)}$$
$$\times \text{(% reduction/100%)} \times \text{(\$/unit)} \times \text{(h/yr)}$$

Recoverable yield loss is the difference between the current yield and the theoretical maximum yield. This is an important distinction since there are usually physical or chemical limits that prevent a manufacturing unit from achieving 100% yield.

Throughput Stake

Throughput improvements can be achieved by avoiding operating conditions that result in downtime due to interlocks, plugging, corrosion, etc., and by operating closer to equipment constraints like flooding in a distillation column that increase the capacity. Improved control in batch operations that results in reduced cycle times also increases throughput. Improved throughput has the biggest impact in a sold-out market in which control improvements leading to increased throughput usually have an overwhelming effect on earnings, a much larger effect that just about anything else:

$$P = \text{(current production rate in units/yr)} \times (\% \text{ increase in throughput}/100\%)$$
$$\times (\% \text{ time sold out}/100\%) \times (\text{earnings in \$/unit of increased production})$$

The percentage of time sold out is the percentage of the time the plant needs to run at maximum production to meet demand. Presumably when the plant is not sold out, there is time to make up for lost production and the production stake is not important.

SUSTAIN THE BENEFITS

After the improvement projects have been implemented, the original team or a subset of it should go back and audit the results to determine how much of the estimated stake has been achieved. The results can be used to refine the estimating techniques for future use and discover any additional needs that were not addressed the first time. Ideally, the plant will institutionalize the process just described so that control improvements will be addressed periodically. If the process is not repeated, chances are the benefits will be lost over time because of changes to the process, equipment deterioration, changing demands, turnover of people, etc.

Performance Metrics

Another useful tool for sustaining the benefits is on-line performance metrics. The purpose of metrics is to discover any poor operation of the control system quickly so that the appropriate action can be taken before large losses can occur. There are two kinds of metrics:

- utilization metrics that indicate when key measurements or control strategies are not being properly utilized
- variability metrics that indicate when the controls are not functioning properly

Figure 18 shows examples of utilization metrics. One metric indicates when an analyzer is out of service and needs to be recalibrated or repaired. Another metric indicates when a key control loop is not being operated in automatic mode. This condition might be caused by poor tuning, a failed transmitter, or just poor design. The metrics are applied only to the areas in which not fully utilizing the measurement or control strategy will result in off-quality product, low yield, reduced rates, or other penalties.

Figure 19 illustrates a sample of a variability metric report. Again, this metric is applied to key variables for which we want to minimize variability. A report of this type contains a lot of information about the health of the controls. For example, if the average does not conform to the limits (that are set to ensure meeting the business goals), it may be just sloppy operation (not adhering to the specified

Utilization Metrics

% Availability of Analyzer = $\dfrac{\text{Time Analyzer is Operating}}{\text{Time Process is Running}}$

% Utility of Control Loop = $\dfrac{\text{Time Control Loop in Automatic}}{\text{Time Process is Running}}$

FIGURE 18   Utilization metrics.

Variability Metrics

Variable	Limits	Mean	S_{tot}	S_{cap}	P_{pk}	C_{pk}
Pressure Column Composition [1]	3 - 6	3.3	0.36	0.03	0.3	3.11
Reactor pH [2]	9 - 11	9.1	0.3	0.25	0.11	0.13
Day Tank Composition [3]	< 10	2.52	4.2	4.1	0.6	0.61
Refining Column Temp [4]	102 - 107	102.1	0.18	0.08	0.23	0.56

FIGURE 19   Variability metrics.

operating conditions) or signal that a special cause of variability has occurred that the controls are not able to compensate for. The standard deviation S_{cap} indicates what should be achievable with perfect control and S_{tot} shows actual standard deviation. If there is a big difference between the two [1], the controls needs attention. Likewise C_{pk} and P_{pk} indicate how far off from world-class performance the plant is operating at that moment. If S_{tot} is approximately equal to S_{cap} but P_{pk} is small [2], it may indicate that the average is too close to one of the limits. If C_{pk} is small, it indicates the process should be modified in some way to eliminate the common cause variability [3] or it may indicate operation too close to the limits [4].

CONCLUSIONS

This section presented a disciplined approach to improving the performance of a manufacturing plant by identifying process control improvements aimed at bringing business value and not just ensuring smooth operation. This approach should not be applied only once but used for continuous improvement on a periodic basis. This is the only way to ensure that the benefits from process control are sustained over the long term.

REFERENCES

1. Shunta, J. P. *Achieving World Class Manufacturing Through Process Control*, Prentice-Hall, Englewood Cliffs, NJ, 1995.
2. Shunta, J.P. "Business metrics in control," Intech, April 1996.

PLANT ANALYSIS, DESIGN, AND TUNING FOR UNIFORM MANUFACTURING

by W. L. Bialkowski*

INTRODUCTION

This section focuses on plant analysis, design, and controller tuning issues that relate to the efficient manufacture of uniform product in continuous process plants, such as paper, plastics, rubber, chemicals, hydrocarbons, pharmaceuticals, food, and others. Joseph P. Shunta, in his book *World Class Manufacturing through Process Control* [1], provides an excellent balanced treatment of the application of process control and where this fits into the big picture of World Class manufacturing. The central theme is the reduction of process variability through better process control, and in some cases, better process design, so that a more uniform, higher-quality product that better conforms to customer specifications can be manufactured more efficiently. These goals link directly with corporate goals of gaining market share, reducing manufacturing costs, and increasing shareholder value. There are many issues involved, both technical and organizational, that must be overcome for a company to become a world class manufacturer. This section embraces Shunta's work completely where it addresses the big issues in process control application. There are, however, a number of key technical issues such as process variability, its origins and sources, time series analysis, model-based controller tuning, and the impact of control valve nonlinearities that are covered in detail in this section. This section is based on two decades of intensive work in process variability reduction in the pulp and paper industry, from which, of necessity, a slightly different focus has emerged. The work has been published both in the control [2] and pulp and paper industry literature [3–5]. In a nutshell, the differences between the chemical industry experience and the paper industry experience can be summed up as follows: slow chemical processes making liquid product on the one hand versus fast hydraulic processes making solid product on the other. In the chemical industry case, the variability caused by the fast control loops and control valve limit cycles has in the past often been ignored, since the dominant process time constants are typically measured in fractions of a hour. Many chemical industry process control practitioners have ignored this fast variability and have focused on the upper-level optimizing controls, such as Dynamic Matrix Control, in an attempt to achieve higher yield and throughput. In the paper industry, the fast lower-level control loops are directly linked with the final product uniformity.

* EnTech Control Engineering Inc., 16 Four Seasons Place, Toronto, Ontario, Canada M9B 6E5.

Ignore these loops, and the process may not be able to manufacture salable product at all. The pulp and paper process control practitioner has had to focus on reducing this fast variability just to keep production going at the budgeted rate. This focus on the fast process variability has brought a number of key techniques into play. These include the following:

1. plant analysis to measure process and product variability by use of time series analysis techniques, plant auditing procedures designed to identify the causes of process variability, an interpretation of the results in both the time and the frequency domains, use of spectral analysis for both diagnostics and design

2. the use of model-based controller tuning such as internal model control (IMC) concepts and Lambda Tuning for both plant design and controller tuning

3. using a tuning strategy to achieve coordinated dynamics of a process area by preselecting the closed-loop time constants for each control loop

4. Understanding the performance-robustness envelope of a control loop.

5. Understanding the impact of actuator nonlinearities on control performance.

6. Understanding the variability propagation pathways through a complex process.

This section sets out to cover this ground at the conceptual level. Theory is presented only to aid in making the concepts easier to understand. References are made to other more detailed works. Pulp and paper process examples are used for illustration, as these data were available to the author. A more detailed treatment of this material is given in Ref. 2. The objective is to learn how to measure and characterize the variability in the final product, how to identify the causes of the variability in the process, how to eliminate these causes where possible, how to tune control loops so that they effectively attenuate variability as best as possible, and to understand how process variability propagates through a process.

PROCESS VARIABILITY

The Process as a Network

All continuous manufacturing plants consist of an interconnected network of reactors, tanks, vessels, and process equipment that interconnect the raw-material supply sources to the final product storages. The process unit operations consist of many different types: reaction kinetics, separation processes such as distillation, hydraulic separation processes such as screening and centrifugal cleaning, hydraulic transport and mechanical operation such as pulp refining, and thermal equilibrium vessels such as boilers. The unit processes can be interconnected in series, in parallel, or in recycle configurations. The process dynamics are often complicated by the presence of long dead times, material or heat integration, nonlinearities, inverse response, and severe interaction. There may be many parallel paths for the variability to travel through the process network. As a result, it is sometimes very difficult to determine precisely how the variability reached a certain point in the process.

There are many control loops in a typical process area—from 20 or 30 to over a 1000. The control loops have nonlinear control valves, they interact with each other, and change their dynamics with operating point. Even though ideally the control loops were designed and installed with the aim of eliminating process variability, they at best can only attenuate variability at certain frequencies. Unfortunately, they also often increase variability at other frequencies because of their natural tendency to cycle and resonate. When loops cycle they can become inadvertent sources of variability in their own right. This tendency to cycle can result from process dead time, aggressive tuning, multiloop interaction, or nonlinearities present in actuators.

Steady-State Plant Design

In almost all cases plants have been designed to meet steady-state design criteria only, with little if any consideration having been given to the dynamics of the process being designed. In the latter stages of plant design, the major equipment has been selected and the lines have been sized. At this point the control valves can be sized and located and the transmitters can also be specified and located. Once this is done, the control loops can be configured by the pairing up of individual transmitters and actuators. Typically before start-up, the control loops are equipped with default tuning parameters, such as a gain of 1.0 and a reset time of 1 min. Next, start-up occurs, and the start-up team works round the clock. In most cases they will have time to tune only those loops that clearly do not work during the start-up phase. Further tuning is seldom done after start-up in most plants. As well, analysis of plant process variability is seldom if ever carried out in the majority of plants.

Economic Opportunity

Yet the raw materials entering the process are laden with variability, control loops cycle and increase variability in automatic mode, and the product uniformity and plant efficiency suffer. The opportunity to increase operating efficiency by use of the techniques described in this section is estimated to represent at least a 5% increase in manufacturing efficiency.

Sources of Variability

The raw materials are never perfectly uniform, hence variability enters the manufacturing process with the raw-material streams. The dynamics of the manufacturing process and control strategy are complex and often inadvertently add to the variability that is already present because of the raw materials. The variability emanates from many point sources such as raw-material streams, cycling loops, rotating equipment, and so on. The net result is that variability is present throughout the process, and in particular in the final product. For the final product the variability is the sum of the effects of all of the point sources, as well as the variability attenuation characteristics of the variability pathways that connect the sources to final product. The variability pathways are dynamically complex. They include the effects of mixing or agitation, control loops, and various combinations of series and parallel interconnections of these effects. Mixing or agitation tends to smooth out fast variability. On the other hand, control loops regulating to a fixed setpoint tend to smooth out slow variability. Unfortunately, control loops can also resonate and amplify certain frequencies.

Operating Constraints, Uptime, Efficiency

Equally important is the potential that the variability sources and pathways may impose operating constraints on the process. For instance, the variability in applied chemical dosage may cause a property such as product strength to vary, which in turn forces the operation to compensate by running the reactor at a slower production rate. In this case, reducing the variability in applied chemical dosage can be directly linked to an increase in production rate. In effect, the variability in chemical dosage has imposed an effective production rate constraint. High variability in fiber delivery to a paper machine headbox will result in a high break frequency on the paper machine. Breaks cause downtime and have a serious impact on manufacturing efficiency. The paper machine operator will run the paper machine at a slower speed in order to keep the occurrence of sheet breaks at a tolerable level. Again, the variability in fiber delivery has imposed a constraint on production rate. As a general rule, process variability can have an impact on manufacturing in two ways:

- High process variability will have an impact on process uptime by causing trips, breaks, plugging, etc.

- High process variability will reduce manufacturing efficiency as a result of
  - lower product yield
  - lower production rate
  - higher energy consumption
  - lower product quality
  - more product downgrade

Solid versus Nonsolid Product

Depending on the industry, the product being made can be a solid product, for instance paper, rubber, or wire, or a nonsolid product such as gasoline, distillates, ethylene oxide, or butane gas. Solid products capture all of the variability that is present in the process and preserve this variability for as long as the product remains in use. The customer will experience the impact of this variability, as this product is used in their secondary manufacturing plants. For example, the paper rolls manufactured in the paper mill will be used to feed the customers' printing presses. If the paper is not very uniform, this can easily cause the printing press to experience paper breaks, hence curtailing production and lowering the efficiency of the customers' operations. Most newspaper pressrooms rate their paper suppliers on the basis of the number of paper breaks that occur per hundred rolls used in the pressroom. Paper manufacturers can be disqualified from supplying paper to a given pressroom once their rating is too high. This is a simple expression of customer dissatisfaction. In contrast, many liquid products are stored in one of the various tanks of a tank farm. Once the production run for a product grade is over, the new product will probably stay in the tank farm for some time before being shipped to the distribution system. In this period of time, the product will likely equilibrate—hence the contents of a given tank will become fairly uniform. However, the tanks that were filled during the grade run will exhibit tank-to-tank variability. Even though the contents of the tanks may not conform to the customer's specification exactly, it is still possible to blend the products of several tanks in order to ensure that the customer's specifications are met for the whole shipment. Adding to this, the final consumer is seldom aware of deviations in the product being consumed. Can the consumer of gasoline tell that the actual octane rating in the last tank of gas was below 97? The answer is no. Clearly, in the manufacture of many liquid or gas products, which involve intermediate storage in a tank farm, the need to focus on variability is not very critical. In contrast, the manufacture of solid product is completely unforgiving—the variability is there to stay, and all variability from very slow to very fast can be measured through testing. Also, the consumer is often critically aware that the variability is present. In the case of paper, pressroom breaks, photocopier jams, or dirt specs present in the sheet are obvious reminders to the consumer that variability is present in the product they have purchased.

Process Example—Paper Making

Before any attempt is made to analyze variability or to tune control loops, it is important to clearly understand the manufacturing process. This understanding should include:

1. the process flow diagram from raw material feed through to final product storage
2. each unit operation including its intended role in the overall process
3. the process dynamics of each unit operation and the impact it should have on process variability.

Finally, the manufacturing goals for both the process and the product should be clearly understood. As well, the specific operating objectives of each unit operation should be clearly understood in the context of how these are supposed to satisfy the overall manufacturing goals. This level of understanding is needed before the dynamic objectives for each unit operation, and each control loop within the overall control strategy can be set.

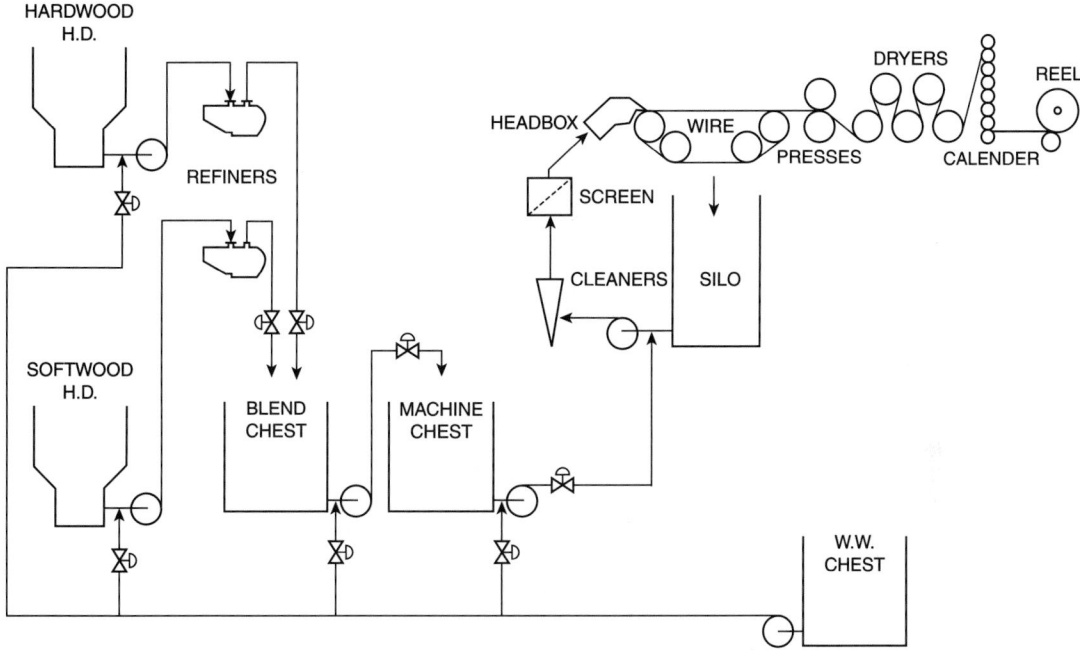

FIGURE 1   A paper machine schematic.

To illustrate these concepts, a paper mill process flow sheet is shown in Fig. 1. The example illustrates a fine paper machine for manufacturing photocopy paper at a production rate of ~500 tons per day. The term fine paper refers to paper for books, photocopiers, laser printers, and other publication grades of paper. The primary raw materials typically include two grades of wood pulp, hardwood, and softwood. In the example, the paper machine is using 70% hardwood pulp and 30% softwood pulp. The hardwood pulp is made from hardwood trees, such as maple and oak. These have a short fiber that allows the manufacture of a smooth paper sheet surface for high-quality printing to be achieved. In Fig. 1 the hardwood fiber, is pumped from the hardwood high-density (HD) chest where the fiber slurry is stored. It is then diluted with water to reduce the consistency, or slurry concentration, before being passed through a rotating machine called a refiner, where the fiber surface area is increased by mechanical cutting action. After the refiner, the pulp is fed to the blend chest, where it is blended with the other raw materials. The other main raw material for fine paper is called softwood pulp. It is made from softwood trees, such as pine or fir. These have a relatively long fiber, which provides sheet strength. Like the hardwood, the softwood pulp is diluted, refined, and pumped into the blend chest. In the blend chest the two ingredients are mixed thoroughly, from where the blended stock is pumped to the machine chest, after further dilution. The machine chest provides a final opportunity to remove process variability through agitation and mixing. From the machine chest the stock is pumped to the paper machine headbox after further dilution, cleaning, and screening have occurred. The headbox is a pressurized vessel that ejects the diluted stock from a rectangular opening, known as the slice, onto a moving fabric mesh known as the wire or the former. The slice and wire are as wide as the paper machine, which can be up to 10 m or 30 ft. The wire is moving at the speed of the paper machine, which can be well over 1000 m/min, or 3300 ft/min. The water drains through the wire mesh, leaving the fiber to form a wet sheet of paper. This sheet is continuously removed from the wire, pressed to remove water, dried by steam-heated dryer cylinders, and compressed to form a smooth finish by means of a calender stack consisting of highly polished steel rolls. The final sheet is wound up on a reel, from where it is further cut into rolls or sheets before shipment to the customer. This has been a very brief description of pulp and paper science; for more details see Ref. 6.

Manufacturing Objectives and Process Dynamics

The manufacturing goals are to make uniform paper to tight tolerances (typically $+/-1\%$) in final properties such as its basis weight or mass per unit area [measured in grams per square meter (gsm)], moisture content, caliper (thickness), color, brightness, and others. Paper properties are measured by on-line sensors also paper samples are tested in the lab. Basis weight is an important final product variable and is useful for illustration purposes.

The manufacturing objectives for each unit operation can be stated as follows:

Refiners: Uniform refining requires uniform mass flow of fiber through each refiner and uniform application of refining energy to the fiber. In turn this requires the pulp flow and consistency fed to the refiner to be as uniform as possible. Any variability that is present in either of these process variables will cause variability in fiber bonding and potentially sheet strength. The refiners are low-volume hydraulic units with a residence time of less than a second. As a consequence of the extremely fast dynamics, the refiners do not offer any ability to attenuate variability.

Blend Chest: The mass ratio of the raw-material streams entering the blend chest should be as constant as possible, as this will determine the fiber composition of the sheet. This requires that the consistencies of both streams be uniform and that the flow controllers for each stream must be adjusted together in unison. The blend chest is agitated and provides an opportunity for process mixing. The residence time is typically 12 min.

Machine Chest: The purpose of the machine chest is to provide a final opportunity to mix the pulp furnish before it is used on the paper machine. The residence time is typically 12 min.

White-Water Silo: This is the tank in which the water, which has drained from the sheet, is collected for reuse. This water contains some fiber. The residence time is $\sim$20 s; however, the silo is not agitated; it may have nonuniform consistency. From a variability standpoint, the silo provides a material recycle path and tends to prolong disturbances that occur in the recycle path. At the discharge of the silo the new pulp stock is mixed with the recycled white water to form the feed to the paper machine headbox. Any variability present in either stream is injected into the headbox.

Headbox Feed: The feed to the headbox is passed through centrifugal cleaners and pressure screens. Both devices are intended to remove dirt, oversized particles, and contaminants. In both cases the units have very short residence times and are essentially flow splitters. They both have a reject flow, which typically runs at $\sim$5% of the feed flow. These units do not offer any potential for attenuation of variability. On the contrary, should their rejects flow rate vary, some of this variability will be passed onto the accepts flow.

Headbox and Wire: The headbox is where the pulp slurry is converted to a sheet product. The headbox velocity is a direct function of the pressure in the headbox. The velocity and the consistency determine the fiber mass in the paper sheet. Since the headbox consistency is very low (less than 1% typically), a large volume of water must drain through the wire. Process variability in this area will drive variability in sheet moisture. The dynamics of the headbox are very fast, since the residency time is less than a second.

Presses, Dryers, Calender: The residence time of the sheet in this section of the paper machine is typically $\sim$20 s. There is no opportunity to reduce variability through mixing, since the fiber in the paper sheet is being permanently set. The variability in water content typically follows the variability in fiber content. The ability to remove water content is provided mechanically in the presses and thermally in the dryers. The dynamics of the dryers are relatively slow, as the thermal mass of the dryer cylinders is large, and the typical thermal time constant is $\sim$2 min. As a result, using the dryers to control the moisture content allows the attenuate of only the very slow components.

Process Control Strategy—Paper Machine Blending

Figure 2 shows a process and control diagram for the blend chest area for the fine paper machine referred to in Fig. 1. The two pulp slurries are pumped out of their respective HD storage chests at $\sim$5% consistency (mass percentage fiber slurry concentration). The hardwood stock is pumped out of the hardwood HD chest and is diluted to 4.5% consistency at the suction of the pump by the addition

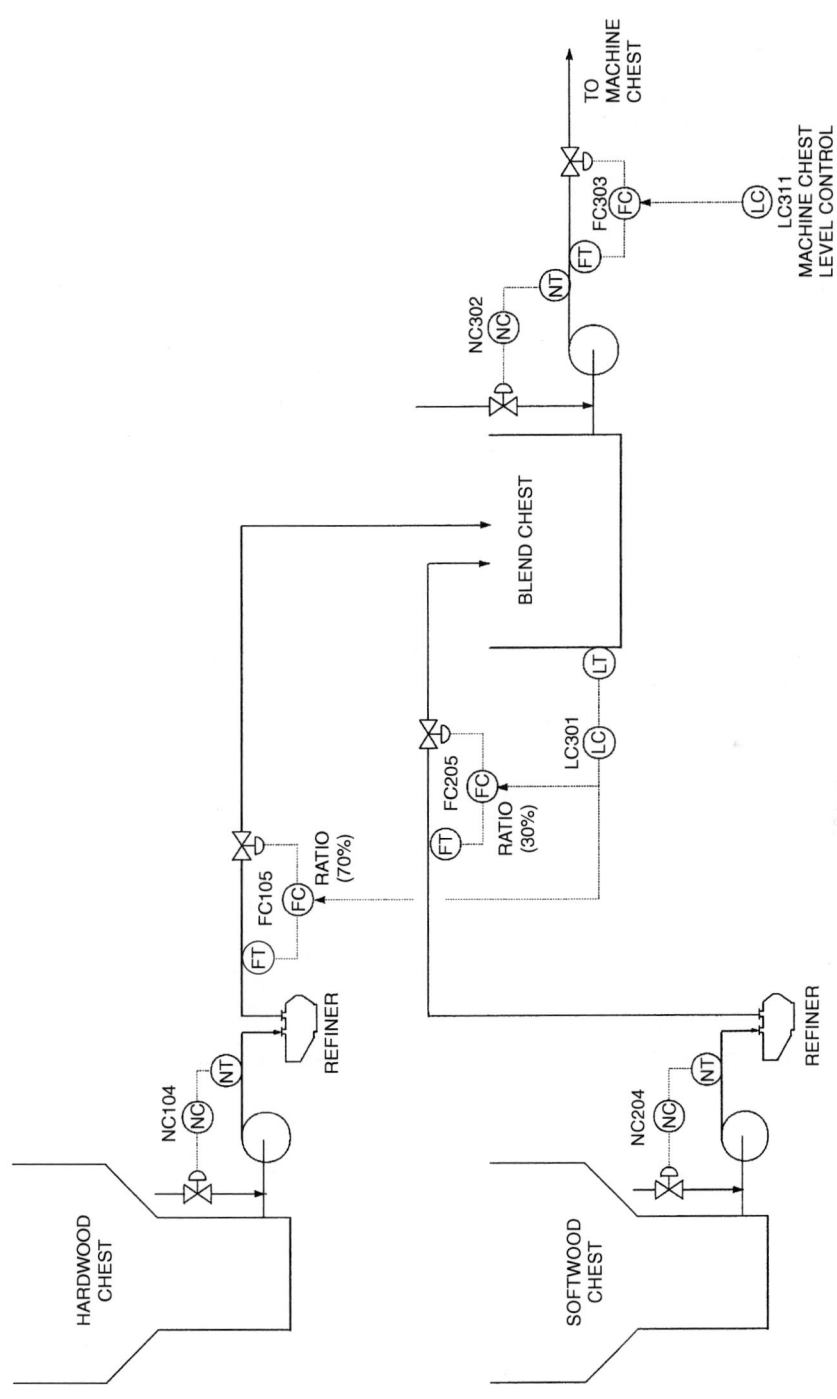

FIGURE 2 The blend chest area of a paper machine.

of dilution water. The dilution water is supplied from a common dilution header (pump and pipe) for this part of the paper machine. In the example of Fig. 2, the dilution water addition is modulated by the consistency control loop, NC-104 [typical loop tag based on Instrument Society of America (ISA) terminology], in order to regulate the consistency to a setpoint of ~4.5%. The consistency sensor is located after the pump and has a measurement transport delay or dead time of 5 s. After consistency control, the hardwood stock is pumped through the hardwood refiner, where the fiber surface area and fiber bonding properties are enhanced. The refining process is sensitive to the mass flow of fiber passing through the refiner. After the refiner, the pulp stream is flow controlled by flow controller FC-105. The softwood line is identical to the hardwood line and includes consistency controller NC-204 and flow controller FC-205. The two flow controllers FC-105 and FC-205 are a part of a cascade control strategy, and their setpoints are adjusted together in order to maintain the blend chest level at the desired setpoint (typically at 70%) while maintaining the desired blend ratio for each stock (70% hardwood and 30% softwood).

Variability Audit Procedure

Process variability audits have been carried out in pulp and paper mills since the early 1980s [2–4]. The process variability audit is intended to:

1. determine the level of variability in the product and to determine the acceptability of the product for the customers' intended purpose, given the variability present

2. determine the level of variability in the process and to determine the how this process variability has an impact on the product variability as well as on the manufacturing efficiency of the process

3. identify the sources for the process variability and in particular those that have an impact on product variability or process manufacturing efficiency

4. if possible eliminate, or recommend how to eliminate, the sources of variability that have been identified.

The procedure involves collecting real-time data while the process is operating. It is advantageous to collect data from final product variables first. This provides a potential signature of variability being generated by upstream variables. It may then possible to identify which upstream variable is causing the variability noticed downstream. As an example, if an oscillation in final product with a period of 1 min has been detected, the hypothesis is that some process variable upstream is cycling at a period of 1 min. Once a potential match has been found and a loop that cycles with a period of ~1 min has been identified, one can prove the hypothesis by altering the behavior of the loop and observing the effect in the final product. For instance, if the cycling loop is placed in manual mode, it is likely that the cycle will stop. Should the oscillation in final product also stop after the loop is placed in manual, then the hypothesis has been proven conclusively—the product oscillation is caused by the loop in question. Then the remaining task is to uncover why the loop has a tendency to cycle. This task is relatively easy, since it involves the loop itself and possibly some adjacent variables.

Variability Examples

Product Variability.   Figure 3 shows variability in final product. The data are presented by a commercially available software package [7] designed for plant process variability analysis. Two data collection runs are shown with basis weight in gsm being measured. The upper plot shows a short-25-s run, while the lower plot shows a long 4.9-h run. The data presented show two key pieces of information about each data run:

- The time series graph shows the variable of interest plotted against time. This represents key historical information that can be used to develop insight and intuition about the dynamic behavior.

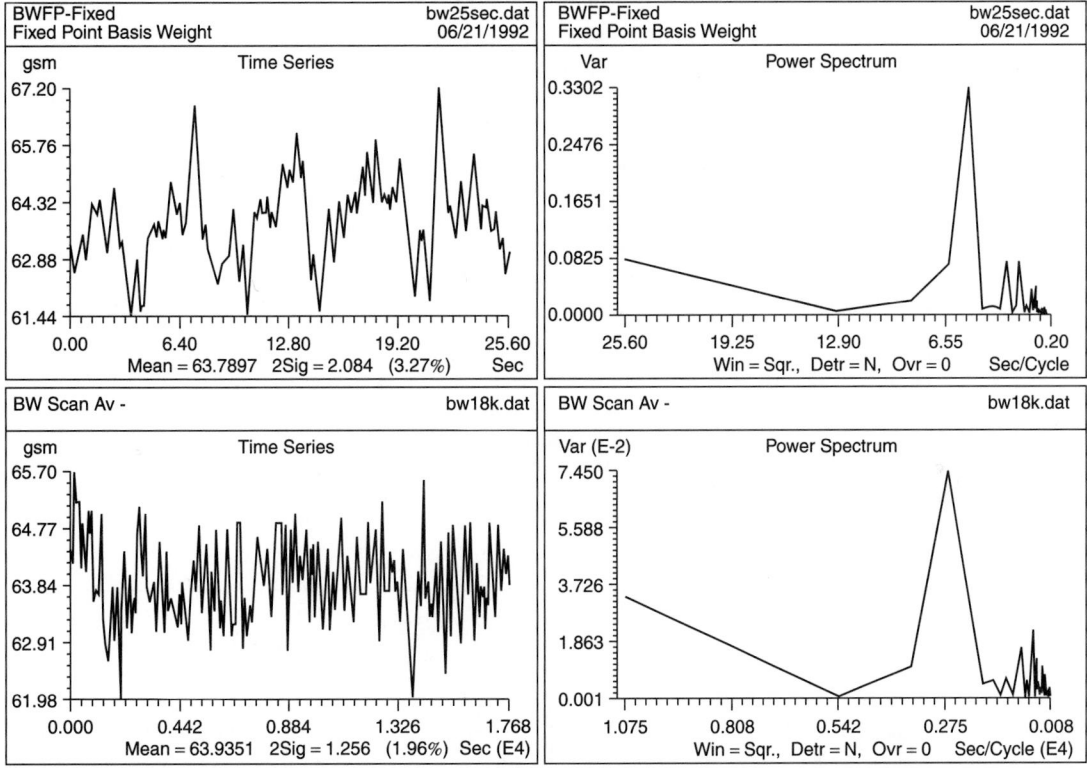

FIGURE 3 Paper machine product weight variability. Upper left – 0.4 minutes, variability = 3.27%, upper right – power spectrum, cycle at 5 seconds, lower left – 4.9 hours, variability = 1.96%, lower right – power spectrum, cycle at 46 minutes.

- The time series statistics are shown below the time series graph. These include
  - the mean value, (e.g., 63.7897 gsm)
  - the two standard deviations (2Sig), which represent the +/−95% confidence limits around the mean value, assuming a normal distribution, (e.g., 2Sig = 1.256 gsm)
  - the two standard deviations (2Sig), expressed as a percentage of the mean value (e.g., 3.27%). This is a useful unitless way of expressing process variability.

The 25-s run collected data at a sample rate of 0.1 s. For this data collection, the basis weight sensor was taken out of its normal mode of scanning the full sheet width and placed in a fixed position on the sheet. The variability in basis weight is 3.27% over 25 s. This is fairly high compared with that of other fine paper machines, hence this should raise concern about the market acceptability of this product. A clear cycle is visible with a period of ~5 s. The likely cause is somewhere in the white-water, cleaner, screen headbox area of the process, since the dynamics of this part of the process are all very fast.

The 4.9-h run collected data at a sample rate of 40 s. For this data collection, the basis weight sensor was its normal mode of scanning the full sheet width. The scan time to traverse the sheet width was 40 s, and each data point collected was an arithmetic average of the basis weight across the sheet. As a result, each data point has had all of the variability faster than the averaging time removed. The variability in scan average basis weight is 1.96% over 4.9 h. This is fairly high compared with that other fine paper machines, the best of which have shown basis weight variability of 1%; hence this should raise concern about the market acceptability of this product. A cycle is clearly visible with a period of ~2750 s or 46 min per cycle. The likely cause is somewhere in the blending area of the process, since the dynamics of this part of the process are comparatively slow.

Process Variability—Typical Examples. Figure 4 shows variability in the two pulp streams entering the blend chest. All the data were collected simultaneously at a sample rate of 2 s for 2.3 h. Figure 4 shows the hardwood consistency (NC-104) and flow (FC-105), as well as the softwood consistency (NC-204) and flow (FC-205). All of the variables appear to cycle with a period of ~46 min per cycle. Both flows appear to be in phase with each other. The consistencies are also in phase, although in the softwood consistency, the cycle is not as pronounced as the others.

Figure 5 shows variability in the blend chest level (LC-301) and the consistency coming out of the blend chest (NC-302). All the data were collected simultaneously at a sample rate of 2 s for 2.3 h. Both the blend chest level and the consistency appear to cycle with a period of 46 min as well. The level cycle appears to be leading the flow cycle by ~90°. This suggests that the level control is the likely cause. The level control is the upper cascade control loop that adjusts the setpoints for the flow controllers FC-105 and FC-205. The cycle in the blend chest consistency appears to be delayed version of the cycle in the blend chest inlet consistencies. The apparent time delay is ~5 min, which is less than the residence time in the blend chest.

When the level controller LC-301 is placed on manual, the 46-min cycle stops everywhere, including the final product. This proves that the 46-min cycle is caused by the blend chest level controller LC-301 and that this is also the source of the cycling of all of the loops.

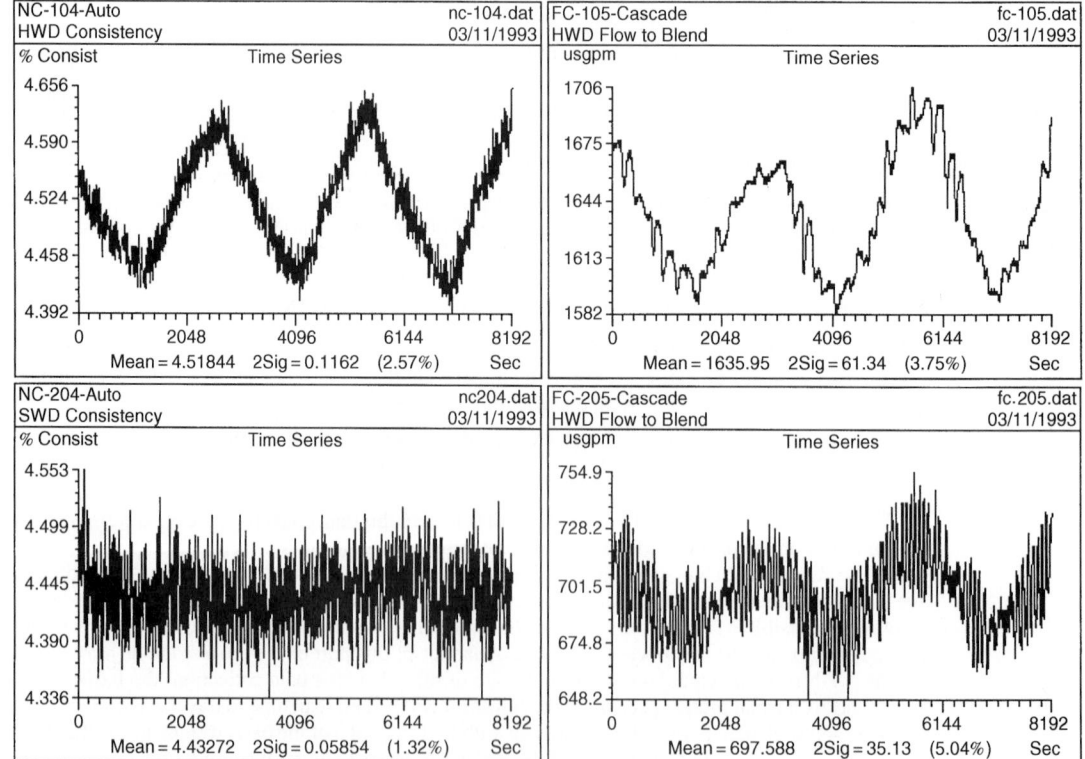

FIGURE 4 Process variability in blend chest consistency and flow variables over 2.3 hours. Upper left – hardwood consistency variability = 2.57%, upper right – hardwood flow variability = 3.75%, lower left – softwood consistency variability = 1.32%, lower right – softwood flow variability = 5.04%.

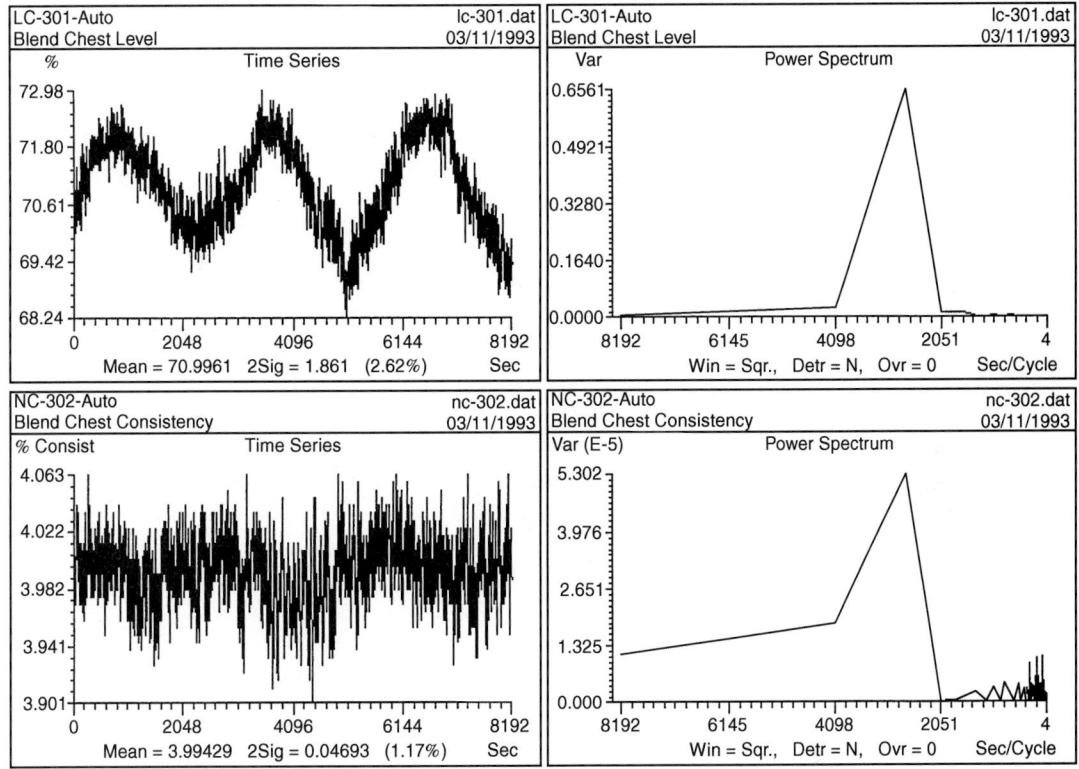

FIGURE 5 Process variability in blend chest consistency and flow variables over 2.3 hours. Left – blend chest level variability = 2.62%, right – softwood flow variability = 1.17%.

Figure 6 shows variability of the screens reject flow in automatic mode. The screens reject flow is a flow controller operating at a fixed-flow setpoint. Both the process measurement and the controller output are shown. This example shows a typical control-valve-induced limit cycle caused by what appears to be ~2.2% dead band in the control valve. The period of the limit cycle is ~1.6 min.

Diagnostic Principles

The variability shown in this example is typical of continuous plants that operate flow, concentration (consistency), and level controllers. Even though the data were taken from a paper mill, the diagnostic principles are applicable universally. The principles used in the diagnostic process depend on the following:

1. A sinusoidal disturbance will be transmitted through a linear system at the same frequency. This means that, by observation of downstream behavior, the upstream disturbance sources can be identified. This is not strictly true if the system is nonlinear, as all processes are; however, in most cases the nonlinearities are not severe enough to invalidate the premise.

2. The amplitude of a sinusoidal signal that has been transmitted from upstream to downstream will depend on the effective coupling gain that exists in the transmission path at the given frequency. This gain may be subject to attenuation at certain frequencies and amplification at others. The net result is that the amplitude of a transmitted variability signal is uncertain and is best determined by carrying out bump tests from source to destination.

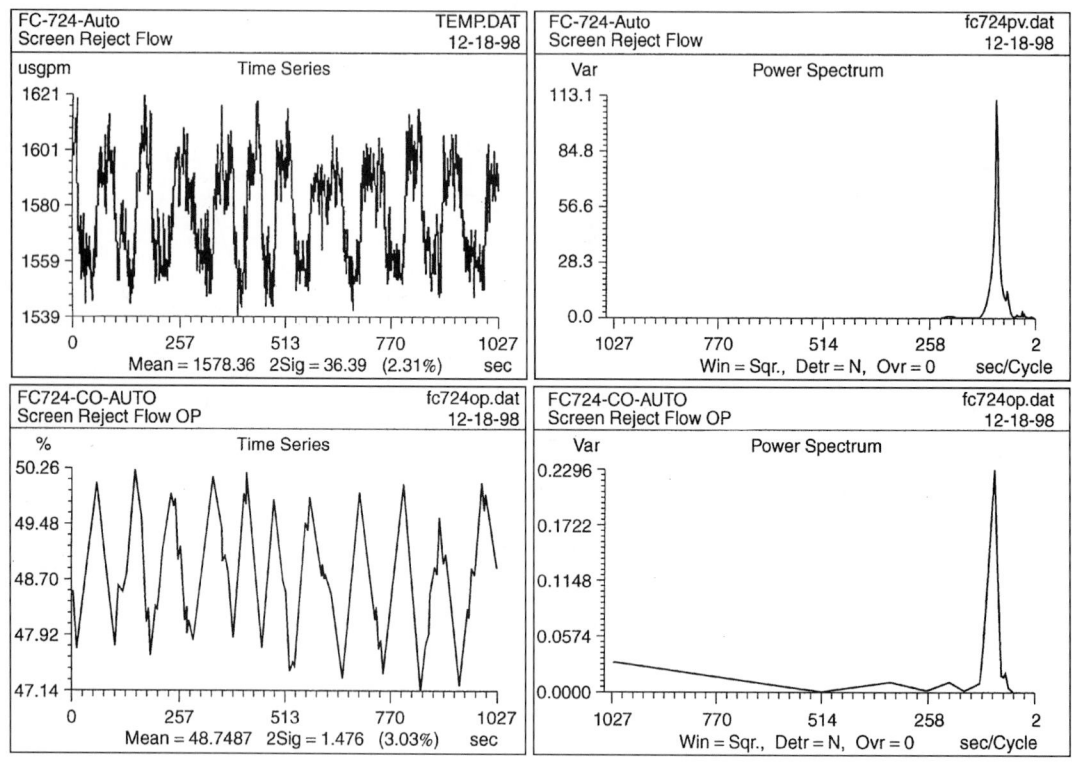

FIGURE 6   Process variability in screen reject flow due to control valve dead band.

TIME SERIES ANALYSIS

Time series analysis involves techniques for analyzing data collected at equal intervals of time [8]. It forms the basis for performing plant variability audits, in which the objective is to correlate the real-time data, give it physical significance, and to ultimately uncover the causes of plant operational problems. Whereas this seems simple enough as an objective, care must be taken to ensure that high-fidelity real-time data are acquired for meaningful analysis to occur, hence avoiding the garbage-in–garbage-out syndrome. In most cases plant data are stochastic or random in nature, and their analysis over a wide frequency spectrum can provide important clues regarding sources of plant variability. However, the amount of data that can be practically analyzed is limited by computing power. For instance, Fourier analysis of a large number of data points (typically more than 32,000) can be unwieldy or impractical on some computers. In turn this means that the sampling rate at which the data are collected may have to vary from very fast to quite slow in order to meet specific analysis requirements. When long-term data are needed to measure slow or low-frequency variability, this may require a sparse sampling interval, which in turn may cause data aliasing to occur.

Time Series Analysis Tools

Plant variability analysis usually involves collecting multiple data channels simultaneously, so that multivariable dynamics and variability can be studied. All of these data should be collected at the same sampling interval and in time synchronism with each other. Most modern plants have distributed-control-system- (DCS-) based control systems, with plantwide data archiving systems in place, and it is tempting to consider that these real-time and historical data are ideal for conducting plant variability audits based on time series analysis. Unfortunately, at the time of writing, this is seldom the case, especially for plants that manufacture a solid product such as pulp and paper mills, in which high-frequency data are important. A detailed treatment of digital data requirements is given in Ref. 9. Some of the more commonly occurring reasons for the unsuitability of these data include:

1. relatively slow sampling of DCS inputs
2. lack of antialiasing filters
3. use of data compression and report-by-exception algorithms, especially for the archived data.

The tools needed to perform effective time series analysis of plant data include:

1. a real-time data acquisition system capable of collecting time-synchronized alias-free plant data at least 10 times faster than the fastest time constants of interest
2. statistical analysis including the calculation of mean, variance, and standard deviation
3. histograms
4. spectral analysis including power spectrum and cumulative spectrum
5. cross-correlation and autocorrelation functions.

Real-Time Data Acquisition

The requirements for high-fidelity real-time data involve ensuring that

1. each sensor or transmitter, which is the source of the real-time data, is correctly installed, calibrated, and is measuring properly
2. the measurement dynamics and internal filtering characteristics of the sensor or transmitter are known and accounted for (the measurement dynamics will alter the apparent dynamics and bandwidth of the resulting information)
3. the input data are sampled with adequate resolution (minimum of 12 bits or 1 part in 4096)
4. the input data are collected and sampled by means of properly designed antialiasing filters to ensure alias-free information.

Sampling Theory—Time Series Data

To understand real-time data acquisition, it is important to grasp a few essential aspects of sampling theory that pertain to in-plant data collection and analysis. Continuous manufacturing plants contain hundreds or thousands of continuous process variables, such as flows, pressures, pH's, concentrations, and temperatures. These variables are continuous or analog in nature, can be described mathematically as continuous functions of time $y(t)$, and can be modeled by differential equations in continuous time. These signals are measured by sensors and transmitters and are then inevitably digitized. The digitization can occur at the DCS input terminals, in a smart transmitter, or in any data acquisition system provided to collect the data. Digitization involves both quantization and sampling. Consider an analog signal $y(t)$ sampled at a given sample interval T_s over a given number of readings N. The result is a discrete time signal y_t, the time series data, which represents a data record NT_s seconds long. The important terms and concepts are the following:

T_s = sampling interval

$T_N = 2T_s$ = Nyquist period. This is the period of the fastest sinusoid that can be represented at a sample interval of T_s (e.g., if $T_s = 1$ s, $T_N = 2$ s).

$f_N = \dfrac{1}{T_N}$ = Nyquist frequency. This is the fastest frequency that can be represented at a sample interval of T_s (e.g., if $T_s = 1$s, $f_N = 0.5$ Hz). The Nyquist frequency is based on Shannon's sampling theorem, which states that the fastest frequency, which can be represented by sampled data, has a period of double the sampling interval.

$\omega_N = \dfrac{2\pi}{T_N}$ = Nyquist angular frequency. This is the period of the fastest angular frequency that can be represented at a sample interval of T_s (e.g., if $T_s = 1$ s, $\omega_N = 3.142$ rad/s). The angular frequency is useful as it is determined by the inverse of a time constant (e.g., a transmitter filter time constant of 2 s corresponds to a filter angular cutoff frequency of 0.5 rad/s, or a cutoff frequency of 0.0796 Hz).

Selecting Sampling Rates

Collecting data for plant analysis requires that sampling rates be chosen carefully. There are two reasons for data collection—analysis of step tests, or "bump tests," for controller tuning purposes and analysis of process variability in order to identify variability sources. When step tests or bump tests, are carried out, the sampling interval should be chosen to be fast compared with the expected process time constant. Ideally, the sample interval should be 10 times faster. In the worst case it should be at least 3 times faster. For a flow loop with a time constant of 3 s, the slowest sampling interval should be 1 s, and ideally it should be 0.3 s. For bump test analysis purposes, additional filters, such as antialiasing filters, are not necessary. These will skew the measured dynamics.

The second reason for data collection is to measure the variability and identify cause and effect. Now the objective is to focus on the frequency content of the signal; hence the antialiasing filters will be needed to ensure signal integrity. The sampling interval will be chosen in order to capture and analyze a significant portion of the frequency spectrum. From a practical point of view there is some maximum number of data points that can be analyzed. A typical number might be 16,000. If very high-frequency data are needed, then the data should be collected at the fastest sample interval available. Suppose this is 50 ms. This means that the length of the data collection run can practically be set at ~800 s, at which point 16,000 data points would have been collected. On the other hand, if the focus is on low-frequency behavior, then slower sampling rates will be needed. Suppose a 10-h run is needed to investigate slow variability. Ten hours represents 36,000 s; hence the sample interval should be 2.25 s in order to not exceed 16,000 data points in the record.

Data Aliasing

Data aliasing is a phenomenon that occurs when a signal containing high-frequency content is sampled too slowly. Figure 7 shows an example of aliasing in which a sinusoidal input signal with a period of 0.85 s (frequency of 1.18 Hz) is sampled once per second. The result is a fictitious, or "aliased," sinusoid with a period of ~6 s. The aliasing phenomenon occurs whenever the frequency content of the signal being sampled is faster than the Nyquist frequency of the sampling operation. In the example shown in Fig. 7, the frequency of the input signal is ~1.18 Hz, while the Nyquist frequency of the 1-s sampler is 0.5 Hz. The resulting alias frequency is always slower than the Nyquist frequency. Data aliasing in a digital control system is potentially dangerous, as the resulting low-frequency

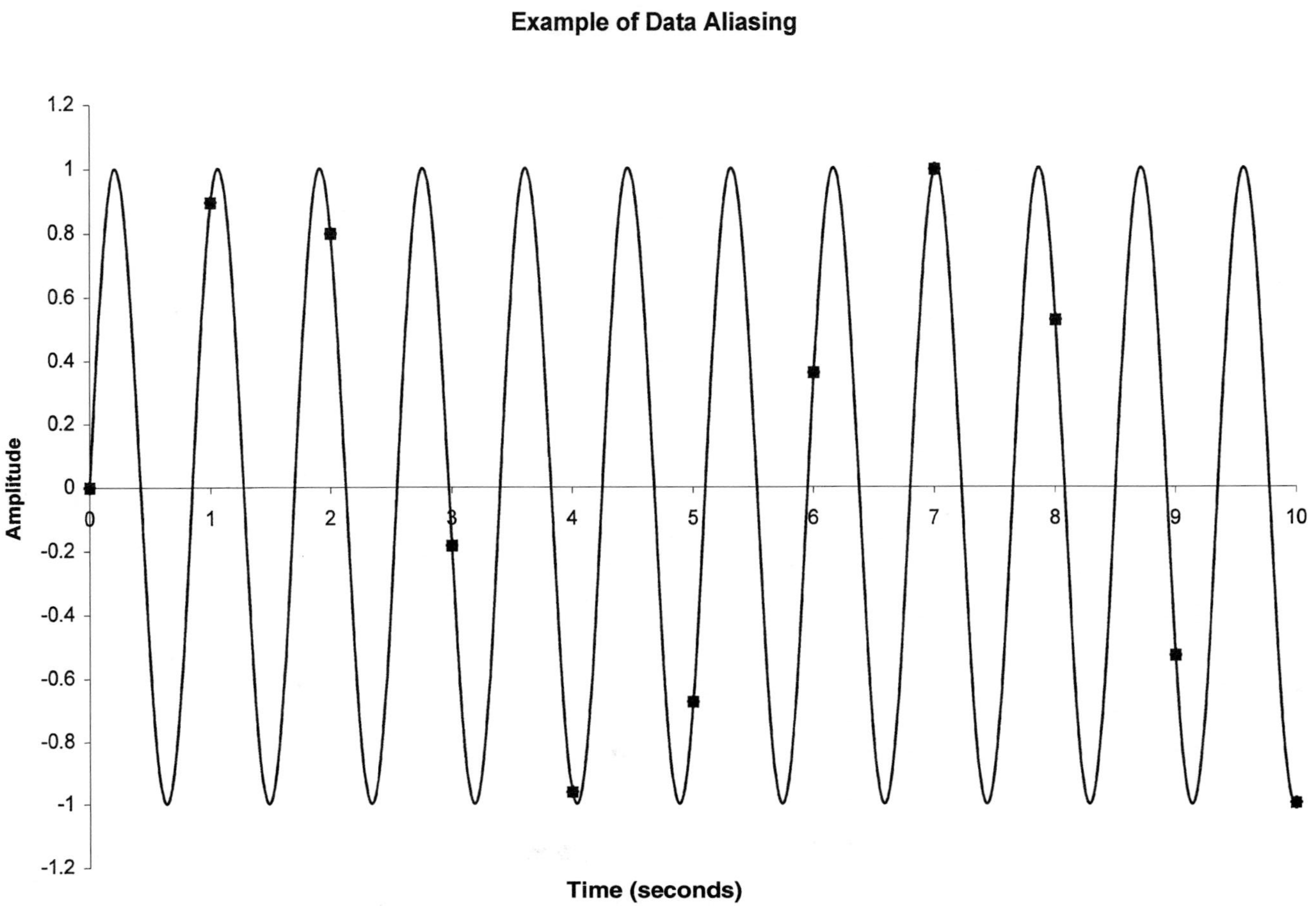

Example of Data Aliasing

FIGURE 7 Example of data aliasing – a 0.85 second input sine wave sampled every second producing an apparent 6 second aliased signal.

signal content will allow the control loop to chase this apparent variability. In turn this will introduce actual, yet unnecessary, variability into the process. The possibility of this happening in a typical DCS installation is real, especially for fast process variables, such as hydraulic pressures. Sampling rates for DCS variables are often chosen in order to keep digital processor loading manageable. Sample rates as slow as several seconds are common. Hydraulic pressures can have measurement time constants faster than a second. Such an example would guarantee that aliasing is going to occur. The presence of aliasing can be detected only by sampling the input signal at progressively faster sampling rates and comparing the results. Each time that the sampling rate is increased, a different result will be seen. Once adequately fast sampling rates are used, then further increases in sampling rate will show the same signal without change.

Antialiasing Filters

In most cases the sampling rate for a digital system is chosen without any knowledge of the frequency content of the signal to be measured. Good signal processing design would prevent aliasing by ensuring that the input signal is filtered before it is sampled. Such a filter is known as an antialiasing filter, and its purpose is to attenuate the signal amplitude at the Nyquist frequency of the sampler to the point that the resulting alias component will be so small that it will not matter. How much attenuation is needed to achieve this goal depends on the level of noise present at the Nyquist frequency, as well as an acceptable threshold. In absolute terms it would be desirable to attenuate any signal alias down to the least-significant bit of the analog–digital (A/D) conversion. If a 12-bit A/D converter is used, this means that attenuation down to 1:4096 or -72 dB is required. In practice it may be argued that the signal-to-noise ratio should also be considered. It can be argued that only -52 dB are needed if the signal-to-noise ratio is 10.

It often happens that signals are sampled several times. The first sampling occurs at the A/D converter, but then the signal is sampled again for other purposes. In a DCS, the first sampling occurs at the A/D converter; however, the controller often runs at a slower rate. To conform to good design practice, a second filter is needed before the signal is used for control. This is seldom the case in most DCSs. Another example involves data collection for time series analysis, as shown in Fig. 8 below. The example is based on a commercially available plant data acquisition package [10, 11]. In this case the analog signal is filtered by an analog filter before it is sampled by the A/D converter. Because the analog filter has a given cutoff frequency, the sampling rate of the A/D converter must be selected to ensure sufficient attenuation at the Nyquist frequency of the filter. After the data are sampled at T_1 seconds, it must be sampled again, in order to allow analysis at another sampling rate T_2. This requires that the signal be passed through another antialiasing filter to allow the second sampler to operate without aliasing. The second antialiasing filter is digital and is implemented in software and is executed at the A/D execution rate. Third-order filters with three equal time constants are used in this example for illustrative purposes only. Higher-order filters would normally be used. Assume that a third-order antialiasing analog filter with a time constant of 0.159 s is used. The transfer function

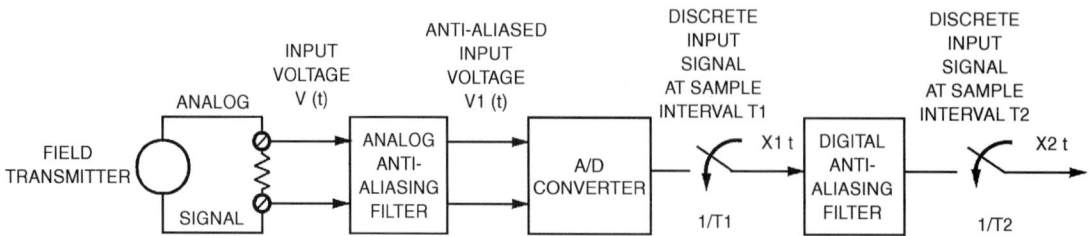

FIGURE 8   Anti-aliasing filter scheme for multiple sampling rates.

for the filter is

$$G_{\text{filter}}(s) = \frac{1}{(\tau s + 1)^3} \quad \text{with} \quad \tau = 0.159 \, \text{s}$$

The cutoff frequency of the filter is

$$\frac{1}{0.159(2\pi)} = 1 \, \text{Hz}$$

Suppose $-60 \, \text{dB}$ of attenuation is needed to achieve adequate antialiasing. The third-order filter achieves $-60 \, \text{dB}$ of attenuation per decade of frequency; hence $-60 \, \text{dB}$ of attenuation will be achieved at a frequency of 10 times 1 Hz, or 10 Hz. Hence 10 Hz should be the Nyquist frequency of the A/D sampler, giving a Nyquist period of 0.1 s, and a sampling interval for the A/D converter of 50 ms.

Suppose data collection at a sample interval of 2 s is desired. Hence the Nyquist period is 4 s and the Nyquist frequency of the second sampler is 0.25 Hz. Suppose $-40 \, \text{dB}$ of attenuation is considered adequate for the second filter, and a third-order filter is to be used for this purpose. A third-order filter achieves $-40 \, \text{dB}$ of attenuation at a frequency 4.5 times the cutoff frequency (corner frequency). Hence the corner frequency must be 0.25 Hz divided by 4.5 or 0.056 Hz. This can be achieved with a time constant of 2.86 s.

Statistical Analysis

Standard statistical analysis of time series data is important. Consider N readings of variable y_t. Important statistical measures are the mean, variance, and standard deviation. These are calculated as follows:

Mean: Mathematically the mean is defined as the expectation of y_t. This is written as

$$E(y_t) = \mu_y \tag{1}$$

The mean value of y_t can be approximated as

$$\bar{Y} = \frac{1}{N} \sum_{t=1}^{N} y_t \tag{2}$$

Variance: The mathematical expectation

$$E(y_t - \mu_y)^2 = \sigma_y^2 \tag{3}$$

The variance of y_t can be approximated as

$$\sigma_y^2 \cong \frac{1}{N} \sum_{t=1}^{N} (y_t - \bar{Y})^2 \tag{4}$$

The variance is expressed in the units of $(y_t)^2$. This is usually thought of as being in power units, to borrow from electrical engineering power, $= V^2/R$. In process variability analysis the concept of variance is important because the amount of variability present, or that which can be reduced, is in variance units.

Standard Deviation: The standard deviation is

$$\sigma_y = \sqrt{\sigma_y^2} \tag{5}$$

Standard deviation is the square root of the variance, and hence is expressed in the units of y_t.

Normal or Gaussian Distribution

An important concept in statistical theory is the central limit theorem, which holds that, as multiple random processes occur, the resulting probability distribution will be normal or Gaussian and will have the familiar bell-shaped curve. The normal distribution is symmetrical about the mean value and is distributed $\pm 3\sigma$ on either side of the mean value. Confidence limits of $\pm 2\sigma$ capture 95% of all readings.

It is common to assume that all probability distributions are normal. In process variability analysis it important to remember that this assumption is not always a good one. For instance, the distribution of a sine wave is bimodal and definitely not normal. When control loops cycle, they tend to appear as noisy sine waves; hence the notion that all variables have a normal distribution is generally not correct.

Two-Sigma.   A common way of expressing variability is by calculating two standard deviations, commonly referred to two-sigma, or 2σ. This has the property that it represents the 95% confidence limit. The units of 2σ are the same as those of y_t. For this reason 2σ is a more natural way of expressing the amount of variability about the mean value. In statistical process control (SPC) terminology, three standard deviations, or 3σ, are used frequently.

Variability—Two-Sigma Percentage of Mean.   Another way of expressing variability that has come into common usage in plant process variability auditing is 2σ expressed as a percentage of the mean value:

$$\% \text{ variability} = \frac{100(2\sigma)}{\text{mean}} \tag{6}$$

This is simply double the coefficient of variation in statistical notation. It has the useful property of being expressed as a percentage, as opposed to a specific set of units. This makes it possible to compare the percent variability of upstream and downstream variables by use of a common basis.

Stochastic Data Structures and Ideal Signals

Stochastic data are in some sense random by nature. In spite of this random nature, stochastic data do have structure [8]. Figure 9 shows four important examples: white noise, white noise plus a sine wave, filtered white noise, also known as autoregressive noise, and integrating-moving-average noise.

White Noise.   White noise is the name given to a signal that is absolutely random. The example shown in Fig. 9 is ideal white noise with a normal distribution. White noise derives its name from white light, which contains all of the colors of the visible spectrum. In the same way, white noise contains equal power at all frequencies. As a result, white noise has a perfectly flat power spectrum. It also has an autocorrelation function that is zero for all lags except lag 0. White noise is the mathematical abstraction of a purely random signal. The concept is used extensively in stochastic signal processing as a source for generating other signals.

White Noise plus a Sine Wave.   This is a simple yet realistic example of a signal that occurs in practice. It often happens that a signal consists of a dominant sine wave with noise added.

Filtered White Noise.   This is another simple yet realistic stochastic signal, which has the appearance of a real signal from a plant. This signal is generated when white noise is passed through a first-order filter. This type of noise is known as autoregressive noise. Passing white noise through a filter of some type is the method used to describe the characteristics of stochastic signals. Such a filter is known as a noise-shaping filter.

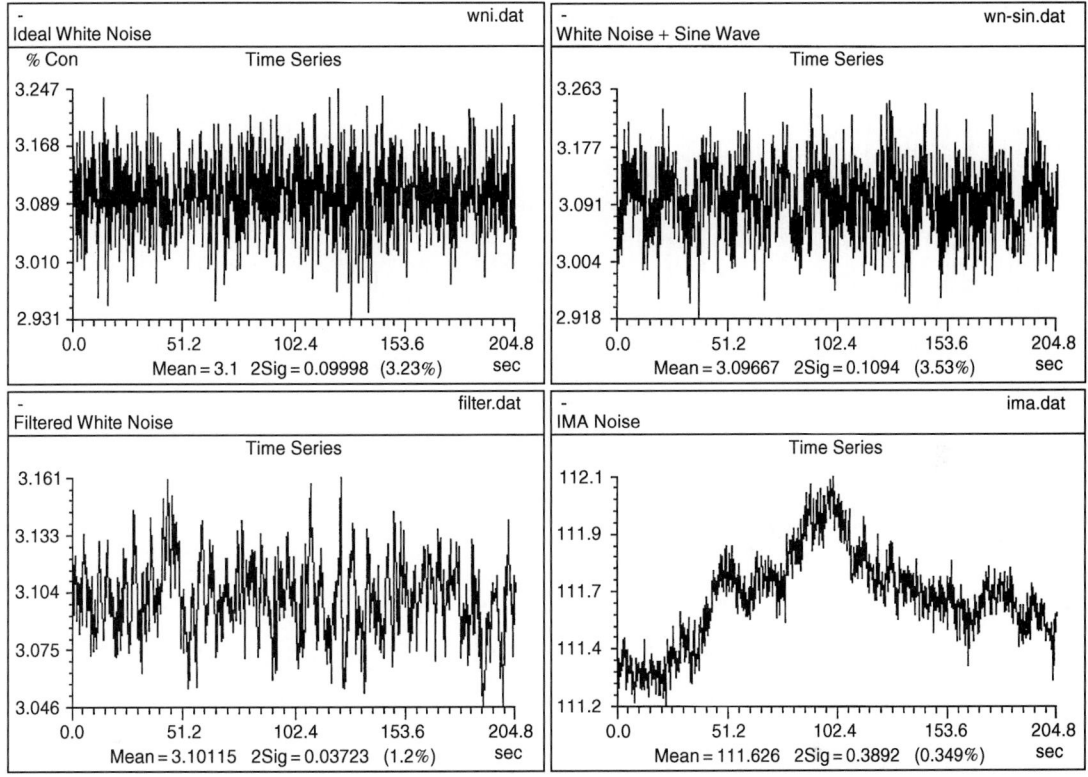

FIGURE 9 Examples of stochastic signals. Upper left: ideal white noise. Upper right: white noise plus a sine wave. Lower left: filtered white noise. Lower left: integrating moving average noise.

Integrating-Moving-Average Noise. This type of noise is characterized by a tendency of a random signal to drift, with white noise superimposed. This type of signal is quite characteristic of normal plant behavior.

Histogram

The histogram is a statistical frequency-distribution plot. Figure 10 shows two examples. The upper plot shows white noise with a nearly normal distribution. The mean value is 3.101. The two-sigma is 0.0999, or 3.22%. The standard deviation is ~0.05, hence, three-sigma is 0.15. As a result a normal distribution should extend from 2.95 through to 3.25. This agrees fairly well with the histogram, which does have a few outliers.

The lower plot of Fig. 10 shows eight and a half cycles of a sine wave. The period is 25 s, and the amplitude is 1.0. The mean is 3.101, and the two-sigma is 0.1417, or 4.47%. The histogram shows a strong bimodal distribution extending from 3.0 up to 3.2. This agrees with the time series plot that extends from 3.0 to 3.2. The bimodal nature of the histogram is due to the nature of the sine wave, which spends most of the time at the extremities of the sinusoid and relatively little time near the mean value. It is interesting to note the two-sigma is 0.1417. The normal interpretation would be that 95% of all readings of this signal would be contained within $+/-0.1417$ and that a further 5% of all readings would be outside this range, assuming a normal distribution. In fact, the sine-wave

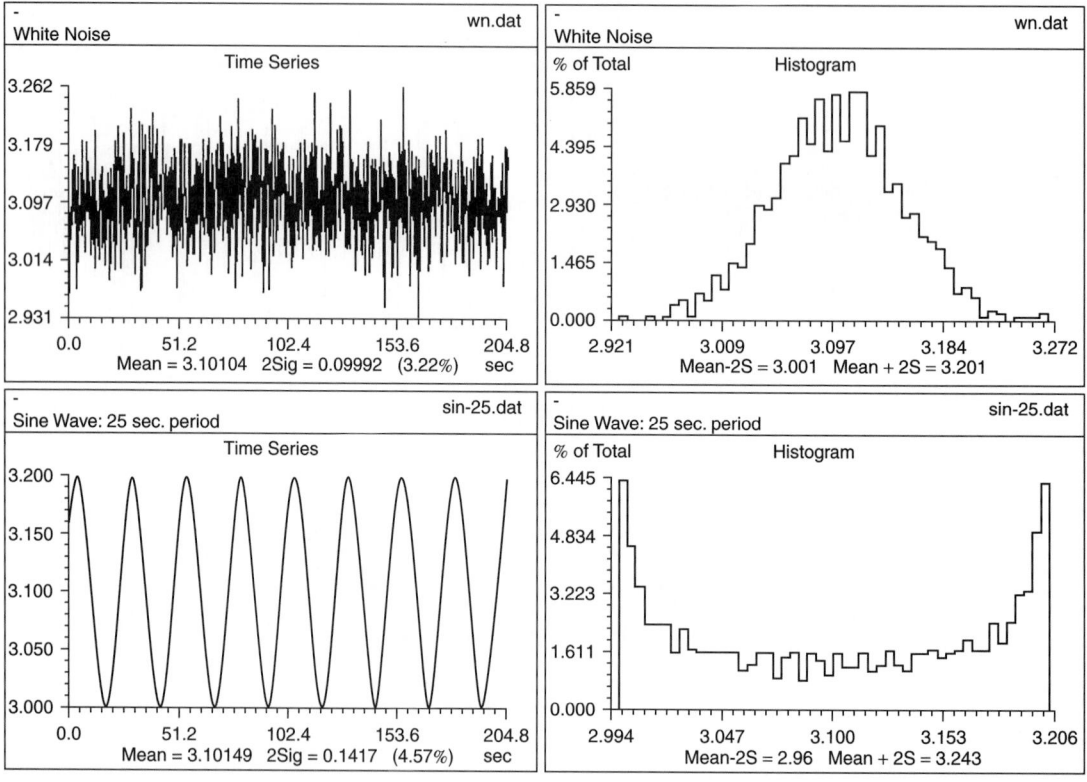

FIGURE 10   Examples of histogram plots: upper left: random noise (white noise) time series, upper right: histogram (normal distribution) lower left: sine wave (25 second period) time series, lower right: histogram (bi-modal).

amplitude is exactly 1.0; hence no readings extend as far as $+/-0.1417$. The two-sigma of 0.1417 can be explained as follows.

The variance of a sine wave is

$$\sigma^2 = \frac{1}{2}A^2,$$

where A is the sine-wave amplitude:

$$A = 1.0, \quad \sigma^2 = 0.5, \quad \sigma = 0.707107, \quad 2\sigma = 1.4142$$

This agrees well with the result shown in Fig. 10.

Spectral Analysis

Spectral analysis is a key component of plant process variability analysis and is used to determine the frequency content of signals. The key techniques include the Fourier transform, the fast Fourier transform (FFT), power spectrum, and cumulative spectrum. It is helpful in diagnosing variability cause-and-effect relationships and also for analyzing control loop behavior. The techniques are based on the work of the French mathematician Jean Baptiste Fourier (1768–1830). The basic idea is that any signal can be expressed as the sum of a series of sine and cosine waves at different frequencies, from very fast to very slow. These fast and slow sine waves are called harmonics. Although Fourier developed the mathematics for the continuous Fourier series—he was analyzing the vibration of violin strings—the discrete Fourier series is the tool needed to analyze time series data. Consider a time series signal y_t containing N equally spaced readings taken every T_s seconds. There are $n = N/2$ harmonics present in the Fourier series y_t, and can be expressed as the discrete Fourier series:

$$y_k = \frac{1}{2}A_0 + \sum_{m=1}^{n-1}[A_m \cos(2\pi f_m T_s k) + B_m \sin(2\pi f_m T_s k)] + \frac{1}{2}A_n(-1)^k \tag{7}$$

for $k = 0, 1, \ldots, N-1$, and there are n harmonics,

$$f_m = \frac{m}{NT_s} \tag{8}$$

for $m = 1, 2, \ldots, n-1$.

The lowest-frequency harmonic is called the fundamental and is

$$f_1 = \frac{1}{NT_s} \tag{9}$$

and has a period of $T_1 = NT_s$. The highest-frequency harmonic is called the Nyquist frequency and is

$$f_N = \frac{1}{2T_s} \tag{10}$$

and has a period of $T_N = 2T_s$. The mth harmonic is given by

$$h_{m,k} = A_m \cos(2\pi f_m T_s k) + B_m \sin(2\pi f_m T_s k) \tag{11}$$

Figure 11 illustrates the concept. Shown is one cycle of a square wave, which is represented by 16 data points. Hence $N = 16$ and $n = 8$. Note that square waves, because of their symmetric nature, contain only odd harmonics. Hence only the first, third, fifth, and seventh harmonics have nonzero values. The A_m and B_m Fourier coefficients are listed in Table 1.

For all 16 values of k in Fig. 11, the sum of the harmonics equals the original data exactly.

It is important to appreciate that the number of harmonics is fixed and is equal to one half of the number data points. The fastest harmonic is the Nyquist, and the frequency is fixed by the sampling interval. The slowest harmonic is the fundamental. This frequency is determined by the length of the data run. All of the other harmonics have fixed frequencies determined by Eq. (8). The Fourier transform treats these fixed frequencies as the bins of a histogram and attempts to represent the power at the actual frequencies as best as it can at bins on either side of the actual frequency. This process is known as leakage between bins.

TABLE 1 Fourier Harmonics and Coefficients

Harmonic	Period	Frequency	A_m	B_m
1	16	0.0625	0.25	1.257
2	8	0.125	0	0
3	5.33	0.188	0.25	0.374
4	4	0.25	0	0
5	3.2	0.313	0.25	0.167
6	2.27	0.375	0	0
7	2.29	0.438	0.25	0.05
8	2	0.5	0	0

Square Wave & 1st, 3rd, 5th, 7th Harmonics

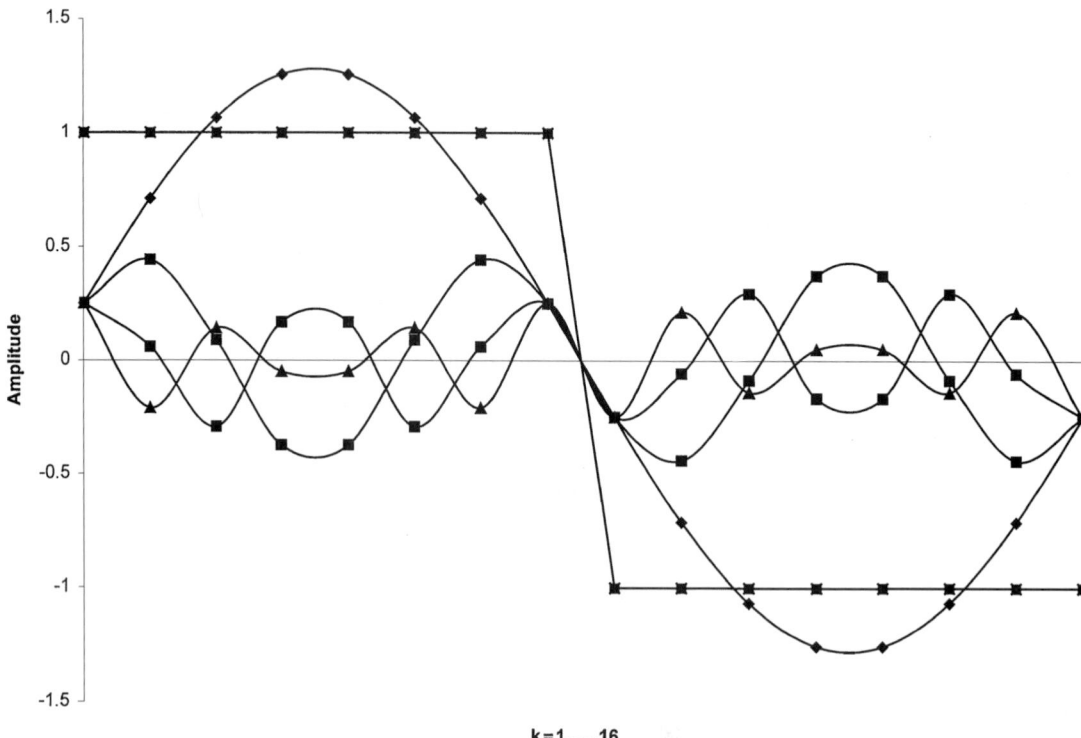

k=1,…,16

FIGURE 11   One cycle of a square wave and its harmonics.

Fourier Transform

The Fourier transform transforms the time series signal into the frequency domain by calculating the *A* and *B* Fourier coefficients for each harmonic, which in turn represent the frequency content of the signal. The *A* and *B* coefficients can also be represented in magnitude and phase notation. In either case, the Fourier transform is a transformation of the amplitude information of the time-domain signal into the frequency domain. Once the coefficients are known, the inverse Fourier transform will allow the reconstruction of the original time series data.

Fast Fourier Transform

The calculation of the Fourier transform is computationally intensive. The number of calculations is proportional to N^2. For a large number of data points the calculation of the Fourier transform can be a slow process computationally. The FFT is a computationally optimized algorithm that calculates the exact transform. The only limitation is that the number of data points must conform to a power of 2; hence, 512, 1024, 2048, 4096, etc. The number of calculations is proportional to N; hence the calculation is much faster. The only complication is if the number of data points is not equal to a power of 2. Most FFT algorithms will simply do the calculation on the basis of the first points that conform to a power of 2. Hence, if 1023 data points are available, the FFT will perform the calculation on the first 512 points. The consequence is that only approximately one half of the data is used. For this reason, most data collection algorithms attempt to collect a binary number of data points. Other algorithms attempt to "pad" the data so that it conforms to a binary number. For instance, if only 1023 data points are available, one additional point is added. Typically the value is set equal to the average for the data run. Naturally, the more padding that is done the more errors that are introduced.

Power Spectrum

The power spectrum is a calculation of the distribution of the variance over the frequency spectrum. The power spectrum is the most useful of the spectral analysis tools and is the most frequently used tool in plant data analysis. The calculation of the power spectrum relates to the Fourier coefficients as follows. For each harmonic, the variance is

$$V_m = \frac{1}{2}(\text{amplitude})^2 = \frac{1}{2}\left(A_m^2 + B_m^2\right) \tag{12}$$

The power spectrum for the square wave of Fig. 11 is shown in Fig. 12. There are 8 data points corresponding to each harmonic. From Parseval's theorem it is known that variance of the time series data is equal to the sum of the variances of all of the harmonics. Hence, the power spectrum is a variance, or power, versus frequency plot and is sometimes known as a periodogram.

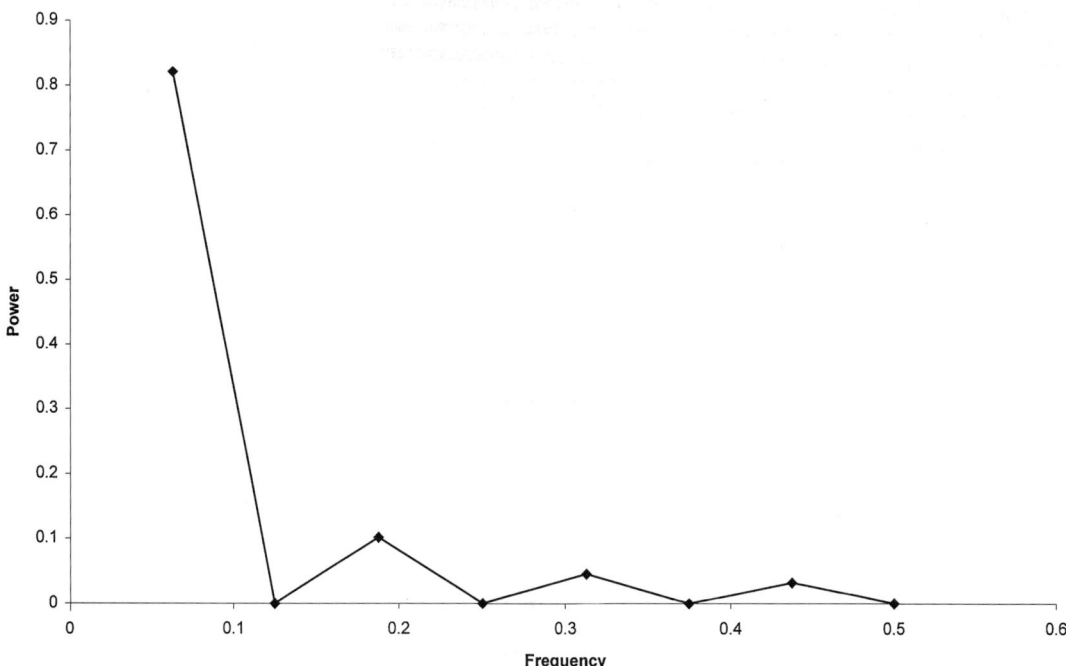

FIGURE 12   Power spectrum of Figure 11 square wave.

Cumulative Spectrum

The cumulative spectrum determines the percentage of the total variance that is occurring at a given frequency. This is a very useful diagnostic tool that helps determine if a cycle at a particular frequency is contributing a significant amount of variance or not. The cumulative spectrum is simply a plot of the percentage variance contribution, from 0% to 100%, as the frequency increases from the fundamental to the Nyquist.

Spectral Analysis Plotting Methods

There are many ways of plotting power spectra, and it is important to identify how each method alters the usefulness of the information. Three methods are shown in Fig. 13.

1. Variance versus frequency (Fig. 13, upper right plot). This is the most common method of plotting spectral information. It is used by most software packages. It is common in certain types of plant analysis, such as vibration analysis. For the purpose of plant process variability analysis this method is the least useful. It tends to deemphasize the low-frequency information, which is often of great interest, while the high-frequency end of the spectrum often contains relatively little information as the sensor has filtered it out.

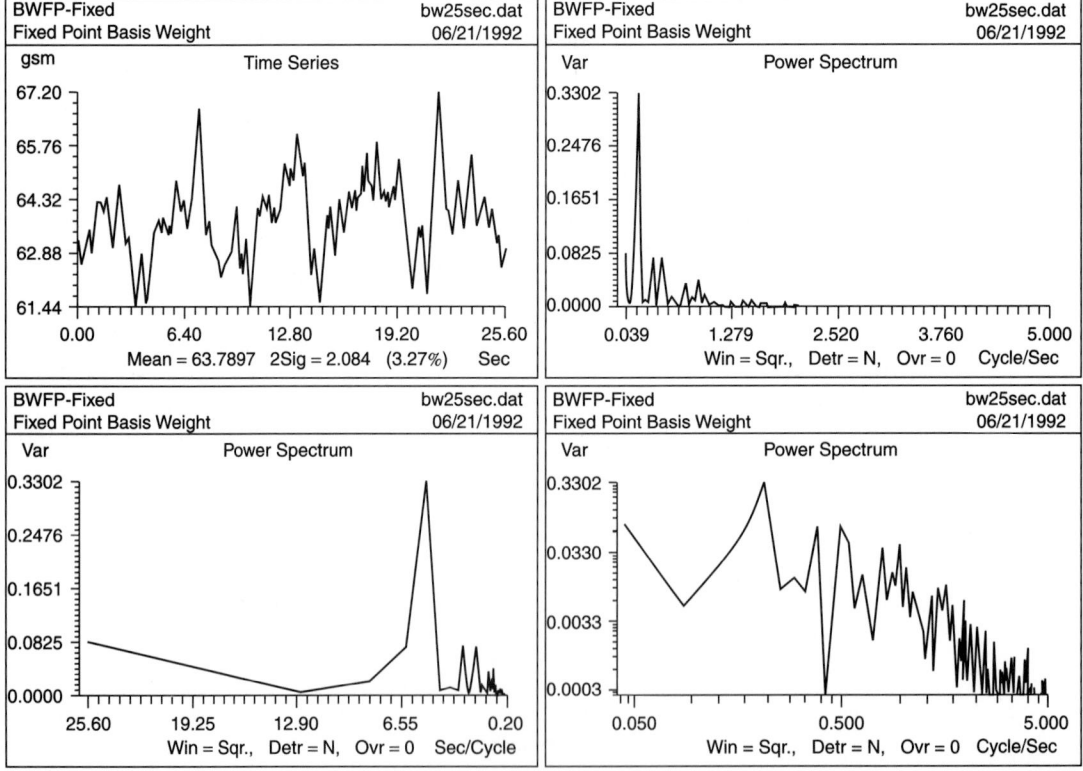

FIGURE 13   Three ways of plotting a power spectrum. Upper right – power vs frequency, lower left – power vs period, lower right – log power vs log frequency.

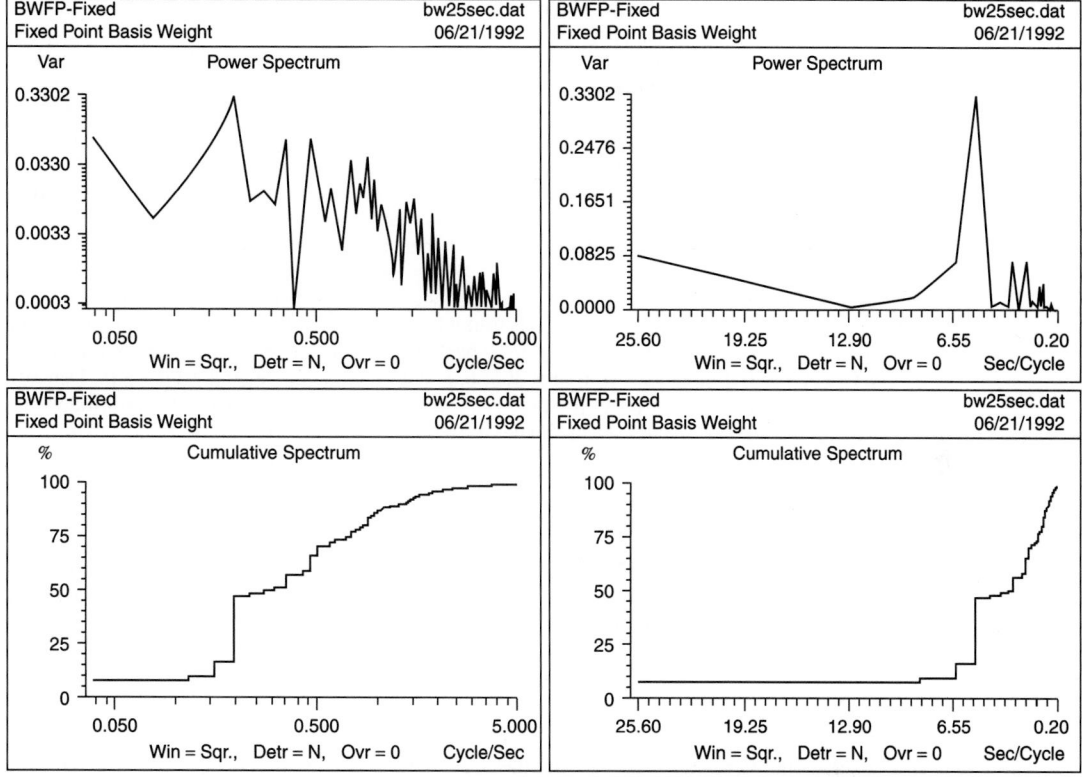

FIGURE 14   Power spectra and cumulative spectra for Figure 13 data. The 6 second cycle represents about 40% of the total variance.

2. Variance versus period. This is an unusual method of plotting and is illustrated in Figs. 3, 5, 6, and 13 (bottom left). This method provides an intuitively obvious interpretation of data that have strong cyclic content. It is useful for human interpretation in a plant production environment, where there may be little formal training in frequency response methods. A disadvantage of the method is that it skews the frequency scale and overemphasizes the low-frequency content.

3. Log variance versus log frequency. This method is illustrated in Fig. 13, bottom right. The method is especially useful for data structure identification and modeling, as the plot has a strong similarity to a Bode plot. In Fig. 13 (lower right) the plot looks very much like a Bode plot with a corner frequency of about ~1 Hz. This would suggest that the data could consist of white noise passed through a first-order filter with a time constant of ~0.159 s.

Figure 14 shows power and cumulative spectra for the same data shown in Fig. 13. The plots include both log–log and period presentations. The 6-s cycle evident in the data represents ~40% of the total variance, as is evident from the cumulative spectrum. If it were possible to remove this cycle completely, the variance would be reduced by 40%, while the standard deviation would be reduced by the square root of this amount, or ~20%.

Spectral Analysis Windowing and Detrending

The Fourier transform attempts to fit harmonics to the time series data exactly. Because the time series data are sampled, and in many cases the data are also nonstationary (their character changes over time), methods are needed to smooth the spectral information and increase the confidence limit for the variance present at any given frequency. Two methods of smoothing are available: windowing and overlapping. Windowing results in a smoothed power spectrum that some windowing techniques achieve by smoothing the time series data before the Fourier transform is taken, while others smooth the power spectrum. Common windowing techniques include square (no windowing applied), Daniel, Parzen, Welch, and Hanning [8]. Another method of smoothing involves breaking up the time series data into overlapping sections, performing spectral analysis of each section independently, and averaging the final results. The overlapping process can be controlled by specifying the number of overlapping segments to be used in the calculations. Overlapping will have an impact on the resulting spectral analysis by shortening the period of the fundamental frequency that is due to the overlapping.

When data contain a tendency to drift, the resulting power spectrum will interpret the drift as a strong low-frequency component. This low-frequency component may "swamp" the remaining spectral information. The detrending option is useful to avoid this tendency. It involves preprocessing the original time series data through a first-order backward difference in order to remove the trend.

The power spectral plots shown in Figs 3, 5, 6, and 14 include the legend "Win=Sqr., Detr=N, Ovr=0." In this time series software package [7], this means that a square window (no windowing) was selected, detrending was not selected, and overlapping segments were not selected either.

Cross-Correlation and Autocorrelation Functions

Time series data can also be analyzed in the time domain, and the two functions that are particularly useful are the autocorrelation and cross-correlation functions. Equation (4) gives the calculation for variance. For two time series variables x_t and y_t, the cross covariance is calculated as follows:

$$C_{xy}^2(k) = \frac{1}{N} \sum_{t=1}^{N-k} (x_t - \bar{X})(y_{t+k} - \bar{Y}) \tag{13}$$

This calculates the correlation between variables x_t and y_t for different time lags k. The cross covariance can be normalized by dividing by the standard deviations of x_t and y_t to produce the cross-correlation function:

$$r_{xy}(k) = \frac{C_{xy}^2(k)}{\sigma_x \sigma_y} \tag{14}$$

The cross-correlation function is a plot of the correlation coefficient between $+1.0$ and -1.0 for different time lags between the data sets. The interpretation is that 1.0 represents perfect positive correlation, 0.0 represents no correlation, and -1 represents perfect negative correlation.

Cross-Correlation Function Use.   The cross-correlation function is useful to provide insight into the relationship between two variables. In particular, it is used to investigate if one variable is related to another through a time delay.

Autocorrelation.   The autocorrelation function for a time series signal y_t is based on the autocovariance function, which is calculated as

$$C_{yy}^2(k) = \frac{1}{N} \sum_{t=1}^{N-k} (y_t - \bar{Y})(y_{t+k} - \bar{Y}) \tag{15}$$

The autocorrelation function then is

$$r_{yy}(k) = \frac{C_{yy}^2(k)}{\sigma_y^2} \tag{16}$$

The autocorrelation function is a plot of the regression coefficient for a time series signal regressed with a time-lagged version of itself. The function varies from $+1.0$ to -1.0 and always has a value of $+1.0$ for a time lag of zero. It represents an effective way to test a signal for degrees of randomness, as a completely random signal (white noise) has an autocorrelation function that is zero for all lags except lag zero. The interpretation of zero correlation really means inside the confidence limits. The confidence limits for an autocorrelation function are calculated as $\pm 2/\sqrt{N}$, where N is the number of data points. Hence for 1024 data points the confidence limits are at $+/-0.0625$.

Autocorrelation and Minimum Variance Control.   The autocorrelation function also serves an important purpose as a control performance benchmark. From minimum variance control theory [12] we know that the lowest variance that a control loop could ever have is if minimum variance control were achieved. Under this condition, the resulting control signal would be perfect white noise, as the controller would remove all nonrandom variability. In the presence of dead time in the loop, perfect white noise could not be achieved, and under this condition the autocorrelation function would be zero for all lags except those less than the dead time plus one sample.

CONTROL LOOP PERFORMANCE FOR UNIFORM MANUFACTURING

How should loops be tuned to enhance, and not hinder, the manufacture of uniform product efficiently? To answer this question the manufacturing requirements for each unit operation in the process must be identified and translated into dynamic performance requirements. A controller tuning strategy must be formulated that translates these into a desired speed of response for each loop and coordinated dynamics for all of the control loops in the plant. At the heart of this concept lie model-based controller tuning methods, such as IMC and Lambda tuning [13–15], all of which require the selection of the closed-loop time constant—often called Lambda, for each loop before the tuning process taking place. Lambda tuning is discussed in the next section. It achieves closed-loop responses that are first order by nature and have a selectable time closed-loop constant. It differs from earlier tuning methodologies such as those of Ziegler and Nichols [16], Cohen and Coon [17], and integral performance criteria methods [18], all of which achieve a closed-loop performance that is in some sense as fast as possible and is also oscillatory. Uniform manufacturing requires two key criteria. First, oscillatory loop tuning is unacceptable as it induces resonance and amplifies variability. Second, the speed of response of control loops should be selected to suit the manufacturing requirements. This requires the selection of the closed-loop time constant for each loop. This affects the robustness of the control loop—the faster the loop, the less robust; this issue is dealt with in the next section. In this section, the focus is on how to determine the closed-loop time constant in such a way that manufacturing is enhanced. The fine paper machine example is used to illustrate this tuning process.

Identifying the Manufacturing Requirements—Fine Paper Machine Example

Consider Fig. 2, which shows the blend chest area of the fine paper machine example. Fine paper is manufactured from two raw materials—hardwood pulp and softwood pulp. Hardwood fibers are short and provide the surface smoothness needed for fine paper grades. Softwood fibers are long and are needed to provide sheet strength. In a typical operation, a 70%-to-30% hardwood-to-softwood blend is required. In the example shown in Fig. 2, hardwood pulp is pumped from the hardwood chest. After consistency control, it is passed through a refiner to improve the fiber bonding properties. It is then flow controlled into the blend chest. Exactly the same procedure is used for the softwood pulp. The blend chest level controller cascades its output to the setpoints of the two flow controllers in order to maintain level while at the same time maintaining the correct blend ratio. The blended pulp is pumped out of the blend chest and is consistency controlled as shown in Fig. 15 before being pumped into the machine chest under both flow and level control. From the machine chest, the pulp is diluted, cleaned,

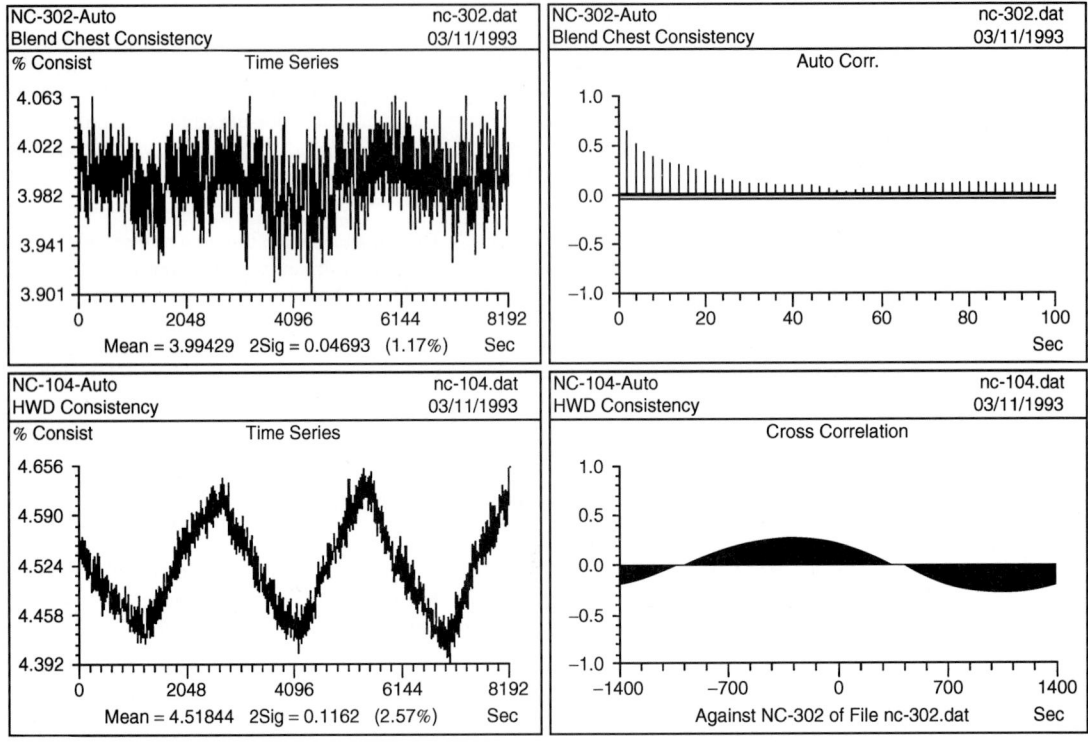

FIGURE 15 Examples of auto and cross correlation functions.

screened, and pumped to the headbox. From the headbox it discharges onto the wire and forms a sheet of paper.

Ziegler–Nichols Tuning

The oldest and best-known tuning method is the one developed by Ziegler and Nichols in 1942 [16]. This method is used for illustration purposes in this example, although any of the earlier methods [17, 18] could also have been used as effectively. If the Ziegler–Nichols tuning method was used to tune the blend chest flows, then the setpoint responses of the hardwood (FC-105) and softwood (FC-205) flow loops might look like that shown in Fig. 16. This shows the response of both loops to a step change in blend chest level controller LC-301 output in manual mode. Both responses have a quarter-amplitude-damped decay ratio, as suggested by the Ziegler–Nichols tuning method. They also have different settling times as a result of differences in open-loop dynamics that are due to differences in pumps, line dimensions, flow meters, and control valves. The proponents of fast oscillatory tuning argue [18] that the positive half cycles of the transient response are averaged out by the negative half cycles. However, in the case of solid product manufacture such as paper, this does not translate into acceptable product. Figure 17 shows the impact of the oscillatory tuning on the fiber blend ratios entering the blend chest. The blend ratios vary by 5% with a period of ~20 s throughout the transient response. Assuming a paper machine speed of 1000 m/min and with downstream mixing neglected (there is no reason to require effective mixing to overcome the effects of badly tuned loops), the Ziegler–Nichols tuning will result in paper with smoothness and strength properties that alternate

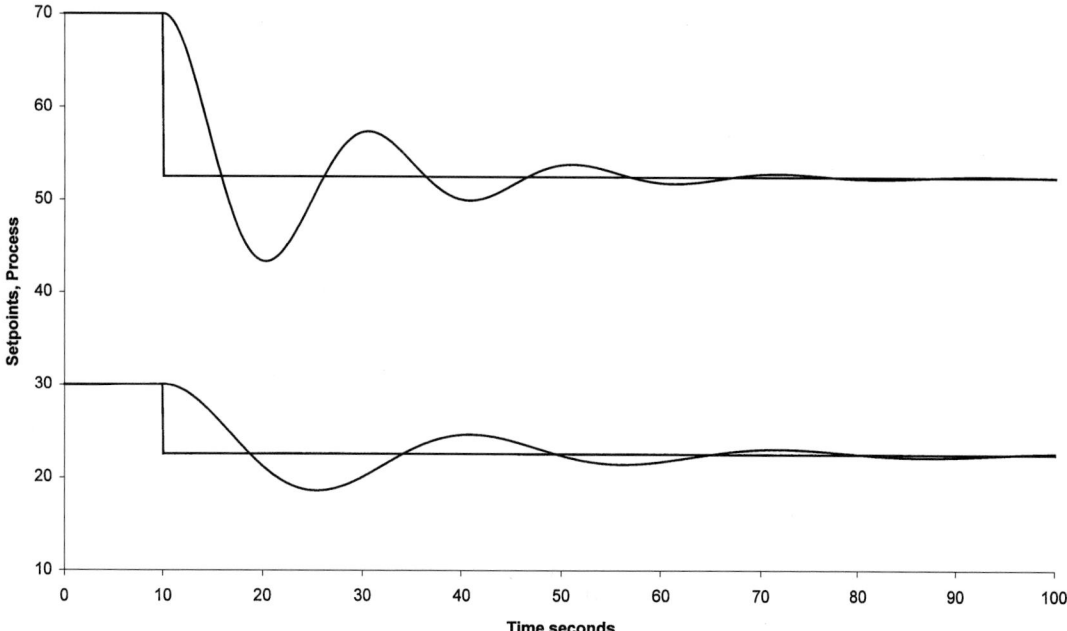

70% Hardwood Flow, 30% Softwood Flow
25% Setpoint Step Changes - Z-N Tuned Loops

FIGURE 16   Blending problem with flow loops tuned using Ziegler-Nichols.

every 170 m. This will cause most of the product to be unsatisfactory as the level controller changing demand will cause the flow loops to cycle continuously.

Clearly, the oscillatory responses and unequal speed of the two loops has resulted in a significant, yet unnecessary disturbance being created in the blend ratio. Oscillatory tuning offers no beneficial effect for uniform manufacturing. This example serves to illustrate the need for coordinated tuning for these two flow loops from a manufacturing standpoint. Clearly, the requirement for the hardwood and softwood flow is to respond in a nonoscillatory manner, with exactly the same speed of response. If both loops were tuned to have first-order responses with equal closed-loop time constants, this would achieve the setpoint change without causing an upset to the blend ratio. This is illustrated in Figs. 18 and 19, which show the impact of Lambda tuning of both loops to a closed-loop time constant of 20 s. The transient lasts ~80 s, while the blend ratio is maintained absolutely constant. This is only one example of how the selection of a specific speed of response for multiple control loops has an impact on uniform manufacturing.

Coordinated Loop Tuning Based on Operational Requirements

The previous example serves to illustrate that the fastest possible speed of response for each loop does not in any way guarantee uniform manufacturing of a product in a continuous plant. In fact the example clearly shows the need to carefully select the desired closed-loop time constant in order to avoid unnecessary disturbances. The paper machine blend chest of Fig. 2 will be used to illustrate other rules for making the control loop speed of response decisions. Collectively these rules can be used to evolve a tuning strategy for uniform manufacturing.

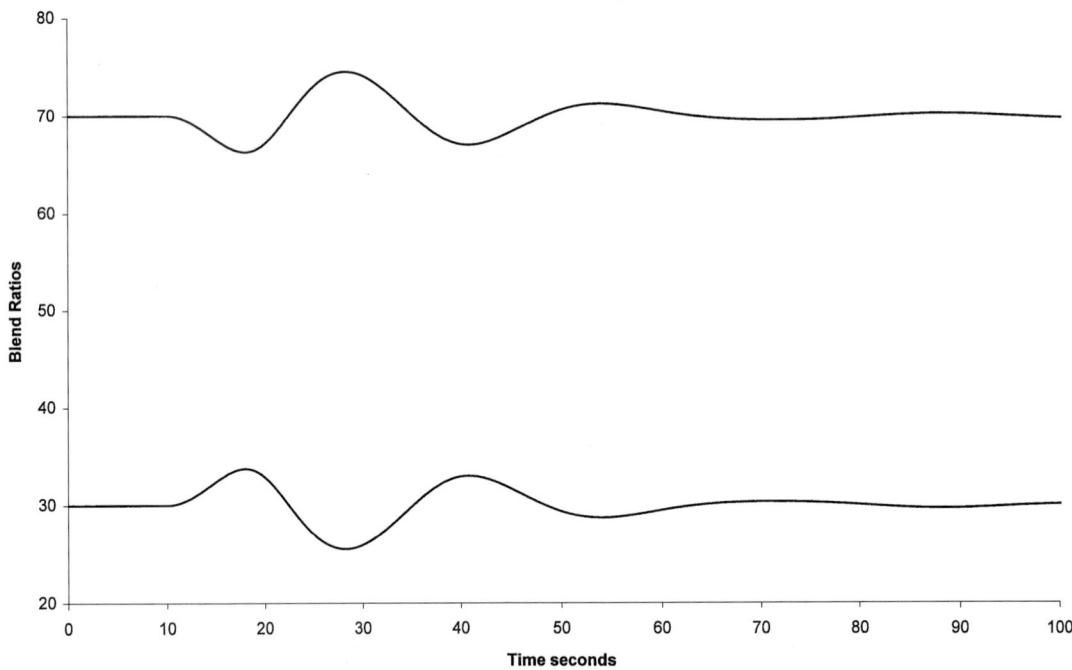

FIGURE 17   Blend ratios vary due to Ziegler-Nichols tuning of flow loops.

Rule of Thumb for Maintaining Constant Ingredient Ratios

The paper machine blend flow example can be generalized to the feed of multiple ingredients to any reactor, which requires the ratio of the ingredients to be maintained constant. Assuming that all ingredient flows are controlled by a common cascade control strategy, all of the flow loops should be tuned to have the same closed-loop time constant.

Rules of Thumb for Process Interaction

Control loops often interact because of process coupling. If both loops are tuned for the same closed-loop time constant they will probably cycle as a result. A good rule of thumb is to select the loop that is most important from an operational point of view and tune it to be as fast as possible, in keeping with the need to maintain robustness and avoid resonance. The next most important of the other interacting loops should then be tuned 5–10 times slower than the loop already tuned. As an example, the consistency loops and the flow loops for both pulp streams are known to interact as a result of process coupling. A change in stock flow immediately results in a need to change dilution water if the consistency is to be maintained constant. On the other hand, a change in consistency will alter the fluid friction in the piping system, and as a result will cause the flow to change. The operational requirement is to deliver a constant mass flow of fiber to the refiners. The refiners are critically important as they determine the degree of fiber bonding and hence the strength of the paper sheet. In view of the need, the consistency loops are probably more important than the flow loops. A reasonable tuning strategy might be to tune the consistency loops to be as fast as possible, subject to

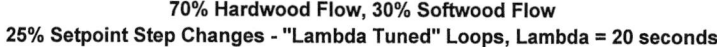

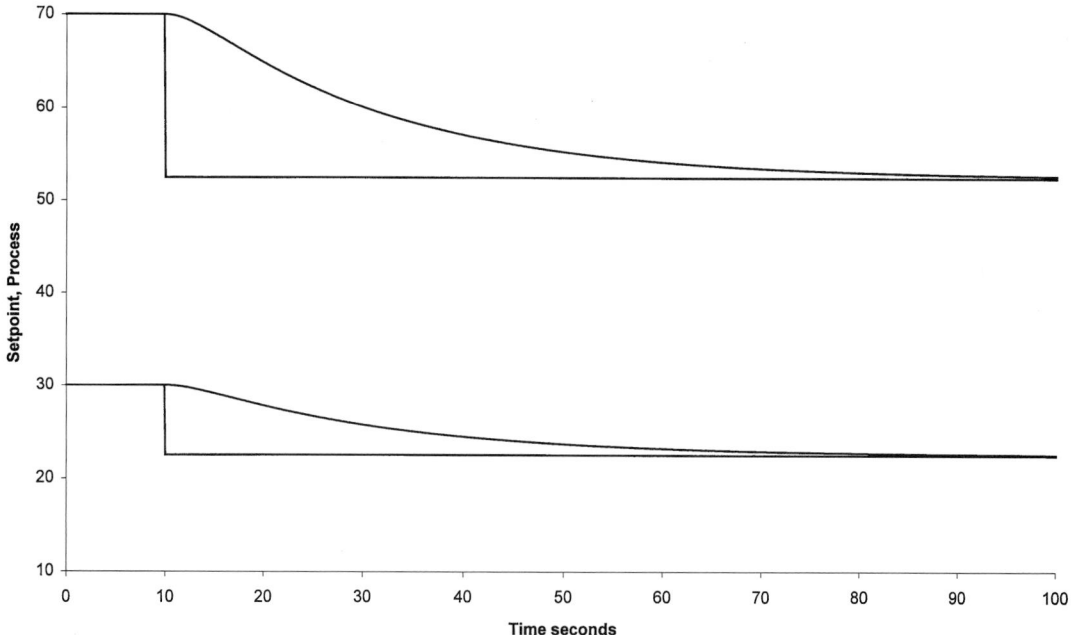

FIGURE 18   Blending problem with flow loops tuned for Lambda = 20 seconds.

considerations about resonance that is due to the dead time present in the consistency loops. In turn, it would be reasonable to tune the flow loops to be 5 or 10 times slower than the consistency loops, so as not to inadvertently upset them dynamically as the flow changes. At the same time, the flow loops must be tuned to have identical speeds of response to maintain the blend ratio.

Rules of Thumb for Cascade Loops

Another way that control loops can interact is through cascade control, with the output of one loop becoming the setpoint of another. In general the 5–10-times rule used for solving process interaction can also be applied to the inner (faster) and outer (slower) cascade loops. When a level controller is the outer loop, the rule should be more stringently set at 10 times faster, because of the sensitivity of level controllers to slow inner loop dynamics. For the paper machine example, the flow loops should be at least 10 times faster than the blend chest level loop.

Rules of Thumb for Buffer Inventory Storage Level Control

The blend chest is a buffer inventory storage, which by design is intended to decouple the downstream manufacturing requirements of the paper machine from the upstream pulp blending process. As such, the blend chest level controller should be tuned as slowly as possible. The blend chest has a residence time of typically 15 min. The operational need for the blend chest level controller is to ensure that the blend chest never overflows and that the level never sinks below some safe level with respect to the agitation zone. The lower limit on the speed of response of the flow loops is at least 10 times faster

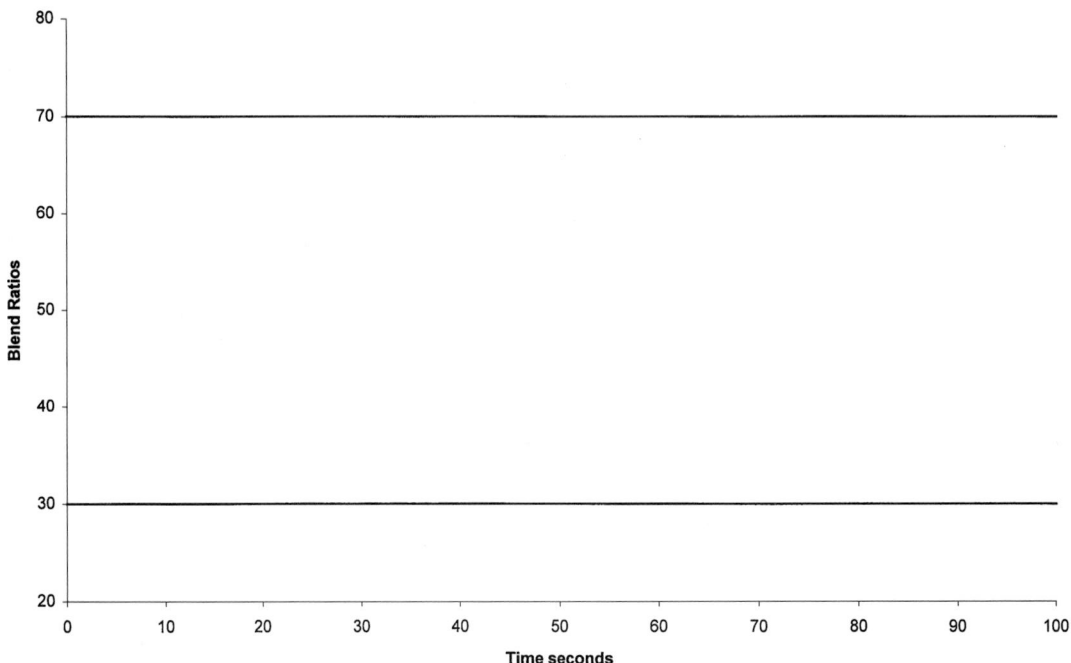

FIGURE 19   Blend ratios are constant due to Lambda tuned flow loops.

than the blend chest level controller. In order to allow the blend chest to work effectively as a buffer inventory, the level control should be tuned as slowly as possible while ensuring that the production rate changes expected on the paper machine will not cause overflow and underflow problems. A simple rule of thumb to achieve this is to set the closed-loop time constant for the level controller to be equal to the residence time of the tank. More detailed considerations should take into account the level setpoint, relative to the total tank volume, as well, as a priori knowledge of the expected downstream flow demand changes that must be accommodated. These considerations should result in a closed-loop time constant selection that could well be longer than the residence time of the tank.

Tuning Rules of Thumb for Uniform Manufacturing—Summary

The tuning requirements illustrated with the blend chest problem above can be generalized as follows:

1. Loops should never be tuned to cycle—control should be smooth.
2. Each loop should be tuned for a specific speed of response as dictated by a plant coordination tuning strategy.
3. The tuning method should be applicable to all loops.
4. The tuning method should identify process dynamics and actuator nonlinearities.
5. Tuning should be robust for all operating conditions.
6. All reactor ingredient feed controls should have equal speeds of response so as not upset the ingredient ratio—feed flows into a blend tank, chemical feed flows, boiler fuel, and air flows.

7. Key loops should be 5–10 times faster than other loops coupled through process interaction

8. Inner cascade loops should be 5–10 times faster than outer loops (10 times faster for level control).

9. Inventory storage tank level controls should be as slow as possible while satisfying production needs.

CONTROL LOOP PERFORMANCE, ROBUSTNESS, AND VARIABILITY ATTENUATION

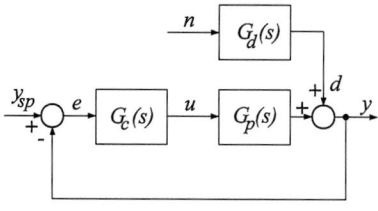

FIGURE 20   Control loop structure.

Before loops in a plant are turned with the hope of reducing process variability, it is important to understand the control loop performance-robustness envelope, as well as the ability of a control loop to attenuate process variability. Figure 20 shows a control loop block diagram in which G_C is the controller transfer function, G_P is the process transfer function, y_{sp} is the setpoint, and y is the measurement. Process disturbances enter the loop at d, which can be modeled as white noise n passed through the disturbance transfer function G_d. This in turn can be shaped to fit realistic conditions. The loop responds simultaneously to both the setpoint y_{sp} and the disturbance d. The setpoint response is given by the forward, or transmission, transfer function $T(s)$:

$$T(s) = \frac{y}{y_{sp}} = \frac{G_C G_P}{1 + G_C G_P} \tag{17}$$

The disturbance or load response is given by the sensitivity function $S(s)$:

$$S(s) = \frac{y}{d} = \frac{1}{1 + G_C G_P} \tag{18}$$

$S(s)$ determines what a control loop can or cannot do, regarding variability attenuation under the assumption of linear dynamics.

Resonance and Bode's Integral

An important property of the sensitivity function that was initially reported by Henrick Bode in 1945 and is known as Bode's integral [19] (also known as the area formula [20] in more modern work) states that

$$\int_0^\infty \log |S(j\omega)| d\omega = 0 \tag{19}$$

for all cases with stable (left half-plane) poles, as long as

$$G_C G_P \neq \frac{k}{s} \tag{20}$$

What Eq. (19) states is that the log of the amplitude ratio (AR) integrates to zero over all frequencies. Practically this means that control loops do not in fact reduce variability; they only redistribute the variability over the frequency spectrum. This is really bad news for achieving variability reduction through control, as the ability to attenuate variability at low frequencies is offset by the tendency of the control loop to resonate and amplify variability at higher frequencies. Inequality (20) provides the only exception to this rule, and offers a ray of hope for variability reduction. Inequality (20) excludes the special in which when the loop transfer function—the product of the controller and the process — results in a pure integrator. Loops that achieve this loop transfer function have the ability to reduce variability at low frequencies without amplification at high frequencies. This can only be achieved in theory by achieving complete pole-zero cancellation, in which the controller cancels out all of the process dynamics, except for one pure integrator. This is precisely what the Lambda tuning technique attempts to achieve.

Lambda Tuning Concept

The term Lambda tuning applies to all controller tuning methods that require the closed-loop time constant to be specified before the tuning is applied [2, 3, 13–15, 21, 22]. The initial concept is attributable to Dahlin [14]. Lambda tuning attempts to achieve a first-order closed-loop response of time constant λ. This can be represented as

$$T(s) = \frac{y}{y_{sp}} = \frac{G_C G_P}{1 + G_C G_P} = \frac{1}{\lambda s + 1} \tag{21}$$

Since the process transfer function can be determined from plant testing, as described below, it follows that

$$G_C = \frac{1}{G_P} \frac{1}{\lambda s} \tag{22}$$

It also follows that, in theory, the loop transfer function would be

$$G_C G_P = \left(\frac{1}{G_P} \frac{1}{\lambda s} \right) G_P = \frac{1}{\lambda s} \tag{23}$$

This does obey the special case of Bode's integral as specified by inequality (20), which allows the control loop to attenuate frequencies slower than the loop cutoff frequency at $1/\lambda$ without amplifying higher frequencies.

Let us consider a first-order process and a PI controller. The process transfer function is

$$G_P = \frac{K_P}{\tau s + 1} \tag{24}$$

When Eq. (22) is applied, the controller type and its parameters can be determined:

$$G_C = \frac{1}{G_P} \frac{1}{\lambda s} = \left(\frac{\tau s + 1}{K_P} \right) \frac{1}{\lambda s} \tag{25}$$

This controller has one integrator and one transfer function zero. Hence it has the form of a PI controller:

$$G_{C_{PI}} = K_C \left(1 + \frac{1}{T_R s} \right) = (K_C) \left(\frac{T_R s + 1}{T_R s} \right) = \left(\frac{\tau}{K_P \lambda} \right) \left[\frac{\tau s + 1}{\tau s} \right] \tag{26}$$

Lambda tuning can hence be achieved by the following PI controller settings:

$$T_R = \tau, \qquad K_C = (1/K_P)(\tau/\lambda) \tag{27}$$

The Lambda tuning concept hence consists of the following steps:

1. identifying the process dynamics
2. using the process dynamics to determine the controller type needed to control this process
3. determining how to calculate the controller settings to achieve the desired λ.

TABLE 2 Process Transfer Functions and Controller Types

Process Model	Controller Type
Pure gain	I or PI
First order	PI
Second-order overdamped	PI or PID
Second-order underdamped	PID
Second order with lead	PI or PID.F
Second order with lead and overshoot	PI or PID.F
Second-order nonminimum phase (wrong-way response)	PI
Integrator	PI or P
Integrator with first-order lag in series	PI or P
Integrator with first-order lead	PI or P
Integrator with nonminimum phase (wrong-way response)	PI or P

Controller Types

Lambda tuning is based on the IMC concept and produces controllers that are nonoscillatory and have first-order responses with a closed-loop time constant λ. The controllers are approximate inverses of the process transfer function. There are many process transfer functions that may occur in practice, and, as a result, many different controllers are involved [2, 3, 13, 15]. One tuning software package [23] allows for 11 process models (all of which may have dead time) and the controllers needed for each of these case, as shown in Table 2.

As illustrated in Table 2, above, the type of process dynamics dictates the controller structure. The PID.F controller referred to consists of a PID controller with a first-order filter in series. This type of controller results from the IMC concept when the process has a transfer function zero, such as a lead.

Industrial Controllers

The problem of tuning is complicated further by the fact that each industrial controller has implemented the PID algorithm in a slightly different way. One tuning software package [23] recognizes over 70 different industrial controllers, each one with its own structure, parameter limits, and idiosyncrasies. For instance, some use gain, others proportional band, some use reset time, others reset rate, some use minutes, other seconds, etc. In general, however, most of the PID forms fall into three main categories: ISA standard form, classical form, and parallel form. It is critically important that the person tuning a control loop be thoroughly familiar with the form of the controller, the meaning of the tuning parameters, and their ranges. Some controllers cannot be tuned with the Lambda tuning concept as a result of parameter limits or internal structure peculiarity, so it is important that these issues be understood in detail. A controller dynamic specification [24] provides more detail.

Common Lambda Tuning Rules

The two most common Lambda tuning rules are listed in Table 3 and apply to first-order and integrator processes with dead time. Collectively, these represent more than 80% of all process dynamics in continuous plants.

Impact of Dead Time

Dead time is the most destabilizing dynamic effect possible. It is important to note that inequality (20) cannot be satisfied in the presence of dead time, as dead time cannot be canceled (even dead-time

TABLE 3 PI Tuning Rules for First-Order and Integrating Dynamics

Process	Process Dynamics	K_C	T_R
First-order plus dead time	$G_P(s) = \dfrac{K_P e^{-sT_d}}{(\tau s + 1)}$	$\dfrac{\tau}{K_P(\lambda + T_d)}$	τ
Integrator plus dead time	$G_P(s) = \dfrac{K_P e^{-sT_d}}{s}$	$\dfrac{2\lambda + T_d}{K_P(\lambda + T_d)^2}$	$2\lambda + T_d$

compensation algorithms do not cancel dead time [22]). Dead time causes resonance according to Bode's integral. When dead time is present, good process design practice should attempt to reduce dead time to a minimum wherever possible.

Let us consider a simple first-order process with dead time and apply Lambda tuning by means of a PI controller. The process is

$$G_P(s) = \frac{K_P e^{-sT_d}}{\tau s + 1} \tag{28}$$

From Table 3, a PI controller is needed for lambda tuning, and the settings are

$$T_R = \tau, \qquad K_C = \frac{\tau}{K_P(\lambda + T_d)}$$

The resulting loop transfer function is

$$G_C G_P = \left[\frac{\tau}{K_P(\lambda + T_d)} \right] \left[\frac{(\tau s + 1)}{\tau s} \right] \frac{K_P e^{-sT_d}}{(\tau s + 1)} = \frac{e^{-sT_d}}{(\lambda + T_d)s} \tag{29}$$

Note that the loop transfer function is not a pure integrator, as required by the Bode integral if resonance is to be avoided. The result is that the control loop with dead time will resonate, and as a result it will amplify high-frequency variability.

Control Loop Robustness and Stability Margins

Control loop robustness refers to the ability of the control loop to deliver consistent and predictable performance. The danger to be avoided is that the loop may become unstable under certain conditions. Unstable means that the loop will oscillate with increasing amplitude, thereby endangering the process. Once the loop has been tuned, the need for robustness results from the fact that the process dynamics may change with time, operating point, or product grade. As the process dynamics change, the process gain, dead time, and time constants may all vary. The loop gain determines the stability of the loop and is the product of the controller gain and the process gain. For example, the slope of the control valve characteristic determines the process gain for a flow loop. The slope of an equal percentage valve characteristic can easily vary by a factor of 5 or more. Similarly, the slope of the titration curve determines the process gain of a pH loop. This can vary by a factor of 10 or more. The degree of robustness of a loop can be measured by the gain margin and phase margin. These are based on the Nyquist stability criterion, which defines a loop to be unstable at frequencies for which the loop gain exceeds 1.0 and the phase shift is 180°.

Gain Margin. The gain margin expresses the ability of the loop to absorb changes in the loop gain resulting from changes in process gain. A gain margin of 4 means that the loop gain can change by a factor 4 before the loop will become unstable. For a flow loop with an equal percentage valve characteristic, a gain margin of 5 or more would be advisable.

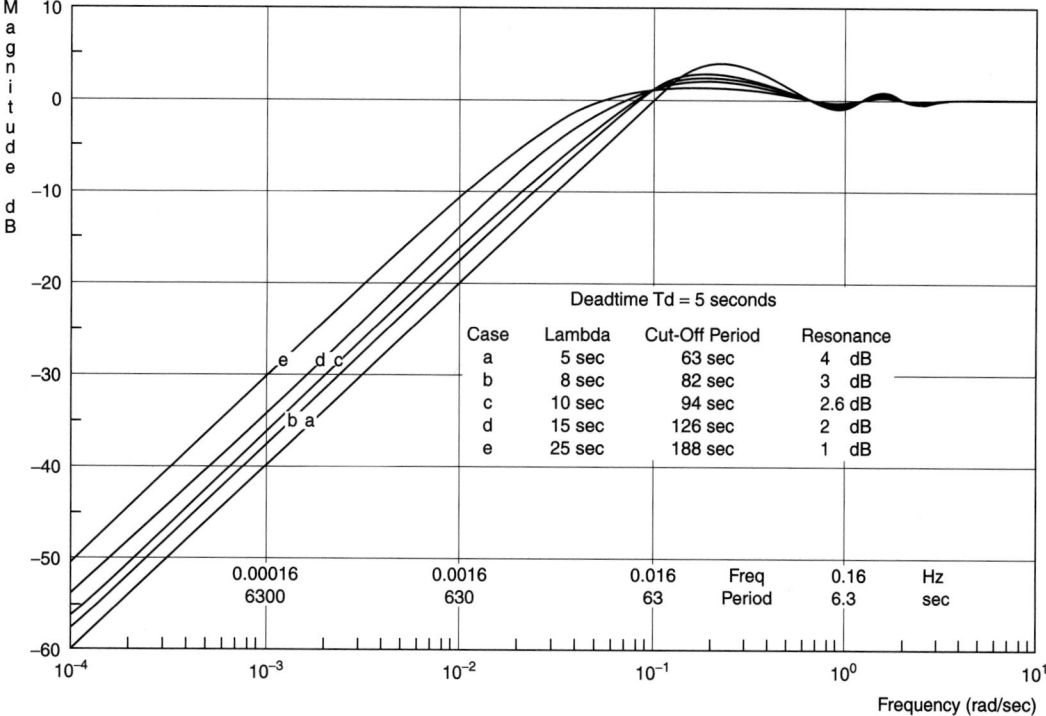

FIGURE 21 Sensitivity function (load frequency response) for loop with deadtime.

Phase Margin. The phase margin expresses the ability of the loop to absorb changes in parameters such as dead time and time constant. A phase margin of 60° means that the combined effect of these changes can cause an additional phase lag of 60° before the loop becomes unstable. Generally loops start to become oscillatory as the phase margin is reduced below 60° or so.

The Control Loop Performance-Robustness Envelope—Speed of Response versus Robustness

The speed of response λ determines the cutoff frequency (or cutoff period), the resonance, and the robustness of the loop. It is a strong function of dead time. Clearly, the faster the speed of response, the greater the resonance and the wider the bandwidth (shorter cutoff period). Figure 21 shows the sensitivity function, or load frequency response Bode plot for a control loop with 5 s of dead time, tuned for λ varying from 5 s (equal to the dead time) through to 25 s. Table 4 shows the speed of response

TABLE 4 Speed of Response versus Robustness and Resonance

Speed of Response λ	Cutoff Period	Gain Margin	Phase Margin	Resonance		
				dB	AR	Amplification (%)
$5T_d$	$12\pi T_d$	9.4	81	1.0	1.12	12
$3T_d$	$8\pi T_d$	6.3	76	2.0	1.26	26
$2T_d$	$6\pi T_d$	4.7	71	2.6	1.35	35
$1.6T_d$	$5.2\pi T_d$	4.1	68	3.0	1.41	41
T_d	$4\pi T_d$	3.1	61	4.0	1.59	59

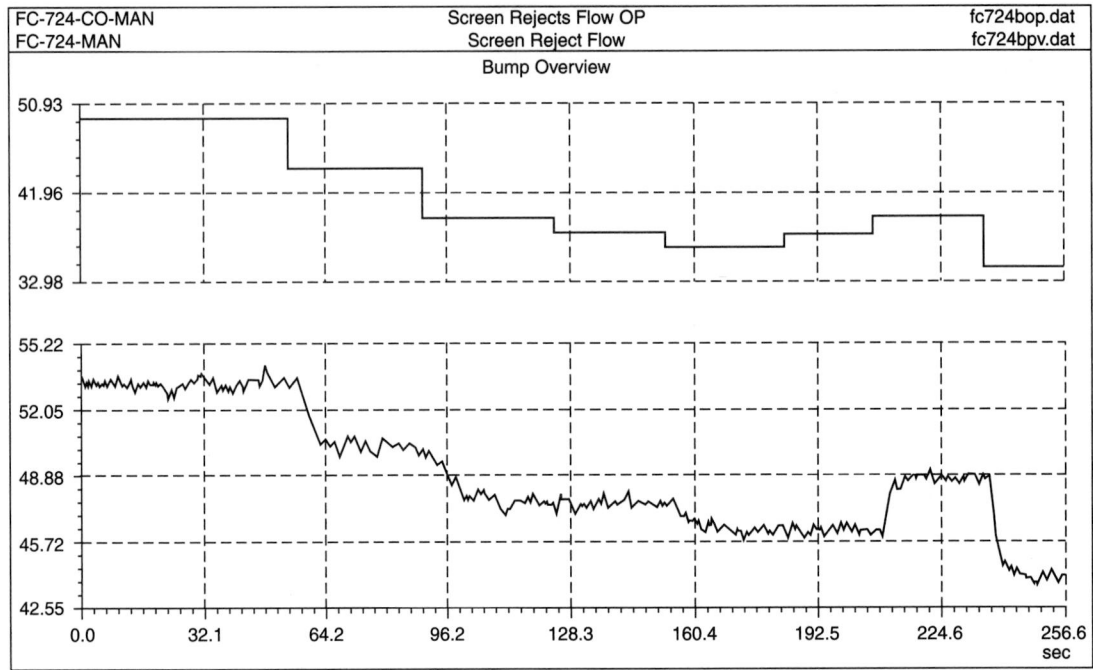

FIGURE 22   Open loop step tests, or bump tests.

in terms of dead-time ratio. It suggests that closed-loop time constants faster than approximately two dead times are probably unacceptable from a robustness point of view. For a closed-loop time constant as fast as two dead times, the gain margin is 4.7 and the phase margin is 71°. These are reasonable values for a loop in an industrial environment. However, faster tuning tends to reduce the margins to values that may be dangerously low. Similar conclusions can be reached from a resonance point of view, since the amplification at the resonant frequency grows to over 40% as the speed of response is increased to a closed-loop time constant of two dead times.

The control loop issue is discussed in more detail in Refs. 2, 21, and 22.

Identifying Plant Dynamics—Open-Loop Step Tests

Lambda tuning requires that the process dynamics be identified. This is best done by placing the loop in manual mode and carrying out a series of step tests, known also as bump tests. Bump tests involve making small step changes in the controller output while collecting data on both the controller output and the process variable. Figure 22 illustrates the procedure. It is important to work very closely with the process operator during this time, as the process is operating and making product for a customer. The operator must at all times be aware of what is happening and should be aware of the potential consequences of each test. From a testing point of view, the objective is to carry out a series of tests— about 10 or more, if possible—that typify the process response. Because the process is nonlinear, the size of the steps should be varied over some practical range from small to large. Of course, the process operator must determine the maximum size of the steps that can be carried out.

Figure 22 shows a series of seven bump tests carried out on a paper machine, Screen Rejects Flow, FC-724, which were collected by a commercial tuner software package [23]. Not all of the bump tests show responses. This is because the control valve actuation nonlinearities limit the valve travel resolution. This aspect is discussed in the next section. The larger bumps appear to produce responses that are first order in character. The identification process is completed by somehow fitting transfer

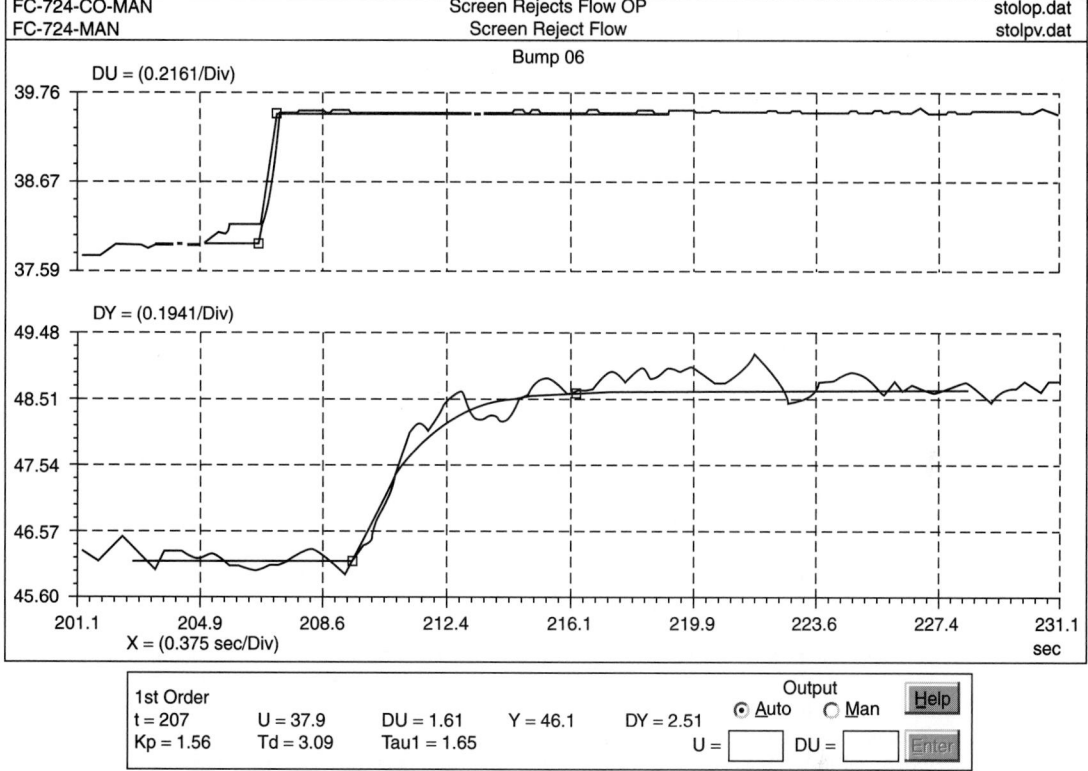

FIGURE 23 Fitting a process transfer function to a bump test.

function models to this data. Figure 23 illustrates how this can be done. In this software package [23], the bump tests are first windowed. This involves breaking the whole data record into sections, or windows, with each one concentrating on the relevant data for each bump. Figure 23 shows the details of bump 06, which occurred at second 207 in the data record. By fitting a transfer function model to each bump, with the user adjusting the goodness of fit, the user generates set of parameters for each bump test. In the case in point, the model being fitted is first order. The process gain is 1.56% of span /% valve. This suggests that the control valve is slightly oversized (the ideal process gain should be 1.0). The dead time is 3.09 s. Ideally, there should be no dead time for an incompressible flow loop. The dead time that is present is most probably due to the control valve actuator and positioner. The time constant is 1.65 s. This is likely due to the damping adjustment in the flow transmitter.

The parameters for each bump will vary. For the seven bumps shown in Fig. 22 the statistics are listed in Table 5.

TABLE 5 Bump Test Parameter Variability

Parameter	Average Value	Parameter Variability (%)
Process gain	0.644	143
Dead time	1.87	100
Time constant	3.33	57

Knowledge of this parameter variability is invaluable in determining the robustness environment in which the loop is being tuned and should be taken into account.

CONTROL ACTUATOR AND TRANSMITTER DEFICIENCIES

Effective process control depends absolutely on the operation of control actuators and transmitters. Some of the more serious process control problems occur in plants as a result of nonlinearities in the final elements and primary elements. When these are severe, they can make effective control impossible.

Control Valve Nonlinearities

Control valve nonlinearities can wreak havoc and create variability where none existed before. Control valve nonlinearities have been identified as the largest single source of variability generation in the pulp and paper industry, based on plant surveys [4]. Consider Fig. 6, which shows a strong cycle of a paper mill screen reject flow loop, FC-724. The flow controller is on local automatic, running to a fixed setpoint of 1578 usgpm. The flow is cycling by +/−2.3%, with a period of ∼100 s. The controller output is cycling by ∼2.5% and is following a sawtooth pattern. This is a classic control-valve-induced limit cycle. A limit cycle is a type of cycle caused by the nonlinear dynamic elements of a control loop. Limit cycling behavior cannot be cured by loop tuning, which can only alter the frequency of the cycle. The amplitude can be corrected only by removing or reducing the offending nonlinearities through maintenance or replacement of the control valve.

The cycle is initiated because the control valve has a tendency to stick once it comes to rest. Unfortunately if the process variable is below setpoint, the controller integral action will slew the controller output in the increasing direction. Eventually the valve input signal will change by an amount big enough to cause the valve to move. In Fig. 6, this amount is ∼2.2%. The valve will then move, but unfortunately it will move too far. In turn, this causes the flow to overshoot the setpoint, and the cycle will start all over again in the opposite direction.

Control valve nonlinearities include dead band and resolution [25]. Dead band is the amount by which the input signal must change in order to cause a reversal of movement of the control valve. Dead band is caused by backlash in linkages, shaft windup in the drive train, and many other potential causes in the actuator and positioner. Resolution [25] is a term used to represent the smallest change that a control valve can execute while traveling in the same direction. It is caused by the differences in static and dynamic friction, sometimes referred to as stick-slip, or stiction. In the case of Fig. 6, the dead band is ∼2.2%. Figure 22 shows a series of open-loop bump tests carried out on this flow loop. There are seven bumps shown. It can be seen that there is no response for bumps 3 and 5. In the case of bump 3, the change in output was −1.5%. The previous bump was −5% in the same direction. There was a response for bump 2. There was no response for bump 3 because the valve was stuck, and the −1.5% step change was not sufficient to overcome static friction. In the case of bump 5, which was +1.4%, there was no response because this was the first bump in the positive direction and the size of the bump was less than the dead band.

Analysis of the seven bump tests showed that the dead time varied from 1.5 s to 3.3 s. This long dead time and its tendency to vary are results of the inability of the positioner to initiate valve motion quickly and in a consistent manner. This valve does not meet the requirement of the EnTech Control Valve Dynamic Specification [26].

Control Valve Dynamic Specification

The EnTech Control Valve Dynamic Specification [26] requires that a control valve be tested in such a way that changes in the input signal actually cause changes to occur in the flow in the pipe. It specifies that the dead band be kept below 1%. It specifies the maximum dead time, speed of response, and overshoot for various valve sizes. It also specifies that the control valve be sized so that the resulting process gain is close to 1.0.

Since this specification was issued in 1994, significant changes have occurred in control valve performance. Many paper companies have adopted the practice of using the specification when ordering control valves. The specification has also been widely used by the control valve manufacturers, who have built control valve test facilities to test valve dynamics with fluid flowing under realistic conditions [27]. As a result of this attention, control valve performance has improved in recent years. The specification has also been used in other industries, such as oil and gas, hydrocarbons, energy, and food. The specification has also drawn the attention of the ISA, which has formed the S75.25 subcommittee that is charged to draft a new control valve standard [25].

In 1988, Version 3.0 of the EnTech Control Valve Dynamic Specification [30] was issued to harmonize terminology and concepts with the emerging ISA S75.25 standard.

Variable-Speed Drives

In recent years electric variable-speed drives have become attractive alternatives on prime movers, such as pumps and fans. In principle, variable-speed drives can be exceptional actuators, being very fast, nearly linear, and highly repeatable. There are, however, a few cautions that must be taken into account.

1. The variable-speed-driven pump or fan can be used to control only one variable. In applications in which multiple streams are to be controlled from a common pressure source, control valves offer the only solution.

2. Variable-speed drives have the potential for response times in the subsecond range. This applies for acceleration, providing the current limit is not exceeded. When the current limit is exceeded, the response becomes velocity limited. Care should be taken in setting up the current limit settings so that for small speed changes needed for good regulation the current limit is not exceeded. The dynamics of deceleration may be governed by the inertia of the rotating machine and can potentially be quite slow. Installing a regenerative breaking system can provide fast dynamics for deceleration.

3. When replacing a control valve with a variable speed drive, one must take care to ensure that shutdown conditions are considered. Shutdown may require a shutoff valve. As well, care must be taken to ensure proper sizing of the pump and pressure drops in the fluid system, which will be significantly different if the control valve is no longer throttled back.

4. In some cases it may be possible to use an existing motor when installing a variable-speed drive. Care should be taken to ensure that this motor will be able to operate reliably in this new environment.

Transmitter Deficiencies

Assuming that the transmitter is correctly installed and has been calibrated, two kinds of dynamic deficiencies may render it unsuitable for effective control. First, it may be installed in such a location that the effective process dead time is too long to allow effective control. Second, there may be one of a number of issues that may degrade the quality of the resulting signal. This is may be true even if the transmitter is a digital smart transmitter. Some of the key issues that govern signal quality are the following.

1. Signal resolution should be at least one part in 4000 or 12 bits. Report by exception should not be used for control purposes. Should it be part of the data transmission design, the resolution should be at least equivalent to 12 bits.

2. Signals should be sampled at an adequately fast rate, which is at least three times faster than the process time constant. Hydraulic flow and pressure loops should be sampled at least every 0.25 s.

3. Nonlinear and adaptive filters should not be used for control.

4. The transmitter should not have latency or dead time greater than one sample.

More detailed recommendations are given in the EnTech Digital Measurement Dynamic Specification for Process Control [9].

INTEGRATED PROCESS DESIGN AND CONTROL— PUTTING IT ALL TOGETHER

Most process plants have been designed with steady-state principles only. The instrumentation is typically chosen after the piping design is complete, and the control loops are configured before start-up. This design process has produced the plants of today, in which process variability is often far in excess of potential, as is typically uncovered when plant variability audits are carried out. Is it possible to design a low-variability plant from the outset? The answer to this question is yes; however, this requires an integration of the process and control design disciplines. How this integration might take place is the subject known today as integrated process and control design, which was initiated by Downs and Doss [28]. Control algorithm design is a very general methodology and is largely based on linear dynamics. When thinking about control loop performance, the engineer pays no attention to the actual behavior of the process. For instance, an important phenomenon in the pulp and paper industry concerns pulp slurries and the transport of fiber in two- or three-phase flow. The physics that govern these phenomena involve the principles of Bernoulli and Reynolds and are very nonlinear. The linear transfer function is a necessary abstraction to allow the control engineer to perform linear control design—the only analysis that can be done well. Yet in the final analysis, a control loop only moves variability from one stream to another, where it is hoped it will be less harmful. Yet the process design has defined the streams without any consideration for the resulting dynamics.

Integrated process and control design must take a broader view of control strategy design. Control provides only high-pass attenuation; hence, it can attenuate only slow variability. Often even this process is flawed, as it is accompanied by the amplification of higher-frequency variability. Process mixing and agitation provide low-pass attenuation of variability. Yet both of these techniques attenuate variability by only so many decibels. Yet the customers demand that absolute variability of the product be within some specified limit in order to meet the manufacturing needs of their processes. Surely the elimination of sources of variability provides the best method to ensure that no variability will be present. These issues are in the domain of process design and control strategy design. Control strategy design does not lend itself to elegant and general analysis. Each process is different and must be understood in its specific detail. Nonlinear dynamic simulation offers a powerful tool to allow detailed analysis of performance trade-offs and is the only available method for investigating the variability impact of different design decisions.

In the blend chest example, for instance, all of the consistency control loops share a common dilution header, fed by a common pump. In the example, this header is not pressure controlled. As the upstream consistency control loops attempt to reduce their variability, they cause pressure variability, which causes consistency variability to be created downstream after the blend chest and the machine chest. One design alternative is provide the blend chest and machine chest consistency loops with their own sources of dilution water so that they will not be effected by the upstream variability. Other process design changes, which may be attractive for the blend chest example, could be (1) to eliminate all flow control valves and use variable frequency pump drives (this will eliminate all control valve nonlinearities); (2) to redesign the existing blend chest to have two compartments, each one being separately agitated) (this will convert the existing agitation from a first-order low-pass filter to a second-order filter and provide a better high-frequency attenuation); (3) to control the pressure of the dilution header by a variable-frequency pump drive. Each of these design alternatives should be evaluated by dynamic simulation before significant funds are committed. The alterations proposed above vary in capital cost from $10,000 to $1,000,0000. Hence the simulation must have high fidelity in its ability to represent the phenomena of importance. In addition, network analysis techniques may be useful in determining how variability spectra may propagate through a given process and control strategy. Changes in process and control strategy design alter these pathways by creating new streams. These concepts are discussed in more detail in Ref. 29.

DEFINING TERMS AND NOMENCLATURE

Basis weight = the paper property of mass per unit area (gsm, lb/3000 sq. ft., etc.)

Bump test = step test

β = process transfer function zero time constant

C = ISA symbol for control

Chest = tank.

consistency = mass percentage of solids or fiber content of a pulp slurry (concentration).

dB = decibel, a method of representing attenuation as $20 \log_{10}$ (AR), or $10 \log_{10}$ (power ratio)

Dead time = time delay

F = ISA symbol for flow

$G_C(s)$ = controller transfer function in the continuous (Laplace) domain

$G_P(s)$ = process transfer function in the continuous (Laplace) domain

HD chest = High-density chest, a large production capacity stock tank with consistency typically in the 10%–15% range with a dilution zone in the bottom

IMA = integrating-moving-average noise structure

ISA tags = ISA tagging convention (e.g., FIC-177 means *F*low *I*ndicating *C*ontroller No. 177)

K_C = controller gain

K_P = process gain

L = ISA symbol for level

Lambda tuning = tuning that requires the user to specify the desired closed–loop time constant λ

λ = the desired closed-loop time constant, usually in seconds

N = ISA symbol for consistency

P = ISA symbol for pressure.

PI = proportional-integral controller: standard or classical form

$$G_C(s) = K_C\left(1 + \frac{1}{T_R s}\right)$$

parallel form

$$G_C(s) = \left(K_C + \frac{K_C}{T_R s}\right)$$

PID = proportional-integral-derivative controller

ISA standard form

$$G_C(s) = K_C\left[1 + \frac{1}{T_R s} + \frac{T_D s}{(\alpha T_D s + 1)}\right]$$

classical form

$$G_C(s) = K_C\left(1 + \frac{1}{T_R s}\right)\left[1 + \frac{T_D s}{(\alpha T_D s + 1)}\right]$$

parallel form

$$G_C(s) = \left[K_C + \frac{K_C}{T_R s} + \frac{K_C T_D s}{(\alpha T_D s + 1)}\right]$$

PID.F = proportional-integral-derivative controller with series filter

Positioner = control valve accessory that acts as a local pneumatic feedback servo

Pulping = the process of removing individual fibers from solid wood

Refiner = a machine with rotating plates used in pulping to cause the wood chips to disintegrate into individual fibers through the mechanical action and in papermaking to fibrilate the fibers to enhance bonding strength

Stock = pulp slurry

T_d = dead time or time delay

T_R = controller reset or integral time/repeat

τ_1, τ_2 = process time constants ($\tau_1 \geq \tau_2$).

REFERENCES

1. Shunta, J., *World Class Manufacturing Using Process Control*, Prentice-Hall, Englewood Cliffs, New Jersey, 1995.

2. Bialkowski, W. L., "Pulp and Paper Process Control," in *CRC Control Handbook,* W. S. Levine, Ed., CRC Press, Boca Raton, Florida, 1996.

3. Bialkowski, W. L., and F. Y. Thomason, "Title of Chapter," in *Process Control Fundamentals for the Pulp & Paper Industry*, N. Sell, Ed., Technical Association, P & P Industries, Atlanta, Georgia, 1995.

4. Bialkowski, W. L., "Dreams Versus Reality: A View From Both Sides of the Gap," Keynote Address, Control Systems '92, Whistler, British Columbia, 1992, published *Pulp Pap. Can.* **94** (11), 1993.

5. Bialkowski, W. L., B.C. Haggman, and S. K. Millette, "Pulp and Paper Process Control Training Since 1984," *Pulp Pap. Can.* **95** (4), 1994.

6. Kocurek, M. J., Series Ed., *Pulp and Paper Manufacture*, Joint Technical Association, P & P Industries and Canadian Pulp and Paper Association textbook, Committee of the Pulp and Paper Industry, 1983–1993, Vols. 1–10.

7. EnTech™, "Analyse" time series analysis software package.

8. Box, G. E. P., and G. M. Jenkins, *Time Series Analysis: Forecasting and Control*, Holden-Day, San Francisco, California, 1976.

9. EnTech™, Digital Measurement Dynamics—Industry Guidelines (Version 1.0, 8/94) (EnTech literature).

10. EnTech™, "Signal Conditioning Module" data acquisition equipment.

11. EnTech™, "Collect" data acquisition software package.

12. Astrom, K. J., *Introduction to Stochastic Control Theory*, Academic, New York, 1970.

13. Chien, I. L., and P. S. Fruehauf, "Consider IMC Tuning to Improve Controller Performance," *Hydrocarbon Processing*, Oct. 1990.

14. Dahlin E. B., "Designing and Tuning Digital Controllers," *Instrum. Control Syst.* **41**(6), 77, 1968.

15. Morari, M., and E., Zafiriou, Robust Process Control, Prentice-Hall, Englewood Cliffs, New Jersey, 1989.

16. Ziegler, J.G., and N.B. Nichols, "Optimum Settings for Automatic Controllers," *Trans. Am. Soc. Mech. Eng.* pp. 759–768, 1942.

17. Cohen, W. C., and G. A. Coon, *Trans. Am. Soc. Mech. Eng.* **75**, 827, 1953.

18. Lopez, A. M., P. W. Murill, and C. L. Smith, "Controller Tuning Relationships Based on Integral Performance Criteria," *Instrum. Technol.* **14**, (11), 57, 1967.

19. Bode H. W., *Network Analysis and Feedback Amplifier Design,* Van Nostrand, Princeton, New Jersey, 1945.

20. Doyle J. C., A. Francis, and A.R. Tannenbaum, *Feedback Control Theory*, Macmillan, New York, 1992.

21. Haggman, B. C., and W. L. Bialkowski, "Performance of Common Feedback Regulators for First-Order and Deadtime Dynamics," *Pulp Pap. Can.* **95**(4), 1994.

22. Bialkowski, W. L., "A Review Of Deadtime Compensator And PI Controller Regulation Performance," *Pulp Pap. Can.* **89**, 1997.

23. EnTech™, "Tuner" Lambda Tuning software package.

24. EnTech™, Automatic Controller Dynamic Specification (Version 1.0, 11/93) (EnTech literature).

25. ANSI/ISA S75.25 Control Valve Response Measurement from Step Reponses Inputs.

26. EnTech™, Control Valve Dynamic Specification (Version 2.1, 3/94) (EnTech literature).

27. Taylor, G., "The Role of Control Valves in Process Performance," presented at the 80th Annual Meeting, Technical Section, Canadian Pulp and Paper Association, Montreal, Canada, 1994.

28. Downs, J. J, and J.E. Doss, "Present Status and Future Needs—A View from North American Industry," presented at the Fourth International Conference on Chemical Process Control, Padre Island, Texas, Feb. 17–22, 1991 (AIChE Proceedings).

29. Tseng, J., W. R. Cluett, and W. L. Bialkowski, "Variability Propagation Through A Stock Preparation System: Implications For Process Control And Process Design," presented at Control Systems '96, Halifax, Nova Scotia, 1996. Published, *Pulp Pap. Can.* **89**(9), T322–T325, 1997.

30. EnTech™, Control Valve Dynamic Specification (Version 3.0, 11/98)(EnTech Literature).

CONTROL VALVE RESPONSE

by Mark T. Coughran*

Control valve response strongly affects the performance of many process control loops. For throttling control, key performance issues are (1) the ability of the control valve to respond consistently to small input changes and (2) the installed loop gain resulting from the match of the valve to the other loop components. The definition of small and the required valve gain depend on the process and other loop components, as shown in Fig. 1. Optimal selection of the control valve requires knowledge of the control objectives of the loop; static characteristics of the process; dynamics of the process, transmitter, and controller sampling; controller tuning practices; source of the disturbances (load or set point); and the frequency distribution of the disturbances. Specification of control performance is distinct from the mechanical configuration requirements typically seen in purchasing documents, such as materials compatibility, seat leakage, pressure class, and end connections. However, the effect of the control valve on process variability should be weighed equally with these other factors, since it may dominate the life cycle cost of the control valve.

In linear mechanical systems, errors between the percentage of input and the percentage of output vary with the history of the input in a predictable manner. Control valves, on the other hand, exhibit nonlinear behavior, i.e., the errors also may vary with amplitude and direction of the input signal changes. The user can specify allowable error in the input–output relationship with several open-loop performance measures. Open-loop testing (also shown in Fig. 1) typically uses simple input signals from a computer—or the existing process controller in manual mode—to the control valve. While open-loop testing does not completely quantify the nonlinearities, it gives some indication whether the control valve will meet the user's requirements for closed-loop control when the input signal is obviously more complex.

The first selection guideline is that the components of the control valve—valve, actuator, positioner, or other accessories—be designed for throttling (not on–off) service and tightly integrated together. This description fits design A in Fig. 2; process responses to 1% changes in the input signal can be detected, despite the presence of process noise. In contrast, design B used a seal design better suited for tight shutoff, an actuator design that added friction and backlash, and a single-stage pneumatic positioner. The process variable data in Fig. 2, recorded with the plant's flow transmitters, show that design B responded poorly to 5% changes and not at all to smaller changes. Automated block valves such as those in design B typically have lower initial cost but greater operating cost because of poorer control.

In most cases, good static performance can be obtained by attention to basic engineering issues such as minimizing friction in the valve and correct sizing of the valve and actuator. However, for the minority of control loops with fast response requirements and fast equipment surrounding the control valve, more complex analysis is needed. This section assumes that sound actuator sizing practices are used, so that friction—not process fluid forces—is the main resisting force to be overcome by the positioner/actuator servoloop.

STATIC PERFORMANCE MEASURES

Positioner test standards [1] define a variety of static performance measures; *static* means the data are recorded after the device has come to rest. The full-scale calibration cycle familiar to instrument engineers provides measurement of hysteresis plus dead band (combined) and linearity, which are

* Research Specialist, R. A. Engel Technical Center, P.O. Box 190, Fisher Controls International, Inc., Marshalltown, Iowa 50158.

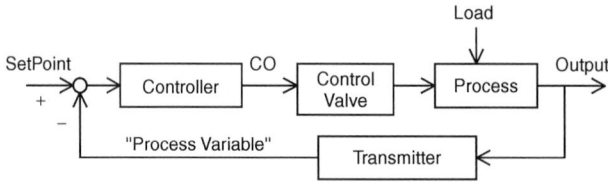

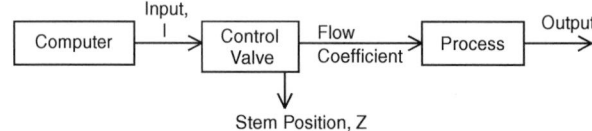

FIGURE 1   Most control valves are used in closed loop control systems. The control valve performance requirements and the appropriate open loop test methods depend on the other components of the loop.

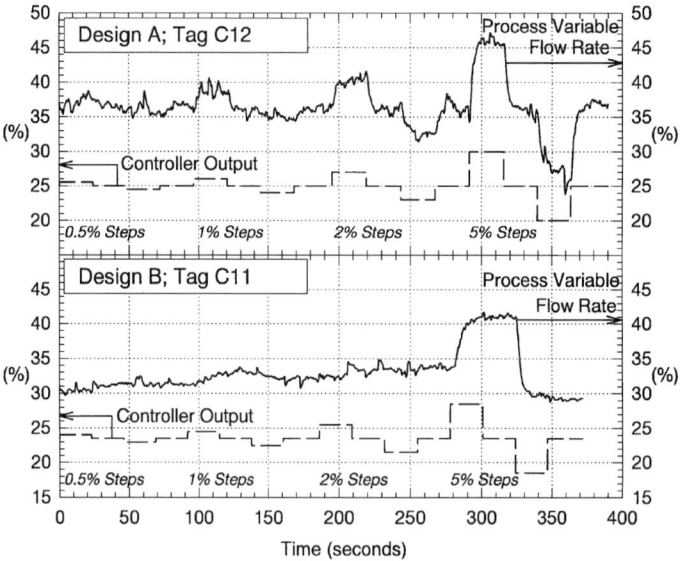

FIGURE 2   Example open loop data from adjacent pulp stock flow control loops in a paper mill; segmented ball valves. Design B used a high-friction seal, a rack-and-pinion actuator, and a low-gain positioner. Measurement time constant for the Process Variable was 2.0 second (magnetic flow tube). Installed dead band was <1% for Design A and ≥5% for Design B.

not the most relevant parameters. Although hysteresis and linearity are relevant for the process transmitter, the control valve plays a different role in the loop. Hysteresis and linearity are smoothly varying errors that accumulate significantly over only large strokes and have an impact on only open-loop systems; for closed-loop control, the discontinuous errors (failure to provide any response over small ranges of input) are more important. Therefore the most useful parameter is dead band, measured with process loading. The benchtop dead band test specified for positioners [1] can be misleading because of discrepancies between actuator position and flow coefficient and discrepancies between benchtop behavior and installed behavior.

Dead band is the range through which an input signal may be varied, following reversal of direction, without initiating an observable change in the output signal [1]. For example, in Fig. 2, design B failed to reverse direction following 5% input reversals; therefore the dead band was greater than or equal to 5%. A dead band of less than 1% can be expected in globe control valves with polytetrafluoroethylene (PTFE) packing, properly sized actuator, and reasonable positioner gain. However, in many process control applications today, the control valve is required to move for signal changes smaller than 0.5%. The inability of some control valves to meet such requirements limits the use of advanced process control strategies, which typically assume negligible dead band.

DYNAMIC PERFORMANCE MEASURES

Dynamic response is the time-dependent response resulting from a time-varying input signal. Possible input signal shapes include pulse, step, ramp, and sinusoid [1]. Step inputs have become popular for installed process testing [2] because of their simplicity. The output signal for dynamic response tests is usually position data measured at the actuator, not the closure member; the location where the positioner receives its feedback is often most convenient. Actuator position data can be misleading for rotary valves [3].

Loop controllers that use proportional plus integral action are most easily tuned when there is a dominant first-order lag in the loop. A dominant lag in the fluid process provides stability, ensures that the transmitter does not hide process events from the controller, and allows the highest controller gain setting—hence best loop response to set point and load disturbances. Relatively slow processes (in which the process dynamics are dominant in the loop) typically include temperature, level, large-volume gas pressure, mixing, and pH. Valve positioners are recommended for such loops [4].

Where the process dynamics are similar to dynamics of other loop components, stability can still be provided by installation of a dominant first-order lag somewhere else in the loop. In the past, a commonly recommended means to accomplish this was to connect an I/P transducer directly to a diaphragm actuator. However, several factors have led to an increasing use of positioners:

1. Popularity of digital controllers with sample intervals of 1 s or longer, resulting in increased focus on static performance and small-amplitude response of the control valve

2. Increased acceptance of springless actuators, which require positioners

3. Increased acceptance of rotary valves with the attendant need for feedback around the nonlinear relationship between actuator pressure and rotation

4. Continued needs for split ranging and higher actuator pressure than commonly available from I/P transducers.

Although a dominant time lag is still desirable to stabilize fast process loops and loops with large dead time, today this can be accomplished with a software lag in a controller output function block or a positioner input filter. Dynamic response requirements for the control valve typically focus on consistent response at small amplitudes and minimizing dead time.

VALVE SIZE AND INHERENT FLOW CHARACTERISTIC

The design of the control loop for stability and best performance requires a dynamic analysis that includes all components [5]. However, dynamic analysis of many generic types of control loops has led to the development of simplified guidelines [6]. Another approach that can prevent many problems—especially for slow loops—is to examine simply the static gain requirement.

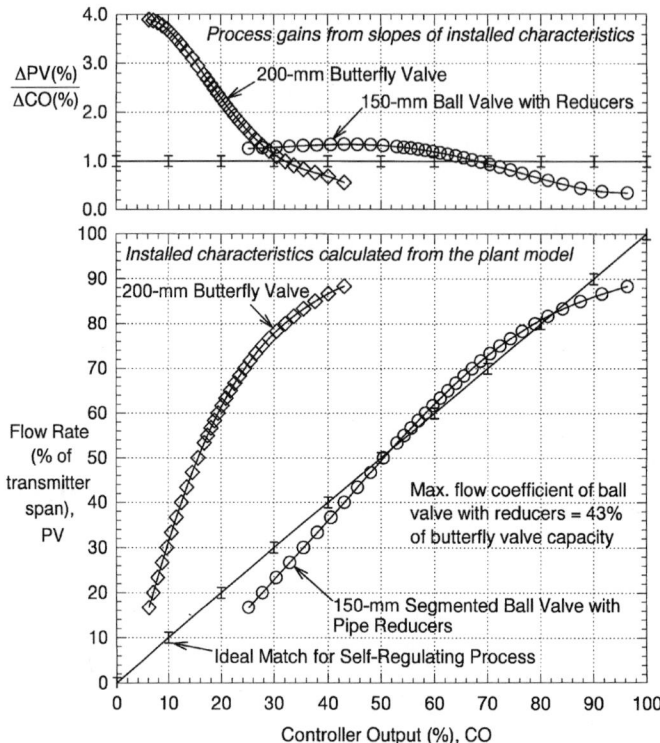

FIGURE 3   Example of variation in open loop static process gain with an incorrectly selected valve; cooling water flow control loop with 200-mm piping in a power plant. The 200-mm butterfly valve originally installed was oversized, whereas the 150-mm ball valve provided a better size and inherent characteristic.

Loop gain is the product of the gains of the controller, control valve, process, and transmitter (Fig. 1). The gains of the process and transmitter are usually fixed by the plant design. Some controllers provide either gain scheduling or an output characterizer block; a static equivalent to the latter is an input characterizer in the positioner (signal-to-position characterizing).[1] However, because the position-to-flow gain of the valve occurs after the signal-to-position dead band, the most important task is selecting the best valve size and best shape of the inherent flow characteristic. High position-to-flow gain multiplies the effect of dead band; for example, if this gain is 4.0 and the dead band is 0.5%, the smallest process variable change that can be made reliably is 2%, regardless of characterization in upstream blocks.

Figure 3 shows an example of a common situation: a greatly oversized valve. In this control loop, unstable behavior occurred at low flow rates. The installed process gain for the 200-mm (8-inch) valve in the 200-mm piping increased by a factor of 6 as the set point moved from high to low, indicating that a fixed-gain controller would give either sluggish response at high flow or excessively reactive response at low flow. The 150-mm (6-inch) ball valve would be a better selection.

Standard control valves are necessarily somewhat oversized because of the availability of a limited number of trim sizes, the tendency of equipment (piping, pump) suppliers to add safety factors, and limited accuracy of process information in the design stage. However, one preventable problem is deliberate oversizing of the valve by the plant owner to allow for future production increases. This is

---

[1] Twenty years ago, feedback cams were recommended for changing the signal-to-position characteristic. However, this method suffers from slope limitations, alteration of the positioner loop gain, and added maintenance issues.

false economy, saving a small amount on future capital costs relative to the continuous adverse effect on present operating costs. Control valves on critical loops should be sized for the present operating conditions and with the most accurate data available for the pump, piping, valves, etc. Globe valves with reduced trim may be the best solution when the user needs reserve capacity for future growth. Software is becoming more readily available for predicting installed characteristics and should see widespread use as plant owners demand that valve size and inherent characteristic be selected to better fit the control loop.

VALVE FLOW STEADINESS

Flow coefficients defined by standards [7] are customarily listed in vendor catalogs under the assumption that they do not vary with time. However, it is possible for a poorly designed valve to exhibit unsteady flow when the trim position is constant [8], and this phenomenon may become significant when the control objectives require small movements of the valve. The flow behavior is best captured by dynamic flow coefficient measurements [9]. Figure 4 gives an example of liquid flow coefficient unsteadiness caused by a particular globe valve (lower graph). The possibility of this behavior is one reason that benchtop stem motion measurements are not a complete method of judging control valve performance.

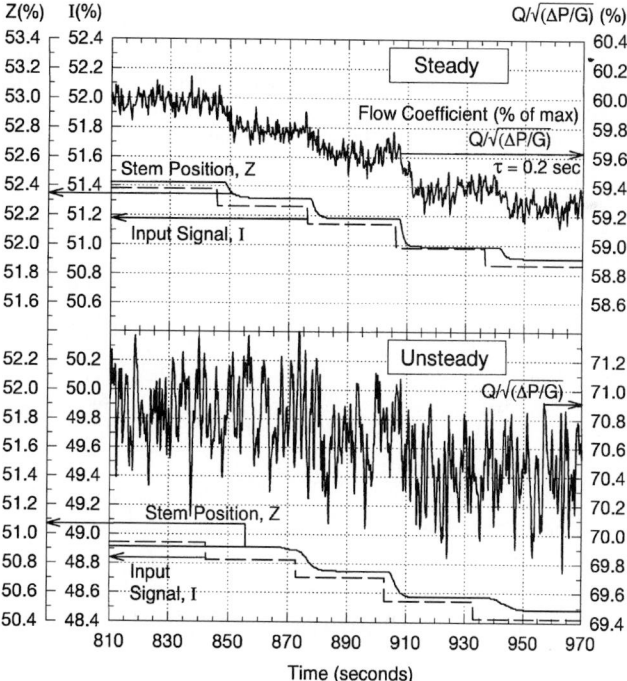

FIGURE 4   Examples of steady and unsteady flow patterns in 50-mm globe valves with advertised linear inherent flow characteristics. Both valves were tested in the same system with a flow coefficient measurement time constant of 0.2 seconds. The unsteady flow pattern created by the valve in the lower graph would limit the performance of fast control systems such as liquid pressure and flow control loops. $Q \equiv$ volumetric flow rate, $\Delta P \equiv$ valve pressure drop, $G \equiv$ specific gravity.

FRICTION AND DRIVE SHAFT DESIGN

High friction in the valve and the actuator increases dead band and degrades the small-amplitude dynamic response. Increased friction on the sealing and guiding surfaces, caused by process temperature, pressure, or fouling, obviously is very specific to the application. In some valve designs, the oft-encountered requirement for tight shutoff adds friction, so that—for critical loops—separation of the control and shutoff functions should be considered (two valves with different purposes). In all cases, attention should be given to the materials, loading, installation, and maintenance of the stem packing. Where temperature extremes would require high-friction packing materials, extension bonnets may be a worthwhile investment. New elastomers with friction characteristics similar to those of PTFE but with higher temperature capability are also becoming available.

Currently no standards exist for defining and measuring friction in control valves. An explanation of the friction measurement typically provided by valve signatures is given by Jackson [10].

It is easy to see how friction opposes stem motion in linear motion valves. A more subtle effect occurs in rotary valves because of finite stiffness of the drive shaft. Friction in the packing, bearings, and seals causes *shaft windup* [3, 10]. Figure 5 shows an example in which actuator position data indicated dead band of less than one step, or < 0.13%, while the flow data proved that actual dead band of the trim (ball) was approximately 0.8%. In the example of Fig. 5, most of the friction was caused by seal-to-ball contact, and the friction increased from 11 to 25 N m after the valve was installed and cycled. This type of friction can be eliminated if the closure element is mounted eccentrically so that it moves out of contact with the seat as the valve begins to open.

The windup form of lost motion is in addition to the more obvious opportunity for *backlash* in rotary actuators that use linkages or in trim components that have more than 1 degree of freedom. This backlash can be prevented by careful design, manufacturing, installation, and maintenance. Many throttling ball valves today use clamped splines to essentially eliminate backlash.

Although some butterfly valves have lower frictional torque than ball valves, they also tend to have smaller-diameter shafts. Then, because torsional stiffness is proportional to the fourth power of

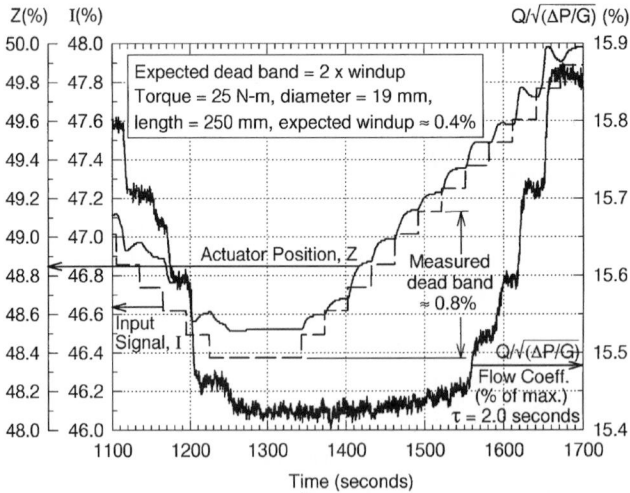

FIGURE 5 Example of shaft windup in a ball valve. From the advertised inherent flow characteristic, the expected flow coefficient change (right) was approximately 0.10% for each 0.13% change in ball position. The flow coefficient measurement time constant was 2.0 seconds in this installation. Note misleading indication of dead band (<0.13%) from actuator position data (left). The benchtop torque of the new valve, before cycling, was only 11 N-m.

diameter, butterfly valves are still susceptible to windup because of friction in packing, bearings, and liners.

SELECTION OF ACTUATOR AND ACCESSORIES

Manufacturers typically size the actuator to overcome (1) shutoff forces and (2) fluid dynamic forces at higher lifts. Friction in the packing is considered only as it affects the ability to close the valve. However, sizing for best control may require a larger actuator for providing either more thrust (relative to the friction) or more actuator volume; the latter allows for higher positioner gain in pneumatic systems. Just as high gain is desirable for the outer process loop in Fig. 1, it is desirable in the positioner servoloop to overcome friction and other nonlinear effects. Gain from proportional action is preferred; integral action in the positioner, combined with friction, tends to cause position cycling.

The positioner design strongly affects dynamic response of pneumatic systems. For small-amplitude response, the positioner is more important than the actuator [11] unless the actuator adds friction. Figure 6 shows an example of small-amplitude response with three positioner designs on one valve and actuator. Positioner A gave dead time approaching 10 s when reversing direction, which would limit the usable controller gain in many control loops. The design of positioner B resulted in much shorter dead time following reversal. This valve and actuator with positioner B would be a good choice for control loops requiring small moves with consistent, fast response.

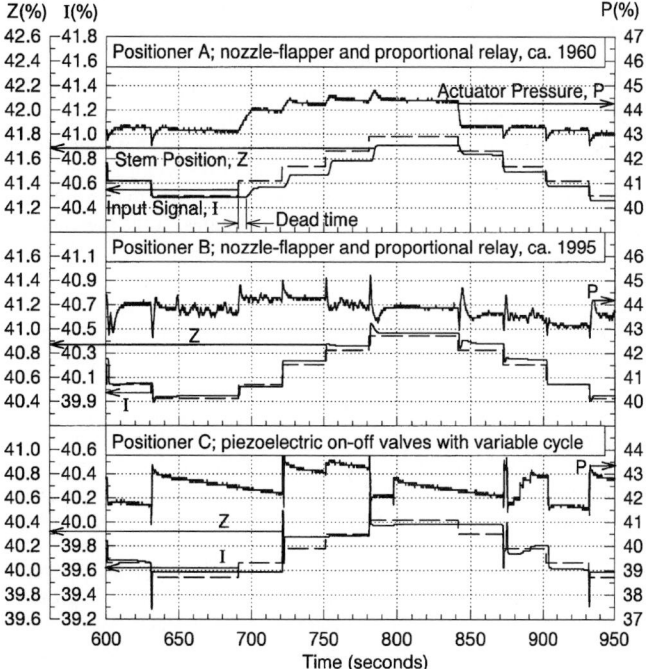

FIGURE 6   Example of positioner design effect on control valve small-amplitude response. Positioners A, B, and C were installed to a globe valve that had steady flow and a diaphragm actuator. Positioner A was an analog two-stage design; Positioner B followed similar design principles but with adjustable gain and digital communication. Positioner C used solenoid valves (supply and exhaust) with a duty cycle controlled by a microprocessor. Actuator pressure is shown in percent of supply.

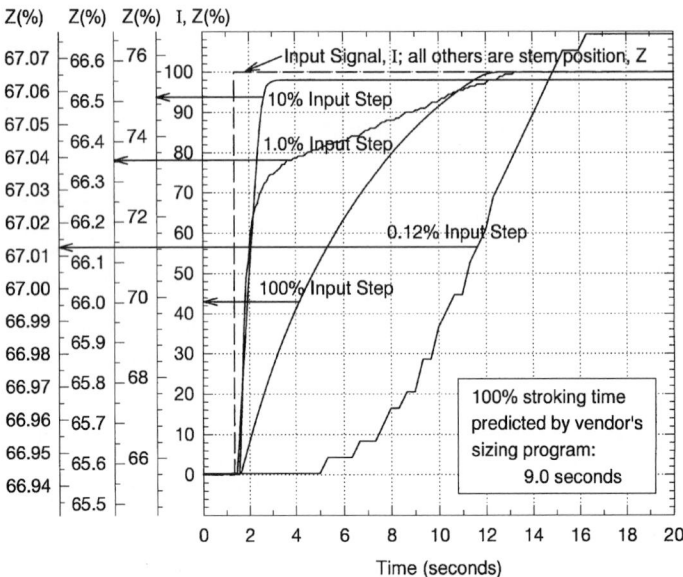

FIGURE 7   Example of amplitude dependence in a control valve with a pneumatic actuator and positioner, illustrating that results from one amplitude cannot be extrapolated to another amplitude.

The conventional design of high-performance positioners uses multistage amplification, typically with signal amplification in a nozzle flapper [12] followed by power amplification in either a poppet-type proportional relay or a spool valve. Within the past 10 years, an alternate scheme has appeared that uses on–off piezoelectric valves whose input pulse width is controlled by a digital algorithm. Normally the bang-bang control scheme includes a dead zone to prevent chattering near the null state. Behavior with this scheme is irregular at small amplitudes, as the positioner C data in Fig. 6 illustrate.

DYNAMIC TEST SIGNAL AMPLITUDE AND SHAPE

Pneumatically actuated control valves tend to respond more slowly as the input amplitude approaches the dead band (small signal changes) and also as the input amplitude approaches relay saturation (large signal changes).[2]   Figure 7 illustrates both effects. The full-range tests, sometimes called stroking time tests, are the only amplitude for which manufacturers make predictions available. In contrast to pneumatics, electric actuators often provide a constant slew rate at amplitudes of 1% to 100%. The user must specify the amplitude for dynamic response tests if appropriate equipment is to be selected based on such tests.

Concerning the shape of the input signal, it is possible for open-loop step inputs to excite overshoot that either will not occur, or will occur to a much lesser extent, in closed-loop application. Figure 1 shows that load disturbances pass through lags in the process, transmitter, and possibly in the controller antialias filter, before arriving at the control valve. Set point changes normally occur slowly from either an outer loop in cascade control or from a supervisory control system; for these cases, tests should use slow input changes. For open-loop testing in a plant environment, the simplicity of step inputs is appealing but the above limitations should be kept in mind.

---

[2] If fast response is desired for very large steps, a volume booster can be used between the positioner and actuator. The booster should use a dead band relay [12], and its action should be limited to large-amplitude input changes.

PROCESS VARIABLE AS THE ULTIMATE OUTPUT

The most useful information arises from a test with process loading, since the most realistic friction occurs when the valve is exposed to the process temperature, pressure, vibration, wear, and possibly fouling. Friction also may change during the break-in period after the valve is first installed; one example is metal-seated ball valves that lap in during service (see Fig. 5 notes and Ref. 10). For most rotary valves, loaded tests that use process data are the only means of measuring true dead band. Even for globe valves, differences in flow behavior can negate the usefulness of benchtop stem motion tests.

The fundamental output of the control valve is the flow coefficient, which can be measured in a laboratory environment and has been used here in Figs. 4 and 5. However, the ultimate function of the final control element is to alter the process variable, and in many plants the process transmitter provides sufficient information to diagnose control valve problems. Open-loop plant testing, in which the responses of the control valve, process, and transmitter are combined, is now widely used for loop tuning and diagnostics. The process data can be recorded at the controller terminations and, for self-regulating processes with adequate signal-to-noise ratio, used to estimate the control valve dead band (e.g., Fig. 2).

SUMMARY: CONTROL VALVE PRACTICES FOR BEST RESPONSE

- Use a control valve designed for throttling service.
- Size the valve trim for present operating conditions; avoid oversizing.
- Select the inherent flow characteristic by calculating the installed static characteristic.
- The valve should provide steady flow when the trim position is held fixed.
- Minimize friction in the valve and actuator (seals, bearings, packing).
- Eliminate lost motion in linkages and maximize stiffness of rotary shafts.
- Select the actuator type and size for best control with the positioner.
- Select high proportional gain in the positioner.
- Test with the appropriate amplitude and shape of input signal for the control loop.
- Judge installed control valve response by change in the process variable.

REFERENCES

1. "Method of Evaluating Performance of Positioners with Analog Input Signals and Pneumatic Output," ANSI/ISA-S75.13-1989.
2. "Control Valve Dynamic Specification," Version 2.1, 1994, EnTech Control Inc., Toronto, Ontario, Canada.
3. Coughran, M. T., "Valves: Testing for Peak Performance," *Intech* **41**(10), p. 58, 1994.
4. Gassman, G. W., "When to Use a Control Valve Positioner," *Control*, Sept. 1989.
5. Lloyd, S. G., and G. D. Anderson, *Industrial Process Control*, Fisher Controls, Publication Stockroom, P.O. Box 190, Marshalltown, Iowa 50158, 1971.
6. Adams, M., "Process Control Valves," in *Process/Industrial Instruments & Controls Handbook*, 4th ed., D. M. Considine, ed., McGraw-Hill, New York, 1993, p. 9.10.
7. " Control Valve Capacity Test Procedure," ANSI/ISA-S75.02-1988.
8. Dvorak, A. D., P. J. Schafbuch, and D. J. Westwater, "Flow Rate Stabilizer for Throttling Valves," U.S. Patent No. 5,765,814, June 16, 1998.
9. Coughran, M. T., "Measuring the Installed Dead Band of Control Valves," *ISA Transactions* **37**, p. 147, 1988.

10. Jackson, R. S., "Friction Effects in Control Valve Performance: Ball Valves," ISA TECH/97, Volume 1, Part 5, p. 67, 1997.

11. Coughran, M. T., "Performance Influences in Globe Control Valves," *Intech* **43**(8), p. 44, 1996.

12. Lytle, R. F., and C.B. Schuder, "Pneumatic Components," Section 42 in *Mechanical Design and Systems Handbook*, 2nd ed., H. A. Rothbart, ed., McGraw-Hill, New York, 1985.

PROCESS IMPACT

by James F. Beall IV*

The effect of process measurements, control valve performance, and controller tuning can have a significant impact on process control. In some audits [1], improving the performance of the control valve and its related components (I/P, actuator, positioner, etc.) significantly decreased the variability of over 50% of the control loops. Case studies presented in this section show that, of the control loops studied, up to 75% need improvements on the instrumentation or control valves in order to meet process control requirements. In most cases, improvements to the instrumentation and control valves must be done before better controller tuning will improve the process control.

A plant program to analyze and correct problems with process measurement instrumentation, control valve response, and controller tuning can significantly reduce process variability and increase process availability. Reductions in process variability of individual control loops 10:1 or more are possible in some cases. This, in turn, can result in a significant reduction in the variability of the process unit (e.g., the analysis of an intermediate or final product).

The economic benefit from these improvements is significant. One benchmark study [2] found that the average improvement of 12 companies with these types of improvements accounted for 25% of the economic benefit from all categories of process control improvements. In addition, good basic control is required for obtaining the potential benefits from advanced control technologies.

PLANT PROGRAM FOR CONTROL OPTIMIZATION

The following is a guide to develop a plant program to decrease process variability and increase process availability by improving controller tuning and the response of instrumentation and control valves.

- Research: Find out if anyone in your company has a control optimization program. Use their experience and track record to help start your program. Consider working as an extension of their group. Start your program based on the success of their program. If there is not anyone in your company with a control optimization program, talk to an industry peer to obtain their insight.

- Select a process analysis system: Consider the following characteristics:

  - Recording capability—number of recording channels, frequency of data sampling, types of signal inputs, communication interface with control system. Although it is convenient to use a communications interface to record data from the control system, this method prevents analysis of all types of control systems and may overlook some common problems. These problems include

---

* Eastman Chemical Company, Longview, Texas.

signal scan time problems, signal aliasing [3], control valve problems, resolution of instrumentation, and others. An analysis system that can record both the actual signals as well as data from a communication interface to the control system would be ideal.

- Accept the required types of signals (pneumatic, electronic, thermocouple, resistance temperature detector (RTD), of position, control valve or final element etc.).
- Suitable for the area electrical classification.
- Analysis techniques—signal editing, simulation, power spectrum analysis, frequency plots, tuning methodology, etc.
- Plan: Develop a schedule of equipment purchases, training, and manpower. Usually it is best to have a plan that is based on an expanding effort. Once the program has some success, the demand for the program will grow.
- Sponsor: Sell your program to a member of your management. Utilize information on the success of similar programs. Make sure you get commitment for consistent, dedicated manpower for the program.
- Training: Obtain good training to use the process analysis system properly and to learn process analysis techniques.
- Partnership: Develop a partnership with an operating department. Work hard to complete successful projects in this department.
- Expand: Use previous successes to expand program to other operating departments. Encourage operating department supervision to share the success in their department with other operating departments.
- Difficult cases: Do not be afraid to tackle control problems that no one else has solved. A more consistent effort and better tools such as a process analyzer may help you solve the problem. If you are successful, it will build credibility. If you are unsuccessful, it does not detract since the problem could not be solved before.
- Unit versus single loop: In general, try to consider as much of the process unit as possible. For example, look at the complete distillation column rather than just the base temperature controller, or maybe look at the whole distillation train or the complete plant.
- Record results: Record before and after results. Try to determine economic benefit to justify and expand the program. A data historian system that collects and stores process data can be very helpful in comparing past and present performance.
- Performance monitor: Develop a program to monitor and maintain the performance improvements. This could be manual system or an automated system utilizing a data management system.

RESULTS OF CONTROL OPTIMIZATION---CASE STUDIES

These case studies are presented to give the reader examples of the magnitude of the benefits, the nature of the instrumentation and control problems encountered, and the value of a graphical presentation of the process performance [7].

Distillation Tower

This example shows how optimization of the instrumentation, controller valves and controller tuning resulted in 6:1 reduction in variability of key control loops on a typical hydrocarbon distillation tower. Figure 1 shows[1] the trends of key control loops on a hydrocarbon distillation tower before improvements were made. The large variability of the D-8 top temperature, trend 2, was the reason

---

[1] All figures (1–20) are from the work of James F. Beall IV, Eastman Chemical Company.

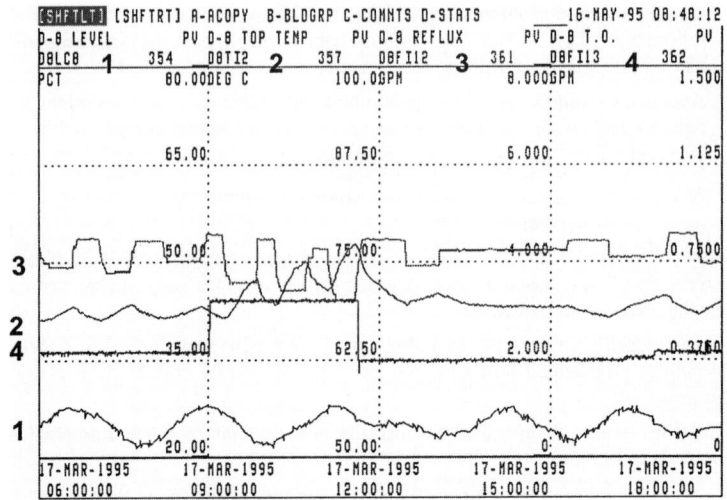

FIGURE 1   Key control loops of a distillation tower, before improvements.

for a control optimization study. Note that the reflux flow, trend number 3, was making square-wave-type step changes that appeared to affect the D-8 top temperature. Also, the column base level was varying in a sinusoidal pattern. Further analysis revealed that the bottom flow out of the column, which was manipulated by the base level controller, also varied in a sinusoidal pattern. The bottom flow was the feed to the next column and therefore created a disturbance in it.

A process analysis system was used to record and analyze the response of the instrumentation and controller valves and to calculate controller tuning constants. Based on the analysis, the following changes were made:

- Positioners were added to two valves.

- Trim size was changed in one valve.

- Tuning constants were changed in four controllers.

Figure 2 show the trends of the loops after the improvements were made. A statistical analysis of these data, shown in Figure 3, reveals approximately a 6:1 reduction in standard deviation and range for the key control loops. Note how well the graphs of the data give a visual quantification of the control improvement.

Pilot Plant Reactor

This example shows how optimization of the instrumentation, controller valves, and controller tuning resulted in up to a 19:1 reduction in the variability of key control loops on a pilot plant reactor. Once again, it was not just retuning the controller that provided the control improvement; it required improvements to the instrumentation and control valves.

Figure 4 shows the trends of key control loops on the pilot plant reactor before improvements were made. The large variability of the reactor temperature (trend 2) and the reactor pressure (trend 1) was the reason for a control optimization study.

The instrumentation and control system was analyzed and the following improvements were made:

- Positioners were installed on two control valves to reduce dead band.

- The pressure control valve was replaced because it was oversized and had excessive dead band.

- Three loops were tuned.

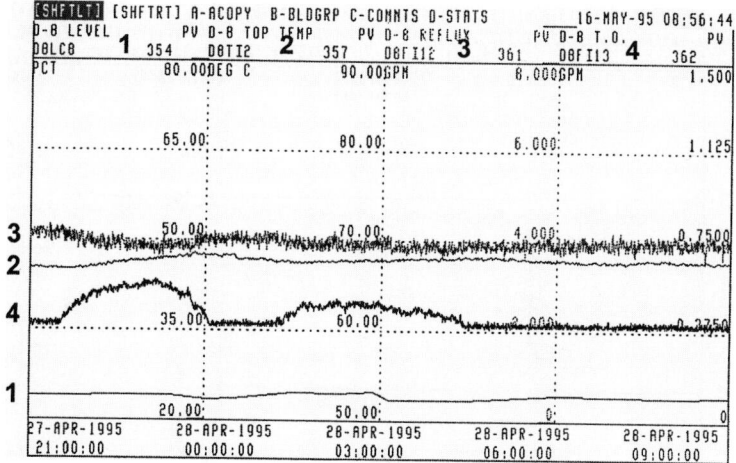

FIGURE 2   Key control loops of a distillation tower, after improvements.

RESULTS:

VARIABLE		BEFORE		AFTER	
		Range	Std Dev	Range	Std Dev
D-8 top temperature	(deg C)	2.8	0.5	0.5	0.09
D-8 mid temperature	(deg C)	9.0	2.3	1.2	0.22
D-8 Bottom flow	(%)	12	na	2	na
(implied by controller output)					
D-8 Distillate flow	Responds as appropriate to remote setpoint from temp controller.				
D-8 reflux flow	Tuned to reflect the changes in take-off flow				

FIGURE 3   Reduction in variability of key loops of a distillation tower.

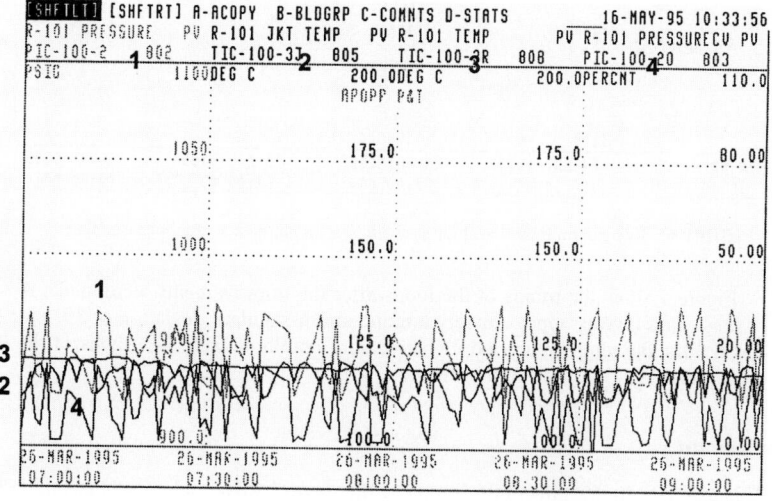

FIGURE 4   Key control loops on a pilot plant reactor before improvements.

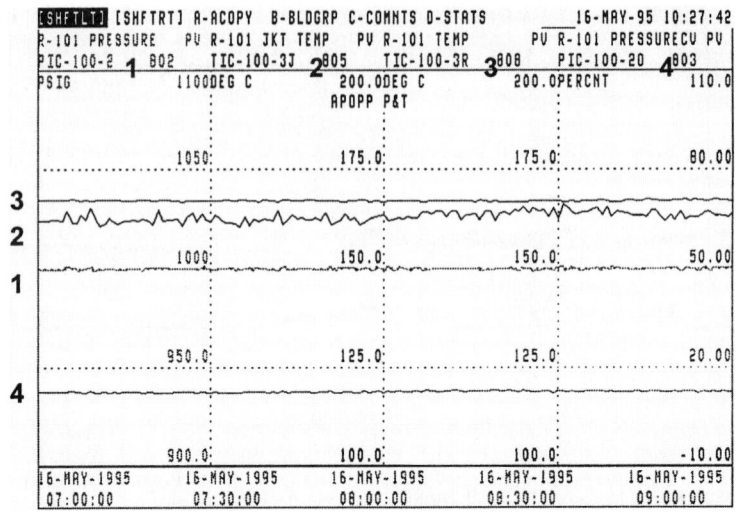

FIGURE 5   Key control loops on a pilot plant reactor after improvements.

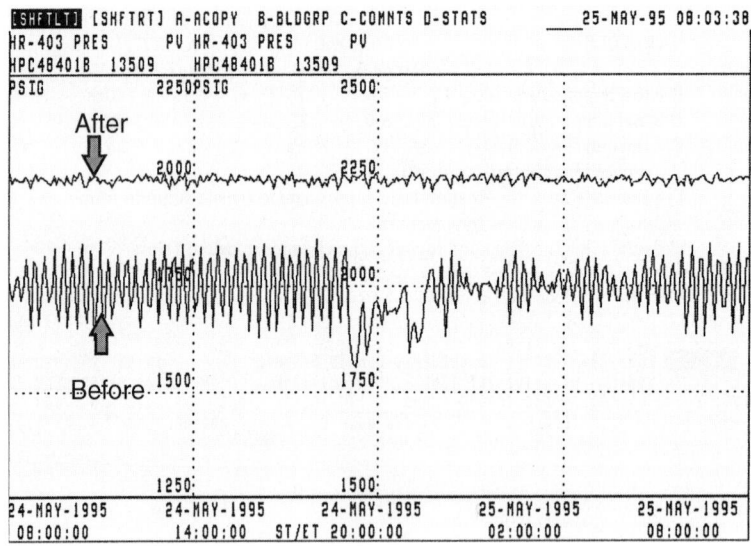

FIGURE 6   Key control loops on a high-pressure reactor before and after improvements.

Figure 5 show the trends of the loops after the improvements were made. A statistical analysis of these data reveals approximately a reduction in standard deviation of 19:1 for the pressure loop and 6:1 for the temperature loop. Once again, note how well the graphs of the data give a visual quantification of the control improvement.

High-Pressure Reactor

This example shows how optimization of the control valve and controller tuning resulted in over a 10:1 reduction in variability of key control loops on a high-pressure reactor. Figure 6 shows the trend

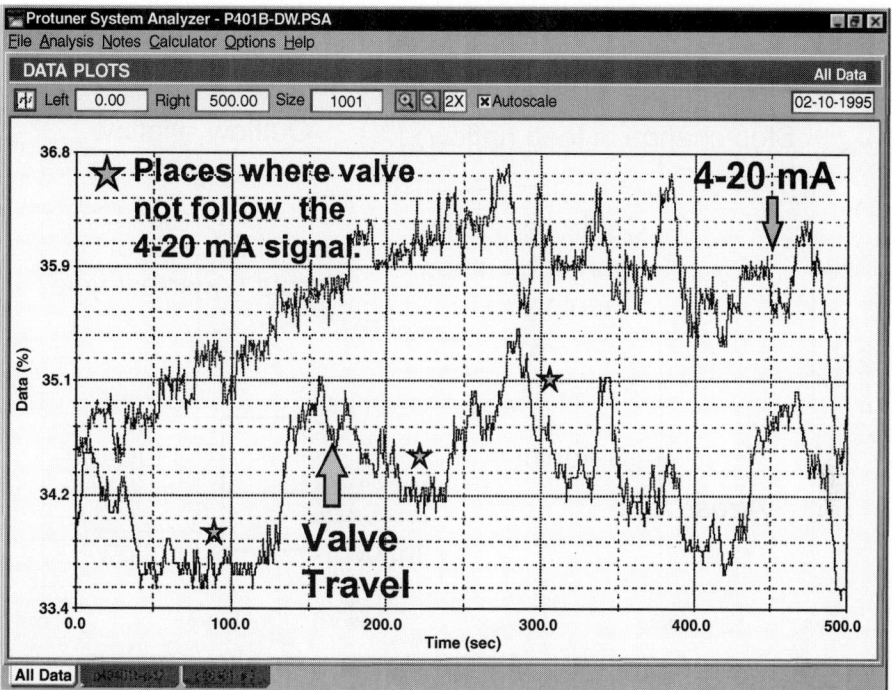

FIGURE 7   Poor performance of the control valve on a high-pressure reactor.

of the loop process variable both before and after the improvements were made. The traces are shifted in time to be able to present both the before and the after improvement data on the same figure (both trends have the same span).

The instrumentation and control system was analyzed, and it was found that poor performance of the control valve was the major source of variability in the control loop. The piston actuator on the control valve was worn, which caused the valve stem to have erratic movement when compared with the control signal to the valve. Figure 7 shows the discrepancy between the signal to the control valve and the actual stem position. The actuator was replaced, the loop was retuned, and the variability was reduced by a factor of 10 or more.

COMMON PROBLEMS

This subsection provides examples of some of the problems with instrumentation, control valves, and controller tuning as well as solutions for these problems. Also, these examples show the nature and characteristics of the problems in an effort to assist in finding other problems.

Improper Tuning on Level Loops

Improper tuning of level loops is a common problem found during control optimization. The typical error is that the integral action on the controller is too fast. Since a level process is usually an integrating process, too much integral action in the controller can cause control instability. Also, it is common for the purpose of a level control loop to allow the level to vary in order to reduce variation of the

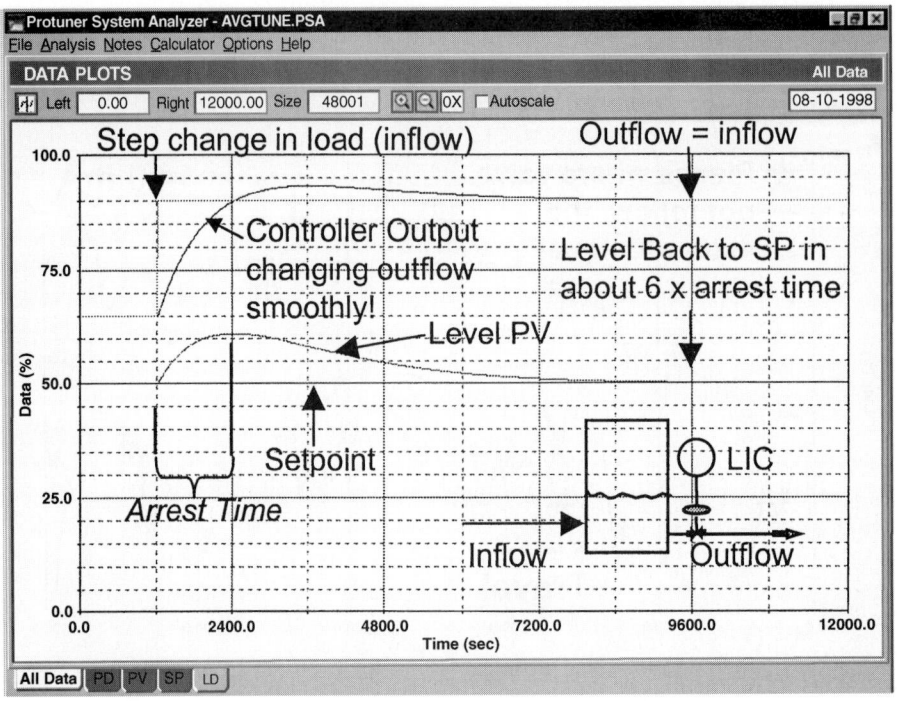

FIGURE 8   Simulation of averaging tuning method for level control.

manipulated flow. However, it is a common problem to find that a level control loop has been tuned such that the level is held tightly and the manipulated variable is varied excessively. In some cases, the tuning is so inappropriate that both the level and the manipulated variable vary excessively.

A solution for this problem is to use a tuning method that results in averaging type response. One such tuning method [4, 5] results in a response to a step change in process load stop or "arrest" the rise or the fall of the process variable (level) in the arrest time and taking approximately 6 times the arrest time to return the process variable back to the setpoint. The selection of a larger arrest time results in more deviation of the process variable from setpoint and less aggressive movement of the manipulated flow (for a given size load disturbance). If the maximum disturbance is known or estimated, the required arrest time can be calculated for an allowable deviation of the process variable. Figure 8 is a simulation of this type of response.

Figure 9 shows the trend of a level control loop before and after the controller tuning was improved. Before the improvement, the level is held close to setpoint by aggressive manipulation of the flow. The flow is the feed to a process and the variability in flow causes process disturbances. After the improvement, the variability of the flow is much less.

Dead Band in Control Valves

Dead band in a control valve can cause control problems. It can cause limit cycling of the process and limit control performance. Dead band creates dead time in the control valve response when the control signal to the valve reverses its direction. Dead band will cause limit cycling of an integrating process.

A new control valve without a positioner can have a dead band of 25% or more because of normal friction forces. A positioner without the integral action reduces the dead band of the control valve by approximately its open-loop gain. For example, if a control valve has a dead band of 15% and the

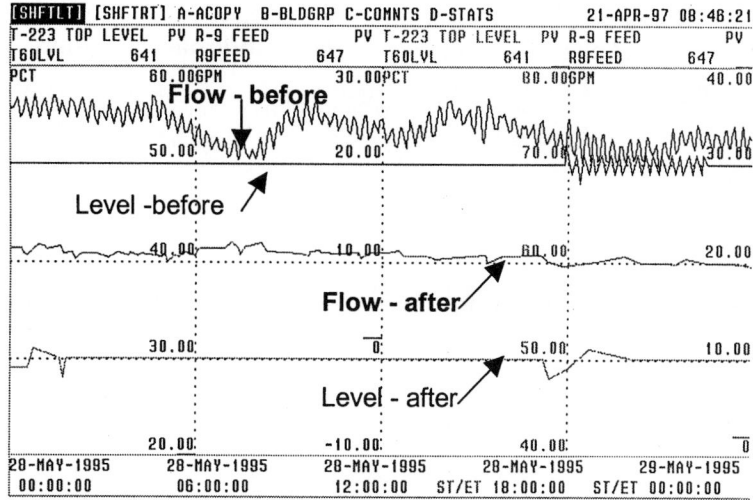

FIGURE 9 Level control before and after averaging tuning is implemented.

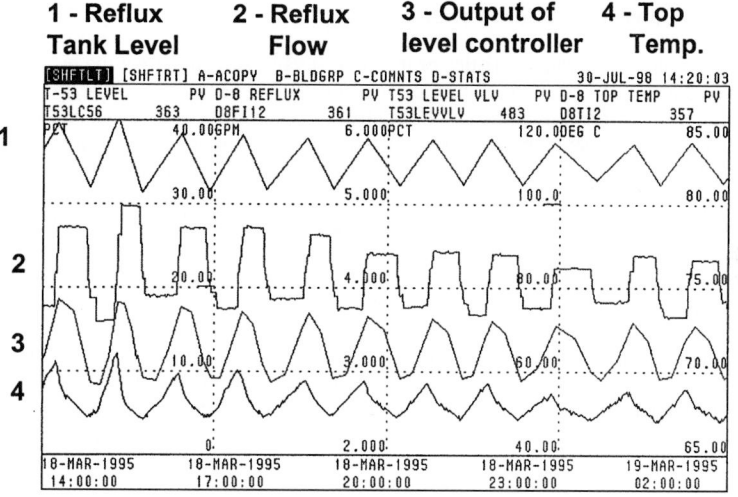

FIGURE 10 Effect of dead band in control valve.

positioner has a gain of 50, the combined dead band of the control valve system will be approximately 0.3%. If the positioner has integral action, the dead band of the control valve system theoretically is 0% but this situation may result in a limit cycle of the valve even if the controller is in manual.

Figure 10 is a trend of control loops associated with the top temperature control of a distillation column. Manipulating the reflux flow controls the reflux tank level. The trend shows the effect of the reflux control valve having approximately 10% dead band. The characteristics of dead band in the control valve on an integrating process such as the reflux tank level is a triangular waveform on the controlled variable and the controller output; and a square wave on the manipulated variable. As shown in Fig. 10, the variation of the reflux flow in a square-wave pattern creates a triangular disturbance on the top temperature on the column.

Figure 11 shows the control after a positioner is added to the reflux control valve to reduce the dead band. The improvement in control is obvious without a statistical analysis.

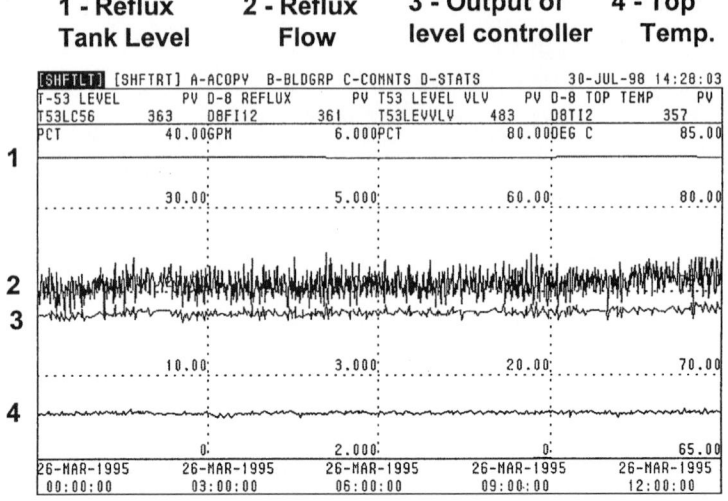

FIGURE 11   Result of adding a positioner to correct dead band in control valve.

Selection of Control Valve Trim Characteristics

The installed flow characteristics of the control valve directly affect the process gain. A variation in the process gain results in a change to the closed-loop response of the control loop. If the change in the process gain is significant, the control loop can become sluggish or, in the opposite direction, unstable. For example, in many heat transfer processes in which the heating medium is throttled, an equal-percentage trim characteristic will help maintain a constant process gain [6]. However, consider the application of a flow control loop with a control valve that has an equal percentage installed flow characteristics. In this application, the process gain of the flow loop can vary by a factor of 10 or more.

The characteristics of the process and control scheme should be considered and the appropriate trim characteristics should be selected to help achieve a constant process gain. If a control valve is installed with the wrong trim characteristics, there are several methods, besides replacing the trim, to modify the valve response. One method is to characterize the output of the controller. If this method is used, loop tuning methods that are based on the output controller should utilize the output of the proportional, integral, and derivative (PID) algorithm, before the characterization function. Another method is to characterize the response of the positioner. Some of the new microprocessor-based smart positioners provide a characterization function to achieve the desired valve response. Reference 6 is an excellent source of information on this subject.

Excessive Dead Time in Control Valve Response

Some positioner and control valve combinations exhibit excessive dead time when the signal to the positioner is reversed by a small amount. The excessive dead time for this type of signal change can create a region of loop instability if the dead time is large compared to the closed-loop time constant. This can create a cycle in the control loop. Figure 12 shows a control valve and positioner with a dead time of 0.4 s when the signal is reversed by a 1% step change. Figure 13 shows the same control valve and positioner with a dead time of 42 s when the signal is reversed by a 0.2% step change. The control loop was tuned based on the response to 1% change in signal; the closed-loop time constant was approximately 3 s. The 42 s of dead time in this small region of signal change causes a region of instability that creates a cycle in the loop. This type of cycle is sometimes called a stick-slip cycle, and the variable dead time is one cause of the cycle. The magnitude of the cycle is limited because, as the output of the size of the change in the controller output

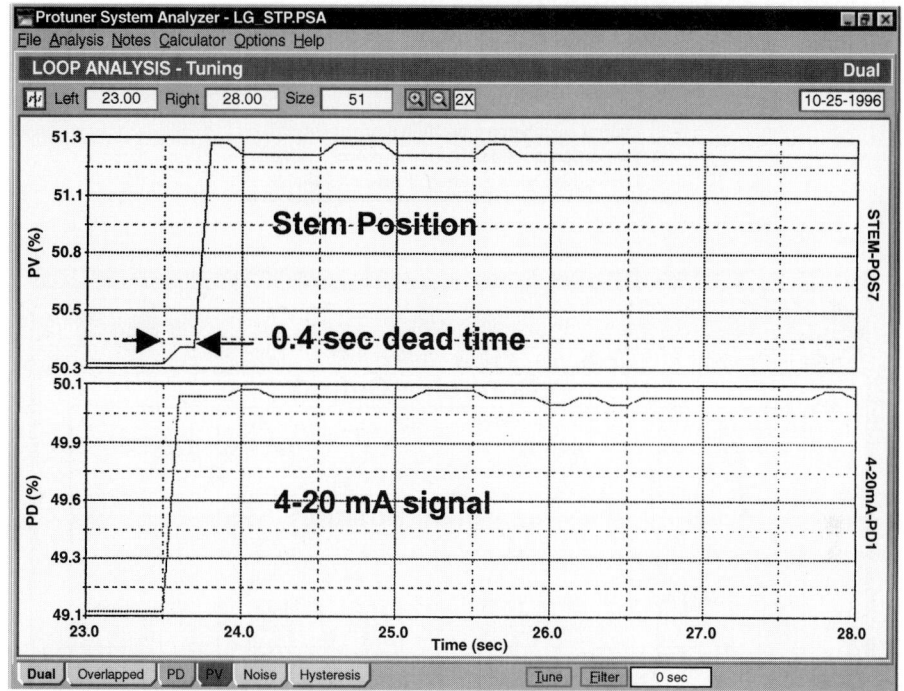

FIGURE 12 Medium-sized (1.0%) step change to positioner and control valve.

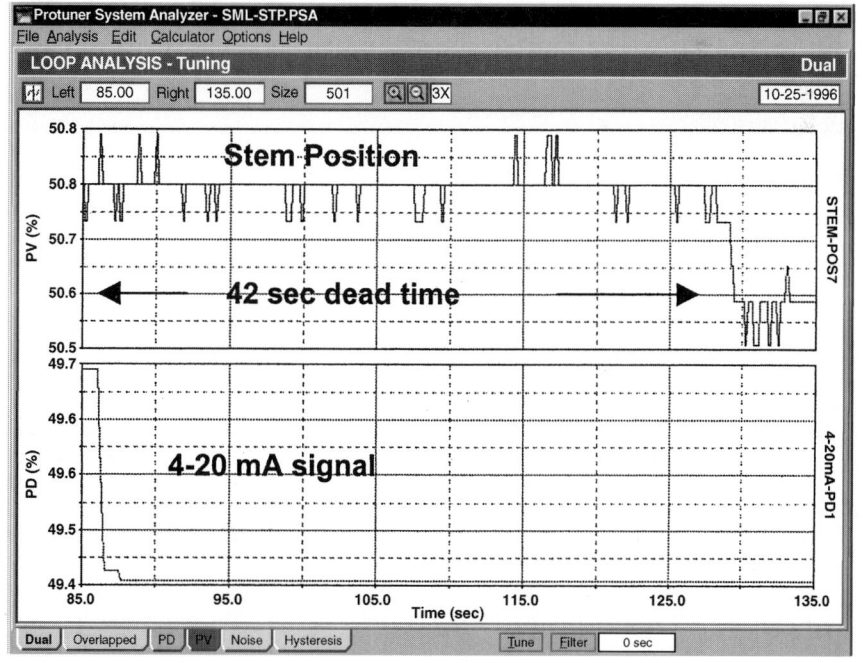

FIGURE 13 Small-sized (0.3%) signal reversal to positioner and control valve.

TABLE 1 PID Implementations

1. Method: ideal, parallel, series
2. Output calculations: positional or velocity
3. Proportional action on error or process variable
4. Derivative action on error or process variable

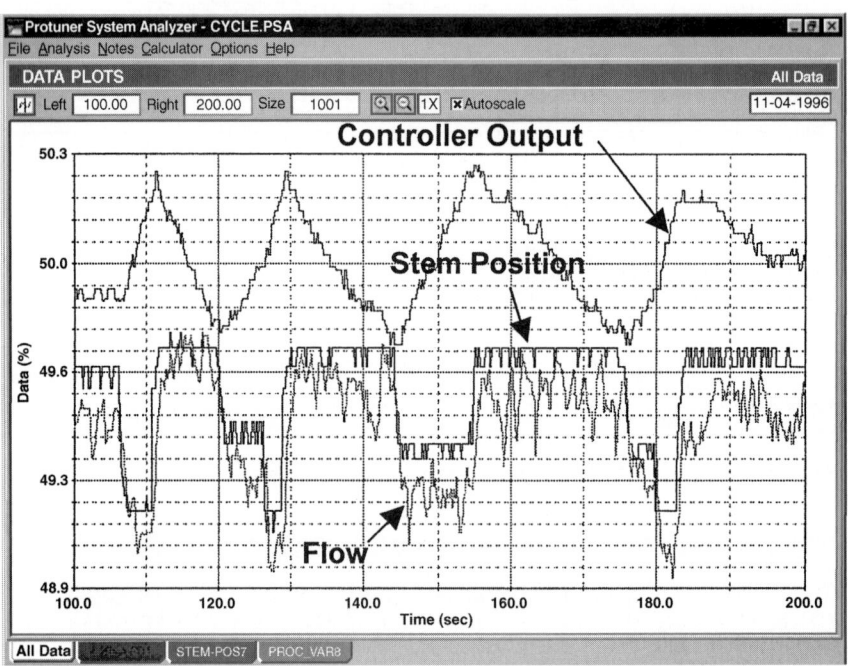

FIGURE 14 Limit cycle in auto when positioner and control valve have large dead time.

increases, the dead time is reduced and the control loop is once again stable. Figure 14 shows this cycle when the loop is in automatic. Figure 15 shows that a different positioner that does not have excessive dead time (same control valve) does not exhibit the limit cycle.

PID Implementation

The PID algorithm can be implemented in several ways. Each method has certain characteristics that may be better suited to particular applications [8]. Table 1 shows the possible combinations.

The most common PID implementation is a series with proportional and integral action on error and derivative action on the process variable. For the same closed-loop control response, the tuning constants must be modified based on the whether the algorithm is series, parallel, or ideal [8].

Implementation of the proportional action on the process variable prevents the output from making a step change when the setpoint is changed. When the setpoint is changed, the integral action will begin to ramp the output to attain the new setpoint. However, the control action for a load change is the same as when the proportional action is on the error. This may be desirable on a level control loop for which it is undesirable to make quick changes to the manipulated variable and it is not critical to get to the new setpoint quickly.

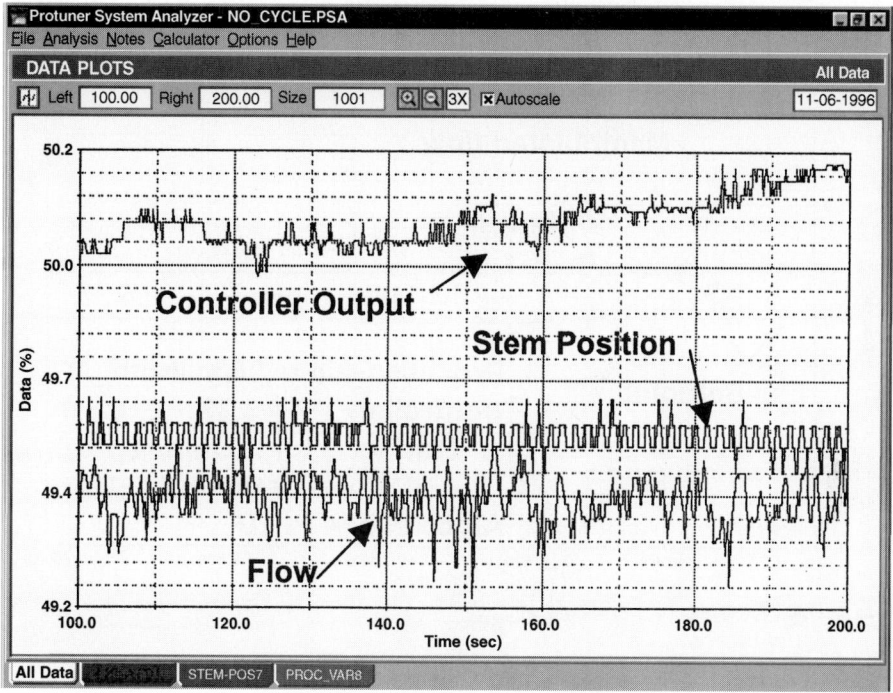

FIGURE 15   No limit cycle in auto when positioner and control valve have small dead time.

Figure 16 shows a simulation of a level control loop with the proportional action on the error. Note the large output movement when the setpoint is changed. Figure 17 is the same process and the same tuning constants but with proportional action on the error. Note the smooth movement of the controller output and how long it takes for the process variable to get to the new setpoint.

Another application for the PID algorithm with the proportional action on the process variable is when the signal filter on the process value is significant compared with the process time constant. Special tuning techniques [3] combined with proper filter selection can be used to better control noisy process value signals.

If the derivative action is applied to the error, a setpoint change causes the controller output spikes in one direction and then spikes back almost immediately. In general, this response is not desirable and the derivative action should be applied to the process variable instead of the error. Figure 18 shows the effect of this algorithm when the setpoint is changed. Figure 19 shows the response to a setpoint change when the derivative action is applied to the process variable rather than the setpoint.

Control Valve Performance Specification

In most control loops the final link back to the process is a control valve. For the control loop to meet the required performance criteria, each component in the loops, including the control valve, must meet performance requirements. As shown in the case studies earlier in this subsection and in other studies [1], the control valve is a common cause of variability in the control loop. This has led to the use of control valve selection and performance specifications [9]. The control valve performance specification should be based on the process dynamics and the process control requirements. This topic is covered in more detail in the Plant Analysis, Valve Response, and Advanced Regulatory Control sections.

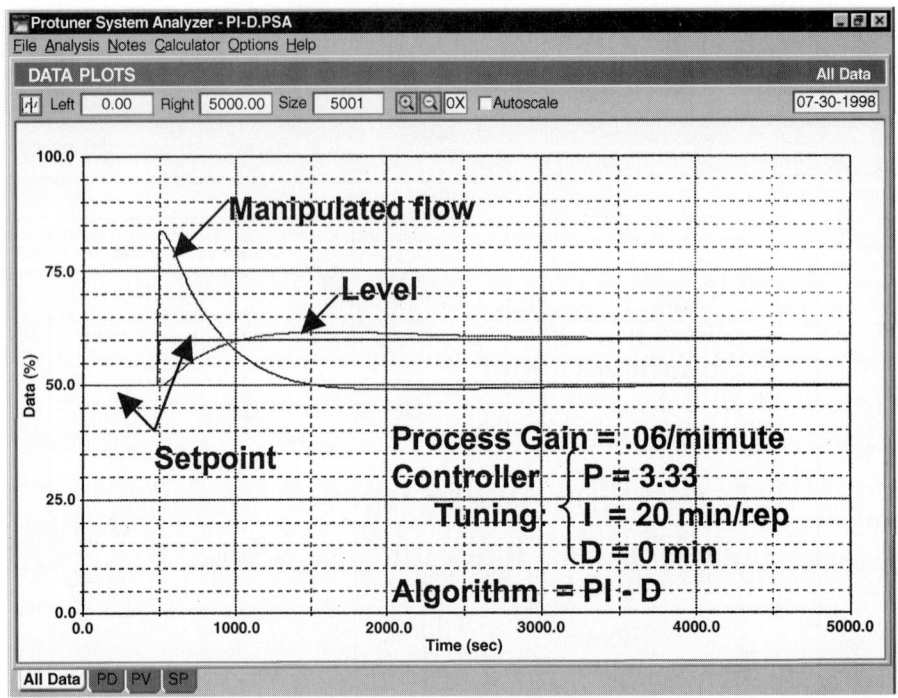

FIGURE 16 Level loop with proportional and integral action on the error.

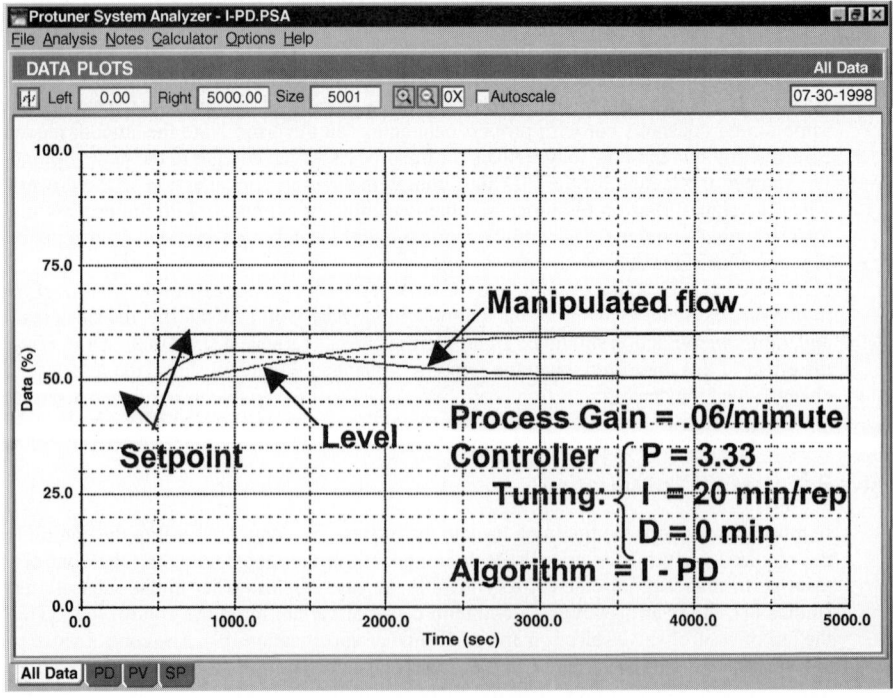

FIGURE 17 Level loop with integral action on the error, proportional and derivative on the process variable.

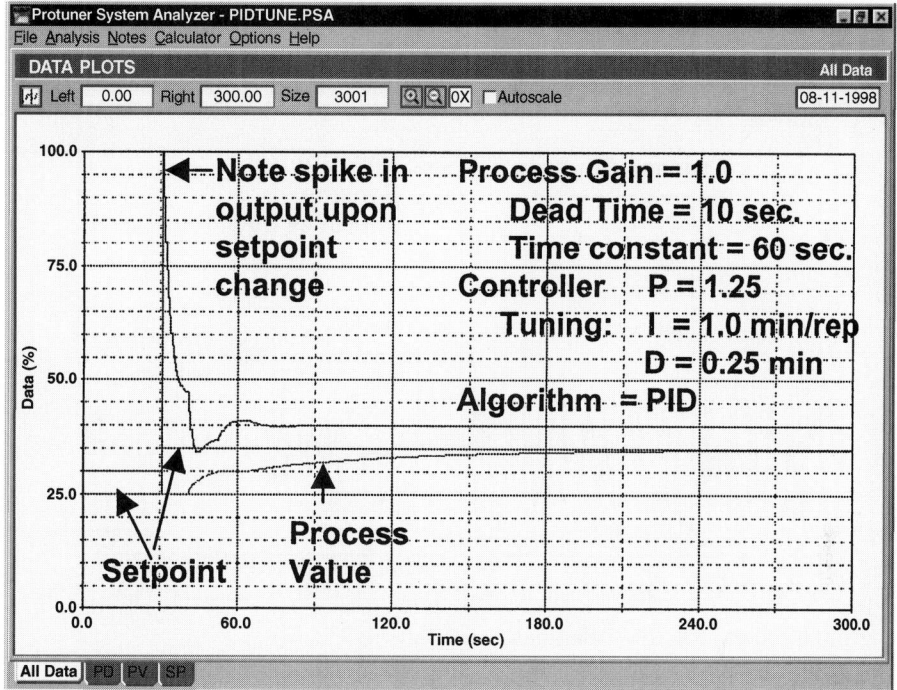

FIGURE 18 Simulated loop with derivative on the error.

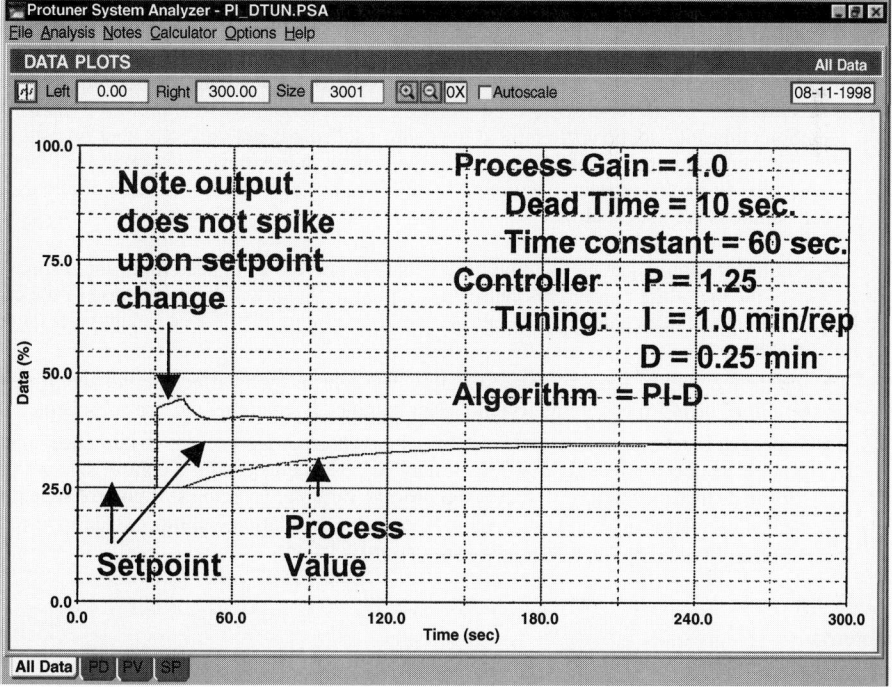

FIGURE 19 Simulated loop with derivative on the process value.

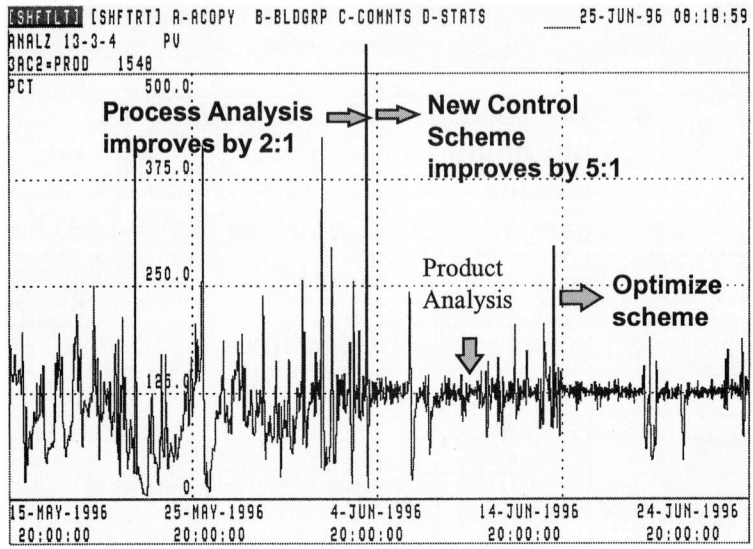

FIGURE 20   New control scheme reduces variation by 5:1

Control Scheme

It is important to choose the correct control scheme to have optimum control. This topic is covered in detail elsewhere in this handbook but deserves to be noted here. Care should be taken in the design phase to select the proper control scheme. Figure 20 shows how improving the control scheme reduced the variability of the product analysis of a distillation process by 5:1. This improvement was made after improvements to the instrumentation and control valves had resulted in a 2:1 reduction in variability.

Sometimes it may be apparent that the control scheme is not correct. Other times it is only after the instrumentation, controller tuning, and the control valves have been optimized that the control scheme is suspected of being incorrect. Some of the potential control scheme problems, symptoms, and solutions are listed below.

- Nonlinear control scheme: Symptoms are that the process gain, dead time, and response time vary over the operating range. This causes a variation in the closed-loop response of the control loop. Solutions include changing the control scheme, changing the process, adaptive tuning, and signal characterization.
- Interactive control loops: Symptoms are when control loops affect by the action of each other. Solutions include decoupling techniques or completely changing the control scheme.
- Poor disturbance rejection: Some control loops correct for process disturbances only after the process has been affected. However, it may be possible to design a control scheme that corrects for the disturbance before it affects the process variable. Examples of control schemes that help correct for disturbances are reflux ratio control on a distillation column, cascade control, pressure–temperature-compensated flow, and feedforward schemes.

Other Problems

The following is a list of other instrumentation and control valve problems that can hinder optimum control.

- Controller scan time: It may be too slow for good control. Generally, the scan time should be 1/5 to 1/10 of the closed-loop response time.
- Signal resolution: Some control systems use a thermocouple or RTD module with a wide temperature range. The actual range of the control point is specified in software but the analog-to-digital conversion is performed over the entire range of the module. This can result in a resolution for the control point that is larger than desired.
- Signal filter: Proper filter techniques should be used [3].
- Smart transmitter response: Make sure that the response time of the smart transmitter is adequate. Find out the complete response (dead time, response time, resolution, etc.) of the transmitter, not just the output update frequency.
- Split range of control valves: Make sure that the split range is correct. Determine if the split range is overlapped, not overlapped, symmetrical, etc.
- Level transmitter calibration: It is common to convert a field-mounted, proportional-only level controller to a level transmitter with a separate controller. While functioning as a level controller, the gain of the controller is included in the calibration. Make sure the level controller is recalibrated properly to function as a level transmitter.

KEY POINTS

- Improvements in process instrumentation, control valve performance, and controller tuning have been found to account for a significant portion of the economic benefit of all control improvement techniques.
- Improvements in process instrumentation, control valve performance, and controller tuning are essential to obtain the full benefit of higher-level control improvement techniques.
- A plant program to optimize the performance of process instrumentation, control valves, and controller tuning will result in less process variability and higher process availability. The key points to implementing a control optimization program are the following.
  - Use the success of others to sell the program.
  - Select a process analysis system with the appropriate features and receive detailed training to fully utilize the system.
  - Partner with an operating department to ensure early successes. Expand the program based on previous successes.
  - Consider as much of the process unit as possible instead of individual loops.
  - Record before and after results and economic benefit to help justify and improve the program.
  - Develop a system to monitor performance to prevent loss of original benefits.
- There is a wide variety of problems with instrumentation, control valves, and controller tuning that can increase process variability. A versatile process analysis system is needed to diagnose and correct many of the problems.
- Control valve performance should be specified based on the process control requirements.
- Control schemes selection may limit control performance. Linearity should be a key consideration in the design of control schemes.

REFERENCES

1. Rinehart, N., and F. Jury, "How Control Valves Impact Process Optimization," *Hydrocarbon Processing*, June 1997.

2. Tolliver, T., "Process Analysis for Improved Operation and Control," Fisher-Rosemount Systems Advanced Control Seminar, 1996.

3. Ender, D. B., *Filter Application Manual*, Techmation, Inc., Tempe, Arizona, 1993.

4. Ender, D. B., *Protuner Application Manual*, Techmation, Inc., Tempe, Arizona, 1993.

5. EnTech Control Engineering Inc., Course PCE-I, "Process Control for Engineers," Toronto, Ontario, Canada.

6. Shinskey, F. G., *Process Control Systems*, McGraw-Hill, New York, 1996.

7. Beall, J. F., "Process Analysis, Diagnostics and Solutions," North Texas ISA Instrumentation and Controls Exhibition, May, 1998.

8. Ender, D. B., *Implementation of PID Algorithms*, Techmation, Inc., Tempe, Arizona, 1993.

9. "Control Valve Dynamic Specification," Version 2.1, 3/94, EnTech Control Engineering Inc., Toronto, Ontario, Canada.

BEST PRACTICES, TOOLS, AND TECHNIQUES TO REDUCE THE MAINTENANCE COSTS OF FIELD INSTRUMENTATION

by Gregory K. McMillan*

BUSINESS IMPACT

Approximately 40% of manufacturing revenues are spent on maintenance according to the U.S. Department of Commerce [1]. For the chemical industry, the percentage can be larger because of extreme pressures and temperatures, highly corrosive fluids, and exposure to the elements. One of the larger maintenance cost categories is instrumentation, and the single biggest subcategory cost is typically calibration. If you take into account that most of the instrument maintenance and the calibration checks are unnecessary, there is a significant opportunity to reduce manufacturing costs. In one large chemical plant, 35% of the instrument checks made for preventative maintenance and 28% of instrument checks from reactive maintenance found no problem [1]. One of the largest chemical companies reports that 65% of calibration checks are unwarranted. If you also consider that most calibration shifts of smart digital instruments result from not using the best selection or installation for the application, calibration should go from being the highest to the lowest subcategory cost.

The exponential increase in technology has led to a corresponding increase in the number of types and models and features of instrumentation. The stocking of spare parts has become an enormous, complex, and costly task to the point where the use of new technologies and manufacturers is discouraged despite performance and/or price advantages.

The other side of the story is that maintenance techniques to date have largely been ineffective in getting at the source of the real instrument problems. Poor measurement and control valve consistency, sensitivity, and reliability reduce process efficiency and capacity. One production unit found that two thirds of the process upsets could be traced to instrument faults [2]. A study of the pulp and paper industry by EnTech revealed that 80% of the loops did more harm than good by increasing process

* Senior Fellow, Solutia Inc., St. Louis, Missouri.

variability. If variability is high, the process has to operate further away from constraints such as equipment limitations and minimum product specifications, which causes lower yields and production rates and more stress on the process equipment. Higher variability and frequency of instrumentation failures mean more violations of constraints from environmental restrictions or interlock settings. The consequence is more waste and less on-stream time. Thus improvements in the performance of field instrumentation can not only reduce maintenance costs but can also reduce the cost of goods and increase fixed cost dilution and revenue.

ENGINEERING PRACTICES

A review of the typical cause of errors and failures for measurements (see Table 1) or for control valves reveals that the performance and reliability of instrumentation is largely determined

TABLE 1 Categories and Classes for a Dynamic Specification for Control Valves (Four classes A–D for each of the four categories 1–4)

1. Minimum Step Classes	2. Maximum Step Classes
Class A: 3.0% +−0.3%	Class A: 5% +−0.5%
Class B: 1.0% +−0.1%	Class B: 10% +−1.0%
Class C: 0.5% +−0.1%	Class C: 20% +−2.0%
Class D: 0.2% +−0.1%	Class D: 50% +−5.0%

Note that the response time is the time for trim (not actuator) to stay within 10% of step or 0.1% span, whichever is largest (overshoot OK if recovery within this offset in response time).

3. Response Time Classes	4. Minimum Positions Classes
Class A: 15 s	Class A: 30%
Class B: 5 s	Class B: 20%
Class C: 2 s	Class C: 10%
Class D: 1 s	Class D: 0%

For example, a DACB class control valve will respond to signals larger than 0.2% and smaller than 5.0% in less than 2 s above a positions of 20%. If you do not care, specify class AAAA—almost any valve will meet it. The idea is to require a control valve to respond by addition of class to valve specification.

- To choose a typical class, use the loop type that actually throttles valve. For cascade loops, you should use the slave loop. Thus, if a column temperature loop sends a remote set point to a distillate flow loop, you should use the classes for a flow loop.
- Vessel or column temperature control => CABC
- Exchanger temperature control => CACC
- Pipeline temperature control => CCCC
- Liquid pressure or flow control => CACC
- Vessel level control => CABC
- Gas or steam pressure control => CBDC
- Compressor surge control => CCCD
- Vessel pH control => DABD
- Pipeline pH control =>DBCD
- Pressure relief => BCCD
- For split range, use class D for last category (minimum throttle position)
- A general-purpose class would be BAAB

during design and construction. The best practices for selection (Table 3) and installation (Table 4) eliminate the source of most maintenance problems and set the stage for developing the confidence to eliminate unnecessary calibrations.

PROBLEMS AND CAUSES

The most common source of problems for pressure measurements is sensing lines and for level measurements is equalization lines. The addition of these long and narrow passages filled with stagnant process fluid is begging for problems. Solids and coating buildup or freezing are likely to plug the lines sooner or later. Plus, the calibration of differential pressure measurement depends on the density of the fluid in the sensing lines, which requires that the state and the composition of the fluid in the lines be constant. If the lines are assumed to be dry, condensate buildup causes a huge error. Similarly, if the lines are assumed filled but the liquid is vaporized or sucked into the vessel or pipeline, the readings become meaningless. Sensing lines are expensive to install and maintain.

The next biggest cause of measurement errors and failures is not enough attention being given to detailing the process conditions and their adverse effect. The coating, fouling, and corrosion rate at different velocities and temperatures should be estimated. The extremes in temperatures and pressures during abnormal operation and special modes of operation, such as defrosting and decontamination, should be identified and taken into account during instrument selection and installation.

Assuming that large case and pneumatic measurement instrumentation with levers, links, bellows, and flapper nozzles is a thing of the past, the final significant source of measurement problems is due to electrical interference or grounding problems caused by poor wiring practices or improper enclosures to protect the terminations. These problems were more predominant in installations completed before cable tray, grounding, and enclosure practices were better defined in the 1970s. However, mistakes are still made, especially when new wires are installed to add a few signals without formal design procedures. Figure 1 shows the erratic signal behavior caused by high signal wiring resistance. The wire was pulled during a control system upgrade and never checked for integrity [3]. Fortunately, there was a redundant pressure transmitter, otherwise the large deviations might have been interpreted as process upsets.

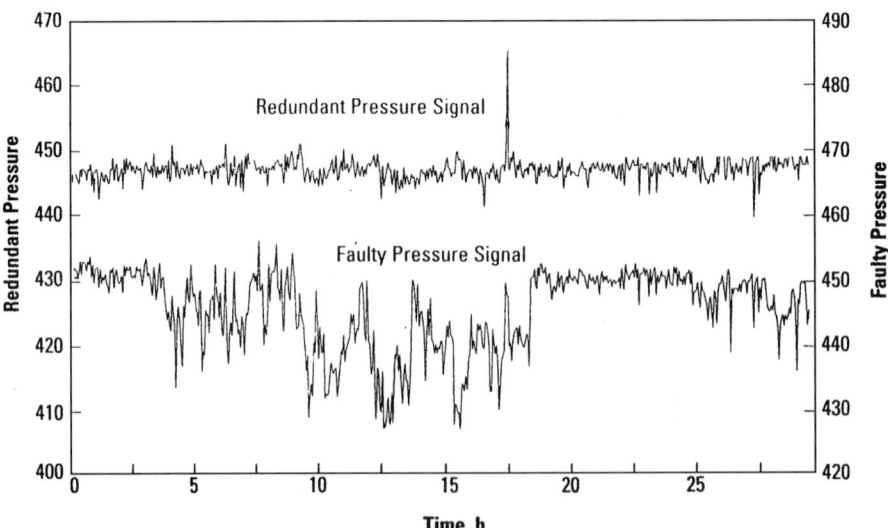

FIGURE 1   Erratic signal behavior caused by a wiring problem [2].

Causes of Measurement Errors and Failures

1. Sensing lines that are plugged or that have liquid when they should be dry or vice versa

2. Sensing elements with excessive coating, fouling, or abrasion

3. Excessive bubbles or solids

4. Sensing elements with deformations, cracks, and holes

5. Low Reynolds numbers

6. Gaskets and O-rings that leak

7. Incorrect materials of construction

8. Sensing, pneumatic, and electronic components affected by process or ambient temperature

9. Moisture on the sensing element or signal connections

10. Electrical interference

11. High connection or wiring resistance

12. Nonrepresentative sensing point

13. Inadequate straight-pipe runs for flow sensor

14. Nozzle flappers that are plugged or fouled

15. Loose feedback linkages and connections

16. Incorrect calibrations

17. Electronic component failures

Unfortunately, control valves are mechanical and use pneumatic components. Their initial performance and degradation rate is coming under closer scrutiny because, next to poor tuning, they are the biggest cause of variability in control loops because of their inability to respond to small changes in controller outputs. Until recently, there were no requirements on a valve specification that the valve actually stroke when asked and no feedback of actual valve position in the control room. A combination of practices on the part of the manufacturer and the user in the past two decades set the stage for the sad situation in which a loop is better off in manual than in automatic. The manufacturer was not asked to make a valve responsive, the user did not know when a valve was sticking or slipping since there was no position feedback in the control room, packing friction increased from the need to reduce packing leaks to meet new environmental regulations, higher sealing friction appeared as users became enamored with tight shutoff and tried to use isolation valves as throttle valves or vice versa, and positioners and actuators became less sensitive from the desire for simpler and cheaper components.

When rotary valves, such as ball and butterfly valves, have high friction and long shafts and just one or two key lock connections between the actuator stem and the valve shaft, the actuator will move, but the ball or disk will not, for small changes in signal that are to twisting of the shaft and play in the connections [3]. Since the feedback of actual valve position is from the actuator stem, the positioner, the user does not know the valve is not moving. Also, the installed valve characteristic becomes very flat at large valve positions to the point where the change in flow per change in stroke approaches zero. Diagnostics and indication of actual valve position in the control room will not show either of these problems. Thus the use of a smart digital positioner is not the final solution. The user must pay attention to valve construction and pressure drops and use a sensitive flow measurement as the ultimate proof that a valve actually changes the flow for small changes in the controller output. Figure 2 shows the response in actuator stem position and flow for various step sizes [4]. It should be noted that the change in controller output per scan typically varies from 0.2% to 1.0% so if the flow does not change for these small steps; the control loop will hunt from reset action and loop variability will be noticeably worse in automatic than in manual.

When the control valve does not respond, the controller continues to increment or decrement its output each scan. The result is a ramp rate of controller output. The time it takes to get beyond the resolution limit sufficiently for the valve to respond is dead time. The rate of change of the loop output

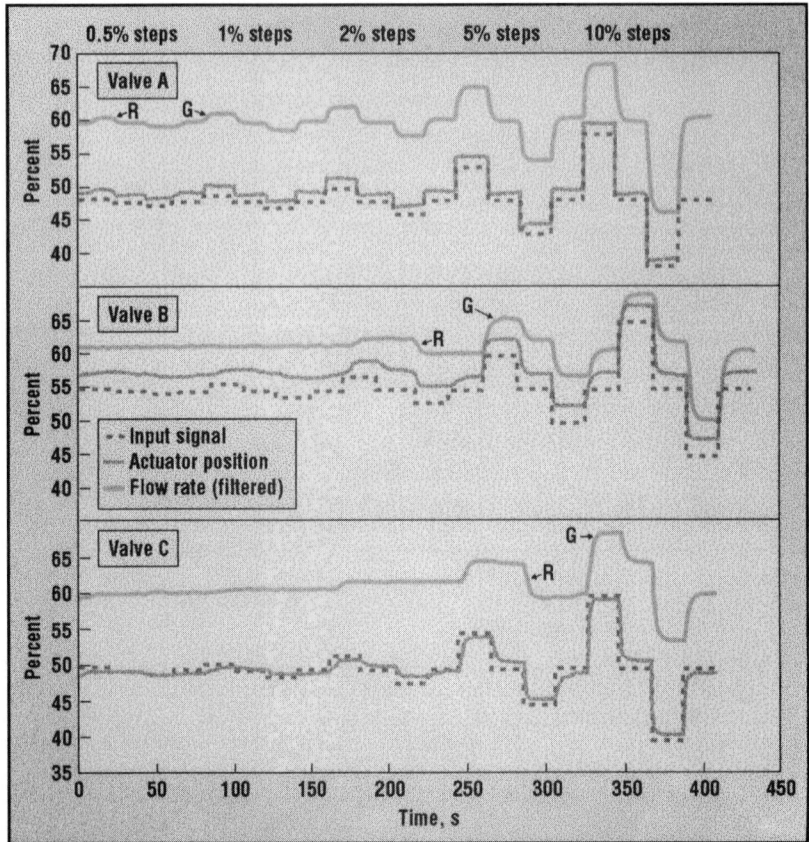

FIGURE 2   Response of a control valve to various step sizes (*R* points to the first reversal of flow and *G* points to the first good response in flow) [4] (*courtesy of Fisher-Rosemount*).

is slower if the loop has low gain and reset action, and the peak error from load upsets is larger because of this additional dead time. If a valve is oversized, the resolution as a percentage of flow and the associated dead time are both larger [5]. The practice of using line sized control valves is expensive from both initial investment and loop variability viewpoints. The permissible stroke range depends on the installed characteristic and the amount of friction near the closed position. Generally the sizing should keep globe valves between 10% and 90%, ball valves between 20% and 80%, and butterfly valves between 25% and 65% open for minimum to maximum flows. The exact ranges depend upon the installed characteristics.

Causes of Control Valve Errors and Failures

1. High trim seating friction for sliding stem valves and high ball, plug, or disk friction for rotary valves
2. High packing friction
3. Loose shaft connections on rotary valves
4. Long shafts or linkages with gaps on rotary valves

5. Single-stage positioners

6. Electrical interference

7. High connection or wiring resistance

8. Nozzle flappers that are plugged or fouled

9. Feedback linkages or connections that are not tight

10. Piston actuators with high sliding friction, wide or worn gear teeth, or yoke slots with play

11. Trim that is plugged or has excessive coating, fouling, abrasion, or erosion

12. Flashing and cavitation

13. Changes in pressure, temperature, and composition

14. Low Reynolds numbers

15. Leaking packing

16. Pneumatic and electronic components affected by process or ambient temperature

17. Incorrect calibrations

18. Pneumatic and electronic component failures

SELECTION

The biggest leverage on maintenance cost is the selection and the installation of instrumentation that has streamlined passages with no restrictions and cavities for stagnant process fluid to accumulate, and utilization of principles of measurements that have low drift, hysteresis, environmentally induced errors, and process-induced errors but high sensitivity. This is true for measurements and control valves. The optimal situation is to have an instrument that is reliable, repeatable, and sensitive. Absolute accuracy is not as important, in that bias errors can be zeroed out for transmitters and corrected by loop reset action for control valves. It is most important that a measurement consistently respond to small changes in the process variable (PV) of interest and that a control valve consistently respond to small changes in the controller output. The requirement for an instrument to be consistent is more inclusive than for it to be repeatable in that it specifies a limit to the variation in the gain, dead time, and time constant of the response.

Field process pressure and temperature switches are generally mechanical in nature and provide no continuous information on their health to the control-room-like transmitters. They are not exercised until needed, which further decreases their reliability. Plant operators are essentially flying blind with these devices, not knowing the value of the PV and never being sure if the device will operate when required.

Orifices are fine as long as you do not need a long-term accuracy of better than 5% and you do not use sensing lines for the differential pressure transmitter. Venturi tubes and flow tubes offer less obstruction, require less upstream diameters of straight run, have a more constant meter coefficient over a wider range of Reynolds numbers, and have a lower permanent pressure drop than orifice meters.

Coriolis, vortex, and magnetic flow meters have minimal or no obstruction to flow and have high consistency and low drift when applied properly. Diaphragm seals and remote heads that can be flush mounted eliminate restrictions and cavities where process fluid can accumulate. Diaphragm seals must be precision filled with hydraulic fluid at a factory or service center whose procedure has been verified to eliminate bubbles. The capillary lengths must be short, shielded from the sun, rain, and other sources of temperature changes, and equal for differential pressure transmitters. Remote heads should use digital signals to prevent the introduction of analog-to-digital (A/D) error into the computation of differential pressure. Also, the absolute pressure should be less than the differential pressure so that the error introduced by separate pressure measurements is not excessive.

Radar devices offer an extremely accurate surface detection and are generally not affected by changes in the vapor or liquid temperature or composition. Nuclear level measurements are non-intrusive and are thus attractive for very nasty fluids and extreme operating temperatures in which nothing else will work. There is a considerable amount of paper work, but this can be contracted out. However, since the level is inferred from the attenuation of radiation by the mass between the source and the detector, nuclear devices depend on the liquid density's being relatively constant. Radar will measure foam level whereas nuclear will not. Radio frequency (RF) admittance level probes can also measure foam. New RF sensors have been developed that have a second probe to measure changes in composition to make the level measurement independent of changes in the electrical properties (e.g., dielectric constant) of the process.

Smart transmitters are available for a slight increase in price. The fact that this investment was smart is evident from the beginning. A plant reported in 1992 that the installation of smart pressure transmitters reduced unplanned maintenance calls by 90% and the average time to diagnose a problem dropped from 2.5 h to 10 min [6]. The estimated savings for 100 transmitters in a more recent implementation of smart transmitters and HART communication protocol was $16,300 for stand-alone HART and $26,050 for HART diagnostics available on the operators' console [7]. The biggest savings of $10,000 and $15,000 was due to the reduction in time required for calibration checks.

Smart transmitters also offer the opportunity to sense, use, and transmit auxiliary PVs. These multivariable transmitters provide more accurate signals by activating alerts or correcting for significant environmental effects and changes in process operating conditions. For example, temperature and pressure sensors integrated into an averaging pitot tube assembly can do pressure and temperature compensation to provide a mass flow measurement for a constant composition that is particularly attractive for large pipelines where the use of Coriolis flow meters is not possible [8]. The addition of temperature and pressure sensors to a Coriolis meter, in combination with the existing mass flow and density measurements, can provide a viscosity measurement. The incorporation of temperature and pressure sensors into a control valve with position feedback can compute flow for a constant composition. The addition of an accelerometer to a photocell can measure vibration that causes false information [9]. Even if the additional information is not used for compensation but is needed in the control room, the savings in wiring and installation is impressive.

The use of digital outputs, such as frequency and pulse count outputs from flow meters, can increase the accuracy of the measurement by more than a factor of 2 by elimination of the digital to analog (D/A) converter in the transmitter and A/D converter in the distributed control system (DCS) or the programable logic controller (PLC). The use of HART frequency outputs for the PV adds a dead time to the loop that could be a concern for antisurge control and some pressure control loops. Fieldbus will reduce this delay by supporting much higher information communication rates.

It is essential to use HART and eventually Fieldbus to get the diagnostics into the control room and part of a system that can document and analyze failures. The use of hand-held interrogators and hand-written reports keeps the knowledge to be gained localized at best. While the savings in paper work for meeting ISO 9000 and FDA requirements is impressive, it is exceeded by the possible improvements from the resolution of process and application problems. Fieldbus will greatly increase the speed and the quantity of diagnostic information available in the control room and enable the interchange of data between instruments to expand the scope of diagnostics. For example, pressure and flow measurements could be cross checked against control valve positions for signal validation before the signals even get to the control room. Fieldbus will dramatically increase the view and scope of instrument diagnostics.

Resistance temperature detectors (RTDs) can respond consistently to changes as small as 0.1°C but this great sensitivity is compromised by not recognizing that each RTD has a slightly different calibration curve and is altogether lost because of A/D error introduced from large spans. Unlike DCS or PLC RTD input cards, the calibration of transmitters can be matched to a sensor, and the spans can be narrowed to cover an individual application. RTDs hold their calibration longer and have greater life expectancy than thermocouples unless there is excessive vibration. While a bare-element RTD may be a couple of seconds slower than a thermocouple, this extra lag is small compared with that of the thermowell and the rest of the temperature loop. Last, three- or four-wire RTD transmitters are important to eliminate the effect of changes in lead wire resistance. Unless the wiring is very short (e.g., less than 20 ft.), a four-wire should be preferentially used over a three-wire assembly. The error of a three-wire RTD for 500 ft of 20-gage wire can be as large as 4.7°F since wires have a resistance tolerance of 10%.

Single straight-tube Coriolis flow meters are better for liquids with solids because the solids will not equally distribute themselves in dual tubes and there are no bends that are particularly susceptible to erosion. The higher excitation energies for thick-wall single straight tubes maintain their accuracy at high solids loading. Also, the resonant frequency is so much higher than other frequencies in the plant that vibration is normally not an issue.

Sliding stem (globe) valves have higher sensitivities, and the actuator stem feedback position more closely represents the final element position than rotary (ball, butterfly, and eccentric plug) valves. Above 6 inches, the economics of rotary valves is compelling, and process conditions such as fouling and solids may dictate a rotary valve. When rotary valves are used, low friction, tight connections, and short large-diameter shafts are essential to minimize gaps (backlash) and twist (shaft windup). Valves designed for on–off service and tight shutoff generally cannot be made into good throttling control valves.

Piston actuators require more maintenance and are less sensitive than diaphragm actuators. However, pistons are less expensive and may be the only alternative for very large control valves. When pistons are used, the emphasis should be on low sliding friction, minimal gap in the slot for scotch yoke, and wear-resistant tight teeth for rack-and-pinion pistons.

A valve response requirement should be added to the control valve specification. This valve dynamic specification must have enough details to ensure meaningful results. It is critical to realize that the manufacturer will naturally choose the step size (10%), the throttle position (45%), the friction (hand tight packing, special lubricants, and/or flow ring), valve size (small), positioner air consumption (high flow spool or relay), and feedback measurement (actuator shaft instead of trim position) that gives the best response time. These test conditions may have nothing to do with the valve actually supplied or the results experienced in the field. The dynamic specification must specify the minimum step, the maximum step, the response time of the trim, and the throttle position expected. The four classes of performance for each of these four categories are shown in Table 1 along with examples of classes for different types of applications. Note that an offset is permitted and that an overshoot is of little consequence if there is a recovery within the response time except where relief valves or rupture disks or instantaneous pressure interlocks might be activated by a pressure excursion. The longest of the response times for the minimum and maximum step size determine the class for the response time category.

For pH and Oxidation Reduction Potential (ORP), either a liquid-filled reference, pressurized to provide a small flow of electrolyte out of the reference junction, or a solid reference is needed to prevent internal contamination of the reference electrode with process ions for high concentrations of salts or acids and bases. The flowing junction provides a more accurate measurement by ensuring a more constant diffusion potential, but it requires refilling.

The life expectancy of pH electrodes is the lowest of all of the common measurements and is highly dependent on operating temperature. Figure 3 shows the average life to be 12 months at 25°C, 6 months at 50°C, 3 months at 75°C, and 1.5 months at 100°C when there are not additional adverse conditions such as abrasion, chemical attack, or dehydration. Caustic solutions above 12 pH increase the degradation of life with rising temperature dramatically. For high temperature and high caustic concentrations, it is essential to limit the exposure by using an actuated assembly to automatically retract the sensor. This is also very effective for coating problems that cannot be prevented by high velocities since the retraction can be followed by an automatic wash of the electrodes with a cleaning solution followed by a flush and soaking of the electrodes. Note that the cleaning solution does not get into the process so it does not have to be chosen to be compatible with the process. However, the retraction and cleaning of electrodes upsets their thermal and ionic equilibrium. So for more moderate conditions, the use of three electrodes and selection of the middle reading is preferred. The middle selection will inherently ignore a single failure of any type and reduces the error from the short-term excursions that are characteristic of pH measurements.

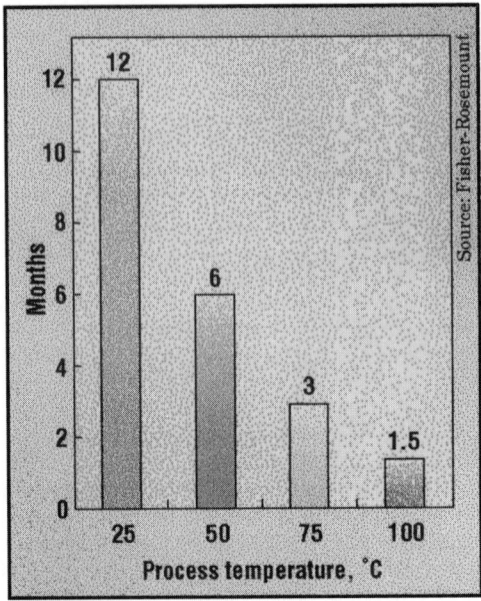

FIGURE 3 Life expectancy of pH electrodes for various process temperatures (*Courtesy of Fisher-Rosemount*).

Best Practices for Instrument Selection

1. Do not use pneumatic transmitters.
2. Avoid sensors with mechanical linkages or bearings that can wear out.
3. Avoid field pressure or temperature switches.
4. Avoid orifice meters.
5. Avoid piston actuators for control valves, especially scotch yoke or rack and pinion with wide teeth.
6. Avoid tight shutoff and automatic block or isolation (on–off) valves as control valves.
7. Use transmitters and positioners with digital rather than analog components.
8. Use smart HART multivariable transmitters and positioners with diagnostics in the control room.
9. Use instrumentation with modular designs to the reduce the time and the cost to replace or overhaul.
10. Use precision filled diaphragm seals or remote heads for differential pressure measurements.
11. Use Coriolis, vortex, and magnetic flow meters.
12. Use radar or nuclear level measurements.
13. Use components that can withstand the extremes of the process and environment.
14. Use sensor matched transmitter calibrations for RTDs.
15. Use three- and preferentially four-wire transmitters for RTDs.
16. If you must use thermocouples, use premium-grade elements and extension wire.
17. Use narrow span transmitters instead of thermocouple or RTD DCS or PLC input cards.
18. Use digital instead of analog output signals for cases for which speed is less important than accuracy.

19. Use level switches (e.g., tuning forks) that ignore coatings and changes in material composition.

20. Use straight single-tube Coriolis or magnetic flow meters for solids.

21. Use diaphragm actuators. If you must use a piston actuator because of valve size, consider a low-friction vane or floating cylinder.

22. Use bellows seals and extension bonnets instead of high-friction graphoil or environmental packing.

23. Use sliding stem (globe) valves. If you must use a rotary valve because of size or plugging, use splined or quadruple keyed shaft connections, no linkages, short shafts, low breakaway torque, and low sealing surface friction.

24. Size the control valve to operate beyond the seating friction and on a portion of the installed valve characteristic where the slope is between 0.25% and 2.5% flow per percentage of stroke.

25. Use pressurized or solid reference electrodes for pH and ORP.

26. For sensors with low life expectancies such as pH, use autoisolation to reduce exposure time or middle selection of three sensors.

INSTALLATION

To date, most installations have concentrated more on accessibility than on reliability. The desire to locate a control valve at a convenient platform level has resulted in long sensing lines, insufficient straight upstream and downstream pipe runs, trapped solids, and longer process dead times from greater transportation delays. The practice is self-fulfilling in that mounting instruments for easy access has increased the maintenance problems to the point where mounting location for such access is deemed essential.

While the need for straight runs upstream and downstream is well recognized for orifice meters, it is sometimes neglected for other instrumentation. Pitot tubes, vortex meters, and thermal gas meters have approximately the same run requirements as orifice meters. Magnetic flow meters and ultrasonic flow meters need approximately half of the run requirements of orifice meters. Ideally, control valves, which are variable orifices, should have approximately the same straight run as an orifice [10]. The manufacturer's test facilities where the control valve characteristics are documented meet or exceed these requirements, but these are rarely adhered to in the field. When the pressure drop across the control valve is small compared with the upstream pressure, which occurs for rotary valves, it is more important; but this is exactly the situation in which it is least likely to be practiced because of the size of the lines. Also ignored is the swage effect on rotary valve, which can reduce the capacity by as much as 40%. Insufficient straight runs cause excessive noise and a loss in consistency for both measurements and valves because of a nonuniform velocity profile, pressure fluctuations, and concentration or temperature gradients. Coriolis meters have no straight run requirements.

Temperature and pH sensing probes should extend into the middle of the pipeline to get the most representative measurement. Both temperature and pH sensors should be at least 10 pipe diameters downstream of a mixer to allow sufficient recombination and dispersion of the flows. Since a thermowell suffers from conduction error where heat flows to or from the tip, depending on the temperature gradient along the thermowell wall, it is also important to ensure that at least 10 diameters are immersed into the process fluid. When an elbow is used for the thermowell installation, most of the insertion length is in the center of the pipeline. Figure 4 shows four different types of installations of thermowells in order of preference to reduce profiling and conduction errors [11].

The most effective and least costly method of keeping probes clean is to increase the liquid velocity to more than 5 feet per second (fps). For pH electrodes, it is desirable to keep the velocity less than 9 fps to reduce wear and noise. For temperature probes (bare elements and thermowells), it is important to keep the velocity less than 30 fps to reduce erosion. The maximum velocity to minimize vibration may be lower, particularly for bare elements or long insertion lengths and RTDs. When velocity does not work for pH probes, jet washer nozzles aimed at the probe tip or autoretractable assemblies with a washing cycle are used [12].

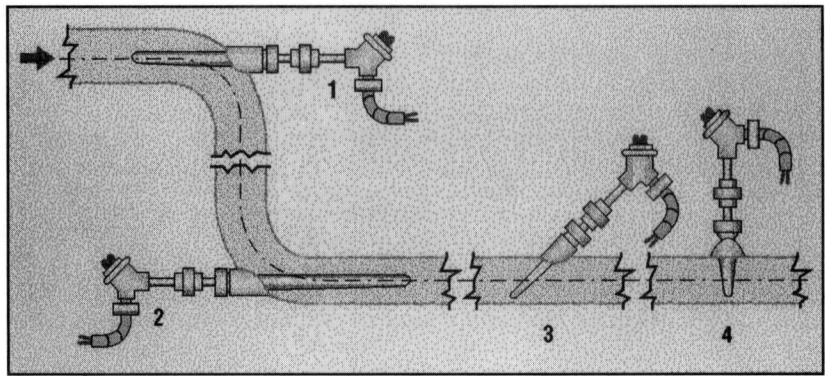

FIGURE 4   Order of thermowell installations to minimize measurement error [11].

It is important that the transportation delay between the control point and the measurement point in the process add less than 10% dead time to the loop so that additional variability is not appreciable. This means that pH electrodes should be mounted approximately 10 to 20 pipe diameters from the discharge of a pump for vessel pH control or a mixer for pipeline pH control. Similar requirements should be used for the control of temperatures at the exit of vessels, desuperheaters, pipeline mixers, and heat exchangers.

For control valve applications for which flashing is possible, a lower temperature and/or high-pressure point in the pipeline, a control valve with a lower recovery coefficient, or a staged pressure reduction can prevent cavitation. Special trim for staged pressure reduction has small graduated passageways prone to plugging, can be as expensive as the valve, and is the epitome of a special part. A combination of orifice plates and valves stroked simultaneously provides a lower total cost of ownership than special trim. When cavitation is inevitable, the control valve should be mounted on the inlet nozzle to a vessel so that cavitation changes to flashing. It is important to note that appreciable damage usually occurs only for water streams.

Best Practices for Instrument Installation

1. Avoid sensing and equalization lines.
2. Avoid sample lines.
3. Close couple and preferably flush mount sensor diaphragms.
4. Use sufficient straight runs upstream and downstream of flow meters, probes, and valves.
5. Preferentially use pumped pipelines sized for 5 to 7 fps for probes to minimize coatings.
6. Mount probes and flow meters in vertical lines to prevent solids accumulation.
7. In vertical lines, use flow up to minimize buildup of condensate, solids, and bubbles.
8. In vertical lines, use flow down to minimize abrasion of pH electrodes.
9. Extend probe tip into the middle of a pipeline to get a representative measurement point.
10. Provide insertion length at least 10 times the outside diameter of thermowells.
11. Use spring-loaded tight-fitted, grounded, sheathed temperature sensors in thermowells.
12. Choose location for probes close to discharge of equipment to be controlled.
13. Choose location for control valves close to entrance of equipment to be controlled.
14. Choose location or size pipe line for probes so velocity is less than 1 fps to reduce abrasion.

15. For vortex and magnetic flow meters make sure the velocity is greater than lower limit (e.g., 1 fps).

16. For magnetic flow meters make sure the conductivity is greater than lower limit (e.g., 1 $\mu\Omega$/cm).

17. Size pipeline and meter to keep the Reynolds number in the range for the best flow meter accuracy.

18. Keep flow meters completely filled with fluid and keep probes completely immersed.

19. Choose location for control valves to prevent cavitation or ensure implosion in vessel vapor space.

20. Use enclosures/fittings that prevent water and corrosion on connections and electronic components.

21. Use separate cable trays and conduit drops to keep instrument signal wiring away from AC wiring.

22. Use separate cables for pulsed, frequency, and switched instrument signals.

23. Use twisted shielded pairs that are properly grounded for instrument wiring.

24. Use autoretractable pH probes for fluids that attack, coat, or dehydrate the glass to limit exposure time.

Even if you have used the best design and implementation practices and there are no instrument-related problems, there are still plenty of equipment and process problems that can lead to unnecessary instrument maintenance requests. Thus reaping the entire potential benefit from better engineering practices requires a change in maintenance practices.

MAINTENANCE PRACTICES

It is better to find the cause than to treat the symptom. Calibration is the technique that tends to be used first when a instrument is suspect because it is the easiest to understand and do and is well documented. Instrumentation instruction manuals have calibration procedures detailed, but troubleshooting practices are either too vague or missing. The emphasis may stem from the days of pneumatic instrumentation when instruments could not hold their calibration for more than 6 months. Now calibration is done in response to application problems or lack of operations confidence. Calibration is at best a temporary fix and often camouflages the real problem. A particularly troublesome practice is the removal of the instrument and calibration in the shop. The instrument is not at the operating or environmental conditions of its application. A calibration may actually introduce errors that will cause a need for calibration again. The result can be calibrations chasing calibrations. This occurs whenever pH electrodes are removed from the process for buffering. There is also the risk of damage and a mistake in the reconnection of the instrument. With reasonably good selections and installations and smart HART transmitters and positioners, most of the maintenance cost of instrumentation can be eliminated by simply not calibrating a measurement or a valve until it is within the wear-out phase of its life expectancy unless the instrument's temperature or pressure limits were exceeded. At this point, it might be better to replace the measurement or overhaul the valve.

Instruments used for process control are continually tested at operating conditions and the results and diagnostics are displayed as part of the normal operation of the loop and the operator interface. In contrast, instruments for safety interlock systems that are exercised only when trip conditions occur and whose health and response at the more extreme operating conditions are not scrutinized have a lower reliability and would benefit from an automated documented on-line testing program.

When the major components that distinguish the primary features of an instrument can be easily connected without special tools, the manufacturer gains from reduced manufacturing costs and the user benefits from reduced maintenance costs. It is a win–win situation. Modular instrumentation, such as the ceramic pressure transmitter shown in Fig. 5, enables a huge number of combinations of ranges, connections, and materials of construction to be assembled within a few hours from a small inventory of spare parts. This means regional service centers can provide same-day delivery for replacements. Next to stopping unnecessary calibrations, the next biggest savings obtainable by best practices is in eliminating plant repair and inventory of component parts.

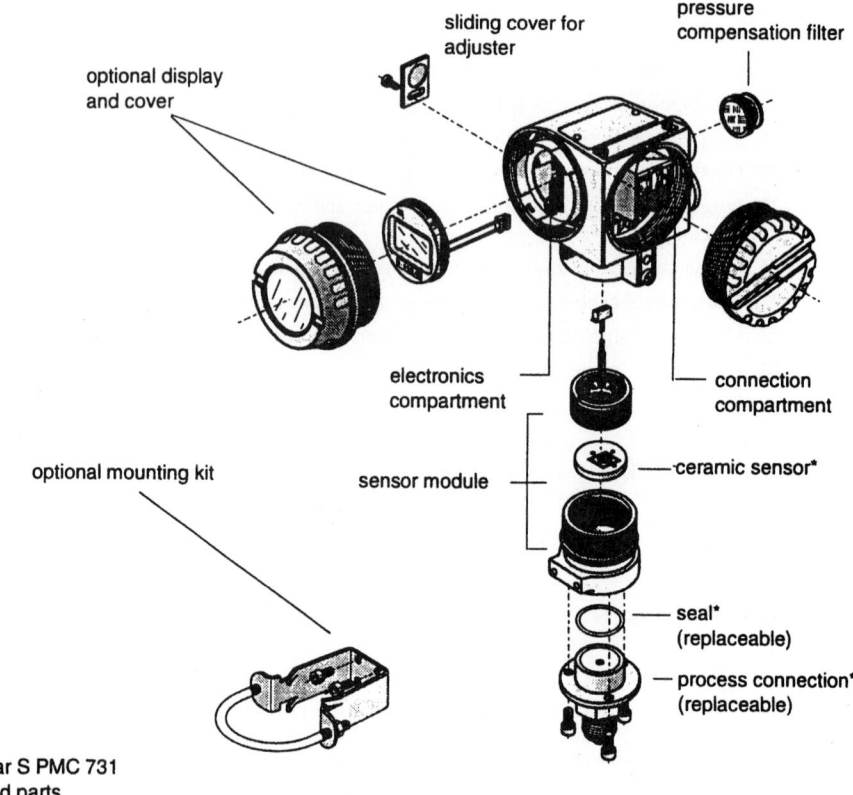

Cerabar S PMC 731
* wetted parts

FIGURE 5   Modular construction of ceramic pressure transmitters to reduce inventory (*courtesy of Endress and Hauser*).

Best Maintenance Practices to Reduce Maintenance Costs

1. Do not calibrate a smart differential pressure transmitter for 2 to 4 years after last calibration.
2. Do not calibrate a coriolis or magnetic flow meter.
3. Do not calibrate a smart four-wire RTD transmitter.
4. Do not calibrate a smart thermocouple transmitter for 5 years after last calibration.
5. Do not calibrate a vortex meter unless the kinematic viscosity permanently changes.
6. Do not calibrate a radar level gage unless the vessel internals change.
7. Do not calibrate a smart nuclear level gage unless the process density permanently changes.
8. Do not calibrate a smart pH transmitter until 2/3 of the electrode life expectancy is reached (e.g., 9 months at 25°C, 4 months at 50°C, 2 months at 75°C, and 1 month at 100°C).
9. Do not calibrate a smart digital positioner unless the valve is overhauled.
10. Do not overhaul a control valve or replace a transmitter until diagnostics indicate a problem.
11. Do not inventory spare parts or repair field instrumentation.
12. Use manufacturer's on-site or regional service centers for replacements and overhauls.

INSTRUMENT-KNOWLEDGE-BASED DIAGNOSTICS

Diagnostics are the key to detecting real problems and eliminating time spent on pseudoproblems. In fact, it is the key to gaining confidence to leave instruments alone until they need to be replaced. However, foolproof diagnostics require the investment of a lot of expertise. A bad diagnostic or a missed diagnostic requires several correctly detected problems to reestablish operations confidence and trust enough to rely on the diagnostics. Since real problems are few and far between when good engineering practices are used, it may take years to recover from a false start.

The diagnostics must be presented near and preferably on the normal operator interface to get the 24-h attention needed for effective use. However, if operators are barraged with false alerts, a valid diagnostic will be ignored. The problem is worse than with false alarms in that operators at first do not believe such expertise is possible, and diagnostic capability, like any new technology, is suspect until proven otherwise.

The investment to ensure reliable diagnostics has been underestimated to date. The investment will be made by manufacturers of instrumentation as an evolution of smart instrumentation to stay competitive if the user recognizes the value of the additional capability. So far, the additional cost of these features is small because the manufacturer can spread the development cost over a large number of units. If instrument-knowledge-based diagnostics can do the job, it is the most cost-effective approach for the user.

One can be misled from the excitement surrounding smart transmitters and think that such transmitters can provide out-of-the-box completely self-sufficient diagnostics and calibration checks. To detect or correct for a drift or span error requires comparing the transmitter output with known signals or values at operating conditions [13, 14]. Some optical devices do this by inserting a known light source and filter in the path. Autoretractable assemblies for pH probes can provide autocalibration by a buffering cycle, but the installation costs of a buffer system are considerable and the calibration is not done at process conditions. A control valve could automatically calibrate its positioner and adjust its tuning (proportional and derivative settings) by comparing the requested and the actual valve positions. Valve manufacturers have built this into their digital positioners as part of an initialization cycle. However, the supposed assessment of drift from comparing the change in transmitter signal from normal operation assumes that the PV is not controlled but is an indicator and that the excursion is not due to changes in process operating points or slow load upsets. To check or correct a calibration requires an internal reference or redundant measurements or inferences from plant-knowledge-based techniques.

A diagnostic check of integrity (e.g., good or bad status) without pinpointing the cause can be obtained from a reference state. For example, a reduction in the noise amplitude of a pressure transmitter compared with the noise during a representative initial 2-day period could be used to determine if the transmitter sensor is coated or the sensing lines or connections are plugged. Transmitters are better suited to do this than expert systems or DCSs because they work on the raw sensor signal before filters, D/A, A/D, and scan times reduce the resolution to noise that is much slower and whose amplitude is altered [13]. Diagnostics to detect the onset of coating will require some periodic adjustment or initialization as operating points or equipment is changed.

Status checks of Coriolis mass flow meters can be made by comparing the excitation energy required for maintaining oscillations, the output amplitude of the sensors, and the resonance frequency of the tubes [13]. Changes in these relative values can be used to detect entrained air or excessive solids.

The status of a tuning fork can be assessed on line by switching to a redundant circuit and drive, as shown in Fig. 6, every second to look for the divergence or nonappearance of the resonant frequency [13]. This essentially complete status check, combined with the tuning fork's low price, simplicity, and ability to ignore nearly all coatings and density changes, makes it an extremely cost-effective level switch. The only adjustment is a threshold specific-gravity adjustment to distinguish between liquid and vapor.

A differential pressure transmitter failure that gets too hot may get noisy before it fails. A diagnostic that keeps track of high temperatures can also help find the source of the problem so that the replacement does not fail [14, 15]. Ideally, the transmitter should keep track of the number, time duration, and magnitude of temperature violations. The same is true for overpressurization.

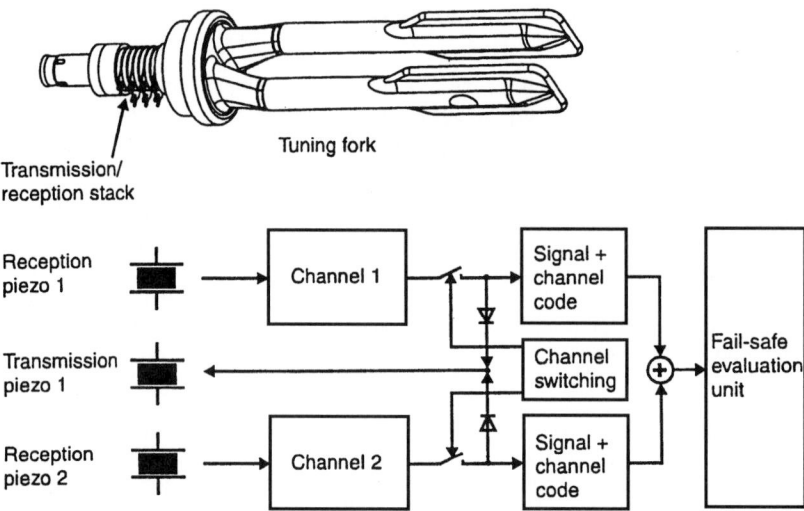

FIGURE 6   Tuning fork design to completely test its integrity every second (*courtesy of endress and Hauser*).

The loss of fill in a differential pressure transmitter can be detected by comparing the measured temperature with the temperature computed from the changes in diaphragm capacitance [13]. Most losses in fill are slow, whereas methods of detection based on response time tests assume a relatively fast loss and a single failure. Since the loss may be over a period of months, concurrent multiple failures are possible and the degradation in response time is not discernible. Of 80 confirmed oil loss failures during a response testing program, none were detected despite 4200 tests [16].

A loop current step response (LCSR) test has been used to determine the in situ response time of thermocouples and RTDs. A small loop current is applied to the sensor leads, and the time for the temperature to rise or decay is used to determine the response time. The method can detect fouling or any other changes in the heat transfer coefficient of the thermowell on line at actual operating conditions [17].

A power spectrum density, combined with data qualification algorithms, can determine the response time of pressure transmitters. The fast Fourier transform (FFT) of noise data of each transmitter is computed to get the power of the fluctuating signal as a function of frequency. The plot of power versus frequency is the power spectrum density [16].

Control valve manufactures have developed extensive diagnostic software. One package that adds measurements of stem position and I/P output and actuator pressures can quantify problems. The capability of the package, including pressure measurements and tests such as valve signature, dynamic error band, step response, drive signal, and output signal tests has also been integrated into onboard advanced diagnostics of a digital positioner so that the tests can be done on command [18]. The results are used to line up spare parts and service center time for repair and reconditioning of control valves during shutdowns or between batches. A savings of 75% in valve maintenance cost was reported by not pulling valves that are fine and by more efficiently correcting actual problems. Sometimes the valve can be fixed without being removed by tightening a bolt or changing a positioner part. Valve diagnostic software has shown the ability to pinpoint the problem. One plant reported a savings of $100,000 per year, which was a 60% reduction in valve maintenance costs. The diagnostic software determined that only 14 of 188 valves scheduled to be pulled during a shutdown for overhaul needed that level of maintenance [19]. There is also the possibility of increased manufacturing revenues from reducing down time.

Another valve manufacturer has developed qualitative diagnostics by noting relative changes in the delay time (dead time) and run time (response time less the delay time) [20]. Depending on the direction of the change in delay time and run time for filling and venting, higher friction, spring failure, air supply problems, air filter contamination, smaller closing force, and leakage in the actuator can be diagnosed.

Some transmitters can be configured to go to a fail-safe signal value following detection of a failure. For decades, it has been possible to specify an upscale or downscale signal following thermocouple burnout (open or high-resistance thermocouple junction). Now smart transmitters have added the capability to hold the last value or to force its signal to a high or low value for a bad PV status that could be caused by a host of diagnosed problems [19]. In practice, signals forced to on-scale values following failure have caused confusion and unforeseen consequences and can actually cause safety issues. To prevent severe process upsets that are due to transmitter failure, hold the controller output at its value from several seconds before the transmitter failure.

Asset effectiveness management software (AEMS) that displays and documents the diagnostics has enabled companies to move to a predictive maintenance mode. The savings during startup alone have been $150 to $200 per device. A pharmaceutical company reported savings of $100 per year per device for a total of $50,000 for 500 devices. A chemical company noted savings of $130 per year per device [19]. Manufacturers are also offering AEMs coupled with on-site loop tuning tools and analysis skills.

Diagnostic Techniques Used or Proposed by Instrument Manufacturers

1. Decrease in sensor signal noise amplitude or frequency to detect plugged, coated, or stuck sensor

2. Shift in sensor noise average to detect drift (highly dependent on numerous assumptions)

3. Change of diaphragm temperature to detect loss of fill fluid in differential pressure transmitters

4. Change in excitation energy, output amplitude of sensor, and resonance frequency of the tubes of Coriolis flow meters to detect changes in the process fluid that affect measurement accuracy

5. Divergence or nonappearance of the resonant frequency of redundant tuning forks to problems with level switches

6. LCSR to determine the in situ response time of thermocouples and RTDs

7. Power spectrum density to determine the response time of pressure transmitters

8. Increase in pH electrode resistance to detect coated or nonimmersed electrodes

9. Divergence of internal redundant reference electrode potentials to detect contamination of fill

10. Capture of pressure or temperature limits violations for almost any type of instrument

11. Totalization of cycles and amplitude to gain access to wear on mechanical components

12. Increase in delay time or run time of small perturbation of valve stroke to detect increase in friction, air supply pressure, clogged air filter, actuator leak, or seat loading

13. Increase in ratio of actuator pressure to drive signal to quantify increase in friction

14. Increase in dynamic error band to detect problem in positioner and valve combination

15. Increase in response time from closed-loop response test response to detect fouled temperature sensors

16. Increase in recovery time of pH measurement after autowashing to detect persistent electrode coating

17. Change in electrode efficiency from autobuffering to detect glass wear, abrasion, and dehydration

18. Fail safe inherently or by autoswitch for bad PV diagnostic

PLANT-KNOWLEDGE-BASED DIAGNOSTICS

Sometimes there are important failure modes that are best determined from material or energy balances or correlation with other instrumentation. However, the user must go into such projects with eyes wide open as to the initial and the continuing costs and commitment and make sure there is some sizable stake. Successful applications in one chemical company occurred when automatic actions were taken by the expert system or neural network that improved yield or capacity and there was strong technical expertise available on site. The greatest broad-based success of neural networks has been in providing on-line intelligent sensors for combustion stack gas analysis. These neural networks have been shown to be more accurate than industrial stack gas analyzers and much less expensive to maintain. Typically, the neural network needs to be retrained once a year by bringing in a validated skid-mounted on-line analysis system. Several states have accepted the results of these neural networks as proof of compliance. It is important that the real-time expert system or neural network have more than just an advisory role in the plant to get the sustained level of attention that it needs.

Neural networks do well when there is a planned test program. For composition measurements, the rental of validated lab or on-line analyzers to generate lots of composition measurements during the test is extremely helpful. Neural networks would benefit from the rigor and the structure of testing used for Constrained Multivariable Predictive Control (CMPC) model identification in which loops are opened (run in manual), unusual operating modes are screened, and several steps in both directions of various sizes are made in all of the manipulated and disturbance variables. Effectively, this is what is done for training and verifying neural networks for combustion analysis. Unfortunately, users have been led to believe that process variables can be modeled from just dumping data from a historian into a neural network and adjusting the inputs, layers, and delays until there is a good fit to the model. This is at best a snapshot in time, since many parameters, such as frictional losses, heat transfer coefficients, and transportation delays, depend on flow [21]. These parameters that are important should be computed as a function of flow before being used as inputs to the neural network. Nonstationary behavior and variables such as level, pressure, composition, and temperature of batch vessels that are the result of the accumulation of mass or energy should be excluded since there is no steady state. Some exceptions are those for which level is determined by a measured gravity discharge flow and pressure is determined by a measured vent flow. Even if integrators could be added to the neural network, the inevitable error accumulation from the slightest measurement error and integration step size would cause the prediction to drift away from reality. For continuous vessels, there may be a steady state, but there are important variable time constants. Neural networks can be used to predict the rate of change of level, gas pressure, temperature, and composition in vessels and the temperature and composition of process equipment for which backmixing is negligible, such as plug flow reactors, static mixers, desuperheaters, stacks, dryers, and heat exchangers. In many cases an equilibrium relationship between temperature, pressure, and composition can also be used to predict one variable from the measurement of another. Also, for sequenced on–off feeds, batch end points can be predicted.

Generic signal validation rules have been developed for real-time expert systems that can determine if common measurement signals are failed (off scale), dead, or have an abnormal rate of change (blipped) or model error high or low (grossly out of calibration). There is also a check for valve saturation because most problems in a control loop are first seen in the valve excursion unless the measurement is completely unresponsive. These rules can be generic in nature and applied to all instrumentation [22].

The model rules in the real-time expert system uses material balances. Energy balances can be used but they are more complex and less generic. For level measurements, the rate of change of level is compared with the net flow in and out of the vessel. These types of checks are done by passing the signal through a large dead-time block so that the new value is compared with a value old enough that the actual level change is much larger than the A/D and measurement noise. If the large difference in age of the two signals used to compute the change is created by a large scan time or calculation interval, the diagnostic has a dead time equal on the average to approximately 1.5 times the calculation interval. This check has been able to determine if the level measurement is not responding or is responding in the wrong direction.

The model for flow measurements is the computed flow through a control valve. This depends on the control valve's having a good positioner and an identified constant installed characteristic. Otherwise, the upstream and the downstream pressures must be measured and used with the inherent characteristic to compute flow by means of the valve sizing equation. The accuracy of this check is approximately 20% at best, mostly because of uncertainties in the valve characteristic and nonideal effects in the valve passageways [22].

Principal component analysis (PCA) can be used to provide a simpler fault identification by creating a new smaller set of uncorrelated variables with orthogonal properties from a much larger set of correlated variables. If you visualize a football in space, the axes of the three orthogonal variables needed to describe the football dimensionally would be rotated so that the long axis and the short axis of the football each lie on the axis of an orthogonal variable. An analogy would be the simplification of instructions to a service person on how to locate an existing window to be repaired from using building number, street name, city, town, floor number, apartment number, room, and wall to just latitude, longitude, and altitude.

In a PCA application to boiler measurements, a sensor validity index (SVI) was created that adds the ability to distinguish abnormal operating conditions from a single sensor fault. False alarms are avoided by application of an exponentially weighted moving average to the SVI to act as a filter. Faulty sensors were replaced with reconstructed values [23]. As more user-friendly software is developed, PCA methods will move from the lab and the university to the plant.

The data reconciliation step in real-time optimization (RTO) can identify instrumentation out of calibration in a few minutes by use of open equations and solving for the measurements with the lowest confidence limits. Since heat transfer coefficients and frictional losses are updated, and the material, component, and energy balances are comprehensive, RTO has a good track record of detecting the degradation of sensors. However, RTO depends on the process's being at steady state, and successful complete runs of the RTO may not occur when the process is upset by a faulty sensor.

The pattern recognition method used in tuning controllers could be used to detect fouled sensors. A coating of just a few millimeters on a pH electrode can cause the loop period to increase by a factor of 5 or more. Also a limit cycle may develop. While the changes are not as dramatic for temperature measurements, the principle is still valid.

A closed-loop performance monitor has been developed that counts the number of loop cycles for which the amplitude or duration is large enough to exceed an integrated absolute error trigger point and be registered as a loop cycle rather than as measurement noise. The cycle total has a forgetting factor that causes the total to decay if the cycling stops. The algorithm is simple enough to be implemented in any controller with some math instruction capability. If the process gain is known, it can be determined whether the cycles are most likely due to a sticky control valve from the controller's tuning settings [24].

Simple statistical measures, such as the process performance and capability indices, can determine how close the control system is performing to its maximum capability and can estimate the benefits from improvements that allow operation closer to constraints. The process capability index is inversely proportional to the mean-square successive difference and the process performance is inversely proportional to the standard deviation. They are both proportional to the spread in specifications (e.g., constraints) [25]. These indices are extremely useful and are simple enough to be installed on line to detect automatically a degradation in the control system.

Finally, a simple diagnostic procedure, whose logic is illustrated in Fig. 7, is helpful in determining whether oscillations are due to measurement noise, tuning, interaction, or a poorly responding control valve. If the oscillations disappear when the loop is in manual, they are due to tuning or a poor valve. If the flow measurement does not change or jumps for a small change in controller output, the valve is sticking or slipping, respectively. If not, the tuning needs to be corrected, based on the type of loop and whether this loop is upsetting other loops. If the oscillations remain with the loop in manual, then they are either measurement noise or caused by the cycling of other loops. An oscillation period of less than 0.1 min is noise from electrical interference, pressure waves, insufficient mixing, or resonance unless there are some loops with unusually fast ultimate periods that are due to a variable speed drive instead of a control valve and a scan time of much less than a second. Slow oscillations are caused by interaction or on–off actions and by erratic action of regulators and steam traps.

Which tool or technique is best depends on the type and the extent of on-site expertise available, the budget, and of course, the nature of the application. Since each technology has its limitations, a combination of methods might be best, such as neural networks embedded in expert systems or trained

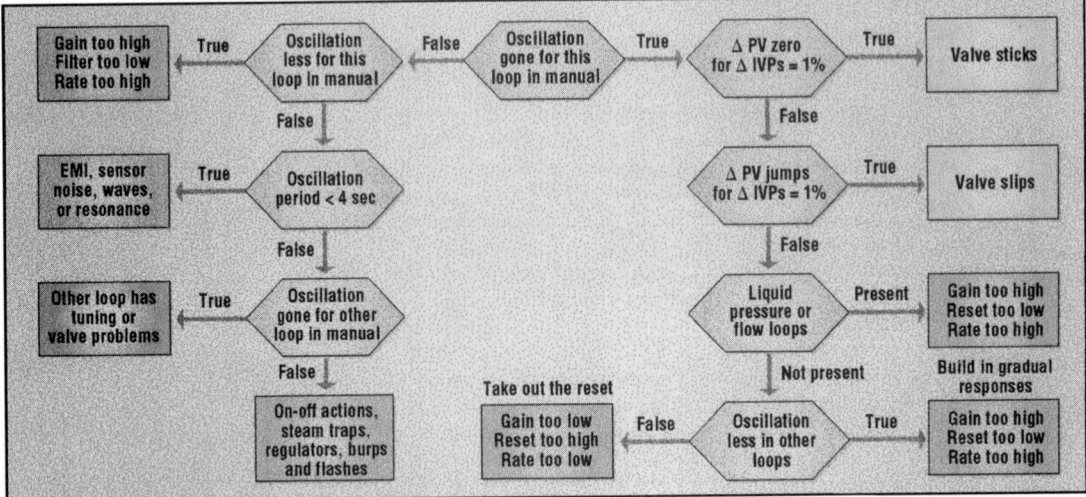

FIGURE 7   Manual troubleshooting to locate the source of an oscillation (IVP refers to the implied valve position or controller output and PV refers to he process variable or controller measurement.

by first-principle models. The dominant method used at a plant site depends on resource requirements and availability.

In general, RTOs are probably the most expensive and require the most outside expertise. Real-time expert systems and neural networks are a close second in cost if the time spent by plant specialists for implementation and maintenance of the system is included. The use of PCA techniques could become less resource intensive as the technology is transferred from academia to industry. The pattern recognition, closed-loop performance monitor, capability indices, and diagnostic procedures are economical but are fairly limited in their ability to pinpoint problems.

Tools and Techniques that Use Plant Knowledge

1. Real-time expert systems for signal validation
2. Real-time neural networks for intelligent sensors
3. PCA for identification of faulty sensors
4. RTO data reconciliation to quantify instrument degradation
5. Pattern recognition to detect sensor fouling
6. Control loop performance monitor to detect control valve sticking
7. Process performance and capability indices to determine control system performance
8. Manual diagnostic methods to track down control loop problems

KEY POINTS

1. The cost of instrument maintenance is a significant portion of the bottom line.
2. The cost of reduced process efficiency and capacity, because of ineffective and reactive mainte-nance, can be greater than the instrument maintenance cost.

3. Most of the maintenance cost to date is unnecessary since it consists largely of calibration checks that are unwarranted or of preventative maintenance that is premature.

4. Most instrument failures are caused by improper selection or installation.

5. Sensitivity, consistency, and reliability are of paramount importance for both measurements and control valves.

6. Sensing lines are the biggest source of maintenance problems.

7. Most smart digital instruments do not need to have their calibration checked.

8. Calibration checks of transmitters are often not possible except through redundancy.

9. Control valves increase loop variability when they cannot respond to small changes in signals.

10. Control valves have the greatest need for diagnostics.

11. Control valves have the most diagnostics implemented.

12. pH needs diagnostics more than any other measurement.

13. pH leads the other measurements in terms of on-line diagnostics of coatings and other problems.

14. Most of the pressure and flow measurement diagnostics implemented to date are for detecting over range or violations of temperature or pressure limits.

15. Calibration checks are done in response to application problems and lack of operations confidence.

16. Instruments can analyze noise before it is filtered or aliased.

17. Diagnostics must be displayed in the control room.

18. Fieldbus will dramatically increase the view and scope of instrument diagnostics.

19. Instrument-based diagnostics are an excellent investment.

20. Plant-knowledge-based diagnostics require significant initial and ongoing internal or external resources.

21. Neural networks and expert systems should have more than an advisory role to ensure ongoing support.

22. Neural networks and expert systems can predict the rate of change of variables that are the integral of mass flow and energy rates (e.g., backmixed volumes) and can predict the actual variable that is the result of a steady-state relationship and a delay (e.g., plug flow volumes and equilibrium relationships).

RULES OF THUMB

1. Engineer applications for the greatest sensitivity, consistency, and reliability.

2. Eliminate any areas of stagnant process fluid.

3. Use instruments that are least affected by the process that holds their calibration the longest.

4. Use smart HART modular digital instrumentation.

5. Phase in Fieldbus when its reliability is established.

6. Select measurements that do not need calibration and do not wear out (e.g., Coriolis meters).

7. Do not install pH electrodes unless you are ready to deal with the short life expectancy.

8. Stop making calibration checks and use field service centers for replacements and overhauls.

9. Use instrument-based diagnostics to develop a predictive maintenance program.

10. Implement on-line calculations of process performance and capability indices.

Most of maintenance is either unnecessary (e.g., calibrations) or reactive, in which action is not taken until there is an actual failure and the damage has been done to the process efficiency and capacity. Preventative maintenance programs suffer from a lack of knowledge as to when a transmitter really does need to be replaced or a control valve overhauled. Most preventative programs are based on time, and consequently most of the preventative maintenance is premature. What is needed is a move to predictive maintenance, in which replacements and overhauls are done based on diagnostics. Good engineering practices can greatly reduce the magnitude of the problems to the point where most of the diagnostics are associated with detecting quality control problems during manufacturing of the instrument or in the wear-out phase of the instrument rather than in application problems. The benefit-to-cost ratio of smart HART transmitters is tremendous because they are much more accurate and reliable and can provide the diagnostics important for predictive maintenance in the control room. These diagnostics are based on many years of instrument knowledge. Sophisticated plant-knowledge-based programs are costly, and their justification and impact on resources should be reviewed.

REFERENCES

1. Diagnostics and Reliability Based Maintenance, 5/8/98, http://www.rosemount.com/products/ams/rbmaint.htm (Internet).
2. Sanders, F. F., "Watch Out for Instrument Errors," *Chem. Eng. Prog.*, 62–66, July 1995.
3. Coughran, M., "Valves: Testing for Peak Performance," *InTech*, 58–61, Nov. 1994.
4. Coughran, M., "Performance Influences in Globe Control Valves," *InTech*, 44–49, Aug. 1996.
5. McMillan, G. K., "Improve Control Valve Response," *Chem. Eng. Prog.*, 76–84, 1995.
6. Feature Focus: Pressure Transmitters and Transducers, *InTech*, 26–29, July 1992.
7. 1996 HART/Fieldbus Investigation Team Tutorial, Monsanto.
8. Schnake, J. B., "Emerging Flow Technology Boosts Accuracy and Savings," *InTech*, 52–56, Jan. 1998.
9. Dierauer, P., "Smart Sensors Offer Increased Functionality," *InTech*, 60–63, May 1998.
10. Luna, S. F., "Installation Practices," in *ISA Handbook of Control Valves*, 2nd ed., Instrument Society of America, Research Triangle Park, North Carolina, 1990, pp. 339–340.
11. McMillan, G. K., and Toarmina C. M., *Advanced Temperature Measurement and Control*, Instrument Society of America, Research Triangle Park, North Carolina, 1995.
12. McMillan, G. K., *pH Measurement and Control*, 2nd ed., Instrument Society of America, Research Triangle Park, North Carolina, 1994.
13. Schneider, G., "Status Monitoring and Self-Calibration of Sensors," Endress and Hauser Report FAR 507E, 1998.
14. Berge, J., "Fieldbus Advances Diagnostics," *InTech*, 52–56, April 1998.
15. Maintenance: Finding the Right Medicine, *Chem. Eng.*, 119–122, May 1998.
16. Weiss, J., "Slow Oil Loss in Pressure Transmitters," *InTech*, 40–43, Oct. 1992.
17. Peterson, K., "Testing Sensors for Accuracy and Speed," *Chem. Eng.*, 131–134, Feb. 1992.
18. FieldVue Instrumentation Fundamentals, Course 1750 Revision B , Fisher-Rosemount Educational Services, 1996.
19. Giovannelli, S., "Controlling Maintenance Costs," *Chem. Process.*, 92–96, May 1998.
20. Kiesbauer, J., and H. Hoffmann, "Improved Process Plant Reliability and Maintenance with Digital Positioners," *Automatisierungstechnische Praxis*, **40**(2), 1998.
21. Shinskey, G. F., "Modeling with Neural Networks—First Principles are More Reliable," *Chem. Process.*, p. 87, June 1998.
22. Mertz, G., "Application of a Real Time Expert System to a Monsanto Process Unit," Process Control Forum of the Chemical Manufacturer's Association, 1986.
23. Dunia, R., et al., "Identification of Faulty Sensors Using Principal Component Analysis," *AICHE J.*, Vol. 42 No. 10, 2797–2812, Oct. 1996.
24. Hagglund, T., "A Control Loop Performance Monitor," *Control Eng. Prac.* **3**(11), 1543–1551, 1995.
25. Shunta, J. P., *Achieving World Class Manufacturing Through Process Control*, Prentice-Hall, Englewood Cliffs, New Jersey, 1995.

NEW DEVELOPMENTS IN ANALYTICAL MEASUREMENTS

by G. K. McMillan*, J. G. Converse**, M. J. Pelletier***, and R. J. Procton****

PERSPECTIVE

The objective of process control is to improve the yield, capacity, and quality of the product. This requires knowledge of the concentrations of key components in various process streams. However, the frequency and the number of laboratory and on-line measurements of these concentrations have steadily declined because of ongoing attrition of analytical chemists, analyzer specialists, and technicians. As a result, many production units are essentially flying blind.

Advanced control systems need concentration measurements in order to truly optimize the process. Neural networks or estimators can utilize conventional measurements of pressure, temperature, and flow to provide virtual analyzers (intelligent sensors), but these models predict steady-state concentrations and ultimately depend upon a concentration somewhere for building and correcting the models.

Thus, as part of any process control improvement, the need for analyzers should be studied. The following compilation of new developments should be valuable in making this assessment, particularly those techniques that require less sample conditioning and recalibration. This means less on-site support and/or opens up the opportunity to measure new components reliably on line.

Many of these newer analyzers utilize technologies that are sophisticated and require extensive engineering and setup. However, often the analyzer either works or does not work and therefore there is less ongoing attention. Also, the manufacturer is willing to provide a turn-key installation and a free plant trial in which the user pays only if the analyzer system performance meets the application requirements.

NEW DEVELOPMENTS IN ANALYZER MEASUREMENTS
by J. G. Converse

Sample Preparation Methods and Hardware

Perhaps the greatest change in process analyzer technology lies in the way we handle the sample rather than in the way we measure chemical concentration. The catch phrase that is fashionable today, total cost of ownership, dictates that we change our present view of process analytical chemistry. We must reduce the hardware, simplify sample preparation, more effectively control the analyzer system, and apply appropriate signal processing.

- Eliminate all hardware possible, including explosion-proof analyzer houses and accessories.
- Extract and operate on a minimum amount of sample to reduce waste and maintenance.
- Simplify programming and controllers required for running the system.
- Utilize the most effective software to process, display, transmit, and store information.

*Senior Fellow, Solutia Inc., St. Louis, Missouri.
**Analytical Services Technologist, Sterling Chemicals, Inc., Texas City, Texas
***Principal Applications Scientist, Spectroscopy Products Group, Kaiser Optical Systems Inc., Ann Arbor, Michigan.
****Corporate Scientist, GAMMA-METRICS, San Diego, California.

The entire concept of design must be a modular one that utilizes standard components to build the analyzer system. Maintenance of individual components must be low skill, and only the necessary pieces can be included in the whole. The individual doing his or her own thing no longer has a place in the industry of commerical analyzers.

There are specific and limited basic functions that must be provided by a module. There are a limited number of circumstances that define a process sample:

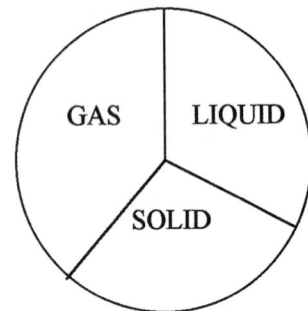

- single-phase gas sample
- gas sample with solid particulates
- gas sample with liquid droplets
- gas sample with liquids and solids
- single-phase liquid sample
- liquid sample with solid particles
- liquid sample with gas bubbles
- liquid sample with gas and solids
- solid samples are a special case

There must therefore be a finite number of components and configurations to handle all cases.

The big change in analyzers has been from single- to multiple-species determination. The danger here lies in the same greed found with wanting to analyze multiple streams with a single analyzer. One wants to save money by manifolding several streams to one detector. Experience has shown that problems with maintenance that are due to complexity can produce an overall loss. Simple reliable systems that encompass no more than is practical still produce the best performance and the lowest total cost of ownership.

It is tempting for the analyzer applications specialist to introduce new measurement technology, and that leads to the many laboratory instruments that may or may not be practical for field use. Some examples will be presented.

Electromagnetic radiation from gamma ray, x-ray, vacuum ultraviolet, ultraviolet (UV), visible, near-infrared (NIR), mid-infrared, far-infrared, microwave, ultrasonic, and sonic regions is utilized to make various measurements to determine chemical concentration. Many ionization phenomena lead to useful analytical measurements as do electrochemical techniques.

It is very important to evaluate the nature of the sample matrix as well as the analyte to be measured before selecting the appropriate measuring device. Selection of the measuring device may in turn have a major effect on the design of the sample preparation system. Choosing a particular analyzer may seem to be clever, but no matter how great the analyzer may be, it has no value unless a representative sample is presented to it. Perhaps the best way to evaluate new on-line analyzers is to discuss assets and "liabilities" with respect to actual applications. The midregions of the electromagnetic spectrum are the more familiar ones used in analytical chemistry. The gamma and vacuum UV at higher energy than the UV and the microwave and the sonic regions at lower energy than the IR are finding new applications.

MULTIPLE-WAVELENGTH NEAR-INFRARED (NIR) ANALYZER

This device has become very popular in the petroleum and the pharmaceutical industries. It utilizes the measurement of intensities of several wavelengths between 700 and 3300 nm.

Assets:

1. much information relatively fast (multiple results per measurement)
2. works on nonvolatile liquids that exclude gas chromatography (GC)
3. remote sensing utilizing fiber optics (decreases sensitivity)

4. reasonably low maintenance depending on method of implementation

5. yields very stable measurements

6. can use long-path-length cell to prevent sample flow restriction

7. can use chemical-resistant window materials

Liabilities:

1. complex mathematical calculation requiring computer and special software

2. time-consuming calibration requiring multiple standards

3. temperature control critical to accurate measurements

4. sample preparation required for removing bubbles and particulates

5. high skill required for setup and maintenance

6. not capable of trace component measurements [<100 parts in 10^6 (ppm)], with a few exceptions

7. relatively high total cost of ownership

A major success has been achieved in monitoring gasoline blends. The blending is done at high speed, which requires very fast response. This sample material is generally clean at the blending site. Cost is small relative compared with the value of the product being on specification. Multiple-wavelength measurement is achieved by several means, as described in Section 6. See the subsection in Section 6 on Wavelength Selection for details that are given for the UV region but that also apply to the IR Region (also in Section 6). This subsection deals with wavelength scanning, multiple filters, and diode array, while Fourier transform infrared (FTIR) is covered in the next subsection.

Scanning the spectrum with a grating (fixed source and detector) was primarily a laboratory technique, but devices have been made rugged enough to utilize them in field on-line measurements. An alternative method uses a fixed source and grating with a series of detectors that are sequentially sampled. These detectors may be individual devices or a diode array of sensors as seen in Fig. 1.

Multiple filter devices can be used for a limited number components, and a wavelength-variable device is utilized if high resolution is not required (~10 cm^{-1}).

The Fourier transform technology has been extended to cover the NIR region, and field-hardened devices are available. Specific applications would favor the technology that fits the specific needs of resolution, sensitivity, stability, skill requirements, and cost.

FOURIER TRANSFORM INFRARED

This device utilizes an interferometer to measure all wavelengths simultaneously in the region of 4800 to 400 cm^{-1}. It can be applied to vapor and liquid samples and produces a lot of information very fast. It requires a fast powerful computer and special software.

Assets:

1. much information relatively fast (multiple results per measurement)

2. works on nonvolatile liquids that exclude GC

3. remote sensing utilizing fiber optics

4. open-path monitoring of vapors in air

Liabilities:

1. complex mathematical calculation requiring computer and special software

2. time-consuming calibration requiring multiple standards

3. moving mirror and optics that may cause high maintenance

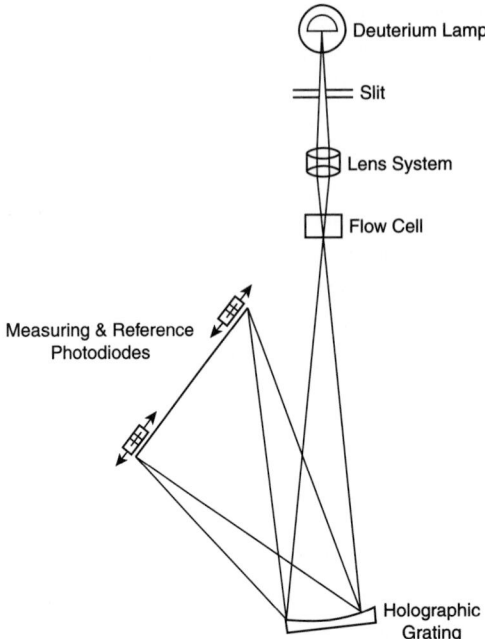

Schematic of a programmable variable wavelength
detector which uses movable photodiodes.

FIGURE 1   Diode array multiple-wavelength NIR analyzer.

4. nonanalyte molecules (e.g., water) may block analyte absorbance

5. high skill required for setup and maintenance

6. relatively high total cost of ownership

FOURIER TRANSFORM SPECTROSCOPY
by K. K. Konrad

The past few years have seen significant changes in the IR market, with a trend from high-performance dispersive instruments to low- and medium-cost FTIR spectrometers. This trend has been due to a significant drop in the price of the FTIR instruments and to an increasing awareness of some fundamental advantages of FT over dispersive instruments. The aim of this discussion is to explain the principles of FTIR and to describe some of the practical considerations in the design and operation of FTIR spectrometers.

Advantages of FTIR

In principle, a well-designed interferometer has several basic advantages over a classical dispersive instrument.

1. Multiplex advantage (Fellgett advantage): All frequencies are measured simultaneously in an interferometer, whereas they are measured successively in a dispersive spectrometer. A complete spectrum can be obtained very rapidly, and many scans can be averaged in the time taken for a single scan of a dispersive spectrometer.

2. Throughput advantage (Jacquinot advantage): For the same resolution the energy throughput in an interferometer can be higher than in a dispersive spectrometer, in which it is restricted by the slits. In combination with the multiplex advantage, this leads to one of the most important features of an FTIR spectrometer: the ability to achieve the same signal-to-noise ratio as a dispersive instrument in a much shorter time.

3. Conne's advantage: The frequency scale of an interferometer is derived from a helium–neon laser that acts as an internal reference for each scan. The frequency of this laser is known very accurately and is very stable. As a result, the frequency calibration of interferometers is much more accurate and has much better long-term stability than the calibration of dispersive instruments. For a broadband source the interference pattern is the sum of the sine waves for all the frequencies present. This interferogram consists of a strong signal at the point where the path difference is zero, falling away rapidly on either side. The customary spectrum showing energy as a function of frequency can be obtained from the interferogram by the mathematical process of Fourier transformation. When no sample is present this gives a single-beam spectrum, the overall shape of which is largely determined by the characteristics of the beam splitter. Normally interferometers operate by first recording this background and then ratioing the spectrum of a sample against this background signal.

Instrument Operation

Because interferometers operate in a single-beam mode, greater overall stability is required than in classical double-beam spectrophotometers. While adequate stability can be achieved by appropriate design, the presence of atmospheric absorption is a major inconvenience. Any change in water or carbon dioxide concentration results in extraneous bands in the final ratioed spectrum. When spectrometers are purged, great care has to be taken to ensure that the degree of purge is the same for sample and background measurements. To avoid this problem, some instruments can be operated in what is called a double-beam mode. This involves taking background and sample interferograms alternately and averaging the two separately during the measurement period. Two single-beam spectra are generated, and the final spectrum is obtained by ratioing these. Variations in atmospheric absorption are similar for both background and sample spectra and are effectively canceled out when the two are ratioed.

Measurement of the interferogram has to start on one side of the point of zero path difference (ZPD) and continue out of the other side to a maximum path difference that depends on the required resolution (Fig. 2). The laser signal cannot be used to identify the point of zero path difference; this is found when the maximum signal in the IR interferogram is located. Although the laser signal precisely monitors chances in path difference, it does not sense the direction in which the mirror is moving

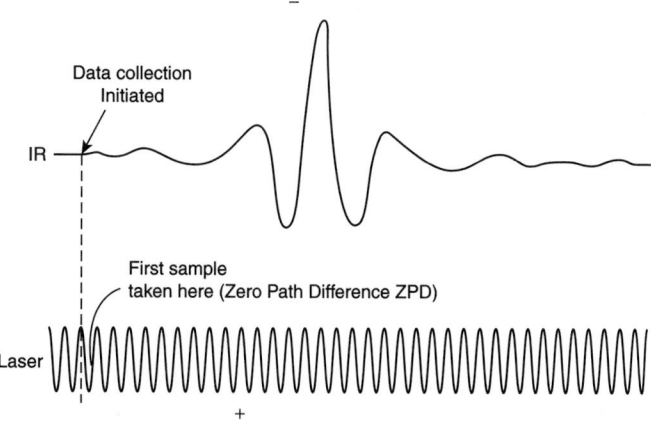

FIGURE 2   Inteferogram.

and thus loses count at the end of each scan. The absolute path difference has to be reestablished for each scan so that successive scans start at the same point. One means of doing this has been to use a white-light source and detector to produce a sharp spike interferogram offset from zero path difference for the IR radiation. This signal is used as a reference point from which to start counting the laser signal. Recently an improved system has been introduced in which two laser detectors are used, from which the path difference can be monitored continuously even when the scan changes direction. This eliminates the need for the white-light interferometer or for any other means of reestablishing the absolute portion for each scan. It also opens up the possibility of scanning the interferometer continuously in both directions. The normal procedure has previously been to scan in one direction only and then fly back rapidly to start the next scan.

The sampling points for measuring the interferogram are derived from the laser signal. If one data point is recorded for each cycle of the laser signal, the data point separation is equal to the laser wavelength of 632.8 nm. The shortest wavelength that could be recognized in the spectrum would then be twice the laser wavelength, that is 1265.6 nm, corresponding to a frequency of 7902 cm^{-1}.

Resolution

The resolution in the final spectrum depends on the maximum optical path difference in the interferogram (the distance the mirror moves). Each wavelength contributes to the interferogram as a sine wave with a separation between successive maxima equal to the wavelength. The Fourier transform of the interferogram is the sum of those frequencies whose contributions fit as an exact number of cycles into the length of the interferogram from zero to the maximum path difference. Two adjacent frequencies will be distinguished if the number of cycles of each in the interferogram differ by one. For two such frequencies $v1$ and $v2$, the number of cycles are $n1$ and $(n1 + 1)$. The maximum path difference is d:

$$v1 = \frac{n1}{d}, \quad v2 = \frac{n1 + 1}{d}, \quad v2 - v1 = \frac{1}{d}$$

That is, the data points in the final spectrum are separated by the inverse of the maximum optical path difference. If the maximum path difference is 1 cm, the final spectrum has data points at intervals of 1 cm^{-1}.

Apodization

The contribution to an interferogram from an infinitely narrow line would be a sine wave that continued indefinitely as the path difference increased. The contribution to an interferogram from any real line with a finite width decreases in amplitude as the path difference increases. To avoid any distortion in measuring the spectrum, the interferogram should be measured out to a path difference where the contribution from the narrowest line has decayed to a negligible value. This is the same criterion used in a classical spectrometer. Apodization also generally reduces noise in the spectrum. The noise in the interferogram comes largely from the detector and affects all the points in the spectrum equally. The spectral information is decreasing as the path difference increases to that apodization that reduces the contribution from those points in the interferogram where the signal-to-noise ratio is lowest. In this respect, apodizing the interferogram is similar to smoothing the frequency spectrum because noise is reduced at the expense of some line broadening.

Phase Correction

An idealized interferometer would produce an interferogram that was exactly symmetrical about the center. In any real instrument the interferogram is not symmetrical as various effects cause phase differences between the contributions at difference frequencies. To obtain an accurate representation of the frequency spectrum, these phase errors must be corrected. The necessary information for doing this is obtained from the central region of the interferogram. The most accurate method of applying the correction involves using a double-sided interferogram, that is, one that is measured for an equal distance on either side of zero path difference. Most instruments use less accurate procedures with single-sided interferograms that are measured for only a short distance on one side of zero path difference and to the maximum path difference on the other. Although measuring a double-sided interferogram requires almost twice as long a time as a single-sided interferogram, it contains effectively as much information as two single-sided scans and so is equally efficient in time.

Trading Rules Relating Resolution, Noise Level, and Measurement Time

Exactly as in a classical spectrometer, it is possible to obtain lower noise levels or reduce measurement time by sacrificing resolution. Generally, spectra are obtained by averaging a number of scans, since the time taken for a single scan is so short. The time taken for a single scan depends on the mirror velocity and on the maximum path difference, which depends on the desired resolution. The mirror velocity depends principally on the type of detector being used. With room-temperature triglycine sulfate (TGS) detectors, the scan should be as slow as possible since the detector works more efficiently, and this offsets the lower number of scans that can be averaged in a given time. Liquid-nitrogen-cooled mercury cadmium telluride (MCT) detectors have a much faster response, and in general the scan should be as fast as possible to optimize the results obtained in a fixed measurement time.

FTIR Interferometer Design [1]

Several different designs have been developed to generate an interferogram. Each designer claims advantage, but the proof is in the performance and maintenance. Figures 3–5 give an overview of current commercial instruments.

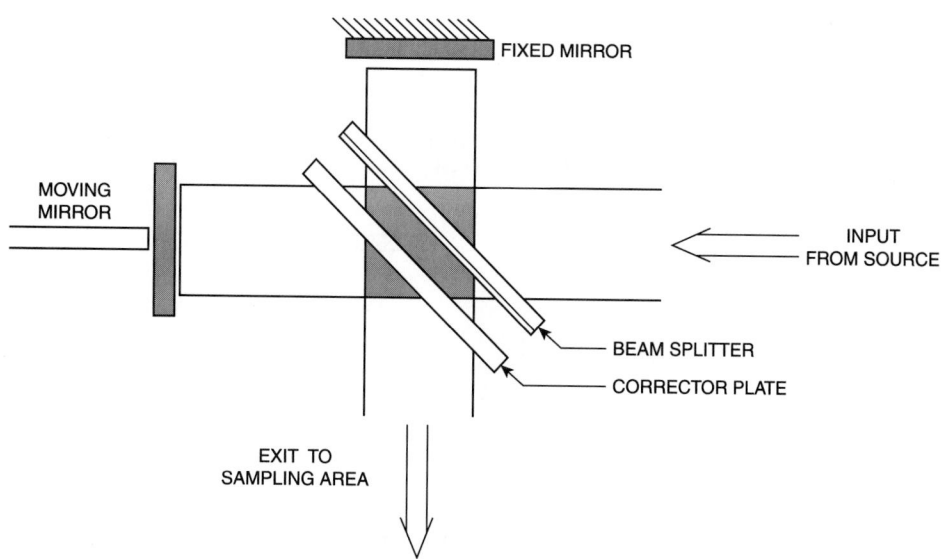

FIGURE 3   Michelson pivoting mirror.

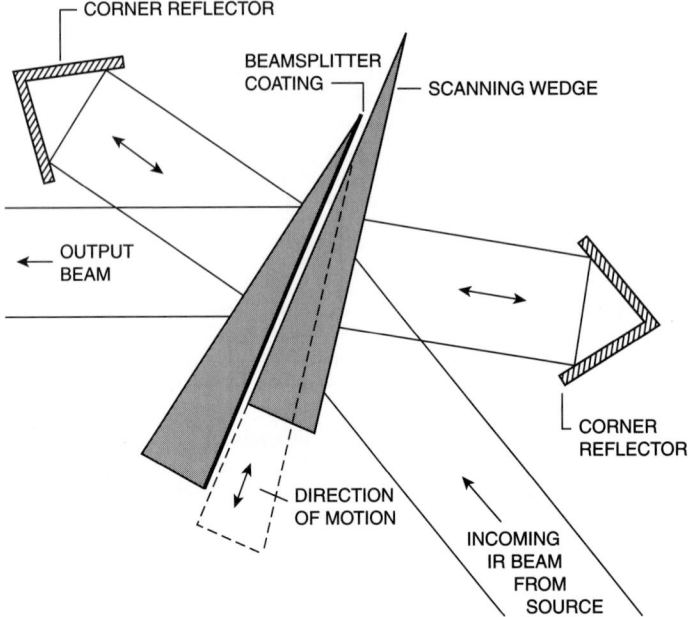

FIGURE 4 Moving wedge.

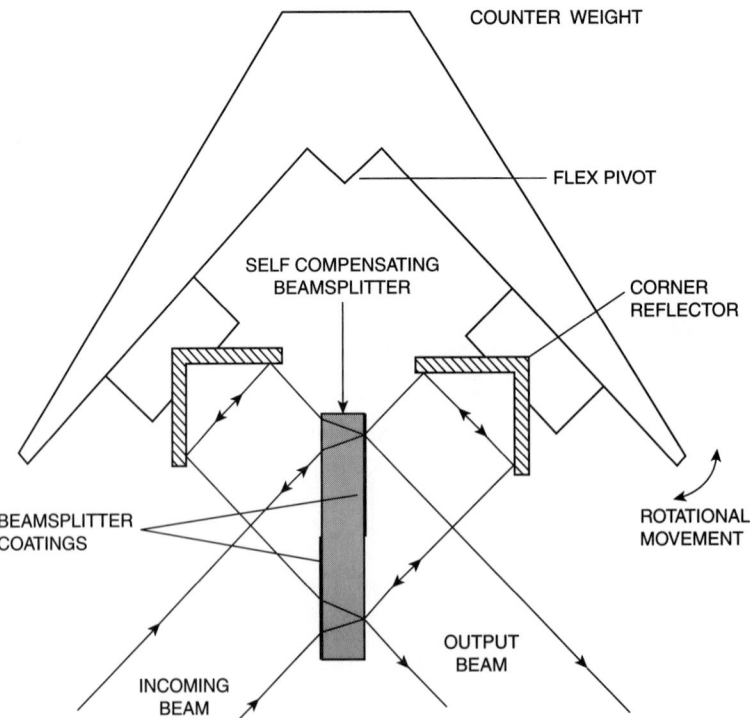

FIGURE 5 Dual moving corner reflector.

THE RESOLUTION–ENERGY–TIME DILEMMA [2]

A common misconception when using a FTIR instruments is, the better the resolution, the better the instrument. This is far from true. The resolution should never be greater than is really necessary in order to distinguish the spectral elements, because the signal is sacrificed [3]. The detection limit for a gas is normally one of the most important parameters to consider when choosing a gas analyzer. In a photoacoustic analyzer the detection limit is, among other things, determined by the signal-to-noise (S/N) ratio, while in a transmission analyzer it is determined by both the S/N ratio and the analyzer's stability.

How does one attain the lowest detection limit and maintain it over a long period of time?

1. Design Considerations
   a. Photoacoustic spectroscopy uses only small amounts of the sample gas and provides good long-term stability.
   b. FTIR spectroscopy provides excellent selectivity and flexibility. The cross-sectional area of the IR beam should be made as large as possible. This makes the price, size, and weight of the analyzer the limiting factors.
   c. Choose an optical frequency resolution that is no better than is absolutely necessary. The S/N ratio is inversely proportional to the resolution at a constant scanning speed.
   d. Since the detector integrates, the chosen scanning speed should be as low as possible. (Both the signal and measuring time are inversely proportional to the scanning speed, while noise is proportional to the square root of the scanning speed.)
2. Averaging
   When a number of spectra are averaged, the S/N ratio improves by the square root of the number of spectra because the noise components are uncorrelated. The cost of averaging the spectra is an increase in measurement time. Therefore it is not practical to average more than approximately 20 spectra.
3. Spectral Integration
   Integrating the measured absorption over a large frequency range, where the desired gas has absorption lines of approximately equal strength, results in the S/N ratio's increasing proportionally by the square root of the bandwidth. This integration results in reduced selectivity, which means a greater chance of interference from other gases. While the factors under Design Considerations are decided by the manufacturer of the analyzer, Averaging and Spectral Integration are decided by the user to optimize the analyzer for the particular application.

Examples

Design Considerations. To illustrate the design considerations for an analyzer, let us assume we want to measure a gas, which has characteristic absorption lines over a range of 40 cm^{-1}, with a measuring time of 100 s.

Example 1. If you design your analyzer to have a resolution of 2.5 cm^{-1} and a scanning time of 4 s (a typical laboratory-type instrument), you can average 25 spectra. Let us assume this gives you a detection limit of 1 ppm with a spectral bandwidth of 40 cm^{-1}.

Example 2. If the scanning speed is kept the same, but the resolution is redesigned to be 10 cm^{-1}, the scanning time will decrease to 1 s (scanning time is inversely proportional to resolution when the scanning speed is constant). 100 spectra can be averaged. The signal will increase 4 times compared with example 1. The noise will decrease 2 times. The S/N ratio will be 8 times greater than in example 1, and a detection limit of 0.125 ppm is achieved.

Example 3. If the scanning speed is reduced 10 times and the resolution is 10 cm^{-1}, the scanning time will be 10 s. Ten spectra can be averaged. The signal will increase 40 times compared with example 1. The noise will decrease 2 times. The S/N ratio will now be 80 times greater than that in example 1, and a detection limit of 0.01 25 ppm is achieved. These examples illustrate how the S/N ratio has been improved 80 times just by designing the instrument properly.

Note: You should bear in mind that, in practice, purging times for typical gas cells far exceed the 4 s scan time given in example 1. Furthermore, a fully resolved single gas line has a bandwidth of the order of 0.1 cm^{-1}. This resolution can only be achieved by big expensive laboratory instruments with liquid-nitrogen-cooled MCT detectors. Lower-resolution room-temperature TGS detectors are less sensitive than the photoacoustic detector. Each type of detector has a parameter called D^*, which measures the S/N ratio associated with the detector and is a function of frequency.

Changes Made by the User

To illustrate how the user can influence the results, let us assume we have measured a gas, with a bandwidth of 40 cm^{-1} and a measuring time of 100 s.

Example A   If the user reduces the optical bandwidth to 10 cm^{-1} to improve selectivity, the signal will decrease 4 times and the noise will decrease 2 times [4(−2)]. This means that the S/N ratio will decrease 2 times (see the User Formula below).

Example B   If the user reduces the measuring time to 10 s to improve response time, the S/N ratio will decrease 410 times since N is reduced 10 times (see the User Formula below).

User Formula:

$$S/N = R[(BN)^{1/2}]   \text{(square root of } B \text{ times } N\text{)}$$

where S/N is the signal-to-noise ratio, N is the number of spectra averaged ($N < 21$), B is the optical integration bandwidth, assuming that gas lines (of equal strength) for the particular gas exist throughout B, and R is a design constant.

THE MEASUREMENT PRINCIPLE

1. The pump draws air from the sampling point through two air filters to flush out the old air in the measurement chamber and fill it with a new air sample.

2. The new air sample is hermetically sealed in the analysis cell by closing the inlet and outlet valves.

3. Broadband light from an IR source is modulated to produce a time-varying signal by passing it through a modified Michelson interferometer. The beam splitter and its two associated platform mirrors are all mounted in fixed positions on a single platform that rotates relative to the base (see Fig. 6) through a very small angle ($\sim$1°) that continuously varies the path length of the beam to produce the interferogram. This represents a single double-sided scan of the sample and takes approximately 13 s. When this optical design is used, the interferometer is highly stable and immune from the effects of external disturbance. A laser diode and detector provide a reference system, which is used to trigger and synchronize sampling of the interferograms produced, the laser signal taking the same path as the IR signal.

4. The IR light reflected from the interferometer output mirror enters the measurement chamber. Here it is selectively absorbed by the gas sample, which causes the temperature of the gas to fluctuate. This produces corresponding pressure fluctuations within the chamber (i.e., an acoustic signal). The frequencies of the acoustic signal depend on the wavelengths of the light absorbed by the gases present in the sample. An intensity detector mounted in the chamber measures the variations in intensity of the IR light. This produces the signal that is used for energy normalization of all measured spectra.

5. The photoacoustic signal is measured by the two microphones mounted on the chamber walls to produce an electrical signal proportional to the amount of absorption that has occurred. The electrical signal is then filtered and sampled (converted from an analog to a digital signal) and Fourier transformed to produce an absorption spectrum. User-defined filter bands determine individual concentration values. These are obtained from the spectrum by integration between the chosen limits for the filter bands and application of a conversion factor for the gas.

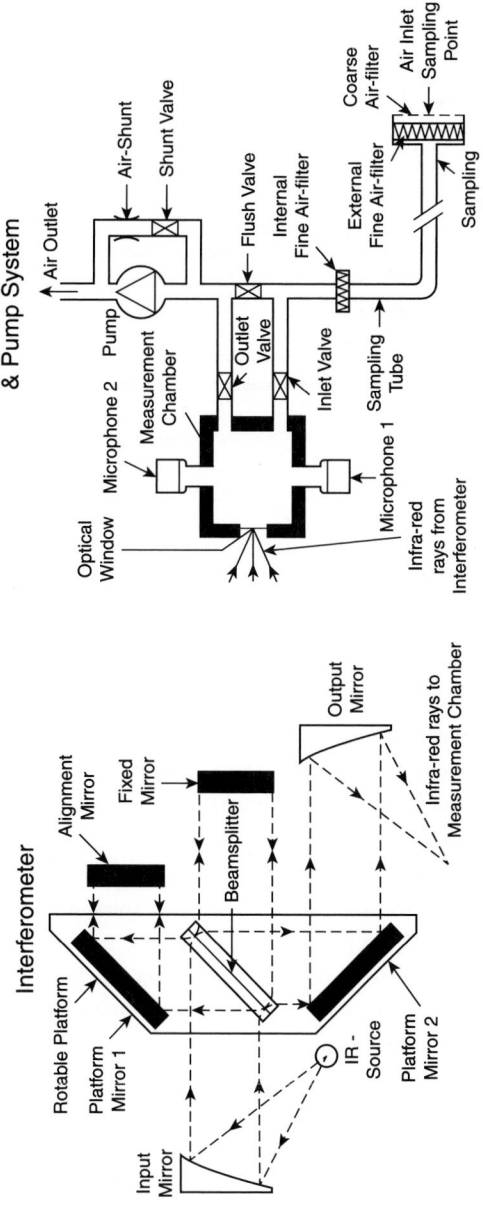

FIGURE 6 Rotating platform.

A Simple Bistatic Configuration

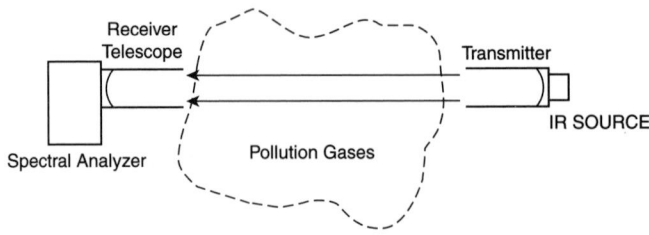

A Unistatic Configuration
(using a single transmitter/receiver telescope and retroreflector array)

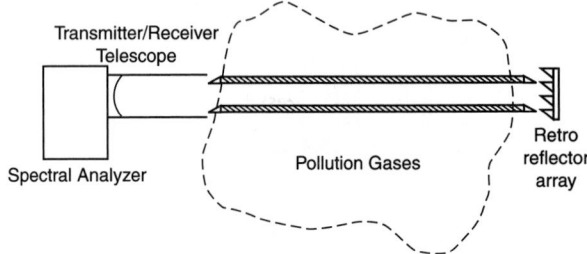

FIGURE 7   Bistatic and unistatic open-path monitoring configurations.

Open-path area monitoring has been extensively evaluated with some successful applications. Both bistatic and unistatic configurations are utilized (see Fig. 7).

The major problems encountered [3] to date with open-path FTIR are

1. acquiring a clean air background reference spectrum

2. fog, rain, snow, and soot shutting down the system

3. temperature, humidity, and vibration corrections to measured spectra

4. wind direction and velocity (meteorology)

5. single line of sight (leaves lots of unmonitored space)

6. signal manipulation software needs improvement (must demonstrate accuracy)

7. hot open IR source (safety)

8. liquid-nitrogen-cooled IR detector (trying mechanical and electrical chillers)

9. component failure and long-term maintenance by hourly personnel

10. defining action for short-term excursions (when to alarm and evacuate?)

11. fence line monitoring not satisfying permit requirements

REFERENCES

1. Doyle, W. M., "Principles and Applications of FTIR," *Process Control Qual.* **2**, p. 11, 1992.

2. Nexo, S. A., "The Resolution–Energy–Time Dilemma," Monograph Series, Gas Products Division, Bruel & Kjaer.

3. Griffiths, P. R., et al., "Low Resolution FTIR," Department of Chemistry, University of Idaho, Moscow, Idaho.

MASS SPECTROMETER
by J. B. Klahn

There are several ways to implement mass spectroscopy:

1. magnetic sector
2. quadruple
3. time of flight
4. ion trap
5. ion cyclotron resonance

Mass spectrometers are generally chosen because they are very fast. They can determine multiple components if the mixture is not too complex without component isolation. As soon as it becomes necessary to separate chemical species by column chromatography, this device loses its speed advantage. The greatest advantage of the mass spectrometer is its ability to identify a molecular species. In process analytical chemistry, one generally knows what species are present and the determination of quantity is the desired result. Scanning the charge/mass spectrum is very useful when monitoring for unknown species that may not be expected. The same is true for scanning spectrophotometers, but the applications for this task are limited somewhat to environmental monitoring. A very valuable asset to mass spectrometer measurements besides the speed, is the low pressure utilized in the measurement process. Sample preparation by use of staged reduction of pressure allows reactive molecules to be isolated before they can react. Reducing the pressure increases the mean free path between molecules and thus their chances of collision and reaction. Species that may be lost on an interactive GC column may survive the transfer from process conditions to high vacuum spectrometer operating environment.

If you require very fast analysis on reactive materials, the cost and complexity of the mass spectrometer system may be justified.

Assets:

1. much information relatively fast (multiple results per measurement)
2. reactive species may be captured without high losses
3. yields very stable measurements

Liabilities:

1. complex mathematical calculation requiring computer and special software
2. vacuum system and ion source may cause high maintenance
3. high skill required for setup and maintenance
4. relatively high total cost of ownership

Quadrupole mass spectrometers are the main process type analyzer. Time-of-flight spectrometers are gaining use because of their fast response and their simplicity. A new device being introduced is the Fourier transform ion cyclotron resonance (FTICR) spectrometer.

Dynamic Mass Analyzers

Quadrupole and FTICR dynamic analyzers rely on either time-varying electric or magnetic fields to effect mass separations. Each has strengths and limitations and finds special areas of application.

Quadrupole Mass Analyzers.   The most widely used mass analyzer today is the quadrupole analyzer or mass filter, as it is often called. As shown in Fig. 8, it consists of four parallel rod-shaped electrodes arranged at the apices of a diamond. Two necessary design conditions are that the rods must be

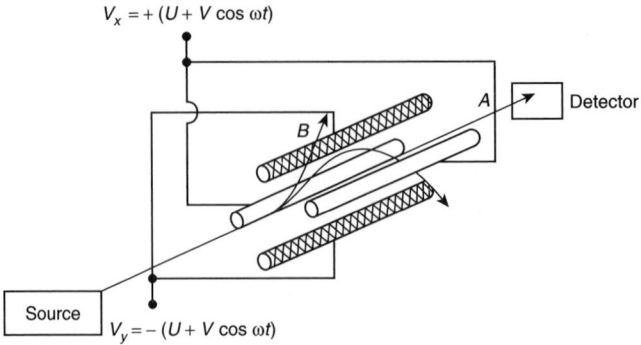

FIGURE 8   Quadrupole mass filter.

precisely machined and aligned so that electric fields with defined gradients are set up, and the usual high vacuum of $10(-7)$ to $10(-8)$ Torr must be maintained in the analyzer to prevent collisions that would change the direction of ions or the focus of the beam.

With a quadrupole analyzer, kinetic-energy differences are inconsequential and ions are extracted from an ion source and focused into a beam by an ion lens at low kinetic energies, 0 to 20 eV. The ion beam moves along the axis of the rod array toward a detector at the far end. The principle of a quadrupole filter is that application of an appropriately varying electrical field to the rods will cause all ions to impact on the rods during transit except those of a particular value of m/z. Ions moving along the z axis through the resulting electric field experience a combination of forces that causes them to undergo oscillations. For any ion oscillatory trajectory to be stable, the dc potential must be less than one-sixth the maximum amplitude of the rf potential: $U < V(\text{rf})/6$. Except for ions of a single value of m/z, their oscillations tend to increase indefinitely in amplitude, causing them to strike a rod. For a given frequency and set of voltages, only ions in a narrow range of values of m/z are able to traverse the analyzer. Representative maximum values for U and $V(\text{rf})$ are 500 V and 3 kV and the rf frequency is commonly between 0.5 and 3 MHz.

How is scanning accomplished'? Theory shows that a linear mass scan with a quadrupole analyzer is straightforward if at a given rf frequency the ratio $U/V(\text{rf})$ is kept constant while both the dc and the rf potentials are increased linearly. Alternatively, the rf frequency can be increased while U/V is kept constant.

Since quadrupole mass filters have both inductance and capacitance, they can be scanned quite rapidly from lowest to highest masses (in a few milliseconds). Their high scanning speed is responsible for their inclusion in many new designs of mass spectrometer. For instance, this advantage makes quadrupole filters especially attractive mass analyzers for high-resolution gas chromatography. They can also mass analyze ions of either polarity without change, since ions of both charges will perform the requisite oscillatory motions.

Resolution can be changed by altering the dc/rf voltage ratio while maintaining a stable voltage ratio that does not exceed the maximum value. Yet resolution also depends in a complicated way on operating parameters. For example, the length of rods affects resolution since it determines the number of oscillations to which ions of a given m/z will be subjected. Further, if a resolution greater than 100 is needed, a trade-off must be worked out between increasing the rf frequency and the use of longer rods.

From a mathematical analysis of the equations of motion for ions, it can be shown that for the entire range of stable ion paths there will be a fixed mass window δm centered around the condition for maximum transmission through the analyzer. Thus all peaks in a quadrupole spectrum will have the same width regardless of ion mass.

The advantages of a quadrupole mass analyzer are low cost, simplicity, capability of rapid scanning, tolerance of high pressures, and ease of control by computer. The analyzer also has high transmission since paths do not depend on ion kinetic energy. The disadvantages of the analyzer relate mainly to the rods. They are difficult to machine precisely and to align well. The rods must also be cleaned periodically to remove deposits; layers that build up are normally nonconducting and become charged as ions have an impact on them. Such charges lead to variation in the quadrupole electric field and loss in analyzer resolution and sensitivity. Sometimes a quadrupole unit with short rods is installed as a prefilter to trap most deposits.

Ion Cyclotron Resonance Mass Analyzer

As discussed in the preceding section, an ion that enters a steady transverse magnetic field with constant velocity will caused to travel in a circular orbit. It can be shown that the angular frequency of circular motion for each ion depends on its value of m/e as well as the magnetic field strength B according to the expression

$$\text{frequency} = v/r = eB/m \tag{1}$$

If an rf field with a set frequency is applied, an ion of the frequency is in resonance with the field, absorbs energy from it, and expands its orbit steadily. To determine a mass spectrum, sample ions are placed in crossed magnetic and rf fields and the magnetic field is increased at a given rf frequency. Ions of different mass are then brought into resonance, and the obsorption of energy at different rf frequencies is monitored by a sensitive device like a marginal oscillator to obtain the mass spectrum. A schematic diagram of this analyzer is shown in Fig. 9. Note that ions enter from the source and that one kind, A, is experiencing resonance and absorbing energy from the rf field. This analyzer has often been used for the study of ion–molecule reactions.

The resolution of all types of mass analyzers is limited by various physical properties, i.e., imperfect vacuum, spatial length, spatial resolution, magnet strength, and field imperfections. In part it was the attempt to reduce the detection time that provided the impetus for the successful development of the Fourier transform mass spectrometer of the ion cyclotron resonance (ICR) type. The resolution of the ICR mass spectrometer is limited by the noise generated through ion impact on the neutral background gases present in the system. However, it has been demonstrated that ICR instruments are capable of the highest resolution of all types of mass spectrometers. For an IRC, there is a gain in resolution for higher-vacuum system. The FTICR technology allows for a small analytical cell volume, which allows lower operating pressures (higher vacuums) than other types of instrument in

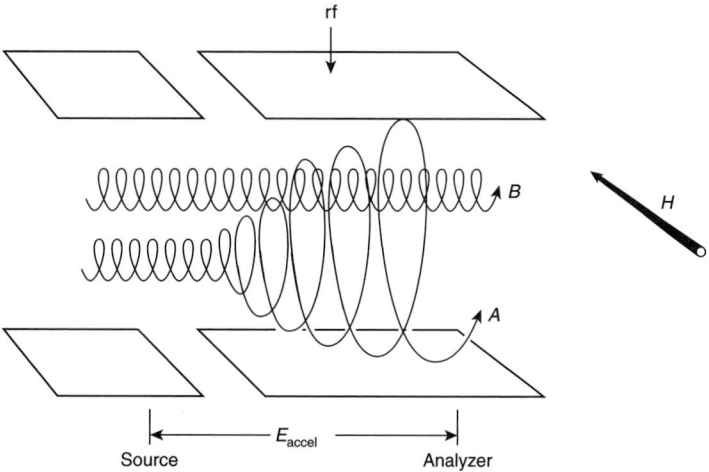

FIGURE 9   Principles of an ion cyclotron resonance spectrometer.

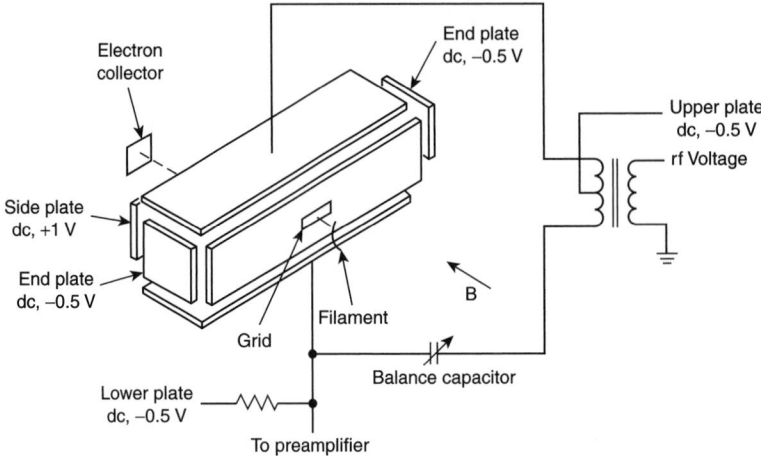

FIGURE 10 Schematic of ICR cell for a Fourier transform mass spectrometer.

general. Additionally, each ion generated may be observed each cycle around the cell (nondestructive measurement), providing a natural gain over other destructive techniques. Modern high-vacuum FTI-CRs can easily trap and observe ions for multiple second time windows. Because the real-time signal generated is an AM- and FM-modulated rf signal composite of all ion frequencies and quantities, a Fourier transform is required for extracting frequency and intensity data into a more useful form.

Fourier Transform Spectrometer. The heart of the Fourier transform mass spectrometer is an ion cyclotron analyzer much like the one illustrated in Fig. 10. The Fourier transform device differs in that all ions from a sample are introduced into the analyzer nearly simultaneously and then excited. The exciting rf signal may be generated by a broadband or a narrow-band generator whose output can be swept quickly (in a 2–5-ms chirp) through the requisite frequency range. The time-domain signal thus generated must be Fourier transformed to yield ICR frequencies. It is Eq. (1) and standardization of frequency against known-mass ions that provide the m/z values in the spectrum. Note that frequency is inversely proportional to m/e. For example, for a magnetic field of 1 T, an m/z 500 ion has a cyclotron frequency of 30.7 kHz, and an m/z 250 ion, a frequency twice that.

What must occur in the ion cyclotron chamber? How can the response of all ions be followed? What changes in instrumentation will make this possible'? A single-cell arrangement that has been used is shown in Fig. 10.

In this cell the several steps in measurements are separated in time. After an initial quench pulse (see below) the sequence of operations is as follows:

1. A sample is admitted.

2. A short electron pulse is generated to cause ionization of the sample.

3. Ions are trapped by a strong steady magnetic field that induces ion cyclotron motion.

4. As soon as the electron pulse ends, a fast rf sweep (~1 ms) coherently excites all ion masses to larger cyclotron orbits.

5. Following the rf sweep, the decay of cyclotron motion induces image currents in a receiver circuit in the cell walls. This time-dependent signal is observed for milliseconds to seconds, the larger time being possible at lower cell pressures.

6. At the end of the measurement cycle a quench pulse expels ions.

Subsequently, the complex decay signal is amplified, phase detected, filtered, digitized, and stored in computer memory. The final step is taking its Fourier transform to shift information from the time to the frequency domain and display of the mass spectrum. To obtain the entire frequency spectrum, the signal bandwidth for data acquisition may be required to be as much as 1 MHz, which means that a large number of bits will be recorded in memory. For example, if $B = 1$ T (1000 G) and the mass range is 15–1200 amu, Eq. (1) shows that the angular frequency extends from 12.5 kHz to 1 MHz.

As is true of other Fourier spectrometric procedures, the Fellgett or multiplex advantage of simultaneous observation of all masses of ions allows either low resolution and very fast scanning or the realization of much higher resolution with a data acquisition time as long perhaps as 1 s. Usually the latter choice is made. The Fellgett advantage is used to increase the resolution as much as possible. Data showing resolutions of 700,000 at m/z values of 80 with a magnetic-field strength of 2 T have been obtained. As the value of m/z increases, resolution decreases. The higher the magnetic field, however, the higher the resolution.

Fourier transform mass analyzers of the ion cyclotron type have the capability of very rapid scanning. The ionization, excitation, and data collection can occur with sufficient speed that scan repetition may approach 10 Hz. The time-of-flight mass analyzer has a potential to produce a complete spectrum at very high rates (where the heaviest ion is the mass 800 amu, a repetition rate of ~11 kHz is possible). Yet time-of-flight spectra are usually obtained by the sampling of only one mass for each pulse of the source. In this case, with an adequate S/N ratio, an acquisition rate of approximately 8–10 Hz can be attained.

The Fourier Transform Ion Cyclotron Resonance Mass Spectrometer (FTICRMS) is able to offer the different aspects of instrumentation in time rather than in space. Since all masses can be scanned simultaneously, the device lends itself to development of time profiles on collision-induced reaction. As mentioned for ion cyclotron mass analyzers, FTICRMS systems require very high vacuum or resolution may be lost.

The sample size for the Applied Automation, Inc. Process FTICRMS analyzer is of the order of micromoles to picomoles of gas. The sample valve is a micromachined piezoelectric device especially designed for minimization of wear and maintenance. The smaple must be clean and dry, with no particulates. GC sample conditioning systems should be adequate. The sample must be vapor at any pressure up to 1–2 atm and temperature up to 300°C.

Applications: ammonia plants, ethylene furnaces, steel mill off-gas, petrochemical plants

Speed: 10 analyses per second maximum

Sensitivity: nominal 0.5 ppm

Mass range: 2 –1000 amu

No scanning; all ions simultaneously; no moving parts

No vaccum pump; sealed 10 (-11) Torr

Sample valve life: 3 years

Estimated cost: $100,000 US

ULTRAVIOLET/VISIBLE ANALYZERS
by J. G. Converse

(Scanning, Filter, Diode Array, Hollow Cathode, and Fourier Transform; see Section 6).

Scanning instruments were considered unreliable in the earlier days of process analytical chemistry because of maintaining component alignment in harsh environments. Today's clean conditioned analyzer shelters have reduced this problem, and more rugged scanning devices have been developed.

There is still a stigma about the conventional scanning spectrophotometer, but they are utilized in some applications. Resolution may be one of the advantages for using such a device. The relative resolutions achieved by the different multiwavelength devices are shown in the following table.

Device	Wavelength Resolution (nm)
Scanning	0.01
Filter	10.0
Diode array	1.0
Fourier transform	0.5

Filter photometers have been the workhorse in the UV/visible area, but the diode array photometer has found some very useful multicomponent applications (CEMS, sloping baseline, etc.). The FTUV device is more of a demonstration that this technique can be utilized in the UV region. It is delicate and expensive and probably cannot be justified in most cases.

Filter Isolation of Discrete Hollow Cathode Lamps [1]

The microprocessor-based BOVAR Western Research, Series 900 photometric analyzer measures the absorbance of UV radiation by a gas sample for up to 6 discrete wavelengths. The concentrations of the components absorbing the light are then determined through application of the ideal gas law in concert with the laws of Bougeer, Beer, and Lambert. The photometric analyzers use two hollow cathode lamps as light sources. Hollow cathode lamps emit discrete lines of intense UV and visible radiation. The narrrow bandwidth of the light sources ensures that a linear response to the analyte concentrations will be observed. The appropriate emission lines from a lamp are isolated through the use of narrow-band interference filters. The emission wavelength of the lamp is completely insensitive to the environmental conditions, being solely determined by the materials used in the lamp construction. This eliminates the possibility of wavelength drift in the spectrometer, thus ensuring that the measurement of the sample absorbance occurs at precisely and accurately known wavelengths. The filter wheel may contain up to six filters, but the number and their wavelengths are application specific. The light sources are pulsed to maximize lamp life.

The outputs of the two light sources are combined by a beam splitter, which also serves to direct one half of the light to a reference photodetector. The other portion of the light beam passes through the sample cell that contains the analyte gas. Dual-beam spectrometers of this design minimize the potential for drift that is due to the light source and provide common-mode noise rejection for data analysis.

Photomultipliers are used as the detectors because of their low inherent noise, dark current and exceptional sensitivity, stability, and linearity of response. A common power supply is used for the two photomultiplier tubes (references and measure) to ensure that any power supply drift or ripple will be canceled out by the common-mode properties of the dual-beam design.

All of the optical components, such as front-sufrace mirrors, are protected from exposure to the analyte gas by quartz windows. The temperature and the pressure of the sample gas are measured simultaneously with measurement of the absorbance signals. This allows the determination of species concentrations in mole percent or parts per million by mole, normalized to standard temperature and pressure.

The operation of the analyzer is controlled by two microprocessors. One, designated the micro-controller, is assigned to the optical bench interface, data conversion, data preprocessing functions, and handling input/output. The other, designated the host controller, is assigned to the front-panel display and keypad control, auxiliary input signal conversion (temperature, pressure, etc.), distributed control system (DCS), and final processing of the data from the microcontroller.

In operation, the analyzer is first zeroed by use of a gas sample (typically nitrogen) that contains no UV-absorbing species. The transmission of zero gas for each wavelength used is determined and stored in memory on the analyzer. When analyte gas is allowed to flow through the sample cell, the transmission of the gas is again determined. The absorbance or optical density of the sample is determined by taking the logarithm of the transmission at each analytical wavelength normalized to that observed for zero gas.

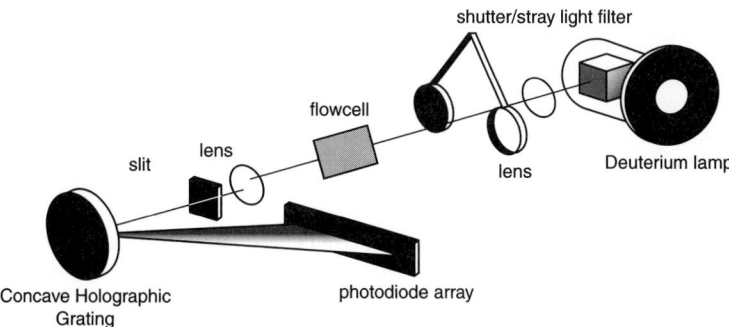

shutter/stray light filter

flowcell

lens

slit

lens

Deuterium lamp

Concave Holographic
Grating

photodiode array

FIGURE 11   Layout of a diode array spectrophotometer.

The concentration of the absorbing species is determined through application of the Beer–Lambert law. Since most chemical species exhibit some overlap in the UV region, multicomponent analysis is performed with linear algebra to extract individual concentration results. The linear response of the analyzer, because of the properties of the hollow cathode lamps, ensures that the matrix analysis is accurate and robust. Instrument bias that is due to variations in transmission on the measure side (change in relative gain of the photodetectors) is eliminated by measuring at one more wavelength than the number of species to be determined.

Diode Array Process Spectrometer [2]

Diode array technology has made it possible to obtain measurements of absorbance at several wavelengths simultaneously without moving the grating or the detector. This makes possible many measurements that were previously difficult, if not impossible, to obtain. The source light is directed through the sample with lenses and stray light is reduced by a blocking filter. It is then directed through a slit onto a concave holographic grating to disperse the different wavelengths. The concave holographic grating combines the functions of producing linear angular dispersion and focusing of the different wavelengths of light at the same point in space while being temperature insensitive. The diode array is positioned so that the individual detectors receive the specified wavelengths of light from the region to be measured. Arrangement of the individual components of the spectrophotometer are shown in Fig. 11.

Solid-state photodiodes have a large dynamic range and are very robust. Proper construction of the instrument produces a resolution of 1 nm, which allows a detailed spectrum of a sample to be obtained. There are 1024 diode elements in a single array that are read simultaneously to produce a scanned spectrum. Light falling on the semiconductor element allows electrons to flow through it, thereby depleting the charge in a capacitor connected across it. The amount of charge needed to recharge the capacitor at regular intervals is proportional to the intensity of the light. Since we have the ability to measure very small quantities of current with high precision, this device provides a very sensitive measurement of absorbance. Each diode is connected to a capacitor and a transistor switch to form a shift register that the computer can query at high speed. A single spectrum can be obtained in 0.1 s, but typically averaging of several scans produces a better spectrum over a time of 0.5 to 1.0 s. The Applied Analytics, Inc., OMA-517 Diode Array Process Spectrophotometer can measure 10-ppmv ammonia (NH_3) with a background of 1000-ppmv SO_2 and 200-ppmv NO_x. This requires low noise, high stability, high resolution, and chemometrics data analysis. Time averaging and wavelength referencing also improve repeatability and accuracy. A stable light source is absolutely necessary since noise from the lamp is often the limiting factor in overall noise performance. Low stray light is also required for high-concentration measurement to provide a wide dynamic range.

REFERENCES

1. Harris, P. and H. Adam, "Design and Practical Application of Process UV Process Analyzers," presented at L'Analyse Industiel '96; Paris, France, 1996.
2. Bardshaw, Y., "OMA-517 Product Description CD," Applied Analytics, Inc., Chestnut Hill, Massachusetts, 1998.

RAMAN SCATTERING EMISSION SPECTROPHOTOMETERS
by M. J. Pelletier

Commercial Raman analyzers are relatively new, although the technology has been around since the early 1930s. The availability of laser sources, diode array detectors, and powerful inexpensive computers has made it practical to move Raman techniques to the field. One major advantage of Raman scattering over IR absorption is the direct application to aqueous solutions. Raman depends on adding or subtracting a small amount of energy (IR region of the spectrum) to a photon of the NIR, visible, or UV region. The light being measured is not attenuated by water and many other solvents. The difference between the source energy and the scattered radiation energy [Stokes($-$) and anti-Stokes($+$)] yields the vibrational information desired.

Assets:

1. can measure vibrational transitions in water
2. can measure absorptions inactive to IR absorption (complements IR data)
3. light source and detector can be isolated from sample (fiber optics)
4. provides fast ($\sim$1–20-s) analysis of multiple components
5. high specificity and non-destructive

Liabilities:

1. complex and delicate (this is changing rapidly)
2. requires computer and calculations
3. lower detection $\sim$0.1%
4. calibration for noncontact applications requiring internal standard
5. high skill required for setup and maintenance
6. relatively high total cost of ownership

Raman Analyzers

The Raman Effect.   A small fraction of the light striking a substance is shifted in wavelength by the molecular vibrations of that substance. This shifted light, called Raman scatter, is used for quantitative and qualitative analysis. Qualitative analysis is based on the wavelength shifts' being different for a wide range of different molecular vibrations. A plot of Raman scattered intensity versus wavelength, called a Raman spectrum, is a fairly unique and interpretable fingerprint for most chemical substances. Quantitative analysis is based on the intensity of the Raman-scattered light being proportional to analyte concentration. Simple spectral band area measurements are often sufficient and robust, but multivariate methods such as partial least squares (PLS) can often extract more information.

Interface to Sample.   A major strength of Raman analyzers for chemical production applications is flexible sampling. Raman probes analyze material at a fixed distance from the end of the probe. Transparent materials between the end of the probe and the sample do not hinder the measurement.

As a result, Raman analysis can be done through quartz or sapphire windows or through closed transparent containers. The sample does not need to be touched, removed from the process stream, or perturbed in any way. The need for slip streams or grab samples is eliminated.

The Raman probe both delivers laser light to the sample and collects Raman scattered light returning from the sample. Since transmission through the sample is not required, both transparent and opaque samples can be analyzed. A fiber-optic cable connects the Raman probe to the Raman instrument. Only one window into the reactor or flowing stream is needed. The transparent window can be as small as 1 mm in diameter. Raman scattered light from a sample is always very weak. It is therefore important to maximize the amount of Raman scattered light that is detected from the sample and minimize all other forms of detected light. External light, such as room lighting, indicator lights, or sunlight, is excluded by the design of the sample interface. This is automatic for measurements done in opaque pipes or reactors, but deserves attention when exposed samples, such as polymers in fiber- or film-drawing lines, are being measured. Laser-induced fluorescence from the sample itself can be a serious problem. The use of NIR laser excitation tends to minimize sample fluorescence. Derivative spectra further reduce baseline fluctuations caused by fluorescence.

Raman Instrumentation.   A schematic diagram of a fully integrated Raman system [1] is shown in Fig. 12. This instrument is contained in a single box measuring 23 in. × 18 in. × 8 in. The only external connections are the 110-V power connection, a serial port to a computer, and the fiber-optic cable to the Raman probehead. Laser light is coupled to the probehead through a single optical fiber. Raman scattered light from the sample is returned to the Raman instrument through a second optical fiber. Those optical fibers can be hundreds of meters long if necessary. Their small core diameters of 50 to 100 μm make long runs of fiber-optic cable reasonably inexpensive. Raman scattered light is dispersed by a spectrograph and detected by a charge-coupled device (CCD) detector. The spectrograph shown in Fig. 12 is an $f/1.8$ axial transmissive spectrograph [2] with a HoloPlex volume transmission grating [1, 3]. This design provides improvements in spectral coverage, optical throughput, and image quality over older, more traditional designs. The spectrograph can accept multiple fiber-optic inputs for simultaneous multichannel analysis.

The performance of the fiber-optic probe is crucial for most chemical production applications. The probe needs to deliver the laser light to the sample, collect Raman scattered light efficiently from the sample, and interface easily to pipes, reactors, sight glasses, or exposed samples. It must also

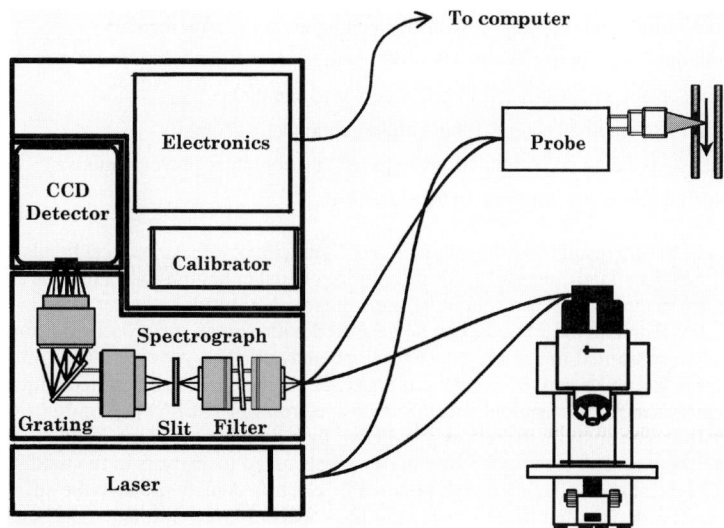

FIGURE 12   Schematic diagram of a Raman analyzer.

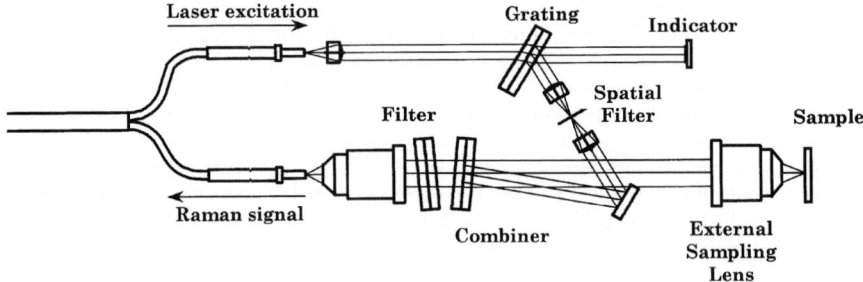

FIGURE 13   Schematic diagram of a Raman fiber-optic probehead.

eliminate and spectral artifacts caused by the optical fibers. An optical diagram of one such probe [1, 4] is shown in Fig. 13.

The combination of a volume holographic transmission grating and a spatial filter in the excitation path of the probehead assembly acts as a monochromator. The monochromator removes Raman and luminescence intensity produced by the laser light traveling through the optical fiber. Volume holographic notch filters [5] reject unshifted laser light before the Raman scattered light is injected into the return fiber. Permanent alignment inside the probehead ensures that the collection optical path and the excitation optical path are always properly aligned, regardless of the external sampling optic used. Numerous external sampling optics for the probehead are available, ranging front simple inversion lenses to research-grade microscopes.

Capabilities and Limitations

1. internal standard needed, do not use absolute Raman intensity
2. effects of background and source fluorescence
3. works with aqueous, opaque, solid, liquid, and gas
4. similarities and differences with IR (techniques are complementary)
5. minimum concentration ~0.1% (1000 ppm)
6. noncontact, nondestructive (optical access to sample)
7. high information content (robust multicomponent, specificity)
8. fast (depending on computations required and computer power available)
9. spatial selectivity and remote measurement

Raman spectra result from the interaction of light with covalent chemical bonds. Materials that do not have covalent bonds, such as a sodium chloride crystal or helium gas, cannot be analyzed directly with Raman spectroscopy. Gases, liquids, and solids such as nitrogen, water, or silicon crystals, which do have covalent chemical bonds, can be measured with Raman spectroscopy. Raman process analyzers usually are limited to sample concentrations above 0.1% by weight. This rule of thumb holds for both the gas phase and the condensed phase. Sensitivity enhancement techniques such as resonance enhancement [6], surface-enhanced Raman spectroscopy (SERS) [7], or multipass sampling [8–10], permit Raman analysis in the low part-per-million range or even less. While useful in the laboratory, these enhancement techniques are often not well suited to analysis in the field.

Chemical bonds having a high polarizable electron density produce the strongest Raman signals. Therefore, double and triple bonds tend to produce a larger Raman signal than the corresponding single bonds. Covalent bonds between heavier atoms tend to produce stronger Raman signals than similar bonds between lighter atoms.

Several properties of the Raman analyzer, the sample interface, and the sample itself can affect the absolute Raman intensity. Often, at least some of these properties cannot be accurately controlled. Quantitative analysis with a Raman analyzer therefore usually uses some type of internal standard. In other words, some part of the Raman spectrum that does not change with the sample property of interest is used to normalize the part of the Raman spectrum that does change with the sample property of interest. For example, the Raman band(s) of the solvent is often used to normalize the Raman bands of the reactants and products of a chemical reaction. In other cases, such as polymerization, Raman bands from a part of a molecule that is not reacting can be used to normalize Raman bands from the same molecule that is reacting. Older laboratory instruments often produced Raman spectra by sequentially measuring Raman intensity at different spectral regions and then combining these measurements into a spectrum. This approach assumed that the Raman signal did not change during the acquisition of the Raman spectrum. The Raman signal from an on-line Raman analyzer may fluctuate rapidly, however, because of bubbles or particles passing through the laser beam. On-line Raman analyzers therefore need to collect all their spectral information simultaneously in order to avoid measurement artifacts caused by changing sample conditions. The internal standard will then correct the Raman spectrum for these types of rapid sample changes.

The weak Raman signal from a sample can easily be overwhelmed by stronger sources of background light. When this happens, the Raman analysis fails. The most common source of background light intensity is laser-induced fluorescence from the sample itself. The fluorescence usually comes from small amounts of impurities in the sample that happen to be highly fluorescent. It is often possible to overcome the fluorescence background problem by changing the laser wavelength. Raman analyzers that use NIR lasers are usually not limited by background fluorescence. Analytical models based on the derivatives of Raman spectra, rather than the Raman spectra themselves, further reduce baseline fluctuations caused by fluorescence. In general the first step in developing a Raman analytical method is to make sure the sample does not fluoresce too strongly. Vendors of Raman analyzers or central analytical groups within major companies can usually provide this preliminary information quickly and easily.

The acquisition of a Raman spectrum usually requires 0.5 s to a few minutes, depending on the sample and the required S/N ratio. The maximum rate at which spectra can be collected, typically 10 to 15 spectra per minute, is usually determined by the time required for getting the spectrum out of the CCD detector and into the computer. The speed of a Raman analyzer can often reveal occurrences in a processes that are not suspected when slower techniques such as gas chromatography are used.

Raman Analyzers are Relatively Expensive. The analyzer may cost from $50,000 to $150,000, and the integration of the analyzer into the production facility can cost as much as the analyzer itself. The main cost of ownership is laser replacement. Helium–neon lasers cost a few thousand dollars and have expected lifetimes in the 30,000- to 50,000-h range (3 to 5 years of continuous operation). Some applications, however, require greater sensitivity, or greater immunity to sample fluorescence than is possible with a helium–neon laser. In these cases more expensive lasers, such as diode-pumped frequency-doubled Nd:YAG lasers or high-power external cavity stabilized diode lasers, are required. The cost of these lasers is currently in the $10,000 to $20,000 range. Their expected lifetimes currently tend to be approximately 10,000 to 15,000 h. Clearly, a Raman analyzer needs to provide significant value to justify its installation.

Examples of Raman Analyzer Applications

Analysis of PCl$_3$ Reactor Material [11, 12]. Monsanto has been using Raman analyzers to continuously monitor reaction vessel composition during PCl$_3$ production. They manufacture PCl$_3$ from elemental phosphorus (P$_4$) and chlorine gas by using PCl$_3$ as a solvent. The heat of reaction is used to distill PCl$_3$ out of the reaction vessel. Unwanted PCl$_5$ is produced when too much chlorine is present. The reactor needs to be periodically shut down because of the buildup of nonvolatile carbonaceous residue from the phosphorus feedstock. At that time all the excess phosphorus needs to be reacted with chlorine, but excessive production of PCl$_5$ needs to be minimized.

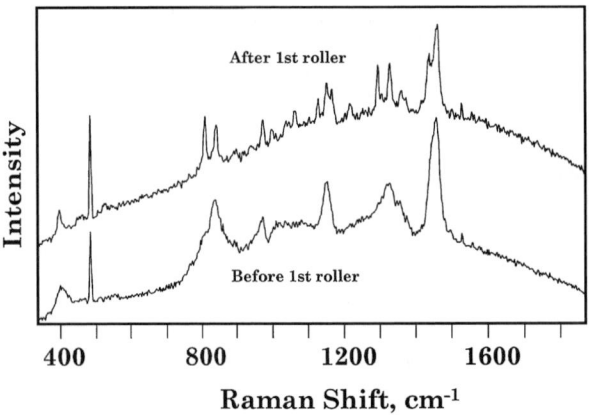

FIGURE 14   Raman spectra of drawn polypropylene film.

The contents of the reactor need to be continuously monitored for safe and efficient production of PCl_3. Monsanto uses Raman probes inside their reactors to eliminate the need for removing a sample of the boiling, toxic, corrosive, pyrophoric liquid from the reactor and transporting it to a laboratory for analysis. The real-time, on-line Raman analysis also provides faster results than off-line methods, allowing more precise control of the reactor. Monsanto uses 785-nm lasers in their Raman analyzers. Shorter wavelengths cause too much fluorescence from the carbonaceous impurities, and longer wavelengths lack the needed sensitivity at acceptable laser power levels.

Polymer Processing [13, 14].   Raman spectroscopy has long been used to measure polymer properties such as residual monomer concentration, crystallinity, orientation, and copolymer composition. Raman analyzers are now making these measurements on operating polymerization reactors, polymer fiber-drawing machines, and fiber-drawing machines. The Raman analyzer provides rapid analytical results without disturbing production or requiring grab samples.

Figure 14 shows Raman spectra collected on polypropylene film at different locations in the drawing process. Before the first roller the polypropylene was mostly amorphous. The polypropylene was mostly crystalline after the first roller, under the drawing conditions used in this application. The large differences among the spectra make the crystallinity determination very robust. The Raman probe in this case was 17 in. away from the region of the file being analyzed. Spectra were collected on line with 532-nm excitation. The sharp band at 482 cm^{-1} is due to fluorescent room lighting.

The crystallinity of poly(ethylene terephthalate) (PET) can also be measured by Raman spectroscopy. Figure 15 shows a calibration curve from a two-factor PLS model that predicts PET density from its Raman spectrum. Laser-induced fluorescence can compromise the accuracy of the Raman analysis, so some form of baseline correction, such as a second-derivative spectrum, is normally used to improve the robustness of the analysis. Raman analyzers that use NIR excitation, rather than visible excitation, are less likely to be influenced by fluorescence.

Dimethyldichlorosilane Distillation [15, 16].   The Dow Corning Company is investigating the benefits of replacing gas chromatographic (GC) analyzers with Raman analyzers. GC analyzers are used to monitor the distillation of dimethyldichlorosilane, the raw material used to make silicones. The GC analyzers require removal and transport of a sample to the analyzer for analysis. The analysis is too slow to detect some of the dynamic processes in the distillation column. In addition, the corrosive nature of the samples leads to frequent analyzer maintenance. The Raman analyzer is much faster, does not require removal and transport of the sample, and is expected to have a lower cost of ownership because of the lower cost of maintenance and consumables. In this application several fiber-optic

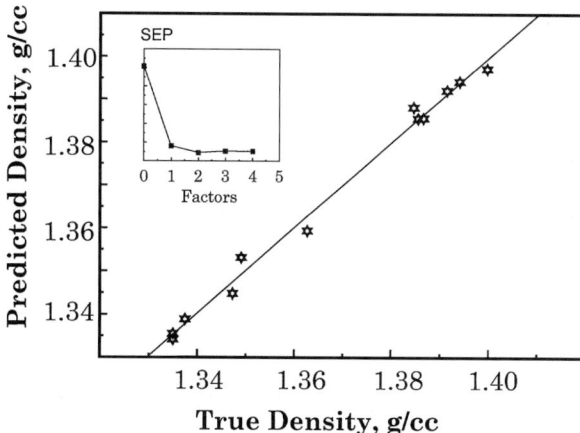

FIGURE 15   Calibration curve for determination of PET density by a Raman analyzer.

probes allow one analyzer to monitor several points in the process, reducing the cost per measurement point.

Oxygen Detection in Closed Containers [17, 18].   The sampling benefits offered by Raman spectroscopy are especially valuable for analyzing gases in closed containers. Raman spectroscopy has long been used to determine the identity of gases trapped in microscopic bubbles in glass or geological samples. More recently it was used to test the integrity of seals in the microaccelerometer chips used to deploy air bags in automobiles. These chips contain a microscopic lever hermetically sealed in a sandwich structure. The cavity containing the lever is only 0.07 mm deep. The integrity of the seal is critical for reliable operation. A Raman analyzer attached to a microscope was able to detect faulty seals rapidly and nondestructively by determining the presence or absence of oxygen in the cavity. The cavities are sealed under nitrogen, and oxygen gas could be present only if the seal had failed. A similar strategy is used by a major pharmaceutical company to test the seal integrity of each vial of one specific product they sell. Their automated Raman system is able to determine vial seal integrity at a rate of several vials per minute.

SUMMARY

Raman analyzers provide rapid, on-line qualitative and quantitative analysis capabilities. Fiber-optic probes allow the analyzer to be hundreds of meters away from the points of measurement. Raman analyzers can usually be added to material production or processing facilities without disturbing normal operation. More detailed information on Raman spectroscopy can be found in recent reviews [19, 20] and books [21–25] on the subject.

REFERENCES

1. Owen, H., D. E. Battey, M. J. Pelletier, and J. B. Slater, "New Spectroscopic Instrument Based on Volume Holographic Optical Elements," in *Practical Holography IX*, S. A. Benton, Ed., *Proc. SPIE* **2406**, 260–267, 1995.

2. Battey, D. E., J. B. Slater, R. Wludyka, H. Owen, D. M. Pallister, and M. D. Morris, "Axial Transmissive $f/1.8$ Imaging Raman Spectrograph with Volume-Phase Holographic Filter and Grating," *Appl. Spectrosc.* **47**, 1913–1919, 1993.

3. Battey, D. E., H. Owen, and J. M. Tedesco, "Spectrograph with Multiplexing of Different Wavelength Regions onto a Single Opto-Electric Detector Array," U.S. Patent 5,442,439, assigned to Kaiser Optical systems, Inc., Ann Arbor, Michigan, 1995.

4. Owen, H., J. M. Tedesco, and J. B. Slater, "Remote Optical Measurement Probe," U.S. Patent 5,377,004, assigned to Kaiser Optical Systems, Inc., Ann Arbor, Michigan, 1995.

5. Tedesco, J. M., H. Owen, D. M. Pallister, and M. D. Morris, "Principles and Spectroscopic Applications of Volume Holographic Optics," *Anal. Chem.* **65**, 441A–449A, 1993.

6. Asher, S., "UV Resonance Raman Spectroscopy for Analytical, Physical, and Biophysical Chemistry Part 1," *Anal. Chem.* **65**, 59A–66A, 1993.

7. Ruperez, A., and J. J. Laserna, "Surface Enhanced Raman Spectroscopy," in *Modern Techniques in Raman Spectroscopy*, J. J. Laserna, Ed., Wiley, New York, 1996.

8. Robinson, J. C., M. Fink, and A. Mihill, "New Vapor Phase Spontaneous Raman Spectrometer," *Rev. Sci. Instrum.* **63**, 3280–3284, 1992.

9. Mitchell, J. R., "Diode Laser Pumped Raman Gas Analysis System with Reflective Hollow Tube Gas Cell," U.S. Patent 5,521,703, assigned to Albion Instruments, Salt Lake City, Utah, May 28, 1996.

10. Che, D., and S. Liu, "Long Capillary Waveguide Raman Cell," U.S. Patent 5,604,587, assigned to World Precision Instruments, Sarasota, Florida, Feb. 18, 1997.

11. Freeman, J. F., D. O. Fisher, and G. J. Gervasio, "FT-Raman on-Line Analysis of PCl_3 Reactor Material," *Appl. Spectrosc.* **47**, 1115–1122, 1993.

12. Gervasio, G. J., and M. J. Pelletier, "On-Line Raman Analysis of PCl_3 Reactor Material," *J. Process Anal. Chem.* **3**, 7–11, 1997.

13. Pelletier, M. J., "Raman Spectroscopy Outside the Laboratory," in *Proceedings of the 43rd Annual ISA Anal. Div. Symposium*, Instrument Society of America, Research Triangle Park, North Carolina, 1998, Vol. 31, pp. 63–72.

14. Everall, N., K. Davis, H. Owen, M. J. Pelletier, and J. Slater, "Density Mapping in Poly(ethylene terephthalate) Using a Fiber-Coupled Raman Microprobe and Partial Least-Squares Calibration," *Appl. Spectrosc.* **50**, 388–393, 1996.

15. Lipp, E. D., and R. L. Grosse, "Analysis of Chlorosilane Process Streams by Raman Spectroscopy," in *Proceedings of the Fifteenth International Conference on Raman Spectroscopy*, by S. A. Asher and P. Stein, Eds., Wiley, New York, 1996, pp. 1092–1093.

16. Lipp, E. D., and R. L. Grosse, "On-Line Monitoring of Chlorosilane Streams by Raman Spectroscopy," *Appl. Spectrosc.* **52**, 42–46, 1998.

17. Weber, W. H., M. Zaini-Fisher, and M. J. Pelletier, "Using Raman Microscopy to Detect Leaks in Micromechanical Silicon Structures," *Appl. Spectrosc.* **51**, 123–129, 1997.

18. Gilbert, A. S., K. W. Hobbs, and P. P. Jobson, "Automated Headspace Analysis for Quality Assurance of Pharmaceutical Vials by Laser Raman Spectroscopy," in *Optical Measurements and Sensors for the Process Industries*, C. Gorecki and R. W. Preater, Eds., Proc. SPIE **2248**, 391–398, 1994.

19. Lyon, L. A., C. D. Keating, A. P. Fox, B. E. Baker, L. He, S. R. Nicewarner, S. P. Mulvaney, and M. J. Natan, "Raman Spectroscopy," *Anal. Chem.* **70**, 341R–361R, 1998.

20. Adar, F., R. Geiger, and J. Noonan, "Raman Spectroscopy for Process/Quality Control," *Appl. Spectrosc. Rev.* **32**, 45–101, 1997.

21. Lin-Vien, D., N. B. Colthup, W. G. Fateley, and J. G. Grasselli, *The Handbook of Infrared and Raman Characteristic Frequencies of Organic Molecules*, Academic, Boston, 1991.

22. Grasselli, J. G., and B. J. Bulkin, *Analytical Raman Spectroscopy*, Wiley, New York, 1991.

23. Turrell, G., and J. Corset, *Raman Microscopy Developments and Applications*, Academic, New York, 1996.

24. Laserna, J. J., *Modern Techniques in Raman Spectroscopy*, Wiley, New York, 1996.

25. Pelletier, M. J., *Analytical Applications of Raman Spectroscopy*, Blackwell Science, Oxford, 1999.

NUCLEAR MAGNETIC RESONANCE
by J. G. Converse

Significant changes in the hardware and technology (permanent magnets, electronics, and computers) required for measuring nuclear magnetic resonance (NMR) have made this laboratory method more amenable to field application. The energy required for making a nuclear process depends on the environment of the atom. This can be very specific for different hydrogen atoms in organic molecules. The chemical shift of the resonant nucleus identifies the type of environment and can provide qualitative information about molecular structure. The signal intensity and time-domain (relaxation) data can be utilized to measure the amount of material present. Computer software can control instrument variable and process measured signal to provide a specified concentration output.

Assets:

1. provides information not available by other means
2. noninvasive and nondestructive
3. no reagents or carrier fluid required

Liabilities:

1. not for trace analysis (<0.5 %)
2. requires computer and calculations
3. requires good temperature control
4. requires skilled maintenance technicians
5. limited commercial suppliers and relatively expensive
6. chemometric calibration required

New Developments in NMR Measurement

1. condensed field magnet technology (small high field permanent magnets 10–60 KG)
2. high-speed data processing capabilities (PCs and workstations)
3. evolution and miniaturization of electronic components
4. reduction of magnet fringe field (reduces interference with other equipment)
5. measurement of relaxation versus chemical shift information
6. sample probe design and materials improvements
7. transmitter–receiver electronics that have more power and resolution
8. computer software that is available to control pulse programming, pulse parameters, and shimming
9. control of temperature and resonant frequency and diagnostics for monitoring
10. chemometrics software PLS and principal component regression (PCR) available for quantification of signal
11. Fourier transform NMR provides faster analysis times

Rules of Thumb

1. Ferromagnetic materials must be kept out of the magnet–probe assembly.
2. Temperature of the magnet and sample must be tightly controlled.
3. Spin frequency must be controlled and locked on free-induction decay resonance.

4. Stopped flow may be required for enhancing S/N ratio.

5. PCR set should contain 20 of more samples for high precision.

Theory of NMR (Conventional versus FTNMR)

Magnetic-field strength and homogeneity determine resolution and dispersion of individual resonant absorption.

14Kgauss

Condensed field magnet:

Autoshimming is the process of adjusting the amount of current flowing to the coils placed within the magnet utilizing PC software to optimize the magnet homogeneity. By electronic shimming, a cylindrical magnetic field of uniform strength is produced, enabling high-resolution determinations to be achieved. Fourier transform converts signal data from the time domain to the frequency domain. Magnetic-field/frequency lock electronics tracks and corrects drift. In NMR spectroscopy, a solid, liquid, or gas sample is subjected to a strong, uniform magnetic field. Simultaneously, the NMR spectrometer applies rf pulses at right angles to the magnetic field. The nuclei in the sample absorb this energy, causing the nuclei to go from parallel to nonparallel orientation (spin) in the magnetic field.

Maximum absorption occurs when the characteristic frequency of a nucleus is the same as the frequency of the rf pulse. The intensity of the absorption is directly proportional to the number of nuclei changing orientation at the specific frequency. Thus peak areas correspond to the numbers of nuclei in a sample and also the absolute concentrations.

A second, more subtle effect is created because each nucleus is also partially shielded from the field by its electrons. This shielding slightly alters the frequency at which the nucleus absorbs rf energy. This phenomenon, called a chemical shift, explains the various regions in the typical NMR spectrum and helps identify the chemical structure of the sample.

Rather than measure the amount of energy absorbed by a sample's nuclei (a difficult proposition), detectors actually measure their energy (or spin) decay. Betwen rf pulses, with the atoms no longer excited, the energy absorbed by each nucleus decays (free-induction decay) and is emitted. A highly sensitive detector measures and records these emissions as a decay pattern that is then transformed into a spectrum.

"Another wonderful advantage of NMR (for tenths-of-a-percent to percent level determinations) is that, in most cases, no calibration is necessary," says James F. Haw, professor of chemistry and director of the Laboratory for Magnetic Resonance and Molecular Science at Texas A&M, College Station, TX. "The NMR signal intensity is directly and fundamentally proportional to the number of protons of a given type. Therefore, the molecular absorptivity is constant for all kinds of protons and all kinds of molecules; it does not change as a result of hydrogen bonding or other kinds of matrix effects that sometimes bedevil the establishment of a calibration procedure." Several actual or potential users of process NMR include Texaco, Amoco, Dow, DuPont, Union Carbide, Vista Chemical, et al. They are making big investments in this technology to supplement their current process analytical chemistry arsenal.

Chemometric analysis is required when the sample spectrum has overlapping peaks.

Applications

1. ideally suited for petrochemicals (hydrocarbons)

2. ethylene and propylene polymerization processes

3. C_4 chemistry-based isomerization units

4. propane in liquid natural gas

5. methyl tertbutyl ether (MTBE) in flowing gasoline streams (also methanol, ethanol, MTAE)

6. physical properties (e.g., viscosity, polymer density, melt index, etc.)

7. measuring black liquor for high percentage solids level and viscosity to optimize heat recovery

8. acid strength (etc.) in sulfuric acid alkylation emulsion stream versus lab titration

FUNDAMENTALS OF THE QUANTITATIVE NMR

by John Edwards*

"Nuclear Magnetic Resonance (NMR) spectroscopy is one of the most powerful experimental methods available for atomic and molecular level structure elucidation. It is a powerful technique in that it is ⋯ noninvasive <and is> used to identify individual compounds, <analyze mixtures>, aid in determining the structures of large molecules ⋯ examine the kinetics of certain reactions, <and predict the physical and chemical properties of materials when those properties are based on the inherent chemical structure of the material>." From *Nuclear Magnetic Resonance Spectroscopy in Environmental Chemistry*, edited by Nanny et al. Oxford Univ. Press, 1997.

The property of nuclear spin, discovered by Rabi and coworkers in 1939, is the fundamental property for observing NMR. By analogy, a spinning nucleus is similar to an electric current flowing in a wire loop. In a wire loop, as charge flows around the loop, it generates an effective bar magnet whose poles are perpendicular to the current flow. Similarly, a nucleus spinning about its axis generates a small magnetic dipole, similar to a bar magnet. Since a dipole is present, a magnetic moment exists for that nucleus. Magnetic moments are defined as the torque felt by the nuclear charge divided by the strength of the applied magnetic field when the nuclear spins are placed in a magnetic field. This torque exists until the net magnetic moment is either aligned with the magnetic field (lower energy state) against the magnetic field (higher energy state) or perturbed from alignment with or against the magnetic field. As there are always slightly more spins aligned with the external magnetic field (lower energy) than aligned against the external magnetic field (higher energy) there is an effective net magnetic moment aligned with the external field under equilibrium (no torque) conditions. The NMR experiment involves the movement of that net nuclear magnetic moment off the axis of alignment with the external magnetic field.

The application of a short (5–15-μ5) low-energy (approximately 1-W) rf pulse perpendicular to the net magnetic moment provides enough torque to move that net magnetic moment off the axis of alignment. Once the pulse is stopped, the bulk magnetic moment relaxes back toward the alignment axis. Chemical information is obtained by observation of the different component relaxation rates of individual magnetic moments. These individual relaxing magnetic moments induce electric currents

---

* Process NMR Associates, LLC (Foxboro), Danbury, Connecticut.

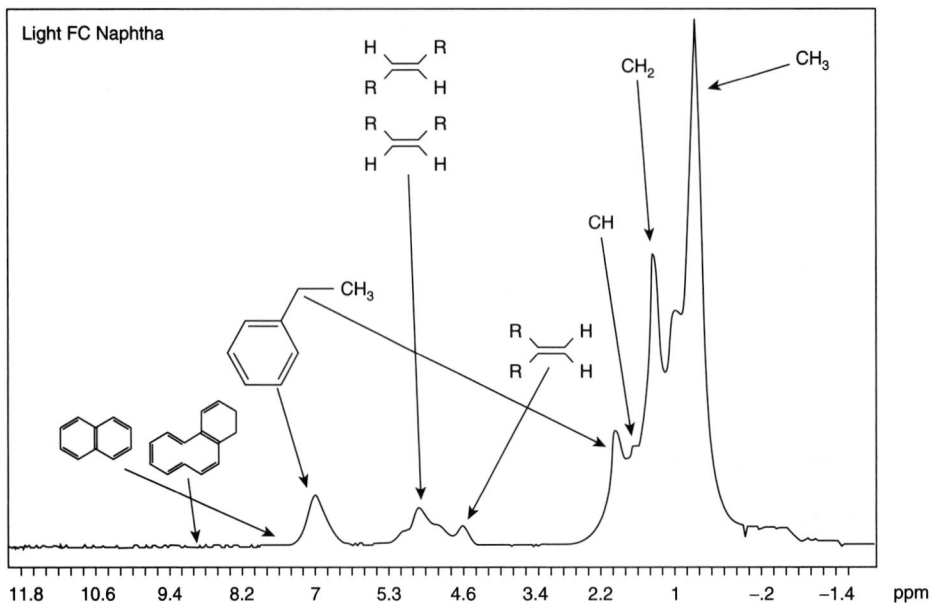

FIGURE 16 NMR spectrum identifying proton environment.

that are picked up in the receiver coil as a free-induction decay. Fourier transformation converts the free-induction decay from the time domain to the frequency domain. Differences in the frequencies of differing nuclear spins are then used to analyze molecular structures, mixtures of molecules, and/or kinetics of spin exchange by NMR. Utilization of NMR information can then be used to correlate NMR spectral information to the chemical and physical properties of the samples that are being analyzed by NMR.

Quantitative analysis is inherent in the NMR experiment. During the rf pulse, nuclei absorb the rf energy, exciting the spins to higher energy states. Unlike any other spectroscopy, this absorption of energy by the nuclei is linearly quantitative across the entire NMR spectral range; absorption of rf energy is dependent on only the number of nuclei present and is not enhanced or dampened by specific molecular environments or chemical functionality. All NMR energy level transitions induced by the rf pulse are equal, in specific contrast to IR transitions, which are dependent on both energy absorption as well as chemical/molecular environment. After the rf pulse, the relaxing nuclei emit rf energy at specific frequencies based on their specific molecular/chemical environment. That energy is detected as an emission signal whose intensity is proportional to the total energy emitted and therefore directly proportional to the number of nuclei excited. The NMR signal intensity contains the quantitative information for the total number of nuclei while the frequency information contains the chemical/molecular structural information (Fig. 16).

For example, for two spins A and B,

$$\text{Total NMR energy} = \text{energy of spin } A + \text{energy of spin } B$$

- if $A = B = 1$, then total NMR energy $= 2$. The integration of peak A must equal the integration of peak B.
- if $A = 1$ and $B = 2$ then total NMR energy $= 3$.

Practically, accurate/quantitative measurement of NMR signals is achieved by

1. generation of precise, square, rf transmitter pulses

2. rapid (low-Q) detection coils with minimal dead times between transmitter pulses and detection

3. high digitization rates and wide dynamic range (A/D converter technology)

Hardware technology developments over the past two decades in all areas affecting NMR (rf technology, computers, solid-state electronics, etc.) have enabled NMR to be a very precise, accurate, repeatable, and reliable technique and technology ready to enter the process environment (see Section 6, Analytical, for application examples).

X-RAY FLUORESCENCE (XRF)
by Tony Harding

New detectors, electronics, and computers have made this technology more field hardened, although it has been utilized for on-line analysis for many years. There is still an argument regarding use of the radioactive x-ray source versus the tube x-ray generator source. Licensing and leakage are of much less concern, but some folks still are afraid of anything bearing a nuclear material sticker. The room-temperature high-resolution detector has not achieved the performance level that had been predicted. There are many detectors that can produce adequate performance without liquid nitrogen cooling.

Multichannel x-ray energy discrimination has seen significant electronics improvements, and software for processing raw x-ray data is readily available.

Assets:

1. flow-through cell provides nondestructive testing

2. excitation source can be selected to optimize efficiency

3. multicomponent measurement of atoms

4. concentration from low ppm to 100%

5. fast (10 to 60 s)

Liabilities:

1. not sensitive to molecular species

2. limited to heavier elements (>aluminum—atomic number 13)

3. not for trace level, and response is element dependent

4. electronics requires excellent shielding (picks up noise spikes)

5. corrections required for matrix interferences

X-Ray Fluorescence

The demand for on-line analyzers capable of compositional determinations in chemical and petrochemical process streams has increased dramatically over recent years. Total control of production plant processes and resources requires the analysis of feed, intermediate, and finished product materials. XRF analysis has made significant progress in taking elemental determinations into the process environment.

There are several general advantages of the XRF technique for process environments compared with other elemental analysis methods. For example, XRF provides rapid, real-time elemental determinations that can trigger immediate process optimization steps. A wide range of components can be studied: aluminum (atomic number 13) to uranium (atomic number 92) inclusive, in concentration ranges of a few ppm to several weight percent, depending on the element and process stream. The technique is continuous, nonintrusive, and nondestructive so that a sample is not altered in any way and oftentimes is not removed from the original process stream. Both aqueous and organic streams

TABLE 1   Sulfur XRF Calibration data

Parameter	Value
Slope[a]	0.268 cps/ppm S
Intercept	0.4 cps
Correlation coefficient	0.998

[a] cps is counts per second.

can be analyzed, and multielement information is provided by XRF simultaneously in many cases. Multiple process streams can be addressed with most XRF analyzer configurations that are capable of Class C Division 1 operation.

This subsection describes two specific applications of on-line XRF analysis: sulfur concentration in fuel and metal catalyst depletion. Issues involved in the successful implementation of the technique for these applications can be applied generally to all other process streams. The measurement theory of XRF is described, followed by a discussion of the instrumentation and configurations of commercially available systems. The final portion of the subsection describes important ancillary aspects of applying XRF to process streams. Examples of those topics include sample conditioning, materials compatibility and handling, and instrument environment. Only energy-dispersive XRF (EDXRF) hardware systems are discussed in this section, for two compelling reasons. First, there exists only one or two commercially available XRF systems based on the wavelength-dispersive (WDXRF) technique. Second, all the other discussions involve x-ray theory, ensuring that the instrumentation is process worthy and that sample handling issues apply equally to WDXRF and EDXRF installations.

Sulfur in Fuel

Environmental regulations have capped the sulfur concentration acceptable in on-road diesel fuels to 500-ppm sulfur. On-line EDXRF is an optimal method for feeding back to the process the sulfur concentration being produced at the refinery or when, in a blending operation, an acceptable sulfur level has been achieved. A slip stream of fuel is provided at approximately 300 ml/min to the analyzer. Materials for plumbing are 316 stainless steel. The sample flow cell is 316 stainless steel. Polymeric flow-cell windows are inappropriate for the diesel stream because of chemical incompatibility and sulfur emission absorption. Therefore a beryllium metal (25-μm-thick) flow-cell window is utilized and x-rays transmitted into and out of the flow cell are only minimally absorbed and beryllium is unaffected by the organic stream. Rhodium anode x-ray tube radiation (7-kV tube voltage) is applied to the sample in the flow cell, and characteristic sulfur radiation is emitted at 2.34 keV. The x-ray detector counts the sulfur x-ray emission over a preset analysis time, usually 5 to 20 min, followed by integration of the sulfur peak (x-ray events or counts per second) in the spectrum. Calibration of the EDXRF is accomplished by linear analysis of sulfur intensity gathered from #2 diesel standards ranging from 300- to 600-ppm sulfur. Data for a typical calibration curve in that sulfur concentration range are listed in Table 1.

Refinery data plotted in Fig. 17 illustrate the XRF-measured trend of the sulfur concentration in the fuel stream in addition to the analysis results of grab samples cocollected and submitted to a laboratory sulfur method. Sulfur accuracy for the EDXRF method is $(+/-)5\%-10\%$.

Monitoring Catalyst Depletion

A second application of EDXRF analysis is monitoring streams to optimize catalyst properties of the process. In general, this technique would apply to determining metal and nonmetal components in process streams. The example used here is the production of terephthalic acid as a precursor to polyester. The catalyst is a mixture of manganese, cobalt, and bromine in acetic acid, xylene, and

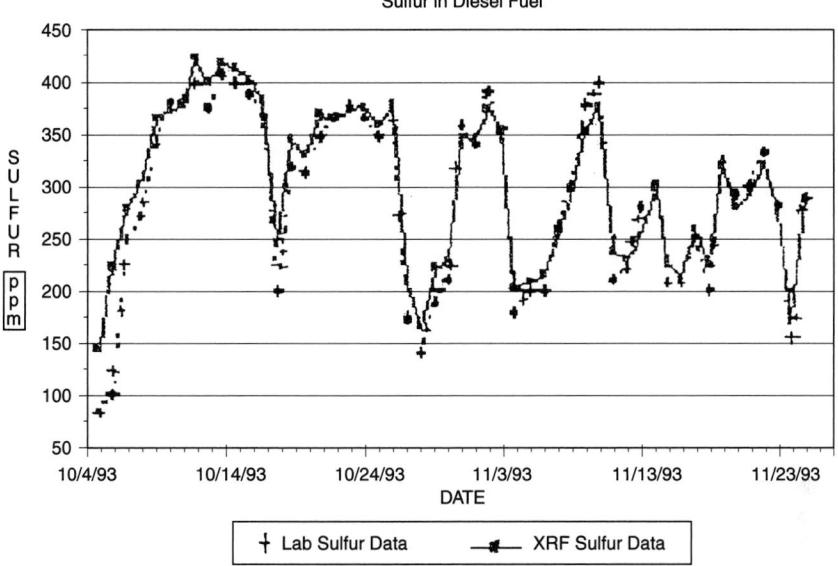

FIGURE 17   Lab sulfur versus on-line XRF sulfur method.

water. Process efficiencies are realized by monitoring catalyst depletion so that components can be recharged to the stream as necessary.

The stream consists of catalyst as well as dissolved product. Before sample introduction to the EDXRF flow cell, stream conditioning is used to reduce stream pressure, filter large particles of product, and adjust the temperature. The last two functions minimize the risk of product precipitation and buildup that could eventually block sample flow to the analyzer. Sample handling materials are 316 stainless steel tubing and flow cell that incorporate a 25-μm polyimide (Kapton, DuPont) flow-cell window. In this case the polyimide provides 35–45 days of continuous use as a flow-cell window.

Systems in use for this application consist of both high- and low-resolution detectors. In the case of high-resolution-detector-based systems, simple peak-above-background regions of interest can be set up for Mn and Co to integrate the x-ray emission peak intensity. For low-resolution-detector-based systems, peak fitting can be used to integrate Mn and Co emission lines; however, they have limited Mn and Co sensivity compared with that of high-resolution detectors. With either detector technology, Br emission is in an interference-free area of the x-ray spectrum. Calibration relies on preformulated mixtures of catalyst with Mn, Co, and Br levels varying in a range of 100–1500 ppm. Typical linear calibration data are listed in Table 2.

Listed in Table 3 are precision data obtained from a high-resolution-detector system in an on-line configuration. Precision better than 2.6% (4-min analysis time) is achieved for all analytes in the stream and accuracy is below 2.5% relative.

TABLE 2   Catalyst Component Calibration Curve Data

Component	Slope (cps/ppm)	Intercept (cps)	Correlation Coefficient
Mn	0.150	0.617	0.999
Co	0.340	0.585	0.999
Br	0.577	1.18	0.999

TABLE 3   Catalyst On-Line XRF Precision and Accuracy Performance

Parameter	Co	Mn	Br
Mean (ppm)	212	402	607
Std. Dev. (ppm)	5.33	10.56	8.20
%RSD	2.5	2.6	1.3
Formulated (ppm)	210	395	595
Relative Accuracy (%)	1.0	1.8	2.1

TABLE 4   Applications of On-Line XRF Spectrometry

Catalyst depletion—Mn, Co, and Br in acetic acid
Plating baths—Fe, Cu, Zn, Pt, and S in acids
Hydrocarbon fuel—S in diesel and gasoline
Crude oil—Ni, S and V in oil

XRF Theory

XRF analysis is a mature analytical technique with the theory of x-ray interaction with matter developed in the early 1900s. Many excellent references are available that describe details of the technique.

The versatility of on-line XRF spectrometry is indicated by the myriad of liquid-sample types effectively analyzed by the technique. Table 4 is a breakdown of areas where process XRF has found extensive use. Advantages of the XRF method compared with those of other elemental analysis techniques include the facts that XRF is nondestructive, requires minimal sample pretreatment, analysis turnaround time is fast (usually 3 to 20 min), and the interpretation of x-ray emission spectra acquired from a sample is straightforward.

PRODUCTION OF X-RAY EMISSION AND ACQUISITION OF XRF SPECTRA

Figure 18 is a schematic of the interacting modules used in combination to produce the x-ray emission spectrum of a sample. The EDXRF spectrometer components diagrammed in Fig. 18 are discussed

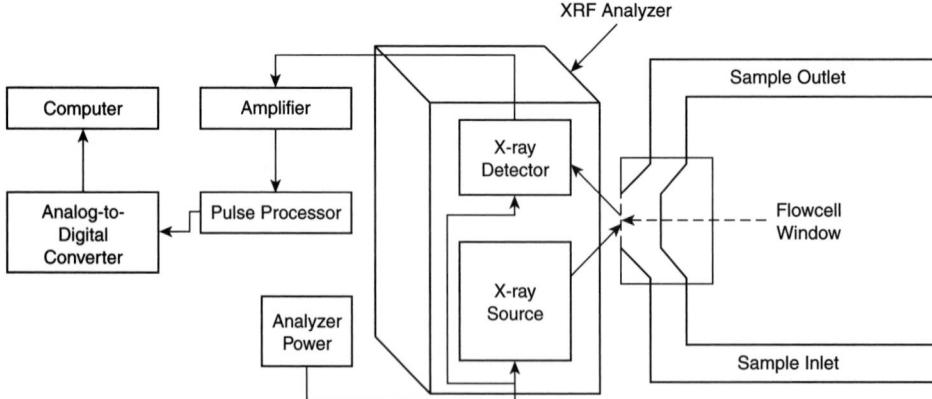

FIGURE 18   Schematic of an on-line XRF spectrometer and stream diagram.

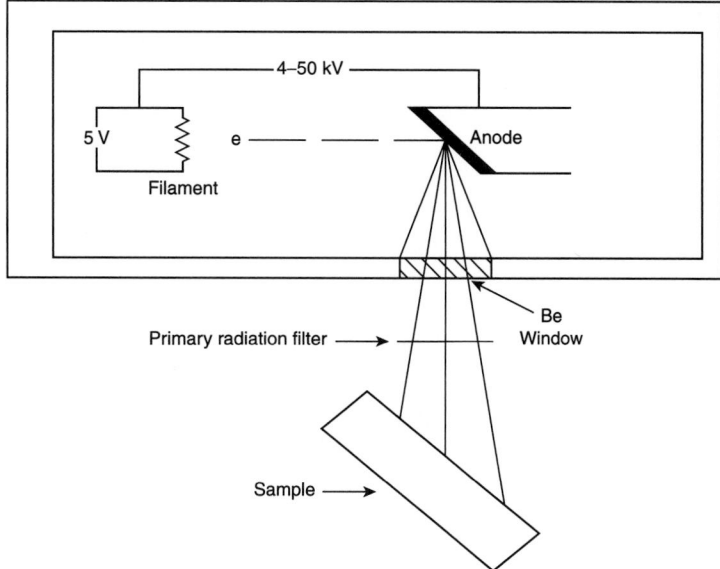

FIGURE 19   X-ray tube and primary radiation filter position.

in the following three subsections:

1. x-ray excitation: x-ray tubes that primary radiation filters or radioisotope sources
2. interaction of x-rays with matter: x-ray absorption followed by x-ray emission
3. x-ray detectors and supporting electronics: generation of the x-ray emission spectrum

X-Ray Excitation

On-line EDXRF spectrometers use one of two types of x-radiation source. They are x-ray tubes coupled with primary radiation filters or sealed radioisotope sources.

X-Ray Tube/Primary Radiation Filters.   The x-ray tube, shown in Fig. 19, is an evacuated glass envelope that comprises

1. filament with fixed or adjustable current control
2. pure metal anode that may be Rh, W, or Ag
3. beryllium window (2 to 5 mils thick) through which the x rays radiate toward the sample

The x-ray tube and high-voltage power supply combination provides a stable, high-intensity x-radiation source. The x-ray tube is electronically interlocked to cabinet doors to prevent accidental exposure to operators. The high-voltage differential between the filament and the anode can be fixed or adjustable (usually 4 to 50 kV) by computer control of the x-ray tube power supply. The highest-energy x-ray output from the tube is equal to the maximum energy imparted to an electron by the accelerating voltage.

This relationship is described by

$$E_{max} \text{ output (keV)} = \text{tube voltage (kV)} \tag{1}$$

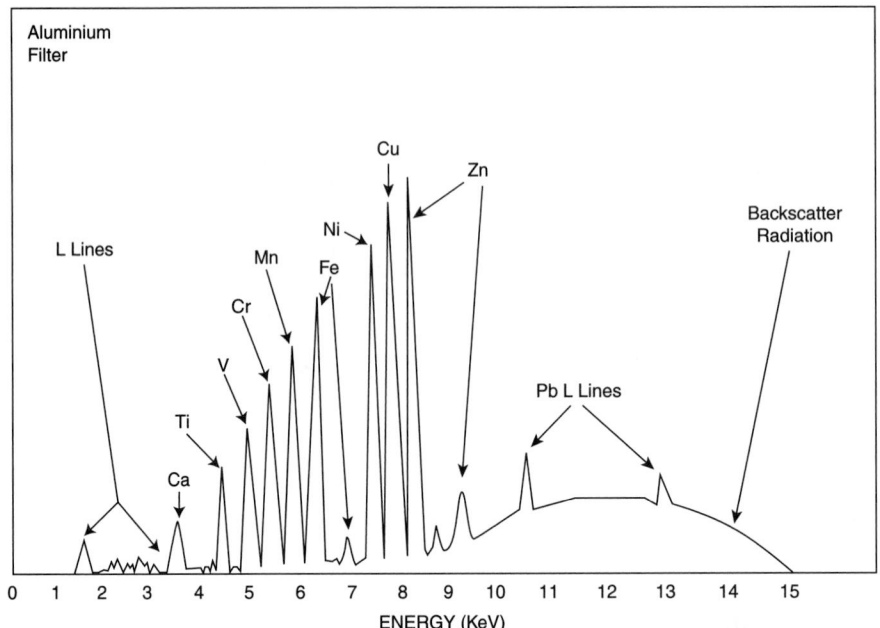

FIGURE 20   X-ray output distributions from an aluminum primary radiation filter. Oil spectrum acquired from a Conostan S-21 standard: 900 ppm of each element.

Two separate processes generate x rays in the tube. First, the electrons strike the anode and gradually decelerate, producing a continuum of x rays. Furthermore, the electron energy imparted to anode atoms may be sufficient to initiate production of characteristic x-ray emission (see the discussion of characteristic x-ray emission below in the subsection on interactor of x rays with matter). As a result of these processes, the output distribution from the x-ray tube is a superposition of intense characteristic anode x rays and continuum, sometimes called bremsstrahlung, x radiation. Direct x-ray tube excitation has high spectral flux output with small power supply requirements (50 W maximum). In practice, x-ray tube voltage is selected in combination with a primary radiation filter to optimize the excitation conditions for the elements of interest in a sample. Primary radiation filters (see Fig. 19), usually thin metal foils, modify the x-ray tube output distribution before striking the sample. Primary filters optimize the x-ray spectrum acquisition conditions in one or more ways:

1. They reduce background beneath element emission lines in a spectral region.

2. They eliminate x-ray tube anode, usually Rh, characteristic line radiation.

3. They transmit x rays of sufficient energy to excite the elements of interest.

Computer automation allows selection of multiple sets of x-ray tube voltage and primary radiation filters to achieve optimum excitation conditions for analytes in complex sample streams. Primary filters are used to optimize emission lines in a particular region of the x-ray spectrum. Figure 20 illustrates a spectrum that has reduced scatter background in the 3.5–6.5-keV energy region with the introduction of a an aluminum filter to modify Rh anode x-ray tube output. The sample analyzed is a multielement oil standard.

Sealed Radioisotopes.   Some unstable isotopes emit x rays after radioactive decay. Radioisotope x-ray sources consist of a radioactive salt sealed in a capsule that includes a window for x-rays to escape. Some on-line XRF spectrometers can be equipped with up to three radioisotope sources mounted on a computer-automated motor. The sources are rotated so that x rays are directed toward

TABLE 5   Common Radioisotopes Used in On-Line XRF Spectrometers

Radioisotope	Half-Life	Emission Line Energies	Typical Source Strength (C_i)
^{55}Fe	2.3 years	5.9 keV	0.05
^{109}Cd	1.3 years	22 keV	0.005
^{241}Am	470 years	Gamma	0.05
^{57}Co	270 days	6,122,145 keV	0.100

the sample. No primary radiation filter is used between the source and the sample, and the source, because of its small size, can be positioned very close to the sample.

The x rays produced by a radioisotope source can be one or more discrete emission lines. This is in contrast to the broad distribution of energies output from the x-ray tube. Table 5 lists the half-lives, emission line energies, and useful elemental analysis regions for each source. Radioisotope sources are smaller and lightweight but provide less output flux than X-ray tubes that require a high voltage power supply. Radioisotope equipped spectrometers can be very small and lightweight for on-line use but sensitivity is reduced compared to X-ray tube excited EDXRF spectrometers.

Interaction of X Rays with Matter

X-Ray Absorption.   EDXRF relies on interactions between x radiation from the primary x-ray source and the atoms in the sample. A prerequisite to x-ray emission is x-ray absorption. Source x rays striking a pure material with intensity I_0 can be absorbed by the material, resulting in an attenuated intensity I that passes through. This relationship is described by

$$I = I_0 \exp[-(m/r)rt] \tag{2}$$

where m/r is the mass absorption coefficient, r is the density (in grams per cubic meter) of the material, and t is the thickness (in centimeters) of the pure material.

The mass absorption coefficient, in units of cubic centimeters per gram, describes the x-ray absorption characteristics of any material. There are two characteristic features of a the mass absorption coefficient: (1) the gradual decrease in m/r with increasing x-ray energy, and (2) the sharp discontinuities corresponding to x-ray absorption. The absorption edge represents the amount of energy required for ejecting an electron from its orbital. For an atom to undergo photoelectric absorption, some fraction of the source x rays incident upon the sample must exceed the absorption edge energy. Photoelectric absorption is illustrated on the atomic scale in Fig. 21. The photoelectron (e- in Fig. 21) is ejected from the electron cloud with some kinetic energy. As shown in Fig. 21, the innermost electron orbitals are designated the K and the L shells. For a particular atom, the energy required for ejecting a K shell photoelectron ($E_{K\,abs}$) is much larger than the $E_{L\,abs}$ for the L shell because the inner K shell electron is held more tightly to the atom. Each atom in the periodic table has unique values of $E_{L\,abs}$ and $E_{K\,abs}$. Photoelectric absorption by the M shell is rarely of interest.

After photoelectric absorption, a vacancy is created in the electron orbital and the atom is in an excited, unstable energy state. To return to the ground state, the excited atom undergoes electronic rearrangement to fill the electron vacancy. The relaxation processes of excited atoms are discussed in the following subsection.

X-Ray Emission.   An electronic rearrangement that produces an x-ray photon is called a fluorescence transition. An x-ray photon emitted by such a transition is designated a characteristic x ray and its energy indicates (1) the atomic number of the emitting atom and (2) the specific fluorescence transition giving rise to the x-ray emission. Subsequent analysis of emitted characteristic x rays is the basis for energy dispersive XRF spectrometry. To date, there are two accepted notation conventions for x-ray transitions: Siegbahn and IUPAC. The IUPAC notation is the most recent and is included in this text in brackets [ ]. Extensive tabulations of known elemental emission line energies are accessed and displayed by the data analysis computer.

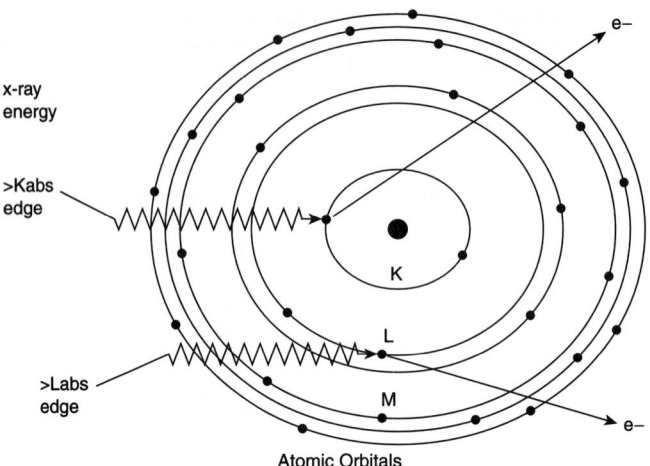

FIGURE 21   X-ray absorption and photoelectron ejection.

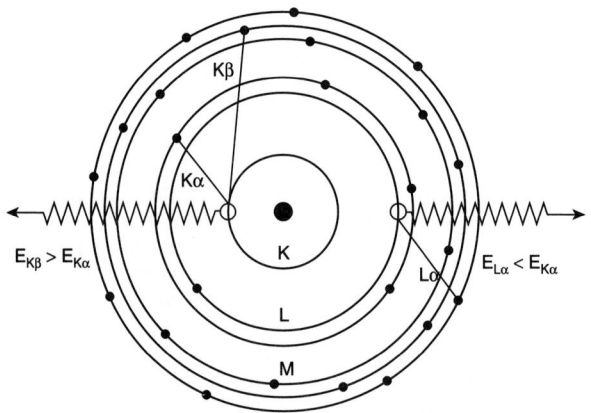

FIGURE 22   XRF transitions.

Consider the electronic transitions shown in Fig. 22. If an electron vacancy occurs in the K shell, either the Ka [K–L$_3$] (from the L shell) transition or the Kb [K–M$_{2,3,5}$] (from the M shell) transition may follow. The Ka [K–L$_3$] transition is approximately 10 times more probable than the Kb [K–M$_{2,3,5}$] transition, and the K emission of an element in the EDXRF spectrum can be identified by a pair of emission lines at an intensity ratio of 10:1. Note the Cu Ka [K–L$_3$] at 8.041 keV Cu Kb [K–M$_{2,3,5}$] at 8.903 keV peaks in Fig. 23, which is the XRF spectrum of a brass sample. The y axis is 20,000 counts full scale.

When an L shell vacancy occurs, electrons from the M or the N shell can fill the vacancy. For an individual atom, photons emitted from L transitions are always less energetic than those produced from K transitions. The pattern of the L emission lines from an element is an intense pair of peaks (Pb La [L$_3$–M$_{4,5}$] at 10.550 keV, Pb Lb [L$_3$–N$_5$] at 12.612 keV) accompanied by other minor emission lines (Fig. 23). Multiple L emission lines arise because there are three separate L shell electron energy levels that can be excited, namely the L(I), L(II) and L(III) absorption edges.

The L emission lines are useful for determination of those elements with atomic number 47 (Ag) and above in the periodic table. Since the x-ray tube maximum output energy is limited to 50 keV

TABLE 6   EDXRF Detectors Used in On-Line Systems

Detector	Energy Range (keV)	Resolution (eV)	Issues
Electrically cooled Si(Li)	1.3–34	160	Heat generation, relatively large size
Mercuric iodide (HgI₂)	2.6–80	270	Small, near room-temperature operation
Si (PIN) diode	1.3–20	200	Small, no cooling, low efficiency at high energy
Gas proportional	1.3–15	500	Small, no cooling, low resolution

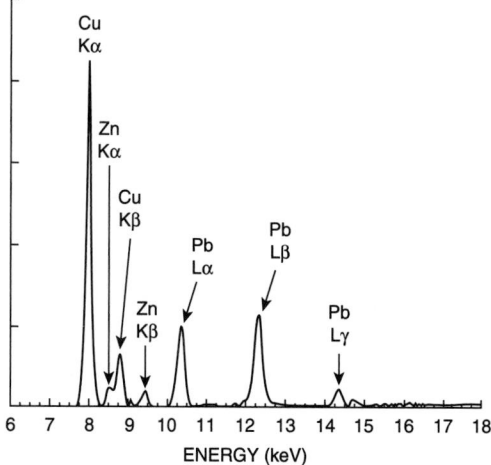

FIGURE 23   X-ray spectrum of a brass sample.

and the extent of the useful spectral range is 0–40 keV, it follows that L line emission must be used for determination of those elements above atomic number 60 (Nd).

The x-ray detector (shown in the block diagram, Fig. 18) in the EDXRF spectrometer analyzes characteristic x-ray photons emitted by fluorescence transitions to determine the x-ray spectrum of a sample. Quantitative analysis by EDXRF relies on the intensity of either L or K emission lines to represent elemental concentrations.

X-Ray Detectors and Supporting Electronics

Two types of x-ray detectors are commonly found in on-line EDXRF systems. The first are solid-state detectors and the second are gas-filled proportional detectors. Table 6 lists four detector systems, the resolution expected and common issues for the incorporating the detector into an on-line system.

Solid-State Detectors.   The solid-state x-ray detector, shown schematically in Fig. 24, can be either a lithium-drifted silicon [Si(Li)] crystal, a HgI₂ crystal, or a Si (PIN) diode. The HgI₂ crystal and the Si (PIN) diode are operated near room temperature, in contrast to the Si(Li) crystal that must be cooled to at least −90°C for operation by use of thermoelectric cooling with the Peltier effect.

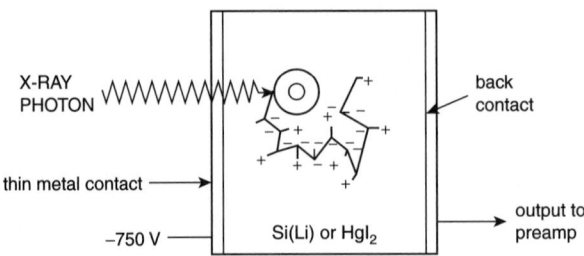

FIGURE 24   Schematic of solid-state on-line XRF detectors.

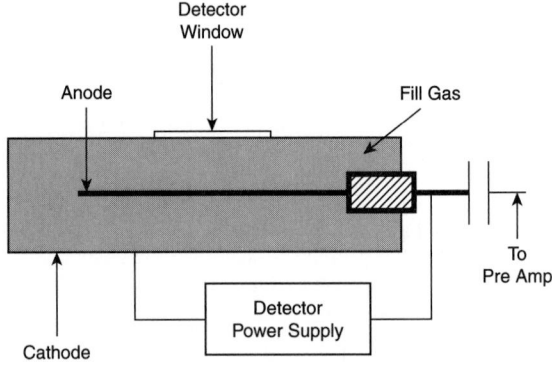

FIGURE 25   Sealed GPD.

The term resolution defines the ability of an x-ray detector to separate x-ray photons very near one another in energy. A smaller full width at half maximum (FWHM) represents improved detector resolution.

The solid-state detector [Si(Li), HgI_2 and Si (PIN) diode] converts x-ray photon energy into a measurable electronic signal. Its operation is as follows: When an x-ray photon enters the crystal, it ionizes atoms in the active area and produces electron ($-$) and hole ($+$) pairs. The number of electron/hole pairs formed is directly proportional to the energy of the x-ray photon that entered the detector. The basis for energy discrimination of x-ray photons emitted from a sample is the proportionality of the detector output signal to the incident x-ray photon energy.

Gas-Proportional Detectors.   Gas-proportional detectors (GPDs) used for on-line EDXRF are filled with a mixture of inert gases. The fill gas depends on the expected energy range required for the EDXRF system. Figure 25 is a schematic of a typical detector system. Internal amplification assists in providing a detector output proportional to the incoming x-ray photon. For simple process streams containing only one or a few x-ray analytes, gas-filled proportional detectors are a good choice. For complex sample streams with many elements or analytes emitting out of the efficient energy range of the GPDs, solid-state, high-resolution detectors are better options.

Regardless of the detector type, detector output signals must first be amplified and then sorted with respect to their magnitude by a pulse-height analyzer (PHA). Components of the PHA, shown in Fig. 18, include an amplifier as well as an analog-to-digital converter. The output from the PHA is the EDXRF emission spectrum and is a plot of the number of x-ray events detected, or counts, on the y axis versus pulse height, or x-ray energy, on the x axis. The EDXRF spectrum shown in Fig. 23 is transmitted to a computer for display and further processing.

The acquisition time selected for collection of the EDXRF emission spectrum depends largely on the goals of a particular analysis method. Precision of an analysis can be improved by acquiring the spectrum for a longer duration, thereby reducing the effect of random counting errors on the overall analytical error. Typical EDXRF spectrum acquisition times can range from a few seconds to 30 min, depending on the application.

On-Line Configurations and Enclosures. Enclosures for on-line XRF installations must address a number of issues. This subsection describes those considerations for an effective commissioning of XRF equipment to the process stream. Virtually all on-line XRF installations are in analyzer shelters or instrumentation areas for maintenance, protection, and sample handling convenience reasons. Modern computer data stations controlling the XRF analyzer are equipped with a modem and communications software for remote diagnostic capabilities. Analyzer shelters are now being equipped with telecommunications capabilities for remote polling of the equipment.

Purged enclosures are effective protection for the XRF analyzer from hazardous environments. Most manufacturers offer explosion-proof X-purged cabinets. Sealed cabinets provide the x-ray modules and support electronics consistent temperatures. Narrow temperature fluctuations assist in the long-term stability of the components. On-line XRF enclosures provide leak detection systems that send external valves into a bypass mode to protect the electronics.

To access multiple sample streams, the XRF analyzer is mounted on a slide that accurately indexes to each flow-cell position. Slides can be very simple for lightweight configurations such as radioisotope–gas-proportional counter systems. Heavier x-ray optic modules require heavy-duty analyzer movement mechanisms. Additionally, when the XRF analyzer can index multiple positions, control samples and calibration standards can be indexed and analyzed automatically.

Flow Cells and Sample Conditioning Issues. Two important aspects of the successful on-line XRF installation are selection of flow-cell materials and sample conditioning. Flow-cell bodies and associated plumbing must be selected with regard to the sample stream temperature and chemical composition. Figure 26 is a schematic diagram of an on-line XRF flow cell. Materials selected most

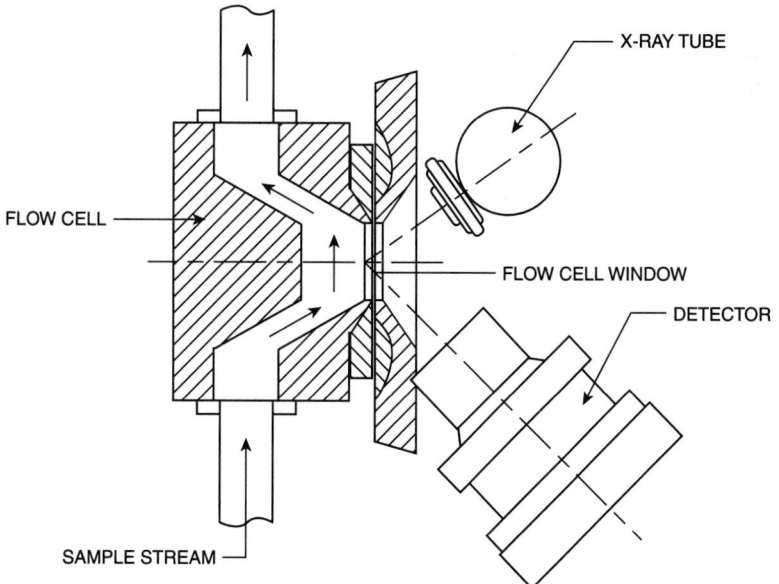

FIGURE 26 Schematic of a XRF on-line analyzer flow cell. Tube and detector, shown for reference, are usually mounted in the horizontal plane.

TABLE 7   Commonly Used On-Line XRF Flow-Cell Windows

Polymer Material	Useful Thickness (μm)	On-Line XRF Compatibility
Polyimide (Kapton)	7.5–50	Good for acids, poor for caustic
Mylar	6	Good for hydrocarbons
Teflon	12–25	Good chemical resistance, poor x-ray transmission
Polypropylene	6–12	Good for most organics, limited strength

often are stainless steel and Teflon. A second important aspect of the flow cell is the choice of window material.

Editor's Note: Other configurations should be considered, depending on the nature of the sample. Mounting the source, detector, and preamplifier above a trough containing a flowing sample may be necessary if solids or gases affect the performance of the vertical flow cell. This configuration also better protects the analyzer hardware. A window material can be placed between the analyzer and the sample if necessary. A blank solution can be periodically introduced to check zero, and a shutter containing the analyte components can be inserted to check span.

The flow-cell window is a critical component because it transmits x rays into and out of the sample and protects the x-ray optics. Window materials are selected based on three criteria:

- chemical and temperature compatibility with the sample stream
- thickness and composition as it has an impact on pressure rating and x-ray absorption
- impurities in the window film and interference with analyte emission measurements

Historically, two basic types of window materials have been used: plastic and beryllium metal. Plastic windows offer the wide range of polymer materials available in varying thickness (6–25 μm). Polymer window films are usually free of impurities that can interfere with the x-ray analysis. Table 7 lists many of the polymer windows selected for on-line XRF equipment. Plastic windows must be replaced periodically (monthly frequencies) and introduce some scheduled maintenance to the x-ray analyzer. The pressure ratings and burst strength of plastic windows are quite low (<30 psi in many cases) and limit the sample stream pressure. Thickness can be increased when measuring higher concentration, and an aluminum screen on the outside of the window will add strength if it does not interfere with the target analyte.

Applications of on-line XRF where the sample stream must be at elevated pressures or temperatures, or, where superior low-energy x-ray transmission is required, require beryllium metal foil windows. Beryllium windows are more expensive than plastic windows but they are rarely replaced and thereby avoid scheduled window maintenance. Disadvantages compared with plastic windows are the toxicity of Be, compatibility of sample streams that can be in contact with the metal, and the impurities in even high-purity Be. Since the window is positioned very near the x-ray focal point, even traces of impurities can cause significant background in the XRF spectrum. For example, high-purity Be can be contaminated with up to 300-ppm iron. This background makes low-level determinations of Fe and Co in a stream more difficult. After evaluation and selection of the proper stream materials and flow-cell window, the sample must be introduced in a controlled manner. Sample conditioning plays a critical role in the success of the on-line XRF installation. In applications that use plastic windows, pressure and temperature reductions are common. Most instrument manufacturers request only small backpressure on the flow-cell window (<10 psi). For Be window applications, pressures up to 100 psi can be used and elevated temperatures (>100°C) can be accommodated. Sample conditioning can also reduce solids in the stream by filtering, maintain an appropriate sample temperature, and/or keep slurry samples adequately mixed. Sample conditioners need to provide reliable flow to the on-line XRF system in the range from 200 to 1000 ml/min. In many cases, the sample conditioning system can return the sample back to the process stream.

CONCLUSIONS

On-line XRF has proven advantages for process stream applications. Specific streams that have been addressed can be generalized to other aqueous or organic streams for the non-destructive, non-invasive EDXRF technique. On-line XRF can determine the elements between aluminum (atomic number 13) and uranium (atomic number 92) and multi-element composition information can be gathered simultaneously. Successful implementation of on-line XRF requires thoughtful selection of an X-ray source and detector appropriate for the complexity and concentration levels of components in the sample stream. These modules must be packaged to protect them from environmental factors such as hazardous vapors and temperature extremes. Multiple sample streams can be addressed if the XRF analyzer is mounted on a movement mechanism to index multiple flowcell positions. Sample stream conditioner design is also critical to the overall success of the installation. Proper selection of final stream temperature and pressure must be considered for reliability and provision should be made to return the sample stream to the process. Lastly, the selection for plumbing and flowcell window materials hinges on their compatibility to the chemical and physical nature of the sample stream. The flowcell window material must also be selected with its X-ray transmission properties in mind with usual choices being polymeric films or beryllium metal windows. After appropriate attention to these issues, the on-line XRF installation can provide many years of low maintenance, high-precision and accuracy compositional analysis.

SONIC AND MICROWAVE ANALYZERS (ULTRA, INFRARED, RESONANCE)
by Joseph Zente & Gregory Gervasio

Advances in sonic measurement technology have provided a nonintrusive type of analyzer that may have a real impact on dirty and hazardous materials monitoring. Long-wavelength radiation can penetrate metal and plastic materials so that the source and the detector can be located externally to the process piping. One company, called Phase Dynamics, offers a free video that explains how the technique works. Absorption of radiation is one way to measure composition, but utilizing phase shifting appears to be another mechanism for monitoring changes in composition.

Microwave absorption has been around a long time, but has been slow in developing as an analytical technique. Monsanto used to sell a microwave analyzer for water in corrosive liquids like acetic acid. The problem was that the system was extremely temperature dependent. A frequency of ~16.8 GHz was selected for water (Fig. 27). This was never commercialized so the analyzer was not a big success. One of the major advantages of this technique was that the inert metal cell containing process material could utilize thick Teflon windows. The method also was unaffected by black or dark-colored materials. The water resonance frequency of 22 GHz provides a temperature-independent absorption with respect to water concentration above ~30°C (Fig. 28). There are several companies today that offer microwave analyzers.

Microwave Spectroscopy

Electromagnetic Wave Interactions with Liquids and Solids

Energy Storage and Velocity Reduction.   When an electromagnetic wave passes through a mixture, it induces an alternating polarization within the material. One can think of this induction as the winding and unwinding of the polar elements of the mixture like small individual springs. The winding action causes the spring to store some of the wave's energy. Later the spring unwinds to return the energy to the wave. This process has the effect of reducing the wave's velocity. At optical frequencies, the refractive index describes this property of the mixture to temporarily store energy and to slow down

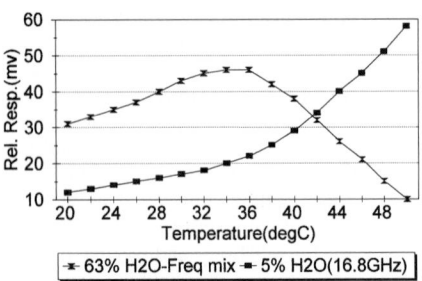

FIGURE 27   Microwave response versus temperature.

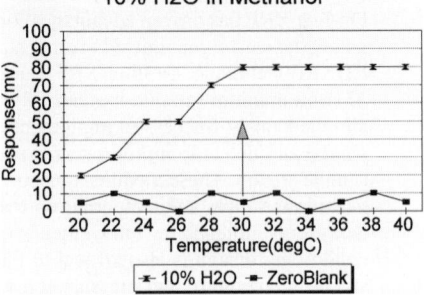

FIGURE 28   Microwave response to water at 22 GHz versus temperature.

the wave:

$$n = V_{vac}/V_{max} \tag{1}$$

where n is the refractive index, V_{vac} is the velocity through a vacuum, which is the speed of light, and V_{mix} is the velocity through the mixture.

At lower frequencies, as in the microwave and radio bands, the dielectric constant that is symbolized by ε' (epsilon prime or sometimes epsilon) describes the velocity reduction effect:

$$\varepsilon' = (V_{vac})^2/(V_{max})^2 \tag{2}$$

Refractive index n and dielectric constant ε' are thus relative ratio measurements like specific gravity. However, instead of water or air being used as a reference, these velocity measurements are given relative to the speed of light in a vacuum.

Energy Loss and Amplitude Reduction.   As the wave winds up the polar element springs, friction causes some of the wave's energy to be lost. This energy is not recovered by the wave since it is converted into heat. As a result, the wave's amplitude progressively declines as it passes through the mixture. At microwave frequencies, the loss factor ε'' (epsilon double prime) describes the capability of the mixture to attenuate the wave.

Energy Storage and Loss for Varying Frequencies.   At low frequencies, the electromagnetic wave slowly winds up the polar elements of the mixture and lets them unwind. This gives the wave enough time to rotate the polar elements completely. However, as frequency increases, the wave attempts to wind and unwind the polar elements at greater speeds. As a result, the wave begins to loose its grip on the larger polar elements because of their higher moments of intertia and spring rates. Thus, as frequency increases, the energy storage and loss effects fall off. Also, the sizes of polar elements measured by the wave can be selected by choosing the appropriate frequency band.

Polar Element Sizes.   Figure 29 shows the different sizes of the polar elements, ranging from the smallest at the electron level to the largest at the grain level.

At the electron level, the electromagnetic wave distorts the electron orbit.

At the ionic level, the wave causes vibration of the bonds that link the atoms to form a molecule. These resonances have energy loss peaks that are used in infrared (IR) and near-infrared (NIR) spectroscopy.

At the orientation level, the wave rotates polar molecules into and out of alignment at the wave's frequency. (Polar molecules have a pair of equal and opposite electric charges separated by a small

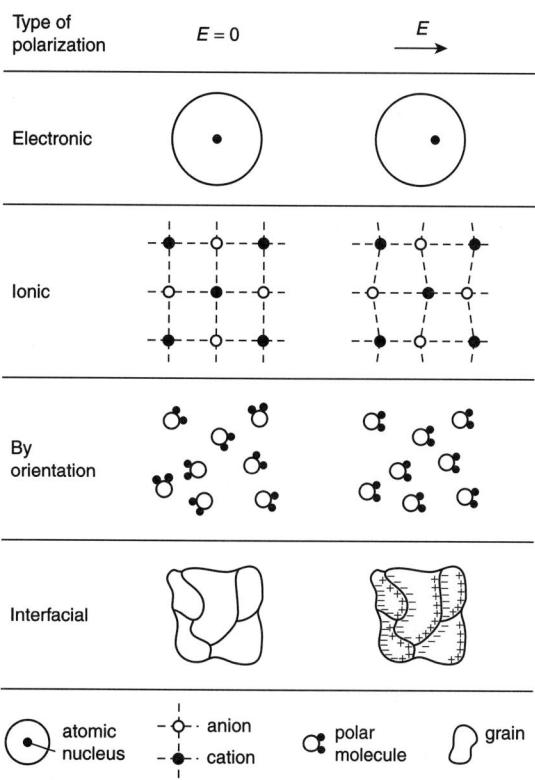

Type of polarization	$E = 0$	$E \longrightarrow$
Electronic		
Ionic		
By orientation		
Interfacial		

atomic nucleus   anion   cation   polar molecule   grain

FIGURE 29   Different sizes of polar elements.

distance). Examples of highly polar molecules are water and sulfuric acid. Examples of semipolar molecules are the alcohols, citric acid, and proteins when they are in a liquid state. Nonpolar molecules, such as oil, fat, sugar, starch, cellulose, and all solids are not rotated. At the interfacial level, the wave polarizes individual water droplets that are held in suspension in an emulsion. This is done by moving the dissolved ions to the water droplet boundaries.

Velocity Reduction versus Frequency.   Because of these differences in polar element size and moment of inertia, the velocity reduction effect (ε') falls off at higher frequencies. Figure 30 shows this effect for an emulsion that contains polar molecules. Notice that rf frequencies see all polar elements, whereas microwave frequencies respond to all elements except interfacial, and NIR and optic frequencies respond only to ionic and electronic effects. (The interfacial and molecular effects are much greater than the ionic and electronic effects because the amount of energy that is stored is based on the distance between the charges. This causes them to have a much larger dipole moment.)

Energy Loss and Amplitude Reduction versus Frequency.   Figure 30 illustrates how the amplitude reduction effect (ε'') varies with frequencies for this same mixture. At 0 frequency, the loss factor is called conductivity and symbolized by σ (sigma). In the rf region, the loss factor declines rapidly with frequency because of the large moment of inertia for ion displacement. The small peak at the interfacial resonance frequency occurs at the midpoint in the ε' interfacial relaxation. In the lower portion of the microwave band, the loss factor increases with frequency caused by the added friction of rotating the polar molecules at a higher frequency. As a result, this is called the dielectric loss factor.

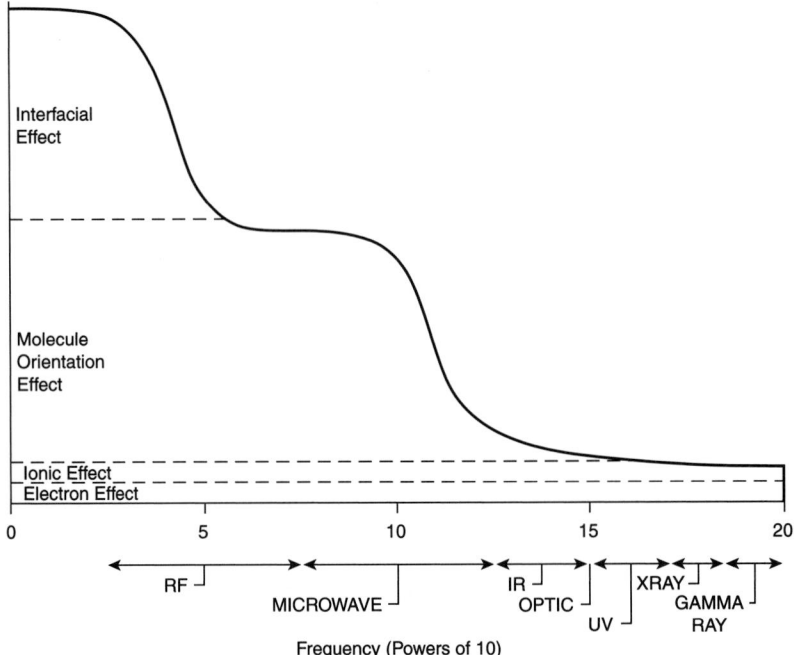

FIGURE 30   Amplitude reduction effect (ε'' versus frequency).

In the higher portion of the microwave band, the loss factor decreases along with the ε' velocity-slow-down effect as the wave looses its grip on the polar molecule, which reduces the friction. The peak at the molecular resonance frequency occurs at the midpoint in the ε' molecular relaxation. The small peaks in the IR region are caused by ionic bond resonance.

Frequency Region Differences.   As shown in Fig. 30, the choice of frequency region selects the size of the mixture's polar elements for interrogation.

Radio Frequencies.   Capacitance probes and other sensors such as Axiomatic's Di Comp sensor and Hewlett-Packard's Colloid Dielectric Probe use the radio frequencies. Because of the high interfacial effect at these frequencies, this region is good for measuring emulsion quality for mixtures with medium to high water content. For moisture and other concentration measurements, however, these frequencies are best used for mixtures with low water content that do not have an interfacial effect.

Microwave Frequencies.   Epsilon's guided microwave spectrometer as well as other instruments from companies such as Berthold, KDC, Kay Ray, Phase Dynamics, and Hewlett-Packard use microwave frequencies. These frequencies are popular for making composition measurements based on the large differences in the molecule rotation effect.

Some of these instruments, including the guided microwave spectrometer, use the lower portion of the microwave band. (The guided microwave spectrometer bandwidth from 200 to 3200 MHz covers the upper portion of VHF section of the rf band and the P, L, and S bands of the microwave spectrum.) For loss factor measurements, this region has the advantage of not only containing information about the low-frequency loss factor (for conductivity and salinity determination), but also information about the high-frequency loss factor (for polar, semipolar, and nonpolar molecule discrimination and for reduction of packing density and pressure effects).

Microwave instruments that only use the mid-to-upper microwave band, (e.g., the X Band around 10 GHz and the K Band around 20 GHz) such as those from Kay Ray, have the disadvantage of containing information about only the high-frequency loss factor. On the other hand, without the loss from salinity, these frequencies can transmit through longer path lengths on conductive fluids.

The mid-to-upper microwave bands have wavelengths of 1 in. (25 mm) and less. Coating has an effect based on its thickness as a percentage of a wavelength. As a result, coating has more of an effect at these frequencies than for the lower microwave and the rf frequencies. In addition, for 1-in. (25-mm) and less wavelengths, scattering occurs from entrained particles 0.1 in. (2.5 mm) or bigger in diameter, which introduces a variety of error sources.

NIR and Optic Frequencies.   The NIR region sees higher-order harmonics of the molecular bond resonances that occur in the mid-IR region. This has the disadvantage of providing a complex array of combination and difference frequencies that make calibration difficult. However, because the loss factor is lower than that of the primary mid-IR resonance, it allows 4- to 25-mm path lengths. The wavelength in the NIR and optic regions is from 0.001 to 0.00001 cm. Since coating effect is proportional to coating thickness as a percentage of wavelength, very small coating thicknesses have a large effect. Changes in material color can also have an effect for some instruments.

Since the wave scattering begins when the diameter of the particle/discontinuity is greater than 10% of the wavelength, scattering is a factor for even very small particles and fine emulsion disconti-nuities. When such scattering takes place, the mode of propogation is called diffused transmission or diffused reflection. This is useful for particle size and emulsion quality measurements. However, for composition measurements, scattering can cause major errors. When particle size or shape changes, it shifts the calibration. When flow rate changes, it changes the scattering pattern because of Doppler frequency shifts to add another error variable. And, when the packing pressure is changed on a compressible slurry, it changes the scattering pattern to add another source of error.

The other alternative for nonclear streams is to use reflection-mode measurements that shorten the path length. This allows sufficient signal reception, but limits the measurement to the mixture's surface, which increases the potential for sampling error. While this reduces the particle size, flow, and pressure effects that are due to scattering since the path length is shorter, they are still a major error variable.

Instrument Bandwidth Differences

Single-Frequency Instruments.   Instruments in this category include capacitance probes, refrac-tometers, and Kay Ray microwave. Such instruments have a lower-cost advantage. Their use is often limited to binary mixtures because of the chances for greater error on more complex mixtures. In addition, these methods have less ability to compensate for other physical effects. For example, an rf/capacitance probe does not have σ and τ information for packing density compensation on granu-lar solids applications or pressure compensation on slurry applications. Meanwhile, a refractometer readout of sugar content in a fruit juice stream has errors when the acid content changes because of the nonbinary nature of the mixture. For the Kay Ray type of amplitude measurement at a single microwave frequency, the instrument cannot tell whether a difference in amplitude is caused by a change in (ε') or (ε'').

Narrow-Bandwidth Instruments.   Instruments in this category include Axiomatics measure of ca-pacitance and conductance in the HF portion of the rf band, phase dynamics measure of oscillator frequency shift and amplitude attenuation in the S band, and KDC measure of harmonic frequency shift and amplitude attenuation in a narrow band anywhere between 0.5 and 10 GHz. With these mea-surements, ε' and ε'' at the narrow frequency of operation provides two measurement variables. Thus, for rf region operation, σ and ε' are measured. For X-Band operation, the dielectric loss factor and (ε') are measured. For S-band operation, (ε'') will change because of either loss factor but the instrument will not be able to distinguish the source of the effect. With this capability, these instruments can measure either one more constituent or one more error variable than a single-frequency instrument, if the correct frequency of operation is selected.

Wide-Bandwidth Instruments. Instruments in the wide-bandwidth category include Epsilon's guided microwave spectrometer, Berthold and Hewlett-Packard microwaves, and NIR. With such a wide spectrum, the guided microwave spectrometer's dual-line or curve-fit software provides three or five electromagnetic variables. When these variables are used, four groups of a mixture's constituents can be measured:

- Polar constituents: liquid water; certain acids such as sulfuric, hydrochloric, and nitric; hydrogen peroxide
- Semipolar constituents: liquid alcohols, certain liquid acids such as amino and citric, and semi-conductive mineral compounds such as metal sulfates
- Nonpolar constituents: oil, fat, sugar, starch, cellulose, fibers, minerals, solids
- Dissolved salts

On mixtures with fewer components, Epsilon's guided microwave spectrometer multiple measured variables compensate for other physical effects. Although not discussed in their literature, Berthold's measure of phase and amplitude response at $\sim$20 frequencies over a wide bandwidth should be able to provide similar capability. Hewlett-Packard laboratory analyzers that can measure phase and amplitude over the full microwave bandwidth have a similar capability.

NIR has a wide variety of harmonic resonance frequencies to choose from. As a result, NIR can measure more constituents than any wideband microwave technology. For example, $\sim$58 constituents are reported to have been measured by NIR instruments to determine the octane number in gasoline.

NIR measurement variables, however, have less orthogonality or independence than the microwave variables. For example, NIR uses the response at the hydrogen bond frequency to measure water content. But salinity changes also effect the total number of hydrogen bonds so that the water and salt concentration effects cannot be separated. With microwave measurements, the salt content effect on σ is clearly separated from the water effect on ε' and τ. In addition, process temperature has an effect on water's hydrogen bonding so that temperature compensation becomes more difficult with NIR than with microwave measurement.

As a result of this orthogonality advantage, microwave wideband technologies measure water content with greater specificity.

Guided Microwave Spectrometry

Guided microwave spectrometry (GMS) is similar to all other electromagnetic techniques in that all of these sensors respond to the wave velocity reduction effect ε' and/or the wave amplitude reduction effect ε''. Such measurements rely on the process materials behavior at the macro or molecular level as opposed to the micro or atomic level. In addition, all of these techniques require calibration against laboratory test results in order to extract the desired quantitative measurements from the process mixture's ε' and/or ε'' properties.

Other microwave approaches assume a free-space environment. As a result, multiple reflection paths and other geometric effects become potential sources of error. The GMS approach is the first microwave instrument to use a waveguide environment with fixed, optimum geometry. At lower frequencies in a waveguide, only the simplest and longest wave still survives at the cutoff region, as shown in Fig. 31. This wave, which is in a stationary oscillation called direct coupling, is larger than the waveguide itself so that it cannot be scattered. Thus it follows a single path from the send antenna through the entire process mixture to the receive antenna.

As a result, the GMS waveguide geometry provides greater precision and repeatability than other microwave techniques wherever particles or discontinuities are present. Also, the single path between GMS antennas eliminates two other sources of error. First, it eliminates the interference from upstream and downstream reflections that may be caused by the changing position of valves, other process equipment, and plant personnel in applications with low signal loss. Second, it eliminates the destructive interference that occurs when multiple out-of-phase signals reach the receive port at the same time to cause signal loss and dropouts that act like resonances.

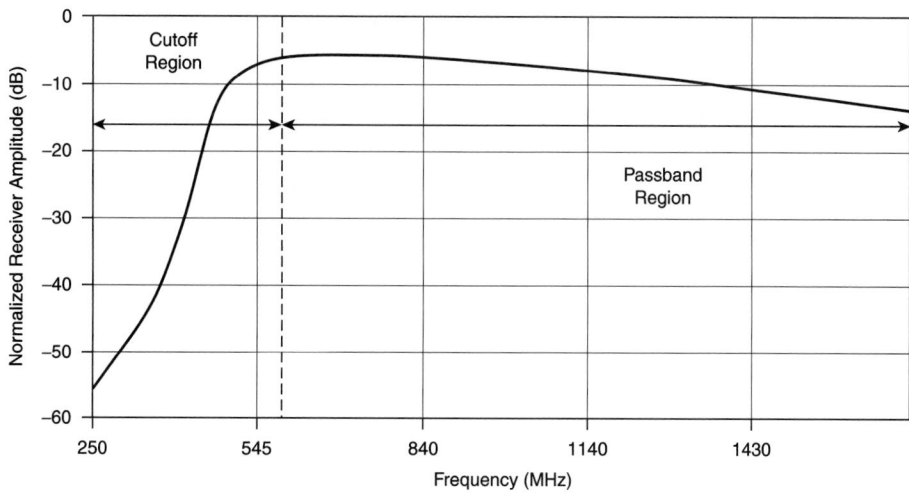

FIGURE 31   GMS direct coupling waveform at cutoff.

GMS spectrometry requires only amplitude measurement to arrive at an accurate determination of the velocity reduction effect ε'. In the past, this velocity measurement (which is sometimes called time-of-flight measurement) has been best achieved in the microwave region by measurement of phase shift. However, phase shift measurement has several disadvantages. For in-line applications, changes in the mixture's flow rate cause Doppler frequency shifts when particles or emulsion discontinuities are present, and this affects the phase measurement. In addition, the phase changes when the ambient temperature around the coax cables linking the sensor to the electronics changes and when the bending of these cables changes.

Epsilon's guided microwave spectrometer transmits microwaves through a mixture at different frequencies, measures the amplitude of the received signal at each frequency, and normalizes this amplitude through the mixture to the amplitude received across a fixed calibration path. With this normalized amplitude versus frequency spectrum, the mixture's ability to cause energy loss ε'' is measured at frequencies ranging from 200 to 3200 MHz.

Figure 32 shows a GMS sensor body that includes a waveguide, transmit and receive antennas, and end pieces to adapt to a given process. The GMS sensor has two, parallel metal surfaces between the send and the receive antennas.

These two metal surfaces form a waveguide that steers the wave front toward the receive antenna. Such a waveguide has the characteristic of transmitting only waves that have a wavelength shorter than twice the distance between the two parallel surfaces, called the a dimension. Larger waves will not fit into the waveguide and are cut off.

Two factors affect wavelength. First, wavelength has an inverse relationship with frequency. As the frequency decreases, the wavelength increases. Second, wavelength has a direct relationship with velocity. As the wave's velocity decreases (by going through a mixture and having its energy temporarily stored by means of the polarization windup effects), its wavelength also decreases. In a waveguide, this has the effect of lowering the cutoff frequency below the cutoff for air.

The general equation for the two wavelength effects is

$$\text{frequency} = \text{velocity}/\text{wavelength} \tag{3}$$

The mixture's velocity reduction effect is described by the equation:

$$\text{velocity} = c/\sqrt{\varepsilon'} \tag{4}$$

where c is the the speed of light (2.998×10^8 m/s) in a vacuum and ε' is the mixture's dielectric constant. Substituting velocity reduction into the equation for wavelength yields

$$\text{frequency} = c/\text{wavelength}(\sqrt{\varepsilon'}) \tag{5}$$

Because the wavelength at cutoff is twice the distance between the waveguide's two parallel surfaces, the frequency at cutoff (fc) is

$$fc = c/2a(\sqrt{\varepsilon'}) \tag{6}$$

where a is the distance separating the waveguide's parallel surfaces and $2a$ is the cutoff wavelength

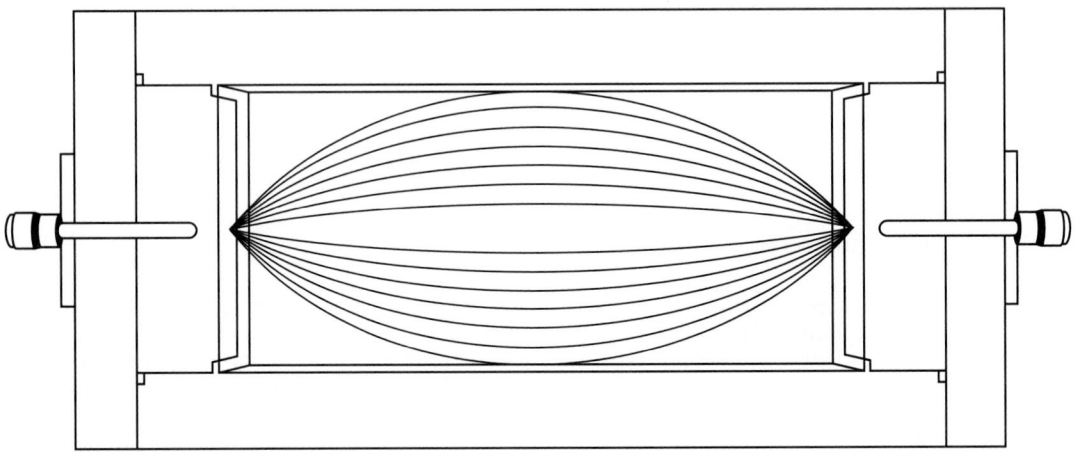

FIGURE 32 Waveguide and antennas.

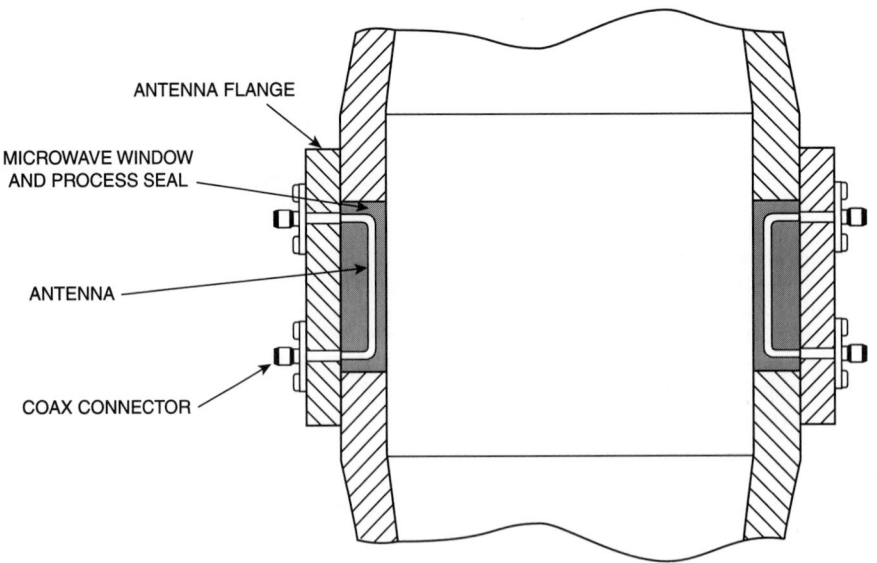

ANTENNA FLANGE

MICROWAVE WINDOW
AND PROCESS SEAL

ANTENNA

COAX CONNECTOR

FIGURE 33 Standard single-antenna assembly.

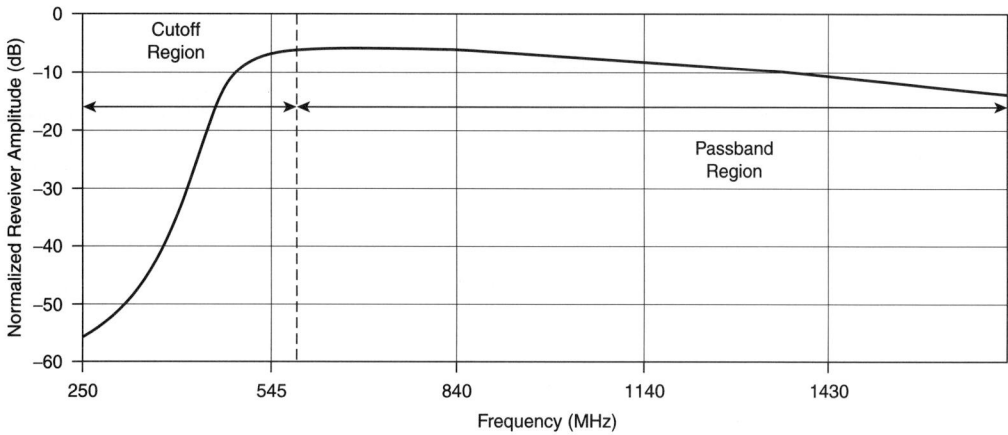

FIGURE 34   Typical GMS spectrum.

Thus, by measuring the cutoff frequency, GMS measures the mixture's ability to store energy and slow down the wave (ε'). By going to different line sizes, the GMS sensor's a dimension varies the ε' to the cutoff frequency relationship.

Interpretation of GMS Spectrums

Figure 34 shows a typical GMS spectrum. Note the location of the cutoff region and the passband region.

Changing [ε'] Effect.   Figure 35 illustrates how the GMS spectrum changes when ε'' holds constant and ε' increases. This would be typical of a case in which water concentration increases in a grain mixture. As shown by the dotted spectrum, such a water increase reduces the cutoff frequency.

Changing Conductivity Effect.   Figure 36 illustrates how the GMS spectrum changes when ε' holds constant while the low-frequency portion of the ε'' response, called conductivity σ, increases. This would be the case with an aqueous mixture that has an increasing salt content. Note that this has two effects on the spectrum: the slope in the rise region decreases and the amplitude throughout the passband region decreases.

Changing the Molecular Relaxation Time Effect.   Figure 37 illustrates how the GMS spectrum changes when ε' and σ hold constant while the high-frequency portion of the ε'' response, called the dielectric loss factor, increases. This has the effect of increasing the slope of rolloff in the passband.
   The relationship between the dielectric loss factor and molecular relaxation time τ depends on whether the passband is below or above the molecular relaxation frequency. For middle to high water applications, the passband falls below this frequency so that the loss factor still increases with frequency. For this case, an increase in molecular relaxation time causes an increase in dielectric loss factor and an increased slope of passband rolloff. This would be the case in which water concentration increases in a mixture with another polar liquid such as an acid that has the same dielectric constant but a lower molecular relaxation time. To a greater extent, the slope of rolloff will increase when water concentration increases in a mixture with a nonpolar constituent that has negligible molecular relaxation time. Thus this slope-of-rolloff loss factor provides another view of the polar, semipolar, and nonpolar concentrations, independent of the ε' energy storage effect.

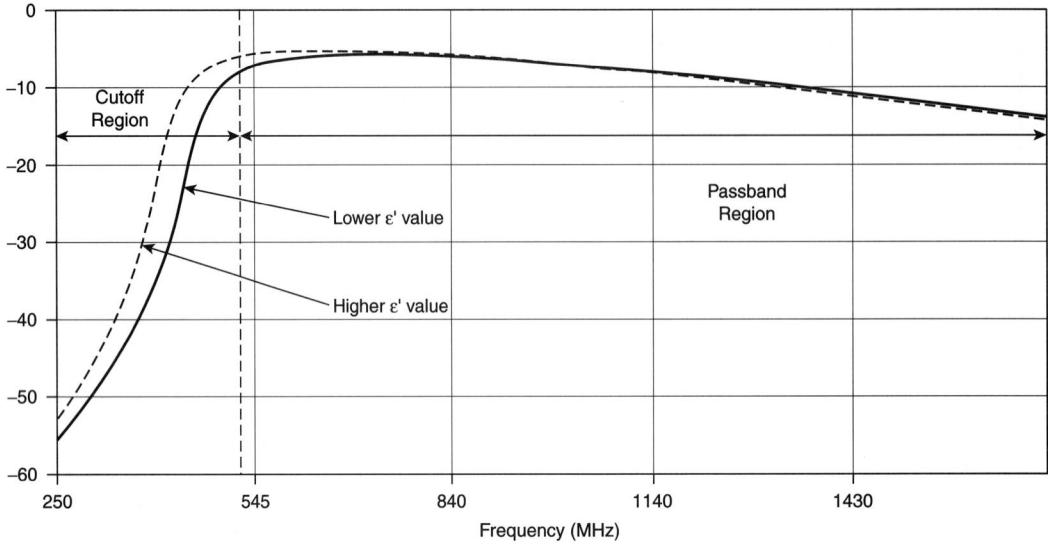

FIGURE 35   Spectrums for changing ε'.

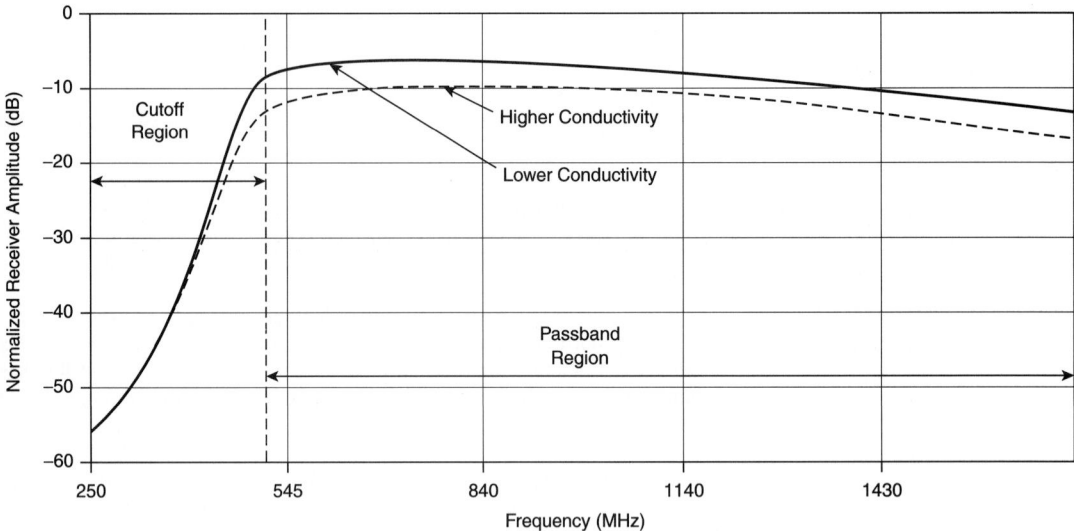

FIGURE 36   GMS spectra for conductivity change.

Typical Software Approach to GMS Spectrum Measurement

Dual-Line-Fit Measurement.   Dual-line-fit software uses a linear regression to draw the best straight line through the cutoff region. Its measured variables are the slope and offset of this cutoff line as well as the cutoff frequency based on where this line intercepts an amplitude setpoint.

Line-fit software also does a linear regression to draw a straight line through the passband region so that its measured variables also include the slope and offset of this passband line.

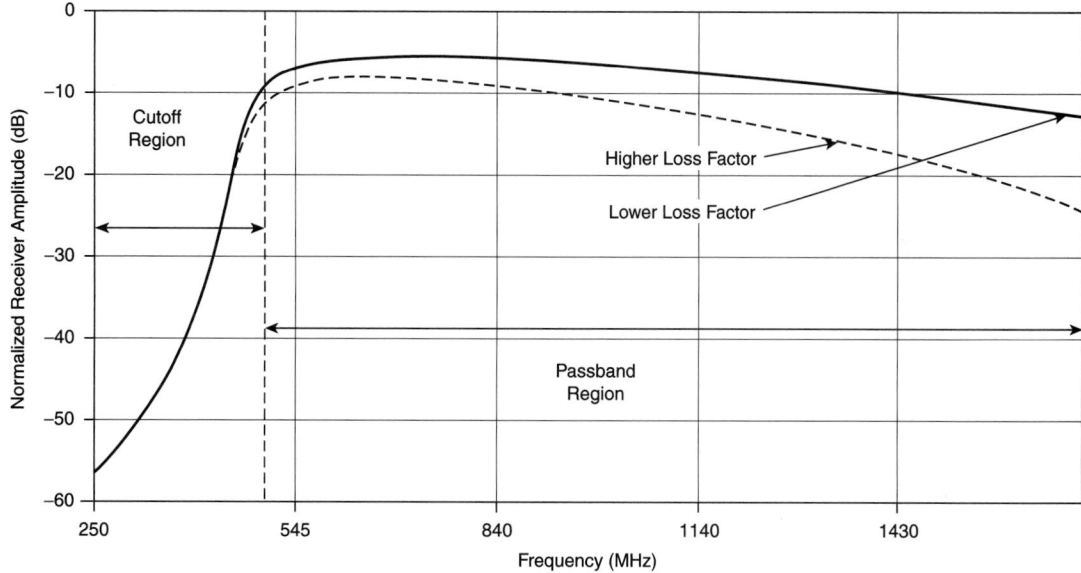

FIGURE 37   Spectrums for high-frequency loss factor change.

The line-fit approach has the advantages of being very fast with updates in less than a second and not requiring a coprocessor. With its five measured variables, it tracks all of the key features of the spectrum so that it can be used on multicomponent mixtures.

Process Effects on Measurement

Entrained Air, Pressure, and Packing-Density Effects.   Entrained air or other gases in liquids or slurries have the effect of lowering ε', σ, and τ because of their nonpolar nonconductive nature. The effect of this air is proportional to the volume of air in the mixture. If this volume remains constant, this effect calibrates out.

In liquid and slurry mixtures, if the process pressure is 25 psi (170 KPa) or greater, the reduced volume of this gas entrainment greatly diminishes any air entrainment effects.

For compressible slurries, the calibration needs to be over the variety of pumping pressure conditions that will be encountered during on-line operation. For binary slurries (e.g., water and a ground-grain batter), such a calibration with the GMS dual-line-fit or curve-fit variables will compensate for a variable pump speed pressure effect. For multicomponent slurries (e.g., meat with water, fat, and protein), pressure also needs to be measured and recorded during calibration to determine the value of an on-line pressure measurement input to the spectrometer for compensation. Often this can be avoided if the measurement output is used only when the pump is on.

For granular-solid applications, the amount of air in between the granular solids varies with particle size. Particle shape has an effect on how uniformly the granules stack in the measurement zone. For in-line granular-solid applications, the use of averaging diminishes the variable stacking effect. Also, calibration with dual-line-fit or curve-fit variables compensates for the variable air concentration effect that is due to variable particle size. Applications with wide changes in particle size, however, may require multiple calibrations for optimum performance.

For at-line granular-solid applications, multiple fill averaging or multiple antenna paths through the mixture reduces the variable stacking effect. Calibration with dual-line-fit or curve-fit variables compensates for the effect of variable air concentration that is due to differences in particle size and packing density.

Temperature and Physical State Effects.   Temperature affects the electrical properties of the mixture proportional to the concentration of polar and semipolar constituents. Meanwhile, nonpolar constituents have less of a temperature. (Although there is less ε' change for oil temperature change, there is still good fc change that is due to the square-law relationship between fc and ε'.) Unless temperature is fixed, it should be measured during calibration and varied over its range to see whether on-line compensation is beneficial. If it does have an effect, the calibration should be temperature compensated. Change of physical state of polar and semipolar constituents has a very large effect on GMS variables. For example, ε' for water is 3.15 when frozen at 0°C; 88 when liquid is at 0°C; 56 when liquid is at 100°C; and 1.0012 as a vapor at 100°C. Water has a low ε' as a solid because the crystalline water molecule is tied to its neighbors with three active bonds instead of the one bond that is active for free water. As a result, it is not free to rotate. Water has a low ε' as a vapor because of the few number of water molecules in the volume and because of the almost frictionless environment. As a result, the few molecules present rotate easily so that little energy is stored.

For these reasons, a measurement point should be selected that has a fixed physical state. For greatest separation between the polar, semipolar, and nonpolar constituents, the polar and semipolar constituents should be liquid.

Coating Buildup Effect.   Coatings have less effect on signals at microwave frequencies than signals in the IR and optic regions because the effect is inversely proportional to wavelength. For example, microwave transmission can occur in spite of nonpolar buildups such as fat, wax, and tar to a depth of over 1 in. (2.5 cm). However, when high resolution is required for narrow spans, such buildups can cause shifts in the calibration's offset. For example, a 3/32-in. (0.25-cm) fat buildup on a meat application can cause a 1% shift in fat concentration readouts.

Therefore coating should be minimized, especially on the process seals that separate the antennas from the process mixture. These seals can be made from any plastic, ceramic, or glass, and they do not have to be visually transparent to allow microwave transmission. The finish of the process seals and the metal portion of the sensor body also can be modified to minimize material buildup.

Also consider the following:

* selecting the optimum velocity and pressure drop through the spectrometer by means of selection of sensor size and upstream/downstream piping
* heat tracing the sensor body
* orienting the piping for vertical flow through the spectrometer instead of horizontal flow

On applications in which the coating buildup reaches a mature thickness and then stabilizes, one can cancel its effect by not gathering calibration data and not operating the instrument until after the coating has reached the mature stage.

Bound versus Free Water Effect.   When water first comes in contact with a dry material, a single layer of the water molecules adheres to all surfaces of the solid. This water is said to be adsorbed. (This is different from the amount of absorbed water, which is the total amount of water that goes into the solid. Absorbed water includes both adsorbed water and the free water in the cavities/capillaries of the solid, which is in contact with only other water molecules.)

Depending on the inner area of the solid, adsorbed water can account for the first 2% to 10% of the mixture's moisture content. Adsorbed water is also called bound water because it has two active hydrogen bonds instead of the one bond active for free water. As a result, bound water does not rotate as well and has less ability to store energy. This makes bound water semipolar with an ε' in between that of ice at 3.15 and free water at 56 to 88. For middle to high water concentrations, the mixture contains a maximum and constant amount of bound water that does not affect the measurement. For very low water concentrations, the calibration will take adsorbed water into account if product temperature is also monitored for on-line compensation. Low water applications with a wide span covering not only low moistures (with only adsorbed water), but also high moistures (with both adsorbed and free water), require a special, nonlinear calibration equation to handle the transition from medium to high ε' sensitivity.

For moisture measurement at low water concentrations, the semipolar bound water provides less ε' contrast to the nonpolar granular solids. However, the GMS approach still has high sensitivity to bound water. The square-law relationship between fc and ε' [see Eq. (6)] provides much higher frequency changes in the low dielectric region, and the amount of frequency change determines GMS sensitivity. (Cutoff frequency measurement resolution is ~ 0.1 MHz.)

Particle Size Effect. When a wave encounters a particle or other discontinuity such as a water droplet in an emulsion, the wave scatters if the diameter of the particle/discontinuity is greater than 10% of the wavelength. This in turn affects both the velocity reduction measurement and the amplitude reduction measurement since varying the path length changes how many polar elements the wave encounters between the send and the receive antennas.

With a fixed mixture composition, this effect can be used to measure particle size or emulsion quality. However, to measure product composition, longer wavelengths 10 times bigger than the particle/discontinuity are needed to be independent of this large potential error variable. GMS uses very large wavelengths relative to the size of the sensor body. For example, in a 3-in. (75-mm) I.D. sensor body, the wavelength at cutoff is twice the a dimension, which is nearly 4 in. (100 mm). As a result, the maximum uniform size of particles or other discontinuities that can flow through the sensor cannot scatter the wave.

Emulsion Phase Effect. The electromagnetic properties of an emulsion shift during a phase change. For example, at low water concentrations, an oil–water emulsion has water droplet surrounded by oil droplets. Such an emulsion has a water internal phase. The emulsion will stay in this phase until the concentration goes from middle water content to high water content, at which time it will switch to a water external phase. The sensitivity of ε' to increasing water content will then shift to a higher value since the water external phase is like free water whereas the water internal phase is like bound water.

NEUTRON ACTIVATION ANALYZERS
by R. J. Proctor

On-line Prompt Gamma Neutron Activation Analyzers

To cut costs, material processing industries are continuously looking for automated solutions to their process control needs. Industrial materials include hydrogenous materials like coat oil and mineral slurries as well as nonhydrogenous materials like cement rocks, mineral ores, and various useful clays like kaolin. One way industry can achieve automation is by the use of on-line analysis of their materials. Prompt gamma neutron activation analysis (PGNAA) is an excellent method for noninvasive on-line analysis of industrial materials

Over the past 10–15 years, instrumentation and computing advances have added prompt gamma neutron activation analyzers to the repertoire of elemental analysis methods for process control A major advantage of PGNAA transducers is that neutrons are a very penetrating radiation that care little for the molecular form, temperature, or physical properties of the process materials being measured. The neutrons reacting with the process materials produce the gamma rays promptly ($\sim 10^{-2}$ s), and most of the gamma rays are even more penetrating than the neutrons. The penetrability and rapid response offers the advantage of very large analysis volumes and no degradation for sample motion. PGNAA analyzers can therefore deliver rapid, sampling free elemental analysis on large-top-size, bulk materials moving at meters per second.

Prompt Gamma Neutron Activation

If a material is bombarded with neutrons, interactions with the nuclei result in the emission of high-energy gamma rays at a variety of energy levels. Two mechanisms predominate in the application of PGNAA to elemental analysis: thermal neutron capture (< 1 eV) and fast neutron inelastic scattering

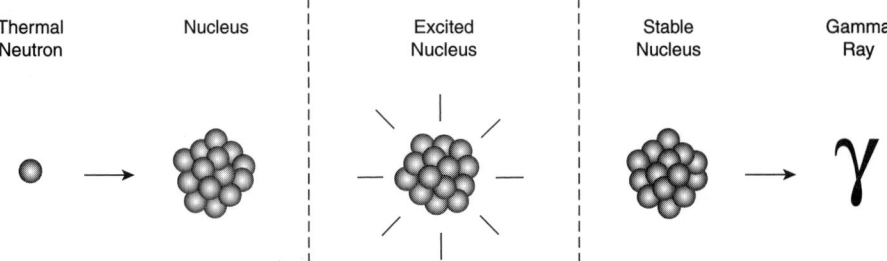

FIGURE 38   PGNAA.

$(>10^5$ eV). Both mechanisms produce high-energy gamma rays, but thermal neutron capture is more efficient and readily achievable with the most inexpensive source of neutrons known, micrograms of the radioactive isotope 252Cf. The fast scattering reactions require a more expensive neutron generator tube. Commercial PGNAA analyzers have predominantly used thermal neutrons to capture gamma rays. The process of thermal neutron capture by a nucleus is shown diagrammatically in Fig. 38.

An example of this reaction is

$$\text{Fe56} + \text{neutron} \Rightarrow \text{Fe57}^* \qquad [^* = \text{excited state}]$$
$$\text{Fe57}^* \Rightarrow \text{Fe57} + \text{gamma rays} \quad [\text{in } 10^{-12} \text{ s}]$$

The prompt gamma radiation is emitted while the neutrons are irradiating the material and happens so quickly that material motion has no effect.

If a material is bombarded with neutrons of a flux $\varnothing(E)$ $n/cm^2/s$, then the reaction rate R_i for N_i nuclei of the element i with absorption cross section $o\hat{}i(E)$ is given by

$$R_i = \text{integral}[N_i o\hat{}i(E)\varnothing(E)dE] \tag{1}$$

If we assume either neutron capture reactions or fast 14-MeV reactions (from a neutron generator), we effectively only have to worry about one main interaction energy. Thus

$$R_i = N_i o\hat{}i\varnothing \tag{2}$$

The reaction rate is the predictor of the observed PGNAA gamma ray fluxes. Thus the observed gamma ray flux is proportional to the number or nuclei in the analysis volume and their absorption cross section.

Signal Processing

PGNAA gamma rays are detected by the deposition of their energy as high-speed electrons within detector materials. The electrons create ionization and hear, which can be detected. The magnitude of the ionization is proportional to the energy of the gamma ray. Examples of detectors include crystalline scintillators and solid-state devices. Scintillators convert the ionization into optical or UV light pulses that can be detected by some form of photodetector, e.g., a photomultiplier tube. Solid-state detectors directly collect the ionization charges. Typically scintillators have poor gamma ray energy resolving power but make up for it in their large size to allow good detection efficiency, whereas solid-state detectors have good resolving power but their maximum sizes are limited. The pulsed gamma ray signals from the detector are typically amplified, processed in analog form, and then digitized to an equivalent gamma ray energy; or they are bash digitized and the processing is performed totally digitally. An analog example of the processing steps is shown in Fig. 39.

The final step is the integration of all the events into a spectrum in the processing computer. An average PGNAA gamma ray spectrum may consist of 8–12 elemental signatures, each consisting

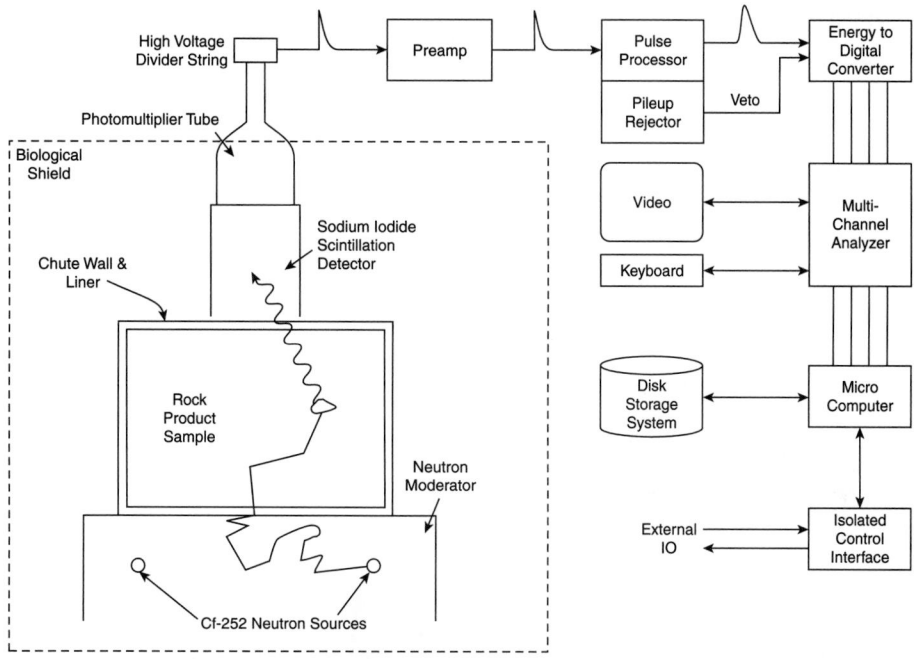

FIGURE 39 Gamma ray event processing chain for a scintillator gamma ray detector.

of 10–200 individual gamma ray lines for a total of several hundred up to thousands of gamma ray lines. In addition, many of the lines are reprocessed into continuum features such that the typical gamma ray spectrum is very confusing, as seen in Fig. 40. To analyze such a spectrum accurately requires a full spectrum analysis technique rather than peak analysis. A commonly used full spectrum method is library least squares [1], which uses instrument responses to pure elements as its library. The library-least-squares method solves the linear matrix equation:

$$Y = XB + e \qquad (3)$$

where Y is a measured spectrum assumed to be composed of known library components held in matrix X with intensities B and random errors e. The best estimate of the spectral intensities b is then the generalized inverse, least-squares solution [2]:

$$b = (X^t X)^{-1} X^t Y \qquad (4)$$

where t represents a matrix transpose and the -1 represents a matrix inverse.

Signal Normalization

All elements in a material emit PGNAA gamma ray fluxes that are proportional to the amount of the element and the average neutron flux. Thus if all the elements are measured, their sum should add to 100% of the material [3]. When such an assumption is valid, then the fractional analysis of each

TABLE 8   Typical Cross-Belt Analyzer Elemental Sensitivity

Sensitivity[a] in weight %	Elements
<0.01%	Cl, Sc, TiNi, Cd, Hg, Srn, Gd, Dy, Ho
0.01–0.1%	S, V, Cr, Mn, Fe, Co, Cu, Rh, Ag, In, Hf, Ir, Au, Nd, Eu, Er, Yb, H_2[b]
0.1–0.3%	N, Na, Al, Si, K, Ca, Ga, Se, Cs, La, W, Re, Os, Pt, Pr, Tm
0.3–1.0%	Li, Be, Mg, P, Zn, As, Mo, Te, I, Ta, Pb, Ce, Tb, Lu, Th, U
1.0–3.0%	C, Ge, Br, Sr, Zr, Ru, Pd, Sb, Tl
>3.0%	He, O, F, Bi

[a] Three-sigma detection limit of 10 min within an elementally simple rock matrix >150 mm thick.
[b] Intrinsically H is <0.01% but it is normally used in neutron shielding, which can greatly degrade the instrumental sensitivity.

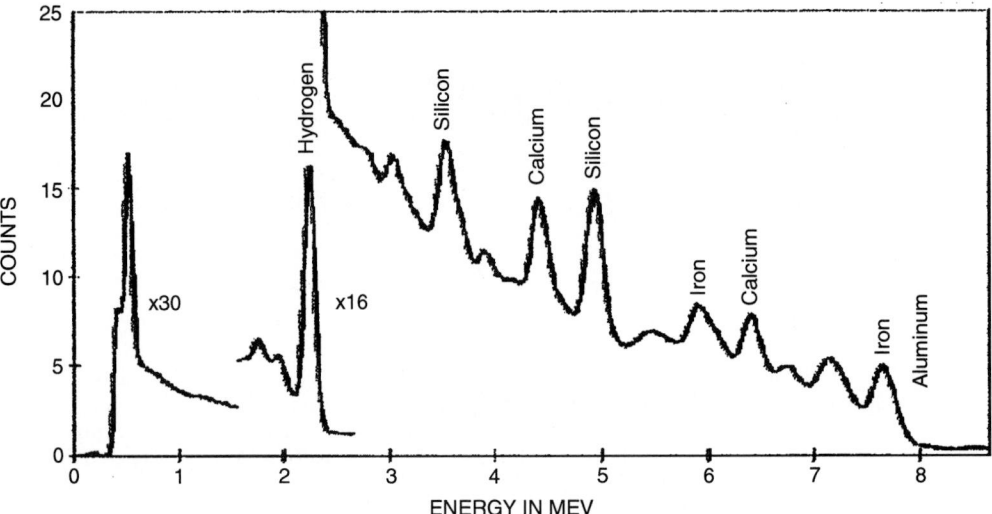

FIGURE 40   Limestone PGNAA gamma ray spectrum from a scintillation detector.

component is given by

$$E_i = \frac{b_i}{\sum_{k=1,\ldots,N} b_k} \tag{5}$$

Where b are the measured signals from Eq. (4) and E are the fractional analysis results.

Sensitivities

Tables of >15,000 PGNAA gamma rays [4] and knowledge of the neutron absorption cross sections allow the deduction of the elemental sensitivities. Such sensitivities for a typical analyzer are given in Table 8. The stronger elements are those with the higher cross sections whilst the weaker are those with not only weaker cross sections but weak gamma-ray production.

FIGURE 41   Cross-belt analyzer analyzing limestone rock on a 1000-mm trough belt at several hundred tonnes per hour.

Instrumentation

The application of PGNAA to real-world analysis requires practical solutions to industrial problems. One solution is the cross-belt analyzer, as shown in Fig. 41, which analyzes a limestone mixture for Ca, Si, AL Fe. Cross-belt analyzers like this can analyze most dry rock materials for important elements. A cross-sectional view through a cross belt, as shown in Fig. 42, displays the two large gamma ray detectors necessary to collect the data for good performance.

The application of such cross-belt analyzers to cement rock analysis can lead to many desirable features.

PGNAA is not limited to dry materials but can analyze high-solid-content slurries. The analyzer in Fig. 43 measures sample stream slurry. A small sample PGNAA analyzer (see Fig. 44) can analyze 48 Kg of rock or coal in plastic buckets, with minimal skilled labor needed.

Examples Plus Sensitivities.

- Matrix effects apply to all elements equally
- Independent of particle size
- Independent of crystalline state

Advantages of PGNAA.

- Very large analysis volume
- Matrix effects apply to all elements equally
- Independent of particle size
- Independent of crystalline state

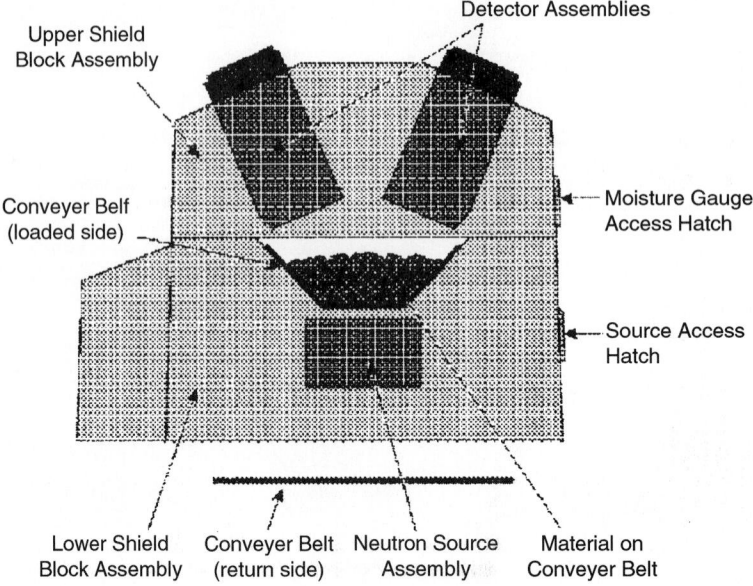

FIGURE 42   Sectional view of a cross-belt analyzer designed for uniform sensitivity.

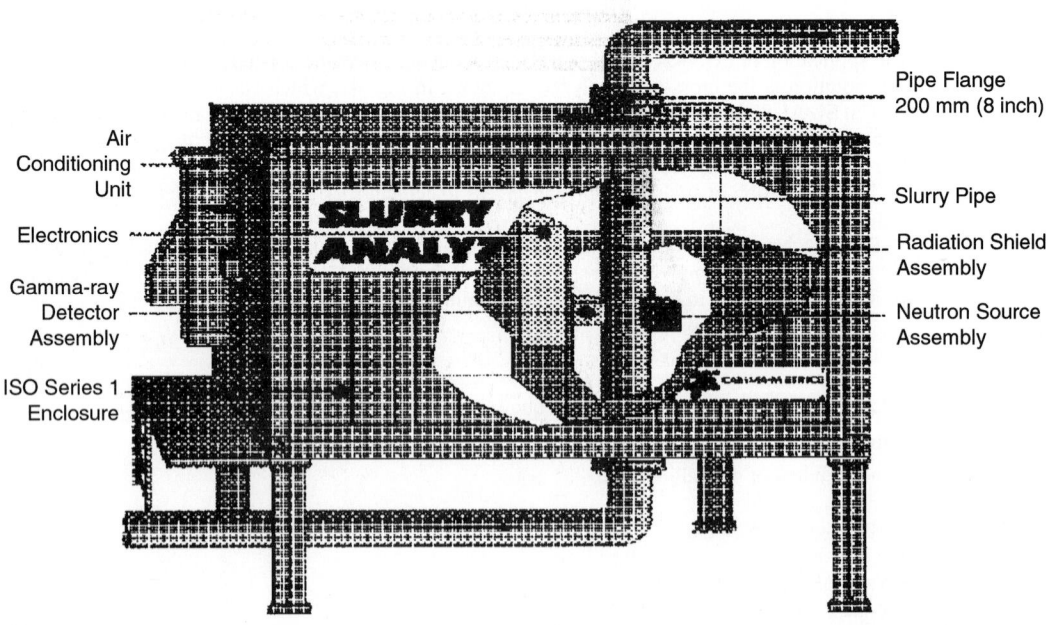

FIGURE 43   Slurry analyzer.

FIGURE 44   Sampled coal PGNAA analyzer can analyze 2–25 tons/h on a dedicated belt.

Calibration Process

Practicalities and Realities.   100% normalization is a simple self-consistent technique, but unfortunately PGNAA does not directly measure *every* element in a sample, preventing this from being an absolute technique. Oxygen is present in most materials but PGNAA, which uses thermal neutrons, has almost no sensitivity to oxygen. For 100% normalization to work, it requires assumptions or predictive models of the molecular state of the detected elements for predicting the undetected elements, e.g., elements are always associated with oxygen by the fixed ratios present in the oxides. In the case of rocks for making cements, the oxide assumption that, e.g., Ca, Si, Fe, Al, Mg, K, Na, and S are in the oxide form of CaO, SiO_2, Fe_2O_3, Al_2O_3, MgO, K_2O, Na_2O, and SO_3 is sufficiently accurate for a calcined analysis. When one element is present in many compounds and the molecular form cannot be predicted, this technique can be inaccurate since the proportion of oxygen cannot be accurately predicted.

REFERENCES

1. Shyu, C. M., R. P. Gardner, and K. Verghese, *Nucl. Geophys.* **7**, p. 241, 1993.
2. Montgomery, D. C., and E. A. Peck, *Introduction to Linear Regression Analysis*, Wiley, New York, 1982, p. 111.
3. Grau, J. A., J. S. Schweitzer, D. V. Ellis, and R. C. Hertzog, *Nucl. Geophys.* **3**, 351–359, 1989.
4. Lone, M. A., R. A. Leavitt, and D. A. Harrison, *At. Data Nucl. Data Tables* **26**, 511–559, 1985.

TABLE 9   Comparison of PGNAA and XRF

Parameter	XRF	PGNAA
Excitation radiation	x rays	Neutrons
Excitation energy	10–100 keV	0–10 MeV
Typical depths of penetration	1–10 mm	100–1000 mm
Penetration limited by	High atomic number and high density	High moisture content and high density
Characteristic radiation	x rays	Gamma rays
Energy of radiation (cement)	1–7 keV	1–11 MeV
Typical escape depths	Submillimeter	200 mm
Matrix effects	Particle size, interelement effects, and crystalline properties	Moisture content and neutron poisons
Detection techniques	Energy and wavelength dispersive	Energy dispersive
Useful count rate per element	$\sim 10^2$–10^4 cps	$\sim 10^2$ cps

ATOMIC EMISSION SPECTROMETERS
by J. G. Converse

EPA On-Line Stack Analyzers for Metals

An inductively coupled plasma (ICP) is an excellent source for exciting chemical elements that subsequently produce radiant emissions. Atomic emission spectroscopy (AES) is a technique for capturing these emissions and sorts them so that chemical analysis of trace level elements such as metals can be performed. The cost of the pure Ar burned at high consumption rates was almost prohibitive for continuos operation until the microtorch was developed. It now appears that this technique may be practical for monitoring elemental emission in stack gas from industrial processes.

A clean water sample (some dissolved salts allowed, but not particulate matter) is aspirated into the plasma to excite the elements in solution. Perhaps the best way to interface the dirty gas sample to the analyzer is by means of a vapor/liquid scrubber. Figure 45 shows a scrubber system utilized to continuously strip and collect target species from stack gases.

Gas is drawn from the stack through the stripper with an air or water eductor. After passing through the stripper, it is sprayed with deionized water from a spray jet. The stripped gas passes on to the eductor. The water flows down through the U–shaped body and up to a standpipe that controls the level in the stripper. It passes on to the sewer. A small diameter tube is placed just below the gas path to extract a sample for the ICP–AES analyzer. This reduces the delay time for the solution to exit at the standpipe. A small-diameter tube produces a higher-velocity flow of sample to the analyzer. A tee fitting at the aspirator of the analyzer allows adequate material to the pulled from the sample stream to meet the needs of the aspirator and bypass some to increase the response time of the sample system.

A switching valve at the aspirator can provide introduction of source deionized water to period-ically check background or protect the aspirator during sample system maintenance. The emission spectrophotometer and computer system can be housed some distance from the stack stripper. Much effort is being made to meet U.S. EPA requirements for this testing.

LIQUID CHROMATOGRAPHY

Liquid Chromatography (LC) is less advanced and higher maintenance than GC, but is having a resurgence in the pharmaceutical industry. Initial attempts at moving LC into the field faired poorly because of unstable columns and large usage of carrier liquid. One cannot vent liquid carrier into the atmosphere as you can He from GC. Stable reliable column materials have been developed, and small-diameter columns have reduced the carrier flow rate significantly. Improved pumps and detectors have

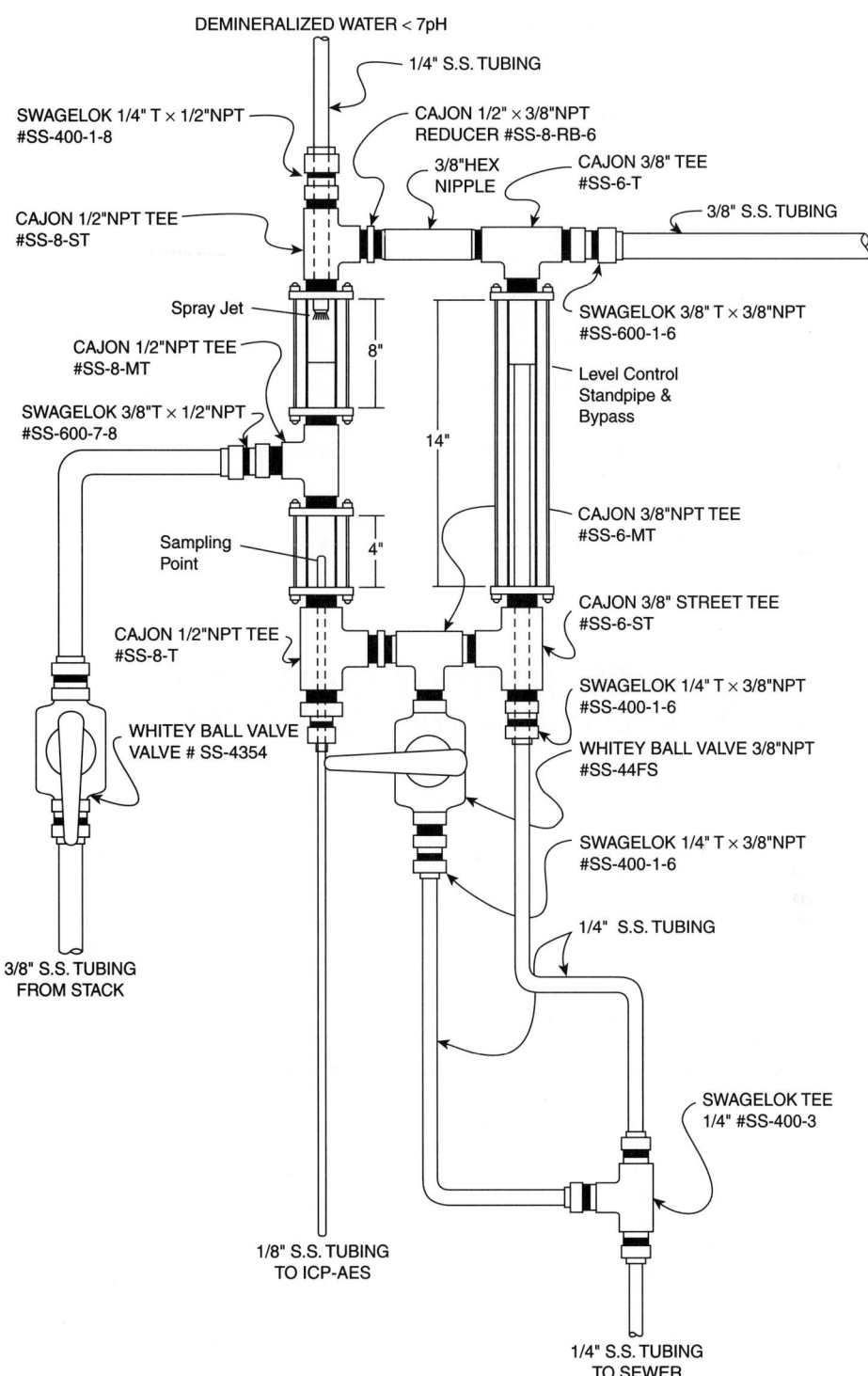

FIGURE 45 Liquid/vapor scrubber for stack gas stripping.

also helped this technology make a comeback. NIR rivals LC for nonvolatile applications, but when time is not critical, separation of components and measurement with simple detectors still has some advantages.

Assets:

1. low detection limits, <1 ppm
2. simple detection device on isolated component
3. multiple components without mathematical calculations

Liabilities:

1. high pressure makes leakage more probable
2. liquid solvent disposal is more difficult than that with GC
3. liquid carrier system is more difficult to operate than that for gases
4. batch process may have long cycle time

THE IMPROVEMENT OF ADVANCED REGULATORY CONTROL SYSTEMS

by G. K. McMillan*

PROBLEMS AND CAUSES

Dead time is the leading cause of control loop problems either directly or by accentuating other causes. A large dead time makes a control loop more vulnerable to nonlinearities, periodic upsets, and poor tuning.

Dead time anywhere in the control loop prevents the proportional-integral-derivative (PID) control algorithm from either seeing or reacting to an upset. The best a PID controller can do is to limit the peak error to how far the process variable deviates in 110% of the dead time if the controller is tuned for quarter-amplitude response (each succeeding peak is $1/4$ the amplitude of the previous peak), as shown in Fig. 1. However, such aggressive tuning is not practical because a 25% or more increase in total dead time or open-loop gain or a 25% or more decrease in the largest time constant can cause instability. The loop is too close to its stability limit. Also, any oscillation in the closed-loop response may be undesirable from a standpoint of causing process variability because of unattenuated cycles in plug flow fluid systems (e.g., pipelines, static mixers, exchangers, extruders, and desuperheaters) or solid systems (yarns, sheets, and conveyors) or periodic upsets to other loops. Thus quarter amplitude is a benchmark and not a practical goal. Typically, the peak error or maximum excursion of the process variable is reached in 150% to 1500% of the total dead time for industrial process control systems. The time to peak in percentage of total dead time is proportional to how much the controller gain is reduced because of tuning practices or to provide additional robustness or decoupling [1].

---

* Senior Fellow, Solutia Inc., St. Louis, Missouri.

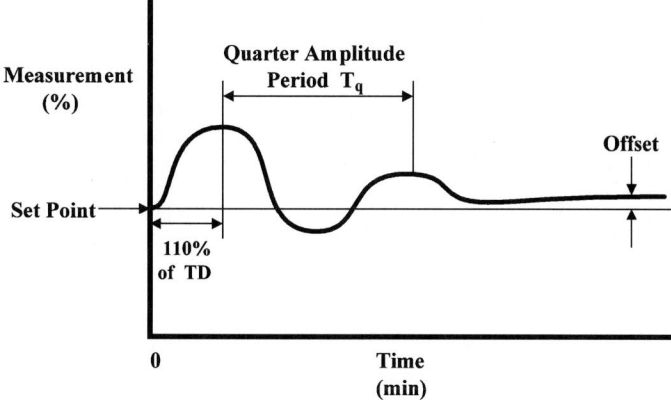

FIGURE 1 Quarter-amplitude benchmark for peak error.

The excursion rate is determined by the time constant of the disturbance and the process. The larger these time constants, the smaller the peak reached in the percentage of the dead time set by the controller tuning.

The total loop dead time is the sum of all the pure dead times and equivalent dead times from all the time constants smaller than the largest time constant in the loop. While the portion of these smaller time constants that causes dead time decreases as they become bigger compared with the largest time constant, all of the small time constants can often be summed up and added to the pure dead times to get the total dead time since it is difficult to find all the sources of dead time.

The biggest source of process dead time is transportation delay, which is the time it takes for a change in composition or temperature to propagate from the final element (e.g., control valve, die, or rotary valve) to the process measurement. For plug flow it is the volume divided by the flow rate, and for yarn, sheet, and conveyor systems it is the length divided by the speed as shown in Eqs. (1) and (2), respectively. For well-mixed volumes (e.g., reactors), the dead time is normally ~0.1 min and can be approximated by Eq. (3). The process dead time for changes in flow and pressure is usually much less than 1 s and therefore negligible. For level loops, the process dead time is the same as that for liquid flow because once the net difference in flows in and out is altered, the change in level is instantaneous [1].

$$\mathrm{TD}_p = V/F \tag{1}$$

(pipelines, static mixers, exchangers, extruders, and desuperheaters),

$$\mathrm{TD}_p = L/S \tag{2}$$

(yarns, sheets, or conveyors),

$$\mathrm{TD}_p = V/(F + F_a) \tag{3}$$

(agitated vessels), where

$F =$ flow rate (gmp)

$F_a =$ agitation rate (gmp)

$L =$ length between final element and sensor (ft)

$S =$ speed (ft/min)

$\mathrm{TD}_p =$ dead time from transportation delay (min)

$V =$ volume between final element and sensor (ft)

The biggest source of dead time in the final element is the control valve resolution limit for fast loops and the control valve dead band for slow loops or loops without positioners. As changes in the controller output approach the resolution limit, particularly for pneumatic positioners, the positioner gain drops and the time required for changing the actuator pressure enough to change the stroke can increase by 2 orders of magnitude (e.g., from 0.5 to 50 s). Before a valve will move it must get out of the dead band. For a signal reversal, the change in signal needed is the full dead band. For a change in the same direction, the change in signal needed can be very roughly approximated as half of the dead band. The dead time for a signal reversal is the dead band divided by the rate of change of the controller output, as defined in Eq. (4). For slow loops, the incremental change in controller output is slower and the time to get out of the dead band is longer. For example, if the valve dead band is 0.5% and a temperature controller output reverses direction and is ramping at the rate of at 0.5% per minute, the additional dead time from dead band is 1 min. When a test is made for an automatic tuning algorithm or a manual tuning procedure, the observed dead time does not include the dead time from the dead band because the change is a step and the size is deliberately chosen to be much larger than the dead band. This additional valve dead time for a load upset to a controller with mostly proportional action as a factor of the observed dead time can be estimated per Eq. (5), where K_x is the factor for how much the controller gain is detuned from the quarter-amplitude benchmark [2]. Note that the dead time becomes very large as the step size approaches $1/2$ of the dead band, which is the resolution limit of the control valve.

$$TD_v = \frac{DB}{\Delta IVP/\Delta t} \qquad (4)$$

$$TD_v = \frac{DB}{K_x(\Delta IVP - DB/2)} * TD_o \qquad (5)$$

where

ΔIVP = change in implied valve position (i.e., controller output) (%)

DB = dead band from valve hysteresis (%)

K_x = fraction controller is detuned from quarter-amplitude response

Δt = change in time (min)

TD_o = dead time seen in open-loop test (min)

TD_v = dead time from control valve dead band (min)

For large control valves (e.g., >4 in.), the actuator size becomes large enough to make the prestroke dead time (time to make an appreciable change in actuator pressure) and stroking time (time to make the total change in actuator pressure) significant (e.g., total response time after exceeding the dead band is 10 to 100 s). Figure 2 shows how the response time of the control valve is affected by the size of the change in controller output and the size of the control valve. The figure does not show (1) that the response is slower in one direction than the other because of the difference in the exhaust and fill rate of the actuator, and (2) the increase in dead band as you drop below 20% for rotary valves and 5% for globe valves because of additional friction from the seal or seat use to reduce leakage in the closed position. If the user or the supplier makes 10% or larger change in valve position, the tests will reveal only the size of the valve actuator and whether it has low friction packing. To determine how well a valve will respond, the step size must be comparable with the normal change in controller output per scan, which is typically from 0.1% to 1% per scan for load upsets. The response to large steps can also be a consideration. If the operator makes a large change in the manual output or the set point or an interlock or a kicker algorithm is activated, the corresponding step in the signal to the control valve can be quite large. For large dampers (e.g., >14 in.), the inertia is so large compared with the friction, an unstable limit cycle can develop for a large step change in signal if the positioner is tuned to respond to small steps. The consequences of a huge oscillating damper are so severe that restrictions (e.g., needle valves) are put in the pneumatic output lines to the positioner. However, this makes the control for normal operation miserable. The better solution is to enforce a velocity limit

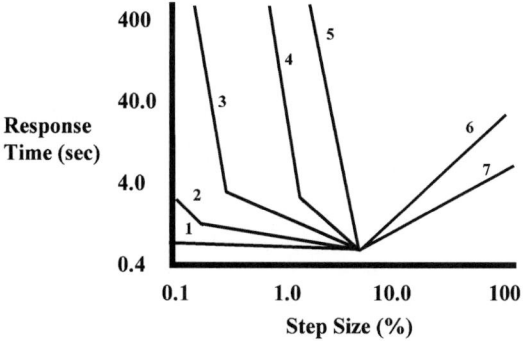

1 - sliding stem valve with diaphragm actuator and digital postioner with pulse width modulated solenoids.
2 - sliding stem valve with diaphragm actuaotr and digital positioner with nozzle flapper
3 - sliding stem valve with diaphragm actuator and pneumatic positioner
4 - rotary valve with piston actuator and digital positioner
5 - rotary valve (tight shutooff) with piston actuator and pneumatic positioner
6 - very large rotary valve (>6") with any type of positioner
7 - sliding stem valve with a digital positioner

FIGURE 2   Effect of size of control valve and the change in controller output on valve response time.

on the controller output in the distributed control system (DCS) or the programmable logic controller (PLC).

To summarize, the response of a control valve depends on the direction of the signal change, the magnitude of the signal change, and the throttle position. If you also consider that the slope of the installed valve characteristic and hence the steady-state gain change with throttle position and the pressure drop that is available to the control valve, you realize that the control valve introduces significant directional, magnitude, and operational nonlinearities that have largely been ignored, despite playing a prominent role in determining the performance of control systems.

Sample transport delay and analyzer cycle time are the sources of the largest dead times in a measurement. The dead time from cycle time varies from one to two dead times, depending on whether the upset arrives just before the next sample or just after the last sample since the result is created at the end of the cycle time. For fast loops (e.g., pressure and flow), the scan time of the DCS is the largest source of dead time in the measurement. In this case, the dead time varies from almost zero to one scan time, depending on whether the upset arrives just before the next scan or just after the last scan since the result is created near the beginning of the scan. This leads to the rule of thumb that the average dead time is 1.5 times the cycle time for analyzers and 0.5 times the scan time for a digital device. However, this rule misleads users into thinking that the dead time is fixed, whereas the variability in dead time is significant and not known. Consequently, if the cycle or scan time is more than 25% of the total loop dead time, the performance of a dead-time compensator is severely degraded. Since the compensator dead time should be set for the minimum dead time, only the time between the reading and the result can be included for compensation. For advance control algorithms that involve complex calculations, it is better if the result is generated and used as soon as the computation is complete rather than waiting to the end of the algorithm cycle. Figure 3 shows how the dead time varies with the relative locations of the upset, result, and reading.

Measurement noise, sensitivity, and analog-to-digital (A/D) converter resolution are considerable sources of dead time that are not discussed in the literature. For a level loop, the time it takes for the level measurement to exceed the sensor sensitivity limit, the DCS input A/D resolution, and get out of the noise band where it can be distinguished by a controller as a true change in level varies anywhere from 1 to 20 min, which is 2–3 orders of magnitude larger than the process dead time. Equations (6)–(9) show that this additional measurement dead time can be estimated as the largest of the sensor sensitivity, A/D resolution, or noise band divided by the rate of change of level, which is

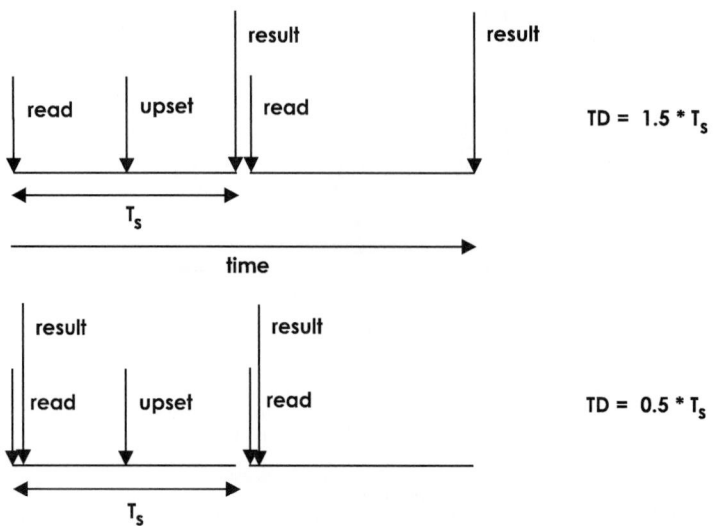

FIGURE 3 Effect of relative location of upset, reading, and result on measurement dead time.

the product of the step size, valve gain, level integrator gain, and measurement gain. The integrator gain is inversely proportional to the product of the cross-sectional area of the vessel and the density of the fluid and is typically quite small. Level rates of change of 0.1% per minute or less are quite common.

$$\text{TD}_m = \frac{\max(E_s + E_r + E_n)}{\Delta \text{PV}/\Delta t} \qquad (6)$$

$$\Delta \text{PV}/\Delta t = \Delta \text{IVP}(K_o) \qquad (7)$$

$$K_o = (K_v)(K_i)(K_m) \qquad (8)$$

$$K_i = 1/(A\rho) \qquad (9)$$

where

A = cross-sectional area of vessel at level measurement (ft^2)

E_n = measurement noise (%)

E_r = A/D resolution (%)

E_s = sensor sensitivity (%)

ΔIVP = change in implied valve position (i.e., controller output) (%)

K_i = integrator gain (ft/lbs/min)

K_m = measurement gain (100% divided by calibration span in feet) (%/ft)

K_o = open loop gain (gain with loop in manual)

K_v = valve gain (slope of the installed flow characteristic) [(lbs/min)/%]

ΔPV = change in process variable (i.e., controller input) (%)

ρ = density of the fluid (lb/ft^3)

TD_m = measurement dead time (min)

Radar level devices have the greatest sensitivity. However, since radar provides a true surface detection, there is more noise from agitation, boiling, foaming, sparging, or recirculation than from a transmitter that responds to the average head across the entire surface. Differential pressure transmitters with ceramic sensors rather than metal diaphragms offer a greater sensitivity since there is no hysteresis (e.g., dead band). While bubblers sense average head, they tend to be noisy, particularly at low bubble rates, because of bubble disengagement from the tip. Floats and displacers see the local changes in surface level and have mechanical components with some backlash. Hence they exhibit greater measurement noise and a lower measurement sensitivity.

For a temperature loop, sensitivity, resolution, and noise also add a large measurement dead time because of the slow response. A typical maximum rate of change for a temperature loop on a vessel is 0.5°C/min. If this is converted to percentage per minute of the temperature measurement span, Eq. (3) can be used to estimate the associated additional dead time. For loops that do not have transmitters but have sensors that are wired directly to a DCS thermocouple or resistance temperature detector (RTD) input cards, the resolution is quite poor because of the large spans associated with the cards. There is also a kick in the controller output because of the step of the A/D converter that seriously limits the amount of gain and rate action that can be used. Equation (10) provides a rough approximation of the kick in the output as a function of span of the measurement calibration and controller scale. Note that the measurement span is the DCS input card span if the sensor is directly wired to the DCS and is the transmitter calibration span if a transmitter is used. It is also useful to realize that RTDs have a much better sensitivity (e.g., 0.002°F) than thermocouples (e.g., 0.1°F) and lower thermal noise [2]:

$$\Delta\%\text{IVP} = (100\%/2^{n-1})(S_m/S_c)K_c(T_d/T_f + 1) \tag{10}$$

where

ΔIVP = change in implied valve position (%) S_c = span of the controller scale (deg)

K_c = controller gain (dimensionless) T_d = derivative (rate) time setting (min)

n = bits in A/D converter (e.g., 12) T_f = derivative filter time setting (min)

S_m = span of the DCS input card or transmitter (deg)

Unlike the level loop, temperature loops with thermowells have a large measurement time constant (e.g., 0.25 to 2 min) and temperature loops on vessels have a large process dead time. These are large because of multiple thermal time constants associated with heat transfer. Many of these thermal time constants can be approximated by the product of the mass and heat capacity divided by the product of overall heat transfer coefficient and area, as shown in Eq. (11). For a vessel with a coil or jacket, Eq. (11) can be used to estimate the time constant for the fluid in the vessel, the metal in the jacket or coils, the fluid in the jacket or coils, the metal in the thermowell, the air in the gap between the thermowell and sensor, and finally, the metal in the sensor sheath. Normally, the liquid volume creates the largest time constant in the loop, and all of the other thermal time constants become equivalent dead time:

$$\text{TC} = \frac{MC}{UA} \tag{11}$$

where

A = overall heat transfer area (ft^2) TC = thermal time constant (min)

C = the heat capacity of material (BTU/lb °F) U = overall heat transfer coefficient [(BTU/min °F)ft^2]

M = the mass of the material (lbs)

The installed characteristic of control valves typically gets so flat above 65% for butterfly valves, 85% for v-notch ball valves, and 95% for globe valves, that the flow may change by less than 2% for the rest of the throttle range. This loss of sensitivity plays through the previously discussed noise, sensitivity, and resolution of measurement to create additional dead time and poor disturbance

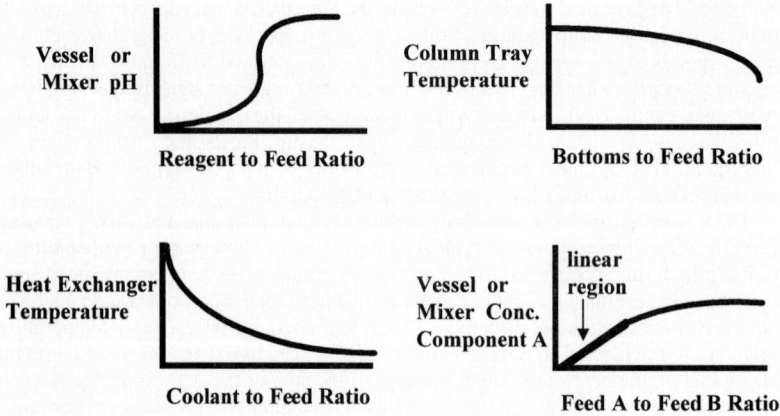

FIGURE 4 Most temperatures and compositions depend on a ratio.

rejection. If the plant is not sold out, the problem can be solved by setting the high valve position limit in the DCS or the PLC to prevent operation on the upper ends of the throttle range. Increasingly, the extra 2% in throughput is needed as old plants are pushed way beyond the original nameplate capacity. Consequently, manufacturing is often reluctant from both a psychological and a practical viewpoint to cut 5% to 35% off of a valve stroke, even if it leads to better loop performance and less process variability. Often these valves are on coolant, air, and steam lines because utility systems have not kept up with plant requirements. Frequently, capital is not available for a new pump, compressor, pipeline, cooling tower, refrigeration unit, boiler, and/or control valve.

Some applications suffer from poor process sensitivity. This occurs in distillation columns in which the change in temperature with composition on the tray is small, in pH systems in which the setpoint is on the extremely flat portion of the titration curve (e.g., below 2 or above 12 pH), and in heat exchangers in which the change in temperature with coolant flow is small because of an insufficient heat transfer coefficient or area. In the first two cases, the trend recordings will look quite nice (e.g., straight lines) even though the actual variability in the composition is quite bad.

The process gain for temperature and composition control loops on most liquid systems is inversely proportional to throughput since the process gain is a function of the ratio of the manipulated flow to the feed flow. Figure 4 illustrates how process variables depend on a flow ratio. The slope of each curve multiplied by the inverse of the feed flow at the operating point is the process gain. This process nonlinearity can be canceled if the temperature or composition loop directly manipulates a control valve with an inherent equal percentage characteristic, which provides a valve gain proportional to flow. However, there are many reasons why the compensation is incomplete and/or undesirable. For example, there are other process nonlinearities (the plot of the process variable versus the ratio is a curve), the installed flow characteristic deviates considerably from the inherent trim characteristic, and a flow loop may be desirable to deal with pressure upsets and valve dead band. For well-mixed vessels, the time constant is also inversely proportional to flow. Since the time constant and process gain have opposite effects on the controller gain, they cancel out. This is not true for plug flow and sheet, yarn, and conveyor lines. Instead of relying on an equal percentage characteristic, it is best to schedule the tuning settings so that the controller gain and reset time (minutes/repeat) are inversely proportional to the feed or the speed.

As a measurement response becomes less repeatable and a control valve response less consistent because of dead band, the gain of the loop becomes more nonlinear because of the variation in the change in the process measurement associated with a change in the controller output. Also, it is important to note that the largest of either the control valve resolution multiplied by the process gain, the measurement repeatability, the sensor sensitivity, or the measurement system resolution sets the lower limit to process variability.

Inverse response is particularly confusing to a controller because the initial response is in the opposite direction of the final response. Since a PID controller does not see past the immediate response, it reacts in the wrong direction to the initial excursion. Inverse response occurs for level control of boiling liquids because of shrink and swell, temperature control of furnaces because of air and fuel cross limits, pressure control of furnaces because of temperature changes from changes in air flow, and any loop in which feedforward signals arrive too soon. In most cases, the initial excursion is faster than the final excursion. In one boiler the initial peak was so much larger than the final peak in drum level that a level trip would occur if the controller did not try to compensate for the initial peak. In this case, the classical solution of a three-element drum level control system, which uses a feedforward signal to change feed water flow in the same direction as steam flow to maintain a material balance, increased the number of drum level trips.

The most difficult type of disturbance for a control loop is a periodic upset. If the upset period is much shorter than the period of the loop, it can be considered to be effectively noise. If the loop reacts to noise, it wears out the control valve packing because of excessive cycling and creates a secondary upset. As the upset period approaches the period of the loop, the loop will increasingly amplify the upset because of resonance. In both of these cases, the loop does more harm than good. A control loop can consistently attenuate only periodic upsets that have a period greater than four times the time to the first peak. The biggest source of periodic upsets are oscillating loops, on–off level and conductivity control, and undersized or sticking steam traps. The major causes of oscillating loops are sticking control valves, interaction, and poor tuning.

Filters are added to attenuate noise. However, the filter adds dead time to the loop. If it becomes the largest time constant in the loop, the trend recording will look smoother and the controller gain can even be increased. The key symptom to watch out for is an increase in the period or settling time of the loop. It is a clue that the real process variability has increased because of a slower reaction to upsets.

Interaction occurs when changes in the manipulated variable of one loop affect another and vice versa. Sometimes is can be the process variable. An example of this is the column sump level and temperature when there is a thermosyphon reboiler because changes in composition change the density and hence the level, and changes in level change the composition by changing the reboiler head and hence boilup.

The tuning settings and the performance of a loop depend on three key variables that describe the open-loop response; total dead time, time constant, and overall gain. Furthermore, various tuning rules and methods excel for particular ranges of the ratio of the dead time to time constant. Also, there are many sources of these variables. Figure 5 summarizes the contributions of the control valve,

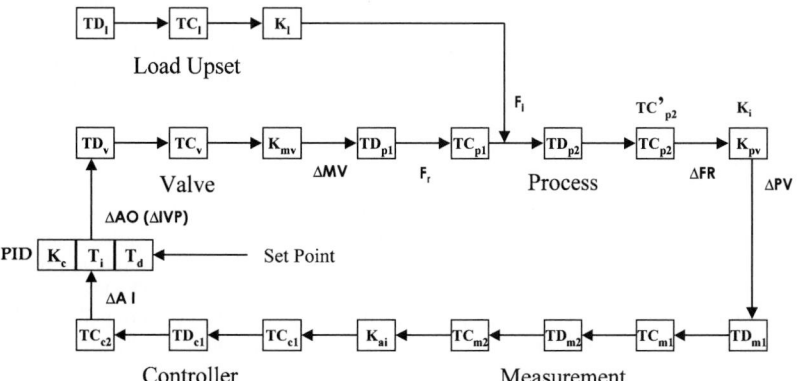

• "Open Loop" - loop is in <u>manual</u> (PID algorithm is suspended)
• "Closed Loop" - loop is in <u>auto</u> (PID algorithm is active)

FIGURE 5   Block diagram detailing the location of dead times, time constants, and open-loop gains.

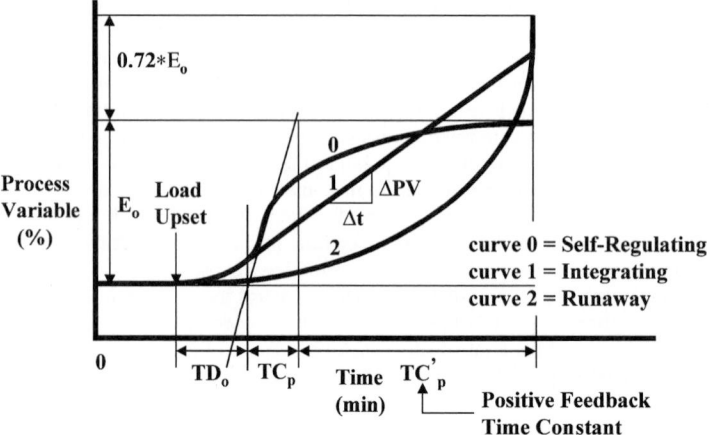

FIGURE 6 The open-loop response for self-regulating, integrating, and runaway processes.

process, and measurement to the three key variables. For ease of approximation, the total dead time can be considered to be the sum of all the pure dead times and small time constants, and the dominant time constant can be considered as the time constant of the loop if it is clearly larger than the rest. The overall gain is the product of the gains associated with the response of the manipulated variable, process, and measurement system. None of the three key variables is constant. It is just a matter of how much do they change. Ideally, the tuning settings should be in a continual state of flux. However, adaptive controllers designed to identify these changes and correct the tuning are always playing catch up. Also, the algorithms to date tend to be too complex to be programmed in a DCS or a PLC and to be set up or maintained by a generalist.

Instead of adaptive controllers, gain scheduling and signal characterization are used because these methods are relatively simple and well understood and provide a preemptive compensation.

There is a tendency for nomenclature and lack of attention to detail to cause erroneous results in the analysis and tuning of controllers. In the literature and practice, the open-loop gain is often referred to as the process gain and no attempt is made to break it up into the contributions from the major components of the loop. Furthermore, the cited gain is often not dimensionless, which means one of the components is missing from the product. For example, a typical mistake is to take the final change in the process variable in engineering units observed on a trend and divide it by the change in controller output to get a steady-state gain. This gain is the product of the gains associated with just the manipulated variable and the process and excludes the measurement gain (the effect of controller scale or sensor nonlinearity). Additionally, the open-loop gain in some books and papers has included the product of the controller gain.

Figure 6 shows how the response of the loop appears if the controller is in manual (open loop) for change in the controller output. It is assumed there are no load upsets to the loop during the test. Normally the response eventually reaches a new steady state and is called a self-regulating (type 0 in Fig. 6). The next most common response is where the process variable ramps away from the set point after the dead time and is called a integrator (type 1 in Fig. 6). The least common and the most dangerous type of response is the one in which the process variable continually accelerates away from set point and is called a runaway (type 2 in Fig. 6). Level loops are the primary examples of loops with a true integrating open-loop response. The next most common example is a conductivity loop for total dissolved solids control. However, pressure and temperature loops with large time constants compared with the dead time appear to the controller to be integrating since control action occurs during the initial excursion. Many tuning methods approximate integrators as processes with large time constants. When a runaway response occurs, it is often isolated to a region of operation. The most common source is an exothermic reaction in which the heat generation rate exceeds the cooling rate. A pH response will appear to be a runaway for an excursion from the flat toward the steepest portion of the titration curve. Loops with integrating and runaway responses require different tuning rules and methods.

To summarize, poor tuning is due to changing dead times, time constants, gains, nomenclature, tuning rules, tuning methods, and insufficient attention to detail, particularly the effect of valve dead band and installed characteristic and measurement resolution and span.

Causes of Poor Performance

1. Large dead time

2. Variable dead time

3. Small process time constant

4. Variable time constant

5. Variable gain

6. Poor sensitivity

7. Poor resolution

8. Insufficient repeatability

9. Inverse response

10. Periodic upsets

11. Noise

12. Interaction

13. Poor tuning

Best Practices

Pair the manipulated and control variables to minimize interaction and maximize leverage. The following simple rules, with examples in parentheses, should be considered in the selection of variables for the overall plant design.

1. The manipulated variable must handle the largest upset (e.g., if the vapor flow is more than 75% of the feed flow in an evaporator, the feed should be manipulated for level control and the discharge flow manipulated for concentration control).

2. The manipulated variable should enclose or inherently compensate for upsets (e.g., steam pressure instead of steam flow should be manipulated for still or reactor temperature control to correct for steam pressure upsets, should maintain a more constant heat transfer surface temperature, and should correct for changes in heat load reflected as changes in condensing rate).

3. The resolution of the manipulated variable multiplied by the open-loop gain must be less than the allowable error (e.g., the resolution of the reagent control valve multiplied by the open-loop gain for the operating point on the titration curve must not exceed the allowable pH error).

4. The sensitivity, resolution, and repeatability of the measurement must be less than the allowable error (e.g., the change in temperature on the tray selected for the allowable change in column composition must be greater than the temperature measurement sensitivity, resolution, and repeatability).

5. The maximum excursion rate multiplied by twice the dead time must be less than the allowable error (e.g., if a change in waste fuel can cause a pressure excursion faster than 1% per second, the exhaust fan speed rather than damper should be manipulated for furnace pressure control).

6. Pressure or level control should be used to prevent an overdetermined system and interaction (e.g., if the feed to coils is flow controlled, the exit flow should be manipulated to control coil pressure).

7. The process variable should inherently maximize capacity while providing tight control (e.g., for maximum capacity, the coolant valve is set wide open and the reactant feed is manipulated to control reactor temperature or pressure).

8. The manipulated variable should not cause inverse response (e.g., if the reactant is cooler than the reactor and the reaction rate is relatively slow, coolant rather than feed should be manipulated for temperature control).

9. A flow must be fixed somewhere in a recycle path, and the makeup should be done by level or pressure control to prevent a divergence of component concentrations (the recycle to a reactor should be set on flow control with a local set point, and the makeup should be manipulated to control the recycle tank level to prevent the composition error from snowballing).

10. The control loops should not interfere with process self-regulation (the steam pressure rather than the flow should be controlled for chlorine vaporizers so that steam rate matches vaporization rate to facilitate self-regulation of the chlorine level by means of changes in the heat transfer area that is submersed).

Specify, locate, and install final elements (e.g., control valves) and measurements to provide a consistent, relatively fast, sensitive, and reliable response. See the subsection on Best Practices, Tools, and Techniques for Maintenance Cost Reduction for more details.

If the disturbance is relatively fast, large, or causes inverse response, and can be measured with a relatively fast and consistent response, use feedforward control. It is particularly important to look for these feedforward opportunities as the dead time in the loop increases. However, you must be certain that the feedforward signal will not arrive too early or too late. Often the timing requirement is overlooked.

Adjust the dead time of the feedforward signal to ensure that the correction does not arrive before the upset. The dead time to be added to the feedforward signal is the time delay from the feedforward measurement of the disturbance to the start of upset to the process variable (TD_u) minus the time delay from the feedforward measurement to the start of the correction to the process variable (TD_c), as defined in Eq. (12). Note that these time delays are often variable and an inverse function of flow. If the time delays cannot be updated, then the largest value of TD_u and the smallest value of TD_c should be used since it is better that the correction arrive late than early.

$$TD_{ff} = TD_u - TD_c \qquad (12)$$

where

TD_c = time delay from the feedforward measurement to the start of the correction in the process variable (min)

TD_u = time delay from the feedforward measurement to the start of the upset in the process variable (min)

TD_{ff} = dead time to be added to the feedforward signal (min)

Adjust the gain of the feedforward signal to ensure that the correction does not exceed the upset. It is better to have a slight undercorrection than overcorrection. If the feedforward measurement is representative of the entire load to the control loop (e.g., single feed), then the feedforward gain is just the ratio of the controller output to the feedforward measurement. If the manipulated variable is a flow loop set point, then it is quite linear and feedforward simplifies to ratio control. If the upset is an oscillation, the amplitude of the feedback controller output to the amplitude of the feedforward signal provides a rough estimate of the feedforward gain. If the relationship is partial or nonlinear, then a step change in the disturbance should be made and feedforward gain estimated as the final change in controller output divided by the change in feedforward signal. If a process simulation provides the change in the process variable for a change in the disturbance variable, this can be multiplied by measurement gain (change in the percentage of measurement for a change in the process variable) and then multiplied by the inverse of the open loop gain to get the feedforward gain (the change in controller output for a change in the disturbance). Note that in every case the feedforward measurement in engineering units is multiplied by the feedforward gain to get the feedforward signal

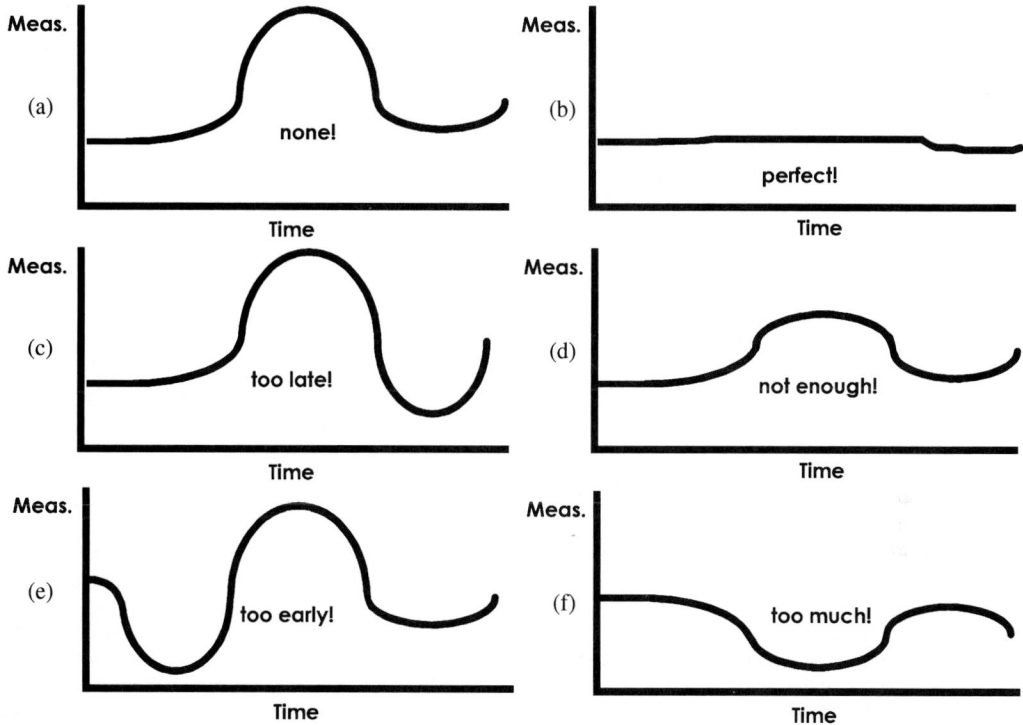

FIGURE 7   The results of feedforward gain and timing on the response to an upset.

in percentage of controller output. If the signal is incorporated into the PID loop, most DCS systems will adjust the contribution from the reset mode to provide a bumpless transfer to feedforward control and automatic. For cascade control, it is advantageous for the feedforward to be active in manual since manual has no real function for the outer (master) loop. In this case, the feedforward signal is converted to engineering units of the inner loop, externally applied to the outer PID loop output after the conversion from percentage to engineering units, and custom logic for bumpless transfer added.

Figures 7 summarize what is observed for various adjustments of the feedforward gain and timing. Figure 7(a) shows the oscillation that results from the upset with just feedback control (no feedforward). Figure 7(b) shows the nearly perfect control that results from the addition of a properly adjusted feedforward signal. Figure 7(c) shows how the second half-cycle amplitude is increased because the feedforward correction arrives after the first peak. Figure 7(d) shows that the first peak is reduced but the oscillation is still noticeable when the feedforward gain is too low. Figure 7(e) shows the inverse response from the feedforward correction affecting the process before the upset. Finally, Fig. 7(f) shows the inversion of the whole oscillation if the timing is right but the feedforward gain is too high.

Theoretically, for most ratio control systems, the feedforward signal should be multiplied by the controller output since the slope is suspected of varying and the intercept is thought to be fixed for a plot of manipulated flow versus feed flow (the manipulated flow should be zero for zero feed flow). However, the use of a feedforward multiplier introduces an undesirable nonlinearity to backmixed vessels, adds more noise at high flows, poses scaling problems, and does not compensate as well for zero shifts in the measurement and dead band in the valve. Therefore, in practice, a feedforward summer is preferably used in which the feedforward signal is either added or subtracted from the controller output, as shown in Fig. 8. However, for plug flow systems and yarn, sheet, and conveyor

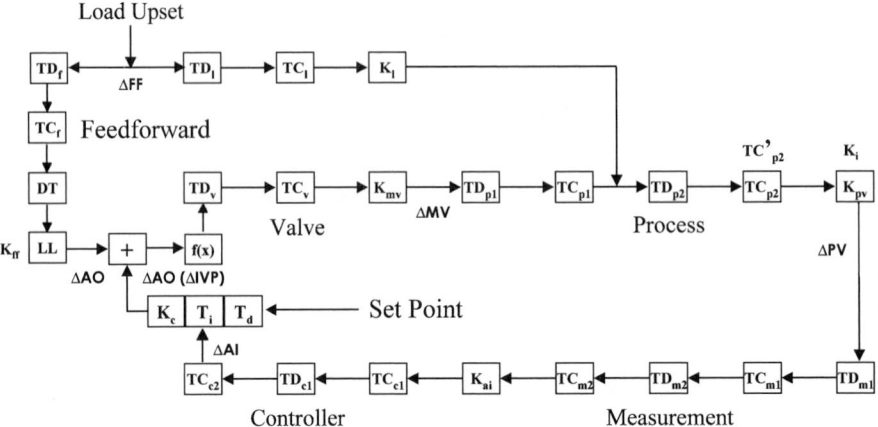

FIGURE 8   Block diagram showing the addition of a feedforward summer.

lines, it might be best to use a feedforward multiplier to eliminate the need for a gain scheduler for start-up and rate changes (the multiplication of the output by the feed or the speed compensates for the process gain that is inversely proportional to throughput).

If the feedforward signal goes to a control valve, the nonlinearity of the installed flow characteristic needs to be compensated for by signal characterization. If the feedback loop benefits from the linearization, then the characterization should be applied, as shown in Fig. 8, to the combined feedforward and feedback signals. The controller output is now a percentage of flow capacity instead of a percentage of stroke for control valve. This can be confusing to the operator and the maintenance technician when the valve is stroked in manual. However, the resolution of the local indicators of valve position are so coarse that often the distinction is not noted. The addition of the display of actual valve position from a HART or Fieldbus positioner to the operator's console may provide the understanding needed. If not, special logic can be added to bypass the signal characterization bumplessly when the control loop is in manual.

Use a velocity limit and/or a filter to keep output fluctuations from measurement within the valve dead band. This practice provides the fastest velocity limit or filter needed to prevent noise from inflicting a secondary disturbance from controller reaction and helps extend packing life. A velocity limit does not add any lag, provided it is set faster than the fastest possible process response. It is preferred for loops with a slow response, such as vessel temperature and temperature, especially to smooth out A/D noise. The velocity limit is normally switched off in manual to facilitate calibration tests and checkout.

If the disturbance is relatively fast, large, and can be enclosed in an inner (secondary or slave) loop with a relatively fast and consistent response, use cascade control. The inner loop can provide a linearization and early correction for upsets. This happens whenever a inner flow loop is added because the flow loop compensates for pressure upsets and the installed valve characteristic. These benefits are also realized when an inner inlet coolant temperature loop is introduced for reactor temperature control because coolant pressure and temperature changes are corrected before they affect the reactor temperature and the process gain (the change in reactor temperature with coolant temperature) is linear.

Tune the inner (secondary or slave) loop with mostly gain action. The purpose of the inner loop is to respond quickly to the demands of the outer loop. Offsets within the inner loop are inconsequential. Users should take note whether the valve positioner, which is an inner valve position loop on the valve, is tuned. It is a proportional-only and, more recently, a proportional-plus-derivative, controller with a gain of 40 or more. However, users have a psychological problem with any persisting offset and feel compelled to add in lots of reset action. A case in point is the most common inner loop: the

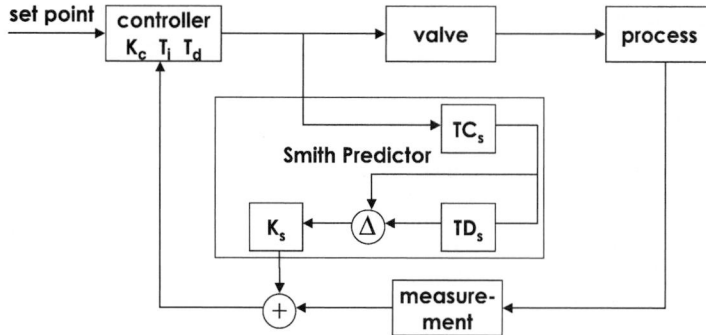

FIGURE 9   Block diagram of a dead-time compensator.

flow loop. Instead of predominantly reset action (e.g., gain = 0.25, reset = 20 repeats/min), the flow loop would benefit from the addition of a filter (e.g., 0.2 min) and the use of more gain action (e.g., gain = 1.0 and reset = 2 repeats/min). For small set-point changes from the outer loop, supervisory loop, or constrained multivariable predictive controller (CMPC), the larger gain helps make sure that the change in valve position is further beyond the resolution limit, which makes the valve response faster and more consistent. The exception is where there are frequent large pressure upsets and a flow feedforward signal is applied to the set point of the flow loop because, in this instance, reset action helps maintain the proper flow ratio. Where high gains are not permissible, the set-point change can be applied as a feedforward signal to improve the set-point response.

Tune the outer (primary or master) loop to be five times slower than the inner loop and tune interacting loops to be five times faster or slower than each other. If a loop must be detuned to be slow enough to reduce interaction, choose the least important loop. Dahlin or lambda tuning provides the most effective method of keeping inner and outer loops and competing loops from fighting by separating the dynamics. In this method, the closed-loop time constant as a multiple of the open-loop time constant is set to be more than five times larger for the slower or less important loop than for the faster or more critical loop.

If the dead time is relatively large and is always known to an accuracy of better than 25%, use a dead-time compensator. Tune the compensator's dead time to be always less than the total loop dead time, the compensator's time constant to be always more than the largest time constant, and the compensator's gain to be always less than the steady-state open-loop gain. As illustrated by Fig. 9, the dead-time compensator computes a corrective signal in percentage of measurement span . One path in the calculation is a model of the loop that uses the previously described three key variables for the open-loop response (dead time, time constant, and steady-state gain). This value is subtracted from the measurement. The other path is a model without the dead time. It is added to the measurement. If the model exactly matches the plant, what the controller sees is a measurement response to changes in its output that has no dead time. It is important to remember that the effects of the valve, process, and measurement must be included in the model. The performance of the dead-time compensator is very sensitive to model miss match. The model dead time must be within 25% of the actual dead time. A larger-than-actual dead time readily causes instability, whereas a smaller-than-actual dead time causes a prolonged choppy peak and slow recovery. Thus the lowest possible actual dead time should be used as the model dead time. While less sensitive to an inaccurate model time constant and gain, it is better to overestimate the former and underestimate the latter to prevent instability. It is interesting that this is the exact opposite of what you want to do for the calculation of the tuning settings for the PID controller.

Tune the reset time (minutes/repeat) of a controller with dead-time compensation to approach but not be less than the sum of the largest time constant plus $^1/_4$ of the uncorrected dead time. If the loop is tuned as it was before the addition of the compensator, the improvement is not noticeable, particularly if a Ziegler–Nichols method was used. This is the source of a lot of confusion in the comparison of the

TABLE 1   Default and Typical PID Settings[a]

Application type	Scan	Gain	Reset	Rate	Tuning
Liquid flow/press	1.0 (0.2→2.0)	0.3 (0.2→0.8)	10 (5→50)	0.0 (0.0→0.02)	⌒
Liquid level	5.0 (1.0→15)	5.0 (0.5→100)*	0.1 (0.0→0.5)[b]	0.0 (0.0→1.0)	CLM
Gas pressure (psig)	0.2 (0.02→1.0)	5.0 (0.5→20)	0.2 (0.1→1.0)	0.05 (0.0→0.5)	CLM
Reactor pH	2.0 (1.0→5.0)	1.0 (0.001→50)	0.5 (0.1→1.0)	0.5 (0.1→2.0)	SCM
Neutralizer pH	2.0 (1.0→5.0)	0.1 (0.001→10)	0.2 (0.1→1.0)	1.2 (0.1→2.0)	SCM
In-line pH	1.0 (0.2→2.0)	0.2 (0.1→0.3)	2 (1→4)	0.0 (0.0→0.05)	⌒
Reactor temperature	5.0 (1.0→15)	5.0 (1.0→50)	0.2 (0.05→0.5)	1.2 (0.5→5.0)	CLM
In-line temperature	2.0 (1.0→5.0)	0.5 (0.2→2.0)	1.0 (0.5→5.0)	0.2 (0.2→1.0)	⌒
Column temperature	5.0 (1.0→15)	0.5 (0.1→10)	0.2 (0.05→0.5)	1.2 (0.5→10)	SCM

[a] Scan in seconds, reset in repeats/minute, and rate in minutes, ⌒, lambda; CLM, closed-loop method; SCM, shortcut method.
[b] An error square algorithm or gain scheduling should be used for level loops with gains <5.

performance of a controller with and without a compensator. The controller needs to take advantage of the degree to which dead time has been eliminated from the reaction to its own action.

If there is inverse response, take a portion of the controller output, send it through a filter, and add or subtract it from the feedback measurement for a reverse- or a direct-acting controller, respectively. This can cancel out most of the inverse response to the controller output. The time constant is set to be approximately $\frac{1}{4}$ of the time to the inverse response peak, and the fraction is the approximately the ratio of the inverse peak to the final change in the process variable. This method does not decrease an inverse response that originates externally to the loop (e.g., load upsets, loop interactions, or early feedforward signals).

Table 1 shows typical scan times and tuning settings for the major types of loops. The first number in front of the parentheses is a suggested download value, and the numbers within the parentheses cover the range of values normally encountered. A tuning setting outside of this range may not be wrong; it may just reflect unusual circumstances such as an exceptionally wide or narrow measurement span or an oversized or very sticky valve. The last column suggests the most appropriate tuning method.

Use the closed-loop method to tune vessel pressure, temperature, and level control loops. This method keeps the loop in automatic, includes the effect of valve dynamics, and forces you to go for the highest gain, all of which are the safest approaches for runaway (e.g., exothermic reactor temperature) and fast integrator loops (e.g., furnace pressure). Normally, the test is concluded before the end of the first cycle of a damped oscillation observed in the controller output. The time between the first peak and the next peak in the opposite direction (half period) is then multiplied by 2 to get an estimate of the whole period. Because of valve nonlinearities, the period for a damped oscillation is generally greater than the ultimate period, but the error leads to less reset and more rate action, which is in the safe direction.

CLOSED-LOOP TUNING METHOD

1. Put the controller in automatic at its normal set point. If it is important not to make big changes, narrow the valve output limits for the test (remember to restore the original limits).

2. Turn off reset and rate action and trend record the controller measurement (PV) and output (IVP).

3. Add a filter to keep output fluctuations within the dead band of the valve from noise.

4. Increase controller gain and momentarily bump the output or the set point to check the response.

5. Stop when the loop starts to oscillate or the gain has reached your comfort limit and note gain setting. For gain settings greater than 1, the oscillation will be more recognizable in the controller output. Make sure the output stays on scale and within the control valve's good throttle range.

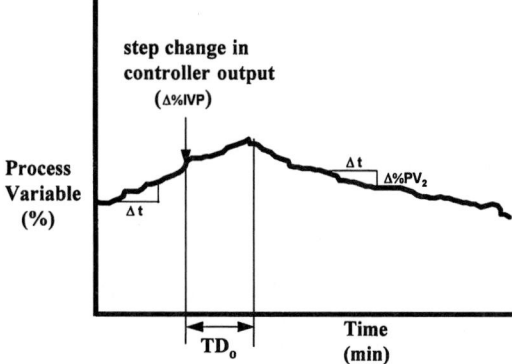

FIGURE 10   Sortcut method for a process that is not at steady state.

6. Reduce the gain until the oscillation just disappears so that the recovery is smooth.

7. If the oscillation period is greater than 0.2 min and is not a square wave (not dead-time dominant), use rate action. If the gain is larger than 10, reset and rate action are not needed. If the manipulated flow will upset other loops, decrease the gain by a factor of 2 or more or use error squared. If a high gain is used, velocity limit the set point and configure the set point to track the PV in manual and DDC so the loop can reopen the control valve from the closed position on start-up.

8. If rate is used, set the rate equal to 1/8 to 1/4 of the period and reset (repeats/minute) to the inverse of the period. If rate is not used, cut the gain by 50%. If the loop is dead-time dominant, increase the reset action to be the inverse of the sum of 1/4 of the dead time and the open-loop time constant.

9. If gains less than 5 are selected for level, add rate if there is no inverse response, decrease the reset (e.g., <0.02 repeats/min), and, most importantly, add a velocity-limited flow feedforward signal.

Use the shortcut method to tune column temperature and vessel pH control loops. This method just needs the dead time and the initial ramp rate or change in ramp rates of the open-loop process response to a bump in the controller output. This method is advantageous for loops with large time constants or small integrator gains. It does not require that the loop be at steady state since the change in ramp rates can be used, as illustrated by Fig. 10. The concept is that these types of loops should be tuned so the reversal of the upset (peak) should occur approximately two dead times after the start of the excursion. Since the controller sees only the initial ramp rate, it should be sufficient for determining the tuning settings.

SHORTCUT TUNING METHOD

1. Adjust the measurement filter to keep the output fluctuations from noise within the valve dead band.

2. Note the size of output change for reaction to typical upsets. With the controller in manual near set point, make a step change of approximately this size in the output, but larger than twice the valve dead band.

3. Note the observed dead time as time to the first change in ramp rate outside of the noise band.

4. Compute the initial new ramp rate of the response during a time period greater than two dead times. If the response was not lined out, compute the change in ramp rates, noting signs.

5. Divide the change in ramp rate by the change in controller output to get the pseudointegrator gain.

6. If the output goes to a control valve, compute the dead time from valve dead band by using Eq. (5).
7. Use Eqs. (13)–(16) to estimate the tuning settings.
8. The test should be repeated for the opposite direction and the most conservative settings used.

$$K_i = [(\Delta\%PV_2/\Delta t) - (\Delta\%PV_1/\Delta t)]/\Delta\%IVP \qquad (13)$$

$$K_c = \frac{0.5}{K_i(TD_o)} \qquad (14)$$

$$T_i = 4.0(TD_v + TD_o) \qquad (15)$$

$$T_d = 1.0(TD_v + TD_o) \qquad (16)$$

where

$\Delta IVP =$ the change in controller output (implied valve position) (%)

$K_c =$ the controller gain (dimensionless)

$K_i =$ pseudointegrator open-loop gain (1/min)

$\Delta PV_1 =$ change in process variable for ramp rate before change in IVP (%)

$\Delta PV_2 =$ change in process variable for ramp rate after change in IVP (%)

$\Delta t =$ change in time (min)

$TD_o =$ dead time seen in open-loop test (min)

$TD_v =$ dead time from control valve dead band (min)

$T_i =$ the integral time setting (min/repeat) (inverse of reset setting)

$T_d =$ derivative (rate) time setting (min)

Use the Dahlin or lambda method to tune flow, pressure, exchanger, desuperheater, static mixer, extruder, yarn, sheet, and conveyor loops. In this method, the closed-loop time constant is set approximately equal to the open-loop time constant to provide a smooth gradual response with much more reset action and less gain action for dead-time-dominant processes than you would get from the Ziegler–Nichols method. Rate action is not used. The method just needs the response time and the open-loop gain.

SIMPLIFIED DAHLIN OR LAMBDA TUNING METHOD

1. Adjust the measurement filter to keep the output fluctuations from noise within the valve dead band.
2. Note the size of output change for reaction to typical upsets. With the controller in manual near set point, make a step change of approximately this size in the output, but larger than twice the valve dead band.
3. Note the observed dead time as the time to the first change in measurement outside of the noise band.
4. Note the open-loop gain (change in % measurement divided by the change in % controller output).
5. Note the response time (the time from the bump to 98% of the final value).
6. The reset setting (repeats/min) is the inverse of $^1/_4$ of the response time [Eq. (17)].

7. The controller gain is $^1/_4$ of the inverse of the open-loop gain [Eq. (18)].

$$T_i = T_{98}/4 = \mathrm{TD}_o/4 + \mathrm{TC} \tag{17}$$

$$K_c = (\Delta \mathrm{IVP}/\Delta\%\mathrm{PV})(1/4) \tag{18}$$

where

$\Delta\mathrm{IVP}$ = the change in controller output (implied valve position) (%)

K_c = the controller gain (dimensionless)

K_i = pseudointegrator open-loop gain (1/min)

$\Delta\%\mathrm{PV}$ = change in process variable (%)

TC = time constant (min)

TD_o = dead time seen in open-loop test (min)

T_i = the integral time setting (min/repeat) (inverse of reset setting)

T_{98} = response time (time to 98% of the final response) (min)

Use optimization and constraint control to help increase process capacity or efficiency. Any equipment or process constraint (operating point limitation) that prevents an increase in the production rate or yield can be the process variable of an override control system if a value indicative of the constraint can be measured or inferred. For multiple constraints, the outputs of override controllers go to low or high signal selector to manipulate a flow or operating condition that will prevent violation of each constraint. The controller that is not chosen has its integral action track the manipulated variable so there is a bumpless transition to selection. Proportional and derivative action determine when a controller is selected. Thus the selection of an integral-only override controller is problematic. The optimization is accomplished by the signal selector's choosing between the override controller outputs and the value for the manipulated feed or operating point that corresponds to the maximum production rate or yield.

When control loops already exist for the process variables that have constraints, the controller output provides the first indication of impending violation. Override controllers are set up to prevent the valve from going beyond its usable throttle position. Normally these valve position controllers are integral only to provide a smooth gradual approach to the optimum operating point.

For exothermic and many batch operations, it is undesirable to use integral action in the controllers. For these applications and to avoid integral-only controllers for valve position constraints, the set point, bias, and gain of a proportional plus derivative controllers are adjusted to provide an allowable offset from lack of integral action. Figure 11 shows the use of an override system to push a batch reactor's capacity.

Push the most aggressive of the documented or practiced constraints values and violation times. Often the stated constraint in not the most aggressive constraint. The constraints need to be constantly challenged by incrementally letting the override system push the plant harder and harder. This means that the override system should be allowed to find the real process limitations.

Rarely considered is the fact that instantaneous violations are not detrimental except where interlocks, relief valves, and runaway reactions can be activated. If an override controller with a new process variable is created that is a function of time when it exceeds the normally considered constraint, the extended constraint provides an important increase in production rate or yield. Each scan when the normally considered constraint is exceeded, the error is multiplied by the scan time and added to the old error. If the constraint is not exceeded, the total error is multiplied by a forgetting factor that is one minus the ratio of the scan time to the supervision time. For a batch reactor, the supervision time is the batch cycle time. Note that this is not averaging or filtering in that the full effect of the newest violation is included. The immediate response to new violations helps establish confidence in the time-based constraint that is the set point of the new override controller.

For loops with supervisory and remote set points, add a filter if necessary to enable a fast setpoint response and use signal characterization to improve the sensitivity and to extend the capacity of these loops. As previously mentioned, the addition of a 0.2-min filter to the flow measurement can allow you to triple the flow controller gain to get the change in controller output beyond the

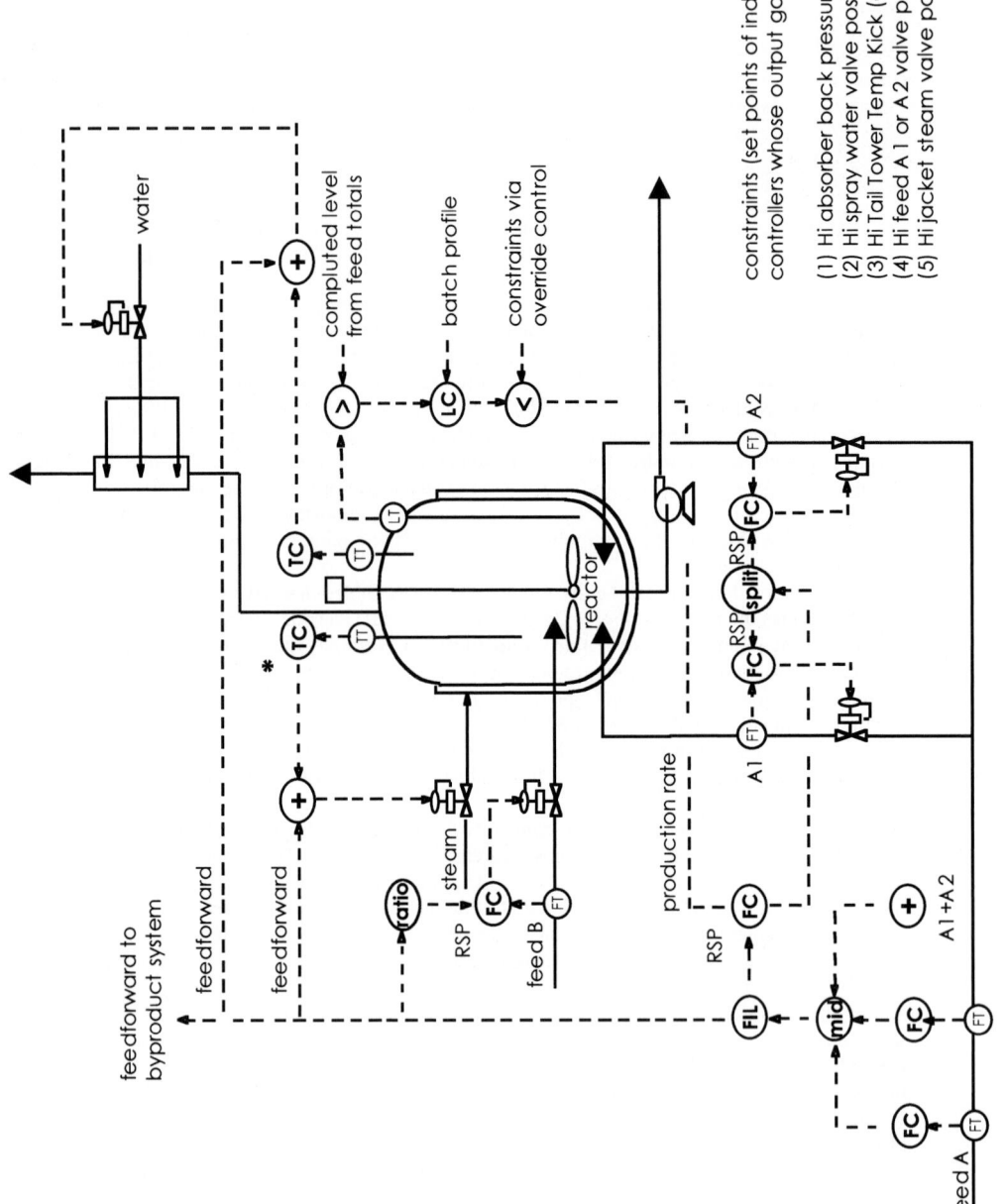

FIGURE 11 Override control to increase batch reactor capacity.

constraints (set points of individual override
controllers whose output goes to low select):

(1) Hi absorber back pressure (PV)
(2) Hi spray water valve position (IVP)
(3) Hi Tail Tower Temp Kick (dPV/dt)
(4) Hi feed A1 or A2 valve position (IVP)
(5) Hi jacket steam valve position (IVP)

resolution limit of the control valve to provide a faster and more consistent response for the master, override, or supervisory loop. Note that it is undesirable to put the filter on the controller output or in a digital positioner because it will slow down the reaction to the set-point change. The addition of signal characterization also provides a more consistent response at the upper flatter end of the installed characteristic by linearization and magnifying signal changes to be beyond the resolution limit. This enables the control system to get the last 2% or so of capacity out of a rotary control valve.

Use a kicker algorithm to get out of undesirable operating regions quickly. No matter how good the override system is, there are situations, possibly because of equipment failures, in which you need to get out of a ditch. When a kicker constraint is violated, the kicker algorithm decrements or increments the signal to a low or high signal selector, respectively, until the process variable has moved sufficiently away from the constraint. A forgetting factor is applied to the kicker signal when deactivated. There normally is a dead band to prevent noise from activating and deactivating the kicker. Besides override control, kickers are used for surge prevention in axial and centrifugal compressors, in which they are called open-loop backups and to prevent RCRA pH violations (waste discharges with a pH above 12 or below 2).

BEST PRACTICES TO IMPROVE PERFORMANCE

1. Pair the manipulated and control variables to minimize interaction and maximize leverage.
2. Specify, locate, and install final elements (e.g., control valves) and measurements to provide a consistent, relatively fast, sensitive, and reliable response.
3. Use a velocity limit and/or a filter to keep output fluctuations within the valve dead band.
4. If the disturbance is relatively fast, large, or causes inverse response, and can be measured with a relatively fast and consistent response, use feedforward control.
5. Adjust the dead time of the feedforward signal to ensure that the correction does not arrive before the upset.
6. Adjust the gain of the feedforward signal to ensure that the correction does not exceed the upset.
7. If the disturbance is relatively fast, large, and can be enclosed in an inner (secondary or slave) loop with a relatively fast and consistent response, use cascade control.
8. Tune the inner (secondary or slave) loop with mostly gain action.
9. Tune the outer (primary or master) loop to be 5 times slower than the inner loop.
10. Tune interacting loops to be 5 times faster or slower than each other.
11. If a loop must be detuned to be slow enough to reduce interaction, choose the least important loop.
12. If the dead time is relatively large and is always known to an accuracy of better than 25%, use a dead-time compensator.
13. Tune the dead time of a dead-time compensator to always be less than the total loop dead time.
14. Tune the time constant of a dead-time compensator to always be more than the largest time constant.
15. Tune the gain of a dead-time compensator to always be less than the steady-state open-loop gain.
16. Tune the reset time (minutes/repeat) of a controller with dead-time compensation to approach but not be less than the sum of the largest time constant plus $1/4$ of the uncorrected dead time.
17. If there is an inverse response, take a portion of the controller output, send it through a filter, and add or subtract it from the feedback measurement for a reverse or direct acting controller, respectively.
18. Use the closed-loop method to tune vessel pressure, temperature, and level control loops.
19. Use the shortcut method to tune column temperature and vessel pH control loops.

20. Use the Dahlin or lambda method to tune flow, pressure, exchanger, desuperheater, static mixer, extruder, yarn, sheet, and conveyor loops.

21. Use optimization and constraint control to help increase process capacity or efficiency.

22. Push the most aggressive of the documented or practiced constraints values and violation times.

23. For loops with supervisory and remote set points, add a filter if necessary for a fast set-point response, and use signal characterization to improve the sensitivity and to extend the capacity of these loops.

24. Use a kicker algorithm to get out of undesirable operating regions quickly.

REFERENCES

1. McMillan, G. K., *Tuning and Control Loop Performance*, 3rd ed., Instrument Society of America, Research Triangle Park, North Carolina, 1996.

2. McMillan, G. K., and C. M. Toarmina, *Advanced Temperature Measurement and Control*, Instrument Society of America, Research Triangle Park, North Carolina, 1995.

MULTIVARIABLE PREDICTIVE CONTROL AND REAL-TIME OPTIMIZATION

by Paul G. Luebbers*

WHAT ARE CONSTRAINED MULTIVARIABLE PREDICTIVE CONTROL AND REAL-TIME OPTIMIZATION?

Discussions of process optimization often focus on the design of a process. A set of operating objectives are specified, including product throughputs, product specifications, and environmental restrictions. Given these objectives, the goal of process optimization is to determine the best arrangement and sizes of equipment to minimize both capital and operating costs.

For existing plants, which may or may not be operating at nameplate capacities, process optimization focuses on an entirely different problem: Given the fixed arrangement and sizes of equipment, the quality and cost of the feedstocks, utilities costs, product specifications, and market demand represented by the existing plant, what are the best operating conditions to produce the most valuable products at the lowest operating costs? This is the objective of real-time optimization (RTO).

The operating constraints of a chemical plant or refinery at any point in time may be a function of feed rate, production targets, ambient conditions, product specifications, or equipment performance. These items and the resulting constraints are all subject to change during normal operating conditions.

* Solutia Inc., Cantonment, Florida.

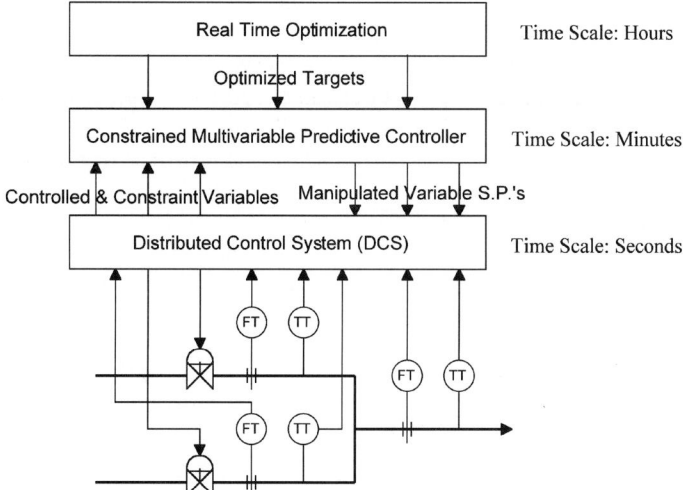

FIGURE 1   Hierarchy of RTO, CMPC, distributed control system and controlled process.

The implementation of a RTO system requires a control system capable of consistently operating the process against these types of constraints. Constrained multivariable predictive control (CMPC) is a multivariable control algorithm that is able to meet these performance requirements.

In addition to operating the process against these constraints, CMPC is effective and efficient for controlling multivariable control systems, inherently compensates for dead-time and process interaction, uses step responses generated from input/output plant test data that are easy to generate, requires no fundamental process modeling, is not compromised by inverse response behavior, and can be made to be robust as part of the design.

RTO and CMPC are implemented in a hierarchy, as shown in Fig. 1. The real-time optimizer runs at a frequency on a scale of hours and generates targets for the constrained multivariable predictive controller that are determined to be a steady-state operating point based on an economic optimization. These targets are passed from the RTO layer down to the CMPC layer in the hierarchy. The constrained multivariable predictive controller runs at a frequency on a scale of minutes and implements the targets sent from the RTO layer such that the process is moved from one steady-state operating point to another while respecting constraints on the process and ensuring a smooth transition. The CMPC accomplishes the implementation of the RTO targets by sending setpoints from the CMPC layer down to the distributed control system (DCS) layer of the hierarchy. The DCS runs at a frequency on a scale of seconds or fractions of a second and either controls the process by manipulating final control elements, such as control valves, directly or through traditional proportional-integral-derivative (PID) controllers.

The three layers of this control hierarchy work together to optimize the operating process by providing different functionality to the overall control system. The DCS provides setpoint following and disturbance rejection capabilities. The CMPC provides for control loop decoupling, predictive capabilities, and constraint handling. The RTO provides economically optimal steady-state targets for the process.

Basic Concepts of Constrained Multivariable Predictive Control

To describe what CMPC is, a simple example will be used. Consider a system to deliver warm water. The system consists of a cold stream of water, a mixing section, a hot water source with a valve for manipulating its flow, and a temperature measurement device, as shown in Fig. 2.

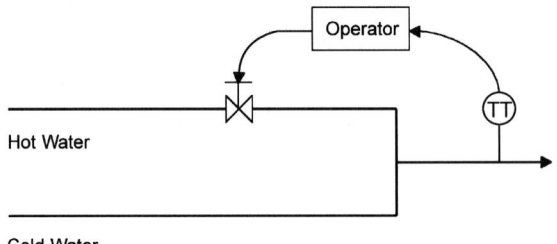

FIGURE 2   Manual Control of a warm water process.

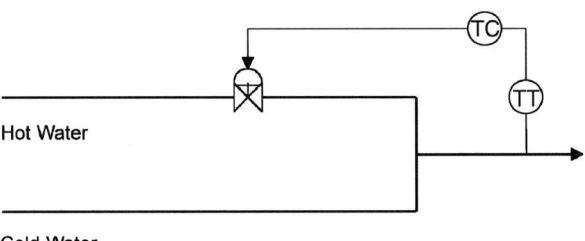

FIGURE 3   Automatic control of a warm water process.

Qualitative Description of Constrained Multivariable Predictive Control.   We make some assumptions:

1. The desired temperature of the delivered warm water does not change often.

2. The supply pressure of the hot water stream remains constant.

3. The temperature of the cold water stream remains constant.

4. The flow rate of the cold water stream remains constant.

5. The temperature of the hot water stream remains constant.

6. The flow rate of warm water delivered is not important.

We could control this process manually by reading the temperature measurement device and adjusting the hot water flow valve manually by trial and error until the desired temperature is reached. If, however, the desired temperature of the delivered warm water changes often (assumption 1 above), we would probably want to control this system with an automatic control system. (Fig. 3) This controller would probably be a PID controller, tuned for setpoint changes. If the supply pressure of the hot water stream was not constant (assumption 2 above), we would probably want to add a flow controller to the hot water stream and cascade the temperature controller to the hot water flow controller (Fig. 4). Again this controller would probably be a PID controller tuned for setpoint changes.

If assumptions 3, 4, and 5 (or any combination of these assumptions) were not true and we could measure the temperatures and flow rates associated with assumptions 3, 4, and 5, we might want to add some feedforward signals to our PID controllers in Fig. 4 or a predictive controller (PC) (Fig. 5). This controller would be implemented at a control level above the regulatory PID controllers and would supply a supervisory setpoint to the hot water flow controller and the cascade between the temperature controller and the hot water flow controller would be opened. In this configuration, the temperature controller would not be used, as its function would be replaced by the PC. The temperature of the warm water is called the controlled variable (CV) and the temperature of the cold water stream, the flow rate of the cold water stream, and the temperature of the hot water stream are called the feedforward variables (FVs).

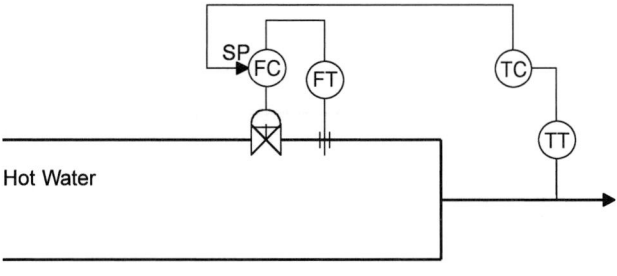

FIGURE 4   Cascade control of a warm water process.

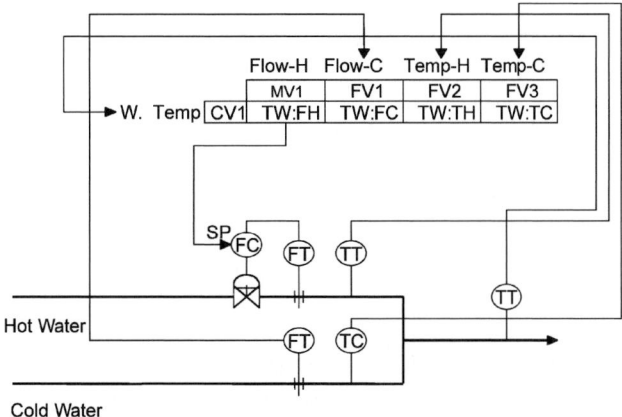

FIGURE 5   Predictive control of a warm water process.

If we wanted to control the amount of warm water to be delivered as well as the temperature of the warm water delivered, we would need an additional controller and control valve on the cold water stream to control the quantity of warm water or we could use a multivariable predictive controller (MPC) (Fig. 6). This MPC is designed to control both the temperature of the warm water and the quantity of the warm water (CVs). The flow of hot water and the flow of cold water are called the manipulated variables (MVs). This controller will control multiple CVs by manipulating multiple MVs, thus the term Multivariable.

Most commercially available MPCs are able to maximize or minimize an engineering objective (often in the form of a cost function). For the example used thus far, we may want to maximize the quantity of warm water to be delivered while maintaining the desired temperature of the water. It is obvious that we will be limited by the quantity of hot or cold water, depending on the temperatures of the hot and the cold streams and the flow rates of these streams. To provide this information to our MPC we might transmit the valve positions of our flow control valves to the controller. These valves (when fully open or closed) constitute constraints on the system and are called constraint variables (XVs). The inclusion of constraints on our MPC would result in a CMPC (Fig. 7).

Characteristics of Constrained Multivariable Predictive Control.   The structure of CMPCs is typically depicted in the form of an input/output matrix that not only includes the MVs, FVs, CVs, and XVs but also describe the relationship between the variables in the form of a dynamic model (Fig. 8). These model relationships and the resulting controller are typically linear. The models can

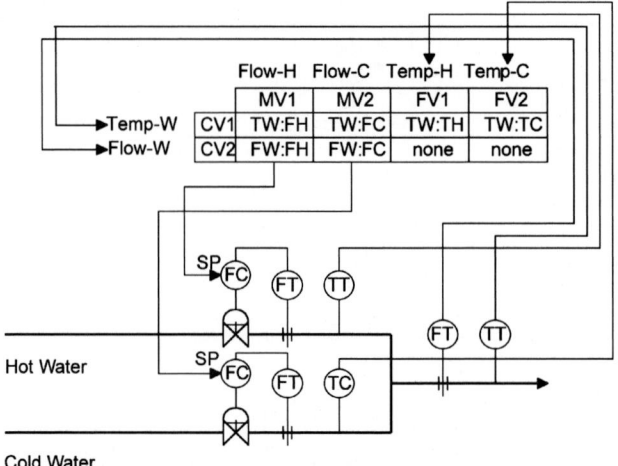

FIGURE 6   MPC of a warm water process.

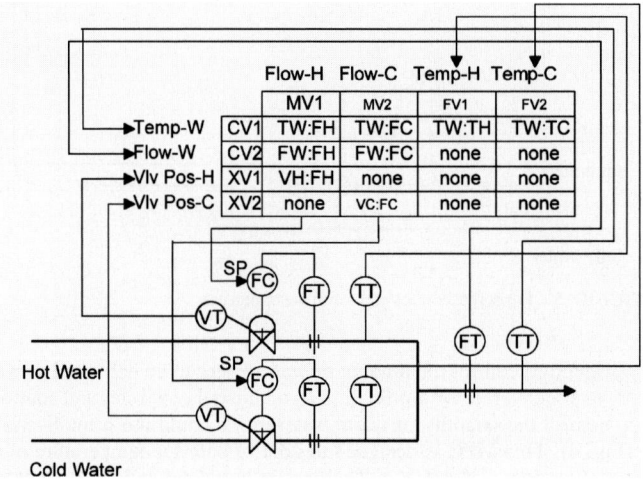

FIGURE 7   CMPC of a warm water process.

take the form of simple low-order parameterized processes with dead times, as state space models, or as discrete step or impulse responses.

As explained in the previous discussion, CMPCs have very desirable characteristics for the control of complex multivariable processes, including the following:

1. predictive feedforward control abilities

2. controller can constrain the movements of the manipulated variables to a maximum or a minimum in the present or any future time interval

3. controller can handle constraints on manipulated variables in the present or any future time

4. controller can handle constraints on controlled variables in the present or any future time

5. drive the process to its economic optimum (for multivariable problems with more inputs than outputs)

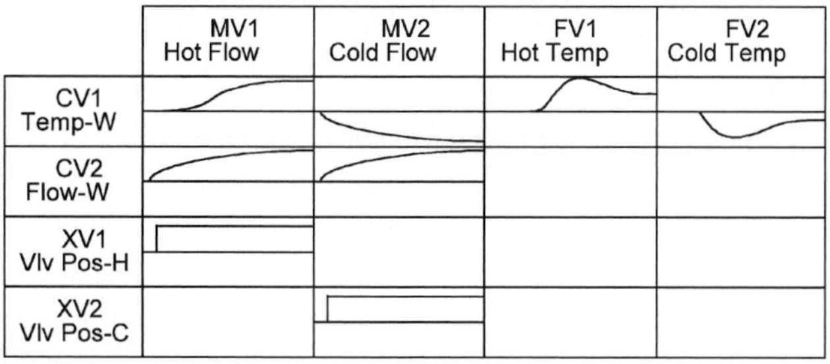

	MV1 Hot Flow	MV2 Cold Flow	FV1 Hot Temp	FV2 Cold Temp
CV1 Temp-W				
CV2 Flow-W				
XV1 Vlv Pos-H				
XV2 Vlv Pos-C				

FIGURE 8   CMPC input/output matrix of a warm water process.

6. allow for any number of feedforward disturbances

7. multivariable decoupling of interacting control objectives

8. handles large number of inputs and outputs

9. accommodates time delays, inverse responses, and long settling times

Basic Concepts of Real-Time Optimization?

RTO uses a detailed steady-state mathematical model of the process in question to find economically optimum operating points for the plant.

The optimization is designed to maximize the profit of the plant by finding the best values for the advanced controller (CMPC) setpoints or targets. This optimization takes the following into account:

1. ambient conditions

2. physical constraints

3. feed-stream availability

4. feed-stream compositions

5. product volumes

6. product qualities

7. process degradation

8. plant configurations

9. thermodynamic, kinetic, and equilibrium relationships

Before the on-line optimization routine can be built, an off-line problem must be developed. This procedure takes the following form:

1. Assemble flow sheets for the plant.

2. Use a commercial process flow sheet simulator to define process configuration.

3. Build a problem to optimize by assembling standard sets of equations and generate physical properties data.

4. Run simulation cases to test reconciliation routines, process constraints, and the objective function to optimize.

5. Assign process tags to problem variables, define the steady-state detection criterion, and the gross error detection criterion.

6. Build on-line problem definition files.

The RTO process follows several steps (Fig. 9):

1. Take plant measurements.
2. Test recent history for steady-state operation.
3. Eliminate bad measurements by using gross error detection.
4. Update model parameters based on process measurements taken from the plant.
5. Calculate new setpoints for the plant to increase profit by optimization of a cost function.
6. Check that the process is still at steady state and that the controls are available.

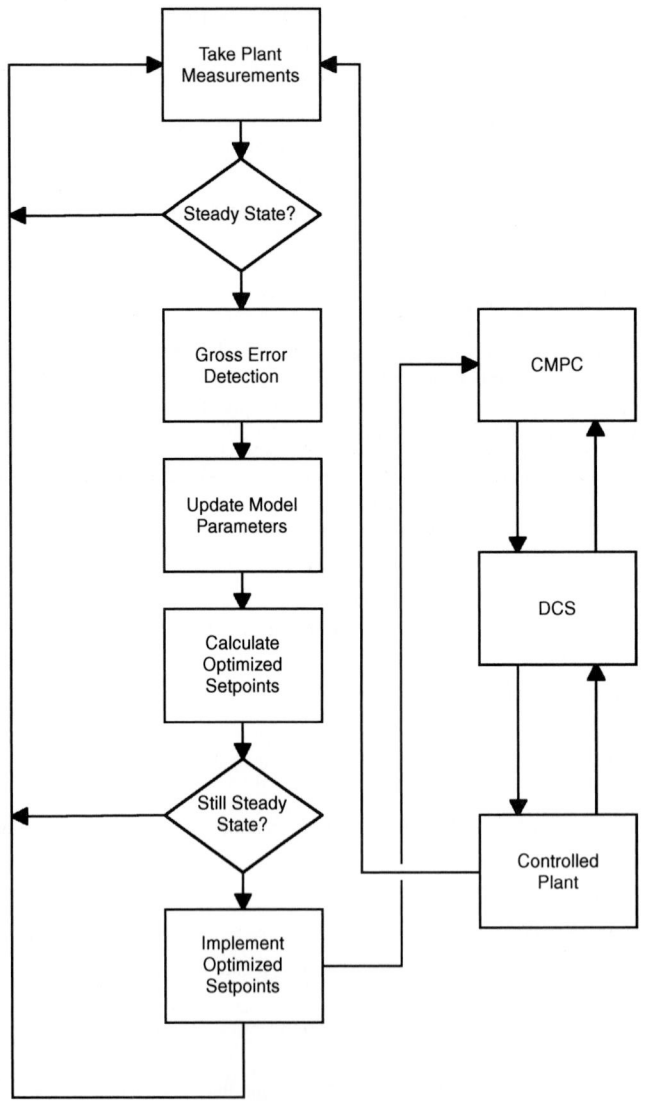

FIGURE 9   One-line process for RTO routine.

7. Implement setpoints.

8. Return to step 1.

MATHEMATICAL FOUNDATIONS OF CONSTRAINED MULTIVARIABLE PREDICTIVE CONTROL AND REAL-TIME OPTIMIZATION

Although there are a large number of ways to formulate the CMPC solution and a large number of enhancements and improvements to the basic technique, what is presented here is typical of the methods that have been used in industry. The original development of CMPC for industrial processes evolved from a technique of representing process dynamics with a set of numerical coefficients [1]. This numerical technique is used to minimize the integral of the error between CVs and target values or setpoints for each of these CV's over a time horizon. The technique assumes that the process can be described or approximated by a set of linear differential equations.

Linear Systems

Processes or systems are defined to be linear if they possess the two properties of superposition and homogeneity [2]. A system that produces an output $y_1(t)$ for an input of $x_1(t)$ and an output of $y_2(t)$ for an input $x_2(t)$ will produce an output $y_1(t) + y_2(t)$ for an input of $x_1(t) + x_2(t)$ if the system possesses the property of superposition (Fig. 10). Similarly, a system that produces an output $y(t)$ for an input of $x(t)$ will produce an output $a[y(t)]$ for an input of $a[x(t)]$ if the system possesses the property of homogeneity (Fig. 11).

A system described by the relation $y = mx + b$ is not linear because it does not satisfy the homogeneity property. A system described by the relation $y = x^2$ is not linear because the superposition property is not satisfied. These systems can be approximated by a linear system about an operating point x_0, y_0 for small changes Δ_x and Δ_y.

If we substitute $x = x_0 + \Delta x$ and $y = y_0 + \Delta y$ into the first of these systems, we have

$$y_0 + \Delta y = m(x_0) + m(\Delta x) + b \tag{1}$$

Therefore we have an equation that describes the operating point where the system was linearized,

$$y_0 = m(x_0) + b \tag{2}$$

and an equation that describes the linearized system,

$$\Delta y = m(\Delta x) \tag{3}$$

A Taylor series expansion of the second relationship, $y = x^2$, about the operating point x_0, y_0 is

$$y = x^2 = (x_0)^2 + d(x^2)/dx[(x - x_0)/1!] + d^2(x^2)/dx[(x - x_0)^2/2!] + \cdots \tag{4}$$

and evaluated at $x = x_0$. The slope at the operating point, $d(x^2)/dx$, is a good approximation to the curve over a small range of $(x - x_0)$, the deviation from the operating point (Fig. 12). Therefore a reasonable approximation of this system would be

$$y = (x_0)^2 + d(x^2)/dx(x - x_0) = y_0 + m(x - x_0) \tag{5}$$

where m is the slope at the operating point. Equation (5) can then be rewritten as the linear equation:

$$(y - y_0) = m(x - x_0) \tag{6}$$

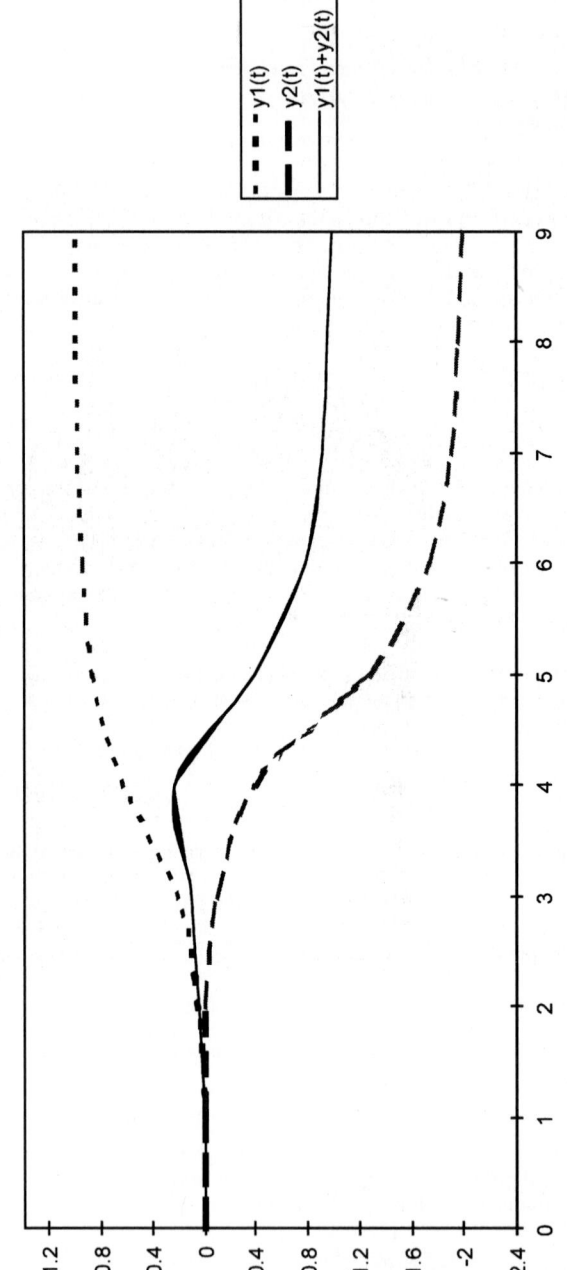

FIGURE 10 Process response for a process that has the property of superposition.

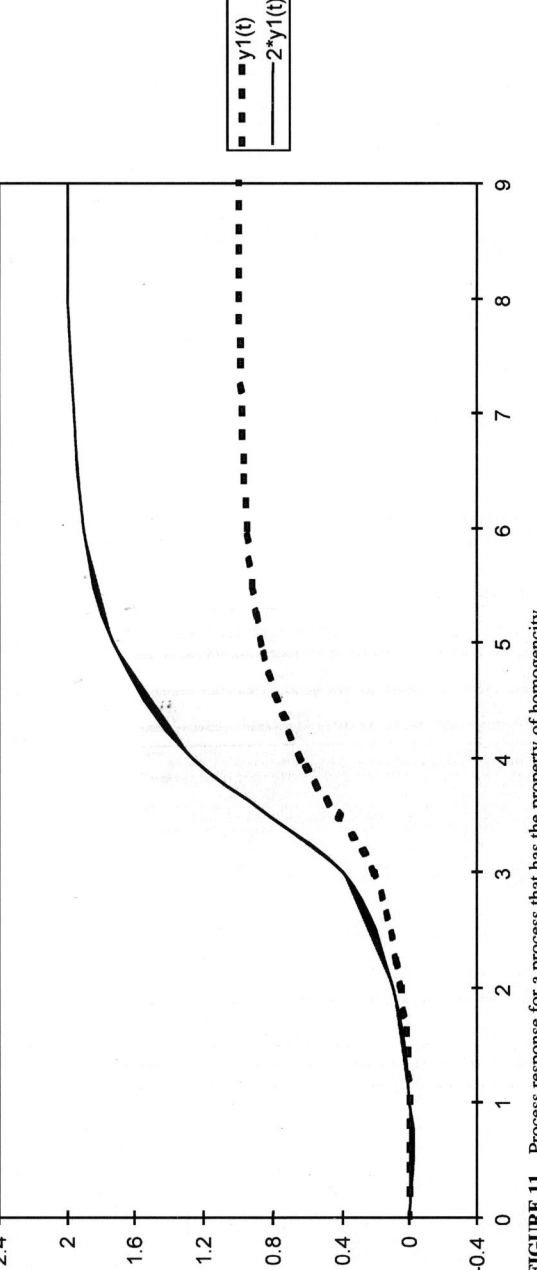

FIGURE 11 Process response for a process that has the property of homogencity.

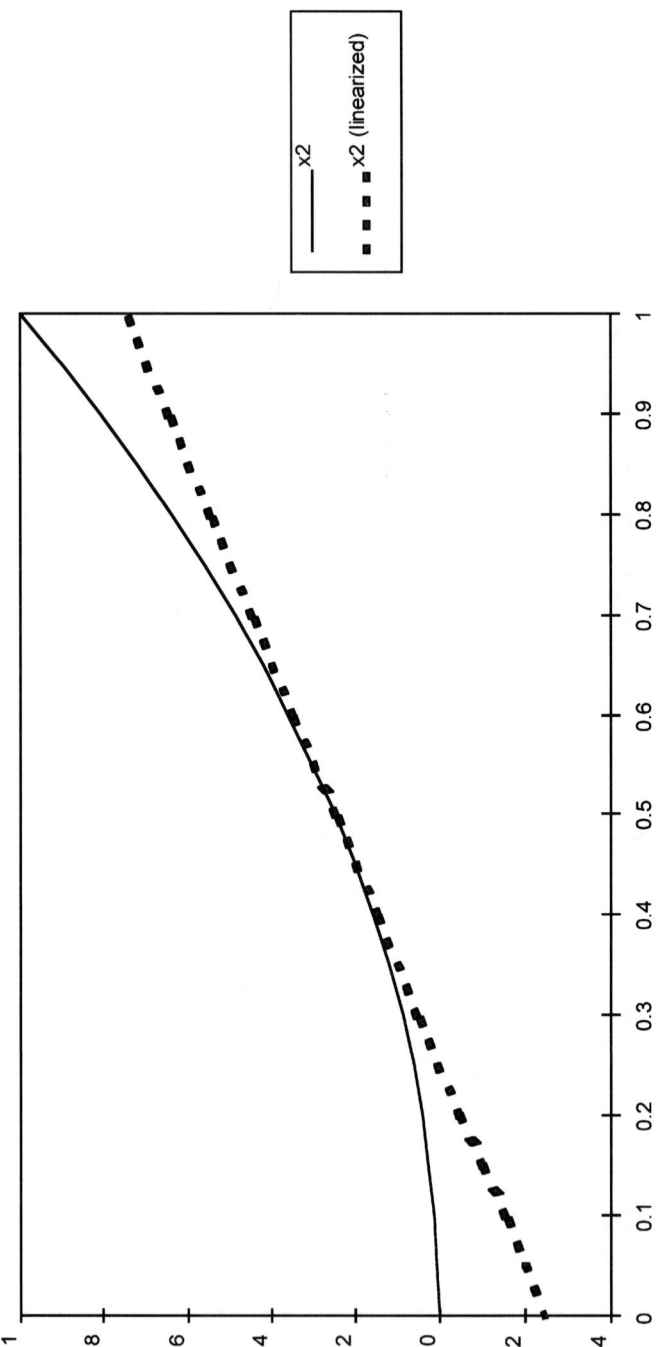

FIGURE 12 Plot of $y = x^2$ with linearization about the point $x = 0.5$.

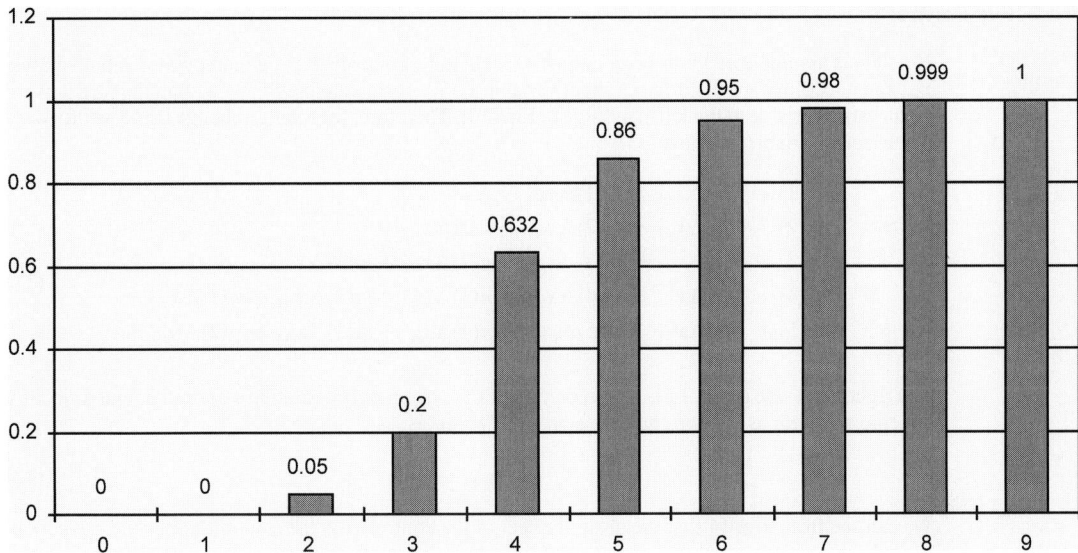

FIGURE 13   Process response represented by a set of numerical coefficients.

or

$$\Delta y = m(\Delta x) \tag{7}$$

Fortunately for many systems in the process industries, this approximation is valid.

Process Representations

These linearized responses can be represented by a set of numerical coefficients (Fig. 13). If we assume that a response y_a to an input x_{a1} at time $(t = 1)$ can be represented by a set of numerical coefficients a, then y_a can be written as

$$\Delta y_a(k) = a(k)\Delta x_{a1} \tag{8}$$

where Δy_a (k) is the change in the output from its initial value to its value at a time k intervals into the future, $a(k)$ is the kth coefficient that represents the response at a time k intervals into the future and Δx_{a1} is the change in the input a at time $t = 1$. If we assume that a second response y_b to a second input x_{b1} can be represented by a second set of numerical coefficients b, then a general expression for the response $y_{(a1 + b1 + \cdots)}$ is

$$\Delta y_{(a1+b1+\cdots)}(k) = a(k)\Delta x_{a1} + b(k)\Delta x_{b1} + \cdots \tag{9}$$

If we assume that the input x_a is a manipulated variable a used to control a controlled variable y_a, where x_a is changed at only discrete intervals, the controller variable y_a can be expressed as

$$\begin{aligned}
\Delta y_a(1) &= a(1)\Delta x_{a1} \\
\Delta y_a(2) &= a(2)\Delta x_{a1} + a(1)\Delta x_{a2} \\
\Delta y_a(3) &= a(3)\Delta x_{a1} + a(2)\Delta x_{a2} + a(1)\Delta x_{a3} \\
\Delta y_a(4) &= a(4)\Delta x_{a1} + a(3)\Delta x_{a2} + a(2)\Delta x_{a3} + a(1)\Delta x_{a4} \\
\Delta y_a(5) &= a(5)\Delta x_{a1} + a(4)\Delta x_{a2} + a(3)\Delta x_{a3} + a(2)\Delta x_{a4} + \cdots
\end{aligned} \tag{10}$$

$$\cdots$$

Predictive Control

Now let us assume that we have an output variable to be controlled y_c, a manipulated input variable x_m, and a feedforward disturbance input variable x_f. We formulate the PC by adding the feedforward disturbance variable to the left-hand side of Eq. (10). Therefore, for a single change in the feedforward disturbance variable, we have

$$
\begin{aligned}
\Delta y_a(1) + f(1)\Delta x_{f1} &= m(1)\Delta x_{m1} \\
\Delta y_a(2) + f(2)\Delta x_{f1} &= m(2)\Delta x_{m1} + m(1)\Delta x_{m2} \\
\Delta y_a(3) + f(3)\Delta x_{f1} &= m(3)\Delta x_{m1} + m(2)\Delta x_{m2} + m(1)\Delta x_{m3} \\
\Delta y_a(4) + f(4)\Delta x_{f1} &= m(4)\Delta x_{m1} + m(3)\Delta x_{m2} + m(2)\Delta x_{m3} + m(1)\Delta x_{m4} \\
\Delta y_a(5) + f(5)\Delta x_{f1} &= m(5)\Delta x_{m1} + m(4)\Delta x_{m2} + m(3)\Delta x_{m3} + m(2)\Delta x_{m4} + \cdots
\end{aligned}
\tag{11}
$$

$$\cdots$$

The right-hand side of Eq. 11 is the prediction of the system $\mathbf{p}$. If we define a desired output trajectory or setpoint vector as $\mathbf{y}_s$, we can define an error vector as

$$
\mathbf{e} = \mathbf{y}_s - \mathbf{p}
\tag{12}
$$

We can see from Fig. 14 that the desired effect of the controller is the setpoint minus the prediction.

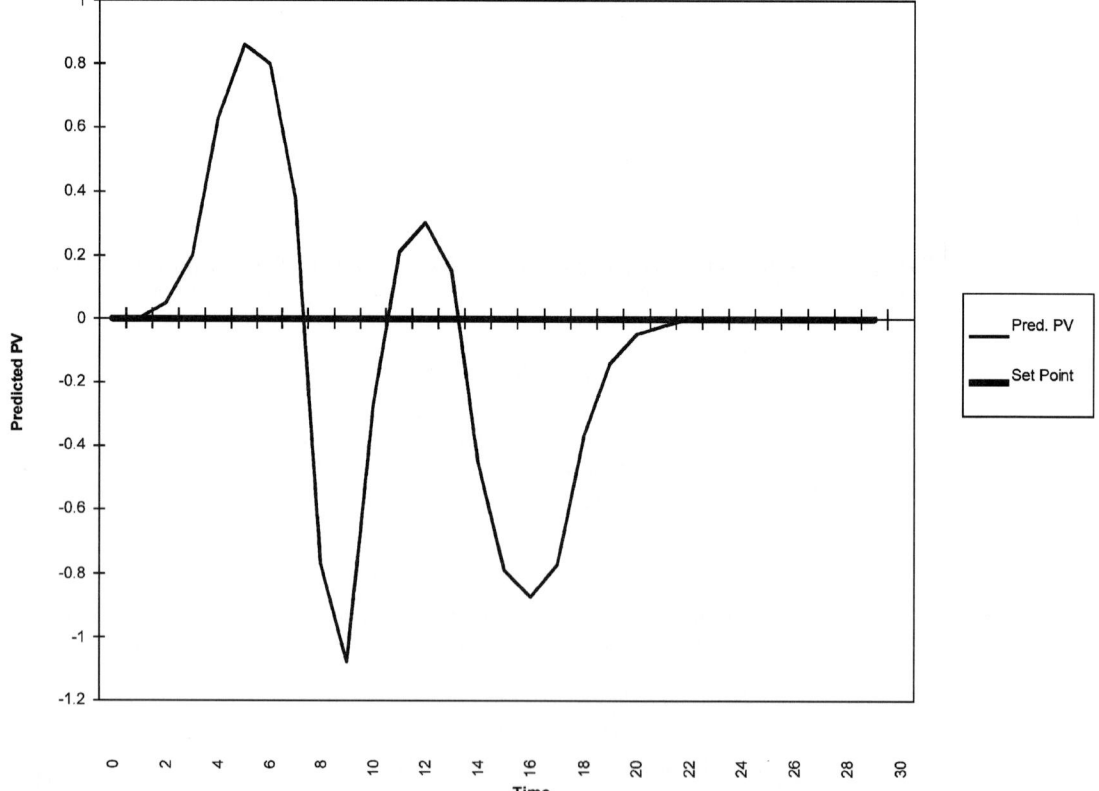

FIGURE 14 Plot of process prediction and process setpoint.

If we define a matrix, M as

$$M = \begin{bmatrix} m(1) & 0 & 0 & 0 & 0 & \cdots \\ m(2) & m(1) & 0 & 0 & 0 & \cdots \\ m(3) & m(2) & m(1) & 0 & 0 & \cdots \\ m(4) & m(3) & m(2) & m(1) & 0 & \cdots \\ m(5) & m(4) & m(3) & m(2) & m(1) & \cdots \end{bmatrix} \tag{13}$$

and define a vector $\mathbf{x}_m$ as

$$\mathbf{x}_m = \begin{array}{c} \Delta x_{m1} \\ \Delta x_{m2} \\ \Delta x_{m3} \\ \Delta x_{m4} \\ \Delta x_{m5} \\ \cdots \end{array} \tag{14}$$

we can express the right-hand side of Eq. (11) as $M_{\mathbf{x}}$, and for the formulation of the PC we have:

$$\mathbf{e} = M\mathbf{x}_m \tag{15}$$

Since the predictions **p** are known (from observing past inputs to the system) and the setpoint vector $\mathbf{y}_s$, is known, the error vector **e** is known. The coefficients $m(k)$ are known and constant; therefore the matrix M is known and constant. The width of the matrix M specifies how many future moves are to be calculated and the height of the matrix M specifies the time horizon over which the controller is to operate.

The set of equations described by Eq. (15) is overdetermined and can be solved with a least-squares criterion. The object of the least-squares criterion is to minimize the residual difference r between the setpoint and the prediction by minimizing the sum of the squared residual errors (Fig. 15):

$$r = (M\mathbf{x}_m - e) \tag{16}$$

The solution to this minimization problem can be found in any linear algebra text [3] and results in

$$\mathbf{x}_m = (M^T * M)^{-1} * A^T \times e \tag{17}$$

This provides the least-squares minimization of the following set of single-input–single-output (SISO) equations.

$$\begin{array}{ccccc} m(1) & 0 & 0 & 0 & 0 \\ m(2) & m(1) & 0 & 0 & 0 \\ m(3) & m(2) & m(1) & 0 & 0 \\ m(4) & m(3) & m(2) & m(1) & 0 \\ m(5) & m(4) & m(3) & m(2) & m(1) \\ m(6) & m(5) & m(4) & m(3) & m(2) \\ m(7) & m(6) & m(5) & m(4) & m(3) \\ m(8) & m(7) & m(6) & m(5) & m(4) \\ m(8) & m(8) & m(7) & m(6) & m(5) \\ m(8) & m(8) & m(8) & m(7) & m(6) \\ m(8) & m(8) & m(8) & m(8) & m(7) \\ m(8) & m(8) & m(8) & m(8) & m(8) \end{array} * \begin{array}{c} \Delta x_{m1} \\ \Delta x_{m2} \\ \Delta x_{m3} \\ \Delta x_{m4} \\ \Delta x_{m5} \\ \\ \\ \\ \\ \\ \\ \end{array} = \begin{array}{c} e_1 \\ e_2 \\ e_3 \\ e_4 \\ e_5 \\ e_6 \\ e_7 \\ e_8 \\ e_9 \\ e_{10} \\ e_{11} \\ e_{12} \end{array} \tag{18}$$

This set of equations describes a problem that has a prediction horizon of 12 periods, a process that has a settling time of 8 periods, and allows the calculation of 4 future moves.

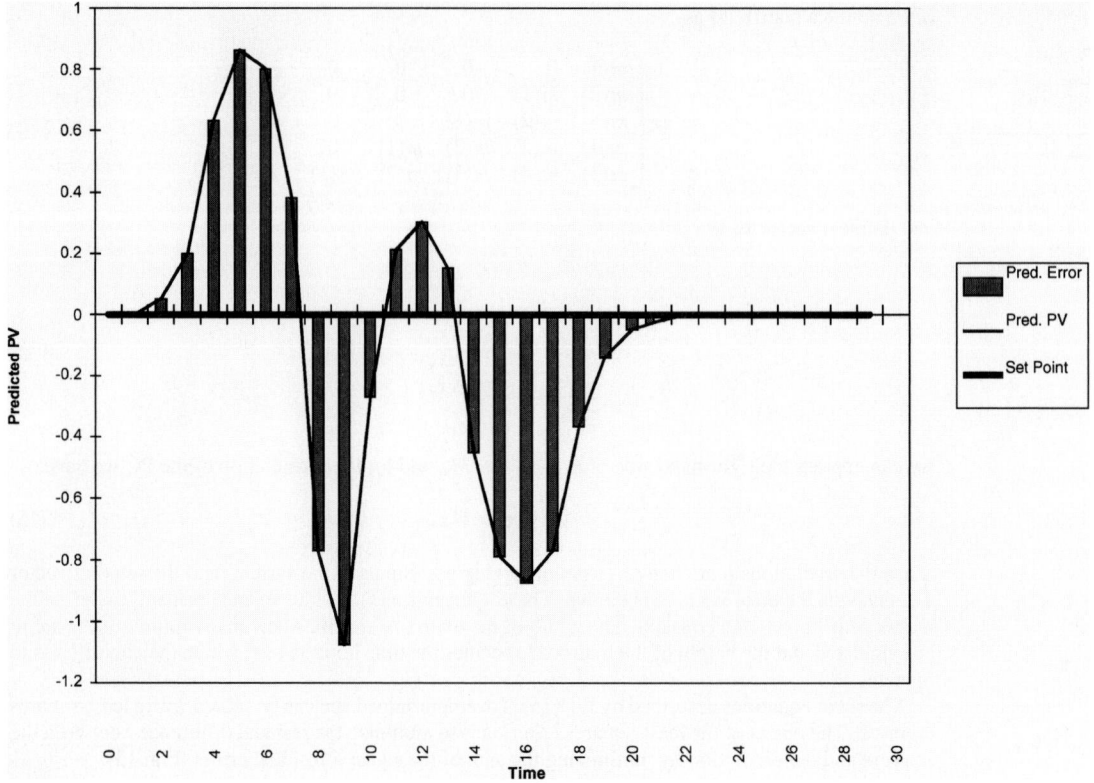

FIGURE 15 Plot of process prediction and residuals between prediction and setpoint of zero.

Move Suppression

This formulation will aggressively determine current and future moves that will minimize the error between the prediction and the setpoint across the entire time horizon. The solution of Eq. (17) will be the optimum solution for minimizing the error. The planned moves are often considered too aggressive, and so it is desired to have a smoother control action even at the expense of having a suboptimum solution to the minimization problem. To solve this, a second objective to minimize the controller movement may be added to the minimization of error criterion. A balance between these two, sometimes conflicting, objectives can be achieved be introduction of an adjustable relative weight, often known as a move suppression (MS) factor. This MS factor can be incorporated into the problem formulation by appending an additional set of equations that tend to minimize the manipulated moves Δx_{mn}. For the case in which five future moves are to be calculated, we would include the following:

$$
\begin{aligned}
k \;*\; \Delta x_{m1} &= 0 \\
k \;*\; \Delta x_{m2} &= 0 \\
k \;*\; \Delta x_{m3} &= 0 \\
k \;*\; \Delta x_{m4} &= 0 \\
k \;*\; \Delta x_{m5} &= 0
\end{aligned}
\tag{19}
$$

This can be expressed in matrix form as

$$
K \;*\; \mathbf{x}_m = \mathbf{z}
\tag{20}
$$

where K is a square diagonal matrix of dimension 5 with the value of the move suppression factor K on the diagonal elements. The vector x_m is the vector of future moves, as previously defined, and $\mathbf{z}$ is a vector of zeros. These equations can be incorporated by appending them to the set of equations described by Eq. (18) as

$$
\begin{array}{|ccccc|}
m(1) & 0 & 0 & 0 & 0 \\
m(2) & m(1) & 0 & 0 & 0 \\
m(3) & m(2) & m(1) & 0 & 0 \\
m(4) & m(3) & m(2) & m(1) & 0 \\
m(5) & m(4) & m(3) & m(2) & m(1) \\
m(6) & m(5) & m(4) & m(3) & m(2) \\
m(7) & m(6) & m(5) & m(4) & m(3) \\
m(8) & m(7) & m(6) & m(5) & m(4) \\
m(8) & m(8) & m(7) & m(6) & m(5) \\
m(8) & m(8) & m(8) & m(7) & m(6) \\
m(8) & m(8) & m(8) & m(8) & m(7) \\
m(8) & m(8) & m(8) & m(8) & m(8) \\
\hline
k & 0 & 0 & 0 & 0 \\
0 & k & 0 & 0 & 0 \\
0 & 0 & k & 0 & 0 \\
0 & 0 & 0 & k & 0 \\
0 & 0 & 0 & 0 & k \\
\end{array}
* \quad
\begin{array}{c}
\Delta x_{m1} \\
\Delta x_{m2} \\
\Delta x_{m3} \\
\Delta x_{m4} \\
\Delta x_{m5} \\
\end{array}
\; = \;
\begin{array}{c}
e_1 \\
e_2 \\
e_3 \\
e_4 \\
e_5 \\
e_6 \\
e_7 \\
e_8 \\
e_9 \\
e_{10} \\
e_{11} \\
e_{12} \\
\hline
0 \\
0 \\
0 \\
0 \\
0 \\
\end{array}
\tag{21}
$$

or

$$
\frac{M}{K} * \; \mathbf{x}_m = \frac{\mathbf{e}}{\mathbf{z}}
\tag{21a}
$$

Extension to the Multivariable Case

For the multivariable 2×2 case we can easily extend Eq. (21a) as

$$
\frac{\begin{array}{c|c} M_1 & N_1 \\ M_2 & N_2 \\ \hline K_1 & Z \\ Z & K_2 \end{array}}{} * \frac{\mathbf{x}_m}{\mathbf{x}_n} = \frac{\begin{array}{c} \mathbf{e}_1 \\ \mathbf{e}_2 \\ \hline \mathbf{z} \\ \mathbf{z} \end{array}}{}
\tag{22}
$$

where M_1 is the model of the response of CV_1 to MV_1, M_2 is the model of the response of CV_2 to MV_1, N_1 is the model of the response of CV_1 to MV_2, and N_2 is the model of the response of CV_2 to MV_2. The vectors $\mathbf{x}_m$ and $\mathbf{x}_n$ are the changes in MV_1 and MV_2, respectively.

The vectors $\mathbf{e}_1$ and $\mathbf{e}_2$ are the error vector between the prediction of CV_1 and the setpoint for CV_1 and the error vector between the prediction of CV_2 and the setpoint for CV_2, respectively. K_1 and K_2 are the diagonal MS matrices for MV_1 and MV_2, respectively. Z is a matrix consisting entirely of zeros with the same dimensions of K_1 and K_2. Control systems with any number of MVs and CVs can be constructed with the same methodology.

For the multivariable case, we may want to ensure that one CV is given more attention than another. This can be done by multiplying the equations associated with a particular CV with a factor that applies a relative importance to it relative to other CVs. This control error weight can be included in Eq. (22) as

$$\frac{\dfrac{w_1(M_1 \mid N_1)}{w_2(M_2 \mid N_2)}}{\dfrac{K_1 \mid Z}{Z \mid K_2}} * \frac{\mathbf{x}_m}{\mathbf{x}_n} = \frac{\dfrac{w_1(\mathbf{e}_1)}{w_2(\mathbf{e}_2)}}{\dfrac{\mathbf{z}}{\mathbf{z}}} \qquad (22a)$$

where w_1 and w_2 are the control error weights for CV_1 and CV_2, respectively.

Constraint Handling and Economic Optimization

In addition to computing the optimum move strategy for controlling a multivariable process, it is often desirable that the solution include a number of constraints on the controller inputs, controller outputs, and other process variables [4]. These constraints may include

1. constraints on the size of movements in the manipulated variables
2. constraints on MVs
3. constraints on controlled variables
4. constraints on other process variables besides explicitly controlled variables

One way to incorporate these constraints and determine a steady-state optimization of the process is to implement a linear program (LP) with linear constraints and a linear objective function. The goal is to find a set of steady-state operating targets (which are to become the targets of the controlled variables of the MPC) that are within the operating limits (process constraints) of the process. If a set of steady state targets as described above exist, then this set represents a feasible solution (FS) to this steady-state optimization problem. If more than one FS exists, then it is possible to use economic cost information to choose the best FS. If it is found that no FS exists because all of the process constraints cannot be satisfied, the least important constraint limit is temporarily relaxed.

Constraints on MVs can be expressed by two equations for each MV:

$$x_{m(ss)} + \Delta x_{m(ss)} \leq x_m \text{ high} \qquad (23a)$$

$$x_{m(ss)} + \Delta x_{m(ss)} \geq x_m \text{ low} \qquad (23b)$$

where $x_{m(ss)}$ represents the value of x_m at an existing steady-state condition and $\Delta x_{m(ss)}$ represents a change in x_m resulting in a new steady state.

For a second MV we would have

$$x_{n(ss)} + \Delta x_{n(ss)} \leq x_n \text{ high} \qquad (24a)$$

$$x_{n(ss)} + \Delta x_{n(ss)} \geq x_n \text{ low} \qquad (24b)$$

Similarly, constraints on controlled variables or other outputs such as other process constraints, can be expressed by two equations for each CV or XV:

$$y_{s(ss)} + \Delta y_{s(ss)} \leq y_s \text{ high} \qquad (25a)$$

$$y_{s(ss)} + \Delta y_{s(ss)} \geq y_s \text{ low} \qquad (25b)$$

For a second CV or XV we would have

$$y_{t(ss)} + \Delta y_{t(ss)} \leq y_t \text{ high} \qquad (26a)$$

$$y_{t(ss)} + \Delta y_{t(ss)} \geq y_t \text{ low} \qquad (26b)$$

Since we control the MVs directly, it is trivial to determine what the steady-state value of the MVs will be after a series of moves that constitute a change from one steady state to another. CVs or XVs are changed or moved by moving the MVs that are associated with them. Since we have a model of the process that includes the steady-state gain of each CV and XV as it relates to the MVs, we can calculate the final steady-state values for the CVs and XVs that result from a series of MV moves. For a two-MV example we would have

$$y_{s(ss)}(\text{final}) = y_{s(ss)}(\text{initial}) + G_{ms} * \Delta x_{m(ss)} + G_{ns} * \Delta x_{n(ss)} \tag{27}$$

where (initial) and (final) refer to the beginning and the end of the move from one steady state to another. From this expression, two constraint equations can be generated:

$$y_{s(ss)}(\text{initial}) + G_{ms} * \Delta x_{m(ss)} + G_{ns} * \Delta x_{n(ss)} \leq y_s \text{ high} \tag{28a}$$

$$y_{s(ss)}(\text{initial}) + G_{ms} * \Delta x_{m(ss)} + G_{ns} * \Delta x_{n(ss)} \geq y_s \text{ low} \tag{28b}$$

For a second CV or XV we would have

$$y_{t(ss)}(\text{initial}) + G_{mt} * \Delta x_{m(ss)} + G_{nt} * \Delta x_{n(ss)} \leq y_t \text{ high} \tag{29a}$$

$$y_{t(ss)}(\text{initial}) + G_{mt} * \Delta x_{m(ss)} + G_{nt} * \Delta x_{n(ss)} \geq y_t \text{ low} \tag{29b}$$

An inspection of relations (23), (24), (28), and (29) shows that, with the steady-state gains from the process models and current prediction of the CVs or XVs, the only information not known is the desired steady-state changes in the MVs.

Relation (28a) can be represented as a constraint line in the space defined by the MVs by first replacing the inequality with an equal sign:

$$y_{s(ss)}(\text{initial}) + G_{ms} * \Delta_{m(ss)} + G_{ns} * \Delta x_{n(ss)} = y_s \text{ high} \tag{30}$$

Equation (30) can now be rearranged as follows:

$$G_{ms} * \Delta x_{m(ss)} + G_{ns} * \Delta x_{n(ss)} = [y_s \text{ high} - y_{s(ss)}(\text{initial})] \tag{31}$$

$$G_{ns} * \Delta x_{n(ss)} = [y_s \text{ high} - y_{s(ss)}(\text{initial})] - G_{ms} * \Delta x_{m(ss)} \tag{32}$$

$$\Delta x_{n(ss)} = [y_s \text{ high} - y_{s(ss)}(\text{initial})]/G_{ns} - (G_{ms}/G_{ns}) * \Delta x_{m(ss)} \tag{33}$$

$$\Delta x_{n(ss)} = (-G_{ms}/G_{ns}) * \Delta x_{m(ss)} + [y_s \text{ high} - y_{s(ss)}(\text{initial})]/G_{ns} \tag{34}$$

Clearly Eq. (34) is the equation of a line with a slope of $(-G_{ms}/G_{ns})$ and an intercept of $[y_s$ high $- y_{s(ss)}$ (initial)$]/G_{ns}$. The other constraints represented by relations (28) and (29) can be represented as lines in the space defined by the MVs in a similar fashion:

$$\Delta x_{n(ss)} = (-G_{ms}/G_{ns}) * \Delta x_{m(ss)} + [y_s \text{ low} - y_{s(ss)}(\text{initial})]/G_{ns} \tag{35}$$

$$\Delta x_{n(ss)} = (-G_{mt}/G_{nt}) * \Delta x_{m(ss)} + [y_t \text{ high} - y_{t(ss)}(\text{initial})]/G_{nt} \tag{36}$$

$$\Delta x_{n(ss)} = (-G_{mt}/G_{nt}) * \Delta x_{m(ss)} + [y_t \text{ low} - y_{t(ss)}(\text{initial})]/G_{nt} \tag{37}$$

For a two-MV problem the constraints represented by relations (23) and (24) can be depicted on a two-dimensional plot representing MV space as shown in Fig. 16. The constraints described by equations (34)–(37) can be added as shown in Fig. 17. Note that the FS for the constraints depicted in Fig. 17 is an area with four vertices. Recall that if more than one FS exists, then it is possible to use economic cost information to choose the best FS. In this case there are an infinite number of solutions;

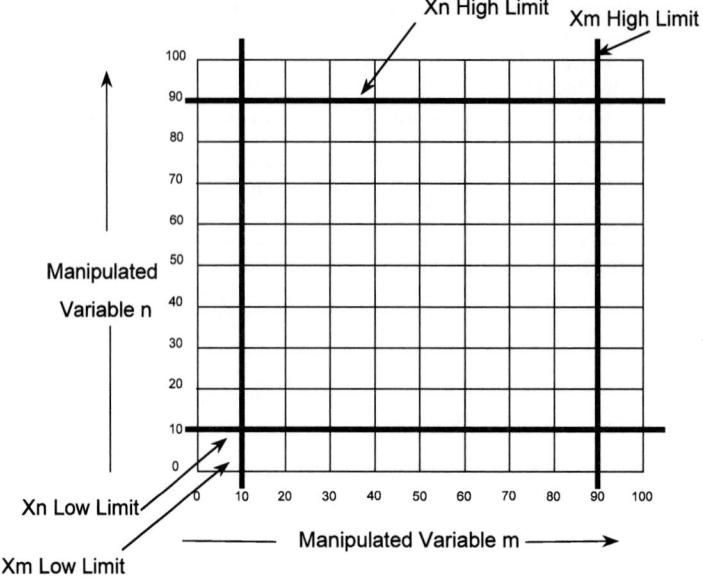

FIGURE 16   MV constraints in MV space.

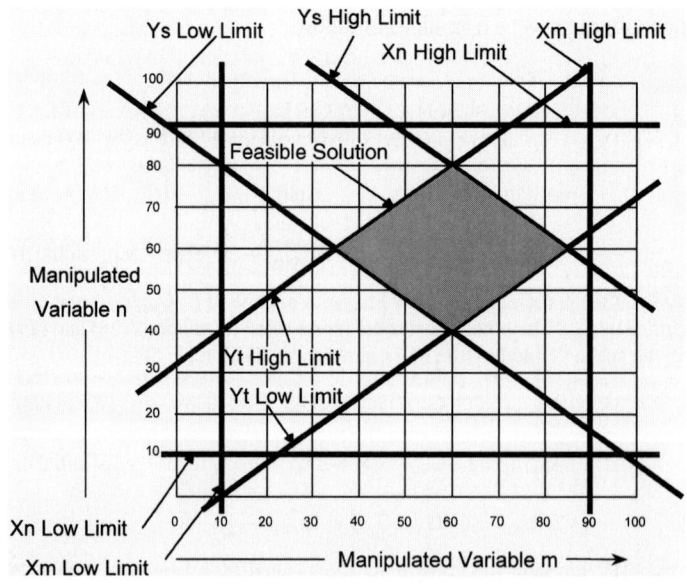

FIGURE 17   Manipulated and controlled variable constraints in MV space with FS on quadrangle.

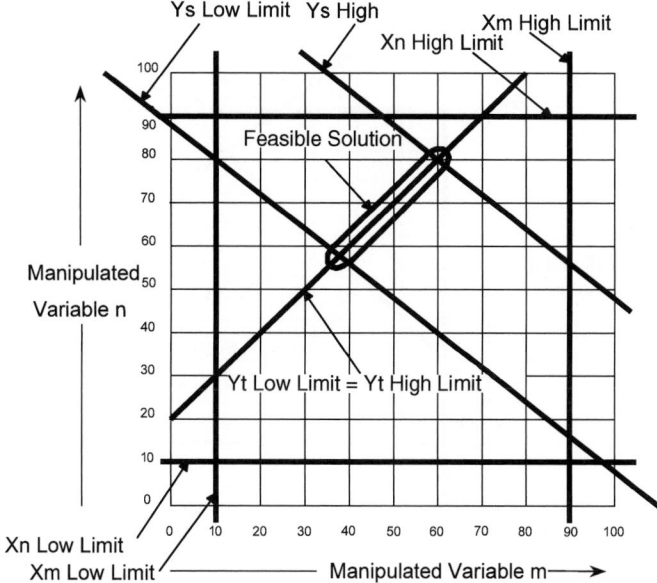

FIGURE 18   Manipulated and controlled variable constraints in MV space with FS on line segment.

however, since the problem is linear, the optimum will exist at one of the vertices. Systematic methods for solving this LP can be found in most texts on optimization.

If the high and low constraints for one of the controlled variables are set equal (this would represent a particular setpoint for the constraint in question), the feasible solution would consist of a line segment, as shown in Fig. 18, and would also have an infinite number of solutions. As before, the optimum will exist at one end or the other of the line segment.

If, as shown in Fig. 19, the high and the low constraints for both of the controlled variables (for the 2 × 2 controller) the FS results in a single point. In this case no economic optimization may be performed, as the degrees of freedom of the problem are zero. The system depicted in Fig. 18 has 1 degree of freedom, and the system depicted in Fig. 17 has 2 degrees of freedom.

If a third controlled variable is added to the problem that has high and low constraints set equal (this is considered a 2 × 3 dimensional control system), as shown in Fig. 20, there will not be a FS. In this case the least important constraint (or setpoint when the high and low constraints are equal) will be relaxed to better accommodate constraints with higher priority.

When the feasible solution results in a region or a line in MV space (not simply a single point, or no feasible solution at all), we can minimize a cost function to optimize the operation of the process. The cost function to be optimized can represent any optimization function that is a linear function of the changes in the MVs (the requirement that the cost function be linear in the MVs allows solution with a LP).

To carry out the optimization, each MV is assigned a cost that describes its effect on the overall cost function. The LP then minimizes the resulting cost function. The LP moves all the MVs within the constraint limits to minimize costs and maximize profits that exist at one of the vertices of the constraints. If there were no constraints, the MVs associated with negative costs (profits) would be maximized and MVs with positive costs would be minimized.

One method for imposing the LP solution on the constrained multivariable controller is to force the sum of all future moves to equal the steady-state changes determined by the LP. We can add additional equations to the previous development to drive each MV to its LP target at steady state.

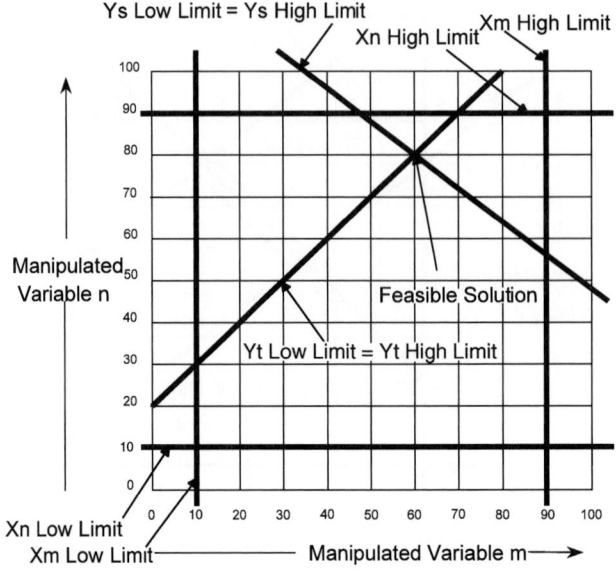

FIGURE 19 Manipulated and controlled variable constraints in MV space with FS at a point.

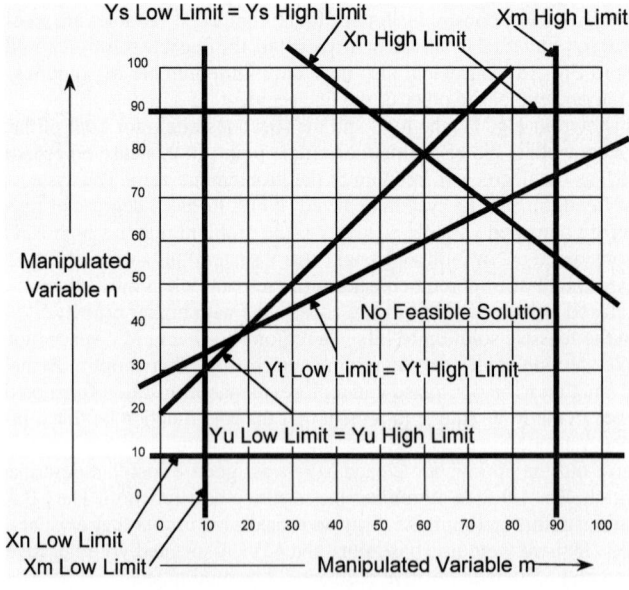

FIGURE 20 Manipulated and controlled variable constraints in MV space with no FS.

Therefore we would require

$$[1 \ 1 \ 1 \ 1 \ 1 \ 1 \ 1 \ 1] \quad * \quad \begin{array}{c} x_{m1} \\ x_{m2} \\ x_{m3} \\ x_{m4} \\ x_{m5} \\ x_{m6} \\ x_{m7} \\ x_{m8} \end{array} \quad = \quad x_{mss} \tag{38}$$

where x_{m1} is the first move in MV_1, x_{m2} is the second move in MV_1, etc., x_{mss} is the total steady-state move in MV_1 as determined by the LP. For two manipulated variables, MV_1 and MV_2, represented by the vectors $\mathbf{x}_m$ and $\mathbf{x}_n$, we have

$$\frac{[1\,1\,1\,1\,0\,0\,0\,0]}{[0\,0\,0\,0\,1\,1\,1\,1]} \quad * \quad \frac{\mathbf{x}_m}{\mathbf{x}_n} \quad = \quad \frac{\mathbf{x}_{mss}}{\mathbf{x}_{nss}} \tag{39}$$

or, incorporated into the previous matrix equations, we have

$$w_q \ * \ \frac{\begin{array}{c|c} w_1(M_1) & N_1) \\ \hline w_2(M_2) & N_2) \\ \hline K_1 & Z \\ \hline Z & K_2 \\ \hline \mathbf{q}_1 & Z \\ \hline Z & \mathbf{q}_2 \end{array}}{} \ * \ \frac{\begin{array}{c} \mathbf{x}_m \\ \hline \mathbf{x}_n \end{array}}{} \ = \ \frac{\begin{array}{c} w_1(\mathbf{e}_1) \\ \hline w_2(\mathbf{e}_2) \\ \hline \mathbf{z} \\ \hline \mathbf{z} \\ \hline \mathbf{x}_{mss} \\ \hline \mathbf{x}_{nss} \end{array}}{} \tag{40}$$

where $\mathbf{q}_1$ and $\mathbf{q}_2$ are vectors of ones, $\mathbf{z}$ is a vector of zeros, and w_q is a tuning parameter representing the importance of the controller following the targets generated by the LP.

JUSTIFICATION OF CONSTRAINED MULTIVARIABLE PREDICTIVE CONTROL AND REAL-TIME OPTIMIZATION

CMPC can be justified for a process when one wants to

1. maximize throughput
2. increase yield of more valuable product(s)
3. minimize energy consumption
4. decouple interacting control systems
5. deal with dead time or complicated dynamics

For an interacting control system, one can determine the applicability of CMPC by considering the flow chart of Fig. 21. If the extent of interaction between variables is low, appropriate pairing of controlled and manipulated variables (using relative gain array or other method) and implementation of multiple SISO controllers may be sufficient. If the interactions are more severe, but the dimension of the interacting system is 2×2, explicit decoupling with override controllers may be the appropriate

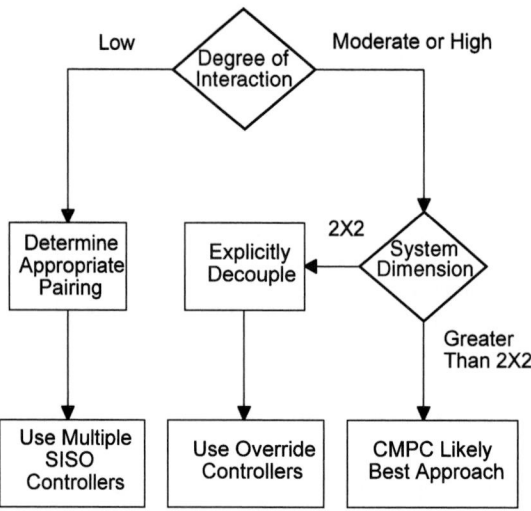

FIGURE 21   Determining the applicability of CMPC for a multivariable process.

approach. However, if the interaction is moderate or high and the dimension of the problem is greater than 2 × 2, then CMPC is typically the best approach.

The ability of the CMPC to deal with dead time or inverse response is also an important consideration. The internal step response model within CMPC allows it to correct for dead time in a manner similar to that of a Smith predictor. The result is that CMPC can be tuned more aggressively than any PID-based strategy on such systems.

RTO can be justified in plants in which there are numerous decision variables. Typically, to justify such a project, historical data are collected and off-line optimization studies are conducted to estimate benefits. As a general rule of thumb, the RTO will push the plant against as many limits as it is given handles. So, for example, if a preliminary analysis has identified 10 setpoints available for optimization and the plant is not currently against 10 limits, in all likelihood it is not at the optimum.

PROCESS MODELING GUIDELINES FOR CONSTRAINED MULTIVARIABLE PREDICTIVE CONTROL

The implementation of CMPC requires the development of input/output models of the process to be controlled. Since these systems are often very large (as large as 30 × 30 systems or larger with numerous process constraints) these models must be built quickly, efficiently, and with high accuracy. There are several characteristics of the modeling process that would be desirable:

1. Models should be built in a reasonable amount of time.
2. The process should not have to be at steady state for the testing of the process to take place.
3. A period of bad data should not invalidate the entire set of test data.
4. The model building procedure should accommodate the movement of all independent variables that are likely to influence the independent or controlled variable.

There are two basic approaches used in building the input/output process models required for CMPC implementations: impulse response modeling and time series analysis.

Impulse Response Modeling

One of the most important and time-consuming aspects of impulse response modeling is determining the time to steady state for each dependent (controlled variable) and independent (MV) pair. During the analysis of the data, the time to steady state (an important design parameter) is approximated and adjusted to determine the best model fit of the data.

From the equations developed previously for a MV x_a, used to control a controlled variable y_a, where x_a is changed at only discrete intervals, the controller variable y_a can be expressed as

$$
\begin{aligned}
\Delta y_a(1) &= a(1)\Delta x_{a1} \\
\Delta y_a(2) &= a(2)\Delta x_{a1} + a(1)\Delta x_{a2} \\
\Delta y_a(3) &= a(3)\Delta x_{a1} + a(2)\Delta x_{a2} + a(1)\Delta x_{a3} \\
\Delta y_a(4) &= a(4)\Delta x_{a1} + a(3)\Delta x_{a2} + a(2)\Delta x_{a3} + a(1)\Delta x_{a4} \\
\Delta y_a(5) &= a(5)\Delta x_{a1} + a(4)\Delta x_{a2} + a(3)\Delta x_{a3} + a(2)\Delta x_{a4} + \cdots
\end{aligned}
\tag{41}
$$

$$\cdots$$

We can rearrange these equations so that the dynamic coefficients are expressed as the independent variables, and since $\Delta y_a(k)$ represents the change in the output from its initial value to its value at a time k intervals into the future from input a, $\Delta y_a(k) = y_{ak} - y_{a0}$, and since Δx_{an} represents the change in the input a at time $t = n$, $\Delta x_{an} = x_{an} - x_{a(n-1)}$ we can express equation set (41) as

$$
\begin{aligned}
y_{a1} - y_{a0} &= (x_{a1} - x_{a0})a(1) \\
y_{a2} - y_{a0} &= (x_{a2} - x_{a1})a(1) + (x_{a1} - x_{a0})a(2) \\
y_{a3} - y_{a0} &= (x_{a3} - x_{a2})a(1) + (x_{a2} - x_{a1})a(2) + (x_{a1} - x_{a0})a(3) \\
y_{a4} - y_{a0} &= (x_{a4} - x_{a3})a(1) + (x_{a3} - x_{a2})a(2) + (x_{a2} - x_{a1})a(3) + (x_{a1} - x_{a0})a(4) \\
y_{a5} - y_{a0} &= (x_{a5} - x_{a4})a(1) + (x_{a4} - x_{a3})a(2) + (x_{a3} - x_{a2})a(3) \\
&\quad + (x_{a2} - x_{a1})a(4) + (x_{a1} - x_{a0})a(5)
\end{aligned}
\tag{42}
$$

$$\cdots$$

If a move is made at time $t = 0$ and the process is sampled at $1/4$ of the settling time, then the effect of this move will be observed in the first four of the equations expressed above, and the contribution of the last term of the fifth equation will essentially be zero, since for any coefficient beyond four settling times [$a(5)$ in this case] will essentially be zero. This further implies that if the system is not at steady state when the testing begins, the first three equations in general will not be valid. For systems for which the process is described by an impulse response of, say, 30 weights, the first 29 equations would not, in general, be valid unless the process was at steady state before the testing of the process.

The difference terms in equation set (42), $(y_{ak} - y_{a0})$ and $[x_{an} - x_{a(n-1)}]$, would typically be sources of error in the analysis of the data set; however, the $[x_{an} - x_{a(n-1)}]$ term is typically representative of regulatory controller setpoint changes and are thus known to a high degree of accuracy. Furthermore the $(y_{ak} - y_{a0})$ term is the change in the output value y from its initial condition, and since the process testing is performed such that the output value exceeds the noise level of the process measurements by six times or more, the $(y_{ak} - y_{a0})$ term will be significantly above the noise level of the measurements.

If there is an unmeasured disturbance on the process, the accumulation term $y_{ak} - y_{a0}$ will contain the effects of the disturbance that will not be accounted for in the $x_{an} - x_{a(n-1)}$ term. Taking the derivative of equation set (42) with respect to the dependent variable will eliminate these accumulations.

The derivative can be found by subtracting sets of equations in equation set (42) as follows:.

$$y_{a1} - y_{a0} = (x_{a1} - x_{a0})a(1)$$

$$\begin{aligned} y_{a2} - y_{a0} &= (x_{a2} - x_{a1})a(1) + (x_{a1} - x_{a0})a(2) \\ -y_{a1} + y_{a0} &= \qquad\qquad\quad - (x_{a1} - x_{a0})a(1) \\ &= \end{aligned}$$

$$y_{a2} - y_{a1} = (x_{a2} - x_{a1})a(1) + (x_{a1} - x_{a0})[a(2) - a(1)]$$

$$\begin{aligned} y_{a3} - y_{a0} &= (x_{a3} - x_{a2})a(1) + (x_{a2} - x_{a1})a(2) + (x_{a1} - x_{a0})a(3) \\ -y_{a2} + y_{a0} &= \qquad\qquad\quad - (x_{a2} - x_{a1})a(1) - (x_{a1} - x_{a0})a(2) \\ &= \end{aligned}$$

$$y_{a3} - y_{a2} = (x_{a3} - x_{a2})a(1) + (x_{a2} - x_{a1})[a(2) - a(1)] + (x_{a1} - x_{a0})[a(3) - a(2)]$$

$$\begin{aligned} y_{a4} - y_{a0} &= (x_{a4} - x_{a3})a(1) + (x_{a3} - x_{a2})a(2) + (x_{a2} - x_{a1})a(3) + (x_{a1} - x_{a0})a(4) \\ -y_{a3} + y_{a0} &= \qquad\qquad\quad - (x_{a3} - x_{a2})a(1) - (x_{a2} - x_{a1})a(2) - (x_{a1} - x_{a0})a(3) \\ &= \end{aligned}$$

$$y_{a4} - y_{a3} = (x_{a4} - x_{a3})a(1) + (x_{a3} - x_{a2})[a(2) - a(1)] + (x_{a2} - x_{a1})[a(3) - a(2)] + (x_{a1} - x_{a0})[a(4) - a(3)] \tag{43}$$

$\cdots$

An impulse response is the derivative of the step response, by definition. Therefore we define the differences in the a coefficients as d, the impulse coefficients:

$$\begin{aligned} d(1) &= a(1) \\ d(2) &= a(2) - a(1) \\ d(3) &= a(3) - a(2) \\ d(4) &= a(4) - a(3) \end{aligned} \tag{44}$$

$\cdots$

then equation set (43) can be rewritten as

$$\begin{aligned} y_{a1} - y_{a0} &= (x_{a1} - x_{a0})d(1) \\ y_{a2} - y_{a1} &= (x_{a2} - x_{a1})d(1) + (x_{a1} - x_{a0})d(2) \\ y_{a3} - y_{a2} &= (x_{a3} - x_{a2})d(1) + (x_{a2} - x_{a1})d(2) + (x_{a1} - x_{a0})d(3) \\ y_{a4} - y_{a3} &= (x_{a4} - x_{a3})d(1) + (x_{a3} - x_{a2})d(2) + (x_{a2} - x_{a1})d(3) + (x_{a1} - x_{a0})d(4) \end{aligned} \tag{45}$$

$\cdots$

For equation set (45) an unmeasured disturbance can still enter the impulse response of the process, but will affect only the equations associated with the change and for one settling time after the disturbance occurs. These equations will be invalid; therefore these equations could be removed from the analysis if an unmeasured disturbance event is recorded. This removal of equations from the analysis is accomplished very easily. If an unknown unmeasureable disturbance occurs, the effect of this disturbance will be imposed on the output of the process response, but will contaminate the data for a period of time equal only to the time to steady state of the disturbance. Although this formulation introduces noise by taking the derivative of the dependent variable, it is typically the preferred form since most industrial processes contain such unmeasured disturbances.

Equation set (45) can be solved in a least-squares sense similar to solving for the dependent variables described above for the formulation of the MPC.

An important design parameter and an important factor in modeling models with this method is the time to steady state. The rigorous approach to modeling the data would be to look at statistical fits for the data in question for each response curve for every possible time to steady state. The statistical analysis would require a numerical coefficient that corresponds to the time interval of collected data. If one considers a process with a 2-h settling time and data are collected every minute, an impulse response model with 300 coefficients might possibly be used to represent the response of the process. This data would conceivably need to be solved for up to 100 time intervals or more. If we consider a system with 20 to 30 independent variables and 20 to 30 dependent variables, the size of the problem and the effort to analyze the data becomes very large. To solve these large problems in a reasonable amount of time, the number of coefficients used to represent the impulse response is limited to 30 to 60 coefficients. To fit the time to steady state into this smaller number of coefficients, the data can be averaged over various time intervals. This averaging procedure changes the basis of the statistical analysis required for verifying the fitness of the model to represent the data. As a result, one can improve the speed of the modeling process by calculating the impulse response for an input/output pairing for various times to steady state and then visually comparing these resulting curves with each other and the raw unfiltered response data. This allows efficient identification for as much as 90% of the models and the rest can be determined with a more rigorous approach.

To summarize, the characteristics of the impulse response modeling method are the following:

1. The process does not need to be at steady state.

2. For systems for which unmeasured disturbances are not a problem, the formulation is particularly robust to noise.

3. For systems for which unmeasured disturbances are a problem, a formulation exists whereby the corrupted data can be easily removed for recorded unmeasured disturbances and that corrupts the data for a time only as long as the time to steady state for the disturbance. The disadvantage is that noise may be introduced into the system as a result of taking the difference of the output response of the system.

4. A statistical measure of model fitness is not generally available.

5. No explicit model form is required to be assumed.

Time Series Analysis

A time series is a set of dependent process response data sampled at a constant time interval and assumed to be affected by an independent variable and one or more unmeasured disturbances. The combined effect of the disturbances on the process is described in terms of a noise model. The object of the analysis is to identify the transfer function or model that describes the input/output relationship and the associated noise model inherent in the time series.

This data analysis method requires that the input to the process and the noise be statistically independent. As a result, the process is deliberately perturbed by applying an input series that is generated by some sufficiently random process [such as pseudorandom binary sequences (PRBSs)]. The resulting process output data along with the input data are then analyzed to find the appropriate process model and noise model of the system.

The actual analysis of these data is beyond the scope of this section, and a number of algorithms are commercially available for performing the analysis. In addition, there are numerous texts available on the theory and methods available for the identification of time series. As a result the following describes the basic underlying concepts associated with this technique [5].

For a sampled data system the sampled data relationship between a process input and the process output can be given as

$$Y(z)/X(z) = z[G_{h0}(s)G_p(s)] \tag{46}$$

where $G_{h0}(s)$ is the Laplace domain transfer function of a zero-order hold, $G_p(s)$ is the Laplace domain transfer function of the process, $Y(z)$ and $X(z)$ are sampled process data output and input observations, respectively, and $z[\cdots]$ is the z transform of the contents of the brackets. This is the sampled data pulse transfer function of a SISO system.

If we assume an overdamped second-order plus dead-time model for the plant, that is, for $G_p(s)$, the application of Eq. (46) results in

$$Y(z)/X(z) = z\{[(1 - e^{-sT})/s] * (K_p e^{-Tds})/[(t_1 S + 1)(t_2 S + 1)]\} \tag{47}$$

In general the time series may not be stationary, such as liquid level systems or other integrating processes. These time series may be transformed to a stationary process by taking the difference one or more times. Typically taking only the first or second difference is sufficient. The resulting system may be assumed to be stationary if the autocorrelations and cross correlations of the input and output $[\gamma_{xx}(k), \gamma_{yy}(k),$ and $\gamma_{xy}(k)]$ damp out quickly.

Typical steps for finding the parameters of the transfer function and noise model that correspond to a input/output time series are as follows:

1. The first step is to identify a likely form of the process model by using various statistical properties of the input/output series. Impulse response weights are used to guess at a likely model form and provide a preliminary estimate of the model parameters.

2. Next the identified model form from step 1 is iteratively fit to the input/output data by using a maximum likelihood routine which minimizes the sum of squares of the residuals of the fit. This routine includes a number of statistical results that can be used to test how well the model form and the estimated parameters fit the input/output data. A noise model is often fit to the residuals during this step.

3. Finally the statistical results from the previous step are used to determine whether the model form identified in step 1 is valid or not. If it is determined that the model form identified in step 1 is not valid, the statistical results may be used to suggest an alternative model form, and the process is returned to step 1, in which a new model form is assumed.

There is evidence that models fitted in this way are superior to the models obtained with the impulse response method discussed previously [6]. The reason cited for this enhanced performance is that the low-order models fitted prevent overfitting the input/output data, thus providing a smoothing effect to the resulting models. The result is that although time series identification typically requires more effort, the models may be obtained with substantially less data and model-based controllers that use these models would be expected to be more stable and robust. This method should therefore be used when the process to be modeled can be represented with a sufficiently simple model form.

To summarize, the characteristics of the time series modeling method are the following:

1. The input/output data do not need to be stationary.
2. The models that result from the time series modeling process are typically low-order models with dead time and can be represented with a small number of parameters.
3. This method typically requires substantially less data to achieve the same degree of accuracy as using the impulse response method [6].
4. A statistical measures of model fitness is available as part of the modeling process.
5. An explicit model form is assumed.

Process Modeling Rules of Thumb

Before starting the response testing of the process in question, several things should be checked out to minimize process and instrument noise and ensure that the value of the test data is maximized:

1. The tuning of the basic regulatory control system should be checked for stability and responsiveness.

2. Check for sticking valves or valves with hysteresis that could result in oscillatory responses to step changes in the regulatory loop setpoint.

3. Filters should be implemented at the regulatory control system level for process measurements which exhibit excessive noise. Filtering at the CMPC level could introduce significant lags in the process models.

4. The span on temperature measurements in the regulatory control system should be checked. If the span is too large, the resolution of the analog-to-digital converter in the regulatory system may cause the measurements to appear noisy.

5. Perform a process pretest to determine approximate dead times, steady-state gains, and settling times for each of the independent/dependent pairs.

6. Check that the data collection system is working properly.

7. Check that the data collection system does not have excessive compression.

Some important guidelines to keep in mind during the planning or actual testing of the process are the following:

1. A log of all unmeasured process changes that affect the dependent controlled variable should be kept by a knowledgable person. This would include instrument calibrations and troubleshooting, removal of equipment for maintenance, starting and stopping pumps, opening bypass valves, changing piping lineups, switching controller modes, changing regulatory controller tuning, etc.

2. When data are analyzed, the data associated with the periods of unmeasured disturbances should be removed.

3. Perform the process testing such that the process outputs are at least six times the measurement noise, but not so large that the response testing violates the linearity assumptions of the process.

4. Monitor the test to ensure that the expected process output is responding to the process input in the expected manner.

5. During the process response testing procedure, valves should be monitored to ensure that they do not become saturated.

6. During the analysis of the data, any unexpected excursion of dependent or independent variables should be investigated.

7. Changes in the independent variable should be made with a maximum frequency of not less than the dead time and not more than approximately $1/2$ of the process lag (1/8 of the process settling time minus the dead time) plus the dead time. In addition to short-duration process changes there should be intermediate- and long-duration process changes. The combination of short, intermediate, and long process changes helps to ensure that good estimates of the process dead time, process dynamics, and process steady-state gain can be made.

CONSTRAINED MULTIVARIABLE PREDICTIVE CONTROLLER TUNING AND CONSTRUCTION GUIDELINES

The tuning and construction of the on-line CMPC involves several considerations. Several characteristics of CMPC influence the design of these controllers:

1. The on-line controller includes a control matrix. The control matrix provides information to the CMPC about the ability to control the dependent variables with specific independent variables. This matrix can be utilized to selectively allow certain manipulated variables to control certain controlled variables. It is often advantageous to have only one manipulated variable be the major controlling factor for a particular controlled variable.

2. A prediction matrix provides information to the CMPC about the dynamic interactions between the dependent and the independent variable pairs. Occasionally one may desire that the CMPC does not know about the effect of one independent variable on a dependent variable. Usually all dynamic relationships that exist between the independent and dependent variables will be expressed in the prediction matrix. In general the control and the prediction matrices are not the same.

3. Tuning parameters available in the CMPC include MS on the manipulated variables and controller error weights on the controlled variables. MS prevents the controller from acting too aggressively, which may result in controller oscillations or even instability. The controller error weights allow one to balance the importance of one control objective over another.

4. Constraint parameters available in the CMPC include error weights on the constraint variables (soft constraints) and absolute as well as move size constraints on the manipulated variables (hard constraints). The constraint error weights constrain the controller to a particular operating region, requiring specific operational parameters to remain within certain boundaries. The absolute move size limits on the manipulated variables limit the manipulated variables to certain values. For instance, minimum or maximum feed rates may be imposed for a particular piece of operating equipment. Move size limits restrict the amount of movement allowed by the manipulated variables during a CMPC control cycle.

The setting of these constraints and limits are typically dictated by operating limits imposed by the operating management, by equipment capabilities, environmental permits, or safety interlocks. The tuning of the various tuning parameters are typically subjective in nature and result from what operations personnel typically expect in terms of performance of the control system. Of course the performance is limited by the process dynamics and the trade-off that ultimately exists as a result of balancing control objectives.

In addition, there are several other design parameters involved in the implementation of CMPC:

1. controller time horizon

2. controller execution frequency

3. number of future moves

REAL-TIME OPTIMIZATION GUIDELINES

A successful RTO project, in general, requires more engineering effort than is true for CMPC. Before proceeding, it is worthwhile to analyze from where the optimization benefits are expected to come. In some cases, maximizing one or two variables (such as charge) may be sufficient to capture most of the benefit. This type of optimization can usually be implemented at the CMPC level though simple constraint pushers.

To justify a more elaborate RTO project, maximizing the plant profit should involve coordinating targets involving multiple CMPC applications. Other considerations include the following:

1. Can the plant profit function be measured on line? How are intermediate streams valued?

2. What constraints will the RTO push the plant against? In between RTO executions is there a fail-safe DCS or CMPC scheme in place to take evasive action if a plant disturbance causes these constraints to be exceeded?

3. Is the scope of the optimization large enough that the RTO solutions will be feasible to implement? For example, there is little point is optimizing a single reactor if all the limitations are in downstream vessels and have not been included.

Although there have been improvements in the efficiency of optimization algorithms, execution time of an RTO is difficult to predict. Users should ensure that they have adequate CPU speed so that significant plant drift does not occur during an execution cycle.

APPLICATIONS OF CONSTRAINED MULTIVARIABLE PREDICTIVE CONTROL AND REAL-TIME OPTIMIZATION

Space does not permit an exhaustive description of applications of CPMP to the refinery and chemical processing industries; however, a short description of some successful implementations has been included to give an idea of the types of savings and improvements that can be realized.

Below are some very brief descriptions of some of the many successful implementations of CMPC. Note that these implementations have been carried out in many countries on many types of petroleum and chemical processes.

1. Hydrocracker Reactor (Sarnia, Ontario, Canada), American Control Conference, June 1987
   Controller has 5 MVs, 2 FVs, and 4 CVs
   Average reduction in energy used by the reactor preheat furnace of 55%
   Savings of approximately $1000/day

2. Catalytic Cracking Unit (Wales, United Kingdom), International Symposium—Advanced Process Supervision and Real-Time Knowledge Based Control, November 1988
   Controller uses 7 MVs, 3 FVs and 8 CVs
   Project payback of 3 months

3. C3C4 Splitter (Sarnia, Ontario, Canada), The National Meeting of The American Institute of Chemical Engineers, April 1988
   Controller uses 2 MVs, 2 FVs, and 2 CVs
   Project benefits of $120,000/year

4. Refrigerated Fractionator (Baytown, Texas), The National Meeting of the Instrument Society of America, October 1989.
   Controller uses 5 MVs, 3 FVs, and 5 CVs
   Average plant ethylene losses to the fuel gas system were reduced by over 50%

5. Plastics Intermediate Process (Mt. Vernon, Indiana), The National Meeting of the Instrument Society of America, October 1989
   Two CMPC controllers
   Controller 1 uses 3 MVs, 2 FVs, and 2 CVs
   Controller 2 uses 1 MV, 2 FVs, and 1 CV
   Resulted in 3% plant capacity increase
   Value to the business of $4000/day

6. Hydrogen Plant (Sarnia, Ontario, Canada), 1989 American Control Conference, June 1989.
   Controller uses 10 MVs, 3 FVs, and 2 CVs
   Economic benefits of $1000/day

7. Ethylene Plant (Channelview, Texas), International Conference on Productivity and Quality in the Hydrocarbon Processing Industry, February 1992
   Controller uses 2 MVs, 1 FV, and 6 CVs
   Project payback in less than 9 months

8. Polyethylene Reactor (Canada), Hydrocarbon Processing, June 1993
   60% reduction in transition time for grade switches
   50% reduction in off-standard product
   15% increase in first through yield

9. Residue Cracking Unit (Australia), Hydrocarbon Processing, October 1993.
   Payback period less than 3 months
   Economic benefits of $2.4 million/year

10. Crude Unit (Denmark), Hydrocarbon Processing, February 1994
    Increased yield of 1.5%
    Increased throughput of 3%
    Decrease in required fuel by 5%

11. Natural Gas Plant (Mexico), Hydrocarbon Processing, April 1994
Payback in less than 1 month
Increase in product revenues by over $2 million/year.

Additional application descriptions can be found in *Hydrocarbon Processing*, September 1993, "Advanced Control Strategies 1993," in which over 100 successful applications are featured.

MAINTENANCE ISSUES OF CONSTRAINED MULTIVARIABLE PREDICTIVE CONTROL AND REAL-TIME OPTIMIZATION

Typically maintenance of CMPC controllers involves only minor tuning of the controller after commissioning. Occasionally, however, other sources of control inaccuracies occur such that more extensive maintenance must be performed. These areas of control inaccuracies include the following:

1. Analyzer reliability: Analyzer outputs are often critical to the operation of the controllers and the need to have high on-stream times. The result of analyzer problems can result in complete off-stream time of the CMPC. One method for minimizing the problems associated with analyzer inaccuracies is to develop inferential measurements based on more reliable process instrumentation. These inferential measurements can replace many analyzer implementations, resulting in high on-stream time and enhanced economic benefits.

2. Instrument reliability: Level, flow, and pressure controllers all need to function correctly for smooth control. Faulty control action at the regulatory level quickly becomes apparent and results in more responsive attention to instrument problems. These instrument failures often result in at least a part of the controller matrix being taken off line. It is obvious that if the controller is to do the job for which it was designed, accurate instruments need to be maintained. This need for instrument maintenance becomes more critical with the implementation of CMPC.

3. Model accuracy: The plant model must be kept up to date with any process modifications. Even minor changes in processes may result in the need to remodel parts of the process. For example, the addition of a digital positioner to the valve of an important manipulated variable may require new models to be built as well as changes to the regulatory control system. If fact, retuning of regulatory control systems may be sufficient to require that new process models be built. This is why it is very important that the regulatory control system be adequately tuned before initial plant testing begins.

DEVELOPMENTS AND FUTURE DIRECTIONS OF CONSTRAINED MULTIVARIABLE PREDICTIVE CONTROL

CMPC as described in the previous sections has proven to be a superior control technology for processes with large dead times, severe interactions, measured disturbances, many process constraints, etc. Despite the improvement in performance provided by CMPC, there have been a large number of shortcomings in the technology, as described by Qin and Badgwell [7]. These shortcomings have resulted in an ongoing effort to improve vendors products. Qin and Badgwell classify the CMPC products into three categories:

1. First generation: This provides for the basic model predictive control features described previously and a way to handle input and output constraints, but the MV moves computed are not necessarily implemented in an optimum fashion.

2. Second generation: This has algorithms that provide a more systematic way to implement input and output constraints. This generation of products uses a quadratic program to ensure that not only does the process get moved to an optimum steady-state operating point but also implements the MV moves in an optimal way.

3. Third generation: These products provide for constraint prioritization, methods for determining and recovering from ill-conditioned solutions, and separate optimizations first for the CVs and then for the MVs, providing sufficient degrees of freedom.

For a recent review of the state of the art of these technologies, the reader is referred to Ref. 7.

Although CMPC products have evolved significantly over the past two decades, there are still a number of areas with shortcomings, some of which are described by Froisy [8]. Some of these shortcomings that are the subject of active research and development include the following:

1. Improved modeling techniques: The models used in CMPC formulations are of fundamental importance for the robustness and performance of the controllers. Some products use finite impulse response (FIR) models that result in overparameterization. These products use smoothness functions and biased regression techniques to improve the resulting models. These are some products that identify process models based on transfer functions or state space models. These formulations are in general superior to FIR models; however, a large degree of engineering judgment is often required for developing accurate models. New systematic methods for developing parametric models is desirable.

2. Improved robustness: Most of the current product offerings use extensive simulation of the controllers to ensure the stability of the resulting tuned systems. Stability guarantees would be very helpful in putting tuned model predictive controllers on line quickly. In addition to stability criteria, a method for dealing with model mismatch or model uncertainty would be helpful for determining the acceptable region of tuning parameters for a given degree of uncertainty, which would result in stable controllers.

3. Nonlinear CMPC: Nonlinear control of multivariable systems has been studied for many years and products are becoming available from both the traditional vendors of CMPC products as well as artificial neural network vendors. Although virtually all processes are essentially nonlinear, engineers have been able to control the majority of these processes with linear controllers alone or in conjunction with linearizing transformations. Despite the success of linear controllers, there are classes of processes that can still benefit from nonlinear controllers (such as polymerization processes).

4. Adaptive CMPC: Many multivariable processes experience significant changes as the process ages (catalysts deactivate, heat transfer surfaces foul, etc.). As a result, the models that describe the process are no longer accurate or in some cases are no longer useful for high-performance control and consequently the controllers are often detuned to maintain stability. Adaptive or self-tuning controllers would be an obvious improvement to current static CMPC controllers for these processes.

REFERENCES

1. Cutler, C. R., and B. L. Ramaker, "Dynamic Matrix Control—A Computer Control Algorithm," presented at the National Meeting of The American Institute of Chemical Engineers, Houston Texas, April 1979.

2. Dorf, R., *Modern Control Systems*, 6th ed., Addison-Wesley, Reading, Massachusetts, 1992.

3. Noble, B., and J. W. Daniel, *Applied Linear Algebra*, 2nd ed., Prentice-Hall, Englewood Cliffs, New Jersey, 1977.

4. Cutler, C. R., J. J. Haydel, and A. M. Morshedi, "An Industrial Perspective On Advanced Control," presented at the National Meeting of The American Institute of Chemical Engineers, Washington, D.C., October 1983.

5. Box, G. E., G. M. Jenkins, and G. C. Reinsel, *Time Series Analysis—Forecasting and Control*, 3rd ed., Prentice-Hall, Englewood Cliffs, New Jersey, 1994.

6. Dayal, B. S., and J. F. MacGregor, "Identification of Finite Impulse Response Models and Robustness Issues," *Ind. Eng. Chem. Res.* **35** (11), 1996.

7. Qin, S. J., and T. A. Badgwell, presented at the meeting on An Overview of Industrial Model Predictive Control Technology, Chemical Process Control - V, Tahoe City, California, Jan. 7–12, 1996.

8. Froisy, J. B., "Model Predictive Control: Past, Present and Future," ISA Trans. **33**: 235–243, 1994.

NEURAL NETWORKS

by Paul G. Luebbers*

WHAT IS AN ARTIFICIAL NEURAL NETWORK?

Artificial neural networks (ANNs) are networks composed of many nonlinear computing nodes. These computing nodes were originally inspired by biological neurons. Since these computing nodes are a rough, approximate model of biological neurons, the networks are often called ANNs to distinguish them explicitly from true neural networks (such as the human brain for example). Since these ANNs should be easy to distinguish from their biological counterparts, they are referred to simply as neural networks (NNs).

NNs, for our purposes, are nonlinear mathematical approximations. These NNs have various characteristics that make them very convenient for modeling functions and processes:

1. NNs can easily be used to develop linear and nonlinear empirical models. It has been shown that NNs can be used to model any nonlinear function to any degree of accuracy, providing enough computing nodes are used. These models, or relationships, may be static or time varying.

2. Input/output data can be used to construct these empirical models. Often the underlying principles that govern a process or plant are not known well enough to construct a first-principle model; however, input/output data sets that describe the process can almost always be collected.

3. The resulting models can often be used to generalize and predict outputs for input sets that have not been used in the construction or training of the NN model.

HISTORICAL DEVELOPMENT

The human brain is arranged in a massively parallel network of neurons, which permit parallel processing of information. This parallel processing capability is one of the characteristics that contributes to the impressive performance of biological NNs. The first major attempt to model these neurons was made by McCulloch and Pitts [1]. They showed that a neuron can be modeled as a threshold device to perform simple logic functions. An early concept of learning between two neurons was presented by Hebb [2], who suggested a simple rule as follows:

> When neuron A and neuron B are simultaneously excited, increase the strength between them.

An extension of this rule to cover both positive and negative excitation is

> Adjust the strength of connection between two neurons A and B an amount proportional to the product of their simultaneous excitation.

This is often called Hebb's rule or Hebbian learning.

Rosenblatt [3] derived a neuronally inspired network called the Perceptron. This network consisted of a single layer of neurons and was used to classify patterns as similar or distinct. It was not long before Minsky and Papert [4] proved that the perceptron could not be used for complex logic functions, including the XOR or two-parity problem.

---

* Solutia Inc., Cantonment, Florida.

The recent progress in NN development can be attributed to many researchers in many disciplines. Rumelhart and McClelland [5] presented the error backpropagation algorithm for training perceptrons of more than one layer. Although the actual inventor of this algorithm is in question [6], its publication revitalized NN research by showing that the error backpropagation algorithm could train multilayered NNs to perform tasks at which the simple perceptron of Rosenblatt had failed.

Despite the impressive results of NN applications, several drawbacks exist. The form of ANNs is inherently model independent; however, the exact architecture for a particular application (number of layers, number of neurons within a layer, types, and number of connections, etc.) is often chosen according to a rule of thumb or other general guideline. In general the convergence of ANNs is very slow, especially as the number of layers, which the error must be backpropagated across, increases. Perhaps one of the greatest disadvantages of ANNs is that they are essentially a black-box approach. The final network that results from a training session cannot be expressed in a tractable form. This is due to the highly nonlinear, cascade computation that the NN architecture represents and the spatial distribution of information within the trained NN.

Although NNs were originally developed to help understand the workings of brain functions, engineers and scientists are interested in exploiting their problem-solving capabilities. As a result of the adaptive nature of NNs, they can adapt and learn the characteristics of input signals. Because of their nonlinear nature, NNs can perform filtering operations that are beyond the capabilities of conventional linear filtering techniques [7]. NNs can be formulated to provide optimal nonlinear discriminant functions for pattern classification. The applications in the areas of signal processing, control, identification, interpolation, function approximation, artificial intelligence, pattern classification, etc., continue to grow at an amazing rate. The promise of very high-speed computation in parallel hardware implementations indicates that the area of NNs will be extremely important to the future of industry. Karayiannis and Venetsanopoulos [6] indicate that the NN industry is expected to be in the billions of dollars in annual U.S. sales by the turn of the century and sustain an annual growth rate of 40% each year in the form of new software, chip designs, computer systems, and support services.

Currently the term NN is used to describe any computing architecture that consists of parallel interconnections of simple neurons. Many of the architectures in common use today bear little resemblance to any current model of the biological neuron, nervous system, or the brain. As a result, the neurons for such a network are commonly referred to as nodes and NNs as AANs or connectionist architectures.

CLASSIFICATION OF ARTIFICIAL NEURAL NETWORKS

Learning mechanisms may be classified in several ways. Rumelhart and McClelland [5] present a classification based on the learning paradigm and how the resulting model is supposed to work. They present four common learning paradigms that are part of most NN processing systems.

The first, called an autoassociator, involves the repeated presentation of a set of patterns to the system. The system is to store these patterns, then later to retrieve the pattern when a part of one of the original patterns or a similar pattern is presented to the system. This is called an autoassociative process in which a degraded or noisy pattern is to be used as a cue to retrieve the original pattern by associating the pattern with itself. These systems are also known as content-addressable or associative memories [8]. Some NNs, such as the Hopfield net, are specifically designed to implement this type of system, while other architectures, such as the multilayer perceptron, require additional postprocessing stages to accomplish the task.

The pattern associator is designed to associate and produce a second pattern when an associated pattern is first presented to the system. In general one wishes to produce an arbitrary output pattern when presented with an arbitrary input pattern. Multilayer perceptrons have been extensively used for this type of system. When these mappings from one arbitrary pattern to another are static and not dependent on a time series, a typical feedforward multilayer perceptron can be used. If the patterns are the result of a dynamic process, a recurrent multilayer perceptron can be used.

In the case of the classification paradigm, there is a fixed set of classes into which an input pattern is to be classified. During a training session, a set of input patterns and their predetermined classifications are imposed on a network such that the system learns the classification problem. Then, during the classification mode, a pattern similar to one of the classes or a distorted pattern that belongs to a particular class is presented. The network is to determine the proper classification of this pattern. This is the typical paradigm for which the perceptron of Rosenblatt and the multilayer perceptron were designed to operate [5].

The regularity detector is supposed to discover statistically important features of the input population of patterns. This is often called self-organization or unsupervised learning. Unlike the classification paradigm, there are no predefined classifications that can be used to supervise the learning of the system. In this case the system is to determine its own feature representation of the input that best represents the population of input patterns. These networks are typically used in applications of vector quantization and clustering operations.

THE MULTILAYER ERROR BACKPROPAGATION PERCEPTRON

For most of the applications encountered in the process industries, the pattern associator paradigm for both static and dynamic association applications is the most important. The network structure that is typically used for implementing these paradigms is the feedforward multilayer perceptron.

The architecture of a single-layer (zero hidden layers) network is shown in Fig. 1. The network inputs enter from the left of the diagram and are directed to the output node. This output-layer node consists of a summation block for summing the incoming input signals, a bias term, and a nonlinear activation function that produces the output of the node. An individual nonlinear node is shown in Fig. 2.

The output of the ith node at the nth layer (output layer) is the summation of the contributions from the nodes in the $(n-1)$th layer (input layer), which may be represented by

$$O_i^{(n)} = f\left[\sum_j w_{ji} O_j^{(n-1)} + \theta_i \right] \tag{1}$$

where f is the nonlinear activation function and θ_i is the bias or threshold for the node (the layer superscript is dropped hereafter for simplicity). The nonlinear activation function is typically some sigmoidal function such as the hyperbolic tangent. The performance of this network is often taken as

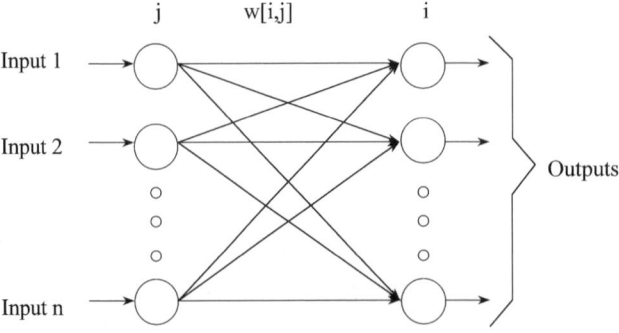

FIGURE 1   Single-layer perceptron.

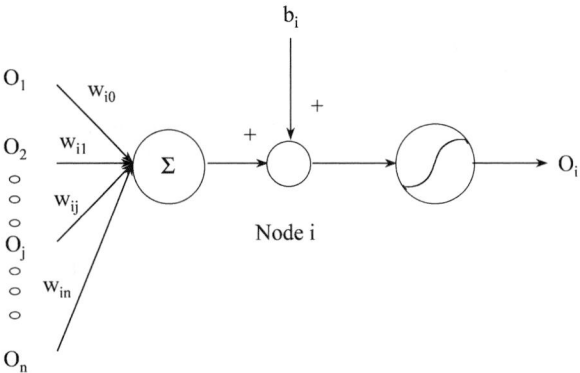

FIGURE 2   Individual NN node.

the squared error at the output:

$$E = \frac{1}{2}\sum_j (T_j - O_j)^2 \tag{2}$$

where T_j is the desired or target for the node j, and O_j is the actual output for node j. The error can be minimized by initializing the connection weights to some random values and adjusting each weight by an amount proportional to the negative of the gradient of the error:

$$\Delta w_{ji} = -\eta \frac{\partial E}{\partial w_{ji}} \tag{3}$$

where h is a positive learning coefficient or step size.

The emergence of the backpropagation algorithm is one of the most influential developments in the field of NNs. This algorithm provides an effective way of training networks of multiple layers. This allows the multilayered NNs to learn complex tasks that were previously unsolvable with single-layer networks.

The architecture of a multilayer (one or more hidden layers) network is shown in Fig. 3. The network inputs enter from the left of the diagram and are distributed to the first hidden-layer nodes. These hidden-layer nodes consist of a summation block for summing the incoming input signals, a bias term, and a nonlinear activation function just like the output nodes of the single-layer network. The output of the ith node at the nth layer is the summation of the contributions from the nodes in the $(n-1)$th layer and is the same as that represented in Eq. (1).

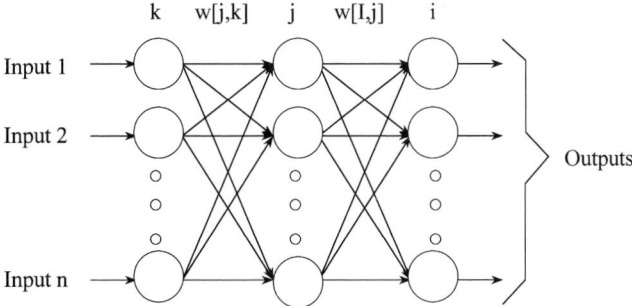

FIGURE 3   Multilayer perceptron.

The backpropagation algorithm uses the same error function given in Eq. 2. The gradient of the error with respect to the hidden-layer weights is estimated by propagating the error at the network output backward to the hidden layers. These error gradient estimates are used to update the weights by

$$\Delta w_{ji} = \eta \delta_j O_i \tag{4}$$

where

$$\delta_j = (T_j - O_j) f_{j'} \tag{5}$$

for the output-layer nodes and

$$\delta_j = \left(\sum_k \delta_k w_{kj} \right) f_{k'} \tag{6}$$

for the hidden-layer nodes. This operation is done layer by layer until all weights back to the input nodes are updated. A detailed explanation of the backpropagation algorithm is not given here. The basic principles of backpropagation can be found in Refs. 5, 9, and others.

APPLICATION CLASSES AND TYPICAL ARCHITECTURES OF MULTILAYER ERROR BACKPROPAGATION PERCEPTRONS

ANNs have been applied to a very large number of scientific and engineering problems. The classes of problems that are typically associated with the applications encountered in the process industries are

1. pattern recognition

2. interpolation and function approximation

3. parameter estimation and system identification

4. control applications

These four application classes are examined in more detail along with typical network architectures which have been used.

Pattern Recognition

Humans possess perceptive powers that allow certain pattern-processing tasks to be performed efficiently. These include the recognition of speech and images in a robust manner despite major variations, distortions, or omissions. Related to this is the ability to retrieve information based on the association of only minimal cues. For example, a glimpse of a person's head can cause us to recall in detail an entire face that is known to us.

It is motivation enough to try to understand the basis of these powers in humans. These traits also motivate engineers to design and develop machines with similar abilities. The perceptron described by Rosenblatt was applied to these areas of research in the 1950s and 1960s. Disappointing results with the perceptron approach led researchers in pattern recognition to concentrate on the mathematical aspects of pattern information processing. For example, researchers interested in building pattern recognition systems placed an emphasis on statistical pattern recognition, while fuzzy logic was used to provide links between pattern recognition research and human behavior.

More recently, ANNs have been revitalized (see the two volumes on parallel distributed processing by Rumelhart and McClelland [5]), and applications of these NN approaches to pattern recognition and the understanding of human perception has received wide attention.

Pattern classification can be viewed as a technique for strictly classifying a set of patterns to a group with similar characteristics or distinct characteristics. A second way to view pattern classification is as a way to determine the probability that a particular pattern belongs to a particular group. Both of these views have been considered in ANN research circles; however, the latter view is more general and useful for examining the field.

The conventional pattern classifier consists of two stages, as described by Lippmann [8]. The first stage computes a similarity score or matching score. The second stage associates the group or the class with the maximum score. The patterns to be classified can be viewed as points in a multidimensional space defined by the degrees of freedom of the input feature measurements. If the input vector has N dimensions and there are M groups or classes, then pattern classification is analogous to partitioning the N dimensional space into M regions. The matching score is generated by an algorithm for each of the M classes. This indicates how closely the input features match some exemplar for each class. This exemplar pattern is that pattern that is most representative of a particular class. In many cases a probabilistic model is used for the generation of input patterns from exemplars and the matching score represents the probability that the input pattern was generated from each of the M exemplars. These probabilistic classifiers assume a priori probability distributions for the input features. Multivariate Gaussian distributions are often assumed, and the parameters of the resulting distributions are subsequently estimated with a supervised training scheme [10].

Adaptive NN classifiers based on multilayer perceptrons have a similar two-stage functionality. Here input values are fed in parallel to the first stage of the NN classifier by means of N input connections. These connections may be binary-valued or continuously valued analog input connections. The first stage computes matching scores and typically outputs these scores over M parallel output connections. The second stage then selects the maximum score and enhances it. The second stage typically has one output for each of the M classes. While in the classification mode, it is expected that only one of the output nodes will be on, while the others are expected to be off.

Preprocessing stages are often added to the neural classifiers. These preprocessors often are modeled after biologically designed sensory systems. For example, preprocessors for image classification have been proposed that are based on the retina of the human eye. Other preprocessor filter banks for speech recognition have been designed as crude analogs of the cochlea of the human ear.

Figure 4 shows a typical multilayer neural network used as a classifier. Note that the two stages (computing a matching score and enhancing the matching score for classification) are often distributed across the network architecture.

Interpolation/Function Approximation

Often one knows the value of a function $f(x)$ at a number of points $x_1, x_2, \cdots x_n$, but an analytical expression for the relationship is not available. The values of the function may be the result of measurements of some physical phenomenon or a complex numerical calculation. Interpolation involves

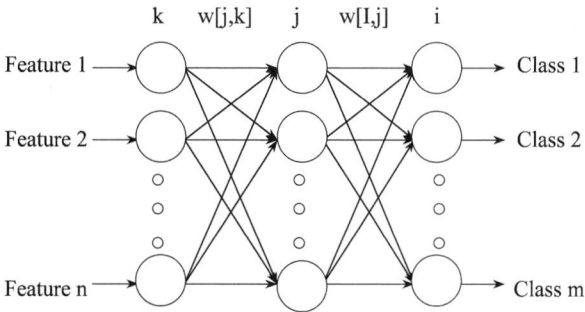

FIGURE 4   ANN for classification.

estimating the value of the function $f(x)$ at arbitrary points x, that lie between the smallest and the largest of the points x_i. To accomplish this task, a model of the function between the points must be generated by some functional form. The most common of the functional forms used are polynomials [11]. Function approximation, which is related to interpolation, involves finding an approximate function of a complex function or relationship that is easy to compute.

The tasks of interpolation and function approximation become much more difficult for multidimensional systems. An added difficulty may arise in cases in which interpolation or function approximation is to be carried out with known points from the system that are nonuniformly sampled.

Many researchers have applied NNs to the task of function approximation. Lapedes and Farber [12], Moody and Darken [13], and Jones et al. [14] have approximated the logistic function $x_{n+1} = 4x_n(1 - x_n)$ by using multilayer perceptrons, radial basis function networks, and the connectionist normalized linear spline (CNLS) net, respectively. The task of predicting the logistic function by use of NNs is simply to learn the quadratic function on the right-hand side of the equation.

Pao [15] used functional link nets to approximate a smooth function of one variable and to synthesize a three-dimensional surface from two two-dimensional Gaussian functions. Jones et al. [16] used the CNLS net to model an example taken from a free-electron laser (FEL) design surface. The most popular network is the feedforward multilayer perception (FFMLP) described by Rumelhart, et al. [17]. It has been shown by Funahashi [18] and Cybenko [19] that a single-hidden-layer network is capable of representing arbitrary continuous mappings from one finite dimensional space to another.

Function approximation can also be accomplished with traditional multiple regression techniques (see Refs. 20 and 21). A major problem in using the regression techniques is that for complex relationships a search for a suitable model can be very difficult. This search involves determining an overall order for the model and deciding which high-order cross terms to include. Pao's functional link net mentioned above suffers from the same shortcoming. In fact, the functional link net is closely related to classical regression.

The NN approaches avoid this problem by using structures, with a large degree of freedom, that are essentially model independent. When using NNs for functional approximation, one chooses a network architecture and, if the desired degree of accuracy is not achieved, adds additional network nodes are until the desired accuracy is met. NNs that adjust the number of nodes in the overall architecture automatically are called ontogenic neural networks. A major problem in using NNs for function approximation is that it is basically a black-box approach. That is, once the network is trained, little can be said about the functional approximation or the correlation between any of the inputs with the output. The resulting network is highly intractable. Finally, the problem of overfitting NNs, resulting in a condition known as brittleness, can be directly related to the high order of the resulting nonlinear NN.

Parameter Estimation and System Identification

The subject of system identification is important in any discussion of adaptive control. The ability of ANNs to model nonlinear functions to any arbitrary precision has been a large influence in their use for nonlinear adaptive control applications. A major portion of ANN research is involved in function approximation and dynamic system modeling.

Parameter estimation is an essential element for pattern recognition, adaptive signal processing, system identification, and adaptive control. The parameter estimation process requires that training data be generated for training the ANNs. These ANN applications of parameter estimation have been classified by Barto [22]:

1. Copying an existing controller: In this case the target outputs for the ANN are the actual control outputs of the expert teacher (existing controller) and the inputs are the same inputs that the existing controller uses for generating the control output (see Fig. 5).

2. Adaptive prediction: Adaptive prediction uses delayed values of a signal to predict the signal at some future time (often in the next time instant). The network to be trained gets the input information from the delayed values of the signal, while the output of the network becomes the signal to be

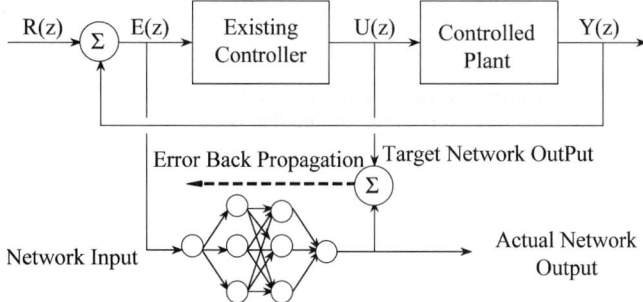

FIGURE 5   Copying an existing controller.

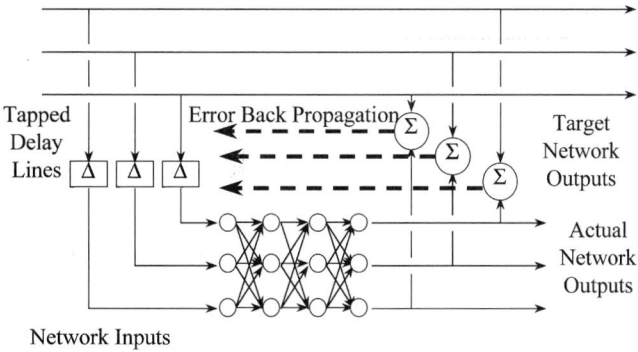

FIGURE 6   Adaptive signal prediction.

predicted (see Fig. 6). The task is to determine the relationship between these inputs and outputs. This approach assumes that a parameterized class of models exists that can represent the relationship between past values of a signal and the signal itself. The scheme illustrated in Fig. 6 is only a very simple case of more general predictive schemes involving cascades of delay units that provide input to the network to be trained.

3. System identification: Training information for system identification can be obtained by observation of the input/output behavior of a plant. The inputs to the ANN may be delayed values of the plant input and plant output (this is especially true for dynamical systems); see Fig. 7. One drawback of this approach is that if the order of the system is large, many past values of the system input(s) are required as ANN inputs. This implies that the input space can be very large for a multiinput high-order system. A comprehensive treatment of using ANNs for system identification can be found in Ref. 23.

4. Differentiating a model: This method identifies a plant model and then uses this model to backpropagate the system error at the plant output back to the input layer of the ANN controller. This permits the ANN controller to learn the plant inverse in an *indirect* manner based exclusively on ANNs (see Fig. 8). A variation of this approach was used by Fujii and Ura [24].

5. Identification of a system inverse: Here, the plant outputs become the ANN inputs and the plant inputs become the ANN outputs or target values (see Fig. 9). A problem with inverse identification arises when different plant inputs generate the same plant output and the mapping is not unique (the plant's inverse is not well defined). In cases in which the plant inverse is well defined or in cases in which the plant inverse can be approximated by more than one network for different operating regions of the plant, adaptive inverse controllers (AICs) can be very effective.

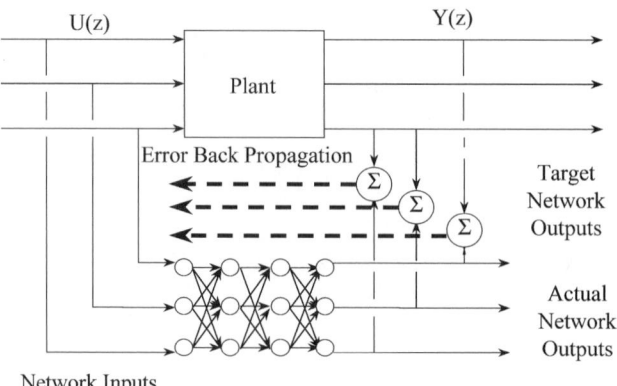

FIGURE 7   NN system identification.

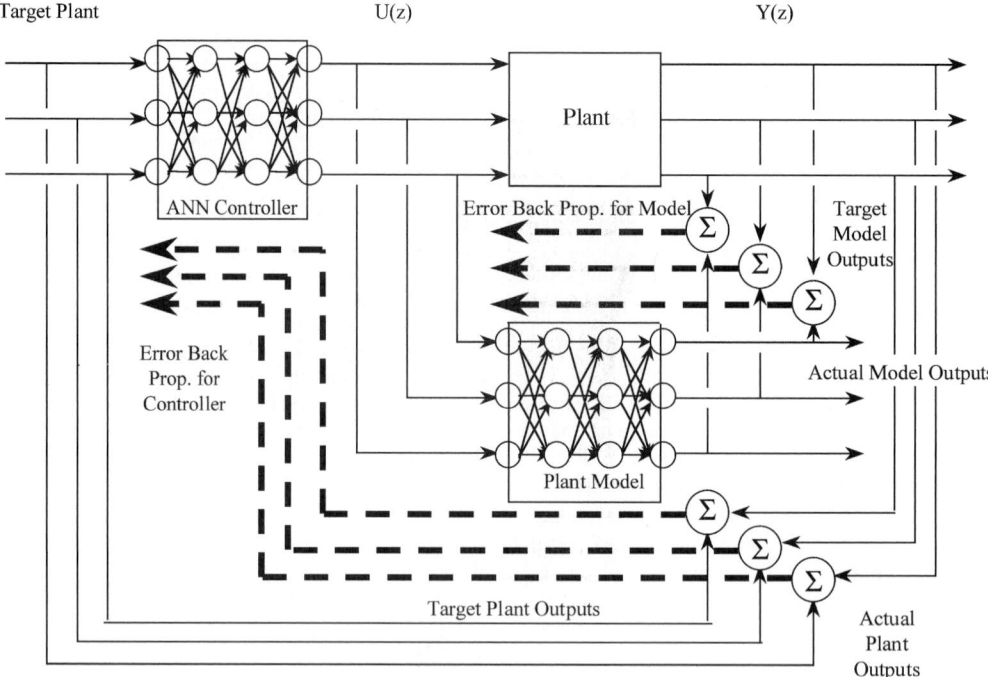

FIGURE 8   Differentiating a model.

All of these techniques are depicted with a multilayer NN. This network architecture is by far the most popular for applications in system identification and control applications. The most common training algorithm for this network is the error backpropagation algorithm described by Rumelhart et al. [17]. This algorithm is developed with gradient descent methods in which the partial derivative of the error function (a quadratic function of the network output) is backpropagated over the hidden layers of the network.

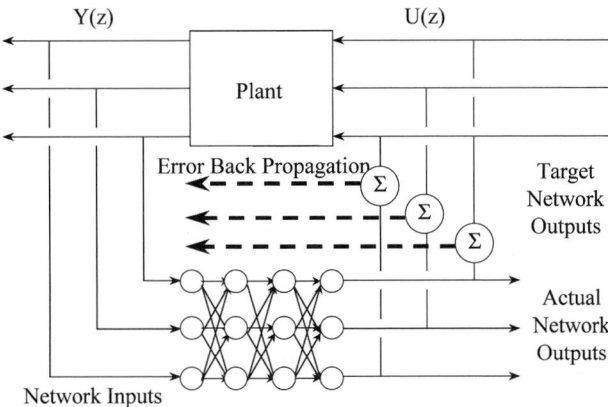

FIGURE 9 Identifying a system inverse.

Control Applications

In the process industries hundreds of parameters must often be tuned continuously to keep processes operating at or near an optimum value. Although the implementation of adaptive control may result in only a slight increase in efficiency, substantial economic savings can often be realized as a result. This fact has always been a motivation for applying adaptive control algorithms to process control. The recent advances in ANNs, and their adaptive characteristics have made the application of ANNs to adaptive control very popular in recent years.

For the most part, the discussion that follows concerns the implementation of multilayer perceptrons, (also called multilayer feedforward networks); however, other network architectures have been used in the adaptive control of dynamical systems, including recurrent or memory model networks. Also, for the most part, these implementations use the error backpropagation learning algorithm, although many other learning algorithms are currently under investigation for use in these applications.

The application of ANN controllers (ANNCs) can be categorized by the type of adaptive control structure in which they are used. Structures that have been used include self-tuning regulators (STRs), AICs, and model reference adaptive controllers (MRACs), among others. For a comprehensive treatment of these controllers in a conventional adaptive framework see Ref. 25.

Another way to classify these controllers is by their use in a direct or an indirect controller structure as that described by Narendra and Parthasarathy [26] (see Fig. 10). In indirect control the parameters of the plant are estimated as the elements of a vector $\hat{p}(k)$ at each instant k and the controller parameters $q(k)$ are estimated assuming that $\hat{p}(k)$ represents the true parameters $p(k)$ of the plant. There is an explicit identification process in this approach. For the direct control approach, the parameters of the controller are directly adjusted to reduce the output error. Both approaches have been used in the implementation of ANNs to control.

Indirect control is obviously more complex to implement than direct control, as can be seen from Fig. 10. In addition it is often required that the inverse plant relationship be determined for accurate tracking control, which can be very difficult or impossible for some systems. While the direct approach is simpler to implement, it is often difficult to estimate the error at the output of the NN controller. This error estimate is required for determining error gradients used for updating the network weights. Since the plant lies between the controller and the output error, the output error does not represent the error at the output of the controller.

Supervised Control. In supervised control, an ANN learns to simulate a human or other controller which is already familiar with controlling the plant (see Fig. 5). This method has been successfully used in the application of ANNs to control the inverted pendulum problem by training the ANN to

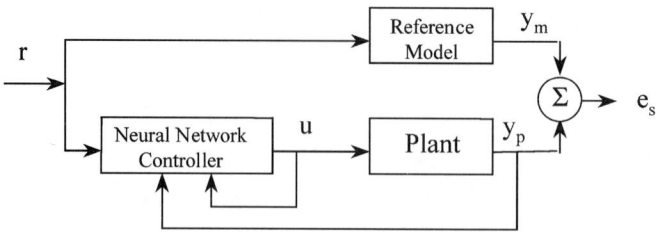

Direct Adaptive Control Using Artificial Neural Networks

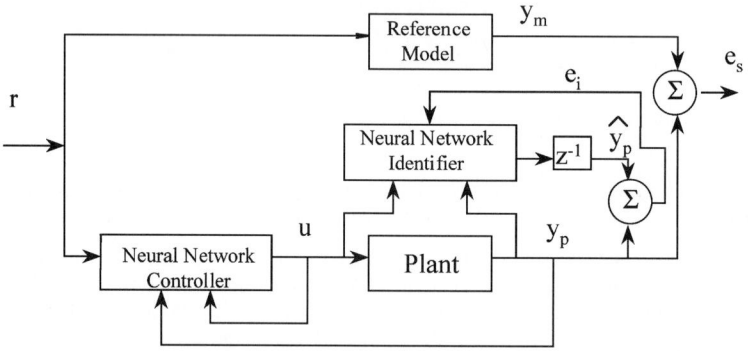

Indirect Adaptive Control Using Artificial Neural Networks

FIGURE 10   Direct and indirect NN controllers.

simulate a human controller [27] and to simulate a simple heuristic controller [28]. Fujii and Ura [24] used an ANN to simulate a simple fuzzy controller as an initial default controller during the application of ANNs to the control of autonomous underwater vehicles. ANNCs may be very useful when it is impractical to use a human controller in a particular environment or when the controller uses system parameters that are easier to measure than those used by the original controller. These controllers can often be useful as default controllers based on ANNs, which may be incorporated into larger overall control systems.

Inverse Control.   In direct inverse control, an ANN is used to learn an inverse mapping, such as learning the static inverse kinematics of a robot manipulator or the inverse of a dynamical plant as in the application of AIC (see Fig. 9). In the case of a learned static mapping, the plant can be controlled in an open-loop fashion. See for example the control of a robot manipulator by Van Der Smagt and Krose [29] or Guez and Ahmad [30] and the control of an autonomous vehicle model by Luebbers and Pandya [31].

AIC [also called inverse model control (IMC)], uses an inverse model of the plant dynamics, which is placed in cascade with the actual plant dynamics, resulting in a transfer function of essentially unit value within the frequency bandwidth of the plant input signal [25]. In the context of AIC, the inverse mapping is continually updated in an iterative manner.

In Ref. 32 the inverse model of the plant was determined directly by use of the plant output/input pairs for training the ANN. Hunt and Sbarbaro [33] use an ANN plant model to generate the input/output pairs used in directly training the plant inverse. Psichogios and Ungar [34] found the plant inverse by using Newton's method on a plant model and then trained the ANN with the resulting inverse mapping. Research is also currently being done in the area of inverting ANNs in an iterative manner (see Refs. 35 and 36).

Differentiating a Model, as described by Barto [22] and depicted in Fig. 8, is also considered to be a form of inverse control. The basic difference is that direct inverse control determines the inverse *directly* while differentiating a model involves finding a model of the plant and then determining the inverse controller in an indirect manner that utilizes the model. ANNs are used as the adaptive controller and are often used as the plant models.

In Ref. 37, an estimate of the partial derivatives of the plant at its operating point is used to further propagate the output error back through to the ANN. Another approach uses an inverse transfer relationship based on the inverse of the system transfer function matrix, and the plant output error signal (see Ref. 38).

Neural Adaptive Control. For neural adaptive control, ANNs are used in place of the linear mappings used in conventional adaptive control techniques. Conventional adaptive control includes designs such as the STRs and MRAC.

STRs are controllers consisting of a parameter estimator, a linear controller, and a block that determines the control parameters for the linear controller based on the estimated parameters [39]. The main disadvantage of this method is that a good prior estimate of the time scale of the process dynamics is required [40]. Currently there are two basic approaches that work well in practice, the pattern recognition approach and the relay feedback autotuner. The pattern recognition method observes the response to a step input or other disturbance and adjusts the tuning of a PID controller based on the pattern of the response. The relay autotuner momentarily switches a feedback relay into the loop, thus causing the loop to oscillate at the ultimate frequency. Once the ultimate frequency is determined, the well-known Ziegler–Nichols (see Ref. 41) or some other tuning criteria are applied. Little work has been done in applying ANNs to this area because a systematic method for tuning linear controllers by use of ANNs has not been developed. One simple application of this technique has been presented in Ref. 42.

Another promising area for the application of ANNs is in the development and implementation of model reference adaptive controllers as described by Widrow and Stearns [25]. Model reference controllers are similar to AICs; however, they involve the use of a reference model to generate a control trajectory that the system is to follow.

For a general treatment of linear and nonlinear plants see Ref. 43 or 32. For a more complete and unified approach to applying ANNs to the adaptive control of nonlinear systems by use of MRAC see Ref. 26.

Neural Network Extension To Linear Optimal Regulators. Iiguni et al. [44] present a method for extending the theory of linear optimal regulators to account for nonlinearities and uncertainties in the linear model of a plant. The classical linear optimal regulator is a full state feedback controller designed to operate a given plant with an acceptable deviation from a reference condition and with acceptable amounts of control action. The relative amounts of these deviations and control actions are specified in a cost function that is to be minimized (optimized). This method produces an optimal feedback gain (based on the cost function) by solving the Riccati equation. It is important to note that this controller is then optimal (in the sense of the cost function) for a given linear system that models the plant. Real plants contain uncertainties that cannot be modeled with a linear system. Therefore the linear optimal regulator is not optimal for the real plant.

These controllers are robust to a certain extent; however, the plant may become unstable if these uncertainties are too large. This method incorporates an ANN to model the plant uncertainties that are not modeled by the linear system equations. The linear model is used to generate the linear optimal regulator gain. A second ANN is used to generate an additional control command that is added to the output of the linear optimal regulator. This second ANN is trained to minimize the same cost function that was minimized in the design of the optimal linear regulator.

APPLICATIONS OF ARTIFICIAL NEURAL NETWORKS

Virtual Sensors

The largest application area for ANNs has been virtual sensors, especially for virtual continuous emissions monitors (VCEMs). It has been recently reported that more than 70% of the applications of ANNs in the process industries has been for this very important application class. NNs have been used for building virtual sensors for over a decade. Virtual sensors, also called virtual analyzers and inferential measurements, provide a process measurement or quality parameter that would normally not be available or not continuously available or available only with hardware that may be expensive to install and maintain. These virtual sensors use other easily obtainable process measurements that are in turn used to train a NN to provide the measurement of interest.

An example of such an application is that in which an important quality parameter is available only by taking samples and after a period of time receiving a lab analysis of the sample. Some action is typically taken if the analysis shows that the quality parameter is out of specification or appears to be heading out of the specification limits as determined by some statistical process control (SPC) analysis. The major problem with this case is that the time between when the sample is taken and the lab analysis is returned may be several hours or longer. As a result, the process may continue to be out of specification limits for some time, producing poor-quality products. The solution would be to have an on-stream analyzer give a continuous indication of the quality parameter so that adjustments could be made to the process in a manual or automatic fashion. Unfortunately, many quality parameters cannot be measured by a continuous analyzer or the analyzer can provide measurements in only a discrete time fashion, often with large amounts of time between readings, or the analyzers may be very expensive or difficult to install and maintain. A NN running in a computer or directly in the distributed control system (DCS) can often provide the desired measurement in a quick, accurate, and cost-effective manner.

The modeling described in the previous paragraph typically falls under one of two basic categories: first-principle modeling methods and empirical modeling methods. First-principle models use fundamental physical properties and the underlying equations that describe the dynamic nature of these processes to model a process or plant. Empirical modeling methods typically involve some sort of data regression to arrive at a mathematical expression for the model of the process or plant.

First-principle models have several disadvantages over empirically determined models:

1. First-principle models, while nonlinear in nature, often do not capture all of the important characteristics of a process. First-principle models, by their very nature, are often only approximate models of the actual process and in fact very detailed models are often too difficult to build in a reasonable amount of time. The approximate models often do not provide accurate indications of the desired measurement.

2. First-principle models often take a relatively long time to compute and converge, and therefore are often not fast enough to provide information to be used in a real-time control system.

3. Often there are no physical properties (or first principles) known that describe a particular process.

Empirical models (especially nonlinear models such as NNs) are often able to overcome all of these first-principle shortcomings.

Virtual Sensor Application. A typical application of using NNs as virtual sensors is the prediction of a melt index in a polyethelene process. The process described by Keeler and Ferguson [45] consists of four low-pressure reactors in series and is used to produce more than 60 different product grades with a melt index ranging from 0.1 to nearly 100. First-principle modeling of this process is very difficult, and the resulting process is highly nonlinear over the operating range. Lab samples are taken every hour but there is a significant time delay between when the melt index is measured versus when the controller manipulation occurs, and, in addition, the lab samples take an additional 2 to 3 h to be analyzed. This application was to build a virtual sensor that provided a melt index (in addition to other quality parameters) on a minute-by-minute basis.

The NN included the monomer and comonomer flow, hydrogen flow, catalyst and cocatalyst flow, and reactor temperatures as network inputs. The model consisted of 27 total input variables with time delays of from 45 to 50 min. The model predicts the melt index to within 10% of any sample analysis and provides the measurement approximately 2 h before the lab analysis is available. In addition this virtual sensor works well for all 60 products produced.

This can be contrasted with a hardware melt index meter that cost approximately $200,000, was hard to maintain, provided measurements that were 45 min downstream, and often provided measurements that were more than 10% off the laboratory analysis. It was decided to use only the virtual sensor based on the NN.

Sensor Validation Application.   Sensor validation is a concept that has been used in the process industries for many years. In its simplest form it may involve no more than a voting scheme between three redundant sensors (this is often done with pH loops) in which the one that is believed (and reported to the control system) is the middle select signal. The idea is that it would be a rare occurrence for two or more sensors to fail simultaneously; therefore, if two or more sensors are working properly, the middle select signal should be close to the correct measurement. Also one can determine that if one of the sensors is reading significantly different from the other two, then it is probably in error and may have failed.

NNs can be used to check the validity of a sensor by comparing its signal with an inferred measurement of the signal produced by the NN. In this case the NN would use various process measurements to infer the measurement of the sensor in question. In the case of a very troublesome instrument, these inferred measurements are often as reliable and less prone to failure, even though the inferred model relies on multiple process measurements to be working correctly.

A discussion of typical applications, implementation issues, and justification of NN sensor validation systems are described by Tay [46].

Virtual Continuous Emission Monitors.   VCEMS, also known as predictive emission monitors, have been a very active application area for NNs. The U.S. Environmental Protection Agency (EPA) requires that plants in the process industries that produce certain amounts and types of pollutants install emissions monitors. In 1993 the EPA (and some states) allowed the use VCEMs as a substitute for CEMs for certain sources. Since that time nearly 100 VCEMs have been installed and certified [47].

An application of a VCEM was described by Keeler and Ferguson [45] and involved modeling the emissions from a 100-MW cogeneration gas turbine. The VCEM provided three outputs, NO_x in parts in 10^6 (ppm), CO in ppm and O_2 in percent readings. The network used nine of the most significant process variables as inputs. The original relative accuracy test audit (RATA) on the turbine was 2%, 3%, and 4% at the low, medium, and high load ranges, respectively. The system has been in operation for more than 2.5 years and has passed all subsequent RATA tests with better than 99% uptime. Eight other VCEMs were installed with similar performance. The savings compared with hardware CEMs is estimated to be nearly $3 million for 15 installations by the customer. The savings come not only from the reduced cost of the initial installation but also from reduced operation and maintenance costs, especially since hardware CEMs often require daily calibration that the VCEMs do not.

Neurocontrollers and Process Optimization

One of the most important advanced control technologies to be exploited over the past two decades has been the application of constrained multivariable predictive control (CMPC). This technology has been implemented in over 2000 implementations and resulted in millions of dollars in reduced energy and raw material costs and enhanced production and yield. One of the fundamental assumptions of most CMPC products is that the process models are assumed to be linear. While this may be a reasonable assumption for plants that operate near a consistent operating point, in other cases the assumption of linearity is clearly not correct. Although CMPC vendors have found ways of accommodating these nonlinear processes, a better approach often is to use a nonlinear controller. Recent developments have allowed the use of NN-based model predictive controllers or neurocontrollers. Additionally vendors of neurocontrollers are integrating real-time optimizers (RTOs) into their products. Traditionally these RTOs have been implemented as an additional layer in the control hierarchy by linear CMPC vendors.

Demoro et al. [48] describe an interesting neurocontroller application to a commercial polypropylene reactor located in Brazil. This application included NNs for property prediction and sensor validation in addition to optimizing multivariable predictive control. The objective of the implementation was to maximize the production of the polypropylene product while minimizing variability in product quality and minimizing transition times.

The neurocontroller was compared with a linear CMPC controller for performance. Both controllers had three manipulated variables (reactor feed flow, modifier feed flow, and catalyst feed flow), three controlled variables (melt flow, solids concentration, and production rate) and two disturbance variables (inert concentration in the feed and reactor temperature). A NN sensor validation system was built for the melt flow sensor and for the neurocontroller implementation in general. It was shown that the performance of the neurocontroller was superior compared with that of the linear CMPC controller for a product transition from a 30 to a 35 melt flow transition. The addition of tuning designed to optimize the economics of the process was demonstrated and shown to maximize the production rate during the same product transition as before while maintaining product specifications.

Additional neurocontroller applications can be found in Refs. 49 and 50. Other process applications that do not specifically include neurocontrollers have also been reported in the literature, for example, applications described by Riddle et al. [51] and Slatsky and Warriner [52].

Other Artificial Neural Network Applications

The following ANN applications were taken from a presentation by Mertz [53]. These selected applications provide a good example of the ways that ANNs have been used in addition to virtual sensors and neurocontrollers. These brief case studies suggest certain rules of thumb and lessons that have been learned as a result of implementing NN applications over the course of several years.

1. Product dryer moisture: An ANN that used several inputs including temperatures, flows, pressures, etc., was designed to predict the product dryer moisture. An estimate of the product moisture out of the dryer was provided by lab samples of the dryer product. The correlation of the resulting model was not particularly good ($R^2 = 0.5$ to 0.6). A couple of observations were made: Sample times for lab analysis were not consistent and did not have accurate time stamps, and the relative humidity and ambient temperature were needed as inputs to the model. Model correlation improved dramatically when a sine wave with a 24-h period was included.

2. The prediction of product yield from reactor conditions: The ANN model used temperatures, flows, pressures, and a calculation to estimate the reactor yield. The actual reactor yield was determined from lab samples. The model correlation was excellent ($R^2 = 1.0$)! Input variables were discarded to determine the most important factor related to yield. Eventually there was only one input left, that being a calculation that was later determined to have the yield as one of its components!

3. Estimate slurry level in a continuous scraped surface crystallizer: The slurry level in question was not able to be measured directly. Operators were able to estimate the level based on observations of various crystallizer temperatures. The ANN model was built with the estimates supplied by the operators. The resulting model correlation was very good ($R^2 = 0.97$). The resulting model was put on line; however, operations were not satisfied with the performance. The model was retrained, and again the resulting model did not provide satisfactory results.

4. Estimation of a critical quality parameter in a flaker: The ANN model was built with 11 inputs and 8-h lab samples. Model correlation of the quality parameter was good ($R^2 = 0.9$). The model was to be put on line after a routine turnaround. The performance of the resulting model was very poor. It was determined that the replacement of water spray nozzles inside the flaker drum resulted in a major shift in the model performance. Subsequently the model was retrained and the resulting model put on line with good performance.

5. Batch process fault detection: The batch process in question had a very critical pH requirement. Often the batch would produce off-specification product because of problems with the pH control of the batch without operations having knowledge of the developing problem. An ANN model of good batch profiles was developed with 8 inputs, including levels, flows, temperature and agitator amplifiers with pH as the output. The model correlation was good ($R^2 = 0.99$). The model verified very well with good batches; however, bad batch data into the ANN quickly showed that something was wrong with the resulting pH of the system. This fault detection allows operations to take action immediately, which saves the batch from going out of specification.

6. ANN supplement to a first-principle model: A first-principle model of an electrolysis reaction was developed to optimize the process. Impurities in ppm concentrations in the process had a significant effect on the process models that could not be accounted for with first principles. Three models were built that related three individual impurities to their individual effects on the process. All ANN models had good correlation ($R^2 = 0.95$ or better). The resulting ANN models were incorporated into the first-principle model with very impressive results, and relationships between one of the impurities and the process were discovered as a result of the modeling process.

7. Neurocontrol of a unit operation by modeling the inverse of a process: A difficult to control waste water distillation column had large variations in feed quantities and feed concentrations. It was very difficult for operations to make correct changes in the manipulated variables of the column to account for the variations in the feed, and it was determined that the relationship was highly nonlinear. The basic idea was to model the inverse of the plant at correct operating conditions for a given feed rate and quality. The initial modeling effort used training data from the actual operation of the column. The resulting model provided poor performance as the model had captured the poor control of the column that had been demonstrated by operations. A first-principle model that used commercially available software was developed to develop "correct steady-state operating conditions" for the column by adjusting the feed quantity and quality and desired column output, then waiting for the model to converge to the proper or correct column-manipulated variables. The ANN was trained with 1300 data sets, which covered the operating conditions of the waste column. The resulting model had a very good correlation ($R^2 = 0.997$). The resulting model was put on line and has performed exceptionally well for several years.

8. End of batch prediction: A batch polymer extrusion was controlled manually and resulted in quality problems and safety considerations, in addition to being labor intensive. It was determined that an ANN model could be built with a set of temperature profiles provided by a resistance temperature detector bundle of various lengths inserted into the bottom of the batch autoclave. Initial results were not too good, and it was determined that the variability in batch extrusion rates was causing inaccuracies in the ANN model. Additional instrumentation was added to the process to control the extrusion process more closely. A new ANN model was developed that had a very good correlation ($R^2 = 0.98$). The resulting model predicts the end of batch times to within $+/- 15$ s. The ANN runs every 2 s.

9. Knowledge discovery: It was desired to shorten the overall time for completing a batch reaction that consisted of numerous reactants with multiple charges. In addition, numerous downstream quality problems were caused be problems with the reactor. A design of experiment (DOE) that used statistical process control techniques was conducted to improve the batch processing time of the reaction with limited results. The DOE examined 10 variables with 22 test batches run and resulted in 3 major recommendations. An ANN model of the process was built with approximately 25 resulting inputs from the process and a key process time interval as the network output. The correlation of the ANN was fair ($R^2 = 0.7$). A posttraining sensitivity analysis of the resulting model confirmed the findings of the original DOE but surprisingly indicated a more important factor contributing to the key process time interval. As a result of the study, process changes were made that resulted in significant batch reaction time reductions and the resulting improvement confirmed the predictions of the ANN.

Additional references are abundant in the literature, including applications to improve the accuracy of first-principle models by Bhat and Martin [54], Thompson and Martin [55], and Sabharwal et al. [56]; applications on process predictions by Morrison and Qin [57]; and an application on batch time prediction by Slusher et al. [58].

SELECTION OF AN ARTIFICIAL NEURAL NETWORK TOOL

In selecting an ANN tool, many considerations of the problem being solved and use of the resulting models become very important. For instance, the problem may involve huge data sets that can be handled by only certain large-scale tools. The resulting model may be used for doing real-time process control and therefore it may be desired to implement the resulting model at the DCS level. If the problem at hand is to do some knowledge discovery of your process, then a good postmodeling analysis becomes very important.

The ANN tools available to engineers cover a spectrum from being very simple (and simple to use), inexpensive (many are free or included with texts on the subject), to very large, complex tools with many features that are very expensive. Commercially available tools in use today include most of the popular features required for most ANN modeling efforts. However, there are still some differences that make one tool a better choice over another for a particular application. These differences include

1. Ease of use: If a tool is too difficult to use or too complex to learn to use in a timely fashion, it is likely that the tool will not be used, regardless of how many features it includes.
2. Cost: The costs of these tools vary considerably, and additional costs are often associated with putting the resulting models on line for control application. Some require that additional mainte-nance fees be paid while others do not (although new versions or later releases are typically not supported).
3. Implementation issues: Some of the resulting models must be implemented on a general-purpose computer and interfaces must be built to support the resulting application, while others may be implemented directly on DCS systems with standard interfaces already in place.

It is clear that the correct tool for the particular application is not always the same tool. Some typical good features that are available with many of the commercially available tools are as follows:

1. Statistical data analysis as a standard part of the data analysis (standard deviation, minimum, maximum, average, etc.)
2. Graphical cutting of "bad" data for preprocessing
3. Many data transformations available (filtering, logarithms, etc.)
4. Can handle huge amounts of data
5. Can do various types of interpolation to get additional data sets
6. Good posttraining analysis
7. Good posttraining what-if examples
8. Additional features available (Sensor Validation, Neuro-Control, Virtual Sensors, etc.)

Some not so good features of some tools include

1. complex features that make the tool difficult to use
2. no efficient way of choosing inputs and delays for the ANN model
3. large tools that can be slow (relative to other tools) to converge and for developing models
4. software that can be very expensive
5. on-line implementations that can be very expensive

6. on-line implementations that are relatively difficult

7. no control over architecture

8. final model is that black box

9. additional features that must be developed by user

The most important characteristics for ANN Tools are, in general,

1. efficient convergence engine

2. good preprocessing of input data

3. good and efficient method for determining model inputs and input delays

4. efficient and quick overall ability to produce ANN models

5. reasonable ease of use

6. reasonable low cost to purchase and put on line

7. good postprocessing analysis capabilities

PRACTICAL GUIDELINES FOR BUILDING ARTIFICIAL NEURAL NETWORKS

Some of the functional characteristics of ANNs are as follows:

1. can store static and dynamic patterns

2. can act as associative memories, associating a particular input pattern with an output pattern

3. can learn and generalize a set of patterns

4. can be used to perform pattern recognition and classification

5. are tolerant to faults in individual processing units caused by algorithmic errors or programming

6. are suitable for handling nonlinear system modeling

7. can handle multivariate data

8. can learn and adapt

As a result, these powerful tools are often used for general regression and model building by engineers who have little experience with either of these complex subjects. In general it is important that the individual building the ANN model have an intimate understanding of the underlying process. Nothing (especially a modeling tool) can take the place of deep process knowledge.

It is also important that the data collected are a true representation of the process or phenomenon to be modeled. For instance, if data are taken from a closed-loop control system, the resulting model will be a representation of the action of the controller in the process.

The limitations of NNs must also be taken into consideration when one is developing ANN applications. For instance, ANNs typically are good at interpolation but poor at extrapolation. While ANNs are capable of being universal approximators, they can overfit an input/output set of training data and as a result give poor generalization of the process being modeled.

While it is not the intent of this effort to describe in detail all of the pertinent issues involved in building ANNs, some general rules of thumb can be listed that cover a lot of the mistakes made by novice or first-time ANN modelers.

1. ANNs need data with accurate time stamps when lab analyses are used for training data sets.

2. ANN needs the correct input variables to get good correlation. For example, when moisture levels in a dryer product are being modeled, the relative humidity often needs to be included in the model.

3. Be careful that input variables are not a function of output variables. Often the person doing the modeling is not the same person who gathered the data. Be sure the inputs are not related to the model output.

4. Need real, ground truth data to model (guesses do not count). If the data are estimates of process parameters, the ANN at best can only estimate the parameter and in general will not be accurate.

5. Watch out for even minor changes to the process. These minor process changes can often significantly degrade the ANN model performance.

6. If you model a poorly controlled process in an attempt to build a neural controller, you will likely have a controller that controls poorly.

7. You need to ensure that the full operating range is represented by the data.

8. Uncontrolled variability in processes can cause model inaccuracies.

While it is clear that the tools available for building ANN applications are powerful and easy to use, they can easily be misused. In general it is important to know the process being modeled, know the data that are being used to develop the model, and know the limitations of the tools being used to develop the models.

RECENT DEVELOPMENTS AND FUTURE DIRECTIONS FOR ARTIFICIAL NEURAL NETWORKS

A number of technologies promise to improve the performance of various ANN applications in the near future. Of these, the following hold promise for improving the applications associated with the process industries:

1. recurrent NN architectures

2. genetic and evolutionary training algorithms

3. fuzzy logic/NN Hybrid Systems

Recurrent Network Architectures

One drawback of the approach described in the section on system identification, which is shown in Fig. 7, is that if the order of the system is large, many past values of the system input(s) are required as ANN inputs. This implies that the input space can be very large for a multiinput, high-order system. Recurrent ANNs are an attempt to address this problem.

ANNs can be grouped into two large classifications: feedforward networks and recurrent networks. In the context of modeling dynamical systems, this implies that either a parallel-series (feedforward training) or parallel (recurrent training) model be used for training, as shown in Fig. 11, which has the functional form of Eqs. (7) and (8), respectively:

$$\hat{y}(k+1) = -\sum_{i=0}^{n-1} a_i y(k-i) + \sum_{j=0}^{m-1} b_j u(k-j) \tag{7}$$

$$\hat{y}(k+1) = -\sum_{i=0}^{n-1} a_i \hat{y}(k-i) + \sum_{j=0}^{m-1} b_j u(k-j) \tag{8}$$

where the left side of each equation is the estimation of the dynamical system. While it has been shown that for the series-parallel model, the estimated parameters will converge to the true parameters as the tracking error approaches zero, the same cannot be said about the parallel model [59].

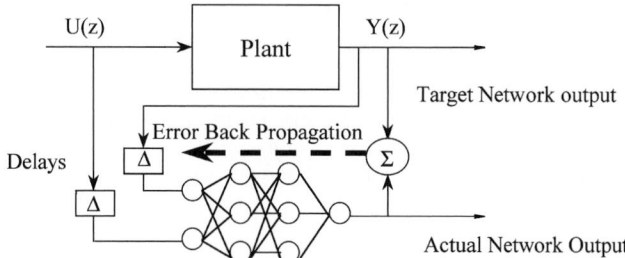

Parallel / Series Model

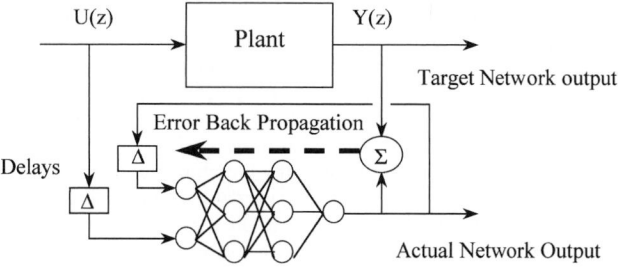

Parallel Model

FIGURE 11   Parallel and series system identification.

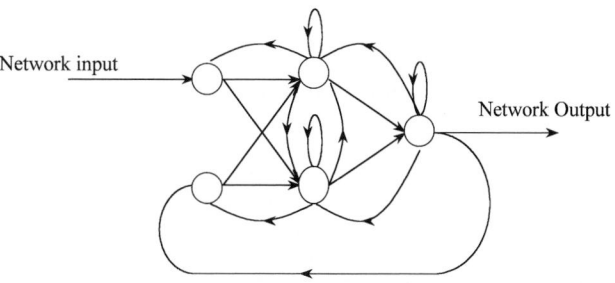

Recurrent Network

FIGURE 12   Recurrent ANN.

Feedforward multilayer NNs feed node outputs (network states) forward to nodes in successive layers. These networks can be used for approximating static mappings without any modifications to the architecture (see Fig. 3). The modeling of dynamical systems requires the use of connections from the output of the system to be modeled to an input node (see Fig. 7). In contrast, what are commonly referred to as recurrent ANNs feed delayed node outputs to other nodes in the current layer (including itself perhaps), or feed back the node output to nodes in preceding layers, or both (see Fig. 12). These networks are dynamical in nature, thereby producing a form of temporal memory. This feature allows the modeling of high-order systems without storing a large number of past values of the system inputs and outputs.

Backpropagation through time (BTT) is a method for training these recurrent NNs. Here the user specifies a utility function or performance measure to be maximized. BTT is used to calculate the derivative of the utility function summed across all future times with respect to the current actions. These derivatives are then used to adapt an ANN that outputs the actions. A major disadvantage of this approach is that the derivatives must be stored for all time steps to be computed, thus requiring large amounts of memory. This algorithm has therefore been used predominantly in off-line training modes. In some cases an approximation is made to this training method, which helps to alleviate the memory constraints. For a brief discussion of the application of recursive NNs see Refs. 23 and 60. Williams [61] focuses on the theory of BTT and popular approximations of this technique.

Genetic and Evolutionary Training

Occasionally one wishes to optimize a performance criterion by using an ANN technique; however, because of a particular controller or system identification structure, a set of input/output training pairs is not available. One method for training the ANN without input/output data is called genetic training. In this technique, weights are updated by subjecting a population of candidate weight vectors to operations described as mutation, crossover, etc., which produces a new generation of weight vectors. By "natural selection," vectors that minimize the performance criterion are retained while those that do not are discarded. Therefore the system tends to converge to a global minimum. The population of these weight vectors evolve over time so as to maximize or minimize some performance measure. Two references to these training methods in the literature are: Refs. 62 and 63.

Evolutionary training is a similar method, except the candidate weight vectors are subjected to a stochastic perturbation, thus producing a much larger population of weight vectors. One determines the best weight vectors by comparing them with a subset of the total population and calculating their ability to minimize the cost function. For a complete treatment of these algorithms see Ref. 64.

Fuzzy Logic/Neural Network Methods

Fuzzy logic controllers (FLCs) have been used in many applications over the past decade, allowing the application of heuristic information to be used in the control of complex systems. The heuristic information in a FLC consists of a set of linguistic IF–THEN rules, comprising the fuzzy logic rule base. Inputs to the FLC are in the form of linguistic variables that belong, in varying degrees, to various fuzzy membership sets. These variables are matched to conditions defined by the individual members of the fuzzy rule base, and the response of each rule is weighted according to the degree of membership of its inputs. Finally the weighted average of the responses of all the rules in the rule base is calculated to generate an appropriate output signal.

Currently there is no systematic way of designing FLCs. A typical approach is to define membership functions and the fuzzy logic rule base by studying and extracting the required information from human operators, existing controllers, or experts in a particular field. The performance of the FLC is then measured, and the rules and membership functions are adjusted until a desired performance is achieved. It is clear that the learning and generalizing characteristics of ANNs may have a substantial impact on the application of these controllers.

Kosko [65] has presented an approach for applying NN techniques to the development of fuzzy system controllers. These can be referred to as neural network/fuzzy logic adaptive controllers. These hybrid controllers show great promise in the area of very complex control problems, where conventional methods are difficult to apply. Typically ANNs are used to adapt the membership functions or adjust the weighted average of the rule base output responses, or both. A number of these types of applications have recently appeared in the literature (see Refs. 66–70).

REFERENCES

1. McCulloch, W. S., and W. Pitts, "A Logical Calculus of the Ideas Imminent in Nervous Activity," *Bull. Math. Biophys.* **5**, 115–133, 1943.

2. Hebb, D.O., *The Organization of Behavior, A Neurophysiological Theory*, Wiley, New York, 1949.

3. Rosenblatt, F., "The perceptron, a probability model for information storage and organization in the brain," *Psychol. Rev.* **62**, 368–408, 1958.

4. Minsky, M., and S. Papert, *Perceptrons: An Introduction to Computational Geometry*, MIT Press, Cambridge, Massachusetts, 1969.

5. Rumelhart, D. E., and J. L. McClelland, Eds., *Parallel Distributed Processing*, MIT Press, Cambridge, Massachusetts, 1988.

6. Karayiannis, N. B., and A. N. Venetsanopoulos, *Artificial Neural Networks*, Klewer Academic, Boston, 1993.

7. Muller, B., and J. Reinhardt, *Neural Networks*, Spinger-Verlag, New York, 1990.

8. Lippmann, R. P., "An Introduction to Computing with Neural Nets," *IEEE ASSP Mag.*, April, pp. 4–21, 1987.

9. Wasserman, P. D., *Neural Computing*, Van Nostrand Reinhold, New York, 1989.

10. Duda, R.O., and P.E. Hart, *Pattern Classification and Scene Analysis*, Wiley, New York, 1973.

11. Press, W. H., S. A. Teukolsky, W.T. Vetterling, and B. P. Flannery, *Numerical Recipes in C: The Art of Scientific Computing*, Cambridge Univ. Press, New York, 1992.

12. Lapedes, A. S., and R. Farber, "Nonlinear Signal Processing using Neural Networks: Prediction and System Modeling," Tech. Rep., Los Alamos National Laboratory, Los Alamos, New Mexico.

13. Moody, J., and C. J. Darken, "Fast Learning in Networks of Locally-Tuned Processing Units," *Neural Comput.* **1**, 281–294, 1989.

14. Jones, R. D., Y. C. Lee, C. W. Barnes, G. W. Flake, K. Lee, P. S. Lewis, and S. Qian, "Function Approximation and Time Series Prediction with Neural Networks," *in Proceedings of the International Joint Conference on Neural Nets*, 1990, Vol. 1, pp. 649–665.

15. Pao, Y.H., *Adaptive Pattern Recognition and Neural Networks,* Addison-Wesley, Reading, Massachusetts, 1989.

16. Jones, R. D., Y. C. Lee, S. Quian, C. W. Barnes, K. R. Bisset, G. M. Bruce, G. W. Flake, K. Lee, L. A. Lee, W. C. Mead, M. K. O'Rourke, I. J. Poli, and L. E. Thode, "Nonlinear Adaptive Networks: A Little Theory, A Few Applications," presented at the meeting on Cognitive Modeling in System Control, Santa Fe, New Mexico, June 10–14, 1990.

17. Rumelhart, D.E., G. E. Hinton, and R. J. Williams, "Learning Internal Representations by Error Propagation," in Ref. 5.

18. Funahashi, K., "On the Approximate Realization of Continuous Mappings by Neural Networks." *Neural Networks* **2**, 183–192, 1989.

19. Cybenko, G., "Approximation by Superpositions of a Sigmoidal Function," *Math. Control, Sig., Syst.* **2**, 303–314.

20. Draper, N., and H. Smith, *Applied Regression Analysis*, Wiley, New York, 1966.

21. Neter, J., W. Wasserman, and M. Kutner, *Applied Linear Regression Models*, Irwin, Homewood, Illinois, 1983.

22. Barto, A. G., "Connectionist Learning for Control: An Overview," in *Neural Networks for Control*, W. T. Miller, R. S. Sutton and P. J. Werbos, Eds., MIT Press, Cambridge, Massachusetts, 1991.

23. Narendra, K. S., and K. Parthasarathy, "Neural Networks and Dynamical Systems Part II: Identification," Rep. 8902, Center for Systems Science, Electrical Engineering, Yale University, New Haven, Connecticut, 1989.

24. Fujii, T., and T. Ura, "Development of Motion Control System for AUV Using Neural Nets," in *Proceedings of the Symposium on AUV Technology*, 1990, pp. 81–86.

25. Widrow, B., and S. D. Stearns, *Adaptive Signal Processing*, Prentice-Hall, Englewood Cliffs, New Jersey, 1985.

26. Narendra, K. S. and K. Parthasarathy, "Neural Networks and Dynamical Systems Part III: Control," Rep. 8909, Center for Systems Science, Electrical Engineering, Yale University, New Haven, Connecticut, 1989.

27. Guez, A., and M. Piovoso, "Custom neurocontroller for a time delay process," in *Proceedings of the 1991 American Control Conference*, Vol. 2, pp. 1592–1596, 1991.

28. Heung, J. Y., and R. J. Mulholland, "Using a neural network as a feedback controller," in *Proceedings of the 32nd Midwest Symposium on Circuits and Systems*, 1989, Vol. 2, pp. 752–755.

29. Van Der Smagt, P. P., and B. J. A. Krose, "A real-time learning neural robot controller," presented at the 1991 International Conference on Artificial Neural Networks, ICANN-91, Espoo, Finland, June 24–28, 1991.

30. Guez, A., and Z. Ahmad, "Solution to the inverse kinematics problem in robotics by neural networks," presented at the International Conference on Neural Networks, March 1988.

31. Luebbers, P. G. and A. S. Pandya, "Video image based neural network guidance system with adaptive view angles for autonomous vehicles," in *Applications of Artificial Neural Networks II*, S.K. Rogers, Ed., Proc. SPIE **1469**, 756–765, 1991, "Neural Network Guidance System for Autonomous Vehicles," in *Progress in Neural Networks*, O. M. Omidvar, Ed., 1991.

32. Levin, E., R. Gewirtzman, and G. Inbar, "Neural network architecture for adaptive system modeling and control," in *Proceedings of the International Conference on Neural Networks*, 1989, Vol. 2, pp. 311–316.

33. Hunt, K. J., and D. Sbarbaro, "Neural networks for nonlinear internal model control," *IEE Proc. D* **138**, 431–438, Sept. 1991.

34. Psichogios, D. C., and L. H. Ungar, "Nonlinear internal model control and model predictive control using neural networks," in *Proceedings of the Fifth IEEE Symposium on Intelligent Control*, IEEE, New York, pp. 1082–1087, 1990.

35. Hoskins, D. A., J. N. Hwang, and J. Vagners, "Iterative inversion of neural networks and its application to adaptive control," *IEEE Trans. Neural Networks*, **3**, 292–301, 1992.

36. Lowe, D., "On the iterative inversion of RBF networks: A statistical interpretation," *Second Intl. Conf. on Artificial Neural Networks*, Institution of Electrical Engineers, Bournemouth International Centre, UK, pp. 29–33, November 18–20, 1991.

37. Psaltis, D., A. Sideris, and A. Yamamura, "A multilayered neural network controller," *IEEE Control Sys. Mag.*, pp. 17–21, April 1988.

38. Chen, V. C., and Y. Pao, "Learning control with neural networks," in *Proceedings of the IEEE Conference on Robotics and Automation*, IEEE, New York, 1989, Vol. 3, pp. 1448–1453.

39. Astrom, K. J., U. Borisson, L. Ljung, and B. Wittenmark, "Theory and Applications of Self-Tuning Regulators," in *Automatica*, Pergamon, New York, 1977, Vol. 13, pp. 457–476.

40. Astrom, K. J., "Toward Intelligent Control," *IEEE Control Syst. Mag.*, April 1989.

41. Shinskey, F. G., *Process Control Systems: Application/Design/Adjustment*, McGraw-Hill, New York, 1979.

42. Claudio, C. C., C. L. Nascimento, and T. Yoneyama, "An auto-tuning controller with supervised learning using neural nets," *Intl. Conf. on Control 91'*, Institution of Electrical Engineers, Heriot Watt University, Edinburgh, UK, Vol. 1, pp. 140–144, March 25–28, 1991.

43. Weerasooriya, S., and M. A. El-Shakawi, "Identification and control of a DC motor using backpropagation neural networks," *IEEE Trans. Energy Conversion,* 1991 **6**, 663–669.

44. Iiguni, Y., H. Sakai, and H. Tokumaru, "A nonlinear regulator design in the presence of system uncertainties using multilayered neural networks," *IEEE Trans. Neural Networks,* **2**, 410–417, 1992.

45. Keeler, James D., and R. Bruce Ferguson, "Commercial Applications of Soft Sensors: The Virtual On-Line Analyzer and the Software CEM," *IFAC Conference Proceedings*, San Francisco, CA, Pergamon, New York June 30–July 5 1996.

46. Tay, M. E., "Keeping Tabs on Plant Energy and Mass Flows," *Chem. Eng.* Vol. X, 82–88, 1996.

47. Keeler, James D. and Paul Reinerman, "Pavilion's Software CEM: A Low-Cost, Reliable Alternative for Enhanced Monitoring," Executive Enhanced Monitoring Conference. (Reprints available from Pavilion Technologies.)

48. Demoro, E. C. Axelrud, D. Johnson, and G. Martin, "Neural Network Modeling and Control of Polypropylene Process," presented at the Society of Plastic Engineers, Polyolefins X International Conference, Houston, Texas, February 23–26, 1997.

49. Keeler, J. D., G. Martin, G. Boe, S. Piche, U. Mathur, and D. Johnson, "The Process Perfector: The Next Step in Multivariable Control and Optimization," Pavilion Technologies, Inc. Technical Report, 1996.

50. Zhao, H., J. Guiver, and C. Klimasauskas, "NeuCOP II: A Nonlinear Multivariable Process Modeling, Control and Optimization Package," presented at the AIChE 1997 Spring National Meeting, Houston, Texas, March 9–13, 1997.

51. Riddle, A. L., N. V. Bhat, and J. R. Hopper, "Neural Networks Help Optimize Solvent Extraction Process in a Lube Oil Plant," paper Supplied by Pavilion Technologies, Inc., no date.

52. Slatsky, M., and G. Warriner, "Boilers: A Hot Application for Neural Nets," *Control*, November, 1997.

53. Mertz, G., "Neural Networks: Not Just For Virtual Sensors," presented at the Fisher-Rosemont Users Group Meeting, Denver, Colorado, 1998.

54. Bhat, N., and G. Martin, "How Hybrid Modeling Approaches Between Neural Networks and Physical Models can Benefit Customers," paper Supplied by Pavilion Technologies, Inc, no date.

55. Thompson, W., and G. Martin, "How Neural Network Modeling Methods Complement Those of Physical Modeling," presented at the NPRA Computer Conference, Atlanta, Georgia, November 11–13, 1996.

56. Sabharwal, Amish; Bhat, Naveen and Wada, Tetsuya, Benefits of Integrating Empirical and Physical Modeling in an Oil Refinery Process, Paper Supplied by Pavilion Technologies, Inc., no Date.

57. Morrison, S., and J. Qin, "Neural Networks for Process Prediction," *ISA Trans.*, Vol. 34, pp. 443–450, October, 1995.

58. Slusher, S., R. L. Bennett, and D. L. Deitz, "Fermentation Application Uses Neural Networks to Predict Batch Time," *Pharmaceut. Eng.* **14** (15), 1994.

59. Narendra, K. S., and A. M. Annaswamy, *Stable Adaptive Systems*, Prentice-Hall, Englewood Cliffs, New Jersey, 1989.

60. Pourboghrat, F., "Neuromorphic Controllers," in *Proceedings of the 28th IEEE Conference on Decision and Control*, IEEE, New York, 1989, Vol. 2, pp. 1748–1749.

61. Williams, R. J., "Adaptive State Representation and Estimation Using Recurrent Connectionist Networks," in *Neural Networks for Control*, W. T. Miller, R. S. Sutton, and P. J. Werbos, Eds., MIT Press, Cambridge, Massachusetts, 1991.

62. Ichikawa, Y., and T. Sawa, "Neural Network Application for Direct Feedback Controllers," *IEEE Trans. Neural Networks* **3**, 224–231, 1992.

63. Fogel, D. B., L. J. Fogel, and V. W. Porto, "Evolving Neural Networks," *Biol. Cybern.* **63**, 487–493, 1990.

64. Fogel, D. B., *System Identification Through Simulated Evolution: A Machine Learning Approach To Modeling*, Ginn Press, Needham Heights, Massachusetts, 1991.

65. Kosko, B., *Neural Networks and Fuzzy Systems*, Prentice-Hall, Englewood Cliffs, New Jersey, 1992.

66. Lin, C. T., and C. S. George, "Neural-Network-Based Fuzzy Logic Control and Decision System," *IEEE Trans. Comput*. **40**, 1320–1336, 1991.

67. Iwata, T., M. Machida, and Y. Toda, "Fuzzy control using neural network techniques," *Intl. Joint Conf. on Neural Networks*, IEEE, San Diego, CA, Vol. 3, pp. 365–370, June 17–21, 1990.

68. Jang, J. S., "Fuzzy Controller Design without Domain Experts," in *Proceedings of the IEEE International Conference on Fuzzy Systems*, IEEE, New York, 1992, pp. 289–296.

69. Wang, B. H., and G. Vachtsevanos, "Fuzzy Associative Memories: Identification and Control of Complex Systems," in *Proceedings of the 5th IEEE International Symposium on Intelligent Control*, IEEE, New York, 1990, pp. 910–915.

70. Berenji, H., "Neural Networks and Fuzzy Logic in Intelligent Control," in *Proceedings of the 5th IEEE International Symposum on Intelligent Control*, IEEE, New York, 1990, pp. 916–920,

SECTION 11
STANDARDS OVERVIEW*

Terry Blevins
Fisher-Rosemount Systems, Inc. Austin, Texas (Fieldbus)

Thomas G. Fisher, P. E.
Lubrizol Corporation, Wickliffe, Ohio (Batch Control)

Paul Gruhn, P. E.
Moore Products, Houston, Texas (Safety Instrumented [Interlock] Systems)

---

* Persons who authored complete articles or subsections of articles, or who otherwise cooperated in an outstanding manner in furnishing information and helpful counsel to the editorial staff.

SAFETY INSTRUMENTED (INTERLOCK) SYSTEMS

by Paul Gruhn*

INTRODUCTION

This article provides an overview of the ISA Standard on Safety Systems (S84), with an emphasis on the layers of protection and the estimates of the availability of different types of instrumentation and controls. The definition of a safety instrumented system is a system designed to respond to conditions of a plant, which may be hazardous in themselves, or if no action were taken could eventually give rise to a hazard. It must generate the correct outputs to prevent the hazard or mitigate the consequences.

The ISA (International Society for Measurement and Control) S84 and IEC (International Electrotechnical Commission) 61508/61511 standards, along with the AIChE CCPS (American Institute of Chemical Engineers, Center for Chemical Process Safety) Guidelines on safety instrumented (interlock) systems, as well as process safety management legislation [1]–[4] are performance oriented, not prescriptive. They do not tell people what technology logic system to use (relay, solid state, or software based), what logic and field device configuration to use (single, dual, or triplicated), or how often to test a system (monthly, quarterly, or yearly). They merely list the performance requirements for the overall system. In other words, the greater the level of risk of the process, the greater the performance needed of the safety instrumented system.

What People Really Want

However, what most people really want is a simple "cookbook" of preplanned solutions. For example, for a refinery, turn to page 35 in standard ABC. There it shows dual sensor, dual logic, simplex valves, yearly test interval, and so on. For an offshore platform, turn to page 63. There it shows. . . . Unfortunately, at this point in time, industry standards are not written this way. The standards do *not* give clear, simple, precise answers. They do *not* mandate technology, level or redundancy, or test intervals.

What the Standards Actually Are

There is a fundamental change in the way industry standards are being written. Standards are moving away from prescriptive standards and toward more performance-oriented requirements. After all, it's relatively easy to be prescriptive about something we have a great deal of experience with (e.g., boilers). The same cannot be said of relatively new and unproven processes. This means each plant will have to decide for itself just what is safe, and each plant will have to decide on how it will determine and document that its systems are, in fact, safe. Unfortunately, these are difficult decisions that few want to make, and fewer still want to put in writing.

DESIGN LIFE CYCLE

Designing a single component may be viewed as a relatively simple matter, one that a single person can handle. Designing any large system, whether it's a car, a computer, an airplane, or a safety instrumented system, however, is typically beyond the ability of any single individual. Large systems

---

* P.E., Moore Products, Houston, Texas.

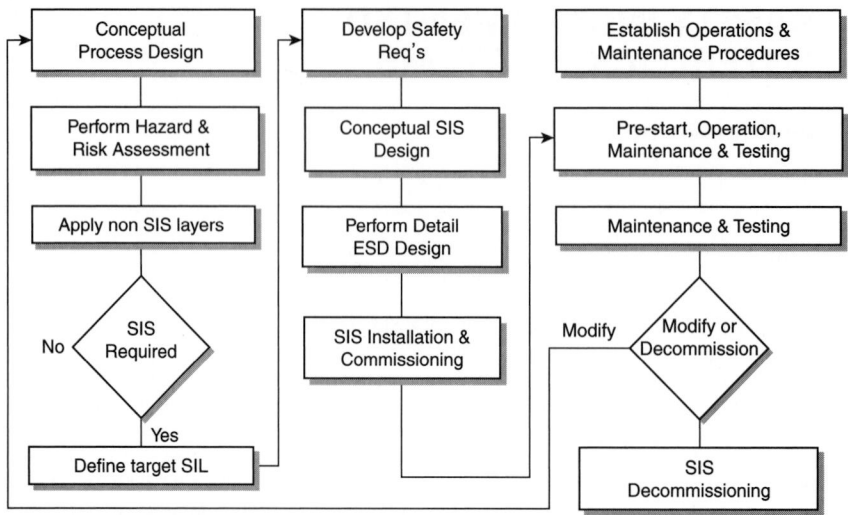

FIGURE 1   ISA S84 design life cycle.

require a multidiscipline team. The control system engineer should not feel that the entire burden of designing a safe plant rests on his or her shoulders alone, because it does not.

Experience has shown that a detailed, systematic, methodical, well-documented design *process* is called for in the design of safety instrumented systems. This starts with a safety review of the process, implementation of other safety layers, systematic analysis, and detailed documentation and procedures. The steps are described in most documents as a safety design life cycle. The intent is to leave a documented, auditable trail, and make sure that nothing is neglected because it has fallen between the inevitable cracks within every organization.

Large systems require a methodical design process. Figure 1 shows the life-cycle steps as described in the ISA S84 standard. This should be considered one example only, as there are variations of the life cycle presented in other industry documents. A company may wish to develop its own variation of the life cycle based upon its unique requirements.

Some will complain that performing all of the life-cycle steps, like all other tasks designed to lower risk, will increase overall costs and result in lower profitability and productivity. One in-depth study in the past, conducted by a group including major engineering societies, 20 industries, and 60 product groups with a combined exposure of over 50 billion hours, concluded that *production increased as safety increased* [5].

Conceptual Process Design

The first step in the life cycle is to develop an understanding of the process, the equipment under control, and the environment (physical, social, political and legal) in sufficient depth to enable the other life-cycle activities to be performed. The goal is to design an inherently safe plant. The activities in this step are generally considered outside the realm of the control system engineer.

Hazard Analysis and Risk Assessment

The next step is to develop an understanding of the risks associated with the process. Risks may impact personnel, production, capital equipment, the environment, company image, and so on. Hazard analysis consists of identifying the hazards. There are numerous techniques one can use (HAZOP, What If, Fault Tree, Checklist, etc.) and numerous texts describing each method [6]–[8]. Risk assessment consists of classifying the risk of the hazards that have been identified in the hazard analysis. Risk is a function of the frequency or probability of an event, and the severity or consequences of the event. Risk assessment can be either qualitative or quantitative. Qualitative assessments subjectively rank the risks from low to high; quantitative assessments, as the name obviously implies, attempt to assign numerical factors to the risk, such as death or accident rates. This is not intended to be the sole responsibility of the control system engineer. There are obviously a number of other disciplines required in order to perform these assessments.

Application of Non-SIS Layers

The goal of process plant design is have a plant that is inherently safe, or one where residual risks can be controlled by the application of noninstrumented safety layers. KISS (keep it simple, stupid) should be an overriding theme.

Is an SIS Required?

If the risks can be controlled to an acceptable level without the application of an instrumented system, then the design process stops (as far as a safety instrumented system is concerned). If the risks cannot be controlled to an acceptable level by the application of noninstrumented layers, then an instrumented system will be required.

Define the Target Safety Integrity Level

The safety system performance should match the level of risk. In other words, the greater the level of process risk, the better the safety system one needs in order to control the risk. This requires identifying the individual risks and assessing their impact.

The most difficult step in the overall process for most organizations seems to be determining the required SIL (safety integrity level). This is not a direct measure of process risk, but rather a measure of the safety system performance required in order to control the risks identified earlier to an acceptable level. The standards describe qualitative methods on how this can be done. One method is outlined in Fig. 2. A potential problem with the qualitative methods is that they are subjective and very often not repeatable. Different organizations may review the same process and each come up with different SIL requirements. There are also quantitative methods available for determining the SIL. A problem for many trying to use quantitative methods, however, is that they must decide on a quantitative target safety goal. Deciding "what is a tolerable death rate" is something few people wish to do, and fewer companies wish to put in writing. One can just imagine an attorney saying, "What do you *mean* you considered it *tolerable* to *kill* four people every 100 million man hours?!"

Develop Safety Requirements Specification

The next step consists of developing the safety requirements specification, essentially the functional logic of the system. This will naturally vary for each system. There is no general, across-the-board recommendation that can be made. For example, if temperature sensor TT2301 exceeds 410°C, then close valves XV5301 and XV5302. Each safety function should have an associated SIL requirement, as well as any reliability requirements if nuisance trips are a concern. One should include *all* operating conditions of the process, from start-up through shutdown, as well as maintenance. (One may find that certain logic conditions conflict during different operating modes of the process.)

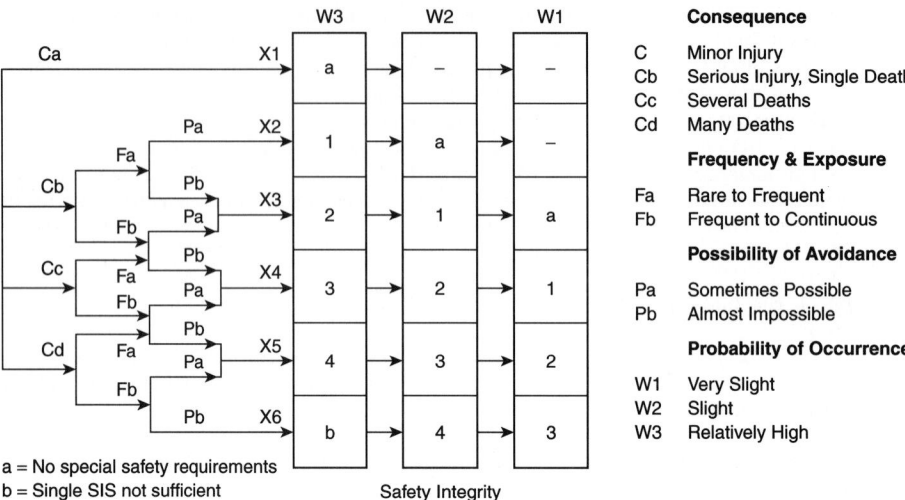

a = No special safety requirements
b = Single SIS not sufficient

Safety Integrity

FIGURE 2   One qualitative method of determining the required SIL.

The system will be programmed and tested according to the logic determined during this step. If an error is made here, it will carry through for the rest of the design. It won't matter how redundant or how often the system is manually tested—it will not work properly when required. These are referred to as systematic, or functional, failures. Using diverse redundant systems, programmed by different people using different languages and tested by an independent team, will not help in this situation, because the functional logic they all based their work on could be in error.

Conceptual SIS Design

One doesn't pick a certain size jet engine for an aircraft based on intuition. One doesn't size a million dollar compressor by gut feel. One doesn't determine the size of pilings required for a bridge by trail and error (at least not any more).

The purpose of this step is to develop an initial design in order to see if it meets the safety requirements and SIL performance requirements. Initially one needs to select a technology, configuration (architecture), test interval, and so on. This pertains to the field devices as well as the logic box. Factors to consider are overall size, budget, complexity, speed of response, communication requirements, interface requirements, method of implementing bypasses, testing, and so on. One can then perform a relatively simple quantitative analysis to see if the proposed system meets the performance requirements [9]–[12], or make a qualitative judgment based on prior experience (although this is obviously harder to substantiate). The intent is to evaluate the system *before* one specifies the solution. Just as it is better to perform a HAZOP *before* you build the plant (it's hard to change the design once it's already been built), it is better to analyze the proposed safety system *before* you specify it, or else how will you know if it meets the performance goal?

Detailed SIS Design

The purpose of this step is to finalize and document the design. Once a design has been chosen, the system must be engineered and built following strict and conservative procedures. This is the only realistic method of preventing design and implementation errors that we know of. The process requires thorough documentation, that is, an auditable trail that someone else may follow for verification purposes.

Installation and Commissioning

This step is to ensure the system is installed per the design and performs per the safety requirements specification. Before a system is shipped from a factory, it must be thoroughly tested for proper operation. If any changes are required, they should be made at the factory, not at the installation site. At installation, the entire system, this time including the field devices, must be checked as well. There should be a detailed installation document outlining each procedure to be carried out. Finished operations should be signed off in writing, showing that each function and operational step has been checked.

Operations and Maintenance

In order to function properly, every system requires periodic maintenance. Not all faults are self-revealing, so *every* safety system *must* be periodically tested in order to make sure it will respond properly to an actual demand. The frequency of inspection and testing will have been determined earlier in the life cycle. All testing must be documented.

Modifications

As process conditions change, it will be necessary to make changes to the safety system. All proposed changes require returning to the appropriate phase of the life-cycle. A change that may be considered minor by one individual may actually have a major impact to the overall process. This can only be realized if the change is thoroughly reviewed by a qualified team. Hindsight has shown that many accidents have been caused by this lack of review [13].

Decommissioning

Decommissioning a system should entail a review to make sure removing the system from service will not impact the process or surrounding units, and that means are available during the decommissioning process to protect the personnel, equipment, and environment.

MULTIPLE INDEPENDENT SAFETY LAYERS

Figure 3 appears in a number of different formats in most all of the standards. It shows how there are various safety layers, some of which are prevention layers, others which are mitigation layers. The basic concept is simple: don't put all your eggs in one basket. (Everything fails; it's just a matter of when.) The more layers there are, the better. In addition, each layer should be as simple as possible, and the failure of one layer should not prevent another layer from performing its intended function. Some refer to this as defense in depth.

Process Plant Design

The process plant itself must be designed with safety in mind. This is why HAZOP (Hazard and Operability Studies) and other reviews are performed, such as fault trees, checklists, what-if, and so on.

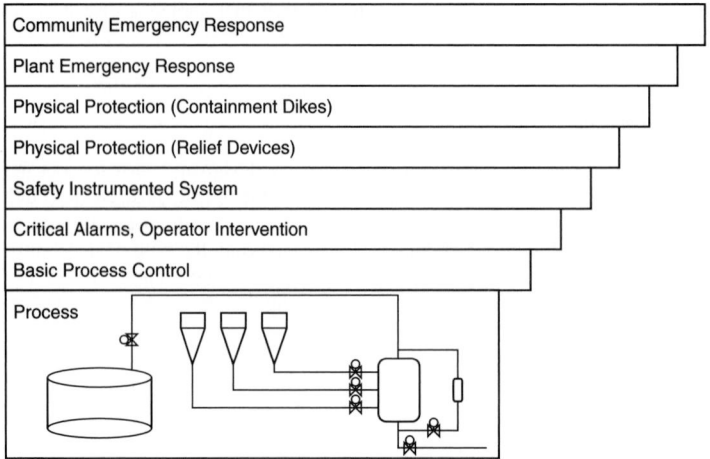

FIGURE 3   Multiple independent safety layers.

A major thrust within the process industry is to design inherently safe plants. Don't design a dangerous plant with the intention of throwing on lots of band aids to fix the problem. Design it so the band aids aren't even necessary. Work with low-pressure designs, low inventories, nonhazardous materials, and so on. Eliminating or reducing hazards often results in a simpler design, which may, in itself, reduce risk. The alternative is to add protective equipment to control hazards, which usually adds complexity.

Process Control System

The process control system is the next layer of safety. It controls the plant for optimum fuel usage, product quality, and so on, and it keeps all variables (e.g., pressure, temperature, level, and flow) within safe bounds. Some are reluctant to consider the process control system a safety layer, yet this author has no such problem, as long as it is not the *only* safety layer.

Alarms and Operators

If the process control system fails to do its function (for any number of reasons), alarms may be used to alert the operators that some form of intervention is required on their part. Alarm and monitoring systems should (1) detect problems as soon as possible, at a level low enough to ensure action can be taken before hazardous states are reached, (2) be independent of the devices they are monitoring (they should not fail if the system they are monitoring fails), (3) add as little complexity as possible, and (4) be easy to maintain, check, and calibrate.

Alarm and monitoring systems constitute the safety layer at which people get actively involved. Operators will usually be required for the simple reason that not everything can be automated. It is essentially impossible for designers to anticipate every possible set of conditions that might occur. Human operators may need to be considered since only they will be flexible and adaptable enough in certain situations. This is a two-edged sword, however, because events not considered in the design stage will no doubt also not be included in operator training either. In contrast, simply blindly following procedures has resulted in accidents [14]–[16]. Deviation from the rules is a hallmark of experienced people, but it is bound to lead to occasional human error and related blame after the fact.

Shutdown/Interlock Systems

If the control system and the operators fail to act, automatic shutdown systems take control. These systems are usually completely separate, with their own final elements. These systems require a higher degree of security to prevent inadvertent changes and tampering, and a greater level of fault diagnostics. The focus of this chapter is on these systems.

Fire and Gas Systems

If the shutdown system fails and an accident ensues, fire and gas systems may be used to mitigate or lessen the consequences of the event. In the U.S., these are traditionally alarm-only systems—they do not take any automatic control actions. Typically, the fire crews must go out and manually put out the fire. Outside the U.S., these systems frequently take some form of control actions, or they may be integrated with the shutdown system.

Containment Systems

If an atmospheric storage tank were to burst, dikes could be available to contain the release. In nuclear power plants, reactor are usually housed in containment buildings to prevent accidental releases. (The Soviet reactor at Chernobyl did not have a containment building, whereas the U.S. reactor at Three Mile Island did.)

Evacuation Procedures

In the event of a catastrophic release, evacuation procedures are used to evacuate plant personnel from the area, and if necessary, even the outside community. While these are procedures only, and not a physical system (apart from sirens), they may still be considered one of the overall safety layers.

SYSTEM TECHNOLOGIES

Logic Systems

There are a number of technologies available for use in shutdown systems—pneumatic, electromechanical relays, solid state, and PLCs (programmable logic controllers). There is no one overall best system; each has advantages and disadvantages. The decision of which system may be best suited for an application will depend upon many factors, such as budget, size, level of risk, flexibility, maintenance, interface and communication requirements, security, and so on.

Pneumatic systems are most suitable for small applications in which there are concern over simplicity, intrinsic safety, and lack of available electrical power.

Relay systems are fairly simple, are relatively inexpensive to purchase, are immune to most forms of EMI/RFI interference, and can be built for many different voltage ranges. They generally do not incorporate any form of interface or communications. Changes to logic require a manual change of documentation. In general, relay systems are usually only used for relatively small applications.

Solid-state systems (hardwired systems that do not incorporate software) are also available. Several of these systems were built specifically for safety applications and include features for testing, bypasses, and communications. Logic changes still require a manual change of documentation. These systems have fallen out of favor with many people because of their high cost, along with the acceptance of software-based systems.

Software-based systems, generally industrial PLCs, offer software flexibility, self-documentation, communications, and higher level interfaces. Unfortunately, many general purpose systems were not designed specifically for safety and do not offer features required for more critical applications (such as effective self-diagnostics). However, certain specialized dual and triplicated systems were developed for more critical applications and have become firmly established in the process industry.

Field Devices

In the process industries more hardware faults occur in the peripheral equipment—that is, the measuring instruments/transmitters and the control valves—than in the logic system itself. The overall reliability of a computerized control system may therefore not be significantly different than a conventional hardwired electrical system [17].

Sensors

Sensors are used to measure process variables, such as temperature, pressure, flow, level, and so on. They may consist of simple pneumatic or electric switches, which change state when a setpoint is reached, or they may contain pneumatic or electric analog transmitters, which give a variable output in relation to the strength or level of the process variable.

Sensors, like any other device, may fail in a number of different manners. They may cause nuisance trips, that is, respond without any change of input signal. They may also fail to respond to an actual change of input conditions. While these are the two failure modes of most concern for safety systems, there are additional failure modes as well, such as leaking, erratic output, and responding at an incorrect level.

Most shutdown systems are designed to be fail safe. This usually means that when power is lost, the safety system makes the process revert to a safe state, which usually means stopping production. (Nuisance trips should be avoided for safety reasons as well. Startup and shutdown modes of operation involve the highest levels of risk.) Thought must be given to how the sensors should respond in order to be fail-safe. This usually means the sensor has normally closed and energized contacts, although this is not always the case. Transmitters can usually be configured to fail upscale or downscale in the event of a failure of the internal electronics. Thought should be given to the failure mode for each type of transmitter. A recommendation that is overall, across the board, and the same for all applications simply cannot be made.

Some measurements may be inferred from other variables. For example, if a system is designed to shutdown as a result of high pressure, it may be helpful to monitor temperature (if, because of the process, an elevated temperature might also imply a high pressure). Special care should be taken when operating sensors at the low end of their ranges, because of the potential low-accuracy problems. For example, a sensor designed to operate at 1,000 psi may not be able to differentiate between 20 and 25 psi.

Technologies.   Discrete switches do not provide any form of diagnostic information. For example, if under normal circumstances a pressure sensor has a closed set of alarm contacts that are designed to open upon low pressure, and the contacts become stuck and cannot open as a result, the sensor has no means to annunciate the failure. The only way to know whether devices such as these are working is to *test* them.

Transmitters, in contrast, provide an analog signal in relation to the input variable, thus indicating in a limited sense whether the device is functioning properly. Any information is better than none. However, if the transmitter output is never monitored by the operators, or if the logic system does not automatically check for "noise" or occasional drifting of the signal, then there really may be no more usable information than a discrete switch. It would be like having a color printer, but only printing black and white documents. The perceived benefit of having the color printer is illusory if one is usuable to take advantage of the desired feature.

Redundancy.   If the failure of any one sensor is of concern (i.e., a nuisance trip or a fail-to-function failure), then redundant or multiple sensors may be used. Ideally, the possibility of two sensors failing at the same time should be very remote. Unfortunately, this does not account for common cause failures, which might impact multiple sensors at the same time. Common cause failures are usually associated with external environmental factors such as heat, vibration, corrosion, and plugging. If multiple sensors are to be used, they should be connected to the process using different taps, so as to avoid common cause plugging failures. Consideration may be given to using different sensors from different manufacturers, or having different maintenance personnel work on the sensors (so as to avoid the possibility of a maintenance technician incorrectly calibrating all the sensors).

Final Elements

Final elements generally have the highest failure rates. They are mechanical devices and subject to harsh process conditions. Safety shutoff valves also suffer from the fact that they are usually open and not activated for long periods of time, except for testing. One of the most common failure modes is that the valve is stuck, or frozen in place.

Valves should be fail safe upon loss of power. This usually requires a spring. A pneumatic or hydraulic valve would require a volume bottle to be fail safe, but the "availability" of the bottles may be too poor to rely on.

Solenoids.   Solenoids are one of the most critical components of final elements. It is important to use a good industrial grade solenoid valve, especially for outdoor use. The valve must be able to withstand high temperatures, including the heat generated by the coil itself. In general, the reliability of solenoids is very low. One of the most common failures is burning out a coil, which causes a false trip. A dual coil would keep the solenoid energized if one coil were to burn out. Solenoids should be tested frequently.

SYSTEM ANALYSIS

What is suitable for SIL 1, for SIL 2, and SIL 3? Things are *not* as intuitively obvious as they may seem. Dual is *not* always better than simplex, and triple is *not* always better than dual. Which technology to use, what level of redundancy, what manual test interval, and what about the field devices?

We do not design nuclear power plants or aircraft by gut feel or intuition. As engineers, we must rely on quantitative evaluations as the basis for our judgments. A quantitative analysis may be imprecise and imperfect, but it nevertheless is a valuable exercise for the following reasons:

1. It provides an early indication of a system's potential to meet the design requirements.

2. It enables one to determine the weak link in the system (and fix it, if necessary).

In order to predict the performance of a system, one needs performance data of all the components. Information is available from user records, vendor records, military-style predictions, and commercially available databases in different industries.

When modeling the performance of a safety system, one needs to consider two failure modes. Safe failures result in nuisance trips and lost production. The preferred term for these failures is the nuisance trip rate (measured in years). Dangerous failures result in hidden failures in which the system will not respond when required. Common terms used to quantify performance in this mode are pfd (probability of failure on demand), RRF (Risk Reduction Factor), and SA (Safety Availability).

There are a number of modeling methods used to predict safety system preformance. The ISA technical report TR84.02 [9] provides an overview of simplified algebraic equations, fault trees, and Markov models. Each method has its pros and cons. No method is more right or wrong than any other, as they are all simplifications and account for different factors. Using such techniques, one can model different technologies, levels of redundancy, test intervals, and field device configurations. One can model systems using a hand calculator or develop spreadsheets or stand-alone programs to automate and simplify the task.

TABLE 1   General System Recommendations

SIL	Sensors	Subsystem Logic	Final Elements
1	simplex switches	relays solid-state systems general purpose PLCs	simplex dumb
2	redundant switches simplex transmitters	relays fail-safe or fully tested solid-state systems certified software-based systems (simplex or redundant)	redundant dumb simplex smart
3	redundant switches redundant transmitters	relays fail-safe or fully tested solid-state systems certified software-based systems (dual or triplicated)	redundant smart

Notes for Table 1

Such tables are by their very nature oversimplifications. It is not possible to show the impact of *all* design features (failure rates, failure mode splits, diagnostic levels, quantities, manual test intervals, common cause factors, etc.) in a single table. Users are urged to perform their own analyses in order to justify their design decisions. The above table should be considered an example only, based on the following assumptions:

1. Separate logic systems are assumed for safety applications. Safety functions should not be performed solely within the BPCS (basic process control system).
2. Field devices are assumed to have an MTBF in both failure modes (safe and dangerous) of 100 years.
3. Simplex transmitters are assumed to have 80% diagnostics; redundant transmitters >95%.
4. Dumb valves offer no self-diagnostics; smart valves are assumed to offer 80% diagnostics.
5. When a consideration is made of solid-state logic systems, only solid-state systems specifically build for safety applications should be considered. These systems are either inherently fail safe (like relays) or offer extensive self-diagnostics.
6. General purpose PLCs are only appropriate for the lowest safety levels. They do not offer effective enough diagnostic levels to meet the higher performance requirements. Check with your vendors for further details.
7. One year manual testing is assumed for all devices. (More frequent testing would offer higher levels of safety performance.)

8. Redundant configurations are assumed to be either 1oo2 or 2oo3. 1oo2 configurations are safe, at the expense of more nuisance trips. 2oo2 configurations are less safe than simplex and should only be used if it can be documented that they meet the overall safety requirements.

9. The above table does not categorize the nuisance trip performance of any of the systems.

KEY POINTS

- follow the steps defined in the safety design life cycle
- if you can't define it, you can't control it
- justify and *document* all of your decisions (i.e., leave an auditable trail)
- the goal is to have an inherently safe process (i.e., one where you don't even need an SIS)
- don't put all of your eggs in one basket (i.e., have multiple, independent safety layers)
- the SIS should be fail safe and/or fault tolerant
- analyze the problem *before* you specify the solution
- all systems *must* be periodically tested
- *never* leave points in bypass during normal operation (or be prepared to suffer the consequences)

RULES OF THUMB

- maximize diagnostics (This is the most critical factor in safety performance.)
- any indication is better than none (e.g., transmitters have advantages over switches, systems should provide indications even signals are in bypass, etc.)
- minimize potential common cause problems
- general purpose PLCs are not suitable for the higher safety integrity levels
- when possible, use independently approved and/or certified components/systems (e.g., FM, TÜV, etc.)

REFERENCES

1. "Application of Safety Instrumented Systems for the Process Industries," ANSI/ISA-S84.01-, 1996.
2. "Functional Safety—Safety Related Systems," IEC Draft Standard 61508, 1997.
3. *Guidelines for Safe Automation of Chemical Processes*, AIChE, CCPS, 1993.
4. CFR Part 1910.119, Process Safety Management of Highly Hazardous Chemicals U.S. Federal Register, February 24, 1992.
5. Leveson, N. G.; *Safeware—System Safety and Computers*, Addison-Wesley, Reading, Mass., 1995.
6. *Guidelines for Hazard Evaluation Procedures,* AIChE CCPS, 1992.
7. *Guidelines for Chemical Process Quantitative Risk Analysis*, AIChE CCPS, 1989.
8. Taylor, J. R., *Risk Analysis for Process Plants, Pipelines and Transport*, E & FN Spon, an Imprint of Chapman & Hall, London, UK, 1994.
9. "Safety Instrumented System (SIS)—Safety Integrity Level (SIL) Evaluation Techniques," ISA Draft Tec. Rep. dTR84.02, 1997.
10. Gruhn, P., "The Evaluation of Safety Instrumented Systems—Tools to Peer Past the Hype," *ISA transactions* Vol. 35, pp. 25–32, 1996.

11. Gruhn, P., "Safety Systems: Where is Your Weak Link?" *InTech*, December 1993.

12. Smith, D. J., *Reliability, Maintainability and Risk*, Butterworth Heinemann, Oxford, UK, 1993.

13. "Out of Control: Why Control Systems Go Wrong and How To Prevent Failure," *Health & Safety Executive* (UK), 1995.

14. Kletz, T. A., *What Went Wrong? Case Histories of Process Plant Disasters*, Gulf Publishing, Houston, TX, 1986.

15. Kletz, T. A., *An Engineer's View Of Human Error*, The Institute of Chemical Engineers, Warwickshire, England, 1985.

16. Kletz, T. A., *Lessons From Disaster—How Organizations Have No Memory and Accidents Recur*, Gulf Publishing, Houston, TX, 1993.

17. Lowe, X., *Measurement and Control*, Vol. 17; p. 317, 1984.

AN OVERVIEW OF THE ISA/IEC FIELDBUS

by Terry Blevins*

INTRODUCTION

The Instrument Society of America (ISA) approved in 1985 the charter of the SP50 committee to develop a digital fieldbus standard. This standard defines a digital, two way, multidrop communication link among intelligent field devices and automation systems. It provides for communications at 31.25 K baud over existing wiring and meets intrinsic safety requirements of the process industry. By being able to multidrop devices off one single pair of wires as shown in Fig. 1, a significant saving in wiring and termination cost of 30–40% can be achieved over a traditional installation. In addition, since devices communicate digitally over the fieldbus, it is possible to obtain or send multiple pieces of information to a fieldbus device, that is, the device is no longer restricted to providing or receiving a single measure or output value. To take maximum advantage of this capability, each fieldbus device supports a standard function block application. Through this function block application, it is possible for fieldbus devices to be used to meet application requirements for measurement, alarm, calculations, and control.

Within the International Electrotechnical Commission (IEC), a working group, SC65C WG6, is responsible for the international fieldbus standard. This working group holds joint meetings with SP50 on the physical layer and communication stack. A separate working group, TC65 WG6, is defining function block architecture, IEC1499, targeted to meet the requirements of both manufacturing and process automation.

The fieldbus standard defined by ISA/IEC ushers in the next generation of control and automation products and systems that represent change and opportunity for the process industry:

- advanced function added to field instruments
- expanded view for the operator

---

* Fisher-Rosemount Systems, Inc. Austin, Texas.

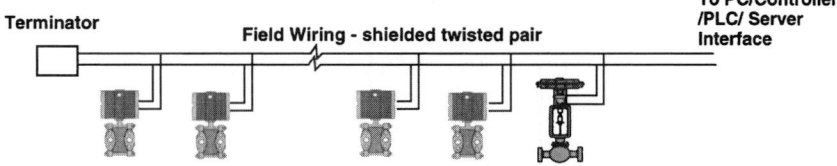

Terminator   Field Wiring - shielded twisted pair   To PC/Controller /PLC/ Server Interface

Transmitters, Analyzers, No/Off and Regulating Valves, Discrete and Low Level I/O Mux, Controllers, Converters, etc

FIGURE 1   Fieldbus segment: multiple devices may be connected to a single fieldbus segment.

- reduced wiring and installation costs
- reduced I/O equipment by one-half or more
- provide increased information flow to enable automation of engineering, maintenance, and support functions

The Fieldbus Foundation (FF) is an independent, nonprofit organization established to support commercialization of the ISA/IEC fieldbus standard. Foundation Fieldbus devices use the IEC physical layer and communication stack. The Function Block Application Process defined by the Fieldbus Foundation specification includes architectural concepts and terminology of IEC1499. Training schools are sponsored by the Foundation to support fieldbus device development. In addition, the Foundation has established test procedures, which are used to certify field devices as interoperable.

PHYSICAL INSTALLATION OF A FIELDBUS SYSTEM

The normal twisted pair with shield wiring used in most plants today may serve as the fieldbus. Thus, in an existing installation, wiring may be reused when Foundation Fieldbus devices are installed. Typically, the fieldbus segments will be connected into a control system through a controller or control system bridge, which is specifically designed to support fieldbus devices, as illustrated in Fig. 2.

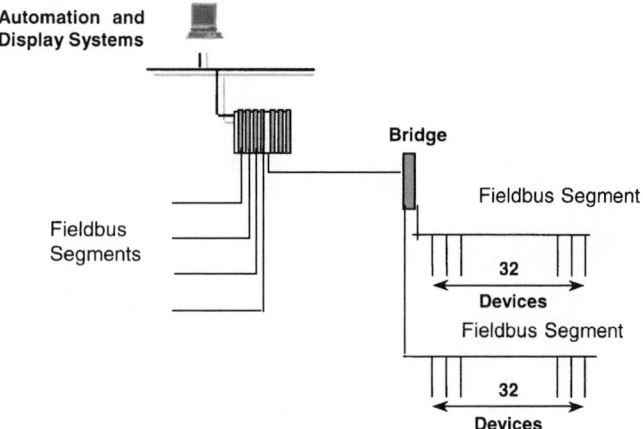

Automation and Display Systems

Bridge

Fieldbus Segment

Fieldbus Segments

32 Devices

Fieldbus Segment

32 Devices

FIGURE 2   System connection: fieldbus segments may be integrated into a control system by using a controller or bridge.

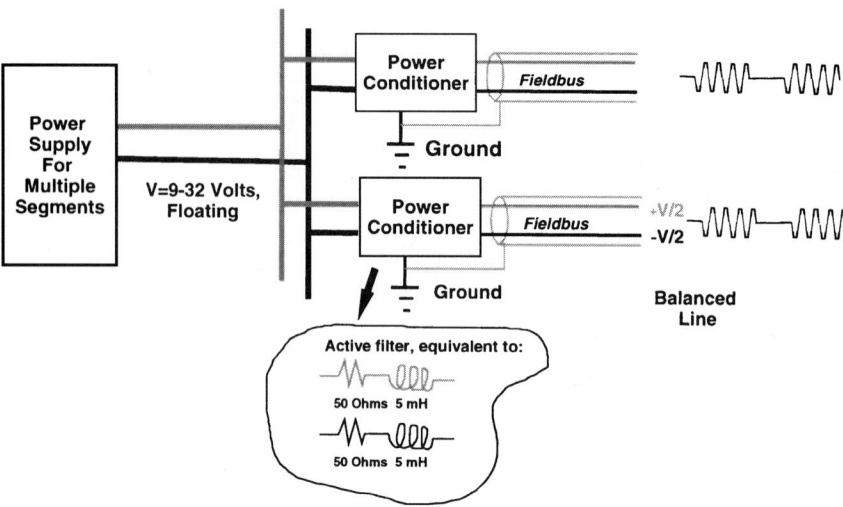

FIGURE 3 Segment Power: some power conditioners may be used with a common bulk power supply.

A maximum of 32 field devices may be connected to a single fieldbus segment. Twelve of the devices on a fieldbus segment may powered from the fieldbus segment, similar to the powering of tradition two-wire transmitters. The power for each fieldbus segment is provided through a power conditioner. The purpose of the power conditioner is to prevent the communication signal from being attenuated by the power supply and to eliminate cross talk between fieldbus segments through a common power supply. Also, the power conditioner provides a balanced conductor for communications; that is, the fieldbus conductors are maintained at $\pm\frac{1}{2}$ the supply voltage with respect to ground. In some cases, a common power supply with floating output may be used to power multiple power conditions, as shown in Fig. 3.

In an intrinsically safe installation, power conditions with built-in barriers are available that support 4–6 fieldbus powered device. An example of an installation that utilizes commercially available power conditions is shown in Fig. 4.

FIGURE 4 Example of a commercial power conditioner.

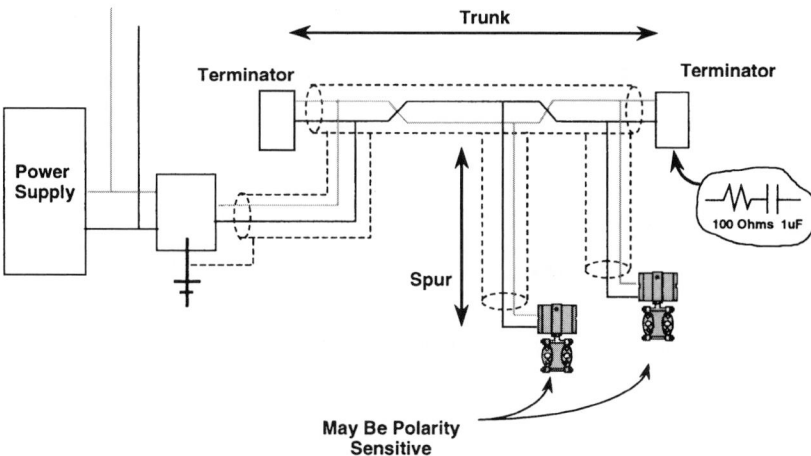

FIGURE 5   Fieldbus Trunk: the main path of a fieldbus segment may have multiple spurs.

The main path of a fieldbus segment is commonly called the truck or home run, as illustrated in Fig. 5. Terminators are required at the far ends of the fieldbus trunk. Branches off the fieldbus trunk, known as spur, may be a maximum of 120 m in length. The total length of the trunk and spurs that make up a fieldbus segment may be up to 1,900 m when an 18-gage single twisted pair with shield wiring is used.

The maximum length of a fieldbus segment will depend on the type of cable used—as detailed in Table 1. Also, the maximum length of a spur will vary with the number of devices on the segment and the spur, as shown in Table 2.

TABLE 1   Maximum Fieldbus Segment Length

Type	Cable Description	Size (mm^2)	Max Length[a] (m)
Type A	Shielded, twisted pair	0.8 (#18AWG)	1,900
Type B	Multitwisted pair, w/shield	0.32 (#22AWG)	1,200
Type C	Multitwisted pair, w/o shield	0.13 (# 26AWG)	400
Type D	Multicore, w/o shield	1.25 (#16AWG)	200

[a] Total of trunk plus spur(s) length, based on IEC-1158-2 and ISA S50.02-1192.

TABLE 2   Maximum Recommended Fieldbus Spur Length[a]

Total devices	Device per Spur		
	1 Device (m)	2 Devices (m)	3 Devices (m)
1–12	120	90	60
13–14	90	60	30
15–18	60	30	1

[a] Based on IEC-1158-2 and ISA S50.02-1192 Part 2, Annex C (informative) notes: these lengths are "recommended" and are not required.

Daisy-Chain Topology

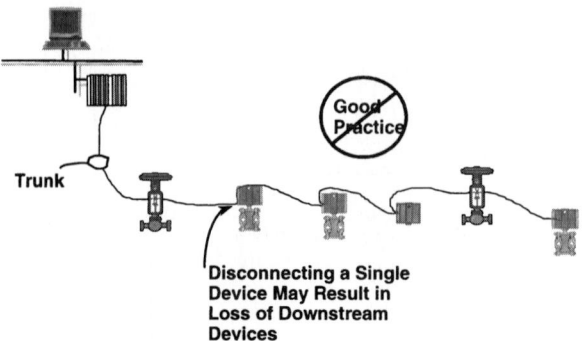

FIGURE 6   Daisy-chain arrangement: it is not a good practice to daisy-chain fieldbus devices on a segment.

In the initial design of a fieldbus segment, one of the critical decisions is what devices will be placed on an individual fieldbus segment. When information is to be exchanged between fieldbus devices for control or calculations, then these devices should be located on the same segment. Naturally, there is a tendency to place as many devices on a fieldbus segment as possible to reduce the overall cost in wiring and fieldbus interface cards. However, depending on the speed with which information must be accessed, some manufacturers will limit the number of devices that may be placed on an individual segment.

Once the decision is made concerning what devices should be on a segment, then a detailed layout of the segment is possible. In the design of the segment, a key decision will be how the devices will be multidropped from the fieldbus segment. Even though it is possible to wire the fieldbus devices in a daisy-chain topology as shown in Fig. 6, this arrangement is not recommended because of the problems this may present in maintenance.

For example, in a daisy-chain arrangement, all power may be lost downstream of a fieldbus device that is removed for maintenance. When individual devices are to be multidropped off the fieldbus segment, then a branch topology, shown in Fig. 7, supports ease in maintenance, checkout, and installation.

Branch Topology

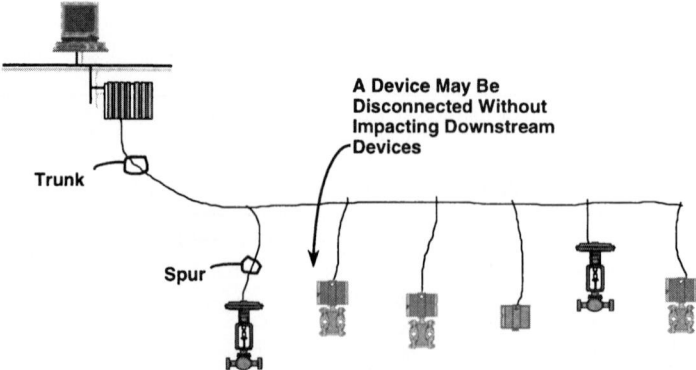

FIGURE 7   Branch topology: this allows a single device to be added or removed without disrupting other devices on the segment.

FIGURE 8   Spur wiring: example using a conduit box and prefabricated T.

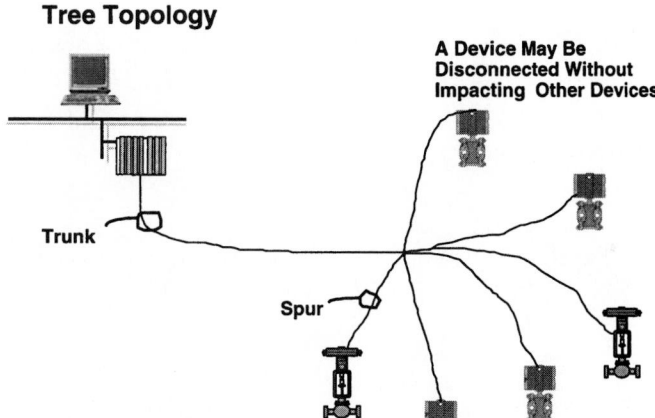

FIGURE 9   Most fieldbus installations wiil utilize a tree topology.

When the branch topology is used, it is possible to add or remove individual devices from the fieldbus segment without disruption to other devices on the segment. A number of commercial products are available to support the installation of the branch topology. When conduit and junction boxes are selected, a commercially available junction board may be installed in the conduit box to facilitate wiring between the main segment and the spur. If quick disconnects are installed, then commercially available T's may be used for the spur connection to the fieldbus trunk. Examples of spur wiring using a condulet and T are shown in Fig. 8.

In a typical installation, multiple fieldbus devices may be located in the same physical area. In this case, the spurs from these devices may be wired together in a junction box to form a tree topology, as shown in Fig. 9.

Traditional terminal strips may be used for wiring inside the junction box. Alternately, commercial DIN rail-mounted termination blocks are available for fieldbus installation. Using such devices will minimize the potential for wiring mistakes. These termination blocks are available with a built-in

FIGURE 10   Fieldbus junction box: pre-fabricated components may reduce installation and checkout time.

FIGURE 11   Example installation in which the field devices were purchased with quick disconnects.

terminator for use at the far end of the segment. Also, prefabricated junction boxes are available for quick disconnect terminations, as shown in Fig. 10.

In some installations, as many as 16 fieldbus devices will be connected to a single fieldbus segment. In such cases, a short along the fieldbus segment could result in the loss of communications with all devices on the segment. To support maintenance, many manufacturers will provide quick disconnects at the fieldbus device to minimize the chance of shorting the segment while adding or removing a field device. Examples of fieldbus device quick disconnects are shown in Fig. 11.

UTILIZING FIELDBUS DEVICES TO MEET APPLICATION REQUIREMENTS

In a fieldbus environment, the user application is defined through the configuration of function blocks. This approach is similar to the configuration of control and monitoring in a distributed control system (DCS) today. However, the fieldbus function blocks support applications, which involve measurement and control, to be distributed between fieldbus devices. Such capability will allow the base level process control and measurement done today in distributed control systems and single-loop digital controllers to be implemented in a fieldbus device. Some of the advantages gained by standardizing the user application are:

- consistent, easy, block-oriented configuration of functions
- distribution and execution of function in field devices from different manufacturers in an integrated, seamless manner
- consistent definition of information that will be communicated and function that will be executed
- avoidance of custom interfaces and cumbersome mapping

Monitoring, calculation, and control functions may be defined by configuring function blocks within a fieldbus device and configuring the connections between function block input and output parameters. Fieldbus devices provide a new level of capability. Fieldbus valve and transmitters may support control and calculations. In addition, auxiliary measurements such as stem position or limit switch status for an on–off valve will often be made available through fieldbus.

The function block application in Foundation Fieldbus devices is designed to allow control, measurement, and calculation functions to be distributed between field devices. For example, a flow measurement might be implemented in a transmitter using the analog input block, AI block. This transmitter publishes its measurement value and its status as shown in Fig. 12. A valve on the fieldbus segment may support a control block, the PID block, which subscribes to the published measurement value. Based on this value and its target setpoint, the PID block may calculate an output required to maintain the setpoint. Within the valve, this output might be used by an output block, the AO block, to adjust the valve.

To successfully distribute control between fieldbus devices without degrading control by communication delay, the scheduling of function block execution and related communications is critical. Thus, as an integral part of the Foundation Fieldbus specification, scheduling of communications in

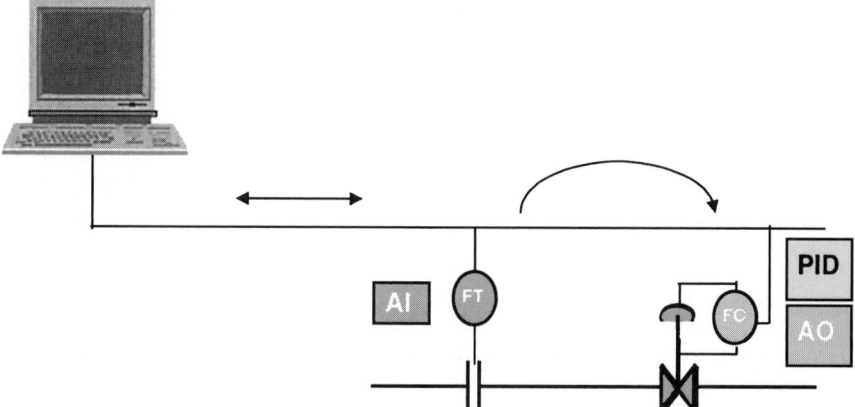

FIGURE 12   Fieldbus communication: support is provided to distribute control and calculations between fieldbus devices.

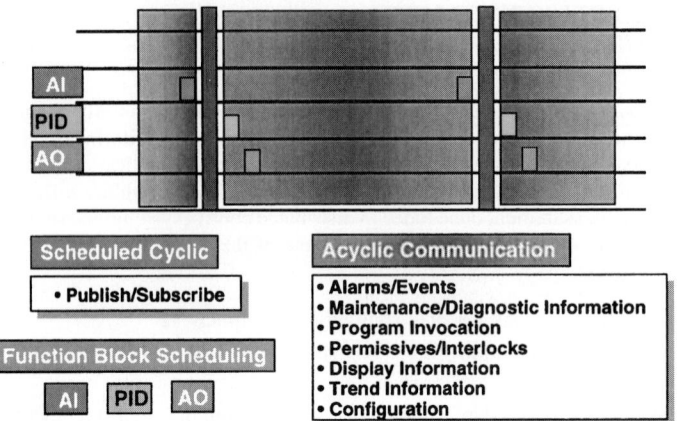

FIGURE 13 Both control-related communication and function block execution may be scheduled to minimize delay.

conjunction with function block execution is defined. A typical schedule for the above example is illustrated in Fig. 13. When no control communications are scheduled, then other information for display, alarm notification, and so on may be communicated.

Function blocks and tools used for process analysis, such as trending of measurement versus time, often assume periodic sampled values. In such cases, it may not be practical to compensate for variation or jitter in the sample time. Thus, fieldbus allows measurements to be sampled on a precisely periodic basis, independent of fieldbus communications. Through the scheduling of block execution, these sampled values may be made available to control blocks with minimal delay.

The low processing power of intrinsically safe devices, combined with the distribution of control between fieldbus devices, imposes some constraints on the design of fieldbus devices. To support the periodic sampling of inputs, a common sense of time is maintained in each device on a fieldbus segment. Each device maintains a schedule of which blocks within the device are to be executed. Based on this, it is possible to schedule communication of block output values between devices for calculations and control.

To obtain an accurate measurement, a device may sample a process measurement at a much faster rate than is needed for control or monitoring requirements. For example, a measurement may be processed 20 times/s, even though it may be used in control only 5 times/s. A transducer block is defined in fieldbus devices to contain the parameters associated with the basic measurement. The processing of the transducer block is defined to be independent of function blocks that reference the transducer block, as shown in Fig. 14. To accommodate these differences in execution rates, software filtering is provided in transducer blocks. Through this filter, the frequency content of the measurement may be matched to the function block execution rate.

As part of input block processing, checks on associated hardware and software are performed. In addition, process alarm detection may be done. On the detection of a change in block status or process alarm condition, an event notification will be generated. Included in this notification will be the time of detection and a substatus that gives further information on the alarm or event.

Ten basic blocks and 19 advanced function blocks have been defined by the function block specification, as shown in Table 3. These blocks allow both analog and discrete monitoring, calculations, and control to be done in field devices. It is estimated that as much as 80% of the DCS and PLC controller functionality may be distribute to the fieldbus devices using this capability.

Using these function blocks, most analog and discrete control problems may be addressed. In Tables 4 and 5, some examples of how Foundation fieldbus function blocks may be applied in typical control applications are shown.

TABLE 3   Function Blocks Defined by Foundation Fieldbus Specification

Basic

discrete input (DI)	P, PD Control (PD)
discrete output (DO)	control selector (CS)
analog input (AI)	manual loader (ML)
analog output (AO)	bias/gain station (BG)
PID, PI, I control (PID)	ratio station (RA)

Advanced

pulse input (PCI)	deadtime (DT)
complex analog output (CAO)	arithmetic (ARITH)
complex discrete output (CDO)	calculate (CALC)
step output PID (STEP)	integrator (INT)
device control (DC)	timer (TIM)
setpoint ramp generator (SPR)	analog alarm (AALM)
splitter (SPLIT)	discrete alarm (DALM)
input selector (ISEL)	analog human interface (AHI)
signal characterizer (CHAR)	discrete human interface (DHI)
lead lag (LL)	

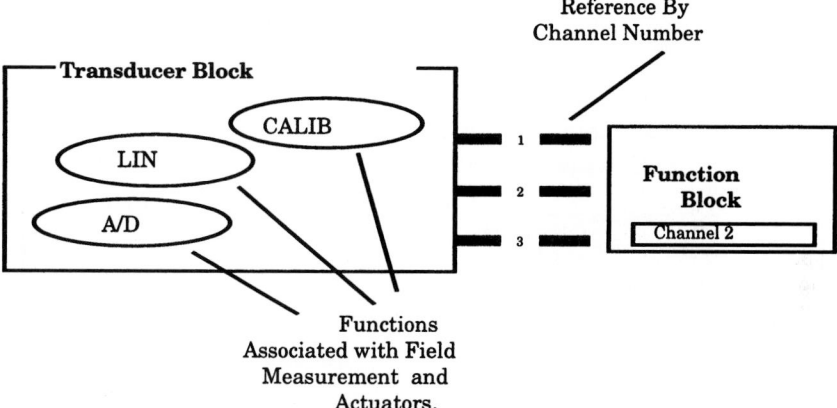

FIGURE 14   Input and output function blocks may reference the field value(s) provided by the transducer block.

Before a field device is certified as interoperable by the foundation, it is put through a series of test to verify that the device conforms to the definitions of the specification. Only if the device supports the block features defined by the function block specification can it earn the interoperable check issue by the Fieldbus Foundation. In purchasing a fieldbus device, then it will be important to verify that the device has passed interoperability testing.

DIAGNOSTIC SUPPORT OF FOUNDATION FIELDBUS DEVICES

The ability to quickly troubleshoot a process and its control and instrumentation system is an important aspect of plant operation. Any time spent identifying the source of a malfunction may contribute to

TABLE 4 Example Applications Addressed by Fieldbus Devices

Control/Measurement Function	Fieldbus Function Blocks Required
Transmitter	**AI**
Regulating Valve	**AO**
Pressure Switch	**DI**
On/Off Valve	**DO**
Manual Loader	**AI** → **ML** → **AO** LT112 HIC112 SC112
Single Loop Control	**AI** → **PID** → **AO** LT108 LIC108 SC108

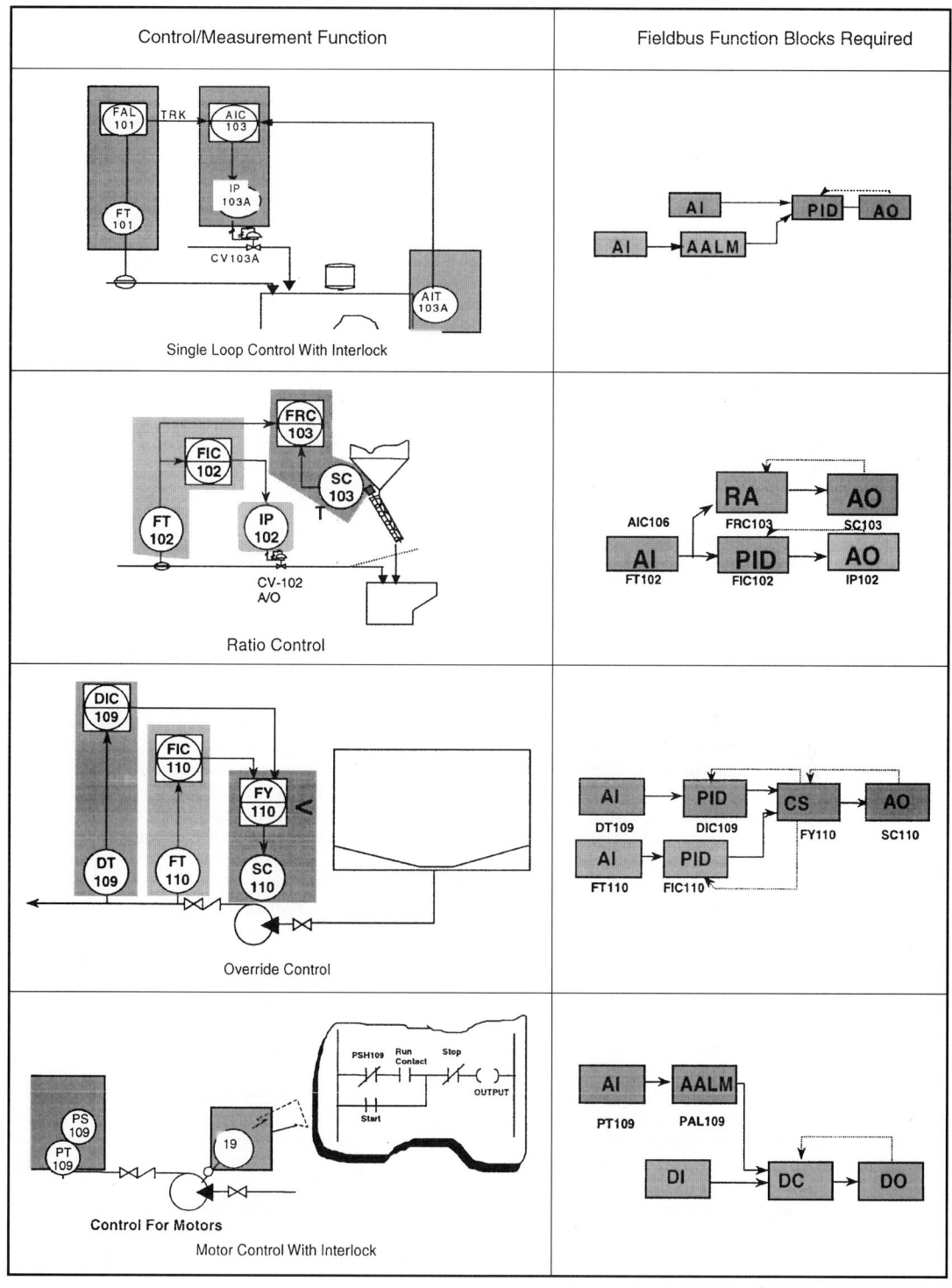

Control/Measurement Function	Fieldbus Function Blocks Required

Single Loop Control With Interlock

Ratio Control

Override Control

Motor Control With Interlock

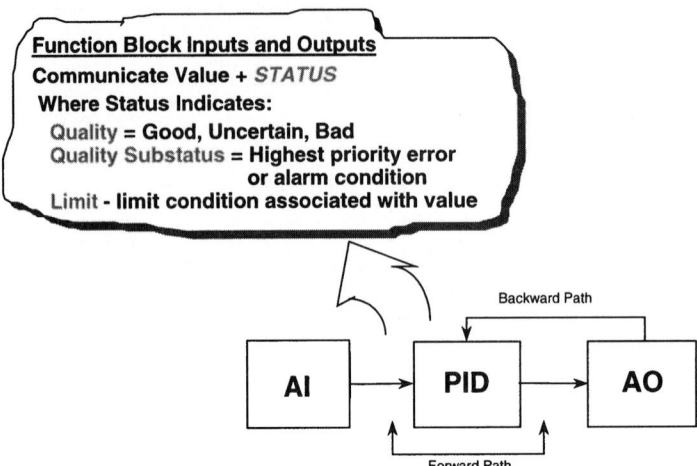

FIGURE 15   Input/Output status: each function block input or output supports a value and status.

process downtime and lost production. In many digital control systems in use today, the process measurement is received as a 4–20 mA signal and the actuator is sent as a 4–20 mA signal. Plant instrument technicians are familiar with the troubleshooting process instrumented in this manner. Within fieldbus devices the signal from the measurement element will be converted directly to its digital representation. The only accessible representation of the measurement will be the digital value communicated over the fieldbus. Thus, the tools and techniques used in calibration, setup, commissioning, and troubleshooting will change. The design of any fieldbus system must address the requirements for process analysis through communicated values.

Features are included in the Fieldbus Foundation Specification for the support of process analysis and troubleshooting field problems. Special consideration is given to signal filtering and sampling to ensure that the resulting digital values adequately represent the true process measurement. Conditions detected by the field device that would impact the validity of the measurement are made available as the status of the digital value. To allow this information in a device to be used, an efficient mechanism is provided to access field device information without disrupting the control distributed between field devices.

As part of input block processing, checks on associated hardware and software are performed. Analog values will be transferred as floating point values in engineering units. The status of an output parameter is calculated by the block to give an explicit indication of the quality of the value: *good, uncertain, or bad* as shown in Fig. 15. A substatus attribute indicates the primary condition, which determines the quality. Also, there is an explicit indication in the status attribute of whether the value is limited high or low. When a block input or output parameter is access for viewing at an operator console, historian, or diagnostic tool, both the value and its associated status may be communicated. The quality information and its substatus may be useful in diagnosing the cause of a bad or uncertain measurement.

A substatus attribute of status indicates the primary condition that determines the quality. The substatus for good quality provides additional information that may be needed for control or monitoring purposes. For example, the good substatus for input blocks is defined to give further information on the status of alarms that are defined for the block. As part of block definition, the processing and propagation of the status is defined.

Communication support is provided for an explicit indication of stale data. Data are defined to be stale if the input value has not been updated since it was last read or if a value has been communicated but the value received is the last (old) value. Within a device, the input parameter status will be

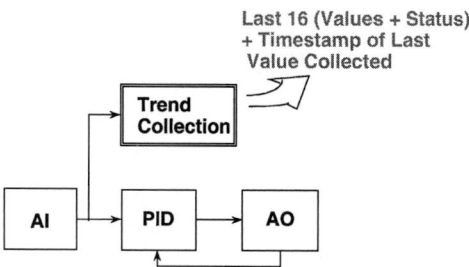

FIGURE 16 Diagnostic informtion may be obtained by using trend objects supported by a fieldbus device.

set to bad—no communication if the value is stale for a specified time. When communications are re-established, then the input will again reflect the status of the output parameter.

When a block input or output parameter is access for viewing at an operator console, historian, or diagnostic tool, both the value and its associated status may be communicated. The quality information and its substatus may be useful in diagnosing the cause of a bad or uncertain measurement. To facilitate the use of status information, the representation and meaning of quality and its associated substatus have been defined in the Foundation Fieldbus Specification.

When information is accessed for process analysis and diagnostics, it may not be practical or technically feasible to communicate individual parameter values as fast as the block executes rate. To address these issues, trend objects are defined in fieldbus devices as illustrated in Fig. 16. At least one trend object is included in each field device. A trend object maintains a collection of 16 parameter samples. Through trend-object configuration, it is possible to specify the input or output parameter to be collected at the collection rate. The block collects both the parameter value and status. Samples rates are an integer multiple of the parameter's block execution rate. Sample values may be averaged if the collection rate is slower than the block execution rate.

When 16 new samples are collected, this information along with the time stamp of the last value collected will be automatically reported to one or more devices. Through this mechanism, the communication load associated with the collection of the parameter value and status may be reduced by as much as a factor of 16. In addition, by collecting samples at the device, the collection of information is not skewed or limited by the communication rate of the fieldbus.

View objects are provided for each function block to facilitate information access. A view object allows multiple parameters to be read or written with a single communication request. Through this mechanism, communications to support updating operator interfaces (etc.) may be done efficiently. The processing of such requests is done when no other communications are scheduled. In addition, process alarm detection may be done in the fieldbus device. On the detection of a change in block status or process alarm condition, an event notification will be generated. Included in this notification will be the time of detection and a substatus, which gives further information on the alarm or event.

CONTROL SYSTEM IMPACT

A new generation of control systems that is designed to effectively utilize fieldbus devices is available from major control system vendors. Such systems are smaller in size physically because of the reduction in the number of I/O card required to interface with field devices. To allow the configuration and calculated parameter values in a fieldbus device to be accessed, each manufacturer will provide a device description (DD) of his or her fieldbus devices written in the device description language (DDL) standardized by the Fieldbus Foundation. Configuration and operator interface stations, which

FIGURE 17   Example installation of a fieldbus controller that is connected to 96 fieldbus devices.

have been designed to utilize device descriptions, will be able to access device function blocks after the device description is loaded. An example of a controller designed for fieldbus is shown in Fig. 17.

Information for process analysis and diagnostics is available in fieldbus devices that are based on the Fieldbus Foundation's specification. To take advantage of this capability, future tools for process analysis, diagnostics, and reporting will utilize more than the measurement values. In particular, these tools will access and make visible:

- alert notification time stamp and substatus detailing alarms or events
- quality and substatus accompanying function block inputs and outputs

To support the analysis of blocks executing at fast rates, future devices for viewing trends and archiving trend data will utilize value and status samples that are collected and reported through trend objects in the field devices.

FIGURE 18   World's first large commercial installation based on the IEC/ISA fieldbus standard.

EXAMPLE INSTALLATIONS: COMMERCIAL FIELDBUS INSTALLATIONS

The first two major commercial fieldbus installations may be used to illustrate the labor and material savings that are achieved by using fieldbus. The world's first commercial installation of Foundation Fieldbus was at a Dow Chemical Company in Alberta, Canada. Mr. David Taylor was the project manager for this historic installation. Initially, a total of 96 fieldbus transmitters were installed at Dow in April, 1997 (shown in Fig. 18). Based on the success of this installation, an additional 480 fieldbus devices were installed in July, 1997.

At the Dow installation, quick disconnects and prefabricated fieldbus junction boxes are used extensively. Dow reported the total project cost as 30% lower than that documented for a similar installation that used traditional instrumentation.

The second major commercial Foundation Fieldbus installation was at ARCO, West Sak, Alaska. Mr. Duane Toavs was the ARCO project manager for the large installation. This project included fieldbus transmitters and valves for 30 oil wells (shown in Fig. 19). The system was commissioned in December, 1997.

FIGURE 19   At the West Sak oil field on Alaska's North Slope, operated by ARCO Alaska, Inc.

ARCO reports the savings associated with the West Sak fieldbus installation as follows:

- 16% reduction in wellhead terminations
- 69% reduction in comparable wiring costs
- 98% reduction in home run wiring
- 83% reduction in instrument commissioning and checkout
- 90% reduced configuration time to add an expansion well
- 92% reduced engineering drawings to add or expand a well
- ~$210,000 material cost savings
- ~$320,000 labor savings
- ~$90,000 engineering savings

Additional long-term benefits are anticipated from the diagnostic and maintenance features of fieldbus devices.

ESTIMATING SAVINGS FROM USING FIELDBUS TECHNOLOGY

By the use of fieldbus technology, savings are possible in the following areas:

1. Reduction of terminations
2. Reduction in number of I/O cards

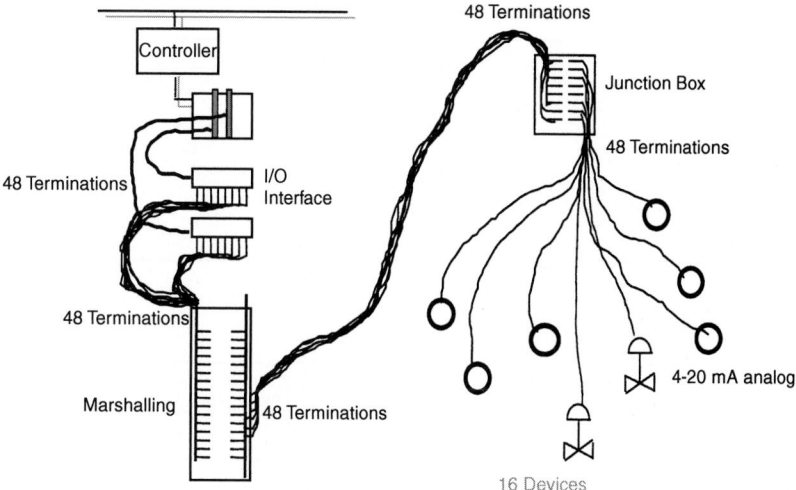

FIGURE 20 Illustration of terminations for a traditional I/O with 16 devices.

3. Reduction in home run wiring

4. Reduction in instrument control room panel space

The magnitude of these cost savings may be determined by analyzing the controls configuration in a manufacturing plant. Each area of cost reduction may be addressed separately.

Reduction of Terminations and Home Run Wiring

In a typical plant situation, each transmitter or valve is connected to a junction box with a single, shielded twisted pair, wire. Since each wire has three terminations, the number of terminations into the junction box is the number of devices multiplied by three (two conductors plus the shield). The junction box is often connected to a marshalling panel with the same number of terminations (number of devices times three), and finally, the marshalling panel is connected to the I/O cards with the same number of terminations. The schematic layout of the terminations in a traditional I/O system with 16 devices is shown in Fig. 20.

In a fieldbus situation, the number of terminations is significantly reduced. The same number of wires is required to connect the devices to the junction box, but one wire (three terminations) connects the junction box to the marshalling panel, and one wire (three terminations) connects the marshalling panel to the H1 fieldbus card, as illustrated in Fig. 21. Assuming the same number (16 devices) of valves and transmitters that are connected to the fieldbus segment, a maximum of 16 wires (48 terminations) can be connected to the fieldbus segment at the junction box, as shown in Table 6.

From a total cost perspective, a tremendous installation cost savings can be achieved with the usage of fieldbus technology. The reduction of 240 terminations to 60 for an installation of 16 devices represents a 75% reduction in termination cost. A significant savings in wiring reduction will result from any fieldbus installation. Using traditional I/O, 16 devices require 16 wires from the junction box to the marshalling panel and 16 wires from the marshalling panel to the I/O card. With fieldbus technology, one wire connects the junction box to the marshalling panel, and one wire connects the marshalling panel to the I/O card. As with a traditional I/O layout of 16 devices, a fieldbus layout of 16 devices requires 16 wires to the junction box. The reduction of 16 wires to 1 wire results in an 83% reduction.

TABLE 6   Wiring Terminations for 16 Devices

Terminations from the . . .	To the . . .	Traditional I/O Terminations	Fieldbus Terminations
16 devices (16 wires, times 3)	junction box	48	48
Junction box	marshalling panel	96	6
Marshalling panel	I/O cards	96	6
Total		240	60

TABLE 7   Card Requirements for Traditional I/O

No. of Measurements	Channels per Card	No. of I/O Cards	Equiv. Fieldbus Cards
64	8	8	1

Reduction in the Number of I/O Cards

Traditional I/O cards consist of digital input, digital output, analog input, and analog output. Each card may, for this example, accommodate up to eight channels. With fieldbus, analog input, analog output, digital input, and digital output function blocks reside in the device, thus eliminating the need for the traditional I/O cards. A fieldbus card, which is used to interface with the fieldbus devices, may, for example, accommodate 64 function blocks from multiple fieldbus devices. To illustrate the impact fieldbus has in the number of I/O cards required, compare a traditional I/O layout of 64. Assuming that two function blocks are supported by each fieldbus transmitter, a single fieldbus connection equates to two analog input channels in a tradional installation. A traditional configuration of 64 analog inputs may require, in some, cases, a total of eight I/O cards. The equivalent information accessed on fieldbus may require only one fieldbus interface card, as shown in the Table 7. This 87% reduction in the number of I/O cards will result in reduced system cost and space requirements for I/O cards.

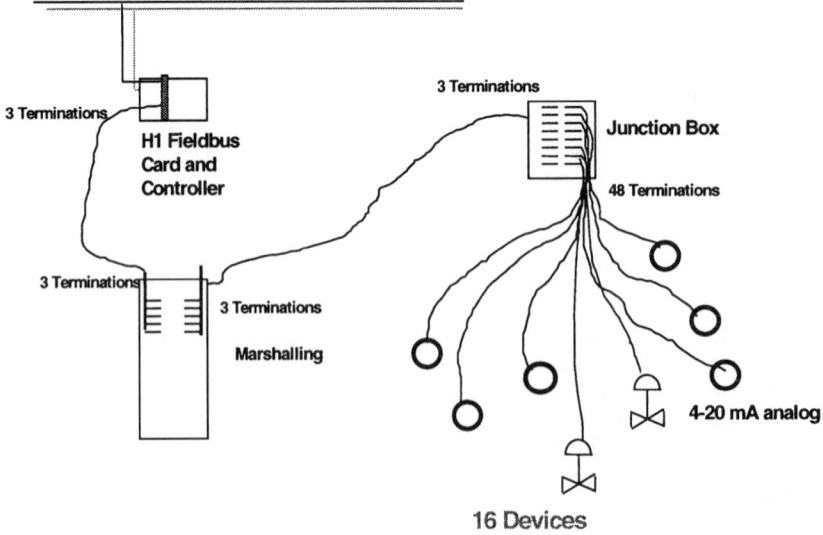

FIGURE 21   Example terminations for 16 fieldbus devices.

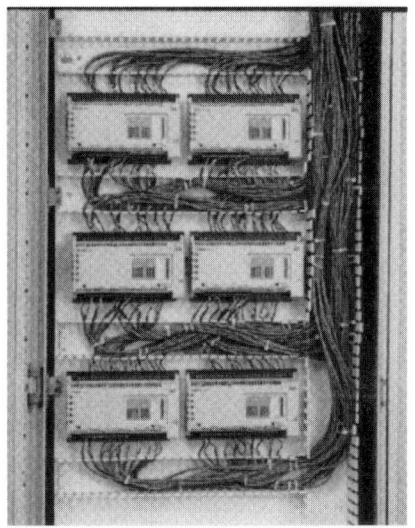

Tradtional I/O Footprint

**Fieldbus controller and
I/O Footprint**

FIGURE 22   System footprint: comparison of traditional to fieldbus systems for the same number of field devices.

Reduction in Instrument Room Space

The reduction of an instrument room for control is directly related to the I/O card reduction previously discussed. A visual comparison of a traditional installation to a fieldbus system with an equivalent I/O capability shows the dramatic reduction in instrument room space, which may be achieved using Founation Fieldbus devices (Fig. 22).

SUMMARY

Fieldbus technology clearly presents the opportunity for cost reductions in the following areas:

- installation labor and material costs of terminations
- installation labor and material costs of wiring
- reduction in the use of components such as I/O cards
- reduction in instrument control room panel space

Additional benefits that have tangible cost savings are in the following areas:

- reduction of start-up and commissioning costs
- integration of plantwide information to provide timely access to information
- automation of plant functions
- reduction of operating and maintenance costs (remote diagnostics, troubleshooting, spares)
- enable improvements in quality and safety
- maximize plant up time

It is too early to have hard numbers on the long-term benefit that the diagnostic and maintenance feature of fieldbus devices will provide these initial Foundation Fieldbus installations. However, the feeling of plants that have installed fieldbus is that these savings will match or exceed the installation savings that fieldbus has provided.

Best Practices in Applying Fieldbus

1. Limit the number of fieldbus devices on a single fieldbus segment to 16. This will help ensure reasonable update rates of dynamic fieldbus parameters in the host interface.

2. Install on a single fieldbus segment no more than 12 fieldbus devices that are powered by the fieldbus segment. For an intrinsically safe installation, limit the maximum to four to six devices based on individual device power consumption.

3. Locate fieldbus devices on a single segment if control or calculations are distributed between these fieldbus devices.

4. Utilize a tree topology whenever possible for connecting fieldbus devices to a segment. Avoid daisy-chaining fieldbus devices since this will cause installation, commissioning, and maintenance problems.

5. Locate field junction boxes within 120 m of fieldbus devices that are to be on a fieldbus segment.

6. Install an approved barrier with the fieldbus power supply when fieldbus devices are to be installed in a classified area and the fieldbus segment is not protected by hard conduit or other approved materials.

7. Limit the fieldbus segment length to 1,900 m when using Type A cable (shielded twisted pair; 0.8 mm^2).

8. Utilize a maximum of 16 function block input and output links between fieldbus devices on a single segment to provide reasonable loop execution rates and display update time at a host.

9. Purchase fieldbus devices that have been certified by the Fieldbus Foundation to be interoperable. Consider the function blocks and diagnostics supported by the fieldbus device when selecting a fieldbus device.

10. Select a control system that is specifically designed to support the interface, configuration, and diagnostics of fieldbus devices. In particular, the configuration system should utilize the DD provided by the device manufacturer to allow all fieldbus block parameters of a device to be accessed.

11. Include the location of power conditioners and terminators on fieldbus segment drawings and verify their installation before initially powering up the segment.

12. Continue shields throughout the fieldbus segment and have only one ground reference—at the source of power.

13. Estimate terminations and a wire reduction of 70–80% when 16 fieldbus devices are installed on each segment.

14. Plan on an 80–90% reduction in the number of I/O cards when 16 devices are included on a fieldbus segment. However, the cost of a fieldbus interface card may be higher than a traditional I/O card.

15. Take advantage of the fact that the rack room space for a complete fieldbus installation with 16 devices per segment will typically require only 30% of the space needed for a traditional installation.

16. Consider the fact that some projects have reported an 80–90% deduction in time for instrumentation drawings and instrument commissioning and checkout time over a traditional installation using fieldbus. However, on your first installation in a plant, plan on some of these savings being offset by additional time spent on fieldbus training.

REFERENCES

1. Blevins, T., "Fieldbus Ushers In A New Era in Process Control," *Plant Services*, September 1998.

2. Blevins, T., J. Duffy, and R. Willems, "DCS Integration of Fieldbus," *ISA Conf. proceedings,* 1996.

3. Blevins, T., and W. Wojsznis, "Fieldbus Support for Process Analysis," *ISA Trans.* Vol. 35, pp. 177–183, 1996.

4. Instrument Society of America, "User Layer Technical Report for the Fieldbus Standard," ISA-TR50.02, Part 9-TR1, Tech. Rep. ISA/SP50-1993-389F, 1993.

5. Furness, H., "Distributed Control Functionality Moves Downstream," *Control Engineering*, December 1993.

6. International Electrotechnical Commission, Technical Committee No. 65, Industrial-Process Measurement and Control, Working Group 6, Committee Draft: Function Blocks for Industrial-Process Measurement and Control Systems, Part 1—General Requirements.

7. Fieldbus Foundation, Fieldbus Specification, Function Block Application Process—Part 1&2, FF-94-890, FF-94-8891.

8. Bialkowski, W. L., and A. D. Weldon, "The Digital Future of Process Control: Possibilities, Limitations, and Ramifications," *Tappi Journal*, Vol. 77, No. 10, October 1994.

9. Wheelis, J. D., and K. Zech, "Benefits Observed During Field Trials of an Interoperable Fieldbus," *ISA Conf. Proceedings*, 1994.

BATCH CONTROL: APPLYING THE S88.01 STANDARD

by Thomas G. Fisher*

INTRODUCTION

The S88.01 standard [1] provides the basis for an object-oriented approach that fits very well with batch control and the automation of batch processes, including the development of objects that can be reused from project to project. Significant savings from applying these standards and guidelines have been demonstrated in many phases of batch control projects. Considerable emphasis is given in the standard to the separation of the recipe procedure (which tells how a batch should be made) from the equipment logic (which actually executes the batch). This separation is one of the major reasons that this standard has been so successful.

S88.01 is entitled Models and Terminology, but it is really a communications standard, although not in the sense of communications protocols over a local area network. S88.01 makes it possible for people to communicate about batch control by using a common language and a common set of models that describe batch control and batch manufacturing. S88.01, *Part 1– Models and Terminology*, was approved in February 1995. S88.01 became a joint ANSI (American National Standards Institute) and ISA (International Society for Measurement and Control) standard in October 1995. IEC 61512 [2] is a corresponding international standard from IEC (the International Electrotechnical Commission).

The NAMUR NE-33 guidelines [3] is often referenced in discussions about batch control. NAMUR is a group of German user companies, and it was founded in 1949 as an association of the measuring and control engineering departments of the chemical industry. Working Group 6, Batch Control, was founded in 1966. The first official attempt to set criteria and guidelines for batch automation and

* Operations Technology Manager, Lubrizol Corporation, Wickliffe, Ohio.

related application software was made by this group under the chairmanship of R. J. Uhlig. In July 1985, they drew up a set of guidelines for batch control software that was entitled "Generation of Control Sequences for Batch Processes with Changing Recipes by Configuration, Using Predefined Functional Modules." This 1985 work formalized the approach of defining recipes by configuration, using predefined software building blocks (programs and data) that reflect and implement the fundamental operations (Grundfunktionen) of a batch process. These guidelines were published as a NAMUR status paper in 1987. The NE-33 guidelines, "Requirements to be Met by Systems for Recipe-Based Operations," were issued in May of 1992.

DEFINITIONS

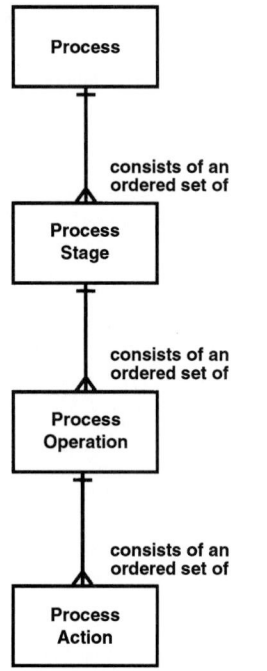

FIGURE 1 Process model. (Copyright © ISA. Reprinted with permission. All rights reserved.)

A Batch Process is a process that leads to the production of finite quantities of material by subjecting quantities of input materials to an ordered set of processing activities over a finite period of time, using one or more pieces of equipment. The processing activities are defined in the various levels of the process model (see Fig. 1). The process model describes what is required by the batch process, and it is a way of organizing the subdivisions of a batch process in a hierarchical fashion.

The process consists of one or more process stages that are organized as an ordered set that can be serial, parallel, or both. A process stage usually results in a planned sequence of chemical or physical changes in the material being processed. An example of a process stage might be polymerize, that is, polymerize vinyl chloride monomer (VCM) into polyvinyl chloride. Each process stage consists of an ordered set of one or more process operations. A process operation usually results in a chemical or physical change in the material being processed. An example of a process operation for the polymerization process stage might be Charge (add demineralized water and add surfactants).

Each process operation can be subdivided into an ordered set of one or more process actions that carry out the processing required by the process operation. Process actions describe minor processing activities that are combined to make up a process operation. A typical process action for the Charge process operation might be Add (the required amount of demineralized water to the reactor).

A *Batch* has two meanings in S88.01. The first meaning of batch is the material that is being produced or that has been produced by a single execution of a batch process. The second meaning of batch is an entity that represents the production of a material at any point in the process. Therefore, batch means both the material made by and during the process and also an entity that represents the production of that material. Batch is used as an abstract contraction of the words "the production of a batch."

Batch control refers to control activities and control functions that provide a means to process finite quantities of input materials by subjecting them to an ordered set of processing activities over a finite period of time, using one or more pieces of equipment.

A *recipe* is the necessary set of information that uniquely defines the production requirements for a specific product. The recipe tells the batch control system and/or the operator how to make product. A recipe usually exists for each intermediate and finished product that is to be produced.

Equipment control is the equipment-specific functionality that provides the actual control capability for an equipment entity, including procedural, basic, and coordination control, and that is not part of the recipe.

An *equipment entity* is a collection of physical processing and control equipment and equipment control grouped together to perform a certain control function or set of control functions.

RECIPES

Recipes provide a way to describe products and how those products are to be produced. A product may be made in many different arrangements of equipment at many different sites. Recipes that are appropriate for one site or set of equipment may not be appropriate for another site or set of equipment. This can result in multiple recipes for a single product. Sufficient structure should be provided in the definition of recipes to allow tracing of the genealogy of any given recipe.

The recipe contains neither scheduling nor equipment control; it contains process-related information for a specific product. This concept allows batch processing equipment to make many different products without having to redefine equipment control for each product. This section discusses the four types of recipes that are covered in the S88.01 standard and the five categories of information that are contained in a recipe.

Recipe Types

Fundamental to the practical application of recipes is the concept that different parts of an enterprise may require information about the manufacture of a product in varying degrees of specificity, because different recipients of the information use it for different purposes. Therefore, more than one type of a recipe is needed in an enterprise. The S88.01 standard defines only the general recipe, site recipe, master recipe, and control recipe (see Fig. 2). General and site recipes are substantially different than master and control recipes. The general and site recipes describe the technique (i.e., how to do it in principle). Master and control recipes describe the task (i.e., how to do it with actual resources).

Whether a particular recipe type actually exists, who generates it and where it is generated will vary from case to case and from enterprise to enterprise. An enterprise may choose not to implement one or more of the recipe types, and, depending on the specific requirements of an enterprise, other recipe types may exist.

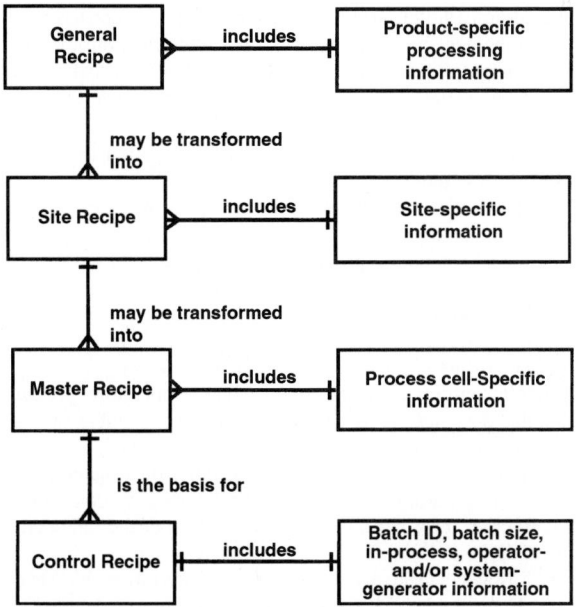

FIGURE 2   Recipe types. (Copyright © ISA. Reprinted with permission. All rights reserved.)

General Recipes.   A general recipe is the parent of all lower-level recipes. A general recipe provides general manufacturing requirements in an equipment-independent format, and it applies to all sites within the enterprise.

Site Recipes.   A site recipe is specific to a particular manufacturing site within the enterprise. A site recipe is derived from a general recipe, and little transformation is typically required. A site recipe is also in an equipment-independent format, and it is generally used for the following:

* planning
* costing
* long-term production scheduling

Master Recipes.   A master recipe is derived from a site recipe, and a major transformation is expected to be required because the master recipe is specific to equipment at the manufacturing site. A master recipe may be targeted at a class of equipment or to specific pieces of equipment. Master recipes contain information that is needed for batch scheduling.

Control Recipes.   A control recipe starts its life as a copy of a master recipe, and it is the recipe that is used to make a batch of product, whether automatically by a control system or manually by an operator. A control recipe is specific to a particular batch of product, and it may change as the batch progresses.

Recipe Information Categories

Recipes contain the following categories of information: header, formula, procedure, equipment requirements and other information. The following discussion provides details regarding these categories.

Header.   The administrative information in the recipe is referred to as the header. Typical header information may include the recipe and product identification, the version number, the originator, the issue date, approvals, status and other administrative information. For example, a site recipe may contain the name and version of the general recipe from which it was created.

Formula.   The formula is a category of recipe information that includes process inputs, process parameters, and process outputs.

A process input is the identification and quantity of a raw material or other resource that is required to make the product. In addition to raw materials that are consumed in the batch process in the manufacture of a product, process inputs may also include energy and other resources (e.g., manpower). Process inputs consist of both the name of the resource and the amount required to make a specific quantity of product. Quantities may be specified as absolute values or as equations based upon other formula parameters or the batch or equipment size. Process inputs may specify allowable substitutions, and they are expressed in the same basic form.

A process parameter details information such as temperature, pressure, or time that is pertinent to the product but that does not fall into the classification of input or output. Process parameters may be used as set points, comparison values, or in conditional logic.

A process output is the identification and quantity of a material and/or energy that is expected to result from one execution of the recipe. These data may detail environmental impact, and they may also contain other information (e.g., specification of the intended outputs in terms of quantity, labeling, and yield).

The types of formula data are distinguished to provide information to different parts of an enterprise, and they have to be available without the clutter of processing details. For example, the list of process inputs may be presented as a condensed list of ingredients for the recipe or as a set of individual ingredients for each appropriate procedural element in a recipe.

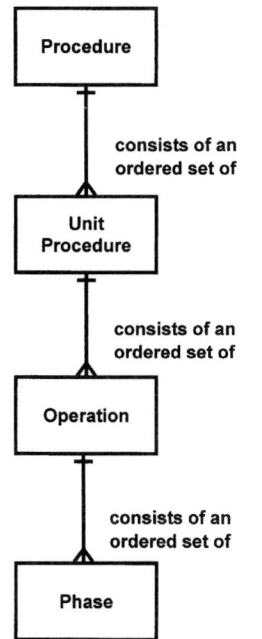

FIGURE 3   Procedural control model.
(Copyright © ISA. Reprinted with permission. All rights reserved.)

Procedure.   The procedure defines the strategy for carrying out a process. The general and site recipe procedures are structured by using the levels described in the process model because these levels allow the process to be described in equipment-independent terms. The master and control recipe procedures are structured by using the procedural elements of the procedural control model (see Fig. 3) because these procedural elements have a relationship to equipment.

The recipe creator is limited to the use of procedural elements that have been, or will be, configured and made available for use in creating a procedure. He or she may use any combination of these procedural elements to define a procedure. Determination of which of these procedural elements that may be part of the procedure is an application specific design decision that is based on many factors, including the capabilities of the controls and the degrees of freedom appropriate for the recipe creator in a given application.

The procedural control model shown in Fig. 3 defines the control that enables the process equipment to provide the process functionality required by the batch process. Procedural control is made up of procedural elements, and they are combined in a hierarchical manner in order to accomplish the task of a complete process as defined by the process model.

A procedure is the highest level in the hierarchy, and it defines the strategy for carrying out a major processing action (e.g., making a batch). A procedure is defined in terms of an ordered set of unit procedures. An example of a procedure is "Make PVC."

A unit procedure consists of an ordered set of operations that cause a contiguous production sequence to take place within a unit. An example of a unit procedure in the Make PVC procedure might be "Polymerize VCM."

An operation is an ordered set of phases that define a major processing sequence that takes the material being processed from one state to another, and a chemical or physical change is usually involved. An example of an operation in the Polymerize VCM unit procedure might be React (Add VCM and catalyst, Heat, and Wait for the reactor pressure to drop).

A phase is the smallest element of procedural control that can accomplish a process oriented task. The intent of the phase is to cause or define a process oriented action, while the logic or set of steps that make up a phase is equipment specific. An example of a phase in the React operation might be Add (Add catalyst).

Equipment Requirements.   Equipment requirements constrain the choice of the equipment that will eventually be used to implement a specific part of the procedure. In the general and site recipes, the equipment requirements are typically described in general terms, such as allowable materials and required processing characteristics. The guidance from and the constraints imposed by equipment requirements will allow the general or site recipe to eventually be used to create a master recipe that targets appropriate equipment. At the master recipe level, the equipment requirements may be expressed in any manner that specifies allowable equipment in process cells. If trains have been defined, then it is possible for the master recipe (and the resulting control recipe) to be based on the equipment of the train rather than the full range of equipment in the process cell. At the control recipe level, the equipment requirements are the same as, or a subset of, the allowable equipment in the master recipe. The control recipe may be used to include specific allocations of process cell equipment, such as Reactor R-501, when this becomes known.

Other Information.   The data that are contained in the Other Information category are usually one of the following:

- recipe-dependent safety comments, but not MSDS (material safety data sheet)
- recipe-dependent compliance comments
- data-collection requirements
- special reporting requirements

EQUIPMENT ENTITIES

This section discusses equipment entities that are formed from the combination of equipment control and physical equipment. This combination results in four equipment entities: process cells, units, equipment modules, and control modules. Guidelines for structuring these equipment entities are also discussed.

When the terms process cell, unit, equipment module, and control module are used, they generally refer to the equipment and its associated equipment control. Whether equipment control in an equipment entity is implemented manually or by way of automation, it is only through the exercise of equipment control that the equipment can produce a batch.

The notion of equipment control being part of an equipment entity is to be understood logically because it is not a statement of the physical implementation of equipment control. However, equipment control for a particular equipment entity must be identifiable.

This interaction of equipment control and physical equipment is described purposely without any reference to language or implementation. The intent is to describe a framework within which equipment control and physical equipment may be defined and discussed.

Equipment Control

Three types of control are defined: basic control, procedural control, and coordination control.

Basic Control.   Basic control is dedicated to establishing and maintaining a specific state of equipment and process, and it includes the following:

- regulatory control
- interlocking
- monitoring
- exception handling
- repetitive discrete or sequential control

Procedural Control.   Procedural control is a characteristic of batch processes, and it is used to direct equipment-oriented actions. Procedural control is based on the procedural control model.

Coordination Control.   Coordination control directs, initiates, and/or modifies the execution of procedural control and the utilization of equipment entities. The following are some examples of coordination control:

- supervising the availability of equipment
- allocating equipment to batches
- arbitrating requests for allocation
- coordinating common resource equipment

More coordination control is to be expected at the process cell level, while little coordination control is typically found at the control module level.

Physical Model

The physical model (see (Fig. 4) describes what equipment is available for the batch process. The physical assets of an enterprise that is involved in batch manufacturing are usually organized in a hierarchical fashion as described by the physical model. Lower-level groupings are combined to form

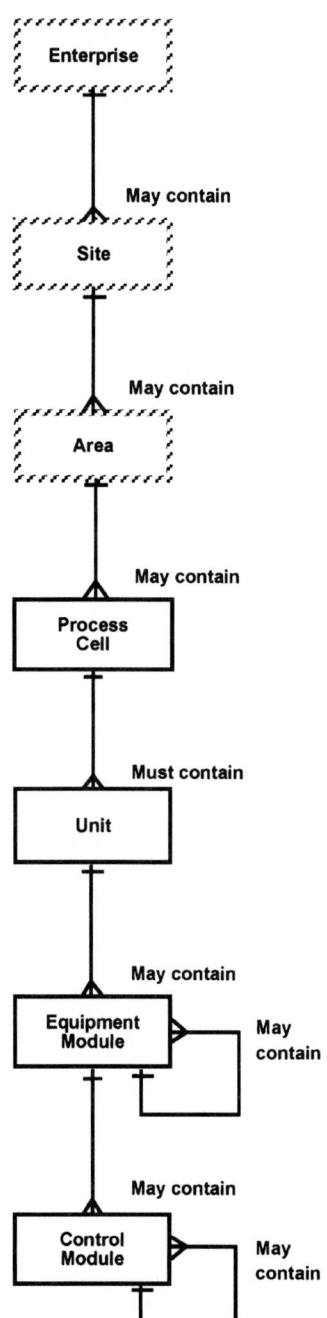

FIGURE 4 Physical model. (Copyright © ISA. Reprinted with permission. All rights reserved.)

higher levels in the hierarchy. In some cases, a grouping within one level may be incorporated into another grouping at that level.

The model has seven levels, starting at the top with an enterprise, a site and an area. These three levels are frequently defined by business considerations, and they are not modeled further in S88.01. The three higher levels are part of the model to properly identify the relationship of the lower-level equipment to the manufacturing enterprise.

The lower four levels of this model refer to specific equipment types. An equipment type in Fig. 4 is a collection of physical processing and control equipment grouped together for a specific purpose. This grouping is usually done to simplify operation of the lower-level equipment by treating it as a single larger piece of equipment.

Equipment entities are defined for the lowest four levels of the physical model (i.e., process cell, unit, equipment module, and control module).

Control Module. A control module is a collection of sensors, actuators, other control modules, and associated processing equipment. A control module acts as a single entity from a control standpoint, and it is the direct connection to the process through its sensors and actuators. A control module cannot execute procedural elements. The following are some examples of control modules:

1. A flow control loop that operates by means of the set point of the controller.
2. An on–off automatic block valve with limit switches that operates by means of the set point (e.g., open and close) of the valve.
3. A header that contains several automatic block valves and that directs flow to different destinations based on a set point to the header.

Equipment Module. An equipment module is a collection of control modules and/or other equipment modules. An equipment module can carry out a finite number of minor processing activities (i.e., phases), and it contains all the necessary processing equipment that is needed to carry out these processing activities. The following are some examples of equipment modules:

1. A weigh tank that is shared by multiple units but that can only be used by one unit at a time.
2. A filter that is a permanent part of a particular unit.
3. An ingredient supply system that is shared by multiple units and that can be used simultaneously by all units.

Control modules and equipment modules are used because these combinations of various instrument functions can be addressed as a single entity. The alternative to these equipment entities is to deal with the individual equipment-oriented checks and actions.

Unit.   A unit is usually centered on a major piece of process equipment, and it frequently operates on or contains the complete batch. Although a unit may operate on or contain only a portion of a batch, it cannot operate on or contain more than one batch at a time.

Units are the primary object for automatic control, and they have a direct relationship with unit procedures and operations. A unit is made up of control modules and/or equipment modules, and there will often be multiple units involved in making a batch. Control modules and equipment modules can exist as:

- permanently included parts of a unit
- temporarily attached parts of a unit
- totally separate from a unit

When control modules and/or equipment modules are not part of a unit, they may be connected to a unit, and then they may be commanded like any other object in the unit.

Process Cell.   A process cell is a logical grouping of equipment that is required for the production of one or more batches. A process cell may contain more than one grouping of equipment that is needed to make a batch. That grouping is referred to as a Train. The equipment that was actually used to execute a batch is referred to as the Path. A process cell frequently contains more than one batch at a time.

Partitioning Equipment Entities

This section discusses the general principles involved in partitioning a process cell into equipment entities that can carry out specified processing activities or equipment-specific actions [4], [5].

The physical process cell design can greatly influence the implementation of batch control. Minor differences in the physical system can dramatically affect the organization of equipment entities and procedural elements.

All control-related sections of the standard assume that the process cell in question (both physical equipment and related control activities) has been subdivided into well-defined equipment entities such as units, equipment modules, and control modules. Effective subdivision of the process cell into well-defined equipment entities is a complex activity, and it is highly dependent on the individual requirements of the specific environment in which the batch process exists. Inconsistent or inappropriate equipment subdivisions can compromise the effectiveness of the modular approach to recipes suggested by this standard, and they may result in a solution that:

- is difficult to support and enhance
- fails to exploit the inherent flexibility of the plant
- requires control system experts to assist process personnel when they are developing new recipes or modifying existing recipes

Subdivision of the process cell requires a clear understanding of the purpose of the process cell's equipment. Such understanding allows the identification of equipment entities that can work together to serve an identifiable processing purpose.

Process Cells.   The subdivision of a process cell usually follows the principles listed below:

1. The function any equipment entity serves in product processing must be clear and unambiguous.
2. The function performed by the equipment entity must be consistent in terms of processing task, and it should be usable for that task no matter what product is being manufactured at a given time.
3. Subordinate equipment entities should be able to execute their task(s) independently and asynchronously because this allows the highest level equipment entity to orchestrate the activities of its subordinates.

4. Interactions between equipment entities should be minimized. While planned interaction is periodically necessary, each equipment entity should perform its functions while influencing the functioning of other equipment entities as little as possible.

5. Equipment entities must have clear boundaries.

6. A consistent basis is required for the definition of equipment entities. An operator subsequently interacting with similar equipment entities should be able to do so naturally and without confusion.

7. Necessary interaction between equipment entities is, insofar as possible, coordinated by equipment entities at the same level or at the next higher level.

8. The process cell may have multiple units and/or equipment modules of the same type that operate in parallel.

9. The number of units in the process cell will not necessarily be the same as the number of vessels in the process cell.

10. The equipment entities that are defined should perform essentially independent tasks.

Units.   The definition of a unit requires knowledge of the major processing activities, as well as the equipment capabilities. The following guidelines apply:

1. One or more major processing activities (i.e., operations), such as reaction or crystallization, may take place in a unit, but only one may be active at a time.

2. Units should be defined such that they operate relatively independently of each other.

3. A unit can operate on only one batch at a time.

4. A unit can operate on all or a part of a batch.

5. A unit may contain equipment modules and control modules, but a unit may not contain another unit.

Equipment Modules.   The definition of an equipment module requires knowledge of specific minor processing activities and equipment capabilities. The following guidelines apply:

1. An equipment module can carry out a finite number of minor processing activities, such as dosing and weighing.

2. These activities are typically centered around a set of process equipment.

3. Collections of control modules can be defined as equipment modules or as control modules. If the collection executes one or more equipment phases, then it is an equipment module.

4. An equipment module may be part of a process cell, a unit, or another equipment module.

Control Modules.   A control module is the lowest level of grouping that operates as a single entity. The following guidelines apply:

1. A control module executes basic control.

2. A control module cannot execute procedural control.

Procedural Control Model/Physical Model/Process Model Relationship

The general relationship between the procedural control model, the physical model, and the process model is illustrated in Fig. 5. This mapping of procedural control with individual equipment provides the processing functionality described in the process model.

The concept of equipment capabilities and usage of these capabilities to accomplish processing tasks is a major point of the S88.01 standard. The procedural control capability of equipment entities is the mechanism that enables this. The procedural control may be entirely defined as part of the equipment entity—or it may be based on procedural information passed on to the equipment entity from the recipe procedure.

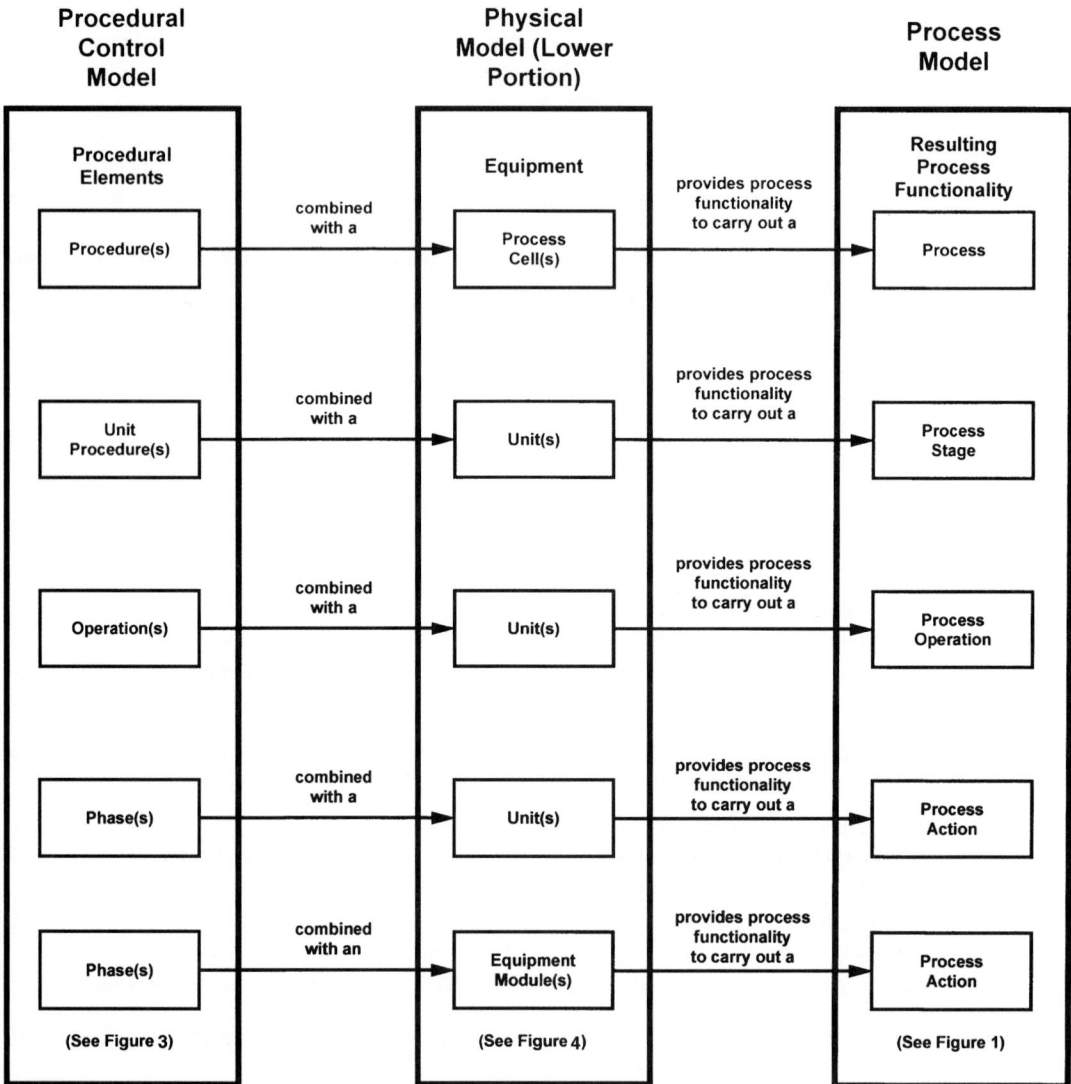

FIGURE 5   Procedural control model/physical model/process model relationship. (Copyright © ISA. Reprinted with permission. All rights reserved.)

RECIPE PROCEDURE/EQUIPMENT CONTROL SEPARATION

This concept of separating the recipe that describes how the batch is to be made from the equipment that is actually used to make the batch results in the following advantages:

1. Improves recipe transportability because the recipe does not contain all the equipment-specific logic.

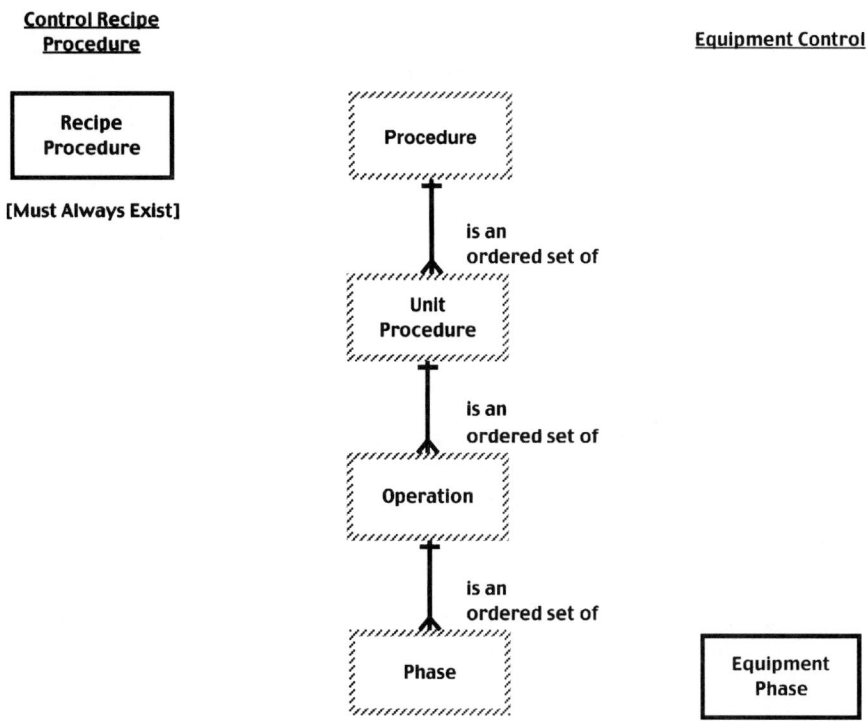

FIGURE 6   Recipe procedure/equipment control separation. (Copyright © ISA. Reprinted with permission. All rights reserved.)

2. Makes the recipe more flexible because it is applicable to a wider range of batch control equipment, even if the batch control equipment operates differently. As an example, a recipe phase that adds an ingredient to a batch doesn't care if the fluid is metered with a flow meter or a weigh tank.

3. Simplifies recipe validation because the equipment logic does not have to be revalidated every time a change is made to a recipe.

4. Makes software modularity feasible. The goals of software modularity can be accomplished by:

   • Making the creation or modification of a recipe easy enough so that control systems experts are not needed to develop and maintain recipes.
   • Using predefined and pretested building blocks (i.e., standard software modules) for implementing equipment control.
   • Making it easy to link the recipe with these building blocks.

Figure 6 shows the separation between the control recipe procedure and equipment control. The control recipe procedure must contain at least one procedural element, which is the recipe procedure. Equipment control must also contain at least one procedural element that provides the linkage needed to operate the physical equipment. For the example described in Fig. 6. this procedural element is assumed to be the equipment phase.

The control recipe procedure might not include recipe unit procedures, recipe operations, and recipe phases. Such a recipe procedure must then be linked (by reference) to an equipment procedure in equipment control if batches are to be executed. Whenever a procedural element (i.e., recipe procedure, recipe unit procedure, recipe operation, or recipe phase) is linked to equipment control, it must exist as that recipe procedural element (such as a recipe operation) and as that equipment procedural element (such as an equipment operation). Whenever recipe phases are used in the control recipe procedure, recipe phases are linked to equipment phases.

When recipe unit procedures, recipe operations, and recipe phases are not used as part of the control recipe procedure, the use of lower-level equipment procedural elements (some or all) as part of equipment control can provide a modular structure to the equipment control.

Control Recipe Procedure/Equipment Control Linking

There must be some method to link the control recipe procedural elements with the equipment procedural elements. This linking is done by associating the recipe procedural elements with equipment procedural elements. In this way, the call for a certain processing function is separated from equipment control, and this enables the same recipe procedural element to use different equipment procedural elements, depending on what equipment the recipe addresses.

An equipment phase may be initiated by things other than the execution of a control recipe (e.g., by the request of another unit or on the request of an operator). The independent execution of a phase may be useful for handling exception conditions, during start-up or maintenance and/or to prepare a unit for production.

If unit procedures, operations, and phases are part of the control recipe procedure, linking (by reference) of the control recipe procedure to equipment control is done at the phase level (see Fig. 7). This drawing applies to one control recipe.

In a batch control system, this linking is accomplished with a phase logic interface. A phase logic interface is a layer of software that provides the interface between the recipe procedure and the corresponding equipment procedural element.

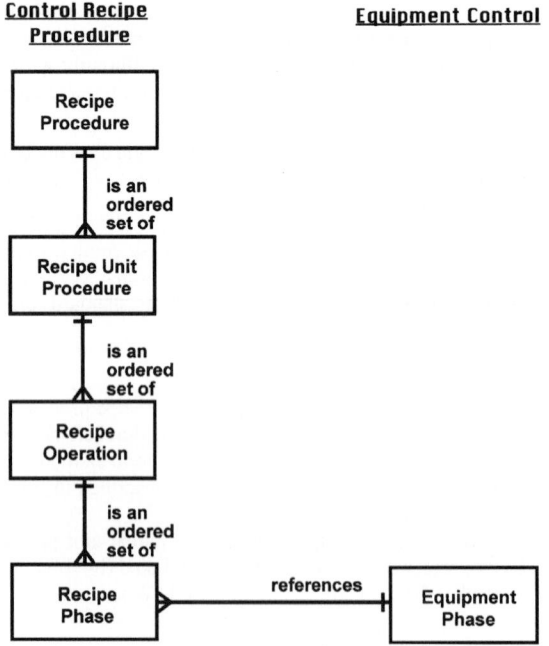

FIGURE 7   Recipe procedure/equipment control linking example.
(Copyright © ISA. Reprinted with permission. All rights reserved.)

Control Recipe/Equipment Procedural Elements

The following are typically associated with recipe procedural elements:

- a description of the functionality required
- formula and other parameter information specific to the procedural element
- equipment requirements specific to the procedural element

The equipment procedural element to be linked typically has the following:

- an identification that can be referenced by the recipe procedural element or a higher-level equipment procedural element
- a description of the functionality that is provided
- variables that can receive the formula and other parameter information from the recipe
- execution logic

In order for a recipe procedural element to be able to reference an equipment procedural element, it must have an identification that enables the element to be correctly linked. In other cases, it must reference or include other recipe procedural elements and a specification of the execution order of those procedural elements.

PROCESS AND CONTROL ENGINEERING

In order for required processing functions to be properly carried out in a batch manufacturing environment, the equipment structure needed, the process functionality, and the exception handling for that equipment have to be fully developed. This requires a coordinated engineering effort that continues from initial definition through the life of the batch processing facility. This section describes the process and control engineering needed for the design of the controls needed to support the recipe hierarchy, the definition of equipment capability, and the development of the functionality required in the procedures to produce a batch.

Process and control engineering is needed at the general and site recipe levels to describe procedures, process stages, process operations, and process actions and at the master recipe level to describe recipe procedures, recipe unit procedures, recipe operations, and recipe phases.

The precise definition of appropriate procedural elements and equipment entities is an iterative process. The dual work process is illustrated in Fig. 8. Considerations affecting one decision process also affect the other. Processing considerations are the primary input to the definition (or selection) of procedural elements that will characterize functionality for associated equipment entities. Since the functionality defined will be affected by the equipment used, equipment considerations must be a secondary input. In the same way, equipment considerations form the primary input and processing considerations form the secondary input when making the definition (or selection) of equipment entities.

Recipes can be constructed by using these procedural elements and specific product information. The equipment entities are arranged into a path that is determined by scheduling and taking into account arbitration constraints. The combination of the results of these activities provides a framework within which a batch of material can be manufactured.

Process and control engineering also includes the development and revision of the equipment phases corresponding to the recipe phases that are used to define the recipe. As far as possible, recipe and equipment phases should be defined such that any reasonable functionality of a unit can be expressed in terms of these phases. They should generally not be tailored to a set of known recipes. Then, new recipes can in most cases be written by using existing recipe phases that reference existing equipment phases. The development and revision of recipe and equipment phases is an ongoing activity that provides ongoing support to the batch manufacturing facilities. This activity is the result of the ongoing drive for continuous improvement and the periodic addition of new process technology.

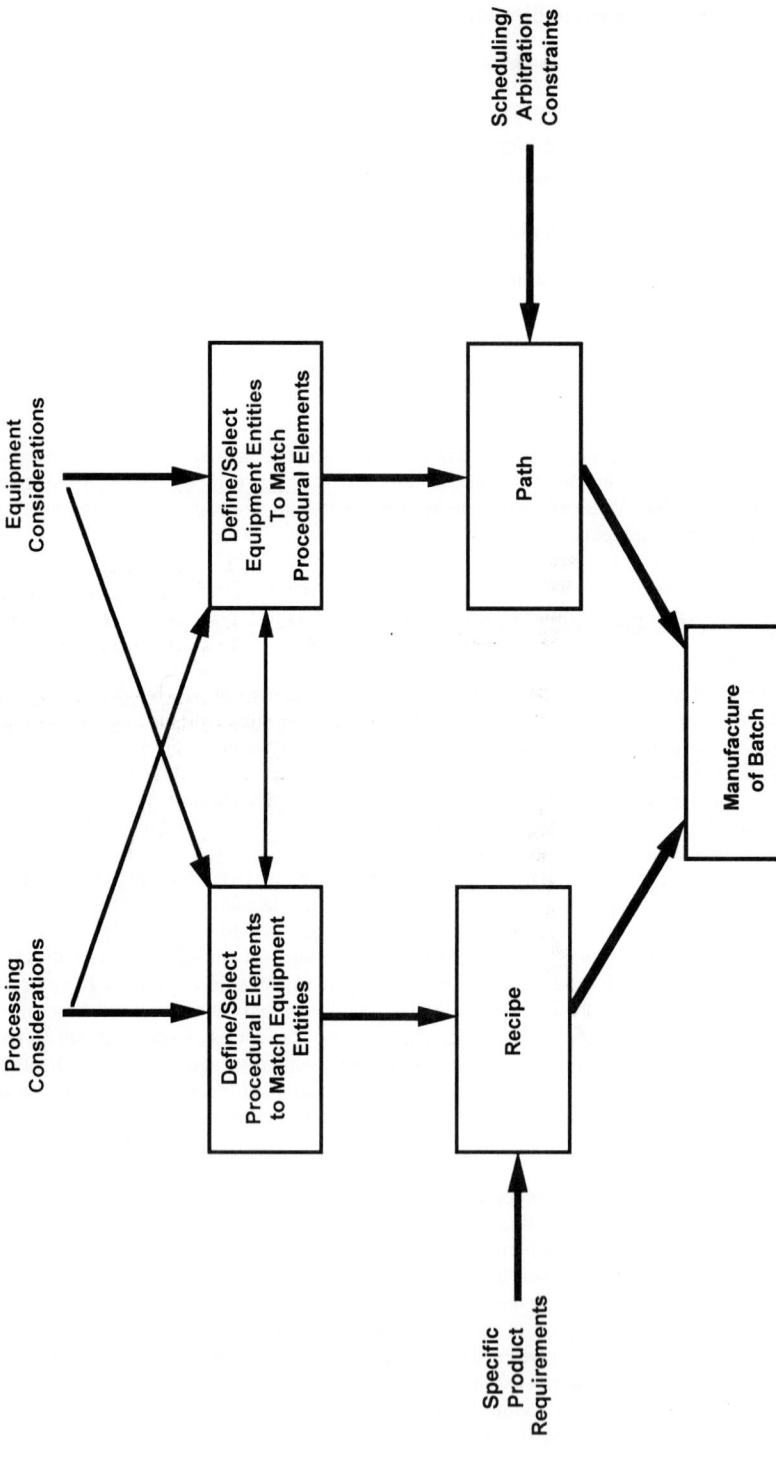

FIGURE 8 Definition of procedural elements and equipment entities. (Copyright © ISA. Reprinted with permission. All rights reserved.)

CONTROL SYSTEM FUNCTIONAL SPECIFICATIONS

The models and terminology that are defined in S88.01 can be used as the basis for developing vendor-neutral or vendor-specific functional specifications for batch control. By the application of object-oriented techniques, reusable modules can be developed that can then be used from project to project.

What Is Needed To Define Batch Control

The following is the information that is needed in order to define the batch control requirements for a particular project:

- how the product is to be made (i.e., recipes)
- what equipment is to be controlled (i.e., equipment entities)
- what control is needed (i.e., control activities and control functions)

Recipes and equipment entities were discussed in previous sections. Control activities and control functions are defined by the control activity model (see Fig. 9). This model provides the basis for defining the control functionality that is required.

The control activity model is the cornerstone of the S88.01 standard. These control activities are the ones that are necessary to manage and control production in a batch manufacturing plant. The

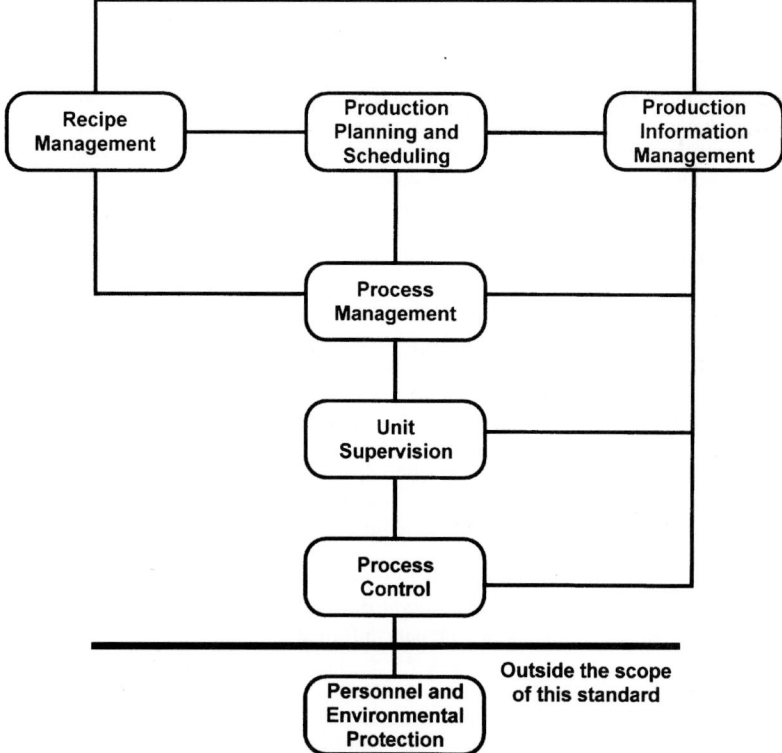

FIGURE 9   Control activity model. (Copyright © ISA. Reprinted with permission. All rights reserved.)

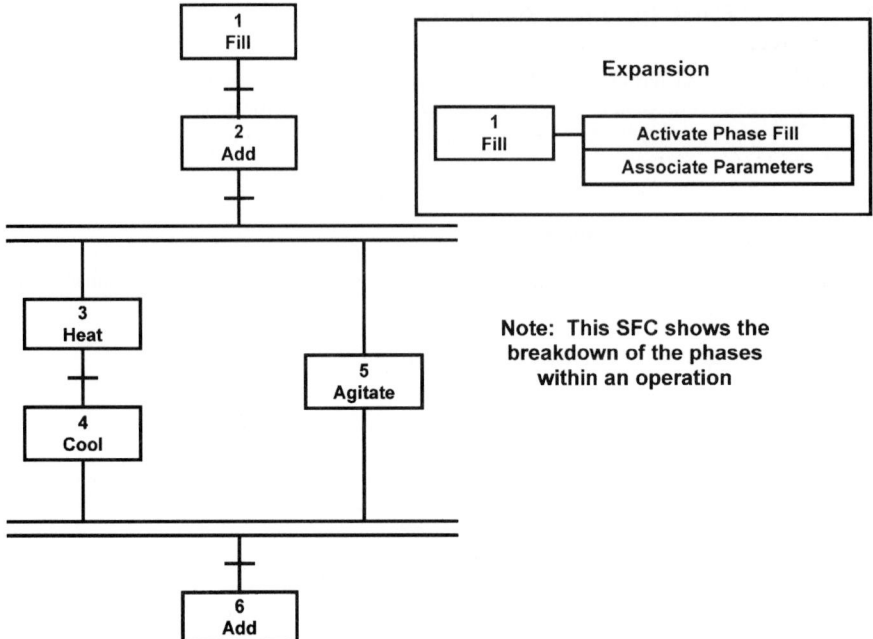

FIGURE 10   Using sequential function charts (SFCs) to depict recipe procedures.

control activity model is used to define what has to be accomplished to do batch control, whether the batch control is done automatically or manually. The control activities shown relate to real needs in a batch manufacturing environment. For example, the need to have control functions (note that control activities are broken down into control functions) that can manage general recipes, manage site recipes, and manage master recipes implies a need for the Recipe Management control activity.

Recipe Management.   For control system functional specifications, the primary recipe of interest is the master recipe. The master recipes that are needed for the project can be developed by using the techniques that are defined in S88.01. The user must decide on the language that is used for recipe development and user presentation. An example using sequential function charts (SFCs) is shown in Fig. 10. This figure shows the breakdown of the phases in a React operation.

Production Planning and Scheduling.   The user must decide how batch scheduling will be handled (i.e., manual or automatic). The format of the batch schedule must be determined in order to minimize any mismatch between the control function that supplies the batch schedule and the control system that uses the batch schedule. In addition, the functional specification should define what information must be fed back to the scheduling control function.

Production Information Management.   A list of the information that must be collected from the control system must be developed, along with the format of that information. In addition, the functional specification should define what summary information must be supplied to higher-level business processes.

Equipment Entities.   The three higher-level control activities that were described above communicate directly with the Process Management control activity. Process Management and the lower-level control activities are typically dealt with as part of Equipment Control.

STANDARDS OVERVIEW **11.51**

Process Management maps to the process cell equipment entity. The process cell and the equipment that is contained within the process cell must be partitioned into equipment entities that correspond to the three lowest levels of the physical model:

- unit
- equipment module
- control module (includes safety interlocks)

The relationships between these equipment entities must also be specified.

The behavior of the equipment entities that were defined above must be specified so that the control functionality that was defined using the Control Activity Model can, in fact, be accomplished. This behavior may be defined by using a combination of the following:

- basic control
- procedural control
- coordination control
- data collection
- operator interface requirements (not discussed here)

Equipment Entity Details

The following are things that must be dealt with as part of the definition of the equipment entities:

- phase logic
- allocation and arbitration
- unit-to-unit synchronization
- modes and states
- data collection
- exception handling

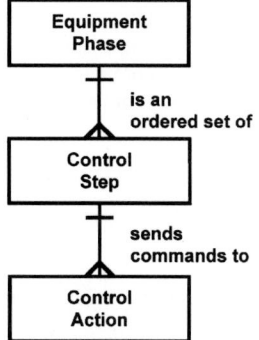

FIGURE 11 Equipment phase subdivisions.

Phase Logic.   Equipment phases are usually broken down into control steps and control actions. The steps and transitions that are described in IEC 60848 or IEC 61131-3 document one method for defining the subdivisions of a phase (see Fig. 11).

The following are some things that control steps can do:

- enable and disable regulating and state-oriented types of control actions
- specify the set points and initial output values of control actions
- set, clear, and change alarms and other limits
- set and change controller constants, controller modes, and types of algorithms
- read process variables (e.g., alarm limits, set points, and controller status)
- send a message to an operator

Some of the categories of control actions are:

- control actions that perform an actuating element algorithm

- control actions that perform a regulating type of control algorithm or a state-oriented type of control algorithm
- control actions that perform arithmetic calculations
- control actions that communicate with people and/or other equipment
- control actions that make decisions and then control the direction and timing of higher-level control components

Allocation and Arbitration.   Allocation is a form of coordination control that assigns a resource to a batch or unit. Allocation determines the routing or the "path" of the batch through the units. This patch may be dynamic (i.e., the path may change from batch to batch) depending on unit availability at the time the batch is executed.

Allocation must deal with common resources. When more than one unit can acquire or request the services of a single resource, that resource is a common resource. Common resources are often present with complex batch processes, and they are often implemented as equipment modules or control modules.

Common resources may be either exclusive-use resources or shared-use resources. When a resource is designated as an exclusive-use resource, only one unit can use that resource at a time (i.e., it is exclusive to that unit). An example is a shared weigh tank that can be used by many reactor units but that can only be used to charge one reactor at a time. This common resource must be allocated properly to avoid unit downtime while the unit is waiting for the use of the resource.

When a resource is designated as a shared-use resource, several units may use the resource at the same time. An example is a raw-material distribution system that is capable of delivering material to more than one unit at a time. If the capacity of the shared-use resource is limited (e.g., a maximum flow capability of a steam header), controls must be put in place to ensure that the resource's capacity is not exceeded. Controls must also be put in place to ensure that one unit does not improperly shut off or deactivate a shared-use resource while other units are using the resource.

Arbitration is a form of coordination control that determines how a resource should be allocated when there are more requests for the resource than can be accommodated at one time. Arbitration is required when there are multiple requestors for a resource. Then some method must be put in place to resolve contention for the resource according to some predetermined algorithm and to provide definitive routing or allocation direction.

Unit-To-Unit Synchronization.   The need for unit-to-unit synchronization is very common when transfers are made between units. This synchronization is usually handled with a Transfer operation in one unit and a Receive operation in the other unit. However, the actual communication between the units occurs at the equipment phase level. At least one commercial system handles transfers as part of the routing or path.

Figure 12 shows the communication between the Transfer operation in an esterification unit and the Initialize and Receive operations in a stripping unit.

Modes and States.   A mode is defined as:

- the manner in which the transitions of sequential functions are carried out within a procedural element or
- the accessibility for manipulating the states of equipment entities manually or by other types of control

The modes that are defined in S88.01 are shown in Table 1.

Users must deal with mode propagation. For an example, a unit procedure changes to the semi-automatic mode. Should all lower-level procedural elements in that unit go to the semi-automatic mode? Propagation may be from higher to lower and vice versa.

TABLE 1 Modes Defined in S88.01

Mode	Behavior	Command
Automatic (procedural control)	The transitions within a procedure are carried out without interruption as appropriate conditions are met.	Operators may pause the progression, but they may not force transitions.
Automatic (basic control)	Equipment entities are manipulated by their control algorithm.	The equipment cannot be manipulated directly by the operator.
Semi-automatic (procedural control only)	Transitions within a procedure are carried out on manual commands as appropriate conditions are fulfilled.	Operators may pause the progression or redirect the execution to an appropriate point. Transitions may not be forced.
Manual (procedural control)	The procedural elements within a procedure are executed in the order specified by an operator.	Operators may pause the progression or force transitions.
Manual (basic control)	Equipment entities are not manipulated by their control algorithm.	Equipment entities may be manipulated directly by the operator.

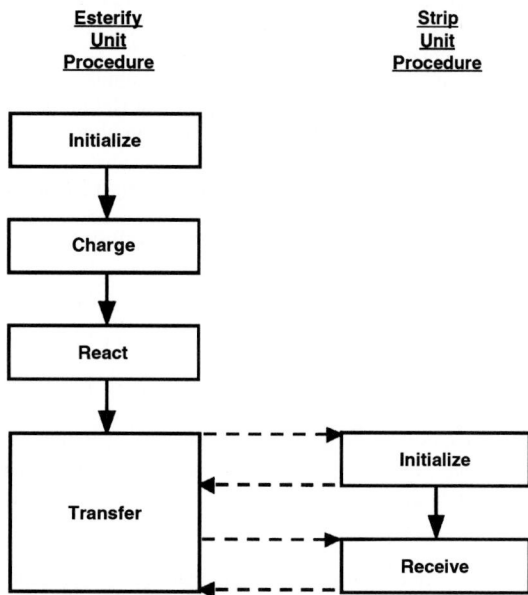

FIGURE 12 Unit-to-unit synchronization example.

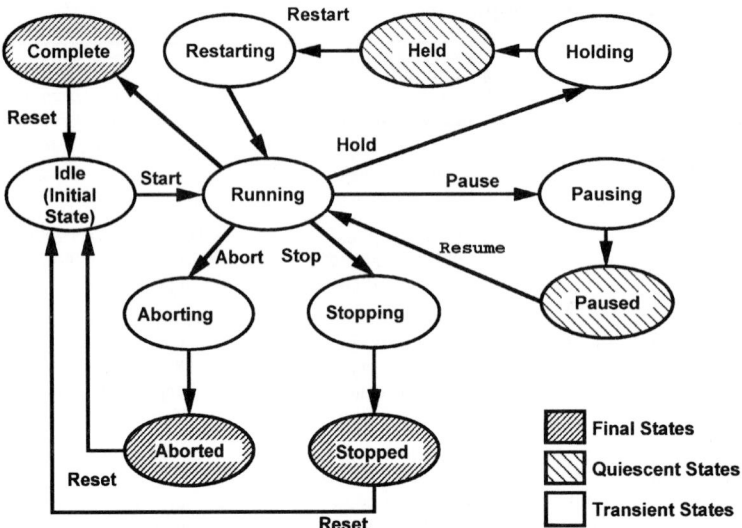

FIGURE 13   State transition diagram. (Adapted from S88.01, ISA.)

State is the condition of an equipment entity or of a procedural element at a given time. Some examples of states for equipment entities are On, Off, Closed, Open, Failed Tripped and Available. Some examples of states for procedural elements are Running, Holding, Paused, Stopped, Aborted, and Complete.

Procedural elements also have commands that move the procedural elements from one state to the next. Some examples of the commands that are applicable to procedural elements are Start, Hold, Pause, Stop and Abort. An example of the relationship between the procedural element states and the procedural element commands is shown in Fig. 13.

Users must also deal with state propagation. For example, a unit procedure moves to the Held state. Should all procedural elements in that unit go to the Held state? Should all procedural elements in *all units* go to the Held state? Propagation may be from higher to lower and vice versa.

Data Collection.   Data collection is present in the lower four elements of the control activity model:

- process management: collect batch and process cell information
- unit supervision: collect batch and unit information
- process control: collect data
- personnel and environmental protection: not specified by S88.01, but collect data

A mechanism must exist to specify what data must be collected. These data must be related to the batch, to the equipment entity, and to the procedural element. Data must be collected on those things that are expected to happen during the course of the batch, but they must also be collected on those things that are not expected to happen. For example, if the recipe procedure is changed during the execution of a batch, data on any new recipe procedural elements and their execution must be collected.

Exception Handling.   An exception is an event that occurs outside the normal or desired behavior. Exceptions can occur at all levels in the control activity model, and they may be part of procedural control, basic control, and coordination control. Exception handling typically accounts for a very large portion of the control definition (i.e., 50–80%).

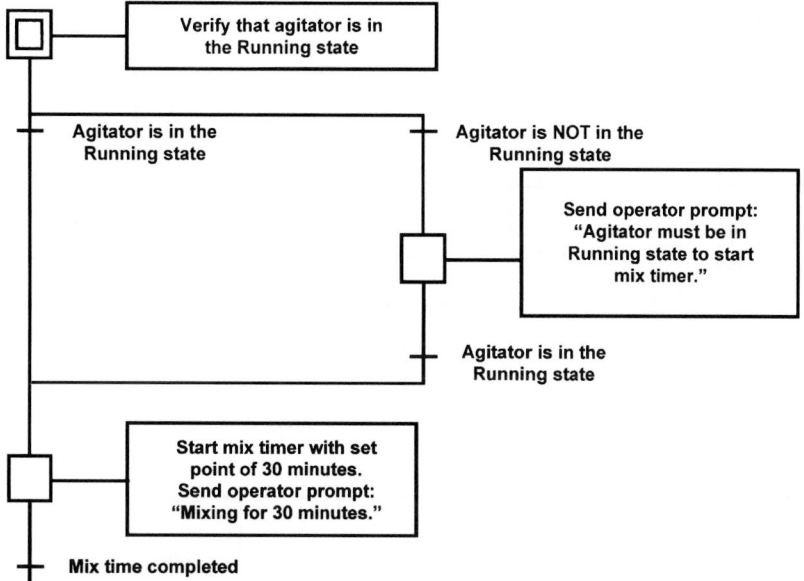

FIGURE 14   Exception handling in phase logic.

An event must be detected, then evaluated, and a response generated. Exceptions may affect the modes and states of equipment entities and procedural elements. For example, high pressure in a reactor could lead to the exception handling function transferring the process to the Stopped state.

The following are some different ways of implementing exception handling:

1. Implement exception handling that changes infrequently in equipment control.
2. Implement exception handling that changes frequently in the recipe.
3. Consider implementing product-related exceptions in equipment control using formula parameters.

If exception handling is implemented in the recipe, it should be as simple as possible and implemented at the recipe phase level, if possible.

Most exception handling is implemented in equipment control. The following are some different ways of implementing exception handling in equipment control:

1. Implement directly in phase logic.
2. Implement in phase logic using states.
3. Implement in control actions.

Figure 14 shows an example in which exception handling is implemented directly in phase logic. This method is effective if the exception handling functions are relatively simple, but it can get very complex if the exception handling functions are complex.

Exception handling is usually best done with states. The user must define what states are needed, but the states that are defined in S88.01 will meet most needs. The user must also define the conditions that cause state transitions (i.e., commands). If the states that are defined in S88.01 are used, then the equipment phase should contain the following logic:

- running logic
- holding logic

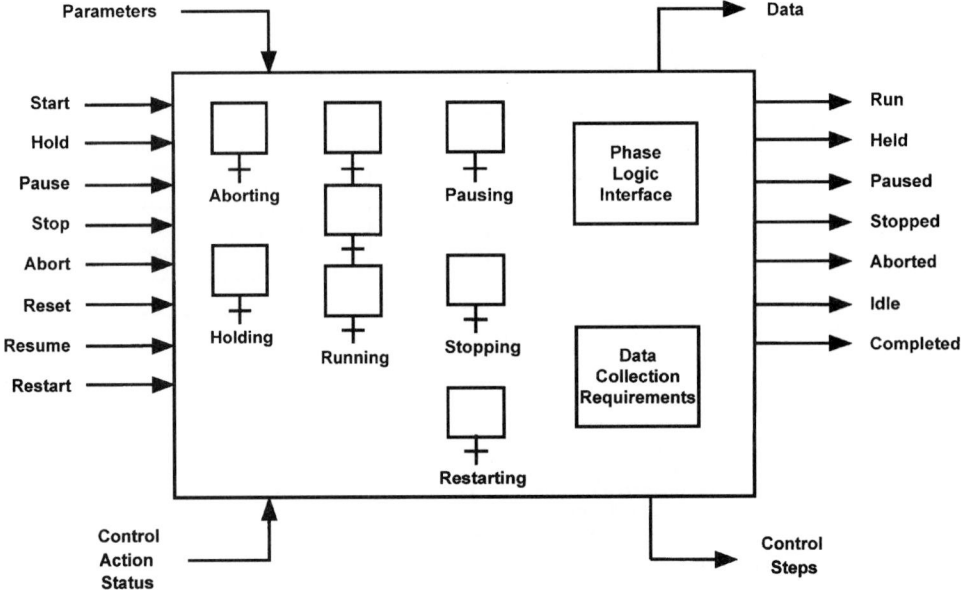

FIGURE 15   Equipment phase object.

- pausing logic
- aborting logic
- stopping logic
- restarting logic

An example of an equipment phase is shown in Fig. 15. Commands come in the left side of the phase block. The current state of the phase is shown in the status signals that come out of the right side of the block. Parameters come from the recipe, and data are fed back to higher-level elements. Commands (control steps) to control actions come out of the phase block on the bottom right side. Feedback on the status of the control actions comes back into the phase block on the bottom left side.

The phase logic interface is shown within the equipment phase. The phase logic interface is responsible for interfacing to the recipe and for enforcing the states of the phase. Data-collection requirements for the equipment phase are also shown. The logic for the various states is included within the equipment phase.

Figure 16 shows an example of exception handling in control actions. A safety interlock and a process interlock are both examples of control actions. When either of these interlocks trip, they can affect the operation of other control actions and phase logic.

SUMMARY

S88.01 is a good tool for improving batch control-related communications between users and vendors and for developing functional specifications. This standard also provides the basis for actual implementation of the batch control system. S88.01 is a models and terminology standard, so many of the implementation details are left up to the user.

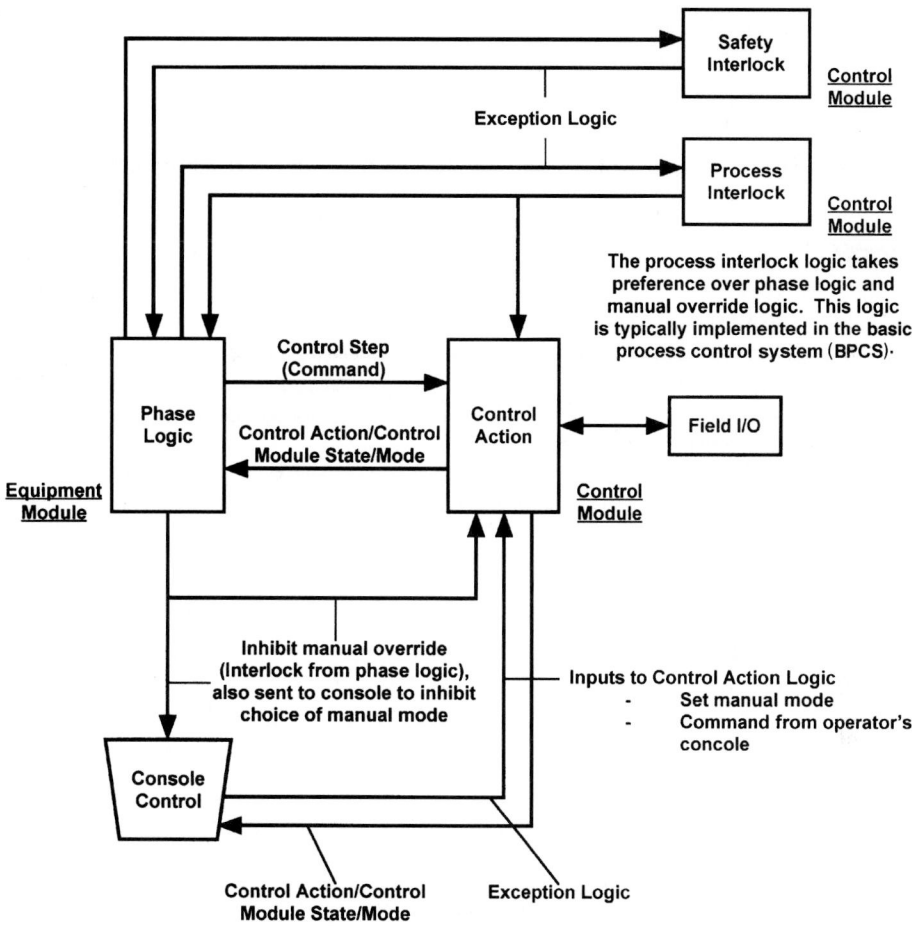

The safety interlock logic takes preference over phase logic, manual override logic and process interlock logic. This logic, when required, is typically implemented in an independent safety interlock system.

The process interlock logic takes preference over phase logic and manual override logic. This logic is typically implemented in the basic process control system (BPCS).

FIGURE 16   Exception handling in control actions.

Key Points

1. The S88.01 standard is ideally suited to a methodolgy for the design of batch control that is object oriented.

2. Logic that changes frequently should be incorporated in the recipe procedure, when possible.

3. Logic that changes infrequently should usually be incorporated in equipment control.

4. Partitioning of process cells requires knowledge of both the equipment in the process cell and the recipes for the products that will be made in the process cell.

5. Exception handling can amount to 50–80% of the total configuration time for a batch project.

6. Reusable software components provide a lot of advantages in a batch control system.

7. Data collection is part of each equipment entity.

Rules of Thumb [6]

1. Understand the process before designing a batch control system for that process.

2. Design the control system before beginning implementation of the control system.

3. Get the end users involved in the design.

4. Exception handling is usually best done in equipment control.

5. The use of states simplifies dealing with exception handling in equipment control.

REFERENCES

1. ANSI/ISA-S88.01-1995, *Batch Control—Part 1: Models and Terminology*, 1995.

2. IEC61512:1997, *Batch Control—Part 1: Models and Terminology*, 1997.

3. NAMUR NE-33, *Requirements To Be Met By Systems for Recipe-Based Operations*, May 19, 1992.

4. Hopkinson, P., and J. Hancock, "A Case History of the Implementation of An S88-Aware Batch Control System," *World Batch Forum*, 1998.

5. Fleming, D. W., and P. E. Schreiber, "Batch Processing Design Example," *World Batch Forum*, 1998.

6. Christie, D. A., "A Methodology for Batch Control Implementation—Real World Lessons," *World Batch Forum*, 1998.

INDEX